AF335638

# Ullmann's Encyclopedia of Industrial Chemistry

Volume A 26

Surgical Materials
to
Thiourea

# Ullmann's Encyclopedia of Industrial Chemistry

Volumes A 1 – A 28: alphabetically arranged articles
Volumes B 1 – B 8: basic knowledge

## Editorial Advisory Board

# Ullmann's Encyclopedia of Industrial Chemistry

Fifth, Completely Revised Edition

Volume A 26:
Surgical Materials to Thiourea

Editors: Barbara Elvers, Stephen Hawkins, William Russey

Editorial Director: Christina Dyllick-Brenzinger
Editorial Assistants: Ilse Bedrich, Reinhilde Gutsche, Ingeborg Harders, Monika Pikart-Müller, Ulrike Winter
Copy Editor: Don Emerson


Library of Congress Card No. 84-25-829

A catalogue record for this book is available from the British Library.

Die Deutsche Bibliothek – CIP-Einheitsaufnahme

**Ullmann's encyclopedia of industrial chemistry** / ed.: Barbara Elvers ... [Ed. advisory board Hans-Jürgen Arpe ...]. — Weinheim ; Basel (Switzerland) ; Cambridge ; New York, NY ; Tokyo : VCH.
 Teilw. executive ed.: Wolfgang Gerhartz
 Bis 4. Aufl. u. d. T.: Ullmanns Encyklopädie der technischen Chemie
 ISBN 3-527-20100-9
NE: Gerhartz, Wolfgang [Hrsg.]; Elvers, Barbara [Hrsg.]; Encyclopedia of industrial chemistry

Vol. A. Alphabetically arranged articles.
 26. Surgical materials to thiourea. – 5., completely rev. ed. – 1995
 ISBN 3-527-20126-2



Distribution

VCH Verlagsgesellschaft, P.O. Box 10 11 61. D-69451 Weinheim (Federal Republic of Germany)

Switzerland: VCH Verlags-AG, P.O. Box, CH-4020 Basel (Switzerland)

Great Britain and Ireland: VCH Publishers (UK) Ltd., 8 Wellington Court, Wellington Street, Cambridge
     CB1 1HZ (Great Britain)

USA and Canada: VCH Publishers, 220 East 23rd Street, New York NY 10010-4606 (USA)

Japan: VCH, Eikow Building, 10-9 Hongo 1-chome, Bunkyo-ku, Tokyo 113, Japan



Production Managers: Claudia Baumann, Peter J. Biel
Cover design: Wolfgang Schmidt
Composition, printing, and bookbinding: Graphischer Betrieb Konrad Triltsch, D-97070 Würzburg
Printed in the Federal Republic of Germany

# Contents



# Cross References

# Symbols and Units

Symbols and units agree with SI standards (for conversion factors see pp. VIII–IX). The following list gives the most important symbols used in the encyclopedia. Articles with many specific units and symbols have a similar list as front matter.

| Symbol | Unit | Physical Quantity |
| --- | --- | --- |
| $a_B$ | | activity of substance B |
| $A_r$ | | relative atomic mass (atomic weight) |
| $A$ | $m^2$ | area |
| $c_B$ | $mol/m^3$, $mol/L$ (M) | concentration of substance B |
| $C$ | C/V | electric capacity |
| $c_p$, $c_v$ | $J\,kg^{-1}K^{-1}$ | specific heat capacity |
| $d$ | cm, m | diameter |
| $d$ | | relative density ($\varrho/\varrho_{water}$) |
| $D$ | $m^2/s$ | diffusion coefficient |
| $D$ | Gy ($= J/kg$) | absorbed dose |
| $e$ | C | elementary charge |
| $E$ | J | energy |
| $E$ | V/m | electric field strength |
| $E$ | V | electromotive force |
| $E_A$ | J | activation energy |
| $f$ | | activity coefficient |
| $F$ | C/mol | Faraday constant |
| $F$ | N | force |
| $g$ | $m/s^2$ | acceleration due to gravity |
| $G$ | J | Gibbs free energy |
| $h$ | m | height |
| $h$ | $W \cdot s^2$ | Planck constant |
| $H$ | J | enthalpy |
| $I$ | A | electric current |
| $I$ | cd | luminous intensity |
| $k$ | (variable) | rate constant of a chemical reaction |
| $k$ | J/K | Boltzmann constant |
| $K$ | (variable) | equilibrium constant |
| $l$ | m | length |
| $m$ | g, kg, t | mass |
| $M_r$ | | relative molecular mass (molecular weight) |
| $n_D^{20}$ | | refractive index (sodium D-line, 20 °C) |
| $n$ | mol | amount of substance |
| $N_A$ | $mol^{-1}$ | Avogadro constant ($6.023 \times 10^{23}$ $mol^{-1}$) |
| $p$ | Pa, bar* | pressure |
| $Q$ | J | quantity of heat |
| $r$ | m | radius |
| $R$ | $J\,K^{-1}mol^{-1}$ | gas constant |
| $R$ | $\Omega$ | electric resistance |
| $S$ | J/K | entropy |
| $t$ | s, min, h, d, month, a | time |
| $t$ | °C | temperature |
| $T$ | K | absolute temperature |
| $u$ | m/s | velocity |

* The official unit of pressure is the pascal (Pa).

Symbols and units (continued from p. VII)

| Symbol | Unit | Physical Quantity |
| --- | --- | --- |
| $U$ | V | electric potential |
| $U$ | J | internal energy |
| $V$ | $m^3$, L, mL | volume |
| $w$ | | mass fraction |
| $W$ | J | work |
| $x_B$ | | mole fraction of substance B |
| $Z$ | | proton number, atomic number |
| $\alpha$ | | cubic expansion coefficient |
| $\alpha$ | $W\,m^{-2}K^{-1}$ | heat-transfer coefficient (heat-transfer number) |
| $\alpha$ | | degree of dissociation of electrolyte |
| $[\alpha]$ | $10^{-2}\deg\,cm^2g^{-1}$ | specific rotation |
| $\eta$ | Pa · s | dynamic viscosity |
| $\theta$ | °C | temperature |
| $\varkappa$ | | $c_p/c_v$ |
| $\lambda$ | $W\,m^{-1}K^{-1}$ | thermal conductivity |
| $\lambda$ | nm, m | wavelength |
| $\mu$ | | chemical potential |
| $\nu$ | Hz, $s^{-1}$ | frequency |
| $\nu$ | $m^2/s$ | kinematic viscosity ($\eta/\varrho$) |
| $\pi$ | Pa | osmotic pressure |
| $\varrho$ | $g/cm^3$ | density |
| $\sigma$ | N/m | surface tension |
| $\tau$ | Pa $(N/m^2)$ | shear stress |
| $\varphi$ | | volume fraction |
| $\chi$ | $Pa^{-1}\,(m^2/N)$ | compressibility |

## Conversion Factors

| SI unit | Non-SI unit | From SI to non-SI multiply by |
| --- | --- | --- |
| *Mass* | | |
| kg | pound (avoirdupois) | 2.205 |
| kg | ton (long) | $9.842 \times 10^{-4}$ |
| kg | ton (short) | $1.102 \times 10^{-3}$ |
| *Volume* | | |
| $m^3$ | cubic inch | $6.102 \times 10^{4}$ |
| $m^3$ | cubic foot | 35.315 |
| $m^3$ | gallon (U.S., liquid) | $2.642 \times 10^{2}$ |
| $m^3$ | gallon (Imperial) | $2.200 \times 10^{2}$ |
| *Temperature* | | |
| °C | °F | °C $\times 1.8 + 32$ |
| *Force* | | |
| N | dyne | $1.0 \times 10^{5}$ |

Conversion factors (continued from p. VIII)

| SI unit | Non-SI unit | From SI to non-SI multiply by |
| --- | --- | --- |
| *Energy, Work* | | |
| J | Btu (int.) | $9.480 \times 10^{-4}$ |
| J | cal (int.) | $2.389 \times 10^{-1}$ |
| J | eV | $6.242 \times 10^{18}$ |
| J | erg | $1.0 \times 10^{7}$ |
| J | kW $\cdot$ h | $2.778 \times 10^{-7}$ |
| J | kp $\cdot$ m | $1.020 \times 10^{-1}$ |
| *Pressure* | | |
| MPa | at | 10.20 |
| MPa | atm | 9.869 |
| MPa | bar | 10 |
| kPa | mbar | 10 |
| kPa | mm Hg | 7.502 |
| kPa | psi | 0.145 |
| kPa | torr | 7.502 |

## Prefixes for Powers of Ten

| | | | |
| --- | --- | --- | --- |
| E (exa) $10^{18}$ | M (mega) $10^{6}$ | d (deci) $10^{-1}$ | n (nano) $10^{-9}$ |
| P (peta) $10^{15}$ | k (kilo) $10^{3}$ | c (centi) $10^{-2}$ | p (pico) $10^{-12}$ |
| T (tera) $10^{12}$ | h (hecto) $10^{2}$ | m (milli) $10^{-3}$ | f (femto) $10^{-15}$ |
| G (giga) $10^{9}$ | da (deca) 10 | $\mu$ (micro) $10^{-6}$ | a (atto) $10^{-18}$ |

## Abbreviations

The following is a list of the abbreviations used in the text. Common terms, the names of publications and institutions, and legal agreements are included along with their full identities. Other abbreviations will be defined wherever they first occur in an article. For further abbreviations, see p. VII (Symbols and Units), p. XIV (Companies and Country Codes in Patent References). The names of periodical publications are abbreviated exactly as done by Chemical Abstracts Service.

| | |
| --- | --- |
| abs. | absolute |
| a.c. | alternating current |
| ACGIH | American Conference of Governmental Industrial Hygienists |
| ACS | American Chemical Society |
| ADI | acceptable daily intake |
| ADN | accord européen relatif au transport international des marchandises dangereuses par voie de navigation interieure (European agreement concerning the international transportation of dangerous goods by inland waterways) |
| ADNR | ADN par le Rhin (regulation concerning the transportation of dangerous goods on the Rhine and all national waterways of the countries concerned) |
| ADP | adenosine 5'-diphosphate |
| ADR | accord européen relatif au transport international des marchandises dangereuses par route (European agreement concerning the international transportation of dangerous goods by road) |
| AEC | Atomic Energy Commission (United States) |
| AIChE | American Institute of Chemical Engineers |

| | |
|---|---|
| AIME | American Institute of Mining, Metallurgical, and Petroleum Engineers |
| AMP | adenosine 5'-monophosphate |
| APhA | American Pharmaceutical Association |
| API | American Petroleum Institute |
| ASTM | American Society for Testing and Materials |
| ATP | adenosine 5'-triphosphate |
| BAM | Bundesanstalt für Material-prüfung (Federal Republic of Germany) |
| BAT | Biologischer Arbeitsstoff-Toleranz-wert (biological tolerance value for a working material, established by MAK Commission, see MAK) |
| Beilstein | Beilstein's Handbook of Organic Chemistry, Springer, Berlin–Heidelberg–New York |
| BET | Brunauer–Emmett–Teller |
| BGA | Bundesgesundheitsamt (Federal Republic of Germany) |
| BGBl. | Bundesgesetzblatt (Federal Republic of Germany) |
| BIOS | British Intelligence Objectives Subcommitee Report (see also FIAT) |
| BOD | biological oxygen demand |
| *bp* | boiling point |
| B.P. | British Pharmacopeia |
| BS | British Standard |
| ca. | circa |
| calcd. | calculated |
| CAS | Chemical Abstracts Service |
| cat. | catalyst, catalyzed |
| cf. | compare |
| CFR | Code of Federal Regulations (United States) |
| Chap. | chapter |
| ChemG | Chemikaliengesetz (Federal Republic of Germany) |
| C.I. | Colour Index |
| CIOS | Combined Intelligence Objectives-Subcommittee Report (see also FIAT) |
| CNS | central nervous system |
| Co. | Company |
| COD | chemical oxygen demand |
| conc. | concentrated |
| const. | constant |
| Corp. | Corporation |
| crit. | critical |

| | |
|---|---|
| CTFA | The Cosmetic, Toiletry and Fragrance Association (United States) |
| DAB 9 | Deutsches Arzneibuch, 9th ed., Deutscher Apotheker-Verlag, Stuttgart 1986 |
| d.c. | direct current |
| decomp. | decompose, decomposition |
| DFG | Deutsche Forschungsgemeinschaft (German Science Foundation) |
| dil. | dilute, diluted |
| DIN | Deutsche Industrie Norm (Federal Republic of Germany) |
| DMF | dimethylformamide |
| DNA | deoxyribonucleic acid |
| DOE | Department of Energy (United States) |
| DOT | Department of Transportation – Materials Transportation Bureau (United States) |
| DTA | differential thermal analysis |
| EC | effective concentration |
| EC | European Community |
| ed. | editor, edition, edited |
| e.g. | for example |
| emf | electromotive force |
| EmS | Emergency Schedule |
| EN | European Standard (European Community) |
| EPA | Environmental Protection Agency (United States) |
| EPR | electron paramagnetic resonance |
| Eq. | equation |
| ESCA | electron spectroscopy for chemical analysis |
| esp. | especially |
| ESR | electron spin resonance |
| Et | ethyl substituent ($-C_2H_5$) |
| et al. | and others |
| etc. | et cetera |
| EVO | Eisenbahnverkehrsordnung (Federal Republic of Germany) |
| exp (...) | $e^{(...)}$, mathematical exponent |
| FAO | Food and Agriculture Organization (United Nations) |
| FDA | Food and Drug Administration (United States) |
| FD & C | Food, Drug and Cosmetic Act (United States) |
| FHSA | Federal Hazardous Substances Act (United States) |
| FIAT | Field Information Agency, Technical (United States reports on the chemical industry in Germany, 1945) |
| Fig. | figure |

| | |
|---|---|
| *fp* | freezing point |
| Friedländer | P. Friedländer, Fortschritte der Teerfarbenfabrikation und verwandter Industriezweige, Vol. 1–25, Springer, Berlin 1888–1942 |
| FT | Fourier transform |
| (g) | gas, gaseous |
| GC | gas chromatography |
| GefStoffV | Gefahrstoffverordnung (regulations in the Federal Republic of Germany concerning hazardous substances) |
| GGVE | Verordnung in der Bundesrepublik Deutschland über die Beförderung gefährlicher Güter mit der Eisenbahn (regulation in the Federal Republic of Germany concerning the transportation of dangerous goods by rail) |
| GGVS | Verordnung in der Bundesrepublik Deutschland über die Beförderung gefährlicher Güter auf der Straße (regulation in the Federal Republic of Germany concerning the transportation of dangerous goods by road) |
| GGVSee | Verordnung in der Bundesrepublik Deutschland über die Beförderung gefährlicher Güter mit Seeschiffen (regulation in the Federal Republic of Germany concerning the transportation of dangerous goods by sea-going vessels) |
| GLC | gas-liquid chromatography |
| Gmelin | Gmelin's Handbook of Inorganic Chemistry, 8th ed., Springer, Berlin–Heidelberg–New York |
| GRAS | generally recognized as safe |
| Hal | halogen substituent ($-F$, $-Cl$, $-Br$, $-I$) |
| Houben-Weyl | Methoden der organischen Chemie, 4th ed., Georg Thieme Verlag, Stuttgart |
| HPLC | high performance liquid chromatography |
| IAEA | International Atomic Energy Agency |
| IARC | International Agency for Research on Cancer, Lyon, France |
| IATA-DGR | International Air Transport Association, Dangerous Goods Regulations |
| ICAO | International Civil Aviation Organization |
| i.e. | that is |
| i.m. | intramuscular |
| IMDG | International Maritime Dangerous Goods Code |
| IMO | Inter-Governmental Maritime Consultive Organization (in the past: IMCO) |
| Inst. | Institute |
| i.p. | intraperitoneal |
| IR | infrared |
| ISO | International Organization for Standardization |
| IUPAC | International Union of Pure and Applied Chemistry |
| i.v. | intravenous |
| Kirk-Othmer | Encyclopedia of Chemical Technology, 3rd ed., J. Wiley & Sons, New York–Chichester–Brisbane–Toronto 1978–1984 |
| (l) | liquid |
| Landolt-Börnstein | Zahlenwerte u. Funktionen aus Physik, Chemie, Astronomie, Geophysik u. Technik, Springer, Heidelberg 1950–1980; Zahlenwerte und Funktionen aus Naturwissenschaften und Technik, Neue Serie, Springer, Heidelberg, since 1961 |
| $LC_{50}$ | lethal concentration for 50 % of the test animals |
| LCLo | lowest published lethal concentration |
| $LD_{50}$ | lethal dose for 50 % of the test animals |
| LDLo | lowest published lethal dose |
| ln | logarithm (base e) |
| LNG | liquefied natural gas |
| log | logarithm (base 10) |
| LPG | liquefied petroleum gas |
| M | mol/L |
| M | metal (in chemical formulas) |
| MAK | Maximale Arbeitsplatz-Konzentration (maximum concentration at the workplace in the Federal Republic of Germany); cf. Deutsche Forschungsgemeinschaft (ed.): Maximale Arbeits- |

| | |
|---|---|
| | platzkonzentrationen (MAK) und Biologische Arbeitsstoff-Toleranz-Werte (BAT), VCH Verlagsgesellschaft, Weinheim (published annually) |
| max. | maximum |
| MCA | Manufacturing Chemists Association (United States) |
| Me | methyl substituent ($-CH_3$) |
| Methodicum Chimicum | Methodicum Chimicum, Georg Thieme Verlag, Stuttgart |
| MFAG | Medical First Aid Guide for Use in Accidents Involving Dangerous Goods |
| MIK | maximale Immissionskonzentration (maximum immission concentration) |
| min. | minimum |
| *mp* | melting point |
| MS | mass spectrum, mass spectrometry |
| NAS | National Academy of Sciences (United States) |
| NASA | National Aeronautics and Space Administration (United States) |
| NBS | National Bureau of Standards (United States) |
| NCTC | National Collection of Type Cultures (United States) |
| NIH | National Institutes of Health (United States) |
| NIOSH | National Institute for Occupational Safety and Health (United States) |
| NMR | nuclear magnetic resonance |
| no. | number |
| NOEL | no observed effect level |
| NRC | Nuclear Regulatory Commission (United States) |
| NRDC | National Research Development Corporation (United States) |
| NSC | National Service Center (United States) |
| NSF | National Science Foundation (United States) |
| NTSB | National Transportation Safety Board (United States) |
| OECD | Organization for Economic Cooperation and Development |
| OSHA | Occupational Safety and Health Administration (United States) |
| p., pp. | page, pages |
| Patty | G. D. Clayton, F. E. Clayton (eds.): Patty's Industrial Hygiene and Toxicology, 3rd ed., Wiley Interscience, New York |
| PB report | Publication Board Report (U.S. Department of Commerce, Scientific and Industrial Reports) |
| PEL | permitted exposure limit |
| Ph | phenyl substituent ($-C_6H_5$) |
| Ph. Eur. | European Pharmacopoeia, 2nd. ed., Council of Europe, Strasbourg 1981 – |
| phr | part per hundred rubber (resin) |
| PNS | peripheral nervous system |
| ppm | parts per million |
| q. v. | which see (quod vide) |
| ref. | refer, reference |
| resp. | respectively |
| $R_f$ | retention factor (TLC) |
| R. H. | relative humidity |
| RID | règlement international concernant le transport des marchandises dangereuses par chemin de fer (international convention concerning the transportation of dangerous goods by rail) |
| RNA | ribonucleic acid |
| R phrase (R-Satz) | risk phrase according to ChemG and GefStoffV (Federal Republic of Germany) |
| rpm | revolutions per minute |
| RTECS | Registry of Toxic Effects of Chemical Substances, edited by the National Institute of Occupational Safety and Health (United States) |
| (s) | solid |
| SAE | Society of Automotive Engineers (United States) |
| s.c. | subcutaneous |
| SI | International System of Units |
| SIMS | secondary ion mass spectrometry |
| S phrase (S-Satz) | safety phrase according to ChemG and GefStoffV (Federal Republic of Germany) |
| STEL | Short Term Exposure Limit (see TLV) |
| STP | standard temperature and pressure ($0°$ C, 101.325 kPa) |
| $T_g$ | glass transition temperature |
| TA Luft | Technische Anleitung zur Reinhaltung der Luft (clean air regulation in Federal Republic of Germany) |
| TA Lärm | Technische Anleitung zum Schutz gegen Lärm (low noise |

| | |
|---|---|
| | regulation in Federal Republic of Germany) |
| TDLo | lowest published toxic dose |
| THF | tetrahydrofuran |
| TLC | thin layer chromatography |
| TLV | Threshold Limit Value (TWA and STEL); published annually by the American Conference of Governmental Industrial Hygienists (ACGIH), Cincinnati, Ohio |
| TOD | total oxygen demand |
| TRK | Technische Richtkonzentration (lowest technically feasible level) |
| TSCA | Toxic Substances Control Act (United States) |
| TÜV | Technischer Überwachungsverein (Technical Control Board of the Federal Republic of Germany) |
| TWA | Time Weighted Average |
| UBA | Umweltbundesamt (Federal Environmental Agency) |
| Ullmann | Ullmanns Encyklopädie der Technischen Chemie, 4th ed., Verlag Chemie, Weinheim 1972–1984; 3rd ed., Urban und Schwarzenberg, München 1951–1970 |
| USAEC | United States Atomic Energy Commission |
| USAN | United States Adopted Names |
| USD | United States Dispensatory |
| USDA | United States Department of Agriculture |
| U.S.P. | United States Pharmacopeia |
| UV | ultraviolet |
| UVV | Unfallverhütungsvorschriften der Berufsgenossenschaft (workplace safety regulations in the Federal Republic of Germany) |
| VbF | Verordnung in der Bundesrepublik Deutschland über die Errichtung und den Betrieb von Anlagen zur Lagerung, Abfüllung und Beförderung brennbarer Flüssigkeiten (regulation in the Federal Republic of Germany concerning the construction and operation of plants for storage, filling, and transportation of flammable liquids; classification according to the flash point of liquids, in accordance with the classification in the United States) |
| VDE | Verband Deutscher Elektroingenieure (Federal Republic of Germany) |
| VDI | Verein Deutscher Ingenieure (Federal Republic of Germany) |
| vol | volume |
| vol. | volume (of a series of books) |
| vs. | versus |
| WHO | World Health Organization (United Nations) |
| Winnacker-Küchler | Chemische Technologie, Carl Hanser Verlag, München |
| wt | weight |
| $ | U.S. dollar, unless otherwise stated |

# Abbreviations for the Names of Frequently Cited Companies

| | | | |
|---|---|---|---|
| Air Products | Air Products and Chemicals | ICI | Imperial Chemical Industries |
| Akzo | Algemene Koninklijke Zout Organon | IFP | Institut Français du Pétrole |
| Alcoa | Aluminum Company of America | INCO | International Nickel Company |
| Allied | Allied Corporation | 3M | Minnesota Mining and Manufacturing Company |
| Amer. Cyanamid | American Cyanamid Company | Mitsubishi Chemical | Mitsubishi Chemical Industries |
| BASF | BASF Aktiengesellschaft | Monsanto | Monsanto Company |
| Bayer | Bayer AG | Nippon Shokubai | Nippon Shokubai Kagaku Kogyo |
| BP | British Petroleum Company | PCUK | Pechiney Ugine Kuhlmann |
| Celanese | Celanese Corporation | PPG | Pittsburg Plate Glass Industries |
| Daicel | Daicel Chemical Industries | Searle | G.D. Searle & Company |
| Dainippon | Dainippon Ink and Chemicals Inc. | SKF | Smith Kline & French Laboratories |
| Dow Chemical | The Dow Chemical Company | SNAM | Societá Nazionale Metandotti |
| DSM | Dutch Staats Mijnen | Sohio | Standard Oil of Ohio |
| Du Pont | E.I. du Pont de Nemours & Company | Stauffer | Stauffer Chemical Company |
| Exxon | Exxon Corporation | Sumitomo | Sumitomo Chemical Company |
| FMC | Food Machinery & Chemical Corporation | Toray | Toray Industries Inc. |
| | | UCB | Union Chimique Belge |
| GAF | General Aniline & Film Corporation | Union Carbide | Union Carbide Corporation |
| W.R. Grace | W.R. Grace & Company | UOP | Universal Oil Products Company |
| Hoechst | Hoechst Aktiengesellschaft | VEBA | Vereinigte Elektrizitäts- und Bergwerks-AG |
| IBM | International Business Machines Corporation | Wacker | Wacker Chemie GmbH |

# Country Codes

The following list contains a selection of standard country codes used in the patent references.

| | | | |
|---|---|---|---|
| AT | Austria | ID | Indonesia |
| AU | Australia | IL | Israel |
| BE | Belgium | IT | Italy |
| BG | Bulgaria | JP | Japan* |
| BR | Brazil | LU | Luxembourg |
| CA | Canada | MA | Morocco |
| CH | Switzerland | NL | Netherlands* |
| CS | Czechoslovakia | NO | Norway |
| DD | German Democratic Republic | NZ | New Zealand |
| DE | Federal Republic of Germany (and Germany before 1949)* | PL | Poland |
| | | PT | Portugal |
| DK | Denmark | SE | Sweden |
| ES | Spain | SU | Soviet Union |
| FI | Finland | US | United States of America |
| FR | France | YU | Yugoslavia |
| GB | United Kingdom | ZA | South Africa |
| GR | Greece | EP | European Patent Office* |
| HU | Hungary | WO | World Intellectual Property Organization |

* For Europe, Federal Republic of Germany, Japan, and the Netherlands, the type of patent is specified: EP (patent), EP-A (application), DE (patent), DE-OS (Offenlegungsschrift), DE-AS (Auslegeschrift), JP (patent), JP-Kokai (Kokai tokkyo koho), NL (patent), and NL-A (application).

# Periodic Table of Elements

1A ("European" group designation according to old IUPAC recommendation)
1  (group designation according to 1985 IUPAC proposal)
IA ("American" group designation, also used by the Chemical Abstracts Service until the end of 1986)

| 1A<br>1<br>IA | 2A<br>2<br>IIA | 3A<br>3<br>IIIB | 4A<br>4<br>IVB | 5A<br>5<br>VB | 6A<br>6<br>VIB | 7A<br>7<br>VIIB | 8<br>8<br>VIII | 8<br>9<br>VIII | 8<br>10<br>VIII | 1B<br>11<br>IB | 2B<br>12<br>IIB | 3B<br>13<br>IIIA | 4B<br>14<br>IVA | 5B<br>15<br>VA | 6B<br>16<br>VIA | 7B<br>17<br>VIIA | 0<br>18<br>VIIIA |
|---|---|---|---|---|---|---|---|---|---|---|---|---|---|---|---|---|---|
| 1.0079<br>$_1$H | | | | | | | | | | | | | | | | | 4.0026<br>$_2$He |
| 6.941<br>$_3$Li | 9.0122<br>$_4$Be | | | | | | | | | | | 10.811<br>$_5$B | 12.011<br>$_6$C | 14.007<br>$_7$N | 15.9994<br>$_8$O | 18.998<br>$_9$F | 20.180<br>$_{10}$Ne |
| 22.990<br>$_{11}$Na | 24.305<br>$_{12}$Mg | | | | | | | | | | | 26.982<br>$_{13}$Al | 28.086<br>$_{14}$Si | 30.974<br>$_{15}$P | 32.066<br>$_{16}$S | 35.453<br>$_{17}$Cl | 39.948<br>$_{18}$Ar |
| 39.098<br>$_{19}$K | 40.078<br>$_{20}$Ca | 44.956<br>$_{21}$Sc | 47.88<br>$_{22}$Ti | 50.942<br>$_{23}$V | 51.996<br>$_{24}$Cr | 54.938<br>$_{25}$Mn | 55.847<br>$_{26}$Fe | 58.933<br>$_{27}$Co | 58.69<br>$_{28}$Ni | 63.546<br>$_{29}$Cu | 65.39<br>$_{30}$Zn | 69.723<br>$_{31}$Ga | 72.61<br>$_{32}$Ge | 74.922<br>$_{33}$As | 78.96<br>$_{34}$Se | 79.904<br>$_{35}$Br | 83.80<br>$_{36}$Kr |
| 85.468<br>$_{37}$Rb | 87.62<br>$_{38}$Sr | 88.906<br>$_{39}$Y | 91.224<br>$_{40}$Zr | 92.906<br>$_{41}$Nb | 95.94<br>$_{42}$Mo | 98.906<br>$_{43}$Tc* | 101.07<br>$_{44}$Ru | 102.91<br>$_{45}$Rh | 106.42<br>$_{46}$Pd | 107.87<br>$_{47}$Ag | 112.41<br>$_{48}$Cd | 114.82<br>$_{49}$In | 118.71<br>$_{50}$Sn | 121.75<br>$_{51}$Sb | 127.60<br>$_{52}$Te | 126.90<br>$_{53}$I | 131.29<br>$_{54}$Xe |
| 132.91<br>$_{55}$Cs | 137.33<br>$_{56}$Ba | | 178.49<br>$_{72}$Hf | 180.95<br>$_{73}$Ta | 183.85<br>$_{74}$W | 186.21<br>$_{75}$Re | 190.2<br>$_{76}$Os | 192.22<br>$_{77}$Ir | 195.08<br>$_{78}$Pt | 196.97<br>$_{79}$Au | 200.59<br>$_{80}$Hg | 204.38<br>$_{81}$Tl | 207.2<br>$_{82}$Pb | 208.98<br>$_{83}$Bi | 208.98<br>$_{84}$Po* | 209.99<br>$_{85}$At* | 222.02<br>$_{86}$Rn* |
| 223.02<br>$_{87}$Fr* | 226.03<br>$_{88}$Ra* | | | | | | | | | | | | | | | | |

| 138.91<br>$_{57}$La | 140.12<br>$_{58}$Ce | 140.91<br>$_{59}$Pr | 144.24<br>$_{60}$Nd | 146.92<br>$_{61}$Pm* | 150.36<br>$_{62}$Sm | 151.97<br>$_{63}$Eu | 157.25<br>$_{64}$Gd | 158.93<br>$_{65}$Tb | 162.50<br>$_{66}$Dy | 164.93<br>$_{67}$Ho | 167.26<br>$_{68}$Er | 168.93<br>$_{69}$Tm | 173.04<br>$_{70}$Yb | 174.97<br>$_{71}$Lu |
|---|---|---|---|---|---|---|---|---|---|---|---|---|---|---|
| 227.03<br>$_{89}$Ac* | 232.04<br>$_{90}$Th* | 231.04<br>$_{91}$Pa* | 238.03<br>$_{92}$U* | 237.05<br>$_{93}$Np* | 244.06<br>$_{94}$Pu* | 243.06<br>$_{95}$Am* | 247.07<br>$_{96}$Cm* | 247.07<br>$_{97}$Bk* | 251.08<br>$_{98}$Cf* | 252.08<br>$_{99}$Es* | 257.10<br>$_{100}$Fm* | 258.10<br>$_{101}$Md* | 259.10<br>$_{102}$No* | 260.11<br>$_{103}$Lr* |

* Elements with unstable isotopes; mass of most important isotope given

# Surgical Materials

DAVID WILLIAMS, Department of Clinical Engineering, University of Liverpool, Liverpool, United Kingdom



## 1. Introduction

Surgery in the late 1900s places heavy demands upon sophisticated technology, ranging from functional imaging equipment to assist in diagnosis, to lasers and robotics to improve accuracy and reliability. One of the most important areas in which technological advances have significantly enhanced surgery involves the use of synthetic materials for implantation in the body to assist in surgical reconstruction and repair.

These surgical materials form an important subset of the more general group of materials referred to as "biomaterials", which are defined as materials intended to interface with biological systems to evaluate, treat, augment, or replace any tissue, organ, or function of the body [1].

Within the broad group of biomaterials, a number of subsets, including dental materials and drug delivery materials, may be distinguished (see → Dental Materials and → Pharmaceutical Dosage Forms). Surgical materials are defined in

Ullmann's Encyclopedia
of Industrial Chemistry, Vol. A 26

the context of this review as biomaterials used for implantation in the body to assist in surgical reconstruction and repair.

Biomaterials are rarely designed specifically for this use. Instead, they are generally materials developed for other applications but which have been found to be appropriate for medical use, often being modified or optimized for these applications. Thus the materials described in Chapter 4 are generally no different in structure and properties to materials used in chemical, aerospace, nuclear, and other industries. It is the matching of the properties of materials to the specific and unique requirements of surgical reconstruction that is the key to biomaterials science.

## 2. Surgical Applications of Materials

In this section the reasons why materials are used surgically and the criteria for materials selection that can be established on the basis of the functional performance required are reviewed. General reviews of the clinical applications of materials may be found in [2]–[6].

### 2.1. General Principles

Biomaterials are implanted in the body for many different reasons but these can be broadly grouped as follows.

**Congenital Deformities.** Children are often born with a physical deformity which results in functional or aesthetic impairment. In some cases, the consequences may be life threatening, as with certain deformities associated with spina bifida, while in other cases they are more psychological, as in the case of facial disfigurement. The treatment and the precise usage of the materials vary from one situation to another.

**Developmental Defects.** A child apparently healthy and normal at birth may develop abnormally, perhaps due to impaired hormonal function, leading to defects that require correction. This usually involves the skeletal system, and procedures such as leg-lengthening and spinal curvature correction are included in this category.

**Trauma.** At any time in life, injuries may be sustained which require surgical intervention as part of the treatment. This may involve the use of materials and devices to assist the body to repair itself, such as sutures to close soft tissue wounds and fracture plates, nails, and screws to facilitate bone repair, or may involve structures that replace or augment damaged tissue, such as artificial skin, bone substitutes, ligament prostheses, and nerve guides.

**Disease.** The majority of diseases can be treated nonsurgically. However, a significant number affect tissues, leading to pain, loss of movement, or loss of vital function, such that surgery becomes essential. Specific examples involving the use of surgical materials in reconstruction include osteo- and rheumatoid arthritis, which may be treated by total joint replacement; atherosclerosis within the vascular system, which can be surgically treated with artificial arteries; and cataracts of the eye, which can be excised and replaced by intraocular lenses. Of great importance here is the order of preference in treatment: (1) nonsurgical methods (e.g., drugs); (2) minimally invasive procedures in which the defect is treated remotely or endoscopically; (3) surgical reconstruction involving the patient's own tissues, such as bone or skin grafts; and (4) reconstruction involving surgical materials. Thus, these latter procedures are only used when other treatments are ineffective or inappropriate.

**Psychological Factors.** There are many occasions in which adults find that their appearance causes them psychological problems. Under such circumstances, even when there is no medical or physical disability and where few others would agree that the appearance requires correcting, surgery, usually minor plastic surgery, may be carried out.

**Tissue Atrophy.** As part of the ageing process, tissues change their shape and consistency. In particular, some tissues lose their form and so augmentation or restoration is desirable. This may relate to loss of bone, especially in osteoporosis.

### 2.2. Orthopedic Surgery

Orthopedic surgery has historically been the most significant user of biomaterials. It is concerned with the musculoskeletal system and the treatment of diseases and disturbances to its components, particularly bones, cartilage, ligaments, muscles, and tendons. It is this system that produces movement and that responds to forces. It is therefore susceptible to injury and to stress-related diseases, the consequences of which are significant since they affect the ability to move. Of equal importance is that any replacement of these tissues has to perform in the same mechanical environment, and attainment of the relevant mechanical performance is not a trivial matter.

### 2.2.1. Bone and Joint Replacement

Several diseases affect the bones and joints but they predominantly include osteo- and rheumatoid arthritis. These conditions affect the tissues of the joints, osteoarthritis being largely a mechanical disruption to the cartilage and underlying bone, while rheumatoid arthritis is an inflammatory disease which causes very painful structural changes to the tissues. Pain relieving drugs and anti-inflammatory agents may be reasonably effective but the diseases are progressive and in many cases, excision of the affected tissue is the only long-term solution.

Since in most cases the disease affects much of the joint, it is usual to resect a considerable amount of tissue and use a total joint replacement prosthesis. In the hip, for example, the first joint to be replaced in this way, it is usual to replace both the head of the femur and the socket within the pelvis (the acetabulum) [7]. In the knee, which is equally affected but more difficult to replace, the lower end of the femur, the upper end of the tibia, and the patella are involved [8].

The main requirements for these joint replacements are the ability to transmit forces under sliding conditions and the ability to be attached to the remaining bone, with durability measured in decades. Combinations of materials that give low friction and high wear resistance for the articulating surfaces are used, for example, a metal or ceramic bearing against ultrahigh molecular mass, high-density polyethylene, high-strength alloys for the main structural components [9], and a system to permit fixation to the bone. In the latter case, this is most often achieved by using a so-called bone cement, generally an in situ curing poly(methyl methacrylate) [10], or sometimes by means of a porous surface which allows ingrowth of bone to achieve

stability [11], or bioactive coating that encourages rapid regeneration of bone at the interface [12].

Joint replacements are not without their problems but are generally able to provide pain-free movement for patients for a reasonable length of time. Whilst at one stage success rates, at least in hips, were high with an expectation of 15–20 years performance, as the age range in which these prostheses are used has widened, particularly as younger and more active patients have been treated, the demands on them have escalated resulting in reduced performance. An increasingly common problem is that the prostheses become loose. The principle causes of this loosening are the tissue reaction to the wear debris produced at the articulating interface [13], and the difficulty of attaining and retaining an acceptable stress distribution in the prosthesis–bone system [14]. It is for these reasons that attempts to enhance wear resistance and fixation with surface modifications and the introduction of composite materials to facilitate load transfer are being carried out (see Chap. 4).

### 2.2.2. Ligament Augmentation and Repair [15]

Ligaments are dense bundles of tissue, mostly composed of the structural protein collagen, that are attached to the bones within a joint and provide stability and constrain movement to acceptable limits. Damage to ligaments severely compromises this stability and therefore restricts the controlled movement of the patient. Ligaments rarely suffer disease, primarily because of their simple construction but they are very susceptible to injury, especially sports injuries. Here it is the ligaments of the knee, especially the cruciate ligaments which are frequently affected. Loss of function of these ligaments makes walking impossible.

Surgical treatment can take several forms. It is possible to attempt reconstruction by total replacement of the damaged ligament with a prosthesis; however, matching the elastic characteristics of the natural ligament and producing a reliable anchorage to the bone are difficult. The alternative is to use an augmentation device in which an implant is added to the damaged tissue and works with it to transmit the forces. Materials used in these types of device include polyester fibers, carbon fibers, and expanded polytetrafluoroethylene (PTFE).

### 2.2.3. Fracture Fixation

Bone possesses an extremely efficient mechanism for self-repair. Thus, when bone fractures, it heals spontaneously such that the newly formed bone is microstructurally indistinguishable from the existing bone. However, the repair only occurs with the appropriate alignment and geometrical arrangement if the fracture surfaces are held together in their correct position, without gross movement, during the healing process, which may take a few months.

With the long bones, such as the femur, tibia, and humerus, correct alignment and stability can be achieved by using a plaster cast. However, there are many occasions were this is not possible, perhaps because there are multiple fragments. In this case, surgical intervention is necessary to achieve the correct juxtaposition of the bones and to hold them in this position. This may be done through the use of various types of plate, nail, screws, or other fixtures. Although there have been a few attempts to use other materials, the vast majority of these devices are made from metals, which must combine the properties of strength, especially fatigue strength, with corrosion resistance and biocompatibility [16].

Because bone is a dynamic tissue that requires the stimulus of mechanical stress for maintenance of its internal structure, problems arise if rigid metal plates are retained within the bone after healing [17]. Ideally, all fracture plates should be removed once healing is complete; because of the slight risks involved with this additional surgical procedure, attention is now on the development of biodegradable fracture plates [18], which is difficult since the requirements for degradability and inherent strength are almost mutually exclusive.

## 2.3. Maxillofacial Surgery

Closely allied to the problems of the skeletal system are those concerning the structural tissues of the head, mostly involving bone but also the associated soft tissue. Here, however, the causes of the problems are a little different, as are the solutions. There are many subspecialities, including oral surgery, craniofacial surgery, orthognathic surgery, and perhaps more conveniently as an all-embracing term, maxillofacial surgery. The most important aspects follow [19].

### 2.3.1. Dental Implants

The area of dental implants lies at the interface between surgery and dentistry as far as biomaterials are concerned. The loss of natural teeth, through either disease or trauma, has, for many years, been compensated by the provision of artificial teeth in the form of bridges or dentures. These essentially provide an esthetic replacement of the crown of the tooth but do not replace the root and its attachment to the bones of the jaws. Whilst they are quite satisfactory in many patients, a number of disadvantages exist including the need to interfere with adjacent teeth when a bridge is used, and the problems of denture stability, and so it has long been the ambition of dentists to use implants that replace teeth in their entirety. This is difficult because the implants contact a number of different types of tissue, including an epithelial barrier, must reside in the microbiologically hostile region of the mouth, and are exposed to a complex stress system without much support. For many years, a very poor success rate reflected this difficulty, but more recently, improvements to materials, designs, and techniques have given far better performance and dental implants now represent a very respectable alternative to traditional dentures [20]–[22]. The materials of choice are titanium alloys and alumina, often coated with a porous surface or bioactive material such as calcium phosphate (see → Dental Materials, **A 8**, pp. 266–267).

### 2.3.2. Maxillofacial Trauma

Fractures of the facial skeleton have traditionally been treated conservatively; in contrast to fractures of the long bones, which are often compound and displaced, those of the face are usually simple without distortion of shape. The requirements for stability are more easily satisfied, for example, by interdental wiring or external halos which consist of a frame secured to the skull by means of pins. However, it is becoming increasingly recognized that better results and far more convenience to the patient can be achieved through the use of internal plating. Small titanium or stainless steel plates for the fixation of mandibular or maxillary fractures are now widely used [23]. The requirements for degradability mentioned in Section 2.2.3. also apply here, and the less severe mechanical environment makes this a more realistic possibility.

### 2.3.3. Reconstructive Maxillofacial and Orthognathic Surgery

The connective tissues of the head and neck also suffer from a variety of diseases and deformities that may require treatment. These range from inflammatory disease of the periodontium, the tissues that support the teeth, to extensive hard and soft tissue tumors. The treatment of advanced periodontitis may involve the implantation of thin strips of a porous material around the root of the tooth, such as expanded PTFE, which facilitate the regeneration of new connective tissue and act as a barrier to infection, or the implantation of artificial bone. The latter approach may also be used in patients who suffer localized osteoporosis, with substantial loss of bone, within mandible or maxilla after tooth extraction.

Tumors of the head and neck present considerable problems of reconstruction [24]. This is partly because prior irradiation treatment used in an attempt to eradicate the tumor severely compromises the ability of the tissue to heal and to accept the presence of an implanted device. If the tumor affects the bone, it is necessary to remove a large amount of that bone, which makes reconstruction by a prosthesis, with attachment to the remaining part of the facial skeleton, very difficult. This is a particularly challenging area of implant surgery.

## 2.4. Cardiovascular Surgery

Diseases of the circulatory system occur as frequently as those of the musculoskeletal system but, by their very nature, are more life threatening than immobilizing. The diseases can affect the heart, the major circulatory system, and the microvasculature, the consequence of which is a perturbation to the flow of blood. Treatment that does not require major intervention is preferred, for example, pharmacological and endoscopic techniques.

### 2.4.1. Heart Valves

The valves of the heart control the flow of blood into and out of the chambers. There are four valves of interest: the mitral, aortic, pulmonary, and tricuspid, and their opening and closing mechanisms slightly differ, as do their

susceptibility to disease. The mitral and aortic valves are the most critical in this respect, suffering from two major modes of malfunction, stenosis and incompetence. *Stenosis* is an obstruction of the valve orifice which disturbs the pattern of blood flow and increases the pressure necessary to achieve adequate flow. At these high pressures, turbulence becomes more significant and considerable stress is placed on the heart. *Incompetence* arises from the inability of the valve to close properly, allowing a retrograde flow back through the valve, again disturbing the flow profile.

When valvular disease has progressed to such an extent that it compromises the ability of the heart to maintain the circulation, it becomes necessary to remove the offending structure and replace it with a prosthesis [25]. This has been possible for over thirty years. The original valves consisted of a ball held within a cage which was sewn into the heart. These worked reasonably well although not without problems and the valve, both with respect to materials and design, has evolved over the years [26]. The common designs today incorporate a flat disc that tilts within a mechanical constraints, or flaps which open and close on a pivot or hinge. Although mechanical failures may occur with such devices, which are usually catastrophic and fatal [27], the main problem arises from the thrombogenicity of the materials used. This means that there is a tendency for the blood to clot on the surface of the valve and, as a consequence, all patients fitted with them have to regularly take anticoagulant drugs.

Because of this, there have been attempts to produce valves that are more compatible with blood and which are able to open and close in a more physiological mode [28]. The development of the so-called bioprosthetic valve that is made of natural tissue, derived usually from the muscle wall of cows hearts or the valves of pigs has ensured. These valves do not require anticoagulants but are susceptible to slow deterioration and calcification, necessitating periodic replacement [29].

### 2.4.2. Vascular Prostheses

Atherosclerosis is a condition in which deposits building up along the inner surface of blood vessels, especially the arteries [30]. The resulting stenosis interrupts blood supply and places the heart under considerable strain. Most arteries can be affected. When the coronary arteries are involved, the blood supply to the heart itself is interrupted, leading to a heart attack. If an artery in the lower leg is affected, for example, the femoropopliteal artery, the result is restricted circulation to the feet and, eventually, gangrene. The deposit that forms involves a collection of cells and proteins which can sometimes be broken up using drugs. More usually, by the time the condition is diagnosed, it is not treatable in this way. It may, however, be amenable to endoscopic techniques, which involve the insertion of a device into the vessel, its guidance to the site of the blockage, and the removal of the deposit, either mechanically, or by laser ablation [31]. If these angioplasty techniques fail, or if the problem recurs, then reconstruction may be the only option. Ideally the vascular surgeon will choose to use a vein or artery graft, i.e., transposing a vessel such as the saphenous vein or internal mammary artery to the critical, damaged site [32]. This works well under most circumstances, and indeed is currently the only option for the replacement of vessels < 4 mm diameter, but it is not always possible to obtain a suitable source and very difficult to find large segments. Under these circumstances, artificial arteries, often incorrectly termed vascular grafts, may be employed [33]. Their design is crucial, as is the material and method of construction chosen for the tubular structure. Without exception, successful artificial arteries employ microporous polymers, either woven or nonwoven, such as expanded PTFE [34] or polyester textiles [35]. Their mechanical characteristics, such as compliance and kink resistance, and their compatibility with blood components are critical.

### 2.4.3. Extracorporeal Circuits

Although external life support systems are not included here within the definition of surgical materials, it is important to recognize that extracorporeal circulation is extremely common. The best examples are the hemodialysis machines, oxygenators, and liver perfusion equipment. In all these cases, blood is removed from the patient and treated by circulation through tubing, over membranes, or through columns (for example, to remove toxic metabolic products). All the concerns about blood flow and

compatibility between blood and materials discussed in relation to implantable devices are also relevant here [36].

### 2.4.4. Cardiac Pacemakers

Conditions other than the mechanics of blood flow may affect the ability of the heart to function correctly, particularly if there is discontinuity in the conduction of stimuli from the nervous system. If there is such a disturbance, either permanant or transient, then an implantable cardiac pacemaker may be used to supply and deliver signals to the heart muscle to control the heartbeat. Pacemakers consist of a package of microelectronics, usually hermetically sealed inside a welded titanium casing, connected to the wall of the heart by an insulated lead, terminating in an electrode [37].

## 2.5. Ophthalmology

The structure of the eye is conceptually very simple, consisting of a collection of tissues and fluids that direct and focus incident light onto the retina, where the light energy is converted into electrical signals that are passed to the brain via the optic nerve. Blindness or partial loss of sight can arise from disruption to the light pathway, an inability of the retina to capture the light with appropriate fidelity, or a failure within the nervous system. Problems of the latter nature can rarely be remedied. However, there are procedures that can be used to reconstitute a light pathway that has been disturbed through disease.

### 2.5.1. Corneal Replacement

A common condition affecting the cornea is glaucoma, where there are inflammatory and structural changes in the tissues which prevent light passing through undisturbed. Glaucoma can be treated by drugs but also by using grafts. Corneal grafting, using corneas derived from cadavers is one of the easiest types of tissue grafting. The cornea, an avascular structure, is far more easily tolerated by the body's immune system. On the rare occasion where corneal grafts are not possible, artificial corneas, the so-called keratoplasty prostheses, typically made of a clear plastic such as poly(methyl methacrylate) (PMMA), are available [38].

### 2.5.2. Intraocular Lenses

Cataracts, areas of the lens which become cloudy, are extremely common. Treatment involves excision of the lens, which leaves the patient with a severe visual defect. This can be compensated by spectacles, but not efficiently, and the treatment of choice is the intraocular lens [39]. This is a poly(methyl methacrylate) lens, usually discoid in shape, which has attachments around the periphery, made from either the same material or an alternative polymer and known as haptics. The plastic lens is placed within the eye at approximately the same location as the extracted lens. Although at one stage it was difficult to achieve a high level of success with these lenses because of the damage caused to the surrounding tissue and the resulting inflammation, improvements to the surface quality and to the implantation technique have resulted in considerable success for this quite simple structure [40].

### 2.5.3. Contact Lenses

Although not strictly implantable materials, contact lenses are usually worn as an alternative to spectacles for correction of vision but may also be worn for medical reasons. Contact lenses must have good optical properties and excellent compatibility with the cornea [41]. The cornea is an avascular tissue which derives its oxygen by diffusion through its external surface. This should be compromised by the placement of a plastic structure ovei the surface. For this reason it is better that the contact lens has high oxygen permeability. It should also produce minimal mechanical irritation to the sensitive corneal surface. Several materials are used for these lenses; hard lenses are usually made of PMMA, while soft lenses may be made of silicone elastomer or a hydrogel such as poly(hydroxyethyl methacrylate) [42].

### 2.5.4. Vitreoretinal Surgery

A far less common cause of visual defect involves damage to the retina or associated tissues at the back of the eye, mainly detached retinas. A variety of materials, mainly silicones, may be used in attempts to physically relocate and reattach the retina [43].

## 2.6. The Nervous System

Intuitively, the brain and nerves appear to be tissues far too sensitive to accommodate invasion by surgical materials. However, there are a variety of circumstances in which foreign materials have been placed within the central nervous system without apparent effect. The difficulty is actually achieving anything profitable as far as performance of the system is concerned. One simple but very important application within the brain is the insertion of a catheter into the lateral ventricles to allow the drainage of cerebrospinal fluid in infants with hydrocephalus, a potentially fatal condition associated with spina bifida [44]. The tolerance of brain tissue to biomaterials is also seen from the use of a variety of substances, including in situ curing acrylics, to seal cranial defects.

### 2.6.1. Nerve Repair and Regeneration

Nerve tissue does have the capacity for repair, but it is limited in extent and efficiency. Repair of peripheral nerve damage can be assisted by nerve guides, small tubular structures placed around the severed ends of a nerve, which allow the nerve tissue to regenerate and reconstitute the nerve pathway. Ideally the materials used should actively encourage this growth although the simple polymers currently employed probably achieve their goal passively [45].

### 2.6.2. Nerve Stimulators in Pain Relief

Chronic, severe pain is extremely difficult to treat. It is known that electrical impulses can be applied to nerves to block the impulses transmitted to the brain, that give rise to pain. The use of implantable pain relief stimulators is relatively new [46]. The same principles of construction described for pacemakers apply.

### 2.6.3. Treatment of Hearing Loss

Sound is conducted to the brain via the ear. The middle ear, involving the three small ossicular bones (stapes, incus, and malleus), conducts sound waves to the tympanic membrane. The disease of otosclerosis causes inflammation and eventual fusion of these bones such that their ability to vibrate is compromised. These bones may be excised and replaced by quite simple structures which can themselves vibrate within the ear to conduct sound [47]. If the inner ear is affected by disease, treatment is far more difficult; indeed restoration of hearing to the profoundly deaf is virtually impossible in the vast majority of cases. Nevertheless, some progress has been made in the conversion of sound waves into electrical signals and delivering these directly to the cochlea [48].

## 2.7. General Surgery

### 2.7.1. Wound Closure and Artificial Skin

Although there is a very strong trend toward minimally invasive surgery, in which incisions are very small, there is still a considerable need for materials and techniques for wound closure. Traditionally, tissues have been closed with sutures. These are monofilament or multifilament threads, used to sew tissues together. They may be permanent, as with polypropylene sutures, or intentionally degradable and resorbable, either natural (catgut) or synthetic (some polyesters). Materials such as silk are used where long-term retention is required but where degradation eventually takes place [49].

Clips, usually metallic, are often used instead of sutures, largely because of greater convenience. In addition, there is much interest in adhesives. The specifications for surgical adhesives are, however, very strict especially in relation to the wet environment and the need to avoid toxic effects associated with a setting adhesive. There is still much progress to be made [50].

Artificial skin is required in situations where there is considerable skin loss, usually resulting from burns. Skin grafting is preferred, but where there is substantial loss of tissue and insufficient donor tissue, a synthetic alternative is desirable. Due to the complexity of skin structure and function this is a difficult task but some progress has been made with the use of a degradable polymer scaffold, infiltrated with epithelial cells from the patient prior to application [51]. This type of structure allows skin to regenerate within the matrix, which then degrades.

### 2.7.2. Soft Tissue Reconstruction

Many surgical procedures used for correction or reconstruction are simple and only involve the placing of a piece of soft plastic material into the body to alter shape or contour. Since these operations are largely cosmetic, they tend to concern the face and female breast. In theory they should be the simplest of procedures and provide the best results. This is not always the case, however, and significant problems have arisen over the use of some of these materials, especially the silicones and polyurethane-coated silicones used for breast prostheses [52]. These problems have considerably reduced public demand for such prostheses.

### 2.7.3. Surgery of the Urinary Tract

The urinary tract, including bladder, urethra, ureter, and associated tissues, suffers a number of diseases, primarily involving incontinence and infection, and a very large number of patients are affected. However, treatment of many of these conditions remains problematic and the use of materials to assist in reconstruction or restoration of function is not well established. It is extremely difficult to use biomaterials within the urinary system, because of the need to regulate urodynamics, to avoid infection, and to resist encrustation. There is a considerable need for prosthetic replacements for these tissues and also for devices to effectively control incontinence [53].

## 2.8. Artificial Organs

Clearly some organs can be transplanted, and although there remain very many logistic and ethical issues, this must be considered the preferable mode of replacement. In many circumstances, however, this is not possible and an artificial device has to be considered. A number of years ago, an implantable artificial heart was used in a few patients but this was not particularly successful. It is unlikely that any of the major organs can be replaced by a synthetic substitute, simply because of their complex and multiple functions. Progress is being made in a limited number of situations in which synthetic materials can be used in conjunction with cells or tissues, derived either from the patient or a donor, producing a hybrid artificial organ. This is undoubtedly an area of high promise for the future.

# 3. Criteria for Materials Selection [54]

A wide variety of functions are required from surgical materials and no single material or group of materials can satisfy all the conditions. It is necessary, therefore, to consider, generically, the requirements of biomaterials and to identify the key elements of materials selection. These can be divided into properties that determine suitability of the device to function, those which determine practical aspects of manufacture and supply, and, most importantly, the features of biocompatibility.

## 3.1. Mechanical and Physical Properties

Although surgical materials are intended to replace, both in structure and function, the diseased or otherwise damaged tissues, the performance required of surgical materials is generally quite limited and largely restricted to space filling or mechanical performance, either structural or fluidic. There are some circumstances where physical properties are critical, and some situations where biological performance is required are emerging.

### 3.1.1. Mechanical Performance

Since the tissues to be replaced vary from the extremely hard enamel to the very soft lens, and from the rigid cortical bone to the very extensible arterial wall, a generalized mechanical performance for surgical materials cannot be specified. However, there are a number of important general points.

*Strength* is always important and it is crucial that stress systems within the material–tissue complex are established and performance specifications, with respect to both yield and fracture strength, determined. Finite element analyses are increasingly being used for this purpose, and the stresses generated within the human body are often surprisingly high. Because the body is a dynamic medium, both viscoelastic and fatigue behavior are crucial. Nevertheless, these mechanical considerations per se should not be serious limiting factors, since, with a few exceptions, there are sufficient materials available, off-the-shelf, to satisfy these needs.

With respect to *elasticity*, the most important aspect is the matching of the elastic constants and deformation behavior of devices to the requirements of the body. Ideally the rigidity, flexibility, compliance, or other characteristics of elastic deformation should enable the device to participate in the transfer of load and movement exactly as if it were the natural component. Failure to do so can have major consequences, both in relation to device function and to the overall effect on the body.

Because *mechanical performance* must be achieved within the chemically hostile environment of the tissue fluids, mechanical–environmental failure modes, such as stress corrosion cracking, corrosion fatigue, stress crazing, stress accelerated macromolecular degradation, and hydrolysis induced delamination of composites, may occur.

### 3.1.2. Electrical and Optical Properties

Electrical and optical properties are the two most significant physical properties. In some situations electrical conductivity is required, as in pacemaker or stimulator electrodes while in other cases, as in their coverings, insulation is necessary. Electrodes delivering signals to tissues are often made of noble metals, particularly platinum group alloys, because of the need for inertness under conditions in which charge has to be transferred across the interface, without the simultaneous transfer of toxic metal ions. Because of inherent difficulties here, there is much interest in other conducting materials such as conducting polymers. Insulation is not normally too much of a problem although the diffusion of body fluids, themselves electrolytes, into polymers has to be taken into account when considering long-term performance.

Optical properties are reasonably straightforward, there being a small range of optically clear amorphous polymers available. Poly(methyl methacrylate) is usually preferred.

## 3.2. Manufacturing and Sterilization

### 3.2.1. Constraints on Fabrication

The very significant advances in manufacturing technology and process engineering imply that there should not be too many problems over the fabrication of medical devices. The one major factor to consider is that no two patients are alike and that ideally many of the devices should be custom-made. This becomes difficult technically since usually prior to an operation the extent of tissue resection is not known and, therefore, the shape of the prosthesis cannot be predetermined. Nowadays, a variety of shapes and sizes have to be made available and hospitals are required to hold a large inventory so that the nearest fit can be used. In the future there will be more emphasis on the use of medical imaging techniques to determine the extent of the disease and predict the shape that will be required, and of CAD/CAM techniques for customized fabrication [55].

### 3.2.2. Market Analyses [56]

One of the main constraints to the supply of both materials and devices is the size and fragmented nature of the market. Whilst it is often thought that the medical device area, like the pharmaceutical industry, commands very high prices, and that markets are very large, in reality it is not quite so simple and many clinical conditions are not sufficiently common for manufacturers to take the risks associated with small production runs. Of even greater significance is the fact that raw materials suppliers usually do not benefit from the high unit prices of medical devices, but carry responsibility for quality. Few are, therefore, willing to supply small amounts of medical-grade materials to this industry.

### 3.2.3. Legal and Regulatory Constraints

There are quite severe regulatory constraints on the use of surgical materials in medical devices. In the United States, the industry is regulated by the Food and Drug Administration (FDA) while the European industry is now coming under the control of the European Community (EC) Medical Devices Legislation. There are stringent requirements concerning biological testing and conformity to certain requirements, especially quality control.

### 3.2.4. Sterilization Techniques [57]

The implantation of a medical device into the body carries with it an inherent risk of infection. Of course any operative procedure is associated with a finite risk but the use of materials placed within tissue considerably enhances that risk. It is extremely important, therefore, that the risks are minimized by appropriate sterilization of the devices. Care has to be taken since all of the available techniques can adversely affect some of the surgical materials. For example, two processes rely on heat to kill bacteria, either by dry heat or with superheated steam (autoclaving), the temperatures involved affecting many plastics. The main alternative is gamma irradiation, whereby the free radicals generated during this process can have significant and deleterious effects on many materials. The third method involves chemical agents, principally ethylene oxide. Whilst effective, it is very difficult to remove all residues from the materials, potentially leading to toxic or irritant effects, and it is likely that this process will be prohibited in the future.

## 3.3. Biocompatibility [58]

The interactions between surgical materials and their host are the most influential factors determining their performance, especially in the long term. Biocompatibility can be defined as the "ability of a material to perform with an appropriate host response in a specific application" [1]. It must be emphasized that the range of materials in use today has been based on the assumption that biomaterials have to be inert and have no effect on the tissues. Whilst this may be appropriate under many circumstances, and while we never wish there to be adverse effects, it is now considered that the total lack of interaction implicit in that concept is not appropriate to the use of devices that are intended to become incorporated into the body, both anatomically and functionally, and that a distinctive and positive interaction is a far better goal. It is this concept that underpins the following discussion and which explains the recent developments in materials and especially their surfaces.

### 3.3.1. General Principles

Biocompatibility is concerned with all of the reactions and effects that take place between a material and the body [59]. It is possible to distinguish several groups or types of reaction in order to identify mechanisms, although all are potentially mutually interdependent. These categories of reaction are determined by the location of the effect.

Firstly, upon immediate exposure of the material to the tissue fluids, whether that be blood, serum, saliva, tears, or extracellular fluid, molecules are adsorbed from the fluid onto the surface. These are usually proteins, of which there are many different types in all body fluids and the characteristics of the adsorbed layer may significantly influence subsequent events. Of particular importance are the selectivity of the process, the changes to the conformational state of the proteins after adsorption, and the dynamics of adsorption and desorption as a multilayer is established. At this point, it is still difficult to identify which material characteristics control these phenomena.

Secondly, there are the effects of the tissues on the material, particularly in relation to corrosion and degradation mechanisms.

Thirdly, and usually of greatest significance, there are the effects of the material on the tissues adjacent to the device; these effects vary from one site to another and with time.

Fourthly, there is the possibility that the interfacial reactions and the systemic transport of reaction products away from the interface can give rise to remote effects, perhaps at the site of deposition or accumulation of these products.

### 3.3.2. Corrosion and Degradation

The environment of the human body is surprisingly hostile to synthetic materials. It is an isotonic saline solution with a variety of additional anions and cations and a range of biological macromolecules, free radicals, and cells, all of which have the potential to be biologically as well as chemically active. For metallic materials, the saline electrolyte ensures a highly corrosive environment [60], although it is known that the proteins present can also influence corrosion processes, especially if the metal concerned is able to bind to proteins [61].

Because of this aggressiveness, only the noble metals and those that are passive under physiological conditions can be used (see Section 4.1).

With ceramics, the behavior varies considerably [62]. Whereas oxide ceramics such as alumina and zirconia are extremely stable, some calcium-based ceramics are used which are degradable over a relatively short period. It is not quite clear whether this is a passive dissolution process or whether specific biological activity is implicated. It is instructive to note, however, that natural coral is currently used as a biomaterial; in the oceans, coral is stable over thousands of years but within the body, it may dissolve in months. Coral mainly consists of calcium carbonate.

Polymeric materials are the most difficult to judge from the degradability point of view. Polymers are normally degraded by physical agencies such as heat, light, and ionizing radiation (which are absent in the body) so that they should be largely protected. However, some are susceptible to degradation by hydrolysis, and active molecular species within tissues, particularly free radicals and enzymes, can participate in degradation mechanisms [63], [64]. There is, therefore, a spectrum of behavior ranging from some homochain, nonhydrolysable, hydrophobic materials such as polytetrafluoroethylene, which appear to be nondegradable, to hydrolysable and rapidly degradable polymers, always heterochain and usually with carboxyl groups such as the aliphatic polyesters [65]. In the middle are a selection of polyamides, polyurethanes, polyesters, and many others which slowly degrade, perhaps over years, within the body. This degradation can affect device performance, for example, the eventual rupture of a polyester artificial artery [66], or affect the tissue response [67].

### 3.3.3. Soft Tissue Biocompatibility

It is still uncertain exactly what device characteristics determine the host response, but some general principles can be described. It is convenient to regard these on a temporal basis and to consider the response as a perturbation to wound healing, repair, and remodeling processes that occur in response to tissue damage [68]. If the surgical implantation of the material into an area of soft tissue is thought of as an injury, the effects of the material itself can be analysed as a superimposition onto the effects of the surgery. Inflammation results whenever tissue is damaged and has to be recognized as an inherently desirable response under these circumstances since it is designed to eliminate the source of the irritation (e.g., bacteria). Inflammation involves changes to the microvascular network and the infiltration of large numbers of cells into the area. If the source of the injury can be eliminated and any debris removed, both being brought about by the activity of these inflammatory cells, then a repair process is initiated, and the inflammatory cells slowly withdraw. In soft tissue, the repair process is normally controlled by fibroblast cells which synthesize collagen, a protein that constitutes scar tissue.

A biomaterial placed at the site of a surgical incision may interfere with either or both inflammatory or repair processes. With a totally chemically inert material, however, these opportunities are minimal and it is likely that inflammation and repair will proceed as normal, the only influence of the material being as a physical barrier to regeneration. Classically, the implantation of a sample of an inert biomaterial into soft tissue results in the formation of a thin layer of fibrous tissue, structurally analogous to scar tissue, that surrounds or encapsulates the implant [69].

If, however, the implant is not inert, which is usually the case, the reactions that take place will influence events. They may aggravate and extend the initial acute inflammatory response. More importantly, the continuation of interfacial reactions means that there remains a persistent stimulus to inflammation resulting in a chronic inflammatory response, merging with a delayed and continuous fibroblastic reaction. Their effects on the performance of the device and the health of the patient vary considerably, but if uncontrolled they are likely to lead to cell and tissue death. Moreover, these interfacial reactions usually initially involve the release of some component from the material into the tissue, either a corrosion or degradation product, or a leachable residue. The development of the inflammatory response, in which large numbers of cells are attracted to the area and activated to release large quantities of superoxide ions and oxidative enzymes, for example, rapidly increases the aggressiveness of the tissue, causing further and enhanced degradation of the material. The process is, therefore, autocatalytic [70].

The precise features of the host response are highly dependent on the individual circumstances. Material characteristics obviously influence events, particularly with respect to bulk and surface chemistry, rates of degradation, and the nature of the degradation products and physical features such as surface finish, size, and shape. Of equal importance are the patient variables, such as location of the device and age, sex, activity, general health, and pharmacological status.

### 3.3.4. Hard Tissue Biocompatibility

Concerning location, one type of tissue, bone, is extremely important, for the host response is a little different and the consequences of enormous importance. The critical aspect is the ability to attain and maintain good apposition and preferably adhesion to bone, especially with joint replacements [71]. If the prosthesis is loose or becomes loose, then it will fail. If the generalized reaction of tissues to materials is, after inflammation and repair, a zone of soft collagenous tissue around the device, and if this were to take place around a material placed within bone, then it would be impossible to obtain the necessary attachment. It is essential for there to be a bone–material interface without an intervening layer of soft tissue. This inherent mechanism of fibrous encapsulation, therefore, has to be avoided or suppressed. This has been largely attempted through the use of materials of maximal inertness, such as titanium, alumina, poly(methyl methacrylate), or polyethylene. Of considerable current interest is the approach involving materials that are positively attractive to bone, the so-called bioactive materials, such as calcium hydroxyapatite [72], which permit preferential formation of bone rather than soft tissue at the interface.

The acquisition of a direct interface with bone is important, but so is its retention. Two factors may contrive to destroy such an interface. Firstly, as bone is a dynamic tissue it is dependent on the appropriate level of mechanical stress for it to maintain its normal structure and architecture. Without stress, bone will resorb. With the wrong type of stress, which results in relative movement between bone and prosthesis, resorption will occur. It is vital to ensure the optimal stress distribution in the bone–prosthesis system in order to maintain the bone; failure to do so, arising from poor materials selection, design, or technique, will result in the prosthesis becoming loose [73]. Secondly, if restimulation of the inflammatory response occurs, this could cause resorption through cellular activity. The main causes here are wear particles generated at the bearing interface in total joints [74]. Depending on their rate of release and the particle size, a significant chronic inflammation can arise, some of the cells involved also having the capability of destroying bone. This is probably the ultimate limitation to the performance of orthopedic prostheses.

### 3.3.5. Blood Compatibility

Just as the interactions between materials and bone control the performance of orthopedic prostheses, so the interactions between materials and blood control the performance of all intravascular and extracorporeal circulatory devices. These are extremely complex since they involve so many different components and mechanisms (see Section 2.4). The principle points are as follows [75].

The initial event when blood flows over a surface is that some plasma proteins are adsorbed onto that surface. The nature of the resulting *protein layer* varies from one material to another and with time. A thin layer of protein on a surface is not inherently undesirable in most circumstances; with the exception of intravascular sensors where the layer may disturb analyte transport. However, it is this protein layer which determines whether blood will clot on the surface, primarily through its effect on clotting cascade proteins and platelets [76]. The former process involves a sequence of enzyme-catalyzed reactions in which various circulating proteins are sequentially transformed into fragments, resulting in conversion of prothrombin into thrombin, which catalyzes the conversion of fibrinogen to fibrin and its subsequent polymerization. For reasons which are not entirely clear, some surfaces are far more likely to activate this sequence of events than others [77].

The second process is *platelet activation* [78]. These cells do not normally interact with the surface of blood vessels; they only do so when the vessel is damaged and then the interaction is very valuable in hemostasis. Unfortunately, the conditions which arise when a foreign surface is exposed to blood are similar to those of a damaged vessel surface, and the platelets react in the same way. A few platelets are attracted to the surface and activated. A series of chemicals are released from these activated platelets attracting vast numbers of other platelets which form an aggregate. The combination of polymerized fibrin and platelets constitutes a blood clot or thrombus, which is both damaging to device performance and compromises the health of the patient.

The selection of blood-contacting biomaterials is determined by the need to minimize this tendency. There are several possible approaches. First, extreme inertness may yield relatively nonthrombogenic surfaces, such as PTFE, although this is no guarantee. The second approach is to use materials which inhibit or prevent these processes from taking place. This usually involves treating a surface with a substance which interferes with the clotting process (e.g., heparin [79]), or with the platelet adhesion process (e.g., prostacyclin [80]). Thirdly, the surfaces can be prepared such that they mimic the natural surface, as with the attachment of phospholipids to polymer surfaces [81], or are partially made of natural components such as surfaces seeded with endothelial cells [82].

Currently, there is no universally blood-compatible material or surface but these approaches provide a promising future.

### 3.3.6. Systemic Biocompatibility [83]

If the products of an interfacial reaction can be transported away from the site, either by passive diffusion through the tissues or actively within lymphatic or vascular systems, there is the possibility that they could promote effects at some point distant from the site of application of the device. There is little evidence of this, but by the very nature of the phenomenon, evidence would be hard to obtain. The greatest fear here is that the products could accumulate in a sensitive area and cause serious effects where there was no obvious causal link to the device; the risk of biomaterial-induced cancer is at the forefront of these concerns [84].

### 3.3.7. Bioactivity

Inertness and lack of biological recognition is not always desirable, and a more active role for the material may be preferable. There are at least two possible mechanisms by which these so-called bioactive materials could operate, either through the inhibition of undesirable events or through the promotion of desirable events. Examples have already been given for bioactivity in the cases where bone adaptation has to be encouraged and where blood interactions have to be controlled. Further examples can be found in Chapter 4.

## 4. Materials

An encyclopedic review of biomaterials is given in [85]

## 4.1. Metals and Alloys

Metals are primarily employed in situations requiring their unique combination of mechanical properties, particularly fatigue strength, toughness, and ductility. Also when electrical conductivity is needed, metals are currently the materials of choice.

Because the mechanical specifications can be met by many alloy systems, considerations of biocompatibility, and especially corrosion resistance and toxicity, are decisive. It is possible to use the corrosion resistance of noble metals, and indeed, platinum group metals are used in electrical applications, gold alloys in dentistry, and silver in a variety of circumstances due to its antibacterial activity. However, these noble metals do not generally have sufficient strength, and so engineering alloys are preferred. These have to be of exceptional corrosion resistance and it is only passive alloys that can be considered. These are based on the passivity of chromium or titanium.

**Steels.** Stainless steels rely on chromium for their passivity and provide good corrosion resistance under many conditions. However, the physiological environment is such that their corrosion resistance is marginal, and only a small group of austenitic stainless steels are used, particularly AISI 316L [86]–[88]. These alloys possess a good combination of properties, but their popularity is declining in view of the enhanced performance of the alternatives discussed below.

**Cobalt-Based Alloys.** For over sixty years, alloys of the Stellite variety have been used in dentistry and surgery. These are alloys primarily of cobalt and chromium, in solid solution, with a variety of other elements to confer specific properties. Initially these alloys, with additions of molybdenum, nickel, and carbon gave a good balance of properties but the very restricted ductility limited the alloys to situations in which they could be cast. Variations in alloying additions then permitted conventional forging operations to be used and now a wide variety of formulations and production technologies provide some extremely good properties [89]. The corrosion resistance is better than that of stainless steels and the excellent hardness makes these alloys the materials of choice for one of the bearing surfaces of many joint replacements. The main drawback lies with the putative toxicity of cobalt and chromium, although the significance is questionable.

**Titanium and its Alloys.** Titanium is the most corrosion-resistant non-noble metal in physiological fluids and is widely used in medical applications. It does not have sufficiently good mechanical properties for it to be used as the pure metal in most circumstances and so, for structural applications, alloys are employed. For many years, the $\alpha + \beta$ alloy Ti–6% Al–4% V has dominated the scene, but more recently, a few others have been introduced. The rationale for this has been a concern over adverse effects of aluminum and vanadium, but again these are speculations rather than objective decisions and there is no evidence of such adverse effects. Indeed, for many applications, the Ti–Al–V alloy possesses the best combination of mechanical, corrosion, and biological properties [90], [91]. Two points about the orthopedic uses of titanium should be noted. First, it has an elastic modulus, ca. one half that of either steel or cobalt alloys and is, therefore, more closely matched to bone with respect to rigidity. Secondly, its wear resistance is considerably inferior to these alternatives and its use as the bearing surface in joints is not recommended.

**Platinum Group Metals.** The inertness of platinum coupled with its excellent electrical properties suggests that it should be very appropriate for electrodes. Because implanted electrodes are subjected to forces within the body, mechanical performance is important, and certain alloys, such as Pt–Ir or Pt–Rh, may be superior [92].

**Other Metals.** There are few opportunities for other metals in the body. As noted above, silver is of interest because of its antibacterial activity [93]. One other possibility is the use of shape-memory alloys for intravascular strents, orthodontic wires, and orthopedic devices, where Nitinol, a nickel–titanium alloy is the most obvious candidate [94].

## 4.2. Ceramics and Glasses

Until a decade ago, the inherent brittleness of ceramics precluded their use in critical structural applications within the body. However, advances in ceramic processing technology have enabled significant improvements to fracture strengths and toughness to be obtained and now ceramics, and their amorphous counterparts the glasses and glass ceramics, are frequently used in these applications. There are three qualities of ceramics with underpin these uses, the inertness of certain oxide ceramics, their hardness, and within a small group, a degree of bioactivity.

**Oxide Ceramics.** Some of the simplest ceramics, those combining one metal with one nonmetal, provide the best properties. In particular, when the nonmetal is oxygen, a small series of extremely inert and hard ceramics may be distinguished. Of these, alumina (aluminum oxide), is the most widely used in surgery [95], [96]. Although there have been a few reports of property changes to alumina on implantation, this is not a significant problem, and the long-term biocompatibility is exceptionally good. The hardness is responsible for the selection of alumina as the bearing surface in many hip replacements. While the strength is reasonably high, its toughness is still only marginally acceptable, and there have been problems of brittle fracture in joints where alumina has been used for both femoral and acetabular components. There has, therefore, been a move toward tougher ceramics, particularly transformation-toughened, partially stabilized zirconia [97].

**Calcium-Based Ceramics.** Calcium compounds are not renowned for their mechanical properties nor for their hardness, but the chemical and structural similarity between some calcium compounds and the mineral phase of bones and teeth is attractive for certain medical purposes. This phase is essentially calcium hydroxyapatite, a hydrated calcium phosphate.

*Calcium Phosphates.* Synthetic calcium phosphates can be placed in contact with bone and stimulate bone regeneration [98]. These cannot be used in a structural capacity, but there are several forms which are of interest. Of the *chemical variations*, the two main forms are calcium hydroxyapatite and tricalcium phosphate. The former is essentially stable, while the latter is rapidly degradable. There are also *structural variations*. The materials can be used in particulate form, whereby a quantity may be placed within defects in bone, in a similar way to that employed for bone grafts. New bone forms around the particles, which effectively act as a scaffold. With the nondegradable calcium hydroxyapatite, the result is a composite structure of biomaterial and new bone; with the degradable tricalcium phosphate, new bone may continue to form as the particles degrade, resulting in total bone restoration. Bone defects in the facial skeleton and the spine may be treated in this way [99], [100].

**Calcium Carbonates.** Some forms of calcium carbonate are also able to demonstrate bioactivity with respect to bone. In particular, coral, a natural form of calcium carbonate, is very effective in stimulating bone ingrowth and regeneration; the coral degrades over months and is wholly replaced by bone [101].

An alternative use is as a coating on a tougher substrate. In particular, hydroxyapatite is frequently applied by plasma spraying to metallic prostheses to encourage adaptation of bone [102].

**Glass Ceramics.** While the oxide ceramics are completely stable and materials such as coral are totally degradable, some ceramics and glass ceramics display intermediate behavior with a degree of surface degradation. These are often referred to as bioglasses. These are complex structures, based on oxides of calcium, silicon, sodium, and phosphorus. Interaction with body fluids causes dissolution of calcium and phosphorus, which appears to permit formation of new bone at the surface, but the reaction is limited by the establishment of a silica-rich layer. These controlled surface-reaction glass ceramics are, therefore, bone-bonding [103] but their inherent brittleness and lack of toughness severely limits their applications.

## 4.3. Polymers

Because metals and ceramics are generally more rigid and hard than tissues, even bone, their use for tissue replacement or augmentation is counterintuitive. Polymers are preferred for these purposes because of their greater similarity to the natural structures. Furthermore, the wide variety of features, ranging from chemical reactivity to mechanical characteristics, that can be developed within the family of polymers, allows far better tailoring of these materials to meet precise specifications.

Polymers have the inherent capacity to display good biocompatibility since the macromolecules themselves are usually noncytotoxic. However, these macromolecules are very rarely used as the sole component of a material, most of which contains a variety of other substances, some desirable additives, other undesirable residues or contaminants, which tend to control biocompatibility. Plasticizers, antioxidants, residual monomers, and catalyst residues, for example, determine the properties. Polymers can be produced in a variety of forms ranging from elastomeric to rigid, cross-linked solids, from viscous resins to high molecular mass thermoplastics, by varying the molecular structure within the same chemical family.

Some of the major groups of biomedical polymers are listed below. These and other polymer families are described more fully elsewhere [104].

**Polyethylene** (→ Polyolefins) is, chemically, the simplest of all polymers. As a homochain polymer it is essentially stable and is suitable for long-term implantation in many situations. Moreover, it is amenable to a variety of production technologies, is relatively inexpensive, and has good mechanical properties, resulting in versatility with applications ranging from catheters to joint replacement. The latter is probably the most important and significant. After unsuccessful attempts with polymers that did not have adequate wear resistance, for example, PTFE, an ultrahigh molecular mass, high-density polyethylene became the material of choice for the acetabular component of hip replacements [105]. Although it does perform adequately in many prostheses, wear does occur, and there is evidence now that if the wear rate exceeds a certain limit, the reaction to the wear debris will be destructive to the bone and will predispose it to failure [106].

**Polypropylene** (→ Polyolefins) is closely related to polyethylene, and has a similar versatility. Its medical uses range from high-strength, nondegradable sutures to the flexible integral hinges used in finger joint replacements.

**Fluoropolymers** (→ Fluoropolymers, Organic). The extreme inertness of some fluoropolymers is responsible for their widespread use, for example, polytetrafluoroethylene (PTFE). Unfortunately, the relatively poor mechanical properties limit their practical applications to soft-tissue reconstruction. There is no record of

degradation of PTFE in the body, and tissue response to solid devices made of this material is minimal. Microporous expanded PTFE (Gore-Tex) is successfully used in small diameter artificial arteries [107], periodontal surgery, and other such applications.

**Poly(methyl methacrylate)** (PMMA) (→ Polymethacrylates) is a hard, brittle polymer that would appear to be unsuitable for most clinical applications. However, as well as being very resistant to degradation and well tolerated by tissues, it does possess two distinct advantages, which underpin many of its uses. Firstly, being an amorphous polymer it is transparent and is, therefore, the material of choice for most ophthalmological devices, especially intraocular lenses and hard contact lenses [108].

Secondly, it can be produced and fabricated under ambient conditions. It can be processed in dental laboratories which is crucial to its use in custom-made dental devices such as dentures and it may also be prepared intraoperatively, explaining its unique position as a bone cement which is made by mixing prepolymerized PMMA with monomeric methylmethacrylate, forming a dough [109]. This can be inserted into bone as a grout to hold a prosthesis firm within the bone. Although there are a number of deficiencies in this situation, including the release of heat associated with the polymerization process, the toxicity of the volatile methylmethacylate, and the poor fracture toughness, no superior bone cement has been developed to date.

PMMA is the best known example of an acrylic polymer used in surgery but there are several derivatives of polyacids such as poly(acrylic acid) that are also employed. In particular, poly(hydroxyethyl methacrylate) is a cross-linked hydrogel that was developed specifically for use in contact lenses with the high oxygen permeability of its water-containing structure [110].

**Polyesters** (→ Polyesters) have a number of features which make them very attractive for medical use. The chemical structure of polyesters is intrinsically hydrolyzable so whether the polymer based on a specific polyester molecule is degraded or not will depend on the accessibility of the carboxyl groups to water.

Aliphatic polyesters based on naturally occurring substances such as lactic acid are *biodegradable*, with a combination of properties that render them highly suitable for transient functions in the body such as sutures or drug delivery systems. Since many polyesters are ideal for fabrication in the form of fibers, these materials, such as polyglycolic acid, polylactic acid, and polydioxanone, can be used as absorbable sutures. The precise molecular structure determines their rate of degradation and since the degradation products are natural and metabolizable, this process is generally well accepted [111].

The fiber-forming character of polyesters enables them to be prepared as *textiles*. Aromatic polyesters such as poly(ethylene terephthalate) (PET), are extensively used as the woven or nonwoven structures required in artificial arteries and ligament regeneration [112].

**Polyurethanes** ($\rightarrow$ Polyurethanes) have had a very variable history of use in medical applications, partly arising from the wide range of materials that can be prepared with the urethane group. Bearing in mind the fact that the precursors of polyurethanes, such as toluene diisocyanate, have well-known toxicities, and the fact that the structures generally are susceptible to degradation, it is not surprising that these do display varying biocompatibility characteristics. Certain polyurethanes, however, have excellent mechanical properties, especially elastomeric properties, which make them very suitable for areas of soft-tissue reconstruction. Polyurethanes also often have excellent blood compatibility. For these reasons, polyurethanes, especially polyether urethanes, have been used in critical blood-contacting applications, such as in the experimental implantable artificial heart [113]. They are also very suitable for catheters and similar devices.

However, their inherent susceptibility to degradation has resulted in a number of difficulties, for example, stress cracking of pacemaker lead encapsulation [114] and the degradation of porous coverings on breast implants [115]. The ideal characteristics, therefore, of elastic properties and blood compatibility needed for small diameter artificial arteries have yet to fully realized in devices of suitable long-term stability.

**Silicones** ($\rightarrow$ Silicones, $\rightarrow$ Inorganic Polymers, A 14, pp. 251 – 252). Although the vast majority of synthetic polymers are based on the carbon – carbon bond, a small number are based on silicon. Linear chains of alternating silicon and oxygen atoms yield the family of siloxanes. Depending on the nature of the side groups attached to the silicon, the molecular size, and the degree of cross-linking, a range of substances may be produced, from oils to elastomers. It has long been thought that the inherent strength of the Si – O bond, coupled with the fact that this bond is not normally found in biological structures and should therefore be immune to enzyme-catalyzed degradation, should render it very stable. Thus, over several decades, substances such as polydimethylsiloxane have enjoyed widespread medical and dental use, ranging from elastomeric dental impressions to finger joint prostheses, catheters, intraocular oils, contact lenses, and a variety of soft tissue augmentation prostheses [116], [117].

However, the biocompatibility of silicones has been questioned in recent years [118] and claims have been made that silicones are capable of causing a wide variety of conditions, primarily autoimmune diseases. There is no clear evidence, but this matter is having a profound influence on the use of such materials in hitherto very common applications such as breast implants [119].

**Natural Biopolymers.** There is a trend toward the use of polymers derived from natural components where in some situations, the polymers themselves may be analogous to those occurring naturally.

Some tissues consist almost entirely of *collagen* and these may be used as biomaterials; one example is the tendon derived from the tails of certain animal species, for example, the kangaroo. Fortunately collagen is only a weakly antigenic protein and is unlikely to invoke an immune response when used in this way. It may be preferable, however, to prepare alternative forms of collagen, for example, by solubilization and reprecipitation. Some collagen products have quite widespread use in plastic surgery, although their stability is still questionable [120].

*Hyaluronic acid* is a polysaccharide that is widely distributed in tissues, especially in synovial fluid. It can be polymerized and modified in many ways resulting in a variety of degradable polymers with a number of potential applications [121].

## 4.4. Carbon

Carbon, as the major elemental constituent of organic molecules, would appear to be intrinsically safe to use within the body. On the basis of this putative inertness and biocompatibility, attempts have been made to use carbon, in several of its forms, as a biomaterial. Its major deficiency is that it is not a structural material and is difficult to fabricate into useful artefacts. As a bulk material it has very limited use; this can be seen by the attempts to use it in one of its few monolithic forms, as glassy carbon, in dental and other applications, the inevitable brittleness resulting in mechanical failures [122].

The interest in carbon is still intense but the forms in which it is used try to obviate these structural deficiencies. For example, it is used very successfully as a thin coating, in the form of pyrolytic graphite, deposited onto tough metallic substrates, for blood-contacting surfaces in heart valves [123]. An alternative carbon surface is the so-called diamond-like coating [124].

An alternative approach has involved the use of carbon fibers, well known for their intrinsic high strength and modulus of elasticity. Some applications have utilized tows or bundles of the fibers themselves, as with their use in artificial ligaments, but the majority are concerned with carbon composites, either carbon–carbon [125], carbon-fiber-reinforced thermosetting resins [126], or carbon-fiber-reinforced thermoplastics such as polyetheretherketone [127]. However, whilst chemically stable, the fibers themselves, by virtue of their tendency to fracture and produce small fragments of cellular dimensions, can be immensely irritating to tissues.

## 4.5. Composites

Attempts have been made to use composites in surgical materials, although up until now with limited success, since problems of biocompatibility have been aggravated through the increasing complexity of the structure.

There are several principles underlying the design of composites, usually mechanical in origin and commonly based upon maximizing fracture strength and optimizing elasticity. Strength is not usually markedly improved by the use of dispersed fibers or particles in a ductile plastic matrix, but it may be possible to achieve modest improvements. There is considerable scope for tailoring the elastic modulus by accurate control of components and their volume fractions. Although restorative dentistry has seen the development of a very successful range of particulate composites of quartz-like phases dispersed in

dimethacrylate resin matrices [128], efforts with composite surgical materials have been far less successful. Apart from the carbon-fiber composites mentioned above, there have been unsuccessful attempts to strengthen bone cement by incorporating short fibers, and to improve the wear and creep resistance of polyethylene [129]. Especially disappointing has been the development of a carbon- or alumina-filled porous PTFE, intended as a prosthetic material to be replaced and incorporated into a variety of tissues [130].

## 4.6. Surface Modifications and Coatings

**General Principles.** Engineering materials used as structural components, with their surfaces modified to optimize biocompatibility is generally quite common. However, the varying requirements for biocompatibility and the need for long-term performance hinder their practical application as witnessed by the diverse and imaginative approaches already attempted.

**Hard Coatings.** Improving the hardness of a surface is thought to have a beneficial effect, not only when that surface is involved in a joint, but also when the release of any component through abrasion with tissue is possible. A variety of ion-plating, ion-assisted deposition, ion-implantation, and other similar techniques may be employed [131].

**Hydrophilic Coatings.** Improved biocompatibility is thought to result if the surface is hydrophilic rather than hydrophobic. Such surfaces may be developed by grafting hydrogels onto polymer surfaces [132]. Better performance is likely when molecular species are grafted, in particular, when substances such as poly-(ethylene glycol) are attached [133].

**Pharmacological Coatings.** In some circumstances it may be more appropriate to seek specific activity rather than passivity. One method would be to attach pharmaceuticals to surfaces, either by immobilizing them on the surface or incorporating them into the surface layer from which they may be released. This is most relevant in the case of blood-compatible surfaces, where anticoagulant, antiplatelet, fibrinolytic, or other agents may be attached to polymers (see Section 3.3.5).

**Biomimetic Surfaces.** The attachment of a layer of natural tissue-like material to engineering substrates may be difficult to imagine. One approach, however, involves biomimicry, the use of molecules that are analogous to components of tissue, which can behave as if it were that tissue. For example, phospholipids are the major molecular species of cell membranes and can be attached to a variety of substrates where, among other things, they substantially reduce the extent of protein adsorption. This is a particularly attractive property for contact lenses [134].

**Cell seeding** of biomaterial surfaces is of considerable interest; here it is often possible to render a surface attractive to a type of cell, for example, the endothelial cells that line blood vessels, and to incubate that material in a culture of cells derived previously from the eventual recipient. If these cells are able to colonize the surface, then a pseudo-natural surface is generated. This technology is still under early stages of development but it is a very attractive solution to biocompatibility problems [135].

**Hybrid Structures and Materials.** Combinations of synthetic materials and natural components, either viable or nonviable, are now in experimental use. For example, the treatment of diabetes may be radically altered by the use of a primitive artificial pancreas, in which islets, the insulin-producing cells, are obtained from donors and encapsulated in membranes that allow the cells to respond to glucose levels and produce insulin, but protect them from the destructive proteins of the immune system [136].

# 5. References


[1] D. F. Williams, J. Black, P. J. Doherty: "Second Consensus Conference on Definitions in Biomaterials," in P. J. Doherty, R. L. Williams, D. F. Williams, A. J. C. Lee (eds.): *Biomaterial-Tissue Interface*, Elsevier, Amsterdam 1992, p. 528.

[2] D. F. Williams: *Implants in Surgery*, Saunders, London 1973.

[3] D. F. Williams, *Physics in Med. Biol.* **25** (1980) 611–636.

[4] J. Black: *Orthopaedic Biomaterials in Research and Practice*, Livingstone, New York 1988.

[5] J. W. Boretos, M. Eden (eds.): *Contemporary Biomaterials*, Noyes Publications, Park Ridge, N.J. 1984.

[6] M. Szycher (ed.): *High Performance Biomaterials*, Technomic Publications, Lancaster, Pa. 1991.

[7] D. S. Hugerford, L. C. Jones, *Clin. Orthop. Rel. Res.* **235** (1988) 12–24.

[8] J. A. Rand, C. D. Dorr (eds.): *Arthroplasty of the Knee*, Aspen, Rockville 1987.

[9] P. Ducheyne, D. Kohn: "Materials for Bone and Joint Replacement," in D. F. Williams (ed.): *Medical and Dental Materials*, VCH Verlagsgesellschaft, Weinheim 1991, chap. 2.

[10] J. Charnley: *Acrylic Cement in Orthopaedic Surgery*, Williams & Wilkins, Baltimore 1970.

[11] R. M. Pilliar, *Clin. Orthop. Rel. Res.* **176** (1983) 42–51.

[12] P. Ducheyne, *J. Biomed. Mater. Res. Appl. Biomat.* **21** (A 2) (1987) 219–236.

[13] H. C. Amstutz, P. Campbell, N. Kossovsky, I. C. Clarke, *Clin. Orth. Rel. Res.* **276** (1992) 7–18.

[14] R. M. Rose, A. S. Litsky, "Biomechanical Considerations in the Loosening of Hip Replacement Prostheses," in D. F. Williams (ed.): *Current Perspectives on Implantable Devices*, vol. 1, JAI Press, London 1989, pp. 1–45.

[15] A. E. Goodship, P. Cooke, *CRC Crit. Rev. Biocompat.* **2** (1986) 303–334.

[16] H. Uhthoff: *Current Concepts of the Internal Fixation of Fractures*, Springer Verlag, Berlin 1980.

[17] G. Heimke, R. J. Kolbe, R. A. Latour, *Mater. in Med.* **3** (1992) 204–211.

[18] O. Bostman et al., *Clin. Orthop. Rel. Res.* **238** (1989) 195–203.

[19] D. F. Williams: "Dental Implants," in R. W. Cahn (ed.): *Encyclopaedia of Materials Science and Engineering*, vol. 1, Suppl., Pergamon Press, Oxford 1988, pp. 115–121.

[20] T. Albrektsson et al., *J. Prosthet. Dent.* **60** (1988) 75–84.

[21] A. Kirsch, K. Ackermann, *Dtsch. Zahnaerztl. Z.* **38** (1983) 106–112.

[22] D. F. Williams: "Materials for Oral and Maxillofacial Surgery," in D. F. Williams (ed.): *Medical and Dental Materials*, VCH Verlagsgesellschaft, Weinheim 1991, pp. 259–283.

[23] J. Breme, E. Steinhauser, G. Paulus, *Biomaterials* **9** (1988) 310–313.

[24] B. F. Conroy, J. E. Bowerman, J. M. Harrison: "Restoration of the Mandible by Implant Prostheses," in D. F. Williams (ed.): *Biocompatibility in Clinical Practice*, vol. II, CRC Press, Boca Raton 1982, pp. 169–182.

[25] E. Bodnar, R. Frater (eds.): *Replacement Cardiac Valves*, Pergamon Press, New York 1991.

[26] D. P. Giddens, A. P. Yoganathan, F. Schoen, *Cardiovasc. Path.* **2** (1993) 167–177.

[27] W. R. Dimitri, B. T. Williams, *J. Cardiovasc. Surg.* **31** (1990) 125–134.

[28] Y. A. Goffin, M. M. Black, P. Lawford: "The Stability and Performance of Bioprosthetic Heart Valves," in D. F. Williams (ed.): *Current Perspectives on Implantable Devices*, vol. 2, JAI Press, London 1990, pp. 65–120.

[29] R. J. Levy, F. J. Schoen, H. C. Anderson, *Biomaterials* **12** (1991) 707–714.

[30] P. Libby, F. Schoen, *Cardiovasc. Path.* **2** (1993) 43–52.

[31] T. Ischinger (ed.): *Practice of Coronary Angioplasty*, Springer Verlag, New York 1986.

[32] R. P. Leather, A. M. Karmody, D. M. Shah, J. D. Corson, *Acta Chir. Scand. Suppl.* **529** (1985) 47–56.

[33] J. C. Stanley (ed.): *Biologic and Synthetic Vascular Prostheses*, Grune and Stratton, New York 1982.

[34] M. A. Golden, S. R. Hanson, T. R. Krikman, P. A. Schneider, A. W. Clowes, *J. Vasc. Surg.* **11** (1990) 838–845.

[35] R. Guidoin, M. King, D. Marceau, *J. Biomed. Mater. Res.* **21** (1987) 65–87.

[36] R. VanHolder, *Clin. Mater.* **10** (1992) 87–132.

[37] S. S. Barold: *Modern Cardiac Pacing,* Futura Publ. Co. Mount Kisco, New York 1985.

[38] P. S. Binder, E. Y. Savala, J. K. Deg, *Cornea* **4** (1983) 119–125.

[39] D. F. Williams: "Materials for Ophthalmology," in D. F. Williams (ed.): *Medical and Dental Materials,* VCH Verlagsgesellschaft, Weinheim 1991, pp. 416–427.

[40] R. Caramazza, P. Versura (eds.): *Biomaterials in Ophthalmology,* Studio ER Congressi, Bologna 1990.

[41] R. W. T. Bowers, B. J. Tighe, *Biomaterials* **8** (1987) 83–88.

[42] O. Wichterle, D. Lim, *Nature* **185** (1960) 117–119.

[43] B. W. McCuen, E. de Juan, M. B. Landers, R. Machemer, *Retina* **5** (1985) 189–205.

[44] A. D. Hockley, P. Collins, J. R. Anderson: "Ventricular Catheters," in D. F. Wiliams (ed.): *Biocompatibility in Clinical Practice,* vol. II, CRC Press, Boca Raton 1982, pp. 151–168.

[45] I. V. Yannas et al., *Trans. Soc. Biomat.* **10** (1987) 6.

[46] D. Fiume, M. Palombi, V. Sciassa, M. Tamorri, *PACE* **12** (1989) 698–710.

[47] J. J. Grote: *Biomaterials on Otology,* Martinus Nijhoff, Boston 1984.

[48] R. S. Tyler, B. C. J. Moore, F. K. Kuk, *J. Speech and Hearing Res.* **32** (1989) 887.

[49] C. C. Chu, *CRC Crit. Rev. Biocompat.* **1** (1985) 261–322.

[50] B. E. Causton: "Medical and Dental Adhesives," in D. F. Williams (ed.): *Medical and Dental Materials,* VCH Verlagsgesellschaft, Weinheim 1991, pp. 285–302.

[51] I. V. Yannas: "Biologically Active Analogs of the Extracellular Matrix," in D. F. Williams (ed.): *Medical and Dental Materials,* VCH Verlagsgesellschaft, Weinheim 1991, chap. 12.

[52] W. Kaiser, G. Biesenbach, U. Stuby, P. Grafinger, J. Zazgornik, *Ann. Rheum. Dis.* **49** (1990) 937–938.

[53] P. Aebischer, M. Goddard, P. M. Galletti, M. Lysaght: "Biomaterials and Artificial Organs," in D. F. Williams (ed.): *Medical and Dental Materials,* VCH Verlagsgesellschaft, Weinheim 1991, pp. 133–178.

[54] D. F. Williams: "Biofunctionality and Biocompatibility," in D. F. Williams (ed.): *Medical and Dental Materials,* VCH Verlagsgesellschaft, Weinheim 1992, pp. 1–30.

[55] N. Jedynakiewicz, N. Martin: *CAD-CAM in Restorative Dentistry,* Liverpool University Press, Liverpool 1993.

[56] M. Szycher: "Markets for Medical Plastics," in M. Szycher (ed.): *High Performance Biomaterials,* Technomic, Lancaster, Pa. 1991, pp. 3–30.

[57] I. Matthews, *Clin. Mater.* **1994**, in press.

[58] D. F. Williams, *Interdisc. Science Rev.* **15** (1990) 20–33.

[59] D. F. Williams, *J. Biomed. Eng.* **11** (1989) 185–192.

[60] D. F. Williams: "Electrochemical Aspects of Corrosion in the Physiological Environment," in *Fundamental Aspects of Biocompatibility,*" vol. 1, CRC Press, Boca Raton 1981, pp. 11–42.

[61] D. F. Williams, *CRC Crit. Rev. Biocompat.* **1** (1985) 1–30.

[62] S. F. Hulbert et al., *J. Biomed. Mater. Res.* **4** (1970) 433–456.

[63] J. P. Santerre, R. S. Labow, G. A. Adams, *J. Biomed. Mater. Res.* **27** (1993) 97–109.

[64] D. F. Williams, S. P. Zhong, *Adv. Mater.* **3** (1991) 623–626.

[65] R. Smith, D. F. Williams, C. Oliver, *J. Biomed. Mater. Res.* **21** (1987) 1149–1166.

[66] E. Vinard et al., *J. Biomed. Mater. Res.* **25** (1991) 499–514.

[67] M. Mohanty, J. A. Hunt, P. J. Doherty, D. Annis, D. F. Williams, *Biomaterials* **13** (1992) 651–656.

[68] D. F. Williams, *J. Mat. Sci.* **22** (1987) 3421–3445.

[69] M. Spector, C. Cease, X. Tong-Li, *CRC Crit. Rev. Biocompat.* **5** (1989) 269–295.

[70] D. F. Williams, *J. Mat. Sci.* **17** (1982) 1233–1246.

[71] T. Albrektsson, *CRC Crit. Rev. Biocompat.* **1** (1984) 53–84.

[72] L. L. Hench, E. C. Ethridge: *Biomaterials; An Interfacial Approach,* Academic Press, New York 1982.

[73] L. Linder, L. Lindberg, A. Carlsson, *Clin. Orth. Rel. Res.* **175** (1983) 93–104.

[74] T. P. Gross, D. W. Lennox, *J. Bone Jt. Surg.* **74 A** (1992) 1096–1101.

[75] R. W. Coleman, *Cardiovasc. Path.* **2** (1993) suppl. 23–31.

[76] T. A. Horbett, *Cardiovasc. Path.* **2** (1993) suppl. 137–148.

[77] G. P. Clagett, "Artificial Devices in Clinical Practice," in R. W. Coleman, J. Hirsh, V. J. Marder, E. W. Salzman (eds.): *Hemostasis and Thrombosis,* Lippincott, Philadelphia 1987, pp. 1348–1366.

[78] J. M. Anderson, K. Kottke-Marchant, *CRC Crit. Rev. Biocompat.* **1** (1985) 111–204.

[79] Y. Ito, *J. Biomater. Appl.* **2** (1987) 235–265.

[80] C. H. Bamford, I. P. Middleton, K. G. Al-Lamee, *J. Biomat. Sci.* **2** (1991) 37–52.

[81] D. Chapman, S. Charles, *Chem. in Brit.* **28** (1992) 253–256.

[82] S. P. Schmidt, W. V. Sharp, M. M. Evancho, S. O. Meerbaum: "Endothelial Cell Seeding of Prosthetic Vascular Grafts," in M. Szycher (ed.): *High Performance Biomaterials,* Technomic, Lancaster, Pa. 1991, pp. 483–496.

[83] D. F. Williams, *Systemic Aspects of Biocompatibility,* vols. I and II, CRC Press, Boca Raton 1983.

[84] R. B. Pedley, G. Meachim, D. F. Williams: "Tumour Induction by Implant Materials," in D. F. Williams (ed.): *Fundamental Aspects of Biocompatibility,* vol. II, CRC Press, Boca Raton 1981, pp. 175–204.

[85] D. F. Williams (ed.): *The Concise Encyclopaedia of Medical and Dental Materials,* Pergamon, Oxford 1991.

[86] E. J. Sutow, S. R. Pollack: "Stainless Steel," in D. F. Williams (ed.): *Biocompatibility of Implant Materials,* vol. I, CRC Press, Boca Raton 1981, pp. 45–98.

[87] A. Cigada, G. Rondelli, B. Vicentini, M. Giacomazzi, A. Roos, *J. Biomed. Mater. Res.* **23** (1989) 1087–1095.

[88] R. M. Pilliar, G. C. Weatherly, *CRC Crit. Rev. Biocompat.* **1** (1986) 371–403.

[89] ASTM, "Standard Specification for Cast Co–Cr Mo alloy for Surgical Implant Applications," ASTM, Philadelphia 1984, F 75–82.

[90] H. A. Luckey, F. Kubli, "Titanium Alloys in Surgical Implants," ASTM, Philadelphia 1983, spec. tech. publ. 796.

[91] D. F. Williams: "Titanium," in D. F. Williams (ed.): *Biocompatibility of Clinical Implant Materials*, vol. I, CRC Press, Boca Raton 1982.

[92] K. R. Cote, R. C. Gill, *Ann. Biomed. Eng.* **15** (1987) 419–426.

[93] R. L. Williams, P. J. Doherty, D. G. Vince, G. J. Grashoff, D. F. Williams, *CRC Crit. Rev. Biocompat.* **5** (1989) 221–244.

[94] L. S. Castleman, S. M. Motzkin, F. P. Alicandri, V. C. Bonawit, A. A. Johnson, *J. Biomed. Mater. Res.* **10** (1976) 695–710.

[95] J. M. Dorlot, P. Christel, A. Meunier, *J. Biomed. Mater. Res. Appl. Biomat.* **23** (1989) 299–310.

[96] H. Mittelmeier, *The Hip* **12** (1984) 146–160.

[97] P. Christel, A. Meumier, M. Heller, *J. Biomed. Mater. Res.* **23** (1989) 45–62.

[98] R. LeGeros, *Clin. Mater.* **14** (1993) 65–88.

[99] N. M. Meenen, J. F. Osborn, M. Dallek, K. Donath, *Mater. in Med.* **3** (1992) 345–351.

[100] H. Alexander, J. R. Parsons, J. L. Ricci, P. K. Bajpai, A. B. Weiss, *CRC Crit. Rev. Biocompat.* **4** (1988) 43–77.

[101] G. Guillemin, J. L. Patat, J. Fournie, M. Chetail, *J. Biomed. Mater. Res.* **21** (1987) 557–567.

[102] B. Koch, J. G. C. Wolke, K. de Groot, *J. Biomed. Mater. Res.* **24** (1990) 655–668.

[103] L. L. Hench, J. Wilson, *Science* **226** (1984) 630–636.

[104] D. F. Williams (ed.): *Biocompatibility of Clinical Implant Materials*, CRC Press, Boca Raton 1982.

[105] P. Eyerer, M. Kurth, H. A. McKellup, T. Mittlmeier, *J. Biomed. Mater. Res.* **21** (1987) 275–291.

[106] R. M. Rose, E. L. Radin, *Clin. Orth. Rel. Res.* **170** (1982) 107–115.

[107] R. B. Rutherford: *Vascular Surgery*, Saunders, Philadelphia 1989.

[108] R. Larsson, G. Selen, H. Bjorklund, P. Fagerholm, *Biomaterials* **10** (1989) 511–516.

[109] R. D. Mulroy, W. H. Harris, *J. Bone Jt. Surg.* **72 B** (1990) 757–760.

[110] B. J. Tighe: "Hydrogels as Biomedical Materials," in R. W. Cahn (ed.): *Encyclopaedia of Materials Science and Engineering*, suppl. I, Pergamon Press, Oxford 1988, pp. 216–221.

[111] M. Vert, S. M. Li, G. Spenlehauer, P. Guerin, *Mater. in Med.* **3** (1992) 432–446.

[112] Y. Marois et al., *Biomaterials* **14** (1993) 255–262.

[113] R. K. Jarvik, *Sci. Amer.* **244** (1981) 74–80.

[114] K. Stokes, *Cardiovasc. Path.* **2** (1993) suppl. 111–119.

[115] H. G. Bruck, *Aesth. Plast. Surg* **14** (1990) 85–86.

[116] J. B. Crawford, *Am. J. Ophthalmol.* **101** (1986) 680–683.

[117] E. E. Frisch, "Polysiloxanes as Biomedical Materials," in R. W. Cahn (ed.): *Encyclopaedia of Materials Science and Engineering*, suppl. I, Pergamon Press, Oxford 1988, pp. 418–427.

[118] N. Kossovsky, J. P. Heggers, M. C. Robson, *CRC Crit. Rev. Biocompat.* **3** (1987) 53–85.

[119] E. E. Sahn, P. D. Garen, R. M. Silver, J. C. Maize, *Arch. Dermatol* **126** (1990) 1198–1202.

[120] E. Sableman, "Biology, Biotechnology and Biocompatibility of Collagen," in D. F. Williams (ed.): *Biocompatibility of Tissue Analogs*, vol. I, CRC Press, Boca Raton 1985, pp. 27–66.

[121] A. Rastrelli, M. Beccaro, F. Biviano, G. Calderini, A. Pastorello: "Hyaluronic Acid Esters, A New Class of Semisynthetic Biopolymers," in G. Heimke, U. Soltesz, A. J. C. Lee (eds.): *Clinical Implant Materials*, Elsevier, Amsterdam 1990, pp. 199–205.

[122] A. D. Haubold, H. S. Shim, J. C. Bokros, "Carbon in Medical Devices," in D. F. Williams (ed.): *Biocompatibility of Clinical Implant Materials*, vol. II, CRC Press, Boca Raton 1982, pp. 3–42.

[123] A. D. Haubold, R. A. Yapp, J. C. Bokros: "Carbons for Biomedical Applications," in M. B. Bever (ed.): *Encyclopaedia of Materials Science and Engineering*, vol. I, Pergamon Press, Oxford, pp. 514–520.

[124] D. F. Williams, *Med. Dev. Technol.* 1993, in press.

[125] P. S. Christel, *CRC Crit. Rev. Biocompat.* **2** (1986) 189–218.

[126] L. Tayton, C. Johnson-Nurse, B. McKibbin, *J. Bone Jt. Surg* **64 B** (1982) 105–111.

[127] D. F. Williams, A. McNamara, R. M. Turner, *J. Mat. Sci. Lett.* **6** (1987) 188–190.

[128] D. C. Watts: "Dental Restorative Materials," in D. F. Williams (ed.): *Medical and Dental Materials*, VCH Verlagsgesellschaft, Weinheim 1991, chap. 6.

[129] H. E. Groth, J. M. Shilling, *J. Orthop. Res.* **1** (1983) 129–135.

[130] B. L. Florine, D. J. Gatto, M. L. Wade, D. E. Waite, *J. Oral Maxillofac. Surg.* **46** (1988) 183–188.

[131] H. A. McKellop, T. V. Rostlund, *J. Biomed. Mater. Res.* **24** (1990) 1413–1426.

[132] B. D. Ratner: "Biomedical Applications of Hydrogels," in D. F. Williams (ed.): *Biocompatibility in Clinical Practice*, vol. II, CRC Press, Boca Raton 1982, pp. 146–175.

[133] E. W. Merrill, E. W. Salzman, *ASAIOJ* **6** (1983) 60–64.

[134] A. A. Durrani, J. A. Hayward, D. Chapman, *Biomaterials* **7** (1986) 121–125.

[135] J. A. Hubbell, S. P. Massia, N. P. Desai, P. D. Drumheller, *Biotechnology* **9** (1991) 568–572.

[136] M. F. A. Goosen, *CRC Crit. Rev. Biocompat.* **3** (1987) 1–24.

**Suspension Polymerization → Polymerization Processes**
**Sutures → Surgical Materials**

# Sweeteners

GERT-WOLFHARD VON RYMON LIPINSKI, Hoechst Aktiengesellschaft, Frankfurt/Main, Federal Republic of Germany

## 1. Introduction

The term "intense sweeteners" is commonly used for substances that are much sweeter than sucrose, and which are either not metabolized in the human body or do not significantly contribute to the energy content of foods and beverages. Intense sweeteners may be produced synthetically or isolated from plants. Sucrose, related carbohydrates, and sugar alcohols should be clearly distinguished from intense sweeteners.

Plants containing sweet compounds have been used for centuries to sweeten foods and beverages, particularly in tropical countries, e.g., the thaumatin-containing plant, *Thaumatococcus daniellii* Benth [1].

The history of intense sweeteners began with the discovery of saccharin by C. FAHLBERG and I. REMSEN [2] in 1878. Fahlberg recognized the scientific and commercial importance of saccharin and took steps to market it; production commenced in 1884. Since then, many other sweet compounds have been discovered, mostly by chance. Only a very few of these compounds have been used as sweeteners in foods and beverages. Several substances have been withdrawn, owing to the emergence of toxicological problems or unsatisfactory properties.

Saccharin enjoyed great commercial success in periods of short sugar supply, e.g., during World Wars I and II. As a consequence, saccharin was considered a cheap and inferior sugar substitute. Taxes were imposed on saccharin to protect the sugar industry. Diabetics were the main consumers.

In the last three decades, the general perception of intense sweeteners has changed. The incidence of obesity, and increasing health-consciousness have paved the way for a more favorable attitude. The increasing importance of industrial food production, the need for systematic recipe development, and the resulting possibilities for large-scale production of products containing intense sweeteners has led to widespread availability of these products. Their main applications are no longer products especially intended for consumption by diabetics or people suffering from other metabolic disorders. Intense sweeteners are now rather used in products aimed at consumers interested in body weight management and a calorie-controlled diet.

The increasing demands on safety of food additives in general, and sweeteners in particular, resulted in a reevaluation and reassessment of the intense sweeteners commonly used until the 1960s. The results provoked an international discussion of their safety, which, in turn, stimulated an intensive search for alternative compounds with improved properties. Two of the sweeteners developed, acesulfame K and aspartame, have found widespread international approval, and others, such as thaumatin and neohesperidin dihydrochalcone are approved and used in some countries. More products are under evaluation (e.g., alitame) or about to enter the market (e.g., sucralose). There continues to be promising market opportunities for sweeteners with good taste, stability, and handling characteristics, but above all, a clean safety record.

## 2. Sensory Properties

### 2.1. Structural Requirements for Sweetness

The generally accepted theory for the phenomenon of sweetness was developed by SHALLENBERGER and ACREE [3]. According to this theory, a molecular system of a proton donor and proton acceptor is necessary, with a distance of ca. 0.3 nm between the groups. It is assumed that a component has a sweet taste whenever this so-called AH – B-system can interact with a complementary system located in the membrane of taste receptor cells.

Changes in the distance between groups, as well as changes in electronic structure influence the occurrence of the sweet taste, and may change the general taste perception, sometimes eliminating sweetness totally, or changing it to bitterness.

The AH – B-system (Fig. 1) is a necessary but not sufficient structural requirement for sweetness, and is therefore not a reliable indicator for its occurrence or absence. Accordingly, the

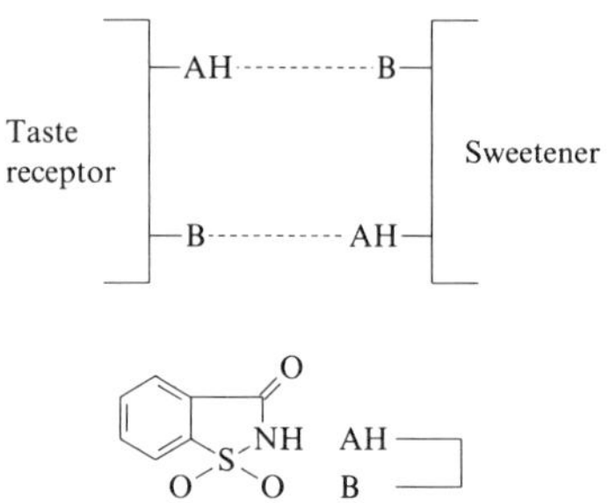


**Figure 1.** Interaction of sweeteners and taste receptors [3] through the AH – B-system of saccharin

molecular theory was extended to include a third binding site that interacts with the sweet receptor by hydrophobic contact; this model allows better correlation between molecular structure and sweetness [4].

On the basis of intensive research on very sweet molecules, the number of potential interaction sites of sweetness receptors with sweetening agents has been extended to eight, two of which must be in contact with sweet molecules to elicit sweetness, and four have to be in contact for extremely intense sweetness [5].

In recent years, more systematic approaches have sought to reveal the molecular structure of sweetness receptor molecules. These included structure–activity investigations in homologous series of sweet substances [6], [7]. Investigations on nitroanilines, sulfamates, oximes, isocumarins, dipeptides, saccharins, acesulfames, chlorosugars, tryptophans, and ureas showed that, for certain substituents, an increase in chain length results in increased sweetness, but a sharp decrease to complete absence of sweetness was observed above the optimum. Similar observations were made for amino acid amides [8].

Superposition of different sweet compounds was used to identify structural requirements for the sweetness receptor molecules [9]. Assuming a peptide-based structure, interaction with the L-aspartyl-L-phenylalanine methyl ester with a helical peptide could be demonstrated, whereas the nonsweet D–D isomer was unable to form hydrogen bonds with this peptide [10].

It is still uncertain whether there is just one sweetness receptor with different binding sites, or several types of sweetness receptor with different response characteristics. It seems probable, however, that specific binding characteristics for different sweet tasting structures have to be assumed. A sweetness receptor model with binding sites for these structures has been suggested [11].

## 2.2. Sweetness Intensity

The sweetness intensity of a substance can be expressed in several ways. The sweetness intensity factor, or sweetness power, is most frequently used. It is the factor by which an equisweet sucrose concentration is more concentrated than a solution of the sweetener. Commonly used values are often based on comparison with 0.1 mol/L sucrose solutions, or solutions of similar concentration. Sweetness intensity factors compared with sucrose are given below:

| Sweetener | Sweetness intensity |
| --- | --- |
| sucrose | 1 |
| sodium cyclamate | 35 |
| glycyrrhizin | 50 |
| stevioside | 160 |
| acesulfame K | 200 |
| aspartame | 200 |
| rebaudioside A | 250 |
| neohesperidin dihydro-<br>    chalcone | 330 |
| sodium saccharin | 450 |
| saccharin | 550 |
| alitame | 2500 |
| thaumatin | 3500 |

Sweetness intensities generally depend on the sweetener concentration and the sucrose concentration used for comparison. Therefore, values other than those given above may apply at lower or higher concentrations. A demonstration of this concentration dependence is given in Figure 2 for some commonly used sweeteners. Esti-

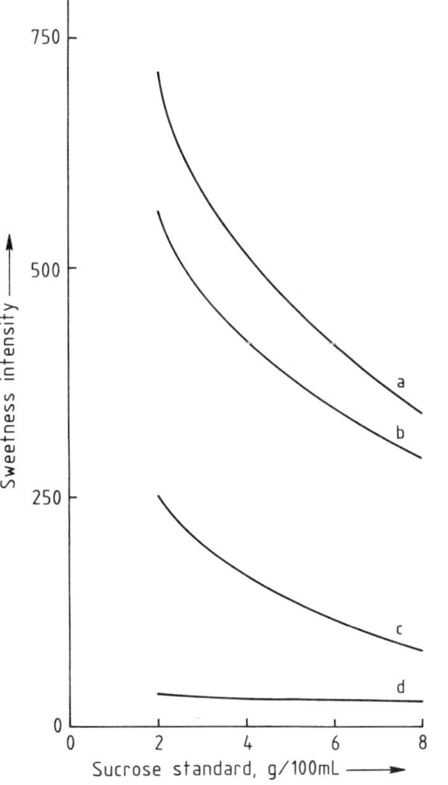


**Figure 2.** Sweetness intensity of saccharin (a), sodium saccharin (b), acesulfame K (c), and sodium cyclamate (d) compared with sucrose

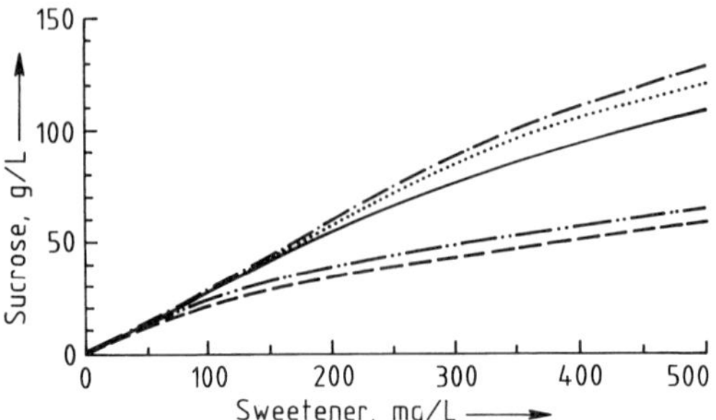

**Figure 3.** Equisweet concentrations of acesulfame (– – –), aspartame (– · · –), sweetener blends and sucrose
Blend ratio: (—) 2:1; (· · ·) 1:2; (– · –) 1:1

mates of use levels in foods and beverages are often possible on the basis of comparison tables of sweetness intensities [12].

Much higher sweetness intensity factors are observed at the taste threshold [12]. These values are sometimes reported, but are of no practical significance.

Apparently, maximal sweetness intensities exist for all sweet substances, above which an increase in concentration does not induce a higher sweetness perception. This property is well illustrated by graphs of equisweet solutions of sweeteners and sucrose (Fig. 3).

Determination of sweetness intensities requires sensory tests; panelists have to be used according to their sensory abilities, and to be adapted to the test procedures. Owing to individual variations in sweetness perception, results have to be evaluated on a statistical basis [13], [14].

### 2.3. Taste Characteristics

The taste quality of sweeteners depends on several characteristics in addition to sweetness. Time–intensity profiles of sweetness perception are important for the assessment of taste quality as well as the occurrence of side-tastes and after-tastes.

Temporal sweetness perception characteristics can be studied with modern sensory techniques [15]–[17]. As sucrose is not only the standard for sweetness intensity assessments, but also the ideal to be matched in other sensory characteristics, the time–intensity profile of sucrose is often considered the target. Lasting sweetness or delayed sweetness onset may impair

the applicability of sweeteners. Time–intensity perception of sweetness is influenced by several substrate factors, e.g., viscosity, temperature, and the presence of taste-modifying compounds. Blending of sweeteners with different time–intensity profiles often results in a mixed profile, closer to that of sucrose.

The bitter aftertaste of saccharin is well known. As sweetness and bitterness are structurally closely related [9], bitter aftertastes are observed not only in some intense sweeteners, but also in a number of sweet carbohydrates. Individual susceptibilities for bitterness perception and taste thresholds vary. Off-taste thresholds as reported in water [18], however, give general information on taste characteristics of sweeteners rather than information on limitations of practical use, as interactions with food constituents may change taste perception substantially. Other off-tastes occurring in some intense sweeteners are menthol- or licorice-like tastes.

Taste modification of intense sweeteners has been widely studied in order to improve their sensory characteristics, especially time–intensity profiles, and reduce aftertastes. A wide variety of substances is claimed to have taste-modifying properties [19].

### 2.4. Synergism

In sweetener blends, synergistic sweetness enhancement is often observed (Fig. 4). Commonly used synergistic sweetener blends are acesulfame–aspartame, aspartame–saccharin, and

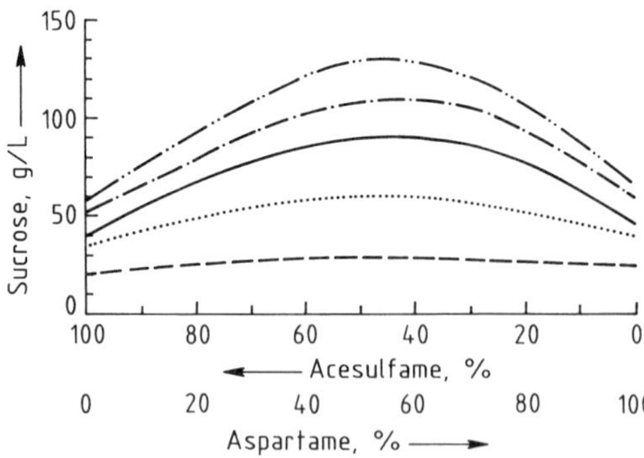

**Figure 4.** Equisweet concentrations of different levels of sucrose and acesulfame–aspartame blends
Sweetener blend, mg/L: (– –) 100; (· · ·) 200; (—) 300; (– · –) 400; (– · · –) 500

cyclamate–saccharin. The apparent intensification of sweetness compared with single sweeteners is not fully understood. Even definitions of the term "synergism" and methods for determination of synergistic effects vary [20]–[22]. For a number of combinations, calculation of equi-sweet concentrations on the basis of sweetness intensity data for single sweeteners seems possible [12]. In these cases synergism has to be regarded as the consequence of nonlinear sweetness response functions. For certain examples, e.g., acesulfame–aspartame blends, sweetness intensities calculated on the basis of the single sweetener response curves are below the observed values.

As time–intensity profiles of sweeteners are different, and often differ from the generally accepted standard sucrose, taste modifications can be achieved by blending; the resulting taste improvement is sometimes referred to as "qualitative synergism" [22].

When saccharin and cyclamate were generally available, blends of these two sweeteners were widely used. With the availability of acesulfame, aspartame, etc., synergism has attracted renewed interest, and blends of acesulfame–aspartame, aspartame–saccharin, and even tertiary or quaternary blends of sweeteners are increasingly used in order to exploit both types of synergism.

## 3. Uses

### 3.1. Foods and Beverages

Foods and beverages are the most important fields of application of intense sweeteners, with calorie reduction being the main goal. Single sweeteners or combinations with other sweet substances, e.g., carbohydrates, or sugar substitutes may be used. Intense sweeteners can be used in diabetic foods and beverages; depending on the type of product, either as single sweetening agents or combined with bulk sugar substitutes suitable for diabetic consumption (isomalt, lactitol, maltitol, mannitol, sorbitol, xylitol, etc.).

Whenever sugar and other sweet carbohydrates are used for their sweet taste, and not for functional reasons, replacement by intense sweeteners is generally simple. Functional sugar uses, such as viscosity enhancement, reduction of water activity, providing bulk, assistance in gelling, and others, cannot be provided by intense sweeteners owing to the minute concentrations used for normal sweetness levels. For foods requiring such functional properties, combinations of intense sweeteners and bulk sugar substitutes are an appropriate alternative whenever sugar or sweet carbohydrates are not to be used.

Beverage uses of intense sweeteners account for more than 50 % of human consumption; sugar replacement by intense sweeteners is simple, as carbohydrates do not play any important functional role in beverages, and the special taste characteristic commonly referred to as "body" can be achieved by addition of small quantities of sweet carbohydrates or viscosity-enhancing agents, such as certain polysaccharides.

Other important applications are fruit-flavored dairy products, desserts, etc.

In products such as pickles, marinated fish, and delicatessen salads, intense sweeteners not only round out the taste and balance the acidity, but also reduce the risk of spoilage (in contrast to sweet carbohydrates, they do not promote the growth of microorganisms).

In a number of applications, functional sweet carbohydrates are required for the specific product characteristics. For these applications, e.g., baked goods, confectionery products, and chewing gum, combinations of intense sweeteners with bulk sugar substitutes or bulking agents are an alternative.

General information on applications is given in product-related brochures and some surveys [23]–[25].

### 3.2. Table-Top Sweeteners

For household use, intense sweeteners are formulated into table-top sweeteners, such as sweetener tablets, powders and spoon-by-spoon products, and liquids.

Sweetener tablets may be produced as effervescent tablets containing a $CO_2$-releasing ingredient and an acid, such as sodium hydrogen carbonate and tartaric acid. Owing to the small amounts equivalent to a teaspoonful of sugar or a sugar cube, excipients, e.g., lactose or sugar alcohols, are often used for noneffervescent tablets.

For granular sweeteners, the same excipients or, if they are not intended for diabetics, glucose or sucrose are often used.

For spoon-by-spoon products, the bulk density of the granular product has to be reduced substantially; maltodextrins are most commonly used. To facilitate application of table-top sweeteners and to bring them close to standard sugar formulations, sugar-reduced cubes containing intense sweeteners with the same sweetness level as normal sugar cubes, and aqueous solutions of intense sweeteners, are also available.

### 3.3. Cosmetics

Several types of cosmetics, especially oral hygiene products, are sweetened to make them more pleasant for consumers. For oral hygiene products, noncariogenic ingredients have to be used. In dentifrices, polyols, such as glycerol or sorbitol, are used as humectants, and also provide some basic sweetness. The desired sweetness level is adjusted with an additional quantity of intense sweetener [26].

### 3.4. Pharmaceuticals

Intense sweeteners are used to mask undesired flavors and tastes of active pharmaceutical ingredients, e.g., bitterness, whenever the pharmaceuticals are intended for use by diabetics. Sweeteners are used in syrups, and soluble tablets and powders.

### 3.5. Feed

Intense sweeteners are used in animal feed to secure balanced nutrition. The main application is in starter feed for early weaning, in which intense sweeteners balance the palatability, and avoid decreased feed consumption immediately after weaning.

### 3.6. Others

Among other applications, tobacco and tobacco-related products are most important. In chewing tobacco, intense sweeteners are added to the flavoring systems. A further application is in cigarette tips, the palatability of which is improved by addition of intense sweeteners.

## 4. General Toxicology and Physiology

### 4.1. Toxicology

Intense sweeteners have to undergo the safety tests generally required for food additives, with long-term studies in at least two species, including a carcinogenicity study, mutagenicity studies in different test systems, studies on reproductive toxicity, pharmacokinetics, and metabolism.

Guidelines for these studies are available on an international and, in some countries, a national basis [27], [28].

### 4.2. Metabolism

Most available intense sweeteners are not metabolized in the human body or they are excreted without metabolic utilization; they are therefore noncalorific. Some, especially peptide-based sweeteners, are digested and metabolized. In spite of this, these sweeteners are virtually noncalorific, as normal use levels are extremely low and the contribution to the nutritive value of foods and beverages is insignificant compared with nutritive bulk sweeteners.

### 4.3. Suitability for Diabetics

Intense sweeteners are metabolized independently of insulin, and therefore suitable for diabetics. Consumption of intense sweeteners in doses equivalent to a normal soft drink serving neither stimulates insulin secretion nor influences blood glucose levels [29].

### 4.4. Dental Effects

The commonly used intense sweeteners are not metabolized, or at least not into acids, by the bacteria of the oral cavity. They are therefore noncariogenic. Inhibitory effects on *Streptococcus mutans*, a caries-causing bacterium, have been demonstrated at sweetener concentrations above normal use levels. Synergistic enhancement of these effects in sweetener blends or blends of sweeteners and fluoride have also been demonstrated [30]–[32].

## 5. Food Legislation

### 5.1. International Regulations and Assessments

On the basis of available safety data, acceptability of intense sweeteners is assessed by several international authorities, especially the Joint Expert Committee on Food Additives of the WHO and FAO (JECFA), and the Scientific Committee for Food of the EU (SCF). JECFA bases a general evaluation on the available safety data. Acceptance for food use is demonstrated by allocation of an acceptable daily intake (ADI) value. The results of the evaluation are published as a toxicological monograph. In parallel, specifications are recommended which are normally used as guidelines for national food legislation. The SCF performs a similar evaluation, having produced a report on sweeteners in foods, and several subsequent publications [33].

Beyond national legislation, an EU directive on sweeteners harmonizes sweetener legislation in the Community, listing approved sweeteners, fields of application, and maximum use levels.

### 5.2. National Food Legislation

Under the food legislation of most countries, intense sweeteners are food additives, requiring approval for food use. Approvals are listed in national regulations, e.g., in the United States [34], the United Kingdom [35], and Germany [36].

## 6. Substances Commonly Used as Sweeteners

### 6.1. Acesulfame K

#### 6.1.1. General Information

Acesulfame K was discovered by CLAUSS and JENSEN during investigations on oxathiazinone dioxides. The sweet taste was found by chance [37], [38]. Several other oxathiazinone dioxides taste sweet, but have less favorable characteristics.

#### 6.1.2. Physical and Chemical Properties

Acesulfame K is the potassium salt of 6-methyl-1,2,3-oxathiazin-4($3H$)-one-2,2-dioxide (also 3,4-dihydro-6-methyl-1,2,3-oxathiazin-4-one-2,2-dioxide) [55589-62-3], $C_4H_4KNO_4S$, $M_r$ 201.2.

Acesulfame K

Acesulfame K does not have a well-defined melting point, and starts decomposing at ca. 225 °C. It is soluble in water (ca. 270 g/L at 20 °C and ca. 1300 g/L at 100 °C). Solubility in most organic solvents is low. Aqueous solutions of acesulfame K are neutral [39].

At pH 3–7, the common range for most foods and beverages, acesulfame K is virtually stable. Losses of a few percent were reported only after prolonged exposure to lower pH [39], [40]. Acesulfame K does not decompose during normal processing conditions of foods and beverages. No decomposition was observed during baking [41], ultra-high-temperature treatment [42], and microwave heating [43]. Hydrolysis in strong acids yields acetone, $CO_2$, and ammonium sulfate [44].

Compared with 3% sucrose solution, the sweetness intensity of acesulfame K is approximately 200 [12]. An off-taste may become perceptible at elevated concentrations [18]. Acesulfame K is often synergistic with other sweeteners [45].

#### 6.1.3. Production

Several synthesis routes for acesulfame K have been described. Potential synthesis starts from halogen sulfonyl isocyanates and keto compounds. Depending on the ketone, oxathiazinone dioxides with different substituents can be produced, some of which are also sweet, but inferior to acesulfame K in taste quality.

One production route (Fig. 5) starts from fluorosulfonyl isocyanate reacted with acetoacetic acid tert-butyl ester to a labile addition compound. On heating, this intermediate releases $CO_2$ and isobutene to form N-(fluorosulfonyl)acetoacetic acid amide; in the presence of bases, this compound cyclizes to the oxathiazinone dioxide ring [37], [46].

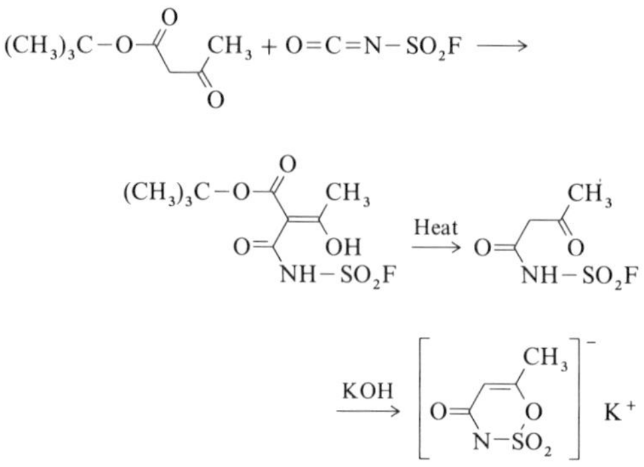

**Figure 5.** Synthesis of acesulfame K [37], [46]

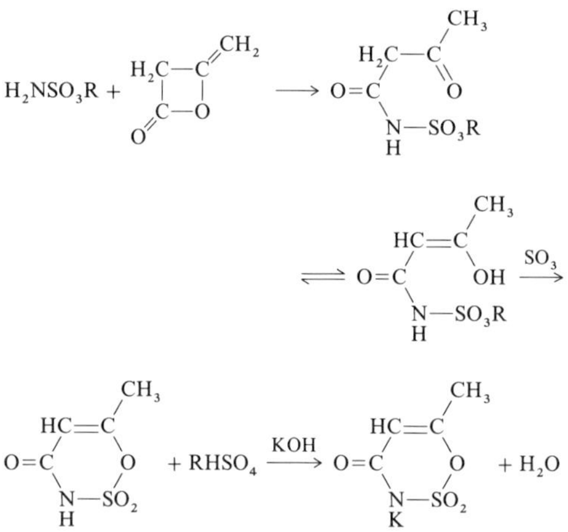

**Figure 6.** Synthesis of acesulfame K: alternative route [41], [48]

An alternative synthesis route (Fig. 6) has been described in which diketene and amidosulfonic acid are reacted to acetoacetamide-$N$-sulfonic acid. In the presence of suitable dehydrating agents, acesulfame acid is formed, which can be neutralized with potassium hydroxide to acesulfame K [47], [48].

### 6.1.4. Specifications and Analysis

For food use, acesulfame K has to conform to specifications laid down in national food legislation; these are often based on the suggestions made by JECFA [49], in which a purity > 99% is required. Specific methods for the analysis of pure acesulfame K are given in these suggestions, and general methods are listed in a guide to the specifications [50].

Although several analytical methods for the quantitative determination of acesulfame K in foods and other products are available, HPLC analysis, allowing simultaneous determination of acesulfame K and other sweeteners, is often preferred [51], [52].

### 6.1.5. Toxicology and Legislation

Acesulfame K has been subjected to the toxicological studies required for food additives. It is devoid of carcinogenicity, chronic toxicity, teratogenicity, and pharmacological side effects. It is absorbed quickly, and excreted at a similar rate, completely unmetabolized. Owing to the rapid excretion, accumulation in the body seems impossible. JECFA allocated an ADI of 0–15 mg/kg [53], as did the FDA [54].

Acesulfame K is approved for food use in most countries where intense sweeteners are used in foods and beverages, and is also approved for use in oral hygiene products and pharmaceuticals.

### 6.1.6. Uses

Acesulfame K is used in all fields of applications of intense sweeteners. Common applications are: table-top sweeteners; beverages; foods, such as dairy products, desserts, bakery products, confectionery, chewing gum, pickles, and marinated fish; oral hygiene products, such as tooth paste and mouthwash; and pharmaceuticals. Owing to its synergistic characteristics, acesulfame K is often used in sweetener blends, and in combination with bulk sweeteners in products requiring good stability, e.g., confectionery or bakery products [23], [40], [55].

## 6.2. Aspartame

### 6.2.1. General Information

The sweet taste of aspartame, $N$-L-$\alpha$-aspartyl-L-phenylalanine methyl ester, was discovered by SCHLATTER [56], [57] in 1965, although the compound had been synthesized earlier. Systematic research on the sensory properties of dipeptide derivatives has revealed many sweet substances [58], [59].

## 6.2.2. Physical and Chemical Properties

Aspartame [22839-47-0], $C_{14}H_{18}N_2O_5$, $M_r$ 294.3 is a white crystalline powder virtually free from odor, with no well-defined melting point. At 196 °C it reacts to 5-benzyl-3,6-dioxo-2-piperazineacetic acid [55102-13-1] [60].

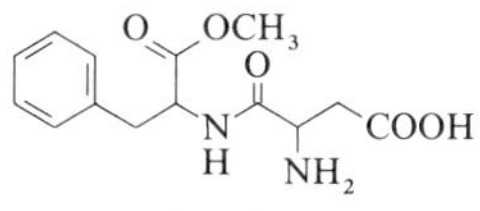

Aspartame

Solubility of aspartame in water depends on pH. At pH 5.2, the isoelectric point, solubility in water is approximately 1 g in 100 mL. With decreasing pH, solubility increases to a maximum value at pH 2.2. Water solubility also increases with temperature. In most organic solvents, the solubility is lower than in water [60].

At elevated temperature, solid aspartame slowly releases methanol to form aspartylphenylalanine and the dioxopiperazine. In aqueous solutions aspartame slowly hydrolyzes to amino acids and methanol; the dioxopiperazine can also be formed. The maximum stability in aqueous solution is at ca. pH 4. The rate of hydrolytic decomposition increases steeply with both increase and decrease of pH [60].

Aspartame stability studies have been carried out, e.g., in aqueous solutions [61], [62], beverages [63], [64], fruit preparations, yoghurt [65], [66], and under conditions favoring Maillard reactions [67]. Coating or encapsulation by cyclodextrins has been suggested to improve its stability [68].

Compared with 3 wt % aqueous sucrose solution, aspartame has a sweetness intensity of ca.

180–200 [12]. Its taste characteristics are generally assessed as good [18]. The sweetness perception can be slightly delayed and lasting [17].

## 6.2.3. Production

Production of aspartame normally starts from L-aspartic acid and L-phenylalanine or L-phenylalanine methyl ester. The production follows common routes of peptide synthesis, as the L-configuration of amino acids has to be retained (essential for sweetness).

Common production routes require protective groups for aspartic acid. The protective groups commonly used in peptide synthesis have been suggested for aspartame synthesis, but simple protection for large-scale production can be obtained by use of formylaspartic acid anhydride. It can be reacted with phenylalanine methyl ester to N-formylaspartame [69], from which the protective group has to be removed [70]. Especially simple production (Fig. 7) is possible from N-formylaspartic acid anhydride and L-phenylalanine. The resulting N-formylaspartylphenylalanine is treated with hydrochloric acid and methanol. The aspartame hydrochloride obtained can be crystallized for purification and transformed into aspartame after addition of a stoichiometric amount of a neutralizing agent [71], [72]. Adjustment of the solvent increases the yield by improving the ratio of α- to β-isomers [73]. Crystallization of α-aspartame hydrohalides is a suitable purification step, as β-hydrohalides are more soluble [74], [75].

A commercially used alternative to chemical peptide synthesis is enzymatic formation of the peptide bond. Some proteinases, e.g., ther-

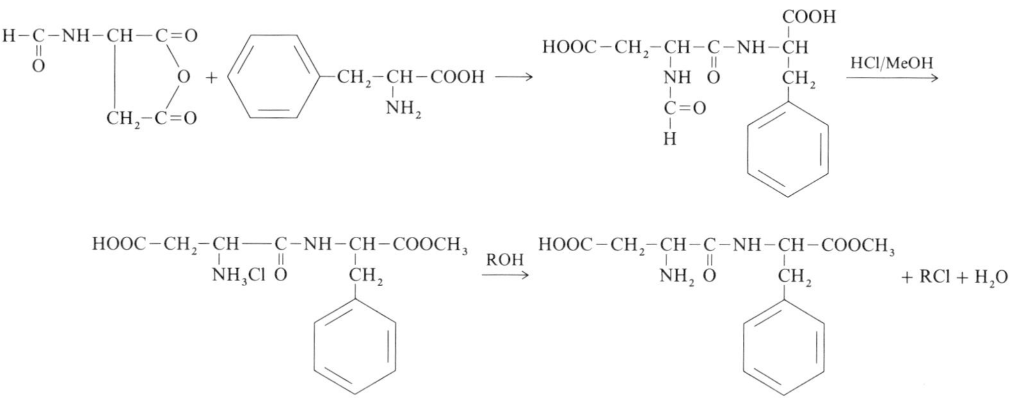

**Figure 7.** Synthesis of aspartame

molysin, are able to catalyze peptide bond formation. Protected or nonprotected aspartic acid anhydride can be used. In contrast to chemical peptide bond formation, L-phenylalanine is selectively used from a racemic phenylalanine, and the remaining D-isomer can be racemized again [76].

### 6.2.4. Specifications and Analysis

National food legislation, for the use of aspartame in foods, is often based on the suggestions made by JECFA [77] in which a purity > 98 % is required. Specific methods for the analysis of pure aspartame are given in these suggestions, and general methods are listed in a guide to the specifications [50]. Aspartame is also listed in the Food Chemicals Codex [78].

Although several analytical methods for the quantitative determination of aspartame in foods and other products are available, HPLC analysis, allowing simultaneous determination of aspartame and other sweeteners, is often preferred [51], [52].

Methods allowing determination of aspartame and its breakdown products are available [61], [63].

### 6.2.5. Toxicology and Legislation

Aspartame is devoid of carcinogenicity, chronic toxicity, and teratogenicity.

Being a dipeptide ester, aspartame is metabolized, with a nutritive value of 17 kJ/g (4 kcal/g); at normal use levels of several hundred milligrams per kilogram, this nutritive value has no practical importance.

JECFA allocated an ADI of 0–40 mg/kg [79], whereas the FDA value is 50 mg/kg [80].

Owing to its phenylalanine content, aspartame should be avoided by persons suffering from phenylketonuria, and a warning on aspartame-containing products is required in many countries.

Aspartame is approved for food use in most countries where intense sweeteners are used.

### 6.2.6. Uses

The sensory characteristics of aspartame allow its use in all common sweetener applications. Limitations are imposed by its susceptibility to hydrolytic decomposition, and limited tempera-

ture stability. The shelf life of aspartame-sweetened products with high water content has to be limited to periods in which no decrease in sweetness can be observed. No limitations exist for low-humidity products, unless they are exposed to high temperatures. Owing to the limited solubility of aspartame, preparations have been developed to facilitate dissolution [81].

In addition to single-sweetener applications, the synergistic properties of aspartame are used in a variety of blends.

Applications of aspartame in foods, beverages, and other products are described in surveys [24], [82].

## 6.3. Cyclamate

### 6.3.1. General Information

The sweet taste of salts of cyclohexylaminosulfonic acid was discovered accidentally by SVEDA in 1937 [83].

### 6.3.2. Physical and Chemical Properties

Cyclamates are salts of cyclohexylaminosulfonic acid (cyclohexylsulfamic acid) [100-88-9], $C_6H_{13}NO_3S$, $M_r$ 179.2. Cyclohexylsulfamic acid melts at 169–170 °C, but its salts do not have well-defined melting points and decompose at elevated temperature. Cyclohexylaminosulfonic acid is less suitable for food use but is approved, like its sodium, potassium, and calcium salts [84], [85].

$$\langle\!\!\!\bigcirc\!\!\!\rangle\text{—NHSO}_3\text{M}$$

Cyclamates (M = e.g., Na, K, ½ Ca)

Sodium cyclamate, $C_6H_{12}NNaO_3S$, $M_r$ 201.2 [139-05-9], is the most important cyclamate sweetener. It forms white crystals, decomposing at ca. 260 °C. Its solubility is 20 wt % in water at room temperature, whereas in most organic solvents solubility is low or extremely low [84], [85].

Calcium cyclamate [5897-16-5], $C_{12}H_{24}CaN_2O_6S_2 \cdot 2 H_2O$, $M_r$ 432.6, is of less importance. At pH 2–10, the stability of cyclamates to hydrolytic decomposition is high, so they are not affected by common food processing conditions [84], [85].

Compared with 0.1 mol/L sucrose solutions, the sweetness intensity of sodium cyclamate is ca. 35 [12]. Aftertastes become perceptible at elevat-

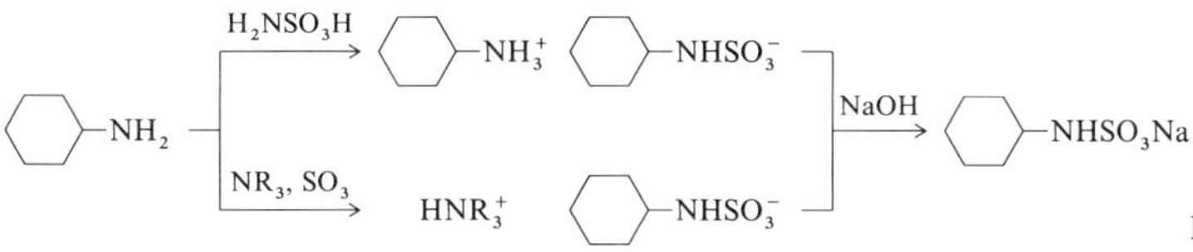

**Figure 8.** Synthesis of cyclamates

ed concentrations only, rendering cyclamate generally superior to saccharin [18].

### 6.3.3. Production

The starting material for cyclamate production is cyclohexylamine, which is reacted with sulfonating agents (Fig. 8) [86]. The originally used chlorosulfonic acid has been replaced by other compounds. In high-boiling-point solvents, cyclohexylamine reacts with sulfamic acid to $N$-cyclohexyl-$N$-cyclohexylammonium sulfamate. Reaction with sodium or calcium hydroxide yields cyclamates and free cyclohexylamine. In a corresponding way, equimolar amounts of cyclohexylamine and tertiary amines react to tri-alkylammonium-$N$-cyclohexyl sulfamates, the respective cyclamates being formed after reaction with sodium or calcium hydroxide. Further possibilities are reactions of cyclohexylamine with sodium methylthiosulfate or sulfamic acid metal salts. Reactions of cyclohexyl compounds with $SO_3$ or sulfuric acid, or of cyclohexylhydroxylamine and $SO_2$ or sulfites, producing cyclamates or cyclohexylsulfamic acid have also been described [87], [88].

### 6.3.4. Specifications and Analysis

The suggestions made by JECFA for cyclamates in foods [89] require a purity > 98 %. Specific methods for the analysis of pure cyclamate are given in these suggestions, and general methods are listed in a guide to the specifications [50].

Owing to the absence of a chromophore for UV detection, analytical methods for cyclamates have to be based on either color changes or other properties. A suitable HPLC method with post-column derivatization for easy detection has been described [90].

### 6.3.5. Toxicology and Legislation

Reevaluations of toxicological data have concluded that cyclamates are devoid of carcino-genicity. Discussion focused on the toxicity of cyclohexylamine, a potential metabolite formed by intestinal bacteria [91]. Current ADI values are based on the toxicity data for cyclohexylamine, and the assumption of a fairly high metabolic rate in the human body. Both JECFA and SCF allocated ADI values of 0–11 mg per kilogram body weight [33], [92].

Whereas some countries approve cyclamate use in a variety of applications, some important sweetener markets, such as the United States, still ban them.

### 6.3.6. Uses

Owing to their good stability, cyclamates are suitable for all applications of intense sweeteners. They can be used in acid products, e.g., soft drinks and pickles, and withstand exposure to elevated temperature as used in the production of canned fruit [85].

The main application of cyclamates, however, is in blends with saccharin in a 10:1 ratio by weight. In these blends, quantitative and qualitative synergistic effects can be used to advantage, rendering them an important sweetener system in countries approving the use of both sweetenes [85].

## 6.4. Saccharin

### 6.4.1. General Information

Saccharin was the first intense sweetener ever introduced. It was discovered in 1878 [2]. Small-scale production commenced in 1884, and it has been available since. For decades, saccharin was the most important intense sweetener.

### 6.4.2. Physical and Chemical Properties

Saccharin, often referred to as $o$-benzoic acid sulfimide, or $o$-sulfobenzoic acid imide, is 1,2-benzisothiazol-3(2$H$)-one-1,1-dioxide [*81-07-2*], $C_7H_5NO_3S$, $M_r$ 183.2. Solid saccharin is a white, crystalline, odorless powder, melting at 228–230 °C. Solubility in water at 20 °C is ca. 2 g/L and increases to ca. 40 g/L at 100 °C. Aqueous solutions of saccharin are acidic, pH ca. 2 for a saturated solution at room temperature. In aqueous solutions of alkali hydroxides or carbonates, saccharin is easily soluble, forming salts [93].

The sodium salt is more important than saccharin itself. Sodium saccharin [6155-57-3], $C_7H_4NNaO_3S \cdot 2\,H_2O$, $M_r$ 241.2, crystallizes with water of crystallization. Commercial preparations, however, may not contain the stoichiometric water content, but a slightly lower value. In contrast to saccharin itself, sodium saccharin is soluble in water. At room temperature, ca. 1000 g can be dissolved in 1 L of water, and more than 3000 g dissolve in 1 L of water at 100 °C [93].

Calcium saccharin, $C_{14}H_8CaN_2O_6S_2 \cdot 3\frac{1}{2}H_2O$, $M_r$ 467.5 [6381-91-5], is of less importance than sodium saccharin. At 20 °C, ca. 370 g can be dissolved in 1 L of water, but at 100 °C solubility is almost as high as for sodium saccharin [93].

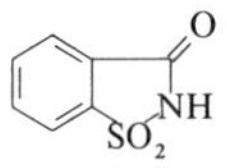

Saccharin

Compared with dilute sucrose solutions, the sweetness intensity of saccharin is ca. 550, whereas the values for its salts are ca. 450 [12]. At medium or elevated use levels, saccharin sweetness is accompanied by a metallic or bitter aftertaste [18]. Several masking agents for this sidetaste have been suggested, and sweetener blends, especially a blend of 1 part sodium saccharin to 9–10 parts sodium cyclamate, are often used [94].

### 6.4.3. Production

In most countries, commercial saccharin is produced by the Remsen–Fahlberg process (Fig. 9), which basically follows the synthesis described by its discoverers [95], [96]. Toluene is used as the starting material and reacted with chlorosulfonic acid to a mixture of isomeric toluene sulfochlorides. In the presence of ammonia, 2-toluene sulfochloride forms 2-toluene sulfonic acid amide which is oxidized under appropriate conditions to saccharin. Potential oxidizing agents are potassium permanganate or

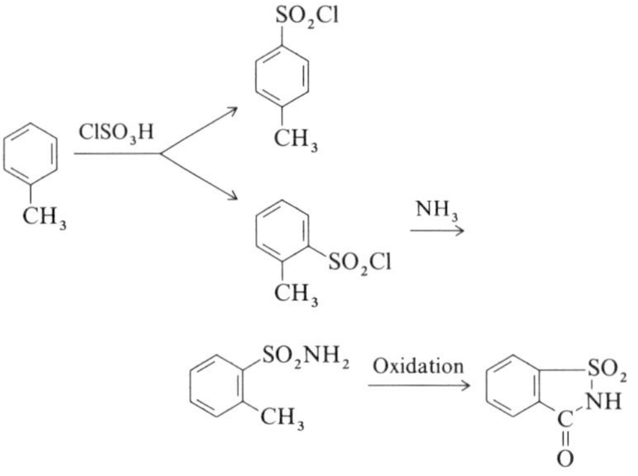

**Figure 9.** Remsen–Fahlberg synthesis of saccharin [96]

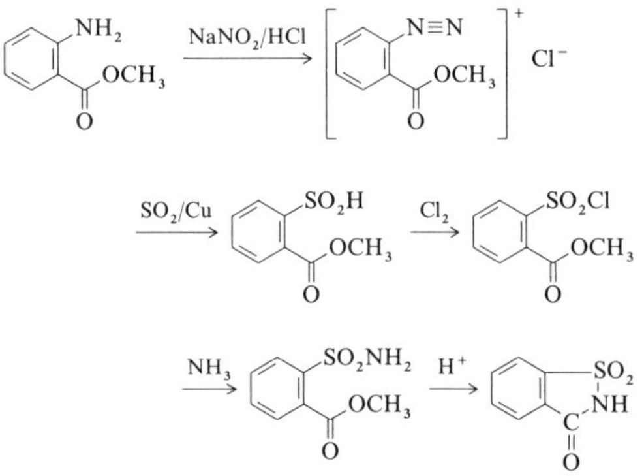

**Figure 10.** Maumee synthesis of saccharin [98]

chromic acid; syntheses not involving oxidizing agents have also been described [97].

In the United States, the Maumee process is used (Fig. 10). This starts from either anthranilic acid or its methyl ester, which can be obtained from phthalic acid anhydride. With sodium nitrite, anthranilic acid forms a diazonium compound, which is reacted with sulfur dioxide or sulfites, and chlorinated to 2-chlorosulfonylbenzoic acid methyl ester. In the presence of ammonia, this ester yields the respective amide, from which, after de-esterification, saccharin is obtained. Reaction with sodium hydroxide or calcium hydroxide forms the respective saccharin salts [98].

### 6.4.4. Specifications and Analysis

The suggestions made by JECFA [99] for saccharin in foods require a purity > 99 %. Specific methods for the analysis of saccharin and saccharin salts are given in these suggestions, and general methods are listed in a guide to the specifications [50]. Saccharin and its sodium and calcium salts are also listed in the Food Chemicals Codex [98].

Although several analytical methods for the quantitative determination of saccharin and sodium saccharin in foods and other products are available, HPLC analysis, allowing simultaneous determination of saccharin, sodium saccharin, and other sweeteners, is often preferred [51], [52].

### 6.4.5. Toxicology and Legislation

Saccharin is not metabolized in the human body. It is without influence on diabetics. Safety studies on saccharin have been repeatedly performed. In long-term feeding studies in rats an increased bladder tumor rate was observed after prolonged ingestion of high doses. This was attributed to a nongenotoxic action of saccharin, or to a co-carcinogenic effect. On the basis of further studies, the relevance of these observations for humans has been questioned. Information on saccharin toxicology has been reviewed [100]. JECFA allocated an ADI of $0-5$ mg/kg [101].

Saccharin use is restricted in several countries, although in most countries it is still available as an intense sweetener. In the United States, a special label is required for saccharin-sweetened products [102].

### 6.4.6. Uses

Saccharin is most frequently used as its sodium salt. Important fields of application are soft drinks, table-top sweeteners, and desserts. For taste reasons, blends with other intense sweeteners, or combinations with reduced sugar levels are preferred wherever such blends are approved. In oral hygiene products, e.g., toothpastes and mouthwashes, saccharin masks undesired tastes of other ingredients [93].

In starter feed for livestock, saccharin is used to avoid reduced feed intake after weaning.

Beside its applications as an intense sweetener, saccharin is used in electrolytic nickel deposition. Addition of saccharin to the nickel salt solutions increases the hardness and brightness of the nickel plate. This effect is apparently specific to saccharin [93].

## 7. Sweeteners of Lesser Importance

### 7.1. Glycyrrhizin

#### 7.1.1. General Information and Properties

Glycyrrhizin is the name for mixed calcium and potassium salts of glycyrrhizic acid which are found in licorice root, *Glycyrrhiza glabra* L., grown in Southern Europe and Asia.

Glycyrrhizic acid [*1405-86-3*], $C_{42}H_{62}O_{16}$, $M_r$ 823.0, is a glycoside, consisting of glycyrrhetic acid, a triterpene of the β-amyrine type, and two β-1,2-glycosidic-linked glucuronic acids. Extracts containing $6-14\%$ glycyrrhizinates are available as licorice. It is used as a flavoring ingredient in confectionery and pharmaceuticals [103].

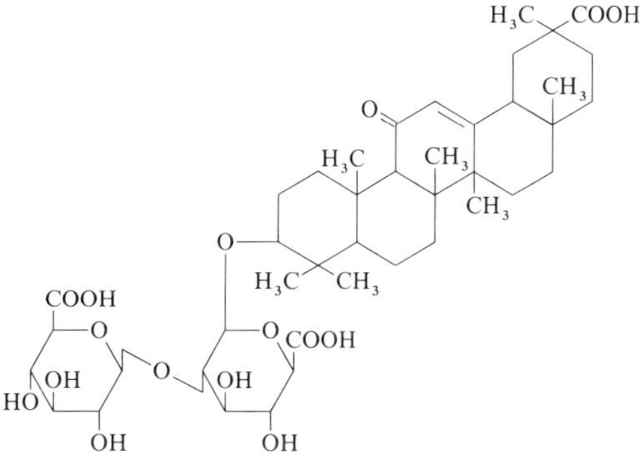

Glycyrrhizic acid

In addition to calcium and potassium glycyrrhizinates, crude ammonium glycyrrhizinate and monoammonium glycyrrhizinate are available. Ammonium glycyrrhizinate dissolves in water above pH 4.5, and in aqueous ethanol, but the solubility of monoammonium glycyrrhizinate in water is fairly low. At elevated temperature, glycyrrhinates are not fully stable to hydrolytic decomposition [103].

Glycyrrhizin is claimed to be $50-100$ times sweeter than sucrose in dilute aqueous solution, whereas for ammonium glycyrrhizinate the sweetness intensity is ca. 50. Glycyrrhizinates are characterized by a delayed sweetness onset, a lingering sweetness, and a pronounced licorice taste; they are therefore considered to be flavorings rather than sweeteners [103].

### 7.1.2. Production

The starting material for the production of glycyrrhizin is dried, cut licorice root, with a glycyrrhizin content of 2–7wt%, depending on its origin. A crude extract is obtained by countercurrent extraction with water. After removal of polysaccharides, glycyrrhizic acid can be precipitated from the crude extract with sulfuric acid. Ammonium glycyrrhizinate is produced by dissolving crude glycyrrhizic acid in aqueous ammonia, and subsequent drying or precipitation with ethanol. Repeated recrystallization yields monoammonium glycyrrhizinate which, in contrast to the dark colored ammonium glycyrrhizinate, is almost colorless [103].

### 7.1.3. Toxicology and Legislation

Glycyrrhizin is characterized by a variety of physiological activities, which discourage the use of large quantities for sweetening purposes. Accordingly, it is not approved as a sweetener in many countries, but can be used as a licorice flavoring agent.

### 7.1.4. Uses

In addition to their flavoring properties, glycyrrhizin and ammonium glycyrrhizinate have flavor-enhancing effects in certain products. Important fields of application are confectionery, soft drinks with higher pH levels, tobacco products, and oral hygiene products [103].

Extracts of licorice root, glycyrrhizin, and glycyrrhizic acid have been suggested for treatment of a number of diseases, e.g., gastric ulcer [104].

## 7.2. Neohesperidin Dihydrochalcone

### 7.2.1. General Information and Properties

Neohesperidin dihydrochalcone, 2′,3,4′,6′-tetrahydroxy - 4 - methoxydihydrochalcone - 4′ - β-neohesperidoside [20702-77-6], $C_{28}H_{36}O_{15}$, $M_r$ 612.6 is the important compound among the sweet dihydrochalcones. A number of related dihydrochalcones, e.g., naringin dihydrochalcone, are without practical importance.

Neohesperidin dihydrochalcone has a melting range of 152–154 °C. In water only ca. 0.5 g/L dissolves at room temperature, but much higher quantities can be dissolved at high temperature and in ethanol (ca. 20 g/L). In alkaline

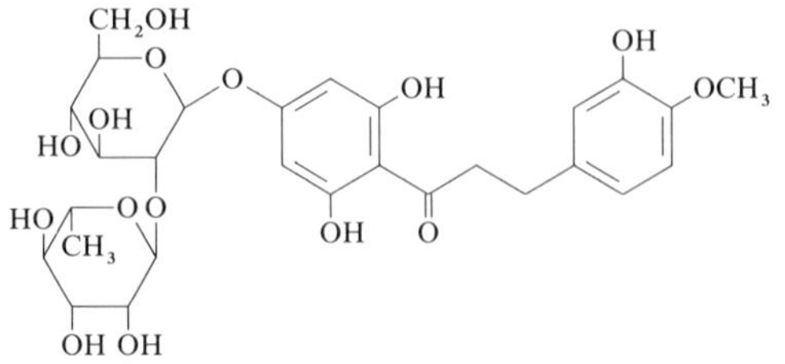

solution, dissolution is favored owing to salt formation [105].

Despite the possibility of hydrolytic cleavage of the glycosidic bond, the stability of neohesperidin dihydrochalcone is acceptable under food processing and storage conditions [105].

Compared with dilute sucrose solutions, the sweetness intensity of neohesperidin dihydrochalcone is ca. 330. The much higher values sometimes reported were determined at the taste threshold. Neohesperidin dihydrochalcone is characterized by a pronounced menthol-like aftertaste which limits its applicability and allows it to make only a partial contribution to the total sweetness of most products [105].

Neohesperidin dihydrochalcone can be analyzed by HPLC [106].

### 7.2.2. Production

Neohesperidin dihydrochalcone is obtained by hydration of neohesperidin under alkaline conditions. This compound can be isolated from the skins of Seville oranges. A further production route starts from naringin (Fig. 11), found in the skins of several other citrus species. Treatment of

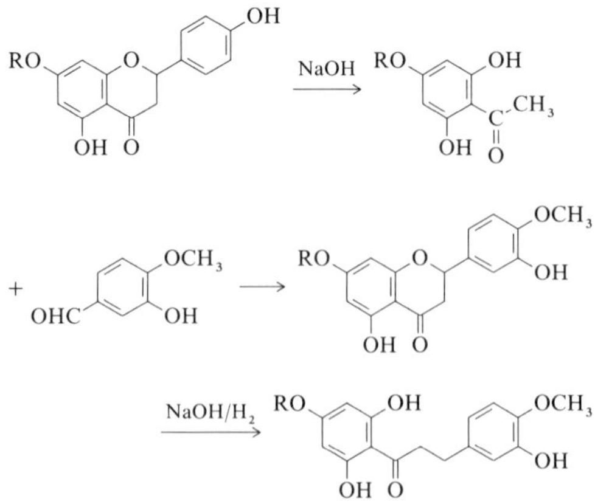

Neohesperidin dihydrochalcone

**Figure 11.** Synthesis of neohesperidin dihydrochalcone, starting from naringin

naringin with alkali produces phloroacetophenone-4′-β-neohesperidoside which is condensed with isovanillin to neohesperidin chalcone. Hydrogenation under alkaline conditions yields neohesperidin dihydrochalcone [107], [108].

### 7.2.3. Toxicology and Legislation

Neohesperidin dihydrochalcone has been tested in a full range of safety studies, and an acceptable daily intake value of $0-5$ mg per kilogram body weight was allocated by SCF [109]. Approval of this sweetener is limited to few countries, e.g., those in which it can be produced from the skin of citrus fruits. Extension of approval seems possible.

### 7.2.4. Uses

The specific taste characteristics of neohesperidin dihydrochalcone limit its use to few applications, or low use levels. It acts synergistically with a number of other sweeteners, e.g., acesulfame and aspartame, and can be used in fairly small quantities in blends with these sweeteners. The main fields of application are certain confectionery products and chewing gum, but small concentrations are suitable for use in beverages and dairy products. In addition to sweetness, neohesperidin dihydrochalcone has flavor-enhancing properties in a variety of foods and other products [105].

## 7.3. Stevioside and Rebaudioside

### 7.3.1. General Information and Properties

Stevioside and other steviol glycosides are constituents of the South American plant *Stevia rebaudiana* Bertoni. It is claimed that the leaves of the plant have been used for centuries to sweeten foods and beverages.

Stevioside is a glycoside consisting of the aglycone steviol, *ent*-13-hydroxykaur-16-en-19-oic acid, and three glucose molecules, two of which are linked to the disaccharide sophorose. Stevioside is therefore named as 13-$O$-β-sophorosylsteviol-18-$O$-β-glucosyl ester [*57817-89-7*], $C_{38}H_{60}O_{18}$, $M_r$ 804.9.

Beside steviol, the leaves of *Stevia rebaudiana* contain a number of other sweet glycosides, the most important of which is rebaudioside A. Re-

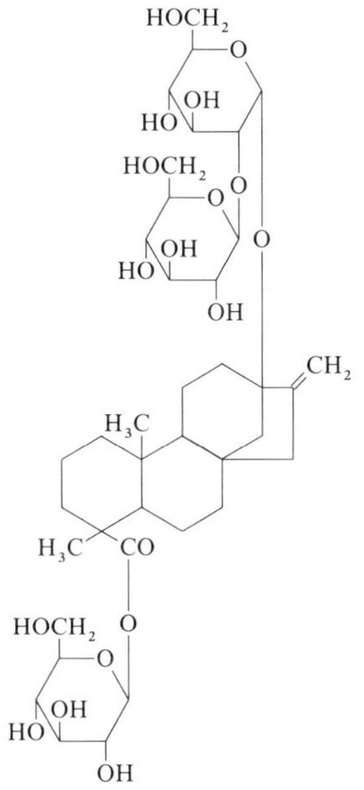

Stevioside

baudioside A is 13-$O$-β-glycopyranosylsteviol-19-$O$-β-glucopyranosyl-(1,3)-sophorosyl ester [*58543-16-1*], $C_{44}H_{70}O_{23}$, $M_r$ 967.0. Further sweet constituents of *Stevia* leaves are other steviol glycosides, such as other rebaudiosides, dulcoside, and steviolbioside.

*Stevia* originates in South America, but is also grown in several Asian countries. Dried *Stevia* leaves contain $2-22$ wt % of stevioside with a mean of ca. 7 %. Rebaudioside levels are in the range $1.5-10$ % [110].

Stevioside melts in the range $196-198\,°C$. The solubility in water is approximately $1.2-1.3$ g/L, only slightly higher than the concentrations necessary for medium or high sweetness. Stevioside is of acceptable stability for food production and storage, e.g., for use in soft drinks. At elevated temperature, the stability decreases with decreasing pH [111].

Compared with dilute sucrose solution, stevioside is ca. $160-170$ times sweeter. The higher sweetness intensity values often reported were determined at the taste threshold. The sweetness of stevioside is accompanied by a licorice-like aftertaste. Side-tastes and other sensory characteristics depend on the decomposition of stevioside products which are often purified extracts containing other plant constituents in addition to stevioside [112].

Rebaudioside A melts in the range $242-244\,°C$. In water, it is more soluble than stevioside. Although the stabilities of steviosides and rebaudioside A are similar, rebaudioside A has been found to decompose on exposure to light [111].

The sweetness intensity of rebaudioside A is approximately one-third higher than that of stevioside, and its taste characteristics are superior [110].

### 7.3.2. Production

Stevioside can be isolated from the leaves of *Stevia* by extraction with water or water–ethanol mixtures. Crude *Stevia* extracts are more or less purified, e.g., by treatment with calcium or magnesium hydroxides or carbonates. Further purification by ion-exchangers, or absorption of undesired ingredients is possible. Stevioside can be obtained from the extracts by precipitation with methanol [112].

Depending on the processing conditions, stevioside products may contain other sweet steviol glycosides, and other constituents. Accordingly, the properties of commercially available stevioside products vary.

Rebaudioside A is isolated from *Stevia* leaves in a similar way to stevioside. The ratio of rebaudioside A to stevioside can be increased by extraction with methanol, and further purification with column chromatography yields rebaudioside-A-rich products. Stevioside can, in addition, be transformed enzymatically to rebaudioside A [110].

Commercial production routes often fail to yield pure stevioside or rebaudioside A.

### 7.3.3. Toxicology and Legislation

Although safety studies on stevioside and rebaudioside have been performed, they have not been accepted internationally, owing to the lack of generally accepted specifications. Accordingly, no ADI has been allocated for these compounds on an international basis. *Stevia* extracts are approved for food use in several South American and Asian countries, but lack approval in Europe and North America [110], [111].

### 7.3.4. Uses

Stevioside, the more important product, is used in table-top sweeteners, confectionery, certain soft drinks, fruit products, and Japanese-style vegetable products in countries approving the sweeteners, generally as stevioside-rich *Stevia* extracts.

## 7.4. Sucralose

### 7.4.1. General Information and Properties

Sucralose is the common name for 1′,6′-dichloro-1′,6-dideoxy-β-D-fructofuranosyl-4-chloro-4-deoxy-α-D-galactopyranoside, sometimes referred to as trichlorogalactosucrose.

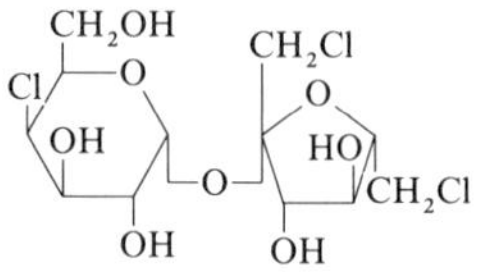

CH₂OH  CH₂Cl
Cl  O  O
OH  HO  CH₂Cl
O
OH  OH
Sucralose

The sweetener was discovered in 1976 [113]. Following the discovery that halogenation increases the sweetness of carbohydrates substantially, several chlorinated and brominated mono- and disaccharides of intense sweetness were found.

Sucralose [56038-13-2], $C_{12}H_{19}O_8Cl_3$, $M_r$ 397.6. The solubility in water is ca. 283 g/L at 20 °C and ca. 90 g/L in ethanol [114]. Sucralose melts at 125 °C with decomposition.

Solid sucralose is not fully stable, and slowly releases HCl, with discoloration. In contrast, aqueous solutions are highly stable, and it is marketed in this form. Owing to its stability over the normal pH range of foods and beverages, it is suitable for all liquid sweetener applications [115].

Compared with dilute sucrose solution, sucralose is ca. 600 times sweeter. The sweetness is perceived with a slight delay and a lasting effect, similar to aspartame, and generally considered of good quality.

### 7.4.2. Production

The starting material for the synthesis of sucralose (see Fig. 12) is sucrose. The production route commonly described blocks the three primary hydroxyl groups as trityl ethers, and transforms the secondary hydroxyl groups to acetates to yield 6,1′,6′-trityl-2,3,4,3′,4′-pentaacetylsucrose. By deblocking the primary hydroxyl groups, 2,3,4,3′,4′-pentaacetylsucrose is obtained.

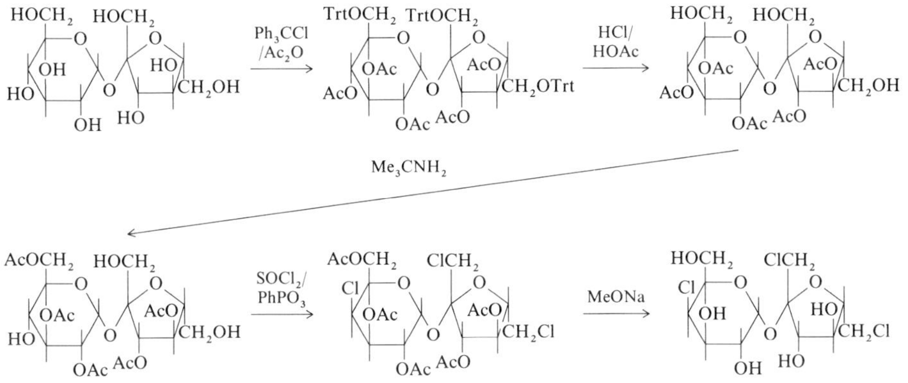

**Figure 12.** Synthesis of sucralose
Trt = $Ph_3C$ = trityl group

Under appropriate conditions, this compound can be transformed to 2,3,6,3',4'-pentaacetylsucrose. Chlorination at the exposed hydroxyl groups gives 4,1',6'-trichloro-4,1',6'-tridesoxygalactosucrose on conversion of the glucose moiety to galactose. Removal of the acetate groups by hydrolysis yields sucralose. Other blocking agents have also been described [116].

### 7.4.3. Toxicology and Legislation

Sucralose has been studied in the test systems commonly used for the evaluation of food additives. JECFA allocated an ADI of 0–15 mg/kg [117]. Sucralose has not yet been endorsed by other international authorities, and still lacks approval in most countries.

### 7.4.4. Uses

Owing to the lack of approval, sucralose uses are limited. Its good stability is considered an incentive for single sweetener use, which appears feasible on the basis of its sensory characteristics, but its synergistic characteristics probably favor its use in blends.

## 7.5. Thaumatin

### 7.5.1. General Information and Properties

Thaumatin is a mixture of structurally related proteins found in the arils of the fruits of *Thaumatococcus daniellii* Benth, a plant originating in West Africa, where it is grown on planta-

tions. The sweet taste of *Thaumatococcus* fruits, locally called Katemfe and traditionally used for sweetening purposes, was first described in the scientific literature in 1855 [1].

Thaumatin consists of several structurally related polypeptides with approximate $M_r$ 21 000 [53850-34-3]. Normally, thaumatin I and II are the major components, but other similar peptides may be present in thaumatin products. The amino acid composition and tertiary structure are known [118], [119].

Thaumatin is highly soluble in water, and solutions containing more than 1000 g of thaumatin in 1 L of water can be prepared [120]. Thaumatin stability data in the literature are not consistent. Good stability below pH 5.5 has been claimed, but limited stability is also reported [120].

Compared with dilute sucrose solutions, thaumatin is approximately 3500 times sweeter The sweetness shows a delayed onset and long persistence, accompanied by a licorice-like side-taste. Thaumatin is often used as a flavor enhancer rather than as a sweetener [120].

### 7.5.2. Production

Thaumatins are extracted from the arils of *Thaumatococcus* with water; removal of solids is by centrifugation and ultrafiltration. Further purification is possible by ion-exchange chromatography [120].

As production of thaumatin in countries growing *Thaumatococcus* is difficult, attempts have been made to produce thaumatin by means of genetically engineered microorganisms. Expression of the thaumatin genes in several microorganisms has been achieved, but no biotechnological production has yet been established [121], [122].

### 7.5.3. Toxicology and Legislation

Thaumatin is digested like other peptide food constituents. No effects of toxicological significance have been observed in safety studies on thaumatin. It has been accepted internationally as a food additive by JECFA [123], with ADI "not specified". It is approved as a sweetener and flavor enhancer in a number of countries.

### 7.5.4. Uses

Although thaumatin was primarily introduced as a sweetener, the temporal sweetness characteristics do not favor its use as a single sweetener, and limit blend use to fractions of the total sweetness.

Flavor-enhancing properties, in contrast, are of some importance; thaumatin is used as a flavor enhancer in confectionery, chewing gum, and similar products [120].

## 8. New Developments

### 8.1. Alitame

Alitame, L-α-aspartyl-$N$-(2,2,4,4-tetramethyl-3-thietanyl)-D-alanine [$80863$-$62$-$3$], $C_{14}H_{24}O_4N_3S \cdot 2\frac{1}{2} H_2O$, $M_r$ 375.3, is an intensely sweet dipeptide derivative [124].

$$HOOC-CH_2-\underset{\underset{NH}{|}}{CH}-\underset{\underset{O}{||}}{C}-NH-\underset{\underset{CH_3}{|}}{CH}-\underset{\underset{O}{||}}{C}-NH-CH\begin{smallmatrix}H_3C\diagdown_{C}\diagup CH_3\\ |\\ S\\ |\\ C\\ H_3C\diagup \diagdown CH_3\end{smallmatrix}$$

Alitame

Alitame melts at 100 °C, solidifies at 102 °C, and starts decomposing at 136–137 °C. At room temperature, ca. 130 g/L can be dissolved in water at pH 5.0, with increasing solubility at higher and lower pH, and higher temperature [125].

Being a dipeptide, alitame is not completely stable, but superior to aspartame at all important pH levels, especially in the neutral range.

Compared with dilute sucrose solutions, its sweetness intensity is > 2500. Sensory properties are similar to aspartame.

Alitame is under safety evaluation and has not been endorsed as a food additive by international authorities, nor yet approved in most countries [125].

### 8.2. Others

Systematic investigations into structure–activity relationships of sweeteners have led to a variety of products with sweetness intensities comparable to the sweetest products already known, or even higher. A number of products have been described, e.g., derivatives of $N$-phenylguanidinoacetic acid and $N$-phenylethanamidinoacetic acid [126], [127], tetrazole derivatives [128], or amide-, glycine- and β-alanine-based sweeteners with heterocyclic substituents [129]. These products are still far from being marketed, although some are apparently under safety investigation for human consumption.

**Monellin** is a polypeptide constituent of the fruits of *Dioscoreophyllum cumminsii* Stapf Diels, with approximate $M_r$ 11 500. The yield from 1 kg of fruit is ca. 3–5 g [130], [131]. Its sweetness intensity is ca. 2000. Owing to its structure, monellin is sensitive to conformational changes caused by heat or hydrolytic decomposition, rendering it unsuitable for normal sweetener use. As monellin is intensely sweet, and has a pleasant taste, studies have been initiated to improve the stability by structural changes, without impairing the good sensory characteristics.

## 9. Substances Formerly Used as Sweeteners

### 9.1. Dulcin

Dulcin is the common name of $N$-(4-ethoxyphenyl)urea [$150$-$69$-$6$], $C_9H_{12}N_2O_2$, $M_r$ 180.2. Dulcin melts at 173 °C and is sparingly soluble in water (ca. 1.25 g/L). Compared with dilute sucrose solutions, it is approximately 250 times sweeter. Sweetness characteristics are good.

$$H_5C_2-O-\!\!\left\langle\!\!\bigcirc\!\!\right\rangle\!\!-NH-CO-NH_2$$

Dulcin

Dulcin was synthesized by reaction of phenetidine hydrochloride with sodium cyanate or urea, and by reaction of 4-phenetidine with phosgene and ammonia [132], [133]. Dulcin achieved some practical importance as a sweetener, despite safety concerns. It was blended with saccharin or sodium saccharin in order to mask the aftertaste of saccharin. As some dulcin metabolites were considered potentially hazardous, it is now banned in most countries.

## 9.2. Others

*Suosan* is the common name for the sodium salt of *N*-(ethyl)-*N*-(4-nitrophenyl)urea, $C_{10}H_{10}N_3NaO_5$, $M_r$ 275.2 [134]. Pure suosan is a yellow crystalline substance ca. 300 times sweeter than sucrose in dilute solutions. It can be obtained by reacting 4-nitrophenyl isocyanate and β-alanine.

For some time, the sodium salt of *2-(4-methoxybenzoyl)benzoic acid*, $C_{15}H_{11}NaO_4$, $M_r$ 278.3 [135], was offered as the product S 23/46. This substance is ca. 150 times sweeter than sucrose, but is bitter at elevated concentrations. Low solubility of the corresponding acid, which is released in acid foods and beverages, limited the applicability. The compound was synthesized from phthalic acid anhydride and anisole in the presence of aluminum chloride.

Several substituted nitroanilines are distinguished by high sweetness intensity. The sweetest compound in this group is *2-propoxy-5-nitroaniline*, $C_9H_{12}N_2O_3$, $M_r$ 196.3, which is ca. 4000 times sweeter than sucrose. Attempts to market this product, as Ultrasüß or P 4000, were unsuccessful, owing to unsatisfactory sensory properties. It was synthesized by reaction of 1-chloro-2,4-dinitrobenzene and sodium propylate, with subsequent partial reduction, or by reaction of 2-nitrophenol with 1-chloropropane, reduction of the intermediate to the amino compound, and subsequent nitration [135].

*Perillatine*, the *E*-oxime of 4-(2-propenyl)-1-cyclohexene carboxaldehyde, $C_{10}H_{15}NO$, $M_r$ 165.2, commonly called perilla aldehyde, was marketed for a time in the United States. The product is ca. 2000 times sweeter than sucrose, but has less than satisfactory sensory characteristics [136].

# 10. References

[1] W. F. Daniell, *Pharm. J.* **14** (1855) 158–160.

[2] C. Fahlberg, I. Remsen, *Ber. Dtsch. Chem. Ges.* **12** (1879) 469–473.

[3] R. S. Shallenberger, T. E. Acree, *Nature (London)* **216** (1967) 480–482.

[4] L. B. Kier, *J. Pharm. Sci.* **61** (1972) 1394–1397.

[5] J. M. Tinti, C. Nofre in D. E. Walters, F. T. Orthoefer, G. E. DuBois (eds.): "Sweeteners, Discovery, Molecular Design, and Chemoreception," *ACS Symp. Ser.* **450** (1991) 206–213.

[6] A. van der Heijden, H. van der Wel, H.-G. Peer, *Chem. Senses* **10** (1985) 57–72.

[7] A. van der Heijden, H. van der Wel, H.-G. Peer, *Chem. Senses* **10** (1985) 73–88.

[8] M. Tamura, I. Shinoda, H. Okai, C. A. Stammer, *J. Agric. Food Chem.* **37** (1989) 737–740.

[9] H.-D. Belitz et al. in G. Charalambous (ed.): *Frontiers of Flavours,* Elsevier, Amsterdam 1988, pp. 49–62.

[10] T. Suami, L. Hough, *Food Chem.* **46** (1993) 235–238.

[11] J. C. Culberson, D. E. Walters in D. E. Walters, F. T. Orthoefer, G. E. DuBois (eds.): "Sweeteners, Discovery, Molecular Design, and Chemoreception," *ACS Symp. Ser.* **450** (1991) 214–223.

[12] K. Hoppe, *Lebensmittelindustrie* **38** (1991) 13–14.

[13] A. Fricker, J. Gutschmidt, E. Prochazka, *Dtsch. Lebensm. Rundsch.* **71** (1975) 138–139.

[14] A. Tunaley, D. M. H. Thomson, J. A. McEwan, *Int. J. Food Sci. Technol.* **22** (1987) 627–635.

[15] D. B. Ott, C. L. Edwards, S. J. Palmer, *J. Food Sci.* **56** (1991) 535–542.

[16] W. E. Lee, R. M. Pangborn, *Food Technol. (Chicago)* **40** (1986) no. 11, 71–78, 82.

[17] A. C. Noble, N. L. Matysiak, S. Bonnans, *Food Technol. (Chicago)* **45** (1991) no. 11, 121–124, 126.

[18] K. Hoppe, B. Gassmann, *Nahrung* **29** (1985) 417–420.

[19] Hoechst, DE-OS 3 331 517, 1982 (G.-W. von Rymon Lipinski).

[20] N. Ayya, H. T. Lawless, *Chem. Senses* **17** (1992) 245–259.

[21] R. A. Frank, S. J. Mize, R. Carter, *Chem. Senses* **14** (1989) 621–632.

[22] G.-W. von Rymon Lipinski, *Int. Food Marketing Technol.* **4** (1990) no. 5, 22–25.

[23] Hoechst, SUNETT®, Product Brochure, Frankfurt 1992.

[24] Nutrasweet AG, NutraSweet, Technical Bulletin, Zug 1987.

[25] G.-W. von Rymon Lipinski, *Int. Food Marketing Technol.* **5** (1991) no. 2, 5, 8–11.

[26] G.-W. von Rymon Lipinski, E. Lück, *Manuf. Chem.* **52** (1981) no. 5, 37.

[27] Principles for the Safety Assessment of Food Additives and Contaminants in Food, *Environ. Health Criter.* **70** (1987).

[28] Toxicological Principles for the Safety Assessment of Direct Food Additives and Color Additives Used in Food, *US Food Drug Adm. Bur. Foods,* Washington 1982.

[29] B. Härtel, H.-J. Graubaum, B. Schneider, *Ernaehr. Umsch.* **40** (1993) 152–155.

[30] S. C. Ziesenitz, G. Siebert, *Caries Res.* **20** (1986) 498–502.

[31] G. Siebert, S. C. Ziesenitz, J. Lotter, *Caries Res.* **21** (1987) 141–148.

[32] A. T. Brown, G. M. Best, *Caries Res.* **22** (1988) 2–6.

[33] Commission of the European Communities: "Sweeteners," *Report of the Scientific Committee for Foods*, no. 16, Luxembourg 1985.

[34] US Code of Federal Regulations, Title 21, parts 172, 180, and 189.

[35] *The Sweeteners in Food Regulations 1983*, SI 1983, no. 1211, London 1983.

[36] Zusatzstoff-Zulassungsverordnung, amendment dated 13.06.1990, BGBl I 1990, 1053–1067.

[37] Hoechst, DE 2001017, 1970 (K. Clauß, H. Jensen).

[38] K. Clauß, H. Jensen, *Angew. Chem.* **85** (1973) 965–973; *Angew. Chem. Int. Ed.* **12** (1973) 869–876.

[39] K. Clauß, E. Lück, G.-W. von Rymon Lipinski, *Z. Lebensm. Unters. Forsch.* **162** (1976) 37–40.

[40] G.-W. von Rymon Lipinski in D. Mayer, F. Kemper (eds.): *Acesulfame K*, Marcel Dekker, New York 1991, pp. 209–225.

[41] C. Klug, G.-W. von Rymon Lipinski, D. Böttger, *Z. Lebensm. Unters. Forsch.* **194** (1992) 476–478.

[42] A. Lotz, C. Klug, G.-W. von Rymon Lipinski, *Int. Z. Lebensmitteltech. Verfahrenstech.* **43** (1992) EFS 21–EFS 23.

[43] M. Korb, B. Kniel, E. Meyer, *Int. Z. Lebensmitteltech. Verfahrenstech.* **43** (1992) 494, 496, 498.

[44] H.-J. Arpe in B. Guggenheim (ed.): *Health and Sugar Substitutes*, Karger, Basel 1979, pp. 178–183.

[45] Hoechst, DE 2628294, 1975 (G.-W. von Rymon Lipinski, E. Lück).

[46] Hoechst, DE-OS 2264235, 1972 (K. Clauß, H. Jensen).

[47] A. Linkies, D. B. Reuschling, *Synthesis* **1990**, 405–406.

[48] Hoechst, EP-A 155634, 1985 (K. Clauß, A. Linkies, D. B. Reuschling).

[49] *Compendium of Food Additive Specifications*, vol. 1, Food and Agriculture Organization of the United Nations, Rome 1992, pp. 9–10.

[50] *Guide to Specifications for General Notices, General Methods, Identification Tests, Test Solutions and Other Reference Materials*, FAO Food and Nutrition Paper no. 5/Rev. 2, Food and Agriculture Organization of the United Nations, Rome 1991.

[51] J. F. Lawrence, C. F. Charbonneau, *J. Assoc. Off. Anal. Chem.* **71** (1988) 934–937.

[52] U. Hagenauer-Hener, C. Frank, U. Hener, A. Mosandl, *Dtsch. Lebensm. Rundsch.* **86** (1990) 348–351.

[53] Evaluation of Certain Food Additives and Contaminants, 27th Report of the Joint FAO/WHO Expert Committee on Food Additives, *WHO Tech. Rep. Ser.* no. **806** (1991) 20–21.

[54] *Fed. Reg.* **53** (1988) 28379–28387.

[55] G.-W. von Rymon Lipinski, B. E. Huddart, *Chem. Ind. (London)* **1983**, 427–433.

[56] R. H. Mazur in L. D. Stegink, L. J. Filer (eds.): *Aspartame*, Marcel Dekker, New York 1984, pp. 3–9.

[57] R. H. Mazur, J. M. Schlatter, A. H. Goldkamp, *J. Am. Chem. Soc.* **91** (1969) 2684–2691.

[58] A. van der Heijden, L. B. P. Brussel, H. G. Peer, *Chem. Senses Flavour* **4** (1979) 141–152.

[59] Y. Miyashita et al., *J. Med. Chem.* **29** (1986) 906–912.

[60] J. N. Bergmann, W. Vetsch in G.-W. von Rymon Lipinski, H. Schiweck (eds.): *Handbuch Süßungsmittel*, Behr's, Hamburg 1991, pp. 425–444.

[61] P. Langguth, R. Alder, P. Merkle, *Pharmazie* **46** (1991) 181–192.

[62] M. N. Tsoubeli, T. P. Labuza, *J. Food Sci.* **56** (1991) 1671–1675.

[63] W.-S. Tsang, M. A. Clarke, F. W. Parrish, *J. Agric. Food Chem.* **33** (1985) 734–738.

[64] K. Noda, T. Iohara, Y. Hirano, H. Hayabuchi, *Kaseigaku Zasshi* **42** (1991) 691–695.

[65] J. W. Fellows, S. W. Chang, W. H. Shazer, *J. Food Sci.* **56** (1991) 689–691.

[66] S. E. Keller, S. S. Newberg, T. M. Krieger, W. H. Shazer, *J. Food Sci.* **56** (1991) 21–23.

[67] L. N. Bell, T. P. Labuza, *J. Food Sci.* **56** (1991) 17–20.

[68] M. E. Brewster, T. Loftsson, J. Baldvinsdottir, N. Bodor, *Int. J. Pharm.* **75** (1991) R 5–R 8.

[69] Ajinomoto, EP-A 227301, 1986 (T. Yukawa et al.).

[70] Farmitalia, BE 900317, 1985 (F. Dallatomasina, R. Ortica, P. Giardino, E. Oppici).

[71] Monsanto, US 5053532, 1990 (J. B. Hill et al.).

[72] Monsanto, EP-A 221878, 1986 (J. S. Tou, B. D. Vineyard).

[73] NutraSweet, WO 9113860, 1991 (Y. Gelman).

[74] Y. Arioshi et al., *Bull. Soc. Chem. Jpn.* **46** (1973) 1893–1895.

[75] Ajinomoto, DE-AS 2152111, 1970 (Y. Arioshi et al.).

[76] K. Oyama, S. Irino, T. Harada, N. Hagi, *Ann. N. Y. Acad. Sci.* **434** (1984) 95–98.

[77] *Compendium of Food Additive Specifications*, vol. 1, Food and Agriculture Organization of the United Nations, Rome 1992, pp. 161–166.

[78] *Food Chemicals Codex*, 3rd ed., National Academy of Sciences, Washington 1981, pp. 28–29.

[79] Toxicological Evalution of Certain Food Additives. 24th Report of the Joint FAO-WHO Expert Committee on Food Additives, *WHO Tech. Rep. Ser.* **653** (1980) 20–21.

[80] *Fed. Reg.* **46** (1981) 38284–38308.

[81] Toyo Soda, EP-A 255092, 1987 (H. Wakamatsu et al.).

[82] B. Homler in L. D. Stegink, L. J. Filer (eds.): *Aspartame*, Marcel Dekker, New York 1984, pp. 247–262.

[83] L. F. Audrieth, M. Sveda, *J. Org. Chem.* **9** (1944) 89–101.

[84] K. M. Beck, *Food Technol. (Chicago)* **11** (1957) 156–158.

[85] L. Kreutzig in G.-W. von Rymon Lipinski, H. Schiweck (eds.): *Handbuch Süßungsmittel*, Behr's, Hamburg 1991, pp. 413–424.

[86] Du Pont, US 2275125, 1940 (L. F. Audrieth, M. Sveda).

[87] Abbott, GB 662800, 1949.

[88] Abbott, GB 669200, 1949.

[89] *Compendium of Food Additive Specifications*, vol. 1, Food and Agriculture Organization of the United Nations, Rome 1992, pp. 283–285, 473–474, 1333–1334.

[90] M. Lehr, W. Schmidt, *Z. Lebensm. Unters. Forsch.* **192** (1991) 335–338.

[91] B. A. Bopp, R. C. Sonders, J. W. Kesterson, *CRC Crit. Rev. Toxicol.* **16** (1986) no. 3, 213–306.

[92] Toxicological Evaluation of Certain Food Additives, 26th Report of the Joint FAO-WHO Expert Committee on Food Additives, *WHO Tech. Rep. Ser.* no. **683** (1982) 27–28.

[93] L. Kreutzig in G.-W. von Rymon Lipinski, H. Schiweck (eds.): *Handbuch Süßungsmittel,* Behr's, Hamburg 1991, pp. 397–412.

[94] H. C. Vincent et al., *J. Am. Pharm. Assoc.* **44** (1955) 442–446.

[95] Fahlberg-List, DE 35 211, 1884.

[96] J. W. Orelup, US 1 601 505, 1921.

[97] Fahlberg-List, DD 237 262, 1984 (B. Müller et al.).

[98] Maumee Development, US 2 667 503, 1951 (O. F. Senn).

[99] *Compendium of Food Additive Specifications,* vol. 1, Food and Agriculture Organization of the United Nations, Rome 1992, pp. 333–334, 1275–1278, 1385–1387.

[100] D. L. Arnold, D. Krewski, L. C. Munro, *Toxicology* **27** (1983) 179–256.

[101] World Health Organization PSC/93.8, Summary and Conclusions of the 41st Joint FAO/WHO Expert Committee on Food Additives Meeting.

[102] *Fed. Reg.* **42** (1977) 62 209–62 211.

[103] M. K. Cook, B. H. Gominger in G. E. Inglett (ed.): *Symposium Sweeteners,* AVI, Westport 1974, pp. 211–215.

[104] M. N. Sela, D. Steinberg in T. H. Grenby (ed.): *Progress in Sweeteners,* Elsevier Appl. Sci., London 1989, pp. 71–96.

[105] A. Bär et al., *Lebensm. Wiss. Technol.* **23** (1990) 371–376.

[106] J. F. Fisher, *J. Agric. Food Chem.* **25** (1977) 682–683.

[107] R. M. Horowitz, B. Gentili, US 3 375 242, 1968.

[108] L. Krbechek et al., *J. Agric. Food Chem.* **16** (1968) 108–112.

[109] Commission of the European Communities, *Report of the Scientific Committee for Food on Sweeteners,* Luxembourg 1987.

[110] B. Crammer, R. Ikan in T. H. Grenby (ed.): *Developments in Sweeteners – 3,* Elsevier Appl. Sci., London 1987, pp. 45–64.

[111] S. S. Chang, J. M. Cook, *J. Agric. Food Chem.* **31** (1983) 409–412.

[112] K. C. Phillips in T. H. Grenby (ed.): *Developments in Sweeteners – 3,* Elsevier Appl. Sci., London 1987, pp. 1–43.

[113] L. Hough, P. S. Phadnis, *Nature (London)* **263** (1976) 800.

[114] M. R. Jenner in T. H. Grenby (ed.): *Progress in Sweeteners,* Elsevier Appl. Sci., London 1989, pp. 121–141.

[115] M. E. Quinlan, M. R. Jenner, *J. Food Sci.* **55** (1990) 244–246.

[116] McNeilab, EP-A 220 907, 1986 (W. Tully, N. M. Vernon, P. A. Walsh).

[117] Evaluation of Certain Food Additives and Contaminants, 37th Report of the Joint FAO/WHO Expert Committee on Food Additives, *WHO Tech. Rep. Ser.* no. **806** (1991) 21–23.

[118] R. B. Iyengar et al., *Eur. J. Biochem.* **96** (1979) 193–204.

[119] A. M. de Vos et al., *Proc. Natl. Acad. Sci. USA* **82** (1985) 1406–1409.

[120] J. D. Higginbotham in C. A. M. Hough, K. J. Parker, A. J. Vlitos (eds.): *Developments in Sweeteners—1,* Applied Science, London 1979, pp. 103–121.

[121] L. Edens et al., *Gene* **18** (1982) 1–12.

[122] N. Overbeeke, *Biotechnol. Ser.* **13** (1989) 305–318.

[123] Toxicological Evaluation of Certain Food Additives, 26th Report of the Joint FAO-WHO Expert Committee on Food Additives, *WHO Tech. Rep. Ser.* no. **733** (1986) 35–36.

[124] Pfizer, US 4 411 925, 1982 (T. M. Brennan, M. E. Hendrick).

[125] Pfizer, Alitame Technical Summary 1987.

[126] Université Claude Bernard, EP-A 195 731, 1986 (C. Nofre, J.-M. Tinti).

[127] Université Claude Bernard, EP-B 289 430, 1988 (C. Nofre, J.-M. Tinti, F. Ouar).

[128] NutraSweet, EP-A 299 533, 1988 (R. Mazur, W. H. Owens, C. A. Klade, D. Madigan, G. W. Muller).

[129] Université Claude Bernard, EP-A 321 369, 1988 (C. Nofre, J. M. Tinti, F. Ouar).

[130] H. van der Wel, *FEBS Lett.* **21** (1972) 88–90.

[131] H. van der Wel, K. Loeve, *FEBS Lett.* **29** (1973) 181–184.

[132] DE 63 485, 1891 (J. Berlinerblau).

[133] W. Kegler, *Seifen Öle Fette Wachse* **77** (1951) 606.

[134] S. Petersen, E. Müller, *Chem. Ber.* **81** (1947) 31–38.

[135] H. P. Kaufmann, D. Schweizer, *Fette Seifen* **55** (1953) 321–324.

[136] F. M. Acton, M. A. Leaffer, S. M. Oliver, H. Stone, *J. Agric. Food Chem.* **18** (1970) 1061–1068.

# Sympatholytics and Sympathomimetics

Sympatholytics and sympathomimetics act on the sympathetic nervous system ($\rightarrow$ Neuropharmacology, **A17**, pp. 138–139; 154). Sympatholytics are blocking (antagonist), sympathomimetics, stimulant (agonist) drugs. They are subdivided into $\alpha_1$, $\alpha_2$, $\beta_1$, and $\beta_2$ agents according to the receptor types that they affect.

**Sympatholytics** are known as adrenergic blockers since they inhibit the stimulatory (adrenergic) effects of norepinephrine (noradrenaline) and epinephrine (adrenaline) by blocking the corresponding receptors.

$\alpha$-Sympatholytics are vasodilatory and they thus lead to a lowering of the blood pressure. $\alpha_1$-Drugs that selectively block the $\alpha_1$-adrenoceptors, such as prazosin [19216-56-9], indoramin [26844-12-2], and urapidil [34661-75-1], are effective in the treatment of hypertension ($\rightarrow$ Blood Pressure Lowering Agents, **A4**, pp. 245–246).

$\beta$-Sympatholytics are a widely used class of drugs that reduce heart rate by blocking cardioaccelerator receptors, and lower blood pressure by a mechanism that is not entirely clear. The most commonly used drug of this class is propranolol [525-66-6], which blocks both $\beta_1$- and $\beta_2$-receptors. It is used in the treatment of coronary heart disease, hypertension, and cardiac arrhythmias and other conditions ($\rightarrow$ Coronary Therapeutics, **A8**, pp. 5–7; $\rightarrow$ Blood Pressure Lowering Agents, **A4**, pp. 247–250; $\rightarrow$ Antiarrhythmic Drugs, **A2**, p. 443, 448). Other important $\beta$-sympatholytics are metoprolol [37350-58-6], nadolol [42200-33-9], pindolol [13523-86-9], timolol [26839-75-8]. Timolol is also used as eye drops in the treatment of glaucoma ($\rightarrow$ Ophthalmological Preparations, **A18**, p. 138). For further information, see $\rightarrow$ $\beta$-Receptor Blocking Agents.

**Sympathomimetics,** also called adrenergic drugs, act on the sympathetic nerves in the same way as norepinephrine and epinephrine ($\rightarrow$ Hormones, **A13**, p. 108). The action of $\alpha$-sympathomimetics corresponds to that of norepinephrine, resulting in vasoconstriction and increased peripheral resistance and blood pressure. They include hypertensive drugs such as norfenefrine [536-21-0].

The stimulation of $\beta_1$-receptors leads to increased heart rate and cardiac output, whereas the stimulation of $\beta_2$-receptors causes relaxation of smooth muscle, e.g., vascular and bronchial. Sympathomimetics stimulating both $\alpha$- and $\beta$-receptors are commonly employed to raise blood pressure ($\rightarrow$ Blood Pressure Increasing Agents, **A4**, pp. 229–232) and there is limited application in the short-term treatment of heart failure (e.g., dopamine [51-61-6], dobutamine [34368-04-2]). The cardiac glycosides and synthetic cardiotonic drugs act by different mechanisms to stimulate the heart (**A5**, pp. 282–284). $\beta_2$-Sympathomimetics (e.g., fenoterol [13392-18-2]) cause bronchodilation and are used in the treatment of asthma ($\rightarrow$ Antiasthmatics, **A2**, pp. 455–461).

Sympathomimetics are also used as mydriatics to dilate the pupil of the eye for diagnostic purposes ($\rightarrow$ Ophthalmological Preparations, **A18**, pp. 141–142) and in the treatment of glaucoma ($\rightarrow$ Ophthalmological Preparations, **A18**, p. 138).

A distinction is made between indirectly acting and centrally acting sympathomimetics. Indirectly acting sympathomimetics, e.g., stimulants such as amphetamines, ephedrine, and cocaine, release norepinephrine from sympathetic nerve endings. Centrally acting $\alpha$-sympathomimetics, such as clonidine [4205-90-7] and methyldopa [555-30-6], cause stimulation of $\alpha_2$-adrenoreceptors in the central nervous system which, in contrast to stimulation of reception in blood vessels, leads to a fall in blood pressure ($\rightarrow$ Blood Pressure Lowering Agents, **A4**, p. 242).

**Synthesis Gas $\rightarrow$ Gas Production**

Ullmann's Encyclopedia
of Industrial Chemistry, Vol. A 26

# Talc

WERNER TUFAR, Fachbereich Geowissenschaften der Philipps-Universität Marburg, Marburg/Lahn, Federal Republic of Germany



## 1. Talc

Talc, an important industrial mineral, forms with pyrophyllite (Chap. 2) a separate group of silicates, the phyllosilicates, which have a mica-like lamellar structure. Talc is a magnesium hydrosilicate, and pyrophyllite an aluminum hydrosilicate. The minerals are very similar crystallographically and in their properties.

Pure talc has the chemical composition $3\,MgO \cdot SiO_2 \cdot H_2O$. Its mineralogical chemical formula

$$\frac{2}{\infty}Mg_3{}^{[6]}\{(OH)[Si_2O_5]\}_2 \quad \text{or}$$
$$\frac{2}{\infty}Mg_3{}^{[6]}[(OH)_2/Si_4O_{10}]$$

clearly represents the layer silicate structure.

The talc lattice (see Fig. 1) is composed of infinite two-dimensional ($\infty$) silicate double layers $\frac{2}{\infty}\{[Si_2O_5]^{2-}\}_2$ or $\frac{2}{\infty}[Si_4O_{10}]^{4-}$ in which the apical O atoms of all of $SiO_4$ tetrahedra within one individual layer point in the same direction, and the OH groups occupy the centers of the hexagons formed by these apical oxygen atoms. In the double layer formed by two individual layers, the oxygen layers formed by the O atoms and the OH groups are located directly opposite to one another, i.e., are directed towards each other. These two single layers within the double layer are linked together by magnesium atoms which are coordinated octahedrally by $4\,O$ and $2\,OH$, the coordination number being symbolized by $Mg^{[6]}$. The double layer ("talc layer") so formed is therefore electrically neutral, and is only weakly bonded to the neighboring double layers by Van der Waals forces. This explains the pronounced cleavage properties of talc in the direction $\{001\}$.

Structurally, pyrophyllite is comparable to talc, which it also resembles in other respects:

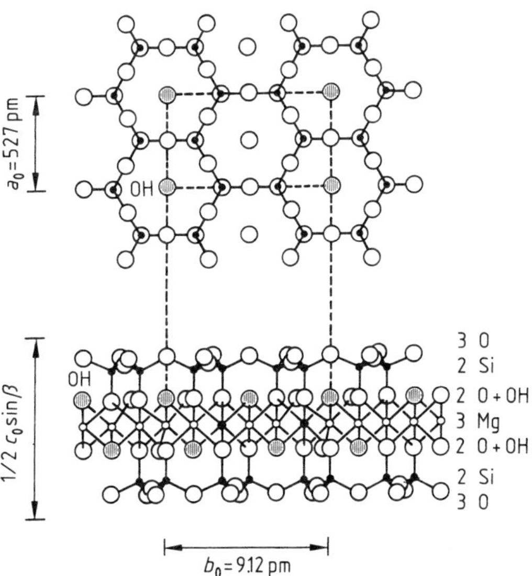



**Figure 1.** Structure of talc
Projection parallel to $c$ on (001) and a plane perpendicular to the $a$ axis; the monoclinic angle $\beta$ is 100° 00′
In pyrophyllite, only the locations shown as open circles are occupied by aluminum (2 Al in place of 3 Mg)

$$\frac{2}{\infty}Al_2{}^{[6]}\{(OH)[Si_2O_5]\}_2 \quad \text{or}$$
$$\frac{2}{\infty}Al_2{}^{[6]}[(OH)_2/Si_4O_{10}]$$

Unlike talc, the two single layers in pyrophyllite are bonded by $Al^{3+}$ instead of $Mg^{2+}$ ("pyrophyllite layer"). All $^3/_3$ octahedral coordination centers in talc are occupied by $Mg^{[6]}$, but in pyrophyllite only $^2/_3$ are occupied by $Al^{[6]}$. Thus, whereas talc shows trioctahedral cation occupation, in pyrophyllite it is dioctahedral.

The pyrophyllite layer and the talc layer form the basis of the mica structure ($\rightarrow$ Mica). In micas, within the pyrophyllite layer (muscovite series) and the talc layer (biotite series), one-fourth of the $Si^{4+}$ ions of the $SiO_4$ tetrahedra are replaced by $Al^{3+}$. Charge balance is achieved by the introduction of 12-coordinate $K^+$ ($K^{[12]}$) between the double layers.

Both minerals form monoclinic crystals with space group $C_{2h}^6 - C2/c$ or $C_s^4 - Cc$. In the most common stacking sequence, $2M_1$, talc has the following lattice constants: $a_o = 527$ pm, $b_o = 912$ pm, $c_o = 1885$ pm, $\beta = 100° \, 00'$.

The axial ratio is $a_o : b_o : c_o = 0.578 : 1 : 2.067$ and $Z = 4$ (number of formula units per unit cell). The interlayer distance $d_{002} = \tfrac{1}{2} c \cdot \sin \beta$ is 928 pm (9 Å layer silicate). If iron is present, these lattice constants can vary slightly.

The $Mg^{2+}$ in talc can be partly replaced by $Fe^{2+}$ or $Ni^{2+}$, giving the following talc varieties:

Minnesotaite: $\overset{2}{\infty}(Fe^{2+}, Mg, H_2)_3[(OH)_2/(Si, Al, Fe^{3+})_4 O_{10}]$

Willemsite: $\overset{2}{\infty}(Ni, Mg)_3[(OH)_2/Si_4 O_{10}]$.

Idiomorphic talc formations as crystals or pseudohexagonal lamellae or platelets are rare. The mineral mainly occurs in lamellar or broad columnar radial aggregates and masses. It commonly occurs in a flaky to compact form (soapstone or steatite). Talc masses intergrown with chlorite are known as potstone.

Talc has a greasy feel due to its very low hardness (Mohs hardness = 1), and cleaves extremely readily parallel to {001}. The density $\rho$ is 2.7–2.8 g/cm³. The optical refraction ($\alpha$, $\beta$, $\gamma$) and birefringence $\Delta$ in a section under the polarizing microscope are as follows:

$\alpha = $ 1.539–1.550
$\beta = $ 1.589–1.594
$\gamma = $ 1.589–1.596

$\Delta = -(0.050 - 0.046)$

The optic axis angle is $2V_\alpha = 0 - 30°$ with distinct dispersion r > V.

Talc is formed under hydrothermal conditions and is a typical mineral of weaker regional metamorphosis (regional dynamo-thermometamorphosis). It often occurs in association with chlorite, serpentine, or magnesite.

The main parent rocks that undergo metamorphic mineral reactions leading to talc formation are either ($\pm$ magnesite-bearing) siliceous dolomites, or olivine- and/or pyroxene-containing ultrabasites. In ultrabasic rocks, talc is often a product of autohydrothermal conversion.

The formation and occurrence of talc can be illustrated by a number of metamorphic mineral reactions that can be reproduced in experimental hydrothermal syntheses. Increasing temperature shifts the reactions towards liberation of water and $CO_2$.

$$3\,CaMg[CO_3]_2 + 4\,SiO_2 + H_2O$$
Dolomite     Quartz
$$\rightleftharpoons Mg_3[(OH)_2/Si_4O_{10}] + 3\,CaCO_3 + 3\,CO_2 \quad (1)$$
Talc     Calcite

$$5\,Mg_3[(OH)_2/Si_4O_{10}] + 6\,CaCO_3 + 4\,SiO_2$$
Talc     Calcite     Quartz
$$\rightleftharpoons 3\,Ca_2Mg_5[(OH,F)/Si_4O_{11}]_2 + 6\,CO_2 + H_2O \quad (2)$$
Tremolite

$$2\,Mg_3[(OH)_2/Si_4O_{10}] + 3\,CaCO_3$$
Talc     Calcite
$$\rightleftharpoons Ca_2Mg_5[(OH,F)/Si_4O_{11}]_2 + CaMg[CO_3]_2$$
Tremolite     Dolomite
$$+ CO_2 + H_2O \quad (3)$$

$$2\,CaMg[CO_3]_2 + Mg_3[(OH)_2/Si_4O_{10}] + 4\,SiO_2$$
Dolomite     Talc     Quartz
$$\rightleftharpoons Ca_2Mg_5[(OH,F)/Si_4O_{11}]_2 + CO_2 \quad (4)$$
Tremolite

$$Mg_3[(OH)_2/Si_4O_{10}] + 5\,CaMg[CO_3]_2$$
Talc     Dolomite
$$\rightleftharpoons 4\,Mg_2[SiO_4] + 5\,CaCO_3 + 5\,CO_2 + H_2O \quad (5)$$
Forsterite     Calcite

$$11\,Mg_3[(OH)_2/Si_4O_{10}] + 10\,CaCO_3$$
Talc     Calcite
$$\rightleftharpoons 5\,Ca_2Mg_5[(OH,F)/Si_4O_{11}]_2$$
Tremolite
$$+ 4\,Mg_2[SiO_4] + 10\,CO_2 + 6\,H_2O \quad (6)$$
Forsterite

$$13\,Mg_3[(OH)_2/Si_4O_{10}] + 10\,CaMg[CO_3]_2$$
Talc     Dolomite
$$\rightleftharpoons 5\,Ca_2Mg_5[(OH,F)/Si_4O_{11}]_2$$
Tremolite
$$+ 12\,Mg_2[SiO_4] + 20\,CO_2 + 8\,H_2O \quad (7)$$
Forsterite

$$3\,MgCO_3 + 4\,SiO_2 + H_2O$$
Magnesite  Quartz
$$\rightleftharpoons Mg_3[(OH)_2/Si_4O_{10}] + 3\,CO_2 \quad (8)$$
Talc

$$Mg_3[(OH)_2/Si_4O_{10}] + H_2O$$
Talc
$$\rightleftharpoons Mg_3[(OH)_4/Si_2O_5] + 2\,SiO_2 \quad (9)$$
Serpentine     Quartz

$$2\,Mg_3[(OH)_4/Si_2O_5] + 3\,CO_2$$
$$\text{Serpentine}$$
$$\rightleftharpoons Mg_3[(OH)_2/Si_4O_{10}] + 3\,MgCO_3 + 3\,H_2O \qquad (10)$$
$$\text{Talc} \qquad \text{Magnesite}$$

$$5\,Mg_3[(OH)_4/Si_2O_5]$$
$$\text{Serpentine}$$
$$\rightleftharpoons 6\,Mg_2[SiO_4] + Mg_3[(OH)_2/Si_4O_{10}] + 9\,H_2O \qquad (11)$$
$$\text{Forsterite} \qquad \text{Talc}$$

$$Mg_3[(OH)_4/Si_2O_5] + 2\,SiO_2$$
$$\text{Serpentine} \qquad \text{Quartz}$$
$$\rightleftharpoons Mg_3[(OH)_2/Si_4O_{10}] + H_2O \qquad (12)$$
$$\text{Talc}$$

$$Mg_3[(OH)_2/Si_4O_{10}] + 5\,MgCO_3$$
$$\text{Talc} \qquad \text{Magnesite}$$
$$\rightleftharpoons 4\,Mg_2[SiO_4] + H_2O + 5\,CO_2 \qquad (13)$$
$$\text{Forsterite}$$

$$4\,SiO_2 + 3\,MgCO_3 + H_2O$$
$$\text{Quartz} \quad \text{Magnesite}$$
$$\rightleftharpoons Mg_3[(OH)_2/Si_4O_{10}] + 3\,CO_2 \qquad (14)$$
$$\text{Talc}$$

$$9\,Mg_3[(OH)_2/Si_4O_{10}] + 4\,Mg_2[SiO_4]$$
$$\text{Talc} \qquad \text{Forsterite}$$
$$\rightleftharpoons 5\,(Mg,Fe)_7[(OH)/Si_4O_{11}]_2 + 4\,H_2O \qquad (15)$$
$$\text{Anthophyllite}$$

$$7\,Mg_3[(OH)_2/Si_4O_{10}]$$
$$\text{Talc}$$
$$\rightleftharpoons 3\,(Mg,Fe)_7[(OH)/Si_4O_{11}]_2 + 4\,SiO_2 + 4\,H_2O \qquad (16)$$
$$\text{Anthophyllite} \qquad \text{Quartz}$$

$$2\,Mg_3[(OH)_2/Si_4O_{10}] + MgCO_3$$
$$\text{Talc} \qquad \text{Magnesite}$$
$$\rightleftharpoons (Mg,Fe)_7[(OH)/Si_4O_{11}]_2 + H_2O + CO_2 \qquad (17)$$
$$\text{Anthophyllite}$$

Assuming that these formulas represent the chemical composition of the parent rock, talc can be formed over a wide temperature range. For example, under isobaric conditions with $X_{CO_2} = 0.3$, the equilibrium temperature for Reaction (1) is 425 °C, while that of Reaction (3) is ca. 50 °C lower ($X_{CO_2}$ is the molar fraction of $CO_2$ in the gas phase, which consists of $CO_2$ and $H_2O$). The upper temperature limit for the existence of talc is set by Reaction (16), which occurs at ca. 700 °C and a water vapor pressure of 1000 bar. Reactions (4) and (7) only occur if magnesite is present with quartz and dolomite in the original rock prior to regional metamorphosis. Reaction (15) requires the presence of $CO_2$.

In commercially important deposits, talc therefore also occurs in association with tremolite, calcite, quartz, and dolomite, as the product of a more intense regional metamorphosis of siliceous dolomites, and with forsterite and anthophyllite due to intense regional metamorphic overprint of ultrabasites. However, talc can also be observed as an authigenic new formation, e.g., in sandy sediments and salt deposits.

Reactions (1)–(17) include a number of minerals associated with talc in its deposits. Other minerals that occur in association with talc include chlorites, mica, actinolite, feldspars, rutile, magnetic iron pyrites, pyrites, magnetite, and hematite. Limonite, a product of the weathering of iron-containing minerals, especially ores, can often intersperse with talc.

Deposits of talc are widely distributed and mined worldwide in regional metamorphic complexes. The important deposits in Europe are in Germany (Göpfersgrün-Thierstein in the Fichtel Mountains of Bavaria, Schwarzenberg on the Saale), in Austria (Rabenwald in Eastern Styria, Lassing, Weisskirchen), Norway (Altermark near Bergen), Finland (Lahnaslampi near Sotkamo, Central Finland; Polvijärvi near Sola), Spain (mainly in the provinces of León, Gerona, Málaga, and Almería), and Italy (Pinerolo/Piedmont, Sardinia). There are also deposits in France (Dept. Ariège), Sweden, and the United Kingdom (Cornwall, Shetland Isles).

Important deposits occur in Canada (Madoc/Ontario and Van Reet/Quebec, which are deep mined, and Penhorwood/Timmins, which is opencast mined), the United States [deposits in 11 states; > 80 % of production from Texas: Allamoore Talc District with Van Horn/Hudspeth County; also Montana: Yellowstone, Beaverhead, Treasure Chert; Vermont: Hammondville, Rainboro; California (trains carry talc from deposits in the south east to Nevada); New York State], Uruguay, Brazil (Paraná, Sao Paulo; Bahia: Serra das Eguas), Argentina, Australia (Three Springs, Mt. Seabrook/Western Australia; Mt. Fitton/South Australia), Japan (Hokkaido, Kyushu), the People's Republic of China (Haicheng), Taiwan, South Korea (Dongyang/Chungju region, other deposits in the Geangju, Yesan, Muju/Jinan regions), India (Rajasthan, Andhra Pradesh, Madhja Pradesh, Tamil Nadu, Bihar), and the former Soviet Union.

European talc production is ca. 500 000–700 000 t/a, of which ca. 100 000 t/a come from Austria and Norway.

As talc and pyrophyllite can often be substituted for each other, they are usually grouped together in production and consumption statistics.

A very important application of massive talc is in the production of ceramics (e.g., refractory castings, high-voltage insulators, burners in acetylene lamps). Here, the important properties of talc are its low hardness (enabling it to be easily cut and shaped), the stability of cut products towards firing, and the very low degree of shrinkage. The steatite porcelains are never pure white in color, but are yellowish to brownish, and have higher strength than normal porcelains.

For many other uses, talc is first ground. Powdered talc is used as a lubricant or polishing agent, in pastels, as a filler in the pigment industry, in cosmetics (face powder), as an additive for soaps, in paper production, as a finishing agent in the textile industry, as a filler and glazing agent in a wide range of industries, as a filler in the plastics and rubber industries, as coating for rubber products to avoid sticking and in asphalt processing.

In Japan, talc in the form of soapstone was formerly used in the food industry for polishing rice. However, this was abandoned after it was shown that asbestos, present as an impurity in soapstone, was causing an increased incidence of stomach cancer in Japan.

Talc competes with kaolin as a filler, but cannot compete with cheaper fillers such as fine silica sand and calcite. In ceramics (steatite ceramics, high strength steatite ceramics), however, talc cannot be replaced by kaolin.

Talc is used wherever fillers with absorption properties are required or where there are very high specifications for color and/or toxicity (e.g., in the dyestuff, cosmetics, and pharmaceutical industries).

Garnierite, $(Ni, Mg)_6[(OH)_8/Si_4O_{10}]$, which was formerly regarded as nickel serpentine or nickel antigorite, is now regarded as a mixture of 7 Å and 10 Å layer silicates, and alternately deposited mineral components. The term garnierite is now restricted to the 9 Å component, which is classed as nickel talc. Together with other nickel hydrosilicates (e.g., with the nickel chlorite schuchardtite), it occurs as an important component of the "New Caledonia" type of deposit, which commercially is a very important type of workable nickel deposit.

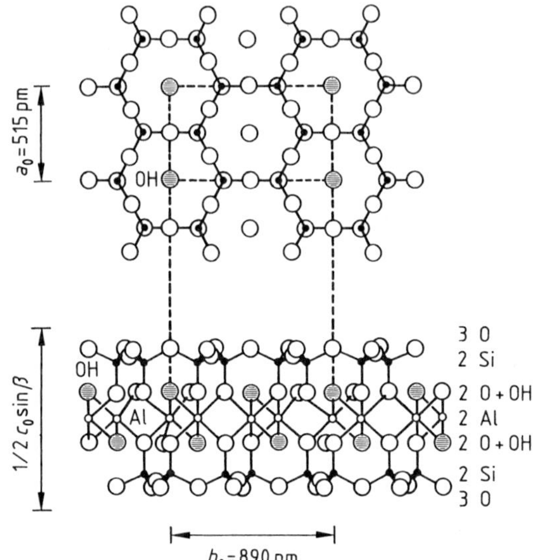



**Figure 2.** Structure of pyrophyllite
Projection parallel to $c$ on (001) and a plane perpendicular to the $a$ axis; the monoclinic angle $\beta$ is 99° 55′

## 2. Pyrophyllite

The important industrial mineral pyrophyllite forms, with talc, a special group of phyllosilicates (layer silicates) with micalike laminar structures. Pyrophyllite is an aluminum hydrosilicate, whereas talc is a magnesium hydrosilicate. Pyrophyllite is similar to talc crystallographically and even more so in its other properties.

The pyrophyllite lattice (see Fig. 2) is an infinite coherent two-dimensional silicate double layer of linked $[SiO_4]$ tetrahedra: $^2_\infty\{[Si_2O_5]^{2-}\}_2$ or $^2_\infty[Si_4O_{10}]^{4-}$, in which all apical O atoms of the $SiO_4$ tetrahedra within one $[SiO_5]$ layer point in the same direction, and the OH groups occupy the centers of the hexagons formed by these oxygen atoms. In the double layer, the oxygen layers formed by the O atoms and OH groups are situated directly opposite to each other, and they are linked together by $Al^{3+}$, which is octahedrally coordinated by 4 O and 2 OH. This coordination number is symbolized by $Al^{[6]}$. The resulting silicate double layer ("pyrophyllite layer") is therefore electrically neutral and is only weakly bonded to the neighboring double layers by Van der Waals forces. This results in the laminar structure and pronounced cleavage of pyrophyllite in the direction {001}. In talc, unlike pyrophyllite, the two single layers are bonded by $Mg^{2+}$ ("talc layer") instead of $Al^{3+}$. Thus, in pyrophyllite only $2/3$ of the octahedral coordination centers are occupied by $Al^{[6]}$, while all $3/3$ in talc are occupied by $Mg^{[6]}$. Thus, whereas pyrophyllite shows dioctahedral cation occupation, in talc it is trioctahedral, like the corresponding charged mica layers, e.g., in muscovite $K^+\{Al_2[(OH)_2/Si_3AlO_{10}]\}^-$.

Pure pyrophyllite has the chemical composition $Al_2O_3 \cdot 4SiO_2 \cdot H_2O$, its mineralogical chemical formula being

$$^2_\infty Al_2^{[6]}\{(OH)[Si_2O_5]\}_2 \quad \text{or}$$
$$^2_\infty Al_2^{[6]}[(OH)_2/Si_4O_{10}]$$

Pyrophyllite forms monoclinic crystals with space group $C_{2h}^6 - C2/c$ or $C_s^4 - Cc$, and has the following lattice constants: $a_o = 515$ pm, $b_o = 892$ pm, $c_o = 1859$ pm, $\beta = 99° 55'$.

The axial ratio is $a_o:b_o:c_o = 0.577:1:2.084$ and $Z = 4$ (number of formula units per unit cell). The interlayer distance $d_{002} = 1/2 c \cdot \sin \beta$ is 916 pm. If iron is present, these lattice constants can vary slightly.

The $Al^{3+}$ in pyrophyllite can be partly replaced by ca. 2 atom% (average) of $Fe^{3+}$. The small amounts of MgO and CaO found in many analyses can be attributed to mechanical contamination of the analytical samples. Chrome pyrophyllite, containing ca. 3 wt% $Cr_2O_3$, also occurs naturally.

Pseudohexagonal lamellar or tabular crystals, developed in the {001} direction, are always formed in coarse radial, random, or fan-shaped crystalline aggregates. Massive pyrophyllite is known as agalmatolite or figure stone (e.g., large deposits in Minas Gerais/Brazil). Pyrophyllite is translucent, silvery or silvery-white, white, yellowish, golden yellow, greenish, or apple green, and has a nacreous luster. The lamellae are flexible and have a slightly greasy feel. Cleavage in the {001} plane is fully developed. The Mohs hardness is $1\frac{1}{2}$ and the density 2.79 $g/cm^3$.

Since water in pyrophyllite is in the form of hydroxide only, strong heating is needed to evolve it, leading to exfoliation and the formation of wormlike shapes. Pyrophyllite is difficult to melt. It is only slowly decomposed by sulfuric acid, but is more readily attacked by hydrofluoric acid.

In thin sections under the polarizing microscope, pyrophyllite is always colorless ($x < y \approx z$). Its optical refraction $\alpha$, $\beta$, $\gamma$ and birefringence $\Delta$ are

$$\alpha = \quad 1.552 - 1.556$$
$$\beta = \quad 1.558 - 1.589$$
$$\gamma = \quad 1.600 - 1.601$$

$$\Delta = -(0.048 - 0.045)$$

The optical axis angle is $2V_\alpha = 53 - 62°$ with weak dispersion $r > V$.

Depending on its origin, pyrophyllite can mainly consist either of soft, fragile aggregates or coarsely radiating lamellar crystals, very similar to lamellar talc, or dense, coarse aggregates with an uneven, matt to lustrous fracture, similar to massive talc.

Pyrophyllite can occur in pneumatolytic formations, where it is formed by the transformation of aluminum-rich silicates (e.g., vein deposits with cassiterite or amblygonite). In hydrothermal regions, pyrophyllite can be formed from feldspar if the alkalis can be removed by hydrothermal solutions, or from kyanite (disthene), andalusite, kaolinite, etc., provided that there is a supply of $SiO_2$. Suitable rocks (such as tuffs and acidic extrusive rocks, e.g., hyalorhyolite) can be (hydrothermally) converted, sometimes completely, into pyrophyllite rock or agalmatolite. The transport of pyrophyllite in hydrothermal solutions leads to its deposition in fissures, where it occurs with quartz. Pyrophyllite also occurs with quartz in nodules and lenses of regional metamorphic rocks (crystalline schists). In regional metamorphic formations, pyrophyllite can be found in the weak epizones.

Important metamorphic mineral reactions which elucidate the occurrence and distribution of pyrophyllite in certain formations include

$$CaAl_2[(OH)_2/Si_2O_7] \cdot H_2O + 2SiO_2 + CO_2$$
$$\text{Lawsonite} \qquad\qquad \text{Quartz}$$
$$\rightleftharpoons CaCO_3 + Al_2[(OH)_2/Si_4O_{10}] + H_2O$$
$$\text{Calcite} \qquad \text{Pyrophyllite}$$

$$Al_4[(OH)_8/Si_4O_{10}] + 4SiO_2$$
$$\text{Kaolinite} \qquad \text{Quartz}$$
$$\rightleftharpoons Al_2[(OH)_2/Si_4O_{10}] + H_2O$$
$$\text{Pyrophyllite}$$

The equilibrium conditions of this metamorphic reaction are (with $P_{H_2O} = P_{total}$):

1000 bar/325 $\pm$ 20 °C;
2000 bar/345 $\pm$ 10 °C;
4000 bar/375 $\pm$ 15 °C.

$$Al_2[(OH)_2/Si_4O_{10}] + CaCO_3$$
Pyrophyllite          Calcite

$$\rightleftharpoons CaAl_2[(OH)_2/Al_2Si_2O_{10}] + 6SiO_2 + H_2O + CO_2$$
Margarite          Quartz

$$CaAl_2[(OH)_2/Al_2Si_2O_{10}] + 4SiO_2$$
Margarite          Quartz

$$\rightleftharpoons Ca[Al_2Si_2O_8] + Al_2[(OH)_2/Si_4O_{10}]$$
Anorthite          Pyrophyllite

$$2Al_2[(OH)_2/Si_4O_{10}] +$$
Pyrophyllite

$$\{Fe_2Al(OH)_6\} \cdot \{Fe_2Al[(OH)_2/Al_2Si_2O_{10}]\}$$
Fe-rich Chlorite

$$\rightleftharpoons 2Fe_2AlAl_3[(OH)_4/O_2/(SiO_4)_2] + 6SiO_2 + 2H_2O$$
Chloritoid          Quartz

$$Al_2[(OH)_2/Si_4O_{10}] \rightleftharpoons AlAl[O/SiO_4] + 3SiO_2 + H_2O$$
Pyrophyllite          Andalusite,    Quartz
Kyanite

Under certain hydrothermal conditions, pyrophyllite can be converted into diaspore ($\alpha$-AlOOH).

Pyrophyllite can be synthesized from neutral or strongly acid colloidal solutions of aluminum hydroxide and $SiO_2$ at high temperatures (420 – 575 °C) and relatively low pressures.

By far the most important reserves are the Roseki deposits ("wax stone", Japanese term for hydrothermally formed pyrophyllite-rich rocks with a waxy appearance) in Japan (Mitsuishi District/Okayama Prefecture, Shokozan District/Hiroshima Prefecture SW of Honshu, Goto District/Nagasaki Prefecture on Goto Island) and the Nab-Suk deposits (Korean term for deposits of similar origin to the Japanese "Roseki") in South Korea (southeast: Wondong, Milyang, Dongnae, and in the far southwest: Wando, Kasado, Sungsan). Other deposits are found in northern Finland (Hirvivaare), in the United States (North Carolina: Robbins, Glendon, Hillsborough, and also California), in Canada (Manuels/Newfoundland), Brazil (Districts of Onca de Pitangui and Matheus Leme in Minas Geraes), India (Madhya Pradesh, Uttar Pradesh), Thailand, Turkey (Malatya Province), and Australia (Pambyula/New South Wales). In the Republic of South Africa (Ottosdal/Lichtenburg District) "wonderstone" is extracted, which contains 90 % pyrophyllite with some epidote, chlorite, and rutile.

As pyrophyllite has very similar properties to talc, its industrial uses are also similar, and the minerals can be substituted for one another.

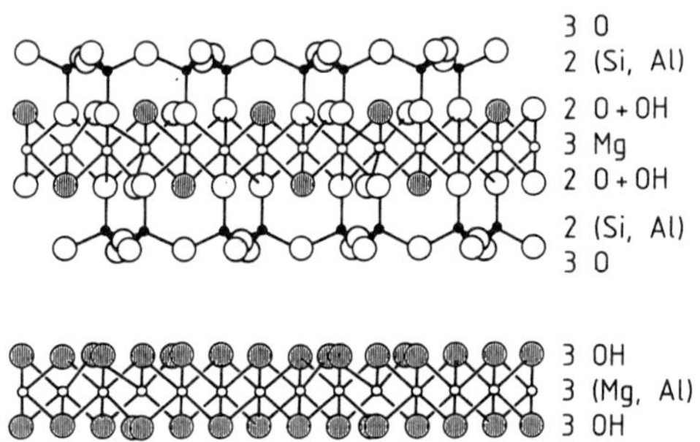

Figure 3. Idealized structure of chlorite
Projection on a plane perpendicular to the *a* axis

Outside Europe (e.g., Japan, South Korea, United States), pyrophyllite is also used as a raw material in the refractory industry. A special application is in diamond synthesis where it is used as a lining material for the ultrahigh pressure cells. In this application it is advantageous that the melting point of pyrophyllite increases with increasing pressure.

## 3. Chlorite

The chlorites ("14 Å chlorites") form a very heterogeneous group, and can sometimes be important rock-forming minerals. Their chemical structure (Fig. 3) shows them to be essentially hydrated Mg–Fe–Al silicates that are free of alkali metals and calcium and are built up of two kinds of oppositely charged layer ions.

Formerly, low-iron orthochlorites were regarded as "true chlorites" and were distinguished from iron-rich leptochlorites. However, orthochlorites are now considered to be those in which $Fe_2O_3 < 4\%$, and leptochlorites those in which $Fe_2O_3 > 4\%$.

In the chlorites, trioctahedral phyllosilicates (layer silicates) are present, consisting of negatively charged mica layers ($\rightarrow$ Mica **A 16**, p. 551, **A 23**, p. 662, 670, 671) alternating with positively charged brucitelike layers. All octahedral sites are occupied by cations (Mg, Al, Fe) forming trioctahedral layer silicates.

As in the other phyllosilicates, the $Si^{4+}$ in chlorites is partly replaced by $Al^{3+}$ in the tetrahedral positions of the mica layer. Charge balance is achieved by the replacement of divalent cations ($Mg^{2+}$, $Fe^{2+}$, $Mn^{2+}$, $Cr^{2+}$, $Ni^{2+}$) by trivalent cations ($Al^{3+}$, $Fe^{3+}$) in the hydroxide layer. This results in a large number of possible chemical variants, and further variations can result from the replacement of $Mg^{2+}$, mainly by $Fe^{2+}$, in the octahedral positions of the talc layer and the brucite layer.

In contrast, the sudoite series of minerals, which have only been known since the early 1960s, are "dioctahedral chlorites". They consist of alternating muscovite and (charged) brucite layers. Charge compensation, necessary because of the partial substitution of Si by Al in the $[SiO_4]$ tetrahedra, results from the introduction of trivalent in place of divalent cations in the brucite-like layer as well as by over-occupation of the octahedral interstices.

With regard to the six octahedrally coordinated cations, the minerals of the chlorite series can also be regarded as mixed crystals of four endmembers, although these are hypothetical, and do not occur in nature in the pure state:

$$Fe_6\ \text{Chlorite}\quad \begin{cases} Fe_3^{2+}[(OH)_2/Si_3AlO_{10}]^- \\ Fe_2^{2+}Fe^{3+}(OH)_6^+ \end{cases}$$

$$Mg_5Al\ \text{Chlorite}\quad \begin{cases} Mg_3[(OH)_2/Si_3AlO_{10}]^- \\ Mg_2Al(OH)_6^+ \end{cases}$$

chemical composition comparable to antigorite

$$Fe_4^{2+}Al_2\ \text{Chlorite}\quad \begin{cases} Fe_2^{2+}Al[(OH)_2/Al_2Si_2O_{10}]^- \\ Fe_2^{2+}Al(OH)_6^+ \end{cases}$$

daphnite

$$Mg_4Al_2\ \text{Chlorite}\quad \begin{cases} Mg_2Al[(OH)_2/Al_2Si_2O_{10}]^- \\ Mg_2Al(OH)_6^+ \end{cases}$$

chemical composition comparable to amesite

However, the two magnesium-rich minerals antigorite and amesite, sometimes described as endmembers, do belong to the serpentine minerals and not to the chlorites. The conversion from the chlorite structure to the serpentine structure only requires the apical O atoms of a $T_2O_5$ (T = Si, Al) layer to be shifted to the other side. If the transformation is only partial, the poorly crystallized 7 Å chlorites or septechlorites result.

The minerals of the chlorite series form monoclinic crystals, usually with the space group $C_{2h}^3 - C2/m$, $Z=4$ (number of formula units per unit cell). However, the number of possible stacking variations is much greater than in the case of micas.

Much simplified, chlorite has the following lattice constants: $a_o \approx 530$ pm (5.3 Å), $b_o \approx 930$ pm (9.3 Å), $c_o \approx 2860$ pm (28.6 Å). The axial ratio is $a:b:c \approx 0.57:1:3.09$, $\beta \approx 97°$. The layer thickness is $d_{002} = 1420$ pm (14 Å layer silicate).

Of the trioctahedral chlorites (chlorites in the narrower sense), several groups ("series") can be distinguished.

The talc chlorites include:

Talcchlorite (hypothetical end member) $\begin{cases} Mg_3[(OH)_2/Si_4O_{10}] \\ Mg_3(OH)_6 \end{cases}$ (same chemical composition as serpentine, $Mg_3[Si_2O_5/(OH)_4]$, which is thermodynamically more stable)

Penninite $\begin{cases} (Mg,Al)_3[(OH)_2/Al_{0.5-0.9}Si_{3.5-3.1}O_{10}] \\ (Mg,Al)_3(OH)_6 \end{cases}$

Clinochlore $\begin{cases} (Mg,Al)_3[(OH)_2/AlSi_3O_{10}] \\ (Mg,Al)_3(OH)_6 \end{cases}$

Sheridanite $\begin{cases} (Mg,Al)_3[(OH)_2/Al_{1.2-1.5}Si_{2.8-2.5}O_{10}] \\ (Mg,Al)_3(OH)_6 \end{cases}$

Rhipidolite $\begin{cases} (Mg,Fe,Al)_3[(OH)_2/Al_{1.2-1.5}Si_{2.8-2.5}O_{10}] \\ (Mg,Al)_3(OH)_6 \end{cases}$

Corundophilite $\begin{cases} (Mg,Fe,Al)_3[(OH)_2/Al_{1.5-2}Si_{2.5-2}O_{10}] \\ (Mg,Al)_3(OH)_6 \end{cases}$

The ferro chlorites are formed from

Brunsvigite $\begin{cases} (Fe^{2+},Mg,Al)_3[(OH)_2/AlSi_3O_{10}] \\ (Fe,Mg,Al)_3(OH)_6 \end{cases}$

Daphnite $\begin{cases} (Fe^{2+},Al)_3[(OH)_2/Al_{1.2-1.5}Si_{2.8-2.5}O_{10}] \\ (Fe,Al)_3(OH)_6 \end{cases}$

Pseudo-thuringite $\begin{cases} (Fe^{2+},Al,Mg)_3[(OH)_2/Al_{1.5-2}Si_{2.5-2}O_{10}] \\ (Fe,Mg,Al)_3(OH)_6 \end{cases}$

The ferro-ferri chlorites include the following: (oxychlorites with partial dehydrogenation:

$$Fe^{2+} + OH^- \longrightarrow Fe^{3+} + O^{2-} + \tfrac{1}{2}H_2)$$

De-lessite $\begin{cases} (Mg,Fe^{2+},Fe^{3+})_3[(O,OH)_2/Al_{0-0.9}Si_{4-3.1}O_{10}] \\ (Mg,Fe^{2+})_3(OH)_6 \end{cases}$

Chamosite $\begin{cases} (Fe^{2+},Fe^{3+})_3[(O,OH)_2/AlSi_3O_{10}] \\ (Fe,Mg)_3(OH)_6 \end{cases}$

Thu-ringite $\begin{cases} (Fe^{2+},Fe^{3+},Al)_3[(O,OH)_2/Al_{1.2-2}Si_{2.8-2}O_{10}] \\ (Mg,Fe,Fe^{3+})_3(OH)_6 \end{cases}$

The following can also be mentioned here:

Pennantite $\begin{cases} (Mn,Al,Fe^{3+})_3[(OH)_2/(Al,Si)Si_3O_{10}] \\ (Mn,Al)_3(OH)_6 \end{cases}$

Gonyerite $\begin{cases} (Mn,Mg,Fe^{3+})_3[(OH)_2/(Si,Fe^{3+})Si_3O_{10}] \\ (Mn,Mg,Fe^{3+})_3(O,OH)_6 \end{cases}$

Chromium chlorites include:

$$\text{Kämmererite} \quad \begin{cases} (Mg,Cr)_{<3}[(OH)_2/AlSi_3O_{10}] \\ (Mg,Cr)_3(OH)_6 \end{cases}$$

$$\text{Kochubeite} \quad \begin{cases} (Mg,Al)_{<3}[(OH)_2/(Cr,Al)Si_3O_{10}] \\ (Mg,Al)_3(OH)_6 \end{cases}$$

Nickel chlorites also occur: schuchardtite is a nickel penninite which contains less nickel than magnesium, whereas in nimite, nickel predominates.

Crystals of chlorite are only occasionally found, and are usually lamellar and in flake or massive form. Well-developed crystals usually have a hexagonal, rhombohedral, or pseudo-hexagonal tabular crystal habit.

The Mohs hardness is ca. 2, and the density (depending on iron content) ca. $2.6-3.3$ g/cm$^3$. Unlike the micas, lamellae of chlorite are not elastically flexible.

Owing to the lamellar twinning commonly present, the crystals appear optically uniaxial. There are several twinning "laws", e.g., the penninite law [twinning plane (001)] or the mica law (twinning plane perpendicular to (001) and parallel to [110]). Also, twinning occurs even in the lattice, leading to polytypes, which have different symmetries and multiple $c$ values (variations in stacking sequence).

Depending on chemical composition, the optical character of the chlorites is either biaxially positive or biaxially negative, and is sometimes different for different wavelengths. The plane of the optical axis is usually (010), and the acute bisectrix is approximately normal to (001).

Chlorites usually have a green color, from which the name is derived. As the iron content increases, the color becomes more intense, increasing from colorless to dark green or brownish green (on oxidation of the iron). Other colors include pink to violet-pink (chromium chlorites) and orange-red (higher Mn content and lower Fe content).

The optic axis angle can vary greatly ($-20°$ to $+70°$), e.g., for clinochlore and grochauite (iron-containing sheridanite) $2V_\gamma = 0-70°$, and for daphnite, bavalite, and thuringite $2V_\alpha = 0-30°$. However, optic axis angles are usually very small, converging to $0°$.

The iron content has a considerable influence on optical refraction. Increasing iron content leads to an increase in optical refraction ($\alpha, \beta, \gamma$), while birefringence $\Delta$ is usually low. Pleochroism is usually strongly developed.

| Mg chlorites: | Penninite | Clinochlore | Sheridanite |
|---|---|---|---|
| $\alpha =$ | 1.562 | 1.581 | 1.594 |
| $\beta =$ | 1.565 | 1.581 | 1.594 |
| $\gamma =$ | 1.565 | 1.586 | 1.606 |
| $\Delta =$ | $-0.003$ | $+0.005$ | $+0.012$ |

Chromium chlorites (Kämmererite, Kochubeite)

| | | | |
|---|---|---|---|
| $\alpha =$ | 1.576 | 1.584 | 1.585 |
| $\beta =$ | 1.576 | 1.584 | 1.590 |
| $\gamma =$ | 1.580 | 1.584 | 1.591 |
| $\Delta =$ | $+0.004$ | $+0.000$ | $-0.006$ |

Mg–Fe$^{2+}$ chlorites:

| | Mg diabantite | Fe rhipidolite | Corundophilite |
|---|---|---|---|
| $\alpha =$ | 1.589 | 1.627 | 1.605 |
| $\beta =$ | 1.595 | 1.627 | 1.608 |
| $\gamma =$ | 1.595 | 1.629 | 1.615 |
| $\Delta =$ | $-0.006$ | $+0.002$ | $+0.010$ |

| Ferro chlorites | Brunsvigite | Aphrosiderite | Daphnite |
|---|---|---|---|
| $\alpha =$ | 1.632 | 1.647 | 1.665 |
| $\beta =$ | 1.638 | 1.651 | 1.675 |
| $\gamma =$ | 1.638 | 1.651 | 1.675 |
| $\Delta =$ | $-0.006$ | $-0.004$ | $-0.010$ |

| Ferro-ferri chlorites | Delessite | Thuringite |
|---|---|---|
| $\alpha =$ | 1.595 | 1.671 |
| $\beta =$ | 1.599 | 1.684 |
| $\gamma =$ | 1.599 | 1.685 |
| $\Delta =$ | $-0.004$ | $-0.014$ |

Anomalous interference colors can often occur. Optically positive chlorites with low iron contents and low refractive index show typical anomalous brown ("leather brown") interference colors, whereas optically negative chlorites with higher iron contents and higher refractive index show anomalous violet to blue ("duck blue" or "mallard blue") interference colors.

If chlorites are heated, they liberate water in two steps, but their structure remains unchanged almost up to complete dehydration. With magnesium-rich chlorites, water is first liberated from the brucite layers, and olivine is formed in the second dehydration step. Further temperature increase leads to the formation of spinel and enstatite.

As a rock-forming mineral, chlorite occurs mainly in low-grade regional metamorphic rocks or epizonal crystalline schists, i.e., in the metamorphites of the greenschist facies (e.g., chlorite schists). Low-iron chlorite often occurs in association with talc. Also, chlorite is widely distributed as a hydrothermal druse mineral. Chlorite is often formed by hydrothermal decomposition and alteration ("chloritization") of various (magnesium-rich) silicates, e.g., from biotite, garnet, hornblende, and augite; and also displaces and causes pseudomorphism of these minerals. Chlorites are also formed by the transformation of ultrabasic rocks (mainly olivine-rich rocks) into serpentinites (serpentinization). Chlorite can also be formed by contact metamorphosis (e.g., of rocks of the albite–epidote–hornfels facies). Clay sediments can also contain chlorite, usually originating by diagenesis.

The iron-rich chlorites, e.g., chamosite (in the Jura mountains from Chamoson in Wallis/ Switzerland) or thuringite (Thuringia/Federal Republic of Germany), which practically correspond to the older leptochlorites, can occur in certain oolitic iron deposits, and can even be the predominant mineral in some places. These formations are due to sedimentary deposition, but now exist in a slightly metamorphosed form.

The nickel chlorite schuchardtite occurs near other nickel hydrosilicates (e.g., with garnierite, hitherto classified as nickel serpentine or nickel antigorite, with the mineralogical chemical formula $(Ni, Mg)_6[(OH)_8/Si_4O_{10}]$, more recently also regarded as mixtures, so that garnierite is then restricted to the 10 Å component as nickel talc) in the "New Caledonia" type nickel deposits, which are of major economic importance.

The chromium chlorites, which are characterized by their pink to peach-blossom or violet-pink colors, occur in chromite deposits. They were formed by the serpentinization of the chromite-containing ultrabasic parent rocks.

Apart from its use as an ore [an important iron ore in former times, and a nickel ore today (→ Nickel **A 17**, p. 157)], chlorite is only of very minor importance as an industrial mineral (e.g., as a filler or pigment vehicle) in spite of its wide distribution and frequent occurrence in laminar form. Its potential as a replacement for other minerals (e.g., talc, mica), which have some similar properties is very limited at present. Chlorite (penninite) is sometimes used in the wallpaper industry. Leuchtenbergite, an almost iron-free variety of clinochlore, together with sericite and quartz, is the most important component of "white schist", a metamorphic rock of which there are deposits, for example, at Aspang (Ausschlag-Zöbern in Lower Austria). Although white schist is known as "Aspang kaolin", it contains no kaolin and is mainly of interest for its sericite content.

## 4. References


H. U. Bambauer, F. Taborszky, H. D. Trochim, W. E. Tröger: *Optische Bestimmung der gesteinsbildenden Minerale*, part 1: Bestimmungstabellen, 4th ed., E. Schweizerbart'sche Verlagsbuchhandlung, Stuttgart 1971.

H.-R. Bosse et al.: *Untersuchungen über Angebot und Nachfrage mineralischer Rohstoffe: XIX Industrieminerale*, Bundesanstalt für Geowissenschaften und Rohstoffe Hannover – Prognos AG Europäisches Zentrum für Angewandte Wirtschaftsforschung Basel - Hannover - Basel 1986.

O. Braitsch, W. E. Tröger: *Optische Bestimmung der gesteinsbildenden Minerale*, part 2: Textband, E. Schweizerbart'sche Verlagsbuchhandlung, Stuttgart 1967.

W. A. Deer, R. A. Howie, J. Zussman: *Rock-Forming Minerals*, vol. 3: Sheet Silicates, Longmans, Green and Co. Ltd., London 1962.

W. A. Deer, R. A. Howie, J. Zussman: *An Introduction to the Rock-Forming Minerals*, 7th ed., Longman Group Ltd., London 1974.

*Ind. Miner.* (London) (1977) no. 6, 19–27.

H. Kittel: *Lehrbuch der Lacke und Beschichtungen*, vol. 2, W. A. Colomb i. d. Heenemann GmbH, Berlin 1974, p. 379.

O. Lückert: *Pigment + Füllstoff Tabellen*, M. u. O. Lükkert, Laatzen 1980.

F. Machatschki: *Spezielle Mineralogie auf geochemischer Grundlage*, Springer-Verlag, Wien 1953.

P. Ramdohr, H. Strunz: *Klockmann's Lehrbuch der Mineralogie*, 16th ed., Enke-Verlag, Stuttgart 1978.

H. J. Rösler: *Lehrbuch der Mineralogie*, 2nd ed., VEB Deutscher Verlag für Grundstoffindustrie, Leipzig 1981.

H. Strunz: *Mineralogische Tabellen*, 7th ed., Akademische Verlagsgesellschaft Geest & Portig KG, Leipzig 1977.

H. G. F. Winkler: *Petrogenesis of Metamorphic Rocks*, 4th ed., Springer-Verlag, New York - Heidelberg - Berlin 1976.

W. Tufar: "Anthophyllit und Talk von Vorau (Oststeiermark)," *Mitteilungsbl. Abt. Mineral. Landesmus. Joanneum* **47** (1979) 37–52, 181–196.

W. Tufar: "Die Vererzung der Ostalpen und Vergleiche mit Typlokalitäten anderer Orogengebiete," *Mitt. Österr. Geol. Ges.* **74/75** (1981/82) 265–306.

W. Tufar: "Die geologischen Grundlagen für den Bergbau in Niederösterreich," in H. Feigel (ed.): *Studien und Forschungen aus dem Niederösterreichischen Institut für Landeskunde*, vol. 10, A. Kusternig (ed.): Bergbau in Niederösterreich, Amt der Niederösterreichischen Landesregierung, Vienna 1987, pp. 1–53.

W. Tufar et al.: "Formation of Magnesite in the Radenthein (Carinthia/Austria) Type Locality," *Monogr. Ser. Miner. Deposits* **28** (1989) 135–171.

# Tall Oil

LARS-HUGO NORLIN, BERGVIK KEMI (retired), Sandarne, Sweden



"Tall" is the Swedish word for pine, and the term tall oil is an American adaptation of "tallolja," a Swedish product name associated with the kraft pulping process ($\rightarrow$ Paper and Pulp, **A 18**, pp. 557–561) and with products from the fractional distillation of crude tall oil.

Pine wood contains an extractable fraction, $\approx 3\%$, consisting of free and combined fatty and rosin acids and nonacidic compounds (tall oil components). In the alkaline pulping process the tall oil components are saponified into a soapy material that, when acidulated, gives a mixture of free rosin and fatty acids and neutral components—crude tall oil. Crude tall oil is refined mainly by vacuum distillation processes to separate the various compounds almost completely into rosin and fatty acid fractions.

The *rosin* part of crude tall oil is a complex mixture of rosin acids, a three-ring fused system with the empirical formula $C_{20}H_{30}O_2$ ($\rightarrow$ Resins, Natural, **A 23**, pp. 81–83). The tall oil *fatty acids*, primarily $C_{18}$ types, are similar in composition to those derived from soybean oil. The *neutral constituents* of tall oil—predominantly esters, sterols, and hydrocarbons—are generally regarded as troublesome "impurities" that impair its utilization and distillation.

Abbreviations of basic tall oil names and common analyses are often used in publications and reports.

Crude tall oil (CTO) [8002-26-4]
Tall oil fatty acids (TOFA) [61790-12-3]
Tall oil rosin (TOR) [8052-10-6]
Distilled tall oil (DTO) [8002-26-4]
Tall oil pitch (TOP) [8016-81-7]
Heads, light ends (H) [65997-03-7]

## 1. Crude Tall Oil as Byproduct and Raw Material

The kraft pulping process was developed late in the 19th century. Some years later it was discovered in Sweden that a product similar to oil or wood tar could be obtained when the soapy material was acidulated with sulfuric acid. The recovery of byproducts from the kraft pulping process is shown in Figure 1.

**Tall Oil Precursor in Pine Wood.** Exploitation of the original tall oil materials in pine wood in the form of products such as tar and pitch dates from centuries BC. Gum and wood rosin are still

Ullmann's Encyclopedia
of Industrial Chemistry, Vol. A 26

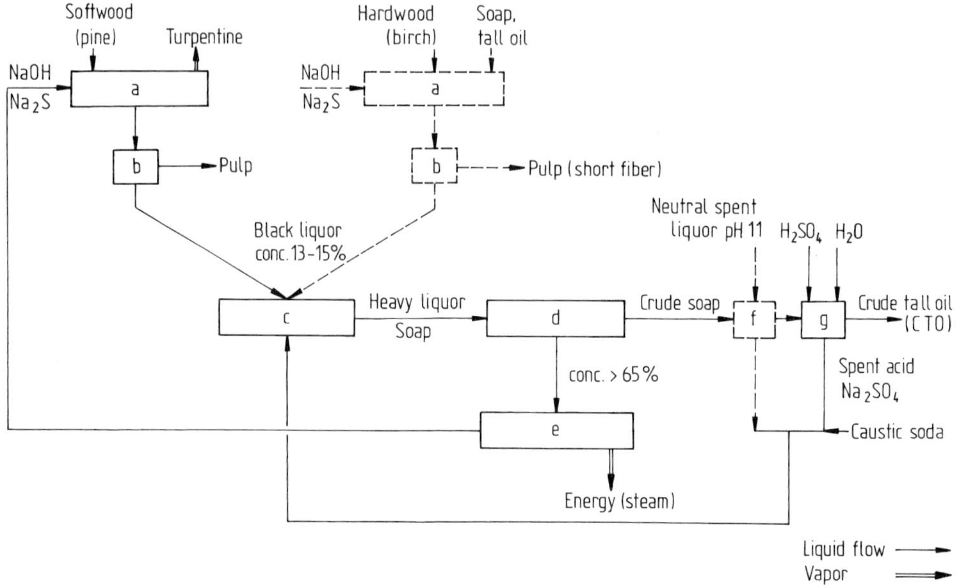


**Figure 1.** Recovery of byproducts from the kraft pulping process
a) Digester; b) Filter; c) Evaporator; d) Settling tank; e) Chemical recovery, energy production; f) Washing; g) Reactor

produced, and compete with the rosin fraction, TOR, derived from CTO.

Products such as gum rosin (residue product, → Resins, Natural, **A 23**, p. 80), and turpentine oil fractions (→ Turpentine) (distillates) obtained by distillation of exuded oleoresin from wounded living trees are still of high volume and importance. However, gum rosin production in the United States has declined drastically from ca. 300 000 t/a in 1930 to almost zero today owing to the lack of affordable labor. Despite this fact, worldwide gum rosin production in the early 1990s has been reported to be as high as 700 000 t/a. The People's Republic of China alone reports 400 000 t/a. Extraction of products such as wood rosin and turpentine fractions from stumps rich in rosin and liquid gum has declined because of the shortage of stumps (→ Resins, Natural, **A 23**, pp. 80–81).

**Replacement of Original Products by Tall Oil.** The production of CTO as a byproduct of the kraft pulping process and the subsequent refining of CTO have supplied important and renewable resources such as tall oil rosins and tall oil fatty acids as two basic chemical raw materials. The decreasing supply of wood and gum rosin in the United States is well compensated by the TOR production, which has increased proportionally to pulp production (see Fig. 2).

**Tall Oil in Wood.** The extractable portion of wood differs chemically, in part, from the compounds found in the CTO product. The chemical changes occurring during various processing steps are shown in Figure 3. Examples of composition of the extract from different wood species are listed in Table 1.

## 1.1. Composition and Properties of Crude Tall Oil

The wide range of compositions and properties of crude tall oil on the market can be explained and understood from the chemistry of tall oil. Many tall oil components are sensitive to chemical reactions and changes in properties and composition. The use of wood species other than pine [e.g., spruce and hard wood (birch)] in the kraft pulping process lowers the yield of CTO and the selection of available fatty and rosin acids. It also changes the composition and properties of the CTO. The quality and yield of CTO from the kraft pulping process consequently depend on many variables such as type and length of wood storage, season of the year, amount of wood residue used, hardwood usage, and type of pulping employed.

Crude tall oil is a dark brown, tarlike product; its odor, influenced by sulfur compo-

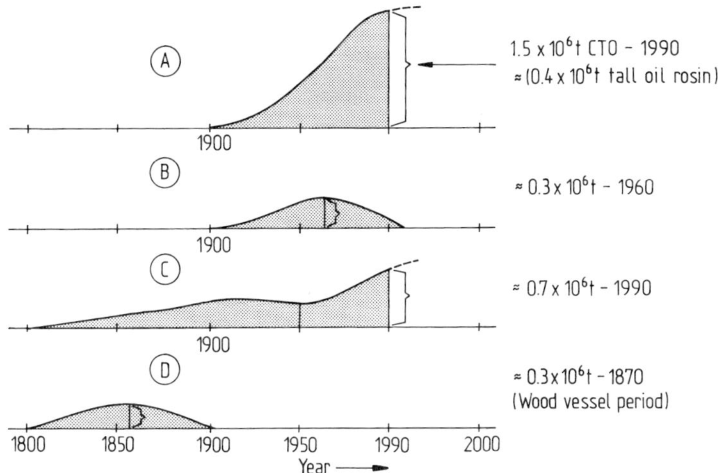



**Figure 2.** Historical development of global production of tall oil materials
A) CTO; B) Wood rosin; C) Gum rosin; D) Tar

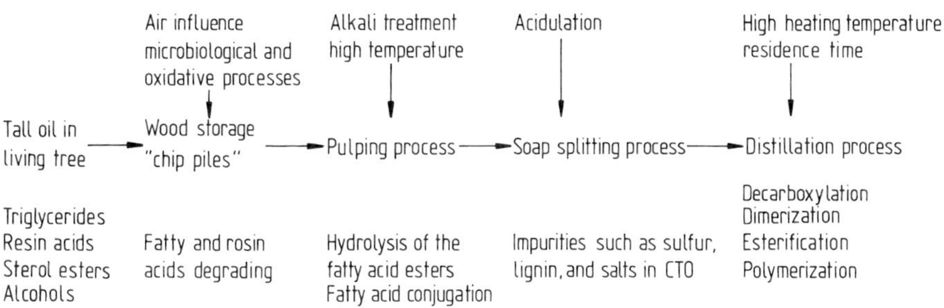



**Figure 3.** Chemical changes in tall oil during processing

**Table 1.** Composition of the extract from different wood species

| Compounds | Pine | Spruce | Birch |
|---|---|---|---|
| Esters, % | 40 | 43 | 58 |
| Free fatty acids, % | 18 | 10 | 6 |
| Free rosin acids, % | 34 | 25 | 0 |
| Total, % | 92 | 78 | 64 |

**Table 2.** Primary constituents of tall oil in different wood species

| | Extractives, % | Fatty acids, kg/t wood | Rosin acids, kg/t wood | Unsap*, kg/t wood |
|---|---|---|---|---|
| Pine | ≈ 3 | 15–20 | 5–10 | 2–5 |
| Spruce | ≈ 1 | 4–8 | 1–2 | 1–3 |
| Birch | ≈ 2 | 12–18 | | 5–10 |

* Unsaponifiables.

nents from pulp processing and soap skimming/acidulation, is unpleasant and penetrating. The three main constituent groups—fatty acids, rosin acids, and neutral materials (unsaps)—vary widely in composition, depending on the wood's origin (see Table 2). The saponification number (SN) of CTO is always higher than the acid number (AN) because of the unsplit esters in the crude oil. The difference increases for lower qualities, indicating less available rosin and fatty acids.

The crude is normally transported and stored at 50–80 °C to avoid crystallization and settling of rosin acids. The viscosity of the crude is variable; it is mainly a function of the rosin content. Viscosity must be measured on a clear, noncrystallized product. Impurities such as lignin, soap residue, water–oil emulsions, sodium salts, and sulfuric acid occur in the product. The amount of these undesirable materials depends on the efficiency of the soap washing and splitting process. Sour components such as sulfuric acid in the separated brine and degassed sour vapors from heated CTO in storage tanks impart the crude

**Table 3.** Typical data and composition of crude tall oil

| Characteristic | | Scandinavia, typical mix | United States, pine | Canada, mixed | France |
|---|---|---|---|---|---|
| Acid number | | 145 | 165 | 140 | 165 |
| Saponification number | | 160 | 172 | 165 | 172 |
| Fatty acids, % | | 45 | 45 | 42 | 40 |
| Palmitic | (16:0)* | 1.5 | 3 | 2 | 2 |
| Stearic | (18:0)* | 0.5 | 1 | 1 | 1 |
| Oleic | (18:1)* | 10 | 20 | 10 | 15 |
| Linoleic | (18:2)* | 17 | 13 | 15 | 11 |
| Pinolenic | (18:3)* | 5 | 1 | 3 | 1 |
| Arachidic | (20:0)* | 1 | 0.5 | 1 | 0.5 |
| Other | | 10 | 6.5 | 10 | 9.5 |
| Rosin acids, % | | 30 | 42 | 30 | 50 |
| Pimaric | | 2 | 3 | 2 | 5 |
| Palustric | | 4 | 7 | 5 | 7 |
| Isopimaric | | 2 | 4 | 4 | 4 |
| Abietic | | 12 | 15 | 10 | 18 |
| Dehydroabietic | | 5 | 4 | 4 | 5 |
| Neoabietic | | 4 | 6 | 3 | 6 |
| Others | | 1 | 3 | 2 | 5 |
| Neutrals, % | | 25 | 13 | 28 | 10 |
| Moisture | | | < 2% | | |
| Ash | | | < 0.2% | | |
| Mineral acid ($H_2SO_4$) | | | < 0.02% | | |
| Sulfur | | | < 0.3% | | |

* The figure before the colon denotes the number of C atoms per molecule, that behind the colon the number of double bonds.

corrosive properties. The physical properties of tall oil (empirical values) are listed below:

| | |
|---|---|
| Boil-up temperature (at 1.33 kPa) | 180 – 270 °C |
| Heat of vaporization | 290 – 330 kJ/kg |
| Specific heat | 2.1 – 2.9 J/g |
| Heat of combustion | – (33 000 – 38 000) kJ/kg |
| Density (at 20 °C) | 950 – 1020 kg/m³ |
| Liquid viscosity (at 70 °C) | 25 – 40 mm²/s |

Typical data and composition of crude tall oil are presented in Table 3.

## 1.2. Recovery of Soap and Conversion to Crude Tall Oil

During the kraft pulping process, CTO precursors are hydrolyzed to soluble sodium soaps in the cooking liquor. The soap and spent cooking liquor (black liquor) are washed and filtered from the pulp and concentrated in multiple-effect evaporators from 13 – 15 % to > 65 % solids. Evaporation increases the solids content so much that the soap—contaminated with black liquor to a greater or lesser extent—precipitates and rises continuously to the surface of the upgraded liquors in skimming tanks connected to the evaporation process. When the soap is skimmed from the top of the black liquor tanks, contamination of the soap with some black liquor is unavoidable. The major part of this black liquor containing about 10 % lignin should be removed. Otherwise, precipitation of the lignin and gypsum (gypsum is formed because of the presence of calcium in the black liquor) will occur during subsequent acidulation. Good decantation or separation of the soap is thus important. A low black liquor content of the soap decreases acid consumption and the amount of toxic hydrogen sulfide and carbon dioxide released during acidulation. In addition, soap present in black liquor and in the heavy black liquor (i.e., concentrated black liquor) decreases the capacity of both the evaporator system and the recovery boiler.

The *splitting* or *acidulation process* is a chemical reaction between the sodium soaps and sulfuric acid. The tall oil soap is converted into free fatty and rosin acids by reaction with dilute sulfuric acid (ca. 30 %) at elevated temperature (ca. 98 °C). The reaction can be summarized as follows:

$$2\ RCOONa + H_2SO_4 \longrightarrow Na_2SO_4 + 2\ RCOOH$$

where R denotes a rosin or fatty acid alkyl group.

Three phases are formed during the reaction: liberated tall oil, a lignin–fiber phase, and the spent acid phase ($Na_2SO_4$, free $H_2SO_4$). The phases can be separated because of their different densities. The splitting process is performed either batchwise or continuously. In both cases the equipment must be made of acid-proof material, able to resist dilute sulfuric acid (spent acid or brine) at 90–100 °C. A pH of 2.5–3.5 is normal for complete reaction with a minimum of equipment corrosion. The Hedemora–Celleco continuous tall oil system is shown in Figure 4.

## 1.3. Upgrading of Soap Extraction (Crude Soap Refining Process)

On a limited scale, crude birch–pine soap is upgraded by extraction of neutrals before acidulation. The crude soap refining (CSR) process developed by HOLMBOM and AVELA in Finland yields not only a cleaner CTO but also commercial compounds such as sitosterols, terpenes, and aliphatic alcohols from the extracted neutrals. This upgrading process has been in operation since 1977 (Kaukas, Finland).

## 2. Upgrading of Crude Tall Oil by Distillation

**History.** In 1930–1950 a number of relatively small distillation plants ($< 10\,000$ t/a) came onstream in Scandinavia and the United States, producing rosin-rich distillates and distilled tall oil. The "rosin oil" fractions were crystallized and centrifuged. The rosin crystals (rosin acid content $\approx 90\%$) were melted and partly saponified.

The amorphous *rosin product* replaced gum and wood rosin in Scandinavia and was used in the Delthirna process for making paper size. The *distilled tall oil fraction* (rosin acid content 20–30%)—in Scandinavia named "såpolja" (soap oil)—was used to make laundry soaps, as gel and solid products.

The introduction of *Linder trays* in fractionation towers was a promising alternative to already existing "dry" distillation (randomly packed) column processes. The first Linder plant came onstream in 1947 at Mo och Domsjö, Sweden as a dry distillation process designed for a rosin fraction introduced at the bottom of the column. The tall oil flow followed a convoluted path through a series of trays, encountering vapors generated in an electrically heated reboiler at the bottom of the column. Later, several Lin-

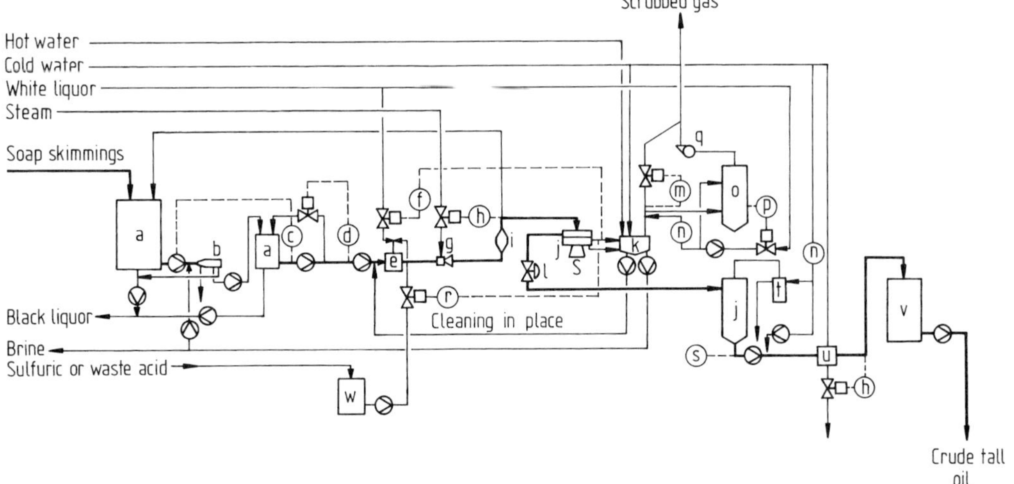


**Figure 4.** Hedemora–Celleco continuous tall oil system
a) Soap tank; b) Decanter–centrifuge; c) Level transmitter; d) Pressure transmitter control; e) Mixing-in piece; f) Density control; g) Direct steam heater; h) Temperature control; i) Reactor; j) Centrifugal separator; k) Brine tank; l) Constant pressure valve; m) Pressure switch; n) Flow meter; o) Gas scrubber; p) Liquid control; q) Fan; r) pH control; s) Temperature transmitter; t) Condenser; u) Tall oil cooler; v) Tall oil tank; w) Acid tank

der plants came onstream, further developed to produce TOFA with a low rosin content.

In 1949, Arizona Chemical and E. R. Badger introduced a *"wet" distillation process* in the United States, based on the experiences and techniques of the petroleum industry. This distillation differed from the Scandinavian process. The use of bubble-cap towers for distillation made the addition of steam necessary for vaporization of the high-boiling tall oil components. To achieve sufficient separation of the fatty and rosin acids, a considerable number of trays is required. To overcome the high static pressure drop in the column, sparge steam must be used to avoid undesirable reactions of the products, caused by too high boil-up temperatures. Initially, the Arizona plant came onstream as a two-pass operation in a two-column system. The feed of the first pass is CTO, that of the second the crude fatty acid fraction from the first pass.

In 1970, Luwa, Switzerland built a plant for Krems Chemie in Austria using Luwa *thin film evaporators* for dehydration, depitching, and heat exchange for boil-up in the fractionation tower (packed columns). This approach minimized the time during which heat-sensitive compounds in tall oil were exposed to elevated temperature.

The introduction of *large-scale structured packing* (Mellapak, Sulzer Chemtech, Switzerland, see Fig. 5) at the Bergvik plant in 1979 improved the efficiency and capacity of the towers to achieve more complete fractionation of the

**Table 4.** Comparison of the capacity of two equal rosin columns

| Characteristic | Pallring/minirings | Mellapak |
|---|---|---|
| Top pressure, kPa | 0.2 | 0.2 |
| Bottom pressure, kPa | 1.6–2.0 | 0.8–1.2 |
| Bottom temperature, °C | 270–280 | 255–265 |
| Capacity, t/a | 40 000 | 80 000 |
| Reflux ratio | 4 | 2 |

vaporized high-boiling compounds in CTO. Worldwide, approximately 15–20 rosin columns and approximately 10–20 fatty acid columns equipped with structured packings are onstream today. A capacity comparison of two equal rosin columns, one of which is equipped with structured packing (Mellapak), is given in Table 4.

## 2.1. Technical and Economic Conditions

Lengthy development work was required from the time CTO was discovered until a scientifically designed distillation process was achieved. Process performance may differ depending on local conditions, CTO quality, capacity demand, operation control equipment, and in-plant experiences, but the technology advances introduced have resulted in plants all over the world that are capable of producing fatty and rosin acid fractions of very high purity and quality. Special arrangements—mainly for the rosin fraction—were necessary because of variations

**Figure 5.** Sulzer's structured packing

in the quality of the CTO. However, these changes enabled the production of fractions with specific composition and properties. In practice, distillation of CTO has been achieved in two ways, both carried out in vacuum and at elevated temperature. The high-boiling components of tall oil can be vaporized either in the presence of superheated steam (wet process) or without adding any steam at all (dry process). In both cases, low pressure is necessary to vaporize the high-boiling compounds. At actual operating temperature the compounds are sensitive to undesirable chemical reactions, depending on heating conditions. In addition, the compounds degrade rapidly in the presence of oxygen. Therefore, the entire apparatus involved in distillation must have the functional security sufficient to meet the demand of steady running.

The "dry" process is more favorable today than the "wet process"; it involves lower energy costs and less air and water pollution. The Swiss technology—rapid heating for boil-up and improved mass transfer at lower pressure drop and lower reflux ratios—for distilling CTO has been successful and used completely or partly in new and upgraded plants all over the world.

Recently, the trend of activities in the tall oil field has been directed toward increased capacity, in order to gain scaleup effects, meet pollution restrictions, and produce tall oil fractions that are suitable for further processing into specific products. To obtain this, computerized control facilities are generally necessary and of great importance. Most of all, a stable operating pressure (vacuum) must be maintained to obtain fractions that meet certain values and limits. Even a small pressure change results in different process temperatures, corresponding to the new pressure. Improved techniques and resistant materials, more suited for treatment of high-boiling and heat-sensitive tall oil compounds, have led to the development of efficient distillation processes. The vapor pressure curves of tall oil components are depicted in Figure 6. The residence time of the product and the temperature differences between the product and the heating medium in different reboilers (see Table 5) influence product quality and yield. Therefore, heating conditions are very important.

As heating medium Dowtherm ($\rightarrow$ Evaporation, **B 3**, pp. 3–23; $\rightarrow$ Hydrocarbons, **A 13**, p. 263) is often used in liquid form. The desired or programmed product temperature and com-

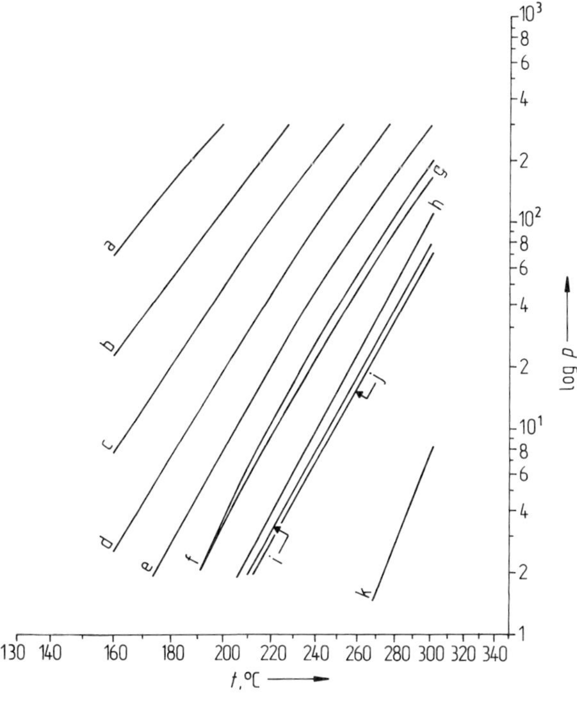



**Figure 6.** Vapor pressure curves of tall oil components ($\log p - 1/t$ curves, $t$ in °C is depicted on $1/t$ axis for convenience)
a) Caprylic acid ($C_8H_{16}O_2$); b) Capric acid ($C_{10}H_{20}O_2$);
c) Lauric acid ($C_{12}H_{24}O_2$); d) Myristic acid ($C_{14}H_{28}O_2$); e) Palmitic acid ($C_{16}H_{32}O_2$); f) Oleic acid ($C_{18}H_{34}O_2$); g) Stearic acid ($C_{18}H_{36}O_2$); h) Arachidic acid ($C_{20}H_{40}O_2$); i) Abietic acid ($C_{20}H_{30}O_2$); j) Behenic acid ($C_{22}H_{44}O_2$); k) Heavy neutrals
(Courtesy of Sulzer Chemtech AG)

**Table 5.** Residence time and $\Delta t$ (temperature difference between heating medium and product) for different types of reboiler

| Type of reboiler | Residence time | $\Delta t$, °C |
|---|---|---|
| Forced circulation | $\approx$ 60 min | $\approx$ 5 |
| Thin-film evaporator with fixed wipers | $\approx$ 5 s | $\approx$ 40 |
| Falling-film evaporator with circulating pump | 1–2 min | $\approx$ 15 |

position can be established with great precision at each heating station by regulating the amount of Dowtherm liquid flow in the loops connected to the head coil.

Dowtherm in vapor form as heating medium is still used mainly in U.S. plants, in a way similar to conventional steam-heated systems.

The principal flow sheet for the tall oil distillation process is, in practice, quite old. A simplified flow sheet of a large-capacity plant (> 100 000 t/a) is given in Figure 7. The Bergvik Kemi tall oil plant is shown in Figure 8.

## 2.2. Process Description

### 2.2.1. Dehydration

CTO is fed directly from the storage tank into the thin-film evaporator (Fig. 7, a) for rapid preheating up to 200 °C under vacuum (3–6 kPa). Inert combustible gases, water ($\approx$ 1 %), low-boiling compounds such as organic heads ($H_1$) and turpentine ($\approx$ 1 %) are distilled during the first passage through the evaporator. This operation makes possible a stable low pressure ($\approx$ 0.3 kPa) in the dry distillation and fractionation of the relatively high-boiling tall oil components that follow.

### 2.2.2. Depitching

The applicability of dry distillation not only is influenced by the CTO composition but is also highly dependent on conditions [e.g., operating pressure and temperature and residence time of the products in the depitcher (Fig. 7, b)].

The vaporization and squeezing of vaporizable products, such as separation of the fatty and rosin acids from native pitch compounds in CTO and from pitch formed by heating, require a high temperature (270–275 °C), even at low pressure (800–1300 Pa), measured at the vapor entrance into the rosin column or the first fractionation tower. The optimal depitching conditions for obtaining maximum yields and purity of rosin and fatty acid fractions are, however, of delicate nature. At elevated temperature undesirable reactions can occur such as esterification of fatty acids (by alcohols contained in the CTO), polymerization, decarboxylation of rosin acids, and isomerization of abietic-type acids. Too high a depitching temperature also leads to decomposition and vaporization of high-boiling compounds (neutrals, fatty acids) that affect the quality of rosin and fatty acid fractions negatively.

### 2.2.3. Fractionation of Vaporized Tall Oil Compounds

To reduce the pressure drop and achieve the lowest possible pressure in the depitching step the top section of the rosin column has a wider diameter than the lower section. Overhead vapor from the depitcher, consisting mainly of rosin and fatty acids but also of nonacidic compounds, is fed to the rosin column (Fig. 7, c) to recover rosin as a bottom fraction (TOR). The rosin

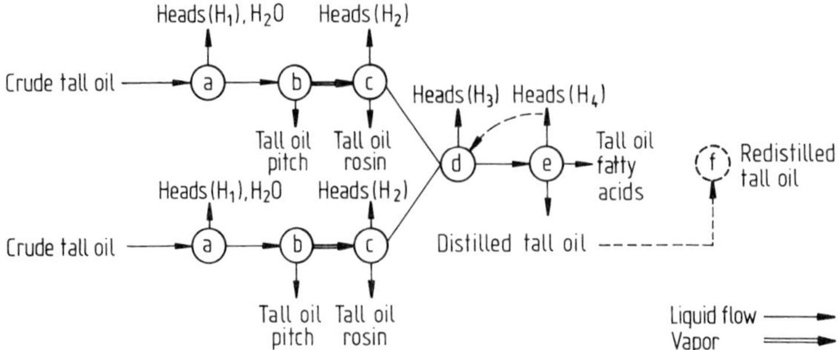


**Figure 7.** Simplified flow sheet for large-capacity plants (> 100 000 t/a)
a) Dehydration; b) Depitcher; c) Rosin column; d) Light ends column; e) Fatty acid column; f) Distilled tall oil columns

acids in TOR are similar in composition to those found in gum and wood rosin. As distillate (side stream from the upper part of the column) the crude fatty acid intermediate fraction, containing about 8 % rosin acids, is taken off and fed to the light ends column (d) to remove light heads (a mixture of neutral compounds and short-chain fatty acids). The bottom fraction, containing mainly $C_{18}$ fatty acids, a smaller portion of rosin acids, and a "minimum" of neutrals, is then fed to the fatty acids column (e) for production of high-quality tall oil fatty acids as a side stream from the top section of the column. From the top of all involved fractionation towers, smaller cuts of head fractions containing mixed neutrals and "low-boiling" ($C_8-C_{16}$) fatty acids are distilled off ($H_1-H_4$).

Distilled tall oil containing most of the rosin acids in the intermediate or crude fatty acid fraction, is usually distilled off as a side stream "middle-boiler" fraction as well as by other means, depending on local technical conditions and feed composition.

The fatty acids in CTO are mainly $C_{18}$ unsaturated acids ($\approx 85\%$) in which a significant proportion of the linoleic acid has been isomerized to the conjugated acid during the alkaline pulping process. The rest of the fatty acids are—if compared to the $C_{18}$ compounds—of "low-boiling" and "high-boiling" character. More than 20 different acids from $C_8$ to $C_{24}$ have been identified, among which palmitic, stearic, eicosanoic, arachidic, and behenic acids are of most interest. The higher-boiling fatty acids are concentrated mainly in the DTO fraction by fractional distillation, but are also contaminants in the rosin acid fraction. Depending on the conditions during pulping and distillation, some abietic types are isomerized to dehydroabietic acid.

Both the tall oil fatty acids and the rosin acids contain two functional groups: double bonds and the carboxyl group, which allow a number of chemical reactions. Some of these occur during processing, depending on CTO composition and external physical and chemical influences. These reactions result in increased pitch yield (by esterification, polymerization, oxidation) or affect the molecule structure of certain sensitive materials (e.g., isomerization of linoleic acid in the fatty acid group). Levopimaric acid is converted to more stable forms, and other isomers of abietic acid are formed.

**Figure 8.** Bergvik Kemi tall oil plant

**Table 6.** Principal composition and yields of tall oil fractions and some empirical volatility data

| Fraction | Yield, % | Acid number | Composition, % | | |
|---|---|---|---|---|---|
| | | | Rosin acids | Fatty acids | Neutrals |
| Head | 5–12 | 70–120 | <0.5 | 30–50 | 40–60 |
| TOFA | 35–45 | 192–197 | <2 | 95–98 | 1–2 |
| DTO | 5–15 | 180–190 | 20–30 | 65–70 | 4–7 |
| TOR | 20–35 | 165–182 | 85–96 | 1–5 | 1–7 |
| TOP | 20–40 | 20–50 | 5–13 | 5–10 | 40–60 |

The principal composition and yields of tall oil fractions are listed in Table 6.

## 3. Energy Recovery

Like many other distillation processes characterized by several steps and elevated temperature (150–300 °C), the energy demand of tall oil distillation is large and costly. Thus, operating the distillation column at low reflux ratios and reusing the energy from the condensation system as heat (i.e., for the distillation process and for supplementary equipment, tanks, etc., belonging to a complete tall oil plant) is very important in terms of economics.

Besides the usual arrangements for heat exchange of hot and cold streams in the process, most of the thermal energy from the condensa-tion system can be transformed into low-pressure steam (see Fig. 9). The quantity of low-pressure steam in principle meets the demands of the tall oil process. It is used mainly in vacuum jets and CTO tanks but also for other heating purposes.

## 4. Security and Construction Material

Acceptable construction material and process equipment that resist the corrosivity of tall oil compounds—particularly fatty acids—ensure reliable operation. Based on many years of "dry" distillation, the Co–Ni–Mo alloy, containing 2.5 % molybdenum (AISI Type 316) was regarded as an acceptable material, combating corrosion generally in the tall oil distillation process. However, as time went on, the corrosion of replacement parts occurred, sometimes followed by unscheduled shutdowns and quite often fires. This resulted in the use of an alloy containing more than 3.5 % molybdenum (AISI Type 317).

Vapor-phase streams are more corrosive than liquid-phase streams, and corrosion increases with increasing temperature. The corrosion rate of alloys with 5–6 % molybdenum is negligible; thus, the use of high-molybdenum alloy in construction of the equipment is essential. Alu-

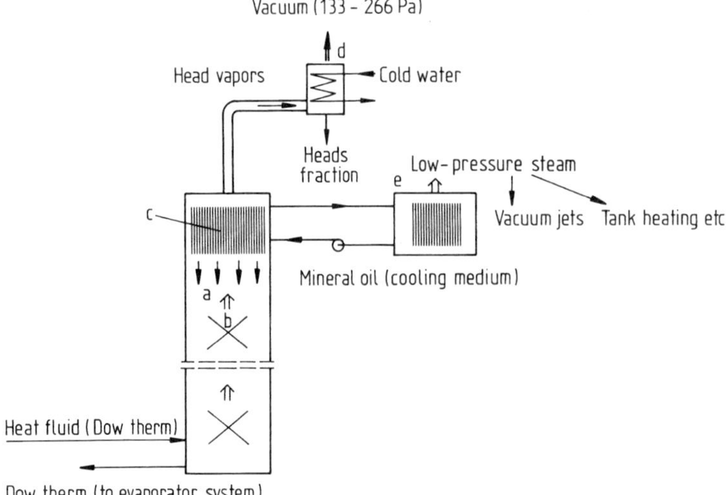


**Figure 9.** Heat recovery system
a) Fractionation tower; b) Flexipack packings (corrugated Al sheet); c) Deflegmator (partial condenser); d) End counter; e) Steam generator station

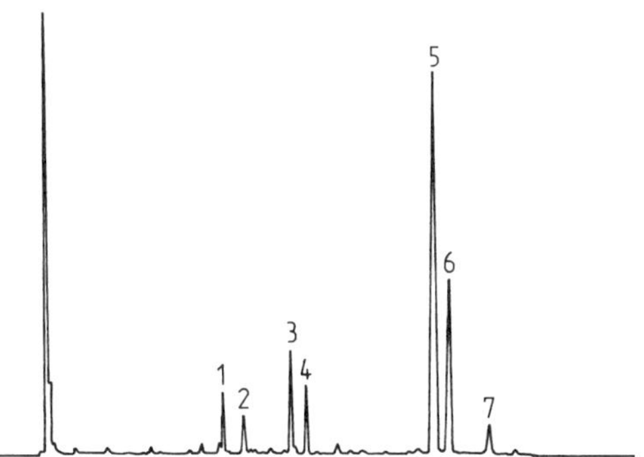

**Figure 10.** Gas chromatogram of tall oil rosin
1. Pimaric acid; 2. Sandaracopimaric acid; 3. Palustric acid; 4. Isopimaric acid; 5. Abietic acid; 6. Dehydroabietic acid; 7. Neoabietic acid

minum has been used for tanks, pipelines, packings, etc., and is still of interest when special strength is not required.

## 5. Analysis

Most of the analytical methods for tall oil are established standards and, as such, collected in ASTM and especially in SCAN (Scandinavian Pulp Paper Testing Committee) publications. About 20 standard methods are available today.

In addition to the standard methods of analyses, gas–liquid chromatography, mass spectroscopy, and gel permeation chromatography are used widely for better understanding of tall oil chemistry. A gas chromatogram of tall oil rosin is shown in Figure 10. These methods are essentially necessary complements to the older standard analyses for further development of the processes and of special or unique products as well. Typical data and composition of tall oil fatty acids and of tall oil rosins are listed in Tables 7 and 8.

## 6. Storage and Transportation

The crude tall oil, melted rosin, and pitch are stored at elevated temperature to reduce viscosity. Air exposure of both fatty and rosin acid fractions results in darkening of the products. Protection with inert gases—$N_2$, $CO_2$, and steam—at elevated temperature is recommended. The fatty and rosin acid fractions are preferably stored and shipped in tanks made of stainless steel or aluminum. When rosin in the melted state is shipped in insulated tank cars, the tem-

**Table 7.** Typical data and composition of some tall oil fatty acids

|  |  | Scandinavia | United States |
|---|---|---|---|
| Acid number |  | 195 | 197 |
| Rosin acids, % |  | 2 | 1 |
| Unsap*, % |  | 2 | 1.5 |
| Iodine value |  | 150 | 130 |
| Color, Gardner |  | 4 | 3 |
| Fatty acids, % |  |  |  |
| Saturated |  | 2 | 2 |
| Oleic | (18:1) | 30 | 48 |
| Linoleic | (18:2) | 44 | 37 |
| Linolenic | (18:3) | 10 | 3 |
| Conjugated | (18:2) | 6 | 6 |
| Other |  | 4 | 2.5 |

* Unsaponifiables.

**Table 8.** Typical data and composition of some rosin types

| | Gum rosin | Wood rosin | Tall oil rosin | | |
|---|---|---|---|---|---|
| | | | United States | Scandinavia | |
| | | | | I | II |
| Acid number | 167 | 157 | 174 | 173 | 180 |
| Rosin acids, % | 90 | 85 | 92 | 90 | 95 |
| Fatty acids, % | | | 2 | 4 | 2 |
| Softening point, °C | 78 | 73 | 75 | 66 | 75 |
| Color, USDA* | | | WW | X | XA |
| Composition of rosin acids fraction, % | | | | | |
| Pimaric | 7 | 7 | 2 | 3 | 3 |
| Palustric | 20 | 8 | 7 | 8 | 7 |
| Isopimaric | 15 | 15 | 8 | 6 | 6 |
| Abietic | 30 | 50 | 40 | 40 | 40 |
| Dehydroabietic | 5 | 8 | 20 | 22 | 23 |
| Neoabietic | 15 | 5 | 3 | 4 | 4 |

* For a definition of USDA color shades, see → Resins, Natural, **A 23**, p. 81.

perature is kept at $> 160\,°C$. Rosin derivatives should be produced in direct connection with the distillation process. By storing rosin in solid form (galvanized drums), additional costs (packaging, handling, remelting, etc.) may arise. In addition, TOR in drums can crystallize easily to become nonbrittle and often difficult to remove. Today, the modified resin products are generally flaked or compressed into tablets and delivered in paper bags.

## 7. Uses

**Uses of CTO.** Most of the world production of CTO ($\approx 1.5 \times 10^6$ t/a) is utilized as raw material in continuous distillation processes, producing high-quality tall oil fatty acid fractions and tall oil rosins as primary products. If the CTO is "low grade" (e.g., the content of valuable components is too low), the product (or the corresponding soap) is used as liquid (or solid) fuel. Since CTO, irrespective of quality, can always be used as a fuel oil, its minimum price is more or less related to actual "fuel oil" prices on the market, including taxes and other associated costs.

**Tall Oil Fractions.** The TOFA and TOR fractions are used widely for further processing into a variety of chemical products, replacing other raw materials such as gum and wood rosin, hydrocarbon resins, and vegetable oils. Improved distillation techniques have led to upgraded fractions of TOFA and TOR that are attractive raw materials for special chemicals. The chemical structure of some fatty and rosin acids makes the primary tall oil fractions suitable as raw material for making intermediate chemicals. Examples are the production of dimeric fatty and rosin acids, polymerized rosin, and maleic and fumaric anhydride (acids)−rosin adducts including their salts and esters ($\rightarrow$ Resins, Natural, **A 23**, pp. 84−86). The dimeric acids are utilized as raw materials to make products used in adhesives, coatings, inks and epoxy resins. Most of the intermediate chemicals are processed in the presence of certain catalysts. An overview of the applications of tall oil fractions is given in Table 9.

# 8. Economic Aspects

The total world production of CTO was in the range of $1.5 \times 10^6$ t in the early 1990s. Most of the available quantities are produced and refined in the United States and Europe. The world tall oil fractionation capacity by region is as follows:

| | |
|---|---|
| United States | 56 % |
| Scandinavia | 21 % |
| Other Western Europe | 7 % |
| Other | 16 % |

A breakdown of world tall oil fractionation capacity by company is given in Figure 11.

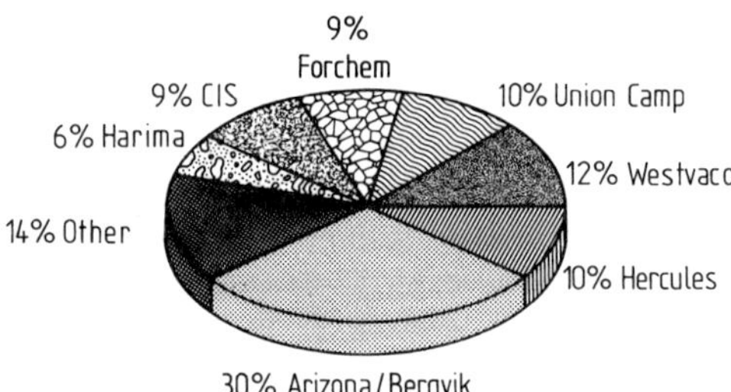


**Figure 11.** World tall oil fractionation capacity by company

**Table 9.** Applications of tall oil fractions

| Composition, uses | Pitch | Rosin | Distilled tall oil | Fatty acids | Heads |
|---|---|---|---|---|---|
| Principal composition | esters, sitosterol, estolides, rosin dimer, free fatty and rosin acids | rosin acids (isomeric diterpene acids, complex mixture), free fatty acids, neutrals, unsaps 3−10%, esters | fatty and rosin acids, neutrals, unsaps, esters < 10% | fatty acids (oleic, linoleic types > 90%), rosin acids (neutrals < 10%) | fatty acids (saturated) |
| As intermediates in miscellaneous processes | mixtures for chemical modification | fortification, Diels−Alder reaction, dimerization (catalytic process), isomerization, esterification, disproportionation | alkyd formation, esterification, saponification, epoxidation | dimerization, isomerization, esterification, alkyd formation, saponification, separation, ethoxylation, decarboxylation, $C_{21}$ diacids | mixtures, crystallization (palmitic acid) |
| Final uses | fuel, binders, coatings, rubber modifier, asphalt, sizing, printing inks, hardboard impregnation | paper size, protective coating, adhesive, ink, rubber, hot melt | soap, coating, flotation, board impregnation | protective coating, soap, ink | fuel, flotation, asphalt emulsifier |

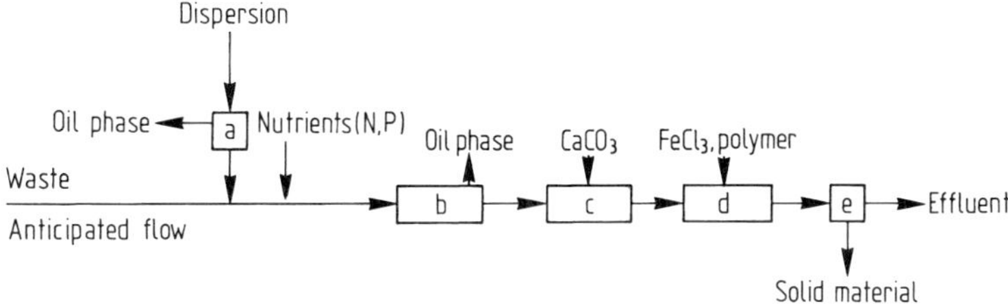


**Figure 12.** Principal waste treatment process
a) Phase separation equipment (Knitmesh filter); b) Oily water skimming basin; c) Aerotin basin (biological treatment); d) Chemical treatment basin; e) Filter, sludge decanter centrifuge
BOD reduction $> 96\%$; COD reduction $> 88\%$

# 9. Toxicology and Environmental Protection

**Toxicology.** The tall oil compounds in living trees, stumps, CTO, fractions, etc., are nonhazardous natural substances. Modified tall oil products and solutions may, however, become hazardous because of the chemicals and solvents used. Many tall oil derivatives such as glycerol and pentaerythritol esters, alkyd resins, dimerized fatty acids, and rosin acids are generally regarded as nonhazardous products.

The toxicity of tall oil components in effluent water to fish is not a function of oxygen demand. Fish may tolerate some higher (allowed) concentration limits of rosin and fatty acids or their sodium soaps in the waste stream. These concentrations may be indicated along with the other conventional parameters (COD, BOD, etc.) of the wastewater.

**Environmental Protection.** During vacuum distillation, some product losses arise, resulting in polluted water and inert gases. The water streams from the various processes are of different compositions but after passing through special oil separation equipment, they are collected for the removal of organic waste constituents by biological, physical, or chemical methods before they are discharged (see Fig. 12). The noncondensed (odor) component of the gases may be disposed of as fuel gas if necessary.

The dry distillation method has the advantage of requiring fewer measures to combat pollution compared to the wet process. In distillation, most of the polluted water comes from the dehydration steps, where the "low-boiling" components (1 %) together with the water in CTO (1 %) are distilled off from the feed before depitching. Additional sources of polluted water are related to the steam jet vacuum systems, plant cleanup wastes, CTO storage tanks, pumps, etc.

# 10. References


[1] International Tall Oil Symposium, Imatra, Finland, June 7–9, 1983, Turku Akademi, Turku 1983, p. 259.
[2] D. F. Zinkel, J. Russel (eds.): *Naval Stores, Production Chemistry Utilization*, Pulp Chem. Assoc., New York 1989, p. 1060.
[3] J. Drew, M. Propst: *Tall Oil*, Pulp Chem. Assoc., New York 1981, p. 199.
[4] E. E. Sweeney, H. G. Arlt, J. Russel: *Tall Oil and Its Uses – II*, Pulp Chem. Assoc., New York 1987, p. 132.
[5] W. Sandermann: *Naturharze, Terpentin, Tallöl*, Springer Verlag, Berlin 1960, p. 483.
[6] P. Knoer, *South. Pulp Pap. Manuf.* **38** (1975) no. 9, 18–22.
[7] *Ullmann*, 4th ed., **22**, 385.
[8] Å. Linder: "Tall Oil Refining," *Ingenioersvetenskapsakad. Handl.* **207** (1952) 56.
[9] *Analytical Procedures for Tall Oil Products*, Pulp Chem. Assoc., New York 1976.
[10] K. Ukkonen, *Kem. Kemi* **18** (1991) nos. 7–8, 603–604.
[11] A. Rütti, *Fette, Seifen Anstrichm.* **88** (1986) Mai, 515–519.


**Tallow → Fats and Fatty Oils**
**Tanning Agents → Leather**

# Tantalum and Tantalum Compounds

KLAUS ANDERSSON, KARLHEINZ REICHERT, RÜDIGER WOLF, H. C. Starck GmbH u. Co. KG, Goslar, Federal Republic of Germany



Tantalum, Ta [7440-25-7], atomic number 73, relative atomic mass 180.948, has the electronic structure $1s^2$, $2s^2p^6$, $3s^2p^6d^{10}$, $4s^2p^6d^{10}f^{14}$, $5s^2p^6d^3$, $6s^2$, which accounts for its principal oxidation states +2 and +5. There are two naturally occurring isotopes: $^{181}$Ta and the weakly radioactive isotope $^{180}$Ta (0.012%, half-life ca. $10^{13}$ years). A large number of artificial radioactive isotopes are also known.

Tantalum was discovered by EKEBERG in 1802 [1] [2]. The discovery and history of tantalum and niobium are closely linked (→ Niobium, **A 17**, p. 251 ff.). The two elements usually occur together.

## 1. Properties

### 1.1. Physical Properties

The physical properties of tantalum are as follows [1]–[7]:

| | |
|---|---|
| Crystal structure | Body-centered cubic |
| Lattice constant | $a = 0.33025$ nm |
| Density | 16.6 g/cm³ |
| Melting point | 2996 °C |
| Boiling point | 5425 °C |
| Heat of fusion | 28.5 kJ/mol |
| Heat of vaporization at 3273 K | 78.1 kJ/mol |
| Specific heat capacity at 20 °C | 25.41 J mol⁻¹ K⁻¹ |
| Linear coefficient of thermal expansion at 20 °C | $6.5 \times 10^{-6}$ K⁻¹ |
| Thermal conductivity at 20 °C | 54.4 Wm⁻¹ K⁻¹ |
| Specific electrical conductivity at 20 °C | 0.081 Ω⁻¹ cm⁻¹ |
| Superconductivity | 4.3 K |
| Temperature coefficient (0–100 °C) | 0.00383 |

The effect of temperature on electrical and thermal conductivity has been accurately determined [3].

The mechanical properties of the metal are strongly dependent on its purity, structure, and crystal defects, as is the case with almost all refractory metals. Even low concentrations of interstitial impurities increase the hardness and reduce the ductility. The yield strength is strongly temperature dependent, the value at 200 °C being ca. 30% of that at 20 °C [3]. The important mechanical properties are as follows [1], [3], [7], [8]:

*Tantalum, soft annealed*

| | |
|---|---|
| Tensile strength | 240 MPa |
| Yield strength (0.2% offset) | 210 MPa |
| Breaking elongation | 40% |
| Vickers hardness | 60–120 HV |
| Modulus of elasticity | 177–186 MPa |

Ullmann's Encyclopedia
of Industrial Chemistry, Vol. A 26

| Ductile–brittle transition temperature | 4 K |
|---|---|
| Poisson's ratio | 0.35 |

*Tantalum, cold formed*

| | |
|---|---|
| Tensile strength | 400–1400 MPa |
| Yield strength (0.2 % offset) | 300–900 MPa |
| Breaking elongation | 2–20 % |
| Vickers hardness | 105–200 HV |
| Recrystallization temperature | 1200–1500 K |

## 1.2. Chemical Properties

Like niobium, tantalum has principal oxidation states $+2$ and $+5$. Below $100\,^\circ C$, tantalum metal is extremely resistant to corrosion by most organic and inorganic acids, with the exception of hydrofluoric acid. This is due to a dense adherent film of tantalum oxide, a characteristic which is utilized in the manufacture of electrolytic capacitors (Section 4.3). Above ca. $300\,^\circ C$, tantalum tends to form oxides, nitrides, hydrides, and carbides. All modern tantalum capacitor powders can be handled in air at temperatures up to ca. $100\,^\circ C$. However, they are classified as flammable solids as they ignite at higher temperatures, forming $Ta_2O_5$.

**Table 1.** Tantalum deposits [11]–[15]

| Location | Reserves, $10^3$ kg $Ta_2O_5$ | Production (1991, estimated), $10^3$ kg $Ta_2O_5$ | Annual production capacity (estimated), $10^3$ kg $Ta_2O_5$ |
|---|---|---|---|
| Australia | 17 000 | 235 | 300 |
| Brazil | 18 000 | 100 | 245 |
| Canada | 3 000 | 80 | 130 |
| Malaysia | 2 000 | 80 | 130 |
| Thailand | 16 000 | 150 | 200 |
| Africa | 10 000 | 25 | 50 |
| Russia | > 20 000 | < 50 | |
| China | 17 000 | 100 | 200 |
| Others | 10 000 | < 50 | < 100 |

## 2. Occurrence

Tantalum is 54th in order of abundance of the elements in the Earth's crust (2.1 g/t). In many deposits, it occurs in association with the much more abundant niobium (24 g/t). The most important tantalum-containing minerals are tantalite, wodginite, microlite (the tantalum-rich end member of the pyrochlore series), and columbite [9]. The typical $Ta_2O_5$ concentration in processable pegmatitic deposits, which represent the principal raw material potential, particularly in Australia, is < ca. 0.1 %. The uranium- and thorium-containing minerals yttrotantalite, strueverite, euxenite, and samarskite are sometimes highly radioactive, which limits their workability [10]. An important potential reserve lies in a belt stretching from China to Indonesia via Thailand and Malaysia, and consists of cassiterite ($SnO_2$) that contains some tantalum and niobium. During tin smelting, the tantalum and niobium become concentrated in the slag ($\rightarrow$ Niobium, **A 17**, p. 253). The important deposits, with estimated current production capacities, are listed in Table 1, and the compositions of some tantalum-containing minerals are given in Table 2.

## 3. Extraction

### 3.1. Extraction and Processing of Ore Concentrates

In the important Australian mines, the tantalum-containing minerals are present in finely divided form in hard pegmatitic rock. The rock is extracted by open-cast mining. Further processing, which is different from that used for the mainly niobium-containing deposits

**Table 2.** Composition of important tantalum-containing minerals [9], [16], [17]

| Mineral | Composition | $Ta_2O_5$ content, % | $Nb_2O_5$ content, % |
|---|---|---|---|
| Tantalite | $(Fe, Mn)Ta_2O_6$ | 42–84 | 2–40 |
| Microlite | $(Na, Ca)(Ta, Nb)_2O_6F$ | 60–70 | 5–10 |
| Columbite | $(Fe, Mn)(Nb, Ta)_2O_6$ | 1–40 | 40–75 |
| Wodginite | $(Ta, Nb, Sn, Mn, Fe, Mn)_{16}O_{32}$ | 45–56 | 3–15 |
| Yttrotantalite | $(Y, U, Ca)(Ta, Fe^{3+})_2O_6$ | 14–27 | 41–56 |
| Fergusonite | $(RE^{3+})(Nb, Ta)O_4{}^*$ | 4–43 | 14–46 |
| Strueverite | $(Ti, Ta, Nb, Fe)_2O_4$ | 7–13 | 9–14 |
| Tapiolite | $(Fe, Mn)(Nb, Ta)_2O_6$ | 40–85 | 8–15 |
| Euxenite | $(Y, Ca, Ce, U, Th)(Nb, Ta, Ti)_2O_6$ | 1–6 | 22–30 |
| Samarskite | $(Fe, Ca, U, Y, Ce)_2(Nb, Ta)_2O_6$ | 15–30 | 40–55 |

* RE = rare earth element.

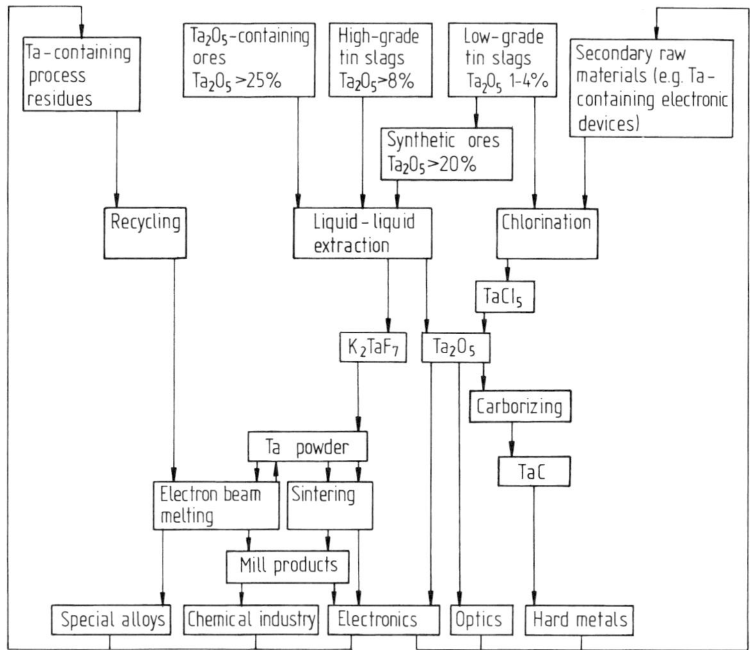


**Figure 1.** Major tantalum processing routes

($\rightarrow$ Niobium, **A 17**, p. 253), is an initial size reduction to $< 15$ mm, followed by grinding in ball or rod mills to $< 1$ mm. Flotation is then carried out, and the soluble constituents are leached out from the concentrates. The latter are then dried, and the magnetic constituents are removed [18].

This ore concentrate, up to 40 % $Ta_2O_5$, is dissolved in concentrated hydrofluoric acid ($\rightarrow$ Niobium, **A 17**, pp. 253–255), and is then separated from the niobium by extraction with methyl isobutyl ketone, so that the tantalum is obtained as $H_2TaF_7$. Potassium salts are added, and potassium heptafluorotantalate, $K_2TaF_7$ [16924-00-8], crystallizes out. This is reduced with sodium to metallic tantalum (Section 4.1).

Major tantalum processing routes are summarized in Figure 1.

## 3.2. Obtaining Ta–Nb Concentrates from Tin Slags

Tantalum-containing tin slags represent an important source of tantalum; e.g., in 1980, ca. 50 % of world demand was recovered from tin slags [19]–[21]. Owing to the low price of tin, its production is in continuous decline, so that the availability of tantalum-containing tin slags is decreasing. Moreover, the tantalum concentration in tin concentrates is also decreasing significantly. The fraction of the world demand for tantalum represented by $Ta_2O_5$ in tin slags is expected to fluctuate around 20 %. The price of tin is not likely to reach a level at which more tantalum-rich tin ores from Thailand and Malaysia can be processed economically.

The tantalum content of these slags fluctuates widely. Bolivian slags do not contain tantalum, but those from Thailand can have tantalum content 15–16 % $Ta_2O_5$. The Ta:Nb ratio varies from 1:0.4 (Thailand) to 1:3 (Nigeria).

The most important grades of slag have the following tantalum and niobium content [19]–[21]:

|  | $Ta_2O_5$, % | $Nb_2O_5$, % |
|---|---|---|
| **High-grade** | | |
| Thailand | 13–16 | 9–10 |
| Zaire | 11 | 10 |
| Australia | 10 | 5–6 |
| | | |
| **Medium grade** | | |
| Thailand | 4–5 | 4 |
| Malaysia | 3–4 | 3–4 |
| Nigeria | 4–5 | 12 |
| South Africa | 6–8 | 8–9 |

Low-grade

| | | |
|---|---|---|
| Malaysia | 0.8–2.5 | 0.8–2.5 |
| Thailand | 0.5–1 | 0.2–0.8 |
| Brazil | 1–2.5 | 1–2.5 |
| Singapore | 1–1.5 | 1–1.5 |

High-concentration slags are used directly in the wet chemical process (dissolution of the slag in hydrofluoric/sulfuric acid, followed by extraction and separation of tantalum and niobium).

From low-concentration slags, Ta/Nb concentrates containing 50–60% $Ta_2O_5/Nb_2O_5$ can be produced, e.g., pyrometallurgically. The Ta/Nb content of these synthetic concentrates is > 99% soluble in hydrofluoric acid, as is the case with natural ores.

In all the concentration processes so far developed, the slag is first melted in an electric furnace with addition of reducing agents, e.g., coke, and fluxing materials. This causes the Ta/Nb to collect in the carbide-containing ferroalloy produced, while some impurities remain in the slag. The solid ferroalloy is produced in relatively large blocks, which must be broken up and ground. The tantalum/niobium content is increased in further process steps, giving a concentrate containing 50–60% oxides of tantalum and niobium.

One method of enriching the tantalum and niobium consists of oxidizing the ferroalloy, which contains the Ta/Nb carbide, with $Fe_2O_3$ (e.g., as hematite). This causes undesired elements to be collected in the oxidic slag, while the Ta–Nb carbide is oxidized only as far as the metals and therefore remains in the metallic phase. In a final reaction stage, this is again oxidized with $Fe_2O_3$ to form a slag that contains tantalum and niobium oxides [22].

The partial oxidation can also be carried out in a single smelting stage using air, oxygen-enriched air, or pure oxygen. Under these conditions, the tantalum is converted to slag much more rapidly than the niobium, enabling the tantalum to be concentrated at the expense of niobium [23]. Ta/Nb carbide-containing ferroalloy can also be treated by wet leaching processes [24]–[26].

## 3.3. Processing Tantalum Scrap

When processing tantalum-containing waste materials, physical and chemical processes are used as much as possible, avoiding processes typically used in ore treatment, i.e., dissolution in hydrofluoric acid and solvent extraction. The tantalum scrap is divided into two categories which differ in difficulty of recycling:

1) Scrap material of the first type can consist of pure unoxidized metallic tantalum, e.g., sintered tantalum powder pellets (anodes) from the manufacture of tantalum electrolytic capacitors, scrap foil, sheet, and wire, or of anodically oxidized sintered tantalum pellets containing 1.5–3% oxygen from capacitor manufacture and oxidized wire scrap arising from this process.

The following processing methods are commonly used for these scrap materials: ingot melting in an electron beam furnace, or conversion to brittle tantalum hydride, grinding, and dehydrogenation at > 600 °C in vacuo or under a protective gas to form metallic tantalum powder [1]. Alternatively, the ground tantalum hydride can be carburized with carbon black to form tantalum carbide [27].

The oxygen content of oxidized tantalum anodes can be reduced by deoxidation with magnesium or calcium to give high-grade scrap [28].

2) Scrap of the second type consists of oxidized tantalum anodes coated with manganese dioxide or with conductive silver, or sometimes welded to nickel conductor wires. This category also includes faulty capacitors, in which the coated and welded tantalum anodes also have an encapsulation of synthetic resin or tinned brass.

Pure oxidized tantalum anodes can be recovered from coated anodes and brass-encapsulated capacitors by successive treatments with nitric and hydrochloric acid. Resin-encapsuled capacitors require complex treatment. These can be coarsely size-reduced in grinding mills, the tantalum being recovered by density separation. Foreign metals are removed by treatment with nitric and hydrochloric acid, leaving a residue of oxidized tantalum anodes in granular form [29].

Another method has been described for the treatment of tantalum anode scrap containing manganese dioxide. This can be directly reduced in an argon/hydrogen plasma and melted to pure tantalum [30]. If the scrap capacitor materials are resin free, the tantalum can be recovered as its pentachloride by chlorination [31].

Some types of tantalum scrap which contain other impurities are roasted to form the oxide, and treated like ore concentrates. These include condensates formed during the melting of tantalum ingots, scrapped alloy materials, impure sawings and turnings, flue dust, and sediments from the wash liquors from the sodium reduction of potassium heptafluorotantalate.

Hard metals that contain tantalum carbide can be attacked by roasting and treating with caustic soda solution [32], anodic oxidation [33], or melting with sodium nitrate [34], converting the tantalum to its oxide for further recovery treatment.

# 4. Metallic Tantalum

## 4.1. Production

Metallic tantalum products (powder, semifinished products, ingots) are produced almost exclusively by reduction of potassium heptafluorotantalate with sodium [35], [36]:

$$K_2TaF_7 + 5\,Na \longrightarrow 2\,KF + 5\,NaF + Ta$$

The strongly exothermic reaction is controlled by adding inert salts (KCl, NaCl, KF, NaF). The reduction can be carried out at constant temperature by controlled addition of one of the reactants, with controlled heat removal. Tantalum powder is recovered by leaching the salts out of the reaction product. Two processes are commonly used to obtain tantalum capacitor powder from this:

1) The sodium-reduced metallic tantalum powder is purified and converted to a compactable and free-flowing product by further processing stages, e.g., high-temperature treatment in vacuo or under protective gas, deoxidation by magnesium [37], or a combination of these two processes [38]. Optimum properties for capacitor production are obtained by the addition of dopants to control sintering of porous pellets [39]–[41].
2) During ingot melting of compressed slugs of tantalum powder in electron beam or electric arc furnaces [42], volatile impurities are driven off, usually in two smelting operations. The ingot can be ground after hydrogena-

tion. The powdered tantalum "hydride" is then dehydrogenated in vacuo [1] and agglomerated by high-temperature treatment.

In addition to the sodium reduction process, other processes have been described, i.e., the molten salt electrolysis of $K_2TaF_7$ with addition of $Ta_2O_5$ or $TaCl_5$, and hydrogen reduction of $TaCl_5$ [1], [43], but these have not achieved economic importance. The carbothermic reduction of $Ta_2O_5$ in high vacuum [1], [4], [44] at ca. 1900 °C leads first to tantalum carbide:

$$Ta_2O_5 + 2\,C \longrightarrow 2\,TaC + 5/2\,O_2$$

This reacts above 2000 °C with residual $Ta_2O_5$ to metallic tantalum:

$$5\,TaC + Ta_2O_5 \longrightarrow 7\,Ta + 5\,CO$$

Because of the high quality standards for particle size and purity required by the producers of tantalum capacitors (Section 4.3), this process is of minor importance.

## 4.2. Semifinished Products

Tantalum metal powder for producing semifinished products is also obtained by reducing $K_2TaF_7$ with sodium. Semifinished products such as sheet, wire, and shaped articles are produced by two different methods:

1) Powders are compacted, and then sintered by passing an electric current at ca. 2500 °C (Coolidge process) or indirectly by resistance or induction heating in vacuo. Tantalum wire for tantalum electrolytic capacitor manufacture is produced by this process. It is advantageous to control the grain size, e.g., by adding dopants [45], to optimize the mechanical properties.
2) Bars produced by powder compression are formed into ingots by electron beam or electric arc melting, and shaped by forging, rolling, extruding, and wire drawing. Intermediate heat treatment in vacuo is usually unnecessary as tantalum does not work harden. After 95 % deformation, a tantalum product has typical hardness 180 HV, after an initial ingot hardness of 80–90 HV [46], [47].

As well as its use for electrolytic capacitors (wire, cans, foil), semifinished tantalum is also

**Table 3.** Typical chemical composition of commercial tantalum semifinished products

|  | Powder metallurgy grade | Electron beam melted grade |
|---|---|---|
| Ta | $\geq 99.9\%$ | $\geq 99.98\%$ |
| Impurity content, ppm | | |
| O | 250 | 100 |
| C | 75 | 30 |
| N | 75 | 30 |
| H | 15 | 10 |
| Nb | 10 | 20 |
| W | 10 | 20 |
| Mo | 5 | 10 |
| Fe | 75 | 10 |
| Cr | 15 | 3 |
| Ni | 20 | 2 |
| Na | 10 | 1 |
| K | 10 | 1 |

used in the chemical industry because of its excellent resistance to corrosion by hot, concentrated mineral acids, with the exception of hydrofluoric acid which strongly attacks tantalum. The passivation of the metal is a result of the dense, adherent oxide film, 1−4 nm thick, which forms spontaneously in air and in acidic media, and which is stable up to ca. 260 °C [48]. Welding of tantalum sheet is mainly carried out by the tungsten−inert gas (TIG) process, which allows complex heat exchangers and intricate heating equipment to be fabricated [49]. Impurity levels in commercial semifinished tantalum products are listed in Table 3.

Because of the high price of tantalum, large vessels and pipework are often merely clad with the metal. Plating of steel with tantalum can be carried out by the explosion method [50], or corrosion protection can be achieved by build-up welding of tantalum [51].

## 4.3. Tantalum Powder for Capacitors

Since the first report that tantalum could be used for electrolytic capacitors [52], the growth of this application has been such that it now accounts for the highest consumption of the metal [53]. The capacitor consists of an anode of tantalum metal powder compressed into a pellet, typically 20−200 mg. This anode is sintered to give electrical continuity and density of ca. 6−9 g/cm$^3$. The electrical connection is provided by a tantalum wire, attached by sintering or welding. The whole of the pore surface of the sintered anode is coated with an amorphous

dielectric layer of $Ta_2O_5$ by anodic oxidation, using an electrolyte of dilute $H_3PO_4/H_2SO_4$, sometimes with the addition of organic compounds [54], [55]. $Ta_2O_5$ has a resistivity of ca. $10^{15}\,\Omega \cdot cm$ [56] and dielectric constant $\varepsilon = 25$. The anode is coated with a semiconducting layer of $MnO_2$ by impregnation with an aqueous solution of manganese nitrate, followed by pyrolysis. This solid electrolyte provides the cathodic contact. Research has been carried out into the effect of the pyrolysis temperature on the $Ta_2O_5$ layer [57] and its dielectric properties [58], and also into the effect of temperature on the capacity [59] and conductivity [60]−[62], and the effect of these on the leakage current of the capacitor [63]. Progressive improvements to powder quality [64] have doubled the specific capacitance of the tantalum metal powder (the product of capacitance and formation voltage per unit mass) [65]. Typical qualities and properties of capacitor grade tantalum powder are given in Table 4.

## 4.4. Other Uses

Tantalum is used in medicine because of its lack of toxicity and very good compatibility with tissue. Reports and discussions have been pub-

**Table 4.** Typical properties of capacitor-grade tantalum powder

| Chemical impurities, ppm | Powder from electron beam melting | | Powder from sodium reduction | |
|---|---|---|---|---|
| | QR 7* | QR 3* | VFI-18 KT** | STA-30 KD*** |
| O | 800 | 1500 | 1800 | 2200 |
| C | 10 | 10 | 35 | 40 |
| N | 20 | 40 | 50 | 70 |
| H | 15 | 15 | 50 | 50 |
| Nb | < 20 | < 20 | < 10 | < 10 |
| W | < 20 | < 20 | < 10 | < 10 |
| Fe | < 10 | < 10 | 25 | 25 |
| Cr | < 5 | < 5 | 10 | 10 |
| Ni | < 5 | < 5 | 20 | 20 |
| Na | < 1 | < 1 | < 2 | < 2 |
| K | < 1 | < 1 | 5 | 10 |
| Specific charge, C V/g (µF V/g) | 3000 | 5000 | 15 000 | 30 000 |
| Application working voltage, V | 50−75 | 35−50 | 25−35 | < 25 |

* H. C. Starck, Newton, United States. ** H. C. Starck − V-tech, Tokyo. *** H. C. Starck, Goslar, Germany.

lished on the suitability of inhaled tantalum powder as a contrast medium in X-ray diagnostics of the throat, trachea, larynx, bronchi, esophagus, and stomach [66]–[68], but the material has been rarely used, mainly for cost reasons. Small tantalum spheres or tantalum wires are used in organs (e.g., the heart), in bone, or in implants as markers when controlling position and function by X rays [69]–[71].

Because of its stable oxide layer, tantalum in the body is completely bioinert. It has been shown in animal experiments that, if broken bones are held in position with tantalum rods, even in septic conditions with corrosive tissue reaction, healing is better than when chromium–nickel steel or niobium are used [72]. Tantalum implants in the jaw have shown good metal–bone contact over many years with no morbid reaction [73]. Tantalum has been largely replaced as a prosthetic implant material by titanium, as this has better strength properties and is sufficiently bioinert [74]. However, tantalum has an important use in clips for rapid occlusion of vessels in surgery, either temporary or permanent, [75]–[77]. In some cases, tantalum mesh and plate can be implanted in damaged areas of the skull [78]–[80].

The possible use of the very dense and ductile tantalum for the manufacture of heavy missiles for armour penetration is under investigation [81].


# 5. Tantalum Alloys

**Tantalum-Based Alloys.** The most common tantalum alloy is Ta–2.5 W, produced by ingot melting. This has higher strength and resistance to deformation than pure tantalum, but still has good cold working and welding properties. It is used for the construction of industrial chemical equipment, especially heat exchangers, pipes, and vessels. Its resistance to hot mineral acids (except hydrofluoric acid) is even somewhat greater than that of pure tantalum [82]. The alloy Ta–7.5 W, produced by powder metallurgy, is harder, and is therefore a preferred material for springs in chemical equipment, e.g., valves. The alloy Ta–10 W, mainly produced by ingot melting, is relatively hard, but still ductile. This, too, is used for special applications in chemical equipment, and in aircraft construction [82], [83]. For cost reasons, tantalum–niobium alloys have

been proposed for chemical plant, but these alloys, Ta–40 Nb [83] and Nb–25 Ta [84], are less resistant to corrosion by hot acids.

The alloys Ta–8 W–2 Hf (T-111) and Ta–9.6 W–2.4 Hf–0.01 C (T-222), which have good machining and high-temperature properties, were developed for use in air- and spacecraft, e.g., in nozzles, and for cladding the interior of combustion chambers and the exterior of rockets [85].

Earlier research led to the development of many other binary and ternary alloys of tantalum with Nb, W, Mo, Hf, and V [86], [87], Hf, W, and Al [88], or Ti and Al [89]. Some of these alloys have improved high-temperature properties and oxidation resistance compared with pure tantalum, but have not achieved industrial importance.

**Superalloys.** In superalloys, tantalum contributes mainly to properties such as high-temperature strength, reduction in fatigue at high temperature, and reduction of corrosion by sea air. Superalloys ($\rightarrow$ High-Temperature Materials, **A 13**, pp. 57–65) are metallic multicomponent materials, with nickel, cobalt, or iron as the main constituent, and additions of tantalum, aluminum, chromium, hafnium, molybdenum, niobium, rhenium, titanium, tungsten, zirconium, etc., developed for use at high operating temperatures and under high mechanical stress. Components, produced by forging, casting, or powder metallurgy, are used in aircraft engines and stationary gas and steam turbines. These conventional superalloys contain up to 9 % tantalum (e.g., TRW-NASA VI A, MAR-M 302) [90]. The manufacture of single-crystal components, e.g., for turbine blades, and the directional solidification of superalloys specially developed for this technique (e.g., Rene 142, with 6.5 % Ta and PWA 1480 with 12 % Ta) are processes of increasing importance [91], [92].

The effect of tantalum in the nickel-based superalloys B-1900 (4 % Ta) and MAR-M 247 (3 % Ta) has been systematically investigated. The favorable high-temperature properties are mainly due to the $\gamma'$-phase of the type $Ni_3(Al, Ti, Ta)$, formed by precipitation from the $\gamma$-phase. Tantalum increases the volume fraction of the $\gamma'$-phase in the $\gamma'/\gamma$ ratio. For the most part, it appears in this phase without changing the partitioning ratios of the other elements. A small part of the tantalum appears in the metal carbide fraction, which improves the high-temperature strength

[93], [94]. As well as the composition of the material or component, a multistage heat treatment is very important for optimization of $\gamma'$-phase precipitation.

# 6. Tantalum Compounds

**Oxides.** Tantalum oxides in lower oxidation states are not industrially important. Tantalum(II) oxide, TaO [*12035-90-4*], is the only lower oxide whose existence has been confirmed. It is produced from tantalum pentoxide by reduction with carbon at 1900 °C, or with hydrogen at 1100 °C. A possible use is for the production of heat-reflecting window glass [95].

No tantalum compound analogous to niobium(IV) oxide, $NbO_2$, is known. However, with tantalum pentoxide, the possibility exists of introducing interstitial tantalum atoms, so that $TaO_x$ compounds can be obtained ($x = 2-2.5$). These oxides have metallic conductivity, but do not form discrete phases.

Tantalum pentoxide, $Ta_2O_5$ [*1314-61-0*] ($M_r$ 441.9, *mp* 1880 °C, density 8.73 g/cm³), occurs in two thermodynamically stable modifications, $\alpha$ and $\beta$. The transition temperature from the orthorhombic $\beta$ modification to the tetragonal $\alpha$ modification is ca. 1360 °C. The existence of an $\varepsilon$ modification, produced hydrothermally from tantalic acid at 300–340 °C, is also known. This $\varepsilon$-$Ta_2O_5$ is isomorphous with $\beta$-$Nb_2O_5$, and is transformed to $\beta$-$Ta_2O_5$ by heating in air above 886 °C [96].

Two processes are used for the manufacture of tantalum pentoxide:

*Wet Chemical Method.* The raw materials (ores and/or tin slag concentrates) are dissolved in hydrofluoric acid at ca. 100 °C. Tantalum and niobium are extracted from this strongly acid aqueous solution, preferably with methyl isobutyl ketone. The impurities that are also extracted, $SiF_6^{2-}$, $FeF_6^{3-}$, $AlF_6^{3-}$, $SbF_6^{2-}$, etc., are washed out with sulfuric acid, and the tantalum and niobium are separated in a multistage operation. Ammonia gas or solution is added to the aqueous $H_2TaF_7$ solution, forming hydrated tantalum oxide, $Ta_2O_5 \cdot n\,H_2O$ [*75397-94-3*]. This is washed, dried, and calcined above 800 °C to form the oxide (→ Niobium, **A 17** pp. 251–264) [97].

*Chloride Process.* The ferroalloy produced by the pyrometallurgical method is chlorinated in an iron chloride–sodium chloride melt at 550–600 °C. The complex salt $NaFeCl_4$ present in the melt acts as a chlorine carrier. The iron loses a chlorine atom and is reduced from the trivalent to the divalent state, and the added chlorine gas causes the iron to be continuously reoxidized to Fe(III). The $TaCl_5/NbCl_5$ mixture is separated by fractional distillation to obtain the chlorides of the two elements in a highly pure state. To obtain the oxide, the ground chlorides are hydrolyzed with steam in a fluidized bed, and then calcined (→ Niobium, **A 17**, p. 251–264).

In the optical industry, tantalum pentoxide is used in lanthanum borate glasses, characterized by their high refractive index and low optical scattering; since the 1980s, niobium pentoxide has been increasingly used for this application, as tantalum pentoxide is 3–5 times as expensive and has twice the density [98].

Tantalum pentoxide, and especially hydrated tantalum oxide $Ta_2O_5 \cdot n\,H_2O$ [*75397-94-3*], is used as an acid catalyst. In its amorphous form, its surface acidity can be as high as $H_0 \leq -8.2$ [99]. It therefore has good catalytic acitivity for alkylation, esterification, Beckmann rearrangement, aldol condensation, etc. [100]–[103].

**Tantalates.** Tantalates can be produced by calcination of oxide mixtures, or by dissolving tantalum pentoxide in molten alkali metal hydroxides or carbonates. Excess alkali metal hydroxides or carbonates leads to the formation of water-soluble tantalate isopolyanions $(H_xTa_6O_{19})^{(8-x)-}$, where $x = 0, 1$, or 2 [104].

The most industrially important tantalates are those of lithium and yttrium. Single crystals of lithium tantalate, $LiTaO_3$ [*12031-66-2*], are pulled from the melt crucible by the Czochralski method [105]. The crystal growth rate for crystals 5–8 cm in diameter is ca. 2–3 mm/h. These ferroelectric single crystals in the unpoled state have domains that are statistically distributed in the crystal. They are poled on passing through the Curie temperature (665–618 °C) in an electric field and then have only a single domain [106], [107]. Single-crystal wafers of $LiTaO_3$ are widely used as surface acoustic wave filters (SAW) for high-frequency applications in communication systems (intermediate-frequency filters in television and video equipment). Also, lithium tantalate is used in combination with lithium niobate in the manufacture of waveguides, frequency doublers (second harmonic generation), optical modulators, and optical switches.

Yttrium tantalate powder, $YTaO_4$ [12143-49-6], produced in a solid state reaction, is applied as an emulsion to a film, and is used in diagnostic medicine as a phosphor for the amplification of X rays. It converts X rays to fluorescent light in very high yield, enabling the X-ray dose received by the patient to be significantly reduced [108].

Barium magnesium tantalate, $Ba_3MgTa_2O_9$ [12231-81-1], and barium zinc tantalate, $Ba_3ZnTa_2O_9$ [12231-88-8], are likely to play an increasingly important role in the future as the base material for microwave resonators for the stabilization of oscillators, or as frequency filters. These dielectric materials, which have the perovskite structure, have high dielectric constant and very low dielectric loss at high frequencies [109]–[112]. These high-quality resonators are especially necessary for the high carrier frequencies in satellite communication.

**Halides.** Tantalum pentafluoride, $TaF_5$ [7783-71-3], ($mp$ 96.8 °C, $bp$ 229.5 °C) is produced by passing fluorine over metallic tantalum, or from tantalum chloride by adding anhydrous hydrogen fluoride [113]–[115]. Both tantalum pentafluoride and niobium pentafluoride are used in the petrochemical industry as isomerization and alkylation catalysts. The fluorides of both elements are also useful as fluorination catalysts for the manufacture of fluorochloro- and fluorinated hydrocarbons [116].

The oxyfluorides $TaOF_3$ [20263-47-2] and $TaO_2F$ [13597-27-8] are also known, but have no industrial importance.

Potassium heptafluorotantalate (potassium tantalum fluoride), $K_2TaF_7$, is an industrially important intermediate in the production of tantalum metal (Section 4.1). It crystallizes in colorless, rhombic needles when $K^+$ ions (e.g., KCl or KF) are added to solutions of tantalum in hydrofluoric acid (e.g., $H_2TaF_7$ solution) from the extraction process. If the acidity is not high enough, undesired oxyfluorides are formed in addition to $K_2TaF_7$.

The solubility of potassium heptafluorotantalate in hydrofluoric acid solution decreases from 60 g/100 ml at the boiling point to < 0.5 g/100 ml at room temperature. As the corresponding stable niobium salt $K_2NbOF_5$ is significantly more soluble, tantalum can also be

separated from niobium by fractional crystallization (Marignac process) [7].

Tantalum pentachloride, $TaCl_5$ [7721-01-9], $mp$ 216 °C, $bp$ 242 °C, forms colorless, strongly hygroscopic, needle-shaped crystals. On an industrial scale, production of this chloride is exclusively by chlorination of metallic tantalum, ferrotantalum, or tantalum metal scrap. This process is mainly used in the production of high-purity oxides [117], [118].

Tantalum pentachloride is soluble in absolute alcohol, with formation of the corresponding alkoxide [119], [121]. Ultrafine tantalum oxide powder can be produced by hydrolysis of this alkoxide [121].

Tantalum pentachloride and tantalum alkoxides are suitable for use in the chemical vapor deposition of tantalum metal or tantalum oxide.

The other pentahalides, $TaBr_5$ [13451-11-1] and $TaI_5$ [14693-81-3], are known, but are of no industrial importance.

**Carbides.** Tantalum carbide, TaC [12070-06-3], is golden-brown, $mp$ 3985 °C (one of the highest known). It is added as powder in the manufacture of hard metals. Tantalum carbide can be produced by the direct carburization of Ta metal by carbon, or preferably by reaction of $Ta_2O_5$ with carbon at ca. 1900 °C. In hard metal mixtures based on tungsten carbide, tantalum carbide is present at concentrations 0.5–10 %. It improves the high-temperature fatigue strength and thermal shock resistance of the cutting tool. For details of its properties, production, and use → Carbides, **A 5**, pp. 69–70.

**Nitrides.** Tantalum nitride, TaN [12033-62-4], is a hard material, like tantalum carbide, with good wear resistance. It is produced by heating tantalum in a pure nitrogen atmosphere at ca. 1100 °C. Tantalum nitride films are used as diffusion barriers in semiconductor technology [122], [123], and as protective coatings for moisture sensors [124]. Tantalum nitride is also sometimes found in small amounts (3–8 %) in cermets (base material Ti(C, N)).

**Borides.** The tantalum–boron phase diagram [125] shows the existence of $Ta_2B$ [12045-26-0], $Ta_3B_2$ [12228-67-0], TaB [12007-07-7], $Ta_3B_4$ [12045-92-0], and $TaB_2$ [12007-35-1]. These hard materials are of little industrial importance. The borides may be produced by heating tantalum and boron in vacuo at ca. 1800 °C, or by electrolyzing a melt of $Ta_2O_5$, $B_2O_3$, CaO, and $CaF_2$ ($\rightarrow$ Boron Carbide, Boron Nitride and Metal Borides **A 4**, pp. 303–307).

**Silicides.** The silicides $TaSi_2$ [*12039-79-1*] and $Ta_5Si_3$ [*12067-56-0*] [126] are produced from the elements by heating in vacuo at 1000–1500 °C. Tantalum silicide targets were formerly used in the semiconductor industry for production of interconnections by sputtering [127], but tantalum silicides are no longer of any industrial importance.

# 7.  Analysis

The analytical chemistry of tantalum is described in detail in several monographs [128]–[131]. Analysis for tantalum as a major or minor component or trace element in raw materials, ferroniobium–tantalum, scrap materials, and alloys must be distinguished from analysis for impurities in tantalum products. The classical method of analyzing for tantalum is by gravimetry after extraction as the fluoro complex on an ion-exchange resin [132]. Low concentrations, i.e., in the µg/g region, are usually determined photometrically [131].

When tantalum is a major or minor component in ores, scrap, etc., it is usually determined by fusion with sodium or lithium borate, followed by X-ray fluorescence analysis ($\geq 0.01\%$ Ta) [133]. Analysis of solutions by emission spectrometry with plasma excitation (ICP-AES) is increasingly used. Tantalum in the ng/ml region can be determined by measurement at 296.5 nm.

Because of the high purity required for electronic components, optical glasses, and superalloys, sensitive multielement methods have been developed for the determination of trace impurities in commercial products, e.g. potassium heptafluorotantalate, metallic tantalum, and tantalum oxide. The first methods were based on OES with vaporization in a carbon d.c. arc [134]. Additionally, many elements can be determined by plasma optical emission spectrometry (DCP-AES, ICP-AES) after dissolution by treatment with hydrofluoric or hydrofluoric/nitric acid under pressure [135]. Alkalis are determined by flame atomic absorption spectrometry, and arsenic, antimony, and other hydride formers by hydride techniques. Phosphorus, as a doping element in tantalum powders, can be very reliably determined photometrically [136]. Carbon and sulfur are determined by combustion analysis, and oxygen and nitrogen by hot extraction with a carrier gas.

As the optical spectrum of the tantalum matrix contains a very large number of lines, there are many possibilities for interference; trace element–matrix separation offers the possibility to overcome the problems [135], [137], [138]. For the final determination, ICP-AES, ICP-mass spectrometry, or neutron activation analysis are used. For trace determination in the sub-ppm region, glow discharge mass spectrometry is now the most sensitive routine method of determination [139].

# 8.  Economic Aspects

Figure 2 shows how the price of tantalum has fluctuated over recent years. The most important tantalum processors have made long-term ore contracts with the large mining concerns [140]. World production of tantalum ores (Table 1) is well below capacity.

World demand for tantalum has changed little since 1970, remaining at ca. 1000 t/a. Forecasts based on increasing tantalum demand for electrolytic capacitors [141] have not been borne-out, as developments in tantalum powders have meant that increasingly less tantalum powder is required for a given capacitor type [142]. However, the number of tantalum capacitors produced has grown continually, from ca. $2 \times 10^9$ in 1975 to ca. $9 \times 10^9$ in 1992 [53]. The most important area of use is for electrolytic capacitors, which account for $> 50\%$ of total world production of tantalum. The use of tantalum carbide as an ad-

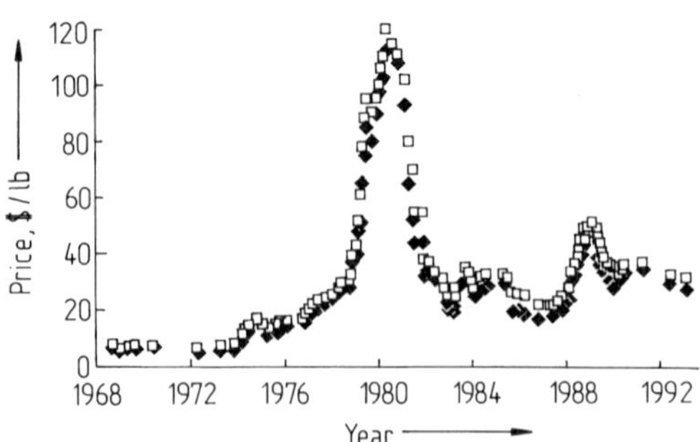


**Figure 2.** Tantalum ore prices
London Metal Bulletin 25/40% $Ta_2O_5$, based on 30% $Ta_2O_5$, cost, insurance, and freight

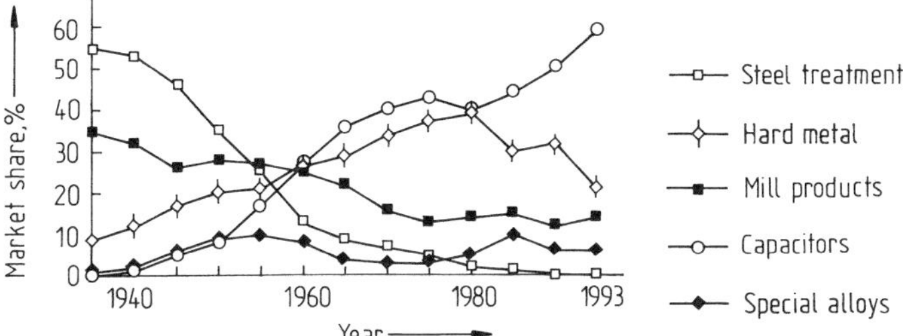


**Figure 3.** Market share of tantalum applications, 1935–1993

ditive in the hard metal industry, which was still important in 1980, has decreased in recent years [17] as TaC is being replaced by chromium and vanadium carbides for cost reasons. Figure 3 illustrates the shifting market shares of the various application areas over recent decades. Applications not included, e.g., $Ta_2O_5$ in optical applications and in X-ray phosphors, account for considerably less than 5% of the market.

## 9. Toxicology

Metallic tantalum is completely nontoxic and bioinert, and allergic reactions are unknown. This has enabled it to be used in X-ray diagnostics and surgery (Section 4.4). Tantalum carbide is not known to have acute toxicity. Tantalum oxide is classified as nontoxic, although animal experiments indicate that the $LD_{50}$ (rat, oral) is 8000 mg/kg (RTECS). For respirable dusts of metallic tantalum, the MAK value is 5 mg/m$^3$. Tantalum chloride has a strongly corrosive and irritant effect, owing to the hydrolytic production of hydrogen chloride. The $LD_{50}$ (rat, oral) is 1900 mg/kg (RTECS). MAK values have not yet been established for tantalum oxide, tantalum carbide, and tantalum chloride. Potassium heptafluorotantalate has toxic properties due to its fluoride content; $LD_{50}$ (rat, oral) is 2500 mg/kg (RTECS), and MAK for dusts is 2.5 mg/m$^3$ total dust (calculated as fluoride) [143].

## 10. References

[1] R. Kieffer, H. Braun: *Vanadin-Niob-Tantal*, Springer Verlag, Berlin – Göttingen – Heidelberg 1963.
[2] *Gmelin*, **50**, part B 1, pp. 1–4.
[3] S. M. Cardonne, P. Kumar, C. A. Michaluk, H. D. Schwartz, *Adv. Mater. Processes* **1992**, no. 9, 16–20.
[4] R. Kieffer, G. Jangg, P. Ettmayer: *Sondermetalle*, Springer Verlag, Berlin – Heidelberg 1971.
[5] O. Kubaschewski: *Atomic Energy Review, Tantalum: Physico-Chemical Properties of its Compounds and Alloys*, International Atomic Energy Agency, Vienna 1972.
[6] W. Köck, P. Paschen, *J. Met.* (1989) no. 10, 33–39.
[7] G. L. Miller: *Tantalum and Niobium*, Butterworths, London 1959, pp. 91–93.
[8] E. Erben, R. Lesser, *Metall (Berlin)* **15** (1961) 679–686.
[9] W. W. Albrecht: "Production, Properties and Application of Tantalum, Niobium and Their Compounds," in P. Möller, P. Cerný, F. Saupe (eds.): *Lanthanides, Tantalum and Niobium*, Springer Verlag, Berlin – Heidelberg 1989, pp. 345–358.
[10] P. Borchers, G. J. Korinek: "Extractive Metallurgy of Tantalum," *AIME Annu. Meet. 110th* **1981**, 97–108.
[11] TIC Tantalum-Niobium International Study Center Brussels, *Int. J. Refract. Hard Met.* **7** (1988) no. 1, 3–7.
[12] L. D. Cunningham: *Mineral Industry Surveys, Columbium (Niobium) and Tantalum in 1993*, U.S. Bureau of Mines, Washington 1994.
[13] R. Gaupp, G. Moteani, P. Möller, *Erzmetall* **36** (1983) 244–251.
[14] H. C. Starck GmbH & Co. KG, Internal Raw Material Availability Survey, Goslar 1993.
[15] TIC Tantalum-Niobium International Study Center Brussels, *TIC Bull.* **62–73** (1990–1993).
[16] T. C. Pool (ed.): *Elements,* Concord Trading Corp., Denver 1992, pp. 12–17.
[17] *The Economics of Tantalum*, 6th ed., Roskill Information Services, London 1986, p. 2.
[18] C. Bird, *Met. Bull. Monthly* **1993**, no. 5, 48–49.
[19] W. Rockenbauer, *Metall (Berlin)* **35** (1981) 584–585.
[20] W. Rockenbauer, *Metall (Berlin)* **38** (1984) 156–159.
[21] J. Bonjer: "Proceedings of the Seminar on Production and Marketing of Associated Heaug Minerals from South-East Asian Tin Deposit in Ipoh/Malaysia and Phuket /Thailand," in W. Gocht (ed.): *Intertechnik* **26** (1985) 179–206.
[22] Kawecki Berylko Ind., US 3721727, 1971 (R. A. Gustison).
[23] Metallurgie-Hoboken-Overpelt, DE 2844914, 1978 (M. C. F. van Hecke, J. Deweck).
[24] Hoboken-lez-Anvers, US 2131350, 1937 (J. P. Leemans).
[25] Kawecki Chemical Comp., US 3447894, 1966 (R. A. Gustison, W. Lawn, F. Gorcyca, J. A. Cenerazzo).
[26] Kawecki Berylko Ind., US 3585024, 1968 (J. A. Cenerazzo, C. E. Mosheim, C. E. Marvasi).

[27] R. Kieffer, F. Benesovsky: *Hartstoffe*, Springer Verlag, Wien 1963.

[28] Western Electric Co., US 3697255, 1972 (W. M. Baldwin, E. O. Fuchs, D. J. Sharp, J. H. Swisher).

[29] H. C. Starck, DE 2133104 C3, 1978 (H. Meyer).

[30] K. Mimura, T. Takahashi, M. Nanjo, *J. Min. Mater. Proc. Inst. Jpn.* **106** (1990) no. 4, 187–192.

[31] N. Sato, T. Takenaka, Y. Wei, M. Nanjo, *Tohoku Daigaku Senko Seiren Kenkyusho Iho* **46** (1990) no. 2, 96–104.

[32] GTE Products Corp., US 4255397, 1981 (B. E. Martin, J. E. Ritsko, H. L. Acla).

[33] G. P. Power, *Chem. Aust.* **46** (1979) no. 7, 303–307.

[34] VEB Kombinat Metallaufbereitung, DD 207932, 1984 (G. Beyer, H. Bernhardt, H. Dam, H. K. Kiesling, W. Tomalik, A. Zimmermann).

[35] NRC Inc., US 2950185, 1960 (E. D. Hellier, G. L. Martin).

[36] NRC Inc., US 4141720, 1979 (H. Vartanian).

[37] H. C. Starck, US 3635693, 1969 (H. J. Friedrichs, H. Meyer).

[38] H. C. Starck, DE 3130392, 1983 (W. W. Albrecht, 33 U. Papp).

[39] Fansteel, US 4356028, 1982 (V. T. Bates).

[40] H. C. Starck, DE 3140248, 1983 (W. W. Albrecht, D. Behrens).

[41] NRC Inc., US 4957541 (T. B. Tripp).

[42] W. W. Albrecht, D. P. Ingalls: "Hochtemperaturreinigung von Tantal und Niob," in Gesellschaft Deutscher Metallhütten- und Bergleute (eds.): *Raffinationsverfahren in der Metallurgie*, Verlag Chemie, Weinheim 1983, pp. 229–242.

[43] *Gmelin* **50**, part A2, pp. 490–491.

[44] C. A. Hempel: *Rare Metals Handbook*, 2nd ed., Reinhold Publ. Corp., London 1961.

[45] Fansteel, US 4859257, 1989 (V. T. Bates, C. Pokross).

[46] E. Raub, E. Röschel, *Z. Metallkd.* **53** (1962) 93–103.

[47] G. Jangg, R. Eck, *Metall* (*Berlin*) **31** (1977) 750.

[48] G. D. Corey, *Proc. Int. Symp. Tantalum 1st* **1978**, 61–82.

[49] F. J. Hunkeler: "Properties of Tantalum for Applications in the Chemical Process Industry," in R. E. Smallwood (ed.): "Refractory Metals and Their Industrial Applications," *ASTM Spec. Tech. Publ.* **849** (1984) 28–49.

[50] M. Hörmann, *Metall* (*Berlin*) **42** (1988) 400–406.

[51] U. Draugelates, B. Bouaifi, H. Steinberg, *Werkst. Korros.* **44** (1993) 269–273.

[52] R. L. Taylor, H. E. Haring, *J. Electrochem. Soc.* **103** (1956) 611–613.

[53] Tantalum-Niobium International Study Center, *TIC Bull.* **73** (1993) 2–3.

[54] M. Fernández, J. Baonza, J. M. Albella, J. M. Martínez-Duart, *Electrocomponent Sci. Technol.* **7** (1981) 205–210.

[55] Sprague Electric Comp., US 4781802, 1988 (E. J. Fresia).

[56] G. P. Klein, N. J. Jaeger, *J. Electrochem. Soc.* **117** (1970) 1483–1494.

[57] T. Kudoh, M. Katoh, M. Watase, *Denki Kagaku oyobi Kogyo Butsuri Kagaku* **40** (1972) 701–705.

[58] D. M. Smyth, G. A. Shirn, T. B. Tripp, *J. Electrochem. Soc.* **110** (1963) 1264–1270.

[59] D. M. Smyth, T. B. Tripp, *J. Elechtrochem. Soc.* **110** (1963) 1271–1276.

[60] D. M. Smyth, G. A. Shirn, T. B. Tripp, *J. Electrochem. Soc.* **111** (1964) 1331–1336.

[61] D. A. Vermilyea, *J. Appl. Phys.* **36** (1965) 3663–3671.

[62] D. A. Vermilyea, *J. Electrochem. Soc.* **112** (1965) 1232–1234.

[63] M. Burnham, Hughs Aircraft Compancy Report, no. M 74-08, Culver City, Cal. 1974.

[64] K. Andersson, H. Naito, T. B. Tripp, *Proc. 6th Europ. Capacitor and Resistor Tech. Symp. CARTS-Europe,* Brugge 1992, pp. 1–6.

[65] Tantalum-Niobium International Study Center Brussels, *TIC Bull.* **69** (1992) 6–8.

[66] D. B. Plone, *J. Am. Osteopath Assoc.* **73** (1974) no. 4, 635–640.

[67] P. J. Friedman, G. M. Tisi, *Radiology* (*Easton, Pa.*) **104** (1972) Sept., 523–535.

[68] E. Kammler, W. Weller, W. T. Ulmer, E. Bruckmann, *Pneumonologie* **146** (1972) 241–249.

[69] G. T. Daughters et al., *J. Thorac. Cardiovasc. Surg.* **104** (1992) no. 4, 1084–1091.

[70] T. Friden, L. Ryd, A. Lindstrand, *Acta Orthop. Scand.* **63** (1992) no. 1, 80–84.

[71] B. Tjornstrand, G. Selvik, N. Eglund, A. Lindstrand: *Arch. Orthop. Trauma Surg.* **99** (1981) no. 2, 73–81.

[72] L. Rabenseifner, W. Küsswetter, P. H. Wünsch, M. Schwab, *Z. Orthop.* **122** (1984) 349–355.

[73] B. Heinrich et al., *Z. Stromatol.* **75** (1977) no. 6, 214–224.

[74] P. Tetsch: *Enossale Implantationen in der Zahnheilkunde,* 2nd ed., Hanser Verlag, München 1991, pp. 46–47.

[75] F. Kylberg, *Acta Chir. Scand.* **141** (1975) no. 3, 242–244.

[76] F. Magistris, *Wien. Klin. Wochenschr.* **86** (1974) no. 8, 225–228.

[77] F. Magistris, *Chirurg* **46** (1975) no. 11, 518–521.

[78] H. Kobayashi et al., *Neurol. Res.* **8** (1986) no. 4, 221–224.

[79] S. A. Wexler, B. R. Frueh, D. C. Musch, M. A. Pachtman, *Ophthalmology* (*Philadelphia*) **92** (1985) no. 5, 671–675.

[80] H. J. Gerhardt, G. Muhler, D. Szdzuy, F. Biedermann, *Zentralbl. Neurochir.* **40** (1979) no. 1, 85–94.

[81] J. Carleone, *TIC Bull.* **71** (1992) 5–10.

[82] R. B. Flanders, *Chem. Eng.* (1979) Dec. 17, 109–110.

[83] M. Schussler, *J. Refr. Hardmet.* **2** (1983) no. 2, 67–70.

[84] A. Robin, M. E. de Almeida, C. A. Nunes, *Corrosion* (*Houston*) **47** (1991) no. 6, 443–448.

[85] *Chem. Eng. News* **42** (1964) no. 38, 47.

[86] F. F. Schmidt, E. S. Bartlett, H. R. Ogden, *Trans. Soc. Min. Eng. AIME* **227** (1963) 856–864.

[87] *Batelle Tech. Rev.* **12** (1963) no. 1, 12–13.

[88] H. R. Babitzke, J. G. Croeni, *Rep. Invest. US Bur. Mines* **7116** (1968) 1–14.

[89] *Batelle Tech. Rev.* **13** (1964) no. 2, 12.

[90] E. F. Bradley: *Superalloys – A Technical Guide,* ASM International, Metals Park, Ohio, 1988, pp. 10–14.

[91] E. W. Ross, K. S. O'Hara: "René 142: A High Strength, Oxidation Resistant DS Turbine Air-Foil Alloy," in: "Superalloys 1992," *Proc. 7th. Int. Symp. Superalloys* **1992**, 257–265.

[92] A. D. Cetel, D. N. Duhl: "Second Generation Columnar Grain Nickel Base Superalloy," in: "Superalloys 1992," *Proc. 7th Int. Symp. Superalloys* **1992**, 287–296.

[93] Z. Meng, G. Sun, M. Li: "The Strengthening Effect of Tantalum in Nickel-base Superalloys," in: "Superalloys 1984," *Proc. 5th Int. Symp. Superalloys* **1984**, 563–572.

[94] G. M. Jankowski, R. W. Heckel, B. J. Pletka, *Metall. Trans. A* **17 A** (1986) 1891–1904.

[95] Central Glass Co., JP 04 243 935, 1992 (N. Takeuchi, T. Ito, M. Takayama, K. Furuyo, H. Nakajima, H. Iida).

[96] F. Izumi, H. Kodama, *J. Less-Common Met.* **63** (1979) 305–307.

[97] H. C. Starck, DE 4 021 207 A 1, 1990 (W. Bludßus, J. Eckert).

[98] T. Ichimura: "Niobiumoxide in Optical Glass Manufacture," in H. Stuart (ed.) *Niobium Proc. Int. Symp.* **1984**, 603–614.

[99] Mitsubishi Chem. Ind., JP 63 051 945, 1986 (H. Wada, T. Ushikubo).

[100] Bayer, DE 4 133 675, 1991 (O. Immel, H. J. Buysch, G. Darsow).

[101] Bayer, EP 433 811, 1991 (O. Immel, H. Waldmann, R. Braden).

[102] Mitsubishi Kasei Corp., JP 01 050 843, 1989 (H. Wada, T. Ushikuba).

[103] Mitsubishi Kasei Corp., JP 63 225 329, 1988 (T. Maki, T. Yokoyama, Y. Sumino).

[104] H. J. Lunk, S. Schönherr, *Z. Chem.* **27** (1987) 157–170.

[105] B. C. Grabmaier, F. Otto, *J. Cryst. Growth* **79** (1986) 682–688.

[106] S. Miyazawa, H. Iwasuki, *J. Cryst. Growth* **10** (1971) 276–278.

[107] S. Miyazawa, H. Iwasuki, *Rev. Electr. Commun. Lab.* **21** (1973) 374–383.

[108] AGFA-Gevaert, EP 0 520 094 A 1, 1992 (D. Philip).

[109] O. Renoult, J. B. Boilot, F. Chaput, *J. Am. Ceram. Soc.* **75** (1992) 3337–3340.

[110] S. Kawashima, M. Nishida, I. Ueda, H. Ouchi, *J. Am. Ceram. Soc.* **66** (1983) 421–423.

[111] D. A. Sagala, S. Nambu, *J. Am. Ceram. Soc.* **75** (1992) 2573–2575.

[112] K. Tochi, *J. Am. Ceram. Soc. Jpn. Int. Ed.* **100** (1992) 1441–1442.

[113] J. K. Gibson, *J. Fluorine Chem.* **55** (1991) no. 3, 299–311.

[114] A. I. Papov, V. F. Sukhoverkov, N. A. Chumaevskii, *Zh. Neorg. Khim.* **35** (1990) 1111–1122.

[115] E. I. Du Pont de Nemours, WO 92/03382, 1990 (M. Nappa).

[116] Asahi Glass Co., WO 9 008 754 A 2 090 809, 1990 (S. Morikawa, S. Samejima, H. Okamoto, K. Ohnishi, S. Tatematsu, T. Tanuma, T. Ohmori).

[117] Ciba, DE 1 066 194, 1957 (W. Scheller).

[118] Ciba, DE 1 056 105, 1956 (F. Kern, W. Schornsein).

[119] D. C. Bradley, B. N. Chakravarti, W. Wardlaw, *J. Chem. Soc.* **52** (1956) 2381–2384.

[120] R. Gut, *Helv. Chim. Acta* **47** (1964) 2262–2278.

[121] N. Sato, M. Nanjo, *High Temp. Mater. Proc.* **8** (1988) 39–46.

[122] M. A. Nicolet, *Thin Solid Films* **52** (1978) 415–443.

[123] J. R. Shappirio, *Solid State Technol.* **28** (1985) 161–166.

[124] E. Jiang, C. Sun, J. Li, Y. Liu, *J. Appl. Phys.* **65** (1989) 1659–1663.

[125] K. I. Portnoi, V. M. Romashov, *Sov. Powder Metall. Met. Ceram. (Engl. Transl.)* **11** (1972) 378–384.

[126] S. P. Murarba, D. B. Fraser, W. S. Lindenberger, A. K. Sinha, *J. Appl. Phys.* **51** (1980) 3241–3245.

[127] K. Hieber, F. Neppl, *Thin Solid Films* **140** (1986) 131–135.

[128] R. Fresenius, G. Jander: *Handbuch der analytischen Chemie*, Part 3, "Quantitative Bestimmungs- und Trennungsmethoden," vol. 5 b, "Elemente der 5. Gruppe, Vanadin, Niob, Tantal," Springer Verlag, Berlin 1957.

[129] R. W. Moshier: *Analytical Chemistry of Niobium and Tantalum*, Pergamon Press, New York 1964.

[130] J. M. Gibalo: *Analytical Chemistry of Niobium and Tantalum*, Ann Arbor, London 1970.

[131] O. G. Koch, G. A. Koch-Dedic: *Handbuch der Spurenanalyse*, Part 2, 2nd ed., Springer Verlag, Berlin – Heidelberg – New York 1974.

[132] S. Kallmann, H. Obertin, R. Liu, *Anal. Chem.* **34** (1962) 609–613.

[133] A. H. Knight, *TIC Bull.* **45** (1964) 7–8.

[134] Autorenkollektiv: *Methoden zur Bestimmung von Spurenanalyse in Niob, Tantal und Wolfram*, VEB Deutscher Verlag für Grundstoffindustrie, Leipzig 1970.

[135] J. Stummeyer, G. Wünsch, *Fresenius J. Anal. Chem.* **342** (1992) 203–206.

[136] O. Hilmer, J. Peters, A. Breustedt, A. Hoppe: "Bestimmung des Phosphors in Refraktärmetallen und deren Verbindungen," in D. Hirschfeld (ed.): *Nichtmetalle in Metallen '90*, Deutsche Gesellschaft für Metallkunde (DGM), Oberursel 1990, pp. 85–92.

[137] V. Krivan et al., *Fresenius J. Anal. Chem.* **341** (1991) 550–554.

[138] R. Caletka, R. Hausbeck, V. Krivan, *J. Radioanal. Nucl. Chem.* **120** (1988) no. 2, 319–333.

[139] N. Jakubowski, D. Stuewer, G. Tölg, *Int. J. Mass Spectrom. Ion Processes* **71** (1986) 183–197.

[140] *Met. Week* **5** (1991) 21.

[141] W. Gocht: *Handbuch der Metallmärkte*, 2nd ed., Springer Verlag, Berlin – Heidelberg – New York – Tokyo 1991, pp. 401–413.

[142] I. Salisbury, *Proc. 5th Europ. Capacitor and Resistor Tech. Symp. CARTS-Europe*, Munich 1991, pp. 81–88.

[143] H. C. Starck GmbH u. Co. KG, Sicherheitsdatenblätter für Tantal-Metall, Tantalcarbid, Tantaloxid, Tantalpentoxid, Tantalchlorid und Kaliumtantalfluorid, Goslar 1994 (in press).

# Tapes, Adhesive

CARL. A. DAHLQUIST, St. Paul, Minnesota, 55144, United States

JOAQUIN DELGADO, RICHARD W. OEHMKE, 3 M Company, St. Paul, Minnesota, 55144, United States



## 1. Introduction

Most pressure-sensitive tapes consist of four basic components: a backing, a low-adhesion backsize (LAB), a primer, and a pressure-sensitive adhesive (PSA) [1], [2]. The *backing* may be a fabric, a tissue, an impregnated paper, a rigid film, a plasticized film, a foamed polymer, a metal foil, or various combinations of films with fabric, tissue, or aligned monofilaments to increase the strength of the backing. The *backsize* may be a release coating to reduce the effort needed to unwind the tape from a roll, a sealant for a porous backing, or both. The *primer* strengthens the bond between the adhesive and the backing, thereby preventing offsetting or telescoping, and the *adhesive* provides tack, and hence the ability of the tape to stick to most surfaces with no more than moderate hand pressure.

The term "pressure-sensitive" connotes more than stickiness. In addition to being aggressively tacky, the adhesive must be sufficiently cohesive for it to be handled with the fingers and removed from most surfaces without leaving a visible residue. The adhesive must do more than merely cling to a surface, but it cannot be excessively soft and sticky [3].

## 2. Historical Background

The earliest tapes sold in commerce were cloth bandaging tapes made by calendering a sticky adhesive mass on to cloth. They were pre-ceded by, and resembled, adhesive plasters. At the turn of the century a cloth insulating tape sometimes called "friction tape" was introduced for wrapping electrical connections. It adhered better to itself than to other surfaces. In the 1920s the need for a tape to make sharp demarcations between two different paint colors in the finishing of two-tone automobile bodies led to the invention of masking tape. This was quickly followed by a transparent tape comprised of cellophane with a transparent adhesive. These developments led to a burgeoning industry in the 1930s, 1940s, and 1950s, where technological advances led to a host of new tapes for specific applications.

## 3. Paper Tapes and PSA-Coated Paper Products

The most representative example is masking tape. Masking tape must be impermeable to the solvents used in paint formulations and conformable, with a degree of "dead stretch," so that when stretched around contours it will not retract and lift from the surface. The creped paper backings which make this conformability possible may be calendered to provide a flatter, thinner tape. To render the tape impervious to paints and lacquers, and resistant to fiber separation or delamination, the backing is impregnated with a fiber-binding composition, typically a moisture-insensitive elastomeric material. The

Ullmann's Encyclopedia
of Industrial Chemistry, Vol. A 26

PSAs most commonly used on masking tapes are blends of natural or synthetic rubbers and tackifying resins, such as rosin. Antioxidants of the nonstaining type extend tape life, and fillers and pigments are often added for cost saving or enhancement of cohesive strength. Cohesive strength may also be increased by cross-linking with sulfur compounds or phenolic resins, or by high-energy radiation. Both saturant and adhesive may be cross-linked.

Masking-tape backsizes are solvent barriers and also serve to reduce the force needed to unwind the tape. Typically they are thermosetting resins overcoated with thin release coatings. The release coatings must provide enough adhesion to coatings sprayed over the tape so that these coatings will not flake off when dry and contaminate subsequent coats of paint.

Special masking tapes have been developed for severe temperature requirements, for weather resistance, and for masking of anodized aluminum and other polished or finely finished surfaces such as stainless steel and poly(methyl methacrylate).

Paper tapes other than masking tapes are used in carton sealing, product bundling, labeling, and surface protection. These tapes typically use flat paper backings which, depending on the end use requirements, may or may not have unification treatments.

Products related to paper tapes are the pressure-sensitive labels and decals. While tapes are typically coated directly, labels and decals are produced by laminating adhesive-coated release liner to various face-stocks, which may be paper or film. Individual labels or decals are then produced by die-cutting the face-stock side as far down as the liner surface and peeling away the "weed" (excess material that lies outside and between the borders of the individual labels). The paper webs used as face-stocks for labels may be filled and glossy, but need not be strengthened with a unifying binder.

Pads of paper with a stripe of specialized PSA coated along one edge are finding wide use for conveying messages, adding instructions to paper documents, and temporary labeling. They can be attached to documents and other surfaces for extended periods of time and removed without damaging the underlying surface or leaving a residue.

Parchment is used as a backing for some tapes, such as the repulpable splicing tape used to join paper webs during the manufacture of paper. Because ordinary PSAs break up and then agglomerate to form "stickies" (small globules of adhesive), it is important that such tapes carry a special PSA which will disperse or dissolve in water under the action of the repulping beaters. "Stickies" cause problems in finished repulped paper by creating unsightly blemishes in the sheet or by causing tearouts when one wrap on a jumbo of paper separates from the one underneath.

## 4. Rigid Film Tapes

Cellophane tape is the oldest rigid film tape, and it is still manufactured, but cellophane itself has been largely replaced by clear rigid films that are less sensitive to moisture and more dimensionally stable. Cellulose acetate tape, so-called office tape, with a matt finish and a long-aging acrylic PSA, is non-yellowing and can be used for mending valuable documents and for a variety of applications requiring durability and dimensional stability. For higher tensile strength some manufacturers choose unplasticized poly (vinyl chloride), polyester, or biaxially oriented polypropylene for the backing of these tapes. However, cellulose acetate tape remains the preference of many consumers because of its ease of cutting on the serrated edge of a tape dispenser.

The original adhesive used on cellophane tape was a blend of natural rubber and rosin or colophony. Because the adhesives produced from natural rubber adhere aggressively to a wide variety of surfaces, it is still widely used, but synthetic substitutes have been found for all components. Synthetic rubbers, including polybutadiene, butadiene–styrene copolymers, polyisobutylene, and block copolymers of styrene and butadiene or styrene and isoprene find use in formulating the PSAs for many tapes. Rosin is often replaced by other tackifying resins, e.g., rosin esters, polyterpenes, coumarone–indene resins, or various resins derived from the $C_5$ and $C_9$ cuts of petroleum.

Polymers which in themselves have the balance of properties requisite for pressure sensitivity without the addition of a tackifying resin include the homopolymers and copolymers of alkyl acrylic esters having alkyl groups of two to eight carbons, or even higher, the corresponding polymers and copolymers of alkyl vinyl ethers, and some of the copolymers of the higher carboxylic acid esters of vinyl alcohol.

Rigid film tapes also require primers, and usually have release coatings. The original primer in cellophane tape was a blend of natural rubber latex and casein, but other primers of the latex–hydrophilic colloid type were developed. For noncellulosic rigid films, solvent-based rubber cements based on natural rubber, polybutadiene, polychloroprene, etc., are often used. The release coatings (LABs) are typically polymers of high molecular mass and low critical surface energy. They are often wax-like, and must be incompatible with the PSA so they will not migrate into its mass and cause deterioration of adhesive performance.

## 5. Conformable Film Tapes

Plasticized poly(vinyl chloride) film tapes were developed as electrical insulating tapes for wrapping electrical connections, replacing the old self-adhering, impregnated-cloth friction tape. The stretchability, conformability, and retractability of the plasticized poly(vinyl chloride) tapes gave them a distinct advantage in producing tight and impervious insulating seals. Backing and adhesive materials can be formulated to produce a range of tapes, some of which perform better at high or low temperature.

The adhesives on electrical tapes are formulated to minimize plasticizer migration from the backing. They must be free of ionic contaminants, resist moisture, and have excellent insulating properties. Many other uses have been found for these tapes (e.g., for floor marking, where different colors are used to identify aisles, safety areas, etc.). High abrasion resistance, stretch and conformability, and resistance to cleaning agents are requisite properties.

Conformable film tapes are used as masking tapes in electroplating operations where resistance to electroplating solutions is required. Gold plating of the connecting fingers on printed-circuit boards requires a special adhesive tape that resists the corrosive chemicals in the plating baths, conforms extremely well to the contours of the circuit elements so plating solution cannot seep into areas where it is not wanted, and is capable of being stripped off without leaving any residue. A PSA made of polydimethylsiloxane and a siloxane resin is often used. Conformable

film tapes with a polyethylene backing and an asphaltic polyisobutylene composition as the PSA are used for corrosion protection of underground pipelines. A cloth-like tape known as "duct tape" finds widespread household use in addition to the use for which it was originally intended, the sealing of air conditioning and heating ducts. On movie sets duct tape is called "gaffers' tape" and is used for holding down extension cords and other items around the set.

Conformable tapes made with a polyethylene backing demonstrate good low-temperature flexibility and low moisture transmission. When the backing is formed of polytetrafluoroethylene, those two properties are combined with resistance to high temperature. For exceptional thermal and chemical resistance, the polytetrafluoroethylene tapes have a PSA derived from polydimethylsiloxane.

## 6. Cloth Tapes

As mentioned in Chapter 2, the earliest pressure-sensitive tapes were made by calendering a pressure-sensitive adhesive onto cloth. These tapes were used as surgical bandages, but have been largely replaced by prefabricated surgical bandages in which the pressure-sensitive tape has a conformable film backing, and by "breathable," hypoallergenic tapes with backings of nonwoven fibers and microporous coatings of synthetic copolymer PSAs. Their adhesive and strength properties are such that in some instances they can even replace sutures. The older cloth or fabric tape has not been completely eliminated, and is still found under the guise of "athletic tape," used by athletes as joint supports.

Cloth tapes find use as electrical insulating tapes. Acetate rayon cloth is commonly used. Other electrical insulating tapes may be formed with a backing of epoxy-saturated glass fiber or fabric made with polyimide fibers. Cloth backings of cotton, or blends of cotton and synthetic fiber, are used for tapes which may be used for insulating but are more commonly employed for bundling and packaging where high tensile strength is needed. They are also used for decorating, coding, or repairing, and are manufactured in various colors. Rayon and silk fabrics are also used as backings for surgical tapes.

## 7. Metal Foil Tapes

Metal foil tapes find application where the unique properties of the backing are essential to performance. PSAs on metal foil tapes, selected for excellent aging, heat resistance, or specific mechanical properties, are usually silicone or acrylic. Low-adhesion backsizes reduce unwinding effort and minimize wrinkling during unwind.

Lead foil tape is an excellent sealing and masking tape, impervious to moisture and resistant to severe weathering conditions (e.g., intense sunlight or salt spray). It is a good barrier to X rays and other high-energy radiation, and, because of its chemical resistance, to electroplating baths or other corrosive solutions.

Aluminum foil tapes are impermeable to water vapor, and resist sunlight and weathering. They serve as heat shields because of their high reflectivity. They are also used to damp vibrations of metal panels. The high tensile modulus of the aluminum backing maximizes the cyclic shearing action on the adhesive between the foil and the vibrating plate. The adhesive is selected to have high hysteretic energy loss, yielding high energy dissipation for effective damping.

## 8. Double-Stick, Transfer, and Foam Tapes

Tapes with PSA coatings on both sides find extensive use for both temporary and permanent bonding. The backings may be unified papers or rigid films. Usually a release liner is incorporated to prevent blocking, but a tape has been perfected which has mutually incompatible PSAs and can be unwound without need for a release liner.

Double-stick foam tapes have been developed for attaching objects to rough surfaces, the high compliance of the foam permitting greater penetration into these surfaces and hence better adhesion. The foams may be polyurethanes, plasticized poly(vinyl chloride), polychloroprene, polyethylene, or acrylic. The tapes are available in various thicknesses, typically 0.8–3.2 mm, and widths of 0.5–20 cm. They have release liners to facilitate unwinding of the rolls. Like label stock, foam tapes are often transfer coated by laminating the foam material to release liners coated with a PSA, and like labels they are sometimes die cut into specific shapes for certain applications. Foam tapes are finding uses which were once the domain of mechanical fasteners, such as attaching side mouldings and emblems to automobile bodies, and installing wall panels or windows on high-rise buildings.

Adhesive transfer tapes consist of a release liner having a coating of unsupported pressure-sensitive adhesive, and serve as a means of dispensing the dry and tacky adhesive as a thin, uniform sheet to the surface of other objects or films. They are basically the same as the adhesive-coated liner mentioned above for use in the lamination coating of label stock, except they are generally converted into narrow rolls or discrete sheets. The tape is applied to a surface and the liner is stripped away, leaving the adhesive attached to the surface. If the surface is part of an object such as a nameplate or the face of a membrane switch, the object can be attached or laminated to another surface. The liner can be left on and the objects shipped off to be attached to something at a remote location, where the liner is stripped just before use. Another typical application is the mounting of photographs in an album.

## 9. Reflective Tapes

These tapes reflect light to provide visibility in poor light conditions. The tapes are composed of metallized glass beads on the surface of acrylic materials, or acrylic materials conformed in geometric patterns, and they are very efficient in reflecting incident light in the direction of the source. Their main use is in roadway applications as pavement marking tapes or as symbols or decals attached to vehicles, obstacles, or pedestrians to increase visibility and safety in traffic.

## 10. Trends

Most current trends are driven by three factors:

1) Cost reduction in tape manufacture
2) Environmental concerns, including solvent reduction/elimination and recyclability
3) Improvements in tape performance and quality

Manufacture of PSA tapes involves the coating of one or more of the four components. Traditionally, the backsize, primer, and adhesive have been coated from organic solvents. Because of the environmental impact of solvent emissions and the rising costs of organic solvents in tape manufacturing, as well as the costs of recovery or disposal of solvents, a transition from solvent coating to solventless methods is occurring. Three technologies are described in research journals, patent publications, and product literature for production of PSA tapes by solventless coating: radiation-curable, hot-melt-coatable, and water-based (latex) materials.

Radiation-curable materials have received the greatest attention in the past few years in terms of numbers of patent applications and journal articles. According to a recent review, over 400 references can be found in the area of PSAs alone from 1976 to 1991 [4]. Radiation-curable materials are either radiation-polymerizable materials that are coated and then irradiated in situ to form the coating, or already formed polymeric resins that are subsequently irradiated to improve certain properties of the coating. Common types of radiation are ultraviolet light, $\gamma$ radiation, or electron beams.

Hot-melt-coatable materials for PSAs are usually introduced into an extruder, softened by heat, compounded with other ingredients (e.g., tackifying resins and stabilizers), and coated onto a tape backing or release liner by a coating die attached to the end of the extruder. Coextrusion technology is being used to prepare multilayer tape backings, liners, and PSA tapes [5]–[8], which opens the way to integrated tape manufacturing.

In attempts to overcome some of the short comings of the early latex systems, special attention has been given to the development of water-based PSAs [9]. Improvement has come in the areas of coating versatility, emulsion stability, foaming, humidity sensitivity, and cohesive strength.

It had been the common wisdom in the tape industry of the 1970s that the use of solvents would be drastically curtailed within a very short period, but that has not occurred. Despite the availability of new technology, solvent-based PSAs are likely to represent about one-third of all PSAs until after the mid-1990s, with hot-melt systems comprising about half, and water-based materials making up most of the remainder. Radiation systems remain slow to catch on, and are likely to represent only ca. 1 % of the total.

## 11. Economic Aspects

Pressure-sensitive adhesive tapes represent a relatively small fraction of the value of all goods produced on a national and even a global basis. While exact figures are unavailable, worldwide sales of pressure-sensitive adhesive-coated tapes, labels, and decals are thought to represent ca. $ 14–16 \times 10^9$ at the manufacturing level. At the retail or consumer level this figure would be roughly twice as large; adhesive tapes represent just under half the total.

The estimated worldwide production of adhesive tapes is $12–13 \times 10^9$ m$^2$. A large portion of this—about half—is tape for packaging and box-sealing applications, the remainder being various office, consumer, and industrial tapes. The United States and Western Europe account for about two-thirds of total production, with Japan and the Asia–Pacific countries accounting for the other third. Japan's share is estimated to be $1.5 \times 10^9$ m$^2$. A little less than half of the total production can be attributed to a small number of international producers that manufacture and sell tape and other products on a global scale. These include Avery Dennison (with headquarters in California, United States), Beiersdorf/tesa (Hamburg, Germany), Nitto Denko and Sekisui Chemical (both in Osaka, Japan), and 3M Company in Minnesota, United States. Over half the world's production of adhesive tapes is due to several hundred smaller firms located in all corners of the earth, including Australia, India, China, Mexico, South Africa, Italy, Singapore, the Philippines, Taiwan, and New Zealand. The principal exporting countries are Germany, Italy, Japan, Taiwan, and the United States. Pressure-sensitive adhesive tapes make a small but valuable contribution to the manufacture, packaging, and sale of goods produced by virtually every industry in the world.

## 12. References

[1] D. Satas: "Adhesive Products in the United States," in D. Satas (ed.): *Handbook of Pressure Sensitive Adhesive Technology*, 2nd ed., Van Nostrand Reinhold, New York 1989, chap. 1.

[2] J. O. Hendricks, C. A. Dahlquist: "Pressure Sensitive Adhesive Tapes," in R. Houwink, G. Salomon (eds.): *Adhesion and Adhesives*, vol. 2, Elsevier, Amsterdam 1967, chap. 17.

[3] C. A. Dahlquist: "Pressure Sensitive Adhesives," in R. L. Patrick (ed.): *Treatise on Adhesion and Adhesives*, Marcel Dekker, New York 1969, chap. 5.
[4] R. W. Oehmke: *RadTech '92 North Am. UV/EB Conf. Expo. Conf. Proc.* **1** (1992) 229–238.
[5] Avery Int., US 4 925 714, 1989 (M. S. Freedman).
[6] Avery Int., US 4 946 532, 1989 (M. S. Freedman).
[7] Minnesota Mining and Manufacturing, US 4 908 278, 1986 (R. H. Bland et al.).
[8] Beiersdorf, DE 3 915 610, 1989 (M. Evers et al.).
[9] D. G. Pierson, J. J. Wilczynski, *Adhes. Age* **33** (1990) no. 9, 52–56.

# Tar and Pitch

GERD COLLIN, DECHEMA e.V., Frankfurt/Main, Federal Republic of Germany (Chaps. 1–6)

HARTMUT HÖKE, Weyl GmbH, Mannheim, Federal Republic of Germany (Chap. 7)



## 1. Origin, Classification, and Industrial Importance of Tars and Pitches

**Origin and Classification.** *Tars* are liquid or semisolid products obtained by thermal decomposition of natural organic materials. *Pitches* are residues of tar distillation or meltable residues directly obtained by distillation of organic substances. Tars and pitches take their names from the source material. Tars are produced in industrial quantities by low-temperature carbonization, coking, or gasification of fossil raw materials or biomass, e.g., coal, lignite, peat, wood, and oil shale. In addition, tars result as higher boiling liquid pyrolysis products from thermal decomposition of organic secondary raw materials and waste products, e.g., pyrolysis residual oils as byproducts of steam cracking of petroleum fractions to alkenes and from pyrolysis of refuse and plastic waste.

Properties and composition of tars and pitches depend not only on the raw material but also on the temperature conditions during thermal treatment. Thus low-temperature tars are produced during low-temperature carbonization or partial gasification below 700 °C. High-temperature tars are generated by coking between 900 and 1300 °C. The differences in the properties and composition of low- and high-temperature tars are especially marked in case of coal tars. Low temperature coal tars contain phenols, hydrogenated aromatics, aliphatic hydrocarbons, and alkenes as their main components, together with small amounts of aromatic hydrocarbons. These compounds are generated from two sources: (1) Coal bitumen, the extractable component of coal, which consists of oxygen-, sulfur-, and nitrogen-containing hydrocarbon compounds, and (2) the bitumen which arises additionally from the insoluble components by cracking of aliphatic C–C, C–O, C–N, and C–S bonds, as well as through the removal of side chains (→ Coal Pyrolysis, **A 7**, pp. 245–246). High-temperature coking of coal generates thermally stable aromatics, which are characteristic of high-temperature coal tars. These heat-stable aromatics arise via the following reaction schemes: Paraffins are cracked pyrolytically and dehydrogenated. The resulting alkenes are converted first into hydrogenated aromatics and finally into aromatics. Alkylphenols are dealkylated and reduced to aromatic hydrocarbons.

*Lignite tars* differ from low-temperature coal tars mainly in an increased content of solid paraffins and a lower amount of cyclic hydrocarbons. The main source of paraffins is the extractable montan wax occurring in lignite. The tars are formed by thermal decomposition of initially present and newly formed "bitumen," similar to the case of coal. Tar obtained by high-temperature coking of lignite is similar to low-temperature lignite tar, as lignite bitumen, in contrast to coal bitumen, is already entirely decomposed and volatilized at lower temperatures.

*Peat tar* properties are determined by the structure of peat, which is a conglomerate of extractable bitumen, humic acid and its salts, and partially decomposed vegetation. In pyrolytic distillation "bitumen" is the source of the bulk of the tar, mainly paraffins. Ketones, methanol, monocarboxylic acids, and phenols arise from the plant components cellulose and lignin ($\rightarrow$ Peat, **A 19**, pp. 25–27).

*Wood tars* are formed by thermal decomposition of the wood components cellulose, hemicelluloses, and lignin. This gives rise to acetic acid, methanol, ketones and phenols but, in contrast to tars from fossil raw materials, almost no hydrocarbons ($\rightarrow$ Charcoal).

**History.** The discovery of *coal tar* goes back to J. R. Glauber, who in 1658 described the formation of "black oleum" during coal retorting. A patent for the production of pitch and tar by coking of coal with condensation of volatiles was granted in England to J. J. Becher and H. Serle in 1681. In the early 1800s coal tar was initially regarded as an annoying byproduct of town-gas and metallurgical-coke production. In 1822 the first industrial tar distilleries were built in Britain. The light oils obtained were used as solvents, the heavy oils as impregnants for building timber and railway sleepers, and the distillation residue, pitch, for the production of carbon black. In 1845 A. W. Hofmann discovered benzene in the light oil used to produce nitrobenzene as a perfume for soaps, named essence of mirbane. H. Garden (1819) had previously discovered naphthalene and J. B. A. Dumas and M. A. Laurent (1832) anthracene in coal tar. With the discovery of synthetic dyes during the second half of the 1800s, demand for these three aromatic hydrocarbons led to coal tar becoming the main source of raw materials for the flourishing organic chemical industry in central Europe. The first German tar distilleries were built in 1842 in Offenbach by E. Sell, and in 1860 in Erkner near Berlin by J. Rütgers, followed by numerous others.

*Lignite tar* was obtained by J. G. Krünitz in 1788 as "rock oil" when retorting central German "earth coal." In 1830 Karl Freiherr von Reichenbach isolated the main component, paraffin, by distillation. The lamp oil, obtained as a byproduct, replaced rapeseed oil, which had been used up to then for lighting. After 1858 low-temperature lignite-carbonization plants were constructed in central Germany for the production of tar-based products and town gas. During the 1930s central German lignite tar was mainly processed by hydrogenation to produce motor fuels.

Peat and wood were carbonized in early times. The Egyptians used *wood tar* for embalming corpses. Solid paraffin for candle production and liquid paraffin oil for lamp oil were obtained from *peat tar* or lignite tar during the 1800s.

**Industrial Importance.** Approximately $15 \times 10^6$ t/a of *coal tar* is produced worldwide, mainly in the form of high-temperature tar as a byproduct of metallurgical coke production by horizontal chamber coking ($\rightarrow$ Coal Pyrolysis, **A 7**, pp. 273–275). Smaller amounts are produced as byproducts of low-temperature carbonization of coal to produce solid smokeless fuel ($\rightarrow$ Coal Pyrolysis, **A 7**, pp. 271–273) and in pressure gasification of coal by the Lurgi process ($\rightarrow$ Coal, Coal Gasification, **A 7**, p. 185). Approximately 100 coal-chemical tar refineries process coal tar to give chemical feedstocks and other aromatic products. Industrial countries with major steel production account for most of the capacity: ca. 50 % in Europe and the CIS, 20 % in North America, and 20 % in South-East Asia, especially China and Japan. In the United Kingdom and India, low-temperature tars are obtained as byproducts of low-temperature carbonization of coal to produce smokeless fuels. Lurgi pressure gasification is a preliminary stage of Fischer–Tropsch synthesis in South Africa.

In contrast to coal tars, *lignite tars* are produced only in a few countries, e.g., as a byproduct of pressure gasification of lignite in the United States (North Dakota), the Czech Republic, Germany (Lausitz), Russia, and China. Lignite is carbonized to produce coke on a minor scale, e.g., in Germany (Rhineland and Lausitz) and India. Normally only the phenolic effluents are processed, whereas the lignite tars are gasified or burnt within the process. The total amount of lignite tar arising is ca. $10^6$ t/a.

*Peat tars* and *wood tars* arise as byproducts in the production of activated carbon and reductive carbon by carbonization of peat and wood ($\rightarrow$ Peat, **A 19**, pp. 25–27, 29–34; $\rightarrow$ Carbon, **A 5**, pp. 126–132). The main producing countries are Russia (peat) and Brazil (wood). The total production of these tars is at most ca. $3 \times 10^6$ t/a.

*Oil-shale tars* are produced by low-temperature carbonization and partial gasification of oil shale, mainly in Estonia (Kiviter process). Approximately $10^6$ t/a are produced in this way ($\rightarrow$ Oil Shale, **A 18**, pp. 108–122). Other processes involving pyrolysis of oil sand and oil shale are still at the pilot-plant stage.

*Pyrolysis residual oils* are tarlike byproducts of the steam cracking of naphtha and gas oil to produce alkenes. Approximately $10 \times 10^6$ t were produced in 1993. Pyrolysis oils with similar compositions are obtained by thermal decomposition of waste plastics. Industrial large-scale tests are being carried out to develop methods of waste-plastics pyrolysis ($\rightarrow$ Plastics, Recycling, **A 21**, pp. 64–68).

# 2. Properties

In general, *tars* are dark or black viscous liquids containing small amounts of emulsified water and dispersed high-carbon solid particles (free carbon, tar sediments) entrained during formation. Organic components may crystallize from high-paraffinic tars at lower temperatures.

*Pitches* are generally dark or black thermoplastic materials with a broad molecular-mass spectrum. By using selective solvents it is possible to separate pitches into fractions with different properties. At temperatures below the softening range most pitches transform into glass-like solids. At higher temperatures polymerization and polycondensation reactions take place, and pitches change partly from liquid crystal intermediate stages into a nonthermoplastic state and finally into coke [5].

The chemical composition of tars and pitches varies widely and depends essentially on the type of raw materials and the temperature conditions during production. In all cases, multicomponent mixtures are produced that contain aromatic and/or aliphatic hydrocarbons as the main components, together with oxygen-, sulfur-, and nitrogen-containing heterocyclic compounds; phenols; amines; and possibly carboxylic acids, ketones, and aliphatic alcohols as minor constituents. It has been estimated that coal tar can contain more than 10 000 different constituents, most of which have boiling points above 500 °C. Multicomponent eutectics can form, especially in the case of condensed aromatics. The liquidviscous nature of pitches and many tar oils arises from strong mutual melting-point depression of the individual components [6].

The main properties of different tars are listed in Table 1. Given approximately equal boiling range and pitch content, the density and carbon content indicate the degree of aromatization of tars. Coke-oven coal tars with average densities of ca. 1.2 g/cm$^3$ and carbon contents $> 90\%$ contain the highest fraction of aromatic hydrocarbons, followed by coal low-temperature gasification tars with densities of ca. 1.1 g/cm$^3$ and carbon contents of ca. 85%. Smaller amounts of aromatics are found in tars produced by low-temperature carbonization of coal and lignite having densities of ca. 1 g/cm$^3$ and carbon contents of max. 85%. Peat and wood tars with carbon contents of only ca. 80% or even $< 70\%$ and varying densities, contain almost no aromatic compounds. Oil-shale tars are hardly or only partially aromatized. However, a high content of aromatic components can be found in pyrolysis residual oils from ethylene/propene plants and from the thermal decomposition of waste plastics if the pyrolysis temperature is at least ca. 700 °C. These pyrolysis residual oils have an aromatic hydrocarbon content of ca. 90% and higher, similar to coke-oven coal tar, whereas pyrolysis oils contain almost no phenols and only small amounts of heterocyclic compounds due to the raw-material basis. The carbon content falls and the hydrogen content rises with decreasing tar aromatization. A sensitive criterion for the "aromaticity" of tars is the naphthalene content, which is ca. 10% in coke-oven coal tars and pyrolysis residual oils, ca. 3% in coal-gasification tars on average, less than 2% in low-temperature coal-carbonization tars, and max. 1% for lignite, peat, and wood tars. With increasing geological age of the raw material (i.e., wood $<$ peat $<$ lignite $<$ coal), the content of phenols and solid paraffins decrease, while the

**Table 1.** Properties and composition of various tars

| | Coal tars | | | Lignite tars | Peat tars | Wood tars | Oil-shale tars | Pyrolysis residual oils from steam cracking of naphtha and gas oil | Pyrolysis residual oils from pyrolysis of plastic waste |
|---|---|---|---|---|---|---|---|---|---|
| | Coke-oven tars | Low-temperature tars | Gasification tars* | | | | | | |
| Density at 20°C, g/cm³ | 1.14–1.25 | 0.96–1.05 | 1.05–1.14 | 0.95–1.05 | 0.94–0.98 | 1.08–1.20 | 0.90–0.97 | 0.96–1.10 | 0.93–1.05 |
| Carbon, % | 90–93 | 83–85 | 84–86 | 81–85 | 78–82 | 60–65 | 78–86 | 89–93,1 | 88–91 |
| Hydrogen, % | 5–6 | 8–9.5 | 6–8 | 9–11 | 8–10 | 6–8 | 9–12 | 6–9 | 8–10 |
| Naphthalene, % | 5–15 | 0–2 | 2–4 | 0–1 | 0–1 | 0–0.5 | | 10–15 | 4–12 |
| Phenols, % | 0.5–5 | 10–45 | 15–25 | 5–30 | 5–25 | 20–40 | 1–30 | 0–trace | trace |
| Bases, % | 0.2–2 | 0.5–2 | 0.5–4 | 0.1–1.5 | 0.5–3 | trace–0.5 | 0.2–9 | 0–trace | 0–trace |
| Solid paraffins, % | 0–trace | 3–15 | 0–5 | 8–20 | 10–30 | trace | 0.5–12 | 0–5 | 0–trace |
| Coking residue **, % | 10–40 | 5–15 | 10–20 | 4–9 | 4–10 | 10–18 | 1–7 | 5–20 | 1–10 |
| Toluene insolubles, % | 2–20 | 0.5–10 | 3–10 | 0.2–1.5 | 2–8 | trace–12 | | trace | trace |
| Ash (800°C), % | <0.5 | <0.2 | <1.5 | <0.2 | <0.2 | <0.2 | | <0.1 | <0.2 |
| Water, % | <5 | <5 | <5 | 5 | 1–10 | 3–10 | | 0–1 | 0.2 |
| Distillation analysis (DIN 1995) | | | | | | | | | |
| Light oil up to 180°C, % | 0.2–2 | 0–4 | 0–10 | 0–2 | 0–3 | 10–20 | 10–15 | 0–10 | 40–70 |
| Middle oil 180–230°C, % | 3–10 | 5–25 | 5–15 | 0–4 | 3–25 | 10–14 | 5–20 | 20–30 | 8–15 |
| Heavy oil 230–270°C, % | 7–15 | 10–20 | 5–20 | 2–18 | 5–20 | 7–12 | 15–35 | 20–30 | 5–15 |
| 270–300°C, % | 3–7 | 8–12 | 5–10 | 5–20 | 10–20 | 5–10 | 0–15 | 10–20 | 3–8 |
| 300°C up to pitch, % | 12–30 | 15–40 | 10–30 | 20–50 | 20–40 | 25–40 | 0–10 | 2–5 | 2–10 |
| Pitch, % | 45–65 | 25–40 | 30–50 | 20–60 | 20–35 | 20–30 | 30–50 | 25–30 | 10–20 |

* From Lurgi fixed-bed pressure gasification of coal with oxygen and steam. ** Conradson method.

**Table 2.** Properties of coke-oven coal tars

|  | Average value | Range of analytical data |
|---|---|---|
| Density, g/cm$^3$ | 1.175 | 1.14–1.20 |
| Water, % | 2.5 | 0.2–5.0 |
| Toluene insolubles, % | 5.50 | 2–12 |
| Quinoline insolubles, % | 2.0 | 0.5–5 |
| Coking value (platinum crucible), % | 14.6 | 10–20 |
| Carbon (waf) *, % | 91.39 | 90–93 |
| Hydrogen (waf), % | 5.25 | 5–6 |
| Nitrogen (waf), % | 0.86 | 0.6–1.2 |
| Oxygen (waf), % | 1.75 | 1.5–2 |
| Sulfur, % | 0.75 | 0.6–1 |
| Chlorine, % | 0.030 | trace–0.1 |
| Ash (800 °C), % | 0.15 | 0.05–0.25 |
| Zinc, % | 0.038 | trace–0.2 |
| Naphthalene, % | 10.0 | 5–13 |
| Distillation analysis (DIN 1995), % |  |  |
| up to 180 °C |  |  |
|   water | 2.5 | 0.2–5 |
|   light oil | 0.9 | 0.2–2 |
| 180–230 °C | 7.5 | 3–10 |
| 230–270 °C | 9.8 | 7–15 |
| 270–300 °C | 4.3 | 3–7 |
| 300 °C up to pitch | 20.1 | 12–30 |
| pitch (softening point 67 °C KS**) | 54.5 | 45–65 |
| distillation loss | 0.5 |  |
| Crude phenols, % | 1.5 | 0.4–5 |
| Phenol, % | 0.5 | 0.08–0.7 |
| Cresols and xylenols, % | 1.0 | 0.3–4.5 |
| Low-boiling pyridine bases (in light oil), % | 0.081 | 0.01–0.2 |
| High boiling pyridine and quinoline bases (in middle and heavy oil), % | 0.065 | 0.2–0.9 |
| Unsaturated compounds (in light oil), % | 0.590 | 0.02–1.0 |

* waf: water- and ash-free. ** KS: Krämer–Sarnow.

"aromaticity" increases. Thus coke-oven coal tars contain on average ca. 2 % phenols and only traces of solid paraffinic hydrocarbons, whereas low-temperature coal-carbonization tars, lignite tars, and peat tars contain up to ca. 30 % phenols and solid paraffins. High coking values and components that are insoluble in toluene normally indicate a high degree of aromatization.

Tar properties are investigated with numerous analytical methods. Table 2 contains the analytical values of a typical coke-oven coal tar. Figure 1 shows a gas chromatogram of a coke-oven tar, and Table 3 lists the most important constituents of coke-oven coal tar. Up to now about 600 compounds have been identified in coal tars; together, they account for ca. 55 % of the tar. The remaining, unidentified compounds are mainly pitch components which decompose before they vaporize.

In general, *pitches* are multicomponent mixtures of higher molecular compounds. *Coal-tar pitch* obtained from coke-oven tar is a mixture of condensed aromatic hydrocarbons and heterocyclic compounds. Figure 2 shows a high-pressure liquid chromatogram (HPLC) of coal-tar pitch. HPLC can analyze ca. 70 % of coal-tar pitch [7]. Gel permeation chromatographic investigations indicate in case of higher molecular pitch fractions that they have molecular weights up to approximately 2500 [8]. The highest molecular components, which are partly insoluble, even in pitch, have higher molecular masses. The higher molecular fractions consist of condensed aromatic hydrocarbons and heterocyclics with insignificant fractions containing methyl- or hydroxyl-substituted compounds. The $^{13}$C NMR spectrum shows that 97 % of the carbon atoms present are aromatic [9], [10]. Higher molecular pitch fractions contain monomeric aromatics and heterocyclic aromatics, as well as oligomeric systems, in some of which aromatics containing nitrogen and other heteroelements are linked by C–C single bonds or bridges such as $CH_2$, O, and S [11], [12]. Very large *peri*-condensed systems arise on thermal treatment of normal pitch, whereby the average molecular mass increases and the solubility of the pitch in solvents decreases. Furthermore, anisotropic liquid crystal structures known as "mesophases" are formed [13], [14].

Properties of coke-oven tar "normal" pitch are listed below:

| | |
|---|---|
| Density at 20 °C | 1.27–1.28 g/cm$^3$ |
| Softening point (Krämer-Sarnow) | 68 ± 3 °C |
| Softening point (Mettler) | 86 ± 3 °C |
| Brittle point (Fraass) | 30–40 °C |
| Elemental analysis: | |
|   Carbon | 92–94 % |
|   Hydrogen | 4–5 % |
|   Nitrogen | 1.0–1.3 % |
|   Sulfur | 0.5–0.8 % |
|   Oxygen | 1–2 % |
| *n*-Hexane solubles | 40–60 % |
| Toluene insolubles | 15–25 % |
| Quinoline insolubles | 3–7 % |
| Coking value (ALCAN) (DIN 51 905) | 43–48 % |
| Ash (800 °C) | < 0.3 % |
|   Iron | 200–300 ppm |
|   Silicon | 100–300 ppm |
|   Vanadium | < 50 ppm |
| Specific heat capacity (140–210 °C) | $C_p/\mathrm{J\ g^{-1}\ K^{-1}} = 1.15 + 0.0037\ T/°C$ |

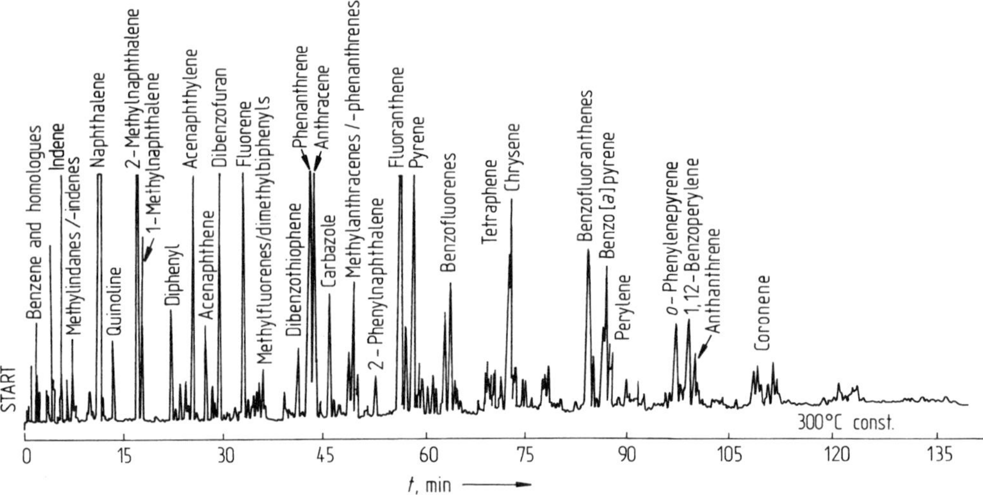



**Figure 1.** Gas chromatogram of a coke-oven coal tar 22 mm glass-capill. CP-sil 5 (inner diameter 0.5 mm), split, 50 °C 2 min, 2 °C/min up to 300 °C const., helium.

For reasons of clarity only selected peaks are shown with compound names. The concentrations of selected components (in wt %) are:

benzene/benzene homologues (1), indene (1), methylindanes/methylindenes (0.8), naphthalene (10), quinoline (0.3), 2-methylnaphthalene (1.5), 1-methylnaphthalene (0.5), diphenyl (0.2), acenaphthylene (2), dibenzothiophene (0.3), phenanthrene (5), anthracene (1.3), carbazole (0.7), methylanthracenes/methylphenanthrenes (1.8), 2-phenylnaphthalene (0.3), fluoranthene (3.3), pyrene (2.1), benzofluorenes (2.3).

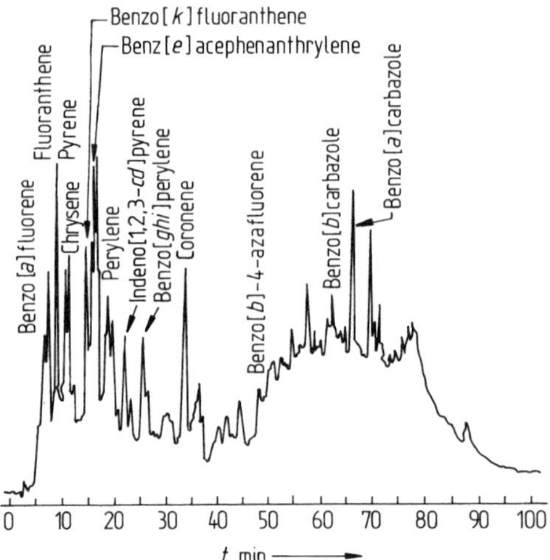



**Figure 2.** High-pressure liquid chromatogram of coal-tar pitch [7]
Nucleosil 5-NO$_2$, 5 µm, 2 columns of 200 mm length and 4 mm inner diameter and pre-column with 20 mm length and 4 mm inner diameter (switched one after the other, with gradient elution (chloroform–$n$-hexane) and UV detection at 300 nm)

| | |
|---|---|
| Coefficient of thermal expansion | 0.00045–0.00056 K$^{-1}$ |
| Thermal conductivity (25–105 °C) | 0.0014 J cm$^{-1}$ sec$^{-1}$ K$^{-1}$ |

| | |
|---|---|
| Surface tension (120–220 °C) | $\gamma/\mathrm{mN\ cm}^{-1} = 18.4 \cdot d_t^4$ ($d_t$ = density at measuring temperature) |
| Dielectric constant (25 °C) | 3.3–3.4 |
| Magnetic susceptibility | $0.69 \times 10^6$ |

## 3. Processing of Coke-Oven Coal Tar

### 3.1. Survey

The refining and processing of coke-oven coal tar preferentially isolates aromatic chemicals by a combination of six basic processes (Figure 3):

1) *Distillation* is used for primary separation of crude tar into six fractions: light oil, carbolic oil, naphthalene oil, wash oil, anthracene oil, and pitch. It is also used for further separation of primary and secondary tar fractions.

2) *Crystallization* is used to produce crystalline condensed aromatic hydrocarbons and nonpolar heterocyclics from distillate fractions (e.g., naphthalene, methylnaphthalenes, acenaphthene, dibenzofuran, fluorene, phenanthrene, anthracene, carbazole, fluoranthene,

**Table 3.** Important constituents of a typical coke-oven coal tar

| Compound | Formula | bp at 1.013 mbar, °C | mp, °C | Average content, % |
|---|---|---|---|---|
| **Hydrocarbons** | | | | |
| Naphthalene | 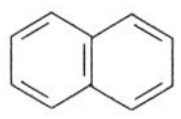 | 217.95 | 80.29 | 10.0 |
| Phenanthrene | 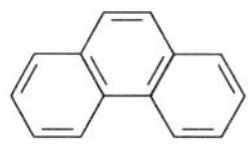 | 338.4 | 100.5 | 4.5 |
| Fluoranthene | 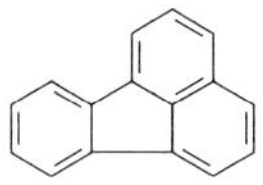 | 383.5 | 111 | 3.0 |
| Acenaphthylene | 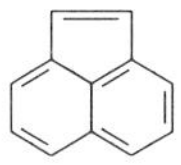 | 270 | 93 | 2.5 |
| Pyrene | 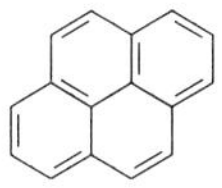 | 393.5 | 150 | 2.0 |
| Fluorene | 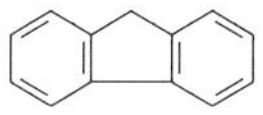 | 298 | 115 | 1.8 |
| 2-Methylnaphthalene | 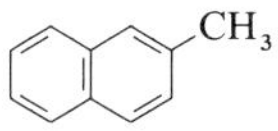 | 243.1 | 34.6 | 1.5 |
| Anthracene | 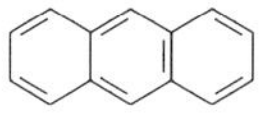 | 339.9 | 218 | 1.3 |
| Chrysene | 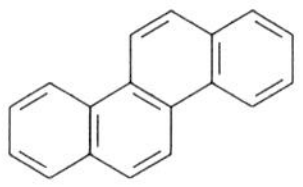 | 441 | 256 | 1.0 |
| Indene | 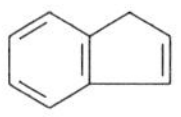 | 182.8 | −1.8 | 1.0 |
| 1-Methylnaphthalene | 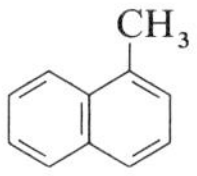 | 244.4 | −30.5 | 0.7 |
| Benzene |  | 80.1 | 5.53 | 0.4 |
| Diphenyl | 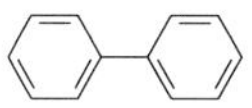 | 255.9 | 71 | 0.4 |
| Acenaphthene | 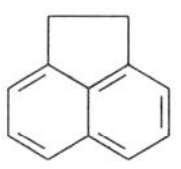 | 277.2 | 95.3 | 0.2 |
| **Heterocyclics** | | | | |
| Dibenzofuran | 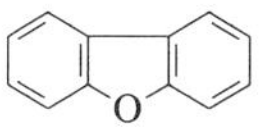 | 287 | 83 | 1.3 |
| Carbazol | 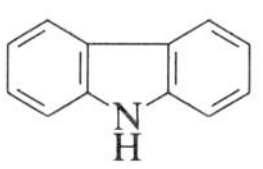 | 354.75 | 245 | 0.9 |

**Table 3.** Important constituents of a typical coke-oven coal tar (continued)

| Compound | Formula | $bp$ at 1.013 mbar, °C | $mp$, °C | Average content, % |
|---|---|---|---|---|
| Dibenzothiophene | 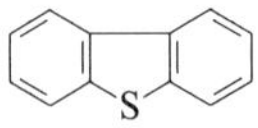 | 331.4 | 97 | 0.4 |
| Benzothiophene | 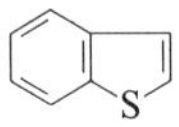 | 219.9 | 31.3 | 0.3 |
| Quinoline | 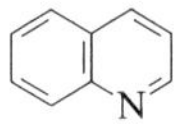 | 273.1 | −15 | 0.3 |
| 7,8-Benzoquinoline | 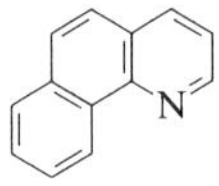 | 340.2 | 52 | 0.2 |
| 2,3-Benzodiphenylenoxide | 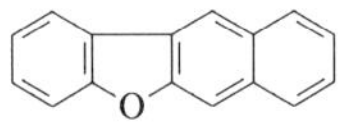 | 394.5 | 208 | 0.2 |
| Indole | 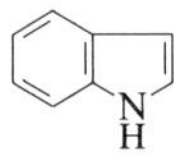 | 254.7 | 52.5 | 0.2 |
| Acridine | 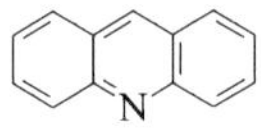 | 243.9 | 111 | 0.1 |
| Isoquinoline | 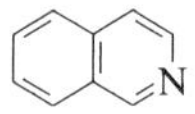 | 243.25 | 26.5 | 0.1 |
| Quinaldine | 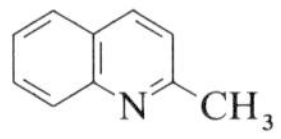 | 245.6 | −2 | 0.1 |
| Phenanthridine | 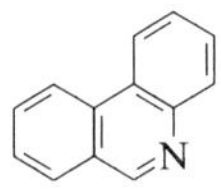 | 349.5 | 107 | 0.1 |
| Pyridine | 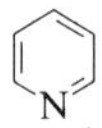 | 115.26 | −41.8 | 0.03 |
| 2-Methylpyridine | 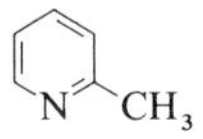 | 129.41 | −66.7 | 0.02 |
| **Phenols** | | | | |
| Phenol | 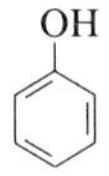 | 181.87 | 40.89 | 0.5 |
| $m$-Cresol | 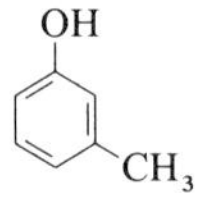 | 292.23 | 12.22 | 0.4 |
| $o$-Cresol | 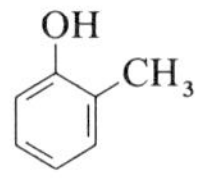 | 191.00 | 30.99 | 0.2 |

**Table 3.** Important constituents of a typical coke-oven coal tar (continued)

| Compound | Formula | $bp$ at 1.013 mbar, °C | $mp$, °C | Average content, % |
|---|---|---|---|---|
| $p$-Cresol | | 201.94 | 34.69 | 0.2 |
| 3,5-Dimethylphenol | | 221.96 | 63.27 | 0.1 |
| 2,4-Dimethylphenol | | 210.93 | 24.54 | 0.1 |

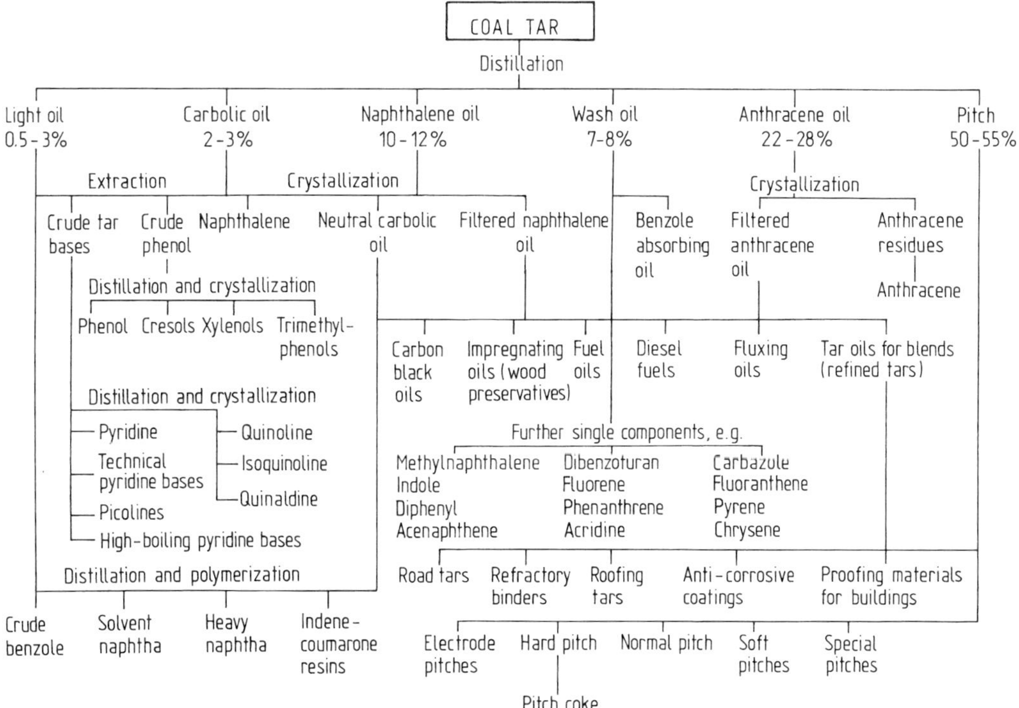



**Figure 3.** Processing of coal tar [1]

and pyrene). The aromatic oils obtained as filtrates are suitable as components for carbon-black oils, impregnating oils, fuel oils, diesel fuels, fluxing oils, and pitch–oil mixtures such as road tars, refractory pitches, anti-corrosion and protective agents for buildings.

3) *Extraction* with alkalis and acids allows the recovery of phenols and polar nitrogeneous tar bases from light oil, carbolic oil, and filtered naphthalene oil.

4) *Catalytic polymerization* with Friedel–Crafts catalysts yields thermoplastic indene–coumarone resins from light oil fractions from

which the phenols and bases have been extracted. The nonpolymerizable component is crude benzole, which together with coke-oven benzol is processed to benzene and benzene homologues; the higher boiling fractions are refined to solvent naphtha and heavy naphtha.

5) *Thermal polymerization* of coal-tar normal pitch yields higher value pitch products such as hard pitch and various special pitches.

6) *Coking* of pitch leads to pitch coke as technically pure carbon which is suitable for further processing into graphite, carbon electrodes, and carburizing agents.

## 3.2. Primary Distillation

Crude coal tar is generally pretreated prior to primary distillation. The emulsified aqueous solution (up to 5%) of ammonia, hydrogen sulfide, and ammonium chloride is removed as far as possible. Suspended fine solids (tar sediments) may also be removed.

The simplest procedure for a partial separation of water is to store crude tar at elevated temperatures. Storage for several days at 60 °C decreases the water content by 1–2%. Due to its lower density the water collects at the top of the tank, thus enabling its periodical discharge. To reduce the tar viscosity and increase the settling rate of water, crude tar can be stored in pressure tanks at 100–200 °C. This pressure dehydration can decrease the water content to ca. 1–1.5%. The settling rate can be further increased by centrifuging at 6000–7000 g. Disk centrifuges with discharge nozzles reduce the water content of crude tar down to ca. 0.5–1%.

The ammonium chloride contained in tar water dissociates into ammonia and hydrogen chloride above 225 °C, causing corrosion. The ammonium chloride content of crude tar can be greatly reduced by extraction with water prior to mechanical tar–water separation. Another method is to mix in an amount of sodium carbonate or sodium hydroxide solution equivalent to the ammonium chloride content. However, during tar distillation nonvolatile sodium chloride is formed and remains in the pitch, increasing its ash content.

Further impurities of crude tar are dispersed fine solids (tar sediments), which mainly consist of entrained fine coke particles. The partial removal of these sediments takes place together with mechanical water separation by sedimentation during storage. It is also possible to obtain a sediment concentrate from the crude tar by two- or three-phase centrifugal separators either with water as third or as second solid phase. Thus, erosion in pumps, pipes, and valves is avoided and the content of inorganic trace contaminants in the pitch primary distillation residue is decreased.

Nowadays, partly dehydrated and pretreated crude tar is distilled continuously in pipe stills and rectifying columns. Numerous distillation processes have been developed. They can be divided into flash processes, bottom-circulation processes, and processes in which these two principles are combined.

In *flash distillation* the tar is heated continuously under pressure, usually in a two-stage pipe still. It is then flashed into a system of rectification columns where the oil components are evaporated and fractionated. The first heating step dehydrates and evaporates the light oil, and in the second step the remaining oil components are evaporated, leaving pitch as the residue.

In *bottom-circulation distillation* the rectifying column bottoms are pumped continuously through pipe stills where they are provided with the thermal energy required for distillation.

These two distillation processes differ in the thermal stress on the tar fractions. In the flash process the oil fractions are subject to temperatures well above their boiling range, whereas the distillation residue pitch is not exposed to high temperatures. In the bottom-circulation process, however, the pitch is subject to high temperatures for long periods, but there is only a low thermal stress on the vaporized oil components. As several coal-tar constituents are unstable under distillation conditions, the two distillation processes differ significantly in the yield and quality of certain tar products.

*Pipe stills* are mostly rectangular or cylindrical, with an inner refractory lining. To maximize the heat utilization of firing gases, pipe stills are divided into radiant and convection zones and are often equipped with air preheaters. The burners of the radiant zone can be operated with gases or liquid fuels of various qualities. The tubes are made of steel; austenitic 18/8/2 chromium nickel molybdenum steels have proved to have especially good corrosion resistance.

As *rectifying columns*, bubble-cap columns, valve tray columns, sieve-plate columns, packed columns, and combinations thereof are used.

Stoneware Raschig rings are used as packing materials. The lower regions of the dehydration columns and flash columns mostly contain simple cascade trays. In zones with high corrosion risk the columns consist of high-grade stainless steel or cast iron, otherwise of mild steel.

To separate high-boiling tar fractions the columns normally operate under vacuum. In some processes steam is injected to lower the evaporation temperature. Heat of condensation and the thermal energy contained in the flowing bottom products are transferred via heat exchangers to the crude tar and its intermediate products to lower the total energy demand.

The *GfT/Koppers* continuous tar distillation process is shown in Figure 4 as an example of flash distillation. In the first step the crude tar is heated to ca. 160 °C in the pipe still and then flashed through an expansion valve into the middle of the atmospheric dehydration column. The pressure before the expansion valve is up to 10 bar. Light oil and water evaporate in the dehydration column, which is equipped with bubble and overflow trays, and are separated into two phases after condensation. Part of the light oil is refluxed to the top of the dehydration column. The dehydrated (topped) tar as column bottom is preheated in a heat exchanger by the hot pitch stream, then it is heated under pressure to 360 °C in the second stage of the pipe still. At this temperature it is flashed in the pitch column at a pressure of 100–130 mbar. The pressure before the expansion nozzle is up to 5 bar. To take

off the high-boiling anthracene oil II above the flash chamber, there is a column, filled with Raschig rings, which is dephlegmated at the top with the lower boiling anthracene oil I. The column tops of the vacuum system are refluxed with a split stream of the oil fraction taken from the bottom of the next column. The bottom product of the wash oil column is directed via a side column to evaporate the remaining naphthalene. This side column is additionally heated by bottom circulation, yielding an improved rectification efficiency. Concentrated ammonium chloride is extracted with water from the bottom product of the wash oil main column before the oil is refluxed to the top of the anthracene oil column. This continuous removal of ammonium chloride prevents clogging and corrosion of the column by the salt, which condenses to form a solid in the temperature range 180–200 °C. The tar oil fractions anthracene oil I, wash oil, and naphthalene oil are separated as column bottoms. Carbolic oil is condensed in the final condenser as the top product of the naphthalene column. The uncondensed gases and vapors are extracted with oil and water before entering the vacuum equipment. Light oil and naphthalene vapors are extracted in the oil scrubber with cooled wash oil. Hydrogen sulfide is removed with water in the water scrubber. The heat consumption of the total plant is ca. 900 MJ/t of wet crude tar.

Figure 5 shows a diagram of the *Rütgers process* for continuous tar distillation using the bot-

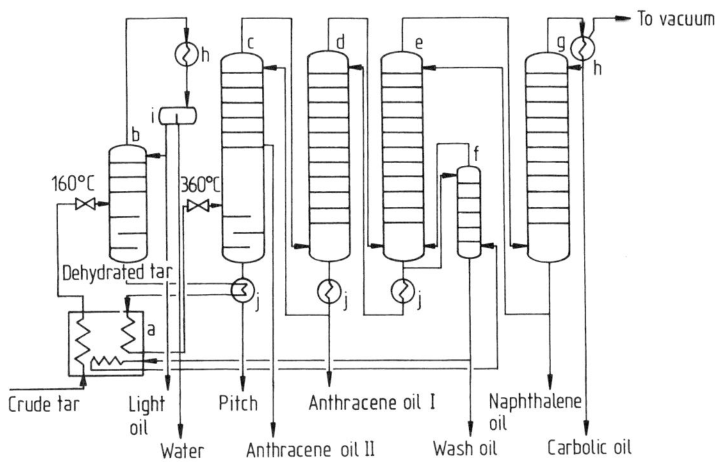


**Figure 4.** Continuous tar distillation with multiflash system (GfT/Koppers process)
a) Pipe still; b) Dehydration column; c) Pitch/anthracene oil II column; d) Anthracene oil I column; e) Wash oil column; f) Wash oil side-stream column; g) Naphthalene oil column; h) Condenser; i) Decanter; j) Heat exchanger

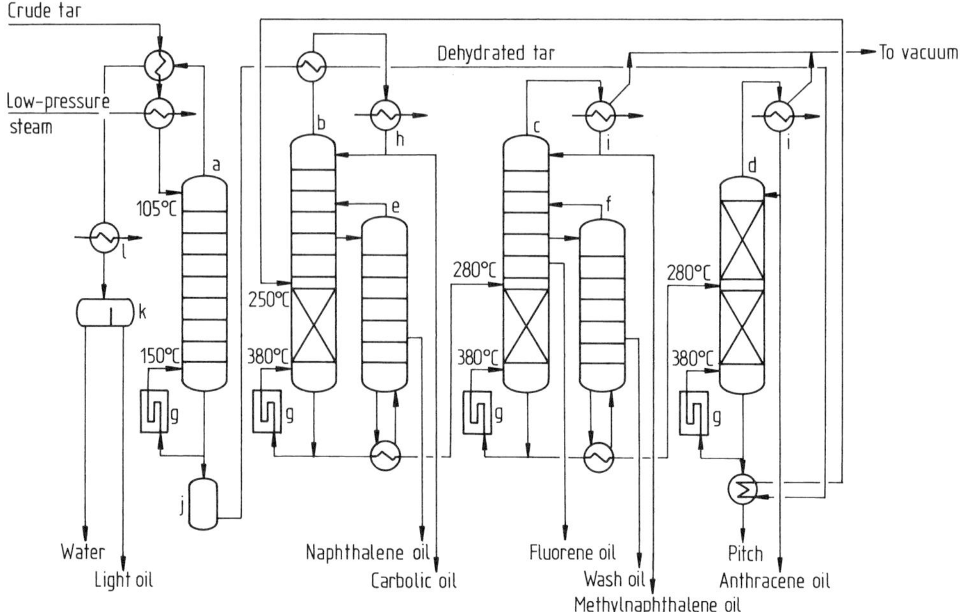



**Figure 5.** Continuous tar distillation (Rütgers process)
a) Dehydration column; b) Carbolic oil column; c) Methylnaphthalene oil column; d) Anthracene oil column;
e) Naphthalene oil side-stream column; f) Wash oil side-stream column; g) Pipe still; h) Condenser; i) Condenser (steam generator); j) Vessel for dehydrated tar (topped tar); k) Decanter; l) Cooler

tom-circulation principle. Each of the main rectifying columns has its own pipe still for bottom circulation to achieve independent control of the column reflux. Condensing water, light oil vapors of the dehydration column, and waste heat vapor preheat the crude tar to 105 °C. Afterwards the crude tar is pumped into a dehydration column which contains cascade trays. The bottoms of the dehydration column are circulated through a pipe still and kept at 150 °C. After intermediate storage the dehydrated topped tar is passed through to heat exchangers where the heat of condensation of carbolic oil and the heat of pitch is used. The topped tar, at ca. 250 °C, enters the middle section of the carbolic oil column, which is operated at atmospheric pressure and is equipped with valve or Perform trays in the lower part and, bubble trays for example, in the upper part. The pipe still, with a discharge temperature of 380 °C, allows adjustment of the reflux ratio up to 16:1. A distillate fraction in the carbolic oil column is cycled to the top of a side-stream column where the naphthalene oil fraction is taken as a vapor from its base. The main-stream bottom product is led via heat exchangers to the methylnaphthalene column, which is run with a top pressure of 250 mbar and which also contains a packing or Perform trays as well as bubble trays. The reflux ratio of this column is up to 17:1. A methylnaphthalene fraction is obtained as top product. The top vapors are condensed in a low-pressure steam generator. A wash oil fraction is taken from a side column and a fluorene fraction is taken as a further side stream of the main column. The bottom product of the methylnaphthalene column is led to the anthracene-oil packed column with a top pressure of 90 mbar. Its top product is anthracene oil, and the bottom product is normal pitch. This column only requires a reflux ratio of 1.5:1. The condensation heat of the anthracene oil is used for steam generation.

Besides the product flows described, other column interconnections are possible. Due to the high reflux ratios in efficient rectifying columns the tar oils are separated into many single fractions. Thus a high-yield extraction of technically pure components can be achieved. For example, the naphthalene oil fraction contains up to 90 % or more of the naphthalene occurring in crude tar at a concentration up to 90 % (crystallizing point up to 75 °C). Despite the increased energy needed for high reflux ratios, the heat requirement of the plant is only ca. 880 MJ/t crude tar due to the heat recovery system.

If the production of coal-tar pitch rather than tar aromatics and oils is the main aim then the primary distillation of crude tar should be combined with a thermal treatment at increased temperature and longer time under pressure. This is the case, for example, in the production of binders or coke for carbon and graphite electrodes.

The *Cherry-T process* of Osaka Gas [15] serves as an example of such a procedure. In this process water and light oil are separated from the crude tar as a first step, usually in a dehydration column. The dehydrated tar is heated in a pipe still to almost 400 °C and afterwards thermally polycondensed in a pressure reactor. After this reactor, a flash column is used for oil/pitch separation. If necessary, the oil can also be treated thermally in a second pressure reactor. Then, after additional flashing, a secondary oil pitch and nonpolymerized oils are obtained. The total yield of primary and secondary pitch is ca. 75 %, compared with 50–55 % for tar distillation without intensified thermal treatment.

Whereas in the Cherry-T process, electrode pitches are the main products of primary processing, pitch coke is the main product when tar distillation is coupled with a *delayed coker* (→ Oil, Oil Refining, **A 18**, pp. 77–78). Some Japanese plants use such a combination (see Section 3.3). Here only light, carbolic, naphthalene, and wash oils are separated by primary distillation. Soft pitch obtained as residue is pumped into the bottom of a combination tower where it is distilled together with coker oil. The bottom product of the combination tower is transferred to the pipe still of the delayed-coking plant where it is heated to almost 500 °C and pumped into the delayed-coking drum (see Section 3.3).

Nondistillation processes for the primary processing of coke-oven coal tar, such as liquid or supercritical extraction and hydrocracking, are presently not used in large-scale production.

## 3.3. Processing of Coal-Tar Pitch

### 3.3.1. Cooling

Approximately 50–55 % of the products of a coal tar primary distillation is a coal-tar normal pitch with a softening point of 68 °C (Krämer–

Sarnow). This is either processed directly in the liquid state or solidified by cooling, which converts the thermoplastic pitch into its glassy state.

Modern pitch-cooling processes operate continuously and use water as cooling medium. The pitch is obtained in granulated form.

The problem of continuous pitch cooling is to obtain a rapid heat transfer despite the relatively low thermal conductivity of pitch. If the granules do not cool completely, undesired agglomeration during storage and transport can occur. For various applications the pitch granulate must be moisture-free; therefore, cooling is followed by drying. Several continuous processes for direct cooling of molten coal-tar pitch with water have been developed.

Figure 6 shows the diagram of the Rütgers pencil-pitch process [16]. Relatively thick strings are formed by pumping pitch through heated nozzles directly into water flowing cocurrently. The water acts as cooling and transport medium. Before the liquid pitch is formed by the nozzles, it is cooled in a heat exchanger that produces low-pressure steam. The nozzles are shaped such that by adjusting the temperature and water/ pitch ratio, a pitch string without internal water inclusion is obtained. The cooling water and the solidified pitch string run together through a cooling tube 20–25 m long. This tube may be bent to shorten the equipment. Each nozzle has its own cooling tube to prevent the warm, thermoplastic pitch strings merging. At the end of the cooling tube the outer skin of the pitch strings at least has cooled and solidified. The strings can be deposited together on a slowly

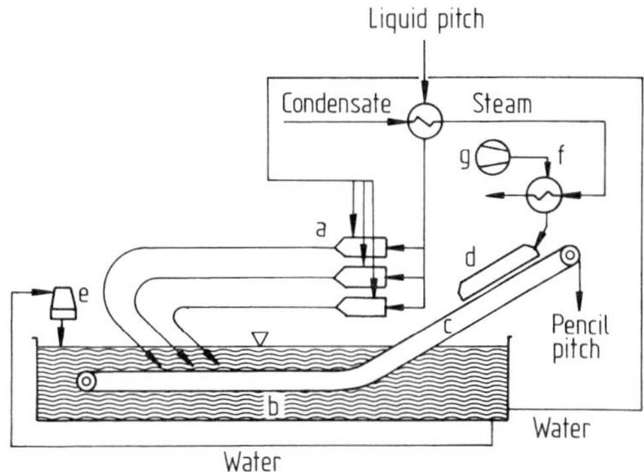


**Figure 6.** Direct cooling of coal-tar pitch (Rütgers pencil-pitch process)
a) Nozzles; b) Cooling water basin; c) Conveyor belt; d) Dryer; e) Cooling facility; f) Air preheating; g) Blower

moving conveyor belt running submerged in a long cooling-water basin. The cooling water is cooled and recycled. At the end of the cooling section the conveyor belt removes the cold, solidified pitch strings from the water. They pass a dryer whose airstream is heated by the low-pressure steam produced in the heat exchanger. The dried pitch strings are air-cooled and then loaded into transport and storage containers. During loading the pitch strings break into short, air-conveyable pieces. This process is also applicable to other pitches with different softening points. The throughput capacity is $1-2$ t per nozzle per hour.

### 3.3.2. Production of Electrode Pitch

Electrode pitches are the most important products derived from coal-tar pitch. They are used as reactive binders, especially for the production of carbon anodes for aluminum production by electrolysis ($\rightarrow$ Aluminum, **A 1**, pp. 463–471) and furthermore as binder and impregnating pitches for the production of graphite electrodes for electric steelmaking ($\rightarrow$ Carbon, **A 5**, pp. 103–120; $\rightarrow$ Steel, **A 25**, pp. 99–102).

Electrode pitches should have good caking and binding properties towards electrode cokes (petroleum cokes, pitch cokes), the solid components of the electrode paste. On heating, electrode pitches should give a high yield of a high-strength coke. Medium and higher molecular pitch fractions are mainly responsible for the caking and binding properties. They are analyzed by solvent fractionation (e.g., quinoline insolubles, QI; toluene insolubles, TI; $\beta$-resins = TI − QI). The volatile matter of the electrode pitch should degas smoothly to prevent cracking during carbonization of the electrode paste. Moreover, electrode pitches should wet electrode coke particles well and should have a desirable viscosity–temperature behavior and their properties should not change significantly on hot storage at temperatures of, e.g., 200–250 °C. The content of ash-forming constituents should be low. Table 4 lists typical characteristics of various electrode pitches. Bench-scale processes have been developed that allow the quality properties of electrode pitches to be determined more precisely. It is now possible to produce test electrodes with electrode coke on a laboratory scale. These electrodes are tested according to various methods [17], [18]. Quality aspects of the test electrodes include high mechanical strength, high electrical conductivity, and low air permeability, as well as low air reactivity at, e.g., 525 °C and reactivity to $CO_2$ at, e.g., 1000 °C [19], [20].

The properties required for electrode pitch are generally not fulfilled by normal pitch from conventional coal-tar primary distillation. The aim of production processes for electrode pitch is to increase the pitch coking value. This is achieved either by thermal condensation and polymerization of pitch components in secondary processes or by gentle distillation of vaporizable pitch components integrated with the tar distillation. In both processes the softening

**Table 4.** Specifications of various electrode pitches from coal tar

| | Electrode pitches for aluminum | | Electrode pitches for electrographite | |
|---|---|---|---|---|
| | Prebaked anodes | Søderberg anodes | Binder pitches | Impregnating pitches |
| Softening point, °C | | | | |
|    Krämer–Sarnow | 88–103 | 108–112 | 90–95 | 68–74 |
|    Mettler | 106–121 | 126–130 | 108–113 | 87–93 |
| Density at 20 °C, g/cm³ | $\geq 1.31$ | $\geq 1.31$ | $\geq 1.31$ | |
| Quinoline insolubles, % | 5.5–13 | $\leq 14$ | 7–12 | $\leq 5$ |
| Toluene insolubles, % | 23–36 | $\geq 32$ | 27–34 | 16–22 |
| β-Resins, % | 18–26 | $\geq 22$ | 20–22 | 11–18 |
| Coking value (Alcan), % | 53–63 | $\geq 57$ | 56–63 | $\geq 43$ |
| Ash (800 °C), % | 0.15–0.35 | 0.15–0.35 | 0.15–0.35 | $\leq 0.53$ |
| Sulfur, % | $\leq 0.7$ | $\leq 0.7$ | $\leq 0.7$ | $\leq 0.7$ |
| Sodium, ppm | $\leq 200$ | $\leq 200$ | $\leq 200$ | |
| Iron, ppm | $\leq 200$ | $\leq 200$ | $\leq 200$ | |
| Silicon, ppm | $\leq 200$ | $\leq 200$ | $\leq 200$ | |
| Distillation up to 360 °C (ASTM D 2569), % | $\leq 3$ | $\leq 3$ | $\leq 3$ | * |

* Flash point $\geq 260$ °C (open cup).

point and the content of medium-molecular β-resins are increased, and in the thermal condensation and polymerization process, the amount of quinoline insolubles rises. These "secondary" quinoline insolubles, generated during thermal retreatment, mainly consist of liquid crystalline carbonaceous mesophases. These mesophases advantageously increase the coking value of total pitch, but at the same time generally decrease its wetting ability. Therefore, in many cases electrode pitches produced without mesophase formation are preferred for electrode production [21].

Figure 7 shows the flowsheet of a thermally gentle electrode-pitch production. The process is part of the coal-tar primary distillation. In this case the coal-tar primary distillation (see Fig. 5) is extended by a flash column. By regulation of temperature and vacuum it is possible to directly control the desired increased softening points and the coking values related therewith.

The Cherry-T process of Osaka Gas (see Chap. 3.2) is a continuous thermal pitch-condensation process forming part of the coal-tar primary distillation. This process gives a high yield of electrode pitch from coal tar in two different qualities. In the first step a primary pitch yield of 55–60 % from coal tar is achieved. The pitch has a softening point of 80–100 °C, 31–38 % toluene insolubles, and 8–14 % quinoline insolubles, thus having a β-resins content of ca. 22 %. In the second step, further thermal treatment of the flash oil from the first reactor gives 40–50 % of a secondary pitch with a soft-

ening point of 70–90 °C, 23–31 % toluene insolubles, and a very low amount of quinoline insolubles (max. 2 %) [22].

Not only the production conditions for electrode pitches but also the choice of parent coal tars is important. The desired pitch components may already have been formed to a large extent in the coal tar. Moreover, the tars have to meet the requirements concerning the content of inorganic contamination, which very often require a separation of these microparticles by centrifugation or extraction. For use as binder pitches for carbon anodes in aluminum production, for example, a low content of zinc and iron compounds is important. Under the conditions of aluminum oxide electrolysis in the melt, elemental metals are formed together with aluminum at the cathode, thus reducing the purity of the aluminum product. Furthermore, the content of alkaline earth and alkali metal compounds should be as low as possible because they catalyze the reaction of $CO_2$ with the carbon anode. In the case of impregnating pitches it may also be necessary to partly remove the highest molecular mass quinoline insoluble α-resins from coal tar, as the impregnating behavior depends upon the content of α-resins and upon their particle size distribution. On the other hand, α- and β-resins considerably contribute to the stability of the pitch coke skeleton formed from electrode pitch, thus giving the mechanical and thermal stability of carbon and graphite electrodes.

For large-scale production of electrodes, bituminous binding agents other than coal-tar

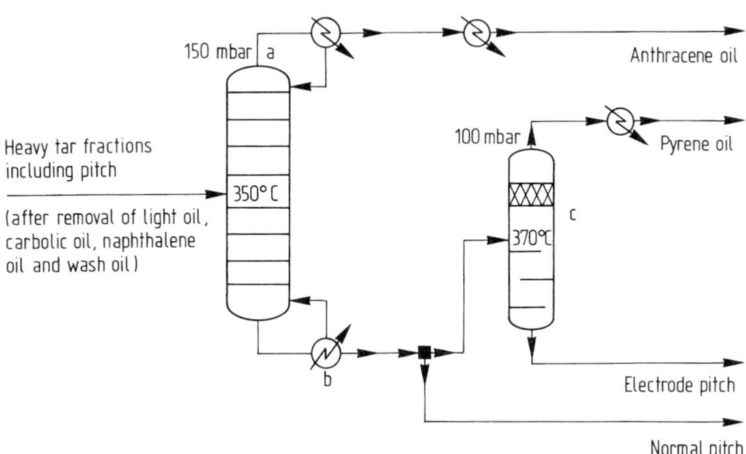


**Figure 7.** Flowsheet of a coal-tar primary distillation with integrated electrode-pitch fabrication (Rütgers process)
a) Pitch/anthracene oil column; b) Reboiler; c) Flash column

pitch have not gained acceptance. The reason is the good coking behavior of coal-tar electrode pitches due to the high content of condensed aromatics and fixed carbon. Coal-tar pitch has the highest carbon content of comparable bituminous binding agents. In contrast to petroleum asphalt and coal extracts, the carbon content of coal-tar pitch is considerably above 90%, on average ca. 93%. The atomic C/H ratio is 1.7–1.8 compared with, for example, 1.3 for coal extract from the solvent refined coal (SRC) process and 0.8 for high-vacuum bitumen with a comparable softening point [23]. It is only possible to produce aromatized petroleum pitches from strongly aromatized residues of catalytic petroleum cracking by thermal treatment. These may serve as substitutes for coal-tar electrode pitches in some situations [24], [25].

Despite the advantages of coal-tar pitch as electrode binder, it does have environmental and toxicological drawbacks. Coal-tar pitch has a high content of cancerogenic polycyclic aromatics, especially benzo[a]pyrene. However, the uptake of harmful polycyclic aromatic hydrocarbons can be avoided if pitch fumes and dusts are not handled or inhaled. Nevertheless, attempts have been made to reduce the benzo[a]pyrene content of electrode pitches. The physical removal of the cancerogenic polycyclic aromatic hydrocarbons by extraction or distillation is technically possible. The softening point of the pitch increases on removal of these low-molecular compounds. Bench-scale experiments showed that it is possible to obtain carbon electrodes with improved technical properties when using such high-melting electrode pitches [26]. In an electrode pitch with an increased melting point of 170 °C, for example, the benzo[a]pyrene content is decreased by > 60%, down to 0.3–0.4%. A further increase of the melting point up to, for instance, 260 °C by a > 99% reduction of the benzo[a]pyrene content to less than 100 or 50 ppm is possible in principle. For practical applications, however, modification of the electrode production process by using a different mixing technology for the electrode masses, or adjustment of the softening point is necessary [27].

### 3.3.3. Production of Pitch Coke

Like calcined petroleum coke, pitch coke is used as electrode coke, the solid component for the production of carbon and graphite electrodes. In addition, technically pure carbon pitch coke is used to produce other carbon and graphite materials, especially for high-temperature and chemically resistant equipment. It is also used as conductive filler in carbon brushes, batteries, plastics, and brake linings; as a carburizing agent in steel and cast-iron production; as adsorption coke for water purification; as an almost ash-free solid reducing agent in metallurgy and chemistry; and for the production of highly isotropic graphite for nuclear reactors. Specifications of different pitch cokes are listed in Table 5.

The most important processes for producing pitch coke are:

1) Discontinuous high-temperature coking of hard pitch in normal coke ovens with gas heating and liquid feeding, used especially in the CIS.
2) Discontinuous coking of soft pitch by the delayed-coking process in Japan and China.
3) Continuous hard-pitch coking in an indirectly heated rotary-drum reactor (still in development).

*Hard pitch* with a softening point > 150 °C (Krämer–Sarnow) can be obtained either from the normal pitch produced by continuous tar distillation by dehydrogenative condensation with atmospheric oxygen (air blowing) or solely by thermal treatment, similar to electrode-pitch production processes.

In the air-blowing process, air is blown through liquid normal pitch at 300 °C in a retort

**Table 5.** Properties of pitch cokes

|  | Regular coke | Needle coke |
|---|---|---|
| Density, g/cm³ | 2.04–2.10 | ≥ 2.12 |
| Microstructure | coarse mosaic or fine isotropic | coarse fibrous |
| Coefficient of linear thermal expansion * $10^{-6}$ K⁻¹ | 2.5–5.5 | < 1 |
| Ash content, % | < 0.5 | < 0.2 |
| Elemental analysis |  |  |
| C, % | ≥ 98 | > 98 |
| S, % | 0.4–0.5 | < 0.3 |
| N, % | 0.5–0.7 | < 0.5 |
| H, % | 0.1–0.2 | 0.1–0.2 |
| Si, % | ca. 0.03 | < 0.03 |
| Fe, % | ca. 0.03 | < 0.03 |
| V, ppm | < 5 | < 5 |

* Measured on graphite rods, 20–200 °C.

directly after coal-tar primary distillation [28]. There is an air distributor at the bottom of the retort, and the air is either sucked or blown through the pitch. The reaction is exothermic above 300 °C and takes place without additional heating. The temperature of the retort contents rises to ca. 400 °C. The main product is a hard pitch which can be processed to coke with a yield of up to 83 %. The byproduct pitch distillate can be converted into a secondary hard pitch by distillative thermal treatment and air blowing. By using a mixture of primary and secondary hard pitch, different pitch-coke qualities can be obtained in the subsequent coking process.

The thermal condensation of normal pitch to hard pitch is usually carried out by continuous treatment of normal pitch in pipe stills under pressure followed by flash distillation to remove vaporizable components. Compared to the air-blowing process, the yield of hard pitch is lower, but the yield of high-boiling aromatic oils is higher.

In discontinuous *pitch coking in horizontal coke ovens* hard pitch is pumped at a temperature of above 300 °C into heated pitch tanks. In the Koppers process [29] several oven blocks, each with, e.g., five pitch-coke ovens, can be fed by these tanks. The hard pitch is pumped continuously via a ring main from the pitch tanks to the oven blocks. The liquid hard pitch is pumped in portions over a period of several hours into the coke ovens from this ring main. The final coking temperature can be up to 1200 °C, and the coking time 14–24 h. The pitch-coke yield is ca. 70 %.

The pitch-coke oil byproduct can be used for secondary-pitch production. By mixing this pitch with the primary pitch, the quality of pitch coke can be varied and the pitch-coke yield increased to 80–85 %. Under these conditions ca. 250 m³/t process gas, containing ca. 80 vol% hydrogen and ca. 12 vol% methane, are generated as a byproduct. This gas can be used for the production of hydrogen, or as low-calorific gas for heating the coke ovens and pitch tanks. With heat requirements of ca. 18 MJ per tonne of hard pitch, about 80 % of the process gas is required. Analogous to metallurgical coke production, after carbonization the pitch coke is pushed mechanically out of the horizontal chamber of the coke oven and cooled in a quenching car by spraying with a calculated amount of water. It is possible to achieve a water content of 1–2 % in spite of the high porosity of pitch coke. Due to the high carbonizing temperature, the pitch coke produced does not require secondary calcination.

The second process for production of technical pitch-coke from coal tar is *delayed coking*, well known from petroleum refining (→ Petroleum, Coke). To produce a pitch coke with a low volatiles content a second calcination step is required. Soft pitch containing high-boiling aromatic oils, obtained by primary distillation of coal tar, is used as feed. The soft pitch is pumped into a combination column, the bottoms of which are fed to the pipe still of the delayed coking plant.

Figure 8 shows the process flow sheet for production of pitch coke from coal-tar soft pitch

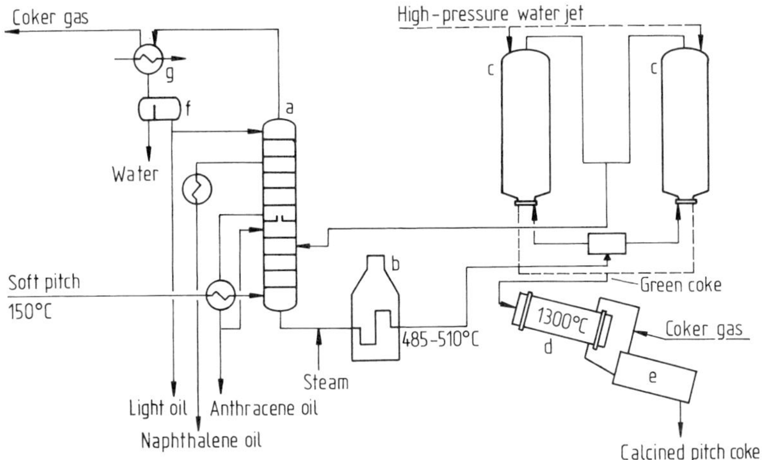


**Figure 8.** Production of pitch coke by delayed coking
a) Combination column; b) Pipe still; c) Coker drums; d) Rotary kiln; e) Rotary cooler; f) Decanter; g) Condenser

by delayed coking using the Lummus process as applied in Japan and China [30]. The soft pitch coming from the coal-tar primary distillation at 150 °C is heated in a heat exchanger by the heavy oil of the combination column and then pumped into the bottom of this column. The coker oil of the delayed-coking plant is also added to the column. Light oil, middle (naphthalene) oil, and heavy (anthracene) oil fractions are taken from the combination column. The bottoms of the column are heated in a pipe still to 485–510 °C by gas or fuel oil. Then they are fed from below into one of two coker drums via a switching valve. In these drums the delayed coking of pitch takes place during a 24 h cycle with a residence time of 12 h at a pressure of, for example, 5 bar. When the first coker drum is full, the feed is switched to the preheated empty coker drum. After switching, the full drum is steamed for 2 h to remove residual coker oil, which is mainly present in the upper part of the coker drum. The oil obtained is recycled to the combination column. After steaming, the coke charge is cooled directly with water. Then the coke is cut out by a high-pressure water jet, which takes 5–7 h. The green coke obtained is dried, broken, and calcined in a hearth furnace or a rotary drum. The pipe still and the calciner can be heated with coker gas arising as the noncondensable top product of the combination column.

Normally the green-coke yield is at least 60 % of the soft pitch fed, with approximately 3 % process gas and 35 % coker oil as byproducts. The combination column yields ca. 10 % light and middle (naphthalene) oil, as well as ca. 25 % heavy (anthracene) oil. The coker gas contains ca. 50 % hydrogen and 50 % methane. The middle oil has a boiling range of 180–310 °C and contains ca. 30 % naphthalene. The heavy oil has a boiling range of 265–400 °C and a composition that differs from that of an anthracene oil from normal continuous tar distillation. Under the conditions of pressure carbonization in the delayed coker, some distillable tar aromatics are preferentially converted to coke. The green coke still has 7.5–9.5 % volatiles, which are removed to a residual content of < 0.5 % by calcination at up to 1300 °C.

Highly anisotropic cokes, known as *needle cokes*, are obtained via intermediate liquid-crystalline mesophases by delayed coking of pitches from which the α-resins (quinoline insolubles) have been largely removed by filtration [31] or extraction [32]. These needle cokes are used for

the production of ultra-high power (UHP) graphite electrodes with high electrical conductivity and low thermal expansion coefficient.

The carbonization of hard pitch in an indirectly heated *rotary drum reactor* [33], [34] is a continuous pitch-coke production process. Hard pitch is pumped into an indirectly fired rotary drum reactor where it is carbonized with a residence time of 0.5–1.5 h at an inner-wall temperature of 500–800 °C. The gases and vapors arising are pumped in countercurrent to the carbonizing pitch, if necessary together with an inert gas. The green-coke yield is 70–75 % of the hard pitch feed. The grained coke has volatile matter content of ca. 3.5 wt % and relatively high strength and density. It is calcined without precooling in a second directly heated rotary drum reactor at up to 1300 °C. The yield of calcined coke is ca. 90 % of the green-coke feed, and 65–70 % of the hard pitch input. The advantages of this continuous process compared with the discontinuous processes of horizontal chamber coking and delayed coking are the possibility to adjust the residence times and heating periods, thus yielding a more homogeneous pitch coke.

### 3.3.4. Production of Special Pitches

*Hard pitches, soft pitches*, and *extractive pitches* produced by solvent extraction with considerably decreased content of α-resins (primary quinoline insolubles) are special pitches that are intermediate products of pitch carbonization. Other special pitches include plasticized pitches for protective coatings, pitches used for pitch–polymer combinations, and road-binding agents which are compatible with polymers and bitumen, as well as the newly developed and produced mesophase pitches and high-melting isotropic pitches for further processing to carbon materials.

Soft pitches with increased coking values are used as *refractory binders*. They are reactive binding agents for the production of dolomite and magnesite bricks used especially for the lining of steel converters. During brick calcination a binding coke skeleton is formed by carbonization of the soft pitch, which considerably increases the compressive strength of the preform and counteracts corrosive attack by slags. Refractory binder pitches with increased coking values are obtained by adjusting hard pitches with high-boiling aromatic oils to give the desired softening point. Also soft-pitch mixtures with phenolic resins or their precursors are used. Special formulations contain less than 50 ppm benzo[*a*]pyrene.

For the production of *anticorrosion protective coatings* (→ Paints and Coatings, **A 18**, p. 423) the plasticity range (difference between softening and brittle point) of 30 K of coal-tar normal pitch is too low. To increase this plasticity range to 100 K or more, hard pitches with a high content of toluene insolubles are adjusted to the desired softening point with high-boiling tar oils, preferably by adding crystalline tar aromatics. A further increase in the plasticity range is obtained by hot-mixing these pitches with extenders such as finely ground coal, minerals, diatomaceous earth, or fly ash. Such plasticized pitches are used for long-term anticorrosion pipe coatings and as sealing compounds. The advantages compared with the bitumen coatings are the excellent resistance of coal-tar pitch to moisture, bacteria, fungi, mineral oils, and gasoline. When pitch coatings are used together with cathodic pipe protection the low water absorption of coal-tar pitch is important. The water content of a 2.5 mm tar-pitch coating is only 0.3–0.6 % after prolonged storage in water. Plasticized pitches without extenders are used as dip-coating agents for protecting cast-iron pipes.

To meet especially high anticorrosion requirements, coal-tar pitches are combined with polymers [35]. Especially suitable for such *pitch–polymer-combinations* are epoxy resins, polyurethanes, and some elastomers (→ Paints and Coatings, **A 18**, pp. 423–424). Normally two-pack systems are produced for epoxy resin combinations. One component contains the non-hardened epoxy compound dissolved in a compatible special coal-tar pitch, and the other contains the hardener for the epoxide compound. After combination of the components the epoxy resin polymerizes and incorporates the pitch components. Usually low molecular-mass liquid epoxy resins are combined with soft coal-tar pitches. After chemical hardening, coatings with high-temperature stability and resistance to chemicals are obtained. Pitch–polyurethane two-pack systems are produced and processed similarly. One component contains the special soft coal-tar pitch and the polyol, and the other the polyisocyanate. In combination with polyurethanes, one-pack systems that cure by reaction with atmospheric moisture are also feasi-

ble. Other polymers and elastomers are mixed directly with the hot liquid pitch, whereby miscibility gaps and the decomposition temperatures of the polymers must be taken into account. It is also possible to add high-boiling or unvaporizable, preferably aromatic homogenizing agents.

By using high-boiling tar aromatics as homogenizing agents, soft coal-tar pitches with a low content of $\alpha$-resins can be combined in any ratio with appropriate, preferably naphthene-based bitumens. Carbobitumen, a highly deformation resistant *road binding agent*, is a blend of soft pitch and hard bitumen, containing 20–30 % of a special pitch. The pitch is enriched with high-boiling crystalline aromatics [36], [37]. These crystalline aromatics cause a separation of microcrystals in the pitch–bitumen matrix when cooled to ca. 100 °C. These microcrystals crosslink the thermoplastic system and harden the binding agent at lower temperatures. Thus carbobitumen is harder at temperatures below 70–80 °C and, therefore, more deformation resistant than the road bitumen usually used. At higher temperatures, however, it is less viscous and, therefore, can be applied at lower temperatures than road bitumen.

Normal pitches, hard pitches, and special pitches with carbon fillers and polymers are suitable as reactive additives for *cocarbonization of weak-coking coal blends* in amounts of 2–10 %. The mechanical strength of the metallurgical coke produced in horizontal chamber-coke ovens increases on addition of an amount of pitch ascertained by laboratory tests [38], [39]. With increasing temperature the pitch wets the coal grains in the coke-oven chamber. It partially dissolves the coal macerals in the softening range of the coal. During carbonization above 450 °C inert macerals and solid leaning agents such as coke dust or petroleum coke are also brought into the generated coke skeleton. Used plastics, for example, can be utilized as additives. These polymers should first be subjected to cothermolysis with pitch, to obtain a resistant reactive additive as liquid phase in the coal-softening range [40].

Further special pitches are fine-grained *foundry pitches* produced by spraying. They have a high softening point and high coking value. They serve as binding agents for the production of foundry molds. Another special pitch is *fiber-pipe pitch* which has a low content of $\alpha$-resins. This is used as impregnating material for cellulose fibers in the production of pitch fiber pipes. *Clay pigeon pitch* is used as a brittle binding agent with increased softening point for clay pigeons used in sport shooting.

*Mesophase pitches* are pitches with a high content of carbonaceous mesophases (see Chap. 2). They are produced by heat treatment of pitches that have been filtered or extracted to remove primary $\alpha$-resins [31]. By heat treatment, mesophase spherules are formed, which can be separated from the isotropic pitch matrix by settling and/or by solvent fractionation. Pitch prefractionation with solvents or supercritical extraction is often carried out before thermal mesophase formation. Two-step thermal pitch treatment is also utilized. In the first step the especially reactive pitch aromatics are separated by condensation to give secondary quinoline insolubles. The second thermal treatment step leads to mesophases with low to medium molecular mass and correspondingly low viscosity, which can be advantageous. Mesophase pitches can also be produced from special petroleum pitches, hydrogenated coal-tar pitches, and pyrolysis residual pitches. By using different processes special properties can be achieved [41]. Mesophase pitches also can be produced by catalytic polycondensation of naphthalene and its homologues and higher condensed polycyclic aromatic hydrocarbons in the presence of homogeneous Friedel–Crafts catalysts. Thus, after 2–17 h at 140–180 °C 80–100 % mesophase pitches with a yield of 50–80 % are formed from naphthalene, 1,1-binaphthylanthracene, and 1,2-benzanthracene and from a 1:1 mixture of naphthalene and phenanthrene by reaction with aluminum chloride/copper(II)chloride [42]. From naphthalene it is possible to produce mesophase pitches by using $HF/BF_3$ catalysts, either by one step at 260–300 °C, or by two steps, the first at 80 °C and the second at 400–480 °C under vacuum or pressure [43]. Instead of naphthalene, 1-methylnaphthalene, 2-methylnaphthalene, or mixtures thereof can also be used. Here mesophase pitches are formed by using $HF/BF_3$ at 260 °C under a pressure of ca. 30 bar in 4–5 h with a yield of 80–90 % [44].

Mesophase pitches and high-melting isotropic pitches are intermediates in the production of new forms of technical carbon such as anisotropic and isotropic carbon fibers, mesocarbon microbeads, and sinterable carbon powders [41]. In their production the special pitches are normally first stabilized by oxidation after pre-forming by melt spinning or fine grinding and are then partly or entirely carbonized. For the production of high-modulus carbon fibers from mesophase pitches various processes have been developed ($\rightarrow$ Fibers, Inorganic, **A 11**, pp. 55–60). Some of these processes have been used commercially. Mesophase microbeads and sinterable carbon powders can be pressed isostatically without additional binding agents and converted by thermal treatment into highly dense isotropic carbon materials [41]. In the last few years isotropic fine-grained carbon materials have found a wide range of applications, as have carbon fiber reinforced carbon materials (CFC) [45]. Widespread areas of application for these carbon materials can be predicted for the future, especially in energy and transport technology. This is due to their low density compared to metals such as steel and aluminum, their mechanical strength over a wide temperature range, and their excellent properties at temperatures up to 4000 °C if protected against oxidation.

### 3.3.5. Pitch–Oil Blends

Coal-tar pitches are highly soluble in various tar oils. For manufacturing of pitch–oil blends (refined tars) the liquid pitch is mixed with tar oils.

*Road tars* based on pure coal tars are blends of normal pitch with middle oils (boiling range 170–270 °C), heavy oils (boiling range 270–300 °C), and anthracene oils (boiling range above 300 °C). The content of normal pitch is between 60 and 80 %. These road tars are suitable for surfacing low-traffic roads. The advantages of road tars for this purpose are good adhesive strength, quick rock wetting due to their content of polar compounds, and their property of gradual surface resinification. The middle and heavy oil components evaporate during setting, leaving as the actual binder a soft pitch consisting of anthracene oil and normal pitch. The setting speed depends on the initial viscosity and the pitch/anthracene oil ratio. Normal setting road tars have a pitch/anthracene oil ratio of ca. 3.5:1. By increasing the ratio of anthracene oil II (boiling range above 350 °C) and anthracene oil I (boiling range up to 350 °C), aging-resistant road tars can be produced. Bitumen-containing road tars include, for example, 15 % asphalt basic distillate bitumen with low oil content; bitumen-rich road tars contain 35–40% bitumen, and *pitch bitumen* contains 70–85% bitumen. These tar–bitumen blends with increased or predominant bitumen components combine the advantages of tar (good setting properties, resistance to mineral oil and fuel, easy processability, and good road grip) with those of bitumen (low temperature dependence of viscosity and wide plasticity range). Carbobitumen is a specially deformation resistant pitch–bitumen blend (see Section 3.3.4).

Besides bitumen, plastics such as poly(vinyl chloride), polystyrene and epoxy resins are used for the modification of road tars. Thermosetting tar–epoxy resin combinations (see Section 3.3.4) are particularly suitable for highly exposed road surfaces and airport runways due to their good wear properties and their resistance to liquid fuels.

*Roofing tars* used as impregnating, coating, and adhesive material for tarred felts and tarred sealing webs are usually blends of pitch and filtered anthracene oil. By using plasticized pitches or by adding extenders the plasticity and temperature stability of roofing tars is improved considerably. Tarred felts produced in this manner have the same temperature stability as bitumen felts. Moreover, they have a high resistance to water, microorganisms, and plant roots.

*Pitch paints* are solutions of coal-tar pitch in highly volatile aromatic solvents such as light and heavy tar-oil naphtha. Normally plasticized pitches are used or extenders are added. Like the plasticized pitches themselves, which are processed while hot, pitch paints have excellent resistance to water, bacteria, and chemicals. Anticorrosion and building-protective coatings are applied, for example, as underwater protection and as protective paints for steel, concrete, and underground brickwork [35].

Effective anticorrosion protective coatings are obtained by combining pitch with polymers. The best-known *pitch combination coatings* are two-pack systems based on soft pitches and epoxy resins (see Section 3.3.4). The coatings are resistant both to low temperatures and high temperatures up to 100 °C. They are resistant to fresh and salt water, moderately strong mineral acids and alkalis (except ammonia), salt solutions, lubricating oils, and nonaromatic motor fuels. Two- and one-pack pitch–polyurethane coatings have a similar range of applications.

## 3.4. Processing of Tar Distillates

This chapter deals with the further processing of primary tar distillates to fractions and the various aromatic oils. The processing to technically pure aromatic compounds is dealt under separate keywords. Only the processing to technically pure phenols and pyridine bases is described in more detail here.

### 3.4.1. Processing of Anthracene Oil

In the primary distillation of coal tar, ca. 20–30% of the crude tar distills as anthracene oil, which at room temperature is a semisolid, green-brown, crystalline material with a boiling range of ca. 300–450 °C. In most primary distillation plants two fractions are produced: the larger fraction being low-boiling anthracene oil I, and as the smaller, high-boiling anthracene oil II. Anthracene oil I contains phenanthrene, anthracene, and carbazole, whereas fluoranthene and pyrene are concentrated in anthracene oil II.

The crude anthracene oil or anthracene oil I is cooled in water-cooled discontinuous or continuous crystallizers and partially crystallized to give a filtered anthracene oil and anthracene residues. Filtered anthracene oil does not crystallize at normal temperature. The discontinuous cooling process is usually used. Anthracene oil at 80–100 °C is pumped from primary distillation into cylindrical vertical stirred coolers. It is cooled over 24–36 h to 20–30 °C by indirect water cooling. Early sedimentation of crystals is prevented by a slowly rotating stirrer (ca. 15 rpm). After cooling the crystal slurry is discharged via a gate through the conical lower part of the cooler and centrifuged with a peeler centrifuge. The crude anthracene oil yields 10–15% anthracene residues and 85–90% filtered anthracene oil. The residues are further processed to anthracene, phenanthrene, and carbazole.

Filtered anthracene oil I and high-boiling anthracene oil II are important blending components for technical *aromatic oils* and pitch–oil mixtures. Aromatic oils with anthracene oil as main component include carbon-black oils, impregnating oils (creosote), medium fuel oils, special diesel fuels, gasometer oils, fluxing oils, naphthalene wash oil and numerous others. Table 6 lists specifications for some of these aromatic tar oils.

**Table 6.** Specifications of various coal-tar oils

| | Impregnating oil for railway sleepers (Deutsche Bundesbahn) | Impregnating oil for telegraph poles (Deutsche Bundespost) | Impregnating oil AWPA (P 1–78) (USA) | Carbon-black oil (light quality) | Carbon-black oil (heavy quality) | Fluxing oil | Light fuel oil (DIN 51 603) | Diesel fuel | Benzole absorbing oil |
|---|---|---|---|---|---|---|---|---|---|
| Density at 20 °C, g/cm³ | 1.04–1.15 | 1.04–1.15 | at 15.5 °C ≥ 1.050 | 1.07–1.1 | 1.1–1.15 | 1.02–1.07 | 1.02–1.10 | 1.04–1.10 | 1.02–1.06 |
| Clear point or free of sediment, °C | ≤ 23 | ≤ 23 | | < 40 | < 50 | 0 | 0 | 0 | < 0 |
| Water content, % | ≤ 1 | ≤ 1 | ≤ 1.5 | ≤ 0.2 | | ≤ 0.5 | ≤ 0.3 non-settleable | ≤ 0.3 | ≤ 0.1 |
| Toluene insolubles, % | traces | traces | ≤ 0.5 | ≤ 0.1 | | | | ≤ 0.1 | |
| Coking value[a], % | | | | ≤ 1 | | | ≤ 1 | ≤ 0.5 | |
| Ash content, % | | | | | | | a) ≤ 0.01[d]<br>b) ≤ 0.04 | ≤ 0.01 | |
| Phenols, % | < 3 | < 2.3 | | | | | | | |
| Sulfur, % | | | | ≤ 0.8 | ≤ 0.8 | | a) ≤ 0.3[d]<br>b) ≤ 0.8 | | |
| Correlation index[b] | | | | ≥ 140 | ≥ 150 | | | | |
| Flash point[c], °C | | | | | | > 55 | > 85 | > 80 | |
| Calorific value | | | | | | | ≥ 37.7 MJ/kg | ≈ 37.7 MJ/kg | |
| Boiling range, % | up to 235 °C: ≤ 10<br>up to 300 °C: 20–40<br>up to 355 °C: 55–75 | up to 235 °C: ≤ 10<br>up to 300 °C: 40–60<br>up to 355 °C: 70–90 | up to 210 °C: ≤ 2.0<br>up to 235 °C: ≤ 12.0<br>up to 270 °C: 10–35<br>up to 315 °C: 40–65<br>up to 355 °C: 60–77<br>residue: 23–40<br>density at 15.5 °C<br>fraction 235–315 °C: ≥ 1.027 g/cm³<br>fraction 315–355 °C: ≥ 1.095 g/cm³ | 200–400 °C<br>up to 200 °C: ≤ 20<br>up to 300 °C: ≥ 60<br>up to 350 °C: ≥ 70<br>up to 400 °C: ≥ 90 | 250–450 °C | 180–360 °C | | up to 250 °C: ≤ 55<br>up to 300 °C: ≤ 60 | 230–300 °C |

[a] Conradson. [b] Bureau of Mines. [c] Pensky–Martens. [d] a) and b) are two types of fuel oil.

Due to their low sulfur content and high aromaticity, coal-tar oils are high-value raw materials for carbon black. *Carbon-black oils* obtained from coal tar have, due to their high content of condensed aromatics, the highest Correlation Index (Bureau of Mines). They are pyrolyzed at ca. 1500 °C by the furnace-black process to yield filler blacks, mostly for rubber, and pigment blacks, mainly for printing inks. Depending on the required quality of the carbon black the yield is 40 – 70 % ( → Carbon, **A5**, pp. 140 – 150).

*Impregnating oils* (creosote) obtained from coal tar are important for long-term wood preservation ( → Wood, Surface Treatment and Preservation). The aromatic components of anthracene oil and other tar-oil components have, when mixed, strong antifungal properties and significant pesticidal properties for wood pests such as the house long horn beetle (*Hylotropes bajulus*), the ship worm (*Toredo navalis*), anobium beetles (*Anobium punctatum*) and termites, and furthermore inhibit dry-rot fungi. Railway sleepers, telegraph poles, pit props, timbers for use in water and other building timbers are protected against decay and insect damage over a long term by vacuum – pressure impregnation with tar oils. Coal-tar impregnating oils are not washed out by water, are noncorrosive, and have high electrical resistance. The service life of such impregnated woods is prolonged from 5 – 15 years to 30 – 50 years. The various types of impregnating oils are blends of filtered anthracene oil, wash oil, and filtered naphthalene oil.

*Fuel oils* based on coal tar have a short, hot flame with high radiation intensity due to their high carbon content. Because of these properties they are used in various metallurgical plants. Coal-tar fuel oils are produced in varying types which differ in boiling range, density, viscosity, and sedimentation behavior. Table 6 lists data for a light fuel oil. Heavy fuel oils for steel works may also contain pitch, which itself is used as extraheavy fuel oil. Coal-tar fuel oils have a minimum calorific value of ca. 38 MJ/kg.

*Diesel fuels* manufactured from coal tar are suitable for the operation of stationary, slowly running diesel motors. They are blends of filtered anthracene oil ( < 40 %) and tar oil boiling below 300 °C ( > 60 %). This diesel fuel contains no active sulfur (copper strip test) or corrosive chlorine ( < 0.01 %) and has, like all coal-tar oils, a minimum calorific value of ca. 38 MJ/kg.

*Fluxing oils* for bitumen are, depending on their quality, blends of low- and high-boiling tar oils. They decrease the softening point and the viscosity of bitumen and, therefore, lower the processing temperature of road bitumen (cutback bitumen).

Other technical aromatic oils with coal-tar anthracene oil as main component are applied as solvents, flotation oils, quenching agents, and plasticizers.

### 3.4.2. Processing of Wash Oil

Wash oil, or Solvay oil, which makes up 7 – 8 % of the crude tar, is obtained by coal-tar primary distillation directly as marketable benzol wash oil. It boils between 230 and 290 °C and contains, besides slight amounts of naphthalene, mainly the naphthalene homologues, indole, diphenyl, acenaphthene, dibenzofuran, fluorene, ca. 2 % phenol homologues, and 4 – 6 % quinoline bases. If suitable rectifying conditions are used, crystals will not settle out from wash oil, even at 0 °C. This serves as a good example for the mutual melting-point depression of the tar aromatics, most of which are solids.

Wash oil can be directly used as benzol wash oil. Wash oil and anthracene oil also serve as blending components for technical aromatic oils such as impregnating oil (creosote), diesel fuels, and light fuel oils. In pitch – oil blends wash oil is the heavy oil component and offers good solubility, fungicidal activity, and slow evaporation. Furthermore, wash oil is a starting material for numerous technically pure tar aromatics such as dibenzofuran ( → Furan and Derivatives, **A12**, pp. 130 – 131); acenaphthene, indene and fluorene ( → Hydrocarbons, **A13**, pp. 265 – 272); indole ( → Indole) and alkylnaphthalenes ( → Naphthalene and Hydronaphthalenes, **A17**, pp. 5 – 6).

### 3.4.3. Processing of Naphthalene Oil

Naphthalene oil is obtained by primary distillation of coal tar, in 10 – 12 % yield based on crude tar. Depending on the separation efficiency of the distillation process the oil fraction, boiling range ca. 210 – 220 °C, has a solidification point of 65 – 75 °C, i.e., a naphthalene content between 73 and 90 %. Besides naphthalene, the oil contains phenols, tar bases, dibenzothiophene, and small amounts of other aromatic hydrocarbons. The naphthalene oil is further processed by crystallization ( → Naphthalene, **A17**, pp. 1–3). Phenol homologues (carbolic oil II) are generally also extracted from the filtered naphthalene oil.

### 3.4.4. Processing of Carbolic Oil

A yield of 2–4% carbolic oil based on crude tar, is obtained by primary distillation of coal tar as a fraction with a boiling range of 180–210 °C (carbolic oil I). It contains benzene homologues, some naphthalene, pyridine bases, and ca. 20–30% phenols. The phenols are a mixture of phenol, cresols, and dimethylphenols (xylenols).

Phenol and its homologues are isolated from carbolic oil I, filtered naphthalene oil (carbolic oil II) with a crude phenol content of 15–25%, light oil containing ca. 5–10% phenols, and possibly the wash oil and the fractions obtained from wash oil components with a total content of high-boiling phenol homologues of ca. 2%. Furthermore, during processing phenols extracted from the water effluents from coke-oven plants are added, either as crude phenol or sodium phenoxide solution. The coal-tar phenols are, apart from some sterically hindered dimethylphenols, readily soluble in dilute sodium hydroxide solution to form sodium phenoxides. For continuous removal of the phenols from phenol-containing oils normally a 8–12% aqueous sodium hydroxide solution is used, which dissolves only slight amounts of hydrocarbons.

Figure 9 shows the processing flow sheet for the isolation of *crude phenol* from tar fractions by extraction with sodium hydroxide solution. The process has four main steps:

1) Oil extraction with dilute sodium hydroxide solution, yielding the crude sodium phenoxide solution
2) Refining the crude sodium phenoxide solution
3) Precipitation of crude phenol from the refined sodium phenoxide solution with carbon dioxide and formation of sodium carbonate solution
4) Regeneration of the sodium hydroxide solution by caustification of the sodium carbonate solution with calcium oxide

The hot carbolic oils from the primary distillation of coal tar and naphthalene crystallization are pumped with a solution of phenols in sodium phenoxide solution, recycled fom the second step, into the mixing chamber of the first extraction step. The oil–alkali mixture is continuously recirculated at ca. 50 °C and kept at a constant level. Part of the mixture is pumped continuously into a separation column from which the heavier crude sodium phenoxide solution is drawn off at the bottom. The carbolic oil is removed at the top. About 4–6% phenol remains in the carbolic

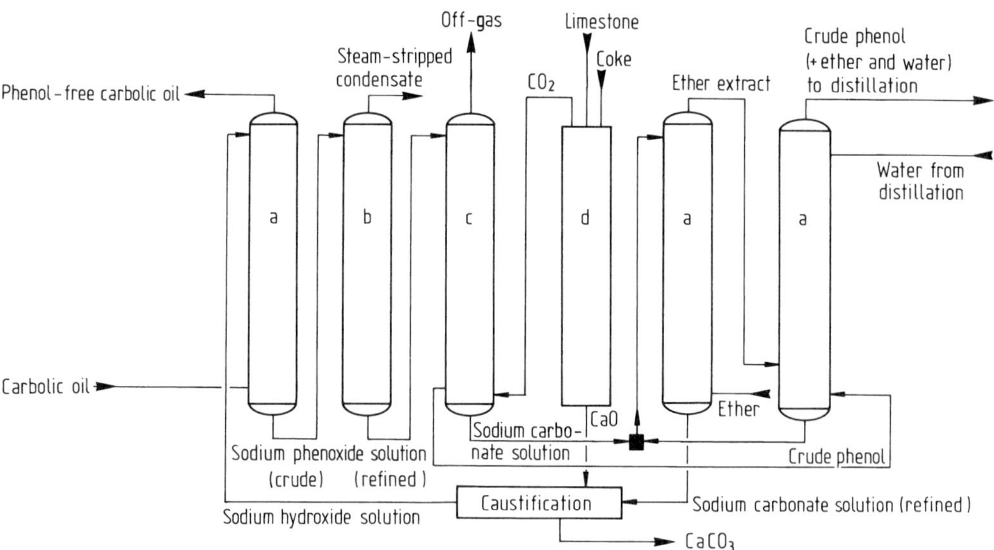


**Figure 9.** Isolation of crude phenol by sodium hydroxide solution extraction
a) Extraction; b) Steam stripping; c) Saturation; d) Lime kiln

oil. In the second extraction step this carbolic oil is extracted analogously to the first step with fresh sodium hydroxide solution to yield a carbolic oil containing less than 0.5 % residual phenols. The solution of phenols in sodium phenoxide solution is returned to wash the fresh carbolic oil in the first step.

The crude sodium phenoxide solution contains approximately 0.5 % dissolved nonphenolic components (hydrocarbons and pyridine bases) which must be removed before the precipitation (saturation or "springing") of the crude phenols. Distillative impurity removal by steam stripping to the clear point is widely used. This takes place in continuously operating packed columns heated at the bottom with direct or indirect steam. To save energy it is favorable to connect several columns in sequence to a multistage evaporator. The vapors taken off at the top of the steam-stripping plant are condensed in a heat exchanger with the crude sodium phenoxide solution and separated into an oil and a water phase. The steam-stripped oil is fed to the light oil processing stage. The dissolved pyridine bases can be extracted from the aqueous phase with benzene.

The hot sodium phenoxide solution (ca. 100 °C) is then pumped to the top of the packed precipitation column (saturation or "springing"). Carbon dioxide is bubbled through the column. The precipitated crude phenol and the sodium carbonate solution formed flow off at the bottom and are separated.

The carbon dioxide used for precipitation of crude phenol from sodium phenoxide solution is normally produced in a lime kiln. This provides kiln gas with a carbon dioxide content of 30–35 % and the calcium oxide required for the caustification of the sodium carbonate solution. The sodium carbonate solution flowing from the precipitation column still contains 3–4 % physically dissolved phenols which can be isolated before caustification by ether extraction. Caustification takes place in stirred vessels at ca. 100 °C; the carbonate solution can be treated without cooling. The suspension of precipitated calcium carbonate in sodium hydroxide solution flowing out of the caustification plant is pre-separated in a sedimentation vessel. For isolation of sodium hydroxide solution the settling calcium carbonate is filtered on a vacuum rotary filter and washed. The regenerated clear sodium hydroxide solution is recycled to the carbolic oil extraction step. The filter cake of the vacuum filter consists of precipitated microcrystalline calcium carbonate with a water content of 40–50 %. It can be used as neutralizing agent, for instance, in the inorganic chemical industry, for flue-gas desulfurization, or as a concrete additive.

The crude phenol from the saturation step (generally 3 parts) and the ether extract from sodium hydroxide solution refining (1 part) is subjected to extraction with water (from dehydration of the crude phenol) in a further column to demineralize the crude phenol. The crude phenol is liberated from ether and water in a continuous distillation plant prior to distillative separation of individual phenols. This process is carried out on a large scale in a plant having a capacity of ca. 30 000 t/a [3], [46], [47].

The isolation of crude phenol by this process requires ca. 550 kg of limestone and 55 kg of coke per tonne water-free crude phenol.

The Lurgi *Phenoraffin process* is the only industrially important process in which crude phenol is extracted by selective solvents.

Figure 10 shows the processing flow sheet for crude phenol isolation by the Phenoraffin process. As selective solvent an aqueous sodium phenoxide solution is used with excellent dissolving power for phenols. The phenoxide solution can be oversaturated to 350 % of the amount of stoichiometrically bound phenol. The phenols are extracted from the sodium phenoxide solution with diisopropyl ether. The plant consists of numerous extraction and distillation columns. The carbolic oil is extracted in two steps in countercurrent flow with the phenoxide solution. Between the two steps the oil is de-based with dilute sulfuric acid. Hydrocarbons and pyridine bases are extracted from the supersaturated solution of phenols in sodium phenoxide solution with, for example, toluene. After de-basing with dilute sulfuric acid and extraction with water the toluene is regenerated by distillation. The distillation bottoms, which contain phenols and hydrocarbons, are returned to the carbolic oil. The phenoxide solution after extraction with toluene and still charged with phenols is stripped with steam to remove the remaining toluene. It is then blown with air to destroy colored components and then extracted with diisopropyl ether. After the ether extraction the ether dissolved in the phenoxide solution is removed by distillation. The regenerated phenoxide solution is recirculated to the oil extraction step. The ether phase, which contains phenoxide solution, is de-based with sulfuric acid, washed with water and sepa-

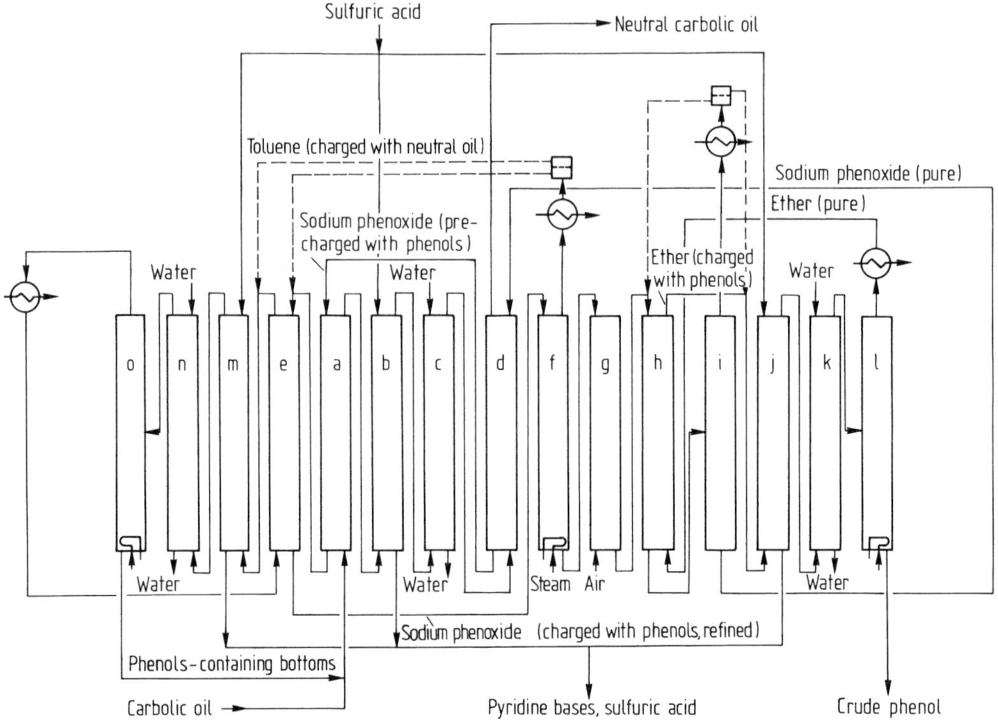


**Figure 10.** Crude phenol isolation (Phenoraffin process)
a) Oil extraction with sodium phenoxide (1st step); b) Oil de-basing by extraction with sulfuric acid; c) Oil extraction with water (neutralization); d) Oil extraction; e) Removal of phenols from the sodium phenoxide solution; f) Steam stripping; g) Air blowing; h) Ether extraction of the refined phenoxide solution of phenols; i) Ether stripping; j) Ether de-basing by extraction with sulfuric acid, k) Ether extraction with water (neutralization); l) Distillation column for separating ether and crude phenol; m) Toluene de-basing by extraction with sulfuric acid; n) Toluene extraction with water (neutralization); o) Toluene distillation

rated by distillation into diisopropyl ether and crude phenol. The ether is recirculated. The crude phenol is purified by distillation.

Energy and chemical requirements for production of one tonne of crude phenol by the Phenoraffin process are 5 t steam, 165 m³ cooling water, 60 kWh electrical energy, 20 kg diisopropyl ether, and 10–200 kg sulfuric acid (depending on the amount of pyridine bases).

The crude phenol extracted from coal-tar carbolic oils by sodium hydroxide solution has the following composition: 15% water, 30% phenol, 12% *o*-cresol, 18% *m*-cresol, 12% *p*-cresol, 8% dimethylphenols (xylenols), 5% trimethylphenols and higher boiling phenols. The quantitative composition of the crude phenol produced in coal-tar refineries varies widely as varying amounts of sodium phenoxide solution or crude phenol from coking plant effluents are used. These mainly contain phenol and cresols but insignificant amounts of higher phe-

nol homologues. Normally crude phenol from sodium hydroxide extraction has an ether content of ca. 2% which is decreased by water washing before distillation to ca. 0.3%.

Figure 11 shows the processing flow sheet of a continuous crude phenol distillation. The first stage is dehydration and ether removal by distillation of the crude phenol, normally under atmospheric pressure. From the dewatered, ether-free crude phenol, a residue containing higher boiling phenol homologues and unvaporizable phenolic pitch is separated in a vacuum stripping column. The stripped vapors are rectified, for instance, in four interconnected tray columns operating under a graduated vacuum of 10–140 mbar and equipped with steam-heated reboilers. The tray columns are preferably made of stainless steel. For example, in the first, 80-tray column, technically pure phenol as top product is obtained with a reflux ratio of, e.g., 12:1. From the bottom stream of this phenol column, a technically pure *o*-cresol is obtained as top product in the second, 65-tray column with a reflux ratio of, e.g., 25:1. From the bottom stream of this *o*-cresol column, a *m/p*-cresol mixture is obtained as top product in the third 45-tray column with a reflux ratio of, e.g., 5:1. The bottom product of this *m/p*-cresol column together with the bottom product of the stripping column is separated in the fourth 45-tray xylenol column into a dimethylphenol fraction and phenolic pitch. The phenolic pitch contains polymerized and oxidized phenols, which can be depolymerized to phenols by distillation with tar oils.

### 3.4.5. Processing of Light Oil

In coal-tar primary distillation, light oil is obtained azeotropically during distillative dehydration. It accounts for 0.5–3 % of the crude tar and has a wide boiling range of ca. 70–200 °C. In addition to the main components, benzene and its homologues, it contains 10–40 % unsaturated, polymerizable aromatic hydrocarbons, 3–10 % phenols and cresols, 2–7 % pyridine bases, and varying amounts of naphthalene. Coal-tar light oil, preferably together with carbolic oil I, is processed by distillation, extraction, and polymerization to obtain benzene and its homologues, pyridine bases, phenols, and indene–coumarone resins.

*Pyridine bases* are obtained from the dephenolized light oil and carbolic oil I, to which the strip oil from the sodium phenoxide solution refining can also be added, by discontinuous or continuous extraction with dilute sulfuric acid and precipitation with ammonia or sodium hydroxide solution.

Figure 12 shows the flow sheet for the recovery of pyridine bases. Extraction is carried out with 25–35 % sulfuric acid. After separation of acidic resins that may form the dissolved hydrocarbons are removed from the acidic solution of pyridine-base sulfates by steam stripping or solvent extraction with benzene. From this refined solution the crude pyridine is precipitated with gaseous or aqueous ammonia. The aqueous phase is separated and concentrated to yield ammonium sulfate.

The composition of crude pyridine bases from coal tar varies widely, depending mainly on the boiling range of the starting oil. The main components of a typical water-free tar base mixture from coal-tar light oil are listed below. The crude pyridine bases contain mainly pyridine and its methyl homologues as well as small amounts of aniline and its methyl homologues.

| | |
|---|---|
| Pyridine | 12 % |
| α-Picoline | 10 % |
| 2,6-Lutidine | 6 % |
| β-Picoline | 6 % |
| γ-Picoline | 8 % |

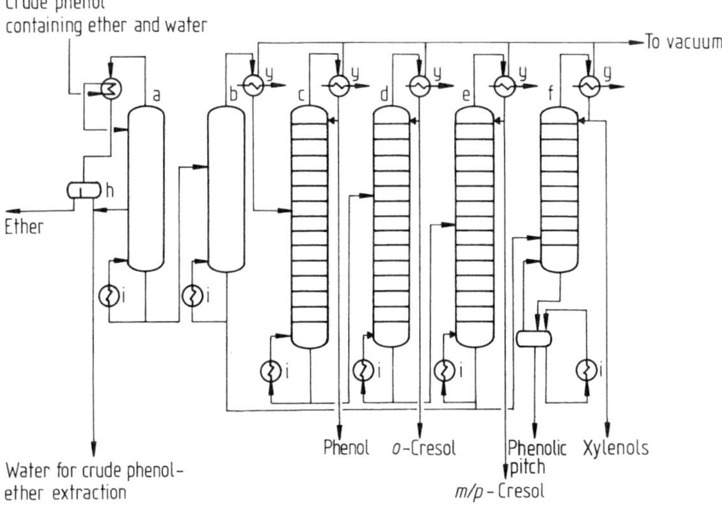


**Figure 11.** Distillation of crude phenol
a) Column for the separation of ether and water; b) Stripping column; c) Phenol column; d) *o*-Cresol column; e) *m/p*-Cresol column; f) Xylenol column; g) Condenser (low-pressure steam production); h) Separator; i) Reboiler

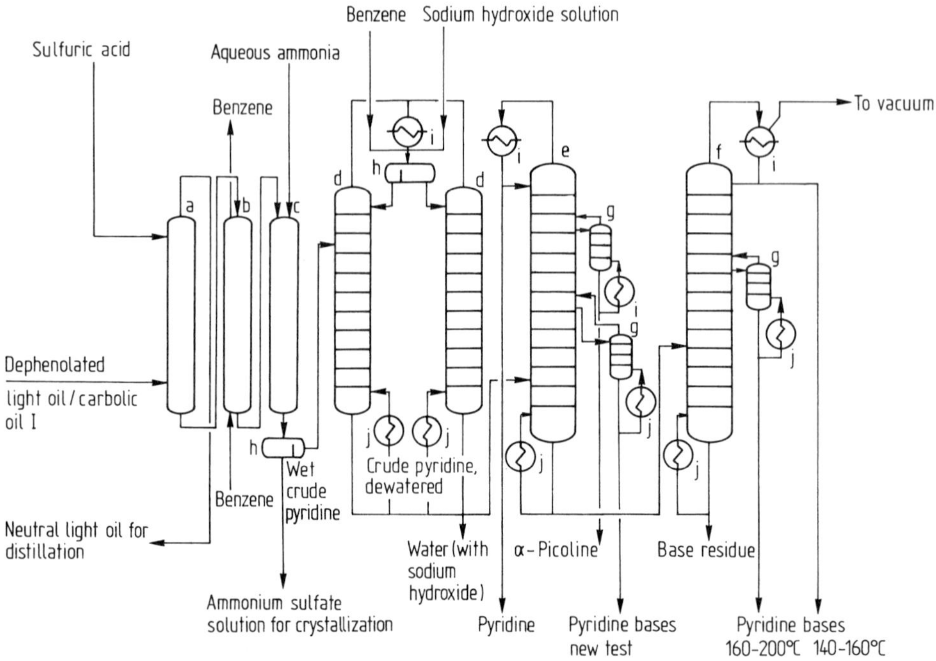


**Figure 12.** Recovery of pyridine bases
a) De-basing; b) Refining of the base sulfates solution; c) Base precipitation; d) Dewatering column system; e) Pyridine main column; f) Vacuum main column; g) Side column; h) Separator; i) Condenser; j) Reboiler

| | |
|---|---|
| 2-Ethylpyridine | 2% |
| 2,5-Lutidine | 4% |
| 2,4-Lutidine | 11% |
| 2,3-Lutidine | 2% |
| 3,5-Lutidine | < 1% |
| 2,4,6- and 2,3,6-Collidine | 9% |
| Aniline | 12% |
| o-Toluidine | 4% |
| m/p-Toluidine | 8% |
| Higher boiling bases | 6% |

The extracted crude pyridine is distilled (Fig. 12). The crude pyridine still contains water after precipitation with ammonia. Dehydration is preferably carried out by azeotropic distillation with benzene in a continuous two-column system. In the first column the water-containing crude pyridine is separated into a water-free bottom product and an azeotropic pyridine–water top product. After condensation and cooling, sodium hydroxide solution or benzene or both are added to the top product in a separator, thus obtaining a predominantly pyridine-containing and water-containing phases. The aqueous phase is pumped to the upper tray of the second column where it is separated into a pyridine-free, aqueous sodium hydroxide solution as bottom product and an azeotropic pyridine–water mix-

ture as top product. The top products of the first and second column are both condensed, cooled, and separated into the above-mentioned phases. The phase that contains the pyridine bases from the separation vessel is fed as reflux to the first column. The dehydrated crude pyridine is separated into pyridine and various homologue fractions by, for example, continuous rectification in a column at atmospheric pressure and a vacuum column with accompanying side columns. In the main column at atmospheric pressure, technically pure pyridine is taken off as top product. An α-picoline fraction and a base fraction ("pyridine bases new test"), with a boiling range of 130–150 °C, consisting of picolines and lutidines, are obtained in a side column. In the vacuum main column the bottom product is split into a pyridine base fraction, boiling range 160–200 °C, as top product and the base residue. A fraction containing collidine and aniline bases, boiling range 160–200 °C, is obtained in a side column. The reboilers are generally heated by pipe stills in heat exchange. By using additional side columns or by redistillation, further splitting of the base fractions is possible. The following main fractions are obtained:

1) Technically pure pyridine, *bp* 115.3 °C
2) Technically pure α-picoline, *bp* 129.4 °C
3) A β/γ-picoline/2,6-lutidine-fraction, boiling range 144.0–145.4 °C
4) A 2-ethylpyridine fraction, boiling range 146–157 °C
5) A lutidine fraction with 2,4- and 2,5-lutidine as main components, boiling range 157–159 °C
6) An intermediate fraction with 2,3-lutidine, 3- and 4-ethylpyridine, and 2-methyl-6-ethylpyridine as main components, boiling range 159–170 °C
7) A collidine fraction with 2,4,6- and 2,3,6-collidine as main components and small amounts of 3,5-lutidine, boiling range of 170–173 °C
8) An aniline fraction, *bp* ca. 184 °C
9) A toluidine fraction with the three toluidine isomers, boiling range 195–205 °C
10) A xylidine fraction with the six isomeric dimethylanilines, boiling range 210–225 °C

The fractions 1–7 contain pyridine and its homologues, and the fractions 8–10 contain aniline and its homologues.

Whereas pyridine and α-picoline are obtained technically pure by distillation alone, isolation of the other nitrogen bases from the distillate fractions is difficult. The majority of the pyridine bases, however, are not isolated in pure form, but are used as distillate fractions or as blends.

The commonly used technical pyridine-base blends, which are mainly employed as solvents, are classified according to boiling range. The "pyridine bases new test" are used to denaturize ethanol. The higher boiling base fractions are important anti-corrosion agents. The isomeric mixture of toluidines serves as absorbant for $SO_2$ and as a vulcanization accelerator. The pyridine homologue fractions can be demethylized catalytically in the presence of hydrogen to give pyridine and lower pyridine homologues.

After the extraction of phenols and pyridine bases, the "neutral oil" obtained from light oil and carbolic oil I is separated by distillation into different fractions. The fraction with a boiling range of 160–185 °C is of special interest as it contains up to 60 % polymerizable unsaturated hydrocarbons, with indene as main component and slight amounts of coumarone, dicyclopentadiene, and methylstyrenes. These compounds are polymerized cationically to indene–coumarone resins with Friedel–Crafts catalysts such as boron trifluoride and its complexes or aluminum chloride (→ Resins, Synthetic, **A 23**, pp. 94–98).

The nonpolymerizable benzene homologues which are distilled from the resins under vacuum after polymerization are combined with the remaining light-oil fractions boiling up to 185 °C, which are free of phenol and organic bases. They are processed as benzol pre-product together with the coke-oven crude benzol after refining with sulfuric acid or catalytical hydrogenation pressure refining to yield technically pure benzene, toluene, and benzene homologues (→ Benzene, **A 3**, pp. 480–481). After extraction of phenols and bases, the light-oil fractions that boil above 185 °C are used as solvents for paints, varnishes, and binders.

## 4. Processing of Low-Temperature Coal Tars

Only in a few locations (e.g., United Kingdom, India, China, and South Africa) are coal tars from low-temperature carbonization and gasification of coal processed by distillative and extractive refining. The aim of the recovery process is mainly the isolation of the phenols (see Table 1 for composition), rather than the isolation of aromatic hydrocarbons.

Figure 13 shows the flow chart for low-temperature tar, light condensate oil, and aqueous ammonia liquor from low-temperature carbonization of highly volatile coal to produce smokeless fuels at ca. 640 °C using the Coalite process [48]–[51]. In this process one tonne of coal yields 70–75 kg of crude coal oil and about 160 L of ammonia liquor. From the crude gas, tar and light condensate oil are obtained by two-stage condensation. Analoguous to coke-oven coal tar, they are dewatered and separated into three to five fractions by continuous primary or multi-stage distillation at atmospheric pressure and under vacuum:

1) Light oil, boiling range ca. 90–160 °C, about 2 % of crude tar, containing roughly 30 % aromatic hydrocarbons, 35 % alkenes, 20 % alkanes + naphthenes, 10–15 % phenols, and 2 % tar bases.
2) Middle oil, boiling range 200–300 °C, ca. 40 % of crude tar, containing roughly 45 % phenols, 25 % aromatic hydrocarbons, 10–15 % alkenes, and 10–15 % alkanes + naphthenes.

3) Upper distillate, boiling range 300–400 °C, ca. 30% of crude tar, containing ca. 65% aromatic hydrocarbons and alkanes. If required they can be separated by further vacuum distillation into a heavy oil and a wax oil with a significant content of solid paraffin hydrocarbons.

4) Low-temperature coal tar pitch as a distillation residue, ca. 25% of crude tar.

Alternatively, the residue from the base of the atmospheric fractionating column, after removal of light and middle oil (reduced crude) can be utilized, for example, as a heavy fuel oil or for making tar–bitumen blends for road construction.

*Low-temperature coal-tar pitch* is composed primarily of a complex mixture of hydrocarbons. It can be produced with softening points within the range of 40–180 °C (ring & ball). With a softening point of 55–60 °C it is used in production of road binders, sometimes in combination with bitumen. It is also used as an ingredient in the material prepared for coating submarine telephone cables and in the preparation of pitch–fiber pipes. The pitch that softens between 73 and 77 °C is also used for the production of pitch–fiber pipes and as a binder in briquette production. Special pitches with softening points of, for example, 82–88 °C and 100–110 °C are obtained by air blowing at elevated temperatures. These products (Florastic 85 and Florastic 100) are used for the production of colored or black pitch–mastic flooring for domestic and business premises and for the manufacture of clay pigeons. These pitch–mastic floors are resistant to oils and fats. The significantly lower content of benzo[a]pyrene in low-temperature pitch compared with pitch obtained from coke-oven coal tar is highly advantageous for these applications.

*Heavy oil* or upper distillate is processed to produce creosote oils with various boiling ranges. These are used as wood preservatives for railway sleepers, telegraph poles, electricity-transmission poles, and timber; fluxing oils for

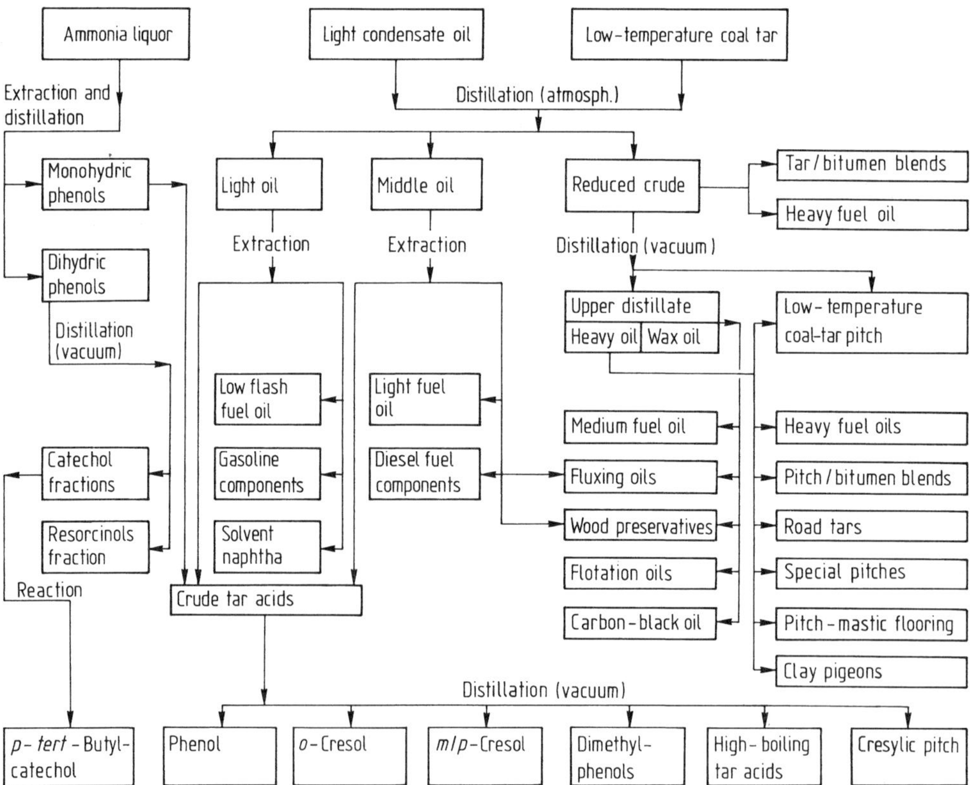


**Figure 13.** Processing of liquid products from low-temperature carbonization (Coalite process)

cutting bitumens and road tars; as carbon-black oil; and as medium industrial fuel oil.

Similar to the processing of coke-oven coal tar phenols are extracted from the *middle oil* continuously with sodium hydroxide solution to form a solution of water-soluble sodium salts known as cresylate. A neutral oil fraction then separates. Neutral oil entrained in the cresylate is removed by extraction with a lighter oil, followed by steam treatment to remove the last traces of oil. The phenolic cresylate is converted to a crude mixture of free phenols (crude tar acids) by acidification with flue gases containing carbon dioxide. Further refining of the neutral middle oil affords low-viscosity wood preservatives, fluxing oils for bitumens and tar blends, as well as components for diesel fuels and light industrial fuel oils.

Crude tar acids are also separated from *light oil* by extraction with sodium hydroxide solution. Distillation and refining of the neutral oil gives low-flash fuel oils, gasoline components, and solvent naphtha. The oil is also used as a light industrial fuel oil.

The aqueous *ammonia liquor* contains ca. 1.5% dissolved phenols, about two-thirds of which are monohydric and one-third dihydric. These phenols are extracted from the aqueous phase with, for example, isobutyl acetate or diisopropyl ether (Lurgi Phenosolvan process, → Gas Production, **A 12**, pp. 277–278). After distillation of the extractant, the crude phenol mixture is separated by continuous distillation into a lower boiling monohydric phenols fraction and a dihydric phenols fraction. The dihydric phenols fraction is separated by further distillation into catechol and resorcinols fractions. Further processing leads to catechol, 4-methylcatechol, a resorcinols fraction, and 2-methylresorcinol, and as a further downstream product, 4-*tert*-butylcatechol.

The monohydric phenol fraction isolated from ammonia liquor and the crude phenol obtained by extraction of light oil and middle oil is dehydrated by distillation and then separated by continuous distillation into the following fractions:

1) Technically pure phenol (8–10% of the crude phenol)
2) Technically pure crude *o*-cresol (ca. 2% of the crude phenol)
3) An *m/p*-cresol fraction containing 50% *m*-cresol (ca. 15–29% of the crude phenol)

4) Various dimethylphenol fractions, for example a fraction with 90% 2,4- and 2,5-dimethylphenol and one with 50% *m/p*-ethylphenol and ca. 50% 3,5-dimethylphenol. These fractions make up ca. 30–35% of the crude phenol.
5) Refined high-boiling tar acids (boiling range ca. 230–305 °C), approx. 20% of the crude phenol.
6) Cresylic pitch (R & B softening point ca. 115 °C), approx. 18% of the crude phenol.

The *high boiling tar acids* contain as main components $C_3$, $C_4$, and $C_5$ alkylphenols as well as 4- and 5-indanols. They are converted to black and white heavy duty disinfectants, soil sterilizers, horticultural fluids, sheep dips, foundry molding resins, agents for ore and coal flotation, engine degreasers and metal cleaners, specialized dye-markers, resins for food and beverage can linings, and germicides for impregnating toilet tissues.

The cresylic pitch is used as pulverized fuel and as a binding agent.

Because of the corrosive properties of dihydric phenols contained in low-temperature tar crude phenol, the distillation plants must be manufactured of high-quality stainless steel.

In the United Kingdom phenols and phenolic fractions from low-temperature tar crude phenols are extracted by the Coalite process. These phenols and phenolic fractions are then processed by chlorination and phosphorylation as well as by reaction with formaldehyde to give phenolic resins.

As an alternative to the chemical processing, crude tar from low-temperature carbonization of coal can be processed directly by catalytic hydrogenation to produce gasoline and diesel fuels [52].

Low-temperature coal tars obtained as byproducts of coal pressure gasification (e.g., by the Lurgi fixed-bed process) are processed similarly to the coal oil from low-temperature carbonization of coal. Attention has to be paid to the increased concentration of aromatic hydrocarbons which is due to higher generation temperatures (see Table 1). In contrast to the crude phenol resulting from low-temperature carbonization of coal, the crude phenol, from water effluents and tar oils, contains almost no dihydric phenols but increased amounts of phenol, *o*-cresol, and *m/p*-cresol.

Since the synthesis gas formed by coal gasification is the desired product, the crude tar formed is preferably recycled to the gasifier, thus giving an increased gas yield.


## 5. Processing of Other Tars and Tarlike Raw Materials

*Lignite tars* and crude phenols are byproducts of lignite carbonization and gasification. Large-scale plants for lignite coking and chemical processing of lignite tars in Eastern Germany [53], were shut down for economic reasons after German reunification in 1991/1992. Lignite tars are obtained in American, Czech, Indian, and German large-scale lignite gasification plants, especially in fixed-bed gasification by the Lurgi process. These tars are either recirculated to the gasifier to increase gas production (similar to coal-gasification tars) or are utilized as heavy fuel oil for steam generation. Generally only the crude phenol extracted from wastewater is chemically processed, analogous to the refining of coal-tar crude phenol to phenol, *o*-cresol and *m*/*p*-cresol, xylenols, and higher boiling phenols (see Section 3.4.4 and Chap. 4). The Dakota Gasification Co. in Bismarck/North Dakota runs such a plant with a phenol capacity of ca. 15 000 t/a.

*Peat tars* are obtained as byproducts with a yield up to 10 % during carbonization of peat to peat coke, which is used as low-ash metallurgical carbon and for production of activated carbons (→ Peat, **A 19**, pp. 25 – 27, 29 – 34). Peat tars are used only as low-sulfur industrial fuel oils. Technically the distillation of peat tar to obtain oil is possible. This distillation product can be extracted to yield crude phenol, containing as its main components phenol, the three isomeric cresols, guaiacol (2-methoxyphenol), 2,4-, 2,3-, and 3,5-dimethylphenol, *p*-ethylphenol, and higher phenol homologues, as well as small amounts of catechol and its homologues. Peat-tar oils can be used to formulate wood preservatives.

*Wood tars*, methanol, and acetic acid are the liquid byproducts of wood carbonization to give activated carbons and wood charcoal for iron smelting (→ Charcoal). In these plants wood tar normally is used directly as a low-sulfur fuel oil. Wood tar can be separated into wood-tar pitch and various tar-oil fractions. The tar oils are suited to the formulation of wood preservatives and

flotation oils. Extraction with sodium hydroxide solution gives wood-tar creosote, which contains phenol, creosols, guaiacol, and other phenol ethers as main components. Due to its disinfectant properties purified beechwood creosote, boiling range 200 – 220 °C, with cresols and guaiacol as main components, serves as a formulating agent for pharmaceutical preparations (e.g., cough mixtures). Furthermore, it is used as auxiliary solution in mercury maximum/minimum thermometers. Wood-tar pitch can be used as putty for the polishing of optical glasses. Crude wood tars are suitable for jute impregnation, serve as binders for briquettes and active carbon, as game deterrents, and as disinfectant formulations in veterinary medicine.

*Oil-shale tars*, also called shale oils, are the main liquid products of oil-shale retorting (→ Oil shale, **A 18**, pp. 110 – 125). Depending on composition (see Table 1), oil-shale tars can be refined by pressure hydrogenation and subsequent distillation to fuel oils, gasoline and diesel fuels, or can be carbonized [54]. Estonian oil-shale tars, rich in crude phenol (cucersit oil, obtained by the Kivitter process), are separated by distillation into various oil fractions, boiling up to 360 °C. They are used for the production of wood preservatives, road binders, and plasticizers [55]. The pitch residue can be carbonized to electrode coke. Phenols are extracted from water effluents and lower boiling distillate fractions and processed to phenolic resins, injection agents for soil compaction, tanning agents, and alkyl resorcins.

*Pyrolysis residual oils* are tarlike byproducts of steam cracking of naphtha and gas oil to produce ethylene/propene (→ Ethylene, **A 10**, pp. 47 – 61; Resins, Synthetic, **A 23**, p. 91). Pyrolysis residual oils consist almost entirely of aromatic hydrocarbons (see Table 1). Due to the relatively high content of unsaturated, polymerizable aromatics they are less stable than coal tars. The main components are naphthalene and its homologues, indene and its homologues, and thermally polymerized pyrolysis residual resin (pyrolysis pitch), obtained as a dark distillation residue. By using modified coal-tar refining processes (distillation, thermal and catalytic polymerization, crystallization) the basic chemicals naphthalene, methylnaphthalenes, aromatic hydrocarbon resins and aromatic oils are obtained from pyrolysis residual oils on a large scale [56], [57]. Pyrolysis residual oils are also used as raw materials for carbon black.

Pyrolysis residual oils from scrap plastics, tyres, etc. are obtained, similarly to petrochemical pyrolysis residual oils from ethylene/propene plants, as tarlike materials. They can be processed to chemical raw materials. If the pyrolysis is carried out at ca. 700 °C in fluidized-bed or rotary-drum reactors, up to 50 % pyrolysis oils, rich in aromatics, are obtained ($\rightarrow$ Plastics, Recycling, **A 21**, pp. 64–68). These pyrolysis oils have a composition similar to pyrolysis residual oils from ethylene/propene plants (see Table 1). As main components they contain benzene, naphthalene and its homologues, styrene, indene and its homologues, and dark pyrolysis residual resins (pyrolysis pitch), which is obtained as distillation residue [58]–[60]. If modified processes of coal-tar refining are applied, pyrolysis oils yield benzene homologues, naphthalene and methylnaphthalenes as well as technical aromatic oils.

# 6. Uses of Tar Products and their Economic Importance

About $30 \times 10^6$ t/a of all kind of tars are recovered, of which ca. $15 \times 10^6$ t/a are processed to basic chemicals and other products. The main raw materials are high-temperature coal tars from coking plants. Low-temperature coal tars from coal carbonization and gasification; lignite tars, oil-shale tars, and pyrolysis residual oils are of lesser importance. Peat tars, wood tars, and pyrolysis oils from scrap are not currently processed on a large scale.

Tars are the basic raw material for:

1) Industrial carbon products: carbon electrodes for aluminum production, graphite for production of electrodes and high-temperature materials and carbon black as a rubber filler and black pigment (ca. $4 \times 10^6$ t/a)
2) Technically pure condensed aromatic hydrocarbons such as naphthalene and its homologues, anthracene, phenanthrene, acenaphthene and pyrene for synthesis of polymers, plasticizers, and dyes (ca. $1 \times 10^6$ t/a)
3) Technically pure aromatic heterocyclics such as carbazole, quinoline, pyridine, and indole for syntheses of dyes, pesticides, and pharmaceuticals (several thousand tonnes per annum)
4) Phenols for synthesis of thermosets, pesticides, disinfectants, and antioxidants (several hundred thousand tonnes per annum)
5) Thermoplastic indene–coumarone resins for the formulation of paints, varnishes, adhesives, and rubber products (ca. $100 \times 10^3$ t/a)
6) Aromatic oils as wood preservatives, solvents, and gas-washing oils ($> 2 \times 10^6$ t/a)

Whereas aliphatic hydrocarbons and mononuclear aromatics presently are mainly produced from petrochemicals, polynuclear aromatics from coal tar cover a high percentage of the total requirements. In the case of naphthalene the fraction arising from coke-oven coal tar is ca. 85 % worldwide. Naphthalene homologues, the polynuclear aromatics anthracene, phenanthrene, acenaphthene and pyrene, and the heterocyclics carbazole and quinoline are obtained exclusively from coal tar. The portion of phenols derived from tar is only 3 % worldwide, whereas in case of phenol homologues it is ca. 40 %. Approximately 20 % of world aromatic hydrocarbon resin production is accounted for by indene–coumarone resins based on coal tar. In the case of binders for carbon and graphite electrodes, coal-tar electrode pitches account for over 90 %. Carbon-black oils from coal tar are a basic raw material for more than a quarter of world carbon-black production ($> 2 \times 10^6$ t/a), and 75 % of oil-based wood preservatives for pressure-impregnation of timber ($> 10^6$ t/a) are based on coal tar.

As discussed above, tars and pitches of different types are used also as industrial fuel oils, motor fuels, binders and coating materials as well as for co-processing together with coal for gasification and coking.

# 7. Toxicology and Ecotoxicology

## 7.1. Toxicology

The acute toxicity of tars and pitches is low, but high doses of polycyclic aromatic hydrocarbons and structurally related heterocyclic systems are known to result in necrosis of the adrenal gland.

Tars or their fractions may cause irritation of the skin and eyes by direct or indirect exposure. The extent of skin irritation varies, depending on the origin of the tar, the exposure time, and involvement of light. High-temperature coal tar has the strongest, and wood tar the weakest impact on skin. Contact with unprotected human skin may lead to dermatitis, inflammation of the sebaceous glands (folliculitis), and acne. Several years of exposure can cause brownish, speckled pigmentation (melanosis) and tar or pitch warts [61, pp. 130–132].

Certain tar constituents are activated by UV light, resulting in intensified irritant skin reactions (phototoxicity) or allergic contact dermatitis (photosensitization); this may occur also on contact with minor quantities of volatiles from tar or tar fractions.

Pharmaceutical formulations of coal tars, especially in combination with exposure to UV light, have been successful in the treatment of psoriasis [62], though after excessive usage of tar ointments, melanomas or squamous-cell carcinomas developed in isolated cases [61, p. 136]. Furthermore, such combined treatment caused marked inhibition of DNA synthesis in vivo in normal and proliferative skin of hairless mice [61, p. 128].

In general, tar constituents seem to be readily available by all exposure routes and distributed throughout the body, but are also rapidly excreted [63]–[66]. On application of crude coal tar (2% solution in petrolatum) to the skin of humans for 8-h periods on two consecutive days evidence was obtained for absorption of polycyclic aromatic hydrocarbons (PAH) [65].

The potency of tumor formation can be attributed to this fraction of the mixture; PAHs are byproducts of pyrolysis of all organic materials and are generated to a major extent above 500 °C. Thus high-temperature coal tar has the highest content of highly condensed aromatic ring systems.

The experimental generation of skin carcinomas by the topical application of tars, tar fractions such as creosote, anthracene oil, and pyrolytic oil [67] of naphtha, and tar pitch is well documented [61], [68]. Case reports and epidemiological studies have provided sufficient experience about the occurrence of skin, scrotum, bladder, and other systemic cancers in exposed workers [61, pp. 121–127], [68, pp. 140–155].

Evidence has accumulated to confirm that occupational inhalation of tar or tar pitch volatiles leads to increased lung and bladder cancer mortality that is strongly associated with duration and intensity of exposure [61, pp. 134–136], [69, pp. 174–176]. There is suggestive ex-

perimental and epidemiological support for PAH causing cancer by the oral route [61], [69].

Inhalation studies with rats exposed to a condensation aerosol from tar pitch alone and in combination with carbon black [70], [71] corroborate the idea that the carcinogenic impact of PAH on the lungs is enhanced by adsorption onto respirable fine particles: After 43 weeks of exposure, the lung tumor rate was 4% with 1.1 mg/m$^3$ aerosol alone ($= 20$ µg/m$^3$ benzo[$a$]pyrene), which was less than in the control with carbon black alone (8–18%). With 2.6 mg/m$^3$ of aerosol ($= 50$ µg/m$^3$ benzo[$a$]pyrene), the tumor rate amounted to 33%, but this concentration in combination with 2 or 6 mg/m$^3$ of carbon black led to 89 and 72%, respectively. The majority of lung tumors were squamous-cell carcinomas. This potentiating effect can be explained by a PAH depot from which carcinogenic material is sustainably released, thus resulting in long-term exposure to the target cells of the lung.

Some 50 coal-tar compounds have been studied for their cancerogenic effect under experimental conditions, mainly on the skin of rats and mice. About 30 proved to be noncarcinogenic, and ca. 20 were positive to varying extent; for status reports see [72], [73]. The strongest carcinogens are found among systems with five or more condensed benzene rings. They are not tumorigenic by themselves, but must be metabolically converted into highly reactive alkylating dihydrodiol epoxides, the most potent of which

Bay-region dihydrodiol
epoxide of benzo[$a$]pyrene

have a so-called bay region. Benzo[$a$]pyrene, the model compound, is one of the strongest carcinogens among the PAHs, and the carcinogenic potency of tar fractions is frequently related to its concentration.

Examples of carcinogenic high-molecular $N$- and $S$-heterocycles are benzologues of acridine, carbazole, and dibenzo[$bd$]thiophene. Apart from PAHs, 2-naphthylamine – found in tar only in very low concentrations – is well known to specifically cause bladder cancer. This was identified primarily among workers manufacturing aniline dyes where appreciable exposure to aromatic amines occurred under former working conditions [74].

## 7.2. Classification and Legislation

According to IARC, there is sufficient evidence for carcinogenicity of coal tars and coal tar pitches to humans (class 1) [68, pp. 174–176]. In the United States and Germany, the classification is A1 (human carcinogen). In Germany, the occupational limit value for benzo[*a*]pyrene in volatiles from pitch and coke processing is set at $5 \, \mu g/m^3$, from other sources at $2 \, \mu g/m^3$ [75]. In the United States the amount of coal-tar/pitch volatiles is determined as benzene-soluble matter recovered from sampled dust. The TLV or MAC (maximal accepted concentration) is $0.2 \, mg/m^3$ [76]. In the European Union, complex coal-derived materials for industrial purposes must be labelled as carcinogens unless a certain limit concentration of benzo[*a*]pyrene (currently $50 \, mg/kg$) as indicator component for PAHs is met [77]. A comparative carcinogenicity study with two high-temperature coal tar oils (benzo[*a*]pyrene $< 30$ and $< 300 \, mg/kg$) is in progress [78].

## 7.3. Ecotoxicology

The aquatic toxicity of tar and tar-borne materials depends primarily on the physical and toxicological properties and availability of their aromatic constituents from the matrix of the mixture. However, the outcome of test procedures is very much determined by the method of preparation of test solutions and the reference basis for the computation of the effective or toxic concentrations. Therefore, the evaluation for aquatic ecotoxicological properties of mixtures which consist of sparingly soluble substances is problematic.

In particular, organic solvents or detergent-like agents used as cosolubilizers, dispersants, or emulsifiers tend to bias the measurement (see below) [79]. The choice and standardization of adequate testing methods continue to be subject of controversy and debate [80].

The highest toxicity ($LC_{50}$ values mostly $< 1 \, mg/L$) [81], [79] was obtained with acetone solutions of high-temperature coal tar creosote.

In a reliable comparative study on fish and daphnia [79], the aqueous creosote extracts yielded 96 h and 48 h $LC_{50}$ values between 2 and 3.5 mg/L (referring to nominal concentration of creosote); on the other hand, about 0.7 mg/L was found when the organic solvent was used. This can be explained by the finding that the composition of the acetone solution was almost identical to that of the original material, while in the aqueous extract – reflecting more realistic conditions – less toxic components with higher water solubility were enriched.

Aqueous coal tar extracts were somewhat less toxic to fish and daphnia, with $LD_{50}$ values ranging from about 7 to 12 mg/L (nominal). This may reflect its lower content of water soluble fractions and lower availability of the toxic constituents compared with creosote. This seems to be in agreement with findings on four tar oils in the bacteria bioluminescence assay with effective 50 % doses from ca. 0.8 to 1.5 mg/L, since these data are comparable to the range of creosote toxicity mentioned above [82].

For tar pitch, the lowest $LC_{50}$ (96 h) for fish is reported to be ca. 100 mg/L and for daphnia $> 100 \, mg/L$ (nominal) [79].

## 8. References


General References

[1] H. G. Franck, G. Collin: *Steinkohlenteer*, Springer Verlag, Berlin 1968.
[2] G. Collin, E. Wolfrum: "Chemierohstoffe aus flüssigen Nebenprodukten der Verkokung, Verschwelung und Vergasung," in Winnacker-Küchler: *Chemische Technologie, Organische Technologie I*, 4th ed., Hanser Verlag, München 1981.
[3] G. Collin, *Erdöl, Kohle Erdgas Petrochem.* **29** (1976) 159–165; **38** (1985) 489–496.
[4] G. Collin, Proc. World Conf. Future Sources Org. Raw Mater. Chemrawn I, Toronto, July 10–13, 1978, Pergamon Press, London 1979, pp. 283–297.

Specific References

[5] M. Zander, G. Collin, *Fuel* **72** (1993) 1281–1285.
[6] H. G. Franck, *Brennst. Chem.* **36** (1955) 12–20.
[7] G. P. Blümer, R. Thoms, M. Zander, *Erdöl, Kohle Erdgas Petrochem.* **31** (1978) 197.
[8] W. Boenigk, M. W. Haenel, M. Zander, *Fuel* **69** (1990) 1226–1232.
[9] H. D. Sauerland, J. W. Stadelhofer, R. Thoms, M. Zander, *Erdöl, Kohle Erdgas Petrochem.* **30** (1977) 215–218.
[10] P. Fischer, J. Stadelhofer, M. Zander, *Fuel* **57** (1978) 345–352.
[11] M. Zander, *Erdöl, Erdgas, Kohle* **105** (1989) 373–377.

[12] M. Zander, *ACS Fuel Chem. Div. Prep.* **34** (1989) 1218–1225.

[13] J. D. Brooks, G. H. Taylor, *Nature (London)* **206** (1965) 697–699.

[14] M. Zander, *Erdöl, Kohle Erdgas Petrochem.* **38** (1985) 496–503.

[15] S. Takeshita, *CEER Chem. Econ. Eng. Rev.* **9** (1977) nos. 2/3, 25–30.

[16] Rütgerswerke, DE 976913, 1954 (H. Krieger, E. Schweym, J. Geller).

[17] A. M. Odok, W. K. Fischer, *Light Met. (Warrendale, Pa.)* **1978**, 269–286.

[18] A. Alscher, W. Gemmeke, F. Alsmeier, W. Boenig, *Light Met. (Warrendale, Pa.)* **1987**, 483–489.

[19] F. Keller, W. K. Fischer, *Light Met. (Warrendale, Pa.)* **1982**, 729–740.

[20] W. K. Fischer, F. Keller, R. Perruchoud, *Light Met. (Warrendale, Pa.)* **1991**, 681–686.

[21] W. Boenig, A. Niehoff, R. Wildförster, *Light Met. (Warrendale, Pa.)* **1991**, 615–619.
N. I. Steward, P. H. Halley, *Light Met.* (Warrendale, Pa.) **1994**, 517–524.

[22] Osaka Gas Co., DE 2121458, 1971 (K. Ueda, J. Kimoto, M. Moritake).

[23] G. Collin, H. Köhler, *Erdöl, Kohle Erdgas Petrochem.* **30** (1977) 257–263.

[24] C. D. Alexander, V. L. Bullough, J. W. Pendley, *Light Met. (Warrendale, Pa.)* **1971**, 365–376.

[25] G. L. Ball, C. R. Gannon, J. W. Newman, *Light Met. (Warrendale, Pa.)* **1972**, 127–140.

[26] W. Boenig, A. Niehoff, R. Wildförster, *Light Met. (Warrendale, Pa.)* **1992**, 581–584.

[27] W. Boenig, J. W. Stadelhofer, Proc. Carbon '92, Essen, June 22–26, 1992, pp. 30–32.

[28] Gesellschaft für Teerverwertung, DE 642437, 1935 (F. Meyer, O. Preiss).

[29] Koppers, DE 705576, 1939 (H. Koppers).

[30] A. J. Gambro, D. T. Shedd, H. W. Wang, T. Yoshida, *Chem. Eng. Prog.* **65** (1969) no. 5, 75–79.

[31] A. Eckert, R. Marrett, J. W. Stadelhofer, *Erdöl, Kohle Erdgas Petrochem.* **38** (1985) 510–514.

[32] I. Moshida, Y. Q. Fei, Y. Korai, K. Fujimoto, R. Yamashita, *Carbon* **27** (1989) 375–380.

[33] Rütgerswerke, DE 3609348, 1986 (M. Morgenstern, C. Bertrand).

[34] G. Collin, M. Zander, Prep. 19th Biennial Conf. Carbon, University Park, PA, June 25–30, 1989, pp. 466–467.

[35] D. Stoye (ed.): *Paints, Coatings and Solvents*, VCH Verlagsgesellschaft, Weinheim 1993, pp. 91–94.

[36] Rütgerswerke, DE 2623574, 1976 (H. Louis, H. Köhler, R. Oberkobusch, G. Collin, V. Potschka).

[37] G. Herion, G. von Mossen, *Strasse Autobahn* **37** (1986) 3–6.

[38] H. Spengler, W. Weskamp, G. Collin, *Cokemaking Int.* **3** (1991) 25–29.

[39] B. Bujnowska, G. Collin, *Cokemaking Int.* **6** (1994) 25–31.

[40] G. Collin, B. Bujnowska, *Carbon* **32** (1994) 547–552; Prep. Carbon '94, Granada, July 5–8, 1994, pp. 80–81. J. Polaczek, G. Collin et al., DGMK-Tagungsber. 9401, Velen, April 18–20, 1994, pp. 241–245.

[41] H. Honda, *Carbon* **26** (1988) 139–156.

[42] I. C. Lewis, Prep. 20th Biennial Conf. Carbon, June 23–28, 1991, Santa Barbara, CA, pp. 156–157.

[43] I. Mochida et al., *Carbon* **28** (1990) 311–319.

[44] Y. Korai et al., *Carbon* **29** (1991) 561–567.

[45] K. J. Hüttinger, *Chemiker-Ztg.* **112** (1988) 355–366.

[46] Rütgerswerke, DE 2919298, 1979 (H. Spengler, D. Pachaly).

[47] Rütgerswerke, DE 3813634, 1988 (J. Talbiersky, K. Stolzenberg, B. Wefringhaus).

[48] W. A. Bristow, *J. Inst. Fuel* **20** (1947) 109–130.

[49] G. S. Pound, *Coke Gas* **21** (1959) 395–401, 446–453, 521–523.

[50] K. R. Payne: Yearbook Coke Oven Managers' Association 1977, pp. 154–164.

[51] A. D. Gregory: private communication.

[52] S. A. Quader, G. R. Hill, *Chem. Ing. Tech.* **43** (1971) 595–596.

[53] *Ullmann*, 4th ed., **22**, 439–441.

[54] R. Rammler, H.-J. Weiss, A. Bussmann, T. Simo, *Chem. Ing. Tech.* **53** (1981) 96–104.

[55] R. E. Joonas, W. M. Jefimow, T. A. Purre, *Chem. Ind.* **34** (1982) 739–740.

[56] Rütgerswerke, DE 1815568, 1968 (O. Wegener, R. Oberkobusch, G. Collin, M. Zander, H. Buffleb).

[57] J. Mostecky, M. Popl, M. Kuras, *Erdöl, Kohle Erdgas Petrochem.* **22** (1969) 388–392.

[58] H. J. Sinn, *Chem. Ing. Tech.* **46** (1974) 579–589.

[59] H. J. Sinn, W. Kaminsky, J. Janning, *Angew. Chem.* **88** (1976) 737–749.

[60] G. Collin, G. Grigoleit, G. P. Bracker, *Chem. Ing. Tech.* **50** (1978) 836–841.

[61] *IARC Monographs on the Evaluation of Carcinogenic Risks to Humans* **35** (1985).

[62] J. S. Comaish, *J. Invest. Dermatol.* **88** (Suppl.) (1987) 61s–64s.

[63] R. Modica et al., *J. Chromatogr.* **24** (1982) 352–355.

[64] C. L. Sanders et al., *J. Environ. Pathol. Toxicol. Oncol.* **7** (1986) 25–34.

[65] J. S. Storer et al., *Arch. Dermatol.* **120** (1984) 874–877.

[66] P. M. Daniel et al., *Nature (London)* **215** (1967) 1142–1146.

[67] C. S. Weil, N. I. Condra, *Am. Ind. Hyg. Assoc. J.* **38** (1977) 730.

[68] *IARC Monographs on the Evaluation of Carcinogenic Risks to Humans* **34** (1984).

[69] *IARC Monographs on the Evaluation of Carcinogenic Risks to Humans*, updating vols. 1–42, Suppl. 7 (1987).

[70] U. Heinrich et al. in U. Mohr (ed.): *Toxic and Carcinogenic Effects of Solid Particles in the Respiratory Tract*, 4th Int. Inhal. Symposium, Hannover, March 1–5, 1993, pp. 57–73.

[71] U. Heinrich in U. Mohr (ed.): *Assessment of Inhalation Hazards*, Springer Verlag, Berlin 1989, pp. 301–313.

[72] *IARC Monographs on the Evaluation of Carcinogenic Risks to Humans* **32** (1983) 95–452.

[73] J. Santodonatro, P. Howard, D. Basu, "Health and Ecological Assessment of Polynuclear Aromatic Hydrocarbons," *J. Environ. Pathol. Toxicol.* **54** (1981) 1–364.

[74] S. Tola, *J. Toxicol. Environ. Health* **6** (1980) 1253–1260.

[75] List of MAK and BAT Values, VCH Verlagsgesellschaft, Weinheim 1993.

[76] Am. Conf. of Government of Industrial Hygienists (ACGIH): *Threshold Limit Values*, 1990/1991.

[77] EEC: Draft Annex Entries for Complex Coal-Derived Substances XI/51793, July 5, 1993 as Supplement of Chemical Substance Directive 67/548/EEC, Annex 1.

[78] Inter. Tar. Association, 1993 (in preparation).
[79] H. Tadokoro et al., *Ecotoxicol. Environ. Saf.* **21** (1991) 57–67.
[80] A. E. Girling et al., *Chemosphere* **24** (1992) 1469–1472.
[81] G. Sundström: "Creosote," in O. Hutzinger (ed.): *The Handbook of Environmental Chemistry*, vol. 3D, Springer Verlag, Berlin 1986, p. 158.
[82] K. G. Steinhäußer et al., *Vom Wasser* **72** (1989) 93.

# Tar Sands

FRANK J. MINK, RICHARD N. HOULIHAN, Energy Resources Conservation Board, Calgary, Alberta, Canada



## 1. Introduction

The existence of the tar sands has been known for centuries. The term "tar sands" is often used to describe the deposits of unconsolidated bituminous sands found in Canada. In that the "bitumen" is a mixture of hydrocarbons and asphaltic or tarry material "oil sands" is a more appropriate term, and is used throughout this article. The sands are exposed along the banks of the Athabasca and Clearwater Rivers in the northern part of the province of Alberta. A number of other large deposits are found in deeper formations of the province (Fig. 1). Although the oil sands were described by fur traders and explorers, it was not until 1819 that geological description was attempted and it took until ca. 1887 before the nature and extent of the resource was assessed by the Geological Survey of Canada. While the potential for bitumen production and conversion to synthetic crude oil was recognized, considerable technological development was required before commercial projects could get underway.

Developments on the recovery of bitumen started in the 1920s, and expanded rapidly during the early 1940s. It declined in the late 1940s when large amounts of conventional oil were discovered in the province. Commercial surface mining began in the late 1960s, and has expanded since then. In situ recovery testing occurred simultaneously and commercial developments got underway in the mid-1980s (see Chap. 3).

The importance of research and development to oil sands exploitation is analogous to the importance of exploration in conventional oil de-

Ullmann's Encyclopedia
of Industrial Chemistry, Vol. A 26

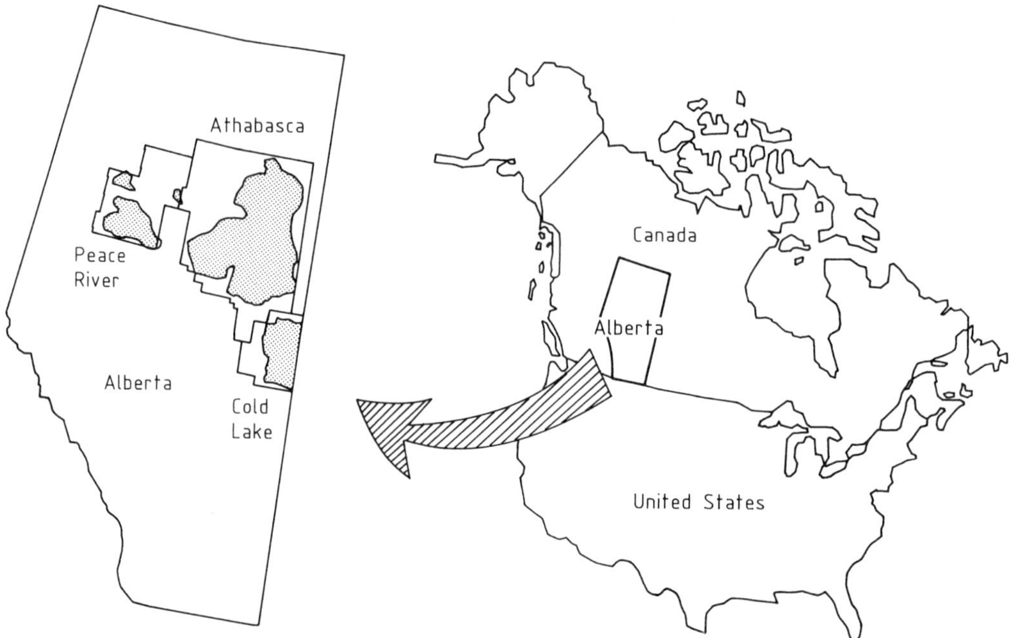

**Figure 1.** Oil sands deposits in western Canada


velopment. New technologies have the potential to change the oil sands industry from a relatively high-cost source of oil to one that is more competitive. The province of Alberta participates directly in that research through the Alberta Oil Sands Technology and Research Authority (AOSTRA). To 1993, AOSTRA and its industrial partners have provided over $ 10^9$ in research and development investment. Most of this is for field pilot testing of in situ production, and mining, extraction, and upgrading technology. These projects are normally carried out on an equal funding basis between AOSTRA and a company or group of companies.

In addition, industry also invests in non-AOSTRA-sponsored research and development. The Alberta Research Council is active in oil sands research, as is the Federal Government of Canada, through the Canada Centre for Mineral and Energy Technology. In total, $ 10^5$ was spent on oil sands research and development in Canada in 1991.

## 1.1. Oil Sands and Crude Bitumen Defined

Oil sands are the combined hydrocarbon and clastic sediment host rocks containing crude bi-

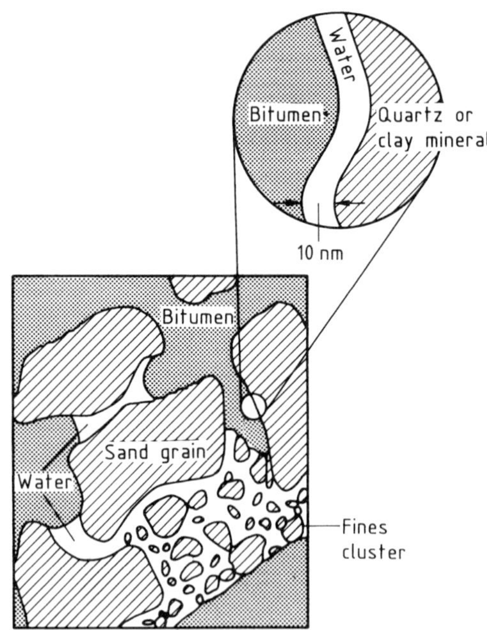


**Figure 2.** Structural model of the Athabasca oil sands [1]

tumen. Typically, they contain 84% solids and 16 wt% crude bitumen and water (Fig. 2). The solids are 90% quartz, with minor amounts of

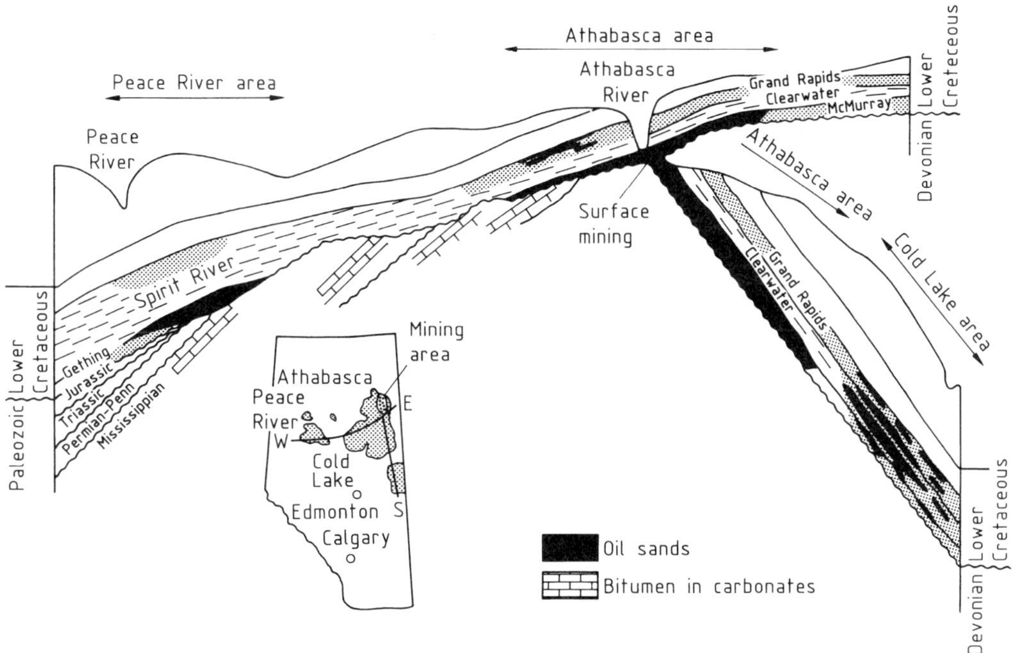



**Figure 3.** Stratigraphic relationships among the oil sands areas [8]

feldspar, muscovite, chert, and clay minerals. Clays are mainly kaolinite, illite, and montmorillonite. The large solid particles are surrounded by a thin film of water containing fine solids; the bitumen occupies the interstitial space.

The Alberta Energy Resources Conservation Board (ERCB) defines crude bitumen as: "A naturally occurring viscous mixture, mainly of hydrocarbons heavier than pentane, which may contain sulphur compounds, and that in its naturally occurring viscous state, will not flow to a well."

Crude bitumen is a complex mixture of organic compounds, containing polymeric asphaltenes and resins of high molecular mass and high levels of nitrogen, oxygen, sulfur, and metals. Typically, it contains sulfur 4.6–5.6 wt %, vanadium 160–300 mg/kg, nickel 60–100 mg/kg [2], and possibly small amounts of dissolved methane and traces of hydrogen sulfide. Crude bitumen is also notable for a deficiency of hydrogen in relation to carbon.

A distinguishing characteristic of crude bitumen is its viscosity. Alberta's crude bitumens are highly viscous, to the point of pseudoplasticity, with natural viscosity $10^4$–$10^6$ MPa · s, and density 960–1030 kg/m$^3$.

## 1.2. Geological Setting

The crude bitumen deposits occur within Lower Cretaceous sands and Upper Devonian and Mississippian carbonates (Figs. 3 and 4) of the western Canadian sedimentary basin, a wedge of sedimentary strata thickening westward from a zero edge along the Canadian Precambrian Shield to the Cordilleran foreland thrust belt [3], [4]. The deposits underlie an area of 86 000 km$^2$, about equal to that of Austria. The ERCB groups the oil sands deposits geographically into areas (Athabasca, Peace River, and Cold Lake) and separates them geologically by formation into deposits [5]; their main geological attributes are presented in Table 1 [6].

The hydrocarbons of the deposits are thought to have originated in the deeper part of the basin to the west, and to have migrated eastward updip, until a reversal in dip was reached in the north-eastern part of the province [7]–[9]. Overlying shales of the Mannville group form the cap-rock for the hydrocarbons. The hydrocarbons were degraded to their present state through the action of microorganisms, water washing, and possibly inorganic oxidation [10].

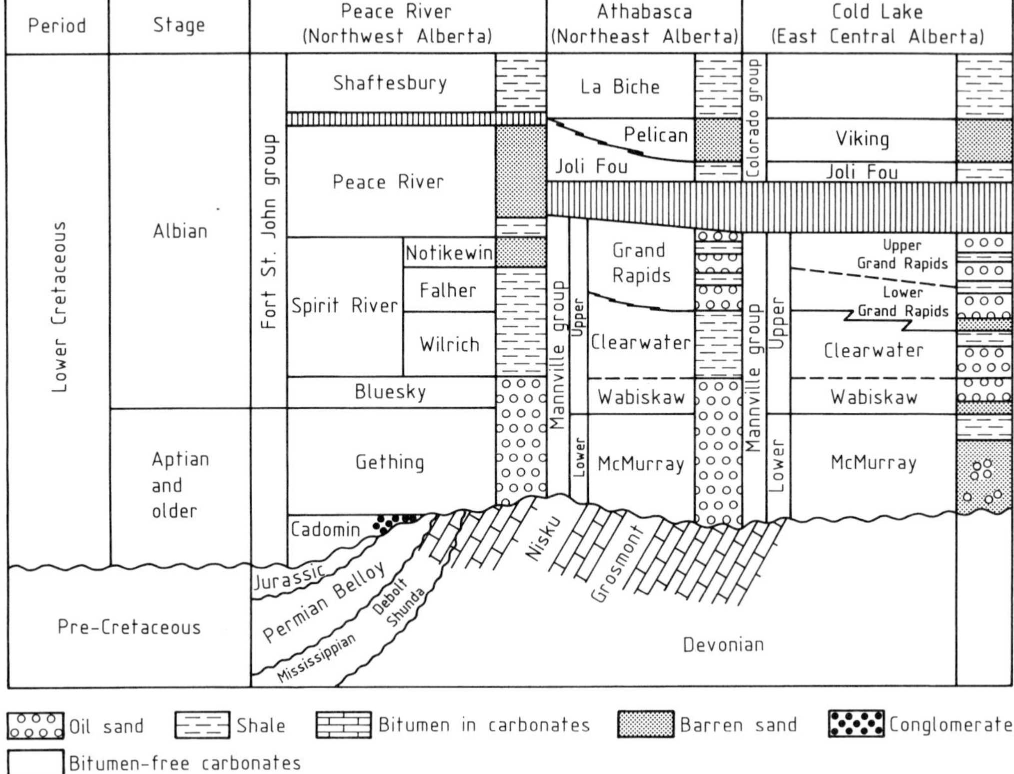



**Figure 4.** Stratigraphic correlation chart for the oil sands areas [8]

### 1.2.1. Athabasca Oil Sands Area

**Grand Rapids Deposit.** The Grand Rapids sands were deposited in a high-energy, wave-dominated shoreline along a low-relief coastal plain. The reservoir sands consist of nearshore, shallow marine, and shore-attached beach complexes, with minor barrier islands and lagoons. The reservoir continuity is excellent parallel to the regional north-east to south-west depositional strike. Vertical reservoir continuity is good, but may be locally interrupted by thin, but extensive calcite or siderite cemented beds, or zones of calcareous concretions [6].

**McMurray Deposit.** The McMurray Formation consists of fluvial and estuarine sediments deposited in a ridge and valley system developed on the underlying Devonian carbonates [11]–[13]. An erosional contact separates the McMurray Formation from the overlying marine sands of the Wabiskaw member (Clearwater Formation) [6].

In the channelized McMurray Formation deposits, the sands are well sorted with little clay content, and are of excellent reservoir quality. The reservoirs are relatively narrow, but owing to vertical stacking, through the amalgamation of channel sands, thicknesses up to 70 m may occur.

The Wabiskaw member sands are marine in character, and generally consist of lower to upper shoreface and offshore sands, silts, and shales. They are generally fine to very fine grained and form laterally extensive, but relatively thin reservoirs [12], [14].

**Grosmont/Nisku Deposits.** Crude bitumen occurs along the eastern, updip edge of the Devonian Grosmont and Nisku Formations that subcrop below the Cretaceous sands in the Athabasca oil sands area [15]. The Grosmont Formation is a large, shallow-marine carbonate platform. Original depositional environments include relatively deep water, through patch reef accumulations within a platform interior setting,

**Table 1.** Generalized geological attributes of the oil sands deposits [6]

| | Depositional environment | Dominant grain minerals | Dominant clay minerals |
|---|---|---|---|
| **Athabasca** | | | |
| Grand Rapids | | | |
| Upper | | | |
| Middle | shoreline and shallow marine | quartz, chert, feldspar, rock fragments | kaolinite, chlorite, illite } oil, sands |
| Lower | | | |
| Wabiskaw/McMurray | | | smectite, illite } |
| Minable | continental to marine shelf | quartz | kaolinite, illite } water, sands |
| In situ | | | |
| | | | |
| **Cold Lake** | | | |
| Grand Rapids | | | |
| Upper | continental to marine shoreline | quartz, rock fragments, feldspar | kaolinite, illite, smectite |
| Lower | | | |
| | | | |
| Clearwater | marine shoreline | quartz, feldspar, rock fragments, altered grains | kaolinite, illite, smectite |
| | | | |
| Wabiskaw | marine shoreline | quartz, glauconite quartz | kaolinite, illite |
| McMurray | continental | | |
| | | | |
| **Peace River** | | | |
| Bluesky | estuarine to shallow marine | quartz, chert, rock fragments | kaolinite, illite |
| Gething | | | |
| | | | |
| **Carbonate deposits** | | | |
| Debolt | open, marine | limestone, dolomite | |
| Shunda | tidal flat to open marine | limestone, dolomite | } not significant |
| Grosmont | tidal flat to open marine | dolomite | |

to restricted-marine shorelines and tidal flats. A vertical profile shows a series of shallowing-upward depositional cycles with excellent lateral continuity, but highly variable vertical reservoir pore fabric [6], [15]. The Nisku Formation is composed of carbonates overlying the Grosmont and separated from it by Ireton Formation shales.

### 1.2.2. Peace River Oil Sands Area

**Bluesky/Gething Deposit.** The Gething Formation consists of continental and marginal marine sands, silts, and shales overlaid by marine sandstones deposited along the trend of a paleovalley developed on the underlying Paleozoic surface. The Bluesky is a thin reworked Gething sand that represents the marine transgression that flooded the Gething coastal plain [16]. The bitumen is trapped in an updip pinchout of the sands against Paleozoic highs on the sub-Cretaceous unconformity [6].

**Belloy Deposit.** The Belloy Formation unconformably underlies the Bluesky/Gething. The reservoir rock was deposited as Permian, shoreline-related, basin margin sands, with sandy dolomites representing proximal, shallow marine facies [17].

**Debolt/Shunda Deposits.** The Debolt and Shunda Formations are Mississippian carbonate formations that unconformably underlie the Belloy. They form part of an overall shallowing-upward succession, representing restricted shelf–peritidal carbonates, siliciclastics, and evaporates [18].

### 1.2.3. Cold Lake Oil Sands Area

**Grand Rapids Deposit.** The Grand Rapids Formation consists of four distinct depositional environments, marine shelf–shore sediments, delta front sediments, bay-fill/delta plain sediments, and distributary channel-fill sediments

[19]. The thickest oil sands are associated with the channel sands, while the marine sands are thinner, but form the most extensive continuous reservoirs [6].

**Clearwater Formation.** The sands of the Clearwater Formation are interpreted as river-dominated deltaic deposits with laterally offset shoreface sands [20]. Owing to the extensive vertical and lateral continuity of the Clearwater sands, they form the main reservoir in this area [21].

**McMurray Deposit.** The McMurray Formation sands in the Cold Lake area were deposited in fluvial and lacustrine environments. The Wabiskaw member sands are the transitional sediments between the underlying continental McMurray Formation and the marine Clearwater Formation [6].

## 1.3. Reserves

The ERCB appraises crude bitumen reserves in each of Alberta's oil sands deposits and updates its appraisal annually. Various categories of reserves are published, ranging from the volume determined to be present based on drilling information (the in-place volume) to the volume which is conservatively expected to be recovered at existing developments (the established reserve). The ERCB also provides a broad estimate of the reserve that could be recovered before activity ceases (the ultimate potential). The in-place volume, established reserves, and ultimate potential are currently $269 \times 10^9$ m³, $482 \times 10^6$ m³, and $49 \times 10^9$ m³, respectively.

### 1.3.1. In-Place Reserves

The oil sands deposits in Alberta contain an estimated in-place volume of $269 \times 10^9$ m³ crude bitumen (Table 2). Initial in-place volumes of crude bitumen in each deposit are determined with different cutoff criteria. Crude bitumen in-place volumes within the Cretaceous sands are determined with a minimum saturation of 3 wt % crude bitumen (approximately 20 % of pore volume), and a minimum saturated zone thickness of 1.5 m; in the carbonate occurrences, they are determined with a minimum bitumen saturation of 30 % of pore volume and porosity 5 %.

### 1.3.2. Minable Reserves

The minable area is defined as that part of the Athabasca Wabiskaw–McMurray deposit

**Table 2.** Reserves and characteristics of the oil sands deposits [5], [6]

| | In-place volume, $10^6$ m³ | Area, km² | Overburden depth, m | Pay thickness, m | Bitumen saturation | | Average porosity, % | Water saturation, % |
|---|---|---|---|---|---|---|---|---|
| | | | | | wt % | Pore, vol % | | |
| **Athabasca** | | | | | | | | |
| Upper Grand Rapids | 4140 | 3340 | 150–450+ | 9 | 6.2 | 55 | 30 | 45 |
| Middle Grand Rapids | 1410 | 1820 | 150–450+ | 5 | 7.7 | 68 | 30 | 32 |
| Lower Grand Rapids | 1220 | 1730 | 150–450+ | 6 | 5.1 | 45 | 30 | 55 |
| Wabiskaw/McMurray | 141 750 | 43 290 | 0–750+ | 35 | 9.2 | 62 | 28 | 30 |
| Nisku | 10 330 | 4990 | 200–800+ | 8 | 5.7 | 63 | 21 | 37 |
| Grosmont | 50 500 | 118 910 | 200–800 | 10 | 5.0 | 75 | 16 | 25 |
| **Cold Lake** | | | | | | | | |
| Upper Grand River | 7400 | 8160 | 300–600 | 6 | 8.1 | 58 | 30 | 42 |
| Lower Grand Rapids | 12 950 | 7500 | 300–600 | 12 | 8.7 | 60 | 31 | 40 |
| Clearwater | 11 050 | 5890 | 300–600 | 15 | 8.9 | 64 | 30 | 36 |
| Wabiskaw/McMurray | 3860 | 6250 | 300–600 | 6 | 5.7 | 51 | 25 | 49 |
| **Peace River** | | | | | | | | |
| Bluesky/Gething | 14 040 | 9760 | 300–700 | 6 | 6.1 | 60 | 23 | 40 |
| Belloy | 282 | 260 | 675–700 | 8 | 7.8 | 64 | 27 | 36 |
| Upper Debolt | 1830 | 1000 | 500–800 | 13 | 5.0 | 61 | 19 | 39 |
| Lower Debolt | 5970 | 2020 | 500–800 | 29 | 5.1 | 67 | 18 | 33 |
| Shunda | 2510 | 1430 | 500–800 | 14 | 5.3 | 52 | 23 | 48 |

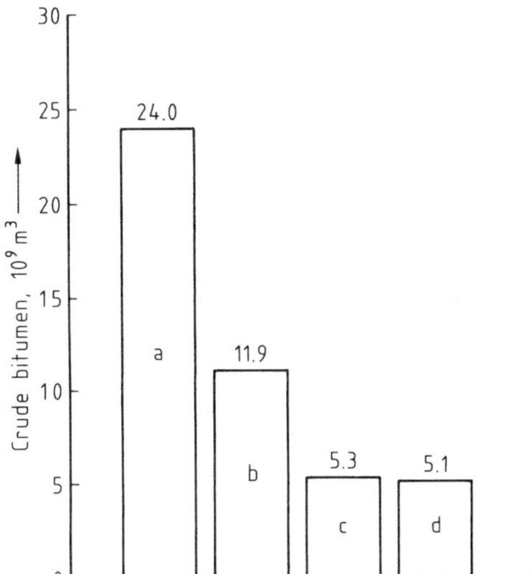

**Figure 5.** Crude bitumen reserves categories within the surface-minable area
a) Initial in-place volume; b) Initial minable in-place volume; c) Initial established minable reserve; d) Remaining established minable reserve

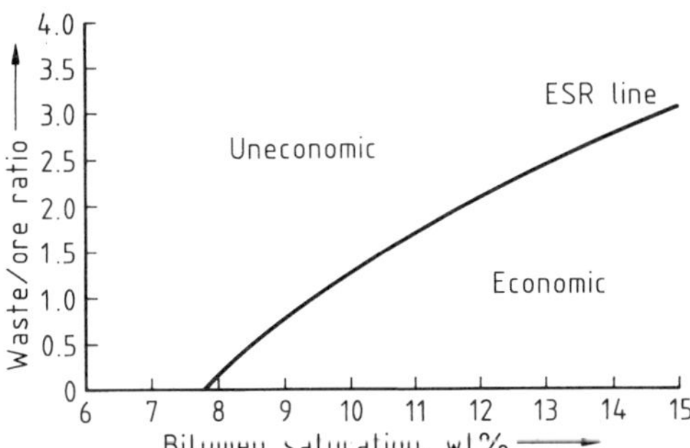



**Figure 6.** Economic stripping ratio relationship

where the overburden thickness generally does not exceed 75 m. This area is ca. 3250 km$^2$, more than 7 % of the total area of the Athabasca oil sands. Approximately $24 \times 10^9$ m$^3$ crude in-place bitumen is contained in this area. The determination of minable reserves from this volume is discussed below and illustrated in Figure 5.

Potentially minable areas are identified by applying economic stripping ratio (ESR) criteria, a minimum saturation of 5 wt % crude bitumen, maximum shale content 45 vol %, and minimum saturated zone thickness 1.5 m. The ESR is a modified version of the more widely used cutoff stripping ratio; it defines a curve of in-creasing strip ratios at increasing ore grades where the area below the curve is economic and that above the curve is uneconomic (Fig. 6). The ESR considers product price, project costs, and bitumen recovery for diferent ore grades to define a pit limit within an ore body, where the net revenue is zero [22].

Within the potentially minable area, the initial, or potentially, minable in-place volume of crude bitumen is calculated as $11.9 \times 10^9$ m$^3$ (see Fig. 5). Allowing for surface facilities such as plant sites, tailing ponds, discard sites, environmental protection corridors along major rivers, and isolated minable areas, assuming a combined mining/extraction recovery factor of 0.82, the resulting in-place minable reserve of crude bitumen is estimated as $5.3 \times 10^9$ m$^3$ (initial established minable reserve). Production to date has reduced this reserve to $5.1 \times 10^9$ m$^3$, the remaining established minable reserve. Technological improvements, better placement of surface facilities in future projects, and improved project economics could increase the estimates. Only a small portion of the initial minable reserve is being developed. The existing Suncor and Syncrude Canada projects have estimated established minable crude bitumen reserves at December 1992 of $82 \times 10^6$ m$^3$ and $352 \times 10^6$ m$^3$.

### 1.3.3. In Situ Reserves

In deeper formations, crude bitumen can be recovered through wells (in situ recovery). Since crude bitumen is not normally mobile in the reservoir, the application of thermal energy is required, and only areas that may be developed for thermal recovery are included in the established reserves. The initial in-place volume for developed areas is based on the expected drainage areas and a regard for the spacing of wells or well clusters. Established reserves are determined based on recovery mechanism, e.g., 20 % recovery is assumed for cyclic steam stimulation projects. For projects with a primary recovery component, the in-place volumes are based on the assumed full development of all project lands not currently developed for thermal recovery. The initial established primary reserves are based on an average primary recovery factor for the particular sand being developed. In the active experimental schemes, an initial established reserve figure of $17.7 \times 10^6$ m$^3$ is set.

The ERCB estimates the remaining established reserves of crude bitumen from the deposits under active development to be $434 \times 10^6$ m³ for surface-minable schemes, and $48.2 \times 10^6$ m³ for in situ schemes. The reserves remaining to be developed are vast. In the surface-minable area the established reserve is $5.1 \times 10^9$ m³.

## 2. Minable Oil Sands Development

Surface mining of oil sands and production of synthetic crude oil is well established in Alberta. The Syncrude and Suncor projects produced approximately 17 % of Canada's oil requirement in 1992. The large reserve base, proven technology, and an established infrastructure all have a positive influence on the viability of oil sands projects, but their development is handicapped by high capital and operating costs and long lead time from project initiation to production.

Figure 7 shows the Syncrude project. Table 3 provides an overview of the quantities of oil sand processed, bitumen recovered, and synthetic

**Figure 7.** Syncrude oil sands project
Dragline mining in foreground; bitumen extraction and upgrading at top right.

**Table 3.** Oil sands mining projects, 1992 production volumes [23]

|                                        | Syncrude | Suncor |
| -------------------------------------- | -------- | ------ |
| Oil sands mined, t/d                   | 341 650  | 110 140 |
| Bitumen in oil sands, wt %             | 10.8     | 11.9   |
| Bitumen production, m³/d               | 33 610   | 11 620 |
| Synthetic crude production, m³/d       | 28 490   | 8720   |
| Coke production, t/d                   | 3990     | 2730   |
| Sulfur production, t/d                 | 1230     | 340    |

crude oil (SCO), sulfur, and coke produced at the Syncrude and Suncor projects in 1992.

### 2.1. Mining Operations

Mining the oil sands and recovering the bitumen involves stripping the overlying overburden, mining the oil sand, and transporting it to the bitumen extraction plant, where hot water washes the bitumen out of the sand. The bitumen is upgraded and refined to produce synthetic crude oil, and the waste sand is pumped to disposal sites. For economic operation, very large volumes of material must be mined, transported, processed, and disposed of continuously. In 1992, the large Syncrude project typically handled over 700 000 t/d of in-place materials to produce 341 650 t/d of mined oil sands; additionally over 263 000 t/d of waste sand product (tailings) was disposed of.

#### 2.1.1. Waste Management

The first steps in the mining operation are tree clearing and muskeg drainage and stripping. Muskeg is a relatively thick, water-saturated peat, composed of black organic material in varying stages of decomposition. It varies in depth from 2 to 6 m and, because of its high organic content, is stored for later use in mine reclamation. Before it can be removed, the muskeg must be drained by an extensive series of ditches.

Overburden stripping is carried out by a truck and shovel fleet. Shovels vary in size and type from 14 m³ hydraulic diesel-powered shovels to 44 m³ cable-electric-powered shovels. Haul fleets vary from 68 to 218 trucks.

Overburden materials are mainly composed of tills underlying glacial lacustrine soils, or a variety of meltwater or outwash alluvial deposits [24]. Most of the overburden material is used to construct waste dumps and containment dykes for the tailings produced from the extraction process. Some of the more granular deposits are used for road construction and the wetter, siltier material is dumped in-pit or within the core of the waste dumps (Fig. 8). The wet overburden soils are dumped into an inner core, formed by constructing a shell of compacted overburden. The waste dumps are 30–55 m high with side slopes ca. 3.3:1–4.4:1. Most have been con-

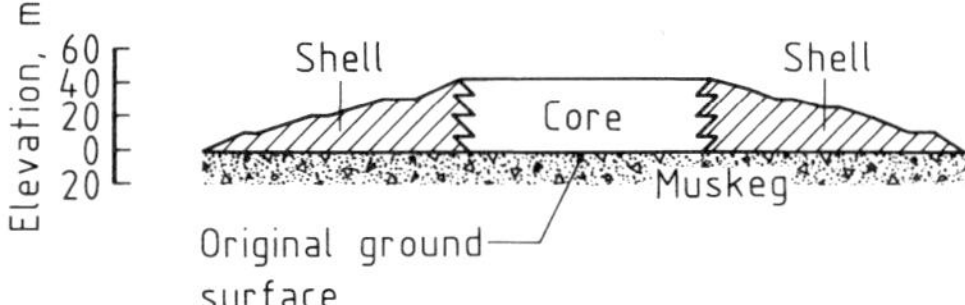



**Figure 8.** Cross section of a waste dump

structed out-of-pit and founded on thick muskeg deposits. Careful attention must be given to foundation conditions and construction rates to prevent high pore pressures developing, and to consolidate the underlying in situ muskeg.

Oversize lumps and clay or siltstone stringers are removed from the ore during the early stages of processing. This material is very wet and difficult to use as a construction material. Some success has been achieved by using it to construct a clay core within a tailings sand dyke [25]. Oversize material not needed for construction purposes is wasted into the mined-out pit areas.

### 2.1.2. Mining

Before mining can begin, the basal water sands in the mining area have to be depressurized. These are discontinuous silty sands $1-12$ m thick. These aquifers pose potential stability hazards to the single-bench mining highwalls, the final pit walls, and the stability of the bucketwheel excavators operating on the final mine bench. A unique characteristic of oil sand basal aquifers is the presence of dissolved gases in the water [26].

The two mining operations currently in operation have developed different mining schemes to manage their respective reserves. Syncrude uses four 70 m$^3$ walking draglines operating from a single bench. The oil sand is spoiled on the operating bench behind the dragline and is later reclaimed by a system of bucketwheel reclaimers and conveyors. The draglines are also capable of spoiling interburden, either directly or via rehandle, into the mined-out pit. Supplementary oil sands feed is supplied via a mobile equipment fleet. It is sized and transferred to the conveyor network via feeder–breakers and double-roll crushers. A dump pocket with approximate capacities of 15 000 t live and 200 000 t dead storage provides surge capacity.

The Athabasca oil sands have been classified as locked sands [27]. Naturally occurring slopes have been observed up to 65 m high at angles of 55°. Owing to the very high strength of the oil sands and the abrasiveness of the sand, blasting is sometimes used to improve productivity and reduce wear on equipment [28]. Blasting is also done for soil stabilization. Steeply dipping clay beds are blasted at Syncrude to reduce the risk of block slide failures; blast densification of loose beach sands has been done at both Suncor and Syncrude [29], [30].

Highwall stability is a safety concern at both oil sands mines, particularly at the Syncrude mine where the dragline operates on a single-bench highwall with 50° slopes and vertical height $> 60$ m [31]. Four basic modes of highwall failure have been identified, largely in the Syncrude operation: slabbing failures, highwall deformations, shallow block slides, and basal failures (Fig. 9). In contrast, the present Suncor mining operation is carried out on three to five benches, each ca. 20 m high.

Both Syncrude and Suncor rely heavily on instrumentation and observation in dealing with highwall performance and design [32]. An important part of this approach is to obtain reliable geological data based on structural boreholes, highwall mapping, and past highwall performance. When a highwall has been designed, suitable instrumentation is installed to monitor the mining operation and the performance of the highwall. Instrumentation includes slope inclinometers, open standpipe piezometers, pneumatic piezometers, and electronic distance measurement systems. Visual monitoring by specially trained engineers also plays an important role.

Highwall monitoring and design at Suncor has not been required to the same extent as at Syncrude. Bucketwheel excavators, and the current truck and shovel system, operating at the base of a 20 m mining face do not require the same degree of control as that needed for a dragline operating at the top of a 60 m single bench. The first stage of recovering oil from the oil sands involves bitumen recovery.

## 2.2. Bitumen Production

Bitumen production operations include a feed surge system, oil sand conditioning, bitumen separation, and froth treatment facilities. The feed surge is required between the mine and extraction to decouple these operations and al-

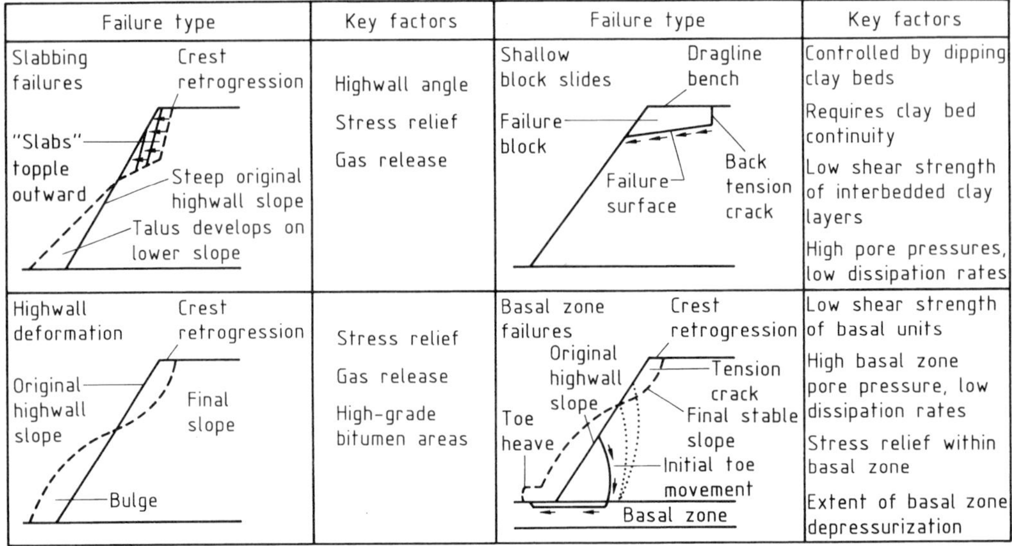


**Figure 9.** Highwall failure modes in oil sands

low either to operate while the other is unavailable for reasons of upset or maintenance. Oil sand is withdrawn from the surge facility through hoppers and apron feeders which feed extraction feed conveyors.

### 2.2.1. Conditioning and Bitumen Froth Separation

The process for bitumen froth production used at commercial operations is based on the hot water extraction process invented by Karl A. Clark in the 1920s [33], [34]; it was developed through pilot testing and first applied commercially in 1967 by Suncor (formerly Great Canadian Oil Sands). The process depends on the water-wetted nature of solids in oil sands.

The Clark hot water extraction process is illustrated in Figure 10. In the first step, oil sand is conditioned by mixing in a rotating drum with hot water, sodium hydroxide, and steam to produce a slurry of ca. 65 % solids at ca. 80 °C. The process involves the addition of mechanical, thermal, and chemical energy. Thermal energy is supplied by hot water and steam, mechanical energy by the rotating and tumbling action in the conditioning drum, and chemical energy results from the generation of surfactants, produced by reaction of sodium hydroxide with organic acids

occurring naturally in the bitumen; occasionally surfactants are available in the oil sands without treatment. Bitumen recovery is highly dependent on surfactant concentration; sodium hydroxide is used to adjust the surfactant concentration and maximize recovery [35], [36]. The conditioning drum action causes the bitumen film which envelops the oil sand particles to rupture, allowing solids and bitumen to separate. Bitumen is also aerated in the process, and separates as a froth from the water–solid slurry in separation vessels. The residence time in the conditioning drum is set to ensure oil sand lumps are broken up and the bitumen is liberated into the slurry.

Variables affecting recovery are: process temperature, water/ore ratio, and residence time. Syncrude's warm water extraction process operates at 50 °C, and a cold water process has been developed, designed to operate at ca. 20 °C. Conditioning drum water/oil ratios in the range 0.2–0.4 are used, depending on the ore grade. Sodium hydroxide use depends on the ore grade and varies from 0 to 0.04 wt % of oil sands feed.

The conditioning drum slurry flows through vibrating screens to remove oversize material, consisting mainly of rocks, clay lumps, and undigested oil sands. The oversize material is washed with hot water to minimize bitumen loss, and conveyed to storage bins that are discharged to trucks and eventual disposal in mined-out pits.

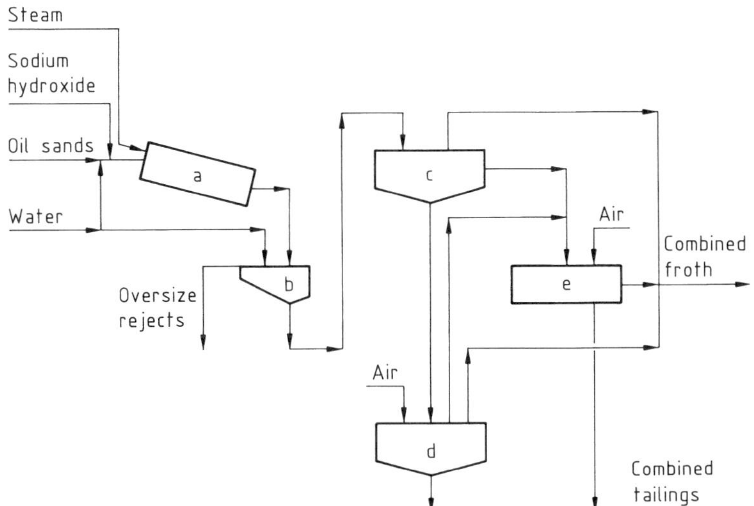

**Figure 10.** Bitumen froth production
a) Conditioning drum; b) Oversize screen and pump box; c) Primary separation vessel; d) Secondary separation vessel;
e) Flotation vessel

The conditioning drum slurry is further diluted with water and pumped to primary separation vessels.

The bitumen floats to the surface of a primary separation vessel as froth where it overflows into launders. The froth, also containing water and solids, is deaerated prior to further processing. Coarse solids settle to the bottom of the primary separation vessel and are withdrawn along with fine solids, some bitumen, and water, either to a tailings oil recovery system, or discharged to a tailings pond. The density of the separation vessel underflow is a key control variable. The tailings oil recovery system is a secondary settling vessel where, through dilution with water and other recycle streams, additional bitumen froth is produced. The underflow from this vessel is pumped to the tailings pond.

Conditions in the middle zone of the primary separation vessel are controlled by upstream variables, such as water and sodium hydroxide addition, and by withdrawing a middlings stream to ensure that the fluid density and viscosity allow aerated bitumen particles to rise to the froth layer [37]. The vessel middlings stream and the middlings stream from the tailings oil recovery circuit are treated in air flotation vessels to produce additional bitumen froth. This froth, of lower quality than primary circuit froth, is then fed to a second froth deaerator. Flotation vessel tailings flow to tailings pump boxes. Bitu-men froth, following deaeration and further cleaning if necessary, is fed to storage or directly to froth treatment.

Syncrude's commercial operation has four similar extraction trains, each with oil sand throughput rate up to 3000 kg/s per train. Suncor has five trains, with throughput rate 1225 kg/s. Sufficient capacity is provided to allow for variations in oil sands feed rates, the water : ore ratios necessary to accommodate changes in oil sands properties and maintenance downtime.

The hot water extraction process is efficient. Bitumen recovery is highly dependent on oil sands feed grade (Fig. 11) [38]–[40]. Key factors are fines content, depositional environment, and bitumen content.

A major concern with the existing process is the large volume of poor settling tailings produced and, as a result, the need for extensive tailings ponds and sand containment areas. Coarse solids in tailings settle rapidly in separate containment areas, or are used to form dykes to contain the remaining fluid tailings. The fluid tailings settle to ca. 30 wt % solids, and must be contained indefinitely or treated. Clear tailings water is recycled for reuse in the extraction plant. Ultimately, fine tailings which do not settle beyond 30 wt % solids in water are disposed of in mined-out pits and reclaimed (see Section 3.4).

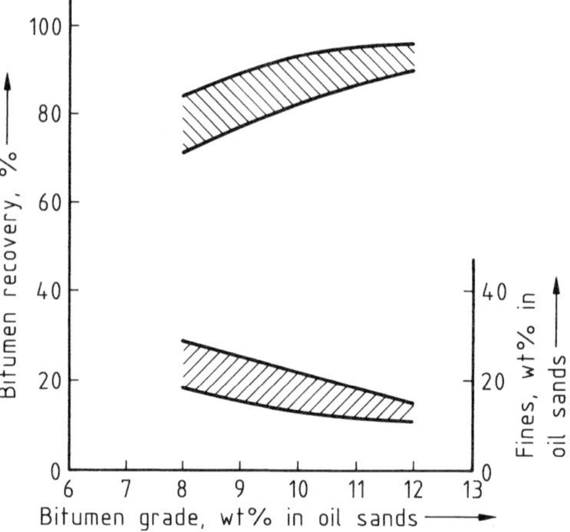

Figure 11. Bitumen recovery and relationship of fines to bitumen grade

### 2.2.2. Froth Treatment

Figure 12 illustrates a commercial froth treatment plant. Syncrude and Suncor processes have two-stage centrifuges of conventional design installed in parallel.

Deaerated bitumen froth containing ca. 8 wt % solids and 40 wt % water is treated by centrifuging to reduce solids to < 0.6 wt % and water to ca. 5 wt %, to satisfy feed requirements for downstream bitumen upgrading units. The froth is diluted with naphtha and centrifuged in two stages at ca. 76 °C. Naphtha recovered from

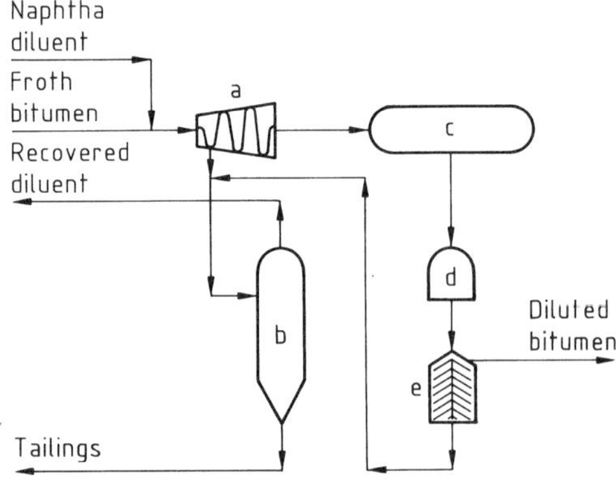



Figure 12. Froth treatment
a) Scroll centrifuge; b) Naphtha recovery unit; c) Disk centrifuge feed drum; d) Filters; e) Disk centrifuge

the upgrading of bitumen is used as diluent at a naphtha/bitumen ratio of 0.8 by weight. The first stage is a scroll-type, horizontal bowl centrifuge operating at 800 rpm (ca. 250 $g$). Coarse solids ($>$ 44 µm) are removed as a cake. The hydrocarbon product containing water and fine solids flows to a second stage disk centrifuge where most of the water and remaining solids are removed. Disk centrifuges operate at 3600 rpm (ca. 2500 $g$). A filter on the feed line to the disk centrifuge removes coarse solids. The diluted bitumen product of the disk centrifuges is pumped to a diluent recovery unit where naphtha is distilled off and recycled. Centrifuge facility tailings, mainly water and solids, contain naphtha, some of which is recovered by a vacuum flash process. Bitumen and naphtha recovery in the treatment process is ca. 99 wt %.

The Syncrude froth treatment plant throughput has been increased by the addition of an inclined plate settler that takes one-third of the feed, and operates in parallel with the centrifuge system [41]. The settler feed diluent : bitumen ratio is 0.5 by weight.

Syncrude has tested a three-stage process using inclined plate gravity settlers with the froth and diluent flowing concurrently. Other research and development has led to a process which, in two stages of mixing and gravity separation at 160 °C, results in a bitumen product of lower water and solids content.

### 2.3. Tailings Disposal

The tailings—sand, clay, water, and bitumen discard from the bitumen production operations—are stored in ponds where the solids are allowed to settle out and the clarified water is recycled to the extraction plant. Containment, handling, and disposal of tailings pose major economic and environmental problems.

In the initial years of operation of an oil sands mine, when recycle water is not yet available owing to the very slow fine tailings dewatering rates, tailings storage is required to hold three times the volume of oil sands for ca. 5 a. This volume increase is called a tailings bulking factor. Later, as recycle water becomes available from the tailings pond, the bulking factor reduces to 1.4. It is this bulking, and the very slow release of water from the fine tailing, that is the focus for much of the ongoing research into tailings activities and final reclamation planning.

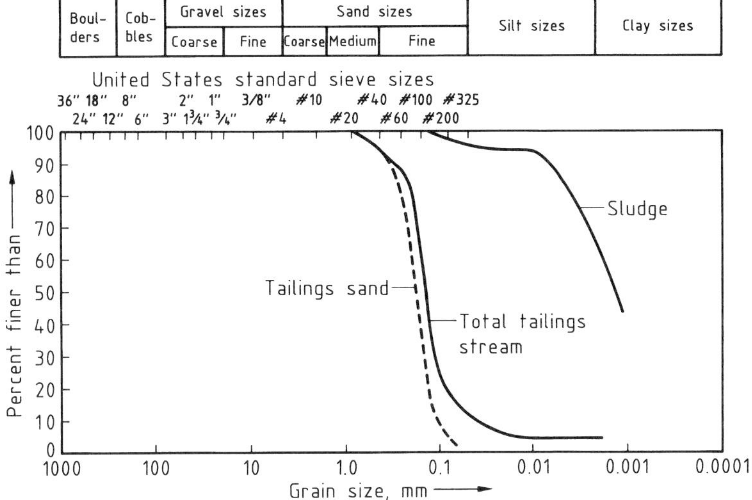


**Figure 13.** Grain size distribution of oil sands tailings

Gradation data for oil sands tailings produced by existing oil sands operations is shown in Figure 13 [42]. While the tailings are relatively clean and coarse overall, a significant slime fraction is present (a mixture of silt, clay, and oil residue).

Poor sedimentation and consolidation characteristics of the slimes have complicated tailings disposal [43]. When the tailings are deposited, the coarse sand fraction settles out almost immediately, while the dispersed silts and clays remain suspended in the water to form a sludge-like material on top of the sand. During dyke construction and beaching operations, ca. 50 % of the fine tailings are captured in the voids of the coarse tailings. The remaining fine tailings enter the pond as a thin slurry (ca. 3–8 wt % solids). The fine tailings consolidate very slowly to a maximum concentration of ca. 35 % solids. The volume of these mature fine tailings and their toxic nature are of major public concern.

Some of the bitumen lost to the pond in the tailings stream is also released into the recycle water layer. This free bitumen either floats to the surface, and in some cases is recovered with booms and vacuums, or collects below the surface to form large suspended pools of bitumen, and is pumped to storage tanks at the shoreline for trucking back to the plant.

Containment of the tailings is also of interest. Tailings dykes have been constructed by a modified method of upstream construction (Fig. 14).

The dyke typically consists of a downstream shell of compacted sand and an upstream supporting beach of sand. The compacted shell and supporting beach are placed by hydraulic construction techniques (Fig. 15).

Beaching is conducted from either a single discharge point or from multiple discharge points along the pipeline (spigotting). Beach slopes range from 2 to 4 %. The discharge point is moved along the dyke at appropriate time intervals to ensure adequate upstream beach support without imposing large stresses on the underwater beaches, thereby triggering a liquefaction failure.

Tailings dyke design and performance monitoring are key components in a safe tailings management program. The observational approach is used as part of this program [44]. As already described in Section 2.1.2, extensive visual and instrumentation monitoring is required. Tailings dyke integrity has been reported as being a function of three major geotechnical design variables: seepage control, foundation movement, and overall dyke stability, as well as the liquefaction potential of loose sands.

Seepage control is important both to minimize groundwater contamination and to ensure structural integrity. Internal drains or filters are installed within the downstream section of the tailings dyke to control the phreatic surface and to militate against potential piping problems.

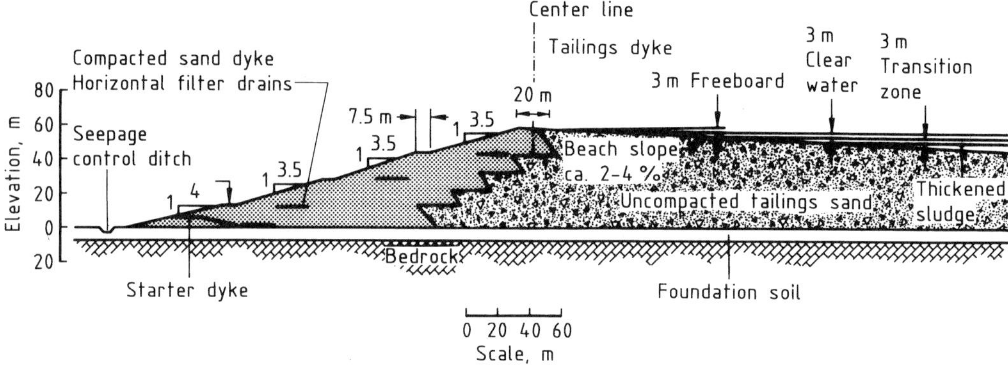


**Figure 14.** Typical dyke section

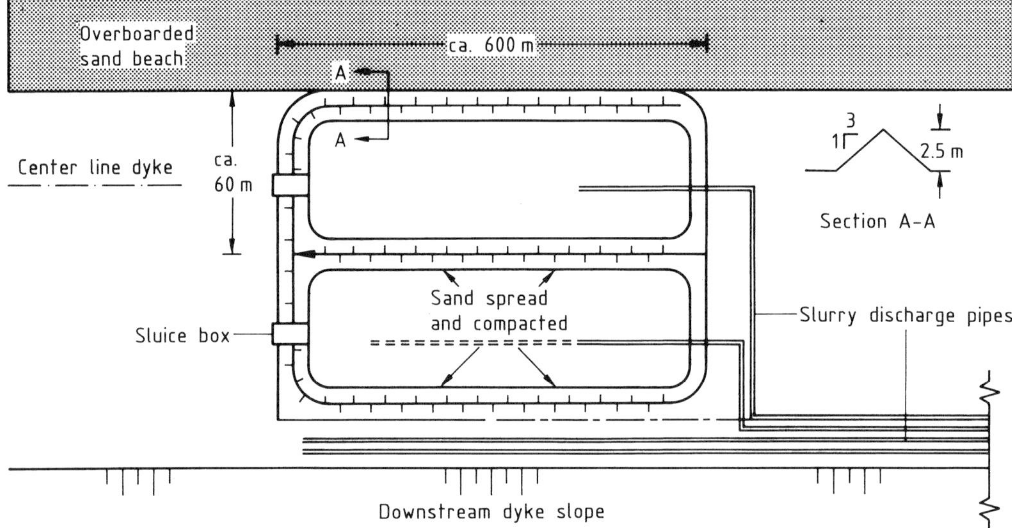


**Figure 15.** Construction of tailings dyke cells

Existing operations have experienced significant foundation movements as a result of tailings dyke construction. Weak presheared clay shales are present under Syncrude's tailings pond [44], and Suncor's first tailings pond was founded on a normally consolidated alluvial silty clay [45]. Several types of remedial measures have been used to protect overall dyke integrity.

Although the Fort McMurray area is in a low seismic risk area, the potential for liquefaction was considered a possibility. Strains induced in the loose beach sands by foundation move-ments and from the additional loading of dyke construction, may be enough to cause liquefaction. Blast densification of loose sands and construction of toe berms has been used successfully on both projects to offset this potential risk.

Both plants are obliged to reclaim the surface of the disturbed areas to levels of productivity similar or better than those prior to development. Because of the large area and volumes, the operators of the plants are advocating an integrated reclamation strategy with a mixture of "dry" and "wet" landscapes.

## 2.4. Bitumen Upgrading

Bitumen produced from oil sands is a heavy viscous material which is difficult to transport and not readily processed in conventional refineries. At oil sands mining schemes in Alberta, upgrading occurs at the bitumen production site to produce "synthetic" crude oil. The characteristics of this partially refined product are described later. The need to refine bitumen depends on economics, integration of mining, bitumen production, and upgrading operations, markets for bitumen or synthetic crude oil, transportation constraints, and the ability of refineries to process bitumen and synthetic crude oil.

The Syncrude and Suncor projects, which commenced operation in the late 1960s and 1970s, are stand-alone, integrated operations comprising mining, bitumen extraction, bitumen upgrading, and utilities. In contrast to the minable oil sands developments, in situ recovery projects (Chap. 3) produce crude bitumen which is blended with a hydrocarbon diluent and pipelined to upgraders, primarily in the United States. Recent trends in the minable oil sands area are toward bitumen production operations at the mine site, and transportation of diluted bitumen to regional or remote upgraders.

The need for upgrading is illustrated by a comparison of the properties of crude bitumen and conventional oil (Table 4). Bitumen has high density, high viscosity, and high pour point. Boiling point data show that bitumen contains little or no hydrocarbon boiling below ca. 195 °C, and approximately 86 % of bitumen boils above 345 °C. The essence of upgrading bitumen is to reduce high-boiling material, to increase the

**Table 4.** Crude properties [46]

|  | Bitumen | Synthetic crude | Conventional crude |
|---|---|---|---|
| °API | 8 | 32 | 41 |
| Sulfur, wt % | 4.8 | 0.08 | 0.2 |
| Nitrogen, wt % | 0.45 | 0.03 | 0.04 |
| Viscosity, $10^6$ m²/s, 40 °C | 3000 | 3.0 | 2.9 |
| Pour point, °C | 18 | − 4.5 | − 6 |
| Metals, mg/kg | 450 | < 1 |  |
| Distillate yield, vol % |  |  |  |
|   IBP–C 5* | 0 | 4 | 3 |
|   C 5–195 °C | 0 | 18 | 36 |
|   195 °C–345 °C | 14 | 47 | 31 |
|   345 °C–560 °C | 86 | 31 | 18 |
|   560 °C + | | 0 | 12 |

* IBP = Initial boiling point. C 5 = Pentane.

hydrogen : carbon ratio, and to remove impurities.

Figure 16 illustrates the bitumen upgrading process. Diluted bitumen product is distilled for diluent recovery. Bitumen is upgraded by a primary upgrading process, and liquid products refined to produce synthetic crude oil. Hydrogen is required by the process. Byproducts are fuel gas, sulfur, and coke. Fuel gas is used in the process, sulfur is sold and, in the absence of on-site use or market, coke is stored.

In the conversion of bitumen to synthetic crude oil, a two-stage upgrading sequence is used. The purpose of the primary step is to convert the high-boiling, high-molecular residue in the bitumen to distillate hydrocarbon. The secondary step involves hydrotreating distillate in desired boiling ranges (→ Oil, Oil Refining, **A 18**,

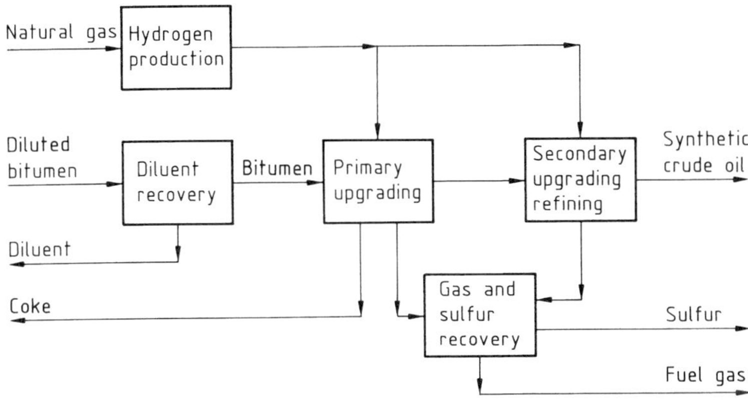


**Figure 16.** Bitumen upgrading

pp. 64–67), and adding hydrogen to saturate hydrocarbons and remove impurities.

Primary processes to upgrade bitumen can be categorized as carbon removal or hydrogen addition. Both types of process increase the hydrogen : carbon ratio compared with crude bitumen. The carbon removal or coking processes involve heating the crude bitumen above 500 °C at near-atmospheric pressure; the high-molecular material cracks to produce distillate hydrocarbons and coke (→ Oil, Oil Refining, **A 18**, pp. 77–79). The coke forms by combination of reactive high-molecular material. Hydrogen addition processes involve reaction of bitumen with hydrogen above 450 °C and pressure > 7 MPa, sometimes in the presence of a catalyst, to produce distillate hydrocarbons and a liquid residuum. The process involves thermal cracking and hydrogen addition to the cracked product.

### 2.4.1. Coking Processes

A wide range of technology is available for primary upgrading, but few processes can be applied commercially for oil sands crude bitumen upgrading. Early developments relied on coking while more recent efforts have turned to hydrocracking to upgrade the bitumen. Coking processes include delayed coking and fluid coking (→ Petroleum Coke, **A 19**, pp. 235–236). In each case the crude bitumen charge is converted to distillate oil and coke. The coker distillate is a partially upgraded material, and is a suitable feed for hydrotreating to a sweet synthetic crude oil. The coke fraction of the product may be used for fuel.

In the delayed coking process, the bitumen is heated to 480–510 °C to initiate the coking reactions. The heated oil is fed to a large drum, where residue is deposited as coke, and more volatile material flows through to a fractionator. Delayed coking is a batch process, and a minimum of two coking drums are required; as coke is formed in the on-line drum, coke is being removed from the off-line drum by steam and water cutting (Fig. 17).

The Syncrude fluid coking technology (developed by Exxon) combines the principle of coking with that of fluid solids processing, as used in fluid catalytic cracking (Fig. 18).

Preheated bitumen feed at ca. 260 °C is introduced into a fluidized reactor. The reactor thermally cracks the feed at ca. 510 °C. Heat for the

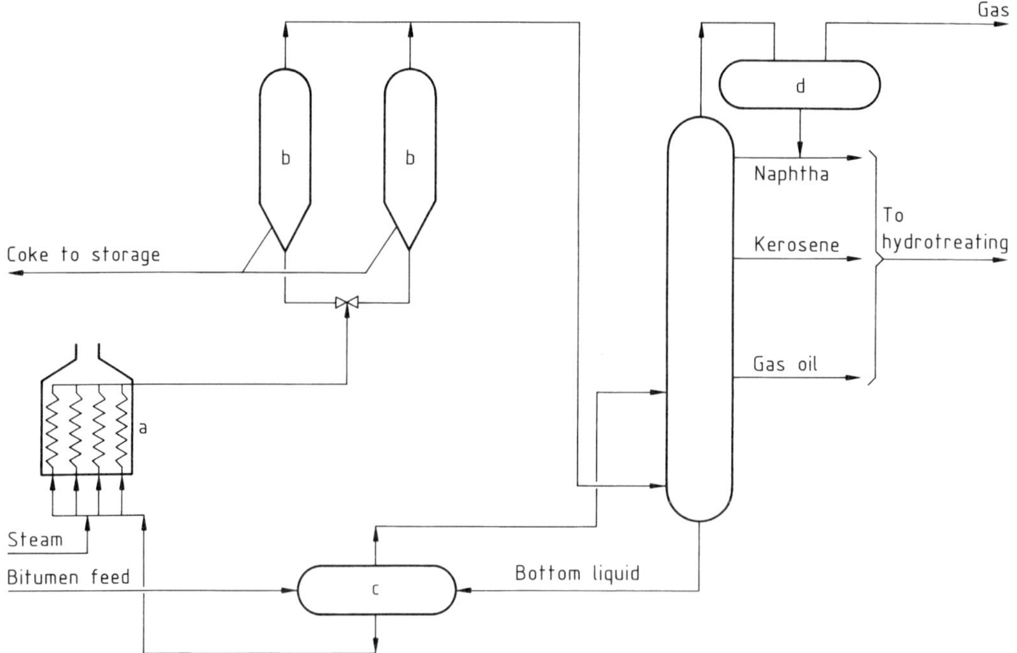


**Figure 17.** Suncor delayed coking
a) Furnace; b) Coke drums; c) Feed drum; d) Condenser

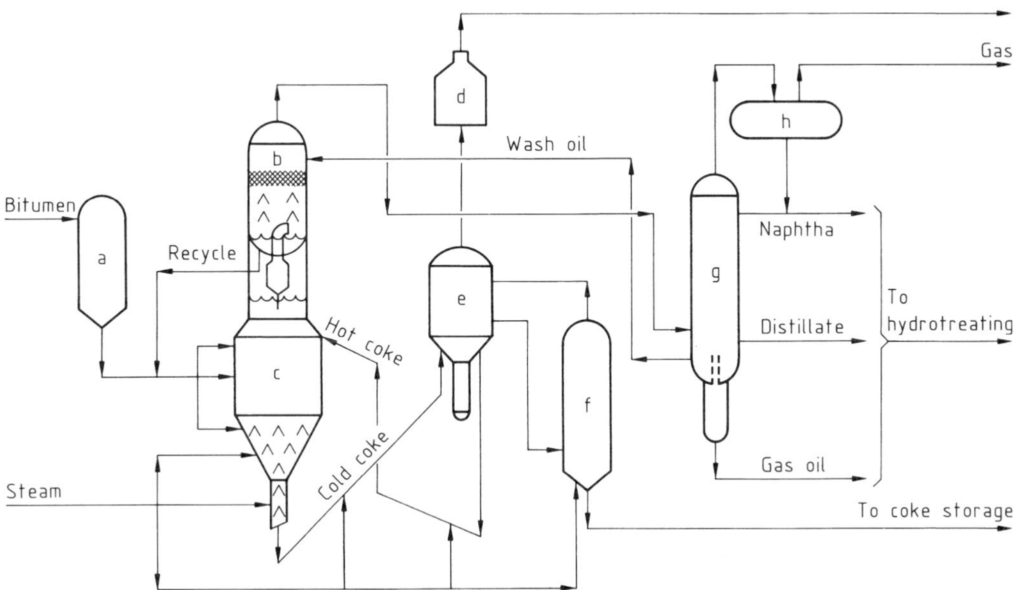



**Figure 18.** Syncrude fluid coking
a) Feed drum; b) Scrubber; c) Reactor; d) CO boiler; e) Coker burner; f) Elutriator; g) Fractionator; h) Condenser

reaction is supplied by recycling hot coke from the coker burner. The cracked vapor is scrubbed with recycled heavy oil and fed to the fractionator. Coke from the reactor is fed to the coker burner, where a portion of the coke is combusted with air to provide heat for the reaction. Recycled coke is used as a heat carrier to the reactor. Excess coke from the burner is removed to storage. Coker burner gas is combusted with supplementary fuel and air to generate steam in a CO boiler. Flue gases are vented through an electrostatic precipitator to remove particulates, and routed to an elevated stack.

The fractionator separates the vapor products from the reactor into gas, naphtha, distillate, and gas oil fractions. The gas and naphtha fractions are further fractionated in the gas recovery system. The gas is reused in the process for fuel. The naphtha and distillate are hydrotreated to reduce sulfur, nitrogen, and aromatic concentrations. The gas oil is hydrotreated to reduce sulfur, nitrogen, and metals and to add hydrogen to hydrogen-deficient aromatic compounds. Hydrotreated product is blended to meet downstream refinery specifications.

Hydrocarbon liquid yield from Syncrude's fluid coker is approximately 81 vol%, and net coke and gas make are each ca. 10 wt%. Syncrude has two fluid cokers, each operating at ca.

$16\,700$ m$^3$/d. Each fluid coker is equipped with a 5700 kg/s CO boiler.

When the Syncrude project was designed in the early 1970s, fluid coking was attractive because of its low capital and operating costs, the existing commercial experience, and the availability of waste heat and fuel gas for use in the extraction and utilities areas of the integrated operation. Disadvantages include relatively low liquid yield, high SO$_2$ emissions, and the excess coke byproduct which, for lack of markets, must be stored. Developments to minimize or eliminate the coke production have been considered, but have not been applied for economic reasons; these include Flexicoking and coke use as a feedstock, for hydrogen production, or fuel.

### 2.4.2. Hydrogen Addition Processes

Hydrogen addition processes are attractive, owing to the higher liquid yields, lower sulfur dioxide emissions, and the absence of unmarketable byproducts such as coke or residuum. Hydrocracking processes are designed to convert up to 95 wt% of the heavier hydrocarbons in bitumen (i.e., material boiling above 524 °C) to lower-boiling material, and to provide product yield > 95 vol%. Bitumen hydrocracking technology is currently applied commercially at the Syncrude project.

At the Syncrude project, an LC-Fining hydrocracker of capacity 8000 m³/d was added to hydrocrack 20 % of Syncrude's bitumen feed upstream of cokers. The LC-Fining technology has ebullated bed reactors designed to convert > 60 % of the 524 °C + content of bitumen. The unconverted portion of the bitumen is routed to the fluid coker for further processing (see Section 2.4.1). This addition provided a high incremental yield with no additional sulfur dioxide emissions, and is the basis for significant additional conversion capacity expansion in the future.

## 3. In Situ Oil Sands Development

The vast majority of crude bitumen from Alberta's oil sands can be recovered only by in situ methods. While commercial in situ recovery has been the focus of greater attention in the past 15 years, considerable research has been going on since the mid-1950s.

In situ thermal processes utilize either methods in which steam is injected into the reservoir from the surface, or methods in which heat is produced in the reservoir by hydrocarbon combustion reactions. Currently, there are three commercial thermal in situ projects in Alberta which inject steam into the reservoir, two projects employing the cyclic steam stimulation process, and one project utilizing a steam drive process. While the combustion process (see Section 5.1.1) appears promising, it has not been applied commercially. Methods in which electrical energy is used to heat the reservoir have also been tested. Operators have also pursued the development of primary production technology in areas previously thought not to be capable of such production. In situ projects typically produce only a bitumen product that is not upgraded in Alberta.

In the Peace River oil sands area, a steam drive in situ project is operated by Shell Canada. In the Cold Lake area, Imperial Oil Resources and Amoco Canada operate projects based on cyclic steam stimulation technology. Other commercial in situ projects are producing on primary. There are also 12 experimental operations currently active. In general, these are considerably smaller than the commercial projects. An illustration of in situ experimental activity over the past 20 years is provided in Figure 19.

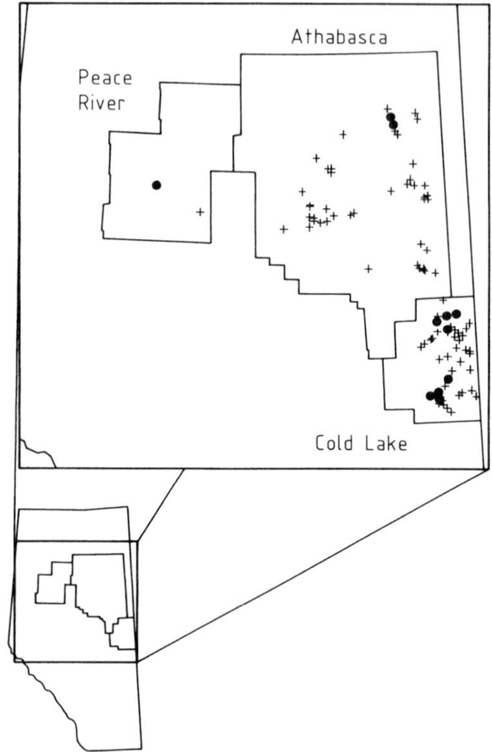


**Figure 19.** Alberta oil sands areas
● = Commercial;   + = Experimental

Owing to significant variations in reservoir characteristics, extensive laboratory experimentation and field testing are required prior to commercial field development of in situ schemes. Experimental pilot schemes apply recovery processes to the geological and petrophysical properties of the reservoirs to identify the operating parameters and conditions that maximize productivity and economics.

Most of Alberta's bitumen, in its natural state, has little or no mobility, owing to a lack of reservoir energy and high viscosity. In situ production requires heat input to the reservoir or the creation of heat in situ to reduce the viscosity of the bitumen (Fig. 20) and the reservoir flow resistance. The bitumen viscosity, and its response to elevated temperature, control the production rate in a recovery process. The application of heat also results in changes in capillary pressure, relative permeability, and wettability. The injection of heat also results in thermal expansion of oil, increased evolution of dissolved gas, and

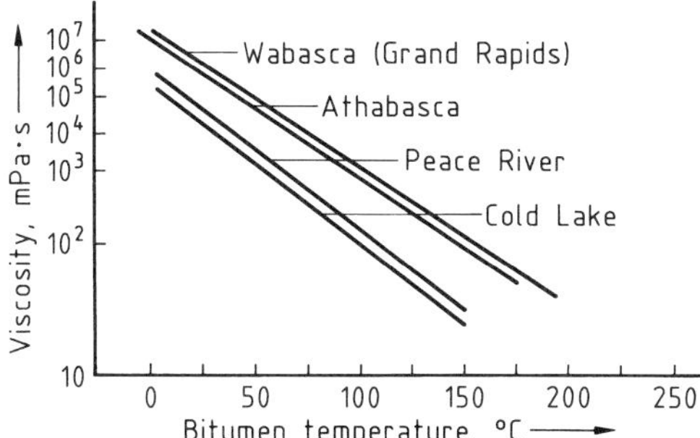

Figure 20. Bitumen viscosity–temperature relationship [47]

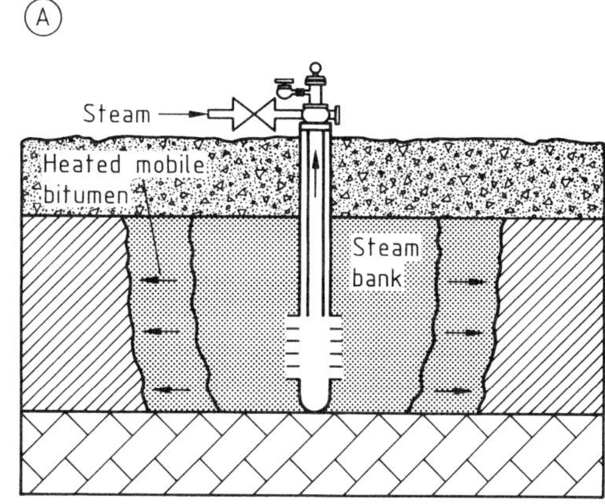



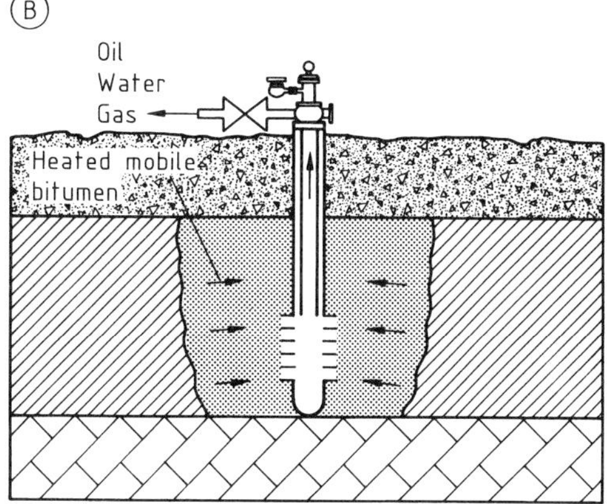



Figure 21. Cyclic steam stimulation
A) Steam injection; B) Bitumen production

changes in bitumen composition through thermal cracking and steam distillation.

## 3.1. Cyclic Steam Stimulation: Cold Lake Project

The cyclic steam stimulation (CSS) or "huff n puff" process widely used in the thermal in situ recovery of Alberta's bitumen, was discovered accidentally in 1959 in the Mene Grande field, Venezuela, when a steam injection well was backflowed to relieve reservoir pressure. The CSS process (Fig. 21) involves steam injection and fluid production in the same wellbore, conducted in a cyclic manner. Each cycle is comprised of four phases; steam injection, soaking, flowback, and pumping. Cycle steam volumes range from 4000 to 20 000 m$^3$ per well, and are injected over 15–75 d, depending on the type of operation. Soaking, for 0–30 d, is the period during which the well relaxes as it waits to be placed on production. The flowback period, up to 45 d, utilizes the reservoir energy to produce fluids. The pumping period, of 75–450 d, utilizes artificial lift to produce fluid until rates become uneconomic, at which time the cycle is repeated.

The CSS process is utilized extensively in the development of the Clearwater member in the Cold Lake oil sands area. One of the commercial projects, Imperial's Cold Lake project, is founded on a strong empirical basis with extensive piloting since 1964. The project has been commercially developed in eight phases, resulting in 1600 wells covering 26 km$^2$. Another two phases are planned to bring bitumen production up to 18 000 m$^3$/d (Fig. 22). Each rectangle represents a pad with 20 wells and area 0.32 km$^2$. Each pair

of phases is served by a central plant, capable of daily processing 3000 m$^3$ of bitumen, generating 12 000 m$^3$ (cold water equivalent) steam, and treating 9000 m$^3$ produced water for reuse.

To reduce surface disturbance, wells are drilled from pads (Fig. 23). The wells are directionally drilled to the reservoir. The figure shows the pattern aspect ratio of 1.7 (assumed length/ width of rectangular drainage area), the well spacing, and the preferential pad orientation. Pad orientation is based on preferential flow directions observed during pilot testing, which is a function of the anisotropic permeability in the reservoir. Figure 24 shows a typical pad.

Initially, the CSS process is managed as a single-well process, however in subsequent cycles the process must be managed as a multiwell process, requiring fieldwide synchronization of

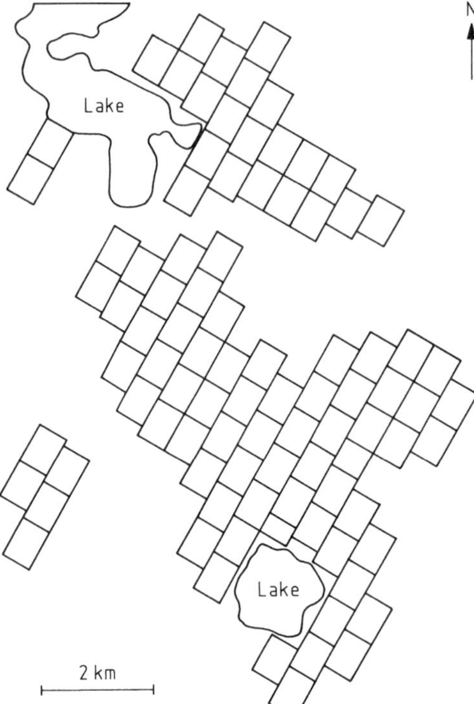

Figure 22. Cold Lake commercial development

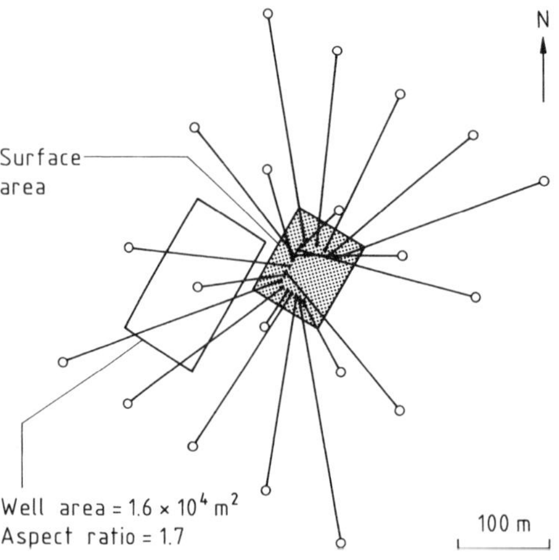

Figure 23. Well trajectories for a typical cyclic steam stimulation pad

Figure 24. Imperial Oil's Cold Lake project, typical well pad

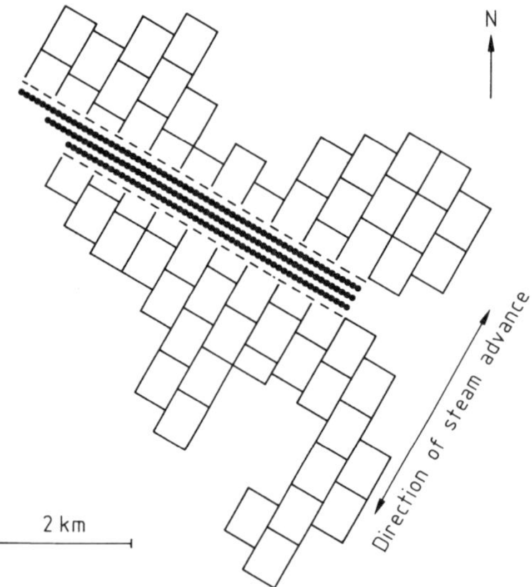

Figure 25. Overlap steaming/soaking of adjacent rows
●●●●●● = Wells receiving steam;  ——— = Wells seaking or shut-in

steaming with overlap steaming of adjacent rows to control unwanted communication [48]. Figure 25 illustrates the configuration of steaming, soaking, and shut-in wells at a given time. As cycle number increases, communication widens and, after about eight cycles, the process becomes uneconomic.

To allow sufficient volumes of steam injection, first-cycle injectivity is achieved by injecting above fracture pressure. In subsequent cycles, initial injectivity can be achieved below fracture pressure; however, as the depleted vapor space is compressed, formation failure is initiated again. Later cycles require increased steam volumes to initiate formation failure. Productivity decreases with increasing cycle number, and once the steam/oil ratio in a given cycle becomes uneconomic the process is discontinued. A typical well has a steam/oil ratio of 3–4 and overall recovery ca. 20%.

## 3.2. Steam Drive: Peace River Project

Another commercial in situ site near Peace River has a steam drive that heats the reservoir via continuous injection of steam through a pattern of injection and production wells. Steam drive or steam floods require hydraulic continuity between wells, and may be preceded by a cyclic steam process to generate this continuity. Steam drive processes usually operate under less severe conditions than cyclic stimulation processes, and may be conducted in continuous or pulsed mode.

Shell Canada developed the pressure cycle steam drive (PCSD) process which is applied in the Peace River area [49], [50]. In this process (Fig. 26), steam is injected into the reservoir without fracturing via an underlying water-saturated sand which is permeable to the steam. This results in a radial heated zone at the base of the reservoir, which mobilizes the overlying bitumen. The PCSD process alternates periods of high steam injection with periods of high production, which causes cycles of pressure buildup and blowdown in the reservoir.

The Shell project, which commenced in 1986, encompasses the operations at a pilot and at 213 wells directionally drilled from eight pads in four well clusters (Fig. 27). Each well cluster comprises 13 injectors and 40 producers arranged in 13 patterns; each pattern has 6 producers surrounding each injector in a hexagonal array. Operations began with 1–3 steam soak cycles at each well, followed by the PCSD. Operations were later modified to a continuous low-pressure steam drive process. Performance in 1992 was stabilized at 1600 m³/d, at injection pressure 3200 kPa. The recovery efficiency to 1992 was 20% and is anticipated to exceed 40% with a steam/oil ratio of 3–4.

## 3.3. Primary Crude Bitumen Production

While crude bitumen recovery from oil sands generally requires thermal stimulation, primary production (pumping wells at natural reservoir temperature) is feasible in some areas. In the southern portion of the Cold Lake area there is

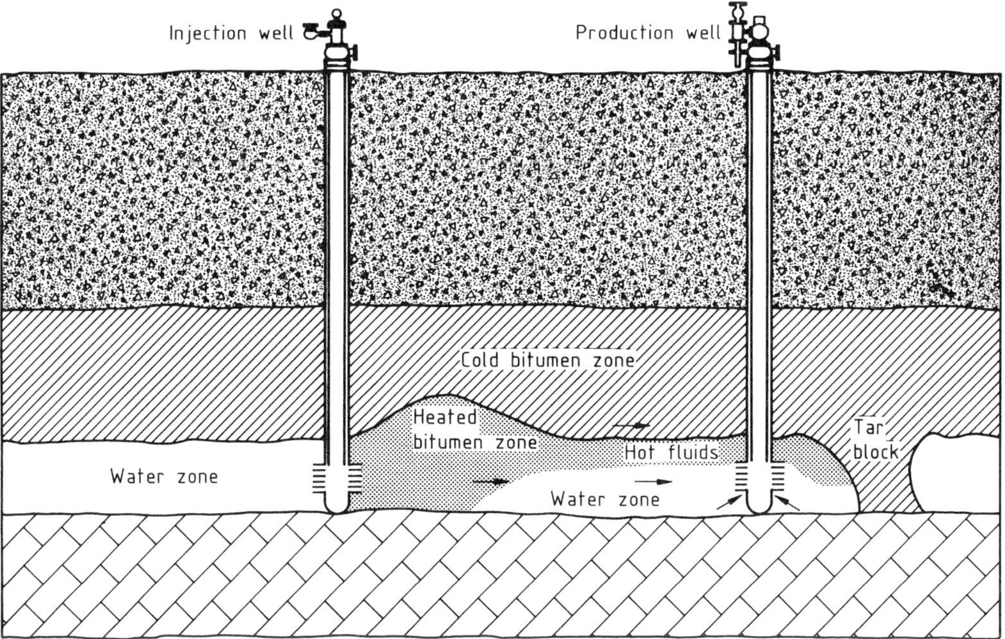


**Figure 26.** Shell Canada's steam drive process

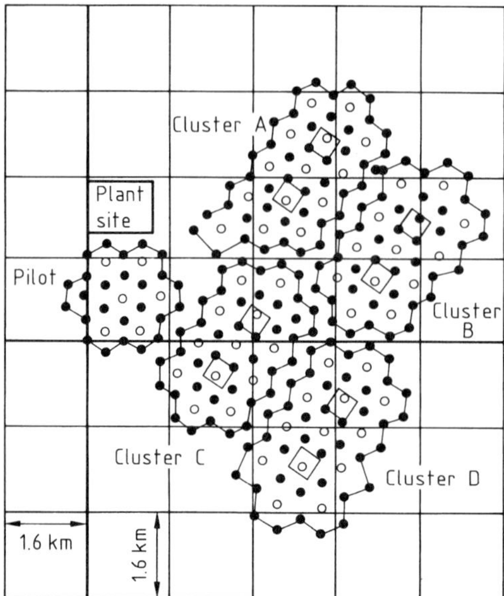

**Figure 27.** Shell Canada's Peace River complex
● = Production well bottom hole location;
o = Injection well bottom hole location;
□ = Pad

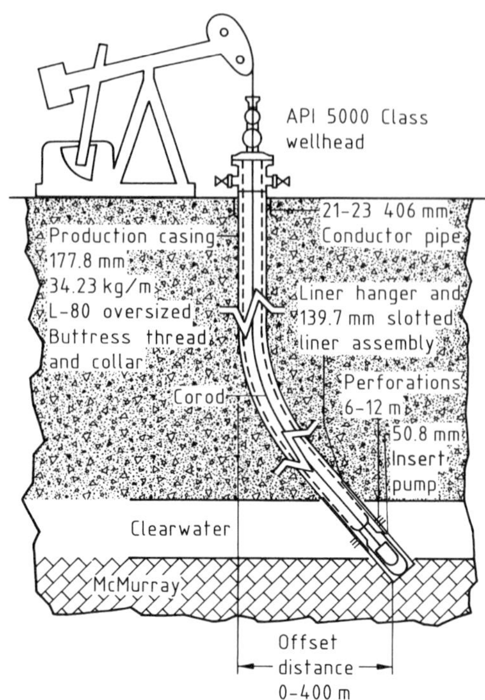


**Figure 28.** Basic well configuration

a transition from deposits of crude bitumen to conventional oil pools. Primary production is enhanced by allowing sand to be produced. The sand production causes wormholes or channels to form in the reservoir which improve bitumen recovery through enhanced drainage radius, grain movement, gas bubble expansion, and continuous pore deblocking [51]. Although the primary reserves estimates are low (2 % for the Cummings sands and 0.1 % for the other Mannville sands) the primary production is significant. In 1992, primary production accounted for 18 % (3700 m$^3$/d) of the bitumen produced from Alberta's oil sands. Progressive cavity rotary pumps are more suited to pumping fluids containing significant amounts of sand (as high as 40 vol %), and in many cases are replacing the more conventional reciprocating action beam pumps.

## 3.4. Technical Issues Affecting In Situ Recovery

### 3.4.1. Casing Design

Thermal casing design must consider possible failure in tension due to thermal stresses upon heating and subsequent cooling. Well casing is designed for the maximum expected temperature and the minimum expected temperature after steaming.

Casing failures were a common problem in the early 1960s when cyclic steaming was first attempted. Most failures occurred at a coupling, usually in the upper part of the wellbore. Casing failures seldom occur now because stronger casing and better cementing practices are employed. Casing and cement integrity are important for environmental protection and production maintenance.

Figure 28 illustrates a typical thermal casing design. A thermal grade casing, along with specially prepared thermal cement, is used with oversized buttress-thread collars to withstand thermal stresses and to resist hydrogen sulfide stress cracking. A special spray coating and thread compound are used to ensure a tight seal at the joints.

### 3.4.2. Gathering and Distribution System

A pipeline gathering system collects production from the satellites for processing in the cen-

**Figure 29.** Imperial Oil's Cold Lake project
Well pad in foreground; pipelines (steam and produced
fluids) to central plant in background

tral plant. Both the production and steam gathering systems are pile-mounted, above-ground pipeline systems with expansion loops (Fig. 29). To minimize surface disturbance, the road and power distribution grid normally follow the same route as the gathering and distribution pipeline system.

### 3.4.3. Production Treating

Field production enters the treating plant at the inlet flowsplitter where free gas is removed (Fig. 30). The fluid stream (bitumen and water) is routed through inlet coolers, to lower the fluid temperature, and allow treating to continue. The fluid proceeds through an inlet separator in which the free water and bitumen are separated by gravity. As the bitumen is cooled, the fluid viscosity tends to increase. At this point diluent is added to reduce viscosity and aid the separation process, and forms up to 30 vol% of the mixed diluent and bitumen stream (dilbit). The remaining water is separated from the dilbit stream in electrostatic dehydrators. The dilbit and oily produced water are then cooled to room temperature in glycol heat exchangers. The dewatered dilbit is degassed in a vertical boot degasser and sent to the dilbit sales tanks. Final product viscosity is controlled by blending diluent in the suction line to the pipeline pump station.

Cooled produced water is treated and reused for the generation of steam in once-through steam generators (Fig. 31). Oily produced water from the production treating facility enters the water treatment plant through a skim tank in which gross oil and suspended solids are removed. Further oil removal takes place in an induced gas flotation (IGF) unit. Effluent from

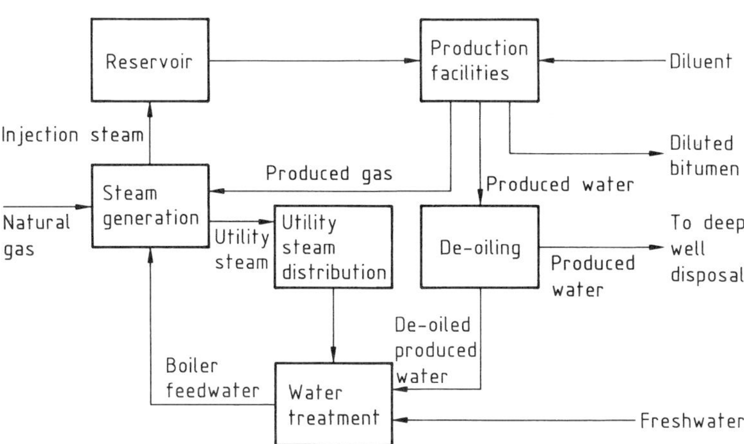


**Figure 30.** Bitumen production by in situ schemes

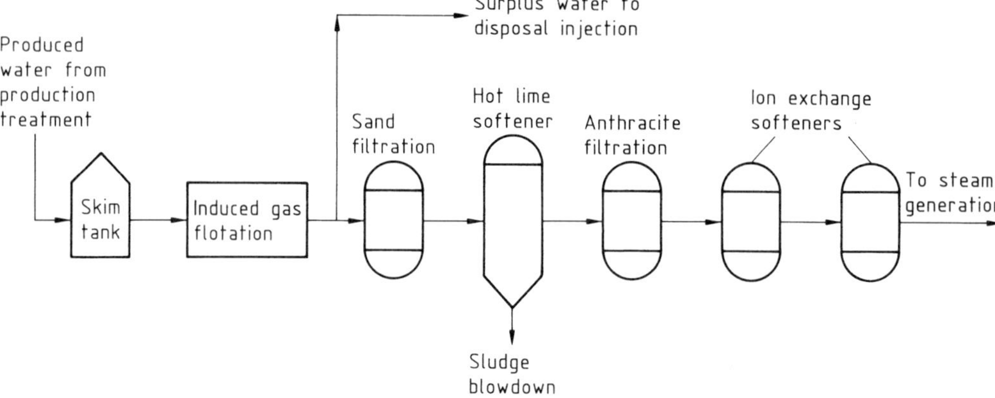



**Figure 31.** Produced water treatment

the IGF flows to the sand filters which remove a portion of the remaining oil and suspended solids, and protect the downstream processes from slugs of oil which may result from an IGF upset. In the hot lime softener, the inlet stream is mixed with lime, magnesium oxide, and coagulant aid. These react with alkali and silica in the water to form insoluble precipitates which are purged to a sludge pit. The hot lime softener effluent then passes through filters to remove any suspended solids. Ion exchange softeners, utilizing a weak acidification resin, are used to remove the residual hardness. Effluent from the ion exchangers flows directly to the steam generators.

Oil recovered through skimming the produced water is collected and recycled through the production treatment process. Emulsions of oil and water that prove difficult to separate are handled by the recycle treater.

### 3.4.4. Oily Waste Disposal

Oily waste typically consists of sands, sludges, and slop oil containing bitumen, generated during oil sands production. Oily waste is disposed of by spreading an oil–sand–gravel mixture onto minor roads (as a dust supressor), by water washing and disposal of "clean sand" in an industrial landfill, landfarming, and by subsurface disposal into a salt cavern (Fig. 32). A salt cavern is created by drilling a well and washing the salt formation, using produced water only. The resulting brine is disposed into an existing water disposal well. This option is expected to

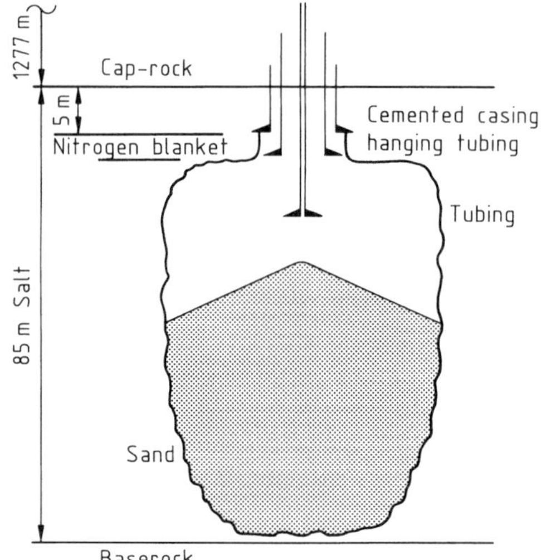



**Figure 32.** Sand disposal cavern

have a long lifetime and to be environmentally and economically sound.

## 4. Economic Development

Alberta's early attempts to develop the oil sands was very sporadic with little or no commercial success. Pilot projects in the 1930s were doomed for a variety of reasons: limited capital, inadequate extraction technology, high development cost, and geographic or climatic impediments. All of these issues had to be addressed and overcome in order to exploit this vast resource; many remain today.

## 4.1. Mining Projects

Alberta's major oil sands deposits for mining are all located north of 56° lattitude. The area is typically forested and much of it is covered with muskeg, a permanently water-saturated soil cover. Northern Alberta is prone to extreme climatic conditions with long and extremely cold winters (average January temperature $-20\,°C$, sometimes $-50\,°C$) and short, hot, humid summers. The infrastructure necessary for industrial development of the area remained undeveloped until the first commercial project in the 1960s. Until then, the region was sparsely populated by a variety of remote indigenous settlements and a small community of 1200 at Fort McMurray.

Development of the Great Canadian Oil Sands Project (now Suncor) in the 1960s was remarkable in many respects, particularly for the long-term vision adopted by the company to scale up and commercialize unproven technology, as well as the corporate risks taken to assure its success. At that time, it was necessary to produce a partially refined product (synthetic crude) to capture a market share. The initial investment in the integrated Suncor project was approximately $\$\,250 \times 10^6$ for synthetic crude oil production of 4930 m$^3$/d. Comparable crude oil typically sold for $\$\,16$/m$^3$ in 1967.

The Syncrude Canada development during the 1970s represented further advances in refining the technology and developing the infrastructure to allow the eventual exploitation of oil sands on a larger scale. It was developed when considerable conventional production was competing for a limited oil market, and crude oil prices in Canada were controlled to less than half their market value. Completed in 1978, the integrated Syncrude project promised to deliver some 20 500 m$^3$ synthetic crude oil per day with an initial capital cost of $\$\,2.3 \times 10^9$. To minimize the corporate risks the project was financed by a partnership of large oil companies and the Alberta and Canadian Governments. Corporate returns were shared through a negotiated fiscal package, and a favorable royalty stream in the early years of production. Alberta's oil sands remain a high-cost source of supply.

Mining of oil sands and production of synthetic crude oil is well established in Alberta. Two large commercial projects deliver ca. 17 % of Canada's oil production (38 000 m$^3$/d). The large reserve base, proven technology, and an established infrastructure all have a positive impact on the commercial viability of oil sands projects, and stable prices should promote further development in the future.

## 4.2. Economic Viability

**Mining Projects.** Development of reserves is hindered by high capital and operating costs and long lead time between project approval and initial production. In 1993, a typical oil sands mining and upgrading project with 13 000 m$^3$/d production was estimated to require investment of $\$\,4-5 \times 10^9$ over a 6–8 year construction period.

New technology and increasing operating efficiency have steadily reduced the cost of bitumen production at Syncrude and Suncor, and costs are close to being competitive with new conventional crude production in Alberta. The unit operating costs of commercial projects are about $\$\,75$/m$^3$, and the projects generate a modest return on investment.

The supply cost of a grassroots project has been estimated at ca. $\$\,150$/m$^3$ (1990 prices). However, significant future cost reductions are expected with the adoption of new mining and extraction technologies.

**In Situ Projects.** In situ recovery of oil sands was tested in the mid-1950s by Shell Canada, Pan American, Mobil Oil, and Petrofina Canada. Early work on steam stimulations and in situ combustion was largely experimental. Higher oil prices in the early 1980s prompted significantly more activity by a variety of firms and technological improvements led to commercial developments.

Although in situ developments are only marginally economic, they offer a number of environmental advantages over mining projects, e.g., limited surface disturbance and no tailings impoundment. Bitumen produced by in situ projects needs to be diluted for transportation by pipeline; it has found a ready market throughout North America.

In situ bitumen production in Alberta was ca. 21 000 m$^3$/d in 1992, including production from nine commercial projects employing thermal and/or primary recovery techniques, and a variety of experimental projects. In the Peace River

**Table 5.** Threshold in situ supply cost for bitumen, 1990 [53]

| Area | Cost, $/m³ |
|---|---|
| **Athabasca** | |
| Steam-assisted gravity drainage | 65 |
| **Cold Lake** | |
| Cyclic steam (superior deposits) | 65 |
| Primary recovery | 50 |
| **Peace River** | |
| Pressure cycle steam drive (superior deposits) | 75 |

area, a steam drive in situ project is operated; in the Cold Lake area, two projects are based on cyclic steam stimulation technology. In 1993, there were 6 other commercial projects producing on primary, while 12 experimental operations were active. Experimental projects are considerably smaller than the commercial projects, and typically operate for less than 5 years, generally producing less than 500 m³/d.

Estimates of threshold bitumen supply costs vary, depending on the source and technology (Table 5), but are typically $ 50–75/m³ [52]. Additional refining costs need to be incurred in order to upgrade bitumen to higher value synthetic crude oil produced at integrated projects. Upgrading is estimated to cost $ 40–60/m³ [54].

Oil sands production is clearly a high-cost source of crude oil at present, and the tenuous commercial viability is likely to require ongoing royalty and fiscal concessions to make projects commercially attractive. Nevertheless, progressive technological improvements and refinements in operating techniques have shown that Alberta's vast reserves are on the frontier of major development.

# 5. Challenges Past and Present

## 5.1. Technology

Technology selection to develop the Suncor project in the 1960s was a major challenge, owing to the limited information available on mining oil sands, extracting the bitumen, and upgrading it. Although experimentation and field testing had occurred, much more work was necessary to modify and apply the available technology. When bucketwheel excavators, used successfully for nearly half a century in the large-scale German lignite mining operations, were applied to oil sands mining in Canadian winter conditions bucketwheel teeth had to be replaced every 18 h. Improved tooth design, blasting and loosening of the oil sands, and revised procedures for operating the bucketwheels during winter eventually solved the problem.

The management of tailings from the extraction of bitumen from oil sands was another huge challenge. Tailings volumes at oil sands mines are enormous compared with general mining experience. The impoundment plan initially prepared by Suncor was inadequate because of the slow fines settling rate. Solutions were found where tailings dykes used the coarse sand from the tailings, and huge impoundments were constructed; however, the challenge of increasing the fine tailings settling rate has yet to be solved.

During the 1970s, while Suncor's focus was on operational problems, Syncrude concentrated on design and construction. Although considerable testing occurred prior to start-up, new challenges arose which required innovations, such as mine depressurization and dewatering ahead of the mine highwall, slope monitoring and stabilization, and improvements to mining and extraction equipment reliability. Syncrude's work on conveyer belting, along with developments by belt manufacturers, greatly improved the operational efficiency of conveyers.

Parallel with this successful start-up at Syncrude, research and development continued on other technology. In bitumen extraction, the hot water process was improved through design modifications to the secondary recovery circuit and addition of the tailings oil recovery. In froth treatment, inclined plate settling has been found to be an effective alternative to the more complex centrifuges, and was added to increase production capability. A naphtha recovery system has also been added which significantly reduces hydrocarbon loss to tailings.

The application of huge fluid cokers at Syncrude has been a remarkable success, bearing in mind the considerable scale-up from any previous installation, and the increase in throughput from the design rate of 11 600 m³/d to the current rate of ca. 17 500 m³/d. The addition of a hydrocracker upstream of the fluid coker is another major innovation by Syncrude, with benefits of increased liquid yield and reduced sulfur emissions.

Challenges and innovations in the in situ recovery area have also been dramatic. Prior to commercial operations in the early 1980s at Imperial Oil's Cold Lake project, various companies conducted extensive in situ tests designed to demonstrate a wide range of in situ recovery techniques. The commercial success of Imperial and Shell is due to the development of steam stimulation processes. Other developments of note are in the area of cluster drilling, the use of slant drilling, horizontal drilling, thermal casing and coupling, and thermal cement. Along with these came a host of production, injection, metering, and control systems unique to the high-temperature, multiphase flows found at in situ operations. The large volumes of water required and produced at steam stimulation schemes led to the development of water reuse technology. The industry developed water treatment and recycle technology which meets environmental and water conservation objectives, and provides energy and cost savings.

Technology development is underway, with emphasis on cost reduction and increased reliability, service factor and product yield. For mining operations, the focus is on materials handling and transportation techniques; for bitumen production, the focus is on a lower-temperature extraction process which fits better with the mining technique. In the upgrading area, the focus is on higher yield and better quality. In the in situ area, combustion techniques have been developed and considerable success has been achieved with the application of horizontal wells which operate at lower pressures and capitalize on gravity drainage.

### 5.1.1. Mining and Bitumen Production Technology

Mining developments have resulted in improvements of truck and shovel capacity and oil sands crushing technology [55]. Mining shovel capacity has increased from the $11.5 \, \text{m}^3$ hydraulic shovel used by Suncor in 1980 to a $46 \, \text{m}^3$ shovel in 1993. Over the same period Suncor and Syncrude have increased truck capacity from 155 to 218 t. Oil sands crushing technology was tested in the early 1980s when a prototype single-roll crusher was tested. This led to the installation of a commercial-scale crusher by Syncrude in the mid-1980s and the installation of a high-speed double-roll crusher in 1991. In 1993, Suncor installed two fixed-plant double-roll crushers rated at $7500 \, \text{t/h}$. Additonally, a low-speed mobile mineral sizer was operated at Suncor in 1990.

As a result of these developments, a truck and shovel mining system is in use at Suncor as a replacement for the early bucketwheel system.

**Hydraulic Transport.** Another major challenge of minable oil sands development is efficient materials handling. Besides the developments in truck and shovel mining, combined with oil sands crushing technology, operators have investigated alternatives, including dredge mining and hydraulic transport of overburden and oil sands. In the mid-1980s, Syncrude and AOSTRA conducted pilot and prototype tests of overburden and oil sands mining, but this technology has not been applied commercially.

In 1988, Syncrude launched a program to develop technology for hydraulic transport of oil sands [56]. The program resulted in the development of an oil sands slurry mixing device, a cyclofeeder, and a process to condition oil sands in a pipeline. This technology, tested at pilot scale in conjunction with a $50 \, ^\circ\text{C}$ bitumen separation process, was found to be feasible. Late in 1993, a commercial test was initiated at an oil sands feed capacity of $1250 \, \text{kg/s}$.

The development of technology to take dry oil sand, slurry it, and pipeline that slurry so that it is conditioned and allows bitumen to be separated, is a major development. This technology, combined with truck and shovel mining, would reduce materials handling costs and energy consumption, and allow remote oil sands reserves to be exploited.

**Alternative Bitumen Extraction.** Commercial operation of the hot water extraction process has shown the need to reduce or eliminate tailings, to increase process energy efficiency, and to reduce water requirement. Technology development in the late 1970s and early 1980s has been assessed in a study commissioned by the ERCB [39]. While the AOSTRA/UMATAC retort process [57] was considered to offer the best long-term commercial potential, its development depended on successful scale-up and proven field demonstration. The commercial development of the Syncrude modified aqueous extraction process was attractive, owing to the relatively low risk involved in scale-up. Unfortunately, tailings disposal requirements for the alternative processes were largely unproven by 1984, and further investigations are ongoing. Recent technology development effort, directed by the ERCB's thrust toward improvement in energy efficiency, tailings management, and environmental acceptability, is focused largely on aqueous processes.

**Minerals Recovery.** The presence of minerals in oil sands and byproducts of processing has been noted, and recovery methods explored. Titanium and zirconium are found in the surface-minable oil sands, and are concentrated in centrifuge tailings streams. Preliminary analysis indicate that tailings would provide sufficient feedstock for large-scale production [52]. Tests indicate that recovery is technically feasible. Tests on fine tailings have also indicated the potential for alumina. Minerals may be recovered from coke, fly ash, and spent catalyst; e.g., vanadium in coke, fly ash, and catalyst from the Syncrude project are estimated at 282 kg/h, 18 kg/h, and 45 kg/h, respectively [58].

### 5.1.2. In Situ Recovery Technology

**In Situ Combustion.** In situ combustion, also known as fireflooding, is an in situ thermal process that has been tested, but not developed commercially [59]. The combustion process generates heat in situ via hydrocarbon combustion reactions.

The process is initiated by igniting the reservoir bitumen. Spontaneous ignition may occur, or ignition may have to be induced, depending on the characteristics of the bitumen. Air, or air enriched with oxygen, is injected continuously to sustain and propagate the firefront away from the injection well toward the production wells. When the firefront moves in the same direction as the injected air the process is "forward" combustion. If air is injected continuously until it reaches the producer and is then ignited, the combustion front moves backward from the producer, "reverse" combustion. The efficiency of the dry combustion process can be improved by alternately injecting water and air to create the wet combustion process. The advantages of wet combustion are that water, because of its greater heat capacity, is better at scavenging heat from the burned zone than air, and that the higher viscosity of the water phase results in a more favorable distribution of the injected air, and a greater sweep efficiency of the hydrocarbons.

The most common method is forward combustion (Fig. 33) [60]. Moving from the injector to the producer, injected air first encounters the burned zone. The combustion front has passed through this zone, leaving only hot, clean sand with air-saturated pores. The air passing through this zone is preheated as it progresses to the combustion front. The heated air passes through the combustion zone itself where it oxidizes the coke left from the thermal cracking of the oil in-place

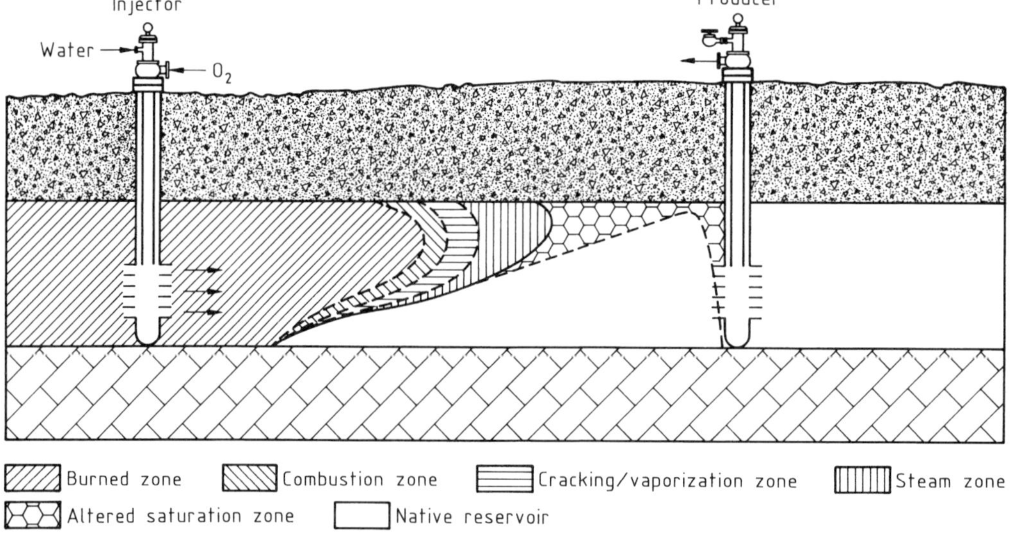


**Figure 33.** Forward in situ combustion

in the cracking/vaporization zone, temperatures are sufficiently high to crack the oil thermally, and to vaporize any hydrocarbons and water. Just beyond the thermal cracking zone, combustion gases, steam, and vaporized light hydrocarbons (steam zone) are transported forward, where vaporized water and hydrocarbons begin to condense (altered saturation zone). The last zone is the native reservoir which has not yet been affected by the advancing fire. The residual gases flow through this zone toward the production wells.

**Horizontal Wells for Thermal Recovery.** The in situ recovery of bitumen by horizontal wells presents new opportunities for efficient recovery and improved project economics. Commercial bitumen recovery currently involves close well spacing and in such cases horizontal wells can be expected to be most effective.

Additionally, horizontal wells improve productivity without coning since the pressure differential is distributed over a larger section of the reservoir [61].

Steam-assisted gravity drainage (SAGD), is a form of the steamflood recovery process that utilizes the effects of gravity and the improved reservoir contact achievable with horizontal wells. Steam is injected continuously, by a vertical or horizontal injector, above the horizontal producer, located at the base of the reservoir. The steam moves upward and sideways (Fig. 34 A) or downward and sideways (Fig. 34 B), and forms a steam chamber [62]. As the steam contacts the interface, it heats the bitumen near the interface. Owing to its reduced viscosity, the bitumen at the interface drains, by gravity, toward the horizontal producer. Although the rate of drainage at any point along the horizontal producer is relatively low, an overall high recovery rate can be achieved by utilizing a long horizontal production well.

Steam-assisted gravity drainage with underground access (SAGDUA) (Fig. 35) is also based on the SAGD process, but these horizontal wells are drilled from a mine shaft and tunnel system in the limestone underlying the bitumen-saturated zone [63]. Drilling costs are reduced and, by injecting steam directly from the mine tunnel, surface and wellbore heat losses are minimized, improving project economics. Highly promising results based on a full-scale field demonstration project indicate that commercial development of the Athabasca bitumen deposits too deeply

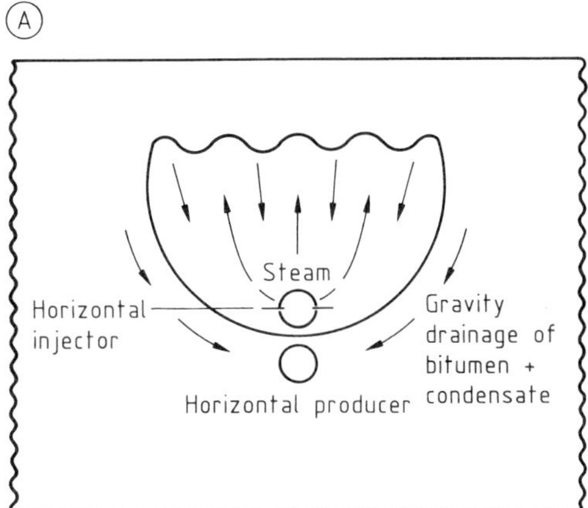


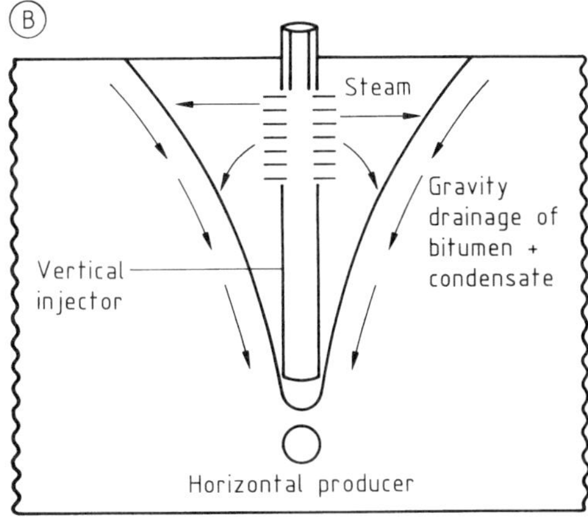


**Figure 34.** Steam-assisted gravity drainage
A) Upward/sideways movement of steam; B) Downward/sideways movement of steam

buried for surface mining is becoming possible through the application of this process.

The heated annulus steam drive (HASD) process, (Fig. 36) is a form of the steam drive process that circulates steam through a horizontal well [64]. This creates a zone of reduced bitumen viscosity, between an injector and producer. The horizontal well is not perforated so there is no fluid communication between the steam in the horizontal conduit and the formation. The zone of reduced bitumen viscosity results in increased bitumen mobility, and allows steamflooding to occur along the horizontal conduit toward the producing well. The injection of drive steam is maintained below fracture pressure so that control of the communication path is not lost.

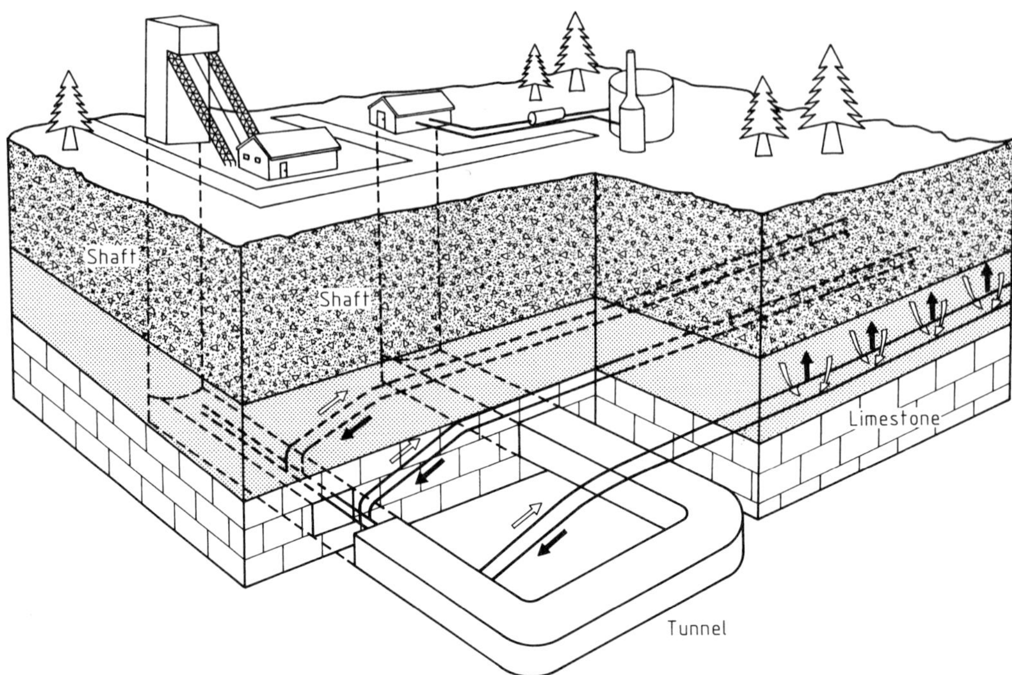



**Figure 35.** AOSTRA's field pilot of the SAGDUA process

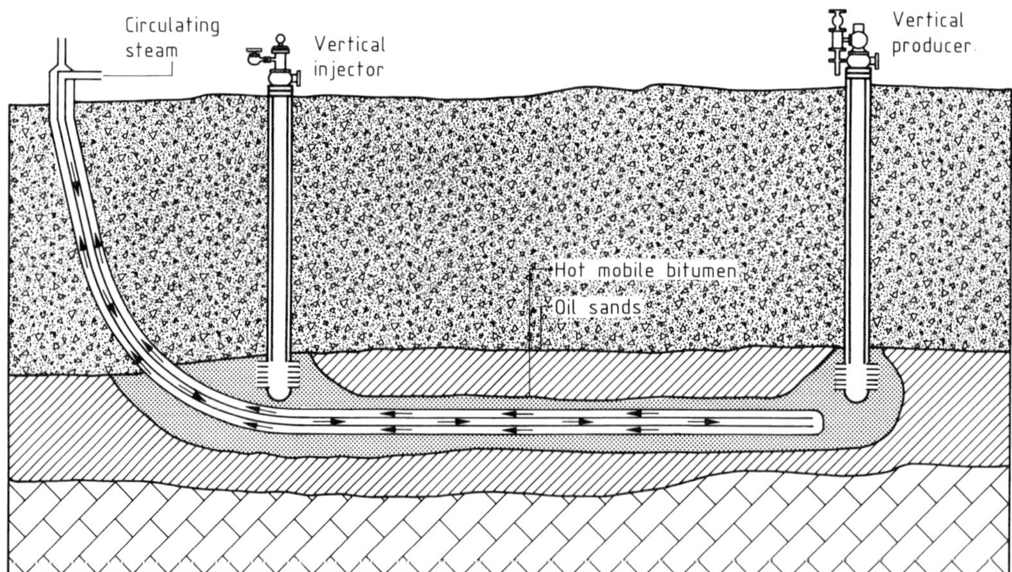



**Figure 36.** Horizontal configuration of HASD process

### 5.1.3. Bitumen Upgrading Technology

Upgrading improvements for oil sands are of interest because ca. 50 % of the cost of producing synthetic crude oil is estimated to be due to upgrading. This, combined with the need for improved product quality to meet marketing and environmental requirements, has resulted in considerable interest and activity on bitumen upgrading and distillate product hydrotreating. Technologies considered include visbreaking, coking, and catalytic hydrotreating and hydrocracking. While progress has been made in visbreaking, coking, and fixed-bed hydrotreating technologies, the major emphasis is on catalytic hydrocracking [65], [66]. Although fixed-bed hydrodesulfurization technology and H-Oil and LC-Fining technologies have been available commercially for some time in a low-conversion mode, these have not been designed and operated commercially in the high-conversion mode. In recent years, slurry phase hydrocracking processes such as Veba Combi cracking (VCC), CANMET, and MICROCAT RC, have been extensively tested on a demonstration scale.

Slurry phase hydrocracking technology is based on German coal hydrogenation technology. The process involves operation at $\geq 450\,°C$ and $27-69$ MPa, to crack residuum in the presence of a finely divided powder, which can be an additive or a catalyst. The powder serves both as a catalyst and as a site where coke can be deposited rather than depositing on process equipment. Such processes convert $> 90$ wt% of the heavier hydrocarbon, and achieve volumetric yields of ca. 105%. The VCC process has been demonstrated on the $475\,m^3/d$ scale in Bottrop, Germany, and the CANMET process has been demonstrated on the $715\,m^3/d$ scale in Montreal, Quebec, Canada. MICROCAT RC is another additive residuum hydrocracking process recently tested on a $1.3\,m^3/d$ pilot scale. The high conversions achieved with these technologies minimize air emissions and the production of unmarketable byproducts such as coke or pitch.

A unique feature of VCC technology is the integration of a fixed-bed hydrotreater with the high-pressure hydrocracking process. This minimizes costs in current operations associated with the pressure and temperature reduction of products of coking or hydrocracking, and reduces capital and operating costs.

## 5.2. Environmental Issues

Environmental issues facing oil sands development are primarily solids waste management, protection and conservation of potable water resources, reducing air emissions, and land reclamation. These issues are driven by more stringent regulatory standards and public expectations.

Fine tailings management and reclamation are the greatest challenges for the minable oil sands sector. Extensive study has identified the fundamental properties of fine tailings which give it an intractable gel-like structure and toxic character [67]. This has indicated ways to reduce fine tailings production, to treat fine tailings, and to reduce their bulk. The fine particles of size $0.1-1.0$ µm in the presence of organic salts produced in the hot water process form the gel, which hinders water mobility. Reduced fines dispersion in a gentler oil sands conditioning process and reduced organic salt generation, by eliminating sodium hydroxide, should reduce fine tailings formation. Alternative extraction processes based on these ideas are being developed.

Reclamation techniques have been proposed for fine tailings, but have not been adopted. The principal options are to produce a dry, nontoxic, trafficable material or to cap fine tailings with water and allow then to detoxify and consolidate over an extended period. In both cases, the tailings would be contained in mined-out pits. Testing has indicated both methods have potential, but further work is required to confirm feasibility and safety. It is expected that, through the development of improved bitumen extraction and/or fine tailings treatment technology, future projects may not produce fine tailings at all.

The existing commercial projects burn the high-sulfur coke byproduct of upgrading bitumen, and consequently have significant $SO_2$ emissions. This and other combustion sources also produce $CO_2$, $NO_x$, and particulates; $CO_2$ is also a byproduct of the production of hydrogen by steam methane reforming. Volatile organic carbons (VOCs) are another source of atmospheric emission of some concern, and various measures are contemplated to reduce their effects. Particulate control in coke combustion operations is achieved by electrostatic precipitators. Studies have indicated the potential to recover $CO_2$ for use in enhanced oil recovery and chemical manufacture.

The need for large quantities of freshwater in oil sands recovery schemes has required maximizing water recycle and, in areas of Alberta experiencing freshwater shortages, the development of alternative sources such as brackish water and freshwater from adjacent water basins. These are feasible, but at added cost.

The disposal of process wastewater is another factor in water use at oil sands schemes. At in situ operations wastewater is disposed of to deep formations. As yet, mining projects wastewater is contained on site; in the long-term, treatment and discharge to surface is likely.

The integrity of wells, and in particular the potential to contaminate potable water aquifers, is a concern. Development of pipe material, couplings, and cement to handle the thermal application has resulted in few failures.

## 5.3. Public Policy

Broad policy issues involving the oil sands industry typically center on fiscal regimes to foster development by private interests, and the establishment of public infrastructure to handle the influx of population to remote or sparsely populated parts of Alberta. The bitumen/synthetic crude industry is characterized by a highly skilled workforce. Commercial development has tended to call for very large unit operations, requiring capital investment in the order of $ 10^9$, and thousands of employees. Economic and social factors play a major role in the pace of development. Given the personnel and technical constraints that may apply, economic issues center largely on the rate of return on invested capital, after allowances for royalties and taxes. Social issues focus largely on infrastructure development and impacts on the social fabric. It is estimated that the cost of infrastructure for a new town requires an expenditure $> \$ 800 \times 10^6$.

The economic benefits to the public and private sectors from large oil sands developments are very significant and usually dominate negotiations of the agreement for a commercial project. The direct social benefits include improvement in the scope and quality of community infrastructure (hospitals, recreation facilities, and transportation systems) although the large transient workforce required to build a project may produce significant dislocation in the community during the construction period.

Current developments have shown that, if the projects are distributed among existing and new communities, long-term community impacts are manageable. Existing institutions find it difficult to accommodate the rapid growth; issues to be considered include:

1) The need for large and advanced municipal financing of infrastructure
2) Tax-sharing arrangements
3) Rights of the indigenous population
4) Training and recruitment of the indigenous population
5) Compensation and relocation of residents affected by the project
6) The availability and affordability of houses
7) Environmental monitoring, changes in land use and land reclamation
8) Provision of adequate social facilities
9) Provision of health care, social services, education, and law enforcement
10) Transportation routes and services

These factors, if not fully resolved in a timely fashion, can constrain the pace of development.

Large bitumen/synthetic oil projects require a permanent workforce ranging from 2100 for in situ schemes to $> 3000$ for mining projects. Many of these positions can be filled by the construction workers involved in developing the site, but special technical skills are also required which are normally acquired through formal secondary education.

Approximately 10 years is required from the project definition phase until plant start-up. Major components of this period are: the $1-2$ years required to obtain regulatory agency and government approvals; ca. $1-1\frac{1}{2}$ years for process design and preparation of a detailed estimate; and a ca. 3-year detailed engineering period. The detailed engineering period overlaps plant project construction, which normally takes ca. 5 years. Start-up operations would likely be conducted over at least a 1-year interval.

A construction workforce of 5000 is required over a 5-year period, with a peak value of 8000–10 000. Following the construction stage, operating staff amount to 2500–3000 per project, and the expected result is a total population increase of 15 000–20 000 per operational plant. This can place a great demand on the infrastructure, and create great need for additional government and private servicing to provide the necessary schools, hospitals, roads, and other services.

While the resource base in Alberta's oil sands is immense, the financial risks and the logistics may place limits on the rate of synthetic oil supply in the foreseeable future.

## 6. References

[1] K. Takamura: "Microscopic Structure of Athabasca Oil Sand," *Can. J. Chem. Eng.* **60** (1982) 538.

[2] O. P. Strausz: "Bitumen and Heavy Oil Chemistry," in L. G. Hepler, C. Hsi (eds.): *AOSTRA Technical Handbook on Oil Sands, Bitumens and Heavy Oils,* AOSTRA, Edmonton, AB, 1989, p. 35.

[3] B. D. Ricketts: *Western Canada Sedimentary Basin, A Case History,* Canadian Society of Petroleum Geologists, Calgary, AB, 1989, pp. 3–8.

[4] J. W. Porter, R. A. Price, R. G. McCrossan, *Philos. Trans. R. Soc. London* **305** (1982) 169–192.

[5] Energy Resources Conservation Board Publications: Oil and Gas Conservation Board: *A Description and Reserve Estimate of the Oil Sands of Alberta,* Calgary, AB, 1963.
H. J. Webber: "The Oil Sands of Alberta," *J. Can. Pet. Technol.* **6** (1967) no. 4, 146–149.
Energy Resources Conservation Board: "Geology and Proved In-Place Reserves of the Cold Lake Oil Sands Deposits," *ERCB Report 73-L-Geol,* Calgary, AB, 1974.
Energy Resources Conservation Board: "Geology and Proved In-Place Reserves of the Peace River Oil Sands Deposits," *ERCB Report 74-R,* Calgary, AB, 1974.
Energy Resources Conservation Board: "Geology and Proved In-Place Reserves of the Wabasca Oil Sands Deposits," *ERCB Report 76-A,* Calgary, AB, 1976.
C. P. Outtrim, R. G. Evans: "Alberta's Oil Sands Reserves and Their Evaluation," *Annu. Tech. Meet. Pet. Soc. CIM* (1977).
Energy Resources Conservation Board: "Crude Bitumen Reserves Atlas," *ERCB ST 91-38,* Calgary, AB, 1990.
Energy Resources Conservation Board: "Alberta Reserves of Crude, Oil, Oil Sands, Gas, Natural Gas Liquids, and Sulphur," *ERCB ST 93-18,* Calgary, AB, 1993.

[6] D. Wightman, B. Rottenfusser, J. Kramers, R. Harrison: "Geology of the Alberta Oil Sands Deposits," in L. G. Hepler, C. Hsi (eds.): *AOSTRA Technical Handbook on Oil Sands, Bitumens and Heavy Oils,* AOSTRA Tech. Pub. Ser. no. 6, Edmonton, AB, 1989, pp. 1–9.

[7] J. A. Masters: "Lower Cretaceous Oil and Gas in Western Canada," in J. A. Masters (ed.): Elmworth – Case Study of a Deep Basin Gas Field, *Mem. Am. Assoc. Pet. Geol.* **38** (1984) 1–33.

[8] S. O. Moshier, D. W. Waples: "Quantative Evaluation of Lower Cretaceous Mannville Group as Source Rock for Alberta's Oil Sands," *AAPG Bull.* **69** (1985) 161–172.

[9] M. J. Ranger, G. Pemberton: "Trapping Mechanism and Time of Oil Migration in the Athabasca Oil Sands Deposit," *Am. Assoc. Pet. Geol. Annu. Conf. Proc.* 1992, 108.

[10] G. Deroo, T. G. Powell, B. Tissot, R. G. McCrossan: "The Origin and Migration of Petroleum in the Western Canadian Sedimentary Basin, Alberta," *Bull. Geol. Surv. Can.* **262** (1977) 114.

[11] R. S. Strobl, L. P. Yuan, W. K. Muwais: "Reservoir Geological Modelling of the McMurray Formation, Alberta," *Bull. Can. Pet. Geol.* **39** (1991) no. 2, 225.

[12] D. A. W. Keith et al.: "Resource Characterization of the McMurray/Wabiskaw Deposit in the Athabasca Central Region of Northeastern Alberta," AOSTRA Tech. Pub. Ser. no. 7, Edmonton, Alberta, 1990.

[13] G. D. Mossop, J. W. Kramers, P. D. Flach, B. A. Rottenfusser: "Geology of Alberta's Oil Sands and Heavy Oil Deposits," *1st UNITAR Conference on Future of Heavy Crude Oils and Tar Sands,* 1981, 197.

[14] J. R. MacGillivray et al.: "Resource Characterization of the McMurray/Wabiskaw Deposit in the Athabasca South Region of Northeastern Alberta," AOSTRA Tech. Pub. Ser. no. 8, Edmonton, AB, 1992.

[15] R. S. Harrison: "The Bitumen – Bearing Paleozoic Carbonate Trend of Northern Alberta," *AAPG Stud. Geol.* **25** (1987) 319.

[16] D. Marion, B. Rottenfusser: "Geology of the Peace River Giant Oil Sands Deposit, Alberta, Canada," *Proc. of Unconventional Hydrocarbon Sources Conference,* St. Petersburg 1992.

[17] J. E. Barclay, F. F. Krause, R. I. Campbell, J. Utting: "Dynamic Casting and Growth Faulting: Dawson Creek Graben Complex, Carboniferous – Permian Peace River Embayment, Western Canada," *Bull. Can. Pet. Geol.* **38 A** (1990) 115–145.

[18] S. C. O'Connell, G. R. Dix, J. E. Barclay: "The Origin, History, and Regional Structural Development of the Peace River Arch, Western Canada," *Bull. Can. Pet. Geol.* **38 A** (1990) 4–24.

[19] B. M. Benyon, S. G. Pemberton: "Resource Characterization of the Lower Cretaceous Grand Rapids Formation: Reservoir Geology and Hydrocarbon Distribution, Cold Lake Oil Sands Area, AB, " *AOSTRA J. Res.* **8** (1992) 108–130.

[20] D. P. Leyden: "Geologic Reservoir Types Challenging the Future Development of the Clearwater Formation at Cold Lake, Alberta, Canada," *Proc. of Unconventional Hydrocarbon Sources Conference,* St. Petersburg 1992.

[21] D. Wightman, T. Berezniuk: "Resource Characterization and Depositional Modelling of the Clearwater Formation, Cold Lake Oil Sands Area, East Central Alberta," in *The Canadian Heavy Oil Association Handbook,* Calgary, AB, 1991, pp. 267–282.

[22] Energy Resources Conservation Board, Alsands Fort McMurray Project, Decision Report D 79-H, Calgary, AB, 1979.

[23] Energy Resources Conservation Board, Oil Sands Annual Statistics, Publication ST 92-43, Calgary, AB, 1992.

[24] E. C. McRoberts: "Waste Management For Surface Mineable Oil Sands Mines," *1st District 5 Meeting, Can. Inst. of Mining and Metallurgy,* Fort McMurray, AB, 1983.

[25] R. Lahaie, W. K. Chan: "Large Dykes of the Suncor Oil Sands Mine," *3rd District 5 Meeting Can. Inst. of Mining and Metallurgy,* Calgary, AB, 1987.

[26] B. Purcell, A. Slawinski: "Basal Aquifer Dewatering Project at the Suncor Oil Sands Mine," *Can. Geotech. Conference,* Regina, Sask., 1987.

[27] M. B. Dusseault, N. R. Morgenstern: "Locked Sands," *J. Eng. Geol.* **12** (1979) 117–131.

[28] R. Doucet, B. Romanchuk: "Drilling and Blasting in Tarsand," *Proc. of the 11th Conference on Explosives and Blasting Techniques,* San Diego, CA, 1985.

[29] A. E. Fair: "Development of Preblasting Techniques to Stabilize Mine Highwall Slopes in Areas of Steeply Dipping Clay Beds," Occupational Health & Safety Division – Mines Inspection Branch – Seminar on Slope Stability, Edmonton, AB, 1985.

[30] C. Fordham, E. C. McRoberts, B. Purcell, P. McLaughlin: "Practical and Theoretical Problems Associated With Blast Densification of Loose Sands," *Can. Geotech. Conference,* Calgary, AB, 1991.

[31] N. R. Morgenstern, A. E. Fair, E. C. McRoberts: "Geotechnical Engineering Beyond Soil Mechanics – A Case Study," 25th Anniversary Special Paper, *Can. Geotech. J.* **25** (1988) 637–661.

[32] A. E. Fair, B. A. Isaac: "Highwall Stability Monitoring at Syncrude Canada Ltd. Open Pit Mining Operation," *Proceedings of the 13th Annual Mining and Metallurgy Industries Symposium and Exhibit*, Salt Lake City, UT, 1985.

[33] K. A. Clark, D. S. Pasternack: "Hot Water Separation of Bitumen from Alberta Bituminous Sands," *Ind. Eng. Chem.* **24** (1932) 1410.

[34] K. A. Clark: "The Hot Water Washing Method for Recovery of Oil from the Alberta Tar Sands," *Can. Oil Gas Ind.* **3** (1950) no. 6, 46.

[35] E. C. Sanford, "Processability of Athabasca Oil Sand: Interrelationship between Oil Sand Fine Solids, Process Aids, Mechanical Energy and Oil Sand Age after Mining," *Can. J. Chem. Eng.* **61** (1983) 554.

[36] L. L. Schramm, R. G. Smith, *AOSTRA J. Res.* **5** (1989) 87.

[37] L. L. Schramm: "The Influence of Suspension Viscosity on Bitumen Rise Velocity and Potential Recovery in the Hot Water Flotation Process for Oil Sands," *JCPT J. Can. Pet. Technol.* **28** (1989) no. 3, 73.

[38] R. N. Houlihan, K. H. Williams: "Recent Enhancement in Mined Oil Sands Bitumen Extraction Technology," *JCPT J. Can. Pet. Technol.* **26** (1987) no. 1, 91.

[39] Dynawest Projects Ltd.: "Oil Sands Bitumen Extraction Process Evaluation," Alberta Energy Resources Conservation Board, Calgary, AB, 1984.

[40] Suncor Bitumen Recovery – Ore Grade/Fines Correlation, private communication, 1993.

[41] R. N. Tipman, R. C. Shaw: "Recent Advances in the Treatment of Oil Sands Froth," *Oil Sands—Our Petroleum Future Conference*, preprint, Edmonton, AB, April 1993.

[42] L. T. Yano, A. E. Fair: "Overview of Hydraulic Construction Techniques Utilised at the Syncrude Canada Limited Tailings Pond," *Specialty Conference on Hydraulic Fill Structures*, Fort Collins, CO, 1988.

[43] M. MacKinnon, A. Sethi: "A Comparison of the Physical and Chemical Properties of the Tailings Ponds at the Syncrude and Suncor Oil Sands Plants," *Oil Sands – Our Petroleum Future Conference*, Edmonton, AB, Apr. 4–7, 1993.

[44] A. E. Fair, G. T. Handford: "Overview of Tailings Dyke Instrumentation Program at Syncrude Canada Limited," *International Symposium on Geotechnical Stability in Surface Mining*, Calgary, AB, 1986.

[45] W. K. Chan, J. Martschuk, E. C. McRoberts: "Performance Review of Tar Island Dyke," *7th Panamerican Conference on Soil Mechanics and Foundation Engineering*, Vancouver, B.C., 1983.

[46] A. W. Hyndman, R. W. Luhning, *JCPT J. Can. Pet. Technol.* **30** (1991) no. 2, 61.

[47] R. W. Luhning, "Heavy Oil, Oil Sands and Enhanced Oil Recovery; Where Will the Technology Breakthrough Come From," *Canadian Heavy Oil Association Conference on Heavy Oil – A New Direction*, Calgary, AB, Dec. 5, 1989.

[48] P. R. Kry: "In Situ Recovery Technology at Cold Lake," *Oil Sands – Our Petroleum Future Conference*, Edmonton, AB, Apr. 4–7, 1993.

[49] L. M. Goobie, H. L. Chang: "The Evolution of a Successful Recovery Scheme for the Peace River Oil Sands, Alberta, Canada," in [48].

[50] M. J. Zatka, R. W. Kangen, J. I. Chambers: "Progress at Shell's Peace River In-Situ Recovery Operation," *AOSTRA Oil Sands 2000 Conference*, Edmonton, AB, March 26–28, 1990.

[51] M. Dusseault: "Cold Production and Enhanced Oil Recovery," *JCPT J. Can. Pet. Technol.* **32** (1993) no. 9.

[52] J. W. Kramers, R. A. S. Brown, *CIM Bull.* **69** (1976) 92.

[53] S. Singh et al.: "Technical and Economic Factors for Establishing Hydrocarbon Resource Cost Thresholds," *Proc. of the 13th World Petroleum Congress*, Buenos Aires 1991.

[54] National Energy Board: Canadian Energy Supply and Demand: 1990–2010, Calgary, June 1991.

[55] A. E. Fair, W. A. Hill, F. R. Payne, B. Bruce: "Economic Comparison of Oil Sand Surface Mining Technologies," *Oil Sands – Our Petroleum Future Conference*, Edmonton, AB, Apr. 4–7, 1993.

[56] W. Maciejewski, R. J. Oxenford: "Hydrotransport – An Enabling Technology for Future Oil Sands Development," in [55].

[57] R. Koszarycz, R. Padamsey, R. M. Ritcey: "Synthetic Crude Oil from Retorted Mined Oil Sands – Economic Facts and Misconceptions," in [55].

[58] Syncrude Application to ERCB, No. 921 321, Energy Resources Conservation Board, Calgary, AB, Sep. 1992.

[59] R. G. Moore et al.: "A Canadian Perspective on In Situ Combustion," *Joint Canada/Romania Heavy Oil Symposium*, Sinaia, Romania, 1993.

[60] R. G. Moore, D. W. Bennion, M. G. Ursenbach: "A Review of In Situ Combustion Mechanisms," *4th UNITAR/UNDP International Conference on Heavy Crude and Tar Sands*, Edmonton, AB, Aug. 7–12, 1988.

[61] R. M. Butler: "The Potential for Horizontal Wells for Petroleum Production," *Annual Technical Meeting of the Petroleum Soceity of CIM*, Calgary, AB, 1988.

[62] K. H. Chung, R. M. Butler: "A Theoretical and Experimental Study of Steam-Assisted Gravity Drainage Process," *4th UNITAR/UNDP International Conference on Heavy Crude and Tar Sands*, Edmonton, AB, Aug. 7–12, 1988.

[63] D. P. Komery, J. C. O'Rourke, J. I. Chambers: "AOSTRA Underground Test Facility UTF Phase B Implications for Commercialization," *Oil Sands–Our Petroleum Future Conference*, Edmonton, AB, Apr. 4–7, 1993.

[64] D. J. Anderson: "The Heated Annulus Steam Drive Process for Immobile Tar Sands," *4th UNITAR/UNDP International Conference on Heavy Crude and Tar Sands*, Edmonton, AB, Aug. 7–12, 1988.

[65] B. L. Schulman, R. L. Dickenson: "Upgrading Heavy Crudes, A Wide Range of Excellent Technologies now Available," *5th UNITAR/INDP International Conference on Heavy Crudes and Tar Sands*, Caracas, Venezuela, 1991, p. 105.

[66] H. D. Sloan, *Hydrocarbon Process.*, Nov. 1991, 99.

[67] D. E. Sheeran: "An Improved Understanding of Fine Tailings Structure and Behavior," Proceedings Fine Tails Symposium, *Oil Sands – Our Petroleum Future Conference*, Edmonton, AB, Apr. 4–7, 1993.

**Tara Gum → Starch and Other Polysaccharides**

# Tartaric Acid

JEAN-MAURICE KASSAIAN, ETS Legré-Mante, Marseilles, France



## 1. Introduction

Tartaric acid is a mixed diacid and dialcohol, empirical formula $C_4H_6O_6$.

It is also known as acidum tartaricum, 2,3-dihydroxysuccinic acid and, according to the official IUPAC nomenclature, 2,3-dihydroxybutanedioic acid. The structural formula is:

$$HOOC-\overset{\displaystyle H}{\underset{\displaystyle OH}{C}}-\overset{\displaystyle OH}{\underset{\displaystyle H}{C}}-COOH$$

The molecule contains two asymmetric carbon atoms; there are four stereoisomeric forms.

**Optically Active Forms.** L(+)-Dextrotartaric acid and D(−)-levotartaric acid rotate the plane of polarization of polarized light. These two forms, which are enantiomers, have identical physical and chemical properties. The optical rotatory powers are equal, but of opposite sign.

**Optically Inactive Forms.** DL-Tartaric acid (racemic tartaric acid), obtained by chemical synthesis, is an equimolar mixture of the two enantiomers; *meso*-tartaric acid (unresolvable tartaric acid) is inactive owing to internal compensation. These two forms have no effect on polarized light. They have the same chemical properties (identical to the active forms), but differ in certain physical properties from each other and the optically active acids.

Tartaric acid, known from antiquity as its acid potassium salt (a sparingly soluble compound which forms a deposit in wine vats), was characterized by the Swedish chemist SCHEELE, who extracted it by decomposing its calcium salt with sulfuric acid. This extraction method is still currently used. BERZELIUS, another illustrious Swedish chemist, established the structural formula in 1830. Between 1848 and 1860, PASTEUR made one of the most fundamental discoveries in organic chemistry while studying the asymmetry and optical activity of salts of tartaric acid, establishing the relationships between racemic tartaric acid and levotartaric and dextrotartaric acids.

PASTEUR showed that, if sodium ammonium tartrate is crystallized below 28 °C, a tetrahydrate is obtained in which half the crystals have supplementary hemihedral faces oriented to the left, and the other half have hemihedral faces oriented to the right. He separated these crystals by hand, using a lens, and showed that some of them yielded a new levorotatory acid, and that the others yielded the already known dextrorotatory acid [1].

L(+)-Dextrotartaric acid is the only natural form and the main form produced commercially.

It is widely distributed in plants, either in its free form or as salts. It is found in many fruits, especially in grapes, where it occurs particularly as the acid potassium salt, and to a small extent as the neutral calcium salt.

During vinification, and as alcoholic fermentation proceeds, the acid potassium tartrate becomes insoluble and crystallizes on the vat walls. This deposit, called tartar, is collected and used as starting material in the production of L(+)-tartaric acid.

## 2. Physical Properties

The physical properties described below refer to natural L(+)-tartaric acid. They are identical to those of D(−)-tartaric acid. A comparison of the characteristic properties of the isomers are given in Table 1 [1]–[6].

Tartaric acid has molecular mass $M_r$ 150.09. It crystallizes in the anhydrous state from solution above 5 °C, as large colorless monoclinic needles, density 1.7598 g/cm$^3$, melting point 169–170 °C. It has a strong acidic taste and is odorless. It is stable in air.

It decomposes on heating above 220 °C, with an odor of caramelized sugar. Tartaric acid is very soluble in water.

At various temperatures the solubility is:

| $t$, °C | 0 | 5 | 10 | 20 |
|---|---|---|---|---|
| g/100 g H$_2$O | 115 | 120 | 125 | 139 |

| $t$, °C | 30 | 40 | 50 | 60 |
|---|---|---|---|---|
| g/100 g H$_2$O | 156 | 176 | 195 | 218 |

| $t$, °C | 70 | 80 | 90 | 100 |
|---|---|---|---|---|
| g/100 g H$_2$O | 244 | 273 | 307 | 343 |

It is also soluble in alcohol: 100 g alcohol dissolves 20.4 g tartaric acid at 18 °C; it is less soluble in ether: 100 g diethyl ether dissolves 0.3 g tartaric acid at 18 °C.

Tartaric acid solutions have a rotatory power that varies according to concentration:

$$[\alpha]_D^{20} = 15.050 - 0.1535\, c$$

where $c$ is the concentration of tartaric acid, in the range 20–50 % wt/vol. At 20 % wt/vol, $[\alpha]_D^{20} = +11.98$ (−11.98 for the D(−)-form).

Its enthalpy of combustion is 1149.9 kJ/mol. Its specific heat capacity (0–100 °C) is 1237 kJ kg$^{-1}$ K$^{-1}$.

Tartaric acid is a diacid. Its electrolytic dissociation constants at 25 °C are $K_1 = 1.17 \cdot 10^{-3}$, $K_2 = 5.0 \cdot 10^{-5}$.

At various temperatures (°C) the relative density $d_4^{15}$ of solutions is as follows:

| 1 | 10 | 20 | 30 | 40 | 50 |
|---|---|---|---|---|---|
| 1.0045 | 1.0469 | 1.0969 | 1.1505 | 1.2078 | 1.2696 |

The boiling points of solutions are: 102.2 °C for a 25 % solution, 106.7 °C for a 50 % solution.

The refractive index at the melting point $n_D^{170}$ is 1.464.

## 3. Optical Isomerism: Configuration

The conventions using the symbols $l$, $d$, L, D to denote the enantiomers can lead to confusion, and an improved system of nomenclature for compounds containing one or more chiral centers was developed by CAHN, INGOLD, and PRELOG in 1960. It uses the symbols $R$ (Latin, *rectus*,

**Table 1.** Physical properties of the isomers of tartaric acid [1]–[6]

|  | L(+), D(−) Tartaric acid | Racemic tartaric acid | *meso*- Tartaric acid |
|---|---|---|---|
| Crystal form | monoclinic prisms | triclinic prisms | tetragonal plates |
| Density, g/cm$^3$ | 1.7598 | 1.788 | 1.666 |
| Melting point, °C (anhydrous) | 169–170 | 206 | 159–160 |
| Solubility, g/100 g H$_2$O (25 °C) | 147 | 25 | 167 |
| Solubility of potassium bitartrate, g/100 g H$_2$O (20 °C) | 0.57 | 0.56 | 13.2 |
| Solubility of calcium salt, g/100 g H$_2$O (20 °C) | 0.026 | 0.004* | 0.034 |
| Water molecules in hydrated Ca salt | 4 | 8 | 3 |

* 25 °C.

right) and *S* (*sinister*, left). This system enables the compounds to be classified without ambiguity [7], [8]:

$$
\begin{array}{ccc}
\text{COOH} & \text{COOH} & \text{COOH} \\
\text{H}-\text{C}-\text{OH} & \text{HO}-\text{C}-\text{H} & \text{H}-\text{C}-\text{OH} \\
\text{HO}-\text{C}-\text{H} & \text{H}-\text{C}-\text{OH} & \text{H}-\text{C}-\text{OH} \\
\text{COOH} & \text{COOH} & \text{COOH} \\
\mathbf{1} & \mathbf{2} & \mathbf{3}
\end{array}
$$

L(+) or (2*R*-3*R*) or [*R*-(*R**,*R**)]tartaric acid [*87-69-4*]    D(−) or (2*S*-3*S*) or [*S*-(*R**,*R**)]tartaric acid [*147-71-7*]    *meso*-tartaric acid [*147-73-9*]

DL-Racemic tartaric acid, an equimolar mixture of (**1**) and (**2**), has CAS No. [*133-37-9*].

## 4. Chemical Properties

**Action of Heat.** Tartaric acid melts between 170 and 180 °C, and isomerizes without loss of water to metatartaric acid. If heating is continued, amorphous anhydrides are obtained, which regenerate tartaric acid when boiled with water. At about 220 °C tartaric acid decomposes, swells, and ignites, leaving a carbonized residue.

**Oxidation.** Tartaric acid is very sensitive to oxidizing agents. The Fenton reaction involving oxidation with hydrogen peroxide in the presence of ferrous sulfate leads to the formation of dioxymaleic acid.

**Reduction.** Reduction with hydriodic acid leads to the formation of succinic acid.

**Complexing Action.** Tartaric acid acts as a complexing agent, preventing the precipitation of heavy metal salts by bases.

**Boiling.** Prolonged boiling of L(+)-tartaric acid solutions in the presence of alkalis (potassium and sodium hydroxides) yields the racemic and meso acids [1], [3].

## 5. Resources and Starting Materials

L(+)-Tartaric acid is the only one of the four isomers to be produced industrially on a large scale. The starting materials are exclusively natural residues derived from vinification. The tartaric acid in these residues exists mainly in the form of potassium bitartrate and, to a lesser extent, calcium tartrate. Depending on their tartaric acid concentration and composition, these materials are classified under four headings [9]−[12].

**Tartar.** This is deposited by crystallization on the walls of wine vats as the ethyl alcohol concentration increases. Tartar contains 80−90 % potassium bitartrate.

**Lees.** These are deposited on the bottom of wine vats and contain 19−38 % potassium bitartrate.

**Dried Lees.** These are lees that have been treated, and contain 55−70 % potassium bitartrate.

**Tartrate.** This is produced by distilleries from wine lees or grape marc (the residue obtained after pressing grapes). After distillation to remove the alcohol, the lees and marc are precipitated with milk of lime, in the form of calcium tartrate.

## 6. Production

Two of the four isomers of tartaric acid are produced industrially, mainly L(+)-tartaric acid, but the racemic acid is also produced in very small amounts by chemical synthesis.

### 6.1. Production of L(+)-Tartaric Acid

Tartaric acid is produced mainly in Spain, France, and Italy. One of the oldest producers is Société Legré-Mante established in Marseilles in 1784. All processes for producing tartaric acid are based on the decomposition of calcium tartrate with sulfuric acid.

Two competing methods have been used to convert the starting materials to calcium tartrate: the Scheurer−Kestner acid process, and the Scheele−Lowitz neutral process, together with the Desfosses variant [9], [13]. Nowadays, only the less costly neutral process is employed.

After grinding, the dried starting material (tartar or lees) passes to a roaster where it is heated for 2 h at 160 °C, to remove the organic material that might interfere in the filtration.

After it has been roasted, the hot product is passed to a reactor where it is diluted with water and neutralized to pH 5 with milk of lime. The reaction temperature is held at 70 °C. The addition of calcium chloride or calcium sulfate in 10% excess with respect to the stoichiometric amount gives a more complete reaction and reduces losses.

The chemical reaction that takes place is:

$$2\,KHC_4H_4O_6 + Ca(OH)_2 + CaCl_2$$
$$\longrightarrow 2\,CaC_4H_4O_6 + 2\,KCl + 2\,H_2O$$

The calcium tartrate is separated from the mother liquor by filtration on a rotary filter and washed.

The tartrate obtained from distilleries enables this first stage to be omitted.

The second stage is the decomposition of the calcium tartrate in aqueous solution with sulfuric acid, which yields tartaric acid solution and insoluble calcium sulfate

$$CaC_4H_4O_6 + H_2SO_4 \longrightarrow H_2C_4H_4O_6 + CaSO_4$$

An excess of sulfuric acid of ca. 5% with respect to tartaric acid is necessary in order to optimize the reaction. The pulp obtained by this decomposition is filtered and washed.

The red tartaric acid solution is recovered at a concentration of ca. 200 g/L. The wash waters are recycled to the decomposition stage and the calcium sulfate is discharged as waste.

The tartaric acid solution is concentrated in vacuum evaporators at 70 °C to a concentration of 650 g/L. This new solution is then concentrated in suitable apparatus under vacuum at 70 °C until crystals appear (at ca. 1300 g/L).

This liquid–crystal mixture is then passed to crystallizers where it is slowly cooled to provide a maximum yield of crystals. After the mixture has been cooled it is dried in centrifuges, where the crystals are separated from the mother liquor. The mother liquor is subjected to repeated evaporation, granulation, and drying cycles to recover the maximum yield of granulated tartaric acid.

The granules obtained in each operation are cooled to give a tartaric acid solution of concentration 650 g/L, which is then decolorized with activated charcoal and chemically purified if excess iron and sulfuric acid are present. This operation is carried out at 70 °C.

This solution is passed through a filter press and becomes almost colorless, and is then concentrated under vacuum until crystallization starts, following which it is passed to crystallizers or cooled to provide maximum yield of crystals.

This mixture is separated by centrifugation. The mother liquor, which still contains tartaric acid, is subjected to repeated evaporation, granulation, and drying cycles before being retreated. The purified granules are dried in an oven at 140 °C and passed to a battery of sieves for separation into different grain sizes, from 2000 µm to < 100 µm (powder). Other processes are sometimes used [14], [15].

Two other production processes have been studied: one involves production of L(+)-tartaric acid by fermentation; the other involves production of racemic DL-tartaric acid by chemical synthesis.

## 6.2. Fermentation Process

Studies and investigations include conversion of glucose by bacteria of the species *Acetobacter suboxydans* [16]; and conversion of *cis*-epoxysuccinic acid and its sodium derivatives by bacteria of the species *Nocardia tartaricans* [17]. Publications and patents relating to other research work are listed in [18].

## 6.3. Chemical Synthesis

Racemic DL-tartaric acid is produced from maleic acid or its salts [19], [20]. Maleic acid in aqueous solution is oxidized by 35% hydrogen peroxide in the presence of potassium tungstate. Epoxysuccinic–tartaric acid is formed, which is hydrolyzed on boiling to racemic tartaric acid. The latter is recovered after cooling, centrifugation, washing, and drying.

Several variants of this process have been studied and published, and numerous patent applications have been filed.

Racemic DL-tartaric acid is produced commercially in a small production unit in South Africa.

## 7. Environmental Protection

In the production of L(+)-tartaric acid, the solid waste is discharged into bins before being sent to authorized dumps. The liquid waste is collected, treated with milk of lime, and filtered before being discharged into the sewage system, where it is retreated by municipal purification plants.

Daily analyses of the COD, suspended matter, and sulfates in the effluent are obligatory. Independent checks are carried out by authorized bodies.

Tartaric acid and its salts are biodegradable compounds.

## 8. Quality Specifications

L(+)-Tartaric acid, 2,3-dihydroxybutanedioic acid ($2R, 3R$) is used mainly in the food and pharmaceutical industries, and therefore has to satisfy extremely stringent quality criteria, defined in the pharmacopoeia of each country. The two principal publications are the European Pharmacopoeia (1986) and the United States Pharmacopoeia (1990), whose requirements are very similar.

Limiting values from the European Pharmacopoeia are:

| | |
|---|---|
| Tartaric acid content | 99.5 – 101 % |
| Loss on drying | ≤ 0.2 % |
| Calcination residue | ≤ 0.1 % |
| Rotatory power in 20 % wt/vol solution | + (12 – 12.8°) |
| Sulfates | ≤ 150 ppm |
| Chlorides | ≤ 100 ppm |
| Calcium | ≤ 200 ppm |
| Heavy metals | ≤ 10 ppm |
| Oxalic acid | ≤ 350 ppm |

## 9. Chemical Analysis

The official analysis of tartaric acid in starting materials is the Goldenberg–Géromont method (1898–1907) as modified by the VIIth Chemistry Congress, London (1909). The accuracy of this analysis is $\pm 0.5\%$. This method enables total tartaric acid to be determined without distinction between the isomers or salts derived [9], [13], [21], [22].

The first stage consists in acidifying the starting material with 5.6 mol/L hydrochloric acid to dissolve all the tartaric acid derivatives (potassium bitartrate, calcium tartrate).

These derivatives are filtered, boiled, and converted to neutral potassium tartrate with a solution of potassium carbonate at 682 g/L. After filtration followed by evaporation on a water bath, the neutral potassium tartrate is reacted with 3.5 mL pure acetic acid to potassium bitartrate. The precipitate is rendered insoluble with 100 mL ethyl alcohol and collected by vacuum filtration, washed with alcohol, taken up and solubilized with boiling water in a porcelain evaporating dish. This solution is brought to the boil and titrated against sodium hydroxide 0.2 mol/L, using neutral litmus paper as indicator.

Other, significantly less accurate methods exist. The most widely used of these is spectrophotometry, whose accuracy is ca. $\pm 2\%$. The reagent used is ammonium metavanadate, which forms a colored complex with tartaric acid derivatives [23], [24]. This method is most suitable for continuous analyses during production.

## 10. Derivatives of L(+)-Tartaric Acid

Tartaric acid forms numerous salts and esters. The commercially best-known compounds are discussed in this Chapter [1]–[3].

**Sodium potassium tartrate** is commonly known as Seignette's salt or Rochelle salt. It was discovered in 1672 by PIERRE SEIGNETTE, a pharmacist working in La Rochelle.

Empirical formula $KNaC_4H_4O_6 \cdot 4\,H_2O$, [6381-59-5], $M_r$ 282.23, density 1.79 g/cm³, $mp$ 70–80 °C, decomposes at 220 °C; solubility (g per 100 mL water) 26 at 0 °C, 66 at 26 °C.

The starting material for its production is tartar with minimum tartaric acid content 68 %. This is first diluted in water or in the mother liquor of a previous batch. It is then neutralized hot with caustic soda to pH 8, decolorized with activated charcoal, and chemically purified before being filtered. The filtrate is evaporated to 42° Bé at 100 °C, and passed to granulators in which Seignette's salt crystallizes on slow cooling. The salt is separated from the mother liquor by centrifugation, accompanied by washing of the granules, and is dried in a rotary furnace and sieved before packaging.

Commercially marketed grain sizes range from 2000 μm to < 250 μm (powder).

**Acid potassium tartrate** (potassium bitartrate, monopotassium tartrate) is commonly called cream of tartar. $KHC_4H_4O_6$ [868-14-4], density 1.96 g/cm³; solubility (g per 100 mL water) 0.57 at 20 °C, 6.1 at 100 °C.

This salt is mainly produced from the mother liquor of Seignette's salt. These decolorized, purified, and filtered solutions are acidified to pH 3.5 with hydrochloric or sulfuric acid. Since cream of tartar is sparingly soluble, it precipitates and is recovered by centrifugation, dried, and ground before being packaged as fine powder.

**Potassium antimonyl tartrate,** also known as tartar emetic, was discovered in 1631 by ADRIEN DE MYNSICHT. $KSbC_4H_4O_7 \cdot \frac{1}{2} H_2O$, [28300-74-5], $M_r$ 333.93, density 2.61 $g/cm^3$; solubility (g per 100 mL water) 8.7 at 25 °C, 35.7 at 100 °C.

The compound is prepared by reacting 3 parts antimony oxide with 4 parts cream of tartar diluted in water. The reaction mixture is heated to 100 °C and held at this temperature for 4 h. After filtration, tartar emetic crystallizes in the cooled solution in a mixer. The granules are extracted by centrifugation. After drying, the granules are pulverized.

**Metatartaric acid** is obtained by heating tartaric acid to 170–180 °C. The conversion takes place without any loss of weight. Two or more molecules react with mutual esterification.

The compound obtained is more or less polymerized. Metatartaric acid has outstanding inhibitory properties with respect to precipitation of tartaric acid salts.

# 11. Uses

L(+)-Tartaric acid and its derivatives are used particularly in the food, pharmaceutical, and viniculture industries. There are also numerous applications in other fields [2], [25], [26].

**L(+)-Tartaric acid** (E 334) is used:

1) In the acidification of wine musts
2) As an acidifier and taste enhancer in sweets, candies, jellies, jams, fruit nectars, ice creams, gelatins, and pastes
3) In fruit, vegetable, or fish preserves, where it acts as a synergistic antioxidant, and also stabilizes the pH, color, taste, and nutritional value
4) In greases and oils, as an antioxidant
5) In the preparation of carbonated beverages
6) In the pharmaceutical industry, as an excipient or carrier for the active principal, helping to correct the basicity
7) Because of its simplicity of use, stability, and high solubility, as an acidification agent for foaming tablets and powders
8) In the cement industry, and particularly in plaster and gypsum, where its retardant action facilitates handling of these materials
9) In polishing and cleaning metals

**Cream of tartar** (E 336) is used:

1) To accelerate the precipitation of tartaric acid salts in wine
2) In the production of chemical yeasts
3) In the production of some drugs, by virtue of its laxative action

**Seignette's salt** (E 337) is used:

1) In electroplating, where it improves deposition and yield
2) In electronics and piezoelectricity
3) As a reducing agent in the silvering of mirrors
4) As a laboratory reagent; it is one of the constituents of Fehling's solution
5) In the production of cigarette paper, as a combustion regulator
6) In the pharmaceutical industry, on account of its laxative action

**Potassium antimonyl tartrate** is used:

1) In small amounts as an expectorant in cough syrups
2) In the treatment of some tropical diseases
3) As a mordant in the textile and leather industries

**Metatartaric acid** is used to inhibit tartrate crystallization in table wines.

**The tartrates of fatty acid monoglycerides and diglycerides** are used as emulsifiers in the bakery industry.

# 12. Economic Aspects

L(+)-Tartaric acid currently has a world market of ca. 30 000 t. The main producers are France, Spain, and Italy; it is also produced in small amounts in Portugal, Argentina and Japan.

The tartaric acid market has remained static over the last decade. Production is entirely dependent on vinicultural raw materials, leading to large fluctuations in the retail price of tartaric acid.

# 13. Physiological and Toxicological Behavior

Tartaric acid, unlike some other organic acids in fruit, is not an intermediate of the Krebs cycle (→ Citric Acid, **A 7**, p. 103), which is a major center of organic synthesis in the human body. It may be ingested in more or less large amounts with foodstuffs such as fruit and wine, where it occurs in the L(+)-form. After oral absorption, about 20 % is excreted in the urine, a large proportion of the remainder undergoing bacterial breakdown in the intestines. No traces are found in the feces. The absorbed fraction is rapidly eliminated from the blood, and is either excreted by the kidneys or oxidized in various tissues.

Long-term toxicity studies in rats have been carried out with L(+) monosodium tartrate. No toxic or carcinogenic effects have been detected. Moreover, no adverse effects on renal function and pathology have been detected at a daily dose of 3 g/kg.

The toxicity of the (DL)-form has been less widely studied. One study has shown that it behaves differently from the L(+)-form. The (DL)-isomer disappears more slowly from the blood and accumulates for a longer period in the kidneys, producing a weight increase in the latter. On account of this nephrotoxic effect, this isomer has not been approved for use in foodstuffs and pharmaceuticals [27]–[30].

# 14. References


[1] V. Grignard: *Traité de chimie organique*, vol. 11, Masson et Cie, Paris 1945, pp. 285–331.

[2] *Merck Index* 1989, pp. 1432–1433.

[3] A. D. Wurtz: *Dictionnaire de Chimie*, vol. 3, Librairie Hachette et Cie, Paris 1882, pp. 198–246.

[4] A. D. Wurtz: *Dictionnaire de Chimie*, vol. 7, Librairie Hachette et Cie, Paris 1908, pp. 663–671.

[5] H. V. Ockermann: *Source Book for Food Scientists*, Avi Publishing, Westport, Conn. 1978, p. 276.

[6] *Dictionary of Organic Compounds*, 4th ed., vol. 5, Eyre and Spottiwoode, Oxford 1965, 2943–2945.

[7] *Vogel's Text Book of Practical Organic Chemistry*, 5th ed., J. Wiley & Sons, New York 1989, pp. 4–7.

[8] J. March: *Advanced Organic Chemistry*, 4th ed., McGraw Hill, New York 1992, pp. 94–114.

[9] J. Ventre: *Dérivés Tartriques de la Vendange*, Librairie Coulet et fils, Montpellier 1909.

[10] C. Cantarelli: "Produits secondaires de la Vinification, Rapport Général," Office International de la Vigne et du Vin, no. 501 (1972) 947–967.

[11] J. Mourgues: "Valorisation des sous produits des distilleries vinicoles," *C.R. Seances Acad. Agric. Fr.* **59** (1973) 475–480.

[12] J. Mourgues, J. Mangenet, *Ind. Aliment. Agric.* **1** (1975) january.

[13] U. Roux: *La Grande Industrie des Acides Organiques*, Librairie Dunod, Paris 1939.

[14] Montecatini, US 3 069 230, 1962 (B. Pescarolo, V. Bianchi).

[15] Orandi et Massera, US 3 114 770, 1963 (L. Dabul).

[16] T. K. Yamada, T. Kodoma, T. Obata, N. Takakashi, *J. Ferment. Technol.* **49** (1971) no. 2, 85–92.
K. Yamada, Y. Minoda, T. Kodama, U. Kotera, US 3 585 109, 1971.
H. K. Bhat, G. N. Quazi, S. K. Chaturvedi, C. L. Chopra, *Res. Ind.* **31** (1986) no. 2, 148–152.
Skowa Chem., Mitsubishi Chem. Ind., JP Kokai 93 163 193, 1993 (N. Ootomo).

[17] Tokuyama Soda, JP Kokai 77 90 690, 1977 (K. Yutani, H. Takesue, K. Fujii, Y. Miura).
Tokuyama Soda, JP Kokai 79 14 917, 1979 (H. Takesue, Y. Miura, M. Shibuya, M. Toyoshima).
Tokuyama Soda., DE-OS 2 605 921, 1976 (Y. Miura et al.).
Takeda Chemical Industries, DE-OS 2 619 311, 1976 (Y. Kamatani et al.).

[18] T. Huang, Y. Qian, *Gongye Weishengwu (Ind. Microbiol.)* **20** (1990) no. 6, 14–17.
Z. Zhang, T. Huang, *Gongye Weishengwu (Ind. Microbiol.)* **20** (1990) no. 2, 7–12, 24.
Miles Laboratory, US 2 314 831, 1943 (I. Kamlet).
Institute of Microbiology, Academy of Sciences, Moldavian S.S.R., SU 514 891, 1976.

[19] J. M. Church, R. Blumberg, *Ind. Eng. Chem.* **43** (1951) no. 8, 1780–1786.

[20] Imperial Chemical Industries, GB 1 442 748, 1976 (F. Lewis, P. A. Rodriguez).
Faming Zhuanli Shenqing Gongkai Shuomingshu, Peoples Republic of China (Xintong Group), 1 050 712, 1991 (X. Wu, Z. Liu et al.).
Faming Zhuanli Shenqing Gongkai Shuomingshu, Peoples Republic of China, 85 100 455, 1986 (Z. Zhang et al.).
Nippon Peroxide, JP Kokai 7 339 437, 1973.
G. M. Grinberg, S. M. Gabrielyan, N. M. Morlyan, M. K. Madroyan, SU 322 043, 1975.
Österreichische Chemische Werke, DE-OS 2 555 699, 1976 (K. Petritsch, P. Korl, F. Pogoriach).
Nippon Peroxide, JP Kokai 7 785 119, 1977 (K. Kazutani et al.).
S. Eisler et al., *Rev. Chim. (Bucarest)* **39** (1988) no. 8, 660–663.
Kawaden Fine Chemicals, JP Kokai 6 372 647, 1988 (Nozue, Moriaki).

[21] Goldenberg, Geromont et Cie, *Z. Anal. Chem.* (1898) 312.

[22] Chem. Fabrik, former Goldenberg, Geromont et Cie, *Z. Anal. Chem.* **47** (1908) 57.

[23] H. Rebelein, *Mitteilungsbl. GDCh Fachgruppe Lebensmittelchem. Gerichtl. Chem.* **24** (1970) 14.

[24] H. Tanner, M. Sandoz, *Schweiz, Z. Obst Weinbau* **108** (1972) 10.

[25] *Pharmaceutical Codex,* vol. 11, Pharmaceutical Press, London 1979, pp. 53, 720, 836, 916.

[26] E. F. James (ed.): *The Extra Pharmacopoeia*, 30th ed., Pharmaceutical Press, London 1993, pp. 39, 900, 904, 1408.

[27] F. P. Underhill et al., *J. Pharmacol. Exp. Ther.* **43** (1931) 359, 381.

[28] FAO/OMS reports nos. 7, 17, 21, 27.

[29] P. Dupuy, "Le Métabolisme de l'Acide Tartrique," *Ann. Technol.* **2** (1960) 139–184.

[30] T. E. Furia: *Handbook of Food Additives.* The Chemical Rubber Co., Boca Raton, Fla. 1968, pp. 261–263.

# Tea

ANITA CROCKER, The Tea Council Limited, London, United Kingdom



## 1. Introduction

Tea is an evergreen plant belonging to the Camellia family and is known as *Camellia sinensis*. It was first discovered ca. 5000 years ago in China by the legendary emperor SHEN NUNG.

Tea is indigenous to both China and India, as the BRUCE brothers, employees of the East India Company, discovered in 1823. Hence today, the tea trade uses the term China or Indian jat to distinguish the origin of the original tea bushes on some of the older estates.

If left to grow wild, tea grows as a tree reaching some 30 m high; thus it was that monkeys used to be trained to pick the leaves and throw them down to people waiting to collect them below.

Today, under cultivation, tea is kept to a height of some 1.5 m (waist height) for easy plucking. The characteristics and flavor of tea vary according to the type of soil, altitude, and climatic conditions of the area in which it grows. Other factors affecting tea's characteristics and flavor are the method by which it is processed—"made" in tea trade terms—and the blending of teas from different areas or countries. Tea offers a wide choice of varieties, more than 1500 blends. There is a tea to suit every occasion.

## 2. History

The early beginnings of tea are wrapped in legend and myth. Circa 2737 B.C. tea drinking was first recorded in China, when SHEN NUNG, whilst touring his empire, sat down beneath a tree as his servant boiled some water. A leaf from the tree dropped into the water and SHEN NUNG decided to try the brew. The tree was a tea tree (*Camellia sinensis*).

Two thousand years later, in the Tang dynasty (906–618 B.C.), tea was China's national drink, and the modern term tea derives from the early Chinese words Tchai, Cha, Tay etc., used to describe both the beverage and the tea leaf. Circa 552 A.D. Buddhist monks are reputed to have taken tea to Japan. They not only cultivated the tea drinking habit, but also planted seeds and pioneered tea cultivation in Japan. The Buddhists were also responsible for developing the Japanese tea ceremony, which is said to be based on the "Code of Practise" written by the Chinese scholar LO YU, circa 900 A.D.

Tea was first mentioned in Europe around 1559 when the Venetian author GIAMBATTISTA RAMUSIO published the stories of a Persian trader, Hajji Mahommed and his tales of tea in China.

Although, the Portuguese were the first to open up trade with the Far East, and Portugal's missionaries are reputed to have brought the tea drinking habit back to Portugal, many of the Portuguese ships were crewed by other nationalities. Thus it was a Dutch navigator, JAN HUGO VAN LINSCHOON, who is reputed to have fired the enthusiasm of Dutch merchants and sea captains to trade with the Far East, and this led to Holland trading in tea with its European neighbors, including Scandinavia and Britain.

Tea is thought to have been brought back to Britain by the East India Company as early as 1633, but the first recorded sale of tea was in 1657 when THOMAS GARWAY sold tea in Garraway's Coffee House in Exchange Alley, London. The first tea auction was held in the early

1700s, and Britain had established herself as the center of the world tea trade by 1790.

## 3. Major Tea Producers

Tea grows in more than 32 countries around the world. Today's main producers are India, China, Sri Lanka, Indonesia, Kenya, Malawi, Tanzania, Zimbabwe, Papua New Guinea, Bangladesh, Mauritius, Rwanda, and the Cameroons. Tea is also grown in South America, Australasia, Eastern Europe, Middle and Far East, but is mostly consumed domestically.

Some teas are seasonal, others not. This depends on altitude, nearness to the equator, and climatic conditions in the producer country. The yield per hectare is also influenced by these natural factors, plus good husbandry on the part of the producer. For example, the harvest weight of plucked tea leaves on an estate can fluctuate between 1600 and 6000 kg per hectare in East Africa or Assam, but is lower in high-altitude areas such as Darjeeling.

India, one of the main tea growers, produces ca. 30% of the world's tea, with over 400 000 hectares under tea cultivation. Teas from India are seasonal and are generally plucked from March until the end of October. The tea producing areas range from low-grown (sea level to 609 m above sea level) to high-grown tea (> 1219 m above sea level). Each growing area produces teas with unique flavor and characteristics, peculiar to a given region. Assam, Darjeeling, and Nilgiri teas are examples, and apart from their use in blended teas, they are also well-known worldwide as speciality teas.

Sri Lanka is the world's largest exporter of tea. Teas from this tropical island retain the name "Ceylon", although the traditional name of Sri Lanka was readopted in 1972 when Ceylon became a Sovereign Republic in the Commonwealth. Ceylon teas have been arriving in the London auction since 1867. Today Sri Lanka has some 240 000 hectares under tea cultivation, and because of its geographical position is able to pluck tea throughout the year.

The west and east side of the island are divided roughly by the central mountain system; thus, due to the natural climatic conditions of the area the western side's tea plucking season alternates with that of the eastern side.

Tea is one of Sri Lanka's main foreign exchange earners, and Ceylon teas are drunk as speciality teas throughout the world, as well as being used for blending. The teas span the whole spectrum of growth from low-grown to high-grown.

Indonesia's tea industry was founded by the Dutch in the 1700s. Teas from there, together with teas from India and Ceylon, dominated the black tea markets of Britain and Europe until 1939 – 40. After World War II Indonesian tea estates were in a sorry state. Wrecked factories and tea bushes reverted to their wild "tree" state were just two of the problems which faced the country.

By 1984, tea exports from Indonesia began to reappear on the tea market. Improvements in tea production, replanting of old estates, and investment in new factory machinery and manufacturing methods has enabled Indonesia to gain an estimated 12% of world tea exports.

Teas from Kenya are probably the best known of the East African teas as Kenya sells her tea as a speciality tea as well as for blending. Kenya's tea history dates back to 1903 when C.S.L. CAINE imported tea seeds from India and began to produce quality tea on a tea farm two acres in size. Other experimental farms followed and by 1928 Kenya was sending tea for sale at the London tea auction. Near to the equator, Kenya is a fertile land with a climate which allows it to grow and pick tea throughout the year.

Malawi pioneered the tea trade in East Africa. Circa 1878 tea farming commenced. However by the 1880s commercial tea production was firmly established in Mulanje. Malawi teas are used mainly for blending and have a bright color and brisk flavor. Today Malawi has a 3.8% share of world tea exports.

Zimbabwe teas, like those from Malawi, are used mainly for blending. They have a 1.1% share of world tea exports, which are important as a foreign exchange earner. The first fully irrigated tea estate was established in Zimbabwe, as the country has an annual rainfall of only 66 cm and tea needs some 127 cm.

The tea industry in Tanzania was originated by the Germans during the reign of WILHELM II, and established itself between the two world wars. Teas in Tanzania are grown at low, medium, and high altitudes, have a bright color and brisk flavor, are exported as a foreign exchange earner, and are used in the West for blending purposes.

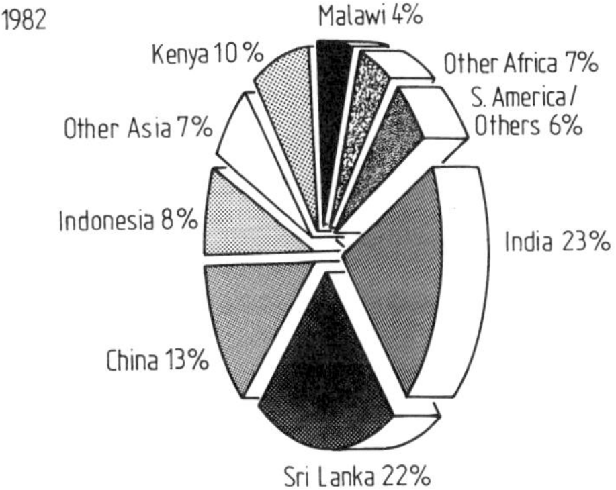


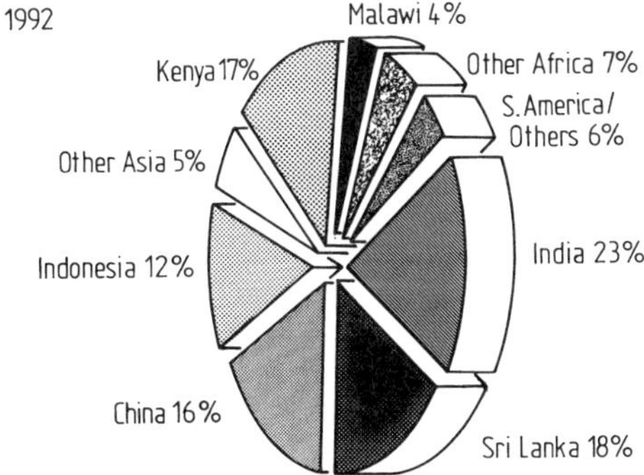


**Figure 1.** World tea exports (courtesy of International Tea Committee)

China is the second largest tea producer after India and the second largest tea exporter after Sri Lanka. The Chinese retain the majority of the tea they grow for home consumption, exporting approximately a quarter of their annual yield. World exports of tea in 1982 and 1992 are shown in Figure 1.

## 4. Growth and Manufacture

Tea is grown on estates or smallholdings. A smallholding can be as small as 0.5 hectares or several hectares. The owner grows tea and sells his leaves to a factory for processing. An estate is a self-contained unit, housing its own factory, schools, hospital, staff houses and gardens, shops, places of worship, reservoir, and guest house.

**Figure 2.** Tea estate

**Figure 3.** Two leaves and a bud

In modern tea cultivation, the bushes are grown from cuttings, or clones. The tender young clones are nurtured and trained into fan-shapes in nursery beds until they are ready for planting out on either a smallholding or estate. On the medium to high altitude areas the bushes are planted around the contours of the hills in rows (Fig. 2), to avoid soil erosion. The bushes are planted approximately 1.5 m apart, with a distance of some 1.0 m between each row, allowing easy access for both the gardeners and pluckers. Each bush is pruned so that it grows in a fan-shape topped by a flat "plucking plateau" roughly 1.5 × 1.0 m in size. Prunings from the bushes are used as mulches and some for domestic firewood. Tea bushes provide a safe haven for wild life and maintain an environmental balance. Only the top two leaves and a bud (Fig. 3) are plucked from the top of the plucking plateau, and tea is a self renewing crop, growing again within 7–14 d of plucking, depending on the altitude at which it is grown.

A skilled plucker can gather between 30 and 35 kg of green tea leaf in a day, sufficient to produce ca. 7.5–9 kg of processed black tea.

Once plucked there are seven stages of processing in tea manufacture: collection, withering, breaking the leaves, fermenting, drying, sorting, and packing. For small holders the leaves are either taken to the factory or collected at a central point from the smallholders, who are paid for their crop.

Figure 4. Orthodox rolling machine

Figure 5. Tea sorting

On an estate the plucked leaves are collected at a central point, each plucker being credited with their own weights of leaf for subsequent payment. The collected leaf is packed into nets and taken to the factory, transported to the top floor and spread thinly on wire racks exposed to the air at 25–30 °C. This is known as withering; the moisture from the leaf evaporates in the warm air leaving the leaf limp. The process can take 10–16 h, depending on the wetness of the leaf.

The withered leaf is then broken by the "orthodox" or "unorthodox" method, so that the enzymes in the veins of the leaf are released, come into contact with air, and are oxidized (known in the tea trade as fermenting). The orthodox process (Fig. 4) rolls the leaves, whereas the unorthodox process, known as "cut, tear and curl" or "rotovane", depending on the type of machinery used, breaks the leaf into smaller particles more suited to modern market needs.

The broken leaf is laid out on trays or in troughs in a cool humid atmosphere for 3–4 h and is periodically turned until all leaf particles are a golden russet color and fermentation is complete.

Next the leaf is dried by a firing process. The leaf is passed slowly through hot air chambers until all the moisture is removed and the leaf turns a dark brown, referred to as "black" in the tea-trade. Then the black tea is sorted by leaf particle size or "grades". The "grades" are Broken Pekoe, the largest particles; fannings, smaller particles which can fall under three or four classifications of decreasing size; and Pekoe Dust, the smallest leaf particles. Other descriptions such as Flowery Orange Pekoe and Golden Flowery Orange Pekoe can be used to describe the leaf appearance. The black tea is sorted by sifting through wire meshes of varying sizes into containers, before being weighed and packed into chests (Fig. 5) or sacks for loading onto pallets. Each chest or sack contains an average of 56 kg of black tea, and a loaded pallet holding 50–100 tea chests or sacks is known as a break.

The above describes the manufacture or "make" of black tea. There are also two other forms of manufacture: green tea and oolong tea.

In green tea manufacture after plucking and collection the tea is left to wither, as with black tea. It is then steamed and rolled before drying. Thus the veins in the leaves are not broken and the enzymes are not oxidized. When green tea is brewed it has a yellowish liquor and the wet leaf is almost entire. Green tea is drunk mostly in China and Japan as well as some parts of South America. A little is drunk in the West as a speciality tea.

Oolong tea is a semi-fermented tea. After withering and rolling, fermentation is stopped after ca. 1 h and the tea dried, then sorted. Oolong is widely drunk in the Far East and is gaining a following in the West as a speciality tea. It is a light liquoring tea with a very delicate flavor.

## 5. Tea Trade

After manufacture the tea is tasted in the factory and a sample airmailed to the estate's brokers in London and national local auction center. Each broker evaluates the tea for quality and price and reports back to the estate, so the tea can be sold to the best advantage. In India there are several auction centers. Sri Lanka has Colombo; East Africa, Mombasa and Limbe; and Indonesia, Jakarta.

Tea destined for the London auction takes 12–16 weeks to arrive at Felixtowe, the main tea port in Britain. From there it is transferred into warehouses under the aegis of the Tea Clearing House. At this stage, samples of the tea are sent to both the selling broker and tea buyers.

The selling broker retastes the tea to ensure it has travelled well, reevalutes it, and decides the price it is expected to bring at auction. All this information is communicated immediately to the estate from which the tea came.

The tea buyers taste the tea for its quality, how much of it is needed for the company's blends and evaluate the tea pricewise from their company's viewpoint. Thus, all parties come to the auction having tasted the teas. After the teas are sold they are transported from the warehouses to the blender and packers factories.

Approximately 25 % of the tea drunk in Britain comes through the London auction, the rest being purchased from auction in the country of origin or sometimes direct from estates.

The rest of the world buys its teas from any of the auctions, on the spot market in Amsterdam, through agents or brokers in the producer countries and London, or direct from the estates. There are no future markets in tea and the price is dominated by supply and demand.

In recent years, with containerization in shipping, an off-shore auction has developed in London, whereby teas that are still on the water en route are sold. Again both sellers and buyers have tasted the teas before they begin their travels, but unlike the landed or national auction, the buyer has to buy the complete container lot, whereas with the landed auction, buyers can split lots and deal across the floor of the auction rooms.

## 6. Tea Market and Consumption

Black tea has the major share of the tea market in terms of production, sales, and amount drunk. Black tea can be roughly divided into popular, national or local blends, and speciality teas. The speciality teas are those which take their name from the area in which they are grown, e.g., Assam, Darjeeling, Ceylon; a blend of teas for a particular time of day such as English Breakfast, Afternoon Tea; teas blended and named after a person (e.g., Earl Grey) or tea (*Camellia sinensis*) which has been flavored with fruit or flower blossoms (e.g., Jasmine tea, Rose Pouchong, or Orange tea). Green teas, Gunpowder being the best known; oolongs; and Lapsang Souchong, with its distinctive smoky aroma and flavor, also come under the speciality tea heading.

Within the popular blend and speciality tea markets there are two main classes: loose leaf tea and teabags.

In Britain tea consumption accounts for 42 per cent of total drinks drunk, excluding tap water. Eighty per cent of the nation drink tea daily and Britain imports annually ca. 16 % of the producer countries' tea exports. In kilograms per capita per annum British consumption during the past decade has remained relatively stable at 2.56. In Eire, over the same period, per capita consumption was 3 kg. The rest of Europe and Scandinavia tend to consume ca. 1 kg of tea per capita per annum, and the United States follows this pattern. Many of the major growing countries such as India, Sri Lanka, and Kenya consume 1 kg per capita, as do Russia and Turkey. In Australasia tea consumption is around the kilogram per capita figure, whereas in Canada the American pattern seems to hold sway with consumption at 0.5 kg or less per capita.

In the Middle East consumption averages ca. 1 kg per capita per annum.

## 7. Tea and Health

Tea is a natural product. It contains no artificial coloring, preservatives, or flavoring and is virtually caloric free if taken without milk or sugar. Tea provides a pleasant way of taking in the fluids the body needs daily for optimum health.

Tea is a rich source of two minerals essential to health: manganese and potassium.

Manganese is essential for bone growth and the body's development. Our bodies need 2–5 mg/d, and 45 % of this can be obtained by drinking five or six cups of tea daily.

Potassium is vital for maintaining a normal heart beat. It is one of the major constituents in living cells; helps balance sodium; enables nerves and muscles to function and regulates fluid levels within the cells. The potassium requirement is 2–3.5 mg/d, and 25 % of this can be provided by five or six cups of tea daily.

Tea contains small amounts of carotene (a vitamin A precursor), thiamin (vitamin B1), riboflavin (vitamin B2), nicotinic acid, pantothenic acid and ascorbic acid (vitamin C).

Tea cools, calms, refreshes, and relaxes. Tea helps digestion of foods and acts as a diuretic.

A cup of tea made with a leaf tea or teabag contains 40 mg of caffeine, and with instant tea, 30 mg. Over 200 scientific papers a year are published on caffeine and there is now general consensus that caffeine has no deleterious effects on the heart, blood cholesterol, or the circulation, and there is no evidence that it gives rise to cancers. The FDA has given caffeine GRAS status, that is, it is generally recognised as safe.

Caffeine is a mild stimulant, which can increase concentration and alertness, accuracy and sensitivity of taste and smell. In high doses ($> 800$ mg/d) it can invoke anxiety and unpleasant gastric sensations. Tea has not been shown to have either effect, possibly due to the fact that to get this amount of caffeine from tea would involve drinking unrealistically large quantities.

Tea also contains polyphenols, termed tannins by the tea trade. The polyphenolic compounds in tea have an important sensory role. They contribute to the color, taste, and refreshing aspect of the drink.

# 8. References


[1] *The Tea Council Annual Reports*; *File on Tea*; *Speciality Tea Leaflets*; *Fact File on Tea*; available from The Tea Council Limited, Sir John Lyon House, 5 High Timber Street, London EC4V 3NJ, United Kingdom.
[2] International Tea Committee: *Annual Statistical Bulletin.*
[3] W. Ukers: *All About Tea*, Vols I and II, Ukers, New York 1935.
[4] D. Forrest *The World Tea Trade*, Woodland Faulkner, Cambridge 1985.
[5] J. Weatherstone *The Pioneers*, Quiller, London 1987.
[6] Flammarion *The Book of Tea*, Flammarion, Paris 1991.
[7] W. Ukers *The Romance of Tea*, Ukers, New York 1935.


**Technetium → Radionuclides**

# Tellurium and Tellurium Compounds

GUY KNOCKAERT, Union Minière, Hoboken, Belgium



Tellurium [*13494-80-9*], Te, atomic number 52, $A_r$ 127.60 $\pm$ 0.03, is a metallic element of group 16 in the periodic table.

Tellurium has 8 natural stable isotopes, and is known to have a total of 21 artificial unstable isotopes. The natural abundance of the stable isotopes is: $^{120}$Te (0.096 %), $^{122}$Te (2.60 %), $^{123}$Te (0.908 %), $^{124}$Te (4.816 %), $^{125}$Te (7.14 %), $^{126}$Te (18.95 %), $^{128}$Te (31.69 %), and $^{130}$Te (33.80 %) [1]. The abundance of tellurium in the Earth's crust is comparable to that of platinum, 0.01 ppm.

Tellurium was discovered in 1782 by MÜLLER VON REICHENSTEIN and named by KLAPROTH in 1798 (Latin, *tellus,* Earth) [2].

## 1. Properties [1], [3]–[12]

**Physical Properties.** Tellurium is a crystalline, bright silver-white metal, which is rather brittle and easily crushed. Tellurium is relatively soft (2.3 Moh, nearly as hard as zinc).

The crystal structure is a three-point helical lattice with hexagonal symmetry, with the helical chains parallel to the hexagonal axis. Many physical and electrical properties, e.g., hardness, thermal expansion coefficient, Hall coefficient, resistivity, and diamagnetic susceptibility are anisotropic. The so-called amorphous phase actually contains very small hexagonal crystals.

Tellurium, a *p*-type semiconductor, demonstrates the phenomenon of piezoelectricity and becomes superconductive at 3.3 K. Some tellurium compounds have excellent thermoelectric properties which are commercially interesting. Trace additives of tellurium to steel, lead, and copper and their alloys have a positive influence on the machinability of tools.

Tellurium has a relatively low melting (723 K) and boiling point (1327 K). On melting, the specific volume increases by ca. 5 %. Tellurium vapor is yellow-gold and consists mainly of $Te_2$ molecules up to 2000 °C.

Some physical properties of tellurium are:

| | |
|---|---|
| Electronic configuration $[Kr]4d^{10}5s^25p^4$ | |
| $A_r$ | 127.60 $\pm$ 0.03 |
| Thermal neutron cross section at 2200 m/s | |
| absorption | $(4.7 \pm 0.1) \times 10^{-28}$ m$^2$/atom |
| scattering | $(5 \pm 0.1) \times 10^{-28}$ m$^2$/atom |
| Crystal structure | hexagonal lattice with trigonal symmetry (A 8) |
| | $a_0 = 0.4457$ nm |
| | $c_0 = 0.5929$ nm |
| | bond angle $(103.2 \pm 0.1)°$ |
| Atomic radius | 0.1285 nm |
| Ionic radius | |
| Te$^{4+}$ | 0.089 nm |
| Te$^{2-}$ | 0.221 nm |
| Atomic volume | $20.42 \times 10^{-6}$ m$^3$/mol |
| Density at 300 K | 6.245 g/cm$^3$ |
| *mp* | 722.6 K |
| *bp* | 1327 K |
| Enthalpy of fusion | 17.489 kJ/mol |
| Enthalpy of vaporization (Te$_2$) | 107.77 kJ/mol |
| Specific heat at 298.15 K | 25.707 J mol$^{-1}$ K$^{-1}$ |
| Entropy at 298.15 K | 49.497 J mol$^{-1}$ K$^{-1}$ |
| Entropy of fusion | 24.201 J mol$^{-1}$ K$^{-1}$ |
| Entropy of vaporization | 36.6 kJ mol$^{-1}$ K$^{-1}$ |
| Coefficient of linear expansion, mean value (anisotropic) | $16.8 \times 10^{-6}$ K$^{-1}$ |
| Vapour pressure $p$ | |
| at 793 K | 0.133 kPa |
| at 923 K | 1.33 kPa |
| at 1111 K | 13.3 kPa |

Ullmann's Encyclopedia
of Industrial Chemistry, Vol. A 26

| | |
|---|---|
| Thermal conductivity at 293 K | $0.060 \ \mathrm{W m^{-1} K^{-1}}$ |
| Surface tension at temperature $T$, $T_{mp}$ = melting point | $0.178 - 2.4 \times 10^{-5}$ $(T - T_{mp}) \mathrm{N/m}$ |
| Electrical resistivity | |
| at 3.3 K | superconducting |
| at 300 K | $9.9 \times 10^{-3} \ \Omega \mathrm{m}$ |
| Standard electrode potentials | |
| $\mathrm{Te} + 2\mathrm{e}^- \longrightarrow \mathrm{Te}^{2-}$ | $-0.92 \ \mathrm{V}$ |
| $\mathrm{Te}^{4+} + 4\mathrm{e}^- \longrightarrow \mathrm{Te}$ | $0.63 \ \mathrm{V}$ |
| Tensile strength | $11.0 \pm 0.25 \ \mathrm{MPa}$ |
| Modulus of elasticity | $4140 \ \mathrm{MPa}$ |

**Chemical Properties.** Metallic tellurium has many properties similar to sulfur and selenium, but it is less reactive, more basic, more metallic, and strongly develops amphoteric properties. When metallic tellurium is heated in air, it burns with a blue-green flame; tellurium powder oxidizes at room temperature.

Tellurium reacts vigorously with halogens at room temperature, cannot be combined directly with sulfur, and reacts with hydrogen at high temperature (920 K). Tellurium does not react with carbon, boron, or nitrogen; it reacts with phosphorus only in a sealed tube heated above 595 K.

Tellurium is insoluble in water, dissolves in concentrated sulfuric and nitric acid, and only slightly in dilute hydrochloric acid. The solubility in caustic alkalis depends on the temperature; tellurium is, however, insoluble in ammonium hydroxide.

When heated in an evacuated ampule, stoichiometric amounts of tellurium and one or more other elements form binary or ternary tellurides. Molten tellurium corrodes iron, copper, and stainless steel.

# 2. Resources and Raw Materials [13]

Tellurium, a relatively rare element with a crustal abundance of 10 μg/kg, is found in close association with sulfur as well as with selenium. Unlike the latter, tellurium does not substitute in sulfide lattices, but forms discrete minerals or microsegregations in the host sulfide mineral [14].

Native tellurium has been observed, but the element is mainly found as a gold and/or silver telluride. The concentration of tellurium minerals in nature is insufficient to allow their economic recovery as principal mining products;

its recovery therefore depends on its concentration during the processing of other nonferrous metals. In copper refining, tellurium accumulates together with precious metals and selenium in the anodic slimes generated by electrolysis. Other primary metals such as zinc, gold, and lead, concentrate tellurium during refining.

Reserves of tellurium are difficult to assess because of the limited knowledge of tellurium content in the copper or other ores from which it is recovered. Estimations of refinery production of tellurium are complicated by the trade in concentrates, blister, and anode copper, as well as other nonferrous residues.

As copper ores are the main sources for tellurium, public statistics are based on the copper industry. The United States Bureau of Mines estimates tellurium resources by applying fixed recovery factors (i.e., 0.065 kg Te per tonne of Cu) and quotes the figures in tonnes [15].

| Continent, country | Tellurium reserves, t |
|---|---|
| North America | |
| United States | 6000 |
| Canada | 2000 |
| Others | 3000 |
| South America | |
| Chile | 6000 |
| Peru | 2000 |
| Others | 1000 |
| Europe | 5000 |
| Africa | |
| Zaire | 2000 |
| Zambia | 2000 |
| Others | 1000 |
| Asia | |
| Philippines | 1000 |
| Others | 1000 |
| Oceania | 2000 |
| World | 34000 |

The results of these estimations do not include tellurium resources included in lead, zinc, or gold reserves.

# 3. Production [16]–[22]

Tellurium can be recovered as a byproduct in the treatment of lead, copper, bismuth, precious metals, and nickel ores, and from sulfuric acid plants. The main source is copper anode slimes, which settle at the bottom of the refining tanks during the electrowinning of 99.9 % pure copper.

**Table 1.** Composition of copper anode slimes: selected examples [23]–[25]

| Element | Typical concentration, wt% | | |
|---|---|---|---|
| | Canadian Copper Refiners | Nippon Mining | Inco |
| Cu | 20.3 | 4.73 | 17 |
| Bi | 0.36 | 1.6 | 0.1 |
| Sb | 0.95 | 0.95 | 0.05 |
| Se | 10.9 | 15.23 | 7 |
| Te | 3.19 | 3.64 | 2 |
| Pb | 8.5 | 6.54 | 1 |
| Au | | 0.99 | 0.1 |
| Ag | 21.3 | 20.58 | 6 |
| As | 1.83 | 1.59 | 0.8 |
| Ni | 0.52 | 0.03 | 26 |

Copper anode slimes contain 0.5–10% tellurium (Table 1).

Tellurium is collected with the precious metals in the anode slimes, and has to be separated from these together with selenium, because they interfere with the separation and refining of the precious metals. Tellurium is mostly present in the form of intermetallic compounds of silver, copper, and sometimes gold [$Ag_2Te$, $Cu_2Te$, and $(Ag, Au)Te_2$]. Pretreatment of the slimes can dissolve the copper which is harmful in the further treatment of the slimes. It is normally followed by roasting or sulfatizing of the decopperized slimes, or a combination of these and alternative methods.

**Pretreatment** consists of dissolving residual copper and tellurium by dilute aerated sulfuric acid or by oxidative pressure leaching with dilute sulfuric acid (normally tankhouse liquid from the copper electrolysis). The first method dissolves 70–80%, the second > 90% of the tellurium. Pressure leaching takes place in an autoclave under an oxygen pressure of 250–350 kPa at 80–160 °C. Depending on the conditions, tellurium is converted to the tetravalent or hexavalent form. Tests have been performed at oxygen pressure up to 1000 kPa whereby copper and tellurium are completely dissolved, together with part of the contained silver and selenium, which, after contact with a reducing agent (e.g., $SO_2$) at atmospheric pressure, reprecipitate [23]. The dissolved tellurium in the filtrate is consequently recovered as copper telluride ($Cu_2Te$) by cementation with copper above 80 °C. This can be performed in a rotating drum containing copper shot or in a fixed-bed reactor filled with copper chippings [26]. Other techniques use copper electrodes suspended in a bath containing the dissolved copper and tellurium sulfate, agitated by vibration or alternating current to the electrode to remove the deposited copper telluride [27], [28]. Figure 1 illustrates a possible pretreatment scheme.

**Roasting.** Pretreated decopperized slimes are roasted with soda ash, whereby tellurium is converted to insoluble sodium tellurite and tellurate:

$$Ag_2Te + Na_2CO_3 + O_2$$
$$\longrightarrow 2\,Ag + Na_2TeO_3 + CO_2 \;\; (400-500\,°C)$$
$$Na_2TeO_3 + \tfrac{1}{2}\,O_2 \longrightarrow Na_2TeO_4 \text{ (insoluble)} \; (> 550\,°C)$$

Normally, complete conversion to hexavalent tellurium is preferred with roasting temperature 550–650 °C. Other processes operate below 400 °C or up to 750 °C, involving different process steps. After water leaching, selenium is recovered as soluble sodium selenate. The residue containing tellurium (as sodium tellurate), lead, and precious metals is either sent to a doré furnace, or treated with sulfuric acid:

$$Na_2TeO_4 + H_2SO_4$$
$$\longrightarrow H_2TeO_4 \text{ (soluble)} + Na_2SO_4 \; (95\,°C)$$

In this processing step, the residue containing lead and precious metals is separated for further treatment. The telluric acid is reduced to tellurium by treatment with hydrochloric acid and sulfur dioxide, or reduced by sodium sulfite to tellurium dioxide, or precipitated as copper telluride by adding copper. If there is no preliminary sulfuric acid treatment, the residue still containing the sodium tellurate is smelted in a doré furnace, where the tellurium is collected in a soda slag leached with water to convert it to sodium tellurite. It is then neutralized with sulfuric acid to tellurium dioxide. Copper telluride and tellurium dioxide can be leached with dilute

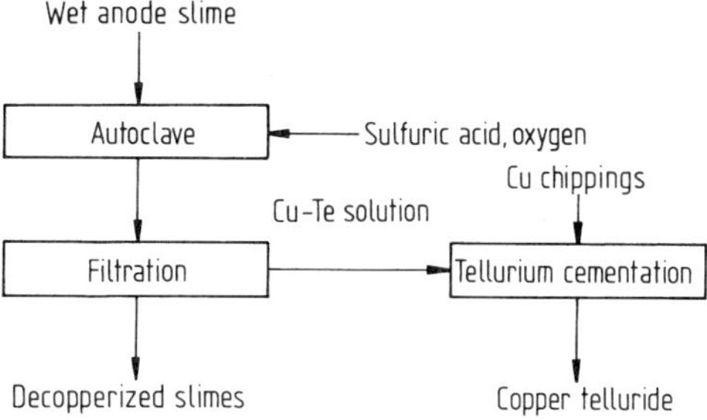


**Figure 1.** Anode slime pretreatment scheme

sodium hydroxide (with aeration in the case of copper telluride) to form a sodium tellurite solution which is further prepared for electrolysis. Figure 2 shows a soda ash roasting process combined with a doré furnace.

**Sulfatizing** takes place in two stages: below 350 °C the sulfuric acid forms metal sulfates with the proper consistency to be roasted above 400 °C in the second stage. In this stage, selenium dioxide is eliminated. For tellurium, the following reactions take place:

$$Ag_2Te + 3 H_2SO_4$$
$$\longrightarrow Ag_2SO_4 + TeSO_3 + SO_2 + 3 H_2O \quad (150-200\,°C)$$
$$3 TeSO_3 \longrightarrow TeO_2 \cdot SO_3 + 2 Te + 2 SO_2 \quad (> 400\,°C)$$
$$TeO_2 \cdot SO_3 \longrightarrow TeO_2 + SO_3 \quad (> 430\,°C)$$

The tellurium dioxide remains in the sulfated slimes, which are leached with water. Part of the

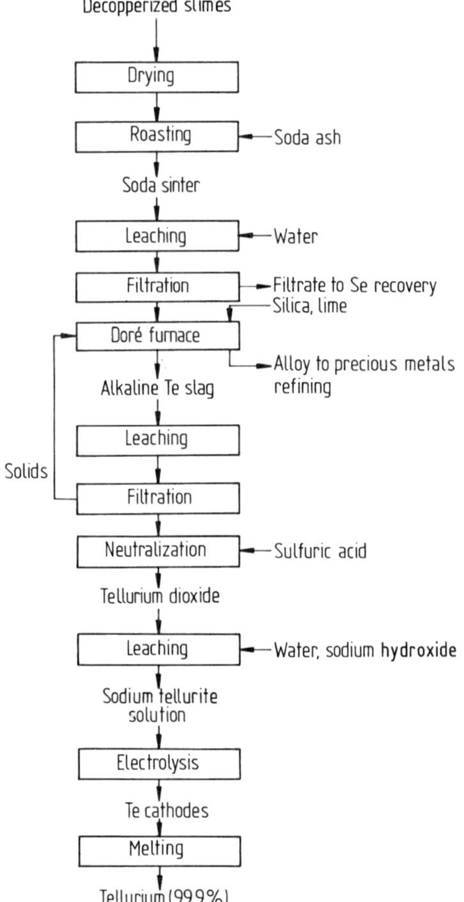


**Figure 2.** Soda ash roasting and doré furnace treatment

tellurium is dissolved, and can be cemented with copper to yield copper telluride. The leach residue can be treated in the doré furnace. Alkaline leaching of the residues yields soluble sodium tellurite, but this process does not recover all the tellurium. Figure 3 shows a sulfatizing roasting process for the recovery of tellurium.

**Other Techniques.** The oxidation of slimes under pressure in alkaline solution can be used to give good separation of tetravalent selenium and hexavalent tellurium after oxidation. Sodium hydroxide concentration is $100-500$ g/L, the temperature is ca. 200 °C, and the oxygen partial pressure is $170-1700$ kPa. Pressure leaching with oxygen or hydrogen peroxide is also possible [29], [30]. Aqueous or dry high-temperature chlorination of slimes has been developed. Dry chlorination being difficult to control, wet chlorination is preferred. The slimes are sparged with chlorine gas or sodium chlorate at ca. 100 °C, and all selenium and tellurium is recovered in the soluble form.

The recovery of tellurium from secondary materials accounts for $10-20\%$ of world tellurium production. International Recoveries of the Philippines started producing tellurium dioxide and tellurium metal from copper telluride raw material in 1984, and other complex residues and waste products are now treated. Treatment of a tellurium tetrachloride residue was described in 1989[31]. Small quantities of thermoelectric scrap are recycled by different tellurium producers.

**Purification.** The following reactions take place during the electrolysis of a tellurite solution:

Anode:    $4 OH^- \longrightarrow 2 H_2O + O_2 + 4 e^-$
Cathode: $TeO_3^{2-} + 3 H_2O + 4 e^- \longrightarrow Te + 6 OH^-$

Typical conditions in an electrolysis cell are: tellurium $100-200$ g/L; free sodium hydroxide 40 g/L; current density 160 A/m²; temperature $40-50$ °C. After drying, melting, and casting in different shapes (or crushing to powder) the commercial tellurium grade is ready for sale. Direct precipitation with zinc is also possible instead of electrowinning.

Further refining consists of vacuum distillation and zone refining, and yields the higher-purity grades $99.99-99.9999\%$. As zone refining is unsatisfactory for some impurities, e.g., aluminum, bismuth, and iron, another process uses chlorination with HCl to tellurium tetrachloride;

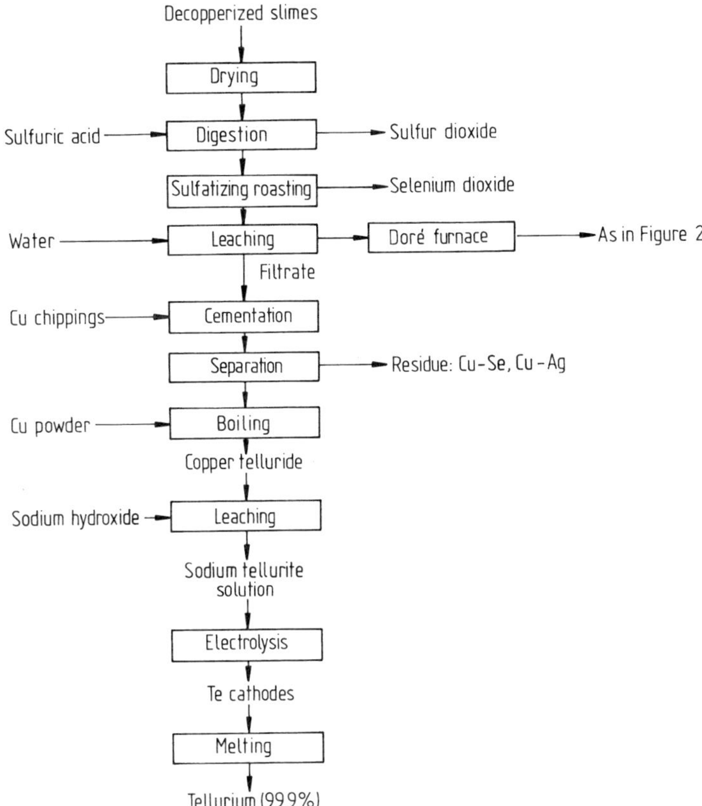


**Figure 3.** Sulfatizing roasting process

distillation and hydrolysis leads to the precipitation of tellurium dioxide of > 99.999 % purity, which after reduction with hydrogen gas gives tellurium of the same purity [32].

## 4. Quality Specifications

Qualities ranging from 99.5 % to 99.9999 % tellurium are commercially available. Impurities may be metallic (As, Na, Ca, Pb, Cu, Ni, Sb, Sn, Se, Fe) and nonmetallic (S, Cl, O).

Commercial grades and forms are: 99.5 – 99.9 % (2N5-3N) powder, sticks, or ingots; 99.99 % (4N) ingots or lumps; 99.999 % (5N) ingots; and 99.9999 (9) % (6N-7N) ingots.

Tellurium dioxide powder is available with purity 99.5 – 99.9 %.

## 5. Analysis (Table 2) [13], [33], [34]

Emission spectrometry (ES) is frequently used for the analysis of standard grade materials

**Table 2.** Detection limits obtained by GFAAS, SSMS, and ES, µg/kg

| Element | GFAAS | SSMS | ES |
| --- | --- | --- | --- |
| Ag | 10 | 1.0 | 100 |
| Al | 50 | 0.8 | 300 |
| Bi | 50 | 2.0 | 100 |
| Cu | 50 | * | 200 |
| Fe | 50 | 1.0 | 200 |
| Mg | 20 | 19 | 100 |
| Ni | 20 | 1.6 | 300 |
| Pb | 50 | 1.0 | 200 |
| Sb | 50 | 1.0 | 300 |
| Sn | 50 | 4.0 | 100 |

* Interference by doubly charged ions.

up to a purity of 99.99 – 99.999 % (with preconcentration if necessary). Atomic absorption spectrometry (AAS) is used for analysis down to sub-ppm levels; this method has, however, no multielement capability. AAS can be adapted with a graphite furnace (GFAAS) and coupled with a Zeeman background correction to allow detection limits at µg/kg levels.

Techniques such as spark source mass spectrometry (SSMS) and glow discharge mass spectrometry (GDMS) offer panoramic elemental capability with low detection limits and maximum selectivity.

# 6. Tellurium Compounds

[5], [11], [35]−[37]

## 6.1. Inorganic Compounds

**Tellurium dioxide**, $TeO_2$ [7446-07-3] is found in the minerals tellurite and paratellurite, which have a rhombic and a tetragonal lattice structure, respectively. It can be formed by burning tellurium in air, or by treating tellurium powder with nitric acid, followed by heating the nitrate:

$$2\,Te + 9\,HNO_3 \longrightarrow 2\,TeO_2 \cdot HNO_3 + 8\,NO_2 + 4\,H_2O$$

The melting point of $TeO_2$ is 1005.8 K (732.6 °C), the clear, dark red liquid oxide vaporizes between 790 and 940 °C. Tellurium dioxide is very soluble in water and is amphoteric, the minimum solubility is at pH 3.8−4.2, the isoelectric point. Tellurium dioxide is an important catalyst in oxidation, hydrogenation, and dehydrogenation processes; it is also used in chalcogenide glasses, and as a vulcanizing agent for rubber. Single crystals of $TeO_2$ are used in acousto-optic detectors and modulators. Other oxides of tellurium are tellurium monoxide, TeO [13451-17-7], tellurium trioxide, $TeO_3$ [13451-18-8], and tellurium pentoxide $Te_2O_5$ [12036-81-6]; $TeO_3$ exists in a yellow α-form and a grey β-form.

**Tellurium Halides.** In tellurium halides, Te has valency 2, 4, or 6. Not all the halides are formed—$TeI_2$ and the $Te_2X_2$ halides do not exist. *Tellurium tetrachloride*, $TeCl_4$ [10026-71-5] forms white hygroscopic needle-like crystals which are soluble in benzene, nitrobenzene, toluene, ethylacetate, and methanol. It is used as a starting material for the synthesis of many organic tellurium compounds, and is also a catalyst for the chlorination of phenol and benzene, and the hydrogenation of acids and esters to alcohols. *Tellurium tetrafluoride*, $TeF_4$ [15192-26-4] forms white hygroscopic needles. It reacts with water, glass, or silica to tellurium dioxide; metal tellurides are formed with copper, silver, gold, and nickel at 185 °C. Platinum is not attacked below 300 °C.

**Hydrogen telluride**, $H_2Te$ [7783-09-7] is a colorless, unstable gas which decomposes above 0 °C and is prepared by electrochemical reduction of tellurium, or by the addition of hydrochloric acid to $Al_2Te_3$. It is used in the production of heavy metal tellurides.

**Tellurous acid**, $H_2TeO_3$ [10049-23-7] is obtained by reaction of alkali tellurites with nitric acid, or by hydrolysis of a tellurium tetrahalide. *Orthotelluric acid*, $H_6TeO_6$ [7803-68-1] is made by oxidizing tellurium or tellurium dioxide. It is used in some chemical processes. *Sodium tellurite*, $Na_2TeO_3$ [10102-20-2] improves the corrosion resistance of electroplated nickel layers. Solutions of *sodium tellurate*, $Na_2TeO_4$ [10101-83-4] containing $Cu^{2+}$ ions are used for black or blue-black coatings on iron, steel, aluminum, and copper.

A large number of *metal tellurides* are known, and many are semiconductors (Table 3). The tellurides are used in infrared detectors ($Cd_xHg_{(1-x)}Te$, PbTe, and $Pb_{(1-x)}Sn_xTe$) and laser diodes (PbTe and $Pb_{(1-x)}Sn_xTe$), thermoelectric elements (PbTe, $Bi_2Te_3$, $Sb_2Te_3$, GeTe), solar cells (CdTe), and γ-ray detectors (CdTe).

**Table 3.** Important tellurides and their properties

| Formula | Crystal type | Density, g/cm³ | Band gap at 300 K, eV | Carrier mobility at 300 K, cm² V⁻¹ s⁻¹ | | Melting point, °C |
|---|---|---|---|---|---|---|
| | | | | $\mu_n$ | $\mu_p$ | |
| CdTe | cubic | 5.86 | 1.5 | 1 000 | 95 | 1092 |
| HgTe | cubic | 8.2 | 0.15 | 25 000 | 350 | 670 |
| PbTe | cubic | 8.16 | 0.29 | 1 800 | 900 | 924 |
| ZnTe | cubic | 5.7 | 2.26 | 530 | 900 | 1238 |
| $Bi_2Te_3$ | hexagonal | 7.86 | 0.15 | 1 300 | 1200 | 588.5 |
| $Sb_2Te_3$ | hexagonal | 6.52 | 0.24 | | 360 | 621.5 |
| GeTe | face-centered cubic | 6.2 | 0.1 | | 180 | 724 |

## 6.2. Organic Compounds (Fig. 4)

The search for organic tellurium compounds and their use in chemical synthesis have developed quickly during the last few decades. In organic compounds tellurium usually has valency 2 or 4. Organotellurium compounds are thermally less stable, and are oxidized more rapidly by air (at room temperature) than organosulfur or organoselenium compounds. The names and formulas of organotellurium compounds are as follows:

| Compound | Formula |
|---|---|
| Tellurol, tellane | $R-Te-H$ |
| Dialkyl(diaryl)(mono,di,tri) tellurides | $R-Te_{1,2,3}-R$ |
| Tellurenyl | $R-Te-X$ |
| Alkyl(aryl)tellurium halides | $R_n-TeX_n$ |
| Tellurinic acid | $RTeOOH$ |
| Dialkyl(diaryl)tellurium oxide | $R_2TeO$ |
| Tellurone | $R_2TeO_2$ |
| Tetraalkyl(tetraaryl) telluride | $R_4Te$ |
| Dialkyl(diaryl)tellurium ketone | $R_2TeCR_2$ |

These compounds have many applications in the pharmaceutical industry. Organic tellurium compounds containing the tellurium isotope $^{123m}Te$ are used as imaging agents for organs. Tellurium diethyldithiocarbamate (TDEC) has been used extensively in the rubber industry, and is also used in copper electroplating. Tellurofulvalenes are organic conductors [38]. Volatile organic tellurium compounds are used in organometallic vapor phase epitaxy for the manufacture of semiconductors.

In organic synthesis, organotellurium compounds are used as mild and highly selective reducing agents, e.g., sodium(hydrogen) telluride, aryltellurols, and tellurides. Other applications are tellurium-mediated formation of anionic species which then react with electrophiles (e.g., Reformatsky-type reactions with sodium telluride), and the preparation of organolithium reagents by lithium/tellurium exchange [40].

# 7. Uses

Consumption of tellurium (in t) according to end use in the western world in 1990 was estimated as follows [41]:

| | |
|---|---|
| Steel additives | 145 |
| Nonferrous metals additives | 55 |
| Chemicals | 45 |
| Electronics | 20 |
| Others | 5 |
| Total | 270 |

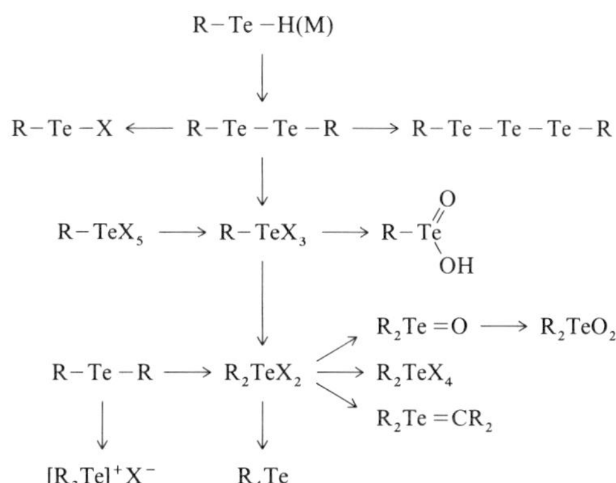

Figure 4. Known types of orgnic tellurium compounds with at least one carbon–tellurium bond (reproduced from [39], with permission)

The machinability of steel can be improved by the addition of small quantities (up to 0.1 %) of tellurium [42]–[45], added during the refining of leaded and/or resulfurized, or low-carbon steel in the form of powder, sticks, or cored wire. Manganese telluride alloys are most frequently used nowadays. Additions to cast iron help control the chill depth of rapidly cooled surfaces and produce hard, wear-resistant top layers.

Tellurium steels and chilled castings are used in the automotive industry, in industrial machinery and tools, in mining, and in railroad equipment.

Many other positive effects of tellurium additions, e.g., deoxidation of liquid steel, refining of the grain size of castings, improvement of the magnetic anisotropy in electrical steels, and improvement of machinability of powder metallurgy steel have been reported.

The use of tellurium in steel could be affected because of decreased consumption of leaded steels owing to environmental constraints.

Traces of tellurium are added to many nonferrous metals [42], [44]–[46]. The tellurium content is mostly less than 1 %. Tellurium–copper (which also contains some phosphorus) reduces cutting resistance, tool wear, heat of cutting, and chip size. It is used in forging, automotive radiators, vacuum applications, and as electrical, motor, and switch parts. Tellurium–copper can replace lead–copper in plumbing, as the latter is under environmental pressure in some countries.

Tellurium lead is used in the sheathing of power and marine cables and in chemical equipment piping (mostly where sulfuric acid is used). In lead acid batteries, tellurium helps strengthen the battery grids. Other nonferrous alloys are tellurium bronze, tellurium–tin, tellurium–chromium, and tellurium–magnesium. Tellurium–cobalt–titanium is used in permanent magnets.

Tellurium is used in combination with other components in many catalysts. The chemical reactions involved are oxidation, ammoxidation, hydrogenation and dehydrogenation, halogenation and dehalogenation, and phenol condensation [47]. The two major industrial applications are the Nitto process: (acrylonitrile ammoxidation) and the Mitsubishi process (tetramethylene glycol synthesis) [48]:

*Nitto:* acrylonitrile ammoxidation catalyst:

$$C_3H_6 + NH_3/air \xrightarrow{\text{Sb–Mo–U–Te}} C_2H_5CN$$

*Mitsubishi:* tetramethylene glycol synthesis
(Ac = acetate):

$$C_4H_6 + HOAc/air$$
$$\xrightarrow{\text{Pd–Te}} AcO–CH_2–CH=CH–CH_2OAc$$
$$\xrightarrow{\text{H}_2/\text{H}_2\text{O}} HO–CH_2–CH_2–CH_2–CH_2–OH$$
$$\xrightarrow{-\text{H}_2\text{O}} (CH_2)_4O \text{ (tetrahydrofuran)}$$

Tellurium is used as a secondary vulcanizing agent in natural, styrene–butadiene, and isobutylene–isoprene rubber [42], [44], [49]. The tellurium rubbers have improved aging and mechanical properties, and better resistance to heat and abrasion. They are used in portable cables, automobile tires, door and window seals, double glazing, and in conveyor belts for special applications. The use of tellurium in rubber has declined over the past few decades, and there is concern over the toxicity of tellurium handling in rubber production, which could further affect its usage.

Other uses of tellurium in chemistry are in heat- and vacuum-stable lubricants for electronics and aerospace applications, metal coatings for silverware, aluminum, and brass, in glass, pigments and fungicides, and in organic derivatives and radioactive isotopes for medicine and biology. Lately, the role of tellurium as a possible regulator in cholesterol synthesis has been studied [50].

In electronics, tellurium is a well-known additive in selenium(-arsenic) photoreceptors [51], [52]. Tellurium has the property of increasing the sensitivity and broadening the spectral response of the photoreceptor. Since organic photoconductors (OPCs) have been taking over the photocopier and laser printer market, starting with the slower machines (up to 20–30 pages per minute), the use of selenium–tellurium alloys is decreasing markedly, especially in the industrialized countries.

Polycrystalline cadmium telluride is used in photovoltaic solar cells [53]. CdTe has a direct band gap of 1.5 eV, which is optimal for solar energy conversion, and a high optical absorption coefficient. Research into higher efficiencies, lower cost, and better reliability has intensified in recent years, and the future of CdTe solar cells looks very promising. By the year 2000, the use of tellurium in solar cells could be more than 50 t/a [54].

Bismuth telluride and lead telluride are thermoelectric materials [55], [56]. Thermocouples of these materials convert electricity into heat (Peltier effect) and heat into electricity (Seebeck effect). In these thermocouples, dopants such as Sb and Se are needed to make *n*- and *p*-type materials.

$Bi_2Te_3$ is mostly used below 200 °C, PbTe at 200–500 °C. The efficiency of a thermoelectric material is expressed by its figure of merit $Z$, a function of the thermal and electrical conductivity and the Seebeck coefficient.

Thermoelectricity has found many applications in picnic boxes, water and food coolers, fiber optic circuits, blood analyzers, and other precision temperature control instruments, remote power generators, solar energy conversion, military equipment (e.g., refrigeration systems in submarines and infrared detectors), and aerospace. Especially the low-tech applications, such as picnic boxes and car drink coolers, have seen a steady rise in sales.

Now that systems based on chlorofluorocarbons (CFCs) are increasingly banned from refrigerators and air-conditioning devices, owing to their detrimental effect on the ozone layer, thermoelectricity could offer a viable alternative, but the present efficiency of the coolers is still too low and the cost too high to suggest a rapid growth in large cooling systems [57].

Optical disks are powerful tools for storing vast quantities of digitized information (up to 1 Gbyte). Different types of write-only and erasable disks exist, and some of their active layers use tellurium alloys which are sputtered onto the disk by vapor deposition techniques from targets containing, e.g., Te–Ge–Sb or Te–Se–Sb [58].

Cadmium-, cadmium zinc-, and cadmium selenium telluride single crystals are used in several electronic applications: γ-ray detectors for medi-

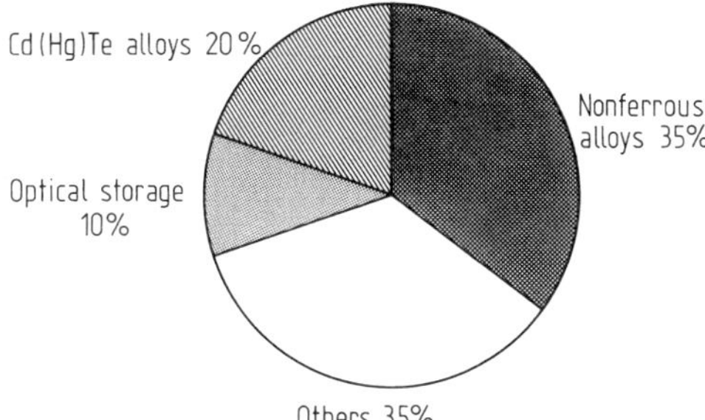


**Figure 5.** Tellurium in scientific publications: main areas of interest in *Chemical Abstracts* (1991)

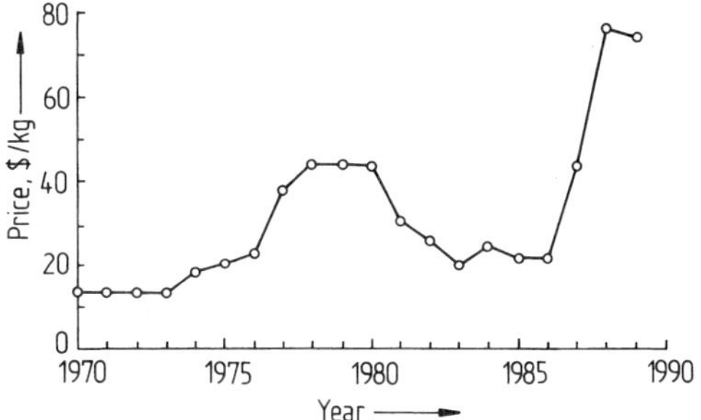


**Figure 6.** Average annual price for standard-grade tellurium (99.5%)

cal diagnostics and dosimetry, nonlinear optics, infrared electro-optic modulators, and photorefractive devices [34], [59]. These crystals are grown by vertical or horizontal Bridgman techniques, or the traveling heater method (THM). They are also used as substrates for epitaxial cadmium mercury telluride ($Cd_xHg_{(1-x)}Te$) thin film detectors for infrared imagers [60], of which lead tin telluride ($Pb_xSn_{(1-x)}Te$) is an example. The detectors operate in the spectral ranges $3-5$ µm and $8-14$ µm. The thin films are grown by methods such as liquid phase epitaxy (LPE), organometallic vapor phase epitaxy (OMVPE), or molecular beam epitaxy (MBE) [34]. The devices are normally cooled to 77 K with liquid nitrogen. The new generations of imagers use two-dimensional focal plane arrays to capture an entire image at once.

Figure 5 shows how the research and development literature focuses on the different uses of tellurium in nonferrous alloys, optical storage, and cadmium- and cadmium mercury tellurides.

## 8. Economic Aspects

World production of tellurium in 1990 has been estimated at 306 t. The estimated production (in t/a) by country and company for 1990 is listed below:

| Japan | |
|---|---|
| Mitsubishi Metal | 22 |
| Mitsui Mining & Smelting | 10 |
| Nippon Mining & Smelting | 12 |
| Sumitomo Metal Mining | 12 |
| Belgium | |
| Union Minière | 90 |
| Canada | |
| Noranda Mines | 15 |
| United States | |
| Asarco | 40 |
| Philippines | |
| Pacific Rare Metal | 35 |
| Soviet Union | 50 |
| Peru | 10 |
| China | 10 |
| *Total* | *306* |

The main producers are in Belgium, the United States, the Commonwealth of Independent States, and the Philippines [44].

Figure 6 shows the evolution of the average annual price for standard-grade tellurium (99.5%) [44]. Tellurium 99.99% is normally $40-50\%$ more expensive, 99.999% is more than double the standard-grade price, and $\geq 99.9999\%$ ranges from 400 to 1200 \$/kg.

The world economic recession of the early 1990s, particularly hitting the steel industry, and cheap material from the Commonwealth of Independent States, led to a price erosion of 50% between 1990 and 1993.

## 9. Toxicology and Occupational Health [61]–[65]

Elemental tellurium is considered to be less toxic than selenium. Organic compounds and reactive tellurides can be a health hazard. Hydrogen telluride ($H_2Te$) and tellurium hexafluoride ($TeF_6$) are both highly toxic, colorless gases. Some $LD_{50}$ values on ingestion of tellurium compounds are listed in Table 4. On the basis of new data for the $LD_{50}$ value for elemental tellurium, the Environmental Protection Agency of the United States has been asked to declassify tellurium as extremely hazardous [65].

**Table 4.** $LD_{50}$ values for different tellurium compounds

| Compound | Animal | Administration | $LD_{50}$, mg/kg |
|---|---|---|---|
| Te | rat, mouse | oral | > 5000 |
| $TeO_2$ | unspecified | oral | > 3000 |
| $Na_2TeO_3$ | white mouse | oral | 20 |
| $Na_2TeO_4$ | white mouse | oral | 165 |

The acute inhalation toxicity value $LC_{50}$ (4 h) for rats is $> 2.42$ g/m$^3$, the maximum attainable concentration [66].

Tellurium intoxication in the working environment can occur through inhalation and ingestion. No irritation of skin or eyes has been reported for elemental tellurium and tellurium dioxide. Exposure to tellurium fumes and dust cause a garlic odor of the breath and malaise, dryness of the mouth and metallic taste, anorexia, occasional nausea, and some other symptoms. Tellurium is accumulated in the liver, kidneys, bones, and neural cells. The elimination of tellurium in breath, sweat, urine, and feces is very slow. The typical garlic odor seems to result from the partial elimination of tellurium through the lungs as dimethyl telluride $(CH_3)_2Te$; it can be suppressed temporarily by administration of vitamin C (ascorbic acid), but the vitamin C may also enhance the toxic effects. Tellurium hexafluoride can cause a bluish-black pigmentation of the skin.

Tellurium intoxication symptoms do not appear when the tellurium concentration in the air is $< 0.01$ mg/m$^3$ and in the urine $< 1$ µg/L. The TLV for tellurium and its compounds is 0.1 mg/m$^3$ (calculated as tellurium). The TLV was established to prevent poisoning, but garlic breath may still occur. There is no specific antidote for tellurium poisoning.

There are no indications of carcinogenic, teratogenic, or mutagenic effects of tellurium or its compounds.

Unlike selenium, tellurium is not an essential trace element for humans or animals. Tellurium and its compounds cannot be used in cosmetics.

# 10. References


[1] *Gmelin*, Suppl. **A 1**, p. 205.
[2] S. C. Carapella, Jr., in W. C. Cooper (ed.): *Tellurium*, Von Nostrand Reinhold Co., New York 1971, p. 1.
[3] *Ullmann*, 4th ed., **22**, 447.
[4] *Handbook of Chemistry and Physics*, 66th ed., The Chemical Rubber Co., Cleveland, Ohio 1985/86.
[5] J. C. Bailar et al.: *Comprehensive Inorganic Chemistry*, 1st ed., Pergamon Press, Oxford 1973, pp. 935–1009.
[6] *Kirk-Othmer*, **22**, 659–661.
[7] S. C. Carapella, Jr.: *Metals Handbook*, 9th ed., vol. **2**, American Society for Metals, Metals Park, Ohio 1979, p. 806.
[8] W. M. Becker, V. A. Johnson, A. Nussbaum in W. C. Cooper (ed.): *Tellurium*, Von Nostrand Reinhold Co., New York 1971, pp. 54–109.
[9] D. M. Chizhikov, V. P. Shchastlivyi: *Tellurium and Tellurides*, Collet's Ltd., London 1970, pp. 1–30.
[10] H. Lumbroso in P. Pascal (ed.): *Compléments au Nouveau Traité de Chimie Minérale*, vol. **8**, Masson, Paris 1977, pp. 17–68.
[11] W. A. Dutton in W. C. Cooper (ed.): *Tellurium*, Von Nostrand Reinhold Co., New York 1971, pp. 110–183.
[12] I. Barin: *Thermochemical Data of Pure Substances*, part II, VCH Verlagsgesellschaft, Weinheim 1989.
[13] M. Caffarey, *Proc. 8th Conf. on Thermoelectric Energy Conversion*, Nancy, France 1989, pp. 40–44.
[14] W. C. Cooper (ed.): *Tellurium*, Von Nostrand Reinhold Co., New York 1971, pp. 1–13.
[15] D. Edelstein in S. C. Carapella, Jr. (ed.): *Proc. 4th Int. Symp. on Uses of Selenium and Tellurium*, Banff, Canada 1989, pp. 4–13.
[16] P. E. Skinner, *Trans. Inst. Min. Metall. Sect. C* **97** (1988) C83–C87.
[17] P. H. Jennings in W. C. Cooper (ed.): *Tellurium*, Von Nostrand Reinhold Co., New York 1971, pp. 14–53.
[18] *Kirk-Othmer*, **22**, 662–663.
[19] R. Bresee, D. Vleeschhouwer, J. Thiriar, *Proc. of the Symp. on Industrial Uses of Selenium and Tellurium*, Toronto, Canada 1980, pp. 31–49.
[20] J. E. Hoffmann, *J. Met.* **41** (1989) no. 7, 32–38.
[21] J. E. Hoffmann, *J. Met.* **42** (1990) no. 8, 50–54.
[22] *Ullmann*, 4th ed., **22**, 448–451.
[23] Noranda Inc., CA 2049276, 1991 (P. L. Claessens, C. W. White).
[24] Nippon Mining Co., JP 60208431, 1985 (T. Abe, Y. Takazawa).
[25] J. E. Dutrizac, T. T. Chen in W. C. Cooper, G. E. Lagos, G. Ugarte (eds.): *Copper 87*, vol. 3, Universidad de Chile, Santiago, Chile 1988, p. 469.
[26] T. Shibasaki, K. Abe, H. Takeuchi, *Hydrometallurgy* **29** (1992) 399–412.
[27] Mitsubishi Metal Corp., Onahama Seiren Co., JP 01079008, 1989 (Y. Sugawara, J. Konishi, S. Hayashi, M. Hayashi).
[28] Mitsubishi Materials Corp., US 5160588, 1992 (T. Shibazaki, K. Abe, H. Takenchi).
[29] Nippon Shin Kinzuko K.K., JP 56084428, 1981.
[30] Interox Chemicals, EP 127357, 1984 (A. Broome).
[31] R. J. Hisshion in S. C. Carapella, Jr. (ed.), *Proc. 4th Int. Symp. on Uses of Selenium and Tellurium*, Banff, Canada 1989, pp. 25–30.
[32] Sumitomo Metal Mining Co., JP 63285106, 1988, JP 63286531, 1988 (M. Yukinobu, J. Oowa).
[33] K. Swenters, J. Verlinden, R. Gijbels, *Fresenius Z. Anal. Chem.* **335** (1989) 900–904.
[34] D. R. Nichols, B. E. Dean, C. J. Johnson in S. C. Carapella, Jr. (ed.): *Proc. 4th. Int. Symp. on Uses of Selenium and Tellurium*, Banff, Canada 1989, pp. 282–299.
[35] *Ullmann*, 4th ed., **22**, 451–452.
[36] D. M. Chizhikov, V. P. Shchastlivyi: *Tellurium and Tellurides*, Collet's Ltd., London 1970, pp. 145–259.

[37] *The Economics of Tellurium 1991*, 4th ed., Roskill Information Services, London 1991, pp. 65–66.

[38] D. O. Cowan et al., *Phosphorus Sulfur Silicon* **67** (1992) 277–294.

[39] K. J. Irgolic in D. Klamann (ed.): "Methods of Organic Chemistry," *Houben-Weyl*, 4th ed., Vol. **E 12 b**, Thieme Verlag Stuttgart 1990, p. XXXIX.

[40] N. Petragnagni, J. V. Comasetto, *Synthesis* **10** (1991) Oct., 793–817, **11** (1991) Nov., 897–919.

[41] E. Hoyne, *The Bulletin of Selenium-Tellurium Development Association, Inc.*, Sept. 1989, pp. 1–4.

[42] *Kirk-Othmer*, **22** 671–676.

[43] E. S. Nachtman in W. C. Cooper (ed.): *Tellurium*, Von Nostrand Reinhold Co., New York 1971, pp. 373–409.

[44] *The Economics of Tellurium 1991*, 4th ed., Roskill Information Services, London 1991, pp. 51–98.

[45] R. J. Raudebaugh, *Proc. of the Symp. on Industrial Uses of Selenium and Tellurium*, Toronto, Canada 1980, pp. 201–221.

[46] P. W. Taubenblat, *Proc. of the Symp. on Industrial Uses of Selenium and Tellurium*, Toronto, Canada 1980, pp. 267–276.

[47] W. C. Cooper (ed.): *Tellurium*, Von Nostrand Reinhold Co., New York 1971, pp. 410–430.

[48] T. A. Koch in S. C. Carapella, Jr. (ed.) *Proc. 4th Int. Symp. on Uses of Selenium and Tellurium*, Banff, Canada 1989, pp. 661–672.

[49] W. R. McWhinnie, J. E. Stuckey, K. G. Karnika Da Silva, *Proc. of the 3rd Int. Symp. on Industrial Uses of Selenium and Tellurium*, Stockholm, Sweden 1984, pp. 206–245.

[50] A. D. Toews, S. Y. Lee, B. Popko, P. Morell, *J. Neurosci. Res.* **26** (1990) 501–507.

[51] L. Cheung, G. M. Foley, P. Fournia, B. E. Springett, *Photogr. Sci. Eng.* **26** (1982) no. 5, 245–249.

[52] R. C. Norton, *Proc. 7th Annual Photoreceptor & Components Conf.*, Santa Barbara, United States 1991.

[53] K. Zweibel: "Toward Low Cost CdTe PV, NREL/TP-413-4841," National Renewable Energy Laboratory, Golden, CO., April 1992.

[54] M. C. King, *Workshop Report: Recycling of Cadmium and Selenium from Photovoltaic Modules and Manufacturing Wastes*, Golden, United States 1992, pp. 90–101.

[55] C. H. Champness in W. C. Cooper (ed.): *Tellurium*, Von Nostrand Reinhold Co., New York 1971, pp. 322–372.

[56] D. P. Burton, *Proc. 9th. Int. Conf. on Thermoelectrics*, Pasadena, United States 1990, pp. 109–123.

[57] P. M. Schicklin, *Proc. 9th Int. Conf. on Thermoelectrics*, Pasadena, United States 1990, pp. 381–395.

[58] M. Hartmann, B. A. Jacobs, *Philips Tech. Rev.* **42** (1985) no. 2, 37–47.

[59] M. Ohmori, Y. Iwase, R. Ohno, *Proc. Symposium F on New Aspects of the Growth, Characterization and Applications of CdTe and Related Cd Rich Alloys of the 1992 E-MRS Spring Conf.*, Strasbourg, France 1992, pp. 283–290.

[60] C. Lucas, *Sens. Actuators A* **25–27** (1991) 147–154.

[61] J. R. Glover, V. Vouk in L. Friberg et al. (eds.): *Handbook on the Toxicity of Metals*, Elsevier/North Holland Biomedical Press, Amsterdam 1979, chap. 35, pp. 587–598.

[62] W. C. Cooper in W. C. Cooper (ed.): *Tellurium*, Von Nostrand Reinhold, New York 1971, pp. 313–321.

[63] *Ullmann*, 4th ed., **22**, 453.

[64] *Kirk Othmer*, 3rd ed., **22**, 666.

[65] *Environmental Outlook*, vol. 3, no. 1, Selenium–Tellurium Development Assoc., Grimbergen, Belgium 1992.

[66] TNO Report V 92.008, TNO-Voeding, Zeist, Holland, Jan. 1992.

**Telomerization → Organometallic Compounds and Homogeneous Catalysis**

**Tequila → Spirits**

# Teratogenic Agents

CHRISTIAN BAEDER, Hoechst AG, Hattersheim, Federal Republic of Germany

Numerous agents are known to be teratogenic in humans. Examples are radiation (therapeutic radiation, nuclear contamination), infections (rubella and cytomegalovirus), maternal metabolic imbalance (alcoholism, diabetes, folic acid deficiency), drugs (androgenic hormons, alkylating agents, diethylstilbestrol, retinoic acid, thalidomide, valproic acid) and environmental chemicals (chlorobiphenyls, lithium, organic mercury) [1]. Typical representatives of teratogenic agents and effects they cause in humans are:

| | |
|---|---|
| Androgenic hormones | adrenal hyperplasia, virilisation of the female genitalia |
| Alkylating agents | degeneration of the embryonic disc and intrauterine death, abortion |
| Diethylstilbestrol | virilisation of the female fetus |
| Retinoic acid | multiple CNS and cardiac malformations, thymic aplasia |
| Thalidomide | limb malformations, phocomelia, ear defects, facial hemangioma, atresia of the esophagus or duodenum, tetralogy of Fallot, renal agenesis, cleft palate |
| Valproic acid | neural tube defects (anencephaly, microcephaly, facial dysmorphology, spina bifida), heart defects, heart ventricular septal defects, postnatal growth deficiency |
| Lithium | cardiovascular defects, Ebstein's anomaly |
| Chlorobiphenyls | dark-brown staining of the skin, exophthalmos, intrauterine growth retardation |

Teratogenicity is a special form of embryotoxicity [2]. It presupposes a detrimental effect on the embryo and on organogenesis. Thus, malformations induced after birth, during the lactation period are to be regarded as postnatal developmental disturbances. For the tolerance classification of an agent, not only its teratogenic effect, but the total embryotoxic potential is of significance. This includes effects on:

1) Gravidity in general
2) Developmental stages of the zygote (the period of time from fertilization of the ovum to implantation in the wall of the uterus), and
3) Intrauterine development (embryogenesis: development of the basic body form and organogenesis; fetogenesis: growth and organ differentiation after the completion of organogenesis)

Forms of embryotoxicity and fetotoxicity are death, disturbance of growth (retardation), teratogenic effects, and induction of minor anomalies. Growth disturbances can be reversible within the uterus or after birth. A distinction is to be made between structural and functional developmental defects. Structural anomalies can be detected macroscopically, microscopically, or by electron microscopy [3]. Functional anomalies are expressed as disturbed organ functions, and enzymatic, endocrine, or immunological changes. Malformations are irreversible. Minor anomalies and variations of organs and the skeleton are characterized by the fact that they do not interfere with the vital functions or the social status of the individual, and can be reversible.

Testing of the embryotoxic potential is carried out preferably on laboratory mammals. As suitable models for humans, they should fulfil the following requirements: maternal–embryonal/fetal unit; similarity in anatomy, placentation, and transplacental passage; and similarity in kinetics, toxicokinetics, metabolism, and sensitivity.

The animals used most often are rats. The advantages of using rats are that results of previous systemic toxicity studies and information

Ullmann's Encyclopedia
of Industrial Chemistry, Vol. A 26

on toxicokinetics and metabolism are generally available. The rabbit is the first species of experimental animal in which the teratogenic effect of thalidomide could be reproduced. Both species have always been widely used for embryotoxicological studies and have proved to be sensitive models. If differences in the toxicokinetics, metabolism, or sensitivity are found between rats or rabbits and humans or if uninterpretable results are obtained, mice and subhuman primates are often used as alternative species. For these animals, extensive knowledge of the spontaneous and agent-induced occurrence of gravidity damage and intrauterine developmental defects is available. In the assessment of test results, lineage specific differences in sensitivity that are species or genetically dependent must be taken into account.

Alternative methods and test models, such as nonmammalian systems in vivo (e.g., chicken, frog, fish or Drosophila embryos, and hydra) as well as primary embryonal cell cultures and cell cultures from mammals and nonmammals, organ cultures (e.g., limb buds), and embryo cultures (pre- and post-implantative) in vitro, concentrate on different and limited test parameters (end points) and do not cover the entire embryonal, fetal, and postnatal development. The metabolism of the mammalian organism can be adopted by introducing a metabolism activating postmitochondrial liver homogenate (S-9 mixture from supernatant after centrifugation at 9000 $g$ of homogenized livers of male rats treated with an enzyme inductor, e.g., Acrolor 1254 plus NADPH) [4]. However, these systems lack above all the maternal–embryonal unit. Therefore, at their present level of development, these systems are suitable only for prescreening, especially of structurally analogous agents, and for mechanistic testing [5]. Intensive work is being done on the development of alternative methods which would lead to decreased use of experimental animals.

The studies are carried out according to national and international test guidelines [6], [7]. The following basic requirements must be complied with before planning a study:

The identification and characterization of the embryotoxic potential of an agent depend on its chemical structure, mechanism of action, and on the dosage–response relationship [8].

Agents can directly or indirectly influence the intrauterine development of the embryo/fetus. Direct actions can influence the development of

cells, tissues, or organs by changing or inhibiting cell proliferation, migration, and differentiation, or can kill the cell. Indirect actions can depend on metabolic or hormonal changes, intoxication of the mother, or disturbance of placenta function. Only a few mechanisms of teratogenesis have been elucidated.

With a few exceptions, the principle of the dosage–response relationship also applies to embryotoxicology/teratology. With increasing dose or concentration of an agent, the severity, variety of effects, and the number of damaged individuals increase. In rare cases, a clear dose–response relationship is not discernible or is concealed by other toxic effects. The determination of the doses or concentrations to be tested is of particular importance. The low dose is based on the maximal or a multiple of the amount expected to be used for daily exposure. With an intermediate dose, the geometric mean of the low and the highest dose selected, the safety margin of the agent should be determined. In the highest dose, the agent should exhibit minimal toxicity for the maternal or embryonal organism. An important purpose of the testing is the determination of the no observed adverse effect level (NOAEL), i.e., the dose or concentration in which no impairment of the physical condition of the mother and embryonal development can be determined. Agents, which are not toxic for the designated species of animal in oral doses of 1000 mg/kg body weight or in concentrations of 5 mg/L of maternal respiration air, or cannot be administered in higher doses or concentrations due to their constitution, should not be tested in smaller or higher amounts because of lacking practical relevance (limit test).

As a rule, route and the frequency of administration should be similar to the exposure of humans in practice. To attain higher concentrations of the agent in the organism, other types of application can also be used.

**Course of Testing.** Adult virginal female animals in heat are mated naturally, a vaginal smear being taken to check for insemination, or artificially inseminated. The day following insemination is called day 1 of gravidity. The animals are treaded daily with the agent to be tested, starting from the implantation of the blastocyst in the uterus (e.g., rat: day 7, rabbit: day 6) and continuing through the organogenesis phase or up to the end of gravidity. The physical condition of the animals is examined every day. Disturbances of gravidity can result in abortion, premature birth, or death of all the embryos.

On the day before the expected birth (e.g., rat: day 21, rabbit: day 29) the animals are killed, a cesarean section is performed, and the uterus contents are examined. The condition of the uterus, the number, appearance, size, and weight of the living and dead fetuses and placentas, as well as the size of the prematurely dead embryonal structures that are in the process of decomposition and resorption are determined. The fetuses are inspected for external anomalies, dissected, and examined for structural anomalies of the inner organs. After exenteration and clearing the animal body in potassium hydroxide solution, the bones are stained with alizarin red S and placed in glycerol. The skeleton of the fetuses is then structurally examined for the state of ossification and for anomalies.

To find functional disorders, the mothers must deliver and raise their offspring. From the duration of gravidity and the course of birth, conclusions can be drawn regarding a possible impairment of gravidity (shortening, extension, and disturbance of parturition) caused by the administration of the agent to be tested. The number, size, behavior, and physical development of the offspring after birth give information about the postnatal effects of intrauterine induced damage.

The parameters to be investigated are the number of offspring that are born alive and dead, suckling, ability to survive, weight development, and physiological development parameters, such as the time of opening of the eyelids, unfolding of the auricles from the head, eruption of the incisors through the jaw, and the beginning of fur growth. Functional tests comprise observations of the free functional behavior including sight and hearing. For behavioral studies, a large number of test methods from the area of psychopharmacology are available. The test program contains, e.g., examinations of the general behavior in the open field, activity, clinging and movement on an inclined plane or movable roll, reflexes, motor system, equilibrium sense, carrying out of given tasks in discrimination tests, as well as learning ability, memory, and ability to relearn. These examinations are effected descriptively or with measuring methods.

With due regard to the dose, sensitivity, comparative kinetics, and metabolism, all the examinations together make possible a careful estimation of the risks involved in the use on or exposure of human beings. After consideration of the criteria mentioned above, the determination of injury to the embryo in animal experiments leads to the contraindication for use in women, except where pregnancy can be definitely excluded. Uncertainty is involved in the assessment of the functional and behavioral tests. Their meaningfulness has not yet been adequately demonstrated in the absence of a cellular (histological) structural substrate.

# References

[1] J. G. Wilson: *Environment and Birth Defects*, Academic Press, New York 1973.
[2] *Ullmann*, 4th ed., **22**, 517–518.
[3] Ch. Baeder, M. Albrecht, "Embryotoxic/Teratogenic Potential of Halothane," *Int. Arch. Occup. Environ. Health* **62** (1990) 263–271.
[4] D. M. Marion, B. N. Ames, "Revised Methods for the Salmonella Mutagenicity Test," *Mutat. Res.* **113** (1983) 173–215.
[5] ECETOC: "Alternative Approaches for the Assessment of Reproductive Toxicity (with Emphasis on Embryotoxicity/Teratogenicity)" *Monograph* **12** (1989).
[6] R. Bass et al.: "Draft Guideline on Detection to Toxicity to Reproduction for Medicinal Products," *Adverse Drug React. Toxicol. Rev.* **9** (1991) 127–141.
[7] OECD: Guidelines for Testing of Chemicals, Section 4: Health Effects, no. 414–416 (1981, 1983).
[8] ECETOC: "Identification and Assessment of the Effects of Chemicals on Reproduction and Development (Reproductive Toxicology)," *Monograph* **5** (1983).

**Terbium → Rare Earth Elements**

# Terephthalic Acid, Dimethyl Terephthalate, and Isophthalic Acid

RICHARD J. SHEEHAN, Amoco Research Center, Amoco Chemical Company, Naperville, Illinois, United States



## 1. Introduction

Terephthalic acid [100-21-0], and isophthalic acid [121-91-5], both $C_8H_6O_4$, have the IUPAC names 1,4- and 1,3-benzenedicarboxylic acid. Dimethyl terephthalate [120-61-6], $C_{10}H_{10}O_4$, is also known as 1,4-benzenedicarboxylic acid dimethyl ester. The acids are produced by oxidation of the methyl groups on the corresponding p-xylene [106-42-3] or m-xylene [108-38-3]. After oxidation to a carboxylic acid, reaction with methanol [67-56-1] gives the methyl ester, dimethyl terephthalate.

Terephthalic acid and dimethyl terephthalate are used to make saturated polyesters with aliphatic diols as the comonomer. Isophthalic acid is used as a feedstock for unsaturated polyesters as well as a comonomer in some saturated products. Structures are as follows:

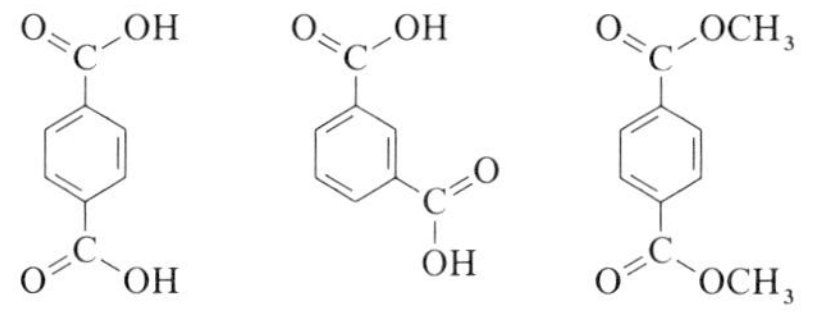

Terephthalic    Isophthalic    Dimethyl
acid            acid           terephthalate

Terephthalic acid came to prominence through the work of WHINFIELD and DICKSON in Britain around 1940 [4]. Earlier work by CAROTHERS and coworkers in the United States established the feasibility of producing high molecular weight linear polyesters by reacting diacids with diols, but they used aliphatic diacids and diols. These made polyesters which were unsuitable to be spun into fibers. WHINFIELD and DICKSON found that symmetrical aromatic diacids yield high-melting, crystalline, and fiber-forming materials; poly(ethylene terephthalate) has since become the largest volume synthetic fiber. Worldwide, terephthalic acid plus dimethyl terephthalate ranked about 25th in tonnage of all chemicals produced in 1992, and about tenth in terms of organic chemicals.

## 2. Physical Properties

Terephthalic acid, $M_r$ 166.13, is available acommercially as a free-flowing powder composed of rounded crystals. It forms needles if recrystallized slowly. Vapor pressure is low: 0.097 kPa at 250 °C, with sublimation at 402 °C and atmospheric pressure. Melting has been reported at 427 °C.

Dimethyl terephthalate, $M_r$ 194.19, melts at 140.6 °C and has sufficient vapor pressure for vacuum distillation. The molten form is preferred commercially, but flakes and briquettes are available when long transport distances are required.

Isophthalic acid, $M_r$ 166.13, melts at 348 °C. Vapor pressure is 0.61 kPa at 250 °C.

Both terephthalic and isophthalic acid are stable, intractable compounds with low solubilities in most solvents; isophthalic acid is 2–5 times more soluble than terephthalic acid in the same solvent. Terephthalic acid solubilities > 10 g per 100 g solvent at room temperature occur with ammonium, potassium, or sodium hydroxide, dimethylformamide, and dimethyl sulfoxide. Tetramethylurea and pyridine each dissolve ca. 7 g per 100 g.

Dimethyl terephthalate is also stable, and more soluble in common organic solvents than the acids. Its main physical properties are a melting point below the point of degradation, and a vapor pressure which allows for purification by distillation.

Other physical properties of terephthalic acid and isophthalic acid are listed in Tables 1 and 2.

Some physical properties of dimethyl terephthalate are listed below:

Solubility, g/100 g solvent
| | |
|---|---|
| Methanol, 25 °C | 1.0 |
| Methanol, 60 °C | 5.7 |
| Ethyl acetate, 25 °C | 3.5 |
| Ethyl acetate, 60 °C | 16.0 |
| Trichloromethane, 25 °C | 10.0 |
| Trichloromethane, 60 °C | 23.0 |
| Benzene, 25 °C | 2.0 |
| Benzene, 60 °C | 14.0 |
| Toluene, 25 °C | 4.3 |
| Toluene, 60 °C | 10.4 |
| Dioxane, 25 °C | 7.5 |
| Dioxane, 60 °C | 26.5 |

Vapor pressure, kPa
| | |
|---|---|
| 140 °C | 1.58 |
| 160 °C | 2.62 |
| 200 °C | 10.88 |
| 250 °C | 42.60 |

| | |
|---|---|
| Density at *mp* | 1.07 g/cm$^3$ |
| Viscosity, 180 °C | 0.0071 Pa · s |
| Viscosity, 200 °C | 0.0060 Pa · s |
| Specific heat, solid | 1.55 J g$^{-1}$ K$^{-1}$ |
| Specific heat, liquid | 1.94 J g$^{-1}$ K$^{-1}$ |
| Heat of fusion | 159.1 kJ/Kg |
| Heat of vaporization | 355.5 kJ/Kg |
| Heat of formation | − 740.2 kJ/mol |
| Heat of combustion, 25 °C | 4660 kJ/mol |
| Flash point, DIN 51 758 | 141 °C |

# 3. Production

*p*-Xylene is the feedstock for all terephthalic acid and dimethyl terephthalate production; *m*-

**Table 1.** Physical properties of terephthalic acid and isophthalic acid

| | Terephthalic | Isophthalic |
|---|---|---|
| Crystal density, g/cm$^3$, 25 °C | 1.58 | 1.53 |
| Specific heat, J g$^{-1}$ K$^{-1}$, to 100 °C | 1.20 | 1.22 |
| Heat of formation, kJ/mol | − 816 | − 803 |
| Heat of combustion, kJ/mol, 25 °C | 3189 | 3203 |
| First dissociation constant | $2.9 \times 10^{-4}$ | $2.4 \times 10^{-4}$ |
| Second dissociation constant | $3.5 \times 10^{-5}$ | $2.5 \times 10^{-5}$ |
| Ignition temperature in air, °C | 680 | 700 |

**Table 2.** Solubility of terephthalic acid and isophthalic acid (g/100 g solvent)

| | 25 °C | 150 °C | 200 °C | 250 °C |
|---|---|---|---|---|
| Terephthalic acid solubility, g/100 g solvent | | | | |
| Water | 0.0017 | 0.24 | 1.7 | 12.6 |
| Methanol | 0.10 | 3.1 | | |
| Acetic acid | 0.013 | 0.38 | 1.5 | 5.7 |
| Isophthalic acid solubility, g/100 g solvent | | | | |
| Water | 0.012 | 2.8 | 25.2 | |
| Methanol | 1.06 | 8.1 | | |
| Acetic acid | 0.23 | 4.3 | 13.8 | 44.5 |

xylene is used for all isophthalic acid. Oxidation catalysts and conditions have been developed which give nearly quantitative oxidation of the methyl groups, leaving the benzene ring virtually untouched. These catalysts are combinations of cobalt, manganese, and bromine, or cobalt with a co-oxidant, e.g., acetaldehyde [75-07-0]. Oxygen is the oxidant in all processes. Acetic acid [64-19-7] is the reaction solvent in all but one process. Given these constant factors, there is only one industrial oxidation process, with different variations, two separate purification processes, and one process which intermixes oxidation and esterification steps.

## 3.1. Amoco Oxidation

About 70 % of the terephthalate feedstock used worldwide is produced with a catalyst system discovered by Scientific Design [5], [6]. Almost 100 % of new plants use this reaction. A separate company, Mid-Century Corporation, was established to market this technology, and subsequently purchased by Amoco Chemical. Amoco developed a commercial process, as did Mitsui Petrochemical, now Mitsui Sekka. Mitsui

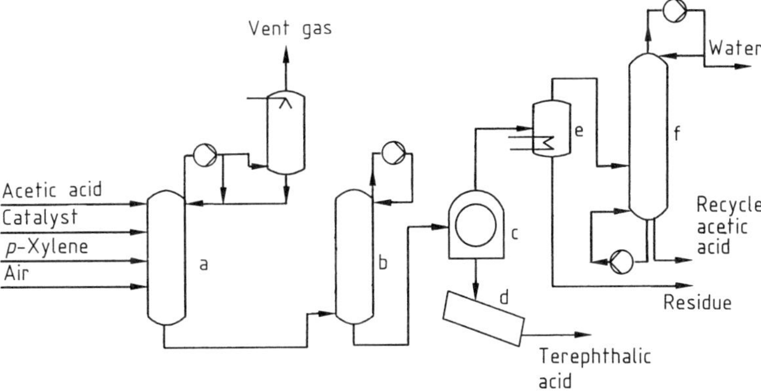



**Figure 1.** Catalytic, liquid-phase oxidation of *p*-xylene to terephthalic acid by the Amoco process
a) Oxidation reactor; b) Surge vessel; c) Filter; d) Dryer; e) Residue still; f) Dehydration column

was an early licensee of Mid-Century. Both Amoco and Mitsui participate in joint-venture companies, and both have licensed the process. Licensees are distributed around the world, and some have relicensed the process to other companies.

A soluble cobalt – manganese – bromine catalyst system is the heart of the process. This yields nearly quantitative oxidation of the *p*-xylene methyl groups with small xylene losses [7]. Acetic acid is the solvent, and oxygen in compressed air is the oxidant. Various salts of cobalt and manganese can be used, and the bromine source can be HBr, NaBr, or tetrabromoethane [*79-27-6*] among others. The highly corrosive bromine – acetic acid environment requires the use of titanium-lined equipment in some parts of the process.

A feed mixture of *p*-xylene, acetic acid, and catalyst is continuously fed to the oxidation reactor (Fig. 1). The feed mixture also contains water, which is a byproduct of the reaction. The reactor is operated at 175–225 °C and 1500–3000 kPa. Compressed air is added to the reactor in excess of stoichiometric requirements to provide measurable oxygen partial pressure and to achieve high *p*-xylene conversion. The reaction is highly exothermic, releasing $2 \times 10^8$ J per kilogram *p*-xylene reacted. Water is also released. The reaction of 1 mol *p*-xylene with 3 mol dioxygen gives 1 mol terephthalic acid and 2 mol water. Only four hydrogen atoms, representing slightly over 2 wt% of the *p*-xylene molecule, are not incorporated in the terephthalic acid.

Owing to the low solubility of terephthalic acid in the solvent, most of it precipitates as it forms. This yields a three-phase system: solid terephthalic acid crystals; solvent with some dissolved terephthalic acid; and vapor consisting of nitrogen, acetic acid, water, and a small amount of oxygen. The heat of reaction is removed by solvent evaporation. A residence time up to 2 h is used. Over 98% of the *p*-xylene is reacted, and the yield to terephthalic acid is > 95 mol%. Small amounts of *p*-xylene and acetic acid are lost, owing to complete oxidation to carbon oxides, and impurities such as oxidation intermediates are present in reactor effluent. The excellent yield and low solvent loss in a single reactor pass account for the near universal selection of this technology for new plants.

The oxidation of the methyl groups occurs in steps, with two intermediates, *p*-toluic acid [*99-94-5*] and 4-formylbenzoic acid [*619-66-9*]. While 4-formylbenzoic acid is the IUPAC name of the intermediate, it is customarily referred to as 4-carboxybenzaldehyde (4-CBA).

*p*-Xylene          *p*-Toluic          4-Formyl-          Terephthalic
                    acid              benzoic acid        acid

4-Formylbenzoic acid is troublesome, owing to its structural similarity to terephthalic acid. It co-crystallizes with terephthalic acid and becomes trapped and inaccessible for completion of the oxidation. Up to 5000 ppm 4-formylbenzoic acid can be present, and this necessitates a purification step to make the terephthalic acid suitable as a feedstock for polyester production.

The slurry is passed from the reactor to one or more surge vessels where the pressure is reduced. Solid terephthalic acid is then recovered by centrifugation or filtration, and the cake is dried and stored prior to purification. This is typically referred to as crude terephthalic acid, but is > 99 % pure.

Vapor from the reactor is condensed in overhead heat exchangers, and the condensate is refluxed to the reactor. Steam is generated by the condensation and is used as a heating source in other parts of the process. Oxygen-depleted gas from the condensers is scrubbed to remove most uncondensed vapors. Similar to the reactor condensate, liquid from centrifuges or filters is sent to solvent recovery. Since the centrifugate or filtrate contains dissolved species, it is first sent to a residue still. Vapor from the still and other vents from throughout the oxidation process are sent to a solvent dehydration tower. The tower removes the water formed in the reaction as the overhead stream, and the acetic acid from the tower bottom is combined with fresh acetic acid to make up for process losses, and returned to the process.

Isophthalic acid is also produced by this process from *m*-xylene. Because isophthalic acid is several times more soluble than terephthalic acid, much less precipitates in the reactor. Consequently, isophthalic acid from this process contains much less 3-formylbenzoic acid, since it tends to stay in solution where complete oxidation can occur. Further purification was not carried out in the past, but a purified grade that is now being produced will become the standard.

## 3.2. Amoco Purification

The purification process developed by Amoco Chemical [8] and used on terephthalic acid from the Amoco oxidation process supplies over 60 % of the terephthalate feedstock for polyester production (Fig. 2).

Crude terephthalic acid is unsuitable as a feedstock for polyester, primarily owing to the 4-formylbenzoic acid impurity concentration. There are also yellow impurities and residual amounts of catalyst metals and bromine. The Amoco purification process removes 4-formylbenzoic acid to < 25 ppm, and also gives a white powder from the slightly yellow feed.

It is necessary to make all impurities accessible to reaction, so the crude terephthalic acid is slurried with water and heated until it dissolves entirely. A solution of at least 15 wt % is obtained, and this requires a temperature ≥ 260 °C. The solution passes to a reactor where hydrogen is added and readily dissolves. The solution is contacted with a carbon-supported palladium catalyst. Reactor pressure is held above the vapor pressure of water to maintain a liquid phase.

The 4-formylbenzoic acid is converted to *p*-toluic acid in the reactor, and some colored impurities are hydrogenated to colorless compounds. The reaction is highly selective; loss of terephthalic acid by carboxylic acid reduction or ring hydrogenation is < 1 %.

After reaction, the solution passes to a series of crystallizers where the pressure is sequentially decreased [9]. This results in a stepped temperature reduction, and crystallization of the terephthalic acid. The more soluble *p*-toluic acid formed in the reactor, and other impurities, remain in the mother liquor. After leaving the final crystallizer, the slurry undergoes centrifugation and/or filtration to yield a wet cake, and the cake is dried to give a free-flowing terephthalic acid powder as the product. Over 98 wt % of the incoming terephthalic acid is recovered as purified product.

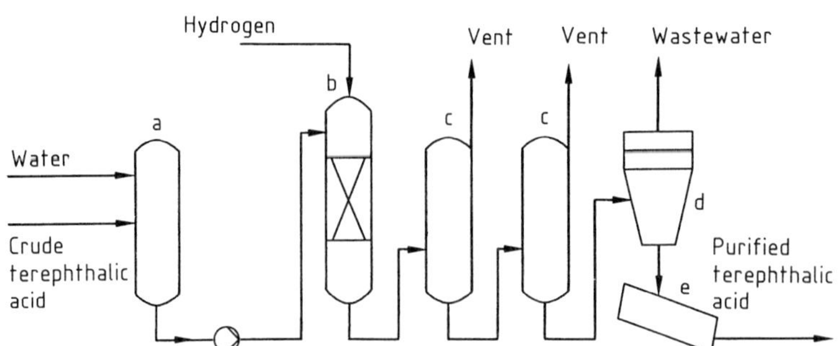


**Figure 2.** Purification of terephthalic acid by the Amoco process
a) Slurry drum; b) Hydrogenation reactor; c) Crystallizers; d) Centrifuge; e) Dryer

As with the Amoco oxidation, this purification process is also used with isophthalic acid.

## 3.3. Multistage Oxidation

Several companies, mostly in Japan [10], [11], have developed processes to reduce the 4-formylbenzoic acid content to 200–300 ppm by more intensive oxidation. A separate purification step is eliminated; the concentration of 4-formylbenzoic acid is low enough for the terephthalic acid to be suitable as a feedstock for some polyester products where high feedstock purity is not critical. The product is often called medium-purity terephthalic acid, and accounts for about 11 % of the terephthalic acid produced. Most of these processes also use the catalyst system discovered by Scientific Design.

Most medium-purity terephthalic acid is produced by Mitsubishi Kasei and its licensees. They have named this product Q-PTA, and it has a typical 4-formylbenzoic acid level of ca. 290 ppm [10]. Mitsubishi has also developed a still more intensive oxidation process where the 4-formylbenzoic acid level is further reduced. The product is called S-QTA.

The oxidation of p-xylene in acetic acid with a cobalt–manganese–bromine catalyst is carried out as in the Amoco oxidation. The slurry is heated to 235–290 °C and oxidized further in another reactor. More catalyst can be added in addition to the temperature increase [12].

Heating gives increased terephthalic acid solubility, and as crystals dissolve, some 4-formylbenzoic acid and colored impurities are released. Although the terephthalic acid is not completely soluble at the higher temperature, the crystals can digest. Digestion is a dynamic equilibrium process wherein crystals constantly dissolve and reform. This increases the release of 4-formylbenzoic acid into solution where oxidation can be completed. While the need for a separate purification process is eliminated, another reactor is needed in the oxidation process. Also, at higher temperature, acetic acid tends to be oxidized to a larger extent to carbon oxides and water [12].

The rest of the Mitsubishi process consists of solid–liquid separation and drying to obtain the powdered product. Acetic acid must be dehydrated and recycled to the process.

Eastman Chemical has also developed a medium-purity terephthalic acid product. The

process does not employ manganese, only cobalt and bromine. Two oxidation stages are used, both at 175–230 °C [13]. Instead of heating between stages to obtain increased solubility, as performed by Mitsubishi, the contents of the first reactor are sent to hydroclones where hot, fresh acetic acid displaces the mother liquor. Samarium may be added to the catalyst. A residence time up to 2 h with air addition provides sufficient digestion of the crystals to yield a 40–270 ppm level of 4-formylbenzoic acid [13]. Downstream recovery of the terephthalic acid crystals by solid–liquid separation and drying must again be performed.

## 3.4. Dynamit-Nobel (Witten) Process

Most dimethyl terephthalate is made by a process first developed by Chemische Werke Witten, with work also being done at a division of Standard Oil of California. Modifications were made by Hercules and Dynamit-Nobel. Hüls Troisdorf currently licenses the process with further modifications [14]. Dimethyl terephthalate is formed in four steps. First, p-xylene is passed through an oxidation reactor, where p-toluic acid is formed. It then passes to an esterification reactor, the second step, where methanol is added to form methyl p-toluate [99-75-2]. The methyl p-toluate is isolated and returned to the oxidation reactor for oxidation to monomethyl terephthalate [1679-64-7], the third step, followed by the fourth step, esterification to dimethyl terephthalate. The sequence is as follows:



The process as licensed by Hüls Troisdorf is illustrated in Figure 3. Fresh and recovered p-

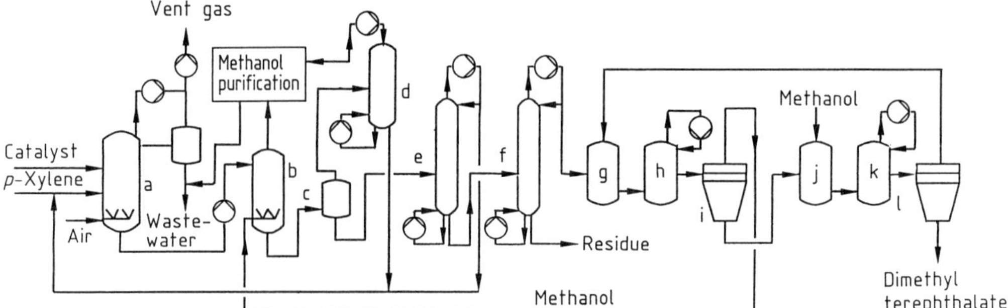


**Figure 3.** Production of dimethyl terephthalate by the Dynamit Nobel process
a) Oxidation reactor; b) Esterifier; c) Expansion vessel; d) Methanol recovery column; e, f) *p*-Methyl toluate and dimethyl terephthalate columns; g, j) Dissolvers; h, k) Crystallizers; i, l) Centrifuges

xylene, along with catalyst (mostly cobalt with some manganese) are combined with methyl *p*-toluate and fed to the liquid-phase oxidation reactor. Because bromine and acetic acid are not used, vessels lined with titanium or other expensive metals are not necessary. Oxygen supplied by compressed air is added at the bottom. Oxidation conditions are 140–180 °C and 500–800 kPa. The heat generated by oxidation is removed by vapors of unreacted *p*-xylene and the water of reaction. Cooling coils in the reactor are used to generate steam. The steam and reactor vapors are condensed and combined to recover *p*-xylene for recycle.

The oxidation effluent is then heated and sent to the esterification reactor, operated at 250 °C and 2500 kPa. Excess vaporized methanol is sparged into the esterifier, where the *p*-toluic acid and monomethyl terephthalate are converted noncatalytically to methyl *p*-toluate and dimethyl terephthalate, respectively. Overhead vapors from the esterification reactor are condensed and fed to a distillation system, where the water from the esterification is separated from methanol, which is recycled.

The remainder of the process separates the dimethyl terephthalate from methanol and methyl *p*-toluate, which are recycled, and residue and wastewater, which go to waste treatment.

The product from the esterifier goes to an expansion vessel. Vapor from this vessel feeds a methanol recovery column, where the methanol overhead goes to methanol recovery, and the methyl *p*-toluate bottoms are recycled to the oxidation reactor. Liquid from the expansion vessel feeds two vacuum distillation columns in series, which yield crude dimethyl terephthalate. The first column recovers more methyl *p*-toluate

overhead for recycle to oxidation, and the bottoms feeds the crude dimethyl terephthalate column, where the product is taken overhead. The bottoms from the dimethyl terephthalate column, containing heavy byproducts and catalyst metals, can be mixed with water from the oxidation, which dissolves the catalyst. The resulting slurry is centrifuged; the catalyst solution is recycled, and the cake is sent to disposal.

The crude dimethyl terephthalate goes through two stages of crystallization. It is slurried with the methanol mother liquor from the second crystallization stage and dissolved by heating. Dimethyl terephthalate is crystallized from this solution on cooling, by flashing the methanol. The dimethyl terephthalate cake is separated by centrifugation, dissolved in fresh methanol, and crystallized in the same way. The wet dimethyl terephthalate cake can be melted and stored molten, or dried and flaked. There is usually a distillation column after the second centrifuge for further purification.

Mother liquor from the first centrifugation is sent to the methanol recovery system. The centrifugation yields a cleaner mother liquor which is combined with methanol recovered from the melting operation for use in dissolving the crude dimethyl terephthalate entering the crystallization section. Side-streams from throughout the process are recycled to appropriate points to maximize yield and to minimize methanol and catalyst use.

## 3.5. Esterification of Terephthalic Acid

Terephthalic acid can be produced and, in a separate process, esterified with methanol to

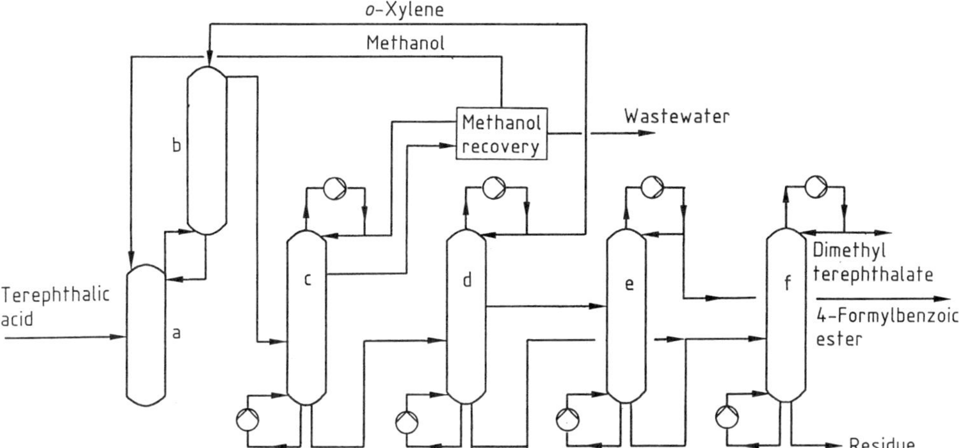



**Figure 4.** Esterification of terephthalic acid and separation of purified dimethyl terephthalate
a) Esterifier; b) o-Xylene scrubber; c) Methanol column; d) o-Xylene recovery column; e) 4-Formylbenzoic ester stripper; f) Purification column

dimethyl terephthalate which is then purified by distillation (Fig. 4) [15], [16]. This process can be used on highly impure terephthalic acid, because of the purification achievable by distillation.

Crude terephthalic acid and excess methanol are mixed and pumped to the esterification reactor. In this example, o-xylene [95-47-6] is used to enhance the subsequent separations. The terephthalic acid is rapidly esterified by the methanol at 250–300 °C without catalysis, although a catalyst can be used. Methanol vapor carries dimethyl terephthalate and o-xylene from the reactor to a column where o-xylene is added to scrub out monomethyl terephthalate and return it to the reactor for completion of the esterification. The vapor contains the dimethyl terephthalate product as well as methanol, o-xylene, water from the esterification, and esterified impurities in the terephthalic acid feed. Several distillation steps are needed to separate out the dimethyl terephthalate and to process the separated streams for recovery of valuable components for recycle.

The reactor overhead vapor first goes to a methanol column where methanol is removed overhead, water and some methanol form a side-draw, and the bottoms contain the dimethyl terephthalate, o-xylene, and impurities.

Next, in the o-xylene recovery column, dimethyl terephthalate purification is started; it operates at 10–20 kPa absolute so that a temperature of 200–230 °C can be used. o-Xylene is removed overhead, the methyl esters of

4-formylbenzoic acid and p-toluic acid are removed in the middle, and dimethyl terephthalate forms the bottoms. A 4-formylbenzoic acid stripping column follows, where the middle stream from the previous column is sent, so that 4-formylbenzoic ester can be removed overhead. Finally, the bottoms from both the o-xylene recovery column and the 4-formylbenzoic acid stripper are sent to a purification column, where the dimethyl terephthalate product is taken overhead.

Methanol from the top of the methanol recovery column is sent to a purification column where the overhead contains low boilers, and the methanol from the bottoms is recycled. The methanol and water from the side-draw of the methanol recovery column are sent to a methanol dehydration column, where the water is removed and the dehydrated methanol recycled.

## 3.6. Hydrolysis of Dimethyl Terephthalate

High-purity terephthalic acid can be produced by hydrolysis of dimethyl terephthalate. Slightly over 2 % of terephthalate feedstock is produced by this route.

Dimethyl terephthalate and recovered monomethyl terephthalate are combined with water from the methanol recovery tower of the dimethyl terephthalate process and heated to

260–280 °C at 4500–5500 kPa in a hydrolysis reactor, where the methyl esters are hydrolyzed. The overhead methanol plus water vapor is returned to the methanol recovery tower for separation. The hydrolysis reactor liquid is sent to a series of crystallizers and cyclones. After cooling to crystallize the terephthalic acid at ca. 200 °C, washing cyclones are used to remove mother liquor which contains monomethyl terephthalate. The cyclone underflow is further cooled to 100 °C, and final crystallization occurs. This slurry is centrifuged and dried to give the final product.

Monomethyl terephthalate from the cyclone overflow is recovered, again by cooling to 100 °C and centrifuging, and the cake is recycled to the hydrolysis reactor.

### 3.7. Alternative and Past Technologies

Oxidation of *p*- or *m*-xylene is also possible with acetic acid solvent and a cobalt catalyst with an acetaldehyde activator. Reaction temperature is 120–140 °C, and as a result the oxidation requires a residence time $\geq$ 2 h. Titanium-lined vessels are not required because bromine is not used. The acetaldehyde is converted to acetic acid as it promotes the reaction. After reaction, the terephthalic or isophthalic acid is recovered by centrifugation or filtration and drying, much like the Amoco process, and acetic acid is recovered for recycle by distillation.

With isophthalic acid, which has a lower concentration of 3-formylbenzoic acid, considerable purification is possible by slurrying in water to ca. 20–25 wt%, heating to 240–260 °C to dissolve, and crystallizing by cooling in batch mode. Recovery of the crystals is again by centrifugation or filtration and drying.

Eastman Chemical in the United States also used an acetaldehyde activator with a cobalt catalyst to produce terephthalic acid. Bromine is now being used in place of acetaldehyde [17].

Mobil Chemical in the United States had a commercial terephthalic acid operation which has since been abandoned [18]. As with the above processes, a cobalt catalyst was used with acetic acid solvent; the activator was 2-butanone [78-93-3]. After reaction, the crude terephthalic acid was leached by adding pure acetic acid and heating to achieve partial solubility; this removed gross impurities. Final purification was by sublimation and catalytic treatment of the vapor. Impure terephthalic acid was vaporized in a steam carrier, and catalyst and hydrogen were added to convert undesirable organic impurities. Subsequent cooling of the stream condensed the terephthalic acid, leaving impurities in the vapor. Product crystals were removed by cyclones.

A process no longer practiced is based on Henkel technology. Starting with phthalic anhydride [85-44-9], the monopotassium and dipotassium *o*-phthalate salts were formed in sequence. The dipotassium salt was isolated from solution by spray drying, and isomerized to dipotassium terephthalate under carbon dioxide at 1000–5000 kPa and 350–450 °C. This salt was dissolved in water and recycled to the start of the process, where terephthalic acid crystals formed during the production of the monopotassium salt. The crystals were recovered by filtration.

Two proposed processes which were never commercialized were by Lummus, using a dinitrile route [19] and by Eastman, using the formation of 1,4-diiodobenzene [624-38-4] with carbonylation to aromatic acids [20].

Toluene [108-88-3] is a potential feedstock for manufacture of terephthalic acid and is cheaper than *p*-xylene. Mitsubishi Gas Chemical has researched a process where a complex between toluene and hydrogen fluoride–boron trifluoride is formed, that can be carbonylated with carbon monoxide to form a *p*-tolualdehyde [104-87-0] complex [21]. After decomposition of the complex, *p*-tolualdehyde can be oxidized in water with a manganese–bromine catalyst system to terephthalic acid. While cheaper toluene is the feedstock and acetic acid is not required, the complexities of handling hydrogen fluoride–boron trifluoride and the need for carbon monoxide add other costs. This process has not been commercialized.

## 4. Environmental Protection

The three chemicals discussed here consist only of carbon, hydrogen, and oxygen, so carbon dioxide and water are the final effluents of oxidative degradation. There are traces of the cobalt–manganese–bromine catalyst in waste streams, however.

Effluents from the terephthalic acid process include water generated during oxidation and water used as the purification solvent. Gaseous emissions are mostly the oxygen-depleted air from the oxidation step. Volumes are large, but the chemicals in the streams can be effectively destroyed and removed.

Water effluents are subjected to aerobic wastewater treatment, where the dissolved species, mostly terephthalic acid, acetic acid, and impurities such as *p*-toluic acid, are oxidized to carbon dioxide and water by the action of bacteria which are acclimated to these chemicals. The bacterial growth is a sludge which can be dried and burned or spread on land. An anaerobic process has been developed to treat the wastewater [22]. Advantages include much less waste sludge production, less utility consumption, and the generation of methane, which can be burned for energy recovery.

Waste gas is scrubbed in process equipment to remove acetic acid vapors for recovery. Trace amounts of other compounds formed in the reactor can be removed by catalytic oxidation, followed by scrubbing to meet the most demanding regulations for process vents [23].

Waste streams from distillation in the dimethyl terephthalate process can be burned for energy recovery.

Another aspect of environmental protection is the recyclability of polyesters. Post-consumer containers can be ground and cleaned, and then extruded and spun into fiber to fill bedding, and quilted clothing. Textile fibers can also be made. The polyester can also be reacted to regenerate the terephthalate feedstock. High-temperature hydrolysis yields terephthalic acid. Reaction with methanol yields dimethyl terephthalate.

## 5. Quality Specifications

Feedstocks for polyester production must be extremely pure; they are among the purest high-volume chemicals sold by industry. If certain impurities are present in high enough concentrations, harmful effects can be measured, e.g., monofunctional compounds can cap the polyester chain and limit molecular mass buildup. Trifunctional carboxylic acids can cause chain branching which leads to undesirable rheological and spinning properties. Colored impurities can be incorporated into the polyester. In particular, 4-formylbenzoic acid limits polyester molecular mass and causes yellowness [24]. The particle size of terephthalic acid determines how

**Table 3.** Specifications and typical analyses of purified terephthalic acid

| Property | Specification | Typical value |
|---|---|---|
| Acid number, mg KOH/g | $675 \pm 2$ | 673–675 |
| Ash, ppm | $\leq 15$ | < 3 |
| Metals,* ppm | $\leq 9$ | < 2 |
| Water, wt % | $\leq 0.2$ | 0.1 |
| 4-Formylbenzoic acid, ppm | $\leq 25$ | 15 ** |
| *p*-Toluic acid, ppm | $125 \pm 45$ | 125 ** |

* Co + Mn + Fe + Cr + Ni + Mo + Ti.
** Medium-purity terephthalic acid typically has 250 ppm 4-formylbenzoic acid and < 50 ppm p-toluic acid.

the powder flows, and the viscosity of the slurry when mixed with 1,2-ethanediol [*107-21-1*].

Owing to the consistent high purity of these products and the different effects of impurities, specifications put limits on specific impurities or color, rather than overall purity (Table 3). Trends in quality specifications are away from maximum or minimum values, toward a target value with an allowable range. This has been instituted to some extent by Amoco Chemical for the *p*-toluic acid content, for example.

In Table 3, acid number, determined by titration, is an overall purity measurement. A perfectly pure sample will have an acid number of 675.5 mg KOH/g, but the low impurity levels make the acid number meaningless as a quantitative indication of purity, and it is being phased out.

Impurities specified include *p*-toluic acid and 4-formylbenzoic acid, residual water, and trace metals, the ash being trace metal oxides. There are also color and particle size measurements. Polyester producers often have specific tests which they prefer for color or particle size.

Medium-purity terephthalic acid specifications address the same impurities, except that 4-formylbenzoic acid concentrations are up to 300 ppm, and residual acetic acid is present.

Typical dimethyl terephthalate specifications are shown in Table 4. Freezing point is an extremely sensitive purity indicator. Since dimethyl

**Table 4.** Specifications and typical analyses of dimethyl terephthalate

| Property | Specification | Typical value |
|---|---|---|
| Freezing point, °C | $\geq 140.62$ | 140.64 |
| Acid number, mg KOH/g | 0.03 | 0.01 |
| Molten color, APHA | $\leq 25$ | 10 |
| Color stability,* APHA | $\leq 25$ | 15 |
| Water, wt % | $\leq 0.02$ | 0.01 |

* 4 h at 175 °C

terephthalate is often shipped and stored in molten form, color measurements are made on the melt. A color stability test can determine the sample's resistance to degradation. This resistance is reduced by the presence of unesterified acid groups, so these residual groups are measured.

Specifications for isophthalic acid are similar to those of terephthalic acid, *m*-toluic acid and 3-formylbenzoic acid being the impurities.

## 6. Storage and Transportation

Terephthalic acid powder is stored in silos and transported in bulk by rail hopper car, hopper truck, or bulk container with a polyethylene liner. Hopper cars and trucks should be dedicated to this service, because of the extreme purity requirements. Shipment is also in 1 t bags, or paper bags containing ca. 25 kg. Isophthalic acid, being a much lower volume chemical, has a higher proportion shipped in bags.

Dimethyl terephthalate is stored molten in insulated heated tanks, and is preferentially shipped molten in insulated rail tank cars or tank trucks. It is also solidified into briquettes or flakes, and shipped in 1 t or 25 kg bags.

The huge production volume of terephthalic acid and dimethyl terephthalate favors the use of bulk shipment with minimized packaging costs. Rail cars are preferred where possible, and they can accommodate up to 90 t. Containers with plastic liners hold 20 t terephthalic acid, and these are the preferred ocean shipping method.

## 7. Uses

Terephthalic acid and dimethyl terephthalate are used almost exclusively to produce saturated polyesters. Poly(ethylene terephthalate), the alternating copolymer of terephthalic acid and 1,2-ethanediol, accounts for > 90% of this use, with a worldwide demand of $12.6 \times 10^6$ t in 1992. Textile and industrial fiber accounted for 75% of this demand. Polyester is the largest volume synthetic fiber. Food and beverage containers accounted for 13%, and constituted the fastest growing segment. Film for audio, video, and photography took 7%.

Other uses are for poly(butylene terephthalate), a high-performance molding resin made by reaction with 1,4-butanediol [74829-49-5], and for special industrial coatings, solvent-free coatings, electrical insulating varnishes, aramid fibers, and adhesives. A small amount of bis(2-ethylhexyl) terephthalate [6422-86-2] is produced as a plasticizer, and some dimethyl terephthalate is ring-hydrogenated to produce the cyclohexane analog, 1,4-cyclohexanedicarboxylic acid [1076-97-7], for specialty polyesters and coatings.

Isophthalic acid is used as a comonomer with terephthalic acid in bottle and specialty resins. It is also a feedstock for coatings and for high-performance unsaturated polyesters, being first reacted with maleic anhydride [108-31-6] and then this resin is cross-linked with styrene [100-42-5].

## 8. Economic Aspects

Together, terephthalic acid and dimethyl terephthalate rank about 25th in tonnage of all chemicals, and about tenth for organic chemicals. The rapid growth of polyester use has led to an above-average growth for these feedstocks, and this is projected to continue as deeper penetration into fiber and container markets is made. The versatility of terephthalate and isophthalate polyesters is projected to drive growth in the foreseeable future.

Capacity for the three types of terephthalate feedstocks since 1980 is given in Table 5. Purified terephthalic acid has grown rapidly, and is projected to supply virtually all the growth for polyesters, about 7% annually to the year 2000. Medium-purity terephthalic acid capacity has grown by construction of a few new plants, and by conversion of some dimethyl terephthalate capacity. Dimethyl terephthalate has shown little change in capacity since 1980, and essentially none is expected in the future.

The Far East contains most of the world's population and most of the garment manufac-

**Table 5.** Capacity of terephthalate feedstocks ($10^3$ t/a)

| Feedstock* | 1980 | 1986 | 1992 |
|---|---|---|---|
| PTA | 2800 | 4000 | 8100 |
| MTA | 420 | 580 | 1300 |
| DMT | 3300 | 3600 | 3300 |

* PTA purified terephthalic acid; MTA medium-purity terephthalic acid; DMT dimethyl terephthalate.

**Table 6.** Geographical distribution of terephthalate production ($10^3$ t/a)

|              | 1980 | 1986 | 1992 |
|--------------|------|------|------|
| North America | 3300 | 3400 | 3600 |
| Europe        | 1500 | 2100 | 2700 |
| Far East      | 1500 | 2300 | 5800 |
| Rest of World |  220 |  380 |  600 |

turing industry. As a result, it is experiencing most of the growth in manufacturing capacity for polyester fibers, and most of the growth for terephthalic acid production (Table 6). Approximately 75 % of the growth in terephthalate feedstocks in 1986 – 1992 was due to purified terephthalic acid plants built in the Far East.

Pricing is often determined in conjunction with long-term contracts, and is most influenced by the cost of $p$-xylene. In 1992, prices were about $ 0.60/kg for terephthalic acid and $ 0.57/kg for dimethyl terephthalate [25]. Because a given mass of terephthalic acid produces 17 % more polyester than the same mass of dimethyl terephthalate, the dimethyl terephthalate price must be adjusted downward, and allowance is made for the value of the methanol generated during transesterification.

The major producer of purified terephthalic acid is Amoco Chemical, with over 17 % of the world's production in 1992. Joint ventures of Amoco, the largest being China American Petrochemical in Taiwan and Samsung Petrochemical in Korea, also produce about 17 %. Other major producers are Mitsui Sekka in Japan and Imperial Chemical Industries in England and Taiwan. China is certain to become a major producer, and large plants are being planned in Southeast Asia. Larger producers of medium-purity terephthalic acid are Mitsubishi Kasei in Japan, and Eastman and DuPont in the United States. Some polyester producers have licensed technology to supply their own feedstock. All licensors are themselves producers of terephthalic acid, or have obtained the technology from a producer.

Major dimethyl terephthalate producers are DuPont, Eastman Chemical, and Hoechst Celanese in the United States, Hoechst in Europe, and Teijin Petrochemical in Japan. There are several small, older plants, mostly in Europe and China.

Isophthalic acid production capacity is $250 \times 10^3$ t/a, ca. 2 % of the terephthalate total. Amoco Chemical is the major producer.

# 9. Toxicology and Occupational Health

Terephthalic acid, dimethyl terephthalate, and isophthalic acid all have low toxicity and cause only mild and reversible irritation to skin, eyes, and the respiratory system.

For oral ingestion of terephthalic acid, $LD_{50}$ values for rats have been reported as 18.8 g/kg [26], and for mice as 6.4 g/kg [27], or $> 5$ g/kg [28]. Dimethyl terephthalate $LD_{50}$ values for rats are 6.5 g/kg [29] and 4.39 g/kg [26]. Some mortality was found at 5 g/kg oral ingestion [30], and with a 28-day oral uptake of 5 % in the feed, ca. 2.5 g/kg. For isophthalic acid, $LD_{50}$ for rats was reported as 10.4 g/kg [26].

Interperitoneal administration of terephthalic acid in rats showed $LD_{50}$ values $> 1.43$ g/kg [28] and 1.9 g/kg [31], with $LD_{100} = 3.2$ g/kg [31]. Mortality was found at 0.8 g/kg in mice and 1.6 g/kg in rats [27]. For dimethyl terephthalate, an $LD_{50}$ of 3.9 g/kg has been reported [29], and for isophthalic acid, 4.3 g/kg [26].

Ingestion of terephthalic acid or dimethyl terephthalate results in rapid distribution and excretion in unchanged form [32]. In rats, at high ingestion rates of about 3 % terephthalic acid or dimethyl terephthalate in the feed, bladder calculi are formed which are mostly calcium terephthalate; these injure the bladder wall and lead to cancer. Calculi cannot form unless the calcium terephthalate solubility is exceeded (threshold effect) [33]. Neither compound has been found to be genotoxic (Ames test) [34].

Inhalation of terephthalic acid dust appears to pose no clear hazards. Inhalation by rats for 6 h/d, 5 d/week, for 4 weeks at 25 mg/m$^3$ produced no fatalities [35]. Dimethyl terephthalate dust at 16.5 and 86.4 mg/m$^3$ for 4 h/d caused no toxicological effects after 58 days [29]. The rats intermittently reacted to the nuisance dust levels at the higher concentration.

Irritation of the skin, eyes, and mucous membranes by any of these chemicals is mild and reversible. In rabbits, the FHSA test for terephthalic acid for eyes gave scores of 14.0/110.0 and 3.5/110.0 at 1 and 24 h after exposure. The FHSA skin test score was 0.4/8.0 [35]. For isophthalic acid, eye and skin irritancy scores were 5.3/110.0 and 0.2/8.0 [36]. Irritation due to dimethyl terephthalate has also been reported to be mild [30].

Terephthalic acid, dimethyl terephthalate, and isophthalic acid can all form dust clouds. As with any flammable substance, an explosion can occur, given proper dust and oxygen concentrations. Reported limits on the explosive region for terephthalic acid dust clouds are minimum dust content of 40 g/m$^3$ and minimum oxygen content 12.4 % at 20 °C. At 150 °C, minimum oxygen is 11.1 % [37]. The maximum concentration of an explosible dust cloud has been calculated as 1400 g/m$^3$ [38].

Dimethyl terephthalate and isophthalic acid have a more stringent oxygen limit. In tests with carbon dioxide diluent, the minimum oxygen content was 12, 14, and 15 % for dimethyl terephthalate, isophthalic acid, and terephthalic acid [39]. The oxygen content is higher for carbon dioxide diluent than for nitrogen, owing to the higher heat capacity of carbon dioxide (the 12.4 % minimum oxygen content reported above for terephthalic acid was for nitrogen diluent).

For molten dimethyl terephthalate, the flash and fire points, determined by the Cleveland open-cup method, are 146 and 155 °C [1].

# 10. References


General References

[1] *Kirk-Othmer*, **18**, 732–777.
[2] P. Raghavendrachar, S. Ramachandran, *Ind. Eng. Chem. Prod. Res. Dev.* **31** (1992) 453–462.
[3] J. E. McIntyre, *Chem. Eng. Monogr.* **15** (1982) 400–444.

Specific References

[4] Calico Printers Assoc, GB 578079, 1941 (J. R. Whinfield, J. T. Dickson).
[5] Mid-Century Corp., US 2833816, 1955 (R. S. Barker, S. A. Soffer).
[6] Mid-Century Corp., US 3089906, 1958 (R. S. Barker, S. A. Soffer).
[7] W. Partenheimer in D. W. Blackburn (ed.): *Catalysis of Organic Reactions,* Marcel Dekker, New York 1990, pp. 321–346.
[8] Standard Oil Company (Indiana), US 3584039, 1967 (D. H. Meyer).
[9] Standard Oil Company (Indiana), US 3931305, 1973 (J. A. Fisher).
[10] K. Matsuzawa, *Chem. Econ. Eng. Rev.* **8** (1976) no. 8, 25–30.
[11] M. Hizikata, *Chem. Econ. Eng. Rev.* **9** (1977) no. 9, 32–38.
[12] Mitsubishi Chem. Ind. Ltd., US 4877900, 1988 (A. Tamaru, Y. Izumisawa).
[13] Eastman Kodak Company, US 4447646, 1983 (G. I. Johnson, J. E. Kiefer).
[14] H. J. Korte, H. Schroeder, A. Schoengen: "The PTA Process of Hüls Troisdorf AG," AIChE Summer National Meeting, Denver, Co. 1988.
[15] E. I. DuPont de Nemours, US 2491660, 1949 (W. F. Gresham).
[16] Standard Oil Company (Indiana), US 2976030, 1957 (D. H. Meyer).
[17] *Eastman News* **43** (1988) no. 14.
[18] H. S. Bryant, C. A. Duval, L. E. McMakin, J. I. Savoca, *Chem. Eng. Prog.* **67** (1971) no. 9, 69–75.
[19] A. P. Gelbein, M. C. Sze, R. T. Whitehead, *CHEMTECH* **3** (1973) 479–483.
[20] Eastman Kodak Company, US 4705890, 1987 (G. R. Steinmetz, M. Rule).
[21] *Chem. Eng. (N.Y.)* **83** (1976) no. 17, 27–28.
[22] S. Shelley, *Chem. Eng. (N.Y.)* **98** (1991) no. 12, 90–93.
[23] T. G. Otchy, K. J. Herbert: "First Large Scale Catalytic Oxidation System for PTA Plant CO and VOC Abatement," *Ann. Air Waste Managem. Assoc. Meet.* **85th 1992**.
[24] W. Berger, D. Dornig, *Faserforsch. Textiltech.* **29** (1978) 256–262.
[25] *Chem. Markt. Rep.* **242** (1992) no. 4, 42.
[26] J. V. Marhold, *Sb. Vys. Toxicologickeho Vysetrien Latid A Priravku* **1972**, 52.
[27] A. E. Moffitt et al., *Am. Ind. Hyg. Assoc. J.* **36** (1975) 633–641.
[28] A. Hoshi, R. Yanal, K. Kuretani, *Chem. Pharm. Bull.* **16** (1968) 1655–1660.
[29] W. J. Krasavage, *Am. Ind. Hyg. Assoc. J.* **34** (1973) 455–462.
[30] Haskel Laboratories, unpublished data, Wilmington, Delaware, USA, 1979.
[31] E. O. Grigas, R. Ruiz, D. M. Aviado, *Toxicol. Appl. Pharmacol.* **18** (1971) 469–486.
[32] A. Hoshi, K. Kuretani, *Chem. Pharm. Bull.* **16** (1968) 131–135.
[33] H. d'A. Heck, *Banbury Rep.* **25** (1987) 233–244.
[34] E. Zeiger, S. Haworth, W. Speck, K. Mortelmans, *EHP Environ. Health Perspect.* **45** (1982) 99–104.
[35] Amoco Chemical Company, Material Safety Data Sheet, Amoco TA-33, Chicago, Ill., 1993.
[36] Amoco Chemical Company, Material Safety Data Sheet, Amoco PIA, Chicago, Ill., 1993.
[37] A. D. Craven, M. G. Foster, *Combust. Flame* **11** (1967) 408–414.
[38] S. Nomur, M. Torimoto, T. Tanake, *Ind. Eng. Chem. Proc. Res. Dev.* **23** (1984) 420–423.
[39] M. Jacobson, J. Nagy, A. R. Cooper, *Rep. Invest. U.S. Bur. Mines* **5971** (1962) 25–26.


**Terpene Resins → Resins, Synthetic**

# Terpenes

MANFRED EGGERSDORFER, BASF Aktiengesellschaft, Ludwigshafen, Germany



## 1. General

### 1.1. Definition and Basic Structures

Terpenes are natural products, whose structures are built up from isoprene units. They are classified according to the number of these units:

| | |
|---|---|
| Monoterpenes | $C_{10}$ |
| Sesquiterpenes | $C_{15}$ |
| Diterpenes | $C_{20}$ |
| Sesterpenes | $C_{25}$ |
| Triterpenes | $C_{30}$ |
| Tetraterpenes | $C_{40}$ |

Terpenes can be acyclic, or mono-, bi-, tri-, tetra-, and pentacyclic with 3-membered to 14-membered rings [1]–[7].

Deviations from the rule occur through rearrangement reactions or degradation of parts of the molecule during biosynthesis. The broader term terpenoids also covers natural degradation products, such as ionones, and natural and synthetic derivatives, e.g., terpene alcohols, aldehydes, ketones, acids, esters, epoxides, and hydrogenation products.

An overview of the most important basic structures demonstrates the structural variety of terpenes [8]. The compounds are normally known by their trivial names, as the systematic nomenclature is frequently less practicable for complicated structures.

Monoterpenes, $C_{10}$

acyclic

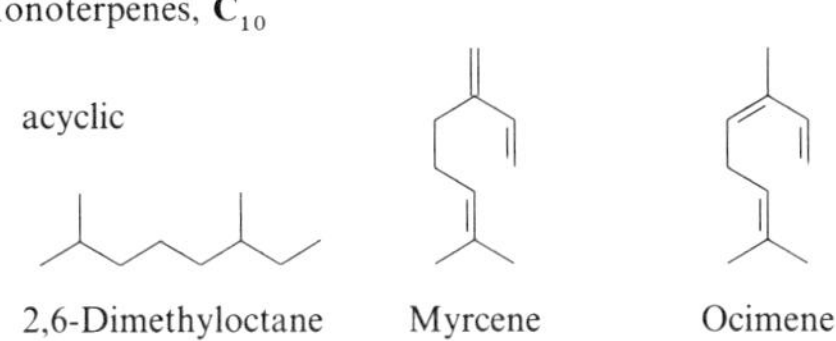




Ullmann's Encyclopedia
of Industrial Chemistry, Vol. A 26

monocyclic

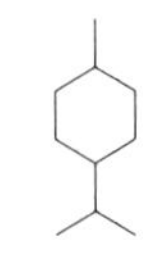

*p*-Menthane

Limone

α-Terpinene

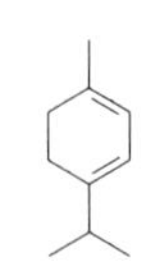

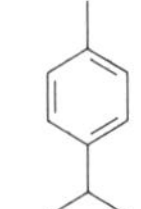

*p*-Cymene

Safrane

bicyclic

Pinane

α-Pinene

Carane

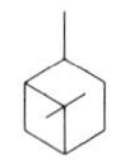

Camphor

Sesquiterpenes, $C_{15}$

acyclic

Farnesane

monocyclic

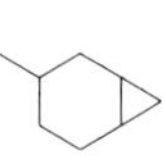

Bisabolane

Germacrane

Humulane

bicyclic

Cadinane

Guaiane

tricyclic

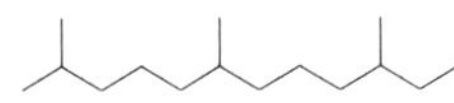

α-Santalane

Patchoulane

Diterpenes, $C_{20}$

acyclic

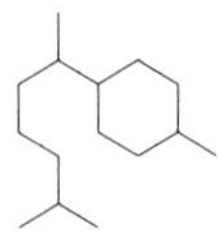

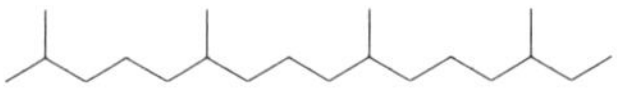

Phytane

monocyclic

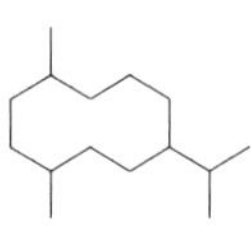

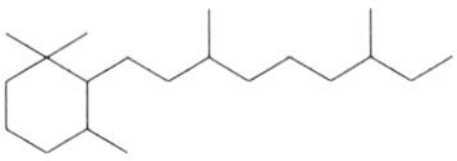

Retinane

bicyclic

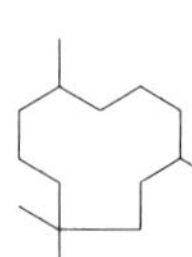

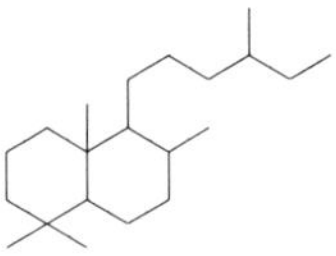

Labdane

tricyclic

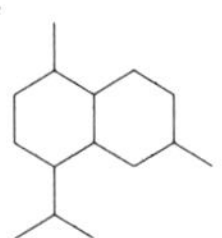

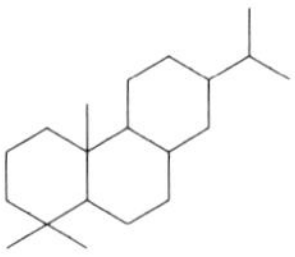

Abietane

tetracyclic

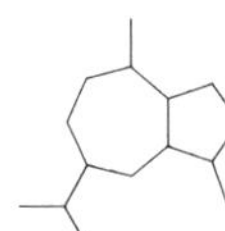

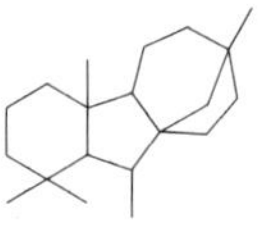

Gibbane

pentacyclic

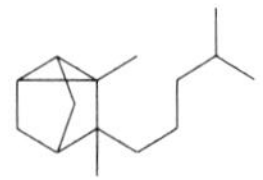

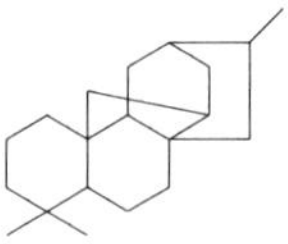

Atisane

Triterpenes, $C_{30}$

acyclic

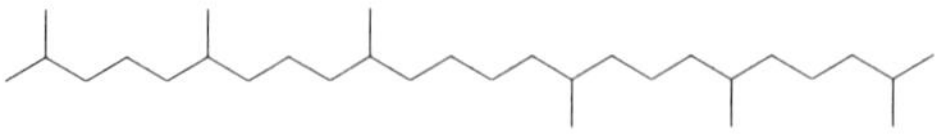

Squalane

tetracyclic

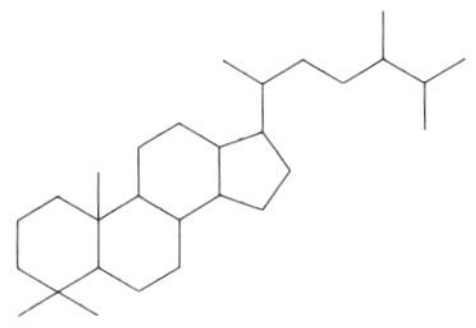

Lanostane

Tetraterpenes, $C_{40}$

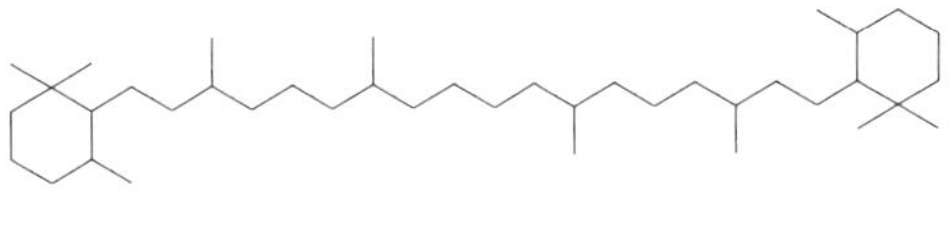

Perhydro-β-carotene

Polyprenes

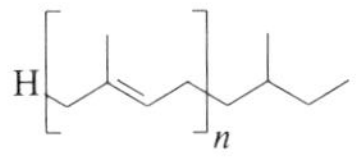

$n = 8-12$ in plants
$n = 16-21$ in animals

## 1.2. Occurrence

Terpenes occur everywhere and in all organisms, in particular in higher plants. Certain components or their combinations are often characteristic of individual types of plant, so terpenes are used for chemotaxonomy [8]. In vegetable raw materials mono- and sesquiterpenes are predominantly found in essential oils [9], [10], sesqui-, di-, and triterpenes in balsams and resins [11], tetraterpenes in pigments (carotenoids) [12], and polyterpenes in latexes [13]. Animal organisms also contain terpenes and terpenoids, which are predominantly incorporated by consumption of plants.

## 1.3. Biosynthesis

The biosynthesis of terpenes was explained by F. LYNEN. Starting from $(3\,R)$-$(+)$-mevalonic

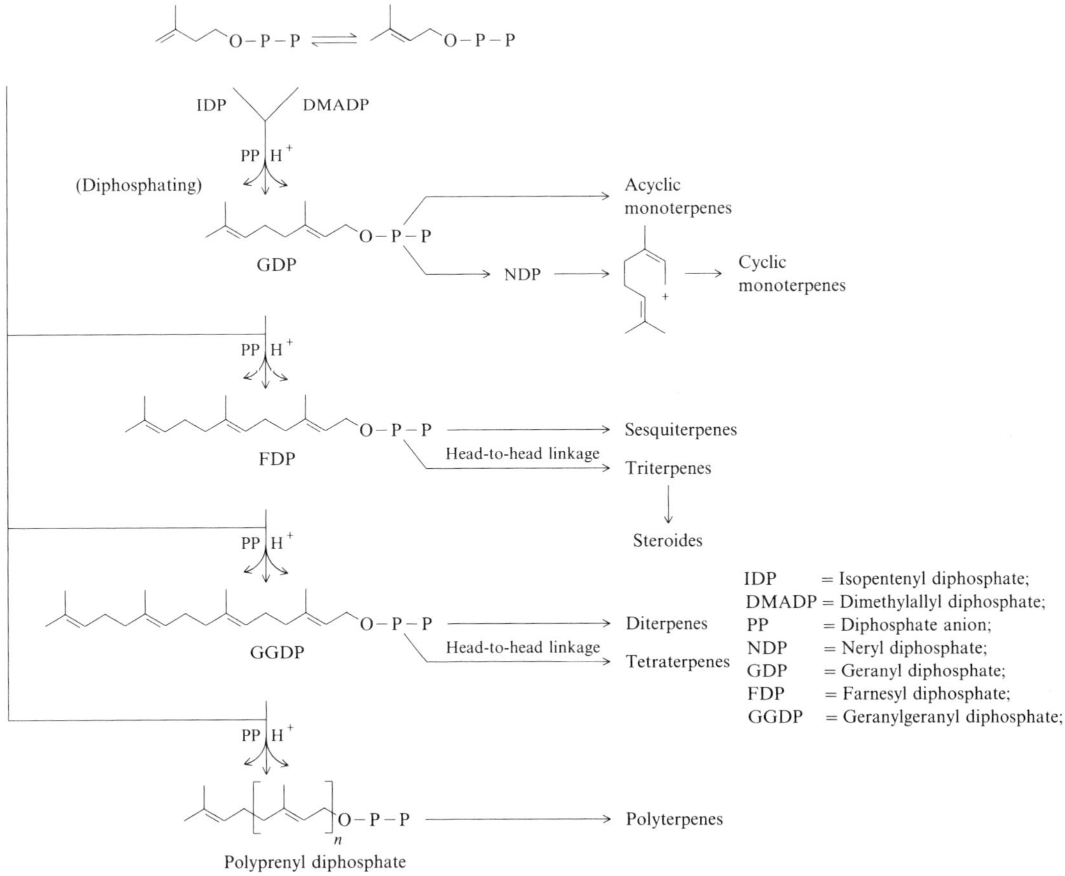

IDP          DMADP

PP $|H^+$

(Diphosphating)

Acyclic
monoterpenes

GDP

NDP

Cyclic
monoterpenes

PP $|H^+$

FDP                      Sesquiterpenes
Head-to-head linkage
Triterpenes

Steroides

PP $|H^+$

GGDP                     Diterpenes
Head-to-head linkage
Tetraterpenes

IDP       = Isopentenyl diphosphate;
DMADP = Dimethylallyl diphosphate;
PP        = Diphosphate anion;
NDP     = Neryl diphosphate;
GDP     = Geranyl diphosphate;
FDP     = Farnesyl diphosphate;
GGDP   = Geranylgeranyl diphosphate;

PP $|H^+$

Polyterpenes

Polyprenyl diphosphate

Scheme 1: Biosynthesis of terpenes

acid ($C_6$), isopentenyl diphosphate (IDP) and dimethylallyl diphosphate (DMADP) [8], [14]–[16] are formed by decarboxylation and isomerization as active $C_5$ building blocks (Scheme 1). (3 *R*)-( + )-Mevalonic acid is formed from three $C_2$ units from acetyl-CoA [8]. Plants synthesize IDP in the chloroplasts, mitochondria, and microsomes; animals synthesize IDP in the liver.

Monoterpenes are formed by stereospecific condensation of IDP with DMADP, whereby a diphosphate anion (PP) is eliminated from DMADP. Geranyl diphosphate is thus formed in a head-to-tail condensation. Neryl diphosphate with the *Z*-double bond is formed when the *S*-proton in isopentenyl diphosphate is eliminated stereospecifically. Neryl diphosphate is also formed from geranyl diphosphate by double bond isomerization.

The hydrolysis of geranyl or neryl diphosphate by a prenolpyrophosphatase gives the monoterpene alcohols geraniol and nerol. The acyclic monoterpene hydrocarbons, such as myrcene and ocimene, are formed by dehydration and isomerization of geraniol. The monoterpene aldehydes geranial and neral are formed from geraniol, and the biosynthesis of cyclic monoterpenes involves the loss of diphosphate from neryl diphosphate. The carbonium ion formed is stabilized by ring closure, sometimes after rearrangement, elimination of a proton, or addition of an anion.

The head-to-tail condensation of IDP with geranyl diphosphate gives farnesyl diphosphate, which rearranges further to give geranylgeranyl diphosphate. A large number of sesqui-, di-, and polyterpenes are formed from farnesyl and geranyl diphosphates in the same way as the monoterpenes. Triterpenes (and hence, steroids) are formed by head-to-head condensation of two $C_{15}$ units (farnesyl diphosphate), and tetraterpenes from two molecules of geranylgeranyl diphosphate.

## 1.4. Biodegradation

Terpenes are degraded by microorganisms, such as Pseudomonas and Aspergillus species [17]. Generally both acyclic and cyclic terpenes are oxidized. Plants are known to be capable of degrading as well as forming terpenes.

In animals acyclic terpenes are degraded by ω-, α- and/or β-oxidation and cyclic terpenes by hydroxylation (elimination as glucuronides).

## 1.5. Importance and Extraction

Since the study of organic chemistry began, it has been intensively concerned with terpenes. Many reactions, such as the Wagner–Meerwein rearrangement, theories of conformational analysis, and the Woodward–Hoffmann rules have their basis in terpene chemistry. The postulation of the isoprene rule by O. WALLACH (1887) and its confirmation and extension by L. RUZICKA (1921, 1953) gave this area of chemistry direction and unity.

Monoterpenes and a few sesquiterpenes are important economically as perfumes and fragrances (→ Perfumes, → Flavors and Fragrances). Depending on the properties, terpenes (mostly essential oils) are obtained from the corresponding plants by steam distillation, extraction, or enfleurage. The pure terpenes are mostly isolated from the extracts by fractional distillation.

As readily available, optically active products, some monoterpenes, such as ( − )-α-pinene and ( − )- and ( + )-limonene are used for the synthesis of other optically active products and reagents for cleaving racemates [18].

Various essential oils and the terpenes isolated from them are still used in pharmaceutical preparations, e.g., turpentine, camomile oil, eucalyptus oil, and camphor. The activity of some plants used in medicine and flavoring is due to their terpene content [19]. Terpenes are also used in the production of synthetic resins [20]–[22], and as solvents or diluting agents for dyes and varnishes [23].

In the following sections only the industrially most important terpenes are dealt with, insofar as they are not part of other articles (→ Flavors and Fragrances, → Turpentines, and → Tall Oil). For natural resins, gibberellic acid (→ Plant Growth Regulators), vitamins A and E (→ Vitamins), triterpenes (→ Steroids), tetraterpenes, and the other terpenes the specialist literature must be consulted.

## 2. Acyclic Monoterpenes

The following acyclic terpenes and terpene derivatives are described elsewhere (→ Flavors and Fragrances):

*Alcohols:* geraniol, nerol, linalool, myrcenol, citronellol, dihydromyrcenol, tetrahydrogeraniol, and tetrahydrolinalool.

*Aldehydes and acetals:* citral, citraldimethylacetal, citronellal, hydroxydihydrocitronellal, hydroxydihydrocitronellal dimethylacetal, and methoxydihydrocitronellal

*Esters and nitriles:* geranyl formate, acetate, isobutyrate, isovalerate, phenylacetate, and propionate; neryl acetate; linalyl acetate, butyrate, formate, isobutyrate, and propionate; lavanduyl acetate; and citronellyl acetate, formate, isobutyrate, isovalerate, propionate, and tiglate.

## 2.1. Myrcene

Myrcene [*123-35-3*], 2-methyl-6-methylene-2,7-octadiene, $C_{10}H_{16}$, $M_r$ 136.23, is a colorless liquid with a pleasant odor.

Some physical properties are listed below:

| | |
|---|---|
| *bp* | 166–168 °C (101.3 kPa) |
| $d^{20}$ | 0.7905 |
| $n_D^{20}$ | 1.4697 |
| *mp* | 50 °C |

Myrcene is very reactive. It polymerizes slowly at room temperature. For this reason it is stored in a cool place from which light and air are excluded. It can be stabilized by the addition of 0.1 % 4-(*tert*-butyl)catechol.

Although myrcene occurs naturally in many organisms, its extraction is uneconomic; it is produced industrially by pyrolysis of β-pinene [24].

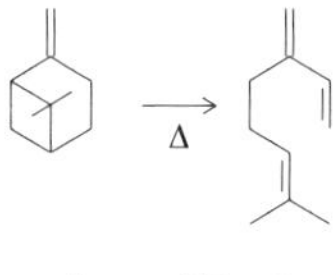

The fragmentation of linalool and linalyl acetate and the catalytic dimerization of isoprene have been described as laboratory methods [25].

Because of its functionality, myrcene is the starting material for a range of industrially important products, e.g., geraniol, nerol, linalool, and isophytol [26]. Addition of hydrogen chloride is industrially important. Depending on reaction conditions, this gives 1-chloro-3,7-dimethyl-2,6-octadiene (geranyl chloride and/or neryl chloride), 3-chloro-3,7-dimethyl-1,6-octadiene (linaloyl chloride), and 7-chloro-7-methyl-3-methylen-1-octene (myrcenyl chloride) [27]. The alcohols are produced from the chlorides by saponification.

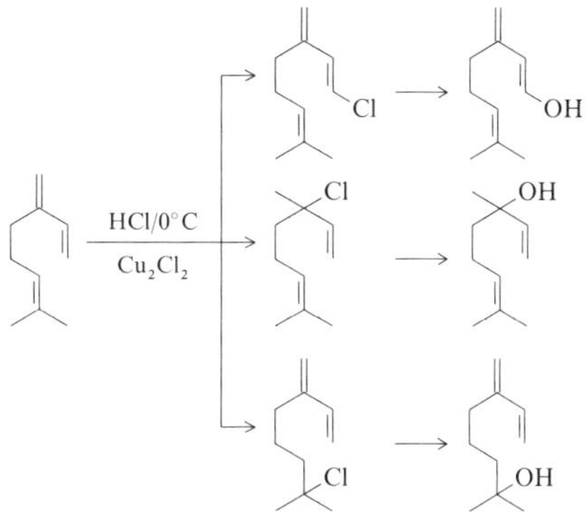

The 1,3-diene system of myrcene reacts with a large number of dienophiles. For example, the maleic anhydride adduct is used to characterize and determine the quality of the product [28].

Isophytol is formed by transition-metal-catalyzed addition of malonic ester and subsequent acidification [29].

In the hydrogenation of myrcene 2,6-dimethyl-2,6-octadiene, 2,6-dimethyl-2,7-octadiene, and 2,6-dimethyloctane are formed, depending on the catalyst and the reaction conditions [30:1H 260, 1EIII 1054].

Besides its main use as an intermediate for the production of terpene alcohols, myrcene is used in the production of terpene polymers [20], terpene–phenol resins [21], and terpene–maleate resins [22]. It can also be used as a solvent or diluting agent for dyes and varnishes [23].

## 2.2. Ocimene

Ocimene [*673-84-7*], (*Z, E*)-2,6-dimethyl-2,5,7-octatriene, $C_{10}H_{16}$, $M_r$ 136.23, is a colorless liquid with a pleasant odor.

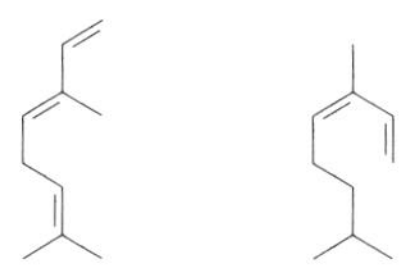

The *Z/E* mixture has the following properties:

| | |
|---|---|
| *bp* | 176–178 °C (101.3 kPa) |
| $d^{20}$ | 0.800 |
| $n_D^{20}$ | 1.4862 |
| *mp* | 50 °C |

Ocimene is very sensitive to oxidation and therefore can be kept for long periods only with the exclusion of air. At elevated temperatures it rearranges to alloocimene [31].

Ocimene is produced by the flash pyrolysis or photochemical isomerization of α-pinene [32].

Ocimene is mainly used as a perfume component. The derivatives ocimenol (2,6-dimethyl-5,7-octadien-2-ol), produced by acid-catalyzed hydration, and the partially and completely hydrogenated products are also important [33], [34].

## 2.3. 2,6-Dimethyl-2,4,6-octatriene

2,6-Dimethyl-2,4,6-octatriene exists as two stereoisomers:

[7216-56-0]
(4E,6Z)-2,6-Dimethyl-
2,4,6-octatriene
(Neoalloocimen)

[3016-19-1]
(4E,6E)-2,6-Dimethyl-
2,4,6-octatriene
(Alloocimen)

(4E,6Z)-2,6-Dimethyl-2,4,6-octatriene, $C_{10}H_{16}$, $M_r$ 136.23, is a colorless liquid with an intense odor, and is sensitive to oxidation.

| | |
|---|---|
| $bp$ | 80 °C (1.9 kPa) |
| $d^{20}$ | 0.8161 |
| $n_D^{20}$ | 1.5437 |

(4E,6Z)-2,6-Dimethyl-2,4,6-octatriene is produced as a mixture with other products by the thermolysis of α-pinene at 350 °C [35], or by heating Z-ocimene to 190 °C [36]. The products are separated by fractional distillation.

(4E,6E)-2,6-Dimethyl-2,4,6-octatriene, $C_{10}H_{16}$, $M_r$ 136.23, has physical properties similar to those of the (4E,6Z)-isomer.

| | |
|---|---|
| $bp$ | 79 °C (1.9 kPa) |
| $d^{20}$ | 0.8106 |
| $n_D^{20}$ | 1.5438 |

(4E,6E)-2,6-Dimethyl-2,4,6-octatriene is formed as a mixture with the (4E,6Z)-isomer and other products in the thermolysis of pinene at 320–330 °C [37], by dehydration of 3,7-dimethyl-3,6-octadien-1-ol [38], and by treating Z-ocimene with base [39].

2,6-Dimethyl-2,4,6-octatriene is used to a small extent in the perfume industry. It is also used as a diluting agent for varnishes and dyes [23], and as a component for terpene polymers [20], [22].

# 3. Monocyclic Monoterpenes

## 3.1. p-Menthane

p-Menthane [99-82-1], 1-isopropyl-4-methyl-cyclohexane, $C_{10}H_{20}$, $M_r$ 140.25, occurs widely as a cis/trans mixture, e.g., in essential oils in the eucalyptus fruit.

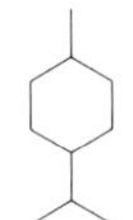

The cis/trans mixture of p-menthane is a colorless liquid with a fennel-like odor.

cis-1-Isopropyl-4-methylcyclohexane

| | |
|---|---|
| $bp$ | 168.8 °C (99.2 kPa) |
| $d^{20}$ | 0.8086 |
| $n_D^{20}$ | 1.4443 |

trans-1-Isopropyl-4-methylcyclohexane

| | |
|---|---|
| $mp$ | −86 °C |
| $bp$ | 168.1 °C (99.2 kPa) |
| $d^{20}$ | 0.7941 |
| $n_D^{20}$ | 1.4369 |

p-Menthane is obtained as a cis/trans mixture by the catalytic hydrogenation of limonene, terpinols, and p-cymene [40]. Raney nickel, platinum, or copper and aluminum oxides are used as catalyst [41].

p-Menthane is predominantly used as a precursor in the production of 1-isopropyl-4-methylcyclohexane hydroperoxide, a catalyst for radical polymerization [42].

## 3.2. α-Terpinene

α-Terpinene [99-85-4], 4-isopropyl-1-methyl-1,3-cyclohexadiene, $C_{10}H_{16}$, $M_r$ 136.23, is the main component of terpinene, the other components being the β- and γ-isomers. It occurs in various essential oils, almost always as a mixture with its isomers.



α-Terpinene is a colorless liquid with a lemon-like odor. It resinifies on prolonged storage.

| | |
|---|---|
| $bp$ | 172.5–173.5 °C |
| $n_D^{20}$ | 1.4780 |

α-Terpinene is obtained by fractional distillation of pine oil [43] and by the thermolysis of α-pinene in the presence of catalysts, such as manganese oxide [44].

α-Terpinene is used in essential oils and as an intermediate for perfumes.

## 3.3. Terpinolene

Terpinolene [*586-62-9*], 4-isopropylidene-1-methyl-1-cyclohexane, $C_{10}H_{16}$, $M_r$ 136.23, occurs in many essential oils and in sulfate turpentine.

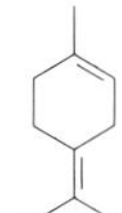

Terpinolene is a colorless liquid with an odor resembling that of turpentine.

| | |
|---|---|
| *bp* | 186 °C |
| $d^{20}$ | 0.8623 |
| $n_D^{20}$ | 1.4861 |

It forms a hydroperoxide on autoxidation [45], [46].

Terpinolene used to be extracted by fractional distillation of wood turpentine [46]. It is now produced by treating α-pinene with aqueous $H_3PO_4$ at 75 °C [47]. It is preferable to distill terpinolene under vacuum to minimize the formation of polymeric products at elevated temperatures.

Terpinolene is mainly used to improve the odor of industrial and household products [48]–[50].

## 3.4. *p*-Cymene

*p*-Cymene [*99-87-6*], 4-(isopropyl)methylbenzene, $C_{10}H_{14}$, $M_r$ 134.21, is the main component of sulfite turpentine and occurs in many essential oils [51].

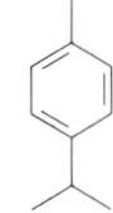

*p*-Cymene is a colorless liquid with an odor typical of aromatic hydrocarbons.

| | |
|---|---|
| *mp* | −67.7 °C |
| *bp* | 177.3 °C |
| $d^{20}$ | 0.8573 |
| $n_D^{20}$ | 1.4906 |

The enthalpy of evaporation is 44.670 J/mol.

*p*-Cymene undergoes catalytic hydrogenation to 1-isopropyl-4-methylcyclohexane [40]. It must be stored in the absence of light and air.

*p*-Cymene is formed during the sulfite leaching of wood [52]. A range of production processes starting from mono- and bicyclic terpenes have also been described. For example, *p*-cymene is obtained in good yields from α-pinene in the presence of copper catalysts [53], [54]. In this process a significant proportion of 1-isopropyl-4-methyl-cyclohexane is also formed, which is dehydrogenated in the presence of palladium catalysts at 260–280 °C to *p*-cymene [55]. The Friedel–Crafts alkylation of toluene with propene is used industrially [56]. Products containing more than one isopropyl group are converted into *p*-cymene with aluminum chloride and hydrochloric acid [57].

*p*-Cymene is used to improve the odor of soaps and as a masking odor for industrial products. Its use as an intermediate for musk perfumes [58], the oxidation to terephthalic acid, and its addition to antiseptic preparations have also been described [59]. *p*-Cymene is also used as a solvent for dyes and varnishes.

## 3.5. 1,8-Cineole

1,8-Cineole [*470-82-6*], 1,8-epoxy-*p*-menthane, eucalyptol, $C_{10}H_{18}O$, $M_r$ 154.25, occurs widely in natural essential oils, mainly eucalyptus oil.

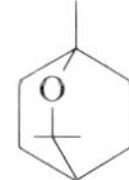

1,8-Cineole is a colorless liquid. Its odor and taste resemble those of camphor.

| | |
|---|---|
| *mp* | 1.55 °C |
| *bp* | 176–176.4 °C |
| $d^{20}$ | 0.9232 |
| $n_D^{20}$ | 1.4575 |

On warming, particularly in the presence of acids, 1,8-cineole forms other terpenes of the *p*-menthane type. It reacts with a range of substances, forming sparingly soluble products.

1,8-Cineole is extracted exclusively from eucalyptus oils with a high 1,8-cineole content. Various processes are used to separate it from the other terpenes, for example, treatment with $H_2SO_4$ in the cold [60], distillation in the presence of phenols, such as cresols or resorcinol

[61], which form loose addition compounds, or by addition to β-naphthol [62]. 1,8-Cineole can also be enriched by rectification. The yield is increased by gasification in the presence of a chromium–nickel catalyst [63].

1,8-Cineole is used as a perfume and fragrance. In the past it was also used as an expectorant and an antiseptic.


# 4. Bicyclic Monoterpenes

The industrially most important bicyclic monoterpenes are subdivided into caranes, pinanes, and bicyclo[2.2.1]heptanes, depending on their basic structure.


## 4.1. 3-Carene

3-Carene [498-15-7], 3,7,7-trimethylbicyclo-[4.1.0]hept-3-ene, $C_{10}H_{16}$, $M_r$ 136.23, occurs in a range of turpentines as the enantiomerically pure compound, but also as the racemate [64], [65].

3-Carene is a colorless compound with a sweet, penetrating odor.

| | |
|---|---|
| $bp$ | 176–176.4 °C |
| $d_4^{20}$ | 0.9232 |
| $n_D^{20}$ | 1.4575 |
| $[\alpha]_D^{25}$ | +17.6° (undiluted) |

It readily undergoes autoxidation and resinification in air. Addition of pyrogallol or resorcinol as stabilizers is recommended [66]. On treatment with peracetic acid in glacial acetic acid and subsequent saponification, 3-carene is oxidized to 3,4-caranediol [67].

Pyrolysis at 300–580 °C in the presence of $Fe^{III}$ oxide on carriers gives $p$-cymene as the main product [68].

3-Carene is obtained exclusively by fractional distillation of certain turpentines [69].

3-Carene is predominantly used in the perfume industry for producing essential oils [70]. It is also an intermediate in perfume synthesis and a precursor for insecticides of the pyrethroid type [71], [72].


## 4.2. Pinane

Pinane [473-55-2], 2,6,6-trimethylnorpinane, 2,6,6-trimethylbicyclo[3.1.1]heptane, $C_{10}H_{18}$, $M_r$ 138.25.

*cis*-(1S)    *trans*-(1S)

Pinane derivatives occur in wood, the leaves of many plants, algae, and insects. Pinane itself apparently does not occur naturally. However, it is formed on hydrogenation of pinane derivatives [73].

*cis*-Pinane (colorless liquid)

| | |
|---|---|
| $bp$ | 167.2–168 °C |
| $d_4^{20}$ | 0.8575 |
| $n_D^{20}$ | 1.4626 |
| $[\alpha]_D^{25}$ | −23.6° (undiluted) |

*trans*-Pinane (colorless liquid)

| | |
|---|---|
| $bp$ | 164 °C |
| $d_4^{20}$ | 0.854 |
| $n_D^{20}$ | 1.4610 |
| $[\alpha]_D^{25}$ | +21.4° (undiluted) |

The oxidation of pinane with air or oxygen gives 2-pinane hydroperoxide or 2-pinanol [74]. In industry, oxidation is carried out in the presence of catalysts [75].

The pyrolysis of $(-)$-(1S)-*cis*-pinane at 500 °C gives a mixture of (3R)-3,7-dimethyl-1,6-octadiene and (1R)-isopropyl-2,3-dimethylcyclopentane [76].

*cis*-Pinane is obtained, together with small quantities of *trans*-pinane, by the catalytic hydrogenation of α- and β-pinene. If $(-)$-α- and $(-)$-β-pinene are hydrogenated, the $(-)$-enantiomer is obtained, and from the $(+)$-pinenes, $(+)$-*cis*-pinane. In industry, α-pinene is predominantly used for the hydrogenation. Raney nickel, Pt, or Pd (and their oxides) on carriers are used as catalysts [77], [78]. The hydrogenation with Raney nickel is carried out at 2–10 MPa at

150 °C. In the presence of halogen compounds *cis*-pinane is obtained predominantly [79].

*trans*-Pinane can be obtained from β-pinene by treatment with sodium borohydride in the presence of boron trifluoride in di(ethylene glycol) and subsequent heating of the reaction mixture in propionic acid [80].

Pinane is used in industry for the production of pinane hydroperoxide [81]. It is also important as an intermediate in the production of 3,7-dimethylocta-1,6-diene, which is used to produce perfumes, such as citronellol, citronellal, and hydroxycitronellal.

## 4.3. 2-Pinane Hydroperoxide

2-Pinane hydroperoxide [*5405-84-5*], 2,6,6-trimethylbicyclo[3.1.1]heptane-2-hydroperoxide, $C_{10}H_{18}O_2$, $M_r$ 170.25, is a colorless liquid, which is readily flammable and insoluble in water.

(−)-(1*R*)

| | |
|---|---|
| *bp* | 48–52 °C (1.3 Pa) |
| $d_4^{20}$ | 1.021 |
| $n_D^{20}$ | 1.4898 |
| $[\alpha]_D^{25}$ | −27.4 ° (undiluted) |

On heating 2-pinane hydroperoxide to > 110 °C 1-((1*R*)-*cis*-3-ethyl-2,2-dimethylcyclobutyl)ethanone is formed [82]. (1*R*,2*S*)-2-pinanol is formed in good yields by catalytic hydrogenation.

2-Pinane hydroperoxide is produced by oxidation of pinane with air or oxygen at 95 °C [83]–[85]. *cis*-Pinane reacts more rapidly than *trans*-pinane [86].

2-Pinane hydroperoxide is used as a radical initiator for polymerization reactions [87], e.g., for the polymerization of diolefins and aromatic vinyl compounds, or for hardening unsaturated polyester resins. It is also an intermediate in the production of perfumes, such as pinanol, linalool, nerol, and geraniol [88].

## 4.4. 2-Pinanol

2-Pinanol [*473-54-1*], 2,6,6-trimethylbicyclo[3.1.1]heptan-2-ol, $C_{10}H_{18}O$, $M_r$ 154.25, is obtained as the *cis* or *trans* isomer, depending on the production process.

*cis*-(−)-(1*R*,2*S*)          *trans*-(−)-(1*R*,2*R*)

*cis*-2-Pinanol forms colorless crystals with a camphor-like odor.

| | |
|---|---|
| *mp* | 78–79.2 °C |
| *bp* | 90–91 °C (1.5 kPa) |
| $[\alpha]_D^{20}$ | −24.5 ° (CHCl$_3$, C = 8 g/L) |

*trans*-2-Pinanol forms colorless needles.

| | |
|---|---|
| *mp* | 58–59 °C |
| *bp* | 88–89 °C (0.12 kPa) |
| $[\alpha]_D^{17}$ | −2.3 ° (ether, 10 g/L) |

The pyrolysis of 2-pinanol at 500 °C gives linalool. (−)-Linalool is formed from (+)-*cis*- and (−)-*trans*-2-pinanol, and (+)-linalool from (−)-*cis*- and (+)-*trans*-2-pinanol. The process is used industrially [89], [90].

*cis*-2-Pinanol is produced industrially by the catalytic hydrogenation of 2-pinane hydroperoxide [91], [92]. Alternatively 2-pinane hydroperoxide can be treated with sodium sulfide in aqueous sodium hydroxide [93] or with sodium methoxide [94]. *cis*-2-Pinanol can also be obtained directly from pinane by air oxidation in the presence of alkalis, such as sodium hydroxide, at 80–100 °C [95].

2-Pinanol is used to produce linalool by pyrolysis [96].

## 4.5. Camphene

Camphene [*5794-03-6*], 2,2-dimethyl-3-methylenenorbornane, $C_{10}H_{16}$, $M_r$ 136.23, occurs in a large number of essential oils in optically active form, both as the *R* and *S* enantiomers.

(+)-(1 *R*)

Camphene is a colorless, crumbly crystalline solid with a camphor-like odor. It has a tendency to sublime.

| | |
|---|---|
| *mp* | 52.5 °C |
| *bp* | 157.8 °C (98.8 kPa) |
| $d_4^{25}$ | 0.84225 |
| $n_D^{54}$ | 1.4564 |
| $[\alpha]_D^{20}$ | +106.8 ° (CHCl$_3$, 44 g/L) |

Camphene is stable in air and light. At elevated temperatures, in the presence of oxygen, it undergoes autoxidation to camphenilone [97]. Peracids attack the double bond, giving camphene oxide [98]. Catalytic hydrogenation gives the saturated hydrocarbon isocamphane [99].

The extraction of camphene by distillation of turpentine is now hardly used in industry. Instead camphene is produced from α-pinene.

(+)-α-Pinene is converted into (+)-bornyl chloride by the action of dry hydrogen chloride in a Wagner–Meerwein rearrangement. Base-catalyzed dehydrohalogenation of the (+)-bornyl chloride gives racemic camphene [100], [101]. The reaction of α-pinene with borophosphoric acid in the gas phase [102], or on $TiO_2$ catalysts [103] has also been described.

Camphene is used to improve the odor of industrial products. It is an intermediate in the production of camphor, isobornyl esters, and the insecticide Toxaphen.

## 5. Acyclic Sesquiterpenes

### 5.1. Farnesene

α-Farnesene
[502-61-4]

β-Farnesene
[18794-84-8]

α-Farnesene, 2,6,10-trimethyl-2,6,9,11-dodecatetraene, $C_{15}H_{24}$, $M_r$ 204.36.

α- and β-Farnesenes and their Z- and E-isomers occur in many essential oils, e.g., in apples [104], citrus fruits [105], and hop oil. α-Farnesene is a natural attractant for the larvae of the codlin moth [106]. It is also found in the gland secretions of certain ants [107].

α- and β-Farnesenes are colorless liquids with fruity odors. They are sensitive to autoxidation

and can be stored for long periods only with the exclusion of light and air.

α-Farnesene
$bp$     98–102 °C (0.5 kPa)
$n_D^{25}$     1.4790

β-Farnesene
$bp$     95–107 °C (0.5 kPa)
$d^{20}$     0.8363
$n_D^{20}$     1.4899

Mixture
$bp$     129–132 °C (1.2 kPa)
$d^{18}$     0.8770
$n_D^{20}$     1.4995

Mixtures of α- and β-farnesenes are obtained by heating farnesol with potassium hydrogen sulfate to 160–170 °C [108]. Other syntheses of the mixture, involving dehydration of farnesol, nerolidol, and isoprene, have also been described [109], [110].

Farnesene is used in small quantities in the fragrance industry.

## 6. Monocyclic Sesquiterpenes

### 6.1. Bisabolene

Bisabolene, $C_{15}H_{24}$, $M_r$ 204.33, occurs in myrrh oil and limett oil. The isomers α-, β-, and γ-bisabolene are known.

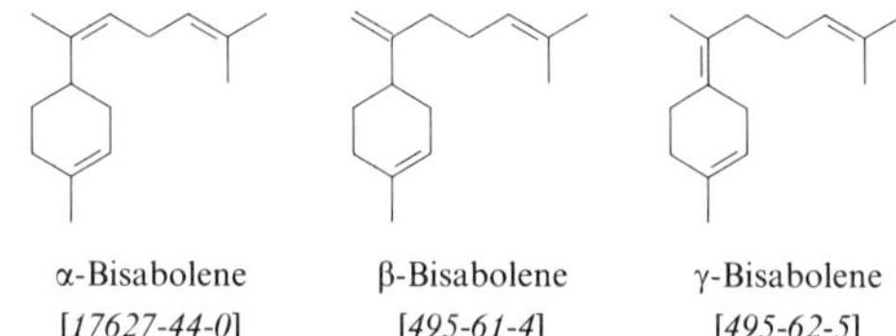

α-Bisabolene
[17627-44-0]

β-Bisabolene
[495-61-4]

γ-Bisabolene
[495-62-5]

α-Bisabolene, (E,Z)-4-(1,5-dimethyl-1,4-hexadienyl)-1-methyl-1-cyclohexane

$bp$     95 °C (0.02 kPa)

β-Bisabolene, 1-methyl-4-(5-methyl-1-methylene-4-hexenyl)-1-cyclohexane

$bp$     148 °C (2.4 kPa)
$n_D^{15}$     1.4893
$[\alpha]_D^{15}$     −67 ° (undiluted)

γ-Bisabolene, (E,Z)-4-(1,5-dimethyl-4-hexenylidene)-1-methyl-1-cyclohexene

$bp$     132 °C (1.4–1.6 kPa)
$n_D^{25}$     1.4928

The individual isomers and their mixtures are colorless liquids with pleasant, balsamic odors.

Isomer mixtures are produced by treating nerolidol and farnesol with acid [111]. Targeted syntheses of α- and β-bisabolene by the Wittig reaction have been described [112]–[114].

Bisabolene mixtures are used in the perfume and fragrance industries.

## 6.2. Bisabolol

Bisabolol [*515-69-5*], $C_{15}H_{26}O$, $M_r$ 222.37, occurs in various plants.

(+)-α-Bisabolol, (+)-2-methyl-6-(4-methyl-3-cyclohexenyl)-6-hepten-2-ol

| | |
|---|---|
| *bp* | 120–122 °C (0.13 kPa) |
| $d_4^{20}$ | 0.9213 |
| $n_D^{20}$ | 1.4919 |
| $[\alpha]_D^{20}$ | + 51.7 ° (undiluted) |

(−)-α-Bisabolol, (−)-2-methyl-6-(4-methyl-3-cyclohexenyl)-6-hepten-2-ol

| | |
|---|---|
| *bp* | 153 °C (1.59 kPa) |
| $d_4^{20}$ | 0.9211 |
| $n_D^{20}$ | 1.4936 |
| $[\alpha]_D$ | − 55.7 ° (undiluted) |

*rac*-Bisabolol, 2-methyl-6-(4-methyl-3-cyclohexenyl)-6-hepten-2-ol

| | |
|---|---|
| *bp* | 157 °C (1.59 kPa) |
| $d_4^{23}$ | 0.9223 |
| $n_D^{23}$ | 1.4917 |

The bisabolols are colorless liquids with slightly flowery odors.

*rac*-Bisabolol is produced by acid-catalyzed cyclization of farnesol or nerolidol [115]. The individual enantiomers are obtained by extraction from the appropriate plants.

Because of their anti-inflammatory and spasmolytic properties [117], both (−)- and *rac*-bisabolol are predominantly used in the cosmetics industry and to a small extent in the pharmaceutical industry.

## 7. Bicyclic Sesquiterpenes

### 7.1. Caryophyllene

Caryophyllene [*87-44-5*], 6,10,10-trimethyl-2-methylenebicyclo[7.2.0]-5-undecene, $C_{15}H_{24}$, $M_r$ 204.36, occurs in many essential oils.

Caryophyllene is a colorless, oily liquid with an odor resembling that of cloves.

| | |
|---|---|
| *bp* | 254–256 °C |
| $d_4^{20}$ | 0.9019 |
| $n_D^{20}$ | 1.4995 |
| $[\alpha]_D^{20}$ | −9.2 ° (undiluted) |

Catalytic hydrogenation on platinum gives tetrahydrocaryophyllene [118], and on palladium a dihydrocaryophyllene [119]. On oxidation with potassium permanganate, a glycol (*mp* 120–120.5 °C) is formed [120], and with perbenzoic acid [121] or peracetic acid, a caryophyllene oxide.

Caryophyllene is best extracted from clove oil. It can also be obtained from other sources, such as certain American pine oil fractions.

Caryophyllene is used as a perfume and fragrance, for example in chewing gum (ca. 200 mg/kg). It is also used as a fixative. It can be used as an intermediate in the synthesis of other perfumes and fragrances [123].

## 8. Tricyclic Sesquiterpenes

### 8.1. Longifolene

Longifolene [*475-20-7*], 3,3,7-trimethyl-8-methylenetricyclo[5.4.0.9-Z9]undecane, $C_{15}H_{24}$, $M_r$ 204.36, occurs in the *d*- and *l*-forms in a range of essential oils. Indian turpentine contains up to 20 % longifolene.

(+)-(1*R*,2*S*,7*S*,9*S*)

Longifolene is a colorless, oily liquid.

| | |
|---|---|
| *bp* | 254–256 °C (93.9 kPa) |
| $d_4^{18}$ | 0.9319 |
| $n_D^{20}$ | 1.5040 |
| $[\alpha]_D$ | + 42.73 ° (undiluted) |

Longifolene forms crystalline products with hydrogen halides [124]. On treatment with bromine followed by *N,N*-dimethylaniline, ω-bromolongifolene is formed, and with nitrogen oxides ω-nitrolongifolene [125].

Longifolene is obtained by distillation, e.g., Indian turpentine.

Longifolene is used as a solvent additive. It has been described as a starting material for the production of perfumes, e.g., by oxidation [126], or by treatment with formic acid [127].

## 9. Acyclic Diterpenes

### 9.1. Phytol

Phytol [150-86-7], (7R,11R)-3,7,11,15-tetramethyl-*trans*-2-hexadecen-1-ol, $C_{20}H_{40}O$, $M_r$ 295.52, is formed by saponification of chlorophyll.

Phytol is a colorless liquid with a slightly flowery odor. The action of acids or heat effects dehydration to phytadiene.

| | |
|---|---|
| $bp$ | 136 °C (1.3 kPa) |
| $d_4^{21}$ | 0.8533 |
| $n_D^{25}$ | 1.4637 |

Phytol can be obtained from natural raw materials by a variety of processes [128]. The industrial synthesis of phytol starts from isophytol. The latter is first treated with formic acid to give phytyl formate, from which $(Z,E)$-phytol (isomer ratio 3:7) is obtained by saponification or transesterification [129].

Phytol is used in the perfume and cosmetics industries [130]. It can also be used, like isophytol, as the starting material for the synthesis of vitamins E and $K_1$ [131]–[133]. The hydrogenation and oxidation products are also known [134].

## 10. Acyclic Triterpenes

### 10.1. Squalene

Squalene [111-02-4], 2,6,10,15,19,23-hexamethyltetracosa-2,*t*-6,*t*-10,*t*-14,*t*-18,*r*-22-hexaene, $C_{30}H_{50}$, $M_r$ 410.73, occurs in liver oils of various species of shark [135], in cod liver train oil [136], in vegetable fats and oils [137], and in human skin fat.

Squalene is a colorless, air-sensitive, almost odorless liquid.

| | |
|---|---|
| $bp$ | 223–226 °C (0.26 kPa) |
| $d_4^{20}$ | 0.8577 |
| $n_D^{25}$ | 1.4961 |

Squalene is extracted from shark liver [138] or synthesized from hexaphenyl-1,4-butanediyldiphosphonium dibromide and 6,10-dimethyl-5,9-undecadien-2-one (geranylacetone, industrial intermediate in the vitamin E synthesis) [139].

Squalene is hydrogenated on platinum or nickel catalysts to dodecahydrosqualene (perhydrosqualene, squalane, cosbiol) [140], which is widely used in cosmetics [141], [142].

## 11. Toxicology

There are excellent reviews of the toxicology of terpenes [143]. The known toxicological test results (Table 1) show that most terpenes have low acute oral toxicity and, in spite of occasionally good skin resorption, low dermal toxicity. Some are irritant to the skin of rabbits in concentrated form, but a 4–16% solutions in vaseline, are generally not irritant to humans. Only 3-carene, which occurs in turpentine, has been shown to have a sensitizing potential in humans and animals.

To investigate detoxification in mammals, a metabolism study on rabbits has been carried out. On oral administration several characteristic oxidation reactions have been detected:

1) Stereoselective oxidation: formation of 3-caren-9-ol, 3-caren-9-carboxylic acid, and 3-caren-9,10-dicarboxylic acid from 3-carene
2) Regioselective oxidation: 3,10-myrcenediol, 1,2-myrcenediol, and 2-hydroxymyrcene-1-carboxylic acid from myrcene
3) Allylic oxidation: verbenol and myrtenic acid from α-pinene, and pinocarveol from β-pinene
4) Homoallylic oxidation: 6-*exo*-hydroxycamphene and 10-hydroxytricyclene from camphene
5) Hydrolysis of epoxides: camphenediol from camphene

With *p*-cymene, three different oxidation products were detected in the first step. The alcohols 2-(4-methylphenyl)-1-propanol and 2-(4-methyl-phenyl)-2-propanol are formed by

**Table 1.** Toxicity data for terpenes

| Substance | Acute oral toxicity (rat), LD$_{50}$, g/kg | Acute oral toxicity (rabbit), LD$_{50}$, g/kg | Skin resorption | IRT* (rat) | Skin irritation (R = rabbit; H = human) | Irritation of rabbit eye mucous membrane | Sensitization (H = human; G = guinea pig) | Specific properties/tests | References |
|---|---|---|---|---|---|---|---|---|---|
| Myrcene | >5 | >5 | + | | R: moderately irritant; H: 4% not irritant | | H: negative | | [145] |
| Ocimene | >5 | >5 | | | R: moderately irritant; H: 5% not irritant | | H: negative | | [145] |
| α-Terpinene | 1.68 | | | | H: 5% not irritant | | H: negative | damage to liver and blood (forms methemoglobin) | [145] |
| Terpinolene | 4.39 ml/kg | >5 | | | R: not irritant; H: 20% not irritant | | H: negative | | [145] |
| p-Cymene | 4.74 | >5 | + | 0/6 after 6 h | H: irritant; R: not irritant; H: 4% not irritant | irritant | H: negative | affects central nervous system (narcosis) | [145], [146] |
| 1,8-Cineole | 2.48 | >5 | | | R: not irritant; H: 16% not irritant | | H: negative | cancerization study in progress | [145], [146] |
| 3-Carene | 4.8 | >5 | | | R: irritant | | H: positive G: positive | | [145] |
| Camphene | >5 | 2.5 | + | | R: slightly irritant; H: 4% not irritant | | H: negative | | [145], [146] |
| Bisabolene | >5 | >5 | | 0/12 after 7 h | R: slightly irritant; H: 10% not irritant | not irritant | H: negative | | |
| Bisabolol | >5 | >5 | | 0/12 after 7 h | R: not irritant | slightly irritant | | | [146] |
| Caryophyllene | >5 | >5 | | | R: irritant; H: 4% not irritant ** | | H: negative | | |
| Phytol | >10 | | | | R: moderately irritant | not irritant | | | [145], [146] |

* IRT: inhalation risk test (rat, test result dependent on toxicity and volatility; 0/6 after 6 h means that after 6 h exposure in an enriched or saturated atmosphere at room temperature no animals died). ** In light petroleum jelly.

the ω- and (ω-1)-oxidation of the isopropyl group, and 4-isopropylbenzyl alcohol by oxidation of the methyl group. The alcohols formed in these different oxidations are either excreted from the organism as their glucuronides or further oxidized [8], [144].

Most of the terpenes described in Table 1 are licensed as food additives by the international bodies responsible (Table 2). This licensing clearly demonstrates the low toxicity of the terpenes concerned.

**Table 2.** Terpenes licensed as food additives

| Substance | Licensing authority |
| --- | --- |
| Myrcene | FDA (GRAS classification), European Council |
| Ocimene | FDA, European Council |
| α-Terpinene | FDA |
| Terpinolene | FDA, European Council |
| p-Cymene | FDA |
| 1,8-Cineole | FDA, European Council |
| Camphene | FDA, European Council |
| Caryophyllene | FDA, European Council |

FDA = Food and Drug Administration;
GRAS = generally recognized as safe.


# 12. References


[1] J. C. Simonsen et al.: *The Terpenes,* 2nd ed., vols. 1–3, Cambridge University Press, New York 1947–1951.

[2] A. A. Newmann: *Chemistry of Terpenes and Terpenoids,* Academic Press, London 1972.
T. K. Devon, A. I Scott: *Handbook of Naturally Occurring Compounds,* vol. II, Terpenes, *Academic Press,* London 1972.

[3] *Terpenoids and Steroids* (A Spezialist Periodical Report) vols 1–6, The Chemical Society, Burlington House, London 1971–1976.

[4] W. Templeton: *An Introduction to the Chemistry of the Terpenoids and Steroids,* Butterworths, London 1969.

[5] W. J. Tayler, A. R. Battersby: *Cyclopentanoid Terpene Derivatives,* Marcel Dekker, New York 1969.

[6] O. Aschan: *Naphthenverbindungen, Terpene und Campherarten,* De Gruyter, Berlin 1929.

[7] J. ApSimon: *The Total Synthesis of Natural Products,* vol. 2, J. Wiley, New York 1973.

[8] O. W. Thiele: *Lipide, Isoprenoide mit Steroiden,* Thieme Verlag, Stuttgart 1979.

[9] E. Gildemeister, F. Hoffmann: *Die ätherischen Öle,* 4th ed., vols. 1–7, Akademie Verlag, Berlin 1960.

[10] E. Guenther: *The Essential Oils,* Van Nostrand, Princeton, N. J., 1949.

[11] W. Sandermann: *Naturharze Terpentinöl Tallöl,* Springer Verlag, Berlin 1960.

[12] O. Isler: *Carotinoide,* Birkhäuser Verlag, Basel 1971.

[13] *Beilstein,* **30** (1938) 64–76.
S. Boström: *Kautschuk-Handbuch,* vols. 1–5, Berliner Union, Stuttgart 1958–1962.
L. Batteman: *The Chemistry and Physics of Rubber-Like Substances,* Maclaren, London 1963.

[14] T. W. Goodwin: *Aspects of Terpenoid Chemistry and Biochemistry,* Academic Press, London 1971.

[15] D. V. Banthopre, B. V. Charlewood, M. J. O. Francis: "The Biosynthesis of Monoterpenes," *Chem. Rev.* **72** (1972) no. 2, 115.

[16] A. Cordell: "Biosynthesis of Sequisterpenes," *Chem. Rev.* **76** (1976) no. 4, 425.

[17] S. G. Cantwell, E. P. Lau, D. S. Watt, R. R. Fall, *J. Bacteriol.* **135** (1978) 324.

[18] H. Siegel, W. Himmele, *Angew. Chem.* **92** (1980) 185.

[19] W. Franke: *Nutzpflanzenkunde,* Thieme Verlag, Stuttgart 1976.

[20] C. D. Ender, *Nav. Stores Rev.* **66** (1956) 10.
W. H. Wiles, *Rubber Plast. Age* **40** (1959) 249.
K. Blaetuer, *Dtsch. Farben Z.* **17** (1963) 34.
H. P. Preuss, *Met. Finish* **62** (1964) no. 8, 63.

[21] *Chem. Age (London)* **63** (1950) 269.
*Prakt. Chem.* **10** (1959) 383.
R. Werner, *Prakt. Chem.* **11** (1960) 331.

[22] J. Schreiber: *Chemie und chemische Technologie der Kunstharze,* Wissenschaftl. Verlagsgesellschaft, Stuttgart 1943, p. 675.
E. R. Littmann, *Ind. Eng Chem.* **28** (1936) 1150.

[23] L. Désalbres, FR 1023339, 1950.

[24] J. M. Mellor, S. Munavalli, *Q. Rev. Chem. Soc.* **18** (1964) 270.

[25] F. Sorm: "Sequisterpenes with Ten-Membered Carbon Rings, A. Review," *J. Agric. Food Chem.* **19,** (1971) 1081.

[26] Van Amerigen-Haebler, Inc., US 2882 323, 1957 (R. Weiss).

[27] Glidden Co., US 3031442, 1958 (R. L. Webb).

[28] T. Sasaki, S. Eguchi, T. Jshi, *J. Org. Chem.* **34** (1969) 3749.

[29] L. A. Goldblatt, S. Palkins, *J. Am. Chem. Soc.* **63** (1941) 3517.
R. L. Burwell, Jr., *J. Am. Chem. Soc.* **73** (1951) 4421.
E. T. Theimer, B. M. Mitzner, *Can. J. Chem.* **42** (1964) no. 41, 959.
E. L. Patton, *Am. Perfum. Essent. Oil Rev.* **56** (1950) 118.

[30] *Beilstein,* 1 H 260, 1 E III, 1054.

[31] J. E. Hawkins, H. G. Hunt, *J. Am. Chem. Soc.* **73** (1951) 5379.
J. E. Hawkins, W. A. Burris, *J. Org. Chem.* **24** (1959) 150.

[32] R. L. Burwell, *J. Am. Chem. Soc.* **73** (1951) 4461.

[33] M. C. J. Enklaar, *Recl. Trav. Chim. Pays-Bas Belg.* **26** (1907) 157, 167.

[34] M. C. J. Enklaar, *Recl. Trav. Chim. Pays-Bas Belg.* **26** (1907) 174; *Chem Zentrabl.* **1907** II, 56; *Ber. Dtsch. Chem. Ges.* **41** (1908) 2084.

[35] K. Alder, A. Dreike, H. Erpenbach, U. Wicker, *Justus Liebigs Ann. Chem.* **609** (1957) 1.

[36] T. Saskaki, S. Eguchi, H. Yamada, *Tetrahedron Lett.* **1971,** 99.

[37] H. Pines, N. E. Hoffmann, V. N. Ipatieff, *J. Am. Chem. Soc.* **76** (1954) 4412.

[38] G. Ohloff, *Chem. Ber.* **90** (1957) 1554.

[39] V. N. Ipatieff, H. Pines, E. E. Meisinger, *J. Am. Chem. Soc.* **71** (1949) 2934.

[40] V. N. Ipatieff, H. Pines, E. E. Meisinger, *J. Am. Chem. Soc.* **71** (1949) 2934.
[41] H. A. Smith, J. F. Fuzek, H. T. Meriwether, *J. Am. Chem. Soc.* **71** (1949) 3765, 3766.
[42] USA Secr. of Agric., US 2 775 578, 1955.
[43] W. Sandermann: *Naturharze: Terpentinöl, Tallöl* Springer Verlag, Berlin 1960, pp. 269–310.
[44] BASF, DE 960 988, 1954.
[45] M. Saito, *Kogyo Kagaku Zasshi* **61** (1958) 326–330.
[46] J. N. Borgein, D. A. Lister, E. J. Lorand. J. E. Reese, *J. Am. Chem. Soc.* **72** (1950) 4591–4596.
[47] Haarmann u. Reimer, DE 411901, 1923 (R. Moelleken).
[48] Monsanto Co., US 4 097 555, 1976.
[49] Mitsubishi Chem. Ind., JP 53 121 891, 1977; *Chem. Abstr.* **90** (1979) 88066 t.
[50] Standard Oil Co., US 4 080 592, 1977.
[51] R. G. Buttery et al., *J. Agric Food Chem.* **22** (1974) 773.
[52] Zellstoffabrik Waldhof, DE 727 475, 1940.
[53] BASF, DE 961979, 1953.
[54] Goodyear Co., EP 77289, 1981 (J. Kuczkowski, L. Wideman).
[55] Hercules Powder Co., US 2 400 012, 1942.
[56] BASF, US 3 555 103, 1967.
[57] Mobil Oil Corp., EP 12514, 1978 (G. T. Burress).
[58] Sumitomo Chemical KK, DE 2910493, 1978.
[59] G. Ohloff: *Riechstoffe und Geruchssinn*, Springer Verlag, Berlin 1990.
[60] Rheinische Kampherfabrik. DE 499 732, 1928.
[61] Newport Ind. Inc., US 2090620, 1936.
[62] Takasago Perfumery Ind. Co. JP 11 72, 1950.
[63] Union Camp Corp., EP 270023 (P. W. D. Mitchell, D. E. Sasser).
[64] B. S. Rao, J. L. Simonsens, *J. Chem. Soc.* **127** (1925) 2494–2499.
[65] N. T. Mirov, *J. Am. Pharm. Assoc. Sci. Ed.* **40** (1951) 410–413.
[66] G. Widmarks, S. G. Blohm, *Acta Chem. Scand. (1947–1973)* **11** (1957) 392–394.
[67] B. A. Arbusow, B. M. Michailow, *J. Prakt. Chem.* **127** (1930) 1–15.
[68] J. Verghese, W. K. Sondhi, B. Bushan, M I. Joshi, *Curr. Sci.* **17** (1948) 359.
[69] B. S. Rao, J. L. Simonsens, *J. Chem. Soc.* **127** (1935) 2494.
[70] S. Arctander: *Perfume and Flavor Chemicals*, Det Hoffensbergske, Etablissement, Kopenhagen 1969.
[71] Shell Oil Co., US 4156692, 1977.
[72] Shell Int. Res., EP 2850, 1977.
[73] D. V. Banthorpe, D. Whittaker, *Chem. Rev.* **66** (1966) 643.
[74] Monsanto US 2097 744, 1934.
[75] SCM Corp., US 4254291, 1981.
[76] J. Tanaka, T. Katagiri, K. Ozawa, *Bull. Chem. Soc. Jpn.* **44** (1971) 130–132.
[77] G. S. Fisher, J. S. Stinson, R. N. Moore, L. A. Goldblatt, *Ind. Eng. Chem.* **47** (1955) 1368–1373.
[78] G. S. Fisher, L. A. Goldblatt, J. Kniel, A. D. Snyder, *Ind. Eng. Chem.* **43** (1951) 671–674.
[79] SCM-Corp., US 4018 842, 1977.
[80] G. Zweifel, H. C. Brown, *J. Am. Chem. Soc.* **86** (1964) 393–397.
[81] *Chem. Eng. News* **60** (1982) no. 11, 5.
[82] G. A. Schmidt, G. S. Fisher, *J. Am. Chem. Soc.* **76** (1954) 5426–5430.
[83] Tagasako, JP 61 200999, 1986.
[84] G. Ohloff, E. Klelin, *Tetrahedron* **18** (1962) 37.
[85] USA Secr. of Agric, US 2735870, 1950.
[86] G. S. Fisher, J. S. Stinson, L. A. Goldblatt, *J. Am. Chem. Soc.* **75** (1953) 3675–3678.
[87] Amer. Cynamid, US 306 1554, 1962.
[88] *Beilstein*, **27**, 5 E IV, 277.
[89] Studiengesellschaft Kohle mbH, DE 1150974, 1961.
[90] G. Ohlof, E. Klein, *Tetrahedron* **18** (1962) 37–42.
[91] Shell, NL 7900588, 1979.
[92] L. A. Shutikova, V. G. Charakew, M. S. Erzhanova, A. K. Alifanova, *Maslo. Zhir. Promst.* **1973**, 23–25.
[93] Cocker Chemical Co., GB 1619649, 1963.
[94] G. A. Schmidt, G. S. Fisher, *J. Am. Chem. Soc.* **81** (1959) 445–448.
[95] Stephan Chemical Co., DE 2 305 363, 1973.
[96] Studiengesellschaft Kohle mbH, FR 1328113, 1963.
[97] Du Pont, *Ind. Chim. Belge Ser.* 2 **11** (1940) 3.
[98] W. Hückel, *Ber. Dtsch. Chem. Ges.* **80** (1947) 41–47.
[99] P. Lipp, *Justus Liebigs Ann. Chem.* **382** (1911) 265–305.
[100] A. Meyer, DE 272562 1911.
[101] V. Hilcken, DE 439 695 1924.
[102] Schering-Kahlbaum, DE 578 569, 1931.
[103] Du Pont, US 2551 795, 1949.
[104] E. F. L. J. Anet, *Aust. J. Chem.* **23** (1970) 2101.
[105] M. G. Moshonas, P. E. Shaw, *J. Agric. Food Chem.* **28** (1980) 680.
[106] O. R. W. Sutherland, R. F. N. Hutchins, *Nature (London)* **239** (1972) 170.
[107] G. W. K. Cavill, P. J. Williams. F. B. Withfield, *Tetrahedron Lett.* **1967**, 2201.
[108] C. Harries, R. Haarmann, *Chem. Ber.* **46** (1913) 1741.
[109] G. Brieger, T. J. Nestrick, C. McKenna, *J. Org. Chem.* **34** (1969) 3789.
[110] G. Brieger, *J. Org. Chem.* **32** (1967) 3720.
[111] L. Ruzicka, E. Capato, *Helv. Chim. Acta* **8** (1925) 259.
[112] F. Delay, G. Ohloff, *Helv. Chim. Acta* **62** (1979) 369.
[113] G. Brieger, T. J. Nestirck, C. McKenna, *J. Org. Chem.* **34** (1969) 3789.
[114] O. P. Vig, S. D. Sharma, P. Kumar, M. L. Sharma, *J. Indian Chem. Soc.* **52** (1975) 614.
[115] E. V. Weber, *Justus Liebigs Ann. Chem.* **238** (1887) 101.
[116] Degussa, DE-OS 2709033, 1978.
[117] M. Holub, V. Heront, F. Sorm, *Cesk. Farm* **4** (1955) 129.
[118] F. W. Semmler, E. W. Mayer, *Ber. Dtsch. Chem. Ges.* **45** (1912) 1384–1394.
[119] Shell, EP 127911, 1983 (A. J. Mulder).
[120] E. Deussen, *Justus Liebigs Ann. Chem.* **388** (1912) 136–165.
[121] H. N. Rydon, *J. Chem. Soc.* **1939**, 537–540.
[122] Soda Shangyo Co. Ltd, JP-Kokai 76 80847, 1975.
[123] BBA, GB 1580184, 1976 (H. R. Ansari, R. Clark, H. R. Wagner).
[124] J. L. Simonsen, *J. Chem. Soc.* **117** (1920) 564–578.
[125] G. R. Dupont, R. Dulou, P. Naffa, G. Ourisson, *Bull. Soc. Chim. Fr.* **1954**, 1075, 1078.
[126] Tagasako Co., JP 04 360848, 1992.
[127] Tosho Corp., JP-Kokai 01 268657, 1989.
[128] E. M. Burdick, US 3248301, 1964.
[129] BASF, DE 28187150, 1978.
[130] H. Ianistyn, DE 1467955, 1964.

[131] B. Staller-Bourdillon, *Ind. Chim. Belge* **35** (1970) no. 1, 13.
[132] Eisai KK, JP-Kokai 88 108, 1976.
[133] Nisshin Flour Milling, JP-Kokai 10 235, 1960.
[134] E. Jellum, L. Eldjarn, K. Try, *Acta Chem. Scand. (1947–1973)* **20** (1966) 2535.
[135] M. Tsujmoto, *Chem. Zentralbl.* **1918** I, 638, 1048.
[136] J. C. Drummond, H. J. Channon, K. H. Coward, *Biochem. J.* **19/23** (1919/1923) 1055/279.
[137] K. Täufel, H. Thaler, H. Schreyegg, *Fettchem. Umsch.* **43** (1936) 26.
[138] M. Hamaya, JP-Kokai 00 3158, 1977.
[139] D. W. Dicker, M. L. Whiting, *J. Chem. Soc.* **1958**, 1994.
[140] J. M. Heilbron, T. P. Hildrich, E. D. Kamm, *J. Chem. Soc.* **1926**, 3135.
[141] H. Janistyk: *Handbuch der Kosmetik und Riechstoffe*, 2nd ed., vol. 1, Hüthig Verlag, Heidelberg 1969, p. 753.
[142] E. S. Lower, *Spec. Chem.* **1** (1981) no. 1, 34.
[143] E. L. J. Opdyke (ed): *Monographs on Fragrance Raw Materials*, Pergamon Press, Oxford 1979.
[144] T. Ishida et al.: "Biotransformations of Terpenoids in Mammals," *Chem. Abstr.* **93** (1980) 8986003.
[146] NIOSH: *Registry of Toxic Effects of Chemical Substances 1979*, II, Washington, D. C., 1979, p. 33.

**Terphenyls → Hydrocarbons**

**Tetrachloroethanes → Chlorinated Hydrocarbons**

**Tetrachloroethylene → Chlorinated Hydrocarbons**

**Tetracycline → Antibiotics**

**Tetraethyllead → Lead Compounds**

**Tetrafluoroethylene Copolymers → Fluoropolymers, Organic**

# Tetrahydrofuran

HERBERT MÜLLER, BASF Aktiengesellschaft (retired), Ludwigshafen, Federal Republic of Germany



## 1. Introduction

Tetrahydrofuran (THF) [*109-99-9*] tetramethylene oxide, oxolane, is a five-membered cyclic ether with wide application in the chemical industry.

The largest proportion of the $1.4 \times 10^5$ t of THF produced worldwide in 1992 was used as a monomer in the manufacture of poly(tetramethylene oxide) (PTMO) also known as poly-(tetramethylene ether glycol) (PTMEG) and polytetrahydrofuran (PTHF), important in the production of thermoplastic polyurethanes, elastic fibers, molded elastomers, and copolyesters or copolyamides. A smaller proportion finds use as a solvent for coatings, adhesives, and special varnishes based, for example, on poly(vinyl chloride) (PVC) or polyurethanes; as an extracting agent; and as the preferred medium for organometallic syntheses.

## 2. Physical Properties

The pure distilled tetrahydrofuran sold commercially is a colorless, clear, volatile, polar liquid with a characteristic acetone-like odor and a minimum purity of 99.9 wt %. THF is miscible in all proportions with water, alcohols, ethers, and many common solvents. Important physical properties include the following [1], [2]:

| | |
|---|---|
| $M_r$ | 72.1 |
| $bp$ | 66 °C |
| $mp$ | −108.5 °C |
| Critical temperature | 267 °C |
| Critical pressure | 5.19 MPa |
| Critical density | 0.322 g/cm³ |
| $d_4^{20}$ | 0.886 |
| $n_D^{20}$ | 1.4073 |
| Specific heat | 1.765 J g⁻¹ K⁻¹ |
| Heat of vaporization (66 °C/101.3 kPa) | 435 J/kg |
| Heat of combustion | −35141 kJ/kg |
| Explosive limits in air (25 °C) | |
|    lower | 1.5 vol % |
|    upper | 12.0 vol % |
| Flash point (Abel–Pensky) | −22 °C |
| Dipole moment | $5.84 \times 10^{-30}$ C m |
| Dielectric constant (20 °C) | 7.6 |

## 3. Chemical Properties

THF is a valuable starting material for a wide range of important reactions. For example, cationic polymerization accompanied by ring opening leads to high molecular mass bifunctional glycol ethers with various chain lengths [3], [4]. These have achieved great economic significance in the preparation of important plastics. Ring-cleavage reactions also constitute the basis for dehydration to butadiene, oxidation to succinic acid, and carboxylation to adipic acid or $\gamma$-valerolactone [5], [6].


Ullmann's Encyclopedia
of Industrial Chemistry, Vol. A 26

THF is the preferred reaction medium for carrying out Grignard reactions or reductions with, for example, $LiAlH_4$. It is also suitable for use as a ligand in coordination complexes, such as those sometimes employed in stereospecific polymerizations [7], [8].

THF reacts readily with oxygen (on contact with air, for example), the principal product being an unstable hydroperoxide [9]. Hydroquinone or 2,6-di-*tert*-butyl-*p*-cresol (BHT) (e.g., 250 mg/kg) can be added to retard peroxide formation. Distillation of peroxide-containing THF increases the peroxide concentration, resulting in a serious risk of explosion, even on a laboratory scale.

# 4. Production

## 4.1. Acetylene/Formaldehyde

A process developed by Reppe in the 1930s was for many years the preferred synthetic route to 1,4-butanediol and THF, and it is still the most common approach in Europe and the United States [10], [11]. The Reppe process involves a reaction between acetylene and formaldehyde to give 2-butyne-1,4-diol, with subsequent hydrogenation to 1,4-butanediol ($\rightarrow$ Butanediols, Butenediol, and Butynediol, **A 4**, pp. 458–459). The saturated diol is very readily cyclized to THF with elimination of water by acid catalysis above 100 °C. Suitable catalysts include inorganic acids, acidic aluminum silicates, and earth or rare-earth oxides.

$$HC \equiv CH + 2 CH_2O \longrightarrow HOCH_2C \equiv CCH_2OH$$
$$\xrightarrow{H_2} HOCH_2CH_2CH_2CH_2OH \xrightarrow[\Delta]{H^+} \underset{O}{\bigcirc}$$
$$\Delta H = -13.4 \, kJ/mol$$

Quantitative conversion and a yield of almost 100 % can be achieved in an atmospheric-pressure, continuous process with aluminum oxide as the catalyst provided the THF is continuously distilled out of the reaction mixture as it forms and pure 1,4-butanediol is fed into the reactor at a rate equal to its rate of consumption. The amount of conversion per unit amount of catalyst is very high. Crude butanediol can be converted into THF very selectively and with little added energy by a medium-pressure process recently described by BASF [12]. The THF–steam mixture obtained from cyclization is first distilled in a rectification column to give the corresponding azeotrope (5.3 wt % water, *bp* 62.3 °C). Treatment with alkali hydroxide [13] and subsequent distillation provides anhydrous THF. The azeotrope problem can also be circumvented industrially by distillation under pressure.

## 4.2. Butadiene Acetoxylation

The Mitsubishi-Kasei Corporation in Japan produces 1,4-butanediol and THF in parallel starting from butadiene [14]. The procedure can be described by the following set of equations:

$$H_2C=CH-CH=CH_2 \xrightarrow[O_2]{AcOH} AcOCH_2CH=CHCH_2OAc$$
$$\xrightarrow{H_2} AcOCH_2CH_2CH_2CH_2OAc$$
$$\xrightarrow{H_2O} AcOCH_2CH_2CH_2CH_2OH + AcOH$$
$$\xrightarrow{H_2O} HOCH_2CH_2CH_2CH_2OH + AcOH \qquad \underset{O}{\bigcirc} + AcOH$$
$$\text{where } Ac = CH_3\overset{O}{\overset{\|}{C}}-$$

Butadiene is oxidized at 3 MPa/80 °C over a palladium–tellurium catalyst with acetic acid and a nitrogen–oxygen mixture to give 1,4-diacetoxy-2-butene. The olefin is hydrogenated to 1,4-diacetoxybutane [15], which can be hydrolyzed to butanediol or THF.

## 4.3. Propylene Oxide Process

Arco recently commenced industrial production of 1,4-butanediol by the following route:

$$H_2\overset{O}{\overset{/ \ \ }{C}}-CH-CH_3 \xrightarrow{Li_3PO_4} H_2C=CH-CH_2OH \xrightarrow[cat.]{H_2/CO}$$
$$HOCH_2CH_2CH_2CHO + HOCH_2\overset{CH_3}{\overset{|}{CH}}-CHO \xrightarrow{H_2}$$
$$HOCH_2CH_2CH_2CH_2OH + HOCH_3\overset{CH_3}{\overset{|}{CH}}-CH_2OH$$

Propylene oxide is isomerized to allyl alcohol by conventional means at 250–300 °C/1 MPa over a trilithium orthophosphate ($Li_3PO_4$) catalyst. The allyl alcohol is then hydroformylated to 4-hydroxybutyraldehyde and the byproduct 3-hydroxy-2-methylpropionaldehyde by a process developed by Kuraray [16]. Subsequent hydrogenation of the aldehydes gives 1,4-butanediol as the main product together with 2-methyl-1,3-propanediol.

## 4.4. Maleic Anhydride Hydrogenation

Because of its structure, maleic anhydride is an attractive precursor for the preparation of butanediol, THF, and $\gamma$-butyrolactone:

This route was originally developed and adapted to industrial scale by Mitsubishi-Kasei [17], but Mitsubishi itself has since abandoned the process and replaced it by the newer butadiene acetoxylation (Section 4.2).

Recently, Davy–McKee Ltd., BP, and Sohio have introduced a similar process for the production of butanediol and THF. Shinwha Petrochemical in South Korea has commissioned a plant based on the McKee process producing 20 000 t/a of butanediol. The first step is a vapor-phase hydrogenation of ethyl maleate [18], and the ratio of THF to butanediol in the product mix can be adjusted by altering the reaction conditions.

Increasing attention is being focused on maleic anhydride as a starting material for THF, and it appears that this approach will become more important than the Reppe process in the medium term [19].

## 4.5. n-Butane – Maleic Anhydride Process

Du Pont has developed a new two-step process for producing THF from n-butane. The process involves oxidation to maleic anhydride with a transported abrasion-resistant oxidation catalyst, which guarantees a high yield (70–75%), with subsequent hydrogenation of aqueous maleic acid solution over a special rhenium-doped palladium catalyst [20]. This method also leads to either butanediol or THF as desired; it has been perfected on a pilot scale and is due to be implemented in 1994 in the Spanish province

of Asturia with an annual production of 45 000 t. The process can be summarized as follows:

## 4.6. Pentosan/Furfural Processes

Because it takes advantage of a renewable raw material, THF production from furfural is currently attracting interest ($\rightarrow$ Biomass Chemicals, **A4**, pp. 102–103). Because of their widespread distribution in agricultural waste products, pentosans are especially promising as starting materials ($\rightarrow$ Furan and Derivatives, **A12**, pp. 122–125) [21]. The production of THF from furfural involves catalytic decarbonylation to furan and hydrogenation of the latter to THF [22].

## 5. Specifications and Analysis

Table 1 illustrates typical quality specifications quoted by a leading European producer of THF. Gas chromatographic (GC) analysis can be conducted with a 30-m WCOT (fused silica) column coated with DB-1 (FID detector, He as carrier gas) using n-octane as an internal standard [2].

## 6. Environmental Protection and Recovery

Based on the considerations presented in the section on toxicology (Section 10), it can be concluded that with proper handling THF constitutes no significant risk with respect to the environment.

**Table 1.** Quality specifications for commercial tetrahydrofuran

| Characteristics | Value | Test method |
| --- | --- | --- |
| THF content | min. 99.9% | GC |
| Water content | max. 0.03% | DIN 51 777, section 21 |
| Hazen color index | max. 10 | DIN/ISO 6271 |

THF is also quite easily recovered. Thus, THF vapors from exhaust gases can be adsorbed on activated charcoal and subsequently desorbed with steam. The efficiency of the process is said to be 98–99%. In the case of high concentrations of THF in air, absorption in water is also a possibility. The THF–water azeotrope (5.3 wt% water, *bp* 62.3 °C) obtained on distillation can be converted into anhydrous THF (< 0.05% water) by distillation at reduced or (preferably) elevated pressure [23].

Extraction of THF from an aqueous mixture with a solvent immiscible with water (e.g., pentane), followed by phase separation and fractional distillation of the organic phase, is another drying technique practiced on an industrial scale. This results in THF containing < 0.1% water [24]. For the separation of THF from water and such organic solvents as toluene by fractional distillation, see [24]. Various methods have been developed for drying on a smaller scale [25], [26]. Regardless which method is selected, great care is required because of the danger of peroxide formation [27].

## 7. Storage and Transport

The handling of THF entails only the usual precautionary measures applicable to other volatile organic compounds, and THF can be stored in the usual steel containers. For transportation through piping, welded pipes are preferable; special attention should be directed toward the seals in pumps and valves.

Like other ethers (e.g., diethyl or diisopropyl ether) THF forms peroxides in air if it has not been stabilized. At a peroxide concentration > 0.1% THF is subject to explosive decomposition on distillation. It is stabilized with 2,6-di-*tert*-butyl-*p*-cresol (BHT). In the absence of a stabilizer THF must be stored in an inert gas atmosphere.

A safety data sheet in accordance with DIN 52900 is available for THF.

For THF the following transport regulations are in force:

| | |
|---|---|
| GGVE/RID | Class 3, no. 3b RN 301 |
| GGVS/ADR | Class 3, no. 3b RN 2301 |
| ADNR | Class 3, no. 3b RN 6301 |

| | |
|---|---|
| IMDG Code | Class 3.1 |
| UN no. | 2056 |
| S phrases | 16–29–33 |
| R phrases | 11–19–36/37 |

## 8. Uses

The most important area of application for THF from the standpoint of quantity and turnover is polymerization with simultaneous ring opening to give poly(tetramethylene oxide) (→ Polyoxyalkylenes, **A21**, pp. 579–589). Available in various molecular masses, this is a key component in the production of important elastic construction materials, thermoplastics and molded elastomers based on polyurethanes, polyesters and polyamides, elastic Spandex fibers, and polyurethane coatings. Trade names for commercial PTMO include Terathane (DuPont), Polytetrahydrofuran (BASF), and Polymeg (Quaker Oats).

THF is also a versatile solvent for natural and synthetic resins as well as PVC. It is used extensively in the varnish and film industries and as a cosolvent for printing inks and adhesives.

## 9. Economic Aspects

The projected annual growth rate for tetrahydrofuran is 4–7% based on anticipated increases in the consumption of polytetrahydrofuran. Major producers of THF are listed in Table 2, and regional production capacity and consumption estimates are provided in Table 3.

**Table 2.** Major THF producers in 1992

| Company | Region | Process | Capacity, $10^3$ t/a |
|---|---|---|---|
| Du Pont | USA | Reppe | 70 |
| BASF | USA, Europe, Japan | Reppe | 40 |
| ARCO | USA | PO/AA* | 8 |
| GAF | USA, Europe | Reppe | 18 |
| Quaker Oats | USA | furfural | 10 |
| Mitsubishi-Kasei | Japan | butadiene | 16 |
| Toso Corp. | Japan | Reppe | 3 |
| Shinwha | Korea | butane | 10 |
| | CIS/China | furfural/Reppe | 10 |

*PO = propylene oxide; AA = allyl alcohol.

**Table 3.** Worldwide production capacity and consumption data by geographical area ($10^3$ t/a)

|  | USA 1990/1995 | Western Europe 1990/1995 | Southeast Asia 1990/1995 | China, CIS 1990 |
|---|---|---|---|---|
| Capacity | 95/133 | 30/50 | 20/40 | 5–10 |
| Consumption | 63/80 | 50/70 | 20/30 | 5–10 |

## 10. Toxicology and Industrial Hygiene

THF shows weak narcotic activity. According to very intensive investigations conducted at the Industrial Hygiene and Pharmacology Institute at BASF it is one of the least toxic solvents. This is in sharp contrast to previous reports, which were based on impure samples. Its true toxicity is evidently not greater than that of acetone.

In the investigations at BASF, THF exposure levels of 1000 ppm for 6 h a day, 5 days a week, over a period of 1 year were tolerated by cats, rabbits, and rats with no associated symptoms of poisoning. Exposure to 3000 ppm daily for 8 h over a course of almost 2 years produced as few symptoms of poisoning in rats as did inhalation over the same length of time of 3000 ppm acetone. Concentrations of 1000 ppm or 3000 ppm are completely intolerable for humans because of severe mucous membrane irritation, and would not therefore be experienced in practice. Even higher concentrations of THF vapor produce a narcotic effect similar to that of ether or acetone. Like alcohol, THF is rapidly metabolized in the living organism.

As a consequence of its universal solvent power, THF dissolves the upper keratin-containing layers of skin and mucous membranes, penetrating very rapidly into deeper parts of the skin. Thus it gives rise to very severe skin and mucous membrane irritation, but it does not display a sensitizing effect. This irritant character is apparently reinforced by peroxides, since THF that contains peroxides due to prolonged storage is a more severe skin irritant. THF that is allowed to penetrate a wound or get under the fingernails causes very unpleasant pains.

Like methanol, THF is resorbed very rapidly by the skin. For this reason, prolonged wetting of the skin with THF must be strictly avoided, and hand protection should be provided against repeated wetting.

The maximum workplace concentration for THF has been fixed at 200 mL/m$^3$ (ppm). This limiting concentration is a function only of local mucous membrane irritation caused by THF vapors; much higher concentrations are required to produce a narcotic effect in humans [28].

## 11. References


[1] A. P. Kudschadker et al.: *Key Chemicals Data Books Furan, Dihydrofuran, Tetrahydrofuran,* Thermodynamics Res. Center, Texas A & M University, College Station, Texas 1978.

[2] BASF, technical report "Tetrahydrofuran," Ludwigshafen 1991.

[3] H. Meerwein et al., *Angew. Chem.* **72** (1960) 927.

[4] P. Dreyfuss: "Polytetrahydrofuran," in *Polymer Monographs*, vol. 1.8, Gordon and Breach Sci. Publ., New York 1982.

[5] W. Reppe: *Neue Entwicklungen auf dem Gebiet der Chemie des Acetylens und Kohlenmonoxyds,* Springer Verlag, Heidelberg 1949.

[6] "Furfural" 660.5020 in: *Chemical Economics Handbook*, Stanford Res. Inst., Menlo Park, Calif. 1968.

[7] A. Wong et al., *J. Am. Chem. Soc.* **102** (1980) no. 13, 4529.

[8] R. G. Gastinger et al., *J. Am. Chem. Soc.* **102** (1980) no. 15, 4959.

[9] R. Criegee et al., *Angew. Chem.* **62** (1950) 120.

[10] J. W. Copenhaver, M. H. Bigelow: *Acetylene and Carbonmonoxide Chemistry,* Reinhold Publ. Co., New York 1949; University Microfilms, Ann Arbor, Mich. 1949.

[11] BASF, DE 1 043 342, 1957 (J. Schneiders); DE 2 930 144, 1979 (H. Müller et al.).

[12] BASF, EP 1 536 680, 1979 (H. Müller, Ch. Palm).

[13] IG Farben, DE 713 565, 1939 (W. Reppe, H.-G. Trieschmann).

[14] Mitsubishi Chem., US 3 922 300, 1975 (T. Onoda et al.).
T. Onoda: "1,4-Butylenglycol and Tetrahydrofuran Production from Butadien," *AIChE Summer National Meeting*, Minneapolis, Minn., Aug. 16–17, 1987.
Kurary, US 3 872 163, 1975 (T. Shimizu, T. Yasui).

[15] Mitsubishi Chem., US 4 010 197, 1977 (J. Toriya, K. Shirago).

[16] Kurary, US 4 567 305, 1986 (T. Shimizu, T. Yasui).
Kurary & Daicel, EP-A 129 802, 1984 (M. Milsuo, M. Shini).

[17] J. Kanetaka et al., *Ind. Eng. Chem.* **62** (1970) 24.
Mitsubishi Chem., GB 1 230 276, 1971. GB 1 344 557, 1974.

[18] Davy McKee Ltd., EP 143 634, 1984 (M. Sherif, K. Turner).
Davy McKee Ltd., GB 2 207 914, 1987 (M. Sherif, K. Turner).

[19] BASF, DE 3 423 447, 1986 (R. Schnabel, H. Weitz); DE 3 726 509, 1989 (R. Fischer, W. Harder, K. Malsch).
Huels, 3 539 151, 1986 (R. Nehring, W. Otte).
UCB, EP 339 012, 1989 (J. L. Dalons, J. Markusel).
General Electric, US 4 652 685, 1987 (W. James, E. Norman).

General Aniline, US 4797382, 1989 (D. Waldo, D. Taylor).
Standard Oil, US 4301077, 1980 (F. A. Pesa, A. M. Graham); US 4810807, 1988 (J. Budge, E. Pederson); US 4827001, 1988 (Th. Attig, A. Graham).
Tonen Corp., EP 373946, 1989 (S. Sadakatsu, I. Hiroyuki); EP 431923, 1991 (S. Sadukatsu, I. Tatsumi).

[20] Du Pont, EP-A 189261, 1986 (R. M. Contractor); US 4442226, 1982 (T. A. Bither); US 4371702, 1982 (T. A. Bither); EP 147219, 1989 (A. Marby, W. Pichard).
[21] C. Godawa et al., *Inf. Chim.* **260** (1985) 171–177.
[22] N. N. Machalaba et al., *Gidroliz. Lesokhim. Prom.-st.* **1988**, no. 5, 19–20; *Chem. Abstr.* 109 151883k.
Quaker Oats, US 3233714, 1963.
Du Pont, US 2846449, 1956 (W. H. Banford, H. M. Manes); US 2374149, 1943 (G. M. Whitman). Quaker Oats, US 3021342, 1958 (H. W. Hemker); GB 855255, 1959.
[23] Du Pont, CA 546591, 1957.
[24] Du Pont, US 2790813, 1953 (P. F. Bente).
[25] Du Pont, US 3074968, 1960 (J. W. Meinherz).
[26] Du Pont, bulletin A-25077, recovery of THF.
[27] B. G. White et al., *Org. Synth.* **46** (1966) 105.
[28] BG Chemie, Merkblatt M038,588, Tetrahydrofuran, Jedermann Verlag Dr. Otto Pfeffer, Heidelberg 1988.
Kühn-Birett: *Merkblätter Gefährlicher Arbeitsstoffe*, 26. Erg.Lfg., 3/85-T08-1, Verlag Moderne Industrie, München 1977.
G. Hommel: *Handbuch der gefährlichen Güter*, 5th ed., Merkblatt 192, Springer Verlag, Berlin 1993.
*Ullmann*, 4th ed., **12**, 21.

**Tetralin → Naphthalene and Hydronaphthalenes**
**Tetramethyllead → Lead Compounds**

# Textile Auxiliaries

KLAUS FISCHER, Hoechst Aktiengesellschaft, Frankfurt, Federal Republic of Germany (Chaps. 1 and 8)

KURT MARQUARDT, Frankfurt, Federal Republic of Germany (Chap. 2)

KASPAR SCHLÜTER, Henkel KGaA, Düsseldorf, Federal Republic of Germany (Chap. 3)

KARLHEINZ GEBERT, BASF Aktiengesellschaft, Ludwigshafen, Federal Republic of Germany (Chap. 4)

ERICH KROMM, BASF Aktiengesellschaft, Ludwigshafen, Federal Republic of Germany (Chap. 5)

VOLKER GIESEN, REINHARD SCHNEIDER, BASF Aktiengesellschaft, Ludwigshafen, Federal Republic of Germany (Chap. 6)

ROSSER LEE WAYLAND, Jr., Hickson DanChem Corporation, Danville, Virginia, United States (Chap. 7)



Ullmann's Encyclopedia
of Industrial Chemistry, Vol. A 26

# 1. Introduction

This article goes beyond the account of classical textile finishing given in the 4th edition of Ullmann's Encyclopedia, and includes nonwoven textiles, carpets, and bonded composites, which are textiles in the wider sense. Technical textiles have become increasingly important in recent decades. A chapter (Chap. 8) has therefore been devoted to the auxiliaries for these materials.

## 1.1. Definitions

The TEGEWA nomenclature is used for the definition and classification of the various textile auxiliaries and finishing agents [1.1]. It is also the basis of the comprehensive former tpi catalogue of products [1.2], and is fairly well adhered to in other international product lists. This nomenclature differentiates between textile auxiliaries and textile chemicals such as acids, alkalis, salts, and solvents used in the textile industry, usually in conjunction with auxiliaries. The other chapters of this article also follow the TEGEWA nomenclature.

When textile auxiliaries and chemicals are used, they become attached to the textile. After they have carried out their auxiliary function, they may be removed at some stage of textile manufacture, or they may remain on the textile, so that the finished product then consists of the fiber, the textile structure, and the auxiliaries, which then function as finishing agents. Textile auxiliaries may be qualitatively classified as follows:

1) Nonpermanent: auxiliaries that are almost completely removed, e.g., by water, washing, or dry cleaning
2) Semipermanent: auxiliaries that are gradually removed from the textile
3) Permanent: auxiliaries that remain on the textile both in use and after treatments such as washing or dry cleaning

Generally, the only distinction made is between "permanent" and "nonpermanent."

Environmental regulations make a further distinction, between substances and preparations [1.3]. According to this, substances are chemical elements or compounds of natural or synthetic origin used as auxiliaries necessary to satisfy market requirements. Preparations consist of two or more substances, and may be mixtures or solutions. Most textile auxiliaries are preparations.

The auxiliary and/or finishing functions of textile auxiliaries vary widely, and depend on the intended use of the textile. Textiles can be classified into three groups based on the demands placed on the finished textile, independent of the type of textile (i.e., whether woven, knitted, nonwoven, carpet, or bonded).

1) Classical textiles for clothing, decoration, and home
2) Technical textiles, mainly for industrial use
3) Textiles used in hygiene, care, and medicine

Textile auxiliaries include not only products that are applied to fibers after they have been made into textiles, but also those that are incorporated during fiber manufacture, e.g., optical brighteners in the spinning mass, modifiers for high-strength viscose, delustering agents, and flame-proofing agents.

## 1.2. Historical Aspects

Even in antiquity, auxiliaries were undoubtedly used in the production of textiles from wool, cotton, silk, and linen to facilitate working with the fiber and to give the finished product the desired appearance. This included soap for washing and fulling, oils for improved smoothness in spinning and weaving, starches to stiffen the textile, and inorganic mordants for fixing colorants. With increasing industrialization, requirements for the processibility of the raw materials and the appearance and wearing properties of the textile product became stricter. Thus, soap-washed wool had to be treated with lubricants prior to spinning, warps had to be sized, and the material had to be treated with finishing agents to give the final commercial product.

The development of modern textile auxiliaries dates back to the turn of the century and is closely associated with the development of surfactants. This was initiated by MERCER and RUNGE, who ca. 1830 observed that oils and fats can be made water soluble by treatment with sulfuric acid. In 1875 KÖCHLIN, STORCK, and WUTH, and, independently, CRUM sulfated castor oil and obtained the important Turkey red oils. By increasing the degree of sulfation, STOCKHAUSEN (1896) obtained the monopole brilliant oils, which were used as auxiliaries in dyeing and finishing due to their stability in hard water and their good dispersant and protective-colloid properties. Surfactants discovered in the following years were often first used as textile auxiliaries before becoming established in other areas.

Some other inventions that illustrate the historical development of textile auxiliaries are given in the following:

1885: M. DE CHARDONNET recommends use of additives such as fats and glycerol for the spinning baths for nitrate-silk fibers.

1904: The Stockhausen company produces solvent-containing preparation from monopole soaps as fat-dissolving soaps.

1905: KNECHT and WITT observe that felting of wool can be prevented by treatment with aqueous chlorine.

1911. LILIENFELD applies for patent for poly fatty acids (condensed castor oil) as a spinning aid for viscose fibers.

1913: GÜNTHER (BASF) develops Nekal A (a diisopropylnaphthalenesulfonate), the first high-activity wetting agent.

1921: Pott & Co. introduce the Nekal-based fat-dissolving soap Neomerpin.

1924: HERRMANN and HAEHNEL produce poly(vinyl alcohol), the first synthetic size and finishing agent. Shortly afterwards, styrene–maleic anhydride and polyacrylate sizes are introduced.

1925: BERTSCH develops the spin brighteners Avirol KM extra and Avirol AM extra by means of an improved sulfonation of oils.

1927: Stockhausen introduces Praestabitol V, a highly sulfonated oil, and recommends it as a wetting agent for wool dying and vat dying.

1928: BERTSCH and SCHRAUT develop the primary alkyl sulfates, which under the name Gardinol (Böhme Fettchemie) become widely accepted as detergents in the textile industry.

1928: Tootal Broadhurst Lee applies for patent for the use of urea–formaldehyde resins as finishing agents for textiles made of cellulose fibers. This initiated the development of permanent finishes for cotton and regenerated cellulose fibers.

1929: DAIMLER and PLATZ (Hoechst) apply for patent for the first fatty acid condensation product (Igepon A), followed in 1930 by Igepon T, (fatty acid methyl tauride), a high-quality detergent for the textile industry.

1930: SCHÖLLER, WITTWER et al. (BASF) develop the ethylene oxide adducts. Wide range of uses, depending on hydrophobic group and poly(glycol ether) chain: emulsifiers (Emulphor), detergents (Leonil, Hostapal), levelling agents (Perigal, Uniperol), softeners and preparations (Soromin, Afilan).

1930: Polyphosphates introduced into textile industry as water softeners.

1933: REED (United States) synthesizes the alkyl sulfonates. Shortly afterwards, large-scale production of these products under the name Mersolates commences in Leuna and Wolfen. Important raw materials for detergents and textile auxiliaries.

1934: Shell introduces the olefin sulfonates (Teepol). Still important detergent raw materials and textile auxiliaries.

1935: IG Farbenindustrie patents production of nitrilotriacetic acid and ethylenediaminetetraacetic acid. Widespread use of these complex formers in the textile industry begins.

1938: BENER applies for patent for a process for the wash-fast permanent shaping of cotton goods with resins.

1947: Rohm and Haas (United States) introduce dimethylolethyleneurea as easy care finish for cotton.

1949: Cox (Du Pont) invents the modificator viscose spinning process in which amines are used as coagulation regulators.

1953: Sun Chemical Corp. introduces dimethyltriazinone as finishing agent.

In the 1950s and 1960s, numerous processes for the manufacture of easy care garments were developed; silicones were introduced as finishing agents and spinning auxiliaries, and fluorochemicals as oil and soil repellants.

The history of textile auxiliaries in more recent decades is linked to the growth of synthetic fiber production, as well as to the successful use of finishing techniques to optimize cellulosic chemical fibers and the most important natural fibers, cotton and wool, to meet modern consumer demands.

Discussion of environment aspects, led by Germany to a great extent, has included comparison of the properties, advantages, and disadvantages of natural and chemical fibers, including fibers such as flax and linen [1.4] and has led to new product development. Superfine fibers (microfibers [1.5]) for clothing and technical textiles required special treatment methods. The first imitation leather made of microfibers was marketed in Japan under the trade name Alcantara in 1970. Microfibers for sports clothing must compete with membranes [1.5].

World production of synthetic fibers increased from 70 000 t in 1950 to $17.2 \times 10^6$ t in 1992 [1.6], the manufacture of the most important fibers, i.e., polyester (PES), polyamide (PA), and polyacrylic (PAN) fibers having been made possible by developments that took place within the short time period 1938–1943. In 1968, the proportion of synthetic fibers used in the Federal Republic of Germany reached 50 % of the total [1.7].

In the 1960s and 1970s, interest was directed towards easy-to-keep properties, e.g., minimum ironing, resistance to soiling and graying, ease of washing and cleaning, wearing comfort, the deliberate modification of the textile character (hand), and the fashion aspects of the finished textiles. Nonwoven materials contributed to the stabilizing function of textiles, e.g., interlining materials, and high volumes of cheap, hygienic nonwoven filling materials were used in the bedroom in place of blankets and feathers. The low density of textiles was increasingly utilized in automobile construction, and this trend will continue in the future. Shaped components made of recycled fiber from textile waste bonded with hardening resins consisting of phenol–formaldehyde (novolac) or melamine–formaldehyde precondensates, or butadiene–styrene–acrylonitrile copolymers in dispersion form provided good handling properties, elegance, practical use, comfort, safety, and ecological advantages in automobiles [1.8].

Textiles with technical applications have become much more important in recent times (Chap. 8) [1.9]. Their importance in ancient times is shown, e.g., by the use of glass fibers by the ancient Egyptians to decorate vases [1.10]. In 1886, C. AUER VON WELSBACH used cotton fibers saturated with lanthanum salts and then ignited to increase the brightness of gas flames [1.11]. The first light bulb of T. A. EDISON contained a carbon fiber filament.

In 1921–1922, animal glue was marketed by the company Röhm, Darmstadt, Germany, for use as a binder for waste-fiber wadding [1.12], which was employed as an upholstery material for mattresses and automobile seats. In 1928, bonded fiber was used by the company Weisweiler and Kalff, Euskirchen, Germany, to make stiffening material for shoes.

The first styrene–butadiene copolymer was synthesized by Bayer in 1929. Natural rubber was used to bond nonwoven material in 1930. In 1936, the company Chicopee, United States, obtained a patent for the chemical bonding of wet and dry formed nonwoven material, and soon afterwards the company Freudenberg, Germany, used natural rubber to bind carded nonwovens [1.13], [1.14].

The trade name Appretan (from *Appretur*, the German for textile finishing) was registered on April 20th, 1934 [1.15]. It denoted a product group poly(vinyl acetate) used for many years to complement and replace natural materials (e.g., unmodified starch, shellac, and animal glue) to give a stiffening finish.

In 1947, the company Freudenberg introduced nonwoven interlining to the market. Malimo technology (sewing technique) was discovered in 1949. The large-scale production of

nonwovens started in 1952. The development of a wide range of special polymer dispersions, mainly acrylate based and self-cross-linking, began in the 1950s connected to the name H. WILHELM, BASF [1.16]. The history of nonwovens in the former German Democratic Republic is described in [1.17].

The development work that led to spunbonded web technology was started in the late 1950s by Du Pont, United States, and Freudenberg, Germany [1.18]. The first spunbonded was marketed by Du Pont in 1965 under the trade name Reemay. Today, geotextiles or impregnated spun PES in roof coverings are typical technical textiles of the spunbonded type.

A patent was applied for in 1943 in the United States for the first tufted-carpet machine, and production started in the early 1950s. Production of tufted carpets began in Germany in 1956. The coating materials, mainly synthetic latex (butadiene–styrene copolymers), made it possible to manufacture these durable, hardwearing carpets, whose textile character and cheapness compared with woven carpets have now become fully accepted by the consumers.

The dates of more recent events in the history of technical textiles are given below:

1954    Lamination (foam rubber–textile)
1955/56 The first European iron-on (transfer) materials made from polymer powders
1960    Rapid growth of flocking technology in Europe (initiated in the 1930s)
1964    Durogan process: latex foam-backed carpets
1966    Discovery of paper coating (wet deposition) of clothing in the United States; production of Corfam synthetic leather (Du Pont)

New methods were necessary for the finishing of double-ribbed knitted fabrics [1.18a].

Research and development also affect testing methods, which must meet the latest requirements. For example, as the result of increased efforts in recycling and rot-proofing biological degradation, a long used test had to be modified (DIN 53933/1, new draft of 1990).

## 1.3. Economic Aspects

In 1992, 112000 t textile auxiliaries were produced in Germany [1.19]. These were textile auxiliaries in the classical sense as defined by TEGEWA and 11 300 t textile dyes and 100 000 t textile chemicals. The total sales of textile auxiliaries in Germany are shown in Table 1.1 [1.20].

According to figures from the European Disposables and Nonwovens Association EDANA [1.21], ca. 40 000 t binders (calculated as 100 % solids) were used in Western Europe in nonwovens. Figures from EDANA are given in Table 1.2 [1.22]. Because not all binder consumers are members of EDANA (e.g., some wadding producers), the quantities quoted may be too low.

**Table 1.1.** Sales of textile auxiliaries in Germany, $10^6$ DM

|          | 1989 | 1990 | 1991 | 1992 |
|----------|------|------|------|------|
| Domestic | 528  | 542  | 507  | 481  |
| Export   | 1218 | 1184 | 1118 | 1107 |

**Table 1.2.** Binders for nonwoven fabrics in Western Europe (tonnes dry weight; EDANA statistics [1.21], [1.22]

| Dispersions | 1980 | 1985 | 1990 | 1993 |
|-------------|------|------|------|------|
| Acrylates, vinyl acetate–acrylate | 16 800 | 18 400 | 15 600 | 20 000 |
| Vinyl acetates including vinyl acetate–ethylene | 2 500 | 5 200 | 5 200 | 8 600 |
| Total of above | 19 300 | 23 600 | 20 800 | 28 600 |
| SBR + NBR* | 8 000 | 9 300 | 9 900 | 9 400 |
| Halogen-containing (flame retardant) | 2 300 | 600 | 700 | 200 |
| Others | 2 100 | 1 600 | 7 600 | 3 000 |
| Grand total | 31 700 | 35 100 | 39 000 | 41 200 |

*SBR = Styrene–butadiene rubber; NBR = Acrylonitrile–butadiene rubber.

**Table 1.3.** World fiber production in 1992

| Fiber | Production | |
|-------|-----------|---|
|       | $10^3$ t | % |
| Raw cotton | 18 115 | 46 |
| Raw wool | 1 676 | 4 |
| Man-made fibers: | | |
|   Synthetic fibers | 17 212 | 43 |
|   Cellulosic chemical fibers | 2 620 | 7 |
| Total | 39 623 | |
| Glass fibers (estimated) | 1 650 | |

Also, not all the needle-felt materials used as floor coverings are included.

According to [1.23] ca. 180 000 t of vinyl and acrylic dispersions, with a sales value of ca. $ 184 × 10^6$, were produced in Western Europe in 1989. The same source quotes consumptions in the United States in 1990 for nonwoven materials as follows: 26 000 t pure acrylate dispersions, 17 000 t vinyl acetate–ethylene copolymer dispersions, 14 500 t butadiene–styrene copolymer dispersions, and 7700 t vinyl acetate–acrylic ester copolymer dispersions.

A rough estimate of latex consumption in Western Europe, which correlates mainly with carpet production, is 250 000 t, calculated as 100 % dry material [1.24]. In the carpet sector, by far the most important materials used are butadiene copolymer dispersions (SBR latex, carboxylated latex).

Consumption of textile auxiliaries, e.g., fiber preparation agents, can be estimated from the figures for fiber production (Table 1.3) [1.6]. Production of nonwoven material in Western Europe shows the following growth pattern [1.25]: 185 600 t in 1980, 450 700 t in 1990 [1.13], 480 000 t in 1991, and 510 000 t in 1992. In 1990, production in the United States was 650 000 t, and in Japan 200 000 t [1.26].

Classifying fiber production by use (Table 1.4) [1.6] reveals that a high proportion of fibers are used in technical textiles: 22 % in the United States, and 18 % in Western Europe. In Japan, the proportion of fiber used for technical textiles is even higher. Technical textiles have become an important growth market [1.27], the highest proportion of the total (23 %) being used in motor vehicles [1.28]. There is increasing use of textiles for roofing and damp courses. Of the total of $228 × 10^6$ m$^2$ in Germany 1992, glass-fiber nonwovens account for ca. $40 × 10^6$ m$^2$. Polyester spunbonds with polymer–bitumen coatings are used in smaller, but growing, amounts. Other

**Table 1.4.** Applications of textile fibers in 1991 ($10^3$ t)

| Use | Chemical fibers | | Cotton | | Wool | | Total | |
|-----|------|------|------|------|------|------|------|------|
|     | USA  | EC   | USA  | EC   | USA  | EC   | USA  | EC   |
| Clothing | 1007 | 1169 | 1470 | 666 | 65 | 300 | 2542 | 2135 |
| Carpets | 1334 | 596 | 9 | 9 | 12 | 73 | 1355 | 677 |
| Other home textiles | 402 | 518 | 582 | 384 | 5 | 34 | 988 | 935 |
| Tires | 128 | 86 | 2 | 2 | | | 130 | 88 |
| Other technical textiles | 1110 | 588 | 150 | 127 | 6 | 6 | 1265 | 721 |
| Total | 3981 | 2956 | 2213 | 1188 | 88 | 413 | 6280 | 4556 |

damp course materials include felt, cardboard, plastics, and metal [1.29].

Table 1.5 shows the growth and recession that have taken place in textile finishing and printing in Germany [1.30]. The worldwide growth in printed textiles (ca. $18\,000 \times 10^6$ m$^2$ in 1980, and ca. $22\,500 \times 10^6$ m$^2$ in 1990) occurred mainly outside Europe [1.31].

## 1.4. Environmental Aspects

As late as the early 1970s, impressive developments in graft polymerization and finishing technologies using solvents were being made [1.7]. From today's viewpoint, such techniques are regarded as ecologically problematic. Potential problems include impurities in wastewater and exhaust gases, and toxic residues on the textile. Since the 1970s, producers of textile auxiliaries, the textile industry, the textile finishing industry, and producers of fibers, nonwovens, carpets, and technical textiles have been confronted by a continually growing number of regulations and restrictions relating to the environmental situation. These concern wastewater, emissions into the atmosphere, the chemical substances themselves [1.3], [1.32], waste [1.33], packaging [1.34], soil and waste disposal [1.35], energy recovery from residual materials [1.36], and the recovery and reuse of used textiles [1.37], unused auxiliaries, residual treatment liquor, etc.

For the year 1992, the Association of German Textile Finishers stated that the cost of environmental protection as a percentage of total finishing costs was 7.2 %. In the Netherlands, the figure was 4.1 %, in Switzerland 2.6 %, in the United Kingdom 3.25 %, and in France 1.5 % [1.30].

Attacks on the textile industry have been mounted by the media in Germany. Various representatives of textile companies have responded [1.38]. Many other publications have attempted to elucidate the environmental situation, which has become very complex [1.39]. Activities of this kind in the United States are described in [1.40], and the environmental situation with respect to polymers in [1.41].

**Wastewater.** The textile finishing industry is one of the largest consumers of water. The auxiliary agents and process steps of the textile finishing industry have been brought into question more and more by technically oriented discussions on wastewater discharge laws in the Federal Republic of Germany [1.42], and the introduction of a wastewater levy in 1981 [1.43]. Stricter threshold values, drafted in Appendix 38 (1993) to the water resources policy act indicate the further restrictions the textile finishing industry will have to deal with (Table 1.6).
Some general requirements are as follows:

Reprocessing and reuse of wash water from the printing plant.

Segregation of dyeing and washing liquors with high dye contents, and of residues from finishing materials.

Precleaning of vessels and hanks without generation of wastewater.

Residues of textile chemicals, dyes, and auxiliaries to be collected separately and reprocessed to recover or dispose of the material.

Recovery of synthetic sizing materials after desizing, followed by reuse, separate disposal, or treatment. Use of synthetic sizing agents that give a COD elimination of 80 % after 7 d.

Avoidance of the use of organic complexing agents (such as EDTA) that are difficult to eliminate biologically.

Surfactants must comply with the law on washing and cleaning agents (and associated regulations) concerning biological degradability.

Separate wastewater streams for desizing, bleaching, printing, dyeing, finishing, coating, laminating, cleaning of vessels and hanks, and antifelting finishing of wool.

Equivalent Austrian regulations are given in [1.44].

Sizing and desizing have attracted the most attention from the outset [1.42], [1.45]–[1.47], as not only sizing agents (starches) that are easily biologically degraded were appearing in wastewater in large quantities, but also microorganism-adapting synthetic sizing agents [poly(vinyl alcohol) [1.48]] and nonadapting sizing agents (polyacrylates, cellulose ethers).

**Table 1.5.** Textile finishing in Germany

|  | 1985 | 1989 | 1990 | 1992 |
|---|---|---|---|---|
| Textile finishing, $10^6$ m$^2$ | 1721 | 1802 | 1799 | 1514 |
| Outher wear, % |  | 35 |  | 33 |
| Lining fabrics, % |  | 16 |  | 17 |
| Underclothes, % |  | 7 |  | 3 |
| Furnishing fabrics, % |  | 6 |  | 15 |
| Home textiles, % |  | 13 |  | 13 |
| Technical fabrics, % |  | 20 |  | 15 |
| Other textiles, % |  | 3 |  | 5 |
| Textile printing, $10^6$ m$^2$ | 477 | 527 | 573 | 502 |
| Clothing (woven), % |  | 44 |  | 47 |
| Clothing (knitted), % |  | 15 |  | 13 |
| Furnishing fabrics, % |  | 12 |  | 14 |
| Home textiles, % |  | 21 |  | 19 |
| Technical textiles, % |  | 4 |  | 4 |
| Other textiles, % |  | 4 |  | 3 |

**Table 1.6.** Wastewater from textile manufacture and finishing: specifications from the draft Appendix to 38th Wastewater Administrative Regulation (selection as of 1993)

| Parameters | Specifications new (old) | Relevance to textile auxiliaries |
|---|---|---|
| Chemical oxygen demand (COD) | 160 mg/L (280 mg/L) | affected by all organic products, COD dependent on C content and product concentration, COD values of 300–2200 mg $O_2$/g product |
| Biological oxygen demand ($BOD_5$) | 25 mg/L (40 mg/L) | wide range, e.g., 5–700 mg/L<br>700 maize starch or alkane sulfonate (60%)<br>150 melamine cross-linker (75%)<br>85 paraffinic hydrophobic agent, cationic (30%)<br>10 silicone hydrophobic agent (30%) |
| Phosphorus, total | 2 mg/L (–) | P-containing flame retardants, phosphates, preparation agents containing phosphoric esters, P-containing sequestrants |
| Ammonium nitrogen | 10 mg/L (5 mg/L) | pH regulator (ammonia), thickening agents, flame retardants, ammonium phosphates, melamine cross-linking agents, amine hydrochloride catalyst, ammonium stearate |
| Total nitrogen in ammonium, nitrite, and nitrate | 20 mg/L | |
| Iron | 3 mg/L (–) | |
| Aluminum | 3 mg/L (–) | flame-retardant filler aluminum hydroxide, precipitating agent aluminum sulfate |
| Sulfite | 1 mg/L (1 mg/L) | reducing agent |
| Total hydrocarbons | 10 mg/L (10 mg/L) | certain thickening agents, defoamers in small quantities |
| Chromium(VI) compounds from use as oxidizing agents for sulfur dyes | none | |
| Chromium(VI) from other sources | 0.1 mg/L | |
| Chlorine-organic carriers | none | |
| Mercury and mercury compounds | none | preservatives |
| Alkylphenol ethoxylates (APEO) from washing and cleaning agents | none | surfactants for dispersing, wetting, and washing |
| High-boiling mineral oil hydrocarbons (initial $bp > 200\,°C$) | none | lubricants |
| Absorbable organic halogen compounds (AOX) | 0.5 mg/L (–) | chlorine- and bromine-containing flame retardants, chlorine-containing bleaching agents |
| Free chlorine | 0.3 mg/L (0.3 mg/L) | chlorine bleaching agents |
| Total aromatics (benzene and homologues) | 0.1 mg/L | |
| Hydrocarbons | 15.0 mg/L | defoamers |
| Sulfide | 1.0 mg/L (0.1 mg/L) | sulfur dyeing |
| Total chromium | 0.5 mg/L (2 mg/L) | |
| Copper | 0.5 mg/L (1 mg/L) | in pigments |
| Nickel | 0.5 mg/L (–) | |
| Zinc | 2.0 mg/L (3.0 mg/L) | Zn catalysts |
| Biologically difficult to degrade complexing agents such as EDTA | to be avoided | sequestering agents, stabilizing agents for coating pastes |
| Toxicity to fish as dilution factor $G_F$ | 2.0 (4.0) | depends on product concentration |

The rate of COD elimination in the presence of activated sludge in a biological clarification plant can be determined in the laboratory by the Zahn–Wellens test [1.49]. If air is introduced into the system with agitation, activated sludge degrades poly(vinyl alcohol) and other sizing agents, and textile auxiliaries in general [1.50]. The Zahn–Wellens test has been introduced as a test method [1.51], [1.52]. The elimination rates determined by this test are applicable to wastewater treatment plants using the mechanical/biological system [1.53]. The COD and BOD were at first the primary consideration to minimize pollution. Later it became necessary to minimize organic halogen compounds or to avoid them completely. These materials are mainly organochlorine carriers and hypochlorite, a bleaching agent, both of which produce high AOX values [1.54]. The AOX contamination caused by the bleaching agent sodium chlorite is considerably less than that caused by hypochlorites [1.55]. A future AOX limit of 3 mg/L for partial streams has been drafted in Germany.

Special collection of residues and excess materials from impregnation baths, from the nip rollers, spraying, coating, and sizing, and the reuse or separate waste disposal of these materials have led to considerable reduction in the pollution of natural water resources.

The decoloration of wastewater (e.g., by ozone or by adsorption) is being further developed [1.56].

**Classification of Water Contaminants.** Textile auxiliaries are divided into various classes according to the hazard to water that they represent [1.57]:

Class 0: Generally harmless to water (e.g., citric acid, calcium carbonate)
Class 1: Slightly harmful to water (e.g., acetic acid, crosslinking agents, polymer dispersions)
Class 2: Harmful to water (e.g., ammonia, formaldehyde, diesel fuel, surfactants)
Class 3: Very harmful to water (e.g., perchloroethylene)

**Emissions to the Atmosphere.** Emissions due to the use of textile auxiliaries in the textile finishing industry, including the use of the finished textile, recycling, and incineration include the following:

| | |
|---|---|
| Application of the product | Odor from the product. MAK values apply, if available, and also TA-Luft [1.58]. |
| Drying, curing, fixing | Volatile and steam-volatile substances are produced. The TA-Luft and 4. BImSchG/apply. |
| Textile is finished | Odor from the textile. Determination of chemical constituents on the textile (e.g., formaldehyde in the chamber test). Tests for ecological labels and special trade names. |
| Use in interior spaces | For example, automobiles [1.59]: odor, fogging [1.60]. |
| Recycling | Odor, soiling. TA-Luft, sometimes MAK values. |
| Incineration with utilization of energy | Volatile materials, combustion products. TA-Luft, etc. |

**Foodstuffs and Commodities.** Certain textiles come into contact with foodstuffs. The laws and regulations that apply to foodstuffs then apply to the textile auxiliaries [1.61]. The commodity regulations in Germany (Bedarfsgegenstände-Verordnung) [1.62] in which a number of EC guidelines have been incorporated, deals with plastics that come into contact with foodstuffs. The permitted monomers and other starting materials are covered by Appendix 3/Part A, while Part B covers provisionally permitted materials.

**Skin Compatibility.** The information given in safety data sheets [1.63] on skin and mucous membrane compatibility (skin and mucous membrane irritation determined on rabbits) are valid for commercial auxiliaries. However, in the case of textiles, the effects of drying and fixing always weaken these effects considerably. The material on the textile is present at low concentration or as a thin film in a solid, hardened state, with inert properties compared with liquid commercial products in ready-to-use form. In spite of the very large number of different textile finishes produced, very little skin damage can be attributed to the direct effect of chemicals on textiles. However, mechanical effects (rubbing) also play a role.

**Seals of Quality and Ecological Labels.** In 1991, the "Gemeinschaft umweltfreundlicher Teppichboden" (GuT) (association of environmentally-friendly carpets) [1.64] was founded, followed in 1992 by the "Verein für verbraucher- und umweltfreundlicher Textilien" (association of consumer-friendly and environmentally-friendly textiles) [1.65]. The seal of quality for environmentally-friendly carpets is granted by Western European carpet manufacturers according to established testing criteria.

The following harmful substances must be absent from floor-coverings bearing the "envi-

**Table 1.7. Qualifying criteria Öko-Tex-Standard 100 (selection as of 1995)**

| | Öko-Tex-Standard | | | | | |
| | 101 | 104 | 107 | 110 | 112 | 114 |
| Criterion | Textiles for clothes except clothes for babies | Textiles for babies | Textile floor coverings | Upholstery fabrics | Bed linen | Home textiles |
|---|---|---|---|---|---|---|
| Organochlorine carriers | n.d.* | n.d. | n.d. | n.d. | n.d. | n.d. |
| No flame-retardent finish | | | | | | |
| No biocidal finish | | | | | | |
| **Colorfastnesses** | | | | | | |
| Water | 3 | | 3 | 3 | 3 | 3 |
| Perspiration (acidic) | 3–4 | | | 3–4 | 3–4 | 3–4 |
| Perspiration (alcaline) | 3–4 | | | 3–4 | 3–4 | 3–4 |
| Abrasion wet | 2–3 | 2–3 | 2–3 | 2–3 | 2–3 | 2–3 |
| Abrasion dry | 4 | 4 | 4 | 4 | 4 | 4 |
| Saliva and perspiration | | yes | | | yes, for babies | yes, for babies |
| **Emissions, mg/m$^3$** | | | | | | |
| Toluene | | | 0.1 | | | |
| Styrene | | | 0.005 | | | |
| 4-Vinyl cyclohexene | | | 0.002 | | | |
| 4-Phenyl cyclohexene | | | 0.03 | | | |
| Butadiene | | | 0.002 | | | |
| Vinyl chloride | | | 0.002 | | | |
| Aromatic hydrocarbons | | | 0.3 | | | |
| Volatile organic compounds | | | 0.5 | | | |
| **Odor** | | | | | | |
| SNV 195651 (Switzerland) | | | 3 | | | |
| Others | | no odor with mildew, heavy benzine, fish, aromatics or odor improver | | | | |
| **pH** | | | | | | |
| Wool, silk | 4.0–7.5 | 4.0–7.5 | 4.0–7.5 | 4.0–7.5 | 4.0–7.5 | 4.0–7.5 |
| Others | 4.8–7.5 | 4.8–7.5 | 4.8–7.5 | 4.8–7.5 | 4.8–7.5 | 4.8–7.5 |
| **Formaldehyde, mg/kg** | | 20 | 300 | 300 | 75 | |
| Skin-tight | 75 | | | | | 75 |
| Skin-distant | | | 300 | | | 300 |
| For babies | | | | | 20 | 20 |
| **Eluable heavy metals, mg/kg**** | | | | | | |
| As | 1.0 | 0.2 | 1.0 | 1.0 | 1.0 (0.2) | 1.0 (0.2) |
| Pb | 1.0 | 0.2 | 1.0 | 1.0 | 1.0 (0.2) | 1.0 (0.2) |
| Cd | 0.1 | 0.1 | 0.1 | 0.1 | 0.1 | 0.1 |
| Cr | 2.0 | 1.0 | 2.0 | 2.0 | 2.0 (1.0) | 2.0 (1.0) |
| Cr (VI) | n.d. | n.d. | n.d. | n.d. | n.d. | n.d. |
| Co | 4.0 | 1.0 | 4.0 | 4.0 | 4.0 (1.0) | 4.0 (1.0) |
| Cu | 50.0 | 25.0 | 50.0 | 50.0 | 50.0 (25.0) | 50.0 (25.0) |
| Ni | 4.0 | 1.0 | 4.0 | 4.0 | 4.0 (1.0) | 4.0 (1.0) |
| Hg | 0.02 | 0.02 | 0.02 | 0.02 | 0.02 | 0.02 |
| **Pesticides and pentachlorophenols, mg/kg**** | | | | | | |
| DDT, DDD, DDE | 1.0 | 0.5 | 1.0 | 1.0 | 1.0 (0.5) | 1.0 (0.5) |
| HCH's | 1.0 | 0.5 | 1.0 | 1.0 | 1.0 (0.5) | 1.0 (0.5) |
| Lindan | 0.5 | 0.25 | 0.5 | 0.5 | 0.5 (0.25) | 0.5 (0.25) |
| Aldrin | 0.2 | 0.1 | 0.2 | 0.2 | 0.2 (0.1) | 0.2 (0.1) |
| Dieldrin | 0.2 | 0.1 | 0.2 | 0.2 | 0.2 (0.1) | 0.2 (0.1) |
| Toxaphen | 0.5 | 0.5 | 0.5 | 0.5 | 0.5 | 0.5 |

**Table 1.7.** (cont).

| | Öko-Tex-Standard | | | | | |
| | 101 | 104 | 107 | 110 | 112 | 114 |
| Criterion | Textiles for clothes except clothes for babies | Textiles for babies | Textile floor coverings | Upholstery fabrics | Bed linen | Home textiles |
|---|---|---|---|---|---|---|
| Heptachlorine, hepta-chlorine epoxide | 0.5 | 0.25 | 0.5 | 0.5 | 0.5 (0.25) | 0.5 (0.25) |
| 2,4-D | 0.1 | 0.1 | 0.1 | 0.1 | 0.1 | 0.1 |
| 2,4,5,-T | 0.05 | 0.05 | 0.05 | 0.05 | 0.05 | 0.05 |
| Pentachlorphenol | 0.5 | 0.05 | 0.5 | 0.5 | 0.5 (0.05) | 0.5 (0.05) |
| Total incl. PCP | 1.0 | 0.5 | 1.0 | 1.0 | 1.0 (0.5) | 1.0 (0.5) |
| **Dyestuffs** | | | | | | |
| Able to release amines of MAK classes IIIA1 and III A2 or dyestuffs may be carcinogenic or allergic | n.d. | n.d. | n.d. | n.d. | n.d. | n.d. |

* n.d. = not detectable.   ** Values in parentheses: textiles for babies.

ronmentally-friendly" label (GuT regulation of 1994):

Pentachlorophenol (PCP)
Formaldehyde
Butadiene
Vinyl chloride (VC)
Volatile chlorofluorocarbons (CFCs)
Asbestos
Dyeing accelerators (carriers)
Azo-dyes whose amino-components are in the Class III A 1 list of MAK values
Vinyl acetate
Pesticides on natural fibers

The seal of quality for textiles tested for harmful substances (MST) conforms with the criteria of German textile companies. The Austrian Textile Research Institute in Vienna and the Hohenstein Research Institute, Germany, mark products "Tested for harmful substances according to Öko-Tex-Standard 100" [1.66]. German and Austrian standards were unified in 1994 (Table 1.7). Formaldehyde, which is the substance most often discussed in the textile field, is included in all these quality seals and ecological labels.

# 2. Auxiliaries for Fiber Production and Processing

## 2.1. Introduction

In the production and processing of fibers, products are used for which the term "auxiliaries" is hardly adequate, as they are in fact virtually indispensable for the production of man-made fibers or for spinning yarns. The products may be grouped according to their area of use, i.e., for

1) Synthetic fibers or natural fibers
2) Primary spinning or secondary spinning

The general term "spinning auxiliaries" includes:

1) Spinning preparations (spin finishes) for fiber production, also known as primary spinning
2) Spinning oils, lubricants, and batching oils for secondary spinning
3) Antistatic and lubricating agents for aftertreatment of fibers in dyeing and finishing, as well as winding oils and lubricating (waxing) agents for yarn and threads

Products used as additives to spinning solutions and spinning melts include modifiers for the production of regenerated cellulose fibers and rayon, delusterants, optical brighteners, coloring pigments or master batches (only for spinning melts), antiaging agents, flame retardants, and incorporated antistatic agents.

In the trade there are numerous terms in use which vary with geographical areas and/or fields of application, very often without any logical coherence. Such terms include spinning preparation, avivage, spin finish, antistatic agent, antistat, fiber finish, spinning aid, winding oil, lubricating agent, lubricant, spinning oil, batching oil.

It has not yet been possible to achieve agreement on an international nomenclature that would enable the subgroups listed above under the general heading "spinning auxiliaries" to be clearly distinguished from one another.

The general term "avivage," which is often met with, should not be used in the field of fiber production as it implies reactivation, although in this case there is no restoration of an original condition [2.1]–[2.5].

## 2.2. Economic Aspects

World production of man-made fibers, which continues to increase, reached ca. $20 \times 10^6$ t in 1992. This required an amount of fiber preparation materials (based on active substance) of ca. 150 000 t, assuming that this was used economically. Losses associated with the production process are included in this estimate [2.6].

Spinning preparations contribute significantly to the quality of man-made fibers and yarns, and their importance is reflected in added value. Modern products have made possible innovations in production techniques and equipment, giving higher primary and secondary production rates. They have also been important in the development of finer types of fiber.

## 2.3. Use of Fiber Preparations

Spinning preparations ensure the best possible results during production and all subsequent processing stages for both staple fibers and filaments. Applied soon after the product leaves the spinneret and forms the fiber, they start to act immediately and, in the case of staple fibers, accompany these through all processing stages, from secondary spinning right up to the finished yarn. They function similarly in the production of filament yarns, whether flat, draw twisted, or textured.

Spinning preparations or spin finishes, lubricants, and batching oils were developed for use in secondary spinning, and ensure optimum running properties in all processing stages of worsted, semiworsted, carded (woollen), and bast fiber spinning mills, and in the production of needle punched nonwovens.

Antistatic and lubricating agents are always used when the spinning preparation has been removed by a previous wet process (e.g., dyeing). In the case of staple fibers, antistatic and lubri-

cating agents must restore these to a condition such that their behavior during processing is again suitable for the requirements of secondary spinning. They also support mechanical processes in the final finishing of the textile surface structure, e.g., raising a nap (teazling). In yarn dyeing, such auxiliaries are used both to facilitate rewinding and to improve processing of the yarn to form woven textiles.

In general, winding oils, lubricating and waxing agents are applied to yarns to improve especially warp and weft knitting including Raschel. With sewing threads, these auxiliaries prevent excessive heating of the needles, and help to increase the sewing speed and prevent thermally induced thread breakages and sewing defects.

Such oils are also used when winding filament yarns onto warp beams. In two-for-one twisting, these oils prevent the yarn defects that can be caused by increased abrasion in the twisting pot.

### 2.3.1. Use of Fiber Preparations in Primary Spinning

There are various methods of applying spinning preparations in primary spinning, depending on the production technology used. These can involve one or more steps. Equipment used includes licking rolls, preparation thread guide tubes, dipping baths, and sprinkling and spraying equipment. Combinations of these methods are also used. However, in all cases, the process is one of forced application [2.7]–[2.12].

**Viscose Rayon Production.** The production of viscose rayon can be continuous or discontinuous. In the continuous process, the filaments successively pass through the various aftertreatment operations, such as washing, carbon disulfide recovery, desulfurization, bleaching, preparation application, and drying. The preparation is applied in a dipping bath [2.11].

In discontinuous operation, the bobbins or spinning cakes are attached to carriers similar to those of a package dyeing apparatus, and run through the various aftertreatment baths in the individual machine sections. Each section carries out a single treatment [2.11].

**Melt-Spun Filament Yarns.** When melt spinning filament yarns from polymers such as polyesters, polyamide 6, polyamide 66, and polypropylene, the spinning preparation is generally applied at the bottom of the spinning duct, whereby two processes are used. In the classical method using licking rolls, the threads are passed tangentially onto a ceramic roll rotating with its lower part in the direction of movement of the threads in a trough filled with the solution or emulsion of the preparation. A regulator ensures that the trough filling is always kept at the same level. In this way, a thin film of the dilute preparation is transferred from the roll as the filament yarn passes over it.

The amount of preparation applied is controlled by varying the rotation speed of the roll or, within limits, the degree of dilution of the preparation [2.7]–[2.11].

To ensure that the film thickness of the preparation remains constant on the application roll, the solutions of the products used must behave as Newtonian fluids.

In the production of yarns spun at high speed (drawing rates >2000 m/min) metering by means of preparation thread guide tubes has advantages. Here, the solution of preparation is fed by a precision toothed-wheel metering pump to a ceramic preparation tube with a fine bore. The thread then quantiatively picks up the amount delivered by the pump. Since the preparation systems are closed, addition of bactericides, germicides, or fungicides is not necessary.

Compared with application by rolls, this process gives considerably improved uniformity of application of the preparation to all yarn packages.

**Staple Fiber Production.** The method of application of spinning preparations to staple fibers differs from that used for filament yarns in that the process takes place in at least two stages. In the case of polyesters, the most important man-made fibers, three stages are generally used. This type of fiber is used here as an example, and the methods used and the role of the preparations applied to it are described below. The most important features also apply to other polymers such as polyamide and polypropylene.

Three processes are used:

1) Licking roll method
2) Immersion bath method
3) Spraying method

The choice of method depends on the fiber production technology and mechanical equipment used. The individual process stages (see Figs. 2.1 and 2.2) include:

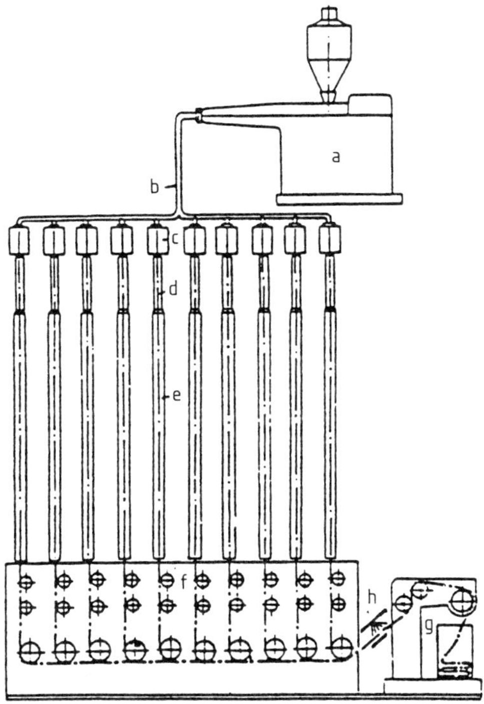


**Figure 2.1.** Schematic diagram of spinning equipment for tow production
a) Extruder; b) Melt feed; c) Spinnerets; d) Blow ducts; e) Spinning ducts; f) Spin finish application rolls; g) Tow deposited in cans; h) Spraying equipment

1) Spinning
2) Preparation application
3) Coiling of the tow (not yet drawn) into cans
4) Withdrawal of the tow from the cans
5) Gathering and dividing of the tows to form separated fiber sheets
6) Immersion bath at the inlet of the drawing line
7) Drawing
8) Preliminary thermal annealing
9) Final preparation
10) Crimping
11) Drying and final thermal setting

Depending on the subsequent processes, the tow of man-made fiber is cut and compressed into bales, or passes as tow to the conversion stage.

The objectives and requirements of the preparation process in the individual process stages of fiber production are as follows:

The first stage of preparation application takes place after cooling of the multifilaments at

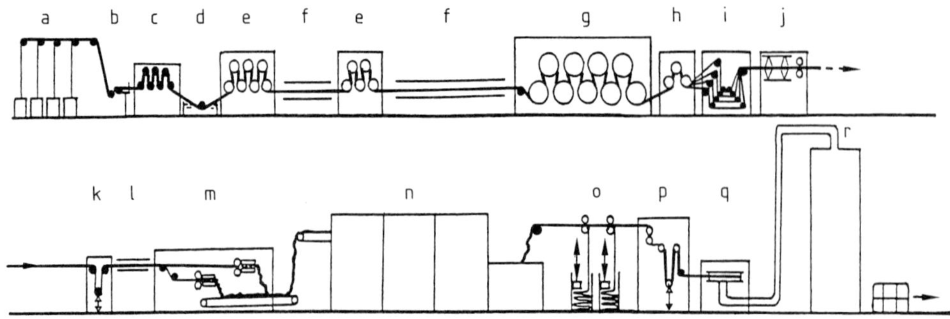

**Figure 2.2.** Schematic diagram of a PES tow production line
a) Can creel; b) Frame and reed; c) Infeed rollers; d) Immersion bath; e) Drawing septet; f) Heating zone for first and second drawing zones; g) Drawing and thermal setting godets; h) Cooling godets; i) Gathering tows; j) Spraying equipment; k) Dancing rolls; l) Steam chanel; m) Stuffer box; n) Dryer; o) Tow coiling; p) Dancing rolls; q) Cutter; r) Baling

the bottom of the spinning duct of each spinning station by tangential contact with licking rolls. A tow is formed by deflecting through 90°, and, before this is coiled into cans, it is again sprayed with preparation solution to increase the pickup of water.

Good wetting is necessary as the fiber must be well set at this stage to ensure that the coiling of the tow into the cans and also the withdrawal at a later stage will take place without problems.

To ensure a stable and uniform condition of each tow coming from the creel, several of them are gathered to form one or more larger sheets, and passed through an immersion bath. This compensates for irregularities caused by variation in the dwell times of the individual tows in the cans before drawing. As drawing is generally carried out wet, electrical charges play no part in this process. However, filament–filament friction is of major importance in the drawing operation. The individual filaments must slide against each other easily, as otherwise the number of capillary, breakages will increase considerably.

Other requirements include very low volatility and good thermal stability of the preparations. Before crimping, application of the final preparation is usually carried out, either in the dipping bath with subsequent squeezing off, or by spraying. As the crimping equipment has a strong squeezing action, the concentration of the bath or the spraying solution must be such that the desired final spin finish pickup is achieved [2.7]–[2.10].

**Viscose Staple Fiber.** With viscose staple fiber, the preparation is applied as the last oper-ation of a series of aftertreatments. Directly after the drawing stage, the tow is cut to the desired staple length, placed in this form on a perforated endless belt, and passed through the aftertreatment line, which is divided into several sections, separated by pairs of squeeze rolls.

Before entering the preparation section, the wet fiber fleece is squeezed off very strongly, sprinkled with preparation solution, and again squeezed off. The preparation solution must be uniformly distributed over the staple fiber. A gentle squeezing off at the exit of the preparation section ensures good uniformity of the applied preparation. The preparation liquor squeezed off is returned to a tank, restored to a usable concentration by adding a stock solution of preparation, and returned to the preparation section for reuse [2.11].

### 2.3.2. Use of Antistatic or Lubricating Agents in Secondary Spinning

In secondary spinning, antistats or lubricants are applied exclusively by spraying aqueous solutions. This is carried out by using spray nozzle systems or, as is still common, e.g., in worsted spinning, with the aid of rotating disks [2.2].

**Cotton and Rotor Spinning Mills.** The mechanical equipment and functioning of these types of spinning systems require the best possible primary preparation of man-made fibers.

In exceptional cases, or in countries with unsuitable climates, the yarn feeding behavior can be improved at the spinning pretreatment stage, i.e., before carding, by spraying on extremely small amounts of antistatic spinning agents.

Cotton contains its own optimal "preparation" in the form of solid cotton wax. Moreover, its histological structure gives good spinning properties.

**Worsted and Semiworsted Spinning.** Lubricants and antistatic agents are a permanent feature of worsted spinning. This is true for wool, man-made fibers, and blends thereof.

The natural wool grease present in the fiber is unsuitable as a lubricant owing to its adhesive properties, and must be removed as far as possible by scouring before processing.

The scouring of the wool not only removes natural grease, grease residues, and other soil, but also creates a suitable substrate to which a lubricant appropriate for the needs of worsted spinning can be applied. Application is carried out by dripping or spraying onto the dried, scoured wool before initial carding.

After the combed top has been produced, it is again lubricated on the draw frame or during blending with man-made fibers on the melangeuse, either by the sandwich method at the infeed of the draw frame or on the delivered top.

In a semiworsted or long staple spinning, the fiber flock is sprayed before carding. Because coarser fibers must be processed here, larger quantities of spinning auxiliaries and water must be applied.

Woolen spinning mills use the largest quantities of lubricants. The lubricant is applied to the fiber flock also before carding.

### 2.3.3. Use of Antistatic and Lubricating Agents in Dyeing

The use of antistatic and lubricating agents is necessary if the original preparation is lost during a dyeing operation. The new agents must be suitable for the changed fiber parameters, since the crimping, surface condition, and stiffness of the fiber have changed during dyeing.

It is therefore not effective to apply the original preparation again. At this stage cationic active single component systems, which are suitable for exhausting processes, should be used.

In procedures for the forced application such as back washing or centrifugal methods, nonionic/anionic systems can also be used, provided that no selective adsorption on the fiber surface takes place. The product used is chosen so as to

enable the frictional and antistatic properties required for the subsequent spinning process to be reestablished [2.13].

## 2.4. Physical Properties

The task of preparations is to modify the interactions of the fiber surfaces with one another and with the various surfaces of fiber guiding devices and processing machines. This requires precise knowledge of their surface behavior and hence of the resulting frictional processes.

A number of theories of surface coating, chemisorption, and intermediate valency forces are described in the literature. These explain how molecules of almost equal molecular mass and equal viscosity can show different frictional behavior. Similar phenomena can also be observed with different substrates [2.2], [2.14]–[2.26].

For example, if a prepared polymer fiber is heated above its glass transition temperature, reorientation of the molecules of the preparation on the fiber surface often takes place, leading to changes in zeta potential and frictional behavior. Moreover, frictional processes lead to partial redistribution of the preparation between the fiber material and the friction partner.

This is influenced by the surface of the solid undergoing friction, its structure, and the tendency to form a surface layer.

The object of a preparation is not to produce the lowest possible friction, but to give the optimum frictional properties that ensure steady and uniform processing.

Important parameters for frictional behavior include the consistency of the preparation (e.g., solid paste or oily liquid), viscosity, yarn speed, and reciprocal applied pressure.

These frictional phenomena are shown in the so-called Striebeck curve (Fig. 2.3). This shows that solid paste systems are less sensitive than oily liquid components. The latter are subject to the shear forces that develop within the liquid film between the fiber and the surface of the solid.

A very important function of a preparation is conductive removal of the electrostatic charge formed by the frictional processes. This is achieved by incorporating polar groups in the preparation molecules.

Formulating a spinning preparation necessitates knowledge of the physical properties of the individual components and of the full range of

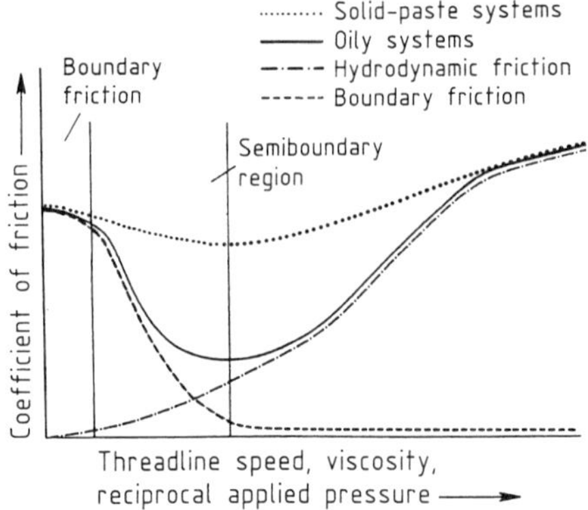



**Figure 2.3.** Striebeck curve for oily and solid-paste systems

requirements for the intended use. The following main properties are required for all areas of use:

1) Physiological harmlessness (e.g., dermatologically) and nontoxic by inhalation if an aerosol is formed
2) Ecological harmlessness, i.e., biologically degradable (>80 % in the OECD simulation test)
3) Antistatic action
4) Good spreading properties on the fiber surface (wetting properties)
5) Resistance to aging, i.e., no tendency towards resinification
6) Easily removable by washing
7) No detrimental effects on later finishing processes
8) No yellowing tendency
9) Good shelf life of the product and good durability after deposition on the substrate
10) Minimum migration into the interior of the fiber and no interaction with the fiber polymer (oligomer retention)

**Requirements for Filament Preparations.** Requirements for *flat yarns* are:

1) Low dynamic friction against solids, e.g., thread guides
2) Good fiber bond (i.e., static fiber–fiber friction for problem-free winding)
3) Low to medium dynamic fiber–fiber friction to avoid fluff formation (capillary breakage) during stretching
4) Good stability of the lubricant film
5) No spattering
6) High thermal stability and low tendency to evaporate, avoiding fume and condensate formation on hot stretching

Requirements for *textured fibers*, especially for high-speed spinning of POY (partially oriented yarn), are:

1) For friction texturing, a high dynamic dry (fiber/solid) friction coefficient is required to impart optimum twist distribution in the friction unit.
2) Little or no snow formation
3) No glazing effect on ceramic disks
4) Compatibility with polyurethane disks (does not cause swelling of the polyurethane material)
5) Residue-free thermal degradation, avoiding soiling problems on heaters
6) Consistent behavior under tension, giving uniform crimping data, extension, strength, and uniform dyeing properties with no sealed-in areas (tight spots)

**Requirements for Staple Fiber Preparations.** In both the production of staple fibers and in processing them in secondary spinning, the requirements of secondary spinning must be considered as well as the criteria listed above. The following are of equal importance:

1) Antistatic behavior
2) Fiber–fiber and fiber–metal friction in relation to the secondary spinning process

Thus, operating frictional conditions can be assigned to individual secondary spinning processes. The friction diagrams described in the literature indicate the operating situation for each spinning system, but vary from one author to another. These apparent discrepancies in the position of the optimum friction conditions are due to differences in measuring methods and their individual evaluation.

## 2.5. Chemical Properties

The basic materials for fiber preparations are oleochemical and petrochemical derivatives [2.12], [2.14], [2.17], [2.20].

Filament preparations are oily liquids formulated from several components which complement each other or have synergistic activity. They are used as aqueous solutions, stable emulsions, or neat oils.

Staple fiber preparations can be liquids, pastes, and waxlike compounds.

Depending on the area of use, antistatic and lubricating agents, as well as those applied after dyeing include both formulations of the filament preparation type and individual compounds of the staple fiber preparations.

As well as providing a purely lubricating or cohesive function, these surface-active compounds can provide all intermediate functions, depending on their chemical structure, and can also have emulsifying and antistatic effects.

The following review includes only a selection of the commonly used compounds.

1) Lubricants based on naturally-occurring materials
a) Mineral
   Oils
   Waxes
b) Vegetable (triglycerides)
   Coconut oil
   Peanut oil

2) Synthetic lubricants
   Esters of higher carboxylic acids with fatty alcohols or polyhydroxylic alcohols
   Polyethers (ethylene oxide based)
   Polyethers (ethylene oxide–propylene oxide based)
   Silicones

3) Emulsifiers and wetting agents
a) Anionic
   Fatty acid soaps with alkali metals or alkanolamines
   Sulfonated vegetable oils
   Sulfosuccinates
   Salts of ethoxylated alkyl phosphates
b) Cationic
   Fatty amines and their salts
   Ethoxylated fatty amines
   Quaternary nitrogen compounds
   Quaternary phosphorus compounds
c) Nonionic
   Glyceryl mono-oleates
   Ethoxylated fatty alcohols
   Ethoxylated fatty acids
   Ethoxylated sorbitan esters
d) Amphoteric
   Aminoacids and their salts
   Betaines
   Sulfobetaines

4) Antistatic agents
a) Anionic
   Alkyl phosphates
   Alkyl phosphonates
   Alkanolamine salts
   Soaps

b) Cationic
   Quaternary ammonium compounds, e.g., chlorides, methylsulfates, ethylsulfates
   Quaternary phosphonium compounds
c) Nonionic
   Ethoxylated fatty acid amides
d) Amphoteric
   Betaines
   Sulfobetaines

5) Additives include corrosion inhibitors, antioxidants, biocides, and solubilizers

*Analysis* is carried out by all relevant analytical methods from, for example, oleochemistry, petrochemistry, and surfactant chemistry [2.3], [2.27].

## 2.6. Fiber Preparations for Special Applications

Fiber preparations for special applications include products used on fibers for hygienic and medical applications. The compounds used must comply with the U.S. Food and Drug Administration (FDA) and the European Community (EC) Regulations, and must meet strict requirements with respect to toxicological and dermatological properties.

## 2.7. Additives for Spinning Solutions, Spinning Baths, and Spinning Melts

These include pigments and master batches for spinning dyes, optical brighteners, delusterants to reduce the luster of man-made fibers, and flame retardants.

In wet spinning processes such as viscose fiber and polyacrylonitrile fiber production, the pigments are added to the viscose or polyacrylonitrile polymer solution. However, in melt spinning, master batches are used. These consist of pigments incorporated in high concentration in a carrier polymer. They are added to the extruder in chip form. The delusterant used is extremely finely divided titanium dioxide, which is added to the spinning solutions or polymer melts.

In the production of viscose fibers and filaments, modifiers are used, especially to improve the wet strength modulus. Without modifiers, only fibers with a pronounced core–shell structure can be produced. The core is largely amorphous and is thus responsible for the low strength. The modifiers delay the saponification of the xanthogenate, and hence lead to the formation of a uniform crystalline structure over the entire cross section of the fiber. The modifiers consist of amine derivatives such as ethoxylated fatty amines, quaternary ammonium bases, and cyclohexylamine. Nitrogen-free products include polyethylene glycols [2.11].

## 2.8. Environmental Aspects

The increased importance attached to environmental protection has led to many compounds that were formerly successfully used being dispensed with. The alkyl aryl ethoxides are an example. Although these are biologically degradable, some metabolites formed have been found to be toxic to fish.

In the case of textiles for domestic use, the products described in this chapter enter the home environment on domestic textiles and, in the case of clothing, come into direct contact with the skin.

Fiber preparations appear in the wastewater from fiber producers or, at a later stage, from textile producers. Owing to increasingly strict environmental regulations, especially in Europe, in future there will be a requirement to use only biologically degradable compounds in fiber preparations.

Some of the fiber preparation material appears in the exhaust air, especially where thermal processes are used during fiber production and textile finishing. The general legislation that regulates emissions from industrial plants also applies to the exhaust air from fiber and textile plants [2.28].

## 2.9. Commercial Products

Some producers and trade names of auxiliaries used in the production and processing of fibers are listed below [2.29]:

| | |
|---|---|
| BASF | Soromin |
| Dr. Th. Böhme | Filapan, Synthesin |
| CHT | Avistat |
| Hansa Textilchemie | Duron |
| Henkel | BK-, LW-Marken, Nopco |
| Hoechst | Afilan, Leomin |
| ICI | Cirrasol |
| Schill & Seilacher | Limanol, Silastol, Syntex |
| Stockhausen | Praelanol, Tallopol |
| Takemoto | Delion |
| Zschimmer & Schwarz | Fasavin |

# 3. Sizing Agents

## 3.1. Introduction

Woven fabrics are two-dimensional bodies consisting of fibers in the form of threads intersecting at right angles (the warp and weft), produced using the shed-forming method. Whereas each weft thread is stressed only briefly as it is placed in position, the warp threads undergo repeated stress during each insertion of the weft and at each change of shed. The warp threads undergo stress in the form of yarn–metal abrasion when the weft thread is pushed by the reed, by yarn–yarn abrasion during the change of shed, and by cyclic stretching processes. The warp threads are normally unable to withstand these extreme stresses, and must therefore be provided with a protective coating – the sizing agent – that adheres to the fiber, forming an abrasion-resistant, elastic film.

With staple fiber yarns, the sizing agents have the task of making the yarn resistant to the frictional processes that take place during weaving. Protruding fibers are caused to adhere to the main body of the yarn, thereby preventing neighboring warp threads from catching or entangling. The overall increase in the tensile strength of the thread of ca. 20 % is of minor importance, but the increase in strength at the weakest points is crucial [3.1]. With filament yarns, the sizing agents cause the individual filaments to stick together, giving a compact thread, drastically reducing the risk of filament breakage. The sizing agent must adhere strongly to the fiber, and its film properties should be largely independent of the climatic conditions, especially atmospheric humidity, and be unaffected by fiber finishes and sizing additives. The elongation of the warp thread should not be reduced by presence of the sizing agent.

After the gray cloth has been woven, the task of the sizing agent is complete. As it would usually have a deleterious effect on subsequent finishing processes, it must be completely removed. Removal is simple in the case of cold-water-soluble sizing agents, but starch products that are insoluble in cold water require preliminary enzymatic or oxidative breakdown before the desizing stage (Section 4.2.2.1.). The removal of the sizing agent may pose special wastewater treatment problems in finishing plants.

A large number of classes of chemical substances are used as sizing agents. They can be divided into two main groups [3.2]:

1) Macromolecular natural products and their derivatives: starches and starch derivatives, carboxymethyl cellulose, galactomannan, and tamarind flour derivatives
2) Synthetic polymers: poly(vinyl alcohol), poly(meth)acrylates, polyester condensates, and polyvinyl compounds

The growth of synthetic fibers and the developments in weaving technology have accelerated the development of synthetic sizing agents. The consumption of sizing agents worldwide is ca. 700 000 t, of which ca. 70 % are starches and starch derivatives, and ca. 15 % are poly(vinyl alcohol). The remainder consists of carboxymethyl cellulose (CMC), poly(meth)acrylates, polyester condensates, and other sizing materials. The areas of use of the various sizing agents are summarized in Table 3.1. General information on sizing agents is given in [3.3], [3.4].

## 3.2. Sizing Methods

As a rule, sizing agents are applied to the warp threads from aqueous liquors (Fig. 3.1). The warp threads pass from the warp beams (a) through the sizing trough where they are dipped in the sizing liquor (b) and impregnated with it. The excess liquor is then squeezed off in the squeezing equipment (d), giving a liquor pickup of 50 – 200 %, depending on the warp parameters (fiber material, spinning method, thread thickness, etc.), sizing machine parameters (speed, squeezing pressure, etc.), and liquor parameters (viscosity, temperature, wetting properties, etc.) [3.5], [3.6]. The warp threads are dried in the cylinder drier, and are then separated in the dry splitting section (f). They then pass through the comb (g), which adjusts the distance between the warp threads and makes them all parallel, after which they are wound onto the weaving beam (h), which is then taken to the loom. There are two methods of applying the size: either all the warp threads are sized simultaneously or a proportion of the warp threads are sized as a warp beam (the so-called single-end sizing process). These warp beams are then assembled in a separate operation to form the complete weaving beam. This single-end sizing process is used only for filament yarns, to prevent damage to the delicate individual filaments on separation of the threads stuck together by the sizing material. Staple fiber yarns are always sized all at the same time. In state-of-the-art technology, high-pressure squeezing off is used, ensuring that the liquor pickup is small, keeping drying costs low [3.7], [3.8].

In the cold sizing process, very coarse or twisted yarns are coated with a small amount of sizing liquor by kiss roller application during

**Table 3.1.** Areas of use of sizing agents

| | Starch | CMC | Galacto-mannan | PVAL* | Polyacrylates | Polyesters | Vinyl acetate copolymers |
|---|---|---|---|---|---|---|---|
| *Staple fiber yarns* | | | | | | | |
| Cotton, linen viscose staple fiber | + | + | + | + | + | | + |
| Wool | + | + | + | + | + | | |
| Polyester – cellulose | + | + | + | + | + | | + |
| Polyester – wool | | + | + | + | + | | |
| Polyester | | + | + | + | + | | |
| Polyamide | | | | + | + | | |
| Polyacrylonitrile | | + | | + | + | | |
| *Filament yarns* | | | | | | | |
| Viscose | | | | + | + | | |
| (Tri)acetate | | | | + | + | | + |
| Polyamide | | | | + | + | | |
| Polyester | | | | + | + | + | |

* PVAL = Poly(vinyl alcohol).

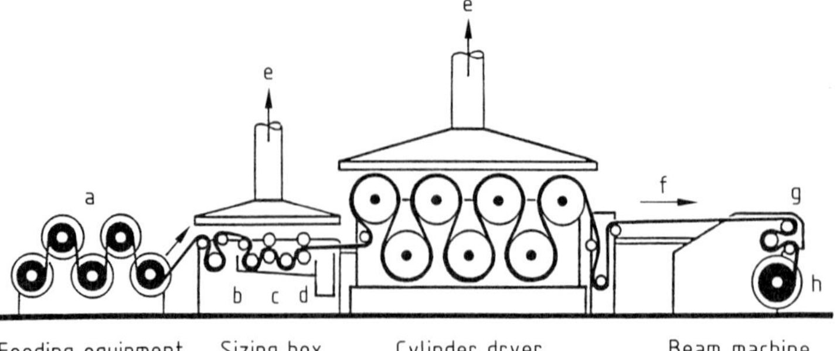



**Figure 3.1.** Schematic diagram of the sizing process
a) Warp beams; b) Sizing liquor; c) Immersion roller; d) Squeezing equipment; e) Steam outlet; f) Dry splitting section; g) Comb; h) Weaving beam

warp beam preparation. The application process is carried out at room temperature, and drying of the yarn is not necessary [3.9], [3.10]. Other alternative sizing techniques, such as solvent sizing [3.11], [3.12], hot melt sizing [3.13]–[3.16], and foam sizing [3.17]–[3.22] have not yet become established.

## 3.3. Sizing Agents Based on Natural Raw Materials

The most important natural raw materials used as sizing agents are starch and cellulose. Other polysaccharides have so far played only a minor role.

### 3.3.1. Starch and Starch Derivatives

Starch and its derivatives are still the most important class of sizing agents with respect to total consumption. This is because of their low price, good sizing effect, and worldwide availability. The raw material basis of this class of sizing materials is naturally occurring starch, a polysaccharide based on $\alpha$-D-glucopyranose. Starch is not a single chemical substance, but is composed of two structurally different polymers: amylose and amylopectin. Amylose consists of chains of glucose units linked by $\alpha$-1,4-glucosidic bonds, whereas amylopectin additionally contains $\alpha$-1,6-glucosidic bonds which cause branching of the polymer chain [3.23]–[3.26] ($\rightarrow$ Starch and Other Polysaccharides, **A 25**, p. 2).

The degree of polymerization of amylose is ca. 100–1000 glucose units, and that of amylopectin, ca. 6000–10^6. Amylopectin is the main constituent of starch, making up 73–86 % of the total, depending on the type of starch.

The most important sizing agents are potato, maize, and tapioca starches. Wheat, rice, and sago starches are also used. The characteristic properties of these starches are determined by the amylose/amylopectin ratio, the degree of polymerization of these two constituents, and the size and fine structure of the starch grain. These parameters determine the swelling and solution behavior, and also the properties of the film (Table 3.2).

Natural starch is insoluble in cold water because of the hydrogen bonds linking parallel polymer chains. The starch is brought into "solution" by heating. The starch grains first absorb

**Table 3.2.** Composition and properties of starches

| Property | Potato starch | Maize starch | Tapioca starch |
|---|---|---|---|
| Grain size, µm | 30 | 15 | 20 |
| Amylose content, % | 21 | 28 | 17 |
| Glucose units per molecule of amylose | ca. 3000 | ca. 8000 | ca. 3000 |
| Starch paste temperature, °C | 60–65 | 75–80 | 65–70 |
| Viscosity of starch paste | very high | medium | high |
| Shear stability | medium to low | low | high |
| Retrogradation | medium to low | high | low |
| Strength and elasticity of film | high | low | high |

water until swelling is at its maximum. Above a certain temperature, characteristic for each type of starch and known as the gelatinization temperature, the starch grains burst and form a gel. The viscosity increases to a maximum, and then decreases asymptotically to a limiting value as the solubilized polymer molecules disperse. Complete solubilization of the individual molecules of a starch grain only occurs above 100 °C. The viscosity value is important in size application, as it has a considerable effect on the amount of liquor pickup.

On storage and with decreasing temperature, starch pastes solidify to a pulpy mass. This retrogradation is caused by stretching of the molecular chains, parallel alignment of the chains, and formation of hydrogen bonds between neighboring chains, with loss of water of hydration [3.26]. This retrogradation has detrimental effects on the sizing agent, leading to poor storage properties, skin formation, formation of deposits on the rollers, and reduced adhesive strength. Therefore, natural starches are increasingly being replaced by starch derivatives.

All modified starches that have lost their original properties are referred to as starch derivatives. These include thin-boiling starches, dextrins, starch esters, and starch ethers (→ Starch and Other Polysaccharides).

Thin-boiling starches are produced by acid hydrolysis or oxidative degradation in aqueous suspension, and dextrins are produced by thermal depolymerization, usually in the presence of acids. They gelatinize at low temperatures, give solutions of low viscosity, and can be dissolved in high concentrations. Furthermore, it is easier to produce a liquor with a predetermined viscosity, and the tendency to retrogradation is considerably reduced.

The starch esters mainly used in sizing materials are those of phosphoric acid (the phosphate starches) and acetic acid (the acetyl starches). These starch derivatives are usually not only esterified, but also depolymerized, giving lower liquor viscosities and decreased retrogradation. In general, they give better sizing effects than thin-boiling starches.

The three most important types of starch ethers are the hydroxyethyl, hydroxypropyl, and carboxymethyl starches, produced by reaction of starch with ethylene oxide, propylene oxide, and chloracetic acid or sodium chloroacetate in the presence of caustic soda, respectively. The degree of substitution of these starch derivatives is generally ca. 0.1 or less. Because of its ionic character, sodium carboxymethyl starch is soluble in cold water and therefore does not require enzymatic desizing. Also, the starch ethers have a better sizing effect than starch derivatives that are simply depolymerized.

The mechanical properties of films cast from solutions of starch or starch derivatives depend on the degree of hydration (which depends on the relative humidity of the atmosphere in the weaving mill) and on the ratio of amylose to amylopectin and type of modified starch.

**Uses.** The main use of starch and starch derivatives is for sizing yarns of pure cotton and its blends with other fibers. For these yarns, a large number of sizing formulations are used which consist either exclusively or principally of starch or starch derivatives. Starch and its derivatives adhere relatively strongly to cotton [3.27], [3.28]. Cotton yarns woven on high-speed looms or yarn blends with a high proportion of synthetic fiber must be sized with sizing formulations that contain additionally carboxymethyl cellulose, poly(vinyl alcohol), or poly(meth)-acrylates to improve the sizing effect. Starches are detected by the blue coloration with iodine, and this reaction is also used for the semiquantitative determination of residual size content [3.29].

**Ecology.** Starch and starch derivatives are in general biologically degradable, and derivatives such as the hydroxypropyl, phosphate, and carboxymethyl starches are also classified as such [3.30], [3.31]. The COD of sizing agents based on starch is ca. 900–1000 mg $O_2$/g starch (or derivative). In comparison with other sizing agents, starch and its derivatives must be applied at ca. twice the amount used for carboxymethyl cellulose or synthetic sizing agents because of the poorer sizing effect, thereby leading to a higher COD in the wastewater. Toxicity towards waterborne organisms such as fish, daphnia, and bacteria is very low ($LC_0 > 100$ mg/L).

### 3.3.2. Sodium Carboxymethyl Cellulose

Unlike starch, cellulose contains the 1,4-β-glucosidic link and has a straight chain ($\rightarrow$ Cellulose, **A 5**, p. 376). Depending on its origin, the degree of polymerization of the macromolecule corresponds to $\bar{P}_\eta = 1000–7000$ glucose units. Because of its chemical uniformity and high degree of crystallinity, cellulose hardly swells in water and is practically insoluble. This prevents the use of the natural product as a sizing agent. To obtain water-soluble cellulose derivatives, the cellulose is etherified or esterified [3.32], [3.33] ($\rightarrow$ Cellulose Ethers, $\rightarrow$ Cellulose Esters).

Of the water-soluble cellulose derivatives, only sodium carboxymethyl cellulose (CMC) is used as a sizing agent. CMC is produced by reacting cellulose with sodium hydroxide and sodium chloroacetate, whereby the cellulose polymer is simultaneously depolymerized. Solubility in water is produced when the degree of substitution (DS) is greater than ca. 0.5. CMC with a DS of ca. 0.7 is usually used for sizing agents. In the production process, large amounts of salts are produced, mainly sodium chloride and sodium glycollate. Unpurified grades of CMC containing up to 40% salt and purified grades with < 2% salt are commercially available. The salt content only slightly affects the rheology of CMC, but because it can lead to considerable corrosion damage in sizing machines and looms, purified CMC grades are preferred. Depending on the method of production and the DS, CMC can contain small amounts of insoluble cellulose fibers. If the content of these materials is too high, deposits may form on the drying cylinders of the sizing machine, which can seriously affect the sizing process. Owing to its chemical affinity to cellulose, purified CMC adheres strongly to cotton. It also adheres to polyester and polyacry-

lonitrile, its performance sometimes equalling that of poly(meth)acrylates and poly(vinyl alcohol) [3.27], [3.28].

On drying CMC solutions, tough, transparent films are formed whose mechanical properties are highly dependent on water content. This in turn depends on the salt content of the film and the relative humidity of the air (24–31% water content at 20 °C and 65% R.H.).

**Uses.** CMC is the preferred additive for increasing the adhesion of starch sizes to cotton and cotton blends. Other important uses include the sizing of cotton yarns for weaving terry toweling and viscose staple fiber yarns. Because CMC is easily removed by washing, fabrics with warp threads sized with CMC can be dyed without a preliminary wash. Owing to the high viscosity of CMC, the concentration range for sizing liquors is limited to ca. 20–100 g/L. CMC is suitable for size recovery processes, whether used on its own or in combination with poly(vinyl alcohol) (see Section 3.4.1) [3.34].

The methods of detecting CMC are much less selective and less practical than the iodine–starch test. CMC can be detected by precipitation with uranyl nitrate, copper (II) sulfate, or cationic surfactants [3.35], [3.36].

**Ecology.** Although CMC is degradable by biological action, its rate of breakdown does not justify its classification as a readily biologically degradable material [3.30]. The COD is ca. 900 mg $O_2$/g CMC. As with starch derivatives, the toxicity of CMC towards waterborne organisms is very low ($LC_0 > 100$ mg/L).

### 3.3.3. Sizing Agents Based on Other Natural Raw Materials

Older sizing agents based on linseed oil [3.37] and bone or leather glue [3.38] are scarcely used today, having been largely replaced by synthetic sizing agents. However, sizing agents based on galactomannans have some importance as biologically degradable alternatives to CMC [3.39], [3.40]. Although galactomannan derivatives from guar or carob flour have long been known as thickening agents for printing, they have not been used as sizing agents. Products of this type are produced from the polysaccharide extracted from cassia beans, which is made up of galactose and mannose monosaccharide units in the ratio

1:5. Solubility in water and ease of removal by washing are produced by etherification to form the hydroxypropyl or carboxymethyl derivatives.

Tamarind seed also appears to be a suitable raw material for sizing agents [3.41], [3.42]. However, like the galactomannan polysaccharides, this raw material has not yet been used in any substantial quantity.

## 3.4. Synthetic Sizing Agents

The growth in the use of synthetic fibers and high-speed looms (projectile and air jet looms) means that the warp yarns must receive optimum sizing treatment, and this has led to an increasing demand for high-quality synthetic sizing agents. The most important of these are poly(vinyl alcohols), poly(meth)acrylates, and polyester condensates.

### 3.4.1. Poly(Vinyl Alcohols)

Poly(vinyl alcohols) have been used as sizing agents for around 60 years, but large-scale use started only in 1960, coinciding with the appearance of the polyester–cotton yarns. They are the most important group of synthetic products used as sizing agents.

Technical-grade poly(vinyl alcohol) can be considered to be formally a copolymer of vinyl alcohol and vinyl acetate:

$$(CH_2-CH)_m- (CH_2-CH)_n -$$
$$\qquad | \qquad\qquad\quad |$$
$$\quad OH \qquad\qquad O \ CO \ CH_3$$

Poly(vinyl alcohol)

As monomeric vinyl alcohol does not exist, poly(vinyl alcohol) is produced by the hydrolysis or deesterification of the poly(vinyl acetate) ($\rightarrow$ Polyvinyl Compounds, Others, **A 21**, pp. 743–747).

The parameters that determine sizing properties are the degree of polymerization (DP) and the degree of hydrolysis. Products are generally characterized by the viscosity of an aqueous 4% solution as a measure of the DP, and by the degree of hydrolysis in mol%. These parameters are controlled via the DP of the starting material poly(vinyl acetate), and conditions during hydrolysis (deesterification). There are two grades: the partly hydrolyzed grade with a degree of hydrolysis of ca. 88%, and the fully hydrolyzed grade with a degree of hydrolysis of ca. 98%.

Maximum solubility in water occurs for a degree of hydrolysis of ca. 88%.

Partially hydrolyzed poly(vinyl alcohol) grades adhere more strongly to nonpolar fibers such as polyester [3.27], [3.28] and also show maximum water solubility. These are therefore preferred for use as sizing agents.

In addition to these conventional poly(vinyl alcohol) grades, copolymers with methyl methacrylate or with comonomers having carboxyl functionality are also used as sizing agents. These have the advantage of better water solubility under the alkaline conditions used for desizing and scouring, and after the thermofixing process applied to fabrics made from polyester-containing yarns [3.43]–[3.45].

The poly(vinyl alcohol) films are tough and elastic. Their mechanical strength, which is unmatched by other synthetic sizing agents, is less sensitive to atmospheric humidity than that of other water-soluble sizing agents. The water pickup of poly(vinyl alcohol) films is 5–10% at 65% R.H., increasing to only 20% at 86% R.H. [3.46].

**Uses.** Poly(vinyl alcohol) has its greatest importance in the sizing of staple fiber yarns. Because of its high efficiency on cellulose and polyester fibers compared with starch and its derivatives [3.27], [3.45], it is used as a high-quality additive in sizing formulations for cotton and especially for polyester blends. Here, as much as 50% of the sizing agent can be saved compared with sizing exclusively with starch and its derivatives. Special areas of use include the sizing of wool where it is generally used in combination with CMC, and of polyacrylonitrile, viscose, acetate, triacetate, and polyamide filament yarns. Although poly(vinyl alcohol) adheres strongly to polyester and has an excellent sizing effect on polyester staple fiber yarns, it is normally not effective on polyester filament yarns.

Poly(vinyl alcohol) is extremely suitable for size recovery processes, whether in a partial recycling operation by means of simple washing techniques or by full recycling using ultrafiltration. Conventional partially or fully hydrolyzed poly(vinyl alcohols) and copolymers are equally suitable [3.45], [3.47]–[3.49]. Poly(vinyl alcohol) can be used alone as a sizing agent or in combination with carboxymethyl cellulose or polyacrylate [3.50].

Several methods for determining poly(vinyl alcohol) are available. Of these, the iodine reaction [3.51], intensified by addition of boric acid [3.52] is the most important, and can also be used for quantitative determination in wastewater [3.46], [3.53].

**Ecology.** Unlike most synthetic polymers, poly(vinyl alcohol) is biologically degradable, the degradability being dependent on the adaptation of the activated sludge microorganisms [3.30], [3.54], [3.55]. Given such adaptation, poly(vinyl alcohol) can be classified as having good biological degradability. The breakdown behavior is independent of the degree of hydrolysis and the viscosity of the poly(vinyl alcohol), which has a COD of ca. 1700 mg $O_2$/g. The aquatic toxicity towards fish, daphnia, and bacteria is $LC_0 > 100$ mg/L [3.30].

### 3.4.2. Poly(meth)acrylates

Poly(meth)acrylates are another important class of synthetic sizing agents. They show considerable variation in chemical structure, and hence have a very wide range of applications ($\rightarrow$ Polyacrylamides and Poly(Acrylic Acids); $\rightarrow$ Polyacrylates; $\rightarrow$ Polymethacrylates).

The important chemical building blocks and the construction scheme for poly(meth)acrylate sizing agents are shown below [3.56]:

$$CH_2{=}CH \qquad CH_2{=}CR^1 \qquad\qquad CH_2{=}CR^1$$
$$\quad|\qquad\qquad\quad|\qquad\qquad\qquad\qquad\quad|$$
$$COOH \qquad COO^-Na^+(NH_4^+Ca^{2+}) \qquad COOR$$

Acrylic acid          Salts                    Esters

$$CH_2{=}CH \qquad CH_2{=}CH \qquad\qquad CH_2{=}C\overset{\textstyle CH_3}{\underset{\textstyle COOH}{\big<}}$$
$$\quad|\qquad\qquad\quad|$$
$$CONH_2 \qquad\quad C{\equiv}N$$

Acrylamide     Acrylonitrile          Methacrylic acid

$$-(CH_2{-}CR^1)_x{-}\,(CH_2{-}CR^1)_y{-}\,(CH_2{-}CR^1)_z{-}$$
$$\qquad|\qquad\qquad\qquad|\qquad\qquad\qquad|$$
$$\quad COONa \qquad\quad COOH \qquad\quad COOR$$

Schematic structure

$$R^1 = H, CH_3$$
$$R = CH_3, C_2H_5, C_3H_7$$

Hydrophilic monomers such as methacrylic acid and its salts and acrylamide give good adhesion to polar fibers, ease of removal by washing, and compatibility with alkalis. Hydrophobic monomers such as the acrylic esters increase the elasticity of the sizing films and improve the ease of removal from the wastewater. (Meth)acrylic acids and their esters increase the hardness of the

films, while acrylonitrile improves their elasticity and abrasion resistance. The (meth)acrylic acid in the polymer is usually present as the sodium or ammonium salt. The preferred base is ammonia, because of the low hygroscopicity of the product. Complete or partial neutralization with alkaline earth bases [MgO, Ca(OH)$_2$] leads to a further reduction in the hygroscopicity, and also avoids the problem with ammonium polyacrylates that they tend to be difficult to remove by washing due to formation of cross-links during thermofixing [3.57].

In recent years, poly(meth)acrylate sizing agents based on methacrylic esters have been developed. These give good sizing effects owing to their high adhesive strength to natural and synthetic fibers and exhibit low sensitivity to atmospheric humidity [3.56].

An example of chemical optimization to enable extreme quality requirements to be met is provided by the sizing agents used for filament yarns for water jet looms:

$$(CH_2{-}CR^1)_x{-}\,(CH_2{-}CR^1)_y{-}\,(CH_2{-}CR^1)_z{-}$$
$$\qquad|\qquad\qquad\qquad|\qquad\qquad\qquad|$$
$$\quad CN \qquad\qquad COOR \qquad\qquad COONH_4$$

$$R^1 = H, CH_3$$
$$R = CH_3, C_2H_5, C_3H_7$$

On drying, ammonia is cleaved from the water-soluble ammonium polyacrylate, converting it to the water-insoluble, almost nonswelling acid form. A polyester or polyamide yarn sized in this way can be woven on water jet looms. The sizing agent can then be washed off under alkaline conditions.

**Uses.** Unlike other staple fiber sizing agents, which are usually supplied as the 100 % solid product, poly(meth)acrylates are usually produced as a 25 % aqueous solution, and have a wide range of uses owing to their wide variability. Like poly(vinyl alcohol), they are used as a high-quality additive in sizing formulations for all cotton yarns, especially for blends of cotton with synthetic fibers, and also for viscose staple fiber, wool, and polyacrylonitrile. Poly(acrylic acid) is the preferred sizing agent for polyamide filament yarns. Products based on acrylic acid copolymers are suitable for viscose, acetate, and triacetate filament yarns, while the hydrophobic poly(meth)acrylates based on (meth)acrylic esters are preferred for sizing polyester filament yarns. The excellent water-solubility of the sizing agents based on (meth)acrylic acid means that the gray cloth can be treated in subsequent processes without first desizing it. Recent research has shown that hydrophobic polyacrylate sizing agents are suitable for recovery and concentration by ultrafiltration [3.50].

**Ecology.** Poly(meth)acrylates are practically biologically nondegradable. Hydrophobic methacrylic acid ester copolymers are adsorbed by the activated sludge, and can be eliminated in this way, while conventional hydrophilic products are eliminated to only a small extent [3.54], [3.56]. COD values of poly(meth)acrylate sizing agents are in the range ca. 1300 – 1700 mg $O_2$/g, based on solids material. Their aquatic toxicity is low, like that of the other sizing agents.

### 3.4.3. Polyester Condensates

With the introduction of the single-end sizing process [3.58], it became possible to apply water-dispersible sizing agents based on polyesters. Today, these are the standard sizing agents for plain polyester filament yarns.

Polyester sizing agents are, in general, polycondensates of aromatic dicarboxylic acids with diols (e.g., ethylene glycol, diethylene glycol, or cyclohexane dimethanol) and sulfonated aromatic dicarboxylic acids (e.g., 5-Na-sulfoisophthalic acid), which provides solubility or dispersibility in water [3.59], [3.60]:

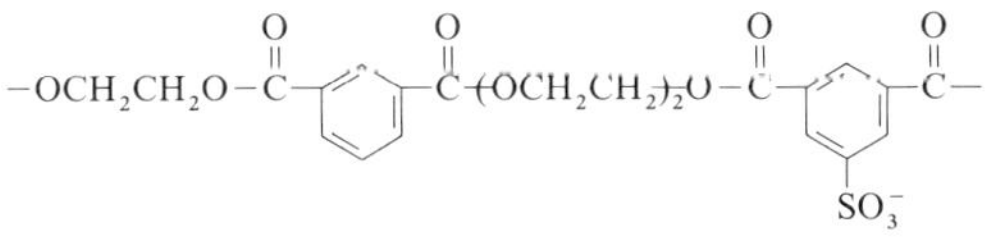

$$-OCH_2CH_2O-\overset{O}{\overset{\|}{C}}-\!\!\bigcirc\!\!-\overset{O}{\overset{\|}{C}}(OCH_2CH_2)_2O-\overset{O}{\overset{\|}{C}}-\!\!\bigcirc\!\!-\overset{O}{\overset{\|}{C}}- \atop SO_3^-$$

The proportion of ionic sodium sulfonate groups is such that the polyester condensates do not form true solutions in water, but dispersions of small particle size.

The properties of the polyester sizing agents can be controlled over a wide range by variation of the monomers, their relative proportions, and the molecular mass. The aim is to produce products that disperse readily in water and give films that are tough, resilient, and do not adhere to each other at high humidities. Polyester sizing agents have a chemical affinity to the polyester fiber and adhere to it strongly [3.27]. The possibility of interaction with the spinning preparation must be taken into account [3.61].

**Uses.** The main use of polyester sizing agents is for sizing plain polyester filament yarns. However, they are also used as a synergistic component to increase the adhesive strength when sizing staple fiber yarns made of polyester blends [3.62]. Note that when polyester sizing agents are washed out of the textile, there is a danger that they may be precipitated by electrolytes, especially polyvalent cations [3.63]. They can be determined on the textile by color reactions with basic dyes [3.36], [3.64], [3.65].

### 3.4.4. Other Synthetic Sizing Agents

Apart from the synthetic sizing agents described above, vinyl acetate copolymers have some practical importance. These consist of copolymers of vinyl acetate with acidic comonomers such as (meth)acrylic acid or maleic half-esters. In their almost universal applicability, they resemble the poly(meth)acrylate class of products. Owing to the tendency of vinyl acetate to hydrolyze, and the fact that the films are relatively weak and tend to stick together, these products are not so widely used as the poly(meth)acrylates. They are used as additives in starch-based sizing formulations for cotton and cotton–polyester blends to increase the adhesion, but also for acetate, triacetate, and polyester filament yarns [3.66].

## 3.5. Sizing Additives

Sizing additives are added to the sizing liquor to aid the weaving process by their softening, smoothing, or antistatic effects. Other such products are used as antifoaming agents or to increase the liquor pickup [3.4], [3.67].

**Viscosity regulators** are used to control the viscosity of starch-based sizing liquors. Complex formation between $Na_2B_4O_7 \cdot 10\,H_2O$ (borax) and the hydroxyl groups of starch increases the viscosity of starch paste [3.4], while urea reduces it [3.68]. Important viscosity regulators include starch-degrading agents that act by hydrolytic or oxidative cleavage of the macromolecules.

Acids are important in the industrial manufacture of starch derivatives, but are not used in the sizing process because of poor reproducibility of the effect. Amylases, which have an enzymatic action, are used in desizing, but are little used to reduce starch viscosity for sizing. The

**Table 3.3.** Sizing agents producers

| Producer | Trade name | Classification |
| --- | --- | --- |
| Abco, USA | Absize, Synthesize | mixed products |
|  | Plasticryl | poly(meth)acrylate |
| Air Products, USA | Airvol | poly(vinyl alcohol), mixed products |
| Allied Colloids, UK | Alcosize | mixed products |
|  | Alcowax | mixed products |
|  | Collasyn | poly(vinyl alcohol) |
|  | Colvinal | vinyl acetate copolymer |
|  | Sincol, Vicol | poly(meth)acrylate |
|  | Tescol | poly(meth)acrylate, polyester |
| Amylum, BE | Resamyl, Supramyl | starch derivative |
| Avebe, NL | Kollotex | starch derivative |
|  | Quicksolan | starch derivative, mixed products |
|  | Solvitose, Retamyl | starch derivative |
| BASF, FRG | Schlichte CA/CB/CE/S | poly(meth)acrylate |
|  | Schlichte T10 | vinyl acetate copolymer |
|  | Schlichte 3204 | polyester |
|  | Textilwachs | sizing fat |
| Blattmann, CH | Noredux, Noresin, Noresol | starch derivative |
|  | Noregum | galactomannan |
| Bozzetto, IT | Wisacet, Wisacryl | poly(meth)acrylate, vinyl acetate copolymer |
|  | Wisester | polyester |
| Du Pont, USA | Elvanol | poly(vinyl alcohol) |
| Eastman, USA | Eastman LB, WD | polyester |
| Emslandstärke, FRG | Emsize | starch derivative, mixed products |
| Henkel, FRG | Quellax | starch derivative |
|  | Horsil | CMC |
|  | Inex | poly(vinyl alcohol) |
|  | Fibrocol, Fibropur, Fibrosint | mixed products |
|  | Molvenin | galactomannan |
|  | Olinor, Grünau Pillenwachs | sizing fat |
| Hercules, USA | Blanose | CMC |
| Hoechst, FRG | Arkofil | galactomannan, polyester |
|  | Leomin | sizing fat |
|  | Tylose | CMC |
|  | Vinarol | poly(vinyl alcohol) |
| Kuraray, JP | Poval | poly(vinyl alcohol) |
| Lamberti, IT | Carbocel | CMC |
|  | Lamcol | polymethacrylate, polyester, vinyl acetate copolymer, mixed products |
|  | Overwax | sizing fat |
| Metsä-Serla, SF | Finnfix | CMC |
| Morton Intern., USA | Polyfilm | polymethacrylate, polyester, vinyl acetate copolymer |
| National Starch, USA | Kofilm, Lubrisize | starch derivative |
| Nippon Gohsei, JP | Gohsenol | poly(vinyl alcohol) |
| Rhône Poulenc, FR | Bevaloid | polymethacrylate, vinyl acetate copolymer |
|  | Gerol | polyester |
| Roquette, FR | Tissalys | starch derivative |
| Seydel, USA | Aztex, Seycofilm | polymethacrylate, polyester, mixed products |
|  | Plystran | mixed products |
|  | Sicosize | polymethacrylate |
| Südstärke, FRG | Sobex, Sobotex | starch derivative |

most important starch degradation agents for this application are oxidizing agents such as peroxodisulfates, peroxosulfates, and other compounds such as chloramine T (sodium *p*-toluenesulfonylchloramide, [*127-65-1*]). These reagents are often marketed as blends with starch, CMC, poly(vinyl alcohol), or sizing fats.

**Sizing fats** are used to improve the dry splitting and weaving behavior of the warp by plasticizing the sizing film, providing a smoothing and antistatic action without reducing the adhesion and abrasion resistance of the sizing agent. At the same time, the emulsifying properties must be good, and washing out of the product must be problem-free [3.69]. Suitable materials include self-emulsifying sizing fats such as sulfated fats and oils, and especially preparations of fatty acid esters with nonionic and anionic emulsifiers. These sizing fats often have antifoaming properties. Antistatic surfactants based on poly(glycol ethers) are added to reduce the static charge.

If it is not necessary to achieve maximum reduction of the friction coefficient, poly(ethylene glycols) with a molecular mass of ca. 4000 can be used as sizing fats. These improve the flexibility of the sizing agent film, facilitate the dry splitting process, and reduce dust formation during weaving by means of their antistatic effect. The most effective means of reducing friction coefficients is after-waxing. For this, the most suitable materials are self-emulsifying paraffin-free waxes based on fatty acids [3.69].

**Wetting Agents.** The liquor pickup during sizing depends partly on the wettability of the yarn. If the necessary size add-on can not be achieved by using high concentrations of sizing agent, a low foam wetting agent must be added. As the adhesive strength of the sizing agent can be greatly reduced by surfactants, the amount added to the sizing liquor must not exceed ca. 1 g/L. Fatty alcohol poly(glycol ethers) with a low degree of ethoxylation are usually employed.

**Defoamers.** If the sizing agents used tend to produce foam [e.g., poly(vinyl alcohol)] or if wetting agents are added, the addition of a defoamer is often necessary. Suitable products can be based on paraffin oils, phosphoric esters, fatty acid esters, or silicone oils. The most effective materials are silicone oils, but even small additions of these can lead to a drastic reduction in the adhesive strength of the sizing agent [3.69].

**Preservatives.** Sizing liquors that are stored for long periods and contain biologically degradable components such as starch or starch derivatives must be protected against degradation by adding fungicides and/or bactericides. Classical preservatives include formaldehyde, formaldehyde-releasing substances, phenol derivatives, and heterocyclic compounds of the isothiazolin type.

## 3.6. Commercial Products

A selection of sizing agents is listed in Table 3.3. Detailed information on these can be obtained from the various catalogues of textile auxiliaries [3.70] and from information leaflets provided by the manufacturers.

# 4. Pretreatment Auxiliaries

## 4.1. Processes and Auxiliaries for the Pretreatment of Textile Materials

Pretreatment processes are used to prepare textile material (greige) for subsequent processes such as dyeing, printing, optical brightening, and finishing.

Precleansing, extraction, and bleaching processes remove foreign materials from the fibers. The reactive groups of the substrate, previously blocked by impurities, are exposed, and the whiteness of the textile is improved.

For gray goods made of natural fibers such as cotton, linen, flax, wool, and silk, the technical task is more difficult than for those made of synthetic or cellulosic chemical fibers. For example, cotton fabric can be accompanied by up to 20 % of materials that can interfere with later processing (sizes, preparations, waxes, hemicelluloses, pectins, seed husks, proteins, and alkaline earth and heavy metal compounds), whereas crude polyester textiles contain only preparations, water-soluble synthetic size, and soil, which can be removed by simple washing processes.

The processes used depend both on the form of the fiber (flocks, yarn, woven or knitted fabrics), and on the machinery available.

Pretreatment processes are substrate-specific. Because of the wide range of chemical reactions and physicochemical processes involved, the preparation of gray goods usually

**Table 4.1.** Pretreatment processes

| Substrate | Process |
|---|---|
| Cotton, linen, polyester–cotton | singeing |
| | enzymatic desizing |
| | oxidative desizing |
| | demineralization |
| | alkaline extraction |
| | bleaching with hydrogen peroxide |
| | bleaching with sodium hypochlorite |
| | bleaching with sodium chlorite |
| | bleaching with reducing agents |
| | caustic soda treatment |
| | mercerization |
| Wool, silk | raw wool washing |
| | carbonization |
| | boiling-off |
| | bleaching with reducing agents |
| | bleaching with hydrogen peroxide |
| Synthetic fibers (PA, PES, PAN) | desizing (water-soluble sizes) |
| | washing |
| | bleaching with sodium chlorite |

consists of a series of part processes. The choice and sequence of these processes mainly depend on the type of material, the quality requirements (which are often specific to the article) and the available machines or equipment (Table 4.1) [4.1].

In the various processes, the auxiliaries have the following tasks:

1) Transport of the washing liquor into the interior of the fiber (air removal, wetting, impregnation)
2) Removal of minerals (dissolution, complexation)
3) Mobilization and removal of foreign materials and reaction products (dispersing, emulsification, complexation, colloid protecting action)

4) Fiber protection (reduction and complexation of heavy metal ions)

Modern auxiliaries are mostly multifunctional synergistic mixtures of highly active individual components; thus, a single product performs several tasks simultaneously. Other requirements include ease of handling, availability in concentrated liquid form, suitability for metering and pumping, stability in washing liquors, low foaming properties under highly turbulent washing conditions, and good storage stability. More recently, demands for minimum wastewater contamination and biological degradability have become increasingly important.

## 4.2. Preparation of Textile Material made of Cellulose Fibers

### 4.2.1. Pretreatment Processes

The pretreatment process depends principally on the gray goods and the object of the subsequent treatment. The nature of the material and the method of treatment determine the choice of equipment (see Table 4.2).

The following methods are used:

1) "Long liquor" treatment
2) Impregnation processes
3) Open-width treatment
4) Treatment in rope form
5) Discontinuous methods
6) Continuous methods

Worldwide, the pretreatment of woven textiles is quantitatively of far greater importance than the finishing of knitted fabrics.

**Table 4.2.** Preparation techniques

| Type of material | Type of treatment | Type of machine |
|---|---|---|
| Flock, cotton wool, combed sliver | discontinuous | bale apparatus |
| Yarn (hank, wound package) | discontinuous | apparatus for bales, hanks, cross-wound packages, warp beams |
| Piece (woven, knitted) | discontinuous in rope form | boiler, bale apparatus, rope apparatus, winch dyeing machine |
| | continuous, in rope form | place of deposit, rope storage, J-box, immersion jigger |
| | discontinuous, in open-width form | pad-roll, pad-batch, jigger, beam, centrifuge |
| | continuous, in open-width form | combination steamer, U-box, roll steamer, conveyor, sieve belt (travelling screen) steamer, roller-bed steamer, loop steamer, pressure HT-steamer, immersion jigger |

**Discontinuous Methods.** The treatment stages for open-width materials include impregnation with the chemicals containing liquor, squeezing off, batching (winding onto a roll; up to 10 000 m per roll), storage at room temperature (cold pad-batch method) or in a hot steam chamber (pad-roll system), and washing in an open-width washing machine. The discontinuous treatment of fabrics in rope form is always carried out in a long bath (see above).

**Continuous Methods** [4.2]–[4.4]. When treating textiles in open-width form the material is impregnated with the treatment liquor, squeezed off, and then steamed in a reaction chamber (steamer). After steaming (2–45 min, depending on the type of steamer, e.g., roller steamer, combination steamer, U-box, conveyor, or open-width J-box, see Fig. 4.1), it is washed in an open-width washing machine. The impregnation equipment, e.g., saturator and squeezing rollers, steamer, and washing machine represent one unit of each treatment stage. In multistage preparation plants, up to three units are arranged in line (desizing, boiling-off, bleaching, see Fig. 4.2). In these plants, the production rate is ca. 60–120 m/min. When the fabric is treated in rope form, the rope is impregnated, squeezed off, and kept in a steam atmosphere in a storage device known as a J-box or boot for a treatment period of 45–60 min, and then washed in a rope scouring machine. Here, too, multistage plants are usual. The production rates are in the range 100–220 m/min.

As crease marks can be formed, especially with polyester–cotton fabrics, the open-width treatment method has become the state-of-the-art technology.

The mercerization of woven fabrics can only be carried out in the open-width state in chain or chainless mercerization plants. Circular knitted materials are usually mercerized in the tubular state.

**Standard Methods for Pretreatment of Cotton and Polyester–Cotton Fabrics** [4.1]. In the pretreatment of cotton and polyester–cotton fabrics, the three-stage process (desizing, boiling off, and bleaching) gives the best and most reproducible results. The continuous treatment of the raw textile in the open-width state ensures the best uniformity.

| Types of steamer | | Dwell time, min |
|---|---|---|
| | Roller steamer | 1–2 |
| | U-box | 10–30 |
| | Combination steamer | 45–60 |
| | Conveyor | 10–30 |

**Figure 4.1.** Types of steamer

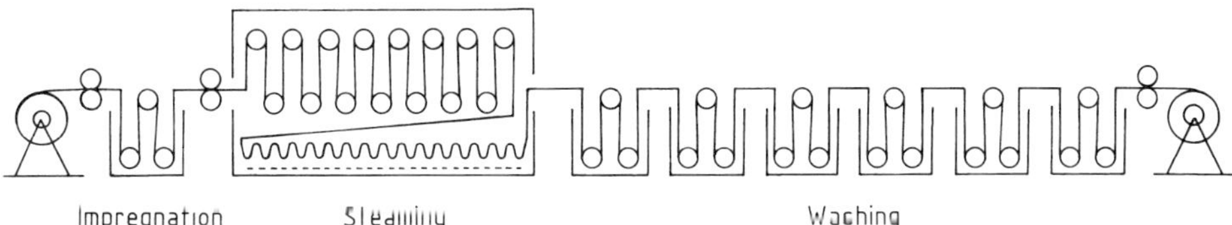


**Figure 4.2.** Continuous plant for the open-width preparation

The mercerization of cotton and cotton blend fabrics is a standard pretreatment process in most plants for improving dye yield and dimensional stability. A continuous desizing process can be achieved by using thermally stable enzymes in a steam desizing process that can be fully integrated into the pretreatment sequence, or by simultaneous oxidative desizing and caustic scouring.

If the pretreatment equipment has only one steamer, the following two-stage processes are often used:

1) Cold bleaching, alkaline extraction
2) Cold bleaching, steam bleaching
3) Enzymatic desizing, bleaching

The "Extraction Cold Bleach" followed by the "Intensive Wash" is mainly carried out in textile finishing plants in which only pad batch equipment and an open-width washer are available. Each reduction in the amount of processing steps leads to a narrowing of the textile-quality safety margins. Special auxiliaries intended to ensure improved safety margins even using shortened methods are being developed. One of their most important tasks is to give more intensive extraction of foreign matter including heavy metal ions (catalysts).

For white articles with brightnesses of > 85 % reflectance (Elrepho, Zeiss), a second bleaching stage is always necessary (hydrogen peroxide, sodium hypochlorite, sodium chlorite).

In the European textile industry, the increasing importance attched to the limits on AOX (adsorbable organic halogens) in wastewater may lead to the complete prohibition of chlorine-containing bleaching agents.

### 4.2.2. Pretreatment Part Processes

#### 4.2.2.1. Desizing

Modern dyeing and printing processes require completely desized fabrics. The most important desizing processes are enzymatic degradation of starch-based size by the impregnation-retention process, and the removal of water-soluble sizes such as carboxymethyl cellulose (CMC), poly(vinyl alcohols) and polyacrylates (see Chap. 3). Other desizing processes depend on the machines available and the production methods. They include

1) Continuous enzymatic desizing with amylases stable under steaming conditions [high-temperature (HT) enzymes]
2) Oxidative desizing with hydrogen peroxide
3) Simultaneous oxidative desizing and boiling-off

Table 4.3 lists examples of desizing agents.

In addition to the size and the sizing auxiliaries, a number of impurities are present on the fabric, which can be partly extracted, during the desizing process. In particular, these include oils, preparations, fats, waxes that melt below 70 °C, hardening agents, and catalysts.

**Table 4.3.** Examples of desizing agents [4.13]

| Product name | Producer | Use | Active substances |
| --- | --- | --- | --- |
| *Enzymes* | | | |
| Aquazym | Novo Nordisk | enzymatic desizing | enzyme |
| Bactosol TK | Sandoz | cold desizing | modified amylase |
| Beisol B 260 | CHT | enzymatic desizing | amylase |
| Biolase TS | Hoechst | enzymatic desizing | amylase |
| Encylase HT | Diamalt | hot desizing | heat-resistant amylase |
| Rapidase fl. | International Biosynthetics | enzymatic desizing | amylase |
| Thermozym | Novo Nordisk | hot desizing | heat-resistant amylase |
| *Oxidative desizing agents* | | | |
| Beisol DO | CHT | oxidative desizing/boiling off | inorganic per-salts |
| Leonil EB | Hoechst | oxidative desizing/boiling off | mixture of peroxisulfate and surfactants |
| Lufibrol DO | BASF | oxidative desizing/boiling off, cold bleaching | stabilized peroxosulfate dispersion |
| Persulfate | Degussa | oxidative desizing/boiling off, cold bleaching | peroxosulfates |

The result of a desizing process depends on

1) Solvation of the polymeric sizing agent (swelling)
2) Removal of hydrophobic impurities
3) Reaction temperature
4) Reaction time
5) Washing conditions (temperature, water supply, bath turbulence)

In practice, optimum conditions hardly ever exist, so that the process only rarely gives a completely desized product. For example, cotton wax and warp wax melt during the earlier singeing process to form a film which prevents the water absorption necessary for the swelling process, and hydrophobic impurities can locally hinder liquor absorption. Furthermore, the machines and the required production rate determine or limit the reaction time. Therefore, the desizing process is usually carried out to completion in the subsequent alkaline boiling-off and bleaching stages. The initial water-insoluble starch degradation products and the residual size are broken down partly hydrolytically and partly oxidatively and removed. Likewise, the extraction of the other foreign materials usually takes place successively during the later process steps.

**Enzymatic Degradation of Starches.** The main component of starch, amylopectin, differs from cellulose in the linkage of the glucose units ($\alpha$-glucosidic 1,4-linkage) and partial branching of the polymer chain.

Bacterial amylase [4.5], pancreas amylase [4.6], [4.7], and malt diastase ($\rightarrow$ Enzymes, **A10**, pp. 390–392) break the $\alpha$-glucosidic bond of starch stereospecifically. The $\beta$-glucosidic bond of cellulose is not attacked. Fission fragments of various molecular masses (oligosaccharides, dextrins, and glucose) and various solubilities are formed.

The stability and activity of the amylases depend on pH [4.8], temperature, and, especially with HT amylases [4.9], on the presence of sodium, potassium, and calcium ions [4.10]. Detergents also affect enzyme activity. The effect of pH and temperature on the activity of a bacterial amylase is shown in Figure 4.3.

**Desizing Auxiliaries.** Inadequate desizing is the most frequently reported pretreatment problem [4.11]. Therefore, the most important auxiliaries are products that deaerate and wet the textile, so making available as much water as

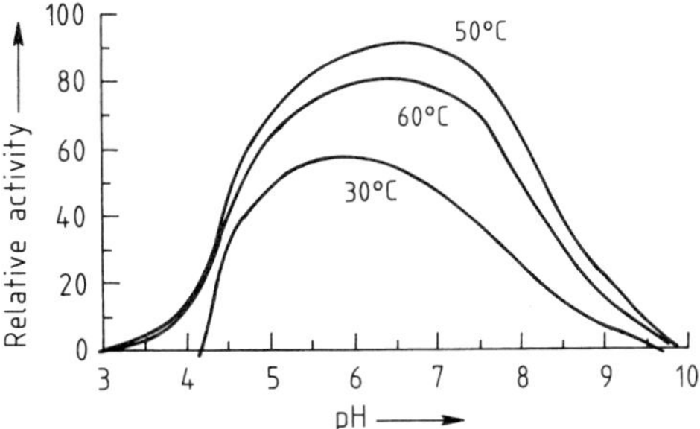

**Figure 4.3.** Activity of bacterial amylase as a function of pH and temperature

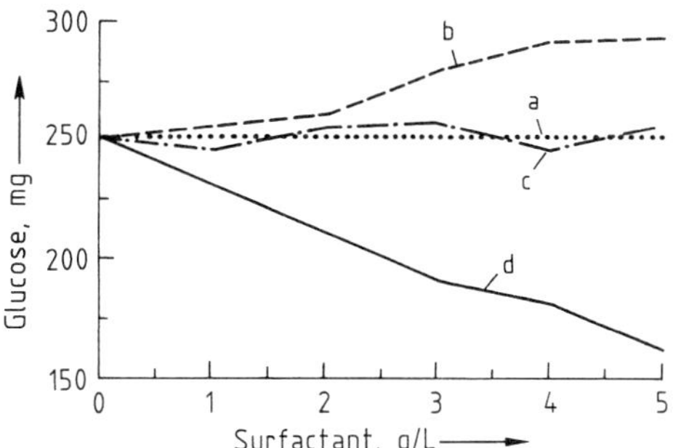

**Figure 4.4.** Effect of surfactants on the degradation of starch to glucose by enzymes
a) No surfactant; b) Octylphenol ethoxylate; c) Mixture of nonionic and anionic surfactants; d) Ester of sulfosuccinic acid

possible for the solvation and swelling of the hydrophilic polymer. This applies both to polysaccharides and synthetic sizes. If the wetting agent acts as an emulsifier, lipophilic impurities are removed. A dispersing action is necessary to disperse soil and water-insoluble size-degradation products. Special complexing agents act to demineralize the textile and extract the heavy metal ions (catalysts). Oxidative desizing agents, mainly peroxosulfates or peroxosulfate-containing auxiliaries are used in alkaline boiling-off and in cold bleaching with hydrogen peroxide. Wetting agents and detergents used with enzymes must not affect the enzyme activity at the concentrations used.

The effect of surfactants on the enzyme activity, measured indirectly as the glucose concentration on the textile, is shown in Figure 4.4.

**Oxidative Degradation of Starch Size.** Oxidative desizing with hydrogen peroxide is carried out as the first stage of a completely continuous preparation process in some Far Eastern countries. Bleach stabilizers and wetting agents are used in addition to the bleaching agent. In Europe, simultaneous oxidative desizing and scouring is carried out, if it is desired to avoid a discontinuous batching process for enzymatic desizing, if the textile contains enzyme poisons (fungicides), or if sizes are present that are difficult to degrade (branched polysaccharides).

For oxidative desizing peroxosulfates and peroxodisulfates are generally used. Oxidative desizing with bromites discussed in the past for some time is not practiced. Oxidative desizing agents, inlcuding hydrogen peroxide, attack both starch and cellulose in the same way because of their chemical similarity.

Damage to the fibers is minimized by suitable choice of formulation and by using short reaction times, so that "mild" starch degradation occurs. The treatment should lead to dispersible rather than water-soluble degradation products. Oxidative partial desizing also takes place when a cold pad-batch bleach is carried out on undesized fabric.

**Detection of Sizing Agents** [4.12]. In practice, the detection of sizing agents on gray cloth is carried out using simple methods:
1) The iodine test for the detection of starches
2) The spot test with potassium dichromate for the detection of poly(vinyl alcohol)
3) Detection of carboxymethyl cellulose by uranyl nitrate
4) Test for soluble sizes by extraction of the textile with water

The completeness of the removal of the starch size in the pretreatment process is tested for by the very sensitive starch-iodine reaction. A scale of blue colorations is available from TEGEWA (Frankfurt/Main), which enables the residual amount of starch on the textile to be estimated.

### 4.2.2.2. Alkaline Extraction

Raw cotton contains a wide range of organic and mineral foreign materials in amounts depending on its origin:

| | |
|---|---|
| Cellulose | 88–96% |
| Pectins | 0.7–1.2% |
| Wax | 0.4–1.0% |
| Proteins | 1.1–1.9% |
| Other organic compounds | 0.5–1.6% |
| Inorganic constituents (as ash) | 0.7–1.6% |

The combustion residues from cotton include alkali metal salts, which do not affect the preparation process, varying amounts of calcium and magnesium phosphate, and aluminum and iron oxides. Recent analyses of cottons show calcium, magnesium, and iron figures that fluctuate over a surprisingly wide range. Also, modern harvesting, processing, and packing plants cause increased iron contents due to abrasion, e.g., of the often rusty steel bands used for binding the cotton bales.

Alkaline boiling-off and the subsequent washing process are aimed at the efficient extraction of the above-mentioned foreign substances, which include low molecular mass cellulose components, size degradation products, residual size, dirt, and dust. Depending on the amount of impurities and the reaction and washing conditions, a weight loss of 3–7% can occur on alkaline extraction [4.14].

The extraction, also known as boiling-off, kier-boiling, or scouring, is carried out either batchwise or continuously by the impregnation/steaming process or by prolonged hot treatment. It is carried out at 90–100 °C under normal pressure. Pressurized high-temperature (HT) steamers for the open-width treatment and kier-boiling vessels for treatment in rope form are still used occasionally.

During boiling-off, intra- and intermolecular hydrogen bonds of cellulose are broken, and the polar hydroxyl groups of the polysaccharide are solvated. Swelling of the fiber takes place, which favors transport of the impurities from the inside to the outside of the fiber. During swelling, the fiber diameter increases by up to 28%, while there is a very small increase in the length of the yarn (ca. 1%). Uniform swelling improves the dyeing properties of the substrate.

The treatment with caustic liquor causes hydrolytic decomposition of leaves, fruit pods, and seed husk residues in the raw cotton, but these are only partially removed. Fats and waxes are also hydrolyzed, the degree of saponification depending on the temperature and treatment time. Sparingly soluble lime soaps are formed from the alkaline earths already present on the gray goods or picked up at a later stage. Pectins, which are mainly located in the cuticle of the cotton fiber and are bound to the cellulose fiber by hydrogen bonds, can also react via their carboxyl groups with alkaline-earth metal ions and form sparingly soluble pectinates, which must also be extracted. Proteins are released from the cotton fibers during boiling-off, which is of particular significance if the cotton cloth is bleached with sodium hypochlorite, a process still operated in many countries. This leads to the formation of chloramines, which are very difficult to remove, have a yellowing effect on the material, and damage the fibers.

The alkali concentration of 40–60 g/L NaOH used in alkaline extraction produces sodium cellulose, which is susceptible to oxidation at high temperatures. The operations are therefore carried out in saturated steam with no air present in the equipment. Where this is not possible, e.g., in the pad-roll chamber and in older conveyors, there is a risk of fiber damage by the formation of oxycellulose and cleavage of the cellulose molecule, resulting in lower DP values (mean degree of polymerization, a measure of the fiber damage, see p. 261).

Alkaline boiling-off does not completely remove all the foreign materials from the cotton. Rather, they are "chemically degraded", and can readily be oxidatively and hydrolytically decomposed and eliminated in the subsequent bleaching stage. Depending on the initial extent of soiling of the textile materials, an amount of lightening is produced which may be adequate for dyeing in medium to dark shades. The absorbency of the treated fabric only becomes fully developed in the bleaching process.

Alternatively, the material can first be bleached (e.g., by hydrogen peroxide at room temperature in the cold pad batch process) and then boiled-off. This gives an extremely good extraction of the degradation products formed in bleaching, and hence good absorbency.

**Auxiliaries for Alkaline Extraction.** In alkaline extraction, the auxiliaries have the following tasks:

1) Thorough wetting of the gray goods by the boiling-off liquor
2) Emulsification of lipophilic impurities
3) Dispersion of insoluble degradation products, soil, and lime soaps
4) Complexation of hardness formers and heavy metal ions
5) Prevention of fiber damage by atmospheric oxygen

These functions of boiling-off auxiliaries are performed by alkali-stable wetting agents/detergents (e.g., alkyl sulfonates), complexing agents such as the sodium salts of nitrilotriacetic acid (NTA), ethylenediaminetetraacetic

acid (EDTA), diethylenetriaminepentaacetic acid (DTPA), gluconic acid, phosphonic acids, special surfactant-free dispersing agents such as polyacrylates and phosphonates, and reducing agents such as sulfite and dithionite. Some examples of these auxiliaries are listed in Table 4.4.

Detailed investigations into the action of the various components of the auxiliaries are described in [4.15], [4.16]. The effects of alkali concentration, temperature, and time on the results of the treatment process are reported in [4.17]. The fiber-protecting caustic extraction has become known as the Lufibrol KB process (BASF) [4.18].

### 4.2.2.3. Bleaching with Hydrogen Peroxide

Gray goods, especially natural fibers, always contain colored foreign matter that cannot always be completely removed by washing and extraction. The whiteness of the textile materials is improved by an oxidative or reductive decomposition of these foreign materials.

Today, the most important oxidative bleaching agent is hydrogen peroxide ($H_2O_2$), although sodium hypochlorite (NaClO) and sodium chlorite ($NaClO_2$) are still important. The redox potentials of these substances under normal circumstances depend very much on the pH. The relatively low redox potential of $H_2O_2$ means that it can be used in cold and hot processes [4.19].

The most common reductive bleaching agent is sodium dithionite ($Na_2S_2O_4$); thiourea dioxide is used only to a small extent. Auxiliaries, including activators, stabilizers, buffer systems, and surfactants, are used to control the bleaching process to avoid damage to the treated gray goods and to develop good absorbency.

Hydrogen peroxide offers both technical and ecological advantages over sodium hypochlorite and sodium chlorite. First, wide range of bleaching processes can be used, including cold pad batch bleaching, bleaching under steaming conditions, impregnation methods, and bleaching processes in a long bath. Second, the decomposition of hydrogen peroxide which takes place during the bleaching reaction forms only water and oxygen.

Hydrogen peroxide is supplied as a 30 or 50 vol% aqueous solution. With a dissociation constant of $1.78 \times 10^{-12}$ at 20 °C, $H_2O_2$ is a very weak acid.

**Table 4.4.** Some auxiliaries used in boiling-off and extraction [4.13]

| Trade name | Manufacturer | Composition |
| --- | --- | --- |
| *Wetting agents/detergents* | | |
| Alcopol KBW | Allied Colloids | mixture of anionic and nonionic surfactants |
| DS 70 | Bozzetto | sulfated ether oxide |
| Lanaryl NA | ICI | alkylphenol ethoxylate and solvent |
| Kieralon B | BASF | anionic and nonionic surfactants, APEO-free* |
| Solvent Scour 880 | ICI | mixture of perchloroethene and surfactants |
| Subitol NA | CHT | modified alkanesulfonate |
| *Surfactant-free extraction agents* | | |
| Invatex SA | Ciba–Geigy | esterified derivatives with water-soluble polymers |
| Ladit 1026 | BK Ladenburg | polyphosphonate and hydroxycarboxylates |
| Lufibrol DK | BASF | mixture of polyacids |
| Lufibrol KB | BASF | organic polyacids and reducing agents |
| Sequion CM 40 | Bozzetto | salt of an acrylic copolymer, phosphate-free |
| *Surfactant-containing extraction auxiliaries* | | |
| Cottoblanc EXT | CHT | APEO-free ethoxylates, anionic surfactants, complexing agents, and dispersants |
| Delinol FE-NS | Dr. Th. Böhme | alkyl polyglycol ethers and complexing agents |
| Kappawet AT 91 | Kapp-Chemie | surfactants and complexing agents |
| Kieralon CD | BASF | anionic and nonionic surfactants with poly-acids, APEO-free |
| Ladit N 26 neu | BK Ladenburg | phosphonate, hydroxycarbonate, and surfactants |
| Sandopan SF | Sandoz | phosphoric esters, sulfonate, and poly-hydroxy hydrocarbon |
| Sevalin 2000 | TC | surfactants and sequestering agents |
| Tanawat BC | Sybron/Tanatex | nonionic and ionic surfactants and complexing agents |

* APEO = Alkyl phenol ethoxolate.

A simple description of the bleaching process is as follows:

$$H-O-O-H \xrightarrow{OH^-} H-O-O^- + H_2O$$
$$\textbf{1}$$

Hydrogen peroxide        Perhydroxy anion

$$H-O-O^- \longrightarrow H-O^- + O$$
$$\textbf{2}$$

Chromophore        Oxirane        Diol

The hydrogen peroxide is activated by the addition of alkali ($OH^-$ ions), which leads to the formation of the perhydroxy anion (**1**). This decomposes into the more stable hydroxyl ion and singlet oxygen (**2**). This "active oxygen" reacts with the double bonds of the chromophores (e.g., carotenoid pigments) that impart the characteristic brownish-yellow color to the raw cotton.

The true mechanism of the bleaching reaction is still unclear. As well as the reaction mechanism involving the perhydroxy anion and singlet oxygen atom, radical mechanisms ($OH^\bullet$) are under discussion [4.20]–[4.22].

In competition with the desired bleaching reaction, hydrogen peroxide can decompose at a comparable rate into water and oxygen with no bleaching action:

$$2\,H_2O_2 \longrightarrow 2\,H_2O + O_2$$

Cleavage to form radicals requires a high activation energy (201 kJ/mol).

$$H_2O_2 \longrightarrow 2\,OH^\bullet$$

Heavy metal compounds and other ill-defined impurities can catalyze the decomposition reaction, which then competes with the bleaching reaction. Cleavage of the hydrogen peroxide to form radicals by the Haber–Weiss mechanism is favored especially by metals whose ions can exhibit several valencies (e.g., iron, manganese, copper, and cobalt) [4.23]. The attack of radicals on the cellulose fibers leads to damage, and can form "catalyst holes" in the cotton fabric [4.24].

Alkaline oxidation of chromophore pigments is accompanied by oxidation of cellulose (Fig. 4.5), with terminal hydroxyl groups being oxidized to aldehydes and carboxylic acids, and inner hydroxyl groups to ketones. Further oxidation eventually leads to cleavage of the cellulose molecule, decreasing the mean degree of polymerization (DP value).

The DP values for crude cotton, determined by viscometry, are 2600–3200 glucose units per cellulose molecule. In practice, DP values of 1600–2200 are accepted for bleached cotton. However, the measured DP values do not give any indication of the nature and number of the various oxidation stages outlined above, or of the quality of the oxycellulose formed. Eisenhut's damage factor $s$ provides a relationship between the DP value of the bleached material and that of the greige material. In practice, the values for $s$ are usually 0.2–0.7. Higher values indicate above-average damage and make it necessary to check the pretreatment process used.

$$s = \frac{1}{\log 2} \cdot \log\left[\frac{2000}{DP_{tx}} - \frac{2000}{DP_x} + 1\right]$$

where $DP_{tx}$ is the DP value after bleaching and $DP_x$ the DP value before bleaching.

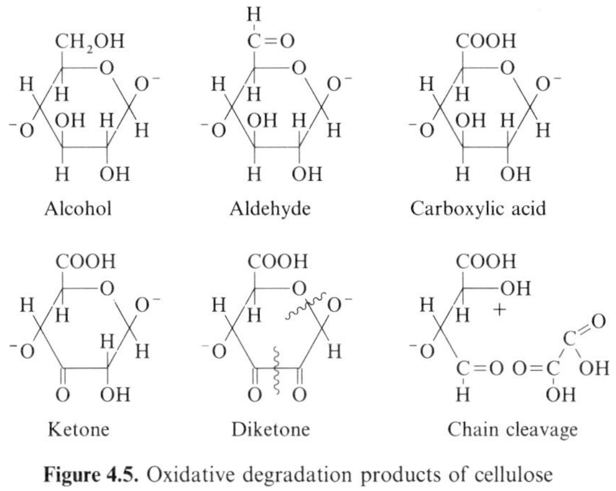

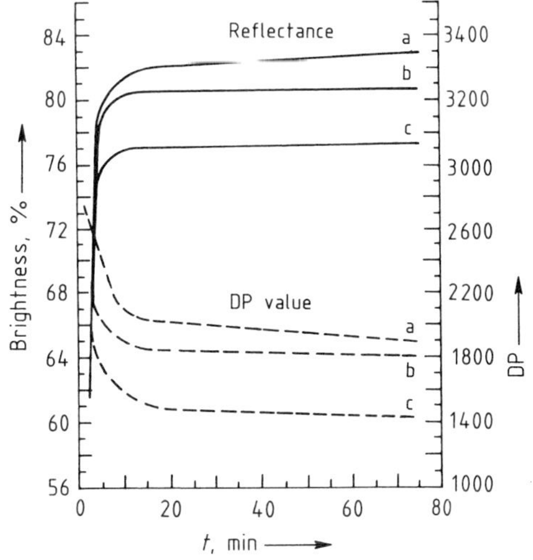


**Figure 4.5.** Oxidative degradation products of cellulose

**Effects of Hydrogen Peroxide Bleaching.** In the course of the oxidative–hydrolytic bleaching reaction, the extraction processes are carried out to completion, and the following effects are achieved:

1) The seed husks are completely removed
2) The natural dyes in the raw cotton are bleached oxidatively or are extracted after chemical degradation
3) Residual size and its degradation products are hydrolyzed, oxidized, and removed
4) Whiteness is achieved with the minimum decrease in the DP value
5) The absorbency of the material is made uniform

The resulting product is suitable for dyeing and printing. To produce completely white articles, a second bleaching process is usually necessary.

The course of a peroxide bleaching under steaming conditions is shown in Figure 4.6, which shows the reflectance–time curves and the degradation of the cellulose as functions of the increase in brightness.

The curves show that

1) The reflectance obtained is a function of the quality of the prior extraction process.
2) The decrease in the DP values correlates with the increase in the reflectance curve, depending on the quality of the prior extraction.



**Figure 4.6.** Peroxide bleaching under steaming conditions a) Desizing, boiling-off, bleaching; b) Boiling-off, bleaching; c) Bleaching

3) Longer steaming times for a given formulation do not affect the bleaching. They only affect the removal of the seed husks and the oxidation of the residual sizing agent.

If the reaction time in the steamer is limited to 10–20 min, and if the absorption of liquor on impregnation of the fabric is only 90–100%, the alkaline boiling-off stage is usually unavoidable. If the liquor pickup is 120–150%, as has recently become technically possible, the alkali stage can usually be dispensed with.

**Auxiliaries for Hydrogen Peroxide Bleaching** (Table 4.5). When bleaching with hydrogen peroxide under alkaline conditions, bleach stabilizers must be used. These inhibit the decomposition of the bleach-active perhydroxy anions and ensure a high oxidation potential over the whole bleaching time. Residual peroxide on the fabric after bleaching in amounts of 15–40% of the original hydrogen peroxide concentration indicates that the bleaching process was satisfactory, and that spontaneous catalytic decomposition has not occurred.

Stabilizers contain magnesium ions and oxidation-stable costabilizers. The positively charged magnesium ions stabilize the perhydroxy anions, while the anions of the costabilizers (e.g., phosphonate ions) form complexes with the heavy metal ions, thereby inactivating their catalytic effect. Thus, direct stabilization is caused by the magnesium ion and indirect stabilization by the costabilizer. As sodium silicate inactivates heavy metal ions, its anion also has an indirect stabilizing action [4.1].

**Table 4.5.** Some auxiliaries used in bleaching [4.13]

| Product | Manufacturer | Composition |
|---|---|---|
| Hydrogen peroxide | Peroxid-Chemie Degussa | $H_2O_2$ |
| *Stabilizers* | | |
| Actiron ASH | Protex | mixture of complexing agents |
| Baystabil KF | Bayer | organic hydroxylic compounds and carboxylic acids |
| Candoxyl ST | Sigma | phosphonic acid derivatives |
| Celidon 2820 | Bozzetto | organic complexing agents and silicates |
| Contavan SNF | CHT | organic phosphorus compounds |
| Cottozon 190 | Rotta | phosphonates |
| Delinol NSR | Dr. Th. Böhme | hydroxycarboxylic acids and surfactants |
| Lamepon 56 | Grünau | protein–fatty acid condensate |
| Lastabil 1020 | BK Ladenburg | phosphonates and organic substances |
| Prestogen D | BASF | salts of organic polyacids |
| Stabicol A | Allied Colloids | acrylic polymers and protective colloids |
| Stabilol S | Henkel | combination of organic and inorganic substances |
| Tannex 15 | Sybron/Tanatex | mixture of organic stabilizers |
| Thorstabilisator BF | Thor | combination of organic and inorganic compounds |
| Tinoclarit G | Ciba | organic metal complex compounds |
| *Co-stabilizers* | | |
| Kieralon CD | BASF | polyacids, anionic and nonionic surfactants, APEO-free* |
| Securon 540 | Henkel | organic phosphorus compounds |
| *Activators for acid bleaching* | | |
| Actiron PHD | Protex | mixture of complexing agents |
| Prestogen SP | BASF | mixture of organic activators |
| *Wetting agents/detergents and deaeration agents* | | |
| Arbyl BF | Grünau | alkanesulfonate with alkyl polyglycol ether |
| Cottoclarin OW | Henkel | combination of anionic and nonionic components |
| Erkantol CW | Bayer | phosphoric esters with emulsifiers |
| Hostapal SF | Hoechst | combination of surfactants |
| Kieralon JET-B konz. | BASF | defoamed nonionic surfactants, APEO-free* |
| Kieralon OLB konz. | BASF | combination of anionic and nonionic components, APEO-free* |
| Subitol L-SN | CHT | synergistic mixtures of surfactants |
| Tanawet PS | Sybron/Tanatex | nonionic and anionic surfactants |

* APEO = Alkyl phenol ethoxolate.

The stabilizers available commercially have various compositions. Apart from the magnesium ions responsible for the direct stabilization, there are complexing agents such as EDTA, DTPA (diethylenetriaminepentaacetic acid), gluconic acid, phosphonic acid, and poly(acrylic acid) derivatives. These so-called organic stabilizers do not contain sodium silicate. The silicate-containing stabilizers include sodium metasilicate in aqueous solution at a concentration of 38–40° Bé. Organic stabilizers containing surfactants are also marketed.

In modern steamers with reaction times up to 30 min, silicate-free bleaching in the presence of organic stabilizers has become established. Stabilization of bleaching liquors with sodium silicate and magnesium ions has the disadvantage that silicate encrustations (scaling) form in the bleaching equipment. These are difficult to remove, and damage the surface of the fabric. Silicate can also be deposited on the fabric, and this spoils the hand (feel) of the product and reduces its absorbency.

Stabilizers must have the following properties:

1) Good stabilizing action
2) Good resistance towards oxidizing agents
3) Prevention of silicate built-up on rollers in the steamer (scaling)
4) Inactivation of the catalysts
5) Good metering and pumping properties

Surfactant-containing stabilizers used with sodium silicate in the bleaching both must have silicate dispersing properties.

Surfactants used in bleaching, have emulsifying, dispersing, and wetting properties which promote the removal of hydrophobic impurities and soil, and assist in the transport of the reaction products formed by the bleaching process. The wetting properties are necessary to correct the absorbency of the pretreated goods and to make it uniform. To meet these requirements, the surfactants used are usually mixtures of anionic compounds such as alkyl sulfonates and alkyl aryl sulfonates with nonionic compounds such as alkylphenol ethoxylates or the biologically degradable fatty alcohol ethoxylates. The products must be stable in the bleaching bath, and must be suitable for metering equipment.

**Bleaching under Weakly Alkaline to Weakly Acidic Conditions.** Bleaching of cellulose fibers with hydrogen peroxide in the pH range of 6.5–8.0 is of some industrial importance. The process is used for goods dyed in the form of yarn when the dyes present change or bleed out in a normal alkaline bleach or are otherwise not bleachfast. The process is also used for blends of cotton with alkali-sensitive fibers such as wool. It is also carried out after bleaching with hypochlorite in order to save water in the subsequent washing process.

Whereas in alkaline bleaching the hydrogen peroxide is so strongly activated by the $OH^-$ ions present that it must be stabilized, in the middle pH range activators are required to give bleaching activity [4.25].

Activators include acetylated amines, acetylated amides, mixed anhydrides of acetic acid with inorganic acids such as phosphoric [4.26] and phosphorous acids, and nitriles [4.27]. The bleaching process (oxidation) presumably does not proceed directly with the hydrogen peroxide molecule, but via an intermediate association compound. Because the pH of the bleaching bath tends to decrease, a buffer must be added.

### 4.2.2.4. Bleaching with Sodium Hypochlorite

In Europe, bleaching with chlorine compounds is in decline for ecological reasons. Outside Europe, it remains highly important as a cheap bleaching method and as a prebleach for cotton with a peroxide bleach to follow [4.28].

Damage to the cotton fibers is minimized if bleaching is carried out in the pH range 9.0–11.5 [4.29]. Precise control of pH and reaction time is therefore necessary. The bleaching process must be carried out at room temperature because of the high reactivity of the hypochlorite.

### 4.2.2.5. Bleaching with Sodium Chlorite

Chlorite bleaching for the pretreatment of cotton, linen, flax, jute, and other cellulosic fibers is carried out by the impregnation–steam process (e.g., pad-roll system) or long bath process. When used in conjunction with hydrogen peroxide bleaching, either before or after the chlorite bleaching, high reflectance is obtained. Textiles with a high catalyst content can be safely bleached, as heavy metal compounds are catalytically inactive in chlorite bleaching [4.30].

The chlorite method has declined in importance for ecological reasons and because of its severe corrosivity, but the main reason for its decline is the introduction of continuous open-width bleaching processes for cotton articles.

### 4.2.2.6. Mercerization and Causticizing

Mercerization is the treatment of cotton under tension with caustic soda solution at 15–25 °C for 25–40 s. The concentrations normally used in practice for woven materials are in the range 20–24% NaOH (26–30° Bé), and for yarn, 24–26% NaOH (30–32° Bé). Treatment with concentrated caustic soda destroys the spiral form of the cellulose with formation of alkali cellulose (alcoholate), which is changed to cellulose hydrate on washing out the caustic soda.

The process has the following beneficial effects on the textile [4.31]:

1) Improved ability to take up dyes
2) Improvement of the dyeability of dead cotton
3) Luster
4) Good dimensional stability of woven fabrics
5) Higher tear strength of yarns

In dry mercerization, the impregnation of the dry fabric can be done on the greige cloth or after any part process, e.g., desizing, alkaline boiling-off, or bleaching, followed by an intermediate drying [4.32].

In the Mercevic process, which is also referred to as dry mercerization, the process is carried out while drying the fabric on the stenter frame [4.33], [4.34]. In wet mercerization, intermediate drying is not carried out for cost reasons. Here, it is essential for effective mercerization that the water in the fabric should be displaced as completely as possible by the concentrated caustic solution.

For effective mercerization, complete penetration (wetting) of the cotton fibers by the concentrated caustic liquor must occur. This is hindered by the hydrophobic substances present on the gray goods, size or size residues, and by the high surface tension of the liquor. Effective mercerization therefore requires the use of wetting agents.

Hot impregnation followed by mercerization at room temperature or with cooling, in the presence of wetting agents, is also used (hot mercerization) [4.35]–[4.37].

In the causticizing process, the material is treated with less concentrated caustic liquor without applying tension. The material is allowed to shrink, whereby the dye absorption is improved. Cotton is treated with 13–16% liquor (18–22° Bé), and viscose staple fiber with 3–5% liquor. Causticizing of viscose staple fiber and polyester–viscose articles gives the material a wool-like hand. A combined caustic scour/causticizing process with 18–22° Bé NaOH under steam conditions is sometimes applied to cotton fabric. Mercerization wetting agents include $C_4$–$C_{12}$ sulfuric half-esters of aliphatic alcohols, alkyl sulfonates from the sulfochlorination of alkanes, and ether alcohols such as butylene glycol monoethyl ether.

Alkanolamines and their mixtures with alkyl sulfonates and ether sulfates are also used. To reduce foaming phosphoric esters of $C_4$–$C_8$ alcohols may be added. The wetting agents for treatment with 13–16% caustic liquor are similar to those used for mercerization. Wetting agents used for the caustic treatment of viscose staple fiber include anionic surfactants or mixed anionic/nonionic wetting agents based on alkyl sulfonate/alkyphenol ethoxylate or alkyl sulfonate/fatty alcohol ethoxylate. These must be stable in the alkaline bath (i.e., they must not separate, forming a creamy layer at the surface). They must have a good wetting action and must not cause foaming. The first wetting agents used in the mercerization of cotton were phenol- and cresol-containing products which were mixed with amines, naphthenic acids, resin acids, and alkylnaphthyl sulfonates. The mercerization wetting agents used today are free of phenol or cresol.

The effectiveness of a mercerization wetting agent is assessed by measuring the shrinkage of a cotton yarn as a function of the wetting time.

Some auxiliaries for mercerization and causticizing are listed in Table 4.6.

## 4.3. Preparation of Woolen Textiles
(→ Wool)

The composition of raw wool varies widely, depending on the amount of impurities and foreign materials. On average, raw wool contains only ca. 50% fiber. The approximate composition is as follows:

1) Wool wax (5–10%), which is insoluble in water, consists of a mixture of fatty acid esters of higher alcohols, free fatty acids, and free higher alcohols

**Table 4.6.** Some auxiliaries used in mercerization and caustic treatment [4.13]

| Trade name | Producer | Compositon |
|---|---|---|
| Floranit 4028 | Henkel | derivative of special alcohols |
| Invadin MC neu | Ciba | preparation from an alkanol sulfonate |
| Leonil GA | Hoechst | derivative of special alcohols |
| Leophen BN | BASF | alkylsulfonate with additives |
| Leophen MC | BASF | derivative of special alcohols |
| Mercerisin OR | Thor | alkyl sulfate |
| Mercerol QW | Sandoz | alkyl sulfate |
| Natron | Bozzetto | sulfated alcohols |
| Subitol MLF | CHT | sulfated alcohols |
| Sultafon MES 200 | Stockhausen | combination of various sulfonates and sulfated fats |

2) Wool grease (2–13%), which is soluble in water, consists of lower carboxylic acids and aldehydes and inorganic constituents
3) Vegetable and mineral foreign materials (5–40%)
4) Fibrous materials (proteins, 20–80%)
5) Water (10–20%)

**Raw Wool Washing.** The first cleaning process applied to the fleece after shearing is washing in the Leviathan wool washing machine. This continuous process removes the wool wax, wool grease, sand, soil, and other impurities from the loose fibrous material. The washing medium, which has a pH of 7–10, contains alkylphenol ethoxylates and fatty alcohol ethoxylates [4.38]. Soaps are still occasionally used.

**Carbonization** is the treatment of wool with acid to remove plant constituents such as burrs, straw, wood fragments, and seeds. In the treatment of spinning mill wastes, any cellulose fibers still present are also removed. The vegetable impurities are mainly decomposed by hydrolysis with sulfuric acid, and can then be easily removed by beating or brushing. If the process is correctly carried out, very little damage to the wool occurs. Carbonization with sulfuric acid consists of the following steps:

1) Impregnation of the fiber with carbonization liquor (3–6% $H_2SO_4$)

2) Concentration of the carbonization chemicals on the wool fiber by heating to 70–90 °C (evaporation of some of the water)
3) "Burning" by dehydration and carbonization of the vegetable matter at 100–115 °C
4) Mechanical removal of the carbonized vegetable impurities
5) Neutralization and removal of the chemicals

The use of wetting agents (Table 4.7) [4.39]–[4.43] such as alkylnaphthalene sulfonates, alkylphenol ethoxylates, and alkyl polyglycol ethers leads to thorough wetting of the gray goods and reduces impregnation time. They also enable the concentration of the acid to be reduced and protect wool fibers [4.40].

The treated material must be completely neutralized to avoid damage to the wool. This is achieved by having an appreciable excess of sodi-

**Table 4.7.** Some auxiliaries used in the carbonization and bleaching of wool [4.13]

| Trade name | Producer | Composition |
|---|---|---|
| *Carbonization auxiliaries* | | |
| Bura-Netzer | Baur & Gäbel | sulfonate mixture with nonionic surfactants |
| Lavotan DSU 100 | CHT | fatty alcohol ethoxylates with sulfonates |
| Leophen RA | BASF | sulfosuccinic acid ester |
| Sultafon TK | Stockhausen | special ethylene oxide condensation product |
| Teca-Rapid S | Dr. Angele | alkane sulfonate mixture |
| *Bleaching auxiliaries* | | |
| Blankit IN | BASF | combination of stabilized reducing agents |
| Blankolen K | Brüggemann | combination of stabilized reducing agents |
| Blankit D | BASF | combination of active reducing agents |
| Blankit IAN | BASF | combination of stabilized reducing agents with optical brighteners |
| Celidon LA | Bozzetto | phosphate derivatives |
| Kappazon B 84 | Kapp-Chemie | alkali polyphosphate |
| Prestogen W fl. | BASF | salts of organic polyacids |
| Tubotex PCA fl. | CHT | inorganic salts, alkali buffers, and complexing agents |

um carbonate in the neutralizing liquor. The surfactants present improve the wetting of the wool fiber [4.43].

As well as sulfuric acid, HCl gas and hydrochloric acid are used for carbonization, especially when recovering wool from rags [4.44]. In the United States, a mixture of aluminum chloride and hydrochloric acid at a treatment temperature of ca. 130 °C is preferred [4.45]. Aluminum chloride gives the best results in the carbonization of wool–synthetic fiber blends [4.46].

**Bleaching Processes.** The colors of washed wool range from cream to almost bleach (e.g., the strongly pigmented fibers of Karakul wool). All types of wool are bleached to give the whiteness necessary for dyeing with pastel tones or to enable the wool to be used for white articles. Bleaching agents are also often used during raw wool washing to brighten the color of the raw wool and to give it a more pleasing appearance.

Oxidative bleaching agents such as hydrogen peroxide and reductive bleaching agents such as sodium dithionite are used. To give a high degree of whiteness and improved lightfastness, two-stage oxidative and reductive bleaching is carried out in a conventional complete bleaching process. This bleaching method leads to high reflectance but is very time-consuming, so that ways of improving productivity are being sought. New bleaching agents can either give more rapid bleaching in existing equipment, or can enable the padding technique, a semicontinuous or continuous process, to be used. Also, several operations can be combined, leading to cost reduction. The following processes are used [4.47]:

1) Conventional wool bleaching in two stages: oxidative and reductive
2) Rapid acid bleaching with hydrogen peroxide in either a discontinuous or a continuous process
3) Reductive bleaching with stabilized sodium dithionite or thiourea dioxide, with or without whiteners
4) Reductive bleaching based on activated sulfinic acid derivatives, enabling fully and partly continuous bleaching with simultaneous whitening to be carried out
5) Oxidative and reductive bleaching in the dyebath to produce clear color shades

Oxidative bleaching of wool with hydrogen peroxide is performed in the weakly alkaline pH range (8–9.5) below 50 °C because of the sensitivity of the wool fibers to alkali. These conditions lead to long reaction times. Sodium pyrophosphate is usually used as a stabilizer. Peracetic acid, produced in the bleaching bath from hydrogen peroxide and acetic acid, is used in the weakly acid range [4.48]. Pure peracetic acid is less effective in the bleaching bath than the combination $H_2O_2/CH_3COOH$ [4.49]. Activation of the hydrogen peroxide can also be achieved with formic acid or formaldehyde, which form performic acid [4.50], [4.51]. Peroxycarboxylic acids formed, e.g., by the addition of Prestogen W (BASF), have a similar effect.

Another special use of activated acid solutions of hydrogen peroxide is in the mordanting of wool and the catalytic mordanting of animal hair. Here, activators that cause decomposition of the $H_2O_2$ include metallic salts and modified metallic salts of iron, cobalt, nickel, manganese, and copper [4.52]–[4.54].

The products used in the reductive bleaching of wool are mainly based on sodium dithionite, though some have thiourea dioxide as their main component. All reductive bleaching agents contain stabilizers (Table 4.7) to prevent the reducing agent from rapidly decomposing, a reaction which is catalyzed by the acidic degradation products of the bleaching process.

The pH is usually kept within the range 7–9 at the start, and 7–5 at the end of the process. Phosphates are used for buffering. Complexing agents are used to combine with calcium and iron ions and prevent the formation of precipitates (carbonate hardness). The bleaching time is determined by the consumption of reducing agent, which is affected by the amount of air dispersed in the circulating liquor [4.53].

When wetting agents and detergents are required, surfactants based on alkylphenol ethoxylates or fatty alcohol ethoxylates, sometimes mixed with alkyl sulfonates or alkylaryl sulfonates, are used.

## 4.4. Ecological Aspects of Fabric Pretreatment [4.55]–[4.62]

The new laws and regulations for the protection of the environment that are being applied in most industrialized countries are having a considerable impact on the textile industry.

In Germany, according to the dangerous substances regulations covering the discharge of

**Table 4.8.** Threshold values for impurities in wastewater

| AOX* | 3 mg/L | Cr | 2 mg/L |
|---|---|---|---|
| VHC** | 1 mg/L | $Cr^{6+}$ | 0.5 mg/L |
| Hydrocarbons | 50 mg/L | Zn | 10 mg/L |
| Cu | 2 mg/L | Sn | 10 mg/L |
| Ni | 2 mg/L | | |

* The AOX threshold value for bleaching with chlorine and antifelting finishing of wool is 8 mg/L (until 31.12.96).
** VHC = volatile halocarbons

**Table 4.9.** Threshold values attainable by current technology

| COD | 160 mg/L | ammoniacal nitrogen | 10 mg/L |
|---|---|---|---|
| $BOD_5$ | 25 mg/L | iron | 3 mg/L |
| Phosphorus total | 2 mg/L | aluminum | 3 mg/L |

**Table 4.10.** Threshold values attainable by current technology for direct and indirect dischargers

| $T_F$ (Dilution factor) 2 | | | |
|---|---|---|---|
| AOX | 0.5 mg/L | Cu | 0.5 mg/L |
| VHC* | 0.1 mg/L | $Cr^{6+}$ | 0.5 mg/L |
| Free chlorine | 0.3 mg/L | Ni | 0.5 mg/L |
| Hydrocarbons | 15 mg/L | Pb | 0.5 mg/L |
| Sulfide | 1 mg/L | Zn | 2 mg/L |
| Sulfite | 1 mg/L | Sn | 2 mg/L |

* VHC = volatile halocarbons

wastewater into open sewage installations, a license is required if the amounts and concentrations exceed prescribed threshold values (Tables 4.8–4.10). According to the recent law on detergents, wetting agents and detergents are deemed to include sizing agents, boiling-off auxiliaries, bleaching stabilizers, complexing agents, softeners, and finishing materials. All these products must be biologically degradable or capable of being biologically eliminated. Since 1989, the TEGEWA Association of textile companies has voluntarily abandoned the use of APEO-containing auxiliaries, as the biological breakdown of these materials produces metabolites that are toxic to fish.

# 5. Textile Dyeing Auxiliaries

**Introduction.** The dyeing of textiles requires the use not only of dyes and chemicals, but also a number of special products known as dyeing auxiliaries [5.6]. These materials may constitute an integral part of the dyeing process (e.g., reducing agents for dyeing with vat dyes) or can lead to technical improvements such as more (level) dyeing, better fastness, etc. [5.7].

The following classification and definitions of dyeing auxiliaries are taken from the TEGEWA brochures [5.8]. Classification is based on product use. The descriptions of the product groups indicate whether they are suitable for the exhaustion dyeing process (batchwise long-liquor dyeing) or for the padding process (→ Textile Dyeing), although many dyeing auxiliaries can be used in either process, including dye solubilizing and hydrotropic agents, dye wetting agents, aftertreatment agents, and reducing agents. Some special padding auxiliaries have been placed in a category of their own (Section 5.9). Levelling agents, carriers, and anticreasing agents are only used in the exhaustion dyeing process.

The dyeing auxiliaries form a very heterogeneous group of chemicals. Though many of them are surfactants, there are many other types, including inorganic compounds, water-soluble polymers and oligomers, polymer dispersions, and solubilizing agents. Most commercial dyeing auxiliaries are preparations containing several components. The individual chemical compounds, usually along with relevant environmental data, are described under separate keywords (e.g., → Polyacrylates, → Polyacrylamides and Poly(Acrylic Acids), → Reduction → Sulfites, Thiosulfates, and Dithionites, → Solvents, → Surfactants).

Consumption of dyeing auxiliaries is estimated at 60–70 % of the consumption of the dyes themselves. In 1992 this corresponded to ca. 280 000 t/a dyeing auxiliarics for a world consumption of 440 000 t/a dyes used for dyeing textiles [5.9].

## 5.1. Dye Solubilizing and Hydrotropic Agents

To dissolve large amounts of a dye in a small amount of water, solubilizing and hydrotropic agents are used to increase its solubility. This is mainly necessary when dye baths are used in padding processes, dye stock solutions, and printing pastes (see Section 6.2.6). The effectiveness of hydrotropic agents is due to their amphiphilic properties. Intermolecular cohesive forces lead to the formation of molecular associates with the substances to be dissolved, and these are water-soluble due to the presence of hydrophilic groups [5.10]. In this respect, the hydrotropic agents resemble surfactants, which also have a solubilizing effect. A difference is that surfactants can increase the solubility of sparingly soluble substances at relatively low concentrations by absorption into micelles [5.1].

Solvents are used in dyeing and printing to wash dye residues from equipment and apparatus. Many auxiliaries used in continuous dyeing contain solvents, hydrotropic agents, and surfactants, not only because of their ability to solubilize dyes, but also to improve the fixing process, produce of fiber swelling, and to provide wetting and foaming properties, etc. (see Section 5.9).

The commercial products supplied for dissolving dyes often contain mixtures of solvents, dispersants, and surfactants. Solvents and hydrotropic agents are needed when dyeing with the following classes of dyes:

**Developing Dyes.** To dissolve naphthols by the cold dissolution process (→ Textile Dyeing, p. 417), ethanol or certain heterocyclic bases (pyridine derivates, *N*-methylpyrrolidone) are used as solvent.

**Phthalogen Dyes.** To dissolve the sparingly soluble aminoisoindolines and other precursors of these dyes, special mixtures of glycols, amines, and surfactants are used (→ Textile Dyeing, pp. 420–421).

**Reactive Dyes.** In the continuous dyeing of cellulose fibers with reactive dyes by the pad dry thermofix process (→ Textile Dyeing, Section 4.1), large amounts of urea (50–100 g/L) are added to the liquor. These increase the solubility of the dye in the pad-dyeing liquor and, during the fixing process, form a melt which allows the dye to diffuse into the fiber. In many cases, this improves the dye yield [5.11]. Dicyandiamide, which has a lower tendency to sublime, can be used instead of urea.

**Acid dyes and metal-complex dyes** often have very low solubility in water, and dye solubilizing agents must be added if they are to be used in pad liquors or printing pastes for wool or polyamide, especially with dark colors. Suitable materials include ethanol, propanol, di- and triglycol, thiodiglycol, various glycol ethers, and nonionic surfactants.

## 5.2. Dye Protecting Agents and Boildown Protecting Agents

Under unfavorable conditions, certain dyes can be changed or destroyed during application. In these cases, special protecting agents are added to the dye baths. Boildown effects occur when dyeing wool and wool–cellulose fiber blends, meaning that the azo dyes used are reductively decomposed by breakdown products from the wool proteins, or undergo changes in shade [5.12]. Dye boildown can be prevented by adding buffering agents and special auxiliaries. Azo dyes can also be reduced when dyeing cellulose fibers with direct dyes at boiling temperatures. These defects can be prevented by precise control of the pH by the addition of buffers and oxidizing agents. Similar defects can occur on dyeing with reactive dyes, which can be modified through reduction by cellulose degradation in the presence of alkali at elevated temperatures. The dye is protected by the mild oxidizing agent sodium 3-nitrobenzenesulfonate [5.13]. This product is also used to protect the dye from reduction when boiling off before bleaching colored piece goods.

When dyeing bleached cotton with reactive dyes, which are very sensitive to oxidation and reduction, even very small amounts of residual hydrogen peroxide must be destroyed by weak reducing agents. Vat dyes of the indanthrone type tend to give cloudy and green tints at dyeing temperatures above 60 °C owing to over-reduction. This effect can be prevented by adding sodium nitrite or glucose [5.14]. Some commercial products are listed in Table 5.1.

## 5.3. Wetting and Deaerating Agents

A fundamental prerequisite for successful dyeing with an aqueous dye bath is complete wetting of the textile. Only then is uniform and level dyeing possible. This is achieved by means of wetting and deaerating agents [5.15], whose use depends on the dyeing process and the nature and condition of the textile [5.1].

**Exhaustion Dyeing Process (Batch Dyeing).** Textiles made from natural fibers such as cotton and wool generally have good wetting properties after pretreatment such as washing, boiling-off, or bleaching. However, it may be economically advantageous to carry out pretreatment and dye-

**Table 5.1.** Some commercial dye protecting agents [5.5]

| Name | Producer | Uses | Composition |
|---|---|---|---|
| Buralit N | Baur & Gäbel, Germany | prevention of boildown of substantive dyes | combination of ammonium salt with aliphatic and aromatic sulfonic acids |
| Degal SBN | Sigma, Germany | antireduction agents | sodium nitroarylsulfonate |
| Dupranin RKM conc. | Thor-Chemie, Germany | reducing agent for single bath bleaching and dyeing with reactive dyes | inorganic salt |
| Intratex BD | Crompton & Knowles, United Kingdom | prevents reduction on boiling off, dyeing, and printing | sodium salt of an aromatic nitrosulfonic acid |
| Ludigol | BASF, Germany | oxidizing agent | sodium 3-nitrobenzenesulfonate |
| Matexil BA-PK | ICI, United Kingdom | for removal of hydrogen peroxide on dyeing with reactive dyes | mixture of inorganic salts |
| Matexil PA-L | ICI, United Kingdom | weak oxidizing agent | sodium 3-nitrobenzenesulfonate |
| Meropan SO2 | CHT, Germany | prevention of boildown of direct and reactive dyes | benzenesulfonic acid derivative |
| Revatol S/SP Powder | Sandoz, Switzerland | prevention of reduction on boiling off, dyeing, and printing | sodium 3-nitrobenezenesulfonate |
| Sustilan N | Bayer, Germany | fiber protecting agent for wool and boildown protecting agents | combination of ammonium salts, aliphatic and aliphatic aromatic sulfonic acids |

ing in a single step. This can be achieved by the use of wetting agents and protective colloids (see Section 5.4.2). Wetting agents are not necessary when batch dyeing synthetic fibers. However, when dyeing wound packages (cross-wound bobbins or warp beams) a deaeration step is required at the beginning of the dyeing process because of the high density and thickness of the textile, irrespective of the type of fiber and its pretreatment. This prevents mechanical deformation of the wound package by the included and absorbed air when the dye liquor is introduced under pressure. The deaeration of wound packages of pretreated natural fibers and of synthetic fibers can usually be carried out by gently displacing the air in a preliminary wetting stage, accelerated by using a deaerating agent. With raw cotton, which is difficult to wet, the use of a deaerating agent is essential.

**Continuous Processes** (see Section 5.9). Continuous dyeing processes are characterized by very short contact times between the pad liquor and the textile when applying the dyes and chemicals.

Inadequate, nonuniform wetting leads to reduced uptake of the dye and impairment of the fabric appearance. Wetting agents must therefore always be added to the impregnation bath when gray cloth, inadequately pretreated textile materials, or any material that is difficult to wet is dyed by the pad dyeing process [5.1], [5.3], [5.16].

The very short contact times in the pad dyeing process therefore require the use of products that cause very fast wetting (rapid wetting agents). Not only is the liquor caused to penetrate the interior of the textile material (capillaries, fiber crossing points) rapidly, but also a continuous film of dye liquor is formed on the surface of the textile. Such wetting agents must maintain their effectiveness not only in neutral solutions, but also in baths containing alkalis and electrolytes, and must ensure that as little foaming as possible occurs in the baths even for high throughputs [5.16], [5.17]. The stability of the pad liquor must not be affected, as even small amounts of slight creaming and dye precipitation on the cloth can lead to spots and marks on the textile. Dye wetting agents differ from pretreatment products as follows:

1) Exhaustion dyeing process: low foam formation, no dye-retention effect
2) Continuous process: no fiber affinity, good wetting action

Wetting and deaerating agents are surface-active substances. They cause spreading of the aqueous liquor over the surface of the fibers, thereby displacing the air [5.18]. In the usual type of surfactant wetting agents, which promote spreading by reducing the surface tension of the liquor, this is associated with foam formation [5.19], which causes severe problems, especially during the removal of air from wound packages. Wetting with the formation of little or no foam can be achieved in two ways. Surfactant mixtures for reducing the surface tension of the liquor can be used in combination with antifoams, or the spreading of the liquor is achieved by increasing the surface tension of the fibers by adsorption of polar, hydrophobic compounds [5.20], which do not form stable foam lamellae. Wetting agents for exhaustion and continuous processes usually contain the following surfactants: alkyl- or alkylaryl sulfates, alkylaryl ethoxylates, polyglycol ether sulfates. Rapid wetting agents contain a branched hydrophobic molecular component (e.g., secondary alkylsulfonates, isooctyl esters of sulfosuccinic acid, and dialkylsulfonimides). Deaerating products are emulsions of hydrophobic phoshoric triesters and mixtures of phosphoric acid partial esters, and mixtures of surfactants with antifoams. Some commercial products are listed in Table 5.2.

## 5.4. Dispersants and Protective Colloids

Insoluble dyes applied in the form of aqueous dispersions are used in a large number of dyeing and printing processes. Dispersants are required to produce the dye preparations for these processes and to stabilize the finely dispersed state during application. Powdered dispersion and vat dyes contain 50–80% of these products, and dyes prepared in liquid form 10–30%. To maintain the stability of the dispersion throughout the dyeing or printing process, additional dispersant is added to the dye bath. Also, during the dyeing of plant fibers (e.g., raw cotton) which have not been pretreated, considerable amounts of impurities in the form of waxes, pectinates, and water hardness are introduced into the dye baths. The dispersants must also prevent these from being precipitated.

Dispersants used in textile finishing can be divided into two classes:

1) Water-soluble oligo- and polyelectrolytes ("protective colloids") [5.4, p. 459]
2) Surfactants

Both have an amphiphilic structure with hydrophilic and hydrophobic components. The most important examples of these compounds are the polysulfonates, polyacrylates, polyvinylsulfonates, and anionic and nonionic surfactants. The activity of these substances is based on the formation of electrostatic and mechanical protective films around the dispersed particles which prevent precipitation and agglomeration.

Specific dispersants are available for dyes, plant fiber impurities, and precipitates due to water hardness.

**Dispersants for Dyes.** Condensation products of aromatic sulfonic acids with formaldehyde and lignosulfonates are widely used. Anionic and

**Table 5.2.** Some commercially available wetting and deaeration agents for dyeing [5.5]

| Name | Producer | Uses | Composition |
| --- | --- | --- | --- |
| Albegal FFA | Ciba-Geigy[a] | deaeration agent | alkylaryl ethoxylates + polysiloxane |
| Arbyl ASN | Grünau[b] | rapid wetting agent | diisooctyl sulfosuccinate |
| Erkantol CW | Bayer[b] | deaeration agent | phosphoric esters + emulsifiers |
| Irgapadol FFU | Ciba-Geigy[a] | pad wetting agent | alkylaryl polyglycol ether sulfate + poly(ethylene glycol) |
| Leophen M | BASF[b] | deaeration agent | phosphoric acid esters + emulsifiers |
| Leophen RBD | BASF[a] | rapid wetting agent | alkylsulfonimide |
| Perenin 4649 | Dr. Th. Böhme[b] | pad wetting agent | phosphoric esters |
| Periwet SL | Dr. Petry[b] | deaertion agent | fatty alcohol ether sulfate |
| Primasol NF | BASF[b] | pad wetting agent | phosphoric esters |
| Rewopol SBD 075 | Rewo[b] | rapid wetting agent | diisooctyl sulfosuccinate |
| Subitol SAN Neu | CHT[b] | wetting agent | phosphoric esters |
| Tanawet PAD | Sybron/Tanatex[c] | pad wetting agent | phosphoric esters |

[a] Switzerland;   [b] Germany;   [c] The Netherlands.

nonionic surfactants are used for special applications.

*Formaldehyde Condensation Products.* Condensation products of β-naphthalenesulfonic acid with formaldehyde [5.22] have been known since 1913 [5.21], and are still used in dye production and dyeing. In addition to these, condensation products of phenol with formaldehyde and sodium sulfite (sulfomethylation products) are now also used as dispersing agents. The degree of condensation depends on the conditions of the reaction with formaldehyde, and can give products with 2–10 aromatic nuclei. However, the structure of this, the most common dispersing agent, has not yet been fully elucidated. Conventional formaldehyde condensation products and the lignosulfonates described below are only ca. 30 % degraded in biological wastewater treatment plants (test method OECD 302 B). However, a formaldehyde condensation product which can be > 70 % eliminated from the wastewater has been commercially available since 1993 [5.23]. This product is used both for finishing Palanil and indanthrene dyes and as a dyeing auxiliary (see Section 5.4.2) under the trade name Setamol E.

*Lignosulfonates.* Products based on natural lignins form a further large group of disperants for finishing and as auxiliaries in the dye bath. They are produced from sulfite pulping liquor or from alkali lignins produced in the kraft pulping process [5.24]. The molecular mass of industrial lignosulfonates, whose structure is still not completely elucidated, lies between 2000 and 100 000 [5.25]. The basic building block is phenylpropane, from which a large number of derivatives can be obtained. The suitability of the lignosulfonates for the various applications depends very much on their purity, degree of sulfonation, number of hydroxyl groups, and molecular mass distribution. The desired properties are obtained by the purification and chemical treatment of the crude products [5.26]. Some undesirable properties of the lignosulfonates include a reducing action on sensitive azo dyes during high temperature dyeing (130 °C) and possible soiling of the fiber due to the dark color of these products.

*Other Compounds.* Other natural and synthetic high molecular mass compounds are used, especially in the production of pigments [5.27]. These include polysaccharides, alginates, cellulose derivates, polyacrylates and polyvinyl compounds.

*Dispersants for cotton impurities and precipitates from water hardness* are polycarboxylates [5.28]–[5.30], which are homopolymers and copolymers of acrylic and maleic acids with relative molecular masses of 1000–10 000. With higher molecular mass, the protective colloid effect becomes a flocculating action [5.28]. These substances have wide application in many processes for dyeing cellulose fibers, as they do not cause foaming or dye retention.

### 5.4.1. Dispersants in Dye Production

Vat and disperse dyes must be finely dispersed in the dye bath (→ Textile Dyeing, pp. 399–421; 434–441). This is achieved by a special treatment (finishing) which follows the synthesis of the dye. Finishing consists of precipitation, filtration, grinding in the presence of dispersants, and drying. Dye finishing has a major influence on the dyeing process. Formaldehyde condensation products and lignosulfonates are widely used in dye finishing [5.31]. They give improved particle size reduction on grinding, good redispersibility of the finished dye, low staining of the textile fibers by the dispersant, and good solubility and temperature stability of the dispersion. To give improved compatibility with synthetic thickeners for printing, dyes for the printing process are often finished with nonionic dispersants.

### 5.4.2. Dispersants for Dyeing (Batch and Continuous Processes)

**Batch Dyeing Processes with Disperse Dyes.** Disperse dyes have very low solubility in water, and exist as suspensions or emulsions even at the high temperatures of the PES (polyester) dyeing process (120–140 °C). They provide the reservoir for a saturated dye solution from which molecularly dispersed dye is absorbed by the fiber [5.32]. Since fineness of this dispersion and its stability during the entire dyeing process are of crucial importance for dye deposition, that dispersants must be used. They prevent a filtering effect of agglomerated dye in package dyeing, and hence prevent unlevel results, spots, and unsatisfactory dye yield and dyefastness. Although all disperse dyes already have a high content of dispersants, 0.5–2 g/L dispersant is usually added to the dye liquor if the thermal and mechanical stress on the dye bath is very intense.

Formaldehyde condensation products and similar compounds, nonionic surfactants, and anionic surface active compounds can be used. As the nonionic surfactants often have a levelling effect as well as a dispersing effect, auxiliaries for dyeing polyester fibers often consist of mixtures of various compounds, and are used both as dispersing agents and levelling agents.

When dyeing secondary acetate fibers and polyamide fibers with disperse dyes, the sulfoaromatic condensation products described above or nonionic surfactants are used as dispersants.

In the aftertreatment of dyed polyester fibers, dispersants help to remove the dye adhering to the surface in an additional, usually reductive washing stage, thereby improving the fastness of the dyeing. Nonionic surfactants in amounts of 0.5–1 g/L are usually added as dispersants and washing agents to the final washing bath.

**Batch Dyeing Processes with Vat and Sulfur Dyes.** Vat and sulfur dyes are converted by reduction into the water-soluble, oxidation-sensitive leuco form which is adsorbed by to the cellulose fiber.

Dispersants are therefore only necessary for process steps in which the pigment has not yet been reduced or has been re-formed by oxidation:

1) In the preimpregnation process (→ Textile Dyeing, Sections 4.3.6, 4.5.4), dispersants prevent destabilization of the dye dispersion.
2) When dyeing in open equipment (jiggers, hank dyeing machines), pigment is continuously formed by atmospheric oxygen. Dispersants prevent agglomeration, and thereby enable rapid revatting to be carried out.
3) In the oxidation stage at the end of the dyeing process, a large amount of pigment is formed in the dye liquor, especially with dark colors. Dispersing agents prevent the agglomeration of these pigment particles.

Dispersing agents used include naphthalenesulfonic acid–formaldehyde condensates, lignosulfonates, sulfonated oils, and protein–fatty acid condensation products, which are used in concentrations of 0.5–2 g/L [5.33], [5.34].

When dyeing raw cotton, the following auxiliaries are also used in concentrations of 1–2 g/L to prevent the agglomeration of impurities and of precipitates due to water hardness: polycar-

boxylates as dispersing agents and moderately strong complexing agents, and pure complexing agents (see Section 5.5) such as phosphonates, EDTA (ethylenediaminetetraacetic acid), and NTA (nitrilotriacetic acid).

**Batch Dyeing Processes with Other Classes of Dye.** When dyeing with developing dyes, dispersants are used during dissolution of the naphthols and preparation of the primary and development baths [5.35]. Suitable products for the primary baths include sulfonated oils, lignosulfonates, protein–fatty acid condensation products, and condensation products of aromatic sulfonic acids. Fatty alcohol ethoxylate is added to the development baths in amounts of 0.5–2 g/L to enable the excess insoluble azo pigments to be easily washed out.

For the same reason, nonionic dispersing agents are added to the development baths when dyeing with leuco esters of vat dyes.

When dyeing raw cotton with water-soluble reactive dyes, dispersants are only needed for the impurities in the raw cotton and the precipitates due to water hardness. Polycarboxylates and polyphosphates are suitable for this.

**Continuous Dyeing with Disperse and Vat Dyes.** In continuous dyeing with insoluble dyes, a dispersing agent is added to the pad liquor to ensure stability of the dye dispersion. This gives a uniform fabric appearance without danger of marks due to agglomerates. In chemical or development baths, dispersants facilitate the redispersion of unfixed dye during final washing. This gives improved rubbing fastness. Almost all dispersing agents can be used in pad baths. Some commercial dispersants and protective colloids are listed in Table 5.3.

### 5.4.3. Testing

Whereas improvement of the fineness of the dispersion is the main criterium when testing dispersants for use in dye production, the property that is tested when the dyes are used in a dye bath is their ability to prevent agglomeration (stabilizing action).

The effectiveness of dispersants is characterized by their effect on dye dispersions, impurities, and precipitates from hard water, whereby a large number of testing methods are used [5.36]–[5.42]:

**Table 5.3.** Some commercial dispersants and protective colloids [5.5]

| Name | Producer | Country |
| --- | --- | --- |
| *Condensed aromatic sulfonates* | | |
| Anogal DA | Baur & Gäbel | Germany |
| Avolan IS | Bayer | Germany |
| Dispersogen P liq. | Hoechst | Germany |
| Irgasol DAM | Ciba-Geigy | Switzerland |
| Matexil DA-N | Zeneca | United Kingdom |
| Setamol E | BASF | Germany |
| Setamol WS | BASF | Germany |
| *Ligninsulfonates* | | |
| Dispergal CN | Impocolor | Germany |
| Lamepon N | Grünau | Germany |
| Reax product series | Westvaco | United States |
| Ultrazine | Borregaards | Norway |
| Serlasol | Serlachius Cop. | Finland |
| *Polycarboxylic acids and mixed products* | | |
| Alcosperse AD | Allied Colloids | United Kingdom |
| Dekol SN | BASF | Germany |
| Dispergal PS | Impocolor | Germany |
| Irgasol CO | Ciba-Geigy | Switzerland |
| Masquol A 340 N | Protex | France |
| Meropan VD | CHT | Germany |
| Sandopur RSK | Sandoz | Switzerland |
| *Protein condensates* | | |
| Adipon TH conc. | Henkel | Germany |
| Lamepon BN | Grünau | Germany |
| Ofnapon ASN | Hoechst | Germany |
| *Sulfonated oils* | | |
| Geipolan EB | Geissler | Germany |
| Monopolbrillant Oil | Stockhausen | Germany |
| Sandozol KB | Sandoz | Switzerland |
| *Ethoxylated products* | | |
| Dispersogen ASN | Hoechst | Germany |
| Lyocol ACN | Sandoz | Switzerland |
| Matexil DN-VL | Zeneca | United Kingdom |
| Ruco Levelling Agent EN | Rudolf-Chemie | Germany |
| Sarabid 200 LL | CHT | Germany |
| Uniperol EL | BASF | Germany |

1) Determination of the particle size distribution of dye dispersions
2) Determination of the filtering properties of dye dispersions
3) Determination of the increase in the solubility of dyes
4) Determination of the antiflocculation effect on accumulations of insoluble dyes and impurities (waxes, pectinates)
5) Calcium carbonate dispersing capacity (CCDC) [5.43]
6) Tests of performance in use, e.g., laboratory dyeing tests or flow tests

## 5.5. Complexing Agents

Water quality is of great importance for the success of the dyeing process. Insoluble impurities, hardness components, and heavy metal salts can cause considerable problems. The presence of alkaline earth and/or heavy metal salts lead to the following problems during dyeing [5.44], [5.45]:

1) The formation of sparingly soluble saltlike compounds with anionic dyes leads to filtering-out problems in package dyeing, levelling problems, and impairment of the resistance to rubbing and washing
2) The formation of stable complexes with dye molecules causes change of the shade, often accompanied by loss of brilliance

Purified and softened water is therefore used in textile finishing. Complexing agents are added to the dye bath to combine with multivalent cations, especially the calcium, magnesium, and iron salts which are carried into the dye liquor by the textile material (e.g., cotton [5.46], [5.47]). The use of these products is also extremely important for washing out unfixed dye after completion of the dyeing stage (see Section 5.10) [5.48], [5.49]. In general the products used are the same as those used in the washing and cleaning processes of textile pretreatment [5.1, p. 995], [5.50].

The use of ion exchangers is not possible as these are by definition insoluble in the treatment solution.

Complexing agents differ with respect to the stability of the metal complex and the specific effect on metal cations. Also, for all products, the stability of the complex is strongly dependent on the pH [5.51].

EDTA (ethylenediaminetetraacetate; → Ethylenediaminetetraacetic Acid and Related Chelating Agents, **A 10**, p. 95), DTPA (diethylenetriaminepentaacetate), NTA (nitrilotriacetate; → Nitriloacetic Acid, **A 17**, p. 377), and derived phosphonic acids are very effective on a wide range of cations including heavy metal ions and those from hard water.

Specific products to combat the effects of water hardness include mild complexing agents such as polyphosphates and various polycarboxylic acids. These also have a dispersing action on precipitates from water hardness (see Section 5.4) [5.43], which the strong complexing agents referred to above do not have.

Specific mild complexing agents for the heavy metal ions of iron, copper, and manganese include various polyhydroxy compounds, e.g., sorbitol, gluconic acid, glucoheptanoic acid, and alkanolamines.

Commercial products often also contain mixtures of various compounds.

With metal complex dyes, the use of weak complexing agents is especially preferable, so that the metal is not removed from the dye molecule. Strong complexing agents are suitable for stripping these dyes, as the metal-free dye is less strongly bound to the fiber.

Detailed information on complexing agents and their use in dyeing can be found in the literature [5.44], [5.45], [5.51]. A list of commercial products is given in Table 5.4.

## 5.6. Levelling Agents

Levelling agents promote uniform distribution of the dye in the textile in the exhaustion dyeing process, so that the dyeing is level, with a uniform shade and depth of color.

Inequalities are caused or intensified by [5.52]:

1) High and variable affinity of the fibers for the dye
2) Imperfect distribution of the liquor in the textile
3) Temperature differences in the textile
4) High and variable affinity of the dye for the fibers

Unsatisfactory levelling can be prevented by dyeing techniques (e.g., improvement of the diffusion of the liquor within the textile, pH control) and by levelling agents.

Levelling agents act mainly by reducing the dyeing rate, increasing the rate of migration of the dye within the textile, and improving the compatibility of dyes.

Levelling agents also have other effects which do not directly influence the dye–fiber interaction, but nevertheless have a favorable effect on dye levelling. These include improvement of the solubility or dispersion stability of the dye, and

**Table 5.4.** Some commercial complexing agents [5.5]

| Name | Producer | Composition |
| --- | --- | --- |
| Actiron GHN 30 | Protex[a] | gluconate derivative |
| Aquamollin BC | Hoechst[b] | EDTA |
| Calgon T neu | BK Ladenburg[b] | polyphosphate |
| Dekol SN | BASF[b] | polycarboxylate |
| Delinol VB | Dr. Th. Böhme[b] | polycarboxylate |
| Heptol NWS | CHT[b] | polycarboxylate + polyphosphate |
| Heptol FEW | CHT[b] | polyhydroxycarboxylic acid |
| Ladiquest P 97 | BK Ladenburg[b] | polycarboxylate + phosphonate + polyphosphate |
| Lamepon KB | Grünau[b] | organic phosphorus compound |
| Masquol EL 40 | Protex[a] | EDTA |
| Periquest AL | Dr. Petry[b] | polycarboxylate |
| Plexene UL | Sybron/Tanatex[c] | polycarboxylate |
| Tetralon A | Allied Colloids[d] | EDTA |
| Trilon FE | BASF[b] | mixed complexing agents |
| Trilon TA | BASF[b] | NTA |
| Trilon TB | BASF[b] | EDTA |
| Verolan NBI | Rudolf-Chemie[b] | polycarboxylate |

[a] France; [b] Germany; [c] The Netherlands; [d] United Kingdom.

prevention of deposition of impurities. Levelling agents often exhibit several effects simultaneously, sometimes because the individual components themselves have these properties, and sometimes because mixed products with complex effects are deliberately produced. Levelling agents that directly influence the dyeing process can be divided into products with an affinity for dyes, and products with an affinity for fibers [5.53].

Products with an affinity for dyes form loosely bound addition compounds with the dyes whose stability is concentration dependent and usually decreases with increasing temperature. Hence, the dye distribution equilibrium between the dye liquor and the fiber is shifted towards the dye liquor. The increased dye concentration in the liquor enables areas of the textile that are dyed to different intensities to be levelled by dye migration.

Effective levelling agents have an affinity for the dye, sufficient to reduce its rate of absorption and/or to promote its migration. Also, differences in the absorption behavior of different dyes can be equalized, so that the dyes are taken up from mixtures at equal rates. However, if the interaction is too selective, this leads to variations in color shade.

There is a series of products with increasingly high dye affinity, so retaining increasing amounts of dye in the liquor. Some of these can be used as stripping agents or for brightening dyed materials.

Auxiliaries with an affinity for dyes are also used for levelling already dyed material.

Levelling agents with an affinity for fibers are absorbed onto the fibers in competition with the dye. This competitive reaction reduces the absorption rate of the dye and promotes migration.

Products with fiber affinity may also be used in some cases (polyamide, wool), to even out differences in the dye affinity of the fiber. The most important levelling agents for the most important fibers and dyes are listed in Table 5.5.

## 5.6.1. Levelling Agents for Dyeing Cellulose Fibers

Cellulose fibers are dyed with reactive dyes, vat dyes, leuco esters of vat dyes, direct dyes, sulfur dyes, developing dyes, and pigments (→ Textile Dyeing, pp. 399–421).

Levelling agents are mainly required for vat dyes, and to a lesser extent for direct dyes. Dyeing auxiliaries that improve levelling are also used for dyeing with the other types of dye, but their effectiveness depends not on directly influencing the dyeing process, but on secondary effects such as wetting and dispersing, protective colloid formation, and increase of solubility.

**Levelling Agents for Vat Dyes.** Because of the high affinity of the reduced leuco form of many vat dyes for the cellulose fiber, the textile is dyed very rapidly and unevenly. Evenly dyed products can be produced by:

Controlling the *concentration of the leuco form.*

The leuco form of the vat dye is not yet formed at the start of the dyeing process (exception: leuco process), but is produced

1) In the course of the dyeing process (semipigmentation process)
2) After a phase in which the vat dye pigment (low affinity for the fiber) is distributed within the textile (prepigmentation process)
3) By controlled metering [5.54], [5.55] of the vat dye to the reducing liquor
4) By controlled metering of the reducing agent to the dispersed vat dye pigment

The dyeing equilibrium is shifted in favor of the liquor by dyeing at a high temperature (80–110 °C). The increased dye concentration in the liquor favors levelling.

*Levelling agents* with an affinity for dyes compete for the cellulose fibers with the leuco form of the vat dye. This retards the absorption process and improves levelling.

The use of levelling agents can also be combined with modification of the process.

The equilibria of the dyeing process and the effect of an auxiliary with fiber affinity are illus-

**Table 5.5.** Types of levelling agent

| Type of fiber | Dye | Composition | Effect |
|---|---|---|---|
| Cellulose | vat dyes | nonionic surfactants | affinity to dye |
| | | polyamide amines | affinity to dye |
| | | polyvinylpyrrolidone | affinity to dye |
| | direct dyes | nonionic surfactants | affinity to dye |
| | | polyvinylpyrrolidone | affinity to dye |
| Wool | acid, metal complex, and reactive dyes | nonionic surfactants | affinity to dye |
| | | weakly cationic ethoxylated products | affinity to dye and fiber |
| | | anionic surfactants | affinity to fiber |
| Polyamide | acid and metal complex dyes | nonionic surfactants | affinity to dye |
| | | cationic surfactants | affinity to dye |
| | | ethoxylated products | affinity to dye |
| | | anionic surfactants | affinity to fiber |
| | disperse dyes | nonionic surfactants | affinity to dye |
| Polyester | disperse dyes | carriers, special products | affinity to fiber |
| | | nonionic surfactants | affinity to dye |
| Polyacrylonitrile | cationic dyes | cationic surfactants and nonsurfactants | affinity to fiber |
| | | anionic surfactants | affinity to dye |

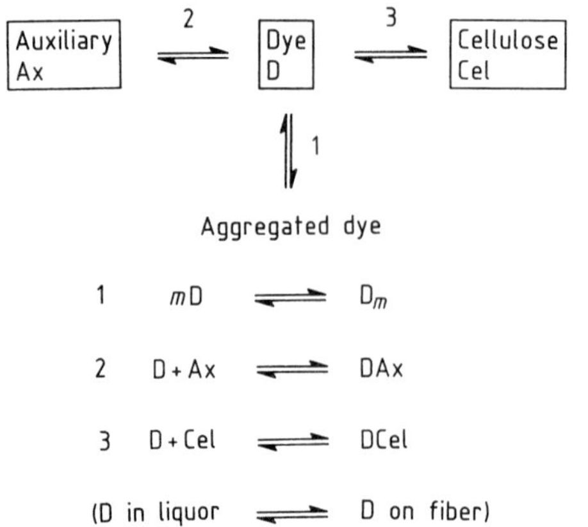



**Figure 5.1.** Equilibria involving a dissolved dye in a system with cellulose

trated schematically in Figure 5.1. The interaction between the levelling agent and the leuco dye is believed to involve the formation of an adduct, which is indicated by the characteristic displacement of the bands in the absorption spectra of the dye [5.56].

The substances with an affinity for the dye that are most suitable for vat dyeing are stable in the strongly alkaline dye liquor. As foam formation seriously affects the dyeing process in modern dyeing equipment, the substances are classified according to their surface active properties [5.57]:

*Surface Active Compounds*

a) Fatty alcohol ethoxylates: mild levelling effect, strong bond to the dye (stripping agent)
b) Fatty amine ethoxylates: mild levelling effect, strong bond to the dye (stripping agent)
c) Alkylbenzimidazolesulfonates: moderate levelling effect, dispersing effect

*Nonsurface Active Compounds*

a) Betaines (*N*-pyridiniumacetic acid): levelling effect, low dye-bonding effect
b) Polyalkyleneamines
c) Oligomeric aminoamides: good levelling and absorption-retarding properties with low, unselective dye-bonding action; the most important levelling agents for vat dyes
d) Polyvinylpyrrolidone: very strong absorption-retarding effect, and very strong dye-bonding action (stripping agent).

**Levelling Agents for Substantive Dyes.** Mixtures of nonionic and anionic surfactants have long been used for this purpose. Nonionic components include ethoxylated fatty alcohols, fatty amines, fatty acids, alkylphenols, or propylene oxide polymers. Anionic components include fatty alcohol sulfates, alkylaryl sulfates, and Turkey red oil.

The task of these products is not so much to prevent too rapid absorption of the dye, as this can be controlled by salt addition, but more to ensure thorough wetting, and to disperse soil, preparations, dye deposits, etc.

Material dyed with substantive dyes can be levelled at a later stage by the use of polyvinylpyrrolidone, which is an auxiliary with an affinity for the dye.

**Levelling Agents for Other Dyes.** Other products referred to as "levelling agents" for dyeing with sulfur dyes or reactive dyes actually consist of wetting and dispersing agents.

Some commercial products are listed in Table 5.6.

### 5.6.2. Levelling Agents for Wool Dyeing

Anionic dyes of the following types are used for dyeing wool: acid dyes, metal-complex dyes, afterchrome dyes, and reactive dyes (→ Textile Dyeing pp. 422–426). Levelling agents for wool dyeing must reduce the rate of absorption of the dye, improve the migration of the dye during the dyeing phase, and improve the uniformity of color shade and color intensity of wool fibers that may have undergone damage (light, segregation, pretreatment, etc.) and may be of various origins. Nonionic, ionic, and amphoteric surfactants are used for these purposes, although their mechanism of action is still not completely understood [5.58].

Nonionic surfactants form a hydrophilic complex with the anionic dye, and thus are products with an affinity for the dye [5.59], [5.60]. The molecules contain a hydrophilic group consisting of long polyglycol ether chains, and a hydrophobic portion, usually fatty alcohols, alkyl phenols, fatty acid alkylolamides, etc.

The ionic surfactants include cationic compounds, some of which are the most commonly used commercial levelling agents for wool dyeing. These have affinity both for fibers and dyes, and only have a levelling effect above the critical micelle concentration [5.61]. Typical compounds of this group include polyglycol ethers of fatty

**Table 5.6.** Some commercially available levelling agents for dyeing cellulose fibers [5.5]

| Name | Producer | Uses | Composition |
|---|---|---|---|
| Albatex OR | Ciba-Geigy[a] | vat dyes | oligomeric amide |
| Albatex PON conc. | Ciba-Geigy[a] | vat and direct dyes | alkylbenzimidazole ethoxylate |
| Albigen A | BASF[b] | vat and direct dyes | polyvinylpyrrolidone |
| Anogal KF | Bauer & Gäbel[b] | vat dyes | polyglycol ether |
| Anogal NK | Bauer & Gäbel[b] | direct dyes | polyglycol ether |
| Cerafil BFA | Dr. Th. Böhme[b] | bleaching + dyeing | phosphoric ester + alkylarylsulfonate |
| Degal CR | Sigma[b] | vat and direct dyes | polyamide |
| Dupramin NFS | Thor-Chemie[b] | vat and direct dyes | polyglycol ether |
| Impegal SR | Impocolor[b] | vat dyes | condensation product |
| Levegal KNS | Bayer[b] | vat and direct dyes | polyglycol ether |
| Lyogen E | Sandoz[b] | vat dyes | oligomeric amide |
| Matexil DN-VL | ICI[c] | vat and direct dyes | polyglycol ether |
| Peregal P | BASF[b] | vat dyes | oligomeric amide |
| Ruco-Egalisierer IPM | Rudolf-Chemie[b] | vat dyes | amine condensation product |
| Sarabid SBF | CHT[b] | direct dyes | polyglycol ether + sulfonates |
| Sarabid VAT | CHT[b] | vat dyes | amine condensation product |
| Solidegal GL | Hoechst[b] | vat dyes | heterocyclic compound |
| Solidegal SR | Hoechst[b] | vat dyes | higher alkylamine |
| Uniperol O | BASF[b] | vat and direct dyes | polyglycol ether |

[a] Switzerland; [b] Germany; [c] United Kingdom.

amines, fatty acid amide amines, and fatty alkyl-polyamines.

Anionic surfactants exhibit fiber affinity. They include alkylnaphthalenesulfonic acids, fatty alkyl sulfates, alkylbenzenesulfonates, alkyl polyglycol ether sulfates, and polycarboxylates [5.62].

Amphoteric products combine the properties of cationic and anionic surfactants, and have an affinity for both dyes and fibers. Their effect can be varied by varying the auxiliary/dye ratio and the length of their polyether chains [5.63]. Typical products include ethoxylated nitrogen-containing fatty alkyl compounds with an anionic group on the nitrogen or at the end of the polyether chain.

Commercially available levelling agents for wool include many kinds of synergistic mixtures of surfactants of various ionic types. Products of universal application are available, as well as mixtures optimized for certain groups of dyes (Table 5.7).

### 5.6.3. Levelling Agents for Dyeing Polyamide Fibers

Polyamide can be dyed with disperse dyes (→ Textile Dyeing, p. 430) and with those classes

**Table 5.7.** Some commercial levelling agents for dyeing wool [5.5]

| Name | Producer | Uses* | Active substances |
|---|---|---|---|
| Albegal A | Ciba-Geigy, Switzerland | e | alkylamine polyglycol ether sulfate |
| Albegal B | Ciba-Geigy, Switzerland | f | ethoxylated fatty acid amide derivative |
| Albegal SET | Ciba-Geigy, Switzerland | e | modified nitrogen-containing fat and alcohol ethoxylates |
| Avolan AV 200% | Bayer, Germany | a, c | alkylarylamine polyglycol ether |
| Avolan W | Bayer, Germany | e | alkyl polyglycol ether |
| Avolan REN | Bayer, Germany | f | alkylarylamine polyglycol ether |
| Avolan UL75 | Bayer, Germany | a–e | alkylamine polyglycol ether sulfate |
| Breviol SCN | Henkel, Germany | a, d | amine condensate |
| Eganal SZ | Hoechst, Germany | a–e | surfactant combination |
| Keriolan FMK | CHT, Germany | e | polyglycol ether derivative |
| Lyogen WD | Sandoz, Switzerland | a–d | fatty amine polyglycol ether |
| Uniperol AC | BASF, Germany | a, b, d | alkylamine polyglycol ether |
| Uniperol KM | BASF, Germany | e | amphoteric product |
| Uniperol SE | BASF, Germany | a, b, d, e | amphoteric product |
| Uniperol W | BASF, Germany | a, c, d, e | polyglycol ether sulfate |

* a = Acid dyes (strongly acid); b = Acid dyes (weakly acid); c = Afterchrome dyes; d = 1:1-Metal complex dyes; e = 1:2-Metal-complex dyes; f = Reactive dyes.

of dyes that are also important for wool dyeing (acid dyes and metal-complex dyes).

*Dyeing with Disperse Dyes.* In long liquor dyeing of polyamide with disperse dyes, dispersants are always added to stabilize the dispersion (see Section 5.4). With rapid acting processes, e.g., the dyeing of hosiery special mixtures of surfactants with a wetting, washing, and dispersing effect are used. These contain polyglycol ethers as dispersants as well as nonionic ethoxylation products and anionic conditioning agents. High molecular mass polyglycol ethers improve stripping of the hosiery products from the patterns.

*Dyeing with Acid Dyes.* When dyeing polyamide fibers with anionic dyes, anionic, nonionic, and cationic surfactants are used, as with the dyeing of wool (see Table 5.7).

Anionic levelling and resist agents with an affinity for fibers are used (see Section 5.14). These are mono- and polysulfonates of high molecular mass aliphatic and aromatic compounds [5.1], [5.64]. They also prevent formation of "stripes" on the dyed polyamide [5.65].

Cationic levelling agents (e.g., fatty amines) are similar or identical to those products used in wool dyeing (see Section 5.6.2).

Nonionic surfactants (ethylene oxide adducts such as ethoxylated fatty alcohols) form a complex with the anionic dye, thus having dye affinity [5.65]–[5.67].

When dyeing polyamide, mixtures of nonionic, ionic, and amphoteric products are also used (see Table 5.8).

When choosing an auxiliary, care must be taken that it does not have a harmful effect on the compatibility of dyes properties [5.68] or lightfastness of dyes [5.69], [5.1].

Uniform absorption of the dye can be achieved by pH control as well as by the use of retarding agents. This is carried out by automatically metering acid into the dye bath, or by adding substances that gradually release acid (see Section 5.6.7).

### 5.6.4. Levelling Agents for Dyeing Polyester Fibers

The dyeing of polyester fibers with disperse dyes by the exhaustion process is carried out either at 100 °C with dyeing accelerants (carriers) (Section 5.7) or at 120–140 °C under pressure (high-temperature process) (→ Textile Dyeing, pp. 436–437).

At boiling point and in the presence of a carrier, a levelling agent is usually unnecessary, as polyester fibers are dyed slowly and uniformly under these conditions.

To give uniform dyeing in the high-temperature process, levelling agents are necessary to prevent excessively rapid dye absorption, equal-

**Table 5.8.** Some commercially available levelling agents for dyeing polyamide fibers [5.5]

| Name | Producer | Uses | Active substances |
|---|---|---|---|
| Alviron CNF | TC, Switzerland | levelling agent, antistriping agent | sulfonates, ethylene oxide condensates |
| C 14 | CHT, Germany | antistriping agent | araliphatic sulfonates and polyglycol ethers |
| Dispergal B liq. | Auschem, Italy | levelling agent | sulfonated naphthalene derivatives |
| Gisapon 1476 | Rotta, Germany | levelling agent | modified fatty amine ethoxylate |
| Levegal FTS | Bayer, Germany | levelling agent | polyglycol ether derivative and sulfonates |
| Lyogen PN | Sandoz, Switzerland | antistriping agent | aromatic sulfonate |
| Perigen HTP | Dr. Petry, Germany | hosiery dyeing | fatty acid polyglycol esters, oleates, terpenes |
| Ruco-Egalisierer PA | Rudolf-Chemie, Germany | levelling agent, antistriping agent | alkylamine ethoxylate |
| Sandogen NH | Sandoz, Switzerland | levelling agent | fatty amine polyglycol ethers |
| Uniperol AC | BASF, Germany | levelling agent | alkylamine polyglycol ether |
| Univadin PA | Ciba-Geigy, Switzerland | levelling agent | combination of anionic surfactants |

ize the different absorption rates of the various dyes, and/or improve the migration of the dye in the high-temperature phase. They must not affect the stability of the dye dispersion, as otherwise filtering effects can severely disturb the process, causing unlevel dyeing and impairment of fastness.

There are three typical groups of levelling agents for the high temperature process: dyeing accelerants (carriers), ethoxylated products, and new special products.

In the high temperature dyeing process, carriers increase the rate of diffusion of the dye in the fiber and in the fiber/liquor equilibrium, thereby improving the migration of the dye [5.70]. They are also suitable for levelling nonuniformly dyed materials [5.71]. However, carriers can increase the exhaustion rate of the dye so much that unlevel dyeing takes place. They should therefore only be added to the dye bath at higher temperatures or in the migration phase. Because they cause swelling of the fiber, carriers reduce stripiness due to nonuniform dye uptake [5.72], [5.73].

The group of carriers used in high temperature dyeing includes the same compounds as are used in dyeing polyester fibers at boiling temperatures (see Section 5.7): halogenated benzenes, halogenated toluenes, 2-phenylphenol, diphenyl, diphenyl ethers, salicylic esters, methylnaphthalene, etc. [5.74]. However, these compounds often have disadvantages such as reduced light-fastness, environmental pollution, excessive fiber swelling, and strong odor. As alternatives to these products with their associated environmental problems, aliphatic carboxylic esters are used [5.75]–[5.77]. However, the carrier effect of these products is inadequate when dyeing at boiling temperatures.

Ethoxylated products have a retarding effect on dyes in the absorption phase, and are therefore used especially in rapid dyeing processes or in shading at high temperature [5.78]–[5.81]. The main members of this group of ethoxylated products are ethoxylated castor oil, stearic acid, alkylphenols, and the sulfuric or phoshoric esters of ethoxylated fatty alcohols or alkylphenols.

Special levelling agents include product mixtures that produce equal rates of absorption of dye mixtures (synchronization) during the heating phase, and improve dye migration in the high-temperature phase. These products usually consist of combinations of alcohols, esters, or ketones of medium chain length ($C_6$–$C_{12}$ or $C_7$–$C_{16}$) with emulsifying systems [5.82]. The effect

of cyclodextrins as levelling agents for dyeing polyester fibers under high temperature conditions is discussed in [5.83], [5.84].

Commercial levelling agents are listed in Table 5.9.

*Carrier-Free Dyeing of Modified Polyester Fibers.* The modified polyester fibers known as NCD (noncarrier dyeable) fibers can be dyed with disperse dyes at boiling temperatures without the addition of carriers ($\rightarrow$ Textile Dyeing, p. 441). To prevent unlevel effects caused by excessively rapid absorption of the dye, levelling agents of the ethoxylated product type are used.

*Dyeing of Polyester Fibers in Alkaline Media* ($\rightarrow$ Textile Dyeing, p. 439). Polyester fibers are also dyed in an alkaline medium to reduce the oligomer problem and to facilitate the process of finishing (desizing and dyeing in one step). Levelling agents of the ethoxylated product type are used for this, and special products are also used to buffer the liquor in the pH range 9–9.5 [5.85].

### 5.6.5. Levelling Agents for Dyeing Polyacrylonitrile Fibers

The cationic dyes used for dyeing anionically modified polyacrylonitrile fibers usually have a very high affinity for the fiber, and diffuse very rapidly into the fiber above the glass transition temperature ($\rightarrow$ Textile Dyeing, Section 9.2.2, pp. 444–447). The ability of these dyes to migrate is low, and unlevel effects due to excessively rapid absorption of the dye are difficult to rectify. The absorption rate can be moderated by controlling the temperature and adding levelling agents (cation active or anion active products), thereby improving the uniformity.

Cationic products known as retarders are of the greatest importance. These can be of the "permanent" or "temporary" type [5.86], the more important of these being the permanent retarders, which have a high affinity for the fiber. These retard dye absorption by attaching themselves to the anionic sites in competition with the dye. For given characteristic parameters of the fiber, dye, and retarder, the amount of retarder required can be calculated for each dye bath formulation [5.87], [5.88]. If the amounts added are too high, the retarder can block the fiber, affecting the dye yield. It must therefore be stripped by an anionic auxiliary to enable further dyeing to be carried out. Commercially available retarders

**Table 5.9.** Some commercial levelling agents for dyeing polyester fibers [5.5]

| Name | Producer | Uses | Active substances |
|---|---|---|---|
| Alviron PES | Allied Colloids, United Kingdom | levelling, dispersing, migration and crease prevention agent | modified phosphoric acid esters |
| Anogal NTH | Baur & Gäbel, Germany | levelling and dispersing agent | ethoxylated compound with anionic part |
| Breviol LC | Henkel, Germany | levelling agent | ester combination |
| Eganal PS | Hoechst, Germany | levelling and dispersing agent | linear polycondensate |
| Egasol WFE | CHT, Germany | levelling agent with washing properties | aromatic esters with ethoxylates |
| Levegal HTC | Bayer, Germany | levelling agent | halogenated aromatic alkyl polyglycol ethers, alkylarylsulfonates |
| Levegal HTN | Bayer, Germany | levelling and dispersing agent | alkylphenol and fatty acid polyglycol ethers |
| Levegal MSF | Bayer, Germany | levelling agent, crease prevention agent | fatty acid glyceryl esters |
| Palegal HT | BASF, Germany | levelling agent | nonionic aliphatic compounds |
| Palegal SFD | BASF, Germany | levelling and migration agent | aromatic esters and emulsifiers |
| Perigen HTN | Dr. Petry, Germany | levelling agent | aromatic esters and emulsifiers |
| Ruco-Egalisierer SBB | Rudolf-Chemie, Germany | levelling agent and stripe formation prevention agent | carboxylic acid alkyl esters, oil emulsifiers, and antifoaming agents |
| Sandogen BM | Sandoz, Switzerland | levelling agent | carboxylic acid esters, emulsifiers |
| Synthapal BM | Dr. Th. Böhme, Germany | levelling and migration agent | aromatic esters and emulsifiers |
| Synthapal HTM | Dr. Th. Böhme, Germany | levelling, migration, and dispersing agent | polyglycol esters, aromatic sulfonate |
| Trevylen | Kreussler, Germany | carrier | $N$-alkylphthalimides |
| Univadin 3flex | Ciba-Geigy, Switzerland | levelling, migration, and dispersion agent | fatty acid esters, aryl polyglycol ethers, polyalcohol |

contain quaternary ammonium salts with long fatty alkyl side chains ($C_{12}$–$C_{14}$ alkyltrimethyl or $C_{10}$–$C_{14}$ alkyldimethylbenzyl chains) [5.89]. Polymeric compounds [5.90 or polyethoxylated amines [5.91] are also often used.

Cationic products with a significantly lower affinity for the fiber than for the dye are marketed as migration auxiliaries. These increase the migration of the dye during the equilibrium phase. They have only a slight effect on the absorption rate, and do not block the fiber [5.92]. Quaternary ammonium salts with aromatic ring systems (e.g., trimethylbenzylammonium chloride), electrolytes (e.g., NaCl, $Na_2SO_4$) or organic salts with larger cations [5.93] increase migration.

Other levelling agents used in PAN dyeing include anionic products with an affinity for the dye (e.g., naphthalenesulfonic acid–formaldehyde condensation products). The possibility of precipitation of dye–auxiliary adducts can be avoided by adding nonionic dispersing agents (ITW process, [5.94]). A disadvantage of the anionic products is that they reduce bath exhaustion.

A list of commercial products is given in Table 5.10.

### 5.6.6. Levelling Agents for Dyeing Other Synthetic Fibers and Fiber Blends

**Acetate Fibers.** When dyeing acetate fibers ($\rightarrow$ Textile Dyeing, p. 442) with disperse dyes, levelling agents with a dispersing action are used (nonionic or anionic surfactants, see Section 5.4). These products often also need to have a detergent action to remove fiber preparations during dyeing and to maintain them in emulsified form.

**Triacetate Fibers.** When dyeing triacetate fibers ($\rightarrow$ Textile Dyeing, p. 442) the same dispersants as are used for dyeing other fibers with

**Table 5.10.** Some commercially available levelling agents for dyeing polyacrylonitrile fibers [5.5]

| Name | Producer | Uses | Active stubstance |
|---|---|---|---|
| Astragal M | Bayer, Germany | migration agent | organic amino compound |
| Atragal PAN | Bayer, Germany | retarder | quaternary ammonium compound |
| Astragal TR | Bayer, Germany | temporary retarder | quaternary ester ammonium compound |
| Basacryl Salz AN | BASF, Germany | migration agent | quaternary ammonium compound |
| Basacryl Salz MX | BASF, Germany | retarder | quaternary ammonium compound |
| Basacryl Salz TR | BASF, Germany | retarder | quaternary ammonium compound |
| Geneucol ART | Dr. Th. Böhme, Germany | retarder | quaternary ammonium compound |
| Geneucol MI | Dr. Th. Böhme, Germany | migration agent, nonblocking | quaternary ammonium compound |
| Geneucol RTM | Dr. Th. Böhme, Germany | retarder | quaternary ammonium compound |
| Osimol MA | Grünau, Germany | migration agent | quaternary ammonium compound |
| Osimol RAC | Grünau, Germany | retarder | quaternary ammonium compound |
| Remartard AC | Hoechst, Germany | retarder | quaternary ammonium compound |
| Remartard T | Hoechst, Germany | retarder | cationic nitrogen compound |
| Retargal AN | Sandoz, Switzerland | retarder, nonblocking | quaternary ammonium compound |
| Sarabid UZ 1290 | CHT, Germany | anionic retarder | ethylene oxide adducts |
| Tinegal B | Ciba-Geigy, Switzerland | retarder | araliphatic quarternary fatty acid amine |
| Tinegal MR | Ciba-Geigy, Switzerland | retarder with migration effect | $n$-tetraalkylammonium halide |
| Tinegal PAC | Ciba-Geigy, Switzerland | retarder | quaternary ammonium compound |

disperse dyes (polyester, polyamide, and acetate), are used as levelling agents. When dyeing at 100 °C, special carriers are often used (see Section 5.7).

**Fiber Blends.** If the various types of fiber in a fiber mixture are dyed successively (→ Textile Dyeing, Chap. 11), the same levelling agents can generally be used as when dyeing the fibers individually. For single-bath dyeing with dyes of differing ionic character, special dispersants must be used to prevent mutual precipitation of the two types of dye, which could have a harmful effect on the levelling and rubbing fastness. These special auxiliaries are mixtures of nonionic and anionic surfactants (e.g., Uniperol KA, BASF).

If the individual components of a fiber blend are not dyed to the same color intensity by a particular dye (e.g., when dyeing wool–polyamide with anionic dyes, or cellulose–polyamide with substantive dyes), a better choice of dye, careful temperature control, and the use of an auxiliary with an affinity for the fiber can reduce the dye affinity of that component of the mixture with the greater affinity. The products used for this are the same as those used as chemical resist agents (see Section 5.14). For tone-in-tone dyeing of polyamide–polyurethane blends, cationic products with an affinity for the dye or ethoxylated fatty amines are used.

A list of commercially available products is given in Table 5.11.

### 5.6.7. pH Regulators (Acid- and Alkali-Releasing Agents)

The pH influences the absorption of anionic dyes onto wool and/or polyamide fibers and the fixing of reactive dyes onto cellulose fibers. By controlling the pH, it is possible to improve dye levelling in the absorption phase, or, when dyeing polyester–cotton blends with disperse and reactive dyes, to control the fixing of the dye.

**Dyeing of Wool and Polyamide.** The level absorption of anionic dyes onto wool and/or polyamide fibers can be controlled by means of the pH of the dye liquor. The dye equilibrium in acids lies on the side of the fiber, and the dye is effectively absorbed onto the fiber. In alkalis, the equilibrium is shifted in favor of the liquor.

Dyeing is started at a pH of 8–8.5, and the pH is then successively reduced by adding acids or acid salts.

This method has the disadvantage that there is a sharp change in pH at the point where the acid is added, leading to a pH gradient in the dye bath.

The use of acid releasing agents based on esters of organic acids [5.95]–[5.100] enables the acid to be liberated continuously on heating the dye bath and hydrolyzing the ester. This causes the pH to decrease at a slow and constant rate in all parts of the dye bath, so that the anionic dye is absorbed in a controlled and uniform manner.

**Dyeing of Cotton and Polyester–Cotton Blends.** The dyeing of mixtures of cotton and polyester fibers by the exhaustion dyeing process in the same dye bath (single bath/two stages or single bath/single stage) is more economic than the process that includes a change of dye liquor [5.101].

The single-bath/single-stage process in which an increase in temperature is used to cause an acid releasing agent present in the liquor to lower the pH is especially advantageous. Fixing of the reactive dye begins in alkaline solution at pH 10–10.5 and 60 °C, and the dyeing conditions for the disperse dye are reached at 130 °C and pH 5–6 after ca. 130 min.

The acid-releasing agents used are salts of $\alpha$-halocarboxylic acids which decompose on heating. Phosphates are also used as buffering agents.

For continuous dyeing with reactive dyes, separate application of the dye and the fixing alkali is used (two baths) to prevent hydrolysis of the dye (pad steam process).

Using the alkali-releasing agent sodium hydrogencarbonate in the dye bath makes the single-bath pad steam process possible, in which the reactive dye is stabilized at room temperature and pH 7–7.5, and is fixed at 100 °C in the steamer. The alkaline fixing conditions are then produced by decomposition, liberating $CO_2$ and giving a pH of 9–9.5.

A list of commercially available products is given in Table 5.12.

### 5.6.8. Testing

To test the effectiveness of levelling agents in the laboratory, the effects of the auxiliary are measured either on individual processes in the dyeing phase, or by assessing any improvement in uniformity produced by addition of the level-

**Table 5.11.** Some commercial levelling agents for dyeing fiber blends [5.5]

| Name | Producer | Uses | Active substances |
|---|---|---|---|
| Dispergal WP | Auschem, Italy | WO/PA | derivatives of sulfonated mixtures |
| Emulsol OS | Auschem, Italy | PA/WO | ethoxylated fatty alcohol |
| Gelonal W | Geissler, Germany | PA/WO | alkylamine polyglycol ether |
| Levegal ER liq. | Bayer, Germany | CEL/PA | aromatic sulfonates |
| Ritandin CD | Baumheier, Germany | CEL/PAN | fatty acid condensate, polyglycol ether |
| Teban ERP | Dr. Th. Böhme, Germany | PA/CEL | modified arylsulfonates |
| Teban WPA | Dr. Th. Böhme, Germany | WO/PA | solvent-containing arylsulfonate mixture |
| Transferin UN | Dr. Th. Böhme, Germany | WO/PES, WO/PAN, PES/PAN | modified alkyl polyglycol ether |
| Uniperol KA | BASF, Germany | PAN/WO | mixture of nonionic surfactants |

WO = Wool; PA = Polyamide; PAN = Polyacrylonitrile; CEL = Cellulose; PES = Polyester.

**Table 5.12.** Some commercially available acid releasing agents [5.5]

| Name | Producer | Active substances |
|---|---|---|
| Emenil DG | Emequimica, Spain | heterocyclic compound |
| Eulysin WP | BASF, Germany | organic ester |
| Levafix Salz PC | Bayer, Germany | aliphatic carboxylic acid, sodium salt |
| Meropan EF | CHT, Germany | special esters |
| Sandacid V | Sandoz, Switzerland | heterocyclic compound |
| Sandacid VS | Sandoz, Switzerland | organic formate |
| Verolan GBK | Rudolf-Chemie, Germany | low molecular mass ester |

ling agent to a system that would otherwise give unlevel results.

**Testing of the Retarding Effect.** To assess the retarding effect of a levelling agent, the progress of absorption of the dye from the dye bath and onto the fiber is observed with and without the auxiliary concerned. During the dyeing process, samples are removed from the dye liquor, and either the amount of dye deposited on the fiber is determined by extraction or by reflectance measurements, or the amount of dye remaining in the liquor is determined colorimetrically or by dyeing more undyed textile in this liquor [5.102]–[5.105]. By using dye combinations, the effect of the dyeing auxiliary on the simultaneous absorption of the dye components can also be determined from any change in the color shade [5.106].

**Testing of the Migration Effect.** The migration effect of a levelling agent leads to levelling of nonuniformly dyed material either during the course of the dyeing process or subsequently. Assessment of the migration effect is carried out in a migration, levelling, or boildown test [5.107]–[5.110]. Dyed material is treated with undyed material under dyeing conditions in a blind dye bath, a bath that contains all the chemicals necessary for dyeing except the dye. The effect of the levelling agent in promoting transfer of the dye from dyed material to accompanying undyed material is then observed [5.111], [5.112]. Testing of the retention effect of the levelling agent (i.e., the reduction in the bath exhaustion) is described in Part 2 of DIN 54 290 [5.113]. This test is important in the assessment of levelling agents.

**Rectification of Unlevel Dyeing.** To test the ability of levelling agents to rectify unlevel dyeing, caused by different dye absorbtion of fibers, special test materials and dyes are used in which the formation of stripes is especially noticeable. The levelling effect of a levelling agent is assessed visually, by reflectance measurements, or by extraction of the absorbed dye [5.108], [5.113]–[5.116].

**Special Levelling Tests.** To test the effectiveness of levelling agents, experimental conditions are often used which give unlevel dyeing without the use of auxiliaries. The extent to which the addition of a levelling agent improves the uniformity of dyeing is then assessed [5.117]–[5.123].

As well as its levelling effect, other effects of a levelling agent are often tested, e.g., wetting effect, dispersing effect, foaming tendency, etc., as these properties can also affect the uniform appearance of a dyed material.

**Chemical Identification and Quantitative Determination of Levelling Agents.** As most levelling agents are surfactants or similar products, the usual methods of surfactant analysis are used [5.1], [5.124].

## 5.7. Dyeing Accelerants (Carriers)

Dyeing accelerants (carriers) are used in the dyeing of synthetic fibers by the exhaustion process to give increased rate of absorption of the disperse dye onto the fiber, more rapid diffusion into the fiber, and better dyeing yield.

The use of carriers enables polyester fibers to be dyed with disperse dyes to a satisfactory intensity even at 100 °C. The carrier dyeing process is therefore of particular importance for blended fibers of polyester and wool, as wool must not be subjected to wet treatment at temperatures significantly above 100 °C (→ Textile Dyeing, pp. 445–446). Carriers need not be used to increase dyeing yield in the high temperature dyeing of pure polyester (→ Textile Dyeing, Chap. 7), but they are often used as levelling agents in this process (see Section 5.6.4.). They are much less used for dyeing triacetate, polyacrylonitrile, and polyamide fibers than for polyester fibers.

Research into carriers has concentrated on their effects on polyester fibers [5.70], [5.125]–[5.129]. It is currently believed that the activity of carriers in accelerating dyeing is mainly located in the fiber region.

Dyeing in the presence of carriers can also bring a number of problems. Most carriers are toxic, contaminate wastewater, and lead to a high COD value [5.130], [5.131]. Carrier dyeing can also cause fiber swelling sufficient to change the technological properties of the fiber. Also, carrier residues remaining on the fiber after dyeing can affect lightfastness and heat setting fastness. Apart from avoiding or minimizing these problems, an efficient carrier must also fulfil the following requirements [5.1]:

1) Maximum effectiveness from a minimum amount of carrier used
2) Independence of the carrier effect from the chemical constitution of the dye
3) Rapid formation of a stable emulsion of the carrier under dyeing conditions
4) Low volatility in steam
5) Ease of removal from the fiber
6) Very little or no odor contamination of the environment
7) With disperse dyes, no increase in soiling of the other fibers present in fiber blends

Commercial products are usually liquid and selfemulsifying, i.e., they contain emulsifiers and sometimes solvents as well as the active substances. Typical carrier formulations contain 60–80% active substance, 10–30% emulsifier, and 0–10% solvent [5.132].

Active substances for dyeing accelerants include the following groups of compounds [5.74], [5.133], [5.134]:

1) Liquid halogenated benzenes
2) Aromatic hydrocarbons
3) Aromatic hydroxy compounds
4) Aromatic alcohols, ketones, carboxylic acids and their esters
5) Alkyl phthalimides
6) Substituted phenylglycols and their esters

**Dyeing Accelerants for Polyester Fibers.** Carriers are mainly used as dyeing accelerants in polyester–wool blends and as levelling agents in the high-temperature dyeing of pure polyester fibers. The most important active substances used in dyeing accelerants for polyester fibers are 1,2-dichlorobenzene; 1,2,4-trichlorobenzene; 2-phenylphenol; diphenyl; diphenyl ether; methyl, butyl, and benzyl benzoate; methyl salicylate; dimethyl phthalate; tetralin; $\alpha$- and $\beta$-methylnaphthalene; phthalic acid $N$-butylimide; and chlorophenoxyethanol.

The composition of a dyeing accelerant is always a compromise. The volatility and toxicity of chlorobenzenes limit their use in closed equipment, and the reduction of lightfastness by naphthalene derivatives, diphenyl, or 2-phenylphenol necessitates subsequent fixing of the material at $> 180\,°C$. Chlorobenzenes, methylnaphthalene, and carboxylic esters are particularly suitable for polyester–wool blends, as they do not increase soiling of the wool by disperse dyes.

The properties of the various types of dyeing accelerant are compared in [5.135]–[5.137]. A method of testing the accelerating effect and dye yield is described in DIN 54 289.

**Dyeing Accelerants for Other Synthetic Fibers.** With triacetate fibers, dyeing accelerants are only necessary to increase the dye yield for very deep tones or with dyes that do not diffuse well. The most commonly used compounds are esters of benzoic acid, salicylic acid, or phthalic acid.

To avoid severe shrinkage of polyacrylonitrile fibers, dyeing is occasionally carried out below the glass transition temperature $T_g$ in the presence of dyeing accelerants. Ethylene and propylene carbonate and particularly benzyloxypropionitrile are suitable.

Certain modified polyacrylonitrile fibers become yellow and delustered or lose strength if they are dyed at $100\,°C$. Such fibers can be dyed at lower temperatures by using dyeing accelerants. Dyeing accelerants for the Verel (modified acrylic) fibers include triisobutyl phoshates.

Poly(vinyl chloride) fibers are normally dyed in the presence of dyeing accelerants (usually benzoic acid esters) with disperse dyes at 60–70 °C.

With polyamide fibers, it is often difficult to obtain a high enough dyeing temperature, especially when dyeing in a jigger. The dyeing properties are therefore improved either by means of pretreatment with a benzyl alcohol–ethanol mixture [5.137], or by the direct addition of benzyl alcohol to the dye bath.

When dyeing aramid fibers, which have poor dyeing properties at boiling temperatures, butyl benzoate, salicylic esters, benzaldehyde, or acetophenone are added as dyeing accelerants.

**Environmental Aspects of Dyeing Accelerants.** The environmental problems presented by many dyeing accelerants include toxicity to humans, fish, and sewage sludge, poor biological degradability, and odor problems [5.130], [5.138], [5.139]. However, as these active substances have an affinity for the fiber, and are ca. 75–90 % adsorbed by the textile material [5.74], [5.130], only the biologically harmless emulsifiers normally remain in the dye bath and subsequently enter the wastewater. The dyeing accelerant (carrier) that remains on the fiber is expelled during the drying or fixing process, so that the waste air cleaning processes control the emission of this material [5.140].

Commercially available dyeing accelerants are listed in Table 5.13.

## 5.8. Anticreasing Agents

Crease formation is one of the most unpleasant faults that can occur during the finishing of piece goods. There is an extensive literature on the mechanism, prevention, and cure of creasing [5.141].

Because the removal of many types of crease is either impossible or extremely difficult, textile finishers always try to prevent them as far as possible [5.142]–[5.147].

Longitudinal creases are always formed when the textile is treated in rope form and with high lengthwise tension. If piece goods are treated in wide form, a crease-free product can be obtained, although the special character of many textile qualities (bulk, handle, fullness, etc.) can only be obtained by finishing in machines that do not produce high tension (winches, overflows, and jets). As well as measures that can be taken

involving processing and mechanical aspects, (preliminary fixing of the material, low treatment temperature, slow cooling after heat treatment, finishing on special machines with low longitudinal tension), the formation of permanent creases can also be minimized by adding special auxiliaries to the treatment bath [5.1], [5.142]–[5.144], [5.148].

The action of anticreasing agents is not fully understood. They probably act as lubricants which improve the sliding of one part of the textile over another, so that creases formed during movement of the textile open up and disappear more easily [5.143], [5.145], [5.146].

The following are the main classes of compounds used:

1) Synthetic products based on fatty acids or their esters, amides, alkylol esters, and alkylolamides also fatty alcohols, usually ethoxylated. Apart from these nonionic compounds, ionic compounds with carboxyl, sulfonic, or phosphoric acid groups can also be used.
2) Products based on lecithin.
3) Products based on water-soluble alkoxylated high molecular mass polyamides.
4) High molecular mass polyethoxylates, polyacrylates, and acrylamide–acrylic acid copolymers.

A list of commercial products is given in Table 5.14. These are often mixed products which also contain dispersing and levelling agents as well as the anticreasing agents. The auxiliaries must not cause foaming, but should if at all possible have an antifoaming effect.

**Table 5.13.** Some commercially available dyeing accelerants (carriers) [5.5]

| Name | Producer | Uses | Active substance |
|---|---|---|---|
| Carolid 250 | Sybron/Tanatex, Netherlands | PES | 2-phenylphenol |
| Dilatin NAN | Sandoz, Switzerland | PES | aromatic hydrocarbons |
| Levalin ACE | Bayer, Germany | PAN | aromatic ethers |
| Levegal DTE | Bayer, Germany | PES, CTA | halogenated aromatics |
| Levegal PEW | Bayer, Germany | PES, CTA | $N$-alkylphthalimide |
| Levegal PTN | Bayer, Germany | PES, CTA | aromatic carboxylic acid esters |
| Matexil CA-MN | Zeneca, United Kingdom | PES | methylnaphthalene |
| Remol HT | Hoechst, Germany | PES | diphenyl |
| Remol LF | Hoechst, Germany | PES | diphenyl/methylnaphthalene |
| Sarapol DLU | CHT, Germany | PES | aromatic esters and hydrocarbons |
| Setavin CA | Zschimmer + Schwarz, Germany | PES | phthalic acid derivatives |
| Tanadel PTN | Sybron/Tanatex, Netherlands | PES | aromatic ester |
| Tanadel TMO | Sybron/Tanatex, Netherlands | CTA | aromatic ester |

PES = Polyester; CTA = Cellulose triacetate; PAN = Polyacrylonitrile.

**Table 5.14.** Some commercially available anticreasing agents [5.5]

| Name | Producer | Active substance |
| --- | --- | --- |
| Breviol LFP | Henkel, Germany | modified phosphoric ester |
| Breviol PAM | Henkel, Germany | polyamide derivative |
| Cibafluid U | Ciba-Geigy, Switzerland | mixed polymerizate based on polyethers |
| Eganal LUB | Henkel, Germany | fatty acid polyglycol ester |
| Hydrocol LFV | Rudolf-Chemie, Germany | ethoxylated caprolactam derivative |
| Lubit LC | Sybron/Tanatex, Netherlands | modified phosphoric ester |
| Osimol LVP | Grünau, Germany | modified phosphoric ester |
| Palatex S | BASF, Germany | polyamide derivative |
| Persoftal L | Bayer, Germany | sulfonated fat |
| Pretolon L | Ciba-Geigy, Switzerland | modified lecithin |
| Rucolin JES | Rudolf-Chemie, Germany | polyacrylamide |
| Sevosoftal U | TC, Switzerland | sulfated fats and lubricating components |
| Tebolan MDF liq. | Dr. Th. Böhme, Germany | phosphoric ester, special emulsifiers |
| Viscavin S 700 | CHT, Germany | modified phosphoric esters |

## 5.9. Auxiliaries for Continuous Dyeing

Continuous and semicontinuous dyeing processes operate according to the following fundamental principle [5.1], (→ Textile Dyeing, pp. 377–390):

1) Padding (impregnation) of the textile with the dye liquor followed by squeezing-off to give a definite amount of absorbed liquor
2) Intermediate drying (not with all processes)
3) Developing and/or fixing of the dye by chemical, hot air, steam, or contact heat treatment
4) Removal of the unfixed dye and auxiliaries by washing

Continuous dyeing processes are used both with water-soluble dyes (substantive, reactive, acid, and cationic dyes) and water-insoluble dyes (disperse and vat dyes).

The following products are used as auxiliaries for continuous dyeing (padding auxiliaries):

1) Dispersants and protective colloids (see Section 5.4), solubilizing agents and hydrotropic agents (see Section 5.1) are used to improve the solubility of the dye or the stability of the dye dispersion
2) Wetting agents (see Section 5.3) and products to increase liquor pickup lead to uniform wetting and formation of a continuous adherent film of liquor on the textile
3) Antimigration agents prevent of migration of the dye during intermediate drying
4) Dye solubilizing agents and hydrotropic compounds, fiber swelling agents, and fixing accelerants (see Section 5.1) improve transfer of the dye into the fiber

5) Antifrosting agents promote dyeing of protruding parts of the fiber

An auxiliary must also have the following properties to ensure problem-free processing without harming the quality of the textile:

1) No marked affinity for the fiber, so that the composition of the pad liquor is kept constant
2) No detrimental effects on handle and fastness properties
3) Very low foaming tendency, although foam formation may sometimes be desirable (antifrosting agents)

The auxiliaries discussed below are used exclusively for continuous dyeing.

### 5.9.1. Antimigration Agents

In all continuous dyeing processes, after padding of the dye there is always the risk of dye and pigment migration due to nonuniform drying. The dye migrates with the water to the more rapidly drying parts of the textile, causing variations in color intensity and imperfect dye penetration. Antimigration agents prevent this dye movement. The action of these products, which takes place during the drying process, limits the growth of the dye particles by causing a kind of adhesion effect and viscosity increase [5.149], [5.150]. These auxiliaries are not often used when dyeing with water-soluble dyes. Highly active products include alginates or high molecular mass acrylic acid or acrylamide copolymers. Other compounds are used to a lesser extent,

**Table 5.15.** Some commercially available auxiliaries for the pad dyeing process [5.5]

| Name | Producer | Uses | Composition |
|---|---|---|---|
| Irgapadol MP | Ciba-Geigy[b] | antimigration agent | poly(acrylic acid) derivative |
| Levalin KAM | Bayer[a] | antimigration agent | polyacrylate |
| Matexil FA-MIV | ICI[c] | antimigration agent | organic polyelectrolyte |
| Primasol AMK | BASF[a] | antimigration agent | acrylic acid copolymer |
| Solidokoll N | Hoechst[a] | antimigration agent | polyacrylamide derivative |
| Arbyl FRO | Grünau[a] | antifrosting agent | fatty acid condensate |
| Emigen DPR | Hoechst[a] | antifrosting agent | polymerization product |
| Lyogen CW | Sandoz[b] | antifrosting agent | alkyl polyglycol ethers |
| Matexil PA-S | ICI[c] | antifrosting agent | poly(ethylene glycol) |
| Primasol SD | BASF[a] | antifrosting agent | preparation of nonionic surfactants |
| Impegal EPR | Impocolor[a] | increase in liquor uptake | polyacrylamide |

[a] Germany; [b] Switzerland; [c] United Kingdom.

e.g., cellulose ethers, starch ethers, carboxymethyl celluloses, modified polysaccharides, and poly(vinyl alcohols) [5.151].

### 5.9.2. Antifrosting Agents

With certain textile articles and thick-pile, parts of the fiber that protrude from the yarn structure are inadequately wetted by the dye liquor, and hence are inadequately dyed. The textile then becomes covered with a gray "veil" (frosting). The effect can be prevented by two methods, in both of which the protruding fibers are held in the liquor [5.152], [5.153]:

1) Foam production during heating of the liquor in dye fixing. Mixtures of nonionic and anionic surfactants are used, sometimes with special volatile antifoaming agents that prevent foam formation during padding.
2) Increasing the viscosity and thickness of the liquor film to prevent the fibers from protruding after squeezing off. This is achieved by using high molecular mass polyacrylates and acrylamide–acrylic acid copolymers.

### 5.9.3. Products to Increase Liquor Uptake

Some continuous and semicontinuous dyeing processes rely on the textiles picking up a large amount of liquor (e.g., the wet-in-wet process for vat dyes → Textile Dyeing, p. 385 and the pad-jig process, → Textile Dyeing, p. 372). High liquor pickup is synonymous with maximum wetting and penetration of the textile, and is obtained by using effective pad-wetting agents (see Section 5.3). A further increase in liquor pickup is achieved by increasing its viscosity, thereby pre-

venting it from draining out of the interstices of the textile material. The effectiveness of this method is strongly dependent on the structure of the textile (open or compact).

Products for increasing liquor pickup include high molecular mass polyacrylates and acrylamide–acrylic acid copolymers. Some commercially available products are listed in Table 5.15.

### 5.10. Aftertreatment Agents to Improve Fastness

Many dyes of interest for textiles do not give adequate wetfastness properties (fastness to water, washing, perspiration, and ironing) or fastness to rubbing and light. Textiles dyed with direct dyes often require aftertreatment to improve fastness, though this also applies to other types of dyed textiles.

Two principal methods are used to improve fastness properties:

1) Unfixed dyes, dye aggregates, or dye degradation products are removed by washing or cleaning with complexing agents [5.48] or various types of surfactants with a dispersing action [5.154], by reductive decomposition [5.155], or by an organic solvent [5.156]. These auxiliaries are not usually classified as aftertreatment agents.
2) Bonding of poorly fixed dye to the fiber. If removal of the unfixed dye by washing does not give the required fastness properties, the poorly fixed dye is bonded to the fiber by an aftertreatment process. The auxiliaries used for this differ in chemical structure and the mechanism by which they act, depending on the type of dye and type of fiber [5.157].

**Aftertreatment of Direct Dyes.** The basic principles of the use of aftertreatment agents are illustrated by the example of dyeing with substantive dyes (→Azo Dyes , **A 3**, pp. 288–291).

*Aftertreatment with Cationic Produts* [5.158], [5.159]. Cationic products improve wetfastness by forming sparingly soluble dye salts with the anionic groups of the dye [5.2]. Aftertreatment agents of this type include quaternary (poly)-ammonium compounds with long hydrocarbon chains, polyamines, or polyethyleneimine derivatives.

*Aftertreatment Agents Based on Formaldehyde Condensation Products, and Formaldehyde-Releasing Products.* The use of these basic condensation products presumably leads to the formation of sparingly soluble adducts with the dye molecules [5.2], [5.160], [5.161]. Formaldehyde-releasing condensation products are particularly effective with dyes that have free hydroxyl or amino groups. The dye molecules probably become linked to the sparingly soluble dye aggregates by methylene bridges. Commercial products of this type include formaldehyde condensation products with amines, polynuclear aromatic phenols, cyanamide, or dicyandiamide.

*Aftertreatment with Metal Salts.* Aftertreatment with copper sulfate or potassium dichromate leads to complexation with certain substantive dyes. These reduce the aqueous solubility of the dye [5.2], [5.162], [5.163].

*Aftertreatment with diazotized bases* can only be used with dyes that have amino and hydroxyl groups in suitable positions, and, for example, couple with diazotized 4-nitroaniline [5.162].

**Aftertreatment of Other Classes of Dye.** Aftertreatment with cationic products can be used with all dyes that contain anionic groups (reactive, sulfur, acid, and metal-complex dyes). The following aftertreatment methods are also used:

*Acid Dyes.* Condensation products from aromatic sulfonic acids are used as well as cationic products, especially with polyamide fibers.

*Sulfur Dyes.* In addition to the use of cationic products to improve wetfastness, dark colored dyed materials must be aftertreated with sodium carbonate or sodium acetate to neutralize the sulfuric acid produced by oxidation of the sulfur dyes on storage, which can damage the fiber.

*Reactive Dyes.* Aftertreatment is usually carried out in two stages:

1) Removal (as thoroughly as economically possible) of the substantively bonded dye hydrolyzate by washing with very soft water. This process is very much improved by the use of complexing agents, which suppress the formation of sparingly soluble alkaline-earth salts of the dye hydrolyzate in incompletely softened water [5.48].
2) Fixing of the residual dye hydrolyzate by a cationic aftertreatment agent.

Some trade names used for these products are listed in Table 5.16.

## 5.11. Reducing Agents

In dyeing, reducing agents are used for the following purpose:

1) Reduction of vat and sulfur dyes to form the leuco compound (which has an affinity for the fiber) when dyeing
2) Aftertreatment for improving the colorfastness of textiles dyed and printed with disperse dyes (see Section 5.10) by the reductive decomposition of dye adhering to the surface

**Table 5.16.** Trade names used for aftertreatment agents to improve wetfastness [5.5]

| Name | Producer | Dyes whose wetfastness properties can be improved |
|---|---|---|
| Alcofix | Allied Colloids, United Kingdom | direct, reactive, and acid dyes |
| Basolan | BASF, Germany | acid and metal-complex dyes |
| Cibatex | Ciba-Geigy, Switzerland | acid dyes |
| Cyclanon | BASF, Germany | reactive and direct dyes |
| Fissatore | Bozzetto, Italy | direct and reactive dyes |
| Impofix | Impocolor, Germany | direct, reactive, and acid dyes |
| Levogen | Bayer, Germany | direct and reactive dyes |
| Matexil | Zeneca, United Kingdom | acid and reactive dyes |
| Mesitol | Bayer, Germany | acid dyes and cationic dyes for polyamide |
| Rewin | CHT, Germany | direct, reactive, sulfur, acid, and metal-complex dyes |
| Sandofix | Sandoz, Switzerland | direct and reactive dyes |
| Solidogen | Cassella, Germany | direct, reactive, and sulfur dyes |
| Tinofix | Ciba-Geigy, Switzerland | reactive and direct dyes |

3) Stripping of defectively dyed textiles (see Section 5.13) by reductive decomposition of the dye
4) In dye protection agents (see Section 5.2), for reducing residual hydrogen peroxide in bleached cotton dyed with reactive dyes

The most important reducing agents are those used in the dyeing of cellulose fibers with vat and sulfur dyes. These reduce water-insoluble dye pigments in the alkaline pH range to form water-soluble leuco compounds with an affinity for the fiber. These are absorbed by the fiber and then fixed by oxidation.

The reducing agents used can be divided into three groups:

1) Sulfur-containing compounds ($\rightarrow$ Sulfites, Thiosulfates, and Dithionites; $\rightarrow$ Sulfinic Acids; $\rightarrow$ Sulfides, Polysulfides, and Sulfanes). These are directly or formally derived from the following parent compounds dithionous acid ($H_2S_2O_4$) and sulfinic acid ($RSO_2H$): sodium dithionite ($Na_2S_2O_4$), hydroxyalkylsulfinates, thiourea dioxide (formamidinesulfinic acid); sulfurous acid ($H_2SO_3$): sodium sulfite, hydroxyalkylsulfonates; hydrogen sulfide ($H_2S$): sodium sulfide, sodium hydrogen sulfide.
2) Organic compounds. These include substances with the $\alpha$-hydroxycarbonyl structure (reductones): glucose, hydroxyacetone.
3) Complex hydrides: $NaBH_4$.

Vat dyes and sulfur dyes pose different requirements on the activity of the reducing agent, and are therefore considered separately.

**Reducing Agents for Vat Dyes.** Vat dyes are anthraquinoid and indigoid dye pigments and are generally more difficult to reduce than sulfur dyes. They were discovered in 1901 by R. Bohn, BASF but their large-scale use only became possible in 1906 when highly reactive sodium dithionite was successfully produced in anhydrous and therefore storable form. Sodium dithionite then became the most important reducing agent in the textile industry.

The mechanism of the heterogeneous reduction reaction has been thoroughly investigated [5.164]–[5.166].

In the dyeing process, sodium dithionite is consumed by reduction of the dye and also by reaction with atmospheric oxygen. It must there-fore either be present from the start in excess or be added continuously. A wide range of techniques have been proposed to reduce these losses [5.167], including the addition of stabilizing additives [5.168]. A number of reducing agents less sensitive to oxygen have also been proposed [5.166]. They include thiourea dioxide and sodium borohydride [5.169], which have not been widely used, mainly for economic reasons, but also because of their low activity.

Following increasing environmental pressures, biologically degradable sulfur-free organic reducing agents have been investigated [5.170]. Hydroxyacetone is significantly more effective than glucose, but cannot match the range of applications of sodium dithionite.

Although many other compounds have been proposed, sodium dithionite is still the dominant reducing agent for vat dyes.

**Reducing Agents for Sulfur Dyes.** The most common reducing agent for sulfur dyes is sodium sulfide ($\rightarrow$ Textile Dyeing, p. 412). Some sulfur dyes can also be reduced with sodium hydrogensulfide. As this has a less alkaline reaction, it is used for dyeing alkali-sensitive regenerated cellulose fibers. Sodium dithionite can cause overreduction of a number of sulfur dyes, leading to changes in color shade.

Sulfide-free reducing agents are being increasingly recommended, mainly for environmental reasons. These include low-sulfur mixtures of organic reducing agents with sodium dithionite or thiourea dioxide [5.171], hydroxyacetone, and glucose.

Some commercially available products are listed in Table 5.17.

## 5.12. Oxidizing Agents

The oxidizing agents employed in dyeing are mostly the same as those used in textile bleaching (see Chap. 4), i.e., hydrogen peroxide, sodium peroxosulfate, sodium peroxoborate, sodium hypochlorite, and sodium chlorite. For special applications, sodium chlorate, sodium nitrite, sodium dichromate, and sodium 3-nitrobenzenesulfonate are also used.

Oxidizing agents are required when dyeing cellulose fibers with vat dyes, leuco esters of vat dyes, and sulfur dyes, and also for the production of aniline black ($\rightarrow$ Textile Dyeing, pp. 408–409; 412–413; 419).

**Table 5.17.** Some commercially available reducing agents [5.5]

| Name | Producer* | Uses | Composition |
|---|---|---|---|
| Bayreduct PRN | Bayer | peroxide removal before reactive dyeing | inorganic sulfur compound |
| Cyclanon R | BASF | cleaning of dyed polyester materials | reducton + dispersing agent |
| Hydrosulfit conc. | BASF | vat dyes | sodium dithionite |
| Lavan RH | Textilcolor | cleaning of dyed polyester materials | reducton |
| Lorinol R | Henkel | cleaning of dyed polyester materials | thiourea dioxide |
| Redoxarin | Rubensperger | sulfur dyes | dextrose |
| Reduktionsmittel F | Degussa | cleaning of dyed polyester materials | thiourea dioxide |
| Rongal HT | BASF | vat dyes, high temperature dyeing | preparation of sodium dithionite |
| Rongal PS | BASF | vat dyes, pad steam | preparation of sodium dithionite |
| Zetesal UR | Zschimmer + Schwarz | cleaning of dyed polyester materials | glucose derivatives |

* in Germany.

When dyeing with vat dyes, the insoluble pigment is regenerated by oxidation. With some dyes and dyeing processes, this can be performed by the oxygen dissolved in the water used for washing, or by hanging the textiles in air. However, it is usually accelerated by adding an oxidizing agent [5.165]. Hydrogen peroxide and sodium 3-nitrobenzenesulfonate are preferred. The latter can also be used in weakly alkaline solution, enabling washing to be minimized. The color shade of the dye can vary to some extent, depending on the oxidation method used [5.165].

Sulfur dyes are in generally easier to reduce and hence more difficult to oxidize than vat dyes. Oxygen alone will not cause oxidation to form the insoluble dye pigment. Oxidation of the leuco compounds of the sulfur dyes can in principle be performed by all the inorganic oxidizing agents mentioned above, although color shade and handle may vary [5.172]. Hydrogen peroxide is preferred for environmental reasons [5.171].

For developing the leuco esters of vat dyes, where hydrolysis of the sulfuric dye ester in acid solution is coupled with oxidation, the usual oxidizing agent is sodium nitrite.

The oxidizing agent used to form aniline black for "single-bath black" and "oxidation black or steam black" is sodium chlorate.

The most commonly used oxidizing agents for dyeing are readily available simple chemical substances. A few such compounds are marketed as textile auxiliaries under trade names. Some of these are listed in Table 5.18.

## 5.13. Stripping Agents

When dyeing and printing, any defective textile products that may sometimes be produced must be rectified for economic reasons. Defects

**Table 5.18.** Some commercially available oxidizing agents [5.5]

| Name | Producer | Composition |
|---|---|---|
| Emectol PAL | Emequimica[a] | NBS* |
| Lamesal P | Grünau[b] | NBS |
| Ludigol Granulat | BASF[b] | NBS |
| Lyoprint RG | Ciba-Geigy[c] | NBS |
| Perlavin OSL | Dr. Petry[b] | NBS |
| Revetol S | Sandoz[c] | NBS |
| Rewin SW | CHT[b] | inorganic oxidizing agent + non-ionic additives |

* NBS = 3-Nitrobenzenesulfonic acid, sodium salt.
[a] Spain; [b] Germany; [c] Switzerland.

may be due to nonuniform dyeing properties of the textile, process errors (incorrect operation of machines, incorrect use of dyes and chemicals), or soiling during transportation of the material. Before the material can be correctly dyed or printed, the original dye must normally be stripped (i.e., removed or chemically destroyed). Depending on the severity of the defect, the dyed material may be simply brightened, (partial dye removal), or the dye may be chemically destroyed by reduction or oxidation, completely removing it from the fiber.

It is often possible to brighten the material while retaining the color shade and colorfastness by treating the defective product with an auxiliary with an affinity for the dye, and the effect can be intensified by repeating the treatment. The auxiliary with an affinity for the dye, which usually has the same composition as a levelling agent (see Section 5.6), shifts the equilibrium between the dye on the fiber and the dye in the liquor in favor of the liquor.

Chemical decomposition of the dye is carried out with the usual reducing agents, oxidizing agents, and discharging agents used in dyeing and printing. These include common chemicals such as sodium dithionite, sodium hypochlorite, hydrogen peroxide, etc., and products marketed under special trade names (see Sections 5.11, and 5.12).

The most important methods of brightening and stripping dyed and printed materials are described below. Precise bath formulations can normally be obtained from the sample books and information sheets of the dye producers. A comprehensive treatment of the subject can be found in [5.173].

**Stripping Dyed Cellulose Fibers.** Materials dyed with direct dyes can be brightened by boiling with an auxiliary with an affinity for the dye in a solution containing sodium carbonate. In the case of dyed materials whose fastness has been improved by cationic products, these must first be removed by treatment with strong solutions of formic acid or with diluted hydrochloric acid. Stripping is performed by chemical decomposition with alkaline sodium dithionite, sodium hypochlorite, or sodium chlorite solution.

Materials dyed with reactive and naphthol dyes cannot be brightened (though in exceptional cases materials dyed with reactive dyes can be brightened in a bath containing strong alkali or acetic acid at 60–85 °C). The method of destroying reactive dyes is similar to that for direct dyes. Materials dyed with naphthol dyes are stripped with alkaline sodium dithionite solution with an added "reduction catalyst" (anthraquinone).

Materials dyed with vat and sulfur dyes can be brightened in a "blind vat" (alkaline sodium dithionite solution) by an auxiliary with an affinity for the dye. Only in exceptional cases is the dye destroyed with sodium hypochlorite solution.

Suitable auxiliaries with an affinity for the dye include fatty amine ethoxylates and especially polyvinylpyrrolidone (PVP).

**Stripping Dyed Wool.** Wool dyed with acid or metal-complex dyes can be brightened by boiling with levelling agents and sodium sulfate in neutral or weakly alkaline baths. Almost all dyed woollens can be reductively stripped with zinc formaldehyde sulfoxylate and formic acid in the presence of a levelling agent with an affinity for the fiber (e.g., fatty amine ethoxylate). The stripping effect can be improved by a neutral treatment with stabilized sodium dithionite (e.g., Blankit IN, BASF). As wool is sensitive to alka-lis and oxidizing agents, it should only be treated in acidic or, if necessary, weakly alkaline baths, and only with reducing agents.

**Stripping Dyed Polyamide.** As polyamide is more resistant to alkalis, materials dyed with anionic dyes can be brightened in solutions of sodium carbonate and a levelling agent with affinity for the dye. Disperse dyes can be brightened with nonionic levelling and dispersing agents, or can be stripped in acid solution by reduction with zinc formaldehyde sulfoxylate or oxidation with sodium chlorite.

**Stripping Dyed Polyester and Triacetate.** Stripping fibers dyed with disperse dyes is very difficult and is only possible at high temperature. These textiles can be brightened by using high concentrations of nonionic levelling agents and carriers, but some fiber damage can occur. The dye can be destroyed by chlorite bleaching with addition of a carrier.

**Stripping Dyed Polyacrylonitrile.** Polyacrylonitrile fibers dyed with disperse dyes can be brightened by nonionic levelling agents, or stripped by sodium chlorite or zinc formaldehyde sulfoxylate. Basic dyes can be substantially brightened by anionic surfactants. Although most basic dyes can be removed by reductive or oxidative stripping by zinc formaldehyde sulfoxylate, sodium chlorite, or sodium hypochlorite, this causes permanent yellowing of the fiber.

**Stripping Dyed Fiber Blends.** Methods suitable for sensitive fibers should be used when stripping fiber blends. For example, a waste textile such as reclaimed wool, which is actually a blend, is treated like pure wool (i.e., with zinc formaldehyde sulfoxylate or stabilized sodium dithionite).

Some commercially available products especially suitable for stripping defectively dyed materials are listed in Table 5.19. For brightening dyed materials, levelling agents can generally also be used (see Section 5.6).

## 5.14. Resist Agents for Dyeing and Printing

In some dyeing operations, resists are employed to prevent dyes from being absorbed on certain parts of the textile, to produce special

**Table 5.19.** Some commercially available products for stripping defectively dyed materials [5.5]

| Name | Producer | Active substances |
| --- | --- | --- |
| Albigen A | BASF, Germany | polyvinylpyrrolidone |
| Anogal KF | Baur & Gäbel, Germany | amine ethoxylates |
| Decrolin | BASF, Germany | zinc formaldehyde sulfoxylate |
| Deflavit ZA | BASF, Germany | stabilized sulfoxylate |
| Emegal A | Emequimica, Spain | polymerization product |
| Gisapon 1430 | Rotta, Germany | alkylamine polyglycol ether sulfate |
| Perlavin A | Dr. Petry, Germany | polyvinylpyrrolidone |
| Ritornal AL | Geissler, Germany | polypyrrole derivative |
| Tinegal W | Ciba-Geigy, Switzerland | alkylamine polyglycol ether |
| Uniperol AC | BASF, Germany | fatty amine polyglycol ether |

effects. In the case of fiber blends, it is used to protect components of the fiber that give a different color shade with certain dyes, so that they can subsequently be dyed with other dyes to give the same color shade as the other components.

Resist areas can be produced mechanically (by tying or by covering selected areas with wax, e.g., batik) or chemically by using resist agents. The use of resist agents enables mixture effects to be produced, or, with fibers of different dye affinity, the tone-in-tone effect (levelling agents, see Section 5.6).

In textile printing (→ Textile Printing), there are many ways of obtaining resist effects. Resists applied before printing are known as preprint resists, and those applied afterwards, as overprint resists. "White" resists completely prevent dyeing, half-tone resists partially prevent it, and, in colored resists, the printing paste that contains the resist also contains a dye which is itself not resisted, but causes coloration where it is applied. Discharge resists contain a discharging agent as well as a resist.

Not only can a textile be made resistant to dyeing, but it can also be made resistant to the acquisition of other surface qualities, e.g., abrasiveness, a crêpe effect (alkalis), or roughness effects. With polyester fibers, resist printing is mainly used in the production of discharge resists, and, on cellulose fibers, reactive resists.

**Resist Agents for Dyeing.** Resist agents to give uniform dyeing of fiber blends are mainly used in the dyeing of wool–cellulose, polyamide–cellulose, and wool–polyamide fibers with acid, metal-complex, and direct dyes.

With wool–cellulose and polyamide–cellulose blends, the cellulose fiber is usually dyed with direct dyes, which, with increasing temperatures and dyeing time, are absorbed onto both the wool and the polyamide. In the presence of anionic resist agents (e.g., polycondensation products of aryl- or alkylarylsulfonic acids with formaldehyde), which have a strong affinity for wool and polyamide and in the acid pH range can be absorbed as a colorless dye on these fibers, wool and polyamide can be wholly or partially protected before dyeing with direct dyes. In the tone-in-tone dyeing of wool–polyamide blends, anionic surfactants (e.g., fatty alcohol sulfonates, alkylarylsulfonates, sulfonated castor oil, etc.) retard absorption of the dye by the polyamide fiber, which has an affinity for the dye, and enable the two types of fibers to be dyed to an equal intensity.

To obtain mixture effects (with wool and polyamide), part of the fiber material is treated with a resist agent, mixed with untreated fiber, and then dyed. Suitable resist agents include the same products as those used in the tone-in-tone dyeing of wool–polyamide blends. In the Sando-Space-R process [5.174], resist effects are produced on wool and polyamide by reducing the dye affinity of fibers by a chemical reaction with a reactive resist agent (e.g., the sodium salt of dichlorotriazine-*p*-sulfanilic acid). A list of commercially available products is given in Table 5.20.

**Resist Agents for Printing.** Resist effects can be obtained by mechanical means or by printing a textile material with a concentrated paste containing a pigment or a resin, followed by dyeing or printing. This type of resist is often combined with one of the many possible chemical resists. Apart from the actual resist agents, the choice of paste plays an important part in resist technology.

In the following examples, some important methods of producing resists for various fibers and classes of dye are described. (The various resist techniques are described in → Textile Printing.)

**Table 5.20.** Some commercially available resist agents [5.5]

| Product | Producer | Uses* | Composition** |
|---|---|---|---|
| Alviron RM | Bayer, Germany | WO/PA, PA/CO | PCAS |
| Erional RF | Ciba-Geigy, Switzerland | WO/PA, PA/CO | PCAS |
| Levegal ER liq. | Bayer, Germany | PA/CO | PCAS |
| Levegal ESR | Bayer, Germany | WO/PA | PCAS |
| Mesitol HWS liq. | Bayer, Germany | WO/PA, WO/CO | PCAS |
| Perigen PAW | Dr. Petry, Germany | WO/PA | PCAS |
| Sandospace R | Sandoz, Switzerland | PA | chlorotriazine derivative |
| Sandospace S | Sandoz, Switzerland | PA | triazine derivative |
| Thiotan TR | Sandoz, Switzerland | PA, WO, CO | fatty amine polyglycol ether |
| Thiotan TRN | Sandoz, Switzerland | PA carpets | fatty amine polyglycol ether |
| Zetesal CRN | Zschimmer + Schwarz, Germany | PA | combination of polyglycol ethers |

* CO = Cotton; PA = Polyamide; WO = Wool. ** PCAS = Polycondensates of aromatic sulfonic acids.

*Cellulose Fibers* [5.175]. In the case of vat dyes, a printing paste containing heavy metal salts (Zn, Mn) and oxidizing agents (e.g., Ludigol, BASF) is first applied by printing. On overdyeing or overprinting with the vat dye, the heavy metal salts prevent dye fixing, partly mechanically (formation of hydroxides), and partly chemically (combination with the alkali). The oxidizing agent simultaneously prevents reductive development of the dye.

Reactive dyes react with the cellulose in an alkaline medium. This reaction can therefore be prevented by preprinting (or, in two-phase printing, by overprinting) acids or acid-releasing materials, producing white resists. Colored resists are produced by including dyes that are fixed in an acid medium (pigment printing, Naphtol AS dyes). Reactive dyes with a β-sulfoethylsulfonic acid reactive group (Remazol and Basilen F dyes) can be resisted by certain sulfite salts (see Section 6.2.7).

As an example of resists for developing dyes, acidic salts [e.g., aluminum sulfate or tin(II) chloride] and a reducing agent can be printed onto a textile treated with Naphtol AS, preventing subsequent coupling with the diazonium component. Colored resists are made from leuco esters of vat dyes, reactive dyes, or pigments.

The formation of phthalocyanine dyes by the condensation of isoindolenines can be prevented by preprinting with certain primary amines (Phtalotrop B), zinc salts, and organic acids.

*Polyamide Fibers* [5.176]. Resist printing of polyamide is used to a limited extent in the carpet sector. Polycondensates of aromatic sulfonic acids are used as resist agents for acid dyes containing sulfonic groups and metal complex dyes.

*Polyester Fibers*. With discharge resists [5.177], the textile is first pad dyed with dischargeable dyes and then dried. The discharge resist printing paste, which contains a reducing agent [e.g., Decrolin (BASF), or Rongalit DP (BASF)] and a white pigment or discharge-resistant dye, is overprinted. In the high-temperature process, the dyes are then discharged and/or fixed in a single step. The discharging agent decomposes the dye before it can diffuse into the fiber. This process can produce very detailed patterns with sharp contours.

*Polyacrylonitrile Fibers*. The discharge resist technique can be used in the printing of cationic dyes onto polyacrylonitrile fibers [5.178].

Some commerical products are listed in Table 5.20.

# 6. Textile Printing Auxiliaries

Textile printing auxiliaries are defined as the noncoloring products used at all stages of the printing process from preparation to the final wash [6.1]. The pronounced diversification of this product group is a consequence of the application of different dye classes [6.2].

In principle a distinction can be made between printing processes based on pigments, which have no affinity to the fiber, and dyes (reactive, vat, disperse, and others), which have such an affinity. The essential auxiliaries for pigment printing are necessary to adhere the pigments permanently to the fiber; they remain on the fiber and give colorfastness. In the case of dyes with an affinity for the fiber, the auxiliaries are generally removed from the fabric in a final wash.

Unlike other printing techniques pigments can be used on almost all types of textile substrates [6.3]. It is therefore not surprising that almost 50% of the world's printed textiles (ca. $19 \times 10^9$ m/a in the early 1990s) are produced by this process [6.4].

The advantages of the application of other dye classes in printing are especially noticeable in higher fastnesses, improved handle, and in enabling more demanding design requirements to be met.

## 6.1. Auxiliaries for Pigment Printing

The idea that "cheap" pigment printing is synonymous with poor quality is today completely unfounded. High-quality pigment printing has now become established as the method of choice for such articles as home textiles, sheeting, apparel, and many others. The great variety of effects from this printing technique (direct printing with pearlescents, spangles etc.; matt-white and matt-colored printing on a colored ground fabric, three-dimensional printing, etc. [6.5]) enables rapid response to fashion trends.

At least in industrialized countries, it is current standard practice to use solvent-free recipes, which, for environmental, safety and cost reasons, have replaced formulations containing solvents, however, these are still used in other parts of the world [6.3].

The recipes contain the thickening system, a binder and, if necessary, other auxiliaries (fixing agents, hand modifiers etc.), which are used either to ensure a problem-free production process or to meet the required functional properties of the finished pigment printed article.

### 6.1.1. Thickening Agents

The choice of the thickening agent determines the rheology of the print paste, and hence affects the results of the printing process in many ways. The thickening system has a considerable influence on the printing result as characterized by evenness, sharpness of outlines, color strength, and brillance, as well as on handfastnesses, consumption of print paste, and running properties.

Synthetic thickeners are widely used today in preference to white spirit-containing emulsion thickenings, which contain up to 700 g white spirit per kilogram print paste. During conden-sation ammonia is liberated, and the polymer is converted to the acid form, generating the slightly acidic pH necessary for cross-linking of the binder and, if used, the fixing agent.

The most commonly used pigment printing thickening agents are liquid, easily pourable preparations of synthetic polymers in a mineral oil. They can be aqueous formulations, completely neutralized with ammonia and with a solids content of ca. 25%, or anhydrous, partially neutralized products with a solids content up to 60%. The latter require the addition of a predetermined amount of ammonia to the print paste. In Europe and North America, completely solvent free granulated solid products are also gaining in importance. With these, pigment printing can be a practically emission-free process.

### 6.1.2. Pigment Printing Binders

The binders used in pigment printing are polymers that form clear, colorless, and relatively soft films, and have very little effect on color shades and handle. The binder film should coat the pigment and provide physical adhesion to the substrate, and so protect the printed pigments from mechanical abrasion (rubbing fastness, washing fastness, etc.). The thickness of the film is ca. 10 μm.

Commercially available binders are aqueous dispersions of polymers (mainly based on acrylic esters, butadiene, and, less commonly, vinyl acetate) with solids contents of ca. 40–50%. The amount used depends on the amount of pigment and the textile substrate, and usually varies within the range 50–150 g per kilogram of print paste.

The surface active substances needed to produce a stable dispersion are responsible for the compatibility of the binder with the other components of the print paste.

The fastness, i.e., the resistance of the binder film to mechanical stress and its swelling tendency in water and organic solvents, must be increased to an acceptable level by the cross-linking reaction of the binder. Whereas the older types of cross-linkable binders reacted with separate fixing agents during the condensation process (hot air at 150 °C), self-cross-linking binders are now well established. These contain reactive groups, usually from copolymerization with monomers such as $N$-methylolacrylamide or

**Table 6.1.** Some commercially available auxiliaries for pigment printing

| Product | Composition | Producer |
| --- | --- | --- |
| *Thickeners* (based on acrylic acid copolymers) | | |
| Acraconz BN | liquid | Bayer |
| Alcoprint PTF | liquid | Allied Colloids |
| Lutexal HEF | liquid | BASF |
| Lutexal P | granules | BASF |
| Polymerconcentrate 475 | liquid | Morton Intern. |
| Texipol 67-5028 | liquid | Scott Bader |
| *Self-cross-linking binders* | | |
| Acramin ALW | acrylate | Bayer |
| Acramin BA | butadiene | Bayer |
| Binder ACM | acrylate | Minerva |
| Helizarin Binder ET | acrylate | BASF |
| Helizarin Binder TW | acrylate | BASF |
| Helizarin Binder UD | acrylate | BASF |
| Imperon Binder MTB | vinyl acetate | Hoechst |
| Polybinder HS 359 | acrylate | Morton Intern. |

similar compounds. Three-dimensional cross-linking of the binder film can then be achieved by acid catalysis under the usual condensation conditions [6.6].

Binder films based on butadiene can age by the action of light and oxygen, i.e., they can become yellow, and their fastness properties can deteriorate. Butadiene binders are therefore not recommended for pigment printed textiles that are continuously exposed to light (curtains, awnings, etc.) [6.7].

Some commercially available products are listed in Table 6.1.

### 6.1.3. Fixing Agents

The level of the wetfastnesses (wet rubbing, home laundry, etc.) achievable simply by the use of a binder is often inadequate for many purposes, especially on smooth fibers such as polyester. A significant improvement in fastness is obtained by the use of an additional fixing agent that strengthens the three-dimensional network of the binder film [6.6]. However, this has a certain disadvantageous effect on the hand.

Melamine–formaldehyde condensates etherified with methanol has proven to be especially suitable, as their reactive groups take part in a cross-linking reaction under the same condensation conditions as those of the self-cross-linking binders.

However, these substances are also the main source of formaldehyde emitted by pigment-printed fabrics [6.8]. The various limit values imposed today, whether by official regulations, by clothing manufacturers, or by consumer-oriented eco-labels, can be complied with by using modern, modified low-formaldehyde fixing agents of similar chemical type.

Some commercially available products are Acrafix MF (Bayer), Helizarin Fixierer S, Helizarin Fixierer LF (BASF), and Tubiprint Fixierer R (CHT).

Alternative fixing agents of different chemical bases such as aziridines or isocyanates are occasionally used, but these should be treated with circumspection as experience of their use is limited, and toxicological hazards cannot be excluded.

### 6.1.4. Hand Modifiers

Hand modifiers are of two types: silicones and fatty acid esters.

The silicones, mainly poly(dimethylsiloxanes), significantly improve the dry rubbing fastness without decreasing the wetfastness. Used in conjunction with fixing agents, these substances make it possible to produce high-quality pigment printed materials. As well as improving the dry rubbing fastness, the poly(dimethylsiloxanes) give a smooth, dry hand to the fabric, which may be necessary if very soft binders with a somewhat sticky hand are applied.

"True" softeners such as dioctyl phthalate and fatty acid esters cause the binder film to be more mobile. This results in a considerably softer hand. Also, effects on the brillance and color strength (mainly positive) and on the fastness (negative, if too much is used) may be noticeable. Some commercially available products are Acramin Weichmacher SI (Bayer), Luprimol SIG, and Luprimol CW (BASF).

### 6.1.5. Emulsifiers

In high- and low-solvent pigment printing, the emulsifier stablizes the solvent (i.e. white spirit) as an oil-in-water (o/w) emulsion, which functions as a thickened paste with the appropriate rheology. Water-in-oil (w/o) emulsions are nowadays very seldom used. In solvent-free pigment printing, the main tasks of the emulsifier are to prevent agglomeration of the pigment, screen blocking, and separation of components of the print paste. At the same time, the fastness properties and the surface printing must not be affected, as sometimes happens with simple wetting agents.

The aryl- and alkyl-substituted polyglycol ethers are suitable substances, giving a reliable pigment printing process with both solvent-containing and solvent-free recipes. The ethylene oxide and propylene oxide derivatives of various alcohols are also used to overcome special problems, e.g., to improve wetting properties, cleaning properties, etc. Some commercially available products are Emulgator W (Bayer), Luprintol PE New and Luprintol MP (BASF), and Solegal W (Hoechst).

### 6.1.6. Other Additives

*Antifoaming agents* are discussed in Section 6.3.

*Acid donors* are used in solvent-based printing to obtain the acid pH required for the fixation. They consist of ammonia salts of inorganic acids (e.g., diammonium phosphate).

*Compounds* are mixtures of several pigment printing additives, and are used to simplify print recipes.

## 6.2. Auxiliaries for Dye Printing

### 6.2.1. Thickening Agents
(→ Textile Printing, pp. 507–512)

Natural thickening agents, both unmodified and chemically modified, are today widely used in printing with dyes, and have largeley replaced emulsion thickenings based on white spirit. Etherified or carboxymethylated starches are often used for printing with disperse dyes, and alginates for reactive printing [6.9]. Recently, synthetic polycarboxylates, specially developed for reactive printing, have become established. These have improved dye fixation yields [6.4a] and are easy to use, swelling rapidly in water, and having good resistance to bacterial attack.

Advantages of the alginates include ease of removal in the after-wash, resulting in a printed fabric with a soft hand, and also low sensitivity of the thickening effect towards electrolytes in the print paste.

### 6.2.2. Oxidizing Agents

**Sodium *m*-Nitrobenzenesulfonate.** In direct reactive printing on cellulose fiber, the addition of sodium *m*-nitrobenzenesulfonate to the print paste limits harmful effects due to chemical reduction of the reactive dyes by the natural thickening agents and by the fabric itself. This oxidizing agent thus promotes smooth and problem-free printing [6.10]. Sodium *m*-nitrobenzenesulfonate is also used in combination with natural thickening agents in direct disperse printing on acetate, triacetate, and polyester fibers [6.11]. It is not recommended for use with polyamide fibers as yellowing can occur.

In the discharge printing of vat dyes on substantive or reactive predyed fabrics, the ground colour can be reductively damaged, both in the printed and unprinted areas. Even traces of print paste containing the reducing-agent, carried from the already printed areas onto the ground color during printing, steaming, or storage of the fabric are sufficient to such damage. To prevent this undesired effect the predyed fabric can be impregnated with a solution of sodium *m*-nitrobenzenesulfonate before printing [6.1, p. 676]. Some commercially available products are Ludigol, Ludigol granules, and Ludigol liquid (BASF), Matexil PA-L (ICI), and Rapidoprint UL 22 (CHT).

**Sodium chlorate** is used only occasionally to prevent reductive dye decomposition when printing disperse dyes.

**Hydrogen peroxide** is used as in dyeing to reoxidize printed vat dyes [6.12].

### 6.2.3. Reducing and Discharging Agents
[6.13]

In textile printing, reducing agents with a wide range of reducing power are needed. They are used for the fixation of vat dyes in direct and discharge printing, for discharge printing on natural and synthetic fibers, and for afterclearing of printed polyester fibers. Sodium dithionite is used in many applications, but for certain printing processes and equipment the reducing agent must be very stable towards atmospheric oxygen, and in these cases the alkali metal salts of the sulfinic acids are preferable. Some commercial products are listed in Table 6.2.

**Sodium dithionite**, $Na_2S_2O_4$ is usually known in textile finishing as "hydrosulfite". It is widely used for the reductive clearing of polyester fibers printed with disperse dyes [6.14]. Hydrosulfite is

**Table 6.2.** Some commercially reducing and discharging agents

| Trade name | Composition | Producer |
|---|---|---|
| Hydrosulfite conc. | sodium dithionite | BASF |
| Rongalit 2 PH-A | sulfinic acid derivative | BASF |
| Rongalit 2 PH-B liquid | sulfonic acid derivative | BASF |
| Rongalit C | sodium hydroxymethane-sulfinate | BASF |
| Rongalit FD liquid | sodium salt of a sulfinic acid derivative | BASF |
| Rongalit ST liquid | sodium salt of a sulfinic acid derivative | BASF |
| Decrolin | zinc hydroxymethane-sulfinate | BASF |
| Rongalit DP | stabilized sulfinic acid derivative | BASF |
| Reduktionsmittel F | formamidinesulfinic acid | Degussa |

a strong reducing agent with limited stability towards atmospheric oxygen. It is therefore unsuitable for direct printing with vat dyes, as severe losses of the reducing agent in the print paste and premature vatting of the dye would occur. It can, however, be used in two-phase printing [6.14], [6.15].

**Rongalit 2 PH-A/Rongalit 2 PH-B liquid** is a combination of reducing agents especially intended for two-phase printing with vat dyes [6.16]. In this process, the reducing agent is not printed directly with the dyes, but is subsequently applied to the fabric together with sodium hydroxide by an additional padding process. In the second phase the dye is then fixed by steaming (ca. 20–40 s) in a flash ager with exclusion of air. Cold solutions of Rongalit 2 PH-A and Rongalit 2 PH-B liquid with sodium hyroxide are less sensitive to atmospheric oxygen than hydrosulfite solutions, and are therefore significantly more stable. Unlike hydrosulfite, the pad liquor of the combination product with sodium hydroxide does not cause premature vatting after padding, and hence does not lead to running of the printed vat dye. The ease of reoxidation of the printed fabric and the relatively low level of impurities in the used wash liquor give improved operational safety.

**Sodium hydroxymethanesulfinate**, sodium formaldehydesulfoxylate, $HOCH_2SO_2Na \cdot H_2O$, Rongalit C, is appreciably more stable in alkaline solution than hydrosulfite or the Rongal-it 2 PH-A/2 PH-B liquid combination. It is used for printing cellulose fibers by various methods.

It can be used in direct printing or two-phase printing with vat dyes [6.16]. Because the steaming time is considerably longer (ca. 4–6 min) compared with Rongalit 2 PH-A/2 PH-B liquid (20–40 s), it is only used in a two-phase process if the ager does not allow rapid dye fixation. The most important application of sodium formaldehydesulfoxylate is in discharge printing [6.14]. Reactive dyes, and to a lesser extent naphthol and direct dyes, are used to produce the discharge ground.

In color discharge printing, the process must be carried out so that no decomposition of the reducing agent occurs before the dye is fixed by steaming. Great care must be taken to exclude moisture on storage and oxygen during steaming [6.17]. For white discharge printing, further auxiliaries are added to ensure a pure white discharge effect.

**Rongalit FD liquid**, a sodium sulfoxylate stabilized with formaldehyde and ammonia, is more stable than sodium formaldehydesulfoxylate. It is used as a reducing agent for direct printing with vat dyes [6.16]. The improved stability of the printed fabric before steaming is especially important for screen printing on tables. The higher stability requires a longer steaming time (ca. 15–20 min). Optimum color yields are obtained by using steam at a higher pressure (1.5 bar, 12–15 min).

**Rongalit ST liquid** is also a sodium sulfoxylate stabilized with formaldehyde and ammonia that is even more stable than Rongalit FD liquid. It is used as a reducing agent in pigment discharge printing. Pigment print pastes which contain Rongalit ST liquid in a slightly alkaline solution are applied to substantive or reactive predyed cellulose fabrics. The decomposition of the ground color takes place during steaming of the printed fabric. A careful choice of the dyes is necessary for this process.

**Zinc hydroxymethanesulfinate**, zinc formaldehydesulfoxylate, $(HOCH_2SO_2)_2Zn$ in aqueous solution has a slightly acid reaction and is stable for several days. Zinc formaldehydesulfoxylate, unlike the sodium salt, is therefore suitable for discharge printing in weakly acid media. Applications are:

1) White and colored discharge printing on acetate, triacetate, polyester, polyamide, wool, and silk [6.18], [6.19]. Only a limited number of discharge-resistant dyes are suitable for printing colored effects.
2) Color discharge printing with pigments on cellulose fibers [6.20], mainly on substantively dyed grounds.

**Rongalit DP** is a stabilized sulfoxylic acid derivative that is suitable for discharge printing on polyester [6.18]. It is mainly used for color discharge resist printing, whereby discharge-resistant disperse dyes are printed together with Rongalit DP. Discharge grounds are unfixed prints or pad dyeings made with dischargeable disperse dyes that are only fixed during printing. On light and medium grounds, white discharge resist printing is also possible, but on dark grounds, zinc formaldehydesulfoxylate should be used, because it results in a purer white effect.

**Formamidinesulfinic acid**, thiourea dioxide, THDO, is relatively stable in acid solution and unstable in alkaline solution. THDO can therefore be used for printing with vat dyes on wool, silk, acetate, and synthetic fibers. Since the application process takes place in the acid range, the dye is absorbed by the fiber in the form of the vat acid [6.1, p. 789]. THDO can also be used for color discharge printing with vat dyes on acetate fabrics. This product is now mainly used for the reductive afterclearing of dyed and printed polyester materials.

**Tin(II) chloride** has several disadvantages, in particular corrosion of the steamer by hydrochloric acid which is liberated during the discharge process. This reducing agent is therefore only used in special applications [6.18], e.g., for colored discharge printing on polyacrylonitrile fibers. In this process tin(II) chloride cannot be replaced by any other discharging agent [6.21].

### 6.2.4. Auxiliaries for Discharge Printing

In discharge printing, auxiliaries are used in addition to the discharge agents. They are not able to decompose the ground dye by themselves, but improve the discharging effect and/or facilitate the fixation of the printed dye used for the colored effet. Auxiliaries especially used in discharge printing include:

**Sulfonated Dimethylphenylbenzylammonium Salt.** This benzylating agent, which is sulfonated in both aromatic rings, enables white discharge prints of indigo and certain vat dyes [6.17]. The dye, after reduction by Rongalit C, is converted by benzylation into a water-soluble, oxidation-resistant derivative that can easily be washed out [6.1, pp. 532, 683, 790]. A typical commercial product is Leukotrop W (BASF).

**Anthraquinone** improves the discharging effect in discharge printing of vat dyes on cellulose fibers dyed with reactive, direct, or naphthol dyes [6.20]. It is also recommended for discharging of materials dyed with indanthrene dyes and indigo. Anthraquinone acts as a reduction catalyst and is converted to anthrahydroquinone as an intermediate. A typical commercial product is Anthraquinone Powder (BASF).

**White pigments** such as zinc oxide, titanium dioxide or barium sulfate can also be added to discharge print pastes. These cover the remaining undischarged ground dyes, and produce a plastic effect in patterns with small motives or fine outlines.

**Optical brighteners** can intensify the white effect in white discharge prints. They are mainly used with polyester fibers.

### 6.2.5. Fixation Auxiliaries

In addition to oxidizing and reducing agents, there are many other auxiliaries that improve the fixation of soluble dyes. The effect of many fixation auxiliaries is to promote solubilization of the dye during the fixation process. In other cases they cause swelling of the fibers.

**Fixation Auxiliaries for Printing on Cellulose Fibers.** *Hydrotopic auxiliaries.* Urea has very good hydrotropic properties and is widely used in printing with reactive dyes [6.22]. During the fixation of the dyes by steaming, only a small amount of solubilizer, in the form of condensed steam, is present. The hydrotropic properties of urea significantly improve the fixation rate of the dye, which is especially important for regenerated cellulose fibers [6.1, p. 781].

When printing with vat dyes and during vat discharge printing, thiodiglycol is often added, since it increases the solubility of the leuco form of the dye and hence improves the fixation rate of certain vat dyes [6.1, p. 781], [6.20].

*Alkalis* such as sodium carbonate or sodium hydrogencarbonate are necessary for the fixation of reactive dyes. Alternatively, the special products Basilen Fixing Agent F-RP (BASF) and Remazol Salt FD (Hoechst), which are sodium salts of a chlorinated carboxylic acid, can be used. These salts act as alkali-releasing agents, i.e., the alkali is only liberated during the dye fixation process (steaming) [6.23].

**Fixation auxiliaries for printing on acetate and polyester fibers** are mainly fatty acid derivatives and products based on polyglycols [6.24], [6.25]. These auxiliaries not only improve the fixation rate, but also simplify the fixation process of the dye. When printing on acetate, the use of fixation auxiliaries enables the fixation time to be reduced from ca. 30 min to 8–12 min [6.26]. With triacetate fibers, suitable auxiliaries enable a dye-fixation in a continuous process at 100 °C with saturated steam instead of the batch pressure steaming process that is otherwise necessary [6.26].

**Fixation Auxiliaries for Printing on Polyacrylonitrile Fibers.** Certain fixation auxiliaries containing nitrile groups have proven to be especially suitable for this printing process. They can be used with cationic dyes and some disperse and metal complex dyes [6.27].

Some examples of commercial fixation auxiliaries are listed in Table 6.3.

## 6.2.6. Dye Solubilizing and Dispersing Agents

The dyes in print pastes must be in a completely dissolved or very finely divided form, otherwise problems could arise during the printing processes, resulting in uneven prints.

Dye solubilizing agents have the task of dissolving the water-soluble dyes during the preparation of the paste. Many polar organic solvents are recommended and used for this purpose, e.g., ethanol, ethylene glycol, ethylene diglycol, butyl glycol, butyl diglycol, glycerine, and thiodiglycol [6.1, p. 781].

Thiodiglycol is widely used in textile printing. It is recommended for solubilizing cationic dyes, acid dyes, metal-complex dyes, substantive dyes, naphthol dyes, and leuco esters of vat dyes. The solubilizer can be directly added to the required

**Table 6.3.** Some commercially available fixing accelerators

| Trade name | Composition | Producer |
| --- | --- | --- |
| *For acetate* | | |
| Luprintan DCA | poly(ethylene glycol) | BASF |
| *For polyester and triacetate* | | |
| Rapidoprint RE | mixture of organic compounds | CHT |
| Luprintan HDF | fatty acid derivative | BASF |
| Tanaprint ASD | modified fatty acid amide | Tanatex |
| Lamefix 680 | ethylene oxide adduct to polyglycol ether | Grünau |
| *For polyacrylonitrile* | | |
| Luprintan PFD | cyanoethylation product | BASF |
| Rapidoprint PAN | organic compound containing nitrile groups | CHT |

**Table 6.4.** Some commercially available dye solubilizing agents

| Trade name | Composition | Producer |
| --- | --- | --- |
| Lyocol BC | aliphatic ether | Sandoz |
| Rapidoprint LM | aliphatic ether | CHT |
| Glyezin A | thiodiglycol | BASF |
| Tinosol G 133 % | thiodiglycol | Ciba-Geigy |

amount of water. With powder dyes, it is often advantageous to paste the dye with the solubilizer, and to form the solution by adding hot water to this mixture.

Dispersing agents can cause insoluble dyes to become finely dispersed during the preparation of the print paste, and stabilize this state of dispersion. The dispersing agents are usually already present in the dye preparations (e.g., of vat and disperse dyes), so that a separate addition to the print paste is not necessary.

Some commercial dye solubilizing agents are listed in Table 6.4.

## 6.2.7. Auxiliaries for Reactive Resist Printing

In reactive resist printing, fixation of the ground (vinylsulfone dyes) is prevented at the printed areas by pre- or overprinting with a resist agent (white resist). If the print paste contains monochlorotriazine dyes that are resistant to the resist agent, this is known as a colored resist print. Special sulfur-containing compounds are used as resist agents [6.28], [6.29].

Some commercial products are BASF Liquid Reactive Resist Agent (BASF) and Cleantex PWC-N (Kyoei Kagaku).

### 6.2.8. Auxiliaries for Burn-out Printing [6.30]

Almost the only textiles used in burn-out printing are woven and knitted goods made of polyester–cellulose and polyamide–cellulose fibers. In both fiber blends, the cellulose component is destroyed by acid hydrolysis, i.e., it is burnt out. Sodium hydrogensulfate or the less aggressive aluminum sulfate in concentrations of 80–150 g/kg printing paste are used for the purpose. Glycerine is added to burn-out pastes to facilitate their removal by washing. The addition of glycerine is necessary to enable a rapid rewetting of the burned-out areas and easy and complete removal of the decomposed cellulose during the washing process. If yellowing occurs or if the residues are difficult to remove, the textile must be bleached.

### 6.2.9. Aftertreatment Agents and Detergents

When printing with dyes that have affinity to the fiber, the fabric must be afterwashed. The goal is to remove the unfixed dye, the thickening agent, and the auxiliaries from the goods, giving the printed product the highest possible brillance and the required fastness properties [6.31].

Detergents with a good dispersing action are necessary, especially with water-insoluble dyes such as vat dyes. Detergents with a dye-solubilizing effect are used particularly for washing materials printed with disperse dyes. With printed polyester, the effect of the detergent is enhanced by treatment with the reducing agent hydrosulfite [6.32]. The addition of polyvinylpyrrolidone to the washing baths of prints made with reactive and direct dyes prevents reabsorption of the dissolved part of the dye. Condensation products of aromatic sulfonic acids with formaldehyde show substantivity to polyamide fibers and are mainly used for the aftertreatment of polyamide textiles printed with acid and metal-complex dyes [6.33]. They prevent staining of the white grounds during washing of the printed textile, and also improve the wetfastness of the printed dyes.

Some commercial products are listed in Table 6.5.

## 6.3. Other Auxiliaries

*Preservatives* such as formaldehyde are added to natural thickeners to prevent bacterial degradation [6.9].

*Coagulation agents* such as borax or aluminum sulfate are sometimes added to the pad liquor for flash-age printing of vat dyes. They prevent bleeding of the dye leuco compound and marking of the printed material on the padding and steaming rollers [6.34].

*Acids* such as citric acid or ammonium sulfate are used to produce mildly acid conditions in print pastes when printing with disperse and cationic dyes.

*Antifoams* prevent the formation of foam during the make-up of the print paste and during the printing process itself. Many such products are available based on, e.g., silicone oils, organic and inorganic esters, aliphatic hydrocarbons, and many other types of compounds.

*"Print oils"* (mineral oils) are sometimes added to print pastes to reduce the friction between the doctor blade and the roller of a roller printing machine.

## 6.4. Textile Printing Adhesives [6.35]

In screen printing, the textile goods are almost always secured by means of an adhesive so that it cannot move during the process.

Water-soluble adhesives are the most common and include those based on natural prod-

**Table 6.5.** Some commercially available detergents and aftertreatment agents

| Trade name | Composition | Producer |
|---|---|---|
| *Detergents* | | |
| Foryl M 8 liquid | mixture of alkyl polyglycol ethers | Henkel |
| Levegal AN liquid | mixture of alkylaryl polyglycol ethers | Bayer |
| Kieralon DB | mixture of substances that are surface active, have an affinity for the dye, or have a dye solubilizing effect | BASF |
| Printoblanc | mixture of fatty amine polyglycol ethers | CHT |
| Rucogen FWK | aliphatic ethylene oxide condensation products | Rudolf |
| *Aftertreatment agents* | | |
| Albigen A | polyvinylpyrrolidone | BASF |
| Sandofix WE | methylolamide | Sandoz |
| Mesitol PS | condensation product of fused aromatic sulfonic acids with formaldehyde | Bayer |

**Table 6.6.** Some commercially available adhesives [6.36]

| Trade name | Composition | Producer |
|---|---|---|
| *Water-soluble adhesives* | | |
| Texogum 3013 | starch ether | Diamalt |
| Vianarol ST | poly(vinyl alcohol) | Hoechst |
| Lubasin RF | polyvinylcaprolactam | BASF |
| *Permanent adhesives* | | |
| Kiwotex TDK | acrylate-based copolymerisate | Kissel & Wolf |
| Lubasin TP | acrylate-based copolymer | BASF |
| Tubigum DK 2 | polymer dispersion | CHT |

ucts such as degraded starch, starch derivatives, or vegetable gums, and synthetic adhesives such as poly(vinyl alcohol) and polyvinylcaprolactam. The latter, which does not contain hydroxyl groups, is used especially with textile goods made of synthetic fibers to which it is difficult to form an adhesive bond.

The water-insoluble adhesives include thermoplastic adhesives, which are acrylate-based water-soluble polymers and have a softening temperature of 50–80 °C. They are applied to the print blanket as a solution in a volatile organic solvent. In the printing process, the coated print blanket is heated by special heating equipment, and the goods are pressed onto this by a roller, causing adhesion. Thermoplastic adhesives are permanent glues, i.e., they enable printed fabrics to be bonded many times to the same adhesive film. As well as thermoplastic adhesives, permanent adhesives with a lower softening range are also used. Unlike the thermoplastic adhesives, these already form a strong bond even at room temperature, so that no heating is required to cause adhesion to the goods. They consist of dispersions of acrylic esters. Examples of commercial adhesives are listed in Table 6.6.


# 7. Finishing Agents

## 7.1. Introduction

Finishing is the name given to those final processes applied to a fabric before it is sold or made into end-use articles, such as garments or home furnishings, which impart desirable end-use properties to it. Finishing may involve both chemical and mechanical treatment. In mechanical treatment, chemical agents are often used to enhance, facilitate, or make durable the effect of the treatment.

Historically three general finishing systems have evolved, based on the processing of the three major natural fibers, cotton, wool, and silk. These general systems differ in the processes and equipment used. The synthetic fibers and their blends with natural fibers are finished in the general system that best fits the fabric type in which they are used. The cotton finishing system predominates and can be assumed in this chapter unless another system is specified.

Cotton is the most widely used textile fiber. World cotton consumption grew at an average of 3.7 % per annum in 1985–1990, from 15.8 $\times 10^6$ t/a to just under $19 \times 10^6$ t/a [7.1]. In 1988 cotton represented ca. 49 % of textile fibers used worldwide, while wool accounted for only ca. 5 %, silk only ca. 0.2 %, rayons ca. 8 % and noncellulosic synthetic fibers ca. 38 %.

In the cotton finishing system, dyeing – either in the yarn or fabric state – normally takes place prior to finishing, but some fabrics may be printed after finishing, especially when pigment prints are used. Finishing may sometimes be carried out on greige fabrics, but in most cases it is preceded by preparation processes such as singing, desizing, bleaching and mercerizing. For most chemical finishes, such as easy care, soil-release, water repellents, and fire retardants, the fabrics must be properly prepared and be free from fiber finishes and contaminants, residual sizes, surfactants, and inorganic salts if optimum results are to be obtained.

Although there have been some attempts to use solvent systems to apply chemical finishes [7.2], most are applied commercially from aqueous systems. The traditional method for applying finishes has been to pass dry fabric through a bath containing the finishing ingredients, then through a pair of squeeze rolls to squeeze out as much of the treating solution as possible, before passing the fabric into a drying device, such as a tenter, loop, or cage dryer. The dryer is often followed by a high-temperature curing device (e.g., an oven) if curing is needed. Wet pickup may vary from ca. 60 % to 100 %, depending upon the composition and construction of the fabric, and the composition and pressure of the squeeze rolls in the conventional pad-squeeze system. Driven initially by attempts to conserve energy in the drying step, but later by economic and environmental considerations, low wet pickup methods of finish application have been proposed [7.3]. These include expression methods, in which the fabric is saturated with a relatively large amount of treating solution, after which the excess is removed by special squeeze rolls, a vacuum extractor [7.4]–[7.6] or air or steam jets, and topical methods, in which a limited amount of the treating solution is applied to the fabric by devices such as kiss-roll applicators (e.g., Triatex MA [7.7]), engraved rolls (e.g., Goller, Sando, Toyoda, and Zimmer), spray applicators (e.g. Burlington, Farmer, Norton and WEKO) and open (e.g., United Merchants and Manufacturing) and closed (i.e., Gaston County–FFT [7.8]) foam applicators [7.9]. With these systems pickups as low as 12 % have been reported on 100 % synthetic fabrics. They are particularly effective on polyester–cotton blend fabrics where pickups of 25–35 % are usual. In addition to the savings due to reduced energy consumption for drying, the low wet pickup methods also provide savings in chemical costs, improved hand, and less migration of finish. On fabrics containing cotton or rayon, sufficient water must be applied with the finishing solution to swell the fibers and allow the finish to penetrate. In cross-linking only the cross-linking agent within the cellulosic fiber is effective. Any material on the surface or within the capillaries of the fabric or on the noncellulosic fibers present is ineffective in producing easy-care and dimensional-stability effects. It is thus essentially wasted and can contribute to migration, hand, odor, and sewability problems. However, decreasing the wet pickup increases the degree of control required for uniform pickup, thus making the control mechanisms for the low wet pickup systems critical to their success [7.10]. For this reason the expression methods, which are somewhat easier to control and tend to give more uniform treatment, are more widely used commercially than the topical methods. Other attempts to conserve energy in finishing have included wet-on-wet finish applications and the elimination of washing after finishing, except in cases where it is absolutely necessary, such as with certain fire retardant finishes.

In most cases agents designed to provide the various properties desired in a given fabric are all applied from a single bath rather than from separate baths. In addition to cross-linking agents and catalysis, finishing baths often also contain wetting agents, softeners, hand builders, soil-release agents, weighters, antistats, anti-crock agents, antifoams, optical brighteners, tints,

odorants, and other additives. When multiple agents are applied from the same bath it is very important that all of the components be compatible and not cause precipitation or interfere with the function of other components.

## 7.2. Easy Care Finishes

Easy care finishes are chemical finishes applied to cellulose-containing fabrics to reduce their propensity for wrinkling in the wet and dry state and to stabilize them against progressive shrinkage during laundering. The terms easy care, wrinkle resistant, wash and wear, no-iron, durable press and permanent press have all been applied to this type of finish and although they may denote different levels of performance, they are often used interchangeably. They function by introducing cross-links between the cellulose molecules of cotton and rayon and by so doing reduce the swellability and extensibility of the fibers but increase their resilience. The cross-linking results in a fabric with a "memory" that tends to return to the state it was in when the cross-links were introduced. Thus if a fabric is smooth and flat when cross-linked it will tend to return to a smooth, flat state after wrinkling or laundering. Likewise if the fabric is creased when cross-linked it will tend to retain the crease through wear and laundering.

### 7.2.1. Cross-Linking Agents

Multifunctional methylol derivatives of urea [57-13-6], substituted ureas, or melamine [108-78-1] produced by reacting formaldehyde with these compounds have been used almost exclusively as the cross-linking agents for commercial easy care finishes. The methylol derivatives of urea itself were the first compounds used as cross-linkers and are still used to some extent, particularly on rayon fabrics and where more sophisticated products may not be readily available. They were first used by Tootal, Broadhurst and Lee around 1926 [7.11]. Dimethylol urea, $N,N'$-bis(hydroxymethyl)urea [140-95-4] is available as a white crystalline material with a formaldehyde to urea ratio of 2 : 1. Products having lower formaldehyde to urea ratios (i.e., 1.4 : 1 to 1.7 : 1) are available in paste form, are more reactive than the dimethylol compound, and tend to form self-condensation polymers as well as cross-links with the cellulose. When methylol ureas are reacted with methanol under acidic conditions some of the methylol groups are converted to methoxymethyl groups accompanied by some polymerization, leading to syrup resins [68071-45-4] that are useful for stiffening fabrics, especially synthetics. All resins derived from unsubstituted urea form chloramides when fabrics containing them are bleached with hypochlorite [7.12]. Upon heating, as in ironing, these chloramides decompose to form hydrogen chloride, which can seriously degrade the fabric. This phenomenon is known as chlorine retention and can be tested for by AATCC (American Association of Textile Chemists and Colorists) Test Method 92 [7.13]. Urea–formaldehyde finishes also tend to release formaldehyde from fabrics treated with them and can evolve "fishy" odors due to the formation of methylamines if not prepared properly or if cured with ammonium salt catalysts.

Methylol derivatives of melamine, which can contain up to six methylol groups per molecule, and particularly those in which the methylol groups have been partially or completely converted into methoxymethyl groups by reaction with methanol have been used in textile finishing, either alone [7.14] or in combination with other cross-linkers [7.15]. A widely used commercial product has been the dimethyl ether of trimethylolmelamine [1852-22-8]. Melamine derivatives tend to yellow on bleaching with hypochlorite, and fabrics treated with them tend to evolve considerable formaldehyde on storage. They are not used extensively today, except for stiffening synthetic fabrics, where the methoxymethyl products are used, and in special finishes such as fire-retardant and rot-resistant fabric finishes.

The methylol derivatives of many substituted ureas have been used commercially as cross-linkers for easy care finishes. These substituted ureas include ethylene and propylene ureas, triazones, and urons. The ethylene and propylene ureas are made by reacting primary ethylene- and propylenediamines with urea at high temperatures, with the evolution of ammonia [7.16]. Ethylene urea reacts with formaldehyde to form 1,3-bis(hydroxymethyl)-2-imidazolidinone (DMEU) [136-84-5]. This product was a widely used cross-linker during the 1950s and 1960s. The dimethylol derivative of propylene urea, 1,3-bis(hydroxymethyl)tetrahydro-2-(1H)-pyrimidinone [3270-74-4], has some advantages over DMEU, such as improved freedom from discoloration and improved durability, but it is more expensive and never gained commercial importance [7.17], [7.18].

Triazones, such as hydroxyethyltriazone, 5-hydroxy-1,3-bis(hydroxymethyl)hexahydro-s-triazin-2-one [1852-21-7], are made by reacting urea with formaldehyde and a primary amine [7.19], [7.20]. The triazones were used extensively as cross-linkers during the 1950s, mostly in combination with other cross-linkers such as methylol ureas where they have a "depressant" effect upon chlorine damage due to their basic character [7.21], [7.22]. They were abandoned because of their tendency to yellow fabrics, generate amine odors, and evolve formaldehyde.

The urons, i.e., tetrahydro-3,5-bis(hydroxymethyl)-4H-1,3,5-oxadiazin-4-one [7327-69-7] are made by reacting 1 mol of urea with 4 mol of formaldehyde and cyclicizing the tetramethylol compound. The urons are usually methylated by reaction with methanol under acidic conditions to form N,N'-bis(methoxymethyl)uron [7388-44-5] [7.23], [7.24] which is used alone or in combination with other products, such as melamine resins.

The early cyclic urea products greatly alleviated the chlorine retention problem but did little to reduce formaldehyde evolution from fabrics treated with them. Formaldehyde evolution became particularly critical when afterwashing was eliminated from the finishing process for cost and energy savings and because of the advent of durable press in the mid-1960s.

Methylol derivatives of carbamates, especially methyl carbamate [598-55-0], and methoxyethyl carbamate [1616-88-2], which can be produced by reacting the appropriate hydroxy compound with urea at elevated temperatures [7.25]–[7.27] have also been used commercially as cross-linkers and were valued for their lack of yellowing and resistance to severe laundering conditions [7.28]; however, they evolved considerable formaldehyde.

Many other compounds which are capable of cross-linking cellulose have been proposed as easy care finishes and some have been run commercially on a small scale, but have since been abandoned for cost, environmental, or deficiency reasons. These include polyepoxy compounds [7.29], polyaziridinyl compounds [7.30], dihydroxyethylsulfone (Ganalok) [7.31], polyacetals [7.32], formaldehyde [7.33], and several dialdehydes [7.34], [7.35].

Methylol derivatives of dihydroxyethylene urea [3720-97-6], produced by reacting glyoxal with urea [7.36], have replaced almost all of the other products formerly used as cross-linkers in easy care finishes. These products first appeared commercially in the early 1960s and were adopted as the agents of choice for permanent press when this type of finish became popular in the mid-1960s because of their lower formaldehyde evolution potential, both in the cured and uncured state, and their stability in the uncured or partially cured state. Post-cured permanent press, in which the cross-linker is left uncured [7.37], [7.38] or only partially cured in the fabric by the finisher and then fully cured in a garment oven after the garment is sewn became very popular in the 1960s, and many garment plants installed garment curing ovens for this purpose. Later it was found that precured fabrics could be reformed on hot head presses to smooth seams and install creases, and use of garment ovens declined. The original glyoxal–urea products were dimethylol dihydroxyethylene ureas (DMDHEU) [1854-26-8]. Later, buffered versions and versions with slightly less than a 2:1 formaldehyde to DHEU ratio appeared, which offered better fabric whiteness with certain catalyst systems and slightly lower formaldehyde evolution from the finished fabric. Later, products with varying degrees of methylation appeared, such as 1,3-dimethoxymethyl DHEU [3001-61-4] and the fully methylated product [4356-60-9]. Commercial products are commonly 25–50% methylated. The methylated products provide lower formaldehyde evolution from fabrics than unmethylated DMDHEU. They are prepared by reacting DMDHEU with methanol at low pH. Hydroxyl-containing compounds with low volatility, such as glycols or glycerin, when added to or reacted with either DMDHEU or methylated DMDHEU provide even lower formaldehyde evolution potential from the treated fabric. Although specific hydroxy compounds, such as the nitroalcohols [7.39], have been promoted from time to time, any nonvolatile hydroxy compound will work, and the efficiency of the compound in reducing formaldehyde evolution is directly related to its hydroxyl-equivalent weight. A patent has issued on DMDHEU reacted with glycols [7.40], but a mixture of DMDHEU with nonvolatile hydroxy compounds is equally effective [7.36] presumably due to reactions that take place on the fabric during curing. Various formaldehyde "scavengers" have been proposed for reducing formaldehyde evolution [7.41]. Products based on nitrogen-containing compounds have been proposed but not used extensively on a commercial basis. Hydroxy and active methylene compounds have also been shown to be effective formaldehyde scavengers [7.41], [7.42] and are used commercially. Methylated DMDHEU-based products in general afford about one-half the formaldehyde evolution potential of the unmethylated products. Products containing nonvolatile hydroxy compounds, such as glycols, have about one-half the formaldehyde evolution potential of the methylated products. Products that have been both methylated and contain a nonvolatile hydroxy compound provide even lower formaldehyde evolution potential. All of the various modifications of the glyoxal–urea type products are in commercial use. The products which have been both methylated and contain a hydroxy compound afford the lowest formaldehyde evolution potential of all commercially available easy care finishes of the methylol type.

In selecting a cross-linker for use in easy care finishing there are many factors to be considered, including toxicity, cost, durability to laundering, chlorine resistance, effect on hand, discoloration potential, odor, formaldehyde-evolution potential, effect on fabric strength, rate of cure, energy of activation, effect on shade of dyes, effect on lightfastness of dyes, and effect on fabric soiling and soil removal. The glyoxal–urea products have satisfactory ratings on all of these factors, which has made them the products of choice for most easy care finishes worldwide. The driving force for achieving lower formaldehyde-evolution potential from finished fabric has been the identification of formaldehyde as a suspected carcinogen and tightened threshold values for formaldehyde in the workplace. Formaldehyde-evolution potential is influenced by many factors, including the nature of the cross-linker, the amount used, the degree of cure, and the pH of the finished fabric. Many studies have been carried out on factors influencing the evolution of formaldehyde from finished fabrics [7.43]–[7.46]. Numerous other studies have been carried out on test methods for formaldehyde-evolution potential [7.47]–[7.52]. AATCC Test Method 112 [7.13] for formaldehyde release gives a value that includes both free formaldehyde and formaldehyde released by hydrolytic degradation of the finish and unbound reactant during the incubation step of the test. Thus it is designed to simulate the most severe conditions of high humidity and elevated temperatures likely to be encountered during the storage of finished fabrics. Other, milder test methods are available, including the Japanese Law 112 Method.

Easy care finishes free of formaldehyde or formaldehyde precursors and thus having no formaldehyde-evolution potential on fabrics have been proposed, but have not achieved significant commercial use [7.53]. Dimethyldihydroxyethylene urea, 4,5-dihydroxy-1,3-dimethyl-2-imidazolidinone [3923-79-3], made by reacting dimethylurea with glyoxal [7.54], has been available commercially but has not been used to any significant extent because of its high cost, low effectiveness, and color and odor problems. Improvements which claim to have overcome these problems have been made [7.55] and there is some commercial use of the improved products.

Polycarboxylic acids such as citric acid [77-92-9] [7.56]–[7.58] and butanetetracarboxylic acid (BTCA) [1703-58-8] [7.59], [7.60] have been proposed as cross-linkers for easy care finishes, with catalysts such as hypophosphite salts being used to form ester cross-links, but these have enjoyed little commercial success because of their high cost and color problems. Zero-formaldehyde systems based on polyacrylics, polyurethane, and silicones have also been explored but have problems with cost, hand, and performance. It appears doubtful that zero-formaldehyde finishes can compete with DMDHEU-type products unless zero-formaldehyde finishes are mandated by government regulation, which currently appears unlikely.

### 7.2.2. Catalysts

Many catalysts have been used to cure the cross-linkers of easy care finishes, but those based on magnesium chloride are most widely used. Almost any acid or acid salt will catalyze the cross-linking reaction, but many such compounds are unsuitable because they are too strong, too weak, or present environmental problems. Zinc salts were formerly used but have been mostly abandoned for environmental reasons. Nitrate salts were used extensively during a period when there was speculation that chloride catalysts might give rise to the known carcinogen bis(chloromethyl) ether [542-88-1], when used with formaldehyde-containing products. This speculation proved unfounded, and chloride cat-

alysts replaced the nitrates again. Magnesium chloride is often used alone, but it can be combined with organic acids such as citric acid, or more acidic salts such as aluminum chloride, to provide "hotter" catalysts which cure the cross-linkers at lower temperatures or in shorter times than pure magnesium chloride. A number of studies have demonstrated the advantages of mixed catalysts [7.61]–[7.64].

Curing conditions vary widely from finishing plant to finishing plant. Some plants use roller ovens while others cure on tenter frames. Curing temperatures normally range from 150 to 200°C, and curing times from 10 s to several minutes with the shorter times being at the higher temperatures vice versa.

### 7.2.3. Testing

The effectiveness of an easy care finish is measured by the washed appearance of the fabric, the washed appearance of creases and seams in a garment, and the shrinkage of the fabric on laundering (AATCC Test Method 123) [7.13]. Since cross-linking tends to embrittle cellulosic fibers, tests that measure the effects of this embrittelement, such as fabric strength, abrasion resistance, frosting, and dusting, are also usually measured in evaluating easy care finished fabrics. Dyefastness properties, such as lightfastness, washfastness, and crocking, may also be tested to measure the effect of the finish on these properties.

### 7.2.4. Hand Builders

Hand builders stiffen fabrics and give them a firmer, bulkier hand. They are all polymeric compounds. Starches from natural products, such as corn, potatoes, and rice, were formerly used extensively as hand builders and are still used on plain-finish fabrics which are not treated with a cross-linker. In easy care finishes on cellulose-containing fabrics, soluble and emulsion types of hand builders are used. The soluble types tend to give a crisp, "papery" hand to the fabric and include poly(vinyl alcohols) and modified starches (e.g., hydroxyethyl starches). These polymers are insolubilized by the cross-linker used in the easy care finish bath and become durable to laundering.

The emulsion polymers tend to give a thicker, "leathery" hand. Polymers that do not discolor on exposure to light, heat, or atmospheric exposure such as poly(vinyl acetate) and acrylate and methacrylate copolymers are the most widely used emulsion polymer hand builders. The stiffness of the polymer and consequently the degree of stiffness imparted to the fabric is related to the glass-transition temperature of the polymer, with polymers having a higher glass transition temperature being stiffer. Polymers with very low glass-transition temperatures that are very soft can contribute to soil pickup or redeposition during laundering. Some copolymers contain reactive groups, such as methylol amide or epoxy groups, which undergo cross-linking on curing, making the polymers more insoluble and resistant to laundering.

Polymeric methylated urea – and melamine – formaldehyde resins are used as hand builders on 100% synthetic fabrics, such as nylon and polyester, where they adhere well and give a crisp resilient hand.

### 7.2.5. Softeners

Softeners are used to improve the hand of woven and knitted fabrics and to improve fiber, yarn, and fabric lubricity. The improved lubricity exhibits itself in improved tear strength, reduced needle cutting during sewing, improved fabric drape, and softer feel. Anionic softeners account for < 10% of the total softener market, with the remainder being almost equally divided between nonionic and cationic products. Many softeners are fatty based. For greatest softening effect the chain length of the fatty group should be about $C_{18}$. The fatty group is connected to a hydrophilic solubilizing group. A great variety of linkages and solubilizing groups are used.

In nonionic softeners the linkage is usually an ester or ether group, and the hydrophilic portion of the molecule may be glycerin, polyoxyethylene, sorbitan, or ethoxylated sorbitan. Nonionic softeners are especially useful where freedom from yellowing is important or where compatibility with anionic components such as optical brighteners is needed. They tend to give surface smoothness to fabrics, without excessive drape. Nonionic softeners are often used as lubricants to assist in napping and other mechanical processes which involve metal – fiber friction.

A wide variety of cationic softeners are available. Quaternary methylammonium compounds containing one or two fatty alkyl groups are used extensively as softeners in home laundering but only to a small extent in fabric finishing. Fatty amides of diamines, such as aminoethylethanolamine [111-41-1], which have been further reacted with acetic acid to form acetate salts have been used as softeners, but these compounds tend to yellow with heat or aging. More complex products formed by reacting the amino amides with ethylene oxide, followed by quaternization with chloromethane or methyl or ethyl sulfate exhibit less tendency to yellow and are widely used. The amino amides may also be cyclicized to form imidazoles before being converted into acid salts or quaternized. Cationic softeners are substantive to cellulosic fibers and can be applied by exhaustion as well as by padding. Cationics are more effective in increasing the drape and limpness of cellulose-containing fabrics than are the nonionics.

Secondary emulsions of low molecular mass oxidized polyethylenes are widely used in finish baths where they function more as lubricants than as hand-modifying agents. They improve the tear strength and abrasion resistance of fabrics and improve sewability but increase seam slippage. Polyethylene emulsions are often blended with cationic softeners.

Dimethylsilicone fluid emulsions are used as softeners and are particularly effective on synthetic fabrics. They impart a unique surface smoothness to fabrics that is different from that given by other softener types. On cellulose-containing fabrics, organo-modified silicones containing epoxy, hydroxy, or amino groups are often used. The aminosilicones are particularly effective softeners but tend to yellow [7.65], [7.66]. They are often blended with fatty cationic softeners, with which they exhibit a synergistic effect.

Anionic softeners consisting of sulfonated oils or fatty esters, often blended with unsulfonated oils, are widely used on plain finish fabrics, such as denims, to facilitate compressive shrinkage of the fabrics. These products act as lubricants as well as wetting agents.

## 7.3. Water Repellents

### 7.3.1. Historical Aspects

The earliest treatments for water repellency consisted of coating fabrics with various substances which are impervious to water, such as natural fats and oils, waxes, pitch, and asphalt. Later, vulcanized natural rubber became an important waterproofing material. Even later, synthetic polymers replaced the natural products as coating materials because of their superior physical properties. The coated rainwear, often known as "slickers", was very uncomfortable on extended use because of its impermeability. Coated fabrics are still used for tents and tarpaulins, but fabrics treated with repellents which do not reduce permeability to air and water vapor have largely replaced them for apparel rainwear. Coating materials used for waterproofing are applied as hot melts, latexes, or solvent solutions.

Durable water repellent finishes that allow air and water-vapor permeability began to be used in the mid-1930s with the introduction of reactive pyridinium type repellents. Other types were developed, and by 1990 fluorochemicals dominated the water-repellent fabric market, with ca. 90 % of the value of all products consumed. The value of all fluorochemicals used in the US textile industry in 1990 was ca. $ 100 \times 10^6$. The fluorochemicals are followed, in order of importance, by silicones, resin-based finishes, wax-metal emulsions, organometallic complexes and reactive quaternary compounds. All of these types were developed many years ago and there have been few new developments in water repellents since.

### 7.3.2. Reactive Quaternary Repellents

Reactive quaternary water repellents were introduced by ICI under the trade name Velan in 1937, and later by DuPont under the trade name Zelan [7.67]. These were the first truly durable water repellents. They are based on stearaminomethylpyridinium chloride [4261-72-7]. They are made by reacting stearamide with paraformaldehyde and pyridine hydrochloride. Commercial products contain not only the quaternary but other byproducts and derivatives such as methylol stearamides and methylenedistearamide. The quaternary compounds react with the hydroxyl groups of cellulose on curing, chemically linking the hydrophobe with the cellulose. The bound hydrophobe then tends to retain other unbound hydrophobic materials onto the fabric [7.68]. The quaternary type repellents are very durable to laundering, even at boiling temperatures. They are also durable to dry clean-

ing. They are dispersible in water and are applied from aqueous baths, followed by drying and curing at 135–205°C. They impart a soft but sometimes waxy hand to fabrics. Reactive quaternary repellents have fallen into disuse because of the necessity of afterwashing fabrics treated with them to remove byproducts such as pyridine salts.

Quarpel, a finish developed by the US Army Natick Laboratories in 1959 [7.69], combines a pyridinium repellent with a fluorochemical to give outstanding water and oil repellency for military garments. It has not been used for civilian garments because of its high cost.

### 7.3.3. Organometallic Repellents

Fatty acids, such as stearic acid, form Werner-type complexes with chromium which can be used as water repellents for natural and synthetic fabrics [7.70]. Commercial products include Quilon C, M, and S (DuPont) and Phobotex CR (Ciba-Geigy). They are neutralized with an amine (e.g., hexamethylenetetramine) or sodium hydroxide and applied by a pad–dry–cure process. Curing at 150–170°C results in maximum repellency. Chrome complexes have fair to good durability to washing and dry cleaning but are sensitive to alkali and some detergents and soaps. They are green in color and can cause a slight color change in the fabric. Aluminum complexes are colorless but are less effective than the chromium complexes. The organometallic complexes, like the pyridinium quaternaries, have fallen into disuse in recent years.

### 7.3.4. Wax-Based Repellents

Early processes involving wax emulsions used soaps as emulsifiers and precipitated the soaps as aluminum salts by treating the fabric with alum or another soluble aluminum salt in a second bath. Later, wax emulsions were combined with aluminum acetate or formate in a one-bath process. Still later, zirconium salts were substituted for the aluminum salts [7.71]. The wax employed is usually a paraffin wax with an emulsifier and a protective colloid such as glue, poly(vinyl alcohol) or an acrylic polymer. The wax–salt repellents are still in use, although their market share has shrunk from its peak in the 1950s. They are inexpensive and give good water repellency, but they have poor durability to laundering or dry cleaning. The zirconium-based products are more durable than the aluminum-based products and are more widely used. Application, consisting of padding and drying without curing, is very simple.

### 7.3.5. Resin-Based Repellents

Resin-based repellents, produced by condensing fatty materials with methylolated melamines and emulsifying the resulting product, became popular in the 1950s. Many variations of the resin-based repellents have been developed and marketed under trade names such as Aerotex 96 and Permel Resin B (American Cyanamid, now Freedom Chemical), Phobotex FTC (Ciba-Geigy), Nalan W and GM (DuPont), and Norane 16 (Sun, now Sequa). Some of the manufacturing processes for these products are quite complex, as exemplified by Phobotex FTC, which is made by a three-step process involving methylated methylol melamine, stearic acid, diglycerides, triethanolamine, monochlorobenzene, and paraffin wax [7.72]. The resin-based repellents produce durable water repellency on a broad range of fibers and fabrics. They are less expensive than the fluorochemical, silicone, or reactive quaternary types. They are applied by the pad – dry – cure process, often together with cross-linking agents. They require an acid catalyst and curing temperatures of up to 175°C for maximum repellency and durability. They are normally applied at a rate of 1 – 4 wt % of repellent solids on the fabric. They produce a hand which is soft but firmer than that given by quaternary or silicone repellents. The resin-based repellents have been used as extenders for the more expenisve fluorochemical repellents.

### 7.3.6. Silicone Repellents

Silicone repellents were introduced in the early 1950s [7.73]. These products are methylhydrogen- and dimethylsiloxane copolymers that react in the presence of catalysts to form cross-linked water-repellent films. The silicones are oils and must be emulsified for application. Typical catalysts are zirconium, tin, or aluminum salts, zinc octoate, tin(II) oleate, copper naphthenate, or metal-free compounds such as epoxy amines or amides. They are usually applied by a pad – dry – cure process and have good fastness on synthetic fabrics. On cellulosics they exhibit good dry-cleaning fastness but poor washfastness. Attempts to increase the washfastness of silicone repellents on cotton have been only moderately successful [7.74]. They are more expensive than the wax – salt repellents but less expensive than the newer fluorochemical finishes. They produce a very soft, silky hand, particularly on synthetics, where they have been found especially useful. Silicone repellents are normally applied at a level of 0.5 – 1.5 wt % solids on the fabric. Curing temperatures of ca. 150°C are usual. Silicone repellents can be coapplied with durable press finishes. In recent years organo-modified silicones with various reactive functional groups have become available which offer the possibility of greater durability; however, these have found more use as softeners than as water repellents.

### 7.3.7. Fluorochemical Repellents

Fluorochemicals were introduced in the late 1950s and have become the most widely used repellents. They are marketed under the trade names Scotchgard (3M), Zepel and Teflon (DuPont), Persistol (BASF), and Repellan (Henkel). They provide both oil and water repellency, due to highly fluorinated alkyl groups. For best repellency at least four fully fluorinated carbon atoms should be present, and the end group should be trifluoromethyl. The fluorochemical repellents used for textiles are mostly copolymers of fluoroalkyl acrylates and methacrylates [7.75]. The comonomers are esters of acrylic and methacrylic acids containing a variety of alkyl and substituted alkyl groups chosen to modify the physical properties of the polymers, improve performance, and reduce cost. Although the most expensive of the water repellents available they have become popular because of the additional oil and soil repellency which they afford. They repel both water and oil because they produce an extremely low energy surface. They are applied by a pad – dry – cure process with the cure being optional but usually used for the benefit of other finishing bath components such as cross-linkers. They firm the hand somewhat but not enough to make the fabrics unsuitable for most uses. Fluorochemical types are often used with "extenders" which may be other repellents such as the quaternary or resin-based types or even materials normally used as softeners [7.76]. With the extenders the amount of fluorochemical required can often be reduced by 50 % or more resulting in greatly improved economics.

### 7.3.8. Testing for Water Repellency

Probably the most widely used test for civilian rainwear is the Spray Test (AATCC Test Method 22) [7.13], in which 250 mL of water is allowed to fall through a spray head onto a taut fabric held at a 45° angle from a height of 6 inches (ca. 15 cm). The pattern of the wetting on the fabric is rated visually against a set of standard photographs. This test measures only the water repellency of the fabric surface and cannot predict the probable rain penetration of water through the fabric. It is, however, especially suitable for mill production control work, because of its speed and simplicity.

In the Impact Penetration Test (AATCC Test Method 42) [7.13] 500 mL of water is sprayed from a height of 24 inches (ca. 61 cm), and evaluation is by measuring the increase in weight of a blotter clamped under the fabric. This test measures the resistance of fabrics to penetration of water by impact, and thus can be used to predict the probable rain penetration resistance of garment fabrics. The results obtained with this test depend on the construction of the fabric as well as its water repellency.

The Rain Test (AATCC Test Method 35) [7.13], another impact penetration test, differs from the Impact Penetration Test in that it can be performed at different intensities. It involves directing a horizontal spray from a column of water 2–8 feet (ca. 60–240 cm) in height onto a fabric backed by a blotter from a distance of 12 inches (ca. 30 cm) for 5 min. Evaluation is by measuring the weight gain of the blotter.

The Hydrostatic Pressure Test (AATCC Test Method 127) [7.13] measures the pressure required to force water through a fabric and is especially suitable for evaluating fabrics which are expected to be used in contact with water such as heavy ducks and tarpaulins.

The Static Absorption Test (AATCC Test Method 21) [7.13] and the Dynamic Absorption Test (AATCC Test Method 70) [7.13] involve immersing the fabric in water under static and dynamic conditions for a specified length of time, blotting the fabric, and weighing the amount of water retained. The results of these tests depend primarily upon the resistance to

wetting of the yarns and fibers in the fabric and not upon the construction of the fabric.

The Bundesmann Test, which is used primarily in Europe, involves showering water on the fabric from 150 cm at a rate of 62–68 mL/min for a specified time, while the underside of the fabric is being rubbed, and weighing the amount of water which passes through the fabric or is absorbed by it.

Rain rooms are used primarily by the military for evaluating garments under severe rain conditions. In these tests water is sprayed onto subjects wearing repellent garments, simulating conditions encountered in rain storms. Such tests evaluate not only the fabrics but garment and seam construction.

## 7.4. Soil-Repellent and Soil-Release Agents

### 7.4.1. Historical Aspects

Soiling and soil removal have always been a problem with textiles, but with the advent of durable press, where polyester–cotton fabrics are treated with large amounts of cross-linking agents and other additives, the problem was accentuated. Durable press fabrics began to be a major factor in the market in 1964, and interest in soil-release finishes increased soon thereafter. Soiling and soil release had been studied extensively before the 1960s, and many research reports were published by workers at Harris Research, ITT, the Cotton Council, the SRRL of the USDA, and others [7.77], [7.78]. Many of the early workers realized that hydrophilic polymers such as starch or CMC improve soil removal, and these polymers were used as a laundry finish to improve soil removal but they were not durable. Deering Milliken announced a soil-release finish based on polymers containing carboxylic acid groups in June 1966, and proposed a test for soil release in December 1966 which later was adapted into AATCC Test Method 130 [7.13]. A patent was issued to Deering Milliken on their process in 1968 [7.79]. ICI offered soil-release agents for polyester fibers under the names Cirrasol and Permalose, first in Europe and then in the United States, during the 1960s [7.80], [7.81]. During the late 1960s there was feverish activity in the laboratories of textile mills and their suppliers to develop soil-release finishes. Rohm and Haas announced SR-488, a carboxyl-containing acrylic emulsion, in July 1967. 3M introduced a fluorochemical soil-release agent, Dual Action Scotchgard, in late 1967. All through 1967 fabric producers were announcing tradenamed soil-release finishes. The trade names did not imply a particular chemical finish, and there was much changing of formulations as soil-release finishing developed. The advertising, promotion and interest in soil-release finishes probably reached a peak in late 1967 or early 1968. It has been estimated that ca. 50% of all durable-press fabrics marketed during 1968 contained a soil-release agent [7.82]. Subsequently, interest in soil-release finishes began to wane, although they are still used on selected fabrics. Some of the factors which contributed to their decline were the high costs of soil-release finishes in relation to their performance, the undesirable side effects exhibited by many of them, such as poor hand, poor compatibility, and poor sewability, the misuse of the finishes, and the fact that the soil-release concept was not valued by consumers. Fabric producers found themselves promoting a concept which the consumer felt should be inherent in any fabric offered for sale. Customers wanted soil release but were unwilling to pay a premium for it. Even after interest in soil-release finishes had peaked, new and improved finishes continued to be offered as research initiated during the peak interest years came to fruition. 3M announced FC-218, an improved fluorochemical soil-release finish, in the spring of 1968 [7.83] and continued to make improvements in its fluorochemical products into the 1980s. When the Deering Milliken patent [7.79] on carboxylated acrylic emulsions was issued in April 1968 questions of infringement arose on many finishes then being used. In July 1968 Rohm and Haas announced RL-528, a water-soluble acrylic polymer soil-release agent, with the claim that it did not infringe the Milliken patent, which included only emulsion polymers. In July 1968 ICI announced Permalose TG, an improved version of Cirrasol PT, which was more durable and required less add-on. Later they introduced Milease T. In 1968 BASF announced Perapret D [7.84] and promised to indemnify all users against suits for patent-infringement by Deering Milliken.

Soil-release finishes based on the caustic treatment of polyester were developed and promoted, particularly by J. P. STEVENS whose trade name was Fantessa [7.85]. It soon became generally recognized that any finish that imparts a hydrophilic surface to textiles would improve soil release and numerous products of various types began to be proposed, patented, and marketed. Although there are dozens of such products, few have gained appreciable commercial significance.

### 7.4.2. Factors Influencing Soil Release; Testing

Some of the factors which influence soil release are:

1) Type of soil
2) Amount of soil
3) Size and shape of fabric sample which influences work applied during laundering
4) Type and amount of detergent
5) Mechanical energy of the wash
6) Temperature of the wash

There is a conflict between the optimum conditions for durable press performance and soil release during laundering. Durable press is best at low temperatures and mild agitation, whereas soil release is most effective at high temperatures and with vigorous mechanical action.

The most widely used test method for soil release is the Oily Stain Release Method (AATCC Test Method 130) [7.13]. In this test a stain is produced on a test specimen by using a weight to force a given amount of the staining substance into the fabric. The stained fabric is then laundered in a prescribed manner and the residual stain rated on a scale from 5 to 1 by comparison with a graduated series of standard stain replicas.

Other tests have been devised to measure facets of the soiling problem, such as redeposition. In redeposition tests, clean samples are laundered with soil or soiled fabrics, and their reflectance is measured before and after the laundry cycle. Fabrics with a propensity for soil redeposition pick up soil in the laundry cycle and show a reduction in reflectance after laundering.

### 7.4.3. Soil-Release Finishes

The most widely used soil-release agents are:

1) Anionic polymers
2) Low molecular weight polyesters
3) Caustic treatments on polyester
4) Fluorochemicals

The anionic polymers are probably the least expensive of these finishes. They may be either emulsions or solutions. They are usually vinyl polymers containing carboxyl groups, although almost any anionic polymer which can be fixed to the fiber surface will work. A variety of acid monomers have been proposed, including acrylic, methacrylic, and maleic acids. Acrylic acid is the most important commercially [7.86]. A nonacidic comonomer is used to improve the flexibility and durability of the polymer. Lower alkyl acrylates such as methyl or ethyl acrylate are common comonomers. At least 15 % of the carboxylic acid monomer is needed in the polymer for practical soil-release performance. Comonomers such as $N$-methylolacrylamide, which can provide limited cross-linking of the polymer when used in small amounts (1–2 %), can add durability to the finish without causing excessive cross-linkage, which can inhibit swelling and decrease the effectiveness of the finish. The anionic polymers are usually supplied in concentrations of 20–30 % solids, and up to 15 wt % of the products as supplied is used on the fabric. They are usually applied in the finishing bath along with other chemicals in pad–dry–cure systems. Being polymers they tend to give a firmer hand to the fabric. Some method of fixing the polymer on the fabric must be incorporated into the finish to obtain durability. This can be achieved through an internal reactive monomer component or by reaction with an external cross-linking agent. The anionic soil-release agents present some compatibility problems, particularly with cationic agents such as softeners. They can sometimes interfere with the catalysis of durable press finishes. Other side effects associated with the anionic polymers are metal-ion pickup during laundering, yellowing, and the loss of fastness of pigment dyes.

The low molecular mass polyester types are used mostly on polyester fabrics, where they are claimed to "co-crystallize" with the polyester fiber component. The polyesters contain large hydrophilic polyglycol segments as recurring segments along with short blocks of poly(ethylene terephthalate) units. They are usually supplied in concentrations of ca. 15 %, and up to 15 wt % of the product as supplied is used on the fabric. The products are usually applied to the fabric by padding and are fixed by heating which causes them to be absorbed into the surface of polyester fibers. Alternatively they can be exhausted onto polyester fibers from baths at elevated temperature under conditions similar to those used for dyeing. The solubility of the products decreases with increasing temperature, which aids their exhaustion onto polyester fibers at elevated temperature. This type of finish is not effective on cellulosic fibers. Finishes of this type sometimes have an adverse effect on the fastness of dyes. Cirrasol PT, Permalose T, and Milease T (ICI) are products of this type.

When polyester fibers are treated with strong alkalis such as sodium or potassium hydroxide the surface of the fiber is hydrolyzed, liberating carboxyl and hydroxyl groups, and the resulting fibers have soil-release properties. Auxiliaries such as quaternary compounds are sometimes used with the caustic treatment to enhance the effect. There is always some weight loss associated with caustic treatment and, if carried to the extreme, severe weakening of the fiber can occur. This weakening has been proposed as a cure for pilling on polyester fabrics. Hydrolysis reduces the size of the polyester fibers and leaves the treated fabrics with a soft silky hand. The soil-release properties of caustic-treated polyester fabrics are durable until the fabrics are heated, as in ironing, after which the soil-release properties are lost. This loss of soil-release properties on heating is thought to be due to the fact that only the surface of the fiber is modified, and that on heating the surface rearranges to bring unhydrolyzed polymer to the surface.

The fluorochemicals used for soil release differ from those used solely for water and oil repellency in that they contain hydrophilic groups as well as perfluoro alkyl groups [7.83]. The hydrophilic groups are usually polyoxyethylene chains. In the dry state the perfluoro alkyl groups are oriented toward the surface, giving water and oil repellency, while the polyoxyethylene chains are coiled. In the presence of water the polyoxyethylene chains are hydrated, uncoil, and become oriented toward the surface, giving good soil-release properties. These dual-action fluorochemicals were first offered by 3M as FC-216 and FC-218. Other versions have been marketed by 3M and other suppliers. Both nonionic and cationic products are available, as are low-flash-point products containing acetone and high-flash-point products without acetone. The products are supplied at concentrations from 15 to 30 %. Application of 0.3–0.4 wt % of active solids on the fabric is recommended. The fluorochemical agents are usually applied with other finish bath ingredients in durable press formulations by a pad–dry–cure system. The fluorochemical types are the most expensive soil-release agents, but they are probably the most widely used types because of their performance, compatibility with durable press finishes and freedom from deleterious side effects.

## 7.5. Flame Retardants

(see also → Flame Retardants)

### 7.5.1. Flammability of Fibers

Most fibers will burn, some more readily than others. The limiting oxygen index (LOI), which measures the lowest concentration of oxygen at which a fiber will support combustion, is a good measure of the inherent flammability of fiber types [7.87]. The LOIs of some common fiber types are as follows:

| | |
|---|---|
| Acrylic | 0.182 |
| Acetate | 0.186 |
| Polypropylene | 0.186 |
| Rayon | 0.197 |
| Cotton | 0.201 |
| Nylon | 0.201 |
| Polyester | 0.206 |
| Wool | 0.252 |
| FR Cotton | 0.270 |
| Nomex | 0.282 |

The concentration of oxygen in air at sea level is ca. 21 % or 0.210.

Judged solely on fire resistance, Nomex would be the fiber of choice, and it has become the fiber which is most often used when fire resistance is the most important property, such as in garments for firefighters, racecar drivers, steelworkers, and certain military personnel. Nomex is quite expensive, however, and it is difficult to obtain in a complete range of colors. It has not found general usage in apparel or industrial fabrics, except where fire resistance is of paramount importance. Other inherent fire resistant synthetic fibers such as Kynol, polybenzimidazole, and Durette fall in the same category as Nomex.

Cotton and rayon can present serious fire hazards, particularly in certain fabric constructions and in certain end uses. Wool, in contrast, is difficult to ignite and burns very slowly. Thermoplastic synthetic fibers, such as nylon, polyester, and polyolefin, are generally less serious fire hazards than the cellulosics because they tend to melt away from a flame, but they present the additional hazard of molten drip.

When blended with cellulosics the thermoplastics represent a greater hazard because the cellulosics form a matrix which retains the molten thermoplastics and prevents them from dripping away from the flame.

Thermal decomposition of fibers, which always precedes flaming, results in gaseous, liquid, tarry, and solid products. As the flammable gases burn, the liquids and tars volatilize, and some of these fractions burn. Other fractions yield a carbonaceous material (char) which does not burn readily. The carbonaceous residue glows until essentially all organic material is consumed and only a fluffy ash remains. Thus there are two stages of burning, flaming and afterglow. The temperature of flaming cellulose is 400–450 °C, whereas the afterglow temperature is ca. 600 °C.

### 7.5.2. Nondurable Flame Retardants

Many water soluble salts render cellulosic fabrics flame retardant, until they are laundered or otherwise exposed to water [7.88]. These are Lewis acids or compounds which give rise to Lewis acids on pyrolysis. Some of the compounds which have been used include diammonium phosphate, boric acid–borax mixtures, aluminum sulfate, ammonium sulfate, sulfamic acid, ammonium sulfamate [7773-06-0], zinc chloride, various phosphate salts, various ammonium salts and combinations thereof. They are applied from aqueous solution by padding or spraying and drying, usually at concentrations of 10–15 %. Phosphoric and boric acids and their salts also prevent or inhibit afterglow, as well as serving as flame retardants. While effective, the nondurable types are removed by laundering and thus are suitable only for fabrics, which are seldom or never laundered and can be retreated whenever laundering is done, such as buntings and some draperies.

### 7.5.3. Semidurable Flame Retardants

Cellulose can be phosphorylated by heating it with a mixture of phosphoric acid and urea [7.89]. The resulting phosphate ester of cellulose is fire retardant, but it picks up anions from laundering to form the sodium, calcium, or magnesium salts, which are not flame retardant. To be effective this finish must be regenerated by reconverting it to the acid or ammonium form after each laundering.

Another semidurable finish which has been proposed is a combination of ammonium phosphate with a dicyandiamide – formaldehyde condensation product, which forms an insoluble resin within cellulosic fibers when applied to the fabric and baked. This finish has a tendency to weaken the fabric if heated above 150°C. Various modifications of this finish have been proposed.

Cyanamide and phosphoric acid (Pyroset CP) also produce semidurable fire retardancy when applied to cellulosic fabrics, dried, and cured [7.90]. Optimum results are obtained when the solution contains ca. 3 mol of cyanamide per mole of phosphoric acid. The higher the solids add-on the greater the fire resistance. At 45% add-on this finish withstands 7 – 10 home laundry cycles, but it is subject to deactivation by sodium, calcium, or magnesium ions from laundry water and must be reactivated by souring and ammonia treatment. The finish causes considerable loss of fabric strength. It has found some commercial use, especially on draperies.

### 7.5.4. Durable Flame Retardants

During the 1960s and 1970s a number of durable flame-retardant finishes were developed after which research on such finishes appeared to decline. Some of the finishes developed, such as APO (tris-aziridinyl phosphine oxide [545-55-1]) [7.91] which was promoted for cotton, and tris-(bromoethyl) phosphate [126-72-7] (TRIS), which was promoted for polyester, are no longer in use because the reactants were found to be carcinogenic. Many of the most successful finishes have been based upon tetrakis (hydroxymethyl) phosphonium chloride [124-64-1] (THPC) or its derivatives. In the Proban process [7.92] THPC is reacted with urea to form a water-soluble compound that is stable in solution. This compound is applied to fabric in a pad – dry process, then the fabric is exposed to ammonia and ammonium hydroxide consecutively to produce an insoluble polymer within the fibers by reaction of the methylol groups of the THPC with urea and ammonia.

Numerous variations of THPC-derived finishes have been proposed. One of these is the THPC-amide process in which THPC is applied with urea, trimethylol melamine and sodium hydroxide by a pad – dry – cure process. A typical formulation is:

| | |
|---|---|
| THPC | 17% |
| Urea | 10% |
| Trimethylolmelamine | 10% |
| Sodium hydroxide | 1.5% |

This finish generally decreases breaking strength by ca. 30%, increases stiffness, and is discolored by chlorine bleaches. It is very durable to laundering and increases wrinkle recovery somewhat.

THPC can be converted to THPOH [512-82-3] by treatment with alcoholic sodium or potassium hydroxide. If cellulosic fabrics are padded with ca. 30% THPOH, dried to 10 – 20% moisture, then treated with ammonia, a fire-resistant fabric results [7.93]. This finish produces little strength loss and little stiffness. Its durability is good, but it produces no improvement in wrinkle recovery and it discolors with chlorine bleaches.

If a methylolmelamine is combined with THPOH and an ammonia treatment, followed by curing, a more efficient finish is obtained which has the lowest cost of the THPC family [7.94]. A typical bath contains ca. 18.5% THPOH and 11.5% trimethylolmelamine. The fabric is padded with the finish, partially dried, exposed to ammonia, and finally cured for ca. 3 min at 150°C. This finish shows little strength loss, only a slight increase in stiffness and has good durability. It produces no increase in wrinkle recovery and yellows with chlorine bleaches.

The THPOH – amide finish is similar to the THPC – amide process except that THPOH is substituted for THPC [7.95]. A typical finishing bath contains 16% THPOH, 10% urea, and 10% trimethylolmelamine. The fabric is padded, dried, and cured at ca. 150°C for 3 min. With this finish about 90% of fabric strength is retained, with a slight increase in stiffness. Durability is very good, wrinkle recovery is greatly increased, and yellowing with chlorine bleach is very slight.

All THPC and THPOH finishes must be afterwashed to remove unreacted materials and byproducts. THPC has been replaced by the sulfate salt THPS [55566-30-8] because of possible toxicity concerns with THPC, which is no longer manufactured.

Permanent fire retardancy on polyester – cotton blends has been reported for a combination of a specially formulated halogen-free phosphonate with the THPC – urea – ammonia process [7.96]. The phosphonate alone is claimed to afford durable flame retardancy on 100%

polyester fabrics when ca. 5% is applied by padding, drying and thermosoling.

Another durable fire retardant which has been used commercially is the methylol derivative of the addition product of a dialkyl phosphite with acrylamide, Pyrovatex CP [*20120-33-6*] (Ciba–Geigy) [7.97]. It is used alone or in combination with trimethylolmelamine, which improves its efficiency and durability. Add-ons of ca. 20–25% are required on cotton. It is applied by a pad–dry–cure process. Treated fabrics have tear and tensile strength losses of 20–30%. The finish is durable to 30–50 launderings and drycleanings.

Specialty resins, such as methylated thiourea–formaldehyde resins, have been used to flameproof nylon, acetate and acrylic fabrics.

### 7.5.5. Flammability Testing

There are numerous tests for the flammability of textiles. Many of these are specialized tests for specific fabric constructions or specific end uses and have been designed to simulate a particular fire hazard situation. For apparel fabrics the most widely used tests are the 45° angle test called for in US Commercial Standard 191-53, which is now a part of the US Flammable Fabrics Act, and a variety of vertical tests. The CS 191-53 test was not designed for fire-resistant textiles, but was designed to screen out those fabrics which presented an unusual flammability risk as items of commerce. In it a flame is presented to the surface of a fabric held at a 45° angle for 1 s, and if it ignites the time required for the flame to spread and burn through a drop-cord thread 5 inches (ca. 13 cm) above the point of ignition is measured. In the vertical tests a strip of fabric is held in a vertical position and a standard flame is applied to its lower edge for a standard time. If the fabric is not totally consumed the char length of the burned area is measured by applying a standard weight to one side of the fabric to tear it through the charred area, after which the length of the char from the bottom of the fabric to the end of the tear is measured. Some variation of this test is most commonly used for testing fire retardant fabrics. The shorter the char, the more fire retardant the fabric.

### 7.6. Wool Finishing

The finishing of wool involves both mechanical and chemical processes. Mechanical finishing to develop the hand, drape, and surface characteristics of the fabric is at least as important as chemical finishing.

Softeners, lubricants and water repellents used on wool are similar to those used on cotton.

### 7.6.1. Setting

Crabbing is normally the first process employed in wool finishing. It is a wool setting process in which latent strains in the fabric from earlier fabric forming processes are relieved to avoid distortions such as cockling, crowsfeet, and uneven shrinkage. The process can be accomplished by hot or boiling water or steam on a crabbing machine with only a wetting agent, or chemical agents can be used. Agents based on monoethanolamine sulfite, monoethanolamine formate, and ammonium sulfamate have been used [7.98]. Mixtures of a reducing agent and an alcohol or an ionic surfactant may also be used.

### 7.6.2. Scouring

The second operation in wool finishing is usually scouring. It is essentially a washing operation in which contaminants applied or picked up during earlier processes are removed. It has been estimated that 70% of all processing faults originate from inadequate scouring [7.99]. Nonionic surfactants, such as alkyl phenol and fatty alcohol ethoxylates are the most widely used scouring agents, although alkyl benzenesulfonates are also used, as are ethoxylated products which have been sulfated, phosphated, or carboxymethylated. Supporting electrolytes such as sodium carbonate, sodium bicarbonate, or sodium sulfate are usually added to the scouring bath.

### 7.6.3. Milling

Milling, fulling, or felting is a process in which wool fibers are progressively entangled by mechanical action to consolidate the fabric structure [7.100]. Wool fibers felt because of the scales on their surface. Milling is usually carried out in rotary milling machines, although other mechanical forms of milling are used. About 150% liquor on the weight of the fabric is used in milling. Three types of liquor are used: grease, acid, or soap. In grease milling, sodium carbonate or carbonate/bicarbonate mixtures are used to form soap from the natural greases and spinning oils. Acid milling at pH 2–4 is used to minimize color bleeding and to increase the rate of milling. Wool felts least at its isoelectric point of pH 4.8, and felting is slow in the pH range of 4–7. In acid milling acid-stable surfactants such as amphoterics and amine oxides must be used. Soaps and synthetic detergents are used in milling to increase lubrication and increase the rate of felting. High-foam surfactants are often preferred for milling. Scouring and milling are sometimes combined. Alkaline milling is used on unscoured fabrics. Increasing the temperature to 40–50°C increases the rate of milling.

### 7.6.4. Carbonizing

Carbonizing is used to remove vegetable matter from wool fabrics. It is based upon the difference in stability to mineral acids shown by animal fibers and vegetable matter. Sulfuric acid is the most common agent used for carbonizing, although other acid substances, such as aluminum chloride, may be used on blends containing fibers which are sensitive to sulfuric acid, such as polyester and acetate. The agent is applied in dilute solution, after which the wool is dried, and baked to degrade the vegetable matter present. After baking, the degraded contaminants are crushed and removed by mechanical action, then the wool is neutralized. Surfactants are added to the acid solutions to promote rapid and uniform wetting. The carbonizing process has been well studied [7.101]–[7.103].

### 7.6.5. Shrink-Resistant Finishes

A number of processes can be used to make wool shrink resistant. One of the processes most used commercially involves subjecting the fibers to a light chlorination, which confers some degree of shrink-proofing, then coating them with a resin such as Hercosett 125, which masks the surface scales [7.104], [7.105]. The chlorination–polymer treatments have been in commercial use since the 1970s and account for about 80 % of all machine-washable wool fabrics. Hercosett 125 (Hercules Chemical Co.) is a polyamide–epichlorohydrin polymer capable of self cross-linking. It is applied at 2 wt % solids on the fabric in a bath adjusted to pH 7.5–8.0 with sodium bicarbonate or ammonia and exhausted by

raising the temperature to 30°C. After treatment, the fabrics or garments are extracted and dried. Nonchlorination processes which utilize polymers alone have been developed. One such process uses Synthappret BAP (Bayer), a reaction product of sodium bisulfite with a prepolymer formed by capping poly(propylene oxide) with an aliphatic diisocyanate, and Impranil DLN (Bayer), an aqueous anionic dispersion of a polyurethane. The polymers are exhausted from a bath containing magnesium chloride at pH 7.5–8.0 by raising the temperature from 30°C to 60°C over 30 min, after which the pH is increased to 9.0 and the temperature maintained at 60°C for 30 min to cure the polymer [7.106]. Other polymer-based products and processes have been proposed for shrink-proofing wool [7.107], [7.108].

### 7.6.6. Mothproofing

Mothproofing of wool is an important chemical process. One of the most widely used mothproofing agents has been polychloro-2-(chloromethylsulfonamide)diphenyl ether (PCSD) marketed as Eulan 933 and Eulan WA New (Bayer) [7.109]. Dieldrin, a general-purpose insecticide, has also been used [7.110]. Because of environmental pressures, mothproofing agents based on synthetic pyrethroid insecticides, such as permethrin, have been introduced [7.109].

## 8. Auxiliaries for Technical Textiles

### 8.1. Technical Textiles—an Overview

**Introduction.** The typical processes of classical textile finishing—sizing, pretreatment, bleaching, dyeing, optical brightening, printing, and finishing—are supplemented by another series of finishing processes in the production of technical textiles: forming (molding), fiber bonding, heat sealing, etc. Because many of the textile auxiliaries and finishing agents [8.1] are the same as those used for finishing classical textiles, it is not always appropriate to refer to special auxiliaries for technical textiles. However, the emphasis is different. The important auxiliaries are not the optical brighteners, the color shades, or the softening agents that give a silky or other hand, but polymer dispersions for fiber bonding or coating, polymer powders for composite materials, finishing agents for barrier textiles, plasticizers for polymers, glass-fiber coupling agents, cross-linking agents, resins, oleophobic and hydrophobic agents, nonslip agents, antistatic agents, flame retardants, thickeners, light stabilizers, etc. Some technical textiles are treated with sizing agents, but the residual amounts are small compared with the size-free nonwovens, floor coverings, moldings, and composites.

**Economic Aspects.** The growth in the importance of technical textiles is described in [8.2]. Technical textiles were first reported separately in German textile statistics in 1988 [8.3]. The fraction of the total output of textile finishing production used for technical textiles was 19 % in 1992 [8.4]. A working group, "Technical Textiles", was set up in 1989 by the association of German textile auxiliary-producing industries (TEGEWA, Frankfurt/Main) [8.5].

Many problems would be technically insoluble without technical textiles, for example, awnings tire cord fabric, tents, filters, barrier textiles, clothing for spray paint rooms and medical clean rooms, coverings for waste tips, textiles for purification processes, upholstery, fishing nets, and safety equipment for building construction [8.6].

**Application Methods.** The auxiliaries are applied by the usual processes, e.g., padding, with nip rollers, spraying, coating and strewing. Wet processes of surface finishing may involve the use of foam. Dry processes are increasingly used, both to apply powders, e.g., for sintering and melting, and in the transfer process.

The well-known conditions for drying and curing are mostly the same as those used in classical textile finishing. To these must be added molding operations and adhesion processes.

**Definition.** Technical textiles cannot be precisely defined. For example, a textile can be made into classical clothing, or, if it is first provided with a coating, into technical clothing. Abrasive-coated fabrics are technical textiles. Hand-made carpets are archetypal classical textiles. Tufted carpets are used in the home, but are also used as technical products in automobiles. A nonwoven padding, e.g. for an anorak lining, is grouped with the clothing textiles, but a similar product, used as a filter, is a technical textile. A coated material can be used for clothing or as a tarpaulin. This duality makes a sharp division between classical and technical textiles impossible. The roof covering or the glove tray in an automobile and the lattice textile in a facade insulating plaster must certainly be grouped among the technical textiles. Medical textiles are often included with the technical textiles, although this can be regarded as inappropriate.

The importance of technical textiles is indicated in Table 8.1 [8.7].

The term technical textiles refers not only to loose stock, yarn, woven, warp-knitted, or knitted materials, and ropes but also includes nonwovens ($\rightarrow$ Nonwoven Fabrics), carpets, and textile floor coverings, ($\rightarrow$ Floor Coverings), shaped textile products, nets, layered yarn structures, bonded textiles, and composites.

**Fibers.** The range of fibers used in the production of technical textiles includes natural and synthetic fibers, fiber blends, organic and inorganic fibers, polyester and glass fibers, low- and high-melting fibers, high-strength and extra-high-strength fibers, chemically modified fibers, primary and recycled fibers, and normal, waste, and special fibers. Recycled fibers, which have been used in technical textiles for a long time, will become increasingly important [8.8].

The composite materials [8.9] produced from certain fibers and fiber–plastics combinations are described in $\rightarrow$ Plastics, General Survey, **A 20**, pp. 644–657; see also $\rightarrow$ Fibers, 4. Synthetic Organic, **A 10**, 1987, pp. 599–600.

**Table 8.1.** Areas of use of textile fibers (%)

| Use | Germany 1992 | | EC 1992 | | United States 1991 | |
|---|---|---|---|---|---|---|
| | Total | Man-made fibers | Total | Man-made fibers | Total | Man-made fibers |
| Clothing textiles | 39 | 29 | 47 | 39 | 40 | 25 |
| Home textiles* | 36 | 38 | 35 | 38 | 37 | 44 |
| Technical textiles | 25 | 33 | 18 | 23 | 22 | 31 |

* Including floor coverings.

## 8.2. Auxiliaries

There are a number of technical textiles that are not given a final treatment with auxiliaries, e.g., geotextiles made from spunbonded polyester. Here, the technical textile is the end product of the production process. However, many other products only receive their desired properties on treatment with chemical products. Such auxiliaries are indispensable, as the completely finished technical textile can only be produced with their aid.

There are three basic requirement profiles for auxiliaries, which apply to the auxiliary itself, to additives used in conjunction with the auxiliary, and to the conditions under which the auxiliary is applied. Auxiliaries can be used alone or in combination with others.

The tasks of the auxiliary include:

1) To achieve and maintain stabilization of the fiber material during use (e.g., in the production of bonded fleeces), to reduce shrinkage of spunbonds, loosening of the textile structure and fraying, to prevent dust formation, to improve edge strength and resistance to cutting, and to maintain shape during heat treatment
2) Permanence of, for example, strength under various types of stress, e.g., boil washing, water, seawater, moisture, steam, climatic conditions, dry cleaning solvents, light, UV, heat, cold, and oxygen
3) To give repellency effects, e.g., hydrophobic and/or oleophobic properties, and to facilitate the removal of impurities, soil, and dust
4) Prevention of aging

A list of auxiliaries for technical textiles with TEGEWA nomenclature [8.10] is given below, chapters of [8.1] in parentheses:

*Auxiliaries used alone or as polymer additive*
Polymer dispersions (3.11, 4.5, 4.5.2, 4.18, 4.19, 4.20)
Polymer powders (4.19)
Polymer solutions
Resins
Cross-linking agents (4.2)
Catalysts (4.4)
Nonslip agents (4.14)
Hydrophobic agents (4.7.1)
Oleophobic agents (4.7.22, 4.7.3)
Flame-proofing agents (4.15)
Antistatic agents (4.6)
Affinity promoters
Antimicrobial agents (4.16)

*Auxiliaries mainly used as additives*
Plasticizers
Thickeners (3.12, 4.18)

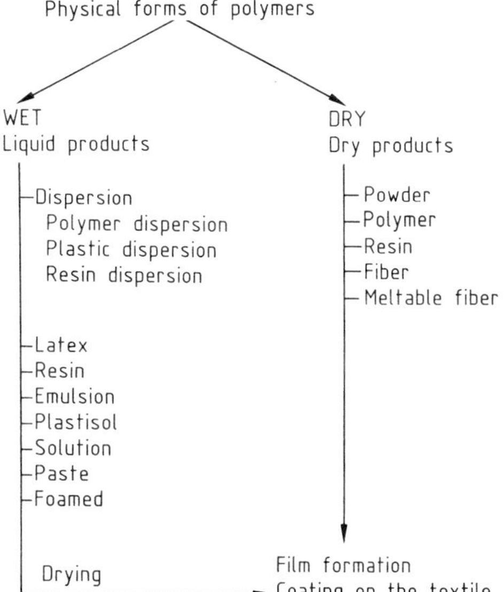


**Figure 8.1.** Physical forms of polymers suitable for application to textiles

Surfactants as emulsifiers (5.1), dispersants (5.3) and foaming agents (5.3)
Wetting agents
Antifoams (5.2)
Light stabilizers
Pigments
Carbon black
Inorganic fillers

**Application** (Fig. 8.1). Polymers are applied to textiles in water (hardly ever in solvents today) or as dry powders. Alternatively, low-melting textile fibers can bond other textile fibers.

### 8.2.1. Polymers

**Polymer Dispersions.** The terms dispersion and latex describe the same form of a polymer [8.11], i.e., finely dispersed water-insoluble polymer particles in water produced by means of dispersants (emulsifiers, protective colloids; see → Disperse Systems and Dispersants). Dispersed polymers are particularly suitable for textile applications. The terms "emulsion", derived from emulsion polymerization (→ Emulsions) [8.12], [8.12a] and "resin" [8.13] are also used. Typical polymer emulsions include the silicone emulsions used to give textiles hydrophobic properties.

The polymer dispersions for technical textiles are a series of products with many different constituents (Fig. 8.2) [8.14].

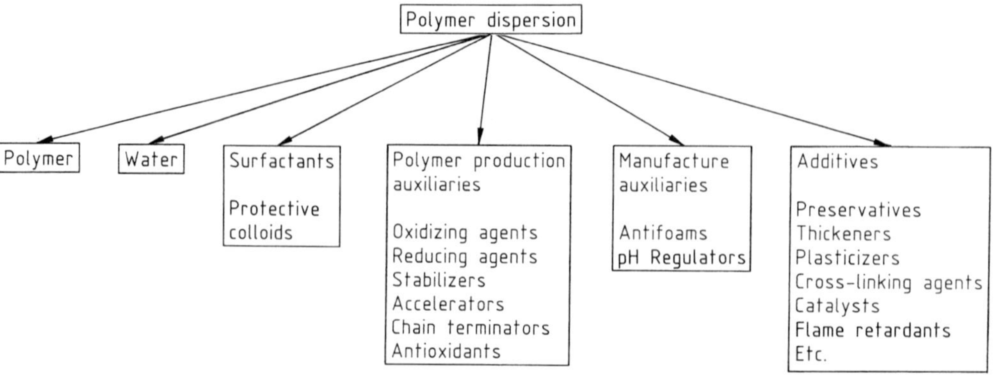


**Figure 8.2.** Components of polymer dispersions

A typical composition of a polymer dispersion [8.15] is:

| 47.0 % Copolymer of | monomer I | 60 % |
| | monomer II | 33 % |
| | monomer III | |
| | (dispersive effect) | 1 % |
| | reactive monomer | 6 % |

2.0 % Emulsifier (ionic or nonionic)
0.7 % Production auxiliaries: reducing/oxidizing agents, buffers, pH regulators
0.1 % Preservative and antifoam
0.2 % Residual monomer
50.0 % Water

A distinction is made between primary and secondary dispersions, the former being obtained directly by polymerization, and the latter by dispersing solid or molten polymers or resins in a separate operation.

*Environmental Aspects.* According to the German chemical regulations [8.16], polymer dispersions are classed as preparations. They do not require special labeling, being classified as a Class 1 (low) hazard to water, and can be eliminated in wastewater purification plants, although not biologically degradable. Their $LD_{50}$ (rat) is generally > 2000 mg/kg. There are no known significant problems associated with the drying or curing of polymers on textiles. For applications covered by the German food regulations, the dispersions recommendation XIV [8.17] is also supplemented by more recent EC guidelines [8.18] which apply to the drying of polymer dispersions on textiles to form polymers.

**Plastisols** [8.19] are polymer powders dispersed in plasticizers; they are described in → Poly(Vinyl Chloride) **A 21**, pp. 734–736.

**Polymer Solutions.** Polymers used in the form of aqueous solutions include sizing agents [poly(vinyl alcohols), cellulose ethers, starches, polymers and copolymers of acrylic acid], protective colloids for the production of polymer dispersions, and some dispersants such as maleic anhydride copolymers, surfactants, ethoxylated alkylphenol, ethoxylated fatty alcohol, some finishing agents based on polyacrylamide, textile print adhesives [poly(vinyl alcohol)], and many thickening agents. Their properties include solubility in water, hydrophilicity, viscosity increase, adhesive effects, ease of removal by washing with water, and hardening of the textile (except for surfactants).

**Polymer Powders.** The use of polymers without water offers significant environmental advantages, although the range of problems that can be solved is more limited than when aqueous systems are used. Emissions are minimized compared with aqueous polymer systems, but control of the addition rate is more difficult.

**Monomers.** Polymer dispersions can be produced by radical polymerization. The ultimate properties of a textile are closely linked with the polymerizable monomers "concealed" in the polymer (see Table 8.2). These monomers therefore contribute to such basic properties as hardness, softness, strength, solubility in water, and oleophobicity. Some copolymers are described in [8.20].

**Auxiliaries in Polymer Manufacture.** Polymers contain the auxiliaries used in their production (see Fig. 8.2). The production of polymer dispersions (latexes) in an aqueous medium

**Table 8.2.** Monomers used in the production of the polymers in polymer dispersions

| Monomer | Glass transition temperature of homopolymer* $T_g$, °C | | | Characteristics of polymer dispersion attributable to monomer |
|---|---|---|---|---|
| | A | B | C | |
| Methyl methacrylate | +107 | +105 | +105 | hardness, limited elongation, strength, brittleness |
| Acrylonitrile | +105 | + 75 | +105 | thermal stability, hardness, permanence, limited elongation, brittleness |
| Styrene | +100** | +105 | +107 | thermal stability, hardness, brittleness, limited elongation |
| Vinyl chloride | + 83 | + 81 | + 77 | hardness, thermoplasticity, heat sealing, yellowing, flame-proofing, strength |
| Vinyl acetate | + 33 | + 29 | + 42 | hardness, thermoplasticity, increased adhesion, limited elongation |
| Dibutyl maleate | 0 to −10*** | | | filling effect |
| Methyl acrylate | | + 8 | + 22 | slight hardening effect |
| Vinyl propionate | − 7 | | + 8 | filling effect, elasticity |
| Vinylidene chloride | − 17 | − 19 | − 18 | hardness, yellowing tendency, flame-proofing, heat sealing |
| Ethyl acrylate | − 24 | − 22 | − 8 | softness, permanence, tearing strength |
| Ethylene | | − 26 | | softness, adhesion, elasticity |
| Ethylhexyl acrylate | − 50 | − 85 | − 58 | softness, adhesion |
| Butyl acrylate | − 55 | − 54 | − 43 | softness, increased adhesion, elasticity |
| Butadiene | − 78 | −102 | −102 | softness, elasticity |
| *Reactive monomers* | | | | |
| Acrylamide | | +165 | +220 | hardness, ability to cross-link via N-methylol groups, limited elongation |
| N-hydroxymethylacrylamide (N-methylolacrylamide) | | | | permanence, limitation of thermal adhesion, limited elongation, hardness |
| N-hydroxymethylmethacrylamide (N-methylolmethacrylamide) | | | | permanence, limitation of thermal adhesion, limited elongation, hardness |
| Acrylic acid | | +103 | +130 | hardness, hydrophilicity, stabilizing, poor wet strength |
| Methacrylic acid | | +230 | +162 | |
| Itaconic acid | | | | |
| Crotonic acid | | | | |
| Hydroxyethyl methacrylate | | + 55 | | hardness, hydrophilicity |
| Glycidyl methacrylate | | + 46 | | hardness, permanence |

* A: various literature sources; B: from [8.12 a]; C: see →Polyacrylates **A 21**, p. 169.  ** Hoechst measurement (DSC).
*** Extrapolated.

normally requires a redox system consisting of peroxides (hydrogen peroxide, organic peroxides) or peroxosulfates, reducing agents (dithionite, bisulfite), polymerization initiators, and dispersing/emulsifying agents. Dispersants have a considerable influence on the ultimate properties of the treated textile, e.g., yellowing, usually due to ammonium salts, or odor from organic peroxides (see Table 8.3).

Dispersants can be protective colloids or surfactants. They emulsify one or more liquid monomers, which usually have low solubility in water, and also disperse the polymer formed in the polymerization [8.21].

**Table 8.3.** Effects of dispersants on a polymer dispersion

*Surfactants*

| | |
|---|---|
| Nonionic | coarser particle size, stability with cationic additives, lower capacity for fillers, reduction of surface tension, hydrophilicity not permanent, no effect on hand |
| Anionic | finer particle size, stability problems with cationic additives, increase in capacity for fillers, improvement of penetration, reduction of surface tension, hydrophilicity not permanent, no effect on hand |
| Nonionic/anionic | hydrophilicity, increase in salt stability and penetration, reduction of surface tension, hydrophilicity not permanent, no effect on hand |
| Cationic (seldom used) | coarser polymer particles, thickening, yellowing, compatibility problems with anionic additives, affinity to cotton, increase of surface tension |
| Amphoteric (seldom used) | stability problems, yellowing |

*Protective colloids*

| | |
|---|---|
| Poly(vinyl alcohol) Cellulose ethers Maleic anhydride copolymers | coarser polymer particles, thickening, reduction of penetration and capacity for fillers, hardening |

The influence of the individual components of a polymer dispersion on the ultimate properties of a textile are listed below:

*Effects due to polymer*

Viscosity
Drying properties
Hand: hardness, softness, full, stiffness
Elasticity, elongation, strength
Tack, adhesion to the textile
Swelling by water and solvents
Resistance to water, washing liquors, and solvents
Thermal stability, moldability
Soiling, graying
Welding properties
Flame-proofing effect
Yellowing, stability to light and UV

*Effects due to dispersant and emulsifier*

Particle size
Stability
Ionicity
Viscosity
Compatibility with additives, salts, and fillers
Spraying properties, foaming properties
Drying behavior

Migration
Color intensification with pigments
Swelling by water and solvents
Soiling
Adhesion/separation from substrate
Wet strength
Hydrophobicity, hydrophilicity, oleophobicity

*Effects due to polymerization additives*

Yellowing

*Effects due to water*

Limitation of stability
Drying properties
Volatility of constituents with water vapor

The surface tension of a polymer dispersion system under conditions of use depends on the surfactants or protective colloids used in its manufacture [8.22]. As a rough guide to determining whether a polymer dispersion can be regarded as hydrophilic or whether it will give poor wetting of the textile, the critical value of the surface tension is ca. 40 mN/m. Finishing agents containing polymer dispersions with a surface tension of $> 40$ mN/m are less hydrophilic.

A dispersing effect is also shown by carboxyl groups in the polymer derived from copolymerized acrylic, methacrylic, or itaconic acid, and from sulfonate groups in sodium vinylsulfonate. Other components with an important effect on application properties are antifoaming agents and thickeners.

**Reactive Monomers (Functional Monomers).** The fundamental properties of the polymers can be improved by including reactive monomers in the polymer chain such that the reactivity of the reactive groups is still present and utilizable. Polymer dispersions can provide a technical textile with many useful properties if such reactive groups are incorporated.

Permanence properties (wet strength and stability towards water, washing processes, solvents, and dry cleaning) are improved by copolymerization with the reactive monomer $N$-hydroxymethylacrylamide (NHMAM) ($N$-methylolacrylamide, NMAM,) [8.23].

Other commonly used reactive groups include the epoxy group of glycidyl methacrylate and the hydroxyl group of hydroxyethyl acrylate.

**Cross-linking** of the polymers applied to the textile is necessary, especially for applications in the technical sector. This enables swelling of the fiber and shrinkage of the textile to be minimized, and high wet and dry strengths to be achieved. Cross-linking of the polymer can be produced by two methods:

1) Externally, by the addition of a cross-linking agent to the polymer, e.g., a *N*-hydroxy-methylmelamine, partially or fully etherified.
2) Internally, by means of a cross-linking component copolymerized into the polymer (self-cross-linking). The polymer on the fiber cross-links in the presence of acid catalysts and heat.

For example, the effect of an increase in the percentage of an internal cross-linking agent (NHMAM) in a polymer is shown by the following comparison of the properties of two polymer dispersions (pH 3.6) based on a vinyl acetate–ethylene copolymer with a film glass transition temperature of $-25\,^\circ$C applied to a cellulose spunlaced nonwoven: Increasing the amount of reactive monomer in the copolymer from 3 to 4 % increases the dry tear strength from 3.7 daN to 4.7 daN, and the wet tear strength from 1.2 daN to 2.0 daN.

The 1 % increase in the percentage of reactive monomer also reduces the water swelling properties of the film and increases its resistance towards hot water and perchloroethylene.

**Pre-cross-linking.** Pre-cross-linked polymers can be produced from bifunctional monomers (dimethacrylic esters of ethylene glycol or of di- and polyglycols, diallyl phthalate, or divinyl benzene) and tetraallyloxyethane, a reaction product of glyoxal and allyl alcohol, which are copolymerized in small amounts during the production process. Pre-cross-linked polymers are sparingly soluble, difficult to disperse in water, less thermoplastic, and have reduced swelling properties in comparison to chemically analogous but not pre-cross-linked polymers.

**Externally and Internally Plastified and Thickened.** Polymers that form brittle films as homopolymers can be softened by copolymerization with a monomer that gives a soft film when polymerized (internal plasticization).

Subsequent addition to the polymer of a plasticizer such as a dialkyl (dibutyl or dioctyl) phthalate or dioctyl sebacate, e.g., to PVC powder or by mixing 5, 10, or 20 % into a polymer dispersion, is known as external plasticization ($\rightarrow$ Plasticizers). With such a product, care must be taken that the plasticizers do not migrate during use of the finished textile, as these can loosen and weaken contiguous materials or coatings.

In a similar dual approach, a monomer with a thickening effect (e.g., acrylic acid) can be incorporated into a copolymer. If this is made alkaline, thickening occurs, and the viscosity undergoes a large increase due to the formation of ammonium, sodium, or potassium salts. Alternatively, the viscosity can be increased by subsequent addition of a thickening agent.

**Sensitization.** Polymer dispersions are colloidal systems (see $\rightarrow$ Colloids, **A 7**, pp. 361 – 362). They are coagulated by temperature increase, the mechanical effects of shear stress, or the addition of salts, acids, alkalis, etc. ($\rightarrow$ Colloids, **A 7**, pp. 362 – 364). Polymer dispersions are normally stable up to ca. 80 $^\circ$C. A dispersion stable up to this temperature can be rendered unstable by using a different dispersing agent during the production process, and/or by means of certain additives, so that controlled coagulation takes place at 50 – 60 $^\circ$C. The benefit of this sensitization technique is that separation into an aqueous phase and a polymer phase occurs at lower temperatures, so that the water evaporates more readily and the coating dries more rapidly because the formation of a film that can retain water under its surface prevented. Delamination, layer formation, and migration are also prevented.

**Migration** [8.24]. Like dyes, polymers can migrate from aqueous phases to the surface of the textile. Dye and polymer migration can be detected by the formation of a layer with intensification of the color. A given amount of pigment can cause stronger or weaker coloration in polymer dispersions whose chemical compositions are similar but which are differently dispersed. In the case of color intensification, the dye migrates to the surface of the polymer, whereas in the other case it becomes distributed over the entire thickness of the coating.

Heat treatment can cause the binder to become concentrated at one side of the textile surface. Also, treatment with washing liquors or solvents can cause binders to swell and then concentrate at the surface. This surface film formation leads to loss of strength in the interior of the textile structure (e.g., a nonwoven) and to hardening at the surface. Also, soil particles that have been picked up by a solvent, e.g., during dry cleaning, can become visible at the surface.

**Monomers for Direct Polymerization on the Fiber.** For some decades, the possibility of directly polymerizing monomers on the fiber to form coatings and to give fiber bonding has been investigated. However, this has so far not been possible in large-scale textile production, whether by initiation with peroxides, cerium compounds, etc., or by the use of UV, or electron beam radiation, or $\gamma$-rays [8.25]. Recently, renewed attempts have been made to use electron beams in textile technology [8.26]. The combined use of styrene and perfluoroacrylate leads to oleophobic and hydrophobic effects [8.27]. The reasons why direct polymerization has so far not been used in practice are as follows:

1) High costs of equipment, shielding, and energy
2) Incomplete polymerization, leaving residual monomers which are difficult to remove from the textile
3) Difficulties in controlling the process to give reproducible results
4) Toxicological problems with monomers in the open conditions of textile production
5) Need for the replacement of cheap monomers by more expensive special monomers of lower volatility
6) Unwanted coloration and graying effects

**Glass Transition Temperature.** Polymers can be characterized by their glass transition temperature $T_g$ [8.28]. Although this parameter was formerly of very little importance in textile applications, it has since become an aid to grading the very many different types of polymers used in textiles [8.29]. For polymer dispersions, $T_g$ is measured on the film produced by drying [8.30], while powders, fibers, and fiber melts can be measured directly. In the case of coated textiles, an amount of sufficient coating material for a $T_g$ determination can be scraped off. However, it is also possible to carry out a direct determination together with the associated woven or nonwoven substrate [8.31]. A typical $T_g$ curve for a polymer coating is shown in Figure 8.3. As well as the $T_g$ of an applied polymer, the $T_g$ and melting point of the fiber substrate can also be determined. Polymer $T_g$ can also be measured in the presence of organic or inorganic flame retardants or fillers provided that these do not themselves have a glass transition range.

In polymer mixtures, the components can only be differentiated if the glass transition temperature ranges are far apart.

The $T_g$ measurements are of interest because the results for films of polyacrylates, poly(vinyl

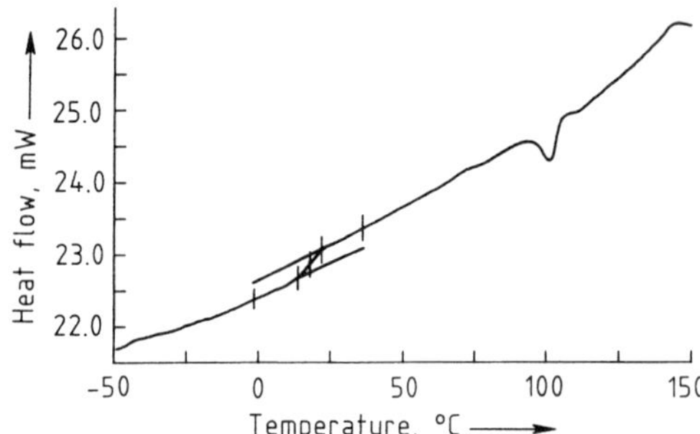

**Figure 8.3.** Glass transition temperature curve ($T_g$ ca. 17 °C) of a methyl methacrylate–ethyl acrylate copolymer dispersion dried on polyester spunbond
Liquor bath 500 g/L, drying 120 °C, curing 5 min at 150 °C.

acetates), analogous copolymers, and styrene and butadiene copolymers show good correlation with the hand of the finished textile. Since low $T_g$ corresponds to soft hand, and vice versa, glass transition temperatures can be ordered to correlate with hand.

For primary or secondary polyurethane dispersions [8.32] ($\rightarrow$ Polyurethanes, **A 21**, pp. 677–680; 706–707), this correlation between hand and $T_g$ does not apply, although a harsher hand is noticeable for very low $T_g$ compared to poly(vinyl acetate) (PVAC) or acrylics. Whereas a polyacrylate dispersion with $T_g = -31$ °C gives a soft hand after impregnation/drying and a soft, elastic, tacky coating, a polyurethane secondary dispersion with the same $T_g$ gives a flexible, relatively hard coating, suitable for coating vertical blinds.

Tackiness in a textile coating means that dust adheres more readily, or unrolling a roll of coated textile is more difficult. For example, a textile coated with a 70/30 styrene/butyl acrylate copolymer dispersion ($T_g$ +50 °C) exhibits no dry soiling in a dust-containing air after 7 d open storage, while a 70/30 butyl acrylate/styrene ($T_g$ −8) coating leads to significant soiling of the textile in a dust-containing atmosphere after 7 d. The $T_g$ also correlates with soiling behavior.

Glass transition temperatures and minimum film-forming temperatures of polymer dispersions are listed in Table 8.4 [8.33].

Various polymers (e.g., those with perfluoro groups) [8.34] and certain surfactants have melting points, and are therefore at least partially crystalline. A nonylphenol ethoxylate with 30 ethylene oxide units has a melting point of ca. 50 °C, measured by DSC. This melting point can

**Table 8.4.** Minimum film-forming temperature (MFT, white point temperature) according to DIN 53787 and glass transition temperature ($T_g$) of polymer dispersions measured by DSC

| Dispersion film | Use | MFT, °C | $T_g$, °C |
|---|---|---|---|
| Polystyrene | stiffening shoe toecaps | >100 | 98 |
| Butadiene–styrene–acrylonitirile copolymer | moldable floor coverings | 90 | 112; −15* |
| Methyl methacrylate–acrylate copolymer, pre-cross-linked | hard finish | 90 | 103 |
| Acrylic acid copolymer | thickener | 50 | 65 |
| Styrene–butyl acrylate copolymer, self-cross-linking | glass textile finish | 60 | 50 |
| Perfluoroacrylate polymer | oleophobizing | | 47; *mp* 74.5 |
| Acrylate copolymer, self-cross-linking | nonwoven binder | 29 | 38 |
| Poly(vinyl acetate) | stiffening agent | 15 | 33 |
| Acrylate copolymer, self-cross-linking | nonwoven binder | 10 | 24 |
| Vinyl acetate–dibutyl maleate copolymer | handle modifier | 12 | 19 |
| Vinyl acetate–ethylene copolymer | carpet backing | 0 | 5 |
| Vinyl acetate–ethylene copolymer | carpet backing | 0 | −6 |
| Butyl acrylate–styrene copolymer, self-cross-linking | glass textile finish | 0 | −8 |
| Vinyl acetate–butyl acrylate copolymer, self-cross-linking | pigment binder | 0 | −12; *mp* 44** |
| Polyurethane | pigment binder | 0 | −30; *mp* 52 |

* Polymer mixture. ** Caused by surfactant; no longer present after heating twice.

also be measured when the surfactant is used as a dispersant (Fig. 8.4).

No correlation is observed between the thermal resistance of a textile material and the $T_g$ with which it is finished. Thus, while one compression molded textile treated with a polymer with a $T_g$ of, e.g., 100 °C, may have a stable shape when the textile is later heat treated at 100 °C, another polymer of different chemical composi-

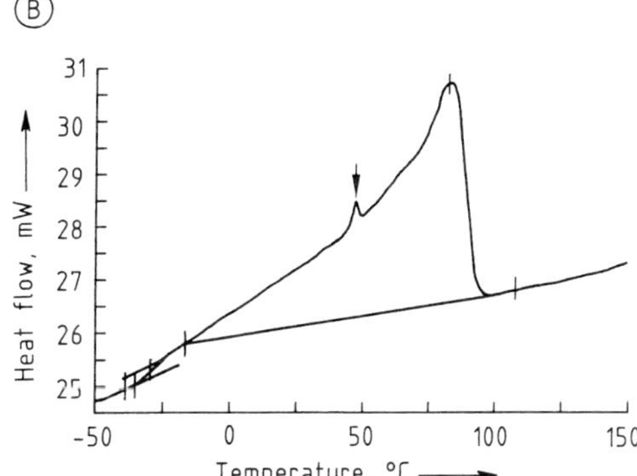

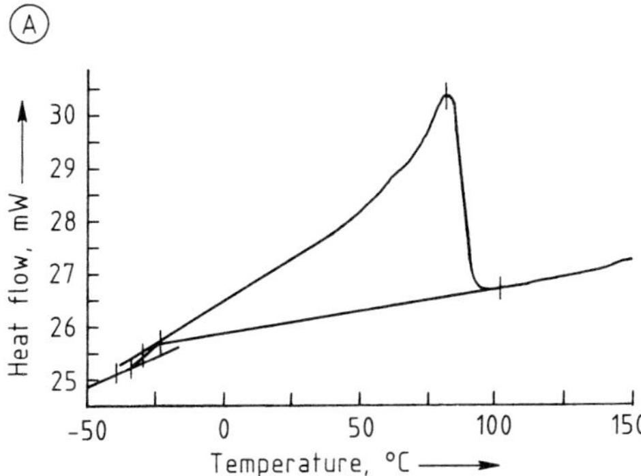

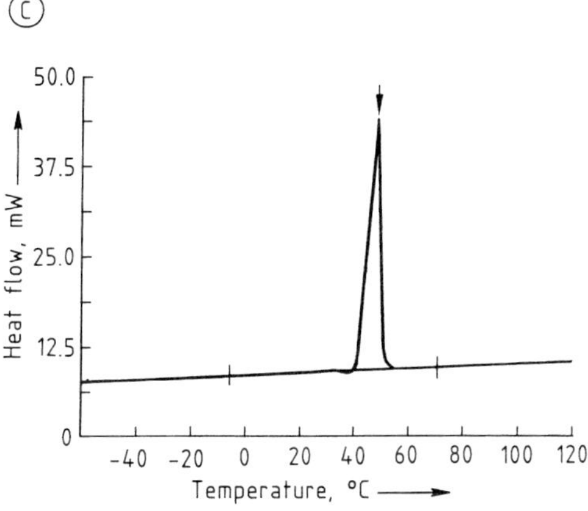

**Figure 8.4.** Glass transition temperature curve of an ethylene–vinyl acetate powder with 2% surfactant (nonylphenol ethoxylate, 30 EO)
A) Polymer without surfactant: $T_g$ − 29.5 °C, peak 81.7 °C; B) Polymer with surfactant on PES spunbond: $T_g$ − 30.6 °C, peaks 47.2 °C and 83.0 °C (PES); C) Surfactant without polymer: peak 49.1 °C

tion but with the same $T_g$ may not give the same thermal stability when applied to the same textile material.

**Uses of polymer dispersions** are as follows:

| | |
|---|---|
| Textile floor coverings | backing of woven and tufted carpets, secondary adhesion backing of tufted carpets impregnation and backing of needle punch |
| Nonwovens | |
|   Nonwoven wadding | spray stiffening |
|   Flat nonwoven | impregnation |
|   Needle felt | coating |
|   Spunbond | bonding |
|   Spunlaced | lamination |
|   Melt blown | heat sealing thermobonding flocking pressing printing |
| Glass textiles | |
|   Woven and nonwoven lattice | impregnation coating finishing |
|   Glass fibers | sizing |
| Woven, knitted fabrics, lattices | impregnation coating finishing printing flocking bonding, laminating hot sealing metallization |
| Ropes, nets | impregnation |
| Deposited materials | coating |
| Textiles bonded to paper, foil, plastics, metal | bonding, adhesion metallization |
| Loose stock | spray stiffening |
| Recycled fibers | impregnation |
| Mineral wool | |
| Flock | bonding in coating |

Binders include nonwoven binders [8.35], pigment, printing binders [8.36], or dye binders, flocking binders [8.37], and binders for abrasives. Coating materials, either smooth or foamed, are used as carpet backings [8.38] or for coating textiles [8.39]. Adhesives [8.40] are used to bond textiles or coated textiles to other materials, or are termed flocking adhesives when they bind flock in the coating materials.

**Technical effects** achieved by auxiliaries include:

1) Surface effects, e.g., gloss, luster, roughness, oil, water, and soil repellency, surface strengthening of waddings, coatings, antipilling, and adhesive effects
2) Penetration of full bath impregnation to give through hardening of moldings, stiffness, water impermeability, and prevention of delamination, etc.
3) Incorporation of fibers or textiles into polymers in the production of fiber-reinforced materials.

As an example, the effect of using various polymer dispersions on various fiber materials can be seen in Table 8.5.

**Suitable Products for Various Technical Textiles.** A list of selected technical textiles and suitable polymer dispersions is given in the following:

| | |
|---|---|
| Polyester spunbond for bituminized roofing | hard thermally stable film formers based on acrylic ester – methacrylic ester or acrylic ester – styrene and/ or acrylonitrile |

**Table 8.5.** Comparative effects of polymer dispersions on the technical textiles polyester spunbond and woven glass textile used for bituminized roofing material (application conditions: padding, drying at 120 °C, curing for 3 min at 170 °C)

| Formulation | $T_g$, °C | Amount | Glass textile | | | Spunbond | | |
|---|---|---|---|---|---|---|---|---|
| | | | Transverse tearing strength, N | Transverse elongation, % | 24-h absorption height, cm | Transverse tearing strength, N | Transverse elongation, % | 24-h absorption height, cm |
| Vinyl acetate – dibutyl maleate copolymer dispersion, 55 %* | 19 | 200 g/L | 1230 | 1.7 | 6.5 | 300 | 50.2 | 0 |
| Starch derivative | | 40 g/L | | | | | | |
| Acrylic ester – styrene copolymer dispersion, 50 %** | 28 | 300 g/L | 890 | 1.7 | 10.5 | 326 | 47.5 | 0 |
| Untreated | | | no stability | no stability | 7.0 | 225 | 770 | 0 |

* Typical formulation for glass textile without hydrophobic agent.    ** Typical formulation for spunbond.

Glass woven textile for bituminized roofing — hard film formers based on vinyl acetate or styrene, methacrylic ester/acrylic ester

Webbing belts — soft film formers based on vinyl acetate–ethylene or acrylic ester

Automobile carpets for final powder coating and molding — medium-hardness film formers based on vinyl acetate–ethylene

Backing for automobile carpets — medium-hardness film formers based on butadiene–styrene

Adhesion — vinyl acetate homopolymers, vinyl acetate copolymers with ethylene or acrylic esters, styrene–butadiene

Pressure-sensitive adhesion — vinyl acetate–ethylene or acrylic ester copolymers with very low $T_g$

Technical articles made by flocking — soft film formers based on acrylic esters, acrylic ester–acrylonitrile, butadiene–acrylonitrile, sometimes with added melamine cross-linking agent

Glass textile wall-coverings — tough elastic film formers based on vinyl acetate–ethylene with flame-proofing components

Glass textile structural lattice — medium hardness film formers based on styrene–acrylic ester or vinyl acetate–ethylene

Milk filters — soft film formers based on acrylic ester or vinyl acetate–ethylene

Technical nonwovens — medium hardness film formers based on vinyl acetate–ethylene, acrylic ester, or acrylic ester–styrene

Hard coating of technical woven textiles — styrene–acrylic ester copolymers, methyl methacrylate ester copolymers

Wadding, quilting — hard film formers and medium hardness binders with no surface tack based on acrylic/methacrylic ester and vinyl acetate–acrylic ester or vinyl acetate–ethylene

Nonwovens for cleaning purposes — soft film formers based on vinyl acetate–ethylene or acrylic ester

Needle felt moldable for automobiles — hard film formers based on butadiene–styrene–acrylonitrile or styrene–acrylic ester

Technical knitted goods — soft film formers based on vinyl acetate–ethylene

Back coating of technical warp-knitted plush — soft film formers based on acrylic ester, acrylic ester–styrene, vinyl acetate–acrylic ester, vinyl acetate–ethylene

Nonwoven filters — soft film formers with surface tack based on vinyl acetate–ethylene, acrylic ester or butadiene–styrene

Sails — combination of hard acrylic/methacrylic ester binders with melamine cross-linking agents

Some practical formulations (each case, application of the auxiliary is followed by drying and curing) are as follows:

Sizing formulation for abrasive fabrics —
4% poly(vinyl alcohol), partly hydrolyzed
4% carboxymethylcellulose, low viscosity
0.3% size plasticizer
0.1% lubricant

Thermobonded polyester nonwoven, slop padded with a rapidly breaking dispersion foam in the nip roller process — 100 parts polyacrylate dispersion 50%, $T_g$ ca. $-35\,°C$

Impregnation of a polyester spunbond to give a hard product for bitumen coating — 500 g/L polyacrylate dispersion 50%, $T_g$ ca. $30\,°C$

Cleaning cloth with soft hand of viscose nonwoven, nip roller process — 100 g/L ethylene–vinyl acetate copolymer dispersion 50%, $T_g\ -18\,°C$

A fiber blend needle felt, nip-rolled, dried, and molded in a press at $170\,°C$ (hard molded articles for automobiles) — 500 g/L copolymer dispersion mixture based on styrene–butadiene–acrylonitrile 50%, $T_g$ ca. $112\,°C$ and $-17\,°C$

Hardening of nonwoven filter of polyester fibers by spraying the binder (hard bonding points are formed) — 250 g/L copolymer dispersion based on vinyl acetate–butyl acrylate 45%, $T_g$ ca. $32\,°C$

Spray on an high-volume nonwoven — 160 g/L polyacrylate dispersion 60%, with $T_g\ 40\,°C$
40 g/L polyacrylate dispersion 60%, with $T_g\ 12\,°C$

Coating for a cotton textile, medium hard hand — 75 parts polyacrylate dispersion 45%, $T_g$ ca. $25\,°C$
25 parts water
15 parts titanium dioxide
0.1 parts pigment
2.5 parts thickener 30%, viscosity 6000 mPa · s

| Soft-elastic coating on woven fabric | as above, with polyacrylate dispersion 50%, $T_g$ − 9 °C |
|---|---|
| Impregnation of vertical blind of Trevira CS, stiffening without reduction of flame-retardant properties | 500 g/L polyurethane secondary dispersion 35%, $T_g$ ca. − 31 °C |
| Coating a fabric for later flocking | 100 parts polyacrylate dispersion 45%, $T_g$ − 15 °C<br>0.2 parts antifoaming agent<br>8 parts N-methoxymethyl-melamine cross-linking agent 50%<br>0.8 parts amine hydrochloride catalyst<br>4.5 parts thickener 30% ammonia to adjust pH to ca. 8, viscosity ca. 6000 mPa · s |
| Finishing of glass textile wall coverings, impregnation | 300 g/L copolymer based on vinyl acetate/ ethylene/vinyl chloride 50%, $T_g$ 15 °C<br>50 g/L modified starch<br>25 g/L flame retardant, e.g., organic bromine compound |
| PVC coating with plastisol on polyester nonwoven | 100 parts PVC powder<br>65 parts plasticizer dioctyl phthalate<br>3 parts coplasticizer<br>1.5 parts stabilizer<br>10 parts chalk<br>5 parts titanium dioxide |
| Hot sealing [8.41]: sprinkling with a sifter, sintering, pressing for 15 s, at 5 bar and 116 °C in the sealing gap, 127 °C in the press | 100 parts copolymer powder based on ethylene–vinyl acetate, mp 87 °C |

**Polymer powders and hot melts** are characterized by their melting behavior (melting point, glass transition temperature, melt index, and melt viscosity; see Table 8.6), adhesive properties, resistance to water and solvents, and the high-temperature strength of adhesive bonds. See → Adhesives, **A 1**, p. 228.

A high melt flow index indicates that the polymer has a low melt viscosity, and melts more readily at a low temperature. The polymer powders used as hot melt adhesives [8.42], [8.44] give a strong adhesive bond which depends on the strength of the textile, the strength of the interface between the textile and the polymer adhesive (adhesion), and the cohesive properties of the film of adhesive (cohesion). It is also dependent on the method of applying the adhesive and the thickness of the textile. Adhesive bonds are also formed between a plastic film (PVC, polyurethane, polyester) and a textile.

Polymer powders are available in various particle sizes: 0–80, 80–200, 200–300, and 300–500 µm.

The polymers with the largest market share in Western Europe as hot melt adhesives are those based on ethylene–vinyl acetate.

Powders in aqueous suspension are also used [8.42]. The process used to disperse the large polymer particles must be carefully chosen, as surfactant-type dispersants can weaken the bond to the textile. This is also true of polymer powders that have been given into polymer dispersions [8.41]. If a powder and a polymer dispersion are used together, the polymer dispersion

**Table 8.6.** Hot melt adhesives of various chemical types

| Composition | | mp | DSC measurements | | Melt flow index (MFI) [8.43], 160 °C/2.16 kp, g/10 min |
|---|---|---|---|---|---|
| | | | Peak*, °C | $T_g$, °C | |
| Polyethylene (LDPE) | | 107 | 109 | | 22 (190 °C) |
| Polyethylene/vinyl acetate, 3% VAC | | 102 | 104<br>104** | | 1.1 |
| | 19% VAC | 87 | | − 30 | 6.3 |
| | 28% VAC | 70 | 52<br>62** | − 30<br>− 34** | 14.5 |
| Polyamide | | 122 | 103<br>77<br>−** | <br><br>+ 23** | 16 |
| Polyamide | | 135 | 127**<br>73** | <br>+ 21** | 16 |
| Polytetrafluoroethylene | | 370–375 | | | |
| Tetrafluoroethylene copolymer | | 350–370 | | | |

* DSC after first heating.   ** DSC after second heating.

can help to give uniform distribution, and the cohesive strength of two so-bonded fabrics is increased.

*Example:* The cohesive strength at 120 °C of a paste coating (100 g/m$^2$) with polyacrylate dispersion, thickener, and alkenesulfonate wetting agent, pressed at 170 °C and 25 bar [8.45] is as follows:

Polymer disperion 50%, $T_g$ 18 °C     187 N/m
  (100 parts)

Polymer dispersion (100 parts)     377 N/m
  + copolyamide powder (30 parts)

Copolyamide powder, $T_g$ 21 °C,     430 N/m
  *mp* 127 °C; 73 °C (100 parts)

Hot melt fibers for textiles are described in [8.46].

A list of *trade names* is given below:

Platamid, Platabond, Rilsan (Atochem, FRG), Levasint (Bayer, FRG), Abifor (Billeter, CH), Riblene (Enichem, I), Isatherm (Fuller, USA), Griltex (Ems, CH), Guttacoll-Klebefolien (Gutacoll, FRG), Hostalen, Hostalit, Hostaflon, Synthacryl (Hoechst), Coathylene (Hoechst-Celanese/Plastlabor, CH), Mybond (Mydrin, GB), Vestamelt, Vestoplast (Hüls, FRG), Permatex (Permatex, FRG), Schaettifix (Schaetti, CH), Sarpifan, Estekoll (Stockhausen, FRG)

**Thermoplastics, Thermosets, and Elastomers** [8.47]. Many polymers applied to textiles have thermoplastic properties. They range from soft elastic coatings ($T_g$ ca. − 15 °C) up to articles with hot strength (e.g., automobile fittings) and $T_g$ ca. 100 °C. Melamine, urea, phenol–formaldehyde precondensates, and epoxy resins are, in contrast, thermosets [8.48].

Phenol–formaldehyde [8.49] and melamine–formaldehyde precondensates are used in powder form or in an aqueous medium to produce shaped, high-strength, heat stable, press molded articles made of textile fibers for the automobile [8.50]. The precondensates, phenol–formaldehyde novolacs, harden at 170–180 °C on the textile to give resin structures. Melamine–formaldehyde precondensates can be used either alone or together with polymer dispersions, but phenol–formaldehyde precondensates are often used alone. Glass transition temperatures in the range 60–90 °C can be measured in the precondensates, but not after the resin has been formed.

Polytetrafluoroethylene is neither a thermoset, nor a thermoplastic, and its stability at high temperatures can be utilized, e.g., in combination with glass fibers. Some thermoplasticity can be obtained by copolymerization [8.51].

**Starch** is also suitable for technical textiles. It is a macromolecular carbohydrate with a high decomposition temperature (> 230 °C) and thermoset properties, stabilizing at high temperatures, having a hardening and thickening effect, and is biologically degradable and compatible with other auxiliaries. Starch and modified starches are therefore standard constituents, e.g., for finishes for glass textile wall coverings, to produce stiffening effects, and to reduce thermoplasticity, often in combination with polymer dispersions [8.52].

**Pretreatment.** Pretreatment processes to optimize textile properties for technical use include not only the classical processes of mercerization, scouring, bleaching, and washing, but also various methods of modifying the surface of fibers [8.53] such as roughening (abrasion [8.54]), treatment of polyester fibers with alkali, etching, treatment with affinity promotors (e.g., for tire cord [8.55]) for fiber-reinforced plastics and composites [8.9], and glass fibers and for metallization [8.56]. Other surface modifications include coatings or sizings for glass fibers [8.57], corona treatment [8.58], plasma treatment [8.59], laser treatment [8.60], irradiation [8.26], and the reaction of fibers with reactive products, e.g., 2-aminoethylsulfuric acid [8.61], which reacts with the hydroxyl groups of cellulose.

The penetration of the finishing agent into the textile is increased by these pretreatment processes, as is the adhesiveness of the bond between the fiber and the dried/fixed auxiliary.

Finally, appropriate pretreatment and selection of textiles is needed to give compatibility with a coating applied by spraying on the back of the textile [8.62], e.g., in an automobile.

The following materials can be used either as independent finishing agents and auxiliaries or as additives [8.63], especially to polymer dispersions [8.64].

### 8.2.2. Cross-Linking Agents

Melamine cross-linking agents [8.65] are preferably used with polymer dispersions. They are produced by hydroxymethylation of melamine in the presence of alkaline or acidic catalysts followed by etherification, e.g., with methanol. The resulting precondensates are cross-linking agents that cure by polycondensation with liberation of water in aqueous solution or in powder form on the textile in the presence of acidic catalysts at elevated temperatures. Precondensates in the form of aqueous solutions are used as the main components of finishing agents for sail fabrics together with a hydrophobic or oleophobic agent, or are used with poly(vinyl alcohols) for stiffening of textiles.

Melamine cross-linking agents in powder form, also water-soluble, have glass transition temperatures between 55 and 70 °C. With some products, the DSC curves show melting peaks in addition to the glass transition. After curing on the textile, neither $T_g$s nor melting peaks are present.

On prolonged storage in aqueous solution melamine cross-linking agents undergo continuous chemical condensation with slow reduction of pH to below the neutral point, especially if they have a low degree of etherification and are therefore very reactive. This can lead to the formation of precipitates at higher ambient temperatures.

These products give solvent-resistant bonds after curing or molding processes.

When melamine cross-linking agents are added to polymer dispersions, they give improved resistance to water, washing, perchloroethylene, and solvents, increased strength and thermal stability, and reduced shrinkage. Alone it has a flame retardant effect due to the nitrogen content.

*Example:* N-methoxymethylmelamine cross-linking agent, partially etherified, in aqueous solution, with a binder, $T_g$ +40 °C, based on acrylic and methacrylic esters on polyester spunbond (180 g/m$^2$). Application by padding, dried at 120 °C, cured for 3 min at 150 °C. Dry addon: ca. 25 % (500 g/L binder 50 %, 20 g/L melamine crosslinking agent 75 %, 5 g/L catalyst). Properties were as follows (values with cross-linking agent in parentheses):

| | |
|---|---|
| Breaking strength at room temperature, N | 585 (594) |
| Elongation at room temperature, % | 39.5 (36.8) |
| Breaking strength at 180 °C, N | 266 (234) |
| Elongation at 180 °C and 100 N load, % | 14.8 (12.3) |
| Absorption height, cm | 0.9 (3.7) |

This example illustrates the reduction in hot strength and hot tensile elongation as a consequence of hardening by the cross-linking agent, and the reduction in the hydrophobic behavior of the binder.

The cross-linking agents used in resin finishes for classical textiles (pp. 302–305) are also used for technical textiles. Internal and external cross-linking are described on p. 321.

**Catalysts** accelerate external and internal cross-links. They are mostly acidic catalysts, i.e., salts such as amine hydrochlorides, magnesium chloride, or ammonium dihydrogenphosphate, and have good solubility in water and aqueous products.

Heavy metal salts such as zinc nitrate or zinc chloride are very effective, but are not used for environmental reasons. Acids such as phosphoric, oxalic, maleic, and citric acid are also highly effective. The low pH values of many polymer dispersions, resulting from their production, are suitable for cross-linking.

Catalysts impair the effect of thickeners.

If polymer dispersions which have been made alkaline (e.g., SBR and NBR dispersions) are used with cross-linking agents with an acid reaction and the appropriate catalysts, the extent of cross-linking, which takes place in the alkaline range, may be decreased; this can be compensated for by increasing the amount of catalyst.

### 8.2.3. Nonslip Agents

One of the main aims in the manufacture of technical textiles is to achieve high mechanical stability, which may not be provided by the fiber material alone. An important aspect of mechanical stability is the prevention of the movement of fibers, fiber systems, or yarns within the textile structure. Nonslip (antisnag) agents are also said to increase the seam strength of the material [8.66].

The stabilization of the textile is mainly due to the prevention of movement at points where threads meet or interlace.

Smooth fibers and open textile structures are the main causes of slipping. Knitted products made of smooth fibers or filaments, woven textiles, lattice woven materials, layered yarn structures, nets, and ropes are finished to give slip resistance.

Nonslip agents for technical textiles include

1) Polymers: polymer dispersions, natural resins, PVC plastisols, polymer powders, poly(vinyl alcohol), starch, and starch derivatives
2) Silica and silica derivatives
3) Metal oxides

Two fundamental principles are utilized: Formation of a film on the fibers that reduces movement of individual fibers, warp threads, and yarn at points of interlacing, or production of a rough surface by silica or metal oxides.

**Polymer Dispersions and Plastisols.** Polymer dispersions with interior cross-linking are less effective against slipping than their chemical analogues without reactive groups. Poly(vinyl alcohol) ($T_g$ ca. 70 °C), poly(vinyl acetate) dispersions ($T_g$ 30–35 °C), or copolymer dispersions with $T_g$ between 15 and +30 °C (hard-film formers) are suitable.

To give a softer hand, polymer dispersions with a $T_g$ range of −5 to −20 °C are used. As a general rule of thumb (although there are exceptions), lower $T_g$ values corespond to softer films and lower slip resistance. In spite of the stronger adhesion of pressure-sensitive adhesives, which form very soft films, these provide very little slip resistance. Moreover, soft films interfere with subsequent calendering processes by causing adhesion, and if many textile layers are stored under their own heavy weight, they can block. This can be avoided by using a harder film former, which improves slip resistance, but often makes the hand undesirably harsher. The addition of cross-linking agents such as *N*-methoxymethyl-melamine compounds to polymer dispersions does not improve slip resistance.

PVC plastisols are used for strengthening lattice woven materials and layered yarn structures because of their softness, elasticity even at relatively low temperatures, and superior cold fracture resistance compared with the above-mentioned polymer dispersion.

Slip can also be reduced by polyurethane dispersions combined with vinyl acetate–ethylene copolymer dispersions.

**Silica and its Derivatives.** Products based on silica are the most typical textile nonslip agents. Silica products dry to form a brittle, powdery material. They are often combined with other textile auxiliaries, especially polymer dispersions, to improve washing stability. When using silica products, their compatibility with other auxiliaries must be tested, as, although the cationic silica nonslip agents are often very effective, many polymer dispersions have anionic character, so that precipitates are likely to be formed.

Some commercial silica-based nonslip agents follow:

| | |
|---|---|
| Aerosil | Degussa, Germany |
| Burasil | Baur Gaebel, Germany |
| Durasol | Henkel, Germany |
| Durifan | Thor, Germany |
| Feran | Rudolf, Germany |
| Fixan | Rotta, Germany |
| Fornax | Pfersee, Germany |
| Hispasil | Hispaño Quimica, Spain |
| Lurapret | BASF, Germany |
| Schiebefest | Dr. Th. Böhme, Germany |
| Sevofast | Textilcolor, Switzerland |
| Silene | Auschem, Italy |
| Syntharesin | Bayer, Germany |
| Sytex | Bozetto, Italy |
| Tubicoat | Chem. Fabrik Tübingen, Germany |

**Structure Combinations.** Very good slip resistance can be produced by bonding or sealing a very thin film, e.g., of polyethylene, to the textile. This allows the textile to stay soft while maximizing slip resistance. Nonwovens are made slip resistant by thermobonding in a thermo-calander.

In bonded fiber products, the slip resistance is a result of the direct bond between the fiber and the polymer produced by pressing.

Slip resistance is often required only temporarily, e.g., if a somewhat unstable material has to withstand mechanical stresses in the course of further processing. For example, a woven glass fiber for use as a wall covering can be impregnated, dried, and cured, so that it will remain stable while in contact with wallpaper paste during application to the wall.

### 8.2.4. Water Repellents [8.67]

Apart from the water repellents used in classical textile finishing (see Section 7.3), ammonium stearate is used as a hydrophobic component of polymer dispersion coating pastes. The pH must be controlled to prevent precipitation of sparingly soluble stearic acid in acid media.

### 8.2.5. Oleophobic Agents

Oil repellency is usually associated with hydrophobicity and soil repellency. Oil, water, and soil repellency are very desirable effects in technical textiles [8.68]. The ability to repel oils, mineral oils, naphtha, diesel fuel, and hydrocarbons, is achieved by the presence of perfluoro groups in polymers.

In principle, an "open" textile can be made oil-proof by sealing it with an oil-repellent fluorine-free polymer that does not swell in oil, e.g., a hydrophilic polymer such as poly(vinyl alcohol). Especially high oil repellency is given by polymer coating with perfluoroalkyl groups ($\geq C_6$). They are usually applied as dispersions and are generally used in combination with polymer dispersions that do not contain perfluoroalkyl groups.

*Example:* A thin, oil-repellent polyester nonwoven is obtained by impregnating (padding process) with a liquor of the following composition:

| | |
|---|---|
| Vinyl acetate–ethylene copolymer dispersion, 50 % ($T_g - 18\,°C$) | 200 g/L |
| Oleophobic agent based on a perfluoromethacrylate, fluorine content 16.5 % ($T_g$ 47 °C, *mp* 75 °C) | 10 g/L |
| Melamine cross-linking agent, 50 % | 25 g/L |
| Amine hydrochloride catalyst | 5 g/L |

Dried at 120 °C and cured for 3 min at 150 °C

Unlike the silicone hydrophobic agents, there are oleophobic–hydrophobic finishes that do not affect the bonding of a hot-sealed solid polymer subsequently applied.

### 8.2.6. Flame Retardants ($\to$ Flame Retardants)

Low-flammability fibers such as aramid, polyester (Trevira CS), acrylonitrile copolymers, PVC, glass, and other inorganic fibers, lose some flame-resistance when treated with finishing agents such as polymer dispersions, hydrophobic agents, and softeners. Some auxiliaries do not affect the flame resistance if applied sparingly. For example, combinations of melamine cross-linking agents and polymer dispersions containing flame retardants are used for finishing glass textile wall coverings.

In addition to giving the flame-proofing effect required regulations [8.69], flame retardants should also exhibit stability towards water,

washing, wiping, chemical cleaning, heat, and light. The following product groups are used for technical textiles:

1) Halogen-containing compounds: PVC pastes, chlorinated rubber, poly(vinyl chloride) and poly(vinylidine chloride) dispersions, vinyl acetate–ethylene–vinyl chloride copolymers, chlorinated alkanes in combination with antimony trioxide, tris(dichloropropyl) phosphate, organic bromine compounds, bromides
2) Inorganic and organic phosphorus compounds: ammonium dihydrogenphosphate

*Examples:* flame-proofed viscose fiber nonwoven:

| | |
|---|---|
| Impregnation with polymer dispersion based on vinyl acetate–ethylene, 50 % ($T_g - 15\,°C$) | 70 parts |
| Diammonium hydrogenphosphate | 30 parts |
| Water | 100 parts |

Guanidine phosphate and ammonium polyphosphate are also combined with polymer dispersions.

3) Nitrogen compounds: polyurethane dispersions for coatings, melamine cross-linking agents
4) Aluminum hydroxide as a filler
5) Microencapsulation of flame-proofing agents [8.70]

### 8.2.7. Antistatic Agents

For technical textiles, antistatic agents are mainly used in combination with polymer dispersions. Table 8.7 gives a brief overview of the antistatic behavior of typical materials used in a coating paste for a carpet backing. The use of

**Table 8.7.** Effectiveness of various antielectrostatic additives to a polymer dispersion applied on a tufted carpet backing Formulation: copolymer dispersion (vinyl acetate–ethylene based, $T_g$ 5 °C), 100 parts, polyacrylate thickener 2 parts, viscosity of coating paste ca. 5000 mPa · s, drying temperature of the carpet 120 °C; dried to form a film at room temperature.

| Coating | Content of antistatic agent (100 %) in coating, % | Surface electrical resistance, $\Omega$ | |
|---|---|---|---|
| | | Film | Carpet |
| Without antistatic agent | | $\times 10^9$ | $2 \times 10^{11}$ |
| With stainless steel fibers (Bekinox LT)[a] | 1 | $< 10^7$ | $< 10^7$ |
| With carbon dispersion, anionic[b] | 2 | $6 \times 10^8$ | $12 \times 10^9$ |
| With potassium formate | 2 | $< 10^7$ | $3 \times 10^9$ |
| With anionic surfactant: secondary alkyl sulfonate[c] | 2 | $< 10^7$ | $3 \times 10^8$ |
| With nonionic surfactant: fatty alcohol 13 EO[c] | 2 | $2 \times 10^8$ | $4 \times 10^{10}$ |
| With antistatic agent Tallopol[d] | 2 | $< 10^7$ | $\times 10^9$ |

[a] Bekaert Deutschland, Bad Homburg, FRG.   [b] Degussa, Kalscheuren, FRG.   [c] Hoechst, Frankfurt/Main, FRG.
[d] Stockhausen, Krefeld, FRG.

**Table 8.8.** Antistatic and hydrophobic properties of polypropylene (PP) and polyamide (PA) spunbonds impregnated by padding (1 bar); polypropylene dried at 120 °C; polyamide dried at 120 °C, cured for 2 min at 150 °C

| Formulation | Surface resistivity at 25% R.H., $\Omega$ | | Water column, first drop, mm | | Oleophobicity (AATCC Test 118, 1966) | |
|---|---|---|---|---|---|---|
| | PP | PA | PP | PA | PP | PA |
| Untreated fabric (basis weight 35 g/m²) | $2 \times 10^{13}$ | $2 \times 10^{13}$ | 30 | 10 | 0 | 0 |
| 20 g/L antistatic agent (solids content 42%) | $10^8$ | $6 \times 10^7$ | 10 | 0 | 0 | 0 |
| 20 g/L antistatic agent 20 g/L oleophobic agent (solids content 27%), F content 16% | $2 \times 10^8$ | $10^9$ | 10 | 30 | 6 | 5 |
| 20 g/L antistatic agent 20 g/L oleophobic agent 50 g/L polymer dispersion, styrene–acrylate, 50%, $T_g$ +28 °C | $10^8$ | $3 \times 10^8$ | 15 | 50 | 6 | 2 |
| 50 g/L polymer dispersion, $T_g$ +28 °C | $10^{13}$ | $3 \times 10^{13}$ | 30 | 20 | 0 | 0 |
| 20 g/L oleophobic agent | $2 \times 10^{12}$ | $10^{13}$ | 55 | 60 | 0 | 0 |

special antistatic fibers is described in the literature [8.71].

The electrostatic charging correlates with the hydrophobic behavior of a material. Typical auxiliaries for producing antielectrostatic finishes are hydrophilic: surfactants, salts, and carbon-black suspensions (which contain a high level of dispersing agents). This effect counteracts the often desired hydrophobicity of the textile. In some cases, metallic conductivity due to metal fibers or metal powder can be used, provided that these are not completely coated with insulating materials. The example of polypropylene spunbond in Table 8.8 demonstrates the difficulty of obtaining good hydrophobicity of a relatively open structure when small amounts of auxiliary are used, while maintaining the character of the textile.

### 8.2.8. Plasticizers (see also Section 7.2.5)

Unlike polymer plasticizers, textile softeners, typically act on the fibers. They modify the hand of a yarn or a woven or knitted material, e.g., from a rough finish to a smooth, flowing, full-bodied, or "crunch" hand.

Problems can be caused by migration of the plasticizer. The migration of plasticizers insoluble in water can loosen applied coatings and soft-en the material. Loss of strength can occur. In the case of a carpet backing coating that contains a plasticizer that migrates readily, this could reach a PVC underlay.

Barrier layers of softener-resistant polymers can help to prevent softeners migrating to the surface.

As plasticizers can have varying degrees of volatility, and can be emitted from the polymer, they can become deposited on car windscreens (fogging) [8.72].

PVC powders used for floor coverings are formulated as pastes [8.19]. A typical formulation (parts by weight) is as follows:

| | |
|---|---|
| PVC powder | 100 |
| Plasticizer (e.g., dioctyl phthalate) | 67 |
| Epoxy plasticizer | 3 |
| Stabilizer | 2 |
| Chalk | 10 |
| Titanium dioxide | 5 |

### 8.2.9. Coupling Agents

A brief overview of coupling agents is given below:

1) Applied to glass fibers to increase the affinity to organic polymers      aminopolysiloxanes

| 2) | Applied to tire cord (viscose) to improve bonding to the rubber | resorcinol – formaldehyde condensation products together with latex or resorcinol with formaldehyde-releasing substances (hexamethylenetetramine or *N*-hydroxymethylmelamines) |
|---|---|---|
| 3) | Applied to polyester and aramid | pretreatment with isocyanate or poly-epoxide |
| 4) | Applied to nonwovens to improve the bond to resins for glass fiber nonwovens | copolymer dispersions based on acrylic esters, cationically emulsified polymers or polymer dispersions with cationic modification of the polymer |

## 8.2.10. Thickeners

The application of auxiliaries and their subsequent effect on the technical textile are strongly influenced by viscosity. High viscosities lead to increased rates of application to the surface and give sharp outlines while low viscosities give good penetration. This is true for the application of the auxiliaries both "flat" or as a foam. The auxiliaries themselves can be of low, medium, or high viscosity, pourable easily or with difficulty, or in paste form. They are often diluted with water before application. Variation of the viscosity is therefore an important method of ensuring that a material can be satisfactorily applied to technical textiles. Practical viscosity ranges (mPa · s) are as follows:

| | |
|---|---|
| Low viscosity | 20–600 |
| Medium viscosity | >600 |
| High viscosity | >6000 |
| Application process | |
| Padding (foularding) | <1000 |
| Nip roller | ca. 2000 |
| Coating | ≥ca. 6000 |
| Printing | >10 000 |

The viscosity of polymer dispersions depends on:

| 1) | The polymer/copolymer | increase in molecular mass of the polymer correlates with increased viscosity |
|---|---|---|

| 2) | Content of reactive monomers in the copolymer | *N*-hydroxymethylacrylamide has a greater tendency to increase viscosity than *N*-methoxymethylacrylamide Carboxyl groups greatly increase viscosity due to salt formation |
|---|---|---|
| 3) | Protective colloids (polyvinyl alcohol, cellulose ethers) | increase the viscosity |
| 4) | Plasticizers | increase the viscosity |
| 5) | Fillers | increase the viscosity at high concentrations |
| 6) | Incompatibility between the polymer dispersion and the additive before the onset of coagulation | different ionicity (anionic/cationic); low shear stability; low tolerance to salt; incompatibility with thickeners |

In the production of coatings, thickeners based on polyacrylic acid are mainly used as low-viscosity dispersions thickened by alkali addition (usually ammonia), or are used as high-viscosity solutions of the ammonium or alkali metal salt. The so-called inverse thickeners based on polyacrylic acid have higher activity per unit solids content [8.73]. Acidic polymer dispersions are usually adjusted to pH > 7 by the addition of ammonia before adding the thickener. With alkaline dispersions (e.g., SBR, and NBR), pH adjustment is unnecessary.

The thickening effect of cellulose ethers is independent of pH. Carboxymethyl, hydroxyethyl, and hydroxymethyl cellulose are used, and are effective in acidic and alkaline media. Their viscosity can be controlled both by the degree of substitution and by their concentration in the liquid medium to be thickened.

The thickening effects of poly(vinyl alcohols) and starches are also exploited (see → Textile Printing, p. 510). Thickeners based on alginate and guar gum used in textile printing are less often used in the technical textile area.

Thickeners are water-soluble polymers that harden films on drying. The special components of inverse thickeners that produce the inverse phase decrease the glass transition temperature of coating materials (Table 8.9).

Thickeners are effective stabilizers for auxiliaries and polymer dispersions that are processed as foams.

Salt additions and catalysts often impair the thickening effect.

**Table 8.9.** Polymer dispersions and thickeners: $T_g$ of film by DSC

| Styrene–acrylate copolymer dispersion 50% | $T_g$, °C |
|---|---|
| Without additive | 26.1 |
| With 2% polyacrylate thickener 30%/ammonia | 26.5 |
| With 2% inverse thickener 30% | 22.5 |
| Polyacrylate thickener/ammonia (as ammonium salt, without dispersion) | none |

### 8.2.11. Surfactants (→ Surfactants)

Surfactants have a wide range of uses:

1) Emulsification, dispersing, dissolution
2) Increasing the mechanical stability (shear stability) of liquid products, and preventing the deposition on rollers
3) Improvement of the spraying properties of auxiliaries
4) Improvement of the stability of auxiliaries towards salts
5) Increasing the amount of filler that can be incorporated in a liquid product
6) Improvement of dyeing yield
7) Production of foamable liquid products
8) Wetting of particles or textiles and improvement of penetration
9) Increasing hydrophilicity and water absorption capacity

They may also affect the volatility of water vapor, and the reduction of the hydrophobicity and oleophobicity of coatings and the strength of adhesive bonds between the textiles and coatings. For example, nonionic surfactants in coating pastes for carpet secondary backings in concentrations of 1% can considerably weaken the bond. If surfactants are added to polymer dispersions in a separate stage, the surfactants can form gels on mixing with water, and these are difficult or impossible to disperse homogeneously in liquid products. Certain surfactants are suitable for sensitizing polymer dispersions. The cloud point can be used as an indicator of their suitability for this use.

The various kinds of foaming behavior which the surfactants in auxiliaries can initiate can be utilized to produce low-foam polymer dispersions. These can then be treated with a surfactant with stronger foaming properties to give a stable foam for an application process.

The effect of surfactants on the surface tension (mN/m) of a polymer dispersion is illustrated by the example of a vinyl acetate–ethylene copolymer dispersion ($T_g$ 5 °C) applied to a viscose fiber nonwoven: no additive, 39.5 mN/m; with 2% anionic emulsifier 37%, 35.3 mN/m; with 1% sulfosuccinate 65%, 28.2 mN/m. Wetting by a water drop occurred after 30 s, 10 s, and 2 s, respectively.

### 8.2.12. Antifoams

Although antifoams are useful for transporting dispersion and in preventing extreme foam formation during application, problems can arise when they are used. The products consist of oils, mineral oils, paraffin oil, high-boiling hydrocarbons, silicone fluids, silicone derivatives, high-boiling esters, and high molecular mass alcohols, and are mainly supplied as emulsions for use in aqueous products or liquors (see → Foams and Foam Control, **A 11**, 1988, pp. 483–488).

Antifoams are used during the manufacture of polymer dispersions to prevent them foaming while they are being charged into other vessels. Excessive foaming can be caused by certain surfactants and sometimes when residual substances are evolved in gaseous form.

When antifoams are added to impregnation baths containing polymer dispersions, they can cause reduction in their wetting properties in subsequent processes, e.g., wetting of an impregnated and dried cellulose fiber nonwoven, or can cause a reduction in the bond strength of a polymer-coated nonwoven. They can also promote the pickup of dry or wet soil.

Antifoams can form agglomerates with polymer dispersions, and these can lead to the formation of deposits on rolls and textiles. The appearance of "fish eyes" in the polymer film formed by the polymer dispersion indicates the nonuniform distribution of an antifoam, and that oily particles have separated.

Antifoams also hinder the formation of foam required for textile application. However, this can be conteracted by using larger quantities of foaming agent (surfactant).

### 8.2.13. Antimicrobial Agents

The presence of microorganisms causes problems of two kinds.

Auxiliaries can contain very large numbers of microorganisms, and if they also contain sulfur-containing compounds (e.g., surfactants with sulfonate groups) hydrogen sulfide can be formed on storage. Auxiliaries generally contain < 1000 microorganisms per milliliter, and antimicrobial agents can deal with this. Microbe counts of $10^6/m\,L$ can cause problems. Product contamination is usually caused by the use of contaminated water in the manufacture of the auxiliary, and this can occur from time to time in the summer months in a central European climate. Contaminated dilution water for the storage tank in the textile plant can also cause or exacerbate the problem. The odor produced can be transferred to the textile during the treatment process. As it is almost impossible to remove the odor at a later stage, strict control of water quality is very important.

Minimization of the residual monomer content of polymer dispersions for ecological reasons can have a detrimental effect on product stability, so that special preservatives must be added. In many countries, formaldehyde/paraformaldehyde is used, which is one of the simplest and most effective methods of preservation.

The second kind of problem is the damage caused to textiles by microbial attack, mostly by mildew. This is due to storage of an auxiliary in moist, warm conditions, sometimes with pressure to the textile, which provides a breeding ground for microbial growth. The remedy is to add antimicrobial agents to treatment liquors, coating pastes, and binders [8.74].

### 8.2.14. Light Stabilizers

Light stabilizers prevent yellowing and embrittlement of polymers by light and ultraviolet radiation [8.75]. Benzotriazoles and benzophenones are examples of such stabilizers, which bond to fibers and protect the body against sunlight [8.76].

The action of light stabilizers depends on the high reactivity of these substances towards the radicals formed when the polymers are subjected to UV radiation.

### 8.2.15. Fillers [8.77]

When inert substances are added to aqueous product systems, they produce an increase in volume, an increase in weight, a reduction in cost, and a matt rather than a transparent appearance, but seldom an improvement in strength properties. Inorganic fillers are used in considerable quantities in combination with polymer dispersions in technical textiles.

Fillers for polymer dispersions and their useful properties are:

| | |
|---|---|
| Carbon black (aqueous dispersion) | black pigment, e.g., for a black-out coating antistatic agent for carpet backing |
| Graphite | antistatic agent |
| Titanium dioxide | white pigment for coatings |
| Calcium carbonate (chalk) | filler for flat and foamed coatings for carpet backings |
| Aluminum hydroxide | flame-resistant component |
| Kaolin | filler with a soft filled structure |
| Sand (fine) | for coating glass textiles |
| Barium sulfate | to give a highly-weighted textile |

When incorporating powdered fillers into liquid products, a preliminary preparation stage in which surfactants or polyphosphates are added is advantageous to ensure that the polymer dispersions have a rapid wetting effect.

When chalk is incorporated into slightly acidic polymer dispersions, the pH must first be raised to $> 7$ by adding ammonia, sodium carbonate solution, or dilute sodium hydroxide solution to prevent liberation of $CO_2$.

A typical formulation (parts by weight) for a first backing coating with a high percentage of filler for tufted carpet is given below:

| | |
|---|---|
| Copolymer dispersion 50%, carboxylated butadiene–styrene, $T_g$ ca. 5 °C | 200 |
| Dispersant | 1 |
| Chalk | 700 |
| Water | 100 |
| Thickener up to 6000 mPa · s | |

For the second backing coating, the proportion of filler used is somewhat less, e.g., ca. 200 – 300%, based on the solid polymer.

Polymer dispersions which dry to form soft films or coatings have a higher capacity for filler than hard-film formers. There is therefore a correlation with $T_g$, provided that these are the usual vinyl, acrylic, or butadiene copolymers.

The special properties of organic fillers such as rubber powder, recycled fibers, and polymer powders are used in a number of important technical textiles.

### 8.2.16. Pigments [8.78]
($\rightarrow$ Pigments, Organic; $\rightarrow$ Pigments, Inorganic)

White, black, or colored pigments are used to color coatings, flock adhesives for the flock

**Table 8.10.** Trade names of selected polymer dispersions, etc.

| Company | Polymer dispersions | Polyurethane, aqueous | Melamine–formaldehyde precondensates | Oleophobic agents | Thickeners |
|---|---|---|---|---|---|
| Air Products, USA | Airflex, Vinac, Flexbond | | | | |
| Allied Colloids, UK | Alcoset, Alcoprint | | | | Viscalex, Alcoprint |
| American Canamid, USA | | | Cymel, Aerotex | | |
| Asahi Glass, JP | Lumiflou | | | Asahiguard | |
| Astro, USA | | | Astromel | | |
| Atochem, IT | Repolem | | | Foraperle | |
| Auschem, IT | Crilat | Rolflex | Permafresh Mel | | |
| BASF, FRG | Acronal, Butofan, Diofan, Lutofan, Perapret, Propiofan, Styrofan, Vinofan, Helizarinbinder | Emuldur | Saduren, Kaurit | Persistol | Collacral, Latekoll, Lutexal |
| Bayer, FRG | Acralen, Acrafloc, Perbunan, Baystal, Acramin | Desmoderm, Impranil | Acrafix | Baygard | Acraconz Verdicker |
| Dr. Th. Böhme, FRG | Atebin, ATB-Appretur, Elastofix, Pretavyl | Elastofix | Elastofix | Pluvion | Verdick-ungsmittel |
| Buna, FRG | Sconatex, Scopacryl, Scovinat, Präwozell, Buna-Latex, Scopadien | | | | |
| Chemische Fabrik Tübingen, FRG | Arristan, Tubicoat, Tubvinyl | Tubicoat | | Tubicoat | Tubivis |
| Chuou Rika, JP | Rikabond | | | | |
| Ciba-Geigy, CH | Dicrylan | Dicrylan | Lyofix | | Lyoprint, FRGicrylan |
| Chemische Fabrik Pfersee, FRG | Vibatex | | Knittex | Oleophobol | |
| Daikin, JP | Polyflon | | | Dicguard | |
| Dainippon Ink, JP | Voncoat, FRGicnal | Hydran | | Noxguard | Dic Thickener |
| Dairen, Taiwan | DA | | | | |
| Daisel, JP | Sdyne | | | | |
| Doverstrand, UK | Revinex | | | | |
| Dow, USA | Dow-Latex | | | | |
| DuPont, USA | Hypalon, Teflon | | | Zepel, Zonyl, Stain Master | |
| Enichem, IT | Ravemul | | | | |
| Dai Yang Chem. Ind, KO | Polysol | | | | |
| FAR, IT | Policril | | Metoxil | | Policril |
| Fuller, USA | Isatex | Isatex | | | |
| Glo-Tex, USA | Glocryl, Reacticryl | | | | |
| B.F. Goodrich, USA | Geon, Hycar | Estane | | | Carbopol |
| Harco Chemical, UK | Emultex, Revacryl, Viking | | | | |
| Hansa Textilchemie FRG | | | | Hansa Phob | |
| Henkel, FRG | Dynapress, Stabiform | | Stabitex | Aversin, Repellan | |

**Table 8.10.** (Continued)

| Company | Polymer dispersions | Polyurethane, aqueous | Melamine–formaldehyde precondensates | Oleophobic agents | Thickeners |
|---|---|---|---|---|---|
| Hickson & Welch, UK | | | | Bradfluor | |
| Hispaño Quimica, ES | Acronet | | | | |
| Hoechst, FRG | Appretan, Imperonbinder, Hostaflon | Appretan | Cassurit | Nuva | Verdicker, Tylose |
| Hüls, FRG | Bunatex, Lipolan, Vestolit | | | | |
| ICI, UK | Fluon | | | Milease | |
| Kanebo NSC, JP | Yodosol, Necrylic | | | | |
| Kapp-Chemie, FRG | Kappatex | | | | |
| Korea Polymer, KO | Koplex | | | | |
| Kurare, JP | Panflex | | | | |
| Lefatex, FRG | Lefasol | Lefasol | | Lefasol | |
| 3M, USA | | | | Scotchgard | |
| Magma, Turkey | Magmalaminator | | | | |
| Mitsui Toatsu, JP | Borron, Polylack | | | | |
| Montedison, IT | Vinavil, Algoflon | | | | |
| Morton, USA | Polybinder | | | | |
| Mydrin, UK | Mydrin | | | | |
| Nasydco, Egypt | Apomul, Poliva | | | | |
| National Starch, USA | Dur-o-set, Dur-o-cryl, Dur-o-cote | | | | |
| Nebopolymers, NL | Neboplast | | | | |
| Nippon Carbide, JP | Nikasol | | | | |
| Organic Kimia, Turkey | Orgaclear | | | | |
| Para Chem., USA | Paranol | | | | |
| Protex, FR | Mulron, Acrymul | | Prox | | Acryron |
| Reichhold, USA | Synthamul, Elvace, Tylac | | | | |
| Revertex, UK | Revacryl, Revertex-Latex | | | | |
| Rohm & Haas, USA | Primal/Rhoplex, Primax | | | | Acrysol |
| Röhm, FRG | Plextol | | | | Rohagit |
| Rotta, FRG | Rotta-Coating, Rotta-Finish | Rotta-Coating | Rottafix | Dipolit | |
| Saiden, JP | Savynol | | | | |
| Sandoz, CH | Melioresin, Ratifix | Meliopromul | Finish MH | Sandofluor | |
| Sapkin, Turkey | Sapakril | | | | |
| Schill & Seilacher, FRG | Texaran, Ukadan | | Ukadan | Evoral | Verdicker |
| Scott Bader, FR | Texicote, Texicryl | | | Texipol | |
| Sequa, USA | Suncryl, Warcoset | | Permafresh | Sequapel | Sequabond |
| Showa High Polym., JP | Polysol | | | | |
| Stahl, NL | Permuthane | Permuthane, Permutex | | | |
| Stockhausen, FRG | Estekoll, Sarpifan | Fixamin | | Estarfin | Mirox |
| Sumitomo, JP | Sumikaflex | | Sumiretz | | |

**Table 8.10.** (Continued)

| Company | Polymer dispersions | Polyurethane, aqueous | Melamine–formaldehyde precondensates | Oleophobic agents | Thickeners |
|---|---|---|---|---|---|
| Synthomer, FRG | Synthomer-Latex | | | | Verdicker |
| Takeda, JP | Crosron | | | | |
| Teikoku, JP | Teisan Resin | | | | |
| Thor, FRG | Rhenappret | | Quecodur | | |
| Toyo Ink, JP | Tocryl | | | | |
| Union Carbide, USA | Ucar | | | | Ucar |
| Union Chemicals, USA | RES | | | | |
| Valchem, USA | Valbond | | | | |
| Vinamul, UK | Vinacryl, Vinamul | | | | |
| Wacker-Chemie, FRG | Vinnapas, Vinnol | | | | |
| Weserland, FRG | Supron | | | Stralin | |
| Zschimmer & Schwarz, FRG | Cefacoll | | | Antidryn | Carbolan |

printing process, and nonwoven binders, and for printing textile surfaces and labels.

Technical textiles are rarely dyed. The compatibility of the pigment dyestuffs with polymer dispersions and the consequent dye yield must be considered. Here, the type and quantity of the dispersing agent used in the production of the polymer dispersion are important.

### 8.2.17. Producers and Trade Names

Table 8.10 lists commercially available polymer dispersion auxiliaries and their producers.

# 9. References


**References for Chapter 1**

**General References**

Gesamtverband der deutschen Textilindustrie, TEGEWA, Fachausschuß Farbstoffe und organische Pigmente im Verband der Chem. Ind. (ed.): "Wissen kleidet – Textilveredlung und was man darüber wissen sollte", brochure, 1991.
TEGEWA-Rechtssammlung, Loseblatt-Sammlung, 1994.
DIN/ISO 9000–9004 (1990), EN 29000–29004, Qualitätsmanagement- und Qualitätssicherungsnormen
9000: Leitfaden zur Auswahl und Anwendung
9001: Modell zur Darlegung der Qualitätssicherung (QS) in Design/Entwicklung, Produktion, Montage, Kundendienst
9002: Modell zur Darlegung der QS in Produktion und Montage
9003: Modell zur Darlegung der QS bei der Endprüfung
9004: Qualitätsmanagement und Elemente eines Qualitätssicherungssystems, Leitfaden.

**Specific References**

[1.1] TEGEWA: *Zur Nomenklatur der Textilhilfsmittel, Leder- und Pelzhilfsmittel, Papierhilfsmittel und Tenside*, Frankfurt am Main, 1987.
[1.2] *Textilhilfsmittel Katalog 1994/95*, Konradin Verlag, Leinfelden 1994.
[1.3] Chemikaliengesetz, Gesetz zum Schutz vor gefährlichen Stoffen, 14.03.1990, BGBl I S. 521, 1. Fassung v. 16.09.1980.
[1.3a] *Ullmann*, 4th ed. **23**, pp. 4–5.
[1.4] A. Schenek et al.: "Einsatzmöglichkeiten von dampfaufgeschlossenen Flachsfasern in der Spinnerei," *Melliand Textilber.* **74** (1993) 465–474.
[1.5] M. Matsui: "Microfasern gestern, heute und morgen," *Chemiefasern/Textilind.* **43/95** (1993) 223–227.
G. G. Feth: "Membran kontra Mikrofaser," *Frankfurter Allgemeine Zeitung*, Dec. 28, 1993.
[1.6] R. Bauer, H. J. Koslowski: *Chemiefaser Lexikon*, Deutscher Fachverlag, Frankfurt am Main 1993.
[1.7] K. Fischer: "Die Entwicklung der Textilveredelung in den Siebziger Jahren," *Dtsch. Färber Kal.* **75** (1971) 97–134.
[1.8] T. Lampe, P. Neumayer, W. Roggenbach: "Alterung von Textilwerkstoffen im Kraftfahrzeug," *Melliand Textilber.* **73** (1992) 402–416.
[1.9] K. Mayer: "Kettengewirkte technische Textilien," *Kettenwirkpraxis* 2/1986, 2/1987, 4/1989, company brochure, Obertshausen.
[1.10] P. A. Koch, *Z. gesamte Textilind.* **56** (1954) 770–774.
[1.11] O. Krätz: *Faszination Chemie*, Callwey Verlag, München 1990.
[1.12] R. Klitzsch: "Von der Polsterwatte zur Isolierwatte," *Vliesst. Nonwoven Int.* **1993**, 92–94.

[1.13] A. Bührler: "Zur Geschichte der Vliesstoffe," *Textilveredlung* **28** (1993) 219–223.

[1.14] P. Böttcher: "Vliesstoffe und Freudenberg," *Int. Nonwovens Bull.* **3** (1992) no. 3, 22–25.

[1.15] K. Fischer: "Moderne Trends mit Appretan-Marken für die Ausrüstung, zum Binden und Beschichten von Textilien," *Melliand Textilber.* **71** (1990) 290–303.

[1.16] H. Wilhelm: "Reaktionsfähige Hochpolymere für die Textilveredlung," *Textilveredlung* **23** (1988) 96–100, 178–181, 225–230, 307–312.

[1.17] P. Böttcher, Vliesstoffindustrie, in den 5 neuen Bundesländern, *Chemiefasern/Textilind.* **41/39** (1991) 273–274.

[1.18] S. R. Malkan, L. C. Wadworth: "Ein Rückblick auf die Spinnvlies-Technologie," *Int. Nonwovens Bull.* **3** (1992) no. 3, 4–14.

[1.18a] C. Wilkens: "Geraschelte Abstandsgewirke," *Melliand Textilber.* **74** (1993) 993–997.

[1.19] Verantwortung für die Zukunft, Enquete-Kommission des Deutschen Bundestages: *Schutz des Menschen und der Umwelt*, Economica Verlag, Bonn 1993, p. 263.

[1.20] TEGEWA, annual reports 1990, 1991, 1992, Frankfurt am Main.

[1.21] EDANA, European Disposables and Nonwovens Assoc., Brussels, reports.

[1.22] K. Fischer: "Aktuelles zur adhäsiven Vliesverfestigung," *Chemiefasern/Textilind.* **41/93** (1991) 800–802.

[1.23] "Der Vinyl- und Acryl-Emulsions-Markt in Europa," *TPI Text. Prax. Int.* **48** (1993) 11.

[1.24] "Binders: Putting it all Together," *Nonwovens World* **3/4** (1990) 41.

[1.25] E. Schramm: "Westeuropäischer Vliesstoffmarkt 1968–1988," *Chemiefasern/Textilind.* **38/90** (1988) 596–597; "Vliesstoffmarkt in WEU," *Chemiefasern/Textilind.* **42/94** (1992) 934.
"Die 25 größten Erzeuger von Vliesstoffen erbringen 80% der weltweiten Produktion im Wert von 5 Milliarden Dollar," *Vliesst. Nonwoven Int.* **5** (1990) 177.
"Westeuropa: mehr Spinnvliese," *Chemiefasern/Textilind.* **41/93** (1991) 173.
Latin America: Nonwovens Factbook and Directory, 1993, Fibers and Binders for Nonwovens, Miller Freeman Books, Gilroy, USA.

[1.26] E. Gehmayr, W. Six: "Viskosefasern für die Nonwoven-Industrie," *Textilveredlung* **28** (1993) 225–227.

[1.27] G. Egbers: "tT – ein Wachstumsmarkt," *Jahrbuch der TI 1986*, Gesamttextil, Frankfurt am Main 1986, pp. 27–32.

[1.28] H.-J. Koslowski: "Aktuelle Markttrends für tT," *Tech. Text.* **35** (1992) T 85.

[1.29] Verband der Dach- und Dichtungsbahnen, Frankfurt am Main, personal communication, 1993.

[1.30] Jahresbericht des Gesamtverbandes der deutschen Textilveredlungsindustrie, Frankfurt am Main 1992.

[1.31] W. Kothe, G. Vogl, J. Winter: " Textildruck – quo vadis?" *Melliand Textilber.* **74** (1993) 1038–1040.

[1.32] Pentachlorphenol-Verbotsverordnung, Dec. 12, 1989, BGBl I, 2235.
Verordnung über gefährliche Stoffe (Gefahrstoff-Verordnung) Aug. 26, 1986, BGBl I, 1470 mit Änderungen 1987, 1989, 1990, 5.6.1991, BGBl I, 1218.

[1.33] Gesetz über die Vermeidung und Entsorgung von Abfällen (Abfallgesetz) Aug. 27, 1986, BGBl I, 1410.

[1.34] Verpackungsverordnung, June 12, 1991, BGBl I, 1234.
EC Guideline for the Classification, Packaging, and Labelling of Hazardous Substances 92/32 EC, June 5, 1992, and Hazardous Preparations 88/379, June 7, 1988.

[1.35] TA-Siedlungsabfall, vom 1. 6. 1993.
K. Finsterwalder, U. Mann: "Emissionsabschätzung für Deponien," *Wasser, Luft und Boden* (1992) 7–8, 70.
"Jeden Tag acht Tonnen Müll am Krankenbett," *Frankfurter Allgemeine Zeitung*, Feb. 14, 1980.

[1.36] A. H. Magloth; "Umweltschutz aktiv: neue Ökotechnik Sparal," *Wirkerei- Strickerei-Tech.* **39** (1989) no. 10.
F. Rüegge: "New Technology for Exhaust Air Cleaning in the Textile Industry," *2nd Int. Conf. on Coating and Laminating*, Zürich, Nov. 9, 1992.

[1.37] P. Knobloch: "Wertstoffe aus Restmengen von Vliesstoffen," *Tech. Text.* **35** (1992) T 104–T 105.
"Recycling von PES-Textilien," *Chemiefasern/Textilind.* **42/94** (1992) 924.
H. Meierkord: "Recycling-Möglichkeiten für PES," *Chemiefasern/Textilind.* **43/95** (1993) 524–526.
H. Borufka: "Nach jeder Messe Ärger mit Einweg-Teppichen," *Frankfurter Allgemeine Zeitung*, March 6, 1993.
Sandler: "Index Award für recyclingfähige Vliesmatten," *Chemiefasern/Textilind.* **43/95** (1993) 523.
R. Weber: "Recycling bei Kraftfahrzeugen," *Konstruktion* **42** (1990) 410–414.
J.-P. Peckstadt: "Polyolefinfaser-Abfallmanagement," *Chemiefasern/Textilind.* **43/95** (1993) 534–536.
E. Taeger et al.: "Chemische Verwertung von Polyamidabfällen," *Chemiefasern/Textilind.* **43/95** (1993) 526–531.
G. Menges, W. Michaeli, M. Bittner: *Recycling von Kunststoffen*, Hanser Verlag, München 1991.
J. M. Marzinkowski: "Reststoffe aus Textilveredlungsprozessen," *Int. Text. Bull.* **39** (1993) no. 2, 29–33.

[1.38] W.-H. Hempel: "Baumwolle – produktökologisches Sorgenkind der TI?," *Textilveredlung* **28** (1993) 168–170.
C. Kummer: "Baumwolle – eine Naturfaser im Clinch zwischen Markt- und Ökologieanspruch," *Textilveredlung* **28** (1993) 170–176.

[1.39] J. M. Marzinkowski: "Umweltschutz in der Textilveredlung," *Melliand Textilber.* **73** (1992) 438–444.
A. Weber: "Textilchemie und Umweltschutz," *Textilveredlung* **27** (1992) 388–391.
U. Baumann, U. Engler, W. Keller, W. Schefer: "Pragmatischer Versuch, Textilchemikalien-Sortimente umweltverträglicher zu gestalten," *Textilveredlung* **27** (1992) 43–45.
K. Fischer: "Technik und Umwelt – ein Bündnis für große Herausforderungen," *Melliand Textilber.* **73** (1992) 90–94.

[1.40] K. L. Kochenderfer: "The Environmental Activities of the US Textile Industry," *2nd Int. Conf. on Coating and Laminating*, Zürich, Nov. 9, 1992.

[1.41] R. Saffert: "PVC Plastisols for Fabric Coating, Environmental Effects," *2nd Int. Conf. on Coating and Laminating*, Zürich, Nov. 9, 1992.

J. G. Kennedy: "Carpet and the Environment," Company brochure, Dow Chemical, Oct. 29, 1991.
K. Fischer: "Binder und Binderadditive – Variationsbreite unter Umweltgesichtspunkten," *TPI Text. Prax. Int.* **48** (1993) 709–714; "Polymer-Dispersionen – Einsatz als Bindemittel f. Vliesstoffe unter Umweltgesichtspunkten." *Textilveredlung* **28** (1993) 213–218.

[1.42] A. Stiebert: "Kann die TI mit dem Abwasssserabgabengesetz überleben?" *Chemiefasern/Textilind.* **24/76** (1974) 575–582.

[1.43] Abwasserabgabengesetz, Sept. 13, 1976, BGBl I, 2721–2726.
E. Fitza; *"Die Bedeutung des Abwasserabgabengesetzes für die TVI"*, Taschenbuch für die Textilindustrie, Schiele + Schön, Berlin 1977.
K. Fischer: "Ab 1981, Abwasserabgabe," *Text. Prax. Int.* **36** (1981) 1245–1248, 1324–1327.

[1.44] Verordnung über die Begrenzung von Abwasseremissionen aus Textilveredlungs- und -behandlungsbetrieben, Sept. 24, 1992, BGBl OE, no. 612, 2813.

[1.45] H. Leitner, W. Schenk: "Konsequenzen aus dem Einsatz und der Rückgewinnung spezieller Polyacrylatschlichten," *Melliand Textilber.* **58** (1977) 537–541.

[1.46] O. Deschler: "Probleme und Möglichkeiten der Schlichte-Rückgewinnung," *Melliand Textilber.* **60 + 61** (1979) 1003–1006, (1980) 42–50.

[1.47] H. Schönberger: "Schlichtmittel und Abwasserbelastung aus Sicht der Wasserbehörde," *Melliand Textilber.* **71** (1990) 429–431.

[1.48] K. Fischer: "Eliminierung von Polyvinylalkohol aus Abwässern" *Melliand Textilber.* **65** (1984) 269–274, 340–345.

[1.49] R. Zahn, H. Wellens, *Chem. Ztg.* **98** (1974) 228–232.

[1.50] K. Fischer: "Eliminierungen von Inhaltsstoffen in Abwässern der Textilveredlung," *Melliand Textilber.* **59** (1978) 487–494, 582–588, 659–669; "Abwasser-Inhaltsstoffe verringern, und vermeiden," *Wirkerei-Strickerei-Tech.* **29** (1979) 492–499. "Abwasser: Fällung, Bio- Abbau, Vorbehandlung," *Chemiefasern/Textilind.* **34/86** (1984) 850–854.

[1.51] DIN 38412/L 25.

[1.52] TEGEWA: "Modifizierter Zahn-Wellens-Test," *Melliand Textilber.* **73** (1992) 755–758.

[1.53] A. Killer, II. Schönberger: "Refraktäre Stoffe im Textilabwasser," *Textilveredlung* **28** (1993) 44–53.

[1.54] G. Schulz, D. Fiebig, H. Herlinger: "Minimierung der Umweltbelastungen bei Prozessen der TVI," *Melliand Textilber.* **74** (1993) 137–143.

[1.55] R. Kleber: "Die Natriumchlorit-Bleiche und ihr ökologisches Umfeld," *Melliand Textilber.* **74** (1993) 395–397.

[1.56] G. Schulz, H. Herlinger, F. U. Gähr, T. Lehr: "Oxidativer Abbau von Farbstoffen durch Ozon," *TPI Text. Prax. Int.* **47** (1992) 1055–1060, 1062.
H. J. Buschmann, C. Carvalho, U. Driessen, E. Schollmeyer: "Die Entfärbung von text. Abwasser durch Bildung von Farbstoff-Einschlußverbindungen," *Textilveredlung* **28** (1993) 176–182.

[1.57] L. Roth: "Selbsteinstufung von Handelsprodukten," *Wassergefährdende Stoffe*, vol. 2, chap. 2.3, ecomed Verlag, Landsberg 1989.

[1.58] TA Luft, Erste Allgemeine Verwaltungsvorschrift zum Bundes-Immissionsschutzgesetz, June 27, 1986, BGBl I 1986, 95.

[1.59] *2. Symposium Emissionen im Fahrzeuginnenraum*, VW, Wolfsburg, Oct. 5, 1992.

H. Freiberg: "Saubere Abluft an Textilmaschinen: Wohin geht der Weg?" Melliand Textilber. **74** (1993) 875–883.

[1.60] DIN 75201, Bestimmung des Foggingverhaltens, 1992.
D. Eisele: "Geruch und Fogging von Automobil-Innenausstattungsmaterialien," *Melliand Textilber.* **68** (1987) 206–215.
P. Ehrler, H. Schreiber, S. Haller: "Emission textiler Automobil-Innenausstattung: Ursachen und Beurteilung des Kurzzeit- und Langzeit-Foggingverhaltens," *Textilveredlung* **29** (1994) 254–260.
P. Hardt, T. Weihrauch: "Prognosen zum Fogging- und Abluftverhalten von Textilhilfsmitteln," *Text. Prax. Int.* **49** (1994) 163–167.

[1.61] Lebensmittel- und Bedarfsgegenständegesetz, D, Aug. 15, 1974, BGBl I, 1945, 1946 and Jan. 22, 1991, BGBl I, 121.
R. Franck: *Kunststoffe im Lebensmittelverkehr*, Loseblattsammlung, 41st, Carl Heymanns Verlag, Köln 1991.

[1.62] Bedarfsgegenstände-Verordnung v. 10.4.1992, BGBl I, 866–902 Anlage 1: verbotene Stoffe. Anlage 3: Zugelassene Monomere und sonstige Ausgangsstoffe.

[1.63] EG-Richtlinie 91/55 v. 22.3.1991, Festlegung von Einzelheiten eines besonderen Informationssystems für gefährliche Zubereitungen.

[1.64] Gemeinschaft umweltfreundlicher Teppichboden e.V., Wuppertal, Satzung v. 14.11.1992, Erweiterung 1994 und *Chemiefaser/Textilind.* **41/93** (1991), 262.

[1.65] Sicherheit beim Kauf von Textilien, Informationsblatt des Vereins für verbraucher- und umweltfreundliche Textilien (seit 3.6.1992), Eschborn, 1993.

[1.66] Öko-Tex Standard 100, Österreichisches Textil-Forschungsinstitut, Wien, No. 5/1992.

**References for Chapter 2**

[2.1] H. Rath: *Lehrbuch der Textilchemie*, 3rd ed., Springer Verlag, Berlin 1972.

[2.2] A. Chwala, V. Anger (eds.): *Handbuch der Textilhilfsmittel*, Verlag Chemie, Weinheim 1977.

[2.3] H. Stache: *Tensid-Taschenbuch*, Carl Hanser Verlag, München 1979.

[2.4] *Handbuch der Werkstoffprüfung*, vol. 5, Springer Verlag, Berlin 1960.

[2.5] B. v. Falken (ed.): *Synthesefasern*, Verlag Chemie, Weinheim 1981.

[2.6] Chemiefasern/Textilind. (ed.): *Man-Made Fiber Year Book 1993*, p. 6.

[2.7] K. Riggert, *Chemiefasern/Textilind.* **27/79** (1977) 1084–1088; *Chemiefasern/Textilind.* **28/80** (1978) 117–122.

[2.8] E. Welfers: "Herstellung von Chemie-Spinnfasern," *4th Symposium für Dozenten des Textilwesen*, Oct. 14, 1976, Karlsruhe, *Chemiefasern/Textilind.* **26/78** (1976) 1079–1096; *Chemiefasern/Textilind.* **27/79** (1977) 42–60.

[2.9] L. Riehl: "Klassische Polyesterfaser-Technologie und Avivierung," *Chemiefasern/Textilind.* **12** 78 1039–1046; *Chemiefasern/Textilind.* **1** 79, (1977) 24–30.

[2.10] R. Kleber: "Avivagen und Avivierungsmethoden bei Chemie-Schnittfasern und -kabeln," *Melliand Textilber.* **58** (1977) 187–194.

[2.11] Z. A. Rogowin: *Chemiefasern*, Georg Thieme Verlag, Stuttgart 1982.

[2.12] E. M. Verenitch, *Khim. Volokna* **6** (1991) 44–47.

[2.13] K. Marquardt: "Reibungs- und griffbeeinflussende Avivagekomponenten in der Flockfärberei," *Reutlingen–Denkendorfer Technologie-Berichte*, Konradin-Verlag Robert Kohlhammer, Leinfelden-Echterdingen 1981, pp. 19–22.

[2.14] R. Kleber, *Chemiefasern/Textilind.* **27/79** (1977) 322–328.

[2.15] R. Kleber, *Melliand Textilber.* **61** (1980) 831–832.

[2.16] H. R. Billica, *Chemiefasern/Textilind.* **27/79** (1977) 328–335.

[2.17] R. Kleber, *Melliand Textilber.* **60** (1979) 263–267.

[2.18] W. Sprenkmann, *Chemiefasern/Textilind.* **6** (1984) 415–419; **7/8** (1984) 489–491.

[2.19] H. Herlinger et al., *TPI Text. Prax. Int.* **9** (1985) 970–972, 977–979; **11** (1987) 529–532, 537.

[2.20] A. Speidel, *Melliand Textilber.* **62** (1981) 912–914.

[2.21] I. S. Olsen, *Text. Res. J.* **39** (1969) 31–44.

[2.22] M. I. Schick, *Text. Res. J.,* **44** (1974) 758–762.

[2.23] M. M. Robins et al., *J. Text. Ins.* **1** (1988) 126–139.

[2.24] H. I. Geursen, *Lenzinger Ber.* **47** (1979) 128–134.

[2.25] A. Schulberger, Ph. D. Thesis, Universität Stuttgart 1983.

[2.26] H. Dunken, *Freiberger Forschungsh. A*, 164 (1960), 35–42.

[2.27] H. König: *Analyse von Tensiden*, Springer Verlag, Berlin 1971.

[2.28] K. Winck, *Chemiefasern/Textilind.* **42/94** (1992) 893–896.

[2.29] *Chemiefasern/Textilind.* **43/95** (1993) 893–897.

**References for Chapter 3**

[3.1] J. Trauter, R. Vialon, B. Ruess, *Chemiefasern/Textilind.* **39/91** (1989) 123.

[3.2] H. Moroff, *Melliand Textilber.* **37** (1956) 678.

[3.3] H. Hacking: *The Sizing of Spun Yarns*, Scientific Era Publ., Stamford 1980.

[3.4] K. Ramaszeder: *Die chemische und mechanische Technologie des Schlichtens*, Steinkopff Verlag, Dresden 1973, p. 91.

[3.5] J. Trauter, H. Bauer, *Melliand Textilber.* **62** (1981) 519, 621.

[3.6] J. Trauter, H. Bauer, B. Rueß, *Melliand Textilber.* **59** (1978) 524.

[3.7] E. Becker, *Melliand Textilber.* **64** (1983) 813.

[3.8] J. Trauter, H. Böttle, K. Götz, *TPI Text. Prax. Int.* **35** (1980) 1200.

[3.9] J. Trauter, W. Wunderlich, H. Böttle, *Melliand Textilber.* **73** (1992) 551, 623.

[3.10] K. Ramaszeder, *Textilbetrieb (Würzburg)* **104** (1986) no. 5, 26.

[3.11] M. Dawson, *TPI Text. Prax. Int.* **37** (1982) 357.

[3.12] A. von Kannen, *Chemiefasern/Textilind.* **27** 79, (1977) 1113.

[3.13] Burlington Ind., US 3 862 475, 1973 (W. F. Illmann, R. C. Malpass, D. M. Conklin).

[3.14] Burlington Ind., WO 80/02035, 1980 (W. F. Illmann, D. M. Conklin, C. H. Karnes, R. I. Malpass).

[3.15] P. Ellis, S. Galuszinsky, *Text. Inst. Ind.* **18** (1980) 268.

[3.16] E. Sonntag, V. Hüttner, *Textiltechnik (Leipzig)* **36** (1986) 305.

[3.17] W. S. Perkins, R. P. Walker, *Text. Chem. Color.* **16** (1984) no. 4, 89.

[3.18] C. G. Namboodri, *Text. Res. J.* **56** (1986) 87.

[3.19] W. Beck, *TPI Text. Prax. Int.* **44** (1989) 623.

[3.20] J. Trauter, R. Vialon, *TPI Text. Prax. Int.* **44** (1989) 370.

[3.21] J. Trauter, *ITB Flächenherstellung* **104** (1986) no. 3, 19.

[3.22] J. Trauter, D. Scholze, *TPI Text. Prax. Int.* **48** (1993) 292.

[3.23] J. A. Radley: *Starch and its Derivatives*, Chapmann & Hall, London 1968.

[3.24] M. W. Rutenberg in R. L. Davidson (ed.): *Handbook of Water-Soluble Gums and Resins*, McGraw-Hill, New York 1980, chap. 22.

[3.25] J. BeMiller in R. L. Whistler, J. N. BeMiller (eds.): *Industrial Gums*, Academic Press, San Diego 1993, p. 579.

[3.26] G. Tegge: *Stärke und Stärkederivate*, Behr's Verlag, Hamburg 1984.

[3.27] J. Trauter, M. Laupichler, *Melliand Textilber.* **57** (1976) 375, 443, 545, 615, 713, 797, 875, 979; **58** (1977) 23, 111.
J. Trauter, H. Bauer, B. Rueß, M. Laupichler, *Textilbetrieb (Würzburg)*, **96** (1978) 46.

[3.28] J. Trauter, *TPI Text. Prax. Int.* **44** (1989) 1297.

[3.29] P. Wurster, G. Schmidt, *Melliand Textilber.* **68** (1987) 581.

[3.30] K. Schlüter, *Melliand Textilber.* **71** (1990) 195.

[3.31] M. Seekamp, *TPI Text. Prax. Int.* **48** (1993) 206.

[3.32] R. L. Feddersen, S. T. Thorp in R. L. Whistler, J. N. BeMiller (eds.): *Industrial Gums*, Academic Press, San Diego 1993, p. 537.

[3.33] G. I. Stelzer, E. D. Klug in R. L. Davidson (ed.): *Handbook of Water-Soluble Gums and Resins*, McGraw-Hill, New York 1980, chap. 4.

[3.34] CMC Warp Size Bull. VC-505, Hercules, Inc., Wilmington, 1979 **39** (1984) 252.

[3.36] U. Denter, E. Schollmeyer, *Melliand Textilber.* **73** (1992) 267.

[3.37] H. M. Ulrich, *Dtsch. Färber Kal.* **52** (1951) 173.

[3.38] H. Moroff, *Melliand Textilber.* **37** (1956) 1304.

[3.39] P. Habereder, M. Denkler, R. Cosse, *Melliand Textilber.* **64** (1983) 526.

[3.40] K. Schlüter, *TPI Text. Prax. Int.* **48** (1993) 408.

[3.41] H. Srivastava, S. Harshe, M. Gharia, G. Mudia, *J. Text.Assoc.* **33** (1972) 139, 148.

[3.42] T. Gerard in R. L. Davidson (ed.): *Handbook of Water-Soluble Gums and Resins*, McGraw-Hill, New York 1980, chap. 23.

[3.43] M. Maruhashi in C. A. Finch (ed.): *Polyvinyl Alcohols – Developments*, Wiley Interscience, London, 1992, p. 158.

[3.44] R. K. Tubbs in C. A. Finch (ed.): *Polyvinyl Alcohols – Developments*, Wiley Interscience, London, p. 361.

[3.45] D. L. Nehrenberg, *Melliand Textilber.* **69** (1988) 171.

[3.46] J. Langer, *Melliand Textilber.* **71** (1990) 666.

[3.47] O. Deschler, *Melliand Textilber.* **64** (1983) 716.

[3.48] J. Trauter, *Textilveredlung* **25** (1990) 27.

[3.49] J. Trauter, *Melliand Textilber.* **71** (1990) 17.

[3.50] H. Leitner, P. Dürrbeck, *TPI Text. Prax. Int.* **48** (1993) 330.

[3.51] D. O. Hummel, F. Scholl: *Atlas der Kunststoffanalyse*, vol. 1, part 1, Hanser Verlag, München, p. 30.

[3.52] W. T. Brown, E. S. Olson, H. J. Keegan, *Am. Dyest. Rep.* **19** (1967) 703.

[3.53] J. M. Finley, *Anal. Chem.* **33** (1961) 1925.
Du Pont, Quantitative Analysis of Elvanol, A 85389, Wilmington 1973.

[3.54] H. Bauer et al., *Melliand Textilber.* **73** (1992) 755.
[3.55] Q. Wheatley, F. Baines, *Text. Chem. Color.* **8** (1976) no. 2, 28.
[3.56] P. Dürrbeck, H. Leitner, *Melliand Textilber.* **72** (1991) 819.
[3.57] W. Schenk, A. Würz, *Melliand Texilber.* **60** (1979) 830.
[3.58] *Melliand Textilber.* **47** (1966) 154.
[3.59] Eastman Kodak, US 3 546 008, 1968 (C. J. Kibler et al.).
[3.60] C. Brun, Y. Girardeau, B. Pointud, D. Roberjot, *Melliand Textilber.* **68** (1987) 175.
[3.61] J. Spijkers, H. Scheer, H. Lorenz, *Melliand Textilber.* **63** (1982) 254.
[3.62] C. Mayfield, *Melliand Textilber.* **71** (1990) 110.
[3.63] M. S. Kellou, C. Pividori, *Melliand Textilber.* **68** (1987) 270.
[3.64] S. Dugal, G. Heidemann, *Melliand Textilber.* **65** (1984) 216.
[3.65] U. Denter, S. Dugal, E. Schollmeyer, *Melliand Textilber.* **66** (1985) 142.
[3.66] V. Heap, *Melliand Textilber.* **61** (1980) 691.
[3.67] H. Chwala, V. Anger: *Handbuch der Textilhilfsmittel*, Verlag Chemie, Weinheim 1977.
[3.68] L. G. Bercsenyi, I. Kovacs, *Melliand Textilber.* **61** (1980) 688.
[3.69] J. Trauter, H. J. Schneider, *Melliand Textilber.* **56** (1975) 869.
[3.70] *Textilhilfsmittelkatalog*, Konradin Verlag, Leinfelden 1991.

**References for Chapter 4**

[4.1] BASF, Techn. Information, TI/T 215, Ludwigshafen, June, 1991.
[4.2.] K. Gebert, *SVCC Symposium*, Zürich 1976, Abstracts, p. 171.
[4.3] S. Y. Kamat, A. K. Prasad, *Colourage* **38** (1991) 33.
[4.4] S. Fornelli, *TPI Text. Prax. Int.* **45** (1990) 1178.
[4.5] R. Delecourt, *Melliand Textilber.* **46** (1956) 721.
      H. Jalke, *Z. Gesamte Textilind.* **62** (1962) 760.
[4.6] V. Windbichler, *Melliand Textilber.* **37** (1956) 297.
[4.7] J. Voss, *Text. Prax.* **11** (1956) 1117.
[4.8] W. Schenk, H. Leitner, *Melliand Textilber.* **59** (1978) 148.
[4.9] P. Habereder, *Dtsch. Färber Kal.* **81** (1977) 66.
[4.10] F. Svoboda, *Z. Gesamte Textilind.* **66** (1964) 757.
[4.11] P. Wurster, *Textilveredlung* **13** (1978) 341.
[4.12] U. Denter, E. Schollmeyer, *Melliand Textilber.* **72** (1991) 267.
[4.13] *Textilhilfsmittelkatalog*, Konradin Verlag, Leinfelden 1991.
[4.14] J. Reicher, E. Csiszar, *Melliand Textilber.* **73** (1992), 427.
[4.15] L. Kollmann, *Melliand Textilber.* **8** (1927) 270.
[4.16] F. S. Gore, G. M. Nabar, *J. Soc. Ind. Res.* **8** (1949) 142.
[4.17] R. Blankenhorn, *Melliand Textilber. Int.* **51** (1970) 1063.
[4.18] U. Kirner, A. Würz, *Melliand Textilber.* **49** (1968) 187.
[4.19] W. Kind: *Die Bleiche von Pflanzenfasern*, 3rd ed., Springer Verlag, Berlin 1932.
[4.20] C. S. Foot, S. Wexler, W. Ando, R. Higgins, *J. Am. Chem. Soc.* **90** (1968) 975.
[4.21] W. G. Steinmiller, D. M. Cates, *Text. Chem. Color.* **8** (1976) 14.

[4.22] J. Dannacher, W. Schlenker, *Textilveredlung* **25**, (1990) 205.
[4.23] A. F. Hollemann, E. Wiberg: *Lehrbuch anorg. Chemie*, 100th ed., De Gruyter, Berlin 1985, p. 467.
[4.24] V. Nevelling, W. Sebb, *Textilbetrieb (Würzburg)* **93** (1975) 30; **94** (1976) 41; **95** (1977) 50.
[4.25] E. Just, *Text. Prax.* **19** (1964) 1015.
      BASF, DE-OS 1 469 608, 1964 (O. Schmidt).
[4.26] Degussa, GB 1 106 732, 1965.
[4.27] U. Kirner, *Melliand Textilber. Int.* **51** (1970) 1069.
[4.28] P. Wurster, F. Conzelmann, *TPI Text. Prax. Int.* **45** (1990) 1269.
[4.29] A. Agster, *Textilveredlung* **1** (1966) 276.
[4.30] W. Sebb, *TPI Text. Prax. Int.* **35** (1980) 1216.
[4.31] S. K. Patel, J. Varghese, *Colourage* **30** (1983) no. 5,3.
[4.32] P. F. Greenwood, *J. Soc. Dyers Colour.* **103** (1987) 342.
[4.33] K. H. Gottschalk, M. Stuhlmiller, *Melliand Textilber.* **61** (1980) 72.
[4.34] G. Prelini, *Textilia* **58** (1982) 33.
[4.35] G. Rösch, *TPI Text. Prax. Int. (Foreign Ed.)* **43** (1988) 847.
[4.36] D. Bechter, A. Roth, *TPI Text. Prax. Int.* **45** (1990) 500.
[4.37] C. Duckworth, L. M. Wrennall, *J. Soc. Dyers Colour.* **93** (1977) 407.
[4.38] H. Rath: *Lehrbuch der Textilchemie*, 2nd ed. Springer Verlag, Heidelberg 1963, p. 242.
[4.39] J. Knott, I. F. Polet, H. Müller, *Textilveredlung* **10** (1975) no. 277.
[4.40] W. G. Crewther, *Proc. Int. Wool. Res. Conf. Australia* vol. E, (1955) p. 408.
      A. E. Davis, E. J. Johnson, L. R. Mizell, *Text. Res. J.* **31** (1961) 825.
[4.41] H. Zahm, E. Hille, *Z. Naturforsch. B: Anorg. Chem. Org. Chem.* **133**, (1958) 824.
      H. Zahn, *J. Soc. Dyers Colour.* **76** (1960) 226.
      E. Hille, *Melliand Textilber.* **40** (1959) 893.
[4.42] M. S. Nossar, M. Chaikin, A. Datyner, *J. Text. Inst.* **64** (1973) 490.
[4.43] H. Zahn, W. Stein, G. Blankenburg, *Text. Res. J.* **37**, (1967), 701.
[4.44] Brändy, Brumann, *Text. Manuf.* **60** (1943) 381.
      I. Park, *J. Soc. Dyers Colour.* **87** (1917) 111.
[4.45] E. Elöd, R. Rudolph, *Z. Gesamte Textilind.* **41** (1938) 295.
[4.46] W. van Bergen, H. R. Mauersberger: *American Wool Handbook*, Textile Book Publ., New York 1938, p. 426.
      G. v. Hornuff, H. Hiller, *Melliand Textilber.* **42** (1961) 684.
[4.47] P. A. Duffield: *Review of Bleaching*, IWS Development Center, Ilkley, West Yorkshire, 1986.
[4.48] K. Prett, *Melliand Textilber.* **39** (1958) 999.
[4.49] P. Ney, *Text. Prax. Int.* **29** (1974), 1392, 1461, 1552, 1557, 1565.
[4.50] BASF, DE-AS 1 272 876, 1961 (O. Schmidt, V. Steinhoff).
      BASF, DE-AS 1 269 991, 1964 (O. Schmidt, K. Rükker).
[4.51] M. Giesen, K. Ziegler, *Melliand Textilber.* **62** (1981) 482.
[4.52] K. Rücker, *Dtsch. Färber Kal.* **75** (1971) 169.
[4.53] K. Reincke: *Taschenbuch für die Textilindustrie*, Fachverlag Schiele und Schön, Berlin 1977, p. 250.
[4.54] S. Heimann, *Melliand Textilber.* **72** (1991) 562.

[4.55] W.-D. Kermer, *Melliand Textilber.* **69** (1988) 586.
[4.56] Abwasserabgabengesetz vom 05.03.1987 und 3. Novelle vom 21.09.90.
[4.57] Klärschlammverordnung v. 25.06.1982 und Novellierungsentwurf v. 31.01.1990.
[4.58] Bundesimmissionsschutzgesetz v. 15.03.1974.
[4.59] Gesetz zum Schutz vor gefährlichen Stoffen v. 22.03.1990.
[4.60] Gefahrstoffverordnung v. 26.08.1986.
[4.61] Gesetz über die Umweltverträglichkeit von Wasch- und Reinigungsmittel v. 05.03.1987.
[4.62] S. Held, *Textilveredlung* **24** (1989) 394.

**References for Chapter 5**

**General References**

[5.1] H. Chwala, V. Anger: *Handbuch der Textilhilfsmittel*, Verlag Chemie, Weinheim 1977.
[5.2] H. Rath: *Lehrbuch der Textilchemie*, 3rd ed. Springer Verlag, Heidelberg 1972.
[5.3] K. Lindner: *Tenside-Textilhilfsmittel-Waschrohstoffe*, 3 vols., Wissenschaftl. Verlagsgesellschaft, Stuttgart 1964–1971.
[5.4] J. Stauff: *Kolloidchemie*, Springer Verlag, Berlin 1960.
[5.5] TPI Text. Prax. Int., TEGEWA (eds.): *Textilhilfsmittel-Katalog 1991*, Konradin Verlag, Leinfelden-Echterdingen 1991.

**Specific References**

[5.6] DIN 54295, 1986.
[5.7] S. Heimann, *Textilveredlung* **19** (1984) 42.
[5.8] TEGEWA: *Zur Nomenklatur der Textilhilfsmittel, Leder- und Pelzhilfsmittel, Papierhilfsmittel und Tenside*, 3rd ed., Frankfurt/Main, May, 1987.
[5.9] Gesamttextil (Deutscher Textil-Verband): *Jahrbuch 1993*, Textilservice und Verlagsgesellschaft, Eichborn 1993.
[5.10] H. Rath, *Tenside* **2** (1965) 1.
C. Matasa, *Chem. Ztg.* **95** (1971) 507.
[5.11] H. Herlinger, D. Fiebig, B. Kastl, *TPI Text. Prax. Int.* **45** (1990) no. 12, 1291–1298.
[5.12] D. Behr: *Taschenbuch der Textilchemie*, VEB Fachbuchverlag, Leipzig 1988, p. 312.
H. Zahn, I. Souren, U. Altenhofen, *Melliand Textilber.* **65** (1984) 467.
[5.13] C. Preston: *The Dyeing of Cellulosic Fibres*, Dyer's Comp. publ. Trust, Bradford 1986, pp. 163, 203.
[5.14] U. Baumgarte, *Melliand Textilber.* **68** (1987) 189.
[5.15] R. H. Traber, *Textilveredlung* **27** (1992) 312.
[5.16] G. Wecker, *Chemiefasern/Textilind.* **27** (1977) 935.
[5.17] B. Hartmark, *Melliand Textilber.* **48** (1967) 1217.
[5.18] H. Herlinger, *Melliand Textilber.* **67** (1986) 807.
[5.19] H. E. Tschakert, *Tenside* **3** (1966) 317.
[5.20] K. Reincke, *Text. Prax. Int.* **28** (1973) 461.
[5.21] BASF, DE 292531, 1913.
[5.22] S. Heimann, K. Dachs, *Melliand Textilber. Int.* **53** (1972) no. 5, 580–586.
[5.23] P. Richter, *Melliand Textilber.* **74** (1993) no. 9, 872–875.
[5.24] K. v. Sarkanen, C. H. Ludwig: *Lignins*, J. Wiley, New York 1971.
[5.25] S. Heimann, *Rev. Prog. Color Relat. Top.* **11** (1981) 1–8.
[5.26] G. Prazak, *Am. Dyes. Rep.* **59** (1970) 44.
[5.27] R. D. Athey, *Tappi* **58** (1975) 55.
[5.28] S. Heimann: *Taschenbuch für Textilindustrie*, Verlag Schiele und Schön, Berlin 1985.
[5.29] U. Strahm, *Textilveredlung* **19** (1984) 123.
[5.30] BASF, DE 2444823, 1974 (H. Wolf, J. Gerendas, E. Wilhelm).
[5.31] G. Prazak, *Colourage* **37** (1990) no. 19, 62–67.
[5.32] J. R. Aspland, *Text. Chem. Color.* **24** (1992) no. 12, 18–23.
[5.33] W. Prenzel, *Chemiefasern/Textilind.* **24/76** (1974) 293.
[5.34] S. Heimann, *Melliand Textilber.* **63** (1982) 885.
[5.35] Hoechst: *Naphtol AS Anwendungsvorschriften 4026*, Frankfurt 1971.
[5.36] F. Jones, *J. Soc. Dyers Colour* **100** (1984) 66–72.
[5.37] H. Brünger, A. Bossmann, E. Schollmeyer, *TPI Text. Prax. Int.* **43** (1988) no. 8, 843–846.
[5.38] T. v. Chambers et al., *J. Soc. Dyers. Colour* **105** (1989) 214–218.
[5.39] H. Leube, H. Uhrig, *Textilveredlung* **9** (1974) 97.
[5.40] J. Skelly, *J. Soc. Dyers Colour* **89** (1973) 349.
[5.41] J. Prikryl, J. Ruzicka, L. Burgert, *J. Soc. Dyers Colour* **95** (1979) 349–351.
[5.42] DIN 53908, 1982.
[5.43] F. H. Richter, E. W. Winkler, R. H. Baur, *JAOCS J. Am. Oil Chem. Soc.* **66** (1989) 1666.
[5.44] J. Han Wood, *Am. Dyest. Rep.* **65** (1976) 32.
[5.45] X. Kowalski, *Am. Dyest. Rep.* **68** (1979) 49.
[5.46] P. Wurster, *TPI Text. Prax. Int.* **41** (1986) 1331.
[5.47] U. Strahm, *Textilveredlung* **19** (1984) 123.
[5.48] Sandoz, DE 2926098, 1980 (J. P. Chavannes, R. Fischer, S. Fornelli, F. Palacin).
[5.49] K. H. Weible, *Melliand Textilber.* **71** (1990), 772.
[5.50] J. F. Leuck, *Text. Chem. Color* **10** (1978) 32–61.
[5.51] B. J. J. Engbers, G. Dierkes, *TPI Text. Prax. Int.* **47** (1992) 557.
[5.52] E. Schönpflug, *TPI Text. Prax. Int.* **30** (1975) 1554; **30** (1975) 1968; **31** (1976) 53.
W. Beckmann, F. Hoffmann, *Chemiefasern/Textilind.* **27/79** (1977) 557.
R. Weingarten, *Melliand Textilber.* **59** (1978) 59.
A. Kretschmer, *Melliand Textilber.* **59** (1978) 823.
F. Hoffmann, *Textilveredlung* **24** (1989) 340; **24** (1989) 381; **25** (1990) 49.
[5.53] U. Baumgarte, *Textilveredlung* **15** (1980) 413.
[5.54] U. Baumgarte, *Textilveredlung* **23** (1988) 241.
[5.55] H. Schlüter, *Textilveredlung* **25** (1990) 218.
[5.56] U. Baumgarte, *Melliand Textilber.* **59** (1978) 311.
[5.57] E. P. Frieser, *SVF-Fachorgan Textilveredl.* **18** (1963) 146.
[5.58] K. Hannemann, *Textilveredlung* **27** (1992) no. 10, 321–325.
[5.59] K. Knopf, U. Denter, E. Schollmeyer, *Tenside Surf. Deterg.* **29** (1992) no. 5, 306–310.
[5.60] D. M. Stevenson, D. G. Duff, D. I. Kirkwood, *J. Soc. Dyers Colour* **97** (1981) no. 13, 13.
[5.61] E. Deniz et al., *DWI Reports 1993*, Aachener Textiltagung, Nov. 1992.
[5.62] Olin Corp., EP 479911, 1991 (T. Hemling, H. Stitzel).
[5.63] J. Cegarra, A. Riva, *J. Soc. Dyers Colour* **104** (1988) 227–233.
[5.64] D. Fiebig, E. Schollmeyer, *TPI Text. Prax. Int.* **37** (1982) 1083–1085.
[5.65] V. Wassileva, K. Stoyanov, M. Duschewa, *Melliand Textilber.* **72** (1991) no. 3, 200.
[5.66] Y. Nemoto, H. Funahashi, *Ind. Eng. Chem. Prod. Res. Dev.* **19** (1980) 136–142.

[5.67] D. Fiebig, H. Herlinger, W. Brennich, *Melliand Textilber.* **74** (1993) no. 11, 1168.

[5.68] F. Hoffmann, *Melliand Textilber.* **59** (1978) 239.

[5.69] I. L. Rush, E. H. Hinton, *Text. Chem. Color* **11** (1979) 26.

[5.70] H. Herlinger, D. Fiebig, S. Koch, D. Schnaitmann, *TPI Text. Prax. Int.* **38** (1983) no. 6, 583.

[5.71] E. S. Lower, *Int. Dyer Text. Printer* **4** (1988) 29–31.

[5.72] TEGEWA, *Melliand Textilber.* **64** (1983) no. 3, 230–232.

[5.73] F. Bartsch, *Dtsch. Färber Kal.* **81** (1977) 245–253.

[5.74] A. Murray, K. Mortimer, *Rev. Prog. Color Relat. Top.* **2** (1971) 67–72.

[5.75] R. Iltscheva, *Textilveredlung* **26** (1993) no. 6, 191.

[5.76] Ciba–Geigy, US 4032291, 1977 (K. A. Dellian).

[5.77] Ciba–Geigy, US 5009668, 1991 (H. Berendt, R. Töpfl).

[5.78] D. Fiebig, H. Herlinger, *TPI Text. Prax. Int.* **38** (1983) no. 8, 785–793.

[5.79] K. Knopf, E. Schollmeyer, *Tenside Surf. Deterg.* **24** (1987) no. 2, 101–106.

[5.80] K. Knopf, E. Schollmeyer, *Tenside Surf. Deterg.* **24** (1987) no. 3, 134–142.

[5.81] D. Fiebig, D. Soltau, *TPI Text. Prax. Int.* **46** (1991) no. 6, 543–546.

[5.82] BASF, DE 2444102, 1978 (M. Daeuble).

[5.83] H.-J. Buschmann, D. Knittel, E. Schollmeyer, *TPI Text. Prax. Int.* **45** (1990) no. 5, 376.

[5.84] Deutsches Textilforschungszentrum Nord-West e.V., DE 4036328, 1991 (D. Knittel, H. J. Buschmann, E. Schollmeyer).

[5.85] H. Imafuku, *J. Soc. Dyers Colour* **109** (1993) no. 11 350–352.

[5.86] I. Komminos et al. US 3716329, 1970.

[5.87] W. Beckmann, *Z. Gesamte Textilind.* **71** (1969) 603.

[5.88] U. Mayer, *Z. Gesamte Textilind.* **71** (1969) 792.

[5.89] Bayer, DE-AS 1154788, 1961 (H. von Brachel, L. Nüßler, E. Degener).

[5.90] S. Shukla, M. Mathur, *J. Soc. Dyers Colour* **109** (1993) 330–333.

[5.91] J. Cegarra, P. Puente, J. M. Fiadeiro, *J. Soc. Dyers Colour* **102** (1986) 274–278.

[5.92] J. Cegarra, P. Puente, B. Castro, *Tinctoria* **85** (1988) 67–74.

[5.93] F. Feichtmayr, S. Heimann, *Z. Gesamte Textilind.* **68** (1966) 509.

[5.94] B. A. Laddy, *Am. Dyest. Rep.* **49** (1960) 272.

[5.95] Sandoz, CH 16126/72, 1972 (S. Fabbri, J. Frauenknecht).

[5.96] Sandoz, DE 2803309, 1978 (E. Hervot, Y. Rene, A. Verdoucq).

[5.97] BASF, DE 2812039, 1980 (M. Daeuble, V. Weberndoerfer, H. Schulze).

[5.98] BP Chemicals, GB 2132641, 1984 (A. Currie).

[5.99] Sandoz, DE 3417780, 1990 (K. H. Weible, J. Palleiro Cardona).

[5.100] BP Chemicals, EP 0442749, 1991 (B. Black).

[5.101] D. Hildebrand, J. Fiegel, *Melliand Textilber.* **64** (1983) 290.

[5.102] J. K. Skelly, *Textilveredlung* **8** (1973) 102.

[5.103] R. Schiffer, B. Bormeister, *Dtsch. Textiltech.* **16** (1965) 264.

[5.104] R. Rokohl, *Tenside* **2** (1965) 76.

[5.105] G. Siegrist, *Text. Rundsch.* **17** (1962) 143.

[5.106] E. Schönpflug, *Melliand Textilber.* **60** (1979) 244.

[5.107] W. Haebler, D. Hildebrand, *Text. Ind. (Moenchen-Gladbach, Ger.)* **72** (1970) 501.

[5.108] J. A. Hughes, H. H. Sumner, B. Taylor, *J. Soc. Dyers Colour* **87** (1971) 463. S. Blackborn, T. L. Dawson, *J. Soc. Dyers Colour* **87** (1971) 473.

[5.109] A. N. Derbyshire, W. P. Mills, J. Shore, *J. Soc. Dyers Colour* **88** (1972) 389.

[5.110] TEGEWA, *Melliand Textilber.* **67** (1986) no. 10 750–751.

[5.111] DIN 53988, 1976.

[5.112] DIN 54290, 1982.

[5.113] J. Carbonell, H. Egli, R. Hasler, *Textilveredlung* **7** (1972) 472.

[5.114] F. Lessinsky, *Dtsch. Färber Kal.* **80** (1976) 69.

[5.115] M. Scheller, H. Steinmüller, *Dtsch. Textiltech.* **18** (1968) 639.

[5.116] W. Beckmann, K. Langheinrich, *Melliand Textilber. Int.* **51** (1970) 316.

[5.117] C. L. Zimmermann, A. L. Cate, *Text. Chem. Color* **4** (1972) 150.

[5.118] E. Schönpflug, *Melliand Textilber.* **60** (1979) 244.

[5.119] H. Egli, *Textilveredlung* **2** (1967) 211.

[5.120] J. Hertig, *Textilveredlung* **1** (1966) 668.

[5.121] J. H. Brooks, *J. Soc. Dyers. Colour* **90** (1974) 158.

[5.122] P. Richter, *Dtsch. Färber Kal.* **83** (1979) 228.

[5.123] R. Zimmermann, *TPI Text. Prax. Int.* **33** (1978) 451.

[5.124] R. Egginger, *Textilbetrieb (Würzburg)* **97** (1979) no. 8, 49.

[5.125] Piedmont Section of AATCC, *Am. Dyest. Rep.* **48** (1959) no. 22; **48** (1959) no. 23, 37.

[5.126] G. Roberts, S. Solanki, *J. Soc. Dyers Colour.* **95** (1979) 226, 427.

[5.127] G. Schreiner, *Textiltechnik (Leipzig)* **28** (1978) 371, 780; **29** (1979) 378; **30** (1980) 192, 257; **31** (1981) 436.

[5.128] W. Beckmann, H. Hamacher-Brieden, *Text. Chem. Color* **5** (1973) 118.

[5.129] K. Thurner, *Text. Ind. (Moenchen-Gladbach, Ger.)* **72** (1970) 872.

[5.130] D. Fiebig, *TPI Text. Prax. Int.* **39** (1984) no. 2, 144.

[5.131] J. S. Gow, *Textilveredlung* **18** (1983) no. 4, 119–125.

[5.132] K. Keller, *Textilveredlung* **13** (1978) 140.

[5.133] K. Jacobs, *Text. Prax. Int.* **28** (1973) 521.

[5.134] G. Weckler, *Text. Prax. Int.* **28** (1973) 335.

[5.135] G. Weckler, *Text. Prax. Int.* **27** (1972); **28** (1973) 335; **28** (1973) 457.

[5.136] K. Jacobs, *Text. Prax. Int.* **28** (1973) 521.

[5.137] A. Bendak, H. Hanna, *Teintex* **44** (1979) 11.

[5.138] P. Medilek, M. Quurke, P. Jablonski, *Melliand Textilber.* **57** (1976) 583.

[5.139] P. B. Simmons, D. R. Branson, R. I. Mollenaar, R. E. Bailey, *Am. Dyes. Rep.* **66** (1977) no. 8, 21.

[5.140] P. Richner, *Textilveredlung* **13** (1978) 134.

[5.141] Deutsches Textilforschungszentrum Nord-West: *1. Forum für Verfahrenstechnik der Textilveredlung,* Krefeld, March 4, 1977.

[5.142] S. Heimann, W. Tobias, *Melliand Textilber. Int.* **55** (1974) no. 10, 893–894.

[5.143] G. Weckler, *Textilbetrieb (Würzburg)* **96** (1978) no. 3, 74–80.

[5.144] K. Dengler, *Wirkerei Strickerei Tech.* **26** (1976) no. 1, 32–33.

[5.145] W. Rüttiger, *Lenzinger Ber.* **45** (1978) 160–171.

[5.146] D. Fiebig, D. Bechter, *TPI Text. Prax. Int.* **31** (1976) 1057, 1199.

[5.147] D. C. Prevorsek, R. H. Butler, G. E. R. Lamb, *Text. Res. J.* **45** (1975).

[5.148] W. H. Hempel, *TPI Text. Prax. Int.* **34** (1979) no. 12, 1626.

[5.149] G. Weckler, *Chemiefasern/Textilind.* **27** (1977) 935.

[5.150] H. Gerber, *Melliand Textilber. Int.* **53** (1972) 335.
H. Lehmann, F. Somm, *Text. Prax. Int.* **28** (1973) no. 1, 52.
S. Heimann, *Z. Gesamte Textilind.* **66** (1964) no. 2, 129.
H. Haas, E. Hilgeroth, *Melliand Textilber. Int.* **51** (1970) 699.
W. Beckmann, R. Kuth, *Melliand Textilber.* **48** (1967) 1441.
J. N. Etters, *Textilveredlung* **8** (1973) no. 3, 187.
J. N. Etters, *Am. Dyest. Rep.* **77** (1988) no. 9, 15.
P. Sarabi, *Text. Chem. Color* **23** (1991) no. 3, 21.

[5.151] S. Heimann, K. Dachs, *Melliand Textilber. Int.* **53** (1972) 580.

[5.152] N. D. Stewart, *J. Soc. Dyers Colour* **89** (1973) no. 7, 2.

[5.153] Casella, DE 2710898, 1977 (H. Beiertz, G. Weckler, W. Weikardt).

[5.154] C. Heinrichs, W. Becker, S. Dugal, E. Schollmeyer, *TPI Text. Prax. Int.* **40** (1985) no. 11, 1218–1222.
G. Siegrist, *Dtsch. Färber Kal.* **74** (1970) 148–158.
R. S. Asquith, A. K. Booth, *Text. Chem. Color.* **3** (1971) no. 3, 99–101.
R. Vogel, *Wirkerei Stirckerei Tech.* **30** (1980) no. 6, 394.

[5.155] S. Heimann, T. Hölken, *Chemiefasern/Textilind.* **30** (1980) no. 11, 898–901.

[5.156] J. Rieker, *Chemiefasern/Textilind.* **27** (1977) no. 12, 1122.
J. Rieker, W. Braun, *Melliand Textilber.* **62** (1981) no. 6, 496–501.

[5.157] C. C. Cook, *Rev. Prog. Color. Relat. Top.* **12** (1982) 73–89.

[5.158] H. Fischer, *Textilveredlung* **25** (1990) no. 2, 54.

[5.159] D. M. Lewis, *J. Soc. Dyers Colour* **108** (1992) nos. 7/8, 317–324.

[5.160] H. D. Pratt, *Am. Dyest. Rep.* **69** (1980) no. 11, 32.

[5.161] C. Oschatz, *Textilveredlung* **9** (1974) no. 9, 442.

[5.162] J. Offenbach, *Dtsch. Färber Kal.* **76** (1972) 185.

[5.163] M. S. Aboul-Fetouh et al. *Indian Text. J.* **88** (1978) no. 8, 73; **89** (1979) no. 2, 111.

[5.164] U. Baumgarte, *Textilveredlung* **4** (1969) 821.

[5.165] U. Baumgarte, *Melliand Textilber.* **68** (1987) 189–276.

[5.166] U. Baumgarte, *Melliand Textilber. Int.* **56** (1975) 228.

[5.167] J. N. Etters, *Am. Dyest. Rep.* **78** (1989) no. 3, 18.

[5.168] R. C. Shah, *Text. Chem. Color* **4** (1972) 268–269.

[5.169] G. L. Medding, *Am. Dyest. Rep.* **69** (1980) 30.

[5.170] E. Marte, *TPI Text. Prax. Int.* **44** (1989) 737.

[5.171] Hoechst: *Die Schwefelfärbung unter erhöhten ökologischen Anforderungen*, Frankfurt 1993.

[5.172] W. Titzka, *TPI Text. Prax. Int.* **33** (1978) 1387.

[5.173] W. H. Hempel, *Int. Text. Bull.* **1978** no. 3, 282; **1979**, no. 2, 243.

[5.174] J. Frauenknecht, D. Schwer, *Textilveredlung* **5** (1970) 912.

[5.175] J. Schwab, *TPI Text. Prax. Int.* **34** (1979) 282.
M. Hückel, *TPI Text. Prax. Int.* **33** (1978) 1095.

[5.176] M. Homuth, *TPI Text. Prax. Int.* **35** (1980) 55.

[5.177] P. Kutschera, G. Vogl, *TPI Text. Prax. Int.* **34** (1979) 173.

[5.178] W. Kühnel, *Melliand Textilber.* **57** (1976) 315.

**References for Chapter 6**

[6.1] H. Rath: *Lehrbuch der Textilchemie* 3rd ed., Springer Verlag, Heidelberg 1972.

[6.2] G. Faulhaber, *Melliand Textilber.* **61** (1980) 1032–1034.

[6.3] a) W. Schwindt, G. Faulhaber, *Rev. Prog. Color. Relat. Top.* **14** (1984) 166–175.
b) F. Carlier, *Ind. Text. (Paris)* **1222** (1991) 68–75.

[6.4] a) W. Kothe, G. Vogl, J. Winkler, *Melliand Textilber.* **74** (1993) no. 10, 1038–1040.
b) Stork Brabant, *Chemiefasern/Textilind.* **41** (1991) 507.
c) Stork Brabant, *Int. Text. Bull.* **36** (1990) no. 4, 15 ff.
d) Stork Brabant, *Melliand Textilber.* **67** (1986) 261–264.

[6.5] a) U. Perkuhn, *Melliand Textilber.* **67** (1986), no. 12 E 362–E 365.
b) K. Dorfner, *TPI Text. Prax. Int.* **37** (1982) no. 10, 1048–1050.
c) U. Perkuhn, *Int. Text. Bull. Dyeing/Printing/Finishing* **3** (1981) 229–236.

[6.6] D. Bechter et al., *TPI Text. Prax. Int.* **44** (1989) no. 3, 290–291.

[6.7] W. Berlenbach, *Int. Dyer Text, Printer Bleacher Finish.* **140** (1968) 693–699.

[6.8] F. Reinert, *Melliand Textilber.* **73** (1992) no. 4, 353–358, E 157–E 160.

[6.9] W. Tiedemann, P. Hülsberg, P. Horlacher, D. Kinast, *TPI Text. Prax. Int.* **47** (1992), 337–338, 343–345.

[6.10] R. Schwäbel, K. Bühler, F. Belde, *Bayer Farben Rev. Sonderh.* **12** (1970) 4–21.

[6.11] G. Meyer, *Text. Prax. Int.* **29** (1974) 484–485.

[6.12] R. Hofstetter, *Textilveredlung* **11** (1976) 186–194.

[6.13] N. Grund, *Melliand Textilber.* **67** (1986) 896–902.

[6.14] E. Feess, *Melliand Textilber.* **45** (1964) 67–70, 172–181, 296–299, 413–427.

[6.15] W. Scheuermann, W. Küppers, *Melliand Textilber.* **46** (1965) 1207–1212, 1333–1338.

[6.16] A. Blum, *Melliand Textilber.* **41** (1960) 587–593.

[6.17] G. Dillmann, *TPI Text. Prax. Int.* **38** (1983) 54–58.

[6.18] P. Kutschera, G. Vogl, *TPI Text. Prax. Int.* **34** (1979) 173–174, 179.

[6.19] K. Roth, *Textilbetrieb (Würzburg)* **98** (1980) no. 9, 50–65.

[6.20] H. Schwab, *TPI Text. Prax. Int.* **34** (1979) 282, 295–297.

[6.21] W. Kühnel, *Melliand Textilber.* **57** (1976) 315–320.

[6.22] H. Herlinger, D. Fiebig, B. Kastl, *TPI Text. Prax. Int.* **45** (1990) 1291–1298.

[6.23] E. Feess, *Text. Prax.* **23** (1968) 335–341.

[6.24] H. Ilg, H. Fischer, *Text. Prax.* **25** (1970) 484–487.

[6.25] H. Ilg, W. Römpp, *Melliand Textilber. Int.* **54** (1973) 82–88, 661–667.

[6.26] C. Sommer, *Melliand Textilber. Int.* **50** (1969) 692–698.

[6.27] a) G. Meyer, *Melliand Textilber. Int.* **50** (1969) 698–702.
b) G. Meyer, H. Dietz, *Text. Prax. Int.* **27** (1972) 294–298.
c) F. Hoch, R. Bäuerle, *Ciba Geigy Rundsch.* **2** (1973) 34–37.

[6.28] H. D. Optiz, *Melliand Textilber.* **71** (1990) 775–782.

[6.29] R. Eisenlohr, *Melliand Textilber.* **70** (1989) 945–947.

[6.30] a) G. Bertolina, *Melliand Textilber. Int.* **51** (1970) 812–815.
b) H. Schwab, *Int. Dyer Text. Printer Bleacher Finish* **146** (1971) 637–640.

[6.31] a) J. Winkler, *TPI Text. Prax. Int.* **34** (1979) 302, 307–309.
b) W. Schwindt, *Melliand Textilber.* **62** (1981) 562–564.

[6.32] C. Heinrichs, W. Becker, S. Dugal, E. Schollmeyer, *TPI Text. Prax. Int.* **40** (1985) 1218–1226.

[6.33] R. Koch, *Melliand Textilber.* **61** (1980) 167–173.

[6.34] E. Feess, *Melliand Textilber.* **45** (1964) 413–427.

[6.35] K.-H. Stukenbrok, *Melliand Textilber.* **72** (1991) 46–49.

[6.36] H. Dahm.: "Verdickungsmittel und Kleber für den Textildruck," 6th ed., *Bayer Farben Rev.*, Sep. 1981 (reprint).

**References for Chapter 7**

[7.1] D. A. Lämmermann, *Melliand Textilber.* **73** (1992) 274.

[7.2] E. Kurz, *Am. Dyest. Rep.* **25** (1974) Jan., 25.

[7.3] H. B. Goldstein, H. W. Smith, *Text. Chem. Color.* **12** (1980) no. 3, 49.

[7.4] R. A. Holser, R. J. Harper, Jr., A. H. Lambert, *Text. Chem. Color.* **18** (1986) no. 11, 29.

[7.5] A. D. Broadbent, *Text. Chem. Color.* **22** (1990) no. 8, 13.

[7.6] A. D. Broadbent, *Text. Chem. Color.* **22** (1990) no. 9, 53.

[7.7] M. Schwemmer, H. Bors, A. Gotz, *Textilveredlung* **10** (1975) 15.

[7.8] Union Carbide Corp., US 4023526, 1977 (D. H. Ashmus, W. W. Rankin, A. T. Walter).

[7.9] G. R. Turner, *Text. Chem. Color.* **17** (1985) no. 10, 30.

[7.10] R. S. Gregorian, *Text. Chem. Color.* **19** (1987) no. 4, 13.

[7.11] Tootal Broadhurst Lee, GB 291473, 1926; GB 291474, 1926 (R. P. Foulds et al.).

[7.12] R. L. Wayland, Jr., et al., *Am. Dyest. Rep.* **49** (1960) no. 24, P843.

[7.13] *AATCC Technical Manual* **70** (1995).

[7.14] J. Bancroft & Sons Co., US 2501857, 1950 (W. R. McIntire).

[7.15] Dan River Mills, Inc., US 2690404, 1954 (M. J. Spangler, R. L. Wayland, Jr.).

[7.16] J. F. Mulvaney, R. L. Evans, *Ind. Eng. Chem.* **40** (1948) 393.

[7.17] Dow Chemical Co., US 2930715, 1960 (W. N. Bakke, W. F. Tousignant).

[7.18] Dan River Mills, Inc., US 3158501, 1964 (R. L. Wayland, Jr.).

[7.19] Dan River Mills, Inc., US 3324062, 1967 (G. S. Y. Poon).

[7.20] E. I. DuPont de Nemours & Co. Inc., US 2304624, 1942 (W. J. Burke).

[7.21] A. M. Paquin, *Angew. Chem.* **60** (1948) 267.

[7.22] R. L. Wayland, Jr., *Text. Res. J.* **29** (1959) 170.

[7.23] H. Kadowaki, *Bull. Chem. Soc. Jpn.* **11** (1936) 248.

[7.24] Sumitomo Chemical Co., US 3089859, 1963 (T. Oshima).

[7.25] Dan River Mills, Inc., US 3524876, 1970 (J. E. Gregson).

[7.26] BASF, US 3639455, 1972 (H. Petersen, K. Renner, H. Diem).

[7.27] BASF, US 4156784, 1979 (H. Petersen, T. Dockner).

[7.28] J. D. Reid, R. M. Reinhardt, J. S. Bruno, *Am. Dyest. Rep.* **54** (1965) 485.

[7.29] R. Steele, *Text. Res. J.* **31** (1961) 257.

[7.30] G. L. Drake, J. D. Guthrie, *Text. Res. J.* **29** (1959) 155.

[7.31] G. Tesoro, P. Linden, S. B. Sello, *Text. Res. J.* **31** (1961) 283.

[7.32] Quaker Chemical Products Corp., US 2785947, 1957 (B. H. Kress).

[7.33] J. T. Marsh, *J. Soc. Dyers Colour.* **75** (1959) 244.

[7.34] Shell Development Corp., US 2826514, 1958 (C. W. Schroeder).

[7.35] Courtalds Ltd., US 3312521, 1967 (J. G. Stenner).

[7.36] BASF, US 2731364, 1956 (B. v. Reibnitz, W. Ruemens, K. Beideck).

[7.37] Korotron Co., Inc., US 2974432, 1961 (W. K. Warnock, F. G. Hubener).

[7.38] H. B. Goldstein, J. M. May, *Text. Res. J.* **34** (1964) 325.

[7.39] International Minerals and Chemical Co., US 4238545, 1980 (J. H. Hunsucker).

[7.40] Sun Chemical Corp., US 4396391, 1983 (B. F. North).

[7.41] B. F. North, *Text. Chem. Color.* **23** (1991) no. 4, 23.

[7.42] C. Tomasino, *Text. Chem. Color.* **16** (1984) no. 12, 33.

[7.43] R. L. Wayland, Jr., L. W. Smith, J. H. Hoffman, *Text. Res. J.* **51** (1981) no. 4, 302.

[7.44] B. A. Kottes Andrews, R. M. Reinhardt, *Text. Chem. Color.* **16** (1984) no. 5, 21.

[7.45] D. Katovic, I. Soljacic, *Text. Res. J.* **58** (1988) no. 9, 552.

[7.46] Y. K. Kamath, R. U. Weber, S. B. Hornby, H.-D. Weigmann, *Text. Res. J.* **55** (1985) 519, 589.

[7.47] E. C. Roberts, A. J. Rossano, Jr., *Text. Chem. Color.* **16** (1984) no. 3, 29.

[7.48] D. M. Pasad, C. L. Cochran, B. A. Kottes Andrews, *Text. Chem. Color.* **21** (1989) no. 6, 13.

[7.49] B. J. Collier, Y. Chen, J. R. Collier, *Text. Chem. Color.* **24** (1992) no. 10, 26.

[7.50] S. H. Yoon, *Text. Chem. Color.* **17** (1985) no. 4, 33.

[7.51] B. A. Kottes Andrews, *Text. Chem. Color.* **19** (1987) no. 10, 19.

[7.52] B. A. Kottes Andrews, R. M. Reinhardt, B. J. Trask-Morrell, *Text. Res. J.* **58** (1988) no. 5, 255.

[7.53] D. Lämmermann, *Melliand Textilber.* **73** (1992) no. 7, 274.

[7.54] J. G. Frick, R. J. Harper, *Text. Res. J.* **51** (1981) 51.

[7.55] BASF, US 5246904, 1993 (P. Hois, T. Simenc).

[7.56] B. A. Kottes Andrews, *Text. Chem. Color.* **22** (1990) no. 9, 63.

[7.57] B. A. Kottes Andrews, *Text. Chem. Color.* **25** (1993) no. 3, 52.

[7.58] H. M. Choi, *Text. Chem. Color.* **25** (1993) no. 5, 19.

[7.59] C. M. Welch, *Text. Res. J.* **58** (1988) 480.

[7.60] C. M. Welch, *Text. Chem. Color.* **23** (1991) no. 3, 29.

[7.61] H. J. Buschmann, E. Schollmeyer, *Textilveredlung* **26** (1991) no. 11, 361.

[7.62] S. N. Pandey, C. R. Raje, *Text. Res. J.* **57** (1987) no. 4, 265.

[7.63] B. A. Kottes Andrews, R. M. Reinhardt, *Text. Res. J.* **52** (1982) 123.

[7.64] C. Chen, *Text. Res. J.* **60** (1990) 669.

[7.65] H. J. Lautenschlager, J. Bindl, K. G. Huhn, *TPI Text. Prax. Int.* **47** (1992) no. 5, 460.

[7.66] J. D. Turner, *Text. Chem. Color.* **20** (1988) no. 5, 36.

[7.67]   ICI, GB 475170, 1937 (A. W. Baldwin, E. E. Walker).
[7.68]   H. A. Schuyten, J. W. Weaver, J. G. Frick, Jr., J. D. Reid, *Text. Res. J.* **22** (1952) 424.
[7.69]   U.S. Dept. of the Army, US 3347812, 1967 (C. G. DeMarco, G. M. Dias).
[7.70]   R. K. Iler, *Ind. Eng. Chem.* **46** (1954) 766.
[7.71]   E. B. Higgins, *Text. Inst. Ind.* **4** (1966) 255.
[7.72]   Ciba, US 2927090, 1960 (A. Hiestand).
[7.73]   W. Noll: *Chemistry and Technology of Silicones*, Academic Press, New York 1968.
[7.74]   F. Fortess, *Ind. Eng. Chem.* **46** (1954) 2325.
[7.75]   3M Company, US 3529995, 1970 (S. Smith, P. O. Sherman).
[7.76]   H. B. Goldstein, *Text. Res. J.* **31** (1961) 377.
[7.77]   J. Compton, W. J. Hart, *Text. Res. J.* **24** (1954) 263.
[7.78]   N. F. Getchell, *Text. Res. J.* **25** (1955) 150.
[7.79]   Deering-Milliken Research Corp., US 3377249, 1968 (F. W. Marco).
[7.80]   ICI Ltd., US 3459590, 1969 (W. M. Corbett, D. Harrison).
[7.81]   D. A. Garrett, D. N. Hartley, *J. Soc. Dyers Colour.* **82** (1966) 252.
[7.82]   *Text. World* **118** (1968) May, 112.
[7.83]   P. O. Sherman, S. Smith, B. Johannessen, *Text. Res. J.* **39** (1969) 449.
[7.84]   H. E. Bille, A. Eckell, G. A. Schmidt, *Text. Chem. Color.* **1** (1969) 600.
[7.85]   J. P. Stevens, US 4008004, 1977 (B. M. Latta, I. E. Pensa).
[7.86]   D. M. Gagarine, *Text. Chem. Color.* **10** (1978) 247.
[7.87]   G. C. Tesoro, C. H. Meiser, Jr., *Text. Res. J.* **40** (1970) 430.
[7.88]   W. A. Reeves, G. L. Drake, Jr., R. M. Perkins: *Fire Resistant Textiles Handbook*, Technomic Publishing Co., Westport, Conn. 1974, p. 45.
[7.89]   F. V. Davis, J. Findlay, E. Rogers, *J. Text. Inst. Trans.* **40** (1949) T839.
[7.90]   S. J. O'Brien, *Text. Res. J.* **38** (1968) 256.
[7.91]   G. L. Drake, J. V. Beninate, J. D. Guthrie, *Am. Dyest. Rep.* **50** (1961) 129.
[7.92]   Albright and Wilson Ltd., US 2983623, 1961 (H. Coates).
[7.93]   J. V. Beninate, E. K. Boylston, G. L. Drake, Jr., W. A. Reeves *Am. Dyest. Rep.* **57** (1968) no. 25, 74.
[7.94]   J. V. Beninate, R. M. Perkins, G. L. Drake, Jr., W. A. Reeves, *Text. Res. J.* **39** (1969) no. 4, 374.
[7.95]   J. V. Beninate, E. K. Boylston, G. L. Drake, Jr., W. A. Reeves, *Text. Res. J.* **38** (1968) no. 3, 267.
[7.96]   N. Flügel, *Melliand Textilber.* **71** (1990) 219.
[7.97]   Ciba Ltd., US 3374292, 1968 (A. C. Zahir).
[7.98]   M. A. White, *Text. Prog.* (1983) no. 2, 13.
[7.99]   W. H. Hempel, *TPI Text. Prax. Int.* **34** (1979) 1626.
[7.100]  B. I. J. Maxon, J. Barritt: *A Survey of Scouring and Milling in the Wool Industry*, Wira, Leeds, UK, June 1950, p. 87.
[7.101]  W. Zhao, M. T. Pailthorpe, *Text. Res. J.* **57** (1987) no. 1, 39.
[7.102]  W. Zhao, M. T. Pailthorpe, *Text. Res. J.* **57** (1987) no. 9, 523.
[7.103]  C. Wang, M. T. Pailthorpe, *Text. Res. J.* **57** (1987) no. 12, 728.
[7.104]  R. H. Earle, R. H. Saunders, L. R. Kangas, *Appl. Polym. Symp.* **18** (1971) 707.
[7.105]  A. delaMaza, J. Z. Parra, L. J. Sanchez, F. Comelles, *Text. Res. J.* **58** (1988) no. 10, 571.
[7.106]  K. R. F. Cockett, R. Kettlewell, D. M. Lewis, P. Smith, *J. Soc. Dyers Colour.* **96** (1980) 214.
[7.107]  A. Bourn et al., *Proc. Int. Wool Text. Res. Conf. 7th* **IV** (1985) 272.
[7.108]  K. Reincke, *Melliand Textilber.* **74** (1993) 408.
[7.109]  R. J. Mayfield, *Text. Prog.* (1982) no. 4, 11.
[7.110]  M. Lipson, J. B. McPhee, *Text. Res. J.* **28** (1958) 679.

**References for Chapter 8**

**General References**

B. Gnauck, P. Fründt: *Einstieg in die Kunststoffchemie*, Hanser Verlag, München 1991.
Schweiz. Verband der Geotextil-Fachleute (SVG): *Das Geotextil Handbuch*, 2nd ed., Vogt-Schild AG, Solothurn 1988.
K. H. Möller: *Markisen-Handbuch*, Verlag Kleffmann, Bochum 1984.
*Kirk-Othmer*, 4th ed., **3**, 931–962; **7**, 29–108; **6**, 595–605, 606–635, 635–661, 661–669; **1**, 445–446, 459, 461; **8**, 905–934.
*Kunststoff-Handbuch*, vol. 10, Hanser Verlag, München 1988, pp. 490–508, 763–775, 731–742, 911–921, 908–910.
M. Peter, H. K. Rouette: *Grundlagen der Textilveredlung*, Deutscher Fachverlag, Frankfurt/Main 1989.
J. Lünenschloß, G. Albrecht: *Vliesstoffe*, Thieme Verlag, Stuttgart 1982.
P. Sroka: *Handbuch der textilen Fixieranlagen*, Hartung-Gorre, Konstanz 1993.
M. Antonietti et a., Kolloide, Company brochure, BASF, Ludwigshafen 1994.

**Specific References**

[8.1]   *Textilhilfsmittel-Katalog* 1994/95, Konradin Verlag, Leinfelden 1994.
[8.2]   G. Egbers: "Technische Textilien – ein Wachstumsmarkt," in Gesamttextil (eds.): *Jahrbuch der Textilindustrie 1986*, Frankfurt 1986, pp. 27–32.
[8.3]   H. Viereck: "Die dritte Säule der Textilproduktion," in Gesamttextil (eds.): *Jahrbuch der Textilindustrie 1988*, Frankfurt 1988, p. 42.
[8.4]   Gesamtverband der dtsch. Textilveredlungsindustrie, *Jahresbericht 1992*, Frankfurt/Main p. 55.
[8.5]   TEGEWA, *Melliand Textilber.* **71** (1990) 315.
[8.6]   K. Mayer: "Kettengewirkte technische Textilien," *Kettenwirk Prax.* **2**/1987, company brochure, Obertshausen.
         P. Böttcher: "Techtextil '93, Neue Anwendungsmöglichkeiten," *Int. Text. Bull.* **39** (1993) no. 3, 20–24.
         S. Anton, Berliner Stoffdruckerei, Fahnenfabrik, "Veredlung von Fahnen," personal communication, 1993.
         K. Mayer, Obertshausen: *Chemiefasern/Textilind.* **36/88** (1986) T4.
         P. Ehrler, K. Maute, "Überlegungen zu einer Gliederung des Arbeitsgebiets Technische Textilien," *Chemiefasern/Textilind.* **36/88** (1986) T102–T105.
[8.7]   H.-J. Koslowski: "Aktuelle Markttrends für technische Textilien," *Techn. Text.* **36** (1993) T148.
         R. Bauer, H.-J. Koslowski: *Chemiefaser Lexikon 1993*, Deutscher Fachverlag, Frankfurt 1993, pp. 238, 239.
[8.8]   E. Otto: "Recycling von thermoplastischen Elastomeren in der Automobilindustrie," *Kunststoffe* **83** (1993) 188–194.

Sächs. Textilforschungsinstitut, Symposium: "Reißfaser '93," Chemnitz, Dec. 1993.

A. Watzel: "Vom Textilabfall zum Nonwoven-Produkt – Nutzen durch Recycling," *Melliand Textilber.* **73** (1992) 397–401, 487–496, 561–563.

T. Lampe, M. Bachor: "Anforderungen an Autotextilien am Beispiel von Sitzbezugstoffen," *Tech. Text.* **36** (1993) T207–T213.

R. J. Ehrig: *Plastics Recycling, Products and Processes*, Hanser Verlag, München 1992.

D. Eisele: "Nadel-Polvliesbeläge für den Automobilbau," Fa. Borgers, company brochure, Bocholt.

D. Eisele: "Nadel-Polvliesbeläge für den Automobilbau," *TPI Text. Prax. Int.* **47** (1992) 723–727.

E. Hufnagl, A.-M. Bartl, R. Arnold: "Möglichkeiten zur direkten Verarbeitung von Kunststoffabfällen zu technischen Texilien," *Melliand Textilber.* **74** (1993) 913, 914.

[8.9] J. Brandt: "Faserverbundwerkstoffe werden überwiegend dort eingesetzt, wo extremer Leichtbau gefragt ist," *Frankfurter Allgemeine Zeitung*, 50 March 1, 1993.

K. Fischer: "Technische Textilien mit eigenem Weltforum in Frankfurt," *TPI Text. Praxis Int.* **41** (1986) 759, 760.

J. C. Howson, R. J. Rymill, R. F. Pinzelli: "Aramidfaser-Laminate für den Bootsbau," *Kunststoffe* **83** (1993) 546–549.

P. Ehrler, M. Hottner, G. Schmeer-Lioe: "Entwicklungstechnische Fragen zu Barriere-Textilien," *Int. Text. Bull.* **39** (1993) no. 4, 6–10.

B. Hinz: "Aufgabe und Funktion der textilen Verstärkung in hochbelasteten Faserverbundwerkstoffen," *Chemiefasern/Textilind.* **41/93** (1991), T34–T38.

R. Kleinholz, C. Lackmann: "Technische Textilien aus Glasfilamentgarnen: Märkte heute," *Chemiefasern/Textilind.* **41/93** (1991) T211–T214.

J. Rellmann, H. Schenck: "Barrieremedian," *Kunststoffe* **82** (1992) 729–738.

*Int. Encycl. Compos.* **3** (1990) 102–126, 172–174, 394–420, 420–445, 458–465, 490–525.

A. R. Bunsell: *Composite Materials,* vol. 2, Elsevier, Amsterdam 1988.

B. Clauß, H. Herlinger: "Anorganische Filamentgarne für technische Textilien und Verbundwerkstoffe," *TPI Text. Prax. Int.* **48** (1993) 905.

P. Artzt et al.: *Die Zukunft der Textilindustrie*, expert Verlag, Ehningen 1993, pp. 133, 149.

[8.10] TEGEWA: *Zur Nomenklatur der Textilhilfsmittel, Leder- und Pelzhilfsmittel, Papierhilfsmittel und Tenside*, Frankfurt 1987.

[8.11] DIN/ISO 1629, Kautschuk und Latices, 1992.

[8.12] a) R. D. Athey: *Emulsions Polymer Technology*, Marcel Dekker, New York 1991.
b) S. M. Milnera, W. Schulz, S. Sisman: "Hydrolysebeständigkeit von Baumwollem," *Melliand Textilber.* **76** (1991) 641–654.
G. Odian: *Principles of Polymerization*, J. Wiley, Chichester 1992.
H. Marschner, G. Mautner: "Abhängigkeit der Verfahrensgestaltung auf das Qualitätsbild von Emulsionspolymerisaten," *Plaste Kautsch.* **33** (1986) 85–87.
H. Warson: "Developments in Emulsion Polymerization," *Polym. Paints Colour J.* **180** (1990) 473–475, 486, 507–512.

G. Markert: "Zur Herstellung von Polymer-Dispersionen," *Angew. makromol. Chem.* **123/124** (1984) 285–306.

[8.13] DIN 55958, Harze, 1988.

[8.14] "Der Vinyl- und Acryl-Emulsions-Markt in Europa," *TPI Text. Prax. Int.* **48** (1993) 11.

[8.15] K. Fischer: "Binder und Binderadditive – Variationsbreite unter Umweltgesichtspunkten," *TPI Text. Prax. Int.* **48** (1993) 709–714.
D. Patil, R. D. Gilbert, R. E. Fornes: "Environmental Effects on Latex Paint Coatings," *J. Appl. Polym. Sci.* **41** (1990) 1641–1650.

[8.16] BGBl I, 1945, 1946 Lebensmittel- und Bedarfsgegenständegesetz, Aug, 15, 1974; BGBl. I, 526, Neufassung v. 8. 7. 1993.
Gesetz zum Schutz vor gefährlichen Stoffen (Chemikaliengesetz), 1990.
7. Änderungsrichtlinie 92/32 EWG zur Richtlinie 67/548/EWG zur Angleichung der Rechts- und Verwaltungsvorschriften für die Einstufung, Verpakkung und Kennzeichnung gefährlicher Stoffe. Amtsblatt der EG No. L154/1 – 29. v. 05.06.1992.
6. Änderung der EG-Richtlinie 67/548 EWG entspricht dem deutschen Chemikaliengesetz von 1980.
K. Fischer: "Polymer-Dispersionen, Einsatz als Bindemittel für Vliesstoffe unter Umweltgesichtspunkten," *Textilveredlung* **28** (1993) 212–129; "Umweltaspekte bei Polymer-Dispersionen auf Basis Vinylacetat/Ethylen und Acrylester," *Textilveredlung* **26** (1991) 354–361.
ISO TC 45, Rubber and Rubber Products, Working Draft II, Head Space Method, 1993.
Bestimmung von leichtflüchtigen KW in Latices, Prüfverfahren TFI No. 8, Teppich-Forschungsinstitut, Aachen 1992.
B. Kolb, P. Pospisil, M. Auer: "Quantitative Headspace Analysis of Solid Samples," *Chromatographie* **19** (1984) 113–122.

[8.17] R. Franck: "Kunststoffe im Lebensmittelverkehr," *Loseblattsammlung*, 41st., XIV. Kunststoffdispersionen, 15.04.1991, Carl Heymanns Verlag, Köln 1991.

[8.18] EC-Guideline 92/39 EC, Plastics Materials and Objects which come into Contact with Foodstuffs, *Off. J. EC* **L168** (1992) 21–29.
EC-Guideline 78/142 EC, Vinyl Chloride Monomer Containing Materials and Objects which come into Contact with Foodstuffs, *Off. J. EC* **L44** (1978) 15–17.
Bedarfsgegenständeverordnung, Bundesgesetzblatt 1992, part I, no. 20, pp. 866–902, Apr. 10, 1992.

[8.19] DIN 53773/1, Prüfung von Farbmitteln in PVC-Pasten (Plastisolen), 1988.
"Beschichtung von Twaron-Aramidgarn-Geweben," *Chemiefasern/Textilind.* **39/91** (1989) T148.
W. R. Bursian: "PVC-Plastisole zur Ausrüstung synthetischer Trägermaterialien," *Plastverarbeiter* **30** (1979) 57–64.
H. Hille: "Herstellen und Eigenschaften von PVC-Pasten," *Kunststoffe* **68** (1978) 735–741.

[8.20] K. Fischer: "Polymer-Dispersionen für tT," *TPI Text. Prax. Int.* **42** (1987) 518–522; *Can. Text. J.* **105** (1988) 22–29.
H. Bille: "Hochpolymere in der modernen Textilausrüstung." *Melliand Textilber.* **65** (1984) 704–709.

[8.21] H.-J. Adler, J. Lichtenbelt, A. Reuvers: "Moderne Methoden zur Charakterisierung von Polymerdispersionen," *Farbe + Lack* **97** (1991) 103–108.
K. Fischer, B. Schwalenstöcker: "Zur Stabilität von Polymer-Dispersionen," *Dtsch. Färber Kal.* **93** (1988) 154–163.

[8.22] K. Fischer: "Tensid-Additive für Textil-Polymerdispersionen," *TPI Text. Prax. Int.* **44** (1989) 503, 506–508, 510, 517.

[8.23] H. Wilhelm: "Reaktionsfähige Hochpolymere für die Textilveredlung," parts 1–4, *Textilveredlung* **23** (1988) 96–100, 178–181, 225–230, 307–312.

[8.24] DIN 53775/3 Farbmittel in PVC – Ausbluten, Migration 1984.

[8.25] F. Molenaar, P. Buijsen, C. Smit: "Haftung elektronenstrahlhärtender Beschichtungen auf Metallen," *Farbe + Lack* **98** (1992) 679–681.
K. Fischer: "Veredlung von textilen Flächengebilden durch Bildung von Polymeren auf der Faser," *Melliand Textilber.* **53** (1972) 692–696, 813–819.
Z. A. Rogovin: *Chemiefasern*, Thieme Verlag, Stuttgart 1982.

[8.26] U. Eisele: "Strahlenchemische Veredlung von Baumwolle," *Melliand Textilber.* **72** (1991) 440–446.
B. Clauß: "Elektronenstrahlinduzierte Polymerisation wäßriger Präpolymer-Emulsionen," *Melliand Textilber.* **72** (1991) 866–869.
P. M. Fletcher, S. V. Nablo: "Environmental Benefits of Electron beam/UV Curing," *2nd Int. Conf. on Coating and Laminating*, Zürich, Nov. 9, 1992.
H. Herlinger, P. Hirt: "Strahlenvernetzte PUR-Fasern," *Chemiefasern/Textilind.* **43/95** (1993), 810.

[8.27] K. Fischer: "Veredlungsverfahren mit polymerisierbaren Monomeren," *Textilveredlung* **8** (1973) 397–412.

[8.28] DIN 51007, Differenzthermoanalyse, 1992.
DIN 51005, Thermische Analyse, Begriffe 1983.
S. Knappe: "Thermische Analyse in der Qualitätssicherung," *Kunststoffe* **82** (1992) 993–998.
E. Kaisersberger, H. Möhler: "DSC an Polymerwerkstoffen," *Netsch-Jahrbücher*, vol. 1, Netsch Gerätebau, Selb 1991.

[8.29] K. Fischer: "Die Glastemperatur," *Textilveredlung* **19** (1984) 146–152.

[8.30] K. Fischer, N. Ruiz, H.-G. Herrel: "Festigkeitsvergleich an Nonwoven und Bindemittelfilmen," *Vliesstoff Nonwoven Int.* **1** (1986) 259, 260.

[8.31] K. Fischer: "Was leisten Polymer-Dispersionen bei der Vliesbindung," *Chemiefasern/Textilind.* **38/90** (1988) 678, 680, 682, 683.
K. Fischer, G. Schwarz, H.-J. Hein: Qualitative Binderanalyse durch TG-Bestimmung, *Taschenbuch für die Textilindustrie*, Fachverlag Schiele & Schön, Berlin 1991, pp. 405–415.
ISO TC 45, Rubber and Rubber Products, Working Draft II, Head Space Method, 1992.
Bestimmung von leichtflüchtigen KW in Latices, Prüfverfahren TFI No. 8, Teppich-Forschungsinstitut, Aachen 1992.
B. Kolb, P. Pospisil, M. Auer: " Quantitative Headspace Analysis of Solid Samples," *Chromatographie* **19** (1984) 113–122.

[8.32] P. Hardt: "Chemie und Eigenschaften von Hydrophilausrüstungen," *TPI Text. Prax. Int.* **45** (1990) 387–392.

R. Van der Beke, L. Dekoninck: "Neuentwicklungen auf dem Gebiet der mikroporösen Beschichtungen," *Melliand Textilber.* **67** (1986) 824–829.
J. Hemmrich: "Polyurethanbeschichtung ohne Lösemittel," *Int. Text. Bull.* **39** (1993) no. 3, 53–56. D. Bechter, H. Herlinger, E. Bader, B. Wittstock: "Die Polymer-Faser-Haftung bei Beschichtungen/PUR-Dispersionen," *Melliand Textilber.* **70** (1989), 952–956.
"Wäßrige Polyurethan-Beschichtungssysteme," *Chemiefasern/Textilind.* **39** 91 (1989) T 149, T 150.
W. Schröer: "Die Beschichtung von Textilien mit Polyurethanen," *Textilveredlung* **22** (1987) 459–467.
K. H. Stukenbrock: "Ironisch koagulierbare PUR-Dispersionen," *Melliand Textilber.* **65** (1984) 756–758.

[8.33] DIN 53787, Prüfung von wäßrigen Kunststoff-Dispersionen; Bestimmung der Mindestfilmbildetemperatur, 1974.

[8.34] K. Fischer, R. Schardt: "Die Glastemperatur II," *Textilveredlung* **20** (1985) 49–52.

[8.35] H. Herlinger, D. Bechter, G. Kurz: "Wirkungsweise von Bindemitteln bei der Vliesverfestigung," *Chemiefasern/Textilind* **31/83** (1981) 936–942.
K. Fischer: "Moderne Trends mit Appretan-Marken für die Ausrüstung, zum Binden und Beschichten von Textilien," *Mellaind Textilber.* **71** (1990) 290–303; "Aktuelles zur adhesiven Vliesverfestigung," *Chemiefasern/Textilind.* **41/93** (1991) 800–802.
K. Fischer, "Bindemittel standen zur Diskussion auf der Index '93", *Vliesstoffe Nonwoven Int.* **8** (1993) 123.
EDANA, Nonwovens, brochure, Brussels 1990.
K. H. Schumacher et al.: "Neue Generation formaldehydfreier Acrylatbinder," *Chemiefasern/Textilind.* **43/95** (1993) 621–624.
Röhm GmbH: "Bindemittel, für Vliesstoffe," *Allg. Vliesstoff-Rep.* **8** (1980) 196–204.
W. Stepanek: "Plextol-Bindemittel für Beschichtungsträger aus PES-Vlies," *Allg. Vliesstoff-Rep.* **18** (1990) 72–76.
"Ethylen-Copolymer-Emulsionen als Binder für Nonwovnes," *Melliand Textilber.* **63** (1982) 894–896.
W. Loy: "Vliesstoffe – eine Übersicht nach Herstellungsart und Verwendungszweck," *Textilveredlung* **28** (1993) 204–212.
"Nonwovens: The Bonded Fabrics," *Nonwoven World* **1990** no. 1, 16–23.

[8.36] U. Eisele: "Färben mit Hilfe von Bindemitteln," *Melliand Textilber.* **72** (1991) 847–851.

[8.37] M. Rasche: "Gibt es einen Zusammenhang zwischen Benetzbarkeit und Haftung?" *Flock* **19** (1993) no. 71, 6, 8, 10–14.
U. Zorll: "Einfluß von Pigmentgehalt und Polarität auf das Haftvermögen von Flock-Klebstoffen zu Substrat und Flock," *Flock* **14** (1987) no. 46, 7–20.
G. P. Bauer: "Qualitätsprüfung beflockter Materialien für die Autoindustrie," *Chemiefasern/Textilind.* **41/93** (1991) T 23–27.
K. Fischer: "Aktuelle Aspekte zu Polymer-Dispersionen, die als Flockklebstoffe verwendet werden," *Flock* **17** (1991) no. 62, 12–17.
V. W. Dimitriew: *Faserpräparation für die elektrostatische Beflockung*, University of Leningrad 1986.

[8.38] A. Stei: "Die Tuftingteppichbeschichtung in der Bewährung," *Chemiefasern/Textilind.* **40/92** (1990) no. 92, 723–726.
K. Fischer, G. Schachtner: "Aspekte der Teppichrückenbeschichtung mit Polymer-Dispersionen," *TPI Text. Prax. Int.* **40** (1985) 990–992, 997, 998, 1092, 1097–1100.

[8.39] S. Schwenkedel: *Textilbeschichtung – Kunstleder-Syntheseleder,* Eder Verlag, brochure, Stuttgart 1970.
DIN 55945, Beschichtungsstoffe, Begriffe, 1988.
K. Fischer, K. Winter: "Textil-Beschichtungen," *Dtsch. Färber Kal.* **89** (1985) 150–170.
R. N. Hildred: "Acrylic Additives for Waterborn Coatings," *Polym. Paint Colour. J.* **180** (1990) 579–583.
O. Lückert: "Bindemittelprüfung in Normung und Literatur," *Farbe + Lack* **98** (1992) 277–280.

[8.40] H. Zander: "Kleben statt Weben," *Kunststoffe* **83** (1993) 100, 101.
H. J. Kauderer: "Verfahrenstechniken der Kontinuierlichen Textillaminierung," *TPI Text. Prax. Int.* **48** (1993) 895–899.
E. Bürkle, G. Rehm, K. Zweig: "Großflächige Bauteile im Urformwerkzeug kaschieren," *Kunststoffe* **82** (1992) 896–901.

[8.41] C. Trebs: "Korrelationen zwischen Polymer-Dispersionen und Polymer-Pulvern," Thesis, Fachhochschule Kaiserslautern 1993.

[8.42] K.-H. Stukenbrock: "Möglichkeiten des Druckens von Vliesen," *Melliand Textilber.* **71** (1990) 303–305.

[8.43] DIN/ISO 1133, Bestimmung des Schmelzindex (MFR) 1993.

[8.44] R. Hinterwaldner: "Schmelzklebstoffe," *Coating* **24** (1991) 110–114.
"Eine breite Palette Hüls-Produkte für textile Anwendungen," *Allg. Vliesstoff-Rep.* **19** (1991) 68–70.
C. Lipman: "Powder Coatings Hope for Legislation," *Eur. Chem. News.* **8** (1993) 28.
"Wer liefert welches Pulver?" *Oberfläche JOT,* Market Survey (1992) 41–44.
E. Lattke, H. Drecker: "Mit Pulver ins Jahr 2000," *Oberfläche + JOT* (1992) no. 9, 34–40.
H. Scholten: "Hot-Melt Adhesives for Textile Laminates," *Symposium Coating,* Zürich, Nov. 9 1992.
H. Jahn: "Verbunde mit Vliesstoffen unter Verwendung von Schmelzklebern," *Melliand Textilber.* **74** (1993) 297–299.

[8.45] G. Schwarz, Hoechst AG, personal communication 1992.

[8.46] V. Baron: "Herstellung, Eigenschaften und Anwendungen von Schmelzklebefasern," *Techn. Text.* **36** (1993) T 200–206.
D. Bechter, G. Kurz, E. Maag, J. Schütz: "Der Einfluß der Kalandrierbedingungen, auf die Festigkeit von PP-Vliesstoffen," *TPI Text. Prax. Int.* **46** (1991) 1236–1240.
"Bindefasern (Produktion Westeuropa)," *Int. Text. Bull.* **39** (1993) 34.

[8.47] DIN 7724, Polymere Werkstoffe, 1993.

[8.48] A. Gardiella: "Duroplaste im Automobilbau," *Kunststoffe* **83** (1993) 57–59.

[8.49] R. Schäfer: "Phenolharze für schalldämpfende Verkleidungselemente im Fahrzeugbau," *Kunstharz Nachr.* (1993) no. 29, 16–19.
DIN ISO 10082, Phenolharze, Begriffe und Prüfverfahren, 1992.

A. Knop, W. Scheib: *Chemistry and Application of Phenolic Resins,* Springer Verlag, Berlin 1979.
A. Gardziella, H. G. Haub: "Phenolharze," in: *Kunststoff-Handbuch,* vol. 10, Hanser Verlag, München 1988, pp. 14–40.

[8.50] R. Kleinholz: "Vorgeformte textile Flächengebilde für Kunststoff-Formteile," *Chemiefasern Textilind.* **40/92** (1990) T 182.
S. Anton, Berliner Stoffdruckerei, Fahnenfabrik, personal communication.

[8.51] Hoechst, *Hostaflon,* Company brochure, Frankfurt 1984.
Hoechst, *Hostafalon TF-Dispersionen,* Frankfurt 1992.
Hoechst, *Fluorkunststoffe und Feuer, Fragen und Antworten,* Frankfurt 1987.
P.-K. Eichhorn: "Hostaflon beschichtetes Glasgewebe in Neue Perspektiven in der Architektur," *Magazin Hoechst,* Frankfurt 1989.

[8.52] G. Tegge: *Stärke und Stärkederivate,* Behrs Verlag, Hamburg 1984.

[8.53] H.-J. Jacobasch, K.-H. Freitag, U. Panzer, K. Grundke: "Charakterisierung und Modifizierung der Oberflächeneigenschaften von Fasern für Verbundwerkstoffe," *Chemiefasern/Textilind.* **41/93** (1991) T 39–T 47.

[8.54] H. Johnen: "Schmirgeln, Rauhen, Scheren," *Textilveredlung* **27** (1992) 348–351.

[8.55] B. von Falkai: *Synthesefasern,* Verlag Chemie, Weinheim 1981, pp. 370–373.

[8.56] "Blei-Akkumulator mit bleibeschichteten Glasfaser-Elektroden," *Frankfurter Allgemeine Zeitung,* Nov. 24, 1993.

[8.57] "Glass Fibers, Fiber Tables According to P. A. Koch 1993," *Techn. Text.* **43** (1993) E 89–E 100.

[8.58] K. W. Gerstenberg: "Sortex, Oberflächenmodifikation von Textilien durch Koronabehandlung," *Allg. Vliesstoff-Rep.* **21** (1993) 28.

[8.59] C. Gruner, B. Rapp, H. J. Zimmermann: "Problemlösungen beim Lackieren von PP-Blends," *Kunststoffe* **82** (1992) 802–806.

[8.60] T. Bahners: "Laser zur Materialbearbeitung in der Textilveredlung," *Textilveredlung* **26** (1991) 319–326.
D. Knittel, E. Schollmeyer: "Haftung von UV-laserbestrahlten Hochleistungsfasern zu Gummi und Epoxidharz," *Techn. Text.* **36** (1993) T 61–T 66.
T. Ryback, D. Knittel, E. Schollmeyer: "Pigmentdruck auf UV-laservorbehandeltem PES-Gewebe," *Melliand Textilber.* **73** (1992) 985–989.

[8.61] H. Herlinger, R. Braun, Institut f. Textilchemie, Universität Stuttgart. DE 4142776 A1, 1991 (H. Herlinger, R. Braun); DE 4219280 A1, 1992 (H. Herlinger, R. Braun).

[8.62] J. Mischke, G. Bagusche: "Hinterspritzen von Textilien, Teppichen und Folien," *Kunststoffe* **81** (1991) 199–203.

[8.63] DIN 55945, Beschichtungsstoffe, Begriffe, 1988.

[8.64] J. K. Rogers et al.: "Chemicals and Additives," *Mod. Plast. Int.* **1993,** Special Report, 33–68.
K. Fischer: "Additive für Dispersionen – Dispersionen als Additive" *Taschenbuch für die Textilindustrie 1987,* Verlag Schiele & Schön, Berlin 1987, pp. 329–343; "Additive für Polymer-Dispersionen: Tenside, Verdicker, Katalysatoren," *Chemiefasern/Textilind.* **36/88** (1986) 920, 923, 924.

[8.65] T. Götze, W. Adam: "Melamin-Formaldehydharze," *Kunststoffe* **80** (1990) 1185–1188.

W. Adam: "Melaminharze," in: *Kunststoff-Handbuch,* vol. 10, Hanser Verlag, München (1988) pp. 41–50.

[8.66] DIN 53868, Bestimmung des Nahtschiebewiderstandes, 1992.

[8.67] A. J. Sabia, G. A. Policello: "Silicon finishes Enhance Nonwoven Substrates," *Nonwoven World* **1990**, no. 6, 42–44.

[8.68] M. Lewin, S. B. Sello: "Chemical Processing of Fibers and Fabrics, Functional Finishes, Repellent Finishes," *Handbook of Fiber Science and Technology,* vol. 2: Marcel Dekker, New York 1984, pp. 144–204.

D. Lämmermann: "Fluorcarbone in der textilen Endausrüstung – Eigenschaften und Anwendungen," *Melliand Textilber.* **72** (1991) 949–954.

D. Lämmermann: "Moderne Fleckschutzausrüstung von Textilien," *Melliand Textilber.* **74** (1993) 883–889.

M. Wilhelm: "Fluorchemikalien für die Textilausrüstung," *Int. Text. Bull.* **39** (1993) 57–60.

R. E. Wiltgen: "Fluorchemikalien," *Textilveredlung* **21** (1986) 384–389.

K. Fischer: "Hygiene und Schmutz aus Sicht der Textilveredlung," *Melliand Textilber. Int.* **54** (1973) 1239–1242.

R. Kleber: "Moderne Aspekte der textilen Oleophob-Ausrüstung," *TPI Text. Prax. Int.* **27** (1972) 499–503.

[8.69] M. Lewin, S. B. Sello: "Chemical Processing of Fibers and Fabrics, Flame Retardancy," *Handbook of Fiber Science and Technology*, vol. 2, Marcel Dekker, New York, 1984 pp. 2–122.

DIN 4102, Brandverhalten von Baustoffen und Bauteilen; Brandklassen, 1988.

[8.70] B. Hanneken, M. Schuirer, A. Schüler, H. K. Rouette: "Moderne flammhemmende Ausrüstung für 'Black-out'-Vorhang-Beschichtung," *Int. Text. Bull.* **39** (1993) no. 3, 38–46.

[8.71] J. Laser, J. Glaser: "Elektrisch ableitende Teppichverbundsysteme," *Melliand Textilber.* **74** (1993) 630–631.

S. Rolfe: "Spezielle Antistatikfaser," *Vliesstoff Nonwoven Int.* **8** (1993), 138, 139.

F. Marchini: "Metallisierte Fasern zum Schutz gegen Elektrostatik und Strahlung," *Chemiefasern/Textilind.* **40/92** (1990) T 164–T 170.

[8.72] DIN 75201, Bestimmung des Fogging-Verhaltens, 1989.

P. Hardt: "Prognosen zum Fogging- und Abluftverhalten von Textilhilfsmitteln," *TPI Text Prax. Int.* **49** (1994) 163–167.

F. Look, T. Lampe, A. M. Bahadir: "Temperaturabhängigkeit des Fogging-Phänomens," *Kunststoffe* **83** (1993) 201–205.

P. Ehrler, H. Schreiber, S. Haller: "Emission textiler Automobil-Innenausstattung," *Textilveredlung* **29** (1994) 254–260.

D. Jachowski, A. C. Poppe: "Fogging-Messungen mit variiertem Verfahren," *Kunststoffe* **82** (1992) 818–820.

DIN EN 665, Elastische Bodenbeläge; Bestimmung der Weichmacherabgabe, 1992.

[8.73] H. Herlinger, U. Erzinger, G. Schulz: "Rheologische Charakterisierung und Qualitätsbeurteilung von Druckpasten und Verdickungen," *TPI Text. Prax. Int.* **46** (1991) 247–252.

[8.74] J. Warmund-Cordelier: "Zuverlässiger Schutz gegen Mikroorganismen," *Textilveredlung* **27** (1992) 395–397.

T. Klinkenberg, W. Stöneberg: "Symprol – umweltfreundliche Markisenstoff-Imprägnierung," *Chemiefasern/Textilind.* **36/88** (1986) 608–610.

K. H. Wallhäußer, K. Fischer; "Die antimikrobielle Ausrüstung von Textilien," *Textilveredlung* **5** (1970) 3–14.

K. Fischer, K. H. Wallhäußer: "Die antimikrobielle Ausrüstung – ein Problem unserer Zeit," *Dtsch. Färber Kal.* **74** (1970) 324–344.

K. H. Wallhäußer, W. Fink: "Konservierung von Polymer-Dispersionen," *Farbe + Lack* **76** (1970) 471; **82** (1976) 108.

K. Fischer: "Hygiene und Schmutz aus Sicht der Textilveredlung," *Melliand Textilber. Int.* **54** (1973) 1239–1242.

[8.75] G. Reinert, V. Misun: "Neuer UV-Absorber für PES-Faserstoffe," *Melliand Textilber.* **74** (1993) 1007–1014.

Hoechst, *HALS-Produkte für hochwirksamen Lichtschutz*, company brochure, Frankfurt 1992.

[8.76] "Textilien mit hohem Sonnenschutz," *Frankfurter Allgemeine Zeitung*, Sept. 24, 1993.

[8.77] Degussa, Pigmente company brochure no. 47: Ruß, Frankfurt 1991.

Omya-Füllstoffe in Latex-Teppichrückenbeschichtungen, Laborbericht no. 6446, Plüss-Staufer, Oftringen, Schweiz, 1988.

"Keramikteilchen eingearbeitet in schützendem Vliesstoff," *Vliesstoff Nonwoven Int.* **7** (1993) 87.

[8.78] DIN 55943, Farbmittel, Begriffe, 1993.

# Textile Dyeing

HERBERT LEUBE, BASF Aktiengesellschaft (retired), Ludwigshafen, Federal Republic of Germany (Chaps. 1, 7, 8, 10, 11, Sections 4.7–4.9)

WILHELM RÜTTIGER, BASF Aktiengesellschaft (retired), Ludwigshafen, Federal Republic of Germany (Chaps. 2, 3)

GERD KÜHNEL, BASF Aktiengesellschaft, Ludwigshafen, Federal Republic of Germany (Section 4.1)

JOACHIM WOLFF, Bayer AG, Leverkusen, Federal Republic of Germany (Section 4.2)

GÜNTHER RUPPERT, MICHAEL SCHMITT, BASF Aktiengesellschaft, Ludwigshafen, Federal Republic of Germany (Sections 4.3, 4.4)

CHRISTIAN HEID, Cassella AG (retired), Frankfurt, Federal Republic of Germany (Section 4.5)

MAX HÜCKEL, Hoechst Aktiengesellschaft (retired), Offenbach, Federal Republic of Germany (Section 4.6)

HANS-JOACHIM FLATH, Dresden, Federal Republic of Germany (Chaps. 5, 6)

WOLFHARD BECKMANN, Bayer AG (retired), Leverkusen, Federal Republic of Germany (Chap. 9)

ROLF BROSSMANN, Bayer AG (retired), Leverkusen, Federal Republic of Germany (Chap. 12)

MANFRED SÖLL, Bayer AG (retired), Leverkusen, Federal Republic of Germany (Chap. 13)

ULRICH SEWEKOW, Bayer AG, Leverkusen, Federal Republic of Germany (Chap. 14)

# 1. History, Economic Importance

## 1.1. Historical Dyeing Methods

The technique of dyeing textiles can be traced back to prehistoric times. Archaeological studies have shown that elaborate chemical–technological dyeing processes were known to all ancient highly civilized cultures. The first written documents containing detailed dyeing instructions are the papyri Leidenenses and Holmiensis dating from the 3rd century A.D. [1.1]. The Plictho of Rosetti is considered to be the oldest printed book on the art of dyeing [1.2]. In preindustrial times, textiles were dyed primarily with plant dyes [1.3], [1.4]–[1.6]. Dyeing with inorganic pigments or the precipitation of metal salts on fibers (mineral dyes), which is still practiced today, was of only secondary importance. Also, dyeing with animal dyes was restricted to special cases. Well known are ancient Tyrian purple (6,6'-dibromoindigo), which is extracted from the Mediterranean purple sea snail, and kermes and cochineal, which contain tetrahydroxyanthraquinone carboxylic acid as the principal dye. These substances are obtained from insects and require a metallic mordant for dyeing.

Dyes isolated from plants, such as indigo from the Indian indigo plant or dyer's woad (→ Indigo and Indigo Colorants) and Turkey red, alizarin (→ Anthraquinone Dyes and Intermediates, **A2**, p. 375), from the root of the madder plant, were of greater significance. Dyeing black with iron salts and tannic acid from the bark of oak was important. Until recently, campeachy wood black, a pyrone dye that occurs in a special Brazilian wood, was used because of its rich hue. Yellow was frequently produced by using dyes of the flavone family, which occur in flowers and roots, e.g., weld (*Reseda luteola* L.), or carotinoids, e.g., crocin from saffron. Dyes from red berries (anthocyans) and the orchella weed (archil) gave dyeings that were less fast. Plant extracts were also used to produce the color green.

Apart from the actual dyeing process, the work of the dyer often included isolation and preparation of dyes from plants that were either imported or grown locally. As a chemical reaction with rather impure raw materials, the dyeing process was complicated and unclear. In fact, dyeing was an art. Indigo was produced by fermentation from plants and had to be applied to the fiber in the reduced form. The reduction step was performed in a fermentation vat with putrefied urine and later with iron sulfate and lime (vitriol vat). Turkey red gave especially fast dyeings as a lake (metal complex) only when the fabric was pretreated with metallic salts. Alum, tin, iron, and copper salts were most commonly used as mordants. The colors obtained depended on the skill and dexterity of the dyer. The old dyeing processes are still being used in the arts and crafts, the substrate being mainly sheep's wool [1.7], [1.8]. For ecological reasons, particularly, the old dyeing processes do not meet the requirements of modern industrial production.

In the 18th century, chemical studies dealt with the nature of color lake formation, the processes involved in vatting, the suitability of individual coloring substances for certain fibers, and making up recipes for particular shades with different dyes. Furthermore, new dyeing and printing techniques were developed in the 18th and early 19th centuries. Apart from providing new color effects, these techniques focused on producing fast dyeings and largely contributed to the upswing in textile manufacture [1.9]. A review of the dyeing methods commonly used at the end of the 19th century is given in [1.10] together with many individual recipes based on publications in *Reimanns Färbezeitung* (1872–1879).

## 1.2. Economic Importance of Textile Dyeing

The worldwide consumption of fibers by the textile industry is shown in Table 1.1 (see also, → Fibers, 1. Survey, **A10**, p. 462). The importance of regenerated cellulose fibers has decreased and that of polyester fibers has clearly increased.

**Table 1.1.** World fiber consumption

| Type of fiber | 1980 | | 1992 | |
|---|---|---|---|---|
| | $10^6$ t | % | $10^6$ t | % |
| Cotton | 14.1 | 47 | 18.7 | 47 |
| Viscose | 3.2 | 11 | 2.4 | 6 |
| Wool | 1.6 | 5 | 1.6 | 4 |
| Polyamide | 3.2 | 11 | 3.9 | 10 |
| Polyester | 5.1 | 17 | 8.3 | 21 |
| Polyacrylonitrile | 2.1 | 7 | 2.2 | 5 |
| Polypropylene | | | 1.8 | 4 |
| Others (silk, linen etc.) | 0.5 | 2 | 1.3 | 3 |

**Table 1.2.** Regional distribution of textile processing, %

| Region | 1976 | 1992 |
|---|---|---|
| Western Europe | 17 | 12 |
| Eastern Europe | 19 | 11 |
| East Asia | 35 | 44 |
| North America | 19 | 16 |
| South America | 6 | 6 |
| Africa, West Asia | 4 | 7 |

Developments since the mid 1970s show that Europe and North America are losing their share of the textile finishing industry and East Asia's share is becoming more significant (Table 1.2). The growing industrialization of China will enhance this trend.

The amount of fiber apportioned to dyeing is closely connected to the consumer and fashion habits of the population. It can be assumed that ca. 75–80 % of the world fiber consumption consists of colored materials; the rest remains undyed or is used for technical articles. Sulfur and reactive dyes for cellulose fibers and disperse dyes especially for polyester fibers are quantitatively in the lead. The estimated world consumption of synthetic textile dyes for dyeing and printing in 1992 (in $10^3$ t) is given below:

| | |
|---|---|
| Indigo | 12 |
| Vat dyes | 26 |
| Reactive dyes | 108 |
| Direct dyes | 45 |
| Naphtols | 19 |
| Sulfur dyes | 100 |
| Cationic dyes | 21 |
| Anionic dyes | 74 |
| Disperse dyes | 102 |
| Pigment preparations | 40 |
| Total | 547 (for comparison 1980: 450) |

More than 80 % of the total consumption of 547 000 t of dye is used for dyeing and the rest for printing.

In textile finishing in Germany (west), the quantitative importance of individual products was shown by an investigation made by the University of Münster [1.11] (Table 1.3).

The methods and processes of the textile finishing industry must adapt to these market requirements. As a result of increasingly shorter yardages, the trend is toward batchwise production and away from the more economical continuous methods. However, cotton and polyester–cotton fabrics as well as certain carpet grades are still dyed in large quantities in continuous processes.

The theoretical principles of dyeing are discussed with individual dye and fiber types.

## 2. Dyeing Technology

### 2.1. General

#### 2.1.1. History

The very first method used to dye clothing was probably with pigments (clay or ocher as the dye and fat or glue as the binder) [2.1], [2.2]. During Old Testament times dyebaths were used, dyeing was performed partly in open lead vessels with fires underneath to heat the liquor [2.3].

For a long period, the main method of dyeing was in open stone or wood vessels in which the yarn was agitated in hank form. The use of a hand-operated rotating mechanism (reel) over the dyebath facilitated the dyeing of fabrics (piece goods).

The use of power driven equipment to move the goods and to drive liquor pumps began in the early 19th century. Also, the first continuous dyeing plants were built at this time, these being

**Table 1.3.** Textile dyeing and finishing in Germany (West) 1992

| Application | Flock (27 000 t), % | Yarn (119 000 t), % | Fabric (1 800 × 10⁶ m²), % | Knitwear (110 000 t), % | Print (600 × 10⁶ m²), % |
|---|---|---|---|---|---|
| Clothing | 11 | | | | 60 |
| Outer clothing | | 42 | 33 | 31 | |
| Lining materials | | | 17 | | |
| Furnishings | 43 | 16 | 28 | | 33 |
| Underwear | | 4 | 3 | 34 | |
| Industrial textiles, especially automotive | 41 | 11 | 15 | 23 | 4 |
| Others | 5 | 27 | 5 | 12 | 3 |

in principle identical to those still in operation today [2.4].

### 2.1.2. The Field of Dyeing Technology

Dyeing technology matches a dyeing process to the goods and the dyeing machines. Dyeing technology also includes laboratory work, necessary not only for the development of formulations (sometimes an extremely difficult task), but also contributing to the development and realization of dyeing processes.

Color measurement for research and production has become an important and quite separate part of dyeing technology. Color measurement is mainly used as an aid in the development of formulations and for the measurement of undesired color variations of the textile material. Another important aspect of dyeing technology is automatic metering. Although the metering of auxiliaries is already widely used, the considerably more difficult metering of dyes is much less widespread.

Dyeing technology is a rapidly expanding area of textile dyeing. Suitable dyeing technology can determine not only the feasibility of manufacturing a new article but also its production costs, and can be essential for the continued existence of some business enterprises.

### 2.1.3. Fundamental Principles of Dyeing

#### 2.1.3.1. Dyeing Systems

In the field of dyeing technology no generally accepted system of classification exists as yet.

Based on the method of introducing the dye into or onto the textile fiber three methods of dyeing can be distinguished:

1) **Exhaustion dyeing**
   Diffusion of the dissolved dye into the fibers (batch and continuous dyeing) [2.5]
2) **Pigment dyeing**
   Deposition of insoluble dye onto the fiber, and fixation with a binder [2.6]
3) **Mass and gel dyeing**
   Incorporation of the dye during production of synthetic fibers

The market share of exhaustion dyeing exceeds that of mass dyeing, which in turn is more important than pigment dyeing.

As exhaustion dyeing is the most important process in terms of market share, the discussion of fundamental principles is limited to this process. Continuous pigment dyeing is mentioned in Section 2.3.

#### 2.1.3.2. Phases of Exhaustion Dyeing

In exhaustion dyeing, a dye which must be at least partially dissolved migrates by diffusion through the surface of the fiber and into its interior. The dye is usually in a liquid medium (the liquor), which also contains the textile material [2.6].

The exhaustion dyeing process can be divided into three phases (Fig. 2.1). Each phase is characterized independently of the other phases as far as possible.

1) **The dyeing phase (exhaustion and absorption)**
   Diffusion of the dye into the fiber (dyeing kinetics)

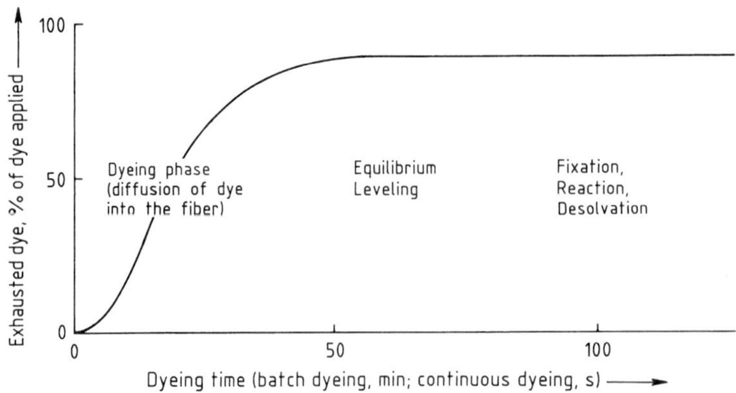


**Figure 2.1.** Phases of exhaustion dyeing

2) **The equilibrium phase**
   The final dye concentration on the fiber (dye yield) is nearly reached; leveling and dye penetration occur at this stage
3) **The dye fixation phase**
   Chemical processes, diffusion or "desolvation" of the dye, fastness improvements

### 2.1.3.3 Dyeing Phase (Dyeing Kinetics)

In this phase, the dye diffuses from the liquor into the textile fiber. In most cases, soft water is used as the liquid medium. Despite intensive research, organic solvents (chlorinated hydrocarbons) have not become established as dyeing media, and air–water foam systems have been even less successful [2.7]. However, some interesting experiments to use supercritical $CO_2$ as liquor have been carried out (see Section 7.2.3) [2.8], [2.9].

There are essentially two techniques available for dyeing the textile material:

1) *Batch Technique:* The liquor and the textile are placed together in a vessel, and the required amount of dye is added.
2) *Continuous Technique:* The dye is dissolved or dispersed in the liquor. A definite quantity of the dye liquor is locally applied to the textile.

In both cases, the dye subsequently diffuses into the textile fibers. Normally, the rate and uniformity of this diffusion are controlled.

The diffusion rate of the dye is normally proportional to its concentration difference between the liquor and the fiber (difference in chemical potential as the driving force for the mass transport, see also Chap. 11) [2.6].

During the exhaustion process, the concentration of the dye in the liquor continuously decreases, so that the concentration difference is also reduced. The dyeing rate therefore continuously slows down and reaches a final value (equilibrium).

The terms "exhaustion" and "absorption" are equivalent in many ways. However, the differences should be noted. Strictly speaking, the *exhaustion* is the decrease in the amount of dye in the bath, because of absorption of the dye by the textile and other processes such as precipitation of the dye. The *absorption* is the deposition of dye in the textile substrate. Absorption thus includes only the dye taken up by the textile material not dye residues or sedimented dye.

In the continuous technique, the term *dye fixation* or *degree of fixation* is often employed. This is the ratio of the dye deposited in the textile substrate to the total amount of dye applied.

In the following, the term exhaustion is used in a general sense, and absorption if the textile substrate is mainly considered.

The characteristic parameter that describes the dyeing properties of a dye in a dyeing system is the rate of dyeing or "strike". The dyeing rate is defined as the quantity of dye absorbed by the textile substrate per unit of time. The *absolute dyeing rate* is expressed in terms of the amount of dye in g/kg or wt% based on weight of fabric (wof). However, the end of the exhaustion process is often of more significance, and hence the rate at which dyeing equilibrium is approached, i.e., the relative dyeing rate. The amount of dye uptake in this case is expressed as percentage of the amount of dye added.

*Example*:

| | |
|---|---|
| Amount of dye added: 1.8% | $= 18$ g/kg goods |
| Absolute dyeing rate | $= 0.4 \text{ g kg}^{-1} \text{min}^{-1}$ |
| | (measured value) |
| Percentage: $(0.4/18) \times 100$ | $= 2.2\%$ of exhausted |
| | dye per minute |
| Relative dyeing rate | $= 2.2\%/\text{min}$ |
| | (calculated) |

If the relative dyeing rate ($2.2\%/\text{min}$) were kept constant, the dyeing time would be $100\% : 2.2\%/\text{min} = 45$ min.

The *instantaneous dyeing rate* can be read off from an exhaustion curve. The slope of the curve gives the value of the instantaneous (relative) dyeing rate, steeper curves indicating higher dyeing rates. Relative dyeing rates in practice fall within two very different ranges:

   $0.5-5\%/\text{min}$ for the batchwise process
   $50-150\%/\text{min}$ for the continuous process
   (i.e., 100% can be absorbed in $< 1$ min.)

What are the causes of these well-established limits? Dwell times in a particular type of dyeing equipment are linked to the relative dyeing rates given above. They are as follows:

   $20-200$ min for the batchwise process
   $0.7-2$ min for the continuous process.

These are the mechanically imposed capabilities of the dyeing techniques available (The semicontinuous process, with a dwell time of several hours, is discussed in Section 2.3.). Although the actual dyeing mechanism remains the same and also many dyes are used in both processes,

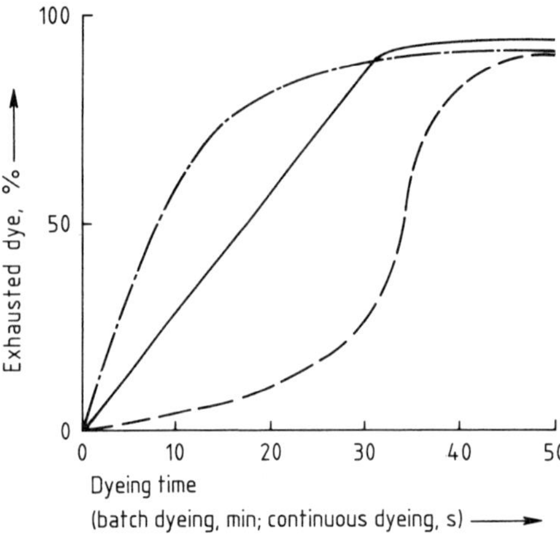



**Figure 2.2.** Kinetics of exhaustion for three different cases
The relative dyeing rate is given by the slope of the curves
— · — · — exponential dyeing curve; —— linear dyeing curve;
— — — S-shaped dyeing curve

significant differences between the two processes exist.

**Batchwise exhaustion dyeing** uses a closed system (dyebath) which contains the textile and the liquor. The dye is added in accordance with the formulation. An intense interaction between the textile and the liquor then occurs. Detailed descriptions of systems are given in Section 2.2.3.

General technical data for batchwise dyeing techniques are given below:

| | |
|---|---|
| Mean dyeing rate | 0.5–5%/min |
| Liquor ratio ("liquor/goods") | 4:1–25:1 (i.e., 4–25 L/kg goods) |
| Dye concentrations (2%, i.e., 20 g dye/kg goods) | 0.8–5g/L |
| Dyeing times (exhaustion) | 20–200 min |

The dyeing rate and hence the dyeing time can be controlled within wide limits: The (relative) dyeing rate increases with increase of dyeing temperature, fiber swelling, and carrier concentration, and with decreasing dye concentration, size of dye molecule (higher diffusion coefficient of smaller molecules), retarder concentration, and often decreasing liquor ratio.

Figure 2.2 illustrates typical exhaustion or absorption curves for batch dyeing. In general, a linear absorption curve gives better leveling.

The batchwise dyeing method shows a tendency to unlevel results. Color correction can be carried out, giving a more reliable dyeing result. Higher dyeing rates are usually associated with poorer leveling properties [2.10], [2.11]. An example of this is the vat dyeing of cellulose fibers. This dyeing usually proceeds so rapidly that an unlevel product results even in modern machines. However, this can be very readily leveled out (see Sections 4.3.6 and 4.4.2).

The batchwise dyeing process has a very large share of the market. It is now a very modern and highly automated technique.

**Continuous Exhaustion Dyeing.** In continuous exhaustion dyeing the dye is first dissolved or dispersed in the liquor, and the dye liquor so produced is locally applied in definite amounts to the textile which continuously passes through. The diffusion transfer of the dye to the textile substrate is controlled by effects on the textile goods, e.g., by temperature and/or auxiliaries.

Detailed descriptions of dyeing systems are given in Section 2.3.

General technical data for the continuous dyeing technique are given below:

| | |
|---|---|
| Mean dyeing rate | 50–150%/min |
| Liquor ratio ("liquor/textile") | 0.4:1–1.2:1 |
| Dye liquor loading | 40–120% |
| Dye concentration (2%) | 17–50 g/L |
| Dyeing times | 0.6–2 min |

(The liquor ratio is given only for comparison with the batchwise technique. This parameter is not normally used.)

If the fibers are hygroscopic, diffusion into the textile frequently occurs immediately after adding the fabric to the dyebath. Sometimes auxiliaries must be used to solubilize the dye and impart its substantivity. Strong heating (e.g., to 100 °C in a steamer) considerably speeds up the process. Some continuous processes also proceed in the cold (cold dwell process, semicontinuous technique).

After the dye has been applied, the textile is often first dried (intermediate drying). During intermediate drying the dye in the liquor migrates with the water into the surface zones of the textile (color intensification by impaired dye penetration, see Chap. 3). If this usually undesired migration is nonuniform, it can lead to unlevelness (at selvedges and ends).

With thermoplastic fibers, true absorption occurs at high temperature by thermosolvation of the dye in the fiber (thermosol process, see Section 7.2.4).

With hygroscopic fibers, developing agents must be applied in aqueous solution. The process then proceeds as described above (see Section 4.3.6). Here, the dyeing rate can be determined partially or entirely by the preliminary reaction, e.g., reduction.

Many proposals have been made aimed at improving the dyeing technique, but none of these has found widespread practical application even after many years. For example, ultrasound is claimed not only to improve the dyeing rate, but also to increase the dyeing yield [2.12].

In the continuous technique, corrections (e.g., of color discrepancies and unlevelness) are very difficult and expensive. However, the conversion of white goods to finished dyed textiles in a single operation is an elegant and hence very attractive process. It is especially important in dyeing big lots (high total yardages).

#### 2.1.3.4. Equilibrium Phase

The equilibrium phase of dyeing follows the exhaustion phase without any sharp transition. It is reached in practice when the dye concentration in the liquor undergoes no further appreciable change [2.6].

Equilibrium is characterized by the terminal degree of exhaustion of the bath or the terminal dye yield as a percentage of the added dye. It is usually in the range 50–99.9 %. The dye still remaining in the liquor increases production and wastewater costs. An important task of dyeing technology is to reduce the amount of this dye.

The dye yield increases with decrease of liquor temperature, liquor ratio, retention effect of the auxiliaries, and dye concentration. The yield increases with increasing salt concentration, substantivity of the dye, affinity of the textile substrate, and build-up properties of the dye.

The most important relationships of the various classes of dye are well known. For example, the distribution coefficient $K_D$ is very useful for characterizing the position of the dye equilibrium [2.13].

If $C_F$ represents the concentration of the dye in the fiber material (solid phase) in g/kg and $C_L$ represents the concentration of the dye dissolved in the liquor (liquid phase) in g/L, the distribution coefficient in L/kg is

$$K_D = \frac{C_F}{C_L}$$

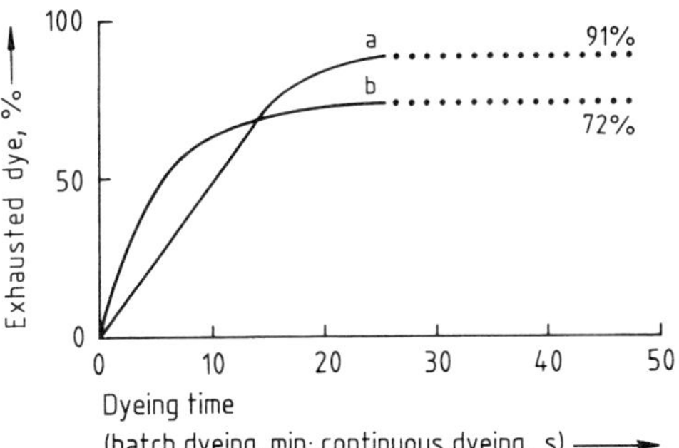


**Figure 2.3.** Dyeing yield as function of dyeing time and temperature
At equilibrium for a liquor ratio of 10:1 a distribution coefficient of 91:0.9 = 100 L/kg is obtained at 60 °C and at 80 °C a distribution coefficient of 72:2.8 = 26 L/kg.
a) Lower temperature; b) Higher temperature
··· = Equilibrium range

The value of $K_D$ can range between 10 L/kg for reactive dyes without additional salt to > 10 000 L/kg for cationic dyes on polyacrylonitrile.

The distribution coefficient is independent of the liquor ratio in a given dyeing process, and is therefore a suitable characteristic number for a dye (substantivity number). However, it is not an invariant, as it depends on the dye concentration (build-up properties of the dye), temperature, pH, addition of salt, and various other parameters (see Chaps. 4–11 for details) (see also Fig. 2.3).

When dyeing equilibrium is reached, the dyeing result is often still unlevel. However, the dye that is not yet fixed can then migrate from the areas that are too dark to the lighter areas. This equalization or leveling usually occurs more rapidly with larger residual amounts of dye in the liquor and at higher temperatures.

Table 2.1 shows that for high distribution coefficients (very high substantivities), the dye yield

**Table 2.1.** Variation in residual dye content (wastewater) for 2 % dye addition, while the difference in dye uptake (yield) is very small; liquor ratio 10:1 (10 L/kg)

| Dyeing run no. | Yield | Residual dye | Distribution coefficient |
|---|---|---|---|
| A | 99.2 % | 0.8 % (16 mg/L) | 1215 L/kg |
| B | 99.6 % | 0.4 % (8 mg/L) | 2430 L/kg |
| Difference | 0.4 % | factor of 2 | |

differences are extremely small ($< 1\%$), although the concentration of residual dye in the liquor differs by a factor of two. Residual dye concentrations of 8–16 mg/L are almost too low for leveling to take place.

With certain leveling auxiliaries, the amount of residual dye (the so-called dye retention) can be very effectively controlled (see Chaps. 4–11 for details).

### 2.1.3.5. Dye Fixation, Improvement of Colorfastness

When the dyeing process has reached equilibrium, the dye is usually still in the outer zones of the textile fibers. It is mobile, and can bleed out again at any time (a process taking place during leveling). The dye must be fixed, a process also known as development, finishing, or aftertreatment.

The treatment required is primarily dependent on the chemistry of the dye. The following fixation methods exist:

1) Fixation of the dye by chemical reaction, e.g., oxidation of soluble dyes to give insoluble pigments, (vat dyes, see Sections 4.3.4, 4.4.2) or substantive bonding to OH groups of the cellulose (reactive dyes, see Section 4.1.1)
2) Fixation by improved deposition within the fiber substrate, e.g., by deeper penetration into the fibers (disperse dyes, see Sections 6.4.1, 9.2, 11.1.2, Chap. 7) or increased aggregation, e.g., by soaping (vat dyes) or "precipitation" of anionic dyes by cationic auxiliaries (see, e.g., Section 4.2.7)

In general, a further washing operation is also required after fixation to completely remove the residual dye and auxiliary agents to achieve satisfactory colorfastness.

### 2.1.3.6. Sources of Further Process Data

The computation of dyeing processes with the aid of process functions and dye data is only slowly becoming established. The most important delaying factor is the complexity of the dyeing processes. Data currently available (e.g., from sample information sheets) are usually presented in the traditional way, and parameters suitable for computer use are not provided.

The sections on individual dyeing processes give some of these data or at least the relevant literature references, e.g., for vat dyes [2.14], [2.15] or reactive dyes [2.16]–[2.19].

Different authors use different concepts and parameters, e.g., [2.16], [2.20]. In order to utilize the information, careful consideration of the system is required, with mathematical conversion of the data if this is at all possible.

Computer programs should be developed that can simulate the dyeing process and depict it on the computer screen, even when the information available is somewhat incomplete. Simulation on the computer screen, however, involves very complex programming [2.21], [2.22].

"Expert systems" (ES) go far beyond data and literature banks [2.23]. These give extremely rapid access to stored technical information (knowledge-based or rule-based systems).

## 2.2. Batchwise Dyeing (Bath Dyeing)

Batchwise dyeing is performed in an essentially closed system in which the textile material, the liquor, and the dye interact over a long period. The essential elements of this dyeing system are the dyebath and the storage of the textile material.

### 2.2.1. Fundamental Principles and Equipment

The term "bath dyeing" describes this dyeing technique more accurately than the term "batch dyeing" more commonly used today. Bath dyeing is a special type of exhaustion dyeing. In the dyeing equipment, the three phases of exhaustion dyeing—exhaustion, establishment of equilibrium, and dye fixation—usually occur in succession.

An important feature is the interaction (movement) of the textile material and the liquor to prevent unwanted concentration differences and consequently unlevelness (Section 2.2.2).

Dyeing equipment in which the textile material remains stationary and the liquor is passed through the textile material, e.g., by pumps, is called *circulating machine*. The most important types of circulating machines are described in Section 2.2.3.

However, if the textile material is moved, e.g., mechanically or hydraulically, the term "*circulating-goods machinery*" is used, even if the liquor is also moved. The most important examples of this are described in Sections 2.2.4. and 2.2.5.

**Table 2.2.** Forms of textile goods and appropriate dyeing equipment

| Type of dyeing equipment | Loose stock (also card sliver and tow) | Yarn (bobbins, hanks) | Textiles (woven, knitted) | Finished garments |
|---|---|---|---|---|
| Circulating machines (Section 2.2.3) | package dyeing machines | bobbin, hank, and warp beam dyeing machines | beam dyeing machines | hosiery dyeing machines |
| Circulating-goods machines (Section 2.2.4) | | hank dyeing machines | winch and jet dyeing machines, jiggers | paddle and drum dyeing machines |

Special types of equipment are described in Section 2.2.6.

Bath dyeing has the great advantage that it can be used at any stage of textile production, i.e., on the fiber, the yarn, the textile, or the finished garment. Approximately since the International Textile Machinery Exhibition (ITMA) 91 there has been a tendency to carry out the dyeing as near to the end of the production chain as possible, i.e., moving in the direction of the finished manufactured garment. A general overview of forms of textile goods with the appropriate dyeing equipment is given in Table 2.2.

### 2.2.2. Theoretical and Technical Fundamental Principles

A particular advantage of bath dyeing is that wet goods can be treated in the dyebath. For example, woven textiles can be (continuously) desized, and can then be charged, while still wet, to the dyeing equipment, avoiding drying costs.

**Absorption of the Dye (Dyeing Kinetics).** Here, the dyeing rate is the dominant parameter (see Section 2.1.3.3). The dye must be absorbed uniformly to give a level result [2.11]. Two basically different strategies can be used to achieve this. These must be optimized, also in quite different ways:

1) Accelerated absorption, giving unlevel dyeing of the goods, followed by accelerated leveling (temperature, auxiliaries, pH). Thus, the absorption phase and the leveling in the dyeing equilibrium are controlled.
2) Retarded absorption (the dyed product is always level, and longer times are required for the absorption phase).

To predict the unlevelness in the absorption phase, the significant dyeing rate can be calculated from the absorption curve [2.10]. The further the dyeing rate of the textile–equipment system lies above the (critical) dyeing rate, the more severe is the unlevelness. Correction of unlevelness is usually carried out by a cumbersome and expensive leveling operation, by bleaching and redyeing, or simply by dyeing black.

Unlevelness is one of the major problems of short dyeing processes, because these require the use of high dyeing rates [2.24].

Since the dyeing rate depends on diffusion, it is controlled by the concentration of the dye in the liquor. A high liquor ratio (low dye concentration) leads to a low dyeing rate. For example, to produce a concentration of red dye in the goods of 2% (20 g per kilogram fiber), the starting concentration of the dye can be 4 g/L (for a liquor ratio 5:1) or 1.33 g/L (for a liquor ratio 15:1).

However, if the system used forms a dye reservoir, the concentration in the liquor remains constant as long as the dye is supplied quickly enough (an "infinite" bath). For example, the dispersed dye particles in polyester dyeing act as a reservoir that maintains a constant concentration of dissolved dye independently of the liquor ratio [2.25].

In general dyeing rate can be controlled by, e.g., heating profiles (time–temperature curves), addition of leveling agents, carriers, and retarders and by pH–time and metering–time profiles.

When these control tools are used, the most favorable shape of the absorption curves shown in Figure 2.2 is linear, S-shaped curves being the most critical. Further details are given in Chapters 4–11.

**Dyeing Equilibrium in Bath Dyeing.** The dyes used for textile dyeing are mostly strongly substantive, i.e., they form a strong bond to the textile substrate. Thus high dye concentration in the fiber $C_F$ is in equilibrium with a low residual dye concentration in the liquor $C_L$. The simplest indicator of substantivity is the distribution coefficient $K_D = C_F/C_L$ (see Section 2.1.3.4). Other substantivity parameters also exist.

This is illustrated by the following examples:

Red R 2%, distribution coefficient = 170 L/kg

| | 5:1 | 15:1 |
|---|---|---|
| Liquor ratio | | |
| Dye concentration in the textile ($C_F$) | 19.44 | 18.38 g/kg |
| Dye concentration in the liquor ($C_L$) | 0.114 | 0.108 g/L |
| Dye yield | 97.2% | 91.9% |
| Dye losses (wastewater) | 2.8% | 8.1% |

Yellow 2G 1%, distribution coefficient = 400 L/kg

| | | |
|---|---|---|
| Dye yield | 98.8% | 96.4% |
| Dye losses (wastewater) | 1.2% | 3.6% |

In a combination dyeing operation, a change to a higher liquor ratio (5:1 → 15:1) would cause more red dye to remain in the liquor (8.1% compared with 3.6%), and would shift the shade towards yellow.

The factors influencing the position of the equilibrium (distribution) (see Section 2.1.3.4) can also be used for control purposes.

The results of a bath dyeing process are easily corrected if necessary. At the end of the process, the dyeing result can still be adjusted (shading) by making further additions. Also, time can be saved by interrupting the exhaustion process early, before a close approach to equilibrium (99.0–99.9%) is achieved. The following typical figures illustrate this:

| Bath exhaustion, % | 64 | 86 | 95 | 98 | 99 |
|---|---|---|---|---|---|
| Dyeing time, min | 20 | 40 | 60 | 80 | 100 |

In this case, the exhaustion process is preferably interrupted after 60 min at 95% exhaustion rather than after 90 or 100 min with virtually complete exhaustion.

If the textile substrate of the same chemical composition is nonuniform, e.g., of various origins, the steps taken to prevent unlevelness must be specifically matched to the chemistry of the fiber and the dye (→ Textile Auxiliaries).

To sum up, the equilibrium phase essentially determines the dyeing result, including the ultimate levelness of a dyed product.

**Dye fixation** is described in Section 2.1.3.5. Specific recommendations for the various classes of dyes and methods of obtaining the required colorfastness are dealt with in Chapters 4–11. Tests for colorfastness are given in Chapter 13.

**Bath Dyeing Technology.** The most important part of the process is the interaction between the textile and the dye liquor, especially as this is influenced by the design of the equipment. It characterizes the types of dyeing equipment, which are described in more detail here.

As some dyeing processes are performed above the boiling point of water, high pressure dyeing equipment has been designed in which dyeing can be carried out up to ca. 140 °C [high temperature (HT) equipment] [2.26]. Since the vapor pressure of water is 2.7 bar at 130 °C and 3.6 bar at 140 °C, the safety valves are usually set at 6–7 bar.

To allow for thermal expansion of the dye liquor (ca. 6% between 20 and 100 °C), either a gas cushion is used in the dyebath or an overflow with a cooler and an open collection vessel which can then also be used to hold the additives (d in Fig. 2.4).

## 2.2.3. Circulating Machines (Stationary Goods, Circulating Liquor)

### 2.2.3.1. Systems and Functions

The textile material, which can be prepared in the form of a roll or a pack, is placed in the apparatus and held there mechanically. The dye liquor is pumped through the goods. A circulating machine for HT conditions is shown in Figure 2.4, including the most important components [2.27].

An important dyeing parameter is the liquor flow rate through the textile goods maintained by the circulation pump (Fig. 2.4f). In modern dyeing apparatus, the flow rate is displayed as a percentage (0–100%), but this percentage usually represents the pump drive motor current and is therefore not a true flow measurement. The conversion of the percentage figure to the absolute liquor flow rate $Q$ in $m^3/h$ and to the specific flow rate $q$ through the textile goods in $L\,kg^{-1}\,min^{-1}$ is explained in Figure 2.5. The specific flow rate is a useful control parameter for the dyeing operation. Dyeing carried out in different types of apparatus (e.g., laboratory scale and full scale) with the same specific flow rate and the same liquor ratio usually gives similar results. This also applies within limits for dyeings in different apparatus with identical percentage figures from the pump display.

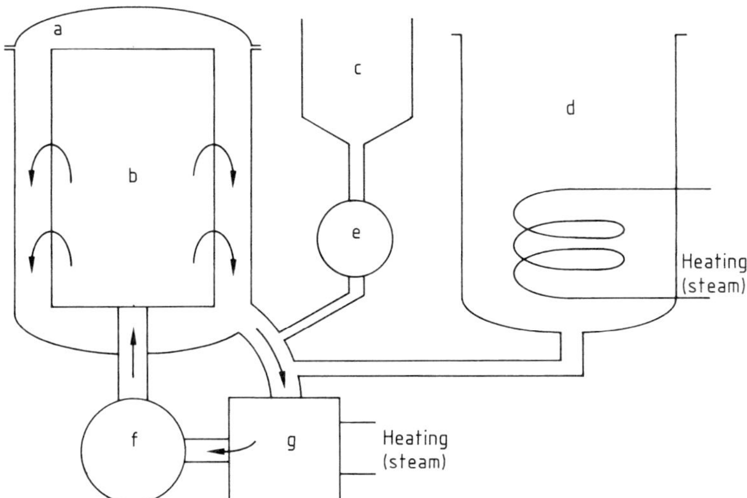

**Figure 2.4.** High temperature dyeing machine
a) Dyeing vessel, pressure-tight (water temperature: 140 °C), liquor volume: ca. 0.5–1 m³ per 100 kg textile goods; b) Textile goods in package form (loose stock), on bobbins, or on a beam (woven, knitted, or warp yarn); c) Mixing vessel for additions to the dyebath (dyes, auxiliaries) (volume ca. 10 % of that of dyeing vessel); d) Auxiliary vessel for preparation of liquor (volume similar to that of dyebath); e) High pressure pump, delivery pressure: ca. 4 bar, delivery rate: ca. 2 m³/h; f) Circulation pump showing flow through the goods (i→o), flow rate: ca. 50–200 m³/h for 100 kg textile goods; g) Heat exchanger (usually steam) for the dyebath (external) and often heating coils in the dyeing vessel (internal)
Valves, device for reversing direction of circulation, and rinsing equipment are not shown.

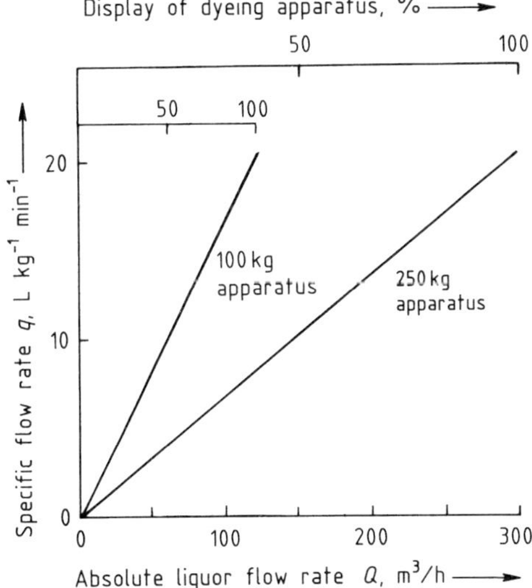



**Figure 2.5.** Characteristic figures for absolute liquor flow rate $Q$ and specific flow rate $q$ through the textile
The specific flow rate $q$ is proportional to the percentage of absolute flow rate shown at the apparatus display; 100 % mostly corresponds to $q = 20$–$30$ L kg⁻¹ min⁻¹.

The dye liquor should be contacted uniformly with the textile goods (no channeling and no extended flow paths, e.g., at the edges).

As the dye liquor flows through the textile goods, the dye exhausts gradually. The liquor flowing away from the textile goods therefore always has a lower dye concentration, and the textile has a lighter color at the outlet than at the dye inlet point. If this concentration difference is kept at $< 1$–$2\%$ by controlling dyeing rate and recirculation rate good leveling results can be expected.

The liquor flow usually has no effect on dyeing shades obtained (the dyeing equilibrium), especially if there is no sedimentation of the dye.

The characteristic curve of the pump is also an important characteristic of every circulating machine (see Fig. 2.6). It is usually nonlinear, and shows the liquor flow rate for a given pressure drop (differential pressure $\Delta P$) across the textile goods [2.28]

$$Q = Q_{max}(1 - \Delta P/P_{max})\, m$$

where $Q$ is the liquor flow rate (m³/h); $Q_{max}$ the liquor flow rate with no flow resistance (m³/h); $P_{max}$ the pressure at the textile goods when liquor flow is blocked (bar); and $m$ the exponent of the pump characteristic curve, e.g., ca. 0.8.

The characteristic curve is primarily determined by the design of the pump [2.28]: *Axial*

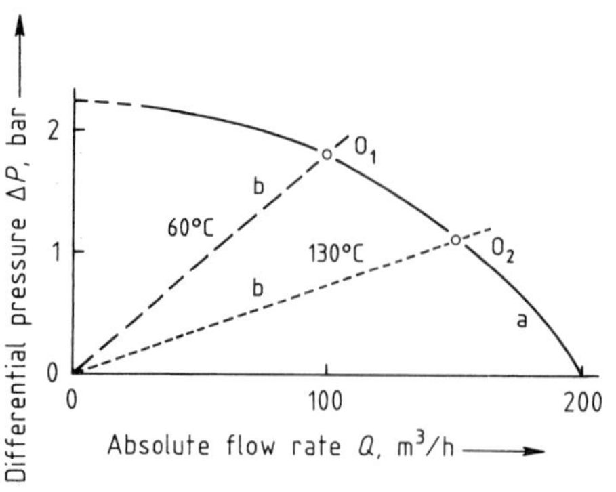

**Figure 2.6.** Pump characteristic curve and flow resistance of textile goods
a) Pump characteristic curve; b) Characteristic lines of flow resistance for two temperatures
$O_1$, $O_2$ = Operating points ($O_1$: $Q = 100$ m$^3$/h, $\Delta P = 1.8$ bar; $O_2$: $Q = 150$ m$^3$/h, $\Delta P = 1.1$ bar)

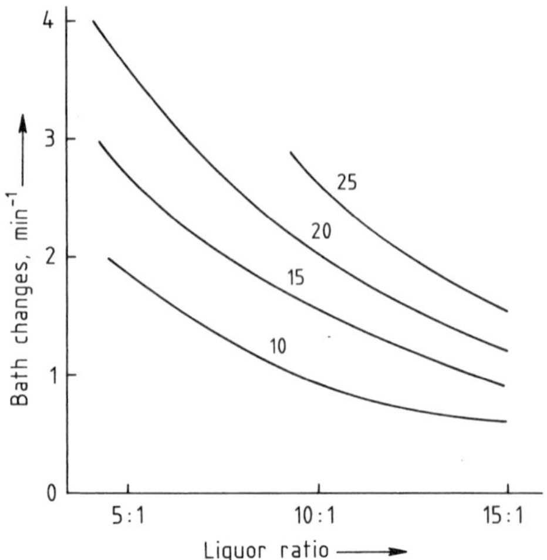

**Figure 2.7.** Frequency of bath changes as a function of liquor ratio at different specific flow rates (10, 15, 20, and 25 L kg$^{-1}$ min$^{-1}$)

*pumps* of the propeller type are characterized by high volumes and low pressures.

In *radial pumps* the liquor is accelerated from the center to the outer edge of the pump, high pressures can be produced and lower volumes are pumped.

Another important parameter is the ratio of diameter of the pump wheel to that of the pump housing.

Secondarily, the resulting characteristic curve also depends on the pump settings, the setting of the throttle and bypass valves, of the pump motor speed (rpm), and of the angle of the propeller blades.

Thus the performance of the pump, the most important part of circulating machines, can be largely controlled by the pump settings.

Equally characteristic of the particular dyeing system is the flow resistance of the textile goods. The absolute value $R_{abs}$ of the flow resistance in mbar m$^{-3}$ h$^{-1}$ is given by

$$R_{abs} = 1000 \times \Delta P/Q$$

(The differential pressure $\Delta P$ is sometimes incorrectly termed flow resistance.)

In normal dyeing conditions, $R_{abs}$ is largely independent of the liquor flow rate.

In the differential pressure–flow diagram, the resistance can normally be represented by a straight line through the origin. This intersects the pump characteristic curve at the so-called operating point O.

The flow resistance often changes during a dyeing operation for reasons other than temperature, e.g., it increases on shrinkage and/or swelling of the textile or if the direction of circulation changes from i → o to o → i (inwards = i, outwards = o).

Increased flow resistance results in a decrease in flow rate and an increasing tendency towards unlevelness. Another useful measure of liquor circulation rate is the number of bath changes per minute (see Fig. 2.7). For normal levelness requirements, the reduction in dye concentration for each bath change should be only 2–3 %.

A comprehensive review of designs of circulating machines produced by the most important manufacturers (ITMA 91) is given in [2.29], mainly in the form of machine drawings.

### 2.2.3.2. Loose Stock Dyeing Machines

A loose stock dyeing machine is used for dyeing of flock, card sliver, and tow.

**Dyeing of Flock.** Flock consists of randomly oriented fiber which is packed (wet) into a cylindrical container with central tube [2.29].

Only the coarser and more stable types of fiber can be dyed as flock without damage. The packing density of flock is in the range of 200–400 g/L. It increases when water and auxiliaries are added. The liquor flow rate is in the region of

$10 \, L \, kg^{-1} \, min^{-1}$, and the liquor ratio is usually in the range $5:1 - 20:1$.

It is difficult to pack and densify the loose stock uniformly, so that unlevelness can easily occur, especially because of the long flow paths in the packed material. However, subsequent spinning (followed by doubling) of the fibers often evens out unlevelness mechanically.

Loose stock dyeing apparatus is usually constructed to hold up to 1 t material, so that not too many dyeing batches are necessary for each batch of spun yarn. Loose stock dyeing comprises ca. 1 % of all bath dyeing.

**Dyeing of Card Sliver.** Card sliver consists of a long strip of any kind of staple fiber, the fibers being mainly parallel. Fiber dyeing is necessary in the production of melange (blended) yarns.

The slivers are placed in dyeing baskets or coiled into bundles (wool "tops" having a maximum weight of 25 kg) [2.29], [2.30]. The slivers are sometimes wound onto a beam for dyeing (see Section 2.2.3.5).

**Dyeing of Tow.** Monofilament material produced during the manufacture of synthetic fibers (primary spinning) is formed into strips (tow), which is packed into large containers or carriers [2.29]. The filled containers are then placed in the dyeing apparatus.

### 2.2.3.3. Package Dyeing Machines (Cross-Wound Packages)

The schematic shown in Figure 2.4 also applies to package dyeing machines, which are usually operated at high temperature. In the (obsolete) overflow cooling system, the liquor that overflows due to expansion collects in the additives (or expansion) vessel. The advantage of this system is the ease of sampling and manual addition of products, and the disadvantage is a high energy consumption, air oxidation of the reducing agents in vat dyeing, and increased foam formation. In modern apparatus, the liquor expansion is taken up by means of a gas cushion.

Package dyeing often allows sampling of the yarn for color evaluation through a pressure lock during dyeing.

The yarns (fine to coarse) are often wound from the spinning bobbins onto dyeing bobbins (metal or plastic). For this, cross winding is prefer-

able (weight of yarn per package: 0.5 – 2.5 kg). The packages are either wound in a conical shape (for easier unwinding of the yarn) and then stacked with separator plates, or are wound cylindrically, and then usually compressed (elastic tubes). Very high package densities (ca. 800 g/L) are obtained with polyester sewing yarns (compressed), and very low values (ca. 200 g/L) with polyacrylonitrile packages (risk of deformation because of thermoplasticity).

The cross-wound packages are held in place in the apparatus by placing them on spindles, and the columns of packages so formed are closed off at the top by a spring-loaded plate. The spindles are metal rods (with a star section) or perforated tubes through which the dye liquor is pumped (i → o and o → i). The spindles are mounted on floor plates which provide both the support and the connection to the liquor supply.

Quantities loaded by this method into package dyeing machines can be large or small. Normal loads for production equipment are between 50 and 500 kg. For larger batches, two pieces of dyeing equipment are joined together (tandem operation).

Occasionally, package dyeing is used with parallel wound yarns (rocket packages, thermofixation packages) or with narrow ribbons instead of yarns (e.g., lace, zippers).

The liquor ratio can range from ca. 4:1 (short liquor ratio dyeing machine) to ca. 15:1 (open winding, older dyeing machine).

The short liquor dyeing machine is usually only partially flooded, and permits liquor circulation only from the inside to the outside (i → o, foaming tendency). The equipment must be specially designed to give a high enough pressure on the suction side of the circulation pump to prevent cavitation (Fig. 2.8).

Higher dyeing rates can often be obtained by reversing the direction of circulation, usually every 3 – 8 min. The time required for this reversal process (dead time) is usually ca. 10 – 100 s.

The techniques used for producing the change in direction of circulation are:

1) Reversal of the direction of the pump (brief stoppage of the electric motor, expensive method)
2) Sliding valve system (pump always running in the same direction, see Fig. 2.9).
3) Adjustment of the pump blades (while motor is running)

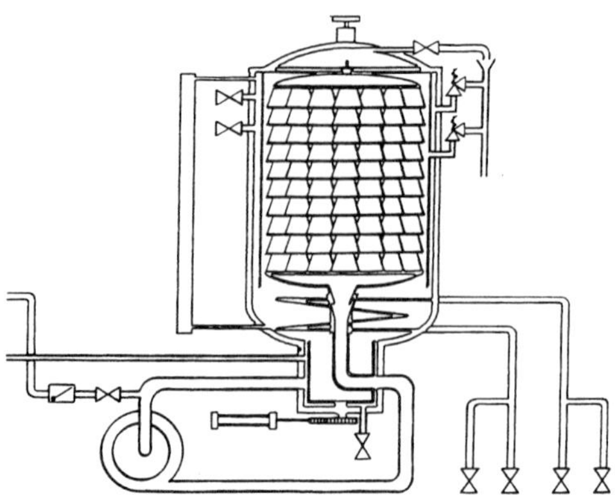

**Figure 2.8.** Short-liquor dyeing machine for cross-wound packages, operating i → o and o → i (eco bloc, courtesy of Thies)
Liquor ratio 3.5:1–6:1.

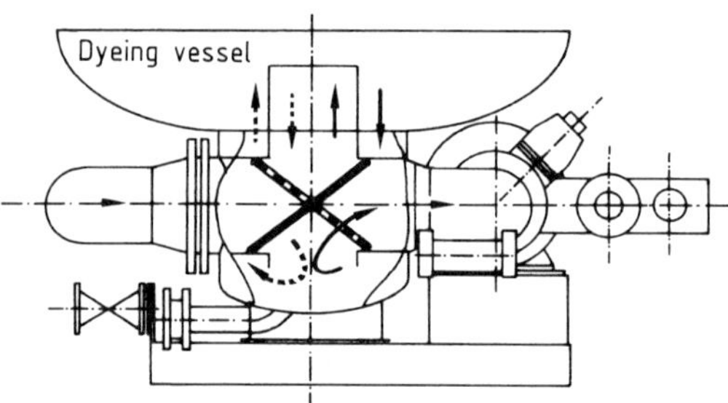



**Figure 2.9.** Special valve system for reversing the direction of flow in a dyeing machine with monodirectional circulation pump (courtesy of Thies)

Auxiliaries can be added to the additives container manually or automatically. The auxiliaries are pumped from here by the pressure pump into the dyebath.

Package dyeing equipment has a high share of the market. Most yarns for color weaving and knitting are dyed by this method.

#### 2.2.3.4. Hank Dyeing Machines

The textile materials, which are usually delicate yarns (high bulk materials, texturized yarns, embroidering yarns) are formed into hanks, i.e., loose coils of yarn without carriers, ca. 0.5–1 m long. The hanks of yarn are hung fairly closely together over poles inside the apparatus, which is usually rectangular in shape (a box) [2.31].

The liquor flows mainly tangentially through the hanks at a high rate. This results in consider-

able tension in the yarn, sometimes leading to flattening of the yarn at the points of suspension (which usually cannot be changed). The liquor ratio is usually 10:1. A bath change occurs every 1–3 min. The differential pressure is only 0.1– 0.2 bar, but the pump must deliver the liquor at a high rate (20–50 L kg$^{-1}$ min$^{-1}$). Axial pumps (propellers) are often used.

Information on hank dyeing apparatus are given in [2.31], and information on ITMA 91 in [2.29].

Hank dyeing is now one of the older processes, and represents a small fraction of yarn dyeing. Circulating-goods machines for hanks of yarn are described in Section 2.2.6.2.

#### 2.2.3.5. Beam Dyeing

In contrast to the apparatus described in the previous sections, the dyeing vessel in beam dyeing is normally arranged horizontally.

**Piece Beams.** In piece dyeing by the beam method, the woven textile (piece), which must not be affected by pressure on the surface (woven, knitted, or nonwoven material), is evenly wound onto a beam (perforated tube) in full width. The interior diameter of the beam is usually 0.3–0.5 m, and the thickness of material on the beam is usually 10–40 cm, depending on the type of fabric and size of the batch. Since the density of the roll of textile on the beam is similar to the density (g/L) measured on a single layer sample of textile, it can be predicted in this way.

Because dyeing with piece beams is so important, calculation methods have been developed based on simple measurements on a single layer sample to calculate the number of layers, the length in meters, the roll diameter, and the outside diameter. Measurement of the flow rate through a sample enables even the flow resistance of the roll of textile to be accurately predicted [2.23]. In practice, the calculation of the liquor–textile interaction is most advanced in the dyeing of piece goods by the beam method.

The direction of flow of the liquor is normally from the inside to the outside of the roll of textile. The ends of the beam are covered with metal sheets (sleeves) before the fabric is wound on, to prevent a short circuit of the liquor.

Although a high liquor flow rate is desirable, a limit is imposed by the resulting high differential pressure, which causes the fabric to "lift" from the beam, often at a differential pressure of only 1 bar. This disturbs the controlled flow of the liquor. A possible preventive measure is to wind the fabric under high tensile load, which, however, is problematical for certain fabrics. Batch sizes can be 500–3000 m. An open (permeable) fabric is usually more suitable than a dense (low permeability) fabric. Also, significant swelling of the fabric reduces its permeability to the liquor (greater pressure buildup).

The piece beam process has a market share of ca. 10 % of circulating machines.

**Warp beam dyeing** is similar to piece beam dyeing. Here, the warp or a part of it is wound onto the beam. The yarns are parallel to each other, and have a relatively high flow resistance as the threads support each other. If the differential pressure becomes too high, a roll of this type cannot lift, but the threads "part".

Warp beam dyeing is mainly important for the process of multicolored weaving with a plane-dyed warp, it accounts for < 10 % of all circulating machines.

### 2.2.4. Circulating-Goods Machines with Textile Storage (Winch Type)

#### 2.2.4.1. System and Functions

Section 2.2.4 describes dyeing machines in which the textile is moved, i.e., it circulates through these types of dyeing machines, and is stored in folds in the dyebath to obtain additional time for dyeing. These dyeing machines have been developed from the winch beck (Section 2.2.4.2).

All the fundamental principles of exhaustion dyeing described in Section 2.1 apply to this dyeing technique, and especially bath dyeing as described in Section 2.2.1.

Dyeing machines with storage devices (rope dyeing machines) consist basically of a means of transporting the textile and a storage zone into which it is transported. Movement of the textile can be achieved

1) Mechanically by a rotating body (winch, i.e. a rotating body with an elliptical section) and electric motor
2) Hydraulically by a jet
3) By a hydraulic lifting pipe and the force of gravity in a propelling system (overflow)
4) Pneumatically (Airflow)

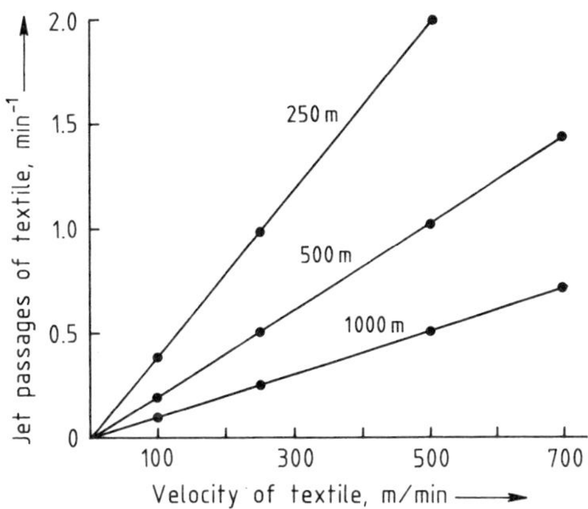


**Figure 2.10.** Jet passages per minute in dyeing machines with textile storage as a function of the velocity of textile (for batch sizes up to 1 t of goods; running speed ca. 70–700 m/min)

In the region of the drive unit, an intensive mass transfer occurs between the free liquor and the liquor in the interstitial volume of the textile, although only very briefly.

In the storage region, the dye has time to diffuse out of the interstitial spaces of the textile and into the interior of the fibers. The depleted liquor in the interstitial volume must then be replenished.

The number of cycles of the textile per minute is usually in the range 0.2–1, with velocities of ca. 30–600 m/min and batch sizes usually < 1000 m. The relationship between these process parameters is shown in Figure 2.10.

If the load in the dyeing machine is too high, the dwell times in the storage region become too long, and unlevelness and crease marks result. A rule of thumb for a level dyeing result is that only 1–3 % of the dye should be extracted for each cycle of the textile.

Liquor ratios range from ca. 3:1 (Airflow machine) to ca. 30:1 up to 40:1 (winch beck).

The greater the number of cycles per minute, the lower is the risk of unlevelness. At one cycle per minute, a dye exhaustion rate of ca. 2 %/min can be used, leading to exhaustion times of ca. 50 min with correct control. The liquor ratio is of greater significance for the running behavior of the fabric and the formation of crease marks than for uniformity between ends.

Loading of the dyeing machine with the fabric requires time. The fabric is loaded in rope form, usually at a speed below the operating speed of the machine.

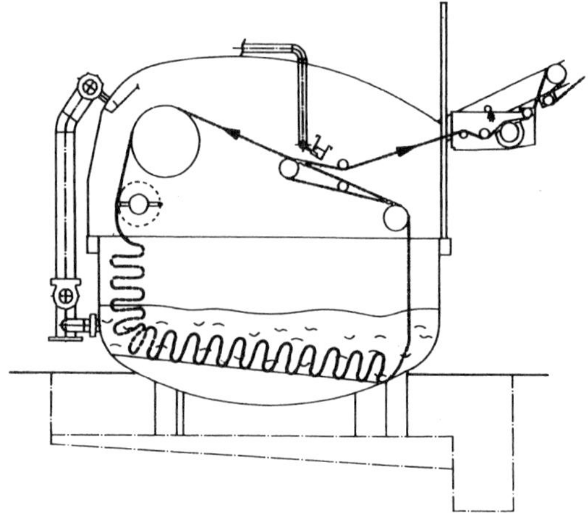

**Figure 2.11.** HT winch beck for dyeing of carpets (Supraflor, courtesy of Brückner)
Working width 4.2 or 5.2 m; capacity max. 1 t of goods; running speed of liquor circulating pump 70 – 100 m/min.

A very comprehensive overview of dyeing machines can be found in the form of machine drawings in [2.29].

### 2.2.4.2. The Winch Beck

In the winch beck, a motor driven roller (usually elliptical, the so-called winch) is located above a dyebath. This winch draws the textile (woven, knitted, or carpet) via a guide roller out of the bath and returns it in folds into the bath. This removes excess liquor, and the liquor from the interstitial space of the textile is interchanged with "outer liquor" see Fig. 2.11. The velocity of the textile is 50 – 100 m/min.

The fabric in a batch is divided into pieces of equal length whose ends are sewn together to form endless loops. Usually, there are up to 10 loops of fabric side by side on the winch, separated by rods or guiding plates. Carpets up to 5 m wide are dyed at full width or are sewn together lengthwise to form a tube (mainly in the United States). The liquor ratio is usually very high (25:1 – 50:1).

Winch dyeing is generally considered to be obsolete. However, it is a very gentle process, a very soft and voluminous fabric being produced. The fabric must not, however, be prone to form crease marks.

Mixing of the liquor in the bath is poor, so that it is difficult to maintain a constant liquor temperature in all parts of the bath and to eliminate differences in dye concentration. Pumps are often used to solve these problems by providing forced circulation of the liquor. Dyeing rates of < 2%/min and long leveling times are often necessary to obtain level dyeing.

The high liquor ratio has a very unfavorable effect on the dye yield (see Section 2.1.3.4).

Some of the winch becks used today are high temperature machines [2.26]. The market share of winch becks for batchwise carpet piece dyeing is very high [2.33], but it is low for the dyeing of woven and knitted textiles.

### 2.2.4.3. Jet Dyeing Machines

In jet dyeing machines, a ring nozzle (jet) is located above the dyebath, the storage area for the fabric. The textile goods are transported through the nozzle, the required liquid flow to propel the fabric hydraulically is produced by a liquor pump. The diameter of the nozzle must match the "thickness" of the rope, i.e., it depends on the goods.

The fabric in the dyebath is moved towards the point where it is lifted up to the nozzle by sliding, a transportation device, or by floating. The liquor ratio with partly flooded jet machines is 10:1 – 20:1, and with fully flooded machines can be 30:1 – 40:1.

Jet dyeing machines are usually constructed for high temperature operation, enabling polyester textiles (knitted or woven) to be dyed satisfactorily. Because of many combinations of the various types of nozzle (including adjustable nozzles) and fabric storage devices, very different types of jet dyeing machines exist [2.34]. An example of an industrial design is shown in Figure 2.12.

The high turbulence in the nozzle as well as the entering of the fabric into the liquor lead to the entrainment of air in the dye liquor, with consequent danger of foam formation. The fabric storage system does not facilitate liquor interchange, because the fabric retains much "old" liquor (ca. 200 – 300%), and the extent of liquor interchange in the storage zone is small. In partially flooded jet machines, a relatively large air space exists, which can cause problems especially with vat dyeing (loss of reducing agent by air oxidation). The fabric velocity is usually in the range of 200 – 400 m/min. The batch size is ca. 200 – 600 m.

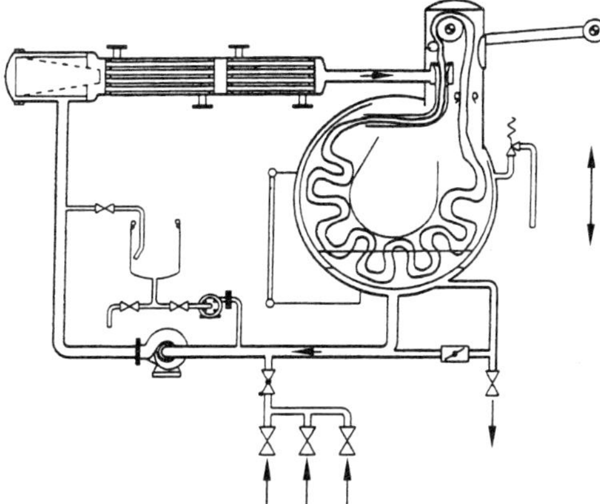

**Figure 2.12.** Schematic of a jet dyeing machine (Colorstar, courtesy of Scholl AG)
Ring nozzle (top left in dyeing vessel); fabric storage (bottom); liquor circulation pump (left); heat exchanger (top left)
Running speed ca. 60–500 m/min.

Although their space–time yield is not very high, jet dyeing machines have now become widely established. They represent the majority of all dyeing machines.

#### 2.2.4.4. Overflow Dyeing Machines

In overflow dyeing machines the textile material (in rope form) is raised by the flow of liquor in a pipe system, and the excess liquor overflows. A winch (usually cylindrical and not motor driven) is situated above the fabric transport point, and the fabric "hangs" over this. A longer length of textile hangs from the exit side of the winch than from the inlet side. Gravitational forces pull the longer length of textile downwards more strongly than the shorter, and the textile is thereby transported very gently [2.13, p. 700, Fig. 5.]. The fabric enters into the storage zone or the pipe system into which the liquor is pumped. In the pipe system, the liquor continually flows past the fabric in cocurrent. Good renewal of the liquor in the storage zone is therefore ensured, which positively affects the liquor interchange with the fabric to some extent (see Fig. 2.13).

The pump must provide the lifting energy to overcome the difference in height (0.1–0.2 bar). The flow resistance of the fabric to the liquor has almost no influence.

Overflow dyeing is suitable for woven and knitted fabrics, especially for delicate materials which justify the cost (tension-sensitive, soft, voluminous fabrics with a structured surface or high bulked yarns). The liquor ratio is ca. 15:1–25:1, the textile velocity is ca. 20–80 m/min, and the pump provides a bath circulation rate of ca. 1–2 circulations/min.

This special type of machine has a considerably lower market share than, e.g., jet dyeing machines.

#### 2.2.4.5. The Air Jet ("Airflow") Dyeing Machine

The "Airflow" dyeing machine has been developed by the company Then. The fabric (in rope form) is transported pneumatically through a nozzle, by air. It then passes into a storage zone which contains a very small amount of free liquor. The dye, auxiliaries, and liberated liquor are atomized and applied to the moving fabric in the form of an aerosol (see Fig. 2.14) [2.35], [2.36]. A powerful fan and a suitable design of

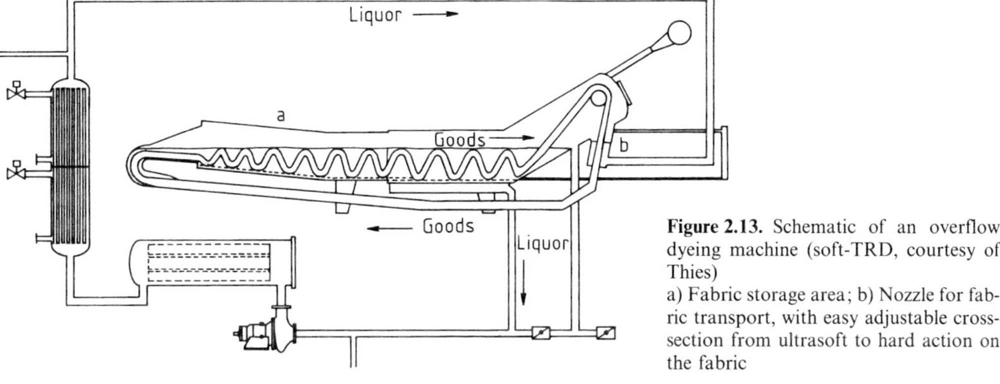



**Figure 2.13.** Schematic of an overflow dyeing machine (soft-TRD, courtesy of Thies)
a) Fabric storage area; b) Nozzle for fabric transport, with easy adjustable cross-section from ultrasoft to hard action on the fabric

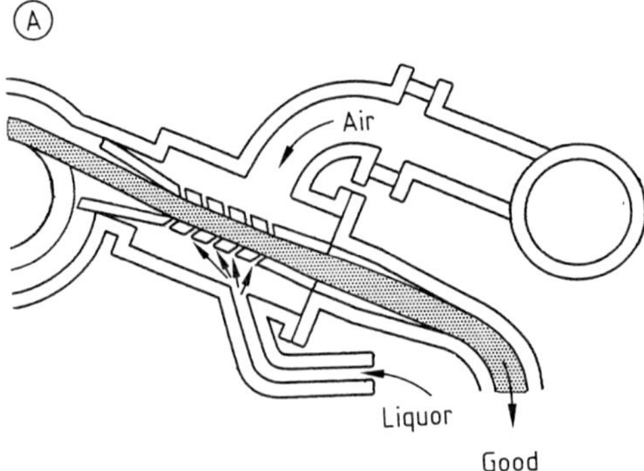

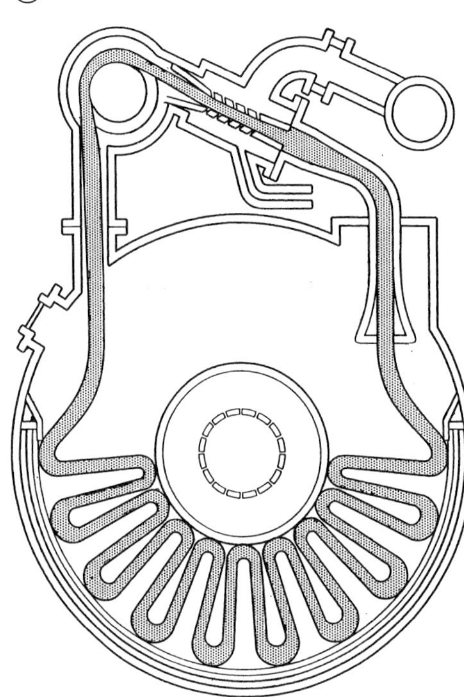



**Figure 2.14.** Air jet dyeing machine, HT type (Then-Airflow AFS, courtesy of Then)
A) Schematic of nozzle system: pressurized air strikes the fabric in an inclined angle and thus propells it; liquor is simultaneously entrained from the nozzles, sprayed and deposited on the goods; B) Schematic of the entire machine

the ring nozzle enable fabric speeds of up to 900 m/min to be achieved without damaging the fabric. The air which propels the fabric is recirculated.

Because of the short liquor ratio applied (2:1–4:1) the dye must have adequate solubility. For a 2% dyed product, dyeing with reactive dye requires 10 g/L dye and 40 g/L sodium chloride.

This dyeing technique is suitable for woven fabrics and stable knitted fabrics. Performance depends greatly on the particular type of textile, however, it is not possible to check in advance whether goods are suitable or not (leveling problems).

The "Airflow" machine was the first air jet dyeing machine to be established on the market. The "Airflow" now has a share of ca. 10% of all jet dyeing machines.

### 2.2.5. The Dyeing Jigger

#### 2.2.5.1. Normal (Direct) Jig Dyeing

The batch of fabric is drawn into the jigger, through a dyebath, and is taken up by a roll on the other side of the bath (Fig. 2.15) [2.13, p. 701, Fig. 6], [2.36]–[2.38].

Generally woven textiles are dyed by the jig process. The textiles must not be sensitive to tension (e.g., texturized warp) or pressure (e.g., bulked material). However, fabrics susceptible to creasing can be dyed, as dyeing is exclusively performed in full width. The jigger is the most suitable dyeing machine for thin and impermeable woven textiles, e.g., lining materials. High temperature jiggers have been available since the 1980s, so that all fiber substrates, from cellulose to polyester garments, can be dyed in these machines.

The fabric can be loaded either dry or wet, e.g., after a continuous desizing operation. The fabric velocity is ca. 50–130 m/min. The liquor ratio is usually 3:1–5:1, i.e., very short. The size of the batch is usually 300–400 kg. So-called jumbo jiggers also exist in which a batch of almost 1 t textile can be dyed.

The fabric passes through the dyebath in which only a limited amount of liquor exchange occurs. A complete passage of a batch is known as a pass. During the time that the fabric is in the bath (ca. 0.5–1 s), the free liquor in the bath is exchanged with the depleted liquor in the textile. The exhaustion of the liquor begins in the wound up batch of fabric, where the fabric remains for a considerably longer time. This can lead to nonuniformity between the ends of the fabric.

During one pass, the dye content of the liquor becomes much depleted. The intensity of dyeing at the ends of the fabric is reduced because of short dwell times, and this must be lev-

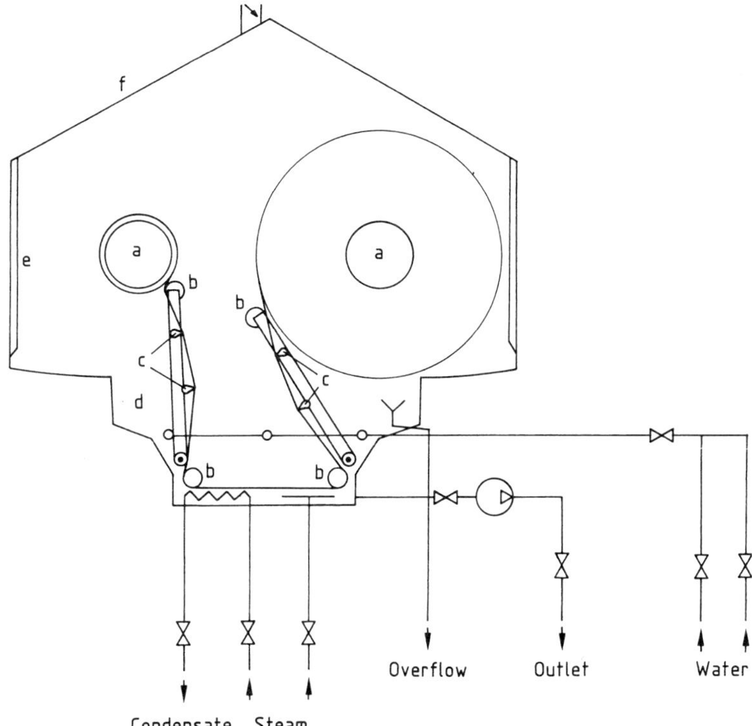

**Figure 2.15.** Schematic of a jigger (jumbo type; courtesy of Henriksen)
a) Main roller; b) Guiding devices; c) Expander (static); d) Basin; e) Window; f) Hood

eled by an increased number of passes. To reduce the unlevelness of the ends, the dye is usually added during the first two passes (1:1). The dyeing process is controlled by the number of passes (exhaustion time, final concentration in the bath) and by means of temperature, pH, salt concentration, etc. (see Sections 2.1.3. and 2.2.1.).

The passes are necessary to achieve intense interchange of the fabric with the liquor and to provide sufficient time for penetration of the dye. The overall dyeing time, determined by the number of passes, varies with the fabric velocity (usually constant) and the length of the batch (usually variable) (see Fig. 2.16).

Minimum dyeing times (exhaustion times) and minimum numbers of passes required to give level dyeing are given in Figure 2.16 [4]. These figures are valid for a batch of 1000 m; for shorter batches (500 m), the number of passes must be doubled; and for batches 2000 m in length, only two passes may be necessary if the end unlevelness remains within limits.

A general problem with jiggers working at atmospheric pressure is to maintain the tempera-

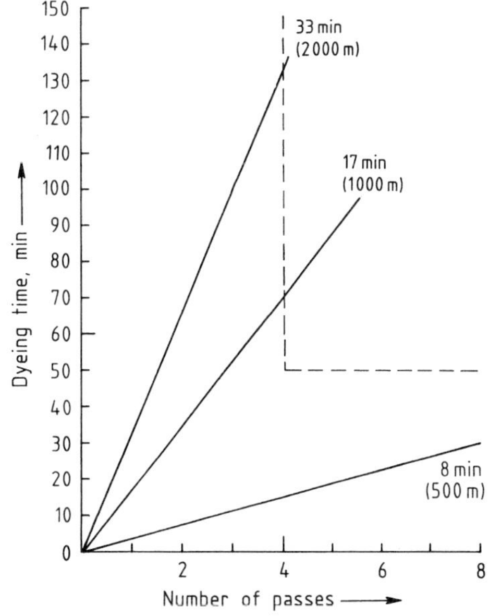


**Figure 2.16.** Dyeing time as a function of number of passes for different batch lengths and pass times
— — — minimum line for good operation

ture constant. The covers should always be kept closed. However, even closed covers have an undesirable cooling effect on the system. The fabric in the batch certainly is at about liquor temperature (70–90 °C), but there is a tangential flow of air caused by the fabric until it reaches the roll, and consequent water evaporation then leads to cooling. In addition, the metal spindle at the center of the roll of fabric abstracts heat, so that the inner layers are cooler. The liquor is heated by direct steam, heating coils in the bath, or external heat exchangers, but, although the liquor is continually circulated, local temperature differences of several degrees Celsius often exist inside the bath. However, even if a pump is installed, the bath cycle times are often insufficient, > 5 min. An air flow on the sides of an open jigger often leads to edge unlevelness.

Jiggers represent a large proportion of dyeing machines used in the industry. In some sectors, they are even in the majority, e.g., for the correction of unlevelness in continuously dyed materials.

### 2.2.5.2. Pad Jig Process

The pad jig process is strictly speaking a semicontinuous method (Section 2.3). However, in accordance with accepted practice, it is included with the batchwise dyeing operations.

In the pad jig process, the dry goods are padded with dye liquor (see Section 2.3.1.1) and are batched up. This batch of fabric is then either hung in the jigger or drawn through in the usual way, corresponding to a single pass. The remainder of the pad liquor is usually added to the jigger (saving of dye, reduction of wastewater load, varying composition causes formulation problems).

The rest of the dyeing process is as outlined in Section 2.2.5.1. However, the minimum number of passes is considerably reduced, usually halved. This results in a considerable reduction in dyeing time for large batches (Fig. 2.16). The minimum number of passes is two using the pad jig process, and the dyeing time required is ca. 60 min for a 2000 m batch, whereas in the normal jig process 5–6 passes are needed with a dyeing time of 100–120 min.

Disadvantages include the problems of dye padding. The fabric to which the dye must be applied is usually sized gray cloth. Smooth fabrics with low interstitial volumes take up < 50 % liquor, this small amount of liquor leads to prob-

lems of solubility of the dyes with dark color shades (> 100 g/L). Other disadvantages are the operating costs of the dye application (smaller pieces, continuous process) and the complicated handling of the goods.

### 2.2.6. Special Bath Dyeing Equipment

These types of dyeing equipment can be circulating machines (e.g., star-shaped suspension equipment) or circulating-goods machines (e.g., vertical star equipment). They have only a small market share of all bath dyeing equipment. However, they play an important role in special areas (e.g., the dyeing of hosiery or velvet).

### 2.2.6.1. Star-Shaped Dyeing Frames

The carriers for the materials are two disk-shaped radial frameworks separated by a spacer tube at a distance equal apart to the width of the goods. On the radial metal rods, set at an angle of ca. 30°, there are small hooks (ca. 1 cm apart) on which the goods (woven or knitted) are hung. These are sometimes also pinned on one side only (usually for pretreatment, allowing free shrinkage). The star framework is an old type of dyeing equipment. The goods form a spiral cylinder and do not touch each other, i.e., dyeing is very gentle.

The *star-shaped hanging frame*, also known as a vertical star, is suspended in the bath. The goods are normally completely immersed.

The star is sometimes moved up and down (high liquor volumes and a liquor ratio of ca. 20:1–30:1) or it can be fixed in the dyebath and the liquor movement can be produced by a circulation pump (low liquor volumes and a liquor ratio of ca. 12:1–25:1). In neither case longitudinal tension occurs in the goods, and any transverse tension is usually due only to their weight.

In *horizontal star suspension systems*, the spacer tube can be rotated, and is located above the dyebath. The star is rotated by a motor. Only part of the star is immersed in the liquor along with the goods (short liquor ratio). The liquor is taken up when the goods are rotated. The spiral shape of the goods ensures that the inner parts are also fully supplied with liquor.

The star with the goods can be very easily transferred into other vessels, after dyeing in a stationary bath, e.g., into an aftertreatment bath or in a star steamer.

The space–time yields are low, so that this dyeing technique, which is now only seldom used, is expensive. However, the very gentle treatment of the goods can justify the process for textiles made from velvet and silk.

### 2.2.6.2. Machines for Dyeing Hanks of Yarn

Spray dyeing machines have been used for a long time for dyeing hanks of yarn, but tubular knitted goods can also be dyed by this method in some cases.

The hanks are placed on a perforated tube. The liquor is then pumped through this tube, sprays out of the perforations onto the goods and runs off. Normally, the hank does not dip into the liquor, so that there is some freedom of choice of liquor volume (liquor ratio can be between 10:1 and 40:1). A carrier arm moves the hank a short distance into a new position from time to time. This helps considerably to avoid flattening of the yarn [2.31].

The Colourdry dyeing machine by MCS is an interesting further development [2.39]. The hanks are placed loosely on triangles of tubing. The liquor is sprayed from the tubing onto the yarn (liquor ratio < 15:1). Centrifuging and drying are carried out in the same equipment without additional handling of the yarn.

The very gentle treatment is advantageous, but the space–time yield is low, so that these machines are usually limited to special applications.

### 2.2.6.3. Paddle Dyeing Machine

This machine, also known as a paddle tank, is used for dyeing finished garments (sweaters, pants and trousers, etc.). The textile goods "float" in the dyebath. The goods are vigorously agitated by

1) Motor driven mechanical paddles.
2) Injection nozzles (tangential and radial) through which the liquor is pumped. This produces the necessary relative movement of the textile and the liquor.

These dyeing machines are also constructed for high temperature operation. They have a large market share of the dyeing of ready-made goods, although this is a relatively small part of the total market [2.40].

**Figure 2.17.** Cabinet dyeing machine for dyeing of hosiery (Amre P, courtesy of Cubotex)

### 2.2.6.4. Rotary Dyeing Machine

The rotary dyeing machine resembles a domestic washing machine. It can contain ca. 50–300 kg goods. It is mainly used for dyeing ready-made garments, and can also be constructed for high temperature operation [2.40], [2.41].

The drum dyeing machine is very suitable for producing wash-out effects. A high proportion of jeans are so treated. Various processes exist, e.g., stone washing (stones and chlorine-based bleach).

### 2.2.6.5. Cabinet Dyeing

The cabinet apparatus resembles a cabinet through which liquor is pumped (Fig. 2.17). If the apparatus is used for dyeing hanks of yarn, the cabinet contains poles that carry the hanks. Stockings and small articles of clothing can also be dyed, these being packed in perforated trays which are loaded into the cabinet. The liquor is circulated by a pump.

The cabinet dyeing machine has the advantage that it is easy to load with the textile goods. It is very important in the mentioned, highly specialized fields.

### 2.2.6.6. Hosiery Dyeing Machines

In these machines, which are specially constructed for dyeing ladies' stockings, the stockings are placed over metal sheets formed into a leg shape, the sheets and stockings are briefly dipped into the vigorously agitated hot liquor, and both are then moved on automatically to the next stage. The dyeing takes place in a standing bath which is continuously replenished [2.40].

### 2.2.7. Automatic Control of Bath Dyeing

#### 2.2.7.1. Aims

The term automatic control (process control) is used very generally here. It includes everything that can be contributed to the operation of a dyeing plant by regulation, control, and data evaluation not involving human intervention options according to the operation manual [2.42].

Automatic control today is expected to give good reliability and repeatability. Standard control systems should be compatible with both types of dyeing machines in a dyehouse (use of specially adapted software).

The requirements on automatic control can be roughly divided into five stages (generations):

1) **Simple process control** (partial automation)
   Temperature–time program, alarm systems
2) **Full automatic control**
   Additional requirements: filling and emptying of the liquor, liquor flow rate control, concentration control (pH, conductivity, etc.), automatic metering, data transmission, operations planning, etc.
3) **Automatic charging and discharging** (robots)
   Preparation of textile goods, charging into dyeing equipment, discharging and transportation after dyeing
4) **Optimization and trouble shooting**
   Evaluation (including statistical methods) of data collected by the automatic control equipment and other measurements, special software and testing equipment [2.20]
5) **Intelligent process development**
   Simulation of a bath dyeing operation on the computer screen including local data from equipment and the dye data (see Section 2.1.3.6)

The stage of today's technology is essentially that stage 2 is complete and stage 3 has started.

On-line control in textile finishing is becoming to be part of the training program for textile engineers [2.43].

Automatic control has now become widely established in bath dyeing. It makes a considerable contribution to the reduction of manpower costs compared with the continuous dyeing process.

With the spread of automatic control the problem arose of precisely defining a dyeing process and representing it in an easily understood form. The tabular form in which formulations have been expressed for a long time is already well adapted for the purpose (see also Section 2.5).

The three usual forms of representation are [2.27], [2.44]:

1) Description of the dyeing process in text form
2) Tables of the control functions required
3) Process diagram with symbols for the individual functions in the apparatus [2.27], [2.45]

The instructions for a dyeing operation are usually rather generalized, so that the programmer has some degrees of freedom, e.g., with respect to liquor draining times during bath changes. Experienced dyers very familiar with automatic control and the dyeing process can often utilize the available degrees of freedom to achieve considerable process optimization. A dyeing diagram is shown in Figure 2.18.

#### 2.2.7.2. Functions of Automatic Control

The automatic control of the charging of dyeing equipment with the textile goods (robot systems) is described in [2.36], [2.46], [2.47]. This can include automatic recognition of the batch number (stage 3 in list provided in Section 2.2.7.1) [2.36].

The basic classification is into (1) direct and (2) indirect effects of the control program on the process (local, decentralized) and (3) an intelligent control of the entire process, which is usually carried out centrally in a computer [2.21], [2.48]

1) *Direct control* of the process elements by the stored program (stages 1 and 2, Section 2.2.7.1), occurs by control of valves and motors or pumps (e.g., flow direction $i \rightarrow o$, $o \rightarrow i$), alarm indicators for product additions, i.e., *process control*.

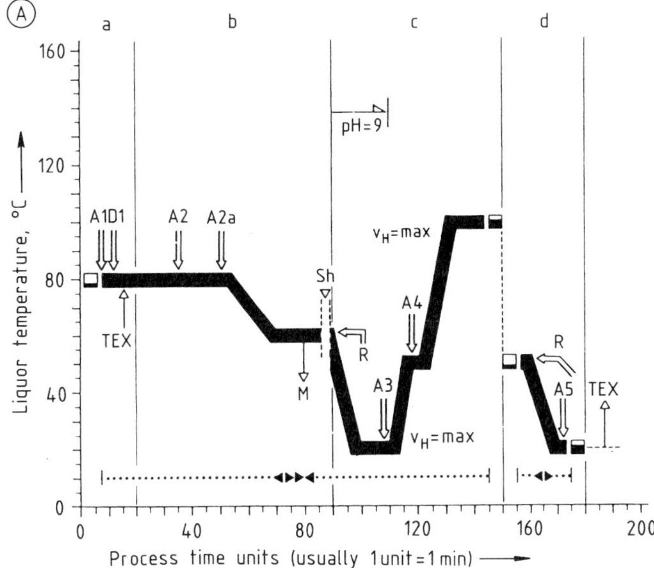


B

| Symbol | Process step | Symbol | Process step |
|---|---|---|---|
| D1 | Dye addition no.1 to dyeing vessel | pH=5 | Proceed as soon as condition indicated (pH=5) has been fulfilled |
| A1 | Addition of auxiliary no.1 to dyeing vessel | Sh | Shading program |
| R | Overflow rinse with rinsing water | C1 | Machine-cleaning program |
| M | Matching-off decision on shading | 3 min | Liquor circulation from inside to outside; 3 min |
| TEX | Entering the textile material into the dyeing vessel | 2 min | Liquor circulation from outside to inside; 2 min |
| | Indirect heating in dyeing vessel | 3 min | Note: Process requires time indicated |

$v_H = \ldots$ °C/min

**Figure 2.18.** Process diagram for dyeing [2.27], [2.44], [2.66, p. 331]
A) Dyeing with vat dyes by the pre-pigmentation process; B) Table of process symbols a) Preparing the dyeing liquor; b) Dyeing; c) Rinsing, oxidizing soaping; d) Rinsing neutralization

2) *Indirect effect* of the stored program on the process (stages 1 and 2). The set points are fed to the local control systems as a function of time. The control systems, which have local intelligence, carry out measurements (e.g., temperature, differential pressure, pH [2.49]), compare with the current set point, and issue the necessary control command to the actuating device, e.g., steam or metering valve, pump which is usually continuously controlled, use of feedforward and closed-loop control, i.e., *program control* (time – temperature program).

3) *Intelligent control* of the entire process (stage 4).

The data of the dyeing operation are usually fed into a central computer. The central computer produces operations planning, usage of the equipment (with or without shading), and batch control.

Data for cost calculations are sometimes also produced. The evaluation of color shading additions, dyeing rate, effects of auxiliaries, unlevelness, etc. is at an early stage of development. The preparation of dyeing programs by the dyehouse with the aid of software is described in, e.g., [2.15].

In all kinds of bath dyeing, i.e., in circulating – goods or circulating – liquor dyeing machines, or special equipment, the requirements are similar. Therefore the automatic control devices are essentially similar for the various tasks and are adapted to the specific tasks by additional modules. The control devices are made compatible with the software. Also, the same kind of personal supervision is applicable to all these control systems.

The requirements of an automatic control system cover a very wide spectrum today. However, a true process control system based on artificial intelligence is still awaited.

### 2.2.7.3. Equipment Requirements

Automatic process control usually has two components:

1) The control computer with the dyeing program and the operator's console (company-specific, Fig. 2.19)
2) A control system with programmable memory, a so-called stored program controller (SPC), constructed from universal low cost components (modules) which carries out individual functions

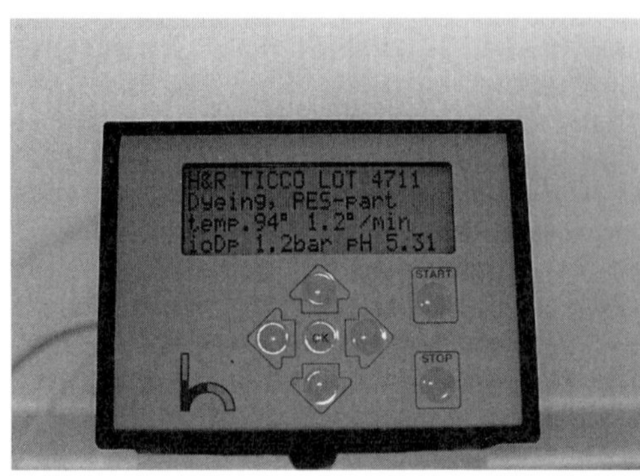


**Figure 2.19.** Local control equipment, connected to the control computer for data transfer and down loading of the dyeing programs [2.49], [2.50] (TICCO equipment, courtesy of Herrig & Rüttiger)

The individual functions are programmed into the SPC, and are called by the control computer as required. Likewise, the different closed-loop control systems are usually formed and backed up by the memory.

The programs stored in the memory of the SPC are completely adapted to the individual parts of the dyeing plant, and cannot be altered by the user.

The control computer contains the actual dyeing program. This is either entered at its console or on a PC, and is then loaded down to the control computer. It is usually specific to the dyeing equipment and not directly transferable. Only the process scheme is transferable from apparatus to apparatus in most cases.

*Important values for closed-loop control*
Liquor temperature (dyebath)
Liquor differential pressure (dyebath)
pH (dyebath, auxiliaries vessel)
Salt concentration

*Important feedforward control functions*
Filling (water, auxiliaries vessel)
Emptying (to drain or recycle)
Flow direction (i →o, o →i)
Metering (linear, nonlinear)
Time – temperature profiles
pH and redox profiles

Some devices are specifically necessary for the dyeing equipment, e.g., cycle detectors for jet and overflow dyeing machines; and passage counters to record the number of passes and devices to measure longitudinal tension for jiggers.

The control computer is often connected to a PC for transfer of process data.

In general, modern equipment for controlling dyeing machinery gives better reliability than good manual control.

Certainly the increase in production over recent years would not have been possible without automatic process control. Moreover, in view of the excellent durability of stainless steel dyeing equipment, the cost of installing or modernizing automatic control equipment can in many cases be justified by the benefits that follow (saving of manpower and gain in reliability).

## 2.3. Continuous and Semicontinuous Dyeing

The continuous dyeing process includes some very different dyeing techniques which can be characterized as follows:

1) The dyeing of fabrics (woven, nonwoven, knitted), which pass continuously through a line of processing units (exhaustion dyeing and pigment/binder dyeing)
2) Dyeing of flock, monofilament yarns, or warp yarns in special dyeing equipment (exhaustion dyeing)
3) Dyeing during the spinning of synthetic fibers (mass dyeing), i.e., during production

*Mass dyeing* is in principle carried out by the fiber producer, > 95 % of polypropylene fibers are dyed in this way. Mass dyeing is not described here, as this dyeing technique mainly lies outside the area of textile finishing.

The continuous dyeing of warp yarn is the main technique used for dyeing with indigo (→ Indigo and Indigo Colorants, **A 14**, p. 153). Continuous dyeing of flock and monofilament yarns is described in Section 2.3.3.

The main topics of Section 2.3 are the continuous and semicontinuous dyeing of fabrics in dyeing lines, using the technique of exhaustion dyeing. Pigment dyeing, which represents a small market volume, is described in [2.51].

Special processing units were developed for continuous dyeing, most of them in the 19th century, and these are assembled to form the dyeing lines [2.4].

Some general rules and relationships apply to the fully continuous and semicontinuous dyeing technique:

1) Fabrics are treated in open width (not in rope form, because crease marks cannot be removed in a continuous system).
2) Fabrics are passed through a processing line at a constant speed of ca. 20–200 m/min. Possible stretching or shrinking of the fabric is compensated for by longitudinal tension sensors and speed controllers. The longitudinal tension on the fabric is in general adjustable. It usually causes a reduction in the width of the goods of several percent and leads to minimization of longitudinal creases [2.36].
3) The running speed of the fabric determines the (usually short) dwell time in the individual treatment units (ca. 5–120 s, depending on the length of fabric in the unit). In semicontinuous processes, i.e., continuous processes in the course of which the goods are batched up, dwell times of several hours are possible [2.52].
4) In all continuous processes, interruptions cause significant staining.
5) Any unevenness in the equipment across the width of the goods leads to unlevel dyeing (streaks and middle-to-edge unlevelness).

### 2.3.1. The Principal Stages of Continuous Dyeing

#### 2.3.1.1. Dye Pickup

In this treatment stage, the dye liquor is taken up (continuously) by the fabric. Two groups of application techniques exist:

1) Direct application of the required quantity of dye liquor to the fabric, i.e., application of the liquor with a metering device. These techniques are of little industrial importance. Application techniques with direct pickup of the dye liquor include: pouring [2.53], kiss padding, spraying, foam application by doctor blade [2.7], and printing (1000-point roll).
2) Immersion of the fabric in the dye liquor. In this case an excess of dye liquor is taken up by the fabric. This is followed by removal of the excess to give the required charge of dye liquor on the fabric. This group of processes is of great industrial importance. Application techniques with immersion of the fabric and removal of the excess liquor include: squeezing off (padder) [2.55], squeegee (Flexnip), suction [2.54], and free runoff [BDA (British Dyer's Association) System].

The *padding* or *pad process* is by far the most important application technique (> 99 % of all continuous production) and is described below.

The *pouring process* is very important in the dyeing of carpets because it can handle articles up to 5 m in width. A small number of fabrics can be uniformly charged with dye liquor by a *squeegee operation*. Fashion effects can be achieved by printing the fabric on both sides in a uniform color (→ Textile Printing). In all continuous techniques, uniform application of the dye liquor can only be achieved on a dry fabric. Therefore, in contrast to the batch dyeing process, the wet fabric must be dried before treatment which increases the costs of the process.

**Padding Process.** In the padding process the dry fabric passes through the pad trough, where it picks up the dye liquor. Good wetting of the textile is therefore important. With suitable auxiliaries it is often possible to pad even gray cloth (→ Textile Auxiliaries). After leaving the pad trough, the fabric is squeezed between rubber rolls. The liquor pickup, i.e., the resulting liquor charge, varies according to the fabric material (40 – 100 %).

The amount of dye liquor picked up (liquor charge or padding effect) strongly depends on the squeeze pressure produced by the two rolls (pad mangle). Figures for the liquor pickup are usually expressed as wt % based on dry fabric (at ambient atmospheric humidity). The curve of the liquor pickup plotted against squeeze pressure is nonlinear (Fig. 2.20).

A low liquor pickup is often desirable to minimize both migration effects and the cost of an interim drying stage. A higher liquor pickup is preferred for vat dyeing of cotton (before steaming), as this gives a smoother fabric appearance.

In the pad dyeing process, the nip of the mangle ensures that the liquor flows into the fabric with high energy, i.e., the liquor is "worked" into the fabric which results in good dye penetration and very little two-sidedness.

Because the padding process is so effective, it has been improved as much as possible.

One of the problems to be solved is that the two rolls, when squeezed together by means of the end bearings, do not provide a uniform pressure over the full width of the fabric because bending takes place. A possible remedy is to fit a "bend tube" inside the body of the (tubular) roll to compensate for most of the nonuniformity (e.g., the Babcock system).

By spreading the squeezing pressure evenly over the width of the fabric, level dyeing across the full width is often ensured and middle-to-edge unlevelness and longitudinal streaking are prevented.

The most important techniques for providing a localized squeeze pressure include:

1) Swimming rolls ("S" pad rolls, Küsters see Fig. 2.21). Variation of the squeeze pressure is achieved by varying the oil pressure [2.4], [2.55].
2) The air cushion control method (the Bicoflex Foulard system of Ramisch–Kleinewefers) in which a tubular squeeze roll body slides on a tube which bears on a stator, by means of air cushions inflated to various extents. This

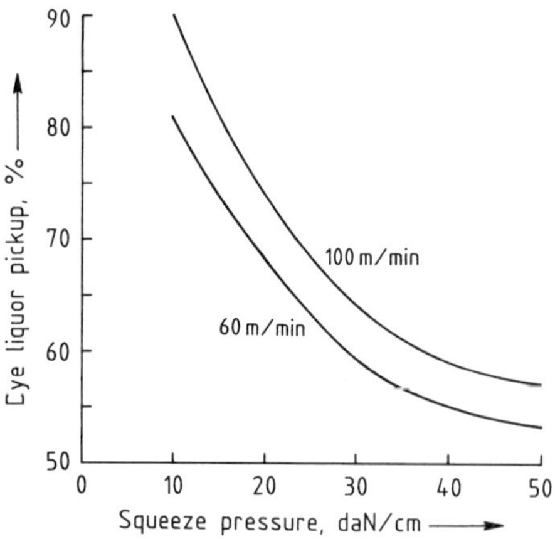



**Figure 2.20.** Effect of squeeze pressure of the pad rolls on dye liquor pickup at different fabric running speeds (60 and 100 m/min)
Values for a "swimming roll" system.

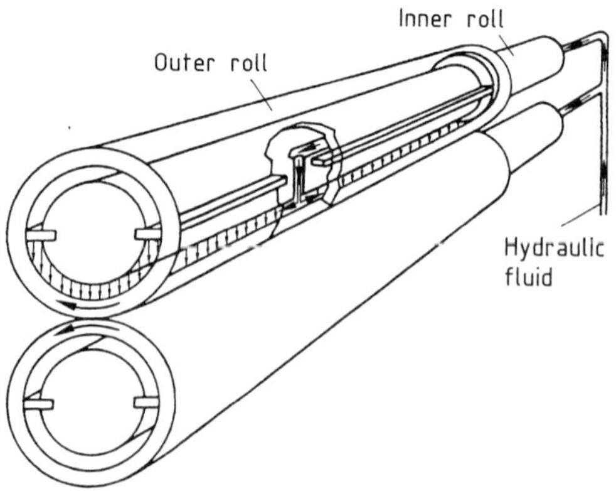



**Figure 2.21.** Swimming rolls (courtesy of Küsters)

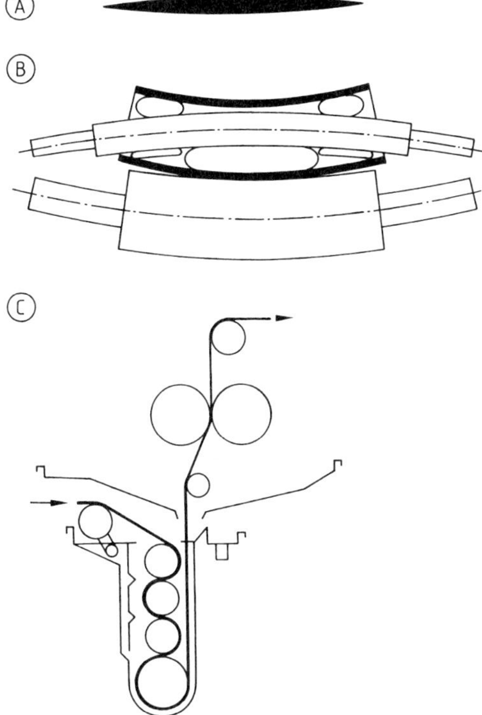

**Figure 2.22.** Pad rolls with dye liquor trough; the squeeze pressure adjustable across the width of the roll by inflated air cushions (supervision) (BICOFLEX system, courtesy of Ramisch–Kleinewefers)
A) Distribution of the squeeze pressure over the width; B) Inflated air cushions; C) Dye liquor trough (cross-section)

enables the squeeze roll to be bent in various ways (see Fig. 2.22).

3) The wave profile of the pad rolls (Matex system, Monfort). Two squeeze rolls whose diameters vary over the width are arranged adjustably in opposition to one another. This enables any desired pressure distribution over the width to be produced [2.56].

4) Sectional rolls (Nipco rolls, Escher–Wyss). The squeeze pressure needed at particular points is obtained by varying the hydrostatic pressure of individual elements ([2.13], p. 702).

As a rough rule of thumb for rubber-coated rolls, increase in the softness of the rubber coating increases both the amount and the uniformity of the liquor pickup (see also Fig. 2.21).

**Constancy of the Dye Liquor.** Uniform liquor pickup is disturbed by the substantivity of the dye. Substantivity reduces color intensity at the end of the batch because of depletion of the dye content of the padding liquor. Possible countermeasures include:

1) Low immersion volumes (rapid attainment of equilibrium)
2) Low temperature, ca. 15 °C (often higher substantivity but lower absorption rate)
3) Control of liquor concentration (e.g., water addition at the start of the batch)

Problems are also caused by instability of the pad liquor which mainly results from:

1) Chemical conversions in the pad liquor, e.g., reaction of reactive dyes with alkali, inactivation of dyes by hydrolysate formation (see Sections 4.1), or reaction of vat dyes with reducing agents (see Sections 4.3, 4.4) [2.57].
2) Gradual precipitation of the dye if it is present at too high a concentration (higher than its solubility), mutual interaction of various dyes used for mixed fabrics (cotton–polyester dyeing), or transfer of sizing or harmful impurities from the gray cloth into the pad liquor.
3) Buildup of foam and lint (stippling due to deposition or resist effects, or printing effects due to small masses of lint).

Pickup of the dye is of great importance in continuous and semicontinuous processes, as defects that occur are extremely difficult to rectify or repair. Color intensity is determined not only by the amount of dye applied, but also by the dye penetration, which determines the "visual dye yield".

A theoretical treatment of pad dyeing is still a long way off. In particular, it cannot be predicted whether a fabric will be dyed lighter or darker at the edges even if squeezing-off is uniform. The dyeing practice is determined by the experience of the specialist, who decides on formulations, settings of the padding equipment, and on on-line measurements during production (see Section 2.3.4).

#### 2.3.1.2. Intermediate Drying

After padding the fabric is often dried (pad dry process). In some cases this is absolutely unavoidable, e.g., when dyeing polyester of polyester mixtures by the thermosol process (see Section 7.1.2). In other cases intermediate drying can be a cost saving, especially with dark color

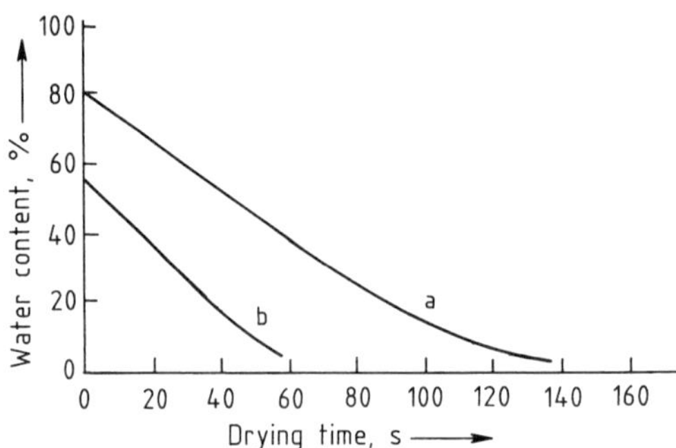

**Figure 2.23.** Drying curve of a hot flue dryer for hygroscopic fiber material
Decrease of water content with increasing drying time at 130 °C for a dryer length of 34 m. The initial linear section corresponds to Stage 1 drying.
a) Cotton, 100%, 200 g/m$^2$, initial moisture content 80%;
b) PES – cotton 67 : 33, 156 g/m$^2$, initial moisture content 55%

shades. During drying the dye migrates to the surface of the fabric, which results in a considerable increase in the "visual dye yield" (up to ca. 30%, see Chap. 11).

In the vat dyeing (see Section 4.3.4), the application of liquor chemicals (reducing agents, alkali) on a predried fabric proceeds better, and, in particular, is less dependent on the material than a wet-in-wet process.

The drying process takes a very characteristic course (see Fig. 2.23) which is largely independent of the type of dryer [2.58], at least as long as the drying intensity is uniform along the length of the equipment. The drying process can be divided into the following stages:

**Stage 1.** Water evaporates from the surface of the fabric in a way largely independent of the nature of the fabric. Migration of the dye occurs, caused by flow of the dye liquor out of the interior of the fabric. The dye contained in the liquor is deposited in the surface zones of the fabric. This process ends when the liquor charge is ca. 30%.

**Stage 2.** Water evaporates from the interstices in the yarns and fibers. Dye deposition occurs in the interior of the fabric where the evaporation occurs. During this stage, there is very little danger of unlevelness due to migration during drying. Dryeing conditions being equal, the drying rate is much lower than that in Stage 1.

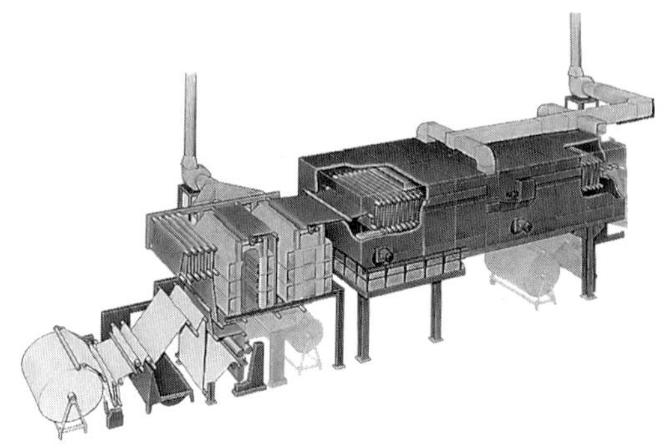

**Figure 2.24.** Intermediate drying
Fabric passage from left to right through: padder, air passage, IR shaft for predrying (VTG/3M, gas heated), and hot flue for final drying.

**Stage 3.** This is only found with hygroscopic fiber materials [e.g., cotton, wool, poly(vinyl acetate)]. Water evaporates from the *interior* of fibers of this type. As some of the water is strongly bonded by hydrogen bonds, the energy requirement for evaporation is higher than in the other stages (further reduced drying rate).

**Drying Equipment.** Two different types of drying equipment are normally used for the intermediate drying of fabrics dyed by a padding process (see Fig. 2.24):

1) *Hot flue dryers* produce a hot air stream (jets, perforated drums), drying occurs by convection. Hot flue dryers have a higher absolute drying rate, but low drying rate per unit of surface area (low energy density).
2) The *infrared (IR) shaft* is a radiation dryer. It is usually only used as predryer before a hot flue dryer to give residual moisture levels of 30 – 25%. Radiation dryers have a lower absolute drying rate, but higher drying rate per unit of surface area (high energy density).

Predrying with a radiation dryer has become widely established, although uniform migration can be disturbed, leading to unlevelness (Stage 1). For pigment dyeing with a binder, a tentering frame is often used for drying before the fixation stage.

**Drying Capacity.** The drying capacity (evaporation rate) can be characterized in various ways, in which the rate can refer to either the dryer or the fabric.

The *drying capacity of a dryer* (absolute drying rate, $DR_{abs}$ in kg/min is the amount of water evaporated per minute. For example in Stage 1 drying, $DR_{abs} = 2.3$ kg/min. An illustrative example is given below:

Fabric goods: 200 g/m², 1.5 m width (water content: 300 g per meter fabric); predrying: water content reduced from 65 % to 30 %.

Amount of water evaporated per meter length of fabric: 35 % of 300 g/m fabric = 105 g/m fabric. Drying capacity of the dryer: $DR = 2.3$ kg/min. Fabric running speed (Stage 1 drying): ($DR_{abs} = 2.3$ kg/min, evaporated water = 0.105 kg/m fabric) = 22 m/min.

The *relative drying rate* $DR_{rel}$ based on the fabric indicates the intensity of drying of the fabric (energy density, fabric stress). The relative drying rate means the decrease in water content of the textile expressed as percentage units per minute. A reduction in the length of the dryer leads to an increase in the energy density for a given reduction in water content and a given fabric running speed.

Extending the above example gives:

| | |
|---|---|
| Length of dryer: | 4.2 m |
| Fabric running speed: | 22 m/min |
| Drying time (dwell time): | 4.2 m/22 m/min = 0.19 min = 11.4 s |
| Moisture reduction (65 % → 30 %): | 35 % |
| Drying rate (relative, based on fabric): | 35/0.19 = 184 %/min  35/11.4 = 3.1 %/s |

Although these figures are useful when fabric changes are involved (light/heavy, sensitive/robust), they have not been widely used in practice. From the figures, it follows that:

$$\text{Fabric throughput} = \frac{200 \text{ g/m}^2 \times 1.5 \text{ m} \times 22 \text{ m/min}}{1000}$$
$$= 6.6 \text{ kg/min}$$

Reduction in water content $= 35\%$ of 6.6 kg/min = 2.3 kg/min. This is a gentle drying operation. For a soundly based assessment, figures for the "energy impact" on the fabric are also important (e.g., length of the drying zone, drying times).

**Levelness.** Both when drying with hot air and with radiation (steam production), differences in flow rate occur between the middle and the edges of the fabric. These can lead to differences in migration behavior (middle-to-edge unlevelness).

**Energy Consumption.** Dryers can be optimized with respect to energy consumption, in particular by prevention of excessive losses due to hot waste air (e.g., optimization to 150 g water vapor per m³ air, or 7 m³ air per kg water) [2.59].

### 2.3.1.3 Dye Fixation

The dye is fixed inside the fiber in the exhaustion dyeing process (see Section 2.1.3.5) or on the surface of the fiber in pigment binding. In the continuous exhaustion dyeing process, the penetration of the dye into the fiber is part of the fixation process. The sometimes complex processes of fixation, which are not sharply separated from each other, can be classified independently of the dyeing technique used as follows:

*Chemical reactions:*

1) Covalent bonding of the dye to the fiber material, e.g., reactive dyes (→ Reactive Dyes)
2) Change in the solvation of the dye molecules, e.g., oxidation of vat or sulfur dyes that have previously been reduced (see Sections 4.3.4, 4.5.3)

*Incorporation into the fiber (desolvation):*

1) Solvation change: dye liquor/fiber material, e.g., disperse dyes (see Chap. 7)
2) Ion pair formation and aggregation, e.g., acid dyes (see Sections 4.8.3, 5.1.2.1) or cationic dyes (see Section 4.8.4) and vat dyes at the soaping stage (see Section 4.3.5)

*Incorporation into the fiber surface zone and binder film:*

Surface fixation of a film of pigments, binders, and cross-linking agents, which sometimes undergo chemical reaction (see Section 4.7).

The *chemical reactions* proceed at a certain rate that determines the fixation time. In the absence of other information (see Chaps. 4–11), it can be assumed that the rate of reaction increases by a factor of 2–3 if the temperature is increased by ca. 10 °C. A reaction that requires ca. 10 h for 95 % conversion at 25 °C can take place in ca. 1 min at 85–95 °C.

*Diffusion processes* proceed slowly in solids (fibers). However, above the glass transition temperature of a synthetic fiber or with very swollen cellulose, diffusion can occur very rapidly (see Section 3.2).

In practice, the dye liquor formulations and range of products are optimized for the various requirements of each dye class, matching to the existing production technology. The most important fixation methods can be divided into two groups.

**Semicontinuous Processes.** *Cold Pad Batch Dyeing.* The fabric is charged with dye liquor by padding wound onto rolls and these batches of cloth are stored for 4–10 h at room temperature. This is important in the case of reactive dyes (see Section 4.1), where the dye–cellulose reaction is brought about by the alkali in the liquor during storage. The reaction rate determines the liquor stability and fixation differences occur if the storage time is too short. The outer layer of the fabric roll is stored shorter time than the inner layer and does not reach the same degree of fixation.

*Pad Roll Dyeing.* The fabric, which is charged with dye liquor by padding, is fed into a chamber where it is wound onto a roll. This is most suitable for cellulose fabrics (see Chap. 4), and is sometimes also used for acetate fabrics with disperse dyes (see Chap. 8).

**Continuous Processes.** Here, the dyed fabric is available for further treatment after passing through the dyeing line (ca. 5 min).

*Steam Treatment (Pad Steam Process).* An additional developing agent is usually applied (by padding or kiss padding) to the fabric after it has been charged with dye and often predried. Examples of developing agents include reducing agents as used in vat (see Section 4.3.4) and sulfur dyeing (see Section 4.5.3) which produce a substantive form of the dye, or alkali and salt for the fixation reaction in reactive dyeing.

Saturated steam at ca. 100 °C for a dwell time of ca. 0.5–2 min is used to achieve fixation or at least a partial fixation reaction.

The most important machinery used for this is the continuous steamer (see Fig. 2.25). This is based on a fabric guide roll system, and has a fabric capacity of ca. 30–60 m. If atmospheric oxygen has to be excluded from the steamer during a dye reduction process, a steam blanket is provided at the inlet point and a water seal at the exit.

Other processes include the "Standfast" dyeing process in which the fabric is passed through molten metal (Wood's metal), heating the fabric in the liquor with the Williams Unit [2.13], and high frequency heating [2.60], [2.61].

*The Thermosol Process* (see Section 7.1.2). The dry fiber, loaded with dyestuff, is heated to around its glass transition temperature. The process is used with thermoplastic fibers, especially polyester, and sometimes also with polyacrylics. The fabric, whose fibers are covered with a dried film of disperse dye and auxiliary, is held at ca. 210–220 °C for 0.5–1 min. This causes the dye to diffuse into the interior of the fiber, although it is not clear whether this transfer takes place mainly by sublimation (gas phase) or whether the dye "dissolved" in the film of auxiliary exhausts from this film. Dye fixation of disperse dyes on polyester fibers as a function of time and temperature is shown in Figure 2.26. The thermosol process is often used for dyeing the polyester component in polyester–cotton blends (see Section 11.1). Dark shades require subsequent washing to remove residual dye on the surface to improve fastness properties.

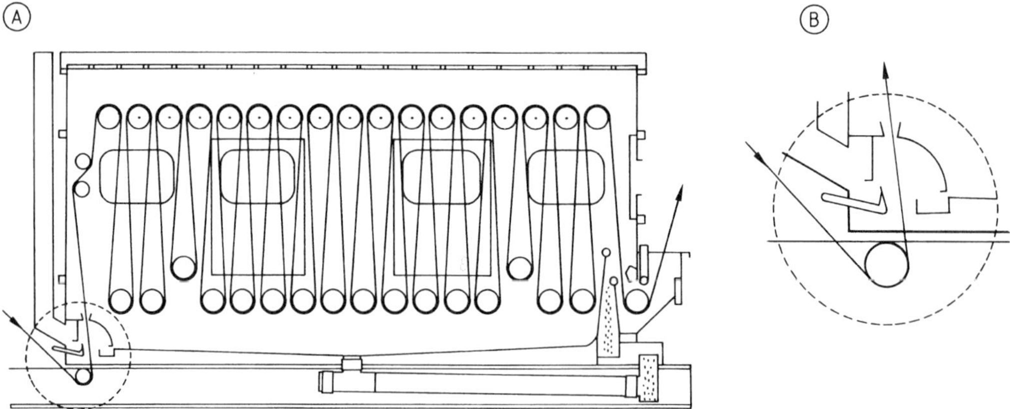

**Figure 2.25.** Continuous steamer; fabrics inlet and water seal specially designed for vat and sulfur dyes (courtesy of Babcock)
A) Entire machine; B) Fabric inlet

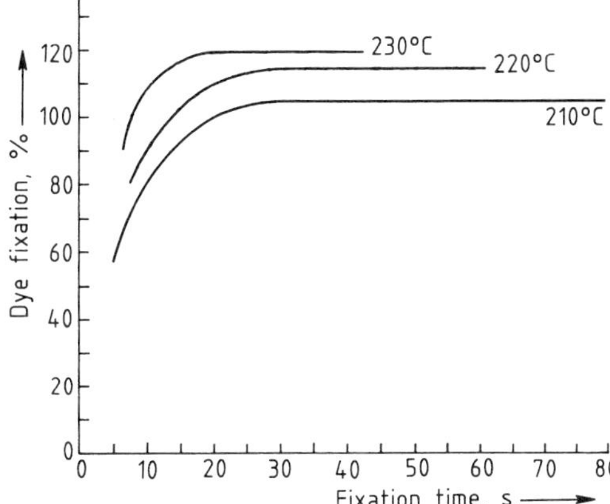

**Figure 2.26.** Fixation of disperse dyes on polyester fibers as a function of time and temperature (Rapid thermosol (RT) unit, courtesy of Fleissner)

*Fixation of the Dye in Pigment Dyeing by Chemical Condensation of the Binder System.*

The combination of dye and (reactive) binder is first applied to the fabric by a pad dry process. The fixation stage is described in [2.6], [2.51], [2.62] and in Section 4.7.

**Fixation by Oxidation.** Dyes that have been chemically reduced before applying them to the fabric must afterwards be reoxidized to achieve fixation (see Sections 4.3.4, 4.5.3).

The oxidation is normally carried out in specially designed washing equipment (see Fig. 2.27). Excess alkali is first washed out of the fabric [2.63]. The most important oxidizing agent used today is hydrogen peroxide [2.50].

A typical example of the oxidizing conditions used in a vat dyeing operation is as follows: hydrogen peroxide (100%) 1 g/L, pH 10–8.5, oxidation time 0.5–1 min, temperature 50–80 °C. If the liquor is too alkaline, this can lead to decomposition of the hydrogen peroxide [2.64], [2.65]. For deeper shades, the oxidation is sometimes completed in the subsequent soaping stage (95 °C), by entrained oxidant.

The dyeing produced is often dependent on the pH of the oxidation liquor.

### 2.3.1.4 Aftertreatment of the Dyed Fabric (Finishing)

Aftertreatment has the following main objectives:

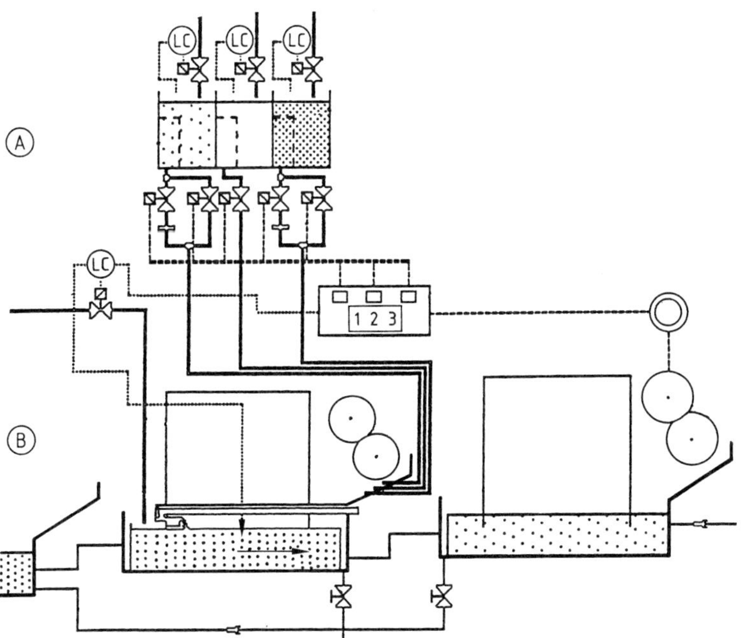

**Figure 2.27.** The oxidation stage in two sections of a washing plant with metering of hydrogen peroxide (sometimes also with acid or buffering agent) (courtesy of Babcock)
A) Metering section with buffering vessels for products and dispensing valves; B) Dyeing line with dosing controller working with running impulses and level control of the oxidation bath (movement of goods from left to right)

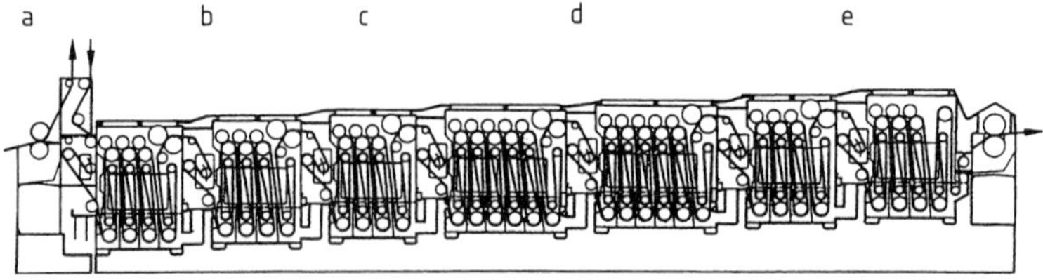

**Figure 2.28.** Continuous open-width washing machine for the aftertreatment of dyed fabrics (equipment especially for vat aftertreatment; courtesy of Benninger)
a) Padder; b) Rinsing; c) Oxidation; d) Soaping, near boil; e) Final rinsing

1) To improve or complete the fixation of the dye
2) To "develop" the shade, i.e., to adjust it to the original pattern card
3) To improve fastness properties
4) To wash out residual chemicals or auxiliaries
5) To improve hand (feel.)

Some process steps that have already been treated in Section 2.3.1.3, e.g., the fixation of most vat or sulfur dyes by oxidation, are often classified as aftertreatment or finishing processes.

Wet-in-wet methods are always used for aftertreatment. Aftertreatment is performed in the continuous open-width washing machine [2.63] (Fig. 2.28) or some of its units.

**Application of Dye Fixation Agents.** During the final wash cationic auxiliaries (to fix substantive dyes) and reactive hydrolysates (to improve wetfastness properties) are applied. The auxiliaries are metered into a section of the washing equipment or a special trough. Good squeezing-off is required before treating the fabric and aftertreatment must be carried out in a standing bath to prevent products from entering the countercurrent washing stream.

**Soaping of the fabric** is carried out in the soaping liquor close to boiling temperature. In the case of reactive dyes the dyeings are rinsed with pure water to remove hydrolysate residues. During vat dyeing soaping occurs in a quasi-standing bath to complete fixation and to develop the color. (It is claimed that soaping also improves rubbing resistance but this is questionable.)

The addition of auxiliaries, methods of improving colorfastness, and the problems that arise are described in Chapters 4–11. For control of dyeings and dyeing results see Chapter 13.

**Reductive Aftertreatment.** Especially in the case of dark shades produced by disperse dyes on polyester-containing fabrics (see Chap. 7), the dye that is adsorbed only on the surface of the fibers must be removed. For this, reducing agents, dispersants, and alkali are added in one or two sections of the washing equipment (standing bath method).

**Neutral Washing.** In this last treatment step product residues (especially alkali) are removed from the fabric [2.63]. During washing the liquor is often neutralized with acid, with pH control.

## 2.3.2. Dyeing Plants

In dyeing plants the individual stages of a dyeing operation are assembled of different machine parts (units) to form a station or production line. Generally the dyeing process can be carried out consistently and at a high production rate using the classes of dye for which the plant was designed, whereas improvisation may be necessary for other classes of dye. In some cases, tasks involving certain fabrics or classes of dye may be impossible.

A continuous dyeing plant must provide for continuous fabric supply including:

1) Synchronous throughput (start-up, production)
2) Adequate longitudinal tension (stretching of the fabric)
3) Maintenance of open width of the fabric (special mechanical devices)
4) Avoidance of longitudinal crease formation
5) Continuous product addition (metering).

Three typical plants are described as an example:

**Thermosol Pad Steam Plant for Polyester – Cotton Blends** (Fig. 2.29). In the following example a mixture of dispersion and vat dyes is first padded on the fabric. The fabric is then dried by

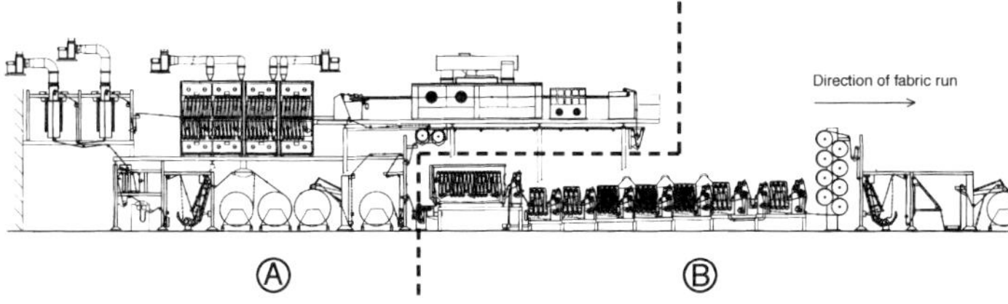

**Figure 2.29.** Thermosol pad steam line (courtesy of Babcock)
A) Disperse dye (thermosol) section: pad application of dye, dryer (IR, hot flue), thermosol unit; B) Vat dyeing section: pad application of chemicals, steamer, rinsing, oxidation, soaping and final washing (neutralization stage) sections

IR shaft and hot flue dryer and subsequently treated in the thermosol unit to fix the disperse dye. This is performed in hot flue or perforated drum equipment. The reducing agent and alkali are then padded on the fabric and steamed. This causes reduction of the vat dye to the leuco form which is absorbed by the cotton, and also reduction of the unfixed disperse dye as a part of the aftertreatment of the polyester component (reductive cleaning).

Aftertreatment of the dyed cotton component is then carried out by:

1) Cooling (to ca. 30 °C, water seal on the steamer)
2) Cold rinsing (for alkali removal, in washing equipment)
3) Oxidation (at ca. 65 °C in washing equipment with metering of additives)
4) Soaping (at ca. 95 °C in special washing equipment)
5) Neutral washing (ca. 60 °C, in 2–3 washing sections)
6) Cooling and final squeezing-off (in squeeze rolls)

**Wet-Steam Vat Dyeing of Cellulose Fabrics** (Fig. 2.30). First the vat dye is padded on the dry fabric. The liquor that contains the reducing agent is then applied to the fabric while it is still wet. The usual intermediate drying is omitted. The dye is then fixed by steaming and oxidation (see Section 4.3.4) [2.66, pp. 319, 357]. The term "wet-steam" here indicates the wet-in-wet dyeing technique and not the condition of the steam.

In this technique, the fabric which is fed into the steamer is charged with a large amount of liquor, so that there is a danger of streaking if the liquor runs back. This depends very much on the fabric. Fabrics with a high capacity for liquor often give no problems.

This dyeing technique only requires a small amount of machinery and little drying energy. The individual process steps are:

1) Dye application (absorbent fabric, padding to give ca. 60–80% pickup)
2) Application of chemicals, i.e., reducing agent and alkali (padding, kiss padding, wet-in-wet, ca. 100% total liquor pickup)
3) Steaming (vat dye steamer, saturated steam, air-free, duration ca. 1–2 min)
4) Cooling (water seal), rinsing, oxidation, and aftertreatment as in the thermosol pad steam plant

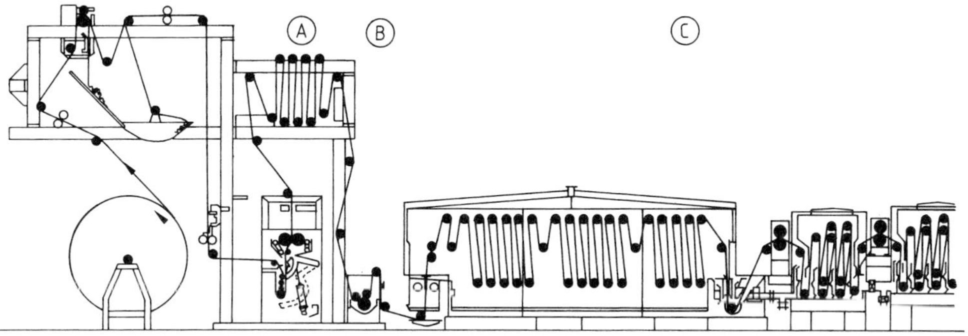

**Figure 2.30.** A wet-steam plant for dyeing with vat dyes (courtesy of Ramisch – Kleinewefers)
A) Padding of liquor (Bicoflex foulard) and air passage; B) Kiss padding of chemicals; C) Steaming and aftertreatment

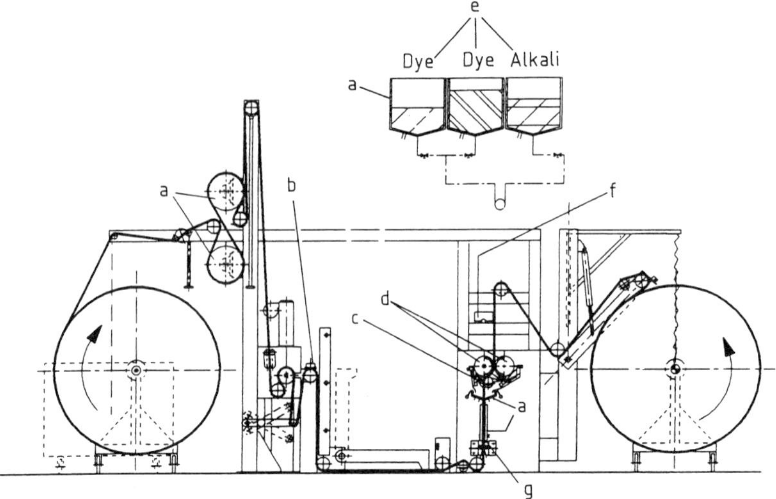



**Figure 2.31.** Cold pad-batch dyeing station (courtesy of Küsters)
Dyestuff uptake is controlled by, e.g., liquor photometric analysis, pickup, initial humidity, and fabric temperature.
a) Cooling, liquor or fabric; b) Measurement of initial moisture content; c) High-speed cleaning; d) Swimming rolls; e) Dye kitchen; f) Measurement of liquor pickup; g) Color measurement, liquor

**Cold Pad-Batch Process for Reactive Dyes on Cellulose Fabrics** (see Fig. 2.31). In this typical semicontinuous process, the short duration steam treatment is replaced by long storage at room temperature.

The dye and alkali (reaction accelerator) are first padded on the fabric [2.67], [2.68]. The fabric is wound onto a roll and stored for ca. 5–10 h in the cold where reactive dye fixation occurs. The fabric is then washed and finished (see Section 4.1).

The individual process steps are:

1) Preparation of the dye liquor: addition of a weighed amount of dye followed by addition of the alkali shortly before use to prevent undesired hydrolysis of the dye in the trough
2) Application of the dye (cold padding at ca. 15 °C to give ca. 55–80 % liquor pickup)
3) Winding (cold, onto a roll)
4) Storage (cold, rotating batch, ca. 5–10 h)
5) Washing cold (ca. 15 °C) and hot (60–90 °C), washing machine or washing beam, preliminary rinsing and soaping
6) Aftertreatment (cold, wet-in-wet, with addition of auxiliaries)

### 2.3.3 Continuous Dyeing of Yarn and Fiber

By far the most important continuous dyeing process for yarn is that for dyeing with indigo [dyeing of warp yarns (slasher, rope lines) for denim material, → Indigo and Indigo Colorants, **A 14**, p. 153]. Dyeing with other dyes (vat and reactive dyes) can be also performed on these plants, the dyes are used alone or as additions to indigo. Dyeing with dyes other than indigo is, however, rather unimportant.

The continuous dyeing of flock is occasionally carried out, e.g., in a "steamer tube" (Seracant equipment for polyacrylic fiber), or by a padding process in high frequency fixation equipment (at ca. 80–95 °C) [2.61].

To obtain space dyeing effects (yarns with local variations in the amount of dye), the yarn can be knitted to tubes and locally printed (vigoureux process, → Textile Printing). After fixation of the dye, the knitted product is unravelled (knit-de-knit process). Yarn is also continuously dyed with local variation of the amount applied, and is then fixed in special yarn storage systems [2.69].

### 2.3.4. Automatic Operation of Continuous Dyeing Plants

The state of automation in the fabric finishing industry has overall reached a high level.

The automation of continuous dyeing plants differs from the automation of bath dyeing (see Section 2.2.7), mainly in that constant conditions are involved—requiring constant adjustments—and that no time programs are required.

Measurement and control equipment exist to a much greater extent in other industries, and the equipment used in fabric finishing represents only a part of the wide range availabe. There is no uniformity in the equipment used.

Computer-integrated manufacturing (CIM) is not (yet) widely used, presumably because the product information available to the dyeing industry is still too generalized and expensive [2.70].

The complex subject of the automation of continuous processes is here split up into two different aspects to facilitate the discussion:

1) Automation of individual items of equipment, mainly to improve their effect on the fabric
2) Automation of assemblies of equipment, mainly to improve their performance and their potential for working together as a large unit

### 2.3.4.1 Important Process Stages and their Automation

Selected process stages—dye application, preparation of baths, drying after padding—are described below in tabular form.

*Characteristic process parameter* (for both the fabric and the process), together with the *measurement and control elements* needed for determining and setting these parameters and for keeping them constant are as follows:

#### Dye Application (Padding of the Dye Liquor)

| Process characteristic parameter | Measurement and control |
|---|---|
| Liquor pickup | control via squeeze pressure (continuously regulated) |
| Middle-to-edge unlevelness | control via variation of squeeze pressure over the roll width |
| Dye concentration in the trough | photometric measurements (transmission) of dye solutions and dispersions |

*Indicators of the amount of dye applied to the fabric:*

| | |
|---|---|
| Dye concentration in the surface zones of the fabric | on-line reflection measurements on the fabric in the visible range |
| Amount of dye liquor on fabric | reflection or absorption of microwaves (cm range) [2.56] or IR [2.71] |
| Total amount of fabric fibers and liquor on the fabric | absorption of β rays [2.56] |
| Volume of dye liquor required per running meter | measurement of liquor added to the pad trough |

A complete list of characteristics of the padding process would contain still more items, e.g., signal generators and receivers to ensure uniform fabric running speed [2.36] and to determine residual moisture on the fabric after predrying, and the absorbency of the fabric. Control of dyeing pickup is only seldom achieved only by means of instrumentation.

Finally, the amount of dye liquor charged is of less interest than the dyeing produced on the fabric. This can easily deviate by $\pm 10\%$ from the figure expected from the amount of dye liquor charged. Here, the distribution of the dye in the fabric material (dye penetration and migration to the fabric surface) is presumably the important factor. Considerable research effort is needed to elucidate this effect and the often puzzling middle-to-edge unlevelness before a truly reliable automation process can be developed [2.72].

**Control of the Treatment Bath (Continuous Washing Units).** Linked to the dyeing processes, sections of open-width washing machines are widely used in important continuous process operations [2.36].

In practice, treatment baths have not yet reached a high level of automation. One reason

| Process characteristic parameter | Measurement and control |
|---|---|
| Bath temperature | temperature control (heating, cooling, water mixing) |
| Bath renewal | control of feed rate (fresh water, countercurrent) or of amount of liquor on the fabric |
| Bath volume | level control (overflow flaps) |
| pH of the liquor | pH controllers, addition of acid, alkali, and buffering agents |
| Concentration of reducing or oxidizing agents | titration with redox probes as indicators or direct measurement with multiprobes to control addition rates [2.50], [2.65] |
| Salt content | conductivity measurement (electrodeless or 2, 3, and 4 electrode systems) and control of dilution or salt addition |
| Concentration profile in the bath | control via partition walls, circulation pumps, or location of feed points |

for this may be the buffering capacity of the large bath volumes employed (ca. $0.5-2$ m³), which nearly levels out concentration changes and automatic control is therefore less important. This is even more significant because dyeing of small fabric batches (pieces) is becoming increasingly important. Simple and reliable metering devices are available which automatically adjust to the fabric running speed (e.g., Polycomat, Babcock).

The addition of acid or alkali is mainly carried out under automatic pH control. However, damage to electrodes (e.g., poisoning) and the presence of nonlinear transition functions in control characteristics still present problems.

Equipment for direct measurement of reducing and oxidizing agents is in the development stage. The most reliable method is still titration.

The main problem for a continuous plant is that a constant dwell time is necessary for a consistent dyeing operation, whether the fabric is light or heavy. Therefore, all weights of fabric pass through the plant at the same speed and differences in their dyeing characteristic must be compensated by other process parameters.

**Drying after Padding (Intermediate Drying)**

| Process characteristic parameter | Measurement and control |
| --- | --- |
| Intermediate and residual moisture content of the fabric | measurement: electrical resistance of the fabric, microwaves (cm range), electrical capacitance, cooling of the drying air, fabric surface temperature |
| Fabric running speed, drying intensity, drying stress on the fabric | measurement: speedometer, values derived from two sources |
| Drying temperature (temperature profile) | measurement: air temperature, fabric surface temperature (conditions suitable for the use of antimigration agents) |
| Energy consumption | water content of exhaust air, e.g., "atmospheric humidity pipes" (Mahlo) |

The following refers to the two-stage intermediate drying process (see Section 3.3.1.2).

It is important that all measurements should be made without contacting the fabric (to avoid streaking effects).

The *predrying stage* (IR radiation dryer) usually leads to a moisture content of $25-35\%$. The moisture content is usually measured indirectly from the surface temperature of the fabric (thermal radiation measuring equipment). In this humidity range, nonuniform migration can occur (middle-to-edge unlevelness, streaks, two-sidedness). The drying is controlled by varying the radiation intensity of the dryer. The fabric running speed must not be changed because this would lead to marking.

*Second drying* by the hot flue process is considerably less intense, and therefore more uniform, with very little danger of migration. The fabric has a longer dwell time in the dryer. The residual moisture content of the dried fabric is usually $0.5-6\%$, and this is often determined from the electrical conductivity of the fabric. The *drying controller* can adjust the temperature of the drying air (heavy fabric requires higher temperatures than light fabric), which takes some time. Alternatively, the heat transfer rate can be varied by varying the rotation rate of the fans, which produces a rapid reaction. Other devices automatically clean the idler rolls that sometimes become smeared with dye liquor (thickened) after the predrying stage.

If the drying is included in a dyeing line, the dye fixation (fixation time) determines the fabric running speed, and the drying equipment must run at a needlessly slow speed even with light fabrics.

However, if the dried fabric is batched up, the dryer can be used efficiently, as the light fabric can be run rapidly, and the heavy fabric more slowly.

### 2.3.4.2. Technology of Automation
(see, → Process Control Engineering, **B 6**)

**General Process Control.** It is first necessary to have local measurement and control equipment at or inside each machine. The measuring instruments (sensors) provide the information (measured data and signals), and the control equipment makes it possible to influence the operation by means of control commands (actuators).

**Local Measurement and Control Equipment.** The measuring instruments (temperature, pressure, fabric running speed, longitudinal tension) are located very close to the point of measurement, and are designed with a direct drive of the display equipment (pointer and scale).

The closed-loop controllers are also constructed on this principle. These have a (mechanical) means of storing the set points. Their control commands mostly go to valves, which are often pneumatically actuated. The necessary controlling action (types of water, countercurrent flow, outflow, etc.) are usually carried out by lever systems or pneumatically.

The advantage of local measurement and control equipment is the low costs, but there are disadvantages:

1) Inadequate supervision and fact that only some of the functions take place automatically lead to faulty adjustment of machines.
2) The investigation and repair of faults can be difficult. Equipment is very individual, and know-how and replacement parts are difficult to obtain. This is also true for calibration.

**Measurement and Control Equipment at a Central Operator's Desk.** The local sensors and associated control instruments are connected to a central operation point. Amplifiers are provided to form a standard signal from the sensor output, and for transmission to the central display equipment.

A disadvantage is that determination of the weight per running meter of fabric is not carried out because this is still very expensive. Thus, quantities associated with the weight per running meter are not available, e.g., the water consumption expressed as L/kg fabric (from the water flow rate), and the moisture content of the fabric in wt % (from the amount of water per unit of area in grams per square meter) or the steam consumption per kilogram fabric.

Some control equipment (actuators) is set at the start of the lot, e.g., liquor flow control or flap valves. Also, the regulators which act on the control equipment are given their set points.

The centralized arrangement is considerably more expensive than the localized equipment, but has the advantage of security of operation.

*Important Advantages:*

1) Control is mainly from the operator's desk, and there is little hand operation of the machines. Change-over to operation with different dyes and fabrics is partly automatic, and uses control programs that can be called up.
2) In trouble shooting, only standard electrical and pneumatic signals are needed for control.

**Process Computer as Control System.** Sensors, regulators, and control equipment produce signals of a type (digital) that can be read by a process computer. Conversely, the computer sends signals (control commands, parameters) that can be understood by the equipment. Also, a PC can undertake a high proportion of the process computing tasks, if suitable software is available (on-line operation) [2.73].

Developments are now in the direction of local control equipment that provides the necessary digital signals or converts the analogue sensor and actuator signals to digital form (field bus operation). At present, many types of field bus are used, leading to unclear and nonstandard operating systems.

Continuous dyeing production lines with process computers sometimes afford the possibility of changing over to a kind of hand operation for controlling the plant and trouble shooting. A complete set of control instruments (usually centralized) exists for hand control of the plant in spite of the costs.

Really useful software tools for trouble shooting are usually in short supply (high programming cost, small number of possible machine combinations).

**Intelligent Process Control.** A computer is now expected to provide also support for the management of the dyeworks. The computer should provide information on:

1) Cost reduction (cost reporting and control)
2) Minimization of reject lots (reproducibility, buffering effect)
3) Shorter treatment time of the goods (optimization of operations planning)
4) Less production of wastewater (independent optimization, washing process)
5) Optimization of product range (amount and range of stored materials, standardization of similar products)
6) Development of new methods, e.g., by calculation of the relationships between dyeing behavior and process parameters

Personal computer programs are only available for some of these areas [2.20], [2.15]. They are usually written by dyeworks staff for particular types of dye. The information required is often only available indirectly (see also Section 2.1.3.6). The simulation of a dyeing process on the computer screen to enable different procedures to be evaluated is still at an early stage of development [2.21], [2.22], [2.25].

## 2.4. Laboratory Dyeing Techniques

### 2.4.1. Objectives

An important task of the laboratory is to carry out preproduction tests, especially of new formulations, to give an indication of the results to be expected [2.74]. The laboratory also helps in maintaining consistency of production. It should also decide whether tolerance ranges should be wide or very narrow. Factors to be considered include dyeing rate, batch time, and cost.

**Important Individual Tasks of the Dyeworks Laboratory**

1) *Determination of Formulations for Production Dyeing.* Searching in the pattern collection, feasibility of producing the color shade on the given substrate, transfer to the production plant.
2) *Shading Formulations.* Determination of dye additions during dyeing.
3) *Testing of Dyed Products.* The various fastness properties are most important. Very specialized methods and equipment are sometimes required (see Chap. 13). Supporting tests of mechanical properties of textiles are also carried out (see Chap. 3).
4) *Dyeing of Samples for Color Measurement* (calibration curves). Preparation of special dyeings of new fabrics supplied by the customer.
5) *Investigation of Defects and Recommendations for their Correction.*
6) *Optimization* (cost minimization, reduction of water consumption, minimization of dye concentration in wastewater, etc., selection of dyes with improved mutual compatibility [2.68]; improved matching of the fastness of lighter or darker shades; statistical analyses).
7) *Reclamation Analyses.* Investigation of samples of rejected materials for quality discrepancies.
8) *New Methods and New Developments.* Development of methods of dyeing and finishing new materials with the available machinery. Ideas for new effects and transformations in the finishing process.

### 2.4.2. Laboratory Dyeing

#### 2.4.2.1. Typical Laboratory Equipment

The amount of textile material available for a laboratory scale dyeing operation is often in the region of a few grams. With such small amounts, the dyeing equipment in the laboratory cannot adequately represent the full-scale plant. The main aim is to match the patterns provided and to produce dyed material for fastness tests.

**Rope Dyeing Machines.** The textile material (hanks of yarn, woven or knitted material) is mounted on a carrier, usually of the double-forked type, and moved up and down in the bath. The dyeing bath is placed in a large and sturdy test tube in a temperature-controlled bath at between room temperature and ca. 95 °C. Rope dyeing machines are a mechanized form of the early method of rope dyeing by hand. A typical example of this is the Ahiba apparatus.

**Beaker Dyeing [High-Temperature (HT) Bomb].** The textile material and liquor are placed in a narrow, cylindrical, pressure-tight steel container with a screw top. The container is placed in a high temperature medium (oil, sand, air, glycol, aluminum block [2.36], or an infrared field [2.75]) where it is kept in continuous motion, so giving good interaction between the fabric and the liquor. Dyeing temperatures of 140 °C can be reached in the HT bomb. This system has found wide application.

**Through Flow Dyeing Equipment.** A certain range of laboratory equipment is available in which the liquor flows through the textile material. This equipment requires a larger quantity of textile. These types of dyeing equipment are used for special tasks, a typical example being the "Colorstar" equipment.

#### 2.4.2.2. Small-Scale Production Equipment

**Batch Dyeing (Bath Dyeing).** To provide a meaningful representation of the dyeing conditions in full-scale equipment, larger textile samples (in the range 100 g to 10 kg) are necessary, which usually make it possible to work with materials of the form used in full-scale production. Dyeing equipment taking samples up to ca. 50 kg in weight covers the size range between the labo-

Figure 2.32. Laboratory Multiflex dyeing machine (courtesy of Then)

ratory and the very small scale production equipment. In general there are laboratory versions of the equipment described in Sections 2.2.3 – 2.2.4 [2.76].

**Dyeing Equipment (Stationary Goods, Circulating Liquor).** Various inserts to the dyeing vessel are available so that yarns (on spools, hanks), flock (compressed into baskets) and piece goods (beams of woven or knitted fabrics) can be dyed.

An example is the Multiflex dyeing machine (Fig. 2.32) which has been specially designed to provide an adaptable model of the production process.

**Dyeing Machines (Circulating-Goods, Stationary/Circulating Liquor).** The most important dyeing machines are the *jigger*, in which the width of the fabric is ca. 40 cm, and the *jet* or *overflow* dyeing machine which can contain between 100 g and several kilograms of fabric. The extent of automation corresponds to that used in production equipment (see Section 2.2) [2.77].

**Continuous and Semicontinuous Dyeing.** The heart of the continuous laboratory dyeing machine is the padding system with a trough for applying the dyeing liquor (see Section 2.3.1.1). This determines the maximum operating width (ca. 30 – 50 cm) and the range of running speeds of the fabric (0.5 – 1.5 m/min). Many different versions of the padding process exist, ranging from the simple cold pad-batch process to a complete dyeing plant [2.31]. A steamer and a dryer (for intermediate drying) [2.78] are often available, the latter also serving as thermosol equipment. Finishing of the dyed product is often still

carried out batchwise in a glass beaker or a bucket, as this does not require space and capital expenditure for operations that can easily be improvised.

In addition to universal dyeing equipment, more specialized equipment is also available, e.g., continuous yarn dyeing equipment for indigo ($\rightarrow$ Indigo and Indigo Colorants, **A 14**, p. 153). This is understandable, as dyeing with indigo accounts for by far the largest volume of dyed products. Here, the dye is applied to the (usually gray) yarn wet-in-wet, which does require special equipment. The Looptex laboratory equipment is of interest here.

### 2.4.3. Laboratory Dyeing Technology

All the principles described in Section 2.1.3 and all the recommendations that do not relate to specific equipment made in Sections 2.2 and 2.3 also apply to laboratory scale dyeing. Some dyeing methods that are special to laboratory dyeing are:

**Dyeing with Exhausted Liquor.** A dyeing operation is carried out on white material in the exhausted liquor. This can also be carried out in dyeing operations that have not reached equilibrium to obtain a shadow dyeing. Information about positions of equilibria, dye compatibility, and dye absorption rates can be obtained in this way. Dyeing with exhausted liquor gives results that provide more information to the dyer than plotted curves, and constitute a descriptive and reliable document.

**The sample collection** may contain a sample similar to the pattern provided, and this sample often has an optimized combination of dyes and reliable fastness properties, so that the relevant formulation can then easily be modified to match the pattern.

**Defect Analysis and Correction.** The successful correction of a lot often confirms the defect analysis. The causes of defects can be very complex. Computer aided methods for defect analysis are being developed (e.g., expert systems). However, systematic instructions have not been published (company know-how).

**Statistical Evaluation.** Statistical topics of interest include: frequencies of defects in returned materials (e.g., whether stains are more common than color discrepancies, effecting repair costs), "transfer factors" enabling colorimetric formulations to be converted to laboratory formulations and then production formulations, test dyeing methods [2.68], distribution of the additives for color shading, substrate, dyeing equipment, and, of course, "substrate calculations", e.g., viscose → cotton [2.79], [2.80].

## 2.5. Techniques of Dispensing Products used in Dyeing

### 2.5.1. Dispensing of Dyes

The dispensing of dyes forms an important part of dyeing technology [2.36], [2.81].

Amounts required for a dyeing operation range from ca. 5 g dye (small batch of 50 kg fabric and a very low dye content, 0.01 %) to ca. 50 kg dye (large batch of 500 kg fabric and dark dyeing, e.g., black, requiring 10 % dye). Hand measurement is much more common than fully automatic metering for dyes.

**Hand Weighing of the Dye.** Hand weighing of the dye still has advantages with which automation is only slowly catching up:

1) Wide range of products
2) Very wide range of addition rates
3) All forms of dye handleable
4) Ease of changeover between products
5) Low capital cost

Disadvantages include high staff requirements, the need for protective measures against dust from the dye, and occasional mistakes in weighing.

**Computer-Aided Dye Weighing.** The balance is linked to a computer. This displays the name of the dye required and the amount to be weighed.

Weighing by robots is also possible in the laboratory. Robots to some extent reproduce the range and scope of hand weighing.

**Automatic Dispensing of Dye.** *Continuous Dispensing of Dyes, Liquors, and Additions.* For liquors with a limited service life it is advantageous to prepare stable masterbatches which are mixed shortly before use. These are almost exclusively used in reactive dyeing.

At the pad trough the dye solution and dilute alkali are delivered by a single metering pump with two heads. The finished liquor being fed into the pad trough via a mixer. In this way, the volume ratio of the two components is kept constant even if the total liquor consumption varies.

Thus, highly reactive dyes can be used in the cold pad-batch process, in which a sufficiently rapid reaction takes place on the fabric while it is stored (see Section 4.1). The very short time for which the dye liquor is stored at room temperature is not long enough to lead to any significant hydrolysis of the dye.

*Batchwise Dispensing of Dyes.* This involves the preparation of measured amounts of dye as required by a formulation. The most important requirements in such a system are:

1) Quantity range ca. 5 g to 50 kg
2) Measuring accuracy ca. 1 – 2 % of amount dispensed
3) Dispensing rate, e.g., 10 formulations per hour

The timing problem is solved by making all the addition times equal, which corresponds to widely different addition rates.

Metering systems are now available which operate at a rate corresponding to the quantity measured out. Two systems exist, in the first the weighing operations take place successively (rough weighing followed by accurate weighing, time-consuming), in the second large and small amounts are metered out simultaneously (using pulsing pumps in parallel, expensive).

The most important types of metering equipment are listed below:

1) *Balances* (often 2 – 3 for various ranges), which give high accuracy, and provide a practicable set of equipment for many or all dyes, whether powdered or liquid.
2) *Metering pumps* (piston or membrane pumps, both with valve problems), in which the accuracy is determined by the stroke, one stroke being the smallest metered amount. The metering rate is adjustable, product distribution is required if several dyes are metered per pump. Pumps are only suitable for liquid dyes and liquid masterbatches.
3) *Flowmeters* in which measurement is performed by direct measurement of flow rate or a counter system (water meter principle). For dispensing, valve switching is necessary between several flowmeters of various sizes to

give accuracy and metering capacity. Flowmeters are only suitable for liquid dye mixtures and masterbatches.

In general, the range of dyes is selected so that the concentrated dyes from the same manufacturer can be mixed and therefore do not require preliminary dilution.

Liquid dyes and masterbatches are used for liquid metering systems. The masterbatches often require special formulation to ensure storage stability (dye sedimentation).

Powdered and granulated materials are still at present only metered by weighing systems. The dye is kept in a container (silo) from which it is fed into a feeding device (vibratory chute).

### 2.5.2. Dispensing of Dye Auxiliaries

Dye auxiliaries, which must be matched very specifically to the formulation, e.g., retarders for dyeing polyacrylic fibers, are usually dispensed like the dye. The products usually have an affinity to the fiber. The quantities of auxiliaries required for the formulation are usually expressed as wt % of the weight of the fabric. Their concentration in the liquor therefore varies as the liquor ratio varies. The metered volume (mL) is given by the following equation:

$$\text{Volume} = \frac{C\,\% \times W_{\text{Tex}} \times 10}{D}$$

where $C\,\%$ is the concentration of the auxiliary in wt %, $W_{\text{Tex}}$ the weight of fabric in kg, and $D$ the density of auxiliary in g/mL. For example, if the concentration of a retarder is 0.7 %, its density 0.93 g/mL, and the weight of fabric 100 kg,

$$\text{Volume} = \frac{0.7 \times 100 \text{ kg} \times 10}{0.93 \text{ g/mL}} = 753 \text{ mL}.$$

The other auxiliaries are usually dispensed with the aid of a liquid metering station. These mainly include products that modify the liquor, e.g., surfactants, dispersants, leveling agents, or complexing agents, which are thus also products with an affinity for the dye. The amounts required in the formulation are usually given in g/L.

The metered volume (mL) is given by the following equation:

$$\text{Volume} = \frac{C_{\text{g}} \times W_{\text{Tex}} \times LR}{D}$$

where $C_{\text{g}}$ is the concentration of the auxiliary in g/L, $W_{\text{Tex}}$ the weight of fabric in kg, $LR$ the liquor ratio in L/kg, and $D$ the density of auxiliary in g/mL.

Powdered products are often converted into stable masterbatches in simple dissolving stations.

The most important measuring devices are metering pumps, flowmeters, and measuring cylinders.

### 2.5.3. Dispensing of Chemicals

The main chemicals used in dyeing are salt, alkali, acid, reducing and buffering agents. These are often required in large quantities, and are stored separately. Salts and liquids (alkaline solution) are metered directly, salts from the salt wagons (mobile containers with feeding equipment), alkaline solutions from the solution wagons (mobile containers with delivery pumps). Stock solutions, e.g., hydrosulfite and alkali (vat dyeing, reductive washing) are prepared in a powder dissolving station. The measured amount of powder is fed into a mixer container by a screw conveyor. The solution of alkali is added, and the powder dissolved. The solution is then delivered through pipework to the point of use. Acid and alkali can be added under the control of a pH meter [2.49]. Alternatively, the addition can follow a predetermined program which specifies the (absolute) amount, the addition time, and the metering characteristics (linear, increasing rate, decreasing rate).

### 2.5.4. Preparation of the Initial Liquor Charge and its Replenishment

The required measured amounts of products must be dispersed, diluted, or mixed, and then added to the dyeing equipment. Details vary according to the dyeing technique.

#### 2.5.4.1. Batch Dyeing

**Dyes.** *Initial Charge.* The measured amounts of dye must be predissolved or prediluted, either at a central location or in situ on or in the dyeing equipment. The volume of the liquor and hence the concentration do not need to be exact. To achieve this, the dyes required for the dyeing operation are predissolved or dispersed in containers fitted with stirring or mixing devices, using a volume of water equal to ca. 10% of the final volume of dye liquor, and this is added to the bath.

Transport of the product is by pumping or gravity in inert pipework, and distribution is by valves or swinging arm distributors [2.82].

*Replenishment.* When dyeing with a standing bath, the dye used must be replenished batchwise before each new batch of fabric. Replenishment is carried out with dye masterbatch solutions and dispersions. The required quantities are withdrawn with pulsed metering pumps from the silo and distributed via valves to the individual dyebaths.

**Auxiliaries and Chemicals.** All the auxiliaries needed for a treatment stage are added to the bath at the same time, either directly or via a central metering and mixing station [2.82].

The quantities and addition of products are usually called up by the dyeing program control system. Good communication between the control computer for the dyeing equipment and the central metering and control computer is necessary.

### 2.5.4.2. Continuous Dyeing

**Dyes.** In *continuous dyeing* of warp yarn with indigo standing baths and special wet-in-wet methods of application and fixation of the dye are used.

A masterbatch of replenishment liquor containing reduced indigo (indigo white, leucoindigo) is metered into the dyebath. The fabric itself in a way controls the replenishment rate via (1) the difference between the amounts of liquor on the fabric entering and fabric leaving the bath and (2) via the evaporation per pass ("skying").

The masterbatch liquor (replenishment liquor), containing ca. 50–70 g indigo, is prepared from indigo (100%), hydrosulfite (excess), and sodium hydroxide solution.

Alternatively, grades of indigo are available (BASF) that already contain the reduced dye. This considerably simplifies preparation of the replenishment liquor (see Section 4.4.2).

In *pad dyeing* the total concentration of dye and auxiliary in the liquor must exactly meet the requirements of the formulation. Therefore the liquor volume must be determined as precisely as possible and the rinsing water must be included in the final volume of liquor. The predicted consumption of liquor is calculated from the weight of the fabric and the pickup. This figure is increased to allow for the liquor volume in the trough (ca. 10–100 L) and a safety factor. Photometric measurements for control of the liquor preparations are under discussion.

The finished batch of liquor is transferred to a storage container near the equipment, and is transferred by pumping as required into the trough.

The dispensing and mixing of liquid additions to the liquor is described in Section 2.5.1.

**Dyeing auxiliaries and chemicals** (e.g., wetting agents and defoamers) are added to the dyebath along with the dyes, and are then applied to the fabric.

Liquid chemicals and stock solutions are prediluted, usually as a part of the metering process, and are added to the liquor.

Powdered products can be weighed out and made up into masterbatches which are then metered to the process. Alternatively, the powder can be dispensed directly at the point of use in the process (usually by screw feeders).

Replenishing solutions are usually made up at the point of addition to the dyeing operation.

## 2.6. Colorimetry
(→Dyes, General Survey, **A9**, pp. 97–105)

Colorimetry means the measurement of the reflection spectrum of a sample and the interpretation of this with the aid of special software. This software calculates color coordinates (CIELAB tristimulus values) and color differences, usually for comparison purposes (patterns/samples). Alternatively, the software can calculate dye formulations based on calibration curves, and can indicate the quality of a calculated formulation. The most important uses of colorimetry are listed below:

In the *laboratory* colorimetry is used to support preparation of formulations and to evaluate the addition of dye for color shade adjustment, for the investigation of defects, and for quality control. In *production* on-line color measurement is used for the control of dye pickup, and as an aid to inspection of the fabric.

Colorimetry has virtually replaced visual assessment by the dyer. However, it cannot adequately take account of secondary effects (luster, dichromaticity, moisture in the sample, nap effects, optical color effects with samples sewn together).

### 2.6.1. Measuring Instruments

The most important part of the instrument is the measuring head. Here, polychromatic light falls on the multilayered sample. The reflected light is spectrally decomposed (ca. 30 wavelengths), and reflectance spectra are obtained.

The mathematical interpretation is performed by the built-in or attached computer, today usually a PC. A good review of the measuring instruments exhibited in ITMA 91 is given in [2.85], capabilities of the latest software are discussed in [2.86]. For on-line measurement, modified instruments are used which enable measurements to be carried out remotely from the fabric in full daylight [2.87], [2.88].

### 2.6.2. Methods of Expressing Colorimetric Results

Color coordinates and color differences are expressed using CIELAB tristimulus values (→ Dyes, General Survey, **A 9**, pp. 99–102).

Color differences (pattern–sample) are calculated from the absolute color coordinates (→ Dyes, General Survey, **A 9**, pp. 103–104). The total color difference (d$E$) gives information on color differences for the user.

Tolerable total pattern–sample color differences are typically within the range ca. 0.8–2.0 (strongly dependent on the type of fabric).

Pass/fail formulae [2.89] enable somewhat more reliable decisions to be made concerning the acceptability of dyed products from a given fabric.

Representation of results in tabular form is the most comprehensive system, but for some purposes is too detailed. Figures usually quoted are the color coordinates $X$, $Y$, and $Z$, the absolute CIELAB values (coordinates):

$a*$ = red–green (+, −)
$b*$ = yellow–blue (+, −)
$L*$ = lightness (light = 100)
$C*$ = chroma
$H*$ = hue (angle of 0° = red, 90° = yellow…),

and the color differences $\Delta E*$, $\Delta H*$, $\Delta C*$, $\Delta L*$, $\Delta a*$, and $\Delta b*$, together with an identification number of the sample.

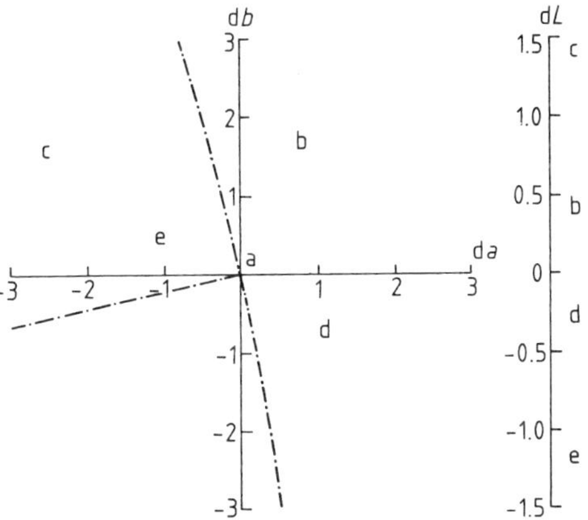


**Figure 2.33.** Graphical representation of color differences (for explanation see text; data MASTER, samples; courtesy of BPI Hohenstein)
a = Pattern; b, c, d, e = Samples

The graphical representation illustrated in Figure 2.33 is widely used, and is almost a standard, despite some problems. The deviation of the sample from the pattern (target) is represented in a field (d$a$, d$b$) and on a scale (d$L$). This gives a very good understanding of the distribution of the samples around the target.

In the graphical representations, neither the color location of the pattern nor the color differences d$E$ between the samples and the pattern (which are very important for many assessments) are usually shown.

The pattern $a$ is in the center of the d$a$, d$b$ field; sample $c$ lies in the sector −d$a$ (more green than the pattern), +d$b$ (more yellow than the pattern); sample $d$ lies in the sector +d$a$ (more red than the pattern), −d$b$ (more blue than the pattern).

Although results from colorimetric comparison measurements are useful for the quality control of the fabric, they are less useful to the dyer for defect analysis and prevention. For defect analysis knowledge of the concentration defects is required; these provide information about the dyeing operation (e.g., unintentional retention of a dye in the bath) [2.91].

Another important application is in the calculation of formulations. With a new pattern, formulations are calculated with the aid of the calibration curves. For this, $K/S$ values (approximately proportional to concentration; $K$ is the coefficient of absorption, $S$ the coefficient of scattering, see Chap. 12) are calculated from the measured reflection values. The $K/S$ curves are then superimposed until the $K/S$ spectrum of the pattern is matched. This enables an estimate to be made of matching, fastness properties, and costs, and a series of laboratory tests are carried out until the pattern is matched with sufficient accuracy.

### 2.6.3. Developments in Colorimetry

The development of colorimetric measuring instruments is proceeding rapidly (e.g., hand-held instruments [2.36], [2.86] and photo-acoustic color measurement).

The development of the interpretation of measurements is remarkably stagnant. After a measurement, the software should be able to provide all possible interpretations, as with data-driven expert systems. Thus, derivation of such information as the characterization of unlevelness, the elucidation of variations in dye properties, and the assessment of migration and two-sidedness is often desirable, but is still only at the development stage.

# 3. Physical Properties of Textiles Important for Dyeing

## 3.1. Classification of Textile Properties

The *physical properties* described here include all the geometrical, mechanical, and optical properties of the textile, together with physicochemical properties such as wetting and swelling.

Information on the following physical properties can be found in the literature:

Thermal and calorific data [3.1]–[3.3]
Amorphous and crystalline regions [3.4]–[3.7]
Frictional behavior [3.8], [3.9]
Hairiness of yarns [3-10]–[3.12]
Conductivity and electrostatic charge [3.13]–[3.15]
Stiffening and strengthening [3.10]
Shrinkage behavior [3.16]
Surface appearance [3.17]–[3.19]

The *chemical properties* of the textile substrate and the functional groups that cause bonding of the dye are described in detail in Chapters 4–11.

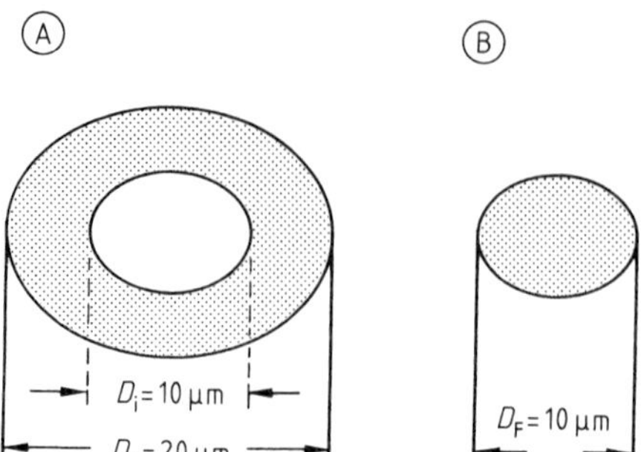


**Figure 3.1.** Cross sections of fibers of various thickness after a standard dyeing process; the shaded zone is homogeneously charged with dye
A) Specific surface area = 154 m²/kg; the colored ring zone represents 50 % of the diameter, but 75 % of the cross-sectional area and of the fiber volume; B) Specific surface area = 310 m²/kg

## 3.2. Fibers [3.20]–[3.24]

(see also, → Fibers, 1. Survey, **A10**, pp. 452–453; → Fibers, 6. Testing and Analysis, **A11**, pp. 72, 79)

The *fineness* (cross section) of the fiber is an important geometrical parameter. The *titer* (mass per 10 km fiber length, unit dtex) is determined from the fineness and the density of a fiber. The specific *surface area* (ca. 50–300 m²/kg) is also important for the dyeing process, indicating the area of the surface into which the dye diffuses and the extent of the dyed layer (ring zone of the fiber impregnated with dye, Fig. 3.1). This determines dye consumption: a thicker fiber takes up dye more slowly because the specific surface area for the entry of dye is smaller. It also consumes less dye (formation of a ring zone [3.25], [3.26]), which partially compensates the lower dyeing rate. This holds, especially for synthetics, when the orientation of the fibers is nearly the same. For a uniform dye concentration in equally thick ring layers of different fibers, the average dye concentration in the thicker fiber is lower, e.g., 75 % in the example given in Figure 3.1. If the transparency of the fibers is not too high the visual impression of both fibers should be the same. Some physical properties of various textile fibers are given in Table 3.1. (The figures listed are mean values of those cited in [3.25], [3.27]. They are reference values for comparisons and calculations.)

Fibers usually show *stress–strain behavior* in which an elastic (reversible) deformation is followed by permanent elongation. Internal

**Table 3.1.** Physical properties of various textile fibers

| Type of fiber | Density, g/cm$^3$ | Diameter, μm | Titer[a], dtex | Surface[b], m$^2$/kg | B[c], g/dtex | EB[d], % |
|---|---|---|---|---|---|---|
| Cotton | 1.52 | 16 | 3 | 160 | 4 | 6 |
| Fine wool | 1.3 | 20 | 4 | 154 | 1.5 | 35 |
| Coarse wool | 1.3 | 60 | 37 | 50 | 1.2 | 35 |
| Silk | 1.3 | 10 | 1 | 310 | 4 | 22 |
| Glass | 2.54 | 8 | 1.3 | 200 | 8 | 2.8 |
| Cellulose acetate (2.5) | 1.32 | 28 | 8 | 110 | 1.2 | 27 |
| Viscose, normal | 1.52 | 26 | 8 | 100 | 2 | 22 |
| Viscose, high strength | 1.52 | 12 | 1.7 | 220 | 3.5 | 15 |
| Polyamide, normal | 1.14 | 27 | 6.5 | 130 | 5.2 | 28 |
| Polyamide, high strength | 1.14 | 30 | | | 6.9 | 19 |
| Polyester filament | 1.38 | 20 | 4.3 | 150 | 4.6 | 26 |
| Polyester staple | 1.38 | 21 | 4.8 | 140 | 3.8 | 42 |
| Polyester, micro | 1.38 | 9 | 0.9 | 322 | | |
| Polyester textile | 1.38 | 15 | 2.5 | 190 | | |
| Polyacrylonitrile (X-51) | 1.17 | 21 | 4.0 | 163 | | 27 |
| Polyacrylonitrile (Acrilan) | 1.35 | 22 | 5.1 | 135 | | 18 |

[a] 1 dtex = 1 g per 10 000 m = 1.11 denier. [b] Specific surface area. [c] Breaking strength of single fiber (dry). [d] Elongation at break (dry).

("frozen") stresses are produced. A partial "melting away" of these permanent stresses leads to (latent) markings. The strongly anisotropic structure of a fiber shows many kinds of reactions to stress (relaxation, retardation). An attempt to formulate the equation of state of a fiber and to describe the transitions between states can be found in [3.28].

Another important parameter for dyeing is *tenacity*, even though the stresses occurring during dying are much lower than this. Tenacity is based on the cross section and the orientation of the fiber and expressed in g/dtex (Table 3.1). The *elongation* of fibers is equally important for certain dyeing techniques. The elongation at break is a useful first approximation (Table 3.1). Generally, the elongation at break is lower for higher-strength fibers, particularly among fibers of the same chemical type.

An important optical property is *light scattering*, which directly affects the depth of color (see also Chap. 12, p. 460). Light scattering increases with delustering of synthetic fibers and decreases with mercerization of cotton (corresponding to an increase in color intensity).

Hygroscopic fibers (e.g., cellulosic fibers or wool) *swell* under the influence of water or alkaline solution and sometimes also undergo structural changes [3.6], [3.29]. This swelling facilitates the diffusion of dyes into the fibers [3.30].

When thermoplastic fibers are below the glass transition temperature range, the dye diffuses very slowly into the fibers (which behave like a solid), but above this temperature, diffusion is very rapid (fiber behavior resembles that of a liquid). In the transition region, diffusion (dyeing rate) increases strongly with increasing temperature.

## 3.3. Yarns

The important properties for the dyeing of yarns are their shrinkage and stretching behavior. These are influenced strongly by the swelling and temperature (below or above the glass transition range). Other physical properties associated with the yarn are determined mainly by the fiber material (Section 3.2), the makeup used for the dyeing process (Section 3.5), and finally, the properties of the woven or knitted article (Section 3.4) that is made of the yarn.

Yarn has by far the highest marketed volume of all dyed goods, due to Indigo dyeing of warp yarn for denim fabrics made of gray cotton. Extreme ring dyeing (thin ring layer) of these yarns is preferred.

## 3.4. Fabrics

Fabrics [e.g., woven products (smooth and pile fabrics), knit products, and nonwovens] represent the final stage of textiles before manufacture into clothing, etc. [3.31]. Woven fabrics account for most of the textiles dyed by continuous dyeing. Around 50 % of fabrics are batch dyed.

The geometrical properties of fabrics are determined mainly by their structure and the type of yarn. The most important of these include mass per square meter (100–1000 g/m²), width (0.8–5 m), and mass per running meter (80–5000 g/m).

A good indication of the structure is obtainable from the thread count (threads per centimeter, warp, weft) or the number of stitches [3.32].

The terms open or dense refer to the structure of a textile, and hence its permeability properties, which are also important. These are indicated by the thickness, interstitial volume and porosity, light transmission properties, resistance to the flow of fluids (air, water), and yarn number and twist.

Also important for good dye penetration is the accessibility of yarn crossing points to the dye and the release of air from the interstitial volume of the fabric during dyeing (prevention of white spots).

Of the mechanical properties, *stress–strain behavior* is the most important [3.6], [3.33]. Longitudinal tension during dyeing causes not only an elongation (reversible, irreversible) in the direction of stress, but also a very characteristic shrinkage in width. This must be corrected at a later stage by stretching to the specified final width [3.16]. Varying warp tensions over the width of the material in the weaving process lead to loose or slack edges [3.34], [3.35]. This results in variable liquor pickup (usually, the higher the warp tension, the lower the pickup).

The *running properties* of the fabric play a very important role in continuous dyeing (e.g., longitudinal creases, ribbed effects, lateral movement, weft displacement) [3.36]. The running behavior cannot be simulated adequately in the laboratory [3.37], [3.38]. Within limits, running properties can be improved by increasing the longitudinal tension and reducing the running speed. With certain products (especially impermeable materials), the wet material is sometimes not in direct contact with the guide rollers and can move laterally (aquaplaning).

The *liquor pickup* after squeezing off and the *liquor-carrying capacity* of a material are other important parameters [3.39]. As a rough guide, dense materials have a lower liquor pickup (40–60 %) than open textiles (70–90 %). The maximum liquor-carrying capacity (with no streaking due to liquor runoff) corresponds

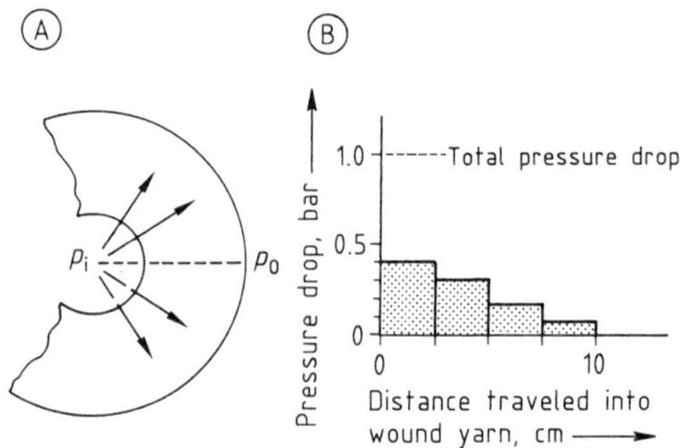

Figure 3.2. Local pressure drop in a cross-wound cylindrical packages during radial passage of dye liquor (density of wound yarn, 300–400 g/L; mass, ca. 2 kg; volume of wound yarn, ca. 5.7 L; i.d., ca. 5 cm; o.d., ca. 25 cm)
A) Cross-wound package, planview; B) Pressure drop across each zone
$p_i$ = inner pressure; $p_o$ = outer pressure

roughly to the interstitial volume of the material (1–3 L/kg).

Of the *optical properties*, the luster of the material and the raising of a nap on its surface are the most important. While the luster (caused by sharp angles of reflection) usually affects only the color measurement, raising a nap often causes a change in color (brightening). Another effect in fiber mixtures is the clear visibility of one type of fiber even though the color intensity difference is small.

## 3.5. Makeup of Textiles for Dyeing

The textile material is usually made up in a rotational symmetrical form for the dyeing machine (rolls of yarn or fabric, packages of flock). During dyeing, the liquor flows radially through the textile material. The greatest resistance to flow and hence the greatest pressure drop occur at the inner parts of the roll. Thus, the inner parts are exposed to the highest stress on the material (ratios of the inner to the outer resistance for cross-wound packages, ca. 5 : 1; for warp beams, ca. 2 : 1, see Fig. 3.2).

An important factor is the uniform and rapid wetting of the material with the dye liquor [3.40]. For this, adhering air must be removed (deaeration) by appropriate pretreatment of the textile (dye-ready material), steaming, or the addition of auxiliaries (wetting agents; → Textile Auxiliaries).

The total resistance to the flow of dye liquor determines the pump adjustment used and the flow rate. This resistance often changes during dyeing because of alternating flow direction of the liquor (greater when flow through the textile or package o → i, smaller when i → o; o = outer part, i = inner part), swelling, temperature changes, and fiber shrinkage, to mention only the more important influences.

When using dyeing machines in which the textiles are stacked during dyeing (winch beck, jet, overflow), crease marks (from the dye) or permanent creases (mechanical) may form [3.41]–[3.44]. The most important methods of preventing these effects are

1) Better presetting of the textile
2) Lower machine loading
3) Better crease control
4) Longer leveling time
5) Slower cooling

Control of the textile throughput rate or cycle time shows whether the material is being transported correctly through the jet equipment (jet diameter, liquor injection rate).

In jiggers, the running problems described in Section 3.4 can occur.

In continuous dyeing yarns [3.3] and woven fabrics [3.4] are used without a special makeup.

# 4. Dyeing of Cellulose Fibers

## 4.1. Dyeing with Reactive Dyes

Reactive dyes are the newest class of dyes for cellulose fibers (→ Reactive Dyes, **A 22**, pp. 651–653). In the dye molecules, a chromophore is combined with one or more functional groups, the so-called anchors, that can react with cellu-

lose. By suitable choice of dyeing conditions, covalent bonds are formed between dye and fiber. More than 100 years after SCHOTTEN and BAUMANN first carried out fundamental investigations into the reaction between alcohols and acyl halides (acylation), ICI introduced the first group of reactive dyes for cellulose fibers, the Procion M dyes (2,4-dichloro-1,3,5-triazine anchor; → Reactive Dyes, **A 22**, p. 651) in 1956, which initiated an intensive development effort. One-third of the dyes used for cellulose fibers today are reactive dyes. They have a better wetfastness than the less expensive direct dyes. The range of available reactive dyes is wide and enables a large number of dyeing techniques to be used. Shades ranging from brilliant to muted can be obtained. Chlorine fastness is slightly poorer than that of vat dyes, as is lightfastness under severe conditions.

### 4.1.1. Fundamentals [4.1]–[4.3]

In general, dyeing with reactive dyes is very similar to the well-known process of direct dyeing. The major difference is that the reactive dye forms a chemical bond to the fiber. Important factors determining the dyeing properties of reactive dyes include the affinity of the dye for the fiber (substantivity), its diffusibility, reactivity, and the stability of the dye–fiber bond. This bond constitutes the fundamental difference between direct and reactive dyes.

The anchor components of reactive dyes are important for both the fixing properties and the wetfastness of the dyed material. More than 300 electrophilic anchors for reactive dyes are described in the patent literature. Of this large number, fewer than ten are currently of practical importance for dyeing cellulose fibers (Fig. 4.1).

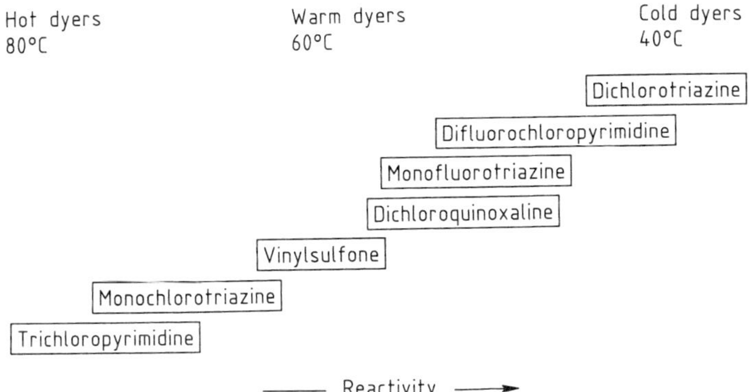


**Figure 4.1.** Reactivities of established anchor systems and optimum dyeing temperatures in exhaustion dyeing [4.8]

**Hydrolysis and Reactivity.** Two possible reaction mechanisms exist: addition and substitution.

Addition anchors include, for example, vinylsulfones

$$\boxed{\text{Dye}}-SO_2-CH=CH_2 + \text{Cellulose}-O^- + H^+ \longrightarrow$$

$$\boxed{\text{Dye}}-SO_2-CH_2-CH_2-O-\text{Cellulose}$$

Substitution anchors include, for example, monochlorotriazines

$$\boxed{\text{Dye}}\!-\!\underset{\substack{\text{triazine ring}}}{\text{(NHR)}}\!-\!Cl + \text{Cellulose}-O^- \longrightarrow$$

$$\boxed{\text{Dye}}\!-\!\underset{\substack{\text{triazine ring}}}{\text{(NHR)}}\!-\!O-\text{Cellulose} + Cl^-$$

As *substitution anchors,* heterocycles with a reduced $\pi$-electron density in the heterocyclic ring are used. The rate of hydrolysis or alcoholysis (i.e., the reaction with the cellulose anion) is influenced by the nature of the heterocyclic nucleus, the substituents, and the leaving groups. In the reaction with the cellulose anion, the substitution anchors listed in Figure 4.1 are bonded by ester linkages. In contrast, cellulose adds to the *vinylsulfone anchor* in a nucleophilic reaction (Michael addition), and an ether link is formed.

Hydrolysis and alcoholysis are competitive processes, both being first-order reactions [4.4], [4.5]. The ratio of the rate constants of the reactions with cellulose and with hydroxide ions (ratio cellulose-$O^-$:$OH^-$) is termed selectivity and can be determined easily in a homogeneous medium by using model substances. For triazine anchors, a decrease in selectivity depending on the leaving group can be observed in the order

$$F > Cl > DABCO > SO_3$$

(DABCO = 1,4-diazabicyclo[2,2,2]octane) [4.6]. However, the differences in selectivities of the established anchor systems are slight.

The preference of the anchor for the fiber is explained by the greater nucleophilicity of cellulose compared to hydroxide. At high pH, the ratio of cellulose-$O^-$:$OH^-$ decreases [4.3], [4.7]. Furthermore, hydrolysis has a higher activation energy than alcoholysis (i.e., it is favored at

higher temperature). As a result, the selectivity with respect to alcoholysis decreases with increasing pH and temperature.

The rate of hydrolysis increases as a rule by a factor of 10 per pH unit (first-order reaction). The rates of both reactions increase by a factor of approximately 2 – 3 per 10 °C. Therefore, for economic reasons (dyeing time) pH and temperature can be varied in practice only within narrow limits. Depending on the optimum dyeing temperature, established anchor systems are classified as cold, warm, and hot dyers (see Fig. 4.1).

In addition to the fastness and application properties the *fixation yield* (i.e., the percentage of dye chemically bound to the cellulose at the end of the dyeing process relative to the total amount of dye employed) is important, particularly for ecological and economic reasons. The fixation yield depends on the chemical nature of the anchor and many other factors [4.9].

**Substantivity.** Substantivity is given by the ratio of the amount of dye in the fiber to the amount in the dye liquor after chemical equilibrium has been reached (without chemical reaction). In analogy to direct and vat dyes, the adsorption properties of reactive dyes can be represented by the Freundlich relation. Substantivity is a necessary precondition for reaction with the fiber. Dye that has not become attached to the fiber is hydrolyzed when the alkali is added and lost to the dyeing process. For monoanchor dyes, high substantivity correlates with attainable fixation yields to a good approximation. Substantivity is influenced by the chemical constitution of the dye and the process parameters. In general, high substantivity is obtained with low dyeing temperature, a low liquor ratio, the addition of electrolytes (e.g., NaCl or $Na_2SO_4$) to the dyebath, and low dye concentration. An extended planar structure of the dye molecule is believed to favor high substantivity.

The dyer must also consider the effect of the substrate. Substantivities for mercerized cotton and regenerated celluloses are higher than those for cotton.

Limits to substantivity exist in practice. If the substantivity is too high, problems can occur with washing off of unfixed dye, wetfastness, or uneven final shade. In continuous processes, tailing can occur (see Section 4.1.2).

**Diffusion Rate.** The diffusion rate is the rate at which dye is transported to the interior of the fiber material. A high diffusion rate leads to rapid establishment of dyeing equilibrium and rapid leveling out of irregularities in the dyeing of the substrate. It also facilitates washing off the hydrolyzed dye. Increased dye substantivity is frequently combined with lower diffusibility. The diffusion rate doubles with a temperature increase of 10–20 °C.

The substrate also has an important influence on diffusion of the dye. The diffusion rate increases with caustic soda treatment or mercerization of cotton. However, with regenerated cellulose fibers, which have a marked skin–core structure, the outer parts of the fiber can act as a diffusion barrier.

**Stability of the Dye–Fiber Bond.** Because of the large variety of reactive dyes, generalizing about colorfastness is difficult. While wetfastness is determined mainly by the anchor system used, most other fastness properties depend on the dye as a whole or the chromophore present. Most reactive dyes are azo or anthraquinone derivatives whose standard of fastness varies greatly. Phthalocyanine, formazan, and triphenyldioxazine derivatives are also very important.

In addition, application conditions [4.10] and finishing processes (→ Textile Auxiliaries, Section 7.1) can affect fastness. Thus, with some resin-finished textiles (dimethylolpropylenurea finish) a decrease in lightfastness is observed.

Significant differences in the stability of the bond toward hydrolysis can be observed in the anchor systems listed in Figure 4.1. The ether bond to cellulose formed by addition anchors (vinylsulfone dyes) is extremely stable to acid hydrolysis. Its instability in alkaline solutions is demonstrated by the poor fastness to hot sodium carbonate solution. In general, the stability of substitution anchors (fluorochloropyrimidines, monochlorotriazines, and monofluorotriazines) is significantly higher in alkaline solution. However, fastness can be reduced considerably by the presence of electron-attracting substituents in the anchor. An example is provided by dichlorotriazines, which are more reactive than monochlorotriazine dyes. The hydroxyl group formed by the second chlorine atom after reaction of the first with cellulose leads to a decreased electron density in the heterocyclic nucleus and to a significant increase in the rate of cleavage of the dye–fiber bond.

One aim of the introduction of heterobifunctional anchor systems has been to combine the good fastness profiles of different anchor systems (→ Reactive Dyes, **A 22**, pp. 653–654).

### 4.1.2. Dyeing Techniques [4.3], [4.12]

Currently, a wide range of reactive dyes of varying constitution is available which are suitable for many different applications. In most cases, dye is fixed to the substrate under alkaline conditions. However, dyes containing phosphonic acid groups (→ Reactive Dyes, **A 22**, p. 653) and quaternary nicotinic acid derivatives (→ Reactive Dyes, **A 22**, p. 652) are exceptions to this rule. Dyes with phosphonate groups are applied in an acid dyeing bath, quaternary nicotinic acid derivatives require neutral conditions. These dyes are applied under weakly acid or neutral dyeing conditions and are used mainly for dyeing polyester–cotton blends.

The suitability of a reactive dye for a particular application depends not only on its reactivity, but also on properties that are determined by its overall structure, such as solubility, substantivity, and diffusion. Special recommendations are therefore made by manufacturers for each type of dye.

Today, ca. 20 % of reactive dyes are used in textile printing, 30 % in pad dyeing, and 50 % in exhaustion dyeing processes.

**Exhaustion Dyeing.** Exhaustion dyeing is a batch process. Loose stock, yarn, or piece goods can be dyed by this method. Equipment consists e.g., of package dyeing equipment, hank dyeing machines, jiggers, winch becks, or dye spraying machines (see Section 2.2.3). The general trend is toward the use of concentrated ("short") liquors (i.e., toward dyeing with the smallest possible volume of dyebath compared to the textile used). With woven textiles, liquor-to-goods ratios of 3 : 1 can be achieved today; with knitted goods, 5 : 1.

In the exhaustion dyeing process, highly or moderately substantive dyes are usually employed. The two main processes are the "all-in" process and the stepwise process. In the *all-in process*, the dye, the salt, and the alkali are all added to the dyeing bath at the start of dyeing. The *stepwise process* consists of an exhaustion and a fixing phase (i.e., alkali is added only after the dye has absorbed to the fiber). The stepwise process has advantages in dyeing materials that are difficult for liquid to penetrate (e.g., dyeing wound packages in a circulating-liquor machine). A preceding exhaustion step aids in preventing an uneven final appearance. The exhaustion step is usually started at 30 °C. Dye and salts are added at the start or portionwise. The salt concentrations employed depend on the substantivity of the dye and the dye concentrations used. For <0.5 % dye based on textile weight, salt concentrations of 10 – 30 g/L are recommended; for less substantive dyes, up to 50 g/L NaCl. For deep shades with dye concentrations of >4 %, salt concentrations of ca. 50 g/L are used; with vinylsulfone dyes having low substantivity, up to 80 g/L. After the dye absorption phase, and sometimes also an additional leveling step (at temperatures sometimes higher than the fixing phase), alkali is added. With hot dyers, a uniform dye – fiber reaction can be obtained by controlled heating. Heating rates are between 1 and 2 °C/min. In the case of cold dyers, fixing can be controlled smoothly by means of pH. Alkali can be added either in portions or by metering. Sodium carbonate or mixtures of sodium carbonate and sodium hydroxide are used almost exclusively today as the fixing alkali. Here, too, the amount of alkali is determined by the reactivity of the system and the depth of shade desired. Cold dyers are applied at lower pH values. The pH is normally adjusted with sodium carbonate. Warm and hot dyers are usually fixed with mixtures of sodium carbonate and sodium hydroxide. Typical concentrations are 5 g/L of sodium carbonate and 1 – 2 mL/L of sodium hydroxide solution (concentration 38° Bé or 440 g/L). The fixing time is 45 – 90 min.

After dyeing, the liquor is usually drained off; the material is rinsed and then washed off with addition of auxiliaries. The washing intensity depends on the substantivity of the dye hydrolysate. The more highly substantive hot dyers require longer washing times or more wash cycles at higher temperature. With alkali-sensitive vinylsulfone dyes soaping at the boil must be carried out in a neutral or weakly acidic liquor, to prevent cleavage of the dye – fiber bond.

**Pad Dyeing Process.** In this so-called impregnating process the dye, additives, and alkali are applied to the textile by dipping followed by squeezing out (see Section 2.3.2). Fixing of the dye on the textile is carried out in a subsequent step.

In these processes, dye and alkali can be applied together or separately. In the *one-bath process*, the so-called pad liquor stability is important. This term denotes the stability of the dye to hydrolysis as a function of time. With increasing reactivity of the dye (vinylsulfone < dichloroquinoxaline < monofluorotriazine < difluorochloropyrimidine < dichlorotriazine dyes), the risk of a loss in color strength with too long a dwell time in the pad box or replenishing tank increases. For this reason, metering equipment is commonly used. Dye and alkali are metered separately into the padder, which means that the dwell time in the pad box can be kept to a minimum. In addition, pad boxes are now constructed so that the liquor volume is as low as possible, and is, on average, replaced within 5 min.

When selecting dyes, the dyer must consider the possibility of so-called tailings. This refers to differences in shade and depth of shade between the beginning and end of the batch of fabric. With highly substantive dyes, the liquor running back into the padder from the squeezing rollers has a lower dye concentration than the initial liquor. This causes a gradual depletion of dye in the liquor. Also, especially with regenerated cellulose fibers, so-called negative tailing occurs. The dye uptake calculated from the liquor pickup is then greater than the actual figure.

In general, pad dyeing can be a semicontinuous or continuous process.

*Semicontinuous Processes.* Semicontinuous retention processes were developed especially for reactive dyes. The *cold pad-batch* process is by far the most important. After the textile has been padded with dye and alkali, the material is either plaited down in a box wagon or rolled up into batches. Fixing takes place during storage. The cold pad-batch process is of special interest from an economic point of view. The amount of alkali and the dwell times required for dye fixation depend on the reactivity and quantity of the dye. With cold dyers, adequate fixation can be achieved with sodium carbonate or bicarbonate; with dichlorotriazine and monofluorotriazine dyes, sodium carbonate concentrations of 2–30 g/L are used. In these systems, ca. 10 g/L of trisodium phosphate can also be used. With less reactive systems, sodium carbonate solution or mixtures of sodium carbonate and sodium hydroxide are employed. If the batches are stored in the open, carbon dioxide from the air can cause a pH decrease and hence a lighter final shade at the selvedges. A high process reliability can be achieved by utilizing the buffering action of waterglass (sodium silicate). In the past, additions of up to 100 mL/L of waterglass (concentration of waterglass solution 38° Bé = 1.35 g/mL) were recommended. For environmental reasons, the trend today is toward lower concentrations or complete avoidance of sodium silicate. By wrapping the batches in plastic film, the negative effects of carbon dioxide can be limited to the actual padding time. If necessary, the solubility of the dyes can be increased by addition of urea (up to 100 g/L). To obtain a level fabric appearance, the fabric must be uniformly absorbent. The maximum possible liquor uptake for cotton (ca. 60–80%) is ca. 10% lower than the figure for regenerated cellulose. To achieve evenness of dyeing by means of higher liquor pickups, auxiliaries that reduce running-down and dripping-out of the liquor can be added. Rotating batches are then used. Nowadays, highly or moderately reactive dyes are used for the cold pad-batch process. For a complete dyeing reaction, 2–8 h is required. However, this time is sometimes extended to 24 h for reasons of work organization. Storage at higher temperature, as in the pad roll process, or the use of accelerators (e.g., 1,4-diazabicyclo[2,2,2]octane) for slow-reacting dyes (e.g., monochlorotriazine anchor) has not widely spread. After dyeing, the material is rinsed and washed off in open-width or rope washing machines. Washing off directly on perforated beams is sometimes employed for economic reasons. The material is then dried.

*Continuous Dyeing Processes.* In continuous dyeing processes, padding, fixing, washing-off, and drying are carried out in one process line. Two different types of processes exist: dry-heat and steaming processes. The *dry-heat process* can be either the single-stage pad dry process or the thermofixation process. Hot dyers are preferred for use in *thermofixation*. Advantages of the thermofixation process include good color yields and covering immature cotton. The dye and alkali are padded at the same time. The preferred alkali is sodium carbonate, used at a concentra-

tion of 10–30 g/L. For less reactive dyes, sodium bicarbonate is recommended. In addition, 20–200 g/L of urea (nowadays because of ecological reasons often cyanoguanidine) is added, which in this process increases the solubility of the dye in the liquor, acts as a solvent for the dye during fixation (*mp* 133 °C), and prevents browning of the cellulose fiber. After pad dyeing, the material is dried, leaving a residual moisture content of 10–15% because further drying can lead to loss of color strength. Subsequent fixation takes place at 150–200 °C, depending on the reactivity of the dye. In most cases, fixation times of 1–2 min are used.

In the *pad dry process*, fixing and drying occur simultaneously. Here, dyes of medium to high reactivity are preferred. The fixing time is 30 s to 5 min.

Of the *steaming processes*, the *pad steam process* is by far the most important. In the first stage, dye is applied by padding. Intermediate drying is then carried out at 110–140 °C. Alkali is applied by a subsequent padding machine. High salt concentrations are used, normally 250 g/L sodium chloride to prevent detachment of the dye and to give a high degree of fixation. The solubility of the dye can be increased by adding urea. Also here, the amount of alkali employed must be adapted to the dyeing system used. For less reactive dyes, up to 50 mL/L of NaOH solution is applied. Dye fixation is carried out with saturated steam at 103 °C for ca. 30–90 s. For dyes that are sensitive to reduction, a weak oxidizing agent can be added to the pad liquor.

In all heat-fixing processes, optimum fixing times must be adhered to. If the times are too short, loss of color strength occurs due to incomplete dye–fiber reactions. With too long a fixing time, cleavage of the dye–fiber bond is a danger, especially for vinylsulfone dyes. Also, in all drying operations, unevenness due to dye migration can occur. This can be prevented by the correct choice of dye combinations, auxiliaries, and drying conditions. After fixation, the material is washed off and dried.

### 4.1.3. Special Processes and Development Trends [4.13], [4.14]

In past years, many attempts have been made to remedy the weak points of the reactive dyeing system as much as possible. Lightfastness can be significantly improved over a wide range of shades by development of effective chromophores (e.g., introduction of the triphenyldioxazine and copper–formazan chromophores in the blue range). However, for demanding applications, especially in the red range, reactive dyes are still inferior to vat and naphthol dyes.

The salt load and discoloration of wastewater from dyeworks using reactive dyes often attract criticism on environmental grounds. With monofunctional dyes (i.e., those having only one anchor per molecule), the fixation yield in exhaustion dyeing processes is in the region of 60%. An important step forward was achieved with the introduction of several anchor groups, giving double- or multiple-anchor dyes. These can be homo- or heterofunctional (mixed-anchor systems). When heterobifunctional dyes are provided with anchors of different reactivity, dyes with a wide range of dyeing properties can be obtained and used in various kinds of dyeing processes (→ Reactive Dyes, **A 22**, pp. 653–654). At the same time, process reliability and dyefastness are improved. Thus far, combination of the triazine and vinylsulfone anchors has had the most important effect.

In terms of dyeing methods, a disproportionate growth in the economical padding process was seen in the 1980s. Nowadays exhaustion dyeing is regaining importance because shorter lengths of textile are being dyed to meet the changing requirements of fashion. In exhaustion dyeing processes, the trend is toward the short-liquor technique. A the same time, attempts are being made to develop new applictions for reactive dyes. One example is the use of reactive dyes in continuous ranges to dye warp yarns (color denim process) [4.11]. These plants were originally designed for dyeing with indigo (→ Indigo and Indigo Colorants, **A 14**, p. 153).

**Special Processes.** In addition, many special processes have been developed. An important aim has been the improvement of fixation yields. The two main techniques are

1) Processes for chemically modifying cellulose fibers
2) Processes in which finishing and dyeing are combined

Introduction of nucleophilic groups (e.g., amino groups) into the cellulose fiber (item 1 above) can significantly increase the degree of exhaustion and fixation yield of reactive dyes. If reactive groups are introduced into cellulose, the possibility arises of bonding not only reactive dyes but also nucleophilic direct dyes covalently to the cellulose.

Of the combined techniques (item 2 above), the system BASF Basazol was the first to be successful commercially. After finishing the cellulose fiber with the trifunctional cross-linking agent 1,3,5-triacroylaminohexahydro-1,3,5-triazine, dyes containing sulfonamide groups were applied with high fixation yields. The Indosol system (Sandoz) is still used today. Here, copper complex dyes are used that can form mixed complexes with aminated resins. The high wetfastness is explained by covalent bonds between the resin and fiber.

However, the practical importance of this technique is still relatively low. Possible reasons for this include the additional cost of applying the dyes, changes in handling of the cellulose fiber, occasionally unsatisfactory fastness properties, and ecological problems.

## 4.2. Dyeing with Direct Dyes

Before the first synthetic dyes were made in 1856, natural dyes of plant or animal origin were used in the dyeing of textiles [4.15]. In most cases, the application of these dyes required time consuming process steps, e.g., mordanting or vatting.

Sun yellow (1883) and Congo red (1884), the first representatives of direct or substantive dyes (→Azo Dyes, **A 3**, pp. 278–290), can be used on cellulose fibers "directly" (i.e., without mordanting). Because of their easy handling a variety of direct dyes has been developed.

### 4.2.1. Applications and Properties

Direct dyes account for 11% (1990) of the worldwide consumption of textile dyes; 75% of this amount is used to dye pure cotton or viscose substrates, and the rest is employed in dyeing mixed fabrics (for use in paper dyeing, see → Paper and Pulp, **A 18**, p. 657).

Substantive dyes were not developed further as intensively as, for example, reactive dyes (→ Reactive Dyes) [4.16]. The main emphasis of research was on the replacement of possibly carcinogenic benzidine dyes (see Chap. 14) and on the improvement of pretreatment and aftertreatment processes.

Only in individual cases do direct dyes achieve the brilliance of reactive dyes. The light-fastnesses of substantive dyes covers the entire scale from 1 to 8, meeting the highest requirements. With increasing depth of color the wet-fastnesses decrease to such an extent that direct dyeings must generally be aftertreated (see Section 4.2.7). Most direct dyes are characterized by excellent affinity to the fiber (i.e., dye liquors contain very little dye at the end of the dyeing process). This is of great importance for ecological reasons.

### 4.2.2. Dyeing Principle

Dyeing with direct dyes is based on the affinity of the dye for the cellulose fiber (substantivity). This affinity depends on the type of chromophore and can be influenced by the choice of dyeing parameters (see Section 4.2.4). Chromophores are azo compounds (→Azo Dyes), stilbenes (→Azo Dyes, **A 3**, p. 282), oxazines (→Azine Dyes, **A 3**, pp. 224–229), or phthalocyanines (→ Phthalocyanines) [4.17]. They always contain solubilizing sulfonic acid groups that are ionized in aqueous solution. The dye molecule is present as the anion. As a result, an electrolyte must be added to the dye liquor because cellulose fibers have a negative surface charge in water, which repels the dye anion. The cations of the electrolyte neutralize the negative charge and favor the aggregation of dye ions on the fiber (salting-out effect). Exothermic adsorption can be described with the help of Freundlich or Langmuir adsorption isotherms [4.18]. After adsorption dye molecules diffuse from the surface through the fiber pores to the amorphous areas of the cellulose. The rate of diffusion can be controlled by the dyeing parameters. Dyes are bound to the fiber by hydrogen bonds or van der Waals forces. Without additional measures (see Section 4.2.7), the strength of these bonds is so low that the dyes can be washed out of the fiber again. At dyeing equilibrium, the rate of absorption is equal to the rate of desorption.

### 4.2.3. Pretreatment of Substrates

Cellulose fibers require careful pretreatment before dyeing. Direct dyes are especially sensitive to differences in fiber affinity [4.16]. Pretreatment involves desizing, bleaching, possibly subsequent mercerization (→ Textile Auxiliaries,

Section 4.2), and increasing the wettability. Other techniques for improving the dyeability of cellulose with anionic dyes (e.g., forming of cations by fixation of quaternary ammonium compounds by reactive systems or by polymerization reactions [4.16 and references cited therein]) have been investigated intensively. Although dye uptake has been successfully increased with these techniques, disadvantages such as shifts of shade, poor fiber penetration, or loss of lightfastness still hinder market acceptance.

### 4.2.4. Dyeing Parameters [4.19]

The dyeing parameters mentioned below are interdependent. The effects of these parameters on the dyeing process are described here.

The *liquor ratio* influences the dye solubility and the strength of the electrolyte effect. Efforts are made to employ lower liquor ratios for ecological, economic, and technical reasons.

The *electrolyte* (usually sodium chloride or sodium sulfate), its concentration, and the speed of addition control the adsorptive behavior of the dyes and the degree of exhaustion.

The *pH value* influences solubility, substantivity of the dyes, and their stability in the dyebath.

*Temperature* generally determines the position of the dyeing equilibrium, which at room temperature is shifted strongly toward adsorption on the fiber. At higher temperature (usually 80–95 °C) the equilibrium is attained more rapidly. The temperature dependence of the maximum degree of exhaustion is dye specific. The same applies to hydrolytic stability, especially in high-temperature processes (see below) and in the presence of carriers in dyeing polyester–cotton mixtures [4.20].

*Dyeing time* is a critical factor for the evenness (levelness) of dyeing. The longer the dyeing time, the more uniform is the result. However, dyeing times should not be prolonged indefinitely because of limited dye stability and economic factors.

*Water quality* (metal-ion content) may be responsible for shifts in shade. Metal-complexing sequestrants are added to avoid this kind of effects.

Other auxiliary agents support the wettability and runnability of the material and improve the evenness of dyeing.

The *leveling capacity* (i.e., ability of dyes to compensate for the unevenness of dyeing) is an important characteristic that can be determined empirically. Unevenness results if dye adsorption occurs to quickly, and can frequently be traced to mistakes made in the addition of dye or salt. The leveling of uneven dye distribution proceeds via continuous desorption of the dye from the fiber and readsorption on a position on the fiber where the dye concentration is lower (migration).

In combination dyeing dyes with similar properties are preferred. To help in the selection of combinable dyes, the Society of Dyers and Colourists has classified direct dyes according to their leveling capacity [4.21], [4.22].

### 4.2.5. Dyeing Techniques

**Exhaustion Process.** In batch dyeing, the preferable mode depends on the type of dyeing equipment, the type of material to be dyed, and the solubility and affinity of the dyes. The dye is made into a paste with small quantity of warm water, diluted with more hot water, and boiled if necessary. It is then filtered and added to the dyebath. The electrolyte is added (if necessary in portions) either during heating of the bath or after attaining the optimum dyeing temperature. The material is dyed for 30 to 45 min and redyed, if necessary, at somewhat lower temperature for a short time. After the dyebath has been drained, the fabric is quickly washed clear with cold water and generally subjected to aftertreatment (see Section 4.2.7).

**High-Temperature Dyeing Process.** With suitable dyes, dyeing can also be performed in a closed apparatus at up to ca. 130 °C (especially for polyester – cotton mixtures). As a result of the high rate of diffusion, high levelness is achieved in short dyeing times, even in the case of poorly penetrable materials. After the high-temperature phase, the dyebath is cooled to 80 – 90 °C and the dye continues to adsorb, resulting in the same depth of color as in normal dyeing processes.

**Pad Processes.** In these continuous or semi-continuous processes, the fabric is first soaked by passing it through a trough filled with dye solution. Excess liquor is subsequently removed by squeezing the fabric between rubberized rolls (padding). The substantive adsorption of dyes during padding is avoided by moving the material rapidly, preferably at ambient temperatures,

and employing minimum amounts of pad liquor. Optimal distribution of the dye is later obtained by either thermal treatment or longer storage.

In *thermal treatment*, the material is passed through a steamer in 1 – 3 min without intermediate drying (pad steam process). Alternatively, it is heated to 80 – 85 °C and rolled up in a closed chamber where it is kept for 2 – 8 h (*pad roll process*).

In the *cold pad-batch process*, the padded material is rolled without heating and stored at room temperature for 8 – 24 h.

The *pad jig process* represents the easiest type of fixation of pad dyeings. The material padded with dye is passed through a salt liquor in a jigger at 80 – 90 °C, thus preventing desorption of the dye. In this process, additional dye can be used for shading (see Section 2.2.5.2).

All the processes mentioned above are followed by quick rinsing with cold water.

### 4.2.6. Special Processes
[4.16 and references cited therein]

An established process is the simultaneous bleaching and dyeing, permitting cost reduction by combining two treatment steps. However, this process can be performed only with selected direct dyes. Dyeing processes with additional energy fields (e.g., ultrasound, magnetic fields, radioactive irradiation, or high-temperature treatment) or with redox systems (persulfate or periodate) have not yet proved useful in practice.

### 4.2.7. Aftertreatment [4.16], [4.23]

As a rule, the wetfastness of direct dyes does not meet the demands of daily use. It can be improved by aftertreatment, giving fabrics that are fairly fast to water and perspiration and, in ideal cases, washable up to 60 °C. The improvement in fastnesses is achieved by reducing the solubility of the anionic dyes. This occurs by enlargement of the molecules via the formation of saltlike compounds with cationic aftertreatment agents. The fixation of the saltlike compound can be enhanced by cross-linking or by formation of covalent bonds to the fiber (→ Textile Auxiliaries, Section 5.10). For certain azo dyes, substantial improvement in lightfastness can be achieved by metal-complex formation (→ Metal-Complex Dyes, A 16, pp. 311 – 312). A combination of these aftertreatments produces especially good fastnesses.

Not only formaldehyde, one of the afore-mentioned cross-linking agents, but also metal complexes are discussed intensively because of the ecological aspects of these compounds. The improvement in fastness by means of diazotization and coupling has decreased in importance because of the resulting changes in shade.

## 4.3. Dyeing with Anthraquinone Vat Dyes

Vat dyes are water-insoluble, organic pigments that are used to dye cotton and other cellulose fibers. The principle of vat dyeing is based on chemical reduction of these dyes to the leuco compounds (which are soluble in aqueous alkali and exhibit fiber affinity), followed by reoxidation to the water-insoluble starting dye within the fiber.

Approximately 10–15 % of cellulose fibers or cellulose-containing fiber blends are dyed with vat dyes. The total consumption of vat dyes is about 22 000 t/a (1993) of commercial products, without indigo.

### 4.3.1. Chemistry of Vat Dyes

Two main groups of vat dyes are distinguished chemically:

1) *Indigoid vat dyes* are derivatives of indigo and thioindigo (→ Indigo and Indigoid Colorants). Leucoindigo compounds have a comparatively low affinity for fibers. Thus, these dyes are used mainly in textile printing rather than dyeing (see Section 4.4).
2) The leuco compounds of *anthraquinoid dyes* have a high affinity for fiber and give especially dyeings that are excellently fast in use and processing.

Vat dyes are dye preparations that consist basically of vattable colored pigment and dispersing agent. *Vattable pigments* are polycyclic quinoid compounds that contain two or more carbonyl groups in a closed system of conjugated double bonds (→ Anthraquinone Dyes and Intermediates) [4.24].

The pigment must be as finely dispersed as possible to achieve a satisfactory vatting rate and uniform dye application in pigmenting processes. For this reason, after the actual synthesis the dyes are wet ground in the presence of selected dispersing agents. The color strength and shade

are then adjusted according to the standards of the intended commercial products (→ Dyes, General Survey, **A 9**, p. 77).

The addition of *dispersing agent* stabilizes the fine dispersion of pigment in the aqueous liquor. Since soluble dispersing agents are discharged entirely into the wastewater, their biodegradability or eliminability in sewage treatment plants is of increasing importance.

The dyeing behavior of vat dyes, i.e., their absorptive behavior and leveling capacity, is determined basically by the substantivity and diffusion of the leuco compounds. The chemical stability and insolubility in water of the dye pigment that is reoxidized in the fiber account for the generally unsurpassed light-, wet-, and weatherfastness of vat dyeings.

The dyeing process involves the following steps:

1) Vatting (reduction) of the dye
2) Absorption by the fiber
3) Oxidation in the fiber
4) Aftertreatment of the dyeing

### 4.3.2. Vatting

Using anthraquinone as an example shows that the dye molecule is converted to the sodium leuco form in alkaline solution by the gain of two electrons (Fig. 4.2).

For a long time, *sodium dithionite* (hydrosulfite) has been the most important reducing agent in vat dyeing. It is inexpensive and, when produced by modern methods, virtually free of heavy metals (Zn, Hg). Its reduction potential ($-970$ mV) is high enough to vat all vat dyes quickly and completely. Studies of the reaction

Pigment
water insoluble
no affinity

Sodium leuco compound
water soluble
high affinity

$$S_2O_4^{2-} + 4\,OH^- \longrightarrow 2\,SO_3^{2-} + 2\,H_2O + 2\,e^-$$

**Figure 4.2.** Conversion of an anthraquinone dye to its leuco form

mechanism have shown that addition of the $S_2O_4^{2-}$ ion to the carbonyl group is the rate-limiting step in the overall reaction [4.25]. Formation of the leuco compound is achieved by the elimination of sulfite by the action of $OH^-$.

The rate of vatting depends not only on the concentration of dye and reducing agent but also on the crystal form, surface, and dispersion of the pigment (i.e., on its finish quality [4.26]). Leuco compounds are soluble in alkali. In the case of anthraquinoid vat dyes, the pH of the vat is about 13. At lower values the risk of vat acid sediments exists. Reduction is usually performed at $50-60\,°C$. At higher temperature, overreduction of certain dyes can occur (i.e., reductive destruction of the dye molecule).

To achieve faultless dyeings, the state of the vat should be monitored [i.e., determination of the concentration of reducing agent by using vat test paper (qualitative), by measuring the redox potential, or by titrimetric methods] because hydrosulfite is consumed by atmospheric oxygen [4.26]. For this reason, less reactive, more stable reducing agents are used in high-temperature dyeing [e.g., *derivatives of sulfoxylic acid* (Rongal HT, BASF)].

The oxidation products of sulfur-containing reducing agents are sulfite and sulfate. In most countries, the concentration of these substances in wastewater must not exceed maximum permissible values stipulated by law.

*Thiourea dioxide* has a stronger reducing effect ($-1100\,mV$) than hydrosulfite. Therefore, with sensitive dyes, a risk of overreduction exists. In addition, the oxidation products of thiourea dioxide contribute to the nitrogen and sulfur contamination of wastewater.

Organic reducing agents such as *hydroxyacetone* (Rongal 5242, BASF) do not contribute to the sulfur contamination of wastewater. However, the reducing effect of hydroxyacetone is much weaker ($-810\,mV$), so it cannot replace hydrosulfite in all applications. For this reason, a combination of hydrosulfite – hydroxyacetone is recommended to lower the sulfur content in wastewater. The use of hydroxyacetone is restricted to closed systems because it forms strong-smelling condensation products in alkaline solution. Newly developed organic reducing agents with improved reducing effects are currently being tested.

Systems for *electrochemical vatting* by means of electric current in the presence of a mediator, an *electron carrier* such as $Fe^{2+}$, or anthrahydroquinones have been described [4.27]. However, these methods have not yet been used in practice.

The vat dyeing system is also supported by auxiliary agents. *Dispersing agents* prevent the agglomeration of pigment particles or the flocculation of slightly soluble components such as vat acids or the impurities in cotton. If the presence of alkaline-earth ions is expected because of the use of hard water, the addition of *complexing agents* is required.

### 4.3.3. Dye Absorption in the Exhaustion Process [4.28], [4.29]

In the dissolved form, the vatted dye is present as the sodium leuco compound, either as a single molecule or as a complex of a few dye molecules [4.30].

In the exhaustion process, the higher the affinity, the more complete is the absorption of the dye by the fiber. The degree of absorption of dyes generally depends on dye concentration, liquor ratio, temperature, and possibly the electrolyte added [4.28]. Sodium salts of leuco vat dyes have a high affinity for the fiber; the exhaustion of the bath is generally between 80 and 90 % in exhaustion dyeing.

In the initial phase of dyeing, vatted dye is absorbed very quickly by the fiber. Extensive exhaustion of the bath is achieved after 10 to 15 min. The dye is then present in a high concentration in the outer regions of the fiber, but its distribution may still be entirely uneven (unlevel). In the subsequent leveling phase, the dye diffuses into the fiber and the fiber becomes dyed uniformly. The more evenly the dye absorbs to the fiber and the stronger the diffusion, the more level is the dyeing. Dyes that are less substantive absorb more slowly and evenly in the initial phase and exhibit more favorable diffusion properties. Diffusion can be improved by increasing the temperature. *Leveling agents* [4.31] can favorably influence absorptive behavior in the initial phase because they themselves possess a high affinity for the dye and release it in a controlled manner during the course of dyeing ($\rightarrow$ Textile Auxiliaries, Section 5.6.1) [4.26], [4.31]. However, a certain amount of dye is retained by the leveling agent, which results in slightly weaker shade.

### 4.3.4. Oxidation

After absorption by the fiber, the leuco dye is converted to the original pigment by oxidation and, in this way, fixed to the fiber. Oxidation can be achieved spontaneously with atmospheric oxygen or by the addition of oxidants, such as hydrogen peroxide, perborate, or 3-nitrobenzenesulfonic acid, to the liquor.

After the dyeing process, the material is usually rinsed with water to remove unfixed dye and decrease the alkalinity of the liquor. Depending on the oxidizing agent, oxidation is carried out at pH 9–12 and 50–60 °C.

The leuco dye present in wastewater is also converted to water-insoluble pigment, which can easily be separated mechanically or adsorbed to clarification sludge. Thus, colorization of the wastewater by vat dyes is negligible.

### 4.3.5. Aftertreatment ("Soaping")

After oxidation, all vat dyeings are subjected to heat treatment in the presence of water. As a rule, treatment is carried out in a weakly alkaline detergent liquor (soap was used in the past) at boiling temperature. The final shade is obtained by this treatment, and the more important fastness ratings (e.g., light- and chlorine-fastness) are improved substantially. Thus, this soap treatment is not restricted to the removal of pigment particles adhering to the surface but is an important individual step in vat dyeing. The processes occurring during soaping are unknown. Crystallization of amorphous dye particles or coarsening of primary particles by aggregation is assumed to occur [4.32]. If soaping takes too long, migration of the aggregates to the surface of the textile can cause deterioration of the rub-fastness.

### 4.3.6. Dyeing Techniques [4.33]

Vat dyeing can be carried out by using various techniques. Batch and continuous dyeing processes can be used. The dyeing properties of vat dyes enable the processes to be adapted to different machines.

**Batch Processes.** In batch processes, the dye must attach to the textile material from the dyebath. Therefore, the affinity and leveling-out properties of the dyes are of special importance. Dyeing is conducted in various circulating-goods machines (e.g., jigger, winch beck, jet, and overflow) and circulating-liquor machines (yarn, piece, and package dyeing machines) (see Section 2.2.4).

Vat dyes are divided into groups according to their affinity for the fiber and the amount of alkali required for dyeing: The following abbreviations are used: IK (I = Indanthren, K = cold), IW (W = warm), and IN (N = normal). The IK dyes have a low affinity for the fiber; they are dyed at 20–30 °C and require little alkali. Salt is added to increase the absorption of these dyes. The IW dyes have a higher affinity than IK dyes. Dyeing is performed at 40–45 °C with more alkali and little or no salt. The IN dyes are highly substantive and applied at 60 °C. They require much alkali but no addition of salt.

Dyeing with vat dyes is performed according to the following processes:

*Leuco Process.* Auxiliaries, dye, sodium hydroxide solution, and reducing agent are added to the dyebath. The textile is introduced into the bath after a vatting time of 10 min. Depending on the depth of shade, dyeing is performed for 30–60 min and completed in the usual manner (rinsing, oxidation, soaping).

*Preimpregnation Process.* At first, the dye liquor contains only the dye as a water-insoluble dispersion. After the temperature is increased and, if necessary, sodium hydroxide solution and salt are added, the pigment partially deposits on the fiber. After the reducing agent is added, vatting occurs rapidly and the dye is absorbed by the fiber. As a result of partial deposition of the dye in the nonsubstantive form, the evenness (levelness) of dyeing is improved. Prepigmentation is conducted in a jigger and is often used in dye packages (cheese, warp, yarn, and piece-dyeing beams).

*Semipigmentation Process.* Semipigmentation is used preferentially for dyeing lighter shades. Dye, sodium hydroxide solution, and reducing agent are added to the dyebath, and dyeing begins immediately at room temperature. Depending on the vatting rate of the dyes, only a small amount of substantive leuco dye is present intially. For this reason, absorption proceeds more slowly and evenly. With increasing dyeing temperature, more and more dye is converted to the leuco form and can be absorbed by the fiber.

*High-Temperature Process.* The leveling-out properties of vat dyes can be improved by increasing dyeing temperature. In the high-temperature process, dyeing is conducted at 90–115 °C.

In bath dyeing, the substrate can be dyed in various processing forms. A distinction is made between loose material (flock, card sliver), yarn (cheese, hank), garments (socks, pullovers, T-shirts, etc.), and fabrics (wovens, knitwear) (see Chap. 3).

**Continuous Processes.** Vat dyes are especially suitable for continuous dyeing because of their clearly defined configuration during the dyeing process (pigment form, leuco form). Dyeing is performed according to the pad steam process or the wet steam process. The continuous process is used almost exclusively for dyeing woven fabrics and to only a small extent for knitwear.

In the *pad steam process*, which is most frequently applied, the textile is impregnated in a padder with the aqueous dye dispersion in the presence of an antimigration agent and a wetting agent, if required. The material is squeezed until it has a specified residual moisture content and then dried. Subsequently, the fabric is passed through a chemical padder, which contains the required amounts of alkali and reducing agent, and is fed immediately to a steamer. In the steamer, the dye is fixed for 45–60 s under saturated steam conditions at a wet-bulb temperature of 99 °C at sea level pressure. After steaming, the material is rinsed, oxidized, and soaped in an open-width washing machine.

To achieve maximum fixation yield and depth of shade, dyes must be very finely dispersed. However, fine dispersion also increases the migration of dye particles during intermediate drying. For this reason, antimigration agents are used (polyacrylates, alginates, etc.), which have an agglomerating effect on the pigment particles and thus restrict their mobility during the drying process [4.34].

The *wet-steam process* is used mainly for voluminous open fabrics with a high liquor-carrying capacity (e.g., toweling, cord, velvet). Unlike the pad steam process, this process does not require drying after dye application in the padder (energy savings).

The continuous dyeing process is especially advantageous economically in dyeing large batches.

## 4.4. Dyeing with Indigo

(→ Indigo and Indigoid Colorants)

### 4.4.1. Chemistry and Historical Development

The chemical constitution of indigo was elucidated in 1883 by ADOLF VON BAYER. The first technically marketable syntheses were developed by HEUMANN and PFLEGER.

As in the case of natural indigo, the dye is made by air oxidation of indoxyl (→ Indigo and Indigoid Colorants, **A 14**, p. 151). Natural and synthetic indigo are absolutely identical chemically.

The demand for indigo depends greatly on fashion; it amounted to about 11 000 t worldwide in 1992.

The indigoid colorant thioindigo and its derivatives are also important vat dyes.

Because of the low affinity of vatted indigoid dyes for the fiber, these dyes are used mainly in textile printing. In comparison with other dyes, indigo occupies a special position because it is used almost exclusively for dyeing warp yarn in the production of blue denim. Being identical to the natural material, indigo has also been used recently in dyeing environmentally friendly clothing.

### 4.4.2. Dyeing Technique

Indigo is available as a powder, granulate, or liquid preparation. The dyeing technique has not changed over the years and, like vat dyeing, involves the following steps (see Section 4.3):

1) Dissolving of the dye by reduction in alkaline medium
2) Dyeing in the vat
3) Oxidation on exposure to air

The first step, dissolving of the dye, can now be omitted by using solutions of pre-reduced indigo (leucoindigo). In this way, about half of the reducing agent (depending on the type of machine) required for dyeing can be saved.

As a result of the relatively low affinity of indigo for the fiber, deep blue shades are attained only if dyeing and oxidation are repeated several times. Thus, in the past, indigo dyeing took several days.

In modern indigo dyeing plants for dyeing warp yarn, several dyebaths and oxidation units (usually up to five units) are arranged in a row for continuous operation [4.35]–[4.37] (see Fig. 4.3). For blue denim articles, pronounced ring dyeing of the yarn is desirable.

There are rope dyeing machines and slashers. In *rope dyeing machines*, 300 to 400 warp threads are combined into a rope and, in this form, passed through the dyeing system. *Slashers* were

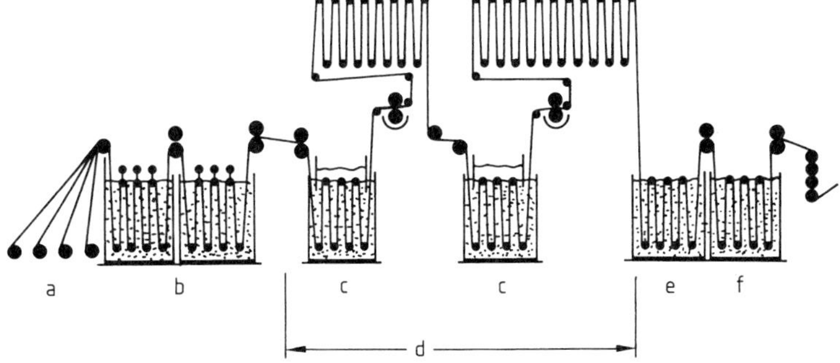

**Figure 4.3.** Schematic of an indigo continuous dyeing machine
a) Ball warps; b) Pretreatment phase; c) Dyebaths and oxidation units, up to five; d) Dyeing phase; e) Aftertreatment phase; f) Drying

developed in the 1970s from sizing machines. In these machines, the warp threads are dyed side by side, parallel to each other [4.38]. Both types of machines have advantages and disadvantages and should be assessed according to the specific requirements (e.g., small or large batches, required fastness of dyeing).

## 4.5. Dyeing with Sulfur Dyes

[4.39]–[4.41] (→ Sulfur Dyes)

The dyeing of cellulosic fibers, especially cotton and its blends with synthetic fibers, is the main field of application of sulfur dyes. In addition, sulfur dyes such as Indocarbon are used in printing these fibers. Sulfur dyes are also used to a limited extent to dye polyamide fibers, silk, leather, paper, and wood.

Although more brilliant sulfur dyes (red and green) have been put on the market in the 1950s and 1960s, the strong point of these dyes still lies in the deeper, muted shades, such as black, dark blue, olive, brown, and green, where their favorable price has its full effect.

With respect to fastness, sulfur dyes are close to vat dyes but not equal to them. They have varying lightfastness that is generally good in the case of the blue, green, olive, and brown shades. The lightfastness in presence of perspiration of sulfur black is very good. The washfastness of sulfur dyes at 60 °C is good to very good, and their fastness to boiling water is moderate to good. Wetfastness can sometimes be improved substantially by aftertreatment of the dyeing with products of the Solidogen IH type. With a few exceptions, sulfur dyeings are not fast to chlorine. Therefore, faulty dyeing can be stripped by treatment with sodium hypochlorite bleach liquor.

A high-grade finishing operation can also improve the wetfastnesses. A decrease in lightfastness, as in the crease resist finishing of materials dyed with other dyes, does not occur; in fact, an improvement is frequently observed.

The main applications according to the textile are:

1) *Piece dyeing:* corduroy and velveteen, twill for working clothes, leisurewear, tarpaulin and rucksack fabrics, coat poplin, and lining materials. The dyeing of the cellulose part of blended fabrics containing polyester is becoming increasingly important.
2) *Yarn dyeing:* sewing thread, warp yarn for denim fabric, yarn for colored woven goods.
3) Dyeing of *flock, card sliver, roving,* etc., for blend yarns with wool and man-made fibers.

### 4.5.1. Types and Mode of Reaction
[4.39], [4.42]–[4.45]

For technical and applicatory reasons, sulfur dyes are available in various modifications, which are classified in the Colour Index under the following general generic names:

**Sulfur Dyes** (C.I. Sulphur). Two types of sulfur dyes exist:

*Amorphous powders,* which are insoluble or partially soluble in water. They must be brought into solution by boiling with sodium sulfide and water, according to the following overall reaction scheme:

$$DS-SD + 2\,S^{2-} \rightleftharpoons 2\,DS^- + S_2^{2-}$$

| Water-insoluble | Reduced |
|---|---|
| sulfur dye | (leuco) dye |

For the sake of simplicity, only one mercapto group has been assumed per dye molecule (D). In strongly caustic alkaline liquors, the following mode of reaction is possible:

$$4\,DS-SD + 2\,S^{2-} + 6\,OH^- \rightarrow 8\,DS^- + S_2O_3^{2-} + 3\,H_2O$$

According to this equation, only one-quarter the amount of reducing agent normally used is required.

With sulfur cooked dyes (i.e., quinonimine sulfur dyes, which are produced by water or solvent reflux thionation), the reducing agent can attack not only the disulfide groups but also the quinonimine group (as in the case of vat dyes).

*Dispersible pigments*, analogous to the Colloisol or Microdisperse grades of vat dyes, are used particularly for pad dyeing.

The dispersible, *partly reduced pigments* carrying the name "Sol" represent a special case. Apart from a dispersing agent, they contain a certain amount of reducing agent.

**Leuco Sulfur Dyes (C.I. Leuco Sulphur).** These ready-for-use, liquid dyes already contain the reducing agent required for dyeing. They must simply be diluted with water before application. Ecologically more favorable, low-sulfide types have been put on the market recently.

**Water-Soluble Sulfur Dyes (C.I. Solubilized Sulphur)** [4.43]. Water-soluble sulfur dyes are available in the form of Bunte salts (*S*-arylthiosulfuric acid salts) of sulfur dyes and can be dissolved in hot water. The addition of alkali and reducing agent gives them an affinity for the fiber:

$$DSSO_3^- + S^{2-} \rightarrow DS^- + S_2O_3^{2-}$$

If NaSH or another reducing agent is used, alkali must be added to trap $H^+$ ions resulting from reduction. In the case of NaSH, $H_2S$ formation is avoided in the following manner:

$$DSSO_3^- + SH^- + CO_3^{2-} \rightarrow DS^- + S_2O_3^{2-} + HCO_3^-$$

The leuco form $DS^-$ is absorbed by the substrate and then oxidized to the fixed dye $DS-SD$. Since in reality several mercapto groups are present

per dye molecule, oxidation causes not only a dimerization but also cross-linking of dye "monomers." For sulfur cooked dyes, the quinonimine group is also involved in the reaction. This is clear from the change in shade that occurs during reduction.

With respect to both application and fastness, a distinction is made between sulfur dyes and sulfur vat dyes. The latter include dyes of the type Hydron Blue (C.I. Vat Blue 43) and Indocarbon (C.I. Sulphur Black 11).

### 4.5.2. Additives to the Dye Bath

**Reducing Agents.** Ready-for-use sulfur dyes contain the reducing agent required for dyeing. Otherwise, the dye must be brought into solution by the addition of a reducing agent or the already water-soluble dye must be made to have an affinity for the fiber.

The reducing agents generally used are sodium sulfide or sodium hydrogensulfide in different forms in the following ratio (parts by weight):

100 parts of crystalline sodium sulfide ($Na_2S \cdot 9\,H_2O$; ca. 30–32 wt % $Na_2S$)

$\cong 50$ parts of concentrated sodium sulfide (ca. 60–62 wt % $Na_2S$)

$\cong 65$ parts of liquid sulfhydrate F 150 (31 wt % NaSH)

$\cong 25$ parts of concentrated sodium sulfhydrate powder (95 wt % NaSH)

The amount of reducing agent required to dissolve or color the sulfur dye ($Na_2S$ or NaSH) is given on sample cards issued by dye producers. In the dye liquor, the amount of concentrated $Na_2S$ should be $\geq 2.5$ g/L.

In aqueous solution, the sulfide ion is largely hydrolyzed according to

$$S^{2-} + H_2O \underset{10\%}{\overset{90\%}{\rightleftharpoons}} SH^- + OH^-$$

The resulting caustic alkaline product is not required at least for dyeing with water-soluble sulfur dyes.

If a combination of NaSH and soda is used, the lower alkalinity offers the following advantages:

1) The dyeing results in a better color yield and can be rinsed clear substantially faster
2) During dyeing and rinsing, the dyeing is much less sensitive to undesirable oxidation effects (bronzing, rubbing-fastness)
3) In the case of regenerated cellulose (viscose), embrittlement of the material is avoided

Sodium dithionite (hydrosulfite) is a proven reducing agent for *sulfur vat dyes*, which is used together with alkali—usually sodium hydroxide solution and occasionally soda.

The color yield of some sulfur dyes is remarkably increased by the addition of hydrosulfite. With blue grades in particular, the addition of hydrosulfite "sharpens" the dye liquor, preventing bronzing. Glucose is employed in one-phase printing with Indocarbon. As a result of possible "overreduction" or destruction of the dye chromophore, hydrosulfite and similar reducing agents can be used to only a limited extent and require a suitable selection of dye.

For the reduction of *water-soluble sulfur dyes,* the aforementioned reducing agents, glucose, and hydroxyacetone are used [4.46].

In the case of *water-soluble sulfur dyes* and *low-sulfide leuco sulfur dyes,* glucose-based combinations of reducing agents (binary reducing agent systems) have recently gained acceptance for ecological, technical, and applicatory reasons [4.47]. The following substances can be considered binary reducing agent components: hydrosulfite, formamidinesulfinic acid, and with limitations, hydroxyacetone. Mixtures of sodium hydroxide solution and soda are used as alkali. The aforementioned reducing agent combinations result in less pollution of wastewater. At the same time, glucose prevents or buffers possible overreduction or destruction of the dye chromophore by the reducing agent component with a higher reduction potential.

Alkali polysulfides are added to *stabilize* sulfur alkaline dye liquors against atmospheric oxygen. In jigger dyeing, dark and bronzed edges or seam marks can be avoided in this way. In the case of winch beck, jet, and continuous dyeing, undesirable oxidation during the absorption phase and rinsing is avoided, improving rubbing-fastness. Problem-free dyeing can be achieved with sulfur vat dyes of the type Hydron Blue by using alkali sulfides and alkali polysulfides, especially on the winch beck.

**Wetting Agents.** A wetting agent or padding auxiliary is usually added. Products based on sulfonated oleic acid amides give good results because they exert a certain leveling effect and confer softness on the material. In comparison, products based on ethoxylates are unsuitable because they have a retarding (equilibrium "dye dissolved in liquor : dye absorbed to fiber" is shifted toward "dye dissolved in liquor"), and sometimes precipitating, effect on the dye.

**Water Softeners.** To exclude the negative effects of alkaline-earth ions on dyeing, sequestrants based on ethylenediaminetetraacetic acid (EDTA) or polyphosphates are used. Additives of this type are required, especially in circulating-liquor dyeing with hardness-sensitive dyes, to prevent unevenness and the splitting of warp beams. Acidic treatment of the material to be dyed (1 – 5 g/L of HCl, HCOOH, or $CH_3COOH$) can often prevent the contamination of wastewater with sequestering agents.

### 4.5.3. Dyeing

**Absorption of Dyes.** In the exhaustion process (in a jigger, circulating-liquor machine or winch beck), the dye generally absorbs at 60 – 110 °C in 30 – 60 min and in the pad steam process, at ca. 102 – 105 °C in 30 – 60 s.

Dyeing with reducing agents is always performed according to the exhaustion process. Pad dyeing processes, such as the pad steam and pad roll process, use a very short liquor ratio (ca. 0.7 : 1). The dye is basically fixed at the place where it has been padded. The fixing yield depends greatly on the liquor ratio. Optimal fixation and dye penetration are achieved with the pad roll process.

**Oxidation of Dyes** [4.39], [4.44], [4.46] – [4.48]. Sulfur colors are finally fixed on the substrate by oxidation. Before this step, the reducing agent, salts, and alkali, as well as the unfixed dye, must be removed by rinsing or washing to attain optimal rubbing-fastness. The following methods of oxidation are common.

*Air Oxidation or Acid Treatment.* The dyeings are subjected to cold and warm rinsing; the last rinse bath should generally be acidified with acetic acid (exception: sulfur black). In general, this process is used only for staple shades because after-oxidation usually occurs on storage, resulting in a change in shade.

*Bichromate – Acetic Acid Oxidation.* The material is treated with ca. 0.5 % potassium or sodium bichromate (based on the weight of the material) at pH 4 – 5 and ca. 70 – 80 °C. Until now, this process was common especially in piece dyeing. In the last years, however, it has been replaced increasingly by other types of oxidation for ecological reasons. In bichromate oxidation the material to be dyed becomes hydrophobic, evidently because of the formation of dye–chromium complexes. Although addition of copper sulfate (1–2 g/L $CuSO_4 \cdot 5\,H_2O$) to the oxidation bath improves the lightfastness of the colors, it is hardly ever applied. This type of oxidation cannot be used for articles to be rubberized.

*Iodate Oxidation.* Iodate oxidation [ca. 0.5 % potassium or sodium iodate (based on the weight of the material) at pH 4.5–5 (acetic acid) and >60 °C] is used only occasionally. In comparison with bichromate–acetic acid oxidation, somewhat brighter shades are obtained.

*Bromate Oxidation.* To attain an adequate rate of oxidation, the addition of vanadate (ca. 0.05 % $NaVO_3 \cdot 4\,H_2O$ with ca. 0.5 % $NaBrO_3$, based on the weight of material) is required. Oxidation proceeds at pH 4–4.5 (acetic acid) and 70 °C.

*Hydrogen Peroxide Oxidation.* With hydrogen peroxide oxidation, bright shades are obtained in the alkaline-to-neutral pH range. However, wetfastness is reduced somewhat compared with other oxidation processes because oxidation with hydrogen peroxide is not specific. Thus, especially if the oxidation is not performed properly [excessively high concentration of $H_2O_2$, excessively high temperature and pH (>50 °C, pH >7), and excessively long oxidation times], overoxidation occurs; i.e., the mercapto groups of the sulfur dyes are oxidized not only to disulfide groups, but also in part to sulfonic acid groups. Wetfastness can be greatly improved by aftertreatment of the dyeings with cationic auxiliaries, which are partly used in combination with oxidizing agents [4.49]. A deterioration of wetfastness can largely be avoided by oxidation in *acetic acid medium* (pH 4–4.5). Since oxidation then proceeds more slowly, the temperature must be increased to 70 °C. The resulting shades are much duller than those obtained by oxidation in the alkaline range. About 1–2 % (based on the weight of the material) of 35 % $H_2O_2$ is used.

*Other oxidizing agents* such as chlorite, bromite, chloramine T, and persulfate are not equally suitable for all dyes and consequently are seldom used. Sodium hypochlorite bleach liquor (hypochlorite) requires careful metering and can be used only with sulfur dyes that are relatively fast to chlorine.

**Alkylation.** With special alkylating agents, DS groups can be converted to the thioether DS–X–SD, where X is, e.g., –CHR–CHR–. In this way, the color is stabilized against oxidative and reductive effects [4.49]. Bi- or polyfunctional alkylating agents are especially effective because they lead to cross-linking and enlargement of the dye molecule. Solidogen IH is an appropriate product that at the same time exhibits substantivity due to its cationic character. About 1–2 %, based on the weight of the material, is used together with soda. In individual cases, the wetfastness of vat dyeing can be achieved by aftertreatment with alkylating agents.

In alkylation, care must be taken that only the mercapto groups and not the quinonimine groups of cooked dyes are alkylated; otherwise these compounds are fixed in their leuco form, which can lead to losses in coloring strength. Thus, in practice, dyeings are subjected to squeezing or suction before aftertreatment so that the leuco vat form is oxidized by air before alkylation. Alternatively, the alkylating agent is used in combination with specifically acting oxidizing agents such as chlorite.

In the case of sulfur black dyeing, cationic alkylating agents can be used not only to improve wetfastness, but also to produce a buffering effect, whereby losses in tensile strength due to fiber damage can be avoided.

**Rinsing or Washing.** After oxidation or alkylation, the dyeing is rinsed or washed at 60–95 °C to remove excess oxidation or alkylating agent and, possibly, detached dye. A soaping process, as in the case of vat dyes, is not common because a decrease in rubbing-fastness occurs and recrystallization cannot take place with the cross-linked sulfur dye molecule.

A disadvantage of sulfur black dyeings (C.I. Sulphur Black 1) is the elimination of sulfuric acid on storage under unfavorable conditions (high temperature and higher humidity) [4.50]. A reaction of this type can even occur after a longer drying period. Therefore, to avoid tendering and losses in tensile strength, sulfur black dyeings must be buffered in the last rinsing bath with soda or sodium acetate. A rather washfast buffering can be achieved by using special cationic auxiliaries, such as dicyandiamide–formaldehyde condensation products and alkylating agents. An improvement can also be obtained by anticipating the elimination of sulfuric acid by means of oxidation with bichromate–acetic acid.

### 4.5.4. Dyeing Techniques
[4.39], [4.41], [4.43]–[4.46], [4.51], [4.52]

**Batch Processes (Exhaustion Processes).** *Jigger Dyeing.* The dyebath is charged with all the chemicals, and the dissolved dye is added in two passages at the desired temperature. The salt is added after the third or fourth passage. After about six passages (or 60 min) and sampling, dyeing is completed by rinsing and oxidation.

*Jigger Dyeing with Preimpregnation.* In the case of water-soluble sulfur dyes, the material can be impregnated at boiling temperature with dye, salt, and wetting agent in the shortest possible liquor ratio. The volume of dyebath is then filled up with water, the reducing agent added, and the dyeing completed as usual. In comparison with the normal jigger process, better penetration is achieved.

*Pad Jig Process.* After the material is padded with dispersed or water-soluble dye and wetting agent (possibly drying, which is generally avoided because of the costs), it is placed in the liquor at the dyeing temperature. This liquor contains the reducing agent, salt, and some padding liquor. The dyeing is completed as usual.

*Winch Beck Dyeing.* The material is pretreated in a bath containing the chemicals and wetting agent at $40-50\,°C$. The dissolved, reduced dye is then added, the bath is brought to dyeing temperature, and salt is added. After $20-30$ min, the material is rinsed clear by the influx of cold water and simultaneous overflow of the liquor. Dyeing is completed as usual.

*Circulating-Liquor Dyeing.* The material is often boiled or subjected to an acidic pretreatment to eliminate hardness-producing substances. The dyebath contains wetting agent, chemicals, and the dissolved, reduced dye. After 10 min of dye liquor circulation at $30-40\,°C$, the dyeing temperature is adjusted, and the process is terminated after about 30 min by rinsing clear and oxidation. The salt is added, in part, at the dyeing temperature.

*Jet Dyeing* [4.53]. Foam control is important in the case of semiflooded jets. After wetting the material (or predyeing the polyester part of polyester–cellulose fiber blends with disperse dyes), reducing agent (e.g., glucose–hydrosulfite) and chemicals are added at ca. $40\,°C$, the dissolved dye being added after about 10 min. If the jet contains a lot of air space compared to liquor, about 1.4 kg of concentrated hydrosulfite and 1.4 L of sodium hydroxide solution $38°$ Bé (ca. 32 wt %) must be added per cubic meter of air. After heating to dyeing temperature, more hydrosulfite is added, if necessary, and the material is dyed for 30 min. Dyeing is completed by rinsing clear, oxidation, and rerinsing.

**Continuous Pad Dyeing Process.** Ready-for-use liquid or water-soluble sulfur dyes and dispersible pigments are employed preferentially in continuous pad dyeing.

As a result of the (varying) substantivity of the dyes in pad processes with reducing agents, tailing can occur. In this case a constant shade is attained only after a certain yardage. This problem can be avoided by taking into account both the feeding liquor factors of the individual dyes and the substrate factors. The recipe of the starting bath and of the feeding liquor can be calculated [4.54].

*Pad Steam Process.* The following variations in the pad steam process are possible: In the *one-bath procedure*, the material is impregnated with dye, reducing agent, and wetting agent at $20-30\,°C$ and subsequently subjected to air-free steaming at $102-105\,°C$ for $30-60$ s. Rinsing is followed by oxidation and rerinsing in a washing machine. The material speed is $40-100$ m/min. Warp for denim articles can be dyed in the same manner.

The *two-bath procedure* (for water-soluble sulfur dyes and dispersible pigments) is preferred especially for mercerized material to obtain a more even appearance. The material is padded at $20-30\,°C$ with dye, wetting agent, and possibly antimigration agent. After intermediate drying, if necessary, it is impregnated with reducing agent, and the process is completed by steaming and aftertreatment, as in the one-bath process.

*Pad Air-Passage Process.* For light to moderate shades, the material can be impregnated with dye, reducing agent, and wetting agent at elevated temperature ($70-80\,°C$). During air passage for ca. 1 min, sufficient fixation is achieved. Dyeing is completed in washing machines as usual.

The *hot pad dry process* is a variant of the pad air-passage process. After impregnation, the material is subjected to hot drying on a drying cylinder. For final fixation of the dye, the material is impregnated with a cationic alkylating agent and redried. This process is used especially in the dying of warp yarns.

**Semicontinuous Padding Processes.** *Pad Roll Process.* After impregnation with the pad liquor, the material is batched up and stored in an air-free steam chamber (dry and wet temperature ca. 98 °C) where it is rotated slowly. After a rotation of about 1 h, dyeing is completed. By this procedure, the dye is almost completely fixed. Excellent penetration is obtained.

**Special Dyeing Processes** [4.39], [4.45], [4.55]. Dyeing with water-soluble sulfur dyes can be performed without reducing agents according to a pad thermofixing process. After padding and drying, the dye is fixed by heat treatment in the presence of acid-donating catalysts, urea, and thiourea. In this way, the water-solubilizing Bunte salt groups are split off. An excellent appearance is obtained in this process.

For denim articles, the warp yarns can be dyed with sulfur vat dyes or water-soluble sulfur dyes and sized in a continuous process. The warp is impregnated with dye, alkali, and reducing agent at 70 °C and subsequently dried and sized (mainly with potato starch). After weaving, the material is treated with a cationic alkylating agent and acetic acid in a padder and then rinsed. This process produces colored and black dyeings in which mainly the outer layer of the fiber is dyed. The stone wash effect can be imitated in this manner. A one-bath variant of this process gives poorer fastness properties, which correspond roughly to those of indigo dyeing [4.56].

### 4.5.5. Combination with Other Dye Groups

Sulfur dyes are combined with other dyes for reasons of shade, fastness, and in the case of polyamide–cellulose blends, tone-in-tone dyeing. Thus, combinations of sulfur vat dyes of the type Hydron Blue with dispersible vat dyes are generally used in the two-bath process. For stretch cord and similar articles made of polyamide–cellulose fibers, metal-complex dyes that are resistant to reducing agents are added to dye the polyamide fibers because only a few sulfur dyes stain polyamide fibers sufficiently compared to cotton fibers.

Water-soluble sulfur dyes can be combined with selected reactive dyes [4.57]. In this manner, more brilliant shades are obtained. Hydrosol Brilliant Red BCL, which is based on mono-chlorotriazine, is a reactive dye of this type.

In jigger dyeing, sulfur colors are usually brightened and shaded in a fresh bath with direct and basic dyes after oxidation.

### 4.5.6. Wastewater [4.58]

Normal wastewater from sulfur dyeing contains sulfides in addition to dye residues and salts. Since sulfides are toxic to aquatic organisms and form hydrogen sulfide with acids, they must be eliminated or their concentration reduced, at least. To avoid these problems, low-sulfide and sulfide-free dyeing methods have increasingly gained acceptance [4.47], [4.59]. Since the presence of chromium in water is restricted because it is ecologically harmful, oxidation with bichromate, which was used widely in the past, has been replaced in many cases by other types of oxidation. According to appendix 38 of the general wastewater regulations, the draft of which should come into force shortly, the presence of chromium(VI) compounds from the use of oxidizing agents for sulfur dyes is not permitted in Germany. In the United States no national standards for wastewaters from the textile industry exist. Minimum requirements are stipulated by the local authorities or the local public treatment plants. Limits for indirect discharge are generally 0.5 ppm for chromium and 5 ppm for sulfides [4.60]. In Japan the national wastewater standards can be adapted by the local prefectures to local requirements. Relevant values for Osaka are given as a typical example: total chromium 2 mg/L, chromium(VI) 0.5 mg/L (see Section 14.2).

**Indirect Discharge.** For indirect discharge, the conditions to be obeyed for components that are not dangerous depend on the municipality and the state. After appendix 38 of the general wastewater regulations comes into force, the requirements for dangerous components stipulated in these regulations must be fulfilled. This means that in textile finishing companies, sulfide values must be decreased to 1 mg/L, as in the case of direct discharge. Wastewater of this type can be treated without problems together with municipal wastewater in municipal sewage treatment plants. However, mixing with acidic wastewater and the resulting formation of hydrogen sulfide in the sewage system must be prevented. Contaminations of this type must be kept as low as possible because $H_2S$ is oxidized microbially to sulfuric acid, which attacks concrete.

Wastewater from dyeworks can be collected in mixing and balancing tanks. Mixing with wastewater from a bleaching plant is of advantage. If possible, the wastewater should be oxidized by aeration or detention or by the addition of oxidizing agents, such as $H_2O_2$ or ozone.

**Direct Discharge.** For official requirements, toxicological, and ecological aspects, see Chapter 14.

In general, wastewater is purified in several stages. *The first stage* is a mixing and balancing tank with aeration, pH regulation, and cooling. During aeration, sulfide and polysulfide react to form thiosulfate. At the same time, the dye becomes insoluble and precipitates with polysulfide sulfur. Oxidation is accelerated by residues of sulfur dyes, which generally make the addition of catalysts such as $Mn^{2+}$ unnecessary. By air or oxygen injection or by the action of air in trickle plants, the sulfides are oxidized in a few hours at room temperature in the presence of the dye residues.

*The second stage* involves chemical treatment by the addition of iron salts alone or together with $Ca(OH)_2$ and flocculation auxiliaries, if necessary. Sulfide and dye residues are eliminated by precipitation and sedimentation.

*The third stage* involves biological treatment, primarily using the activated sludge process with the addition of oxygen (usually air, seldom pure $O_2$). The activated sludge is separated in a secondary tank.

In the Katox process [4.59], [4.61], the main purification is performed in the Katox stage by means of accelerated oxidation in the presence of activated carbon. This is followed by a final purification, which involves chemical precipitation with iron sulfate, calcium hydroxide, and flocculation auxiliaries.

If no sulfides at all are allowed to be discharged into the wastewater, water-soluble sulfur dyes with sulfide-free reducing agents are an alternative [4.47].

## 4.6. Dyeing with Naphtol AS Dyes

An azo dye is formed from a Naphtol AS coupling component and a diazotized base (developing component) under suitable conditions ($\rightarrow$Azo Dyes, **A 3**, pp. 304–314). This coupling can be carried out on the fiber itself. If components without solubilizing groups are used, the dye formed is an insoluble pigment, and washfast dyeing is achieved. This possibility was exploited quite early. A cotton fabric was first impregnated, for example, with a solution of sodium β-naphtholate and, after drying, was passed through a solution of diazotized *p*-nitroaniline to give an attractive deep red dyeing with marked fastness (Para Red). A large number of modified naphthols and suitable diazotizable bases (fast bases) were marketed by the then I.G. Farbenindustrie (now Hoechst and Bayer). By employing similar methods, these components can be used to produce dyeing and printing with excellent fastness in the most widely differing shades. The classic Turkey Red (Alizarin Red) has been replaced by the combinations Naphtol AS-TR/Fast Red TR base and Naphtol AS-ITR/Fast Red ITR base, in particular, which have good fastness. Besides bases that need to be diazotized before use in dyeings, diazotized diazonium compounds in stabilized form have also been marketed as fast color salts ($\rightarrow$Azo Dyes, **A 3**, p. 309). Similar ranges later came on the market from most of the other large dye producers ($\rightarrow$Azo Dyes, **A 3**, p. 305).

Nearly all cellulosic fibers can be dyed in every processing state by using Naphtol AS combinations. Other dyes can rarely give the same depth of color and shade. Handling these combinations is safe and relatively simple. Therefore, Naphtol AS dyes have been introduced into the textile industry on a wide basis. Colorations with outstanding fastness have been achieved, which in some cases are of the same standard as those using Indanthren.

**Dyeing Process.** In the *impregnating process* the cotton fiber is treated first with Naphtol AS, which is dissolved by boiling with caustic soda and a protective colloid (*hot solution process*) (for details of additives, see $\rightarrow$ Textile Auxiliaries). Most Naphtol AS types can also be solubilized readily with alcohol and caustic soda by using the *cold solution process*. The concentrated sodium naphtholate solution is added to the impregnating bath. For certain naphthols the addition of formaldehyde to stabilize the naphthol on the fiber (air stability) is also necessary; otherwise, coupling capacity is impaired. By addition of sodium chloride a rather substantive attachment of the naphthol to the fiber is achieved. On dyeing from a long liquor a lower temperature is used for impregnating the fabric, followed by rinsing and development. The use of a padding machine requires a higher temperature and, in general, intermediate drying.

Hank yarn is impregnated in a long liquor, for example, in a trough or on hank yarn dyeing machines. Wound yarn can be impregnated in the usual mechanical dyeing machines. Piece dyeing is carried out on the winch beck and the jigger or, preferably, continuously on the padding machine.

For *developing*, the base must first be diazotized in the cold, if necessary with the addition of ice, by using a solution of sodium nitrite in hydrochloric acid. Alternatively the base and the nitrite are made into a paste and stirred into the hydrochloric acid solution. To increase the coupling capacity, excess mineral acid must be neutralized or buffered by the addition of sodium acetate, disodium phosphate, or similar compounds. The textile impregnated with naphthol is passed through the cold developing bath containing the diazotized base so that the dye is formed rapidly. In dyeing machines the developing solution circulates through the stationary textile. Alkali carried over from the padding bath must be neutralized by the addition of an alkali-binding agent (e.g., acetic acid, aluminum sulfate) to the developing bath to prevent decomposition of the diazonium salt.

The use of naphthols and bases in solution was introduced in the 1970s. These formulations are safer and simpler to apply. Diazotization of the liquid bases generally does not require the addition of ice. After development, the textile is rinsed and subjected to careful *aftertreatment* in which excess color lake is removed from the fiber surface. Otherwise, it severely impairs the rubbing-fastness of the dyeing. In addition, through the change in pigment crystal structure occurring during aftertreatment, a stable final shade with enhanced fastness is achieved.

The possibility of producing fast, substantive, and bright dyeings on the textile from two different components has led to a large number of special processes and articles, characteristic of different geographical areas. Naphtol AS combinations that can be used at low temperature are extremely suitable for dyeing cellulose fiber materials that are partially coated with wax. After dyeing the wax is washed out by boiling. This technique is still popular as batik, mainly in Southeast Asia.

The use of Naphtol AS dyes for the production of denim articles is of industrial and commercial interest. The cotton warps are simultaneously sized and padded with a coupling component. After weaving the impregnation is developed with a fast salt or a diazotized base and subjected to the usual aftertreatment. The process was introduced as the warp sizing padding process. Another variant is known as the warp sizing and dyeing process in which the warp is padded, dried, and developed immediately with a diazotized base.

## 4.7. Dyeing with Pigments

Being insoluble products, pigments can only be fixed to the surface of the fiber. For pigment coloring, very fine (particle size $\leq 0.5$ $\mu$m) inorganic or organic pigments are used in a nonionic preparation ($\rightarrow$ Pigments, Inorganic; $\rightarrow$ Pigments, Organic). They are fixed with the help of a binder. The binder is an aqueous dispersion of cross-linkable mixed polymers (copolymers or polymer blends, basis polyacrylate, polystyrene, polyurethane). A film, in which the pigment particles are embedded, is formed on the fiber. Cross-linking occurs during heating. The binder film must adhere firmly to the substrate without the material becoming too stiff or sticky. The type and amount of binder used depend on the amount of pigment and determine not only the feel, but also the rubfastness and wetfastness of the dyeing [4.62]. Usually, only light shades are produced by pigment dyeing. Common trade names for pigment–binder combinations are Acramin (Bayer) and Helizarin (BASF). Recent experiments have shown that polyurethane-based binders (Perapret PU, BASF) form elastic films that can bind large amounts of pigment, making deep coloration possible.

Since pigments show no substantivity for textile fibers, piece goods are impregnated continuously with a liquor that contains the pigment, the binder, an antimigration agent, a cross-linking agent if necessary, an acid donor, and a softener. The pieces are dried at 90–120 °C and fixed at 160–180 °C. No afterwash is required. A resin finish can be carried out at the same time (e.g., a crease-resistant and antishrink finish, hydrophobing or a soil release finish).

Nowadays, a washout effect is produced with pigments in an exhaustion process. In this process, cotton is impregnated with a cationic product so that pigments can attach. A reduced amount of binder is applied and fixation is carried out at low temperature (120–130 °C). The disadvantage of the process is that a lot of pigment is lost, which contaminates the wastewater [4.63].

Pigment dyeing offers the advantages of universal applicability to all fibers (including glass fibers). It is a single-stage process, even for fiber mixtures (i.e., an energy-saving process). Aftertreatment is not required. Raw cotton is covered well. The resistance to light is very good; the disadvantages are limited depth of color, stiffer feel of the product, and limits to the rubfastness and wetfastness.

Pigment dyeing is used commonly for heavy textiles (e.g., canvas), light printing grounds, dress materials, shirting, bed linen, and furnishing articles.

## 4.8. Other Dyeing Methods

### 4.8.1. Dyeing with Leuco Esters of Vat Dyes

The leuco esters of vat dyes are anthraquinoid or indigoid vat dyes that have been made water soluble by reduction and esterification of the hydroxyl groups with sulfuric acid (→ Leuco Esters of Vat Dyes).

After application to the fiber, the ester is hydrolyzed, usually with sulfuric acid at room temperature or slightly elevated temperature (up to 70 °C), and the original vat dye is recovered by oxidation (e.g., with sodium nitrite).

The advantages of dyeing with the leuco esters of vat dyes are not only good levelness (evenness) and penetration but also the excellent fastness of vat dyes. The low affinity of the leuco ester for the cotton fiber and the relatively high dye costs are obstacles to wide application. These dyes are used mainly for dyeing high-quality articles of cellulose fibers in light colors and polyester–cellulose blends to moderate depths of color [4.64]–[4.66, p. 568].

**Batch Dyeing.** Because of the low substantivity of leuco esters of vat dyes for cellulose, batch processes (exhaustion processes) are rather unimportant. The usual systems are used, e.g., jigger and winch beck for piece goods and package dyeing machine for fabric and yarn. One- or two-bath processes exist. In the *two-bath mode* of dyeing, the dye is first allowed to strike (bot-

toming) in the presence of salt in a weakly alkaline medium (e.g., on the jigger). The liquor is then removed, and the color is developed with a liquor containing sulfuric acid and nitrite at room temperature. If high-temperature development is required (70 °C), nitrite is added to the bottoming bath. The *one-bath mode* is usually used in winch becks; i.e., the same bath is used for both bottoming and development. Packages are subjected to one-bath or two-bath dyeing, depending on the affinity of the dye.

**Continuous Processes** [4.67]. Continuous processes yield more favorable dye yields. The most important processes are

1) Processes without intermediate drying: padding (dye, soda, nitrite), air passage, development (dipping at 70 °C in acid), air passage, and aftertreatment
2) Processes with intermediate drying: padding (dye, soda, nitrite), drying, development (dipping at 70 °C in acid), air passage, and aftertreatment (this process gives a higher fixing yield than the first method)
3) Pad steam development process without intermediate drying: padding (dye, soda, nitrite), steaming (1 min, saturated steam, 103 °C), development (dipping at 70 °C in acid), air passage, and aftertreatment

As a result of their high price, low substantivity, and toxicological problems during production, the importance of the leuco esters of vat dyes is decreasing. Remaining producers and trade names include Hoechst (Anthrasol Blue IBC), Ciech (Helasol), and Mitsui (Mikethren). Dyeing with leuco esters of vat dyes can be replaced easily by pigment coloring, by reactive or vat dyes.

### 4.8.2. Dyeing with Mordant Dyes
[4.66, p. 578], [4.68, p. 649]

The fiber is first treated with metallic salts (i.e., mordanted). Highly adhesive, basic metal compounds are formed on the fiber. These compounds are capable of producing poorly soluble colored complexes (lakes) with certain azo and anthraquinone derivatives. Alizarin (1,2-dihydroxyanthraquinone) is the best-known anthraquinone derivative. It used to be isolated from the root of the madder plant but has now been replaced by the synthetic product. Suitable azo dyes contain, e.g., hydroxyl or carboxyl groups in the *o*-position to the azo group on one or both of the aromatic nuclei. The shade of the dyeing depends on the type of metallic mordant used. Alizarin and aluminum–calcium salts produce the well-known Turkey red.

The mode of operation in alizarin dyeing is relatively tedious. The old Turkey red dyeing process required at least 10 working steps: boiling, treatment with rancid olive oil, leaching, treatment with a tanning agent, mordanting with alumina, dyeing in hard water, reviving, acidification, and soaping. In the new process, treatment with Turkey red oil (sulforicinoleic acid) has replaced the oiling step. The light- and wetfastnesses obtained are excellent.

In the past, this method was used to dye, e.g., cambric and bunting. Today, easier methods using developing dyes and vat dyes have replaced this process.

### 4.8.3. Dyeing with Acid Dyes [4.66, p. 580]

Acid dyes (i.e., azo or anthraquinone dyes that are made water soluble by sulfonic acid groups) were occasionally used for dyeing cotton; however, they are no longer important. They do not have sufficient substantivity for cellulose. Only those dyes are usable that can form a metal complex on the fiber when applied together with a metallic salt. Excess heavy metal contaminates wastewater.

Acid dyes can attach directly to vegetable hard fibers (jute, sisal) by salt formation because the fiber companion substances contain basic groups.

### 4.8.4. Dyeing with Basic Dyes
[4.66, p. 580], [4.68, p. 548]

Basic dyes (→ Cationic Dyes) show no substantivity for cellulose. Therefore, pretreatment of the fiber with tannic acid (which contains phenolic OH groups) is required. The tannin mordant is insolubilized with antimony salts (tartar emetic). Synthetic products (Katanol type) are also suitable. Thus, the fiber is dyeable in a weakly acidic medium, a saltlike bond being formed with the acidic phenolic hydroxyl groups. Basic dyes attach to bast fibers without a mordant. Suitable dyes are to be found in the azo, diphenylmethane, and triphenylmethane series, and among thiazine, azine, oxazine, thiazole,

and quinoline derivatives. Bright colors are obtained at low cost with rhodamine, auramine, fuchsin, or methylene blue, among others. Since the fastness, especially lightfastness, is poor, this method is of little significance.

### 4.8.5. Dyeing with Mineral Dyes
[4.66, p. 584], [4.68, p. 520], [4.69, p. 395]

In a pad process, inorganic metallic salts can be applied to cotton and then converted in an alkaline medium to the corresponding oxide by treatment with steam. Brown and khaki shades (mineral khaki) are obtained with chromium and iron salts. This dyeing is inexpensive, lightfast, and weather- and rotproof. The disadvantages are wastewater contamination and definite hardening of the fabric. This method is still important in some countries for dyeing tarpaulin and uniform materials.

### 4.8.6. Dyeing with Oxidation Dyes
[4.68, pp. 437–439]

In the application of oxidation dyes, aromatic amines form insoluble polyazine derivatives in the fiber (→Azine Dyes). Synthesis proceeds in several steps in a hydrochloric acid medium by oxidation (e.g., with dichromate). For the chief representative of this group, aniline black, the chromophore consists of dibenzopyran rings. Oxidation dyes are rapidly decreasing in importance because aniline and other aromatic amines as well as the bichromate used for oxidation are toxicologically hazardous. The colors produced by aniline black are characterized by a full bluish black shade and excellent fastness. Since they are easily reserved, they are still used occasionally for printing grounds. For individual processes, see [4.70].

### 4.8.7. Dyeing with Phthalogen Dyes

As with oxidation dyes, the actual dye in the phthalogen process (Bayer)—the insoluble phthalocyanine pigment—is formed in the fiber itself (→ Phthalocyanines, **A 20**, pp. 234–235). Precursors are involved in dyeing. A distinction is made between two processes:

*Low Molecular Phthalogen Developer.* Aminoiminoindolenine or its derivatives are applied together with heavy-metal donors (preferentially

Cu or Ni) by padding. The pigment is formed on heating (150 °C) in the presence of weak reducing agents (e.g., glycols). This is followed by after-treatment with hydrochloric acid, with the addition of sodium nitrite to remove secondary products. The most important aminoimino-indolenine is: C.I. Ingrain Blue 2:1, 74160 (trade names, e.g., Phthalogen Brilliant Blue IF3GM, Bayer or Sumilogen Brilliant Blue, Sumitomo).

*High Molecular Phthalogen Developer.* Poly-isoindolenines, which are complexes of heavy metal and indolenine, are developed with reducing agents in a wet treatment. The polyisoindolenine complexes (trade names, e.g., Phthalogen Brilliant Blue IF3GK, Cu complex or Phthalogen Turquoise IFBK, Ni complex) are fiber affinitive and can be applied by using the exhaustion process. Their substantivity is, however, low but can be increased by pretreating the cellulose with anionic products (e.g., Phthalofix FN). Reductive development is performed in an alkaline medium with hydrosulfite. For details, see [4.70], [4.71].

Dyeings made by phthalogen developers are characterized by high brilliance and excellent fastness. Thus, they are suitable for washfast and weatherproof articles. No danger of photochromism exists if a finishing operation is performed with synthetic resin products.

### 4.8.8. Dyeing with Coupling and Diazotization Dyes [4.66, pp. 547–548], [4.69, p. 286]

Direct dyes (see Section 4.2) that contain aromatic amino groups can be diazotized on the fiber after dyeing and then coupled with a "developer"—a phenol, naphthol, or aromatic amine. Wetfastness, in particular, is improved by enlargement of the molecule and, naturally, the shade changes as well (→Azo Dyes, **A 3**, p. 290).

Conversely, water-soluble, substantive azo dyes, which carry amino or hydroxyl groups capable of coupling, can also be used followed by aftertreatment with a diazonium compound (usually diazotized 4-nitroaniline as a stabilized diazonium salt). The resulting polyazo dye shows excellent wetfastness (see Section 4.6).

### 4.9. Other Plant Fibers
(→ Cellulose, **A 5**, pp. 398–400)

Hard and bast fibers, not, like cotton, isolated from seed hairs but from parts of stems and leaves, can also be colored with direct or reactive dyes. The noncellulosic portion of these fibers is frequently so large that even simple acid and metal-complex dyes attach. Basic dyes can also be used to produce colors that are brilliant but have poor fastness [4.69, pp. 433–4.34], [4.72, p. 534]. Hard fibers, such as jute and coir (the fibrous outer layer of coconut), have a strong self-color that is not lightfast. For this reason, they affect the quality of all types of dyeing. Bleaching does not help because the fibers become yellow again when exposed to light.

## 5. Dyeing of Wool and Silk

### 5.1. Dyeing of Wool

The dyeing properties of wool (→ Wool) vary depending on the origin, breed, age, food, season, and habitat of the sheep. Even fleece from a single sheep reveals pronounced differences in quality. In fact, differences are also found within a single fiber due to biological and environmental factors (tippy dyeing). By means of the specialized sorting of wool provenances and fleece components, batches are put together that largely correspond with each other with regard to fineness and to processing and application properties.

Apart from shorn wool, certain amounts of hide and slipe wool from slaughtered animals are available on the market. While hide wool is equal to shorn wool, slipe wool exhibits alkali damage.

### 5.1.1. Principles (see also, → Dyes, General Survey, **A 9**, pp. 78–79)

Wool fiber is composed of protein filaments and consists mainly of keratin with a very complex structure. The amino groups of keratin are very important for the dyeing process. The number of basic groups titratable with acid is 850 μmol per gram of wool fiber. In the acidic and neutral range, carboxyl groups are present largely in the undissociated state.

The number of ionic groups in the fiber depends on the pH of the medium. Keratin is most stable when the number of negative and positive ions is identical. This *isoionic region* differs somewhat from the isoelectric region of charge neutrality. This means that charges of adsorbed ions add to the ion charges of the fiber. According to

ELÖD, the *isoelectric point* of wool is pH 4.9 at room temperature [5.1].

The *morphological structure* of the fiber determines the pathway that dyes take during dyeing and is of decisive importance for the rate and extent of dye uptake. In the past, the dye was assumed to have to penetrate the scaly layer on the fiber surface. The epicuticle and the exocuticle were regarded as barriers because of their hydrophobic character. This idea was supported by the fact that oxidative or chlorinating degradation of epicuticle and exocuticle considerably facilitates dye uptake. Oxidative or chlorinating degradation is still used in the preparation of wool materials to be printed.

In contrast, LEEDER [5.2], [5.3] and LEAVER [5.4] showed that the penetration of dyes into the fiber proceeds through the intercellular regions. The cell membrane complex is regarded as a "solvent" for hydrophobic textile chemicals [5.5]. The strong swelling capacity of the intercellular cement is important for the dyeing process [5.6]–[5.8]. Only subsequently are the sulfur-rich keratins also penetrated by the dye molecules thus, determining the position of the dyeing equilibrium.

**Bonding Forces between Dye and Fiber.** *Ionic Bonds.* Dye anions can participate in ionic interactions with fibers that possess cationic groups. However, the formation of ionic bonds is not sufficient to explain dye binding because compounds that can dissociate are cleaved in the presence of water.

*Secondary bonds* (dispersion and polar bonds, hydrogen bridges) are additionally formed between dye and fiber. Close proximity between the two is a prerequisite for bond formation. However, this is counteracted by the hydration spheres of the dye and of wool keratin. On approach, these spheres are disturbed, especially at higher temperature, and common hydration spheres are formed ("iceberg structures") [5.9]. The entropy of the water molecules involved is increased in this process (hydrophobic bonding).

Dye binding can be regarded as an ion exchange in which the ionic bond is supported by secondary bonds between the dye and the polymer [5.10]. ZOLLINGER termed it an "Einweisungsfunktion" that is based on the far-reaching potential of ionic groups ("pilot ions").

Coordinate and covalent bonds can be superposed on secondary and ionic bonds. This aspect is discussed in connection with chrome, metal-complex, and reactive dyes.

## 5.1.2. Dye Classes and Dyeing Processes

### 5.1.2.1. Acid Dyes

Acid dyes consist of simple chromophoric systems (→Azo Dyes, **A 3**, pp. 263–270, →Anthraquinone Dyes and Intermediates, **A 2**, pp. 408–410), which are made water soluble by the introduction of sulfonic acid groups. Dissociation gives rise to dye anions that interact with the ammonium groups of the fiber. These groups are formed from amino groups in the presence of acid, which explains the name of this class of dyes.

From a coloristic standpoint, acid dyes are subdivided according to their affinity. Affinity increases from *leveling dyes* and weakly acidic absorbing (moderately leveling) *milling dyes* to neutral absorbing (poorly leveling) *super-milling dyes* (→Azo Dyes, **A 3**, p. 264). This sequence approximates the increase in size of the dye molecule. The presence of aliphatic groups in the dye molecule contributes to a substantial increase in binding to wool, converting leveling dyes to types that are fast to fulling. The sulfonic acid groups determine not only the number of possible ionic bonds to the fiber, but also hydration, which counteracts binding.

Dyes with similar absorptive behavior should be selected for combined application. The combination values for acid dyes established originally for polyamide were later transferred to wool [5.11]–[5.13].

In dyeing, *salt additives* exhibit a retarding and leveling effect. Higher concentrations of sulfate ions are assumed to compete with dye anions for ammonium groups of the fiber. Weakening of the electrostatic attraction between dye and fiber due to coulombic forces is more likely; this effect decreases with increasing pH. Instead, salt additives exert an aggregating effect on dye molecules, and increased attachment can occur [5.14].

**Dyeing Processes.** Dyeing instructions for different types of acid wool dyes vary, especially with regard to the pH range used (see Table 5.1). The greater the affinity for the fiber, the more strongly must the ionic binding component be repressed [i.e., the pH value at the beginning of dyeing must be higher (see Table 5.1)].

**Table 5.1.** Dyeing instructions for acid wool dyes

| Addition | Leveling dyes | Milling dyes | Super-milling dyes |
|---|---|---|---|
| $Na_2SO_4$, cryst. | 2–4 g/L | 2–4 g/L | 2–4 g/L |
| Acid donor | 1–2 g/L $H_2SO_4$ (96 %)* | 1–3 g/L acetic acid (30 %) | 1–3 g/L $(NH_4)_2SO_4$ |
| pH value | 2.5–3 | 4–5 | 6–7 |
| Leveling agent | | 0.5 g/L | 0.5 g/L |
| Final addition | | 1 g/L $H_2SO_4^*$ | 1–2 g/L acetic acid (30 %) |

* Or formic acid.

The starting temperature is 60 °C for leveling dyes, 50 °C for milling dyes, and 30 °C for super-milling dyes. Usually, the acid is allowed to react with the fiber for 10 min; then salt is added followed, after another 10 min, by the dissolved dye. After another 10 min, the system is heated for 30–45 min until a final temperature of 95 °C is reached and then further dyed at 95 °C for 45–90 min. Addition of acid toward the end of the dyeing process completes bath exhaustion. Warm and cold rinsing follows. With selected dyes dyeing can be performed at 80 °C. Milling dyes require the addition of a leveling agent (→ Textile Auxiliaries, Section 5.6.2).

Trichromic dyeing can be conducted with leveling dyes. In the case of milling dyes, the desired shade is better achieved by selection of a dye with similar hue and shading.

*Improvement in Fastness.* The wetfastnesses of chlorinated wool that has been finished with synthetic resin (polyamide–epichlorohydrin or polyurethane) (superwash wool) can be increased by means of methylol amide compounds. Both fastness and antifelting finish can be improved through the application of a polyquaternary compound (i.e., a compound with several quaternary groups) [5.15] [Basolan F (BASF), Sandofix L (Sandoz)]. Anionic condensation products can form a barriere at the surface of the fiber and thus dimish the bleeding of anionic dyes (see Section 6.4.2) [Mesitol HWS (Bayer)].

### 5.1.2.2. Chrome Dyes

Chrome dyes (→Azo Dyes, **A 3**, pp. 270–271) refer to selected acid dyes that form complexes with chromium ions. During complex for-

mation a strong bathochromic shift of shade occurs. As a result of the superposition of several excited states, a marked dulling of the hue is observed.

Complex formation occurs during the dyeing process in a strongly acidic medium under the participation of electron donors (ligands) from the chromophore and the fiber [5.16], [5.17]. Chromium, the central atom of the complex, acts as a link between dye and fiber. This results in a very strong bond, which is reflected in the excellent fastnesses obtained. The binding of chromium occurs by substitution of the H in $-COOH$ or $-OH$ groups and via lone electron pairs from $-\overset{|}{C}O$, $-NH_2$ or $-N=N-$ groups (dative bonds). Thus, the dyes must contain suitable functional groups:

1) Monofunctional: salicylic acid or alizarin type
2) Bifunctional: *o,o'*-dihydroxyazo groups

Chromium(III) with the coordination number 6 acts as the central atom. It is formed from dichromate, which is reduced by the fiber. Strong acids have an activating effect on this process [5.18]; the reducing effect of wool can also be enhanced by organic acids (tartaric, lactic, or formic acid) [5.19]. Thiosulfate also acts as a reducing agent, increasing the rate and degree of conversion in chroming [5.20] and decreasing fiber damage [5.19]. Lowering the temperature from boiling temperature to 90 °C contributes to fiber protection [5.21]. In the past, the amount of dichromate required was 50 %, based on the dye used, but no less than 0.25 % or more than 2.5 %, based on the fiber. According to Ciba–Geigy, the amount of potassium dichromate used is a standard 0.2 + 0.15 × the amount of dye in percent. In the meantime, considerable reductions of potassium dichromate content have been found to be possible [5.22], [5.23]. Individual chrome factors are given in pattern cards.

Auxiliary agents cause the disaggregation of chrome dyes and form adducts with them. These adducts break down only at boiling temperature. In this way, the number of free amino groups available for complex formation is decreased and less acid is required. The dyeing becomes more level, and "tippy dyeing" is reduced. Ethoxylated fatty alcohols, alkylphenols, and fatty amines are suitable auxiliaries.

**Application.** *Afterchroming Method (Chrome-Developing Dyes).* For dyeing the liquor is prepared with 1–2 g/L of formic acid (50%) (pH 3.5–3.8), 0–2 g/L of calcined sodium sulfate, and 1–2 g/L of wool protectant. The process is started at 40 °C, and dissolved dye is added after 10 min. The system is heated within 30 min, and dyeing is performed at 90 °C for 30–45 min. If the exhaustion of the bath is too low, 0.5–1.0 g/L of formic acid is added and dyeing is continued for 15 min.

For subsequent *chroming* the bath is cooled to 70 °C, followed by the addition of potassium dichromate and heating. Temperature is maintained at 90–100 °C for 30–45 min. After 15 min, 5–10 g/L of sodium sulfate is added to detach the saltlike bound chromate from the wool and make it accessible to complex formation (Bayer). Other variants use reducing and complexing agents to support the chemical reactions [5.24]. Finally, neutralization is carried out with ammonia (pH 8).

*One-Bath Chroming Method (Metachrome Process).* Complex formation must be preceded by dye diffusion. For this reason, the release of Cr(VI) must be delayed.

The dye liquor is prepared with 3–8% of metachrome mordant (a mixture of sodium chromate and ammonium sulfate) and 2.5–5 g/L of crystalline sodium sulfate. After a prerun of 10 min at 40–50 °C, the system is heated slowly and dyed at boiling temperature for 45–90 min. Shortly before dyeing is completed, the bath is exhausted by the addition of 1–2 g/L of 30% acetic acid.

The boiling time is shortest in the afterchroming method, and the most level and fast colors are produced. Poor shading possibilities represent a disadvantage. The one-bath chroming method is most important.

The following auxiliaries are used: Albegal W (Ciba–Geigy), Avolan AV (Bayer), Lyogen MS, WD, Lyocol CR (Sandoz), Syntegal V7 (Ostacolor), Uniperol O (BASF).

### 5.1.2.3. 1:1 Metal-Complex Dyes (→ Metal-Complex Dyes, A 16, pp. 303, 305–307)

Metal-complex dyes are chemically very similar to chrome dyes. The risk of fiber damage is reduced because the complex is formed in dye production.

Dyeing is conducted in a strong sulfuric acid medium at pH 1.9–2.2. Amino groups in the fiber are converted to the ammonium form, and ionic bonds to the dye anion are formed. Under these conditions, amino groups are not available as ligands. Only with increasing pH during rinsing can they be included in the complex in exchange for aquo ligands [5.25]. The addition of auxiliaries (alkanolethoxylates) allows a decrease in the amount of acid added (pH 2.5–3) because they form addition compounds with the dye and thus repress complex formation. The dyeing process is slowed down and made uniform. With synergistic amphoteric mixtures of auxiliary agents, dyeing can be performed at even higher pH (3.5–4) [5.26].

**Dyeing Process.** Depending on the liquor ratio, the bath is adjusted with 2–6 g/L of sulfuric acid (96%) to pH 1.9–2.2 (pH 2.5 in the presence of 1–2 g/L of auxiliary agents). After addition of 5–10 g/L of calcined sodium sulfate, the material is treated with the liquor for 10 min at 40–50 °C; then the dissolved dye is added. Ten minutes later, the system is heated up in 30–45 min and dyeing is carried out at boiling temperature for 90 min; 1–2 mL/L of ammonia (25%) or 2–3 g/L of sodium acetate can be added to the last rinsing bath. A lowering of the temperature to 80 °C is possible in the presence of an ethoxylated fatty amine (pH 1.9–2.2).

**Dye Ranges.** Chromolan (Ostacolor), Inochrom (Zeneca), Neolan (Ciba–Geigy), Palatinecht (BASF).

**Auxiliaries.** Albegal NF, Albegal Plus (Ciba–Geigy); Avolan S, SCN (Bayer); Lyogen WD (Sandoz); Syntegal V7 (Ostacolor); Uniperol O and SE (BASF).

### 5.1.2.4. 1:2 Metal-Complex Dyes (→ Metal-Complex Dyes, A 16, pp. 304–305, 308–311)

Metal complexes are formed in a molar ratio of 1:2 in the weakly acidic pH range [i.e., two chromophores are coordinated to one central atom (Cr or Co)] [5.27]. This atom is located between the two chromophores, which are usually arranged perpendicular to each other (Drew–Pfitzner complexes, → Metal-Complex Dyes, A 16, p. 301). The high affinity for the fiber is based on the large size of the dye molecules; their compact, often almost spherical shape; and their negative charge [5.28].

To achieve sufficient evenness (levelness), the incorporation of sulfonic acid groups into the dye molecules was initially avoided so additional ionic interactions between fiber and dye could be excluded. Instead, water solubility was achieved by incorporation of methylsulfone ($-SO_2-CH_3$) [5.29] or sulfonamide ($-SO_2-NH_2$) groups.

Only after 1960 did dyes containing sulfonic acid groups also find application. They have the advantages of inexpensive production, higher yield, and cold solubility. In general, the wetfastness is somewhat higher than that of dyes containing methylsulfone and sulfonamide groups. Unfortunately, dyes containing two sulfonic acid groups, in particular, are susceptible to nonlevel dyeing. To avoid tippy dyeing, auxiliary agents, especially ethoxylated fatty amines, are added [5.30], which form adducts with the dye. These adducts break down at higher temperature. The leveling effect is increased by addition of Glauber's salt.

Since the 1:2 dye complexes are formed in a weakly acidic medium, they can be applied to wool under the same conditions. This enables a gentle mode of dyeing, which implies neither an oxidative attack by Cr(VI) nor a hydrolytic attack by sulfuric acid.

**Dyeing Process.** Ammonium sulfate (1–2 g/L) or ammonium acetate (2–4 g/L) (pH 5.5) is added to the liquor. After a prerun of 10 min at 30–50 °C, dissolved dye is added. The system is heated in 30–60 min and dyed for 30–60 min at boiling temperature.

In the presence of auxiliaries (1–2 g/L) dyeing is performed with the addition of 1–3 g/L of acetic acid (30%) (pH 4.5–5) and 1–2 g/L of calcined sodium sulfate. Acid dosing using control and monitoring systems is also possible [5.31], [5.32]. After rinsing, 1–2 g/L of formic acid is used for acidification to improve the feel and wetfastness.

*Auxiliary Agents.* Albegal A, SET SW (Ciba–Geigy); Avolan IL, IS, IWN, UL75 (Bayer); Eganal SZ (Hoechst); Lyogen FN, MS (Sandoz); Remol U (Hoechst); Uniperol SE, W (BASF); Unisol WL (Zeneca); Wofalansalz EM (Wolfen).

### 5.1.2.5. Reactive Dyes (→ Reactive Dyes)

Reactive dyes for wool produce brilliant colors with good fastness. They are not identical to reactive dyes for cellulose fibers because the reactivity of the amino groups in the wool is considerably higher than that of the hydroxyl groups in cellulose. To achieve level dyeing on wool, the reactivity of the dye must be reduced and an auxiliary agent added.

| Reactive anchor groups | Dye ranges |
|---|---|
| *N*-Methyltaurine ethylsulfone- | Hostalan E, Procilan E |
| β-Sulfatoethylsulfone- | Remalan |
| Acrylamide-, chloroacetyl- | Procilan |
| α-Bromoacrylamide- | Lanasol, selected Lanaset |
| 2,4-Difluoro-5-chloropyrimidyl- | Drimalan F |

Wool contains many reactive groups—amino, imino, and hydroxyl groups being the most important [5.33]. Reactions occur in a weakly acidic medium (pH 3–5) and include nucleophilic substitution of leaving groups (usually Cl, F, and rarely, sulfonate or ammonium groups) or addition reactions to polarized aliphatic double bonds (see Section 4.1).

The first two anchor groups mentioned (*N*-methyltaurine ethylsulfone and β-sulfatoethylsulfone) have the advantage that their functional groups are masked at the start of the process; consequently, no premature reaction occurs below boiling temperature [5.34]. At the same time, level dyeing is made possible by the increased solubility. Drimalan and Lanasol dyes are bifunctional and can have a cross-linking effect on wool [5.9], [5.35].

*Influence of pH.* In the acidic range, ionic bonds are formed between dye and fiber, with the dye still being capable of migration. At pH 5, covalent binding to the fiber predominates.

**Dyeing Process.** Auxiliary agent (1–2 g/L) is added to the liquor, and the pH adjusted to 3–4 with formic acid or 1–3 g/L of acetic acid. The process is started at 40 °C, and dissolved dye is added after 10 min. After 20–30 min, the pH is adjusted to 5–6 with sodium dihydrogenphosphate. Dyeing is conducted at boiling temperature for 1 h.

To eliminate hydrolyzed dye, an aftertreatment is performed for 15 min at 80 °C with the addition of 1.5 g/L of ammonia (pH 8.5–9.0). The last rinsing bath is weakly acidified.

*Auxiliary Agents.* Albegal B (Ciba–Geigy), Avolan REN (Bayer), Eganal GES (Hoechst), Lyogen FN (Sandoz).

### 5.1.2.6. Vat Dyes, Leuco Esters of Vat Dyes

In the past, vat dyes played an important role in dyeing wool. Indigo (→ Indigo and Indigo Colorants) was considered to be "king of dyes" because of its unsurpassed fastness on wool. Until the mid-1950s, navy cloth was dyed with indigo. The application was problematic with regard to the negative effect of reducing agents and alkali on wool. Today, indigo, thioindigo, and their derivatives have been replaced by classes of dyes that are easier to handle.

### 5.1.3. Technology of Dyeing (see Chap. 2)

Wool is dyed in equal amounts in the form of flock and slubbing. There is a trend toward piece and yarn dyeing [5.36]. The overflow can take the place of winch and beam dyeing machine, and jigger if the material has previously been made shrink proof [5.37]. High-quality, cool wool articles that are woven from fine yarns require especially gentle treatment.

Wool dyeing is usually conducted near boiling temperature. The fiber suffers clear degradation at this temperature. In some cases, the temperature can be lowered to 80–90 °C. An increase in dyeing temperature has proved necessary when poorly leveling dyes are applied. However, 108 °C is generally accepted as the upper temperature limit. A further increase in temperature can be required only in dyeing polyester–wool mixtures by the high-temperature process (see Section 11.2). In this case, the upper limit is 120 °C. Formaldehyde is added to the liquor, which leads to formation of methylene bridges in the fiber and stabilizes the keratin to hydrolytic degradation [5.38], [5.39].

The fields of application of various classes of wool dyes are based on the absorptive and leveling capacity [5.40]:

| | |
|---|---|
| <1/1 standard depth | disulfonic acid dyes |
| 1/1–3/1 standard depth | monosulfonic acid dyes |
| >3/1 standard depth | 1:2 metal-complex dyes |

*Continuous processes* are frequently used to dye slubbing. Here, a thickening agent based on etherified guar or locust bean gum, a special auxiliary agent, and an acid or acid donor are added to the pad liquor. Chromium trifluoride is generally used as the chroming agent. The material is steamed for 15–60 min after padding.

### 5.1.4. Properties of Dyeings

Among the aforementioned classes of dyes, reactive and 1:2 metal-complex dyes lead with respect to tonnage, followed by 1:1 metal-complex, chrome, and acid dyes [5.41]. This distribution among the different classes of dyes depends not only on the required level of fastness, but also on the necessary brilliance. Brilliant shades are produced by acid and reactive dyes, whereas chrome dyes dominate in the case of black.

Among the chrome and metal-complex dyes, 1:2 metal-complex dyes are easiest to use, and corrections of shade (shading) are possible. Superwash articles require a high level of fastness, which can be achieved with reactive dyes. The fastness of dyeings with various classes of wool dyes at 1/1 standard depth of shade is given in Table 5.2.

Excellent fastness to milling and potting is achieved with chrome and metal-complex dyes.

Damage of the wool during dyeing is due to action of acid, oxidizing agent (chromate), water, heat, and mechanical stress. Exposure to boiling water for a longer period results in cleavage of cystine cross-links. New bridges are formed, such as lysinoalanine, which can contribute to crease fixation. Thus, woolen fabrics should be setted before dyeing. Hydrolytic degradation increases with increasing distance from the isoelectric point [5.42]. The sodium sul-

**Table 5.2.** Fastness of wool dyeings at 1/1 standard depth of shade

| Class of dye | Washfastness (40 °C) | Fastness to perspiration (acidic) | Fastness to perspiration (alkaline) | Lightfastness |
|---|---|---|---|---|
| Acid dyes | 4–5 | 2–4 | 2–4 | 3–6 |
| Milling dyes | 4–5 | 4–5 | 4–5 | 5–6 |
| Super-milling dyes | 4–5 | 4–5 | 2–5 | 4–6 |
| Chrome dyes | 4–5* | 4–5 | 4–5 | 6–8 |
| 1:1 Metal-complex dyes | 4–5* | 4–5 | 4–5 | 5–7 |
| 1:2 Metal-complex dyes | 4–5* | 4–5 | 4–5 | 6–7 |
| Reactive dyes | 4–5* | 4–5 | 3–5 | 4–6 |

* At 60 °C.

fate present increases hydrolysis, which is closely connected to a deterioration of mechanical properties [5.43]. The 1:2 metal-complex dyes are considered most favorable with respect to maintenance of feel, elasticity, and strength.

Chlorinated wool exhibits a different coloristic behavior. Elimination of the scaly layer facilitates both dye uptake and release. The wetfastness is much lower, so reactive dyes should be used. In contrast, antifelting treatments with synthetic resin seldom lead to changes in coloristic behavior.

### 5.1.5. Dyeing of Wool–Polyamide Blends

The strength of woolen articles is increased by the addition of 5–20% of polyamide (PA) fibers. Stretch fabrics made of wool weft and PA textured yarn warp have gained importance as ski tricot. Both types of fiber are, in principle, dyeable with the same classes of dyes. However, PA is more accessible to the dye and is, consequently, dyed more deeply than wool in the case of light colors. Only when the limited ionic binding capacity of the $NH_3^+$ groups in PA is exhausted does the higher binding capacity of wool become noticeable so that, for deep shades, the wool component is darker. For light shades, the different affinity can be adjusted by using fiber-affinitive retarding agents, such as aromatic sulfonates, which partly reserve the PA component. The level of auxiliary agents depends on the depth of color and is 0.5–4%. Monosulfonated acid dyes are the most suitable.

**Dyeing Process.** Dyeing is performed in the presence of 1–2 g/L of acetic acid (60%) and 2–6 g/L of Glauber's salt for 2 h at 98 °C. Fewer auxiliary agents are added for PA 66 than for PA 6 [5.44].

Dark colors cannot be achieved with acid dyes because of the limited dye affinity of the material. In these cases, 1:2 metal-complex dyes are used.

Auxiliary Agents. Basopal NA (BASF); Edolan A extra (Bayer); Effektan ST (Ostacolor); Endolan PAW (Hoechst); Erional NW, NWS, PW, RF (Ciba–Geigy); Intrasol CWN (Crompton); Mesitol HWS, liquid (Bayer, for 1:2 metal complex dyes).

## 5.2. Dyeing of Silk

Silk (→ Silk) accounts for 0.2% (61 000 t in 1988–1989) of total fiber production. Neverthe-

less, because of its special properties, the importance of silk should not be underestimated, especially for ladies' wear and men's shirts, jackets, ties, and scarves.

### 5.2.1. Fiber Structure and Dyeing Behavior

Silk fibroin consists of 18 different amino acids. As with wool, amino groups are important for the uptake of ionic dyes. At 230 µmol per gram of fiber the number of amino groups is considerably lower than in wool.

The isoionic point of silk is at pH 5.0. The stability of silk is considerably lower than that of wool because silk has no cystine cross-links. Even under mild conditions, e.g., at pH 4.0 and 85 °C, hydrolytic degradation is observed. Therefore, acid must be added continuously during the dyeing process and not at once. For this reason, dyeing is frequently conducted at 70–80 °C, but no higher than 90 °C. In this way, damage to the surface structure and wrinkle and crease formation are minimized [5.45].

As a result of its fineness, light is strongly reflected from the surface of silk. Thus, considerably more dye is required to achieve the required shade than with other materials.

### 5.2.2. Classes of Dyes

**Direct Dyes.** As a result of their good fastness properties, direct dyes are frequently used on silk.

*Dyeing Process.* Dyeing is conducted in the neutral pH range or with the addition of 1–3 g/L of acetic acid (30%) and 2–5 g/L of sodium sulfate. The process is started at 30–40 °C; the system is then heated up within 30–45 min and dyeing is continued at 90 °C for 30–45 min. As leveling agent, 1/3–1/5 of the degumming liquor (→ Silk, **A 24**, pp. 101–102), which has been made weakly acidic with acetic acid, can be added.

**Acid Dyes.** Acid dyes are the most important dyes for silk.

*Dyeing Process.* Dyeing is conducted with the addition of 1–4 g/L of acetic acid (30%) or 1–2 g/L of ammonium sulfate (pH 4–5.5). After starting at 30–40 °C, the system is heated up within 30–45 min and dyeing is continued at 70–85 °C for 45–60 min. Dyeing in the soap bath used for degumming (pH 8–8.5) with the addition of sodium sulfate offers better fiber protection.

**Metal-Complex Dyes.** In principle, after-chroming dyes can be applied to silk [5.46]. They are, however, of little importance. In contrast to wool, *1:1 metal-complex dyes* can be dyed in a weakly acidic medium at 90 °C and produce excellent fastness values.

Best suited for the dyeing of silk are the *1:2 metal-complex dyes* because they can be applied in a weakly acidic medium and produce good fastness.

*Dyeing Process.* Dyeing is performed with the addition of 2–5 g/L of ammonium acetate or 2–3 g/L of ammonium sulfate and 0.5–1 % of leveling agent. After a prerun at 40 °C for 15 min, dissolved dye is added; the system is heated up within 30–40 min and dyeing is continued at 80–95 °C for 45–60 min.

**Improvement in Fastness.** Colors produced with acid, direct, and metal-complex dyes can be aftertreated with 8 % of tannic acid and 4 % of acetic acid (30 %) at 35–40 °C for 60 min. Subsequently, a fresh bath with 4 % of potassium antimony(III) oxide tartrate at 20–25 °C is used without intermediate rinsing.

**Reactive Dyes.** Reactive dyes are applied to silk only when brilliant shades are needed and when the colorfastness produced with acid dyes does not meet the necessary requirements [5.41], [5.47]–[5.50].

*Dyeing Process.* Dyeing is performed with the addition of 10–40 g/L of calcined sodium sulfate, half of which is added after 15 min and the other half after 30 min at 30 °C. The system is heated up to 50–70 °C in 30 min, and after another 15 min, 2 g/L of soda is added. Dyeing is continued for 40 min. An afterwash at 80 °C increases wetfastness.

**Other Classes of Dyes.** *Developing dyes* can be used for dyeing silk [5.51], [5.52]. *Vat* and *leuco esters of vat dyes*, as well as *cationic dyes*, are also used but are less important.

### 5.2.3. Technology of Dyeing

For the pretreatment of silk (steeping, soupling, degumming, hardening), see [5.50], [5.53].

The dyeing of yarn hanks is frequently conducted on spray or package dyeing machines.

The winch, beam and overflow are preferred for fabrics (see Section 2.2.3).

**Aftertreatment.** To produce the typical scroopy feel of silk, weighting [5.52] and reviving are carried out with 1–2 g/L of formic, acetic, lactic, or citric acid.

# 6. Dyeing of Polyamide Fibers

(see also, → Polyamides; → Fiber, 4. Synthetic Organic, **A 10**, pp. 569–579)

## 6.1. Chemical Structure

**Polyamide 6** [*25038-54-4*] **and polyamide 66** [*9011-55-6*] contain, on average, 150 and 75 monomer units, respectively, linked together to give a macromolecule. One chain end consists of an amino group, which can be present in the free state or in the acylated form. Acylation results from reaction with chain transfer agents, which can be added during polymer formation to obtain a uniform degree of polymerization (DP).

Amino groups are of special importance for dyeing because they form ammonium groups in an acidic dyebath by addition of protons. The number of amino groups is of the order of 25 μmol per gram of polymer in the case of PA 6 and 30 μmol per gram of polymer in the case of PA 66.

The reason for the limited dye uptake lies in the comparatively low number of amino groups in comparison to wool. The depth of color achieved on PA 6 is somewhat less than that on PA 66.

If the pH of the liquor falls below 2.4, more acid is taken up by the fiber because the amide groups are protonated under these conditions [6.1]:

$$-NH-CO- + H^+ \longrightarrow -NH_2^+-CO-$$

However, this is of no interest in dyeing because dye that is absorbed below pH 2.4 is less tightly bound, and fiber damage occurs at low pH.

**Modified types of polyamide** (→ Fibers, 4. Synthetic Organic, **A 10**, pp. 576–577) contain a varying number of amino groups [6.2]. The dyeing characteristics of PA fibers as a function of the content of amino groups (μmol per gram of polymer) are given below

| | |
|---|---|
| Weak dyeing | 15–20 |
| Normal dyeing | 35–45 |
| Deep dyeing | 60–70 |
| Ultradeep dyeing | ≥ 80 |

In addition, dyeability with cationic dyes is achieved by the incorporation of sulfonic acid groups (basic dyeable fibers). During finishing, a change in dye affinity can be attained by local reaction with fiber-reactive products, such as chlorotriazine derivatives to block part of the amino groups (Sandospace process, Sandoz).

## 6.2. Supermolecular Structure

(see also, → Fibers, 2. Structure, **A 10**, pp. 486–493)

The state of order of the polymer in the fiber is of decisive importance for its dyeing behavior. The degree of crystallinity and degree of orientation are regarded as determining quantities. However, the amorphous remainder is of greater importance in dyeing because the chain segments in this part have a mobility that permits penetration of swell water and dye molecules. The system can be looked at as an uniaxial stretched dynamic network with the ordered regions as knot points [6.3].

The amorphous regions are present in the glassy state at lower temperature, and the mobility of chain segments begins only above the glass transition temperature $T_g$. In the presence of water, $T_g$ is below the dyeing temperature for PA and, thus, has no effect on the course of dyeing.

The fiber exhibits certain structural differences in the radial direction (skin–core structure); the more compact surface layer in PA 66 obstructs dye uptake slightly [6.4]–[6.6]. Technological operations change the fibrous structure. Stretching leads to increased orientation and impeded dye uptake [6.4]. While classical drawn yarn has a monoclinic α-form, fast-spinning yarn has an unstable pseudohexagonal γ-form. However, this unstable form is converted to the more stable α–γ mixed structure during prewash [6.3], [6.7].

Thermal treatments result in changes in structure. Dyeing behavior is not changed essentially by *hydrosetting*, but the structure is loosened [6.8]. Thus, the size of the dye molecules has a smaller effect on dye uptake. Free shrinkage of the material results in a faster rate of dyeing [6.9]. In comparison, the critical range of absorp-

tion is shifted upward by 15 K on heat setting. The required penetration time is almost doubled. In *texturing*, the internal forces acting during torsion of the thread change the inner structure so greatly that dye molecules can diffuse more easily into the fiber [6.10]. Any unevenness in operations that changes the supermolecular structure of the polymer leads to disturbances of dye uptake ("stripiness" or "Barré effect"). In general, PA 6 dyes more evenly than PA 66 because it has a more open structure.

## 6.3. Interactions Between Dye and Fiber

The binding of dyes to polyamide is based on various interactions.

**Secondary Bonds.** The chromophoric system of the dye, with its aromatic rings (delocalized π-electron systems), easily polarizable azo groups (lone electron pairs), and other donor and acceptor groups, forms many polar bonds, dispersion interactions, and hydrogen bridges with the polymer.

These secondary bonds are superimposed on all other bonds. The larger the dye molecules and the fewer the hydrophilic groups carried by the dye, the greater is the superposition.

**Ionic Bonds.** Acid dyes are present in the fiber in ionic form [6.3]. ELÖD and coworkers were the first to detect a stoichiometric relationship between the content of amino end groups in the fiber and dye uptake [6.11]. This can be ascribed to electron neutrality. Dye anions are able to displace counterions of the ammonium groups in the fiber because they are supported by numerous secondary bonds (ion-exchange mechanism). In the presence of water, ionic bonds exert no binding effect but rather a "Einweisungsfunktion" (pilot function), according to ZOLLINGER [6.12], [6.13].

Additional dye uptake beyond the ionic binding capacity is possible only in the form of purely secondary adsorption ("overdyeing"). Dissociation and hydration counteract this effect. The dyeings obtained by overdyeing have poor wetfastness. The amount of these weakly bound dyes can be determined under conditions in which ionization of the amino groups does not occur ("neutral absorptive capacity" according to HOFFMANN).

The simultaneous action of ionic bonds and secondary bonds determines the degree of dye uptake in equilibrium. This is frequently reflected in a superposition of the Langmuir and Nernst isotherms [6.14], [6.15].

The amount of dye bound at equilibrium decreases with increasing number of sulfonic acid groups in the dye molecule due to increasing hydration. For this reason, mono- and polysulfonated acid dyes should not be used together in combination dyeing.

**Coordinate and Covalent Bonds.** Amino groups of the fiber can participate as ligands in the coordinate bonding of 1:1 metal-complex dyes. This is, however, of little practical importance. In addition, they can form covalent bonds to reactive dyes.

## 6.4. Dyes

### 6.4.1. Disperse Dyes ($\rightarrow$ Disperse Dyes)

More than 50% of disperse dyes are simple azo compounds, about 25% are anthraquinones, and the rest are methine, nitro, and naphthoquinone dyes. They were originally developed for dyeing acetate fibers. With the emergence of polyamides, they also proved useful for polyamide fibers.

Disperse dyes are finely dispersed products that have a solubility in the bath of about 0.1 g/L. The solubilities of polyamide dyes are somewhat higher than those of polyester dyes. The shades of colors obtained with disperse dyes depend on the substrate. Thus, in comparison with acetate or poly(ethylene terephthalate) (PETP), many shades on PA are bathochromically shifted (orange $\rightarrow$ red, red $\rightarrow$ violet). This is attributed to interaction of the amide groups with the chromophores. The colors are generally vivid, except for the red shades.

As a result of the level dyeing of even structurally nonuniform material, disperse dyes are applied especially in dyeing lighter shades. Wetfastness deteriorates with increasing depth of color.

**Dyeing Process.** A dispersing agent (0.25 – 2 g/L) is added to the liquor, and the pH is adjusted to 5 with acetic acid. The process is started with the material at 40 °C and followed by addition of dye after 5 min. After another 5 min, the system is heated up within 30 min, and dyeing is conducted in 60 min near boiling temperature.

**Dye Ranges.** Artisil (Sandoz), Cibacet (Ciba – Geigy), Dispersol (Zeneca), Celliton (BASF), Ostacet (Ostacolor), Resolin P (Bayer), Samaron (Hoechst).

**Auxiliary Agents.** Eganal SZ (Hoechst), Levegal HTN (Bayer), Uniperol EL (BASF), Univadin DP (Ciba Geigy).

### 6.4.2. Acid Dyes

Acid dyes are derived from mono- and disazo compounds ($\rightarrow$ Azo Dyes, **A 3**, pp. 273 – 274) (primarily yellow-red, dark blue, black) and from anthraquinones ($\rightarrow$ Anthraquinone Dyes and Intermediates, **A 2**, pp. 408 – 410) (brilliant blue grades). Through the introduction of sulfonic acid groups, these compounds become water soluble and, at the same time, acquire the ability to form dye anions by dissociation in aqueous medium. The anions are capable of undergoing electrostatic interactions with cationic groups in the fiber. Thus, acid must be added to the dyeing process to convert amino end groups of the fiber to cationic groups, which explains the name of this class of dyes.

The binding of these dyes to the fiber is essentially supported by secondary valence forces. Affinity to the fiber can be increased by introducing additional substituents into the dye molecule that do not change the character of the chromophore [6.16]. The affinity increases along the series: olefinic < aliphatic < aromatic < cycloaliphatic.

Small chromophores (monoazo or anthraquinone) can be connected to one another via a binding link in such a way that no conjugation between the two can occur and—in contrast to direct dyes—the rotatability of the linked parts is maintained. So the affinity to the fiber is increased without changing the color. In larger molecules a similar effect is achieved by a steric hindrance of coplanarity within the chromophore system (benzidine dyes with sulfonic acid groups in $o,o'$-position).

The secondary binding of dye to PA is counteracted by hydratable sulfonic acid groups, which explains the large spectrum of dyeing properties. A classification according to technical application is given below:

1) *Monosulfonated Acid Dyes.* Leveling dyes and neutral absorbing dyes. These grades can easily be combined with each other.

2) *Disulfonated Acid Dyes.* These dyes produce dyeings with good wetfastness. They are used as self-colorants but are not recommended for trichrome dyeing [6.17].

The dyeing behavior of acid dyes can be described with the help of the combination values $K$ (see Chap. 12), which are derived from combination dyeings at 70, 80, and 90 °C [6.18], [6.19]. In the case of matching shades from dyeing and subsequent dyeing in the same bath the combination values of the dyes involved are equated. Monosulfonated acid dyes lie in the range $K = 1-5$, and polysulfonated acid dyes have still higher values. The higher this number, the greater are the contrasts. The combination values of dyes from different dye manufacturers cannot be compared [6.17], [6.20], [6.21].

The pH of the liquor is important for level dyeing, for exhaustion of the bath, for wastewater contamination, and for the effort involved in rewashing after dyeing. With increasing dye affinity, the ionic interaction in the initial phase must be repressed to achieve uniform absorption, i.e., the liquor must be neutral at the start of the dyeing process. Only in the course of dyeing is the liquor slowly acidified by one of the following methods [6.22]:

1) Acid is added at the beginning of dyeing or continuously, preferably by using pH-controlling instruments: Telon ST process (Bayer), Dosacid process (Ciba–Geigy). Special care is required in the neutral range because small amounts of acid give large shifts in pH value.
2) Acid donors release acid during the dyeing process. Substances used are ammonium sulfate, which liberates $NH_3$ during dyeing; sodium pyrophosphate ($Na_4P_2O_7$), which is converted to the acidic disodium hydrogenphosphate [6.23]; esters of organic acids; and heterocyclic compounds, which slowly release organic acids when exposed to heat.

In the case of low-affinity acid dyes, addition of acid toward the end of dyeing is recommended for full exhaustion of the dyebath.

The classical temperature control is more favorable than pH control for regulation of the dyeing process [6.23]. In the case of pH control, absorption of ca. 60 % of the dye is quick and nonlevel, and only the last 50 – 20 % is absorbed in a controlled manner.

In the acidic range, *electrolytes* retard liquor exhaustion [6.24]. The competition of sulfate anions with the dye for dye-affinitive positions is responsible for this effect (see Section 5.1.2.1).

*Auxiliary agents* improve leveling. In the case of fiber-affinitive agents, the sum of the dye anions and auxiliary agent anions should not exceed the binding capacity of the fiber to avoid blocking effects [6.25].

**Aftertreatment.** The wetfastness of dyeings produced with acid dyes is often unsatisfactory. In contrast to wool where the hydrophobic scaly layer hinders the escape of dye molecules from the fiber during washing, PA does not have a barrier of this kind. However, a barrier can be produced by aftertreatment.

In older processes, tannic acid (polygalloylglucose) and potassium antimony(III) oxide tartrate (tartar emetic) are used that form a 1:1 adduct on the material via hydrogen bridges [6.26]. This finishing process is expensive and said to be carcinogenic.

*Synthanes.* The synthetic tanning agents (synthanes) represent a more favorable alternative. They are high molecular mass condensation products of aromatic sulfonic acids with formaldehyde [6.27], [6.28] or condensation products of phenol, cresol, catechol, and naphthol with formaldehyde, which are made water soluble by reaction with bisulfite.

These products deposit on the fiber surface and, because of an accumulation of negatively charged sulfonate anions, form an electrostatic barrier that counteracts the release of dye anions from the fiber. In accord with this is the observation that a highly negative electrokinetic (zeta) potential is formed [6.29].

This treatment mainly increases the contact fastness (e.g., to water, perspiration, and seawater) of acid and metal-complex dyes and, to a lesser extent, their washfastness. The fastness to chlorinated water in swimming pools is hardly improved. Synthanes also have a stain-repelling effect on floor coverings [6.29].

Aftertreatment with synthanes has a number of disadvantages. Synthanes can reduce the adhesive strength of coatings or laminates [6.30]. In addition, they are poorly resistant to dry heat, afterfixing, and steam; treatment with synthanes results in slight changes in shade, reduced lightfastness, and a hardening of the feel. In the presence of cationic softeners, the fastness to rubbing is decreased. Synthanes are frequently applied despite their disadvantages.

The product Fadex CL (Sandoz) was developed specially to improve the fastness of bathing suits to chlorinated water. However, it cannot be used for printed articles with a white ground or bleached material because of its brown self-color.

**Dyeing Processes.** *Dyeing with Temperature Control.* The pH of the dyebath is adjusted to about 4.5–7, and 0.2–2 g/L of dyeing auxiliary is added. The process is started cold with the material and allowed to prerun for 20 min. Dissolved dye is added; the system is heated up in 30–45 min and dyeing is continued for 20 min near boiling temperature.

*Dyeing with pH Control.* Dye is dissolved in the liquor, the pH is adjusted to 8.5–10 with ammonia, and the auxiliary agent is added as described above. The material is placed in the boiling liquor, and acid is added slowly. The final pH of 4.5 is attained after 45–60 min. (Alternatively, 0.25–1 g/L of a pH regulator can be added.)

*Aftertreatment.* The exhausted bath or fresh liquor is adjusted to pH 4.5 with formic acid, and 1–2 % (per weight of fiber) of synthane is added. Aftertreatment is started at 40 °C, the bath is heated to 70–80 °C, and the material is treated for 20–30 min; the material is then rinsed.

*Auxiliary Agents for pH Control.* Eulysin WP (BASF), Sandacid V, VS (Sandoz).

*Synthanes.* Cibatex PA (Ciba–Geigy), Mesitol NBS (Bayer), Matexil FA-SN (Zeneca), Nylofixan P (Sandoz).

### 6.4.3. 1:2 Metal-Complex Dyes

The classical 1:2 metal-complex dyes (→ Metal-Complex Dyes, **A 16**, pp. 308–311) are acid dyes that are coordinately bonded to transition metals, especially chromium and cobalt, in the molar ratio of 1:2. These dyes are negatively charged and can enter into ionic bonding with the fiber. The amino groups of the fiber are not included in the complex. In contrast to acid dyes, the water solubility of classical 1:2 metal-complex dyes is not based on the presence of sulfonic acid groups. Rather, it is due to sulfonamide and methylsulfone groups. Only after 1960 were dyes containing sulfonic groups put on the market (see Section 5.1.2.4). Their high affinity for the fiber is caused by secondary bonds with the large molecules that are frequent-

ly almost spherical. As a result of their low water solubility, the traditional 1:2 metal-complex dyes are not as suitable for dyeing PA as for wool. Dyes containing sulfonic acid groups behave more favorably and, at the same time, produce better wetfastness. Their high fastness level makes them suitable for articles that are exposed to special use, such as floor coverings and car seat covers. The tendency of these dyes to mark structural differences in the material is a disadvantage. For this reason, they are recommended only for dark shades. The disulfonated acid dyes give good light- and wetfastness on PA and are frequently used in PA dyeing despite the poorer build-up properties. The high affinity of 1:2 metal-complex dyes makes the addition of amphoteric or nonionogenic leveling agents necessary.

The absorption of dyes by PA increases with decreasing pH. This tendency is especially pronounced in the case of disulfonic acid dyes. They are dyed in the presence of acetic acid. Dyeing with dyes free of sulfonic acid groups depends less on the charge of the fiber and more on the extent of formation of free complex acid, which absorbs more easily than the anion [6.31].

**Dyeing Process.** Dye and 0.5–1 g/L of auxiliary agent are added to the liquor. It is then made weakly acidic by the addition of 0.5–2 g/L of ammonium sulfate and 0.5–1 g/L of acetic acid (60 %), if necessary. High-affinity dyes can be used in a neutral or weakly alkaline medium. The process is started at 30–40 °C, followed by heating up within 30–60 min, and dyeing at boiling temperature for 30–60 min. A prerun with auxiliary agent at boiling temperature reduces the stripiness of the color.

### 6.4.4. Reactive Dyes

In principle, the reactive dyes used for wool are also suitable for PA (see Section 5.1.2.5). Covalent binding of dyes to the terminal amino groups produces dyeings that have excellent wet- and rubfastness (almost without exception grade 5). The lightfastness corresponds to that of analogous acidic dyes (3–6). The depths of color attainable are limited by the number of end groups and are higher for PA 6 than for PA 66.

**Dyeing Process.** The dyeing bath is made weakly acidic (pH 4.5–5). The process is started at 20–45 °C, followed by heating with a rate of 1 K/min and dyeing at near boiling temperature for 30–90 min. Aftertreatment (soaping) is performed with 0.5 g/L of a nonionic surfactant and 1 g/L of sodium bicarbonate or ammonia at 95 °C for 20 min.

## 6.5. Technology

**Dyeing Material.** The dye-binding capacity of a PA material can be characterized by comparison with a reference material. For this purpose, the ratio $v$ of the dye uptake of the test ($m_F$) and the standard ($m_{F, standard}$) materials in the initial phase is determined (region of linear increase)

$$v = m_F / m_{F, standard}$$

Values between 0.2 and 5 are obtained.

A saturation value can be determined by a comparison of the dyeings at pH 4.7, the amount of dye taken up being given as the fiber sum number in $S_F$ units (1 $S_F$ = 25 mmol/kg) [6.21]. The values lie between 1 and 3. The exhausted bath can be used again, e.g., for a blue color by addition of a yellow dye and the shade compared with a chart [6.32].

**Pretreatment (Washing, Prefixing).** Before dyeing, fabrics generally must be prefixed to compensate for material-related differences in affinity and to reduce the sensitivity to creasing during the dyeing process. Prefixing can be performed as *thermofixing on the frame* [15–20 s at 190 °C (PA 6) or 200–230 °C (PA 66)] or as *hot-water fixing* (45 min at 130 °C).

**Dyeing of Special Materials.** The material to be dyed is available in various forms. Exceptional features in dyeing exist for the following articles.

*Stockings, tights, and socks* are dyed either in drum-dyeing machines or on automatic apparatus (for stockings). In these machines the stockings or tights are drawn over aluminum forms, sprayed with dye liquor, and fixed.

*Belts for technical purposes (safety belts)* can be dyed most efficiently according to the pad steam process in continuous systems.

*Floor Coverings.* In the piece finishing of carpets, stringent demands are made on the evenness of the shade. Dyeing on special, open-width winch becks (4–5-m width) is frequently performed at 90 °C according to the pH shift process, starting at pH 8. Within 30–45 min, the pH is lowered to 4.5–6.5, depending on the depth of color; pH regulators can also be used [6.24], [6.33].

In continuous dyeing of floor coverings thickening agents are added to the pad liquors to prevent dye migration. Application is carried out with the help of a minimal application process or by using foamed liquors. The liquor uptake is between 120 and 500% [6.34], [6.35]. The next step involves steaming. Frothing auxiliaries are added to prevent dye from being washed out of the fiber tips by the formed condensate (frosting effect). Differential dyeing materials require a specific selection of dyes, auxiliaries, and pH values [6.36].

## 6.6. Properties of Dyed Materials

Dyeing with acid dyes can result in a certain degradation of the polymer [6.37]–[6.39]. The degree of hydrolysis depends on the concentration of the dye and the number of sulfonic acid groups per dye molecule [6.40].

In the case of crash and crinkle-look articles, thermal treatment under high-temperature conditions at 110–125 °C is frequently combined with dyeing in jet dyeing systems. In this process, loss of fiber strength can occur. This fiber damage can be largely avoided by adding 0.1–0.2 g/L of a hydroxylamine derivative [6.17].

The fastness of colors produced by various classes of dyes is shown in Table 6.1.

The values for PA 66 are slightly better than those for PA 6. Improvements can be obtained

**Table 6.1.** Fastness of polyamide dyeings at 1/1 standard depth of shade

| Class of dye | Washfast-ness (40 °C) | Fastness to perspiration (acidic) | Fastness to perspiration (alkaline) | Light-fast-ness |
|---|---|---|---|---|
| Disperse dyes* | 4–5 | 4–5 | 4–5 | 4–5 |
| Acid dyes | 3–5 | 2–5 | | 4–6 |
| 1:2 Metal-complex dyes | 3–5** | 4–5 | 4–5 | 6–8 |
| Reactive dyes | 4–5 | 4–5 | 4–5 | 4–6 |

* At 1/6 standard depth of shade. ** At 60 °C.

by aftertreatment with synthanes. The lightfastness of dyeings obtained with acid dyes on PA is slightly lower than on wool. Resistance to the chlorinated water of swimming pools plays a special role for bathing suits. PA 66 is usually superior to PA 6 in this respect.

## 6.7. Selection of Dyes

Very light shades are obtained on PA with *disperse dyes* because these dyes exhibit a good leveling capacity and are best able to compensate for material-dependent differences in affinity. *Acid dyes*, which also have a relatively good leveling capacity, are used for moderate depths of shade. Their fastness is usually sufficient, especially following aftertreatment with synthanes. The depth of shade has a limit of ca. 1/1 standard depth. The *1:2 metal-complex dyes*, which permit any depth of shade and give excellent fastness, are used for all dark and sober shades. They are not suitable for light shades because they have a tendency to nonlevel absorption and to mark structure-related stripiness. *Reactive dyes* are used for brilliant colors if the wetfastness of acid dyes is inadequate. They, too, have a tendency toward stripiness.

## 6.8. Toxicological and Physiological Aspects

Polyamide textiles often come in direct contact with the skin, especially underwear and stockings. For some time occasional allergic reactions have been observed to occur with close-fitting textiles because of their low moisture withdrawing capacity. However, only recently has the toxicological risk from the dyes employed received increased attention [6.41]. The amount of dye residues remaining on materials subjected to the all-in drum dyeing process, which is still occasionally used in the stocking industry, is much greater than after package

dyeing with intermediate rinsing. Thus, in the first case, the potential risk of skin damage is somewhat increased but disappears after the material has been washed.

## 7. Dyeing of Polyester Fibers

For information on polyester (PES) fibers, see → Fibers, 1. Survey, **A 10**, pp. 464–473 (properties); → Fibers, 2. Structure (form, structure); → Fibers, 3. General Production Technology; → Fibers, 4. Synthetic Organic, **A 10**, pp. 579–606 (chemistry and production). For properties, see also [7.1].

PES fiber is quantitatively the most important synthetic fiber (see Table 7.1). Its inexpensive production from petrochemical raw materials and excellent textile properties alone and in combination with natural fibers guarantee PES fibers universal applicability. Starting with tire cord, technical fabrics for tarpaulin and seat belts, carpeting, and furnishing fabrics, uses also include clothing, especially mixed with wool for suiting and trouser materials and mixed with cotton for shirts, raincoats, trousers, and casual wear. Pure PES fibers, especially in textured form, are employed in the knitwear sector.

PES fibers are hydrophobic, thus water-soluble dyes do not attach. In contrast, PES fibers can be dyed easily with water-insoluble, small molecular dyes originally developed for dyeing cellulose acetate. Since the preferred dyeing medium is an aqueous liquor, the poorly water-soluble dyes must be dispersed before application (→ Disperse Dyes).

### 7.1. General

#### 7.1.1. Dyeing in Aqueous Liquor

Since the fundamental work of MEYER and KARTASCHOFF in 1925 [7.2] and the studies of

**Table 7.1.** World polyester fiber production

| | Amount, $10^6$ t | | | Percentage (without polyolefin) of | |
|---|---|---|---|---|---|
| Year | Total | Staple fiber | Filament | Total fiber production | Synthetic fiber production |
| 1970 | 1.6 | 0.9 | 0.7 | 8 | 33 |
| 1980 | 5.1 | 3.0 | 2.1 | 17 | 48 |
| 1990 | 8.7 | 4.7 | 3.9 | 23 | 58 |
| 1992 | 8.3 | | | 21 | 58 |

VICKERSTAFF [7.3] in the beginning of the 1950s, a large number of studies have been published on the kinetics and thermodynamics of dyeing cellulose acetate and synthetic fibers with disperse dyes in aqueous dyebaths. For a survey, see [7.4, pp. 67, pp. 147]. Disperse dyes as well as acid and basic dyes must be molecularly dispersed in the dyebath; i.e., the dye must be dissolved in the aqueous bath before it can adsorb to the fiber surface and then diffuse into the fiber.

Thus, the form in which the disperse dye is present in the dye liquor is decisive for the dyeing process. The water solubility and rate of dissolution are influenced by many factors including [7.5]–[7.7]:

1) Temperature
2) Particle size
3) Crystal modification
4) Crystal growth
5) Melting behavior
6) Type of dispersion and dispersing agent used (fineness, stability, agglomeration, aggregation of dispersion)
7) Presence of salts
8) Behavior in dye mixtures

The water solubility of pure disperse dyes is a few milligrams per liter and increases strongly with temperature. It is also increased manifold by dispersing agents [7.5], [7.8].

The state of the dye in the fiber is often compared to a solid solution. Thus, the thermodynamic dyeing equilibrium of a disperse dye between water and fiber follows the Nernst distribution law:

$$\left(\frac{C_F}{C_L}\right)_{t=\infty} = K$$

At constant temperature, the distribution coefficient $K$ [i.e., the ratio of the concentration of dye dissolved in the fiber ($C_F$) and in the liquor ($C_L$) at equilibrium] is constant.

At normal dyeing temperatures the rate of dissolution of dispersed dyes in the dye liquor is assumed to be high, so that $C_L$ is constant as long as undissolved dye is present. Although not always valid in practice, the motion of the liquor is also assumed to be high enough so that molecularly dissolved dye is always transported to the fiber surface in sufficient amounts and the adsorption of dye proceeds rapidly. Then, diffusion in the fiber is the step that determines the rate of dyeing. The amount of dye $g$ that diffuses in a certain time follows Fick's first law:

$$g = -\frac{D}{h} \cdot (C_L - C_F)$$

The diffusion coefficient $D$ is $10^{-10}$ to $10^{-12}$ cm$^2$/s and can be increased only by an increase in temperature. The diffusion path $h$ is predetermined by fiber geometry; the concentration gradient $C_L - C_F$, by $C_L$.

Many experiments were carried out to develop an equation for the kinetics of dyeing. However, only approximations result because of the complexity of the fiber state and of the processes occurring in the dye liquor. The following phenomena are of special interest:

1) The polyester fiber contains varying proportions of crystalline and amorphous regions, depending on the degree of drawing and the fixing state. Only regions that are amorphous above the glass transition temperature are accessible to dye diffusion. During dyeing, the amorphous portion can change as a function of stress and shrinkage. In addition, the dyeing medium (water) and the dye already diffused into the fiber change the fiber structure and, thus, the dyeability.

2) In the dyebath, the dye goes through a large number of stages, from solid particle to a single molecule, which interact with each other. These include

Dissolving in the liquor
Possibly melting
Adsorption of dispersing agent
Association to dispersing agent
Incorporation into dispersing agent micelles
Diffusion through the laminar liquor layer on the fiber surface
Adsorption to the fiber
Agglomeration, aggregation, crystal growth
Mutual influencing of dyes, isomorphism of different dyes (formation of mixed crystals with different solution behavior than pure components)
Chemical changes (e.g., reductive destruction)

3) The levelness of dyeing – the basic requirement of coloring – is endangered by differences in temperature and concentration. In practice, such differences are unavoidable because of irregularities in the velocity and direction of the liquor flow at various positions in the dyeing aggregate and gradual depletion of dye in the liquor during passage through the dyebath. Hence, a constant sup-

ply of dye at the fiber surface is not guaranteed, and diffusion no longer determines the rate of dyeing.

For this reason, the practical dyeing process cannot (yet) be described by simple theoretical models. Despite many careful studies of model systems and individual processes, which permit a semiquantitative interpretation of individual steps under standardized conditions, dyeing results still cannot be mathematically predicted unambiguously. Thus, a dyeing process must be elaborated from dyeing experiments. Recipes are drawn up on the basis of trial dyeings with the help of color measurements.

When elaborating a dyeing process in practice, it is important to realize that, with regard to the rate of absorption, the system consisting of hydrophobic fiber and disperse dye occupies an intermediate position compared to other classes of dyes. The rate of diffusion of disperse dyes into hydrophobic fibers is moderately temperature dependent. Absorption can be easily regulated by controlling the temperature. Furthermore, this intermediate position between fast and slowly diffusing dyes is characterized by adequate migration capacity of the dyes, especially at high temperature. Thus, unlevel dyeings can be leveled out within manageable periods of time. Both methods (temperature control during exhaustion and high temperature leveling) are basically available for dyeing with disperse dyes:

a) As a result of rapid absorption of the dyes, the dyeing at the end of the exhaust phase is uneven and becomes level through subsequent adequate migration at high temperature.

b) The dyeing at the end of the exhaust phase is largely level due to controlled absorption with a bath exhaustion rate that is below a previously determined limit. The subsequent high-temperature phase serves only to ensure good penetration.

Irrespective of the method selected, a series of basic rules with regard to the time requirement, safety, and quality can be deduced from knowledge of the optimization of the absorption phase (method b). These are

1) In one formulation, dyes should be combined that, in the amounts applied, have exhaustion characteristics that are as similar as possible.

2) While the dyeing apparatus is being heated, the main exhaust range of the dyes, which

covers about 30 °C, should be passed through sufficiently slowly.

3) The heating rate in the critical temperature range should be adjusted to the type and design of the textile and to liquor circulation or the circulation rate of the material [7.9], [7.10].

To follow the first of these rules, information about the combinability of dyes is obtained from dye suppliers. The fact that the temperature range in which the main amount of dye exhausts depends on the amount of dye must also be taken into account. Moreover, dye producers offer premixed disperse dyes whose components are selected in such a manner that the mixture absorbs evenly at relatively low temperature and, consequently, makes fewer demands on the dispersion stability (e.g., [7.11]). At the same time, the melting behavior of mixtures (usually melting point depression) should not be overlooked [7.12]. When using mixed dyes, differences in the fastness of the components must be accepted. Trade names of combination dyes include Foron RD (Sandoz), Resolin K (Bayer), Dianix UN-SA (Mitsubishi-Hoechst), and Sumikaron RPD (Sumitomo).

For the second and third rules, empirical values are used to elaborate apparatus-specific control programs that guarantee optimization of the dyeing process with regard to the time course and dyeing levelness. Balancing the critical dyeing rate of the apparatus system ($V_{CRIT}$) with the significant dyeing rate of the formulation ($V_{SIG}$) (see Section 2.1) is helpful. Programs of this type are marketed as fast dyeing processes [7.13].

The dyeing process can be also controlled with dyeing accelerants (carriers) (see Section 7.2). Carriers are aromatic compounds that exert a swelling effect on PES fibers and a dissolving effect on disperse dyes. In the presence of carriers, as in the case of special leveling agents developed for PES dyeing, theoretical elucidation of the dyeing process becomes even more complicated. Only models are available at present.

### 7.1.2. Thermosol Process

In dyeing PES fibers, the dyeing time can be shortened significantly by increasing dyeing temperature. The rate of dyeing increases exponentially above the glass transition temperature; a temperature increase of 10 °C doubles the rate. For this reason, at ca. 200 °C fast colored fibers can be obtained with disperse dyes within seconds. Especially for PES–cellulose mixtures, this property is used to dye PES fibers continuously in a pad-fix process, which was introduced by Du Pont as the thermosol process [7.4, pp. 191–202], [7.14, pp. 141–142] (see Section 11.1). Application of the dye from an aqueous dispersion is followed by drying and fixing.

During drying, the padded dye liquor on the fiber surface can migrate to positions with a higher rate of drying. This danger is counteracted with an antimigration aid (→ Textile Auxiliaries, Section 5.9.1) [7.15]. To obtain a dye distribution on the fiber surface that is as homogeneous as possible, a uniform liquor film must be produced on the fiber. For this purpose, additives are used that wet the fiber and, at the same time, have a dissolving effect on dyes (e.g., ethoxylated fatty alcohols or fatty acids).

After drying, the dye is present on the fiber surface embedded in a film of dyeing aids and dispersing agents. In the case of PES–cellulose fiber blends, the dye is also present on the surface of the cellulose fibers. At fixing temperatures of 210 °C, most disperse dyes have melted or are dissolved in the liquid film of auxiliary agents. Three processes overlap in actual dye fixation: (1) heating, (2) dissolving of the dye in the fiber surface layer, and (3) diffusion. Disperse dyes diffuse into the fiber only in the molecular form. Sorption on the fiber proceeds rapidly. Buildup of a high dye concentration on the fiber surface layer and diffusion are competing steps that depend on the prevailing temperature and determine the rate of dyeing.

In heat fixation of disperse dyes, suitable heat transfer media are hot air, superheated steam, contact heat, and radiant heat. Hot air and contact heat are used most commonly. Steam has a swelling effect on the fiber. If superheated steam is used for fixation, the temperature can be lowered in comparison with hot air, the fixing times being the same. This is an advantage especially with articles made of textured fibers.

In the case of PES–cellulose fiber blends, the largest part of the liquor is absorbed by the cellulose fiber during drying, because of its higher absorption capacity. Nevertheless, in the subsequent thermofixing step, disperse dyes are almost completely fixed to the PES fiber. Transfer from the cellulose to the PES fiber occurs both by migration via direct fiber contact points and through the gas phase. The type of transfer apparently depends on the sublimation behavior of the dyes. Dyes that are resistant to sublimation migrate by direct contact of the fibers, whereas those that are not prefer the gas phase. Transfer through the gas phase can lead to more level dyeing, but with easily subliming dyes a risk exists of contaminating dyeing equipment and incurring dye losses [7.16].

## 7.2. Dyeing Processes for Polyester Fibers

PES fibers are almost exclusively colored with disperse dyes. For some time, diazotization dyes were used for dark shades. They are coupled to the fiber with suitable components (e.g., 2-hydroxynaphthoic acid). Although fast colors are obtained, the dyeing process is time consuming and susceptible to failure, and dyeing reproducibility is poor. Vat dyes are also occasionally used for PES dyeing. Small molecular thioindigo derivatives, in particular, diffuse into PES fibers if they are applied according to the thermosol process.

### 7.2.1. Requirements to Be Met by Dyes and Auxiliaries

A large variety of disperse dyes are available for coloring PES fibers. For the trade names of dyes recommended for PES fibers, see → Disperse Dyes, **A 8**, p. 574. The Colour Index provides information on dyes made by various producers that correspond chemically with each other. In the case of disperse dyes, however, chemically identical dyes can exhibit marked differences not only in coloring strength and shade, but also in their preparation and finish. Especially in modern dyeing processes, exacting standards are set for fine dispersion and dispersion stability. Disperse dyes are supplied as powder and liquid products. Powdered dyes contain a high percentage of dispersing agent, which is required to maintain a very fine dispersion when drying a dye that has been ground and dispersed in an aqueous slurry. Liquid products contain much less dispersing agent. Depending on concentration, the yields obtained from liquid products in various dyeing processes can deviate from those given by powders. The advantage of liquid products is that they are easier to handle in the preparation of large amounts of dye liquor.

Among the number of trade names available, the dyer selects dyes in accordance with the required properties and with their suitability for certain dyeing techniques. A strict standard is applied to the lightfastness of articles for car seat covers, furnishings, etc., and to the wetfastness of textiles that are washed in commercial laundries. Fastness to thermofixing is important for all articles that are heat-set, pleated, or resin finished after dyeing. Thermal stability plays an important role in the selection of suitable products for thermosol dyeing. Easily subliming dyes contaminate the fixing aggregate. Many dye producers divide their dyes into groups according to thermal stability (e.g., ICI in Class A to D and Sandoz in Class S, SE, and E). Choosing the proper dye for a certain dyeing process is not easy. For this reason, dyes are also classified into groups according to their rate of dyeing and the main temperature range in which they exhaust [7.17]. A combination of dyes with similar absorption characteristics quickly leads to level dyeing. To determine combinability, the American Association of Textile Chemists and Colorists (AATCC) recommends an easy-to-perform immersion test [7.18].

Another classification is based on covering up material differences in textured PES fibers. Dyes that have a low fastness to thermofixing often hardly mark differences in texture; they dye at relatively low temperature and are suited to carrier dyeing processes. Dyes that are very fast to thermofixing diffuse slowly into the fiber and must be dyed at high temperature. With a carrier, the color yield of these dyes is unsatisfactory at boiling temperature; texture and fixing differences in the fiber are frequently marked.

According to the recommendation of the Society of Dyers and Colourists, four tests are used to characterize the dyeing properties of disperse dyes [7.19]:

1)  Migration test for leveling capacity
2)  Building up capacity to deep shades
3)  Exhaustion test to determine the temperature range in which the main amount of dye attaches
4)  Diffusion test (BASF Combi Test [7.20]) for rate of diffusion in the fiber

Many dyeworks illustrate the coloring behavior of dyes by absorption curves (e.g., [7.21]).

For dyeing aids and carriers especially suited to dyeing PES fibers, → Textile Auxiliaries, Sections 5.4–5.7.

### 7.2.2. Dyeing from Aqueous Dye Baths

Articles made of pure PES fibers are dyed almost exclusively in exhaustion processes. Disperse dyes are most stable at a pH of 4–5, acetic acid being used preferentially for pH adjustment. A dispersing agent is also added to the dyebath (→ Textile Auxiliaries, Section 5.4). Dyes sensitive to heavy metals (e.g., anthraquinoid red products) are dyed in the presence of a complexing agent [ethylenediaminetetraacetic acid (EDTA) type].

**Dyeing under High-Temperature Conditions.** Because of the slow diffusion of disperse dyes at boiling temperature, PES fibers are dyed whenever possible at 125–135 °C under pressure (HT). The dyeing aggregates required for this purpose are available in most dyeworks. To develop a course of dyeing that is as quick and efficient as possible, studies have been performed on optimizing the temperature–time program, using dyes together that have similar exhaustion properties and adapting the circulation of the liquor or goods to the type of material to be dyed (see Section 7.1.1). In favorable cases, the entire coloring cycle can be shortened to less than 60 min in this way. Attempts are being made to approach the same goal by using complex dye mixtures [7.22]–[7.24].

Efforts are being made to make HT dyeing more economical by reducing the amount of liquor substantially (see Section 2.2.1–2.2.3).

The evenness of dye absorption is improved by dyeing in the presence of a leveling agent (→ Textile Auxiliaries, Section 5.6). Carriers are also used as HT leveling agents. They are, however, technically outdated and ecologically harmful. For this reason, the carrier dyeing technique is recommended only when dyeing temperatures of 130 °C are impossible (e.g., with PES–wool blends).

**Dyeing below 100 °C.** The highest possible dyeing temperature is of advantage for coloring PES fibers. However, if no pressure-stable dyeing aggregates are available, dyeing can be carried out at atmospheric pressure with the addition of dyeing accelerants. In addition, all conditions must allow the theoretically attainable boiling temperature to be approached as closely as possible under local conditions (i.e., by using closed or at least covered dyeing equipment). At the same time, this also prevents inhalation problems due to the carriers, which are generally steam volatile and can be ecologically harmful (→ Textile Auxiliaries, Section 5.7). Carriers must be accurately dosed; too much or too little can result in decreased yield. Since they can also affect fastness, especially lightfastness, carriers must be removed after dyeing by hot rinsing and washing or by drying at high temperature or heat setting.

### 7.2.3. Special Dyeing Processes

For reasons of levelness, very light shades are sometimes dyed without the use of pressure and without a carrier. Only those disperse dyes that diffuse sufficiently quickly into PES fibers even at 100 °C are suited to this purpose (e.g., C.I. Disperse Yellow 7, Orange 1, 33, Red 4, 11, 60, Blue 56, 81). The dyeing equipment must be as tightly closed as possible, and top heating in vats and jiggers is advantageous. Dyeing time is extended to several hours. Since, in a process of this type, a portion of the dye adheres to the fiber surface without being fixed, dyeing must be followed by a reductive clearing step.

**Dyeing from Foam.** For saving energy, attempts were made to do entirely without an aqueous dye liquor and to apply disperse dyes to PES articles from a foam medium. Problems can arise with regard to the levelness of dyeing (Aqualuft dyeing machines from the company Gaston County). When considering the energy advantages of the foam dyeing technique, it is important that although water is saved in the dyeing process, considerable amounts of water are required e.g., for pretreatment and aftercleaning. For continuous foam application, see Section 7.2.4.

**Air as Transport Medium.** A clear economic advantage is observed when air is used as the transport medium for the textile material (Then Air Jet, see Section 2.2.4.5). Here, efforts are made to prevent foam formation. The small amount of aqueous dye liquor present serves as the medium for dyes and auxiliaries.

**Dyeing in Alkaline Medium.** Oligomeric fiber components migrate out from the fiber during dyeing and can cause problems due to formation of agglomerates with dyes and deposition on the textile material or in the dyeing equipment. The cyclic trimer of ethylene terephthalate is especially harmful because of its low water solubility ($< 0.5$ mg/L at 100 °C). The oligomer problem is counteracted by dyeing in alkaline medium. In this process, oligomers leaving the fiber are hydrolyzed to water-soluble components before their deposition. To achieve this the pH is adjusted to ca. 10.5 with NaOH at the start of dyeing; it then decreases gradually to about 7.5. Alkalistable dyes must be used (e.g., Dianix SPH, UPH, Hoechst–Mitsubishi). Many azo blue products are unsuitable for this process. The reductive alkaline aftertreatment can be omitted.

**Dyeing from Organic Solvents.** Although dyeing from organic solvents in which disperse dyes are highly soluble has been studied carefully, this method has not achieved practical importance. Chlorinated hydrocarbons, in particular, have been recommended as a medium for continuous or batch dyeing of acetate and PES fibers. For continuous dyeing, dye application from either chlorinated hydrocarbons or an aqueous liquor with fixation in solvent vapor is possible. This method is not favored for ecological and toxicological reasons.

**Dyeing from Supercritical Carbon Dioxide.** According to studies performed in the beginning of the 1990s, dyeing from supercritical $CO_2$ is possible because of its adequate solubilizing power for disperse dyes [7.25], [7.26]. Temperatures around 130 °C and pressures $> 20$ MPa are used. Short dyeing times, good levelness, and low disposal costs are expected. The $CO_2$ is recovered. The equipment is naturally very complex.

**Dyeing via the Gas Phase.** Dyeing with dye vapor in the vacuum or in a gas stream as the medium, has also been investigated. Although advantageous from an energy standpoint, the problems encountered on a practical scale have not yet been solved [7.27].

### 7.2.4. Continuous and Semicontinuous Dyeing Processes

**Thermosol Process.** The most important continuous dyeing process for PES fibers is the thermosol process. It is applied primarily to PES–cellulose fiber blends (see Section 11.1).

The thermosol process consists of four individual steps:

1) Padding of the dye liquor on the fabric
2) Drying
3) Fixing of the dyes in the fiber
4) Aftertreatment

The four individual steps usually follow one another in one pass. Systems also exist that include also overdyeing of the cellulose component (e.g., by using the pad steam technique; see Section 2.3.2).

The quality of thermosol dyeing is assessed according to its levelness over large yardages. For good quality, uniform pretreatment of the material and a very uniformly operating thermosol system are required.

*Pretreatment* ensures, above all, that the material is free of burls and wrinkles, contains no dirt or preparation stains, and is uniformly wettable. It includes careful singeing, washing, and bleaching.

In *padding*, a long dipping section is used so that the liquor penetrates the material completely and is applied evenly over the width of the material. Wetting agents are avoided if possible because they foam, and strong foam formation interferes with the process. Silicone-based foam depressants can negatively affect the stability of the dispersion. The pH of the pad liquor should not exceed 6–7 because disperse dyes can be sensitive to alkali. A padding auxiliary, e.g., a polyacrylate or an alginate thickener (→ Textile Auxiliaries, Section 5.9), is used to prevent migration of dye liquor on the fiber surface during drying. Migration can cause two-sidedness of the material, as well as darker- or lighter-colored fiber ends (frosting). Furthermore, dye migration is counteracted by low liquor application (concentrated pad liquor, high squeezing pressure in the padder), low drying temperature, and low air circulation. *Drying* is carried out contact free up to a residual moisture of 20–30 % and then on any equipment available.

The *fixing time* and temperature depend on the aggregate, the dyes used, and the depth of shade (see Table 7.2 for guide values). For equipment requirements, see Section 2.3.1.

In one formulation, dyes with similar rates of diffusion should be used as much as possible. Very quickly diffusing dyes sublime easily, contaminate the fixing aggregate, and produce dyeings that have a low fastness to thermofixation. The color yield of very slowly diffusing dyes is sensitive to temperature variations.

After the dyes are fixed, dyes and auxiliaries that adhere superficially must be washed out, if

**Table 7.2.** Fixing times and temperatures in different dyeing aggregates

| Dyeing aggregate | Temperature*, °C | | Time, s |
|---|---|---|---|
| | Dyes with average rate of diffusion | Dyes with low rate of diffusion | |
| Hot flue, frame | 200–215 | 215–225 | 60–30 |
| Suction drum | 200–215 | 215–225 | 45–20 |
| Cylinder fixing | 215 | 220–225 | 30–15 |

* Lower dyeing temperature corresponds to longer dyeing time.

necessary by an alkaline reductive treatment. In the case of PES–cellulose fiber blends, this washing can be combined conveniently with afterdyeing of the cellulose component, e.g., simply by overdyeing with vat dyes (see Section 11.1). For exact instructions for use of the thermosol process, see [7.21, pp. 122–131].

**Pad Roll Process** (Section 2.3.2). Disperse dyes are fixed only incompletely to PES fibers under the conditions of the pad roll process (padding, retention in saturated vapor atmosphere). Only light colors can be produced economically with adequate fastness. Carrier additions are not useful because very large amounts would be required. In the pad roll process, rapidly diffusing dyes must be used at any rate.

**Continuous Dye Application in Foam.** Dyeing from foam (Section 7.2.3) has also been studied for continuous dyeing [7.28]. Since a slightly inhomogeneous application of the foam becomes visible immediately in a color difference, the consistency, pore size, and disintegration time of the foam must be kept as constant as possible and the articles must be carefully precleaned [7.29]. Equipment for foam formation and for continuous application has been developed. The continuous foam dyeing technique offers advantages for dyeing floor coverings by greatly reducing liquor application. However, it is not an attractive method for dyeing PES piece goods.

**Continuous Dye Fixation with Microwaves.** The moist fabric padded with dye liquor is irradiated with microwaves in the presence of vapor. The addition of urea or carrier is recommended for rapid diffusion of the dyes [7.30]–[7.32].

### 7.2.5. Dyeing of Microfibers

Fibers with a fineness of less than 1 dtex are called microfibers. Articles made of microfibers are characterized by a low weight, soft feel, weather resistance, and good breathing activity. The large fiber surface leads to rapid dye absorption and, thus, to the risk of unevenness. For physical reasons, the colors also appear to be lighter than on normal fibers. Therefore, greater amounts of dye are required to achieve a required shade. The large fiber surface and the great amounts of dye used also lead to lower light- and wetfastness. Preliminary tests are useful [7.33], [7.34].

PES fibers that have been subjected to a surface peeling process with alkali behave similarly to microfibers [7.35].

### 7.2.6. Dyeing of Modified Polyester Fibers

Since disperse dyes diffuse very slowly into PES fibers, efforts are made to increase the rate of dye strike by chemical or physical alteration of the fiber. The fiber is also modified to reduce the pilling tendency, to increase shrinkage and elasticity, and to reduce flammability ($\rightarrow$ Fibers, 4. Synthetic Organic, **A 10**, p. 587). Such modified fibers exhibit improved dye receptivity. Fibers with improved dyeability can be dyed with disperse dyes at boiling temperature without a carrier or with basic dyes when they are modified with acidic components (5-sulfoisophthalic acid). Fibers of this type are used if dyeing cannot be carried out easily above 100 °C (e.g., in the case of floor coverings, articles made of PES–wool blends, stretch materials, and cord). Strongly crimped PES bicomponent fibers are produced for special purposes. These fibers are normally also dyeable at the boil and without a carrier [7.36]–[7.40].

Modified PES fibers are usually more sensitive to hydrolysis than normal fibers. This must be taken into account in finishing. The lightfastness of the dyeings is often lower than on normal fibers. Thus, dyes for coloring carpeting, upholstery, and drapery must be carefully selected. When dyeing, one should take into account the fact that dyes start exhausting at low temperatures (ca. 60 °C) and the dyebath is exhausted after a short time, so problems with levelness may arise.

*Anionically modified PES fibers* not only exhibit affinity for cationic dyes but also are more easily dyeable with disperse dyes than normal fibers. They are dyed with basic dyes up to 115 °C in the presence of Glauber's salt to protect the fibers from hydrolysis. If the dyeing aggregates available do not allow dyeing temperatures of 110 °C, a carrier is employed to give deep shades on anionically modified fibers.

The joint use of anionically modified and normal PES types can be exploited for differential dye effects (joint application of different classes of dyes that differ in their affinity to the fiber components).

### 7.3. Aftertreatment

In the case of light and medium PES dyeings, as in the coloring of other fibers, the material need only be rinsed thoroughly or soaped after dyeing. In the case of dark shades, nonfixed dye components are held firmly on the fiber surface and decrease fastness. Oligoesters (oligomers) leaving the fiber can also interfere with spinning or winding because they deposit on yarn guiding devices. Therefore, loose material and yarn, in particular, must frequently be subjected to reductive alkaline clearing after dyeing. For this purpose, "hydrosulfite" (sodium dithionite) is used in the presence of an emulsifying washing agent.

In the case of loose stock, reductive aftertreatment also improves spinnability because the fiber becomes smoother. Good spinnability is especially pronounced when organic reducing agents are used for aftertreatment (Cyclanon, BASF, $\rightarrow$ Textile Auxiliaries, Section 5.10).

To improve the lightfastness of dyes used to color car fittings, benzophenone or triazole compounds (e.g., Cibafast, Ciba) are often employed in the dyebath or as aftertreatment [7.41].

For partial and complete stripping of faulty dyeings, $\rightarrow$ Textile Auxiliaries, Section 5.13.

## 8. Dyeing of Cellulose Acetate

For properties of cellulose acetate fibers, see $\rightarrow$ Fibers, 1. Survey, **A 10**, pp. 464–472; for chemistry and production, see $\rightarrow$ Cellulose Esters, **A 5**, pp. 438–444, 447–448. Cellulose 2.5-acetate (CA) and cellulose triacetate (CT) are preferentially processed as filaments. Because of their silklike feel, dull gloss, and pleasant wearing quality, they are especially popular for dress and blouse materials, scarves, and linings. In the textile sector, the importance of cellulose acetate is decreasing: world consumption in 1970 was 354 000 t, in 1980 338 000 t, and in 1992 232 000 t. Cellulose acetates are being increasingly replaced by synthetic polyamide and polyester fibers. In contrast to the other regenerated cellulose fibers, CA and CT are hydrophobic. Therefore, they have little in common with viscose or cuprammonium rayon from the viewpoint of dyeing. As hydrophobic fibers, they can be dyed with disperse dyes (→ Disperse Dyes). The dyeability of CA and CT with disperse dyes is very similar to that of PES and other synthetic fibers. The rules described for dyeing PES fibers with disperse dyes (see Chap. 7) basically apply to CA and CT as well. Tests have been drawn up by the Society of Dyers and Colourists for the selection of suitable dyes [8.1]. Trade names of disperse dyes available for CA and CT are, e.g., Artisil (Sandoz), Celliton (BASF), Cibacet (Ciba – Geigy), Dispersol (ICI), Dianix (Mitsubishi – Hoechst), and Serisol (Yorkshire).

In addition, the ranges of disperse dyes intended for PES fibers contain numerous products that are very well suited to CT in particular.

The choice of dye depends, on the one hand, on the desired use and production fastness and, on the other hand, on the suitability for certain dyeing techniques. Dyes that produce especially washfast dyeings require higher dyeing temperatures for level absorption. An important criterion for the selection of dyes for CA can be fastness to exhaust gases (see Section 8.3). Articles made of CT are subjected to thermal treatment in the course of finishing. The fastness to thermofixing or pleating of the disperse dyes recommended for CT is usually at least as good as on PES fibers.

## 8.1. Dyeing Processes for Cellulose Acetate [8.2], [8.3]

Theoretical aspects of dyeing CA and CT are described in [8.4, pp. 74 – 92]. Cellulose acetate is dyed by the exhaustion method with disperse dyes in the presence of a nonionic or anionic dispersing and leveling agent in a weakly acidic bath (pH 5 – 6). A series of less wetfast disperse dyes are taken up by acetate at temperatures as low as 50 – 60 °C. For this reason, the process must be started at a correspondingly low temperature to obtain level dyeings. Dyeing is normally done at 80 – 85 °C. Dyes that are fast to wet treatments require temperatures up to 90 °C. If CA is dyed at boiling temperature, the fibers may become delustered.

Piece goods made of CA are usually dyed in a jigger, less often in a star frame or winch beck with gentle material guidance. The piece *beam* is suitable for warp materials. Yarn was previously dyed in hanks. Since suitable winding machines are available, CA can also be dyed on cylindrical cheeses. Woven CA fabrics are occasionally dyed semicontinuously in a pad roll system (Artos) at 80 – 90 °C (see Section 2.2.3). Dyeing auxiliaries are subsequently washed out.

A dyeing process with developing dyes, which was used in the past for various shades, is still being employed for black. Although this process produces dyeings that are especially fast to wet treatments, it is more tedious than dyeing with direct disperse dyes. The process starts with the application of an amino-group-containing diazotizable disperse dye [e.g., Cellitazol (BASF), Cibacet-Diazo (Ciba – Geigy)], which is diazotized on the fiber and then coupled with a naphthol [e.g., 2-hydroxynaphthoic acid (Developer ON)]. Thus, a large-molecular, insoluble, wetfast dye is formed in the fiber.

## 8.2. Dyeing Processes for Cellulose Triacetate [8.3]

In terms of dyeing and finishing, CT is more similar to purely synthetic fibers than CA. It can be permanently pleated. For stress relaxation, articles made of CT, like those made of PES, are heat set (thermofixed) after dyeing [8.4, pp. 92 – 100]. Triacetate, like PES, can be dyed according to the thermosol process (see Section 7.1.2).

The absorption of disperse dyes from a long liquor by CT fibers is slower than CA and somewhat faster than PES fibers. With a few disperse dyes, acceptable dyeings are obtained on CT even at boiling temperature. As a rule, however, dyeing is conducted at 120 °C, especially to obtain dyeings with adequate fastness properties. If this is not possible, a dyeing accelerant (based on butyl benzoate or butyl salicylate) is used. As in the case of CA, CT is dyed in a weakly acidic liquor (pH 5 – 6) in the presence of a leveling dispersing agent. The lower the dyeing temperature, the more must leveling be supported with a suitable auxiliary agent.

Like CA, CT can also be dyed with developing dyes. Since CT is less easily accessible to ionic products than CA, the dyeing process is altered here. Dyeing is begun with the diazotizable azo dye, the developer is then applied at a higher temperature, and diazotization is performed in the cold. The diazonium salt formed couples immediately with the developer present in the fiber.

Dyeing equipment suitable for CA is also suited to dyeing CT. In accordance with the higher dyeing temperature used for CT, pressure dyeing aggregates are preferred (e.g., piece beam dyeing machines, jet dyeing machines, or HT winch becks; see Sections 2.2–2.3).

CT can be textured like PES. Textured yarns made of both fibers behave similarly in textile finishing.

For information on the dyeing of cellulose acetate and triacetate in mixtures with other fibers, see [8.3].

## 8.3. Aftertreatment, S-Finish

CA and CT absorb nitrogen oxides and other acidic, oxidizing exhaust and industrial waste gases from the air. These gases can destroy sensitive dyes in the fiber so that the shade changes. Products that are not resistant to gas fading are found especially among the red and blue anthraquinone dyes ($\rightarrow$ Anthraquinone Dyes and Intermediates, **A 2**, p. 401). If dyeing cannot be done with dyes that are fast to gas fading, the fiber can be protected to a certain extent by aftertreatment with cationic products that undergo saltlike bonding with the mostly acidic gases.

Articles made of CT are subjected to heat setting (thermofixing) after dyeing to make them dimensionally stable.

In the case of CT, effects similar to those produced by thermofixing are obtained by alkali treatment at higher temperature (S-finish). The S-finish results in superficial saponification of the fiber, making the fiber surface hydrophilic, which in turn corresponds to a permanent antistatic finish. The resistance to gas fading of dyeings is also improved. CT with an S-finish is dyeable with the usual direct dyes, the depth of shade depending on the degree of S-finishing.

# 9. Dyeing of Acrylic (Polyacrylonitrile and Modacrylic) Fibers

## 9.1. Economic Aspects

In 1991, worldwide production of acrylic fibers amounted to $2.31 \times 10^6$ t. Thus, these represent the third most important synthetic fibers after PES and PA. About 702 000 t were produced in Western Europe, 350 000 t in Japan, and 206 000 t in the United States. Fiber production decreased from 1988 to 1991, with an average annual rate of $-2.3\%$ [9.1].

The main advantages of acrylic fibers are their similarity to wool in feel, heat retention, and processing, as well as good dyeability. In Western Europe and the United States, ca. 60 % are used for clothes and 30 % for household textiles [9.1]. The proportion used for clothes worldwide is considerably larger.

Dyeing is performed mainly on yarns by exhaustion processes (batchwise). However, the tendency toward gel dyeing is increasing. The proportions of different dye application processes in 1989 and of various textile makeups are given below (in percent):

Proportion of application processes
| | |
|---|---|
| Exhaustion process | 72 |
| Gel dyeing | 13 (increasing) |
| Continuous dyeing | 9 (decreasing) |
| Printing | 4 |
| Mass dyeing | 2 (decreasing) |

Proportion of textile makeup for coloration
| | |
|---|---|
| Yarn | 50 |
| Stock, cable, top | 22 |
| Gel | 13 |
| Piece (dyeing) | 9 |
| Printing | 4 |
| Mass | 2 |

Total dye consumption is estimated at ca. 24 000 t/a.

## 9.2. Principles

For methods of production and properties of acrylic fibers see, $\rightarrow$ Fibers, 4. Synthetic Organic, **A 10**, pp. 629–642. The polymerizates used initially, which were made of 100 % acrylonitrile without suitable comonomers, gave fibers with a high glass transition temperature $T_g$ ($\rightarrow$ Fibers, 4. Synthetic Organic, **A 10**, p. 629) and with insufficient dye-binding possibilities. Fibers of this type could be dyed only in light shades with *disperse dyes* or with *acid dyes* according to the complicated copper ion process [9.2]. Today, anionic comonomers are used, and anionic groups are incorporated into the chain ends of the fiber molecules. In this way, $T_g$ is lowered and anionic groups are available that can act as dye sites for cationic dyes. Thus, acrylic fibers are reliably and economically dyeable.

### 9.2.1. Disperse Dyes

Disperse dyes ($\rightarrow$ Disperse Dyes) can be used to produce light to medium deep shades. The dyeing mechanism and process correspond to those used on polyester fibers (see Chap. 7). However, dyeing can be performed at $< 100\,°C$. Addition of carriers is not required. The good migration properties of disperse dyes result in problem-free level dyeing.

### 9.2.2. Cationic Dyes

Cationic dyes ($\rightarrow$ Cationic Dyes) are used predominantly for acrylic fibers [9.3] because virtually any depth of color can be obtained and they are dyeable with high brilliancy and good fastness. In choosing the dyeing conditions, the dyeing mechanism and relevant properties of the dyes and fibers should be taken into account. This is relatively easy because despite the great variety of fibers and dyes, uniform and—in comparison with other dyeing processes—simple laws are valid for dyeing with cationic dyes. A detailed description is presented in [9.4], and a discussion with a literature survey of the more recent developments in certain areas (e.g., pore and surface structure, fiber modifications, thermal yellowing, effect of carriers and solvents during dyeing, and transfer print) in [9.5]. The physical chemistry of dyeing in general is presented in [9.6], [9.7].

The dyeing process with cationic dyes can be described by the following model. The dye rapidly absorbs to the surface of the fiber from the aqueous solution in the dyebath until saturation occurs. Above $T_g$, dye diffuses into the fiber. This diffusion is usually rate determining for the entire process. Because of their considerably higher affinity, dye cations displace an equivalent amount of cations ($H^+$, $Na^+$) that were previously present at the anionic groups in the fiber

(ion exchange). This process determines the dye distribution in the dyeing equilibrum [9.8]–[9.11] and can be described by the Donnan theory [9.12], [9.13].

In acrylic dyeing, the dyeing (or absorption) rate $v$ is especially important. Because of the high affinity of cationic dye for the fiber, their migration capacity is poor. Thus, dyeing must be sufficiently level from the start; this requires the control of $v$, which can be reliably achieved only if the effects of relevant parameters are known. Since $v$ is determined by the diffusion of dye in the fiber, the following relation is valid according to the laws of diffusion [9.14] (see Section 2.1.3)

$$v \sim \Delta C \cdot D \tag{9.1}$$

where $\Delta C$ is the concentration gradient and $D$ is the average, apparent diffusion coefficient of dye cations in the fiber. The parameter $\Delta C$ depends on the dye concentration on the fiber surface $C_s$. The amount of dye $C_F(t)$ that has diffused into the fiber after time $t$ follows a $\sqrt{t}$ law [9.15]:

$$C_F(t) \sim C_s \sqrt{D t} \tag{9.2}$$

Except when only a small number of dye cations are available (for production of very light dyeings), the fiber surface is almost saturated during dyeing, i.e., $C_s$ is approximately constant until most of the dye is exhausted. At constant dyeing temperature, $D$ can also be regarded as approximately constant. Thus

$$C_F(t) \sim \sqrt{t} \tag{9.2a}$$

To a good approximation, this is valid until ca. 70 % of the dye supplied is absorbed (bath exhaustion BE = 70 %). Thus, some rules exist concerning the dependence of dyeing time and rate of dyeing on the concentration $p$ of supplied dye [9.16]. For a given temperature or temperature program:

1) The dyeing time required increases with $p^2$ ($p$ is normally given in percent based on fiber mass).
2) The relative (based on $p$) rate of dyeing $v_{rel} = dp/dt$ (percentage of dye per minute) is inversely proportional to $p^2$.
3) The absolute rate of dyeing $v_{abs} = dC_F/dt$ (milligrams per minute) is independent of $p$.

Thus, the lighter the color, the faster is the absorption and the greater the danger of non-levelness if absorption is not controlled by a suitable temperature program or retarders. Furthermore, Equation (9.2) denotes the following. Uneven surface charging with dye cations leads to differences in $C_F$ (i.e., nonlevelness). Surface charging occurs rapidly and can hardly be influenced by temperature guidance. In the case of light colors, when the number of dye cations provided is insufficient to saturate the surface, retarders ($\rightarrow$ Textile Auxiliaries, Section 5.6.5) must be applied to achieve this purpose; only then can uniform surface charging be expected.

The *temperature dependence* of $v$ is especially pronounced in acrylic dyeing [9.8], [9.16], [9.17]. Depending on the type of fiber, $T_g$ is in the range of 70–80 °C; above $T_g$ is the absorption range. In the absorption range, $D$ and, according to Equation (9.2), $v$ increase steeply with $T$ in accordance with the Williams–Landel–Ferry (WLF) equation derived from the theory of chain segment motion [9.17]–[9.19]. The Arrhenius equation is also valid to a usable approximation and is more practical because it contains only one constant, the activation energy $E$ [9.6]. The values of $E$ for $v$ are 250–300 kJ/mol [9.8], [9.16], the lower values being for wet spun fibers. This corresponds to a doubling of $v$ for a temperature increase of only 2.5–3 °C. (In the case of disperse dyes on PES or acid dyes on PA a doubling of $v$ is observed at a temperature increase of ca. 5 and 10 °C, respectively.) These values are practically independent of the choice of dye, depth of color, and mode of dyeing [9.16]. The steep increase of $D$ and $v$ with temperature is predominantly caused by changes in the fiber.

Apart from $T$ and the amount of dye applied (in the case of $v_{rel}$), $v$ also depends on the properties of individual dyes and the type of fiber. To take this into account, in practice, especially in systematic dyeing processes, these effects are separated and described in terms of dye- and fiber-specific characteristic values. For instance, the *dyeing times of dyes under standard conditions* and the *rate of dyeing* $V$ of the fiber have been introduced for this purpose [9.16].

The degree of occupation by dye cations of anionic dye sites present in the fiber is decisive for the dyeing equilibrium, in practice the final exhaustion of the dyebath. At constant temperature (maximum dyeing temperature), dye distribution follows a Langmuir isotherm [9.20], [9.21].

$$C_F = \frac{s \cdot C_L}{s/a + C_L} \qquad (9.3)$$

where $C_L$ is the dye concentration in the dyebath, $s$ is a saturation concentration, and $a$ is essentially the affinity of the dye. The value $s$ denotes the maximal possible dye uptake by the fiber for very high $p$. In this case, all dye sites are occupied, the number of dye cations $n_{cat}$ on the fiber equals the number of anionic dye sites of the fiber $n_s$. [Virtually all of the anionic dye sites of the fiber are accessible to dye cations under dyeing conditions (i.e., above $T_g$)]. If $n_{cat} < n_s$, the supplied dye is exhausted almost completely because of the normally high affinity; if $n_{cat} > n_s$, the excess remains in the bath. This relationship is stoichiometrically valid and usually independent of dyeing conditions. Thus, the position of the dyeing equilibrium can be calculated for each individual dyeing process. For this purpose, fiber- and dye-specific values derived from molar quantities are used. The following have proved useful [9.16], [9.22]:

1) The *fiber saturation value* $S_F$, a measure of the concentration $N$ (in millimoles per kilogram of fiber) of anionic dye sites in the fiber:

$$S_F = N/25 \qquad (9.4)$$

2) The *saturation factor* $f$ of a cationic dye having a molecular mass $M_r$ and a net dye content $e$ (grams of pure dye per gram of commercial dye):

$$f = 400 \, e/M_r \qquad (9.5)$$

The *saturation concentration* $p_s$ (percent dye, based on the mass of textile material) of an individual dye (with $n_{cat} = n_s$) is

$$p_s = S_F/f \qquad (9.6)$$

A combination of several dyes (1, 2, 3,...) exhausts almost completely if

$$(p \cdot f)_1 + (p \cdot f)_2 + (p \cdot f)_3 + \ldots < S_F \qquad (9.7)$$

In the case of PAC types with strongly anionic (acidic) dye sites, $S_F$ increases slightly with increasing pH (by ca. 10 % from pH 4 to 6.5). In the case of PAC types with weakly anionic dye sites (carboxyl groups), the increase is considerably greater [9.21].

Another important quantity is the *relative saturation* of the fiber in the dyeing equilibrium [9.16]

$$S_{rel} = \Sigma(p \cdot f)/S_F \qquad (9.8)$$

The higher the $S_{rel}$ (other parameters being equal), the better is the levelness of dyeing during absorption. If $S_{rel}$ is about equal in different dyeings and certain other conditions are met (compatibility, use of retarder, proper temperature guidance), a similar course of the dyeing process with regard to the rate of dyeing, bath exhaustion, and levelness can be expected.

**Retarders and Auxiliaries.**

*Cationic Retarders.* With light and moderately deep shades, the amount of dye applied is often not enough for an adequate $S_{rel}$. Sometimes even saturating the fiber surface is not sufficient. To meet the danger of nonlevelness in such cases, cationic retarders are used ($\rightarrow$ Textile Auxiliaries, Section 5.6.5) [9.23], [9.24]. These are colorless cationic compounds that absorb like dyes and compete with dyes for acidic groups in the fiber and on the fiber surface. They are usually quaternary amines with an aromatic substituent and an aliphatic chain. The longer this chain, the greater is the affinity for the fiber [9.25], [9.26].

These compounds are used mainly to *increase* $S_{rel}$, thereby improving levelness. The increase can be calculated in accordance with Equations (9.5)–(9.8), the retarder being assigned a saturation factor according to Equation (9.5) and treated formally like a dye [9.16]. This calculation is important to guarantee the desired effect and prevent $S_{rel}$ from becoming too large (close to 1 or higher). Dye absorption would then be incomplete (blocking of anionic groups).

Depending on the affinity and amount applied, cationic retarders partially displace the dye from the fiber surface. $C_s$ (dye) and the concentration gradient of the dye become smaller and remain constant longer in the course of dyeing [9.27]. For this reason, at constant temperature $T$, the *rate of dyeing is lowered* by the retarders. However, this effect does not justify the use of retarders because it would be obtainable at a lower cost by dyeing at lower temperature.

*Cationic Softeners.* Depending on their affinity and penetrating ability, some cationic softeners—when used in the dyebath—act like cationic retarders and contribute to $S_{rel}$. Thus, less cationic retarder is required. However, their softening effect is lower than when applied in aftertreatment at temperatures below $T_g$.

*Migration Aids.* Colorless cationic auxiliary agents having a considerably lower affinity for the fiber than dyes are available as migrators (migration aids) ($\rightarrow$ Textile Auxiliaries, Section 5.6.5). They can be used in larger molar amounts than cationic retarders without danger of blocking. They influence $S_{rel}$ and levelness during absorption only slightly, but they promote migration and help compensate for unevenness in the fiber [9.28].

*Electrolytes.* The addition of electrolytes has a similar effect [9.24]. In accordance with their much lower affinity compared to migration aids, this effect is weaker, based on the number of cations used. However, since this number is generally much higher than in the case of migrators comparable effects are obtained.

*Polycationic retarders* [i.e., compounds (usually polymeric) with numerous cationic groups] have been recommended as leveling agents [9.29]. They occupy the fiber surface without penetrating into the fiber and counteract dye adsorption, thereby reducing the rate of dyeing [9.30], [9.31]. Depending on their adsorption characteristics, they also decrease the migration capacity and the bath exhaustion in the dyeing equilibrium. However, a favorable effect on leveling has not been found. (The rate of dyeing can be influenced better through a suitable temperature program.)

*Anionic Retarders.* In some cases, anionic auxiliaries are also used as retarders ($\rightarrow$ Textile Auxiliaries, Section 5.6.5) [9.32]. These contain two or more anionic groups (sulfonic acid groups) per molecule, forming a 1:1 addition compound with the cationic dye; that is soluble in the dye liquor. To ensure this solubility, at least two anionic groups must be available from the start for each dye cation. The minimum amount of anionic retarder required can be calculated via $f$ values according to Equation (9.5) [9.4]. In the course of dyeing a growing excess of auxiliary anions is formed, and the concentration of free dye cations in the liquor falls below that required for saturation of the fiber surface. Dyeing rate and bath exhaustion at dyeing equilibrium are decreased.

Anionic retarders promote the migration of cationic dyes. They reduce the danger of precipitations when anionic dyes are used in addition to cationic dyes (for fiber blends). They influence the compatibility of dyes (see below). The higher the affinity of the cationic dyes, the greater is their effect. The main disadvantages are the lower bath exhaustion and poorer reproducibility.

**Compatibility of Cationic Dyes.** Cationic dyes influence each other during absorption [9.9],

[9.11]. If several dyes exhaust tone-in-tone from one bath (i.e., the depth of color, but not the shade, changes in the course of absorption), the dyes are referred to as compatible. In the case of cationic dyes on PAC, the property of being compatible is dye specific and also applies to dyeing under other conditions and in other shades. It depends on the product $D \cdot A$ ($A$ = dye affinity) and can be described by the compatibility value $K$ (see Chap. 12) [9.16], [9.33], [9.34].

In acrylic dyeing, combinations of dyes with the same $K$ exhibit considerable advantages in level dyeing. In combinations with different $K$, dyes with a lower $K$ absorb before those with a higher $K$. Cationic retarders can also be assigned $K$ values [9.4], [9.35]. The $K$ values change with the use of anionic retarders.

## 9.3. Exhaustion Process

Normal PAC textile material is dyed as cable or stock in package dyeing, as yarn in hanks or packages (cross wound, cylindrical, or conical) (320–350 g/L, direction of liquor inside–outside), as pieces on beam, overflow, paddel (knitwear), or drums (socks) (see Section 2.2). High-bulk (HB) yarns are applied as muffs (cylindrical bobbins without tubes, without intermediate plates) [9.36]. Strongly soiled textile material is prewashed, preferably in a weakly alkaline wash.

The principles of recipe and process operation are described in [9.27]. Because of the poor migration properties, dye absorption must be as level as possible from the start. Correspondingly, attention should be paid to uniform packaging, package density, flow or movement of textile material, and controlled machine conditions.

As far as possible, dyes with the same $K$ are selected (the largest choice of dyes and most suitable retarders are available at $K = 3$). In cases where sufficient levelness during absorption is critical, dyes with a relatively low affinity (with high $K$) are recommended. These dyes migrate somewhat better than normal dyes [9.37]. The dyeing formula is drawn up with the assistance of a color measuring system (see Chap. 12) or according to available documentation, and the amount of retarder is adjusted to give a favorable $S_{rel}$. The dyebath is charged with the dissolved dye and all the other additives (see Table 9.1) at 60–70 °C, and the textile material is added. After a mixing time of ca. 5 min, dyeing is conducted

**Table 9.1.** Additives to the dyeing bath, general formulation

| Additive | Light hues | Medium hues | Deep hues |
|---|---|---|---|
| Nonionic dispersant, g/L | 0.2–0.5 | 0.5 | 0.5 |
| Acetic acid, final pH | 3.6 | 3.6 | 4.5 |
| Sodium acetate anhydrous, g/L |  |  | 0.5 |
| Sodium sulfate anhydrous, g/L | 5 | 5 |  |
| Cationic retarder*, % | 1.5–3 | 0.3–1.5 |  |

* Retarder with $K \approx 3$, $f \approx 0.5$, in percent based on mass of material; higher amounts are required for textiles with higher $S_F$, at critical requirements with respect to levelness, dye combintions with different $K$ values or $K \leq 2$, and for retarders with $K > 3$ (see Section 9.2.2).

**Table 9.2.** Temperature range of controlled heating (general method) ($T_{start} - T_{end}$) and residence time ($t_{max}$) at maximum temperature $T_{max}$

| | $S_{rel}$ | | |
|---|---|---|---|
| | Low | Medium | High |
| $T_{start}$, °C | 75 | 80 | 85 |
| $T_{end}$, °C | 95 | 98–100 | 98–100 |
| $t_{max}$, min | | | |
| (at $T_{max}$ = 98 °C) | 20 | 40 | 60 |
| (at $T_{max}$ = 102 °C) | 15 | 30 | 40 |

by controlling the temperature, as shown in Tables 9.2 and 9.3. This is followed by cooling (ca. 0.3 °C/min to 60 °C), rinsing, and aftertreatment.

The details given in Tables 9.1–9.3 take into account in a general way (same conditions for larger groups of dyeings) the principles explained in Section 9.2.2—especially the need to achieve sufficiently level dyeing from the start and to

**Table 9.3.** Heating rate (general method)

| Temperature range* | Heating rate, °C/min | | |
|---|---|---|---|
| | Stock, top | Yarn (pack.) | Hank, piece |
| < $T_{start}$ (until absorption of dye starts) | quick | quick | quick |
| $T_{start} - T_{end}$ (usually linear) | 0.5 | 0.3–0.4 | 0.25 |
| $T_{end} - T_{max}$ (until absorption of dye is complete) | quick | quick | quick |

* Numerical data for the temperature can be obtained from systematic optimization processes.

have the shortest possible dyeing time. The liquor movement or circulating speed of the material can be taken into account [9.38]–[9.40]. As a rule, satisfactory results are obtained in this manner.

Dyeing can be performed more economically and reliably according to systematic methods for the optimization of formulations and dyeing conditions [9.16], [9.27], [9.41]. Here, dyes with the same $K$ are selected and the following are determined for each dyeing:

1)  The amount of retarder that leads to the most favorable relative saturation $S_{rel}$
2)  The temperatures at which exhaustion starts or ends (the critical temperature range outside which rapid heating will not cause unlevelness, see Section 2.2.1)
3)  The most favorable rate of heating in this temperature range
4)  The dwelling time at the maximum (final) dyeing temperature required for penetration

In following this method, the rules mentioned in Section 9.2.2, the properties (characteristic values) of the fibers and the dyes, and the conditions of the dyeing machine and the article are systematically taken into account. This optimization can be performed with the help of tables [9.41] or with a computer [9.42].

Substantial advantages are obtained by:

1)  Appropriate amounts of auxiliary agents applied
2)  Shorter dyeing times
3)  More protection against nonlevelness
4)  Better programmability
5)  Better reproducibility
6)  Saving of laboratory work
7)  Help in searching for errors

## 9.4. Dyeing of Special Fiber Types According to the Exhaustion Process

**High-Bulk Material.** The bulking of high-bulk material proceeds either as a pretreatment (continuous for yarn, batchwise and usually with steam for yarn and knitwear) or in the dyebath before dye absorption at ca. 80 °C for 5–10 min without liquor movement [9.36]. To avoid nonlevelness in this process, retarders are employed (high $S_{rel}$) to ensure that very little dye absorbs during the bulking time. The formulas given in Section 9.2.2 or systematic dyeing processes

(Section 9.3) are used. With HB bicomponent fibers, a high $S_{rel}$ can interfere with the reversible shrinkage. For this reason, anionic retarders and dyes with $K \geq 4$ are preferred [9.37].

**Pore Fibers.** For the surface and pore structure of PAC fibers, see [9.5]. As a result of their greater light scattering, pore fibers (type Dunova, light fibers with high water absorption capacity) require up to three times more dye than normal fibers to achieve a given shade [9.43]. This relation depends on the angle of observation. Navy and black shades are hardly dyeable; for dyeing other deep hues, an ammoniacal reductive aftercleaning step is required. The diffusion of dyes proceeds mainly in the pores of the fiber [9.44], [9.45], changing the $K$ values of the dyes. With the exception of a higher softener requirement, the rules of normal PAC dyeing apply.

**Microfibers.** PAC microfibers (single titer below 1 dtex) behave differently in some respects from normal acrylic fibers [9.46]. First, more dye is required for the desired shade in each case because of the greater light scattering of the fiber. This requirement increases more than two-fold with increasing depth of color (similar to pore fibers). For this reason, the fiber saturation $S_F$ is usually adjusted to a higher value by the fiber producer. The characteristic dyeing rate $V$ of the fiber is also higher. Corresponding to these increased values, more cationic retarder is required and the starting temperatures are lower. Above $S_{rel} \approx 0.75$, the rub- and wetfastness can be lower; reductive aftercleaning is sometimes necessary. Some dyes exhibit increased sensitivity to boiling down (lower dyebath stability) when the pH clearly deviates from 4.5. For this reason, dye producers recommend a special selection of dyes for microfibers. In the dyeing of stock, top, or yarn, one should take into account the fact that the liquor flow is lower and its uniformness is more problematic than in the case of normal acrylic fibers. For this reason, packaging and winding must be especially even and the rate of heating should be slower if possible. In piece dyeing, longitudinal tension should be kept as low as possible and creasing must be avoided. Cooling to < 60 °C should not proceed more than twice as fast as heating.

**Modacrylic Fibers** (→ Fibers, 4. Synthetic Organic, **A 10**, p. 639). Modacrylic fibers containing more than 20 % of vinyl chloride or vinylidene chloride as comonomer are used because of their reduced flammability. Depending on the comonomers, various types of modacrylic fibers exist that differ especially in the range of dyeing temperatures. In all cases, the critical temperature range is greater (25–50 °C) than with normal fibers (15 °C), and absorption starts in most cases at lower temperature. Modacrylic fibers generally exhibit increased plasticity and a strong tendency to shrink at higher dyeing temperature. The fiber types and their dyeing properties change frequently. The lightfastness on these fibers is considerably poorer than on normal PAC fiber, making a corresponding dye selection necessary [9.47], [9.48]. Dyeing is performed at pH 4.5 with acetic acid, 0.5 wt% of sodium acetate, and 1–2 wt% of nonionogenic dispersing agent. Depending on the type of fiber, carriers or cationic retarders and different temperatures are recommended. Examples of fiber types are given in Table 9.4. The relustering required in the past after dyeing is usually no longer necessary.

Modacrylic fibers can also be dyed with *disperse dyes* (similar to normal PAC fibers, but the fastness to sublimation is poor).

## 9.5. Continuous Processes

Stock, cable, and top can be dyed on special machines according to the *pad steam process* (see Section 2.3.2) [9.49]–[9.51], preferably with pressurized steam of clearly more than 100 °C to

**Table 9.4.** Dyeing properties and recommended range of dyeing temperature for different types of modacrylic fiber

| Type of fiber (producer) | $S_F^a$ | $V^b$ | Auxiliary | Temperature range, °C |
|---|---|---|---|---|
| Dynel[c] (Union Carbide) | 0.9 | 1.0 | carrier | 40–98 |
| Kanekalon S (Kanegafuchi) | 2.2 | 1.2 | migrator carrier (deep hues) | 60–98 |
| Teklan (Courtaulds) | 4.3 | 2.4 | cationic retarder | 60–98 |
| Velicren FRS (Snia) | 4.1 | 2.5 | cationic retarder | 80–98 |
| Verel (Eastman) | 0.8 | 1.0 | carrier | 30–70 |

[a] $S_F$ = Fiber saturation value (see Section 9.2.2).
[b] $V$ = Characteristic value for the rate of dyeing of a fiber (see Section 9.2.2). [c] Several types with partially different properties.

obtain short fixing times. To avoid drying and condensed water spots, the steam should not be either superheated or wet. A typical pad liquor contains acetic acid (pH 4.5) and dye solvent in addition to especially steam-resistant, readily water-soluble cationic dyes (usually liquid brands) [9.50], [9.52]. *Space dyeing* can be regarded as a variant of this mode of dyeing. A new suggestion involves simultaneous bulking and dyeing in a continuous process [9.53].

The *pad steam process with fixing by saturated steam* at 100–102 °C is used for piece goods, especially upholstery material (velour). This requires longer fixing times (10–20 min). Rapidly diffusing cationic dyes and dye solvents, which also exhibit a carrier effect (→ Textile Auxiliaries, Section 5.6.5) [9.5], [9.54], [9.55], are required. If substantive dyes are used simultaneously (for cotton backs), the dye solvent should also have a good dispersing action.

Dry heat fixing (*thermosol process*, see Section 2.3.2) is unimportant for PAC dyeing because the fiber tends to yellow and harden in this process and cationic dyes are fixed slowly and incompletely [9.5], [9.50]. *Transfer printing* is also not suitable without either elaborate pretreatment or special swelling agents or solvents [9.5]. Disperse dyes produce only very light shades.

Dyeing wet spun PAC fibers during the production process in the *gel state* is especially economical. This occurs after the solvent has been washed out and before the usual follow-up processes, especially stretching and drying [9.56]–[9.59]. The gel absorbs cationic dyes very rapidly (within seconds or a few minutes). There is almost no difference from normal PAC dyeing with regard to saturation, shade, and fastness. The dyes are required in large amounts. Thus, liquid brands, which are miscible without the occurrence of precipitation, are preferably used. Mixtures, even diluted, must be stable in storage (premixability).

The *mass dyeing* of PAC is important only for highly weather-resistant, special articles (awnings and tents).

## 10. Dyeing of Other Synthetic Fibers

### 10.1. Poly(Vinyl Chloride) Fibers

Poly(vinyl chloride) (PVC) fibers (→ Fibers, 4. Synthetic Organic, **A 10**, pp. 642–645) are characterized especially by their flame retardance. For this reason, they are used, e.g., for airplane seat covers, quilt fillings, and purely technical purposes.

PVC fibers are dyed preferably with *disperse dyes* (→ Disperse Dyes) [10.1, p. 404], [10.2, pp. 611]. As with modacrylic fibers, excessively high temperatures must not be used because of shrinkage of the PVC fiber. Hence, some fibers are dyed at 60–65 °C with dyeing accelerants. Other PVC fibers can be dyed at 100 °C without a carrier and a few even at 110 °C. Dyes must be selected with regard to the lightfastness desired.

*Cationic dyes* in combination with an anionic auxiliary can also be used to achieve average depths of shade. Here, good lightfastness is achieved with selected dyes.

Similar to certain modacrylic fibers, some PVC fibers, such as Leavil, also lose their luster on dyeing. It can be recovered by relustering (dry heat at 110–130 °C).

## 10.2. Elastomeric Fibers

Elastomeric fibers made of polyurethane (→ Fibers, 4. Synthetic Organic, **A 10**, pp. 609–615) are contained in most stretch articles and also in fashion materials and knitted fabrics [10.3]. Mixtures with polyamide are used for corsetry, orthopedic support stockings, panty hoses, and swim or sports wear. Articles made of cotton–elastomeric fibers and PES–cotton–elastometric fibers are also very common. The companion fiber is often spun around the elastomeric thread, which makes separate dyeing of the polyurethane fiber unnecessary. This is substantiated by the usually moderate fastness of most dyes on polyurethane fibers.

Polyurethane fibers can be dyed with acid dyes (→Anthraquinone Dyes and Intermediates, **A 2**, p. 408; →Azo Dyes, **A 3**, p. 263), metal-complex dyes (→Azo Dyes, **A 3**, p. 272; → Metal-Complex Dyes), chrome dyes (→Azo Dyes, **A 3**, pp. 270–271, and disperse dyes (→ Disperse Dyes). In general, higher dyeing temperatures, especially below pH 4 and above pH 9, are not permissible because the fibers are degraded (i.e., they lose their elasticity and strength). Efforts are made to circumvent the difficulties arising due to the inadequate dye receptivity of polyurethane fibers with the help of fiber-affinitive cationic dyeing aids in combination with ethoxylated fatty alcohols or fatty amines [10.2, pp. 572–574].

Light shades can be made tone-on-tone on polyamide–polyurethane mixtures with disperse dyes at 95–98 °C and pH 6–7. However, the wetfastness of these dyeings on polyurethanes is lower than on polyamide. The mode of dyeing with acid and metal-complex dyes corresponds to that of polyamide dyeing at boiling temperature and pH 5–6 (see Section 6.5.2) [10.4]. If polyamide–polyurethane mixtures are dyed with chrome dyes, the polyurethane fibers lose elasticity and the articles acquire a brittle feel.

Because of the temperature sensitivity of polyurethane fibers, mixtures of elastomeric and polyester fibers must be dyed with small molecular, rapidly diffusing disperse dyes in 30 min at 120 °C according to the HT process [10.5]. Modified PES fibers that are dyeable at 100 °C without a carrier are often used in mixtures with elastomeric fibers.

Fiber blends that contain cotton in addition to elastomeric fibers can be dyed with vat dyes, sulfur dyes, combinations of reactive and disperse dyes, and combinations of substantive and acid dyes [10.1, pp. 342–349].

In all dyeing processes for elastomeric fibers, dyeing equipment that permits low-strain guidance of the material and the lowest possible thermal stress are important.

## 10.3. Polypropylene Fibers

Polypropylene is an inexpensive raw material that produces fibers with a series of desirable properties such as dimensional stability, low weight, low water absorption, resistance to chemicals, and rotproofness (→ Fibers, 4. Synthetic Organic, **A 10**, pp. 615–623).

Consisting of pure aliphatic hydrocarbons, these fibers have no groups with a special dye affinity that can specifically interact with dyes normally used for textile dyeing. Lipophilic dyes, which are employed for dyeing mineral oil, have sufficient affinity for polypropylene so that they were occasionally recommended for dyeing carpet backing materials. However, they produce dyeings of such poor fastness that they are not useable for textile purposes. Hence, mass dyeing with pigment formulations is the commercial-scale method of choice for coloring unmodified polypropylene fibers (→ Fibers, 4. Synthetic Organic, **A 10**, p. 620) [10.6].

For wide application, fibers should be dyeable in processes commonly used in textile dyeing. For this reason, numerous attempts have been made to modify polypropylene to make it dyeable from aqueous liquor. For surveys, see [10.7]–[10.9], [10.10, pp. 671–675], [10.11].

In all dyeing processes, the low melting point of the polypropylene fiber must be taken into account. The fiber becomes plastic at 130 °C, and treatment temperatures must be kept below 120 °C because of shrinkage.

**Modified Polypropylene Fibers.** Polypropylene, which contains nickel compounds for light stabilization, can be dyed with dyes that are capable of forming nickel complexes. Nonfading, wetfast dyeings are produced in mostly dull shades with water-insoluble, chelating dyes [e.g., Altcolene (Althouse)].

The polypropylene fibers that were on the market for some time were modified by basic components and could be dyed with the acid dyes commonly used for wool and polyamide fibers. To obtain acceptable lightfastness, pretreatment with an acid or an alkylsulfonate was required before dyeing.

In a third method, the polypropylene structure is loosened by a polyether or a polyamide, making the fiber accessible to disperse dyes. The depth of shade and fastness properties of these dyes are sufficient for many textile purposes including floor coverings.

Preliminary tests are required in any case. The type of modification can be determined easily by two test dyeings. If an acid dye is absorbed by the fiber, the modification must be basic. A test with a disperse dye capable of complex formation shows if nickel is present in the fiber. Polypropylene fibers that contain nickel are dyed brown with C.I. Disperse Red 91, and differently modified fibers are dyed pink.

## 10.4. Poly(Vinyl Alcohol) (PVA) Fibers

The importance of PVA fibers for the textile industry is decreasing greatly. The cotton-like feel, high strength, low moisture absorption, good rot resistance, and weatherproofness make these fibers suitable for sewing thread and for canvas and other articles used outdoors (→ Fibers, 4. Synthetic Organic, **A 10**, pp. 645–649). The fiber is water soluble after spinning from an aqueous solution. It is subsequently cross-linked with formaldehyde (acetalization). The higher the degree of cross-linking, the more

difficult it becomes to dye the fiber because dyeability depends on the number of free hydroxyl groups.

Although PVA fibers are resistant to dry heat (exposure to 150 °C reduces the strength by only 10 %), the fibers harden. In moist heat, steam, or an aqueous dyebath, the temperature should not exceed 100 °C to prevent greater shrinkage and hardening of the fiber. In general, PVA fibers need not be bleached; optical brightening is sufficient for white materials. PVA fibers resemble cellulose and polyamide in their dyeing properties. *Disperse dyes* are suitable only for light shades because of poor light- and wetfastness. Although the same applies to *direct dyes*, their wetfastness can be improved by the usual aftertreatment. The color yield in the case of direct dyes depends on the number of sulfonate groups in the dye molecule; it should not be too high. Deep shades can also be obtained [10.1, pp. 405–406].

*Cationic, reactive*, and *sulfur dyes* are taken up to PVA fibers and exhibit varying fastnesses. The most suitable are *vat dyes* (e.g., Indanthren brands) and poorly water-soluble *metal-complex dyes* (e.g., Avilon, Ciba–Geigy; Vialon, BASF). With *vat dyes*, IW or IN dyeing is carried out (see Section 4.3) depending on the individual dye. To improve color yield, the dyeing temperature should be increased to 85 °C. Nevertheless, really deep shades are difficult to achieve. In dyeing, the fact should be taken into account that although vat dyes level easily in the leuco form, they are not or are only incompletely revattable after oxidation so belated leveling becomes almost impossible. Vat dyes produce a fastness with PVA fibers that is comparable to or better than that obtained with cotton.

With *weakly acidic metal-complex dyes*, dyeing is performed at pH 4 and boiling temperature in the presence of a leveling agent, e.g., an ethoxylated fatty alcohol sulfate (Uniperol W, BASF). The higher the temperature, the better is the color yield. The leveling capacity is very good. Very fast dyeings are produced that are also suitable for weatherproof articles.

## 10.5. Aramid Fibers

Polyamides made from aromatic amines and dicarboxylic acids are called aramids (→ High Performance Fibers, **A 13**, pp. 3–8). In their textile application properties they are similar to polyamide and polyester fibers. They have the advantage of being highly heat resistant and flame retardant.

For this reason, aramid fibers are used for protective clothing. However, the relatively high cost and moderate lightfastness of the dyeings prevent further application in the textile sector.

The aramid fiber most commonly used for textile purposes is Nomex (Du Pont). Nomex must be heat set by steaming (at 120 or 90 °C, depending on the type) before it is wet finished. Steaming at excessively high temperature reduces dyeability. Thorough washing is also required for level dyeing. *Cationic dyes* are used exclusively for coloring Nomex. Suitable products are, e.g., C.I. Basic Yellow 15, 21, 23, 25, 29, 49, 53, 79, Basic Orange 22, 35, Basic Red 29, 46, Basic Blue 41, 54, and Basic Green 6. A lightfastness of 3–4 is achieved, with Basic Green 6 giving 5–6. To attain a good color yield, dyeing is carried out at pH 3–4 and 120 °C for 2 h. In addition, a carrier is used, preferably benzyl alcohol in amounts of 10–40 g/L, depending on the depth of color. The usual carriers can also be employed. Exhaustion of the dyebath is improved by the addition of 25 g/L of sodium nitrate. After dyeing, thorough afterwashing is required with the addition of sodium hydrogensulfite, if necessary, to remove superficially adhering dye [10.10, pp. 675–676].

In the case of faulty dyeing, the dye can be largely stripped by treatment at 120 °C with benzyl alcohol or with a carrier and a retarder which has a leveling effect in the presence of hydrosulfite in an alkaline liquor.

Other very strong or high-temperature-resistant fibers [e.g., poly(tetrafluoroethylene) or carbon fibers] cannot be dyed with the methods usually employed in textile dyeing.

# 11. Dyeing of Fiber Blends

For dyeing mixtures of wool and polyamide fibers, see Section 5.1.5.

From time immemorial the properties of textiles have been varied by mixing fiber types. Blends that are rather unimportant today (e.g., half linen, a mixture of cotton and linen, and half wool, a mixture of wool and cellulose fibers) were widely employed in the past. For the dyeing of these blends, see [11.1]. Efforts are currently made to combine the favorable technological properties of synthetic fibers with the pleasant feel and wearing comfort of natural fibers. Of the worldwide consumption of PES fibers, 55–60 % is used in blends with cellulose fibers or wool. About 40 % of polyamide is used in blends, and for polyacrylonitrile fibers, 50 % is used especially in blends with wool for knitwear. Thus, the introduction of man-made fibers has resulted in rapid expansion of the mixed processing of various types of fibers. Of the large number of such blends, some have proved very worthwhile, especially PES–cellulose blends.

Since efforts are made to place the dyeing step as close as possible to the end of the finishing process, the problem of coloring fiber blends arises frequently. In the joint dyeing of two components, a compromise often must be made with regard to the dyeing process, tone-on-tone dyeing or fastness of the finished product.

For this reason, not every fiber blend is advantageous from the point of view of finishing, and satisfactory dyeing methods still have to be found for some blends (e.g., PES–cellulose acetate).

## 11.1. Polyester–Cellulose Blends

The most important fiber blend consists of polyester and cellulose fibers. A large part of the entire production of PES, ca. 45 %, is used to make this mixture. Polyester–cellulose blends are used for all types of clothing textiles and for bed linen. A considerable amount is consumed by the knitwear sector. The cellulose component is usually cotton, but viscose staple fibers, modal fibers (→ Cellulose, **A 5**, pp. 407–408), and occasionally linen are also used. Modal fibers are often employed in knitwear because of their luster. The preferred mixing ratio is PES:cellulose 67:33 and, for textiles worn close to the skin, 50:50 to 20:80.

Articles produced from a mixture of cellulose and PES have more versatile application than those made of either cellulose or PES fibers alone. Cellulose with its hydrophilicity, its ability to transport moisture, and its low electrical charging contributes to wearing comfort. The PES fiber improves tear and abrasion resistance, increases dimensional stability, and makes washing and care easier.

### 11.1.1. General Finishing Information

For detailed information on the dyeing and finishing of PES–cellulose blends, see [11.2, pp. 209–289], [11.3, pp. 453–457], [11.4, pp. 592–600].

As in the case of cellulose articles, the dyeability of PES–cellulose mixtures is influenced greatly by the methods used for pretreatment, the boiling and bleaching of the cellulose component, and the heat setting of the PES component. Pretreatment methods depend especially on the type of cellulose fiber employed. In addition to the shades to be dyed and the scheduled dyeing method, tests must also be conducted to determine the required degrees of whiteness and hydrophilicity (→ Textile Auxiliaries, Section 4.2).

In dyeing PES–cellulose mixtures, disperse dyes are used for the PES component. The cellulose portion can be dyed with practically all classes of dyes suited to cellulose (see Chap. 4). The selection of suitable dyes depends not only on the equipment available, the desired shade, and the batch size, but also on economic factors, with the required fastness taken into account.

Thus, *reactive dyes* are used preferentially for all types of clothing textiles, but *vat dyes* are employed where extreme demands are made on washfastness or weatherproofness.

Dyeing with mixed dyes that were first put on the market by BASF is especially easy. Examples of suitable mixtures of disperse and vat dyes are Cottestren (BASF) and Teracoton (Ciba–Geigy). The mixed dyes can be applied in a single bath via the exhaustion or continuous process. The mixed dyes are formulated in such a way that the same shade is obtained on both fibers, which facilitates shade matching [11.5].

In dyeing PES and cellulose components with two dye classes, the staining of one fiber by the dye used for the other fiber should be taken into account. *Disperse dyes* stain cellulose fibers only slightly. The soiling can be removed easily by washing or, in stubborn cases, by alkaline reductive treatment, which is required anyway in vat dyeing. Most of the dyes used for cellulose stain PES only slightly or not at all. Dyes such as selected vat dyes that are fixed to PES by a special application process (see Section 7.2) and sulfur dyes, which often dye PES deeply, exhibit good fastness in use and produce a shade that corresponds with the tint on cellulose.

In the case of light shades, a slight staining of the partner fiber can give an attractive appearance to the entire article. Therefore, for pastel shades, depending on the article and its design, only one component may be dyed, provided that the staining on the other is fast enough. *Leuco*

*esters of vat dyes* are also suitable for the simultaneous fast dyeing of PES and cellulose in pastel shades. The mode of dyeing is simple (see Section 4.8.1). To obtain good fastness on the PES component, drying should be carried out at 170–180 °C. *Pigment dyeing* is also commonly used for light shades (see Section 4.7).

For matching off the two fibers, the components often must be inspected separately. Fiber blends often look better being dyed tone-on-tone if the glossy fiber, PES in this case, is kept somewhat darker. To inspect the dyeing of the components, the cellulose in PES–cellulose blends can be dissolved out by treatment with sulfuric acid (72%, 45 min at room temperature or 5 min at 70 °C) or orthophosphoric acid (85%, 3–5 min at 80–90 °C), which is facilitated by ultrasonic treatment. Furthermore, small samples of the individual components in a presentation similar to the material to be dyed can be dyed for inspection together with the fiber mixture.

### 11.1.2. Dyeing Processes

Since safe and efficient continuous dyeing methods are well known and large yardages are often dyed in the same color, PES–cellulose mixtures are frequently dyed in continuous processes. Nevertheless, the batch dyeing is of major importance for dyeing yarn as well as piece goods, especially in the knitwear sector.

### 11.1.2.1. Dyeing with Disperse and Vat Dyes or Appropriate Mixed Dyes

*Batch Dyeing Process.* In batchwise dyeing, a single bath is used in two stages. A suitable choice of dyes produces dyeings with unsurpassed fastness to washing (e.g., for work clothes) and thermofixing (e.g., for colored weaving yarn). Mixed dyes are usually prepared in such a way that PES–cellulose blends mixtures of ca. 80:20 to 50:50 can be dyed in the same shade. The PES component is dyed at ca. 130 °C without a carrier since carriers can hold back vat dyes in the dyebath (for mode of dyeing, see Section 7.2.2). During the HT step, the cellulose portion is uniformly covered with vat dye. After the PES component is dyed, the bath is cooled to 80 °C and made alkaline, the vat dye is vatted with hydrosulfite, and dyeing is completed according to the rules for vat dyeing (see Section 4.3.6). The disperse dye attached to the cellulose portion is removed by means of the vat so that a reductive clearing step to improve fastness becomes unnecessary. The mode of operation in individual dyeing aggregates is described in Section 2.2.

*Continuous Dyeing Process.* For PES–cellulose blends, continuous dyeing with disperse and vat dyes is especially safe, easy, and reproducible and produces very fast dyeings. Migration of the dye pigment during drying is a frequent source of faults. In PES–cellulose blends, the cellulose fiber is the absorptive component and counteracts uninhibited migration. In addition, active migration inhibitors (e.g., polyacrylates), which can be washed out easily later, are available. In continuous dyeing, the materials must be carefully pretreated to guarantee rapid wetting and uniform absorption of liquor (→ Textile Auxiliaries, Section 4.2).

After dye application, the fabric is uniformly and gently dried and the disperse dye is fixed to the PES component by the thermosol process (see Section 7.2.3). Subsequently, the vat dye on the cellulose fiber portion is developed either continuously according to the pad steam process or batchwise in a jigger (see Section 4.3.6). *Jigger development* is performed according to the pad jig process for vat dyes and is advantageous with respect to penetration into nap containing or densely woven materials. The most elegant method is *pad steam development*, which frequently follows the thermosol process in one pass. In pad steam development, the fabric is padded with reducing agent and alkali, and the vat dye is fixed by steaming in saturated steam (103–105 °C dry thermometer). Finely dispersed vat dyes must be used in this process because of the short vatting time. The mode of operation with individual dyeing aggregates is described in Section 2.3.2.

### 11.1.2.2. Dyeing with Disperse and Reactive Dyes

Reactive dyes are often used to dye the cellulose portion of PES–cellulose blends because the coloring possibilities of reactive dyes are especially diverse and a large number of relatively easy application processes produce dyeings that are sufficiently fast for the clothing sector and, frequently, for household textiles.

*Batch Processes.* Since the worldwide tendency is toward smaller batch sizes, PES–cellulose blends are dyed mainly batchwise by the exhaustion method.

The *two-bath method* is a safe mode of dyeing that results in reproducible dyeings. The PES component is usually predyed with disperse dyes, and the cellulose portion is topped with reactive dyes after reductive intermediate clearing, if necessary.

In a *one-bath two-stage process*, the bath contains all the dyes from the start. After the PES component is dyed at 130 °C, the bath is cooled to ca. 80 °C, and alkali is added, resulting in the fixation of the reactive dyes to the cellulose component. A careful selection of dyes is required because the reactive dyes must be stable at pH 4–5 and 130 °C.

In a *reverse process*, the cellulose component is first predyed at alkaline pH, the pH is then lowered, and the PES component is dyed at 130 °C. In this case, alkali-stable disperse dyes must be chosen if the disperse dye is added at the start. In addition, large amounts of salt in the bath can interfere with dispersion stability. Hence, especially fast exhausting disperse dyes are used.

Alkali dosing processes with controlled addition of alkali are of great importance for dyeing the cellulose component, particularly when vinylsulfone anchor dyes are used. The control program can also allow a progressive increase in pH. Thus, the rate of absorption, which is responsible for leveling, and the rate of fixation are optimized. Occasionally, the pH is controlled by cleavable alkali-releasing chemicals.

Of little importance is a *thermosol exhaustion process* in which the PES component is predyed by the thermosol method and the cellulose portion is then topped in a bath dyeing process. Also of little significance is a *pad batch HT process* in which disperse dye, reactive dye, and chemicals are applied together in the padder. The reactive dye is then fixed in a pad batch process, and the disperse dye is fixed in an HT step. Here, too, alkali-stable disperse dyes are required.

*Continuous Processes.* The most important continuous dyeing methods are the thermosol pad steam process and the thermosol pad batch process. In all continuous processes, the substantivity of the reactive dyes should be taken into account (i.e., the reactive dye can exhaust from the pad liquor or accumulate to a higher or lesser extent even at room temperature as a function of the pH value and salt content).

In the *thermosol pad batch process*, a semi-continuous process, alkali-sensitive disperse dyes can be used if the alkali is applied after the thermosol process. Conversely, if the thermosol process is carried out after the pad batch step, the pH must be lowered before the thermosol process so that the disperse dye is not destroyed.

The *thermosol pad steam process* is especially important. Disperse and reactive dyes are padded together, and alkali and salt are padded after the thermosol passage. If a steamer is not available, the reactive dye can be fixed continuously by means of "alkali shock" in a roller vat with a relatively high alkali concentration.

Although the *thermosol thermofixing process* is especially easy and economical, it is less easily reproducible and, therefore, of little importance. In this process, disperse and reactive dyes are padded together with alkali and fixed by dry heat at ca. 210 °C. This is followed by washing in an open-width washing machine and soaping. Disperse dyes used for this process must be alkali-stable, and the dispersing agent contained in the dye must not react with the reactive dye. Especially suitable reactive dyes are hot dyers that are stable and poorly substantive in the pad liquor.

However, none of the one-step processes with disperse and reactive dyes achieves the safety of operation of the thermosol pad steam process with disperse and vat or sulfur dyes.

### 11.1.2.3. Dyeing of the Cellulose Component with Other Dyes

*Sulfur Dyes.* The thermosol pad steam process is the primary method for application of disperse and sulfur dyes. However, the sulfur dyes that are used preferentially in the ready-for-dyeing soluble form, e.g., Cassulfon (Hoechst) and Sodyesul (Southern Dyestuff), and are specially inexpensive for muted shades, must be applied after the thermosol step because of their incompatibility with disperse dyes.

*Leuco Esters of Vat Dyes.* If a thermosol passage is provided under conditions that cleave the ester and the vat acid can diffuse into the PES fiber, leuco esters of vat dyes can produce up to medium shades with a high fastness level without the additional use of disperse dyes (ATE process, Hoechst).

*Small Molecular Vat Dyes.* Only one class of dyes is needed when small molecular vat dyes are used. Fast dyeings are obtained in lighter shades.

Padding and the thermosol process are, if necessary, followed by vatting and normal aftertreatment to improve fastness on the cellulose component.

*Direct Dyes.* If no great fastness demands are made on the dyed article, PES–cellulose blends can be dyed in a one-bath, one-step or two-step process with disperse and direct dyes. The direct dyes selected here must be stable under HT dyeing conditions. Dyes whose wetfastness can be improved by aftertreatment are of advantage. The subsequent resin finishing must also be taken into account in dye selection.

In a two-bath, two-step dyeing process, other dyes suited to cellulose fibers may also be used for PES–cellulose blends. For instance, in spite of its tediousness, *Naphtol AS dyeing* is still employed, especially for wetfast red shades (see Section 4.6), and the *phthalogen process* is used for turquoise (see Section 4.8.7). *Pigment dyeing* is also applied to PES–cellulose blends (see Section 4.7).

## 11.2. Polyester–Wool Blends

Mixtures of PES fibers and wool are used widely, especially as woven goods and knitwear in the outerwear sector. They offer the advantages of dimensional stability, easy care, and fashionable design possibilities. PES is generally used as staple fiber and frequently as a low-pilling type [11.6]. The ratio PES:wool 55:45 is widely employed.

In methods used for pretreating and dyeing PES–wool blends, the relatively high sensitivity of wool must be taken into account. Difficulties are circumvented by dyeing the components separately as loose stock or slubbing (see Chaps. 5 and 7). Especially fast dyeings are obtained in this manner. However, quick changes in fashion and short-term planning frequently do not permit separate dyeing.

### 11.2.1. Dyes

In dyeing PES–wool mixtures, disperse dyes are used for the PES component and acid or metal-complex dyes for the wool. Disperse dyes can soil wool to a great extent. Since they produce poorly fast dyeings on wool, the dyes selected must stain wool as slightly as possible or must be easily removable by a washing step, which may be reductive if necessary. Frequently used dyes are C.I. Disperse Yellow 23, 54, 64, C.I. Disperse Orange 30, 33, C.I. Disperse Red 50, 60, 73, 91, 167, 179, and C.I. Disperse Blue 56, 73, 87. Premixed dyes consisting of disperse and wool dyes are occasionally available, e.g., Forosyn (Sandoz).

Wool is stained less at high temperatures and longer dyeing times because the disperse dye that was initially on the wool component migrates to the PES fiber. Carrier-free dyeable PES fibers should preferably be employed; these fibers are easily dyeable even at boiling temperature (see Section 7.2.6). The nature of the wool also plays a role, fine wool and reclaimed wool being more strongly soiled.

## 11.2.2. Dyeing Processes

For practical information about the dyeing of PES–wool mixtures, see [11.2, pp. 291–328], [11.7]. Wool cannot be dyed at the high temperatures normal in the HT dyeing of PES fibers or PES–cellulose blends. The dyeing time should also be as short as possible so that the wool is not damaged. The following dyeing processes are usually employed:

1)  At boiling temperature with a carrier
2)  At 103–106 °C with little carrier
3)  At 110–115 °C with the addition of formaldehyde as a wool protectant and without a carrier (HT dyeing)

The significance of HT dyeing has increased in more recent times.

The dyeing of PES–wool blends is carried out in one or two baths. In the *one-bath process* preferred in practice, PES and wool are dyed simultaneously at pH 4–5, which is favorable for wool. Reductive aftertreatment is possible only if the wool dyes used can withstand it.

The *two-bath process* is applied for deep shades and for stringent fastness requirements. Disperse dyes, which soil wool to a somewhat greater extent, can also be used in this case. The first step involves the use of disperse dyes for predyeing the PES part, followed by reductive intermediate clearing and wool dyeing. However, while dyeing the wool component, dye must always be expected to detach from PES and retint the wool.

Prebleached wool becomes somewhat yellow on dyeing. To obtain clear pastel shades bleach and dye are used preferably in the same bath. Mildly bleaching reducing agents (especially stabilized bisulfite or hydroxylamine salts) can be employed because they do not affect the dyes used.

Continuous or semicontinuous dyeing processes are uncommon because dry heat treatment (thermosol) hardens wool. Besides, application of the wool dye requires dyeing in aqueous liquor or dye fixation by steaming.

## 11.2.3. Auxiliaries

At dyeing temperatures > 100 °C, formaldehyde (5 % on weight of fabric) is added to protect the wool. Complexing agents should be used with care in the dyebath if metal-complex dyes are present. Ethoxylated fatty amines or ethoxylated fatty alkyl sulfates are suitable leveling dyeing aids. Condensation products of naphthalenesulfonic acid and formaldehyde, which hardly foam, have proved especially useful dispersing agents. A leveling agent is required primarily if tippy-dyeing grades of wool are present.

In the selection of a carrier, technical as well as ecological factors must be considered (→ Textile Auxiliaries, Section 5.5). As in the dyeing of pure PES fibers, the amount of carrier to be applied is determined by the type of carrier, the dyes, the type of fiber, the dyeing temperature, the depth of shade, the liquor ratio, and the dyeing aggregate.

If possible, yarn is dyed at 106 °C or still higher temperatures because, in this way, the production fastness (e.g., fastness to setting) required for further processing is easier to guarantee. This is possible because thermally stable disperse dyes that exhaust only at higher temperatures are used.

To separately inspect the PES component of PES–wool mixtures, the wool can be dissolved out with sodium hydroxide (5 % solution, boil for 5 min) or a chlorine bleach (50 g/L active chlorine, 5 min at 50 °C).

**Aftertreatment.** After dyeing, an afterwash is always indicated to remove the disperse dye attached to the wool. The materials are treated with an ethoxylated fatty amine in a weakly acidic liquor at 60 °C. Reductive aftertreatment is possible only in the case of appropriately stable wool dyes.

## 11.3. Polyamide – Cellulose Blends

Polyamide (PA) fibers are included in articles made of cellulose fibers to improve dimensional stability, ease of care, and durability. Examples of articles include sportswear and knitwear, corduroy fabrics, or plush articles in which a PA pile is often anchored to a cellulose fabric base. Different dyeing methods are described in [11.3, pp. 433–437], [11.4, p. 570].

**Dyeing Processes.** In dyeing PA–cellulose mixtures, consideration should be given to the fact that PA fibers are more or less strongly dyed by all dyes normally used for cellulose. The shades on the PA and cellulose components often do not correspond in color strength and in shade. Their fastness can also vary. Since good color matching between the PA and cellulose components is usually required, a careful dye selection must be made and the dyeing process adapted appropriately.

If dyeing is carried out in the presence of 1 % sodium dihydrogenphosphate without salt at boiling temperature, solid shades from light to moderate can be obtained with high-leveling *direct dyes* (e.g., C.I. Direct Yellow 44, Direct Red 81, and Direct Black 51). The choice of suitable dyes is greater if dyeing is performed in a single bath with *direct and disperse dyes* at pH 8 in the presence of a leveling agent and a resist for the PA fiber ($\rightarrow$ Textile Auxiliaries, Section 5.6.3) to prevent the direct dyes to exhaust mainly on the PA component.

If, instead of disperse dyes, wetfast *acid or metal-complex dyes* are used for the PA component, union shades having good lightfastness and moderate wetfastness can be obtained. Dyeing is carried out at pH 5–8, depending on the depth of shade and the type of acid dye used. The dyeing temperature is raised to boiling to guarantee good leveling and firm fixation of the wetfast acid dyes in the PA fiber.

Dyeings of the highest wetfastness are obtained on PA–cellulose blends with *vat dyes*. The coloring strength on the PA component depends on the type of PA fiber and on dyeing conditions (the depth of shade on PA 6 is deeper than on PA 66). The lightfastness of some vat dyes can be lower on PA than on cellulose fibers. The advantage of using vat dyes is that in a single bath with selected products, dyeings of the same shade are obtained on both components. For instance, C.I. Vat. O. 26, V.B. 4 and 14, V. Br. 55, V. Bl. 9 and 25 are widely applicable (in the exhaust and pad steam process). The dye yield on the PA component increases with increasing temperature. With a dyeing time of 60 min, the most favorable temperature for the dyes mentioned above is 80–90 °C. The additives to the dyebath and the dyeing processes are the same as those used for dyeing cellulose fibers with vat dyes (see Section 4.3).

Clear shades with very good fastness are obtained on PA–cellulose blends with *reactive dyes*. Like vat dyes, the depth of shade of reactive dyes depends relatively strongly on the type of PA and structural differences. Dyeing is carried out in a three-step process with appropriately selected products. First, the reactive dyes in a weakly acidic liquor are allowed to absorb to the PA component. Salt is then added to improve the yield on the cellulose component. Finally, the liquor is made alkaline for reaction with the cellulose fiber. Dyes (e.g., with monochlorotriazine anchor) that dye PA from a neutral liquor in the presence of salt are applied in a two-step process, as in the case of cellulose. In the reversal of this dyeing process, the cellulose component is dyed first at alkaline pH, followed by neutralization with acid, and the PA component is then covered at elevated temperature.

Dyeing processes with reactive dyes are described in Section 4.1.

## 11.4. Polyacrylonitrile – Cellulose Blends

Polyacrylonitrile (PAC)–cellulose blends are used for household textiles and imitation fur. In these plush and fur materials, the pile consists of PAC fibers while the back is made of cotton or cellulose regenerated fibers. The wool-like character, favorable price, and easy care of PAC fibers combine with the high water-absorbing capacity of cellulose fibers. Furthermore, PAC–cellulose mixtures are used for leisure wear, sport stockings, drapery, and table linen. The percentage of PAC in the mixtures varies between 30 and 80 %. For a survey of dyeing methods, see [11.3, pp. 474–476] and [11.4, pp. 608–610].

Continuous dyeing of PAC–cotton plush with *cationic* and *direct dyes* according to the pad steam process plays an important role. The choice of dyes must take into account liquor stability, reservation of PAC or cellulose fiber, and solubility. Precipitation of cationic and anionic dyes present in the pad liquor at relatively high concentrations cannot be avoided solely by dye selection. Suitable auxiliary systems have been developed. Differently charged dyes are kept in solution separated from each other in two phases by the combination of anionic and nonionogenic surfactants. With the help of fixing accelerators, good penetration of PAC fibers can be achieved in 10–15 min with saturated steam at 98–100 °C.

With *cationic and vat dyes*, dyeing can be conducted batchwise according to the pigment process in one bath and two steps. Here, the PAC component is dyed with cationic dyes and the cellulose portion is simultaneously pigmented. After being cooled to 70 °C (maximum) and vatted, the Indanthren dye is fixed (see also Section 4.3). Another possibility is to complete the dyeing of the PAC component first with cationic dyes and then add the vat dye to the same bath after cooling.

PAC–cellulose mixtures can also be dyed in a one-bath, two-step process with *reactive and cationic dyes*. Here, too, the PAC component is predyed at pH 5 with cationic dyes, and after the pH is raised, the cellulose component is covered with reactive dyes. Process safety is obviously increased when a two-bath process is employed. For the dyeing of yarn, see [11.8].

## 11.5. Polyacrylonitrile – Wool Blends

No other synthetic fiber achieves the wool-like character of PAC fibers. For this reason, PAC fibers are especially suited to the production of blends with wool for applications in which the woolly character is indispensable. The high water-absorbing capacity of wool and the high strength of PAC fibers complement each other and confer good wearability on articles made of this mixture. Woven and knitted fabrics possess a greater dimensional stability and mechanical stressability than corresponding wool textiles. The very low water-absorbing capacity of the PAC fiber does not have such an unpleasant effect as in textiles made of PAC alone, and the felting tendency is lower than that of wool. The blending ratio of PAC to wool varies from about 20:80 to 80:20. The blends are preferred for outer clothing, especially knitwear (e.g., pullovers and socks), and for household textiles.

**Dyeing Processes.** In principle, three possibilities exist for dyeing PAC–wool mixtures [11.3, pp. 473–474], [11.4, p. 610]:

1) The two-bath method, i.e., dyeing of wool and PAC in succession in two baths
2) The one-bath, two-step method, i.e., dyeing of wool and PAC in succession in the same bath
3) The one-bath, one-step method, i.e., dyeing of wool and PAC simultaneously

The one-bath, one-step method is preferred for economic reasons. However, problems arise due to the simultaneous presence of cationic and anionic dyes in the dyebath, with the danger of mutual precipitation. Precipitation can be prevented by dispersing auxiliaries and precise dye selection. Very deep shades (e.g., black) are normally dyed in a one-bath, two-step process.

Although only cationic dyes may be used for dyeing the PAC component, metal-complex dyes, chrome dyes, or reactive dyes can be used for wool.

In the case of 1:1 metal-complex dyes, almost no danger of precipitations exists because of intramolecular compensation of the negative charge by the positively charged metal atom. The pH is adjusted to 2.5–3 (see also Section 5.1.4). Most cationic dyes can be applied at this low pH value. Only the auxiliaries and the acid are added while the dyebath is being heated. After the initial dyeing temperature (ca. 70 °C) is reached, the cationic dyes are added first. After balancing concentration differences (ca. 5 to 10 min), the anionic dyes are then added. A pH of 4.5–5.5 is maintained for acid, chrome, reactive, and 1:2 metal-complex dyes.

In the choice of dyes, attention must be paid not only to the prevention of precipitation but also to the reserve of PAC fibers and wool. Cationic dyes attach first to wool, producing colors of very poor fastness, and then boil over onto the PAC fiber. Even if well-reserving dyes are selected, dyeing must be conducted for a sufficiently long time (60–90 min) until the boilover is complete, in order to obtain the absolutely essential good wool reserve.

Apart from chemical problems (precipitation, reserve), the mechanical properties of the materials must also be taken into account. Knitwear made of PAC–wool must not be exposed to excessively high tensile stress during dyeing.

## 11.6. Blends of Synthetic Fibers

No fiber exists that has not at least on a trial basis been combined with every other fiber to attain new fashionable or technical effects (see survey in [11.3, pp. 420–421]). From a dyers' standpoint, these mixtures do not always make sense. Practically all synthetic fibers can be dyed with disperse dyes. Thus, the affinities of these dyes for the fiber components and their fastness properties should always be taken into account in dyeing blends of synthetic fibers [11.9]. In the selection of a dyeing process, the individual components must be considered. With blends of *PA and PES fibers*, for example, one should note that although disperse dyes have a higher affinity for PES than for PA fibers, the rate of dyeing on PES at boiling temperature in the absence of dyeing accelerants is so low that the PA component is usually dyed to a deeper and, frequently, in a different shade. If the same shades are to be produced on a blend of this type, a small selection of disperse dyes is used at the highest possible temperature under exactly stipulated dyeing conditions. To obtain better fastness on the PA component, the disperse dye can then be reductively eliminated to a large extent and the material redyed with acid or metal-complex dyes [11.2, pp. 359–363].

A frequently processed mixture (e.g., for outerwear or furnishing fabrics), consists of *PES and PAC fibers*. No basic difficulties are involved in dyeing, both union shades (tone-on-tone dyeing) and bicolor dyeing being possible. Selected disperse and cationic dyes are used for dyeing. Although disperse dyes color PAC more weakly than PES, the shade produced is the same. In the case of bicolor dyeing, the PAC component must always be kept darker. A dyeing temperature of 106 °C should not be exceeded because of the PAC component. Thus, carriers must be used as a rule. In one-bath dyeing, attention must be paid to the compatibility of the various dyebath additives with each other. Cationic retarders are not used in one-bath dyeing [11.2, pp. 346–355], [11.4, pp. 601–602].

Problems with the compatibility of the dyes used can also arise in dyeing mixtures of *PA and PAC fibers*. Mixtures of this type are occasionally produced for skiing articles and for wool–PA imitations. The cationic dyes used for the PAC component do not dye normal PA fibers. Acid and metal-complex dyes leave PAC fibers practically white. Sulfo group containing 1:1 metal-complex dyes possess a zwitterionic character and, therefore, can be applied in low concentration in the same bath with cationic dyes, without the occurrence of precipitation. In this manner,

one-bath dyeing can be carried out. The stability of the bath can be improved by addition of a nonionic auxiliary (e.g., an ethoxylated fatty alcohol). Cationic retarders should be excluded in all cases. Dyeing is accomplished most safely in two steps: first with cationic dyes, because they exhaust almost quantitatively, and then with 1:1 or 1:2 metal-complex dyes for the PA component. Disperse dyes can naturally be used as well. However, a lower wetfastness level must be expected, and a distinct amount will absorb to the PAC component [11.3, pp. 471–473]. For the dyeing of mixtures containing elastomeric fibers, see Section 10.2.

# 12. Characterization of Textile Dyes

This chapter gives a description of those properties that are of general importance for dyes or groups of dyes. Such a description is based on experimentally determined quality data.

**Color Strength, Hue, Chroma** ($\rightarrow$ Dyes, General Survey, **A 9**, pp. 99–105). The dominating characteristics of a dye are its coloring properties. These are assessed by preparing a dyed test sample whose color is evaluated. This must, in principle, always be done by the human eye because color perception, being a subjective sense impression, is not accessible to direct measurement. However, with the aid of colorimetry, which can be regarded as a simulation of the visual perception of color, this visual perception can be represented by measurable quantities. Since colorimetry is an objective method and is therefore more accurate and reproducible than subjective visual assessment, it is very widely used today. Color is a three-dimensional quantity and must therefore be expressed by a set of three numbers (color coordinates). In practice, these are typically the values of color strength, hue, and chroma ($\rightarrow$ Dyes, General Survey, **A 9**, pp. 102–103) [12.1].

Color can be described with any accuracy only in a comparative way (i.e., in terms of deviations from a standard). Although the three-dimensional color coordinates can be determined by colorimetry (as absolute values), this is in practice not sufficient to provide an exact specification. In the case of a dye, the reference is a standard dye sample provided by the producer against which all further supplies are controlled and standardized. To be used as a basis for assessment the standard sample must be extremely stable.

To test a dye, dyed test specimen are prepared from both the sample dye and the standard under conditions that are as identical as possible and then assessed visually or colorimetrically. In practice, color strength is calculated from various color strength values agreed upon by the producer and the customer. A widely used method is a determination based on the weighted sum of the $K/S$ values ($K$ is the coefficient of absorption, $S$ the coefficient of scatter), which is used in standard specifications [12.2] ($\rightarrow$ Dyes, General Survey, **A 9**, p. 104). The relative color strength is calculated as the ratio of the color strength values for the sample dye and the standard expressed as a percentage.

Dyers can in practice compensate for differences in color strength by adjusting the concentration. Therefore, the hue and chroma of a dyeing are always judged against a standard dyeing of equivalent color strength. This is carried out by visual assessment in which the colorist estimates the remaining differences in hue and chroma for a strength-corrected dyeing. In a colorimetric evaluation, the residual color difference is determined by a mathematical correction to give equivalence in color strength. The result is expressed in CIELAB color difference units (DIN 6174) and is usually separated into components for hue ($\Delta H^*$) and chroma ($\Delta C^*$) [12.2], [12.3] ($\rightarrow$ Dyes, General Survey, **A 9**, pp. 103–104).

In practice, color strength is often determined by the simpler and more easily reproducible measurement of the extinction of the dye solution. However, this does not in all cases agree with the values determined by dyeing because the dye is not always absorbed quantitatively by the fiber [12.4].

**Solution and Dispersion Behavior.** For the dyeing process in aqueous liquor, the dye must have adequate solubility or dispersibility. In general, good solubility is necessary for good application properties. If the solubility is poor (i.e., if any of the dye is present in the dye liquor in the form of undissolved particles), local coloration (specks), spots, uneven effects, and poor fastness can be produced, leading to serious defects and costly complaints.

In addition to solubility properties, the stability of the liquor during preparation and use (i.e., the *solution stability*) is important. For reac-

tive dyes applied to cellulose, the solution stability in the presence of electrolytes is important (*electrolyte stability*). To test for this, the dye is dissolved under defined conditions of temperature, time, and concentration, and is filtered after a storage time or after addition of electrolytes. The filter should then be free of residues. To give consistent and comparable results, the values of various parameters (temperature, etc.) should be kept constant and should be specified in standards. For example, the temperatures for hot and cold solubility are set at 90 °C and 25 °C, respectively (ISO/DIS 105-Z07, ISO/DIS 105-Z08, and ISO/DIS 105-Z09) [12.5].

For *disperse dyes*, the dispersion properties are of similar importance, but they are not easy to describe because of the complex processes and unstable dispersion states. Therefore, a large number of test methods exist, the results of which are usually limited in their application. A critical review can be found in [12.6].

**Dust.** The dust produced from dyes is of great importance for many reasons, including possible health hazards and intense coloration, leading to severe soiling, that can be caused by very small amounts of dust particles. The content of fine dust (1–10 μm) must be reduced by adhesive bonding or prevented by the use of special formulations (granulation) of the dye. Fine dust is especially troublesome because of the stability of its suspensions in air, and it can constitute a health hazard when inhaled.

The amount of dust depends not only on the dust behavior of the dye, but also on the nature and intensity of handling operations that generate dust [12.7]. Therefore, the dusting behavior of a product is significant only for comparison with other products and cannot be used to predict dust exposure. In a widely used test method, the dust produced when a defined amount of dye is allowed to fall is determined quantitatively or qualitatively [12.8]. The dust is first collected on a filter, and assessment occurs by visual comparison of the dust-laden filter with a gray scale divided into five levels (ISO 105-A03) or by quantitative gravimetric or photometric determination of the quantity of dust.

**Compatibility of Basic Dyes for Acrylic Fibers.** Basic dyes usually have limited migration ability. Therefore, with combination (trichromic) dyeing processes, the dyes are chosen so that they are absorbed at as uniform a rate as possible to prevent local color differences that would otherwise occur readily.

To characterize combination behavior, dyes are divided into five groups numbered 1 to 5 (compatibility values). Higher numbers correspond to a lower bath exhaustion rate. The compatibility of a dye is determined by producing samples of dyed materials in which the test dye is combined with five standard dyes whose compatibility values are known (AATCC Test Method 141–1987; ISO/DIS 105-Z03) [12.9].

# 13. Testing of Colorfastness [13.1]

In the 1980s most of the standard methods for testing colorfastness were revised, and some new methods were introduced.

## 13.1. Organization

Questions of standardization of test methods and of the availability of apparatus and reference materials used in testing colorfastness are treated in Germany by the Working group 511 of the Normenausschuß Materialprüfung at DIN. Working group 511 is called the "Deutsche Echtheitskommission" (DEK) and was founded in 1911. About 70 DIN standards exist describing test methods for colorfastness.

In the European Colourfastness Establishment (ECE), Germany is represented by DEK. Proposals of new test methods to the ISO are prepared by ECE.

In the ISO, Technical committee 38 (Textiles), Subcommittee 1 (TC 38/SC1) is engaged with questions of colorfastness testing, color measurement, and dye characterization. Until now 16 meetings of ISO/TC 38/SC1 have taken place.

Since 1978 the test methods have been published as *International Standard ISO 105*; each separate part is characterized by code numbers, for instance, ISO 105-B02 for testing the "Colour fastness to artificial light: Xenon arc lamp fading test." Since 1978, International Standard 105, which contains more than 70 different test procedures, has been revised, and the different parts have been given code numbers. These standards are published in English, French, and Russian.

The latest, very important development is the edition of *European Standards (EN)*, which are published in English, French, and German. For testing colorfastness, these European standards correspond completely with ISO 105, which became EN 20105 with the same code numbers (i.e., ISO 105-B02 becomes EN 20105-B02). Since the editorial work takes considerable time, publishing the European standards will take several years. At the moment, very few EN denominations exist. When the European standards are published the national standards (e.g., DIN standards) are withdrawn.

## 13.2. Standard Test Methods

The following summary for each test method gives the denomination of the international standard and the title and denomination of the German DIN standard given.

### 13.2.1. Group A: General Principles

**ISO 105-A01: General Principles of Testing** (= DIN 54000-A01: Grundlagen für die Festlegung und Durchführung der Prüfungen und für die Bewertung der Prüfergebnisse). This part of ISO 105 provides general information about methods for testing the colorfastness of textiles for the guidance of users. The uses and limitations of the methods are pointed out, several terms are defined, an outline of the form of the methods is given, and the contents of the clauses constituting the methods are discussed briefly.

"By colour fastness is meant the resistance of the colour of textiles to the different agencies to which these materials may be exposed during manufacture and their subsequent use. The change of colour and staining of undyed adjacent fabrics are assessed as fastness ratings."

The methods may be used not only for assessing the colorfastness of textiles but also for assessing the colorfastness of dyes. Conditions in the tests have been chosen to correspond closely to the treatments usually employed in manufacture and to the conditions of ordinary use. At the same time, they have been kept as simple and reproducible as possible.

**ISO 105-A02: Grey scale for assessing change in colour** (= DIN 54001-A02: Herstellung und Handhabung des Graumaßstabes zur Bewertung der Änderung der Farbe) describes the gray scale for determining changes in the color of textiles in colorfastness tests, and its use. A precise colorimetric specification of the scale is given.

**ISO 105-A03: Grey scale for assessing staining** (= DIN 54002-A03: Herstellung und Handhabung des Graumaßstabes zur Bewertung des Anblutens) describes the gray scale for determining staining of adjacent fabrics in colorfastness tests, and its use. A precise colorimetric specification of the scale is given.

**ISO 105-A04: Method for the Instrumental Assessment of the Degree of Staining of Adjacent Fabrics** (DIN standard in preparation). This part of ISO 105 specifies an instrumental method for assessing the degree of staining of adjacent fabrics in any fastness test, as an alternative to the visual method.

### 13.2.2. Group B: Colour Fastness to Light and Weathering

**ISO 105-B01: Colour Fastness to Light: Daylight** (= DIN 54003-B01: Bestimmung der Lichtechtheit von Färbungen und Drucken mit Tageslicht). A specimen of the textile is exposed to daylight under prescribed conditions, including protection from rain, along with eight dyed blue wool references. Colorfastness is assessed by comparing the change in color of the specimen with that of the references. Five different methods are explained. In an annex, general information on colorfastness to light is given.

**ISO 105-B02: Colour Fastness to Artificial Light: Xenon Arc Fading Lamp Test** (= DIN 54004-B02: Bestimmung der Lichtechtheit von Färbungen und Drucken mit Xenonbogenlicht). A specimen of the textile is exposed to artificial light under prescribed conditions, along with blue wool references. Two different sets of blue wool references exist, which are not interchangeable. Colorfastness is assessed by comparing the change in color of the specimen with that of the references used. For white (bleached or optically brightened) textiles, the fastness is assessed by comparing the change of whiteness with that of the references used. Five different methods are described.

**ISO 105-B03: Colour Fastness to Weathering: Outdoor Exposure** (= DIN 54070-B03: Bestimmung der Wetterechtheit von Färbungen und Drucken in freier Atmosphäre). Specimens of the textile are exposed under specified conditions in the open air without any protection from weathering. At the same time and place, eight dyed wool references are exposed to daylight, but they are protected from rain, snow, etc., by a sheet of glass. Fastness is assessed by comparing the change in color of the specimen with that of the references. Two different methods are described.

The wide variations in conditions under which outdoor exposures are usually carried out make replicate exposures at different times of the year desirable. The most reliable indication of weathering fastness is obtained by taking the mean of the assessment of several exposures.

**ISO 105-B04: Colour Fastness to Weathering: Xenon Arc** (DIN 54071-B03: Bestimmung der Wetterechtheit von Färbungen und Drucken durch künstliche Bewetterung in Xenonbogenlicht). Specimens of the textile are exposed under prescribed conditions to light from a xenon arc lamp and to water spray. At the same time, eight dyed wool references are exposed to the same light source but are protected from water spray by a sheet of window glass. Fastness is assessed by comparing the change in color of the specimen with that of the references. Two different methods are described. This test procedure can also be used for detection of wet light-sensitive textiles.

**ISO 105-B05: Detection and Assessment of Photochromism** (= DIN 54069-B05: Verfahren zur Erkennung und Bestimmung der Photochromie von Färbungen und Drucken). A specimen of the textile is exposed to light of high intensity for a time much shorter than that necessary to cause a permanent change. The change in color of the specimen is assessed immediately after exposure, by using the gray scale. The specimen is then stored in the dark and assessed again.

### 13.2.3. Group C: Colour Fastness to Washing and Laundering

**ISO 105-C01: Colour Fastness to Washing: Test 1** [= DIN/EN 20105-C01: Bestimmung der Waschechtheit von Färbungen und Drucken Wäsche 40 °C (Waschprüfung ISO 105-C01)]. This part of ISO 105 specifies Test no. 1 of a series of five washing tests that have been established to investigate the fastness to washing of colored textiles. The five tests cover the ranges of washing procedures from mild to severe.

**Table 13.1.** Conditions of colorfastness to washing tests according to ISO

| ISO Designation | Test no. | Corresponding DIN/EN designation | Time, min | Temperature, °C |
|---|---|---|---|---|
| ISO 105-C01 | test 1 | (DIN/EN 20105-C01) | 30 | 40 |
| ISO 105-C02 | test 2 | (DIN/EN 20105-C02) | 45 | 50 |
| ISO 105-C03 | test 3 | (DIN/EN 20105-C03) | 30 | 60 |
| ISO 105-C04 | test 4 | (DIN/EN 20105-C04) | 30 | 95 |
| ISO 105-C05 | test 5 | (DIN/EN 20105-C05) | 240 | 95 |

The five tests are designed to determine the effect of washing only on the colorfastness of the textile. The method is not intended to reflect the result of the comprehensive laundering procedure. A specimen of the textile in contact with one or two specified adjacent fabrics is agitated mechanically under specified conditions of time and temperature in a soap solution, then rinsed, and dried. The change in color and the staining of the adjacent fabric(s) are assessed with the gray scales.

The description of the test procedure for test 1 applies to the four other tests; all five are distinguished by the specified conditions of time and temperature as stated in Table 13.1.

**ISO 105-C06: Colour Fastness to Domestic and Commercial Laundering** ( = DIN 54017-C06: Prüfung der Waschechtheit von Färbungen und Drucken bei der Haushaltswäsche und bei dem gewerblichen Waschen). The test methods in this part of ISO 105 are intended to reflect the effect of comprehensive laundering by either domestic or commercial procedures, as distinct from the washing test methods given in ISO 105-C01 to C05.

A specimen of the textile in contact with specified fabric or fabrics is laundered, rinsed, and dried. Specimens are laundered under appropriate conditions of temperature (40, 50, 60, 70, and 95 °C), duration (30 or 45 min), alkalinity (adjustment of pH to pH 10.5 or no adjustment), bleaching (none or 1 g/L sodium perborate, none or 15 mg/L available chlorine), and abrasive action such that the result is obtained in a conveniently short time. The abrasive action is

accomplished by the use of a low liquor ratio and an appropriate number of steel balls (10, 25, 50, or 100). The change in color of the specimen and the staining of the adjacent fabric(s) are assessed with the gray scales.

### 13.2.4. Group D: Colour Fastness to Dry Cleaning

**ISO 105-D01: Colour Fastness to Dry Cleaning** ( = DIN 54024-D01: Bestimmung der Trockenreinigungsechtheit von Färbungen und Drucken). A specimen of the textile in contact with a cotton fabric bag together with noncorrodible steel disks is agitated in perchloroethylene, then squeezed or centrifuged, and dried in hot air. The change in color of the specimen is assessed with the gray scale. Subsequently, the coloration of the solvent is assessed by comparing the filtered solvent with unused solvent by transmitted light, via the gray scale for assessing staining.

**ISO 105-D02: Colour Fastness to Rubbing: Organic Solvents** (DIN standard in preparation). A specimen of the textile is rubbed with cotton cloth impregnated with solvent. The change in color of the specimen and the staining of the cotton rubbing cloth are assessed with the gray scales.

### 13.2.5. Group E: Colour Fastness to Aqueous Agencies

**ISO 105-E01: Colour Fastness to Water** ( = DIN 54006-E01: Bestimmung der Wasserechtheit von Färbungen und Drucken). A specimen of the textile in contact with one or two specified adjacent fabrics is immersed in water, drained, and placed between two plates under a specified pressure in a test device for 4 h at 37 °C. The specimen and the adjacent fabric(s) are dried separately. The change in color of the specimen and the staining of the adjacent fabric(s) are assessed with the gray scales.

**ISO 105-E02: Colour Fastness to Sea Water** ( = DIN 54007-E02: Bestimmung der Meerwasserechtheit von Färbungen und Drucken). This test is performed in the same manner as colorfastness to water but a 30-g/L aqueous sodium chloride solution is used instead of water.

**ISO 105-E03: Colour Fastness to Chlorinated Water (Swimming-Bath Water)** (=DIN 54019-E03: Bestimmung der Farbechtheit von Färbungen und Drucken gegenüber gechlortem Wasser). A specimen of the textile is treated with a weak aqueous chlorine solution of a given concentration and dried. The change in color of the specimen is assessed with the gray scale.

Three alternative test conditions are specified: Active chlorine concentrations of 50 mg/L and 100 mg/L are intended for swimwear. An active chlorine concentration of 20 mg/L is intended for accessories such as beach robes and towels.

**ISO 105-E04: Colour Fastness to Perspiration** (=DIN 54020-E04: Bestimmung der Schweißechtheit von Färbungen und Drucken). Specimens of the textile in contact with adjacent fabrics are treated in two different solutions containing histidine, drained, and placed between two plates under a specified pressure in a test device for 4 h at 37 °C. The specimens and the adjacent fabrics are dried separately. The change in color of each specimen and the staining of the adjacent fabrics are assessed with the gray scales.

**ISO 105-E05: Colour Fastness to Spotting: Acid** (=DIN 54028-E05: Bestimmung der Säureechtheit von Färbungen und Drucken). Drops of a solution of acid (acetic acid, sulfuric acid, tartaric acid) are placed on the specimen, whose surface is rubbed gently with a glass rod to ensure penetration. The changes in color of the textile, while it is still wet and after drying, are assessed with the gray scale.

**ISO 105-E06: Colour Fastness to Spotting: Alkali** (=DIN 54030-E06: Bestimmung der Alkaliechtheit von Färbungen und Drucken). Drops of a solution of sodium carbonate are placed on the specimen, whose surface is rubbed gently with a glass rod to ensure penetration. The change in color of the textile is assessed with the gray scale.

**ISO 105-E07: Colour Fastness to Spotting: Water** (=DIN 54008-E07: Bestimmung der Wassertropfenechtheit von Färbungen und Drucken). Drops of water are worked into the specimen with a glass rod, so that a spot approximately 20 mm in diameter is formed. The change of color of the textile is assessed with the gray scale after 2 min and after drying.

**ISO 105-E08: Colour Fastness to Water: Hot Water** (=DIN 54047-E08: Bestimmung der Heißwasserechtheit von Färbungen und Drucken). A specimen of the textile in contact with adjacent fabrics is rolled around a glass rod, treated with slightly acidified hot water (pH 6 at 70 °C), and dried. The change in color of the specimen and the staining of the adjacent fabrics are assessed with the gray scales. The method is applicable mainly to wool and to textiles containing wool.

**ISO 105-E09: Colour Fastness to Potting** (=DIN 54048-E09: Bestimmung der Pottingechtheit von Färbungen und Drucken). A specimen of the textile between adjacent fabrics is rolled around a glass rod and treated with boiling water. The specimen and adjacent fabrics are dried separately. The change in color of the specimen and the staining of the adjacent fabrics are assessed with the gray scales.

**ISO 105-E10: Colour Fastness to Decatizing** (=DIN 54054-E10: Bestimmung der Dekaturechtheit von Färbungen und Drucken). A specimen of the textile is wrapped around a perforated cylinder, and steam is passed through it for 15 min. The change in color of the dried specimen is assessed with the gray scale. Correct handling of the method is controlled by use of a test-control specimen tested under identical conditions. Two tests, mild and severe, are given.

**ISO 105-E11: Colour Fastness to Steaming** (=DIN 54058-E11: Bestimmung der Dämpfechtheit unter atmosphärischem Druck von Färbungen und Drucken). A specimen of the textile in contact with specified adjacent fabrics is rolled into a cylinder and placed for 30 min in the neck of a flask containing boiling water. The staining of the adjacent fabrics is assessed with the gray scale.

**ISO 105-E12: Colour Fastness to Milling: Alkaline Milling** [=DIN 54041-E12: Bestimmung der alkalischen Walkechtheit (schwere Beanspruchung)]. A specimen of the textile in contact with one or two specified adjacent fabrics is milled at 40 °C in a jar containing 50 steel balls and a solution of soap and sodium carbonate. The severity of action is controlled by means of a test-control dyeing milled separately in the same way. After rinsing and drying, the change in color and the staining of the adjacent fabrics are assessed with the gray scale.

**ISO 105-E13: Colour Fastness to Acid-Felting: Severe** [DIN 54043-E13: Bestimmung der sauren Walkechtheit von Färbungen und Drucken (schwere Beanspruchung)]. A specimen of the textile in contact with adjacent fabrics is milled in solutions of acetic or sulfuric acid as used under severe conditions in the acid-felting process, rinsed, and dried. The change in color of the specimen and the staining of the adjacent fabrics are assessed with the gray scales.

**ISO 105-E14: Colour Fastness to Acid-Felting: Mild** [=DIN 54042-E14: Bestimmung der sauren Walkechtheit (leichte Beanspruchung)]. A specimen of the textile in contact with adjacent fabrics is milled in mineral acid solutions at 60 °C, as used under mild felting conditions in the hatmaking and felt industries, rinsed, and dried. The change in color of the specimen and the staining of the adjacent fabrics are assessed with the gray scales.

### 13.2.6. Group G: Colour Fastness to Atmospheric Contaminants

**ISO 105-G01: Colour Fastness to Nitrogen Oxides** (=DIN 54025-G01: Bestimmung der Stickoxidechtheit von Färbungen und Drucken). Specimens of textiles are exposed to nitrogen oxides in a closed container until either one (dyes with poor color fastness to nitrogen) or three (dyes with good color fastness to nitrogen) test-control specimens exposed simultaneously with the test specimens have changed color to a predetermined extent. The change in color of the textile is assessed with the gray scale.

**ISO 105-G02: Colour Fastness to Burnt Gas Fumes** (no DIN standard issued up to now). A specimen of the textile and the test-control fabric are exposed simultaneously to oxides of nitrogen from burnt gas fumes until the test-control shows a change in color corresponding to that of the standard of fading. The change in color of the specimen is assessed with the gray scale. If no color change is observed in the specimen after one exposure period or cycle, exposure may be continued either for a specified number of periods or for the number of periods required to produce a specified amount of color change in the specimen.

**ISO 105-G03: Colour Fastness to Ozone in the Atmosphere** (no DIN standard issued up to now). A specimen and a swatch of test-control sample are exposed simultaneously to ozone in an atmosphere at ambient temperature and relative humidities not exceeding 65 %, until the control sample shows a color change corresponding to that of a standard of fading. This exposure period constitutes one cycle. The cycles are repeated either until the specimen shows a definite color change or for a prescribed number of cycles.

### 13.2.7. Group J: Measurement of Colour and Colour Differences

**ISO 105-J01: Measurement of Colour and Colour Differences** (no DIN standard). This part of ISO 105 selects from several options published by the CIE those best suited to the needs of the textile industry whenever the difference in color between two specimens has to be quantified.

**ISO 105-J02: Method for the Instrumental Assessment of Whiteness** (no DIN standard). This part of ISO 105 specifies a method intended for quantifying the whiteness of textiles, including fluorescent materials.

### 13.2.8. Group N: Colour Fastness to Bleaching Agencies

**ISO 105-N01: Colour Fastness to Bleaching: Hypochlorite** [=DIN 54035-N01: Bestimmung der Hypchlorit-Bleichechtheit (schwere Beanspruchung)]. A specimen of the textile is agitated in a solution of 2 g/L sodium hypochlorite or lithium hypochlorite, rinsed in water, agitated in a hydrogen peroxide solution or sodium hydrogensulfite solution, rinsed, and dried. The change in color of the specimen is assessed with the gray scale.

**ISO 105-N02: Colour Fastness to Bleaching: Peroxide** (=DIN 54033-N02: Bestimmung der Peroxid-Bleichechtheit von Färbungen und Drucken). A specimen of the textile in contact with adjacent fabrics is immersed in the bleaching solution, containing sodium peroxide in concentrations commonly used in textile bleaching processes, rinsed, and dried. Four different bleaching baths are described. The change in color and the staining are assessed with the gray scales.

**ISO 105-N03: Colour Fastness to Bleaching: Sodium Chlorite: Mild** [= DIN 54036-N03: Bestimmung der Chlorit-Bleichechtheit von Färbungen und Drucken (leichte Beanspruchung)]. A specimen of the textile in contact with specified adjacent fabrics is treated in a sodium chlorite solution (1 g/L, pH 3.5 at 80 °C for 1 h), rinsed, and dried. The change in color and the staining of the adjacent fabrics are assessed with the gray scales.

**ISO 105-N04: Colour Fastness to Bleaching: Sodium Chlorite: Severe** [= DIN 54037-N04: Bestimmung der Chlorit-Bleichechtheit von Färbungen und Drucken (schwere Beanspruchung)]. A specimen of the textile in contact with specified adjacent fabrics is treated in a sodium chlorite solution (2.5 g/L, pH 3.5 at 80 °C for 1 h), rinsed, and dried. The change in color of the specimen and the staining of the adjacent fabrics are assessed with the gray scales.

**ISO 105-N05: Colour Fastness to Stoving** (= DIN 54038-N05: Bestimmung der Schwefelechtheit von Färbungen und Drucken). A specimen of the textile, containing its own mass of soap solution, and a test-control specimen are exposed in an atmosphere containing sulfur dioxide. The change in color of the specimen and the staining of the adjacent fabrics are assessed with the gray scales.

### 13.2.9. Group P: Colour Fastness to Heat Treatments

**ISO 105-P01: Colour Fastness to Dry Heat (Excluding Pressing)** (= DIN 54060-P01: Bestimmung der Trockenhitzefixier- und Trockenhitzeplissierechtheit von Färbungen und Drukken). A specimen of the textile in contact with specified adjacent fabrics is heated by close contact with a medium that is brought to the required temperature (30 s at 150, 180, or 210 °C). The change in color of the specimen and the staining of the adjacent fabrics are assessed with the gray scales.

**ISO 105-P02: Colour Fastness to Pleating: Steam Pleating** (= DIN 54059-P02: Bestimmung der Dämpfechtheit mit Überdruck von Färbungen und Drucken). A specimen of the textile in contact with specified adjacent fabrics is steamed under pressure (135, 170, 270 kPa, for

5, 10, or 20 min) and dried. The change in color and the staining of the adjacent fabrics are assessed with the gray scales.

### 13.2.10. Group S: Colour Fastness to Vulcanization

**ISO 105-S01: Colour Fastness to Vulcanization: Hot Air** (= DIN 54065-S01: Bestimmung der Heißluftvulkanisierechtheit von Färbungen und Drucken). A specimen of the textile in air is in direct contact with an (initially) unvulcanized rubber compound. The change in color of the specimen is assessed with the gray scale.

**ISO 105-S02: Colour Fastness to Vulcanization: Sulfur Monochloride** (= DIN 54064-S02: Bestimmung der Kaltvulkanisierechtheit von Färbungen und Drucken). A specimen of the textile is exposed to sulfur monochloride vapor under conditions usually occuring during the cold vulcanization of rubber. The change in color of the specimen is assessed with the gray scale before and after neutralizing with aqueous ammonia.

**ISO 105-S03: Colour Fastness to Vulcanization: Open Steam** (= DIN 54066-S03: Bestimmung der Heißdampfvulkanisierechtheit von Färbungen und Drucken). A specimen of the textile is heated in live steam in direct contact with an (initially) unvulcanized rubber compound—the textile material being wrapped in either

1) Sheeting impermeable to steam and water (Method A) or
2) Undyed bleached cotton cloth, but ensuring that live steam is not prevented from infiltration into the specimen (Method B)

The change in color of the specimen and the staining of the adjacent fabric are assessed with the gray scale.

### 13.2.11. Group X: Colour Fastness to Miscellaneous Agencies

**ISO 105-X01: Colour Fastness to Carbonizing: Aluminium Chloride** (= DIN 54044-X01: Bestimmung der Karbonisierechtheit von Färbungen und Drucken mit Aluminiumchlorid). A specimen impregnated with aluminum chloride solution is dried, baked (15 min at

115 °C), rinsed, and neutralized. The method is applicable mainly to wool and textiles containing wool. The changes in color after rinsing, neutralizing, and drying are assessed with the gray scale.

**ISO 105-X02: Colour Fastness to Carbonizing: Sulphuric Acid** ( = DIN 54045-X02: Bestimmung der Karbonisierechtheit von Färbungen und Drucken mit Schwefelsäure). A specimen impregnated with sulfuric acid solution is dried, baked (15 min at 105 °C), rinsed, and neutralized. The method is applicable mainly to wool and textiles containing wool. The changes in color after rinsing, neutralizing, and drying are assessed with the gray scale.

**ISO 105-X04: Colour Fastness to Mercerizing** ( = DIN 54039-X04: Bestimmung der Mercerisierechtheit von Färbungen und Drucken). A specimen of the textile in contact with a specified cotton adjacent fabric is treated with a 300 g/L sodium hydroxide solution (5 min at 20 °C), rinsed with 70 °C water and subsequently with cold water, acidified, rinsed again, and dried. The change in color of the specimen and the staining of the cotton adjacent fabric are assessed with the gray scales. Completely resistant specimens may show an apparent increase in depth of color. In such cases an assessment of 5 cannot be given; the fastness number is marked by an asterisk, which must be explained in a footnote.

**ISO 105-X05: Colour Fastness to Organic Solvents** ( = DIN 54023-X05: Bestimmung der Lösemittelechtheit von Färbungen und Drucken). A specimen of the textile in contact with adjacent fabrics is agitated in the solvent and dried. The change in color of the specimen and the staining of the adjacent fabrics are assessed with the gray scales.

**ISO 105-X06: Colour Fastness to Soda Boiling** ( = DIN 54031-X06: Bestimmung der Sodakochechtheit von Färbungen und Drucken). A specimen of textile between specified undyed cloths is rolled around a glass rod and treated with boiling sodium carbonate solution for 1 h with and without the addition of a reduction inhibitor. The composite specimen is rinsed and dried. The change in color and the staining of the undyed cloths are assessed with the gray scales.

**ISO 105-X07: Colour Fastness to Cross-Dyeing: Wool** ( = DIN 54049-X07: Bestimmung der Überfärbechtheit von Färbungen und Drucken). Specimens of the textile in contact with adjacent fabrics are treated in different types of wool dyebaths, but without any dye. The change in color of the specimen and the staining of the adjacent fabrics are assessed with the gray scales.

**ISO 105-X08: Colour Fastness to Degumming** ( = DIN 54055-X08: Bestimmung der Entbastungsechtheit von Färbungen und Drucken). A specimen of the textile in contact with adjacent fabrics is treated with a soap solution, similar to those used in degumming raw silk, then rinsed, and dried. The change in color of the specimen and the staining of the adjacent fabrics are assessed with the gray scales.

**ISO-105-X09: Colour Fastness to Formaldehyde** ( = DIN 54061-X09: Bestimmung der Formaldehydechtheit). A specimen of the textile is exposed in a closed container to the action of gaseous formaldehyde (24 h at 20 °C). The change in color of the specimen is assessed with the gray scale.

**ISO 105-X10: Assessment of Migration of Textile Colours into Polyvinyl Chloride Coatings** ( = DIN 54072-X10: Bestimmung der Echtheit von Färbungen und Drucken gegenüber PVC-Weichmachern). A specimen of a textile impregnated with plasticizer is brought into contact with a white-pigmented poly(vinyl chloride) film and kept under pressure for 3.5 h at 80 °C. Then the specimen and excess plasticizer are removed from the film, and the staining of the film is assessed with the gray scale.

**ISO 105-X11: Colour Fastness to Hot Pressing** ( = DIN 54022-X11: Bestimmung der Bügelechtheit von Färbungen und Drucken). Three methods are described; in each method, hot pressing is performed for 15 s at 110, 150, or 200 °C:

1) *Dry Pressing.* A dry specimen is pressed with a heating device at a specified temperature and pressure.
2) *Damp Pressing.* A dry specimen is covered with a wet cotton adjacent fabric and pressed with a heating device at a specified temperature and pressure.

3) *Wet Pressing*. The upper surface of a wet specimen is covered with a wet cotton adjacent fabric and pressed with a heating device at a specified temperature and pressure.

After hot pressing, the change in color of the specimen and the staining of the adjacent fabric are assessed with the gray scales immediately and again after a period of exposure to air.

**ISO 105-X12: Colour Fastness to Rubbing** (= DIN 54021-X12: Bestimmung der Reibechtheit von Färbungen und Drucken). Specimens of the textile are rubbed with a dry rubbing cloth and with a wet rubbing cloth in a suitable testing device. Two alternative sizes of rubbing finger are specified. The staining of the rubbing cloths is assessed with the gray scale.

**ISO 105-X13: Colour Fastness of Wool Dyes to Processes using Chemical Means for Creasing, Pleating, and Setting** (no DIN standard). A specimen of the textile, treated with the chemical solution, is placed in contact with specified adjacent fabrics and subjected to steam pressing. A comparison specimen, not treated with the chemical solution, is steam pressed simultaneously. The specimens are dried, and any differences between the colors of the two specimens and the staining of the adjacent fabrics are assessed with the gray scales.

**ISO 105-X14: Colour Fastness to Acid Chlorination of Wool: Sodium Dichloroisocyanurate** (no DIN standard). A specimen of the textile is treated in a formic acid buffer solution to which solutions of sodium dichloroisocyanurate and sodium hydrogensulfite are added successively, and is then rinsed and dried. The change in color of the specimen is assessed with the gray scale.

### 13.2.12. Group Z: Colourant Characteristics

**ISO 105-Z01: Colour Fastness to Metals in the Dyebath: Chromium Salts** (= DIN 54052-Z01): Bestimmung der Farbechtheit gegenüber Chrom-VI-Verbindungen im Färbebad). The difference in color between dyeings made with and without potassium dichromate is assessed with the gray scale.

**ISO 105-Z02: Colour Fastness to Metals in the Dyebath: Iron and Copper** (= DIN 54053-Z02: Bestimmung der Farbechtheit gegenüber Eisen(III)- und Kupfer(II)-Ionen im Färbebad). The difference in color between dyeings made in the presence and in the absence of salts of the metal is assessed with the gray scale.

## 14. Legal Aspects, Toxicology, and Ecology

The textile finishing industry is one of the branches of industry with the highest specific water consumption. According to the German Association of Textile Finishers (Deutscher Textilveredlungsverband, TVI), the German textile finishing industry consumed ca. $65 \times 10^6$ m$^3$ of water in 1992, which is 0.6% of the total ca. $11 \times 10^9$ m$^3$ water requirement of the mining and processing industries combined. Of this quantity, 88.4% is discharged as wastewater, and 11.6% represents evaporation losses. The average water consumption was 146 m$^3$ per tonne of fiber material. About 90% of the raw water is self-produced and costs, on average, 0.50 DM/m$^3$; wastewater costs are 2.35 (range 0.63–5.35) DM/m$^3$. Together, the fresh water and wastewater costs average 5.2% of the turnover (only finishing work) [14.1].

### 14.1. Water Consumption

The water consumption of the textile finishing industry depends on the finishing process and method, type of fiber, form, and the machines and equipment used. A survey of the amounts of water required in terms of textiles and dyeing equipment is presented in Tables 14.1 and 14.2 [14.2].

In the past, water consumption has been reduced perceptibly by the use of water-saving equipment and continuous processes.

### 14.2. Legal Regulations

As a result of the intensive use of water in the textile finishing industry, the regulations for

**Table 14.1.** Water consumption (including pretreatment)

| Textiles | Consumption, m$^3$ per tonne fiber material |
|---|---|
| Cotton fabric | 80–240 |
| Cotton woven goods | 70–180 |
| Woolen fabric | 100–250 |
| Polyacrylic fabric | 10–70 |

**Table 14.2.** Liquor ratio of various dyeing equipments

| Dyeing equipment | Liquor ratio |
| --- | --- |
| Star dyeing apparatus | 100:1 |
| Winch beck | 10:1–30:1 |
| Jigger | 3:1–5:1 |
| Beam dyeing machine | 8:1–10:1 |
| Overflow | 4:1–10:1 |
| Padder | 0.6:1–0.8:1 |

wastewater discharge are most important. A distinction must be made between wastewater that is discharged into a municipal sewage system (indirect discharge) and wastewater that is discharged into surface waters (after purification, if necessary). The demands to be met in the second case are greater.

**Germany and Austria.** The regulations for wastewater that is discharged into surface waters are stipulated in the German Water Resources Act (Wasserhaushaltsgesetz, WHG) [14.3] and particularly in the decrees and administrative directions given in §7a. The requirements for discharge are regulated in a branch-specific manner through the ordinance on the source of wastewater [14.4] and the general administrative regulations [14.5] with the corresponding appendices. For wastewater that comes largely from the textile finishing industry, the 38th administrative regulation from September 5, 1984, is still valid (end of 1994) (Table 14.3), but it should be replaced shortly by Appendix 38 of the general administrative regulations [14.7].

Apart from the minimum requirements for wastewater discharge in keeping with the state of the art and generally accepted technological conventions (Table 14.3), general demands are also made on the use of certain chemicals, treatment of partial streams, concentrates, and residual amounts of materials used. About 95 % of the German textile finishers are indirect dischargers. Their wastewater discharge is controlled not on-

**Table 14.3.** Minimum requirements for wastewater direct discharge in Germany

| Parameter | 38. VwV[a] from 1984[b] | Appendix 38, generally accepted technological conventions | Appendix 38, state of the art[c] |
| --- | --- | --- | --- |
| Filterable substances, mg/L | 40 | | |
| Settleable substances, mg/L | 0.3 | | |
| COD, mg/L | 280 | 160 | |
| $BOD_5$, mg/L | 40 | 25 | |
| Phosphorus, total, mg/L | | 2 | |
| Ammonium nitrogen, mg/L | 5 | 10 | |
| Nitrogen from nitrate, nitrite and ammonium, mg/L | | 20 | |
| Iron, mg/L | | 3 | |
| Aluminum, mg/L | | 3 | |
| Sulfite, mg/L | 1 | 1 | |
| Hydrocarbons, mg/L | 10 | 10 | |
| Fish toxicity as dilution factor | 4 | | 2 |
| Absorbable organic halogens (AOX), mg/L | | | 0.5 |
| Free chlorine, mg/L | | | 0.3 |
| Aromatics (benzene and homologues), mg/L | | | 0.1 |
| Sulfide, mg/L | 0.1 | | 1 |
| Chromium(VI), mg/L | | | 0.1 |
| Chromium, total, mg/L | 2 | | 0.5 |
| Copper, mg/L | 1 | | 0.5 |
| Nickel, mg/L | | | 0.5 |
| Zinc, mg/L | 3 | | 2 |
| Color (translucent color value) | | | |
|    436 nm (yellow) | | | $7\,m^{-1}$ |
|    525 nm (red) | | | $5\,m^{-1}$ |
|    620 nm (blue) | | | $3\,m^{-1}$ |

[a] VwV = Administrative regulation. [b] 2-h mixed sample. [c] Qualified random sample or 2-h mixed sample.

ly by state water laws, but also by local regulations of the concerned sewage treatment plants. The amended §7a of the WHG represents a reform because the same minimum requirements must be met for dangerous substances by direct and indirect dischargers [14.8]. (This refers to substances that are assessed as dangerous because of the fear of toxicity, longevity, capability of accumulation, or a carcinogenic, teratogenic, or mutagenic effect.) A similar decree exists in Austria for direct and indirect dischargers [14.9].

**England and Wales.** In England and Wales, the Water Law of 1980 [14.10] gave rise to the National River Authorities (NRA) that establish and enforce quality standards for waters [14.11], [14.12].

**United States.** In the United States, no uniform national or state wastewater standards exist. As of now, industry-specific guidelines established by the EPA do not exist for the textile industry. The minimum requirements are determined by public-owned treatment plants or local officials.

**Japan.** In Japan, the environmental authorities determine the national wastewater standards, which can, however, be raised according to the special situation of local prefectures. A selection of Japanese wastewater standards for direct discharge is given below [14.13]

| | |
|---|---|
| pH (inland waters) | 5.8–8.6 |
| BOD, COD | 160 mg/L |
| Settleable substances | 200 mg/L |
| Copper | 3 mg/L |
| Chromium, total | 2 mg/L |
| Chromium(VI) | 0.5 mg/L |

| | |
|---|---|
| Zinc | 5 mg/L |
| Iron | 10 mg/L |
| Fluoride | 15 mg/L |
| Nitrogen | 120 mg/L |
| Phosphorus | 16 mg/L |
| Tetrachloroethylene | 0.1 mg/L |

A compilation of the requirements to be met for the discharge of some important substances into public-owned treatment plants in various countries is given in Table 14.4 [14.12], [14.14].

## 14.3. Components of Textile Wastewater

The composition of textile wastewater depends to a large extent on the dyed substrates, dyes, textile auxiliaries, and chemicals used, which makes the statement of average values impossible. As a rule, untreated textile wastewater has an alkaline pH, a higher temperature, high conductivity, and a poorer $BOD_5$:COD ratio (i.e., lower degradability) compared to domestic wastewater. Typical $BOD_5$:COD values lie between 1:2.5 and 1:5 [14.10]. The organic load of the wastewater originates largely from pretreatment of the materials—$BOD_5$ values of up to 140 kg $O_2$ per tonne of material being obtained for desizing and 210 kg/t in the case of alkaline boiling out of polyester–cotton blend fabrics [14.16], [14.17]. Wastewater from dyehouses, which work primarily with sulfur or vat dyes, have high COD values because of the organic or inorganic reducing agents (glucose, hydroxyacetone, sodium dithionite) used. An analysis of the wastewater from a cotton finisher is shown in Table 14.5.

The contaminants, which are controlled on discharge into the public sewage system or into

**Table 14.4.** Requirements for indirect discharge (selection)

| Parameter | Belgium | Denmark | Switzerland | Austria | United States* | Japan (Osaka)** |
|---|---|---|---|---|---|---|
| pH | 5.5–10 | 5.5–10 | 5.5–10 | 6.5–9.5 | 6–9 | 5–8 |
| Temperature, °C | 45 | 30–35 | 60 | 40 | 50 | 40 |
| AOX, mg/L | | | 0.1 | 0.5 | | |
| Sulfate, mg/L | 2000 | | 300 | 200 | | |
| Chlorine, mg/L | | 0.5 | 3 | 1 | | |
| Chromium(VI), mg/L | 0.3 | 2 | 0.5 | 0.1 | | 0.5 |
| Total chromium, mg/L | 4 | | 2 | 1 | 0.5 | 2 |
| Copper, mg/L | 0.5 | 1 | 1 | 0.5 | 0.5 | 3 |
| Nickel, mg/L | 1 | 1 | 2 | | | |
| Zinc, mg/L | 5 | 1–2 | 2 | 2 | 1 | |

* Typical example [14.14]. ** [14.15].

**Table 14.5.** Typical wastewater of a cotton finisher [14.1]

| Parameter | Average | Minimum | Maximum |
|---|---|---|---|
| pH | | 8.5 | 10.3 |
| Conductivity, mS/m | 650 | 420 | 1400 |
| Temperature, °C | 27 | 25 | 38 |
| COD, mg/L | 650 | 420 | 1400 |
| $BOD_5$, mg/L | 180 | 80 | 500 |
| Phosphate (as P), mg/L | 50 | 26 | 80 |
| Sulfate, mg/L | 810 | 750 | 1050 |
| Ammonium (as N), mg/L | 0.7 | 0.6 | 1.0 |
| Hydrocarbons, mg/L | 5 | 3 | 15 |
| Chloride, mg/L | 800 | 400 | 1500 |
| AOX, mg/L | 0.8 | 0.5 | 1.2 |
| Settleable substances, mg/L | 10 | 1 | 150 |
| Copper, mg/L | < 0.1 | < 0.1 | 0.6 |
| Chromium, mg/L | < 0.01 | 0.015 | 0.034 |

surface water, can have many different sources (Table 14.6) [14.14], [14.16], [14.18]–[14.20]. Apart from metal complexes, the heavy-metal content in dyes is usually in the range of a few micrograms per gram and is due partially to the catalysts used in production [14.14].

Depending on the coloristic class, dyes are fixed in various quantities (see below) [14.2]:

| | |
|---|---|
| Direct dyes | 70–95 % |
| Vat dyes | 80–95 % |
| Sulfur dyes | 60–70 % |
| Reactive dyes | 50–95 % |
| Disperse dyes | 80–92 % |
| Acid dyes | 80–93 % |
| Afterchroming dyes | 98–99 % |
| Cationic (basic) dyes | 97–98 % |

Nevertheless, their share (ca. 1–2 %) of the organic wastewater load is very low [14.21].

## 14.4. Reduction of Wastewater Load by Special Preventive Measures

If certain requirements are not fulfilled, the wastewater must be treated or the discharge of the contaminants in question must be lowered or completely eliminated. As a rule, preventive measures are more economical than subsequent contaminant elimination. The discharge of a number of the contaminants listed in Table 14.6 can be reduced or eliminated by process changes

**Table 14.6.** Important sources of contaminants in wastewater from dyehouses

| Parameter, contaminant | Source |
|---|---|
| Alkaline pH value | dyeing with reactive, vat, and sulfur dyes; mercerization, boiling-off, bleaching |
| Acidic pH value | dyeing with basic, acidic, and disperse dyes; easy care finish |
| Color | reactive and sulfur dyes |
| Heavy metals | |
|   Chromium | metal-complex dyes, oxidant |
|   Cobalt | metal-complex dyes |
|   Copper | metal-complex dyes, pigments, improvers of lightfastness |
|   Nickel | metal-complex dyes |
|   Zinc | cationic dyes, resist and reducing agents, biocides, catalysts, water piping |
|   Antimony | flame retardant |
|   Tin | biocides |
| Halogenated hydrocarbons | detergents, degreasers, carrier, chlorine bleaching |
| AOX | dyes, carrier, chlorine bleaching, preservatives, mothproofers, wool chlorination |
| Mineral oils | yarn preparations, spinning oils, defoamers, dedusting agents, print thickening |
| Phosphorus, phosphates | buffers, sequestrants, flame retardants |
| Sulfide | sulfur dyes |
| Neutral salts | reactive and substantive dyeing |

or exchange of products. Residual color from reactive dyes can be reduced by using dyes with a higher fixing yield (i.e., with more affinitive, bifunctional, or bireactive dyes [14.22]). Higher yields are also achieved by a shorter liquor ratio or by using continuous or semicontinuous processes instead of exhaustion processes [14.23]. Another possibility is the use of higher salt concentrations, which, however, stand in the way of the demand for lower electrolyte concentrations in wastewater. From an economic viewpoint, optimization of the formulation can be achieved by means of suitable computer programs [14.24]. In the case of copper-containing metal-complex dyes, these methods result in the lowering of an increased copper concentration. The complete avoidance of metal complexes, which are employed for some violet, blue, navy blue, and black shades in direct and reactive dyes, results in marked losses in lightfastness. Turquoise and brilliant green shades can be produced only with copper or nickel phthalocyanine complex dyes.

In padding processes, a general problem is the *unused dye liquor* in the pad-trough, piping, and dissolving tank. Better utilization of dye liqour can be achieved by using padders with a smaller content [14.25] or by reusing the remaining liquor in other dyeing processes, which requires colorimetric determination of the correct formulation in the laboratory [14.26]. With reactive dyes, however, reuse is limited because the liquors are not stable for long in the presence of fixing alkali.

In the wastewater from wool dyeworks, increased *chromium concentrations* are frequently encountered. They are caused by chromium metal-complex dyes and/or by the chroming of chrome dyes. To lower chromium concentration, optimized chroming methods have been developed. In fact, chromium concentrations in residual liquors have been reduced from 20 – 150 mg/L in the case of traditional dyeing to 1 – 6 mg/L. Thus, the concentrations lie below those obtained on black dyeing with 1:2 metal-complex dyes, which can have up to 13 mg/L of chromium [14.27], [14.28]. Optimized chroming methods include the International Wool Secretariat (IWS) low-temperature chrome dyeing method at 90 °C with the addition of thiosulfate, the Bayer Glauber's salt technique with dye-specific chrome factors, and the Sandoz method with minimum chrome factors and a special auxiliary agent.

The *bleaching of cotton* with hypochlorite is an important source of absorbable organic halogen (AOX). In laboratory experiments with raw cotton, the AOX detected in the bleaching liquor was up to 30 mg/L and, in the case of desized materials, up to 80 mg/L. In addition, up to 11 mg/L of chloroform can be formed by a haloform reaction during hypochlorite bleaching under standard conditions. In contrast, bleaching with sodium chlorite or hydrogen peroxide gives ca. 10 or less than 0.5 mg/L AOX, respectively [14.29], [14.30]. The transition to bleaching with hydrogen peroxide can reduce both AOX content and free chlorine in the wastewater.

*Phosphorus* enters wastewater largely via buffer substances or sequestering agents. Although some countries have already installed stages for phosphate elimination in sewage treatment plants, a reduction is desirable because of the possibility of eutrophication of surface waters. Apart from the use of phosphorus-free substances, an alternative to phosphorus compounds as buffers is dosing with acids or bases via automatic dosing equipment. As sequestering agents and peroxide stabilizers in peroxide bleaching, polyhydroxy carbonates, aminocarboxylates [e.g., ethylenediaminetetraacetate (EDTA) or nitrilotriacetate (NTA)], phosphonates or modified polyacrylates are used instead of phosphorus compounds [14.31]. However, EDTA should not be used because it is poorly biodegradable and can remobilize heavy metals from the sediment. The use of sequestering agents can be almost avoided by performing an extraction step with formic or hydrochloric acid before bleaching, eliminating hardness and heavy metals from the material.

In reactive dyeing, the use of *salt* can be avoided when continuous or semicontinuous processes are used instead of exhaustion processes. In the case of reactive dyes that have a low salt sensitivity, the amount of salt can be lowered without yield losses [14.24]. *Sulfites and sulfates* are also formed in the oxidation of sodium dithionite (hydrosulfite), which is used both as a stripping agent for faulty dyeing and for reduction of vat dyes. Thiourea dioxide or reductones such as hydroxyacetone are alternatives to hydrosulfite. The disadvantages of both chemicals are overreduction in the case of vat dyes of the indanthrone type and changes in shade [14.32]. Hydroxyacetone, which is easily degradable, is being used increasingly in indigo dyeing. *Sulfide* formed when dyeing with sulfur dyes must be oxidized before discharge. The use of low-sulfur, prereduced dyes, which are reduced by glucose, lowers the electrolyte load and avoids the required sulfide oxidation [14.26].

*Multiple use of exhausted dye liquors* is usually restricted to dyes with high exhaustion yield and exhibit low interactions with the auxiliaries remaining in the liquor. Examples are the dyeing of cotton with indigo or sulfur dyes and the dyeing of cotton velvet with direct dyes. Polyamide can be dyed with acid dyes, acrylic fibers with cationic dyes, and polyester with disperse dyes in a standing bath [14.21], [14.32]. Multiple use of the residual liquors from finishing processes should also be aimed at. However, the possible staining of the padding liquors with dyes, the correct analysis of present composition, and the storage stability can cause problems [14.33]. The advantages of reuse are reduced organic load of the wastewater and savings of water, heating energy, and, in part, auxiliary agents and chemicals. Water consumption can also be decreased by using dyeing equipment with shorter liquor ratios [14.34] (see Section 14.1), optimized process operation (e.g., washing in reactive dyeing [14.35]), and a series of other measures [14.21].

## 14.5. Special Methods of Wastewater Treatment

The usual methods of wastewater treatment can also be applied to the textile finishing industry. A special case is, however, the removal of dyes. Apart from paper dyeing, the presence of dyes in wastewater to this extent is specific to the textile industry.

**Biological Treatment.** As mentioned in Section 14.3, textile wastewater is frequently poorly biodegradable. This is of minor importance if textile wastewater represents a relatively small part of the total wastewater of a municipal biological sewage treatment plant. Nevertheless, textile wastewater can be subjected satisfactorily to biological treatment even if it is not mixed with municipal wastewater. Because of the different textile processing program, textile wastewater has a complex and rapidly changing composition. For this reason, preceding equalization of composition in adequately dimensioned balancing tanks is required for optimal purification. In addition, a certain temperature compensation is achieved, extreme pH values are avoided by internal neutralization, etc. The balancing tanks also eliminate the need for design of the treatment plant based on maximum wastewater values, serve charge of the biological treatment plant on work-free weekends, and represent safety stages in case of breakdown. The COD value of wastewater has been reported to be reduced almost 50 %, and the mineral oil concentration by 70 %, by appropriate design of the balancing tank, introduction of adapted microorganisms, and additional aeration [14.36]. The pilot plant and the operation of a one-stage aerated biological sewage treatment plant run by the finisher of cotton and blended fabrics are described in [14.37]. Adaptation of microorganisms in the one-stage plant was shown to considerably improve treatment efficiency in comparison with the pilot plant and the results at the start of operation. Biological treatment plants can be improved by combination with preceding

or subsequent precipitation–flocculation [14.38], [14.39] and addition of activated carbon or cheaper brown-coal cokes in the activated sludge part [14.40]–[14.42]. Pilot experiments on the decolorization of the outflow from a biological sewage treatment plant using an activated carbon fixed-bed reactor show a bacterial degradation of the dye, which occurs in addition to purely adsorptive processes [14.43]. Highly loaded wastewater (e.g., partial streams from pretreatment and desizing), is very suitable for anaerobic treatment. Partial streams from dyeworks are not easily degradable anaerobically, especially in the presence of better degradable substrates [14.44], [14.45].

**Precipitation – Flocculation.** Today, chemical precipitation–flocculation belongs to the generally accepted technological conventions and is used for extensive wastewater treatment in many branches of industry. The advantages are increased elimination of contaminants, easy combinability with other processes, and relative insensitivity of these processes to substances that are toxic for bacteria and to fluctuations in wastewater composition [14.46]. The usual inorganic precipitating agents are lime, aluminum and iron salts [e.g., iron chloride sulfate, iron(II) and iron(III) chloride]. Instead of direct dosing into the wastewater, electrochemical production of the metal cations or their hydroxides by corresponding sacrificial electrodes is possible. Apart from reducing the contaminant load (COD, BOD, heavy metals, and mineral oils), dyes are also eliminated, at least partially. These dyes include vat, disperse, sulfur, direct, and acid dyes, as well as pigments [14.47]–[14.50]. Reactive dyes are not adequately removed by the usual inorganic precipitants. For this purpose, polycationic precipitating agents are used. With dyes containing sulfonic acid groups, these agents form compounds that are poorly soluble in water and flocculate [14.16], [14.23], [14.51]–[14.53]. In both cases, the sludge can be separated by flotation or sedimentation.

**Oxidation.** The disadvantage of precipitation and flocculation is the production of large amounts of sludge, which must be treated and disposed of. Chemical processes, especially oxidation and reduction, usually do not have this disadvantage. The oxidizing agents used in practice or in research projects for the treatment of textile wastewater (especially for the removal of residual color) are chlorine [14.42] or hydrogen peroxide, the latter usually together with iron(II) sulfate (Fenton's reagent) [14.54]–[14.56], ozone [14.57]–[14.59], or UV irradiation [14.60], [14.61]. A combination of ozone and UV irradiation also exists. The disadvantage of chlorination is the formation of organic chloro compounds. Fenton's reagent effectively destroys water-soluble dyes. However, it also nonspecifically oxidizes organic pollutants expressed as COD, leading to a high consumption of peroxide and correspondingly high costs. Ozone reacts specifically with dyes, especially at neutral and acidic pH. Thus, other water components such as surfactants and thickeners are hardly attacked. Wastewater treated with oxidizing agents shows better degradability and a lower toxicity for bacteria. The reaction products of ozone treatment include saturated and unsaturated carboxylic acids (e.g., oxalic, malonic, maleic, and phthalic acids) [14.62]. For the special treatment of wastewater from sulfur dyeing, see Section 4.5.6.

**Reduction.** Sodium dithionite is the agent of choice for reductive decolorization. Under suitable reaction conditions, not only azo dyes but also dyes based on other chemical structures (e.g., anthraquinone or formazane dyes) [14.16], [14.63] can be permanently reduced. Treatment with iron(II) sulfate–lime, which uses the reduction potential of divalent iron, represents a combination of reduction and precipitation. This stage precedes the biological wastewater treatment plant [14.64]. Divalent iron ions can be produced electrochemically on activated iron electrodes, which can also reduce hexavalent chromium from wool dyeing [14.28]. Both direct and indirect cathodic dye reduction [i.e., via an iron(II)–iron(III)–triethanolamine redox system] is possible by electrolysis with higher current strengths [14.65]. In purely chemical methods of treatment, which proceed without the use of precipitants, no elimination of heavy metals (e.g., from metal-complex dyes) occurs.

**Adsorption.** The most important adsorbent in wastewater treatment is activated carbon. A distinction is made between the stirring process with powdered carbon and the percolation process with granulated carbon [14.10], [14.21], [14.42], [14.66]. The former has already been mentioned in connection with biological treatment. Besides the actual adsorption, biological processes (i.e., combinations of aerobic and anaerobic degrada-

tion) also play an important role. Apart from water-soluble dyes, polar substances are preferentially adsorbed to activated carbon. As a result of the limited adsorption capacity and high cost, activated carbon should be used preferably for the treatment of less loaded wastewater (e.g., in fine purification for reuse or as safety filters). Cheaper adsorbents under investigation are coking coal, peat, kieselguhr, activated alumina, clay, chitin, and other natural substances. Synthetic polymers and ion exchangers are suited to the separation of polar and nonpolar water components. As with activated carbon, periodic regeneration is required [14.42]. For the selective separation of heavy-metal cations, AOX, and dyes by polymeric diaza crown ethers, cyclodextrins, calixarenes, and curcubituril, see [14.67], [14.68].

**Membrane Separation.** The separation of contaminants by using membrane techniques (e.g., reverse osmosis, nano- or ultrafiltration) results in no final decrease in contamination. Concentrates with the separated contaminants must be treated further by using known techniques, while the clean permeates can be discharged or reused in the plant [14.69], [14.70]. Membranes that are applied include those made of cellulose acetate, polyamides, and polysulfones, and the robust, dynamic membranes made of zirconium(IV) hydroxide–polyacrylates on porous ceramic or steel carriers [14.42].

**Neutralization.** Textile wastewater must be neutralized since the pH is frequently more alkaline than permitted for discharge. Apart from strong acids such as sulfuric or hydrochloric acid, $CO_2$ is also well suited to neutralization and can be used in the pure state [14.71] or as a component of flue gas [14.72]. In comparison with mineral acids, the advantages of neutralization with $CO_2$ are lower chemical costs, less material corrosion, no salting, and more safety in dosing.

## 14.6. Recycling of Water and Valuable Substances

**Recycling of Water.** Because of high costs for treatment or discharge of industrial and wastewater, reuse of water occasionally appears reasonable. The multiple use of dye liquors was mentioned in Section 14.4. Similarly, slightly loaded rinsing baths can be reused in many cases without previous treatment. For universal applicability, however, purification, usually a combination of several processes, is required. Some of these are given below:

1) *Weakly loaded wastewater*
   Powdered activated carbon
   Polyaluminum chloride precipitation and sedimentation
   Multilayer filter [14.73]
2) *Total wastewater*
   Charcoal-supported biology in stirred loop reactor
   Adsorption in stirred loop reactor on pulverized brown coal
   Polyaluminum chloride precipitation and sedimentation
   Fixed-bed filtration with granulated brown coal [14.41], [14.74]
3) *Total wastewater, mainly from reactive dyeing*
   Precipitation with cationic precipitating agents and flotation
   Fixed-bed filter with sand and brown coal [14.75]
4) *Wastewater from polyester dyeing*
   Electrocoagulation
   Filtration [14.49]
5) *Total wastewater*
   Precipitation – flocculation – filtration
   Nanofiltration [14.76]
6) *Total wastewater*
   Ozonization
   Precipitation with calcium chloride and sedimentation
   Electrolysis
   Treatment with powdered activated carbon
   Fixed-bed filtration [14.77]
7) *Total wastewater*
   Distillation or evaporation [14.78], [14.79]

**Recovery of Valuable Substances.** Unused dyes, textile auxiliaries, and chemicals are valuable materials that can be recovered and reused, saving considerable costs. The energy content of hot baths can be used to heat raw water by means of heat exchangers. A lowering of temperature of dyebaths is often required anyway to maintain discharge limits [14.80]. The recovery of synthetic size by ultrafiltration is becoming increasingly important. Not only the value of the size, but also the reduction of wastewater loading with poorly degradable agents should be taken into account [14.81]. The ultrafiltration plant amor-

tizes in less than two years. Diluted sodium hydroxide solution, which is produced in the washing of mercerized fabrics, is purified by ultrafiltration before it is concentrated by evaporation [14.82]. The recovery of indigo from rinsing baths is also successfully carried out by ultrafiltration [14.81], [14.82]. Dyes and auxiliaries can be eliminated from the rinsing baths and concentrated with hot water- and pH-resistant dynamic membranes. Reuse is then possible after combination with residual dye liquors and correction of the chemical concentrations [14.83]. Salts, which have a concentration of up to 100 g/L in exhaustion dyeing with reactive dyes, can be largely recovered if the salt-containing residual liquor and the first rinsing bath, if necessary, are subjected to ultrafiltration. (The less loaded rinsing baths are discharged into the sewage system without treatment.) The amortization time for the investment of these plants is less than three years because of the recovery of the salt [14.69], [14.84].

## 14.7. Ecotoxicology of Textile Dyes

**Biological Degradability.** In view of their high stability to a number of physical and chemical influences, dyes have a similarly high stability to microorganisms. Although indications of mineralization have been found in some investigations [14.85], dyes are regarded as nondegradable during their relatively short detention time in normal *aerobic biological wastewater treatment plants* [14.86]. For some dyes (e.g., C.I. Basic Violet 1 and 3 [14.87], the azo dyes C.I. Acid Orange 12 and 20, and the triphenylmethane dye C.I. Acid Red 88 [14.88]), degradation by adapted bacteria has been detected. Dyes can also be biodegraded on biofilters or biofilms (i.e., packing with a bacterial layer) [14.89], [14.90]. The most important route of elimination in wastewater treatment plants is adsorption on clarification sludge, the extent depending on the constitution of the dye [14.85], [14.86], [14.91], [14.92]. Microbial degradation is possible under *anaerobic conditions* (e.g., in digestion tanks, soil, or sediments). This applies not only to azo dyes, in which reductive cleavage of the azo bond occurs, but also to other classes of dyes [14.93]. The aromatic amines expected from the anaerobic degradation of azo dyes have been detected. Although the amines are poorly degradable anaerobically, aerobic degradation occurs readily [14.94]. From a model landfill with sludge from sewage treatment plants containing mostly textile wastewater, neither dyes nor the corresponding amine components could be eluted [14.95].

**Toxicity for Aquatic Organisms.** Dyes have a low toxicity for aquatic organisms. Of 200 compounds tested, only 18 had a harmful effect on wastewater bacteria at concentrations of < 100 mg/L. These were mostly cationic dyes [14.96]. However, these dyes exhaust very well in the dyebath, are easily adsorbed by biosludge, and thus present no problems in wastewater treatment plant effluents. With respect to algae and fish toxicity, cationic dyes are usually the culprits [14.10], [14.97]. Of more than 3000 tested commercial dyes, 59 % were nontoxic to fish at a concentration of > 100 mg/L, 27 % had an $LC_{50}$ between 10 and 100 mg/L, and only 3 % were toxic at concentrations < 10 mg/L [14.98]. Dye concentrations in surface waters are estimated at the most at $1-10$ µg/L (i.e., three to five orders of magnitude lower than the acute fish toxicity) [14.99]. Bioaccumulation in fish does not occur with water-soluble dyes and is practically negligible for disperse dyes [14.100].

**Toxicity for Warm-Blooded Animals and Occupational Health.** In general, dyes have a low toxicity for *warm-blooded animals*. Of 4500 tested commercial products, 82 % had an $LD_{50}$ (rat, oral) of > 5000 mg/kg and < 1 % had a value of < 250 mg/kg [14.101]. According to the ChemG, only 13 known textile dyes are considered toxic ($LD_{50}$ < 200 mg/kg oral and $LC_{50}$ < 25 mg/m³ inhalation), several of which are no longer produced [14.102]. Studies on rats to determine the subchronic toxicity of eight commercial products with an $LD_{50}$ > 2000 mg/kg showed no measurable effects in the case of 22 daily doses during 28 days of 1000 mg/kg each [14.103]. A carcinogenic effect of 13 textile dyes, mainly azo dyes, has been detected in animal experiments [14.103]–[14.105]. In metabolism, aromatic amines may be formed by the enzymatic reductive cleavage of azo dyes. According to a recommendation of the German MAK Commission (1988), azo dyes that release carcinogenic arylamines of the MAK groups III A 1 and III A 2 under reductive conditions should be treated like the released amines themselves. In the GefStoffV [14.106], this recommendation led to consequences with regard to classification, labeling, packaging, and contact in the factory. The important representatives involved are dyes based on benzidine and its congeners. The members of the Ecological and Toxicological Association of the Dyestuff and Pigment Industry (ETAD) have agreed voluntarily to neither produce nor market benzidine dyes. Up to now the use of benzidine dyes and the importation of textiles dyed with them have been banned only in Sweden. In India, the production of benzidine dyes has been banned since February 1993. In Germany, the sale of textiles dyed with dyes that are able to cleave under formation of carcinogenic amines and can come into contact with skin will be banned from January 1996 on (2nd and 3rd Ordinance for Amendment of Consumer Goods Ordinance from July 15, 1994, BGBl. 1670–1671, and from December 16, 1994, BGBl. 3836). Benzidine and its metabolites could be detected in the urine of workers who had contact with benzidine dyes and had inhaled their dust. In plants with a high standard of industrial hygiene, the results were negative [14.107].

In the case of fast dyeing—since the migration of dyes on and through the skin is estimated at most at a few micrograms—the risk of endangering health is very low. An exception is the risk of allergies, which exists for a few dyes. In comparison with worldwide dye consumption, the number of skin allergies due to dyes is negligibly small [14.108]. The dyes with increased allergic potential are essentially those listed below [14.109]–[14.111].

| Colour Index | Dye class |
| --- | --- |
| Disperse Yellow 1, 9 | nitro dye |
| Disperse Yellow 39, 49 | methine dye |
| Disperse Yellow 54, 64 | quinoline dye |
| Disperse Orange 1, 3, 76 | azo dye |
| Disperse Red 11, 15 | anthraquinone dye |
| Disperse Red 1, 17 | azo dye |
| Disperse Blue 1, 3, 7, 26, 35 | anthraquinone dye |
| Disperse Blue 102, 106, 124 | azo dye |
| Acid Violet 17 | azo dye |

In textile dyehouses, especially in dyeing kitchens, additional exposure to inhalable dust should be considered. Apart from skin irritation, respiratory or nasal problems and itching of the eyes can occur, especially due to reactive dyes. Remedial action can be taken by means of good occupational hygiene and dust-free working conditions [14.112], [14.113], [14.114].

# 15. References

References for chapter 1

[1.1] H. Vogler, *Textilveredlung* **27** (1992) 352–358.
[1.2] S. M. Edelstein, H. C. Borghetty: *The Plictho of Gioanventura Rosetti*, The MIT Press, Cambridge–London 1969.
[1.3] *Ullmann*, 4th ed. **11**, 99–134.
[1.4] H. Schweppe: *Handbuch der Naturfarbstoffe, Vorkommen, Verwendung, Nachweis*, Ecomed Verlag, Landsberg 1993.
[1.5] *Kirk-Othmer*, 3rd ed., **8**, 351–373.
[1.6] F. Brunello: *The Art of Dyeing in the History of Mankind*, Neri Pozza, Vicenza 1973, English Translation by Phoenix Dyeworks, Cleveland, OH.
[1.7] G. Fieler: *Farben aus der Natur, Eine Sammlung alter und neuer Färberezepte für das Färben auf Wolle, Seide, Baumwolle und Leinen*, M. H. Schaper, Hannover 1978.
[1.8] R. J. Adrosko: *Natural Dyes and Home Dyeing, a Practical Guide with over 150 Recipes*, Dover Publ. Inc., New York 1968.
[1.9] E. Bemiss: *The Dyers Companion*, vol. 2, New London 1806, 3rd. enlarged ed. New York 1973.
C. G. Gilroy: *A Practical Treatise on Dyeing and Calico-printing*, 2nd ed., New York 1846.
[1.10] P. Richter, *Wirkerei- Strickereitech.* **40** (1990) no. 3 to **41** (1991) no. 10.
[1.11] Gesamtverband der deutschen Textilveredlungsindustrie, TVI-Verband e.V.: Annual Report 1992.

References for Chapter 2

[2.1] Arbeitgeberkreis Gesamttextil: *Textilveredelung, Färben–Aus der Geschichte der Färberei*, Frankfurt 1990, pp. 1.01–1.02.
[2.2] R. Germer: *Die Textilfärberei und die Verwendung gefärbter Textilien im alten Ägypten*, Verlag Otto Harrasowitz, Wiesbaden 1993, p. 150.
[2.3] C. Hütz: "Textiltechnik zur Zeit des alten Testaments," *Textilveredlung* **28** (1993) 59–67.
[2.4] A. Kretschmer: "Textilveredlungsmaschinen (gestern), heute und morgen," *Melliand Textilber.* **64** (1983) 676–680.
[2.5] M. Peter, H. K. Rouette: *Grundlagen der Textilveredlung, Handbuch der Technologie, Verfahren, Maschinen*, 13th ed., Deutscher Fachverlag, Frankfurt 1989.
[2.6] H. Zollinger: *Color Chemistry*, Verlag Chemie, Weinheim 1987.
[2.7] J. Skoufis: "Low-Liquor Ratio Foam Processing," *Am. Dyest. Rep.* **68** (1979) 20–23.
[2.8] W. Saus, D. Knittel, E. Schollmeyer: "Färben aus überkritischem Kohlendioxid – eine Alternative zur HT-Färbung von Polyester," *Text. Prax. Int.* **47** (1992) 1052–1054.
[2.9] W. Saus, S. Hoger, D. Knittel, E. Schollmeyer: "Färben aus überkritischem Kohlendioxid, Dispersionsfarbstoffe und Baumwollgewebe," *Textilveredlung* **28** (1993) 38–40.
[2.10] W. Rüttiger, J. Ehlert: "Die kritische Färbegeschwindigkeit, eine verfahrenstechnische Kenngröße für die zeitsparende und egale Färbung auf dem Apparat," *Text. Prax.* **27** (1972) 609–616.
[2.11] H. Leube, W. Rüttiger: "Egalität und Färbesicherheit bei unterschiedlich geführten Badfärbungen von Polyesterfasern," *Melliand Textilber.* **59** (1978) 836–842.
[2.12] B. Smith, G. Mcintosh, S. Shanding: "Ultrasound – A Novel Dying Accelerant," *Am. Dyest. Rep.* **77** (1988) no. 10, 15–20.
[2.13] *Ullmann*, 4th ed., **22**, 635–716.
[2.14] U. Baumgarte: "Das Aufziehverhalten der Indanthrenfarbstoffe als wichtige Grundlage für die Kurzzeitfärbetechnik," *Melliand Textilber.* **55** (1974) 953–962.
[2.15] U. Nahr, W. von Bistram: "Grundlagen für die EDV-gestützte Chemikalienberechnung zum Färben mit Indanthren-Farbstoffen in teil- oder vollgefluteten geschlossenen Apparaten," *Text. Prax. Int.* **46** (1991) 978–983.
[2.16] H. Gerber: "Kennwerte zur Charakterisierung des färberischen Verhaltens von Reaktivfarbstoffen," *Textilveredlung* **28** (1993) 405–411.
[2.17] P. S. Collishaw, D. A. S. Philips, M. J. Bradbury: "Controlled Coloration: A Success Strategy for the Dyeing of Cellulosic Fibers with Reactive Dyes," *J. Soc. Dyers Colour.* **109** (1993) 284–292.
[2.18] W. Schrott, K. Bacher: "Kreativität führt zu Spitzenleistung. Wo stehen wir im Heißausziehsegment mit Reaktivfarbstoffen?" *Text. Prax. Int.* **47** (1992) 1145–1150.
[2.19] F. Hoffmann, H. Schubert, J. Fiegel: "Einsatz von Rechnern zur Optimierung von Färberezepten und Verfahren," *Text. Prax. Int.* **47** (1992) 233–237.
[2.20] U. Baumgarte: "Zustand und Verhalten von Farbstoffen beim Färben am Beispiel der Leukoküpenfarbstoffe," *Textilveredlung* **15** (1980) 413–425.
[2.21] W. Rüttiger: "Prozeßgrößen und Kostenfaktoren in der Veredlung am Bildschirm verfolgen," *Text. Prax.* **38** (1983) 1216–1217.
[2.22] W. Rüttiger: "Process Simulation and On-Line Data – A Step Towards Microprocessor Control of Textile Processing," *1985 AATCC Technical Programm Book*, 1985, pp. 234–252.
[2.23] W. Rüttiger: "Expertensysteme – das künftige Instrument technischer Führungskräfte," *Textilveredelung* **23** (1988) 127–134, 159–162, 199–203.
[2.24] H. Wolf, W. Rüttiger: "Möglichkeiten und Grenzen der Kurzzeit-Färbetechniken," *Melliand Textilber.* **55** (1974) 876–879.
[2.25] W. Rüttiger: "Prozess-Simulation am Beispiel der Polyesterfärbung," *Chemiefasern/Textilind.* **40/99** (1990) 446.
[2.26] E. Saas, W. Franke: "Welche Vorteile bietet der Einsatz einer Hochtemperatur-Haspelkufe," *Chemiefasern* **17** (1967) 218–219.
[2.27] P. Toldrian, W. Rüttiger: "Verfahrenssymbole als Voraussetzung für die automatisch gesteuerte Färbung mit Apparaten," *Melliand Textilber.* **55** (1974) 973–978.
[2.28] A. Kretschmer: "Welchen Pumpentyp für den Färbapparat von morgen," *Text. Prax.* **25** (1970) 39.
[2.29] H. U. von der Eltz: "ITMA 1991 Färbeapparate und -maschinen für das Ausziehfärben," *Text. Prax. Int.* **47** (1992) 154–162, 246–251.
[2.30] G. Rordorf: *Neues großes Handbuch der Textilkunde*, Fachbuchverlag Dr. Pfanneberg, Giessen 1956, pp. 271–272.
[2.31] J. Ehlert: "Maschinen und Apparate für die Textilfärberei," *Deutscher Färberkalender 1970*, Deutscher Fachverlag, Frankfurt 1969, pp. 89–147.

[2.32] A. Kretschmer: "Die Durchströmungsrechnung als Optimierungshilfe in der Apparatefärberei," *Textilveredlung* **15** (1980) 50–55.

[2.33] H. Beiertz: "Färben und Drucken textiler Bodenbeläge aus Trevira CF- und CS-Fasern," *Chemiefasern/Textilind.* **42/94** (1992) 64–67.

[2.34] MCS: "Färbemaschinen der Baureihe Lang-Horn erweitert," *Text. Prax. Int.* **48** (1993) 995.

[2.35] THEN: "Airflow noch wirtschaftlicher," *Text. Prax. Int.* **47** (1992) 1171–1172.

[2.36] H. K. Rouette: "ITMA 91: Highlights, Trends und Zukunft – Perspektiven in der technischen Entwicklung von Färberei, Druckerei und Veredlung," *Chemiefasern/Textilind.* **42/94** (1992) 57–61.

[2.37] Küsters: "Universal-Jigger für Vorbehandlung, Färberei und Spezialbehandlungen," *Text. Prax. Int.* **48** (1993) 248.

[2.38] Henriksen: "Hochgeschwindigkeitssteuerung für Jigger bringt 30% Zeitersparnis, soft-set-Trockner jetzt auch kontinuierlich arbeitend," *Text. Prax. Int.* **48** (1993) 131–132.

[2.39] MCS: "Neue Strangfärbemaschine," *Melliand Textilber.* **74** (1993) 566.

[2.40] "ADR's Annual Buyers' Guide to Garment Dyeing/Washing Machinery and Stone/Chemical Suppliers," *Am. Dyest. Rep.* **82** (1993) no. 5, 38–45.

[2.41] Thies: "Trommelfärbeapparat für Fertigkonfektion," *Melliand Textilber.* **67** (1986) 723.

[2.42] J. Park: "Developments in Dyeing Process Control," *Rev. Prog. Color. Relat. Top.* **15** (1985) 1–5.

[2.43] *On-Line-Steuerungen in der Textilveredlung*, lecture held at Fachhochschule Reutlingen.

[2.44] W. Rüttiger, P. Toldrian: "Einheitliche Verfahrensdarstellung zur Automatisierung und Optimierung der chargenweisen Färbung," *Text. Prax.* **27** (1972) 674–678, 719–722.

[2.45] Sepa: "Offene Prozessleittechnik für die Textilveredlung," *Text. Prax. Int.* **47** (1992) 674.

[2.46] J. H. Heetjans: "Handhabungs- und Transportsysteme in der Garnfärberei," *Chemiefasern/Textilindustrie* **39/91** (1989).

[2.47] G. Hofmann: "Automatisches Beschicken von Färbeapparaten und Garnveredlung," *Chemiefasern/Textilind.* **42/94** (1992) 61–63.

[2.48] J. Ehlert: "Rationalisierung in der Textilfärberei," *Deutscher Färberkalender 1971*, Deutscher Fachverlag, Frankfurt 1970.

[2.49] E. Voss: "Erfahrungen mit pH-Automaten für die Färbung von Stück- und Konfektionsware, sehr hohe Ausziehgrade für Polyamid und Wolle in der Badfärbung," *Melliand Textilber.* **74** (1993) 564–566.

[2.50] K. Tiedemann, G. Grüninger, W. Rüttiger: "Methoden zur On-Line-Überwachung der Wasserstoffperoxidkonzentration für die Bleiche und Küpenfärbung," *Deutscher Färberkalender 1993*, Deutscher Fachverlag, Frankfurt 1992, pp. 140–172.

[2.51] H. Fuchs: "Erkenntnisse beim einbadigen Pigmentfärben und Ausrüsten," *Text. Prax. Int.* **49** (1994) 507–510.

[2.52] W. Kambach: "Kontinuierliches Aufwickeln von Textilbahnen zu Großkaulen," *Text. Prax. Int.* **48** (1993) 36–38.

[2.53] A. Günther: "Geringere Abwasserbelastung durch neue Konstruktionen bei Teppich-, Färbe- und Druckanlagen," *Text. Prax. Int.* **48** (1992) 848–852.

[2.54] Farmer Norton: "Neues Vakuum-Flottenauftragssystem," *Text. Prax. Int.* **48** (1993) 129–130.

[2.55] J. Polster: "Foulardtechnik – tema con variazioni," *Melliand Textilber.* **74** (1993) 1173–1174.

[2.56] K. van Wersch: "Kontrollierte Applikation von Färbeflotten," *Textilveredlung* **28** (1993) 218–223.

[2.57] P. Salzmann, W. Schindler: "Die gemeinsame Applikation von Küpenfarbstoff und Reduktionsmittel beim Kontinuefärben ohne Zwischentrocknung," *Melliand Textilber.* **74** (1993) 525–529.

[2.58] W. Rüttiger, W. Rümens, G. Burkhardt, H. Petersen: "Übersicht der industriellen Verfahrenstechniken der Hochveredlung von Zellulose," *Melliand Textilber.* **11** (1972) 1278–1289.

[2.59] H. Paulsen: "Energiebilanzen in der Textilveredlung," *Melliand Textilber.* **74** (1993) 550–556.

[2.60] S. M. Burkinshaw, W. J. Marshall: "Continuous Dyeing of Piece Goods Using Radiofrequency Heating," part I, *J. Soc. Dyers Colour.* **102** (1986) 263–268.

[2.61] P. Lennox-Kerr: *Int. Dyer Text. Printer Bleacher Finish.* **171** (1986) 20–21.

[2.62] U. Einsele: "Neuere Methoden zum Färben von Polypropylen – 1. Mitteilung: Färben mit Hilfe von Bindemitteln," *Melliand Textilber.* **72** (1991) 847–851.

[2.63] W. Rüttiger: "Das Ausnutzen des Waschwirkungsgrades – eine Einführung in das Beherrschen von Waschprozessen," *Text. Prax.* **34** (1979) 1380–1387, 1544–1551, 1629–1643.

[2.64] W. Rüttiger, U. Kirner: "Die Geschwindigkeit des Wasserstoffperoxide-Eigenzerfalls als wichtige verfahrenstechnische Kenngröße in der Textilveredlung," *Text. Prax.* **25** (1970) 21–30.

[2.65] D. W. Ravensberger, R. B. M. Holweg: "Monitoring and Control of Hydrogen Peroxide Dosing," *J. Soc. Dyers Colour.* **109** (1993) 72–75.

[2.66] BASF: *Ratgeber Cellulosefasern, Schlichten, Vorbehandeln, Färben*, Verlag BASF AG, Ludwigshafen 1977, p. 319, 357.

[2.67] K. van Wersch: "Anlagen für das Färben nach dem Pad Batch-Verfahren," *Melliand Textilber.* **73** (1992) 431–432.

[2.68] U. Denter, E. Schollmeyer: "Verfahrensanalyse und Entwicklung einer Testfärbemethode für die Rezeptierung der KKV-Reaktivfärbung von Baumwollgeweben (AIF 7997)," *Melliand Textilber.* **74** (1993) 251.

[2.69] Suderba: "Space-dyeing-Maschine für sechs Farben," *Text. Prax. Int.* **48** (1993) 1006.

[2.70] Carbonell: "Praxiserfahrungen beim Versuch, automatische Prozeß-Steuerungen auf die Erfordernisse einer computerintegrierten Fertigung umzugestalten," *Text. Prax. Int.* **47** (1992) 468–472.

[2.71] Qualico: "Schnellste kontinuierliche Feuchtekontrollen," *Text. Prax.* **48** (1993) 247.

[2.72] J. Rieker, T. Guschlbauer: "Farbabläufe in der Kontinue-Färberei und Materialdichteunterschiede in Geweben – Kann die On-Line-Farbmessung helfen?" *Melliand Textilber.* **74** (1993) 1256–1260.

[2.73] W. Rüttiger: "Meßwerte im Computer für die Veredlungspraxis," *Text. Prax.* **41** (1986) 555–562, 776–781.

[2.74] O. Annen, H. Gerber: "Probleme bei der Übertragung von Laborergebnissen in die Praxis," *Melliand Textilber.* **75** (1994) 406–411.

[2.75] Roaches: "Neue infrarotbeheizte Laborfärbemaschinen entwickelt," *Text. Prax. Int.* **48** (1993) 131.

[2.76] "ITMA 91 – Labor- und Musteranlagen für die Textilveredlung," *Melliand Textilber.* **73** (1992) 180–183.

[2.77] J. Oegarra: "On-Line-Qualitätskontrolle bei der Ausziehfärbung," *Melliand Textilber.* **74** (1993) 538–549.

[2.78] K. Stommel: "Einsatz von Laborgeräten in der Textilveredlungsindustrie," *Melliand Textilber.* **45** (1964) 283–286.

[2.79] O. Annen, H. Gerber, B. Seuthe: "Das Färbeverhalten von Viskose- und Modalfasern im Vergleich zu Baumwolle," *J. Soc. Dyers Colour.* **108** (1992) 215–261.

[2.80] E. Floegel: *Statistik in BASIC*, Hofacker-Verlag, Holzkirchen 1986.

[2.81] M. Warmes: "The Central Dye Preparation Station – A Necessity for an Efficient Process Cycle in the Dyehouse," *International Textile-Bulletin* **2/82** (1982).

[2.82] Then: "Das Farbküchensystem AMC-CKM," *Text. Prax. Int.* **48** (1993) 1009.

[2.83] H. P. Locher, M. Firmann: "Verbesserte Produktionssicherheit in der Foulardfärberei durch photometrische Kontrolle der Farbansätze," *Textilveredlung* **26** (1991) 393–398.

[2.84] J. Rieker, U. Scherschel-Gaedke: "Zur Problematik der photometrischen Bestimmung von Farbstoffkonzentrationen in Färbebändern," *Text. Prax. Int.* **48** (1993) 980–982.

[2.85] J. Rieker: "Labor- und Farbmessungen," *Melliand Textilber.* **73** (1992) 258–261.

[2.86] Datacolor: "Integrierte Farbdatenverarbeitung," *Chemiefasern/Textilind.* **43/95** (1993) 70.

[2.87] K. van Wersch: "On-Line-Farbmessung in der Kontinuefärberei aus der Sicht des Maschinenbauers," *ITB Veredlung* **36** (1990) no. 2, 21–42.

[2.88] W. Pape: "Farbmessung in der Produktion," *Melliand Textilber.* **73** (1992) 152–155.

[2.89] J. Rieker, E. Furthmüller: "Untersuchung der farbmetrischen Abmusterung mit Pass/Fail-Formeln," *Text. Prax. Int.* **42** (1987) 422–423.

[2.90] J. Rieker, K. Hilscher: "Über die visuelle Relevanz von Farbstoffkonzentrationsunterschieden auf gefärbten Textilien," *Text. Prax. Int.* **49** (1994) 499–501.

**References for Chapter 3**

[3.1] B. Wulfhorst, K. Meier: "Untersuchungen an einem kurzen Hochtemperaturheizer," *Chemiefasern/Textilind.* **43** (1993) no. 95, 40–45.

[3.2] *Melliand Textilber.* **74** (1993) 204.

[3.3] Y. G. Bryant: "Neue Fasern als temperaturgesteuerte Wärmefluß-Sperre," *Tech. Text.* **35** (1992) T 88–T 89.

[3.4] J. Gacén, D. Cayuela, J. Maillo, T. Rodriguez: "Einfluß der thermischen Vorbehandlung auf die Feinstruktur von thermofixiertem Polyester," *Melliand Textilber.* **74** (1993) 797–801.

[3.5] H.-J. Flath, C. Scholz: "Charakterisierung zugänglicher Strukturanteile in Cellulosefasern," *Melliand Textilber.* **74** (1993) 219–221.

[3.6] W. Rüttiger: "Mechanische Zustandsänderungen von Baumwolltextilien bei der Alkalikaltbehandlung, part 1: Grunderscheinungen der mechanischen

Vorgänge und ihre Deutungsmöglichkeiten über Faserstrukturmodelle," *Melliand Textilber. Int.* **51** (1970) 1449–1465.

[3.7] R. V. Meyer, E. Haug, G. Spilgies: "Elastane – Chemie, Eigenschaften, Einsatzgebiete," *Melliand Textilber.* **74** (1993) 194–198.

[3.8] H. G. Dejong: "Yarn-to-yarn Friction in Relation to some Properties of Fiber Materials," *Text. Res. J.* **63** (1993) 14–18.

[3.9] G. Bühler: "Garnreibwerte. Ein altes, jedoch immer noch nicht ganz bewältigtes Problem in der Praxis der Maschenwarenherstellung," *TPI Text. Prax. Int.* **48** (1993) 108–113.

[3.10] P. Offermann, G. Putzger: "Qualitätsanalyse im Garnveredlungs- und textilen Fertigungsprozeß," *Melliand Textilber.* **74** (1993) 237–242.

[3.11] U. Scholze, J. Trauter: "Auswirkungen der Haarigkeit von Rohgarnen und der Fasereigenschaften auf das Webverhalten geschlichteter Kettgarne (Part 1)," *TPI Text. Prax. Int.* **48** (1993) 773–779.

[3.12] J. P. Rust, S. Peykamian: "Yarn Hairiness and the Process of Winding," *Text. Res. J.* **62** (1992) 685–689.

[3.13] *TPI Text. Prax. Int.* **48** (1993) 122.

[3.14] W. Löbel: "Weiterentwicklungen der Restfeuchtemessung von Textilien nach dem konduktometrischen Verfahren," *Melliand Textilber.* **73** (1992) 771.

[3.15] W. Rüttiger: "Regel- und Steuerprobleme bei der kontinuierlichen Textilausrüstung," *Melliand Textilber.* **47** (1966) 552–559, 680–690.

[3.16] W. Rüttiger, G. Schmidt: "Krumpf und Verformung, zwei wesentliche Faktoren zur Gütesicherung von Maschenware," *Melliand Textilber.* **61** (1980) 526–532.

[3.17] A. Eickmeier, U. Beller, D. Knittel, E. Schollmeyer: "Untersuchungen von Polymeren im oberflächennahen Bereich mit dem Verfahren der Akustomikroskopie," *TPI Text. Prax. Int.* **48** (1993) 534–536.

[3.18] L. Andrews: "Charakterisierung von Texturen geschmirgelter Textilgewebe mit Hilfe der Statistik dritter Ordnung," *Melliand Textilber.* **74** (1993) 231–233.

[3.19] A. Lehnen: "Untersuchung der Zusammenhänge zwischen Ergebnissen der subjektiven Beurteilung von Teppichoberseitenstrukturen sowie deren Veränderungszustände und optoelektronisch meßbaren Kennwerten," *TPI Text. Prax. Int.* **48** (1993) 104.

[3.20] R. H. Peters: *Textile Chemistry*, vol. III, The Physical Chemistry of Dyeing, Chapt. 2, Importance of Physical Properties of Fibers, Elsevier, Amsterdam – Oxford – New York 1975, pp. 23–26.

[3.21] M. Peter, H.-K. Rouette: *Grundlagen der Textilveredlung, Handbuch der Technologie, Verfahren, Maschinen*, 13th ed., Deutscher Fachverlag, Frankfurt 1989, Chap. 5.1.

[3.22] W. Albrecht: "Welche Bedeutung haben die physikalischen Eigenschaften von Cellulosefasern für ihr Färbeverhalten," *Melliand Textilber. Int.* **51** (1970) 1437–1448.

[3.23] H. Rath: *Lehrbuch der Textilchemie*, Chap. I.A. Chemie und Physik der Cellulose und ihrer Begleitstoffe, 3rd ed., Springer Verlag, Berlin 1972.

[3.24] W. Bubeth (ed.): *Textile Faserstoffe – Beschaffenheit und Eigenschaften*, Springer Verlag, Heidelberg 1993.

[3.25] P. Richter: "Färben von alkalisierten Polyesterfasern und Polyester-Microfasern," *Chemiefasern/ Textilind.* **41/93** (1991) 1118–1125.

[3.26] H. Gerber: "Kennwerte zur Charakterisierung des färberischen Verhaltens von Reaktivfarbstoffen," *Textilveredlung* **28** (1993) 405–411.

[3.27] M. Harris: *Handbook of Textile Fibers*, Textile Book Publ., New York 1954.

[3.28] W. Rüttiger: "Eine Zustandsgleichung für Cellulosefasern als Erfassungsmöglichkeit für Längung, Krumpf, Knitterbildung und -erholung," *Textilveredlung* **2** (1967) 428–435.

[3.29] D. Fiebig, H. Herlinger, B. Kastl: "Strukturveränderungen bei Cellulosematerialien durch Netz-/ Trockenprozesse," *TPI Text. Prax. Int.* **48** (1993) 789–794.

[3.30] P. R. Brady: "Diffusion of Dyes in Natural Fibers," *Rev. Prog. Color. Relat. Top.* **22** (1992) 58–78.

[3.31] B. Reichstädter, O. Pajgrt: "Struktur und Eigenschaften von Chemiefasern und Eigenschaften der fertigen Textilien," *Chemiefasern Text. Anwendungstech.* **20** (1970) 877–884.

[3.32] P. Ehrler: "Präzisionsgewebe: Notwendigkeit verfahrenstechnischer Modifikationen," *Tech. Text.* **35** (1992) T94–T97.

[3.33] L. Vangheuwe: "Methode zur Bestimmung des dynamischen Moduls von Garnen," *Melliand Textilber.* **74** (1993) 717–718.

[3.34] H. Grams: "Neues Websystem," *Melliand Textilber.* **74** (1993) 198–201.

[3.35] J. Rieker, W. Braun: "Kontinuierliche Messung von Kante/Mitte/Kante-Unterschieden an Webware nach dem Durchströmungsstaudruck-Prinzip," *Melliand Textilber.* **68** (1987) 749–752.

[3.36] D. H. Abrahams: "Reducing Cracks Creases Chafes," *Am. Dyest. Rep.* **63** (1974) no. 9, 32–36.

[3.37] G. Heidemann, H.-J. Berndt: "Effektivtemperatur und Effektivspannung, zwei Meßgrößen zur absoluten Bestimmung des Fixierzustandes von Synthesefasern," *Melliand Textilber.* **57** (1976) 485.

[3.38] G. Heidemann, H.-J. Berndt: "Die substrat- und prozeßspezifischen Größen der Fixierung von Synthesefasern, eine Analyse am Beispiel der Thermofixierung von Polyesterfasern," *Chemiefasern/ Textilind.* **24/76** (1974) 46.

[3.39] A. Kretschmer: "Grenzwerte der Flottenbeladung von flächigem Textilgut beim Foulardieren," *Text. Prax.* **25** (1970) 217–218.

[3.40] E. D. Dresden: "Transparenzmessung – eine Möglichkeit zur Kontrolle von Veredlungsvorgängen," *TPI Text. Prax. Int.* **48** (1993) 230–231.

[3.41] W. Rüttiger: "Ein plausibler Mechanismus für das Entstehen von Falten in Ablage-Färbeaggregaten," *Lenzinger Ber.* **45** (1978) 160–171.

[3.42] Deutsches Textilforschungszentrum Nord-West e. V.: "Lauf- und Liegefalten," Krefeld 1977.

[3.43] *Z. Gesamte Textilind.* **58** (1956) 504–505.

[3.44] Bayer-Faser-Institut, company brochure, Bildung und Fixierung von Falten – Dralon, VII f 1, Leverkusen, Feb. 1967.

References for chapter 4

[4.1] K. Venkataraman: *The Chemistry of Synthetic Dyes*, vol. 6: "Reactive Dyes," Academic Press, New York 1972.

[4.2] D. Hildebrand, *Text. Prax. Int.* **25** (1970) 292–296, 351–354, 428–430, 492–495, 621–624; **26** (1971) 45–47, 431–433, 564–567.

[4.3] M. R. Fox, H. H. Sumner in C. Preston (ed.): *The Dyeing of Cellulose Fibers*, Dyers' Company Publications Trust, Bradford 1986, pp. 142–195.

[4.4] P. Rys, H. Zollinger, *Helv. Chim. Acta* **49** (1966) 749–754.

[4.5] J. P. Luttringer, P. Dussy, *Melliand Textilber.* **62** (1981) 84–94.

[4.6] A. H. M. Renfrew, J. A. Taylor, *J. Soc. Dyers Colour.* **105** (1989) 441–445.

[4.7] H. H. Sumner, T. Vickerstaff, *Melliand Textilber.* **42** (1961) 1161–1166.

[4.8] U. Ruf, W. B. Egger, *Textilveredlung* **13** (1978) 304–308.

[4.9] S. Abeta, K. Akahori, U. Meyer, H. Zollinger, *J. Soc. Dyers Colour.* **107** (1991) 12–19.

[4.10] M. Matsui, U. Meyer, H. Zollinger, *J. Soc. Dyers Colour.* **104** (1988) 425–431.

[4.11] BASF, DE-OS 4 216 591, 1992. (J. Schulze et al.).

[4.12] M. Peter, H. K. Rouette: *Grundlagen der Textilveredlung*, 13th ed., Deutscher Fachverlag, Frankfurt a. M. 1989, pp. 510–516.

[4.13] A. H. M. Renfrew, J. A. Taylor, *Rev. Prog. Color. Relat. Top.* **20** (1990) 1–9.

[4.14] D. M. Lewis, *J. Soc. Dyers Colour.* **109** (1993) 357–364.

[4.15] N. S. Knaggs, *Am. Dyest. Rep.* **81** (1992) no. 11, 109–111.

[4.16] J. Shore, *Rev. Prog. Color. Relat. Top.* **21** (1991) 23–42.

[4.17] J. R. Aspland, *Text. Chem. Color.* **23** (1991) no. 12, 30–32.

[4.18] J. J. Porter, *Text. Chem. Color.* **25** (1993) no. 2, 27–37.

[4.19] J. R. Aspland, *Text. Chem. Color.* **23** (1991) no. 11, 41–45.

[4.20] M. I. Khalil, A. Bendak, S. S. Aggour, *Am. Dyest. Rep.* **70** (1981) no. 10, 31.

[4.21] D. R. Lemin, E. J. Vickers, T. Vickerstaff, *J. Soc. Dyers Colour.* **62** (1946) 132.

[4.22] Report of the Commitee on Direct Dyes, *J. Soc. Dyers Colour.* **62** (1946) 280–285; **64** (1948) 145–146.

[4.23] J. Varghese, N. Bhattacharyya, A. S. Sahasrabudhe, *Colourage* **36** (1989) no. 3, 16–31.

[4.24] F. Weiss: *Die Küpenfarbstoffe*, Springer Verlag, Wien 1953.
M. R. Vox: *Vat Dyes and Vat Dyeing*, Chapman and Hall, London 1948.

[4.25] U. Baumgarte, *Textilveredlung* **4** (1969) 821–832.

[4.26] U. Baumgarte, *Textilveredlung* **2** (1967) 896–907.

[4.27] T. Bechtold, E. Burtscher, D. Gmeiner, O. Bobleter, *Melliand Textilber.* **72** (1991) 50–54.
T. Bechtold, E. Burtscher, A. Amann, O. Bobleter, *Angew. Chem.* **104** (1992) 1046–1047.

[4.28] U. Baumgarte, *Melliand Textilber. Int.* **55** (1974) 953–962.

[4.29] T. Vickerstaff: *The Physical Chemistry of Dyeing*, Oliver & Bayd, London 1954.

[4.30] U. Baumgarte, *Textilveredlung* **15** (1980) 413–424.

[4.31] U. Baumgarte, *Melliand Textilber.* **59** (1978) 311–319.

[4.32] E. I. Valko, *J. Am. Chem. Soc.* **59** (1941) 1433–1437.

[4.33] BASF-Ratgeber Cellulosefasern: Schlichten, Vorbehandeln, Färben, Ludwigshafen 1977.

[4.34] U. Baumgarte, H. Schlüter, *Melliand Textilber.* **62** (1981) 555–561.

[4.35] P. Richter: "Kontinuefärben von Baumwollkettgarn mit Indigo," *Textilveredlung* **10** (1975) no. 8, 312–316.

[4.36] BASF, Continuous Dyeing of Cotton Warp Yarn with Indigo, Technical Information, Ludwigshafen 1985.

[4.37] J. A. Greer, G. R. Turner: "Indigo Denims: the Practical Side," *Text. Chem. Color.* **15** (1983) no. 6, 101–107.

[4.38] L. Haas: *Indigo-Färberei: Verfahrens- und maschinentechnische Lösungen*, Int. Text. Bull. Veredlung **2** (1990) 45–50.

[4.39] C. Heid, K. Holoubek, R. Klein: "100 Jahre Schwefelfarbstoffe," *Melliand Textilber.* **54** (1973) 1314–1327.

[4.40] E. Krusche: "Die Bedeutung der Schwefelfarbstoffe heute und in der Zukunft," *Text. Prax.* **23** (1968) 342–344.

[4.41] D. G. Orton: "Sulfur Dyes," in O. K. Venkataraman (ed.): *The Chemistry of Synthetic Dyes*, vol. VII, Academic Press, New York 1974, pp. 26–33.

[4.42] *Colour Index*, vol. 3, pp. 3649–3704 (1971), vol. 4, pp. 4475–4501 (1971), vol. 6 (Revised Third Edition), pp. 6375–6385 (1975).

[4.43] C. Heid: "Chemismus und Anwendung der wasserlöslichen Schwefelfarbstoffe," *Melliand Textilber.* **45** (1964) 648–652, 833.

[4.44] W. Prenzel: "Der heutige Stand der Schwefelfärberei," *Chemiefasern Textind.* **24/76** (1974) 293–302.

[4.45] C. Heid: "Neue Anwendungsmöglichkeiten von Hydrosol- und Hydrosol-Licht-Farbstoffen beim kontinuierlichen und halbkontinuierlichen Färben cellulosehaltiger Stückware," *Z. Gesamte Textilind.* **70** (1968) 626–630.

[4.46] W. Titzka: "Das Färben mit Schwefelfarbstoffen unter besonderer Berücksichtigung der Reduktions- und Oxidationsmittel," *TPI Text. Prax. Int.* **33** (1978) 1387–1390.

[4.47] M. Hähnke: "Die Schwefelfärbung in ökologisch anspruchsvoller Umwelt, Internat. Veredler-Jahrbuch," *Dtsch. Färber-Kal.* **96** (1992) 154–156.

[4.48] L. Tigler: "Sulfur Dye Oxidation," *Am. Dyest. Rep.* **61** (1972) no. 9, 46, 107.

[4.49] C. Heid: "Die Verbesserung der Wasch- und Peroxidwaschechtheit von Färbungen mit Schwefel- und Schwefelküpen-Farbstoffen durch Solidurit IH," *Melliand Textilber. Int.* **51** (1970) 322–325.
C. Heid: "Theorie und Praxis der Verbesserung der Wasch- und Peroxidwaschechtheit von Färbungen mit Schwefel- und Schwefel-Küpenfarbstoffen," *Melliand Textilber.* **59** (1978) 247–254.

[4.50] C. Heid: "Die Anwendung von Fibradurit FS bei Schwefelschwarzfärbungen zur Vermeidung von Faserschädigungen während des Trocknens und Lagerns." *Z. Gesamte Textilind.* **67** (1965) 364–367.

[4.51] C. Heid: "Hydron-Blau, der Indigo des 20. Jahrhunderts," *TPI Text. Prax. Int.* **33** (1978) 285–287.
C. Klopfstock: "Färben von Frottiermaterialien mit Immedial-, Hydrosol-, Cassulfon-, Hydron-Blau- und Indanthren-Farbstoffen," *TPI Text. Prax. Int.* **32** (1977) 1119–1121, 1230–1232.
W. Titzka: "Einsatz von Schwefelfarbstoffen in der Cord-Färberei," *Melliand Textilber.* **60** (1979) 254–256.

[4.52] G. Krauzpaul: "Die Schwefel- und Schwefelküpenfarbstoffe in der Ausziehfärberei aus der Sicht des Praktikers," *TPI Text. Prax. Int.* **42** (1987) 140–142.

[4.53] H. M. Tobin: "Jet Dyeing with Sulfur and Vat Colours," *Am. Dyest. Rep.* **68** (1979) no. 9, 26, 28.

[4.54] H. Bernhardt: "Welche Problematik wird durch die Substantivität bei Klotzprozessen mit Schwefelfarbstoffen ausgelöst? Gibt es dafür Problemlösungen?" *Int. Text. Bull.* **33** (1987) 5–6, 8–9, 12, 16, 19.
H. Bernhardt: "Der Einsatz des Taschencomputers bei der Rezepterstellung in der Kontinuefärberei mit Schwefelfarbstoffen," *Int. Text. Bull.* **33** (1987) 46, 50, 54, 57.

[4.55] C. Heid: "Neue Fixiermöglichkeiten von wasserlöslichen Schwefelfarbstoffen," *Z. Gesamte Textilind.* **69** (1967) 106–108.
Cassella Farbwerke, DE 1 619 403, 1967 (C. Heid, W. Gunzert).

[4.56] R. Klein: "Phänomen Denim," *Textilveredlung* **10** (1975) 112–117.

[4.57] Cassella Farbwerke, DE 1 276 596, 1963 (M. Stuhlmiller, G. Böttiger, C. Heid).

[4.58] "Probleme und Problemlösungen textiler Abwässer," *Chemiefasern Textilind.* **28/80** (1978) no. 1, 85–89.
J. M. Marzinkowski, W. Keiler: "Konzepte zur Vermeidung von Abwasserbelastungen in einem Textilveredlungsbetrieb," *Chemiefasern Textilind.* **41/93** (1991) 895–901.

[4.59] B. Kerres, G. Valk: "Verfahren zur Beseitigung von Geruchsbelästigungen aus dem öffentlichen Kanalnetz bei Einleitung von Färbereiabwässern mit Schwefelfarbstoffen," *Melliand Textilber.* **66** (1985) 588–590.

[4.60] B. Smith, J. Rucker: "Water and Textile Wet Processing–Part II," *Am. Dyest. Rep.* **76** (1987) 68–78.

[4.61] C. Oehme: "Abwasserklärung nach dem Katox-F-System," *TPI Text. Prax. Int.* **33** (1978) 1076–1077.
B. Höke: "Die Reinigungsleistung des Katox F-Verfahrens," *Textilbetrieb (Würzburg)* **96** (1978) no. 6, 92–94; nos. 7, 8, 60–62.
G. Wysocki: "Katox-Fällungsverfahren für die Reinigung von Abwässern der Textilveredlungsindustrie: Technologie und Wirtschaftlichkeit," *Chemiefasern Textilind.* **26/78** (1976) 353–356.

[4.62] W. Schwindt, G. Faulhaber, A. J. Moore, *Rev. Prog. Color. Relat. Top.* **2** (1971) 33.

[4.63] T. Lever, *J. Soc. Dyers Colour.* **108** (1992) 477–478.

[4.64] H. U. von der Eltz, H. Walbrecht, *Dtsch. Färber-Kal.* **77** (1973) 79–129.

[4.65] H. U. von der Eltz, *Textilveredlung* **7** (1972) 692–697.

[4.66] H. Rath: *Lehrbuch der Textilchemie*, 3rd ed., Springer Verlag, Berlin 1972.

[4.67] U. Baumgarte, *Rev. Prog. Color. Relat. Top.* **5** (1974) 17–32.

[4.68] R. H. Peters: *Textile Chemistry*, vol. 3, Elsevier, Amsterdam 1975.

[4.69] A. Schäffer: *Handbuch der Färberei*, vol. 1, Konradin-Verlag R. Kohlhammer, Stuttgart 1949.

[4.70] *Ullmann*, 4th ed., **22**, p. 656.

[4.71] H. Vollmann: "Phthalogen Dyestuffs," in K. Venkataraman (ed.): *The Chemistry of Synthetic Dyes*, vol. 5, Academic Press, New York 1971, p. 283.

[4.72] M. Peter, H. K. Rouette, *Grundlagen der Textilveredlung*, 13th ed., Deutscher Fachverlag, Frankfurt 1989.

References for chapter 5

[5.1] E. Elöd, H. Reutter, *Melliand Textilber.* **19** (1938) 67–72.
[5.2] J. D. Leeder, J. A. Rippon, F. E. Rothery, I. W. Stapleton, *Proc. Int. Wool Text. Res. Conf. 7th* **1985**, Tokyo, vol. V, 99–108.
[5.3] J. D. Leeder: "The Cell Membrane Complex and its Influence on the Properties of the Wool Fibre," *Wool Sci. Rev.* **3** (1987) 3–35.
[5.4] V. Sideris, L. A. Holt, I. H. Leaver, *J. Soc. Dyers Colour.* **106** (1990) 131–135.
[5.5] H. Zahn, *Proc. Int. Wool Text. Res. Conf. 6th* **1980**, Pretoria, suppl. p. 1.
[5.6] M.-L. Klotz, U. Altenhofen, H. Zahn, *Textilveredlung* **21** (1986) 119–121.
[5.7] H. Zahn, *Lenzinger Ber.* **42** (1977) 19–34.
[5.8] Forschungsprojekt des DWI: Interne Lipide der Wolle: Zusammensetzung, Veränderung und Bedeutung bei Vergilbung und bei Veredlungsprozessen von Wolle (AIF-Nr. 7160).
[5.9] D. Mäusezahl, *Textilveredlung* **5** (1970) 839–845.
[5.10] H. Zollinger, G. Back, B. Milicevic, A. N. Roseira, *Melliand Textilber.* **42** (1961) 73–80.
[5.11] F. Hoffmann, *Rev. Prog. Color. Relat. Top.* **18** (1988) 56–64.
[5.12] F. Hoffmann, W. Langmann, H. Elberzhager, M. Schnee, *Int. Text. Bull. Färberei/Druckerei/-Ausrüstung* no. 3 (1978), 205, 206, 211, 212, 217, 218.
[5.13] H. Flensberg, W. Mosimann, H. Salathé, *Melliand Textilber.* **65** (1984) 472–477.
[5.14] D. Hildebrand, H. Zahn, *SVF Fachorgan Textilveredl.* **13** (1958) 376–385.
[5.15] K. Reincke, *Melliand Textilber.* **74** (1993) 408–417.
[5.16] W. Ender, A. Müller, *Melliand Textilber.* **19** (1938) 182–183.
[5.17] F. R. Hartley, *J. Soc. Dyers Colour.* **85** (1969) 66–71.
[5.18] O. K. Dobozy, *Am. Dyest. Rep.* **62** (1973) no. 3, 36–48, 76–77.
[5.19] D. M. Lewis, G. Yan, *J. Soc. Dyers Colour.* **109** (1993) 193–197.
[5.20] P. Spinacci, N. C. Gaccio, 12th IFVTCC Congr., Budapest 1981, pp. 10–13.
[5.21] P. A. Duffield, R. D. D. Holt, *Textilveredlung* **24** (1989) 40–45.
[5.22] L. Benisek, *Textilveredlung* **12** (1977) 406–415; *J. Soc. Dyers Colour.* **94** (1978) 101–105.
[5.23] G. Meier, *TPI Text. Prax. Int.* **31** (1976) 898–901.
[5.24] A. C. Welham, *J. Soc. Dyers Colour.* **102** (1986) 126–131.
[5.25] I. D. Rattee, *J. Soc. Dyers Colour.* **69** (1953) 288–295.
[5.26] D. de Meulemeester, I. Hammers, W. Mosimann, Aachener Textiltagung 1989; *Melliand Textilber.* **71** (1990) 69.
[5.27] F. Beffa, G. Back, *Rev. Prog. Color. Relat. Top.* **14** (1984) 33–42.
[5.28] G. Schetty, W. Kuster, *Helv. Chim. Acta* **44** (1961) 2193.
H. Pfitzner, *Melliand Textilber.* **35** (1954) 649–651.
[5.29] G. Schetty, *J. Soc. Dyers Colour.* **71** (1955) 705.
[5.30] J. A. Bone, *Schriftenr. Dtsch. Wollforschungsinst. Tech. Hochsch. Aachen* **93** (1984) 170–178.
[5.31] W. Mosimann, *Textilveredlung* **17** (1982) 289–295.
[5.32] C. Mancheno, H.-W. Hensen, I. Souren, H.-K. Rouette, *Melliand Textilber.* **71** (1990) 875–882.
[5.33] G. v. Hornuff, H. J. Flath, *Faserforsch. Textiltech.* **12** (1961) 559–567.
[5.34] D. M. Lewis, *Melliand Textilber.* **67** (1986) 717–723.
[5.35] D. M. Lewis, *J. Soc. Dyers Colour.* **98** (1982) 165–175.
[5.36] K. Hannemann, F. Grüner, *Int. Text. Bull. Veredlung* **38** (1992) no. IV, 22–31.
[5.37] H. Angelmahr, F. Bartsch, *TPI Text. Prax. Int.* **47** (1992) 455–456.
[5.38] A. Würz, *Melliand Textilber.* **42** (1961) 439–444.
[5.39] H. Baumann, *Textilveredlung* **14** (1979) 515–522.
[5.40] D. Schwer, H. Ritter, K. Zesiger, *Textilveredlung* **16** (1981) 479–484.
[5.41] H. Flensberg, W. Mosimann, H. Salathé, *Schriftenr. Dtsch. Wollforschungsinst. Tech. Hochsch. Aachen* **93** (1984) 150–169.
[5.42] D. M. Lewis, *Rev. Prog. Color. Relat. Top.* **19** (1989) 49–56.
[5.43] W. Mosimann, *Am. Dyest. Rep.* **80** (1991) no. 3, 26, 28, 30, 32, 34, 62.
[5.44] F. Blum, *Bayer Farben Rev.* **7** (1964) 1–9.
[5.45] R. Hofstetter, *Melliand Textilber.* **72** (1991) 366–373.
[5.46] F. Shimizu, K. Jyoko, I. Sakaguchi, *Sen'i Gakkaishi* **32** (1976) 388–393.[5.48] H. Putze, *TPI Text. Prax. Int.* **39** (1984) 1051–1054.
[5.47] H. Putze, *TPI Text. Prax. Int.* **39** (1984) 1051–1054.
[5.48] A. Liddiard, *J. Soc. Dyers Colour.* **99** (1983) 56–59.
[5.49] R. Rohrer, *Textilveredlung* **20** (1985) 85–87.
[5.50] K. Y. Chu, J. R. Provost, *Rev. Prog. Color. Relat. Top.* **17** (1987) 22–28.
[5.51] A. Kremer, *TPI Text. Prax. Int.* **47** (1987) 746–757.
[5.52] A. Kremer, H. Kremer, *Melliand Textilber.* **73** (1992) 580–582.
[5.53] M. I. Guljarani, *Rev. Prog. Color. Relat. Top.* **22** (1992) 79–89.


References for chapter 6

[6.1] R. H. Peters, *J. Soc. Dyers Colour.* **61** (1945) 95–100.
[6.2] A. Anton, *Text. Chem. Color.* **13** (1981) 46–50.
[6.3] G. Kühnel, H. J. Flath, R. Gärtner, *Melliand Textilber.* **72** (1991) 288–290; 360–362.
[6.4] J. A. Coates, V. Ellard, I. D. Rattee, *J. Soc. Dyers Colour.* **96** (1980) 14–18.
[6.5] I. D. Rattee, S. So, *Mater. Sci. Monogr.* **21** (1984) 575–584.
[6.6] B. Catiore, R. Hagege, *Bull. Sci. Inst. Text. Fr.* **2** (1973) 209.
[6.7] G. Kühnel, Dissertation, TU Dresden 1989.
[6.8] F. Hoffmann, *Chemiefasern/Textilind.* **31/83** (1981) 762–768.
[6.9] B. von Falkai, H. Wilsing, *TPI Text. Prax. Int.* **33** (1978) 1324–1330.
[6.10] D. Dornig, Dissertation, TU Dresden 1974.
[6.11] E. Elöd, T. Schachowsky, *Melliand Textilber.* **23** (1942) 437–440.
[6.12] T. Iijima, H. Zollinger, *Teintex* **32** (1967) 245–263.
[6.13] H. Zollinger, G. Back, B. Milicevic, A. N. Roseira, *Melliand Textilber.* **42** (1961) 73–80.
[6.14] E. Sada, H. Kumazawa, T. Ando, *J. Soc. Dyers Colour.* **98** (1982) 121–125.
[6.15] V. S. Praptowidodo, K. Hamada, T. Iijima, *Angew. Makromol. Chem.* **144** (1986) 159–165.
[6.16] J. Meybeck, P. Galafassi, *Appl. Polym. Symp.* **18** (1971) 463–472.
[6.17] D. Schwer, *Textilveredlung* **23** (1988) 296–301.

[6.18] W. Beckmann, F. Hoffmann, H.-G. Otten, *Melliand Textilber. Int.* **54** (1973) 641–646; *J. Soc. Dyers Colour.* **88** (1972) 354–360.

[6.19] W. Langmann, *Melliand Textilber. Int.* **54** (1973) 654–659.

[6.20] H. Beiertz, *Am. Dyest. Rep.* **68** (1979) no. 6, 22–26.

[6.21] F. Hoffmann, *Rev. Prog. Color. Relat. Top.* **18** (1988) 56–64.

[6.22] B. C. Burdell, *Rev. Prog. Color. Relat. Top.* **13** (1983) 41–49.

[6.23] O. Annen, J. Carbonell, *Textilveredlung* **15** (1980) 296–302.

[6.24] H. Scheidegger, H. Flensberg, R. Bauhofer, *Textilveredlung* **13** (1978) 302–304.

[6.25] J. R. Aspland, *Text. Chem. Color.* **25** (1993) 19–23.

[6.26] J. Shore, *J. Soc. Dyers Colour.* **87** (1971) 37–48.

[6.27] J. Guthrie, *J. Appl. Polym. Sci.* **27** (1982) 2567–2575.

[6.28] C. C. Cook, *Rev. Prog. Color. Relat. Top.* **12** (1982) 73–89.

[6.29] T. F. Cooke, H. D. Weigmann, *Rev. Prog. Color. Relat. Top.* **20** (1990) 10–18.

[6.30] O. Annen, *Textilveredlung* **19** (1984) 8–11.

[6.31] H. Egli, *Textilveredlung* **2** (1967) 856–864.

[6.32] Bayer, Musterkarte Telon-S-Verfahren Sp 502, Leverkusen.

[6.33] R. Weber, *Melliand Textilber.* **58** (1977) 48–51.

[6.34] W. Hartmann, *Am. Dyest. Rep.* **69** (1980) no. 6, 21–22.

[6.35] M. Mitter, *Chemiefasern/Text. Anwendungstech.* **31** (1981) 55–56, 558.

[6.36] I. Strelecky, *Ciba Rundsch.* **1968**, no. 2, 40–44.

[6.37] W. R. Remington, E. K. Gladding, *J. Am. Chem. Soc.* **72** (1953) 2553.

[6.38] M. Greenhalgh, A. Johnson, R. H. Peters, *J. Soc. Dyers Colour.* **78** (1962) 315–321.

[6.39] N. J. Bhatt, E. H. Daruwalla, *Textile Res. J.* **34** (1964) 435–444.

[6.40] G. Valk, G. Stein, A. El Bendak, *Melliand Textilber.* **57** (1976) 166; **58** (1977) 580–582.
G. Valk, G. Stein, *Melliand Textilber.* **59** (1978) 593–594.

[6.41] D. Dornig, H. Sehm, *Melliand Textilber.* **74** (1993) 212–215.


References for chapter 7


[7.1] Faserstoff-Tabellen according to P. A. Koch, *Chemiefasern/Textilind.* **43** (1993) no. 6, 508–522.

[7.2] K. H. Meyer, *Melliand Textilber.* **6** (1925) 737–739; V. Kartaschoff, *Helv. Chim. Acta* **8** (1925) 928–942; **9** (1926) 152–173.

[7.3] T. Vickerstaff: *The Physical Chemistry of Dyeing,* Oliver & Boyd, London 1954.

[7.4] R. H. Peters: *Textile Chemistry,* vol. 3: The Physical Chemistry of Dyeing, Elsevier, Amsterdam 1975.

[7.5] H. Braun: "Particle Size and Solubility of Disperse Dyes," *Rev. Prog. Color. Relat. Top.* **13** (1983) 62–72.

[7.6] H. Leube, *Text. Chem. Color.* **10** (1978) no. 2, 32–39; 39–46.

[7.7] F. Jones, *J. Soc. Dyers Colour.* **100** (1984) no. 2, 66–72.

[7.8] G. L. Baughman, T. A. Perenich, *Text. Chem. Color.* **21** (1989) no. 1, 33–37.

[7.9] H. Leube, W. Rüttiger, *Melliand Textilber.* **59** (1978) 836–842.

[7.10] T. M. Baldwinson, *Rev. Prog. Color. Relat. Top.* **15** (1985) 6–14.

[7.11] H. Gerber, *Melliand Textilber.* **69** (1988) 195–200.

[7.12] A. Keil, H. Noack, H. J. Flath, *Textiltechnik* **40** (1990) 264–267, 318–320, 382–384, 432–437.

[7.13] R. Turner, *Text. Chem. Color.* **15** (1983) no. 9, 180/61–188/69.

[7.14] D. M. Nunn (ed.): *The Dyeing of Synthetic Polymer and Acetate Fibres,* Dyers Comp. Publ. Trust, Bradford 1979.

[7.15] F. Somm, R. Buser, *Int. Dyer Printer* **170** (1985) no. 12, 6–7.

[7.16] H. Gerber, *Textilveredlung* **8** (1973) 449–456.

[7.17] A. Maier, AATCC Int. Dyeing Symp., Washington 1977, pp. 66–69.

[7.18] R. A. Walsh, *Text. Chem. Color.* **7** (1975) no. 10, 184/35–187/38.

[7.19] Disperse Dye Comm., *J. Soc. Dyers Colour.* **93** (1977) 228–237.

[7.20] M. Hammoudeh, E. Schönpflug, *Melliand Textilber.* **52** (1971) 1063–1068.

[7.21] BASF Manual: Dyeing and Finishing of Polyester Fibers B 363 e, pp. 384–501, 1975.

[7.22] J. Park, *Rev. Prog. Color. Relat. Top.* **15** (1985) 26–27; T. M. Baldwinson, 6–14.

[7.23] G. R. Turner, *Text. Chem. Color.* **15** (1983) no. 9, 61–69.

[7.24] E. Podogrocka, *Melliand Textilber.* **70** (1989) 198–199.

[7.25] K. Poulaikis et al., *Chemiefasern/Textilind.* **41** (1991) 142–147.

[7.26] W. Saus, D. Knittel, E. Schollmeyer, *TPI Text. Prax. Int.* **48** (1993) 32–36; *Text. Res. J.* **63** (1993) no. 3, 135–142.

[7.27] F. Jones, J. Kraska, *J. Soc. Dyers Colour.* **82** (1960) 333–338.

[7.28] Ch. Guth, *Textilveredlung* **14** (1979) 270–274.

[7.29] AATCC Palmetto Section, *Text. Chem. Color.* **11** (1979) no. 12, 270/23–282/35.

[7.30] Nippon Kayaku, JP 59 066 585, 1982 (N. Mitsuro).

[7.31] J.-H. Chiao-Cheng, B. M. Reagan, *Text. Chem. Color.* **15** (1983) 12–19.

[7.32] C. A. Li, *Chem. Abstr.* **100** (1984) no. 24, 193 415.

[7.33] W. Griesser, H. Tiefenbacher, *Textilveredlung* **28** (1993) no. 4, 88–96.

[7.34] M. Laufer, *Textilveredlung* **28** (1993) no. 4, 96–101.

[7.35] P. Richter, *Chemiefasern/Textilind.* **41/93** (1991) no. 9, 1118–1125, E 129–133.

[7.36] W. Beckmann, H. Hamacher, *Chemiefasern + Text. Anwendungstech. Text. Ind.* **23** (1973) no. 5, 436–439.

[7.37] C. Renard, *Melliand Textilber.* **54** (1973) 1328–1335.

[7.38] J. Cegarra, F. J. Carrion, P. Puente, *Melliand Textilber.* **65** (1984) 405–409.

[7.39] C. Grardel, *Ind. Text.* (Paris) **1155** (1985) no. 5, 519–520.

[7.40] H. Beiertz, *Chemiefasern/Textilind.* **42/94** (1992) no. 1, 64–67, E 1–3; H. Beiertz, K. H. Röstermundt, *TPI Text. Prax. Int.* **48** (1993) no. 2, 135–137.

[7.41] G. Reinert, V. Misun, *Melliand Textilber. Int.* **74** (1993) 1007–1014.

References for chapter 8

[8.1] Report of Committee: "Dyeing Properties of Disperse Dyes, I Cellulose Acetate," *J. Soc. Dyers Colour.* **80** (1964) 237–242; "II Cellulose Triacetate," **81** (1965) 209–210.

[8.2] H. Leube, *Z. Gesamte Textilind.* **70** (1968) no. 4 to **71** (1969) no. 2; English version see: Dyeing and Finishing of Acetate and Triacetate and their Blends with Other Fibres, BASF Manual S 388 e.

[8.3] D. Blackburn in D. M. Nunn (ed.): *The Dyeing of Synthetic Polymer and Acetate Fibres*, Comp. Publ. Trust, Bradford 1979, pp. 76–128.

[8.4] R. H. Peters: "The Physical Chemistry of Dyeing," *Textile Chemistry*, vol. 3, Elsevier, Amsterdam 1975.

References for chapter 9

[9.1] Stanford Research Inst.: Chemical Economics Handbook, Fibers 5 433 000 D, Menlo Park, Sept. 1993.
Comité International de la Rayonne et des Fibres Synthetiques, Bruxelles, Brochures 1993.

[9.2] T. A. Field, *Am. Dyest. Rep.* **41** (1952) 475.

[9.3] R. Raue, *Rev. Prog. Color. Relat. Top.* **14** (1984) 187.

[9.4] W. Beckmann in D. M. Nunn (ed.): *The Dyeing of Synthetic Polymer and Acetate Fibres*, The Dyer's Company, Bradford 1979, p. 359 ff.

[9.5] I. Holme, *Rev. Prog. Color. Relat. Top.* **13** (1983) 10.

[9.6] F. Jones, *Rev. Prog. Color. Relat. Top.* **4** (1973) 64.

[9.7] H. Sumner in A. Johnson (ed.): *The Theory of Colouration of Textiles*, chap. 2, Society of Dyers and Colourists, Bradford 1989.

[9.8] O. Glenz, W. Beckmann, *Melliand Textilber.* **38** (1957) 296, 783, 1152.

[9.9] W. Beckmann, *J. Soc. Dyers Colour.* **77** (1961) 616.

[9.10] S. Rosenbaum, *Text. Res. J.* **33** (1963) 899.

[9.11] U. Mayer, W. Ender, A. Würz, *Melliand Textilber.* **47** (1966) 653, 772.

[9.12] G. Alberghina, M. Amato, S. Fisichella, *J. Soc. Dyers Colour.* **105** (1989) 163.

[9.13] H. Zollinger, *Text. Res. J.* **59** (1989) 264.

[9.14] P. R. Brady, *Rev. Prog. Color. Relat. Top.* **22** (1992) 58.

[9.15] A. V. Hill, *Proc. R. Soc. London B* **104** (1928) 39, 65.

[9.16] W. Beckmann, *11th Int. Textile Seminar*, Kingston, Ontario, 1968, Book of Papers, p. 97; *Z. Gesamte Textilind.* **71** (1969) 603.

[9.17] S. Rosenbaum, *J. Polym. Sci. Part A* **3** (1965) 1949.

[9.18] M. Williams, R. Landel, J. Ferry, *J. Am. Chem. Soc.* **77** (1955) 3701.

[9.19] R. Asquith, H. Blair, N. Spence, *Text. Res. J.* **47** (1977) 446; *J. Soc. Dyers Colour.* **94** (1978) 49.

[9.20] S. Rosenbaum, *Text. Res. J.* **34** (1964) 159.

[9.21] D. Balmforth, C. Bowers, T. Guion, *J. Soc. Dyers Colour.* **80** (1964) 577.

[9.22] D. G. Evans, G. J. Bent: "Basic Dyes on Acrylic Fibres" *J. Soc. Dyers Colour.* **89** (1973) 292.

[9.23] J. Neary, R. Thomas, *Am. Dyest. Rep.* **46** (1957) 625.

[9.24] I. Holme, *Rev. Prog. Color. Relat. Top.* **1** (1970) 31.

[9.25] S. Cohen, A. Endler, *Am. Dyest. Rep.* **46** (1957) 625; **47** (1958) 325.

[9.26] S. Heimann, F. Feichtmayr, *Z. Gesamte Textilind.* **68** (1966) 509.

[9.27] F. Hoffmann, *Melliand Textilber.* **59** (1978) 239.

[9.28] C. Zimmermann, A. Cate, *Am. Dyest. Rep.* **83** (1972) 150.

[9.29] Bayer, CA 793 184, 1968 (W. Beckmann, J. Nentwig, H. Rudolf, J. Schneider).

[9.30] M. Dullaghan, A. Ultee, *Text. Chem. Color.* **8** (1976) 22.

[9.31] S. Shukla, M. Mathur, *J. Soc. Dyers Colour.* **109** (1993) 330.

[9.32] J. Leddy, *Am. Dyest. Rep.* **49** (1960) 272.

[9.33] D. G. Evans, C. J. Bent: "Basic Dyes on Acrylic Fibres Comm." *J. Soc. Dyers Colour.* **88** (1972) 220.

[9.34] F. Hoffmann, *Rev. Prog. Color. Relat. Top.* **18** (1988) 56.

[9.35] AATCC Metropolitan Sect., *Text. Chem. Color.* **8** (1976) 165.

[9.36] J. Park, J. Shore, *J. Soc. Dyers Colour.* **97** (1981) 223. J. Park: *A Practical Introduction to Yarn Dyeing*, Society of Dyers and Colourists, Bradford 1980.

[9.37] W. Biedermann, *Rev. Prog. Color. Relat. Top.* **10** (1979) 1.

[9.38] F. Hoffmann, M. Schnee, H. Schubert, *Melliand Textilber.* **65** (1984) 478.

[9.39] F. Hoffmann, *Text. Chem. Color.* **22** (1990) 11.

[9.40] A. Kretschmer, *TPI Text. Prax. Int.* **44** (1989) 1098, 1316; **45** (1990) 225; **46** (1991) 1220.

[9.41] W. Beckmann, K. Jakobs, *Bayer Farben Rev.* **16** (1968) 1.

[9.42] F. Hoffmann, H. Schubert, J. Fiegel, *TPI Text. Prax. Int.* **47** (1992) 233.

[9.43] P. Dorsch, H. Wilsing, K.-H. Peters, *Melliand Textilber.* **62** (1981) 188.

[9.44] R. Rohner, H. Zollinger, *Text. Res. J.* **56** (1986) 1.

[9.45] U. Meyer, J. Zhang, H. Zollinger, *Textilveredlung* **19** (1984) 39.

[9.46] U. Reinehr, G. Häfner, A. Nogaj, *Chemiefasern/Textilind.* **35** (1985) no. 87, 588.

[9.47] G. Müller, *Textilveredlung* **10** (1975) 438.

[9.48] L. Witt, P. Toldrian, *Chemiefasern/Textilind.* **28** (1978) no. 80, 469.

[9.49] J. Shore, *Rev. Prog. Color. Relat. Top.* **10** (1979) 33.

[9.50] W. Beckmann, W. Langmann, *Dtsch. Textiltech.* **18** (1968) 243.

[9.51] H. Gerber, H. Lehmann, F. Somm, *Melliand Textilber. Int.* **54** (1973) 77.

[9.52] L. Kostova, R. Iltscheva, R. Detscheva, *Textilveredlung* **27** (1993) 398.

[9.53] I. Hardalov, J. Mikhailova, *J. Soc. Dyers Colour.* **109** (1993) 369.

[9.54] R. Weber, *Text. Prax.* **21** (1966) 670.

[9.55] W. Ingamells, *J. Soc. Dyers Colour.* **96** (1980) 332.

[9.56] P. Ackroyd, *Rev. Prog. Color. Relat. Top.* **5** (1974) 86.

[9.57] R. Detscheva, R. Iltscheva, K. Dimov, *Textilveredlung* **9** (1974) 312; **10** (1975) 318.

[9.58] A. Läpple, A. Schneider, *Textilveredlung* **10** (1975) 63.

[9.59] I. Holme, *Rev. Prog. Color. Relat. Top.* **7** (1976) 1.

References for chapter 10

[10.1] D. M. Nunn (ed.): *The Dyeing of Synthetic-polymer and Acetate Fibres*, Dyers Comp. Publ. Trust, Bradford 1979.

[10.2] M. Peter, H. K. Rouette: *Grundlagen der Textilveredlung*, 13th ed., Deutscher Fachverlag, Frankfurt a. M. 1989.

[10.3] H. W. Partridge, *Rev. Prog. Color. Relat. Top.* **6** (1975) 56.

[10.4] BASF: technical information TI/T 138e, Ludwigshafen 1986.
[10.5] BASF: technical information TI/T 151e, Ludwigshafen 1985.
[10.6] G. Biehler, *Chemiefasern/Textilind.* **29/81** (1979) 848–853.
[10.7] H. Leube, *Melliand Textilber.* **46** (1965) 743–749.
[10.8] J. Shore, *Rev. Prog. Color. Relat. Top.* **6** (1975) 7–12.
[10.9] J. Shore in [10.1], pp. 393–403.
[10.10] R. H. Peters: *Textile Chemistry*, vol. 3: The Physical Chemistry of Dyeing, Elsevier, Amsterdam 1975.
[10.11] B. D. Gupta, A. K. Mukherjee, *Rev. Prog. Color. Relat. Top.* **19** (1989) 7–19.

References for chapter 11

[11.1] *Ullmann*, 3rd ed. **7**, 46–54.
[11.2] BASF Manual Dyeing and Finishing of Polyester Fibres B 363e, 1975.
[11.3] J. Shore in D. M. Nunn (ed.): *The Dyeing of Synthetic-polymer and Acetate Fibres*, Dyers Comp. Publ. Trust, Bradford 1979.
[11.4] M. Peter, H. K. Rouette: *Grundlagen der Textilveredlung*, 13th ed., Deutscher Fachverlag, Frankfurt a. M. 1989.
[11.5] BASF Manual Dyeing and Finishing of Cellulose Fibres B 375e, pp. 383–414, 1975.
[11.6] K. H. Rostermundt, *TPI Text. Prax. Int.* **47** (1992) no. 7, 649–654; and XIX–XXIII.
[11.7] S. M. Doughty, *Rev. Prog. Color. Relat. Top.* **16** (1986) 25–38.
[11.8] W. Haertl, *Textilveredlung* **24** (1989) no. 6, 214–218.
[11.9] D. Blackburn, *Int. Dyer Text. Printer, Bleacher Finish.* **171** (1986) no. 1, 13–14.

References for chapter 12

[12.1] A. Berger-Schunn: *Praktische Farbmessung*, Muster Schmidt, Göttingen 1991.
[12.2] ISO/CD 787-26 (Draft).
[12.3] W. Baumann et al., *Melliand Textilber.* **67** (1986) 562–566; *J. Soc. Dyers Colour.* **103** (1987) 100–105.
[12.4] R. Broßmann et al., *Melliand Textilber.* **67** (1986) 499–502; *J. Soc. Dyers Colour.* **103** (1987) 38–42.
[12.5] A. Berger-Schunn et al., *Melliand Textilber.* **67** (1986) 638–639, 716–717, 812–813; *J. Soc. Dyers Colour.* **103** (1987) 138–139, 140–141, 272–274.
[12.6] T. v. Chambers et al., *Melliand Textilber.* **69** (1988) 755–758; *J. Soc. Dyers Colour.* **105** (1989) 214–218.
[12.7] Brit. Occup. Hyg. Soc., Technical Guide no. 4, part 1 and 2 (1985).
[12.8] A. Berger-Schunn et al., *Melliand Textilber.* **70** (1989) 690–693; *J. Soc. Dyers Colour.* **107** (1991) 270–273.
[12.9] *J. Soc. Dyers Colour.* **88** (1972) 220–222.

References for chapter 13

[13.1] P. J. Smith, *Rev. Prog. Color. Relat. Top.* **24** (1994) 31–40.

References for chapter 14

[14.1] TVI-Verband: Water/Wastewater Investigation 1992 (Association of the Textile Finishing Industry), Frankfurt a. M. 1993.
[14.2] Arbeitsbericht der ATV-Arbeitsgruppe Textilveredlungsindustrie (Working Report of the ATV Working Group Textile Finishing Industry), *Korresp. Abwasser* **36** (1989) 1074–1084.
[14.3] Gesetz zur Ordnung des Wasserhaushaltes, Wasserhaushaltsgesetz, Neufassung vom 12.2.1990, BGBl. I (Water Resources Act, revised February 12, 1990, Federal Law Gazette), p. 205
[14.4] Abwasserherkunftsverordnung vom 3.7.1987, geändert am 27.5.1991, BGBl. I. (Wastewater Origin Decree, July 3, 1987, revised May 27, 1991, Federal Law Gazette), p. 1197.
[14.5] Allgemeine Rahmenabwasserverwaltungsvorschrift über Mindestanforderungen an das Einleiten von Abwasser in Gewässer, Neufassung vom 29.10.1992, GMBl 1992 (General Legal Framework on Minimum Requirements on Discharge of Wastewater into Surface Waters, revised October 29, 1992, Joint Ministerial Law Gazette 1992), pp. 1265–1266.
[14.6] 38. Abwasserverwaltungsvorschrift über Mindestanforderungen an das Einleiten von Abwasser in Gewässer (Textilherstellung) vom 5.9.1984, GMBl. [38th Wastewater Ordinance on Minimum Requirements for the Discharge of Wastewater into Surface Waters (Textile Processing), September 5, 1984, Joint Ministerial Law Gazette] 1984, p. 348.
[14.7] Bundesratsdrucksache 181/93 (Printed Matter of the German Bundesrat), pp. 13–21.
[14.8] 612. Verordnung des Bundesministers für Land- und Forstwirtschaft über die Begrenzung von Abwasseremissionen aus Textilveredlungs- und Behandlungsbetrieben, BGBl. 207. Stück v. 24.9.1992 (612. Ordinance of the Federal Minister for Agriculture and Forestry on Limitations of Discharge from Textile Finishing and Processing Plants, Federal Law Gazette, September 24, 1992), pp. 2828–2835.
[14.9] The Water Act (London. IIMSO).
[14.10] I. G. Laing, *Rev. Prog. Color. Relat. Top.* **21** (1991) 56–71.
[14.11] P. Cooper, *J. Soc. Dyers Colour.* **108** (1992) 176–182.
[14.12] H. Reetz, *Melliand Textilber.* **72** (1991) 932–933.
[14.13] The Government of Japan, Environment Agency: Annual Book, Quality of the Environment in Japan in 1992.
[14.14] B. Smith, J. Rucker, *Am. Dyest. Rep.* **76** (1987) 68–78.
[14.15] Sewerage Bureau of Osaka Municipal Government: The Sewerage Law and Regulations for Control of Water Quality, Osaka 1990.
[14.16] U. Sewekow, *Melliand Textilber.* **70** (1989) 589–596.
[14.17] U. Rössner, *Textiltechnik (Leipzig)* **38** (1988) 661–665.
[14.18] B. Smith, *Am. Dyest. Rep.* **78** (1989) 26–32.
[14.19] W.-D. Kermer, *Melliand Textilber.* **69** (1988) 586–590.
[14.20] B. M. Müller, *Rev. Prog. Color. Relat. Top.* **22** (1992) 14–21.

[14.21] J. Park, J. Shore, *J. Soc. Dyers Colour.* **100** (1984) 383–399.

[14.22] J. Wolff, H. Henk, *Textilveredlung* **25** (1990) 213–218.

[14.23] W. Beckmann, U. Sewekow, *TPI Text. Prax. Int.* **46** (1991) 346–348, 445–449.

[14.24] F. Hoffmann, H. Schubert, J. Fiegel, *TPI Text. Prax. Int.* **47** (1992) 233–237.

[14.25] R. Puk, D. Sedlak, *TPI Text. Prax. Int.* **47** (1992) 238–245.

[14.26] J. M. Marzinkowski, *Chemiefasern/Textilind.* **41/93** (1991) 895–901.

[14.27] P. A. Duffield, K.-H. Hoppen, *Melliand Textilber.* **68** (1987) 195–202.

[14.28] P. A. Duffield, R. R. D. Holt, J. R. Smith, *Melliand Textilber.* **72** (1991) 938–942.

[14.29] W. Sebb, *TPI Text. Prax. Int.* **44** (1989) 841–843.

[14.30] F. Conzelmann, P. Wurster, A. Zahn, *TPI Text. Prax. Int.* **44** (1989) 644–649.

[14.31] G. Rösch, *TPI Text. Prax. Int.* **45** (1990) 495–499.

[14.32] W. Beckmann, J. Pflug, *TPI Text. Prax. Int.* **38** (1983) 160, 165–168.

[14.33] R. Teichmann, *Melliand Textilber.* **74** (1993) 900–902.

[14.34] H.-U. von der Eltz, *TPI Text. Prax. Int.* **45** (1990) 727–731.

[14.35] D. Fiebig, D. Soltau, *TPI Text. Prax. Int.* **43** (1988) 644, 649–650; **44** (1989) 533–535; **44** (1989) 1124–1126.

[14.36] D. Frahne, S. Koscielski, *Melliand Textilber.* **68** (1987) 594–596.

[14.37] N. Athanasopoulos, *Melliand Textilber.* **71** (1990) 619–628.

[14.38] U. Sewekow, G. Diesterweg, *Textilveredlung* **26** (1991) 142–146.

[14.39] A. Wilking, D. Frahne, *Melliand Textilber.* **74** (1993) 897–900.

[14.40] C. Oehme, *Melliand Textilber.* **67** (1986) 582–588.

[14.41] J. Janitza, S. Koscielski, H. Schnabel, *TPI Text. Prax. Int.* **46** (1991) 1216–1219.

[14.42] P. J. Halliday, S. Beszedits, *Can. Text. J.* **103** (1986) 78–84.

[14.43] E. Thomanetz, D. Bardtke, E. Köhler, *GWF, Gas-Wasserfach: Wasser/Abwasser* **128** (1987) 432–441, 475–481.

[14.44] M. Seekamp, *Textilveredlung* **25** (1990) 125–129.

[14.45] W. Marte, W. Keller, *Textilveredlung* **26** (1991) 224–230.

[14.46] G. von Hagel, *Korresp. Abwasser* **33** (1986) 908–915.

[14.47] T. R. Demmin, K. D. Uhrich, *Am. Dyest. Rep.* **77** (1988) 13–14, 17–18, 32.

[14.48] W. C. Tincher, *Text. Chem. Color.* **21** (1989) 33–35.

[14.49] A. E. Wilcock, S. P. Hay, *Can. Text. J.* **108** (1991) 37–44.

[14.50] A. E. Wilcock, M. Brewster, W. C. Tincher, *Am. Dyest. Rep.* **81** (1992) 15–22.

[14.51] G. Schulz, D. Fiebig, H. Herlinger, *Textilveredlung* **23** (1988) 1–4.

[14.52] G. Schulz, H. Herlinger, G. Mayer, *Textilveredlung* **25** (1990) 23–27.

[14.53] A. Hövelmann, S. C. Bidinger, A. Linder, *TPI Text. Prax. Int.* **48** (1993) 507–509.

[14.54] K. H. Gregor, *Melliand Textilber.* **71** (1990) 976–979.

[14.55] U. Sewekow, *Melliand Textilber.* **74** (1993) 153–157.

[14.56] D. L. Michelsen, W. W. Powell, R. M. Woodby, L. L. Fulk, AATCC Technical Conference, Book of Papers, Atlanta 1992, pp. 135–150.

[14.57] J. M. Green, C. Sokol, *Am. Dyest. Rep.* **74** (1985) 50–51, 67.

[14.58] M. Kolb, B. Müller, B. Funke, *Vom Wasser* **69** (1987) 217–223.

[14.59] J. Carriére, J. P. Jones, A. D. Broadbent, AATCC Technical Conference, Book of Papers, Atlanta 1992, pp. 231–236.

[14.60] M. Pitroff, K. H. Gregor, *Melliand Textilber.* **73** (1992) 526–529.

[14.61] H. Reissig, T. Jentsch, R. Fischer, *Gewässerschutz, Wasser, Abwasser* **125** (1991) 275–304.

[14.62] G. Schulz, H. Herlinger, F. U. Gähr, T. Lehr, *TPI Text. Prax. Int.* **47** (1992) 1055–1062.

[14.63] M. Kolb, P. Korger, B. Funke, *Melliand Textilber.* **69** (1988) 286–287.

[14.64] M. Kolb, B. Funke, A. Baur, N. Peschen, *Korresp. Abwasser* **32** (1985) 986–987.

[14.65] E. Burtscher et al., *Melliand Textilber.* **74** (1993) 903–907.

[14.66] J. J. Porter, *Am. Dyest. Rep.* **61** (1972) 24–27.

[14.67] H.-J. Buschmann, E. Schollmeyer, *Melliand Textilber.* **72** (1991) 543–544.

[14.68] H.-J. Buschmann, *Chemiefasern/Textilind.* **42/94** (1992) 408–411.

[14.69] D. Tegtmeyer, *Melliand Textilber.* **74** (1993) 148–151.

[14.70] U. Sewekow, AATCC Conference, Book of Papers, Montreal 1993, pp. 235–246.

[14.71] W. Peukert, *Seifen, Öle, Fette, Wachse* **114** (1988) 474–477.

[14.72] *Text. Horizons* **1987**, 58–59.

[14.73] K. Roennefahrt, *Melliand Textilber.* **70** (1989) 203–206.

[14.74] J. Küßner, J. Janitza, S. Koscielski, *TPI Text. Prax. Int.* **47** (1992) 736–741.

[14.75] P. Wragg, *J. Soc. Dyers Colour.* **109** (1993) 280–282.

[14.76] S. N. Gaeta, U. Fedele, *Desalination* **83** (1991) 183–194.

[14.77] G. M. Elgal, *Text. Chem. Color.* **18** (1986) 15–20.

[14.78] F. Richartz, *TPI Text. Prax. Int.* **46** (1991) 567–572.

[14.79] W. Hess, *Bild Wiss.* **29** (1992) 84–88.

[14.80] E. Spachmann, E. Tillig, *TPI Text. Prax. Int.* **45** (1990) 129–130.

[14.81] J. Trauter, *Melliand Textilber.* **74** (1993) 559–562.

[14.82] J. J. Porter, *Text. Chem. Color.* **22** (1990) 21–25.

[14.83] D. Kuiper, C. A. Brandon, H. G. Spencer, G. L. Gaddys, *TPI Text. Prax. Int.* **40** (1985) 1124–1126.

[14.84] A. Ersvell, C. J. Brouckaert, C. A. Buckley, *Desalination* **70** (1988) 157–167.

[14.85] G. M. Shaul, T. J. Holdsworth, C. R. Dempsey, K. A. Dostal, *Chemosphere* **22** (1991) 107–119.

[14.86] U. Pagga, D. Brown, *Chemosphere* **15** (1986) 479–491.

[14.87] G. B. Michaels, D. L. Lewis, *Environ. Toxicol. Chem.* **5** (1986) 161–166.

[14.88] T. Ogawa, C. Yatome, E. Idaka, H. Kamiya, *J. Soc. Dyers Colour.* **102** (1986) 12–14.

[14.89] K. Ackermann, D. Frahne, *Melliand Textilber.* **63** (1982) 66–69.

[14.90] C. Harmer, P. Bishop, E. Holder, P. V. Scarpino, *Proc. Ind. Waste Conf.* **46th**, Lewis Publishers, Chelsea, Michigan 1992 pp. 217–228.

[14.91] G. M. Shaul, C. R. Dempsey, K. A. Dostal, R. J. Liebermann, *Proc. Ind. Waste Conf.* **41st**, publ. 1987, 603–611.

[14.92] H. R. Hitz, W. Huber, R. H. Reed, *J. Soc. Dyers Colour.* **94** (1978) 71–76.

[14.93] D. Brown, P. Laboureur, *Chemosphere* **12** (1983) 397–404.

[14.94] D. Brown, B. Hamburger, *Chemosphere* **16** (1987) 1539–1559.

[14.95] W. C. Tincher: *Dyes in the Environment: Dyeing Wastes in Landfill*, Final Report, Georgia Institute of Technology, Atlanta 1988.

[14.96] D. Brown, H. R. Hitz, L. Schäfer, *Chemosphere* **10** (1981) 245–261.

[14.97] R. Anliker: "Organic Colorants" in M. Richardson (ed.): *Toxic Hazard Assessment of Chemicals*, The Royal Society of Chemists, London 1986, pp. 167–187.

[14.98] E. A. Clarke, R. Anliker, *Rev. Prog. Color. Relat. Top.* **14** (1984) 84–89.

[14.99] D. Brown, *Ecotoxicol. Environ. Saf.* **13** (1987) 139–147.

[14.100] R. Anliker, E. A. Clarke, P. Moser, *Chemosphere* **10** (1981) 263–274.

[14.101] R. Anliker, *Textilveredlung* **16** (1981) 431–438.

[14.102] K. H. Leist, *Ecotoxicol. Environ. Saf.* **6** (1982) 457–463.

[14.103] R. A. Moll, *Melliand Textilber.* **72** (1991) 836–840.

[14.104] *IARC Monographs on the Evaluation on the Carcinogenic Risks to Humans* **48** (1990) 139–147.

[14.105] U. S. Dep. of Health and Human Serv., National Toxicology Program, Technical Report Series no. 285 (1992) 11–13.

[14.106] GefStoffV vom 26.8.1986, in der Fassung vom 26.10.1993, BGBl. I (GefStoffV, August 26, 1986, version October 26, 1993, Federal Law Gazette), 1782–1810.

[14.107] R. Anliker, D. Steinle, *Textilveredlung* **25** (1990) 42–49.

[14.108] O. P. Hornstein, *Melliand Textilber.* **70** (1989) 222–227.

[14.109] K. L. Hatch, *Text. Res. J.* **54** (1984) 644–682, 721–732.

[14.110] B. M. Hausen, K. H. Schulz, *DMW Dtsch. Med. Wochenschr.* **109** (1984) 1469–1475.

[14.111] B. M. Hausen, A. Kleinheinz, H. Mensing, *Allergo J.* **2** (1993) 13–16.

[14.112] J. M. Wattie, *J. Soc. Dyers Colour.* **103** (1987) 304–307.

[14.113] G. A. Heath, W. Dyson, *Text. Chem. Color.* **22** (1990) 25–31.

[14.114] P. Rosenthal, *Int. Text. Bull.* **36** (1990) 58–66.

# Textile Printing

ROBERT KOCH (retired), JOHANN HEINRICH NORDMEYER (Section 5.10), Bayer AG, Leverkusen,
Federal Republic of Germany
This contribution is based on the article Textildruck in Ullmann's, 4th ed., written by JOACHIM ZAHN,
JOHANNES EIBL, WERNER KÜHNEL, HELMUT DAHM, ROBERT KOCH, RICHARD SCHWAEBEL, and WILHELM
BERLENBACH.

# 1. History

Early testaments to printed textiles have survived only because of the dry climate of Egypt. These include a cylindrical printing block 4.4 cm in height with embossed patterns on both flat surfaces; a linen tunic for a child of ca. 3 years, printed by the resist technique with a white diamond pattern on a light-blue ground dyed with indigo (woad); and fragments of directly printed woven textiles, including a two-color printed fabric in red and black. These items, discovered by FORRER in 1894 in the necropolis of Achmim (the Panopolis of antiquity), are Coptic remains from the period between the 5th and 10th centuries A.D. The gap between these particular items and the true beginnings of the localized dyeing of textiles, which lie much farther back in time, is unlikely ever to be closed. Substantial amounts of material have simply rotted away. There are also no known written reports with the exception of a passage in PLINY THE ELDER (23 or 24 to 79 A.D.), where he describes a process used in Egypt that later came to Europe via India. Printing with solutions of metallic salts (mordanting) was followed by dyeing of the fabric in a hot bath of madder (alizarin). Apart from the vat dyes Tyrian purple and indigo, almost all natural dyes are of the mordant type, and they produce truly dyed textiles only when combined with metallic salts, forming the so-called colored "lakes." Some dyes produce very different colors with different metals.

In the Middle Ages, most of the textile printing in Europe occurred in monasteries. The few records surviving from this epoch also specify the dyes used. Almost all were the same inorganic pigments employed by painters; e.g., black: soot from carbonized vines; white: white lead (basic lead carbonate); yellow to brown: various types of ocher; green: verdigris (mixtures of basic copper(II) acetates); blue: azurite (basic copper carbonate) and powdered indigo; red: red ocher, red lead (lead oxide), and cinnabar (mercury sulfide). All of these were bonded with linseed oil in the printing process. Printing with such oil colors was practiced for many centuries, and it is rightly regarded as the precursor of today's *pigment-printing* process.

After the 15th century, textile printing also developed outside the walls of the monasteries. It gradually became an independent trade, and reached industrial scale in certain countries as early as the 18th century. A powerful impulse came from India, with its rich textile tradition. This increased interest in printing brought remarkable achievements, and not only in the use of color. The first process that did not use oil colors was *indigo blue printing*, established in Holland in the 16th century. In principle, this was the same technique that had been used for the child's tunic in Achmim/Panopolis: printing a pattern with a paste (a liquid wax resist in the case of the Copts) by means of a printing block, dyeing the result in a cold indigo vat, and washing out the printed resist to reveal a pattern that had stayed white during dyeing. In the 17th century, GEORG NEUHOFER brought indigo blue printing from Holland to Augsburg, where the first German calico printing works was built by his brother, JEREMIAS NEUHOFER, in 1689.

In 1759, JOHANN HEINRICH SCHÜLE began a manufacturing operation that at its peak employed 3500 workers. In addition to indigo blue printing, *direct printing* and hand coloring with supplementary dyes such as alizarin soon were of major importance as well. White and color *discharges* also took their place in company with the conventional printing process. Printing was even initiated on the basis of engraved copper plates instead of wooden patterns.

CHRISTOPH PHILIPP OBERKAMPF, who was born in Weissenburg near Ansbach in 1738, moved in 1758 to Paris and soon thereafter (1760) founded the most important textile printing works of the time at Jouy, near Versailles. "Toiles de Jouy" gained a worldwide reputation, and became the absolute embodiment of artistic textile printing. Indeed, LOUIS XVI elevated OBERKAMPF to the nobility and designated his operation a "Manufacture Royale."

The technique of *chlorine bleaching* was practiced at an early date in Jouy, and wooden patterns were replaced by firmly bonded copper plates of arbitrary size, permitting the printing of large designs. Printing with copper plates required use of a *press* for the first time, because the printing paste was to be transferred to the fabric not from a raised area but rather from the recesses in the plate. A simple printing machine with a roller bearing a raised pattern was also part of the technical arsenal at Jouy.

First reports of a continuous *rotary press* appeared at the turn of the 18th century. This wooden device was operated by a hand crank and included two rollers, one above the other. The top roller bore the raised design, to which colored printing paste had to be applied manually. The lower roller pressed the moving fabric against the upper roller.

ALOIS SENEFELDER (1771–1834), the inventor of lithography, failed to achieve significant success with this particular process, so he turned toward calico printing, but his invention of an improved rotary printing machine came too late. By 1783, the Scotsman THOMAS BELL, a printer of copper engravings, had already patented a machine in which engraved copper rollers continuously printed a multicolored pattern onto a fabric. He was thus the true inventor of *cylinder printing*. Present-day roller printing machines developed on the basis of his design. By contrast, rotary printing with a roller containing a raised pattern was not destined for a great future.

In 1834, the French machinist PERROT constructed a flatbed printing machine that quickly came into widespread use, and was in fact called the *perrotine* after its inventor. A rather complicated mechanism coated the printing surface with paste and pressed it against the fabric, subsequently causing the fabric to advance to the next printing position (i.e., by one pattern-repeat unit). The printing blocks extended across the entire width of the fabric. The significant advantages of the perrotine over hand printing included great accuracy, production of a more uniform print thanks to the mechanical transport feature, multicolor printing during a single pass of the fabric, and higher production rates.

An ancient, artistically very ambitious, and still much used Japanese printing technique, the Yuzen process, is named for its inventor, the painter YUZENSAI MIYASAKI (1654–1736). It was this that later developed into the *screen printing* process. A Yuzen screen stencil consisted of several layers of paper, glued together, into which a pattern was cut. The protective paper regions were joined together with hairs (later: silk threads) to keep the edges of the stencil from bending up. Eventually the threads were replaced by silk gauze, then by woven screens made of chemical and metallic fibers. The function of the paper was ultimately assumed by a coating of paint. Starting in 1926, the screen-printing process also became established industrially in Europe. Further development led to the modern continuous rotary-screen printing process based on hollow, cylindrical patterns made of thin nickel foil, which has become a more economical way of producing prints than roller printing.

Since the mid-1960s the *transfer process* has become increasingly important. Here, the pattern is first printed onto an intermediate carrier (usually paper), after which it is transferred to the fabric.

Total world production of printed textiles is now estimated to be ca. $24 \times 10^9$ m$^2$/year.

## 2. Introduction

**Methods of Textile Printing.** Ornamental patterns consisting of one or several colors can be reproduced on fabrics by the local application of sharply defined areas of dye preparations (printing pastes) either to undyed or previously dyed cloth.

If in the fixation process that follows application of such a printing paste there is destruction of a dye applied previously, the procedure is known as *discharge printing*. Use of a discharge printing paste that also contains its own discharge-resistant and fixable dye is known as *color discharge printing*.

*Resist printing* is actually the oldest of the printing processes. In this case a special printing paste is applied to certain areas of the fabric to prevent dye fixation. In the case of a physical resist, penetration of the dye liquor is inhibited by a wax-induced hydrophobicity, or by application of a difficultly wettable thickener with a high solids content. With a chemical resist, dye fixation is prevented by a chemical reaction. Resists are most easily applied in a *preprinting* step; i.e., the resist paste is first printed and dried, after which the entire fabric is exposed to dye in a continuous dyeing device, for example by dipping the fabric in the dye liquor, squeezing it between rollers, and drying it in a continuous dryer (the Foulard or *padding* process). Overprinting resists can also be applied, but only if dye already present in the previously dyed and dried fabric is still in its unfixed state, as in the case of oxidation dyes (aniline black goods) and other developing dyes. A resist appropriately described as an "intermediate printing resist" can be prepared with naphthol dyes that are subject to two-stage application. Here, a fabric previously treated with a naphtholate is printed with a resist that prevents the normal coupling with a stabilized diazonium salt to form an azo dye.

*Color resists* contain a dye that is compatible with the resist agent. *Discharge resists* constitute a special case. The difference between this process and normal discharge printing is that here a dyed fabric is printed in a dry but still unfixed state. Synthetic fibers are especially well-suited to this process, since after padding and drying the deposited dye resides mainly on the surface of the fibers, leaving it much more readily accessible to the discharge chemicals than if fixation has occurred in the interior of the fibers. Preprinted resists are often based on the combination of a physical resist with a chemical resist.

A more recent approach is the "wet-in-wet" printing of a resist, leading to what are known as "African prints." In practice, the resist is overprinted with a surface-covering dye in a continuous "wet-in-wet" process. The contours produced by this type of resist are not as sharp as those obtained either with dried, preprinted resists or with discharge printing.

Direct printing, resist printing, and discharge printing differ with respect to the mechanism of action of the various chemicals, as well as in the dyes used, but all can be conducted by any of the standard printing methods (printing blocks, rollers, etc.), and either by hand or mechanically.

**Dye Application and Fixation.** There are significant differences between absorption of a printed dye and absorption of a dye applied from a dye bath. In the latter case, dye is often present in the dye bath at low concentration, and the process is allowed to proceed to the point of true equilibrium between dissolved and absorbed dye. The position of this equilibrium depends on the substrate (fiber), the dye, the temperature and pressure, and sometimes also the presence of additives (e.g., salts). However, the end of the process is associated with a well-defined final state.

Dye availability from a printing paste is characterized by high dye concentration in the medium and a printing thickener that inhibits dye diffusion. The extent of dye transfer is determined in part by the printing method used (e.g., the depth of engraving). The printing thickener is responsible for limiting capillary flow between the threads of the fabric, as a result of which dye penetration is much reduced into cellulosic fibers, which are subject to swelling. Except in the case of carpet printing, smearing of the dye as a result of transporting the print over guide rollers during drying is a potential problem. Evaporation of the water leads to an increase in the dye concentration and ultimate restoration of the dye to a solid state. As a result, soluble dyes diffuse to some extent into fibers that exhibit swelling properties under the influence of a concentration gradient.

The actual fixation step is responsible for transporting into the fibers as much as possible of the dye that has been retained in the dried thickener. This is especially important with dyes that do not dissolve in the printing paste but are converted into a soluble form only under the influence of chemicals introduced during the fixation process. In the great majority of cases, fixation is completed by treatment with water vapor (steaming), in the course of which steam condenses on the printed material, flows into the thickener, heats the fabric, and provides a dye-diffusion transport medium in the form of the water that swells the thickener.

The distribution of dye between fiber and thickener is determined in part by the "retaining power" of the thickener. The thickener is often composed of polysaccharides, and thus competes with cellulose in the fibers. For this reason the extent to which a thickener is subject to removal by washing is especially important. This depends not only on the swelling properties and cold solubilities of the various modified starches, celluloses, or mucilages present, but also on the nature of the drying and fixation processes employed. Direct fixation with hot air or high-temperature steam may make the removal of a thickener very difficult.

The color intensity of a print depends very much on the thickener used. The dye-distribution ratio between fiber and thickener is important, but so is the possibility of chemical reactions between the thickener and the dye (reactive dyes, metal-containing complex dyes, developing dyes of the Phtalogen type) and between the thickener and various auxiliaries.

The behavior of thickeners toward dyes and printing auxiliaries depends in turn on the textile substrate: synthetic fibers often act quite differently from cellulosic fibers in this respect.

**Printing Auxiliaries.** As a general rule, the auxiliaries used for printing are the same as those used in dyeing with a dye bath (→ Textile Auxiliaries). They can be grouped into various categories as follows:

*Humectants*; e.g., urea, glycerine, glycols
*Solvents*; alcohols, ethers, esters, polyglycols
*Fiber swelling agents*; cyanoethylformamide
*Wetting agents*; nonionic, cationic, anionic
*Reducing agents*; e.g., sodium oxymethanesulfinate (Rongalit C, BASF), sodium dithionite (Hydrosulfit, BASF), formamidinesulfinic acid, tin(II) chloride ($SnCl_2$), boranates, glucose
*Oxidizing agents*; e.g., hydrogen peroxide, sodium *m*-nitrobenzenesulfonate (Ludigol, BASF)
*Carriers*; e.g., cresotinic acid methyl ester (Levegal PN, Bayer), trichlorobenzene, *n*-butylphthalimide in combination with other phthalimides (Levegal PEW, Bayer), methylnaphthalene (Remol TRM, Hoechst)
*Retarders*; e.g., derivatives of quaternary amines (with basic dyes), but also levelling agents
*Resist agents*; e.g., zinc oxide, alkalis, amines, complexing agents (Trilon A and Trilon B, BASF)
*Metal complexes*; copper or nickel salts of sarcosine or hydroxyethylsarcosine
*Acids*; acetic acid, tartaric acid, citric acid, lactic acid, glycolic acid, formic acid

Thickening agents, discussed in detail in Chapter 4, constitute a special group of printing auxiliaries. There also exist various special auxiliaries, which frequently consist of mixtures; examples include emulsifiers, amides, glycol ethers, and aminoalcohols (Levasol, Bayer). The use of certain once-common auxiliaries (e.g., petroleum ether or naphtha for emulsion printing) has recently been severely restricted as a consequence of environmental legislation.

**Dyes.** Regulations related to workplace safety have led to the almost complete disappearance of certain dyes and processes, including aniline black and other black oxidation dyes based on aromatic amines. As a result, dyes applicable to cellulose fibers are now limited to a few main groups. The commercially most important categories are the *vat dyes* (for color printing and color discharge printing), *reactive dyes*, and *pigments* together with their associated binder systems.

More detailed information on dyes is provided in Chapters 5–7.

**Pretreatment.** The appropriate pretreatment of textiles prior to printing is crucial to the success of the process. This includes singeing, washing, bleaching, etc. (→ Textile Auxiliaries; see also Chapters 5–7 of the present article). Pretreatment is usually carried out today on the full-width fabric, usually with continuously operating equipment ("continuous lines"). Cellulosic fibers in woven form should be bleached and sometimes also mercerized before printing to increase the brilliance and intensity of the colors. Slivers, yarns, and knitted goods also require pretreatment prior to printing, typically in the form of cleaning and processing in such a way as to increase wettability. It is essential that spinning preparations and sizes be removed if one hopes to achieve an adequate degree of whiteness. Removal of fiber components is sometimes necessary as well, such as lignin in the case of bast, stalk, and leaf fibers, and "silk gum" (sericin) in the case of silk. The dye-absorption properties of wool are considerably improved by chlorination (see Sec. 7.1).

Particularly with synthetic fibers, but also cellulose fibers, it is necessary that a fabric's shape be stabilized, for example by preshrinking, removal of tension introduced during weaving, stabilization of the weave structure, and straightening of warps and wefts in the "thread direction." Tentering frames of various designs are used for this purpose.

Pretreatment also includes the application of certain preparations designed to improve dye absorption (e.g., urea and wetting agents), or weak oxidizing agents may be applied to a "ground color" (dyed fabric) to prevent damage to the dye by "tarnishing" (carryover of a discharge printing paste to the unprinted surface), especially with light-colored fabrics. Such preparations are applied by padding or "slop padding" (one-sided application with a roller that rotates half-immersed in the pretreatment solution while at the same time wiping against the fabric), after which the treated fabric is dried.

The only case in which an unprepared fabric is printed is when a nap resist is to be applied. This type of resist prevents the nap from being raised by a napping machine, producing a unique type of pattern.

## 3. Textile Printing Technology

### 3.1. Classification

The following list summarizes the various techniques available for printing on textiles:

1) *Manual Methods* (* indicates a historical method)
   Raised printing forms (relief printing): hand or block printing*, yarn printing*, warp printing*
   Recessed printing forms (gravure printing): copperplate printing (of calico)*
   Stencils: spray printing, screen printing
   Special processes like those often associated with tie dyeing, wax batik, or the Golgas process* (mechanical resists produced by squeezing between plates)
2) *Automatic and Semiautomatic Printing Processes*
   Raised printing forms: perrotines*, relief roller printing*, hank yarn printing, vigoureux (melange) printing (especially for carpets), Stalwart – Pickering, space-dyeing system (yarn and warp printing), Stalwart system (warp printing)
   Recessed printing forms: roller printing, space dyeing (knit – deknit process)
   Stencils: flat-screen printing, rotary-screen printing
   Carpet printing: BDA process (flat-screen printing with vacuum), Aljaba carpet printing*
   Special printing processes: carpet jet-spray printing, deep-dye process (special printing

equipment), transfer printing, STAR printing, Orbis printing

The emphasis in the discussion that follows will be on automatic and semiautomatic methods.

## 3.2. Preparation of a Master Pattern for the Printing Form

A master pattern is produced originally as a drawing, which should be as accurate as possible consistent with the capabilities of the proposed printing process. Single-color or multicolor patterns must be incorporated several times into the actual printing form (roller or screen) to permit single-pass printing onto a surface. The repeating portion of the pattern is known as the *pattern-repeat*. The length of the pattern-repeat in the warp direction defines the pattern-repeat height, that in the weft direction the pattern-repeat width. The pattern repeat may involve no offset (1/1) or it may be offset longitudinally to the extent of 1/4, 1/3, or 1/2. Each color incorporated into a pattern drawing requires a separate printing form. (Normally, the pattern designer is careful to ensure that the number of colors required for color separation—and therefore the number of printing forms—is kept to a minimum.) Proper placement of the colors as each is printed in sequence by the printing forms is referred to as the *registration* [3.9].

Several methods exist for transferring a pattern drawing to a printing form. The basic procedure is the same as that used to make printing forms for book production (→ Imaging Technology), where an intermediate step involves reproducing the color-separated design on a transparent film or photographic substrate (→ Photography) of a size appropriate to the printing form. Transfer to the printing form itself occurs by a copying process adapted to the nature of the form (roller or screen). This is described in conjunction with printing processes themselves in Section 3.3.

*Color separations* can be prepared manually, by a photographic method based on color filters, or with electronically controlled color-separation equipment (textile engraving machines such as those produced by Hell in Kiel, Germany, or by Scitex in Israel, which operate on a scanning principle). Scanners of this type, utilized in conjunction with a computer and a data base, open

the way to color signal processing, addition, pattern reflection and offset, color reversal, scale modification, contour tracing, electronic rastering, etc. This makes it possible to prepare a color separation with a repeating pattern of exactly the desired size. Transfer to a printing form is usually accomplished with the aid of photographic film, although a process involving direct control of an engraving machine (e.g., of the type manufactured by Hell) has also become well established. The resulting engraving is very shallow (0.03 – 0.05 mm), but it can be used for printing plastic films (e.g., wood-grain patterns for furniture) as well as transfer papers. In recent years digitalized patterns have also been transferred by means of a laser beam to undercoated templates. In the process, the laser beam burns its way through the undercoat to expose selected portions of the screen constituting the basis of the nickel template.

## 3.3. Relief Printing

**Block Printing.** The pattern form in this case consists of several layers of wood glued together to form blocks ca. 7–8 cm thick such that the grain in the various layers runs in alternating directions. Areas where printing should not occur are carved out on the basis of a traced drawing. Strips cut from a sheet of brass are used to reinforce the contours, and any required dots are fashioned from brass pegs. Large printing areas may be covered with a felt made from wool fleece to ensure greater uniformity of dye application.

Manual printing is carried out on long felt-covered tables over which rubber sheets have been stretched. For difficult (multicolored) patterns, the fabric is held in place by an adhesive. The printing block is wetted with dye from a trough (chassis). This trough is a box filled with thickener (false color) in which floats a wooden frame with a rubber sheet stretched over it. The frame in turn contains a felt cloth saturated with printing dye. This cloth constitutes the "stamping cushion" from which dye is transferred to the raised portions of a printing block placed upon it. Printing is accomplished by laying the prepared block on the fabric and striking it with a wooden mallet (at the corners as well as in the center). This manual process still finds use today, although mainly in the fine arts field.

Attempted mechanization of this particular printing process led to development of the *per-*

*rotine*, named after its inventor PERROT. A perrotine operates in a repetitive mode: a printing block that has already been applied over the entire width of the fabric is lifted mechanically, fresh dye is applied to the block from a roller covered with felt and impregnated with dye, the fabric is moved by the equivalent of one pattern repeat and then secured again to the padded area, and the printing block is lowered once more. Machines of this type have been designed for printing up to 6 colors, although they are rarely encountered today.

**Relief Printing.** The once widely distributed relief printing machines have suffered a similar fate, although they were still used in carpet printing until a few years ago, and modified versions play a role in this application even today (see Sec. 3.6.1).

## 3.4. Gravure Printing

A feature shared by all recessed-form (gravure) printing processes is application of dye to the entire surface of the printing roller. A doctor blade is used to remove excess dye from the surface, leaving it only in the depressions (engraved areas). The textile subject to printing is supported on a somewhat deformable surface and the roller is then applied to the fabric at relatively high pressure, causing the cloth surface to be pressed into the engraved areas where it picks up dye.

### 3.4.1. Roller Printing

Roller printing is a process (Fig. 1) that is declining steadily in importance.

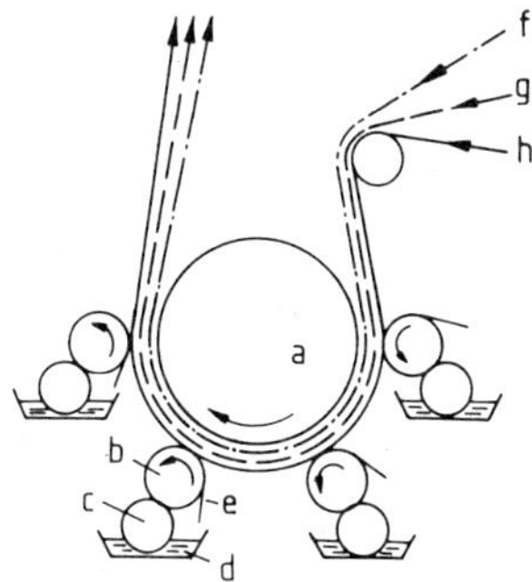


**Figure 1.** Four-color roller printing machine
a) Impression cylinder; b) Printing roller; c) Dye roller;
d) Dye box; e) Doctor blades; f) Undercloth; g) Back gray;
h) Textile

Whereas most manual printing techniques originated in Asia and the Far East, roller printing is a European invention (BELL, 1785, in Lancashire). With only minor improvements, such printing machines continued to operate until the 1950s with virtually no new design features. Subsequent changes have included automated positioning of the pattern repeat, separate control of the doctor blades independent of the rate of fabric transport, somewhat later the introduction of split bearings for the printing rollers, and, finally, the use of hydropneumatic pressure in place of weights or springs. Use of the undercloth (an endless blanket) for driving the drying system (*mansard*) was abandoned in favor of an independent drive mechanism for the dryers and folders. Furthermore, washing systems were introduced for the undercloth and the backing cloth ("back gray"), simplifying the separation especially of those textiles printed by the pigment process from the undercloth and the back gray.

The older type of machinery, used mainly for printing staple-cotton fabrics with a maximum width of 90 cm, has been almost completely displaced by mechanized screen-printing systems (see Sec. 3.4.2), although modern (Saueressig) roller-printing machines have been installed for printing knitted fabrics ca. 160 cm in width. The classical design with a central press has in this case been abandoned in favor of the so-called portal printing machine based on a frame construction in which each printing roller is directly opposed by its own impression cylinder. The frame (portal) consists of two segments joined along one side by hinges. The front part can be opened like a door, facilitating dye changes and minimizing expensive down time.

**Printing Rollers.** Hollow copper cylinders constitute the heart of the roller-printing process. Such a cylinder may have a weight of 40 – 70 kg depending on its size. A slightly conical spindle is pressed into the hollow center of the roller, which has a matching conical shape. Also mounted on the spindle are a pattern repeat wheel, a drive wheel for the dye-application roller, and the necessary bearings.

In newer machines, the continuous spindle has been replaced by a pair of cones pressed pneumatically into the hollow cylinder from opposite ends. These shortened half-axles, each with a cone that can be disengaged pneumatically, remain permanently set in the bearings. Continuous spindles had the disadvantage of being very heavy (80 kg or more), which made manipulation of the printing roller and installation and removal of the spindle both difficult and time-consuming.

The crude roller barrel, already corresponding approximately to the appropriate dimensions, is ground and polished prior to engraving to produce the smoothest possible surface. This is an essential prerequisite for gravure printing, because the doctor blade must be capable of wiping the roller completely free of excess dye. Small defects, such as scratches, inevitably lead to marks on the finished textile.

**Engraving.** The printing roller can be engraved with various types of cutting instruments (graving chisels, hooks), either by hand or mechanically. Embossing or etching processes are applicable as well.

The latest development is the optoelectronically or magnetoelectronically controlled "helioklischograph" engraving machine developed by Hell, (→ Imaging Technology, **A 13**, pp. 632–635), although the very shallow engraving that results with this technique is so far used mainly for printing "transfer papers," and only rarely for textile printing.

**Photogravure.** Color intensities captured as gray levels on a transparent film can be transferred to a copper roller coated with a photosensitive resist (formerly chrome gelatine) as one way of reproducing color shades, but the only technically feasible approach involves the use of a half-tone *screen* (→ Imaging Technology). To produce such an autotype, the color-separation film (negative) of interest is placed directly over an appropriate screen film, and both films are then wrapped together around the coated roller. Light scattered at the edges of the dots constituting the half-tone screen causes these dots to be reproduced on the roller, but with varying sizes. Relatively dark areas of the negative produce dots of a size equal to or even somewhat larger than those of the screen, as there is only a small amount of scattering, while a bright region leads to small dots on the photosensitive-resist coating. After removal by washing of all photosensitive resist that has not been fixed by exposure to light, an engraving is prepared by an etching process, which leads ultimately to the desired autotype.

When an engraved roller is to be prepared for gravure printing, a screen (e.g., a cross or hexagonal screen) is first placed over the coated roller and this is exposed, after which the color separation is set in place. Hardening of the photosensitive resist within the screen apertures takes place to an extent that depends on the local "color intensity" of the separation. A selective etchant is then used, one which penetrates only slowly or not at all into those parts of the coating that have been more strongly illuminated. "Development" with iron chloride leads to an engraving in which the various screen dots, all of which are the same size, have been etched to different depths.

This type of engraving produces what are known as "true halftones." Obviously such a photogravure process could also be used to create printing surfaces with a constant color depth over the full width of the fabric, but other methods are normally employed when the screen pattern itself is the only thing to be copied (e.g., the Szigeti process, or use of an Agfa contour film) [3.10], [3.11].

An engraving suitable for textile printing is always deeper and coarser than one intended for paper printing. When printing onto a textile, interlacing points formed by the threads themselves constitute a second raster pattern, so that higher pressures and larger dye volumes are required to achieve a printed result similar to that obtained with paper. In fact, the two printing techniques are complementary, which is also the case, for example, with the transfer-printing process.

**Aftertreatment and Regeneration of the Rollers.** For all practical applications today the engraved copper rollers are plated with chromium, because the copper of the rollers is much too susceptible to damage (e.g., by the doctor blades). A thin chromium layer (maximum depth 0.04 mm) is applied by electroplating after polishing and degreasing of the roller, and the cleaned rollers are then polished a second time. This hard chromium coating considerably prolongs the useful life of a roller.

It is self-evident that roller printing involves considerable cost, not only in the form of the printing machine, but also through preparation of the printing rollers. These costs can only be justified if the output will be large and color and design changes are limited. For this reason, mechanized screen printing and rotary-screen printing have become the established methods for short production runs. Indeed, rotary-screen printing has recently become popular for long runs as well, in part because of the width of the textiles currently being printed.

### 3.4.2. Screen Printing (Stencil Printing)

As noted previously (Chap. 1), the first stencils were made in Japan of reinforced paper. These devices permitted dye to be applied to the textile underneath through openings that corresponded to the proposed pattern. At a later date a woven screen was used instead, with those parts of the screen where no printing was to occur being simply covered over. Metallic and plastic stencils were also developed for dye application with a spray pistol, through which appropriate patterns were cut, machined, or, in some cases, etched. Thus, a sheet of zinc might be coated with paint wherever printing was to be prevented, permitting the remainder to be etched away. Shading effects were achieved manually in the course of spraying by the use of low-viscosity printing inks. Such techniques are seldom practiced today except occasionally for disguising the joins between large repeating patterns.

A number of different screen-printing methods of ever-increasing importance have evolved from these relatively primitive approaches. The most widely practiced today include:

1) **Flat-screen printing on tables**, where the screens are laid along rails to facilitate alignment of the repeat patterns. The tables are sometimes subject to heating, although in other respects the process is similar to that described in Section 3.3. The dye is applied by hand.

2) **Screen printing with a wheeled stencil carriage** that can be moved on rails from one repeat position to the next, where it stops and is lowered so dye can be applied by hand with the aid of a weighted double squeegee.

3) **Screen printing with a fully automatic screen carriage** that moves the screen and stops it at the pattern-repeat point, where the squeegee is activated by an electric motor. Rails that support the equipment on both sides also carry hand-operated or electrically driven carriages for applying adhesive to the textile and anchoring it in place, as well as a washing carriage to wash off and wipe away excess glue.

Some hand-printing establishments still dry the fabric passively by hanging it over the tables. Mechanical drying devices are described in Section 3.8.

4) **Mechanized screen printing with stationary screens mounted in a frame.** A screen frame performs the function of lifting the screens while the textile, which is glued to a moving endless rubber belt, is advanced to the pattern-repeat point, after which the screens can again be lowered. The screens are wiped with a squeegee while stationary. Squeegees of many different types are used, including wipers, rollers, and magnetic rollers. With a magnetic roller system a magnetic bar located beneath the undercloth is used to move the roller over the screen. The roller pressure and the amount of dye applied are functions of the thickness of the "magnetic squeegee" (a cylindrical iron bar) and the magnitude of the magnetic force, which can be adjusted stepwise. Double rollers and box or slit squeegees are used for printing carpets. These are especially well-suited to printing many colors and pressing the dye deep into the pile.

Printing machines for this purpose consist basically of a facility for moving the textile, a gluing device that causes liquid glue to be applied by a squeegee to the undercloth, and the printing equipment itself. Undercloths precoated with thermoplastic glues are sometimes used instead, in which case the heated textile is squeezed by a roller when it first meets the rubber-coated transport belt, causing glue in the immediate vicinity of the textile to soften and instantly adhere. With permanent adhesives the textile can be fixed simply by placing it in position and pressing it down. Printing occurs with all the stencils on the conveyor belt at once, although the printed design is complete only after it has passed the last screen. The rubber-coated undercloth is then directed downward over a guide roller, washed with water and rotating brushes, wiped to remove excess rinse water, and sent back to the gluing device.

The textile, which has been pulled away from the undercloth, is fed directly to a drying chamber. One type of printing machine manufactured by the Buser company incorporates an ingenious design that permits transport through the dryer to be continuous even though the printing process itself remains discontinuous. This prevents potential damage in the drying chambers due to heat that might accumulate in stationary carrier rollers. Alternatively, drying in a chamber must be preceded by hot-air jet drying.

Automatic screen printing involves the least set-up time, the lowest screen costs, and the smallest loss of printing dye, which in turn minimizes contamination of the waste water that results from washing of the screens.

**Rotary-Screen Printing.** This is currently the most efficient textile-printing process available. Its immediate precursor was a Portuguese printing machine (Aljaba) that for the first time incorporated cylindrical screens. Dye was applied from the interior of the screens by a dye tube and a squeegee that pressed the dye through the stencil openings onto the textile.

The true breakthrough was achieved with development of a rotary-screen printing machine in which the textile was fixed by adhesive to an endless rubber transport belt, permitting cloth to pass continuously under a set of rotary screens arranged in relatively close proximity to one another. The devices used for feeding the fabric, applying adhesive, and removing and drying the printed goods correspond essentially to those used with automatic flat-screen printing, but less space is required for a rotary-screen printing machine (Fig. 2).

The production rate achieved in rotary-screen printing equals or even exceeds that in roller printing. The short distance between the screens and the very brief dwell time experienced by the textile under the screens eliminates adhesion problems even with fabric made of synthetic fibers, such as polyamide, whereas flat-screen printing can be complicated by partial detachment of the fabric from the undercloth, leading to bubble formation.

The availability of automatic equipment for both flat-screen and rotary-screen printing makes it possible to use one or the other technique selectively depending on the circumstances. Thus, rotary-screen devices permit trouble-free printing of longitudinal stripes in the warp direction, for example, whereas with flat-screen printing all the places where it was necessary to lift the frames tend to remain visible. On the other hand, flat-screen printing machines are in many respects more versatile (e.g., squeegee operation can be repeated as often as necessary for application of more dye and increased dye penetration).

**Stencil Preparation.** Flat screens consist of frames that have been covered with a gauze. The frame material may be wood, but today it is more likely to be welded metal tube. Adhesive is used to attach to the frame a stretched prestressed gauze made of monofilament or multifilament artificial fiber.

While the desired pattern might be created by manual application of paint to those areas of the gauze where printing is to be prevented, the process is today accomplished almost exclusively by first coating the entire screen with a light-hardening photosensitive resist. Color separations (film diapositives) are then placed over the coated screens and illuminated; in the case of small repeating patterns a "copy-adding machine" may be used to expose the same screen several times as an alternative way of reproducing a complete design. Because of the high degree of mechanical stress to which it will be subjected, a hardened photo-resist is strengthened by relacquering and fixation after washing out of any unhardened screen coating and drying. Rotary screens for "printing cylinders" can be prepared by any of the following methods:

1) Fabrication of a thin-walled nickel cylinder and subsequent etching of the pattern.
2) Preparation of a cylindrical screen, perforated over its entire surface, which can be processed by the photo-resist method, just like a flat screen; alternatively, a screen might be coated over its entire surface and a pattern then produced by partially burning out the paint with a laser (under computer control based on digitalized design data).

**Figure 2.** Rotary screen-printing machine (Johannes Zimmer Co., Kufstein, Austria)

3) The direct electrolytic method. Here the color separation, together with a half-tone screen, is illuminated on a cylinder that has been coated with photoresist. After dissolution of the unfixed photoresist and drying, the cylinder is connected up as the cathode in an electrolytic bath. A nickel coating is then applied, varying in thickness from a few hundredths to a few tenths of a millimeter depending on the size of the screen. This results in a thin-walled cylinder bearing the pattern. Unlike coated screens, rotary screens of this type cannot later be given a new pattern. On the other hand, there is also a limit to the number of times a fairly resistant paint coating can be removed from a conventional screen without damaging it.

It should be noted that the electrolytic method is capable of producing smaller apertures separated by less metal (e.g., 78–100 apertures per cm), opening the way to subtle color gradations and reduced dye consumption (for transfer printing onto paper, for example).

### 3.4.3. Economic Developments in Gravure Printing

Penetration figures for the textile-printing machinery market in Germany and their evolution with time are presented in Table 1 based on data developed by the Research Center for the General and Textile Market Economy at the University of Münster. Current estimated mean capacities (1000 $m^2$/a) for these facilities excluding carpet printing are as follows:

Roller printing                1233
Rotary-screen printing         3031
Flat-screen printing            620

**Table 1.** Number of textile printing machines in the Federal Republic of Germany

|      | Roller printing | Rotary-screen printing | Flat-screen printing (mechanized) | Flat-screen printing (manual)* |
|------|------|------|------|------|
| 1960 | 320  |      |      |        |
| 1968 | 202  | 22   | 112  | 13 000 |
| 1972 | 135  | 73   | 114  | 7 500  |
| 1980 | 81   | 117  | 94   | 3 265  |
| 1988 | 29   | 124  | 59   |        |

* Data in meters of overall table length

## 3.5. Transfer Printing [3.12], [3.13]

In transfer printing, the pattern is first created on an intermediate carrier and then transferred from there to the textile. The dye may be fixed either in the course of the transfer step or in a subsequent fixation and finishing process.

Transfer processes were of relatively little importance until the mid-1960s, but this situation changed rapidly with development of dry-transfer printing processes, especially the Sublistatic process (heat transfer on polyester; see Sec. 3.5.2), and transfer printing is today involved in the production of ca. 700–800 × $10^6$ $m^2$ of textile per year.

### 3.5.1. Wet-Transfer Printing

The oldest of the transfer-printing processes is the *Sark process* [3.14], also known as the "Orbis" process. A pattern is first generated on a metal roller using a printing-dye mix in which the dyes and auxiliaries are present as a soapy, plastic mass. The pattern is then transferred by rolling it onto a textile previously prepared with a solvent/water mixture. The transfer step is followed by steaming and washing. Only a limited length of fabric can be printed with a single roller.

This process is capable of reproducing patterns with a unique range of colors and elegance, but interest in it has been limited to high-fashion applications. Although the process can in principle be used on all types of fibers, it has so far seen service only for silk and artificial silk.

In the *Fastran* and *Dew-Print* processes [3.15], and most recently in the Danish "*Cotton-Art*" process [3.16], paper is used as an intermediate carrier onto which textile dyes have been printed by paper-printing methods. Moist fabric that has been impregnated with auxiliaries is then pressed onto the paper and heated to 100–105 °C. The resulting steam atmosphere effects transfer of the dye. This is followed by a washing process that may be less intensive than would be required in conventional textile printing because a large part of the unfixed dye and thickener remains on the paper. This technique is again applicable to all types of fibers, including wool and cotton. Since the patterns are printed by paper-printing machines, they are usually limited to four colors.

Neither of the wet-transfer processes described has yet become significant from the standpoint of production volume.

### 3.5.2. Heat-Transfer Printing

**STAR Printing** [3.17]. The *STAR* process was developed in 1950 in Italy. Paper is again printed with a solvent and a paste containing the dyes, but in contrast to the Fastran and Dew-print processes, special thermoplastic binders on specially prepared papers are used to create what amounts to a type of transfer picture.

Dye is transferred to the textile by squeezing the fabric and the paper together between rollers heated to 120–150 °C, causing the film to soften and the entire pattern to be transferred to the fabric. The dyes are then fixed by steaming. The thermoplastic film has of course been transferred as well, so it must be removed by chemical cleaning as part of the finishing process. The STAR method is applicable to all classes of dyes with the exception of developing, oxidation, and pigment dyes. It is so far practiced only by a single company.

**Thermachrome Process.** The *Thermachrome* process is a variation of the STAR process, but one based on pigment printing. The thermoplastic binder in this case is so formulated that it acts as a satisfactory binder on the textile as well. No cleaning step is required subsequent to transfer. The Thermachrome process is of some importance in label printing, and for printing emblems containing photographically precise patterns on finished items of clothing (e.g., T-shirts).

**Sublistatic Process** [3.18], [3.19]. Of all the transfer-printing processes for textiles, the *Sublistatic* process is the only one that is important from a quantitative standpoint, accounting for well over 90 % of the $700–800 \times 10^6$ m²/year of transfer-printed textiles. As with the STAR process, the requisite papers are prepared mainly by four-color printing with paper-printing machines, but thermoplastic binders are avoided. The only coloring materials used are disperse dyes that sublime unchanged in the range 180–230 °C.

The paper and fabric are brought into contact for 60–20 s at 200–220 °C with the aid of presses or rollers, causing only the dye to be transferred from paper to textile. No aftertreatment is required.

The process is restricted almost exclusively to polyesters and polyester blends, because the fibers must be capable of effectively absorbing a disperse dye, and they must not be overly thermoplastic in the range 200–220 °C.

*Transfer Papers.* The stock used in preparing the transfer papers should have a weight of 60–80 g/m², and it must be coated on the printing side with a thin layer of starch (ca. 3 g/m²). Paper that is to be printed on paper-printing presses that accept pastes based on alcohol or solvents need not be as absorbent as the paper used with aqueous pastes on roller-printing or rotary screen-printing equipment. In the case of aqueous pastes, the paper should accept 60–80 g of water per square meter in the course of 60 s, and it must display "good transparency" when held up to the light. Cloudy papers cause uneven printing.

The papers are subject to processing by all the various paper-printing techniques (→ Imaging Technology).

Gravure printing is well-suited to the production of very fine half-tones, shadings, and color sequences. Printing is from roll to roll.

Flexographic printing, another gravure process, is appropriate for less demanding designs characterized by large, uniform surfaces. Here again, printing is from roll to roll.

Offset printing is capable of very fine half-tone reproduction, but only from one sheet to another.

Screen printing is of particular interest because of the low cost of preparing the corresponding stencils. Printing can be either from roll to roll or from sheet to sheet.

Transfer papers can also be printed with textile-printing equipment. Thus roller printing (Sec. 3.4.1) opens the way to very fine half-tone images and shadings, just like gravure printing. Rotary-screen printing (Sec. 3.4.2) is a less expensive approach to reproducing simpler designs. Printing in both cases is from roll to roll with aqueous pastes.

*Dyes.* The dyes employed are selected disperse dyes with fairly low relative molecular masses. These are supplied by the dye manufacturers in various forms suitable for the preparation either of paper-printing inks based on solvents (alcohol or toluene) or water-based printing pastes.

Sublimation and solubility characteristics of the various dyes must be well-adapted to the substrate in question in order to ensure that color shades will remain constant despite variations in the transfer temperature. The depth of penetration into the substrate can also be controlled by suitable choice of the dye.

*Substrates.* Polyesters are the most important substrates because of their good colorfastness characteristics and low thermoplasticity. Over 90 % of thermal printing occurs on woven and knitted fabrics made of this fiber. The remaining 10 % represents mainly polyester – wool mixtures or mixtures of polyester with other synthetics. Other types of synthetic fiber are of minor importance.

Triacetate and acetate fibers tend to be deformed rather severely during the transfer process. They also become somewhat harsh to the touch, and their fastness properties are only moderate. Polyamide fibers can be imprinted to give intense and brilliant colors, but the resulting wetfastness is so poor as to be adequate only in exceptional cases.

There has been no lack of attempts to extend dry-transfer processes based on sublimable disperse dyes to natural fibers, especially cotton. One suggestion involved partial esterification or etherification of the cotton, and many preparations of this type have been described. However, the fundamental problem has still not been solved, because optimization of the dye yield is not accompanied by an improvement in fastness toward moisture and light. It has therefore not proven possible to establish the Sublistatic process in the broad area of polyester – cotton blends, though some of the resulting products might be acceptable for reasons of fashion.

*Nontextile Substrates.* Apart from textiles, plastic films and other flat materials are also subject to printing by the dry-transfer process. Prolonged storage of such products may be accompanied by migration phenomena, however, since disperse dyes are subject to progressive diffusion. With fibers, on the other hand, such diffusion is practically invisible due to the limits imposed by small fiber cross-sections.

*Transfer Machines.* Presses are used for printing single items by the transfer process, whereas rollers are more suitable for long runs of fabric.

Apart from the usual equipment, which results in contact times of 30 s to 4 min at temperatures of 180 – 230 °C, special machines have been designed that take advantage of a vacuum [3.20] to reduce the transfer temperature and time, which also helps to preserve the structure of the textile.

## 3.6. Special Printing Methods

### 3.6.1. Carpet Printing
(→ Floor Coverings and [3.12])

**Jet Printing of Carpets.** Many equipment manufacturers have attempted to perfect processes for the spraying of printing dyes through independently controllable jets [3.21]–[3.25]. The potential advantages are:

1) A printing technique that avoids direct contact with the product
2) Good pile penetration due to the high kinetic energy of the air-borne dye, an effect that cannot be achieved with other printing processes
3) Use of low viscosity printing pastes that are easy to fix and wash out
4) The cost of producing printing forms is eliminated, costs that are not inconsiderable in the case of carpet printing due to the size of the stencils

The technical and electronic challenge is formidable, however. There are at present two ways of carrying out a jet-printing process.

In the first approach, a separate fixed jet holder is assigned to each color. A "spray line" is thus created on the moving textile that advances in the warp direction. The jets must be mounted so close to each other that opening all the jets results in complete coverage. By controlling each individual jet, lines of color of various lengths are produced, permitting the reproduction of any desired pattern (Millitron, Deering Milliken, USA).

In the second method, the jets are mounted on a sliding frame that can itself be moved in the direction of the warp while the fabric remains stationary during the spraying process (discontinuous process). The number of jets employed can thereby be considerably reduced, but with a correspondingly lower rate of production. It would be possible in principle to operate with only one jet assigned to each color, and to mount all the jets on a single sliding frame. In practice, jet-printing machines have been constructed in which each color has its own sliding frame, whereas with others all the colors (eight, for example) are sprayed from a single frame. The number of available fabric colors is of necessity limited, as with any printing processes not based on the trichromic system (Chromotronic or Chromojet, J. Zimmer, Klagenfurt).

Jet-printing machines differ from one another in another respect as well: the method used for controlling the dye jets. The first approach entails jets producing a continuous stream of dye that can either be directed onto the textile or diverted into a channel for return to the dye-storage tank. In the second (Chromotronic) method the dye stream is switched on and off by a magnetically operated needle valve.

The application of jet printing is limited by distortion that occurs as a result of contact between the moving textile and the jet, or, in the Chromotronic process, the moving jets and the stationary textile. A further limitation (though more in the sense of cost) arises from the number of jets that must be installed close to each other. For textile-printing purposes it is currently easy to incorporate $5-7$ jets per centimeter. This arrangement in turn limits the minimum size of the individual points of color to ca. $2 \times 2 = 4$ mm$^2$ based on the minimum width of the corresponding lines and their minimum length as established by the speed of the on/off mechanism. This is roughly equivalent to the pattern potential provided by the tufts of a woven carpet. Nevertheless, it is to be anticipated that there will be further advances in the spray technique, which is already highly developed for such purposes as printing addresses for the delivery of newspapers. Further progress is especially welcome since, except for flat screens, the production of stencils—especially electrolytically produced metal screens—is very costly and consumes a great deal of energy. "Ink-jet devices" that avoid the use of stencils were demonstrated successfully at the 1991 Hannover International Exhibition of Textile Machinery for producing patterns on both paper and textiles (Stork, Brabant), which is to say that the process is by no means restricted to carpets.

**Deep-Dye Process** (Bigelow, USA). The printing element in this case is a shallow plastic tank that permits all the colors required for a design to be applied simultaneously. Individual dyes within the tank are separated by divisions, and each is fed by a separate injector. The carpet is led through the system with the pile side down. Printing requires that the dye container be raised and pressed against the pile (discontinuous operation).

**BDA System** (Bradford Dyers Association, UK). In this screen-printing process based on a double squeegee system, dye penetration is enhanced by a vacuum applied to a perforated plate that acts as an undercloth beneath the carpet (discontinuous process).

**Stalwart–Pickering System** (UK). This is a relief printing process involving rollers to which pieces of foam corresponding to the pattern shape are attached by means of an adhesive. A low-viscosity dye solution is picked up from a tank (continuous process).

All three of the above printing processes are normally carried out before the carpet backing is applied.

**Yarn-Printing Machines.** Special yarn-printing machines have been developed for yarns intended for tufted carpets. An attempt is made whenever possible to apply the dye in an irregular way to prevent the formation of "rivers" (undesirable repeats) on the resulting tufted carpet. With the Laing Space-Dyeing machine (Pickering, UK) the yarn is passed between a pair of rollers, each with a relief printing profile (usually a bar pattern advancing in the longitudinal direction with respect to the rollers). The lower roller runs in a trough of dye, and the upper roller has the larger diameter of the two. Yarn is imprinted only when bars on the two rolls meet or partially overlap. This produces a printed stripe of irregular width on the yarn bundle.

**Stalwart System.** This system operates with two smooth rollers, with an upper roller that can be moved vertically. Dyed bands of varying width are produced by electrically raising and lowering the upper cylinder at irregular intervals.

**Knit–Deknit Process.** The knit–deknit process begins with previously knitted tubes made of a texturable material. These are then imprinted on both sides with deeply engraved rollers and subsequently fixed and washed. Once the yarn has been again unravelled it is ready to be fed into a tufting machine.

Other techniques exist as well for dyeing yarns in a partial and irregular way, but these do not involve printing.

**Warp Printing.** Warp threads arranged parallel to each other and containing "auxiliary wefts" introduced at various intervals are subject to printing with rollers or rotary-screen printing machines. Here, too, a problem is the width of the textile: warp threads cannot be placed parallel to each other in sufficiently close contact to prevent excessive spread in the case of a carpet pile printed with standard equipment. Such carpet can of course also be printed intact by the flat- or rotary-screen process.

### 3.6.2. Other Processes [3.26]

**Brush Printing.** Dye is applied in this case with brushes rather than squeegees (e.g., in the printing of flags). This results in extremely good penetration, corresponding essentially to a completely dyed product. Brush printing can also be used in making imitation fur and other "long-pile materials." However, most challenges of this type are today met by mechanical means (e.g., spray printing) or through use of special printing dyes.

**Melange Printing of Slivers.** Slivers are printed with relief-type printing rollers equipped with diagonally arranged channels (S- or Z-shaped). The width of the printed stripes determines the color shade of the blend (melange) ultimately produced when the fibers are later combed. Printed slivers are not dried, but are instead placed in wire baskets and wet steamed in a round steamer for ca. $1-2$ h in the case of wool. Polyester slivers are steamed under pressure.

**Yarn Printing.** This technique is known as either *pearl* or *flame* printing depending on the size of the printed pattern. The yarns are imprinted by two relief rollers bearing axial stripes or "discs" running in precise opposition to each other, providing both pressure and back-pressure. Each roller has its own device for dye application. The hank, which is stretched between two beams, is passed through the pair of rollers, causing perhaps one-third of the total length to be printed. An equivalent amount of hank is then transferred from one beam to the other and printed once again, the process being repeated until the entire hank of yarn has been printed. The printed hanks are hung up, dried, and fixed—usually by steaming, but sometimes by hot air, as in the case of pigment printing.

**Flock Printing** [3.27]–[3.29]. Here the textile is first coated with an adhesive and immediately "sprinkled" with "flocks" $0.3-3$ mm in length in the presence of an electrostatic field (20 000–60 000 V). These flocks consist of fibers cut to a constant length and usually made of pretreated viscose. In such an electrostatic field the flocks tend to orient themselves in a single direction, and they are impelled like small spears into the adhesive. The potential acceleration of the flocks is limited by the maximum achievable voltage. The distance between the poles must also not be too great, since air exerts a retarding effect such that further increases in the voltage eventually result in no further acceleration, and may in fact cause spark discharge.

Loose flock dust on parts not subject to printing can be removed by reversing the electrostatic field, or by heating the textile after fixation with concurrent suction.

A flocking apparatus of this type can be constructed as a flock-printing carriage (Wirth), which permits its use in conjunction with screen-printing tables. Special devices also exist for combined printing and flocking (Menzel). "Multicolor" flocking (Hofer, Octen/Switzerland) takes advantage of several discrete printing and flocking units. Intermediate drying of the adhesive and vacuum extraction of excess dyed flocks occur between the individual units.

Flock printing is highly dependent on current fashion. Whereas flock covering of complete surfaces has become firmly established, flock patterns are less popular today than formerly.

**The Printing of T-Shirts.** In recent years there has been a great demand for printed T-shirts, prepared, for example, on a modern version of a carousel printing machine, one in which both the carrier and the garments rotate in a circular fashion underneath the stencils. The process is discontinuous, and individual T-shirts are positioned and removed by hand (Variprint KSD II system from Maag and Schenk).

**Batik Printing.** Batik in its original form was not in fact a printing process, but today there are several ways of achieving this resist effect. For example, the penetration of dye liquor can be prevented by tying (*tie dyeing*) or by sewing such that a resist effect is achieved. A related technique is the *bandanna* process, which emphasizes dyeing that occurs under cold or only slightly warm conditions over a short period of time. A similar mechanical resist system, known as the *Golgas* process, involves several layers of textile pressed together between metal plates bearing a relief pattern that are immersed in the dye liquor.

In the classical batik process, wax resists are poured onto a fabric according to a predetermined pattern. It is also common to imprint the molten wax with the aid of stamps made of brass or wood. Unprinted areas are then dyed under conditions that will not cause the wax to melt. The principal dye is often indigo. This process of resist application and dyeing can be repeated many times, whereby the wax resist is "boiled off" after each step.

Batik imitations are produced with the aid of various resist-printing pastes, and on the basis of several printing processes. The one most closely resembling the original batik process entails application of a liquid wax, for example by heated printing rollers. Veined designs result from breaks in the wax layer. Resists are also often printed wet-in-wet, and batik imitations can be produced by printing appropriate resists and bases onto naphtholized fabric (see also Chap. 2 and Sections 5.3 and 5.5.3).

## 3.7. Two-Phase Printing

Two-phase printing refers not to a specific printing technique, but to the separate application of dye and fixing agent. The dye is printed and dried, and the fixing agent is applied later, after which the textile is then steamed. This separation was an essential one with *vat dyes* (see Sec. 5.5.2) due to instability of the associated reducing agents during storage prior to steaming. Production rate also plays a role. Printing machines are capable of operating much more rapidly than the steamers used in the Rongalit–potash process, which entails 8–10 min of steaming time. Separation of the process into two phases makes it possible to use rapid fixing auxiliaries such as sodium dithionite ("Hydro-sulfit," BASF), thereby reducing the steaming time to 30 s–1 min at higher steaming temperatures and at the same time opening the way to greater consistency and efficiency. Two-phase printing also provides the solution to a special problem: the reuse of residual dye. In the Rongalit–potash process, printing-dye residues from the body of the roller machines usually contain excess dye from the doctor blade together with metal ions resulting from chemical or electrochemical reaction. These ions act as oxidation catalysts, and can accelerate decomposition of the Rongalit C to such an extent that no fixing occurs on steaming. It can therefore be dangerous to mix such dye residues into fresh dye.

Analogous two-phase fixation with reduced steaming time is employed with *reactive dyes* (see Sec. 5.10.7). An energy-saving variant of two-phase dyeing is the slop-padding process, also used in conjunction with reactive dyes, which helps avoid the risk of staining by highly concentrated alkali or sodium silicate solutions (osmotic effect; see Sec. 5.10).

Other related processes include the application of dye in two stages (e.g., base application and naphtholate printing; see Sec. 5.8), the resist process with and without intermediate drying, and conversion printing.

**Conversion Printing.** This process, which is used only with cellulose fibers, is a combination process in which overprinted patterns are created by the selective application of a fixation auxiliary (e.g., Rongalit C) that fixes one dye and decomposes another, while in those places where no fixing agent is applied all that occurs is fixation of the second dye.

The most prevalent combination is that of a vat, leucoester, or coupling dye with an oxidation, basic, or reactive dye. The prerequisite for the success of this dyeing process is a difference in behavior on the part of the two dyes in question with respect to the chemicals applied (fixing agents, discharge agents). For environmental reasons, such processes are now mainly of historical interest.

## 3.8. Drying

**Drying Prior to Fixation.** To prevent smearing of the print as it passes over the guide rollers, drying is commenced in such a way that the rollers come into contact only with the back of the textile, or in some cases the material is kept in suspension by heated air emerging from jets on both sides. Drying equipment is usually coupled with the associated printing machine. Various drying methods are practiced, including ones based on hot air or the combustion gases from burners, direct contact with the fabric (cylinder drying), and radiative heating, but the best method found so far utilizes hot air drying in a chamber ("mansard"), preferably one fitted with air jets (Fig. 3). Care must be taken with the use of exhaust gases, because many dyes are sensitive to sulfur dioxide and nitrous gases.

The recent introduction of dye categories and substrates whose color tones are easily altered

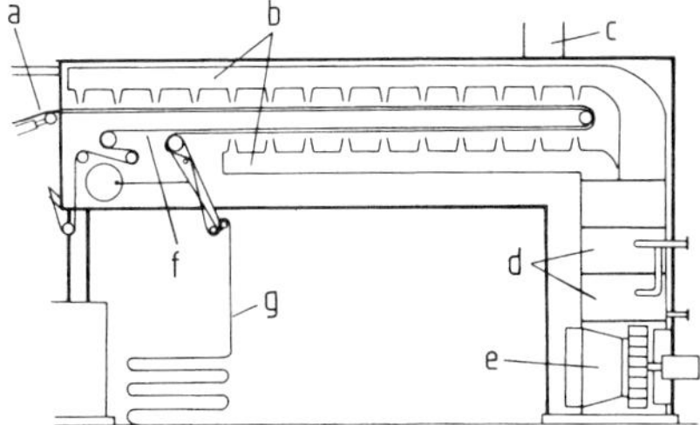

**Figure 3.** Jet-drying chamber
a) Textile inlet; b) Hot-air pipes with jets; c) Exhaust-air extraction; d) Heat exchangers; e) Air recirculation fan; f) Transport belt; g) Textile exit

has required the development of special drying procedures. For example, residual moisture has become an important control characteristic. Also, there must be provisions for uniform air velocity, and the ratio of inlet air to exit air must be precisely controlled.

The drying of textiles made of synthetic fibers brings with it additional problems, since the corresponding fibers are almost incapable of absorbing water except in capillary spaces that the fibers create. For these and other reasons, cylinder drying machines are rarely used in conjunction with modern printing machines.

**Fixation by Dry Heating.** This technique was introduced for use with certain types of dyes and for pigment printing on cellulosic textiles. For most sublimable disperse dyes the high temperatures of the thermosol process (Sec. 6.5.4) lead to better fixation yields with polyester fabrics than can be achieved through the steaming method. Sublimation fixing also works well with the transfer-printing process in the temperature range 190 – 205 °C.

A suitable facility for dry fixation consists of tentering frames, jet dryers, various condensing devices, and high-temperature cylinder arrays heated by gas, oil, or electricity rather than steam. Economical operation is possible because of the much shorter fixation time.

## 3.9. Steaming and Other Fixation Processes

The standard fixation method for dyes on cellulosic fibers is steaming, and in some cases (e.g., with vat dyes) this is the only feasible approach (see Sec. 5.5). The required steamer ca-

pacity depends on the rate at which prints are produced, the type of dye, the printing method (flat-screen printing, rotary-screen printing, or roller printing), and the type of fabric.

Steaming is carried out with saturated or even "wet" steam (i.e., steam containing condensate). With certain dyes (vat dyes) and certain substrates (e.g., wool), the use of saturated steam is preferred. For this purpose, the steam is usually "moistened" by passing it into a tank of water (indirect steam). After heating the water in the tank, a stream of saturated steam passes directly through the vessel, picking up water droplets as it goes, and then enters the steamer from below via a sieve tray. Additional "direct" steam may be introduced through jets. Once all the air has been displaced, the steamer is ready for operation. In general, air inside a continuously operated steamer is vented by way of the same slots through which fabric enters and leaves. The absence of air can only be ensured if excess steam is always present. Steam is lighter than air, so the air should be displaced from above in a downward direction. This is the principle underlying the Arioli-type steamer, which is completely open on the underside and can be operated with either saturated or superheated steam. Fabric is transported through the steamer by motor-driven rollers.

**Continuous Steamers.** The *rapid steamer* system of Mather and Platt includes two rows of rollers between which the textile is moved backward and forward, with only the upper rollers normally being driven. The lid and the inlet box may be heated to prevent drops of water from forming, since these could produce "water stains" on the fabric.

In the *spiral steamer* (Krostewitz) system the cloth is transported by guide rollers that come into contact only with the back of the fabric, and which have a "broken carrying surface" to facilitate keeping the cloth at full width (Fig. 4). The rollers are so arranged that the cloth traverses a

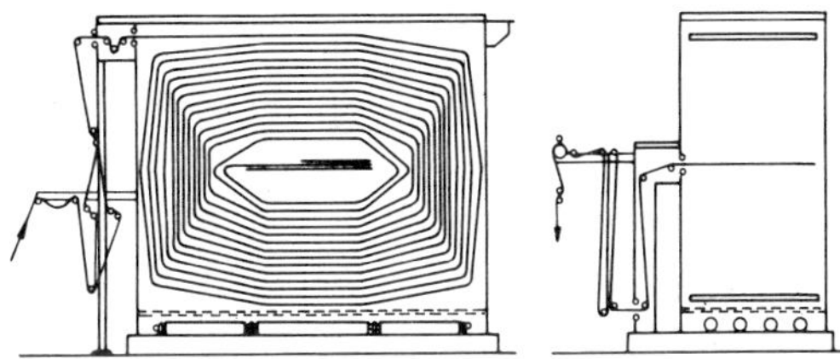

**Figure 4.** Krostewitz steamer

spiral path inward toward the center of the chamber where it is deflected by a guide roller, the so-called "sword," set at a 45° angle to the direction of motion. This directs the cloth out through the middle of the side wall.

*Hanging* or *loop* dryers have only a single upper row of driven rollers. The cloth is first formed into loops by a mechanical looper, and the resulting loops are advanced by a chain of rotating bars. The rollers turn in a synchronized way, but the rate of transport is kept low to prevent the formation of marks at the support surfaces. This type of steamer is exceptionally versatile compared with those described earlier, and is becoming more and more firmly established. Increasing demands with respect to reproducibility of color shades have made it necessary also to incorporate measurement and control devices (from various sources) to ensure constant moisture contents for both cloth and steam.

*Two-phase steamers* (for use in the two-phase printing process, Sec. 3.7) are designed for rapid textile throughput ( $\leq 120$ m/min) and short fixation times (arched steamers, portal steamers, lightning steamers).

**Batch Steamers.** With a *star steamer*, the cloth, sometimes together with a back gray, is hung in spiral fashion on a star-shaped stand, the cloth being "pinned" in place by outward-facing needle bars. This star, complete with the suspended cloth, is then introduced into the steamer. Loading may be through an open lid, through a "door" (cabinet steamers), or from below into a raised "bell" (bell steamers). The steamer itself is a cylinder with a double casing. The bell steamer is a cylinder closed only at the top which is latched to a floor plate after the bell has been lowered.

Steam enters the steamer in most cases via the double casing through openings near the top, passing out through flaps in the floor. However, with a bell steamer (Dedeko, Dupuis), the steam flows through the support columns of the star to the underside of the heated lid, with a steam outlet in the floor.

Most steamers are capable of operating either with or without a sump. Bell steamers, which operate above atmospheric pressure (1 – 2.5 bar), are exceptionally well-suited to fixing dyes on synthetic fibers, especially polyester. A further advantage is that the steaming time can be varied between wide limits. Nevertheless, the number of such steamers in use is decreasing steadily due to their low productivity.

**Other Fixation Processes.** *Wet developing* is possible only with a small number of dyes. The process depends on the action of chemicals in an aqueous bath, usually heated, and it is conducted in the first compartment of a full-width washing machine. Dyes that can be fixed in this way include the leuco vat dye esters (Sec. 5.7), the coupling dyes (Sec. 5.8), and the Phtalogen complex dyes (Sec. 5.9).

Aftertreatments for dyed textiles depend on the dyes in question, and may consist of oxidative or sometimes reductive steps as well as passage through acids or alkalis.

## 3.10. Washing Operations

The washing process is even more important with respect to a print than it is for a conventionally dyed textile, because dyes in different concentrations—sometimes also of different dye types with differing fastness properties—may be present together on a white background. Often, the final colors and fastness properties are established only after the washing process.

Washing devices and washing machines are designed to suit the character of particular fabrics. Woven textiles are washed whenever possible in full-width washing machines, though sometimes rope or winch-back machines may be used. Washing of a printed textile is intended to accomplish removal of thickeners, unfixed dye, and auxiliaries. Unfixed dyes may be either soluble or insoluble, but both cases present the danger of staining or bleeding.

Washing begins with a rinsing process to loosen or swell the thickener. The duration of the swelling effect can be improved by "skying" or laying the fabric over chutes, which is often combined with spraying of the cloth. The overall effectiveness of the wash can be greatly increased by inducing transverse motion of the fabric with beater rollers, turbinators, or vibrators as the material moves progressively forward in the longitudinal direction.

Full-width washing machines are divided into several sections. At the end of each section, squeezing equipment is incorporated to prevent dirty wash liquor from being carried forward. Such squeezing equipment greatly increases the effectiveness of the washing process. Washing is accomplished mainly in countercurrent flow with addition of a calculated amount of feed liquor.

The use of perforated drums with eccentrically mounted bearings results in longitudinal and transverse motion of the wash liquor together with good washing action (Vibrotex, E. Küsters, Krefeld, FRG). A perforated drum with wash liquor flowing through the holes is useful generally for both washing and drying purposes (Fleissner).

The design of newer washing machines is directed increasingly toward conservation of water and energy. Examples include the Mäander full-width washing machine and the Rotojet, both manufactured by E. Küsters and both operating on the countercurrent principle, where the cloth moves upward from below and is steeped and squeezed off many times, while the wash liquor flows downward from above.

Before a washed textile is dried it is often dewatered by centrifugation, squeezing off, application of a vacuum (through a slit), or a blow-through with air or steam. Drying with the aid of equipment like that described in Section 3.8, preferably on tentering frames, is almost always combined with the finishing process.

# 4. Thickening Agents and Thickeners in Textile Printing

## 4.1. Introduction

The composition of a textile printing paste is determined by the printing method (direct, discharge, or resist printing), the substrate, the application method, and the fixation method, not by the type of dye used. Moreover, all printing pastes contain not only dyes and chemicals, but also thickeners.

*Thickeners* are those components of a printing paste that impart a printable, pasty consistency to an otherwise aqueous liquor, and they consist mainly of thickening agents. Thickeners supplied in concentrated form for storage purposes are known as *thickener solutions.*

*Thickening agents* in the narrower sense are substances in powder or granular form (sometimes supplied as low- or high-viscosity "concentrates" or "clears") that display marked swelling properties in water. Such materials form colloidal systems with water, thereby increasing to a

greater or lesser extent the viscosity of a dye or chemical solution [4.1], [4.2].

Printing paste normally contains 40–70% thickener solution. Depending on the solids content, this typically corresponds to 2.5–10% (16% maximum) thickening agent [4.3].

Thickening systems also include emulsions composed of emulsifying agents, water, and heavy naphtha [4.2], [4.4]. The addition of such an emulsion to a swollen colloidal thickening agent produces what is known as a "semiemulsion." However, the use of heavy naphtha emulsions is now decreasing rapidly, mainly for environmental reasons.

### 4.1.1. Purpose and Effect of a Thickener

A thickener confers the required viscosity on an aqueous liquor containing dyes or chemicals to ensure that it can be printed and pumped, producing on the one hand good flow and levelling properties, but at the same time keeping the liquor "short" (i.e., immobile) so that the dye will stay where it should based on the design, permitting the production of sharp contours. In case of an interruption of the mechanical printing operation it is also important that the dye mixture not flow out from engraved areas or screens.

Most thickeners used in textile printing have structural viscous properties; i. e., the apparent viscosity of the mixture changes with changes in mechanical stress (shear stress). Some thickeners also have a flow limit, which is to say that their viscosity in a stationary state is infinitely high, and flow will commence only above some critical value of shear stress. Viscosity requirements differ greatly for the various operations associated with the printing process (e.g., pumping, doctoring) as well as for different printing rates and printing equipment, so it is essential that a printer be familiar with the rheological properties of the various thickeners. The manufacturers of thickening agents provide this information in the form of flow or viscosity curves [4.2], [4.5].

However, the effect of the thickener in a printing paste is not limited to providing some desired set of viscosity or rheological properties, thereby permitting particular effects to be achieved, such as sharp contours, surface levelling, or through-printing of a textile substrate. The thickener must also serve as a protective colloid for the various dyes and chemicals dissolved in the paste, and it must ensure the most complete dye transfer possible from the paste to the substrate. Thickeners are also responsible for regulating the moisture balance during the fixation process, which affects the dye yield [4.6]. Furthermore, dried thickener provides a protective film that prevents the loss of unfixed dye by frictional effects after the textile leaves the drying chamber and before it undergoes dye fixation, a process that does not always follow immediately. This film-forming property sometimes proves inadequate with extremely low-solids thickeners, such as those prepared from the so-called synthetic thickening agents, in which case the further addition of a film former is recommended [4.7]. On the other hand, the film of thickener must not be so brittle that it flakes off prior to dye fixation, a risk that is always present when printing on hydrophobic, smooth fabrics. The choice of a thickening agent is also influenced by the substrate [4.8], [4.9]. It is seldom possible to achieve all the desired effects with a single thickening agent. In practice, several thickening agents are often therefore mixed together, or a ready-mixed combination of thickening agents is utilized.

### 4.1.2. Demands Imposed upon Thickening Agents and Thickeners

The purposes of thickening agents and thickeners as listed in Section 4.1.1 require that they display a number of specific characteristics:

1) Their rheological properties must match the intended use.
2) They must be available in a standardized form at reasonable cost. The most commonly used thickening agents are modified natural products, the availability of which depends upon weather conditions, cultivation potential, and labor content. Unlike finishing agents, thickeners add no value to the textile, but are simply washed out after use. The cost must therefore be as low as possible.
3) Thickening agents and the thickeners prepared from them must have good storage characteristics. Most "natural" thickening agents as they are supplied contain 5–10 % water, and during prolonged storage under "dry" conditions or in a prepared composition the effects of hydrolysis or microbial attack may lead to a gradual loss of thickening

power. The addition of preservatives by manufacturers or users can only delay such decomposition, not prevent it. Furthermore, a thickener solution with a high solids content tends to dry out and acquire a surface crust. Residues of such a crust remaining in a finished dye-containing printing paste might lead to defective "skittery" printing.
4) Thickening agents must be easy to use. Thickening agents are now supplied almost exclusively in cold-soluble form [4.10]. Nevertheless, users often consider it advantageous to mix many thickeners with warm water, or to heat the mixture later [4.11]. Mixing methods commonly employed today (i.e., dispersion in water, perhaps by injection pumping, followed by agitation with high-speed stirrers) require free-flowing, nondusting, nonlumping powders or granules. Therefore, some thickening agents are modified chemically so that they only begin to swell slowly some time after wetting ("delayed swelling," "dispersible"), or only after the addition of acid or alkali [4.12].
5) The finished, fully swollen thickener must be clean and free of large, undissolved swollen particles. Solid impurities could cause the doctor blade of a roller machine to produce marks, and swollen particles might also block the openings of flat or rotary screens. In either case the result is the same: defective printing. The thickener solutions—or better: the printing dyes prepared from them— should therefore be filtered on a vacuum-operated passing machine prior to use. The diameter of the apertures of the corresponding filter sieve should be ca. 15 % smaller than the diameter of the printing-screen apertures.
6) The thickeners must be compatible with all dyes and chemicals that are to be used in the printing pastes, must not react with them, and should themselves have no color or be uniformly colored to prevent problems with spotty or unlevel printing. An important role is also played by ionic properties of a thickener (e.g., possible interaction between an anionic thickener and cationic dyes), as well as by the pH and the electrolyte content of thickeners or the dyes. On the other hand, the incompatibility of some thickeners with alkali, salts of polyvalent metals, and sodium borate (borax) is used to advantage in certain processes (e.g., thickener coagulation in the two-phase printing process).

7) During dye fixation by steam or dry heating, or during wet development of a dye, a good thickening agent should be displaced as completely as possible by the substrate to ensure maximum dye yield, and it should itself adhere or "crab" only slightly to the textile. A thickener that can easily be washed out permits aftertreatment processes to be kept brief, resulting in savings of time and money. Stained white backgrounds attributable to inefficient removal of residual dye and thickener, and printed areas that are harsh to the touch, result in "second-rate" products [4.13].

8) Thickeners should be environmentally friendly. Quite apart from the continually increasing cost of heavy naphtha, environmental considerations and legal restrictions in some countries place severe limits on the use of hydrocarbons in emulsion thickeners. Other types of thickening agents pass completely into the wastewater when a print is washed, so problems associated with wastewater contamination are becoming increasingly important [4.10], [4.14]. This applies both to "natural" polysaccharide thickeners and "synthetic" polymer thickeners [4.15].

## 4.2. Types of Thickening Agents

Especially in manufacturers' information sheets, thickening agents are classified according to the raw materials used in their preparation. The standard categories include alginates, seed derivatives (e.g., guar gum), starch degradation products, and the so-called natural, semisynthetic, and fully synthetic thickening agents. However, this classification system says nothing about the most important characteristics of the products. The characteristics of the starting materials, once considered unchangeable, have over the course of time been dramatically altered by improved processing. The guar flour available ten years ago has nothing in common with a modern highly depolymerized guar ether other than the starting material.

It would therefore be better to classify thickening agents according to chemical criteria; i.e., on the basis of the chemical constitution and ionic charge of the various dissolved long-chain polymers [4.16]. However, this information is not always available, especially for combinations of thickening agents, and it provides little insight into technical printing characteristics of the products. The practitioner therefore seldom looks for a thickening agent of a particular type (whereby exceptions only prove the rule!), but rather follows the recommendations provided by the producer or a product information sheet, determining the true suitability of the product for a particular application through preliminary tests.

**Inorganic products** are almost never used as thickening agents, but they are added as *fillers* that provide a thickening and water-absorbing effect in conjunction with true thickeners or thickening agents. Examples include zinc oxide, organically modified silicates (Bentone EW/ Kronos Titan; → Clays, **A 7**, pp. 116–118, 131–133), and hydrophilic silicic acid (Aerosil/Degussa) [4.17] (→ Silica, **A 23**, pp. 635–642).

**Proteins** also have no importance as thickening agents. *Albumen* was formerly used as an adhesive with a thickening effect in printing pastes for white pigments in direct and discharge printing, but it is now almost never encountered. However, proteins are present in certain polysaccharides, especially the galactomannans (see below), which are processed into carob-flour thickeners. Seedling residues in galactomannan also contain protein, which can have a harmful effect in carob thickeners [4.18].

### 4.2.1. Emulsions

Emulsion thickeners consist of water, nonionic or anionic emulsifiers, and heavy naphtha from a particular boiling range. Their use is very much on the decline worldwide for both cost and environmental reasons. The "water-in-oil" emulsion system was very popular until the late 1980s for pigment printing in the United States primarily because of the high printing rates it permitted, but "oil-in-water" and so-called "semiemulsion" thickeners for printing with reactive, direct, and disperse dyes are now the most widely used systems.

Semiemulsions are prepared by mixing emulsions with thickeners based on swollen polysaccharides, or heavy naphtha may be emulsified into a thickener that already contains emulsifier. The advantages of emulsion printing are good color brilliance, ease of washing out, and minimal influence on the feel of the textile, and these cannot easily be obtained using other thickening systems.

### 4.2.2. Nonionic Unmodified Polysaccharides
(→ Starch and Other Polysaccharides)

Wheat, corn (maize), and rice starch, once popular constituents of thickeners, are now rarely used because of their lack of cold-solubility. The same applies to gum arabic and gum tragacanth, which once were standard products in the printing industry, but today are no longer used because of their high cost and the long preparation times involved.

Unmodified finely ground guar flour is still used in carpet printing processes if it can be obtained at an unusually low price.

### 4.2.3. Nonionic Modified Polysaccharides

All polysaccharides can be hydrolyzed by acids, alkalis, oxidizing agents, enzymes, and the influence of high temperature. This degradation process significantly alters the properties of the raw material, especially its flow characteristics.

**Starch Products.** Examples include the hot-water-soluble heat-treated starches ("British gum"; trade names: Diatex, Diamalt) and the cold-water-soluble heat-treated swelling starches (trade names: Diatex SL, Diamalt). These thickeners are introduced at concentrations of ca. 30–50% and are very resistant to alkalis. They are used, for instance, in alkali-based pastes for crepe production, and are also added as color-intensifying components, such as in acrylic printing [4.19].

**Nonionic Starch Ethers and Starch Esters** (→ Starch and Other Polysaccharides, **A 25**, pp. 17–20). These are seldom used alone, but sometimes serve as thickeners in combination formulations. They are more often used as adhesives for woven cellulose (trade names: Solvitose H and H4, Avebe; Texogum 3013, Diamalt).

**Gum Resin.** The sparingly soluble gum resins are degraded by boiling under pressure. After purification and drying they are marketed in flake or granular form as cold-water soluble *crystal gum* or "industrial gum" (e.g., Cordofan). Solutions (ca. 30%) of these products display low viscosity, almost Newtonian flow characteristics, and good film-forming properties, but for reasons of cost they are used only as combination thickeners or in high-quality applications, such as outline printing (trade names: Karagum Super, Diamalt; Lamegum, Chem. Fabrik Grünau; Nafka Kristallgummi, Avebe).

**Galactomannans** (→ Starch and Other Polysaccharides, **A 25**, pp. 54–57). The galactomannan group includes carob flour, which was formerly used extensively in textile printing but is now avoided for reasons of cost. Because of the better availability, good swelling characteristics, and cold-water-solubility of guar flour, this material is now modified in such a way that it satisfies the demands of textile printing [4.20].

The cold-water-solubilities and hydration rates of nonionic guar derivatives have been improved by hydrolysis or reaction with ethylene oxide or propylene oxide. In contrast to anionic guar ethers (Sec. 4.2.4), the fully etherified products do not react with heavy-metal salts.

Guar derivatives, including the anionic varieties, are marketed under the following trade names, where various grades are distinguished only by added letters: Ago-Gum, Ago Chemicals; Diagum, Diamalt; Guaranate, SFC; Indalca, Cesalpinia; Lameprint, Chem. Fabrik Grünau; Lamberti Print, Fratelli Lamberti; Meypro Gum, Meyhall Chemical; Polygum, Polymer Industries; Polyprint, Polygal; Prisulon, Chem. Fabrik Tübingen; Solvitose, Avebe.

**Cellulose Derivatives.** Methyl, ethyl, hydroxyethyl, and hydroxypropyl derivatives of cellulose are increasingly recommended as thickener components [4.21], but they are seldom used alone. They once played a very important role as stabilizing components for emulsion thickeners containing naphtha.

Trade names include Cellosize, Union Carbide; Fine Gum, Dai-Ichi; Natrosol MR, Hercules; Tylose H and MH, Hoechst; Walocel HT, Wolff Walsrode.

### 4.2.4. Anionic Polysaccharides

**Alginates** (→ Starch and Other Polysaccharides, **A 25**, pp. 34–40). The alkali-metal salts of alginic acid, especially sodium alginates obtained from species of brown algae, have in the last 35 years become extremely important as thickening agents, especially since the introduction of reactive dyes [4.22], [4.23]. Advantages of the sodium alginate thickeners include good water-solubility (facilitating preparation of the thickeners) and ease of removal by washing after fixation of a dye, even at high temperatures. Magnesium and ammonium alginates are also used, the latter for reactive dyes fixed under acidic conditions.

The viscosity of an alginate thickener depends on the origin and species of the algae as well as the corresponding climatic and growth conditions. Degradation reactions occurring during use may lead to reduced viscosity. High-viscosity, medium-viscosity, and low-viscosity alginates are used as thickener solutions in the concentration range 3.5–10%. Calcium salts may alter the flow characteristics of alginate thickeners, but this effect can be prevented by the addition of sequestering agents.

Alginates form elastic films on drying, and it is this property that is exploited when they are used as components in thickening combinations, as for printing on polyesters. Low-viscosity sodium alginates and high-viscosity magnesium alginates are also applicable alone in this case. Unfortunately the cost of alginates has risen considerably in recent years because of cost-intensive methods involved in their production, so attempts are now being made everywhere to find suitable substitutes. Trade names include Alginate, SFC; Cecalgum, Sanofi; Dialgin, Diamalt; Keltex, Kelco; Lamitex, Protan; Manutex, Kelco.

**Carboxymethylated Polysaccharides.** These anionic products, utilized as their sodium salts, have become extremely important in textile printing. They are obtained by reaction of the alcoholic hydroxyl groups of starch, galactomannans, or cellulose with chloroacetic acid, followed by neutralization [4.6]. The products contain a variable excess of sodium chloride.

*Anionic starch ethers* are usually introduced at a concentration of 10% or as combination thickeners in conjunction with certain classes of dyes in African prints. They display alkaline properties, and therefore have only limited application with acidic printing pastes. Such thickeners coagulate in the presence of salts of trivalent metals. Trade names include Emprint CE, Emslandstärke; Monagum W, Diamalt; Solvitose C5, Avebe.

*Anionic etherified seed flours*, mainly based on guar, are supplied in various viscosity grades, and have many uses as thickeners, both alone and in mixtures. They too usually have an alkaline reaction, and the pH should therefore be adjusted so that the mixture is neutral or acidic

after swelling is complete. Very highly substituted grades are recommended for printing with reactive dyes. The trade names are typically identical to those assigned to nonionic etherified seed flours (see above), but with different letters added as distinguishing marks.

*Carboxymethylcelluloses* and their sodium salts are thickening agents with good film forming characteristics. They are seldom used alone, but are advantageous as components in thickener combinations.

Trade names include CMC Cellulose Gum, Hercules; Courlose, British Celanese; Tylose C600, Hoechst; Verdickung V extra Super 2, Chem. Fabrik Grünau; Walocel CRT, Wolff Walsrode.

**Xanthan.** This is a so-called biopolysaccharide (→ Starch and Other Polysaccharides, **A 25**, pp. 51–54) produced by the controlled fermentation of glucose. Its relative molecular mass is reported to be $2 \times 10^6$. Aqueous xanthan solutions have unusual plastic flow properties. When subjected to shear stress their viscosity decreases much more rapidly than is the case with other polysaccharide thickeners once the flow limit has been exceeded. These characteristics are exploited in the printing of thick-pile carpet products.

Incompatibility with cationic dyes can be reduced by special additives introduced into the printing paste [4.24]. Such thickeners are relatively stable toward acids, alkalis, and salts, and combining them with seed-flour products leads to a synergistic increase in viscosity.

Trade names include Kelzan, Kelco; Rhodopol, Rhône-Poulenc.

### 4.2.5. Cationic Polysaccharides

Products in this category are of little practical importance. The reaction of guar with 2,3-epoxypropyltrimethylammonium chloride has been described [4.25]. Such thickeners have been recommended for achieving special effects.

### 4.2.6. Synthetic Polymeric Thickening Agents

The term "synthetic" thickening agents was formerly interpreted to include compounds like acetylcellulose, nitrocellulose, alkylcellulose, hydroxyethylcellulose, polyvinyl alcohol, chlorinated rubber, and the poly(acrylic acids). Of these polymers, only the poly(acrylic acids) are important as fully synthetic thickeners today [→Polyacrylamides and Poly(Acrylic Acids), **A 21**, pp. 151–152].

Other products are obtained by the copolymerization of maleic anhydride and, for example, ethylene, styrene, or vinylmethyl ether with polyfunctional monomers (e.g., divinylbenzene). Aqueous emulsion copolymers of 20–40 % (meth)acrylic acid, olefinic compounds (e.g., ethylacrylate), and up to ca. 2 % polyfunctional monomers are of some importance as well [4.26].

All these thickening agents are long-chain polymers bearing carboxy groups subject to some measure of cross-linking. They swell considerably in water upon neutralization (thickening by swelling), and form high-viscosity gels.

These thickeners are most effective in neutral or slightly alkaline solutions. The commercial products are either already neutralized so that they swell immediately upon addition to water, or else swelling is induced after mixing with water by the introduction of a precisely determined amount of alkali (ammonia, sodium hydroxide, or an amine, depending on the application). The commercial products are supplied either in powdered form or as special liquid preparations, the so-called "concentrates" or "clears" [4.27].

The preparation time for a thickener of this type (i.e., the hydration time for a polymer thickener) is significantly shorter than the swelling time for a polysaccharide thickener [4.28].

The structural viscosity of a prepared thickener is usually higher than that of a polysaccharide thickener. A large decrease in viscosity under shear permits the use of pastes with relatively high static viscosities. Process interruptions thus lead to little or no "running" (drop formation), although at the moment of printing-paste application a corresponding high shear stress causes the mixture to become so thin that it passes without difficulty through very fine screen apertures, whereupon it again becomes highly viscous on the textile. It is therefore possible to use these thickeners for printing very fine outlines.

Polymeric thickeners were first introduced into pigment printing starting about 1967. It thereby became possible to reduce or completely eliminate the use of heavy naphtha in printing emulsions. In later years, the unique rheological properties of synthetic thickeners were exploited for printing with acid and metal-complex dyes, resulting in improved penetration with such voluminous materials as upholstery and carpet fab-

ric [4.29]. However, another feature of polymer thickeners must also be taken into account, namely their sensitivity toward electrolytes. Negative charges on the thickener molecules cause them to become extended and react strongly with added electrolytes present in the form of water hardness, anionic dyes, or setting-up salts. The observed viscosity decreases more or less drastically depending on the type and quantity of electrolyte, and may collapse completely. This has led in recent years to the release of special "nonionic" grades of dyes with a low electrolyte content, but so far only in the area of disperse dyes. Completely synthetic polymer thickening agents for printing on polyester textiles or transfer papers have to date been used only to a very limited extent [4.26], [4.30], [4.31].

Trade names include Acraconz, Bayer; Acramin Concentrate, Miles; Aqua Hue Clear, Blackman Uhler; Clear, Minerva; Hostatherm Verdicker, Hoechst; Lutexal, BASF; Lyoprint, Ciba-Geigy; Tanaclear, Charles S. Tanner.

## 5. Printing on Cellulose

(For pigment printing see Chapter 8)

### 5.1. Pretreatment

(See also → Textile Auxiliaries, pp. 253–267)

The quality of a printed product depends not only on the properties of the printing paste and the dye but also on properties associated with the substrate, so careful pretreatment of textiles is always essential. This applies especially to weaves and knits made of cellulose fibers, because these are heavily contaminated on the one hand with natural substances (e.g., seed-shell residues and waxy substances in the case of cotton), but also with sizing agents and preparations (both cotton and regenerated cellulose). The surfaces of some articles may also have been modified by mechanical treatment. For example, textiles with plain surfaces may be singed (with gas, or occasionally an electrically heated plate) in an attempt to remove protruding fibers and loose fiber dust that might stand in the way of a smooth surface print.

Alternatively, textile materials intended for the production of printed fabrics with a "nap," as well as flannels and similar articles, are "raised" on one or both sides with the rotating wire cards of a raising or napping machine. This may be undertaken either before or after printing. For example, if blurring of the pattern is to be avoided the material is usually raised before printing and again briefly at the conclusion of the various finishing steps. The "nap" of surface prints can be leveled by a cropping machine. Similar effects can be achieved by abrading the material. Fiber dust must in any case be completely removed by air extraction to prevent complications during the printing process.

All soil is removed during the first wet treatment (desizing and/or washing). In the case of cotton this was formerly accomplished with a boiling-off or keir-boiling process in special pressure vessels, but the process has now been considerably simplified through the use of modern apparatus (coiling machines, machines for storage and reaction, and perforated-drum washing units) together with special textile auxiliaries. The next step is usually alkali treatment or mercerization ($\rightarrow$ Textile Auxiliaries, p. 264) to improve the material's dye absorption characteristics and luster. Major improvements in the feel (smooth, silky) are achievable especially with fine knitted materials.

Many cellulose textiles in their natural state or subsequent to manufacture display a yellowish to brownish color (linen), so a bleaching step is usually added, involving hydrogen peroxide, sodium hypochlorite, sodium chlorite, or a combination process ($\rightarrow$ Textile Auxiliaries, pp. 259–264). A pure white color is essential not only for optimum brilliance of the color tones, but also for attractive white-discharge and white-resist effects. Optical brighteners (fluorescent whitening agents) are therefore very often incorporated into the formulations for bleaches, white-discharge printing pastes, and white resists.

A flow diagram outlining the entire operation from gray cloth to finished material is provided in Figure 5.

*Viscose fibers* (filament and staple), including the high-modulus fibers ($\rightarrow$ Cellulose, **A 5**, pp. 401–413), are in principle treated like cotton, but always under milder conditions to allow for their greater sensitivity to mechanical and chemical effects. After desizing and/or washing they may be bleached, but only if necessary, and then nearly always exclusively with hydrogen peroxide (rarely with sodium chlorite). Their dye-absorption properties can be improved by treatment with sodium hydroxide solution ($< 8°$ Bé $= 5.2\%$, $d = 1.06$). After careful rins-

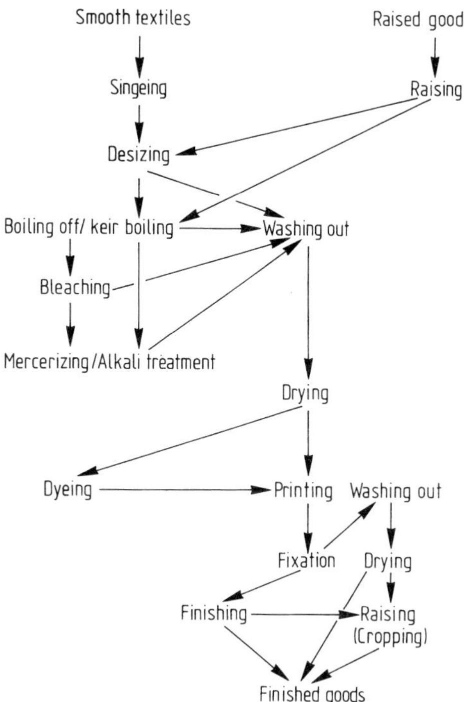


**Figure 5.** Overall process for the pretreatment and printing of cellulose textiles as it applies to both smooth and raised goods

ing and sometimes acidification they are dried, but without the use of high temperature (max. 150 °C) to avoid keratinization of the fibers and any diminution in swelling properties. The greater sensitivity of such regenerated celluloses is also relevant to all subsequent processes as well. For example, with strongly acid or alkaline printing pastes the risk of damage in this case is greater than with cotton. Brittle, dried films of thickening agent can lead to creases and folds, especially with fabrics made of fine or highly twisted yarns, such as crepe articles. The thickening agent must therefore be selected very carefully. Especially delicate, fine-quality fabrics of this sort are often wound into rolls after printing (preferably along with the back gray if the printed paste has a tendency to rub off!), and they are folded and washed in small batches after fixation.

*Knitted goods* are manufactured in both warp and tubular form, and they must be processed with special equipment even during pretreatment that precedes printing. Tubular goods may be processed as tubes on the production line or they may first be cut. Despite edge gumming, an unusable edge remains on both sides in the latter case, and this must be cut off after drying immediately subsequent to passage through the tentering frame.

In the treatment of *cellulose blends*, especially cotton–viscose staple blends or, to a much smaller extent, blends containing linen, the differing natures of the various fiber components necessitates compromises. If, for example, a linen component must be very white, as in the case of table linens, increased damage of fibers derived from the cotton component is unavoidable. On the other hand, if maximum strength is to be maintained with a cotton–rayon blend it may be necessary to accept the presence of seed-shell residues in the fabric, since boiling-off under pressure would necessarily be ruled out [5.1]–[5.4].

## 5.2. Printing Methods: Classification and Significance

Specific fiber materials and dye types are the subject of mutual interactions, and it is these interactions that determine the optimum composition of a printing paste, as well as the nature and sequence of fixation and aftertreatment steps. These matters are discussed here not in the order of their importance, but rather in the order of their historical development, although the most important aspects are consistently emphasized. Each class of dye is treated separately under the classical headings of direct, discharge, and resist printing.

An impression of the relative importance of the various classes of dyes can be gained from Figure 6, in which the overlap of economic factors and style trends is noteworthy [5.5]. Figure 6 also shows that pigment dyes are by far the most important, and that a small number of dye categories account for most of the total demand. Substantive and acid dyes represent barely 1 % of total dye consumption, and cationic and discharge dyes have almost completely disappeared from cellulose printing.

## 5.3. Substantive (Direct) and Acid Dyes

These two classes of dyes differ less in their chemistry (both consist mainly of azoic dyes carrying acidic groups that confer solubility in water) than in their differing affinities for the various fibers (→ Dyes, General Survey). Here they are discussed together because they are usually used together in printing on cellulose, and their subtle differences complement each other. Although their fastness is in most cases only moderate, those dyes are still occasionally used for inexpensive goods because of their favorable price. Also, further savings are possible through suitable choice of particular dyes and simplification of the process.

**Direct Printing.** A printing paste is prepared by dissolving the dyes in hot water to which is added urea and a solvent (ethylene glycol, thioethylene glycol; sometimes glycerine or a similar substance), and this solution is stirred into a thickener that is easily removed by washing. Small amounts of oxidizing agents are added to prevent reduction effects due to the fiber and/or the thickening agent.

Although many substantive and acid dyes reach maximum dye yield and fastness only after fixation with saturated steam for 30–60 min, the steaming time can be reduced to 15 min without incurring significant disadvantages by careful choice of the dye.

If the maximum possible fastness is required it is desirable not only to use a long steaming time, but also to treat the printed textile with auxiliaries that improve wetfastness (cationic quaternary polyammonium compounds) in the course of the final wash. The function of these agents can be interpreted as the linking together of several dye molecules in the fiber so that they become less soluble and also acquire such an awkward shape that they are much less able to diffuse out of the fiber. One negative side-effect must also be tolerated, however: a slight reduction in the lightfastness of prints so treated.

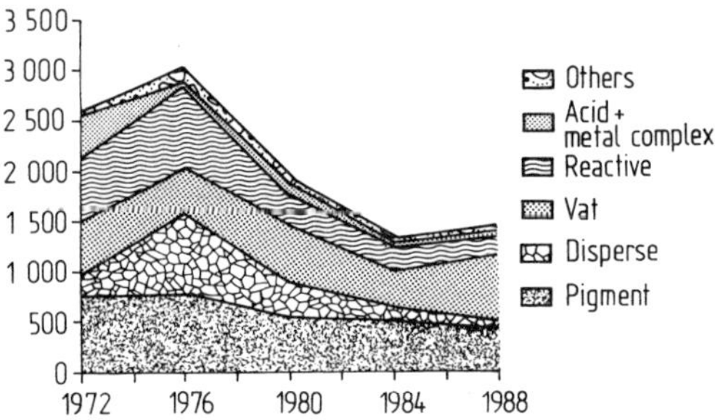


**Figure 6.** Consumption (t/a) of various classes of dyes by the textile-printing industry in the Federal Republic of Germany, 1972–1988 [5.5]

**Discharge Printing.** As fashions change in textile printing, there often occur periods in which discharge printing plays an important role. Substantive dyes with azo bridges subject to facile reductive cleavage are attractive candidates for use as dischargeable ground dyes, whereas acid dyes are rarely used for this purpose on cellulose. Sodium hydroxymethanesulfinate serves as the reducing agent (e.g., Rongalit C, BASF, → Textile Auxiliaries, p. 297), either alone or in combination with discharge auxiliaries such as anthraquinone (a reduction catalyst) or quaternary ammonium compounds (Leukotrop W), depending on the intensity of the color and the ease of discharging the dye. Whiteness effects are improved only in exceptional cases by the addition of alkali, but humectants and wetting agents are advantageous, as are white pigments (preferably zinc oxide or titanium dioxide) or optical brighteners added to improve appearance. Less satisfactory discharge dyeing results are usually obtained with regenerated celluloses, and the fact that there is a risk of fiber damage means that storage of the material (e.g., overnight, either before or after steaming) should be avoided if at all possible. The storage properties of the printed material depend very much on its moisture content and the relative humidity of the atmosphere, and less on ambient temperature.

Special care is required with substantively dyed grounds, whose discharging properties are subject to harm by treatments designed to improve wetfastness. The extent to which such effects are observed is very much a function of the auxiliary used, and it may lead to inferior results (pale or muddy colors) even with colored vat discharge dyeing, especially in the case of bright colors. The cause is not only reduced solubility of the cleavage products of the dyes, which are therefore more difficult to wash out, but also a change in wetting behavior and, in the case of colored vat discharge dyes, the formation of adducts between dyes and auxiliaries that in turn display poor diffusion characteristics under alkaline conditions. In such cases it is preferable to delay the fastness-improvement process until the final wash bath subsequent to printing. Previous washing temperatures should therefore not exceed 30–40 °C at most. Also, the wash-bath liquor should be changed sufficiently frequently to prevent accumulation of dissolved ground dye in the wash liquor and thus avoid any danger of bleeding onto the discharge pattern.

Another difficulty is the possibility of "overdischarging" the ground. In roller printing, rough roller surfaces or damage to the rollers or doctor blades can sometimes cause discharge paste—albeit in very small amounts—to appear outside the intended pattern on the textile, leading to a slight overdischarge of the overall dyed product, or fine lines appearing in the wrong places. One possible countermeasure is treatment of the fabric with oxidizing and/or acidic chemicals such as hydrogen peroxide or sodium *m*-nitrobenzenesulfonate (e.g., Ludigol, BASF), alone or in combination with nonvolatile organic acids, preferably glycolic or lactic acid. This "Ludigolization" can be accomplished by dipping or slop padding either before or after printing. The appropriate concentration depends on the amount of liquor applied, and is typically 5–15 g/L of oxidizing agent and 3–8 g/L of acid (based on 100 % acid).

**Resist Printing.** This technique is today restricted almost exclusively to batik articles, and involves substantive dyes containing groups subject to coupling (i.e., capable of reacting with diazotized bases or diazonium salts, known as "fast color salts"). Cotton fabric to which a wax pattern has been applied is introduced at room temperature into the cold dye solution, subsequently dried in air, coupled with the diazonium salt, and then rinsed. The wax is finally removed by boiling-off in water [5.6]–[5.8].

## 5.4. Cationic and Mordant Dyes

These two classes of dyes (→ Dyes, General Survey; → Cationic Dyes) which are characterized by a very wide variety of chemical groups, are now almost never used in cellulose printing for the following reasons. Cationic dyes alone have insufficient affinity for cellulose, and therefore require fixing with a mordant (a tannin; i.e., a synthetic tanning agent such as Katanol ON or Mesitol) as well as aftertreatment with an antimony salt. Even then the observed colorfastness properties are only moderate, with lightfastness being especially poor. A further disadvantage is that the corresponding printing pastes can be stored only for a very limited time.

Mordant dyes lead once again to environmental problems. The unavoidable use of heavy metal salts, especially chromium salts, is a powerful argument against their application. Also, the colorfastness of mordant dyes (especially wetfastness) is usually only moderate.

Finally, both of these dye categories mandate special thickening agents and dye-specific formulations. All these disadvantages outweigh the favorable economics of the intensely colored and even brilliant representatives of these two dye groups.

## 5.5. Vat Dyes

Vat dyes ($\rightarrow$ Dyes, General Survey) are not only the oldest and most versatile dyes used for printing on cellulose, but also among the most colorfast of all dyes. They are still widely used today in various direct and discharge printing processes, though seldom in resist printing. They are often extremely fast toward moisture, light, and weathering, and are used to great advantage for decorative fabrics, upholstery, and window shades. These characteristics apply with respect to all types of cellulose fibers, including cotton, regenerated cellulose, and linen [5.5].

The dyes are supplied as water-insoluble pigments, and are also printed in this form ($\rightarrow$ Indigo and Indigoid Colorants; Anthraquinone Dyes and Intermediates; $\rightarrow$ Pigments, Organic). They are then converted into the corresponding alkali-leuco compounds ("vats") by the action of alkali and a reducing agent in an atmosphere saturated with water vapor ($\rightarrow$ Leuco Esters of Vat Dyes). This provides substantivity toward cellulose, permitting the dyes to be absorbed by the fiber. There, treatment with oxidizing agents (air in the simplest case) in the course of wet treatment (washing) in weakly acidic medium causes them to be reprecipitated as finely dispersed pigments. Dye pigment deposited in the fibers in this way exhibits extremely good fastness properties.

Nevertheless, the reoxidation process must be carried to completion so that no leuco dye will redissolve in the alkaline soap-containing bath and/or bleed onto the white fabric. Also, the subsequent boiling with alkali must remove not only pigment dye adhering to the surface (to improve rubbing resistance), but also ensure that the vat pigment is converted into its final crystalline (crystallite) form. In very many cases it is only in this way that optimum fastness and brilliance and the correct color shade on printing can be achieved with vat dyes [5.9].

Not all vat dyes that produce good results in normal dyeing are equally well-suited to textile printing. This depends partly on the substantivity of the leuco compounds, which must diffuse as completely as possible into the cellulose fibers during the limited steaming period, but partly on the redox potentials of both dye and reducing agent. Finally, optimum vatting and fixation conditions should be adhered to for each dye with respect to quantities of alkali and reducing agent, time, and temperature.

However, since several very different color shades (and hence different dyes) must ordinarily be fixed simultaneously in a given printed design, the conditions selected necessarily represent a compromise. Under circumstances where, for example, a high alkali content or a longer steaming time might otherwise be desirable, these conditions might in fact lead to the hydrolysis of other dyes present, resulting in changes in shade or a reduction in color intensity. Similar considerations apply in the reverse case as well: use of too little alkali and insufficient steaming can cause rearrangements to dye forms that are more difficult to reoxidize [5.10].

Apart from careful selection of the dyes, this problem can be combated in two ways ($\rightarrow$ Dyes, General Survey):

1) Use of fine or extremely fine pigment-dye powders with increased surface area, permitting considerably faster vatting and hence more rapid dissolution of the dye.
2) Additives introduced by the dye manufacturer to increase the extent of dispersion and provide stabilization and homogenization so that cellulose fibers can be dyed over a wider range of tolerance. With liquid products (pastes), other components are added that confer the maximum possible protection against sedimentation, drying, and frost damage. Especially printing pastes intended for the sulfoxylate printing process (see below) often contain hydrotropic and hygroscopic chemicals along with reduction catalysts. A single dye from a given manufacturer is thus frequently supplied in three or more forms (powder, granules, pastes).

### 5.5.1. All-In (Sulfoxylate) Printing Processes

The term "sulfoxylate" refers to sodium formaldehyde sulfoxylate = sodium hydroxymethanesulfinate [*149-44-0*], C.I. Reducing

Agent 2, marketed, for example, as Rongalit C (→ Textile Auxiliaries, p. 297).

Of the various sulfoxylate printing processes, the "Rongalit C–potash process" developed by W. SIEBER in 1908 has become the most important. It is equally suitable for direct printing or for producing color-discharge effects. It is still widely used today for printing cellulose with vat dyes, although many precautionary measures are necessary, a circumstance that has led in recent years to an increase in the importance of the "two-phase printing process" (Sec. 3.7).

In the Rongalit C–potash process, also known as the "all-in process," the printing paste contains not only the dye but also all the chemicals necessary for fixation. For storage of the paste the dye is often converted to a reduced state, resulting in better dye yields for dyes with medium or dark colors. With light colors, however, decomposition by the reducing agent may lead to an unacceptable loss of color intensity. In the interest of good reproducibility it is therefore advisable to carry out printing tests prior to large-scale use of such a paste.

Manufacturers' recommendations should be followed when selecting a particular *dye*. Products with very soluble leuco forms are preferable, as these will be rapidly and completely reduced and will provide good substantivity with respect to cellulose fibers. Both of these properties depend on molecular constitution, since dyes with relatively low molecular mass are the most soluble, and indigoid dyes are more readily reduced than anthraquinoid dyes, although the latter usually have better fastness properties.

At one time, only paste grades were normally available commercially, but today many users take advantage of the availability of very fine granules with a low dust content. With care, these can be mixed directly into stock pastes using high-speed stirrers to produce speck-free products.

The *reducing agent*, almost always sodium hydroxymethanesulfinate, reduces dye in the printing paste rather slowly at pH 10 and temperatures up to 55 °C. At higher temperatures the reaction proceeds more rapidly, reaching a maximum at 100 °C, which corresponds to the steaming conditions during fixation. If a print is stored before steaming, however, Rongalit C in the printed surface oxidizes relatively rapidly under the influence of high humidity and high temperatures (though hot, dry air is relatively harmless). Complete reduction and fixation of the printed dye is then no longer possible. Storage prior to steaming should therefore be kept to a minimum ( < 6 h).

Disodium iminobismethanesulfinate, first marketed in 1959 under the name Rongalit FD [*8065-69-8*] (BASF), is a more stable sulfinate. However, not all dyes useful for printing can be completely fixed with this substance. Also the steaming time must be extended for maximum dye yield, or else the process must be carried out slightly above atmospheric pressure in a star steamer. Rongalit ST liquid (BASF) is a product with many applications that has more recently been introduced for this purpose (→ Textile Auxiliaries, p. 297) [5.11]–[5.13].

The *alkali-metal carbonate* in printing pastes or thickener stock is usually potassium carbonate (potash), because this is more readily soluble and also more hygroscopic than sodium carbonate. Maximum fixation yield and brilliance are achieved by fixing fabrics printed with vat dyes under saturated-steam conditions at 100–102 °C. As a result of the properties noted above, potash increases the tolerance range for steam superheating. On the other hand, sodium carbonate is not only cheaper, but also effective at considerably lower concentration. However, it has the disadvantage of crystallizing out during prolonged paste storage, increasing the risk of damage to printing rollers and screens (especially rotary screens). Only mixtures of the two carbonates are therefore recommended. Sodium hydroxide is applicable as the source of alkali only in special cases (prereduction processes), since it causes hydrolysis of many of the dyes before the required fixation times have elapsed. Sometimes, especially with blue dye varieties, aliphatic bases are used as the alkali (e.g., triethanolamine) as a way of improving color intensity and purity.

The *thickening agents* (Sec. 4.2) must be stable toward all the chemicals used, especially the alkali, and toward electrolytes in general. Starch/British gum/tragacanth combinations were once very popular, but these have today been replaced by cold-soluble, frequently premixed formulations of starch ethers with seed-flour derivatives and/or special alginates or gums.

Other *auxiliaries* used for printing vat dyes include hygroscopic and hydrotropic substances (glycerine, glycols, glycol ethers, thioglycols, and urea), and sometimes printing oils and defoamers. Whereas oils act in a purely mechanical way to improve printing properties, doctor-blade operation, and the levelness of printed surfaces, the effect of the first-named agents is delayed until the steam-fixation stage, at which point they promote dispersion, dissolution, color intensification, and stabilization, also reducing the negative effect of (slightly) superheated steam by encouraging condensation of the steam on the printed surfaces. Excessively superheated steam (> ca. 106 °C) leads inevitably to the deposition of weak, muddy colors.

Several blue anthraquinoid dyes print at their full brilliance only if smaller amounts of reducing agent are used and triethanolamine is added. The same effect can be achieved by stirring 3–5 g/kg of sodium nitrite or sodium chlorate into the standard mix.

**Steaming.** After printing and drying (whereby the latter must be rapid and above all complete), the fabric is often cooled by air blowing or by passing it over a cooling cylinder to improve its storage properties prior to steaming. Possible effects of unfavorable weather conditions can be excluded by air conditioning (e.g., silica-gel drying).

Steaming with continuous or star steamers using air-free saturated steam at 100–102 °C requires 5–15 min. Regenerated celluloses usually demand longer steaming times. Temperatures in the steamer must be carefully controlled to prevent damage from overheating due to the effect of exothermic processes, including liberation of heat of swelling of the fabric (depending on the initial moisture content), heat of solution for certain chemicals in the condensate, heat of condensation, and the decomposition of reducing agents. It is safer (though more expensive) to achieve the necessary control by maintaining a high flow rate of steam rather than by spraying water into the steamer, as this expedient leads to frequent droplet formation. With water spraying there is also the danger of air leakage, best prevented by a slight overpressure in the steamer.

**Aftertreatment.** After steaming, the printed fabric must not be stored for too long prior to washing because reducing agent residues may continue to decompose exothermically, leading to heat build-up in the stacked material and defective dyeing or even browning of the cellulose. If a delay of several hours is anticipated before the wet aftertreatment the fabric should be cooled with air ("skying") to oxidize at least some of the excess reducing agent [5.14].

Washing of the printed material, whether at full width or in rope form, begins with a thorough rinsing in cold water. After this, reoxidation is carried out with hydrogen peroxide in the presence of a small amount of acetic acid at 50–60 °C. A soap treatment with sodium carbonate at the boiling point should be initiated only after this process is complete.

### 5.5.2. Two-Phase Printing Processes
(See also Section 3.7)

The storage stability of unsteamed fabric from the all-in process is poor, and there are sometimes problems with color reproducibility (prereduction). Also, fiber damage is possible if drying is inadequate, and the results are sensitive to fluctuations in the steaming conditions. This led at an early date to the development of alternative processes (E. PFEFFER, 1926, "Colloresin process"), which became well-established only after about 1954 in conjunction with the construction of special steamers and the development of suitable reducing agents.

In the two-phase processes of today (mainly "flash-aging" methods), dye and chemicals are applied in two separate steps. In the first step only the dye and thickener (the "solid phases") are applied, thereby avoiding all stability problems. The chemicals (reducing agent and alkali, sometimes together with other auxiliaries) are added in a second step in the form of an aqueous solution (the "liquid phase") by padding. The fabric is then immediately steamed; i.e., the dye is fixed by exposure to an air-free atmosphere of steam for 15–60 s. Provided the fabric is wet, no damage is possible during fixation due to heat arising from condensation, enthalpy of reaction, or superheated steam, because the surface of the textile is under saturated steam conditions. For the same reason the steaming temperature also plays a relatively unimportant role. However, it is necessary to ensure that the relationship between the weight of the textile, the amount of water applied during padding, the heat caused by condensation, and the extent of the condensate is such that it excludes possible running of the printed pattern. Considerable help in this respect is provided by addition to the padding liquor of substances for coagulating the thickener. In some cases the added alkali is sufficient, but borax is often added when seed-flour derivatives are used, and aluminum sulfate is introduced with alginates. This measure has the further advantage that it helps prevent marking of the printed product by the guide rollers, a problem that can also be overcome mechanically by only allowing the guide rollers to contact the back of the cloth.

Unfortunately, thickener coagulants may hinder diffusion of leuco compounds into the fiber and thereby decrease fastness toward washing. For this reason it is common practice today to employ mixtures of coagulable products with thickening agents that cannot be coagulated, such as starch or, more recently, starch ethers [5.15].

**Dye Fixation.** Fixation is accomplished by a combination of padding and steaming. Many variations on the process are possible depending on the equipment available. Cloth is either run continuously through the developing bath or it is slop-padded with the liquor on one side only prior to being squeezed off by passage between a pair of rubber rollers or a combination of rubber and steel rollers. The fabric is then fed into the steam chamber through which it is guided by rollers contacting the reverse side only. The steam chamber can be of the tower, channel, vault, spiral, or hanging type, with the latter being used for knitted goods. Depending on the fabric load and the production rate, long- or short-duration steaming methods may be used, in each case with the appropriate formulation and choice of reducing agent (see Table 2).

"Hydrosulfite" (sodium dithionite [7631-94-9], sodium hyposulfite, C.I. Reducing Agent 1) is quite sensitive to oxygen, so air leakage must be prevented between the squeezing-off step and the steam inlet (40 cm maximum) when this agent is employed, and steaming times must be kept very short (20–40 s). Sodium formaldehyde sulfoxylate (sodium hydroxymethanesulfinate, C.I. Reducing Agent 2) is used instead in the long-duration steaming method (3–15 min). A substance superior to hydrosulfite for the short-duration process is Rongalit 2 PH (BASF), which is also a stabilized derivative of hydroxymethanesulfinic acid (C.I. Reducing Agent 13). This substance displays good liquor stability and does not reduce dissolved vat dyes in the temperature range 20–30 °C, thus eliminating the possibility of color changes in the ground. It also permits more thorough wetting of dense fabrics through inclusion of an intermediate air-blowing stage.

The alkali used is normally sodium hydroxide, sometimes mixed with sodium carbonate or potassium carbonate. The most likely further additives apart from coagulation agents are alkali-resistant wetting agents.

If aluminum sulfate is used instead of borax for the coagulation of alginates it is important that it be predissolved in water and then reacted with sodium hydroxide solution to form sodium aluminate, where aluminum hydroxide represents an intermediate stage.

Immediately after padding with the developing liquor (70–80 % uptake of liquor with a concentration equivalent to that listed in Table 2), the wet cloth is steamed. A slight overpressure in the steamer produces an outward flow of steam through the cloth inlet and outlet slots, ensuring the absence of air in the steamer.

The steaming temperature is usually in the range 120–130 °C, but may fluctuate within wide limits depending on the cloth. It is important that the cloth not be allowed to dry out at high temperature, because the leuco compound of the dye would then diffuse back, and sensitive dyes might partially hydrolyze.

**Washing and Reoxidation after Printing.** This is usually carried out continuously in a single production line to prevent marking of the moist cloth during storage. The method is in principle analogous to that used in the all-in process. Thus, the fabric is first thoroughly sprayed in a chute or J-box (although perforated drums with spray tubes and/or water recirculation are often used today), preferably with water from the last rinsing bath of the washing machine. It is then reoxidized by air or added hydrogen peroxide, and finally boiled with soap and sodium carbonate solution [5.16].

**Table 2.** Typical formulations for padded developing liquors

| | Long-duration steaming | Short-duration steaming | |
|---|---|---|---|
| Water* | 700 mL | 700 mL | 700 mL |
| Sodium hydroxide | 40 g | 60 g | 60 g |
| Borax | 0–10 g | 0–10 g | 0–10 g |
| Rongalit C | 80–100 g | | |
| Hydrosulfite (conc.) | | 80–100 g | |
| Rongalit 2PH | | | 80–100 g |

* Additional water is added for a total batch weight of 1000 g

### 5.5.3. Discharge Printing Processes

Vat dyes are very seldom used as discharge grounds, and then only for white discharging of light to medium color shades, because the discharge pastes required are so aggressive that color-discharge printing is practically ruled out because of instability. However, vat dyes have long been used for the illumination of other ground colors with good discharging properties, since the same reducing agents that cleave the ground color (substantive dyes, insoluble azo dyes, reactive dyes) in washable products are also suitable for fixing a vat dye by the sulfoxylate process (see Sec. 5.5.1). However, a successful application requires careful choice of the dyes, and this is true of both the ground dye and the color-discharge dye. Otherwise, although the color might well be discharged, it is possible that the vat dye applied simultaneously would not be fixed, or that fixation would be achieved on printing, but the available reducing power might no longer be sufficient to discharge the underlying ground color. This effect depends on the redox potentials of the dyes employed.

Auxiliaries for increasing discharge effects include anthraquinone, Leukotrop W (BASF), the calcium salt of a disulfonic acid of dimethylphenylbenzylammonium chloride, alkalis (sodium carbonate, potassium carbonate, sodium hydroxide solution), and hydrotropic agents or dispersants. However, the tendency of these substances to attack the color-discharge dye and/or the cloth represents a disadvantage, because it results in poor paste stability, poor storage stability, unsatisfactory reproducibility, and the risk of fiber damage. For example, augmented discharge pastes of this type are not at all suitable for regenerated cellulose [5.17].

In discharge printing, aqueous solutions of the weak oxidizing agent Ludigol (BASF, → Textile Auxiliaries, p. 296) are applied to the fabric by padding either alone or in combination with nonvolatile organic acids, especially glycolic or lactic acid, using a Foulard or special padding machine. The fabric is subsequently dried. This precaution, sometimes delayed until after the printing step, affects the decomposition of any Rongalit introduced through unintentional application of white or color discharges caused by defective rollers, screens, or doctor blades, as well as discharge halos (aureoles) that might be present.

The risk of defective printing is especially severe if errors are not immediately visible, appearing only after steaming. The extra precaution is therefore taken of treating with Ludigol or adding optical brighteners (5–20 g per kilogram of printing paste) that fluoresce under UV light, requiring that special lamps be mounted on the printing machines [5.14].

Examples of typical formulations for white discharge pastes are provided in Table 3.

The formulations for color-discharge pastes are identical to those used in the all-in process (Sec. 5.5.1), although depending on the ease with which the ground is discharged, 50–150 g of Rongalit C may be added per kilogram of printing paste.

After printing and drying, the cloth must be steamed as quickly as possible, because the dual role of the sodium formaldehyde sulfoxylate doubles the harmful effects anticipated during storage. The steaming time, other steaming conditions, and the conditions applicable to the post-printing wash are comparable to those in direct printing (see Sec. 5.5.1.). It is essential that the colorfastness of the ground color at the proposed washing temperatures be thoroughly tested.

**Table 3.** Typical formulations for white discharge pastes

| Ground dye | Vat dye | | Substantive dye | Insoluble azoic dye |
|---|---|---|---|---|
| Thickener* | 450 g | 350 g | 450 g | 450 g |
| Zinc oxide 1:1 | 150 g | 100 g | 100 g | 100 g |
| Rongalit C | 240 g | 240 g | 150 g | 200 g |
| Leukotrop W (conc.) | 60 g | 160 g | | 80 g |
| Ludigol | 30 g | 30 g | | |
| Sodium hydroxide (solution, 38° Bé) | | 180 g | | |
| Potassium carbonate | | | 0–90 g | 60 g |
| Anthraquinone | | | | 20 g |
| Water or thickener to 1000 g | | | | |

* Gum resin/British gum mixtures or alkali-resistant seed-flour ether thickeners

### 5.5.4. Resist Printing Processes

Neither the mechanical (paste resist) nor the chemical resist process is now used with vat dyes. This is not only a consequence of the relative complexity and uncertainty of the operation, but also the fact that the chemicals required (heavy-metal salts) should be avoided for environmental reasons. This leaves only the use of wax resists with indigo as a continuation of the Indonesian folk tradition of batik printing, as well as certain techniques associated with African printing [5.8], [5.18].

## 5.6. Sulfur Dyes (→ Sulfur Dyes)

Apart from black grades, sulfur dyes play almost no part in textile printing because of their dull colors and only modest fastness. They behave in the dyeing process like vat dyes, except that they can be converted to the vat form with such weak reducing agents as sodium sulfide or glucose. However, the black versions (Indocarbon, Hoechst, → Sulfur Dyes, **A 25**, p. 614) can be used in the two-phase printing process alongside vat dyes without any change in formulation.

## 5.7. Leuco Esters of Vat Dyes
(→ Leuco Esters of Vat Dyes)

*Leuco esters* of vat dyes are the water-soluble sodium or potassium salts of sulfuric acid esters of the leuco vat dyes. Their importance in printing has diminished greatly, and they are now used only in special cases, if for no other reason than cost. For example, color shades that could not be obtained by coupling reactions of diazotized bases or stabilized diazonium salts were formerly reproduced by direct printing onto Naphtol AS grounds. Thus red naphthols (AS, AS-OL, AS-D, AS-PH, AS-BO, etc.) were in this way supplemented by yellow, green, and brilliant blue. Today reactive and Phtalogen dyes are used for this purpose in preference to leuco vat-dye esters.

## 5.8. Azoic Dyes Produced by Coupling
(→Azo Dyes)

The production of insoluble azoic dyes directly on a piece of fabric is one of the oldest dyeing techniques. The coupling component 2-hydroxy-3-naphthoic acid anilide [*92-77-3*], Naphtol AS, was introduced in 1911 (→Azo Dyes), and it is still used as such or in the form of one of its many derivatives. The *two-stage* process requires that a "naphtholated" fabric be printed with a diazotized base, usually a diazotized aromatic amine. For this reason the process is also referred to (incorrectly) as "base printing." A naphthol preparation includes the naphthol itself, sodium hydroxide solution, alcohol, and a dispersing agent that prevents the precipitation of naphthol when the concentrated solution is diluted. The naphthol preparation is applied by padding and drying. It is somewhat sensitive to atmospheric moisture, light, and carbon dioxide.

The diazonium salts obtained by diazotizing a base are rather unstable, and must be printed immediately. Apart from the diazotizing solution, the corresponding printing pastes contain sodium acetate, acetic acid, and a suitable printing-paste thickener. The commercial process is simplified today by extensive use of stabilized diazonium salts. These are complex double salts of diazonium ions with, for example, zinc chloride, 1,5-naphthalenedisulfonic acid, or tetrafluoroboric acid (→ Diazo Compounds).

Coupling to form the dye, starting with either a diazotized solution or stabilized diazonium salts, is carried out immediately after printing so that after drying the printed article can be washed to remove unreacted naphthol; it may be, however, that other dyes added to produce specific color shades will require an additional fixing step. Supplementation in this sense is frequently necessary, because not all shades can be obtained with naphthol preparations alone. Missing colors include yellow (except for Naphtol AS-G and other yellow naphthols), turquoise, and a brilliant blue with the clarity of a phthalocyanine dye. Printing on Naphtol AS and AS-G, supplemented with selected reactive dyes, Phtalogen dyes, and pigment dyes, and with the aid of sophisticated resist techniques, is also still extremely important in African printing. Otherwise, however, the technique has declined today to the point of insignificance. The same applies to the single-step application of mixtures of diazo and coupling components (e.g., the Rapidecht, Rapidazol, and Rapidogen dyes).

This displacement in favor of reactive and pigment dyes is somewhat regretable from the standpoint of fastness properties. Mixtures of the type described were of considerable importance in direct and reserve printing with aniline black and Phtalogen dyes. Following disappearance of aniline black articles, with which the best color resists were always obtained (e.g., using Rapidogen dyes), one of the last areas of application for this type of dye has probably been lost. Apart from ecological and toxicological considerations, the following causes can be identified:

1) Interference from so-called cross-coupling in the event of overprinting. Thus, when two dyes consisting of different naphthols as well as different diazo compounds are combined, instead of the desired two dyes, four dye combinations result, with different colors and sometimes differing fastness properties. A predictable result is possible only if a match is present for at least one of the components.
2) The stability of the printing paste depends on stabilization of the diazonium salt, and is in most cases only modest.

## 5.9. Phthalocyanine Developing Dyes
[5.19], [5.20]

Insoluble phthalocyanine–metal complexes (→ Phthalocyanines) can be caused to form on the fibers starting with an isoindolenine and a heavy-metal compound, both of which must be present in the printing paste:

$$R^1 \quad R^2 \qquad + M^{2+},\ -NH_3,\ \text{Heat, Auxiliaries}$$

1-Amino-3-imino-isoindolenine

The substituents $R^1$ and $R^2$ may also be joined in the form of a ring.

Such a printing paste also contains other components designed to improve the solubility of the isoindolenines, suppress hydrolysis, and ensure the liberation of ammonia on heating. The only metals used in practice are copper, cobalt, and nickel. A range of phthalocyanine developing dyes is available from Bayer under the trade name Phtalogen.

**Direct Printing.** Phtalogen dyes are mixed into a paste with solvents, usually consisting of glycols, glycol ethers, aminoalcohols, and amides. To this paste is added an ammoniacal solution of a copper or nickel complex, and the resulting solution or fine dispersion is mixed into a previously prepared thickener. Thickeners for this application must be compatible with the solvents, and must also remain unaffected by the metal ions. Practical advice is available from manufacturers of the thickening agents.

The print must be dried immediately and completely in a batch process, where the endpoint is recognizable by a color change in the direction of the final color. The printed pattern is at first almost colorless, but this changes upon drying at $> 80\,°C$ with appropriate air circulation as a consequence of complex formation, even though this initial color is not yet that of the final pigment. Fixation can be accomplished by steaming for 5–8 min at 100–102 °C, or by hot-air treatment for 5–3 min at 135–145 °C. Inadequately dried prints lose color through hydrolysis in the course of steaming. If the drying-chamber capacity is insufficient to produce consistent drying results, the cloth can be further dried with drying cylinders.

Excess heavy metal should then be removed in an acid treatment (10 g/L HCl, 70–73 °C, 30 s) in order to prevent subsequent damage by peroxide or chlorine from the washing or bleaching agents. The thickener and any dye pigment deposited loosely on the surface are removed by rinsing and boiling-off. Ionic copper can also be removed by a Trilon B bath (1–2 g/L or more, depending on water hardness) if auxiliary dyes happen not to be compatible with a strong-acid treatment.

**Combinations with Other Dyes.** Phtalogen dyes can be printed together with reactive dyes, azoic coupling dyes, leuco-vat esters (Anthrasol and Indigosol), and pigment dyes. With other classes of dyes there is often a problem of incompatibility with the aftertreatment required for Phtalogen dyes, or else the fastness properties do not match.

When printing with reactive dyes, acidification is carried out with 5–10 g of formic acid instead of the usual hydrochloric acid. The boiling-off and rinsing processes must also be adjusted to suit the unique demands of reactive dyes.

Combination of Phtalogen dyes with vat dyes is possible in the two-phase printing process (Sec.

3.7) provided the fabric is dried sufficiently vigorously after printing to ensure proper development of the Phtalogen dye.

Phtalogen dyes are especially important for the printing of cellulosic fibers. In fiber blends (e.g., polyester and cotton), a 50/50 mix usually represents the limit for use of these dyes alone. Mixing Phtalogen and disperse dyes in a paste can open the way to printing other fiber blends. The Phtalogen dyes are so exceptionally colorfast that it is possible in this case to implement washing procedures capable of removing the disperse dye from the cotton constituent.

**Printing with Complex Dyes** (→ Metal Complex Dyes). The initially formed metal-containing complex is a precursor to the actual phthalocyanine. Early examples include Phtalogen Brilliant Blue IF3GK and Phtalogen Turquoise IFBK. These display low solubility in water, and are therefore more resistant to hydrolysis than other Phtalogen developing dyes. Their solubility in mixtures of diglycol ethers and tertiary aminoalcohols (triethanolamine) can be described as adequate.

The complex dyes are developed in the usual way by steaming, hot air treatment, or wet development with reducing agents. Development with reducing agents can be used for direct or two-phase printing in conjunction with vat dyes, because the dye color remains unaffected.

Because of the large size of the molecules (or association compounds) constituting the Phtalogen complex dyes, together with low solubility in water, their use is limited largely to suitably pretreated cotton.

The high proportion of solvent mandates a careful choice of thickening agents. Apart from the classical starch–tragacanth thickeners the only option currently available is starch ethers etherified to a specific degree.

*Phtalogen Blue IBN* is the most important complex dye from a commercial standpoint. It is a cobalt complex that has been rendered water-soluble by the presence of a basic group. Fixation causes this group to be cleaved, and the resulting cobalt phthalocyanine is deposited on and in the fibers. In addition to thermal fixation methods, reductive treatment can also be used to produce the cobalt phthalocyanine pigment, which has a clear, dark blue color.

The solubility of Phtalogen Blue IBN in dilute acetic acid is high, so no additional auxiliary is required. The importance of the product lies in

direct printing and printing on naphtholized textiles, where the acidic printing mixture also permits admixture with stabilized diazonium salts or even diazotized bases. The dark green, brown, or black colors so obtained are both intense and fast.

A typical formulation for 1000 g of printing paste combines 30 g of Phtalogen Blue IBN, 10 g of 60% acetic acid (up to 30 g for an alkaline Naphtol AS preparation), ca. 500 g of thickener, and 460 g of water or additional thickener. Storage properties are improved by the use of lactic acid in place of acetic acid.

Polyacrylonitrile–cotton blends can be printed with combinations of basic dyes and Phtalogen Blue IBN, and polyester–cotton blends with combinations of disperse dyes and Phtalogen Blue IBN.

A chemical resist procedure is very difficult to conduct with Phtalogen Blue IBN, but it is possible to work with either a preprinted paste resist (with intermediate drying) or a wet-in-wet resist (Chap. 2). Resists with reactive dyes can also be obtained with acidic printing pastes.

## 5.10. Reactive Dyes [5.21]–[5.29]
(→ Reactive Dyes)

### 5.10.1. Introduction

There are two basic reasons for the steadily increasing importance of reactive dyes in textile printing:

1) The brilliance and wide range of available colors
2) The ease with which the dyes are applied

Other advantages include stability of the fiber–dye bond, facile removal of unfixed hydrolyzate by washing (i.e., low substantivity), a high fixation yield, and good build-up properties, as well as good stability with respect to the printing paste.

Based on this profile of characteristics the monochlorotriazine (MCT) reactive dyes have established themselves extremely well in textile printing applications. Sulfatoethylsulfone (SES) dyes are used for special two-phase applications. All the other types of reactive dyes, developed mainly for dyebath application, are of minor importance in direct printing.

## 5.10.2. Reaction Mechanism

Reaction occurs primarily in the amorphous region of the cellulose. Depending on the type of reactive group present in the dye, the result is a more ether- or ester-like bond to the fibers. Aliphatic reactive groups substitute preferentially at the C-2 hydroxyl group, forming ether links, whereas triazine derivatives form esters at C-6 hydroxyl groups of the cellulose.

The basic equation is as follows:

$$\mathrm{Chr-R-X + HO-Cell \longrightarrow Chr-R-O-Cell + HX}$$

where $\mathrm{Chr}$ = chromophore, $\mathrm{Cell}$ = cellulose, $\mathrm{R}$ = supporting (bridging) group, and $\mathrm{X}$ = leaving group

Thus,

**MCT:**

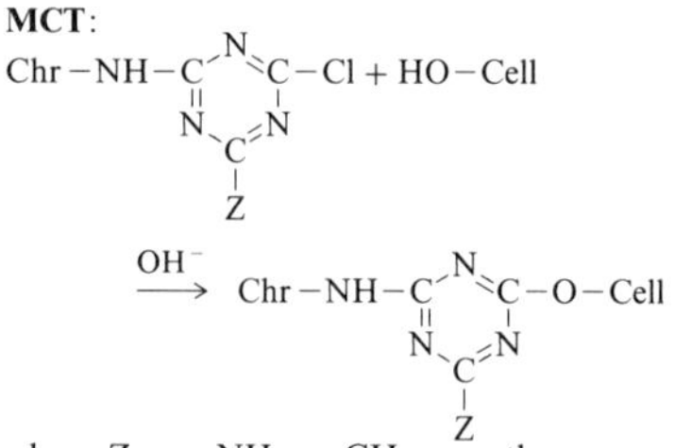

$$\mathrm{Chr-NH-C}\begin{smallmatrix}\mathrm{N}\\ \\ \mathrm{N}\end{smallmatrix}\mathrm{C-Cl + HO-Cell}$$

$$\xrightarrow{\;\mathrm{OH^-}\;}\; \mathrm{Chr-NH-C}\cdots\mathrm{C-O-Cell}$$

where $\mathrm{Z} = -\mathrm{NH_2},\ -\mathrm{CH_3}$, or other groups

**SES:**

$$\mathrm{Chr-SO_2-CH_2-CH_2-OSO_3H} \xrightarrow{\;\mathrm{OH^-}\;} \mathrm{Chr-SO_2-CH=CH_2}$$

$$\mathrm{Chr-SO_2-CH=CH_2 + HO-Cell}$$

$$\xrightarrow{\;\mathrm{SES}\;} \mathrm{Chr-SO_2-CH_2-CH_2-O-Cell}$$

SES dyes must first be converted to the vinyl-sulfone form in alkaline medium before they can participate in an addition reaction with cellulose.

## 5.10.3. Important Reactive Dyes

The reactive dyes most important in textile printing include:

*MCT Dyes:* Basilen P, Cibacron P, Drimaren P, Kayacion P, Levafix PN, Procion P
*SES Dyes:* Remazol, Sumifix

Reactive dyes for textile printing are chosen on the basis of reaction rate, fixation properties, and stability of the corresponding pastes. They react with cellulose in the presence of an alkali such as sodium carbonate or sodium hydrogencarbonate. Application conditions depend upon the alkali used, the reactivity of the dye, and the fixation temperature. Reactive dyes react with hydroxyl compounds generally, so hydrolysis by water competes with the fiber–dye reaction, and a certain amount of hydrolysis is unavoidable. For this reason, moderately reactive dyes are the only ones used in the one-phase fixation processes for textile printing. More highly reactive ranges are in turn applicable only in the two-phase processes (see Section 5.10.7).

## 5.10.4. Pretreatment of the Fabric

Achieving a fiber–dye bond and intense, brilliant, clear colors requires that the fabric first undergo a thorough pretreatment. It should be completely desized, bleached, and given good wetting properties. Mercerization is recommended with cotton, and regenerated cellulose should be treated with alkali. Viscose staple fiber should be treated with 3 % sodium hydroxide at room temperature under tension-free conditions. The mercerized or alkali-treated material should be well rinsed, but not subjected to acid treatment. Further pretreatment with urea and/or alkali should be avoided. Although such a treatment might sometimes lead to an increase in color yield, serious problems can develop in other directions. Fabrics so treated may yellow in the course of steaming, and the moisture–temperature equilibrium of a steamer can be upset by the presence of large quantities of urea. Urea removes large amounts of moisture from the steamer, causing the steam to become superheated. The temperature then rises, and after only a few meters of production there is a noticeable decrease in color (or fixation) yield. This can be remedied only by the introduction of excess steam. Large quantities of urea also present an environmental problem. (For details of the conditioning process see Sec. 5.10.6.)

## 5.10.5. Printing Pastes

**Thickening Agents.** Reactive dyes may also react with thickening agents if these are similar in structure to cellulose. Starches, starch derivatives, tragacanth, and similar materials are therefore unsuitable as thickeners, causing the fabric to become harsh to the touch and producing poor fastnesses. Commercial grades of alginate are used instead, specifically sodium salts of alginic acid obtained from seaweed. High-viscosity sodium alginates are preferred for reasons of cost. Low- and medium-viscosity sodium alginates offer the best dye leveling and dye yields, especially with regenerated cellulose. Since sodium alginates are natural products they are subject to biological degradation, with a consequent loss of viscosity on prolonged storage even in the powdered state. In the colloidal state (i.e., in aqueous medium) a viscosity decrease due to bacterial attack occurs within a few days, which makes it necessary that preservatives be added to the printing paste. Sodium alginate swells in water, but forms precipitates with the substances responsible for water hardness. For this reason the swelling process (often incorrectly referred to as "dissolving") should be permitted to proceed only in soft water. For added safety, a sequestering agent (polyphosphate) is added to the system to combine with and mask any alkaline-earth or heavy-metal ions.

Apart from sodium alginate, so-called *half-emulsions* of the oil-in-water type are often used, but their importance is diminishing for environmental reasons. Half-emulsions are widely employed in markets where cost and environmental considerations are secondary because they result in high degrees of brilliance, color intensity, and dye leveling, and they are easily removed from the print by washing.

In the Asiatic world, carboxymethylcellulose (CMC) has also been used successfully. Suitable grades of CMC must have a very high degree of substitution ($> 2$); i.e., all the reactive hydroxyl groups should be inactivated by substitution.

Synthetic thickeners have been utilized for some time as well. These are sodium salts of poly(acrylic acids), and they have the advantages of producing very brilliant and intensely colored printed results, being virtually immune to bacterial attack, and existing as standardized industrial products available in ample quantity. Synthetic thickeners are sensitive to the presence of electrolytes, however, which cause the viscosity of the paste to decrease. The concentration of thickener must therefore be varied in accordance with the dye concentration. Also, synthetics may confer a harsh "hand" (feel) because they are difficult to wash out. Synthetic thickeners are not soluble in water, but merely swell, so a relatively high degree of agitation is required in the washing equipment to remove them without leaving a residue. It is therefore preferable to use synthetic thickeners in combination with sodium alginates.

**Urea.** Reaction between a reactive dye and cellulose occurs in a liquid phase, which permits the dye to diffuse from the printing paste into the fibers. The function of urea can be regarded as that of "binding" the water; i.e., it causes a certain amount of steam to condense on the print due to urea's negative heat of solution. The resulting aqueous urea phase, in which the dye is dissolved, functions as a dyeing medium. Steam fixation requires the use of 50–80 g/kg of urea for cotton and 100–150 g/kg for viscose rayon. Hot-air fixation is suitable only for cotton, and it requires larger amounts of urea. However, the high nitrogen content of urea makes it a threat to the environment. Urea passes through a sewage treatment plant unchanged, and may contribute to hypertrophy in surface waters. For this reason urea must be introduced at the lowest concentration possible (see also Section 5.10.6.).

**Alkali.** The reaction of a reactive dye with fibers occurs only in an alkaline medium at a rate that is a function of both pH and temperature. Roughly speaking, the reaction rate increases by a factor of ca. 10 with an increase of 1 pH unit or 20 °C. As the reaction rate increases, the rate of hydrolysis by water (the competitive reaction) also increases, which is equivalent to a reduction in the stability of the printing paste. With highly reactive dyes (SES, DCC, FCP, DFT, etc.) applied by the one-phase method, only weak bases (sodium hydrogencarbonate, sodium trichloroacetate, etc.) are used. For dyes of medium reactivity (e.g., MCT) sodium carbonate is appropriate, offering a printing-paste stability of 3–4 weeks at ca. 20 °C. If MCT dyes are used with sodium hydrogencarbonate a yield reduction of ca. 10 % must be anticipated with intense colors. Application of pastes free of alkali is described in Section 5.10.7.

**Oxidizing Agents.** Reactive dyes are more or less sensitive to reduction during steaming depending on the nature of the chromophoric group. The effects of reduction can be countered by adding weak oxidizing agents, such as derivatives of benzenesulfonic acid (Ludigol, BASF). Light colors are especially sensitive to steam fixation. The risk of reduction is generally lower with hot-air fixation.

**Preparation of the Printing Pastes.** Most reactive dyes are quite soluble, so reliable and rapid incorporation into a printing paste can be accomplished by either the premixing or the sprinkling method. In the latter approach the powdered dye is fed slowly with rapid stirring into a mixture containing all the chemicals and thickeners necessary for fixation. Several manufacturers also supply reactive dyes as liquid formulations that are very easy to handle and are suitable for processing in direct dispensing systems known as automated dye "kitchens."

Typical formulations for 1000 g of paste are as follows:

1) Sodium alginate system: water, 50–150 g of urea, 10 g of Ludigol, 15 g of sodium carbonate (25 g of sodium hydrogencarbonate), 450 g of sodium alginate (4%), and dye
2) Semiemulsion: water and chemicals as above, 150 g of sodium alginate (4%), 10 g of emulsifier, and 400 g of kerosene
3) Synthetic thickener: water and chemicals as above, 150 g of sodium alginate (4%), and 40–70 g of synthetic thickener (dispersion)

All these formulations apply to single-phase fixation, the so-called "all-in method."

### 5.10.6. Fixation (One-Phase Method)

The fixation rate is determined principally by the reactive group. Interaction also takes place between the supporting group and the chromophore, however, and the diffusion rate of the individual dye plays a part as well, so reactive dyes may display different reaction rates with the same reactive system. It is well known that yellow (azoic) dyes of low molecular mass are fixed more rapidly than the bulky turquoise (phthalocyanine) types. The fixation time must therefore be adapted to the slowest type of dye used. With MCT reactive dyes in a sodium carbonate medium, exposure to saturated steam at 102 °C for 8–10 min or hot air at 160 °C for 3 min is sufficient. Highly reactive systems could be fixed in less time, but these are seldom utilized in all-in pastes because the storage characteristics of such pastes are unsatisfactory. In practice, reactive-dye prints are usually treated with saturated steam in short-loop steamers for 8 min. Hot air is less often used because it is suitable only for cotton and less satisfactory for regenerated fibers (low yield). Regenerated cellulose requires for optimum dye fixation a relatively high moisture content, and this must be achieved during the fixation process. The problems associated with addition of large amounts of urea have been described in Sec. 5.10.4. In practice, printed goods are often conditioned by external application of moisture immediately prior to steaming. This can be accomplished by spraying according to the Weko rotor principle, foam application by the Stork system, or by contacting the back of the fabric with a wet cloth, the water content of which is controlled (e.g., with equipment of the type supplied by Zimmer). In order to ensure the greatest possible reproducibility, moisturizing equipment of this sort should be subject to hygrometric monitoring and control.

### 5.10.7. Two-Phase Application

Reactive dyes with high reactivity are applied in alkali-free printing pastes because of their low stability. Such pastes consist only of sodium alginate, oxidizing agent, and dye; urea is not required. After printing and drying, the alkali is applied as a separate second step (second phase). Application may occur by one-side "blotch" or two-sided "pad" application of the alkaline liquor. The print is then fixed without intermediate drying in a single operation, through a brief steaming process (20–40 s). Shock fixation by passage through a heated (< 90 °C) alkaline bath is also possible, as is cold-pad batch application with swelling over the course of several hours at room temperature. In all cases, there must be a high electrolyte concentration in the alkaline liquor so that the dye will not bleed, but rather be "pressed" onto the fibers. Suitable additives for this purpose include sodium sulfate or sodium silicate (which also acts as an alkali).

Typical alkali baths are based on 180 g/L of sodium sulfate, 150 g/L of sodium carbonate, 50 g/L of potassium carbonate, and 100 ml/L of NaOH, $d = 1.36$ (38° Bé); the alternative is 800 ml/L of sodium silicate, $d = 1.53$ (50° Bé) and 200 ml/L of NaOH, $d = 1.36$ (38° Bé).

MCT reactive dyes are less suitable for two-phase application. Good color intensities are obtained with regenerated cellulose, but the touch of the fabric may suffer. A steaming step prior to alkali application may prove beneficial in this respect, although there is no theoretical explanation for the phenomenon. The two-phase printing process is questionable from an environmental standpoint because of the large amount of water required for the final wash.

### 5.10.8. Washing of the Prints

The washing of textiles printed with reactive dyes is less problematic the lower the substantivity of the dyes, as this reduces the likelihood that dye will be absorbed from the wash liquor onto the white ground. If the dyes are chosen properly, thorough cold rinsing is enough to remove most of the thickener and dye. This is followed by a boiling step. Detergents are unnecessary with a sufficiently high bath temperature ($> 90\,°C$). Because of the incompatibility of alginate thickeners with hard water, soft water must be used for rinsing, and a complexing agent must be added in the boiling step. Prints treated with cationic aftertreatment agents are subject to the risk of mordanting effects, in which dye bleeds or is absorbed out of the wash liquor onto the white ground, which is covered with cationic substances. This is especially true of prints sewn onto monocolor articles and subjected to wet storage.

### 5.10.9. Fastness

Apart from the colorfastness associated with a given article under normal use, there is one unusual fastness phenomenon associated specifically with prints based on reactive dyes: under unfavorable conditions, the fiber – dye bond may undergo "splitting" (cleavage), permitting dye to bleed onto the white ground. This splitting may be thermally induced (thermal hydrolysis, thermal cracking), or it may occur in an acidic or alkaline medium (acid or alkaline hydrolysis). MCT dyes are sensitive in the acidic range, SES dyes in the alkaline range. Dye splitting can be initiated by excessive drying, excessive acidification during washing, the use of inappropriate catalysts at the finishing stage, condensation at temperatures that are too high, or storage in an acidic atmosphere.

### 5.10.10. Special Printing Processes

**Reactive Discharge Processes.** Discharge printing on a reactive dye ground is of great importance worldwide. Vat dyes are used to produce the printed colors because the reducing agent (sodium formaldehyde sulfoxylate) and alkali (potassium carbonate) required for their development act as the discharge medium for the reactive dye ground. Discharge paste, with or without vat dye, is first printed onto the reactive dye ground, dried and then steamed immediately in the absence of air.

Typical discharge paste: 100 g/kg of $TiO_2$ (white) or vat dye, 150 g/kg of Rongalit C, 90 g/kg of potassium carbonate, and thickener (seed-flour derivative).

The ground is usually dyed with SES dye, which is especially sensitive to alkali and is easily and completely discharged without any reaction residue being left on the fabric. Dischargeable reactive dyes with a fiber – dye bond that is more stable under alkaline conditions may leave behind decomposition products that lead to subsequent oxidative yellowing. Such dyes are usually split off reductively at the azo bridge.

**Resist Printing.** *"Pigment-under-reactive"* resists with the usual organic acids (tartaric and citric) have never become well established because of poor application properties associated with the pigment paste (leading to premature cross-linking of the binder) and unsatisfactory resist effect. However, the use of an organic acid water-soluble only in the form of its ammonium salt (e.g., a pyrazolonecarboxylic acid derivative) permits optimal effects to be achieved. A typical system consists of a normal pigment paste to which is added ca. 20 – 40 g/kg of seed-flour derivative, 30 – 40 g/kg of 25 % ammonia, and 40 – 60 g/kg of Levalin 46040 (Bayer) as the resist agent. This paste is preprinted and then overprinted without intermediate drying ("wet-in-wet") using a normal alkaline reactive paste (MCT or SES), after which the fabric is dried, fixed with hot air for 3 min at 160 °C, and finished in the usual way.

*Reactive/reactive* resists can be obtained by blocking SES reactive dyes with stabilized sodium hydrogensulfite compounds (BASF resists) under MCT reactive dyes. A 30 – 40 g/kg of BASF resist is added to the MCT printing paste (standard formulation), which contains sodium carbonate. This paste is preprinted, and, without intermediate drying, wet-in-wet overprinted with an SES printing paste (sodium hydrogencarbonate, standard formulation). It is then dried, fixed in saturated steam, and finished in the usual way.

**Crepe Effects.** Crepe articles can be prepared by overprinting cotton—which has not been subjected to mercerization or alkali treatment—with a solution of sodium hydroxide. The fabric shrinks in the areas printed with sodium hydroxide, thus producing the desired crepe effect. The selected dye must be characterized by a strong fiber – dye bond in alkaline medium (unlike SES, for example).

## 6. Printing on Synthetic Fibers

Pigment printing is discussed in Chapter 8. Thickeners mentioned in this chapter are discussed in Chapter 4; for other relevant substances → Textile Auxiliaries.

Fabrics based on regenerated cellulose are printed with the same dyes used for cotton (see Chap. 5). However, dyes must be carefully selected for use with fibers derived from cellulose esters (→ Cellulose Esters) and synthetic polymers (→ Fibers, 4. Synthetic Organic), and it is sometimes necessary that dyes be developed specifically for a particular situation.

Compared with natural fibers, man-made fibers based on cellulose esters and synthetic polymers create special dyeing problems associated mainly with the constitution of the fiber material. Other factors are important as well, however, including delustering of the spinning material, fiber cross-section, the spinning processes themselves, and processing to produce filament or staple fiber yarn. The thermoplastic character of synthetic fibers in this group also limits the maximum treatment temperature in processes such as fixation.

Unlike natural fibers, man-made fibers (with few exceptions) are relatively hydrophobic, so dyeing is generally more difficult.

The nature of the textile surface to be printed (woven, knitted, or tufted) also has a significant effect on properties. For example, a closely woven fabric made of polyamide filament behaves like "umbrella silk" and is very hydrophobic. On the other hand, the same polyamide raw material can be used to produce carpets that pick up so much moisture during the printing process that an intermediate drying stage is unnecessary prior to dye fixation by steaming, and no running is observed with pattern contours. This effect is less unusual than it may appear. After all, a rubber sheet is extremely hydrophobic, whereas a rubber sponge can absorb many times its own weight of water.

### 6.1. 2,5-Acetate

Developed about 1925, 2,5-acetate was the first of the fibers to differ significantly in its physico-chemical behavior from the natural fibers used exclusively until that time. Nevertheless, this pioneering material has now been largely replaced by other man-made substances.

Acetate in both woven and knitted form is used for dresses, blouses, linings, and scarves. Its current share of the total volume of printed products is significantly less than 1%.

**Pretreatment.** It is important to note that acetate fiber is sensitive to the action of alkali at temperatures above 85 °C. Partial fiber saponification occurs under these conditions, leading to loss of luster and strength.

Acetate fibers are today sized only with water-soluble preparations, which means that they need only be prewashed at 60 °C with a solution of an anionic detergent (ca. 1 g/L), such as Levapon TH (alkylsulfonate in combination with polyglycol ethers).

The fiber is usually sufficiently white from the outset, so it is seldom bleached. If brightening is necessary (→ Optical Brighteners), the usual choice is a dispersion brightener, such as Blankophor DCR (Bayer) or Hostalux AR (Hoechst). This is applied after the tentering process or after steam padding. The final white shade first becomes apparent following steaming of the printed product.

**Printing Processes.** Printing can be carried out either on tables or with roller or rotary-screen machines. Depending on its quality, the fabric may be glued directly to the undercloth or provided with a back gray.

**Dyes** [6.1] **and Printing Pastes.** Woven and knitted acetate fabrics are today printed almost exclusively with select *disperse dyes* (→ Disperse Dyes). Liquid preparations are preferred, since these are easy to incorporate into printing pastes. Predispersion is unnecessary.

The dyes are selected according to their color shade, lightfastness, wetfastness, sublimationfastness, and fabricationfastness. The relative importance of the various criteria differs from one finished article and fiber to another.

Some dye manufacturers have applied special designations to disperse dyes particularly well-suited to acetate. Examples include the Celliton (BASF) and Cibacet dyes (Ciba-Geigy).

**Fixation.** Disperse dyes (and sometimes cationic dye as well) are fixed by treatment with saturated steam in a star or hanging-loop steamer at 102 °C for 20–40 min. The steam should not be too moist, since the pattern might otherwise run.

**Aftertreatment, Fastness Properties.** The fixed prints are thoroughly rinsed, washed with an anionic detergent at 40–50 °C, and again rinsed until the water is clear.

The fastness of disperse and cationic dyes is only modest. What is known as "fumefastness" can be a problem with some anthraquinoid dyes, but it is subject to improvement by addition of GFN Inhibitor (BASF)—an organic base [6.2]—to the printing paste (see Sec. 6.2).

**Discharge Printing.** Textiles dyed with selected disperse and cationic dyes can be used as a ground for discharge printing. The best white discharge results are obtained with Decroline (zinc formaldehyde sulfoxylate [24887-06-7], → Textile Auxiliaries, pp. 297–298). With color discharge dyeing, tin(II) chloride is used as the discharging agent because of the wider range of dyes resistant to its effects. Processes are also known based on a tautomer of thiourea dioxide (formamidine sulfinic acid [1758-73-2], Manofast).

**Other Classes of Dyes.** *Acid dyes* and *metal-complex dyes* are subject to fixation without steaming. Instead, advantage is taken of the swelling of 2,5-acetate fibers in solvents. A typical 1000-g printing paste formulation consists of 10–60 g of dye, 30 g of urea, 280–330 g of water, 130 g of a solvent mixture, and 500 g of thickener solution. The solvent mixture is made up of 30 g of Glyezin A (thiodiethylene glycol, [111-48-8]), 50 g of benzyl alcohol, and 50 g of ethyl alcohol, whereas the thickener solution (1000 g) contains 40–50 g of 2-hydroxyethyl methyl cellulose [9032-42-2] (Tylose MH), 600 g of ethyl alcohol, 280–290 g of water, and 70 g of ammonium rhodanide [1762-95-4].

Fixation occurs during drying at room temperature or with warm air. Because of its high alcohol content, the printing paste must be kept away from open flames, and the prints may not be dried with infrared radiation. In the final washing process, alcohol previously introduced as a swelling agent is diluted and removed. Dye remaining on the fiber displays a relatively high degree of fastness. White grounds are also not subject to bleeding. However, the brilliance of the resulting prints is somewhat inferior to that obtained with disperse dyes.

**Blends with Other Fibers.** Acetate blends are sometimes produced for reasons of fashion, including the blends with polyamide known as Rhodialon, which are printed with disperse and cationic dyes just like pure acetate. In the case of dark colors (navy blue, wine-red, black) the final wash is complicated by the presence of the polyamide, and the white ground is very easily stained. Nevertheless, very fashionable discharge prints have been achieved with precisely this fiber blend. The ground is dyed with selected disperse dyes, whereas cationic dyes are used for the color-discharge dye system, with a tin(II) salt as the discharge agent.

## 6.2. Triacetate

Triacetate is more resistant to alkalis relative to 2,5-acetate, and it also has less tendency to crease. It occupied an important position in the 1960s in the production of printed fabrics for fashionable clothing. At that time up to 10 % of the prints produced in certain industrialized countries were made of triacetate. Today, it has been largely replaced by polyester, though a small amount is still used in blends with polyamide (e.g., under the name Tricelon).

Triacetate fiber resembles acetate fiber in several of its mechanical and chemical properties, but it is more like polyester in its absorption capacity with respect to disperse dyes.

**Pretreatment.** *Woven* fabrics are first washed with ca. 1 g/L Levapon AN or TH liquid at ca. 60–80 °C, usually in mildly alkaline medium, and subsequently rinsed, all under tension-free conditions if possible. The fabric is then given an "S-finish" whereby a surface soaping produces a delustering effect, also serving to improve the feel, increase fastness toward exhaust gas for subsequently applied dyes, and provide a wash-resistant antielectrostatic finish. However, not all fabrics are subjected in practice to an S-finish because it can reduce their capacity to absorb disperse dyes. This problem does not arise if the dye is fixed in a pressurized steamer (saturated steam at 2.5 bar), but use of a rapid or high-temperature steamer results in a significant loss in color intensity.

Triacetate knits processed on a large scale are washed in full-width low-tension facilities. Temperatures close to the boiling point are required for stabilization. On a smaller scale, book-form or high-temperature equipment is appropriate. The optimum fixation temperature is 130 °C.

Knitted fabrics do not normally receive an S-finish.

**Printing Processes.** Printing is carried out on tables or with roller, rotary-screen, or flat-screen printing machines (see Sections 3.3 and 3.4.2). Especially with discontinuous table and flat-screen printing, the textile must be held firmly to the table by adhesives applied directly or adhesives applied to a previously attached back gray.

**Dyes** [6.3], [6.4] **and Printing Pastes.** Nearly all printing on triacetate is accomplished with specially selected disperse dyes, available from all the major dye producers.

*Fixation Accelerators* (→ Textile Auxiliaries, pp. 298 – 299). The range of available color-intensifying fixation accelerators (auxiliaries) is greater with triacetate than polyester. Even thiourea is effective with some dyes. Various manufacturers offer fixation auxiliaries, but their effectiveness must always be tested as a function of fabric, pretreatment, dye, and fixation process. It should also be noted that many such products fail to dry completely on the fabric, so a film of printing paste may smear if too much fixation accelerator is used. Examples of fixation auxiliaries include Luprintan HDF and PFD (BASF), and Sandotherm ACS liquid (Sandoz) [6.2] (→ Textile Auxiliaries).

*Fastness toward Nitrogen Oxides ("Fume Fading").* This can be an important consideration with both triacetate and 2,5-acetate. A number of disperse dyes on these fibers display more or less marked color change on exposure to nitrous gases like those present in the off-gas from industry and coal combustion plants. Fastness toward nitrogen oxides can be tested by the method prescribed in DIN 54025 and ISO 105601, and it can be improved if necessary by addition to the printing paste of inhibitors, such as GFN Inhibitor.

**Fixation.** Disperse dyes printed on triacetate can be fixed by various methods:

1) Continuous steaming with saturated steam at 100 – 102 °C for 30 – 45 min. Under these "mild" conditions, addition of appropriate fixation auxiliaries considerably improves the color intensity of many dyes.

2) Steaming under pressure in a bell or kettle steamer for 20 – 30 min with saturated steam at 2.5 bar (ca. 127 °C). Optimum yields are obtained without a fixation accelerator, and

the presence or absence of an S-finish has no effect on the results. The disadvantage is the need for batch operation.

3) Continuous fixation with superheated steam (HT-steam fixation) and a steaming time of 6 – 8 min at 185 – 165 °C. Experience shows that temperatures must be somewhat higher in this case than for the same dyes on polyester. Only moderate dye yields are obtained in the absence of fixation accelerators. Disperse dyes on fabric to which an S-finish has been applied cannot be adequately fixed by this method. Also, the high temperatures involved require use of dyes that are stable to sublimation.

4) Continuous thermosol fixing with hot air at 190 – 200 °C for 60 – 45 s. The comments above concerning HT-steam fixation apply here as well.

**Aftertreatment.** Care is necessary with the final wash after steaming to discourage dye from bleeding onto the white ground. Unfortunately, no way has yet been found to prevent this completely. It is therefore recommended that the fabric first be rinsed cold, then warm, and finally washed at ca. 60 °C. As soon as the washing bath becomes contaminated with dye it must be changed immediately. In order to help keep the ground white it is recommended that small amounts of hydrosulfite (1 – 2 g/L) and sodium hydroxide solution (1 – 2 cm$^3$/L) be added to the wash baths. The bath temperature should not exceed 50 °C to prevent attack on dye already fixed in the triacetate fiber, which is more strongly hydrophilic than polyester.

After washing, the fabric is rinsed thoroughly both warm and cold, perhaps neutralized, and then dried.

**Discharges and Resists.** At the time that triacetate was important in textile printing, discharge and resist printing were rarely employed (for reasons of fashion), so no appropriate process of this type was ever developed. Nevertheless, the processes described for polyester should in principle be applicable.

**Blends Containing Other Fibers.** The only triacetate blends available are combinations with polyamide. Marketed under the trade name Tricelon, these can be printed satisfactorily with select disperse dyes. Although such a print can be fixed with high-pressure steam, continuous steamers based on saturated steam are usually employed instead. The use of HT-steamers is not recommended for these articles, especially if the fabric has been subjected to an S-finish.

The same chemicals are used as fixation agents here as with triacetate itself. Excess fixing agent leads to running of the pattern. Optimum amounts depend on the construction of the fabric, and should be ascertained on the basis of preliminary tests. Fixation in high-pressure steam (20 min, 2.5 bar) avoids the need for a fixing agent.

Special care is required when washing the steamed fabric. Especially when large areas have been printed in dark colors there is a danger that the polyamide fraction of the unprinted surface will be stained with excess disperse dye. One should therefore use a long liquor ratio in the first rinse, and the water should be changed rapidly. Alternatively, the fabric might be thoroughly sprayed beforehand.

## 6.3. Polyamide [6.5], [6.6]

Dyeing characteristics of polyamide fibers are described elsewhere ($\rightarrow$ Textile Dyeing, pp. 428–434).

The printing of fabrics made of polyamide first became important with the development of dimensionally stable warp-knitted fabrics made from filament yarns. Woven and knitted fabrics based on polyamide 6 (made from $\varepsilon$-caprolactam) and polyamide 66 (from hexamethylenediamine and adipic acid) are especially suitable for printing, whereas other polyamides (e.g., polyamide 11, from $\omega$-aminoundecanoic acid) play no role whatsoever because of their poor dye-absorption properties.

For some years, so-called nonelastic knitted goods (porous Perlon or Nyltest) attracted the greatest interest. These were followed by articles with a velour effect. Fabrics of this type are of little importance today, at least in the textile-printing industry of Central Europe. Other products have retained their importance, however, including polyamide locknit fabrics, polyamide taffeta in the form of "umbrella silk," and tubular knitted goods for texturized swimwear. The largest area of current application is the carpet sector.

### 6.3.1. Pretreatment

Woven and knitted fabrics made of polyamide fiber must be subjected to several pretreatment steps prior to printing.

Two processes are common with knitted shirt materials and locknit fabrics. Thus, if tentering frames are to be used in the thermofixation process (hot air fixation), a tension-free jigger is required for prewashing and optical brightening. If an HT (high temperature) facility is available, prewashing, thermofixation, and (if necessary) optical brightening are carried out as a single-bath process in the appropriate HT equipment (hot water or hydrofixation).

In the first process, which is also well-suited to woven polyamides, the raw goods are washed in the jigger at 40 °C and thermofixed on the tentering frame (polyamide 6 at 192 °C, polyamide 66 at 210 °C). Grades that have not been brightened during spinning are brightened in the jigger and dried on the tentering frame.

In the case of hydrofixation in an HT facility, the fabric is prefixed on a tentering frame at 160–170 °C, wound onto the perforated beam, washed in the HT apparatus, brightened if necessary, thermofixed (nonelastic knitted goods made of polyamide 6 at 120 °C, Nyltest at 130 °C), and then dried on the tentering frame and pressed.

Thermofixation in an HT facility is especially recommended for knitted shirt fabrics made of polyamide. This leads to more stable fixation, but may also contribute to the appearance of dreaded moiré (optical) effects. The latter are most reliably countered by careful winding onto a beam as a precursor to thorough prefixing. In the absence of prewashing, however, a prefixing can easily lead to graying as a result of diffusion into the fibers of spinning lubricant and other impurities. The best defense against this problem is a full-width wash on the tentering frame prior to prefixing.

A similar pretreatment is appropriate for polyamide velours. The velour effect is produced by a raising machine, preferably after washing and perhaps optical brightening, but before thermofixing on a tentering frame and printing. The process causes the fabric to decrease in width by as much as 50 %, although the length remains virtually unchanged. Printing in advance is impossible due to subsequent pattern distortion.

Optical brighteners may impair the lightfastness of a printed product. For this reason polyamides that have been brightened at the spinning stage are always preferred. If a polyamide must be brightened later in the process the only auxiliaries that can be recommended are those that display the greatest lightfastness (e.g., Blankophor CL or CLE).

*Washing Agents for Precleaning.* Both nonionic and anionic detergents are used for washing the various polyamide products to remove pretreatment materials and soil. Alkylarylpolyglycol ethers are the most widely used of the nonionic detergents. Phosphates are the preferred alkalis for washing, since these help to preserve a soft feel, unlike sodium carbonate, which makes the fabric harsher.

### 6.3.2. Printing Processes

Pretreated woven and knitted polyamide fabrics are printed by roller and screen-printing machines, as well as on tables. Polyamide velour knits are preferentially printed on roller and rotary-screen printing machines. Texturized polyamide tubular knitted goods, polyamide warp-knitted velour, and polyamide locknits, which display high elasticity and poor dimensional stability, are printed almost exclusively on screen-printing tables and machines. They must be passed through the machines by the shortest possible route with as little tension as possible to prevent distortion. In many cases it is useful to attach the fabric to a cotton back gray.

Nonelastic warp-knitted products such as Nyltest or porous Perlon are less problematic, since they acquire a special "fringe" that confers dimensional stability.

If polyamide knits are glued directly onto a rubber undercloth, wet expansion of the polyamide fibers can become a problem. These fibers may expand by as much as 4 % when moist, so a heavy first coating with dye can very easily lead to the formation of bubbles and wrinkles, preventing the immediate pursuit of subsequent printing steps. Knitted goods behave more satisfactorily if their width is extended by ca. 4 % during the last pass through the tentering frames, and if they are glued to the screen-printing table at this width. Such fabrics remain firmly fixed to the table even when printing multicolored patterns that cover the surface completely, and printing proceeds without problems.

### 6.3.3. Dyes and Printing Pastes

Direct printing is accomplished with selected metal-complex and disperse dyes, as well as reactive (Sec. 6.3.7) and occasionally substantive dyes. Apart from the required high degree of fastness, dyes are selected according to their color, solubility, and stability.

The solvents used with all types of dyes are thiodiethylene glycol [*111-48-8*] (Glyezin A) and urea. Urea, which can be replaced by thiourea, increases both the dye solubility and the color intensity of the print. Maximum concentrations of thiodiethylene glycol and thiourea quoted in pattern books must not be greatly exceeded, since otherwise the printed patterns may easily run. Other solubilizing agents include diethyleneglycol monobutyl ether [*112-34-5*] (Glyezin BC) and, in special cases, cyclohexanol.

Ammonium sulfate is added to stabilize the pH between 5 and 3, the region in which polyamide fibers display optimum dye-absorption characteristics. This is especially important with acid dyes, whereas metal-complex dyes are sometimes printed without such an addition.

Because polyamide fibers absorb moisture to a relatively small extent compared with cellulose, the situation is different with respect to thickening agents. In the case of polyamides, water soluble dyes together with a larger amount of solvent must be firmly retained on a less absorbent material during the steaming process, but without any running taking place. A satisfactory thickener must therefore have a high solids content.

Medium- to low-viscosity thickeners, or mixtures thereof, are preferred. High-viscosity products are unsuitable. It is particularly difficult to produce a uniformly printed coating on a closely woven fabric like umbrella silk.

The thickeners with the most satisfactory properties are etherified seed-flour and guar products in combination with degraded starches. The solids content should not be less than 12–16 %.

### 6.3.4. Fixation

After drying, the prints are fixed in steam. For woven and nonelastic knitted products, a continuous or star steamer is appropriate. Printed Perlon warp velours, Nylon velours, polyamide locknits, and polyamide texturized tubular knitted goods are fixed in a star or kettle steamer or a hanging loop steamer. In continuous steamers of the Mather & Platt or Krostewitz type they tend to become curled and stained. Minimum steaming conditions for maximum dye yield are 20–30 min steaming time at ca. 102 °C. In all cases, shorter steaming times lead to lighter color shades.

The best fixing results are obtained in a star or kettle steamer at 1.5 bar. With approximately the same steaming times, the resulting colors are ca. 10–20 % more intense.

### 6.3.5. Aftertreatment [6.7]

Aftertreatment processes for polyamide include washing and measures designed to improve fastness. Washing removes excess dye, dye-paste thickener, and chemical additives from the fabric. Constituents of the printing pastes are sometimes difficult to eliminate by washing, and any residues left behind can easily lead to bleeding onto the white ground. The extent of unfixed dye can vary greatly, and depends upon the extent of the coated area, the amount of dye applied, the weight of the fabric per square meter, and the nature of the paste thickener.

Fabric made of polyamide 6 is capable of taking up more dye than polyamide 66 under equivalent conditions. The dye substantivity of both polyamides can be improved considerably by increasing the steam pressure and temperature during the dye-fixation process, which means that the extent of unfixed dye is also a function of conditions.

Unfixed dye can be removed by cold rinsing, washing at 40–45 °C, and rinsing again.

Bleeding onto a white ground can be prevented by addition to the rinse of certain auxiliaries (2–3 %), usually condensation products of formaldehyde with aromatic sulfonic acids (Mesitol PS or NBS, Erional NWS; → Textile Auxiliaries, p. 300), and adjustment to pH 3 with formic acid. Such additives block the amino groups of the polyamide, preventing any further absorption of dye. Under these conditions the temperature can be increased to 40–45 °C.

Very heavy fabrics, such as upholstery or velour articles containing spandex, can be washed in a bath containing sodium carbonate and Levapon MR (alkylammoniumpolyglycol ether).

*Improvement of Fastness.* A considerable increase in fastness to moisture can be obtained by further aftertreatment with Mesitol PS or NBS (6 % based on the weight of textile) or Erional NWS at 70–80 °C and a pH of 3, obtained by adding formic acid. The maximum increase in wetfastness requires a treatment time of 15–20 min. While fastness to light and rubbing are not adversely affected, the recommended procedure may lead to a harsher feel, although this can be compensated for with softeners.

### 6.3.6. Discharges and Resists [6.8]

*White Discharging with Decrolin.* The most suitable ground dyes are selected acid and Isolan (1:2 metal-complex) dyes. Some disperse dyes are also used. Although a number of substantive dyes are also very suitable for white discharging, the dyeing of piece goods has been found to be difficult with these materials.

The white effects obtained with Decrolin tend to yellow to a varying extent over the course of time.

Dyed textiles are sometimes given an aftertreatment with Mesitol PS or NBS to improve wetfastness, preferably subsequent to printing and discharging.

*Color-Discharge Printing with Vat Dyes.* For color discharging with selected vat dyes on grounds dyed with acid and metal-complex dyes, Rongalit C and Rongalit DS (→ Textile Auxiliaries) are the preferred reducing agents. Vat dyes with good fixing properties are suitable. In addition to a few anthraquinoids, indigoid and thioindigoid dyes are particularly worthy of note.

It is advantageous to carry out reoxidation of the color-discharge prints at the pH of acetic acid. Reoxidation with hydrogen peroxide and ammonia generally leads to more or less severe changes in the color of the ground.

*Discharging with tin(II) chloride* is the most important of the processes used in conjunction with polyamides. It is applicable to both white and color discharge printing, although the color intensity of the ground must be limited to ensure good white effects.

Prior to treatment with Mesitol PS/NBS or Erional NWS (see above), the print must always be thoroughly rinsed to remove any tin chloride residues. Certain concentrations of tin chloride and Mesitol or similar substances result in precipitates that may have a detrimental effect on fabric feel and the clarity of color shades.

The resist process is rarely applied to polyamides.

### 6.3.7. Reactive Dyes [6.9]

Brilliant colors stable to a boiling wash can be obtained with selected reactive dyes. Reactive dyes are absorbed by polyamide fibers even from mildly acidic printing pastes, subsequently undergoing reaction at pH 5–7. The resulting covalent bonds are very stable toward both alkali and acid. Because of this stable dye–fiber bond, textiles printed with suitable reactive dyes display very good fastness properties.

Suitable thickening agents include high- and medium-viscosity alginates and guar ethers, but not starch-containing thickeners.

*Fixation.* The dye can be fixed by steaming in a continuous steamer for 10 min; a star steamer is of course also appropriate, but with a maximum overpressure of 1.5 bar.

*Final Wash.* Prints are first rinsed thoroughly with a cold solution containing 2 g/L of sodium carbonate (to give a pH of 9) and, if necessary, a complexing agent to remove water hardness. Alternatively, however, the fabric can be passed immediately through a hot alkaline rinse bath and then washed with a liquor containing 0.5–1 g/L of Levapon TH at ca. 80–90 °C. After rinsing both hot and cold, a little acetic acid is added to the final cold rinse to effect neutralization. The washing process must completely remove any unfixed residual dye, since this might otherwise discolor the ground in the acidic medium.

Unlike cotton, polyamide is always washed at a temperature near the boiling point, because unfixed reactive dye on the polyamide surface is retained by a salt-like bond analogous to that of an acid dye, whereas residual reactive dye on cotton is held only absorptively to the cellulose fibers.

### 6.3.8. Polyamide Blends

**Polyamide–Cotton.** Blends of polyamide with cotton or viscose staple fiber are so far of minor importance in textile printing because of serious problems—except in the case of pigment printing—of printing in such a way as to produce adequate wetfastness. Moreover, pigment printing is subject to coloristic limits with respect to color intensity and overprint potential.

**Blends with Other Synthetic Fibers** [6.10]. Mixtures with spandex (elasthan) marketed under the trade names polyamide/Lycra and polyamide/Dorlastan are important in the production of swimwear. Because these fabrics are elastic, and also because of the dye-uptake potential of the polyurethane-based spandex, fibers, special procedures must be followed during pretreatment. For example, the fabric is usually treated with Mesitol PS (0.5–1 % based on the weight of the fabric) even before printing (→ Textile Auxiliaries, p. 300) in order to keep the ground white during the wash that follows printing.

Obtaining good tone-in-tone dyeing results requires careful selection of the dye. Acid and metal-complex dyes are used in a formulation analogous to that employed with polyamide printing. Fixing and final washing are also carried out in the same way as with polyamide.

## 6.4. Polyacrylonitrile

Polyacrylonitriles have a character very similar to that of wool. Their market share with respect to total fiber consumption is small: in textile printing, only ca. 1–2 %. Japan and Italy are the two countries where this fiber is printed extensively, especially for women's outer garments. In Germany it is in steady demand for decorative articles and upholstery, whereas the market for women's outer garments is very much governed by fashion.

The polyacrylonitrile fibers from individual manufacturers vary in their coloristic and technological characteristics. This is partly due to a varying content of acidic groups in the polymer (sulfonic, sulfato, and carboxyl groups), which are essential for fixing cationic dyes. The spinning process (dry or wet spinning) also has an influence on properties. It is even possible to induce an affinity for anionic dyes by incorporating cationic compounds into the polymer.

The flame resistance of polyacrylonitrile is increased by the incorporation of a high proportion (> 15 %) of copolymerized vinyl chloride (Modacrylic fibers; → Fibers, 4. Synthetic Organic, **A 10**, p. 639).

### 6.4.1. Pretreatment

It is important to note with respect to pretreatment that there is always a need to remove the sizing agents (most of which are water soluble) from woven goods, as well as water-insoluble paraffinic compounds from knits. Pretreatment of high-bulked (HB) grades of yarn should lead to an increase in volume, whereas filament and normal yarns are more likely to be stretched, thus requiring that different procedures be followed.

**Woven Goods** [6.11], [6.12]. In general, fabrics with a warp containing polyacrylonitrile filament are sized with water-soluble products, but those in which the warp consists of polyacrylonitrile staple fiber are sized with starch-containing preparations. Starch-containing sizes must be eliminated with appropriate enzymatic agents during pretreatment because unhydrolyzed starches cause hardening of the fabric.

A subsequent wash to remove soil, residual preparations, and other foreign substances is carried out with the aid of anionic or nonionic detergents.

The combined washing–shrinking process for all polyacrylonitrile piece goods is effectively carried out in stainless-steel Mezzera full-width washing and alkali-treatment equipment. Continuous-rope washing machines are suitable only for grades that are not overly susceptible to stretching. Stretching by the reeling machine can be prevented by winding the fabric onto the reel cold. The reel is then stopped, and the bath is heated almost to boiling. To desize the fabric the reel is run briefly a few more times during heating, after which the fabric is allowed to cool for 20–30 min with the reel stationary in the machine. It is then run over the reel for a further 10 min for straightening purposes, followed by acid treatment and rinsing.

Polyacrylonitrile piece goods pretreated by one of the above methods are cold rinsed, dewatered by spinning or suction, dried on tentering frames at 110–140 °C, and smoothed. The fabric is provided with an oppertunity to shrink on the tentering frames by stretching it to slightly more than its wet width and overfeeding by ca. 5 %.

**Knitted Goods** [6.13]. In the 1970s, polyacrylonitrile plain knit (jersey) was produced on a large scale. Pretreatment in this case differs in several respects from that described for woven textiles. Removal of the preparation (paraffinic) and soil requires that the fabric be washed. Suitable detergents include anionic (Levapon TH) and nonionic (Diadavin, polyglycol ether) agents, most of which also contain fat solvents.

Cationic auxiliaries should be avoided, since these adhere readily to the fiber surface, impairing fixation of the dye. It is recommended that the last rinse be adjusted to pH 5–6 with acetic acid, because alkaline media may lead to yellowing of the fiber on steaming, as well as to decomposition of the dye.

Fabrics with severely crimped edges are pretreated in the tubular state, and are cut after spin drying. In order to prevent excessive longitudinal tension and the formation of longitudinal creases, both washing and shrinking are carried out on low-tension wool-reeling machines using a long liquor. Creasing and folds can be avoided only by cooling extremely slowly.

After rinsing and spinning briefly the fabric is dried on a tentering frame at 130–150 °C using maximum possible overfeed, after which it is wound onto a reel. The edges should be glued and cut. Minimum ventilation should be ensured in the tentering frames.

The running behavior of the fabric on a rotary-screen printing machine and in continuous hanging-loop steamers can be improved by padding the knits with an alginate thickener at low concentration after washing and intermediate drying, and then leveling them on the tentering frame.

Whether stretched or placed on a printing table, the fabric must correspond precisely to the intended length, width, weight per square meter, mesh number, and wale count as anticipated for the finished product. After printing, steaming, washing, and brightening these parameters associated with the final stretched product must remain absolutely unchanged, since the printed pattern would otherwise be distorted.

### 6.4.2. Printing Processes

The pretreated fabric can be printed on roller printing machines or flat- or rotary-screen presses.

Especially with roller and flat-screen printing it is useful to glue the knitted fabric to a cotton back gray. Only gluing machines with drying cylinders are suitable for this purpose, because the strength of the adhesive bond between the fabric and the back gray will be inadequate unless it is dried immediately. Suitable adhesives include gum–Cordofan 1:1 (see *gum resins* in Sec. 4.2.3), dextrin–water 1:1, and 15–20 % poly(vinyl alcohol). Partial replacement of water in the solvent by alcohol leads to more rapid drying of the adhesive film on the undercloth. Heavy fabrics can also be glued directly onto the rubber printing blanket or undercloth without use of a back gray, although the rubber blanket must be dried completely before the glue is applied. With a screen-printing machine the adhesive-bond strength can be improved considerably by pressing a dummy screen onto the fabric and then drying in a drying device.

### 6.4.3. Dyes and Printing Pastes

Cationic dyes are of greatest importance for printing on polyacrylonitriles because of their high brilliance, spreading power, and considerably better lightfastness relative to other kinds of fibers [6.14]–[6.16].

The dyes in question (Astrazon, Basacryl, Maxilon, and Remacryl) include the classical triphenylmethane and oxazine dyes along with almost all the other known categories; e.g., anthraquinone, azo, naphthalimide, methine, and cyclammonium dyes (→ Cationic Dyes).

In contrast to the localized charges on sulfonate groups in anionic dyes, the positive charges of cationic dyes are often not localized, instead moving around between several basic groups. Moreover, the relative basicities of these groups may differ greatly. This leads to differences in migration, fixation rate, and fixation yield from one dye to another, as well as differences in the effect of pH on these properties and the formation of more or less sparingly soluble salts with anionic constituents from the printing pastes. For example, dyes with a localized cationic group are fixed more slowly than those in which the charge is not localized.

Polyacrylonitrile fibers also differ with respect to their dye-absorption properties and dyeing rates, so reproducible results can be expected only if the type of fiber, the nature of the fabric, the pretreatment, the paste formulation, and the fixing conditions are all held constant [6.17].

Suitable *thickeners* are based on materials with high solid contents that give clear pattern outlines and optimum color intensities, in conjunction with substances that form an elastic film after drying of the printed product [6.18], thereby suppressing flaking when the textile is plaited down and transported.

If large areas are to be printed and exceptionally uniform coloring is desirable, leveling can be achieved by the addition of carboxymethylcellulose, which acts on the cationic dye like an anionic thickener.

Similar leveling effects can be achieved by adding other anionically active materials to the printing pastes. Suitable additives include anionic dispersing agents, such as the condensation products of naphthalenesulfonic acids with formaldehyde. Alginate thickeners behave similarly. In all cases, however, changes in the shade and intensity of the colors must be anticipated.

### 6.4.4. Fixation

**Fixation with High-Pressure Steam.** Over the entire range of basic dyes, the best dye yields are obtained by fixation with pressurized steam at 1.2–1.5 bar for 30 min. Certain dyes of the azo type are subject to reductive decomposition at temperatures above 115–120 °C, so steam temperatures corresponding to the pressures indicated (104 °C and 110 °C, respectively) should not be significantly exceeded. In special cases involving sensitivity of the printing dyes to steaming, a 30 % solution of sodium chlorate should be added at the rate of 5–10 g/kg shortly before printing. A discontinuous mode of operation is not recommended.

**Continuous Steam Fixation at 102 °C for 30 min.** Rapid steamers of the Mather & Platt or Krostewitz type are inappropriate for fixing light woven fabrics. The corresponding steaming temperatures lie within the softening range of polyacrylonitrile fibers, and tensile stress produces distortions of the fabric that are very difficult to correct. Hardening of the goods is also very common.

Continuous steaming at 102 °C in saturated steam became possible with the introduction of hanging-loop steamers. However, selection of the dyes must take these fixing conditions into account. With deep shades, the addition of Luprintan PFD [*69071-73-4*] is necessary.

*Considerations Related to Steaming.* It is difficult to overstate the importance of constant steaming conditions in fixing water-soluble dyes on synthetic fibers. Unlike cotton and wool, synthetic fibers have only a very low moisture content after drying, so it is almost impossible to compensate for fluctuations in the moisture content of the steam. If the steam is too wet, colors begin to run. If it is too dry, extreme cases may result in a dewatering effect rather than the desired extensive swelling of the thickener. Ion exchange between a dye and the fibers requires the presence of water so that the dye can dissociate. The temperature in the steamer must therefore be monitored carefully throughout the fixing process.

**Hot-Steam and Hot-Air Fixation** [6.19]. Fixation processes of this type have been proposed for cationic dyes on polyacrylonitrile, but they have not been widely used.

### 6.4.5. Aftertreatment

Proper aftertreatment is very important for assuring good rubbingfastness, brilliant colors, and a satisfactory feel. The fabric is first rinsed with cold water containing ammonia or sodium carbonate, washed at 40–60 °C with an anionic detergent and hydrosulfite, and then rinsed and washed in fresh liquor at ca. 60–70 °C. If the thickeners used contain starch that has not yet been hydrolyzed, it is wise to include in the first wash bath an enzymic desizing agent.

After washing, the textile is dewatered by spinning or suction and then dried. The final finish is advantageously applied by padding, followed by tentering. The padding liquor might, for example, contain 5–10 g/L of softening agent at pH 5 (citric or acetic acid). The fabric would then be dried at 100–110 °C on a tentering frame and subsequently pressed.

The feel of polyacrylonitrile textiles can be influenced in any desired direction by appropriate softening agents.

### 6.4.6. Discharges and Resists [6.8], [6.20]

Discharge printing methods are of some importance for polyacrylonitrile fabric intended for use in clothing. Polyacrylonitrile is usually printed in the form of a knit. This means that resist methods are inapplicable because they entail a preliminary padding process. In practice, especially in Italy, color-discharge printing is carried out mainly on dyed or printed grounds with the aid of tin(II) salts.

*White discharge* effects can be produced with either Decroline or tin(II) chloride. The resulting effects are subject to brown discoloration on storage, however, hence these methods are of no practical significance. The alternative is overprinting with a white pigment.

*Color-discharge printing* also takes advantage of tin(II) chloride. The fabric is in this case dyed with selected cationic dyes at the lowest possible temperature ( < 100 °C). Printing pastes for color-discharge printing contain selected discharge-resistant cationic dyes, sometimes in conjunction with metal-complex dyes, as well as 40–80 g of tin(II) chloride per kilogram of printing paste. The amount of tin salt used depends on the color intensity of the ground that is to be discharged.

### 6.4.7. Other Classes of Dyes

Apart from cationic dyes, selected disperse dyes are also used for printing polyacrylonitrile. Thus, sharp pattern outlines can be obtained by mixing an anionically dispersed disperse black with a cationic black. It is also sometimes preferable to use select disperse dyes for bright colors, since these are subject to better leveling. Metal-complex dyes can be used if there is a need for especially light, pale colors with very good lightfastness.

Some types of polyacrylonitrile are modified to make them suitable specifically for dyeing with acid and metal-complex dyes.

### 6.4.8. Blends with Other Fibers

**Polyacrylonitrile–Cellulose.** Blends of polyacrylonitrile fibers with cellulosic fibers have been the subject of experimental printing tests with combinations of cationic and reactive dyes. However, fixation by steaming under acidic conditions followed by an "alkali-shock" treatment for the reactive dyes leads to a risk of yellowing of the polyacrylonitrile.

**Polyacrylonitrile–Wool.** Wool blends are printed with selected metal-complex and cationic dyes. The fabric must afterward be very carefully washed to ensure adequate wetfastness. For this reason, interest has developed in mixtures containing grades of polyacrylonitrile that can be acid dyed, since it is easier to print when only one class of dye is involved.

**Polyacrylonitrile Blended with Other Synthetic Fibers.** Apart from blends with polyester (Sec. 6.5.8), only the blends with PVC and acrylonitrile–vinyl chloride copolymers (Modacryl fibers) are of interest, specifically because of their flame-resistant properties. These are printed with selected disperse and cationic dyes.

## 6.5. Polyester

The introduction of HT steamers first made it possible for disperse dyes to be fixed to polyester fibers in a continuous process. As a result, polyester rapidly became the most important synthetic fiber for textile printing. Between 1975 and 1990, ca. $2 \times 10^9$ m$^2$ (300 000 t) were printed annually, ca. 25–35 % by transfer printing and the remainder by the classical direct, discharge, and resist techniques. Nearly all these prints go to the clothing sector, with minor amounts for curtains, upholstery, and decorative fabrics.

Polyethylene terephthalate accounts for well over 90 % of the market; other printed polyester fabrics include polydimethylolcyclohexane terephthalate and fibers modified by the addition of anionic compounds to permit their use with cationic dyes.

### 6.5.1. Pretreatment

**Woven Fabrics** [6.21], [6.22]. These are produced from staple-fiber or continuous-filament yarns. The sizing agents employed are water-soluble, so enzymes are not required for desizing. Precleaning therefore consists only of washing with nonionic or (better) anionic detergents (Diadavin or Levapon TH) at 40–80 °C in the presence of sodium carbonate or trisodium phosphate. Higher temperatures may lead to creasing. If nonionic detergents are used, it is important that their cloud point not be exceeded; otherwise, detergent becomes concentrated at the fiber surface, leading to serious printing problems. Colorfastness under conditions of use may also be impaired.

**Knits.** Unlike woven materials, knits are usually produced from continuous-filament yarn, which might be either smooth or texturized. To facilitate processing on the knitting machines such yarns are treated with preparations based on paraffinic substances melting in the range 40–80 °C. These must be removed completely prior to printing. The residual moisture content should not exceed 0.2 % on fabric ready for printing. The precleaning technique resembles that used for woven fabrics, but with the addition of fat solvents, usually chlorinated hydrocarbons.

Care must be exercised in all high-temperature wash processes to avoid excessive longitudinal stress on the fabric, since this can encourage crease formation. The problem can be combated with the aid of softening agents. These comments apply as well to polyester microfilaments (i.e., fibers with a fineness < 1 dtex or 0.1 g/1000 m) and articles made from them.

**Heat Setting of Woven and Knitted Goods.** Achieving dimensional stability and crease-resistance requires that woven or knitted goods be set by hot air on a tentering frame subsequent to precleaning. Drying and setting can be combined, but the fabric must be completely dry before entering the setting zone. Setting should be accomplished at a temperature ca. 10–20 °C above the highest temperature to which the fabric will be exposed during dye fixation or other aftertreatments. The usual setting conditions for woven fabrics are 20–30 s at 190–210 °C. For articles made of texturized yarn the appropriate setting temperature is somewhat lower: ca. 150–180 °C. Because of anticipated shrinkage—usually 5–7 % in both the warp and weft directions—the fabric is stretched to ca. 5–7 % less than its wet width, and a corresponding amount of overfeed is used. Once it leaves the hot-air zone the fabric must be thoroughly cooled.

The dye-absorption properties of polyester are influenced by this hot-air treatment, with dye uptake by nontexturized materials increasing as setting temperature increases. This effect is not observed with goods made of texturized yarn, however, because they are exposed to high temperature during texturizing.

If the fabric is later to be embossed it is preferable to use a setting temperature ca. 20–30 °C lower.

**Alkali Treatment.** Better fabric feel and increased elasticity can be obtained by an alkali treatment with a 2–3 % solution of sodium carbonate for 15–20 min at the boiling point. Depending on the delicacy of the fabric, which is usually made from filament yarn, the process is conducted on a winch-back machine, a jigger, or a star-dyeing device. This treatment causes polyester molecules on the fiber surface to be hydrolytically cleaved and partially eliminated. The resulting loss of material loosens the weave structure and produces a soft, silky feel.

The optimal duration and temperature for alkali treatment depend on the desired effect. The weight loss is normally 3–6 %. After alkali treatment the fabric is acidified with acetic acid, rinsed, and dried.

## 6.5.2. Printing Processes

Woven or knitted fabrics are printed by screen processes on tables or with flat or rotary machines, as well as by the roller-printing method.

The fabric can be glued either to a back gray or directly to the rubber blanket depending upon the fineness of the material and the intricacy of the proposed design.

As with cellulose fibers, use of a back gray is uncommon with roller printing machines.

In the screen-printing process, heavy-grade textiles (e.g., decorative materials and jersey) are also glued directly to the rubber blanket without a back gray, but knitted goods sometimes cause major problems, especially with flat-film printing. Experience shows that thermoplastic adhesives give the best results, consisting for example of poly(vinyl acetate), poly(vinyl chloride), and an organic solvent. They are used principally with screen-printing machines (flat or rotary). In every case it is important that the adhesive be applied as uniformly and as thinly as possible to prevent mechanical resist effects with the printing paste.

If a back gray seems unavoidable, as is often true with fine textiles made of polyester filament yarn, the fabric is bonded to the cotton back cloth with gum – Cordofan (1:1) and/or dextrin – water 1:1, or with 20% poly(vinyl alcohol). Only machines equipped with a drying cylinder are suitable for gluing, because immediate drying is essential to ensure adequate strength in the adhesive bond between the polyester and the cotton back gray.

## 6.5.3. Dyes and Printing Pastes

Only disperse dyes are used for direct printing on unmodified polyester, although cationic dyes are applicable to anionically modified polyester (Sec. 6.5.7).

**Disperse Dyes** [6.23]–[6.26]. A choice based on several criteria is necessary from among the large number of available dyes when printing on polyester.

*Lightfastness.* The tolerance range is wider for printing than it is for solution dyeing, because colorations based on deep navy blues, clarets, and greens always include lighter nu-

ances as well. For decorative articles in pale colors a light-fastness value of 4 – 5 on the German fastness scale is permitted, and for women's outer garments a value of 3.

*Wetfastness.* Good wet-fastness properties can be obtained with disperse dyes on polyester if recommended printing and aftertreatment instructions are observed. The strongly hydrophobic character of the fiber prevents penetration of wash liquors and perspiration into the fibers, thereby preventing removal of dye. A significant deterioration in wetfastness is observed especially toward perspiration and water if excessive amounts of nonionic products are utilized in the form of detergents, carriers, fixing accelerators, softening agents, or antistatic agents, because they tend to remain on the fiber surface. Heat treatment > 140 °C then causes dye to migrate to the fiber surfaces on which these water-soluble nonionic products have become concentrated (thermomigration). Subsequent testing of fastness toward water or perspiration often shows a decrease of ca. 2 – 3 fastness units. An especially critical — but altogether justified — test is a comparison against polyamide present as a companion fiber, because polyamide accepts disperse dye considerably more readily than polyester at low temperatures.

*Fastness toward Processing.* Whereas dyes are today usually fixed by HT steam at 165 – 180 °C with a steaming time of 8 – 6 min, it was once common to use high-pressure steam (2.5 bar, 125 °C, 20 – 30 min). In such a saturated steam atmosphere, azoic dyes subject to reduction may undergo color changes. Addition of oxidizing agents does not always ensure protection from this effect.

In order to prevent discoloration of a white ground it is important that there be no dye sublimation during dye fixation with HT steam or thermosol treatment (40 – 60 s, 190 – 210 °C). This is one of the important criteria by which the dyes should be selected.

Another important criterion relates to the behavior of a dye in the final wash. The only suitable dyes are those that show no affinity for the fiber under washing conditions, or are destroyed in the wash bath by the reductive washing process.

Applicable dyes are marketed under the trade names Dispersol, Foron, Palanil, Resolin, Terasil, and Samaron. Fabrics made from polyester microfilament do not require increased dye concentrations (in contrast to the situation in vat dyeing), because the unique characteristics of the fabric result in a higher concentration of printed dye at the fabric surface. However, the achievable light-fastness and wet-fastness properties are in this case inferior by ca. 1 fastness unit.

**Printing Pastes.** The requirements with respect to stability, adhesive properties, elasticity, and ease of wash removal cannot be met by one thickening agent alone. Obtaining printed products with optimal pattern sharpness requires the use of high-solids thickeners such as starch ethers and dextrin (British gum), but thickeners of this type lead to a brittle film that can easily break and flake. On the other hand, pastes based on low-solids thickeners, such as low-viscosity alginates or seed-flour derivatives, produce an elastic film that is easily removed by washing. Unfortunately, pastes of this type tend to run, especially with high-pressure steam fixing on a finely woven fabric. In spite of this, they are currently the most commonly used pastes for polyester printing, sometimes in combination with starch and cellulose ethers.

In addition to the thickening agent, the thickener solution may contain such auxiliaries as defoamers and/or printing oils to improve the performance of doctor blades or squeegees. It is more important, however, to add a nonvolatile acid or acid donor. Many disperse dyes are susceptible to chemical change in the presence of alkaline thickeners, especially in the course of high-pressure fixing. This can be countered by adding monosodium phosphate, which has the advantage over a number of organic acids that it has little tendency to attack nickel screens. It is also more compatible with alginates than is an organic acid. In the case of other types of dyes the reductive effect of steam or thickener can lead to color variation, especially with light shades. Here the addition to each kilogram of printing paste of ca. 5 g of an oxidizing agent such as sodium chlorate or sodium nitrobenzene-sulfonate is indicated.

Following the successful use of *synthetic thickening agents* in pigment printing based on ethylene–maleic anhydride polymers and poly-(acrylic acid), their use in other sectors of textile printing has been investigated. Promising formulations are so far known only for printing disperse dyes, apart from special applications in carpet printing involving acid dyes [6.27]–[6.32]. It has since been shown that their use is limited to exceptional situations, however, especially because solvent aftertreatment has proven to be a failure.

### 6.5.4. Fixation [6.33]

Because polyester fibers are strongly hydrophobic they have considerably less swelling capacity than, for example, cellulose. Therefore, a certain amount of effort is necessary to open the fibers to such an extent that dye is able to diffuse into them sufficiently. Appropriate measures include steaming at elevated pressure, steaming with superheated steam, setting with dry heat at ca. 200 °C, and the use of suitable fixation auxiliaries.

The dyeing and fixation of disperse dyes on polyester are purely physical solution processes, where the fiber represents a "solid" solvent for the dye. The capacity of polyester fibers to absorb dye is highly temperature dependent, and is subject to influence by added carriers, also designated in this case as *fixation accelerators* (→ Textile Auxiliaries, p. 299). However, such chemicals may have a negative influence on the fastness properties of certain dyes in the finished print. The specific auxiliaries required are a function of the fixation process.

**Fixation with Saturated Steam** (30 min, 102 °C). Under normal steaming conditions in the absence of carriers, only edge-zone dyeing is obtained, with light to medium color intensities and maximum dye yields of 20–50 %.

Because of the deficiencies associated with this fixation process it is of virtually no significance, even with the use of specially selected dyes and large amounts of carrier. For polyester printing it can therefore be regarded as a solution only in an emergency.

**Fixation with High-Pressure Saturated Steam** (ca. 30 min, 2.5 bar). The dye-absorption capacity of polyester is significantly enhanced by fixation in a kettle steamer under pressure. The increase is proportional to the steam pressure, reaching a color intensity corresponding to essentially 100 % dye fixation at 3.5–4.0 bar, although by this point the dyes begin to sublime. The high dye yields obtainable at high steam pressures cannot be duplicated by longer steaming times at lower pressures. Very good results are achieved at 2.5 – 3.0 bar with a steaming time of 20 – 30 min. At 2.5 bar without addition of a carrier the dye yield is 60 – 90 % depending on the dye used. Addition of a carrier increases the yield only slightly, and is therefore not justified.

**Fixation by HT Steam** [6.34] – [6.37] (7 min, 175 °C). This is becoming an increasingly important method because of the growing use of hanging-loop steamers, and it can be regarded as the best fixation process for printing on polyester.

The process is conducted with continuously operating high-temperature hanging-loop steamers and superheated steam at 170 – 180 °C. These steamers differ from thermofixing facilities (e.g., tentering frames) in their high capacity (200 – 400 m), permitting exposure of the fabric to the requisite high temperatures for 5 – 10 min at maximum throughput.

Because steam is used as the heating medium instead of air, the film or thickener swells at the start of the heating phase (20 – 100 °C), with the water evaporating again on further heating to 170 – 180 °C. As a result, thickeners are less likely to be burned in, the fabric remains softer, and dyes can diffuse more easily from the film into the fibers.

Such a high fixation temperature means that only dyes very resistant to sublimation are applicable. In addition to sublimation or thermofixation stabilities (as established by the DIN method and quoted in pattern books), the sublimation resistance of a print in its unfixed state in steam must also be taken into account. Therefore, when changing to HT steamers, dyes previously used with a high-pressure steam process must be tested once again for fastness in order to avoid the production of prints that are defective due to sublimation onto a white ground. The risk of staining the ground depends on the fixation temperature, the fixation time, and the circulation pattern of the heat-transfer medium.

Thickeners with a high solids content are unsatisfactory for this process because they burn in and harden (albeit to a somewhat lesser extent than in the thermofixation process), and they are therefore difficult to remove by subsequent washing. Good results are obtained with mixtures of alginates and starch ethers in the ratio 4:1 to 3:1. Seed-flour products are also occasionally utilized, either alone or in combination with starch ethers. Pastes containing alginate thickeners should also contain 1 – 3 g/kg of a polyphosphate to facilitate removal by washing and to improve compatibility with acid donors and carriers.

With medium and dark colors the dye yield can be increased significantly by a fixation auxiliary (usually an ethoxylated substance), such as Levegal PEW or Samaron HT Fixer (see also → Textile Auxiliaries, p. 300) [6.25], [6.36].

**Thermosol Fixation** (dry heat, 1 min at 200 °C). In addition to the discontinuous high-pressure steam-fixing process and the continuous HT process, hot-air fixation, which is also continuous, is of some importance. The print is in this case passed through a high-performance tentering frame or a condensation apparatus. The best results are obtained at 200 – 220 °C, but above 200 °C the disperse dyes involved must be exceptionally resistant to sublimation, which greatly limits one's choice.

A corresponding dwell time must be determined for each hot-air temperature. With a high-performance tentering frame this time is typically 40 – 50 s. At 200 °C, the mean dye yield without a carrier is 50 – 70 %. Color intensity can be increased considerably by adding Levegal PEW. In general, fixation by dry heat, because of the high temperature, is suitable only for woven fabrics made of untexturized yarn.

### 6.5.5. Aftertreatment

**Washing** [6.38]. After printing with disperse dyes and fixing by one of the processes described above, the fabric is thoroughly rinsed (warm or cold) and then treated with an alkaline reducing agent (2 g/L of NaOH, 2 g/L of hydrosulfite, and 1 g/L of Levegal HTN). At a bath temperature of 40 – 50 °C, most of the thickener and unfixed dye is either dissolved or reductively decomposed. Because of the low temperature, neither dye nor decomposition products are absorbed onto a white ground.

The first wash bath is followed, sometimes after an intermediate rinse, by a second and third wash at 70 – 80 °C with lower concentrations of chemicals. These are in turn followed by warm and cold rinses under acidic conditions, and drying at 110 – 130 °C under very low tension.

**Finishing with Fabric Conditioners.** These agents further improve the feel, volume, and softness of what is essentially already an "easy-care" and crease-resistant fabric. Treatment with the appropriate substances is possible either in the last rinsing bath or by padding prior to final treatment on a tentering frame.

**Antielectrostatic Finishing.** A certain amount of antielectrostatic effect is conferred by fabric conditioners, although it is not permanent. To achieve a lasting antielectrostatic finish that does not produce a softening effect one can apply 5 – 15 g/L of Statexan PAN (an ester of phosphoric acid) to the fabric by padding. Statexan PAN and Persoftal SWA (which contains silicone) are often applied in combination. For more information on antielectrostatic agents.

### 6.5.6. Discharges and Resists
[6.8], [6.20], [6.39], [6.40]

Polyester is increasingly being printed by the discharge and resist methods in cases where the patterns have very fine designs or registration problems arise due to short pattern repeats. Another motivation is the desire for fashionable coloristic effects that would be difficult to achieve by direct printing. Three methods are available:

1) True discharge printing on a dyed ground
2) Discharge resist on an unfixed ground
3) Ordinary resist on an unfixed ground.

In general, discharge and resist processes always require much stricter production control and more process development than direct printing. In each case, the optimum formulation and operating procedure must be established through preliminary tests carried out by the printer. It is especially important to determine the proper quantities of reducing or resist agents relative to the ground color intensity and the amount of dye applied.

**True Discharge Printing on a Dyed Ground.** If Decrolin or tin(II) chloride is used with a hydroxydiphenyl carrier as the discharge intensifier (Tanalid HP), the result is a white effect that at first appears acceptable. However, because of the hydrophobic nature of polyester, the discharge process usually attacks only the surface of the dyed fabric. Light falling on dye or chemical residues remaining in the fibers and not removable by washing leads to yellowing over the course of time. Moreover, dye diffuses from the

interior of the fibers to the surface, resulting in a color change, especially with dark ground colors. For these reasons, neither white discharging nor color discharging can be recommended.

**Discharge Resists with Tin(II) Chloride.** In the discharge-resist process, the ground that is to be discharged is applied by padding or printing, and then simply dried. The temperature of the fabric must not be allowed to rise above 90 °C to prevent premature fixation of the dye.

In a second step, paste and a discharge agent are printed onto the dried ground, after which the ground and printed pattern are fixed simultaneously in a third step.

The discharge resist/tin(II) chloride combination is used mainly for coloristic reasons. On one hand it provides access to a wide range of dischargeable dyes, but at the same time there is also a broad choice of printing dyes that display good resistance to tin(II) chloride.

The amount of reducing agent required is a function of color intensity and the fixing process. In the two-stage fixing method (presteaming at 100 °C followed by poststeaming for 6 – 7 min at 170 °C), 60 – 100 g of tin salt – water 1 : 1 is usually sufficient, whereas the single-stage process (6 – 10 min of HT steam at 170 °C) requires not only the presence of a polyglycol in the discharge paste but also approximately twice as much tin salt.

One disadvantage of tin(II) chloride is its corrosive effect, attributable to hydrochloric acid liberated during fixing but also present in the paste.

A high-viscosity thickener should be used for printing the dischargeable ground so that the printed surface will remain as elastic as possible after drying.

Fixing processes are described below.

**Discharge Resists with Decrolin or Rongalit DS.** The same process steps are involved with these products as with tin(II) chloride. They offer the advantage of no risk of corrosion. However, one disadvantage is modest stability and reproducibility of the printing dyes. Especially in the red range, light shades of disperse dyes are less stable toward Decrolin and Rongalit DS than toward tin(II) chloride.

Dye in the dischargeable ground is not yet fixed to the fiber, so a relatively small amount of Decrolin (30 – 50 g/kg) is sufficient to cause its decomposition. On the other hand, there are a number of disperse dyes that are stable toward such quantities of Decrolin if their concentration is sufficiently high. Light-colored dilute dyes are not particularly stable even toward small amounts of Decrolin. Since the amount of reducing agent required depends on the intensity of the ground, it may under some circumstances happen that the concentration of Decrolin normally recommended (50 g/kg) will prove insufficient, depending on the fabric and the dye that is to be discharged. However, increasing this concentration severely limits the number of Decrolin-resistant disperse dyes, which are required for color printing. Therefore, in the interest of greater security, Rongalit DS or even tin(II) chloride is preferred in practice.

Methods of fixing discharge reserves with tin(II) chloride, Decrolin, or Rongalit DS include the following:

1)  10 min in a star steamer at 102 °C followed by 20 min at 2.5 bar
2)  20 min in a star steamer at 2.5 bar
3)  8 min in a hanging-loop steamer at 170 – 175 °C with HT steam
4)  8 min in a hanging-loop steamer with saturated steam at 102 °C, followed by 8 min in a hanging-loop steamer at 175 °C with HT steam.

Subsequent to the fixation process the fabric is rinsed and then washed at 60 – 80 °C in the presence of sodium hydroxide solution, hydrosulfite, and Levegal HTN.

**Alkali Resists.** The ground, consisting in this case of special disperse dyes that can be hydrolyzed by alkali, is first applied and then gently dried (maximum temperature 100 °C).

Onto this ground are printed the pastes, which contain sodium carbonate, sodium hydroxide or sodium silicate solution, and disperse dyes that offer good resistance toward alkali.

This process offers environmental advantages, and costs for both the dye and the auxiliaries are low. Disadvantages include sensitivity of the ground to unduly high drying and steaming temperatures, lack of sharpness of the printed pattern, and fiber damage resulting from the application of alkali to thin monofilament fabrics that may already have undergone an earlier alkali treatment.

**Crimped Effects.** These can be achieved with polyester by applying a printing paste containing ca. 400 g/kg of Tumescal PH (*p*-phenylphenol) and then fixing the printed product.

### 6.5.7. Direct Printing on Polyesters with an Affinity for Basic Dyes

Polyester fibers modified by the incorporation of anionic (acidic) groups have not so far achieved great importance in textile printing. They can be dyed with basic (Astrazon) dyes.

Pretreatment before printing is analogous to that for ordinary polyesters, and the procedure for printing with basic dyes is no different from that for printing on polyacrylonitrile (Sec. 6.4.2).

*Fixation.* The dye yield depends on steaming conditions. Steaming is best conducted in a star steamer for 30 min at a maximum pressure of 1.5 bar, because higher pressures can have harmful effects on the dye. However, it may prove necessary to print disperse and basic dyes together in a single design. The disperse dye must then be steamed at ca. 2.0 bar to produce a better dye yield. In this case, dyes sensitive to reduction must absolutely be avoided. Addition to the printing paste of the fixing agent Luprintan PFD can be omitted.

HT fixation for 10 – 6 min at 150 – 175 °C gives unsatisfactory results.

Selected Astrazon dyes can also be fixed very successfully in unpressurized saturated steam (hanging-loop steamers).

Appropriate *thickeners* include seed-flour and guar products, either alone or in combination with such cold-soluble starch derivatives as Diatex SL or Printex S, or proprietary mixtures with this type of formulation (e.g., Diaprint 1024 P).

*Aftertreatment* of the fixed print is usually similar to that applicable to polyacrylonitrile; i.e., cold rinsing, washing at 60 °C with 1 g/L of Levapon TH and 1 g/L of hydrosulfite, and rinsing under acidic conditions.

### 6.5.8. Blends of Polyester with Other Fibers

**Polyester – Cellulose.** These are wide-spread in the form of mixed cotton blends, which are printed at a rate of $4 \times 10^9$ m$^2$/a. Blends with viscose staple fiber are produced in much smaller quantity, and are printed mainly with pigment dyes (see Chap. 8). Only if pigment-printed fabrics are unsatisfactory with respect to feel and rubbingfastness are alternative—and much more complicated—processes used, as described below [6.4], [6.41]–[6.45].

*Mixtures of disperse and reactive dyes* can be prepared by the printer himself, but it is better to use mixtures supplied by the dye manufacturers. Apart from dyes, the corresponding printing pastes contain urea, Ludigol (sodium nitrobenzenesulfonate), and a mild alkali (e.g., sodium hydrogencarbonate). Alginates must be used as the thickening agents.

The print is fixed by the thermosol process (1 min, 200 °C) or better with hot steam in a hanging-loop steamer (6 min, 175 °C).

Careful attention must be directed toward the final wash, especially if the design contains white effects next to large areas of dark ground, and for this reason special washing procedures and test methods have been developed. For example, the *acetone test* can quickly show whether unfixed dye has been completely removed from the textile. In this test a small piece of the printed and washed fabric is shaken with acetone in a test tube. Problem-free washing is indicated if the acetone remains virtually colorless.

Selected *disperse dyes* can be fixed on the cotton in the presence of swelling agents [6.46]. The dyes in this case must be soluble in the swelling agents (usually polyglycols or their esters and ethers), at least at elevated temperature. The swelling agents are added to the printing pastes at the rate of 10–15%, and their role is to displace water from the micelles of the cotton during the drying and fixing process. If the swelling agents are also solvents for the dye, solubilized dye is transported simultaneously into the interior of the cotton fibers. If the solvent partially volatilizes at the fixing temperature ($> 200$ °C), the dye is precipitated. Precipitation in the cotton increases during cooling after fixation, and is essentially complete after dilution of the swelling agent and solvent in the final washing process.

Because disperse dyes also dye the polyester fibers, both types of fiber can be dyed with a single dye group using this method. This approach was introduced by DuPont in 1970 as the Dybln process. The Cellestren dyes of BASF work on the same principle, but both have in the meantime been superceded.

*Burn-out Textiles.* Whereas discharges and resists on polyester–cotton are still in the laboratory stage despite much research, the so-called burn-out process has already achieved considerable importance. The process is based on the fact that cellulose fibers can be destroyed by heating with strong acids or their salts.

Core-spun yarns (i.e., polyester filament yarns that have been surrounded during the spinning process by cotton or viscose staple fiber) are printed with a printing paste that contains strongly acidic metal salts and disperse dyes. They are then subjected to a heat treatment (carbonization) that destroys the cellulose fibers and at the same time fixes the disperse dyes on the polyester component.

The carbonized cellulose is removed by a combination aftertreatment consisting of mechanical action followed by washing, or simply an extremely intense wash.

The dye yield of the selected disperse dyes from this strongly acidic paste is always lower than from pastes free of sulfuric acid salts, and the loss when the salt is aluminum sulfate is greater than with sodium hydrogensulfate.

An important variant of this technique is used in the production of Indian *saris*. Blended yarns of polyester and cotton are woven into fabric and dyed or printed with disperse dyes. The cloth is then treated with 70% sulfuric acid to remove the cotton component, leaving a typical sheer sari fabric.

**Polyester–Wool.** These blends are printed with combinations of disperse dyes with selected acid and metal-complex dyes [6.47]. Fixation is carried out with HT steamers (8 min, 165 °C), although higher dye yields are possible through two-stage fixation with saturated steam (30 min, 100 °C) followed by HT steam (8 min, 165 °C).

**Polyester Blended with Other Synthetic Fibers.** Blends of this type put in an appearance from time to time, although they have so far not achieved commercial importance. Blends of polyester and polyacrylonitrile fibers are perhaps worth special mention. These are printed with select disperse dyes, either alone or in combination with cationic dyes.

## 6.6. Other Synthetics

This group accounts for $< 1\%$ of printed textiles.

*Polypropylene Fiber.* Base-modified polypropylene, which is printed with selected acid and metal-complex dyes, and nickel-containing polypropylene have so far attracted only passing interest in the carpet-printing sector. In the case of the nickel-modified fibers, the dye forms a complex with nickel located both on and in the fibers, thus producing a true dyeing effect [6.48], [6.49].

Other proposed methods of printing polypropylene (e.g., through prior chlorination of the fiber) have not led to practical applications [6.50].

*Polyurethane* fibers are processed together with polyamide in the production of swimwear material. Printing methods are described in Section 6.3.8.

Special polyamide fibers with high strength and good flame resistance have been introduced to the market under the name "Nomex" by Du-Pont. These have hitherto proven difficult to dye, but new carrier-free grades that are subject to dyeing are now of interest for military clothing and tents (camouflage). The development of suitable printing techniques is currently in progress.


# 7. Printing on Protein Fibers

The only protein fibers of importance from the standpoint of printing are wool and silk, termed "animal" fibers because of their origin, (→ Silk, → Wool). These two fibers, together with cotton, formed almost the sole basis for clothing in the past. Today, competition from the comparatively inexpensive synthetic fibers (which unlike natural fibers are also continuously available, and with a consistent level of quality) is the main reason for the steadily decreasing importance of natural products as a fraction of total fiber consumption. Nevertheless, the quality of the natural materials cannot be equalled in many respects, and they are assured of a secure future, especially for high-quality articles.

In the case of wool, its polypeptide backbones bear long side chains, which are very tangled, whereas with silk the polypeptide chains are extended and have only short side chains. Another important characteristic of wool is the fact that the polypeptide chains are linked by the sulfur bridges of cystine.

The simultaneous presence of amino, hydroxyl, and carboxyl groups makes it possible for wool and silk to be dyed by acid, basic, reactive, and the usual substantive dyes. Metal-complex formation is also possible (preferably with chromium, but also copper) in which both principal and secondary valencies of the fibers and dyes play a role.

Several other characteristics are equally important with respect to dyeing and printing of protein fibers. Wool is very sensitive to alkali, but very stable toward dilute acid. On the other hand, silk tends to be more sensitive to acid and less subject to damage by dilute alkali. Both protein fibers are susceptible to abrasive damage, and both display a yellowing tendency above 120–130 °C. High temperature leads to deformation, embrittlement, and irreversible impairment of swelling, and hence of dyeing potential. The felting tendency of wool necessitates special attention before and after printing, or the provision of a special antifelting finish.


## 7.1. Pretreatment

For information on the washing, desizing, crabbing ("burning in"), and bleaching of wool → Textile Auxiliaries, pp. 264–266; → Wool. *Wool chlorination* and the *degumming of silk* are especially important processes with respect to printing.

**Chlorination of Wool.** The acid chlorination of wool, formerly carried out only with dilute hydrochloric or sulfuric acid (first bath), followed by treatment with chlorinated lime or sodium hypochlorite (second bath), alters the scaly surface of the wool fiber such that the material is more easily wetted and exhibits better swelling properties so that it will in turn absorb more dye. In the process, the scales are caused to lie flat on the hard outer layer. The wool also tends not to felt so severely, and it acquires a slight luster. It is extremely difficult to achieve completely uniform chlorination, however, and there is always the risk that "chlorination marks" will form. For this reason two improved processes were developed. In the first, nitrogen-containing auxiliaries are added to the chlorination bath. These combine with the liberated chlorine to form chloramines that react with the wool slowly and gradually (e.g., the Melafix process of Ciba-Geigy). In the second process the chlorinating agent is the less reactive dichloroisocyanuric acid [2782-57-2] or its potassium [2244-21-5] or sodium [2893-78-9] salt (e.g., Basolan DC, BASF) (→ Chloramines).

All the processes described, correctly implemented, produce such a dramatic improvement in the color intensity and brilliance of printed wool—irrespective of the dye—that almost all printed woolen piece goods or knits are subjected to chlorination.

**Antifelting Finish.** The creation of a nonfelting finish (e.g., by the IWS-Hercosett process) so that wool can be washed in a washing machine is almost always carried out on the combed sliver. This produces on the chlorinated fibers a very thin film of resin that has no detrimental effect on typical wool properties but ensures simultaneously both good dye-absorption capacity and nonfelting behavior. Yarn, woven fabrics, and knitted goods produced from wool of this type are printed preferentially with very colorfast reactive dyes to satisfy the demand for good washing properties along with colorfastness.

**Degumming of Natural Silk.** The double fiber emerging from a silk spinneret consists of fibroin covered with a second secretion known as *sericin* (*silk gum*). Woven fabrics produced from silk of this type are matt, rough, and harsh to the touch. They are therefore usually subjected to degumming; i.e., they are carefully boiled with a 4% solution of hard soap either as piece goods in a star-dyeing machine or, more often, in a tension-free full-width washing machine or at an earlier stage in rope form ($\rightarrow$ Silk, **A 24**, pp. 101–102). The resulting colloidal solution of sericin in the wash liquor, known as "degumming liquor," is used in silk dyeing as an excellent protective colloid and levelling agent. The following alternative degumming method is also effective: a solution consisting of 5 g/L of a nonylphenolpolyglycol ether (e.g., Levapon TH, Bayer), 3 g/L of hexametaphosphate, and 2 g/L of sodium carbonate (soda ash) is used to prepare both a degumming bath and a second processing bath. The fabric is treated for 1 h at 90–95 °C in each of these baths. After rinsing and drying, the material retains its silky luster, displays good dye-absorption properties, and can in principle be printed exactly like wool.


## 7.2. Direct Printing

### 7.2.1. Printing with Acid Dyes

In principle, both basic and substantive dyes can be used for printing on wool and silk, but in practice, acid dyes are mainly used, as the others, though often giving excellent brilliance, have unsatisfactory fastness properties. Even with acid dyes it is necessary to restrict the selection to substances that combine brilliant results with acceptable lightfastness and wetfastness properties with respect to the articles in question.

**Printing Paste.** Dissolving the dye requires the use of urea or thiourea, solubilizing agents such as thiodiethylene glycol, and hot water. In rare cases, other solvents (e.g., polyglycol ethers) or dispersing agents are added to prevent thickening of the printing paste on prolonged storage and the possibility of stippled printed effects. Glycerine, which is hygroscopic, is recommended only for wool printing, where it is able to compensate for the harmful effects of slightly superheated steam introduced during fixing, whereas with silk, significant prolongation of drying brings the risk of smearing.

*Thickening agents* used today include seed-flour derivatives (mainly guar products) either alone or in combination with cold-soluble British gum, as opposed to the formerly widely used tragacanth and its mixtures with British gum and/or gum arabic. High-solids thickeners are preferred for fine effects, sharp pattern contours. etc., whereas low-solids products with their better leveling properties and reduced risk of fold and crease formation are preferred for larger printed areas. One of the most important prerequisites for the viability of a thickening agent is ease of removal by washing. The printing pastes also contain an acid donor to promote dye fixation (e.g., ammonium sulfate, tartrate, or oxalate), sometimes also acetic or glycolic acid, and small amounts of sodium chlorate to counteract the reductive action of wool and certain thickening agents in the course of steaming (occasionally essential with "steam sensitive" dyes!). Defoamers and printing oils are usually also necessary for producing smooth, sharp prints.

Wool that has been poorly or inadequately chlorinated or is extremely voluminous may profit from the use of a coacervate former (e.g., Levalin VKUN, Bayer, an alkylarylpolyglycol ether) to prevent "frosting effects." This auxiliary causes dye to be absorbed uniformly during steaming, even at the outer tips of the fibers. If silicone defoamers are introduced simultaneously this effect is negated, therefore high-boiling alcohols are used as defoamers instead (ca. 5 g/kg).

**Fixation.** Fixation of the dye on wool or silk usually requires relatively prolonged steaming (i.e., 30–60 min), and the most brilliant and colorfast prints are obtained only with saturated-steam fixation at 100–102 °C. Excessive drying of the fabric after printing, especially in the case of wool, must therefore be strictly avoided. For this reason wool is sometimes sprayed prior to steaming to ensure that the steam is saturated.

**Aftertreatment.** The final washing process is carried out on winch-back equipment, although one of the various types of tricot washing machine can also be used, depending on the batch size. Just like silk, wool should be treated as carefully as possible—i.e., with minimum mechanical stress and at a maximum temperature of 40 °C (0.5–1 g/L of an anionic synthetic detergent). To prevent bleeding into unprinted areas and the formation of marks, products based on the condensation of high molecular mass aromatic sulfonic acids with formaldehyde may also be introduced (Erional NWS, Ciba-Geigy; Mesitol NWS, Bayer, etc.). These lead to significant improvement in the wetfastness properties of wool if applied for ca. 20 min at the level of 5–6 % of the textile weight at 60 °C in a bath acidified with acetic or formic acid.

Similar improvements in fastness properties are produced by certain polyammonium compounds (e.g., Levogen, Bayer; Tinofix, Sandoz) if all anionic auxiliaries are thoroughly rinsed out after the final wash (to prevent the formation of spots) and the fabric is then treated in a bath containing 2–3 g/L of one of the auxiliaries for ca. 20 min at 35–40 °C.

These inexpensive and very effective processes unfortunately have the disadvantage that cationic aftertreatment may lead to reduced lightfastness, especially with pale colors, and color changes (dulling effects) may be detectable in rare cases.

### 7.2.2. Printing with Metal-Complex Dyes

Both 1:1 and 2:1 metal-complex dyes give a higher degree of colorfastness—especially lightfastness—than acid dyes, but they have the disadvantage of producing rather subdued, dull colors. There are almost no differences in printing applications between metal-complex and acid dyes, except that it is better to print the former in the absence of acids or acidic compounds. The

acid pH range decreases the stability of the paste and encourages a tendency toward dye agglomeration (stippling effects); in addition, one also frequently obtains prints in which leveling of the printed dye is inferior. Leveling problems are in any case common with this type of dye in full-covered design applications.

The dissolution process, the choice of paste-thickening agent, fixation, and the final wash are also identical for both types of dye. This leads to a great many combination possibilities, although printing must be accomplished without acids or acid donors. The use of auxiliaries in aftertreatment processes is also analogous to the situation with acid dyes.

### 7.2.3. Printing with Reactive Dyes

Unlike the dyes described previously, reactive dyes (→ Reactive Dyes) do not form salt-like bonds to protein fibers; instead, true chemical bonds are formed in an acidic medium (pH 3–5) at 80–100 °C with $-SH$, $=NH$, and $-NH_2$ groups from the polypeptide. This results not only in dyes with good fastness properties, but also very brilliant products, thereby significantly enriching the market. The high demand for goods that carry the "woolmark," together with the high serviceability associated with articles produced from these fibers, justify the additional costs entailed in use of these dyes.

From the point of view of printing techniques, there is again no difference between reactive dyes and acid dyes with respect to printing-paste formulation or thickening agents, except that formic acid is often included in the reactive-dye pastes to ensure acidity. In general, the steaming times for fixation can even be reduced—to 10–20 min (saturated steam at 100–102 °C)—but it should be noted in terms of the final wash that a high degree of fastness will be achieved only if dye that has not reacted with the fiber is removed completely. To achieve this end it is necessary after the rinsing bath (to which a small amount of a synthetic tanning agent such as Erional NWS or Mesitol is also normally added) that the fabric be washed in several baths with progressively increasing temperatures (40 °C, 60 °C, 80 °C) using 2 g/L of disodium phosphate (with ammonia added to a pH of 8) and an anionic detergent.

This has two consequences. The energetic washing process does not permit the concurrent use of any but reactive dyes in the color scheme, because these would be subject to serious bleeding. Also, this type of wet aftertreatment is possible only with wool that has received a good antifelting finish; i.e., it is essentially restricted to chlorinated or Hercosett-treated wool.

With silk, which is more resistant to alkali, an alternative procedure is available at alkaline pH with sodium hydrogencarbonate and alginate thickeners. The advantage of this approach is that it often gives a more brilliant result with a better dye yield. The disadvantage is a sometimes inferior print with designs that overlap. No differences are apparent in fastness. (Very detailed patterns can in any case be obtained with pigment dyes without affecting the feel of the fabric.)

### 7.2.4. Printing with Chrome Dyes

The conversion of chrome dyes ($\rightarrow$ Azo Dyes, **A 3**, pp. 270–271) into chromium complexes during steaming leads to very colorfast prints. However, this technique is restricted almost exclusively to melange printing (Sec. 3.6.2), because the range of usable dyes is limited, and the printed colors on clothing materials are not very attractive. Chromium(III) salts (usually chromium fluoride, more rarely the acetate) are added to the printing paste together with formic acid (occasionally oxalic acid), which has a delaying effect on complex formation, thereby preventing premature formation of the chromium pigment. This technique is being used less and less frequently, however, largely for environmental reasons.

After printing, the sliver is steamed for 60–90 min without intermediate drying, after which it is rinsed, washed, dried, and combed.

### 7.3. Discharge Printing

Although the discharge-printing technique on cellulosic materials was almost completely forgotten for almost 15 years, wool and silk for high-fashion articles have always been dyed (albeit in small quantity) with dischargeable substantive or acid dyes and then printed with discharge-resistant acid or substantive dyes. The usual discharging agents are sodium or zinc formaldehyde sulfoxylate (Rongalit C or Decrolin, BASF). A wide range of additives is used in the interest of achieving more or less perfect results. The production of these discharge-dyed articles is difficult, because success depends on many conditions, which are often subject to change. The fabric, its structure, the pretreatment, the degree of whiteness, the color intensity of the ground, the particular combination of dyes, the discharge formulation, the type and amount of paste applied, and the steaming and final wash conditions—all these contribute to the result (as measured, for example, by fastness properties, brilliance, and yellowing tendency). For this reason, only general information can be provided with respect to formulations, which must be adjusted from case to case to suit local conditions.

Special attention must be directed toward the risk of fiber damage through attack on the protein fibers by decomposition products from the reducing agents. Thiourea dioxide (formamidinesulfonic acid) is a mild reducing agent, but its use as a discharging agent on protein fibers has the disadvantage that its poor solubility limits the choice of dischargeable dyes. It is therefore applicable only in special cases.

To ensure good fastness properties, especially with reactive dyes, the dyed fabric must be thoroughly washed before printing. This is also beneficial with respect to the desired discharge effect, and it facilitates the final washing stage. Dischargeable grounds on silk should display a neutral reaction; dye effects achieved with the aid of added acid require that the fabric be neutralized with ammonia and rinsed very carefully.

**Color-Discharge Printing.** The demand for efficiency has led to a reduction in the choice of dyes offered by manufacturers, and the number of dyes available for special applications is decreasing steadily. For this reason it has become increasingly difficult to put together a satisfactory palette of discharge-resistant dyes for wool and silk.

In principle, the following approaches are possible:

1) Color-discharge printing with vat dyes according to the sulfoxylate procedure, using dyes that can be fixed with a small amount of alkali, which means accepting a certain amount of unavoidable fiber damage. This path is only very rarely taken.
2) Color-discharge printing with basic dyes, using tin(II) chloride as the reducing agent. In

spite of a high degree of brilliance and an adequate selection of dischargeable grounds, this method is seldom if ever used because of poor fastness to light and moisture.

3) Color-discharge printing with selected discharge-resistant pigment dyes using soft binders. It is recommended that suitable softening agents be added, with zinc formaldehyde sulfoxylate as the reducing agent. Otherwise, fixation may be unsatisfactory, and the differing swelling properties of protein fibers and binder film may lead to extremely poor rubbingfastness.

4) Color-discharge printing with selected discharge-resistant substantive and acid dyes, where sodium formaldehyde sulfoxylate or zinc formaldehyde sulfoxylate is the discharging agent. Although the choice of dyes in this case is relatively limited, it has proven to be the best method.

## 7.4. Printing on Blends Containing Protein Fibers

Printing of blends of wool with silk, or of one of the two with cellulose, has become a peripheral activity since the emergence of blends with synthetic fibers, although even these are not of great importance in textile printing (see Chap. 6).

**Wool – Silk.** The wool fraction is pretreated like wool (Sec. 7.1), though without chlorination, and the silk fraction is degummed (if necessary) at 90 °C for 1 h in one or two mildly alkaline baths containing hard soap (5 g/L) as well as hexametaphosphate (2 g/L) and disodium phosphate (3 g/L). Silk is sensitive to chlorine, so any bleaching must be accomplished with hydrogen peroxide. The choice of dyes and the formulations are exactly as with pure woven wool products. This is also true for aftertreatment of the print and the production of discharge-printed articles.

**Wool – Cellulose.** Blends of wool with cellulose ("half-wool") are pretreated according to the nature and ratio of the blend, always bearing in mind the sensitivity of wool to alkali. Chlorination can normally be a consideration only if wool is the predominant fiber. Here, too, hydrogen peroxide is the preferred bleaching agent.

Most of the dyes used are substantive, supplemented by certain neutral-dyeing acid and metal-complex dyes. The formulation again corresponds to that recommended for printing on wool alone; additives intended to improve fastness (e.g., cationic quaternary polyammonium compounds) must not be omitted from the aftertreatment process. In the case of half-wool products various special aftertreatment agents have been developed.

Textiles printed with selected reactive dyes display a higher degree of fastness. In most cases these can also be fixed using shorter steaming times. It is essential, however, that the formulations contain only ca. 10 g/kg of sodium hydrogencarbonate, and the two fibers must be intimately mixed in the fabric structure; otherwise, fiber damage and crimping effects must be anticipated due to localized shrinkage.

White- and color-discharge printing can be carried out as described in Section 7.3, and if there is a high proportion of cellulose in the fabric it is also possible to use vat dyes that can be fixed with small amounts of alkali.

**Silk – Cellulose.** Grades known as "half-silk" are pretreated like pure silk and printed like half-wool. However, the lower sensitivity of silk to alkali means that direct or discharge printing with easily fixed vat dyes is more feasible. Otherwise, the same recommendations apply as those given for half-wool.

# 8. Pigment Printing

"Pigment printing" refers to producing printed textiles by the application of colored pigments with the aid of pigment binders, softening agents, flow modifiers, defoamers, and special thickeners.

The colored pigments consist mainly of organic products of the azoquinacridone, dioxazine, and phthalocyanine types, to mention only a few examples. These are supplemented by carbon black and inorganic white pigments, mainly titanium dioxide, and sometimes with iron oxides for brown shades.

All these pigments are used in textile printing in the form of 25 – 50 % dispersions reduced in size mechanically (by grinding) to the optimum particle-size distribution and mixed with various additives (emulsifiers, dispersing agents, evaporation inhibitors, and preservatives).

The resulting viscosity properties should be as stable as possible, and it is important to strive for a consistency that facilitates pumping and metering so that the preparations can be processed by automatic metering equipment. Special effects sometimes call for the use of copper and aluminum alloys in powdered form ("bronze printing"), basic dyes modified with synthetic resins ("luminescent pigments"), or mica that has been coated with titanium dioxide ("nacreous effects"). For further details → Pigments, Inorganic; → Pigments, Organic.

# 9. References


References for Chapter 1

[1.1] W. H. v. Kurrer: *Geschichte der Zeugdruckerei*, Nürnberg 1844.

[1.2] R. Forrer: *Die Kunst des Zeugdrucks vom Mittelalter bis zur Empirezeit*, Straßburg 1898.

[1.3] H. Züblin: *40 Jahre Kattundruck*, v. Berchtold & Gommeringer, Singen 1927.

[1.4] A. Bollinger: *Ein Beitrag zur Entwicklung des europäischen Textildrucks*, Springer Verlag, Wien 1950.

[1.5] E. Bindewald, K. Kasper: *Bunter Traum auf gewebtem Grund*, G. Westermann, Braunschweig 1950.

[1.6] J. Zahn: *Das Abenteuer Farbe – Eine Kulturgeschichte*, Hoffmann und Campe, Hamburg 1983.

[1.7] "Uralter Zeugdruck," *Bayer Farben Rev.* **1–8** (1962–1964).

[1.8] L. W. C. Miles: *Textile Printing*, Dyers Company Publications Trust, London 1981.

References for Chapters 2 and 3

General References

[3.1] G. Georgievics, R. Haller, L. Lichtenstein: *Handbuch des Zeugdrucks*, vols. I–III, Akad. Verlagsgesellschaft, Leipzig 1930.

[3.2] L. Diserens: *Fortschritte in der Anwendung der Farbstoffe*, Birkhäuser, Basel 1949.

[3.3] H. Rath: *Lehrbuch der Textilchemie einschl. der textilchemischen Technologie*, 3rd ed., Springer Verlag, Berlin Göttingen Heidelberg New York 1972.

[3.4] W. Bernard: *Appretur der Textilien (mit Waschmaschinen, Trocknungseinrichtungen, Spannrahmen usw.)*, 2nd ed., Springer Verlag, Berlin 1967.

[3.5] W. Bernard: *Praxis des Bleichens und Färbens*, Springer Verlag, Berlin 1966.

[3.6] W. Bernard: *Druck von Textilien*, Springer Verlag, Berlin Göttingen Heidelberg New York 1969.

[3.7] K. Schmidt: *Der Textildruck*, Dr. Spohr Verlag, Wuppertal 1973.

[3.8] K. Spitzner: *Textildruck*, 3rd revised ed., VEB Fachbuchverlag, Leipzig 1980.

Specific References

[3.9] H. G. Weissgerber, F. Weissgerber: *Walzengravur und Schablonenherstellung*, Melliand Verlag, Heidelberg 1960.

[3.10] *Bayer Farben Rev.* **14** (1968) 28.

[3.11] J. Eibl, *Melliand Textilber.* **57** (1976) 407.

[3.12] E. Bobec, *Melliand Textilber. Int.* **56** (1975) nos. 10–12; *Melliand Textilber.* **59** (1978) no. 9; **61** (1980) no. 7 (literature references).

[3.13] N. L. Moore, *J. Soc. Dyers Colour.* **90** (1974) 318–325 (literature references).

[3.14] Sark, DRP 614 786, 1933.

[3.15] Fastran, GB 1 284 824, 1971.

[3.16] E. Küsters: "Reactiv Transferdruckverfahren für Cellulosefasern," *Int. Textil Service* **1** (1992) 63.

[3.17] Star, DE 35 509 D/8 n, 1949.

[3.18] N. de Plasse, FR 1 223 330, 1958. Ciba, GB 1 221 126, 1968.

[3.19] Sublistatic, DE-OS 2 458 660, 1973.

[3.20] M. Fox, *J. Soc. Dyers Colour.* **89** (1973) no. 12, 474–485.

[3.21] A. Kershaw, *Textilveredlung* **11** (1976) 262.

[3.22] T. L. Dawson, B. P. Roberts, *J. Soc. Dyers Colour.* **93** (1977) 439.

[3.23] J. Eibl, *Chemiefasern/Textilind.* **27** (1977) 636–645.

[3.24] J. Eibl, *Melliand Textilber.* **62** (1981) 565–567.

[3.25] T. L. Dawson: "Jet Printing," *Rev. Prog. Color. Relat. Top.* **22** (1992) no. 12, 22.

[3.26] K. H. Fluss: "Space Dyeing," *Bayer Farben Rev.* **26** (1975) 19. J. Wirtz: "Teppichdruck," *Bayer Farben Rev.* **30** (1978) 47.

[3.27] *Bayer Farben Rev.*, special issue 2 (Beflockung) (1963).

[3.28] *Melliand Textilber.* **38** (1957) 669.

[3.29] P. Weber, *Text. Faserstofftech.* **1956**, no. 3, 129, 130.

References for Chapter 4

[4.1] R. Müller, W. Schwindt, *Melliand Textilber.* **49** (1968) 1325–1330; *Melliand Textilber. Int.* **53** (1972) 201–203.

[4.2] O. Schlösser, *Bayer Farben Rev.* **17/18** (1969).

[4.3] H. Dahm, *Bayer Farben Rev.* **6** (1981) Sept., reprint Verdickungsmittel und Kleber.

[4.4] H. Barth, *Bayer Farben Rev.* **15/16** (1968).

[4.5] W. Berlenbach, *Melliand Textilber. Int.* **53** (1977) 207–210.

[4.6] P. Habereder, F. Bayerlein in A. Chwala, V. Anger (eds.): *Handbuch der Textilhilfsmittel*, Verlag Chemie, Weinheim 1977, p. 621.

[4.7] W. Schwindt, *Melliand Textilber.* **58** (1977) 1009–13.

[4.8] U. Karsunky, M. Thissen, *Melliand Textilber. Int.* **53** (1972) 577–79.

[4.9] F. R. Alsberg, P. R. Dawson, *Textilveredlung* **8** (1973) 365–370.

[4.10] F. Bayerlein, *Melliand Textilber.* **58** (1977) 1017–1020.

[4.11] E. Heym, *Melliand Textilber. Int.* **53** (1972) 213–214.

[4.12] G. Barnhardt, *Text. Chem. Color.* **11** (1979) no. 10, 32–45.

[4.13] M. Thissen, *Melliand Textilber. Int.* **53** (1972) no. 5, 573–576.

[4.14] F. Bayerlein, P. Habereder, *Text. Prax. Int.* **29** (1974) 1411–1413.

[4.15] P. Habereder, *Melliand Textilber.* **61** (1980) 165–166.

[4.16] F. Bayerlein, L. K. Schwörzer, *Melliand Textilber. Int.* **53** (1972) 204–206.

[4.17] K. Roth, *Textilbetrieb (Würzburg)* **98** (1980) no. 9, 50–65.

[4.18] Chem. Fabrik Grünau, Druckverdickungsmittel, company brochure.

[4.19] U. Karsunky, *Chemiefasern/Textilind.* **25** (1975) no. 2, 159–161.

[4.20] J. Rosenbalm, J. Shelso, *Text. Chem. Color.* **11** (1979) no. 10, 28–31.

[4.21] N. S. Volkonskaja et al., *Faserforsch. Textiltech.* **12** (1975) 321–323.

[4.22] H. Schulzen, *Z. Gesamte Textilind.* **1958**, nos. 6, 8, 11.

[4.23] *Ciba-Rundsch.* **1** (1969) 19–34.

[4.24] J. S. Racciato, *Text. Chem. Color.* **11** (1979) no. 2, 31–35.

[4.25] Meyhall, Guar- und Guar-Derivate, bulletin.

[4.26] W. Kühnel, *Bayer Farben Rev.* (special issue) **18** (1979).

[4.27] E. G. Hochberg, *Text. Chem. Color.* **11** (1979) no. 5, 27–28.

[4.28] L. T. Holst Jr., *Text. Chem. Color.* **11** (1979) no. 3, 17–19.

[4.29] T. G. Perkins, *Text. Chem. Color.* **11** (1979) no. 10, 21–25, 44.

[4.30] R. Hofstetter, G. Robert, *Textilveredlung* **14** (1979) no. 2, 51–56.

[4.31] W. Badertscher, *Textilveredlung* **14** (1979) no. 7, 279–82.

References for Chapter 5

[5.1] W. Kothe: "Aktuelle Möglichkeiten der Druckwarenvorbehandlung," *TPI Text. Prax. Int.* **34** (1979) 52.

[5.2] D. Bassing, W. Küppers: "Was bieten die BASF-Vorbehandlungssysteme dem Textildruck?" *TPI Text. Prax. Int.* **34** (1979) 58.

[5.3] BASF-Ratgeber, Zellulosefasern, B 375 d, Ludwigshafen 1977.

[5.4] W. Guth, A. Brüggemann: "Kontinuierliche Verfahren zum Bleichen und Weißtönen von Geweben und Gewirken in Verweilgeräten," *Bayer Farben Rev.* (special issue) **15** (1975) 97.

[5.5] T. Mandt, B. Brüssow: Der Stoffdruck in der Bundesrepublik Deutschland. (Forschungsstelle f. allg. Textil-Marktwirtschaft), Universität Münster 1980.

[5.6] K. Schmidt: "Textildruck," in O. Spohr, E. Wagner (eds.): *Handbuch für Textilingenieure und Textilpraktiker*, 3rd ed., Dr. Spohr Verlag, Wuppertal-Barmen 1951.

[5.7] H. E. Fiertz-David, E. Merian: *Abriß der chemischen Technologie der Textilfasern*, Birkhäuser, Basel 1948, p. 240.

[5.8] IG-Farbenindustrie, Ratgeber für das Bedrucken von Baumwolle und andere Fasern pflanzlichen Ursprungs, 1934.

[5.9] A. Schaeffer: "Die Oxydation der Leukoverbindung zum Küpenfarbstoff," *Melliand Textilber.* **38** (1957) 428.

[5.10] A. Schaeffer: "Vorgänge beim Verküpen von Küpenfarbstoffen," *Melliand Textilber.* **36** (1955) 1033.

[5.11] U. Baumgarte: "Über Reaktionen von Reduktionsmitteln bei der Küpenfärberei," *Textilveredlung* **2** (1967) 896.

[5.12] N. Grund: "Redoxprozesse im Ätzdruck auf Textilien," *Melliand Textilber.* **67** (1986) no. 12, 896.

[5.13] BASF, Rongalit ST flüssig, technical report no. 10, Ludwigshafen 1989.

[5.14] E. Fees: "Der Ätzdruck auf Cellulose," *TPI Text. Prax. Int.* **31** (1976) 161.

[5.15] H. Ilg, V. Krentz: "Untersuchungen zur Ermittlung der günstigsten Einsatzbedingungen für das Zweiphasendruckverfahren." *Text. Prax.* **20** (1965) 324.

[5.16] E. Fees: "Der Zweiphasendruck mit Küpenfarbstoffen," *Melliand Textilber.* **45** (1964) 45.

[5.17] U. Blum: "Ätz- und Reservedruck," *Textilveredlung* **15** (1980) 323.

[5.18] U. Altmann: "Der Afrikadruck—Vita einer interessanten Entwicklung," *Bayer Farben Rev.* **14** (1968) 1.

[5.19] F. Baumann et al., *Angew. Chem.* **68** (1956) 133–150.

[5.20] J. Eibl, *Melliand Textilber.* **39** (1958) 522–527, 660–663, 772–773; **45** (1964) 789–796; *Bayer Farben Rev.* **24** (1974) 47–57 and special issue no. 5 (Rapidogen-Farbstoffe auf neuen Wegen); *Bayer Color.* **10** (1960) 21.

[5.21] D. Hildebrand, R. Schwaebel: "LEVAFIX P – Neue faserreaktive Farbstoffe für Cellulose," *Text. Prax.* **22** (1967) 796–808.

[5.22] D. Hildebrand: "Zum Mechanismus des Färbens mit Reaktivfarbstoffen," *Text. Prax.* **25**, **26** (1970) H 5–H 8.

[5.23] L. Schmidt: "Probleme mit Reaktivfarbstoffen aus der Sicht des Druckereipraktikers," *Textilveredlung* **13** (1978) no. 8, 293–298.

[5.24] H. U. von der Eltz: "Die Chemie der Remazol-Farbstoffe," *Melliand Textilber.* **63** (1982) no. 11, 798–801.

[5.25] H. Herlinger et al.: "Rheolog. Charakterisierung und Qualitätsbeurteilung von Druckpasten und Verdickungen," *TPI Text. Prax. Int.* **42** (1987) no. 10, 1231–1241.

[5.26] H. Beckstein et al.: "Verringerung der Umweltbelastung in der Textildruckerei/Feuchtigkeitsauftrag v.d. HS-Daempfer," ITB Veredlung 1/92.

[5.27] K. H. Blank, V. Wuertz: "Remazol-Farbstoffe, die Voraussetzung für anspruchsvolle, vielseitige Bunteffekte modischer Drucke," *Melliand Textilber.* **64** (1983) no. 10, 752–757.

[5.28] K. H. Blank, "Remazol-Farbstoffe drucken und kalt fixieren," *TPI Text. Prax. Int.* **39** (1984) 249–251.

[5.29] H. Gutjahr, F. v. Vlissingen, EP 0 167 711, 1989.

References for Chapter 6

[6.1] C. Sommer, *Melliand Textilber.* **50** (1969) 692–698.

[6.2] *Textilhilfsmittel-Katalog 1981/82*, Konradin Verlag, Leinfelden-Echterdingen.

[6.3] L. T. Holst, *Text. Chem. Color.* **7** (1975) 154–156.

[6.4] H. J. Manderla, *Chemiefasern* **15** (1965) no. 2, 96–107.

[6.5] H. J. Manderla, *Melliand Textilber.* **43** (1962) 1310–1316.

[6.6] H. J. Manderla, *Bayer Farben Rev.* **8** (1964) 32–38.

[6.7] R. Koch, *Bayer Farben Rev.* **13** (1967) 23–29.

[6.8] W. Kühnel, *Melliand Textilber.* **57** (1976) 315–320.

[6.9] W. Kühnel, *Bayer Farben Rev.* **20** (1971) 27–32.

[6.10] H. Dahm, *Bayer Farben Rev.* **12** (1967) 26–30.

[6.11] H. J. Manderla, *Bayer Farben Rev.* **4** (1963) 24–31.

[6.12] H. J. Manderla, *Bayer Farben Rev.* **5** (1963) 38–51.

[6.13] H. Gutjahr, H. Dahm, *Melliand Textilber.* **10** (1968) 1207–1210.

[6.14] G. Meyer, *Melliand Textilber.* **50** (1969) 698–702.

[6.15] W. Kühnel, *Melliand Textilber. Int.* **51** (1970) 325–330.

[6.16] R. Hofstetter, *Teintex* **44** (1979) 691–696.

[6.17] F. G. Rebske, *Melliand Textilber. Int.* **54** (1973) 519–520.

[6.18] U. Karsunsky, *Chemiefasern/Textilind.* **25** (1975) 728–731.

[6.19] G. Meyer, *Melliand Textilber. Int.* **51** (1970) 1341–1344.

[6.20] Mitsubishi, *JTN* **295** (1979) 77–86.

[6.21] J. Winkler, *Melliand Textilber. Int.* **56** (1975) 728–731.

[6.22] H. J. Manderla, *Bayer Farben Rev.* **7** (1964) 16–30.

[6.23] C. Sommer, G. Vogl, *Melliand Textilber. Int.* **53** (1972) 440–445.

[6.24] F. R. Alsberg, *Chemiefasern/Textilind.* **23** (1973) 439–444.

[6.25] R. Rafael, *Melliand Textilber. Int.* **55** (1974) 361–364.

[6.26] J. F. Dawson: "The Structure and Properties of Disperse Dyes in Polyester Coloration," *J. Soc. Dyers Colour.* **99** (1983) nos. 7/8, 183.

[6.27] R. Hofstetter, G. Robert, *Textilveredlung* **14** (1979) 51–56.

[6.28] W. Schwindt, *TPI Text. Prax. Int.* **34** (1979) 278–281.

[6.29] J. Skoufis, *Text. Chem. Color.* **11** (1979) 279–282.

[6.30] W. Badertscher, *Textilveredlung* **14** (1979) 106.

[6.31] L. T. Holst, *Text. Chem. Color.* **11** (1979) 53–55.

[6.32] W. Kühnel, *Bayer Farben Rev.* (special issue) **18** (1979) 21–27.

[6.33] H. P. Schoepflin, *Text. Chem. Color.* **10** (1978) 225–229.

[6.34] R. Hofstetter, *Text. Chem. Color.* **5** (1973) 172–180.

[6.35] R. Hofstetter, *Textilveredlung* **11** (1976) 186–194.

[6.36] J. Eibl, *Melliand Textilber.* **57** (1976) 663–667.

[6.37] L. T. Holst, *Am. Dyest. Rep.* **65** (1976) 47.

[6.38] J. Winkler, *TPI Text. Prax. Int.* **34** (1979) 302–309.

[6.39] D. Brierley, J. R. Provost: "The Use of Disperse Dyes Containing Diester Groups to Produce Discharge Effects," *J. Soc. Dyers Colour.* **99** (1983) no. 12, 358.

[6.40] D. Knittel, W. Günther, E. Schollmeyer: "Ätzdruck auf laservorbehandelten Polyestergeweben," *TPI Text. Prax. Int.* **45** (1990) no. 1, 46.

[6.41] W. Kühnel, *Bayer Farben Rev.* **13** (1967) 16–22.

[6.42] R. Schwaebel, *Bayer Farben Rev.* **24** (1974) 38–46.

[6.43] G. Gabry, *Teintex* **40** (1975) 267–274.

[6.44] B. Glover, *Am. Dyest. Rep.* **67** (1978) 47.

[6.45] E. Fees, *Melliand Textilber.* **60** (1979) 595–605.

[6.46] F. Miksovsky, A. Blum, *TPI Text. Prax. Int.* **34** (1979) 152–154, 167–168.

[6.47] P. R. Brady, *J. Soc. Dyers Colour.* **95** (1979) 302–344.

[6.48] R R Hyness, *Text. Chem. Color.* **2** (1970) 25–28.

[6.49] G. Buzzoni, *Rev. Gen. Caoutch. Plast.* **52** (1975) 459–462.

[6.50] A. Agster, *Melliand Textilber. Int.* **56** (1975) 470–472.

References for Chapter 7

[7.1] J. Lösch: *Fachwörterbuch Textil*, Verlag J. Lösch, Frankfurt/M. 1975.

[7.2] Knecht-Fothergill: *The Principles and Practice of Textile Printing*, 4th ed., Griffin, London 1952.

[7.3] H. E. Fiertz-David, E. Merian in [5.7].

[7.4] D. Hildebrand: "Strukturchemie der natürlichen Polyamidfasern, *Z. Gesamte Textilind.* **71** (1969) 274–280, 311–317.

[7.5] K. Reincke: Vorbehandlung von Wolle für den Druck, *Text. Prax.* **25** (1970) 419 ff.

[7.6] C. Frommelt: "Bleichen und Weißtönen von Wolle," *Bayer Farben Rev.* **25** (1975) 80 ff.

[7.7] J. Lewis: "Neuere Entwicklung bei Superwaschwolle," *Textilveredlung* **11** (1976) 214 ff.

[7.8] H. Heiz: "Vorbehandlung von Wolle und Wollmischgeweben," *Textilveredlung* **13** (1978) 205 ff.

[7.9] W. Schefer: "Nachweis und Beurteilung chemischer Schäden an Wolle," *Textilveredlung* **13** (1978) 55–57.

[7.10] W. Steger: "Die Naturseide, ihre heutige Veredlung und Ausrüstung," *Text. Prax.* **24** (1969) 379–383.

[7.11] R. Koch: "Bedrucken von Halbwolle," *Bayer Farben Rev.* **33** (1981) 32–41.

[7.12] B. Kramer: "Am seidenen Faden," *Bayer Farben Rev.* **29** (1977) 3–10.

[7.13] P. Bakker, A. Johnson: "The Application of Monochlorotriazinyl Reactive Dyes to Silk," *J. Soc. Dyers Colour.* **89** (1973) 203–208.

[7.14] V. Bell: "Recent Developments in Wool Printing," *J. Soc. Dyers Colour.* **104** (1988) no. 4, 159.

[7.15] A. R. Czerny: "Textildruck auf Naturseide," *Melliand Textilber.* **69** (1988) no. 6, 438.

[7.16] K. Y. Chu, J. R. Provost: "The Dyeing and Printing of Silk Fabrics," *Rev. Prog. Color. Relat. Top.* **17** (1987) 23.

[7.17] H. Thomas et al.: "Anfärbetest zur Überprüfung der Reproduzierbarkeit in der Vorbehandlung von Wollgeweben für den Textildruck," *Melliand Textilber.* **70** (1989) no. 5, 352.

References for Chapter 8

General References

[8.1] P. Wengraf: "Sammelbericht Pigmentdruck in den Kriegsjahren," *Text. Rundsch.* **2** (1947) 125–131.

[8.2] K. Reinartz: "Das ACRAMIN F-Verfahren," *Melliand Textilber.* **33** (1952) 626–628.

[8.3] N. S. Cassel: "New Horizons for Pigment Printing," *Am. Dyest. Rep.* **49** (1960) 184–189.

[8.4] W. Schwind: "Grundlagen und Anwendung der Pigmentfärbe- und Drucktechnik," *Melliand Textilber.* **45** (1964) 529–533, 668–672.

[8.5] M. Glastra, F. van Lamoen: "Zusammenhänge zwischen rheologischen Eigenschaften und Verhalten von Druckpasten beim Rouleauxdruck," *Melliand Textilber.* **46** (1965) 1339–1346.

[8.6] W. Berlenbach: "New Pigment Printing Techniques," *Int. Dyer Text. Printer Bleacher Finish.* **140** (1968) 693–699.

[8.7] W. Schwindt: "Der Pigmentdruck – Bedeutung und Entwicklungstendenzen," *Melliand Textilber.* **50** (1969) 670–672.

[8.8] W. Berlenbach: "Dispersionsverdickungen im Textildruck," *Melliand Textilber. Int.* **53** (1972) 207–210.

[8.9] P. Rasche: "Benzinfreier Druck mit Pigment," *Textilveredlung* **10** (1973) 31–34.

[8.10] U. Perkuhn: "Bronzepigmente im Pigmentdruck," *Int. Text. Bull.* **1981**, 229–236.

# Textile Technology

BURKHARD WULFHORST (Chap. 1), OLIVER MAETSCHKE (Chap. 2), MARKUS OSTERLOH (Chap. 3), ALEXANDER BÜSGEN (Chap. 4), KLAUS-PETER WEBER (Chap. 5), Institut für Textiltechnik, Rheinisch-Westfälische Technische Hochschule, Aachen, Federal Republic of Germany
The entire topic was coordinated by BURKHARD WULFHORST



## 1. General Introduction

The inclusive term *textiles* encompasses items in the following categories:

1) Textile fibers
2) Textile fabrics
3) Goods manufactured from the above

Textile fibers (→ Fibers, 1. Survey) also fall into three categories:

**Natural Fibers.** Natural fibers of vegetable origin include cotton (→ Cotton) and bast fibers (i.e., jute, flax, and hemp; → Cellulose, **A5**, pp. 398–399). The important natural animal fibers are silk (→ Silk) and wool (→ Wool). An example of a natural mineral fiber is asbestos (→ Asbestos).

**Man-Made Fibers Produced from Natural and Synthetic Polymers.** Among the fibers made from natural polymers are viscose ($\rightarrow$ Cellulose, **A5**, pp. 401–413), cupro ($\rightarrow$ Cellulose, **A5**, pp. 413–415), acetate ($\rightarrow$ Cellulose Esters, **A5**, pp. 438–445), alginate, and diene. Man-made fibers from synthetic polymers include polyamide ($\rightarrow$ Fibers, 4. Synthetic Organic, **A10**, 569–579; $\rightarrow$ Polyamides), polyester ($\rightarrow$ Fibers, 4. Synthetic Organic, **A10**, 579–609; $\rightarrow$ Polyesters), acrylic ($\rightarrow$ Fibers, 4. Synthetic Organic, **A10**, pp. 629–642), and elastane (polyurethane, also known as spandex; $\rightarrow$ Fibers, 4. Synthetic Organic, **A10**, pp. 609–615).

**Other Man-Made Fibers.** These are based on inorganic materials, and include carbon fibers ($\rightarrow$ Fibers, 5. Synthetic Inorganic, **A11**, pp. 42–63), metal fibers ($\rightarrow$ Fibers, 5. Synthetic Inorganic, **A11**, pp. 37–42), glass fibers, and spinning paper.

Worldwide production of fibers is estimated at $40 \times 10^6$ t/a, of which 50 % represents cotton.

Fibrous raw materials can be divided into spinnable fibers of limited length (cotton, wool, and man-made fibers that have been cut to a particular length) and so-called endless fibers with essentially unlimited lengths (filaments and filament yarns). Such fibers lend themselves to direct processing into nonwoven fabrics and felts. Spun yarns can be produced from the appropriate fibers by means of various spinning processes. These and the filament yarns are linear in structure, and are used primarily for the manufacture of woven fabrics, braids, knits, and hosiery fabrics. Materials based on other types of surfaces have been developed for special purposes. Thus, bobbinet ("bobbinet lace") presents an open surface comprised of two distinct warp-thread systems, one weft-thread system, and a bobbinet-thread system running diagonally. Fabrics of this type are used for making curtains. Nets are used primarily in the fishing industry. Laces provide surfaces that are both open and transparent. Other materials useful especially in industrial applications include *warp-knitted multiaxial layers* (WIMAG) and *stitch-bonded fabrics* (NVG).

In the case of tufted fabrics, multiple needles are normally used to introduce new threads into a preexisting fabric backing. This results in a series of loops that may subsequently be cut. Fabrics of this type are often used as carpeting material.

Once manufactured, textile fabrics are subject to various types of finishing, including dyeing ($\rightarrow$ Textile Dyeing), printing ($\rightarrow$ Textile Printing), and both wet and dry processing ($\rightarrow$ Textile Auxiliaries). Final make-up leads to products in three major marketing categories: apparel, domestic fabrics and linens, and industrial products. Roughly 50 % of total textile output is directed toward the apparel sector, with each of the others garnering ca. 25 %. The use of industrial fabrics is increasing rapidly, especially in the industrialized nations of the world.

This article is intended to provide an overview of major processes used in converting essentially one-dimensional fibers and yarns into various types of two- and three-dimensional surfaces for further utilization on the basis of the four fundamental technologies spinning (Chap. 2), weaving (Chap. 3), braiding (Chap. 4), and knitting (Chap. 5).

# 2. Spinning

## 2.1. Introduction

The technique of forming staple-fiber yarn includes all the process steps involved in production of staple-fiber yarn from a fibrous raw material. The fibrous raw materials themselves can be either natural products, such as cotton, wool, hemp, and flax, or synthetic materials, such as viscose, polyester, polyamide, or polyacrylic. The work of a spinning mill begins with preparation of the fibers for the actual spinning process, their transformation into a cohesive continuous structure, and the creation of fiber units suitable for further processing.

The basic principle of yarn formation has remained essentially unchanged since the early days of spinning. Hand spinning with simple whorls made of bone, wood, or clay was practiced in the New Stone Age as early as ca. 4000 B.C. By the 13th century A.D. simple hand spinning wheels were already in use in Europe. In 1764, JAMES HARGREAVES constructed the first fully mechanized spinning machine with several spinning positions, and it was this that formed the basis for introduction of industrial spinning into Europe.

Many different spinning processes are practiced depending on the field of application (Table 1), with the cotton-spinning process being the most common. Total annual world production of fiber amounts to about $40 \times 10^6$ t

**Table 1.** Spinning processes

| | Cotton spinning mill (see also Table 2) | | Worsted yarn spinning mill | Carded yarn spinning mill | Semi-worsted yarn spinning mill |
|---|---|---|---|---|---|
| | Ring spinning | OE* Rotor spinning | | | |
| Raw materials | cotton, man-made fibers | | wool, man-made fibers | wool, man-made fibers, cotton | synthetic fibers (wool) |
| Fiber length | $\leq 50$ mm | | $\leq 80-200$ mm | $\leq 120$ mm | $80-200$ mm |
| Spinning-machine set-up | blowroom machines → cards → drawing frames → flyers → ring-spinning frames → spooling frames; rotor spinning | | preparatory machines → card sets → combs → drawing machines gill boxes → flyers/apron frames → ring-spinning frames | preparatory machines → card sets → tape divider/ rubber condensor → carded wool ring-spinning frames; mules; carded wool rotor spinning machines | preparatory machines → card sets → drawing frames → flyers/rubbing frame → ring-spinning frames; covering machines; OE* rotor spinning machines |
| Yarn gauge | 1000–20 tex (carded) 5 tex (combed) | 1000–12 tex | 84–10 tex | 1000–50 tex | 10 000–25 tex |
| Yarn type | firm, hairy (carded) smooth (combed) | less firm, smooth with wrapped fibers | smooth, shiny | less firm, bulky, voluminous, high stretch | |
| Applications | underwear, shirts, blouses, handkerchiefs, sport clothes, home furnishing fabrics, table linens, domestic items | | outerwear | outerwear (cloth), home furnishing fabrics, industrial goods (filters) | Floor coverings, home furnishing fabrics (tapestry, upholstery), technical textiles |

**Table 1.** (Continued)

| | Cotton spinning mill (see also Table 2) | | Worsted yarn spinning mill | Carded yarn spinning mill | Semi-worsted yarn spinning mill |
| --- | --- | --- | --- | --- | --- |
| | Ring spinning | OE* Rotor spinning | | | |
| Raw materials | flax | | hemp | jute | |
| Fiber length | ultimate fibers 25–30 mm<br>bast fibers 200–1500 mm | | ultimate fibers 15–25 mm<br>bast fibers 1000–2500 mm | ultimate fibers 2–3 mm<br>bast fibers 1000–2500 mm | |
| Spinning-machine set-up | heckling machines<br>(industrial fibers 100–600 mm)<br>gill boxes<br>direct spinning / flyers<br>flyer-/ring-spinning frames | | heckling machines<br>gill boxes/<br>drawing frames<br>fine gill spinning frames / flyer-/ring-spinning frames | heckling machines<br>(bast fibers 200–250 mm)<br>breakers<br>jute-spinning frames | |
| Yarn gauge | 580–7.2 tex | spinning limit 8.4 tex | 1700–50 tex | dependent upon fiber type;<br>84 tex to > 840 tex | |
| Yarn type | irregular, smooth, shiny, firm | | irregular | high strength | |
| Applications | woven fabrics<br>knitted goods (blended)<br>nets<br>women's/men's outerwear<br>household fabrics (upholstery, linens)<br>technical textiles (tarpaulins, sails)<br>twisted yarns | | technical textiles (straps, drive belts)<br>carpet backing<br>twisted yarns | technical textiles (straps, hose linings)<br>upholstery fabric, carpets<br>garments (only in blends) | |

* OE = Open end.

**Table 2.** Fiber production in 1000 t (1991)

|  | Staple fiber | Filament yarn | Total |
|---|---|---|---|
| Cotton | 20 793 |  | 20 793 |
| Country of origin |  |  |  |
|   USA | 3 835 |  | 3 835 |
|   Russia | 2 491 |  | 2 491 |
|   China | 5 675 |  | 5 675 |
|   Pakistan | 2 676 |  | 2 176 |
|   India | 2 023 |  | 2 023 |
| Wool | 3 010 |  | 3 010 |
| Silk (raw silk) |  |  | 76 |
| Man-made fibers | 9 658 | 8 014 | 17 673 |
|   Natural origin | 1 682 | 833 | 2 515 |
|   Synthetic | 7 976 | 7 181 | 15 158 |

(Table 2), where cotton accounts for $20.7 \times 10^6$ t (i.e., ca. 50 % of total fiber production). Given that many man-made fibers consist of a combination of filament yarns and staple fibers, and that the corresponding staple fibers are processed in part using cotton-spinning technology, then it is clearly this process that predominates. For this reason the cotton-spinning process is the only one that will be described here.

The cotton-spinning process – also known as the short-staple or three-cylinder process – is suited to all types of fibers with lengths ranging up to ca. 40 mm, and it is very flexible with regard to properties and fields of application for the resulting yarns. Staple-fiber yarns made according to the cotton-spinning process are further processed into fabrics, knits, and braided materials for use as apparel, domestic fabrics and linens, and industrial products.

## 2.2. Steps in the Cotton-Spinning Process

The processing of fiber into yarn is divided into several stages, as indicated in Scheme 1. Depending on the desired yarn properties and the nature of the fibrous raw materials, various mechanical arrangements and spinning methods are employed. In each case it is necessary to establish an optimum compromise between ideal yarn characteristics and economical production.

The matrix in Scheme 1 underscores the main functions associated with the various process steps. The process step *mixing* covers both mixing of a nonhomogeneous fiber substrate and the dosed blending of fibers of differing origin. *Opening* refers to the breaking up of compressed

fiber packets to yield individual fibers. *Cleaning* refers to the removal of interfering particles, such as wood or leaf components and fiber knots ("naps") that cannot be opened. *Parallelization* refers to alignment (insofar as possible) of all the fibers in a single direction. *Drawing* refers to the extension ("drafting") of an oriented collection of fibers. The purpose of drawing is to promote thorough mixing and blending of multiple raw materials as well as further homogenization of sliver weight. This is facilitated by the doubling of several slivers prior to actual drawing.

## 2.3. Harvesting and Ginning

Almost all cotton is now harvested mechanically in accordance with either the "stripper" or the "picker" principle. In *stripping*, all plant components are removed simultaneously by the action of rotating brush rolls. A *picker* incorporates rotating spindle fingers that separate cotton fibers selectively. The raw cotton is then freed of coarse impurities and subsequently separated from seeds in a process known as ginning. Finally, the fiber material is compacted into bales weighing ca. 250 kg each for trade or direct transport to a spinning mill.

The most important cotton-producing countries are listed in Table 2. Cotton bales are traded throughout the world at a price (ca. 2–6 DM/kg) that depends on quality as well as market demand. The most important quality characteristics of cotton include fiber length, fiber fineness, fiber strength, dirt content, and color.

## 2.4. Preparation for Spinning

Two examples of fiber preparation systems are illustrated in Figure 1. The fiber is transported pneumatically from one machine to the next, with a throughput in the range 500–800 kg/h depending on mechanical design.

The specific tasks associated with spinning preparation include:

1) Opening of the bales
2) Separation of interfering particles (refuse)
3) Dedusting
4) Opening of fiber flocks
5) Homogenization of the raw material
6) Mixing of various raw materials

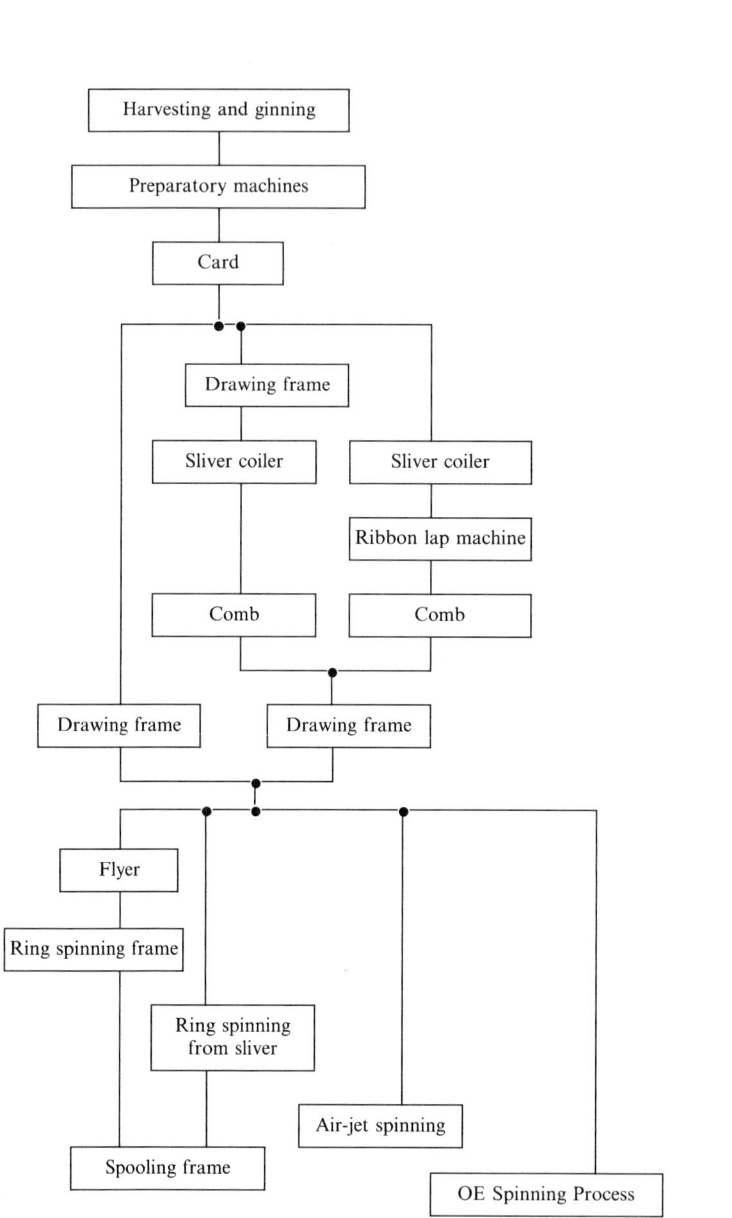

**Scheme 1.** Processing of fiber into yarn

**Bale Opening.** Bale opening refers to breaking up of the bales into individual flocks (see Fig. 1). The nature and intensity of the opening process has a decisive influence on all subsequent processing steps. The more finely the cotton is opened, the more interfering particles will be found on the surface, from which they can more readily be eliminated. In modern bale-opening systems, up to 80 bales are placed side-by-side on the floor and then stripped into layers by a pro-

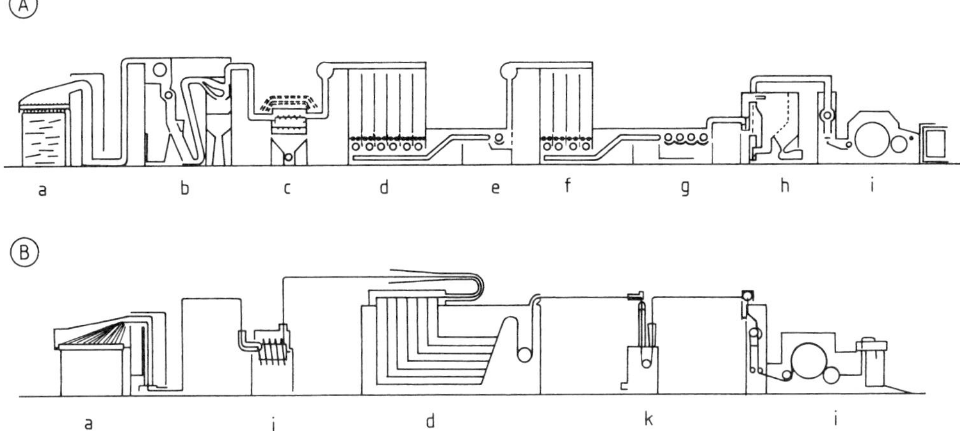

**Figure 1.** Examples of fiber preparation systems offered by two different manufacturers
A) Trützschler; B) Rieter Spinning Systems
a) Bale opener; b) Particle separator; c) Twin-roll cleaner; d) Six-fold mixer; e) Single-roll intensive cleaner; f) Four-fold mixer; g) Quadruple-roll intensive cleaner; h) Dedusting device; i) Card; j) Free-beating cleaner; k) Intensive cleaner

grammable bale opener. Opening into fiber flocks is accomplished mechanically with the aid of milling rolls, after which the flocks are transported further in a stream of air with concurrent dedusting.

**Opening, Cleaning.** The main functions of opening and cleaning machines are additional breaking up of the fiber flocks, elimination of interfering particles (refuse), and dedusting of the cotton. Two basic principles are utilized in accomplishing this task.

With a *free-beating* system, flocks are caught in free flight and accelerated by the working components (beaters) of the cleaner. Interfering particles are separated on grids through the combined effects of centrifugal force and gravity. Such machines may incorporate one or more rolls for this purpose, as illustrated by the example in Figure 2 (see also Fig. 1).

In the case of *constrained beating* the flocks are held between two feed rolls, or between a feed plate and a feed roll, while the beating elements perform their function. Depending on the degree of opening, nose-beaters, spikes, or saw teeth are used as the beating elements (Fig. 3). As a result of the restraining effect, the extent of the opening achieved with constrained beating is greater, but the process is also more aggressive than free beating. Cleaning is accomplished with grids, knives, and exhaust hoods.

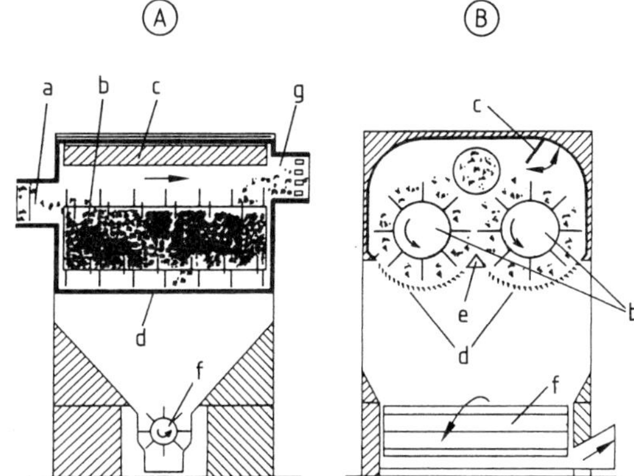

**Figure 2.** Schematic diagram of a twin-roll cleaner
A) Cross section; B) End view
a) Fiber inlet with guide plate; b) Cleaning rolls;
c) Flashing; d) Grid; e) Guide; f) Rotary vane sluice;
g) Fiber discharge port with fresh-air inlet

**Mixing.** Mixing comprises two functions: blending and homogenization. *Blending* refers to the establishment of a definite quantitative relationship between various raw-material partners. This is accomplished through one or more of the following processes:

1) Manual deposition of the appropriate components (imprecise)
2) Automatic bale opening
3) Use of weighing-hopper feeders (batch) and continuous dosing devices inside the preparatory machines

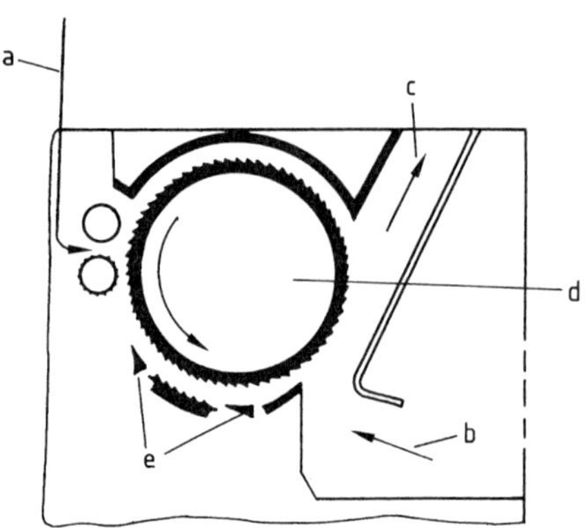

**Figure 3.** Schematic diagram of a sawtooth cleaner
a) Fiber inlet; b) Air inlet; c) Fiber discharge in a stream of air; d) Beating roll; e) Knives (grids) for particle separation

The purpose of *homogenization* is to achieve a distribution of the various components as homogeneous as possible in the final product. With the *mixing-chamber* principle, the maximum available space is filled in horizontal layers and subsequently processed vertically from one side. Multiple mixers (Fig. 4) consist of several independent chambers arranged one behind the other. In this case a distinction can be made between two basic working principles: flocks may be fed in all at once and processed at different times, or fibers that are fed in at different times may be processed simultaneously.

**Carding.** The last machine required in spinning preparation is the card [Fig. 5; see also Fig. 1 (i)]. The functions of a card are

1)  Elimination of interfering particles and short fibers
2)  Opening of the flocks into individual fibers
3)  Parallelization of the fibers
4)  Mixing
5)  Drawing
6)  Slivering
7)  Sliver coiling

The various working elements inside a card are equipped with steel "clothings" (for details see Fig. 5). The fibers themselves are brought between sets of card tips arranged both opposite and adjacent to one another.

Carding (i.e., the separation and parallelization of individual fibers) is a result of relative motion occurring between clothings on the several working elements. If two working elements subject to motion relative to each other are outfitted with card tips inclined in opposite directions, the system is arranged in what is called the *carding position* (Fig. 6). The intensity of the actual carding action is a function of the speed differential between the carding elements and the angles of inclination of the corresponding clothings. If the working elements are instead inclined in the same direction, fibers pass from one fitting to the other, and the system is said to be arranged in the receiving or transfer position.

Homogenized fibrous material is introduced into the card (Fig. 5) via a pneumatic flock feeder (a). The material is then fed to the liker-in (c), a roll covered with saw teeth, via the feed device (b). Here the flocks are also opened as a consequence of a difference in peripheral speeds. Simultaneously, 70–75 % of all interfering parti-

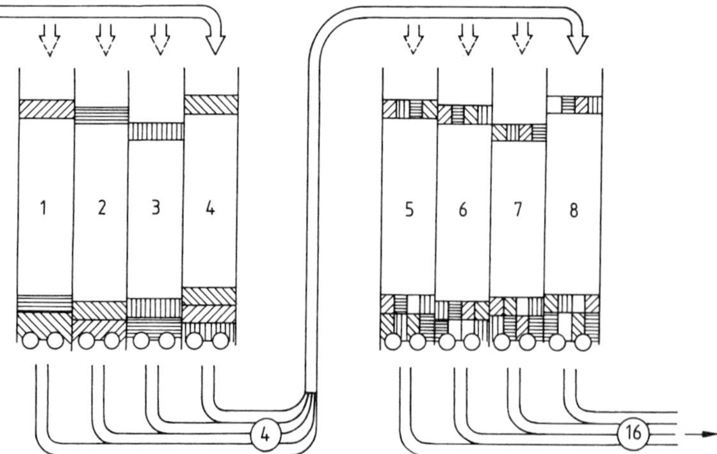

**Figure 4.** Multiple mixer

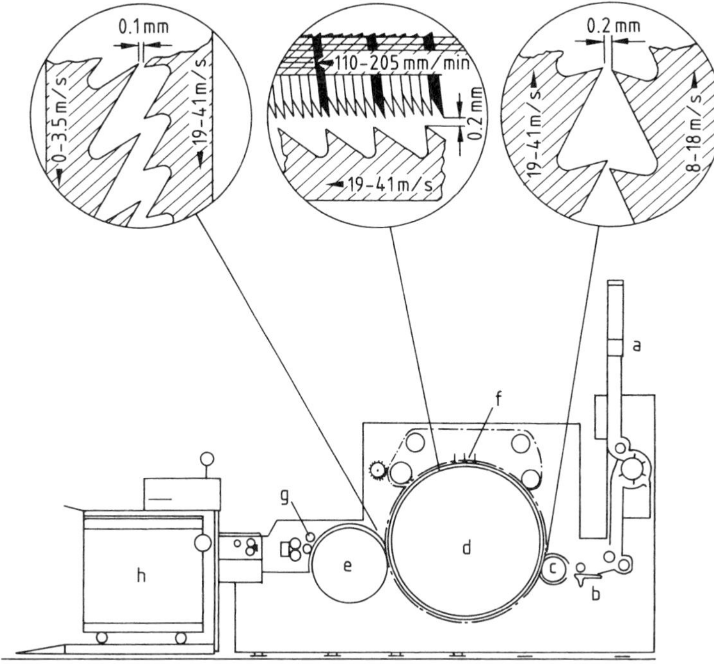



**Figure 5.** Schematic diagram of a card
a) Pneumatic feeder; b) Feed device; c) Liker-in; d) Tambour or drum; e) Doffer; f) Card flats; g) Stripper rolls; h) Can coiler

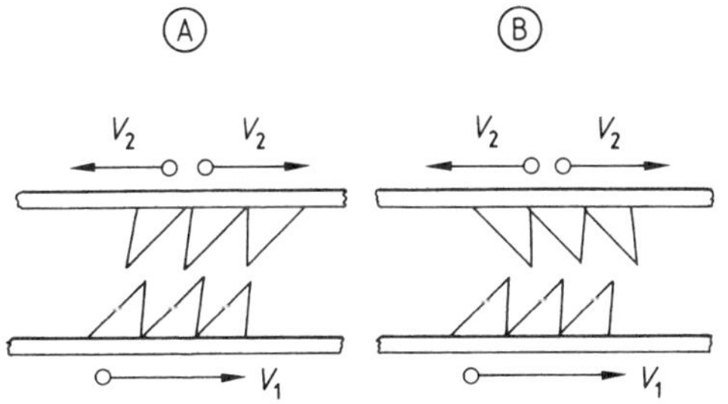



**Figure 6.** Operating principle of a card (schematic)
A) Carding zone; B) Transfer zone
$v_1, v_2$ = velocities of the working elements, where $v_1 \gg v_2$

cles accumulating at the card are eliminated at the liker-in with the aid of knives. The principal operating component of a card is the tambour (d). The surface of a tambour is covered with more than $4 \times 10^6$ card tips, and the tambour moves with a peripheral speed of ca. 26 m/s. Between the liker-in and the tambour is a carding zone, which ensures further separation of the fibers. However, most of the carding effect takes place between the tambour and the card flats (f),

which are arranged in a carding position relative to each other. It is here that separation into individual fibers is achieved, along with fiber parallelization and the simultaneous elimination of particles and short fibers. Processing of manmade fibers is accomplished with fixed flat elements. Modern cards often incorporate additional fixed carding bars and dirt separators in front of and behind the card flats. Even though the region between the tambour and the doffer (e) constitutes another carding zone, fibers are nevertheless transferred from the tambour to the more slowly moving doffer as a result of the angle of inclination of the fittings. The observed transfer behavior is described by a *transfer factor*, which specifies the fraction of the fibers on the tambour that is expected to pass to the doffer with each revolution of the tambour. Interfering particles and dust are eliminated from fibrous material remaining on the tambour by means of grids and knives located beneath the tambour itself.

The condensed fiber web (fleece) is removed from the doffer by means of stripper rolls (g). Remaining interfering particles are pulverized by two squeezing rolls before the fleece is formed into a sliver. A coiler is then used to deposit the sliver in cycloidal form into a revolving can (h). Cards are provided with short- and long-term control devices to compensate for fluctuations in the mass of the sliver.

## 2.5. Spinning Methods

### 2.5.1. Drawing

After carding, the fiber slivers must be doubled and drawn. *Doubling*, a process in which several slivers are combined and subsequently drawn, results in a homogenization of the sliver. A regulating device is often used to reduce fluctuations in the sliver mass. This usually involves mechanical scanning of the incoming slivers, with the feed rate controlled as a function of the measured sliver mass. In the case of slivers of differing origin, doubling can also be used to achieve blending.

A drawing frame (Fig. 7) generally consists of a can creel (a), a framework for the drive and control units, the *drafting* (drawing) *unit* (b), and a can coiler with an automatic changer (c). The drafting unit comprises two or three pairs of rolls arranged one behind the other. A continuous fiber band is drafted between the "nip lines" because the peripheral velocities of successive cylinder pairs increase in the direction of transport. Modern drawing frames are equipped with three pairs of cylinders, resulting in two drafting zones. A light preliminary draft (ca. 1.05-fold) in

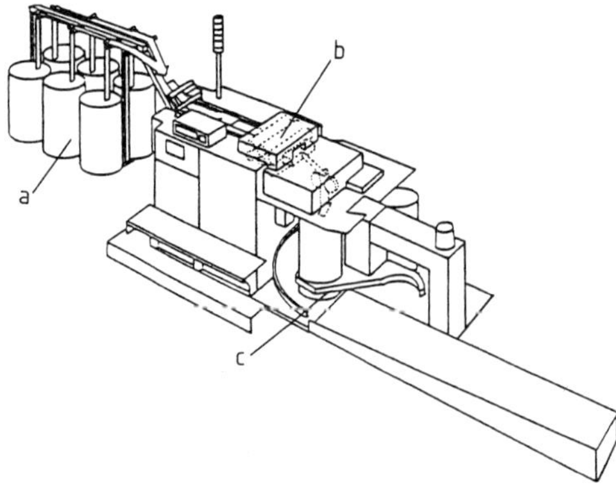



**Figure 7.** Schematic representation of a drawing (drafting) frame
a) Can creel; b) Drafting unit; c) Can changer

the pre-drafting zone is generally followed by ca. 6–8-fold principal draft. The drafts in the individual zones must be multiplied in order to compute an overall draft. The stretched band (sliver) is deposited in cans in cycloidal form. Most spinning processes combine two drawing passages in sequence to achieve effective homogenization, parallelization, and mixing. A typical sliver exit speed is 500–1000 m/min.

### 2.5.2. Combing

The combing process is designed to remove short fibers, impurities, and naps, permitting the spinning of fine yarns and an improvement in the yarn quality. The percentage by weight of combed-out material ("noil") is subject to adjustment. Depending on the raw material, noil percentages between 5 and 30 are common.

The fibrous material must be specially prepared before being fed to the combing machine (see Scheme 1). On a *comber lap machine*, several parallel slivers are combined into a single lap, which is then introduced into the combing machine. Successful removal of fiber "hooks" requires taking the orientation of the hooks into account, because the working principle of the combing machine is such that it is most effective with "guide hooks" – those in which the hook is located at the forward end of the fiber relative to the direction of travel. A fiber hook at the rear of a fiber is called a "trailing hook." The ratio of guide to trailing hooks in a card sliver is typically ca. 15:65. For effective hook removal there must be an even number of machine passages between the card and the combing machine.

Combing is a batch process. Inside the combing machine (Fig. 8) the fiber lap (a) is fed to the feed rolls (b) via a lap trough. These rolls transport the fleece between the top nipper (c) and the bottom nipper (d) into the vicinity of the combing cylinder (e). The two sets of nippers hold the fibers while the combing cylinder combs the fiber tuft. The nippers then open and move toward the detaching rolls (f). The detaching rolls move the slubbing (h) back so that the combed-out fiber tuft can be returned to the slubbing. Subsequently, the detaching rolls rotate in the direction of transport and pull the fibers of the combed-out tuft out of the fleece. A top comb (g) then pierces the material to comb the back end of the fiber tuft, after which the nippers return to their original position and a new combing cycle begins. A

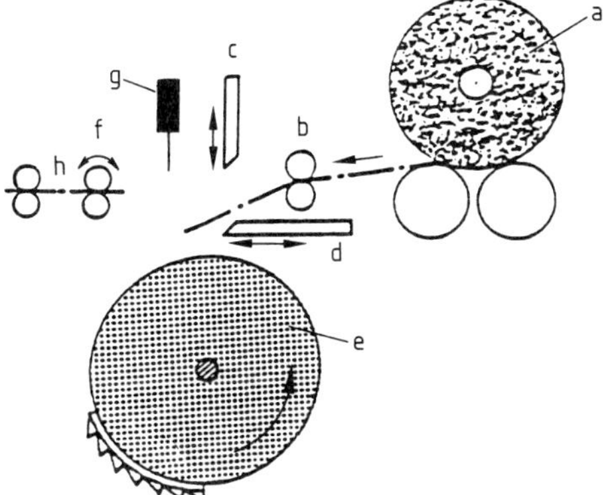

**Figure 8.** Operating principle of a combing machine
a) Lap; b) Feed rolls; c) Top nipper; d) Bottom nipper;
e) Combing cylinder; f) Detaching rolls; g) Top comb;
h) Slubbing (combed sliver)

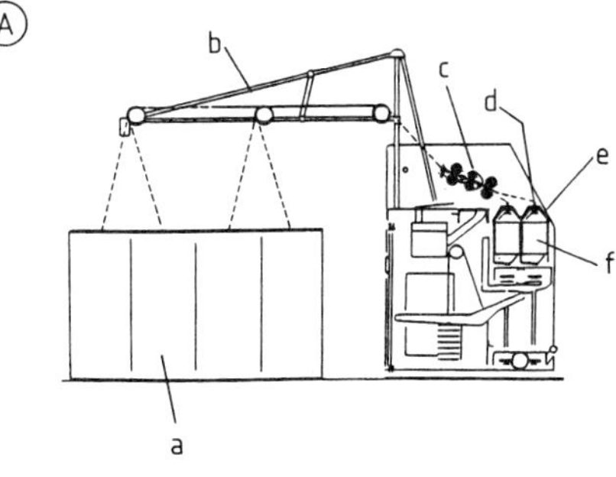

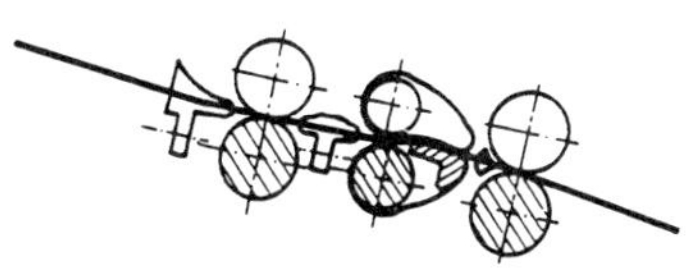

**Figure 9.** Typical arrangement of a flyer
A) Overview; B) Detailed sketch of the drafting unit
a) Can creel; b) Framework; c) Drafting unit;
d) Opening in the top of the flyer leg; e) Flyer leg;
f) Flyer bobbin

rate of 300–400 combing cycles per minute is normal.

### 2.5.3. Conventional Spinning

Conventional spinning refers to the combination of a flyer, a ring-spinning frame, and a spooling frame (see Scheme 1). Today, about 80 % of all staple-fiber yarn is spun according to this process. In an unconventional spinning process the three operations are combined in a single machine.

**Flyer.** The functions of the flyer (or "roving machine") are further drafting of the sliver, coupled with the introduction of protection against faulty drafting in the form of a protective twist (Fig. 9). The twist must be adequate to prevent faulty drafting, but it must also be sufficiently limited to permit a slubbing to be drafted up to the desired final yarn fineness ("gauge") in the drafting unit of the ring-spinning frame.

Slivers are first drafted in the flyer drafting unit (c), which is usually a 3-roll, 2-apron system (see Fig. 9). Aprons are used to guide the fibers into the drafting zone. Drafts as high as ca. 50-fold can be achieved with flyer drafting units. From the drafting unit the drafted sliver ("flyer roving") is transported to the flyer leg (e) via an opening in the top of the flyer leg (d). The sliver exits through an opening at the bottom end of the flyer leg and is led around a spring finger to the bobbin (f). With each revolution of the flyer

the roving acquires one twist. Winding occurs owing to a leading or lagging of the bobbin with respect to the flyer leg. The vertical displacement required for smooth winding is attended to by the bobbin. Since the flyer operates at a constant feed rate, the vertical displacement and rotational speed of the bobbin are subject to constant adjustment as a function of the spool diameter, thereby ensuring uniform twisting and winding. The current velocity limit for flyers is 1300–1500 rpm.

As a result of technical limitations and the fact that it cannot easily be automated, efforts have long been underway to find a replacement for the flyer. Unconventional spinning processes have already made flyers partially obsolete. Certain ring-spinning frames also permit the spinning of slivers directly, although their share of the market is still very small.

**Ring-Spinning Frames.** In order for yarn to be produced from slubbing, further drafting is required up to the desired yarn gauge, and the appropriate degree of twist must be introduced. All of this occurs on the ring-spinning frame (Fig. 10), after which the finished yarn must be wound.

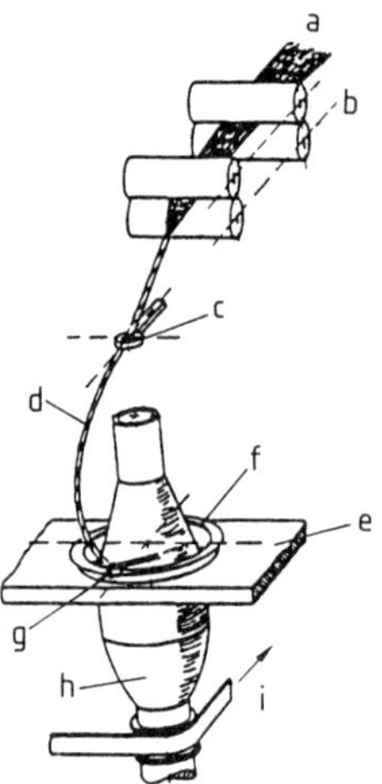

**Figure 10.** Principle of the ring-spinning process
a) Slubbings;  b) Drafting unit;  c) Thread guide;
d) Rotating thread balloon;  e) Ring rail;  f) Ring;
g) Traveler;  h) Spindle;  i) Spindle drive

As shown in Figure 10, slubbings (a) are fed from a creel to a double-apron drafting unit (b) where they are drafted to the desired yarn gauge. The thread passes through a thread guide (c) mounted concentric to the spinning axis and then through the traveler (g) to the spindle (h). The traveler is dragged along by rotation of the spindle, causing it to rotate on the ring (f), whereby the yarn is twisted once with each rotation of the traveler. This twist is propagated up to the spinning triangle. The traveler lags behind the spindle, facilitating winding of the yarn. The extent of the twist $T$ applied to the yarn can be adjusted by altering the ratio of delivery speed to rotational speed of the spindle according to the relationship:

$$T = n_T/f$$
$$n_T = n_{Spi} - f/(d \cdot \pi)$$

where $f$ is the feed rate at the drafting-unit exit, $d$ is the current spindle diameter, $n_T$ is the traveler speed, and $n_{Spi}$ the spindle speed.

The stroke displacement required for satisfactory winding is achieved by means of the ring rail (e). The performance limit for ring spinning is determined by the maximum traveler speed, which is currently ca. 40 m/s, producing local temperatures as high as 450 °C. With higher traveler speeds it is impossible to provide adequate dissipation of the heat generated, resulting in potential damage to the traveler. In the case of

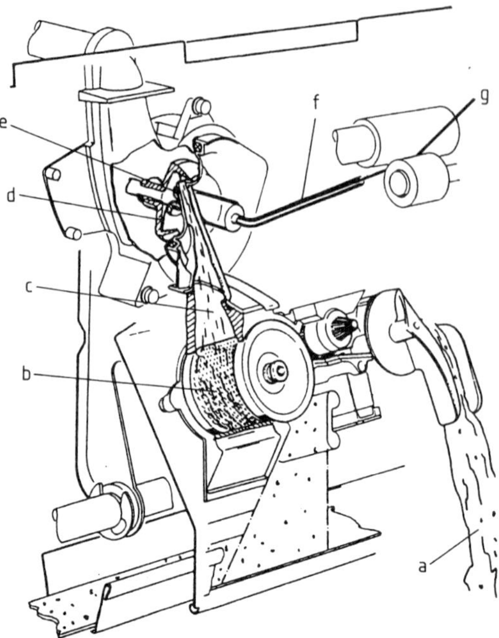

**Figure 11.** Open-end (OE) rotor spinning unit
a) Sliver;  b) Opening roller;  c) Fiber-guiding channel;
d) Rotor;  e) Navel;  f) Doff tube;  g) Yarn

man-made fibers, such heat build-up may also lead to spots of melted fiber in the yarn.

Thread forces in the thread balloon (d) can be reduced with the aid of *balloon-constricting rings*, which diminish the magnitude of centrifugal forces acting on the yarn. Automatic bobbin changing has also become a standard feature of modern ring-spinning frames. Although automatic devices have been developed for repairing breaks in the yarn, this task is still largely performed manually.

### 2.5.4. Unconventional Spinning

**Open-End (OE) Rotor Spinning.** Carded sliver, drawn sliver, and combed sliver can all serve as starting materials for OE rotor spinning. In the spinning unit itself (Fig. 11) the sliver (a) is fed to an opening roller (b) via a feed roll and a feed plate. The surface of the opening roller rotates at 20–30 m/s, and leads to very intensive combing and separation of the fibers. Refuse particles are eliminated below the opening roller, whereas the fibers themselves are removed in a stream of air and accelerated further in a conical fiber-guiding channel (c). The fiber-guiding channel ends in a rotor (d) that revolves with a velocity of up to 130 000 rpm. Centrifugal force causes the fibers to slide along the rotor wall and collect as a ring in the rotor groove. The OE rotor-spinning process is called an "open-end" process because the sliver is broken up into individual fibers immediately prior to yarn formation. Since the fibers are accelerated throughout the process, nearly all fibers collecting in the groove assume an extended configuration. Spinning is accomplished with the help of a previously finished piece of yarn that is fed into the rotor through the doff (discharge) tube (f) and the navel (e). As this piece of yarn rotates in the rotor groove, fibers lying in the groove become bound to its end. Subsequent successive withdrawal of the yarn through the navel and the doff tube results in a continuous spinning process. Yarns can be spun by this method with a fineness ("count") down to 12 tex (where "tex" = g/1000 m).

One technological disadvantage of the OE rotor-spinning process is the presence of "wrapped fibers." Fibers fed onto the previously twisted end of yarn in the rotor groove wrap themselves in such a way that one end of each fiber is oriented in the direction of the yarn twist, whereas the other end wraps around the yarn in the opposite direction (Fig. 12). Wrapped fibers make rotor yarns unsuitable for use in cutpile

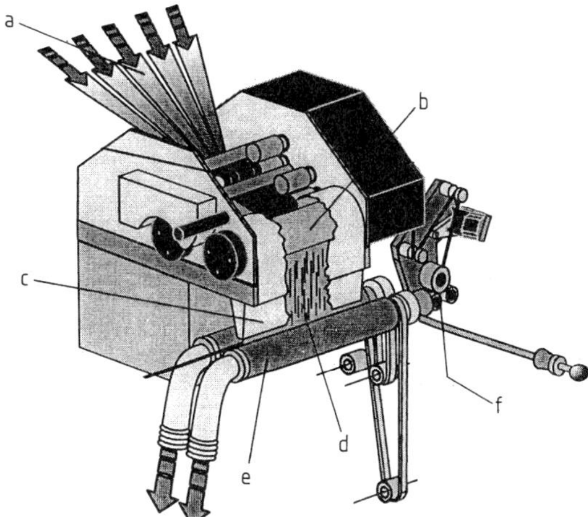

**Figure 13.** Friction spinning unit
a) Slivers;   b) Opening unit;   c) Guide channel;
d) Formation zone;   e) Friction elements;   f) Winding unit;

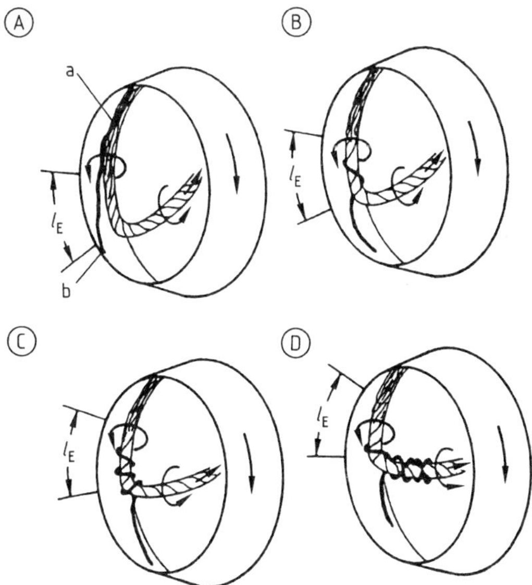

**Figure 12.** Steps (A–D) in the formation of a wrapped fiber
a) Fiber head; b) Fiber end
$l_E$ = length of yarn-forming zone

materials (e.g., carpets and velvet) because they interfere with opening of the pile threads.

As a result of their structure, OE rotor yarns are not as strong as comparable ring yarns. However, the OE rotor-spinning process is more economical, because it avoids the use of both flyers and spooling machines, and an OE rotor-spinning machine can be fully automated.

**OE Friction Spinning.** Another spinning process based on the open-end principle is OE friction spinning (Fig. 13). This process starts with several slivers (a) that are broken up into individual fibers by an opening unit (b). The fibers are subsequently transported in a stream of air through a guide channel (c) into the yarn-formation zone (d). The frictional forces required for conferring a twist are generated by friction elements (e) whose surfaces move perpendicular to the direction of yarn withdrawal. In most cases this entails two perforated, cylindrical friction drums that rotate in a common direction. Fibers are fed into the wedge-shaped gap between the two perforated rolls. Reduced pressure in the region of the wedge ensures that the fibers will be sucked through the perforations and onto the rolls themselves. Normal and frictional forces arising in the process cause the fibers to rotate and bind to the open end of the yarn.

OE friction yarns offer relatively little strength, so the process has gained only limited acceptance, and then only for yarns in the coarse

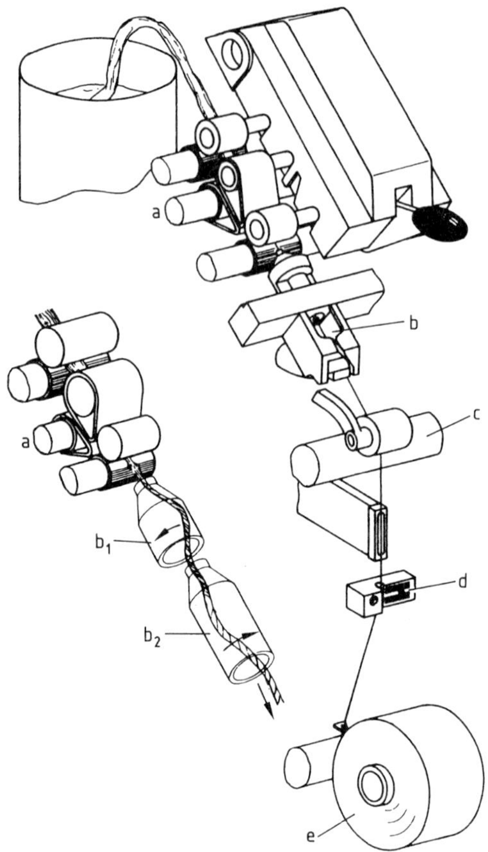

**Figure 14.** Schematic diagram of an air-jet spinning device
a) Double-apron drafting unit; b) Air jets; c) Discharge
rolls; d) Yarn cleaner; e) Spool

range. However, it is suitable for the production of hybrid yarns for industrial applications.

**Air-Jet Spinning.** Air-jet spinning (Fig. 14) is a process that is becoming more and more important. Here the take-off sliver is first drafted to the desired yarn count by means of a drafting unit (a). Two air-jets (b1, b2) are located immediately behind the exit of the drafting unit. Air flowing tangentially through the twist jet (b2) confers upon the thread a "false twist" that is propagated as far as the exit of the drafting unit. This temporary twist is confined to a definite region, and it dissipates as the yarn leaves the false-twist zone. The other jet (b1) has the task of spreading out individual fibers that may protrude from the thread created by the false twist. This is accomplished with the help of air flowing through tangential holes arranged diagonally relative to the direction of yarn transport. Once the false twist has been eliminated through yarn withdrawal, the resulting yarn core is completely free of twist. Fiber ends that have been spread apart by the second jet wrap themselves around the core as the false twist dissipates. Axial force exerted in the course of this wrapping around the yarn core causes yarn compression as a result of frictional contact.

Very fine yarns (down to ca. 10 tex) displaying good uniformity can be spun with the air-jet process. Good yarn characteristics depend upon a fibrous raw material that is fine, strong, and as uniform in length as possible. For this reason the process is best suited to man-made fibers or mixed yarns based on a combination of cotton and man-made fibers. Fine yarns can be produced very economically in this way.

Other spinning processes are known in addition to those described, but none has yet achieved commercial significance.

## 2.6. Automation

All the machines used in preparation for spinning, including the cards, are fully automated, and fibers are transported pneumatically from one machine to the next. The card produces a carded sliver, which is deposited in cans, and these cans then become available for subsequent transport.

Slivers are transported from the card via the drawing frame or the combing machine to the flyer either in cans or as laps. Different forms of transport have been adopted in different mills, but much of the work is still accomplished by hand.

In the case of a conventional spinning line (flyer, ring-spinning frame, winding machine) there exists a fully automated approach that is increasingly being adopted. Unconventional spinning systems are already fully automated. Online quality assurance is usually introduced simultaneously with automation.

## 2.7. Recycling

The recycling of textile raw materials has a long and exemplary history in both the industrial and private sectors. For example, one company in Bocholt (Federal Republic of Germany) has been recycling textiles for more than 130 years.

The Italian city of Prato near Florence is also famous in this respect. Initially, a shortage of raw materials, especially natural materials, always constituted a major incentive.

The total quantity of new textiles introduced into the German market in 1992 is estimated at $2.1 \times 10^6$ t. It must be assumed that roughly the same amount of material is subject annually to disposal. Recycling in this context may take various forms, including *material recycling, chemical recycling* (recovery of chemical starting materials), and *energy recycling* (recovery of energy consumed in production).

Until now, material recycling has been the most common type of recycling in the textile industry, but in the future chemical and energy recycling will necessarily increase in importance. In the case of material recycling, various methods are available, depending in part on the intended purpose of the textiles with respect to the three categories apparel, domestic fabrics and linens, and industrial products. In the case of old textiles there exists the possibility of reuse in the form of second-hand materials. Old-clothes collection represents a second way of disposing of waste textiles. Clothes accumulated in this way are sorted in special facilities into items that are still usable, or from which useful fabric can be recovered, and waste. A third possibility is direct disposal as trash. In this case, fibers can sometimes be recovered through tearing processes, providing materials that can be used as secondary raw materials.

The incorporation of secondary raw materials into carded yarn has a long tradition. Recycling in a cotton-spinning operation can be accomplished either offline or online. In the offline approach, production waste is collected, prepared for use, and returned to the process as a secondary raw material. In the online process, prepared fibers are reintroduced into the same spinning procedure from which they were originally removed in the form of production waste.

# 3. Weaving

## 3.1. Introduction

The technique of weaving has been in use for about 5000 years as a way of producing textile fabrics. Weaving is recognized worldwide as the dominant technology for this purpose. World

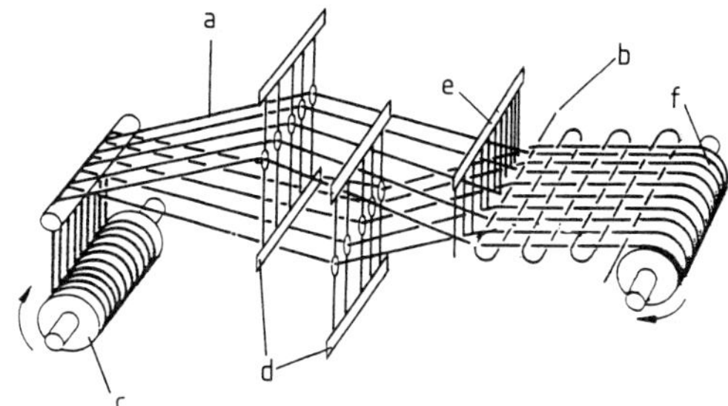

**Figure 15.** Schematic representation of a loom
a) Warp threads; b) Weft thread; c) Warp beam; d) Heald frames; e) Reed; f) Cloth beam

output of woven fabrics amounts to ca. $30 \times 10^6$ t/a [21], [22]. Fields of application for woven fabrics include apparel (especially men's and ladies' garments), table and bed linens, domestic fabrics (carpeting, upholstery, drapery), and industrial products.

The economic importance of industrial textiles is increasing. These now account for about 25 % of all woven fabrics, and include reinforced composites commonly used in light construction as well as geotextiles for road, tunnel, and dam construction. Industrial textiles are also used as protective or filtering materials, and in medical applications.

A woven fabric consists of two or more thread systems that cross each other at right angles, known as the *warp* and the *weft* (see Fig. 15). The warp threads (a) run in the direction of production, and are wound onto a warp beam (c) in the desired width and spacing (ends per inch). Alternate raising and lowering of the heald frames (d) causes warp threads that are supported by these frames to form a passageway (*shed*), through which the weft thread (b) can be inserted. After weft insertion the weft thread is beaten firmly against the edge of the fabric by a reed (e). The relative positions of warp threads are then changed, binding the newly inserted weft thread in its place.

## 3.2. Steps in the Weaving Process

A summary of the process steps from ring-spun yarn to loom-state fabric is provided in Figure 16. Here a distinction is made between preparation for weaving and the weaving process itself.

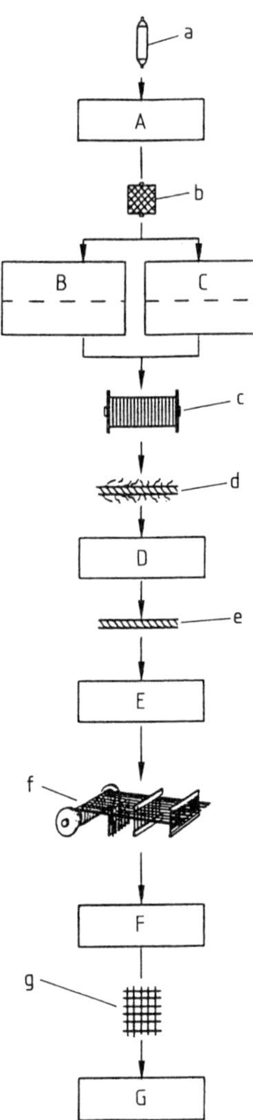

**Figure 16.** Steps in the weaving process
A) Winding;   B) Warping;   C) Sectional   warping;
D) Sizing; E) Drawing-in; F) Weaving; G) Inspection
a) Cops;   b) Crosswound   package;   c) Warp   beam;
d) Unsized   yarn;   e) Sized   yarn;   f) Drawn-in   warp;
g) Woven cloth

Preparation for weaving consists of several operations: winding (A), warping (B) or sectional warping (C), sizing (D), and drawing-in (E). Yarns produced on a ring-spinning frame are supplied as spindles (cops), which must always be rewound in the form of cross-wound bobbins. OE rotor-spinning machines, OE friction-spinning machines, and air-jet spinning machines all deliver cross-wound bobbins directly. Yarn supplies for the warp and the weft are prepared in the *winding* process. Warp threads are then processed into a warp beam either by *warping* or by *sectional warping*. In the subsequent sizing operation the outer surface of the yarn is smoothed by application of a sizing agent. This is intended to reduce friction as well as thread damage during the succeeding weaving process. The fact that the warp is drawn in at several places – by the warp-stop motion, the heald frames, and the reed – minimizes the time during which the weaving machine (*loom*) is in disuse in the course of resetting. If twisted yarn happens to be required, twisting would also be included as one of the steps in weaving preparation.

In the actual weaving process (F in Fig. 16), the two thread systems (warp and weft) are linked together by interlacing (see Fig. 15). The nature of the crossing is called the *weave* or *fabric construction*, and it is this that determines the mechanical and technological properties of a fabric. In the fabric inspection step (G), any flaws in the fabric are detected and recorded. This is followed by textile finishing and make-up.

## 3.3. Weaving Preparation

### 3.3.1. Winding

The purpose of winding (Fig. 17) is to maximize efficiency in the subsequent processes. Winding ensures both the availability of an adequate supply of thread and optimal unwinding characteristics. An integral feature of the winding process is an improvement in yarn quality through the removal of defects (e.g., thick or thin regions).

With respect to spools, a distinction is made between those with parallel winding and cross-wound packages (Fig. 18). Cross-wound packages are used for weaving because of their superior unwinding properties. A spool with parallel winding has a higher package density, and is usually unwound tangentially.

**Random Winding.** The characteristic feature of random winding (Fig. 19 A) is peripheral drive of the tube with the aid of a grooved drum or a smooth friction roll and a moving thread guide (Fig. 17). Thread is thereby wound onto the spool at a constant helix angle $\alpha$ (Fig. 19; see also Fig. 18 A). At the same time, the number of windings per thread-guide pass, known as the winding ratio $W$, decreases with increasing spool

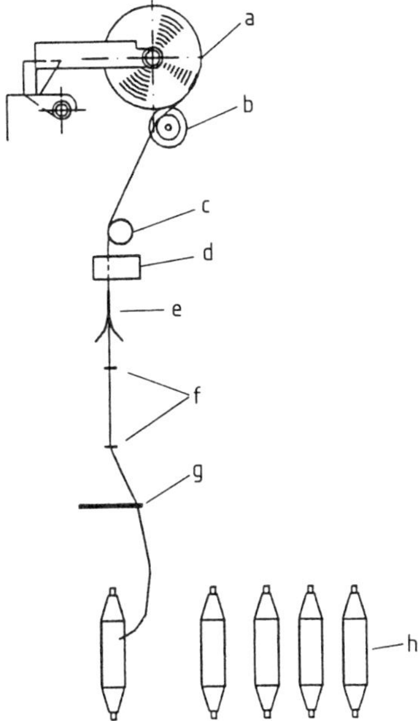

**Figure 17.** Schematic representation of winding
a) Crosswound package; b) Grooved drum; c) Broken-thread stop-motion; d) Thread clearer; e) Thread brake; f) Thread guides; g) Balloon restrictor; h) Cops

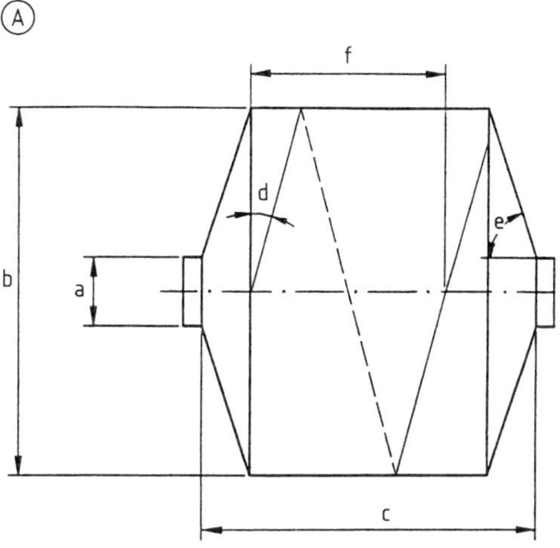

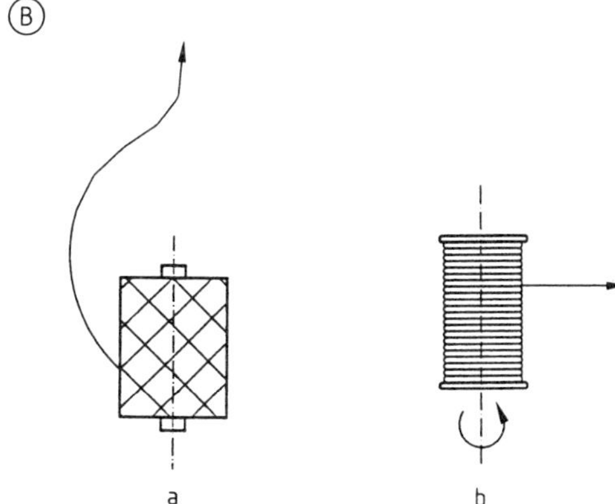

**Figure 18.** Factors associated with yarn spools
A) Spool characteristics; a) Tube diameter; b) Spool diameter; c) Maximum thread-guide traverse; d) Helix angle $\alpha$; e) Slope gradient; f) Pitch
B) Unwinding behavior; a) Crosswound package with axial unwinding; b) Parallel-wound spool with tangential unwinding

diameter. If the winding ratio is an integer, individual layers of thread are wound closely next to each other or on top of each other. This is called patterned or ribbon winding. It may lead to uneven dyeing as a result of variations in the density, as well as thread-layer slippage during thread removal because of insufficient crossing of the layers. Winding machines are therefore equipped with special devices to counteract such "ribboning" by introducing periodic slippage between the drive and the spool.

The advantages of random winding include spool stability, large packages, and mechanical simplicity. The disadvantages are ribboning and difficult unwinding.

**Precision Winding.** In precision winding (Fig. 19 B), the drive impulse is applied to the axis of the spool, resulting in a constant winding ratio. In this case the helix angle $\alpha$ decreases with increasing diameter (see Fig. 19 B). Another characteristic of this type of winding is increasing yarn-winding speed at constant rotational velocity of the spool. Stepwise precision winding is an attempt to combine the advantages of both random and precision winding. Here the winding ratio is changed in a stepwise manner.

The advantages of precision winding are an absence of ribboning and favorable unwinding characteristics. The disadvantages are spool fragility, small spool size, varying thread density, and expensive machinery.

### 3.3.2. Twisting

Twist is a general term applicable to all textiles made by the twisting together of simple yarns or ply yarns. In each twisting step, several

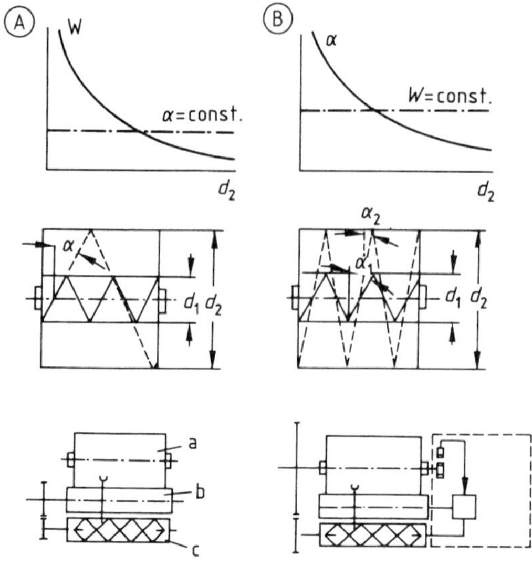

**Figure 19.** Comparison between different types of windings
A) Random winding; B) Precision winding
a) Spool; b) Peripheral drive; c) Grooved drum
$d_1$ = tube diameter; $d_2$ = spool diameter; $\alpha$ = helix angle;
$W$ = winding ratio

starting threads are twisted together to form a single ply yarn (Fig. 20). Distinctions are made between one-stage and multistage twisting (Fig. 20 C, D, E), as well as between Z and S twisting (Fig. 20 A, B). Multistage twist yarns are especially important in the production of cord twist for automotive tires and conveyor belts, including yarns made of polyester, polyamide, or aramide, as well as rayon or steel cord.

A new twist is usually applied in the direction opposite to that of the preceding operation. Successive twisting in the same direction leads to very hard twisted yarns characterized by low extensibility.

The process of twisting serves as a convenient way of increasing strength, improving stretch behavior, reducing unevenness, and achieving special surface and color effects. Ply yarns with special surface effects are widely used in the clothing sector and for home-furnishing fabrics, serving to enliven the appearance of the product. Such materials are made by combining several different take-off yarns. It is also possible to vary periodically the individual yarn-delivery speeds, permitting the introduction of such effects as knops, loops, snarls, or knots.

Twist yarns are made on ring-twisting frames, cablers, double-twisting frames, or Tritec twisters (Fig. 21). These machines differ according to the number of twists applied in a single operation. The importance of the ring-twisting process (Fig. 21 A) is declining, and the double-twist process is now standard [23]. Two spools placed on top of each other are fed simultaneously to the twisting frame. Alternatively, a "doubling bobbin" might be utilized, where the term *doubling* refers to assembly of the threads prior to twisting. In the case of a cabler (Fig. 21 B), the pre-twisted yarn need not be subjected to addi-

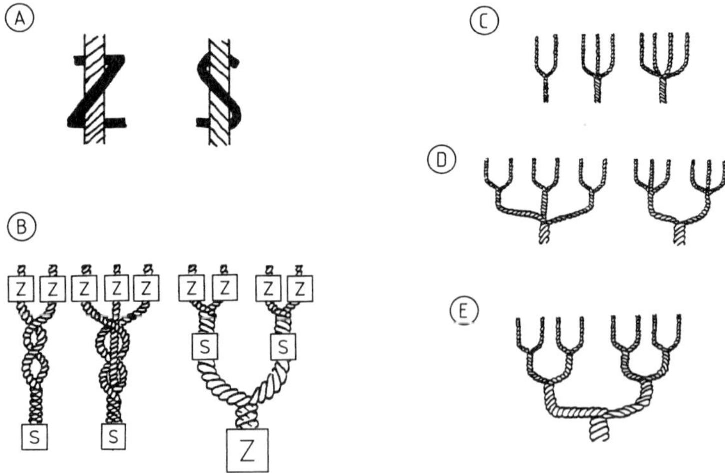

**Figure 20.** Examples of various types of twists
A) Twist direction (Z or S); B) Twist-direction sequences; C) One-stage twist; D) Two-stage twist; E) Three-stage twist

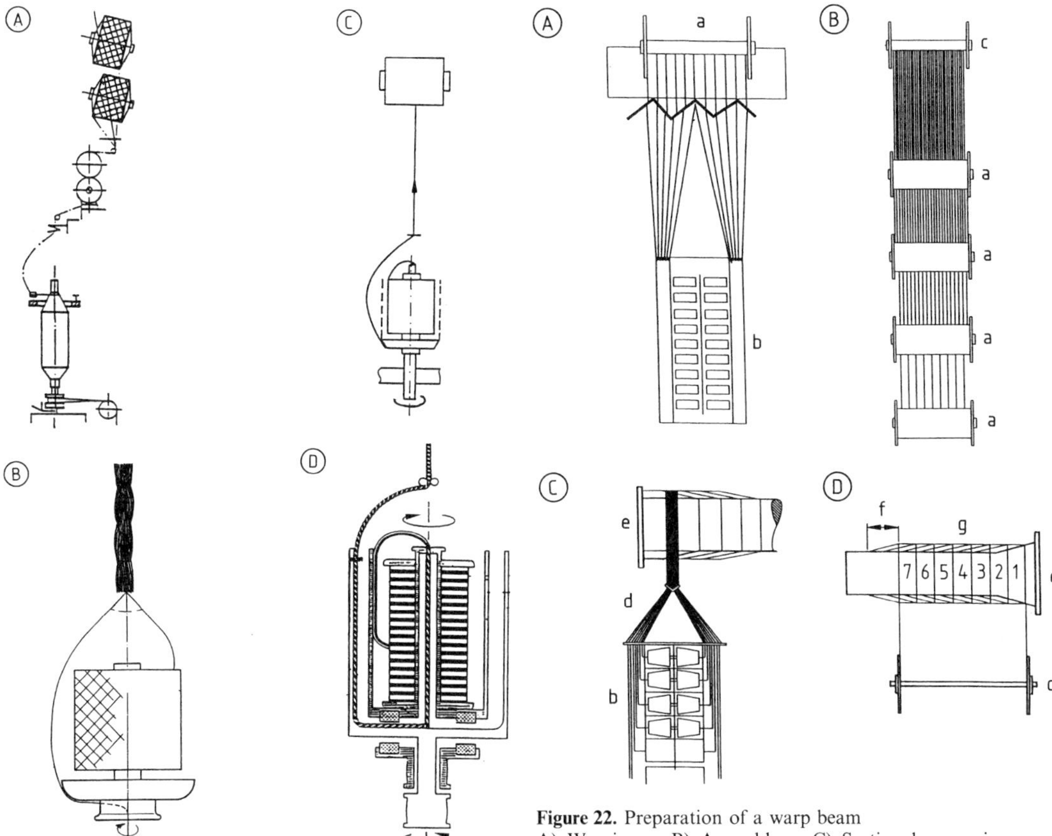

Figure 21. Twisting frames
A) Ring-twisting frame; B) Cabler; C) Double-twisting frame; D) Tritec-twister

Figure 22. Preparation of a warp beam
A) Warping; B) Assembly; C) Sectional warping; D) Rebeaming
a) Sectional warp beam; b) Creels; c) Warp beam; d) Wraithe; e) Conical warping drum; f) Warp layer pitch; g) Sections of warped threads

tional twist. With the double-twist process, two twists are obtained for every spindle rotation (Fig. 21 C), whereas three twists are achieved in the Tritec twister (Fig. 21 D).

### 3.3.3. Preparation of the Warp Beam

A loom requires that there be available a warp system with a defined number of threads, all of which have the same length and thread tension. A distinction can be made between four different production processes with respect to the warp:

1) Weaving from the creel
2) Direct beaming
3) Warping
4) Sectional warping

In the first two processes the number of warp threads is limited by the capacity of the creel, a disadvantage that is avoided with warping or sectional warping.

**Warping.** In warping, a portion of the total number of required threads is first wound onto a sectional beam [(a) in Fig. 22 A) across the entire width of the warp. As many as about 16 sectional beams are then assembled into a warp beam [(c) in Fig. 22 B]. This step is often combined with the sizing process. The process just described offers only limited patterning possibilities, so it is best suited to large lots with simple patterns.

**Sectional Warping.** In the sectional warping process (Fig. 22 C, D), narrow bands of warp threads (g) with the required end spacing and the desired length are wound side by side onto a conical warping drum (e). The conical design prevents slippage of the individual warp sections. As many sections are laid down as required by the desired width. Once sectional warping is complete the warp is rebeamed from the drum to a warp beam (c). The presence of individual sections means that patterning possibilities are virtually unlimited. Therefore this process is appropriate for smaller lots and more complicated warp patterns.

### 3.3.4. Sizing

Sizing refers to a special pretreatment of the warp [24]. A size is a substance that has the function of improving the mechanical properties and loadability of the thread without at the same time reducing its elasticity (→ Textile Auxiliaries, p. 244). Demands placed on a sizing material relate to its behavior during weaving as well as the potential for desizing and recycling. Sizing processes can be classified according to the method of application [25]. The most important categories are:

1) Hot-melt sizing
2) Cold sizing
3) Dry sizing
4) Wet sizing

    a) Conventional wet sizing, which involves the application of a water-soluble size followed by a drying process

    b) Solvent sizing, where dissolved size is deposited on the warp by evaporation of a non-aqueous solvent

5) Warp waxing

Sizing materials themselves also fall into several groups:

1) Starch sizes
2) Carboxymethylcellulose (CMC) sizes based on cellulose ethers
3) Protein sizes
4) Synthetic sizes

Starch sizes are used most commonly because they are both inexpensive and environmentally safe. Sizing processes can be further subdivided according to the nature of the size-application procedure (Fig. 23 A – D).

A sized fabric must be subjected to a *desizing* process prior to the finishing stage. Desizing has a decisive effect on the wastewater load in textile production. Two possibilities exist for reducing this load: use of biodegradable sizing materials, or size recovery by ultrafiltration or evaporation.

## 3.4. Weaving Methods

### 3.4.1. Fabric Construction

As noted previously, a woven fabric is formed by the interlacing of two thread systems, the warp and the weft, and the mode and manner

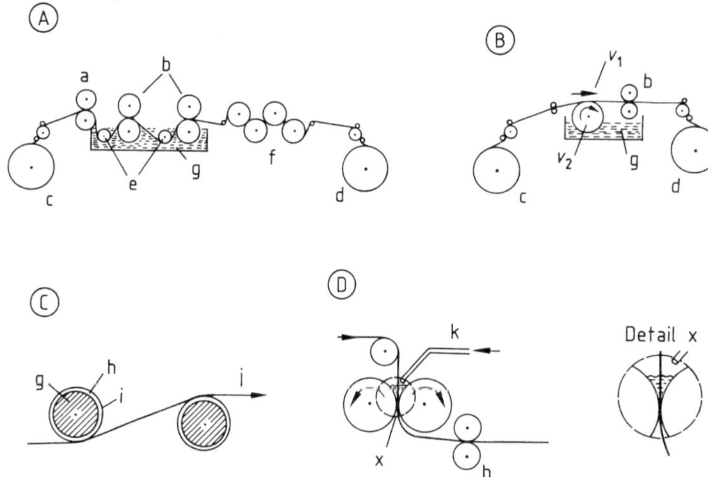

**Figure 23.** Processes for size application
A) Trough sizing; B) Tangential size application; C) Sinter-roll sizing; D) Foam sizing
a) Feed rolls; b) Squeeze rolls; c) Unsized goods; d) Warp beam; e) Immersion rolls; f) Cylindrical dryer; g) Sizing agent; h) Porous steel cylinder; i) Size film; j) Spun yarn; k) Foam introduction
$v_1$ = fabric velocity; $v_2$ = roll velocity

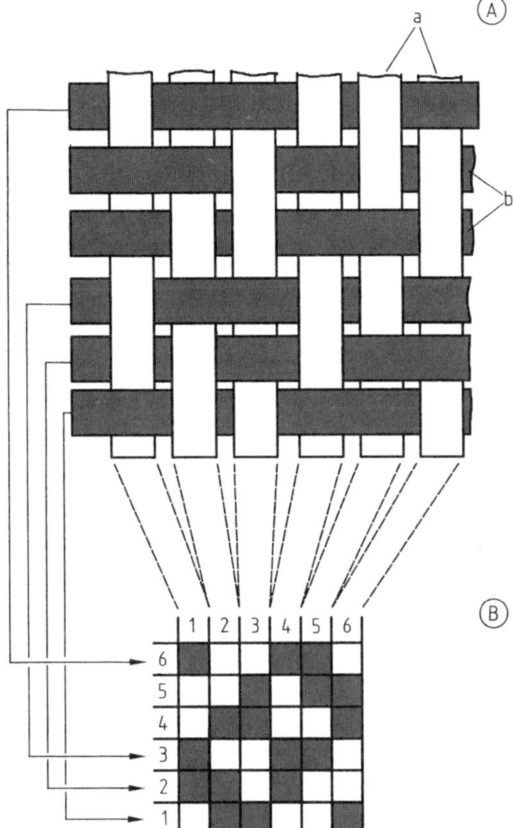

Figure 24. Diagrammatic representation of a particular weave
A) Fabric sample; B) Weave design
a) Warp threads; b) Weft threads

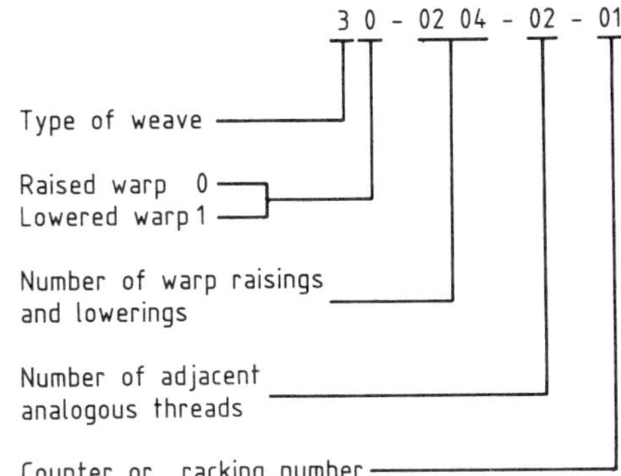



Figure 25. Example of a pattern drafting

of thread crossing is called the weave or fabric construction. Fabric design can be influenced in many ways by choice of a particular type of weave.

**Rapport.** Ordinarily, a weave repeats itself in both the transverse and longitudinal directions. The smallest unit which, when duplicated, gives rise to the overall fabric pattern is called the weave rapport. This can in turn be subdivided into a warp rapport and a weft rapport.

**Floating.** "Floating" refers to the extent of free, unbound thread between two crossing points.

**Weave Design.** The weave design is a graphic representation of a weave pattern, reproduced on squared design paper. A particular square on the paper is filled if at this crossing point the warp thread is raised and the weft thread is lowered (see Fig. 24).

**Pattern Drafting.** A fabric's weave can also be represented by a pattern drafting (see Fig. 25). This is a code based on the type of weave, information related to warp raising and lowering, the number of adjacent comparable threads, and the racking number (see below). An international standard for pattern drafting is stipulated in ISO 9354. The first number in a pattern drafting specifies the base weave (1 for plain weave, 2 for twill weave, or 3 for satin weave; see below).

**Plain Weave.** Plain weave represents the simplest interlacing pattern for warp and weft threads, but also the tightest. Plain weave and its simple modifications can be achieved with the use of only two heald frames. Four, six, or more heald frames are used for very closely woven fabrics with a high fiber density. This weave and its simple modifications are illustrated in Figure 26.

**Twill Weave.** Twill weaves are characterized by ribs that run diagonally across the fabric. A distinction is made between Z- and S-rib twill depending on the orientation of the rib. A Z-rib results from a consistent displacement of the first warp thread in the "upper right" direction. The height of the shift is indicated by the racking number. Examples of basic twill weaves are shown in Figure 27.

**Satin/Sateen Weave.** Satin/sateen fabrics display a closed, smooth, and dense appearance. A

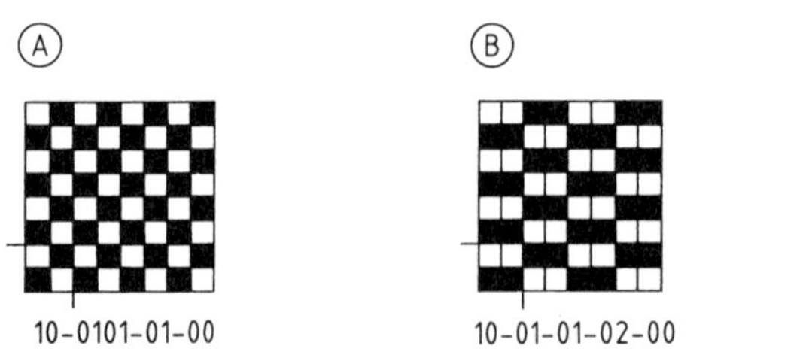

Ⓐ 10-0101-01-00    Ⓑ 10-01-01-02-00    Ⓒ 10-0202-02-00

**Figure 26.** Plain weave and its simple modifications
A) Plain weave; B) Weft-rip fabric; C) Hopsack (panama) weave

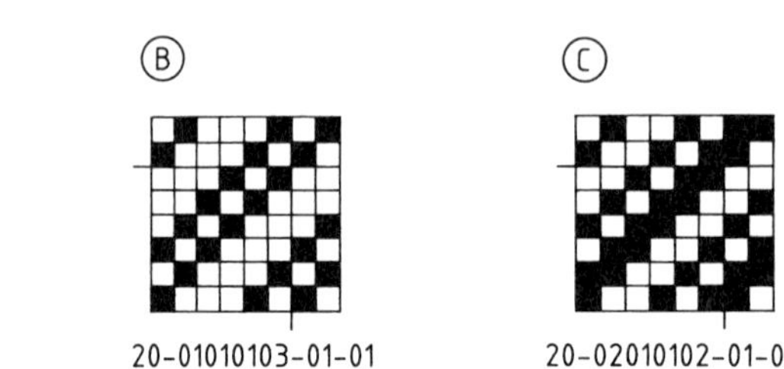

Ⓐ 20-0103-01-01    Ⓑ 20-01010103-01-01    Ⓒ 20-02010102-01-01

**Figure 27.** Twill weave

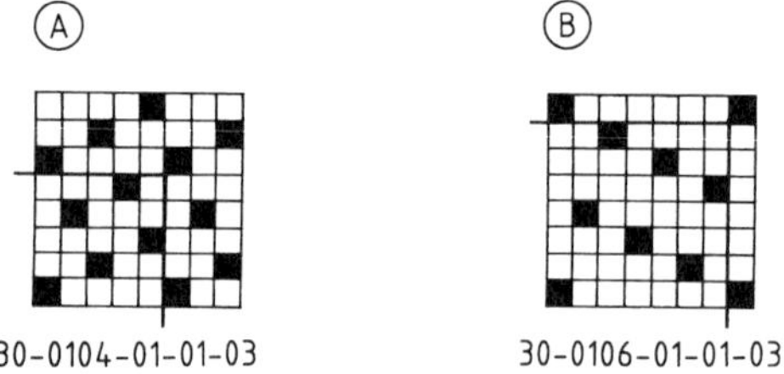

Ⓐ 30-0104-01-01-03    Ⓑ 30-0106-01-01-03

**Figure 28.** Satin/sateen weave, illustrated with a counter of three
A) 5-End (3); B) 7-End (3)

feature of this weave is the fact that none of the uniformly distributed weave points (see Fig. 28) touch each other. The distance from one weave point to the next on the following warp thread is called the *counter* (or move number). Permitted move numbers $i, j$ may not have a common integral divisor, and they must also meet the following supplementary conditions:

$$n = i + j; \; i \neq j; \; i, j \gg 1; \; i, j \, \varepsilon \, IN$$

In a satin weave the warp raisings predominate, whereas in a sateen weave it is the warp lowerings that predominate.

### 3.4.2. Loom Technology

Looms are distinguished according to the nature of the weft-insertion process (Section 3.4.2.3) or the shedding system (Section 3.4.2.1) used. The *shedding system* is what makes possible a selective raising and lowering of the warp threads for insertion of the weft thread. *Weft insertion* refers to the process by which a weft thread is transported through the opened shed. The basic structure of a modern loom is illustrated in Figure 29.

#### 3.4.2.1. Shedding Systems

The task of the shedding elements is to separate the warp threads into an upper shed and a lower shed (Fig. 30, c and d). The weft thread is inserted between these two warp positions. Subsequent changes in warp positions result in binding of the weft threads. The threads are carried and guided in heddles. Various types of shedding mechanisms are described below.

**Eccentric Dobbies.** This shedding system is characterized by a direct, rigid connection between the heald frames and a cam shaft. A particular series of movements repeats itself with

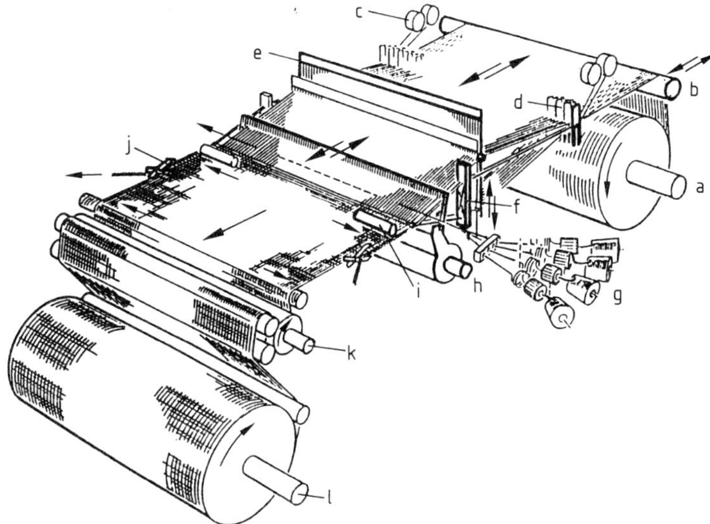

**Figure 29.** Schematic illustration of a modern power loom
a) Warp beam; b) Back rest; c) Right-hand thread spools; d) Warp stop-motion; e) Shedding; f) Selvage formation (in this case, leno weave); g) Weft insertion; h) Reed; i) Temple; j) Selvage cutter; k) Fabric take-up roller; l) Fabric roller

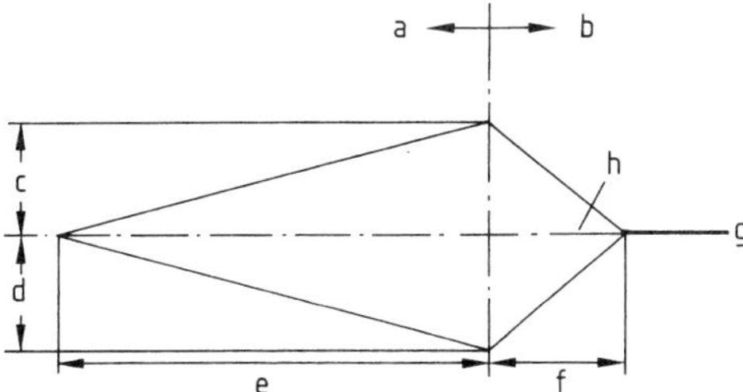

**Figure 30.** Shed geometry
a) Back shed; b) Front shed; c) Height of the upper shed; d) Height of the lower shed; e) Length of the back shed; f) Length of the front shed; g) Fabric; h) Closed-shed position

each rotation of the cam shaft (Fig. 31 A). Eccentric dobbies are compatible with very high weft-insertion frequencies, but they have the disadvantage of a small weft rapport, and a great deal of effort is involved in resetting the loom for a different weave. Eccentric machines are used mainly for mass-produced fabrics.

**Dobbies.** With a dobby, the position to which each heald frame should be moved is ascertained from an information carrier (punch card, computer software) prior to every weft insertion (Fig. 31 B). Theoretically, therefore, the weft re-

peat pattern is unlimited in variety. A change in weave is achieved simply by introducing or activating a new information carrier. A *double-lift* dobby inserts two weft threads per rotation of the main drive shaft. Double-lift dobbies may be based on the Hattersley, positive, or rotary principle.

**Jacquard Looms.** Jacquard looms (Fig. 31 C) permit individual control over each warp thread via a set of harness cords (j). Any desired type of pattern can be produced with this process. A selection mechanism establishes whether or not the knife box (a) takes with it the hook (c) attached to a particular heddle. A selection occurs via needles (f) or electromagnets.

### 3.4.2.2. Warp Let-Off Systems

The warp let-off (Fig. 32) has the functions of ensuring the availability of an adequate supply of warp materials and keeping the warp-thread tension as constant as possible. Together with the movable loom back rest [(g); see also Fig. 29], the warp let-off is expected to help compensate for any thread elongation caused by shedding [26].

The principle of *passive* warp let-off is equivalent to applying a brake on the warp beam.

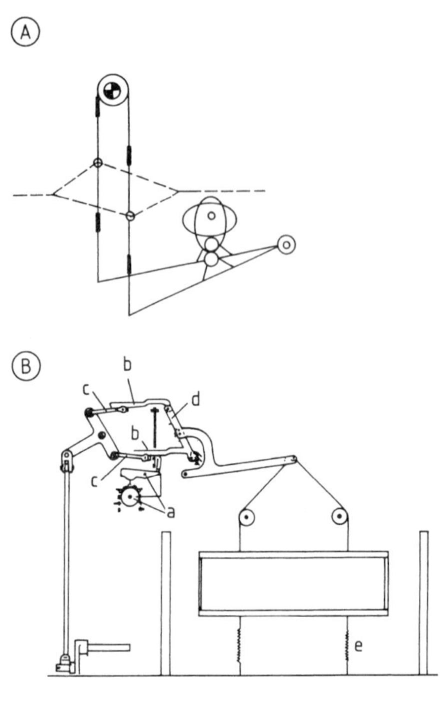

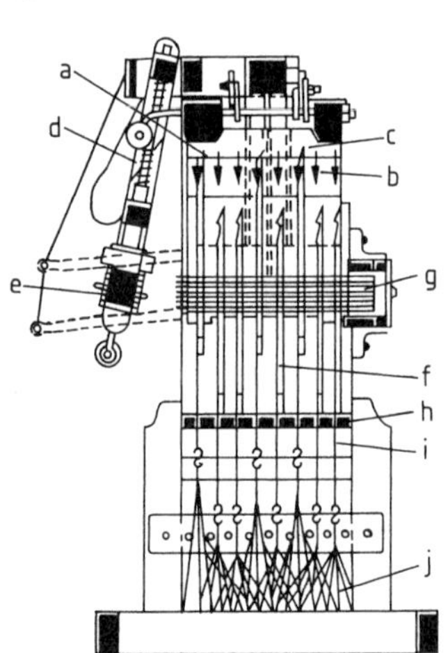

**Figure 31.** Shedding systems
A) Eccentric dobby
B) Dobby: a) Control unit; b) Knives; c) Drawbars;
d) Balance arm; e) Spring
C) Jacquard loom: a) Knife box; b) Knives; c) Hook;
d) Prism batten; e) Pattern cylinder; f) Needles;
g) Compression spring; h) Bottom board; i) Cable;
j) Harness cord

*Active* warp let-off implies that the warp beam itself is driven. The system can be so adjusted as to maintain constant fabric withdrawal as well as

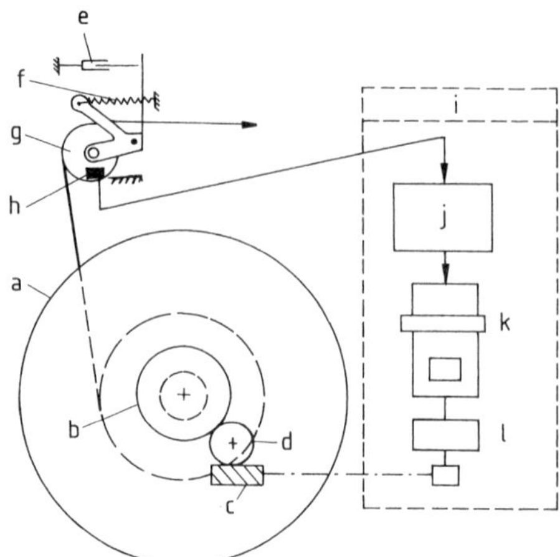

**Figure 32.** Schematic diagram of a warp let-off control system
a) Warp beam; b) Warp-beam gear; c) Worm; d) Worm gear; e) Damper; f) Tension spring; g) Back rest; h) Position sensor; i) Let-off device; j) Controller; k) Motor; l) Drive

either a constant distance between adjacent weft threads (negative fabric withdrawal) or a constant center-to-center weft-thread distance.

### 3.4.2.3. Weft-Insertion Principles

A distinction is made between conventional weft insertion with a shuttle, unconventional or shuttleless systems (missile, rapier, and jet looms), and multiphase weft-insertion systems. Unconventional looms always insert an end of the thread. The inserted weft yarn is subsequently tightened and separated from large, fixed packages, usually cones or cheeses. With a conventional shuttle loom, however, one long continuous thread is inserted repeatedly after being bent around as necessary at the edge of the fabric. A traditional bound edge (selvage) is formed only by a shuttle loom.

The weft-insertion efficiency of a loom is expressed in m/min, obtained by multiplying the rotational velocity by the fabric width. The progressive development in weft-insertion efficiencies as a function of time is shown in Figure 33 for various insertion systems. As a result of the very high and variable unwinding rates from crosswound packages, weft storage systems are used as prewinding devices.

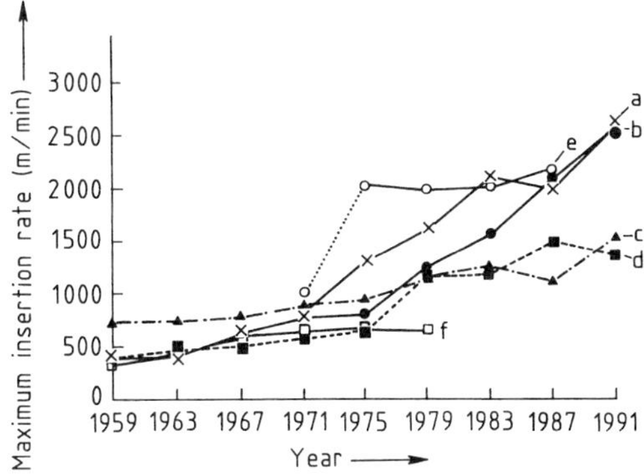

**Figure 33.** Development of weft-insertion efficiencies as a function of time and insertion mechanism, based on presentations at the annual International Textile Machinery Exhibit (Internationale Textilmaschinen-Ausstellung, ITMA) [27]
a) Water; b) Air; c) Missile; d) Rapier; e) Multiphase; f) Shuttle

**Shuttle Looms.** In this type of loom (Fig. 34 A), a shuttle, which contains a spool wound with yarn (the weft spool), is pushed through the shed by a beating device. With each weft insertion, the entire mass of the shuttle (b) and the weft spool must be accelerated and subsequently brought to rest.

**Missile Looms.** With a missile loom (Fig. 34 B) the weft thread is clamped to the end of a projectile (the "missile") that is shot through the shed. The weft is then tightened and separated from the external weft supply. The missile itself must later be restored to its original position externally, which means that there are always several missiles in circulation.

**Rapier Looms.** Looms of the rapier type (Fig. 34 C) are representative of form-closed weft-insertion systems. The head of one rapier (the giver) takes a thread end from the weft bobbin and carries it into the middle of the shed where the thread is then passed along to a second, opposing rapier (the taker). Weft insertion of this type is subject to control and regulation throughout the insertion process. The rapier insertion system is very flexible, and suitable for sensitive materials as well as both fine and coarse fabrics.

**Jet Looms.** In the case of a jet loom (Fig. 34 D), a distinction must be made on the basis of the insertion medium between air-jet and water-jet devices. In other words, the weft thread may be transported through the shed by means of either air pressure or a water jet. A main jet (e)

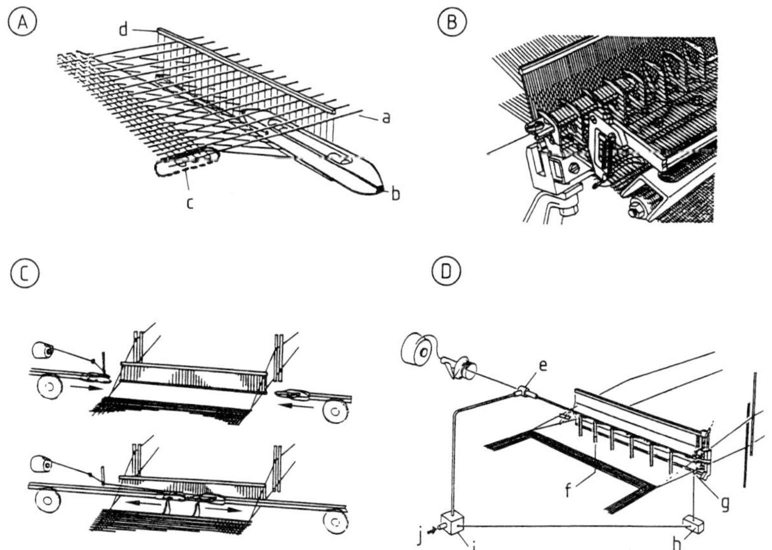

**Figure 34.** Weft-thread insertion techniques
A) Shuttle; B) Missile; C) Rapier; D) Air jet
a) Warp; b) Shuttle and shuttle spool (filling bobbin); c) Traditional selvage; d) Reed; e) Main jet; f) Echelon jets; g) Weft stop-motion device; h) Microprocessor; i) Pressure-regulating valve; j) Compressed air

is supported by echelon jets (f) located at the reed. Jet-weaving machines offer the highest weft-insertion efficiencies, but they are limited in flexibility with respect to both material and fabric width.

**Multiphase Weaving Machines.** In a multiphase weft insertion system, several weft threads are inserted simultaneously. A series-shed loom may have several sheds arranged sequentially in the warp direction, each with its own weft-insertion device. By contrast, with a wave-shed weaving machine the warp is shedded with a wave-like motion to permit passage of a succession of "shuttles," each of which lays in its own length of filling. The beat-up must also be accomplished in a wave-like manner, realized by a rotary slay. Multiphase weaving has not gained broad acceptance because of poor fabric quality and lack of flexibility.

#### 3.4.2.4. Selvage Formation

A firm, non-fraying selvage is required to permit such further processing as textile finishing. With all unconventional weft-insertion processes it is necessary that the weft threads be specially tied at the edge. If possible, the selvage should retain the same thickness and length as the crude fabric. A satisfactory selvage can be achieved in several ways, including:

1) Insertion of an auxiliary thread (*inserted* selvage)
2) Binding of the weft-thread ends (*tuck-in* selvage)
3) Twisting of the warp threads (*leno* selvage)
4) Gluing or welding of the edges

#### 3.4.3. Fabric Defects

**Fabric Defects in the Weft Direction.** Restarting a loom can lead to variations in the distance between individual weft threads. Depending on the origin of the problem, a distinction is made between flaws of the *starting, beating, regulating,* and *shed* type. The extent of observable defects is a critical function of the behavior of the warp let-off device. Defects attributable to improper weft tension or weave errors are less common.

**Fabric Defects in the Warp Direction.** Fabric defects in the warp direction can be a consequence of weave errors, incorrect drawing in the

heald frames or the reed, or varying tension on the warp threads. Flaws in the selvage also fall in this category.

**Fabric Defects Unrelated to Weave Direction.** Such defects include soiling, interwoven foreign matter, holes, and tears. Defects in this class are usually a result of operator carelessness or incorrect machine settings.

#### 3.4.4. Special Weaving Methods

Special weaving methods include the multiphase weft-insertion looms described in Section 3.4.2.3 and systems adapted to the production of special fabrics.

**Circular Looms.** The circular loom is a particular type of multiphase loom. Here the shed moves in a circle, and the loom is in fact similar in appearance to a circular knitting machine (see Section 5.5.2). Such a shed is formed by circulating cams and needles.

**Terry Looms.** A terry loom (Fig. 35) processes two warps simultaneously, a base warp (a) and a pile warp (b), into a single fabric. The base warp is stretched tightly, while the pile warp is very loosely stretched. A group of three or more

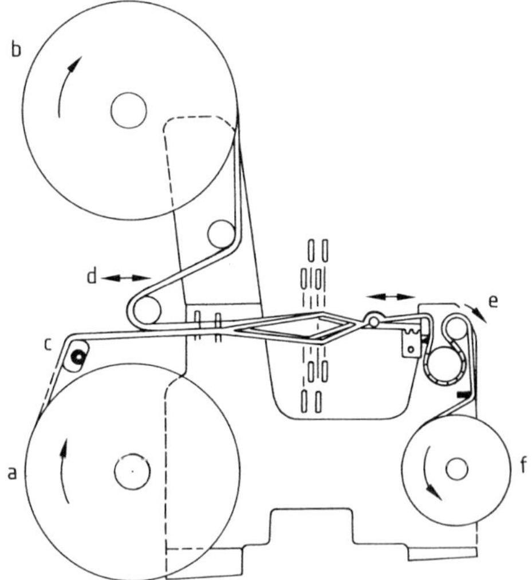

**Figure 35.** Cross-section of a terry loom
a) Base warp; b) Pile warp; c) Tension beam for the base warp; d) Tension beam for the pile warp; e) Fabric takeoff; f) Fabric roller

wefts is inserted one after the other and then beaten as a group. During beating, the weft threads slip over the tightly stretched base warp, thereby pushing up the pile warp and forming loops. The resulting loops are sometimes cut, producing a cut pile.

**Velour and Plush Looms.** With machines of this type, loops or naps in the pile warp are introduced with the aid of pile wires. As the pile wires are withdrawn, all the loops are cut with a knife, again producing a cut pile.

**Carpet Looms.** Double-velvet or double-plush carpet fabric is produced by machines in this category. The pile is achieved by attaching the upper fabric only loosely to the lower fabric via pile threads. These pile threads are subsequently cut with a knife after weft beating.

Large amounts of material are also produced on the basis of the *tufting process* (especially wall-to-wall carpeting). In this case each pile thread is drawn in a needle of a special type of sewing machine, and the pile loops are stitched together into a base fabric. The resulting loops can either be left closed (looped carpeting), or they can be cut (cut-pile carpeting). The back of the carpet is coated in order to fix the loops.

**Shaping Looms.** Almost any type of surface can be achieved by varying the distance between weave points in both the weft and warp directions (Fig. 36). A special shaping loom for this purpose has been developed at the Institute for Textile Technology of the Aachen University of Technology (Federal Republic of Germany).

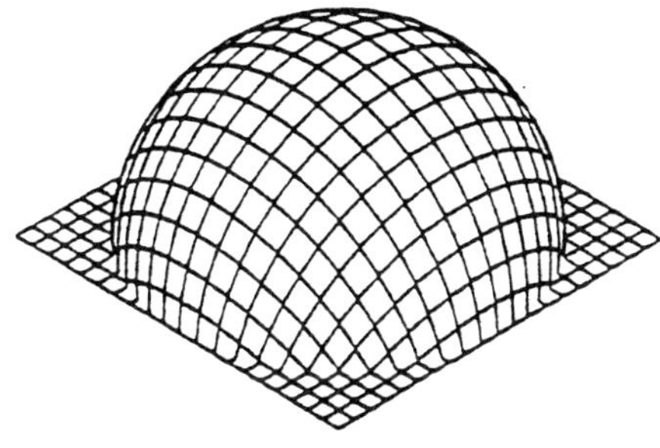

**Figure 36.** Three-dimensionally shaped fabric

# 4. Braiding

## 4.1. Introduction

Machine-made braids exist in many different forms, including cords, tubes, band-like flat products, and three-dimensional objects. All have in common a diagonal thread-crossing and the principle of independent thread motion upon which their production is based.

The first report of a mechanical braider dates back to 1748 when an Englishman, THOMAS WALFORD, received a patent (No. 638) for a device designed for the creation of braids. The first braider made of iron instead of wood was built in 1767 by a mechanic from Barmen (Germany) by the name of BOCKMÜHL [31]. Most of the early braided products were shoelaces and shoe straps, but mechanical braiding was gradually extended to other fields of application. These include articles of clothing, household fabrics, and increasingly, industrial braids, such as sheathing or sealing cords. Research results and recent equipment developments suggest that, in the future, braided three-dimensional structural elements may be useful in light construction.

## 4.2. Definition of a Braid

The term "braiding" refers to the continuous twisting of a set of threads or equivalent elements in the direction of propagation. The resulting textile product consists of a single system of threads twisted diagonally [32]. Uncrossed threads called "cores" or "standing ends" may also be included in the braid. According to DIN 60000, braids (Fig. 37) are defined as "flat or three-dimensional products with a regular thread density and a closed appearance whose braid (lace) threads cross over in a direction diagonal to the edges of the product."

## 4.3. Production of Braids

### 4.3.1. Principles of the Mechanical Braiding Technique

Machine braiding has its origin in the hand-braiding technique, which requires that the material to be braided consist of at least three thread packets. In the course of braiding, the left hand repeatedly moves strands from the upper left to

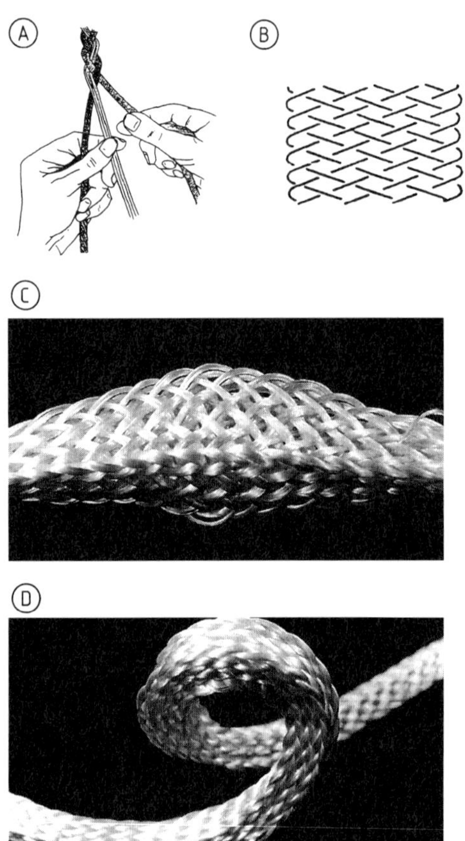

Figure 37. The technique of braiding
A) Hand braiding ("right rib"); B) General form of a braid; C), D) Examples of braided material

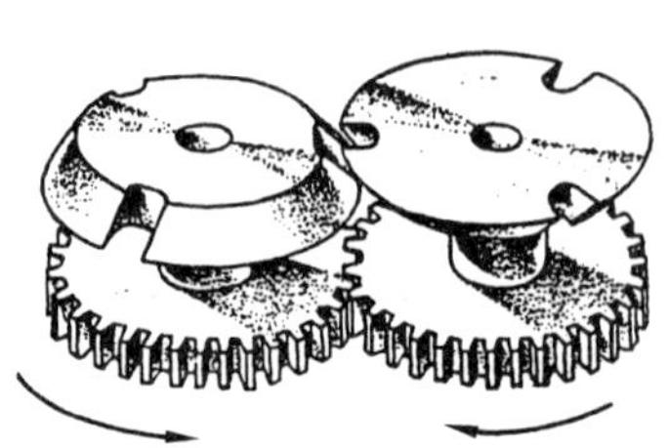

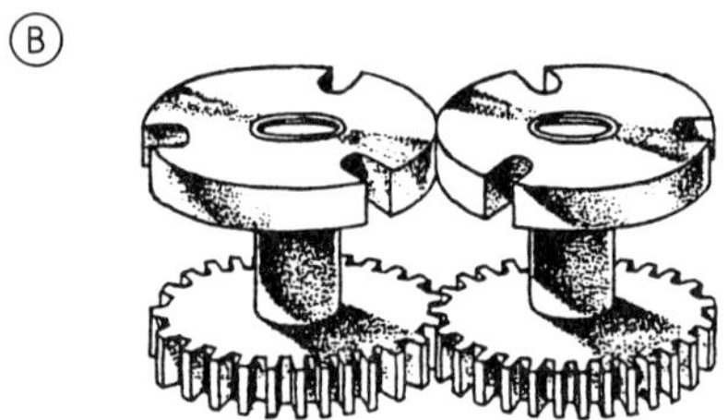

**Figure 38.** Horngears
A) Gliding system; B) Carried system

the lower right position. The resulting strand arrangement is known as a "left rib." If the right hand is used to move strands from the upper right to the lower left, what results is a "right rib" (Fig. 37 A).

During plaiting of the thread strands each hand executes a rotational motion about the central thread. In creating a right rib the left hand moves counterclockwise, while the right hand moves clockwise. Mechanical braiding is accomplished similarly by means of *horngears* rotating in counterclockwise and clockwise fashion (Fig. 38), simultaneously carrying the thread strands around each other just as would the hands. Notches are distributed along the periphery of the horngear. Each notch is capable of accepting a spool carrier, the *braiding carrier*, and transporting it as the horngear revolves. Be-

neath the horngears is a surface into which have been cut guide grooves known as *races*. It is through these races that the lace butts are led. The races also establish paths for the braiding carriers during rotation of the horngears.

### 4.3.2. Functional Elements of a Braider

The basic design of a braider is illustrated in Figure 39 (see also Fig. 37 B).

**Horngears.** With few exceptions, braids are always produced with the aid of rotating horngears. In most braiders these horngears are arranged in a circular pattern. The differences between braiders for ribbon and for round products are clarified in Figure 40. In a ribbon machine the braiding carriers are not passed along to the next horngear at the point of reversal, but are instead returned, and in a direction opposite to that of the previous motion.

From a design standpoint, a distinction must be made between two horngear systems (Fig. 41) [33]. The *gliding* (Barmer) system relies upon a guide table mounted above the horngears. The upper flange at the base of the braiding-carrier foot (the lace butt) secures the carrier to the guide table as the entire assembly glides along the race. The lower part of the lace butt projects

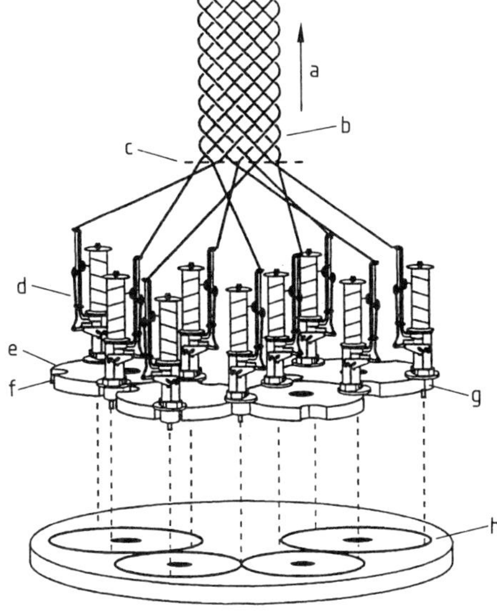

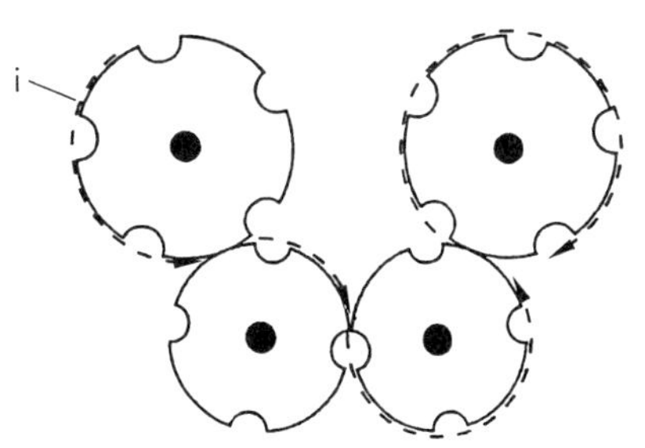

**Figure 39.** Schematic representation of a braiding machine
a) Direction of propagation; b) Braided strand; c) Braiding point; d) Braiding carrier; e) Horngear; f) Notch; g) Sliding shoe; h) Race; i) Path followed by the braiding carrier

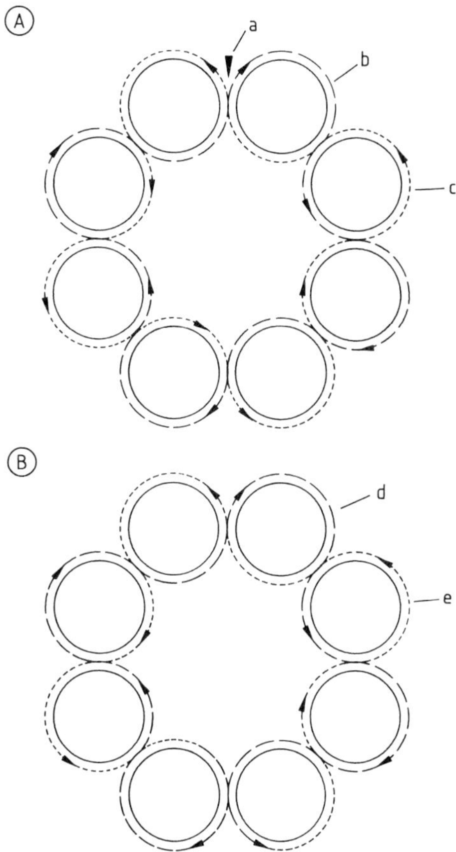

**Figure 40.** Horngear arrangements for (A) a ribbon braider and (B) a round braider
a) Point of reversal; b) Forward course; c) Reverse course; d) First braiding-carrier series; e) Second braiding-carrier series

into the notches. Continuous friction associated with the gliding system requires intensive lubrication, and severely limits the operating speed of the machine.

The *carried* system uses horngears located above the guide table. Here, the lace butt surrounds the notch, which means that the braiding carrier is carried along by the horngear. Only the lower portion of the braiding carrier, the sliding shoe, is subject to friction as the braiding carrier passes along through the race. Considerably higher horngear rotation rates can therefore be achieved relative to a machine based on the gliding system.

The size of the horngears is frequently specified as one characteristic of a braider. The diameter of the horngear usually corresponds to the distance between the axes of adjacent horngears, and is called the "pitch." Braiders typically have pitches in the range of a few centimeters to 1.5 m.

**Braiding Carriers.** As a rule, the yarn spools themselves are not guided along the notches, being confined instead to the braiding carriers, a measure intended to ensure proper uninterrupted unspooling. The braiding carrier plays an important role in the braiding process. Its most important functions are:

1) Mating with an interchangeable spool
2) Controlling the spooling-off process as the braiding carrier moves

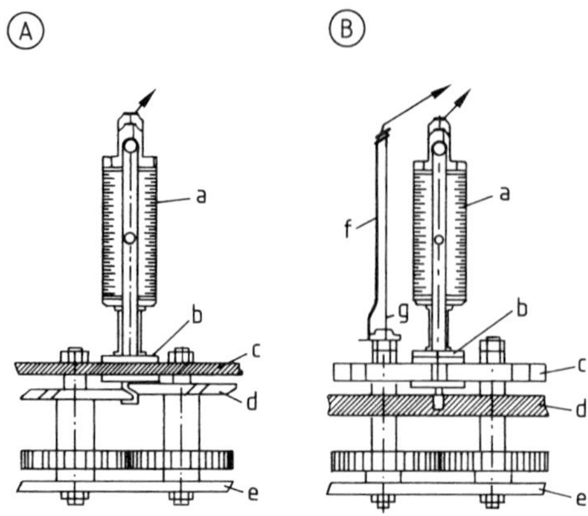

**Figure 41.** Schematic illustrations of two types of braiding systems
A) Gliding (Barmer) system: a) Braiding carrier with braid spool (bobbin); b) Lace butt; c) Upper plate; d) Impeller; e) Lower plate
B) Carried system: a) Braiding carrier with braid spool (bobbin); b) Lace butt; c) Impeller; d) Upper plate; e) Lower plate; f) Standing-end spring; g) Standing end

3) Maintaining an adjustable level of thread tension
4) Interrupting the machine in the event of thread breakage

Braiding carriers come in many different sizes and shapes. The thread spool is usually joined such that its axis is aligned vertically, although a horizontal spool alignment is preferred in the case of a material that is extremely sensitive to bending.

During the braiding process the braiding carrier follows a wave-like course. The changing distance between the carrier and the *braiding point*, where the threads are compacted to form a braid, must be compensated for by a special thread-tightening device, consisting normally of thread-tightening weights or spring-loaded compensator levers. Tension on the thread holds an interrupter located on the braiding carrier in an upright position. If the spool becomes empty, or if the thread breaks, this interrupter falls in such a way as to activate a mechanical or electromagnetic control that shuts down the machine.

A special device, usually a pin or a catch rod, prevents unintentional unwinding of the thread (as a consequence of the rotary motion of the braiding carrier) by locking the grooved edge of the spool. Spool release to permit renewed unwinding of the thread is usually controlled by tension. A conventional braiding carrier with the

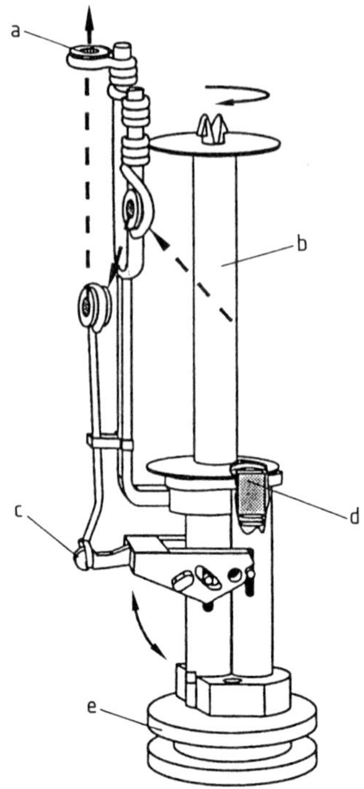

**Figure 42.** Schematic representation of a braiding carrier with faller arm and catch rod
a) Thread eyelet; b) Spool; c) Lever carrier; d) Catch rod serving as a spool brake; e) Lace butt

functions described is illustrated in Figure 42. The lever carrier (c) is lifted by tension of the thread acting against the force of a spring. This pushes down on a pin that pulls a catch rod (d) out of the spool. The spool can rotate to release thread only when this catch rod is completely extracted.

**Races and Track Switches.** It is the race that controls the path of the braiding carrier. Most often this consists of a set of grooves cut in a guide table into which the sliding shoes of the braiding carrier are in turn inserted. When a braiding carrier is pushed from one horngear into the operating zone of an adjacent horngear, the sliding shoe must pass over a race-crossing between the two gears. One alternative to such a crossing is a track switch that blocks the race. Track switches can be controlled by means of cogged repeat rolls or by a Jacquard device. A more recent development is the electronically controlled pneumatic track switch, used especially in braiders for the production of three-dimensional braids [36].

**Material Take-Off.** Above the center of the machine, threads are compacted at the so-called braiding point to form a braid (see Fig. 39). The rate of take-off must be adjusted to match the braiding rate of the braiding carriers. This rate influences both the thread angle in the braid and the extent of thread compacting. Rapid take-off results in a sharp crossing angle and a relatively loose braid. A slow rate of take-off leads to an obtuse crossing angle and a firm, compacted thread structure.

A three-roll take-off unit is frequently employed for ribbon and lacing. The band-like braid is wound around three pressure rolls, the speeds of which can be regulated either with change gears or electronically. Braids that are flexurally rigid or sensitive to bending are withdrawn via a disk with a special adhesive coating.

**Standing Ends.** It is not essential that all the threads of a braid participate in the braiding process on braiding carriers. Individual threads are often introduced in unbraided form, typically as a way of filling hollow spaces in the braid or adding elasticity. Such unbraided threads may constitute either standing ends or cores.

Standing ends are threads that are fed through the bored centers of the horngears and incorporated individually into the braid as the braiding carriers pass around the horngears. This technique is often utilized to make elastic tape, where prestressed rubber threads are introduced in the form of standing ends to add elasticity. Inelastic standing ends are used to provide additional support to the interlacing, or for patterning.

Standing ends introduced through the center of the braiding machine itself are called *cores*. A sheath is created around a single core by all the circulating braiding carriers simultaneously. Filling materials, filling threads, insulating materials, wood and cork forms, electrical cables, lead wires, lead balls, and tubes have all been used as cores in braided products.

### 4.3.3. Mechanical Systems

**Horngear Systems.** The most common type of machine for making braids uses horngears to pick up the braiding carriers, push them through a race, and pass them on to a neighboring horngear (see Figures 39 and 40). The number, arrangement, and notch patterns of the

horngears determine the types of braid that can be produced. Braiders for flat and round braided products usually entail a circular horngear arrangement, whereas packing braiders use square horngear arrays. There are a great many variants as well, combining, for example, circular and flat braiders with individual track switches for shifting the races [34], [35].

*Three-dimensional rotational braiding* is a product of one of the most recent developments in horngear machines. Such a system can be used to produce industrial three-dimensional braids in the form of components [36], [37]. Track switches direct the braiding carriers on individual tracks through a complex array of horngears. This technique permits the introduction not only of cores and standing ends, but also shorter *zero-degree threads*, and for any desired period of time. For this purpose, individual horngears are selectively disengaged at various junctures. Computerized simulation and operational control programs are required to ensure collision-free braiding and to control the activities of the machine. This techniques makes it possible to produce braids with various cross-sectional shapes.

**Horn's System.** Braiders operating according to the Horn system avoid the use of horngears. The general principle is illustrated in Figure 43. Threads are run from the inner spool circle directly into the braid. Threads from an outer spool circle are led over and under the inner circle with the aid of planetary wheels and guide levers. These machines are characterized by a very high production rate, but they also take up a great deal of space. The spools remain the same distance from the braiding point throughout

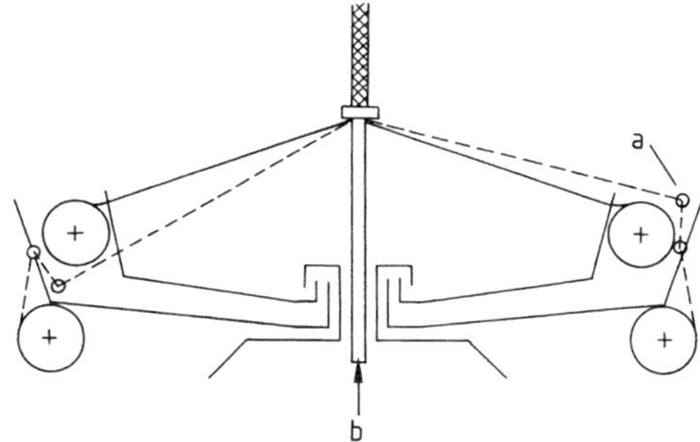

**Figure 43.** Schematic representation of a braider based on the Horn system
a) Planetary wheel or guide lever; b) Core

**Figure 44.** A Cartesian braider
a) Framework; b) Table; c) Material take-off; d) Braid eyelet; e) Braiding carrier; f) Pneumatic cylinder

their rotation, which eliminates the thread compensation required with horngear machines. Horn machines are especially suitable for braiding around a core made of a material that is sensitive to bending, such as wire.

**Cartesian 3-D Machines.** To date, Cartesian braiders have been investigated only on a laboratory scale for the production of industrial three-dimensional braids. The braiding carriers in this case are pushed back and forth on linear tracks, for example by pneumatic cylinders, as a way of producing braid profiles with a constant cross-section. The nature of a Cartesian braider is illustrated in Figure 44.

## 4.4. Interlacing Techniques

A braid interlacing describes the course of a particular thread and its crossing path with respect to threads coming from the opposite direction. The nature of the interlacing depends on the number of notches in the corresponding horngear and the distribution of the braiding carriers among these notches. Depending on the number of notches and the braiding-carrier distribution, a given thread will cross over one or more threads from the opposite direction during each half-rotation of the horngear.

The most common types of braid interlacing are single braid (each thread crosses over and under one opposing thread), double braid (thread crosses over and under two opposing threads), and triple braid (thread crosses over and under three opposing threads).

The interlacing can be altered by changing the notch and braiding-carrier occupation (or fitting). The *notch fitting* establishes which notches are occupied by braiding carriers and which remain empty. The *braiding-carrier fitting* indicates which braiding carriers are to be provided with spools. A braiding carrier lacking thread is called a "missing" or "empty" spool.

## 4.5. Materials Used in Braiding

The choice of materials for braiding depends on the intended application. Demands are placed on the braiding yarn on one hand by the tensile force of the thread, which can lead to frequent thread breaks in larger machines, but also by deflection and friction points in the braiding carrier. Very brittle industrial fibers are most subject to damage by small radii and frequent deflections.

Cotton is often used for making shoelaces and also tapes and ribbons because of its good wet-tear resistance. Viscose and wool are used for trim materials and borders because of the surface effects they are capable of producing. The tear resistance of dyed wool is very low, so wool yarns intended for braiding are mixed with polyamide. Polyamide and polyester are both found in a wide range of applications.

Linen can be regarded as a "rediscovered" natural raw material, and it is being increasingly braided into trimmings. The processing of the corresponding yarns is problematic, however, and it requires air humidification. Polypropylene is braided into ropes and cords. Glass, carbon, and aramide fibers are processed into packing braids and braided three-dimensional structures.

## 4.6. Types of Braids and Their Applications

Braids are classified on the basis of a numerical system and, in most cases, a more detailed description of the character of the product. The numbering begins with the number of braiding carriers required for manufacture, followed by an *interlacing index* that indicates the "braidedness" of the product. For example, a ribbon made from nine braiding carriers with a quadruple-braided interlacing is described as "9/4."

Since this system provides no clear description of the braid itself, a descriptor is often added to clarify the material's character, use, form, or pattern. Examples are [31]:

| | |
|---|---|
| 9/4 | ribbon |
| 32/2 | herringbone |
| 44/2 | shoelace |
| 33/3 | curved ribbon |

### 4.6.1. Ribbon Braids

The terms "heald" refers to a band-like, flat braid. Examples of this type of braid are bobbinet, tailor's braid, elastic ribbon, and high-profile ribbon. Braiding in this category most often serves as a trim or border for outerwear. Compared with a woven band, braided ribbon is especially soft, and it can be sewn in tight curves without creasing. Elastic ribbons are made by including elastomers in the braid. As a rule, double-braided interlacing is preferred for tight enclosure of the rubber thread. Ric-rac braid and bow braids are used as trimmings for folk dress and children's clothes.

### 4.6.2. Round Braids

Round braids are also referred to as round or hollow *cords*. They include all braided articles with a circular or approximately circular cross-section, in contrast to flat braid. The appropriate interlacing pattern can usually be obtained without special control over the braiding carriers. Examples and applications of round braids include cords and laces for trimmings and for the shoe industry; drawstrings for curtains and underwear, rubber cords for the haberdashery, clothing, and sportswear industries; and cables, tubes, and ropes.

### 4.6.3. Packing Braids

Originally, packing braids were classified along with round braids. However, the corresponding cross-sections tend no longer to be round, but rather square (see Fig. 37 C). The horngears in this case are arranged not in a circular pattern, but as a surface array below the braiding point, so the threads are also braided through the center of the cross-section. Packing braids require $2 \times 2$, $3 \times 3$, or $4 \times 4$ horngear arrangements, and the races and braiding-carrier tracks are not adjustable. The square cross-section of such a braid is constant, with edges $\leq 30$ mm in length.

The main application of packing braids is as stuffing-box sealings [34]. Because of their great flexibility packing braids are especially effective for narrow-radius seals against high pressures. Pumps and propelling shafts are examples of devices that often entail the use of such seals.

### 4.6.4. Three-Dimensional Braids

Braided three-dimensional (3-D) structures are particularly valuable in light construction (see Fig. 44) [36], [37]. In combination with appropriate synthetic materials, 3-D braids can be effectively processed into ribs, stringers, or patterns. The diagonal fiber alignment in the braid is advantageous because it provides stability against torsional loads. Development of the corresponding mechanical technology and subsequent processing methods is still in its early stages. Progress to date coupled with the potential for great flexibility in the production of braids similar to construction parts suggest a promising future for the use of braids as structural components in many areas of technology.

# 5. Knitting

## 5.1. Introduction

Knitted material has been produced manually for thousands of years. The Egyptians, Incas, and other ancient advanced civilizations accomplished their knitting with rods. Indeed, it was not until 1589 that the Reverend WILLIAM LEE, a theologian from Calverton, England, invented a hand frame, thereby making it possible for the first time to mechanize the knitting process in the form of *weft knitting*. This device was capable of producing about 16 times the output of a hand knitter. In 1775 a knitter by the name of JOSIAH CRANE from Nottingham, England, patented the first *warp* knitting frame, and in 1863 another theologian, ISAAC W. LAMB, an American Baptist, invented the first *flat* knitting machine.

The production of knitted goods has increased tremendously in the final decades of the twentieth century with the increasing use of synthetic fibers, because these materials are ideally suited to processing on knitting and hosiery machines. Recently developed CAD-supported devices for manufacturing knitted fabrics provide extremely high output rates and the potential for a wide variety of patterns. Current equipment offers 500 000 times the output capacity of a hand knitter. Knitted materials are used today for all types of clothing, household fabrics, and industrial textiles.

## 5.2. Classification of Knitted Fabrics

Knitted stitches consist of thread loops that hang inside one another. Each stitch has upper and lower interlacing points, and consists of a *head, shanks,* and *feet.* A distinction is made between stitch loops (which have only lower interlacing points) and thread loops (with no interlacing points at all), as shown in Figure 45 A. A row of stitches aligned side-by-side forms a stitch course (SC), whereas stitches arranged one above the other (i.e., in the direction of production) constitute a wale (W), as illustrated in Figure 45 B.

The quality of a knitted fabric can be characterized in terms of two parameters:

Density of the stitch course = (no. of SC)/(unit length)
Density of the wale = (no. of W)/(unit length)

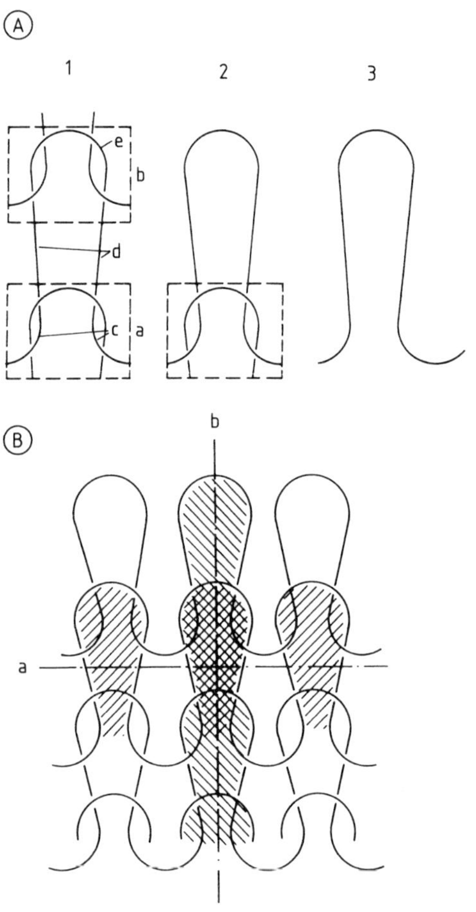

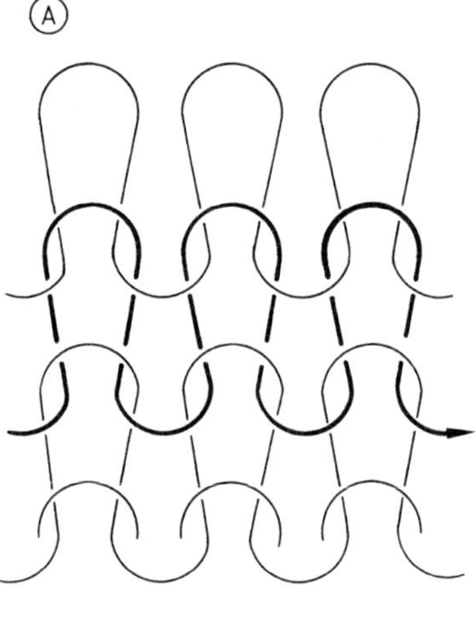

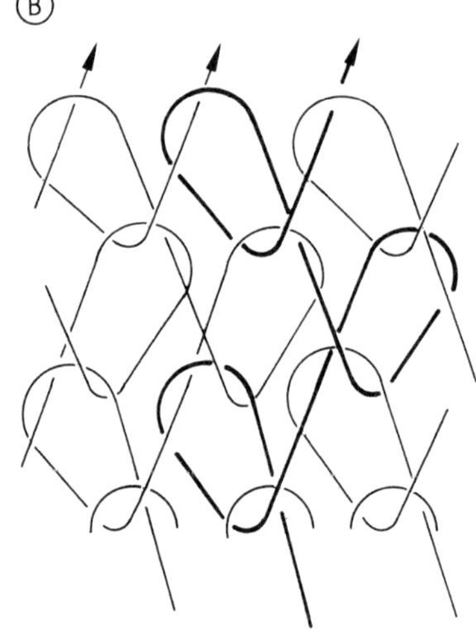

**Figure 45.** The basic characteristics and terminology of knitted stitches
A) Stitches and loops: 1) Knitted stitch; 2) Stitch loop; 3) Thread loop
a) Lower interlacing point; b) Upper interlacing point; c) Feet; d) Shanks; e) Head
B) Stitch combinations: a) Stitch course; b) Wale

**Figure 46.** The two fundamental types of thread path in knitted work
A) Single-thread knitware; B) Warp-thread knitware

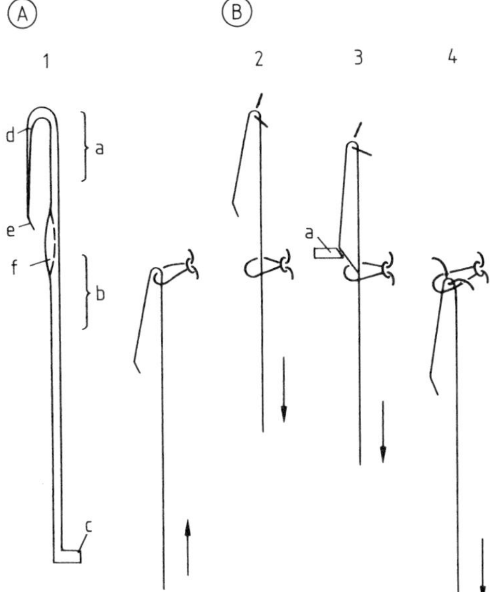

**Figure 47.** Knitting with a bearded (or spring) needle
A) Bearded needle: a) Head; b) Shaft; c) Foot; d) Hook; e) Tip; f) Groove
B) Principle of stitch formation with a bearded needle: 1) Enclosing; 2) Thread laying; 3) Pressing and casting on; 4) Knocking over; a) Press

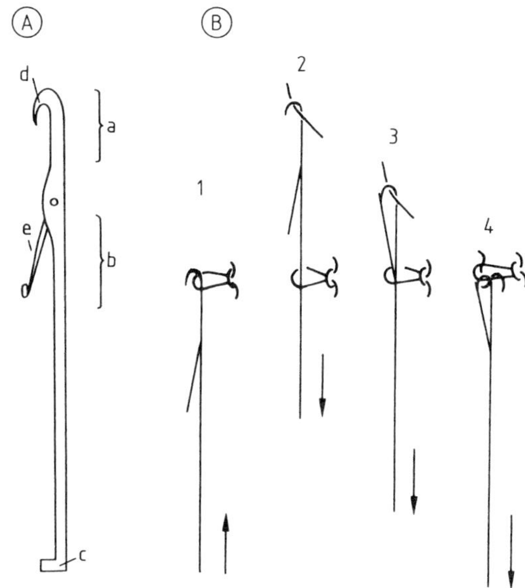

**Figure 48.** Knitting with a latch needle
A) Latch needle: a) Head; b) Shaft; c) Foot; d) Hook; e) Latch
B) Principle of stitch formation with a latch needle: 1) Enclosing; 2) Thread laying; 3) Casting on; 4) Knocking over

In principle, all knitted materials can also be classified according to the nature of the thread take-off as either single-thread (one thread crosswise) or warp-thread (with warp threads lengthwise) knits (Fig. 46).

## 5.3. Technology of Stitch Formation

Most modern knitting and hosiery machines utilize bearded, latch, or compound needles.

The *bearded needle* (Fig. 47 A), invented in 1589 by LEE, consists of an extended hook (d) with a tip (e), a shaft (b) containing a groove (f), and a foot (c). At the start of stitch formation (Fig. 47 B), with the stitch loop already in the hook (1), the needle is pushed forward ("enclosing the loop"), and a thread is placed in the hook (2). The needle is subsequently retracted (3), and a press (or presser bar) pushes the tip into the groove ("pressing") so that a stitch loop can be cast onto the tip ("casting on"). The thread is then pulled through to form a stitch loop (4), and the stitch loop is knocked over the hook, producing a new stitch ("knock-over" or "casting off").

Bearded needles are currently used in both weft knitting and warp knitting machines.

The *latch needle* (Fig. 48 A), invented by TOWNSEND in 1847, has a short hook (d), a hinged latch (e) in the shaft (b), and a foot (c). Stitch formation (Fig. 48 B) starts with expulsion of the needle (1). The stitch loop slides out of the hook, over the latch, and onto the shaft of the needle (enclosing the loop). A thread is then laid into the hook (2) and the needle is pulled back with the foot. During this process, the stitch loop closes the hook by rotating the latch (3). It then slides onto the latch ("casting on") and is subsequently cast off over the hook (4). The stitch loop thereby gives rise to a new stitch, and the thread produces a new stitch loop. The latch needle is used in both standard knitting machines and warp knitting machines.

The *compound* or *slider needle* (Fig. 49 A), invented by WILCOMB in 1924, consists of a hook with shaft and groove and, in addition a separate slider (d). The slider is subject to independent motion in the shaft in such a way that it opens and closes the hook. In forming a stitch (Fig. 49 B), the needle is pushed forward with the slider retracted (1), and the stitch loop is moved

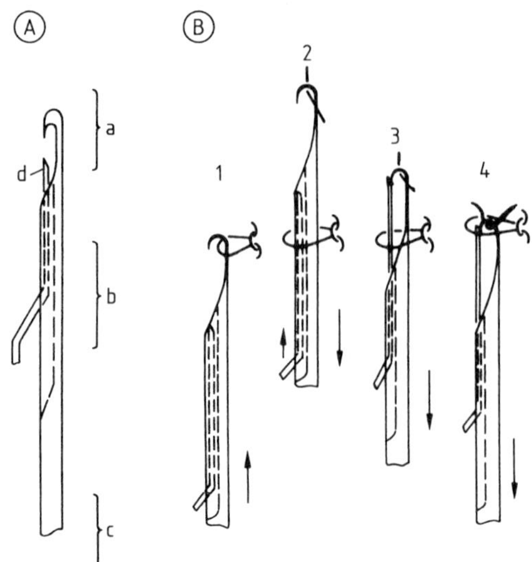

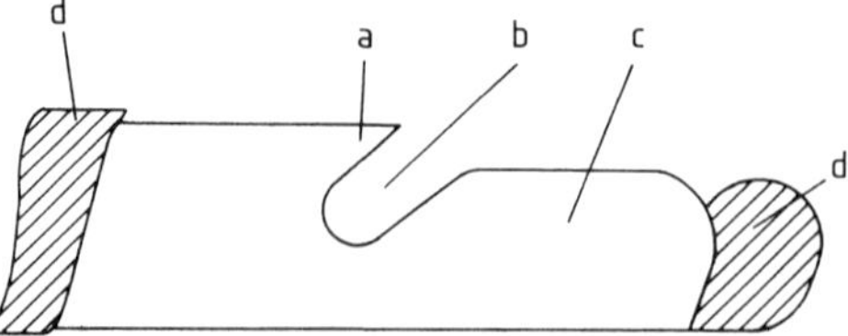

**Figure 50.** The characteristics of a sinker
a) Nose; b) Throat (enclosing the loop); c) Shaft (casting on and knocking over); d) Unit

**Figure 49.** Knitting with a compound needle
A) Compound needle: a) Head; b) Shaft; c) Foot; d) Slider
B) Principle of stitch formation with a compound needle: 1) Enclosing; 2) Thread laying; 3) Casting on; 4) Knocking over

off the hook onto the shaft (enclosing the loop). After thread laying (2), the needle and the slider are moved independently (3) in such a way that the hook is closed and the stitch loop is able to slip onto the slider (casting on). Finally, the stitch loop is pulled over the hook (knocking over), a new stitch is formed, and the thread produces a new loop (4). Slider needles are used in warp knitting machines as well as standard knitting machines (although not extensively at the present time).

To further ensure secure and satisfactory stitch formation, small shaped-steel rods (sinkers) are used in knitting and hosiery machines to improve needle performance, especially in the clearing, casting-on, and casting-off operations (Fig. 50). With respect to the criteria needle mobility and thread take-off, a distinction is made in the production of knitted materials between standard knitting, weft knitting, and warp knitting processes.

The *standard knitting process* (LAMB, 1863), illustrated in Figure 51 A, is characterized by independently moving needles (usually latch needles) that accomplish the various phases of stitch formation successively, and are served with one stretched thread (single-thread take-

off). Stitches are also formed from the thread sequentially. Depending on the arrangement of the needles, a distinction can also be made between *flat-bed* and *circular* standard knitting machines.

In the *weft knitting process* (LEE, 1589), shown in Figure 51 B, all the needles (bearded needles) move in concert (the principle of weft knitting), again with a single thread presented to the hook side of the needles. Unless it is highly elastic, a single stretched thread cannot be simultaneously transformed into stitches by several needles all moving at the same time, so the thread must first be preformed into a series of loops ("sinking the loops"). This is followed by simultaneous stitch formation, with all the needles moving together. In contrast to a flat weft knitting machine, stitch formation on a circular weft knitting machine (with needles arranged in a circle and moved all at once) occurs in a sequential fashion. This process is currently of little importance.

The *warp knitting process* (CRANE, 1775), illustrated in Figure 51 C, also uses bearded, latch, or compound needles in a planar arrangement and all moving together. The threads in the thread chains are usually processed in the longitudinal direction (warp-thread take-off). Threads from the thread chains are transferred in front of and behind the needles with the help of guide bars equipped with thread guides. At the same time, the threads are formed into stitches by needles all moving in concert.

## 5.4. Structural Elements and Structures of Knitted Fabrics

Knitted materials consist of various structural elements: stitches, loops, wefts, floatings, and filler threads. The primary structural ele-

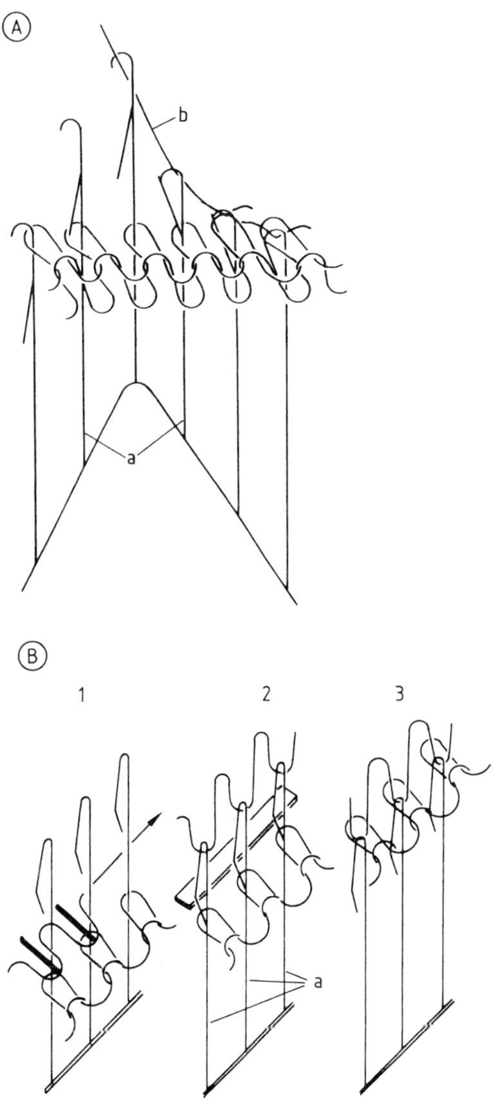

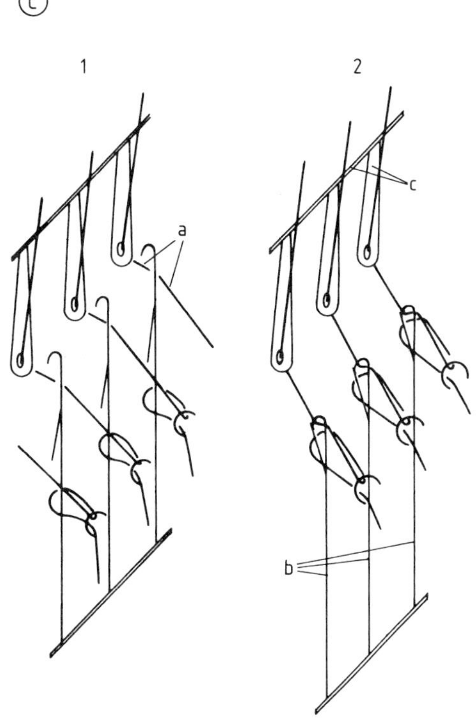

**Figure 51.** Schematic illustrations of the principles underlying various knitting processes
A) The standard knitting process: a) Independently driven latch needles; b) Single-thread feed
B) The weft knitting process: 1) Thread laying and loop sinking; 2) Pressing and casting on; 3) Knocking over; a) Collectively driven bearded needles
C) The warp knitting process: 1) Laying of the warp thread; 2) Stitch formation; a) Warp-thread placement in front of and behind the needles; b) Latch needles moved simultaneously; c) Guide bar with thread guides (guide needles)

ment of any knitted material is the stitch, although other structural elements may be present as well depending on the technology in question, the structural group (see below), and the pattern.

A stitch (Fig. 52 A) can be either open or closed. The head and the shank of the stitch constitute the needle loop, whereas the foot forms the sinker loop. A distinction is also made between the two sides of a stitch: the "face loop" (or *face stitch*, R) and the "reverse loop" (*reverse stitch*, L), recognizable by the form of the lower interlace point (Fig. 52 B). On the face side, the feet are underneath the head of the preceding

stitch, whereas on the reverse side they are above it. This distinctive feature allows one to divide knitted materials into three structural groups:

1) Single-face knits (RL)
2) Double-face knits (RR)
3) Purl knits (LL)

The basic structure of a *single-face* fabric (Fig. 53 A) presents on one face of the fabric the face-loop sides of the stitches, and on the other only the reverse-loop sides. A characteristic feature is a common direction of passage for all the

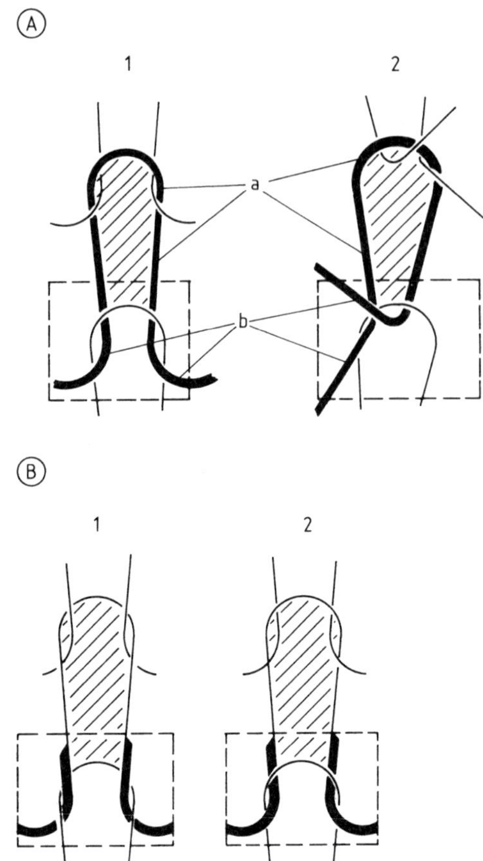

Figure 52. Stitches and the notion of "sidedness"
A) Two basic stitch types: 1) Open stitch (no foot crossing); 2) Closed stitch (foot crossing); a) Needle loop (head and shank); b) Sinker loop (foot bonding)
B) "Sidedness" with respect to a stitch: 1) Face loop ("R"); 2) Reverse loop ("L")

stitches, possible only with a single-needle system (RL technique).

In *double-face* fabrics with the RR structure (Fig. 53 B), the face and reverse sides of stitches alternate along a given stitch course. Nevertheless, only the face sides of the stitches are visible on either face of the fabric because the perpendicular elasticity of the material draws the wales together, concealing the reverse wales. Warp knits of the RR type also display only face stitches on both sides of the fabric. Two needle systems with two directions of passage are required for producing RR knits (the RR technique). Turning off one of the needle systems on an RR machine permits the production of RL materials, and LL materials (described below) can also be produced

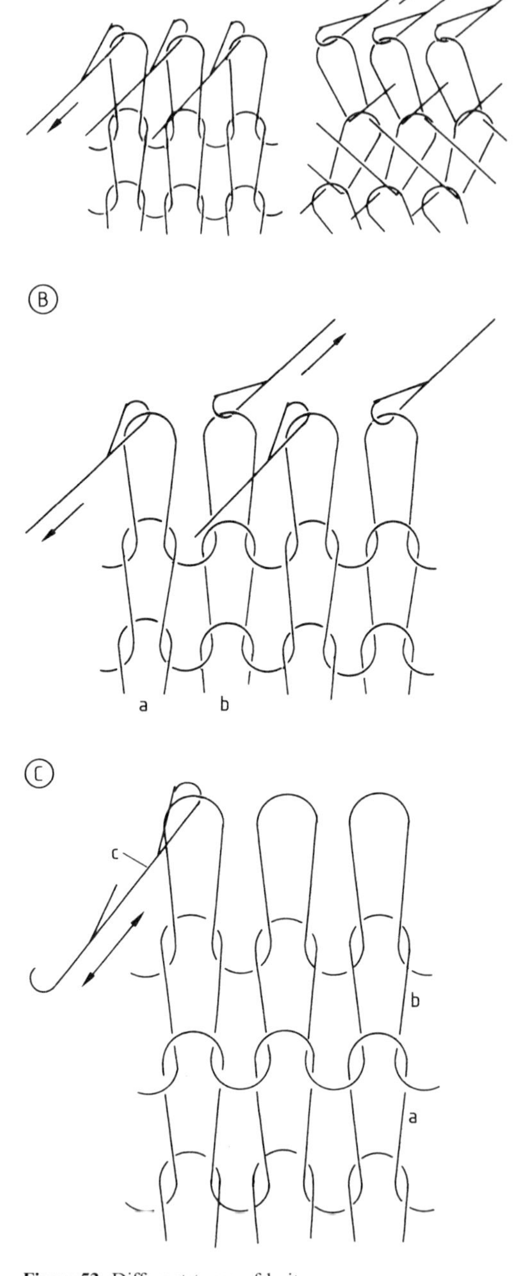

Figure 53. Different types of knits
A) 1) Single-face knitted fabric (face loop); 2) Single-face warp-knitted fabric (reverse loop)
B) Double-face knitted fabric: a) Face wale; b) Reverse wale
C) Purl-stitch knitted fabric: a) Course of face stitches; b) Course of reverse stitches; c) Double-latch needle

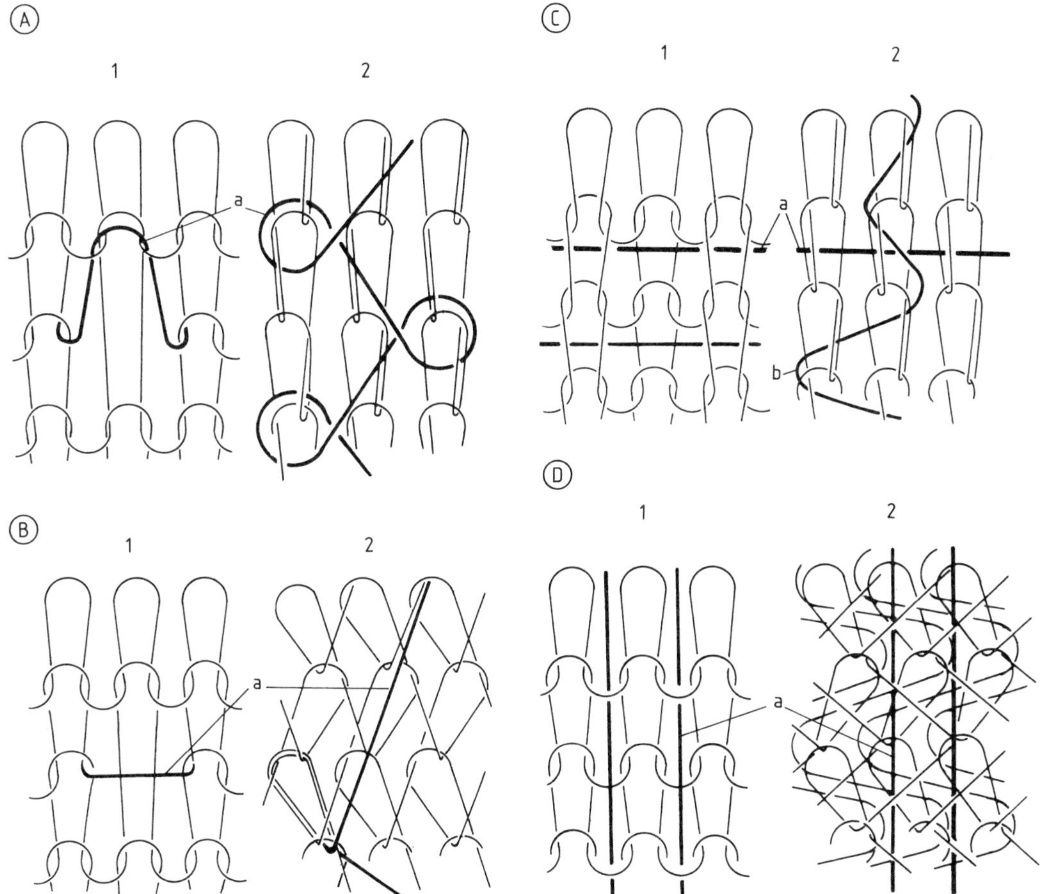

**Figure 54.** Loops, floatings, wefts, and filler threads
A) Loops: 1) RL Single-face knitted fabric (reverse loop); 2) RL Warp-knitted fabric (reverse loop); a) Loops
B) Floatings: 1) RL Single-face knitted fabric (reverse loop); 2) RL Warp-knitted fabric (reverse loop); a) Floatings
C) Wefts: 1) RR Single-face knitted fabric; 2) RL Warp-knitted fabric; a) Full weft; b) Partial weft
D) Filler threads: 1) RL Single-face knitted fabric (reverse loop); 2) RL Warp-knitted fabric (reverse loop); a) Filler thread

provided stitch transfer is possible (see Fig. 58 in Sec. 5.5).

The basic structure of an LL knit (Fig. 53 C) exhibits alternating face and reverse stitches in the wale. The longitudinal elastic behavior of these materials results in the appearance of mainly reverse stitches on both faces of the fabric. Double-latch needles are used in the production of such knits. These needles, activated by selectors, work alternately in different needle beds, causing the direction of passage to be repeatedly reversed (LL technique). Properly sorting the needles in advance in one or two needles beds makes it possible to use an LL device for making

RL and RR knits as well. Technology of this type exists, but it is of little commercial significance.

The following structural elements are incorporated or tied into stitches or other structural elements to establish desired patterning or structural properties of the knitted fabric:

1) A loop (Fig. 54 A) consists of a thread that has been placed in the needle hook along with the stitch-forming thread, and which is bound in place by the stitches.
2) A floating (Fig. 54 B) is a length of thread in a single-thread knit that skips at least one wale, and may skip successive stitch courses.

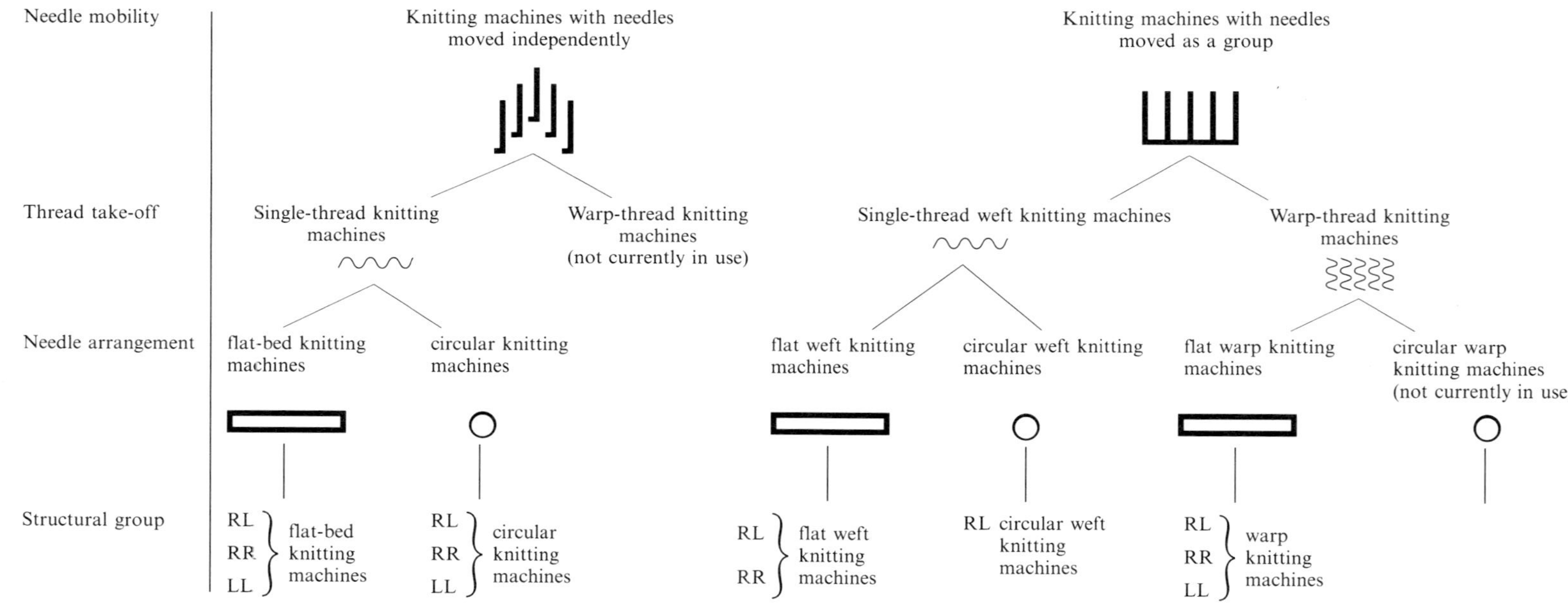


**Scheme 2.** Classification of the machines capable of producing knitted goods

In a warp knit this length of thread must skip at least one stitch course, and may extend over the wales.

3) A weft (Fig. 54 C) is a thread laid in the direction of the course. In warp knits, a distinction is made between a partial weft and a full weft.
4) A filler thread (Fig. 54 D) is a thread laid in the direction of the wale. Single-face knits exist with standing threads, but they are rare.

## 5.5. Classification of Stitch-Forming Machines

Scheme 2 illustrates classification of the currently most important types of knitting machines based on the corresponding needle mobility, thread take-off, needle arrangement, and structural group. These include:

1) RR and LL flat-bed knitting machines
2) RL, RR, and LL circular knitting machines
3) RL flat weft knitting machines (Cotton system)
4) RL and RR warp knitting machines

### 5.5.1. Flat-Bed Knitting Machines

Flat-bed knitting machines are used mainly for outerwear – sweaters, vests, and dresses – with colored or textured patterns. Flat-bed knitting technology has developed tremendously in the last two decades with the help of electronic controls and computer-aided design (the CAD technique). Electronic control and design programs have made it possible to go beyond the limits previously imposed by mechanical control and patterning, and the potential for diversity is

unlikely ever to be exhausted by knitting designers.

The knitting zone of an RR flat-bed knitting machine like that shown in Figure 55 contains two flat needle beds mounted in a roof-like arrangement (Fig. 56 A). Latch needles are independently driven out of channels in these beds and subsequently retracted by the action of moving cams. For this purpose, each cam (Fig. 56 B) consists of driving and retracting segments that

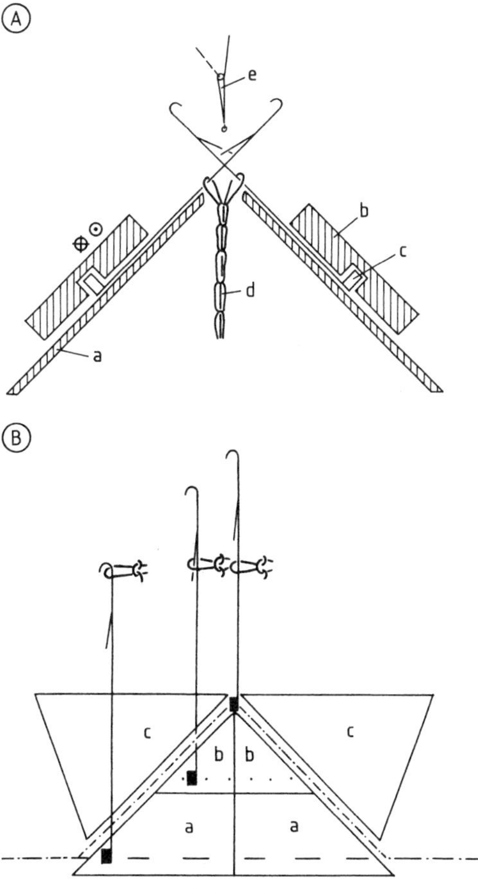

**Figure 56.** Characteristic features of an RR flat knitting machine
A) Knitting zone: a) Needle bed; b) Cam; c) Needle butt in the cam track; d) RR knit; e) Thread guide
B) Knitting cam: a) Lower driving segments; b) Upper driving segments; c) Retracting segments;
—·—·—·— Needle path for knitting (stitch)
········ Needle path for capturing (loop)
-------- Needle path when not knitting (RL, floating)
⊗ ⊙    (here and elsewhere) symbolizes motion perpendicular to the plane of the paper

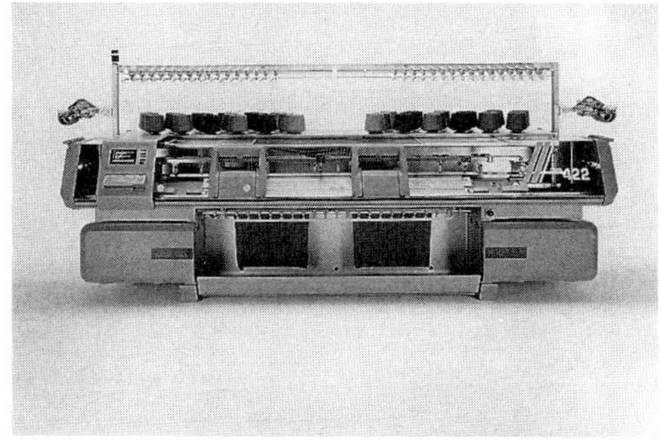

**Figure 55.** Electronically controlled flat-bed knitting machine

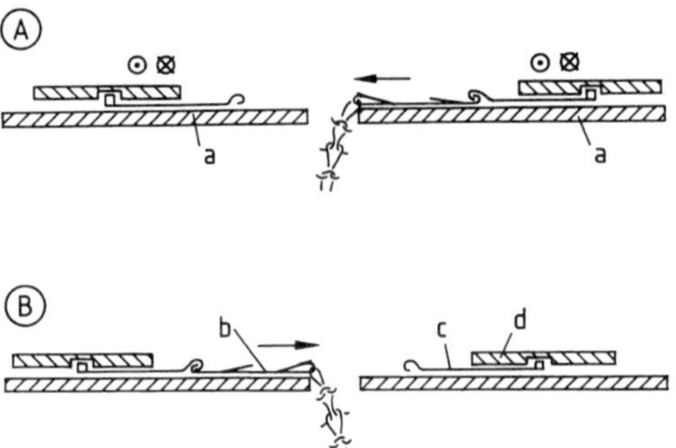

Figure 57. Schematic diagram of the knitting zone of an LL flat-bed knitting machine with needle transfer
A) Stitch formation in the right needle bed; B) Stitch formation in the left needle bed
a) Needle beds; b) Double-latch needle; c) Selector (clavette); d) Clavette cam

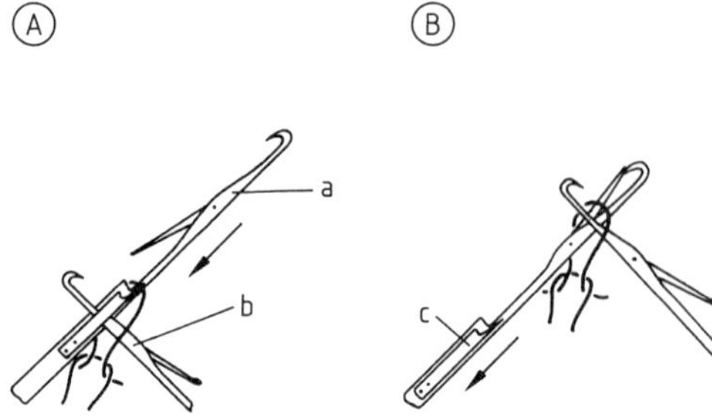

Figure 58. The stitch-transfer method
A) Presentation of a loop by needle (a); B) Capture of the loop by needle (b)
a) Donor latch needle; b) Acceptor latch needle; c) Loop-spreader spring for stitch transfer

constitute a channel for guiding the needle foot during stitch formation.

In the knitting zone of an LL flat-bed knitting machine (Fig. 57) are two flat needle beds mounted horizontally with their channels aligned opposite to each other, facilitating stitch transfer from one double-latch needle to another. Push rods (selectors) are required to move the needles, and these are in turn controlled by cams. Thus, as a consequence of the sliding motion, knitting may either be confined to one of the two needle beds or the growing strand can be transferred from one needle bed to the other. Depending on the direction of passage of a particular needle, a face stitch or a reverse stitch will appear in the knitted material.

With the help of the *Jacquard technique* (individual needle selection), a particular needle can be caused independently to knit (stitch), tuck (loop), or not knit (floating or RL; three-way technique) in each knitting process, using either the RR or the LL bonding technique (Fig. 56 B).

With the RR technique it is possible to transfer any stitch from a needle on one needle bed to one on the other. With this stitch-transfer method (Fig. 58) one is able not only to create fully fashioned parts by adding to or subtracting from the edge of the knit, but also to develop textural designs and patterns similar to those obtained with the LL technique. RR technology is compatible as well with structures of the RL and LL structural groups, either through the stitch-

transfer method or by selectively disengaging one needle system.

Implementation of the Jacquard technique in the LL structural group facilitates needle transfers of every type, and, depending on the direction of transfer, formation of either face or reverse stitches in the knitted material. Patterning based on the "RL–RR–LL" technique, the structural elements "stitch-loop-floating", a stitch-and-needle transfer technique, or some process involving thread changing, needle-bed shifting, etc., is accomplished with modern, electronically controlled flat-bed knitting machines equipped for integral CAD programming (Fig. 59). Pattern input may involve a scanner, a video camera, a manual drawing, or keyboard commands. Pattern development or the modification and generation of an appropriate control

Figure 59. Example of a CAD system for a flat knitting machine

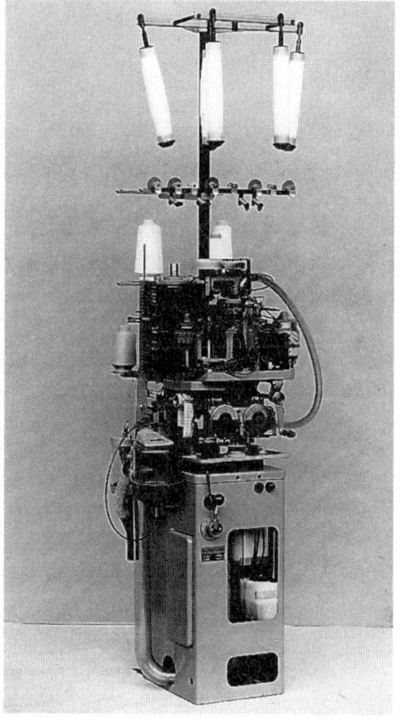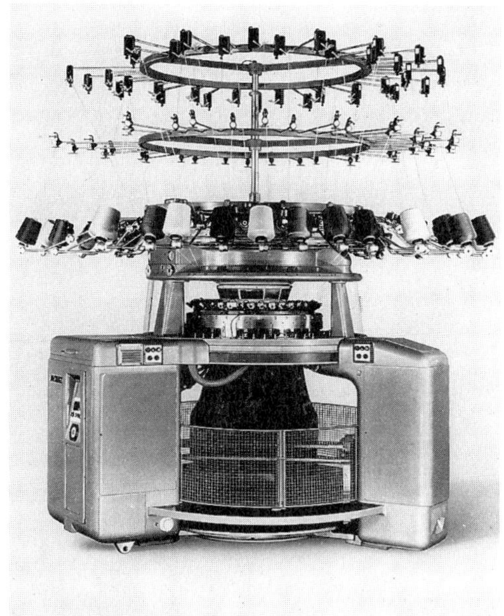

**Figure 60.** Circular knitting machines
A) Small device; B) Large, electronically controlled machine

program can be achieved quite quickly with most CAD systems. Prior to knitting, the complete knitting program for a particular product is transferred either online or via a punched tape or cassette to the flat-bed knitting machine's own computer.

The fineness of a knitted fabric depends on both the yarn strength and the machine gauge, which indicates the number of needles per inch in the needle bed (1 inch = 25.4 mm), where the gauge "E 10" means ten needles per inch. "Spacing" refers to the distance between two needles in the same needle bed. Thus, a machine might also be described as having "12-gauge spacing," meaning twelve needles to the inch (equivalent to "E 12").

### 5.5.2. Circular Knitting Machines

Circular knitting machines are characterized by a very high fabric output ($\leq 10^6$ stitches/min), and they are commonly used in the production of apparel, domestic fabrics, and industrial products. A basic distinction can be made between small (diameter < 165 mm; Fig. 60 A) and large (diameter > 165 mm; Fig. 60 B) circular knitting machines.

RL circular knitting machines (Fig. 61 A) utilize a rotating needle cylinder mounted in the knitting zone, in the channels of which latch needles (cylinder needles) also move vertically. The needle feet slide in the channels of fixed cylinder cams (Fig. 61 B), and can produce several stitches during the course of a single revolution. Each knitting point is known as a system. Up to ca. 150 such systems may be arranged along the periphery of a single circular knitting machine.

In addition to a cylinder and the cylinder cams, an RR circular knitting machine also possesses a dial (Fig. 62 A), in the channels of which are radially movable latch needles (dial needles) driven by dial cams. The dial cams are located above the rotating dial, and serve to apply the appropriate knitting motions to the dial needles. If the dial and cylinder needles are staggered, all the needles can be caused to knit simultaneously (the *dial technique*). If they are directly opposite each other in a head-to-head arrangement the needles in a given system must work together in a criss-cross fashion. Thus, only after two cycles

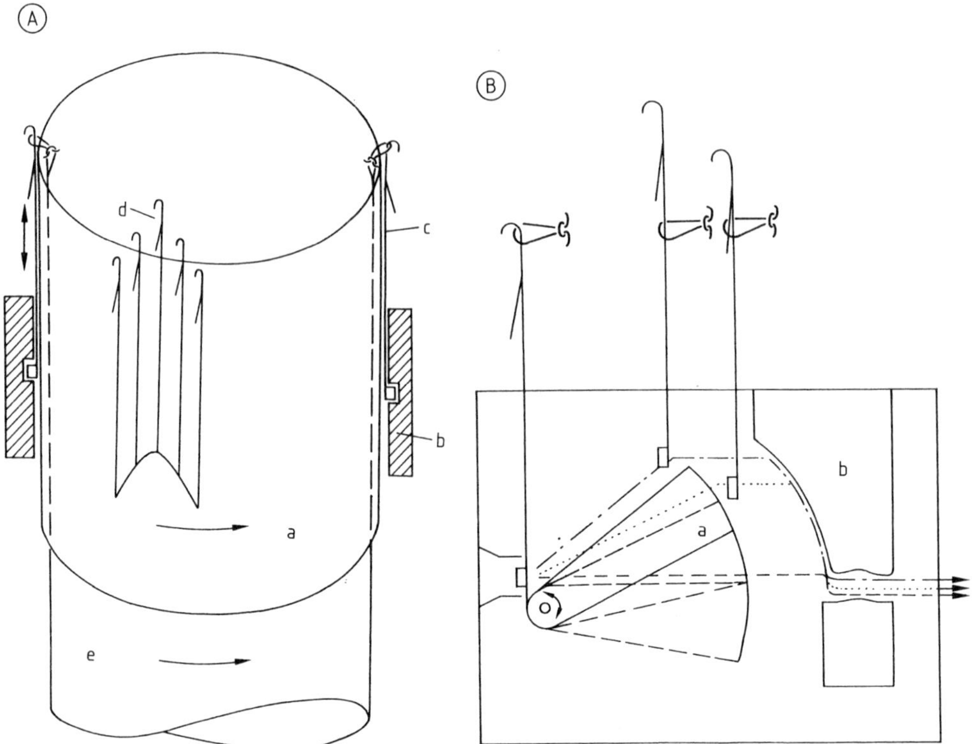

**Figure 61.** Schematic illustration of the RL circular knitting technique
A) General overview: a) Cylinder; b) Cylinder cam; c) Cylinder needle; d) Knitting point; e) RL knitted fabric
B) Detailed view of a cylinder cam: a) Driving segment; b) Retracting segment
–·–·–·– Needle path (cam track) for knitting (stitch)
·········· Needle path for capturing (loop)
– – – – – Needle path when not knitting (RL, floating)

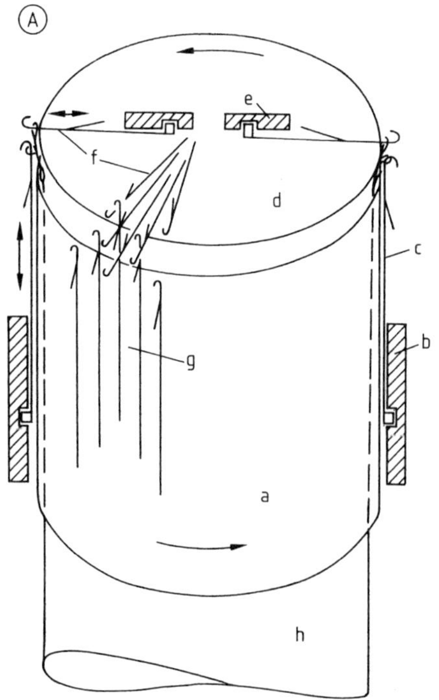

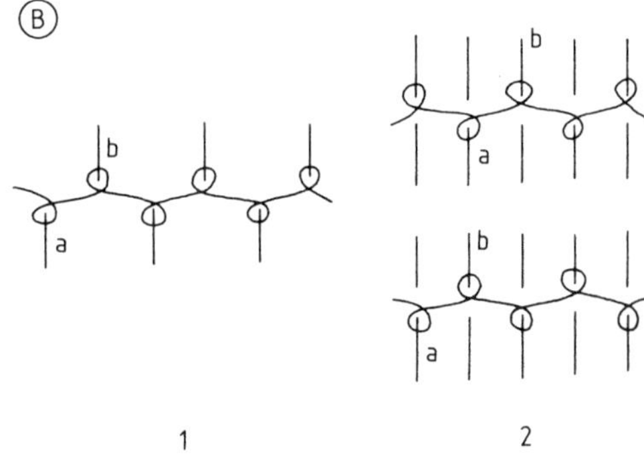

**Figure 62.** Schematic illustration of the RR circular knitting process
A) General overview: a) Cylinder; b) Cylinder cam; c) Cylinder needle; d) Dial; e) Dial cam; f) Dial needles; g) Knitting point; h) RR knitted fabric
B) Path of the thread, alternative systems: 1) Rib technique; 2) Interlock technique; a) Knitting by a cylinder needle; b) Knitting by a dial needle

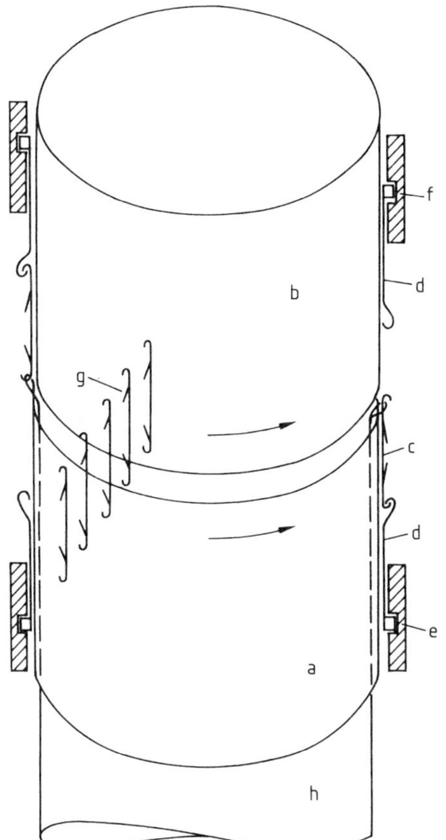

**Figure 63.** Schematic illustration of the LL circular knitting technique
a) Lower cylinder; b) Upper cylinder; c) Double-latch needle; d) Selector; e) Lower selector cam; f) Upper selector cam; g) Needle transfer; h) LL knitted fabric

will all the needles have knitted, leading to two RR knits that are interlocked (*interlock technique*; Fig. 62 B).

LL circular knitting machines (Fig. 63) have two rotating needle cylinders mounted one above the other. The needle channels are so aligned that double-latch needles activated by selectors (clavettes) can not only knit but also transfer thread from cylinder to cylinder. These machines can produce RL, RR, or LL structures depending on the needle arrangement.

With circular knitting, selection of the needles plays an important role in pattern development. Needles with feet that differ (high- or low-butt needles, long- or short-butt needles) make it possible to activate certain needle groups

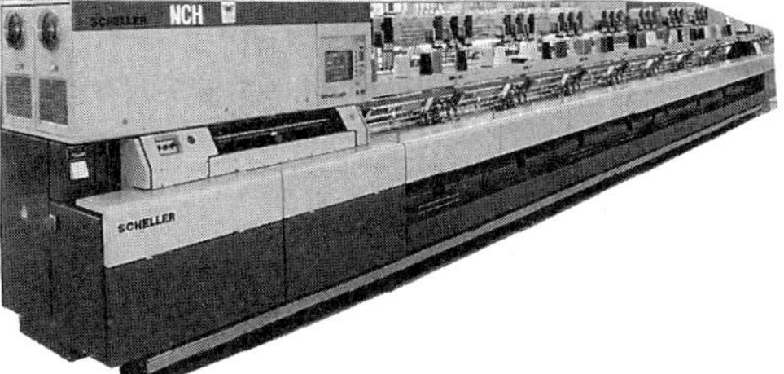

**Figure 64.** Electronically controlled flat weft knitting machine

selectively, opening the way to a limited amount of patterning. Only with the Jacquard technique and fully independent needle control, however, is there complete freedom of patterning. Jacquard needle selection in the cylinder can occur mechanically based on such devices as pattern wheels or pattern drums (peg drums), or the process may be driven electronically via electromagnets. This permits any desired cylinder needle to be designated as "knitting" (stitching), "tucking" (forming a loop), or "nonknitting" (floating).

All the design and process-engineering advantages of the CAD technique can be exploited with electronically controlled circular knitting machines. The gauge designations applied to circular knitting machines correspond to those of flat-bed knitting machines.

### 5.5.3. Weft Knitting Machines

Weft knitting machines are currently constructed with linear or circular needle arrangements, and they almost always employ the RL technique (single-needle system). Circular RL weft knitting machines are of almost no significance, whereas the flat RL weft knitting machine (Fig. 64), known as the Cotton machine after its inventor, W. COTTON (1864), is of some importance for fully fashioned fitted garments like sweaters and dresses. Such fashion forming is achieved by means of stitch transfer within the needle system (narrowing attachment).

The knitting elements (Fig. 65) include bearded needles (a) arranged linearly in a needle bar subject to both horizontal and vertical motion. Loop-forming sinkers (b) are also utilized to form thread delivered by the thread guide (d)

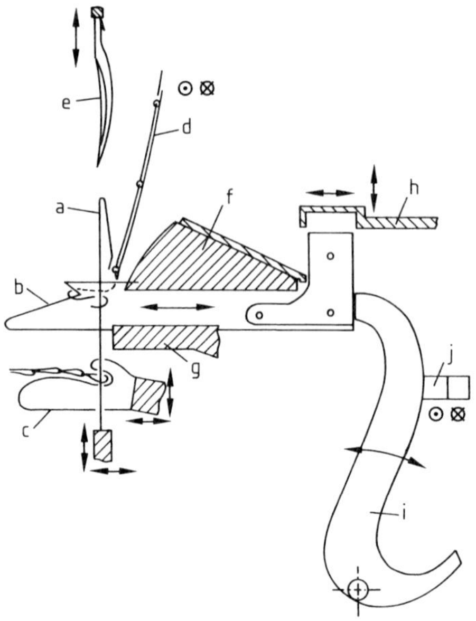

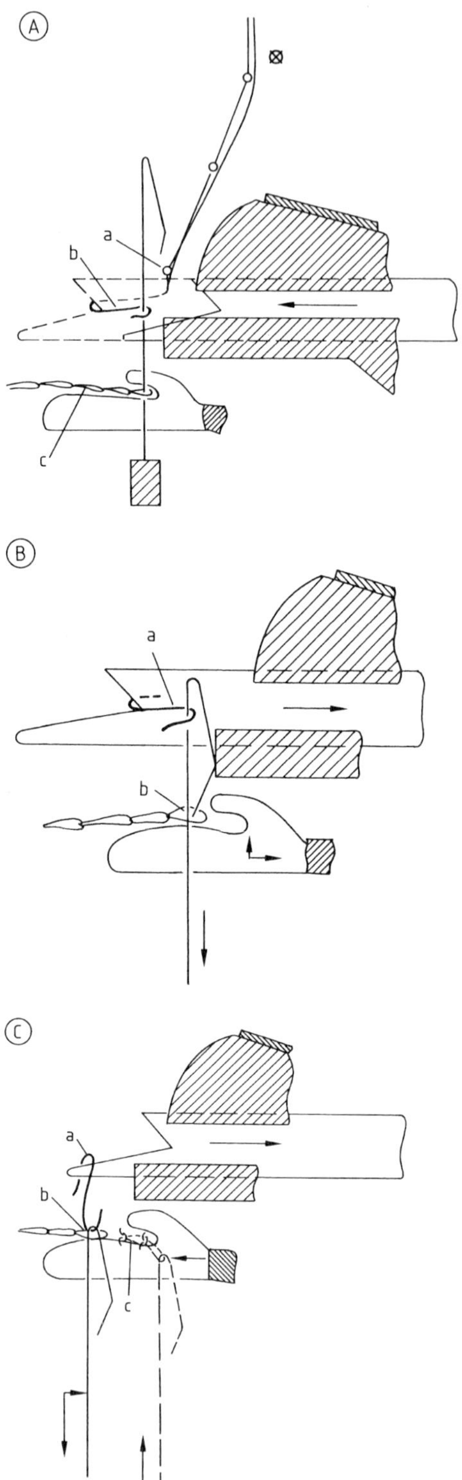

**Figure 65.** Schematic illustration of the key components in a flat RL weft knitting machine
a) Bearded needle; b) Loop-forming sinker; c) Web holder; d) Thread guide; e) Narrowing needle; f) Sinker head, upper part; g) Sinker head, lower part (press); h) Sinker catchbar; i) Jack; j) Slurcock

into successive loops, requiring an independent drive system based on a slurcock or culier cam (j) and jacks (i). The sinkers are activated simultaneously and intermittently by a sinker catchbar (h) as a way of releasing previously formed loops for subsequent stitch formation. Other knitting elements are the hold-down/knock-over sinkers or web holders (c) that facilitate needle operations during stitch formation (holding down the loop, casting-on, casting-off) as well as narrowing needles (e) arranged above the bearded needles. These narrowing needles transfer the stitches to the side for forming and patterning purposes.

Stitch formation (Fig. 66 A–C) begins with laying of the thread. Loop-forming sinkers are pushed one after another into the needle lanes,

**Figure 66.** Loop formation with a weft knitting machine
A) Thread laying and loop forming: a) Laying of the thread; b) Sinking the loop; c) RL knitted fabric
B) Pressing and enclosing: a) Thread loop; b) Stitch loop
C) Casting off and clearing: a) Thread loop; b) Stitch loop; c) Stitch

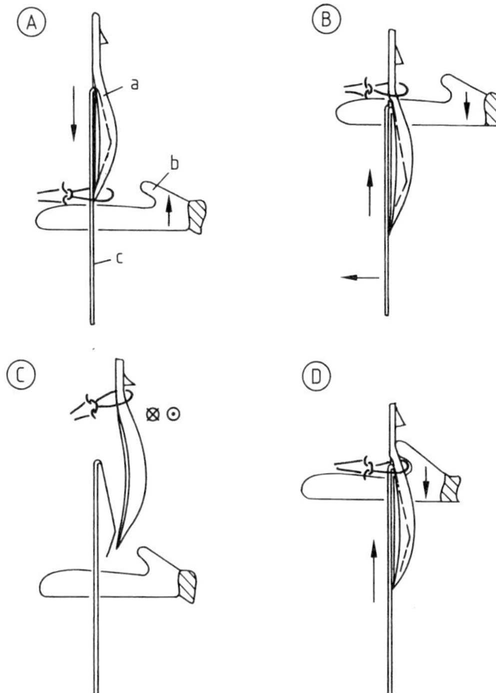

**Figure 67.** Stitch transfer with the aid of a narrowing needle
A), B) Capture of a stitch loop by the narrowing needle;
C) Sideways motion of the narrowing needle; D) Transfer
of the stitch loop to the adjacent needle
a) Narrowing needle; b) Sinker; c) Bearded (spring)
needle

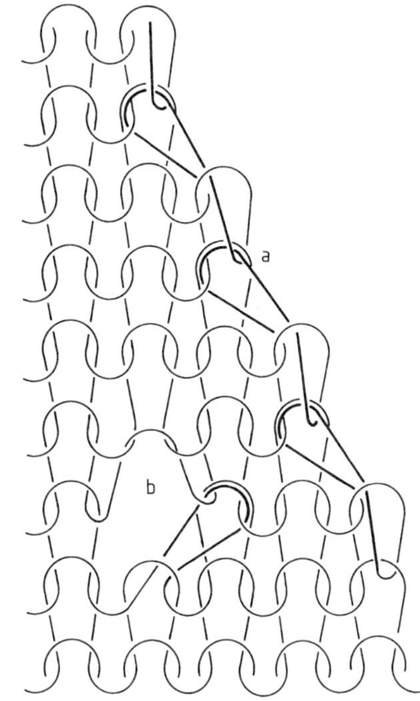

**Figure 68.** The use of "transferred stitches" for forming
and patterning
a) Fully fashioned forming (narrowing); b) Patterning
(eyelet pattern)

forming the thread into a row of loops (loop
sinking). Subsequently, all the needles simultaneously move downward and in front of the
presser bars (pressing), thereby closing them in
such a way that the stitch loops are pushed onto
the tips of the needles (casting-on) and can be
thrown over the hooks (casting-off). The sinkers
grasp the stitch loops in their grooves (enclosing)
and the needles rise, initiating a new stitch-forming process. For forming and patterning, stitches
must be transferred sideways with the help of a
narrowing machine. For this purpose, the narrowing needles are lowered onto the bearded
needles (Fig. 67) where they capture the stitch
loops, move them sideways, and deposit them on
the adjacent needles. Stitch transfer can be used
both for narrowing purposes and to produce an
eyelet pattern in the material (Fig. 68). Color
patterning (ringlet, intarsia, and plated patterns)

can also be achieved by changing or reposition-
ing the thread guide.

The fineness of a Cotton machine is expressed by a *gauge* (gg), which specifies the
number of needles in a span of 1.5 inches (e.g.,
21 gg = 21 needles/1.5 inch).

### 5.5.4. Warp Knitting Machines

Warp knitting machines use needles that
move together and are arranged in a planar array. These machines are equipped with guide
bars, each with its own thread guides (guide
needles) and thus able to move many warp
threads simultaneously in front of and behind
the needles (warp-thread take-off; see Fig. 70).
Sinkers facilitate the needle operation in these
processes as well. Warp knitting machines offer
extremely high output rates of up to ca. $50 \times 10^6$
stitches per minute.

A distinction is made between automatic warp knitting machines (Fig. 69 A), Raschel machines (Fig. 69 B), and crochet galloon machines (Fig. 69 C). The latter are used primarily in the manufacture of ribbons. Their mode of operation is very similar to that of the Raschel machines, and for this reason only the automatic warp and Raschel machines will be described in detail.

Automatic warp knitting machines are usually equipped with compound needles, whereas Raschel machines have latch needles. The angle between the thread chain and the fabric is ca. 90° in an automatic warp knitting machine and ca. 170° in a Raschel machine. The force applied to the needles is thus considerably higher with automatic warp knitting machines than with Raschel machines. Since stitch size in the knitted material is a function of this force, automatic machines are used for fine articles, and Raschel machines are used for coarser products (apparel, domestic fabrics, industrial products).

The process of stitch formation with an automatic warp knitting machine (Fig. 70 A–D) starts with ascent of the needle (enclosing the loop). The guide bars have already completed their lateral motion (thread laying) on the back side of the needle (underlap). They first swing with their guide needles into the needle lanes, move sideways on the hook side (overlap), and then swing back once again. The needles next drop, and in the subsequent casting-on and casting-off processes create a stitch. The knitted material is guided by sinkers in all phases of stitch formation.

The course of the thread ("thread laying") can be seen from a *loop-structure diagram* (Fig. 71). The sinker loops are formed by underlappings, and the needle stitches result from "swinging in," "overlapping," and "swinging out." Depending on the nature of the lappings, various basic structures can be formed, which can be represented by *lapping diagrams* (Fig. 72) showing a top view of the course of the needles in successive stitch-forming processes as seen by the knitter. The illustrated course of the thread shows the resulting structure in a simplified form (pattern notation). In commercial knitted materials, two or more basic structures are combined with each other and with other structural elements as required (Fig. 73).

Lateral movement of the guide bars (i.e., the overlap and the underlap) is accomplished by

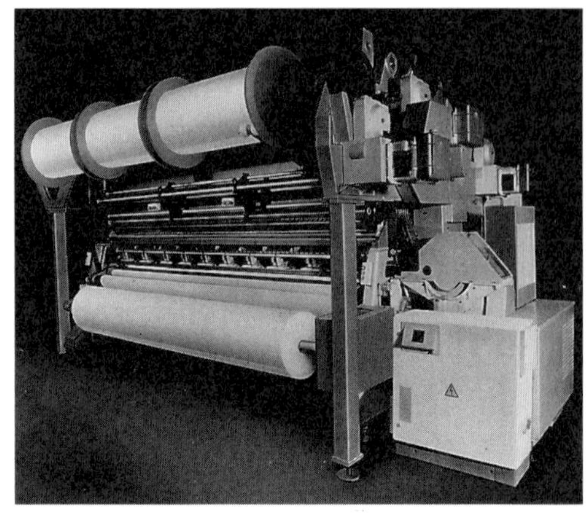

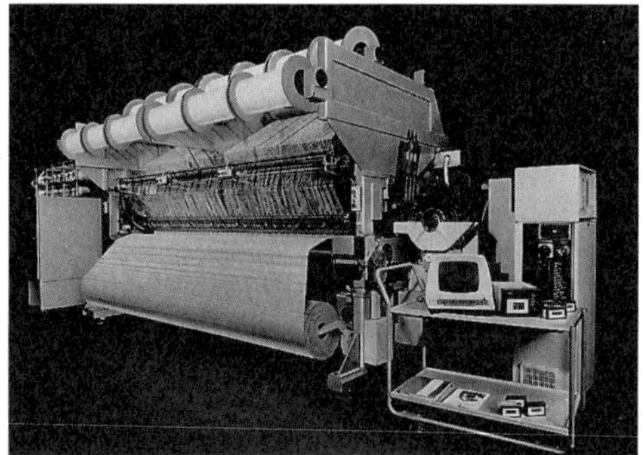

**Figure 69.** Examples of warp knitting machines
A) Automatic warp knitting machine; B) Raschel knitting machine; C) Crochet galloon knitting machine

pattern gears operated either mechanically via chain-link drums or pattern wheels, or else electromechanically.

The stitch-forming process with an RL Raschel machine (Fig. 74 A–D) also begins with enclosing the loop. Web holders (or "hold-down

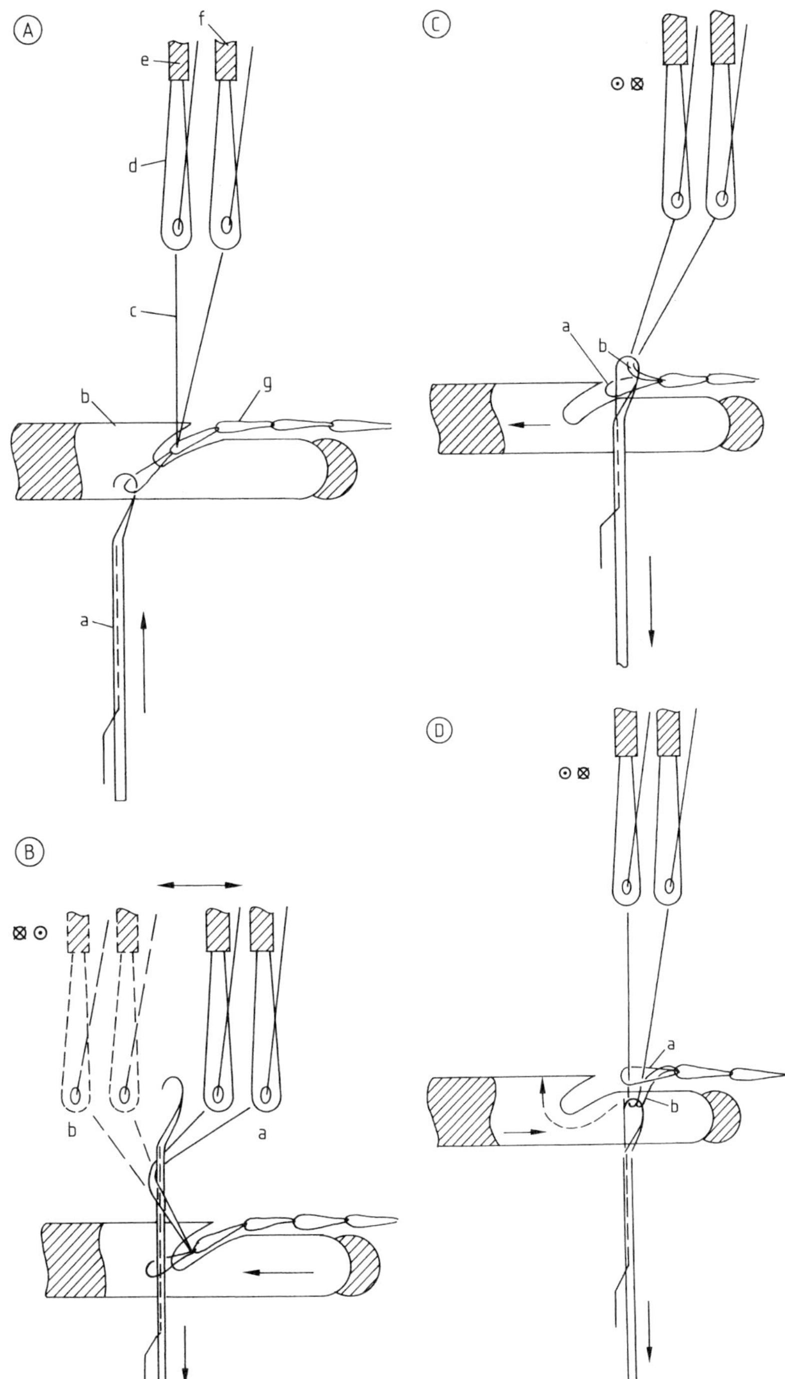

**Figure 70.** Stitch formation with an automatic warp knitting machine
A) Enclosing the loop: a) Compound needle; b) Enclosing/casting-off sinker; c) Warp thread; d) Guide needle; e) Guide bar $G_1$; f) Guide bar $G_2$; g) Warp knitted fabric
B) Overlap: a) Underlap position; b) Overlap position
C) Casting on: a) Stitch loop; b) Thread loop
D) Knocking-over: a) Stitch; b) Stitch loop

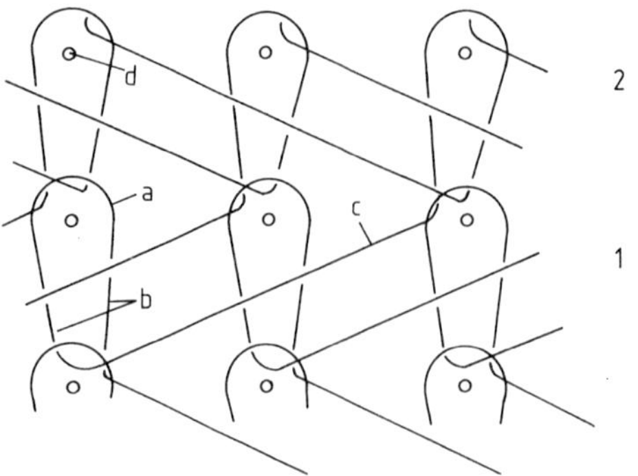

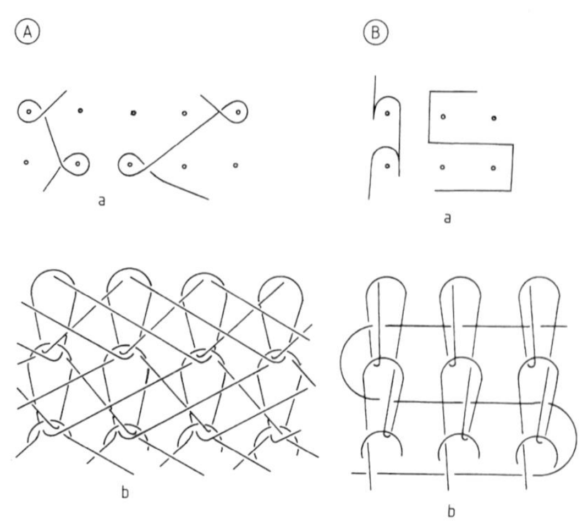

**Figure 71.** Loop-structure diagram for the product of a warp knitting machine
Process of forming (1) the first course and (2) the second course
a) Stitch head (overlap); b) Stitch shank (swinging in and out); c) Stitch foot (underlap); d) Needle
Note: (a) and (b) needle loop; (c) sinker loop

**Figure 73.** Combined stitch formations
A) Combination of closed two-and-one and closed one-and-one lapping movements; B) Combination of open-lap pillar stitch with inlay
a) Lapping diagrams; b) Loop structures

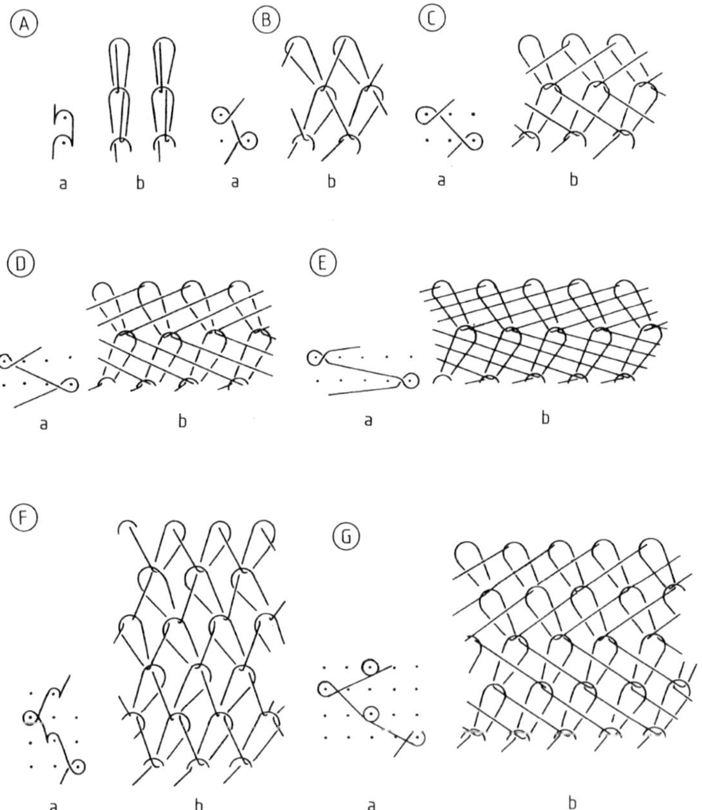

**Figure 72.** Lapping diagrams (stitch bindings)
A) Open-lap pillar stitch; B) Closed one-and-one lapping movement (tricot lap); C) Closed two-and-one lapping movement; D) Closed three-and-one lapping movement; E) Closed four-and-one lapping movement; F) Two-row open atlas lapping movements; G) Two-row floating closed atlas lapping movements with closed reverse
a) Lapping diagrams; b) Loop structures

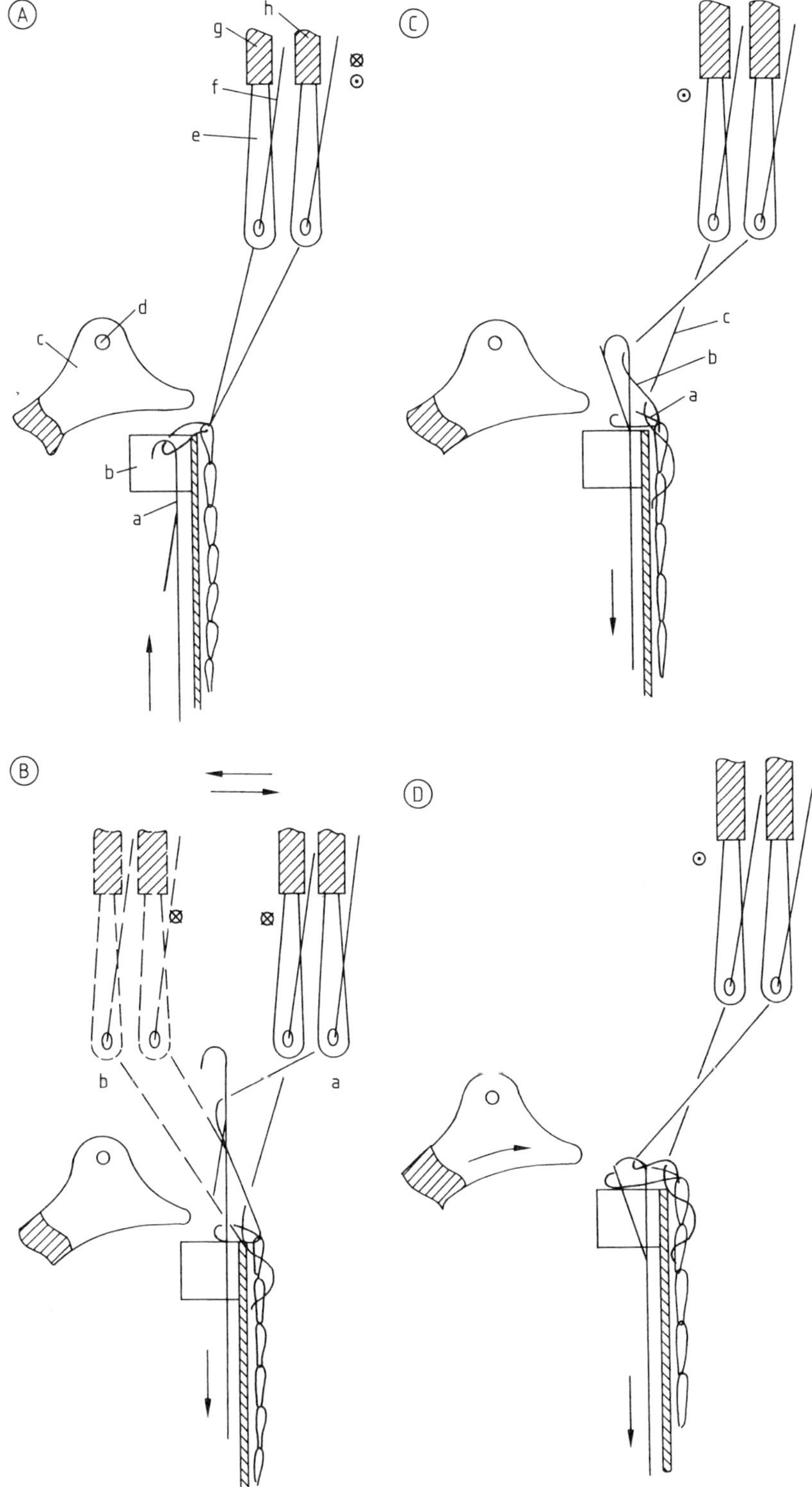

**Figure 74.** The stitch-formation process with an RL Raschel machine
A) Enclosing the loop: a) Latch needle; b) Knock-over bit; c) Web holder; d) Latch-stop wire; e) Guide needle; f) Warp thread; g) Guide bar $G_2$; h) Guide bar $G_1$
B) Overlap: a) Underlap position (e.g., for weft in guide bar $G_2$); b) Overlap position (e.g., for pillar in guide bar $G_1$)
C) Casting on: a) Stitch loop (open pillar stitch); b) Thread loop; c) Inlay lap
D) Casting off (knocking over)

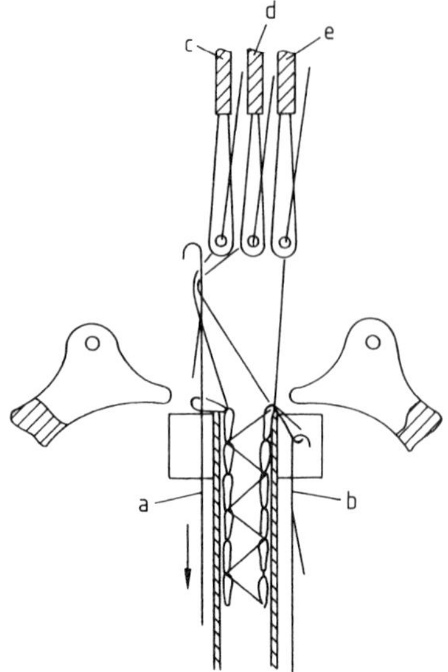

**Figure 75.** Schematic diagram of a double-needle-bar Raschel machine (RR Raschel machine)
a) Needle bar $N_1$; b) Needle bar $N_2$; c) Guide bar for $N_1$;
d) Guide bar for $N_1$ and $N_2$; e) Guide bar for $N_2$

sinkers") (c) support the stitch loops while the latch needles (a) ascend. The underlaps are completed, the guide bars (g, h) swing into the needle lanes, move around the needle hooks (overlap), and then swing back. The needles subsequently descend to form stitches (casting-on and casting-off).

One technique typical of Raschel machines, "pillar stitch with inlay," is illustrated in Fig. 73 B. The weft requires only underlaps, and the pillar stitch only overlaps.

Many patterns can be produced by appropriate laying of the thread: through use of various basic structures, structural elements, independent guide-needle displacements within a guide bar on the basis of the Jacquard technique, colored thread draws in the guide bars, weft insertions, etc.

RR Raschel machines (Fig. 75) operate with two vertically mounted needle bars. Stitch-forming processes are carried out in succession at the two needle bars. This mode of operation permits the production of tube-like warp-knitted fabric (packaging materials, clothing) or RR warp-knitted goods, which when cut open result in two plush or pile fabrics.

The fineness of automatic warp knitting machines is expressed on the basis of one English inch, that of Raschel machines on two English inches. Thus, with an automatic warp knitting machine a gauge of "E 28" means 28 needles per inch; for a Raschel machine the gauge "ER 36" means 36 needles per two inches.

# 6. References


**General References for Chapter 2**

[1] I. Hormes, M. Weber, B. Wulfhorst: *Faserstoff-Tabelle nach P.-A. Koch; Verarbeitung von Baumwolle*, RWTH Aachen 1991.

[2] B. J. Naarding: "Neue Konzepte bei Baumwollanbau-, Ernte- und Entkörnungsverfahren und ihre Bedeutung für die Spinnereien," *Int. Text. Bull. ITS Spinnerei* **2** (1982) 187, 188, 193, 194, 199, 200, 215.

[3] H. H. Perkins, Jr., J. D. Bargeron III: "Einflußfaktoren auf den Kurzfasergehalt von Baumwolle," *Textilbetrieb (Würzburg)* **100** (1982) no. 9, 21–28.

[4] W. Klein: *Manual of Textile Technology, Short-staple Spinning Series*, vol. 2: A Practical Guide to Opening and Carding, The Textile Institute, Manchester 1987.

[5] K. Gilhaus: "Ein Beitrag zum Mechanismus der Baumwollreinigung," Dissertation, RWTH Aachen 1981.

[6] B. Wolf: "Die Karde im Zeichen der modernen Spinnerei," *Int. Text. Bull. ITS Garnherstellung* **34** (1988) no. 3, 37, 40, 41, 44.

[7] "Research on Tandem Carding," parts I–IV, *Text. Top.* **17** (1989) no. 7, 1–3; no. 8, 1–4; no. 9, 1–3; no. 10, 1–2.

[8] W. Klein: *Manual of Textile Technology, Short-staple Spinning Series*, vol. 1: The Technology of Short-staple Spinning, The Textile Institute, Manchester 1987.

[9] M. Frey: "Steigerung von Qualität und Produktivität durch den Einsatz einer modernen Baumwollkämmerei," *Melliand Textilber.* **68** (1987) 157–163.

[10] W. Klein: *Manual of Textile Technology, Short-staple Spinning Series*, vol. 3: A Practical Guide to Combing and Drawing, The Textile Institute, Manchester 1987.

[11] W. Klein: *Manual of Textile Technology, Short-staple Spinning Series*, vol. 4: A Practical Guide to Ring Spinning, The Textile Institute, Manchester 1987.

[12] E. Dyson: *Rotor Spinning – Technical and Economic Aspects*, The Textile Trade Press, Stockport 1975.

[13] R. Nield: *Open-end Spinning*, The Textile Institute, Manchester 1975.

[14] D. E. Henshaw: *Self-twist Yarn*, Merrow Publ. Co., Watford 1971.

[15] W. Klein: *Manual of Textile Technology, Short-staple Spinning Series*, vol. 5: New Spinning Systems, The Textile Institute, Manchester 1993.

[16] L. Hunter: "The Production and Properties of Staplefiber Yarns Made by Recently Developed Techniques," *Text. Prog.* **10** (1978) nos. 1/2, 1–163.

[17] H. Deussen: *Rotor Spinning Technology*, Schlafhorst Inc., Charlotte, N.C. 1993.

[18] J. Möller: "Transportautomation in der Baumwollkämmerei," *4. Reutlinger Ringspinnkolloquium*, Reutlingen 1989.

**General References for Chapter 3**

[19] P. R. Lord, M. H. Mohamed: *Weaving: Conversion of Yarn to Fabric*, Merrow Publishing Co., Watford 1963.

[20] J. J. Vincent: *Shuttleless Looms*, The Textile Institute, Manchester 1980.

**Specific References for Chapter 3**

[21] R. Cocchi: "La Tessitura a Negli Anni 1990," *Tecnol. Tessile* **1985**, no. 4, 24–36.
[22] W. Loy (ed.): *Taschenbuch für die Textilindustrie 1993*, Fachverlag Schiele & Schön, Berlin 1993.
[23] DIN ISO 9947, 11.92: Textilmaschinen und Zubehör: Doppeldraht-Zwirnmaschinen. Begriffe.
[24] J. Trauter, W. Wunderlich, H. Böttle: "Untersuchungen zum Kaltschlichten von Spinnfasergarnen," part I: "Grundlagen: Der Einfluß von Parametern auf den Beschlichtungsgrad," *Melliand Textilber.* **73** (1992) 551–554, 623–626.
[25] L. Simon, M. Hübner: *Vorbereitungstechnik für die Weberei, Wirkerei und Strickerei*, Springer-Verlag, Berlin 1983.
[26] S. Schlichter: "Der Einfluß der einzelnen Maschinenelemente auf die Bewegungs- und Kraftverläufe in Kette und Schuß an Hochleistungswebmaschinen," Dissertation, RWTH Aachen, Aachen 1987.
[27] H. Weinsdörfer: "ITMA 1991: Weberei und Webereivorwerk," *Chemiefasern/Textilind.* **42**/94 (1992) 110, 112–118.

**General References for Chapter 4**

[28] H. Engels: *Handbuch der Schmaltextilien. Teil 1: Maschinen und Verfahren zur Erzeugung konventioneller Geflechte*, Verlag der Fachhochschule Niederrhein, Institut für Textil- und Bekleidungswesen, Mönchengladbach 1994.
[29] H. Engels: *Handbuch der Schmaltextilien. Teil 2: Maschinen und Verfahren zur Erzeugung von Flechtprodukten mit speziellen physikalischen und chemischen Anforderungen*, Verlag der Fachhochschule Niederrhein, Institut für Textil- und Bekleidungswesen, Mönchengladbach 1994.
[30] F. K. Ko, C. M. Pastore, A. A. Head: *Handbook of Industrial Braiding*, Atkins & Pearce, Covington, Kentucky 1989.

**Specific References for Chapter 4**

[31] Arbeitgeberkreis Gesamttextil (eds.): *Ausbildungsmittel Unterrichtshilfen: Textiltechnik Maschinengeflechte*, Frankfurt 1981.

[32] F. Goseberg: Meisterkursus Flechterei, Lehrgangsunterlagen der Textilen Abendschule, Wuppertal, Oct. 1, 1967–Jan. 15, 1970.
[33] B. Lepperhoff: *Die Flechterei*, Eugen G. Leuze Verlag, Saalgau/Wttbg. 1953.
[34] H.-A. Lange: "Die moderne Flechttechnik und ihre Anwendung in zukunftsorientierten Industriezweigen," *Band. Flechtind.* **24** (1987) 20–29.
[35] H. Kruse: "Schneller höhere Leistung – 125 Jahre Flechtmaschinenbau – vom Barmer System zur Flechtmaschine HP," *Band Flechtind.* **24** (1987) 72–76.
[36] B. Wulfhorst, A. Büsgen, J. Brandt, H.-F. Siegling: "Anwendung der Flechttechnik für 3-D-verstärkte Verbundwerkstoffe," *TPI Text. Prax. Int.* **48** (1993) 394–400.
[37] A. Büsgen, B. Wulfhorst, J. Brandt, H.-F. Siegling: "Struktur und Eigenschaften von dreidimensionalen Geflechten zur Herstellung von FVW," *Tech. Text.* **36** (1993) T153–T154, T156.

**General References for Chapter 5**

[38] C. Iyer, B. Mammel, W. Schäch: *Rundstricken*, Meisenbach GmbH, Bamberg 1991.
[39] E. Lesykova: *Fachwörterbuch der Maschenwarenproduktion*, Meisenbach GmbH, Bamberg 1991.
[40] S. Raz: "Warp Knitting Production," *Melliand Textilber.* **68** (1987).
[41] S. Raz: *Flat Knitting*, Meisenbach GmbH, Bamberg 1991.
[42] D. Markert: *Maschen-Abc*, Deutscher Fachverlag, Frankfurt 1990.
[43] K. P. Weber: *Die Maschenbindungen der Kettenwirkerei*, Verlag Werkgemeinschaft Karl Mayer, Obertshausen 1966.
[44] K. P. Weber: *Wirkerei und Strickerei*, Deutscher Fachverlag, Frankfurt 1987.
[45] H. Wignall: *Hosiery Technology*, National Knitted Outerwear Association, New York 1968.
[46] A. Reisfeld: *Warp Knit Engineering*, National Knitted Outerwear Association, New York 1966.
[47] D. S. Paling: *Warp Knitting Technology*, Columbine Press, Manchester and London 1965.
[48] E. Lesykova: *Technical Dictionary for Knitwear and Hosiery Production*, Meisenbach GmbH, Bamberg 1991.
[49] K. P. Weber: *An Introduction to the Stitch Formations in Warp Knitting*, Employees Association, Karl Mayer e. V., Obertshausen 1966.

# Thallium and Thallium Compounds

HEINRICH MICKE, Sachtleben Chemie, Duisburg, Federal Republic of Germany (Chaps. 1–5)
HANS UWE WOLF, Universität Ulm, Federal Republic of Germany (Chap. 6)



## 1. Thallium Metal [1]–[18]

Thallium [7440-28-0], Tl, $A_r$ 204.39, atomic number 81, is in group 13 of the Periodic Table, and has electronic structure $1s^2$, $2s^2p^6$, $3s^2p^6d^{10}$, $4s^2p^6d^{10}f^{14}$, $5s^2p^6d^{10}$, $6s^2p$ [19]. It is a dense, fairly reactive metal, the fresh surface having a bluish-white luster. Thallium is softer than lead, but harder than indium, with a low mp 303 °C.

There are 20 known isotopes of thallium, of which two, atomic numbers 203 and 205, are relatively stable. The isotope $^{204}$Tl is used for materials testing (Section 1.5).

The only thallium-containing substances of commercial importance are the metal itself and thallium sulfate. As thallium compounds are prohibited for insect pest control in the United States and parts of Europe, methods of extraction have not been improved during the last 10 years. Several European producers have stopped production of thallium and its compounds.

## 1.1. Historical

Thallium was discovered in 1861 by the Englishman CROOKES and in 1862 by the Frenchman LAMY, probably independently. During spectrographic examination of a lead chamber slime, CROOKES found a green line that could not be ascribed to any known element. He was convinced that he had discovered a new element, and gave it the name thallium after the Greek word θαλλός, a young green branch. LAMY must have been the first person to isolate the element and establish its metallic character. A heated dispute over priority developed between the two investigators [20].

## 1.2. Properties

In many of its physical properties, thallium resembles its neighbor, lead. Its chemical behavior combines the properties of the alkali metals with those of silver, mercury, lead, and aluminum. Thallium is highly toxic (see Chap. 6).

**Chemical Properties.** Thallium is a very reactive metal. It oxidizes slowly in air, even at ca. 20 °C (more rapidly on heating), to thallium(I) oxide and thallium(III) oxide. Hydrogen peroxide and ozone oxidize the metal to thallium(III) oxide. In the presence of water, a hydroxide is formed. Thallium is therefore stored under petroleum spirit, glycerine, or deaerated water. On prolonged storage under water with free access of air, metallic thallium becomes covered with large crystals of thallium(I) carbonate, and the water becomes saturated with this salt. To protect against oxidation, rods of thallium are coated with paraffin.

Thallium combines with fluorine, chlorine, and bromine at room temperature, and with

Ullmann's Encyclopedia
of Industrial Chemistry, Vol. A 26

iodine, sulfur, phosphorus, selenium, and tellurium on heating. Nitrogen, carbon, and molecular hydrogen do not react with the metal, which does not dissolve in liquid ammonia. The metal is dissolved slowly by hydrochloric and dilute sulfuric acids, but more rapidly by nitric and concentrated sulfuric acids. Its slow reaction with hydrofluoric acid is noteworthy.

In the series of electrode potentials of cations, thallium has a similar normal potential to that of indium ($Tl/Tl^+$ $-0.335$ V, $In/In^{3+}$ $-0.34$ V), lying between cadmium ($Cd/Cd^{2+}$ $-0.402$ V) and tin ($Sn/Sn^{2+}$ $-0.140$ V). Solutions of thallium(III) salts are not stable toward metallic thallium, reacting almost quantitatively:

$$Tl^{3+} + 2\,Tl \longrightarrow 3\,Tl^+$$

The electrode potential of $Tl^+/Tl^{3+}$ is $+1.21$ V.

**Physical Properties.** Thallium is a soft metal (Brinell hardness 29.4 $N/mm^2$, the value for lead being 39.2 $N/mm^2$). Thallium exists in two allotropic crystal forms: $\alpha$-thallium, hexagonal close-packed, stable at room temperature; and $\beta$-thallium, body-centered cubic, stable above 226 °C. A volume increase of 3.23% takes place on solidification. On bending, the metal emits a creaking sound similar to the "cry" of tin. It has good cold forming properties and low strength. It can be melted and worked like lead. Physical properties are as follows [21], [22]:

| | |
|---|---|
| Atomic number | 81 |
| Isotope abundance | |
| $^{203}$Tl | 29.52% |
| $^{205}$Tl | 70.48% |
| Density at 20 °C | 11.85 g/cm³ |
| Atomic volume | 17.26 cm³/mol |
| Melting point | 303.6 °C |
| Boiling point | 1457 °C |
| Latent heat of fusion | 30 J/g |
| Latent heat of vaporization | |
| at normal boiling point | 800 J/g |
| Vapor pressure of melt, Torr | $\log p = \dfrac{-9300}{T}$ $-0.892 \log T + 11.1$ |
| Thermal expansion of melt | 3.1–4.3 %/°C |
| Electrical resistivity | |
| 2.38 K | superconducting |
| 0 °C | $18 \times 10^{-6}\ \Omega \cdot cm$ |
| Specific magnetic susceptibility | |
| Solid thallium | $-0.165 \times 10^{-6}$ cm³/g |
| $\beta$-Thallium | $-0.158 \times 10^{-6}$ cm³/g |
| Molten thallium | $-0.22\ \times 10^{-6}$ cm³/g |
| Surface tension | 4.01 mN/cm |
| Tensile strength | 9.8 kg/mm² |
| Relative elongation | 135 % |
| Brinell hardness | 3 kg/mm² |
| Mohs hardness | 1.2 |

Like mercury, lead, indium, etc., thallium can become superconducting, the average upper temperature limit being ca. 2.38 K [23]. Some compounds, e.g., $Tl_3Bi_5$, $Tl_2Hg_5$, and $Tl_7Sb$ can also become superconducting [24]. It is especially valuable as a superconductor in ceramic materials at higher temperatures than are possible with other metals [25].

Vessels for containing liquid thallium can be made of Fe, W, Ta, Mo, Nb, and Co, in decreasing order of suitability.

## 1.3. Occurrence

Very variable figures are quoted for the concentration of thallium in the Earth's crust (1–3 ppm). It is said to be the 58th most abundant element [26]. In its occurrence, thallium shows a dual character. As a chalcophilic element, it occurs typically, though in low concentration, with the heavy metals that occur in sulfidic ores, usually associated with cadmium, mercury, indium, and germanium. Much more frequently, thallium accompanies the alkali metals potassium, cesium, and rubidium, where its concentration is also low, e.g., in carnallite, sylvine, leucite, lepidolite, feldspar, mica, etc. It is found in seawater, some mineral waters, and in native sulfur [1]. It also occurs in true thallium minerals, of which the following are the most common:

| | |
|---|---|
| Lorandite | $TlAsS_2$ |
| Pirotpaulite | $TlFe_2S_3$ |
| Chalcothallite | $Cu_3TlS_2$ |
| Vrbaite | $Hg_3Tl_4As_8Sb_2S_{20}$ |
| Hutchinsonite | $(Pb,Tl)_2(Cu,Ag)As_5S_{10}$ |
| Bukovite | $Cu_{3+x}Tl_2FeS_{4-x}$ |
| Wallisite | $PbTlCuAs_2S_5$ |
| Hatchite | $PbTlAgAs_2S_5$ |
| Crookesite | $(Cu,Tl,Ag)_2Se$ |
| Avicennite | $7\,Tl_2O_3 \cdot Fe_2O_3$ |

The thallium content of these minerals is 16–85%. For many years, only five true thallium minerals were known. Because of their extreme rarity, these are of no importance for the extraction of the metal [27], [28].

## 1.4. Extraction

Although the amounts of thallium in ores, including those of lead and zinc, are small compared with the amounts in salts, rock, etc., only the former play a significant role in the winning of the metal. They can be economically worked only if the thallium becomes concentrated in the side products in the course of processing the ore to obtain the main metal, and if other metals are present in this side product to an extent that justifies their extraction. Only a small part of the thallium that occurs naturally is therefore available for extraction.

Thallium is present only as a trace element in nonferrous metal concentrates. Only in a few smelting works does the thallium become sufficiently concentrated in particular process steps for extraction to be practical.

The following companies are thallium producers: Vieille Montagne, Belgium; Cimkent, Russia; Toho Zinc, Japan; Cerro de Pasco, Peru; Preussag, Germany.

Thallium produced as a byproduct of the extraction of nonferrous metals is of only minor importance, both quantitatively and economically.

The following concentration processes can be used for the treatment of solid raw materials:

leaching and precipitation
volatilization
special processes

If the raw materials are in solution, the thallium is precipitated by cementation, and is then processed like a solid raw material.

**Leaching and Precipitation.** The simplest process is leaching with water without pretreatment. With airborne dusts from lead or zinc smelting works, or cementation slimes, a yield of 85–90 % of the thallium can be obtained. The thallium-containing liquor, which usually also contains cadmium and zinc, is filtered, precipitated with zinc amalgam, and the relatively pure metal is obtained by an electrolytic process (Fig. 1) [31].

There are many methods of pretreating the starting material before leaching. These include:

oxidative roasting at 500 °C
sulfatization at 175–250 °C
alkaline pretreatment

Leaching of the raw material after pretreatment can extract 50–60 % of the thallium. Flue dusts that contain bismuth, copper, zinc, iron, arsenic, cadmium, thallium, selenium, etc., are given an alkaline pretreatment (fusion with caustic soda and lead sulfide). The alkaline slag so

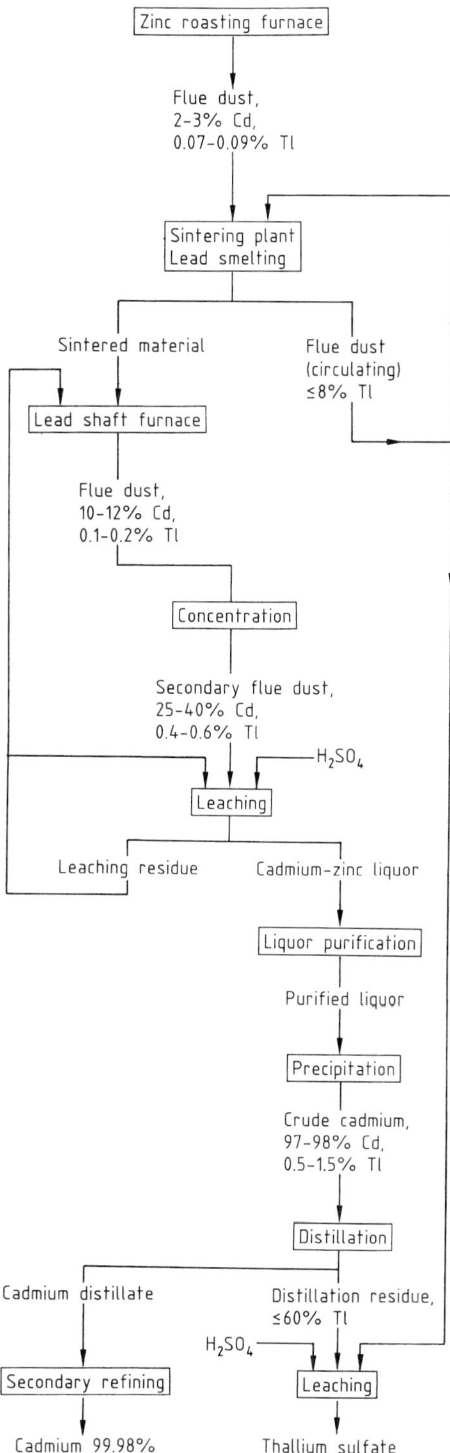


**Figure 1.** Simplified flow diagram of a plant for the combined recovery of cadmium and thallium sulfate from lead and zinc raw materials [29], [30]

formed is granulated with hot water. On cooling, sodium and thallium arsenates are precipitated, and processed to obtain the thallium [32].

The thallium is usually precipitated and recovered as a sparingly soluble salt (cementation thallium and thallium amalgam) from liquors produced in a variety of concentration methods, with and without pretreatment, and also from liquors that are themselves the raw materials.

Processing and refining of the precipitates obtained by cementation at the Duisburg copper refinery [33], [34] was discontinued in 1983.

**Volatilization.** In this process, crude arsenic containing 96 % $As_2O_3$ and 0.2 % Tl is treated by adding 5 % $H_2SO_4$ and 5 % lime and heating to 430 °C, causing 98 % of the arsenic trioxide to be volatilized while 99 % of the thallium remains in the residue. Sodium chloride is added, and the thallium is volatilized at 800 °C [33]. The volatilization process also gives good results with materials of complex composition from lead smelting. The secondary flue dust can be reprocessed. The most well-known process was that of the former company Zinkhütte Magdeburg [34].

**Electrorefining.** Electrorefining is operated in one or several stages, or as amalgam electrolysis. Depending on the purity of the starting metal and the operating conditions, thallium cathodes with variable thallium content (ca. 99.9 – 99.99 % and above) are produced [35] – [37]. Metal of the highest purity can be obtained by multistage electrorefining, in which a perchlorate electrolyte containing 40 – 70 g/L $TlClO_4$ and 60 – 120 g/L $NaClO_4$ with small amounts of organic additives is used in the final stage. The current density is 0.3 – 0.6 A/dm$^2$. The sum of all the impurities in the metal is < 1 g/t, i.e., the thallium produced has a purity of 99.9999 % [38].

Wet metallurgical refining is normally combined with the concentration process. The high solubility of thallium sulfate in water is used when lead and other metals that form insoluble sulfates are precipitated. The insolubility of thallium sulfide in alkaline solvents is used to separate it from the group 1 metals, and its solubility in dilute acids is used to separate it from silver, mercury, bismuth, copper, arsenic, antimony, cadmium, etc. The solution is treated with hydrogen sulfide, or preferably with freshly precipitated thallium sulfide, so that an easily handled thallium sulfate precipitate is produced at the end of all operations [39]. Iron and arsenic can

be removed after oxidation with air followed by neutralization, e.g., by adding zinc oxide. The high solubility of thallium carbonate in water is used to separate it from zinc, iron, cadmium, and other impurities, usually in combination with a concentration process. The low solubility of thallium chromate is then used for both concentration and refining.

Very pure metal can also be obtained by fractional dissolution of the impure metal or by fractional precipitation, mainly by fractional cementation of the impurities and the thallium from impure solutions. Impure thallium metal or thallium amalgam can be used as the starting material. When thallium amalgam is heated with 70 – 100 % sulfuric acid to boiling, the thallium dissolves first. If dissolution is interrupted after almost all the thallium, but none of the impurities, has gone into solution, the solution contains pure thallium sulfate which can be the raw material for metal winning [40].

To obtain compounds of the highest purity for use as intermediates for producing very high-purity metal, several unusual reactions have been proposed. These include precipitation of the thallium as cyclopentadienyl thallium [41], [42], or extraction of thallium(III) chloride, $TlCl_3$, from solution in hydrochloric acid (6 – 7 mol/L) with ether. First, however, the iron must be converted to the divalent form, and the heavy metals removed. Almost complete removal of all impurities takes place by taking the impure starting substance into solution in hydrochloric, hydrobromic, or hydroiodic acid, and extracting the thallium complex with isopropyl ether at acid concentration 0.1 – 4 mol/L. The metal obtained by electrolysis of the purified compound has purity 99.9999 – 99.99999 % [43].

During production of an ultrapure product from the already relatively pure metal by crystallization (especially zone melting and crystal pulling from the melt), the various impurities behave very differently. As well as the distribution coefficients, the operating conditions (mainly the migration rate of the zones and the crystallization rate) have a great influence on the refining. However, many other factors have an influence, e.g., the rotation rate during zone melting, and the rotation rate of the seed crystal during crystal pulling from the melt. Secondary processes are also important, e.g., volatilization and oxidation of certain impurities. The behavior of some of the more common potential impurities for a given combination of operating conditions can make it difficult to obtain very pure end products.

Zone melting is generally carried out in graphite boats. The ingots used are 150–180 mm long, with a ca. 15 mm molten zone, the process being carried out in an atmosphere of purified nitrogen. With a zone migration rate of 0.5–1.0 mm/min and five passes, the copper content in most of the ingot can be reduced from 50 g/t in the starting metal to 2–7 g/t, and, under the same conditions, the silver content in approximately half the length of the ingot is reduced from 3 to < 0.3 g/t. By pulling crystals from the melt in vacuo (ca. $10^{-2}$ Pa, $10^{-4}$ mbar), crystals 100–220 mm long and 8–10 mm diameter can be obtained. With a crystallization rate of 0.4–0.5 mm/min, the manganese content is 0.2 g/t in the lower part and 0.1 g/t in the middle part of the crystal. After a second recrystallization, 0.05 g/t manganese is found in the middle and upper part.

Not all impurities can be removed by crystallization, e.g., lead and iron, and there is a group of impurities that can only be removed by secondary processes. These include mercury, tin, and sulfur [44].

In Germany, thallium is produced solely by Preussag; annual production capacity is 4–5 t of 99.4–99.64 % thallium [45]. Thallium is supplied as rods and granules in "4 nines" to "5 nines" quality (i.e., 99.99–99.999 %). The catalogue also includes "pure" monovalent thallium compounds, including sulfate, bromide, chloride, and iodide. Other compounds can be produced on request.

## 1.5. Uses

No large application potential for thallium has so far been found. The unalloyed metal is unsuitable for direct use, owing to its unfavorable mechanical properties and marked tendency to oxidize. The radioactive isotope $^{204}$Tl is a β-emitter with half-life ca. 3.5 years, and is used as a radiation source in materials testing, principally for thickness measurements on metals and nonmetals. It has been discovered that $^{205}$Tl has the most constant atomic vibrations so far observed. This property has led to its use in atomic clocks.

These special applications require extremely pure grades. The very pure metal supplied by Unterharzer Berg- und Hüttenwerke contains > 99.999 % Tl, i.e., the sum of all the impurities does not exceed 10 g/t [46]. American Smelting and Refining supply "high-purity thallium", with the same purity. The following figures are quoted for the impurities [47]: Cu < 1 g/t; Pb 1 g/t; Fe < 1 g/t; Ag, Sn, Mg, and Si spectrographically undetectable, but < 1 g/t; other elements spectrographically undetectable.

## 2. Thallium Alloys

Thallium readily forms alloys with most other metals. There is complete mutual insolubility with iron, and limited solubility in the liquid state with copper, aluminum, zinc, arsenic, manganese, and nickel. The melting points of thallium–lead alloys are higher than those of the pure metals, the maximum, 380 °C, being reached at 37.8 % Pb. There is a resemblance to the two-component systems of lead, especially in alloys of thallium with copper, aluminum, zinc, and nickel. Gold, silver, cadmium, and tin form simple eutectics with thallium. Thallium also forms binary alloys with antimony, barium, calcium, cerium, cobalt, germanium, indium, lanthanum, lithium, magnesium, strontium, tellurium, and bismuth.

Alkali metals, alkaline earth metals, some metals of the rare earth group (e.g., lanthanum, cerium, praseodymium), and mercury form intermetallic compounds with thallium, some of which have unusually high melting points, e.g., CeTl (1240 °C) and PrTl (ca. 1150 °C). The compound with mercury, $Hg_5Tl_2$, has *mp* 14.5 °C and forms eutectics with thallium and mercury: $Tl–Hg_5Tl_2$ with 40.45 % Tl melts at 0.6 °C and $Hg–Hg_5Tl_2$ with 8.5 % Tl at − 58.4 °C.

Of the systems with three or more components, the combinations with bismuth, lead, and cadmium have been much investigated. These have very low melting points over wide ranges, and in most cases form ternary or quaternary eutectics. For example, the melting point of the Bi–Pb–Tl eutectic with 11.5 % Tl, 55.2 % Bi, and 33.3 % Pb is 90.8 °C, and the melting point of the quaternary Bi–Cd–Pb–Tl eutectic with 8.9 % Tl, 44.3 % Bi, 35.8 % Pb, and 11.0 % Cd is 81 °C [9].

The ternary alloys Tl–Pb–Bi, Tl–Al–Ag, In–Hg–Tl, Sn–Cd–Tl, Bi–Sn–Tl, and Bi–Cd–Tl are used as semiconductors or in ceramic compounds [48], [17].

The quaternary alloys Tl–Sn–Cd–Bi, Pb–Sn–Bi–Tl, and Pb–Cd–Bi–Tl are known [17].

Thallium, like tin, cadmium, lead, etc., has good wear resistance when used in bearings for steel shafts. As additions of thallium also increase the deformation resistance, breaking strength, yield, strength, and hardness of lead and its alloys, thallium-containing alloys are frequently recommended for bearings [49]. The alloy 72 % Pb, 15 % Sb, 5 % Sn, and 8 % Tl is superior to the best tin-based bearing metals with respect to yield strength and breaking strength. Bearing metals based on Pb–Tl–Cd with high cadmium content can have compositions in the range 0.01–5 % Tl, 5–40 % Cd, the rest being Pb [47]. However, thallium has since been replaced by indium as an alloy constituent [45].

Addition of thallium to gold, silver, and copper contacts in the electronics industry (e.g., 0.05–20 % Tl added to copper alloys) reduces the tendency to sticking [45].

In solid-state rectifiers, especially selenium rectifiers, addition of thallium increases the blocking resistance [50].

Addition of thallium can reduce the melting point of mercury from $-39\,°C$ to ca. $-59\,°C$. One use for this low-melting Hg–Tl alloy is for low-temperature thermometers [51].

The increase in the melting point of lead by addition of thallium is used in special types of electrical fuses and in solder (thallium content $\leq 20\,\%$) where a higher melting point is required [5], [6].

## 3. Thallium Compounds

Thallium can be monovalent or trivalent, the stability of the $6s$ electron pair in the electronic structure making thallium preferentially monovalent in inorganic compounds. Thallium(I) compounds have a certain similarity to alkali metal compounds, e.g., high solubility of the sulfate, carbonate, and hydroxide in water. However, they also show similarities to compounds of silver, mercury, and lead, e.g., low aqueous solubility of the chloride, bromide, iodide, and sulfide.

### 3.1. Individual Compounds

**Thallium(I) oxide** [*1314-12-1*], $Tl_2O$, is formed by heating thallium in air at $< 100\,°C$, or by dehydration of TlOH. It is reddish-black to black, volatile in air at higher temperatures, being converted to $Tl_2O_3$, and is very hygroscopic, forming a colorless, alkaline solution of TlOH with water. The enthalpy of formation is given by:

$$2\,Tl + {}^{1}/_{2}\,O_2 \longrightarrow Tl_2O$$
$$\Delta H_f = -180.8 \pm 0.8\ \text{kJ/mol}\ (-425.5 \pm 2.1\ \text{kJ/kg})$$

With $B_2O_3$, borates, or mixtures of borates and fluorides of alkali and alkaline earth metals, $Tl_2O$ forms stable, clear, homogeneous melts which can be used for the electrolytic production of metallic thallium.

**Thallium(III) oxide** [*1314-32-5*], $Tl_2O_3$, is more stable, and can be formed by heating $Tl_2O$ to $> 100\,°C$, treating thallium(III) salts with KOH or $NH_3$, or treating thallium(I) salts with oxidizing agents, e.g., TlCl with NaOCl. It is brown to black, forms cubic crystals with density $(4-21\,°C)$ 9.65 g/cm³, *mp* $717 \pm 5\,°C$, and begins to sublime above $500\,°C$ with partial decomposition:

$$Tl_2O_3 \rightleftharpoons Tl_2O + O_2$$

It can be reduced to $Tl_2O$, and even to the metal, by strongly heating with hydrogen, carbon, or carbon dioxide. It reacts very vigorously with sulfur and hydrogen sulfide, even at room temperature, doubtless in accordance with the equations:

$$2\,Tl_2O_3 + 5\,S \longrightarrow 2\,Tl_2S + 3\,SO_2$$
$$3\,Tl_2O_3 + 5\,H_2S \longrightarrow 3\,Tl_2S + 5\,H_2O + 2\,SO_2$$

Treatment of an aqueous suspension with sulfur dioxide leads to $Tl_2SO_4$:

$$Tl_2O_3 + 2\,SO_2 \longrightarrow Tl_2SO_4 + SO_3$$

It is sparingly soluble in water, but does not form a hydroxide. It dissolves in hydrochloric, sulfuric, and nitric acids with formation of hygroscopic, unstable thallium(III) salts, from which it is precipitated unchanged by alkali. It is also dissolved by acetic and tartaric acids. With oxalic acid, the sparingly soluble oxalate is formed, which decomposes on heating:

$$Tl_2C_2O_4 \longrightarrow 2\,Tl + 2\,CO_2$$

**Thallium(I) hydroxide** [*12026-06-1*], TlOH (92.32 % Tl), is formed on heating metallic thallium with water in the presence of air, or dissolving $Tl_2O$ in water. It is white, becoming dark gray on exposure to light. Its enthalpy of formation is given by:

$$Tl + {}^1/_2 O_2 + {}^1/_2 H_2 \longrightarrow TlOH$$
$$\Delta H_f = -240.4 \text{ kJ/mol } (-1085.8 \text{ kJ/kg})$$

It dissociates to $Tl_2O$ and $H_2O$ on heating above 139 °C.

It is soluble in water, forming a colorless markedly alkaline solution which rapidly absorbs $O_2$ from the air, with formation of $Tl_2CO_3$, and rapidly oxidizes, forming $Tl_2O_3$. The aqueous solution is a stronger base than NaOH, attacking glass very strongly on heating. Sodium sulfide precipitates $Tl_2S$:

$$2 \, TlOH + Na_2S \longrightarrow Tl_2S + 2 \, NaOH$$

**Thallium(I) chloride** [7791-12-0], TlCl, is formed when HCl or alkali metal chlorides are added to neutral thallium(I) salt solutions. The precipitated product is sparingly soluble in water, and is of some metallurgical importance. It is white, becoming gray-brown to blackish-brown on exposure to light. It forms cubic crystals, density 7.00 – 7.05 g/cm³, *mp* 427 °C, *bp* 806 °C, initial sublimation temperature ca. 360 °C, specific heat capacity $c_p$ (2.8 – 45.3 °C) $= 0.221 \text{ J g}^{-1} \text{ K}^{-1}$. The enthalpy of formation is given by:

$$Tl + {}^1/_2 Cl_2 \longrightarrow TlCl$$
$$\Delta H_f = -203.7 \text{ kJ/mol } (-847.3 \text{ kJ/kg})$$

No thermal decomposition takes place at the melting point. On heating in a stream of chlorine, it is first converted to $3 \, TlCl \cdot TlCl_3$, and then to $TlCl_3$. On melting with $NaHSO_4$, it is converted to $Tl_2SO_4$:

$$2 \, TlCl + NaHSO_4 \longrightarrow Tl_2SO_4 + NaCl + HCl$$

Metallic thallium is liberated on melting with alkali metal cyanides, or zinc:

$$2 \, TlCl + Zn \rightleftharpoons ZnCl_2 + 2 \, Tl$$

It is very sparingly soluble in water, 0.29 g dissolving in 100 g $H_2O$ at 15.6 °C, and 2.41 g at 99.35 °C. It reacts with a dilute solution of $Na_2SO_4$ to form $Tl_2SO_4$:

$$2 \, TlCl + Na_2SO_4 \longrightarrow Tl_2SO_4 + 2 \, NaCl$$

On heating with $H_2SO_4$, $Tl_2SO_4$ or $TlHSO_4$ are formed (see above). $HNO_3$ reacts in the cold

to form $2 \, TlCl \cdot TlCl_3$, and, on heating, the very soluble $TlCl_3$ (with some $TlNO_3$), in accordance with the approximate equation:

$$9 \, TlCl + 8 \, HNO_3$$
$$\longrightarrow 3 \, TlCl_3 + 6 \, TlNO_3 + 2 \, NO + 4 \, H_2O$$

On treatment with a 20% solution of $Na_2CO_3$, $Tl_2CO_3$ is formed, which is converted to TlCl on addition of NaCl:

$$2 \, TlCl + Na_2CO_3 \rightleftharpoons Tl_2CO_3 + 2 \, NaCl$$

It reacts with ammonium sulfide to form $Tl_2S$:

$$2 \, TlCl + (NH_4)_2S \longrightarrow Tl_2S + 2 \, NH_4Cl$$

and with sodium hypochlorite to form $Tl_2O_3$:

$$2 \, TlCl + 3 \, NaOCl \longrightarrow Tl_2O_3 + 3 \, NaCl + Cl_2$$

With $CdCl_2$, TlCl forms a sparingly soluble double salt, $TlCl \cdot CdCl_2$, which decomposes again on boiling with water.

**Thallium(I) iodide** [7790-30-9], TlI, is formed by direct combination of the elements or by precipitation from thallium(I) salt solutions by addition of alkali metal iodide. There are two modifications: a yellow rhombic form stable at room temperature, and a red cubic form stable above ca. 165 °C. The yellow form has density 7.1 g/cm³ and the red form 7.45 g/cm³, *mp* 440 °C, *bp* 824 °C. Sublimation starts at 355 °C. The specific heat capacity in $\text{J g}^{-1} \text{ K}^{-1}$ is given by $c_p = 0.152 \, T + 3.43$. The enthalpy of formation is given by

$$Tl + I \longrightarrow TlI \qquad \Delta H_f = -126.28 \text{ J/mol } (-381.2 \text{ J/kg})$$

Metallic thallium is liberated on melting with alkali metal cyanides. The following reaction takes place with sodium sulfide:

$$2 \, TlI + Na_2S \longrightarrow TlS + 2 \, NaI$$

Its solubility in water is considerably less than that of TlCl, but is increased by addition of KOH. It is not attacked by sulfuric and hydrochloric acids, but nitric acid (dilute or concentrated) reacts with liberation of iodine.

**Thallium(I) fluoride** [7789-27-7], TlF, is very soluble in water; unlike the other thallium(I) halides, the solution is alkaline.

**Thallium(I) sulfide** [1314-97-2], $Tl_2S$, is formed on melting together thallium and sulfur (along with $Tl_2S_5$, depending on the mixing ratio). It is also precipitated from thallium(I) salt solutions on adding $(NH_4)_2S$, or an amorphous precipitate is formed when $H_2S$ is added to very weakly acid solutions. This becomes crystalline on heating to 150–200 °C with a solution of $(NH_4)_2S$. From thallium(III) salt solutions, a mixture of $Tl_2S$ and sulfur is precipitated. $Tl_2S$ is also formed when a solution of $Tl_2SO_4$ is heated with $Na_2S_2O_3$. It is black, and the crystals have a blue metallic luster. Its density is 8.4 g/cm³, and its melting point 448 °C. It vaporizes on heating above 300 °C, at first unchanged, but decomposing at higher temperatures (e.g., 900 °C). The enthalpy of formation is given by:

$$2\,Tl + S \longrightarrow Tl_2S \quad \Delta H_f = -90.6 \text{ kJ/mol } (-205.4 \text{ kJ/kg})$$

Precipitated $Tl_2S$ is readily oxidized in air to $Tl_2SO_4$. It is sparingly soluble in water and almost insoluble in alkaline solvents, but dissolves in boiling mineral acids (dilute $H_2SO_4$ and $HNO_3$). Reaction with fuming nitric acid leads to the sulfate:

$$3\,Tl_2S + 8\,HNO_3 \longrightarrow 3\,Tl_2SO_4 + 8\,NO + 4\,H_2O$$

The other sulfides ($Tl_2S_3$, $Tl_2S_5$) become increasingly insoluble as the sulfur content increases, and are converted to $Tl_2S$ with liberation of sulfur on heating.

**Thallium(I) sulfate** [7446-18-6], $Tl_2SO_4$ (80.97 % Tl), is the most important starting material for thallium metal winning. Depending on its method of manufacture, it forms colorless, very lustrous rhombic crystals or a snow-white crystalline powder, density 6.765 g/cm³, mp 632 °C. It vaporizes unchanged at white heat. Above 900 °C, its vapor pressure increases rapidly. During the roasting of zinc concentrates (900 °C) and the sintering of lead concentrates (1000 °C), a considerable amount of thallium can volatilize as its sulfate. The effect of temperature on the vapor pressure can be approximately represented by:

$$\log p = -\frac{9300}{T} - 0.892 \log T + 11.1$$

where $p$ is in Torr (1 Torr = 133 Pa); $T$ is in K.
The enthalpy of formation is given by:

$$2\,Tl + 2\,O_2 + S \text{ (rhombic)} \longrightarrow Tl_2SO_4$$
$$\Delta H_f = -933.9 \text{ kJ/mol } (-1849 \text{ kJ/kg})$$

It is very soluble in water, e.g., 100 g boiling water dissolves 18.45 g $Tl_2SO_4$. The solution is hardly hydrolyzed, and is therefore neutral. Addition of hydrochloric acid or alkali metal chlorides to the solution precipitates sparingly soluble TlCl, though never completely:

$$Tl_2SO_4 + 2\,HCl \rightleftharpoons 2\,TlCl + H_2SO_4$$

Potassium iodide addition can give complete precipitation:

$$Tl_2SO_4 + 2\,KI \longrightarrow 2\,TlI + K_2SO_4$$

Addition of sodium thiosulfate at the boiling point produces $Tl_2S$:

$$3\,Tl_2SO_4 + 4\,Na_2S_2O_3$$
$$\longrightarrow 3\,Tl_2S + 4\,Na_2SO_4 + 4\,SO_2$$

With excess sodium carbonate, partial hydrolysis of the $Tl_2CO_3$ formed, and oxidation of the TlOH, produces a black precipitate. Addition of $Na_2Cr_2O_7$ or $Na_2CrO_4$ produces sparingly soluble $Tl_2CrO_4$, which can be redissolved to $Tl_2SO_4$ and $CrSO_4$ by treatment with dilute $H_2SO_4$ and $NaHSO_4$. Treatment of $Tl_2SO_4$ with $H_2SO_4$ or treatment of TlCl with excess $H_2SO_4$ gives thallium(I) hydrogen sulfate, $TlHSO_4$, or $Tl_2SO_4 \cdot H_2SO_4$, which melts at 115–120 °C and is converted to neutral $Tl_2SO_4$ with liberation of $SO_3$ by strong heating.

**Thallium(I) nitrate** [10102-45-1], $TlNO_3$, is formed when metallic thallium or $Tl_2CO_3$ is dissolved in $HNO_3$. It is white, and forms rhombic crystals (a cubic modification is also known), density 5.55 g/cm³, mp 206 °C. It melts without decomposition, but decomposition begins at ca. 300 °C, and is complete at 450 °C:

$$2\,TlNO_3 \rightleftharpoons Tl_2O_3 + N_2O_3$$

The enthalpy of formation is given by:

$$Tl + \tfrac{1}{2}\,N_2 + \tfrac{3}{2}\,O_2 \longrightarrow TlNO_3$$
$$\Delta H_f = -243.3 \text{ kJ/mol } (-913.4 \text{ kJ/kg})$$

It is very soluble in water, the solubility increasing very rapidly with temperature, 100 g $H_2O$ dissolving 7.93 g $TlNO_3$ at 15.4 °C and 593.93 g at the boiling point. Its solutions are neutral. $H_2SO_4$ reacts with $TlNO_3$ to form $Tl_2SO_4$.

**Thallium(I) carbonate** [6533-73-9], is formed by the action of $CO_2$ on TlOH, or by reacting thallium(II) salt solutions with an alkali or alkaline earth hydroxide, e.g., $Ba(OH)_2$, while passing $CO_2$ into the solution.

It is white, and forms monoclinic crystals, density 7.164 $g/cm^3$, *mp* 272 °C, decomposing at elevated temperatures. The $Tl_2O$ thus formed is soluble in molten $Tl_2CO_3$, and is converted to $Tl_2O_3$ on prolonged heating.

The enthalpy of formation is given by:

$$2\,Tl + C + {}^3/_2\,O_2 \longrightarrow Tl_2CO_3$$
$$\Delta H_f = -699.45 \text{ kJ/mol } (-1492.1 \text{ kJ/kg})$$

It is fairly soluble in hot water (27.78 g $Tl_2CO_3$ dissolves in 100 g $H_2O$ at 100 °C). On dilution, rapid hydrolysis takes place with formation of TlOH, so that the solution is alkaline. It is rapidly decomposed by acids.

**Thallium metasilicate** [*29209-99-2*], $Tl_2SiO_3$, is obtained by treatment of thallium(I) hydroxide with sodium metasilicate, $Na_2SiO_3$, dissolved in a solution of $NaNO_3$. The orthosilicate, $Tl_4SiO_4$, is prepared by reacting thallium(I) hydroxide with sodium orthosilicate, $Na_4SiO_4$. Both silicates are water soluble. Solid vitreous thallium silicate has a high refractive index.

The total quantity of thallium compounds in Germany does not exceed 500 kg (based on thallium metal), and is distributed in laboratories, universities, and research institutes.


## 3.2. Glasses and Single Crystals

Glasses composed of the three-component systems As–Tl–S and As–Tl–Se may be prepared by melting the elements together. They are very mobile liquids, even below 400 °C. Viscosity as low as 3 Pa · s has been observed at 250 °C for a composition of, e.g., 25 % As, 45 % Tl, and 30 % S. The glasses are chemically very stable, and insoluble in dilute acids. They are very slightly attacked by alkalis. The electrical resistivity is ca. $10^6$–$10^{18}$ $\Omega \cdot$ cm. The glasses have good wetting properties toward many metals, and are extremely impervious. They are therefore suitable for hermetically sealing sensitive electrical equipment (capacitors, resistors, semiconductors, etc.) to protect them from moisture and atmospheric effects. The glasses can be vaporized in vacuo and directly condensed onto the article to be coated [52].

The properties of photochromic glass can be improved by the addition of 0.2–4 % thallium (as TlCl added to AgCl).

Single crystals of three-component systems, e.g., $Tl_3VS_4$, $Tl_3NbS_4$, and $Tl_3PSe_4$, are used in acoustical–optical measuring equipment, e.g., laser modulators [53].

In the manufacture of quartz single crystals, Tl compounds can provide wavelength selection (optical windows) [54].


## 3.3. Uses

Since the 1970s, it has not been permitted to use thallium compounds for insect pest control in Europe and the United States, so that they are now sold only for special applications. Thallium compounds have also been proposed for impregnating wood and leather to kill fungal spores and bacteria, and for the protection of textiles from attack by moths. A solution of 0.2 g thallium(III) sulfate in 1 L water is suitable both for wood impregnation and seed treatment.

In the manufacture of ethylene oxide, catalysts containing 1 mmol Tl per kilogram catalyst have certain advantages [55].

When recycling polyurethanes at 120–200 °C, $Tl_2O_3$ 0.01 mol/L and catalysts are added to the solutions [56].

Not only colloidal thallium metal, but also its compounds, especially the benzoate, oleate, and amyl alcoholate have been proposed as antiknock agents for internal combustion engines. These compounds are added to the fuel as insoluble suspensions or soluble compounds, or are introduced into the cylinders in other ways [57].

Binary mixed crystals of thallium halides, especially TlBr–TlI and TlCl–TlBr, are highly transparent in the infrared. They are used for the manufacture of plates, prisms, and lenses for infrared instruments [8].

Addition of thallium to the filling materials of high-pressure mercury vapor lamps improves light emission considerably. If thallium iodide is used in place of the metal, provided that certain operating conditions are observed, the spectrum can be modified, i.e., the shade can be corrected [58]. Thallium salts are also added to fluorescent materials as activators.

Thallium(I) sulfide is one of the most important photosemiconductors, and is used in the manufacture of photocells. These are mainly constructed as barrier layer photoconductive cells, i.e., the resistance of the cell is reduced by absorption of light. The cells are especially sensitive in the long wavelength visible and near infrared, and are superior to selenium cells at low radiation intensities [59].

## 3.4. Environmental Aspects

As the natural concentration of thallium in the Earth's crust is 1 – 3 ppm, it does not constitute a danger to groundwater or rivers (Rhine water, 2 ppb). The most important sources of air pollution are smelting and roasting plants for iron and nonferrous metals, and lignite power stations. Over 95 % of the Tl is retained in cyclones and hot electrostatic filters [18]. The total thallium content of all the lignite used in Germany is 1.7 t/a. Of this, 95 % is retained in the ash, and only 5 % is released to the atmosphere (at an exhaust gas concentration $< 1 \ \mu g/m^3$).

## 4. Analysis

The quantitative determination of thallium is carried out with the aid of the grass-green coloration of the flame, and the spectral line at 535.1 nm, which is easily distinguished from the barium band at 534.7 nm by its sharpness and intensity. There are several wet chemical methods: precipitation of the monovalent ion by hydrochloric acid; the very sensitive reaction with potassium iodide to bright yellow thallium iodide (minimum detectable Tl concentration 1 in $3 \times 10^5$); precipitation from neutral solution as the yellow chromate, $Tl_2CrO_4$; or, the most sensitive detection method, precipitation with thionalide as a lemon-yellow complex (concentration limit 1 in $10^7$).

Quantitative gravimetric determination is by precipitation as thallium(I) iodide or chromate. The reaction:

$$Tl^{3+} + 3 \ KI \longrightarrow TlI + I_2 + 3 \ K^+$$

is used for volumetric determination, the liberated iodine being titrated against thiosulfate after addition of starch.

The Blumenthal permanganate precipitation is used to concentrate traces of thallium, ether being used to extract $Tl^{3+}$ ($Tl^+$ cannot be extracted quantitatively with ether).

For the determination of thallium in the microgram and nanogram regions, atomic absorption spectrometry is used, by flame atomic absorption or flameless techniques. Optimal limits of detection are:

solid samples (ores, earth samples, dusts, etc.), 0.5 ppm
aqueous samples, 0.001 mg/L
biological materials, 0.004 ppm [60]

Field desorption mass spectrometry (FD-MS) enables detection and quantitative determination of metallic cations from biological samples and environmental samples in the picogram range, without a separate conversion to ash. With this new technique for the determination of trace and ultratrace quantities, the limits of detection for thallium are ca. $10^{-11}$ g. The extreme specificity of FD-MS makes elaborate sample preparation unnecessary, except for the tissue and cell components of samples, which must be homogenized and centrifuged; the time required per analysis is only ca. 20 min. The sample size need be only ca. 1 mg [61].

## 5. Economic Aspects

The limited demand can be fully met by suppliers, as the total world demand in 1988 was only ca. 6 – 8 t. No increase is expected for either the metal or its compounds. Annual production in 1975 was 15 t. The amount of thallium available annually from raw materials used in metal refineries is currently estimated as $> 600$ t/a.

The price of the metal has increased tenfold since 1975, and is currently ca. 170 US $/kg. The price of the 99.999 % metal is ca. 350 US $/kg [62].

## 6. Toxicology and Occupational Health [64] – [66]

Thallium is a well-known, highly effective poison. Intoxications occurred in former times due to its use as a medicinal agent for many diseases such as venereal disease, dysentery, and tuberculosis. Another main source of accidental, suicidal, and homicidal exposures was its use as a rodenticide which, however, has been banned during the last decades in many countries. Thallium may be accumulated in mushrooms.

The toxicological significance of thallium is mainly restricted to some inorganic and organic salts of $Tl^+$ such as TlCl, $Tl_2SO_4$ (used as a rodenticide [67]), and thallium acetate. Intoxications by elemental thallium are comparatively rare. There is some evidence that, analogously to mercury, $Tl^{3+}$ can be converted in the environment to methylated compounds by microbial action [68], and thus represents a potential cumulative hazard.

**Toxicokinetics.** $Tl^+$ ions are readily absorbed from the gastrointestinal tract and distributed into various tissues, resulting in high concentrations in kidneys, myocardium, testes, salivary glands, intestine, skeletal muscle, thyroid and adrenal glands. Uptake of thallium-containing dusts via the respiratory tract results in increased thallium toxicity. Transdermal absorption is also possible, especially after the use of thallium-containing ointments. Thallium ions are able to cross the placental barrier and can produce nail abnormalities and alopecia in fetuses exposed during the last trimester [69]. Thallium ions do not undergo metabolic reactions comparable to those of organic xenobiotics (e.g., hydroxylation and conjugation reactions).

Excretion of $Tl^+$ ions after strong acute intoxication mainly occurs via the feces. Due to pronounced enterohepatic circulation, the excretion half-life varies widely in the range 1.7 – 30 d. Only 3 % of the entire body burden is excreted per day in the urine [64], [70].

The mechanism of toxicity is not completely understood. There is evidence for interaction of $Tl^+$ ions with thiol groups, resulting in reactions with a variety of proteins and in inhibition of enzymes such as succinic dehydrogenase. The exchange of the activating $K^+$ ion by the nonactivating $Tl^+$ ion leads to inhibition of $Na^+$, $K^+$-ATPase. Furthermore, $Tl^+$ is stored in axonal mitochondria, resulting in their destruction [71].

**Acute and Chronic Toxicity.** The most prominent and characteristic features of acute, subchronic, and chronic thallotoxicosis are seen in the nervous system, skin, and cardiovascular system. Concerning the clinical presentation, the time course for the development of symptoms can be divided into four stages [65], [72]:

1) *During 3 – 4 h* onset of predominantly gastrointestinal symptoms such as nausea, vomiting, and diarrhea occurs. In severe cases, hematemesis may occur.

2) *Within hours to days* various severe symptoms of the CNS (e.g., disorientation, coma, convulsions, psychosis, cerebral edema with central respiratory failure), PNS (combined motor and sensory neuropathy, including severe hyperesthesia of palms and soles), and ANS (tachycardia, hypertension, fever, salivation, and sweating) develop. Symptoms of the cardiorespiratory system, which may predominate in severe cases, include myocardial necrosis with dysrhythmia and pump failure, hypotension, and bradycardia. Skin symptoms include blue-gray lines on gums, dark pigmentation around the hair roots, and acne. The ophthalmologic symptoms are optic neuritis and ophthalmoplegias.

3) *During 2 – 4 weeks* the skin becomes dry and scaly, and white stripes across the nails (Mees's lines), and scalp and facial hair loss appear as characteristic signs.

4) *During several months* various CNS and PNS abnormalities such as ataxia, tremor, foot drop, and memory loss may persist.

**Genotoxicity.** Chromosome aberrations and an increased frequency of DNA breaks were found in embryonic cell cultures. After chronic thallium exposure, precancerous lesions in the female genital tract occurred in mice [64], [73].

**Reproductive Toxicity.** The reproductive system is highly susceptible to thallium toxicity; in humans some of the highest concentrations have been found in the testes after Tl poisoning [64]. Thallium has been shown to produce teratogenic effects in chick embryos, including achondroplasia, leg bone curvature, parrot beak deformity, microcephaly, and reduced fetal size; the results of teratological investigations in rats, mice, and cats are conflicting [73].

**Toxicological Data and Analysis.** The average lethal dose of thallium salts is ca. 1 g for adults (ca. 10 – 15 mg/kg body weight). Low blood thallium levels are only an indication for an exposure, but do not allow conclusions on the degree of an intoxication to be drawn. However, levels above 300 µg/L blood or 100 µg/L urine indicate severe intoxication (0.3 µg/L urine is the average level in unexposed persons). Moreover, the incorporation of thallium into scalp hair is of analytical significance, 7 – 15 ng/g hair being the normal range.

**Treatment.** Decontamination of the gastrointestinal tract such as by use of syrup of ipecac, lavage, serial doses of charcoal, and use of cathartics are useful during the first few hours after ingestion. Since thallium occurs as a monovalent cation, chelating agents used for the enhancement of excretion such as dithiocarb, $CaNa_2EDTA$, BAL (British anti-Lewisite, dimercaprol, 2,3-dimercaptopropanole), and penicillamine are not effective. The combination of hemodialysis (protecting the kidneys) and potassium diuresis has been recommended in the case of severe poisoning. However, hemoperfusion clearance seems to be superior to hemodialysis clearance.

Prussian blue, potassium ferricyanoferrate, in daily doses up to 24 g (combined with a mild cathartic) may be used to interrupt the marked enterohepatic circulation by exchanging $Tl^+$ for $K^+$. The resulting compound cannot be absorbed and therefore causes only minor side effects [71]. Moreover, it does not liberate substantial amounts of cyanide.

**Occupational Health.** The MAK value is 0.1 mg/m$^3$ soluble Tl compounds measured as total dust. In the United States and the United Kingdom the TLV-TWA is 0.1 mg/m$^3$.

Occupational diseases caused by thallium are included in the list of the German Berufskrankheitenverordnung (Occupation Disease Regulations) [74].

Since thallium tends to accumulate, persons exposed at the workplace should be monitored adequately even in cases of low chronic exposure.

# 7. References


General References

[1] *Gmelin*, Thallium, System no. 38.
*Ullmann*, 3rd ed., **17**, 299.
[2] J. De Ment, H. C. Dake: *Rarer Metals*, Temple Press, London 1949.
[3] N. Lowitzki: Neueres Schrifttum über Thallium aus den Jahren 1938–1948, *Z. Erzbergbau Metallhüttenwes.* **3** (1950) 201.
[4] H. E. Howe, A. A. Smith, Jr.: "Properties and Uses of Thallium," *J. Electrochem. Soc.* **97** (1950) 167 C.
[5] L. Sunderson: "Thallium," *Can. Min. J.* **65** (1944) 624.
[6] W. H. Waggaman, G. G. Heffner, E. A. Gee: "Thallium," US Dept. of the Interior, Bureau of Mines Information Circular 7553, 1950.
[7] V. Tafel: *Lehrbuch der Metallhüttenkunde*, vol. 2, Hirzel, Leipzig 1953.
[8] D. E. Eilertsen: "Thallium, Mineral Facts and Problems," US Dept. of the Interior, Bureau of Mines, Bulletin 585, 1960.
[9] C. A. Hampel: *Rare Metals Handbook*, Reinhold Publ., New York 1961.
[10] W. Schreiter: *Seltene Metalle*, vol. 3, VEB Deutscher Verlag für Grundstoffindustrie, Leipzig 1962.
[11] R. Kleinert: "Thallium, ein seltenes Metall, ein Begleitmetall," *Erzmetall* **16** (1963) 67.
[12] R. Kleinert: "Die Verwendung von Thallium," *Erzmetall* **18** (1965) 363.
[13] A. G. Lee: *The Chemistry of Thallium*, International Idens. Inc., Philadelphia 1974.
[14] H. R. Babitzke: "Thallium, Mineral Facts and Problems," US Dept. of the Interior, Bureau of Mines, Bulletin 667, 1975.
[15] W. W. Stanzo, M. B. Tschernenko: *Die chemischen Elemente Antimon–Wismut, Bausteine der Erde*, vol. 3, Verlag MIR, Moskau, Urania Verlag, Leipzig 1976.
[16] R. J. De Filippo: "Thallium, Commodity Data Summeries," US Dept. of the Interior, Bureau of Mines, 1977.
[17] J. C. Smith, B. L. Carson: *Trace Metals in the Environment*, vol. 1, Ann Arbor Sci. Publ., Ann Arbor, Mich. 1977.
[18] Abfallwirtschaft Forschungsbericht 10301327 des Bundesministers des Innern, Federal Republic of Germany.

Specific References

[19] F. A. Cotton, G. Wilkinson: *Anorganische Chemie*, 2nd ed., Verlag Chemie, Weinheim 1970.
[20] E. Pilgrim: *Entdeckung der Elemente*, Mondus, Stuttgart 1950.
[21] W. Schreiter: *Seltene Metalle*, VEB Deutscher Verlag für Grundstoffindustrie, Leipzig 1960.
[22] J. Fischer: "Die Dampfdruckkurve des Thalliums," Festschrift der TH Breslau, Verlag Korn, Breslau 1935.
[23] W. Meissner, H. Franz, H. Westerhoff, *Ann. Phys. (Leipzig)* **13** (1932) 555.
[24] C. J. Smithells: *Metals Reference Book*, Butterworth, London 1962.
[25] *Am. Ceram. Soc. Bull.* **68** (1989) no. 5, 1070.
[26] B. Mason: *Principles of Geochemistry*, Wiley, New York 1958.
[27] P. Ramdohr: *Die Erzmineralien und ihre Verwachsungen*, Akademie Verlag, Berlin 1960.
[28] K. H. Wedepohl: *Handbuch der Geochemie*, Element 81, vol. 2, no. 3, Springer Verlag, Berlin 1972.
[29] J. Feiser, *Erzmetall* **15** (1962) 578.
[30] J. Feiser, personal communication, 1963.
[31] M. T. Koslovskij et al., *Tr. Inst. Khim. Nauk Akad. Nauk Kaz. SSR* **3** (1958) 5.
[32] Z. A. Serikov, S. M. Anisimov, *Izv. Vyss. Uchebn. Zaved. Chern. Metall.* **3** (1960) no. 6, 65.
[33] I. D. Prater, D. Schlain, S. F. Ravitz, US Dept. of the Interior, Bureau of Mines, report of investigations no. 4900, 1952.
[34] W. Langner, A. Göbel, *Erzmetall* **3** (1950) 370.
[35] W. Kangro, F. Weingärtner, *Erzmetall* **11** (1958) 70.
[36] L. G. Plechanov, *Izv. Akad. Nauk Kaz. SSR Razdel. Metallurgii* **4** (1957) 38.
[37] A. Gäumann, *Schweiz. Arch.* **21** (1955) 337.
[38] A. A. Sokol, L. F. Kozin, *Ukr. Khim. Zh. Kiev* **25** (1959) no. 2, 249.
[39] A. R. Powell, The Institution of Mining and Metallurgy, London 1950.
[40] Duisburger Kupferhütte, DE 954236, 1954.
[41] H. Meister, *Angew. Chem.* **69** (1957) 533.

[42] Chem. Werke Hüls, DE 942989, 1954.

[43] Licentia Patentverwaltungs-GmbH, DE 1045999, 1957.

[44] T. J. Darvojd, V. N. Vigdorovic, N. A. Iordanskaja, *Izv. Akad. Nauk SSSR Otd. Tekh. Nauk Metall. Topl.* **3** (1961) 55.

[45] P. R. Mallory & Co., US 2180845, 1939.

[46] F. Ensslin, *Erzmetall* **15** (1962) 419.

[47] Unterharzer Berg- und Hüttenwerke, DE 740780, 1942.

[48] *Chem. Abstr.* **68** (1968) 98239m.

[49] A. E. Roach, C. L. Goodzeit, P. A. Totta, *Nature (London)* **172** (1953) 301.

[50] A. Hoffman, F. Rose, *Z. Phys.* **136** (1953/54) 152.

[51] H. Moser, *Phys. Z.* **37** (1936) 885.

[52] S. S. Flaschen, *J. Am. Ceram. Soc.* **42** (1959) 450; **43** (1960) 168, 247.

[53] Westinghouse Electric Corp., US 3929970, 1975; 3929976, 1975.

[54] Schafer, *Proc. SPIE Int. Soc. Opt. Eng.* **285** (1981) 183.

[55] US 4267073, 1981.

[56] JP 8199244, 1981.

[57] Ethyl Corp., US 3328440, 1967.

[58] D. A. Larson et al., *Illum. Eng. (N.Y.)* Preprint no. 29 (1962).

[59] W. Brugel: *Physik und Technik der Ultrarotstrahlung,* Vincentz, Hannover 1961.

[60] H. R. Schulten, C. Achenbach, U. Bahr, F. Kochler, R. Ziskoven, *Angew. Chem.* **91** (1979) no. 11, 944.

[61] C. Achenbach et al., *J. Toxicol. Environ. Health* **6** (1980) 519–528.

[61] M. Sager, G. Tölg: "Spurenanalytik des Thalliums," in: *Analytiker-Taschenbuch*, vol. 4, Springer Verlag, Berlin 1984, pp. 443–466.

[62] *Am. Ceram. Soc. Bull.* **69** (1990) no. 5, 885.

[63] L. Manzo, E. Sabbioni: "Thallium," in H. G. Seiler, H. Sigel, A. Sigel (eds.): *Handbook on Toxicity of Inorganic Compounds*, Marcel Dekker, New York 1988.

[64] M. J. Ellenhorn, D. G. Barceloux: *Medical Toxicology, Diagnosis and Treatment of Human Poisoning*, Elsevier, New York 1988.

[65] S. Moeschlin: *Klinik und Therapie der Vergiftungen*, 7. ed., Georg Thieme, Stuttgart 1986.

[66] H. U. Wolf: "Thallium" in H. U. Wolf (ed.): *HAGERs Handbuch der Pharmazeutischen Praxis*, 5. ed., vol. 3: Gifte, Springer-Verlag, Heidelberg 1992.

[67] W. P. Ridley, L. J. Dizikes, J. M. Wood, *Science* **197** (1977) 329.

[68] S. Moeschlin, *Clin. Toxicol.* **17** (1980) 133–146.

[69] W. Stevens et al., *Int. J. Clin. Pharmacol. Ther. Toxicol.* **10** (1974) 1–22.

[70] P. S. Spencer et al., *J. Cell. Biol.* **58** (1973) 79–95.

[71] F. H. Lovejoy, *Clin. Toxicol. Rev.* **4** (1982) 1–2.

[72] E. Sabbioni, L. Manzo in L. Manzo (ed.): *Advances in Neurotoxicology*, Pergamon Press, Oxford 1980.

[73] Berufskrankheitenverordnung, BK Nr. 1106: Erkrankungen durch Thallium oder seine Verbindungen.


**Thermite Process → Welding and Cutting**

# Thermoelectricity

VOLKER TEGEDER, Siemens AG, Erlangen, Federal Republic of Germany

CHARLOTTE TEGEDER, Universität Erlangen, Federal Republic of Germany



## 1. Introduction

Thermoelectric effects result from interactions between thermal and electrical transport processes in electrical conductors in either the solid or the liquid state. For example, in an open circuit made up of two unlike electrically conducting materials that form two junctions, a thermoelectric emf (voltage) can be measured when the junctions are at different temperatures (Fig. 1). This behavior is known as the Seebeck effect.

Thermoelectric phenomena fall into two categories, homogeneous and inhomogeneous. *Inhomogeneous thermoelectric effects* include the Seebeck effect (or the thermoelectric effect in the narrow sense, see Chap. 2) and the Peltier effect (see Chap. 3). *Homogeneous thermoelectric effects* (see Chap. 4) include the Thomson, Benedicks, Bridgman (or internal Peltier effect), and phonon-drag effects (Peltier effect of the lattice). The two classes differ in that inhomogeneous effects occur only when unlike materials are combined.

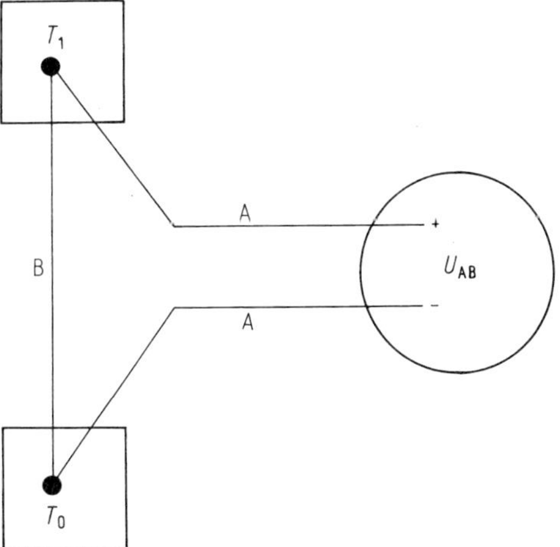


**Figure 1.** A thermoelectric element consisting of two unlike materials A and B; if $S_{AB} > 0$, the potential $U_{AB}$ is present when the junctions are at temperatures $T_1$ and $T_0$ $(T_1 > T_0)$

One well-known application of the Seebeck effect is the thermocouple, which transforms thermal to electrical energy and can thus be used to measure temperature (see Section 2.5.2).

Ullmann's Encyclopedia
of Industrial Chemistry, Vol. A 26

The *Peltier effect* is practically the inverse of the Seebeck effect. If a current flows through the circuit already described, heat is reversibly absorbed or liberated at the junctions. The *Thomson effect* is the reversible absorption or liberation of heat when a current flows through a single, homogeneous conductor over whose length a temperature gradient is maintained.

To a good approximation, these relatively weak thermoelectric effects can be described by linear equations. Three coefficients occur in these equations: the absolute thermoelectric power (thermopower) $S$, the Peltier coefficient $\pi$, and the Thomson coefficient $\mu$. Each of these has a well-defined value for a homogeneous conductor at a constant temperature. The three coefficients are linked by the Kelvin (Thomson) relations so that if one is known at all temperatures, the other two can be calculated. For ease of measurement, the thermopower is usually determined experimentally. The electrical resistivity, thermal conductivity, and thermopower sufficiently describe the electron-transport properties of a conductor.

The principal applications of thermoelectricity are in the study of interactions between electrons and lattice vibrations, the detection of impurities in solids, the measurement of temperature, and cooling and heating.

## 2. Seebeck Effect

In 1822, Thomas Johann Seebeck, working with a system of copper and bismuth as diagrammed in Figure 1, discovered that a voltage is generated in a circuit comprising two unlike conductors between whose junctions a temperature gradient is present [1]–[4].

Figure 2 shows the thermoelectric force $U_{AB}$, plotted versus temperature for several metals (conductor B in Fig. 1). Iron is the reference material (conductor A) in each case.

### 2.1. Seebeck Coefficient

The thermo-emf is a function of the materials in contact and the temperature difference $\Delta T$ between the junctions. If the temperature difference $\Delta T$ is small, so that $S(T)$ is virtually constant, the voltage is described by

$$U_{AB} = S_{AB}\,\Delta T \tag{1}$$

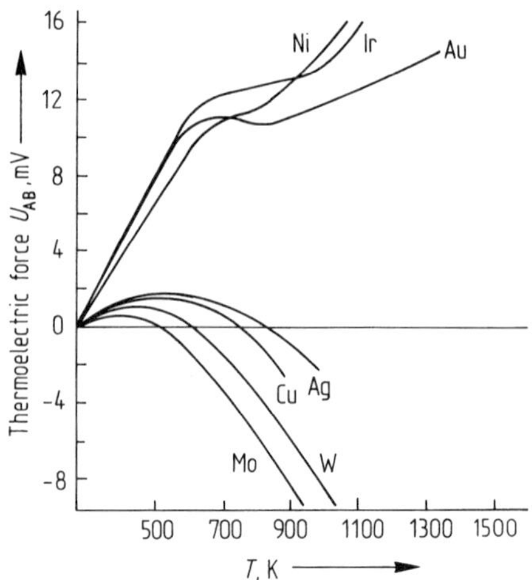


**Figure 2.** Thermoelectric force $U_{AB}$ in couples consisting of various metals (B) and an iron reference, plotted versus the temperature $T$ of one junction; the other junction is at $T = 273$ K [5]

The constant of proportionality $S_{AB}$ is called the differential thermoelectric force or the relative Seebeck coefficient [5]–[12]. $S_{AB}$ is given by the Seebeck coefficients $S'_x$ of the two materials:

$$S_{AB} = S'_A - S'_B$$

The Seebeck coefficient is thus a bulk property of the material, with units $\mu$V/K; its sign depends on the type of charge carrier and the way in which the carriers interact with the lattice. The thermo-emf is independent of the shape of the materials used and the detailed temperature distribution between the junctions. The thermo-emf thus has its source in the materials themselves rather than in the contact, but it can be measured only when unlike materials are in contact, for example, at the junctions of a thermocouple.

The Seebeck effect must be distinguished from the Volta effect. The junction voltage seen in the latter is due to differences in the electron work functions of the two materials and is present even when there is no temperature difference. The thermo-emf is caused by redistributions of charge carriers due to the temperature gradient.

Phenomenologically, all thermoelectric materials (metals, alloys, semiconductors) can be ranked in a thermoelectric series by the value of the thermo-emf against a reference such as copper. One (soldered or brazed) junction of the

**Table 1.** Thermoelectric series for 100 K temperature difference (thermo-emf values in mV) [13]

| Metal $x$ at 273 K | Platinum (373 K) vs. | Bismuth (373 K) vs. | Copper (373 K) vs. |
|---|---|---|---|
| Bi | −7.0 | 0.0 | −8 |
| Constantan* | −3.4 | −4.1 | −4.1 |
| Ni | −1.5 | +5.5 | −2.2 |
| Pd | −0.3 | +6.7 | −1.0 |
| Pt | 0.0 | +7.0 | −0.8 |
| Hg | 0.0 | +7.0 | −0.8 |
| C | +0.2 | +7.2 | −0.6 |
| Al | +0.4 | +7.4 | −0.4 |
| Pb | +0.4 | +7.5 | −0.4 |
| Sn | +0.4 | +7.5 | −0.3 |
| Manganin ** | +0.6 | +7.6 | −0.1 |
| Ir | +0.7 | +7.7 | −0.1 |
| Rh | +0.7 | +7.7 | −0.1 |
| Zn | +0.7 | +7.7 | 0.0 |
| Ag | +0.7 | +7.7 | 0.0 |
| Au | +0.7 | +7.7 | 0.0 |
| W | +0.8 | +7.8 | +0.1 |
| Fe | +1.8 | +8.8 | +1.0 |
| Si | +45 | +52 | +44 |
| Te | +50 | +57 | +49 |
| Se |  | +85 | +98 |

* Alloy of 54% Cu, 45% Ni, 1% Mn. ** Alloy of 86% Cu, 12% Mn, 2% Ni.

thermocouple is held at 273 K. In a thermocouple consisting of two elements from this thermoelectric series (Table 1), if the junction is heated, the element standing lower in the series takes on a positive potential, while the element standing higher takes on a negative potential.

## 2.2. Thermoelectric Power

When larger temperature differences are involved, the differential thermoelectric force must be regarded as a function of temperature given by

$$S_{AB}(T) = dU_{AB}/dT \qquad (2)$$

(Eq. 1 holds for small $\Delta T$). The function $S_{AB}(T)$ is commonly referred to as the thermoelectric power. Figure 3 shows how the thermoelectric power varies with temperature for several materials against copper. The curves are simpler in form than the ones for the thermo-emf (Fig. 2). The thermoelectric power for metals at higher temperatures is generally a linear function of temperature; transition metals exhibit more complicated behavior. The temperature dependence is weaker for semiconductors at higher temperatures.

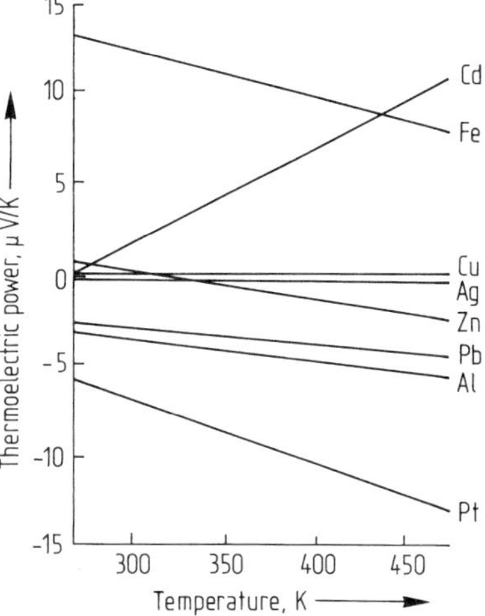


**Figure 3.** Thermoelectric power $S_{AB}$ as a function of temperature for some metals against copper [5]

Accordingly, the $x$ component of the electric field in a homogeneous conductor is, in general, given by

$$E_x = S_{AB}(T)\, dT/dx$$

If the function $S_{AB}(T)$ is known, the voltage $U_{AB}$ can be obtained by integrating this equation

$$U_{AB}(T_0,T_1) = \int_{T_0}^{T_1} S_{AB}(T)\, dT$$

## 2.3. Determination of Absolute Thermoelectric Power

To determine the absolute thermoelectric power (thermopower) $S_A$ of a material, the material can be formed into a thermocouple with a superconductor. The thermo-emf of the superconductor below its transition temperature is 0 V, so that the measured thermo-emf is entirely due to the normally conducting part of the couple, since $S_{AB} = S_A - S_B$. At present, however, this method is restricted to temperatures of $\leq 24$ K. At higher temperature (higher than the transition temperature of superconductors), lead (because of its low thermo-emf) or platinum (because it can be prepared in high purity) is used as the thermoelectric reference.

The thermopower $S_A(T) = dU_A/dT$ (in µV/K) for a given conductor with one end at 0 K and the other at the test temperature must be distin-

**Table 2.** Absolute thermoelectric powers ($\mu$V/K) for some metals [14], [15]

| T, K | Pb | Cu | Ag | Au | Pt | Pd | W | Mo |
|---|---|---|---|---|---|---|---|---|
| 100 | $-0.58$ | 1.19 | 0.73 | 0.82 | 4.29 | 2.00 | | |
| 200 | $-0.83$ | 1.29 | 1.05 | 1.34 | $-1.27$ | $-4.85$ | | |
| 273 | | 1.70 | 1.38 | 1.79 | $-4.45$ | $-9.00$ | 0.13 | 4.71 |
| 300 | $-1.05$ | 1.84 | 1.51 | 1.94 | $-5.28$ | $-9.99$ | 1.07 | 5.57 |
| 400 | | 2.34 | 2.08 | 2.46 | $-7.83$ | $-13.00$ | 4.44 | 8.52 |
| 500 | | 2.83 | 2.82 | 2.86 | $-9.89$ | $-16.03$ | 7.53 | 11.12 |
| 600 | | 3.33 | 3.72 | 3.18 | $-11.66$ | $-19.06$ | 10.29 | 13.27 |
| 700 | | 3.83 | 4.72 | 3.43 | $-13.31$ | $-22.09$ | 12.66 | 14.94 |
| 800 | | 4.34 | 5.77 | 3.63 | $-14.88$ | $-25.12$ | 14.65 | 16.13 |
| 900 | | 4.85 | 6.85 | 3.77 | $-16.39$ | $-28.15$ | 16.28 | 16.86 |
| 1000 | | 5.36 | 7.95 | 3.85 | $-17.86$ | $-31.18$ | 17.57 | 17.16 |
| 1100 | | 5.88 | 9.06 | 3.88 | $-19.29$ | $-34.21$ | 18.53 | 17.08 |
| 1200 | | 6.40 | 10.15 | 3.86 | $-20.69$ | $-37.24$ | 19.18 | 16.65 |
| 1300 | | 6.91 | | 3.78 | $-22.06$ | $-40.27$ | 19.53 | 15.92 |
| 1400 | | | | | $-23.41$ | $-43.30$ | 19.60 | 14.94 |
| 1600 | | | | | $-26.06$ | $-49.36$ | 18.97 | 12.42 |
| 1800 | | | | | $-28.66$ | $-55.42$ | 17.41 | 9.52 |
| 2000 | | | | | $-31.23$ | $-61.48$ | 15.05 | 6.67 |
| 2200 | | | | | | | 12.01 | 4.30 |
| 2400 | | | | | | | 8.39 | 2.87 |

guished from the Seebeck coefficient (also in $\mu$V/K) $S_A$, which is the thermo-emf of a thermocouple in which a platinum reference forms the second leg and one junction is held at a constant 273 K. The thermopower (Table 2) depends on just one temperature. The sign convention states that the polarity is as shown in Figure 1 when $S_{AB} = S_A - S_B > 0$ ($T_1 > T_0$).

## 2.4. Source of Thermo-emf

The experimentally determinable thermo-emf is present because the heat flux due to a temperature difference in a metal or semiconductor is carried partly by phonons (quantized lattice vibrations) and partly by charge carriers, which transport energy along with their charge. In metals, only electrons (charge carriers) with energies near the Fermi level $E_0$ (Fig. 4) contribute to the conductivity. The electron kinetic energy distribution depends on temperature. High-energy electrons from the hot end and low-energy electrons from the cold end diffuse along the conductor at different velocities because of different interactions with the lattice, so that a net current flows until the voltage resulting from this nonuniform distribution stops any further net current flow. The Fermi level also varies with temperature.

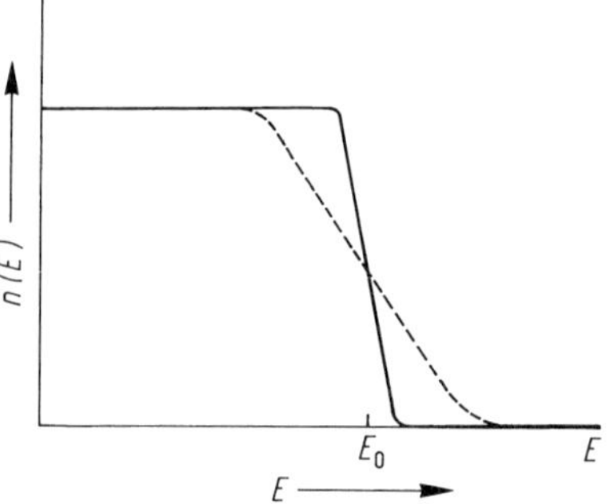

**Figure 4.** Electron-density distribution $n(E)$ versus electron energy $E$ near the Fermi level $E_0$ for two temperatures The dashed curve represents the distribution at the higher temperature

### 2.4.1. Interaction with the Lattice

The strength of interactions with the lattice and phonons is important in determining the thermo-emf generated. Especially at very low temperature, the phonon-drag effect (Gurevich effect) increases the thermopower, which is given by

$$S = S_d + S_g$$

where $S_d$ is the electron diffusion component and $S_g$ is the phonon-drag component. This additional thermopower contribution occurs when elec-

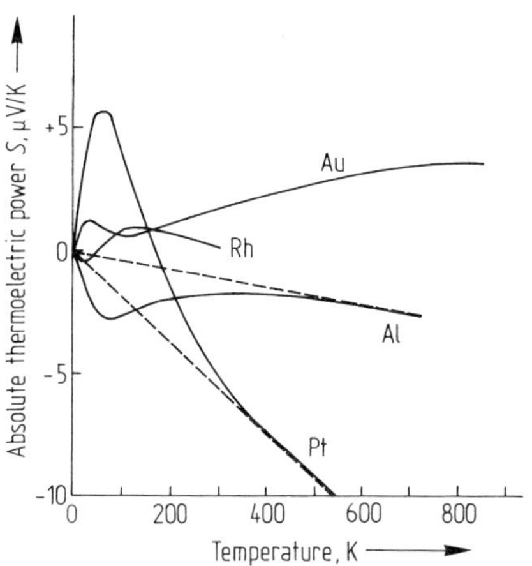

Figure 5. Increase in absolute thermoelectric power at low temperature due to phonon drag [16]

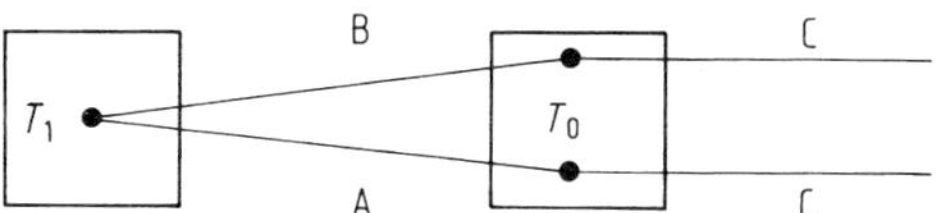

Figure 6. Block diagram of a thermocouple made of metals A, B, and C and used for temperature measurement

trons are entrained by phonons [16] moving from the hot to the cold portion of the conductor because of the temperature difference. The interaction of electrons with these phonons decreases as the temperature increases, since phonon–phonon interactions come to dominate at higher temperature. For this reason, the phonon-drag effect cannot be employed in practical situations (i.e., at higher temperature; see Fig. 5).

### 2.4.2. Effect of Charge Carriers

Thermopower is a function of the carrier concentration. In semiconducting materials, the thermopower increases as the electron or hole density declines; it is thus inversely proportional to the dopant concentration, because the mean carrier energy as a function of temperature increases faster, as carrier concentration decreases. Semiconductors, with their much lower carrier density, thus have much higher thermopowers than metals. The thermopower $S$ can range up to 1 mV/K for semiconductors; it is commonly a few μV/K for metals.

### 2.5. Examples

### 2.5.1. Thermoelectric Magnet

Even though thermo-emf values are very small, relatively large currents occur in some thermocouples if the resistance of the system is low enough. The following experiment demon-

strates these high thermoelectric currents [13]: A copper wire A (diameter 10 mm) is bent into a loop; one end is placed in a beaker of ice water, while the other end is heated over a candle flame. The circuit is completed by two sections B of Constantan brazed to the copper. A thermo-emf of ca. $4.2 \times 10^{-3}$ V is produced. The thermoelectric current of some 40 A can be detected by its magnetic action on two pieces of iron $C_1$ and $C_2$ above and below the loop; $C_2$ can be loaded additionally by a weight of several kilograms.

### 2.5.2. Thermocouples

Thermocouples used for temperature measurement (Fig. 6) are the most important application of the Seebeck effect. A thermocouple is a conducting circuit made up of two or more unlike materials with the junctions held at unequal temperatures [17]–[24]. One of the junctions is kept at a reference temperature (usually melting ice in water, 273 K), while the other takes on the temperature to be measured.

As a consequence of the Seebeck effect, a thermo-emf is generated in the circuit, causing a thermoelectric current to flow. This current is measured with appropriate instruments; the result can be converted to a temperature through use of a table. Many combinations of materials are used in thermocouples in order to cover various temperature ranges. Material pairs for thermocouples and ranges of temperature measured are listed below:

| | |
|---|---|
| Copper–Constantan | 73–673 K |
| Iron–Constantan | up to 1073 K |
| Platinum vs. platinum–10 % rhodium | up to 1573 K |
| Nickel vs. chromium-nickel | up to 1373 K |
| Constantan vs. chromium–nickel | up to 1073 K |
| Chromel (alloy of ca. 90 % Ni, ca. 9 % Cr, and minor amounts of Fe, C, Mn, Cu, Si) vs. Alumel (Ni–Al–Mn) | up to 1473 K |

| | |
|---|---|
| $Bi_2Te_3$ | 273–573 K |
| $Bi_2Te_{2.4}Se_{0.6}$ | 273–573 K |
| $Pb_{1-x}Sn_xTe$ ($x = 0-0.25$) | 423–873 K |
| $Si_{1-x}Ge_x$ ($x = 0.15-0.3$) | 873–1473 K |

Because the sign of the thermopower in semiconductors depends on whether the charge is carried by electrons ($-$) or holes ($+$), a thermocouple made up of semiconductors will have the highest efficiency if both legs consist of different conducting materials, i.e., of $p$-type and $n$-type conductors. Despite the higher thermo-emf obtained with semiconductors, most thermocouples are fabricated from metals because these can be produced with reproducible results at lower cost and can more easily be formed into a variety of shapes.

The producer of a thermocouple must take account of the fact that thermoelectric effects occur even in like materials when the slightest variation in chemical composition exists or when mechanical stresses are imposed (i.e., the Fermi energy changes). For example, a thermocouple made of annealed brass wire and hard drawn brass wire will behave like a thermocouple made of unlike metals; a thermoelectric effect will also be seen if a wire under elastic stress is paired with an unstressed wire or if crystals of the same metal are joined with differently oriented faces coming together at the junction.

### 2.5.3. Determination of Conduction Type

The conduction type of a semiconducting material can be determined by contacting the material with two metal tips, one of which is heated. The polarity of the cold tip will indicate the predominant type of charge carrier.

### 2.5.4. Radiation Measurement

Thermocouples can also be employed for the sensitive measurement of microwave pulse power, as thermoelectric transducers, and as thermoelectric radiation detectors (theoretical detection threshold $3 \times 10^{-11}$ W, practical threshold $3 \times 10^{-10}$ W).

### 2.5.5. Thermoelectric Generator (Seebeck Element)

A thermoelectric generator or heat engine utilizing the Seebeck effect can be made in a configuration similar to a thermocouple (Fig. 7).

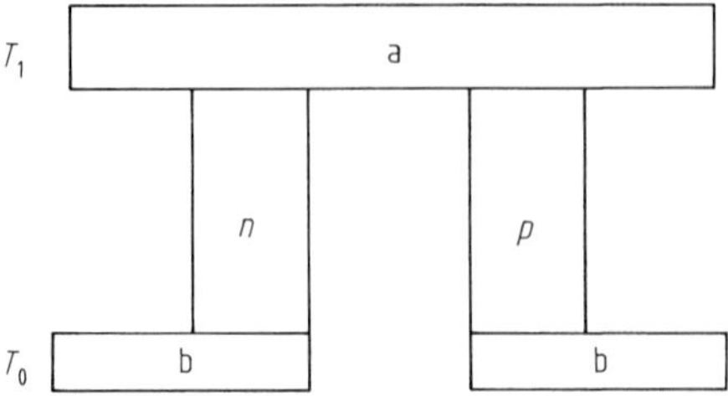


**Figure 7.** Block diagram of a thermoelectric generator with heavily doped connecting plate (a), metallic contacts (b), and $n$- and $p$-type semiconductors

Here, however, the two legs composed of unlike materials are joined by a plate made of a third material. Each leg has a contact. The plate joining the legs is heated relative to the contacts, and electrical energy can be removed at the contacts [24], [25]. A thermoelectric generator combined with a heat source (e.g., the decay of radioactive materials, solar radiation, or combustion) can generate energy for isolated stations such as lighthouses, navigation buoys, remote weather stations, offshore drilling platforms, and man-made satellites.

An alloy of 70 % silicon and 30 % germanium has proved suitable for the legs of a thermoelectric generator.

Generators of this type connected in cascade (thermally parallel, electrically series) offer markedly higher yield, especially if generators of different materials are used in which each material is held at its most efficient temperature. For multistage cascade circuits, an efficiency of 22 % ($\Delta T = 1000$ K) appears possible, and 10–15 % efficiencies have been attained; the corresponding value for single-stage circuits is 7 % ($\Delta T = 800$ K). An output of 1.7 W has been obtained at $T_1 = 1273$ K ($T_2 =$ room temperature) with a generator having arm measurements of 6 mm × 6 mm × 6 mm (Table 3).

**Table 3.** Parameters of important semiconductors for thermoelectric generators [26], [27]

| Material | Temperature range, K | Conduction type | $(z^* T)_{max}$ |
|---|---|---|---|
| $Bi_2Te_3$ | 273–573 | $p$ | 0.6 |
| $Bi_2Te_{2.4}Se_{0.6}$ | 273–573 | $n$ | 0.7 |
| $Pb_{1-x}Sn_xTe$ ($x = 0-0.25$) | 423–873 | $n$ | 1.1 |
| | | $p$ | 1.0 |
| $Si_{1-x}Ge_x$ ($x = 0.15-0.3$) | 873–1473 | $n$ | 1.0 |
| | | $p$ | 0.7 |

* $z$ = Figure of merit (see Section 3.4).

# 3. Peltier Effect

When a current flows through a circuit made up of two unlike conductors, one junction becomes warm and the other cold. This phenomenon was discovered by the clockmaker and physicist JEAN CHARLES ATHANASE PELTIER in 1834. Working with an antimony–bismuth couple, Peltier was able to freeze a drop of water [7]–[9], [12], [28], [29]. If the direction of the current is reversed, the effect is also reversed so that the first junction is cooled and the second heated. The heat released (or absorbed) is directly proportional to the current $I$.

## 3.1. Peltier Coefficient

The Peltier coefficient $\pi_{AB}$ is defined as the quantity of heat generated per unit current and second at the junction of materials A and B. It is measured in volts and is by convention positive if that junction becomes warm at which a positive current flows from A to B. The Peltier coefficient does not depend on the current $I$ or the shape and size of the conductors, and the effect is reversible. The coefficient $\pi$ depends solely on the materials in contact and the junction temperature; it is a measure of the mean energy of the charge carriers in the material. Clearly if

$$\pi_{AB}(T) = \pi_A(T) - \pi_B(T)$$

where $\pi_A$ and $\pi_B$ are the Peltier coefficients of materials A and B, the coefficient is a bulk property and can be defined for a single conductor.

The cause of the Peltier effect lies in the energy transported by charge carriers when a current flows. The Peltier heat $Q_\pi$ (i.e., the heat liberated or absorbed at both junctions when a current flows through the circuit) is given by

$$Q_\pi = \pi_{AB} I t$$

where the product of current $I$ and time $t$ is the quantity of charge transported through the junction.

The Peltier effect is superimposed by simultaneous Joule heating and Thomson heating or cooling, so the Peltier coefficient can be difficultly measured with accuracy. Accordingly, it is commonly determined by applying the Kelvin relations (see Section 4.3) to the thermopower. Table 4 lists some Peltier coefficients determined by calorimetry.

**Table 4.** Peltier coefficients $\pi_{AB}$ for various material pairs [30]

| Fe–Constantan | | Cu–Ni | | Pb–Constantan | |
|---|---|---|---|---|---|
| $T$, K | $\pi_{AB}$, mV | $T$, K | $\pi_{AB}$, mV | $T$, K | $\pi_{AB}$, mV |
| 273 | 13 | 292 | 8.0 | 293 | 8.7 |
| 293 | 15 | 368 | 9.0 | 383 | 11.8 |
| 403 | 19 | 478 | 10.3 | 508 | 16.0 |
| 513 | 26 | 563 | 8.6 | 578 | 18.7 |
| 593 | 34 | 613 | 8.0 | 633 | 20.6 |
| 833 | 52 | 718 | 10.0 | 713 | 23.4 |

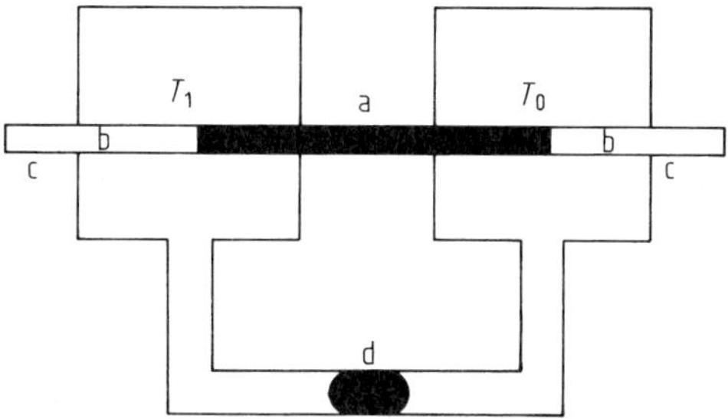


**Figure 8.** EDLUND's experimental demonstration of the Peltier effect with two unlike metals [31]
a) Antimony; b) Bismuth; c) Electrical contacts; d) Liquid drop
$T_1$, $T_0$ = temperatures (unequal) in the chambers

## 3.2. Demonstration of the Peltier Effect

Despite the superimposed Joule effect, the Peltier effect can be demonstrated experimentally. One setup (Fig. 8) has a piece of antimony (a) between two pieces of bismuth (b), with the two junctions sealed into two vessels. The vessels are joined by a U-shaped glass tube with a drop of liquid in it. When a current flows, the liquid drop moves as the junctions are heated and cooled [13].

This effect can be demonstrated at high temperature by brazing together an iron and a Constantan wire (each ca. 12 cm long and 0.3–0.5 mm in diameter) in such a way that the junction is no bigger than the wires. The wires are placed in an evacuated glass tube, and a current is passed through them until they become incandescent. Depending on the direction of the current, the junction will glow more or less brightly than the wires.

## 3.3. Peltier Module

The Peltier module (Fig. 9), a specially designed semiconducting thermocouple utilizing

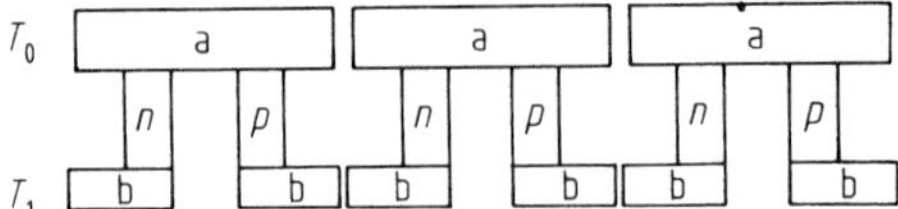

**Figure 9.** Block diagram of a Peltier module
a) Metal bridge; b) Contact
$n$, $p$ = Semiconductors; $T_0$ = Lower temperature; $T_1$ = Higher temperature

the Peltier effect, can be employed for either cooling or heating [32], [33]. The two arms of one element are joined by a piece of metal, which gives off or absorbs heat when a current is passed through the element.

## 3.4. Optimization of the Peltier Element

The selection of materials must take into account the facts that thermal conduction opposes the Peltier effect and that Joule heat is produced because the materials used always have finite electrical resistances. The semiconducting materials must therefore have the best possible electrical conductivity $\sigma$ and poor thermal conductivity $\varkappa$. Suitable materials are Bi–Sb–Te mixed crystals for low temperature and Si–Ge mixed crystals (oxidation resistant) for high temperature. Commonly, materials differing in conduction type are combined. For good cooling performance, the figure of merit $z = S^2\sigma/\varkappa$ must be made as large as possible; $z$ is the ratio of effective power (proportional to $S^2\sigma$) to thermal conductivity $\varkappa$.

Figure 10 shows how to locate the expected optimum point. Because the thermal conductivity $\varkappa$ is the sum of thermal conduction by electri-

cal charge carriers (electrons or holes) $\varkappa_C$ and by the lattice $\varkappa_L$, and because $\varkappa_C$ (like the conductivity $\sigma$) increases with carrier density while thermopower $S$ decreases, an optimum must exist. An optimum for both parameters is located at a free carrier concentration of ca. $10^{19}$ cm$^{-3}$. In semiconductors, however, the lattice conductivity $\varkappa_L$ is usually the crucial parameter for thermal conductivity and varies over three orders of magnitude. The best thermoelectric materials are therefore selected in accordance with this parameter $\varkappa_L$.

The best efficiency $[z(T) \cdot T]_{\max} \approx 1$ is obtained for a few highly doped semiconducting alloys with narrow band gaps. The maximum possible temperature difference $\Delta T_{\max}$ between the hot and cold junctions of a single stage is $\Delta T_{\max} = (\frac{1}{2}) z T_{\mathrm{cold}}^2$; the efficiency for such a stage approaches zero. Accordingly, little heat is transported, but for some applications, heat transport is not needed. This temperature difference can be increased by a suitable cascade arrangement, but no appreciable gain in efficiency results. The best thermoelectric materials, Pb–Te and Bi–Te, exhibit $z$ values of $2$–$4 \times 10^{-3}$ K$^{-1}$ at their best temperatures.

Temperature drops as great as 150 °C can be achieved through the choice of materials and cascading of such elements. Single elements normally have maximum temperature differences of 50–70 K. Table 5 gives the thermoelectric properties of some semiconductors.

## 3.5. Applications

The Peltier element is used to cool sensitive solid-state lasers, the specimen stage in microscopy, infrared detectors, and other sensitive electronic (semiconductor) components that must operate at higher ambient temperatures (e.g., in computers). Other uses are cooling of oil-vapor traps in vacuum pumps, of isothermal microcalorimeters, and of medical devices for cryosurgery and cryotherapy. Small, battery-powered units can be automatically held within a narrow band around 0 °C (reference temperature) for long periods of time. Multistage devices find use as cooling batteries (electrically parallel/ thermally serial configuration or cooling cascade in thermal series).

The advantages of Peltier cooling are higher efficiencies at low cooling power and small $\Delta T$, no rotating parts (hence virtually no wear), small

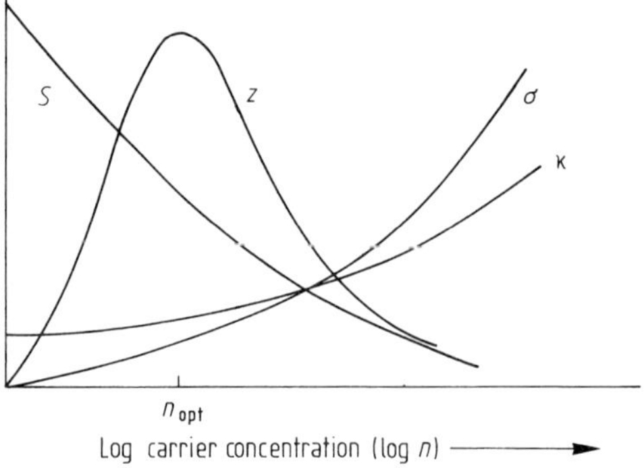


**Figure 10.** Optimal charge-carrier concentration for a Peltier element as a function of $S$, $\sigma$, and $\varkappa$ [25]

**Table 5.** Thermoelectric properties of some semiconductors for Peltier elements [25], [34]

| Material | $E_g$, eV | Type | $\varkappa_L$, W m$^{-1}$ K$^{-1}$ | $z_{max}$, $10^{-3}$ K$^{-1}$ | $T$ for $z_{max}$, K | Maximum temperature, K |
|---|---|---|---|---|---|---|
| Bi$_2$Te$_3$ | 0.15 | $n$ | 1.6 | 2.8 | 300 | 450 |
| Bi$_2$Te$_3$ | 0.15 | $p$ | 1.6 | 2.2 | 300 | 450 |
| BiSb$_4$Te$_{7.5}$ | | $p$ | 1.0 | 3.3 | 300 | 450 |
| Bi$_2$Te$_2$Se | 0.3 | $n$ | 1.3 | 2.3 | 300 | 600 |
| PbTe | 0.3 | $n,p$ | 2 | 1.2 | 300 | 900 |
| GeTe (Bi doped) | | $p$ | 1.6 | 1.6 | 800 | 900 |
| ZnSb | 0.6 | $p$ | 17 | 1.2 | 500 | 600 |
| AgSbTe$_2$ | 0.6 | $p$ | 0.6 | 1.6 | 700 | 900 |
| InAs (P doped) | 0.45 | $n$ | 6.5 | | 900 | 1100 |
| Cu$_3$Te$_2$S | | | 1.2 | 1.5 | 1100 | |
| 70 % Si–30 % Ge | | $n$ | 5 | 0.9 | 1100 | 1200 |
| 70 % Si–30 % Ge | | $p$ | 5 | 0.7 | 100 | 1200 |
| 88 % Bi–12 % Sb | | $n$ | | 6.0 | 80 | |
| 88 % Bi–12 % Sb (magnetic field) | | $n$ | | 8.6 | 100 | |

size, ease of control, and low probability of malfunction.

## 3.6. Other Effects

Phenomena related to the Peltier effect (electrothermal inhomogeneity effect) include the Bridgman effect (or internal Peltier effect), the electron-drag effect (Gurevich effect or lattice Peltier effect), and the transverse Peltier effect.

The *Bridgman effect* occurs when current flows in anisotropic materials; heat is produced when the current changes direction relative to the crystal axes [35]–[38]. In the *electron-drag effect*, an electron current generates an additional flow of heat through the electron–phonon interaction, which is observed chiefly at very low temperature. The *transverse Peltier effect* occurs in noncubic conducting crystals. The formation of a transverse temperature gradient and the associated longitudinal thermo-emf cause the d.c. resistance to be greater than the a.c. resistance (Lenard effct) since no temperature gradient is present if the frequency is sufficiently high [39].

## 4. Thomson Effect

In 1854, WILLIAM THOMSON (later LORD KELVIN) postulated a reversible process of heat production when a current flows in a homogeneous conductor along which a temperature gradient is imposed [40]–[43]. This phenomenon, characterized by the Thomson coefficient $\mu$, is of little practical value but had great theoretical sig-

nificance, because it was essential for a complete thermodynamic theory of thermoelectricity.

If a temperature gradient $dT/dx$ is present in a homogeneous conductor carrying a current $I$ (Fig. 11), a quantity of heat $dQ_T = (\mu I \, dT/dx) \, dx$ per second is developed over an element of length $dx$ (in addition to the Joule heat). This heat is absorbed or liberated, depending on the sign of $\mu$ and the direction of the current relative to the temperature gradient. If the positive current flows in the direction of rising temperature and $\mu > 0$, heat is released. In Figure 11, therefore, if $T_1$ is held constant, the temperature $T_2$ is changed by $\Delta T$ as a result of the transported heat.

The Thomson coefficient is positive for silver and gold, and negative for iron, platinum, zinc, nickel, and Constantan (Table 6). The reason for the effect is that charge carriers flowing through the temperature gradient must exchange energy with the lattice in order to maintain local equilibrium, so that heat is created or absorbed.

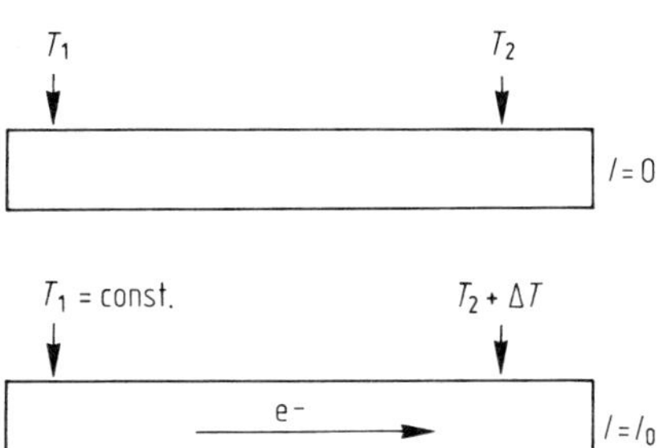


**Figure 11.** Additional heat flux produced when current flows in a conductor with a temperature gradient

**Table 6.** Thomson coefficients for some metals [30]

| Pt | | Cu | | Pb | |
|---|---|---|---|---|---|
| $T$, K | $\mu$, µV/K | $T$, K | $\mu$, µV/K | $T$, K | $\mu$, µV/K |
| 203 | $-9.6$ | 70 | $-0.26$ | 120 | $-0.19$ |
| 233 | $-9.3$ | 90 | $-0.49$ | 150 | $-0.30$ |
| 273 | $-9.1$ | 120 | $-0.26$ | 200 | $-0.45$ |
| 313 | $-9.0$ | 170 | $+0.47$ | 300 | $-0.68$ |
| 353 | $-9.0$ | 220 | $+0.96$ | 400 | $-0.85$ |
| 393 | $-9.2$ | 300 | $+1.52$ | | |

In contrast to the inhomogeneous thermoelectric effects (Seebeck and Peltier effects), the Thomson effect is homogeneous. In other words, it can be observed in a circuit composed of a single (uniform) material; junctions between unlike conductors are not required.

## 4.1. Demonstration of the Thomson Effect

The Thomson effect can be demonstrated in the following way [13]: Wires of equal length and cross section, connected to a current source, are arranged vertically with their free ends dipping into a bath of mercury. The mercury not only completes the circuit but also acts as a cold reservoir. If a current is passed through the wires so as to make them incandescent, the Thomson heat released or absorbed is visualized as a difference in maximum brightness in different regions of the wires, because the current flows with the temperature gradient in one case and against it in the other.

An inverse Thomson effect, in which a varying temperature gradient produces a thermo-emf, must be postulated on grounds of symmetry [44]. This effect has not, however, been observed. The Thomson effect and the inverse Thomson effect are second-order thermoelectric effects.

## 4.2. Other Homogeneous Thermoelectric Effects

Other homogeneous thermoelectric effects include the two *Benedicks effects* [4], [45]–[48]: (1) A thermo-emf is produced in a homogeneous conductor if a very strong temperature gradient exists along the conductor. (2) A very slight rise in temperature occurs at a restriction in a homogeneous conductor when a current flows through the material.

## 4.3. Kelvin Relations

The Kelvin (or Kelvin–Onsager) relations link the thermopower $S$, the Peltier coefficient $\pi$, and the Thomson coefficient $\mu$ [40]–[43]:

First Kelvin relation: $\qquad \pi = ST$
Second Kelvin relation: $\qquad \mu = T\, \mathrm{d}S/\mathrm{d}T$

Thus, if one thermoelectric coefficient is completely known, the other two can be calculated with these equations. The second relation implies that the integral

$$S_\mathrm{A} = \int_0^T \frac{\mathrm{d}\mu_\mathrm{A}}{\mathrm{d}T}\, \mathrm{d}T$$

makes possible the determination of absolute $S_\mathrm{A}$ on a single conductor, provided the Thomson coefficient is known. Here $S(T = 0)$ has been set equal to zero in accordance with the third law of thermodynamics. Because these measurements are very difficult, they have been performed for only a few substances, such as lead and platinum, which can thus serve as references [49].

## 5. References


[1] T. J. Seebeck, *Abh. Kgl. Akad. Wiss. Berlin* **23** (1822) 265.
[2] T. J. Seebeck, *Ann. Phys. (Leipzig)* **73** (1823) no. 1, 430.
[3] T. J. Seebeck, *Ann. Phys. (Leipzig)* **6** (1826) no. 2, 1–20, 133–160.
[4] C. A. F. Benedicks, *C.R. Hebd. Seances Acad. Sci.* **167** (1918) 296.
[5] F. J. Blatt, P. A. Schroeder (eds.): *Thermoelectricity in Metallic Conductors*, Plenum Press, New York–London 1978.
[6] D. D. Pollock: *Thermoelectricity: Theory, Thermometry, Tool*, ASTM STP 852, Am. Soc. Test. Mater., Philadelphia, PA, 1985.
[7] D. D. Pollock: *Physical Properties of Materials for Engineers*, vols. I, II, III, CRC Press, Boca Raton, FL, 1982.
[8] R. R. Heikes, R. W. Ure, Jr.: *Thermoelectricity; Science and Engineering*, Interscience, New York 1961.
[9] R. D. Barnard: *Thermoelectricity in Metals and Alloys*, Taylor & Francis, London 1972.
[10] F. J. Blatt et al.: *Thermoelectric Power of Metals*, Plenum Press, New York–London 1976.
[11] T. C. Harman, H. M. Honig: *Thermoelectric and Thermomagnetic Effects and Applications*, McGraw-Hill, New York 1967.
[12] D. K. C. MacDonald: *Thermoelectricity: An Introduction to the Principles*, J. Wiley & Sons, New York 1962.
[13] L. Bergmann, C. Schaefer: *Lehrbuch der Experimentalphysik*, vol. 4, De Gruyter, Berlin 1981.

[14] R. B. Roberts, *Philos. Mag.* **36** (1977) 91.

[15] N. Cusack, P. Kendall, *Proc. Phys. Soc. London* **72** (1958) 89.

[16] C. Herring: "The Role of Low-frequency Phonons in Thermoelectricity and Thermal Conductivity," *Proc. Int. Coll.,* Garmisch-Partenkirchen 1956, Vieweg, Braunschweig 1958, p. 184.

[17] A. F. Joffé: *Semiconductor Thermoelements and Thermoelectric Cooling*, Infosearch, London 1957.

[18] A. F. Joffé: *Halbleiter-Thermoelemente*, Akademie-Verlag Berlin 1957.

[19] R. P. Benedict (ed.): *Manual on the Use of Thermocouples in Temperature Measurement*, ASTM STP 740 B, Am. Soc. Test. Mater., Philadelphia, PA, 1981.

[20] R. P. Benedict: *Fundamentals of Temperature, Pressure and Flow Measurements*, 3rd ed., Wiley, New York 1960.

[21] N. A. Burley et al.: "The Nicrosil ·Versus Nisil Thermocouple: Properties and Thermoelectric Reference Data," *NBS Monogr. (U.S.)* **161** (1978).

[22] R. L. Powell et al.: "Thermocouple Reference Tables Based on IPTS-68," *NBS Monogr. (U.S.)* **125** (1974).

[23] American Institute of Physics: *Temperature, its Measurement and Control in Science and Industry*, Reinhold Publ. Co., New York, vol. 1, 1941, vol. 2, 1955.

[24] P. Gerthsen: "Thermoelektrische Anwendung von Halbleitern," *Z. Angew. Phys.* **13** (1961) no. 9, 435–442.

[25] D. M. Rowe: "Thermoelectric Power Generation," *Proc. Inst. Electr. Eng.* **125** (1978) (11R), 1113–1136.

[26] H. Preier et al., *Appl. Phys.* **12** (1977) 277.

[27] A. S. Ochotin, A. S. Puskarskij, R. P. Borovikova, V. A. Simonov: "Metody izmerenija charakteristik termoelektriceskich materialov i preobrazovatelej," Moskau: Izd. Nauka 1974.

[28] J. C. A. Peltier, *Ann. Chim. Phys.* **56** (1834) 371.

[29] P. W. Bridgman: *Thermodynamics of Electrical Phenomena in Metals*, Dover, New York 1960.

[30] *Landolt-Börnstein* **II/6**, 59.

[31] P. O. Gehlhoff, E. Justi, M. Kohler, *Abh. Braunschw. Wiss. Ges.* **2** (1950) 149.

[32] H. J. Goldsmid: *Thermoelectric Refrigeration*, Plenum Press, New York 1964.

[33] R. W. Ure: "Thermoelectric Effects in III–V Compounds," in R. K. Willardson, A. C. Beer (eds.), *Semiconductors and Semimetals*, vol. 8, Academic Press, New York–London 1972, pp. 67–102.

[34] R. Wolfe: "Magnetothermoelectricity," *Sci. Am.* **210** (1964) no. 6, 70–82.

[35] P. W. Bridgman, *Proc. Natl. Acad. Sci. USA* **11** (1925) 608.

[36] P. W. Bridgman, *Proc. Am. Acad. Arts Sci.* **61** (1925) 101.

[37] P. W. Bridgman, *Phys. Rev.* **39** (1932) 702.

[38] P. W. Bridgman, *Phys. Rev.* **42** (1932) 858.

[39] P. E. A. Lenard, *Wiedem. Ann. Phys. (Leipzig)* **39** (1890) 28.

[40] W. Thomson, *Math. Phys. Papers* **2** (1882) 192.

[41] W. Thomson, *Philos. Trans. R. Soc. London* **146** (1856) 663.

[42] W. Thomson, *Math. Phys. Papers* **2** (1883) 207.

[43] W. Thomson, *Math. Phys. Papers* **1** (1882) 232, 266.

[44] E. Justi: *Leitfähigkeit und Leitungsmechanismen fester Stoffe*, Vandenhoek & Ruprecht, Göttingen 1948.

[45] C. A. F. Benedicks, *Ann. Phys. (Leipzig)* **55** (1918) no. 1, 103.

[46] C. A. F. Benedicks, *Ann. Phys. (Leipzig)* **62** (1920) 185.

[47] C. A. F. Benedicks, *Math. Astron. Phys.* **23A** (1933) no. 27, 1.

[48] C. A. F. Benedicks, *Math. Astron. Phys.* **24A** (1933) no. 1, 7.

[49] J. W. Christian et al.: "Thermoelectricity at Low Temperatures: VI. A Redetermination of the Absolute Scale of Thermo-electric Power of Lead," *Proc. R. Soc. London A* **245** (1958) 213–221.

# Thermoplastic Elastomers

*A general introduction into Thermoplastic Elastomers is given in Rubber, 3. Synthetic, Chap. 7.*

SABET ABDOU-SABET, Advanced Elastomer Systems, Akron, Ohio 44334, United States (Chap. 1)

HANS-GEORG WUSSOW, Bayer AG, Dormagen, Federal Republic of Germany (Chap. 2)

LARRY M. RYAN, LAWRENCE PLUMMER, E. I. Du Pont de Nemours and Co., Inc., Engineering Polymers, Wilmington, Delaware 19880, United States (Chap. 3)

DIDIER JUDAS, Elf Atochem, Centre d'Application de Levallois, Levallois, France (Chap. 4)

HANS F. VERMEIRE, Shell Research SA, Louvain-La-Neuve, Belgium (Chap. 5)



Ullmann's Encyclopedia
of Industrial Chemistry, Vol. A 26

Abbreviations for polymers and monomers used in this article

| | |
|---|---|
| 2G | ethylene glycol |
| 2GT | dimethylene terephthalate, ethylene terephthalate |
| 3G | 1,3-propanediol |
| 4G | 1,4-butanediol |
| 4GT | tetramethylene terephthalate, butylene terephthalate |
| 5G | 1,5-pentanediol |
| 6G | 1,6-hexanediol |
| 10G | 1,10-decanediol |
| ABS | acrylonitrile – butadiene – styrene copolymer |
| BR | butadiene rubber |
| DOP | dioctyl phthalate |
| DVA | dynamically vulcanized alloys |
| EOPPG | poly(propylene glycol) capped with ethylene oxide |
| EPDM | ethylene – propene – diene terpolymer |
| EPM | ethylene – propene copolymer |
| EPR | ethylene – propene rubber |
| EVA | ethylene – vinyl acetate copolymer |
| $H_{12}$-MDI | 4,4′-dicyclohexylmethane diisocyanate |
| HDI | hexamethylene diisocyanate |
| HIPS | high-impact polystyrene |
| IR | isoprene rubber |
| MDI | 4,4′-methylene diphenyl diisocyanate |
| NDI | 1,5-naphthalene diisocyanate |
| NR | natural rubber |
| PAE | thermoplastic polyamide elastomer |
| PB | polybutadiene |
| PC | polycarbonate |
| PEO | poly(ethylene glycol); poly(ethylene oxide) |
| PETP | poly(ethylene terephthalate) |
| PP | polypropylene |
| PPE | poly(phenylene ether) |
| PS | polystyrene |
| PTMO | poly(tetramethylene oxide) |
| PVC | poly(vinyl chloride) |
| RTPO | reactor thermoplastic polyolefins |
| S–E/B–S, SEBS | styrene – ethylene/butene – styrene copolymer |
| S–E/P–S, SEPS | styrene – ethylene/propene – styrene copolymer |
| SBC | styrenic block copolymers |
| SBS | styrene – butadiene – styrene block copolymer |
| SIS | styrene – isoprene – styrene block copolymer |
| T | terephthalic acid |
| TODI | 3,3′-tolidine-4,4′-diisocyanate |
| TPE | thermoplastic elastomer |
| TPNR | thermoplastic natural rubber |
| TPO | thermoplastic polyolefin elastomer |
| TPR | thermoplastic rubber |
| TPU | thermoplastic polyurethane elastomer |
| TPV | thermoplastic vulcanizates |
| VA | vinyl acetate |

Other abbreviations used

| | |
|---|---|
| BMC | bulk molding compound |
| MFR | melt flow rate |
| SAFT | shear adhesion failure temperature |
| SMC | sheet molding compound |
| T.S. | tensile strength |
| U.E. | ultimate elongation |
| UL | Underwriters Laboratories code |

# 1. Thermoplastic Polyolefin Elastomers

## 1.1. Introduction and Definition

The significant growth of thermoplastic elastomers in the marketplace since the 1970s can be attributed to the family of thermoplastic polyolefin elastomers because of the wide range of products available, their favorable price – performance relationship, and their acceptance by the automotive industry. Historically, Uniroyal Chemical Company was the first company in 1972 to introduce a thermoplastic polyolefin elastomer to the market.

In 1971, W. K. FISHER [1.1], [1.2] filed patent applications on thermoplastic elastomer (TPE) blends based on rubbers that are copolymers of monolefinic monomers—such as ethylene – propene copolymer (EPM) or ethylene – propene – diene terpolymer (EPDM) (→ Rubber, 3. Synthetic, **A 23**, pp. 283–288)—with crystalline polyolefin resin, e.g., polypropylene (PP). FISHER employed partial vulcanization of the rubber component to maintain the thermoplastic processibility of the blend by limiting the amount of peroxide. Cross-linking was performed while mixing under dynamic shear, a process known as dynamic vulcanization. However, the dynamic vulcanization process and the first EPM–PP blend were discovered independently by GESSLER et al. [1.3] (Exxon) and HOLZER et al. [1.4] (Hercules) in 1958 and 1961, respectively. Two types of thermoplastic rubber (TPR) were produced. One type consisted of simple blends of EPDM and PP; the other contained *partially vulcanized rubber* in the blend. In 1972–1976, most major EPDM producers commercialized products based on simple blends. In general, the TPR products were elastomeric according to DIN 7724 and ISO 1382-1982 definitions, yet these products were not seriously considered for true rubber replacement applications.

Significant improvement in the properties of these blends was achieved in 1975 by CORAN et al. [1.5] by *fully vulcanizing the rubber phase* under dynamic shear without affecting the thermoplasticity of the blend. This discovery was advanced by ABDOU-SABET et al. [1.6] in 1977 by the use of phenolic curatives to achieve further improvement in rubber-like properties and processing characteristics.

Therefore, the field of thermoplastic elastomers based on polyolefin rubber–plastic blends grew along two distinctly different product lines or classes: one class consists of a simple blend of $\alpha$-olefin rubber in a crystalline polyolefin resin and classically meets the definition of thermoplastic polyolefin (TPO). In the production of the other class, the rubber component in the TPO is subjected to dynamic vulcanization to generate a TPE with true rubber-like properties. This class is called thermoplastic vulcanizates (TPVs) or dynamically vulcanized alloys (DVAs).

These two classes of materials share a common basic composition. The plastic phase is a crystalline polyolefin such as isotactic and syndiotactic polypropylene, propene copolymer, high-density polyethylene, low-density polyethylene, linear low-density polyethylene, ethylene–vinyl acetate copolymer, ethylene–methacrylate copolymer, or ethylene–ethyl acrylate copolymer. The rubber phase may consist of EPM, EPDM, isobutene–isoprene copolymer (butyl rubber), halobutyl rubber, butyl rubber terpolymer with $\alpha$-methylstyrene, etc. (For a description of component polymers, see → Polyolefins; see → Rubber, 3. Synthetic).

## 1.2. Thermoplastic Polyolefin Blends

The first thermoplastic polyolefins were *mechanical blends* of polyolefin polymers. The most common types of TPO are composed of polypropylene and ethylene–propene rubber (EPR). Advances in Ziegler–Natta catalysis and reactor design have allowed the preparation of such blends in the polymerization reactor from

ethylene and propene monomers. These are commonly called *reactor thermoplastic polyolefins* (RTPO). Commercial demand for TPO and RTPO has been particularly strong in the automobile bumper application, especially in Europe and Japan. Such dominance has overshadowed the opportunities for TPO in other market sectors. Advanced compounding technology is being used to generate TPO compounds specific to the application; such technology can also be applied to RTPO.

**Mechanically prepared blends** are produced by using either batch or continuous mixing techniques. The most common batch mixers are internal intensive mixers, such as the Banbury. Continuous mixing techniques can be performed in either single- or twin-screw extruders. Mixing procedures can be extremely varied because of the wide range of TPO recipes in use, as well as the mechanical characteristics of individual process lines.

**Reactor thermoplastic polyolefins** are the newest variety of TPO. The development of new Ziegler–Natta and metallocene catalysts enabled copolymerization of an elastomeric soft component inside a shell or skin of crystalline PP. This has been accomplished by the *Cataloy process* [1.7], which is based on a two-stage reactor. The first stage consists of a tubular loop reactor that produces crystalline polypropylene. In the second tubular reactor, ethylene and other olefinic monomers are introduced, and ethylene copolymerizes with propene to form the ethylene–propene copolymer rubber inside the crystalline homopolymer PP produced in the first reactor; this minimizes reactor fouling. The formation of a continually growing polymer skin and porous particle provides a reaction bed within which the copolymer is produced. This advanced technology in Ziegler–Natta catalysis has been referred to as *Reactor Granules Technology* [1.8] because the growing polymer particle itself has become the polymerization reactor.

## 1.2.1. Morphology

Thermoplastic polyolefins possess primarily a two-phase morphology. At least one phase—the thermoplastic component of the blend—is continuous. The second, or rubber, phase can be either co-continuous or dispersed, depending on the composition of the blend. The mixing conditions, surface energy, viscosity, and molecular

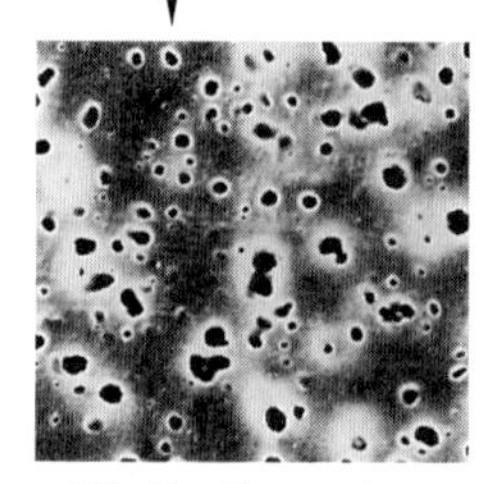



**Figure 1.1.** Morphology of EPDM – PP thermoplastic polyolefins containing 0 and 3 mol % compatibilizer

masses of the two polymers determine at what polymer ratio the two phases are co-continuous [1.9], [1.10], since at extreme concentrations (i.e., a 10:90 or 90:10 ratio) the minor component is expected to be the dispersed phase.

LOHSE [1.11] describes the morphology of the simple blend as unstable because the polymers are not miscible, especially at a 50:50 polymer ratio. The rubber phase, particularly when present as the dispersed phase, tends to agglomerate or coalesce. Such phase growth is undesirable and would lead to changes in properties and performance as a function of mixing and processing. To stabilize against phase growth, additives or *compatibilizers* are usually added. The preferred compatibilizer is a graft or block copolymer of isotactic PP and EPR. One approach has been described in [1.12]. The compatibilizer is generated from EPDM containing vinylnorbornene as the diene; the pendant double bond is then used to graft to the PP. The addition of 3 % of a compatibilizer to the blend improves adhesion between the phases and leads to finer dispersion of the phases. This is demonstrated by the smaller size of the rubber phase and the considerably decreased phase growth (Fig. 1.1).

### 1.2.2. Elastomer Component, Soft Domain

Either ethylene – propene or ethylene – propene – diene rubber is suitable for these blends. The double bonds in the EPDM are not used as chemically active sites for curing. However, they provide considerably wider choices in raw material selection and undoubtedly affect the green strength properties of the elastomer by changing the number of branch points in the rubber molecule.

EPR rubbers contain 45 – 85 mol % ethylene, 55 – 15 mol % propene, and 0 – 10 mol % diene. EPDM with an ethylene content > 75 mol % exhibits crystallinity that can be examined by differential scanning calorimetry or $^{13}C$ NMR by measuring the ethylene sequence index [1.13]. The net effect of ethylene crystallinity is that it greatly strengthens the rubber and, consequently, the blend.

### 1.2.3. Plastic Component, Hard Domain

In most commercial TPO, the hard domain is isotactic polypropylene homopolymer or a copolymer (random or block) with a small amount of ethylene as comonomer. These resins are semicrystalline. The homopolymer has an *mp* of 155 – 165 °C, while the random copolymer has an *mp* of 145 – 155 °C. Ethylene decreases the rigidity of the polymer and improves the low-temperature properties at a sacrifice of the upper performance temperature.

### 1.2.4. Compounds, Trade Names

The two main components of a TPO blend are ethylene – propene rubber and polypropylene. Additional materials, such as other elastomers, processing oils, antioxidants, reinforcing and nonreinforcing fillers, colorants, and other additives to enhance the adhesion or the paintability of the product, are often used. Typical TPO compounds and their properties are listed in Table 1.1, and the suppliers and trade names of such materials are listed in Table 1.2.

Automotive and electrical applications are the major markets for TPO. Automotive constitutes by far the largest market because of the

**Table 1.1.** Typical TPO compositions and properties

| Components, properties | TPO type | | |
|---|---|---|---|
| | 1 | 2 | 3 |
| EPDM, parts by weight | 20 | 57 | 85 |
| PP, parts by weight | 80 | 43 | 15 |
| $CaCO_3$, parts by weight | 60 | 8 | 8 |
| Irganox 1010, parts by weight | 0.2 | 0.2 | 0.2 |
| Tinuvin 327, parts by weight | | 0.15 | 0.15 |
| N-110 black, parts by weight | 3.0 | | |
| Calcium stearate, parts by weight | 0.10 | | 0.10 |
| Process oil, parts by weight | | 7.0 | 64.00 |
| T.S.,* MPa | 16.55 | 13.50 | 5.50 |
| U.E., **% | 200 | 740 | 750 |
| Flexural modulus B, MPa | 1240 | 415 | |
| Shore hardness | 62 D | 54 D | 65 A |
| Gardner impact (at $-29\,°C$), J | 20 | 36 + | |

* T.S. = Tensile strength. ** U.E. = Ultimate elongation.

**Table 1.2.** Suppliers of TPO

| Trade Name | Producer | Location |
|---|---|---|
| Deflex | D & S Plastic International | Grand Prairie, TX, United States |
| ETA, RTA | Republic Plastics/ Himont | Lansing, MI, United States |
| Ferroflex | Ferro | Stryker, OH, United States |
| | | Houston, TX, United States |
| Ferrolene TPE | Ferro | Rotterdam, The Netherlands |
| Kelburon | DSM | Beek, The Netherlands |
| Keltan TP | DSM | Beek, The Netherlands |
| Polytrope | A. Schulman | Akron, OH, United States |
| | | Orange, TX, United States |
| Telcar | Teknor Apex | Pawtucket, RI, United States |
| | | Brownsville, TN, United States |
| Vestolen EM | Chem. Werke Hüls | Marl, Germany |
| Vestoprene | Chem. Werke Hüls | Marl, Germany |
| Vistaflex | Advanced Elastomer Systems | Brussels, Belgium |
| Vitacom TPO | British Vita | Manchester, United Kingdom |
| Milastomer | Mitsui Petrochemical | Chiba, Japan |
| Thermoran | Mitsubishi/JSR | Yokkaichi, Japan |
| P.E.R. | Tokoyama Soda | Japan |

excellent weatherability and low cost of TPOs. Some examples are bumper covers, interior trim, fender extensions, etc.

TPOs have good electrical properties in combination with good ozone resistance, which makes these materials a good choice for low voltage wire and cable applications where hot tension set at 200 °C is not a requirement. Such applications can be found in automotive booster cable, flexible cord, and welding cable insulation.

## 1.3. Thermoplastic Vulcanizates

### 1.3.1. Dynamic Vulcanization

Dynamic vulcanization was discovered by GESSLER et al. [1.3]. In an attempt to improve the impact properties of PP they partially vulcanized halobutyl rubber in a polypropylene matrix under continuous mixing. Dynamic curing generates the same necessary cross-links or three-dimensional polymer structures as static curing. In dynamic curing, however, these structures are generated in small rubber particles dispersed in the uncross-linked thermoplastic polymer matrix as a microgel. The formation of such dispersed-phase morphology depends on establishing a high degree of cure (CORAN et al. [1.5]). The rubber phase is fully vulcanized in a plastic matrix (thermoplastic vulcanizate). The rubber particles are in the order of $1-2$ μm in size. Fully curing the rubber phase has led to significant improvement in the material properties including

1) Higher ultimate tensile strength
2) Lower tension and compression set
3) Better property retention at elevated temperature
4) Greater resistance to attack and swelling by fluids
5) Greater flexural fatigue resistance
6) More consistent processibility

Partially cross-linked blends give modest improvement [1.1], [1.2].

The effect of the size of the rubber particles and the degree of cure on material properties can be seen in Figures 1.2 and 1.3. By proper choice of the curatives, S. ABDOU-SABET et al. [1.6] demonstrated that rubber-like properties can be improved beyond those achieved by CORAN. Improvements in compression set, oil resistance, and the processing characteristics of the material were achieved by applying phenolic curatives

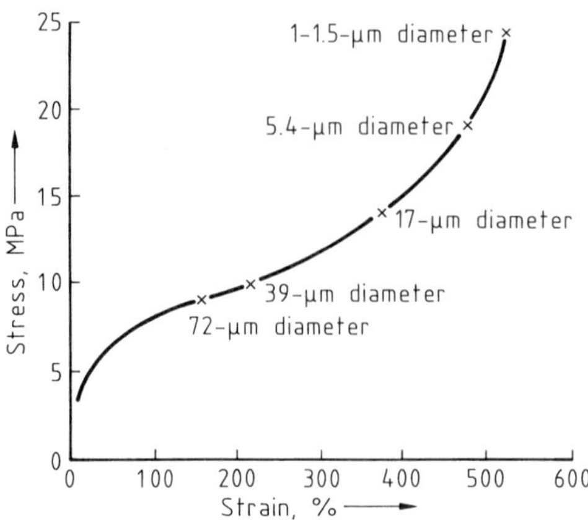



**Figure 1.2.** Effect of particle size on stress–strain relationship

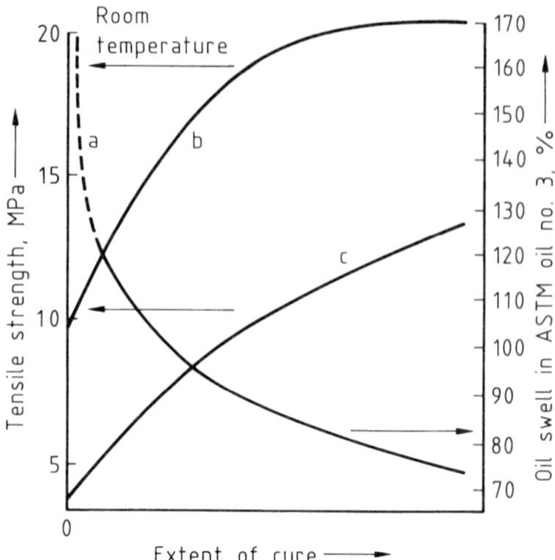



**Figure 1.3.** Effect of the degree of cure on mechanical properties
a) Percentage of swell in ASTM oil no. 3; b) Tensile strength at 100 °C; c) Tensile strength at room temperature

(e.g., dimethyloloctylphenol). These improvements are beyond those expected from the state of cure. They probably are due to in situ formation of a graft copolymer between EPDM and PP that optimizes the interfacial adhesion between the rubber particle and the polypropylene matrix, thus allowing for better elastomeric properties. The graft copolymer could have been generated through the functionalization of PP with dimethyloloctylphenol, which then reacts with EPDM [1.14].

More importantly, significant improvement in processing was achieved over that obtained with the sulfur cure, as shown in Table 1.3. The use of dimethyloloctylphenol as curative allowed the generation of very soft TPV (as soft as 45 Shore A, with compression set approaching that of a thermoset rubber) while maintaining excellent thermoplastic processing. The poor processing characteristics of sulfur-cured TPV are due to phase growth of the dispersed rubber particles. The polysulfidic bonds generated during vulcanization undergo sulfur exchange reactions

leading to significant coalescing of the rubber particles.

### 1.3.2. Production and Morphology

Melt mixing of the two polymers prior to dynamic vulcanization is the first requirement for a good TPV. The phase morphology of the uncross-linked blend will be influenced by the same variables discussed for TPO. S. ABDOU-SABET et al. [1.10] demonstrated for an 80:20 EPDM–PP blend that PP is the dispersed phase in an EPDM matrix. During dynamic vulcanization of such a blend, EPDM and PP have to undergo a phase inversion to maintain the thermoplasticity of the blend. In the initial stages of dynamic curing, two co-continuous phases are generated, and as the degree of cross-linking in-

**Table 1.3.** Effect of cross-linking agent on properties of EPDM rubber–PP TPV

| Property | Cross-linking agent | | | |
| --- | --- | --- | --- | --- |
| | None | Sulfur | Dimethylolalkylphenol | Peroxide |
| Hardness Shore D | 36 | 43 | 44 | 39 |
| T.S.,[a] MPa | 4.96 | 24.3 | 25.6 | 15.9 |
| $M_{100}$,[b] MPa | 4.83 | 8.00 | 9.72 | 8.07 |
| U.E.,[c] % | 190 | 530 | 350 | 450 |
| ASTM oil no. 3 swell, % | disintegrated | 194 | 109 | 225 |
| Comp. set,[d] % (22 h at 100 °C) | 91 | 43 | 24 | 32 |
| Processing characteristics[e] | + | − | + + + | + |

[a] T.S. = tensile strength. [b] $M_{100}$ = modulus at 100% elongation. [c] U.E. = ultimate elongation. [d] Comp. set = compression set. [e] + = good; − = poor; + + + = excellent.

creases during mixing, the continuous rubber phase becomes increasingly elongated and then breaks up into polymer droplets. As these droplets form, PP becomes the continuous phase. This can be seen in the electron micrograph (Fig. 1.4), in which the EPDM phase appears white and the PP phase is black. Photograph 1.4 B shows that both phases are co-continuous.

Reactor TPO can also be used as a raw material to make a TPV [1.15]. Reactor TPO, however, cannot yet be produced with double bonds or other functionality in the rubber molecules. Therefore, cross-linking of the rubber phase can be performed only with peroxide. Peroxide curing is carried out while the necessary precautions are taken to protect polypropylene from chain scission and unzipping by the peroxy radical. Protection can be achieved via the incorporation of compounds in the blend that preferentially undergo reaction with the peroxides without affecting PP [1.16]. Such compounds include polyisobutylene, polybutadiene, and polybutene. Even then, achieving full cure is quite difficult, and only partial curing of the rubber phase is usually accomplished.

To achieve the best improvement in properties of TPV, certain conditions must be satisfied:

1) The higher the crystallinity of the plastic component, the better are the mechanical and elastomeric properties
2) The smaller the difference in critical surface tension between the rubber and the plastic, the smaller is the size of the dispersed rubber phase
3) The lower the entanglement spacing of the rubber chain, the better are the properties of TPV

### 1.3.3. Types

The most common types of TPV are those derived from EPDM–PP [1.17]. Other types can be produced from different rubber–plastic polymer pairs. The discussion in this section is limited to those containing at least one polyolefin polymer.

*Natural rubber (NR)–polyolefin* TPVs [1.18] have been thoroughly investigated in an attempt to develop commercial products that take advantage of the attributes of natural rubber. Most of the research work was focused on *polypropylene-based material*. These NR–PP TPV thermoplastic natural rubbers (TPNRs) exhibited significant disadvantages compared to their EPDM counterparts and were not acceptable. Drawbacks stemmed from the poor thermal stability and the generation of excessive odor from the natural rubber component on processing at higher temperature. Again, partially cross-linked products exhibited co-continuous morphology, while those cross-linked to 95 % or higher of the natural rubber phase showed dispersed-phase morphology. Properties of this type of TPV are listed in Table 1.4.

*Natural rubber–polyethylene-based compositions* [1.19] have also been studied extensively. These TPVs should possess good low-temperature flexibility and elastic recovery for low-temperature applications.

In 1988, R. C. PUYDAK et al. [1.20] introduced *butyl rubber-based TPV* for applications requiring impermeability to oxygen and water vapor transmission. Such products are useful in sports bladders and blood vial stoppers.

Polymers with significantly different critical surface tension [1.21] can still be mixed together with the help of a compatibilizer to improve wetting characteristics or interfacial adhesion [1.22] between the two phases, thus allowing for the generation of preferred morphology prior to dynamic vulcanization. The application of compatibilization technology allowed the mixing of polar rubbers (e.g., *nitrile rubber* [1.23], [1.24] and *acrylate rubber* [1.25]) with polypropylene, followed by dynamic vulcanization to improve the properties as much as possible. Figure 5 illustrates the effect of the compatibilizer (maleated polypropylene) on polypropylene–ethylene–acrylate rubber TPV.

Only a small amount of compatibilizer is needed for compatibilization. The function of the compatibilizer is analogous to that of a surfactant that emulsifies and stabilizes a water–oil mixture. Effective compatibilizers are generated by forming a graft or block copolymer; the structures of the blocks must be similar to the phases to be compatibilized. In that case, one block of the copolymer is soluble in the polyolefin resin, while the other is soluble in the polar rubber phase. Graft formation must be controlled to achieve the desired interfacial adhesion while preserving a dispersed phase morphology. Generation of primary chemical bonds between the two phases could lead to an interpenetrating network morphology. Such networks could result in thermoset characteristics and the loss of the thermoplastic processing advantage.

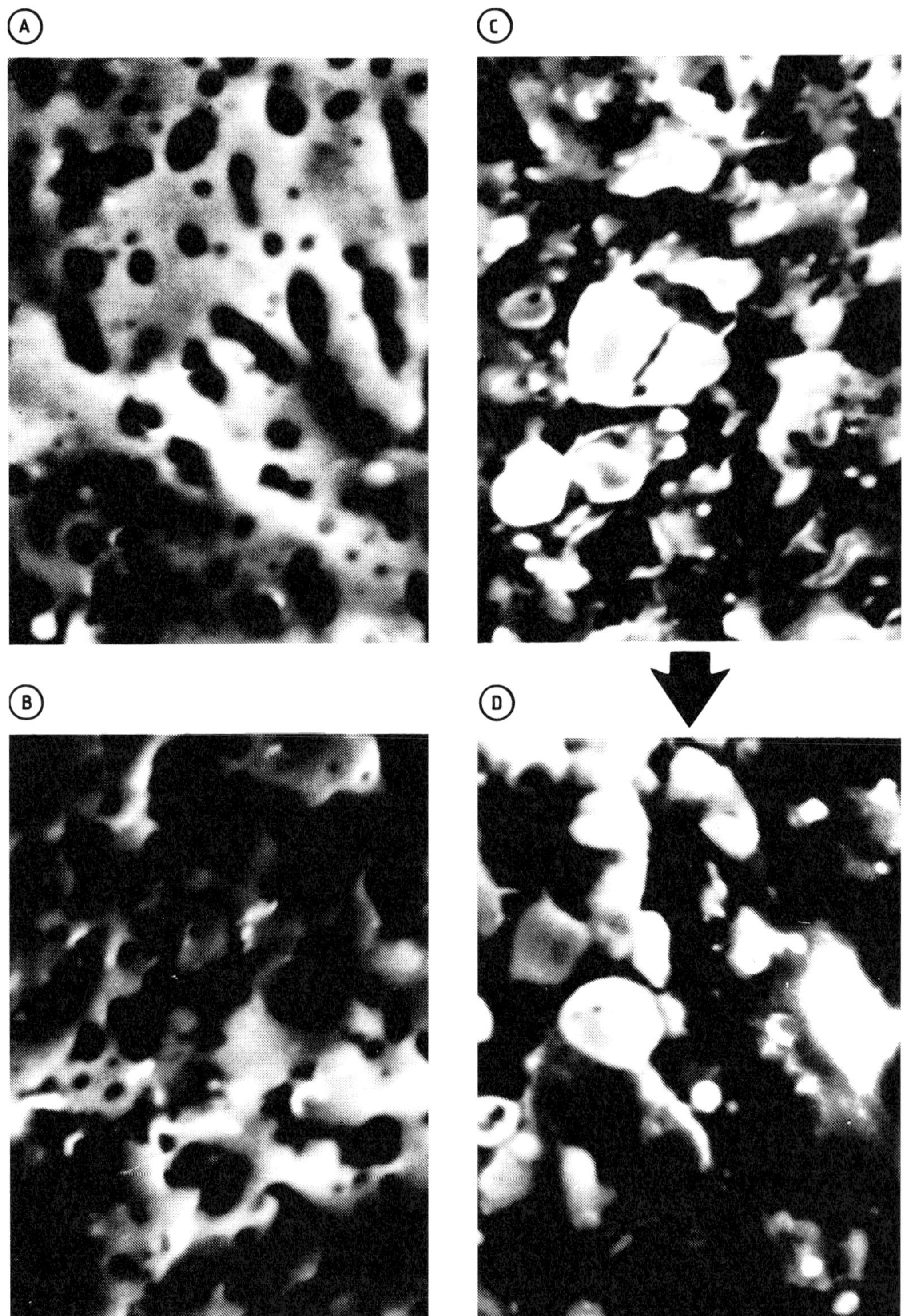

**Figure 1.4.** Electron micrograph of an 80:20 EPDM–PP thermoplastic vulcanizate
A) Unvulcanized blend; B) Partially vulcanized; C) Fully vulcanized; D) Fully vulcanized and diluted to 20:80 EPDM–PP

**Table 1.4.** Typical properties of thermoplastic natural rubbers

| Property | Rubber type | | | | |
|---|---|---|---|---|---|
| | 1 | 2 | 3 | 4 | 5 |
| Shore A hardness | 50 | 60 | 70 | 80 | 90 |
| $M_{100}$,[a] MPa | 3.1 | 4.1 | 5.3 | 6.3 | 8.2 |
| T.S.,[b] MPa | 6.5 | 8.8 | 11.2 | 13.2 | 15.3 |
| U.E. at break,[c] % | 285 | 315 | 340 | 370 | 405 |
| Tear strength, die C (kN/m) | 22 | 31 | 38 | 43 | 51 |
| Tension set, % | 9 | 10 | 14 | 18 | 23 |
| Comp. set,[d] % | | | | | |
|   72 h, 23 °C | 26 | 28 | 31 | 34 | 39 |
|   24 h, 70 °C | 30 | 35 | 36 | 41 | 55 |
|   22 h, 100 °C | 38 | 42 | 43 | 47 | 57 |
| Volume swell after 7 d in | | | | | |
|   ASTM oil no. 1, 23 °C | 14 | 9 | 9 | 7 | |
|   ASTM oil no. 2, 23 °C | 19 | 13 | 12 | 9 | |
|   ASTM oil no. 3, 23 °C | 71 | 53 | 47 | 35 | |
|   ASTM oil no. 1, 100 °C | 101 | 80 | 67 | 61 | |
|   ASTM oil no. 2, 100 °C | 151 | 123 | 108 | 82 | |
|   ASTM oil no. 3, 100 °C | 190 | 164 | 139 | 116 | |

[a] $M_{100}$ = modulus at 100 % elongation. [b] T.S. = tensile strength. [c] U.E. = ultimate elongation. [d] Comp. set = compression set.

### 1.3.4. Processing

Thermoplastic vulcanizates or block copolymers that are made of soft and hard blocks are processed or fabricated by the same techniques normally used for rigid plastics. This enables the fabrication of elastomeric articles with improved efficiency and economy over those of a conventional thermoset rubber. Scrap from TPV can be recycled readily, whereas thermoset rubber scrap is generally discarded, resulting in a sizeable cost penalty. The thermoplastic nature of TPV enables fabrication by methods unavailable to thermoset rubbers. These include blow molding, thermoforming, and heat welding. Thermoplastic vulcanizates have extremely non-Newtonian flow properties, and their viscosity is highly sen-

sitive to shear [1.26]. At low shear rates, they flow very poorly; at high shear rates, however, they are quite fluid, enabling rapid injection molding cycles.

### 1.3.5. Performance and Product Positioning

ASTM D-2000 attempts to classify rubber products according to their performances in dry heat (dry heat service temperature) and ASTM oil no. 3 (see also → Rubber, 3. Synthetic, p. 244, Fig. 1). Conventional thermoset rubber occupies certain regions on a performance chart. Such classification is, however, insufficient because other characteristics such as dynamic properties, stress relaxation, barrier properties, UV resistance, adhesion, and colorability, also play an important role in material selection. Nevertheless, according to the ASTM D-2000 classification, EPDM–PP TPVs have performance characteristics superior to chloroprene (Neoprene) rubber. The properties of TPO, however, are considerably less elastomeric, and TPOs occupy a lower-performance category (Fig. 1.6).

### 1.3.6. Uses

Since their introduction in 1981, TPVs have found thousands of commercial uses worldwide. Generally speaking, these materials have been used in most rubber applications with the exception of tires. Thermoplastic vulcanizates are employed, e.g., in automotive and construction applications, in foods, hardware, appliances, business machines, mechanical rubber goods, and sporting goods, as well as electrical and medical applications. The uses can vary from the syringe stoppers in medical syringes to the rack-and-pinion boots on the steering mechanism of a front-wheel-drive automobile.

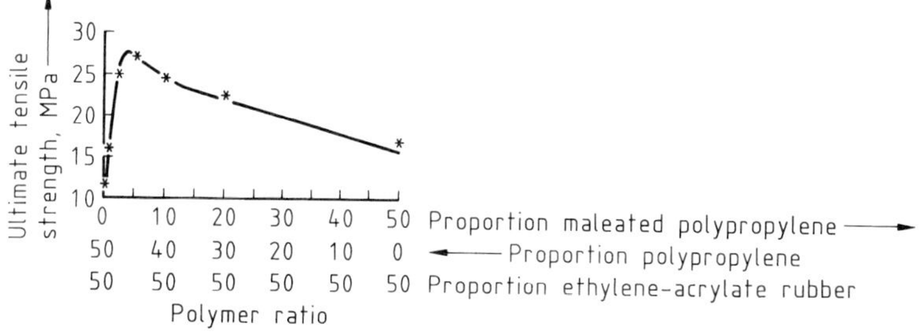


**Figure 1.5.** Effect of compatibilizer (maleated polypropylene) on ultimate tensile strength of polypropylene–ethylene–acrylate rubber TPV

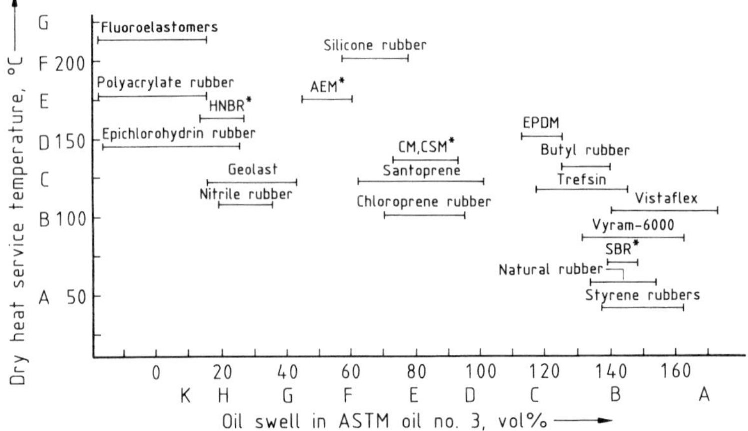



**Figure 1.6.** Classification of elastomers according to heat and oil performance
*AEM = ethyl acrylate–ethylene copolymer; CM = chlorinated polyethylene; CSM = chlorosulfonated polyethylene; HNBR = hydrogenated acrylonitrile–butadiene rubber; SBR = styrene–butadiene rubber; Geolast, Santoprene, Trefsin, Vistaflex, and Viram-6000 are registered trade names of commercial TPVs (see Table 1.5)

### 1.3.7. Commercial Products and Trade Names

Monsanto led the development of thermoplastic vulcanizates and commercialized the first TPV (Santoprene) in 1981. As the FISHER patents were approaching expiration, considerable research and marketing activities occurred in the United States and Europe. Currently, a number of products are on the market (see Table 1.5). Only fully vulcanized materials achieve the full benefit of dynamic vulcanization.

## 2. Thermoplastic Polyurethane Elastomers

(Polyurethane Rubbers, see → Rubbers, 3. Synthetic, pp. 310–312; see also → Polyurethanes)

### 2.1. Introduction

Thermoplastic polyurethanes (TPUs) are a group of elastic materials whose common structural feature is the urethane group $-O-CO-NH-$ arising from the reaction between diisocyanates and long- and short-chain diols. The long-chain polyester or polyether diols form flexible soft segments in the temperature range of use, while the short-chain diols and diisocyanates form rigid or hard segments. Since soft and hard segments are arranged alternately along the molecular chain, TPUs belong to the $[AB]_n$-type block copolymers. Intermolecular interactions between the hard segments lead to the formation of glassy or paracrystalline hard domains, which, as physical cross-linking points, can be reversibly melted during thermoplastic processing. Variation in the ratios of soft and hard segments and monomers permits the production of materials in the modulus range of ca. 5–700 MPa, corresponding to ca. 70 Shore A to 70 Shore D. Because of their very good mechanical properties, especially extraordinary resistance to abrasion, and efficient thermoplastic processibility, TPUs have found a wide range of applications since their introduction to the market at the beginning of the 1960s. Further information on the production and properties of TPUs can be found in [2.1]–[2.4].

**Table 1.5.** Suppliers of thermoplastic vulcanizates

| Trade name | Supplier | Polymer type | Degree of cure |
|---|---|---|---|
| Santoprene[a] | AES | EPDM–PP | full |
| Trefsin[b] | AES | butyl-PP | full |
| Dytron[c] | AES | EPDM–PE | full |
| Sarlink 3000 | DSM | EPDM–PP | partial |
| Sarlink 2000 | DSM | butyl–PP | partial |
| Hifax-XL | Himont | EPR–PP | partial |
| Vyram[d] 9000 | AES | EPDM–PP | partial |
| Vyram[d] 6000 | AES | natural rubber–PP | full |
| Milastomer | Mitsui | EPDM–PP | partial |
| Sumitomo TPE | Sumitomo | EPDM–PP | partial |

[a] Santoprene is a registered trademark of Monsanto Company, licensed to Advanced Elastomer Systems, L.P. [b] Trefsin is a registered trademark of Advanced Elastomer Systems, L.P. [c] Dytron is a registered trademark of Advanced Elastomer Systems, L.P. [d] Vyram is a registered trademark of Advanced Elastomer Systems, L.P.

## 2.2. Raw Materials

(→ Isocyanates, Organic; → Polyesters; → Polyoxyalkylenes)

The raw materials used in TPUs are diisocyanates, short-chain diols with molecular masses of 61 –ca. 600, and long-chain polyester and polyether diols with average molecular masses between 600 and 4000, so-called polyols.

**Diisocyanates.** 4,4′-Diisocyanatodiphenylmethane (4,4′-methylenediphenyl diisocyanate, MDI) [*101-68-8*] is industrially by far the most important diisocyanate. It can be used to produce materials that offer the most favorable compromise between price, thermoplastic processibility, and mechanical properties. Other *aromatic diisocyanates* used are 3,3′-dimethyl-4,4′-biphenyl diisocyanate (3,3′-tolidene-4,4′-diisocyanate, TODI) [2.5] and 1,5-naphthalene diisocyanate (NDI) [*3173-72-6*] [2.6]. Processing of TPUs based on these products is more difficult because of their high melting points. Among the *aliphatic diisocyanates* the following are important: 4,4′-dicyclohexylmethane diisocyanate ($H_{12}$-MDI) [*5124-30-1*] (as the isomeric mixture or the *trans,trans*-product) [2.7] and 1,6-diisocyanatohexane (hexamethylene diisocyanate, HDI) [*822-06-0*] for the production of materials for medicinal technology. The aliphatic diisocyanates are much less reactive toward alcohols than MDI [2.8].

**Short-Chain Diols.** The most important short-chain diol used as a chain extender is 1,4-butanediol. For TPUs with improved dimensional stability at elevated temperature, 1,4-bis(2-hydroxyethoxy)benzene [hydroquinone bis(2-hydroxyethyl) ether] [*104-38-1*] is also used. The use of diol mixtures (e.g., with ethanediol or 1,6-hexanediol) gives products with a broader processing range [2.9].

**Polyols.** *Polyester Polyols.* The most important polyols are polyadipates of short-chain diols with two to ten carbon atoms. Butanediol, hexanediol, ethanediol, and mixtures of these diols are most commonly used. The resistance to hydrolysis of the polyol and TPU increases as the molecular mass of the diol increases. Polycaprolactones and polycarbonate diols with their good to excellent resistance to hydrolysis are also employed. The most widely used products have average molecular masses of 1000–2000. As polycondensation products, polyadipates have a broad molecular mass distribution [2.10]. They frequently contain unreactive ring esters formed as byproducts and the starting glycols. For TPU production the acid number of the ester polyols should be as low as possible, preferably < 1 mg KOH/g, because carboxyl groups react with isocyanates and impair the resistance to hydrolysis.

*Polyether Polyols* (→ Polyoxyalkylenes). Poly(tetramethylene oxides) are the most important polyether polyols for the production of TPUs. Poly(propylene oxides) [poly(propylene glycols)] and copolymers with ethylene oxide give products with poorer properties. The molecular mass distribution of the polyether polyols follows the Poisson distribution and is thus considerably narrower than that of the corresponding polyesters. Poly(propylene oxides) have predominantly secondary hydroxyl groups as terminal units and are therefore considerably less reactive than other polyols. The functionality of poly(propylene oxides) is generally less than 2 because of side reactions occurring during their production, particularly when polymers of higher molecular mass are produced. Recently, almost exactly bifunctional products have come on the market [2.11], [2.12].

**Additives.** Internal release agents, such as esters of montanic acid (octacosanoic acid), silicones, or amide waxes in quantities up to 2 %, are used as *processing aids*. TPUs based on polyesters can be protected effectively against *hydrolytic degradation* by the addition of carbodiimides [2.13]. Protection against *microbial degradation,* for example on prolonged contact with soil, is also possible. Sterically hindered phenols and amines are used as *antioxidants,* mainly to protect against discoloration and degradation, particularly in the case of polyether TPUs at the high production and processing temperatures often required. Yellowing because of the effect of light can be slowed considerably through the use of *light stabilizers* such as benzophenones or benzotriazoles [2.14].

*Plasticizers* are used only to achieve hardness levels below ca. 70 Shore A [2.15]–[2.17]. Short glass fibers are used mainly to *reinforce* TPUs. Many additives, such as dyes, antistatic agents, and flame retardants, are supplied as concentrates.

## 2.3. Production

**General.** The industrial production of TPUs is carried out virtually exclusively in continuous, solvent-free processes [2.18]. According to Flory statistics [2.10] the required high molecular masses are achieved only at high conversions and an essentially stoichiometric ratio of NCO and OH groups. At an NCO:OH ratio < ca. 0.97, products with the required properties are generally not obtained [2.19]. Too high an excess of NCO can lead to cross-linking through allophanate and biuret formation even after long periods [2.20]. A prerequisite for maintaining exact stoichiometric ratios is the use of raw materials that are as anhydrous as possible since water can also consume NCO. Prepolymer and one-shot processes can be used to carry out the reaction (→ Polyurethanes, **A 21**, pp. 682–684). In the one-shot process, OH- and NCO-functional reactants are mixed with each other in one step. In the prepolymer process, a prepolymer with terminal NCO groups is first formed from a long-chain polyol and diisocyanate. This prepolymer is then mixed with the chain extender. Reactivity differences (e.g., between polyether polyols and butanediol) can be overcome with this process. The process used affects the segment length distribution and thus the morphology of the final product [2.21], [2.22].

**Belt Process** [2.23]. Liquid, possibly prepolymerized raw materials are mixed continuously in a mixing head and poured onto a continuous belt, which passes through an oven heated to ca. 70–130 °C, where the reaction proceeds further. The congealed product is ground to irregular pellets. It can be melted in extruders, possibly along with additives, and formed into regular pellets. According to a modification of this process published in 1982 [2.24] the reaction is carried out to a desired degree of conversion, generally > 80%, on the belt, and the congealed mixture is then fed immediately into an extruder where the reaction is continued and the mixture is homogenized and pelletized. In the belt process the temperature and residence times in the heated zone can be adapted to the requirements of the particular product.

**Reaction Extruder Process.** A twin-screw extruder serves as the reactor [2.25]–[2.28]. The screws are conrotating and can be set to the required shearing ratios by a combination of kneading and transporting parts. Temperature is controlled by using barrel sections that can be heated separately. The reactants and additives can be charged at different locations. In this way a continuous prepolymer or one-shot process is possible without additional equipment. The bulk temperatures are generally > 200 °C. The product is extruded continuously and pelletized immediately after the head or after passage through a cooling area. As in the belt process, subsequent sieving and drying processes occur. In the reaction extruder process, finished pellets can be obtained from starting materials with the addition of all additives (including glass fiber or other thermoplastics) in one process step [2.29].

**Comparison of Products.** Because of different shearing and temperature conditions, the type of process used significantly affects the morphology and thus the processing properties of the products. Starting from identical raw materials more opaque products are formed in the belt process with larger hard segment domains that melt at higher temperature, whereas the reaction extruder process gives more transparent, lower-melting products.

## 2.4. Properties

**Morphology** [2.1], [2.3], [2.30], [2.31]. The physical properties of TPUs are determined by their polyphase structure consisting of soft segments with very low glass transition temperatures, hard segments with high melting ranges, and mixed phases. This structure arises from the incompatibility of the hard and soft segments, which, on cooling from the melt during production and processing, leads to partial or complete separation of the phases that are bonded covalently with each other. The extent of separation and the phase structure are determined mainly by the stoichiometric composition of the polymer; the statistical distribution of the hard and soft segments [2.21]; the structure of the polyol [2.32], the diisocyanate and the chain extender, and the thermal history. The latter results from the kinetics of the separation and crystallization processes, and from the cleavage and transurethanization during production, processing, and possible tempering of the product [2.33]–[2.36]. In polyethers, phase segregation is more pronounced than in polyesters and increases as the molecular mass of the polyol increases. Many

**Table 2.1.** Physical properties of TPUs with different hardness based on polyester polyols (example Desmopan, Bayer)

| Property | Test standard (DIN) | Desmopan | | | |
|---|---|---|---|---|---|
| | | 385 | 392 | 359 | 472 |
| Density, $mg/m^3$ | 53 479 | 1.20 | 1.21 | 1.23 | 1.24 |
| Hardness, Shore A/D | 53 505 | 86/32 | 92/40 | 97/59 | 98/73 |
| Tensile stress at break, MPa | 53 504 | 40 | 45 | 50 | 50 |
| Elongation at break, % | 53 504 | 450 | 450 | 350 | 250 |
| Tensile stress, MPa | 53 504 | | | | |
|   10 % elongation | | 1.5 | 3.1 | 10.5 | 28 |
|   20 % elongation | | 2.5 | 4.7 | 13.0 | 29 |
|   100 % elongation | | 6.0 | 9.0 | 20.0 | 31 |
| Tear strength, kN/m | 53 515 | 70 | 80 | 120 | 190 |
| Flexural strength, MPa | 53 452 | 1.4 | 3 | 10 | 32 |
| Abrasion, $mm^3$ | 53 516 | 30 | 25 | 35 | 30 |
| Compression set, % | | | | | |
|   70 h at 22 °C | 53 517 | 30 | 25 | 30 | |
|   24 h at 70 °C | | 55 | 41 | 60 | |

investigations of the structure [2.34], [2.37], phase interactions [2.38], [2.39], and structure – property relationships of TPUs have been carried out, mainly on model substances [2.40] – [2.42]. Macroscopically, the soft segments determine the elastic ductility, the low-temperature properties, and the basic chemical properties, while the hard segments are important for dimensional stability at high temperature and the processing properties.

**Technical Properties.** By varying the proportion of the hard segment the modulus of elasticity can be varied over a range of ca. 5 – 700 MPa without the use of plasticizers or reinforcing agents. Products containing no [2.8] or only very small proportions of soft segments can no longer be regarded as thermoplastic elastomers [2.43]. Table 2.1 gives an overview of some properties of different types of product.

High extension with very low residual extension, high tensile and tear strength, high moduli of elasticity at a given Shore hardness, and excellent abrasion resistance are common to all TPUs. Polyester TPUs are superior to polyether products with regard to strength and resistance to oxidation or swelling by oils, fats, and water, while they are inferior with regard to resistance to hydrolysis or bacteria and impact resistance at low temperature. Hybrid types from polyester and polyether polyols combine the advantages of both types to a certain extent [2.44], [2.45].

Unstabilized TPUs based on MDI turn yellow through the action of light, initially without the impairment of mechanical properties. The lower temperatures of use are determined by the glass transition temperature of the soft phase ($-20$ to $-60$ °C), and the upper ones by the onset of softening of the hard domains. Depending on the hardness, softening starts at ca. 80 – 100 °C (see → Polyurethanes, **A 21**, p. 703, Fig. 34). Hard products can be kept at up to ca. 120 °C for short periods. TPUs are soluble in polar, aprotic solvents such as $N,N$-dimethylformamide, methyl ethyl ketone, or $N$-methylpyrrolidone.

## 2.5. Processing and Use [2.1], [2.3], [2.46]

TPUs can be processed by all methods available for thermoplastics, such as injection molding, extrusion, blow molding, sintering, and calendering, although not all types are suited to every processing method. Generally only products with hardnesses up to ca. 50 Shore D can be extruded.

TPUs are used in all those areas where their high strength and excellent resistance to abrasion, oil, and many other solvents are important. The large number of uses includes bases for shoe heels, soles for sports shoes, screens for the classification of minerals, rollers, tool parts, toothed belts, gaskets, electrical plugs, coupling elements, ear tags for animals, parts for car bodies, bearing shells, ski boots, seals, cable sheathings, tubing of all types, blown films, profiles, coatings, and medical equipment such as tubing and catheters. TPUs are also used for modification of the impact resistance of polyacetals and as blend components for polycarbonates and acrylonitrile – butadiene – styrene (ABS) polymers [2.47], [2.48].

**Table 2.2.** Important producers of TPU and trade names

| Region | Producer | Trade Name |
|---|---|---|
| Europe | Bayer | Desmopan |
| | Elastogran (BASF) | Elastollan |
| | B. F. Goodrich | Estane |
| | Larim | Laripur |
| | Morton | Irogran |
| United States | Dow Chemical | Pellethane |
| | B. F. Goodrich | Estane |
| | Miles (Bayer) | Texin |
| | Morton Thiokol | Morthane |
| | Wyandotte (BASF) | Elastollan |
| Japan | Dainichi Senka | Leathermin |
| | Dai Nippon Ink | Pandex |
| | Nihon Miractran | Miractran |
| | Takeda-BASF | Elastollan |

## 2.6. Recycling

Faulty batches formed in production can frequently be reprocessed or used for less demanding applications. The processor can regranulate sprues or faulty parts and feed them back to the production process without impairing properties. Recycling articles at the end of their service life appears to be economically viable when large quantities of the same type of materials can be collected. Investigations have been carried out on car body parts and ski boots [2.49].

## 2.7. Economic Aspects

TPUs are produced in Europe, the United States, and Japan by the three large producers BASF, Bayer, and B. F. Goodrich (quantities of ca. 20000 t/a) and a number of medium and small producers (Table 2.2).

Almost all producers offer materials of differing hardness based on different polyols within several product ranges. Some highly specialized producers can also be found, particularly in products for medical applications or for further internal processing. At the beginning of the 1990s the estimated market volume was ca. 100000 t/a at somewhat higher capacities. The regional distribution has a peak in Europe (45%), followed by North America (25%), Japan (15%), and Southeast Asia (ca. 8%). The chances of growth are estimated to be positive in view of the possibilities for replacing rubber and plasticized thermoplastics by TPUs and the versatility of this class of materials.

## 2.8. Toxicology and Occupational Health

Since the production process involves polyaddition in a melt, very small quantities of byproducts are formed. Correctly dimensioned and efficiently functioning air extraction allows the permitted workplace concentrations for diisocyanates to be adhered to with certainty by producers and processors. MAK and TLV (TWA) for MDI are 0.005 ppm; an STEL value has not been set (0.02 ppm for other diisocyanates). Other components are less critical. Generally neither aromatic amines nor diisocyanates can be extracted from properly produced TPU articles. If only the allowable raw materials are used, TPUs comply with health authority guidelines for contact with foods (FDA: CFR 175.300, 175.105, 177.1680; BGA: Recommendation XXXIX). Because of their good compatibility with blood and tissue, some products also fulfill the requirements of the U.S. Pharmacopoeia XXI, Class 6 [2.50].

# 3. Thermoplastic Copolyester Elastomers

## 3.1. Introduction

Thermoplastic copolyester elastomers are segmented block copolymers, consisting of repeating, high-melting, rigid "hard" blocks and amorphous, flexible "soft" blocks, which have a very low glass transition temperature. These polymers exhibit a remarkable combination of properties as a result of microphase separation of the hard block (in the form of crystalline domains) from the matrix of the amorphous soft blocks. The crystalline domains impart excellent mechanical properties (tensile strength, toughness, resistance to creep, good retention of properties at elevated temperature, etc.) and outstanding chemical resistance to the polymer. Concurrently, the amorphous domains provide exceptional flexibility and fatigue resistance over a broad temperature range. By varying the ratio of hard to soft blocks, products ranging from relatively rigid plastics to very flexible elastomers can be produced.

Copolyester elastomers are typically produced by condensation of an aromatic dicarboxylic acid or ester with a low molecular mass aliphatic diol and a polyalkylene ether glycol (molecular mass generally between 1000 and 4000) [3.1]–[3.3]. Reaction of the aromatic dicarboxylic acid with the diol leads to the crystalline hard segment phase, while the soft segment is the product of the diacid (or diester) and the long-chain glycol. Other combinations are possible, however (see Sections 3.4.3, 3.4.4).

## 3.2. Raw Materials

Polyester elastomers are produced from aromatic dicarboxylic acids or diesters, polyoxyalkylenes, and an excess of short-chain diols [3.1]. The most common raw materials follow:

**Dicarboxylic Acids.** *Terephthalic Acid* [*100-21-0*]. For production and properties, see → Terephthalic Acid and Dimethyl Terephthalate.
*2,6-Naphthalenedicarboxylic Acid* [*1141-38-4*]. For production, see → Carboxylic Acids, Aromatic, **A 5**, p. 257.

**Polyoxyalkylenes.** *Poly(tetramethylene oxide)*; *poly(propylene glycol)* [*25322-69-4*]; *poly(ethylene glycol)* [*25322-68-3*]; for production and properties, see → Polyoxyalkylenes.

**Short-Chain Diols.** *1,4-Butanediol* [*513-85-9*]. For production and properties, see → Butanediols, Butenediol, and Butynediol, **A 4**, pp. 457–459.
*Ethylene glycol* [*107-21-1*]. For production and properties, see → Ethylene Glycol.

## 3.3. Production
(see also → Polyesters, **A 21**, pp. 233–237)

Polyester elastomers are manufactured by a *multistage melt polymerization process*. Soluble titanium alkoxides are generally used as catalyst. In the first stage of the process, all reactants are mixed and heated at atmospheric pressure to 190–220 °C to form a low molecular mass prepolymer. In the latter stages the prepolymer is converted to high molecular mass polymer by increasing the temperature to 250–260 °C while lowering the pressure to < 133 Pa to distill off the excess short-chain diol. The degree of poly-

merization is monitored by the viscosity change of the molten reaction mass by measuring the torque on the reactor agitator. To protect the polyether ester from oxidative degradation during synthesis and in end use applications, polymerization is usually carried out in the presence of an antioxidant. Hindered phenols such as Irganox 1098 (Ciba–Geigy) and Ethanox 330 (Ethyl Corp) or an aromatic secondary amine [i.e., Naugard 445 (Uniroyal)] are examples of antioxidants that are added during the preparation stage of the prepolymer. At the end of polymerization, the polymer is extruded, quenched, and cut into pellets. Predominantly hydroxyl and carboxylic acid end groups are produced by this process. The acid end groups are formed by the thermal cleavage of polymer chains or by degradation of a butanol end group with simultaneous elimination of tetrahydrofuran.

Both batch and continuous processes produce materials of sufficient viscosity to be fabricated by either extrusion or injection molding. However, a polymer with ultrahigh viscosity is required for blow molding operations and in extrusion processes where very precise dimensions are required. Generally, the polymer viscosities required for these methods of processing cannot be obtained by a melt condensation process. One technique used to manufacture ultrahigh-viscosity copolyester elastomers is *solid-phase polymerization* [3.4]–[3.7]. In this process the solid polymer is heated to just below its melting point under reduced pressure. This leads to the reaction of acid and hydroxyl end groups or of two hydroxyl end groups with elimination of water or short-chain glycols. They diffuse out of the polymer, and the molecular mass increases. A second method of increasing viscosity is by polymer chain extension in a postcompounding step. In numerous examples [3.8]–[3.10], diisocyanates have been used as chain extenders to increase the viscosity of low and moderate molecular mass polyesters through the reaction of the polyester hydroxyl end group to form a urethane.

Thermoplastic copolyester elastomers are stable for an indefinite period of time. However, they will absorb moisture, reaching equilibrium with existing atmospheric conditions very rapidly. Since the ester group hydrolyzes at processing temperature, these products must be dried thoroughly prior to fabrication. The moisture content should not exceed 0.1 wt % when molding or extruding [3.11].

## 3.4. Microstructure and Composition of Segments

### 3.4.1. Short-Chain Ester Units (Hard Segments)

The hard segment employed in the polyester elastomer strongly influences the load-bearing capabilities and processing characteristics of the polymer. Because of its high melting point, rapid crystallization, and low cost, the predominant hard segment in commercial use today is tetramethylene terephthalate (4 GT), an ester of 1,4-butanediol (4 G) and terephthalic acid (T). Hard segments based on ethylene glycol (2 G) and terephthalic acid (2 GT) are high melting and yield polymers with excellent physical properties. However, crystallization of this hard segment is very slow, requiring 36 h to completely develop the physical properties of the polymer. This limitation greatly reduces the usefulness of 2 GT thermoplastic elastomers in injection molding and extrusion applications [3.12]. WOLFE [3.13] has prepared a series of copolyester elastomers with a wide range of short-chain diols and compared their properties (Table 3.1). The properties of the elastomers resulting from 2 G, 3 G, (1,3-propanediol), and 4 G are very similar except for tensile and tear strength. Hard segments based on diols with five, six, and ten carbon atoms do not form high-melting crystals, and as a result, copolyesters made with these diols generally have inferior properties.

Substituting a cycloaliphatic glycol, 1,4-cyclohexanedimethanol, for 1,4-butanediol produces a high-melting polyether ester elastomer that crystallizes somewhat slower than one containing tetramethylene terephthalate but rapidly enough to be blow or injection molded. Because of their slower crystallization characteristics, finished parts made of these elastomers are clear [3.14], [3.15].

Naphthalenedicarboxylic acid, in place of terephthalic acid, in combination with 1,4-butanediol provides products with improved thermal properties compared to their 4 GT counterparts [3.16]. Some mechanical properties such as tear strength are also improved. A major limitation in using naphthalenedicarboxylic acid in commercial thermoplastic copolyester elastomers, however, is its high price.

### 3.4.2. Long-Chain Polyether Soft Segments

The long-chain polyether soft segments are formed by reaction of poly(ether glycols) and diacids. The predominant poly(ether glycols) used in commercial thermoplastic copolyester elastomers today are poly(tetramethylene oxide) (PTMO) and poly(propylene glycol) end-capped with ethylene oxide (EOPPG). In general, the chemical resistance and certain physical properties of thermoplastic elastomers, such as tear strength, are influenced by the poly(ether glycol) segment. For example, products based on PTMO have much higher tensile and tear strengths and significantly better resistance toward oxidative degradation than polymers made with EOPPG. Melting point, modulus, and hardness are generally independent of the type of polyether employed.

Because of their polarity, polymers made using poly(ethylene glycol) (PEO) show the greatest water swell, yet the least swell in nonpolar oils [3.17]. These products have high moisture vapor transmission rates.

A comparison of typical properties of thermoplastic polyester elastomers containing soft segments based on PTMO, EOPPG, or PEO is shown in Table 3.2 [3.18].

**Table 3.1.** Properties of 50% alkylene terephthalate–PTMO terephthalate copolymer as a function of diol structure [3.13]

| Properties | Diol | | | | | |
|---|---|---|---|---|---|---|
| | 2 G | 3 G | 4 G | 5 G | 6 G | 10 G |
| Melting point, °C | 224 | 198 | 189 | 106 | 122 | 106 |
| Shore D hardness | 46 | 48 | 48 | 32 | 33 | 35 |
| Stress at 100% elongation, MPa | 11.4 | 11.7 | 11.7 | 4.5 | 5.2 | 6.1 |
| Tensile strength, MPa | 45.5 | 22.8 | 48.4 | 15.4 | 13.5 | 15.5 |
| Elongation, % | 675 | 660 | 755 | 880 | 750 | 640 |
| Tear strength, kN/m | 42 | 15 | 48 | 25 | 18 | 7.8 |

**Table 3.2.** Properties of 57% 4 GT polyether elastomers as functions of polyoxyalkylene structure [3.18]

| Properties | Polyoxyalkylene | | |
|---|---|---|---|
| | PTMO | EOPPG | PEO |
| Nominal molecular mass | 1000 | 1000 | 1000 |
| Stress at 100% elongation, MPa | 13.4 | 11.7 | 13.2 |
| Tensile strength, MPa | 47.2 | 27.2 | 42.6 |
| Elongation, % | 660 | 675 | 705 |
| Tear strength, kN/m | 62 | 23 | 31 |
| Volume swell, % [3.14] | | | |
|   Water (3 d/24 °C) | 0.5 | 1.0 | 17.0 |
|   ASTM oil no. 3 (7 d/100 °C) | 13.0 | 12.0 | 4.0 |

Through a novel set of reactions, polyethers end-capped with amino groups [3.19] also have been utilized to manufacture copolyester elastomers. These polyoxyalkylenamines are first converted to acid-terminated polyethers by reaction with benzene-1,2,4-tricarboxylic acid, followed by copolymerization with terephthalic acid and a short-chain diol. Since the backbone of the polyoxyalkylenamines is based on ethylene and propylene oxides, the resulting thermoplastic elastomers have properties similar to EOPPG-based products [3.20].

### 3.4.3. Hydrocarbon Soft Segments

Unsaturated straight-chain aliphatic carboxylic acids can be dimerized to form high molecular mass dicarboxylic acids, generally referred to as dimer acids [3.21]. When purified, these dibasic hydrocarbons have been used in place of poly(ether glycols) as soft segments in the preparation of thermoplastic copolyester elastomers. HOESCHELE [3.22] prepared several of these polymers and demonstrated the striking similarities with their polyether counterparts.

One major difference is that products based on dimer acid exhibit poor low-temperature properties because of their high glass transition temperatures. However, due to the hydrocarbon nature of the backbone, these polymers have much better heat aging characteristics and resistance to UV degradation. McCREADY [3.23] has noted that the heat aging of thermoplastic polyether esters can be improved by using a mixed soft segment of dimer acid and poly(ether glycol).

### 3.4.4. Polyester Soft Segments

The use of low molecular mass polyester glycols as soft segments gives thermoplastic copolyester elastomers with improved heat aging characteristics and stability toward UV degradation compared to their polyether counterparts. However, the preparation of these products presents special problems because of the randomization that can occur between the hard and soft ester segments during synthesis. This negatively affects formation of blocks, and the resulting polymers are low melting with poor mechanical properties [3.24].

One approach to solving this problem is to react preformed hard and soft polyester segments in the presence of a titanate catalyst until the reaction mass becomes clear. At that point the catalyst is deactivated by addition of a phosphorus compound [3.25]. On further processing into end use articles, very little ester exchange occurs between the hard and soft blocks, and as a result, parts with good mechanical properties are obtained. Generally, better results are obtained by chemical coupling of the blocks with diisocyanates [3.26] or acylating agents such as diacyl-bis-(N-lactams). A detailed description of diisocyanate coupling between poly(tetramethylene terephthalate) and various aliphatic polyester diols has been published by VAN BERKEL [3.24]. After deactivating the catalyst used for formation of 4 GT, coupling is carried out by using either toluene diisocyanate or methylene diisocyanate. A slight excess of isocyanate is used to obtain optimum mechanical properties. These thermoplastic polyester ester products show improved heat aging and UV resistance, but also have poor hydrolytic stability and stiffen much faster than polyether ester elastomers at low temperature.

## 3.5. Properties

**Physical and Chemical Properties.** Polymers prepared by the method described in Section 3.3 have the molecular mass distribution expected from a condensation polymerization. A number-average molecular mass in the range of 20 000 25 000 can be obtained. The two-phase hard–soft segment structure of these segmented polyesters has been demonstrated clearly by differential scanning calorimetry [3.27]. By varying the ratio of these two segments, thermoplastic polyesters ranging from soft, transparent rubber-like elastomers to semirigid plastics can be prepared. Melting points ranging from 140 to 220 °C have been reported for polymers containing 20–80 wt % 4 GT [3.28]. As expected, the glass transition temperatures of this same series of polymers increase from − 70 to 25 °C.

Thermoplastic copolyester elastomers have good to excellent resistance to hydrocarbons. Products containing > 50 % 4 GT are particularly well-suited for contact with oil, grease, or hydraulic fluids. These polymers offer better resistance to hot hydrocarbons than nitrile rubber.

**Resistant to**
Nonpolar fluids
  Oil
  Fuels
  Hydraulic fluids
Polar fluids at room temperature
  Glycols
  High molecular mass alcohols
  Weak acids and bases

**Not resistant to**
Hot glycols
Hot alcohols
Strong acids and bases
Amines

Resistance to polar fluids, including water, acids, bases, amines, and glycols, is a function of both the composition of the polymer and the temperature of use. At $> 70\,°C$, polar fluids attack the polymer.

**Mechanical Properties.** The predominant hard segment in use today for thermoplastic copolyester elastomers is poly(tetramethylene terephthalate). WOLFE [3.13] has shown that a hard segment content of 30 to 80 wt % produces a wide range of useful products (Table 3.3). A complete summary of mechanical properties for these elastomers has been published [3.29].

## 3.6. Blends

The low melt viscosity and good processing characteristics of thermoplastic elastomers facilitate their use in polymer blends to give products that are otherwise not easily obtainable. In many instances the properties of these blends are an average of the properties of the individual components, for example, the combination of

**Table 3.3.** Properties of tetramethylene terephthalate–PTMO terephthalate copolymer as a function of tetramethylene terephthalate content [3.13]

| Properties | Percentage 4 GT | | | | |
|---|---|---|---|---|---|
| | 30 | 40 | 50 | 57.4 | 80 |
| Melting point, °C | 152 | 172 | 189 | 197 | 212 |
| Shore D hardness | 34 | 44 | 48 | 54 | 70 |
| Stress at 100% elongation, MPa | 5.9 | 8.3 | 11.7 | 13.4 | 24.4 |
| Tensile strength, MPa | 24.6 | 31.0 | 48.4 | 46.7 | 54.5 |
| Elongation, % | 900 | 830 | 755 | 660 | 535 |
| Tear strength, kN/m | 13 | 26 | 48 | 63 | 108 |

poly(vinyl chloride) (PVC) and a polyester based on 4 GT and PTMO [3.30], [3.31]. The addition of 10–30 % of thermoplastic polyester elastomer provides PVC with improved heat resistance, low-temperature flexibility, and resistance to oils and fuels. Another important advantage is that the polyester elastomer cannot be extracted from the blend, which extends the service life of PVC molded parts.

In other instances, blends offer improved physical properties that are not obtainable by direct synthesis. Blending poly(butylene terephthalate) with a thermoplastic copolyester elastomer containing 58 wt % hard segment provides a polymer with improved low-temperature impact resistance compared to a directly synthesized polymer of identical composition [3.30]. The improvement in low-temperature properties is said to result from the incompatibility of the components of the resultant blend.

High-melting, crystalline thermoplastic copolyester elastomers can be softened (thus improving their low-temperature properties) by blending with an amorphous or very low melting polyester made from PTMO, 4 G, and a diacid [3.32]. Since the melting point of the resultant blend is similar to that of the unblended crystalline thermoplastic, evidence that the two components of the blend are not miscible is good. Soft thermoplastic copolyester elastomers can also be prepared by blending the polyester with ethylene copolymers. SHIH [3.33] reports blends with EPDM that retain the melting point of the polyester yet exhibit reduced stress–strain properties. In an extension of this work, blends of polyesters have been prepared with ethylene–carboxylic acid copolymers that had been partially neutralized [3.34]. This novel combination of polymers provides blends with enhanced melt viscosity, which greatly improves the processibility by blow molding and film extrusion of the thermoplastic copolyester elastomers.

Very efficient mixing is required to achieve good disperson of polyester homopolymers and thermoplastic copolyester elastomers because of their incompatibility [3.30]. If the materials being blended differ greatly in polarity, homogeneous blends can sometimes not be achieved unless *compatibilizers* are employed. These compatibilizers generally contain reactive functional groups to assist in dispersing the components of the blend. In two separate examples, polycarbonate resins having improved flexibility and impact strength were obtained by blending with copolyester elastomers. To obtain optimum properties, polystyrene grafted with organosiloxanes or styrene – butadiene copolymerized with acrylic acid derivatives has been used to compatibilize the blend components [3.35], [3.36].

## 3.7. Uses

Thermoplastic copolyester elastomers are available in grades ranging from soft, rubber-like to semirigid plastics. This provides customers with products covering a wide spectrum of load-bearing characteristics, flexibility, and low-temperature properties. These materials are simple to process and can be made into high-performance products by a variety of polymer-forming processes, including injection molding, extrusion, blow molding, rotational molding, and melt casting. All thermoplastic copolyester elastomers have sharp, well-defined melting points, crystallize rapidly, and have good melt stability.

Grades having a Shore D hardness of 50 – 80 have excellent chemical resistance, which makes them suitable replacements for specialty rubber such as nitrile and epichlorohydrin rubbers in hose and tubing applications. They can be extruded into long, continuous lengths with relatively low wall thickness, thereby providing lightweight constructions. Compared to competing flexible polyamide compounds, no plasticizer is required, so the material remains flexible over long periods of time. Overall, hose construction costs using thermoplastic copolyester elastomers are comparable to those using conventional rubber. This is due to the less expensive plastics processing techniques involved and the fact that thermoplastic copolyester elastomers require no postcuring [3.29].

The good impact strength and toughness, dimensional stability, and good paintability of thermoplastic copolyester elastomers have been utilized in automotive applications to replace metal in exterior automotive bumpers and front and rear end spoilers. The use of thermoplastic copolyester elastomers is not limited, however, to the exterior of automobiles. Overmolded interior parts such as door handles having good "feel" properties also have been commercialized [3.37].

In a very demanding application requiring excellent flexural fatigue resistance, thermoplastic polyesters have replaced polychloroprene rubber in constant-velocity joint boot bellows for front-wheel-drive automobiles. In addition to the necessary flexibility, this application also requires a product with good thermal properties, high tear resistance, very good low-temperature properties, and resistance to lubricating greases [3.38].

Thermoplastic copolyester elastomers containing either EOPPG or PEO soft segments absorb significant quantities of water. Because of this characteristic, thin films produced from these materials have high moisture vapor transmission rates. In addition to their use in laminated films for moisture-proof sports and rain wear [3.39], [3.40], these materials have also found application in the medical field [3.41]. By varying the EOPPG or PEO content, a wide range of moisture vapor transmission rates can be obtained. These products are suitable for medical wound care treatment, transdermal patches, and surgical drapes [3.41].

## 3.8. Producers, Trade Names

Thermoplastic copolyester elastomers were originally introduced by DuPont in 1972 under the trade name Hytrel and are manufactured by DuPont in the United States and Luxembourg, and by DuPont – Toray in Japan. Other U.S. producers include General Electric (trade name Lomod), Hoechst – Celanese (trade name Riteflex), and Eastman Chemical Products (trade name Ecdel). Both copolyether esters and copolyester esters are manufactured by DSM in the Netherlands, under the trade name Arnitel, and by Toboyo in Japan, under the trade name Pelprene. Pibiflex is manufactured in Europe by Montedison. More recent manufacturers entering this field include two South Korean producers, Kolon Industries (Kopel) and Cheil Synthetics (Esrel).

# 4. Thermoplastic Polyamide Elastomers

## 4.1. Introduction

Thermoplastic polyamide elastomers (PAEs) are segmented block copolymers—the hard blocks consisting of aliphatic polyamides, the soft segments of aliphatic polyethers. Both segments are linked by ester or amide groups.

Polyamide thermoplastic elastomers were first introduced in Western Europe in the early 1980s by Ato-Chimie (today Elf Atochem) under the trade name Pebax. DELLENS [4.1] had studied the synthesis and properties of this family of TPEs in the 1970s. In polymers of the Pebax type, the same polymer chain can contain different types of polyamides (PA 6, PA 12, PA 11, PA 66, etc.) and all kinds of polyethers or co-polyether glycols derived from THF, and from propylene or ethylene oxide [4.2]. Hard and soft segments are linked by ester bonds.

Another approach [4.3] has been used by Hüls to obtain a PAE family marketed under the trade name Vestamid. These TPEs are produced from lactam 12, dodecanedioic acid, and poly(tetramethylene oxide).

The hard segments of thermoplastic polyamide elastomers produced by EMS (Grilamid ELY and Grilon ELX) are also based on lactam 12.

More recently, at the end of the 1980s, Ube began the development of a new type of PAE, the link between the polyamide and the polyether moiety being an amide group. To achieve this, Ube uses polyethers with amino end groups, especially oligomers produced by Huntsman (trade name Jeffamine).

## 4.2. Production

### 4.2.1. Synthesis of Pebax

Ato-Chimie [4.2], [4.4], [4.5] produces Pebax by a two-stage process. In the first step, an oligoamide with carboxylic end groups ("dicarboxylic oligoamide") is synthesized by polycondensation of a lactam, an amino acid, or a mixture of diacid and diamine (or its salt) with a diacid, according to the following scheme:

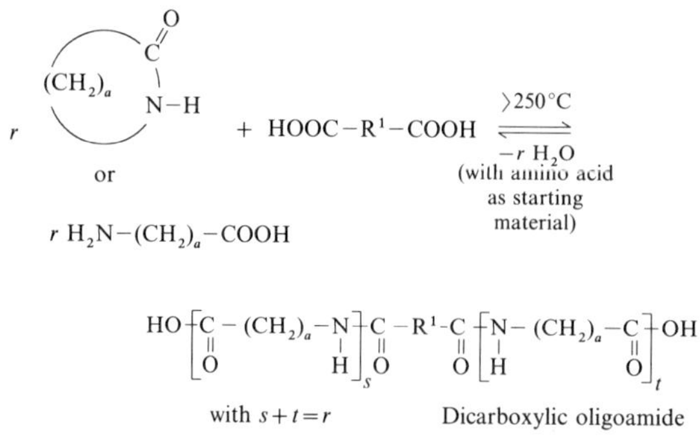

$$r \quad \underset{\text{(CH}_2)_a}{\overset{\text{C}=\text{O}}{\diagdown \text{N}-\text{H}}} \quad + \quad \text{HOOC}-\text{R}^1-\text{COOH} \quad \underset{-r\,\text{H}_2\text{O}}{\overset{>250\,°\text{C}}{\rightleftharpoons}}$$

or

(with amino acid as starting material)

$$r\,\text{H}_2\text{N}-(\text{CH}_2)_a-\text{COOH}$$

$$\text{HO}\left[\overset{\text{C}-(\text{CH}_2)_a-\text{N}}{\underset{\text{O} \qquad\qquad\quad \text{H}}{\|}}\right]_s \overset{\text{C}-\text{R}^1-\text{C}}{\underset{\text{O} \qquad \text{O}}{}}\left[\overset{\text{N}-(\text{CH}_2)_a-\text{C}}{\underset{\text{H} \qquad\qquad\quad \text{O}}{}}\right]_t\text{OH}$$

with $s + t = r$            Dicarboxylic oligoamide

The molecular mass of the oligomer depends on the value of $r$ and the nature of the monomers. In the second step, the dicarboxylic oligoamide is condensed with the polyether glycol at lower temperature (200–250 °C) under reduced pressure ($< 133$ Pa) in the presence of titanium or zirconium catalysts.

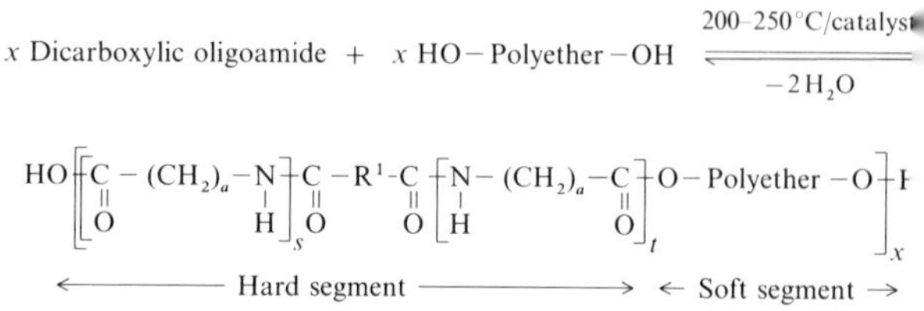

$$x\,\text{Dicarboxylic oligoamide} \ + \ x\,\text{HO}-\text{Polyether}-\text{OH} \ \underset{-2\,\text{H}_2\text{O}}{\overset{200\text{–}250\,°\text{C/catalyst}}{\rightleftharpoons}}$$

$$\text{HO}\left[\left[\overset{\text{C}-(\text{CH}_2)_a-\text{N}}{\underset{\text{O} \qquad\qquad\quad \text{H}}{\|}}\right]_s\overset{\text{C}-\text{R}^1-\text{C}}{\underset{\text{O} \qquad \text{O}}{}}\left[\overset{\text{N}-(\text{CH}_2)_a-\text{C}}{\underset{\text{H} \qquad\qquad\quad \text{O}}{}}\right]_t\text{O}-\text{Polyether}-\text{O}\right]_x$$

$\longleftarrow$ Hard segment $\longrightarrow$   $\leftarrow$ Soft segment $\rightarrow$

The properties of Pebax can be modified by

1) The type of polyamide incorporated, which determines melting point, chemical resistance, and density of the block copolymer.
2) The type of polyether used, which influences hydrophilic and antistatic properties and chemical resistance.
3) The length (or molecular mass) of the polyamide block, which is important for the melting behavior.
4) The ratio of polyamide and polyether segments, which determines the flexibility and hardness of the copolymer. By varying the content of polyether between 10 and 90%, Pebax with different flexibilities can be obtained using the same types of polyamide and polyether blocks.

### 4.2.2. Synthesis of Other Thermoplastic Polyamide Elastomers

**Hüls Process.** In the Hüls process, lactam 12, dodecanedioic acid, and poly(tetramethylene oxide) (PTMO) are used as raw materials. All starting materials are charged to the reactor at once, and polycondensation is carried out in the presence of water [4.3]. During the first step, the reactor is kept at high temperature ($> 250$ °C) to convert the lactam into the corresponding amino acid. Subsequently, the pressure is reduced, water is distilled off, and the molecular mass of the copolymer increases by polyesterification and polyamidation. The copolymers obtained have a statistical distribution of hard and soft segments over the chain [4.3].

**Ube Process.** In 1983, Ube patented several processes for the production of PAEs. Two different synthetic approaches have been described:

1) One-step process: Monomers that build up the polyamide phase and polyethers with amino end groups are polycondensed simultaneously [4.6].

2) Two-step process: After preparation of an oligoamide with carboxyl end groups, the di-amine polyether is polycondensed with this oligomer [4.7]. The possibility of using a mixture of an aliphatic diamine and a polyether has also been described [4.8]. This type of PAE has the general formula:

$$\text{H-O}\left[\text{C} - \text{Polyamide} - \text{C} - \text{N} - \text{Polyether} - \text{N}\right]_x \text{H}$$

Hard segment        Soft segment

The special feature of Ube-PAE is that a dimer acid (derived from a fatty acid) is incorporated in the polyamide chain, linking two amino acid units and acting as a chain limiter.


## 4.3. Properties

### 4.3.1. Properties of PAEs

Thermal transitions in each of the two phases can be detected by differential scanning calorimetry (DSC). Figure 4.1 depicts DSC thermograms of PAEs—prepared from oligoamide 12 (PA 12) and poly(tetramethylene oxide)—that differ in the molecular mass of the polyether segments. Figure 4.1 shows the following:

1) The melting point of the polyamide units is not modified by changing the molecular mass of the polyether segments

2) The melting point of the polyether units disappears when their molecular mass decreases below a certain value ($M_r < 1000$)

3) The crystallinity of each phase changes with the ratio of polyether segments to polyamide segments

**Pebax.** The main characteristics of Pebax grades are [4.9]:

1) High flexibility and impact resistance at low temperature (high resilience)

2) Favorable dynamic properties (no break in alternating flexion–relaxation tests, low hysteresis)

3) Easy processing

The properties of Pebax are listed in Table 4.1. Figure 4.2 shows the effect of temperature

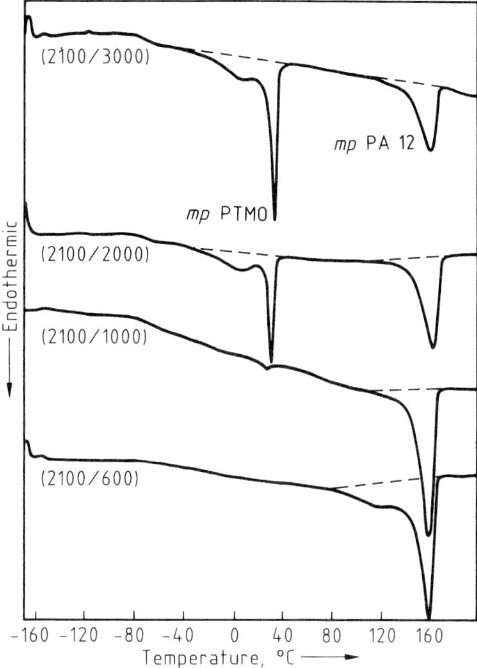


**Figure 4.1.** DSC thermograms of PAEs produced from oligoamide 12 and poly(tetramethylene oxide) [4.10]
The figures denote the molecular mass ratio of PA 12 to PTMO segments.

on the flexural modulus of elasticity of Pebax grades. Pebax retains almost all of its flexibility at low temperature and maintains its excellent mechanical properties down to $-60\,°C$.

**Ube-PAE.** Properties of PAE produced by Ube are given in Table 4.2. The Shore D hardness can vary between 42 and 71. In comparison with Pebax grades, Ube products have poorer mechanical properties, especially a lower tensile strength (see Tables 4.1 and 4.2).

**Vestamid.** Mechanical properties of Vestamid grades are listed in Table 4.3. Hüls offers only PAEs with Shore D hardness in the range of 40–60.

### 4.3.2. Comparison with Thermoplastic Copolyester and Thermoplastic Polyurethane Elastomers

The density of thermoplastic polyamide elastomers (density of Pebax grades of series 33 1.01–1.15 g/mL) is lower than that of TPUs (1.10–1.24 g/mL) and of thermoplastic copolyester elastomers (1.17–1.25 g/mL).

**Table 4.1.** Properties of Pebax grades [4.9]

| Grade* | Shore D hardness | Density, g/mL | Melting point, °C | Tensile strength at break, MPa | Elongation at break, % | Modulus of elasticity in flexure, MPa | Impact strength (IZOD) notched at −40 °C, J/m |
|---|---|---|---|---|---|---|---|
| 2533 SN 01 | 25 | 1.01 | 133.5 | 29 | 790 | 15 | no break |
| 3533 SN 01 | 33 | 1.01 | 143.5 | 30 | 670 | 25 | no break |
| 4033 SN 01 | 42 | 1.01 | 160 | 36 | 450 | 84 | no break |
| 5533 SN 01 | 55 | 1.01 | 159 | 48 | 455 | 170 | no break |
| 6333 SN 01 | 63 | 1.01 | 169 | 55 | 410 | 300 | 100 |
| 7033 SN 01 | 69 | 1.02 | 172 | 52 | 440 | 430 | 50 |
| 4011 MA 01 | 42 | 1.14 | 204 | 32 | 700 | 85 | 90 |
| 5562 MA 01 | 55 | 1.06 | 120 | 51 | 510 | 210 | 90 |

* In Pebax nomenclature, the first dwo digits correspond to the hardness (Shore D scale). The other two refer to the series. The first letter gives information about applications (S = all uses, M = molding, E = extrusion), and the second one stabilization (A = no additives, N = UV stabilized, D = UV stabilized and mold release additive, T = heat stabilized). The last two numbers refer to fillers (0 = without) and formula codes.

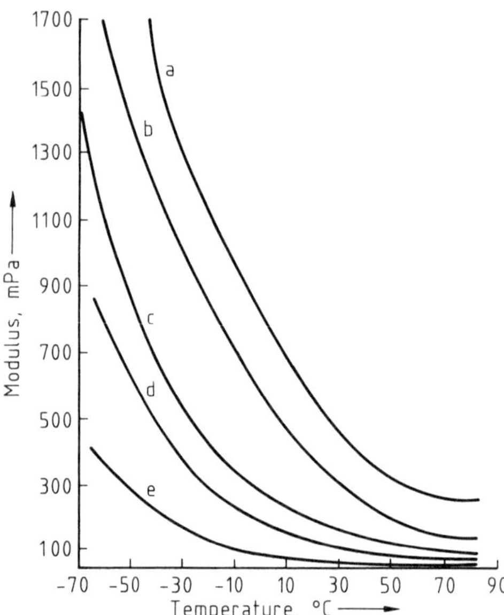

**Figure 4.2.** Flexural modulus of elasticity vs. temperature for the Pebax 33 series [4.9]
a) Pebax 70D; b) Pebax 63D; c) Pebax 55D; d) Pebax 40D; Pebax 35D; e) Pebax 25D

**Table 4.2.** Properties of Ube-PAE [4.11]

| Grade | Shore D hardness | Melting point, °C | Tensile strength at break, MPa | Elongation at break, % |
|---|---|---|---|---|
| PAE 1203 U1 | 71 | 171 | 25 | >300 |
| PAE 1201 U1 | 65 | 165 | 19 | >300 |
| PAE 1200 U1 | 56 | 154 | 9 | >300 |
| PAE 1202 U1 | 42 | 137 | 5.6 | >300 |

**Table 4.3.** Properties of Vestamid grades [4.12]

| Grade | Shore D hardness | Modulus of elasticity in traction, N/mm² | Tensile strength at yield, MPa | Impact strength (IZOD) notched at −30 °C, J/m |
|---|---|---|---|---|
| E62M-S3 | 62 | 360 | 24 | 80 |
| E55M-S3 | 55 | 260 | | 210 |
| E47M-S3 | 47 | 170 | | no break |
| E40-S3 | 40 | 120 | | no break |

Figure 4.3 shows the effect of temperature on the torsional modulus of rigidity of Pebax with Shore D hardness 55, compared to a polyurethane and a polyether–ester elastomer with the same hardness. The freedom of design offered by the chemical structure of PAE bridges

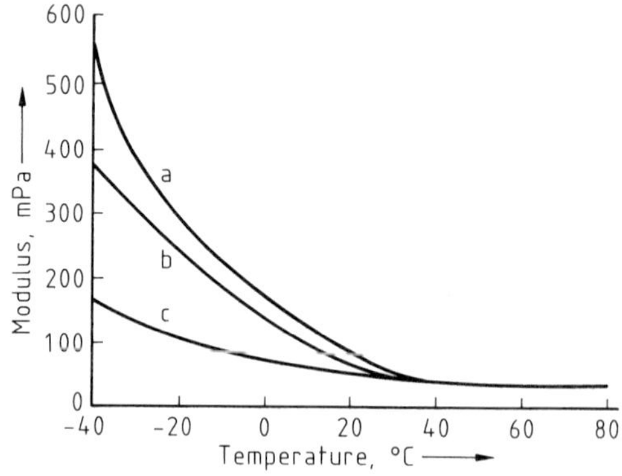

**Figure 4.3.** Torsional modulus of rigidity vs. temperature of various thermoplastic elastomers with same Shore hardness (Shore D 55) [4.9]
a) TPU; b) Thermoplastic polyether–ester elastomer; c) Pebax

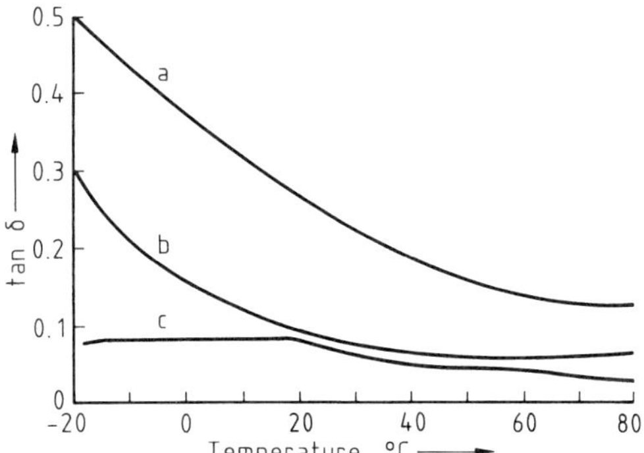

Figure 4.4. Comparative cyclic compression test of various thermoplastic elastomers with same Shore hardness (Shore D 40) [4.9]
Deformation 0.1 %, frequency 500 Hz
a) TPU; b) Thermoplastic copolyester elastomer; c) Pebax

the gap between polyamides and rubbers (hardness of PAEs ranging from Shore D 75 to Shore A 60).

The variation of $\tan\delta$ with temperature for Pebax, a thermoplastic copolyester elastomer, and a TPU during a cyclic compression test is shown in Figure 4.4. The curve shows that the $\tan\delta$ of thermoplastic copolyester and thermoplastic polyurethane elastomers is higher at lower temperature than that of a Pebax with the same hardness (Shore D 40). During cyclic loading, thermal energy (proportional to $\tan\delta$) is dissipated in the material. The resulting increase in temperature changes the vibration-damping properties of the polymer. Because of its small variation of $\tan\delta$ with temperature, Pebax is better suited to applications involving cyclic loading.

## 4.4. Uses

Like other TPEs, thermoplastic polyamide elastomers are used in applications such as sporting goods, hoses, belts, cable jacketing, and automotive components.

## 4.5. Economic Aspects, Producers, Trade Names

In 1991, world consumption of PAEs was as follows [4.13]

| United States | 800 t |
|---|---|
| Western Europe | 5000 t |
| Japan | 600 t |
| Total | 6400 t |

Producers and trade names of PAEs are listed in Table 4.4

# 5. Styrenic Block Copolymers

## 5.1. Introduction

Thermoplastic elastomers known as styrenic block copolymers (SBC) have emerged from developments and research in the organometallic-initiated polymerization of diene monomers in hydrocarbon solvents ($\rightarrow$ Rubber, 3. Synthetic, **A 23**, p. 269–273). Block copolymers currently available on the market are essentially based on styrene, butadiene, and isoprene monomers. The first commercial quantities were produced by Shell in 1964 in its production plant in Torrance,

Table 4.4. Producers, trade names, and structures of commercially available thermoplastic elastomers

| Producer | Trade name | Hard segment | Soft segment |
|---|---|---|---|
| Elf Atochem* | Pebax | PA 12, PA 6 | poly(tetramethylene oxide) |
| Atochem North America** | | | poly(oxypropylene oxide) |
| | | | poly(oxyethylene oxide) |
| EMS Chemie* | Grilamid ELY | PA 12 | poly(tetramethylene oxide) |
| EMS American Grilon** | Grilon ELX | | |
| Hüls | Vestamid | PA 12 | poly(tetramethylene oxide) |
| Ube | Ube-PAE | aliphatic polyamide | polyether with amino end groups |
| Mitsubishi Kasei | Novamid PAE | polyamide | polyether |

* in Western Europe. ** in the United States.

Calif., and marketed under the trade name Kraton [5.1]. Worldwide capacity for styrenic block copolymers in 1993 was estimated at 500 000 t/a.

Styrenic block copolymers combine rubber performance with thermoplastic processibility on the basis of their specific structure. Anionic polymerization allows the selective polymerization of butadiene or isoprene and styrene in such a way that the resulting polymer is essentially a polybutadiene or polyisoprene molecule tipped at both ends with a polystyrene molecule [5.2]. Such a structure is commonly identified as ABA, where A represents the plastic and B the elastomer. More specific notations used are SBS or SIS polymers. Styrenic block copolymers with saturated rubber blocks are produced from SIS or SBS precursors by selective hydrogenation. This process converts the polyisoprene into poly(ethylene–propene) and the polybutadiene into poly(ethylene–butene). To provide rubber performance, the volume fraction of plastic (polystyrene) in these structures must be < 50%, and commercial hydrogenated grades usually contain around 30%. Newer variants of the styrene–ethylene/butene–styrene S–E/B–S or styreneethylene/propenestyrene S–E/P–S polymers are functionalized with maleic anhydride or other reactive comonomers. Functionalization levels are between 0.2 and 5% and the reactive group is grafted onto the rubber mid block [5.3].

## 5.2. Synthesis

The first step in the production of SBCs is activation of purified styrene in a hydrocarbon solvent with butyllithium and subsequent anionic polymerization. The molecular mass of the first polystyrene block is determined by the butyllithium:styrene ratio (see Fig. 5.1). The formed living polymer, PS–Li, is then reacted with purified butadiene to form the second block. When all the diene has been converted, two routes can be followed to complete the formation of SBC:

1) Addition of a coupling agent to produce linear or branched copolymers
2) Additon of a second amount of styrene monomer to form the third block

Technically feasible coupling agents include di-, tri-, tetra-, or multifunctional compounds that allow formation of a multitude of different polymer structures [5.4].

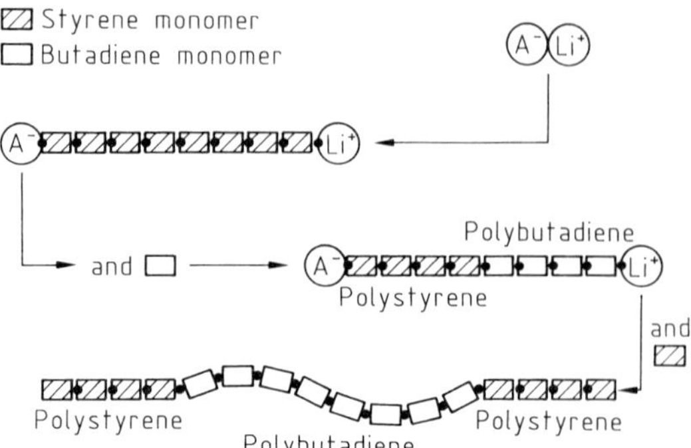


**Figure 5.1.** Formation of block copolymers by anionic polymerization (A⁻ = butyl anion)

The procedure for production of the precursor for *saturated block polymers* is the same. To ensure a rubbery mid block after hydrogenation in the S–E/B–S polymer, a modifier is added during polymerization of the polybutadiene mid block to increase its vinyl content (1,2-content) (→ Rubber, 3. Synthetic, pp. 273–274) to approximately 40% which will ensure optimum elastic properties after hydrogenation. This allows optimization of the ratio of ethylene and butene units in the hydrogenated polymer. After preparation of the precursor, polymer solution is transferred to the hydrogenation section where the double bonds of the polydiene are reacted with hydrogen in the presence of a catalyst.

*Functionalized saturated block copolymers* are made by extruder grafting of an S–E/B–S or S–E/P–S molecule. The polymer is contacted in an extruder with a peroxide to create free radical sites along the rubber mid block. These react in the extrusion step with a compound containing the functional group and a group able to react with the free radical sites (e.g., maleic anhydride). The process requires careful optimization of the amount of peroxide and reaction conditions to minimize polymer degradation.

The result of the polymerization process is a solution of polymer in hydrocarbon solvent. At this stage, additives such as stabilizers, antioxidants, or plasticizers can be added to the solution which is subsequently converted into solid polymer during finishing. The following systems are used industrially:

**Direct Desolventizing.** The polymer solution (cement) is introduced directly into a special vacuum extruder; the solvent is evaporated into a recovery system and the extruded polymer melt is face-cut and cooled at the die head of the extruder to form compact transparent pellets.

**Coagulation with Steam.** Cement is pumped to a vessel where steam is passed into the solution. The resulting coagulate separates from the solvent, and the polymer–water mixture is dried to remove water. The usual form resulting from the coagulation–drying route is referred to as crumb.

## 5.3. Properties

Thermoplastic rubbers based on styrenic block copolymer can be considered polybutadiene [or polyisoprene, poly(ethylene–butene), or poly(ethylene–propene)] tipped with polystyrene end blocks. Polybutadiene (butadiene rubber, BR) and polystyrene of sufficiently high molecular mass are incompatible, which forms the basis for the performance of these thermoplastic rubbers. At temperatures above the softening point of polystyrene the SBC molecule is quite mobile and can be melt processed by, e.g., extrusion or injection molding. On cooling, the PS end blocks separate from the BR mid blocks, and two distinct phases are formed. The separate, discrete phase of the PS domains acts both as a cross-linking material and as a reinforcing filler analogous to the sulfur and carbon black in vulcanized rubber. Figure 5.2 shows an SBC where the PS domains hold together a rubbery BR network. The tensile strength (Table 5.1) and resilience data (Table 5.2) illustrate the truly rubbery behavior of a linear SBS polymer with a polystyrene content of 30 %.

**Table 5.1.** Comparison of tensile strengths of BR and SBS

| Polymer | Tensile strength, MPa |
| --- | --- |
| Unvulcanized BR | 0.2–0.5 |
| Vulcanized BR | 4.0–5.0 |
| Vulcanized + carbon black reinforced BR | 5.0–20.0 |
| Unvulcanized pure SBS | 30.0–35.0 |

**Table 5.2.** Resilience (Lüpke-rebound) of different polymers

| Polymer | Resilience, % |
| --- | --- |
| Unvulcanized SBS | 65 |
| Vulcanized BR | 65 |
| Unvulcanized SIS | 60 |
| Vulcanized NR/IR | 60 |
| Vulcanized SBR | 50 |
| Vulcanized EPDM | 45 |
| Unvulcanized EVA (28 % VA) | 40 |

The *molecular structure* of an SBC can be varied by changing the proportion of PS in the base polymer. Furthermore the total molecular mass can be varied, but practical limits exist to the processability in the melt. Figure 5.3 shows the effect of altering polystyrene content in a series of linear polymers. Polymers can be made that range from very low strength at low (10 %) PS content, through rubbery materials of about 30 % PS, to those with 80 % PS content that behave much like toughened polystyrene. SBC polymers can be linear or branched; branched types are usually higher in molecular mass than linear ones and have a higher melt viscosity at equal strength levels. Furthermore, a minimum molecular mass of the end block exists below

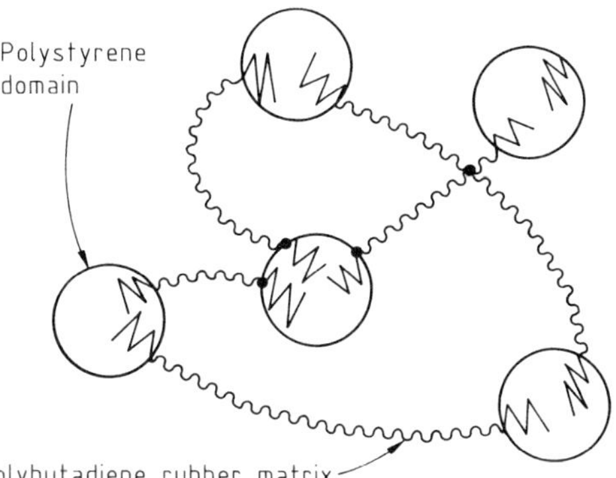


**Figure 5.2.** Schematic phase structure of an SBS block copolymer

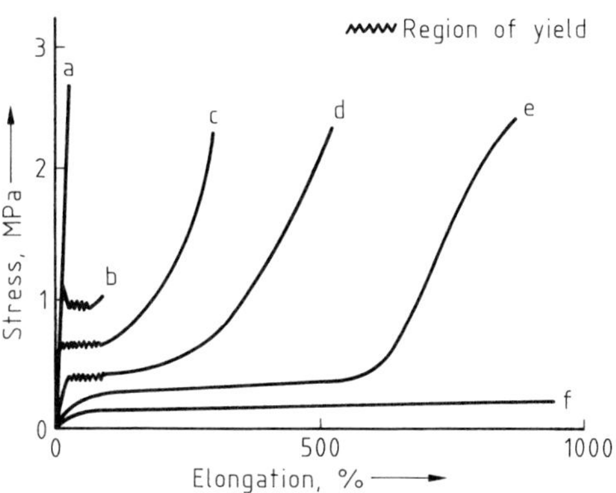


**Figure 5.3.** Stress–strain curves for various SBS copolymers
a) 80 % styrene; b) 65 % styrene; c) 50 % styrene; d) 40 % styrene; e) 30 % styrene; f) 10 % styrene

**Table 5.3.** Properties of styrenic block copolymer structures with different PS block and total molecular mass

| Polymer structure | Molecular mass | | Viscosity | Strength |
| --- | --- | --- | --- | --- |
| | PS block | Total | | |
| Linear | 12 000 | 80 000 | low | high |
| Branched | 12 000 | 160 000 | medium | high |
| Branched | 6000 | 80 000 | very low | low |

which no phase separation occurs, no domain structure forms, and hence unsatisfactory properties are obtained [5.5]. A comparison of properties of polymers differing in the molecular mass of the PS blocks and in total molecular mass is given in Table 5.3.

The *processing properties* of SBCs can be derived from their basic chemical structure. The molecular mass of commercially available grades varies from about 75 000 to 400 000. Their behavior in the molten state is similar to that of polyethylene and polystyrene, for example. However, even in the molten state the tendency to form PS domains persists, especially for those types with a saturated rubber mid block. *Melt processing SBCs* at higher shear rates leads to lower apparent viscosity; this effect is relatively independent of temperature (see Fig. 5.4). Reprocessing of SBCs at < 200 °C has little effect on their properties; saturated grades can be processed and reprocessed at up to 280 °C.

*Solution processing* of SBCs is possible, provided that solvents of sufficiently high solubility are used to dissolve the PS end blocks. Solvents with solubility parameters between 8 and 10 are usually employed. Because of the easy solubility of SBCs, adhesive, sealant, and coating systems can be designed that have low viscosity and high solids content. A consequence of their good solubility in organic solvents is that the oil and solvent resistance of parts based on SBC is limited. However, their resistance to dilute aqueous solutions of the most commonly encountered chemicals is generally satisfactory.

The *service performance* of SBCs and the products derived from them is also related to their basic chemical structure. Because of their moderate molecular mass and their physical and reversible cross-linking, they flow under pressure and exhibit creep. Block copolymers with *BR or IR mid blocks* are unsaturated and sensitive to aging. Exposure to elevated temperature can result in polymer degradation with simultaneous decrease in viscosity and tensile strength. With the exception of black compounds, exposure to UV light for any length of time will result in discoloration and surface cracking, which becomes more pronounced on prolonged exposure. These block copolymers are also susceptible to ozone attack, and in shaped products at points with local stresses, microcracks develop on exposure to atmosphere containing ozone. The use of antioxidants, UV stabilizers, and antiozonants reduces these degradation effects to workable limits in end uses.

The *hydrogenated S−E/B−S or S−E/P−S types*, in contrast, have a saturated rubber mid block and are therefore highly resistant to degradation by oxidation, ozone, and UV light. They can be processed over a much wider processing temperature range.

## 5.4. Uses

Because of their rubbery characteristics (especially at low temperature), high strength, compoundability, and thermoplastic nature, SBCs are used in a variety of applications as highlighted in Figure 5.5.

On the left-hand side of Figure 5.5 the list includes outlets that use SBC as an additive or modifier for other materials. The polymers are delivered to the end user who will incorporate them in his existing process. The right-hand column lists applications that utilize a compounded product with performance tailored to the specific end use and plastic conversion tech-

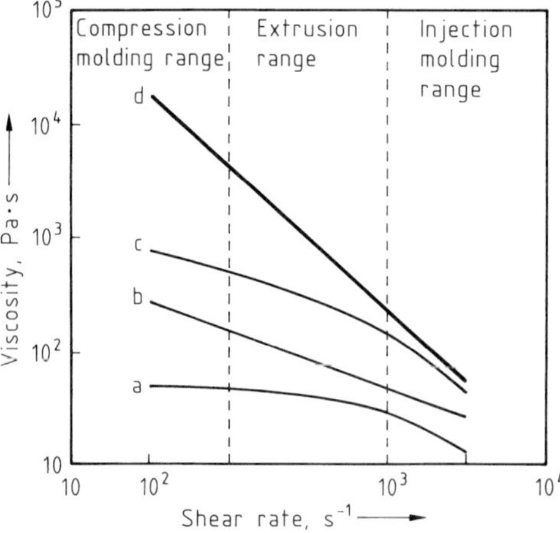


**Figure 5.4.** Viscosity–shear rate relationship at 200 °C
a) Clear polystyrene; b) SBS; c) polypropylene melt flow rate (MFR) = 5; d) S−E/B−S

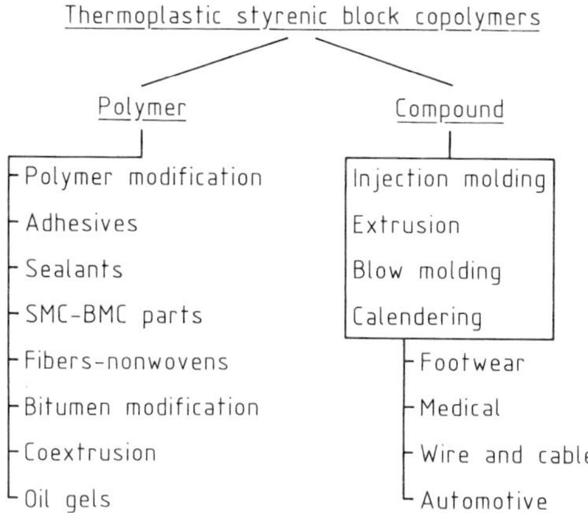


**Figure 5.5.** Applications of styrenic block copolymers SMC = sheet molding compounds; BMC = bulk molding compounds

nology. A comprehensive review of the applications of thermoplastic rubber is given in [5.6], [5.7].

### 5.4.1. Adhesives, Sealants, and Coatings

Adhesive properties are determined primarily by the styrenic block copolymer type chosen for a typical application. SBS polymers with high cohesive strength (high polystyrene content) are commonly used in contact adhesives, whereas SIS polymers with low polystyrene content that can be tackified easily are preferred for pressure-sensitive adhesives. For applications in hot melt form, low molecular mass polymers with a high melt flow rate are selected. When high UV resistance is required, hydrogenated block copolymers should be considered as base polymer. Furthermore, adhesive properties are also determined by the nature and the concentration of the various resins and plasticizers used and their degree of compatibility with the respective mid or end block phase of the block copolymer [5.8].

*Pressure-sensitive adhesives* are materials that remain permanently tacky and adhere instantaneously to solid surfaces on the application of light pressure. Major applications are tapes and labels. Tapes require high shear strength and holding power, while labels require less shear strength but good convertibility. For general-purpose solvent-based pressure-sensitive adhesives, SIS block copolymers are combined with resins that are compatible with the mid block, such as wood rosins, polyterpenes, or synthetic

hydrocarbon resins. A typical composition is shown in Table 5.4, formulation no. 1. The formulation for an adhesive used for permanent labels is given in Table 5.4, formulation no. 2. Pressure-sensitive adhesives for peelable labels have a relatively high cohesive strength and a low peel strength (Table 5.4, no. 3). Labels that are applied under cold conditions must have a low $T_g$, and formulations usually contain low amounts of high-melting resin but large proportions of plasticizer (Table 5.4, no. 4).

In outdoor applications where the adhesive must withstand UV light, saturated block copolymers are employed. Pressure-sensitive adhesives with very high temperature resistance can be obtained by UV or electron beam curing ($\rightarrow$ Radiation Chemistry, **A 22**, pp. 479–484). A special polymer developed by Shell for this application is Kraton D-1320X. The shear adhesion failure temperature (SAFT) of a pressure-sensitive adhesive based on this polymer is increased from 80 °C to ca. 140 °C by either UV or electron beam curing [5.9].

*Contact adhesives* are applied as liquid onto one or two substrates; after a certain time to allow partial evaporation of the solvent (open time) the coated substrates are pressed together to form a strong bond. SBCs are ideal for these types of adhesives because they offer high cohesive strength without vulcanization and are readily soluble in many solvents. For some applica-

**Table 5.4.** Pressure-sensitive adhesive formulations and properties

| Composition, property | Formulation number | | | |
|---|---|---|---|---|
| | 1 | 2 | 3 | 4 |
| *Parts by weight* | | | | |
| SIS polymer* | 100 | 100 | 100 | 100 |
| Piccotac 95 | 90 | | 25 | |
| Escorez 1310 | | 120 | | 90 |
| Hercures C 10 | 10 | 50 | 125 | |
| Wingtack 10 | | | | 40 |
| Irganox 1010 | 1 | | | 3 |
| ZDBC | | 3 | 3 | |
| *Properties* | | | | |
| Melt viscosity (175 °C), Pa · s | | 20 | | 60 |
| Rolling ball tack, cm | 2 | 3 | 3 | 1 |
| Probe tack, N | 10 | >10 | | 10 |
| Loop tack, N/25 mm | 25 | 30 | 8 | 27 |
| 180° peel adhesion, N/25 mm | 17 | 15 | 5 | 16 |
| SAFT, °C | 115 | 100 | 80 | 100 |
| $T_g$ (Fox), °C | −20 | −19 | −33 | −29 |

* Kraton D-1107 from Shell Chemicals.

tions, the contact adhesive is allowed to dry completely to a tack-free nonblocking adhesive film. Exposure to heat reactivates the adhesive, and a firm bond forms under pressure. Large surfaces are commonly coated by spraying with a uniform adhesive layer (e.g., on flexible foams used in the furniture and automotive industries). Because of their molecular structures, SBC-based adhesives combine high strength with low viscosity and ease of spraying. A variety of melt adhesives are used in the construction of hygienic products such as disposable diapers, sanitary napkins, etc. These adhesives can be applied by spray or slot-die coating techniques. Selection of the styrenic block copolymer depends on the properties required and the application technique used.

Styrenic block copolymers with *saturated rubber mid block* are ideally suited as base polymers for both hot-melt and solvent-based sealants that require resistance to long-term degradation and weathering [5.10], [5.11].

In several applications, SBCs are used to formulate *protective coatings* based on solvent systems, but the trend is toward solvent-free systems, for example, in chemical milling (→ Imaging Technology, **A 13**, pp. 649–652).

### 5.4.2. Modification of Bitumen

The addition of SBCs to bitumen increases its upper and decreases its lower operating service temperature, which results in substantial performance improvement in applications such as roofing felts, waterproofing membranes, and asphalt pavement. Selective development of SBS polymers has enabled bitumen to maintain its strength and integrity at elevated temperature without flowing and to remain flexible and ductile at temperatures well below the freezing point [5.12]. Moreover, SBCs greatly enhance the lifetime of bitumen-based products (e.g., roofing felts). This is achieved by development of a three-dimensional network. To optimize the effect of SBS addition, bitumen with the right degree of compatibility must be selected to ensure effective formation of the elastomer network. For roofing felts, SBS addition levels are commonly between 8 and 12%. Blending into the bitumen is carried out at 170–190 °C. For fast dissolution, especially when low-shear paddle mixers are employed, SBS polymer is ground to a powder with particle

size of ca. 1000 µm. For most applications, high-shear mixers are used with regular polymer crumb [5.13], [5.14].

The use of polymer-modified bitumen for asphalt pavements is growing rapidly and offers large potentials for SBC in the future. The SBS addition levels for road applications are lower than for roofing: typically 3–7%. In addition to improved deformation and flexibility, laboratory data clearly demonstrate that the durability of asphalt roads is increased by addition of SBS polymer [5.15]. Many test roads have been installed that are being monitored to gain experience and evidence for performance under practical conditions. For documentation of the most important practical experiences so far, see [5.16]–[5.22].

### 5.4.3. Compounded Products

As mentioned earlier, end use parts that are made via established plastic processing techniques do not employ pure SBC polymers but rather — like traditional vulcanized rubber — compounded and end use-tailored compositions. Depending on their requirements, such compounds have hardnesses in the range from 35 Shore A to as high as 55 Shore D. The actual SBC content can be as low as 15 wt% but is usually between 25 and 40 wt%. SIS block copolymers are rarely used as base polymer for compounds. The main classes of compounding ingredients include the following:

**Mineral oil** is used as plasticizer to increase flow and reduce cost. Paraffinic or naphthenic oils are used that compatibilize with the rubber mid block. Aromatic oils or polar plasticizers such as dioctyl phthalate (DOP) are avoided because they strongly plasticize the end blocks and reduce strength.

**Polystyrenes** are important compounding ingredients to increase hardness, strength, and wear resistance, particularly in SBS polymers. Both clear and high-impact polystyrenes are used.

**Polyolefins**, particularly polypropylene, are important ingredients in S–E/B–S and S–E/P–S block copolymers.

**Fillers**, mainly cheapeners and nonreinforcing types, are used to decrease cost and increase hardness. They also increase density, which is exploited in formulations for sound-deadening applications.

**Additives** such as antioxidants, UV stabilizers, and antiozonants, as well as additives that induce special effects, can be used. Examples are pigments, surface-active slip agents, blowing agents, and antistatics.

Compounding is carried out at elevated temperature and at shear conditions that result in a fine, stable dispersion of the various ingredients. Compounding equipment originates from both the rubber industry with two-roll mills, Banbury-type internal mixers, and Farrel continuous mixers, and the plastic processing industry with mainly single- and double-screw compounding extruders. Modern compounding plants are fully instrumented robots in which a twin-screw corotating compounder with a length:diameter ratio ($L:D$) up to 42 is the key unit. Screw types with feeding, kneading, homogenization, and venting sections are designed for the specific compounding operation. Ingredients are added either as preblended mixture or via separate weigh feeders and oil injection pumps. The granulating section at the end of the extruder is important because it is responsible for formation of the homogeneous, free-flowing pellets required in subsequent plastics conversion equipment. Depending on line capacity and product characteristics (i.e., soft or hard), die plates with knife cutters, water-ring pelletizers, underwater pelletizers, or a strand die with water cooling bath and rotating knife granulator, can be used.

Detailed compositions and selection of base polymers have been published by SBC suppliers in the form of technical bulletins or formulating guides. Compounding know-how and formulations based on SEBS or SEPS polymers is rather specific [5.23]. Table 5.5 lists properties of typical SBC-based compounds tailored for selected end uses. The largest volume of SBCs in compounded form is consumed by the footwear industry. Here, SBS-based compounds are established mainly as high-quality soling materials. An important aspect is the multitude of colors that can be combined in shoe design. Compounding aims to optimize the end products with respect to melt flow rate (for ease of processing), high abrasion resistance (for durability), and good resistance to repeated flexing. Hardness for most typical soling materials falls within the Shore A 55 to 75 range.

**Table 5.5.** Properties of compounded SBCs

| Property | Reference no*, base SBC polymer | | | | |
|---|---|---|---|---|---|
| | 1<br>SEBS | 2<br>SBS | 3<br>SEBS | 4<br>SEBS | 5<br>SEBS |
| MFR (E or G), dg/min | 50 G | 35 E | | 3 G | 17 G |
| Density, g/mL | 0.91 | 1.0 | 1.94 | 0.91 | 1.01 |
| Hardness, Shore A/D | 45 A | 65 A | 36 D | 60 A | 82 A |
| Tensile strength, MPa | 5.0 | 7.6 | 4.8 | 7.5 | 17.3 |
| Elongation, % | 610 | 650 | 250 | 480 | 670 |
| DIN abrasion, mm³ | | 180 | | | |
| Haze, % | 7 | | | | |
| UL 62**, Class 34 rating, °C | | | | | 105 |
| Compression set 22 h at 100 °C, % | | | | 42 | |

* 1 = Transparent medical molding; 2 = soling material; 3 = automotive sound deadening; 4 = high-temperature performance; 5 = electrical wire insulation. ** UL = Underwriters Laboratories code.

Compounds designed for use in medical devices and articles intended for food contact are also important outlets and take advantage of the high purity of polymers made by anionic polymerization. SBS- and SEBS-based compounds which meet the requirements set in the United States by the FDA and in Germany by the BGA, are used for syringes, tubing, bottle nipples, stoppers, i.v. bag systems, etc. Contact of SBS and SEBS with fatty food is restricted.

Thermoplastic rubber products can be recycled, and scrap, trimmings, and off-specification parts can to a certain extent be reintroduced in the manufacturing stream.

### 5.4.4. Modification of Plastics

Most plastics are brittle and lack impact resistance, particularly below 0 °C. Rubber is added to achieve

1) Increased impact strength (Izod, Charpy, FWIS, Dart, etc.) especially at low temperature
2) Increased environmental stress corrosion resistance

The addition of a rubber to a thermoplastic also changes properties such as stiffness, tensile strength, yield modulus, surface friction, softening point, clarity, electrical properties, vapor transmission, and gloss.

With many plastics, an impact resistance improving rubber fraction can be incorporated in the polymerization reaction. Examples are propene–ethylene copolymers, high-impact polystyrene (HIPS), and ABS. Nevertheless, styrenic block copolymers are used as melt-blended component in plastics that do not allow reactor modification or to obtain impact performance in excess of that achievable in the reactor. Polystyrenes are often compounded with flame-retarding additives and then require the addition of SBS polymer to compensate for the loss in impact strength due to addition of the flame retardant [5.24]. Engineering plastics require processing temperatures beyond the operating range of SBS. Here, saturated SEBS and SEPS block polymers are used, especially as modifiers in polycarbonate (PC) and poly(phenylene ether) (PPE). The more polar engineering plastics such as polyamides and polyesters can be adequately impact modified by SEBS or SEPS functionalized with maleic anhydride [5.25].

Styrenic block copolymers are also capable of compatibilizing different and dissimilar plastics into useful and workable blends. Addition of SBS or SEBS to a blend of polystyrene and polypropylene or polyethylene, for example, creates a ductile material that can be extruded or molded without the delamination or phase separation ecountered in the absence of block polymer addition [5.26]. Similar results are obtained when SEBS is used in blends with polycarbonate and polypropylene, as well as polyethylene–poly(ethylene terephthalate) (PETP) blends. Functionalized block polymers are good compatibilizers for, e.g., polyamides and polypropylenes. SBCs can also be used in thermoset resin systems, particularly as additives for sheet molding or bulk molding processes with unsaturated polyesters. Moldings modified with specifically developed SBS or SEBS polymers to control shrinkage show high-quality surface appearance [5.27], [5.28].

The use of SBCs as compatibilizers for mixed plastic streams is becoming more and more important with respect to waste management and upgrading and compatibilization of used plastics [5.29].

# 6. References


References for Chapter 1

[1.1] Uniroyal, US 3758643, 1979 (W. K. Fisher).
[1.2] Uniroyal, US 3806558, 1974 (W. K. Fisher).
[1.3] Exxon, US 3037954, 1962 (A. M. Gessler, W. H. Haslett).
[1.4] Hercules, US 3262992, 1966 (R. Holzer, O. Taunus, K. Mehnert).
[1.5] Monsanto, US 4130535, 1978 (A. Y. Coran, B. Das, R. P. Patel).
[1.6] Monsanto, US 4311628, 1982 (S. Abdou-Sabet, M. A. Fath).
[1.7] SRI Int. Report no. 104 A, Thermoplastic Elastomers, June 1993.
[1.8] P. Galli, J. C. Haylock: "Polyolefins VII International Conference," Society of Plastics Engineers, Houston, Texas, Feb. 24–27, 1991.
[1.9] J. Yin et al., *Sci. Sin. Ser. B* (*Engl. Ed.*) **29** (1986) no. 12, 1233.
[1.10] S. Abdou-Sabet, R. P. Patel, *Rubber Chem. Technol.* **64** (1991) 769.
[1.11] D. M. Lohse, *Annu. Tech. Conf. Soc. Plast. Eng.* **43rd** (1985) 301.
[1.12] D. M. Lohse, S. Datta, E. N. Kresge, *Macromolecules* **24** (1991) 561.
[1.13] B. F. Goodrich, US 4036912, 1977 (J. Sticharczuk).
[1.14] A. Y. Coran, R. P. Patel, D. Williams-Headd, *Rubber Chem. Technol.* **58** (1985) 1014.
[1.15] Himont, US 5196462, 1993 (D. A. Berta).
[1.16] Mitsui, US 4212787, 1980 (A. Matsuda, S. Shimizu, S. Abe).
[1.17] C. P. Rader, S. Abdou-Sabet in S. K. De, A. K. Bhowmick (eds.): *Thermoplastic Elastomers from Rubber-Plastic Blends,* Ellis Horwood, Chichester, England, 1990, p. 159.
[1.18] D. J. Elliott in S. K. De, A. K. Bhowmick (eds.): *Thermoplastic Elastomers from Rubber-Plastic Blends,* Ellis Horwood, Chichester, England, 1990, p. 102.
[1.19] N. R. Choudhury, P. P. De, A. K. Bhowmick in S. K. De, A. K. Bhowmick (eds.): *Thermoplastic Elastomers from Rubber-Plastic Blends,* Ellis Horwood, Chichester, England, 1990, p. 71.
[1.20] R. C. Puydak, D. R. Hazelton, *Plast. Eng.* **44** (1988) no. 9, 38.
[1.21] W. A. Zigman, *Adv. Chem. Ser.* **43** (1964) 1.
[1.22] S. Wu: *Polymer Interface Adhesion,* Marcel Dekker, New York 1982.
[1.23] A. Y. Coran, R. P. Patel, *Rubber Chem. Technol.* **56** (1983) 1045.
[1.24] S. Abdou-Sabet, Y. L. Wang, E. F. Chu, *Rubber Plast. News* **1985**, Nov. 4, 20.
[1.25] Monsanto, US 4654402, 1987 (R. P. Patel).
[1.26] L. A. Goettler, J. R. Richwine, F. J. Wille, *Rubber Chem. Technol.* **55** (1982) 1448.

References for Chapter 2

[2.1] G. Oertel (ed.): *Polyurethane Handbook,* 2nd ed., Hanser Publ., München – Wien – New York 1993.
[2.2] C. S. Schollenberger in A. K. Bhowmick, H. L. Stephens (eds.): *Handbook of Elastomers,* Marcel Dekker, New York – Basel 1988.
[2.3] W. Meckel, W. Goyert, W. Wieder in N. R. Legge, G. Holden, H. E. Schroeder (eds.): *Thermoplastic Elastomers,* Hanser Publ., Munich – Vienna – New York 1987.

[2.4] E. C. Ma in B. M. Walker, C. P. Rader (eds.): *Handbook of Thermoplastic Elastomers,* 2nd ed., Van Nostrand Reinhold, New York 1988.

[2.5] Upjohn, US 3 899 467, 1974 (H. W. Bonk, T. M. Shah).

[2.6] Bayer, DE 3 329 775, 1983 (W. Goyert).

[2.7] H. Hespe et al., *J. Appl. Polym. Sci.* **44** (1992) 2029 – 2035.

[2.8] J. H. Saunders, K. C. Frisch: *Polyurethanes,* part I: "Chemistry," Interscience, New York 1962.

[2.9] Bayer, EP 4393, 1978 (B. Quiring, H. G. Niederdellmann, W. Goyert, H. Wagner).

[2.10] P. J. Flory: *Principles of Polymer Chemistry,* Cornell University Press, Ithaca, N.Y., 1953.

[2.11] Akademie der Wissenschaften der DDR, DD 148 956, 1981 (G. Behrendt, H. U. Schimpfle, W. Lehmann); *Chem. Abstr.* **96** (1982) 21 105.

[2.12] C. P. Smith, J. W. Reisch, J. M. O'Connor, *J. Elastomers Plast.* **24** (1992) 306 – 322.

[2.13] W. Neumann, P. Fischer, *Angew. Chem.* **24** (1962) 806.

[2.14] F. Gugumus in R. Gächter, H. Müller (eds.): *Plastics Additives Handbook,* Hanser Verlag, München – Wien 1989.

[2.15] BASF, EP 0 134 455, 1984 (G. Zeitler, G. Bittner, K. Faehndrich, H. M. Rombrecht).

[2.16] BASF Schwarzheide, DD 300 900, 1987 – 1992 (R. Krech et al.).

[2.17] Elastogran, DE 41 12 805, 1991 (A. Chlosta, F. Lehrich, J. Sadlowski, L. Thil).

[2.18] U. Barth in Verein Dt. Ingenieure (eds.): *Polymerreaktionen und reaktives Aufbereiten in kontinuierlichen Maschinen,* VDI-Verlag, Düsseldorf 1988.

[2.19] C. S. Schollenberger, K. Dinbergs, *J. Elastoplast.* **5** (1973) 222; **7** (1975) 65.

[2.20] H. J. Maaß et al., *Plaste Kautsch.* **34** (1987) 251 – 254.

[2.21] J. A. Miller et al., *Macromolecules* **18** (1985) 32 – 44.

[2.22] S. Abouzaar, G. L. Wilkes, *J. Appl. Polym. Sci.* **29** (1984) 2695 – 2711.

[2.23] ICI, GB 1 057 018, 1964 (J. P. Brown, G. Trappe).

[2.24] BASF, DE 32 24 324, 1982 (G. Zeitler et al.).

[2.25] Mobay, US 3 233 025, 1966 (B. F. Frye, K. A. Piggot, J. H. Saunders).

[2.26] Upjohn, US 3 642 964, 1969 (K. W. Rausch, Jr., T. R. McClellan).

[2.27] Bayer, DE 2 302 564, 1973 (R. M. Erdmenger et al.).

[2.28] A. Bouilloux, C. W. Macosko, T. Kotnour, *Ind. Eng. Chem. Prod. Res. Dev.* **30** (1991) 2431 – 2436.

[2.29] Bayer, DE 2 854 406, 1978 (W. Goyert et al.).

[2.30] C. Li, J. G. Homan, R. A. Phillips, S. L. Cooper in K. C. Frisch, Jr. (ed.): *Recent Developments in Polyurethanes and Interpenetrating Polymer Networks,* Technomic, Lancaster – Basel 1988.

[2.31] Z. Petrovic, J. Ferguson, *Prog. Polym. Sci.* **16** (1991) 695 – 836.

[2.32] C. G. Seefried, Jr., J. V. Koleske, F. E. Critchfield, *J. Appl. Polym. Sci.* **19** (1975) 2493 – 2502, 2503 – 2513, 3185 – 3191.

[2.33] W. P. Yang, C. W. Macosko, S. T. Wellinghoff, *Polymer* **27** (1986) 1235 – 1240.

[2.34] L. Born, H. Hespe, J. Crone, K. H. Wolf, *Colloid. Polym. Sci.* **260** (1982) 819 – 828.

[2.35] G. Pohl, D. Joel, H. Goering, H. E. Carius, *Plaste Kautsch.* **40** (1993) 357 – 362.

[2.36] *Houben-Weyl,* E 20, 534.

[2.37] J. Foks, I. Naumann, G. H. Michler, *Angew. Makromol. Chem.* **189** (1991) 63 – 76.

[2.38] S. Hwang, D. J. Hemker, S. L. Cooper, *Macromolecules* **17** (1984) 307 – 315.

[2.39] L. M. Leung, J. T. Koberstein, *Macromolecules* **19** (1986) 706 – 713.

[2.40] J. W. C. Van Bogart, P. E. Gibson, S. L. Cooper, *J. Polym. Sci. Polym. Phys. Ed.* **21** (1983) 65 – 95.

[2.41] C. P. Christenson et al., *J. Polym. Sci. Polym. Phys. Ed.* **24** (1986) 1401 – 1439.

[2.42] C. D. Eisenbach, T. Heinemann, A. Ribbe, E. Stadler, *Angew. Makromol. Chem.* **202/203** (1992) 221 – 241.

[2.43] Dow Chemical, EP 80 031, 1983 (D. J. Goldwasser, K. Onder).

[2.44] B. F. Goodrich, DE 1 720 843, 1967 (E. G. Kolycheck).

[2.45] Bayer, DE 1 940 181, 1969 (E. Meisert, A. Awater, C. Muehlhausen, U. J. Doebereiner).

[2.46] W. Goyert, H. Hespe, *Kunststoffe* **68** (1978) 2 – 8.

[2.47] E. C. Ma, *Rubber World* **199** (1989) 30 – 35.

[2.48] A. T. Chen et al., *Elastomerics* **122** (1990) no. 9, 19 – 22.

[2.49] H. G. Hoppe, *Plastverarbeiter* **44** (1993) no. 9, 75 – 80; no. 10, 40 – 44.

[2.50] S. Gogolewski, *Colloid Polym. Sci.* **267** (1989) 757 – 785.

References for Chapter 3

[3.1] DuPont, US 3 651 014, 1972 (W. K. Witsiepe).

[3.2] DuPont, US 3 763 109, 1973 (W. K. Witsiepe).

[3.3] DuPont, US 3 766 146, 1973 (W. K. Witsiepe).

[3.4] L. H. Buxbaum, *J. Appl. Polym. Sci.* **35** (1979) 59 – 66.

[3.5] DuPont, US 3 801 547, 1974 (G. K. Hoeschele).

[3.6] BASF, US 4 056 514, 1977 (H. Strehler).

[3.7] General Electric, US 4 732 948, 1988 (R. J. McCready, J. A. Tyrell).

[3.8] Kuraray, JP 61 81 419, 1985 (M. Ishiguro, H. Hira).

[3.9] DuPont, US 5 004 748, 1991 (G. Huynh-Ba).

[3.10] DuPont US 3 726 569, 1973 (G. K. Hoeschele).

[3.11] DuPont: "Hytrel," Product Bulletin H-33430-1, 1994, p. 6.

[3.12] G. K. Hoeschele, W. K. Witsiepe, *Angew. Makromol. Chem.* **29/30** (1973) 267.

[3.13] J. R. Wolfe, Jr., *ACS Adv. Chem. Ser.* **176** (1979) 129.

[3.14] Eastman Kodak, US 4 256 860, 1981 (B. Davis, R. B. Barbee, H. R. Musser).

[3.15] General Electric, US 4 711 947, 1987 (R. J. McCready, J. A. Tyrell).

[3.16] DuPont, US 3 775 374, 1973 (J. R. Wolfe, Jr.).

[3.17] DuPont, US 3 784 540, 1974 (G. K. Hoeschele).

[3.18] J. R. Wolfe, Jr., *Rubber Chem. Technol.* **50** (1977) 688.

[3.19] Texaco Chemical Co.: Polyoxyalkyleneamines, product bulletin, 1991.

[3.20] General Electric, US 4 544 734, 1985; US 4 556 705, 1985 (R. J. McCready).

[3.21] E. C. Leonard, *Dimer Acids,* **1975**, 1.

[3.22] DuPont, US 3 954 689, 1976 (G. K. Hoeschele).

[3.23] General Electric, US 4 594 377, 1986 (R. J. McCready).

[3.24] R. W. M. van Berkel, S. A. G. de Graaf, F. J. Huntjens, C. M. F. Vrouenraets: Development Series, Developments in Block Copolymers-1, p. 261.

[3.25] Teijin, BE 834 004, 1975.

[3.26] Goodyear, US 3 446 778, 1969 (R. Ca. Waller, M. H. Keck).

[3.27] R. J. Cella, *J. Polym. Sci. Symp.* **42** (1973) no. 2, 727.
[3.28] R. J. Cella, Polymer Symposia No. 42, Helsinki 1972.
[3.29] S. C. Wells in B. Walker (ed.): *Handbook of Thermoplastic Elastomers,* Chap. 4.
[3.30] M. Brown, *Rubber Ind. (London)* **9** (1975) 102.
[3.31] DuPont, US 3718715, 1973 (R. W. Crawford, W. K. Witsiepe).
[3.32] DuPont, US 3917743, 1975 (H. E. Schroeder, J. R. Wolf, Jr.).
[3.33] DuPont, US 3963802, 1976 (C. Shih).
[3.34] DuPont, US 4010222, 1977 (C. Shih).
[3.35] General Electric, US 4939205, 1989 (Nan-I Liu).
[3.36] General Electric, US 4814380, 1990 (Nan-I Liu, R. J. McCready).
[3.37] M. Veldstra: *Thermoplastic Elastomers-II*, Rapra Technology Ltd., London 1989.
[3.38] H. Reinhardt, M. Negri, *Thermoplastic Elastomers-II*, Rapra Technology Ltd., London 1989.
[3.39] Teijin, JP 50-35623, 1976.
[3.40] AKZO, US 4493870, 1985 (C. M. F. Vrouenraets).
[3.41] M. H. Horn, K. P. Schodt, G. J. Ostapchenko, *Annu. Tech. Conf. Soc. Plast. Eng.* 1992.

References for Chapter 4

[4.1] G. Deleens, PhD Thesis, University of Rouen, 25.01.1975.
[4.2] Ato-Chimie, FR 2273021, 1974 (G. Deleens, P. Foy, C. Jungblut).
[4.3] Chemische Werke Hüls, FR 2384810, 1978 (K. Burzin et al.).
[4.4] Ato-Chimie, FR 2466478, 1976 (G. Deleens).
[4.5] Ato-Chimie, FR 2364982, 1976 (G. Deleens, P. Foy).
[4.6] Ube Industries, JP 59133224, 1983 (H. Okamoto, Y. Okushita).
[4.7] Ube Industries, JP 59131628, 1983 (H. Okamoto Y. Okushita).
[4.8] Ube Industries, JP 59193923, 1983 (Okamoto Hidemasa, Okushita Yogi).
[4.9] Elf Atochem S.A., Technical brochure on Pebax grades, Paris – La Défense, 1994.
[4.10] Atochem S.A., Internal report on Cerdato, Serquigny, 1986.
[4.11] Ube Industries, Technical brochure on Ube Polyamide Elastomer, Tokyo, 16.09.1993.
[4.12] Chemische Werke Hüls, Technical brochure on Vestamid grades, Marl, 1990.
[4.13] G. M. Bohlmann, J. Wakim: "Thermoplastic Elastomers," SRI International – Economics Program no. 104A (1993) 3–1.

References for chapter 5

[5.1] N. R. Legge: "Thermoplastic Elastomers Based on Three-Block Polymers," *Chemtech.* **13** (1983) 630.
[5.2] Shell Oil Co., US 3149182, 1964 (L. M. Porter).
[5.3] R. J. Ceresa: *Block and Graft Copolymers,* Butterworth, Washington, D.C., 1962.
[5.4] P. Dreyfuss, L. J. Fetters, D. R. Hansen, *Rubber Chem. Technol.* **53** (1980) 728.
[5.5] E. V. Gouinlock, R. S. Porter, *Polym. Eng. Sci.* **17** (1977) 535.
[5.6] B. M. Walker, C. P. Rader (eds.): Handbook of Thermoplastic Elastomers, 2nd ed., Van Nostrand Reinhold, New York 1987.
[5.7] A. D. Thorn: *Thermoplastic Rubbers – A Review of Current Information,* RAPRA, Shawsbury, UK, 1980.
[5.8] D. J. St. Clair, *Rubber Chem. Technol.* **55** (1982) 208.
[5.9] J. R. Erickson, *Rubber Plast. News* **1985**, Sept. 9, 16.
[5.10] G. Holden, S. S. Chin, Paper to the Adhesives and Sealants Conference, Washington, D.C., March 1986.
[5.11] G. Holden, Paper to the Adhesives and Sealants Council Seminar, Chicago, Il., Oct. 1982.
[5.12] E. J. van Beem, P. Brasser, *J. Inst. Pet.* **59** (1973) 91.
[5.13] Shell Int. Chem. Co., Bulletin TR-18, London 1984.
[5.14] *The Shell Bitumen Handbook,* Shell Bitumen UK Ltd., London 1990.
[5.15] J. H. Collins, W. J. Mikols, *60th Meeting of Asphalt Paving Technologists,* San Antonio, Tx., Feb. 1985.
[5.16] D. M. Colwill, M. E. Daines: "Progress in the Trials of Pervious Macadam," *Highways,* Jan. 1989, p. 15.
[5.17] A. Dinnen: "Bitumen Thermoplastic Rubber Blends in Road Applications," 3rd Eurobitumen Symposium, The Hague, Sep. 1985.
[5.18] M. J. W. Downes, R. C. Koole, E. A. Mulder, W. E. Graham: "Some Proven New Binders and Their Cost Effectiveness," *7th AAPA Int. Asphalt Conf.,* Brisbane, Aug. 1988.
[5.19] K. H. Kolb, T. Pallay, J. P. Serfass, O. Ruud: "Asphalt for Better Roads," Symposium, E.A.P.A., Copenhagen 1986.
[5.20] L. Laitinen: "Rubberised Bitumen in Road Surfacing Applications," *Asfaltti* **43** (1988) 18.
[5.21] J. P. Marchand: "15 Years of Experience in Thick Bitumen Pavements," Petrol. Inf. **1987**, Nov., 62.
[5.22] S. H. Carpenter, T. van Dam: "Modified and Unmodified Asphalt Concrete Mixtures," *Transp. Res. Rec.* **1115** (1987) p. 62–74.
[5.23] H. F. Vermeire, *Kautsch. Gummi Kunstst.* **43** (1990) 11.
[5.24] A. L. Bull, G. Holden, *J. Elastomers Plast.* **9** (1981) 281.
[5.25] Shell Int. Res. Co., EP 0215501, 1982 (W. P. Gergen, R. G. Lutz, M. K. Martin).
[5.26] D. W. Bartlett, D. R. Paul, J. W. Barlow, *Mod. Plast.* **58** (1981) no. 12, 60.
[5.27] C. L. Willis, W. M. Halper, D. L. Handlin, Jr., *Polym. Plast. Technol. Eng.* **23** (1984) no. 2, 207.
[5.28] Shell Int. Chem. Co., Technical Bulletin TR.3.6, London 1991.
[5.29] J. Schneider, Paper presented at Recycle '92, Davos, Apr. 1992.

# Thermosets

WALDEMAR SCHÖNTHALER, Bakelite Gesellschaft mbH, Iserlohn-Letmathe, Federal Republic of Germany



## 1. Introduction

Typical of thermosetting plastics (thermosets) is that they are formed in the processing mold from moderately large molecular building blocks (reactive precursors) by chemical reaction. This contrasts with thermoplastics, where the finished high-molecular material is merely physically remelted in the mold. That also explains the main difference between these two kinds of plastics: thermosets no longer melt and are processed at temperatures far below their application temperatures; thermoplastics can be remelted and must therefore always be processed at temperatures far above their temperatures of use.

In German, the thermosets were formerly known as *Pressmassen* (compression-molding compounds) since they were processed by compression molding. Thermoplastics were known as *Spritzgussmassen* (injection molding compounds) since they were processed by injection molding. Since the mid-1960s thermosets have also been processible on injection molding machines, so the term *Formmasse* (molding compound) has been introduced whereby a distinction is made between curable and noncurable molding compounds (thermosets and thermoplastics, respectively).

The classical thermosets, which mainly are discussed here, are based on phenolic, urea, or melamine resins. These were followed by others based on unsaturated polyester, diallyl phthalate, epoxy resin, or silicone resin. Polyurethanes (→ Polyurethanes) and polyimides also can be thermosets. Strictly speaking, even cross-linked thermoplastics can be regarded as thermosets.

A comprehensive survey of the wide field of the thermosets is provided in [1]. Also discussed there are future developments based on nadimide- (polyimide), maleimide- and acetylene-terminated thermosetting polymers.

**History** [2]. Compression-molding compounds were the first fully synthetic plastics. Resins were manufactured from phenol and formaldehyde according to the patents of L. H. BAEKELAND as early as 1910 and, after mixing with fillers, cured to plastics by applying heat and pressure. Phenolic resin compression molding materials were initially based on phenolic resols (one-step resins), and later (in the 1920s) on novolacs ("two-step resins") (see also → Phenolic Resins). The use of novolacs with curing by hexamethylenetetramine was an important development (fast-curing molding compounds), since it also improved the manufacture and the storability.

In the late 1920s, the urea resin compression molding compounds came onto the market and made white and colored light-resistant compression moldings possible. At the end of the 1930s melamine resin compression molding compounds were introduced, which made good water resistance possible.

Compression-molding compounds based on unsaturated polyester resins (see → Polyester Resins, Unsaturated) and epoxy resins (see → Epoxy Resins) became widespread after World War II, the polyester resin compression-molding compounds, at first being only wet-compounded. Today, dry, free-flowing polyester resin compression molding compounds also have gained major importance.

Molding compounds are generally unshaped products which can be processed by pressure and temperature to form moldings.

In the thermosetting plastics, because of the chemical curing process, the composition of the molded material differs from that of the molding compound. In the case of thermoplastics, the finished plastic is processed, and only remelting takes place in the mold.

**Classification.** From the beginning, thermosetting molding compounds were compounded from binders and fillers. This was done to remove the brittleness from the moldings, to economize on resin, and to broaden the range of applications of the materials. An extremely wide variety of materials are used as fillers. The term filler can refer to nonreinforcing and reinforcing fillers.

Fillers can be classified into inorganic and organic, into powder, fibrous, and lamellar types (see Table 1) [3].

These fillers can be combined with various binders. It is also possible to compound combinations of binders with one filler or combinations of fillers with a single binder. Finally, there is also the possibility of combining mixtures of several binders with the mixture of several fillers.

The binder (resin) can be manufactured from diverse molecular building blocks and by various reaction mechanisms (see Table 2) [4]. Today, phenolic resins, urea resins, melamine resins, unsaturated polyester resins, and epoxy resins are widely used. Resins based on diallyl phthalate or silicones are less important.

The curing or cross-linking process in the mold can proceed by other chemical reactions than the manufacture of the resin itself. Polycondensation, polymerization, or polyaddition can be involved. In case of polycondensation, low-molecular substances (e.g., water) are split off, whose vapor pressure at processing temperature must be compensated for.

The properties of the cured molded material are the sum of the properties of the resins and of the fillers. Thus for a given filler, the typical binder properties, and for a given binder typical

**Table 1.** Fillers for thermoset molding compounds

| | Particle form | Example | | Advantages | | Disadvantages |
|---|---|---|---|---|---|---|
| Inorganic | powder<br>short fibers<br>long fibers<br>platelets | stone powder<br>mineral fiber<br>glass fiber<br>mica | high thermal stability | uniform distribution<br>high mechanical values | high density | low mechanical values orientation |
| Organic | powder<br>short fibers<br>long fibers<br>platelets | wood flour<br>cellulose<br>textile threads<br>macerated fabric | low density | uniform distribution<br>high mechanical values | low thermal stability | low mechanical values orientation |

**Table 2.** Resins for thermoset molding compounds

| | Symbol | Resin structural units | Resin-forming reaction | Cross-linking reaction | Reaction product split off on cross-linking |
|---|---|---|---|---|---|
| Phenoplastic | PF | phenol and formaldehyde | polycondensation | polycondensation | water |
| Aminoplastic | UF | urea and formaldehyde | polycondensation | polycondensation | water |
| | MF | melamine and formaldehyde | polycondensation | polycondensation | water |
| Unsaturated polyester | UP | unsaturated dibasic acid and diol | polycondensation | polymerization | none |
| Epoxide | EP | bisphenol and epichlorohydrin | polycondensation | polyaddition with diamine or dibasic acid | none |
| Diallyl phthalate | DAP | monomeric diallyl phthalate | polymerization | polymerization | none |

filler properties are observed. Therefore, resin-determined and filler-determined properties must be differentiated.

**Standardization.** For the standardization and monitoring of thermosetting molding compounds in Germany, the Technical Association of Manufacturers and Processors of Plastics Molding Compounds (Technische Vereinigung der Hersteller und Verarbeiter von Kunststoff-Formmassen) has been established to promote cooperation between raw material manufacturers and processors. Type tables with minimum requirements for several important properties of the molded materials have been drawn up. These properties are tested on defined test specimens according to DIN specifications. The tables, also specify the binder and filler for each type. The most important DIN standards are DIN 7708 (phenolics and aminoplastics) and DIN 16911 (polyester resin molding compounds). For epoxy resin molding compounds, DIN 16912 has been withdrawn owing to the rapid development in this field.

Beyond the type values listed in the DIN specification tables, a further series of properties of the molding compounds is defined. In the case of the phenolics, this is information on the base resin, resin content, and color. For example, a molding material of Type 31 might carry the designation 31-1449, where the first two digits indicate the type according to DIN 7708. The third digit indicates phenolics as base resin, the fourth defines a minimum resin content and the last two the color:

```
3 = 35% resin content
4 = 40% resin content
5 = 45% resin content
6 = 50% resin content
7 = 55% resin content
8 = 60% resin content
00-09 = white, yellow, natural
10-19 = light brown to dark brown
20-29 = red to mahogany
30-35 = green
35-39 = blue
40-49 = grey to black
```

An additional definition is sometimes appended to the type designation (e.g., 51.5):

```
.5 = electrically high-quality
.7 = suitable for tableware for eating and drinking
.8 = content of volatile acids < 0.18%
.9 = ammonia free
```

Attempts to develop a system analogous to DIN 7708 for the quality assurance of thermoplastics failed. This was partly due to the difficul-ty of producing standard test specimens by injection molding. To offer designers the possibilities of comparing the diverse plastics, in Germany the four largest plastics manufacturers, BASF, Bayer, Hoechst, and Hüls, have created the CAMPUS System (Computer-Aided Material Preselection by Uniform Standards), which has been joined by more than 20 international chemical companies. Since then thermoset manufacturers have also decided to use this system on grounds of rationalization, especially since the single-point values originally used for quality assurance are now supplemented by curve diagrams.

The CAMPUS System is offered in the form of diskettes free of charge by the participating firms and can be run on IBM-compatible PCs.

In the course of internationalization, DIN standards have already in many cases been replaced or supplemented by ISO or EN standards.

# 2. Manufacture

Despite the great variety of possible thermosetting molding compounds, only a few processes are used in their manufacture. The binder (resin) must be converted to the liquid state so that good impregnation of the fillers is possible. This is referred to as the fusion process if liquefaction is carried out by melting the resin, and as the liquid resin process if the resin is already liquid at room temperature or is used in solution in water or organic solvents [3].

## 2.1. Fusion Process

In the fusion process (Fig. 1) a premix is produced from finely ground solid resin, fillers, colorants, lubricants, and release agents, etc. Examples of recipes (in wt%) follow:

*Type 31 (PF)*

| | |
|---|---|
| Phenol novolac | 43 |
| Hexamethylenetetramine | 7 |
| Wood flour | 45 |
| Magnesium oxide | 1 |
| Zinc stearate | 2 |
| Dyes | 2 |

*Type 150 (MF)*

| | |
|---|---|
| Solid melamine resin | 50 |
| Accelerator | 0.5 |
| Wood flour | 45 |
| Zinc stearate | 1.5 |
| Dyes | 3 |

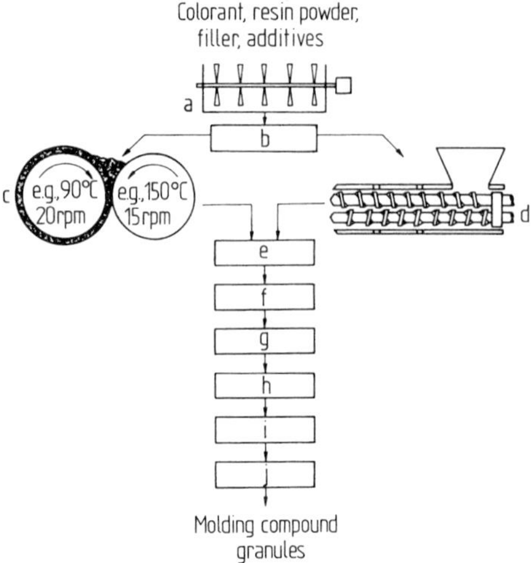


**Figure 1.** Compounding of thermoset molding compounds by the fusion process
a) Premixer; b) Premix; c) Rolls; d) Screw kneader; e) Cooling; f) Precomminution; g) Grinding; h) Sieving; i) Classifying; j) Mixing

The premix is then compounded discontinuously or continuously on roll mills or extruders to form the thermoset molding compound. The roll mills have two contrarotating steel rolls, up to ca. 3 m long, heated by steam, high-pressure water, oil, or electric (for smaller rolls), and rotating at different speeds. Their roll gap, a few millimeters, is variable. The rolls differ in temperature (e.g., 90 °C for the front and 150 °C for the back roll), so that on feeding the premix a layer (sheet) forms on the (cooler) rolls covering it completely. On passage of the sheet through the narrow gap between the rolls, the material is kneaded and mixed as a result of the different roll speeds. In addition, a bank forms on the roll gap rotating contrary to the front roll. On the roll mill, the fillers are impregnated with the binder, the material is homogenized, and in the case of polycondensation compounds (phenolics and aminoplastics) also precross-linked.

The precross-linking is continued until a thermoset molding compound has been formed that the processor can still process in his specific mold with the forming pressure available to him. This enables increased savings of energy and time to be made at the processor's works.

In addition, extensive precondensation also improves the properties of the molded material, such as surface gloss, electrical properties, shrinkage and post-shrinkage behavior. Compounding on roll mills can be carried out discontinuously or continuously. In the discontinuous process, when the sheets have reached the required degree of precondensation they are cut open along the roll and pulled off. In the continuous process, the premix is fed continuously in the middle or on one side of the roll, and the sheet formed is cut off in narrow continuous strips at the roll edge.

In the extrusion process, single- or twin-screw machines or planetary screw extruders are used. In this case the premix is usually processed in large batch premixers.

Processing on the extruders is continuous. In both the roll-mill and extruder processes, the sheet or lumps formed are cooled, giving a grindable product. From this, a granular molding compound of defined grain size is obtained by breaking, milling, and sieving.

For the fusion process, powder or short-fiber fillers such as wood flour, stone powders, and glass fibers, are best suited. Long-fiber fillers are broken down by the high shear forces in the roll gap. Materials manufactured by the hot melt process include phenolic resin molding compounds (PF) of Types 11 (stone powder), 31 (wood flour) according to DIN 7708; the melamine–formaldehyde (MF) and melamine phenol (MP) Types 150 (wood flour), 180 (wood flour), and 155 (stone powder); and the polyester resin Types (UP) 802 and 804 (short glass fibers) according to DIN 16911. The UP compounds are not precured since hardening by polymerization takes place only in the molding tools.

## 2.2. Liquid Resin Process

The liquid resin process (Fig. 2) is used mainly for compounding thermosetting molding materials that contain long-fiber or macerate fillers (coarse-structured materials). Kneaders (e.g., sigma kneaders, which bakeries also use) or edge mills are used as mixing units. The compounded materials can be removed in the dough state after mixing and impregnation. Examples are Types 801 and 803 (DIN 16911), which are compounded with liquid, styrene-containing unsaturated polyester resins and contain long glass fibers.

The compounds can also be dried and further condensed, as in the case of the phenolic of Type 71 (textile fibers) or 74 (textile macerate) or

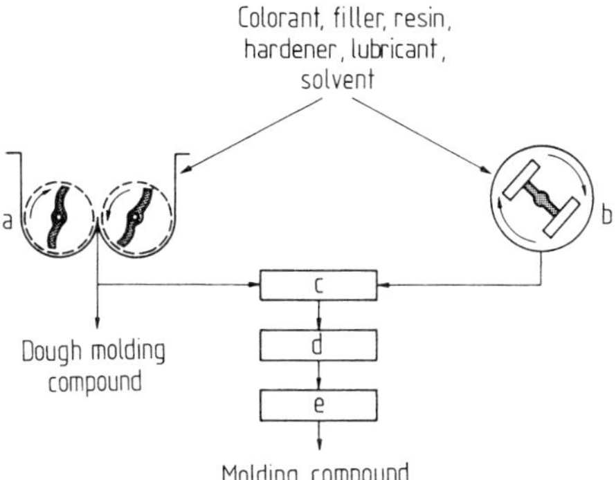



**Figure 2.** Compounding of thermoset molding compounds by the liquid resin process
a) Kneader; b) Edge mill; c) Drying; d) Cooling; e) Coarse grinding

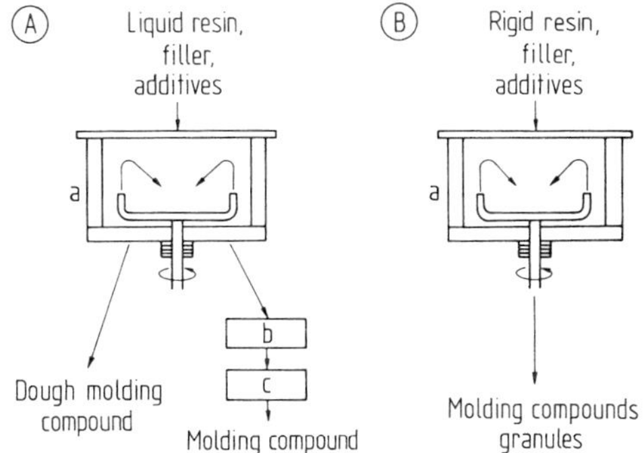



**Figure 3.** Compounding of thermoset molding compounds in a turbomixer
A) Wet mixing and impregnation; B) Dry mixing and impregnation
a) Turbomixer; b) Drying; c) Coarse grinding

the aminoplastic Types 153 (textile fibers) and 154 (textile macerate) (DIN 7708).

Discontinuous tray driers or continuous-flow driers can be used for drying and further condensation. By suitable choice of temperature and drying time, the flow–cure behavior of phenolics and aminoplastics can be adapted to the requirements of the processor. When using solvent-containing resins, however, care must be taken that prior to further condensation the solvent can escape from the molding compound particles. If solvent residues are trapped because drying process has not been carried out properly, surface defects can occur on molding.

The liquid resin process is mainly used for the mild impregnation of long-fiber fillers, but it also allows low-resin molding compounds to be produced if the liquid resin is highly diluted by solvent.

### 2.3. Other Processes

A number of special methods exist. For example, a turbomixer (Fig. 3), which permits very short mixing times, can be used in the liquid resin process. Turbomixers are also suitable for the hot melt process since the large amount of mechanical energy introduced can lead to fusion of the finely ground resin. In addition to impregnation of the filler, the mixture is agglomerated to dust-free granules. As a result, subsequent grinding can be dispensed with, and particularly free-flowing granules are obtained. In addition, recycling of dust and oversize grain to the process, which is otherwise necessary for granule production, is unnecessary.

## 3. Commercial Forms and Manufacturers

Thermoset molding compunds are supplied in diverse forms, from dust-fine material through various grain sizes up to briquetted grades, depending on the mechanical demands made on the molding to be manufactured. Thus, a material compounded with coarse-structured filler cannot be supplied in a fineground form without loss of strength. Fine grinds are therefore useful mainly in recipes containing powdered filler or with short-fiber fillers.

Conventional size reduction by crushers and mills (e.g., hammer mills, pinned disk mills, cutting mills etc.) of the brittle sheets or clods obtained in the production process gives fairly wide grain-size distributions. By sieving out the dust and oversize grain, narrow grain-size ranges can be obtained fixed, but with simultaneous reduction of the production capacity. Oversize grain must be returned to the mills, and the dust component is usually added to the premix and therefore passes once more through the production plant. A reasonable grain size therefore usually comprises the range from 0.2 to 1.0 mm grain diameter. Such a range of grain sizes has a good flowability and does not tend to demix, even during long-distance transport. Removing the very fine fractions on vibrating sieves is sometimes problematic because the very fine dust adheres strongly to the grains of molding compound, and because in the subsequent mixing processes the brittle granules rapidly form dust by abrasion. By careful control of the process, however, almost dust-free granules can be produced. Although these modern low-pollution granules are more convenient for the processor, the older commercial forms with a high dust content (up to 30%) gave moldings with better mechanical properties and surface quality.

With coarser fillers, the upper limit of grain size is extended to ca. 8 mm, whereby the flowability deteriorates. If high-strength materials with good flowability are to be produced, special granulation processes are necessary, such as pelletizing or compression granulation. For thermoset molding compounds with high fiber content the latter is generally carried out on extruders, giving "sausage-like" granulates with diameters of 1–6 mm, depending on the filler.

Molding compounds with particularly coarse fillers, such as those produced by wet compounding of macerates (e.g., Type 74), are also supplied as briquettes, weighing up to ca. 1 kg. A special granulation method is used for the aminoplastic Types 131 and 152. These cellulose-filled molding compounds are usually produced in undyed form and are subsequently dyed in ball mills. This procedure is useful because of the variety of shades of color required for these molding materials. Very fine dust is formed in the ball mills and must subsequently be recompacted in, e.g., tabletting machines, pelletizers, extruders, or pan granulators. Depending on the method used, compaction may be followed by grinding and sieving.

**Manufacturers.** An extensive compilation of the manufacturers and processors of thermoset molding compounds appears in the journal *Kunststoffe*, usually in the February issue.

Some important manufacturers of thermoset molding materials in Europe are:

Bakelite GmbH, Iserlohn Lemathe, Germany
Phönix Gummiwerke AG, Hamburg, Germany
Dr. F. Raschig GmbH, Ludwigshafen, Germany
Resart Ihm AG, Mainz, Germany
Süd-West Chemie GmbH, Neu-Ulm, Germany
Perstorp, Sweden
Vynckier, Belgium

## 4. Storage and Transport

Since the molding compounds discussed here cure on warming they should preferably be stored in cool rooms. The storage temperature depends on the type of material.

Phenolics and aminoplastics should preferably not be stored above room temperature.

Phenolics based on novolac–hexamethylenetetramine (the majority) can be kept in this way for years. The storability of phenolics based on resols, and of aminoplastics is lower. Since these polycondensation molding compounds readily absorb or lose water, depending on the ambient humidity, the relative humidity of the store rooms also plays an important role. A relative humidity of ca. 50% is recommended.

Packaging of molding compounds in multilayer paper sacks with polyethylene liners protects against the uptake or release of moisture for several months. Changes in humidity in phenolics and aminoplastics affect the flow–cure behavior of the molding compounds, in particular.

Polyester resin molding compounds also should be stored below room temperature. The storage life is then at least 6 months. Modern styrene-free, free-flowing polyester compounds achieve storabilities comparable to those of phenolics.

In earlier times epoxy resin molding compounds were stored in refrigerated rooms, and sometimes shipped in cooled vessels. However, for modern epoxy resin molding compounds this is no longer necessary.

Thermoset molding compounds are increasingly transported and stored in containers. This guarantees the exclusion of moisture and permits processing in closed systems on modern injection molding machines. The molding compound can thus be conveyed via suction-pneumatic or mechanical conveyor systems, without dust problems, into the feed hopper of the injection molding machine.

The containers range in capacity from 500 kg to ca. 2000 kg and can be made of metal, cardboard, or plastic. Cardboard containers are used as disposable packaging. They are less moisture-proof and more easily damaged. Containers made of steel sheet or aluminum have considerable weight when empty, and therefore cause additional costs for return freight. However, they are undoubtedly the safest packaging and protect the contents even under highly unfavorable weather conditions.

## 5. Properties

### 5.1. Uncured Molding Compounds

For the uncured molding compound the most important property is the processing behavior. The following properties are tested [4], [5]:

1) Flowability
2) Grain size (or grain-size distribution)
3) Bulk density
4) Filling factor
5) Flow–cure behavior

The *flowability* (DIN 53 492) of granular plastic is usually assessed in funnels of defined shape or with automatic metering devices. A simple method is determination of the angle of repose.

The *grain size* (DIN 53 477) is measured on sieving machines, the residues on the test sieves being weighed. The measured values can be evaluated either directly or by the Rosin–Rammler method, in which the measured values are plotted on a double-logarithmic scale. A straight line is obtained, whose slope allows the uniformity of the grain distribution to be inferred. An advantage of this method is that deviation of the curve from the straight line indicates a fault in the sieving plant.

The *bulk density* (DIN 53 468) is of importance for volumetric metering. First the weight of loose material that can be accommodated in 1000 mL is determined and then divided by the volume to give the bulk density. Since volume metering is often employed in the processing of thermoset molding compounds, the maintenance of the bulk density from delivery to delivery is very important.

The *filling factor* is the ratio of the relative density of the molded material to the bulk density of a molding compound. It plays a role when the transfer chamber of a mold is limited in size. Then, a certain minimum bulk density must be used.

The *flow–cure behavior* is the most important processing property of a thermosetting molding compound. Contrary to thermoplastics, not only does the viscosity of thermosets decrease on heating, it also rises again with the onset of curing (Fig. 4). The viscosity curve is thus U-shaped, the falling branch being determined by the physical process of heating, and the rising

leg by the chemical curing reaction. The more steeply the left-hand branch falls, the more readily the molding compound softens. For compression molding, the softening behavior is the most important property of the thermoset molding compound. The middle part of the U-shaped curve is a measure of the lowest possible viscosity and determines the lowest possible processing pressure. The width of the curve and the steepness of the rising leg are measures of the curing rate.

There is no test method that covers the whole curve. Generally two complementary tests are employed.

In the plates method (Fig. 5) a defined amount of molding compound is pressed flat between two heated plates at a given pressure. The width or height of the "cake" is then used as a measure of the flowability. The drawback of this process is that the specific pressure changes continually during the deformation.

In the beaker method (DIN 53 465, Fig. 6) a molding in the shape of a beaker is produced at a given pressure and a given temperature. The time from the lowering of the ram onto the material to complete sealing of the compression mold is measured. This "beaker time" is a measure of the softening and flow behavior of the material. In addition, the stiffness on demolding is tested. The beaker, still hot, is loaded on its top edge with a defined force (e.g., 10 N) and compressed.

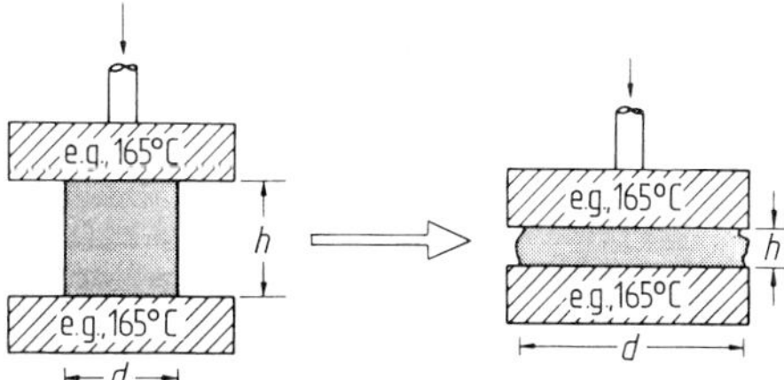



**Figure 5.** Testing of flow behavior by the plates method

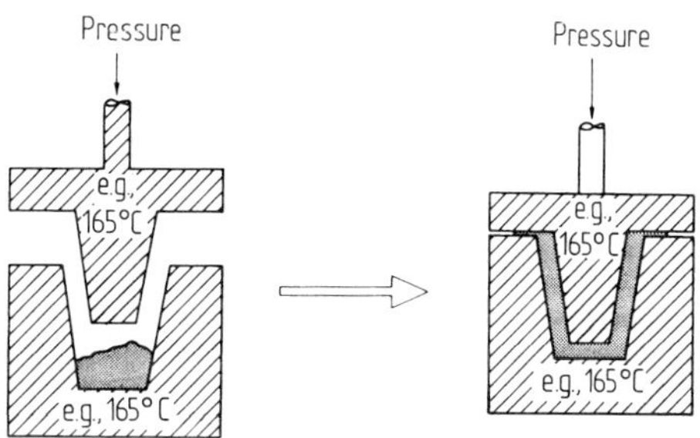



**Figure 6.** Testing of flow behavior by the beaker method

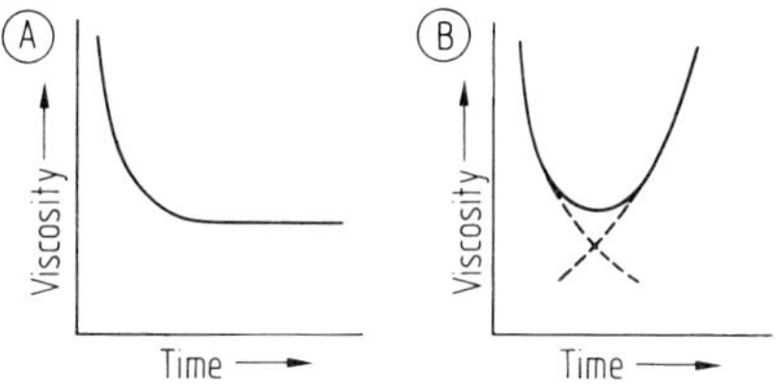



**Figure 4.** Flow-cure behavior of thermoplastics (A) and of thermosets (B) on heating to processing temperature

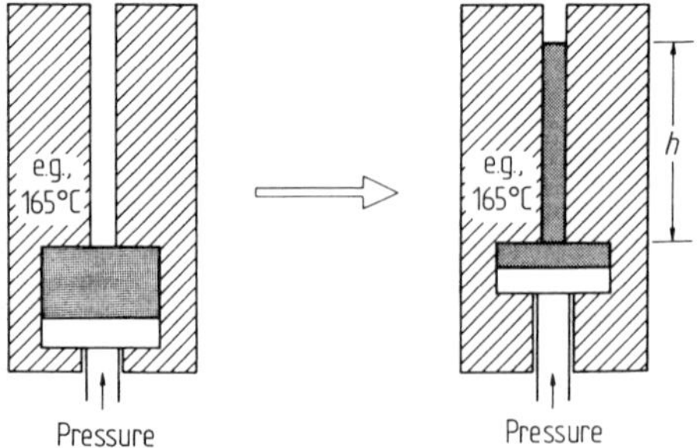


**Figure 7.** Testing of flow behavior by the rod method

It undergoes a deformation that depends on the degree of cure. At a given curing time and curing temperature, the compression in millimeters is a measure of the stiffness on demolding. If several beakers with increasing curing time are tested, the curing rate of a molding material can be determined. In the American version of the beaker method (ASTM D-731) minimum pressure for the outflow of the beaker is measured. Here, therefore, several pressings are for time deter.

In the rod method (Fig. 7) a tablet of the material is pressed from a heated chamber under defined pressure into a heated flow channel. The length of the rod formed is a measure of the flowability. Recording the growth of the rod with time gives softening curves. If the tablet is left in the heated chamber for various lengths of time before it is pressed into the flow channel, and the length of the rod obtained is plotted against the time, curing curves are obtained. This Rossi–Peakes flow test is only suitable for short-fiber molding compounds. For coarse-structured materials, rods of larger diameter are used.

A variation of the rod method is the spiral test, in which the rod is spirally wound, allowing long rods to be accommodated in a relatively small mold. These long rods are required for testing thermoset molding compounds that flow particularly readily. Thermosets of this kind, also known as low-pressure materials, are used in the electronics industry for encapsulating sensitive components. Epoxy resin molding materials in particular are tested by this method in EMMI spirals (EMMI: Epoxide Molding Material Institute).

In the kneader method (Fig. 8) the forces developed at the kneading blades when the molding compound is kneaded are measured (e.g., with a Brabender plastograph). If a defined amount of the molding material is introduced into the small heated kneader, the expenditure of

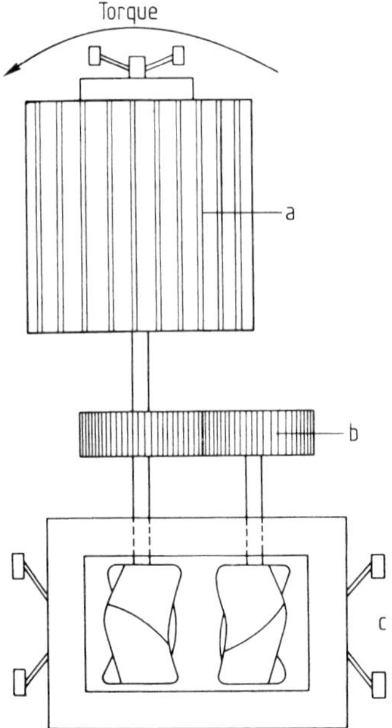


**Figure 8.** Testing of flow behavior with a Brabender plastograph
a) Motor; b) Gearing; c) Motor kneader

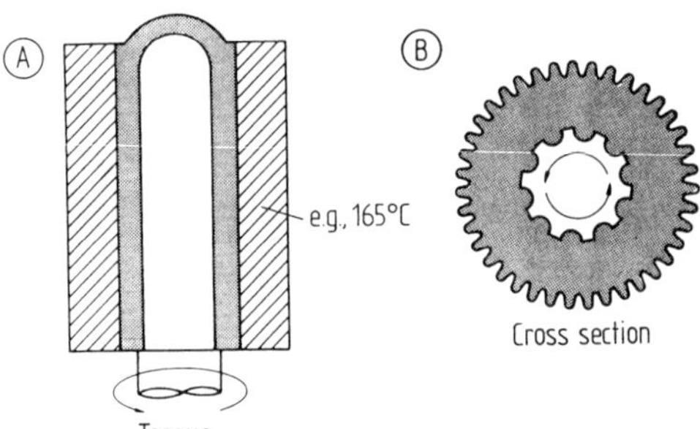


**Figure 9.** Testing of flow behavior by the Kanavec Method

energy at first falls on melting and then, as curing begins, rises again. Finally the material becomes crumbly, and the expenditure of energy falls again. The quantity measured is the torque developed by the driving motor, which gives a U-shaped curve. However, this test does not measure the entire flow–cure behavior, since the material crumbles before it is completely cured.

The rotary viscometer or Kanavec method (Fig. 9) is the most precise test of flow–cure behavior. A test specimen resembling a bicycle hand-grip is molded in a special mold, and during the curing process the male mold is rotated slowly (e.g., 1 revolution in 20 min) relative to the mold cavity. The quantity measured is the torque required for uniform rotation or the shearing stress generated. This method records very accurately the lower part and the rising branch of the U-shaped viscosity curve of thermoset molding compounds.

A special form of the test is the Philips method, in which a mushroom-shaped ram is pressed into a cylindrical mold cavity. The ram has slits, through which the molding compound can escape upwards during the deformation process. The quantity measured is the loss of material or the thickness of the cured residual material.

## 5.2. Cured Moldings

The properties of cured moldings are always a combination of the binder properties and the filler properties. A survey is provided in DIN 7708 for phenolics and aminoplastics, and in DIN 16911 for unsaturated polyester resins.

**Materials Based on Phenolic Resins.** Phenolic resin molding compounds yield moldings with small susceptibility to cracking and small aftershrinkage (Table 3). They are mostly of low flammability, but generally have a low tracking resistance. They are not light stable, and slowly discolor under the influence of light. They are therefore manufactured only in black, brown, or muted red and green shades. Aminoplastics are fully light stable and can be supplied in light and colored shades.

**Materials Based on Urea, Melamine, and Melamine–Phenolic Resins.** These molded materials have particularly low flammability and, in combination with suitable fillers, exhibit very high tracking resistance (Table 4), especially in the case of melamine resins. However, both the urea–formaldehyde (UF) and melamine–formaldehyde (MF) aminoplastics have relatively high aftershrinkage values. In unfavorable cases this can lead to susceptibility to cracking and to distortion at elevated temperatures.

The aminoplastic–phenophlastic (MP) molded materials (Types 180–183), as a result of the small proportion of plasticizing phenolic resin, behave more advantageously. With pale shades, however, only a very small proportion of phenolic resin is possible.

**Materials Based on Polyester Resins.** Polyester resin molding compounds (UP) have extremely small aftershrinkage (Table 5). They can be pale or colored and are light stable, resistant to tracking, and dimensionally stable. Thus they exhibit a combination of the advantageous properties of phenolics and aminoplastics, without their drawbacks. At first, UP materials were produced from styrene-containing unsaturated polyester resins with a relatively high proportion of glass fibers. Owing to their moist, fibrous structure, they were also known as "sauerkraut compounds". They yielded moldings of relatively uneven structure and were therefore mainly used for engineering components.

Improved surface quality was achieved by so-called low-profile additives (thermoplastic additives), which made possible the use of UP molding materials for high-quality visible components.

The use of monomer-free solid UP resins was a further decisive breakthrough. With them, per-

**Table 3.** Properties of phenolic moldings (DIN 7708)*

| Type | Filler | Density, g/cm³ | Bending strength, N/mm² | Impact strength, kJ/m² | Notch impact strength, kJ/m² | Martens temperature, °C | Surface resistance, comparative figure | Water absorption, mg |
|---|---|---|---|---|---|---|---|---|
| 11.5 | stone powder | 1.8 | 50 | 3.5 | 1.3 | 150 | 10 | 45 |
| 12 | mineral fiber | 1.8 | 50 | 3.5 | 2 | 150 | 8 | 60 |
| 13 | mica | 1.9 | 50 | – | 2 | 150 | 10 | 20 |
| 31 | wood flour | 1.4 | 70 | 6 | 1.5 | 125 | 8 | 150 |
| 85 | wood flour + cellulose | 1.4 | 70 | 5 | 2.5 | 125 | 7 | 200 |
| 83 | wood flour + textile fibers | 1.4 | 60 | 5 | 3.5 | 125 | 8 | 180 |
| 84 | chopped cotton cloth + cellulose | 1.4 | 60 | 6 | 6 | 125 | 8 | 150 |
| 51 | cellulose | 1.4 | 60 | 5 | 3.5 | 125 | 7 | 300 |
| 71 | textile fibers | 1.4 | 60 | 6 | 6 | 125 | 7 | 250 |
| 74 | chopped cotton cloth | 1.4 | 60 | 12 | 12 | 125 | 7 | 300 |

* Minimum values, except for water absorption (maximum value).

**Table 4.** Properties of aminoplastic and aminoplastic–phenolic moldings (DIN 7708, minimum values)

| Type | Filler | Base resin | Density, g/cm$^3$ | Bending strength, N/mm$^2$ | Impact strength, kJ/m$^2$ | Notch impact strength, kJ/m$^2$ | Martens tempera-ture, °C | Surface resistance, compara-tive figure | Water absorption*, mg | Tracking resistance |
|---|---|---|---|---|---|---|---|---|---|---|
| 131 | cellulose | urea | 1.5 | 80 | 6.5 | 1.5 | 100 | 10 | 300 | 600 |
| 150 | wood flour | melamine | 1.5 | 70 | 6 | 1.5 | 120 | 10 | 250 | 600 |
| 152 | cellulose | melamine | 1.5 | 80 | 7 | 1.5 | 120 | 10 | 200 | 600 |
| 153 | textile fibers | melamine | 1.5 | 60 | 5 | 3.5 | 125 | 9 | 300 | 600 |
| 154 | chopped cotton cloth | melamine | 1.5 | 60 | 6 | 6 | 125 | 8 | 300 | 600 |
| 155 | stone powder | melamine | 2.0 | 40 | 2.5 | 1 | 130 | 8 | 200 | 600 |
| 156 | mineral filler (or fibers) | melamine | 1.8 | 50 | 3.5 | 2 | 140 | 8 | 200 | 600 |
| 157 | mineral filler (or fibers) + wood flour | melamine | 1.7 | 60 | 4.5 | 1.5 | 140 | 9 | 200 | 600 |
| 180 | wood flour | melamine/ phenol | 1.5 | 80 | 6 | 1.5 | 120 | 10 | 180 | 175 |
| 181.5 | cellulose | melamine/ phenol | 1.6 | 80 | 7 | 1.5 | 120 | 10 | 150 | 600 |
| 182 | stone powder + wood flour | melamine/ phenol | 1.6 | 70 | 4 | 1.2 | 120 | 10 | 120 | 600 |
| 183 | stone powder + cellulose | melamine/ phenol | 1.6 | 70 | 5 | 1.5 | 120 | 10 | 120 | 600 |

* Maximum values.

**Table 5.** Properties of polyester resin molding (DIN 16911, minimum values)

| Type | Filler | Form supplied | Bending strength, N/mm$^2$ | Impact strength, kJ/m$^2$ | Notch impact strength, kJ/m$^2$ | Martens tempera-ture, °C | Surface resistance, comparative figure | Water absorption*, mg |
|---|---|---|---|---|---|---|---|---|
| 801 | glass fiber long + stone powder | moist, semidry and dry | 60 | 22 | 22 | 125 | 10 | 100 |
| 802 | glass fiber short + stone powder | dry, free-flowing | 55 | 4.5 | 3 | 140 | 12 | 45 |
| 803 | glass fiber long + stone powder | moist, semidry and dry | 60 | 22 | 22 | 125 | 10 | 100 |
| 804 | glass fiber short + stone powder | dry, free-flowing | 55 | 4.5 | 3 | 140 | 12 | 45 |

* Maximum values.

fect surfaces without sink marks, such as are required for domestic appliances, can be made. Such UP molding materials are dry, free-flowing, and low-polluting and can be processed on conventional automatic machines like phenolics and aminoplastics. However, unlike the latter, the polyester resin molding materials do not cure by a polycondensation process but by polymerization. Since with this curing process no low-molecular condensation products such as water are split off, polyester moldings can be removed from the mold sooner and molded at lower pressures. The curing process is initiated by the addition of peroxides. The mold temperature necessary for the curing is adjusted to suit the peroxide used. An advantage of the polyester resins is that their synthesis from diols and diacids as well as the chain length between cross-links can be varied within a wide range. A further interesting property of styrene-containing polyester resins is that their viscosity is increased greatly by adding magnesium oxide. This enables the production of UP molding compounds that are resin-rich but still highly viscous. The production of prepregs (DIN 16913) makes use of this effect (→ Reinforced Plastics, **A23,** p. 60).

Dry, free-flowing polyester resin molding compounds containing organic fibers (e.g., cellulose) instead of glass fibers have gained acceptance in recent years, particularly in the domestic appliance sector.

**Materials Based on Epoxy Resins.** Epoxy resin molding (EP) were classified under DIN 16912 (Types 870–872). However, this field is undergoing changes in the binders used, so that some older types have been withdrawn. However, DIN 16912 enables a survey to be made of the basic properties of EP materials (Table 6). With these resins the number of possible combinations is even greater than with polyesters. Epoxy resins can be cured by three different methods: basically with diamines, acidically with diacids or acid anhydrides, and catalytically with Lewis acids. Epoxy resin molding compounds of the older Types 870–872 tend to yellow; therefore, only muted, dark colorations are possible. Hence they are used only for engineering components. Formerly, the storability of EP materials was poor, and cooled containers and cold rooms were necessary. Meanwhile, however, it has become possible to produce types with good storability and considerably improved light stability [6], although the light stability does not reach that of the UP molding compounds. The advantage of the epoxy resin molding compounds lies in their combination of good physical properties, freedom from overcuring, even at very high curing temperatures, and the very low processing shrinkage and aftershrinkage. Epoxy resin moldings therefore have a particularly low tendency to thermal stress cracking. Owing to their chemical composition they have a very good resistance to water and chemicals.

In addition, epoxy resin moldings have the advantage of retaining their electrical properties even under unfavorable climatic conditions. They permit formulations that are processible at low pressure (low-pressure compounds). The adhesion to metallic inserts is very good.

**Miscellaneous Binders.** For particularly demanding electrical requirements, diallyl phthalate molding compounds are used, with a prepolymer from diallyl phthalate as binder. The fillers can be impregnated with the powdery prepolymer by the melt fusion process. Peroxides are suitable for curing, as with the UP resins. Molded materials on this basis have outstanding electrical values, especially in moist atmosphere, and meet U.S. Military Specifications, above all in aircraft construction.

Silicone resin molding compounds satisfy still higher electrical and thermal requirements. These materials play a certain part in electronics for sheathing semiconductors. They have a very high temperature stability, even over long periods of time, but are very expensive.

Cold molding compounds based on special phenolic resins have only limited storability. They have gained some commercial importance because they permit the manufacture tracking-resistant PF moldings. Cold molding compounds (DIN 7708, part 14) are compounded wet in kneaders and molded cold. The moldings are subsequently cured pressureless with step-

**Table 6.** Properties of epoxy resin moldings (DIN 16912)*

| Type | Filler | Base resin | Density, $g/cm^3$, ca. | Bending strength, $N/mm^2$ | Impact strength, $kJ/m^2$ | Notch impact strength, $kJ/m^2$ | Martens temperature, °C | Surface resistance, comparative figure | Water absorption, mg |
|------|--------|-----------|---------|----------|---------|--------------|----------|-------------|-------------|
| 870 | inorganic granular stone powder | epoxy | 1.9 | 50 | 5 | 1.5 | 110 | 12 | 30 |
| 871 | inorganic short-fibrous glass fibers | epoxy | 1.9 | 80 | 8 | 3 | 120 | 12 | 30 |
| 872 | inorganic long-fibrous glass fibers | epoxy | 1.9 | 90 | 15 | 15 | 125 | 12 | 30 |

* Minimum values, except for water absorption (maximum value).

wise increase of temperature over a fairly long period of time (up to several days). The surface quality and dimensional accuracy of such moldings do not, however, approach the high standards of conventional moldings.

**Test Methods for Molded Materials.** Most of the test methods used for testing molded materials are defined in DIN specifications. For thermoset plastics, the test specimens are produced by compression molding. This eliminates the filler-orientation effects which invariably occur in injection molding. Studies are available that allow the directional dependence of the properties of injection molded test specimens, to be estimated, especially the shrinkage behavior [7].

The most important DIN test methods are:

| | |
|---|---|
| Bending strength | DIN 53 452 |
| Impact and notch impact strength | DIN 53 453 |
| Compressive strength | DIN 53 454 |
| Modulus of elasticity | DIN 53 457 |
| Ball indentation hardness | DIN 53 456 |
| Martens heat resistance | DIN 53 458 |
| Linear expansion coefficient | DIN 53 752 |
| Thermal conductivity | DIN 52 612 |
| Incandescence resistance | DIN 53 459 |
| Surface resistance | DIN 53 482 |
| Electrical volume resistivity | DIN 53 482 |
| Dielectric loss factor | DIN 53 483 |
| Relative permittivity | DIN 53 483 |
| Dielectric strength | DIN 53 481 |
| Tracking resistance | DIN 53 480 |
| Water absorption | DIN 53 495 |

In addition there is also a series of special test methods laid down by firms or associations, and test specifications for thermoset plastics from ISO, European Standards EN, and the UL (Underwriter's Laboratories).

Users' requirements for comparability of very different plastics have been generally accepted, which led to revision of DIN 7708. This began with a harmonization of the dimensions and production of the test specimen and was completed with the new edition of DIN 7708, part 8 (February 1992).

The new test specimen is of a multipurpose type, originally proposed for thermoplastics, and adapted to the special requirements for the processing of thermosets. Dyes for producing the test specimens by injection molding and compression molding have been developed, and comparisons made between the test values obtained and the previous values.

# 6. Processing and Aftertreatment

Thermoset molding compounds are hot molded and cured under pressure, sometimes with the addition of hardeners. The usual processes are compression molding, injection molding, transfer molding, injection–compression molding, and extrusion. For a detailed description of the processing of molding compounds to give cured moldings, see [4].

Moldings must generally be deflashed. This can be performed manually by filling, grinding, punching, sawing, abrasive blasting, or brushing. For small moldings or large numbers, mechanical processes are used, such as tumble deflashing, abrasive blasting, or brushing. Sawing, milling, turning, and drilling are also possible [8]. Tools with carbide cutting edges are usually used. Detailed information on steel quality, cutting speed, and tool quality can be found in [8].

If necessary, the moldings can be polished with polishing wheels or in drums. Thermoset moldings can be joined by detachable joints (e.g., screw or clamp joints) or nondetachable joints such as rivets and adhesive bonds [9].

Thermoset moldings are sometimes finished by painting, flocking, hot-embossing, printing, or metallizing by electrodeposition or vacuum deposition.

# 7. Uses

The main applications for moldings made from thermoset molding compounds are electrical engineering, domestic appliances, mechanical engineering, the automotive industry, and the sanitary sector. Thermoset molding materials also have versatile uses as bottle seals, in precision engineering, and in office machine manufacture. Typical moldings in the electrical engineering sector are collectors for electric motors, slip rings, contacts, strip terminals, distributor box cases, switches, plugs, and other connectors. Collectors also play a major role in the automotive industry, as well as ignition distributors, spark plug sockets, control pistons for power brakes, bushes, and axial face seals. In the household sector, thermoset moldings are used for high-quality camping ware, ashtrays, handles for irons, pot and pan handles, grill and toaster components, lamps, and cutlery. In the sanitary sector, high-quality toilet fittings are manufactured in many colors from thermoset molding compounds. They are used for a great variety of cosmetic containers (including metallized ones). In mechanical engineering, they are suitable for highly stressed bearings, including self-lubricating bearings, and for coverings, rollers, reels and operating controls.

## 8. Economic Aspects

The "thermosets" discussed here (thermoset molding compounds) constitute only a narrow area of the thermosetting plastics. They contain 20–60% of thermosetting binders, but such binders are also constituents of a wide variety of other products. Production of curable free-flowing molding compounds is summarized in Table 7.

For the phenolic resins, which make up ca. 2% of foundry molding compounds and ca. 60% of electrical laminates, a total production volume of material for the Federal Republic of Germany in 1988 of almost $3 \times 10^6$ t can be estimated [10]. The production of thermoset molding compounds based on phenolic resin, on the other hand, is much lower, Table 8 [11].

Market growth for thermoset molding compounds is estimated at 3–4% for the next few years. In automobile manufacture even 6–7% may be reached. Here the thermosets compete mainly with metals. Quantitatively, the largest fields of application are electrical engineering, followed by domestic appliances and tableware, and sanitary products. These considerations do not include the large field of polyester dough molding compounds (bulk molding compounds, BMC) and prepregs (sheet molding compounds, SMC).

## 9. Environmental Protection and Recycling [12]

Safety data sheets (DIN 52900) are available for all thermoset molding compounds. They give information on storage and handling, transport, fire protection, toxicology, ecology, waste disposal, etc.

Phenolic moldings that contain > 1% free phenol must be labelled as harmful ($X_n$). However, by the 1970s manufacturers had already undercut this 1% limit by using low-phenol resins. Most phenolic molding compounds therefore do not require such labelling. According to all transport regulations, molding compounds are not deemed to be hazardous materials.

In connection with processing, further questions arise such as which gases and vapors are formed, and in what amounts? For phenolic molding compounds, extensive studies on this subject are available. During processing, phenol, formaldehyde, and ammonia are released, in addition to water vapor. The quantities have been determined by using an enclosed mold, extracting the gases, and absorbing them in appropriate solutions. The following approximate values were obtained:

Phenol, ca. 5 mg per kg molding compound
Formaldehyde, ca. 2–4 mg per kg molding compound
Ammonia, ca. 20–40 mg per kg molding compound

The MAK values for these substances :
Phenol 5 ppm (19 mg/m$^3$)
Formaldehyde, 0.5 ppm (0.6 mg/m$^3$)
Ammonia, 50 ppm (35 mg/m$^3$)
can be maintained by using extraction hoods.

During processing of aminoplastics, formaldehyde quantities of up to ca. 25 mg/m$^3$ have been observed, and demands on the extraction equipment are therefore greater.

Thermoset molding compounds based on phenolic resin, melamine resin, and urea resins contain as binder polycondensation products

**Table 7.** Production of curable, free-flowing molding compounds

| Region | Production, 10$^3$ t | | | |
|---|---|---|---|---|
| | 1979 | 1982 | 1985 | 1988 |
| Federal Republic of Germany * | 58 | 55 | 59 | 60 |
| Western Europe | 210 | 190 | 215 | 207 |
| United States | 189 | 115 | 135 | 133 |
| Japan | 120 | 115 | 114 | 103 |

* Without urea molding materials, which are no longer produced in Germany.

**Table 8.** Production of phenolic molding compounds

| Region | Production, 10$^3$ t | | | | |
|---|---|---|---|---|---|
| | 1987 | 1988 | 1989 | 1990 | 1991 |
| Japan | 59.2 | 64.9 | 67.3 | 65.2 | 64.8 |
| Korea | 5.0 | 5.4 | 5.8 | 5.8 | 6.0 |
| South-East Asia | 16.2 | 17.8 | 25.0 | 26.0 | 28.0 |
| United States | 89.5 | 89.7 | 79.9 | 76.3 | 70.8 |
| Western Europe | 75.5 | 74.0 | 72.0 | 73.0 | 67.0 |
| Germany* | 29.0 | 31.0 | 32.0 | 32.0 | 31.5 |

* Included in data of Western Europe.

produced using formaldehyde. However, owing to polycondensation formaldehyde is not present in the free form. Curing converts the resin molecules via reactive centers derived from formaldehyde into an almost inert network, which gives the molded materials their resistance to temperature and aggressive media. With correct curing, the regeneration of formaldehyde in the molding is possible only to a minor extent in the case of the aminoplastic molding compounds (UF, MF and MP). However, it is observed only under extreme thermal or hydrolytic conditions. In moldings based on phenolic resin, such a release of formaldehyde is practically immeasurable.

No formaldehyde is used in the synthesis of unsaturated polyester resins. For special applications, polyester molding compounds are modified with up to 5 % melamine resin. In these cases small quantities of formaldehyde can escape during the processing.

During the processing of styrene-containing unsaturated polyester dough molding compounds, styrene may be released. Meeting the MAK value for styrene of 20 ppm (85 mg/m$^3$) requires elaborate ventilation systems. Producers of such UP molding compounds are therefore attempting to reduce the styrene content of their products.

A few years ago asbestos was still a popular reinforcing fiber for thermosets. Although the fibers are enveloped by resin in the process, in compounding and aftertreatment of the moldings exposure of the fibers cannot be avoided. Such asbestos fibers are classified as carcinogenic in certain geometries (particles with a length of $> 5$ µm, and a diameter of $< 3$ µm with a length-to-diameter ratio $> 3:1$). Substitutes have since been found and new asbestos-containing molding materials are no longer marketed. The Bakelite company, for example, ceased production of asbestos-containing molding compound at the beginning of 1991.

**Recycling** of thermoset molding compounds was carried out as early as the 1930s, but was not economic at that time, since the wastes had to be very finely ground to avoid surface defects in the new moldings. There was also a risk of entraining impurities. For controlled, quality-assured moldings in Germany the addition of reground material is only allowed if the processor undertakes the inspection necessary to safeguard quality.

As early as the 1970s, studies were carried out worldwide to facilitate the addition of reground material. The conclusion was that the addition of 10–20 % of reground material is possible without significantly impairing the test values.

A similar particle recycling process is used for sheet molding compounds. Raw material manufacturers and processors have established a company for this purpose (Ercom, Karlsruhe).

The problem of damage to thermoplastics on recycling does not occur with thermosets, since the reground material is used as a filler. In the finely ground state, this reground material is a particularly high-quality filler since it is impregnated.

The regenerate can be added either batchwise in mixing vessels or continuously in injection molding machines. The particle size of the regenerate should be ca. 0.1–0.4 mm. Too coarsely ground material leads to surface and textural defects.

**Incineration.** Since most thermoset molding compounds are free of halogens, disposal by incineration is also possible. The calorific values of such wastes of 12500–21000 kJ/kg are relatively high. However, thermoset wastes are not accepted by all refuse incineration plants. It is occasionally possible to use them as additives in cement manufacture to economize on oil and coal.

**Dumping.** Because of their insolubility, cured molded materials can be dumped on landfills for domestic waste. Uncured wastes are regarded as pollutants.

In Germany an active environmental study group for thermosets has existed for several years [13].

A comprehensive survey of the problems of plastics recycling is given in [14].

# 10.  References

[1] S. H. Goodman: *Handbook of Thermosets*, Noyes Publications, Park Ridge, N.J., 1986.
[2] "Duroplaste," *Kunstoff-Handbuch*, vol. X, Hanser, München 1988.
[3] W. Schönthaler, *Maschinenmarkt* **77** (1971) 2227–2231.
[4] W. Schönthaler: *Verarbeiten Härtbarer Kunststoffe*, Taschenbuch T 40, VDI-Verlag, Düsseldorf 1973.
[5] W. Bauer, W. Woebcken: "Verarbeitung duroplastischer Formmassen," *Kunststoffverarbeitung*, vol. 3, Hanser, München 1973.
[6] P. Tschanz, *Kunstst. Plast.* **25** (1978) 1.

[7] W. Schönthaler, K. Niemann, *Maschinenmarkt* **82** (1976) 1638–1641 and **83** (1977) 229–230.

[8] H. Zickel: "Das spanabhebende Bearbeiten der Kunststoffe," *Kunststoffverarbeitung*, vol. 4, Hanser, München 1963.

[9] W. Schönthaler, J. Tetz, *Maschinenmarkt* **81** (1975) 257–259.

[10] W. Schönthaler, U. Braun, *Kunststoffe* **80** (1990) 1165 ff.

[11] M. Kraemer, *Plast. Eur. Kunstst.* Oct. 1992, p. 493.

[12] W. Schönthaler: "Duroplaste Zukunft von Anfang an," Script, Würzburg 1993.

[13] Technische Vereinigung der Hersteller u. Verarbeiter duroplastischer Kunststoff-Formmassen, Würzburg, brochure.

[14] H. Breuer, E. Dolfen, *Kunststoff-Recycling Handbuch 1992*, A. Hüthig Verlag, Heidelberg.

**Thiamine → Vitamins**

# Thin Films

GEORG WAHL, Technische Universität Braunschweig, Institut für Oberflächentechnik, Braunschweig, Federal Republic of Germany (Section 1.1)

PAUL B. DAVIES, University of Cambridge, Department of Chemistry, Cambridge, United Kingdom (Sections 1.2 and 2.6.2)

ROINTAN F. BUNSHAH, University of California, Department of Materials Science and Engineering, Los Angeles, United States (Section 1.3)

BRUCE A. JOYCE, University of London, Imperial College of Science, Technology and Medicine, London, United Kingdom (Section 1.4)

COLIN D. BAIN, University of Oxford, Physical Chemistry Laboratory, Oxford, United Kingdom (Section 1.5)

GERHARD WEGNER, MARKUS REMMERS, Max-Planck-Institut für Polymerforschung, Mainz, Federal Republic of Germany (Sections 1.6 and 2.6.1)

FRANCIS C. WALSH, University of Portsmouth, Department of Chemistry, Portsmouth, United Kingdom (Sections 1.7 and 2.4)

KONRAD HIEBER, Fraunhofer Gesellschaft, Institut für Festkörpertechnologie, München, Federal Republic of Germany (Section 2.1)

JAN-ERIC SUNDGREN, Linköping University, Department of Physics and Measurement Technology, Linköping, Sweden (Section 2.2)

PETER K. BACHMANN, Philips GmbH, Research Laboratories, Aachen, Federal Republic of Germany (Section 2.3)

SHINTARO MIYAZAWA, NTT LSI Laboratories, Kanagawa, Japan (Section 2.5)

ALFRED THELEN, Ingenieurbüro Thelen für optische Interferenzschichten, Schmitten-Seelenberg, Federal Republic of Germany (Section 2.7)

HEINER STRATHMANN, University of Twente, Faculty of Chemical Technology, Enschede, The Netherlands (Section 2.8)



Ullmann's Encyclopedia
of Industrial Chemistry, Vol. A 26

# 1.  Preparation of Thin Films

## 1.1.  Chemical Vapor Deposition Processes

Chemical vapor deposition (CVD) produces layers from the gas phase with thicknesses in the range ca 0.1–10 µm, by means of a chemical reaction. Typical pressure ranges are different from those used in physical vapor deposition ($< 1$ hPa); ca 1–50 hPa for low-pressure CVD (LPCVD), 50 to $< 1000$ hPa for medium-pressure CVD, and atmospheric pressure for atmospheric CVD (ACVD) [1.1]–[1.5]. Accordingly, the Knudsen number $\lambda/d$ (the ratio of the mean free path of molecules in the gas phase to a typical reactor dimension) is small, and diffusion and convection processes are the main transport mechanisms. Reynolds numbers are in the range 1–1000, so laminar gas flow is the normal flow mode and buoyancy effects are important, especially at higher pressure [1.4].

In contrast to chemical transport processes, CVD is an open system with a gas inlet and outlet. The equipment (Fig. 1) has the following main parts: a gasification system (a, b, c, j), a chemical reactor with volume of the order of cm$^3$ or m$^3$, which contains the samples or workpieces to be coated (d, e, f), and an exhaust system for the reacted gas (g, h, i). Precursor compounds can be inorganic (e.g., halides, hydrides, carbonyls) or organic (alkyls, β-diketonates, alkoxides, or organometallics [1.6], [1.7]). Almost all

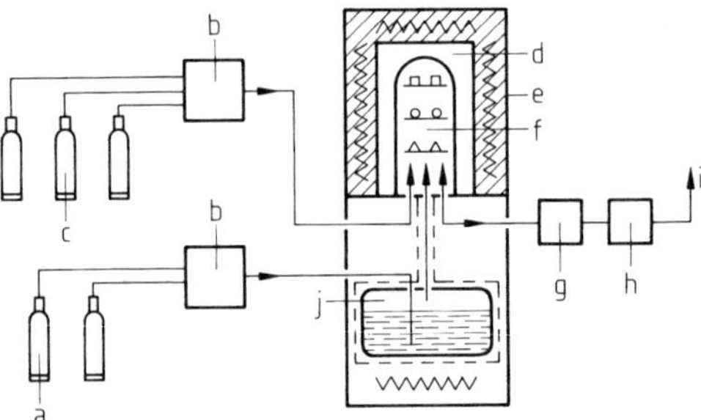

**Figure 1.** CVD system for deposition on tools (e.g., TiN, TiC)
a) Gases; b) Gas control unit; c) Gases for evaporation unit or precursor gases; d) Interspace to remove reactor from the heating system for cooling; e) Heating system; f) Deposition reactor (hot-wall reactor); g) Filter; h) Pump; i) Ventilation, j) Evaporator (liquid or solid)

compounds can be deposited by CVD, provided the material can be dissolved in the gas phase. Usually, the possibility of depositing material from a gas mixture and the expected phase composition can be calculated by thermodynamic equilibrium calculations, minimizing the Gibbs free energy. These estimates are especially suitable for CVD processes at high deposition temperatures (e.g., SiC deposition at ca. 1000 °C [1.8]). At low temperature, kinetic effects are often important, and the deposition of metastable phases (e.g., amorphous SiO$_2$ phases) is possible.

The structure of the coating is the result of several steps: adsorption–desorption processes, surface reactions, surface diffusion, nucleation, growth of the critical nuclei, layer formation, and aging processes in the layer during deposition. The result ist often the formation of preferred orientations of the deposited layer on crystalline and amorphous substrates, depending on the deposition conditions. On monocrystalline substrates, epitaxy is possible if the lattice misfit is not too large [1.9].

There are a number of methods to activate the chemical reactions necessary for deposition.

**Thermal CVD.** The surface or homogeneous reactions necessary for layer formation can be thermally activated (thermal or conventional CVD) by heating the reactor, including the samples (hot-wall reactor, Fig. 1) or by heating the samples only (cold-wall processes) to the deposition temperature $T_d$ (Table 1). These processes have the "ideal" temperature dependence shown in Figure 2. At medium temperature, in the so-called transport-controlled regime, transport in the gas phase determines the deposition rate, and the temperature dependence is very smooth, proportional to $T^v$ with $v = 0.7-1.0$. This range can be used only for deposition on simple geometries, if layers of constant thicknesses are required, because the deposition depends critically on gas flow conditions. Deposition rates are determined by the concentration of precursor molecules in the gas phase and by the gas flow. In stagnation flow reactors, deposition rates up to 100 µm/min can be reached, as shown for tungsten deposition with a precursor content in the hydrogen gas phase of ca. 30% $WF_6$ [1.11].

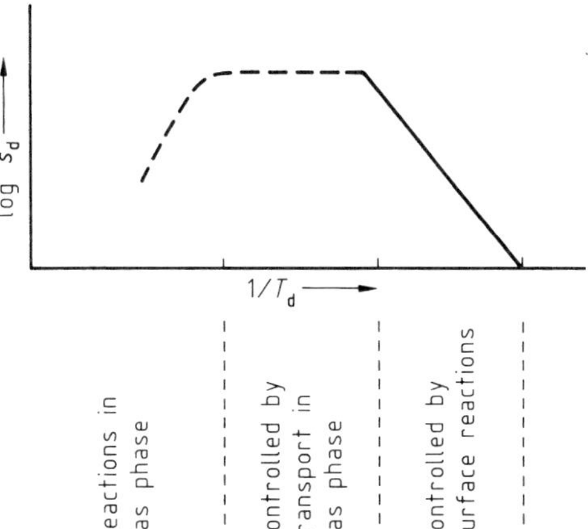



**Figure 2.** Temperature dependence of an "ideal" CVD process

At low temperature, the deposition rate is given by an Arrhenius law:

$$\dot{s} = k \exp(-E/RT) \tag{1}$$

where $k$ is a constant, $E$ is the activation energy, and $R$ the gas constant. In surface reactions, $T$ is equal to the deposition temperature $T_d$. If the rate-determining step is in the gas phase, it is difficult to determine the reaction temperature if the reactor is not isothermal. In spite of this, $T_d$ is usually estimated from Equation (1). Typical activation energies are given in Table 1. At high temperature, the deposition rate often decreases because homogeneous reactions and nucleation in the gas phase reduce transport of precursor molecules to the surface. A more detailed de-

Table 1. Examples of thermal CVD processes

| Deposited material | Type of reaction | Reaction | Deposition temperature, °C | Activation energy, kJ/mol | |
|---|---|---|---|---|---|
| W | pyrolysis | $WCl_6 \rightarrow W + \ldots$ | >1000 | | |
| W | reduction | $WCl_6 + 3H_2 \rightarrow W + \ldots$ | >800 | 627 | [1.10] |
| W | reduction | $WF_6 + 3H_2 \rightarrow W + \ldots$ | >400 | $60 \pm 10$ | [1.11] |
| Si | pyrolysis | $SiH_4 \rightarrow Si + \ldots$ | >500 | $150 \pm 10$ | [1.12] |
| Si | reduction | $SiCl_4 + 2H_2 \rightarrow Si + \ldots$ | >1000 | $150 \pm 10$ | [1.13] |
| $SiO_2$ | oxidation | $SiH_4 + \ldots O_2 \rightarrow SiO_2 + \ldots$ | >400 | | |
| SiC | displacement | $SiCl_4 + \ldots CH_4 \rightarrow SiC + \ldots$ | >900 | | |
| Ge | disproportionation | $2GeI_2 \rightarrow Ge + GeI_4$ | >600 | | |
| TiC | disproportionation | $TiCl_4 + \ldots CH_4 \rightarrow TiC + \ldots$ | >900 | | |
| Ni | pyrolysis | $Ni(CO)_4 \rightarrow Ni + \ldots$ | >150 | | |
| $Y_1Ba_2Cu_3O_{7-\delta}$ | oxidation | $1\,Y\,(dpm)_3 + 2\,Ba(dpm)_2 + 3\,Cu(dpm^*)_2$ $+ \ldots O_2 \rightarrow Y_1Ba_2Cu_3O_{7-\delta}$ | >600 | $82 \pm 5$ | [1.14] |

* dpm = Dipivaloyl methanoat

scription of the dependence of deposition on pressure, temperature, gas composition, and gas flow is possible by the methods of chemical kinetics and the theory of catalysis including fluid dynamics [1.1]–[1.5]. Thus the temperature dependence is often more complicated than shown in Figure 1.

The activation of the chemical reaction at the surface can also be achieved by laser irradiation. In this case, localized deposition with lateral dimensions down to a few micrometers is possible. This laser CVD process can be mainly thermal or mainly photochemical (photo CVD), and allows patterning of surfaces by laser-direct write. The fabrication of three-dimensional microobjects by laser CVD has been demonstrated [1.15].

Practical deposition rates (0.1–1 µm/min) are mostly a compromise between the coating structure and the throwing power (the ability to produce layers of constant thickness on the surface of the workpiece). Thermal CVD processes are used industrially for the production of hard coatings (TiC, TiN, $Al_2O_3$), and in the electronics industry (Si, $SiO_2$, $Si_3N_4$, III–V compounds).

**Plasma CVD.** Very often the deposition temperature of conventional CVD is too high; deposition methods have been developed to decrease the deposition temperature, e.g., plasma-activated deposition processes (PACVD), also called plasma-enhanced CVD (PECVD). A plasma is an electrically neutral gas in which neutral species, ions, and electrons are all present. Plasmas are ignited by d.c. or a.c. in the frequency range MHz to GHz. ln d.c. (continuous or pulsed) discharges, the sample is the cathode and the discharge is ignited between the cathode and the anode. The cathode surface is coated and at the same time bombarded with positive ions with energy in the range of a few hundred volts, and current density 1–10 mA/cm². If the sample to be coated is insulating, an a.c. discharge is necessary to avoid charging effects. The different mobil-

ities of ions and electrons produce a space-charge layer on the electrodes, with a potential drop similar to that in a d.c. plasma. The voltage can be further increased by changing the ratio of the electrode areas, smaller areas being associated with higher voltages (approximately proportional to $(A_1/A_2)^v$, where $A_1$ and $A_2$ are the areas of the the the electrodes, and $v = 1$–4 [1.3], [1.16]. In plasmas used for CVD, the ratio of charged to neutral species can be as low as $10^{-6}$–$10^{-5}$ [1.3], [1.17] (weakly ionized plasmas). The total pressure is $10$–$10^3$ Pa. This low pressure leads to few collisions in the gas phase before species reach the walls or recombine. Therefore, no complete temperature equilibration of neutral species, ions, and electrons is formed. Neutral species and ions have temperatures in the range $kT = 0.1$ eV (where $k$ is the Boltzmann constant, 0.1 eV = 1160 K); electron temperatures are in the range 1–10 eV. The role of the plasma can be twofold:

1) To change the gas phase composition and thus change the reaction system by formation of intermediates, radicals, excited molecules, and ions. Plasmas for this application can also be produced remote from the sample position (Fig. 3 A).
2) To modify the gas phase and deposition conditions. This is achieved by ignition of the gas discharge directly above the deposition surface (Fig. 3 B). Ion bombardment through the space-charge layer changes the surface processes and hence the structure of the coating, usually improving adhesion to the substrate.

Plasma CVD processes are used for the production of hard coatings on steels [1.17] and in the electronics industry ($Si_3N_4$).

## 1.2. Plasma Polymerization

Plasma polymerization refers to the formation of polymeric films on surfaces in proximity to a low-pressure glow discharge in an organic or

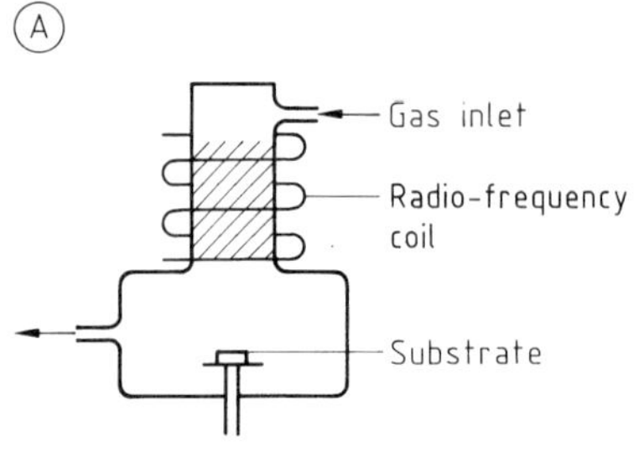



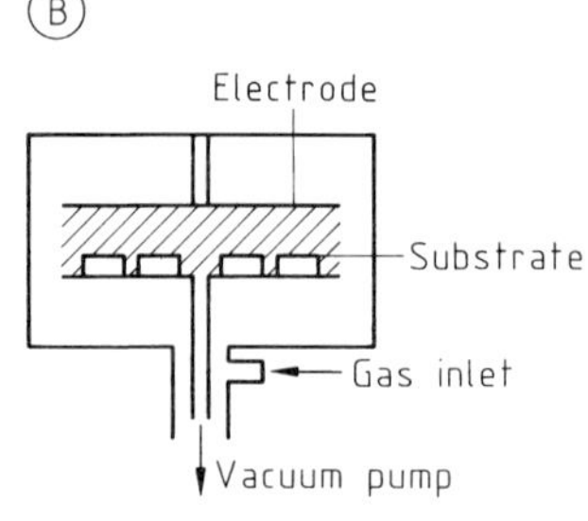



**Figure 3.** Basic configurations of plasma reactors for CVD A) Remote plasma reactor (inductively coupled); B) Parallel-plate reactor (capacitively coupled)

organometallic vapor. Among the unique properties of these films are: controlled thickness up to ca. 2 µm, good adhesion to a variety of substrates, and variable physical properties such as hardness. The useful characteristics of such films has led to their widespread use in applications as diverse as protective coatings and osmotic membranes. Some of the features of plasma polymerization are similar to those found in plasma-enhanced chemical vapor deposition (PECVD). Several plasma parameters are known to influence the polymerization process, including pressure, flow rate, power and frequency of the discharge, and reactor geometry.

A variety of reactor geometries have been used, including those with internal or external electrodes. Reactors with internal electrodes usually have a parallel plate design in which one electrode is grounded (onto which the substrate is placed) and the other is fed with capacitively coupled radiofrequency (r.f.) power. Tubular reactors are often inductively coupled.

A typical bell jar reactor with horizontal, parallel plate electrodes is shown in Figure 4. It has been used [1.18] to prepare protective films

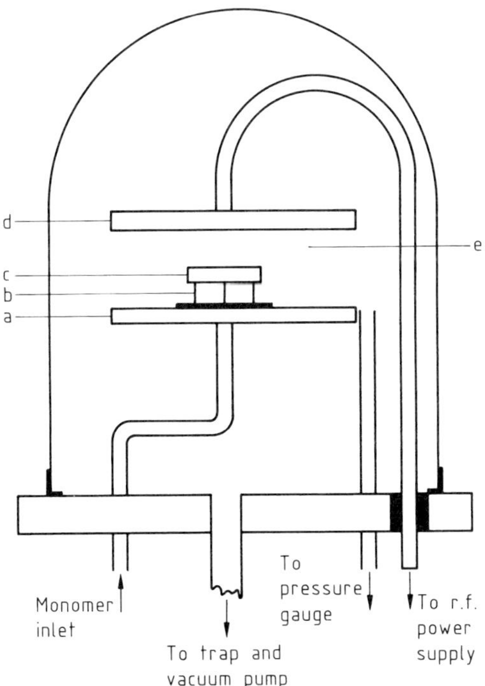


**Figure 4.** A parallel plate capacitively coupled plasma reactor [1.18]
a) Ground electrode; b) Support pedestal; c) Optical crystal; d) Hot electrode, e) Plasma

of fluorinated polymers from tetrafluoroethylene (TFE) and chlorotrifluoroethylene (CTFE) monomers. Plasma reactors must have good vacuum integrity; in this example it was better than 1 Pa. The pressure in the operating reactor was ca. 10 Pa. Typical deposition rates over a 5-cm diameter substrate were $7.9 \times 10^{-6}\,\mathrm{g\,cm^{-2}\,min^{-1}}$ (48 nm min$^{-1}$) with 20 W r.f. power, 13 Pa TFE, and 13 Pa argon. Deposition time for a 1-µm coating on NaCl substrate was ca. 40 min.

The control and measurement of pressure in a reactive plasma presents several problems. The pressure in the reactor can change significantly when the discharge is turned on, even at constant flow rate, owing to changes in the concentration and nature of the plasma species. The ideal device for measuring the pressure in the plasma must be immune from the electrical effects of the plasma, and not disturb the plasma chemistry. A pressure transducer which is independent of the nature of the plasma species is usually selected. At constant flow rate and current density, the deposition rate increases linearly up to some limiting pressure. Maximum reactor pressures which have been used are ca. 500 Pa, but pressures below 250 Pa are usually employed.

Closely allied to the gas pressure is the flow rate $F$. The dependence of deposition rate on flow rate at constant pressure is as follows for ethane plasma polymerization. At high flow rates deposition tends to decrease with increasing flow rate, owing to decreasing residence time. Conversely, the longer residence times at low flow rates enable more of the monomer to react, and under these conditions deposition rate increases with flow rate. While this general behavior is repeated at different pressures, an increase in pressure leads to a decrease in deposition rate, owing to a lower in electron density [1.19].

Increasing the power density increases the deposition rate, as might be expected from the increased electron density. However, the plasma chemistry also changes which directly affects the rate of polymerization. Power dissipation depends uniquely on the reactor geometry and on the monomer under study.

To facilitate comparisons of one plasma with another YASUDA [1.20] has used a composite parameter $W/FM$, where $M$ is the molecular mass of the monomer. The relevant dimensions are $(W/FM) \times 1.34 \times 10^9$ J/kg, where $W$ is the discharge power in Watts and $F$ is in cm$^3$ (STP)/min. The most useful form of this parameter is at the critical power value $W_c$, defined as the power required to fill the reactor with the glow discharge. Above this power level, polymerization is independent of power and the pressure remains constant as the power is increased. The slope of plots of $W_c/FM$ versus flow rate are approximately the same for different monomers containing the same functional group(s). The slope is highest for saturated hydrocarbons and $W_c/FM$ is almost independent of flow rate for compounds with triple bonds or aromatic structure. The slope correlates with the quantity of $H_2$ produced in the plasma and is highest in the degradation of noncyclic saturated hydrocarbons. This has led to general classification [1.20] of hydrocarbons into three categories: compounds containing triple bonds in addition to aromatic structure; compounds containing double bonds and/or cyclic structure; noncyclic saturated hydrocarbons. This classification assumes unchanged reactor geometry.

Plasma polymerization discharges have been operated from d.c. to microwave frequencies (2.45 GHz). Probably the most commonly used frequency is 13.56 MHz. The disadvantage of d.c. plasmas is that extensive polymerization at the electrodes changes the discharge conditions. At high frequencies (13.56 MHz) the energetic charged species in the plasma are contained within the plasma itself. Polymer films which might be sensitive to highly energetic species are therefore best formed at high radiofrequencies.

The mechanism of plasma polymerization is not fully understood, but includes both homogeneous and heterogeneous reactions. A characteristic of plasma polymers is the presence of large concentrations of free radicals. This implies that free radicals are important in the polymerization process. More conventional polymerizations do not involve free radicals, which are normally quenched in the termination step. A free radical mechanism which explains the preponderance of these species is the rapid step growth principle. In RSGP, mono- or biradicals generated in the

plasma can add to further monomer or other plasma molecules to form a new radical species, which can itself add further monomer or, if inactive, can be reactivated in the plasma and reenter the cycle [1.20]. In this scheme no distinction is made between gas phase or surface reactions.

Unlike conventional condensed-phase polymerization, which requires the monomer to contain one or more unsaturated functional groups, no such restriction applies to plasma polymerization and almost any organic molecule with a vapor pressure of a few hundred Pascals is amenable to this treatment. Monomers used include hydrocarbons, fluorocarbons, silane and other silicon compounds (siloxanes and silazines), and organometallic species.

## 1.3. Physical Vapor Deposition (PVD) Processes

The basic PVD processes fall into two general categories: sputtering and evaporation. The application of PVD techniques ranges from decorative to high-temperature superconducting films. The thickness of the deposits can vary from $10^{-10}$ to $10^{-3}$ m. Very high deposition rates (25 µm/s) have been achieved with the advent of electron-beam-heated sources. A very large number of inorganic materials—metals, alloys, compounds, and mixtures as well as some organic materials—can be deposited by PVD technologies. "Ion plating" is a hybrid PVD process, i.e., an atomic film deposition process in which the substrate surface and/or the depositing film is subjected to a flux of high-energy particles sufficient to cause changes in the interfacial region between the film and the substrate, as well as in the properties of the deposited film compared with a nonbombarded film. These changes may be in the adhesion of the film to the substrate, film morphology, density, or stress. The source of the depositing species can be evaporation, sputtering, gases, or vapors (Table 2).

**Table 2.** Classification of PVD processes

|  | Metals/alloys | | Compounds | |
|---|---|---|---|---|
| Process | Basic | Hybrid | Basic | Hybrid |
| Conventional | evaporation deposition sputter deposition | ion plating sputter ion plating | direct evaporation direct sputtering | |
| Plasma-assisted (PAPVD) | | | activated reactive evaporation (ARE) reactive sputtering (RS) | reactive ion plating (RIP) biased RS (BRS) |

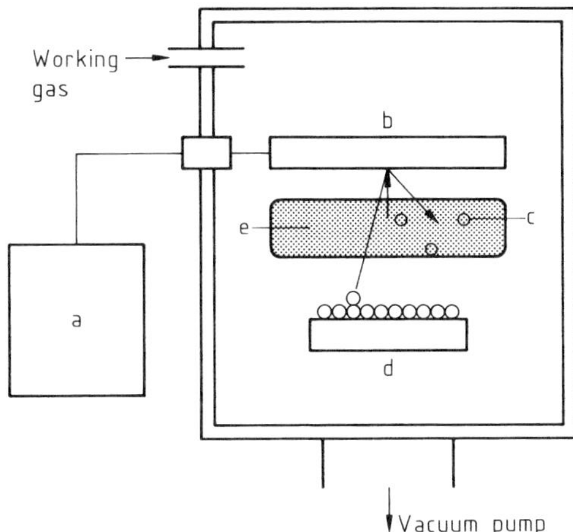



**Figure 5.** The sputter deposition process
a) Power supply; b) Cathode (target); c) Ion; d) Substrate;
e) Plasma

### 1.3.1. Sputter Deposition

Sputtering (Fig. 5) is momentum transfer from an indirect energetic projectile, such as an ion, to a solid or liquid target, resulting in the ejection of surface atoms or molecules. The target and the substrate are placed in the vacuum chamber and evacuated to a pressure of $10^{-2}$–$10^{-5}$ Pa. The target (cathode) is connected to a negative voltage supply and the substrate generally faces the target. A glow discharge is initiated after an inert gas (usually argon) is introduced into the evacuated chamber. Typical working pressure is 2.5–20 Pa.

**Planar Diode Glow Discharge Sputter Deposition.** This is the simplest sputtering system, consisting of the cathode (target) and anode facing each other. The substrates are placed on the anode. The target, which is usually water cooled, performs two functions: as the source of coating material and as the electrode sustaining the glow discharge. The distance between cathode and anode is usually ca. 5–10 cm.

The deposition rate is mainly determined by the power density at the target surface, size of the erosion area, source–substrate distance, source material, and working gas pressure. Some of these factors are interrelated, e.g., pressure and power density; optimal operation is obtained by controlling the parameters to get the maximum power flux which can be applied to the target without causing cracking, sublimation, or melt-

ing. The maximum power limit can be increased if the cooling rate of the target is increased by designing the coolant flow channels properly and improving the thermal conductance between the target and the target backing plate.

Even though planar diode glow discharge sputter deposition techniques are widely used, owing to their simplicity and the relative ease of fabrication of targets for a wide range of materials, they have several disadvantages: low deposition rate, substrate heating due to the bombardment of high-energy particles, and relatively small deposition surface areas.

**Magnetron Sputter Deposition.** By appyling magnetic fields to the diode sputtering process, the ionization efficiency can be greatly increased. In the magnetron sputter deposition process, an applied magnetic field parallel to the cathode surface forms electron traps and restricts the primary electron motion to the vicinity of the cathode. The magnetic field strength is of the order of $10^{-2}$ T and therefore it can influence the plasma electrons, but not the ions. The electrons trapped on a given field line can advance across the magnetic field to the anode or walls via collisions (mostly with gas atoms). Therefore, their chances of being lost to the walls or anode without collisions are very small. Because of the higher efficiency of this ionization mechanism, the process can be operated at ca. 0.15 Pa with high current densities at low voltages, thus providing high sputtering rates.

There are several configurations of the magnetron sputter deposition technology. Figures 6–8 show the cylindrical magnetron, the planar magnetron, and the S-gun magnetron. The cylindrical magnetron is very useful to prepare uni-

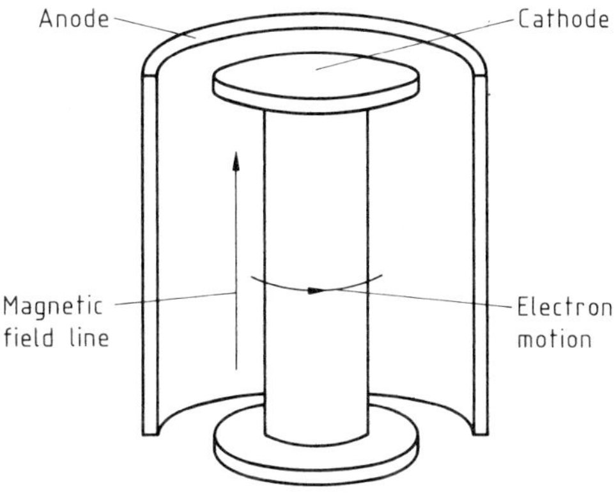



**Figure 6.** A cylindrical magnetron sputtering source

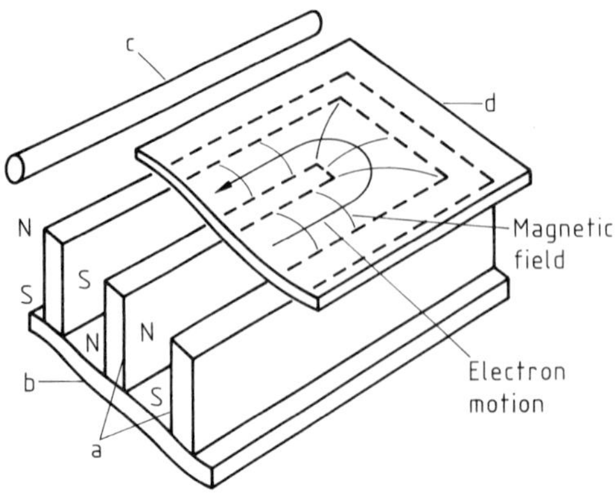

**Figure 7.** Planar magnetron sputtering sources
a) Permanent magnets; b) Pole piece; c) Anode; d) Cathode

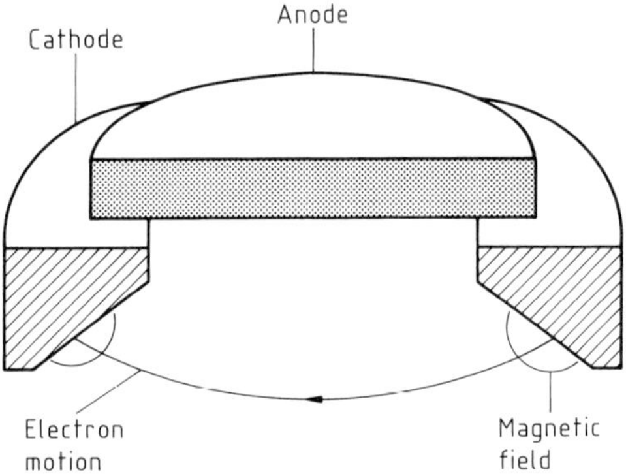

**Figure 8.** S-gun magnetron sputtering source

form coatings over large areas, since long cathodes are employed. Furthermore, the cylindrical hollow magnetron technique is effective for coating complex-shaped objects. The cylindrical post magnetron can be used to avoid substrate bombardment by energetic particles, thus preventing heating of the substrate. Metallic and dielectric films can be deposited at higher deposition rates by planar magnetron sputtering than by diode sputtering. To deposit films on thermally sensitive substrates, such as electronic devices, the S-gun magnetron can be used, since this technique allows good isolation of the substrate from the plasma.

The balanced magnetron was developed primarily for metallization in microelectronic applications where electron/ion bombardment of the growing film is to be avoided. However, for metallurgical applications, ion bombardment of the film is desirable for the changes it produces in its structure and properties. Often, a separate ion gun is incorporated into the system to ion bombard the growing film. In recent years, the development of the unbalanced magnetron has resulted in a practical way to carry out ion bombardment without the extra ion gun. Figure 9 shows the balanced and unbalanced magnetrons. The latter incorporates a controlled leak into the magnetic trap, permitting magnetic field lines as well as electrons and ions to stream toward the substrate. A plasma is thus present near the negatively biased substrate from which positive ions can bombard the growing film.

Another variation is ion beam sputtering where ions of controlled energy from an ion source impinge on a target. The advantages are excellent adhesion; a resulting high-purity deposit because of the low operational pressure (ca. 0.015 Pa); and very low substrate heating effects since the substrate is not in contact with the plasma, as is the case for diode sputtering. If a second ion source is used to ion bombard the film, the process is called dual ion beam sputter-

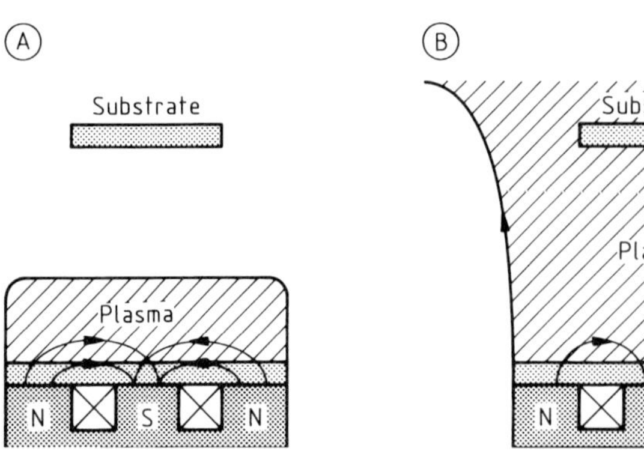

**Figure 9.** A) Balanced magnetron;
B) Unbalanced magnetron

ing or ion-beam-assisted deposition. Optionally, an evaporation source can be used instead of the sputtering source to generate the vapor species.

**Radiofrequency Sputter Deposition.** Deposition can be performed at considerably lower pressures, e.g., 0.5–2 Pa, in r.f. sputtering compared with the planar d.c. discharge, since electrons oscillating at high frequency can obtain sufficient energy to cause ionizing collisions, and the number of electrons lost (without collisions) can be reduced.

Radiofrequency sputtering is widely used to deposit various kinds of conducting, semiconducting, or insulating coatings despite the complexity of the r.f. power source. This technique can also be applied to magnetron sputtering sources.

## 1.3.2. Evaporation Deposition

In the evaporation process, vapors are produced from a material located in a source which is heated by various methods (Fig. 10). The system consists of an electron-beam-heated evaporation source to vaporize the desired material. The substrates are located at an appropriate distance, facing the evaporation source. Resistance, induction, cathodic and anodic arc, electron beam, or lasers are the possible heart sources. The substrate can be heated and/or biased to the desired potential with a d.c./r.f. power supply. Evaporation is carried out in vacuum at $10^{-3}$–$10^{-8}$ Pa. In this pressure range, the mean free path is very large ($5 \times 10^2$–$10^7$ cm) compared with the source–substrate distance. Hence the evaporated atoms essentially undergo collisionless, line-of-sight transport prior to condensation on the substrate, thus leading to thickness buildup directly above the source and decreasing steeply away from it. At a source–substrate distance of 20 cm, the vapor pressure should be ca. 1.5 Pa in order to get reasonable deposition rates (ca. 0.1 µm/min). The source temperature should be adjusted to give this vapor pressure. Specialized evaporation techniques include flash evaporation, laser or electron beam ablation, cathodic arc evaporation (including filtered arc), and anodic arc evaporation.

In the flash evaporation process, pellets of alloy are dropped onto a very hot strip and are vaporized completely, thus maintaining the composition of the alloy in the deposit. It works very well for elements with high vapor pressures. A newer version of flash evaporation is pulsed laser or electron beam ablation where a high energy pulse vaporizes a small chunk of the target material, be it alloy or compound, thus retaining target composition in the deposit.

Cathodic arc evaporation which evaporates small volumes of the cathode (target) may also be used. However, the problem of droplet ejection from the target causing major flaws in the deposited film precludes the application of this process for thin film deposition. One way around the droplet problem is the use of a filtered arc where the ion stream from the cathodic arc plume is extracted by an electromagnetic field and the ions are condensed to deposit a film.

The anodic arc is similar to the cathodic arc except that the anode material evaporates. There is no particulate emission with the anodic arc process. It is as yet a laboratory-scale process.

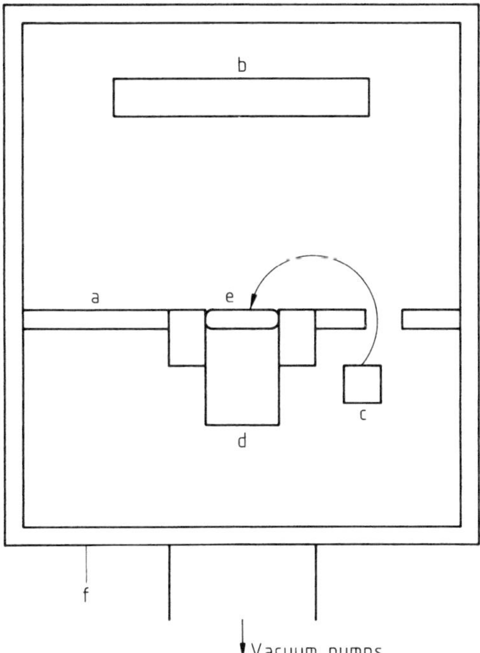


**Figure 10.** Evaporation/deposition process with an electron beam source
a) Separator plate; b) Substrate(s); c) Electron beam gun; d) Ingot rod; e) Molten pool; f) Vacuum chamber

## 1.3.3. Deposition of Metals, Alloys, and Compounds

The great versatility of the PVD processes is their ability to deposit a very large number of materials, including metals, alloys, semiconductors, superconductors, polymers, and to fabricate composites of various types (particulate, fibrous, or laminate).

**Single-Element Deposition.** This can be carried out by evaporation or sputter deposition processes. The deposition rate depends on the process and process parameters.

**Alloy Deposition.** Alloys consist of two or more components, which have different vapor pressures and hence different evaporation rates. As a result, the composition of the vapor phase and the deposit are constantly varying. The following methods have been used to obtain alloy films with stoichiometry close to the source composition by evaporation-based techniques.

1) Co-evaporation or co-sputtering using multiple sources: this technique involves simultaneous co-evaporation of the constitutive elements of the alloy. The composition of the deposited film is controlled by adjusting the evaporation/sputtering rate of the respective elements. In elaborate systems, separate deposition rate monitors are used with feedback networks to control the deposition rate from each source independently.
2) Evaporation from a single source: this technique involves evaporation of an alloy using a rod-fed electron beam source. Evaporation operates under steady-state conditions where the composition and volume of the liquid pool on the top of a solid rod are kept constant [1.21].
3) Flash evaporation: in this process, pellets of the alloy are dropped onto a very hot strip and are vaporized completely, thus maintaining the composition of the alloy in the deposit. It works very well for elements with high vapor pressure.
4) Sputter deposition from an alloy target.
5) Sputter deposition from a segmented target where the segments consist of each of the two components of the alloy and the ratio in the target of each element is inversely proportional to their sputtering yields.
6) Laser ablation from an alloy target.

**Deposition of Compounds.** Deposition of compounds can be performed in two ways: direct evaporation/sputtering of a compound target, where the composition of the target is the same as that of the compound to be deposited; and reactive evaporation, where the elements of the compound are evaporated/sputtered and react with the reactive gas to form the compound [1.21], [1.22].

## 1.4. Thin Film Growth by Molecular Beam Epitaxy (MBE)

### 1.4.1. Introduction [1.23]–[1.26]

Film growth by molecular beam epitaxy (MBE) is based on the formation of thermal, neutral, collision-free molecular (or atomic) beams which interact under ultra-high-vacuum (UHV) conditions with a heated single crystal substrate to form a thin epitaxial film. The beams are formed by effusion from Knudsen sources and the condition for molecular flow is that the mean free path of the atoms or molecules of the beam material contained in the source is greater than the diameter of the effusion orifice. Beams can be formed from elements or compounds, which may be solid, liquid, or gaseous at room temperature, and the design of the Knudsen source is critically dependent on the state of matter involved. It will also influence the pumping system used to achieve UHV.

MBE has been used to prepare films of a wide range of materials, including metals and alloys, semiconductors, superconductors (especially high critical temperature oxides), alkali halides, and other insulators, but by far the most important application is to semiconductor films, with some emphasis on III–V compounds and alloys.

The major advantage of MBE over other methods of thin film formation is that the growth processes can be controlled at the atomic level. This is achieved by the application of a range of in situ diagnostic techniques which can operate in the UHV environment.

### 1.4.2. The Configuration of MBE Growth Systems

The essential elements of an MBE system are illustrated in Figure 11. It comprises a group of Knudsen cells whose detailed design depends on the nature and state of the source material, a heated stage on which the substrate can be mounted, and a mechanical shutter interposed between each source and the substrate, whose function is to intercept the beam and effectively

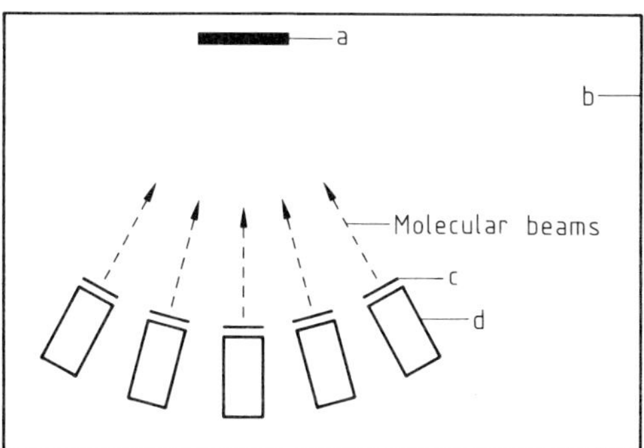



**Figure 11.** The basic arrangement of an MBE system
a) Monocrystal substrate; b) Ultrahigh vacuum chamber; c) Shutters; d) Knudsen sources (effusion cells)

provide a switching capability. All components are contained within a UHV chamber.

Despite this apparent simplicity, implementation of the process requires a complex system, as illustrated in Figure 12. The important components are:

1) The UHV growth chamber, with associated pumping facilities
2) A vacuum interlock arrangement which enables substrates to be inserted and removed from the growth chamber without breaking the vacuum
3) A heated, rotatable substrate holder
4) A set of effusion cells and shutters, the operation of which is computer controlled

**Process Environment.** In general, systems are constructed from stainless steel, but all components should be capable of withstanding bakeout to 200 °C for outgassing. To achieve a base pressure in the growth chamber $< 10^{-8}$ Pa, the primary pumping for solid source systems is usually provided by a combination of sorption and ion pumps, perhaps with additional Ti sublimation and closed-cycle He cryopumps. Where a high gas throughput is involved, which is the case with gaseous or volatile liquid sources, trapped oil diffusion pumps, or more usually turbomolecular pumps are used instead of ion pumping. Secondary pumping takes the form of liquid-nitrogen-cooled cryopanels surrounding as much of the growth environment as possible and is used in all systems except those in which a toxic or pyrophoric condensible (at $> 77$ K) source gas is involved (e.g., $SiH_4$).

**Vacuum Interlock and Substrate Preparation Chamber.** It is critical that UHV conditions are maintained in the growth chamber as substrates are introduced and removed. A vacuum interlock system is therefore used in which the substrates are mounted on a movable holder, transferred from an entry lock into a preparation chamber, and finally into the growth chamber via a series of gate valves. The preparation chamber is used for the initial cleaning of the substrate surface, the final stage being carried out in the growth chamber. It is important that the substrate surface is clean at the atomic level and this

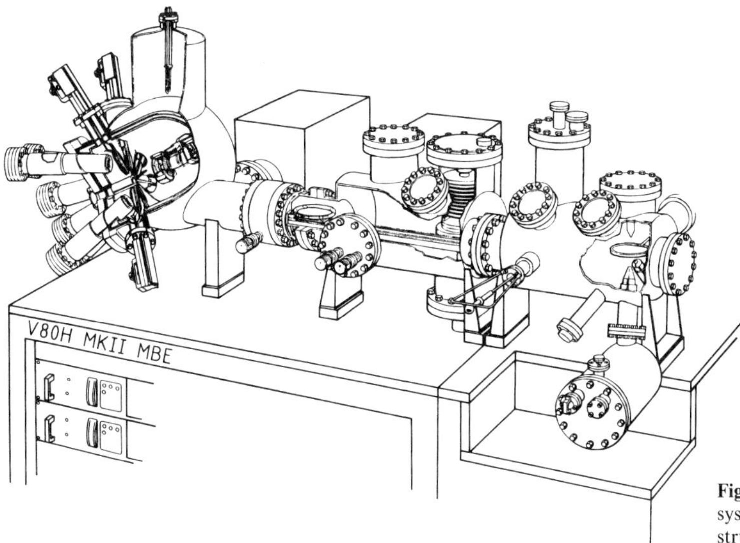



**Figure 12.** A solid source MBE system (VG Semicon, Fisons Instruments)

is usually achieved by ex situ etching and in situ high-temperature treatment to desorb impurities, especially oxides.

**Substrate Holder.** Semiconductor substrates are typically circular wafers ca. 75–100 mm diameter and 0.25 mm thick. The function of the substrate holder is to provide a carriage for the wafers and to heat them uniformly. Growth temperatures are very dependent on materials, but are typically in the range 200–700 °C, most usually 500–700 °C. The holder is usually constructed from a refractory material block (most frequently Mo) heated by resistance or radiation from behind. In addition it must be capable of rotation within the UHV growth chamber at a speed comparable to the rate of growth of a monolayer of material, typically 1 s, so the rotation speed is ca. 1 Hz. The incident beams are neutral, and even with collimation they are necessarily divergent, so uniformity of thickness and composition over a reasonable area ($> 1$ cm$^2$) can be achieved only with substrate rotation. Substrate temperatures are usually measured with a calibrated optical pyrometer, but a thermocouple may also be incorporated in the holder block.

**Beam Sources.** Beam formation is at the heart of the MBE technique and the material from which the beam is to be formed determines the source design and also influences the choice of pumping system. In fact, MBE has become the generic term for a number of closely related techniques which are defined by the nature of the source material (Table 3).

*Solid Sources.* A typical design is illustrated in Figure 13. For low-vapor-pressure materials, such as Ga and Al, the operating temperature may reach 1200 °C. The crucible must therefore be refractory, of high purity, and chemically inert; the preferred material is pyrolytic boron nitride (PBN). The heater is a refractrory metal (usually W or Ta) wound noninductively and supported on PBN insulators; the cell is surrounded by Ta foil radiation shields. The flux is determined and regulated by the cell temperature, so measurement and control are crucial. Measurement is based on an optimally positioned thermocouple, usually W–Re (5 % and 26 % Re) and control is established by proportional integral derivative (PID) controllers and thermocouple feedback, with computer control of the deposition sequence so that temperature–time profiles for each cell can be programmed, together with shutter sequencing and substrate temperature.

The flux $J$ at a distance $d$ (cm) from the cell is given by:

$$J = (1.118 \times 10^{22}) \frac{pA}{d^2 (MT)^{1/2}}$$

$$\text{atoms (molecules) cm}^{-2} \text{s}^{-1}$$

**Table 3.** Processes covered by the generic term MBE, the source materials involved, and some applications

| Process | Source materials | Applications |
| --- | --- | --- |
| Molecular beam epitaxy (MBE) | generic name, but now used for solid source systems, e.g., metals, Si, Ge, and elemental group 13 and 15 sources | magnetic films, silicon, III–V compounds, e.g., GaAs, (Al, Ga)As |
| Metal-organic molecular beam epitaxy (MOMBE) | group 13 organometallic compounds, e.g., triethyl gallium, trimethyl indium, (mostly volatile liquids); elemental group 15 sources | III–V compounds, especially GaAs and $(In_xGa_{1-x})$As |
| Chemical beam epitaxy (CBE) | group 13 organometallic compounds; group 15 hydrides ($PH_3$ and $AsH_3$) | III–V compounds and alloys, especially P-containing materials, e.g., $(In_xGa_{1-x})$P |
| Solid source molecular beam epitaxy (SSMBE) | usually applied only to Si and Ge elemental sources with electron beam evaporation | Si and SiGe alloys |
| Ga source molecular beam epitaxy (GSMBE) | applied mainly to Si and Ge growth from their gaseous hydrides ($SiH_4$, $Si_2H_6$, $GeH_4$) | Si and SiGe alloys |
| Atomic layer epitaxy (ALE); migration-enhanced epitaxy (MEE); growth interruption (GI) | a range of methods characterized by pulsed sources, usually elemental, and applied to III–V compounds and alloys | most III–V compounds and alloys |

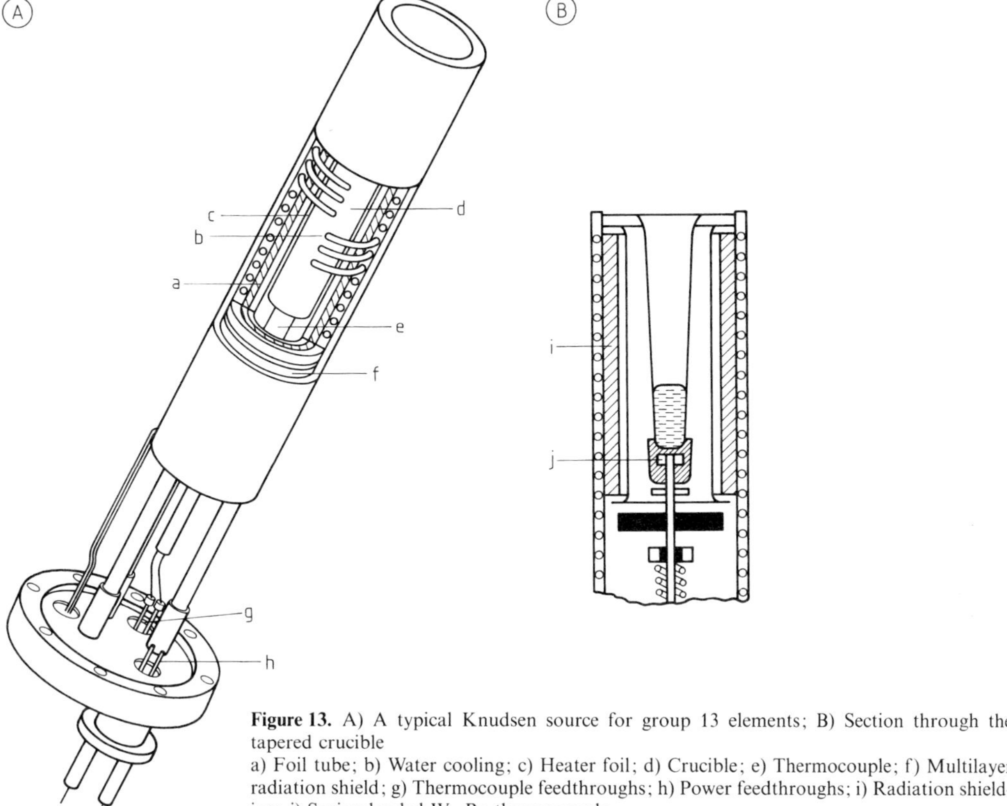


**Figure 13.** A) A typical Knudsen source for group 13 elements; B) Section through the tapered crucible
a) Foil tube; b) Water cooling; c) Heater foil; d) Crucible; e) Thermocouple; f) Multilayer radiation shield; g) Thermocouple feedthroughs; h) Power feedthroughs; i) Radiation shielding; j) Spring-loaded W–Re thermocouple

where $p$ is the pressure in Torr (1 Torr $\simeq$ 133 Pa) of material of molecular mass $M$ in the cell, $A$ is the area of the orifice, and $T$ the cell temperature in K. Typical fluxes at the substrate surface are $10^{14}$–$10^{15}$ atoms (molecules) $cm^{-2}s^{-1}$.

Group 15 elements (group numbering according to IUPAC recommendations) evaporate as tetramers ($P_4, As_4, Sb_4$), and molecular beams from a conventional Knudsen source are consequently also tetrameric. There is evidence, however, that certain electrical properties of films of III–V compounds may be improved by the use of dimeric group 15 species and these can be produced with a special source known as a cracker cell. In essence this is two cells in tandem; a beam of tetramers is formed in the first by normal evaporation from the element and this beam is transmitted into a second cell, referred to as a "cracker," which is maintained at a higher temperature and fitted with baffles to maximize the number of collisions of tetramer molecules with heated surfaces. Under these conditions, the te-

tramers dissociate to form dimers and the resultant beam can contain up to 100 % of the dimerized species, depending on the temperature of the cracker region.

Although it employs a solid source, silicon deposition from elemental silicon is an exception, since it is referred to as Si MBE (or SSMBE), but is not strictly a molecular beam technique. It relies on the evaporation of Si from a Si boule, the central region of which is heated to ca. 2000 °C by a focused electron beam. This source temperature is necessary to obtain an adequate vapor pressure, but molten Si (*mp* ca. 1420 °C) is extremely corrosive and can only realistically be contained by solid Si. The arrangement is to mount the Si boule in a water-cooled copper hearth, which allows a solid outer region to be maintained, even though the central part is molten. The rest of the system is the same as for conventional MBE.

*Gas and Volatile Liquid Sources.* Although solid sources have provided the basis for MBE technology, volatile liquid and gaseous sources have important areas of application. In compound semiconductor technology, organometallic precursors such as trimethyl gallium (TMGa) or triisopropyl gallium (TiPGa) which in general are volatile liquids, are frequently used as the cation source while the gaseous group 15 hydrides ($AsH_3$, $PH_3$) can provide an anion source. They have the advantage of continuous replenishment from outside the UHV system, and operate at or just above room temperature, so potentially generate less contamination than a high-temperature solid source. This low-temperature concept is even more important for Si technology, where the very high-temperature electron beam evaporation process is being superseded by the use of gaseous hydrides ($SiH_4$, $Si_2H_6$). With both volatile liquids and gases, fluxes are normally controlled by mass flow meters, even though it is ultimately the pressure behind the cell orifice which determines the magnitude of the flux.

An additional feature in the growth of III–V compounds from group 15 hydrides is the necessity to dissociate (crack) them before they impinge on the substrate. The technology is similar to that used to produce dimer beams from group 15 elemental solid sources, but in this case the reason for predissociation is simply that the hydrides are too stable thermally to allow growth to occur directly.

The use of nonsolid sources has generated a terminology appropriate to each type and for reference to the literature it is important to define them, as indicated in Table 3.

*Pulsed (Noncontinuous) Sources.* Several modifications to the process of continuous deposition, particularly of III–V compounds and alloys, have been introduced, all based on interruption of growth. The modifications were introduced either to produce more perfect interfaces between layers of different composition or to grow material with superior electrical and optical properties at lower temperature. In the simplest version of growth interruption, the shutter of the group 13 element source, but not that of the group 15, is closed briefly (up to 120 s) to allow the surface to smoothen before deposition of the next layer is started.

Atomic layer epitaxy (ALE) and migration-enhanced epitaxy (MEE) involve periodic interruptions of both fluxes during growth; ALE is based on an alternate supply of constituent elements to the substrate surface, regulated so that a complete monolayer is provided with each pulse. The adsorption of group 15 molecules is self-regulated by the group 13 atoms in the surface, but the latter can accumulate as the free element if the amount per pulse is too large. The MEE process also involves the alternating supply of cation and anion fluxes, and it is claimed that by reducing the surface concentration of the group 15 element during the impingement of the group 13 atoms, their surface migration rate is enhanced, resulting in smoother, higher quality growth at lower temperatures.

### 1.4.3. Monitoring Facilities

The vacuum environment of MBE growth allows a number of powerful diagnostic techniques to be used to monitor various stages of the process. The vacuum quality can be checked with a mass spectrometer to obtain quantitative information on the partial pressures of the background gases; characteristic patterns are observed which may indicate the presence of small leaks, or other contamination, e.g., an oxidized $As_4$ source. The total pressure in the system can be measured with a Bayert–Alpert ion gauge, and a second such gauge placed in the substrate position can be used to quantify the flux of atoms from the growth cells, manifest as a rise and fall in pressure when the shutter in front of each cell is opened and closed. The flux of dopant atoms is usually too small to be detected in this way, and calibration of these cells is performed indirectly, e.g., by measuring the electrical properties of doped layers.

A technique which can provide valuable information on the MBE growth process is reflection high-energy electron diffraction (RHEED). As shown in Figure 14 a beam of electrons of energy ca. $10-25$ keV is incident at glancing angle (ca. $0.5-4°$) on the substrate, and the diffracted beams strike a phosphor screen on the opposite side of the growth chamber. The diffraction patterns produced are related to the surface structure of the substrate (or deposited layer) and can be used to detect the structure of the surface during substrate cleaning, the presence of three-dimensional growth modes (rather than the desired monolayer-by-monolayer deposition), and the occurrence of modifications to the configuration of the surface atoms, the so-called surface reconstructions. In addition, monitoring

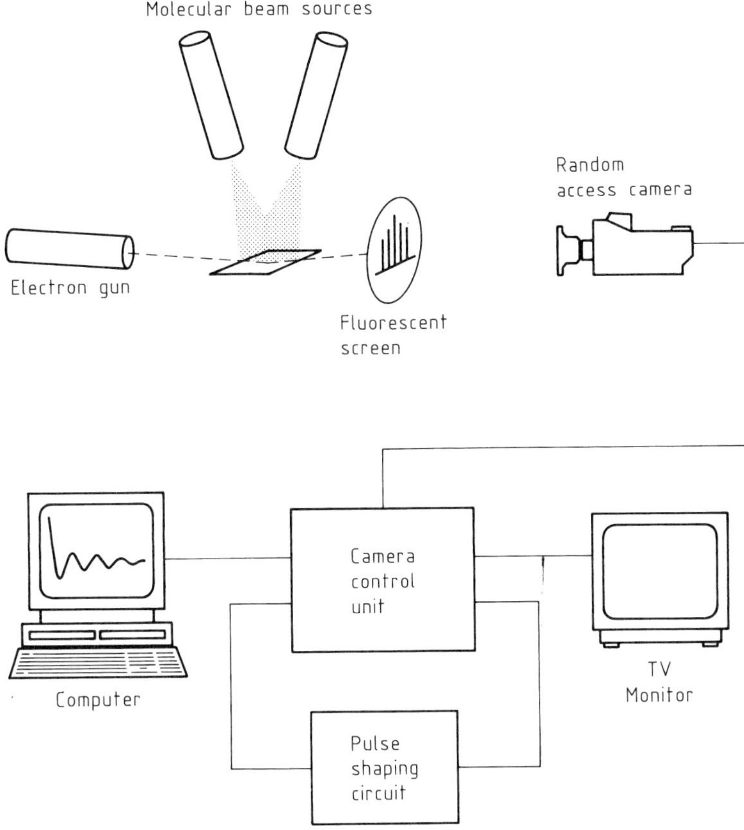



**Figure 14.** A RHEED system used for monitoring MBE growth

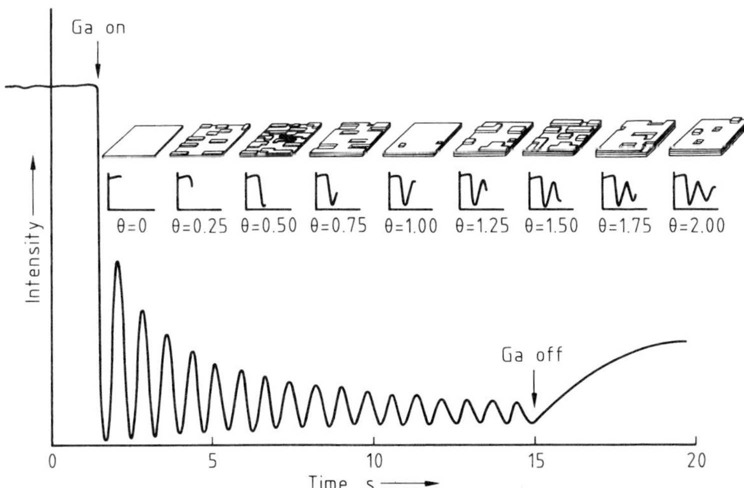



**Figure 15.** Two-dimensional, monolayer-by-monolayer-growth and the corresponding RHEED intensity modulation
$\Theta$ is the coverage in monolayers.

the intensity of the specularly reflected beam allows a very accurate determination of the growth rate. This is possible because the deposition of each monolayer occurs by a "nucleation plus coalescence" process which results in an alternating roughening and smoothing of the surface (Fig. 15) which produces a corresponding intensity modulation of the specularly reflected beam at various stages in the growth. Since electrons are scattered out of the specular beam as the surface roughens, the intensity oscillates with a period equal to the time taken to deposit one monolayer of material; thus, not only can the number of individual monolayers deposited be counted, to give a very precise measure of thickness, but by timing the oscillations, an accurate measure of the growth rate is obtained.

## 1.5. Thin Films by Adsorption Processes

The outermost nanometer of a solid governs many interfacial poperties, including wettability, friction, and electrochemical reactivity. Very thin films of foreign materials on the surface may alter these properties. Consequently, a widely used method of controlling interfacial chemistry is to adsorb a monolayer of a surface-active agent (surfactant) onto the surface. Surfactants are organic molecules, normally adsorbed from solution, although vapor-phase adsorption is also possible.

Adsorbed monolayers are classed as physisorbed or chemisorbed, according to the nature of the interactions attaching the molecules to the surface. Although this division is useful, the distinction between chemisorption and physisorption is not sharp. Physisorbed films are stabilized only by van der Waals or dipolar interactions. Since the interaction with the surface is weak, the monolayers do not normally retain their integrity when the solid is removed from the adsorption solution. A practical example is the use of detergents to enhance the wettability of hydrophobic surfaces. In polymeric adsorbates, the relative weakness of the interaction with the surface is compensated by multiple interactions involving different parts of the polymer chain. The suppression of the growth of wax crystals in diesel fuel is one important use of physisorbed polymers.

In chemisorbed films, a covalent or ionic interaction exists between the surfactant and the solid. This specific interaction with the solid increases the stability of the film and helps prevent multilayer formation, both of which lower the amount of additive required. Surfactants used in aqueous solution frequently also contain one or more long hydrocarbon chains. Hydrophobic interactions between the alkyl chains and water favor aggregation of the surfactant in solution and at a surface; van der Waals interactions be-

tween the hydrocarbon chains in the monolayer contribute additional stability to the adsorbed film.

In commercial applications, mixtures of surfactants are invariably used and the optimum adsorption conditions are developed empirically. The factors governing the choice of surfactant may be illustrated by the practical problem of controlling wettability. In the recovery of ore by flotation, surfactants are required that adsorb selectively to ore particles and make their surfaces hydrophobic. Aqueous solutions of sodium ethylxanthate ($C_2H_5OCS_2^-Na^+$) are commonly used for sulfide ores. It is believed that the monolayer is stabilized by a polar covalent bond between the mineral surface and the sulfur atoms of the surfactant. Metal oxides are more ionic than sulfides. In the flotation of oxide minerals, electrostatic interactions between the surface and an ionic surfactant are exploited. However, ionic interactions alone are insufficient for the formation of stable monolayers, and long hydrocarbon chains are also required. For example, dodecylammonium acetate is used in the flotation of negatively charged oxides, such as goethite and quartz. Conversely, anionic surfactants are used at low pH when the mineral surface is positively charged. Similarities in the chemical structure of the surfactant head group and the solid can be exploited, e.g., sodium stearate is used to make calcite hydrophobic. Typical surfactant concentrations are $10^{-5}$–$10^{-3}$ mol/L.

Surfactants can be deposited from dispersions as well as solutions. Double-chained cationic surfactants, such as $[CH_3(CH_2)_{17}]_2(CH_3)_2N^+Cl^-$, are widely used as fabric conditioners, even though they are virtually insoluble in water. These molecules are present in solution as lamellar mesophases and are adsorbed onto the surface as a result of electrostatic interactions.

Adsorption of monolayers from nonaqueous media requires surfactants that are soluble in organic solvents. A classical example is the adsorption of stearic acid from oil onto iron. The resulting densely packed monolayer acts as an efficient lubricant. Oil-soluble corrosion inhibitors, such as long-chain imidazolines, are highly efficient, even when present in monolayer quantities on the surface of steel.

The silanes are another important class of surface modifiers. Silanes, such as $NH_2(CH_2)_3Si(OCH_3)_3$ (APS), are used as coupling agents to promote adhesion of polymeric resins to mineral surfaces [1.27]. Hydrolysis of the methoxy groups yields free silanols which condense with hydroxyl groups on the surface to form strong covalent Si–O–Si bonds. The terminal functional group is free to couple to the resin: e.g., the amino group of APS can react with polyamide resins. Reactions between silanols strengthen the film by cross-linking, but also lead to polymerization, so the resulting thin films are rarely monomolecular—ten monolayers is a more typical thickness. Silane coupling agents can be applied, either from dilute aqueous solutions or as a minor component of a resin.

### 1.5.1. Self-Assembled Monolayers

Self-assembled monolayers (SAM) are densely packed monolayers formed by adsorption; the term is usually restricted to structurally well-defined monolayers that can be removed from the adsorption medium and characterized by a range of analytical techniques. The molecular architecture and physical properties of SAMs can be varied in a controlled way, in contrast to most of the systems used in commercial processes. Two types are most important: alkanethiols on metals and alkyltrichlorosilanes on oxides.

Long-chain alkanethiols of the generic structure $X(CH_2)_nSH$ adsorb on gold, silver, platinum, and copper (as well as on gallium arsenide and probably other materials) through the formation of a metal thiolate [1.28]. The chain length $n$ is typically $10-21$ and the tail group X can be any functional group chemically compatible with a thiol. The monolayers are stabilized by a strong metal–sulfur bond and by van der Waals interactions between all-*trans* hydrocarbon chains. The tilt of the chains from the surface normal depends on the lattice parameters of the solid and varies from substrate to substrate: 30° on Au (111), 12° on Ag (111), and 57° on GaAs (100).

A SAM on gold can be prepared by immersing a freshly prepared gold mirror in a $10^{-3}$ mol/L solution of a thiol in degassed ethanol (Fig. 16). Adsorption is largely complete within the first few seconds, but a slow period of consolidation follows and many hours of adsorption may be needed before the limiting structure is reached. Multiple functional groups can be incorporated into SAMs by coadsorption of two or more thiols. The composition of a mixed

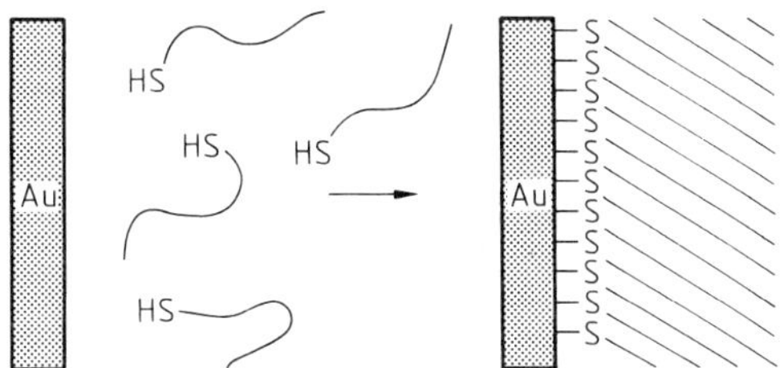

**Figure 16.** Formation of a self-assembled monolayer of an alkanethiol on gold

monolayer does not in general reflect the composition of the adsorption solution and must be determined separately. In the study of interfacial processes such as wettability, surface reactivity, and electron transfer, SAMs of thiols on gold have been extensively exploited as model systems. Practical applications as electrochemical sensors and monolayer resists in microlithography are also being developed.

Alkyltrichlorosilanes $(X(CH_2)_nSiCl_3)$ are similar to the silane coupling agents, but less prone to multilayer formation [1.29]. Elimination of HCl leads to the formation of siloxane bonds to oxide surfaces, in particular to silica. A substrate is typically immersed in a 0.1 % solution of the trichlorosilane in a 3:1 hexadecane/ $CCl_4$ mixture at room temperature for 2 min, although the adsorption time depends critically on the amount of water present. SAMs of alkyltrichlorosilanes are densely packed and the hydrocarbon chains are oriented approximately perpendicular to the surface. One particularly versatile molecule is octadecyltrichlorosilane (OTS). An OTS monolayer on glass is 2.5 nm thick, chemically and thermally stable, highly hydrophobic, and resistant to chemical contamination.

### 1.5.2. Self-Assembled Multilayers

Multilayers can be built up in a controlled fashion by sequential adsorption of self-assembled monolayers. These multilayers are being developed principally for applications in nonlinear optics and optoelectronics, which require oriented thin films of molecules with large nonlinear susceptibilites. More than 25 layers have been built up by the following route at Eastman Kodak (Fig. 17).

An alternative approach, developed at the University of Texas and at AT&T Bell Laboratories, uses layers of alkyldiphosphonates

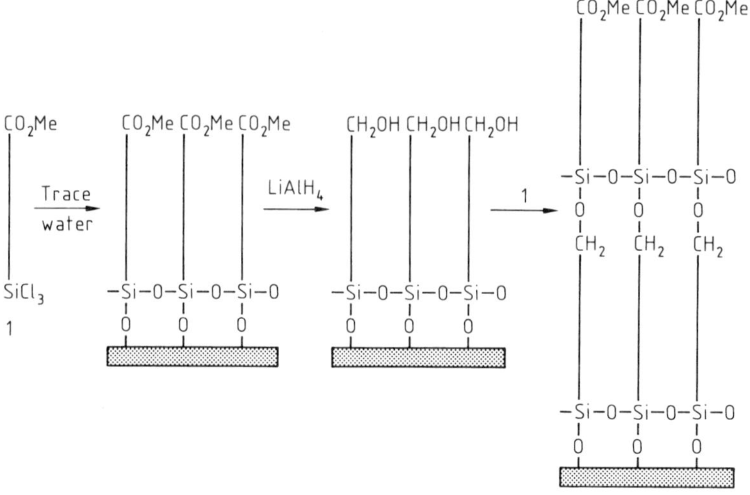

**Figure 17.** Synthesis of organized multilayers of alkyltrichlorosilanes [1.29]

($^{-2}O_3P(CH_2)_nPO_3^{2-}$) held together by sheets of zirconium ions ($Zr^{4+}$) [1.30].

## 1.6. Langmuir–Blodgett and Related Techniques

Monomolecular films have been a subject of scientific interest for more then a hundred years, but it was not until 1917 that LANGMUIR introduced the film balance for the study of molecular films on water surfaces [1.31], the basis of today's experimental and theoretical concepts. He demonstrated that fatty acids of different chain lengths form insoluble monolayers at the air–water interface and that they occupy the same cross-sectional area. Therefore they are oriented perpendicular to the water surface. The transfer of the monomolecular film from the water surface onto a solid substrate provides a method of preparation which is now called the Langmuir–Blodgett (LB) technique, first published by BLODGETT in 1935 [1.32].

The LB technique is commonly used to produce highly ordered ultrathin layered films of monomeric or polymeric materials. The term "LB multilayers" is frequently used to describe the result of the assembly technique. The LB technique offers a number of advantages:

1) Film thickness can be controlled directly by the number of monolayers transferred.
2) The orientation of the molecules in each layer can be precisely controlled.
3) The molecular species in each layer can be varied at will. This is of particular significance with regard to the bottom and the top layers of a multilayered assembly. The bottom layer is responsible for the interaction between film and substrate whereas the top layer determines the surface characteristics, e.g. wetting.
4) Control of the molecular interactions within the film allows optimization of properties such as conductivity or transparency.

Classical substances for the LB technique are amphiphilic molecules consisting of a long alkyl chain with a hydrophilic head group. Films are prepared by first depositing the amphiphilic material, dissolved in a volatile solvent immiscible with water, on the surface of water (subphase) contained in a trough. The subphase temperature is generally controlled by a thermostat. When the solvent has evaporated, the organic molecules may be compressed by a floating barrier to form a floating "two-dimensional" solid. Owing to their hydrophobic head groups, individual molecules align themselves during this process. Once the monolayer is compressed at a suitable constant surface pressure, the molecules can be transferred onto a hydrophilic substrate by dipping the substrate slowly into the subphase [1.33], [1.29]. Deionized water or dilute aqueous solutions of electrolytes have mainly served as subphase, although mercury, hydrocarbons, or glycerol have also been studied [1.34].

Depending on the substance and the substrate, transfer may occur during either the downstroke or the upstroke, or during both (Fig. 18). Three arrangements of amphiphiles in

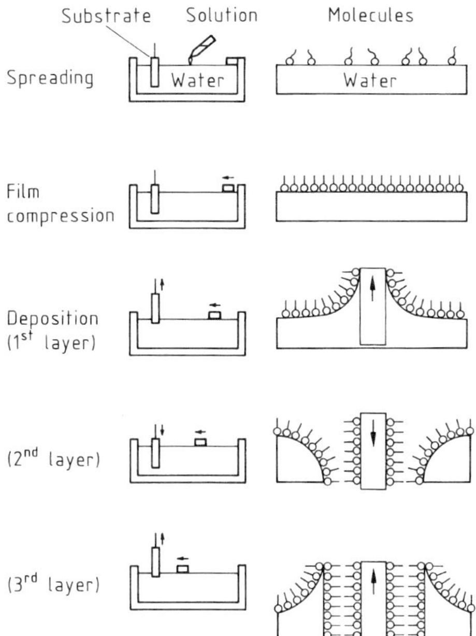

Figure 18. Spreading, compression, and deposition of amphiphilic molecules on the subphase

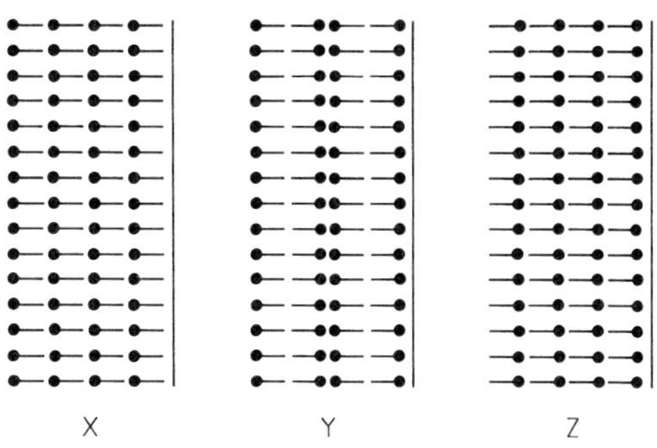



Figure 19. Structure of X-, Y-, Z-type amphiphilic multilayers

the multilayers (X-, Y- and Z-films) are possible (Fig. 19).

During compression, the monolayer undergoes a number of phase transformations. The different phases are in some respects two-dimensional analogs of gases, liquids, and solids, and can be identified by measuring the surface pressure of the monolayer as a function of the area per molecule. Figure 20 shows isotherms of an

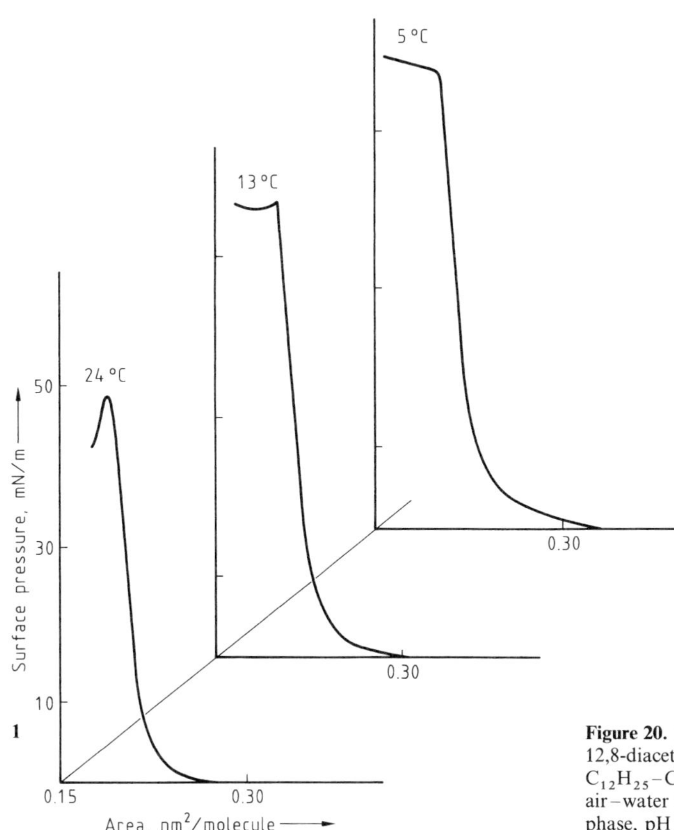

Figure 20. Surface pressure–area isotherms of 12,8-diacetylene monocarboxylic acid, $C_{12}H_{25}-C\equiv C-C\equiv C-(CH_2)_8-COOH$, at the air–water interface on $10^{-3}$ mol/L $CdCl_2$ subphase, pH 6.1 at three temperatures

amphiphilic diacetylene monocarbonic acid at three different temperatures [1.35].

For the 5 °C isotherm, at an area of 0.33 nm² per molecule the first phase transition occurs to form the "liquid expanded" state. In this phase, the monolayer is coherent, except that the molecules occupy a larger area than in the "liquid condensed" state, which is formed at ca. 0.25 nm² per molecule. In the condensed phase, the molecules are closely packed and uniformly oriented. If further pressure is applied, the monolayer collapses and a sharp decrease in the pressure is observed. Frequently, amphiphilic substances show different condensed phases which differ by the mode of packing in two dimensions, depending on the applied surface pressure and temperature. Phase diagrams have been proposed which show the influence of temperature and chain length on the existence range of certain phases [1.36]. A specific state of packing of the molecules on the substrate can be obtained by choosing suitable transfer conditions (temperature, surface pressure, etc.).

**Molecular Design.** Several factors affect the suitability of a substance for the formation of a monolayer:

Solubility
Volatility
Interaction between molecules
Interaction with the subphase
Chemical stability under spreading
and transfer conditions

The deposition of the floating monolayer can be hindered if the affinity of the film to the substrate is too weak. The formation of multilayers is impossible if the substance lacks affinity to itself. Molecules which are able to form monolayers at the air–water interface can be divided into three classes:

1) Low molecular mass substances
2) Amphiphilic polymers
3) Polymers with a rod-like backbone and flexible side chains (hairy rods)

### 1.6.1. Low Molecular Mass Substances

Most of these substances with high affinity to the water subphase consist of a hydrophilic head group and at least one long alkyl chain. The classical film-forming substances are fatty acids [1.37], their derivates and salts, but a number of other substances have been found to form LB layers (Table 4).

Hydrophilic head groups can be formed by nearly all kinds of polar functional groups. The amphiphiles often behave like the corresponding fatty acids.

Aromatic compounds, some of which contain heteroatoms, as well as anthracene derivatives, have been modified by polar groups to form amphiphile head groups. Liquid crystalline (mesogenic) compounds inherently form layered structures; their incorporation increases the thermal and mechanical stability of a two-dimensional fluid phase. Properties of LB films such as selective binding to particular surfaces or conductivity (Section 2.6.1) can be achieved by attaching long alkyl chains to the suitable functional moieties; LB layers containing dyes and ionophores have been prepared. On the other hand, ferrocene is an example of a compound which forms LB layers, if polar substituents are attached.

Lipids are the matrix-forming ingredients of biological membranes and are therefore an important class of LB layer materials. These molecules often show complicated behavior at the air–water interface, due to the presence of two hydrophobic tails.

Porphyrins and phthalocyanins are examples of more complex LB layer molecules. They can carry a variety of functional groups, some of which deviate from the normal requirement of having well-defined head and tail groups. This deviation is reflected in their monolayer properties; the molecules form multilayer stacks in which the structure and stability of the film is governed by molecular packing rather then by hydrophilic–hydrophobic interactions.

For some amphiphiles it may be necessary to add certain ions to the subphase in order to ensure good deposition behavior by salt formation. For example, bivalent ions, e.g., $Cd^{2+}$, in the subphase are sometimes needed to achieve transfer of fatty acids from the water surface onto solid substrates. They are then deposited as Cd salts.

Monolayers can also be built up by a mixture of different molecules; LB films containing molecules which do not themselves form stable films can be prepared from a mixture with a suitable partner.

### 1.6.2. Polymeric Amphiphilic LB Layers

Thermal and structural stability is limited by the melting temperature of the compounds.

**Table 4.** Amphiphiles forming Langmuir–Blodgett films

| | Hydrophobic | Hydrophilic |
|---|---|---|
| Fatty acids and long-chain derivatives | $C_nH_{2n+1}$, $n = 12-22$ | $-COOH$<br>$-CH_2OH$<br>$-CN$<br>$-CONH_2$<br>$-NHCOCH_3$<br>$-CH=NOH$ |
| Aromatic and long-chain compounds | $C_{16}H_{33}-C_4H_6$—$OH$<br>$C_{18}H_{37}$—$O-C_4H_6-NH_2$<br>$C_4H_9$—(anthracene)—$CH_2OH$<br>$C_{12}H_{25}$—$N(CH_3)$—(phenyl)—$N=N$—(phenyl with $NO_2$)—$COOH$ | |
| Others | $C_{18}H_{37}$—$OCH_2$—(18 crown 6)<br>Ferrocene—$C_{18}H_{36}$—$CONH_2$ | |
| Lipids | 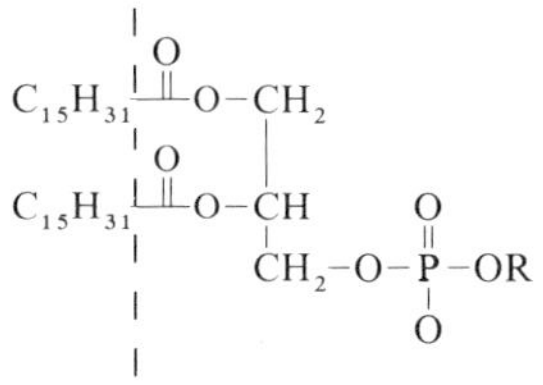 $C_{15}H_{31}\overset{O}{\overset{\|}{C}}-O-CH_2$<br>$C_{15}H_{31}\overset{O}{\overset{\|}{C}}-O-CH$<br>$CH_2-O-\overset{O}{\underset{O}{\overset{\|}{\underset{\|}{P}}}}-OR$ | |
| Porphyrins and phthalocyanins | 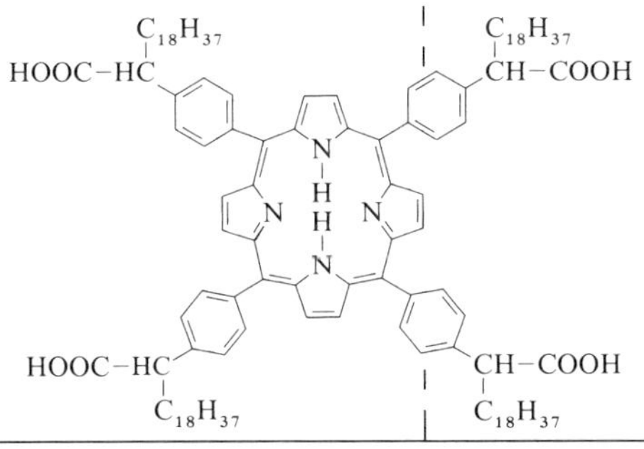 (tetraphenylporphyrin with $C_{18}H_{37}$/$HOOC-HC$ and $CH-COOH$/$C_{18}H_{37}$ substituents) | 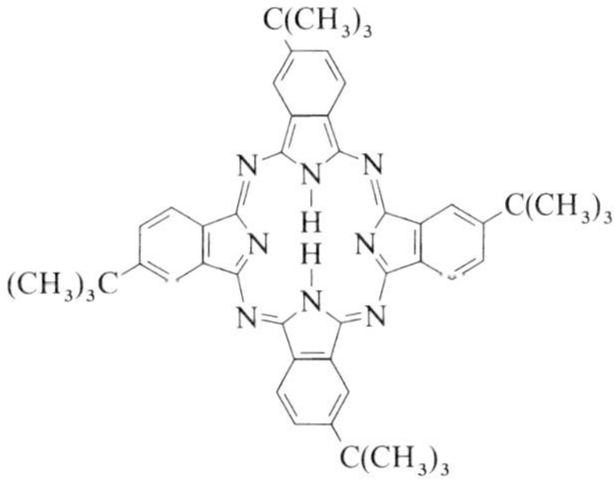 (tetra-tert-butyl phthalocyanine, $C(CH_3)_3$ substituents) |

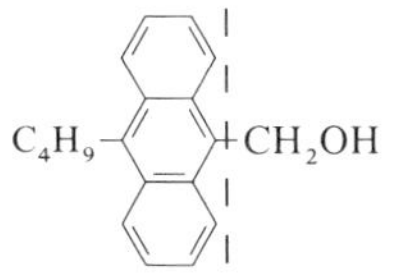

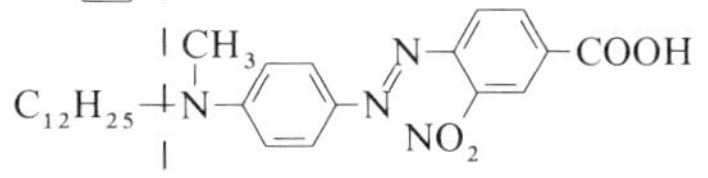

Thermal desorption occurs readily above the melting temperature. The stability can be enhanced by using polymers. Such polymers can be obtained from amphiphilic monomers by in situ polymerization of monolayers or by spreading preformed amphiphilic polymers. The functional groups active toward polymerization of amphiphilic monomers are typically diacetylenes [1.38], acrylates, methacrylates, or cinnamic acid derivatives (Table 5).

Polymerization can be performed on the water surface or in the deposited film. It is generally initiated by irradiation or by the addition of radical-forming substances which can be excited directly or via energy transfer from sensitizers.

The topochemical polymerization of diacetylenes proceeds within LB layers with retention of the crystalline order and the layer structure. The active group to be polymerized can be located on either end of the monomer: near the hydrophilic end or in the hydrophobic part. Oxiranes undergo very little structural change during their ring-opening polymerization.

**Table 5.** Polymerizable amphiphilic monomers forming Langmuir–Blodgett films

| | Hydrophobic | Hydrophilic |
|---|---|---|
| Amphiphiles containing double bonds | $C_nH_{2n+1}$, $n = 10{-}22$ | 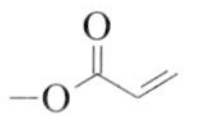 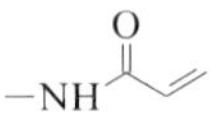 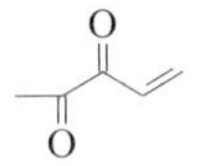 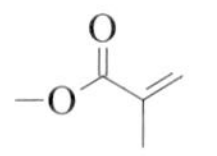 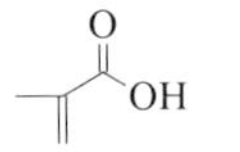 [acrylate, acrylamide, vinyl ketone, methacrylate, methacrylic acid, and cinnamate derivative $C_nH_{2n+1}{-}O{-}C_6H_4{-}CH{=}CH{-}COOH$ structures] |
| Diacetylenic amphiphiles | $CH_3(CH_2)_n{-}C{\equiv}C{-}C{\equiv}C{-}(CH_2)_m{+}$ | [carboxyl group] |
| Oxiranes | $C_nH_{2n+1}{-}HC{-}CH{-}(CH_2)_n{+}$ (epoxide) | [carboxyl group] |

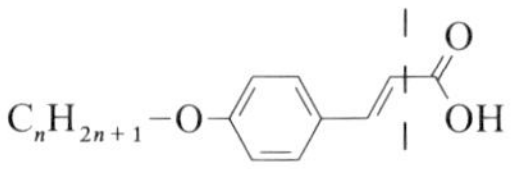

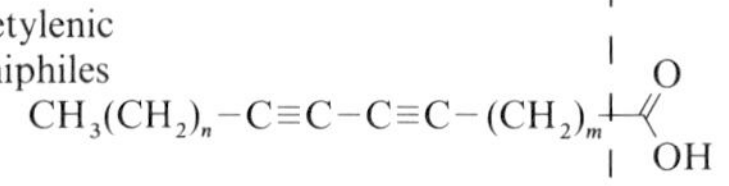

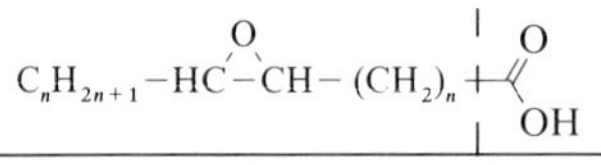

In polycondensation, the constituent molecules join together with the elimination of another molecule. One of the few examples in the present context is the reaction of a monolayer of octadecanal $C_{17}H_{35}CHO$ with a preformed polymer (poly-L-lysine) which is initially dissolved in the subphase.

Preformed amphiphilic polymers have alternating hydrophilic and hydrophobic segments in different parts of the polymer. Spacer groups can be used to decouple the motions of the main chain from that of the amphiphilic side groups. Polymer analogous reactions, if carried out in the deposited multilayer system, may be used to prepare thin films of insoluble polymers, e.g., poly(imides) and poly(p-phenylenevinylenes), on solid substrates.

### 1.6.3. Hairy Rod Polymers

Rod-like polymers can not be classified according to the rules developed for the behavior of amphiphilic compounds at the air–water interface. They exhibit cylindrical symmetry, with low-polarity segments distributed equally over the whole molecule (Fig. 21). These polymers lie flat on the surface. It is essential to surround the rod-like polymers with conformationally mobile side chains. This provides fluidity of the monolayer at the air–water interface and screens direct backbone interactions. The alkyl chains act as chemically attached solvent molecules. Transferring monolayers of hairy rods onto solid substrates aligns the stiff polymer backbones parallel to the dipping direction, producing highly ordered multilayers (Fig. 22) [1.39].

Typical examples of such polymers are phthalocyaninato-(poly)siloxanes (PcPS), several polyglutamates in the α-helical conformation, cellulose alkylethers, specifically substituted polysilanes, and poly(p-phenylenes).

**Film Deposition.** Figure 23 is a schematic diagram of a typical LB trough. The bath (a) is usually made of PTFE or PTFE-coated material. A moving barrier (b) allows control of the pressure applied to the monolayer. The pressure-control device receives information from a pressure sensor (e) on the water surface. A motor (d) is used to lower and raise the solid substrate (e). The trough is temperature controlled.

Two methods can be employed to measure surface pressure. In the Wilhelmy technique, the surface tension is measured by weighing a hydrophilic plate which is dipped through the monolayer into the aqueous subphase. In the Langmuir method, the difference in surface pressure is measured between the pure air–water surface and the surface covered with the film.

The transfer of monolayers requires a constant surface pressure. Thus, during transfer, the barrier moves continuously toward the substrate.

**Other Types of LB Trough.** The circular trough (Fig. 24) [1.40] has a simple reliable motor drive for the barriers (linear troughs require special provision to convert the circular motion of the motor to a linear one). The multicompartment circular trough has two independent drive barriers enclosing the monomolecular film, so that molecules can be compressed as well as being transported from one compartment to another. This allows, for example, a change of subphase while the trough is in use. The texture of the films transferred from the trough reflect the

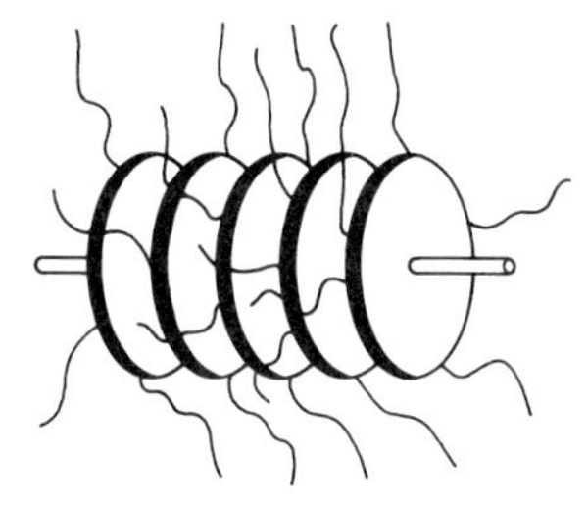


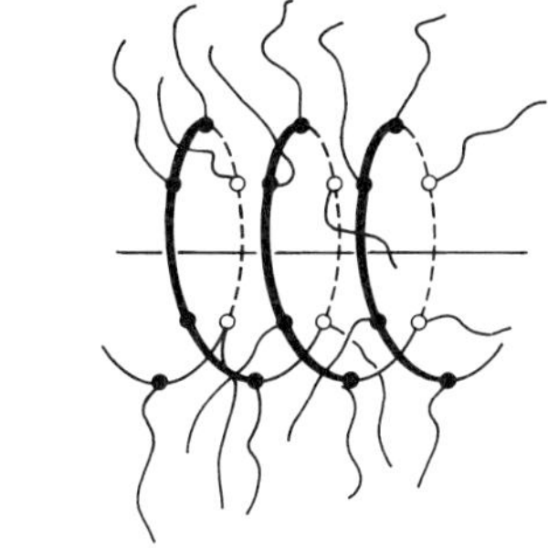


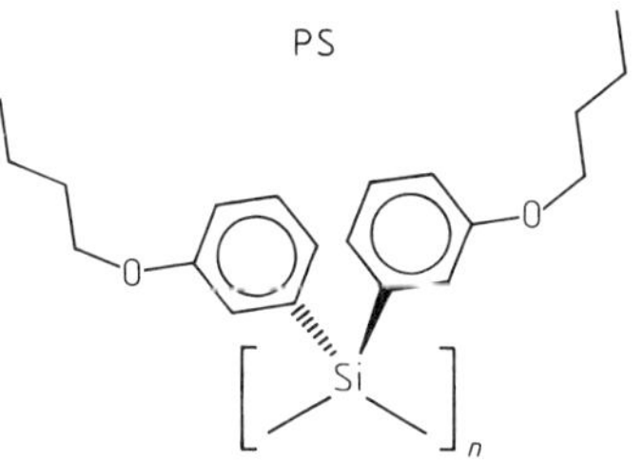


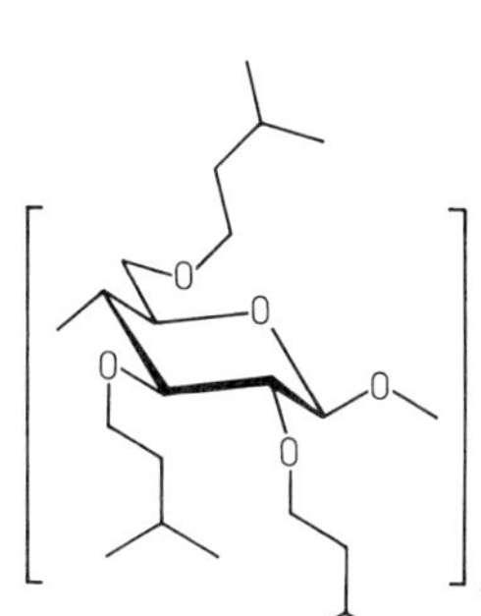


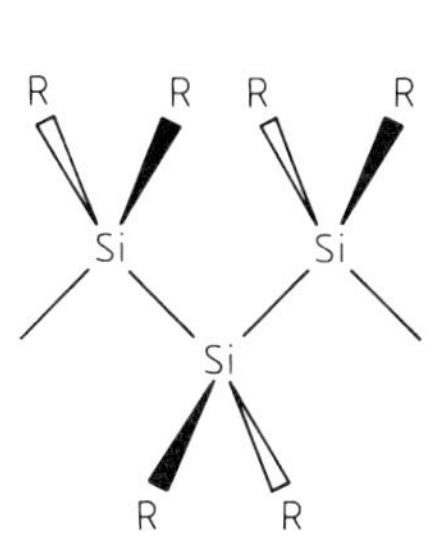


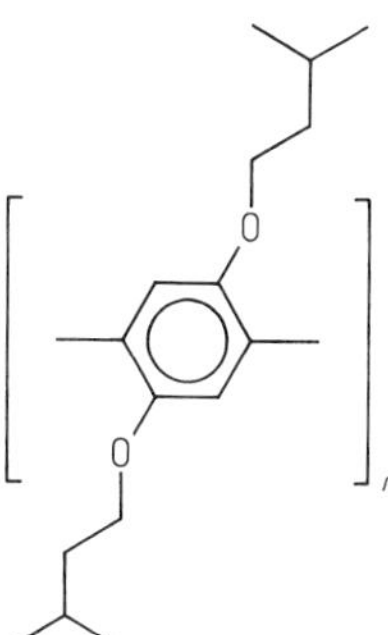


**Figure 21.** Typical hairy rod molecules
PcPS = phthalocyaninato-(poly)siloxanes; PG = polyglutamate; PS = polysilane; IPC = Isopentyl cellulose; PPP = poly-(p-phenylene)

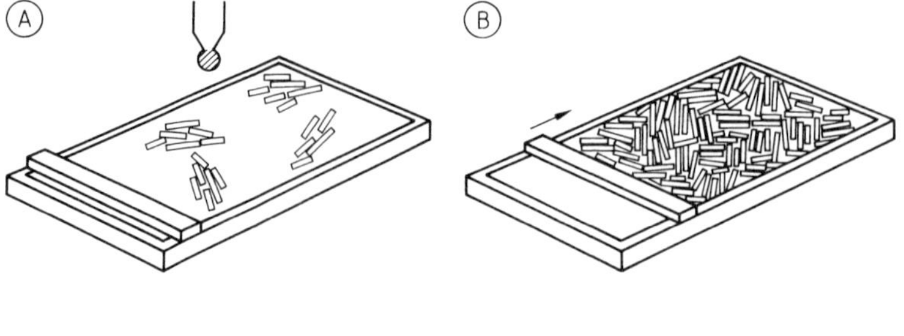

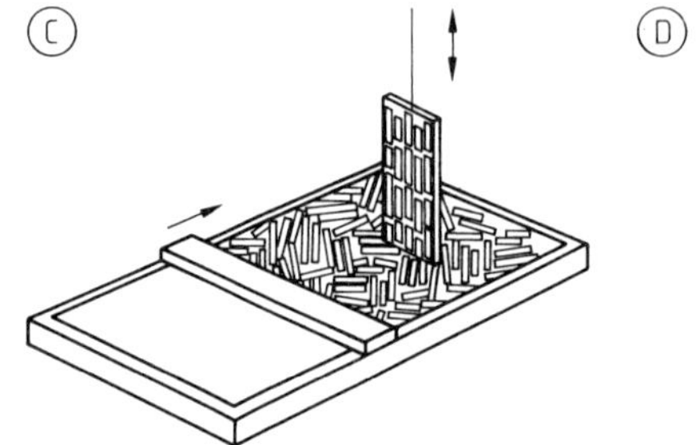

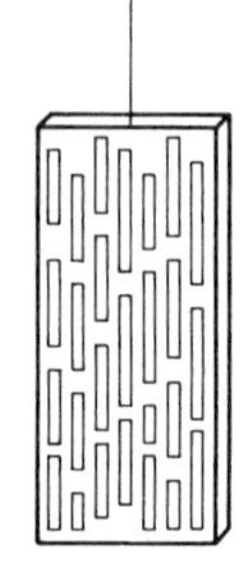

**Figure 22.** Deposition of hairy rod molecules
A) Spreading; B) Compression;
C) Transfer; D) Monolayer structure

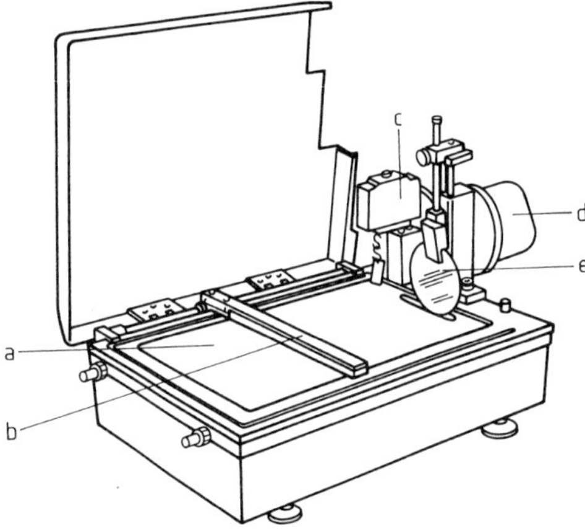

**Figure 23.** A linear LB (Langmuir–Blodgett) trough (Nima Technology, Coventry, United Kingdom)
a) Water basin; b) Moving barrier; c) Pressure sensor; d) Motor; e) Substrate

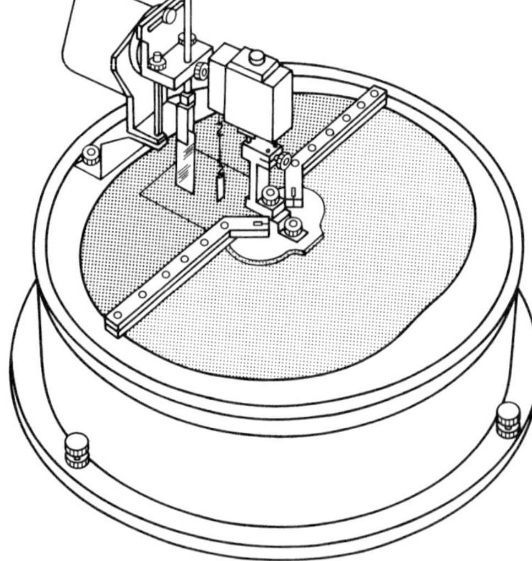

**Figure 24.** Circular LB trough (Nima Technology, Coventry, United Kingdom)

state of order of the monolayer at the air–water interface. The texture is strongly influenced by the flow conditions of the monolayer in compression and transfer. Since the flow processes depend on the trough design, different textures may be obtained by using different designs.

In the constant-perimeter barrier trough [1.41] some of the difficulties encountered with conventional compression systems can be avoided by enclosing the monolayer within a continuous, flexible plastic barrier. Usually, a PTFE band is stretched around four vertical glass spools and a rod. The rod can be rotated to take up the slack as the area is changed.

Double linear troughs [1.33] facilitate deposition of alternating layers. The substrate can be handed over automatically from one trough to another above or below the water level, which allows preparation of a variety of periodically or aperiodically stacked multilayer systems.

All troughs described so far are specifically designed for small-scale production of LB films (typical areas $1-20$ cm$^2$). The number of layers on a substrate is limited by the subphase area and substrate size. For deposition of larger numbers of layers, continuous troughs have been constructed. In principle, the Langmuir time sequence (spreading, compressing, deposition) is replaced by a sequence distributed in space; all the elementary operations take place simultaneously in different compartments. Several different troughs have been developed.

In a trough with a steadily flowing subphase (Fig. 25) [1.42], the horizontal trough is replaced by a flow channel. Water as subphase is circulated to generate a steady laminar flow film on the ramp and thereby compresses the molecules on the surface to form a monolayer. The amphiphile can be continuously spread and deposited.

A roller trough (Fig. 26) [1.43] is divided into four compartments connected by rollers which transport amphiphilic molecules according to the direction of rotation. In the first compartment, the amphiphile is continuously spread by a microelectrovalve and slowly transported into the second compartment where a constant surface pressure of a few mN/m is maintained. The speed of the second roller transporting the molecules into the third compartment, where deposition takes place, is adjusted according to the amount of film removed by a dipping substrate. If the surface pressure exceeds the desired value,

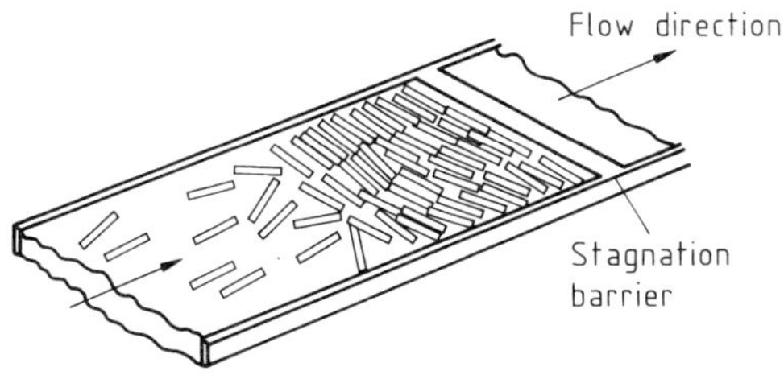


**Figure 25.** Monolayer compression of rod-type molecules in a continuously working LB trough with a steadily flowing subphase

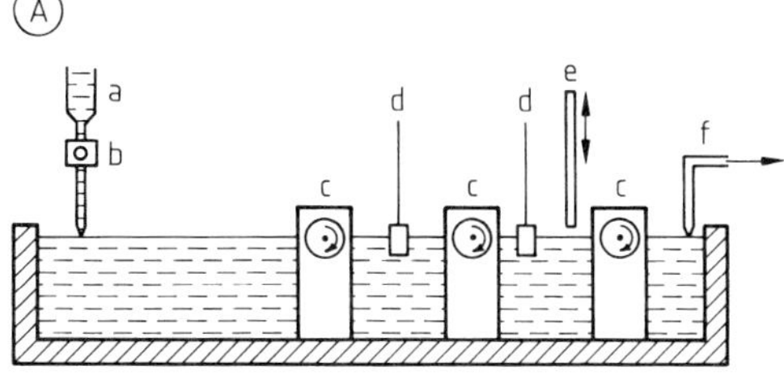


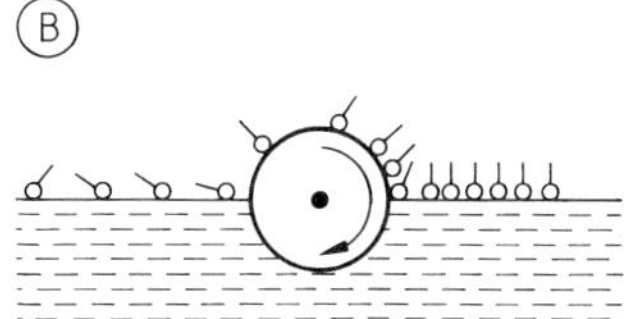


**Figure 26.** A) Continuously working LB trough with rollers
a) Reservoir; b) Microelectrovalve; c) Rollers; d) Pressure sensor; e) Substrate; f) Capillary tube
B) Transport of amphiphilic molecules

an auxiliary roller is actuated to take the excess film to a fourth compartment, where it is removed via a capillary tube.

Another possibility of building LB multilayer structures is the horizontal lifting method (Schäfer's method) [1.44], introduced by LANGMUIR and SCHÄFER in 1938. Schäfer's method is useful for deposition of very rigid films, such as those located in the two-dimensional solid region of the surface–pressure–area diagram (see Fig. 20). A flat substrate is placed horizontally on the compressed monolayer. When this substrate is lifted and separated from the water surface, the monolayer is transferred onto the substrate. In theory, the molecules should maintain the same orientation; in practice, this is not always the case. The molecules near the outer rim of the substrate tend to lose their orientation.

A large number of experimental conditions have to be taken into account for the deposition of LB layers: control of atmosphere, temperature, and humidity around the trough is recommended. The LB trough is usually installed in a laminar flow box to ensure a dust-free environment. The trough should stand absolutely shock free. Since the spread substance has an extremely large surface, special care has to be taken to avoid chemical reactions with the atmosphere or photodegradation.

The water is usually deionized through a Millipore system that contains a final filter to remove bacteria. It is particularly important to keep the subphase free of all surface-active substances. Ionic content and pH must be carefully controlled.

A variety of solids can be used as substrates: glass, quartz, or silicon wafers are commonly used, but in principle most metals or semiconductors can serve as substrates, depending on the desired application of the film. Hydrophobization may be necessary. Cleaning is essential; even trace impurities may adversely affect the deposition process.

## 1.7. Electrochemical Deposition Processes [1.45]–[1.70]

### 1.7.1. Scope

Electrodeposition is the process of producing a thin layer on a substrate, under controlled conditions, via an electrochemical reaction. The choise of substrate and coating may encompass a large number of single metals and alloys. The substrate may be polymeric, ceramic, or a composite. The coating may be a single metal, an alloy, a polymer, a ceramic, or a metal–polymer or metal–ceramic composite. In practice, the choice is severely restricted by technical and economic constraints and there is a trend toward relatively cheap substrates (ferrous, zinc, aluminum, or copper-based alloys) as bulk constructional materials. The most common case of electrodeposition is the d.c. electroplating of a metallic coating onto a metal substrate, the workpiece being made the cathode of the electrolytic cell. However, the field of electrodeposition is much wider, including the production of anodic films (e.g., metal oxides or polymers) and electrophoretic or electropolymerized deposits.

The objective of an electrodeposition process is to produce a deposit which adheres well to the substrate and which has the required mechanical, chemical, and physical properties. It is essential that the process is predictable and reproducible. On the other hand, many metals may (by modification of the bath and plating conditions) be deposited with different properties. For this reason it is not possible to define a single set of conditions for electroplating each metal. Electrolyte composition, current density, temperature, etc. depend on the deposit properties required. Electrochemical techniques are widely used in surface finishing to produce thin layers (typically 1–100 µm) with particular engineering, chemical, corrosion resistance, decorative, or electronics properties.

### 1.7.2 Electrochemical Deposition of Metals and Alloys

The deposition of metal onto a substrate can occur by three electrochemical routes: electrodeposition, electroless plating, and immersion plating. These all involve reduction of a dissolved metal ion, $M^{z+}$, via electron gain to form the metal M:

$$M^{z+} + ze^- \longrightarrow M$$

The three deposition processes are differentiated by the source of the reducing electrons.

The most commonly used technique is electrodeposition or electroplating. An externally applied current provides the electrons. An electroplating bath consists of an anode and a cathode in an electrolyte, usually an aqueous solution of a salt of the metal to be plated. The item to be plated is made the cathode and reduction of ions to metal occurs here. Anodes are made either of an inert conductor, e.g., platinum or graphite, or of the metal which is being plated. Oxidation occurs at the anode. Inert anodes evolve oxygen as current flows. Anodes of the plated metal are dissolved. The thickness is mainly determined by the deposition time. Electrodeposits can be up to 100 µm thick, but thinner layers are much more common.

Electroless plating does not utilize an external power supply; instead, metal ions are reduced by chemical agents in solution and deposition is controlled so that plating occurs only on the substrate. This is achieved by an activation pretreatment of the workpiece which acts to catalyze the reduction or as a nucleation site for film growth. Metal is deposited on both conducting and nonconducting surfaces. The rate of electroless plating is considerably slower than electroplating. It is only used to deposit thin layers, typically 0.25–1 µm.

Immersion plating involves plating of a base metal substrate by reaction of that metal with the salts of a more noble metal, via reduction and deposition of the dissolved ions, and oxidation of the surface of the substrate. This technique has limited application, owing to restrictions on the types of metal that can be plated, and the limited thickness of deposits.

**Table 6.** Properties and applications of electrodeposited metals

| Property | Metals and alloys used | Example of use |
| --- | --- | --- |
| Corrosion resistance | Cr, Ni, Sn, Au, Zn, Rh, Cd | corrosion protection, food cans |
| Buildup of material on surface | Cr, Fe, Ni | reclamation of worn parts |
| Decorative appeal | Cr, Au, Ag, Pt, Ni, 70:30 Cu–Zn | jewelry, cutlery, general decoration |
| Wear resistance | Cr, Ni, Fe, Sn, Ru, Pd | rollers, pistons, bearings, contacts, switches |
| Hardness | Cr, Ru, Os | molds, presses |
| Reflectivity (optical or thermal) | Cr, Rh, Au | lamps, projectors, aerospace screens and visors |
| Electrical conductivity | Cu, Ag, Au | printed circuit boards, antennae, bus bars, cables |
| Oil retention | Cu, 65:35 Sn–Ni | hydraulic systems, lubrication |
| Solderability | Ni, Sn, Cd, 60:40 Sn–Pb | printed circuit boards, electrical contacts |
| Low contact resistance | Ag, Au, Rh, Pd, Sn, 80:20 Pd–Ni | electrical contacts |

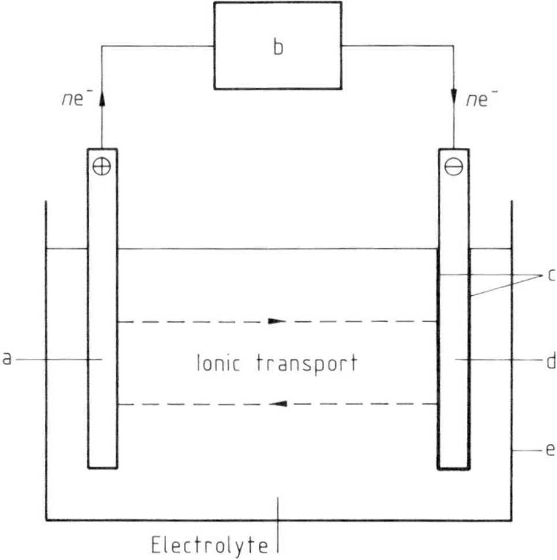


**Figure 27.** Principle of electroplating
a) Anode; b) Power supply, d.c.; c) Electroplated metal layer; d) Cathode (workpiece); e) Electroplating tank
The electrolyte in the plating bath consists of: water, source of metal ions, conducting salts, buffer, and additives.

**Electroplating.** Alloys and composites can also be deposited by electrolysis. The uses of electroplated coatings include engineering, electronic, and decorative applications. The properties and applications of commonly electrodeposited metallic coatings are summarized in Table 6. The essential components of the electroplating process are shown in Figure 27.

The source of metal ions for the cathode reduction reaction may be a complex cation or an anion, rather than the simple metal cation. Examples include $Zn(CN)_4^{2-}$ and $Sn(OH)_4^{2-}$ in cyanide–zinc and alkaline–tin plating baths. The mass of metal deposited is directly related to the charge transferred:

$$w = \frac{\varphi M Q}{z F}$$

where $w$ is the mass of electroplated metal, $M$ the molar mass of metal, $Q$ the electrical charge, $\varphi$ the cathode current efficiency for metal deposition ($< 1$), $z$ the cation charge, and $F$ is the Faraday constant, 96 485 C/mol.

Most electroplating processes are carried out at constant current. If uniform deposition is assumed, the average rate of deposition per unit area is:

$$\frac{w}{tA} = \frac{\varphi M I}{z F}$$

where $A$ is the electrode area, $t$ the time for deposition, and $I$ the current. The rate of thickness development can be expressed as:

$$\frac{x}{t} = \frac{\varphi M I}{\varrho z F A} = \frac{\varphi M i}{\varrho z F}$$

where $x$ is the deposit thickness, $A$ the electrode area, and $\varrho$ the density of the deposited metal. The current efficiency depends on the process used. Electroplating baths are usually designed to have a high efficiency, $0.9 < \varphi < 1$, but some baths, e.g., chromic acid–chromium, are operated at much lower cathode efficiencies, typically $0.1 < \varphi < 0.2$. The most common secondary reaction at the cathode is the evolution of hydrogen which creates safety/environmental problems and the possibility of hydrogen damage to the plated deposit via embrittlement or blinding of the surface.

The throwing power of a bath is a measure of how evenly the workpiece is plated. It is assessed by comparing the plating thickness of areas at different distances from the anode. A high throwing power leads to even coverage all over the workpiece. This is important when coating items with recesses or other surface features. The main factors which affect throwing power are bath composition, temperature, and current density. A conductive electrolyte aids current flow to all parts of the cathode, minimizing the effects of increased resistance with increased distance.

If there is a high concentration of metal ions in the bath, mass transport considerations should not affect the throwing power. An even current density leads to even plating across the workpiece, so the metal plating is affected by the conductivity of the substrate surface. How evenly a metal is plated can be enhanced by auxiliary anodes (usually inert), to help distribute current to inaccessible areas.

Where possible the anode is made of the same metal as that being plated. This dissolves as plating occurs. To maintain a constant metal ion concentration, the rate of dissolution at the anode must be the same as the rate of deposition at the cathode. The current efficiency of the two electrodes must therefore be similar. It is important to minimize production of insoluble sludge as the anode dissolves (from impurities in the metal or particles left over from fabrication). Incorporation of impurities in the plate can affect the smoothness, conductivity, and contact resistance of the metal. The structure of the anode can also be important. A coarse-grain anode can dissolve unevenly, leading to fluctuations in metal ion concentration. The best anodes have a nonporous rolled structure and dissolve steadily at steady current.

Before electroplating, the surface must be cleaned to remove dirt, grease, and surface oxide. This is essential to give good adhesion between the metal plate and the substrate.

**Mechanical cleaning** methods include abrasion, shot blasting, and polishing. These remove scale and some dirt, but further treatment is usually necessary.

**Solvent cleaning** is used to remove residual oils and greases from the workpiece. This allows the surface to be better wetted by water-based electrolytes, but will not remove oxide layers.

**Alkaline cleaners** remove grease, soils, and scale from the metal surface. They consist of a mixture of alkali salts to saponify oil and grease to a water-soluble form, detergents to aid wetting by lowering surface tension at the metal–soil–solution interfaces, sequestering agents such as sodium gluconate to prevent precipitation of metal ions as insoluble hydroxides or carbonates, and a water softener to prevent hard water forming scale or scum on the surface. Alkaline cleaning is often carried out electrolytically. Cleaning is enhanced because the evolution of gas bubbles at the metal–solution interface dislodges soiling and transports fresh cleaner to the site.

**Acid pickling** is used to remove impurities by chemical attack and to stabilize impurities by ionic dissolution. An acid pickle removes oxides and scale from the surface of a metal and is often used after alkaline cleaning. A number of acids are used, either individually or as a mixture. These include nitric, sulfuric, hydrochloric, hydrofluoric, and phosphoric acids at concentration 2–85 %, depending on the metal and finish needed. Acid pickles are usually heated to 50–80 °C. As in the case of alkaline cleaners, electrolytic cleaning is often used to improve pickling.

When the plating process starts, the metal is deposited onto a surface that has a different structure from the plate. This is due to it being a different type of metal, to surface impurities, or to the surface having been machined. This can lead to poor adhesion between the base and the plated layer. Surface preparation such as the techniques outlined above improves adhesion by removing impurities and oxides, but there may still be some difference in structure between the substrate and the plate. Adhesion between these layers is often enhanced by deposition of a thin "strike" plate of metal before the main deposit is applied.

A strike plate is electrodeposited from a low-concentration solution. Such solutions have relatively low throwing power and efficiency. At low current density, a slow plating rate leads to good adhesion of the metal onto the substrate. The strike solution is used to build a good basis upon which to plate from the high-concentration main tank.

Plating baths are a mixture of metal salts, electrolytes, and trace additives which modify the plating characteristics. The precise makeup of commercial baths and chemicals is usually secret, but general functions of each component are well documented.

The metal ion to be plated is kept at high concentration, usually 1–3 mol/L. It may be either the simple hydrated ion or a complex. If the deposition becomes mass transport controlled, owing to a lack of metal ion at the cathode surface, plating quality is impaired because plating is now favored at sites on the workpiece that are easily accessible to the solution, and the lack of metal ions available for reduction promotes electrochemical side reactions. For this reason the concentration of metal ion is kept high.

Items of relatively simple shape are usually coated with solutions of the simple hydrated ion, derived from a salt. This is because the higher throwing power of complex ions is not needed. There are exceptions to this, e.g., when the metal being deposited is difficult to dissolve. To electroplate complicated shapes higher throwing power is needed, utilizing a complexed metal ion electrolyte or auxiliary anodes.

To maximize the conductivity of the bath, high concentrations of electrolyte are added; acidic, alkaline, and neutral electrolytes are used, e.g., $H_2SO_4$, KCN, NaOH, NaF, KF, NaCl, and $NH_4Cl$.

Complexing agents are added to make the deposition potential more negative, if it is necessary to prevent spontaneous reaction between the plating ion and the cathode material. For example, plating copper onto iron, the copper ions in solution attack the metal surface, dissolving iron:

$$Cu^{2+} + Fe \longrightarrow Cu + Fe^{2+}$$

This chemical reaction will lead to a very poor deposit, both porous and badly adherent.

Complexing agents can modify the redox potential of the $Cu^{2+}/Cu$ couple so that it is more negative than the $Fe^{2+}/Fe$ couple and the above reaction is no longer favorable thermodynamically. Common complexants are cyanide ($CN^-$), pyrophosphate ($P_2O_7^{4-}$), and hydroxide ($OH^-$). Complexing agents can also prevent passivation of dissolving anodes which maintains current efficiency.

The structure, morphology, and properties of the electroplate can be modified by low concentrations of organic molecules. The exact identity and concentration of additives for specific baths is kept secret by commercial suppliers. There are four general types of additive.

**Brighteners** produce a visually brighter or more shiny surface finish, by reducing the microscopic roughness of the deposit. The electrical properties are unaffected and brighteners are used when decorative appearance is important. Relatively high concentrations ($g/dm^3$) are used.

**Levelers** act by adsorbtion at points on the plating surface where there would be rapid deposition of metal. The adsorbed molecules reduce the rate of electron transfer, slowing the plating rate at these points and making plating across the whole workpiece more even. Many additives act as both brighteners and levelers.

**Structure modifiers** act by changing the electrodeposit structure to optimize particular properties of the plate. They are used when the plating is required to have certain mechanical properties, e.g., wear resistance for electrical contacts.

**Wetting agents** facilitate the release of gas from the cathode surface. Inclusion of hydrogen can cause the deposit to become brittle and prone to cracking.

Some tyical electroplating conditions are summarized in Table 7.

Metals are usually electroplated to a thickness of $1-100\ \mu m$. Deposits below $1\ \mu m$ are generally too porous, and a plating thickness greater than $100\ \mu m$ is usually ruled out by considerations of cost and time, although chromium plating may give thickness up to $500\ \mu m$, in special cases. Plating thickness for a particular metal is determined by the current, time, area, and electrode efficiency. For any electroplate, the thickness can be controlled by adjusting the current and time of plating. Most plating baths are current controlled and thus operated at known current densities.

The quality of the plating is affected by the rate of growth of the metal. If the plate is build up slowly, the microstructure is well defined, with layered, ridged, or block formation. If the plating rate is increased, a less dense, polycrystalline solid results. At high plating rates, nodules and dendrites of metal form. At the maximum deposition rate the metal goes down as a powder which does not bond to the surface. The change in plating characteristics with rate is mainly due to mass transport considerations. When the metal is deposited slowly, there is time for metal ions to diffuse to the cathode surface and there is always a high concentration of metal ions available for reduction. Since deposition is slow, the metal atoms can form organized, well-ordered structures. When the rate is increased, the concentration of metal ions at the cathode is lower. Despite the short-range order in the structure the faster growth rate does not allow long-range lattices to form (polycrystalline). At high

**Table 7.** Typical process conditions for electroplating single metals [1.49]

| Metal | Electrolyte composition, g/dm$^3$ | Temperature, °C | Current, mA/cm$^2$ | Current efficiency, % | Additives | Anode | Characteristics |
|---|---|---|---|---|---|---|---|
| Cu | $CuSO_4$ (200–250) $H_2SO_4$ (25–50) | 20–40 | 20–50 | 95–99 | dextrin, gelatin, S-containing brighteners, sulfonic acids | P-containing rolled Cu | not suitable for iron and iron alloy substrates; high current, but low throwing power; additives essential |
|  | CuCN (40–50) KCN (20–30) $K_2CO_3$ (10) | 40–70 | 10–40 | 60–90 | $Na_2SO_3$ | $O_2$-free, high-conductivity Cu | good throwing power; deposits adhere well and are bright; used as under-coat; cyanide decomposes at anode |
| Ni | $NiSO_4$ (250) $NiCl_2$ (45) $H_3BO_3$ (30) pH 4–5 | 40–70 | 20–50 | 95 | coumarin, saccharin, benzenesulfonamide acetylene derivatives | Ni pellets or pieces | $H_2$ evolved with deposition; additives depend on purpose of deposit; medium throwing power |
|  | Ni sulfamate (600) $NiCl_2$ (5) $H_3BO_3$ (40) pH 4 | 50–60 | 50–400 | 98 | not essential; naphthalene-1,3,6-trisulfonic acid | Ni pellets or pieces | good throwing power; some $H_2$ with additives, mirror finish at 400 mA/cm$^2$ |
| Ag | $KAg(CN)_2$ (40–60) KCN (80–100) $K_2CO_3$ (10) | 20–30 | 3–10 | 99 | S-containing brighteners | Ag | good throwing power; functional or decorative |
| Sn | $SnSO_4$ (40–60) | 20–30 | 10–30 | 90–95 | phenol, cresol | Sn | poor throwing power; additives essential |
|  | $Na_2SnO_3$ (50–100) NaOH (8–18) | 60–80 | 10–20 | 70 |  | Sn (control current) | good throwing power, but uses twice charge ($Sn^{4+}$) |
|  | $SnCl_2$ (75) NaF (37) KF (50) NaCl (45) pH 2.5 | 65 | 50–200 | 90 | SCN-disulfononaphthalinic acid | Sn | reasonable throwing power; more stable than $SnSO_4$; rarely used |
| Zn | $NH_4ZnCl_3$ (200–300) $NH_4Cl$ (60–120) pH 5 | 25 | 10–40 | 98 | dextrin, organic additives | Zn | poor throwing power; additives essential |

deposition rates, the metal ions arrive quickly at the cathode, and do not form ordered structures. The metal ion concentration at the surface becomes much lower than that of the bulk electrolyte, promoting the growth of nodules and dendrites since plating is favored further from the plane of the workpiece surface where there is a higher concentration of metal ions. The formation of dendrites is further favored since they cause an increase in surface area which reduces the overall current density.

The limiting factor in deposition rate is diffusion-controlled mass transport of the metal ion. At a particular temperature and electrolyte concentration, this has a maximum value (maximum rate of deposition). Deposits laid down this fast have very little structure and form powders which are easily removed by physical agitation. For a typical bath, the limiting current density is ca. 70 mA/cm$^2$. To ensure good adhesion, the current is kept at a much lower level. The main way of increasing plating thickness is to lengthen the plating time.

**Electroless Deposition.** In electroless deposition, the substrate acts as a catalyst for a spontaneous redox reaction. No external current is required and the process takes place by immersion of the workpiece in the plating bath. It is essential, however, that the substrate is pretreated so that deposition of metal takes place over the entire surface. Under constant process conditions, deposition takes place at a constant rate and the deposits tend to be microcrystalline, with the alloying element (e.g., P or B in nickel deposits) in solid solution. Despite the relatively high cost of electroless plating (compared with electrolytic deposition), the technique offers the advantages of a very uniform deposit thickness, even on workpieces with complex shapes. In Ni–P deposits, heat treatment (say 400 °C for 1 h) after deposition can increase the hardness (typically from 500 to ca. 1000 VHN) owing to crystallization of an Ni$_3$P dispersion throughout the coating. In addition to a metal salt and a suitable reducing agent, electroless plating solutions are formulated with additives such as complexants, stabilizers, pH buffers, accelerators, and brighteners to provide stable electrolyte solutions and the required deposit characteristics.

The most common electroless deposits are copper (to produce conductive tracking on printed circuit boards) and nickel (as a corrosion-resistant and engineering coating). Copper deposi-

**Table 8.** Standard electrode potentials of reactions in electroless and immersion deposition

| Reversible electrode process | Electrode potential*, V |
|---|---|
| $Ag^+ + e^- = Ag$ | $+0.799$ |
| $Cu^{2+} + 2e^- = Cu$ | $+0.337$ |
| $HCOOH + 2H^+ + 2e^- = HCHO + H_2O$ | $+0.056$** |
| $Ni^{2+} + 2e^- = Ni$ | $-0.250$ |
| $Fe^{2+} + 2e^- = Fe$ | $-0.440$ |
| $H_2PO_3^- + 2H^+ + 2e^- = H_2PO_2^- + H_2O$ | $-0.500$ |
| $Zn^{2+} + 2e^- = Zn$ | $-0.763$ |
| $HCOOH + 2H^+ + 2e^- = HCHO + H_2O$ | $-1.07$*** |

* Relative to standard hydrogen electrode at 25 °C. ** At pH 0. *** At pH 14.

tion is commonly carried out from an alkaline bath containing formaldehyde as the reducing agent while nickel baths commonly contain hypophosphite (Ni–P deposits) or less commonly borohydride (Ni–B). Typically, Ni–P deposits contain 3–15 wt % Ni.

Oxidation of the reducing agent must have an electrode potential which is more negative than that of the metal deposition process (Table 8). In electroless copper deposition from a bath containing formaldehyde, the simplified reactions are deposition of copper by reduction of its ions at cathodic sites:

$$Cu^{2+} + 2e^- \longrightarrow Cu$$

accompanied by reduction of formaldehyde to formic acid at anodic sites:

$$HCHO + H_2O \longrightarrow 2H^+ + 2e^- + HCOOH$$

The overall redox process on the metal surface is:

$$Cu^{2+} + HCHO + H_2O \longrightarrow Cu + 2H^+ + HCOOH$$

In electroless deposition from baths containing hypophosphite, the simplified reactions are as follows. The anodic reaction is oxidation of hypophosphite to orthophosphite:

$$H_2PO_2^- + H_2O \longrightarrow H_2PO_3^- + 2H^+ + 2e^-$$

and the cathodic reaction is deposition of nickel:

$$Ni^{2+} + 2e^- \longrightarrow Ni$$

The overall redox process on the metal surface is:

$$H_2PO_2^- + H_2O + Ni^{2+} \longrightarrow H_2PO_3^- + 2H^+ + Ni$$

It is important to note that:

1) The substrate metal and the deposited metal must support the electrode process in a catalytic manner
2) In the case of noncatalytic surfaces, the electroless process may be initiated by preliminary deposition of a catalytic metal (as in the case of plastics) or brief electrolytic deposition of the coating metal by passing a current for a short period
3) The process must be operated so as to avoid spontaneous decomposition of the electrolyte in the bath or on the tank surfaces
4) A pH decrease accompanies the overall process
5) The reducing agent is depleted and its oxidation product accumulates
6) The source of metal, i.e., $Ni^{2+}$, declines in concentration as the bath is used
7) In practice, the deposit is an alloy with a significant phosphorous level ($< 15\%$), showing that the above reactions are greatly oversimplified.

Points (4), (5) and (6) are addressed by planned makeup additions to the electrolyte.

By suitable formulation of the bath, electroless deposition can produce single metals, or metals containing a nonmetal component (e.g., P or B). In addition, special baths are available for the deposition of composite "inclusion" coatings, e.g. Ni–PTFE (as a nonstick coating) or Ni–SiC (as a wear-resistant coating), the inert particles being introduced to the electrolyte by agitation and/or surfactant dispersion techniques.

**Immersion Deposition.** If steel is dipped in a solution of copper sulfate, a deposit of copper is spontaneously formed on the surface of the steel. This reaction can be explained with reference to the electrochemical series (Table 8). The more positive (noble) redox system is reduced (deposition) while the more electronegative (base) is oxidized (dissolution).

The surface of the steel consists of a mixed electrode comprised of a mosaic of anodic and cathodic sites; the following reactions take place as the standard electrode potential for copper is more positive than that of iron.

Anodic site:

$$Fe \longrightarrow Fe^{2+} + 2e^- \quad +0.44\ V$$

Cathodic site:

$$Cu^{2+} + 2e^- \longrightarrow Cu \quad +0.34\ V$$

Overall mixed-electrode reaction:

$$Fe + Cu^{2+} \longrightarrow Fe^{2+} + Cu \quad +0.78\ V$$

The deposition of copper continues until almost the entire surface of the steel is coated. At this point there are insufficient anode sites left to dissolve (oxidize) and supply electrons for the cathodic reaction, and deposition ceases. Copper is usually deposited to a thickness of ca. 1 µm. Deposits tend to be very porous and adhere poorly to the substrate. The copper solution becomes contaminated in a relatively short time with ions of the base metal. These considerations limit the applications of immersion plating. It is normally used to produce deposits of silver, gold, or palladium on base metals. Examples include the deposition of gold onto nickel-clad printed circuit boards and the metallization of printed circuit plastics with palladium by deposition onto a tin layer.

### 1.7.3. Electrophoretic Coatings

Electrophoresis is the migration of charged organic molecules to electrodes of the opposite charge. At the electrode, the particles accumulate and deposition follows rapidly. In the organic coating of automobile bodies, prior to the finishing coats of paint, electrophoretic coating is also known as electropriming, electropainting, or electrocoating. It is used to deposit films 1–20 µm thick, the majority of applications being in the range 4–10 µm. These films have high corrosion resistance and an even coverage over the whole workpiece.

There are two types of electropriming: anodic and cathodic. Both use a water-based primer solution as the electrolyte. Water is favored as the solvent because it is cheap, harmless, and effluent treatment is straightforward.

In anodic coating, the article to be coated is given a positive charge. A water-insoluble, organic acid resin is dispersed in water by the action of a base to form colloidal-size anions:

$$\begin{array}{ccc}
\text{OH} & & \text{O}^- \\
| & & | \\
R-C & + KOH \longrightarrow & R-C + K^+ + H_2O \\
\| & & \| \\
\text{O} & & \text{O} \\
\text{insoluble (1)} & & \text{soluble}
\end{array}$$

The anions are attracted to the anodic workpiece where water is electrolyzed to $H^+$ and $O_2$ by removal of electrons:

$$2H_2O \longrightarrow 4H^+ + O_2 + 4e^-$$

The protons produced neutralize the resin, which is deposited as an insoluble, resistive film of (1).

The chemistry of the more commonly used cathodic electrocoating is similar in many ways. Acid is used to form colloidal cations from insoluble aminated resins:

$$\underset{\text{insoluble (2)}}{\overset{\displaystyle R}{\underset{\displaystyle R}{R-N:}} + R-COOH} \longrightarrow \underset{\text{soluble}}{\overset{\displaystyle R}{\underset{\displaystyle R}{R-N^+-H}} + RCOO^-}$$

As these ions collect at the cathode, electrolysis of water produces hydroxyl ions and hydrogen:

$$4\,H_2O + 4e^- \longrightarrow 2\,H_2 + 4OH^-$$

The resin reacts with the hydroxyl ions to deposit on the surface as an insoluble, resistive layer of (2).

Additives can be mixed with the resins to modify their properties. Examples are metals, metal oxides, carbon, sulfur, and cellulose. These are added as colloidal-sized particles which gain a positive charge by adsorption of hydroxonium ions in the case of cathodic electropainting. Large particles can migrate under the influence of a few charged ions.

Electropriming is usually performed onto a metal substrate. The quality of the coverage is affected by the preparation of the metal surface. Conventional cleaning and degreasing techniques are often used, but the adhesion of some primers is enhanced if the base metal has been phosphated prior to electrocoating.

The high throwing power of electropainting is due to the resistive character of the paints. Once a layer has formed, this electrically insulates the surface, preventing further deposition. Coverage on less heavily coated areas is favored.

Following electrocoating, the work is rinsed and cured. Curing removes solvents and irreversibly alters some resins, improving coating finish. The technique is efficient in terms of energy and materials. This has led to a number of applications in large-scale processes, notably the priming of large metal parts such as car bodies prior to further spray coating. Electropainting is also used to coat coil stock, pipes, sheet steel constructions, and cans.

Commercial electropriming tanks use refinements such as ion-selective membranes around the counter-electrode to prevent diffusion of unwanted ions, and thus control paint pH (Fig. 28). The anodes in such systems are usually made of stainless steel, typically with a surface area about one-quarter that of a car body. The body is immersed in the paint, large tanks are required. Direct current at 50–500 V is commonly used to electrodeposit the primer layer. The requirement of large-scale processing equipment leads to electrocoating being used where a large number of parts need to be coated quickly and cheaply.

Electrophoretic techniques may also be used to codeposit polymer or ceramic particles within an electrodeposited metallic film. The inert particles (e.g., PTFE, alumina, silicon carbide, or diamond) can be dispersed in the electrolyte by high-stability (usually fluorocarbon-based) surfactants. The choice of surfactants and their concentration are critical in determining deposit

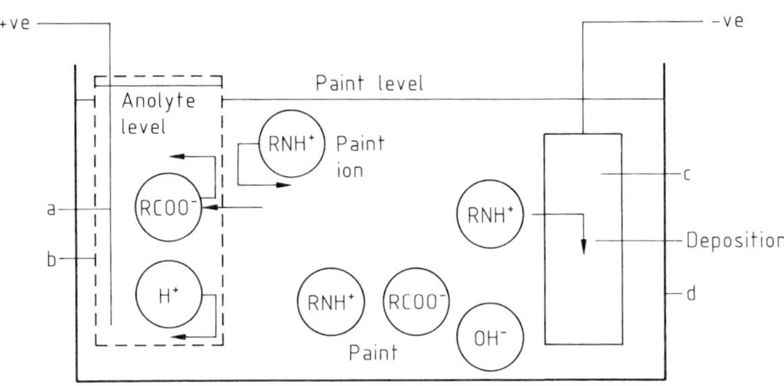


**Figure 28.** Cathodic priming of car bodies
a) Anode plate; b) Ion-selective membrane; c) Car body; d) Paint tank
An ion-selective membrane prevents diffusion of $H^+$ and controls pH.

properties and composition. Examples of the resultant "inclusion coatings" are Ni–PTFE, Co–SiC, and Ni–SiC.

### 1.7.4. Trends in Electrodeposition

Electroplating is vital in almost every branch of industry where the technique can provide a versatile and economic method of applying a thin (1 – 75 μm) metallic coating. However, it is essential to consider the effect of the shape, texture, and material composition of the workpiece at the design stage. Of the many modern developments in electrodeposition technology, several trends may be highlighted:

1) Electroplating of alloys of controlled composition, e.g., Ni–Fe or Pd–Ni as cheaper alternatives to single-metal coatings
2) Codeposition of metals with either polymers or ceramics to provide unique wear-resistant or "self-lubricating" coatings, e.g. Ni–PTFE, Ni–diamond, Co–WC, Zn–alumina
3) Electroless (zero-current) plating of an increasingly wide range of single metals, alloys, and composites, e.g., electroless Ni–P–PTFE
4) Low-temperature nonaqueous electrolytes for deposition of refractory metals, e.g., Ti, Ta, and Nb
5) Selective use of special, programmed current waveforms, e.g., periodic current reversal (rather than the normal steady direct current) in order to produce thick, dense, pore-free coatings, particularly for demanding electronic applications and severe wear-resistant applications
6) Selective, high-speed, automated electroplating processes for reel-to-reel plating of electronics materials
7) Local enhancement of deposition rates by use of laser irradiation, jetted electrolyte flow, as well as by acoustic, mechanical, or ultrasonic vibration
8) Development of polymer coatings based on electrochemical deposition from a monomer in the electrolyte, e.g., polypyrroles and polythiophenes whose conductivity can vary over many orders of magnitude via variation in monomer, electrolyte, and deposition conditions
9) Continued developments in electrophoretic deposition of polymers and ceramics, particularly in the finishing of car bodies

**Journals and Periodicals.** The electrochemistry of electrochemical deposition reactions is considered in the following journals:

Journal of Applied Electrochemistry
Journal of Electroanalytical Electrochemistry
Journal of the Electrochemical Society
Electrochimica Acta
Soviet Electrochemistry
Bulletin of Electrochemistry

Journals devoted to aspects of the surface finishing of metals include:

Plating and Surface Finishing
Metal Finishing
Transactions of the Institute of Metal Finishing
Galvanotechnik
Oberfläche-Surface
Metal Finishing (India)

Other journals relevant to the electrochemical deposition of thin films include:

Surface Technology
Thin Solid Films
Synthetic Metals

# 2. Thin Films as Materials

For a discussion of corrosion-resistant coatings → Steel, **A 25**, pp. 176–179, → 8. Corrosion, **B 1**, pp. **8**-60–**8**-61.

## 2.1. Coatings for Electronic Applications

### 2.1.1. Introduction

The world market of semiconductor-related products increased from $\$45 \times 10^9$ in 1988 to $\$74 \times 10^9$ in 1993. The corresponding market for deposition equipment increased from $\$0.9 \times 10^9$ in 1989 to $\$1.5 \times 10^9$ in 1992. The worldwide hybrid market was estimated at $\$5.5 \times 10^9$ in 1990.

As microelectronics pushes the deposition processes toward shorter cycle time, higher reproducibility, uniformity/conformality, and lower cost, deposition processes have become highly sophisticated. For integrated circuit (IC) assembly, thin film multichip modules are of increasing importance [2.1], but are not discussed here.

### 2.1.2. Hybrid Technology

One of the oldest applications of thin films in electronics are carbon resistors deposited on cyclindrical ceramic substrates by pyrolysis of a mixture of hydrocarbons at 400–600 °C. Because of the high negative temperature coefficient of resistivity (TCR) and the high tolerance of the resistance ($\pm 10$–$20\%$) metal resistors (base materials: NiCr, TaN, or CrSi) are now dominant [2.2]. In R–C networks the negative TCR of the resistor is compensated by the positive TCR of the capacitor [2.3].

Thin film hybrids are at present only a small portion of the hybrid market, but their share is likely to grow because large-scale integration for multipole ICs requires thin film memories, microprocessors, logic devices, semicustom, linear and discrete components, and also because higher pin-in/pin-out counts for dense modules are now achievable.

### 2.1.3. Microsystems

Combining micromachining technology of Si [2.4] with data collection (sensors) and microelectronic data processing has created a new class of products called microsystems (intelligent sensors).

**Infrared sensors** are made of HgCdTe or GaAs. For HgCdTe deposition, a lot of effort has been put into vapor-phase deposition, because of the performance-limiting defects which occur with epilayers. The range of detectable wavelengths can reach 8–12 µm.

**Optical sensors/detectors** consist of Si deposited by evaporation, sputtering or CVD. Applications include scanning, colorimetry, position sensing, fiber optic sensing, multiplexing, and bar-code reading.

**Gas sensors** are made of doped Si or polymers. Applications include smoke detectors, detection of trace quantities of toxic gases, and combustion control for car engines.

**Pressure sensors** employ NiCr, Si, piezo-ceramics (e.g., ZnO), GaP, or heterostructures. They are used in strain gauges, pressure sensing, acceleration sensing, and microphones.

### 2.1.4. Data Storage

**Thermal print heads** [2.5] are fabricated from highly resistive, highly stable compounds with low TCR, e.g., CrSiO, TaN, or HfN as heating elements; $SiO_2$ or $Si_3N_4$ for wear protection. They are used in facsimile machines, bar-code printers, industrial printers and dye-transfer printers. For bubble-jet print heads and magnetic record heads see [2.6], [2.7].

**Magneto-optic (MO) storage.** In MO storage [2.8] the Kerr effect is used for data storage and reading. The materials used are generally amorphous alloys of rare earths and transition metals (Fe, Co) deposited by sputtering. These amorphous materials do not suffer from "grain noise".

### 2.1.5. Microelectronics

The two most important physical processes in microelectronics materials preparation do not involve deposition. Diffusion/ion implantation is used for the local, electrical activation of the bulk Si, and thermal oxidation produces the high-performance gate oxides [2.9]. Nevertheless, for a 16 megabit dynamic random access memory, about 20 deposition processes are needed (two-level metallization with W contact hole-filling). Before depositing the first metallization layer [sputtered Al–Si (1 wt%) or Al–Si (1 wt%)–Cu (0.5–2 wt%)] deposition temperatures above 450 °C may be applied.

**High-Temperature processes.** Applications include insulation (capacitor dielectrics):

$$SiH_2Cl_2 + 2\,N_2O \xrightarrow{\;910\,°C\;} SiO_2 + 2\,HCl + 2\,N_2$$

oxidation masking (capacitor dielectrics):

$$3\,SiH_2Cl_2 + 7\,NH_3 \xrightarrow[\;40\,Pa\;]{\;750\,°C\;}$$
$$Si_3N_4 + 3\,HCl + 6\,H_2 + 3\,NH_4Cl$$

and formation of conformal insulation layers, spacers, etch-stops, and protection against dopant out-diffusion:

$$Si(C_2H_5O)_4 \text{ at } 30\text{–}70\,°C \xrightarrow{\;700\,°C\;}$$
$$\text{(TEOS)} \qquad SiO_2 + 2\,H_2O + 4\,C_2H_4$$

**Medium-Temperature Processes.** Applications include the use of BPSG (Boron–phosphorus–silicate glass) for insulation, smoothing of edges, and planarization of topography by a 850–900 °C anneal/flow step:

$$Si(C_2H_5O)_4 + B(OCH_3)_3 + PH_3 + (O_2)$$
$$\xrightarrow{\;630\text{–}680\,°C\;} BPSG + \text{decomposition products}$$

This process is hard to control under production conditions and new processes have been developed [2.10].

Another application is the production of conduction lines:

$$SiH_4 \xrightarrow[\text{40 Pa}]{630\,°C} Si + 2H_2$$

This CVD process is important in semiconductor technology, and has influenced the development of equipment (tube furnaces, low-pressure CVD). The process gives high-purity deposited polycrystalline Si and high throughput (ca. 100 150-mm wafers per hour) and gives outstanding uniformity and conformality of the film thickness.

The resistivity of doped poly-Si is too high for some applications, and a new class of materials has been introduced (refractory disilicides: $TiSi_2$, $MoSi_2$, $TaSi_2$, $WSi_2$) which are in equilibrium with Si at higher temperatures. The films are sputtered from compound or mosaic targets.

**Metallization Processes** [2.11], [2.12]. For the electrical connections between activated areas, an alloy of Al–Si (0.5–2 %) is sputter deposited. The low-cost Al reduces the native silicon oxide at ca. 450 °C, leading to low ohmic contact resistances. The Si content in the alloy avoids pronounced out-diffusion of the bulk Si into the Al conduction line. After cooling, the Si alloy component forms lightly doped precipitates, preferably on the bulk Si, with average diameter 0.3–1.5 µm.

For sub-micrometer technology, TiN or Ti–W–N layers are applied to reduce Si precipitates in the contact holes. Since these barrier layers have excessive contact resistance to Si, an additional contact layer (e.g., 50 nm sputtered Ti) is necessary. A further problem is that the step coverage of the sputtered films is no longer sufficient. Therefore CVD processes have been developed for TiN deposition [2.13]:

$$Ti[N(CH_3)_2]_4 + 2H^* \xrightarrow[\text{40 Pa}]{430\,°C}$$
$$TiN + 3C_2H_6 + 2CH_4 + \tfrac{3}{2}N_2$$

(where $H^*$ is produced by microwave irradiation) and for contact hole filling with W [2.14]:

$$WF_6 + 3H_2 \xrightarrow[\text{8 kPa}]{430\,°C} W + 6HF$$

**Passivation.** *Low-temperature plasma CVD processes* include:

$$3SiH_4 + 4NH_3 \xrightarrow[\text{30 Pa}]{300\,°C} Si_3N_4 + 12H_2$$

for compressive stress $10^8 - 10^9$ Pa, and:

$$3SiH_4 + 2N_2 \xrightarrow[\text{0.5 kPa}]{390\,°C} Si_3N_4 + 6H_2$$

for compressive stress $5 \times 10^7$ Pa. This is used for protection against alkali and $H_2O$ diffusion.

In *multilevel metallization*, the Al–Si–Cu conduction line levels have to be separated by an intermetal insulating layer deposited by PECVD below 430 °C.

$$Si(C_2H_5O)_4 \xrightarrow[\text{1.3 kPa}]{400\,°C} SiO_2 + \text{decomposition products}$$

For further developments see [2.15].

### 2.1.6. Semiconducting Films

The basic process for *thin film transistors* (TFTs) is the PECVD deposition of amorphous Si (a-Si) at 300–400 °C, starting from $SiH_4$. Many substrates are in use, e.g., Si, GaAs, glass, quartz, and alumina. The breakthrough was reached when doping of the a-Si by a simple codecomposition of *p*- and *n*-type dopant gas species (e.g., $PH_3$, $B_2H_6$) became possible. Applications include television, logic circuits for image sensors (CCDs), printer/copiers, and active matrix liquid crystal displays.

Polycrystalline Si TFTs seem likely to become an important competitor with a-Si for displays, because the drive electronics can be fabricated directly on the glass substrate during the formation of the TFT array.

In the field of *photovoltaics*, a-Si solar cells and GaAs, InP, CdTe, and $CuInSe_2$ cells are fabricated. GaAs and InP cells are attractive for use in space because of their high efficiency and radiation resistance. The efficiency of solar cells is limited by recombination of electrons at the transparent conductive oxide–*p*-layer interface; thin film materials such as ZnO, $SnO_x$ and indium oxide–tin oxide (ITO) are under investigation [2.16].

*Optoelectronic devices* [2.17] are made from $LiNbO_3$ or polymers [2.18]. Applications include phase modulators, amplitude modulators, splitters, polarizers, and switches.

GaAs and GaAlAs [2.19] can also be used for detectors, waveguide components, and high-speed electronic circuitry on the same substrate. Thin film deposition processes promise to play a major part in integrated optics, e.g., semiconductor planar waveguides could be used to integrate several optical components with associated waveguides on a single substrate.

For the use of semiconductor superlattices grown from III–V, II–VI, and IV–VI compounds for quantum wells see [2.20].

### 2.1.7. Ferroelectrics [2.21]

Ferroelectrics can be made from: perovskite compounds, e.g., $BaTiO_3$, $KNbO_3$, $K(Ta,Nb)O_3$, $PbTiO_3$, $Pb(Zr,Ti)O_3$; or non-perovskite compounds, e.g., $LiNbO_3$, $Bi_4Ti_3O_{12}$, $(Sr,Ba)Nb_2O_6$.

The breakthrough for this multifunctional class of materials was an improvement of the deposition process [2.22]. For optical applications of ferroelectric films, the system technologies are not completely mature. Several alternative materials and device technologies exist. For electronic applications, ferroelectrics have commercial potential, e.g., in nonvolatile memories and dynamic random access memories, microactuators, thin film capacitors, and surface acoustic wave devices.

## 2.2. Hard and Decorative Nitride and Carbide Coatings

### 2.2.1. Introduction

Transition metal nitrides and carbides exhibit characteristics of covalent, ionic, and metallic bonding. They have high thermal and electrical conductivities while the covalent nature gives rise to high mechanical microhardnesses, typically 20–30 GPa [2.23], [2.24]. They exhibit a metallic luster, and for some compounds (e.g., TiN, VN, ZrN, TaC) interband transitions locate the reflectance edge in the middle of the visible region, giving a golden color. Most of the compounds are also completely soluble in each other, making it possible to tailor the properties by depositing carbonitrides, alloy nitrides, or alloy carbides [2.23], [2.25]. The combination of high thermal and electrical conductivities, hardness, and attractive color have resulted in extensive use as both wear-protective and decorative coatings. Nitride thin films are also used as diffusion barriers in integrated circuits.

### 2.2.2. Growth Techniques

Nitride and carbide coatings are grown by chemical vapor deposition (CVD) or ion-assisted physical vapor deposition (PVD). In the former, substrate temperatures $> 900\,°C$ are typical; in the latter, growth temperatures of $300–550\,°C$ are common. For acceptable quality of coatings grown by PVD at $300–550\,°C$, ion bombardment is necessary during growth, usually by plasma-assisted techniques, e.g., magnetron sputtering, arc evaporation, or ion plating [2.25]. Owing to the nonequilibrium nature of PVD techniques, film microstructure and morphology, stress levels, hardness, adhesion, etc. depend strongly on the deposition conditions, and a large spread in properties is common for coatings prepared in different equipment [2.24]. The most systematic correlations between deposition parameters, microstructures, and properties are found for TiN coatings.

### 2.2.3. Coating Microstructures

The detailed microstructure of the coatings depends strongly on growth parameters, but under most conditions columnar microstructures are formed. Films grown by PVD at $< 500\,°C$ are low-density coatings with a pronounced columnar microstructure [2.26]. The columns extend from the substrate surface to the top surface (Fig. 29). The surface is typically rough with faceted columns. For dense films, temperatures significantly higher than $500\,°C$ are required, but even in this case the microstructure is columnar. By using ion irradiation during growth, dense films can be grown, even below $500\,°C$. Ion bombardment enhances adatom mobility, forward sputtering, and recoil implantation, leading to densification [2.27]. The energy required to obtain dense films depends on growth conditions, but for typical growth rates of the order of µm/h and ion current densities in the $mA/cm^2$ range, ion energies of $50–100$ eV are required [2.25]–[2.28].

As the ion energy is increased above that required for creating dense films, ion-induced defects (point defect clusters, small dislocation loops, sputtering gas atoms) become trapped within the film. This can lead to continuous renucleation during growth and interruption of the columnar microstructure (Fig. 30). A relatively smooth surface is obtained. As the microstructure changes from a porous columnar structure to a dense noncolumnar structure, the stress in the coatings changes from tensile to compressive. The compressive stress usually in-

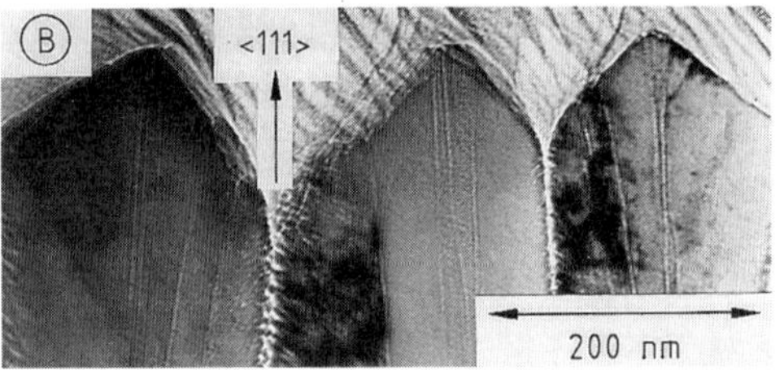


**Figure 29.** Cross-sectional transmission electron micrograph of a Ti$_{0.5}$Al$_{0.5}$N film grown by reactive magnetron sputtering on a stainless steel substrate
Grown with the substrate grounded, i.e., no intentional ion bombardment.
A) Overview micrograph; B) Higher magnification micrograph of the film surface showing faceted top surface and intercolumnar porosity

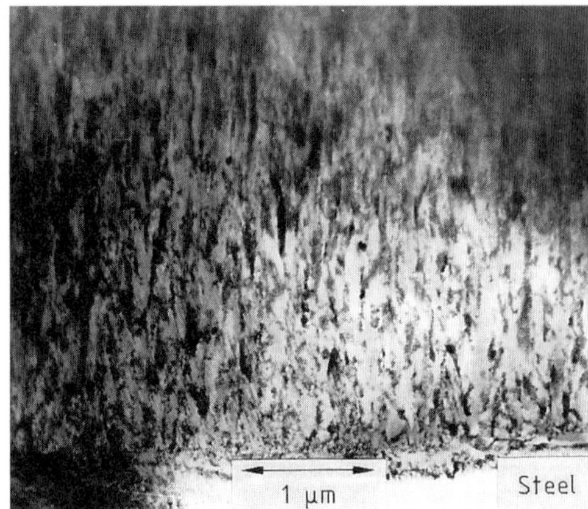


**Figure 30.** Cross-sectional transmission electron micrograph of a Ti$_{0.5}$Al$_{0.5}$N film grown by reactive magnetron sputtering on a stainless steel substrate
Grown with substrate at a negative potential of 150 V, i.e., intentional ion bombardment.

creases to a saturation value as the ion energy is increased [2.28]. Compressive stress levels as high as 10 GPa can be obtained. Thermal stresses due to different thermal expansion coefficients of substrate and coating are usually smaller than growth-induced stresses, unless temperatures significantly above 500 °C are used. The different microstructures are reflected in the physical properties of the coatings. For TiN hardness values of 10–40 GPa are reported [2.24] (the single-crystal bulk value for TiN is 20 GPa [2.23]). The lowest values are obtained for low-density films, the highest values only for coatings with high compressive stress levels. For optical or decorative purposes, low-density films with rough surfaces are not acceptable since their properties change with time and they are very sensitive to fingerprints. For most applications, dense, moderately compressively stressed coatings are required.

### 2.2.4. Adhesion

Since coated tools are subjected to very large forces, the adhesion strength between the coating and the substrate has to be very high. Although the available empirical results do not dictate a unique mechanism for adhesion-enhancing processes, some general requirements for good adhesion can be noted. The most important is to eliminate contamination of the substrate surface prior to growth. During high-temperature growth (e.g., CVD) organic contaminants are readily desorbed from the surface. High temperature also produces interdiffused graded zones between film and substrate, with strong metallurgical bonding, and thus a high adhesion strength (provided no brittle phase is formed). For low-temperature growth, in situ sputter etching in a noble gas (most commonly Ar) plasma is necessary [2.29]. In addition to removing surface contaminants, sputter etching avoids low nucleation density, since lateral growth may result in interfacial voids. During low-temperature PVD growth, graded interfacial zones can be formed by depositing interfacial layers. For TiN coatings, it is common to use Ti interlayers to promote adhesion [2.29], [2.30]. Collisional mixing by ion bombardment can also be used to create graded interfaces, and thus higher adhesion strength.

### 2.2.5. Hard Coatings

**Nitrides.** The most used and studied material for hard and decorative coatings is TiN [2.24], [2.31]. Stoichiometric bulk TiN has a hardness of 20 GPa and a golden color. However, owing to the wide stability range ($0.6 < N/Ti < 1.2$) its properties vary with composition [2.23]. The color can range from that of metallic Ti to almost brown. The different microstructures obtainable give coatings with a hardness range of 10–40 GPa. The majority of commercially used TiN films have compositions close to stoichiometric, hardness 20–27 GPa, and compressive stresses of a few GPa. For N/Ti ratios below 0.6, films consisting of both the NaCl-structure $\delta$-TiN phase and the $\alpha$-Ti phase are reported. Only in a few cases has the tetragonal $\varepsilon$-Ti$_2$N phase been reported in films grown from the vapor phase [2.32]. These usually have high compressive stress values and they have not yet found any commercial use. The exact phase composition is given by the kinetic conditions during growth.

Other common nitrides are the group 4 nitrides ZrN and HfN, both of which have properties similar to those of TiN. Bulk HfN has been reported to have a higher hot-hardness than TiN, and has been claimed to be beneficial in high-temperature applications, but so far HfN films have not been used in commercial wear-protective or decorative applications.

Except for CrN, transition metal nitrides from groups 5 and 6 are not used for wear-protective or decorative purposes.

**Alloy Nitrides.** Most transition metal nitrides are completely soluble in each other, making it possible to vary the properties of the coatings over a large range. The most studied alloy nitride is Ti$_{1-x}$Al$_x$N [2.33]. Even though the NaCl-structure TiN and the wurtzite-structure AlN are not mutually soluble, metastable Ti$_{1-x}$Al$_x$N films with NaCl structure can be grown for $0.4 < x < 0.6$, depending on the growth conditions [2.34]. For growth of such metastable phases it is necessary to use strong kinetic limitations, through low growth temperatures, high

growth rates, or ion-induced collisional mixing during growth. The metastable Ti$_{1-x}$Al$_x$N phase decomposes above 900 °C to TiN and AlN [2.35].

Ti$_{0.5}$Al$_{0.5}$N has a much higher oxidation resistance than TiN, owing to the formation of a protective Al-rich oxide layer, benificial for high-temperature applications [2.35]. The oxidation temperature for Ti$_{0.5}$Al$_{0.5}$N and a few other hard coatings are as follows [2.25]:

| Coating | Oxidation temperature °C |
|---|---|
| TiC | 400 |
| TiN | 550 |
| CrN | 700 |
| Ti$_{0.5}$Al$_{0.5}$N | 750 |
| Cr$_{0.5}$Al$_{0.5}$N | 910 |

The possibilities of using alloy nitrides to increase the oxidation temperature is clearly demonstrated. The structural and mechanical properties of Ti$_{1-x}$Al$_x$N are similar to those of TiN. The electrical and optical properties depend on the Al content (the color changes from golden to dark violet as $x$ is increased from 0 to 0.5).

**Carbides.** The hard carbide coatings used resemble those of nitride hard coatings, i.e., carbides from group 4 (TiC, ZrC, and HfC) dominate. Also, B and Si carbides have been studied for wear-protective applications. The structures of the transition metal carbides resemble those of the corresponding nitrides; all monocarbides in groups 4 and 5 have a B1-NaCl structure. The hardness of the carbide is in general higher than that of the corresponding nitride (more pronounced covalent bonding) [2.23], [2.24]. Compared with nitrides the growth process is more complex (possibility of C precipitation as the C/metal ratio approaches 1, caused by the complex cracking patterns and chemisorption behavior of the hydrocarbon precursors). As free C precipitates, the film structure changes drastically to porous films with very small grain sizes [2.24].

**Carbonitrides.** Since most transition metal carbides and nitrides are completely soluble in each other, carbonitrides are sometimes used. As in the case of alloy nitrides, the properties can be varied continuously from those of the carbide to those of the nitride. Only TiCN coatings have been studied to any significant degree [2.24].

**Multilayers.** By using multilayers of different materials it is possible to take advantage of the

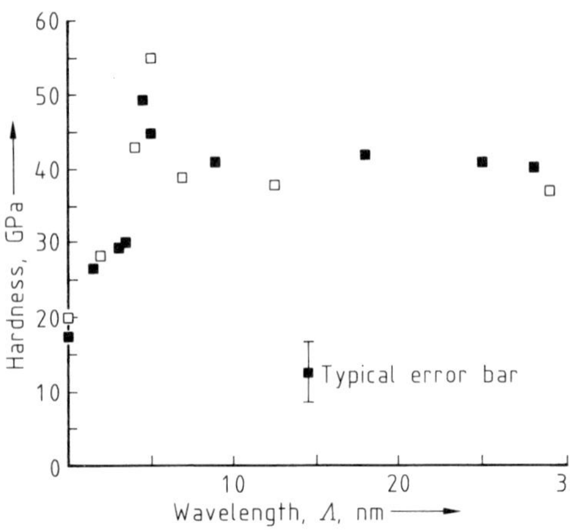

**Figure 31.** Vickers hardness of TiN/VN (open squares) and TiN/NbN (full squares) multilayers as a function of modulation wavelength [2.43]
Total film thicknesses 2.5–3 µm in both cases; hardness values for $\Lambda = 0$ nm correspond to the TiVN and TiNbN alloy values.

characteristic features of each material and to obtain synergetic effects [2.36], [2.37]. The main reasons for using multilayers of nitrides and carbides are to facilitate strong adhesion, to obtain wear-protective films with low chemical reactivity and low friction, and to increase strength and hardness. Multilayers of the type $TiN/Al_2O_3/TiC$ deposited on cemented carbide substrates by CVD are used routinely in order to optimize adhesion and wear properties on cutting tool inserts [2.31].

By depositing nitride multilayer films with individual layer thicknesses of a few nanometers, significantly increased hardness values have been reported for films with total thicknesses of a few micrometers [2.36]–[2.43]. Examples of systems with hardness more than twice that of the individual materials include TiN/VN [2.38], TiN/NbN [2.39], and $TiN/V_{0.6}Nb_{0.4}N$ [2.40]. Figure 31 shows the variation in hardness for TiN/VN and TiN/NbN multilayers as a function of the multilayer wavelength $\Lambda$ (the sum of the individual layer thicknesses). The enhanced hardness can be explained by hindrance of dislocation motion across the layer interfaces (different dislocation line energies in the two materials) and by coherency strain [2.40]. Metal/nitride multilayers such as Ti/TiN, Hf/HfN, W/WN [2.42], and Ti/BN [2.43] have very high hardness values.

## 2.3. Carbon Films

(General references [2.44]–[2.49])

### 2.3.1. Carbon Phases: Composition and Stability

Diamond and graphite are well-defined, long-range-ordered crystalline allotropes of carbon, that occur naturally. Structure, properties, and quality of vapor-deposited thin film variants are best judged by comparison with their natural equivalents. Other carbon phases are less well defined. They do not occur naturally, but are grown as thin films by chemical and physical vapor deposition (CVD and PVD) methods. Composition and properties of these phases depend strongly on the preparation conditions. Resulting differences are manifested by a confusingly long list of names: amorphous hydrogenated carbon (a-C:H), amorphous carbon (a-C), ion-beam-deposited carbon (i-C), mass-selected ion-beam-deposited amorphous carbon (MISB a-C), tetrahedral amorphous carbon (ta-C), and diamond-like carbon (DLC). DLC is also frequently used for the whole class of amorphous carbon phases. They may contain considerable quantities of hydrogen, varying amounts of $sp^3$- and $sp^2$-hybridized carbon atoms. The ternary phase diagram (Fig. 32) is a convenient way to display the compositional differences [2.50], [2.51], to correlate them with different properties, and to outline the relation between the various carbon phases. The $sp^2$-vertex represents graphite, a hydrogen-free crystalline material

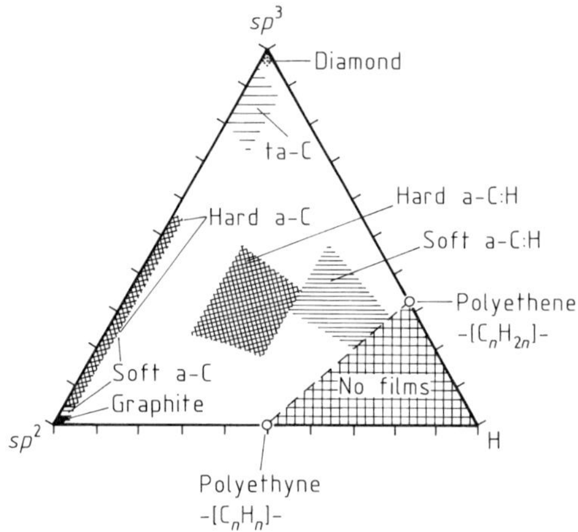


**Figure 32.** Ternary $sp^3$-C/$sp^2$-C/H diagram for carbon films ([2.50], [2.51], modified)

that consists of aromatic carbon layers built from threefold-coordinated $sp^2$-carbon atoms. The $sp^3$-vertex corresponds to diamond, hydrogen-free, cubic crystals formed by exclusively covalently bound, fourfold-coordinated $sp^3$-carbon tetrahedra. Noncrystalline, hydrogen-free carbon phases, called either a-C or, with more than 80 % $sp^3$-carbon, ta-C, are located on the $sp^2-sp^3$ side line. Films with a C–C skeleton are feasible only up to a certain level of hydrogen in the solid. At higher H concentrations films do not form (Fig. 32, dashed line [2.51]). The compositional coordinates of the various carbon thin films in Figure 32 determine whether their physical properties are closer to diamond (high $sp^3$-C, low H content), to graphite (high $sp^2$-C, low H content), or more like polymers (high H content).

Of all the carbon phases mentioned, only graphite is thermodynamically stable. All other carbon phases are metastable, and, although reasonably stable for many applications at room temperature, convert to graphitic material on heating in an inert atmosphere. While a-C:H and a-C change their structure at 300–600 °C [2.52], ta-C seems to tolerate more than 1000 °C [2.53]. Diamond graphitizes rapidly at temperatures > 1500 °C [2.45], [2.49]. Composition, structure, and stability define the physical and chemical properties of the films and, along with the deposition temperature, define their application areas.

## 2.3.2. Properties and Applications of Diamond Films

There is no other material with such a combination of properties as diamond. It is the hardest of all materials. At room temperature, it has the highest thermal conductivity. Diamond is transparent from the UV to the microwave region of the spectrum (except for a small gap between 3 and 6 µm). It is an excellent insulator, but can be doped and converted into a low-resistivity semiconductor. Diamond is chemically inert at room temperature and, owing to its high atom density, is extremely rigid. The properties of the best vapor-grown diamond films are identical with those of natural diamond. Significant differences between natural and vapor-grown material result from the polycrystalline structure of the latter. Polycrystalline diamond films usually exhibit a columnar growth morphology with crystal grains of 0.1 to > 100 µm. The grain boundaries tend to be decorated with contaminants and, depending on the deposition conditions and process, anything from black films with considerable amounts of non-diamond carbon phases, other impurities (Si, O, etc.) and several percent hydrogen, to transparent material with less contamination than natural diamond, is feasible. Table 9 lists the physical properties of diamond, in comparison with some competing materials and possible applications of vapor-grown diamond films.

**Mechanical Properties and Applications.** The hardness and wear resistance of diamond and its resulting application as a material for tools and abrasives is common knowledge. Diamond CVD permits growth of polycrystalline films of at least comparable hardness and wear properties, but free of the cobalt binder that is required when compacting high-pressure diamond powder into tool bits. Binder-free, polycrystalline films have an advantage over single crystals. Cleavage planes are absent, because of the random orientation of crystallites. Grain boundaries prevent crack propagation and the material is tougher and mechanically isotropic. The Young's modulus of CVD diamond is identical to that of single crystals: 1050 GPa. The rupture strength of 10–20 mm diameter CVD diamond plates with thicknesses of 150–300 µm is ca. 1000 ± 200 MPa, a factor of two lower than for single crystals. The fracture toughness is 6 MPa m$^{0.5}$, higher than the 3.5–5 MPa m$^{0.5}$ value reported for single crystals. The test performance of thick CVD diamond plates is hardly affected by the surface morphology of top-quality material, i.e., unpolished disks show the same values as polished material (for mechanical data see [2.54], [2.55]).

For films with thicknesses < 50 µm, the phase purity and hence all film properties depend strongly on the thickness. Vapor-grown, free-standing (substrate-free) diamond plates can be used as single-point tools or, if brazed onto a tool insert, can replace high-pressure synthetic polycrystalline diamond compacts, i.e., in turning, milling, drawing, or bonding tools [2.56]. Diamond can also be directly deposited onto various tool materials, e.g., tungsten carbide or SiAlON inserts. With appropriate surface treatment, the adhesion of such films is sufficient for practical applications. Coated tools require a minimum of diamond. A < 10 µm thick film is usually suffi-

**Table 9.** Properties and applications of diamond films

| Property | | Comparison with competitors | Possible applications |
|---|---|---|---|
| Indentation hardness, kg/cm² | 10 000 | β-BN < 4500; SiC < 4000; WC = 2190; steel = 800 | drill bits, polishing material, cutting tools |
| Coefficient of friction | 0.05–0.1 (only in air) | PTFE ≈0.05 | ultra-low-wear coatings |
| Young's modulus, GPa | 1050 | 2× the value of alumina; 5× that of steel | stiff membranes for lithography masks, lightweight diaphragms for audio devices |
| Sound propagation velocity, km/s | 18.2 | 1.6× the value of alumina | |
| Chemical inertness | oxidation at T > 600 °C | at room temperature, resistant to all acids, bases and solvents | coatings for reactor vessels |
| Range of high transmittance, µm | 0.22–3 and >6 | in the IR orders of magnitude lower than most other materials | UV–vis–IR windows and coatings, microwave windows, X-ray windows (low $Z$) |
| Refractive index | 2.41 | 1.6× the value of silica | |
| Band gap, eV | 5.45 (negative electron affinity) | 1.1 for Si; 1.43 for GaAs; 3.0 for β-SiC | high-power/high-temperature electronics, e.g., thermistors, transistors, particle detectors, cold cathodes for displays, vacuum microelectronics |
| Electron/hole mobility, $cm^2 V^{-1} s^{-1}$ | 1900/1600 | 1500/600 for Si; 8500/400 for GaAs | |
| Dielectric constant | 5.5 | 11 for Si; 12.5 for GaAs | |
| Thermal conductivity, $Wm^{-1} K^{-1}$ | 1300–2200 | highest at room temperature; 4× the value of Cu or Ag | heat sinks, heat spreaders |
| Thermal expansion coefficient, $K^{-1}$ | $0.8 \times 10^{-6}$ | at room temperature, similar to silica ($0.57 \times 10^{-6}$) | thermally stable substrates, e.g., for X-ray lithography masks |

cient. They perform as well or better than conventional diamond tools, without their size limitations and often at a lower price [2.56]. Owing to the complex shapes required, directly deposited diamond layers are specially suitable for drilling tools [2.56]. The excellent mechanical properties are also required for other diamond applications, e.g., as high-pressure or vacuum windows.

**Optical Properties and Applications.** Type IIa natural diamond is optically transparent from its band edge at 0.220 µm to ca. 3 µm. The lattice vibrations of diamond are excited between 3 and 6 µm, hence diamond is weakly absorbing in this region. At higher wavelengths it transmits electromagnetic radiation up to the microwave region of the spectrum [2.45], [2.49]. Figure 33 sketches the optical transmission spectra of diamond and other common optical window materials, e.g., LiF (UV windows), silica (visible), and CsI (IR) [2.57]. Transmission losses of 31 % for

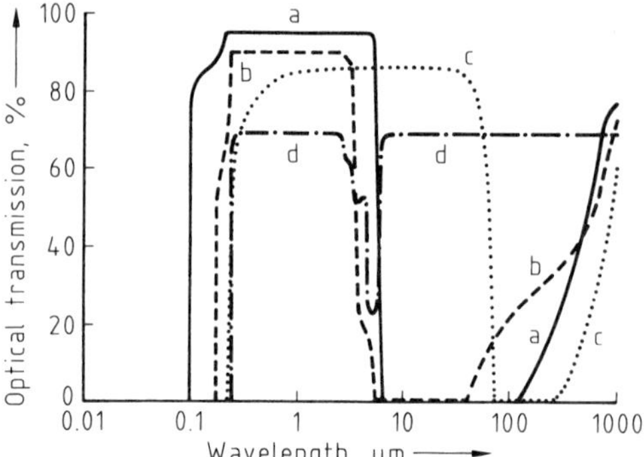


**Figure 33.** Optical transmission spectra for diamond and other optical window materials [2.57]
a) LiF; b) Silica; c) CsI; d) Diamond

diamond are due to reflections at two diamond–air interfaces (refractive index ca. 2.41). The best quality CVD diamond films are now approaching the characteristics of type IIa natural diamond [2.58]. Problems are related to:

1) Excess absorption between 220 and 350 nm, owing to nitrogen incorporation in the solid
2) Excess absorption between 220 nm and 3 µm, owing to non-diamond carbon phases
3) Scattering losses due to the surface roughness of the polycrystalline structure (affected wavelength region depends on grain size)

While excess absorption can be avoided by contamination-free deposition procedures, scattering at the rough growth surface is inherent to polycrystalline material. Either by polishing (for thick diamond plates) or by deposition of very fine-grain material (suitable for thin films and membranes), scattering in the visible region can be minimized. For top-quality material, the grain boundaries inside thick polished plates hardly affect the transmission properties.

Vapor-grown diamond films and plates can be used as windows for electromagnetic radiation of any kind. With its unique combination of mechanical strength, chemical inertness, high transparency, and low atomic number, diamond is particularly useful in harsh environments and as a protective optical coating for sensitive materials such as IR optics.

**Thermal Properties and Applications.** At room temperature, diamond is the best thermal conductor available; at 20 °C, values of 2000–2200 $Wm^{-1}K^{-1}$ were measured for type IIa natural single crystals—four times the value of copper or silver. The thermal conductivity of diamond peaks at ca. 65 K with values around 17000 $Wm^{-1}K^{-1}$. Between room temperature and 500 °C, the thermal conductivity $\kappa$ decreases, following the empirical formula [2.59]:

$$\kappa = 27350\, T^{-1.26}$$

Top-quality material has conductivity ca. 700 $Wm^{-1}K^{-1}$ at 500 °C, still twice the value of copper or silver (Fig. 34). The transport of thermal energy in diamond is a phonon-coupled phenomenon. Despite the polycrystalline nature and the numerous grain boundaries that could disturb phonon propagation, CVD diamond films with the thermal properties of the best natural single crystals have been produced [2.58]–[2.61]. Film thicknesses > 100 µm with grains larger than 20 µm are required to obtain values above 1800 $Wm^{-1}K^{-1}$. Thin films suffer from contamination by excess non-diamond carbon phases and from phonon scattering at the grain

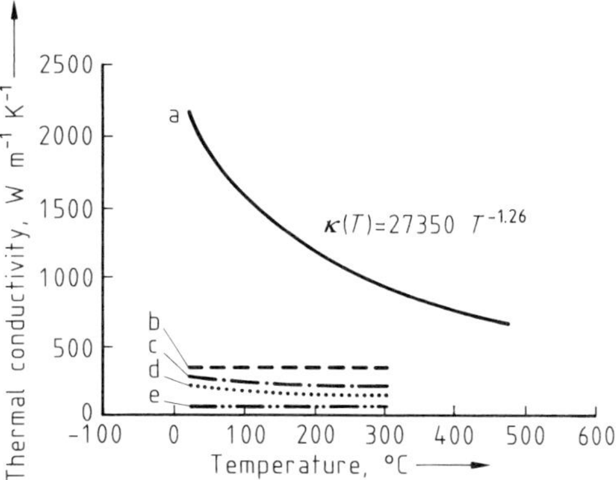


**Figure 34.** Temperature dependence of the thermal conductivity of diamond and other commonly used heat spreader materials
a) Type IIa natural and best quality CVD diamond; b) Cu; c) BeO; d) AlN; e) $Al_2O_3$

boundaries of numerous small ($<$ 5 µm) crystals. For 4 µm thick films, values of ca. 700 $Wm^{-1}K^{-1}$ are feasible [2.61]. Isotopically enriched $^{12}C$-diamond single crystals as well as films exhibit a thermal conductivity 20–50% higher than that of natural (99% $^{12}C$) material [2.62].

In addition to its extremely high thermal conductivity, diamond has a very low thermal expansion coefficient, $0.8 \times 10^{-6}\,K^{-1}$ at room temperature. However, this value applies only to single crystals and top-quality thick CVD diamond plates. Non-diamond carbon contamination in thin films may increase the values to $> 2 \times 10^{-6}\,K^{-1}$.

The application of high-thermal-conductivity diamond as a highly efficient heat spreader for high-power electronic devices and high-thermal-load laser windows is well established. CVD films make it possible to overcome the cost and size limitations imposed by the use of natural single crystals and may compete in applications currently served by poisonous BeO or AlN.

**Electrical Properties and Applications.** Pure diamond is electrically insulating with resistivity $> 10^{11}\,\Omega cm$ and breakdown electrical field $10^7$ V/cm, which is similar to silica [2.45], [2.49]. The values for polycrystalline films are lower [2.57] and depend on the thickness. Again, grain boundary decoration with non-diamond carbon phases poses a problem and only ultra-pure CVD diamond approaches the breakdown voltage values of single crystals. The resistivity of diamond can be modified either by UV illumination or by doping with boron. For both natural boron-doped crystals and vapor-grown films, resistivities $< 0.1$ to $100\ \Omega\,\mathrm{cm}$ are feasible. Boron-doped diamond is a $p$-type semiconductor with activation energy ca. $0.38\ \mathrm{eV}$ [2.57], [2.63]. The feasibility of true $n$-type doping of diamond with significantly low resistivity is still under discussion. Phosphorous, nitrogen, or lithium are possible $n$-type dopants. Implantation of $C^+$ may also lead to $n$-type conductivity. In all cases, the material is still highly resistive and $p\!-\!n$ junctions are not yet feasible. The carrier mobilities in undoped bulk diamond are $2200\ \mathrm{cm^2\,V^{-1}\,s^{-1}}$ for electrons and $1600\ \mathrm{cm^2\,V^{-1}\,s^{-1}}$ for holes [2.45], [2.49]. The polycrystalline nature of CVD diamond severely affects the mobilities. For randomly oriented boron-doped material, carrier mobilities $< 50\ \mathrm{cm^2\,V^{-1}\,s^{-1}}$ are common. At room temperature, highly oriented (100)-textured boron-doped films exhibit values of $300-400\ \mathrm{cm^2\,V^{-1}\,s^{-1}}$ and for homoepitaxially grown boron-doped films $800-1000\ \mathrm{cm^2\,V^{-1}\,s^{-1}}$ has been reported [2.63]. Device operation field-effect transistor, FET at $300-600\ ^\circ\mathrm{C}$ is feasible. For (100)-oriented boron-doped films, the carrier concentration increases from $10^9\ \mathrm{cm^{-3}}$ at room temperature to $10^{16}\ \mathrm{cm^{-3}}$ at $500\ ^\circ\mathrm{C}$. Unfortunately, the hole mobility was found to drop from $350\ \mathrm{cm^2\,V^{-1}\,s^{-1}}$ at room temperature to $120\ \mathrm{cm^2\,V^{-1}\,s^{-1}}$ at $500\ ^\circ\mathrm{C}$ [2.63]. The properties of CVD diamond now allow the construction of the first high-power, high-temperature electronic devices (e.g., transistors, diodes, and sensors). Tremendous progress has been made over the past decade: prototype devices operating at $500\ ^\circ\mathrm{C}$ are available and the first logic circuits [2.64] have been demonstrated. However, to implement a diamond-based semiconductor technology still needs a lot of work.

Diamond is not only interesting as an active device and sensor material, but has unusual electronic surface properties which might be exploited in cold cathodes. The (111) surface of natural boron-doped diamond was found to exhibit "negative electron affinity" (NEA) [2.65]. This means that the energy level of the conduction band of diamond is slightly higher than that of electrons in vacuo. NEA materials emit electrons once they are excited across the band gap ($5.45\ \mathrm{eV}$ for diamond) by, e.g., UV light or high electric fields into the conduction band. Diamond might, therefore, serve as a cold electron source for various applications, e.g., flat panel displays. The NEA properties depend strongly on the surface state of the diamond; H- and HO-terminated surfaces emit electrons while reconstructed diamond surfaces show no emission. CVD diamond films display NEA properties and large-area electron-emitting surfaces seem feasible.

### 2.3.3. Properties and Applications of a-C:H, a-C, and ta-C Films

Amorphous carbon phases are also called diamond-like carbon (DLC) films, referring to the hardness of some of these materials. Hardness, and the fact that the films are carbon-based, are about the only relationship between diamond and some members of the DLC family. Figure 32 illustrates that DLC usually consists of a mixture of $sp^2$- and $sp^3$-hybridized carbon and, depending on the preparation conditions, may contain up to 65 atom % hydrogen. The properties of the different DLC types depend strongly on the composition. They range from polymeric, soft, transparent, and insulating, to hard, dark, and conducting. In all cases the films are amorphous and the properties are isotropic.

Figure 35 correlates the hydrogen content, density, optical band gap, and hardness of DLC films grown under different deposition conditions, i.e., different bias voltages.

**Mechanical and Chemical Properties and Applications.** The hardness, low friction coefficient, and wear resistance are the most useful properties of some members of the DLC family. If deposited properly, a-C:H films can reach a density of $2.0\ \mathrm{g/cm^3}$ and a hardness of $40\ \mathrm{GPa}$ (diamond: $100\ \mathrm{GPa}$). The values for soft, hydrogen-rich a-C:H films and soft, $sp^2$-C-rich a-C are ca. $2-5\ \mathrm{GPa}$. With $90-100\ \mathrm{GPa}$ and densities of $2.5-3.1\ \mathrm{g/cm^3}$, mass-selected, ion-beam-deposited hard a-C are closest to diamond (see also Fig. 32). Unfortunately, the thickness of these films is limited by the high internal compressive stress [2.53].

Friction coefficients $< 0.1$ are common for DLC. Like that of diamond, the coefficient of friction is very sensitive to humidity, but unlike diamond, it is reduced in dry air or vacuum.

Chemically, DLC films are inert to most acids and bases at room temperature. At higher temperatures they are attacked. Moreover (see

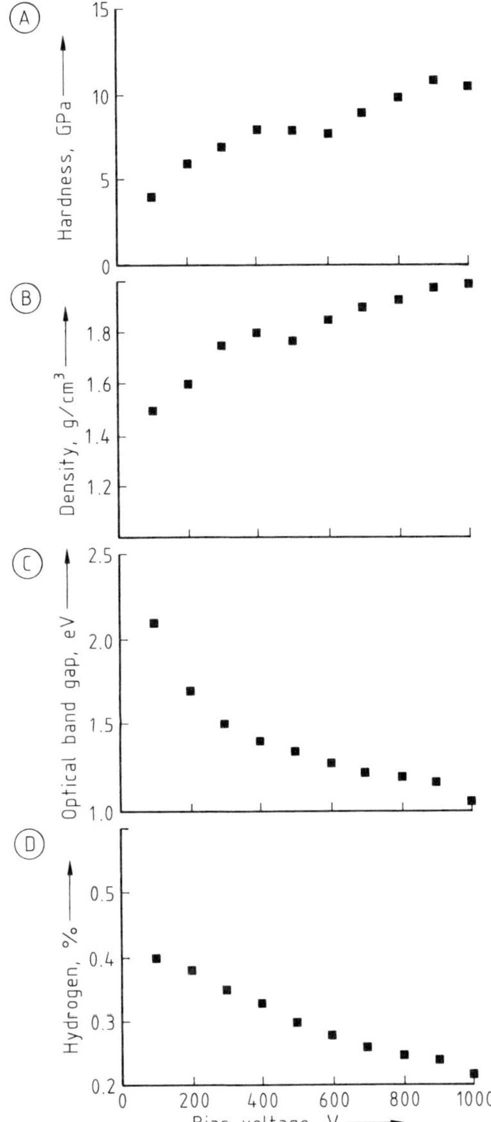

**Figure 35.** Correlation of A) Hardness; B) Density; C) Optical band gap; and D) Hydrogen content of DLC films grown from benzene by r.f. plasma CVD at different bias voltages [2.51]

tics, sunglasses, magnetic recording disks, or metal film recording tapes. Although the properties are inferior to diamond, the low-temperature growth option is a big advantage of DLC films over diamond for such applications [2.67], [2.68].

**Biomedical Properties and Applications.** Carbon has been used for many biomedical applications. Tissue can adhere to it and a protein layer that forms on its surface prevents the formation of blood clots. The ideal solution for biomedical implants would be a combination of the mechanical strength of metals and the biocompatibility of carbon. DLC films deposited onto metal implants seem to provide this combination [2.67], [2.68].

**Optical and Electrical Properties and Applications.** The optical properties of DLC films are also highly dependent on the $sp^2/sp^3$ ratio and the hydrogen content. Films with a high hydrogen content are transparent in the visible with optical band gaps of up to 4 eV. Hard a-C and a-C:H films are less transparent with gaps of 1–2.5 eV. With a gap of 0.4–0.8 eV, soft, graphite-like a-C films are hardly transparent. The optical properties of DLC are often exploited in conjuction with the mechanical properties, e.g., in protective coatings for optical components [2.51], [2.66], [2.68].

Electronically, all DLC films are best described as composite materials, consisting of conductive, graphitic islands embedded in an insulating matrix. Like diamond, ta-C films show photoconductivity, however, white light is sufficient to change its conductance by two orders of magnitude at 125 K [2.53]. Doping with nitrogen changes this effect, and also effects the electrical conductivity of the material. Pure ta-C is a good insulator ($10^{-9}\,\Omega\text{cm}$), N-doped ta-C is more conductive [2.53]. Doping of DLC films has been investigated, but device applications of DLC films, if any, are still in their infancy.

the discussion of stability in Section 2.3.1) they change their structure and ultimately turn into graphite films [2.51], [2.53], [2.66], [2.67].

Because of their excellent mechanical and chemical properties and the fact that it is possible to deposit DLC films at room temperature, they are widely used as protective, low-wear, low-friction, or antireflection coatings on, e.g., IR op-

## 2.4. Metallic Films [2.69]–[2.81]

### 2.4.1. Introduction

Techniques for the preparation of thin films include: electrodeposition, hot dipping, cementation, diffusion, flame spraying, plasma spraying, laser coating, roll bonding, powder rolling, mechanical impingement, explosive bonding, chemical vapor deposition, physical vapor deposition, sputtering, and ion implantation. The properties of metallic coatings depend on the technique used and on the processing conditions; the last decade has seen the increasing use of dry techniques, such as vapor-phase deposition, which are becoming competitive with wet processing techniques in some areas of surface finishing.

Thin metallic films find extensive use in the electronics and engineering industries, as coatings on a substrate or as detached films (e.g., electroforming of foil or mesh). By selection of deposition technique and process conditions, it is possible to produce films in the range 0.1 – 20 µm of materials which include compact single metals and alloys, composite metal – polymer and metal – ceramic layers, conductive polymers, and solid-state ceramics and dispersed coatings of catalysts on a porous substrate (Table 10).

### 2.4.2. Electronics Applications

Thin metallic films may be used to alter the electrical properties of polymer, ceramic, and metal substrates, or the substrate may be viewed as a support for the production of a specific film. Examples include the metallization of laminates, such as glass-reinforced polyester and epoxy materials to produce printed circuit board tracks, switch contacts, and connectors. Following cata-

**Table 10.** Thin film materials and their applications

| Material | Typical film thickness, µm | Deposition technique | Applications |
| --- | --- | --- | --- |
| Zinc | 75–100, 7–20, 20–40 | hot dipping diffusion electroplating | corrosion protection of steel |
| Chromium | 10–120 | electroplating | corrosion- and wear-resistant coatings on steel |
| Copper | 2–10 | electroless plating, electroplating | conductive tracking |
| Gold | 1–5 | electroplating | electronic contacts |
| Copper, nickel, iron, Ni–Co | 10–1500 | electroforming | metal foils and flat meshes; audio and video disk stampers; waveguides; electric razor "foils" and liquid food filter meshes; molds for plastics; magnetic display boards (Ni and Fe) |
| Nickel–phosphorus | 5–30 | electroless plating | bearing and valve surfaces |
| Nickel–boron | 5–30 | electroless plating | high-temperature weat-resistant coatings |
| Dispersed platinum | (1–10 g/m²) | electroplating, chemical deposition | fuel cell electrodes, electrolytic conductivity probes |
| Tin–lead | 5–10 | electroplating | solderable tracks on printed circuit boards |
| Cobalt–nickel–gold | 2–10 | electroplating | edge connectors on printed circuit boards |
| Nickel–silicon carbide | 5–30 | electroplating, electroless deposition | abrasive wheels and cutting tools |
| Nickel–PTFE | 5–30 | electroplating, electroless deposition | self-lubricating, non-stick coatings |
| Titanium nitride, titanium carbide | 4–8 | chemical vapor deposition, plasma deposition | wear-resistant cutting tools |
| Alumina | 15–25 | anodizing | corrosion-resistant coatings on aluminum |
| Tantalum oxide | 1–8 | anodizing | high-stability capacitors |
| Molybdenum disulfide | 4–10 | chemical vapor deposition | lubrication of internal combustion engine surfaces |

**Table 11.** Conductive inks and their applications

| Substrate | Metal(s) | Applications |
|---|---|---|
| Ceramics | gold, silver, copper, silver–palladium blend | conductive tracking; capacitors |
| Printed circuit boards | silver, copper | conductive support for tracking |
| Flexible polymers | silver | conductive tracking; keyboards; switches; RFI* shielding; medical sensors; communications antennae |
|  | copper | flexible printed circuits |
|  | nickel | RFI shielding; magnetic shielding; environmental protection |
|  | silver–carbon | environmental protection; chemical sensors; resistors; keyboards |
|  | carbon–rubber | force-sensitive resistors |
|  | silver–nickel | conductive tracking; magnetic conductors; phone cards |
|  | iron oxides | magnetic cards, "smart" cards |

* RFI = radiofrequency interference.

lytic activation of the surface by a thin palladium deposit, copper may be deposited on a plastic printed circuit board laminate by electroless plating or by screen printing of copper-containing inks. The film may be thickened by electroplating of copper to produce conductive tracks. For edge connectors, further electrodeposition of nickel, then gold, is common, while solderability can be provided by the electrodeposition of tin–lead alloys. Porosity and adhesion problems require care in the formulation of pretreatment sequences and rinses between coating stages.

Conductive inks are cost effective and versatile for the formation of conductive tracks on printed circuit boards and for other applications (Table 11). They consist of a conductive pigment (metal flakes or particles), a binder (polymer resin), and a solvent or dispersant. The pattern is formed by forcing the liquid ink through a masked mesh or screen. The ink is then cured (usually by heat) to remove the solvent. The resin contracts, forcing the pigment particles together to produce a conductive coating. After curing, the coating thickness is usually 3–12 µm, although thicker coatings are possible. Conductive inks are printed onto a range of substrates, principally ceramics, printed circuit boards, and flexible plastics (e.g., polyester or polycarbonate). It is possible to electroplate metals onto some cured inks to increase conductivity and durability. By blending metal pigments it is possible to combine the properties of the constituent metals.

A variety of gas-phase techniques can be used to produce special materials. Reactive sputtering can produce compound semiconductor coatings of CdS by sputtering from a Cd target in an Ar + $H_2S$ atmosphere. The ability of sputtering to control coating composition makes it a useful technique in the electronics industry (metallization of aluminum alloy microcircuits, refractory metal microcircuits, optically transparent conducting electrodes, piezoelectric transducers, optically addressed memory devices, thin film resistors and capacitors, video disks, and thin film lasers). Sputtering can also produce multilayer stacks of dielectric materials such as $TiO_2$, $Ta_2O_5$, $Al_2O_3$, and $SiO_2$, which are harder and denser than evaporated layers.

### 2.4.3. Engineering Coatings

As seen in Section 1.7.1, a wide range of single metals, alloys, and composites can be electrodeposited or electrolessly plated from aqueous solution. Examples include electroless nickel–phosphorous coatings for wear resistance on valves and corrosion protection, and chromium coatings for wear resistance under arduous conditions (hydraulic cylinders, gun barrels). Special applications include the use of Ni–PTFE inclusion deposits for carburetor butterfly valve spindles and the use of Co–WC as a wear-resistant coating for cutting tools.

In dry processing techniques, flame and plasma spraying can produce wear-resistant layers of materials such as stainless steel alloys, WC, $Si_3N_4$, SiC, and $Al_2O_3$. Sputtering, in combination with heating of the substrate (or the use of a reactive atmosphere) can produce hard, wear-resistant coatings such as titanium nitride. A 3-mm TiN film with hardness > 2000 Knoop/

mm² can be produced for high-temperature hobs. Solid lubricants can greatly decrease wear, e.g., sputtered films of $MoS_2$ can have a coefficient of friction of ca. 0.04 against steel, with lifetime more than 100 times that of brushed-on $MoS_2$ films.

For some metals, a number of coating technologies are available and the choice depends on the environment and service lifetime, as well as cost. In corrosion protection, sacrificial coatings of zinc on steel may be produced by electroplating, hot-dip galvanizing, or high-temperature diffusion (Sheradising). Hot-dipped coatings have thickness 20–200 µm, the usual range being 70–100 µm. A series of Fe–Zn alloy layers are produced, the top layer consisting of pure zinc if a low-temperature process is used. The result is excellent adhesion and reasonable resistance to abrasion; this coating process may be chosen when arduous conditions are involved, a long lifetime is sought, or if no further coatings (e.g., paints) are to be applied. Zinc diffusion coatings are usually restricted to smaller workpieces, e.g., fasteners, metal brackets. The surface of the steel is converted to a film of iron–zinc alloys, in a batch process involving zinc powder at elevated temperature, with typical coating thickness 7–25 µm. The coating has reasonable abrasion resistance and uniformity, and is used when no further anticorrosion coatings are applied. For large-scale or continuous coating (e.g., steel coil or sheet) or for substrates that are deformed by high-temperature processes, electroplating of zinc (or an alloy with higher corrosion resistance, e.g., Zn–Ni, Zn–Fe, Zn–Co) is often used, as in car body panels. In electroplating, the films tend to have less toughness and resistance to abrasion, owing to the absence of intermetallic, alloy layers.

Zinc, aluminum, magnesium, and iron alloy surfaces can be chemically modified by a conversion coating to produce, e.g., oxides (anodizing of aluminum), chromates, and phosphates. Such treatments can significantly improve corrosion resistance and, in the case of phosphating, lubrication conditions can also be improved.

For noble coatings, e.g., electroless nickel, electroplated nickel, or chromium, the coating must have low porosity, as pitting of the underlying steel can occur at coating defects. The stress distribution and cracking pattern in chromium-plated layers is important in long-term performance, both for corrosion protection and tribology.

Gas-phase techniques have specialist uses in corrosion protection, e.g., ion plating to coat aircraft and spacecraft steel and titanium fasteners with aluminum. The aluminum layers have good uniformity, adhesion, strength, and mechanical properties, and have, in some cases, replaced cadmium electroplated layers (which tended to suffer from hydrogen embrittlement and stress cracking). Ion plating of aluminum is also used to coat thin-walled stainless steel heat exchanger tubes and interconnecting sections. Other applications include Cr, $Al_2O_3$, $Si_3N_4$, Cd, Ti, C, and Ta on steel for medical, biochemical, and industrial chemical applications.

### 2.4.4. Dispersed Catalytic Coatings

The majority of metallic films are compact, uniform-thickness layers on a nonporous substrate, but an important class of coatings are produced by dispersing a precious metal onto a porous substrate, to produce a catalytically active, high surface area coating which uses a minimum of the costly coating material. Examples include:

1) Deposition of platinum or platinum alloys onto porous carbon paper, carbon cloth, or ionically conductive polymer membrane to produce porous gas diffusion electrodes. These are used in aqueous electrolyte fuel cells, speciality batteries using gaseous reactant (e.g., the aluminum–air cell), in electrochemical gas sensors, and as a hydrogen anode. The electrode surface must remain electrically conductive and act as a three-phase boundary at the interface of gases, liquid electrolyte, and solid catalyst, and it is common to control the wetting properties of the electrode to avoid flooding of the structure, by the use of PTFE which also acts as a binder.
2) The deposition of platinum-group metals and alloys onto porous metal and ceramic substrates to produce catalytic converters for car exhausts. The ability to recycle the expensive catalyst is an important consideration.

In both of these applications, there is a trend toward lower loadings of the precious metals. In fuel cell electrodes for phosphoric acid $H_2/O_2$ cells, the loading of platinum metal on carbon has decreased toward $\leq 1$ mg/cm². Techniques to produce dispersed coatings include combinations of chemical impregnation, electrodeposition, and thermal treatment. More recently, sputtered and other vapor-phase-technology coatings have been considered, although they are much more costly than chemical impregnation followed by heat treatment.

### 2.4.5. Developments and Trends

The range of thin metallic films continues to increase as more advanced techniques are implemented and new alloy, polymer, and composite coatings are explored:

1) Increasing use of vapor-deposition techniques, particularly chemical vapor deposition and ion implantation, to produce very thin, nonporous layers for the electronics and engineering industries
2) Introduction of multilayer coatings by a number of different techniques or the alteration of process conditions during the application of a single technique
3) Increasing use of polymer, alloy, and composite coatings
4) Development of conductive polymer films with electrical conductivity ranging from that of an insulator to about one-tenth the value of some metals
5) Increasing need to use thin films of precious metal catalysts on a less costly, engineered substrate (fuel cell electrodes, specialized gas sensors, and car catalytic converters)
6) Development of cost-effective, environmentally attractive techniques for selective deposition of thin films (ion plating, laser or jet electroplating, and screen printing by jet techniques)
7) Formation of thin diffusion coatings by ion implantation

## 2.5. Superconductive Coatings

Since the discovery of high-temperature oxide superconductors [2.82], coatings (deposition and/or thin film growth) of several cuprates with a superconducting transition temperature $T_c$ above liquid nitrogen temperature, 77.3 K, have been developed. Among the many high-$T_c$ oxide materials synthesized to date, representative thin film materials for practical use are $LnBa_2Cu_3O_{7-\delta}$ (where Ln is a rare-earth element other than Pr, Ce, and Tb) with $T_c \geq 90$ K, and $Bi_2Sr_2Ca_{n-1}Cu_nO_{2n+4}$ with $T_c$ ca. 20 K for $n = 1$, ca. 80 K for $n = 2$, and ca. 110 K for $n = 3$. Deposition of $LnBa_2Cu_3O_{7-\delta}$

thin films is well established by different deposition techniques, while that of the $Bi_2Sr_2Ca_{n-1}Cu_nO_{2n+4}$ system for $n = 2$ and 3 is currently being investigated. Thin film deposition of $Tl_2Ba_2Ca_{n-1}Cu_nO_{2n+4}$ ($n = 1$, 2, 3) is limited because of the toxicity of Tl. A recent development is thin film deposition of $HgBa_2CaCu_2O_{6+\delta}$ by reactive r.f. magnetron sputtering, followed by oxygenation annealing. The film shows a relatively sharp transition up to ca. 120 K [2.83].

Typical methods for in situ deposition of high-$T_c$ cuprate thin films are physical deposition, e.g., r.f. sputtering, pulsed excimer laser ablation or pulsed laser deposition (PLD), and reactive coevaporation, sometimes referred to as molecular beam epitaxy (MBE). Radiofrequency magnetron sputtering was first used for deposition or epitaxial growth of $YBa_2Cu_3O_{7-\delta}$ thin films on MgO and $SrTiO_3$ substrates; PLD with ArF ($\lambda = 193$ nm) and KrF ($\lambda = 248$ nm) excimer lasers is a promising method for $LnBa_2Cu_3O_{7-\delta}$ thin films with $T_c > 90$ K [2.84]. It can now produce in situ deposited multilayered films with different targets installed in a deposition chamber. Figure 36 shows a typical transmission electron micrograph of a $YBa_2Cu_3O_{7-\delta}/PrGaO_3/YBa_2Cu_3O_{7-\delta}/PrGaO_3$ quadrilayered film grown in situ on a $SrTiO_3$ substrate. The great advantage of MBE is its ability to control deposition sequences by observing oscillations of the reflection high-energy electron diffraction (RHEED) intensity [2.85], which enables "layer-by-layer" deposition of constituents and fabrication of superlattices, e.g., $PrBa_2Cu_3O_x/YBa_2Cu_3O_x$.

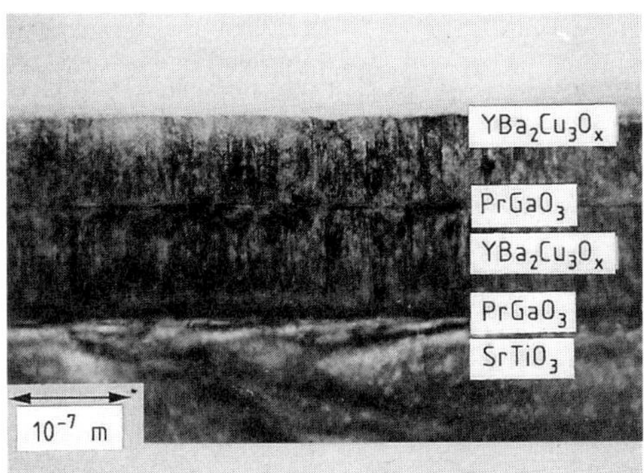


**Figure 36.** Transmission electron micrograph of a $YBa_2Cu_3O_{7-\delta}/PrGaO_3/YBa_2Cu_3O_{7-\delta}/PrGaO_3$ quadrilayered structure on a $SrTiO_3$ substrate deposited in situ by pulsed laser deposition

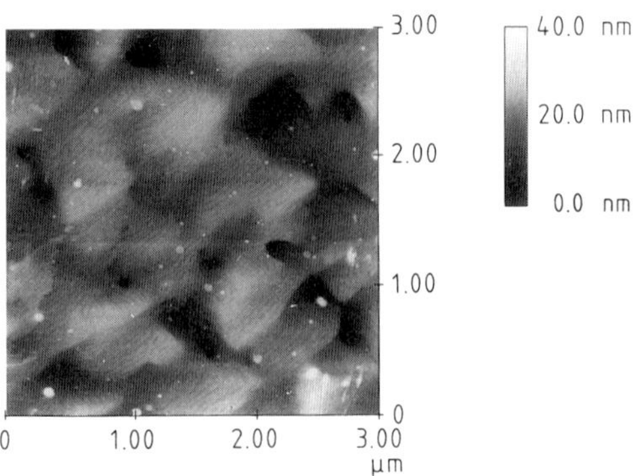

**Figure 37.** Spiral growth islands of $c$-axis-oriented YBa$_2$Cu$_3$O$_{7-\delta}$ thin film revealed by atomic force microscopy
Terraces with step height ca. $12 \times 10^{-10}$ m.

Most YBa$_2$Cu$_3$O$_{7-\delta}$ films grown by physical deposition contain spiral growth islands [2.86] originating from screw dislocations caused by relatively high supersaturation (Fig. 37). Surface roughness is of the order of 0.01 µm.

Physical depositions require low supersaturation of cuprate-generating species, oxygen pressure control, and choice of epitaxial substrate materials. For supersaturation, the most advanced techniques are metallorganic chemical vapor deposition (MOCVD) for Bi systems, and liquid-phase epitaxy (LPE) for LnBa$_2$Cu$_3$O$_{7-\delta}$ systems. Current MOCVD research has obtained high-quality $c$-axis-oriented YBa$_2$Cu$_3$O$_{7-\delta}$ thin films with $T_c$ of ca. 90 K and Bi$_2$Sr$_2$CaCu$_2$O$_6$ ($n = 1$, 2212 phase) thin films with a $T_c$ ca. 80 K.

Liquid-phase epitaxy is the most advanced technology for LnBa$_2$Cu$_3$O$_{7-\delta}$ films [2.87]. The complex phase diagram of LnBa$_2$Cu$_3$O$_x$ gives a relatively high epigrowth temperature, 900–970 °C, and as-grown films do not show a superconducting transition, owing to oxygen deficiency; post-growth annealing in oxygen is necessary. Growth is conducted at minimal supersaturation.

Deposition of LnBa$_2$Cu$_3$O$_{7-\delta}$ thin films, specifically YBa$_2$Cu$_3$O$_{7-\delta}$, has been thoroughly investigated for different deposition techniques. To obtain superconducting YBa$_2$Cu$_3$O$_{7-\delta}$ thin films in an as-deposited state, the relationship between oxygen partial pressure during deposition and substrate temperature has been established for different physical deposition techniques (Fig. 38). Oxygen partial pressure during deposition has to be kept close to the critical

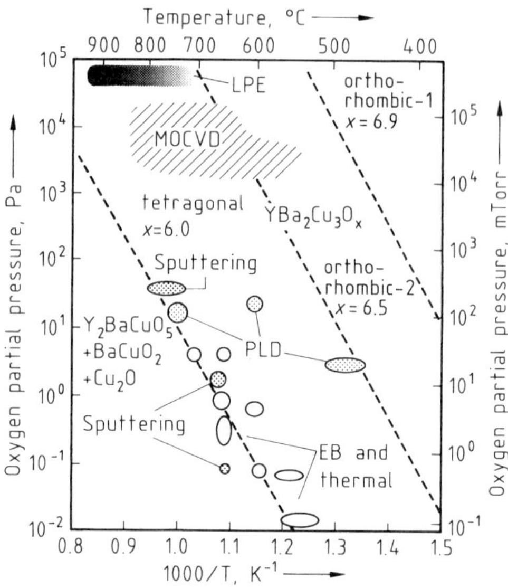

**Figure 38.** Equilibrium oxygen partial pressure and deposition temperature of YBa$_2$Cu$_3$O$_{7-\delta}$ thin films [2.88]
LPE = Liquid-phase epitaxy; MOCVD = Metal-organic chemical vapor deposition; PLD = Pulsed laser deposition; EB = Electron beam

oxygen pressure at the decomposition of bulk YBa$_2$Cu$_3$O$_6$ (nonsuperconducting tetragonal phase). Activation processes to enhance oxidization involve ozone, oxygen plasma, and atomic oxygen radicals. Electron beam deposition involves post-deposition annealing in oxygen.

For epitaxial thin films of LnBa$_2$Cu$_3$O$_{7-\delta}$, the choice of substrate materials becomes more important, especially for active device applications. Figure 39 [2.89] shows thermal expansivities of different substrate materials. LaAlO$_3$ seems practical for microwave applications because of its relatively low dielectric constant [2.90], while NdGaO$_3$ is suitable for $c$-axis-oriented epitaxial thin films because of its good lattice matching [2.91].

Of current interest is the growth of $a$-axis-oriented YBa$_2$Cu$_3$O$_x$ thin films, because of the anisotropy in coherent length ($\xi_{a/b} = 30-40 \times 10^{-10}$ m and $\xi_c = 5-10 \times 10^{-10}$ m). This takes advantage of a Josephson coupling junction across an $a$-axis SIS (superconductor–insulator–superconductor) structure. Many workers have reported $a$-axis-oriented YBa$_2$Cu$_3$O$_{7-\delta}$ film growth in which the deposition temperature is lowered by ca. 100 °C from the usual $c$-axis film deposition temperature. However, this gives $a$-axis films with 90° domains, $T_c$ ca. 20 K. Mean-

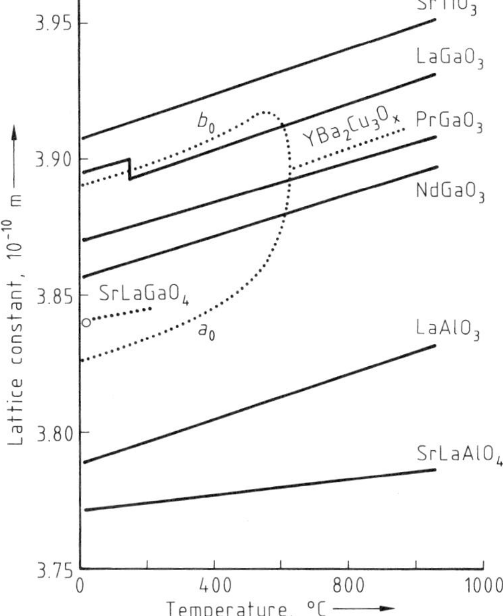

**Figure 39.** Lattice constants and thermal expansivities of new substrate materials
Data from $YBa_2Cu_3O_{7-\delta}$ for comparison [2.89].

while, $c$-axis in-plane-aligned, $a$-axis-oriented $YBa_2Cu_3O_x$ thin films have been grown on (100)-oriented tetragonal $SrLaGaO_4$ and $Nd_2CuO_4$ substrates with a $K_2NiF_4$-type crystal structure. A growth model called "atomic graphoepitaxy" has been proposed [2.92], where the atomic geometry of a (100)-oriented face of tetragonal $K_2NiF_4$-type materials is considered. Films grow epitaxially with perfectly in-plane-aligned $c$- and $b$-axes, along both of which $T_c = 90$ K.

When $YBa_2Cu_3O_x$ films are deposited on semiconducting Si substrates, Si ions diffuse easily into the films, and $T_c$ degrades to ca. 70 K or less. As a result, a variety of oxide materials are being investigated as a buffer layer between the film and Si substrate in order to suppress interdiffusion. One such buffer material is yttrium-stabilized $ZrO_2$ (YSZ).

## 2.6. Polymer Films

### 2.6.1. Langmuir–Blodgett Layers of Polymer Films as Materials

The Langmuir–Blodgett (LB) technique (see Section 1.6.) is a powerful method for the preparation of ultrathin organic films of defined thick-

ness and molecularly controlled internal structure. It consists of the following essential steps:

1) Spreading molecules on a water surface
2) Compressing them to create a two-dimensional condensed layer
3) Transfer of the layer onto a solid substrate to yield mono- or multilayers

The thickness of the films can be varied, in multiples of the monolayer thickness, between a few nanometers (one monolayer) to a few micrometers (several thousand layers). Polymeric LB layers [2.93] can be obtained by: in situ polymerization of multilayers of amphiphilic monomers (Fig. 40 A); spreading and transfer of

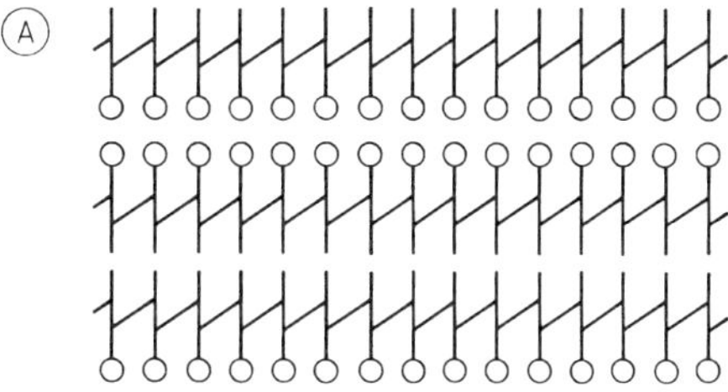

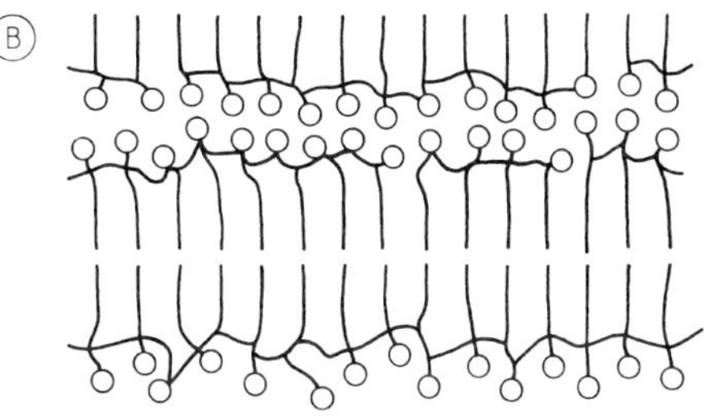

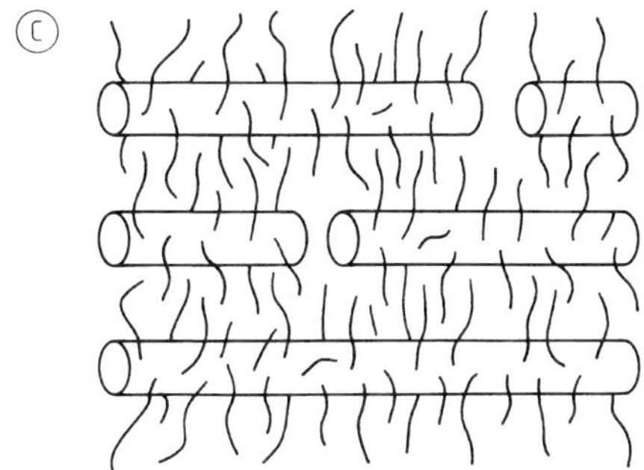

**Figure 40.** Architecture of polymeric LB layers
A) Layers prepared from solid-state polymerization of amphiphilic monomers (the polymerization can be carried out either within the monolayer prior to transfer, or within the multilayer assembly); B) Layers prepared from preformed polymers; C) Layers composed of rod-like main chains

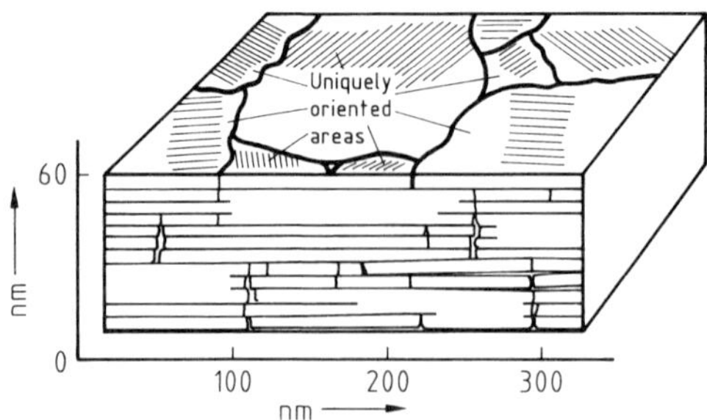

**Figure 41.** Domain structure in multilayer systems

preformed amphiphilic polymers (Fig. 40 B); spreading and transfer of hairy rod polymers (Fig. 40 C).

Individual layers of amphiphilic monomers or polymers are generally composed of two-dimensional crystallites (Fig. 41). The intercrystallite grain boundaries contribute significantly to the properties of LB films. This can result in losses by scattering in such applications as waveguides or cladding layers, and can play a major role in defining transport properties, e.g., in diffusion barriers or electrical insulators.

In multilayers of rod-like macromolecules with flexible side chains (hairy rods), the polymer backbones are oriented parallel to one another and are maintained at a fixed separation by the side chains, which form a continuous matrix (Fig. 42). LB layers of hairy rod polymers exhibit good mechanical stability and high optical transparency, and may be considered as molecular composites. The stability of hairy rod polymer films can be increased by cross-linking functional groups attached to the side chains [2.94].

**Characterization Methods** (Table 12) [2.95]– [2.97]. *Film Thickness.* If the thickness of an indi-

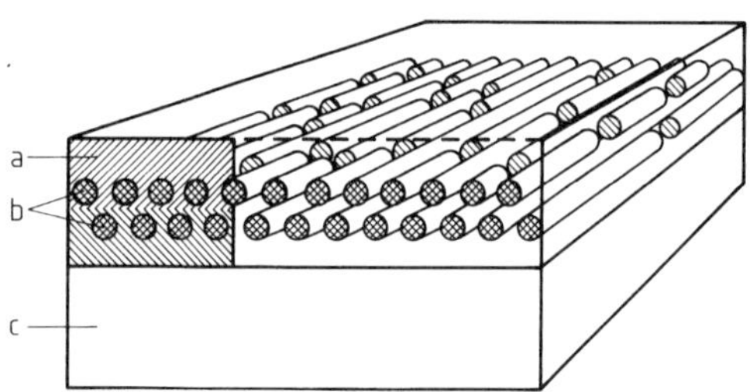

**Figure 42.** Arrangement of hairy rod molecules in a multilayer system
a) Hydrocarbon; b) Rigid rod polymer backbone; c) Hydrophobic substrate

vidual monolayer is known, counting the number of deposition cycles is the simplest method to assess the thickness of the film. The thickness can be determined by a number of experimental techniques. Optical interference techniques yield the "optical thickness," defined as the product of the metric thickness $d$ and the refractive index $n$; knowledge of the refractive index enables evaluation of the metric thickness. The birefringent character of most LB films is a complicating factor.

Phase-shift interferometry [2.98] provides a three-dimensional profile of the film. Study of the profile edge enables the film thickness to be determined. This technique requires that a thin metal film be evaporated on top of the LB layers in order to obtain a reflecting surface. The thickness profile can also be determined by mechanical contact methods and, in the case of only a few layers, by atomic force microscopy (AFM).

In ellipsometry, the change in polarization of light reflected from a thin film on top of a metal surface is analyzed to yield accurate information on the optical thickness. Ellipsometry is also a sensitive technique for detecting minor changes during the lateral organization of LB films on the subphase, and has been used to investigate the properties of monomolecular films of fatty acids on both water and mercury subphases.

Surface plasmon spectroscopy [2.99] is a further method for optical thickness analysis. Surface plasmons are electromagnetic surface waves propagating from a thin metal film into a dielectric medium. Plasma oscillations in the metal film are excited by polarized light from a laser, which couples into the film at a particular angle. The resonance condition for a given system depends on the refractive index and the film thickness. When the LB film is thick enough to be used as a waveguide, the surface plasmon and waveguide modes can be detected simultaneously. Analysis of the waveguide modes gives exact values of the film refractive index and thickness.

The most detailed and accurate information on the thickness of multilayer systems comes from grazing incidence X-ray or neutron reflectometry. The absolute layer thickness can be derived from analysis of the characteristic interference patterns (Kiessig fringes). The thickness of the individual monolayers and the surface roughness of the film can also be obtained.

*Surface Properties* [2.95]. Contact angle measurements of various liquids placed on the LB

**Table 12.** Techniques for the investigation of LB films

|  | Radiation used | Contrast by | Typical information |
|---|---|---|---|
| Optical interference microscopy | light | optical phase shift | optical thickness, topography of surface |
| Ellipsometry | light | refractive index difference | optical thickness |
| Surface plasmon resonance spectroscopy | light | optical phase shift | optical thickness (metric thickness) |
| X-ray reflectometry | X rays | electron density difference | electron density profile along layer normal, metric thickness, roughness |
| Neutron reflectometry | neutrons | selective deuteration | scattering length profile along layer normal, metric thickness, roughness |
| Contact angle | | surface tension | wetting behavior |
| Adhesion test | | | adhesion forces |
| Infrared (IR)/Raman spectroscopies | light | IR/Raman absorption | functional group composition and orientation |
| Transmission electron microscopy (TEM) | electrons | electron density difference in plane | crystal structure, domain structure and texture |
| Low-energy electron diffraction (LEED) | electrons | electron density difference | crystal structure of surface layers |
| Atomic force microscopy (AFM) | | height profile or surface tribology | surface topography |
| Scanning tunneling microscopy (STM) | charge carriers | conductivity differences | chemical composition periodicity of groups |
| X-ray photoelectron spectroscopy (XPS) | X rays | | elemental surface composition |

layer can be used to determine surface energy and wetting properties. The adhesion of the first monolayer to the underlying substrate, and of subsequent monolayers to each other, can be determined only qualitatively. A standard adhesive tape is applied to the surface of the film and then detached. Comparison of the film before and after this procedure gives a measure of the amount of film removed.

Grazing incidence infrared (IR) reflection by mono- and multilayers on a metal substrate and polarized IR transmission are powerful tools for the investigation of the orientation of molecular moieties in the LB film. These techniques are sensitive enough to investigate individual monolayers [2.98].

Raman spectroscopy gives similar information about the structure, orientation, and conformation of the components of thin organic layers. Because of the weak signals, an enhancement mechanism is usually required for the direct observation of LB films. A dramatically increased signal can be produced by waveguide Raman spectroscopy (WRS) and surface plasmon enhanced Raman spectroscopy.

Transmission electron microscopy (TEM) as well as low-energy electron diffraction (LEED) have been used to investigate LB layers. Under optimum conditions single polymer molecules can be observed.

Various scanning near-field microscopies (AFM, STM, SEM, etc.) have been used to gain insight into the surface properties of mono- and multilayer systems.

In X-ray photoelectron spectroscopy (XPS) or electron spectroscopy for chemical analysis (ESCA), the surface is exposed to monochromatic X rays and the energy distribution of electrons emitted gives information on the chemical composition, environment, and binding of atoms at the surface.

Elastic constants of LB assemblies can be determined by Brillouin spectroscopy [2.100] or by a quartz crystal resonator [2.101].

**Table 13.** Properties and applications of LB films

|  | Properties | LB films | Applications |
|---|---|---|---|
| Mechanical properties | stability | low | |
| Optical properties | birefringence | often observed | optical filters |
| | nonlinear optics ($\chi^2/\chi^3$) | observed in films of special composition | optical switches, frequency doubling |
| Electrical properties | conductivity | up to $10^{-2}-10$ S/cm of special components | interconnection in microelectronic circuits |
| | resistance | good insulating behavior, low electric loss, e.g., in arachidic acid | capacitors |
| | photoconductivity | up to 1% quantum efficiency in special cases | |
| | electroluminescence | blue light emitted from devices based on aromatics | displays |
| | pyroelectricity | observed in noncentrosymmetric structures | infrared TV cameras |
| | piezoelectricity | observed in noncentrosymmetric structures | |
| Surface properties | lubrication | drastic reduction of static friction by a few monolayers of fatty acids | magnetic storage |
| | wettability | polarity can be tailored | surface modification of substrates |
| | surface anisotropy | order parameter high, monodomain structure possible | liquid crystal alignment in displays |
| | biocompatibility | possible tailoring of surface | artificial implants, sensors |
| Miscellaneous | permeability | selective permeability often observed | membranes for gas separation, reversed osmosis, ultrafiltration |
| | chemical reactivity | composition dependent | microlithography, positive and negative resists |

**Properties of LB Multilayers and Applications** (Table 13) [2.95], [2.96], [2.102]. Polymeric LB films are much more stable than layers composed of monomers, owing to covalent bonding in the film plane. However, the fragility of these systems remains a problem in many applications. The mechanical properties of mono- and multilayers are generally anisotropic, and may be strongly influenced by structural defects.

*Optical and Optoelectronic Properties* [2.103], [2.104]. Birefringence and dichroism are frequently observed in LB layers. Thus, optical filters and devices for optical applications can be prepared. Optical anisotropy can be induced through the introduction of mesogenic side groups. The optical properties of the film may be varied by thermal treatment or the application of an external electrical field. LB assemblies may be used as waveguides; the best attenuation of guided light reported has been ca. 2 dB/cm.

Nonlinear optical effects have received particular attention. To produce $\chi^2$ effects (frequency doubling, electrooptical switching, etc.)

noncentrosymmetric LB films of hemicyanines, phenylhydrazones, or similar polarizable dyes are necessary. Values of the $\chi^2$ parameter ca. 10–100 times the value of the commonly used material $LiNbO_3$ have been reported. For $\chi^3$ effects (frequency tripling, phase conjugation, etc.) highly polarizable conjugated $\pi$-electron systems with donor–acceptor moieties can be used.

As discussed previously (Fig. 41), the presence of cracks and crystalline domains of sizes in the range of the optical wavelengths must be avoided, and the thermal stability must be enhanced, in order to produce LB films of suitable optic and stability.

*Electric and Dielectric Properties* [2.105]. Electrical measurements on LB layers frequently prove difficult, owing to the inherently fragile nature of the specimens. Various effects have been reported which depend on the chemical nature of the constituents and the state of order of the LB assemblies. For example, tunneling barriers have been studied where an LB film was sandwiched between two metal electrodes. LB layers composed of "organic metals" or semiconducting materials have also been reported. The behavior of LB films as conductors and insulators can be optimized by tailoring the molecular components of the film.

Ultrathin organic films between metal contacts have an enormous charge storage capability, which suggests an application as the dielectric layer in simple capacitors.

LB films on semiconductors can be used for various applications, including metal–insulator–semiconductor diodes, Schottky barriers, or field effect transistors. The properties of LB layers in terms of photoelectric transport and as the active layer in electroluminescent devices have been investigated.

Pyroelectricity and piezoelectricity are observed in noncentrosymmetric LB multilayer systems. Owing to the small heat capacity of the film, the pyroelectric response is very fast. Pyroelectric detector systems for thermal radiation based on LB films could be used in fast infrared TV cameras.

*Surface Properties* [2.102]. The presence of a single monolayer on any substrate can drastically reduce static friction. This effect can be used to reduce frictional forces between different surfaces in mechanical contact. For example, magnetic mass storage media require very thin, but very stable lubricant layers to avoid wear during transit of the tape across the magnetic head. Good results have been achieved with a few layers of fatty acids on the tape, which also serve as an encapsulant to protect the magnetic media.

The wettability of a specific surface can be profoundly influenced by coverage with only one or a few monolayers, e.g., turning a hydrophobic surface into a hydrophilic one. This offers the possiblity of tailoring the surface properties of different substrates. In the field of liquid crystal displays, orientation layers are commonly used to align liquid crystals parallel to the surface. LB films with uniaxial or biaxial orientation, prepared by carefully controlled deposition procedures, are contenders for liquid crystal alignment layers. Excellent results have been reported for $\omega$-tricosenic acid films, polymerized by an electron beam. With azobenzol-dye-containing monolayers, a change of orientation of the liquid crystals from parallel to homeotropic can be induced by photoinduced *cis–trans* isomerization.

The LB technique allows control of surface charge densities, surface potential, degree of hydrophobicity, and type and concentration of functional groups on the surface. It has therefore been proposed to use this to produce coatings on medical devices and, in particular, on implants. To date, only model systems have been developed.

*Permeability* [2.95]. The architecture of LB multilayers of amphiphilic compounds in many ways mimics biological membranes, suggesting possible applications in specific separation tasks, such as ultrafiltration, reverse osmosis, and gas separation. In these situations LB films with high selectivity are deposited on a porous, mechanically stable support. For example, eighteen monolayers of hexacosa-10, 12-diyonic acid on a porous polypropylene support reduce the gas flow of methane by a factor of 30 compared with the uncoated substrate. Structural defects in LB films deposited directly onto the porous support often generate leakage problems. These can be resolved by a composite technique. A very thin, nonporous, but highly permeable layer is installed between the LB film and the porous membrane, which dramatically reduces the number of defects in the film.

*Chemical Variation of LB Multilayers* [2.106]. The polymerization of an LB layer composed of amphiphilic monomers can be initiated locally by UV or high-energy radiation. Resist structures may thus be prepared with a resolution better than 100 nm. This is of importance for integrated circuit technology, where submicrometer resolution lithography is a crucial element of process technology. Both positive and negative resists have been successfully used; simple fatty acid salts show high resolution as positive resists, subliming under irradiation. Enhanced sensitivities are more easily achieved with negative resists, in which case polymerization is initiated by an electron beam. Several types of polymerizable materials have been investigated. The topochemical polymerization of diacetylenic amphiphiles and of $\omega$-tricosenoic acid show promising results. Transfer of these ultrafine structures to solid substrates, however, requires improvement of the resistance of the LB film to plasma etching. Polysiloxanes and polysilanes are good candidates for this task.

**LB Films as Sensors** (Table 14) [2.96]. The properties of LB films as described above suggest many possible applications to sensory devices. The extremely high ratio of surface area

**Table 14.** LB films as sensors

|  | Sensitive absorbing LB layer on top of | Physical parameter changes by absorption of analyte |
| --- | --- | --- |
| Field-effect transistor devices | semiconductor | dielectric properties |
| Optical sensors | optical waveguide | optical spectrum |
| Surface plasmon resonance devices | sandwiched between metal surface and prism | refractive index, film thickness |
| Ellipsometric sensors | metal surface | refractive index, film thickness |
| Acoustoelectric sensors | oscillating quartz | mass of film |

to bulk volume of LB films prevents bulk diffusion, and yields fast response and recovery times. The ordered surface allows a maximum of reactive species per unit surface area, and offers the potential of high selectivities via appropriate molecular design. In particular, exploitation of the intrinsic selectivity of biological systems (e.g., enzyme–substrate, antibody–antigen, and receptor–transmitter) can be incorporated into the design.

A wide range of sensors based upon LB films have been investigated. Chemically sensitive field-effect transistor (ChemFET) sensors detect chemical changes in a medium which is in contact with a metal oxide–semiconductor FET. Changes in this medium cause an alternation in the surface potential within an LB layer placed on top of the FET device. Profound effects on the measurable parameters of a device are noted with even minute changes in the dielectric properties of the interfacial region.

Changes in surface conductivity induced by the interaction of specific gases with semiconducting LB films have formed the basis for the construction of gas sensors.

Optical sensors detect changes in the optical spectrum of an LB film resulting from the adsorption of an analyte. As an example, the fluorescence of porphyrin LB layers can be quantitatively quenched through the absorption of gases such as $NO_2$ and HCl. Other optical methods include plasmon resonance or ellipsometric detection of changes in the refractive index or surface thickness when a substrate interacts with the surface layer.

Acoustoelectric sensors are based on the principle that the resonant frequency of a vibrating body is altered when the surface coverage is changed. Thus, the resonance frequency of a piezoelectric quartz crystal coated with LB layers of $\omega$-tricosenoic acid is changed on adsorption of hydrogen sulfide.

### 2.6.2. Films by Plasma Polymerization

Films formed by plasma polymerization differ from normal polymer material in several respects. A wide range of structure can be fabricated by modifying the plasma conditions (pressure, flow rate, and power). A map delineating different material regions may be constructed using these parameters. In ethylene, for example, it is possible to produce soft and hard films, powder and oil, as well as mixtures of these [2.107]. It is believed that powder is formed when the reaction occurs primarily in the gas phase, and is facilitated by long residence times, high power, and reactive monomers. Both IR and NMR have been used to investigate polymer film structure. Models of the structure of the oil and film products from plasma-polymerized ethylene are shown in Figure 43 [2.108]. A second feature of plasma polymerization is that almost any monomer can be used: organic, organometallic, or containing silicon. The monomer need not contain unsaturated groups. In plasma-polymerized material, very few monomer units remain in a regular repeating pattern as in conventional polymerization. The morphology of plasma films has been investigated by X-ray scattering und electron microscopy (e.g., powder particles in plasma-polymerized ethylene on a chromium substrate observed by transmission electron microscopy [2.107]).

Plasma films contain significant quantities of long-lived free radicals. In plasma-polymerized styrene, one in twenty polymer molecules is a radical site. The radicals are formed either directly in the energetic plasma, and then incorporated in the growing film, or by reactive plasma species or plasma radiation impinging on surfaces. Because the polymer molecules are strongly cross-linked (Fig. 43), the radicals are immobile and long lived. They are shorter lived if the film is

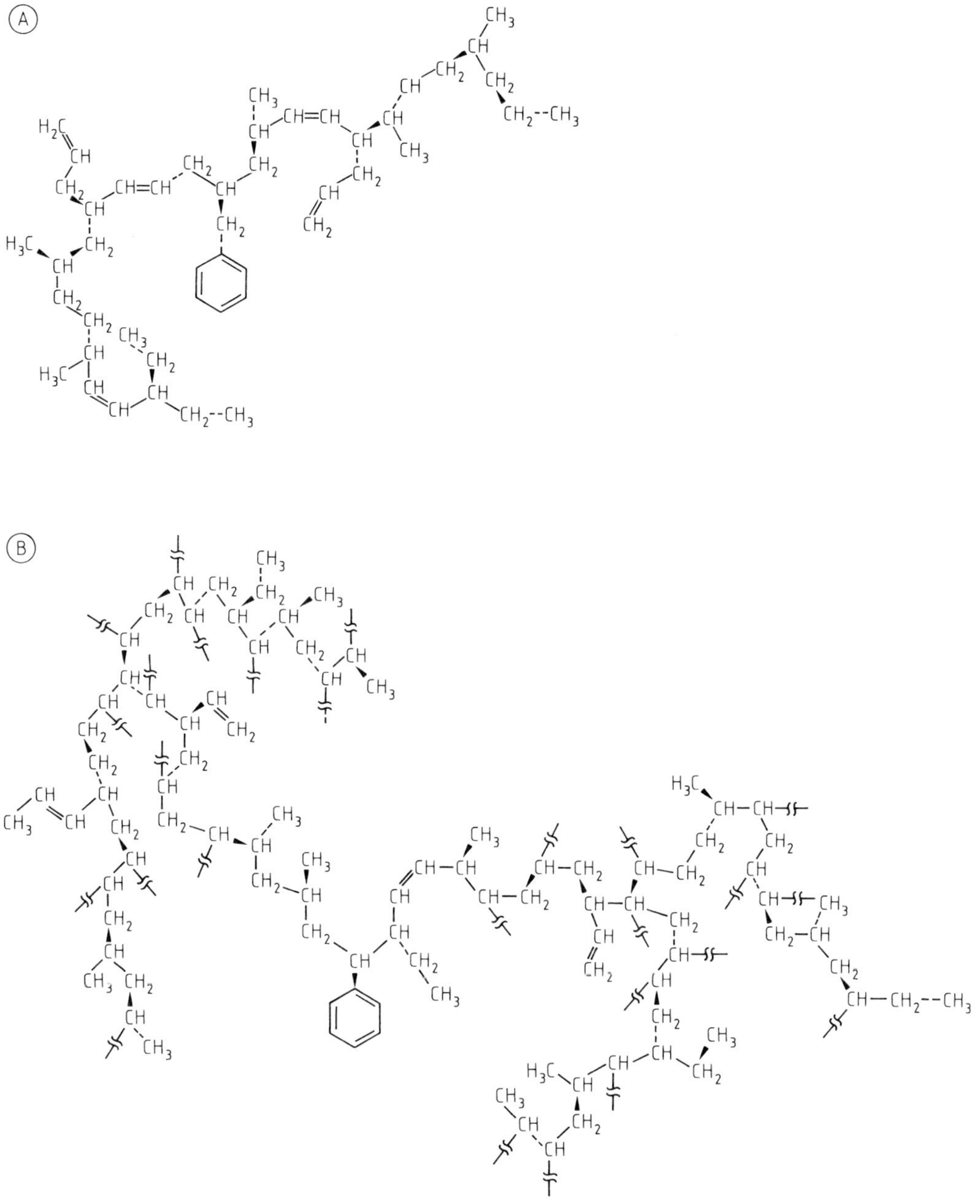

**Figure 43.** Plasma-polymerized material from ethylene monomer [2.108]
A) Oil model; B) Film model

exposed to the atmosphere, when they react with water vapor or $O_2$ (appearance of the characteristic IR bands of OH and CO in exposed film). If a high free radical density is undesirable, it can be reduced by heating or by subsequent exposure to a hydrogen plasma.

Another important characteristic of plasma-polymerized film is its impermeability. This is due in part to cross-linking. The mechanism of permeation in these films has been considered by YASUDA [2.109] who classified plasma films as intermediate between two ideal cases: solution–

diffusion or molecular sieve mechanisms. Plasma films often show extremely low permeability to water, e.g., a methane polymerized film is two orders of magnitude less permeable than conventional polymeric material [2.109].

An important feature of polymerized plasma films is the presence of compressive intrinsic stress, the molecular origin of which is believed to be the high nonequilibrium density of the films due to their highly cross-linked structure. Intrinsic stress has been widely studied because of its implications for film stability. Methods of alleviating it include annealing at $100-300\,°C$. Intrinsic stress is not observed for all polymer films, and films formed from closely related monomers often exhibit quite different stress characteristics. Internal stress is generally larger under slower deposition conditions and increases with film thickness, eventually leading to cracking. Better barrier and adhesion properties are to be found in thinner films. The amount of stress also depends on the substrate and its thermal properties relative to the plasma film deposited on it [2.109].

The electrical properties of plasma films have received much attention. Initially it was believed that they had great potential for forming thin dielectric coatings, but there are some serious drawbacks. Their dielectric properties are susceptible to aging, primarily due to the reactive nature of the film when exposed to air. This behavior is often irreproducible from one film to another, even from the same batch. A second problem is high dielectric loss. Both features are believed to originate from the free radicals in the films, and attempts have been made to eliminate them by heat treatment in vacuo. The effects of $O_2$ and humidity on conduction properties have also been recognised. The effect of $O_2$ on plasma polymer from styrene monomer is an irreversible decrease in conductivity by three orders of magnitude. Photoconduction and bistability have also been investigated.

Potential applications of plasma films which seek to exploit their special properties have been reported [2.110]. The synthesis of osmotic membranes from plasma polymer deposited on porous substrate has produced materials which can efficiently separate salt solutions. Salt rejection as high as $98\%$ has been repeated with nitrogen-containing polymers at fluxes of $320\,\mathrm{L\,m^{-2}\,d^{-1}}$. Permselective membranes for separating gas mixtures have been applied to $H_2/CH_4$, $H_2/CO_2$, and $He/CO_2$ mixtures.

The adhesion properties of the films on a number of substrates and their pinhole-free, cross-linked structure have led to their use as barrier coatings. They have been used to protect metals and moisture-sensitive optical components such as alkali halide crystals. For the latter, fluorinated ethylene films have been used which also give the coated components antireflection properties, which can be varied by changing the monomer and its thickness. Abrasion-resistant coatings can be made from hard films and several silicon-containing monomers have been used.

## 2.7. Optical Coatings

### 2.7.1. Introduction

The characteristic colors of most substances have their origin in selective molecular absorption (chemical colors). Other colors are a result of physical structure. These physical colors have the advantage over chemical colors that they do not fade.

Some physical colors found in nature are iridescent, interference colors, e.g., peacock feathers. "Interference colors are the purest and most brilliant colors known to man. They cannot be matched by even the brightest pigment colors in depth and intensity" [2.111].

Optical coatings are artificial thin films. Although they can be used to color objects, they are normally deposited on optical elements to modify spectral reflection and transmission. Their uses range from the UV to the far IR. The number of individual films can be up to several hundred. Since each additional thin film adds another interface (the origin of an additional partial reflection, contributing to the interference) very high reflections, very low reflections, and steep transitions can be accomplished.

### 2.7.2. Theory

The primary elements of thin film interference are the reflections and the spacings of the interfaces. For nonabsorbing films, this can be related to the indices of refraction $n$ and the optical thicknesses $nd$ of the individual films. The reflectance $R$ of a nonabsorbing optical coating consisting of $m$ individual layers is given by

$$R = 1 - \frac{4n_0 n_s}{2 + (n_0/n_s)M_{11}^2 + n_0 n_s M_{12}^2 + M_{21}^2/n_0 n_s + (n_s/n_0)M_{22}^2}$$

where $n_0$ is the refractive index of the surrounding medium (air), $n_s$ that of the substrate, and $M_{11}$, $M_{12}$, $M_{21}$, $M_{22}$ are the elements of the characteristic matrix of the multilayer:

$$M = \begin{bmatrix} M_{11} & iM_{12} \\ iM_{21} & M_{22} \end{bmatrix} = \prod_{v=1}^{m} \begin{bmatrix} \cos \Phi_v & \dfrac{i \sin \Phi_v}{n_v} \\ in_v \sin \Phi_v & \cos \Phi_v \end{bmatrix}$$

where $n_v$ is the refractive index,

$\Phi_v = \dfrac{2\pi}{\lambda} n_v d_v \cos \alpha_v$ the phase thickness, $\lambda$ the wavelength, $d_v$ the physical thickness of the $v$th layer, and $\alpha_v$ the angle of incidence at the $(v + 1)$th interface.

The "design" process of fitting the spectral reflectance of a multilayer to a prescribed spectral curve generally follows one of two methods:

1) The designer is in charge and the computer is used merely as a calculator (split-filter synthesis [2.112], equivalent layers [2.113], and Chebyshev prototypes [2.114]).
2) The computer is in charge and the decisions of the designer are reduced to a minimum (refinement [2.115] and computer synthesis: comprehensive search [2.116], flip-flop [2.117], and inverse Fourier transforms [2.118]). Several excellent programs are commercially available [2.119].

### 2.7.3. Deposition Methods

Four deposition methods are used commercially:

1) High-vacuum evaporation from boats or with electron guns [2.120]
2) Dipping [2.121]

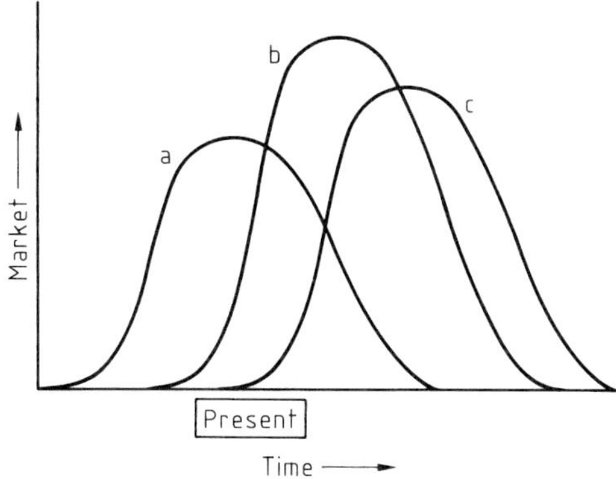



**Figure 44.** Past and projected market share of deposition methods for optical coatings
a) Evaporation/dipping; b) Sputtering/ion plating/ion-assisted evaporation; c) Plasma CVD, PICVD/PECVD

3) Energetic deposition processes [2.122] (d.c. magnetron sputtering, ion plating, and ion-assisted evaporation)
4) Chemical vapor deposition (low-pressure CVD [2.123], plasma-enhanced CVD [2.124], and plasma-impulse CVD [2.125])

Figure 44 estimates the share and trend of each of these methods in world manufacture of optical coatings (total market $\$1.1 \times 10^9$ [2.126]).

### 2.7.4. Coating Materials

Table 15 lists typical coating materials. Mixed oxides are also feasible, but the proportions must allow congruent evaporation, with energetic deposition processes, silicon and aluminum nitrides are also possible.

**Table 15.** Refractive indices and ranges of transparency for typical coating materials [2.127]

| Material | Wavelength, µm | | | | | Range of transparency, µm |
|---|---|---|---|---|---|---|
| | 0.35 | 0.55 | 1 | 1–8 | 10 | |
| Ge | | | | 4.20 | 4.20 | 1.7–25 |
| Si | | | 3.90 | 3.42 | | 1–9 |
| TiO$_2$ | | 2.32 | 2.20 | | | 0.4–6 |
| ZnS | | 2.36 | 2.27 | | | 0.4–14 |
| ZrO$_2$ | 2.15 | 2.05 | 2.00 | | | 0.3 |
| Ta$_2$O$_5$ | 2.31 | 2.16 | 2.09 | | | 0.3 |
| SiO | | 2.00 | 1.90 | 1.85 | | 0.5–8 |
| Al$_2$O$_3$ | 1.66 | 1.63 | 1.60 | | | 0.15–9 |
| Si$_2$O$_3$ | | 1.57 | 1.55 | | | 0.3–8 |
| SiO$_2$ | 1.48 | 1.46 | 1.45 | | | 0.16–8 |
| MgF$_2$ | 1.39 | 1.38 | | 1.36 | | 0.13–10 |
| Cryolite | | 1.30–1.35 | | | | 0.13–9 |

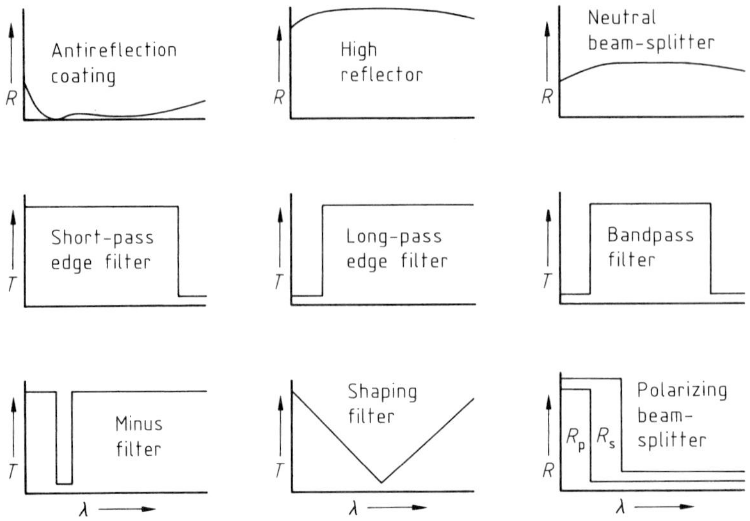



**Figure 45.** Characteristics of classical optical interference coatings

The quality of optical coatings depends on the preparation of the coating materials. Most materials come in 1–2 mm granules and are vacuum sintered. Often, small additions of foreign material reduce crystallization and increase homogeneity.

### 2.7.5. Classical Coatings

The characteristics of Figure 45 can be generated from the UV (0.2 µm) to the IR (50 µm) range of the spectrum. Since different coating materials are necessary in different spectral regions (Table 15), the quality of match to the ideal curve depends on wavelength.

In the visible (0.4–0.7 µm) the following performance figures are possible:

1) Reflectances as low as 0.01 % for narrow regions and < 0.3 % over the whole visible spectrum
2) Reflectances as high as 99.99 % for narrow regions and 99.8 % over the whole visible spectrum
3) Transition from low (< 5 %) to high reflectance (> 95 %) within an interval $< 0.01\,\lambda$
4) Narrow bandpass filters with bandwith as small as $10^{-10}$ m
5) Minus filters with bandwidth ca. $0.01\,\lambda$

### 2.7.6. Nonclassical Coatings

Of increasing importance are optical coatings which, in addition to being part of the interference system, serve an electrical or magnetic function:

1) Transparent conductive coatings: indium–tin oxide conducts electricity well, but is optically transparent [2.128]. Surface resistance of a few $\Omega/m^2$ at 90 % transmittance can be accomplished.
2) Electrochromic coatings: colorless amorphous $WO_3$, on injection of electrons and charge-compensating positive ions, transforms to blue hydrogen tungsten bronze [2.129].
3) Electroluminescent coatings: Mn-doped ZnS films, deposited at ca. 225 °C, emit light in an electrical field [2.130].
4) Magneto-optical films: sputtered alloys of Tb, Fe, and Co show a large Kerr rotation, and can be used for high-efficiency magneto-optical recording [2.131].
5) Diamond-like films: under certain conditions, carbon films can assume a diamond or diamond-like structure [2.132].

### 2.7.7. Applications

Optical coatings have the following four effects:

1) Improved performance of optical components
2) Adaptability to most needs
3) Protection (often harder than the substrate)
4) Solution of otherwise intractable problems

They play key roles in the following areas:

1) Displays: cathode ray tube contrast enhancement, r.f. interference shielding, transparent electrodes for liquid crystal displays, liquid crystal light valves for high-definition television, head-up displays, shop window antireflection coatings

2) Photography/video: antireflection coatings for lenses, mirrors, filters, color film processors, enlarging heads, professional and amateur television cameras

3) Eyeglasses: antireflection coatings, sunglasses, photochromic glasses, high-efficiency sunglasses, frames, laser goggles, welding goggles, electrochromic glasses

4) Lighting: regenerative lamps; low-height, wide-angle street lamps; shopping display lights; traffic/railroad lights; movie projectors; disco lights; stage lighting; dental/surgical lighting

5) Instruments/lasers: antireflection coatings, beam dividers, mirrors, filters, variable filters, Raman spectroscopy filters, fluorescence microscopy filters, laser end mirrors, laser windows, high-power polarizers, laser gyro coatings

6) Photocopiers/data storage: high reflectors for photocopiers, compact disk players, optical data disks, magneto-optical data disks, optical read/write heads

7) Security: burglar alarms; anticounterfeiting for documents, licenses, and currency

8) Buildings: summer films, winter films, antireflection coatings, "smart" windows

9) Automobiles: rear-view mirrors, scratch-resistant coatings for plastic windows, heat-reflecting windows, plastic headlights

10) Optical communications: antireflection coatings on laser ends, narrow-band filters, wavelength-division multiplexers

11) Integrated optics: waveguide coatings, antireflection coatings, integrated filters

12) Home appliances: oven windows, oven wall heater reflectors, optical pot heaters

13) Applied art: jewelry, iridescent color effects, optical art

14) Flexible substrates: architectural window films, cathode ray tube contrast enhancement films, touchscreens, inexpensive solar reflectors, iridescent packaging, interference ink, lightweight mirrors, optically enhanced packaging

## 2.8. Membranes [2.133]–[2.135]

($\rightarrow$ Membranes and Membrane Separation Processes)

### 2.8.1. Introduction

Membranes consist of thin solid or liquid films that act as barriers for the transport of molecular components. Synthetic membranes are widely used in applications such as water desalination and gas separation or the concentration and purification of food and chemical products, and notably in medical devices, such as artificial kidneys or drug delivery systems. In energy conversion and storage systems, e.g., batteries, fuel cells, and electrolyzers, membranes also play an important role.

### 2.8.2. Fundamentals

Membranes used in various applications differ in their functions, structures, and materials. In general terms, a membrane is defined as a solid or liquid, homogeneous or heterogeneous barrier that separates two phases and restricts the transport of different components from one phase to the other in a specific way. To separate molecular mixtures, membranes must have different permeabilities for different components. For high mass transport rates per unit area, membranes should be as thin as possible; they are usually thin films or hollow fibers. To obtain a transport through a membrane, driving forces are required, i.e., differences in the electrochemical potentials of the various components such as differences in concentration, hydrostatic pressure, or electrical potential between the two phases separated by the membrane. Transport of a component is determined by its concentration and mobility in the membrane. Mobility is determined by the hydrodynamic resistance in pressure-driven processes, and by diffusivity of the component in the membrane matrix in concentration gradient- or electrical potential-driven processes. Concentration in the membrane is determined by solubility in homogeneous membranes, by exclusion due to size in porous membranes, or due to electrical charge in ion-exchange membranes. Concentration and mobility of certain components in a membrane may be facilitated by functional groups incorporated in the membrane matrix (facilitated transport).

**Table 16.** Synthetic membranes

| Structure | Material | Function |
|---|---|---|
| Porous films, hollow fibers | polymers, ceramics, metals | gas separation micro- and ultrafiltration, dialysis |
| Homogeneous films, hollow fibers | polymers, metals | reverse osmosis, gas separation, pervaporation |
| Homogeneous or porous films with fixed ions | polymers, ceramics | electrolysis, electrodialysis, dialysis |
| Asymmetric porous or homogeneous films | polymers, ceramics | pressure-driven membrane processes |
| Thin film composite membranes | polymers, ceramics | pressure-driven membrane processes |
| Liquid films with selective carriers | organic liquids | selective removal of ions, gases, etc. |

### 2.8.3. Membrane Structures and their Preparation

**Classification.** Although synthetic membranes vary widely in structure, function, and material, they can be classified into typical groups (Table 16). This classification is rather arbitrary and many structures used today may fit none or more than one of these classes.

**Preparation.** The techniques used for manufacturing the various membrane structures are very different. For the preparation of microporous membranes from ceramics or polymers, pressing and sintering of a fine powder is used. Irradiation of a thin film followed by acid or alkaline etching or the precipitation of polymer from a solution by a non-solvent are also used to prepare porous membranes. Homogeneous membranes are prepared by extrusion of a metal or a polymer into a thin film or by casting polymer solution as a thin film and evaporating the solvent. Asymmetric structures are generally made by phase inversion techniques and composite membranes by dip-coating or interfacial polymerization. Techniques for making liquid membranes include the formation of emulsions for the preparation of unsupported liquid membranes and the filling of porous structures with liquid for the preparation of supported liquid membranes.

**Processes.** Membrane separation processes can differ significantly with regard to the membranes, the applied driving forces, the areas of application, and the industrial relevance (Table 17).

**Applications.** Synthetic membranes are used not only in separation processes, but also in med-

**Table 17.** Technical relevant membrane separation processes

| Separation process | Membrane type | Driving force | Application |
|---|---|---|---|
| Microfiltration | symmetric porous membranes | hydrostatic pressure 10–500 kPa | sterile filtration, clarification |
| Ultrafiltration | asymmetric porous membranes | hydrostatic pressure 0.1–1 MPa | filtration of macromolecular solutions |
| Reverse osmosis | asymmetric homogeneous membranes | hydrostatic pressure 2–10 MPa | desalination of saline solutions |
| Dialysis | symmetric porous membranes | concentration gradient | separation of microsolutes from macromolecular mixtures |
| Electrodialysis | ion-exchange membranes | electrical potential gradient | desalination of ion-containing solutions |
| Electrolysis | ion-exchange membranes | electrical potential gradient | chlor-alkaline production |
| Gas and vapor separation | asymmetric homogeneous membranes | vapor pressure gradient | separation of gases and vapors |
| Pervaporation | asymmetric homogeneous membranes | vapor pressure gradient | separation of azeotropic mixtures |

**Table 18.** Technically relevant applications of synthetic membranes

| Separation process | Controlled-release systems | Membrane reactors, immunoisolation | Energy conversion and storage |
|---|---|---|---|
| Microfiltration | therapeutic drug delivery devices | enzyme and catalytic reactors | battery separators fuel cell and electrolyzer separators |
| Ultrafiltration | artificial organs | biosensors | |
| Reverse osmosis | sustained-release of pesticides | immunoprotection of cell implants | solid electrolytes |
| Dialysis | | | |
| Electrodialysis | | | |
| Gas Separation | | | |
| Pervaporation | | | |

ical devices as barriers for the controlled release of drugs and in energy-conversion systems as ion-permeable separators (Table 18). For large-scale applications in the chemical and food industry or in environmental protection, separation processes, such as micro- and ultrafiltration or gas separation are most important, but commercially the medical uses of membranes, e.g., in artificial kidneys, are equally interesting.

# 3. References


**References for Chapter 1**

[1.1] K. Jensen, G. W. Cullen,: XII Int. CVD Conference, Int. CVD Conferences, Electrochemical Soc., Pennington 1993.

[1.2] T. Mäntilä: European CVD Conferences, last conf.: IX Europ. CVD Conference. *J. Phys. Colloq.* IV C3 (1993).

[1.3] M. L. Hitchman, K. Jensen (eds): *Chemical Vapor Deposition, Principles and Applications,* Academic Press, London 1993.

[1.4] C. E. Morosanu: *Thin Films by CVD*, Elsevier, Amsterdam 1990.

[1.5] C. F. Powell, J. H. Oxley, J. M. Blocher (eds): *Vapor Deposition*, Wiley, New York 1966.

[1.6] D. M. Schleich: "CVD: A Chemical Approach to Ceramic Materials," *J. Phys. Colloq.* C5 (1989) Suppl. to no. 5, vol. 50, 961.

[1.7] T. F. Kuech, *Mater Sci. Rep.* **2** (1987) 1.

[1.8] Y. Wang, M. Sasaki, T. Goto, T. Hirai, *J. Mater Sci.* **25** (1990) 4607.

[1.9] E. A. Fitzgerald, *Mater Sci. Rep.* **7** (1991) 91.

[1.10] W. A. Bryant, *J. Electrochem. Soc.* **125** (1978) 1534. G. Wahl, P. Batzies in G. F. Wakefield, J. M. Blocher (eds): *Proc. IV CVD Conference,* Princeton, N.J. 1973, p. 425.

[1.11] G. Wahl, P. Batzies, in G. F. Wakefield, J. M. Blocher (eds): *Proc. IV Int. CVD Conference*, Electrochemical Society, Princeton, N.J. 1973, p. 363.

[1.12] W. A. P. Claassen, J. M. Bloem, W. G. J. N. Valkenberg, C. H. J. v. d. Brekel, *J. Cryst. Growth* **57** (1982) 259.

[1.13] J. M. Bloem, L. J. Giling in E. Kaldis (ed): *Current Topics in Mat. Sci,* vol. 1, North Holland, 1978, p. 147.

[1.14] F. Schmaderer, R. Huber, H. Oetzmann, G. Wahl, *Appl. Surf. Sci.* **46** (1990) 53.

[1.15] O. Lehmann, M. Stuke, *Materials Letter* **21** (1994) 131–136. O. Lehmann, M. Stuke, *Appl. Phys.* **A53** (1991) 343–345.

[1.16] B. Chapman; *Glow Discharge Processes*, Wiley, New York 1980.

[1.17] B. Rother, J. Vetter: *Plasmabeschichtungsverfahren*, Verlag Grundstoffindustrie, Leipzig 1992.

[1.18] J. R. Hollahan, T. Wydeven, C. C. Johnson, *Appl. Opt.* **13** (1974) 1844–1849.

[1.19] R. J. Jensen, A. T. Bell, D. S. Soong, *Plasma Chem. Plasma Process.* **3** (1983) 163–192.

[1.20] H. Yasuda: *Plasma Polymerization*, Academic Press, Orlando 1985.

[1.21] R. F. Bunshah (ed.): *Deposition Technologies for Films and Coatings*, 2nd ed., Noyes Publications, 1994.

[1.22] J. L. Vossen, W. Kern (eds.): *Thin Film Processes*, Academic Press, London 1991.

[1.23] E. H. C. Parker (ed.): *The Technology and Physics of Molecular Beam Epitaxy*, Plenum Publ., New York 1985.

[1.24] E. Kasper, J. C. Bean (eds.): *Silicon Molecular Beam Epitaxy*, CRC Press, Boca Raton, FL 1988.

[1.25] M. A. Herman, H. Sitter: "Molecular Beam Epitaxy, Fundamental and Current Status," *Springer Ser. Mater. Sci.* **7** (1989).

[1.26] B. A. Joyce, D. D. Vvedensky, C. T. Foxon: "Growth Mechanisms in MBE and CBE of III–V Compounds," in S. Mahajan (ed.): *Handbook on Semiconductors*, vol. 3a, North Holland, 1994, p. 275.

[1.27] E. P. Plueddemann: *Silane Coupling Agents*, 2nd ed., Plenum Publ., New York 1991.

[1.28] L. H. Dubois, R. G. Nuzzo, *Annu. Rev. Phys. Chem.* **43** (1992) 437–463.

[1.29] A. Ulman: *An Introduction to Ultrathin Organic Films*, Academic Press, San Diego 1991.

[1.30] M. L. Shilling et al., *Langmuir* **9** (1993) 2156–2160.

[1.31] I. Langmuir, *J. Am. Chem. Soc.* **39** (1917) 1848.

[1.32] K. B. Blodgett, *J. Am. Chem. Soc.* **57** (1935) 1007.

[1.33] G. Roberts: *Langmuir-Blodgett Films*, Plenum Press, New York 1990.

[1.34] A. H. Ellison, *J. Phys. Chem.* **66** (1962) 1867.

[1.35] B. Tieke, G. Lieser, G. Wegner, *J. Polym. Sci. Polym. Chem. Ed.* **17** (1979) 1631.

[1.36] A. M. Bibo, I. R. Peterson, *Adv. Mater.* **2** (1988) 309.

[1.37] H. Ringsdorf, B. Schlarb, J. Venzmer, *Angew. Chem.* **100** (1988) 117.

[1.38] G. L. Gaines: *Insoluble Monolayers at Liquid-Gas Interfaces*, Interscience, New York 1966.

[1.39] G. Wegner, *Thin Solid Films* **216** (1992) 105.

[1.40] J. Boyle III, A. J. Mantone, *Colloids Surf.* **4** (1982) 77–85.

[1.41] D. B. Ziversmit, *J. Colloid Sci.* **18** (1963) 794.

[1.42] C. Kurthen, W. Nitsch, *Adv. Mater.* **3** (1991) 445.

[1.43] A. Barraud, M. Vandevyver, *Thin Solid Films* **99** (1983) 221.

[1.44] I. Langmuir, J. V. Schäfer, *J. Am. Chem. Soc.* **57** (1938) 1007.

[1.45] F. A. Lowenheim (ed.): *Modern Electroplating*, 3rd ed., J. Wiley & Sons, New York 1974.

[1.46] W. H. Safranek (ed.): *The Properties of Electrodeposited Metals and Alloys: A Handbook*, 2nd ed., American Electroplaters and Surface Finishing Society, Orlando, FL, 1986.

[1.47] J. J. Durney (ed.): *Electroplating Engineering Handbook*, 4th ed., Van Nostrand Reinhold, New York 1984.

[1.48] *The Canning Handbook: Surface Finishing Technology*, 23rd ed., W. Canning, Birmingham 1982.

[1.49] D. Pletcher, F. C. Walsh: *Industrial Electrochemistry*, 2nd ed., Chapman & Hall, London 1991, chap. 8, p. 358.

[1.50] D. R. Gabe: *Principles of Metal Surface Treatment and Protection*, 3rd ed., Merlin Books, London 1993.

[1.51] H. Silman, G. Isserlis, A. V. Averill: *Protective and Decorative Coatings for Metals*, Finishing Publications, Teddington, UK, 1978.

[1.52] R. Suchentrunk: *Metallising of Plastics – A Handbook of Theory and Practice*, Stevenage, Hertfordshire, UK, 1993.

[1.53] R. Weiner: *Electroplating of Plastics*, Finishing Publications, Teddington, UK, 1977.

[1.54] G. Lorin: *Phosphating of Metals*, Finishing Publications, Teddington, UK, 1974.

[1.55] W. Rausch: *The Phosphating of Metals*, Finishing Publications, Teddington, UK, 1990.

[1.56] W. Riedel: *Electroless Nickel Plating*, Finishing Publications, Teddington, UK, 1991.

[1.57] R. Weiner, A. Walmsley: *Chromium Plating*, Finishing Publications, Teddington, UK, 1980.

[1.58] H. Geduld: *Zinc Plating*, Finishing Publications, Stevenage, UK, 1988.

[1.59] S. Wernick, R. Pinner, P. Sheasby: *The Surface Treatment and Finishing of Aluminium and its Alloys*, 5th ed., ASM International, Metals Park, OH, 1987.

[1.60] A. T. Kuhn (ed.): *Techniques in Electrochemistry, Corrosion and Metal Finishing – A Handbook*, J. Wiley & Sons, Chichester, UK, 1987.

[1.61] R. L. Yeates: *Electropainting*, 2nd ed., Robert Draper, Teddington, UK, 1970.

[1.62] C. F. Simpson, M. Whittaker (eds.): *Electrophoretic Techniques*, Academic Press, London 1983.

[1.63] W. Machu: *Handbook of Electropainting Technology*, Electrochemical Publications, Hatch End, UK, 1978.

[1.64] A. Brenner: *Electrodeposition of Alloys: Principles and Practice*, Academic Press, New York 1963.

[1.65] J. K. Dennis, T. E. Such: *Nickel and Chromium Plating*, Butterworths, London 1993.

[1.66] J. W. Price: *Tin and Tin-Alloy Plating*, Electrochemical Publications, Ayr, Scotland, 1983.

[1.67] E. Raub, K. Muller: *Fundamentals of Metal Deposition*, Elsevier, Amsterdam 1967.

[1.68] D. R. Gabe (ed.): *Coatings for Corrosion Protection: A Guide for Production Engineers*, Institute of Production Engineers, London 1983.

[1.69] V. E. Carter (ed.): *Corrosion Testing for Metal Finishing*, Butterworths, London 1982.

[1.70] K. N. Stafford, P. K. Datta, C. G. Coogan (eds.): *Coatings and Surface Treatment for Corrosion and Wear Resistance*, Ellis Horwood, Chichester, UK, 1984.

**References for Chapter 2**

[2.1] P. Lall, S. Bhagath: "An Overview of Multichip Moduls," *Solid State Technol.* **36** (1993) Sept, 65–67, 70, 72, 74, 76.

[2.2] P. L. Kirby: "Applications of Resistive Thin Films in Electronics," *Thin Solid Films* **50** (1978) 211–221.

[2.3] W. Worobey: "Thin Film Technology Used in Bell System Telecommunication Circuits," *Electrocomponent Sci. Technol.* **8** (1981) 3–8.

[2.4] F. Krull, H.-E. Endres: "Spürnasen," *Tech. Rundsch.* **75** (1993) no. 18, 28–34.

[2.5] F.-W. Drees et al.: "Model of The Thermal Transfer Process," *Proc. 4th Int. Congress on Advances in Non-Impact-Printing Tech*, SPSE, New Orleans, 1988, pp. 442–447.

[2.6] E. Unger: "Die Erzeugung dünner Schichten," *Chem. unserer Zeit* **25** (1991) no. 3, 149.

[2.7] J. Bond: "The Incredible Shrinking Disk Drive," *Solid State Technol.* **36** (1993) Sept., 39–45.

[2.8] R. J. Gambino: "Optical Storage Disk Technology," *MRS Bull.*, April (1990) 20–23.

[2.9] K. Hieber, H. Körner, H. Treichel: "Chemical Vapour Deposition of Oxide and Metal Films for VLSI Applications," *Thin Solid Films* **181** (1989) 77.

[2.10] H. Treichel, Th. Kruck: "Molecular Engineering in Semiconductor Technology: Borphosphorous-Silicatglass by Decomposition of a Monomolecular Precursor," *J. Phys. Colloq.* C5 **50** (1989) no. 6, 747–756.

[2.11] D. B. Fraser in S. M. Sze (ed.): *VLSI Technology*, Mc Graw Hill, New York 1984, pp. 347–383.

[2.12] H. Joswig, W. Pamler: "Stoichiometry Effects in TiN Diffusion Barriers," *Thin Solid Films* **221** (1992) 228–232.

[2.13] A. Intemann, H. Körner, F. Koch: "Film Properties of CVD TiN Deposited with Organometallic Precursors," *J. Electrochem. Soc.* **140** (1993) 3215–3222.

[2.14] H. Körner et al.: "A Low Cost, High Throughput Tungsten Plug Contact Metallization with Improved Al Electromigration," *Proc. of the 10th Int. VLSI Multilevel Interconnection Conf. (VMIC) 1993*, VIMIC Catalogue No: 93 IS MIC-102.

[2.15] O. Spindler, B. Neureither: "In Situ Planarization of Intermetal Dielectrics: Process Steps, Degree of Planarization and Film Properties," *Thin Solid Films* **175** (1989) 67–72.

[2.16] H. C. Weller, R. H. Mauch, G. H. Bauer: "Studies on the Interfaces Between Novel Type ZnO and p-a-SiC:H in 1.5 eV a-SiGe:H Pin Diodes in Comparison to SnOx and ITO," *Solar Energy Mater.* **27** (1992) 217–231.

[2.17] M. Kawachi: "Silica Waveguides on Silicon and their Applications to Integrated-Optic Components," *Opt. Quantum Electron.* **22** (1990) 391–416.

[2.18] R. T. Chen: "Polymer-based Passive and Active Guided Wave Devices and Their Applications," *SPIE, Critical Review on Integrated Optics and Optoelectronic*, vol. CR (1993) 298–335.

[2.19] H. Heinecke, E. Veuhoff: "Evaluation of III–V Growth Technologies for Optoelectronic Applications," *Mater. Sci. Eng.* **B 21** (1993) 120–129.

[2.20] L. Esaki: "A Bird's-Eye View on the Evolution of Semiconductor Superlattices and Quantum Wells," *IEEE J. Quantum Electron.* **QE-22** (1986) no. 9, 1611–1624.

[2.21] S. L. Schwartz, V. E. Wood: "Ferroelectric Thin Films," *Condens. Matter News* **1** (1992) no. 5, 4–13.

[2.22] R. Bruchhaus: "Thin Films for Ferroelectric Devices," *Ferroelectrics* **133** (1992) 73–78.

[2.23] L. E. Toth: *Transition Metal Carbides and Nitrides*, Academic Press, New York 1967.

[2.24] J.-E. Sundgren, H. T. G. Hentzell, *J. Vac. Sci. Technol. A* **4** (1986) 2259–2279.

[2.25] J. Musil, J. Vyskocil, S. Kadlec in M. H. Francombe, J. L. Vossen (eds.): *Physics of Thin Films Mechanic and Dielectric Properties*, vol. 17, Academic Press, San Diego 1993.

[2.26] L. Hultman et al., *Thin Solid Films* **205** (1993) 153–164.

[2.27] J. E. Greene, S. A. Barnett, J.-E. Sundgren, A. Rockett in A. Itoh (ed.): *Ion Assisted Deposition*, Elsevier, Amsterdam 1988, chap. 5.

[2.28] S. J. Bull, A. M. Jones, A. R. McCabe, *Surf. Coat. Technol.* **54/55** (1992) 173–179.

[2.29] K. A. Pischow et al., *Surf. Coat. Technol.* **58** (1993) 163–172.

[2.30] U. Helmersson et al., *J. Vac. Sci. Technol. A* **3** (1985) 308–315.

[2.31] H. E. Rebenne, D. G. Bhat, *Surf. Coat. Technol.* **63** (1994) 1–13.

[2.32] M. K. Hibbs, J.-E. Sundgren, B. E. Jacobsson, B.-O. Johansson, *Acta Metall.* **33** (1985) 797–803.

[2.33] W.-D. Münz, *J. Vac. Sci. Technol. A* **4** (1986) 2717.

[2.34] U. Wahlström et al., *Thin Solid Films* **235** (1993) 62–70.

[2.35] D. McIntyre et al., *J. Appl. Phys.* **67** (1990) 1542–1553.

[2.36] J.-E. Sundgren et al., *Thin Solid Films* **193/194** (1990) 818–831.

[2.37] S. A. Barnett in M. H. Francombe, J. L. Vossen (eds.): *Physics of Thin Films Mechanic and Dielectric Properties*, vol. 17, Academic Press, San Diego 1993.

[2.38] U. Helmersson et al., *J. Appl. Phys.* **62** (1987) 481–484.

[2.39] M. Shinn, L. Hultman, S. A. Barnett, *J. Mater. Res.* **7** (1992) 901–911.

[2.40] P. B. Mirkarimi et al., *J. Mater. Res.* **9** (1994) 1456–1467.

[2.41] X. Chu et al., *J. Vac. Sci. Technol. A* **10** (1992) 1604–1609.

[2.42] K. K. Shih, D. B. Dove, *Appl. Phys. Lett.* **61** (1992) 654–656.

[2.43] T. Friesen et al., *Surf. Coat. Technol.* **48** (1991) 169–174.

[2.44] A. Lettington, J. W. Steeds (eds.): *Thin Film Diamond*, Chapman & Hall, London 1994, p. 117.

[2.45] J. E. Field (ed.): *The Properties of Natural and Synthetic Diamond*, Academic Press, London 1992.

[2.46] J. Wilks, E. Wilks (eds.): *Properties and Applications of Diamond*, Butterworth-Heinemann, Oxford 1991.

[2.47] *Diamond and Related Materials*, Journal of the Science and Technology of Diamond and Related Materials, Elsevier Science, Lausanne.

[2.48] *Proc. IEEE* **79** (1991), special issue on "Large Band Gap Electronic Materials and Components."

[2.49] C. A. Davies (ed.): "Properties of Diamond," *EMIS Data Rev. Ser.*, no. 9, INSPEC, Institution of Electrical Engineers, London 1994.

[2.50] P. Reinke, W. Jacob, W. Möller, *J. Appl. Phys.* **74** (1993) 1354.

[2.51] J. Robertson, *Diamond Related Mat.* **3** (1994) 361.

[2.52] B. Dischler, A. Bubenzer, P. Koidl, *Solid State Commun.* **48** (1983) 105.

[2.53] D. R. McKenzie et al., *Diamond Related Mat.* **3** (1994) 353.

[2.54] R. S. Sussmann et al., *Diamond Related Mat.* **3** (1994) 303.

[2.55] T. J. Valentine et al., *Diamond Related Mat.* **3** (1994) 1168.

[2.56] B. Lux, R. Haubner in A. Lettington, J. W. Steeds (eds.): *Thin Film Diamond*, Chapman & Hall, London 1994, p. 127.

[2.57] M. W. Geis, A. A. Tamor in G. L. Trigg (ed.): *Encyclopedia of Applied Physics*, vol. 5, VCH Verlagsgesellschaft, Weinheim 1993.

[2.58] C. H. Wort, C. G. Sweeney, M. A. Cooper, G. A. Scarsbrook, R. S. Sussmann, *Diamond Related Mat.* **3** (1994) 1158.

[2.59] M. Seal in A. Lettington, J. W. Steeds (eds.): *Thin Film Diamond*, Chapman & Hall, London 1994, p. 143.

[2.60] J. E. Graebner et al., *Diamond Related Mat.* **2** (1993) 1059.

[2.61] P. K. Bachmann et al., *Diamond Related Mat.* **4** (1995) 820.

[2.62] T. Anthony, W. F. Banholzer, *Diamond Related Mat.* **1** (1992) 717.

[2.63] P. Bade: "The Business and Technical Outlook for CVD Diamond and Diamond-Like Carbon," *Proc. 2nd Int. Gorham Conference*, Monterey, CA, Jan. 31–Feb. 2, 1994.

[2.64] B. Fox, presented at "Diamond Films 94," Il Ciocco, Italy, accepted for publication in Diamond and Related Mat. 1995.

[2.65] F. J. Himpsel, J. A. Knapp, J. A. Van Vechten, D. E. Eastman, *Phys. Rev. B* **20** (1979) 624.

[2.66] J. Robertson in A. Lettington, J. W. Steeds (eds.): *Thin Film Diamond*, Chapman & Hall, London 1994, p. 107.

[2.67] A. Lettington in A. Lettington, J. W. Steeds (eds.): *Thin Film Diamond*, Chapman & Hall, London 1994, p. 117.

[2.68] A. Matthews, S. S. Esklidsen, *Diamond Related Mat.* **3** (1994) 902.

[2.69] W. G. Wood (ed.): *Metals & Materials Handbook*, vol. 5, American Society for Metals, Metals Park, OH, 1992.

[2.70] C. Bocking, B. Cameron: "The Use of High Speed Selective Jet Electrodeposition of Gold for the Plating of Connectors," *Trans. Inst. Met. Finish.* **72** (1994) 33.

[2.71] W. Riedel: *Electroless Nickel Plating*, Finishing Publications, Teddington, UK, 1991.

[2.72] J. D. Greenwood: *Hard Chromium Plating*, 3rd ed., Portcullis Press, Redhill, UK, 1981.

[2.73] C. F. Coombs, Jr. (ed.): *Printed Circuit Handbook*, 2nd ed., McGraw-Hill, New York 1979.

[2.74] J. A. Scarlett: *The Multilayer Printed Circuit Handbook*, Electrochemical Publications, Ayr, Scotland, 1985.

[2.75] A. Kempster: "The Principles and Applications of Chemical Vapour Deposition," *Trans. Inst. Met. Finish.* **70** (1992) 68–75.

[2.76] Y. Sun, T. Bell: "Combined Plasma Nitriding and PVD Treatments," *Trans. Inst. Met. Finish.* **70** (1992) 38–44.

[2.77] M. S. Wilson, S. Gottesfeld: "Thin-Film Catalyst Layers for Polymer Electrolyte Fuel Cell Electrodes," *J. Appl. Electrochem.* **22** (1992) 1–7.

[2.78] A. K. Shukla, K. V. Ramesh, A. M. Kannan: "Fuel Cells: Problems and Prospects," *Proc. Indian Acad. Sci. (Chem. Sci.)* **97** (1986) 513–528.

[2.79] A. J. Appleby, F. R. Foulkes (eds.): *Fuel Cell Handbook*, Van Nostrand Reinhold, New York 1989.

[2.80] D. Pletcher, F. C. Walsh: *Industrial Electrochemistry*, 2nd ed., Blackies, Glasgow 1993, chap. 8, p. 358.

[2.81] D. R. Gabe: *Principles of Metal Surface Treatment and Protection*, 3rd ed., Merlin Books, London 1993.

[2.82] J. G. Bednorz, K. A. Muller, *Z. Phys. B: Condens. Matter* **64** (1986) 189–193.

[2.83] Y. Q. Wang et al., *Appl. Phys. Lett.* **63** (1993) 3084–3086.

[2.84] T. Venkatesan et al., *Appl. Phys. Lett.* **53** (1988) 243–245.

[2.85] T. Terashima et al., *Jpn. J. Appl. Phys. Part I* **27** (1988) 91–93.

[2.86] D. G. Schlom et al., *Z. Phys. B: Condens. Matter* **86** (1992) 163–175.

[2.87] C. Klemenz, H. J. Scheel, *J. Cryst. Growth* **129** (1993) 421–428.

[2.88] R. H. Hammond, R. Bormann, *Physica B+C* **162/164 B+C** (1989) 703–704.

[2.89] J. D. Jorgensen et al., *Phys. Rev. B: Condens. Matter* **36** (1987) 3608–3616.

[2.90] R. W. Simon et al., *Appl. Phys. Lett.* **53** (1988) 2677–2679.

[2.91] M. Sasaura, S. Miyazawa, M. Mukaida, *J. Appl. Phys.* **68** (1990) 3643–3644.

[2.92] S. Miyazawa, M. Mukaida, *Appl. Phys. Lett.* **64** (1994) 2160.

[2.93] G. Wegner, *Thin Solid Films* **216** (1992) 105.

[2.94] G. Wegner, *Mol. Cryst. Liq. Cryst.* **235** (1993) 1.

[2.95] G. Roberts: *Langmuir-Blodgett Films*, Plenum Press, New York 1990.

[2.96] A. Ulman: *An Introduction to Ultrathin Organic Films*, Academic Press, Boston 1991.

[2.97] M. Stamm in: *Macromolecules: Synthesis, Order and Advanced Properties; Advances in Polymer Science*, vol. 100, Springer Verlag, Berlin 1992, p. 357.

[2.98] C. Bubeck, D. Holtkamp, *Adv. Mater.* **3** (1991) 32.

[2.99] E. F. Aust, S. Ito, M. Sawodny, W. Knoll, *Trends Polym. Sci.* **2** (1994) 313.

[2.100] M. Nieto-Vesperinas, J. C. Dainty: *Brillouin Scattering from Thin Films*, Elsevier Publ., Amsterdam 1990, p. 305.

[2.101] D. Johannsmann, K. Mathauer, G. Wegner, W. Knoll, *Phys. Rev. B: Condens. Matter* **46** (1992) 1088.

[2.102] H. Fuchs, H. Ohst, W. Prass, *Adv. Mater.* **3** (1991) 10.

[2.103] J. D. Swalen, *J. Mol. Elec.* **2** (1986) 155.

[2.104] G. Khanarian, *Thin Solid Films* **152** (1987) 265.

[2.105] P. S. Vincett, G. Roberts, *Thin Solid Films* **68** (1980) 135.

[2.106] K. Ogawa, H. Tamura, M. Sasago, T. Ishihara, *Proc. SPIE Int. Soc. Opt. Eng.* **771** (1987) 77.

[2.107] M. Shen, A. T. Bell: "Plasma Polymerization," *ACS Symp. Ser.* **108** (1979) 1–33.

[2.108] J. M. Tibbitt, M. Shen, A. T. Bell, *J. Macromol. Sci. Chem.* **A 10** (1976) 1623–1648.

[2.109] H. Yasuda, *Plasma Polymerization*, Academic Press, Orlando 1985.

[2.110] M. N. Morosoff in R. d'Agostino (ed.): *Plasma Deposition, Treatment and Etching of Polymers*, Academic Press, Boston 1990.

[2.111] H. Simon: *The Splendor of Iridescence*, Dodd, Mead & Company, New York 1971.

[2.112] S. D. Smith: "Design of Multilayer Filters by Considering Two Effective Interfaces," *J. Opt. Soc. Am.* **48** (1958) 43–50.

[2.113] L. I. Epstein: "The Design of Optical Filters," *J. Opt. Soc. Am.* **42** (1952) 806–810.

[2.114] L. Young: "Synthesis of Multiple Antireflection Films Over a Prescribed Frequency Band," *J. Opt. Soc. Am.* **51** (1961) 967–974.

[2.115] P. Baumeister: "Design of Multilayer Filters by Successive Approximations," *J. Opt. Soc. Am.* **48** (1958) 955–958.

[2.116] J. A. Dobrowolski: "Completely Automatic Synthesis of Optical Thin Film Systems," *Appl. Opt.* **4** (1965) 937–946.

[2.117] W. H. Southwell: "Coating Design Using Very Thin High- and Low-index Layers," *Appl. Opt.* **24** (1985) 457–460.

[2.118] J. A. Dobrowolski, D. Lowe: "Optical Thin Film Synthesis Program Based on the Use of Fourier Transforms," *Appl. Opt.* **17** (1978) 3039–3050.

[2.119] P. Baumeister: "Computer Software for Optical Coatings," *Photonics Spectra* **22** (1988) no. 9, 143–148.

[2.120] H. A. Macleod: *Thin-film Optical Filters*, 2nd ed., Adam Hilger Ltd., Bristol 1986, pp. 357–445.

[2.121] H. Schroeder: "Oxide Layers Deposited from Organic Solutions," in G. Hass, R. E. Thun (eds.): *Physics of Thin Films*, vol. 5, Academic Press, New York 1969, pp. 87–141.

[2.122] P. J. Martin, R. P. Netterfield: "Optical Films Produced by Ionbased Techniques," in E. Wolf (ed.), *Prog. Opt.* **13** (1986) 113–182.

[2.123] D. Z. Rogers: "Manufacture of Optical Interference Coatings by Low Pressure Chemical Vapor Deposition," *Proc. SPIE Int. Soc. Opt. Eng.* **1168** (1989) 19–24.

[2.124] W. Klug, R. Schneider, A. Zöller: "Plasma Enhanced CVD Hard Coating for Opthalmic Lenses," *SPIE* **1323** Optical Thin Films III, New Developments (1990) 88–97.

[2.125] J. Segner: "Plasma Impulse Chemical Vapour Deposition – a Novel Technique for the Production of High Power Laser Mirrors," *Mat. Sci. Eng. A* **140** (1991) 733–740.

[2.126] R. W. Bryant: "Coating Growth Accelerates Through the Decade and Beyond," *Laser Focus World* **28** (1992) Oct., 117–118.

[2.127] A. Thelen: *Design of Optical Interference Coatings*, McGraw-Hill, New York 1989, p. 243.

[2.128] P. Nath, R. F. Bunshah: "Preparation of $In_2O_3$ and Tin-doped $In_2O_3$ Films by a Novel Activated Reactive Evaporation Technique," *Thin Solid Films* **69** (1980) 63–68.

[2.129] F. G. K. Baucke, B. Metz, J. Zauner: "Elektro-chrome Schichtsysteme mit variierbaren optischen Eigenschaften," *Phys. Unserer Zeit* **18** (1987) no. 1, 21–28.

[2.130] C. King: "Thin Film Electroluminescent Displays," *Society of Automotive Engineers, West Coast International Meeting and Exposition*, Portland, OR, Aug. 4–8, 1985, Paper 851461.

[2.131] E. Schultheiss et al.: "Production Technology for Magneto-optic Data Storage Media," *Solid State Technol.* **31** (1988) March, 107–112.

[2.132] J. M. Pinneo, L. C. Conner: "CVD Diamond: New Material, New Applications," *Photonics Spectra* **23** (1989) Oct., 123–126.

[2.133] W. S. Wiston Ho, K. K. Sirkar: *Membrane Handbook*, Van Nostrand Reinhold, New York 1992.

[2.134] S. T. Hwang, K. Kammermeyer: "Membranes in Separations," *Techniques in Chemistry*, vol. VII, J. Wiley & Sons, New York, 1965.

[2.135] R. E. Kesting: *Synthetic Polymeric Membranes*, J. Wiley & Sons, New York, 1985.

# Thiocyanates and Isothiocyanates, Organic

FRANK ROMANOWSKI, HERBERT KLENK, Degussa AG, Frankfurt, Federal Republic of Germany



## 1. Organic Thiocyanates

### 1.1. Introduction

Organic thiocyanates are esters of thiocyanic acid. In the general formula $R-S-C\equiv N$, the group R can be aliphatic, alicyclic, aromatic, or heterocyclic. So far, no organic thiocyanates have been detected in plants or animals.

Special applications have been discovered and developed for only a few synthetic thiocyanates. Certain organic thiocyanates are used as vulcanization accelerators, others have been used as preservatives because of their biocidal and fungicidal properties. Small quantities are produced as fine chemicals.

### 1.2. Physical Properties [1]

The low molecular mass alkyl thiocyanates are colorless oils with an odor reminiscent of leeks. They are insoluble in water, but soluble in ethanol and ether, for example. Some physical properties are listed in Table 1.

### 1.3. Chemical Properties

Thiocyanates isomerize on warming to isothiocyanates. Allylic and tertiary thiocyanates isomerize rapidly, so some thiocyanates can be converted quantitatively to isothiocyanates [2]. In the case of methyl thiocyanate, it is possible to

**Table 1.** Physical properties of some thiocyanates

| Name, formula | $M_r$ | CAS | Freezing point, °C | Boiling point, °C (kPa) | Density, g/cm³ |
|---|---|---|---|---|---|
| Methyl thiocyanate, $CH_3SCN$ | 73.11 | [556-64-9] | −51 | 132 (101.3) | 1.074 |
| Ethyl thiocyanate, $C_2H_5SCN$ | 87.14 | [542-90-5] | −85.5 | 145.5 (101.9) | 1.011 |
| 3-Thiocyanatopropyltriethoxysilane, $(C_2H_5O)_3Si(CH_2)_3SCN$ | 263.42 | [34708-08-2] | −80 | 95 (0.013) | 1.000 |
| 2-(2-Butoxyethoxy)ethyl thiocyanate, $C_4H_9OC_2H_5OC_2H_5SCN$ | 205.28 | [112-56-1] | | 120−122 (0.266) | 1.016 |
| Methylenebis(thiocyanate), $NCSCH_2SCN$ | 130.18 | [6317-18-6] | 103−104 | | |

Ullmann's Encyclopedia
of Industrial Chemistry, Vol. A 26

produce methyl isothiocyanate on an industrial scale by isomerization (Section 2.4).

Reducing agents cleave thiocyanates to thiols and hydrogen cyanide.

$$RSCN + [H] \longrightarrow RSH + HCN$$

The reduction can be carried out with Sn/HCl, LiAlH$_4$, or sodium in liquid ammonia.

Organic thiocyanates are converted to sulfonic acids by strong oxidizing agents:

$$R-SCN + [O] \xrightarrow{H_2O} RSO_3H$$

The most commonly used oxidizing agent is nitric acid, but alkaline hypochlorite or hydrogen peroxide can also be used. Acid hydrolysis of organic thiocyanates leads to thiols via thiocarbamates:

$$R-SCN \xrightarrow[(H^+)]{H_2O} R-S-\underset{NH_2}{\overset{O}{\diagdown\!\!\!\diagup}} \xrightarrow[(H^+)]{H_2O} RSH + CO_2 + NH_3$$

Alkaline hydrolysis gives the corresponding disulfides in good yields [3]:

$$2\,RSCN + 2\,OH^- \longrightarrow RSSR + OCN^- + CN^- + H_2O$$

Because of its polarity, the $C\equiv N$ bond can undergo addition reactions. Intramolecular reactions for the synthesis of heterocycles are particularly important [4], [5].

## 1.4. Production

An important method for the production of alkyl thiocyanates is substitution of a halogen in an alkyl halide by the thiocyanate anion. The most frequently used solvents are ethanol, water, and DMF [4], [6]. As the thiocyanate anion is ambidentate, the reaction can give mixtures of thiocyanates and isothiocyanates, or can proceed selectively to either isomer [7]:

$$R-X \underset{SCN^-}{\overset{SCN^-}{\rightrightarrows}} \begin{matrix} R-SCN \\ R-NCS \end{matrix}$$

$$X = halogen$$

The selectivity of the reaction depends on the solvent, temperature, and thiocyanate concentration. Methylenebis(thiocyanate) [8]–[10] and methyl thiocyanate [11], [12], for example, can be produced by this method.

Organic thiocyanates isomerize partially or completely to the corresponding isothiocyanates during the production process, depending on the type of compound. For example, only isothiocyanates can be produced from β,γ-unsaturated alkyl halides.

Aromatic thiocyanates can be produced by reaction of arenes with dicyanodisulfane. Addition of a Lewis acid increases the electrophilicity of the dicyanodisulfane so much that even unreactive aromatics react:

$$R-C_6H_5 + NC-S-S-CN \longrightarrow R-C_6H_4-SCN + HSCN$$

A further route for the production of aromatic thiocyanates involves reaction of diazonium compounds with salts of thiocyanic acid in the presence of copper salts:

$$ArN_2{}^+X^- + MSCN \xrightarrow{Cu\ salt} ArSCN + N_2 + MX$$
$$Ar = aryl$$

A valuable extension of the possibilities for producing organic thiocyanates is the reaction of thiols with cyanogen chloride [13], particularly as no isomeric isothiocyanate can be formed.

$$RSH + ClCN \longrightarrow RSCN + HCl$$

Other methods exist for the production of organic thiocyanates, but these are not of industrial interest.

Thiocyanic acid adds to epoxides, forming 2-hydroxyalkyl thiocyanates [14], [15]:

$$\overset{O}{\triangle} + HSCN \longrightarrow HO{\frown}SCN$$

Addition of dicyanodisulfane to olefins results in quantitative formation of dithiocyanatoalkanes; the reaction is used to detect double bonds analytically in fat chemistry [16]. Radical initiators or sunlight accelerate the reaction, so a radical addition mechanism is probable.

Reaction of alkylating agents, e.g., dialkyl sulfates or alkyl sulfonates, with salts of thiocyanic acid gives organic thiocyanates [17], [18]:

$$R^1-SO_2OR^2 + MSCN \longrightarrow R^2SCN + R^1-SO_2OM$$

An overview of production methods for thiocyanates can be found in reference [6].

## 1.5. Use and Commercial Names

Industrial applications of organic thiocyanates have been developed for only a few compounds.

Methyl thiocyanate is isomerized to methyl isothiocyanate on an industrial scale (Section 2.4). A small number of compounds are used as insecticides. The compounds known as Lethanes [19] are used as contact insecticides for flies, mosquitoes, and household pests. Lethane 384 [*112-56-1*] has the formula $C_4H_9OC_2H_4OC_2H_4SCN$; Lethane 60 [*301-11-1*] has the formula $C_{11}H_{23}COOC_2H_4SCN$.

Lethane 60 (2-thiocyanatoethyl dodecanoate) is tolerated better by plants than Lethane 384 (2-(2-butoxyethoxy)ethyl thiocyanate) and can thus be used for combatting greenfly.

Methylenebis(thiocyanate) (MBT) and 2-(thiocyanatomethylthio)benzothiazole [*21564-17-0*] [20] exhibit biocidal properties. Both compounds are supplied ready for use in widely differing formulations under the name Busan. The quantities required for use as biocides are predicted to rise [21].

MBT hinders the propagation of slime-forming bacteria and fungi. As Busan 110, it is preferably used in refrigerator circulation systems and in papermills as a 10% solution in inert solvents [22]. To lower the COD in the circulation water in papermills MTB is used as a stable aqueous dispersion [23]. MTB is employed as a fumigant for counteracting fungal growth [24]. Other applications of methylenebis(thiocyanate) are described in the literature [25]–[28].

2-(Thiocyanatomethylthio)benzothiazole has been known as a pesticide for a long time [29]. It is sold as a solution in an inert solvent under the name Busan. It is used as a broad-spectrum bactericide und fungicide in various areas. It is thus possible to preserve water-based or organic-solvent-based dyes with Busan 30 [30]. All types of wood are protected from fungi and insects by a coating of the substance. Attack of leather by mold during processing can be prevented by treatment with 2-(thiocyanatomethylthio)benzothiazole [31], [32]. An advantage of the compound over pentachlorophenol is its better environmental compatibility.

3-Thiocyanatopropyltriethoxysilane (Si 264), $(C_2H_5O)_3SiCH_2CH_2SCN$, is used as a reinforcing additive in molding compounds which can be vulcanized by heating with peroxides. The molding compounds are more readily processed, and the polymer obtained after vulcanization has markedly improved properties [33].

## 1.6. Toxicology

In the case of alkyl thiocyanates, the toxicity in rodents after oral and subcutaneous administration decreases with increasing molecular mass. The higher homologs have an irritant, sometimes corrosive effect on the skin and mucous membranes. They cause irreversible damage to the outer epithelium, death of tissue (necrosis), and defect formation (ulceration), with subsequent hyperplastic reconstruction of the outer epithelium. The systemic toxicity is high (Table 2). Acute poisoning leads to a high degree of circulatory disturbance in the inner organs, with increases in circulation, bleeding, and possibly death.

# 2. Organic Isothiocyanates

## 2.1. Introduction

Organic isothiocyanates are esters of isothiocyanic acid. In the general formula $R-N=C=S$, the group R can be aliphatic, alicyclic, aromatic, acyl, or heterocyclic.

**Table 2.** Toxicological data for important organic thiocyanates

| Name | RTECS | Administration route | Species | $LD_{50}$, mg/kg |
|---|---|---|---|---|
| Methyl thiocyanate | XL 1575000 | oral | rat | 60 |
| | | intravenous | mouse | 18 |
| Methylenebis(thiocyanate) | XL 1560000 | oral | rat | 161 |
| | | intravenous | mouse | 3.6 |
| 2-(2-Butoxyethoxy)ethyl thiocyanate | XK 8400000 | oral | rat | 90 |
| | | intravenous | mouse | 56 |
| 2-(Thiocyanatomethylthio)benzothiazole | XK 8150000 | oral | rat | 1590 |
| | | dermal | rabbit | 642 |

Organic isothiocyanates do not occur naturally in the free form. They are liberated from glucosinolates in plant tissues by enzymatic degradation. Their pungent odor and sharp, mustard-like taste gave these compounds the name mustard oils, by analogy with the ethereal oils of black mustard [34]. Allyl isothiocyanates are liberated from the glucosinolate sinigrin of black mustard by myrosinase [35], [36]. Methyl isothiocyanate is formed by the enzymatic degradation of glucocapparin, which is contained in the caper plant.

Glucocapparin

Many other organic isothiocyanates with, e.g., sulfur-containing or aromatic groups have become known as natural products through the degradation of the corresponding glucosinolates.

Aliphatic and aromatic isothiocyanates are important starting materials for the production of thiourea derivatives, substituted thiosemicarbazides, heterocycles, and other organic intermediates. Some isothiocyanates have nematocidal, bactericidal, and fungicidal properties. Allyl ($\rightarrow$ Flavors and Fragrances, **A 11**, p. 154), methoxycarbonyl, and methyl isothiocyanate are produced on an industrial scale.

## 2.2. Physical Properties

Most aliphatic and aromatic isothiocyanates are colorless oils with a pungent odor. Methyl isothiocyanate ($\rightarrow$ Nematocides, **A 17**, p. 129) is a solid at room temperature. Many isothiocyanates are not only relatively stable to water in the cold, but can also be steam distilled. Aliphatic isothiocyanates with the general formula $R-N=C=S$, in which R is an unbranched or branched alkyl group with 2–6 carbon atoms, distil azeotropically with water at 80–120°C. The boiling points of the pure compounds lie in the range 120–200 °C.

Organic isothiocyanates are readily soluble in most common organic solvents.

Table 3 shows the most important physical properties of some isothiocyanates (for further data on isothiocyanates see reference [1]).

## 2.3. Chemical Properties

The chemical properties of organic isothiocyanates are similar to those of organic isocyanates, but the latter are generally more reactive [37]. Most reactions of isothiocyanates involve addition of a nucleophile to the $-N=C=S$ group. Some examples of nucleophilic addition are given below [1], [38]:

$$R-N=C=S \ + \ NH_3 \ \longrightarrow \ R-NH-C\underset{NH_2}{\overset{S}{\big|}}$$

Thiourea derivative

$$+ \ HNHR \ \longrightarrow \ R-NH-C\underset{NHR}{\overset{S}{\big|}}$$

Thiourea derivative

$$+ \ HNR_2 \ \longrightarrow \ R-NH-C\underset{NR_2}{\overset{S}{\big|}}$$

Thiourea derivative

$$+ \ H_2NCN \ \xrightarrow{NaOH} \ R-NH-C\underset{N-CN}{\overset{SNa}{\big|}}$$

*N*-Cyano-*N'*-methylisothioureas

$$+ \ H_2N-NH_2 \ \longrightarrow \ R-NH-C\underset{NH-NH_2}{\overset{S}{\big|}}$$

Thiosemicarbazides

$$+ \ HCN \ \longrightarrow \ R-NH-C\underset{CN}{\overset{S}{\big|}}$$

Thiooxalic acid amide nitriles

$$+ \ HOR \ \longrightarrow \ R-NH-C\underset{OR}{\overset{S}{\big|}}$$

Alkyl esters of thiocarbamidic acid

$$+ \ \xrightarrow[2)\ H_2O]{1)\ Al(C_2H_5)_3} \ R-NH-C\underset{C_2H_5}{\overset{S}{\big|}}$$

Thiopropanoic acid amides

Isothiocyanates have a very important function in the synthesis of heterocycles. The large number of reactions which have been discovered is notable, and with them access to new compounds has become possible [39]–[42].

Interest in the development of corresponding synthetic processes was directed towards five-

**Table 3.** Physical data for some isothiocyanates

| Name | $M_r$ | CAS | Freezing point, °C | Boiling point, °C (kPa) | Density, g/cm$^3$ |
|---|---|---|---|---|---|
| Methyl isothiocyanate, $CH_3NCS$ | 73.11 | [556-61-6] | 35.9 | 118.7 (100.9) | 1.0691 (37°C) |
| Allyl isothiocyanate, $CH_2=CHCH_2NCS$ | 99.15 | [57-06-7] | −102.5 | 151.8 (100.7) | 1.015 (20°C) |
| Phenyl isothiocyanate, $C_6H_5NCS$ | 135.18 | [103-72-0] | −21 | 221 (100.5) | 1.138 (15.5°C) |
| Vinyl isothiocyanate, $CH_2=CHNCS$ | 85.12 | [1520-22-5] | | 46 (10.0) | 1.018 (15°C) |
| Methoxycarbonyl isothiocyanate, $CH_3OCONCS$ | 117.10 | [35266-49-0] | | 30 (0.12) | 1.152 (15°C) |

and six-ring heterocycles, but four- and seven-ring systems can also be synthesized, such as benzimidazoles, benzothiazoles, and other bicyclic heterocycles.

Amino acids can be converted to thiohydantoin derivatives [43]:

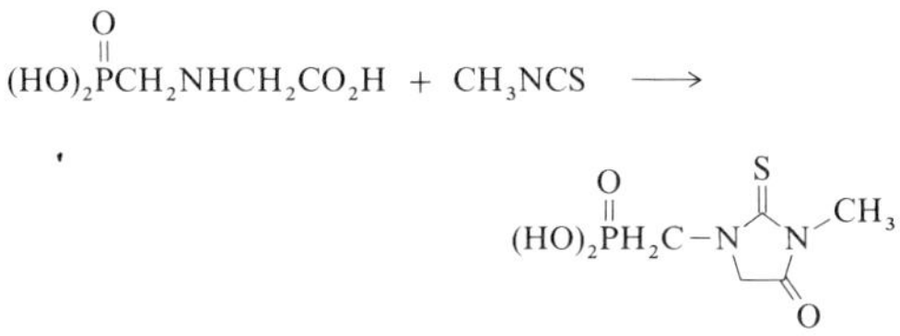

$$(HO)_2PCH_2NHCH_2CO_2H + CH_3NCS \longrightarrow$$

Acyl isothiocyanates react with 1,3-diketones to 5-acyl-4$H$-1,3-oxazine-4-thiones [44]:

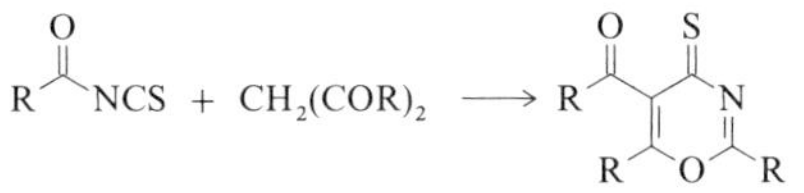

$$R-CO-NCS + CH_2(COR)_2 \longrightarrow$$

Anthranilic acid reacts with methyl isothiocyanate to 3-$N$-methyl-2-thioxotetrahydroquinoxalin-4-one [45]:

The isothiocyanate group $-N=C=S$ can undergo cycloadditions with suitable reaction partners. Methyl isothiocyanate reacts with sodium azide to 5-mercapto-4-methyl-1,2,3,4-tetrazole [46]:

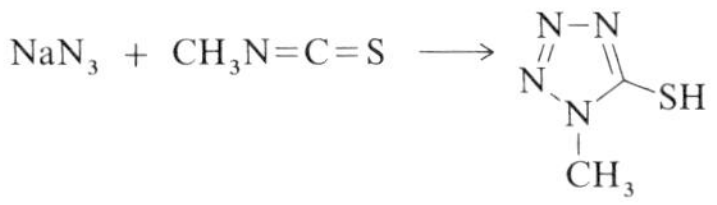

$$NaN_3 + CH_3N=C=S \longrightarrow$$

## 2.4. Production

A large number of processes are known for the production of isothiocyanates [39], [41], [47], [48]. The structure desired and sometimes the other functional groups on the organic skeleton determine the choice of production process.

The most important production methods are:

1) Reaction of alkyl and acyl halides with salts of thiocyanic acid [49], [50]
2) Isomerization of esters of thiocyanic acid [51], [52]
3) Reaction of dithiocarbamates ($\rightarrow$ Dithiocarbamic Acid and Derivatives, **A 9**, pp. 12–13) with: hydrogen peroxide [53], [54], cyanogen chloride [55], cyanuric chloride [56], chloroformates [57], alkali hypochlorite or alkali chlorite [58], phosgene [59], heavy metal salts [60], atmospheric oxygen [61], [62], phosphorus oxychloride [63]
4) Reaction of primary amines with thiophosgene [64]
5) Reaction of thiourea derivatives with mineral acids, or acetic anhydride, or by heating [39]
6) Addition of thiocyanic acid to unsaturated compounds [65]

For vinyl isothiocyanates, special production methods are usually necessary [48].

Methyl isothiocyanate, which is industrially important, is produced by two different processes:

1) Rearrangement of methyl thiocyanate
2) Oxidation of sodium $N$-methyldithiocarbamate with hydrogen peroxide

The isomerization of methyl thiocyanate begins above 100 °C. The reaction is catalyzed by salts, e.g., $ZnCl_2$, alkali-metal thiocyanate [2].

$$CH_3SCN \longrightarrow CH_3N=C=S$$

A patent specification by Morton Chemical [52] describes a multistep continuous process for the isomerization. The temperature of the reaction, which requires a long induction period, is 130–160 °C. In the course of reaction, a mixture is formed, which promotes the isomerization autocatalytically and favors the shift in the equilibrium. The methyl isothiocyanate formed is continuously removed from the vapor phase together with methyl thiocyanate. This mixture is subjected to fractional distillation and the methyl thiocyanate is recycled into the isomerization. The methyl isothiocyanate removed is replaced by fresh methyl thiocyanate. The side products formed (sulfides, disulfides) are removed periodically from the liquid phase. It is possible to convert up to 90% of the methyl thiocyanate into isothiocyanate by this process.

A patent specification by Stauffer [51], describes the isomerization of methyl thiocyanate in the gas phase at 300–400°C on a fixed-bed catalyst. The carrier (silica gel, activated charcoal) is impregnated with catalytically active salts, e.g., alkali thiocyanate, $ZnCl_2$. The vapor phase at the head of the reactor contains up to 55% methyl isothiocyanate. It is purified by fractional distillation and the recovered methyl thiocyanate is recycled.

The production of methyl isothiocyanate from sodium methyldithiocarbamate and hydrogen peroxide has been developed on a large industrial scale [53], [54]. The formation of methyl isothiocyanate takes place via several steps:

$$2CH_3NH-\overset{\overset{S}{\|}}{C}-SNa + H_2O_2 \longrightarrow$$

$$\left[ CH_3NH\overset{\overset{S}{\|}}{C}-S-S-\overset{\overset{S}{\|}}{C}NHCH_3 \right] + 2NaOH$$

$$CH_3HN-\overset{\overset{S}{\|}}{C}-S-S-\overset{\overset{S}{\|}}{C}-NHCH_3$$

$$\longrightarrow 2CH_3-N=C=S + S + H_2S$$

$$H_2S + 4H_2O_2 \longrightarrow H_2SO_4 + 4H_2O$$

The reaction can be represented by the overall equation:

$$2CH_3NH\overset{\overset{S}{\|}}{C}-SNa + 5H_2O_2$$

$$\longrightarrow 2CH_3-N=C=S + S + Na_2SO_4 + 6H_2O$$

With smaller quantities of hydrogen peroxide, sodium thiosulfate is predominantly formed as the side-product:

$$2CH_3NH\overset{\overset{S}{\|}}{C}-SNa + 4H_2O_2$$

$$\longrightarrow 2CH_3-N=C=S + Na_2S_2O_3 + 5H_2O$$

Industrial-scale production is as follows: ca. 40% aqueous sodium methyldithiocarbamate solution (pH 10–12) is reacted with 30–40% hydrogen peroxide at ca. 100°C. The pH during the reaction is 3–4. The reaction mixture is heated so that the methyl isothiocyanate immediately distils as an azeotrope with water. Methyl isothiocyanate (ca. 5–7%) is dissolved in the water that distils over. This can easily be separated by a short distillation. A very pure product is formed in ca. 85% yield. The sulfur precipitated during the reaction can be separated readily from the wastewater, which is readily biodegradable. The small quantity of waste-gas formed is incinerated.

Allyl isothiocyanate is produced by reaction of allyl chloride with alkali rhodanides in the two-phase system water/1,2-dichloroethane [66].

All commercial methoxycarbonyl isothiocyanate is produced in situ from methyl chloroformate and potassium thiocyanate in acetone. It is then converted immediately to thiophanate methyl (Section 2.8.2) [67].

Vinyl isothiocyanate can be produced in 63% yield in a two-step reaction. 2-Aminobromoethane is reacted with thiophosgene in ether. The 2-bromoethyl isothiocyanate obtained after work-up and distillation is converted into vinyl isothiocyanate by elimination of hydrogen bromide [68].

## 2.5. Quality Specifications and Analysis

The processing of industrially produced organic isothiocyanates to higher-value products and their use in pharmaceuticals requires high purity. The quality of the product is indicated by the $-N=C=S$ content. To determine this, the product is treated with excess dibutylamine in dioxan or chlorobenzene/methanol, and the excess amine is titrated [69], [70]. Gas–liquid chromatography gives a detailed overview of the side-product spectrum.

## 2.6. Storage and Transport

Methyl isothiocyanate is generally stored in tanks or in 200-kg steel drums. The product is stable to storage, provided it is free of water. The containers should be stored and handled tightly closed at normal temperature in a dry, well-ventilated area, because the product has a severely irritant effect on the skin and mucous membranes and is toxic [71].

Methyl thiocyanate: UN no. 2477, EINECS 209-132-5, IMDG 6.1 PG II, RID/ADR 6.1 no. 20b.

## 2.7. Commercial Names

Methyl isothiocyanate is sold worldwide by Degussa (Germany) and Morton Chemical (United States).

In formulations, it is sold under the names Trapex and Di-Trapex by Schering. Di-Trapex is sold by Nor-Am under the name Vorlex in the United States. Osmose sells methyl isothiocyanate as MITC-FUME in glass and aluminum ampules. Allyl isothiocyanate is sold in Germany by Haarman & Reimer and in the United Kingdom by Rhone-Poulenc.

## 2.8. Uses

### 2.8.1. Methyl Isothiocyanate

Consumption of methyl isothiocyanate is currently < 2000 t/a; 80 % is used in agrochemicals, ca. 15 % in pharmaceuticals, and ca. 5 % in other applications.

**Agrochemicals.** Methyl isothiocyanate is used directly as a 20 % formulation in an inert solvent (Vorlex, Trapex, Di-Trapex) for protecting soils against fungi and nematodes.

1,3,4-Thiadiazole derivatives are known for their herbicidal properties. Methyl isothiocyanate is used for their synthesis. The most important examples are Spike (Eli Lilly, United States) [72], Ustilan (Bayer, Germany) [73], and Erbotan (Ciba-Geigy, Switzerland) [74].

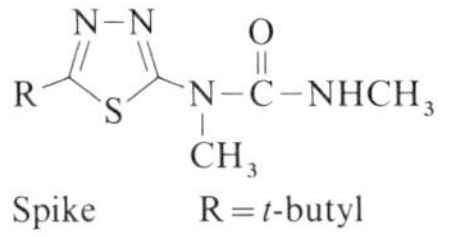

Spike       R = $t$-butyl
Ustilan     R = ethylsulfonyl
Erbotan     R = trifluormethyl

**Pharmaceuticals.** Methyl isothiocyanate is an important basic chemical for the synthesis of the $H_2$-blockers ranitidine (Zantac, Glaxo) [75], [76] and cimetidine (Tagamet, Smith Kline & French) [77].

5-Mercapto-1-methyltetrazole is obtained by treating methyl isothiocyanate with sodium azide. It is a side-group in the cefalosporin antibiotics Latamoxef [78] and Cefamandol [79].

**Other Uses.** Methyl isothiocyanate is very effective in the protection of wood against rotting [80] and premature decay.

A special application of decay prevention for telegraph poles in the United States has been developed by Osmose Wood Preserving. Four holes, on average, are bored at an angle of 45° in the poles. Ampules of methyl isothiocyanate (MITC-FUME) are inserted into the holes and the latter are closed with wood plugs. The methyl isothiocyanate diffuses slowly into the wood. After ca. 30 d the substance can be detected at 60 – 80 cm from the opening of the ampule in dry Douglas pine [81]. The moisture content of the wood determines the rate of migration, the effectiveness, and also the time at which further treatment is required. The decision as to whether a secondary impregnation with MITC-FUME is necessary is made on inspection of the poles after 6 years [82]. Effectiveness is estimated to last 12 years, so that the second inspection cycle is within the expected period of activity. The lifetime of the telegraph poles is expected to be doubled by subsequent treatment with MITC-FUME [83].

### 2.8.2. Other Organic Isothiocyanates

Allyl isothiocyanate is mainly used in the sugar industry because of its fungicidal properties. It protects sugar beet from fungi during storage [84]. Allyl isothiocyanate is also known as an insecticide and an antimicrobial agent, and is used in perfumes. It is used for combatting the cabbage maggot fly *H. brassicae* (→ Insect Control, **A 14**, p. 311).

2-(Hydroxyethyl)allylthiourea, produced from allyl isothiocyanate, has been used for producing light-sensitive paper. Methoxycarbonyl isothiocyanate [67], [85] is treated in situ with *o*-phenylenediamine to give thiophanate methyl (→ Fungicides, Agricultural, **A 12**, p. 99). Thiophanate methyl [*23564-05-8*] is a systemic fungicide, sold under the names Topsin M (Nippon Soda, Japan), Mildothane (May & Baker, United Kingdom), and Cycosin (American Cyanamid, United States). In 1980, United States consumption of thiophanate methyl was 150 t.

## 2.9. Toxicology and Industrial Hygiene

As vapors, organic isothiocyanates have a severely irritant effect on the skin, eyes, and respiratory tract. Direct contact with the body leads to severe damage to the areas affected. Skin resorption is possible. Oral administration leads to general disturbance of function, cramp, and unconsciousness. Acute toxicity after oral administration in animals is generally high (Table 4).

**Methyl Isothiocyanate** [86]–[88]. The vapor has a severely irritant effect on the skin and mucous membranes of the eyes and respiratory tract, and gives rise to a burning pain. On dermal contact, first- or second-degree acid burns are possible. Without immediate treatment, corneal opacity of the eyes is also possible. After swallowing, burning, pain, vomiting, cramp, unconsciousness, complete loss of reflexes, reduction in blood pressure, and death can occur. Methyl isothiocyanate can damage the liver and kidneys and cause lung edema.

**Allyl Isothiocyanate** [89]. Local and systemic toxic effects are very similar to those of methyl isothiocyanate.

## 2.10. Safety Measures and Processing Advice

In all work with organic isothiocyanates a protective mask with combination filter B, thick protective clothing, and safety glasses must be worn. Work areas must be well ventilated and, because sight can be impaired in the event of an accident, straight-line, free escape routes must be provided. Apparatus and opened vessels must be extracted while filling and emptying. Vapors form explosive mixtures with air, so ignition sources must be excluded.

Methyl isothiocyanate which has leaked must be covered with cold water, and the crusts formed added in portions to 10 % aqueous ammonia, and disposed of separately. Spilled allyl isothiocyanate should be covered with activated charcoal and the latter disposed of carefully.

## 2.11. Ecological Aspects

Organic isothiocyanates, particularly methyl and allyl isothiocyanates, represent a hazard to bodies of water. Measures should be taken to avoid contamination of lakes, streams, and groundwater, under all circumstances.

## 3. References


[1] E. E. Reid: *Organic Chemistry of Bivalent Sulfur*, vol. VI, Chem. Publ., New York 1966.
[2] A. Fava in N. Kharasch, Cal. Y. Meyers (eds.): *The Chemistry of Organic Sulfur Compounds*, vol. 2, Pergamon Press, New York 1966, pp. 73–91.
[3] E. Hoggart, W. A. Sexton, *J. Chem. Soc.* **1947**, 815.
[4] R. G. Guy in S. Patai (ed.): *The Chemistry of Cyanates and their Thio Derivatives*, part 2, J. Wiley & Sons, Chichester 1977, pp. 819–878.
[5] B. Schulze, M. Mühlstädt, *Z. Chem.* **19** (1979) 41–48.
[6] *Houben/Weyl*, **E4**, 941–969.
[7] D. Knoke, K. Kottke, R. Pohloudek-Fabini, *Pharmazie* **28** (1973) 574, 617; **32** (1977) 195–215.
[8] Sumitomo Chem., DE 1 232 574, 1967 (T. Kawanami, G. Suzukamo).
[9] Nalco Chem., DE 1 668 494, 1967 (J. Matt, E. W. Hunter, L. A. Goretta).
[10] Kinkai Kagaku, JP-Kokai 60 072 858, 1985 (K. Shinya, H. Seuda, S. Kojima).
[11] Morton, DE 1 295 549, 1966 (J. T. Venerable, J. Miyashiro, A. W. Seiling).
[12] Bayer, DE 1 183 903, 1965 (K. Goliasch, E. Grigat, R. Pütter).
[13] Degussa, DE 1 270 553, 1966 (Ch. Kosel).
[14] H. D. Vogelsang, T. Wagner-Janregg, R. Rebling, *Justus Liebigs Ann. Chem.* **569** (1950) 183.
[15] F. G. Weber, H. Ließert, R. Radeglia, *Pharmazie* **35** (1980) 330.
[16] H. P. Kaufmann: *Analyse der Fette und Fettprodukte*, vol. I, Springer Verlag, Berlin 1958, p. 570.
[17] Dow Chemical, US 2 939 875, 1958 (U. D. Floria).
[18] E. J. Corey, R. B. Mitra, *J. Am. Chem. Soc.* **84** (1962) 2938.


**Table 4.** Toxicological data for organic isothiocyanates

| Name | RTECS | Administration route | Species | LD$_{50}$, mg/kg |
|---|---|---|---|---|
| Methyl isothiocyanate | PA 9625000 | oral | rat | 67 |
| | | dermal | rabbit | 174 |
| | | inhalation | rat | 29.6 |
| Allyl isothiocyanate | NX 8225000 | oral | rat | 112 |
| | | dermal | rabbit | 88 |
| Phenyl isothiocyanate | NX 9275000 | oral | mouse | 87 |

[19] K. H. Büchel in R. Wegler (ed.): *Chemie der Pflanzenschutz- und Schädlingsbekämpfungsmittel*, vol. 1, Springer Verlag, Berlin 1970, pp. 457–459.

[20] Buckman Laboratories, US 1 129 575, 1966.

[21] Kline & Company, Biocides-Industry Report, Investext, March 26, 1990, pp. 1–4.

[22] Buckman Laboratories, Product Data Sheet, Memphis, Tenn., July 1981.

[23] W. D. Henkels, A. Schenker, J. F. Schuetz, *Wochenbl. Papierfabr.* **112** (1984) no. 23/24, 869.

[24] Yamato Chem. Ind., JP-Kokai 59 010 530, 1984 (A. W. Chester, Chu Yang Feng).

[25] Ichikawa Gosei Kagaku, JP-Kokai 03 007 205, 1991 (Y. Igarashi, T. Tsunoda, K. Keisake, R. Imai).

[26] Lester Technologies Corp., US 4 975 109, 1990 (L. A. Friedman, R. F. McFarlin).

[27] T. C. Presuell, D. D. Nicholas, *For. Prod. J.* **40** (1990) 57–61.

[28] W. R. Grace and Co., EP 282 203, 1988 (D. Clarkson, R. P. Cliffort).

[29] T. Egli, E. Sturm in R. Wegler (ed.): *Chemie der Pflanzenschutz- und Schädlingsbekämpfungsmittel*, vol. 6, Springer Verlag, Berlin 1981, p. 380.

[30] Buckman Laboratories, product information "BUSAN 30L".

[31] H. Gattner, W. Lindner, H.-U. Neuber, *Leder* **39** (1988) 66.

[32] W. M. Fowler, A. E. Russel, C. G. Whiteley, *J. Am. Leather Chem. Assoc.* **85** (1990) 243.

[33] Degussa, DE 4 100 217, 1992 (U. Görl, S. Wolff).

[34] A. W. Hofmann, *Ber. Dtsch. Chem. Ges.* **1** (1868) 25.

[35] F. Hoffmann, *Chem. Unserer Zeit* **12** (1978) 182.

[36] A. Kjaer in N. Kharasch (ed.): *Organic Sulfur Compounds*, vol. 1, Pergamon Press, New York 1961, p. 409.

[37] *Houben/Weyl*, **IX**, 768.

[38] S. J. Assony in N. Kharasch (ed.): *Organic Sulfur Compounds*, vol. 1, Pergamon Press, New York 1961, p. 326.

[39] S. Sharma, *Sulfur Rep.* **8** (1989) 327.

[40] A. K. Mukerjee, R. Ashare, *Chem. Rev.* **91** (1991) 1.

[41] M. Avalos et al., *Heterocycles* **33** (1992) 973–1010.

[42] G. Giesselmann, K. Huthmacher, H. Klenk, F. Romanowski, *Chem. Ztg.* **114** (1990) 215–224.

[43] Schering, DE 3 427 794, 1986 (F. Blume et al.).

[44] H. Dehne, P. Krey, *J. prakt. Chem.* **324** (1982) 915.

[45] Ch. H. Chan, *Heterocycles* **26** (1987) 3193.

[46] H. W. Altland, *J. Org. Chem.* **41** (1976) 3395.

[47] *Houben/Weyl*, **E4**, 834–883.

[48] K. Schulze, B. Schulze, Ch. Richter, *Z. Chem.* **29** (1989) 41.

[49] Ciba-Geigy, DE 3 504 016, 1985 (M. Böger, J. Drabek).

[50] B. S. Drach, E. P. Sviridov, A. V. Kirsanov, *Zh. Org. Khim.* **8** (1972) 1872.

[51] Stauffer Chem., US 2 954 393, 1960 (J. N. Haimsohn, G. E. Lukes).

[52] Morton Chem., DE 1 253 265, 1962 (J. T. Venerable, J. Miyashiro, P. L. Weyna).

[53] Degussa, DE 2 105 473, 1971 (G. Giesselmann, K. Günther).

[54] Degussa, DE 4 001 020, 1990 (G. Giesselmann, W. Schwarze).

[55] Degussa, DE 2 603 508, 1976 (G. Giesselmann, R. Vanheertum, G. Schreyer).

[56] Degussa, DE 1 935 302, 1969 (W. Schwarze, W. Weigert).

[57] Bayer, DE 1 178 423, 1963 (R. Heusch).

[58] Bayer, DE 952 084, 1956 (E. Schmidt, R. Schnegg, F. Zaller, F. Moosmüller).

[59] Schering, DE 1 068 250, 1957 (H. Werres).

[60] G. Losse, H. Weddige, *Justus Liebigs Ann. Chem.* **636** (1960) 144.

[61] Stauffer Chem., US 4 713 467, 1987 (J. E. Telschow, D. A. Bright).

[62] Story Chemical, US 3 923 852, 1975 (A. G. Zeiler, H. Babad).

[63] DD 43 996, 1965 (D. Martin, E. Beyer, H. Gross).

[64] Bayer, DE 1 172 257, 1963 (E. Mühlbauer, E. Meisert).

[65] M. S. Kharasch, E. M. May, F. R. Mayo, *J. Am. Chem. Soc.* **59** (1937) 1580.

[66] IG Farbenindustrie, DE 705 561, 1941 (W. Flemming, H. Buchholz).

[67] May & Baker, DE 2 129 960, 1982 (D. A. Eichler et al.).

[68] Dow Chemical, US 2 757 190, 1956 (G. D. Jones, R. L. Zimmerman).

[69] A. G. Williamson, *Analyst (London)* **77** (1952) 372.

[70] M. Ottnad, N. A. Jenny, C.-H. Röder, *Anal. Methods Pestic. Plant Growth Regul.* **10** (1976) 563–573.

[71] Degussa, DIN Sicherheitsdatenblatt "Methylisothiocyanat," Hanau, 1991.

[72] Air Products and Chemicals, DE 2 100 057, 1971 (T. Cebalo, J. F. Alderman).

[73] Bayer, DE 1 816 568, 1968 (C. Metzger, H. Hack).

[74] Bayer, DE 1 770 467, 1968 (D. Rücker, C. Metzger, L. Eue, H. Hack).

[75] Glaxo Group, GB 2 160 204, 1985 (J. F. Seager, R. Dansey).

[76] Union Quimico Farmazeutika, ES 2 003 781, 1988 (I. Lopez Molina, A. D. Coto).

[77] Smith Kline & French, US 3 894 151, 1975 (J. W. Black, M. E. Parsons).

[78] Eli Lilly, US 4 452 778, 1984 (L. G. Brier).

[79] Eli Lilly, US 3 641 021, 1972 (Ch. W. Ryan).

[80] J. J. Morrell, M. E. Corden, *For. Prod. J.* **36** (1986) 26.

[81] J. N. R. Ruddick: "Fumigant Movement in Canadian Wood Species," The International Research Group on Wood Preservation, *15th Annual Meeting*, Stockholm, Sweden 1984.

[82] J. J. Morrell: "The Use of Fumigants for Controlling Decay of Wood: A Review of Their Efficacy and Safety," International Research Group on Wood Preservation, *20th Annual Meeting*, Lappeenranta, Finland, 22–26 May 1989.

[83] J. J. Morrell: *Market Potential for Safer Fumigants for Wood Preservation*, Oregon State University, Corvallis 1986.

[84] I. Leifertova, M. Lisa, J. Stanek, N. Hejtmankova, *Folia Pharm. (Prague)* **5** (1983) 43.

[85] Nippon Soda, DE 1 930 540, 1969 (T. Noguchi et al.).

[86] Industrieverband Agrar: *Wirkstoffe in Pflanzenschutz- und Schädlingsbekämpfungsmittel: physikalisch-chemische und toxikologische Daten*, 2nd ed., BLV Verlag, München 1990, p. 284.

[87] Kühn/Birett: *Merkblätter gefährliche Arbeitsstoffe*, Methylisothiocyanat M 45, ecomed Verlag, Landsberg 1992.

[88] Neue Datenblätter für gefährliche Arbeitsstoffe nach der Gefahrstoffverordnung, Methylisothiocyanat Datenblatt 1032, Weka Verlag, Augsburg, 1992.

[89] Kühn-Birett: *Merkblätter gefährliche Arbeitsstoffe*, Allylisothiocyanat A 106, ecomed Verlag, Landsberg, 1992.

# Thiocyanates, Inorganic

THEO H. J. VAN HOEK, Akzo Nobel Chemicals, Köln, Federal Republic of Germany



## 1. Introduction

Thiocyanates are the salts of thiocyanic acid, $H-S-C\equiv N$. The older name, rhodanides, is still used in many countries. Although HSCN, which is unimportant industrially, is readily decomposed, thiocyanates are stable both in solution and in the crystalline state. Some chemical reactions of thiocyanates are similar to those of halides. One of their characteristic reactions is the blood red coloration they give with the iron(III) ion. Pure thiocyanate solutions are titrated using the Volhard method in nitric acid against 0.1 N silver nitrate with iron alum as indicator.

The older production methods for thiocyanates were based on the use of coke oven gas, which used to be in plentiful supply. These very economical methods have become less important over recent decades because of higher quality requirements for the final product and the availability of suitable alternative processes.

Thiocyanates, in particular the sodium and ammonium salts, are mainly used as raw materials or auxiliaries in fiber production, for agricultural products, in photography, in the chemical industry, and in building. The realization that thiocyanates play an important role in many biochemical processes in animals and humans [1] has recently been increasingly exploited in the production of personal hygiene products and in the foodstuffs and pharmaceutical industries.

## 2. Thiocyanates

### 2.1. Ammonium Thiocyanate

**Properties.** Ammonium thiocyanate [*1762-95-4*], $NH_4SCN$, $M_r$ 76.12, $\varrho$ 1.305 g/cm$^3$, is a white, hygroscopic salt which forms stable monoclinic crystals below 90–92 °C [2], [3]. At higher temperatures it is converted to the rhombic form. The *mp* of the pure salt is 149 °C. If it is heated to above 100 °C it partially rearranges to the isomeric thiourea.

$$NH_4SCN \longrightarrow S=C\begin{smallmatrix} \diagup NH_2 \\ \diagdown NH_2 \end{smallmatrix}$$

The degree of rearrangement depends on the temperature and the heating time. Lower temperatures and longer heating times give the highest proportion of thiourea [4]:

| | |
|---|---|
| 200 h at 122 °C | 31.7 mol % $SC(NH_2)_2$ |
| 10–15 h at 150 °C | 28.0 mol % $SC(NH_2)_2$ |

If the melt is further heated to 180–200 °C, guanidinium thiocyanate is formed [5].

$$2\,NH_4SCN \longrightarrow [H_2N=C(NH_2)_2]SCN + H_2S$$

Here, $CS_2$, $NH_3$, and – via the intermediate melamine – a compound with the formula $C_6H_3N_9$ are formed as byproducts.

Ammonium thiocyanate is readily soluble in many polar solvents. The solubilities in 100 g solvent are for liquid ammonia (25 °C) 312 g, for

Ullmann's Encyclopedia
of Industrial Chemistry, Vol. A 26

liquid sulfur dioxide (0 °C) 46.8 g, and for acetonitrile (18 °C) 7.5 g. Ammonium thiocyanate is readily soluble in ethanol and acetone.

The density of aqueous $NH_4SCN$ solutions (in kg/L) is as follows:

| | |
|---|---|
| 15% $NH_4SCN$ | 1.033 |
| 30% $NH_4SCN$ | 1.068 |
| 45% $NH_4SCN$ | 1.103 |
| 60% $NH_4SCN$ | 1.140 |

and the solubility at various temperatures (in wt%):

| | |
|---|---|
| 0 °C | 55.0 |
| 30 °C | 66.8 |
| 60 °C | 77.8 |
| 90 °C | 87.5 |

The boiling point of a 30% solution in water is 105 °C, of a 50% solution 110 °C, and of an 80% solution 140 °C.

The enthalpy of solution in water is 23.86 kJ/mol. By mixing 100 g water with 130 g ammonium thiocyanate a temperature decrease of 18 °C can be achieved (cooling baths). As the salt of a weak base ammonium thiocyanate is weakly acidic in aqueous solution. The pH value of a 5% solution is between 4.8 and 5.0.

Since small quantities of iron in ammonium thiocyanate which has been exposed to light can cause the typical red coloration, attention must be paid to the chemical resistance of vessels used for storage or processing. Aqueous ammonium thiocyanate solutions and crystals severely corrode mild and stainless steels up to and including the grade 1.4301. The stainless steels 1.4571 (UNS S31635) and 1.4404 (UNS S31603) have limited resistance. Only austenitic steels with a molybdenum content of over 4% or a high nickel content, such as 1.4449/1.4439 (UNS S31726), 1.4539 (UNS N08904), 2.4858 (UNS N08825) and 1.4563 (UNS N08028), have corrosion rates of < 10 µm per year up to 80 °C in concentrated ammonium thiocyanate solution.

Glass, titanium, and rubber or baked enamel linings have also proved to be suitable materials. Plastics, such as polypropylene, poly(vinyl chloride), polyvinylidene fluoride, and PTFE, and epoxy, polyester and vinyl ester resins also show good chemical resistance towards ammonium thiocyanate. Tests have shown that small changes in the chemical compositions of the solutions arising, for example, from differences in origin or pH cause changes in the corrosion behavior with metals.

**Production.** Ammonium thiocyanate is obtained from $CS_2$ and $NH_3$ or from the reaction of hydrogen cyanide or cyanides with sulfur and/or polysulfides.

*Production from Ammonia and Carbon Disulfide.* This continuous, pressure-free process is carried out in two steps [6]. In the first step $CS_2$ and ammonia gas are reacted in the presence of water over activated charcoal at ca. 50 °C in an interface reactor to give ammonium dithiocarbamate.

$$CS_2 + 2NH_3 \longrightarrow \left[ S=C \begin{matrix} S \\ \\ NH_2 \end{matrix} \right]^- NH_4^+$$

The latter is decomposed in the second step at 95–100 °C on activated charcoal.

$$\left[ S=C \begin{matrix} S \\ \\ NH_2 \end{matrix} \right]^- NH_4^+ \longrightarrow NH_4SCN + H_2S$$

If the $NH_3$ and $CS_2$ used are of high purity, the 35–45% ammonium thiocyanate solution formed can be marketed after concentration without further purification. The crystals, which are produced by vacuum evaporation and centrifugation, are of higher purity. In this process, apart from $H_2S$ and excess $CS_2$, no other waste substances are formed. After $CS_2$ and $H_2S$ have been separated, the former can be recycled and the latter can be used, for example, for sulfur production in the Claus process. There is also a one-step variant of this process in which the reaction of $NH_3$ with $CS_2$ and decomposition take place at ca. 115 °C and 5 bar [7].

*Production from Hydrogen Cyanide contained in Coke Oven Gas.* Cooled coke oven gas, which has been freed from tar, contains HCN, $NH_3$, and $H_2S$. $NH_3$ and $H_2S$ are reacted with an aqueous slurry of sulfur in a circulating gas scrubber forming $(NH_4)_2S$, which dissolves the elemental sulfur to form ammonium polysulfide. At the same time the HCN combines with the $NH_3$ dissolving in the washing liquid to give $NH_4CN$, which then reacts with the polysulfide to give ammonium thiocyanate and ammonium sulfide.

$$(NH_4)_2S_2 + NH_4CN \longrightarrow NH_4SCN + (NH_4)_2S$$

The ammonium sulfide is distilled off. A 25–50% ammonium thiocyanate solution remains, which, after purification, is further processed as in the other processes [5], [8], [9].

*Production from Hydrogen Cyanide, Aqueous Ammonia, and Sulfur/Ammonium Polysulfide.* Hydrogen cyanide is neutralized with aqueous ammonia, and the ammonium cyanide formed is treated with elemental sulfur and/or ammonium polysulfide with warming to give ammonium thiocyanate [10].

**Commercial Forms and Specifications.** Ammonium thiocyanate is generally packed in 25 kg (or 50 lb) sacks with polyethylene liners. It is also transported as a 50 or 55% solution in tank cars or containers of 1000 kg capacity or plastic drums of 200–250 kg capacity. The bulk density of the crystalline product is 600–700 g/L. Ammonium thiocyanate is not a hazardous substance under the terms of the transport regulations and, because the corrosion rates measured on aluminum and steel (St 37) are less than 6.25 mm/a, aqueous ammonium thiocyanate solutions also need not be declared as hazardous substances for air and sea transport. Specifications for ammonium thiocyanate solutions and crystals are listed in Table 1.

**Uses.** Most of the ammonium thiocyanate is produced as an intermediate in the production of

**Table 1.** Typical specifications for $NH_4SCN$

| Property | Solution (e.g., 50% in water) | Crystals |
|---|---|---|
| Appearance | colorless, clear liquid | crystalline, white |
| $NH_4SCN$ content | 49–51% | min. 99.0% in dry mass |
| Water content | | max. 1.0% |
| pH | 4.5–6.5 | 4.0–6.0 (5% solution) |
| Iron | max. 10 mg/kg | max. 2.0 mg/kg |
| Iodine consumption | max. 20 mg I/100 g | max. 10 mg I/100 g |
| Sulfides | max. 1500 mg/kg | max. 800 mg/kg |
| Chloride | | max. 50 mg/kg |
| Sulfate | | max. 200 mg/kg |
| Density (20 °C) | 1.113–1.118 g/ml | |
| APHA number | max. 80 | |

other inorganic thiocyanates. It is also used as the starting material for compounds such as aminothiazoles [11].

Large quantities of ammonium thiocyanate are used in agriculture. It is used to combat worms in cattle and, as a mixture with aminotriazole, reinforces the activity of the latter as a broad-spectrum herbicide. Ammonium thiocyanate is also used as a raw material for the production of other herbicides, such as diallates, triallates and tricyclazoles. It is also used as a raw material for fungicides, such as methyl isothiocyanate.

Ammonium thiocyanate is extensively used in photography as an accelerator with sodium thiosulfate in fixing baths, where it removes undeveloped silver ($Ag^+$). Addition of ammonium thiocyanate to electrolytic cells in the production of ammonium peroxodisulfate, $(NH_4)_2S_2O_8$ considerably increases the yield.

Ammonium thiocyanate is used as a solvent for cellulose [12]–[14], which can also be spun [15]. In the textile industry ammonium thiocyanate is widely used to improve desired properties:

1) In dyeing [16] and improving the quality [17], [18] of cotton
2) For improving the resistance of silk to creasing, light, and wear and tear [19], [20]
3) For lowering the absorption capacity of marks on nylon floor coverings [21]
4) For softening jute fibers [22]
5) For the heat-stabilization of polyester fibers [23]
6) In printing [24] or dyeing [25] fabrics

In the steel industry ammonium thiocyanate is used for coating [26], [27] and as a corrosion inhibitor against acidic gases [28]. It is used as an additive for silver [29] and tin [30] plating baths and for etching aluminum [31].

Because of the neutron-absorbing properties of hafnium, only hafnium-free zirconium can be used in nuclear technology. Following their extraction, Zr and Hf are dissolved in aqueous ammonium thiocyanate so that the $Hf(SCN)_4$ complex can subsequently be extracted with an organic solvent.

Ammonium thiocyanate can also be used as a component in the following applications:

1) For energy storage in combination with heat pumps [32]–[35]
2) In vulcanization [36], [37]

3) For the production of fire-extinguishing agents [38]
4) In absorbents for ammonia [39] or steam [40]

## 2.2. Sodium Thiocyanate

Sodium thiocyanate [*540-72-7*], NaSCN, $M_r$ 81.08, and its dihydrate [*1702-40-5*], NaSCN · 2 H$_2$O, $M_r$ 117.11, form white crystals. The hygroscopic NaSCN crystal is orthorhombic [41]. An α- and a β-form of monoclinic NaSCN · 2 H$_2$O are known [42]. The density of NaSCN · 2 H$_2$O is 1.56 g/cm$^3$. On recrystallization the crystals have a severe tendency to agglomerate if residual moisture content is above ca. 0.3–0.5%, particularly in the case of high-purity NaSCN. Anhydrous NaSCN melts at 287 °C and the dihydrate dissolves in its own water of crystallization at 30.4 °C.

The solubility of NaSCN in water (in wt%) at various temperatures is:

| | |
|---|---|
| 30.4 °C | 63.2 |
| 60 °C | 65.2 |
| 90 °C | 67.3 |

and the density (in kg/L):

| | |
|---|---|
| 40 wt% NaSCN | 1.228 |
| 50 wt% NaSCN | 1.298 |
| 55 wt% NaSCN | 1.333 |

The solubility increases sharply up to the triple point (30.4 °C) and thereafter only gradually. The *bp* of a 20% solution in water is 107 °C, of a 40% solution 115 °C, and of an 80% solution 140 °C. The solubility of sodium thiocyanate in methanol increases from 26–35% over the temperature range 16–52.5 °C, that in ethanol from 15.5–19.6% at 19–71 °C, and that in acetone from 6.4–17.5% at 18.8–56 °C. Sodium thiocyanate is less corrosive than ammonium thiocyanate. Besides the materials named under ammonium thiocyanate (see Section 2.1), aluminum and 1.4571 steel (UNS S31635) are also resistant.

**Production.** Sodium thiocyanate can be obtained by using the process described for ammonium thiocyanate from CS$_2$ and NH$_3$, but substituting 1 mol NH$_3$ with 1 mol NaOH (50% solution) [43].

$$CS_2 + 3\,NH_3 + NaOH \longrightarrow NaSCN + (NH_4)_2S + H_2O$$

If ammonium thiocyanate is available and it is possible to use the ammonia formed, sodium thiocyanate can be produced by treating ammonium thiocyanate with sodium hydroxide. This process can also be carried out continuously in a column.

$$NH_4SCN + NaOH \longrightarrow NaSCN + NH_3 + H_2O$$

In another process sodium cyanide formed by neutralizing HCN with NaOH, is treated with rhombic sulfur and/or sodium polysulfide to give sodium thiocyanate [44], [45]. The sulfur can also be added as a suspension in ethanol [46].

Sodium thiocyanate is also formed when wastewater from desulfurization plants for coke-oven gas is treated with NH$_3$, NaOH, and Ca(OH)$_2$ [47].

Like ammonium thiocyanate, sodium thiocyanate is formed as an aqueous solution which may be evaporated after purification to higher concentrations or to give crystals. If the evaporation to form crystals takes place above 30.4 °C the anhydrous salt is formed, and below this temperature the dihydrate.

**Commercial Forms and Specifications.** Sodium thiocyanate is packed as the anhydrous salt and as the dihydrate in 25 (or 50) kg sacks with polyethylene liners. The 50 or 55% solutions are dispatched in tank cars or containers of 1000 kg capacity or in 250 kg plastic drums. Sodium thiocyanate is not classified as a hazardous substance in transport regulations, and because the corrosion rates measured on aluminum and steel (St 37) are less than 6.25 mm/a, aqueous sodium thiocyanate solutions need not be declared as hazardous substances for air and sea transport. For the specifications for sodium thiocyanate solutions and crystal form, see Table 2.

**Uses.** Much of the sodium thiocyanate produced is used in polyacrylonitrile fiber production. Since the decomposition of polyacrylonitrile begins below its melting point, spinning the melt is not possible. However, polyacrylonitrile can be processed to give fibers when dissolved in aqueous sodium thiocyanate solution in the presence of additives. In a similar process polyacrylonitrile is dissolved in an organic solvent.

Sodium thiocyanate is used in the chemical industry as raw material for the production of organic isothiocyanates [48], such as methyl isothiocyanate, other monoalkyl isothiocyanates,

**Table 2.** Typical specifications for NaSCN

| Property | Solution (e.g., 50% in water) | anhydrous crystals |
|---|---|---|
| Appearance | colorless, clear liquid | crystalline, white |
| NaSCN content | 49–51% | min. 99.0% in dry mass |
| Water content | | max. 3.0% |
| pH | 7.5–9.5 | 5.5–7.5 (5% solution) |
| Iron | max. 1.0 mg/kg | max. 2.0 mg/kg |
| Iodine consumption | | max. 20 mg I/100 g |
| Sulfides | max. 80 mg/kg | max. 15 mg/kg |
| Chloride | | max. 100 mg/kg |
| Sulfate | | max. 200 mg/kg |
| Ammonium (as $NH_3$) | max. 300 mg/kg | max. 200 mg/kg |
| Density (20 °C) | 1.291–1.305 g/ml | |
| APHA number | max. 40 | |

or thiosemicarbazides. Some of these products are used as pesticides. Fungicides, such as methylene bisthiocyanate (MBT) and 2-thiocyanomethylthiobenzothiazole (TCMTB) are also produced from sodium thiocyanate. Like ammonium thiocyanate, sodium thiocyanate is used in the production of sodium peroxodisulfate ($Na_2S_2O_8$). In many countries sodium thiocyanate has replaced calcium chloride as the additive for reducing the hardening time of concrete. However, as the corrosive action of thiocyanate on the iron in reinforced concrete cannot be fully explained, there are restrictions in the case of concrete with high-quality requirements and prestressed concrete.

Alkali metal thiocyanates are used in the production of solid cell batteries [49], [50], and in the production of dyes for liquid crystal indicators [51].

Because of complex formation with $Fe^{3+}$ ions, thiocyanates are used as laboratory reagents. The blood red color of $Fe(SCN)_3$ in water is used in oil fields to detect the underground connections between different drillholes. Sodium thiocyanate is also mixed with oil field explosives as a stabilizer. Furthermore, thiocyanates are used in the production of ink for ink-jet printers [52], [53].

In the pharmaceutical industry thiocyanate has been used as an antihypertensive, but the high concentrations required gave rise to side effects. Thiocyanates are currently used as raw materials for antithrombolytics [54] and antitumor drugs [55], and as components of hair growth preparations. In hair dyeing, thiocyanates reduce the duration of the treatment [56], [57].

In recent decades sodium thiocyanate has increased in importance as a preservative. In the lacto-peroxidase system thiocyanate, together with peroxides, forms hypothiocyanite ($OSCN^-$), which exhibits pronounced antibacterial activity. Fresh cows' milk is protected from immediate degradation by the natural presence of thiocyanate (4–6 ppm) and peroxides. In countries where cooling the milk before processing cannot be guaranteed, a further ca. 10 ppm sodium thiocyanate can be added as a preservative together with the oxidizing agent sodium carbonate peroxohydrate ($2\,Na_2CO_3 \cdot 3\,H_2O_2$) [58], [59]. As this method seems to be effective and free of side effects, it is being increasingly used as a means of preservation in the cosmetics industry.

## 2.3. Potassium Thiocyanate

Potassium thiocyanate [*333-20-0*], KSCN, $M_r$ 97.18, forms colorless, prismatic crystals with *mp* 175 °C [60], [61]. At 20 °C it has a density of 1.89 g/cm³. The solubility in water at 20 °C is 20.8 g/100 g. It is readily soluble in ethanol.

Potassium thiocyanate is produced by the same processes as sodium thiocyanate, but using KOH instead of NaOH. Potassium thiocyanate is used for some of the same purposes as sodium thiocyanate, but to a lesser extent because of its higher price. Potassium thiocyanate is particularly important as an intermediate and a final product in the pharmaceuticals industry. It is added to toothpastes and mouthwashes to approximately double the quantity of thiocyanate which is naturally present in saliva. In the lacto-peroxidase system it increases the antibacterial activity in combination with peroxides [62]–[64]. Halides behave similarly, but, unlike thiocyanates, they also cause undesired side effects.

Although the properties of potassium and sodium thiocyanates are comparable, the potassium salt is more corrosive. At the boiling point potassium thiocyanate solutions cause pitting in

**Table 3.** Properties of calcium thiocyanate tetrahydrate and barium thiocyanate trihydrate

| Property | Ca(CSN)$_2$ · 4 H$_2$O | Ba(SCN)$_2$ · 3 H$_2$O |
|---|---|---|
| $M_r$ | 228.31 | 307.58 |
| mp, °C | ca. 59 | ca. 70 |
| Solubility, g/100 g H$_2$O | ca. 1000 g | ca. 240 g |

aluminum and attack 1.4571 stainless steel (UNS S31635) to the same extent as ammonium thiocyanate. In laboratory tests only 1.4577 and 1.4438 steels (UNS S31703) have corrosion rates of < 10 µm per year. With mild steels potassium and sodium thiocyanates exhibit comparable corrosion rates, which are clearly exceeded by that of ammonium thiocyanate.

## 2.4. Alkaline Earth Thiocyanates

Of the known alkaline earth thiocyanates only calcium and barium thiocyanates are important industrially. Some properties of these two compounds are given in Table 3.

Besides Ca(SCN)$_2$ · 4 H$_2$O, di- and trihydrates are known.

Calcium and barium thiocyanates are obtained by treating aqueous ammonium thiocyanate solutions with the corresponding alkaline earth hydroxides.

$$M(OH)_2 + 2\,NH_4SCN \longrightarrow M(SCN)_2 + 2\,NH_4OH$$

After the ammonia has been distilled off, the salt solutions are evaporated under reduced pressure, and the salt crystals are obtained by centrifugation. They then have an assay of 97–98 % based on the hydrate. The yield is almost quantitative.

Calcium and barium thiocyanates [and also strontium thiocyanate Sr(SCN)$_2$] can also be produced from NH$_4$SCN, NH$_3$, and CS$_2$ with the corresponding metal hydroxide [43]. Calcium thiocyanate can also be obtained from the wastewater produced in the desulfurization of coke-oven gas, together with NH$_3$ and Ca(OH)$_2$ [65].

Calcium and barium thiocyanates are used in the chemical and pharmaceutical industries. Calcium thiocyanate is used for the production of special materials in the building industry. Barium thiocyanate is used in printing and dyeing textiles.

## 2.5. Other Thiocyanates

Of the large number of other known thiocyanates, only copper(I), lead(II), and aluminum thiocyanates are important industrially.

**Copper(I) thiocyanate** [*18223-42-2*], CuSCN, $M_r$ 121.62, is a yellowish-white powder with mp 1084 °C and $\varrho$ 2.843 g/cm$^3$ at 20 °C. It is sparingly soluble in water (5 mg/L). It is soluble in alkali thiocyanate solutions and ether, but not in ethanol. It forms a deep blue copper–ammonia complex in aqueous ammonia.

The industrial production of CuSCN starts from ammonium or sodium thiocyanate. Aqueous solutions of the thiocyanate are warmed with copper sulfate solution. SO$_2$ or alkali metal sulfites and hydrogensulfites can be used to reduce the copper(II) thiocyanate formed initially.

$$2\,NaSCN + 2\,CuSO_4 + Na_2SO_3 + H_2O$$
$$\longrightarrow 2\,CuSCN + 2\,NaHSO_4 + Na_2SO_4$$

The CuSCN, which is precipitated quantitatively, is separated from the dissolved sodium salts by decantation and centrifugation, washed until sulfate-free, and dried [66], [67].

CuSCN is used in the production of antifouling agents for protecting the underwater surfaces of ships against vegetation. Anti-fouling paints based on tributyltin (TBT) are only allowed to be used for ships more than 25 m long.

**Lead(II) thiocyanate** [*592-87-0*], Pb(SCN)$_2$, $M_r$ 323.38, forms white to pale yellowish, monoclinic crystals with mp 190–195 °C (decomp.) and $\varrho$ 3.82 g/cm$^3$ at 20 °C. The solubility in water at 20 °C is ca. 0.05 g/100 g, and increases on warming. Lead(II) thiocyanate dissolves in nitric acid, aqueous ammonia, and alkali metal thiocyanate solutions.

Lead(II) thiocyanate is produced by treating water-soluble lead salts (e.g., lead nitrate) with alkali metal thiocyanate solutions [68].

**Lead(II) hydroxythiocyanate**, Pb(OH)(SCN), can be obtained by boiling a slurry of Pb(SCN)$_2$ in water or by the action of basic lead(II) acetate hydroxide on Pb(SCN)$_2$ [69].

Lead(II) thiocyanate is used for producing fuses in the explosives industry and, together with antimony sulfide and potassium chlorate, for producing phosphorus-free matches.

**Aluminum thiocyanate**, $Al(SCN)_3$, $M_r$ 201.21, can be produced as a ca. 30% aqueous solution by treating an aluminum sulfate solution with barium or calcium thiocyanate. After separation of the alkaline earth sulfate, the $Al(SCN)_3$ solution can only be evaporated under reduced pressure to a concentration of ca. 30%. Above this concentration the $Al(SCN)_3$ begins to decompose. $Al(SCN)_3$ solutions are used as mordants in alizarin printing in the textile industry.

## 3. Thiocyanic Acid

**Properties.** Thiocyanic acid, formerly also known as hydrogen rhodanide, HSCN, $M_r$ 59.09, *mp* 5 °C, is a colorless, very volatile liquid. It is readily soluble in water and behaves as a strong mineral acid. In very dilute (max. 5%) aqueous solutions thiocyanic acid is relatively stable up to 20 °C. At higher temperatures or at higher concentrations it polymerizes with loss of hydrogen cyanide, forming 5-imino-1,2,4-dithiazolidine-2-thione.

$$3\,HSCN \longrightarrow HCN + \underset{\underset{H}{N}}{\overset{S-S}{HN=C\diagup\diagdown C=S}}$$

**Production.** Anhydrous thiocyanic acid is obtained as white crystals by passing dry $H_2S$ into dry $Hg(CN)_2$. Pure, gaseous thiocyanic acid can be produced by heating a mixture of KSCN and $KHSO_4$ in $H_2$ under reduced pressure (ca. 60 mbar) [69]. Aqueous solutions of thiocyanic acid are obtained by decomposing a 25% ammonium thiocyanate solution with dilute sulfuric acid.

$$NH_4SCN + H_2SO_4 \longrightarrow HSCN + NH_4HSO_4$$

The yield is almost quantitative when stoichiometric quantities of the starting materials are used. The thiocyanic acid liberated is distilled at ca. 30 mbar and collected in water. The industrial and economic importance of free thiocyanic acid is low.

## 4. Economic Aspects

The annual worldwide demand for thiocyanates is currently ca. 15 000 t, based on 100% crystalline product. As industry mainly uses sodium and ammonium thiocyanates, these account for by far the greatest part of the overall requirement. About 20% of the thiocyanate is used in North America and 35% in Western Europe.

Around 30–40% of the thiocyanate is used in fiber production. The second largest consumer (ca. 20%) is the production of agricultural auxiliaries. About 15% is used in photography. Of the many other uses, application as a reagent for extracting metals such as copper, molybdenum, nickel, lead, zinc, and zirconium from ores is important.

The world market for thiocyanates appears to be relatively stable to slightly recessive. Decreases in the use in agricultural products in the early 1990s were only partially compensated for by a slight increase in the requirement for the photography industry.

The largest producers in order of importance are Western Europe (Germany and the United Kingdom), South East Asia (Japan and China), and the United States.

## 5. Toxicology

Thiocyanates are only slightly toxic. Most inorganic thiocyanates have been assigned to toxicity classes 3–5 according to the Swiss toxicity law. In animal experiments the lethal dose $(LD_{50})$ has been found to be 50–15000 mg/kg [70]. Ammonium thiocyanate with an $LD_{50}$ of 500 mg/kg was assigned to toxicity class 3, and sodium thiocyanate with $LD_{50}$ 764 mg/kg to class 4. The NOEL for rainbow trout for $NH_4SCN$ is 100 mg/L, and the $EC_{10}$ value for bacteria (*Ps. putia*) is 8 g/L. Although resorbed thiocyanate is biologically degradable, degradation can become problematic at high concentrations ($> 36$ mg $SCN^-/L\ H_2O$).

On handling thiocyanates skin contact should be avoided, as prolonged contact can lead to skin irritation and rashes, and thiocyanates are readily resorbed.

The action of strong acids on thiocyanates can liberate toxic hydrogen cyanide. Ammonium and sodium thiocyanates have been assigned to water hazard class 1 (WGK 1) in Germany (September 1994).

# 6. References

[1] W. Weuffen: *Medizinische und biologische Bedeutung der Thiocyanate (Rhodanide)*, VEB Verlag Volk und Gesundheit, Berlin 1982.

[2] J. W. Bats, P. Coppens, *Acta Crystallogr. Sect. B: Struct. Crystallogr. Cryst. Chem.* **B 33** (1977) 1542–1548.

[3] V. E. Zavodnik, Z. V. Zvonkova, G. S. Zhdanov, E. G. Mirevich, *Kristallografiya* **17** (1972) 107–110.

[4] W. Klempt, *Chem. Tech. (Leipzig)* **15** (1963) 1.

[5] W. Klempt, *Brennst. Chem.* **30** (1949) 149, 155.

[6] Enka Glanzstoff, DE 1297088, 1967 (H. Maegerlein, H. Rupp, G. Meyer).

[7] Akzo, DE 3508404, 1986 (G. Meyer, G. Sudheimer, H. Zengel, H. Grothaus).

[8] Sumikin Coke and Chemicals, JP 87-113753, 1987 (T. Sato, K. Takeda).

[9] Scientific Research Institute of Coal Chemistry, SU 87-4171977, 1987 (V. Markov, T. Batyeva, V. Melikentsova).

[10] Sumikinkako, JP 79-164355, 1979.

[11] BASF, DE 3227329, 1984 (G. Seybold).

[12] J. Cho, S. Hudson, J. Cuculo, *J. Polym. Sci., Polym. Lett. Ed.* **27** (1989) no. 8, 1699–1719.

[13] B. Lukanoff, H. Schleicher, B. Philipp, *Cellul. Chem. Technol.* **17** (1983) no. 6, 593–600.

[14] K. Yang, M. Theil, J. Cuculo, *ACS Symp. Ser.* **384** (1989) 156–183.

[15] North Carolina State University, US 4750939, 1988 (J. Cuculo, H. Theil, K. Yang, Y. Chen).

[16] US Dept. of Agriculture, US 171626, 1981 (E. Blanchard).

[17] T. Calamari, D. Thibodeaux, *Text. Chem. Color.* **20** (1988) no. 12, 13–16.

[18] H. Bober, J. Cuculo, P. Tucker, *J. Polym. Sci. Polym. Chem. Ed.* **25** (1987) no. 8, 2025–2032.

[19] H. Ban, JP 01020379, 1989.

[20] Kanebo, DE 3814450, 1988 (T. Fuse, A. Yamamoto, J. Sano).

[21] Allied Signal, WO 8902949, 1989 (D. Hangey et al.).

[22] Indian Jute Ind. Res., IN 152988, 1984 (P. Bhattacharjee, U. Ghosh, M. Masumdar, A. Ray).

[23] Allied, US 4348314, 1982 (S. Lazarus, R. Lofquist).

[24] BASF, DE 3401500, 1985 (A. Blum, N. Grund, S. Schreiner, G. Treiber, N. Zimmermann).

[25] Osrodek Badawczo, PL 116165, 1982 (M. Gralinski et al.).

[26] Nippon Steel, DE 3902457, 1989 (K. Saito, Y. Miyauchi, T. Murata, Y. Shindo).

[27] Wiener Brückenbau- und Eisenkonstr., AT 364218, 1981 (H. Willmitzer).

[28] Dow Chem., US 4446119, 1984 (M. DuPart, B. Oakes, D. Cringle).

[29] Schering, US 3915718, 1975 (R. Ludwig, K. Fuchs).

[30] Kemiska, SE 428805, 1983 (I. Fjell).

[31] Hunt, NL 7303505, 1974 (A. Philip).

[32] J. Ruiter, Rep. ETN-88-91614, Order no. N88-17108, NTIS Fr (1987) 156.

[33] Y. Liu, *Proc. Int. Soc. Energy Convers. 23rd* **4** (1988) 249–252.

[34] M. Jeday, P. Le Goff, *Int. J. Refrig.* **11** (1988) no. 3, 164–172.

[35] I. Fujiwara, M. Ebe, M. Sato, *Kagaku Kogaku* **51** (1987) no. 4, 302–304.

[36] VEB Transportgummi, DD 216944, 1985 (J. Hopfe, A. Nechwatal).

[37] Degussa, US 3947436, 1976 (G. Rocktaeschel et al.).

[38] Amalgamated Chem., US 4816186, 1989 (F. Acitelli).

[39] D. Erickson, US 4784783, 1988.

[40] D. Erickson, WO 8810290, 1988.

[41] J. W. Bats, P. Coppens, Å. Kvick, *Acta Crystallogr. Sect B: Struct. Crystallogr. Cryst. Chem.* **B 33** (1977) 1534–1542.

[42] K. Mereitner, A. Preisinger, *Z. Kristallogr.* **169** (1984) 95–107.

[43] Enka Glanzstoff, DE 67-1592339, 1967 (H. Rupp, H. Maegerlein, G. Meyer).

[44] Bergwerksverband, DE 73-2329894, 1973 (G. Huck, N. Schaefer).

[45] Mitsui Toatsu Chem., JP 76-15771, 1976 (N. Koike, Y. Hatazaki, M. Omura).

[46] Degussa, DE 85-3514408, 1985 (E. Bilger).

[47] Sumikin Coke and Chem., JP 87-113752, 1987 (T. Sato, K. Takeda).

[48] Am. Cyanamid, US 4778921, 1988 (M. Lewellyn, S. Wang, P. Strydom).

[49] Exxon Research and Engineering, US 4190706, 1980 (B. Rao, B. Silbernagel).

[50] Inst. of Phys. and Chem. Research, JP 61256573, 1986 (A. Yamada, J. Shigehara, Y. Kurata, K. Yonahara).

[51] Casio Computer, JP 01146960, 1989 (H. Aoki, K. Kodera, M. Hikasa, T. Okugawa).

[52] A. Dick, GB 1543703, 1979 (K. Hwang, D. Zabiak).

[53] Dainippon Toryo, Matsushita Electric, JP 59053564, 1984.

[54] Wellcome Foundation, EP 304311, 1989 (H. Berger, S. Pizzo).

[55] Engelhard, EP 98133, 1984 (A. Amundsen, E. Stern).

[56] Kintex, NL 7409228, 1976.

[57] Bucaria, US 3823231, 1974.

[58] L. Björck, O. Claesson, W. Schulthess, *Milchwissenschaft* **34** (1979) 726.

[59] H. Korhonen, *World Anim. Rev.* **35** (1980) 23.

[60] S. Yamamoto, M. Sakuno, Y. Shinnaka, *J. Phys. Soc. Jpn.* **56** (1987) 4393–4399.

[61] D. Cookson, M. Elcombe, T. Finlayson, *Mat. Sci. Forum* **27** (1988) 113–116.

[62] O. Poulson, WO 8802600, 1988.

[63] State University of New York, US 461848, 1986 (J. Pollock, T. McNamara).

[64] Laclede Professional Prod., CA 1167381, 1984 (M. Pellico, R. Montgomery).

[65] Sumikon Coke and Chem., JP 87-113751, 1987 (T. Sato, K. Takeda).

[66] Hokko Chem., JP 58036921, 1983.

[67] I. Krut'ko, *Koks Khim.* **1988**, no. 3, 31–32.

[68] Du Pont, US 1977440, 1935.

[69] H. Williams: *Cyanogen Compounds*, 2nd ed., Arnold, London 1948, pp. 257–316.

[70] L. Roth, M. Daunderer: *Giftliste*, Verlag Moderne Industrie, München 1979.

**Thioethers → Thiols and Organic Sulfides**

**Thioglycolic Acid → Mercaptoacetic Acid**

**Thioindigo → Indigo and Indigo Colorants**

# Thiols and Organic Sulfides

KATHRIN-MARIA ROY, Langenfeld, Federal Republic of Germany

The author would like to thank Bayer AG for undertaking the literature research for this article



## 1. Aliphatic Thiols, Sulfides, Disulfides, and Polysulfides

### 1.1. Introduction

Aliphatic thiols, R–SH, were formerly known as mercaptans because of their affinity for mercury compounds (Latin: *corpus mercurium aptans*). They can be regarded as organic derivatives of hydrogen sulfide or thio analogs of alcohols.

According to IUPAC nomenclature the names of aliphatic thiols are constructed by adding the ending "thiol" to the name of the corresponding alkane (e.g., ethanethiol). If substituents with higher priorities are present, the prefix "mercapto" is used instead (e.g., mercaptoacetic acid).

Thiols are found in a wide range of natural materials. Low molecular mass thiols are responsible for the aromas of various foods, such as cheese, milk, coffee, cabbage, and bread. They are probably liberated during processing through enzymatic cleavage of sulfides, disulfides, and other sulfur-containing substances [12]. Thiols can also be detected in a number of plants [13]. Methanethiol occurs in radishes and various alliums, such as onions, leeks, and garlic. The characteristic odor of fresh grapefruit juice is due to the presence of 1-*p*-menthene-8-thiol in high dilution [14]. Asparagus contains 3-mercapto- and 3,3′-dimercaptoisobutyric acids [15]. Butanethiol and 2-methylbutanethiol have been found in the defense secretions of skunks [16]. Low molecular mass alkanethiols are formed in the degradation of biological material during putrefaction [17], and consequently are found in varying quantities in most crude petroleum, depending on its origin [18].

L-Cysteine, one of the natural amino acids found in proteins and a substance that participates in an important reversible redox system with the disulfide cystine, is of considerable biochemical interest, as is coenzyme A (CoA−SH), whose acetyl derivative (CoA−S−Ac) is a key participant in the metabolism of carbohydrates, fats, and proteins.

In 1834 W. C. ZEISE achieved the first thiol synthesis by preparing ethanethiol from barium hydrogensulfide and calcium ethyl sulfate [19].

## 1.2. Properties of Aliphatic Thiols

### 1.2.1. Physical Properties

With the exception of methanethiol, which is gaseous at room temperature, aliphatic thiols with up to 16 carbon atoms are colorless or yellowish liquids. They are characterized by extremely offensive odors, perceptible even at very high dilution. The odor threshold for ethanethiol is less than 0.001 ppm, and that for the $S$-enantiomer of 1-$p$-menthene-8-thiol is only $2 \times 10^{-8}$ ppm [13]. The odor sensitivity is highly concentration dependent, however, particularly for the lower thiols. At very low concentrations the $C_1$ to $C_4$ thiols have an odor similar to that of coal gas. That of other low molecular mass thiols has been described as pleasantly fresh or fruity, but at higher concentrations these compounds have extremely nauseous odors.

The bond energy of the S−H bond (339 kJ/mol) is lower than that of the O−H bond (462 kJ/mol), so hydrogen bonding to sulfur is also much weaker than to oxygen. This in turn means that thiols are less highly associated than the corresponding alcohols, as manifested in higher volatilities and lower boiling points for simple aliphatic thiols compared with their oxygen counterparts. This boiling point difference diminishes as the chain length increases, however, and is eventually reversed. Thus, from 1-octanethiol onward the straight-chain thiols have higher boiling points than the corresponding alcohols (Fig. 1).

The physical properties of important alkanethiols are summarized in Table 1. The high

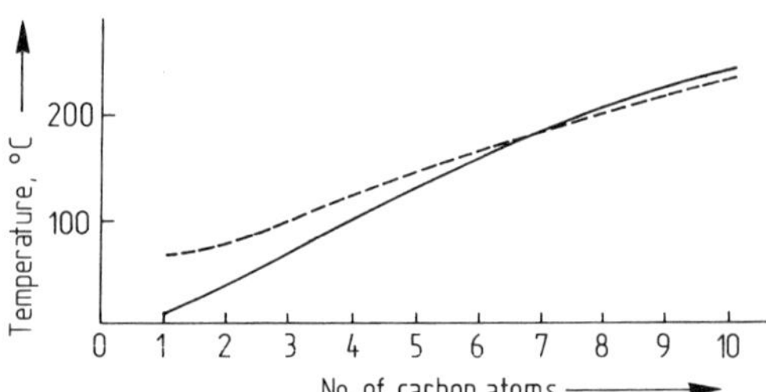


**Figure 1.** Comparison of boiling points for $n$-alcohols and $n$-thiols containing 1−10 carbon atoms
—— RSH　---- ROH

**Table 1.** Physical data for selected aliphatic thiols

| Thiol | CAS registry no. | $M_r$ | $bp$, °C/kPa | $d_4^{20}$ | $n_D^{20}$ | Flash point, °C |
|---|---|---|---|---|---|---|
| $CH_3-SH$ | [74-93-1] | 48.11 | 6.2 | 0.8665 | | |
| $CH_3CH_2-SH$ | [75-08-1] | 62.13 | 35.0 | 0.8391 | 1.43105 | −17 |
| $CH_3-(CH_2)_2-SH$ | [107-03-9] | 76.16 | 67.8 | 0.8415 | 1.43832 | −20 |
| $(CH_3)_3C-SH$ | [75-66-1] | 90.19 | 64.2 | 0.8002 | 1.4232 | −24 |
| $CH_3-(CH_2)_3-SH$ | [109-79-5] | 90.19 | 98.4 | 0.8416 | 1.4430 | 12 |
| $CH_3-(CH_2)_5-SH$ | [111-31-9] | 118.24 | 152.6 | 0.838 | 1.4497 | 20 |
| ⬡−SH | [1569-69-3] | 116.23 | 158 | 0.950 | 1.4921 | 43 |
| $CH_3-(CH_2)_7-SH$ | [111-88-6] | 146.30 | 199.1 | 0.8433 | 1.4540 | 68 |
| $(CH_3)_3C-CH_2-\overset{\underset{\mid}{CH_3}}{\underset{\mid}{C}}-SH$<br>$CH_3$ | [141-59-3] | 146.30 | 155.5 | 0.845 | 1.456 | |
| $CH_3-(CH_2)_{11}-SH$ | [112-55-0] | 202.41 | 142−5/2 | 0.8450 | 1.4589 | |
| $tert.-C_{12}H_{25}-SH$ (isomeric mixture) | [25103-58-6] | 202.41 | ca. 233 | 0.86 | 1.467 | |

flammability of thiols containing up to six carbon atoms is particularly noteworthy. All the corresponding flash points lie below room temperature.

Unlike their oxygen analogs, aliphatic thiols are only slightly soluble in water, but they are readily soluble in hydrocarbons, ethers, and alcohols. Ethane-, propane-, and butanethiols form azeotropes with aliphatic hydrocarbons, but not with benzene or toluene.

Aliphatic thiols are only weakly acidic, but they are significantly more acidic than the corresponding alcohols, which explains the ready solubility of low molecular mass alkanethiols in dilute base. However, the solubility decreases sharply with increasing molecular weight.

$$R-SH + Na^+OH^- \longrightarrow R-S^-Na^+ + H_2O$$

The resulting thiolate salts, $RS^-Na^+$, are less basic but more strongly nucleophilic than the equivalent alkoxide salts.

### 1.2.2. Chemical Properties

**Salt Formation.** Aliphatic thiols form thiolates (formerly known as mercaptides) with numerous metal cations. Many of these salts are sparingly soluble, so their formation can be used for the qualitative and quantitative determination of thiols. Many mercury thiolates, $Hg(SR)_2$, have been used for the identification of thiols because of their defined melting points. Silver thiolates have also been used for the qualitative and quantitative detection of thiols. The formation of lead thiolate constitutes the basis for the "doctor test," a method of determining the thiol content of crude petroleum. Heavy metals, in particular mercury and cadmium, can be removed from wastewater with SH-group-containing resins or ion exchangers [20].

Thiolate formation is also utilized in the treatment of heavy-metal poisoning (copper, lead, mercury, zinc, cobalt, and gold). The thiol groups of essential enzymes that have been blocked by such metal ions are liberated again with the aid of the more strongly chelating thiols. The water-soluble substances 2,3-dimercapto-propane-1-sulfonic acid [4076-02-2] and dimercaptosuccinic acid [304-55-2] are being increasingly used for this complexation reaction, alongside 2,3-dimercapto-1-propanol [59-52-9], also known as dimercaprol or British Anti-Lewisite (BAL), and penicillamine [52-67-5],

which have been known for a long time [21]. Mercaptoacrylic acids are also said to be suitable antidotes for lead and nickel poisoning [22]. Zinc and cadmium salts of cylcohexanethiol can be used as catalysts, crop-protection agents, and polymer stabilizers [23]. Certain gold thiolates are used in the treatment of arthritis. Organic antimony thiolates can be used as passivators for catalysts employed in the cracking of hydrocarbons [24]. Thiolate-coordinated transition metals play an important role in many physiological processes [25].

**Oxidation.** Thiols are readily converted into the corresponding disulfides by mild oxidizing agents.

$$2R-SH \underset{\text{Red.}}{\overset{\text{Ox.}}{\rightleftharpoons}} R-S-S-R$$

This facile oxidizability is the basis of quantitative thiol determination, and it is used industrially for sweetening crude-petroleum distillates. In this process the thiol-containing acidic fractions are oxidized with atmospheric oxygen under basic conditions in the presence of such catalysts as cobalt phthalocyanine monosulfonate [26] or EDTA complexes of copper(II) [27] or cobalt(II) [28]. Many other reagents can also be used for oxidizing thiols to disulfides (see Section 1.6.3). The thiol–disulfide redox system also plays an important role in a range of biological processes (glutathione, cysteine–cystine) [29].

Thiols and disulfides are oxidized to sulfonic acids by stronger oxidizing agents via a series of isolable intermediates. These oxidizing agents include nitric acid, potassium permanganate, chlorine gas [30], a mixture of hydrogen peroxide and aqueous hydrochloric acid [31], and atmospheric oxygen in the presence of catalytic amounts of dimethylsulfoxide and hydrogen halide [32].

**Addition Reactions.** Being strong nucleophiles, aliphatic thiols add to multiple bonds to give thioethers. This reaction, which is also used for the synthesis of unsymmetrical sulfides (see Section 1.6.2), produces a normal Markovnikov adduct ($\rightarrow$ Alcohols, Aliphatic, **A 1**, pp. 285–286) under ionic reaction conditions and an anti-Markovnikov adduct in the presence of radical-forming agents.

$$R^1-CH=CH_2 + R^2-SH \underset{\substack{h\cdot\nu \text{ or} \\ \text{peroxide}}}{\overset{H^+}{\longrightarrow}} \begin{array}{l} \overset{S-R^2}{\underset{|}{R^1-CH-CH_3}} \\ R^1-CH_2-CH_2-S-R^2 \end{array}$$

Thiols add to carbonyl compounds in the same way as alcohols. With aldehydes or ketones the products are thioacetals or thioketals [33], and with carboxylic acids and their derivatives they are *S*-esters of thiocarboxylic acids [34]. Other transformations of thiols are described in [35].

## 1.3. Production of Aliphatic Thiols

### 1.3.1. From Alcohols

The reaction of hydrogen sulfide with alcohols is used mainly for industrial syntheses of low molecular mass thiols.

$$ROH + H_2S \longrightarrow R-SH + H_2O$$

Such reactions are carried out in the gas phase > 300 °C, and generally require a catalyst. The thorium oxide [36] once used for this purpose has now largely been replaced by basic catalysts on a carrier of basic aluminum oxide. These catalysts favor thiol formation, whereas disulfides are the main products in acidic medium [37]. Promoters for thiol formation include a range of metal oxides, potassium and sodium tungstates, and heteropolyacids and their salts [37]–[40]. Thioethers and ethers may form as byproducts, along with olefins (except in the case of methanol).

The method described is used mainly for the industrial production of *methanethiol* [38], [41], although ethanethiol is also produced by this route [42]. The method is not suitable for primary thiols with more than three carbon atoms because of the small boiling-point differences between the products and the starting alcohols (see Fig. 1).

Tertiary thiols can be produced in the liquid phase by reacting alcohols with hydrogen sulfide in the presence of sulfuric acid. This method leads to *triphenylmethanethiol* [*3695-77-0*] in 75 % yield [43], and is used industrially to produce *2-methyl-2-propanethiol* (*tert*-butanethiol) [44].

The reaction of alcohols with phosphorus pentasulfide is frequently employed for this purpose as well, and has been patented for alkanethiols [45]. $C_4$ to $C_{16}$ alcohols are converted in this way into dialkyl dithiophosphates, which give the corresponding thiols on acid hydrolysis.

A more recent method uses Lawesson's reagent [*19172-47-5*] for the conversion of alcohols into thiols [46].



### 1.3.2. From Alkenes

A second industrially important process for aliphatic thiol production is the addition of hydrogen sulfide to alkenes. In principle, both Markovnikov (**I**) and anti-Markovnikov (**II**) adducts can result from this reaction, but formation of one isomer or the other can be favored by the use of suitable catalysts.



The Markovnikov product (**I**) is formed preferentially under ionic conditions in the presence of oxides, sulfides, acids, or elemental sulfur. Branched alkenes react more rapidly than straight-chain or terminal alkenes. Secondary thiols are prepared by the addition of hydrogen sulfide to linear alkenes in the presence of acid catalysts [47]. The analogous reaction of branched alkenes to give tertiary thiols is catalyzed by aluminum chloride, Al/Si oxides [48], zeolites [49], or cation exchange resins [50].

This method is exploited for the industrial production of a variety of thiols. Reaction is carried out in either the gas or liquid phase depending on the catalyst used. The yield of thiol increases with excess hydrogen sulfide (3 – 10 mol per mole of alkene), high pressure ($\leq$ 7 MPa), and relatively low reaction temperature. Dialkyl sulfides are formed as byproducts. *Ethanethiol* is prepared from ethylene and hydrogen sulfide in the gas phase over aluminum oxide catalysts containing potassium tungstate [51] or nickel and tungsten oxides [52]. *Cyclohexanethiol* is obtained with cyclohexene and hydrogen sulfide in the gas phase over nickel sulfide on silica gel [53]. The reaction of cyclohexene with hydrogen sulfide over cobalt molybdate has also been described [54]. The industrial production of "*tert*-dodecyl mercaptan," a mixture of isomeric thiols, starts from oligomers of isobutylene or propylene. In the presence of hydrogen sulfide and a boron trifluoride catalyst, "*tert*-dodecyl mercaptan" can be prepared from triisobutylene at $-40\,°C$ [41] and from propylene tetramer at $60\,°C$ [55]. Aluminum chloride [41], ethyl aluminum dichloride [56], a mixture of boron trifluoride with phosphoric acid and a low molecular weight alcohol [57], and sodium oxide on zeolite [58] have also been described as catalysts. The industrial production of *2,4,4-trimethyl-2-pentanethiol* is similar to that of *tert*-dodecyl mercaptan. Terpenes are also used as industrial starting materials in this reaction. Thus, 1-*p*-menthene-8-thiol [*71159-90-5*], a component of the grapefruit aroma, is obtained selectively and in high yield by the addition of hydrogen sulfide to limonene in the presence of inorganic Lewis acid catalysts [59].

Electron-withdrawing groups, such as the carbonyl, cyano, nitro, carboxyl, or ester functions, activate alkenes toward nucleophilic addition of hydrogen sulfide. Basic catalysts are particularly suitable for these reactions, including amines [60], anion exchange resins [61], and magnesium oxide [61].

$$CH_2{=}CH{-}X + H_2S \xrightarrow{\text{base}} HS{-}CH_2{-}CH_2{-}X$$

$$X = -CO{-}CH_3,\ -CN,\ -NO_2,\ -COOH,\ -COOR$$

Formation of the anti-Markovnikov product (**II**) results from a radical addition of hydrogen sulfide to alkenes ($\rightarrow$ Hydrogen Sulfide, **A13**, p. 471). In industry this reaction is usually initiated with UV light ($\lambda = 253.7$ nm) [62], [63]. Longer wavelength radiation ($\lambda = 300-400$ nm) can also be used in the presence of photosensitizers [62]–[64]. Among the chemical radical initiators, azobisisobutyronitrile is particularly suitable [65]. This can also be used in the presence of accelerators, such as organic phosphines or phosphites [66]. Peroxides, on the other hand, can oxidize the hydrogen sulfide to elemental sulfur, which then acts as a catalyst for the ionic reaction pathway (see above) [67].

The method described is used mainly for the industrial production of primary thiols. The formation of dialkyl sulfides as byproducts is minimized by ensuring that excess hydrogen sulfide is present. Thus, *ethanethiol* is obtained from ethylene and hydrogen sulfide in the liquid phase in the presence of azobisisobutyronitrile and tributylphosphine [68]. *1-Dodecanethiol* is prepared by photochemically catalzyed addition of hydrogen sulfide to 1-dodecene. The olefin is reacted with a five-fold excess of hydrogen sulfide in the presence of $\alpha,\alpha$-diethoxyacetophenone as initiator at $45\,°C$ and 3 MPa [63]. The industrial preparation of *1-octanethiol* is similar to that of 1-dodecanethiol.

### 1.3.3. Other Production Methods

Other methods for the industrial-scale preparation of thiols include the reductive thiolation of carbonyl compounds, epoxide ring-opening with hydrogen sulfide, and nucleophilic substitution of alkyl halides or tosylates.

Aldehydes and ketones are reduced to thiols by hydrogen sulfide and hydrogen in the presence of metal sulfides [69]. This process is used for the industrial production of cyclohexanethiol [70].

$$\overset{\textstyle\diagdown}{\underset{\textstyle\diagup}{}}C{=}O + H_2 + H_2S \xrightarrow{\text{cat.}} \overset{\textstyle\diagdown}{\underset{\textstyle\diagup}{}}CH{-}SH + H_2O$$

Base-catalyzed cleavage of epoxides with hydrogen sulfide gives mercaptoalcohols. Symmetrical bis(1-hydroxyalkyl) sulfides are formed as byproducts. Inorganic bases, amines and anion exchange resins, zeolites [71], or guanidine [72] can be used as catalysts. In industry, preparation of *mercaptoethanol* [*60-24-2*] by addition of $H_2S$ to ethylene oxide is catalyzed by either a cation exchange resin [73] or the byproduct thiodiglycol [74].

$$H_2C{-}CH{-}R + H_2S \xrightarrow{\text{base}} \underset{HS\quad OH}{H_2C{-}CH{-}R}$$

The reaction of alkyl halides or tosylates with metal hydrogensulfides is frequently used to prepare thiols on a laboratory scale [75].

$$R{-}X + M{-}SH \xrightarrow[-MX]{} R{-}SH$$

$$X = Cl,\ Br,\ I,\ -O\text{-tosyl}$$

Alcohols are the usual solvents. The competing formation of sulfide can be suppressed by the presence of hydrogen sulfide or sulfuric acid. Sulfide formation is also strongly inhibited in the reaction of alkyl halides with a mixture of hydrogen sulfide and ammonia or alkylamines [76]. $C_2$ to $C_{12}$ alkanethiols form in essentially quantitative yield from the corresponding bromides and sodium hydrogensulfide in dimethylsulfoxide [77]. The nucleophilic substitution reaction is used industrially to produce *1-dodecanethiol* from 1-chloro- or 1-bromododecane [78].

Substitution of –SH for chlorine occurs particularly readily if the process is activated by the presence of electron-withdrawing groups (e.g., $C=O$, $C\equiv N$) or heteroatoms (O, S, Si) in the $\alpha$-, $\beta$-, or $\gamma$-positions [79]. This method is used to prepare mercaptocarboxylic acids and thiols containing additional functional groups [79], [80].

$$Cl-(CH_2)_n-COOH \xrightarrow[\text{2) HCl}]{\text{1) NaSH}} HS-(CH_2)_n-COOH$$

$$Cl-CH_2CH_2-O-R \xrightarrow{\text{NaSH}} HS-CH_2CH_2-O-R$$

Besides the processes commonly used in industry for thiol production, a number of generally applicable laboratory methods are also available, such as the reaction of an alkyl halide with thiourea, thiosulfate, or dithio- or trithiocarbonate. Details of these methods can be found in the general literature.

## 1.4. Construction Materials, Storage, and Transport

Manufacturing plant and transport facilities for thiols are preferentially made from (coated) normal steel, stainless steel, aluminum, or copper-free alloys. Thiols in containers made of carbon steel should be handled dry and under a blanket of protective gas to prevent the formation of pyrophoric iron–sulfur complexes. Copper and copper-containing alloys are not suitable for use because they are rapidly attacked by thiols.

Stringent demands apply to the impermeability of thiol piping systems. Seals are normally based on asbestos or fluoroplastics.

Regulations in force with respect to the transport of industrially important thiols are summarized in Table 2.

**Table 2.** Transport restrictions applicable to aliphatic thiols

| Thiol | ADR/RID | | IDMG Code | |
|---|---|---|---|---|
| | Class | Number | | |
| $CH_3-SH$ | 2 | 3bt | 2 | UN 1064 |
| $C_2H_5-SH$ | 3 | 18b | 3.1/II | UN 2363 |
| $n\text{-}C_4H_9-SH$ | 3 | 3b | 3.2/II | UN 2347 |
| $t\text{-}C_4H_9-SH$ | 3 | 3b | 3.2/II | UN 1228 |
| cyclo-$C_6H_{11}-SH$ | 3 | 31c | 3.3/III | UN 3054 |
| $n\text{-}C_8H_{17}-SH$ | 3 | 32c | | |
| $n\text{-}C_{12}H_{25}-SH$ | 3 | 32c | | |
| | 6.1 | 21c | | |
| $t\text{-}C_{12}H_{25}-SH$ | 6.1 | 21c | | |

## 1.5. Uses of Aliphatic Thiols

Many aliphatic thiols are important starting materials for the synthesis of crop-protection agents (especially thiophosphoric acid esters) [81], pharmaceuticals, and polysulfides (see Section 1.6.4). They are also widely used as polymerization regulators in rubber and plastics manufacture. *Methanethiol* is important as a starting material for the industrial production of the amino acid DL-methionine, used as an animal-feed additive. Methanethiol is also used as a gas odorizer, as are other low molecular mass thiols (in particular, ethanethiol). For applications of thiolates and chelate-forming thiols see Section 1.2.2.

Examples, drawn mainly from the recent extensive patent literature, are provided below.

**Pharmaceuticals.** Mercaptoamino acids and their derivatives display a broad spectrum of activity. *N*-Isobutyryl- and *N*-pivaloylcysteine are said to exhibit improved bioavailability compared with *N*-acetylcysteine [*616-91-1*], which is used as a mucolytic [82].

Vanadyl–cysteine complexes can be used for oral treatment of insulin-dependent diabetes [83], and substituted cysteinamides for the localized treatment of inflammatory skin diseases [84]. Certain mercaptoamino acids, such as the 2*S*-enantiomer of 1-(3-mercapto-2-methylpropionyl)-L-proline (Captopril [*62571-86-2*]), act as antihypertensives [85].

Other mercaptoamino acid derivatives have been described as analgetics [86] and antiarthritic agents [87].

Diaminodithiols are chelating agents, and thus useful in radioactive diagnosis [88]. Thiol-substituted dioxolanes [89] and cyclohexenes [90] are mucolytics. Mercapto-substituted propanamides exhibit psychotropic [91] and analgesic activity [91], [92].

**Agricultural Chemicals.** Aminodithiols have been described as effective low-toxicity insecticides [93]. Bismuth dithiolates can be used to combat coccidiosis in poultry and as insecticides [94].

Thiol-substituted 2-propanols are fungicides with low toxicity [95]. Bis(dithiobenzyl)nickel complexes absorb IR radiation and can be used as plant-growth regulators [96].

**Other Uses.** *N*-Substituted 1-amino-2-propylmercaptans are used as antioxidants in lubricating oils [97]. 4-Mercaptobutyramides can act as reducing agents in the permanent waving of hair. They are said to be less toxic and accompanied by less noxious odors than mercaptoacetic acid [98]. Mercaptoalkyl esters of acrylic and methacrylic acids are used as binders for metallic fillings for teeth [99]. Many thiol-substituted terpenes [100], esters [101], ketones [102], and mercaptoalkyloxathiolanes and -oxathianes [103] have been patented as fragrances and aroma substances (e.g., for foods, cosmetics, or pharmaceuticals).

## 1.6. Alkyl Sulfides, Disulfides, and Polysulfides

### 1.6.1. Physical and Chemical Properties

*Sulfides* ($R^1-S-R^2$) are the sulfur analogs of ethers, and are thus derivatives of hydrogen sulfide. They were referred to previously as thioethers or sulfanes. Alkyl *disulfides* ($R^1-S-S-R^2$) and *polysulfides* ($R^1-S_x-R^2$) are derived from di- and polysulfanes. Symmetrical sulfides contain R groups that are identical ($R^1 = R^2$), whereas unsymmetrical ones have different R groups ($R^1 \neq R^2$). Sulfides are named in the same way as ethers (e.g., diethyl sulfide, methyl phenyl sulfide). Di- and polysulfides are named either as sulfides or as sulfane derivatives (e.g., dimethyl disulfide, diethyltrisulfane). Most low molecular mass alkyl sulfides have intense, unpleasant odors. Organic sulfides and disulfides are widely distributed in natural materials. For example, they occur (together with thiols) in onions and garlic [104], and have been found to be the source of various well-known aromas, including those of asparagus and coffee. In polecats, mink, and skunks such compounds act as pheromones [105]. The cyclic polysulfide lenthionine confers a characteristic odor on cooked mutton and the shiitake mushroom [106]. Many disulfides are components of biologically active substances (e.g., lipoic acid, cystine, holomycin) and physiologically important peptides (e.g., insulin, oxytocin). Here the activity is associated directly with the presence of the disulfide group. Aliphatic sulfides are also present in crude petroleum, accounting for up to 45 % of the total sulfur, but disulfides are virtually absent [107].

Aliphatic sulfides, disulfides, and polysulfides are colorless to yellow or orange liquids. Most are insoluble in water, but readily soluble in ether, acetone, chloroform, and ethanol. Physical data for a few representative compounds are provided in Table 3.

Alkyl sulfides, like their oxygen analogs, are very weak bases. Because of nucleophilicity attributable to the free electron pair on sulfur, they form stable addition compounds with halogens and salts of heavy metals. Aliphatic sulfides also form crystalline trialkylsulfonium salts with alkyl halides or dialkyl sulfates.

**Table 3.** Physical data for selected alkyl sulfides, disulfides, and polysulfides

| Sulfide | CAS registry no. | $M_r$ | $bp$, °C/kPa | $d_4^{20}$ | $n_D^{20}$ | Flash point, °C |
|---|---|---|---|---|---|---|
| $(CH_3)_2S$ | [75-18-3] | 62.13 | 37.3 | 0.846 | 1.4355 | −36 |
| $(CH_3CH_2)_2S$ | [352-93-2] | 90.19 | 92.1 | 0.8362 | 1.4430 | −9 |
| $(CH_2=CH-CH_2)_2S$ | [592-88-1] | 114.21 | 138 | 0.887 | 1.4889 | 46 |
| $[CH_3(CH_2)_2]_2S$ | [111-47-7] | 118.24 | 142 | 0.838 | 1.4487 | 28 |
| $[CH_3(CH_2)_3]_2S$ | [544-40-1] | 146.30 | 188 | 0.838 | 1.4530 | 76 |
| $[CH_3CH_2CH(CH_3)]_2S$ | [626-26-6] | 146.30 | 165 | 0.839 | 1.4500 | 39 |
| $[(CH_3)_3C]_2S$ | [107-47-1] | 146.30 | 149 | 0.815 | 1.4506 | 48 |
| $[CH_3(CH_2)_5]_2S$ | [6294-31-1] | 202.40 | 230 | 0.849 | 1.4587 | >110 |
| $(CH_3)_2S_2$ | [624-92-0] | 94.20 | 109 | 1.046 | 1.5253 | 24 |
| $(CH_3CH_2)_2S_2$ | [110-81-6] | 122.25 | 152 | 0.993 | 1.5073 | 40 |
| $(CH_2=CH-CH_2)_2S_2$ | [2179-57-9] | 146.28 | 187 | 1.008 | 1.5410 | 62 |
| $[CH_3(CH_2)_3]_2S_2$ | [629-45-8] | 178.36 | 226 | 0.9382 | 1.4926 | 93 |
| $(CH_3)_2S_3$ | [3658-80-8] | 126.27 | 42.5/0.4 | 1.2013 | 1.6012 | |
| $(CH_3)_2S_4$ | [5756-24-1] | 158.33 | 59/0.1 | 1.3065 | 1.6612 | |
| $(CH_3)_2S_5$ | [7330-31-6] | 190.39 | 70/0.01 | | | |

$$R^1\!\!-\!\!S + R^3\!-\!X \longrightarrow \overset{R^1}{\underset{R^2}{}}\!\!\overset{+}{S}\!-\!R^3\ X^-$$

$$X = \text{halogen},\ O-SO_2-O-R^3$$

Depending on the reaction conditions, oxidation of sulfides leads to sulfones or sulfoxides [107]. Sulfonic acids are formed under forcing conditions, and these are then further oxidized to sulfuric acid [108].

$$R-S-R \xrightarrow{[O]} R-\overset{O}{\overset{\|}{S}}-R \xrightarrow{[O]} R-\overset{O}{\underset{\|}{\overset{\|}{S}}}-R \xrightarrow{[O]}$$

$$R-SO_3H \xrightarrow{[O]} H_2SO_4$$

Sulfides can be reduced to hydrocarbons or, depending on the reducing agent and the substrate, thiols or hydrogen sulfide.

$$R-S-R \xrightarrow{\text{Red.}} R-H + R-SH \xrightarrow{\text{Red.}} 2R-H + H_2S$$

The sulfide contraction reaction, a desulfurization with simultaneous C−C bond formation, is important in organic synthesis [109].

Reaction conditions determine whether disulfides are oxidized to thiosulfinic or thiosulfonic acid esters or to sulfonic acids.

$$R^1\text{-}S\text{-}S\text{-}R^2 \longrightarrow$$

(with R−C(=O)OOH → R$^1$−S(=O)−S−R$^2$, thiosulfinic acid ester; with Ph−C(=O)OOH → R$^1$−S(=O)$_2$−S−R$^2$, thiosulfonic acid ester; with O$_2$/H$_2$O → R$^1$−SO$_3$H + R$^2$−SO$_3$H)

The reduction of disulfides to thiols is a reversible reaction of great physiological importance. Hydrogen, zinc in the presence of acid, complex hydrides, or sodium are generally selected as the reducing agents [110]. Disulfides can be reduced selectively with sodium hydrogentelluride in good yield under very mild conditions in the presence of other functional groups [111]. Organic polysulfides can also be reduced to thiols.

### 1.6.2. Production of Alkyl Sulfides

Many of the procedures for the synthesis of dialkyl sulfides correspond to those for aliphatic

thiols, including the alkylation of hydrogen sulfide or alkali metal sulfides with alcohols, alkenes, or alkyl halides to give symmetrical sulfides (Section 1.3). *Dimethyl sulfide*, the industrially most important sulfide, is produced by treating hydrogen sulfide with excess methanol over an aluminum oxide catalyst. Unsymmetrical sulfides are formed in a similar way from thiols and their salts [112], [113], but they can also be obtained by alkylation of thiols with acidic C−H compounds [114]. The alkylation of thiols with alkyl halides in the presence of K$_2$CO$_3$ in DMF occurs particularly rapidly, and in almost quantitative yield [115]. Unsymmetrical sulfides can also be made by exchange reactions involving symmetrical sulfides and thiols, catalyzed by phosphotungstic or phosphomolybdic acids [116].

$$R^1-S-R^1 + R^2-SH \xrightarrow{\text{cat.}} R^1-S-R^2 + R^1-SH$$

Cleavage of an epoxide ring with a thiol proceeds analogously to the cleavage with hydrogen sulfide (Section 1.3.3). Bases (such as sodium hydroxide or triethylamine [117]), metal salts (e.g., lithium perchlorate [118]), and acids [119] have all been used as catalysts. The resulting β-hydroxythioethers are important intermediates for a large number of other products.

Alkylthio-substituted carboxylic acids and their thioesters can be prepared from lactones by treatment with thiols in the presence of an acid catalyst at 100−350 °C [120].

$$\text{(lactone)} + R-SH \longrightarrow R-S-(CH_2)_3-COOH +$$

$$R-S-(CH_2)_3-\overset{O}{\overset{\|}{C}}\overset{}{\underset{S-R}{}}$$

The reductive thiolation of carbonyl compounds (Section 1.3.3) can also be carried out with thiols instead of hydrogen sulfide. The borane−pyridine complex in trifluoracetic acid [121] or treatment with boron trifluoride monohydrate followed by triethylsilane [122] can be used to accomplish the reduction.

$$\overset{R^1}{\underset{R^2}{}}\!\!C{=}O + R^3-SH \xrightarrow{\text{cat.}} \overset{R^1}{\underset{R^2}{}}\!\!CH-S-R^3$$

α-Substituted dialkyl sulfides can be prepared from sulfoxides by means of the Pummerer rearrangement (→Acetic Anhydride, **A1**, p. 68). The deoxygenation of sulfoxides to sulfides is a particularly important approach with respect to asymmetric syntheses. Several methods are known for achieving this reduction [123]. The systems sodium iodide–titanium(IV) chloride [124] and sodium iodide–sulfonic acid [125] have recently been described as especially efficient and mild reagents. The selective reduction of various functionalized sulfoxides can be effected using Lawesson's reagent [126] (see Section 1.3.1).

### 1.6.3. Production of Alkyl Disulfides

Dialkyl disulfides are often formed as byproducts in the synthesis of sulfides. However, the most important synthetic method is the mild oxidation of thiols. In industry this oxidation is carried out in aqueous alkaline solution using air or oxygen. The reaction is catalyzed by radical initiators (copper, cobalt, nickel, or iron salts, or UV light) [127]. Even high molecular mass disulfides can be prepared in this way [128]. Since in this case water-insolubility of the products can be a disadvantage, the reaction is sometimes carried out in a two-phase mixture in the presence of phase-transfer catalysts [129]. For the oxidation of tertiary thiols a catalyst system consisting of alkali-metal and/or alkaline-earth hydroxides and cobalt molybdate is particularly suitable [130]. The oxidation of thiols with atmospheric oxygen is also exploited industrially for sweetening crude petroleum distillates (see Section 1.2.2).

Other potential oxidizing agents include hydrogen peroxide, halogens (in particular, iodine) [131], and potassium hexacyanoferrate(III) [132]. The fire hazard is lowered and the yield of dialkyl disulfide increased using carbon tetrachloride in aqueous alkali as the oxidizing agent together with a phase-transfer catalyst [133].

Interest has recently been shown in the oxidation of thiols with nitrogen oxides. Whereas symmetrical sulfides are obtained in good yields with nitric oxide and nitrogen dioxide [134], dinitrogen tetroxide leads first to thionitrites [135], which may then react with an additional equivalent of thiol to give symmetrical or unsymmetrical disulfides [136].

$$R^1-SH \xrightarrow{N_2O_4} R^1-SNO \xrightarrow{R^2-SH} R^1-S-S-R^2$$

The analogous coupling of thiols with iron(III) nitrate on a bentonite carrier to give symmetrical disulfides takes place under very mild conditions [137].

An important method for the production of unsymmetrical disulfides is the thioalkylation of thiols with sulfenyl halides [138].

$$R^1-SH + R^2-S-Cl \longrightarrow R^1-S-S-R^2$$

### 1.6.4. Production of Alkyl Polysulfides

Alkyl polysulfides are produced industrially by treating thiols with sulfur in the presence of bases (alkali-metal hydroxides or tertiary amines) [139]. Mixtures of various polysulfides are frequently formed in this reaction, so the isolation, purification, and characterization of defined compounds can be difficult. Higher molecular mass dialkyl trisulfides can be obtained in very good yields from the corresponding thiols and sulfur with magnesium oxide as the base at 30–100 °C [140]. The reaction of thiols with electrolytically formed $S_x^{2+}$ gives trisulfides in 53–74% yields [141]. Alkyl polysulfides containing an average of five sulfur atoms can be prepared at room temperature in the presence of thioxanthates as catalysts [142].

The reaction of dihaloalkanes with alkali polysulfides gives polymeric polysulfides with the general formula $-[R-S_x-]_n$. These polysulfide rubbers (thioplastics) are widely used in industry because of their resistance toward chemicals and light and their impermeability toward gases ($\rightarrow$ Dental Materials, **A8**, pp. 287–288; $\rightarrow$ Rubber, 3. Synthetic, **A23**, pp. 312–314; $\rightarrow$ Sealing Materials, **A23**, pp. 504–506).

### 1.6.5. Uses of Alkyl Sulfides, Disulfides, and Polysulfides

The most important industrial application of aliphatic sulfides and disulfides is as intermediates in the production of pharmaceuticals, crop-protection agents, and a wide range of other substances [143]. Alkyl sulfide groups themselves are also present in the active components of many pharmaceuticals and crop-protection agents, especially insecticides [144]. *Dimethyl sulfide*, the industrially most important sulfide, is widely used in the form of the borane complex $(CH_3)_2S \cdot BH_3$ [*13292-87-0*] for the reduction of a variety of functional groups [145]. Dimethyl sulfide is also the starting material for production of the solvent dimethyl sulfoxide (DMSO; $\rightarrow$ Sulfones and Sulfoxides, **A25**, pp. 492–494).

Organic disulfides have a variety of applications as well. L-Cystine [56-89-3] and lipoic acid [1077-28-7] are therapeutic agents for the liver. Holomycin and the pharmaceutically active component in garlic, di(1-propenyl) disulfide [146], exhibit antibacterial activity. Dialkyl disulfides with 2–8 carbon atoms per alkyl residue can be used as insecticides and acaricides (→ Acaricides, **A 1**, p. 21), or to combat algae [133]. Higher molecular mass disulfides are used as additives in the plastics industry for chain transfer during polymerization, and as antioxidants (→ Oxidation, **A 18**, p. 289). Di- and polysulfides are used as high-pressure additives for lubricating oils.

# 2. Aromatic Thiols, Sulfides, Disulfides, and Polysulfides

## 2.1. Physical and Chemical Properties

Aromatic thiols, Ar–SH, are sulfur analogs of phenols. This relationship is also manifested in many names, where the prefix thio is added to the trivial name of the corresponding phenol (e.g., thiosalicylic acid, thiocresol). According to IUPAC rules, aromatic thiols are supposed to be referred to as mercaptoarenes (e.g., mercaptoaniline) or arenethiols (e.g., 2-naphthalenethiol).

Aromatic thiols differ only slightly from alkanethiols with respect to acidity, salt formation, odor, and solubility. Many thiophenols are volatile in steam. Table 4 provides an overview of physical properties for several important arenethiols.

**Table 4.** Physical data for several thiophenols

| R | CAS registry no. | $M_r$ | bp, °C/kPa | mp, °C | Flash point, °C |
|---|---|---|---|---|---|
| H | [108-98-5] | 110.18 | 169 | −15 | 50 |
| 4-CH$_3$ | [106-45-6] | 124.21 | 195 | 43 | 68 |
| 4-Cl | [106-54-7] | 144.62 | 206 | 50 | >110 |
| 2,5-Cl$_2$ | [5858-18-4] | 179.07 | 132/3.9 | 28 | >110 |
| 2,6-Cl$_2$ | [24966-39-0] | 179.07 | | 49 | >110 |
| 2-OH | [637-89-8] | 126.18 | 150/3.3 | 34 | >110 |
| 2-O-Me | [7217-59-6] | 140.20 | 99/1.1 | | 95 |
| 2-NH$_2$ | [137-07-5] | 125.19 | 71/0.03 | 20 | 79 |
| 2-COOH | [147-93-3] | 154.19 | | 167 | |

Unlike aliphatic thiols, arenethiols occur only to a very limited extent in natural materials. One reason for this may be the more facile oxidation of aromatic thiols, which are converted into disulfides simply on standing in air. The powerful nucleophilicity of arenethiols is exemplified primarily by substitution and addition reactions, most of which lead to thioethers. Such reactions correspond largely to those of aliphatic thiols (see Section 1.2.2). Unlike phenol derivatives, derivatives of aromatic thiols undergo a number of rearrangements. A review of the reactions of aromatic thiols can be found in [11].

## 2.2. Production of Aromatic Thiols

### 2.2.1. By Reduction

The reduction of sulfonyl chlorides, which are readily accessible from the corresponding aromatic systems by sulfochlorination, is one of the most important methods for producing arenethiols.

Many different reducing agents can be employed. The use of zinc or tin in dilute acid is still important in industry [147]. For example, *thiophenol* [148], *halothiophenols* [149], or *aromatic dithiols* [150] continue to be produced by this method. The resulting levels of ecologically harmful metal salts can be considerably reduced by the use of such catalysts as lead [151] or various harmless metal salts [152].

Sulfonyl chlorides can be hydrogenated directly in the presence of metal sulfide catalysts to the corresponding thiophenols at 150–275 °C and a pressure of 3–17 MPa [153]. This process is used to produce *p-thiocresol*. The preparation of *2,5-dichlorothiophenol* by pressurized hydrogenation of the corresponding sulfonyl chloride over a platinum catalyst is similar [154].

Another commonly used reducing agent is red phosphorus in the presence of hydrogen iodide in aqueous [155] or glacial acetic acid solution [156]. Two-step reduction of a sulfonyl chloride is also practiced on an industrial scale, as in the preparation of thiophenol. Depending on the reaction conditions, sodium arenesulfinates or disulfides are formed initially by sulfite reduction, and these are subsequently converted into thiophenols by pressurized hydrogenation in the presence of metal sulfides [157].

$$Ar-SO_2Cl \xrightarrow[\text{NaOH}]{Na_2SO_3} Ar-SO_2^-Na^+ \longrightarrow Ar-S-S-R$$

$$Ar-SH$$

(with $H_2$, cat. reductions leading to $Ar-SH$)

The reduction of aromatic sulfonic acids with triphenylphosphine in the presence of a reaction accelerator has also been described [158].

The tendency of aromatic thiols to undergo oxidation means that the actual products are often disulfides, so reduction of the latter and the corresponding polysulfides is frequently important. This indirect route to arenethiols often leads to better yields than the direct synthesis despite the additional reaction step. Many reducing agents are suitable for the reductive cleavage of disulfides. Pressurized hydrogenation in the presence of a catalyst is used in industry [159]. For halosubstituted compounds the preferred catalysts are Raney cobalt or palladium on charcoal [160]. Other reducing agents include metallic aluminum in liquid ammonia with inorganic halogen as the catalyst [161], zinc and acid in the presence of another metal [162], lithium aluminum hydride [163], hydrogen sulfide [164], and alkali-metal sulfides [165]. Functionalized aromatic disulfides are reduced selectively and under mild conditions using lithium (tri-*tert*-butoxy)aluminum hydride, where halogen, nitro, and carboxyl groups are spared from attack [166]. High yields of arenethiols are also obtained in the presence of a wide range of functional groups with a phosphine reagent bonded to a polymeric carrier [167]. For other reducing agents, particularly those used in the laboratory, see [2].

### 2.2.2. By Thiolation

A nucleophilic exchange between aromatically bonded halogen and the thiol group occurs under mild conditions only in the presence of other strongly electron withdrawing substituents, such as nitro or sulfonyl groups [168]. Either sodium hydrogensulfide or sodium disulfide is used as the thiolating agent. Poor yields and selectivities are often observed with sodium sulfide, particularly with weakly activated haloaromatics. Chloronitrobenzenes can be converted into the corresponding thiophenols either with retention [165], [169] or simultaneous reduction of the nitro group [170], depending on the reaction conditions.

Polychlorinated benzenes also react with sodium hydrogensulfide to give the corresponding thiophenols [171].

The thiolation of haloaromatics is often used in the industrial preparation of arenethiols. *Pentachlorothiophenol* is obtained from hexachlorobenzene by treatment with sodium sulfide and sulfur in methanol, or with sodium hydrogensulfide [172], [173]. Various halothiophenols can be produced by a similar method in such polar solvents as DMF, sulfolane, dimethylsulfoxide (DMSO), and *N*-methylpyrrolidone [174]. Mercaptoanilines are formed from halonitrobenzenes with simultaneous reduction of the nitro group [175]. Aromatic dithiols can be obtained in a particularly pure state and high yield from polyhalogenated aromatics by reaction with hydrogen sulfide and sulfur in the presence of iron [176]. Sodium alkanethiolates can be used for the thiolation of nonactivated aryl halides. The aryl alkyl sulfide formed initially is converted into the desired aromatic thiol by acid hydrolysis [177]. Nucleophilic substitution of an aryl iodide with thiourea in the presence of transition metal complexes leads to *S*-arylisothiuronium salts, which can by hydrolyzed to arenethiols [178].

Another thiolating agent, one that is also suitable for the production of alkoxythiophenols, is disulfur dichloride. Disulfides are formed initially, and these are then reduced as described above (see Section 2.2.1). Various metal salts can serve as catalysts for this thiolation [179]. Similarly, monohalogenated benzenes can be converted into halothiols via polysulfides [180].

Other preparative methods applicable to arenethiols include the catalytic reaction of phenols with hydrogen sulfide in the gas phase [181] and reductive cleavage of aryl alkyl sulfides [182].

## 2.3. Uses of Aromatic Thiols

The most important industrial applications of arenethiols are as intermediates in the production of pharmaceuticals, agrochemicals [183], dyes, and pigments, and as chemicals for the electronics industry. Thiophenols are used in the rubber industry as polymerization regulators, plasticizers, and stabilizers [184], as well as for rubber reclamation. Aromatic dithiols and mercaptoanilines are used in the plastics industry as monomers and modifiers [185].

Thiophenolamides act as collagenase inhibitors, and can be used for treating arthritic conditions [186]. Thiophenolate-containing preparations can dissolve seals and coatings based on polysulfides [187].

Methyl and ethyl 2-mercaptobenzoates have been patented as scents [188]. Substituted mercaptoanilines are flotation agents for lead and zinc ores [189]. Polythiols can be used as optical materials in plastics [190], and for the production of lightweight reversible electrodes with high theoretical energy density [191]. The use of mercaptophenyl-substituted pyrrolinones as herbicides and growth regulators for crop-protection purposes has been described [192].

## 2.4. Production of Aryl Sulfides, Disulfides, and Polysulfides

The production processes for aromatic sulfides, disulfides, and polysulfides correspond largely to those for the related aliphatic compounds. Symmetrical aryl sulfides can be obtained by arylation of hydrogen sulfide or its salts with alkyl halides [193]. Similarly, unsymmetrical or mixed (aliphatic–aromatic) sulfides are formed from thiols or thiolates on treatment with alkyl or aryl halides [194]. The addition of thiols to multiple bonds [195] and the reaction of sulfenyl chlorides with arenes or phenols [196] also lead to unsymmetrical sulfides. Sulfides are formed as well in the cleavage of disulfides with phenols [197] or aromatic amines [198].

## 2.5. Uses of Aryl Sulfides, Disulfides, and Polysulfides

Aryl sulfides are used mainly as intermediates in the production of biologically active substances and dyes, and as monomers in the plas-

tics industry. The applications mentioned below are derived in most cases from the recent extensive patent literature.

Tetrasul and Mercaptodimethur are examples of commercial insecticides. Other aryl sulfides have been introduced as fungicides [199], herbicides [200], microbial growth inhibitors [201], acaricides and nematocides [202], miticides [203], and gametocides [204]. Aryl thioethers are components of color developers in recording materials [205] and of the photosensitive elements in photocopiers and printers [206]. Certain monomeric aryl sulfides can be polymerized to materials with interesting optical properties, such as transparency or a high refractive index [207]. In the plastics industry arylthio compounds can be used as hardeners [208], antioxidants [209], or heat and light stabilizers [210]. Other aryl sulfides function as lubricating-agent additives [211], hair dyes [212], and flotation agents [213]. Aromatic sulfides are used in the pharmaceutcial sector for treating immunological diseases [214], skin diseases [215], and asthma [216], and also as muscle relaxants [217].

Aromatic disulfides have been patented as stabilizers for polycarbonate plastics [218], corrosion inhibitors in transport systems for crude petroleum and natural gas [219] and antioxidants for use in lubricating agents [220].

Aryl polysulfides in rubber act simultaneously as antioxidants and cross-linking agents [221].

Poly(*p*-arylenesulfides) are industrially important as thermoplastic polycondensates with good mechanical and thermal characteristics [222].

## 3. Heterocyclic Thiols, Sulfides, and Disulfides

### 3.1. Introduction

Heterocyclic thiols and their derivatives are widely distributed in nature. For example, the eggs of such marine invertebrates as sea urchins, starfish, and squid contain relatively high concentrations of mercaptohistidine and the corresponding methyl derivatives, known as "ovothiols" [223]. These compounds function as biological antioxidants in the fertilized eggs [224].

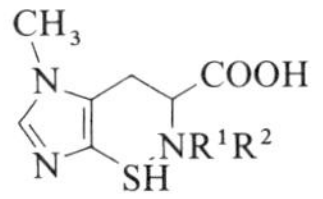

| | $R^1$ | $R^2$ | Ovothiol |
|---|---|---|---|
| | H | H | A |
| | $CH_3$ | H | B |
| | $CH_3$ | $CH_3$ | C |

2-Methyl-3-furanthiol and its disulfide and *S*-methyl derivatives have all been identified as aroma components of cooked meat [225], and they have also been incorporated into synthetic flavorings [226]. 2-Methyl-3-furanthiol has been shown to be present in the aromas of roasted coffee [227] and canned tuna [228]. *S*-Substituted furans, thiophenes, thiazoles, and 3-thiazolines contribute to the odor of yeast extracts [229]. Heterocyclic sulfides occur in cytokinins of the bacteria *Escherichia coli* [230] and in the phallo-toxins, toxic substances from the "death cup" mushroom *Amanita phalloides* [231].

## 3.2. Physical and Chemical Properties

Heterocyclic thiols are generally colorless or pale yellow solids with relatively faint character-istic odors. Table 5 provides an overview of the physical properties of several heterocyclic thiols.

Just like the corresponding aliphatic or aro-matic compounds, heterocyclic thiols form salts with many metals, some of which have therapeu-tic applications (see Section 3.4).

Nitrogen-containing heterocyclic thiols are subject to the same tautomeric equilibrium be-tween thiol and thione forms exhibited by the analogous oxygen species.

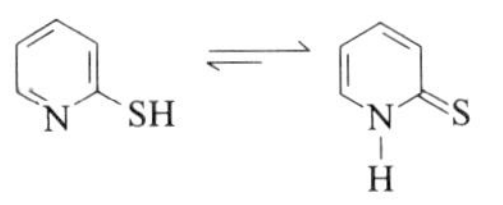

Spectroscopic studies [232] have shown that this equilibrium is displaced in the direction of the thione form. Reactions of heterocyclic thiols can take place from either the thiol or the thione form.

The mild oxidation of heterocyclic thiols leads to disulfides, just as with the corresponding aliphatic and aromatic systems. Catalytic oxida-tion of 2-mercapto-1,3-benzothiazol with atmo-spheric air is not practical, so the corresponding disulfide is prepared industrially by oxidation with sodium chlorate [233].

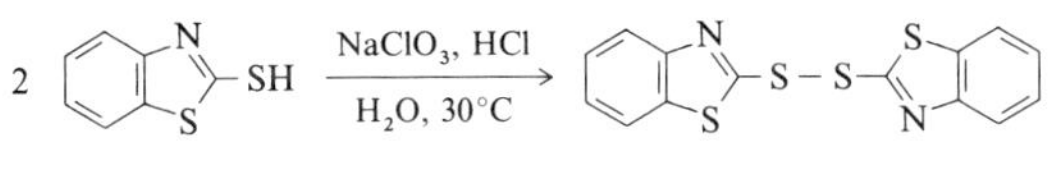

**Table 5.** Physical data for selected heterocyclic thiols

| Formula, name | CAS registry no. | $M_r$ | $mp$, °C |
|---|---|---|---|
| 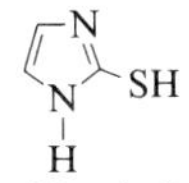 Imidazole-2-thiol | [872-35-5] | 100.14 | 228 – 231 |
| 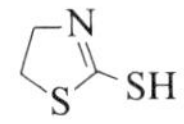 2-Thiazoline-2-thiol | [96-53-7] | 119.21 | 105 – 107 |
| 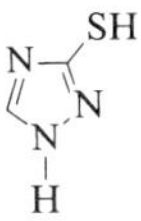 [1*H*]-1,2,4-Triazole-3-thiol | [3179-31-5] | 101.13 | 221 – 224 |
| 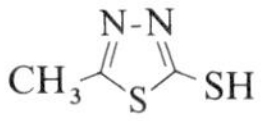 5-Methyl-1,3,4-thiadiazole-2-thiol | [29490-19-5] | 132.21 | 188 – 189 |
| 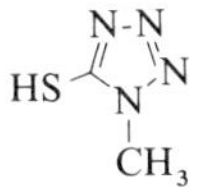 1-Methyl-[1*H*]-tetrazole-5-thiol | [13183-79-4] | 116.15 | 125 – 128 |
| 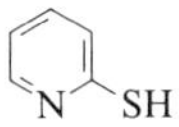 Pyridine-2-thiol | [2637-34-5] | 111.17 | 128 – 130 |
| 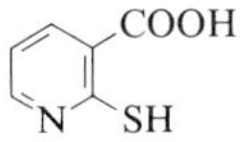 2-Mercaptonicotinic acid | [38521-46-9] | 155.18 | 270 (decomp.) |
| 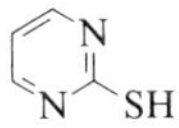 Pyrimidine-2-thiol | [1450-85-7] | 112.15 | 230 (decomp.) |
| 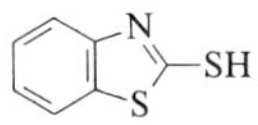 Benzothiazole-2-thiol | [149-30-4] | 167.25 | 182 |
| 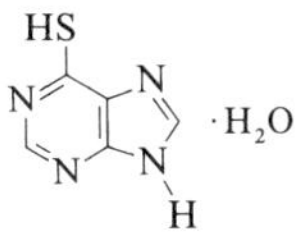 6-Purinethiol hydrate | [6112-76-1] | 170.19 | >300 |
| 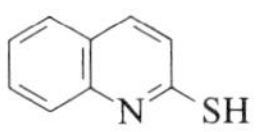 2-Quinolinethiol | [2637-37-8] | 161.23 | 174 – 176 |
| 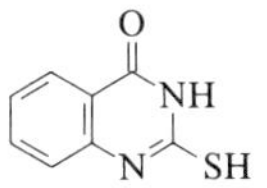 2-Mercapto-4[3*H*]-quinazolinone | [13906-09-7] | 178.21 | >300 |

Stronger oxidizing agents like potassium permanganate or potassium peroxide transform heterocyclic thiols into sulfonic acids [234]. Treatment of 3-mercapto-1,2,4-triazole or 2-mercaptopyridinoimidazole with nitric acid leads to oxidative desulfurization [4].

However, a more widespread desulfurization method for heterocyclic thiols is the reductive approach, in which Raney nickel [235] or sodium borohydride [236] is usually the reagent of choice.

Oxidation of heterocyclic thiols with hypochlorite in the presence of excess amine results in the corresponding sulfenamides [237], and disulfides are subject to similar treatment [238].

Heterocyclic thiols or sulfides are easily transformed into amino or hydrazo derivatives by substitution [239].

$$Het-S-R + H_2N-R' \xrightarrow[-HS-R]{} Het-NH-R'$$

Moreover, heterocyclic thiols are subject just like their aliphatic and aromatic counterparts to such transformations as etherification, esterification, or rearrangement.


## 3.3. Production

A wide variety of methods is available for the preparation of heterocyclic thiols. Thus, a thiol function can either be introduced by substitution into an existing heterocyclic system, or else the heterocycle itself may be constructed by cyclization of a sulfur-containing precursor.


### 3.3.1. Heterocyclic Thiols by Substitution

Generally speaking, heterocycles undergo substitution to thiols under substitution conditions similar to those applicable to activated ben-

zene derivatives. One frequently applied method is the thiolation of a halo-substituted heterocycle, taking advantage of a wide range of thiolation reagents. Particularly suitable in this context are sodium or potassium hydrogensulfide [240] and thiourea [241]. Other possibilities include elemental sulfur [242], thiosulfate [243], alkali thiocyanates, and thiocarboxylic acid derivatives.

Heteroaromatic organomagnesium or organolithium compounds react with sulfur to give thiolates, hydrolysis of which leads to the corresponding thiols. This method permits the preparation of mercaptothiophenes in yields up to 70 % [244].

Similarly, the mercapto function can be introduced into $N$-protected 2-aminopyridines with dibenzylsulfide [245].

Just like aryl amines, heteroaromatic amines are subject to conversion by way of diazonium salts into the corresponding thiols [246].

Heterocycles bearing oxygen substituents can also be transformed into thiols. Suitable reagents include especially phosphorus pentasulfide [247], hydrogen sulfide [248], and the Lawesson reagent [249], just as in the case of alkanethiols (see Section 1.3.1).

Another reaction that has been used to prepare heteroaromatic thiols is the reduction of a sulfonic acid with iodine and triphenylphosphine [250].


### 3.3.2. Heterocyclic Thiols by Cyclization

Cyclocondensations with such sulfur-containing species as carbon disulfide, thiourea, or thiocyanates are especially important reactions for the synthesis of heterocyclic thiols.

**Carbon Disulfide.** 2-Mercapto-1,3-benzothiazoles are prepared on an industrial scale from anilines by cyclocondensation with carbon disulfide and sulfur. One reaction of this type exploited industrially, that of $o$-aminothiophenols, occurs even at low temperature [251].

$o$-Phenylenediamines react analogously to 2-mercaptobenzimidazoles, whereby potassium xanthogenate can be utilized in place of carbon disulfide if so desired [252].

1,2-Diaminoalkanes are transformed by carbon disulfide into 2-mercaptoimidazolines [253].

The cyclization of 2-aminobenzonitriles with carbon disulfide or potassium xanthogenate leads to mercapto-substituted quinazolines [254].

Cycloaddition of carbon disulfide to sodium azide produces the sodium salt of 5-mercapto-1,2,3,4-thiatriazole [255].

**Thiourea.** Condensation of carbonyl compounds with thiourea derivatives represents another important approach to the synthesis of heterocyclic thiols. Thus, cyclocondensation of $\alpha$-hydroxyketones with thiourea results in 2-mercaptoimidazoles [256].

The analogous conversion of $\beta$-dicarbonyl compounds to 2-mercaptopyrimidines has been subject to especially widespread utilization [257].

The usual starting materials are $\beta$-diketones or $\beta$-ketoaldehydes along with the corresponding acetals and acetoacetic or malonic ester derivatives. The latter lead to thiobarbiturates, which are used as short-duration narcotics.

Thiourea reacts with two equivalents of an aldehyde and one equivalent of amine to provide mercapto-substituted 1,3,5-triazines [258].

Thiosemicarbazide undergoes a reaction similar to that of thiourea, and can be transformed with aliphatic or aromatic carboxylic acids into 3-mercapto-1,2,4-triazoles [259].

R: H, alkyl, aryl

**Thiocyanates.** Potassium thiocyanate is a frequently employed reagent that lends itself to cyclization with aminocarbonyl compounds to 2-mercaptoimidazoles [260].

Isothiocyanates react with acylhydrazides in the presence of sodium alcoholate to give 3-mercapto-1,2,4-triazoles [261].

$$R^1-N=C=S + H_2N-NH-CO-R^2 \xrightarrow{NaOC_2H_5} R^1-N$$

Other cyclization processes are described in [262].

### 3.3.3. Heterocyclic Sulfides and Disulfides

The synthetic methods applicable to heterocyclic sulfides and disulfides correspond largely to those for aliphatic and aromatic compounds (see Sections 1.6 and 2.4). Thus, heteroaromatic sulfides can be prepared just like their aromatic analogues by heteroarylation of such inorganic compounds as sodium sulfide or dichlorosulfane [263].

Unsymmetrical sulfides can be obtained from heteroaryl halides by treatment with sodium alkanethiolates or thiols [264]. Good yields are also observed in the palladium-catalyzed reaction of heteroaromatic halogen compounds with organotin sulfides [265].

Sulfenyl chlorides can be transformed into unsymmetrical sulfides with alkenes [266] or arenes [267].

4,5-Dinitroimidazoles are converted in high yield to thioethers through substitution of a nitro group [268].

For the preparation of heterocyclic disulfides see the methods described in Section 1.6.3.

### 3.4. Uses of Heterocyclic Thiols, Sulfides, and Disulfides

Numerous heterocyclic thiols and their sulfur derivatives have found application as intermediates in the synthesis of pharmaceuticals, crop-protection agents, and dyes. Additional applications are described in the sections that follow.

### 3.4.1. Pharmaceuticals

Many heterocyclic thiols and sulfides are among the active ingredients in commercially available drugs. Table 6 presents a few examples.

A wide range of such substances with antibiotic activity has been developed in the last few years, including systems with skeletons that differ from those of penicillin and cephalosporin. A review of the recent patent literature is incorporated into [274]. The description that follows provides a few examples of the broad spectrum of pharmaceutical activity associated with heterocyclic thiols and their derivatives.

Mercapto- (or alkylthio-) triazoles [275], imidazoles [276], imidazolines [277], and pyridines [278] inhibit the activity of dopamine-β-hydroxylases. 7-Thiosubstituted oxoquinolinecarboxylic acids [279], 2-amino-3-mercaptopyridines [280],

**Table 6.** Heterocyclic thiols and sulfides used as pharmaceuticals

| Thiol/Sulfide | Application | Ref. |
|---|---|---|
| 2-Thiobarbituric acid | anaesthetic narcotic hypnotic sedative | [269] |
| 6-Thiopurine | cytostatic agent antileukemic | [270] |
| 2-Thiouracil | thyreostatic | [271] |
| Cephalosporin | antibiotic | [272] |
| 2-Pyridinethiol-*N*-oxide | Na salt antimycotic, Zn salt antiseborrhoeic, disulfide fungicide, bactericide | [273] |

and thiopyridine derivatives of 5-nitro-2-vinyl-furan [281] display antibacterial activity, and the latter are also fungicides. Analgesic, antipyretic, and anti-inflammatory activity has been report-ed for certain thiopyridine derivatives [282], 2-mercaptoimidazoles [283], 3-amino-4-mercap-to-6-methylpyridazines [284], and 3-mercapto-triazoles [285]. Mercapto-substituted imidazoles [286] and triazoles [287] have been patented as medicinal fungicides. 2-Mercaptobenzoxazoles can be used for both prophylaxis and treatment in the case of liver diseases [288]. Various hetero-cyclic thiols and sulfides inhibit the secretion of stomach acid, and can be utilized in ulcer treat-ment [289]. Mercaptopyrrolidine derivatives have been described as cough-prevention agents [290]. Mercaptopyrazolo-pyrimidines and -tri-azines represent potential schistosomicides [291]. Furanyl and pyranyl sulfides display immune-controlling characteristics [292]. Thio-substitut-ed pyridazinones [293], imidazoles [294], and other heterocycles [295] display antiallergic and antiasthmatic activity. The gold salt of thioglu-cose shows in vitro antiviral activity toward the immune-deficiency virus HIV-1 [296].

### 3.4.2. Agricultural Chemicals

Heterocyclic thiols and sulfides are included as crop-protection agents in numerous commer-cial products. For example, several herbicides contain methylthio-1,3,5-triazines [297].

| R | Commercial product |
| --- | --- |
| $- i$-Pr | Semeron |
| $- (CH_2)_3 - O - CH_3$ | Gesaran |
| $-CH_2 - CH_3$ | Ingran |

Other effective herbicides of considerable eco-nomic significance are the 3-methylthio-1,2,4-tri-azine-5(4$H$)-ones, which inhibit photosynthesis [298].

| R | Commercial product | Ref. |
| --- | --- | --- |
| $-NH_2$ | Sencor | [299] |
| $-N=CH-CH(CH_3)CH_3$ | Tantizon | [300] |

Mercapto-substituted pyrazoles [301] and imida-zole-5-carboxylic acids [302] as well as 5-mercap-topyridine-3-carboxylic acid derivatives [303] have also been reported to show herbicidal activ-ity.

A broad spectrum of activity has been report-ed in the extensive patent literature with respect to 1,2-pyridazinones containing thio sub-stituents. For example, they can be used not only as pesticides in veterinary-medicine and crop-protection applications [304], but also as plant fungicides [305]. Other compounds effective as pesticides are derivatives of 1,2,5-thiadiazol-3-thiol [306] and heteroaryl vinyl sulfides [307]. 1-Oxopyridyl disulfides can be utilized as feed supplements in the raising of poultry [308].

### 3.4.3. Other Uses

Heterocyclic thiols and their salts are em-ployed in the photographic sector as stabilizers and as protection against fogging. Patents have been issued in this context covering pyrazoles and triazoles [309], thiadiazoles [310], pyrroli-dines [311], and benzimidazoles [312] with free mercapto groups, but also heterocyclic sulfides [313]. Other photographic auxiliaries contain mercaptopyridines [314].

2-Mercaptobenzothiazole and its derivatives have been used to protect copper and copper alloys against corrosion. Corrosion-inhibiting properties have also been reported for 2-mercap-topyridines [315], 4,5-bismercapto-1,2-dithiol-3-one and its salts [316], and certain pyridine thioethers [317].

A few heterocyclic thiols are effective as sta-bilizers for microbicidal 3-isothiazolones [318].

Mercaptobenzimidazole and its derivatives have proven valuable in preventing the aging of rubber [319]. 2-Mercaptobenzothiazole is an important vulcanization catalyst in the rubber industry. Other compounds suggested for this purpose include mercapto-substituted triazine [320] and pyrimidine [321] derivatives. Deriva-tives of 2,5-dimercaptothiadiazole [322] as well as aminobis- and aminotris(alkylthio)triazines have been suggested as stabilizers for natural and synthetic rubber [323].

Heterocyclic anthraquinone thioethers have been described as dyes for liquid crystals [324]. Thioethers [325] and disulfides [326] of furans and thiophenes are cited in the recent patent lit-erature as flavoring food additives.

2-Mercaptobenzothiazole plays a role in analysis as a reagent for cadmium as well as for the determination of copper, lead, bismuth, sil-ver, mercury, thallium, gold, platinum, and iridi-um. 4-Amino-3-hydrazino-5-mercapto-1,2,4-triazole (Purpald, Aldrich) is a reagent specific for aldehydes [327]. Other analytical applications of heterocyclic thiols and sulfides are discussed in [328].

Cosmetic applications include the use of 2-mercaptopyridine to prevent the formation of melanin in the skin [329], and of certain heterocyclic thiols as active ingredients in deodorants [330].

# 4. Toxicology [331]–[335]

Most thiols are classed as moderately toxic compounds. Since many thiols and sulfides are readily detectable on the basis of their odors at levels far below the danger point, these odors serve to provide a warning. Exceptions include 2,4,4-trimethyl-2-pentanethiol (*tert*-octanethiol), cyclohexanethiol, and thiophenol, which are significantly more toxic than short-chain thiols. Toxicity decreases with longer-chain thiols.

The SH group plays a significant role in biological metabolism (e.g., in metabolite transfer), and for this reason thiols may exhibit either inhibitory or accelerating effects on metabolic processes. A comprehensive treatment of the metabolic, biochemical, and toxicological aspects of xenobiotics containing thiol or sulfide functions is presented in [333].

The discussion that follows constitutes a brief description of the toxic properties of a few representative thiols, sulfides, and disulfides.

*Methanethiol* (Methyl Mercaptan). The greatest risk with respect to this compound involves inhalation, and the corresponding toxic properties are comparable to those of hydrogen sulfide. Low concentrations lead to irritation of the eyes and mucous membranes, headache, nausea, and vomiting. Irritation of the respiratory tract may lead to edema in the lungs. Higher concentrations result in increased blood pressure and pulse rate, coma, severe anemia, and death. Inflammation of the skin (dermatitis) is observed to accompany chronic exposure to low concentrations. MAK: 0.5 ppm (1 mg/m$^3$); TLV: 0.5 ppm.

*Ethanethiol* (ethyl mercaptan) has shown moderately toxic effects in animal experiments in conjunction with inhalation, oral ingestion, or intraperitoneal application. Low concentrations (4 ppm) lead in the case of humans to irritation of the skin and eyes, headache, and nausea, and

respiratory rate also decreases at an exposure level of 50 ppm. Extended exposure to 4 ppm of ethanethiol results in a decrease in odor sensitivity. MAK: 0.5 ppm (1 mg/m$^3$); TLV: 0.5 ppm.

*n-Propanethiol* (propyl mercaptan) has been shown in animal studies to display little toxicity upon oral administration, and it is essentially nontoxic with respect to inhalation. Its effect as an irritant to mammalian skin and eyes is also low. On the other hand, human subjects residing adjacent to a field treated with the pesticide Ethioprop (Mocap) displayed health impairments that persisted for as long as six weeks as a result of the release of propanethiol. The symptoms described included headache, sore throat, diarrhea, and irritation of the eyes, as well as hay fever and asthma attacks [336].

*Butanethiols.* The inhalation of 50–500 ppm of *n-butanethiol* over a period of one hour led in the case of seven human subjects to modest to severe signs of poisoning, including such symptoms as weakness, increased respiratory rate, pain in the neck, nausea, and depression of the central nervous system (diminished reaction capability). MAK: 0.5 ppm (1.5 mg/m$^3$); TLV: 0.5 ppm. *tert-Butanethiol* has been shown in animal experiments to be somewhat less toxic than the straight-chain isomer. Both butyl thiols are eye irritants. MAK and TLV values for these compounds have not yet been established.

*Cyclohexanethiol* causes eye irritation and severe skin damage. Animal experiments (rats) involving oral ingestion have revealed a low LD$_{50}$ of 558 mg/kg.

*tert-Octanethiol* has been found in animal studies to be toxic by oral ingestion, inhalation, or intraperitoneal application, and moderately toxic upon resorption through the skin. The lowest lethal dose (intraperitoneal) is reported for rats at 11 mg/kg.

*Dodecanethiols.* The toxicity of *n-dodecanethiol* is low. Chronic exposure has been observed to produce skin irritation and sensitization. In the case of *tert-dodecanethiol* (isomeric mixture) animal studies based on oral administration produced a very low LD$_{50}$ of 309 mg/kg. Skin and eye irritation was also noted.

*Thiophenol* (phenyl mercaptan) has a severe irritating effect on the skin and eyes, and can lead to acute dermatitis. Headache and dizziness may accompany working with the compound. Animal studies have shown that chronic inhalation of high concentrations may cause damage to the liver, lungs, and kidneys. No MAK value has yet been established; TLV: 0.5 ppm. Thiophenol as well as *o-* and *p-aminothiophenols* cause the transformation of oxyhemoglobin into methemoglobin in human erythrocytes [337]. *p-Chlorothiophenol* and *pentachlorothiophenol* cause severe eye irritation, and the former may be carcinogenic.

*Alkyl sulfides* are generally described as moderately toxic compounds that may lead to hemolytic anemia and allergic dermatitis. *Dimethyl sulfide* is responsible for skin irritation and severe eye irritation. Certain diaryl sulfides [e.g., 2,2′-thiobis(4,6-dichlorophenol) and 4,4′-thiodianiline] are suspected of being carcinogenic.

*2-Mercaptobenzothiazole* has a sensitizing effect, and with chronic exposure (e.g., through the use of rubber gloves) it may induce skin reactions [338]. *2-Mercapto-1-methylimidazole* displays teratogenic effects in humans, and it is suspected of being a carcinogen. *6-Mercaptopurine derivatives* influence cell division, and have been employed as cytostatic agents. They have been found to be teratogenic in animal studies. *Methylthio-substituted triazines*, which are used as herbicides, are only slightly toxic. The corresponding $LD_{50}$ or $LC_{50}$ values derived from animal studies are above the level of 1000 mg/kg [339]. Similar values have been established for a methylthiotriazinone that is the active ingredient in the herbicide Sencor [340].

The professional literature should be consulted regarding the diverse toxic side effects of the many pharmacologically active heterocyclic thiols and sulfides.

# 5. References


**General References**

[1] *Ullmann*, 4th ed., **23**, 175.
[2] *Houben-Weyl*, **E11**, 32, 129; **9**, 55.
[3] N. Kharasch (ed.): *Organic Sulfur Compounds*, Pergamon Press, New York 1961–1966.
[4] G. C. Barrett in: *Comprehensive Organic Chemistry*, vol. 3, Pergamon Press, Oxford 1979, p. 3.
[5] J. L. Wardell in S. Patai (ed.): *The Chemistry of the Thiol Group*, Wiley-Interscience, London 1974.
[6] *Kirk-Othmer*, **22**, 946.
[7] H. Goldwhite in: *Rodd's Chemistry of Carbon Compounds*, 2nd ed., vol. IB, Elsevier Publ., Amsterdam 1965, p. 73.
[8] D. W. Brown in: *Rodd's Chemistry of Carbon Compounds*, 2nd ed., vol. IB, 2nd Suppl., Elsevier Publ., Amsterdam 1991, p. 344.
[9] A. Ohno in S. Oae (ed.): *Organic Chemistry of Sulfur*, Plenum Press, New York 1977, p. 119.
[10] S. R. Sandler, W. Karo in H. H. Wasserman (ed.): *Organic Functional Group Preparations*, 2nd ed., vol. 1, Academic Press, New York 1983.
[11] *Rodd's Chemistry of Carbon Compounds*, Elsevier Publ., Amsterdam, 2nd ed., vol. 3A, 1971, p. 421; Suppl. 3A, 1983, p. 241.

**Specific References**

[12] A. J. Virtanen, *Angew. Chem.* **74** (1962) 374; *Angew. Chem. Int. Ed. Engl.* **1** (1962) 199.
[13] A. E. Johnson, H. E. Nursten, A. A. Williams, *Chem. Ind. (London)* **1971**, 556.
[14] E. Demole, P. Enggist, G. Ohloff, *Helv. Chim. Acta* **65** (1982) 1785.
[15] H. Yanagawa et al., *Tetrahedron Lett.* **13** (1972) 2549–2552.
[16] E. E. Beach, A. White, *J. Biol. Chem.* **127** (1939) 87.
[17] T. Janowski, J. Gawlik, S. Zimnal, *VDI Ber.* **226** (1975) 123.
[18] E. H. Phelps, *Proc. World Pet. Congr. 7th.*, **9** (1967) 201.
[19] W. C. Zeise, *J. Prakt. Chem.* **1** (1834) 257.
[20] C.-D. S. Lee, W. H. Daly, *Adv. Polym. Sci.* **15** (1974) 123.
[21] H. V. Aposhion, *Annu. Rev. Pharmacol. Toxicol.* **23** (1983) 193; L. Magos, *Br. J. Pharmacol.* **56** (1976) 479.
[22] B. L. Sharma et al., *Jpn. J. Pharmacol.* **45** (1987) 295–302.
[23] Pädagogische Hochschule, N. K. Krupskaja, DD 276481, 1990 (K. Andras, H. R. Hoppe, R. Sziburies, H. Selzer).
[24] Sakai Chem. Ind., US 4873351, 1989 (K. Fujita, T. Wachi, Y. Ikeda).
[25] B. Krebs, G. Henkel, *Angew. Chem.* **103** (1991) 785–804. J. H. Dawson, M. Sono, *Chem. Rev.* **87** (1987) 1255–1276.
[26] UOP, US 4490246, 1983 (T. A. Verachtert); US 4498977, 1983 (R. R. Frame).
[27] Co. Française de Raffinage, EP 91845, 1983 (C. Marty, P. Engelhard).
[28] Inst. Française du Petrole, FR 2560889, 1984 (P. H. Bigeard, J. P. Franck, P. Ganne, H. Mimoun).
[29] Y. M. Torchensky: *Sulfur in Proteins*, Pergamon Press, Oxford 1981. H. Sies, *Naturwissenschaften* **76** (1989) 57.
[30] Monsanto, US 4966731, 1990 (Y. Chou).
[31] Pennwalt, EP 313939, 1988 (A. Husain, G. A. Wheaton).
[32] Atochem North America, EP 424616, 1990 (A. Husain, G. A. Wheaton).
[33] E. Campaigne in N. Kharasch (ed.): *Organic Sulfur Compounds*, vol. 1, Pergamon Press, New York 1961, chap. 14.
[34] S. Ahmad, J. Iqbal, *Tetrahedron Lett.* **27** (1986) 3791. T. Imamoto, M. Kodera, M. Yokoyama, *Synthesis* **1982**, 134. T. Kunieda, Y. Abe, M. Hirobe, *Chem. Lett.* **1981**, 1427.
[35] *Kirk-Othmer*, **22**, 953. D. J. R. Massy, *Synthesis* **1987**, 589.
[36] P. Sabatier, A. Mailhe, *C. R. Acad. Sci.* **150** (1910) 1217. R. C. Kramer, E. E. Reid, *J. Am. Chem. Soc.* **43** (1921) 880.

[37] H. O. Folkins, E. L. Miller, *Proc. Am. Pet. Inst. Sect. 3* **42** (1962) 188.

[38] H. O. Folkins, E. L. Miller, *Ind. Eng. Chem. Prod. Res. Dev.* **1** (1962) 271.
A. V. Mahkina et al., *React. Kinet. Catal. Lett.* **36** (1988) 159.

[39] Pennwalt, US 3035097, 1962 (T. E. Deger, B. Buchholz, R. H. Goshorn).

[40] O. Weisser, S. Landa: *Sulphide Catalysts, their Properties and Applications,* Pergamon Press, Oxford 1973, p. 268.

[41] P. Bapsères, *Chim. Ind. (Paris)* **90** (1963) 358.

[42] Toyo Chem. JP 50040511, 1973 (S. Ohno, R. Orita, S. Akita).

[43] M. P. Balfe, J. Kenyon, C. E. Scarle, *J. Chem. Soc.* **1950**, 3309.

[44] Mitsui Toatsu, JA 51048606, 1974 (T. Hayakawa, T. Ichikawa).

[45] Albright & Wilson, GB 917921, 1963 (H. Coates, P. A. T. Hoye).

[46] T. Nishio, *J. Chem. Soc., Chem. Commun.* **1989**, 205.

[47] Societé Nationale Elf Aquitaine (Production), Demande brevet français 8801879, 1988 (E. Arretz).

[48] Pennwalt, US 2951875, 1960 (B. Loev, R. H. Goshorn).

[49] Pennwalt, US 4102931, 1978 (B. Buchholz).

[50] Societé Nationale Elf Aquitaine (Production), EP 101356, 1982 (E. Arretz, A. Mirassou, C. Landoussy, P. Augé).

[51] Pennsalt Chem., NL 142406, 1963.

[52] Societé Nationale Elf Aquitaine (Production), FR 2330674, 1975 (E. Arretz, A. Pfister).

[53] Seitetsu, JP 79132551, 1978 (M. Kawamura et al.).

[54] Phillips Petroleum, US 3963785, 1975 (D. H. Kubicek).

[55] J. F. Frantz, K. I. Glass, *Chem. Eng. Prog.* **59** (1963) 68.

[56] M. A. Korshunov, SU 518489, 1974; *Chem. Abstr.* **85** (1977) 159410.

[57] Phillips Petroleum, US 3214386, 1965 (R. D. Franz, P. F. Warner).

[58] Pennwalt, BE 866426, 1977 (B. Buchholz).

[59] Ogawa, JP-Kokai 63201162, 1987 (H. Masuda, H. Kikuiri, S. Mihara).

[60] R. Dahlbom, *Acta Chem. Scand.* **5** (1951) 690.
Soda Aromatic Co., JP 1052711, 1987 (T. Shirakawa, K. Kinoshita, K. Suzuki, T. Eto).

[61] Phillips, EP 208323, 1985 (J. S. Roberts).

[62] Phillips, US 3050452, 1962 (R. P. Louthan).

[63] Pennwalt, US 4140604, 1978 (D. A. Dimmig).

[64] Shell, US 2411983, 1946 (W. E. Vaughan, F. F. Rust).
Societé Nationale Elf Aquitaine (Production), EP 5400, 1979 (J. Ollivier, G. Souloumiac, J. Suberlucq).

[65] Du Pont, US 2551813, 1949 (P. S. Pinkney).

[66] Stauffer, US 3689568, 1969 (R. J. Eletto).

[67] F. W. Stacey, J. F. Harris, *Org. React. (N.Y.)* **13** (1963) 150.

[68] Nippon Oils & Fats, US 3459809, 1969 (S. Ishida, T. Yoshida).
Stauffer, BE 750555, 1969 (D. J. Martin, E. D. Weil).

[69] M. W. Farlow, W. A. Lazier, F. K. Signaigo, *Ind. Eng. Chem.* **42** (1950) 2547.
Phillips Petroleum, US 3994980, 1976 (D. H. Kubicek).

[70] Seitetsu, JP 79135747, 1979 (M. Kawamura et al.).

[71] Pennwalt, US 4281202, 1980 (B. Buchholz, C. B. Welsh, H. C. Henry).

[72] Societé Nationale Elf Aquitaine (Production), EP 274934, 1988 (E. Arretz, P. Augé, A. Mirassou, C. Landoussy).

[73] Uniroyal, CA 882159, 1971 (G. S. Pande).

[74] Societé Nationale Elf Aquitaine (Production), US 4507505, 1985 (E. Arretz).

[75] L. M. Ellis, E. E. Reid, *J. Am. Chem. Soc.* **54** (1932) 1674.

[76] J. E. Bittell, J. L. Speier, *J. Org. Chem.* **43** (1978) 1687.

[77] A. M. Vasil'tov, B. A. Trofimov, S. V. Amosova, *Zh. Org. Khim.* **19** (1983) 1339; *Chem. Abstr.* **99** (1983) 70162x.

[78] Dow Corning, BE 834487, 1974 (J. L. Speier).
A. G. Koscova et al., *Zh. Prikl. Khim. (Leningrad)* **38** (1965) 170.

[79] Akzo, US 3927085, 1974 (A. G. Zenkel, A. Toth, H. Nagerlein, G. Meyer).

[80] Stauffer, US 4740623, 1988 (J. B. Heather).
Societé Nationale Elf Aquitaine (Production). Demande brevet français 8804693, 1988 (Y. Labat).

[81] Crown Zellerbach, US 3326980, 1964 (D. W. Goheen).

[82] Draco, EP 317540, 1987 (A. R. Hallberg, P. A. S. Tunek).

[83] Panmedica, EP 305264, 1987 (R. Lazaro, G. Cros, J. H. McNeil, J. J. Serrano).

[84] L. R. Morgan, US 4827016, 1985.

[85] Schering, EP 355784, 1988 (M. F. Czarniecki, L. S. Lehmann).
Societé Civile Bioprojet, EP 419327, 1989 (J. C. Plaquevent et al.).
Squibb and Sons, EP 361365, 1988 (N. G. Delaney).

[86] Ono, EP 341081, 1988 (M. Kawamura, Y. Arai, H. Aishita).

[87] Santen, EP 326326, 1988 (T. K. Morita, S. Mita, T. Iso, Y. Kawashima).
Beecham, EP 322184, 1987 (R. E. Markwell, S. A. Smith, L. M. Gaster).

[88] Nihon Medi-Physics, EP 322876, 1987 (H. Yamauchi et al.).
NeoRx, EP 344724, 1988 (A. R. Fritzberg, A. Srinivasan, D. S. Jones, D. W. Wilkening).
Du Pont, EP 279417, 1987 (P. L. Bergstein, E. H. Cheeseman, A. D. Watson).

[89] Proter, EP 288871, 1987 (P. Braga).

[90] Camillo Corvi, EP 343367, 1988 (P. Braga).

[91] Roussel-Uclaf, EP 280627, 1987 (J. P. Vevert, J. C. Gasc, F. Delevallee, F. Petit).

[92] Dainippon, JP 1149763, 1987 (T. Mimura et al.).

[93] Takeda, JP 64000063, 1986 (H. Uneme, H. Mitsuders, T. Uekado).

[94] Nitto, JP 63225391, 1986 (H. Imazaki et al.).

[95] Takeda, EP 421210, 1989 (K. Itoh, K. Okonogi).

[96] Midori, JP 1061492, 1987 (H. Nakasumi, T. Kitao, M. Oizumi).

[97] M. A. Allakhverdiev, *Neftekhimiya* **30** (1990) 570; *Chem. Abstr.* **114** (1990) 26893r.

[98] L'Oreal, EP 368763, 1988 (J. Maignan, M. Colin, G. Lang).

[99] Kurakay, JP 3246360, 1987 (M. Kawashima, I. Komura, J. Yamauchi).

[100] Firmenich, EP 54847, 1982 (E. P. Demole, P. Engist).

Givaudan, US 3 896 175, 1972 (D. Helmlinger et al.).

Fritzsche Dodge, GB 1 546 283, 1979 (B. J. Willis, F. Fischetti, D. I. Lerner).

International Flavors and Fragrances, EP 369 668, 1988 (R. M. Boden, J. A. McGhie).

BASF K & F, EP 418 680, 1989 (P. A. Christenson, R. G. Eilerman, P. J. Riker, B. J. Drake).

[101] Soda Aromatic, JP 1 052 711, 1987 (T. Shirakawa, K. Kinoshita, K. Suzuki, T. Eto).

[102] Ogawa, JP 1 102 019, 1987 (H. Masuda, H. Kikuiri, Y. Shishido, S. Mihara).

[103] General Foods, US 4 496 600, 1985 (T. H. Parliment).

[104] E. Block, *Spektrum Wissensch.* **1985**, 66.

[105] H. Schildknecht et al., *Angew. Chem.* **88** (1976) 228.

[106] *Römpp Chemie Lexikon*, 9th ed., Thieme Verlag, Stuttgart 1992, p. 2484.

[107] E. B. Krein in S. Patai, Z. Rappoport (eds.): *The Chemistry of Sulphur-Containing Functional Groups*, Suppl. S, Wiley & Sons, New York 1993, p. 984.

[108] E. Block in S. Patai (ed.): *The Chemistry of Ethers, Crown Ethers, Hydroxyl Groups and Their Sulphur Analogues*, part 1, Suppl. E, Wiley & Sons, New York 1980, p. 539.

[109] S. Blechert, *Nachr. Chem. Tech. Lab.* **28** (1980) 577.

[110] S. Oae, H. Togo, *Kagaku (Kyoto)* **38** (1983) 48; *Chem. Abstr.* **98** (1983) 197543g.

[111] F. Kong, X. Zhou, *Synth. Commun.* **19** (1989) 3143.

[112] Kuraray, JP 61 155 365, 1986 (T. Onishi, S. Suzuki, M. Shiono, Y. Fujita).

[113] AS Urals Bashkir, SU 1 574 595, 1988 (U. M. Dzhemilev, A. G. Ibragimov, A. B. Morozov, E. G. Galkin); *Chem. Abstr.* **114** (1991) 5808b.

[114] Humboldt-Universität Berlin, DD 266 350, 1987 (E. Wenschuh et al.).

[115] J. M. Khurana, P. K. Sahoo, *Synth. Commun.* **22** (1992) 1691.

[116] Phillips Petroleum, US 4 837 192, 1989 (J. S. Roberts).

[117] M. E. Peach in S. Patai (ed.): *The Chemistry of the Thiol Group*, Wiley, New York 1974, p. 771.

E. J. Corey et al., *J. Am. Chem. Soc.* **102** (1980) 1436, 3663.

[118] P. Crotti et al., *Synlett* **1992**, 303.

[119] Nippon Mining, JP 4 352 764, 1992 (T. Katagiri, K. Furuhashi).

[120] Atochem, EP 446 661, 1991 (B. Buchholz, R. B. Hager, M. J. Lindstrom).

[121] Y. Kikugawa, *Chem. Lett.* **1981**, 1157.

Y. Kikugawa, JP 5 824 557, 1983; *Chem. Abstr.* **99** (1983) 22095.

[122] G. A. Olah, Q. Wang, N. J. Trivedi, G. K. S. Prakash, *Synthesis* **1992**, 465.

[123] Reviews: a) J. Drabowicz, T. Numata, S. Oae, *Org. Prep. Proced. Int.* **9** (1977) 63.

b) M. Madesclaire, *Tetrahedron* **44** (1988) 6537.

[124] R. Balicki, *Synthesis* **1991**, 155.

[125] J. Drabowicz, B. Dudzinski, M. Mikolajczyk, *Synlett* **1992**, 252.

[126] H. Bartsch, T. Erker, *Tetrahedron Lett.* **33** (1992) 199.

[127] T. J. Wallace et al., *Ind. Eng. Chem. Process Des. Dev.* **3** (1964) 237.

[128] Phillips Petroleum, EP 316 942, 1989 (S. H. Presnall).

[129] Shell, EP 288 104, 1988 (E. M. Van Kruchten, C. M. J. Leenaars).

[130] Phillips Petroleum, US 4 277 623, 1981 (D. H. Kubicek).

[131] M. Goodrow et al., *Tetrahedron Lett.* **23** (1982) 3231.

J. Nakayama et al. *Sulfur Lett.* **1** (1982) 25.

J. Drabowicz, M. Mikolajczyk, *Synthesis* **1980**, 32.

[132] F. S. Jørgensen, J. P. Snyder, *J. Org. Chem.* **45** (1980) 1015.

[133] Organic Chem. Tech. Inst., SU 1 712 358, 1992 (O. E. Orlov, L. V. Kaabak, B. A. Kusnetsov, Y. I. Baranov).

[134] A. Pryor, D. F. Church, C. K. Govindan, G. Crank, *J. Org. Chem.* **47** (1982) 156.

[135] Waku Pure Chem. Ind., JP 59 172 456, 1984 (S. Daikiyo, Y. Kin, D. Fukushima).

[136] Waku Pure Chem. Ind., JP 59 172 457, 1984 (S. Daikiyo, Y. Kin, D. Fukushima).

S. Oae et al., *J. Chem. Soc., Perkin Trans. 1* **1978**, 913.

[137] A. Cornélis, N. Depaye, A. Gerstmans, P. Laszlo, *Tetrahedron Lett.* **24** (1983) 3103.

[138] L. Field, W. S. Hanley, I. McVeigh, *J. Org. Chem.* **36** (1971) 2735.

Bayer, DE 3 014 717, 1981 (R. Schubart, U. Eholzer, T. Kempermann, E. Roos).

[139] Rhein Chemie, EP 25 944, 1979 (K. Morche, K. Nuetzel, M. Sauerbier, K. Schilling).

[140] Nippon Soda, WO 83/2 773, 1983 (Y. Koyama, H. Nakazawa, T. Yamashita, H. Moriyama).

[141] Electricité de France, FR 2 669 042, 1992 (G. Le Guillanton, T. Do Quang, J. Simonet).

[142] Societé Nationale Elf Aquitaine (Production), EP 356 318, 1990 (Y. Vallee, Y. Labat).

[143] Nippon Soda, WO 90/543, 1990 (S. Tazaki et al.).

[144] Squibb, US 4 888 424, 1989 (J. E. Sundeen, D. Floyd, V. G. Lee).

Troponwerke, DE 3 012 000, 1981 (H. Jacobi, W. Opitz).

[145] H. C. Brown, S. U. Kulkarni, *J. Org. Chem.* **44** (1979) 2417, 2422.

[146] Nippon Miktron, JP 1 117 856, 1989 (T. Hiramitsu).

[147] R. Adams, C. S. Marvel, *Org. Synth. Coll.* **1** (1964) 504.

[148] Nippon Soda, JA 3 075 165, 1961.

[149] Sumitomo Seika, JP 5 186 418, 1991 (M. Suzuki, N. Kitagishi, A. Fujisawa, J. Ookawa).

[150] Yaroslav Polytechnic Inst., SU 1 421 736, 1987 (E. M. Alov et al.); *Chem. Abstr.* **110** (1988) 192428.

[151] Jenapharm, DD 251 343, 1986 (G. Langbein, H. J. Siemann, S. Berbig, A. Bocker).

Sumitomo Seika, JP 3 153 664, 1989 (T. Morishita, M. Sato, T. Yagi, K. Kaneda).

[152] Sumitomo Seika, JP 3 170 456, 1989 (T. Morishita, M. Sato, N. Kitagishi, A. Onoe).

[153] K. Itabashi, *Yuki Gosei Kagaku Kyokaishi* **19** (1961) 601; *Chem. Abstr.* **55** (1961) 23412.

[154] Ciba-Geigy, EP 2 755, 1978 (H. Thies, F. von Kaenel).

[155] F. Vögtle, R. G. Lichtenthaler, M. Zuber, *Chem. Ber.* **106** (1973) 719.

Bayer, DE 1 643 385, 1967 (H. Leuchs, H. W. Hammen).

[156] A. W. Wagner, *Chem. Ber.* **99** (1951) 375.

J. Morgenstern, R. Bayer, *Z. Chem.* **8** (1968) 106.

[157] Uniroyal, NL 6 611 430, 1967.

[158] Waku Pure Chem. Ind., JP 63 054 355, 1989 (S. Daikyo, K. Fujimori, H. Togo).

[159] O. Weisser, S. Landa: *Sulphide Catalysts, their Properties and Applications*, Pergamon Press, Oxford 1973, p. 193.
Seitetsu, EP 474259, 1991 (M. Kawamura et al.).

[160] Bayer, FR 2008330, 1970.

[161] Nippon Kasei, JP 62209057, 1986 (M. Saito, R. Sato).

[162] Konica, JP 62226957, 1986 (S. Sugita, K. Masuda, S. Nakagawa).

[163] W. Rundel, *Chem. Ber.* **101** (1968) 2956.

[164] Phillips Petroleum, US 4005149, 1977 (D. H. Kubicek).

[165] C. C. Price, G. W. Stacy, *J. Am. Chem. Soc.* **68** (1946) 498.

[166] S. Krishnamurthy. D. Aimino, *J. Org. Chem.* **54** (1989) 4458.

[167] R. A. Amos, S. M. Fawcett, *J. Org. Chem.* **49** (1984) 2637.

[168] Bayer, DE 2903505, 1979 (H. Hagemann, E. Klauke, G. M. Petruck).
F. G. Bordwell, H. M. Anderson, *J. Am. Chem. Soc.* **75** (1953) 6019.

[169] Y. Takikawa, S. Takizawa, *Nippon Kagaku Kaishi* **1972**, 756; *Chem. Abstr.* **77** (1972) 5081.

[170] S. K. Jain, D. Chandra, R. L. Mital, *Chem. Ind. (London)* **1969**, 989.

[171] K. R. Langgille, M. E. Peach, *J. Fluorine Chem.* **1** (1972) 407.

[172] Bayer, US 3560573, 1971 (M. Blazejak, J. Haydn).

[173] ICI, DE 2347462, 1972 (D. M. Dawson, D. J. Harper).

[174] Sumitomo, JP 4182463, 1990 (M. Suzuki, H. Hata, M. Yoshikawa, S. Oe).

[175] Cassella, EP 409009, 1990 (W. Bauer, W. Steckelber).
Hoechst, DE 2127898, 1971 (S. Planker, K. Baessler).
Bayer, DE 1670760, 1966.

[176] Mitsui Toatsu, JP 5117225, 1991 (T. Kobayashi et al.).
Ricoh, JP 4074160, 1990 (R. Momiyama, K. Nagai).

[177] Phillips Petroleum, US 4978796, 1990 (J. E. Shaw).
J. E. Shaw, *J. Org. Chem.* **56** (1991) 3728.

[178] K. Takagi, *Chem. Lett.* **1985**, 1307.
Nippon Chem. Ind., JP 62201862, 1986 (K. Takagi).

[179] Konishiroku Photo, JP 62036354, 1985 (S. Sugita, K. Masadu, S. Nakagawa).
Konishiroku Photo, JP 62067063, 1985 (S. Sugita, K. Masuda, S. Nakagawa).

[180] Sumitomo, JP 5140086, 1992 (M. Suzuki, N. Kitagishi, S. Kimura, A. Fujisawa).

[181] Monsanto, US 4088698, 1978 (N. A. Fishel, D. A. Gross).
Mitsubishi Petrochem., JP 8036409, 1980 (A. Sakurada, N. Hirowatari).

[182] W. E. Truce, J. J. Breiter, *J. Am. Chem. Soc.* **84** (1962) 1621.
A. Ferretti, *Org. Synth. Coll.* **5** (1973) 419.

[183] Sumitomo, AU 8814024, 1988 (T. Haga, S. Hashimoto, E. Nagano, R. Yoshida).

[184] Soc. France Hoechst, EP 368697, 1989 (Y. Christidis).

[185] Mitsubishi Gas Chem., JP 2075656, 1988 (K. Okabe, S. Ametani, S. Murabayashi).

[186] Beecham, EP 358305, 1990 (I. Hughes).

[187] Commonwealth of Australia, WO 87/01724, 1986 (W. Mazurek).

[188] Naarden, EP 219901, 1986 (P. J. Devalois, H. J. Wille, H. L. Vandenheuv).

[189] Consiglio Nat. Ricerche, DE 3601286, 1986 (G. Bornengo, F. Carlini, A. Marabini, V. Alesse).

[190] Showa Denko, JP 3212430, 1990 (T. Arakawa, N. Minorikawa, S. Maruyama, H. Takoshi).
Mitsui Toatsu, JP 63165392, 1986 (K. Sasagawa, M. Imai).

[191] Matsushita, EP 569037, 1993 (Y. Sato, T. Sotomura, K. Takeyama, H. Uemachi).

[192] Bayer, EP 408891, 1989 (H. L. Elbe et al.).

[193] Bayer, EP 302321, 1988 (H. Hagemann, K. Sasse, R. Fischer).

[194] Ciba-Geigy, EP 258190, 1987 (C. Frey, B. Dill).
BASF, DE 1593769, 1967 (H. Eilingsfeld, H. Scheuermann).

[195] Hoechst, DE 4131355, 1991 (J. Dannheim).
Chisso, JP 3287573, 1990 (T. Onishi).

[196] Nisso Petrochem Ind., JP 62077361, 1985 (M. Hoyko, K. Kokoma, T. Hida).
Fuji Photo Film, JP 2286655, 1989 (T. Kawagishi, Y. Shimada, H. Tamai).

[197] DuPont, US 4792633, 1986 (P. W. Wojtkowski).
Societé Nationale Elf Aquitaine, EP 318394, 1988 (C. Vottero, Y. Labat, J. M. Poirier).

[198] Ethyl Corp., US 4670597, 1985 (P. F. Ranken, R. L. Favis).
Ethyl Corp., WO 92/15560, 1992 (D. E. Ballhoff, J. F. Ballhoff).

[199] Shell, EP 411720, 1990 (A. C. G. Gray, T. W. Naisby, S. J. Turner, T. M. Machin).
Sumitomo, JP 5039257, 1991 (S. Tsuyunashi, K. Kumagai, N. Nabeshima, M. Kato).
BASF, EP 379098, 1990 (H. Wingert et al.).

[200] Stauffer, ICI, EP 249150, 1987 (C. G. Knudsen).
Nihon Bayer, JP 5117109, 1991 (T. Goshima et al.).
BASF, EP 402751, 1990 (U. J. Vogelbacher et al.).
ICI, GB 2225014, 1989 (D. Cartwright).
Rhone-Poulenc, EP 351332, 1989 (P. Desbordes, M. Euvrard).

[201] Dow Chem., WO 91/11503, 1991 (C. J. Deford et al.).

[202] Hoechst, EP 336199, 1989 (M. Kern et al.).
Kureha, EP 255935, 1987 (S. Kato, T. Hayaoka).

[203] Sumitomo, JP 5229905, 1992 (T. Takagaki, H. Kishida, K. Umeda, T. Ishiwatari).

[204] Plant Science Res. Inst., SU 1632393, 1980 (M. A. Fedin, T. K. Alsing, T. A. Kuznetsova).

[205] Ricoh, JP 62011676, 1985 (K. Taniguchi, H. Furuya).
Fuji Photo Film, DE 3704627, 1987 (M. Takashima, M. Satomura, K. Ichikawa).

[206] ICI, EP 427404, 1990 (P. Mistry, P. Patel).

[207] Mitsubishi, EP 424144, 1990 (Y. Tamura, F. Watari).
Toray, Green Cross, JP 2289622 (K. Kato, C. Ohama).

[208] Bayer, EP 570798, 1993 (K. Hentschel, U. Walter).
Ethyl Corp., US 4757119, 1987 (P. L. Wiggins, G. G. Knapp).

[209] Sumitomo, JP 4031489, 1990 (H. Hayami).
Bayer, DE 3938422, 1989 (B. Kohler, K. Stahlke, A. Sommer, K. Sommer).

[210] Ciba-Geigy, EP 307358, 1988 (H. O. Wirth, H. H. Friedrich).

[211] Ciba-Geigy, EP 224442, 1986 (H. R. Meier, G. Knobloch).

Ethyl Petroleum, US 4946610, 1989 (W. Y. Lam, G. P. Liesen).
Lubrizol, WO 88/09804, 1988 (M. F. Salomon).

[212] Clairol, Bristol Myers, US 4799934, 1987 (M. I. Lim, J. S. Anderson).

[213] Phillips Petroleum, US 5132008, 1991 (M. D. Bishop, J. E. Shaw).

[214] SmithKline Beecham, EP 327306, 1989 (J. G. Gleason et al.).
Searle, EP 293895, 1988 (R. A. Mueller, R. A. Partis).
Ciba-Geigy, EP 228045, 1986 (A. Vonspreche et al.).
Lilly, GB 2184121, 1986 (S. R. Baker, A. Todd).
Kanegafuchi, EP 211363, 1986 (T. Shiraishi et al.).

[215] BASF, EP 386452, 1990 (B. Janssen, H. H. Wuest).

[216] Green Cross, EP 120426, 1988 (Y. Naito et al.).

[217] Taisho, JP 4145062, 1990 (K. Tomizawa et al.).

[218] Bayer, Miles, US 5214078, 1992 (G. Fennhoft et al.).

[219] Kennelley, CA 2058266, 1991 (K. J. Kennelley et al.).

[220] Lubrizol, US 4894174, 1988 (M. F. Salomon).

[221] Hüls, DE 3941002, 1989 (G. Horpel, K. H. Nordsiek).

[222] Bayer, EP 23313, 1979 (K. Idel, J. Merten).
Hay, US 5239052, 1990 (A. S. Hay, Z. Y. Wang).

[223] F. Rossi, G. Nardi, A. Palumbo, G. Prota, *Comp. Biochem. Physiol. B Comp. Biochem.* **80B** (1985) 843.
E. Turner, R. Klevit, L. J. Hager, B. M. Shapiro, *Biochemistry* **26** (1987) 4028.

[224] E. Turner, L. J. Hager, J. Lisa, B. M. Shapiro, *Science (Washington D.C.)* **242** (1988) 939.
T. P. Holler, P. B. Hopkins, *J. Am. Chem. Soc.* **110** (1988) 4837.

[225] U. Gasser, W. Grosch, *Z. Lebensm. Unters. Forsch.* **190** (1990) 3.
G. MacLeod, J. Ames, *Chem. Ind. (London)* **1986**, 175.

[226] U. Gasser, W. Grosch, *Z. Lebensm. Unters. Forsch.* **190** (1990) 511.

[227] J. Blank, A. Sen, W. Grosch, *Z. Lebensm. Unters. Forsch.* **195** (1992) 239.

[228] D. A. Withycombe, C. J. Mussinan, *J. Food Sci.* **53** (1988) 658, 660.

[229] P. Werkhoff et al., *Chem. Mikrobiol. Technol. Lebensm.* **13** (1991) 30.

[230] W. J. Burrows et al., *J. Biochem.* **9** (1970) 1867.

[231] T. Wieland et al., *Liebigs Ann. Chem.* **12** (1981) 2318.

[232] A. R. Katritzky, J. M. Lagowski in A. R. Katritzky (ed.): *Advances in Heterocyclic Chemistry*, vol. 1, Academic Press, New York 1963, p. 396.
A. R. Katritzky, *Chimia* **24** (1970) 134.

[233] Bayer, DE 1108692, 1958 (E. Kracht, F. Eichler, J. Farana).

[234] Y. H. Kim, G. H. You, *Bull. Korean Chem. Soc.* **2** (1981) 122.

[235] G. Pettit, E. E. van Tamelen, *Org. React. (N.Y.)* **12** (1962) 356.
Y. Nakayama et al., *J. Heterocycl. Chem.* **18** (1981) 631.

[236] Y. Sanemitsu et al., *J. Heterocycl. Chem.* **18** (1981) 1053.
S. D. Aster, S. S. Yang, G. D. Berger, *Org. Prep. Proced. Int.* **23** (1991) 658.

[237] Monsanto, FR 1549002, 1969 (J. J. d'Amico).
D. I. Banks, P. Wiseman, *J. Appl. Chem.* **18** (1968) 202.

E. Kühle: *The Chemistry of Sulfenic Acids*, G. Thieme Verlag, Stuttgart 1972.
Monsanto, DE 2201989, 1972 (J. J. d'Amico).

[238] M. D. Bentley et al., *J. Chem. Soc., Chem. Commun.* **1971**, 1625.

[239] E. C. Taylor, W. A. Ehrhart, *J. Am. Chem. Soc.* **82** (1960) 3138.
H. Hartmann, S. Scheithauer, *J. Prakt. Chem.* **311** (1969) 827.

[240] Nissan Chem. Ind., JP 62004272, 1987 (T. Sato, K. Morimoto, S. Yamamoto).
Nissan Chem. Ind., JP 61109777, 1986 (T. Ogura, Y. Kawamura, H. Suzuki).
Ricoh, JP 3048661, 1989 (K. Nagai).
V. Konecny, S. Kovac, *Chem. Pap.* **41** (1987) 827.

[241] Ishihara Sangyo Kaisha, JP 63203632, 1987 (Y. Tsujii, H. Uenishi, T. Kimura).
Ishihara Sangyo Kaisha, JP 1275562, 1988 (R. Nasu, S. Kimura, A. Mori, R. Koto).

[242] T. Kamiyama, S. Enomoto, M. Inoue, *Chem. Pharm. Bull.* **33** (1985) 5184.

[243] Ricoh, JP 4049282, 1990 (K. Nagai, R. Momyama).

[244] E. Jones, I. M. Moodie, *Org. Synth.* **50** (1970) 104.
J. Cymerman-Craig, J. W. Loder, *Org. Synth. coll. vol.* **V** (1973) 667.

[245] Bayer, DE 3545124, 1985 (A. Klausener, G. Fengler).

[246] C. Corral, J. Lissavetzky, A. S. Alvarez-Insua, A. M. Valdeolmillos, *Org. Prep. Proced. Int.* **17** (1985) 163.
Takeda Chem. Ind., EP 220695, 1987 (I. Aoki, T. Kuragano, N. Okajima, Y. Okada).

[247] R. Gompper, W. Elser, *Org. Synth. coll. vol.* **V** (1973) 780.
R. F. Langler et al., *Can. J. Chem.* **49** (1971) 481.
SmithKline Beckman, EP 308159, 1987 (L. I. Kruse, S. T. Ross).

[248] S. D. Aster, S. S. Yang, G. D. Berger, *Org. Prep. Proced. Int.* **23** (1991) 658.

[249] S.-O. Lawesson, *Nouv. J. Chim.* **4** (1980) 47.
K. Clausen, S.-O. Lawesson, *Bull. Soc. Chim. Belg.* **88** (1979) 305.
Poli Ind. Chim., EP 515995, 1992 (S. Poli, G. Coppi, G. Signorelli).

[250] Waku Pure Chem. Ind., JP 1230557, 1988 (S. Daikyo, K. Fujimori, H. Togo).

[251] C. Ferri: *Reaktionen der organischen Synthese*, G. Thieme Verlag, Stuttgart 1978.

[252] J. A. van Allan, B. D. Deacon, *Org. Synth. coll. vol.* **IV** (1963) 569.

[253] C. F. H. Allen, C. O. Edens, J. A. van Allan, *Org. Synth. coll. vol.* **III** (1955) 394.
SmithKline Beckman, WO 89/6126, 1989 (L. I. Kruse, T. B. Leonard, S. T. Ross).

[254] E. C. Taylor, R. N. Warrener, A. McKillop, *Angew. Chem.* **78** (1966) 333.
H. J. Kabbe, *Synthesis* **1972**, 268.

[255] E. Lieber, E. Ofterdahl, C. N. R. Rao, *J. Org. Chem.* **28** (1963) 194.
C. Cristophersen, *Acta Chem. Scand.* **25** (1971) 1162.

[256] Taisho Pharmaceut. Co., JP 2083372, 1988 (Y. Yoshikawa et al.).

[257] H.-C. Chiang, J.-R. Ko, *Heterocycles* **19** (1982) 529.
Dynamit Nobel, EP 46856, 1980 (H. Peeters, W. Vogt).
H. Petersen, *Synthesis* **1973**, 258.
D. G. Crosby, R. B. Berthold, H. E. Johnson, *Org. Synth. coll. vol.* **V** (1973) 703.
H. M. Foster, H. R. Snyder, *Org. Synth. coll. vol.* **IV** (1963) 638.

[258] H. Petersen, *Synthesis* **1973**, 251.

[259] C. Ainsworth, *Org. Synth. coll. vol.* **V** (1973) 1070.

[260] SmithKline Beckman, US 4 798 843, 1987 (L. I. Kruse).
Ciba-Geigy, Janssen Pharmaceutica, EP 277 387, 1986 (G. R. E. Van Lommen et al.).
Janssen Pharmaceutica, EP 275 603, 1986 (W. R. Lutz, W. G. Verschueren, H. Fischer, G. R. E. Van Lommen).

[261] SmithKline Beckman, EP 359 505, 1988 (L. I. Kruse, J. A. Finkelstein).
SmithKline Beckman, EP 323 737, 1978 (L. I. Kruse, J. A. Finkelstein).

[262] Akad. Wissenschaft DDR, DD 252 372, 1986 (A. Rumler et al.).
N. Okajima, Y. Okada, *Synthesis* **1989**, 398.
Technische Universität Dresden, DD 224 320, 1984 (L. Jakisch, H. Viola, R. Mayer).
Karl-Marx-Universität Leipzig, DD 276 283, 1988 (D. Briel et al.).

[263] M. J. Broadhurst, R. Grigg, A. W. Johnson, *J. Chem. Soc., Perkin Trans. 1* **1972**, 1124.

[264] R. A. Rossi, S. M. Palacios, *J. Org. Chem.* **46** (1981) 5300.
Taiho Yakuhin, JP 56 100 777, 1980.
G. M. Kulkarni, V. D. Patil, *J. Indian Chem. Soc.* **67** (1990) 693.
Ube Ind., EP 283 271, 1988 (H. Yoshioka et al.).
J. Matsumoto et al., *Chem. Pharm. Bull.* **38** (1990) 2190.

[265] C. Jixiang, G. T. Crisp, *Synth. Commun.* **22** (1992) 683.

[266] S. Z. Ivin, V. K. Promonenkov, *Zh. Obshch. Khim.* **37** (1967) 489; *Chem. Abstr.* **68** (1968) 114701.

[267] R. D. Schütz, W. L. Fredericks, *J. Org. Chem.* **27** (1962) 1301.

[268] BASF, EP 293 742, 1988 (H. Köhler, T. Dockner).

[269] Abbott, US 2 876 225, 1956 (G. H. Donnison).
E. Miller et al., *J. Am. Chem. Soc.* **58** (1936) 1090.
G. Renzoni, *Fac. Farm.* **36** (1965) 42.
M. Steinmann et al., *J. Med. Chem.* **16** (1973) 1354.
Lilly, US 2 561 689, 1949.

[270] Burroughs Wellcome, US 3 056 785, 1960 (G. B. Elion, G. H. Hitching).
P. Nuhn et al., *Pharmazie* **34** (1979) 285.

[271] A. Kleemann, J. Engel: *Pharmazeutische Wirkstoffe*, vol. 5, G. Thieme Verlag, Stuttgart 1982.
D. A. McGimty, W. G. Bywater, *J. Pharmacol.* **84** (1945) 342.
Parke & Davis, US 2 153 711, 1934.
Ciba, US 2 666 764, 1951 (A. E. Lanzilotti, A. C. Shabica, J. B. Ziegler).
G. W. Anderson et al., *J. Am. Chem. Soc.* **67** (1945) 2197.

[272] P. Actor et al., *J. Antibiot.* **28** (1975) 471.
Eli Lilly, US 3 928 592, 1974 (J. M. Greene, J. M. Indelicato).
Bristol-Myers, US 3 796 717, 1972 (R. Smith, L. Cheney).
Takeda Chem. Ind., DE 2 715 385, 1977 (M. Ochiai, A. Morimoto, Y. H. Matsushita).
Warner-Lambert, EP 15 771, 1979 (T. H. Haskell, D. Schweiss, T. F. Mich, T. P. Culbertson).
Sankyo, DE 2 455 884, 1974 (H. Nakao et al.).
H. Nakao, *J. Antibiot.* **32** (1979) 320.
Eli Lilly, US 3 943 126, 1973 (C. W. Ryan, W. B. Blanchard).

[273] Olin Mathieson, US 2 745 826, 1953 (M. A. Dolliver, S. Semnoff).
Olin Mathieson, GB 761 171, 1956.

[274] Fujisawa, EP 272 456, 1987 (M. Murata et al.).
Sanraku-Ocean, JP 60 163 882, 1984 (T. Yoshioka et al.).
Rhône-Poulenc, EP 248 703, 1987 (J. C. Barriere, C. Cotrel, J. M. Paris).
Schering, EP 238 285, 1987 (M. P. Kirkup, A. S. Boland).
Fujisawa, EP 111 281, 1983 (T. Takao et al.).
Meiji Seika Kaisha, EP 209 751, 1986 (T. Tsuruoka et al.).
Banyu, JP 58 154 588, 1982.
Bristol-Myers, DE 3 312 533, 1983 (C. U. Kim, P. F. Misco, Jr.).
Hoffmann-LaRoche, US 4 348 518, 1978 (M. Montavon, R. Reiner).
Asahi, EP 68 403, 1982 (J. Nishikido, E. Kodama, M. Shibukawa).

[275] SmithKline Beckman, EP 323 737, 1987 (L. I. Kruse, J. A. Finkelstein).
SmithKline Beckman, EP 359 505, 1988 (L. I. Kruse, J. A. Finkelstein).

[276] SmithKline Beckman, US 4 798 843, 1987 (L. I. Kruse).
SmithKline Beckman, EP 371 730, 1988 (L. I. Kruse, S. T. Ross).
SmithKline Beckman, US 4 749 717, 1987 (L. I. Kruse).

[277] SmithKline Beckman, WO 89/6126, 1987 (L. I. Kruse, T. B. Leonard, S. T. Ross).

[278] SmithKline Beckman, EP 308 159, 1987 (L. I. Kruse, S. T. Ross).

[279] J. Matsumoto et al, *Chem. Pharm. Bull.* **38** (1990) 2190.

[280] Bayer, DE 3 545 124, 1985 (A. Klausener, G. Fengler).

[281] D. Vegh, J. Kovac, M. Kriz, CS 223 450, 1982; *Chem. Abstr.* **105** (1986) 226324g.

[282] Degussa, EP 149 088, 1984 (G. Scheffler et al.).

[283] Taisho, JP 2 083 372, 1988 (Y. Yoshikawa et al.).

[284] R. Bluth, *Pharmazie* **37** (1982) 285, 441.
R. Bluth, DD 150 150, 1980.

[285] Warner-Lambert, US 4 962 119, 1989 (D. H. Boschelli et al.).

[286] Sumitomo, EP 54 974, 1981 (I. Saji).

[287] Pfizer, EP 91 309, 1983 (K. Richardson, P. J. Whittle).

[288] Yamanouchi, JP 62 039 583, 1985 (H. Iwamoto, A. Tanaka, K. Nishikiori, H. Ikadai).

[289] Wyeth, John & Brothers, US 4 374 992 (R. Crossley).
Sankyo, JP 3 014 566, 1989 (A. Yoshida, K. Oda, K. Tabada).
Ohta, JP 60 218 375, 1984 (Y. Hasegawa et al.).

[290] Proter, EP 184 861, 1985 (L. Dall'Asta, G. Coppi, M. E. Scevola).
Poli Ind. Chim., EP 515 995, 1992 (S. Poli, G. Coppi, G. Signorelli).

[291] M. R. H. Elmoghayar, *Arch. Pharm. (Weinheim, Ger.)* **316** (1983) 697.

[292] Taiho, JP 56 100 777, 1981.

[293] Nissan Chem. Ind., JP 61 260 018, 1985 (M. Mutsuto, K. Tanigawa, K. Shikada, R. Sakota).

[294] SmithKline Beckman, EP 168 950, 1986 (W. E. Bondinell, D. T. Hill, B. M. Weichman).

[295] Yamanouchi, EP 181 779, 1985 (M. Kiyoshi, M. Toshiyasu, H. Hiromu, T. Kenichi).

[296] T. Okada, B. K. Patterson, S. Q. Ye, M. E. Gurney, *Virology* **192** (1993) 631.

[297] Ciba-Geigy, CH 337019, 1955.
Geigy, DE-OS 1186070, 1963.
H. Gysin, J. R. Geigy, *Chem. Ind. (London)* **1962**, 1393.
Geigy, US 3207756, 1963 (E. Knusli, W. Stammbach).

[298] W. Draber, K. Dichoré, K. H. Büchel, *Naturwissenschaften* **55** (1968) 446.

[299] Bayer, US 3671523, 1970 (K. Westphal, W. Meiser, L. Eue, H. Hack).
Bayer, US 4058526, 1976 (W. Merz, G. Schümmer).

[300] Bayer, DE-OS 2407144, 1974 (R. R. Schmidt, L. Eue, C. Metzger, K. Dichoré).

[301] Sankyo, AU 515639, 1981 (T. Konotsune, K. Kawakubo).

[302] Janssen Pharmaceutica, Ciba-Geigy, EP 273531, 1986 (W. R. Lutz, G. R. E. Van Lommen, V. Sipido, W. G. Verschueren).
Janssen Pharmaceutica, Ciba-Geigy, EP 289066, 1987 (M. F. L. De Bruyn, G. R. E. Van Lommen, W. R. Lutz).

[303] Monsanto, JP 62273955, 1987 (M. G. Dolson, K. F. Lee, K. L. Spear).

[304] Sumitomo, JP 1180882, 1988 (H. Tomioka, H. Fujimoto, I. Fujimoto).
Ube, EP 283271, 1988 (H. Yoshioka et al.).
Nissan Chem. Ind., JP 1211569, 1988 (Y. Nakajima, K. Hirata, M. Kudo).
Nissan Chem. Ind., EP 183212, 1985 (T. Ogura et al.).
Nissan Chem. Ind., EP 199281, 1986 (T. Ogura et al.).
Nissan Chem. Ind., EP 232825, 1987 (Y. Nakajima et al.).
Hoechst, DE 3733220, 1987 (G. Salbeck et al.).
BASF, DE 3742266, 1987 (J. Leyendecker et al.).
Sumitomo, JP 1197479, 1988 (H. Tomioka, H. Fujimoto).

[305] Sankyo, EP 89650, 1983 (H. Takeshiba, T. Kinoto, T. Jojima).
Nissan Chem. Ind., WO 9208715, 1991 (Y. Nakajima, J. Watanabe, Y. Hirohara, T. Mita).

[306] Schering, DE 3822371, 1988 (D. Huebl, E. A. Pieroh, H. Joppien, D. Baumert).

[307] Bayer, DE 3216416, 1982 (U. Heinemann, P. E. Frohberger, W. Brandes).

[308] SmithKline Beckman, EP 104836, 1983 (G. K. Menon, W. J. Sanders).

[309] Konica, Nissan Chem. Ind., JP 63223635, 1987 (H. Sakamoto, M. Fujiwara, I. Sakata, S. Yamamoto).

[310] Olin, US 4252962, 1980 (E. F. Rothgery).

[311] Chugai Shashin Yakuhin, JP 62178247, 1986 (M. Ito, N. Watanabe, H. Mukai).

[312] Fuji Photo Film, GB 2083243, 1980 (K. Shiba, S. Nakao, T. Toyama).

[313] Agfa-Gevaert, EP 40771, 1980 (E. Ranz, H.-D. Schütz, J. W. Lohmann).
Fuji Photo Film, JP 57026848, 1980 (I. Itoh, T. Toyoda, M. Yamada).
Eastman Kodak, EP 54414, 1980 (H. W. Altland, D. D. F. Shiao).

[314] Fuji Photo Film, JP 1292339, 1988 (M. Goto, K. Morimoto).
Mitsubishi Paper Mills, JP 4012349, 1990 (T. Miura, H. Seiyama, Y. Tsubakii).

[315] F. Zucchi et al., *Werkst. Korros.* **44** (1993) 264.

[316] Technische Universität Dresden, DD 246322, 1986 (G. Reinhard et al.).

[317] Ciba-Geigy, EP 330613, 1989 (E. Phillips, R. C. Wasson).

[318] Rohm & Haas, US 5210094, 1991 (P. F. D. Reeve).

[319] IG Farben, DE 557138, 1931 (M. Bögemann).
DuPont, US 3895025, 1973 (A. L. Goodman).
Bayer, DE 2703313, 1977 (R. Schubart, K. Wagner, K. F. Zenner).
Bayer, DE 3008159, 1980 (R. Schubart, G. P. Langner).

[320] Sankyo, JP 53027641, 1976 (Y. Nakamura, K. Mori, A. Umehara).
Degussa, DE 2850338, 1978 (K. Hentschel, F. Bittner).
Bayer, Degussa, DE 2109244, 1971 (U. Eholzer, T. Kempermann).
H. Ahne et al., *Kautsch. Gummi Kunstst.* **28** (1975) 135.

[321] T. Kempermann, *Bayer-Mitt. Gummi-Ind.* **50** (1978) 29; **51** (1979) 25.

[322] Toyo Linoleum, JP 56026943, 1979 (K. Miyahara, T. Matsukura, I. Niwa).
Vanderbilt, US 4704426, 1986 (L. A. Doe, Jr.).

[323] Geigy, DE 1470832, 1961; *Chem. Abstr.* **62** (1965) 11834.
Mitsubishi Chem. Ind., JP 56129249, 1980 (A. Sugio et al.).

[324] Bayer, DE 3314467, 1983 (M. Blunck, U. Claussen, F. W. Kroeck, R. Neeff).

[325] Haarman und Reimer, DE 3831980, 1988 (R. Emberger et al.).
Haarman und Reimer, DE 4016536, 1990 (R. Emberger et al.).

[326] Givaudan-Roure, WO 93/07134, 1992 (U. Huber).

[327] R. G. Dickinson, N. W. Jacobson, *J. Chem. Soc., Chem. Commun.* **1970**, 1719.
H. Brandl, *Prax. Naturwiss. Chem.* **40** (1991) 25.

[328] U. Schmidt, G. Pfleiderer, F. Bartkowiak, *Anal. Biochem.* **138** (1984) 217.
B. Narayana, M. R. Gajendragad, *Asian J. Chem.* **5** (1993) 121.

[329] Sansei Pharmaceut. Co., JP 1228908, 1988 (Y. Oyama).

[330] Procter & Gamble, EP 483426, 1990 (R. Chatterjee).

[331] R. J. Lewis, Sr. (ed.): *Sax's Dangerous Properties of Industrial Materials*, 8th ed., vol. III, Van Nostrand Reinhold, New York 1992.

[332] E. E. Sandmeyer in G. D. Clayton, F. E. Clayton (eds.): *Patty's Industrial Hygiene and Toxicology*, 3rd ed., vol. 2A, Wiley-Interscience, New York 1981, p. 2061.

[333] L. A. Damani (ed.): *Sulphur-Containing Drugs and Related Organic Compounds*, Ellis Horwood Limited, Chichester 1989.

[334] *Threshold Limit Values for Chemical Substances and Physical Agents in the Workroom Environment*, American Conference of Governmental Industrial Hygienists (ACGIH), 1978.

[335] E. J. Fairchild, H. F. Stokinger, *Am. Ind. Hyg. Assoc. J.* **19** (1958) 171.

[336] R. G. Ames, J. W. Stratton, *Arch. Environ. Health* **46** (1991) 213.

[337] P. Amrolia, S. G. Sullivan, A. Stern, R. Munday, *J. Appl. Toxicol.* **9** (1989) 113.

[338] H. T. H. Wilson, *Br. J. Dermatol.* **81** (1969) 175.
A. A. Fischer, *J. Dermatol. Surg.* **1** (1975) 63.
[339] Industrieverband Pflanzenschutz e. V. (ed.): *Wirkstoffe in Pflanzenschutz- und Schädlingsbekämpfungsmitteln*, Pressehaus Bintz-Verlag, Offenbach 1982.
[340] E. Löser, G. Kimmerle, *Pflanzenschutz-Nachr. Bayer* **25** (1972) 186.

**Thionyl Chloride → Sulfur Halides**

# Thiophene

JOHN WESLEY PRATT, Synthetic Chemicals Limited, Wolverhampton, West Midlands, United Kingdom



The extensive literature on thiophenes is indicative of the commercial and research interest in these five-membered heterocyclics. Some hundreds of patents appear each year, many considering thiophene derivatives as alternative products to, e.g., a parent benzenoid molecule.

The thiophene-containing molecule must possess cost-effective advantages or uniqueness. As well as a means of producing an alternative "me too" effect chemical, the inherently greater chemical sensitivity of the thiophene ring compared to that of benzene may also improve biodegradability or metabolic characteristics.

The process to manufacture thiophene (and its homologs), is capable of unlimited expansion, with consequential lower unit costs, but it remains a fine-chemical business (ca. 1000 t/a).

This article concentrates on the products of current commerce. Chapter 5 is estimated to cover over 90% of the present volume; the remainder consists of 25 or more minor, but highly efficacious products, mainly pharmaceuticals, together with several which are still in the very early stages of development.

## 1. Physical and Chemical Properties

The physical characteristics of thiophene (thiofurfuran) are closer to benzene than any other five-membered heterocyclic. This enables it to replace a benzenoid moiety in the whole spectrum of aromatic chemicals [1] and, as illustrated in Chapter 5, sometimes to introduce worthwhile advantages over the analogous benzenoid compounds.

Some physical and physicochemical properties of thiophene are listed below:

| | |
|---|---|
| Boiling point, 100 kPa | 84.16 °C |
| Melting point | −38.30 °C |
| Density | |
| $d_4^4$ | 1.0873 |
| $d_4^{20}$ | 1.06485 |
| $d_4^{25}$ | 1.05887 |
| Refractive index | |
| $n_D^{20}$ | 1.52890 |
| $n_D^{25}$ | 1.52572 |
| Viscosity | |
| at 20 °C | 0.662 mPa · s |
| at 25 °C | 0.621 mPa · s |
| Vapor pressure | |
| at 20 °C | 8.4 kPa |
| at 25 °C | 10.61 kPa |
| at 60 °C | 45.32 kPa |
| at 110 °C | 143.24 kPa |
| at 130 °C | 270.07 kPa |
| Enthalpy of vaporization | 32.483 kJ/mol |
| Critical temperature | 317 °C |
| Critical pressure | 4.53 MPa |
| Dipole moment | 0.53−0.63 D |
| Dielectric constant | 2.76 |
| Flash point | −6.7 °C |
| Enthalpy of combustion | −2807 kJ/mol |
| Enthalpy of formation | 82.13 kJ/mol |

The numbering of the thiophene ring is orthodox:

Nomenclature can be summarized as

2-Thenyl          2-Thienyl          2-Theonyl


Ullmann's Encyclopedia
of Industrial Chemistry, Vol. A 26

The usefulness of thiophene and its derivatives lies in its chemical sensitivity. It is a $\pi$-electron-rich aromatic (Fig. 1); the literature illustrates the ease of electrophilic substitution in the 2- and 5-positions (Fig. 2). As in other five-membered heterocyclics, the S atom of the thiophene ring plays its part in the formation of cationic reaction intermediates. Thus, the reactivity of thiophene is more akin to that of phenol and aniline than to the comparatively unreactive benzene ring.

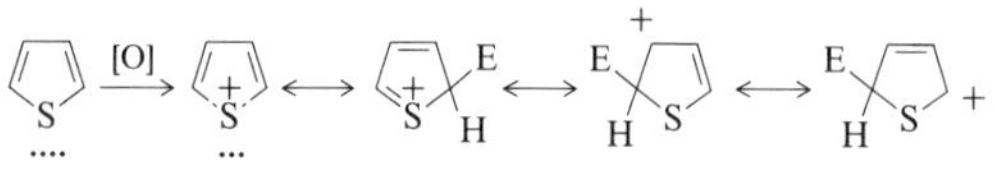

**Figure 1.** Delocalization of $\pi$-electrons in the thiophene ring

With the 2-position occupied by an *o/p* or *m*-directing group, a second group is substituted in the 5-position of the thiophene ring. With such a directing group in the 3-position, mixed isomers are formed.

As can be seen from Figure 2, many 2-substituted derivatives can be produced by orthodox and comparatively mild organic chemical reactions. However, decomposition sometimes interferes with yield expectations, e.g., reduction of an $-NO_2$ group to an $-NH_2$ group ruptures the thiophene ring under normal benzenoid conditions.

While 2-substitution is practically 100 %, single percentage amounts of the 3-isomer may be formed. Routes and reaction conditions to limit this effect need to be optimized.

The order of reactivity toward electrophilic substitution is:

Pyrrole $\gg$ thiophene $>$ furan $\gg$ benzene e.g., bromination of thiophene is $1.25 \times 10^7$ faster than that of benzene; chlorination $1.7 \times 10^7$ faster.

Irradiation causes thiophene to react with $H_2O$ [14]:

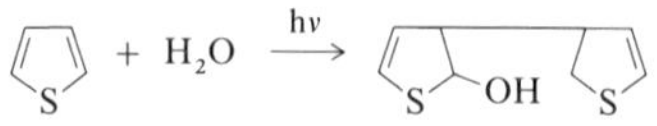

Heating thiophene does not yield $C_2H_2$, but bithenyl [14]:

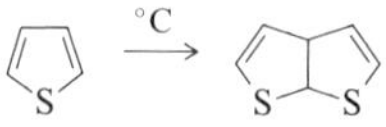

## 2. Resources and Raw Materials

Thiophene occurs in the light hydrocarbon fraction distilled from coal tar. The proportion of thiophene depends on the type of coal, the temperature of carbonization, and the type of retort. Typical quantities in a tar-derived benzol stream are 0.4–1.4 wt % [15].

Traditionally, thiophene is removed from coal tar benzol by washing with $H_2SO_4$, more rarely by solvent extraction, to produce a sweeter, marketable product.

The very large volumes of light hydrocarbons produced from coal carbonization plants represent a considerable quantity of associated thiophene, but the quality demands of the principal markets for thiophene require a purity which cannot be obtained economically from coal tar.

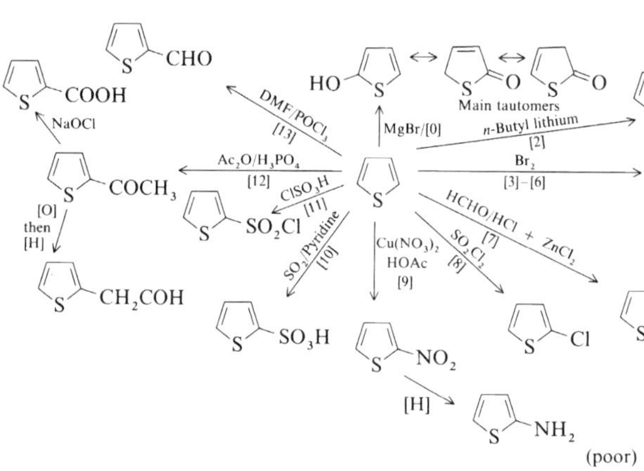

**Figure 2.** Electrophilic substitution reactions of thiophene

Commercialization of thiophene for the fine chemicals intermediate market requires a synthetic high-purity source. This was initiated in the 1960s in the United States by Pennsalt/Pennwalt, now Atochem. Midland-Yorkshire Tar Distillers/Synthetic Chemicals first produced synthetic thiophene on a commercial scale in Europe in 1970.

A key factor in the commercial production of pure thiophene and its homologs is the specificity and cleanliness of the processes used. Vapor phase (VP) cyclization methods are the processes of choice.

The original VP process, using furan and $H_2S$ as the sulfur source [16], is no longer preferred. Synthetic Chemicals operates its own VP process [17], using $CS_2$ plus a selected $C_4$ skeleton. A similar VP process is patented by Elf Aquitaine [18].

Other non-VP processes may be described as laboratory methods and include: (1) syntheses from $C_4$ unsaturates, using NHS, $Li_2S$, $SO_2$, and S as sulfur source; (2) succinic salts and $\gamma$-diketones with, e.g., $P_4S_7$ or $P_2S_5$. Such processes result in very complex reaction products unsuitable for large-scale production.

## 3. Production

Since the current international market for thiophene is < 1000 t/a, the type of VP installation required for its manufacture is much smaller than a typical petrochemical unit. However, the process is capable of expansion to match any required demand.

Tight operating temperatures of 500 – 530 °C, catalyst performance, and regeneration plus overall control, particularly with respect to envi-

ronmental requirements, are key factors in the successful operation of such a unit (Fig. 3).

Typical feedstock portfolio is shown in Figure 4.

The advent of a range of commercially available alkylthiophenes, including those substituted in the difficult 3-, (4-) position, resulted in an alternative raw material to synthesize a range of 3-, (4-) side-chain thiophenes, other than from, e.g., 3-bromothiophene [19].

## 4. Specifications, Transportation, Health and Safety

Data for thiophene, together with its 2-methyl, 3-methyl, 2-acetyl, and 2-bromo derivatives are given in Table 1.

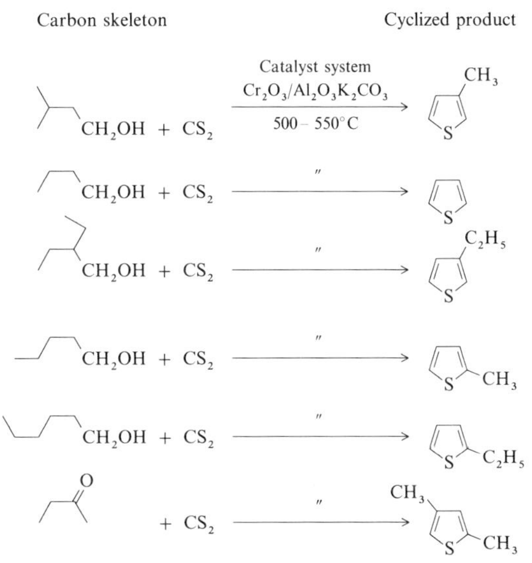

Figure 4. Feedstock for thiophene and its derivatives

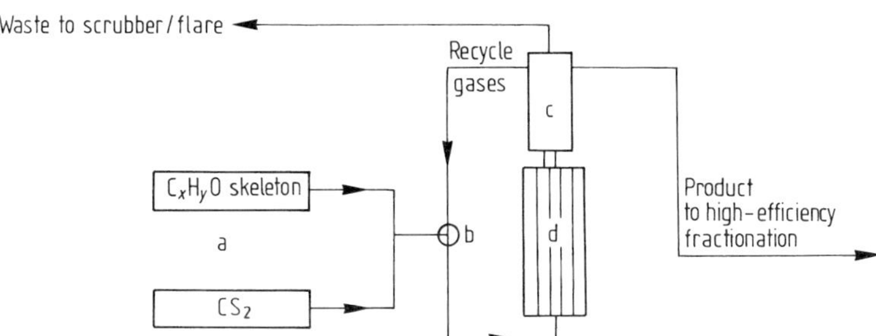

Figure 3. Thiophene production
a) Raw materials tank farm; b) Mixer/flow control; c) Condenser; d) Multitube, fixed-bed VP (vapor phase) reactor

**Table 1.** Specifications, transportation, health and safety of thiophene and derivatives

|  | Thiophene | 2-Methyl-thiophene | 3-Methyl-thiophene | 2-Acetyl-thiophene | 2-Bromo-thiophene |
|---|---|---|---|---|---|
| **Specification** | | | | | |
| CAS No. | [110-02-1] | [554-14-3] | [616-44-4] | [88-15-3] | [1003-09-4] |
| Appearance | clear liquid | clear liquid | clear liquid | yellow liquid ($\geq 9°$) | clear/darkening |
| Odor | aromatic | aromatic | aromatic | | mild pungent |
| Purity (GLC), wt% | 99 | 98 | 98 | 99 | 98 |
| Boiling point, °C | 83–85 (95%) | 113 | 115 | 213–214 (95%) | 153–154 |
| Flash point, °C | −7 | 15.5 | 15.5 | 96 | |
| Miscibility with water | almost immiscible | almost immiscible | almost immiscible | slightly miscible | immiscible |
| **Transportation** | | | | | |
| Tarrif code | 29340 100 | 293490 900 | 293490 900 | 293490 900 | 293490 900 |
| Classification | class 3 highly flammable | class 3 highly flammable/irritant | class 3 highly flammable/irritant | class 6.1 toxic | class 6.1 toxic |
| Packaging group | II | II | II | II | II |
| IMDG page No. | 3284 | 3230 | 3230 | 6231 | 6231 |
| UN (identification) | 2414 | 1993 NOS | 1993 NOS | 2810 NOS | 2810 NOS |
| ADR | item. 3°(b) | item. 3°(b) | item. 3°(b) | item 21° | item. 20°(b) |
| **Health and Safety*** | | | | | |
| R phrases | R11 R21/22 R38 R52/53 | R11 R36/37/38 | R11 R26/37/38 | R23 R24 | R23/24/25 R36/37/38 |
| S phrases | S9/16/33/43 | S9/16/33/43 | S9/16/33/43 | S18/24/30/41/44 | S28/36/37/39/44 |

* Risk and Safety phrases as legislated by the European Commission and stated fully in United Kingdom. Statutory Instruments 1982 No. 1496, Health and Safety Executive, Sudbury, Suffolk, United Kingdom.

# 5. Uses

Successful commercial derivatization is governed by the physical and chemical characteristics of the basic thiophene nucleus, founded on an economic source of guaranteed quality.

By far the greatest volume of simple thiophenes are produced by means of VP cyclizations (see Chapter 3). Derivatives of these homologs account for practically all the growth of this business in the last three decades, in areas as diverse as pharmaceuticals, veterinary drugs, agrochemicals, to dyes, photography, and polymers.

## 5.1. Oxygenated Derivatives

The oxygenated derivatives of thiophene form the largest group of the intermediates market, 70–75% of the world total.

**Thiophene 2-aldehyde** (T2A), $C_5H_4OS$ $M_r$ 112.15, $bp$ 198 °C, is prepared as shown in Figure 2 [13]. It was the first oxygenated derivative to reach a market of several hundred t/a, about two decades ago.

It is used to produce a range of antihistamines, including methapyrilene (**1**) [20], methaphenilene (**2**) [21], and thenalidine (**3**) [22]. However, this usage has practically disappeared.

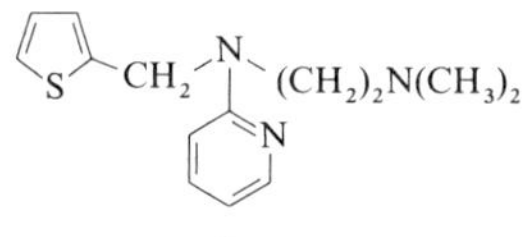

S—CH₂—N—(CH₂)₂N(CH₃)₂ / N

**1**

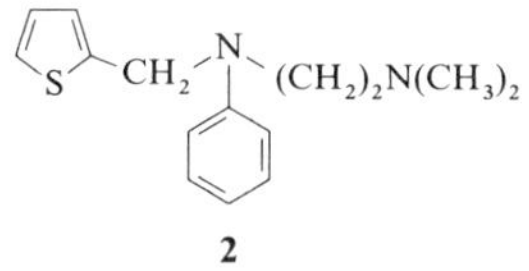

S—CH₂—N—(CH₂)₂N(CH₃)₂

**2**

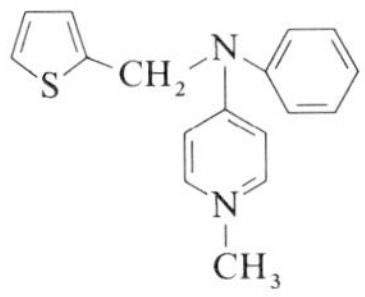

**3**

The anthelmintic pyrantel (**4**) [23] (→Anthelmintics, **A 2**, p. 340) is a major market for T2A, enhanced by new formulations and the development of the medicinal use beyond the original veterinary market.

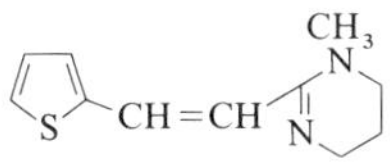

**4**

Other pharmaceuticals containing T2A are azosemide (**5**) [24] a diuretic, and teniposide (**6**) [25] an antineoplastic.

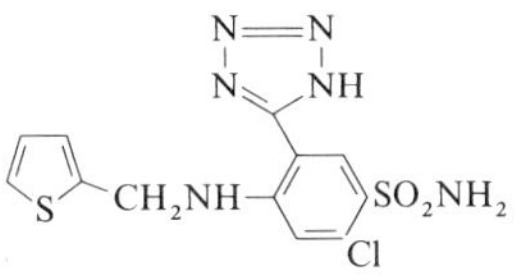

**5**

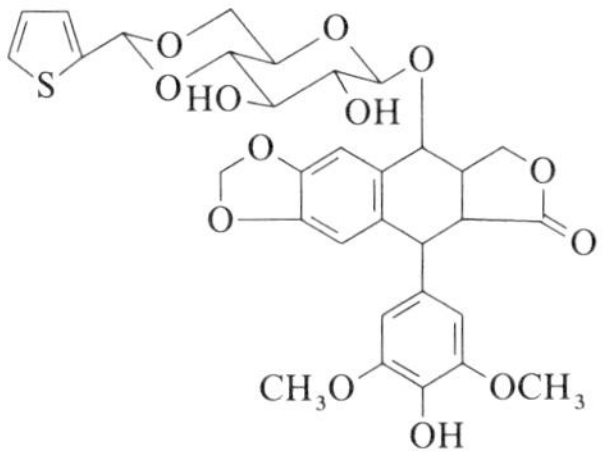

**6**

**Thiophene 2-carboxylic acid**, T2CA, $C_5H_4O_2S$, $M_r$ 128.15, $bp$ 260 °C, $mp$ −130 °C, is prepared via oxidation of 2-acetylthiophene (see Fig. 2).

The sodium salt of T2CA is used to treat influenza (trophires) [26]. Other salts, e.g., Li, have minor medicinal uses. A US Patent [27] describes the use of the Na salt as a lubricating oil thickener.

Suprofen (**7**) [28] a nonsteroidal anti-inflammatory (NSAI), and tienilic acid (ticrynafen) (**8**) [29], a potent diuretic, were both very large prospective markets for T2CA or its acid chloride, but were almost completely withdrawn in the 1980s. It is interesting to compare the success of Roussel Hoechst's NSAI tiaprofenic acid (**9**) in this market.

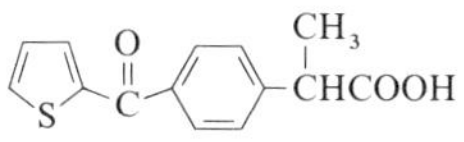

**7**

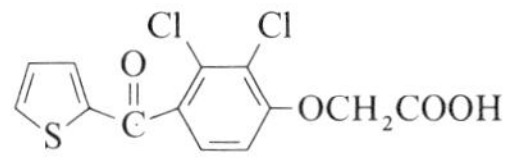

**8**

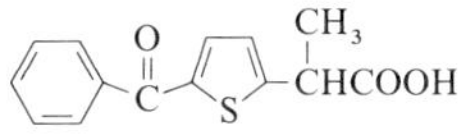

**9**

The 5-nitro derivative of T2CA is used to produce nifurzide, an intestinal antibacterial.

Pfizer's Tenidap anti-inflammatory (**10**) [30] is a developing drug which uses the acid chloride of T2CA as the starting intermediate.

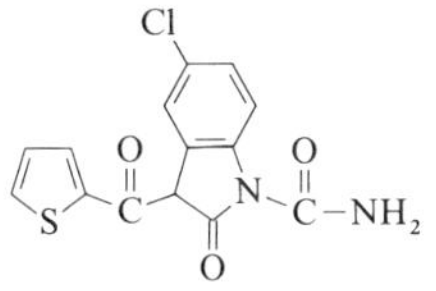

**10**

**Thiophene 2-acetic acid**, T2AA, $C_6H_6O_2S$, $M_r$ 142.18. The use of 7-aminocephalosporanic acid (7ACA), following the classical work by the National Research and Development Council in the 1950s [31] to produce a family of semisynthetic cephalosporins, resulted in important first-generation antibiotics (→Antibiotics, **A 2**, p. 476). Cephaloridine (**11**) [32] and cephalothin [33] were the major products licensed from this work. The second-generation cefoxitin (**12**) [34] came a few years later.

T2AA, or its acid chloride, is used to form the 7-side-chain of these antibiotics. Many processes have been designed to synthesize T2AA (Cl), including chlormethylation of thiophene until the late 1980s in the United States, and starting from T2A in the FAMSO process in Japan.

However, these processes are now superseded, and the reduction of 2-acetylthiophene followed by oxidation is the process of choice.

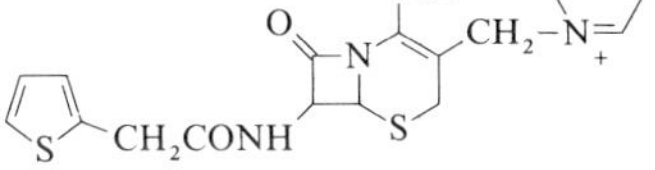

**11**

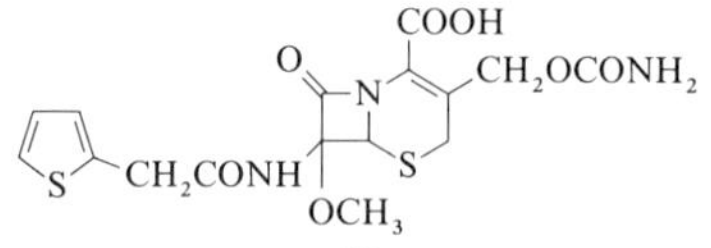

Ketotifen fumarate (**13**) [35] is an antiasthma drug (→Antiallergic Agents, **A 2**, p. 429) which is thought to possess prophylactic properties, and this market is growing.

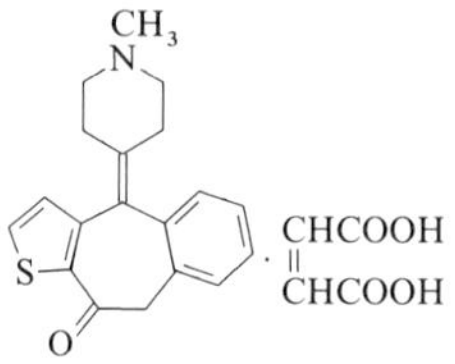

**13**

Pizotyline (**14**) [36] an antimigraine agent, is synthesized from T2AA (Cl).

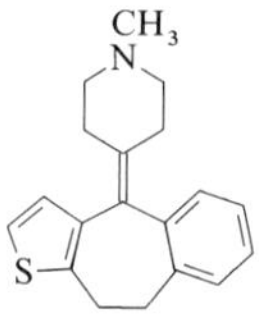

**14**

**2-Acetylthiophene** (2AT) $C_6H_6OS$, $bp$ 214 °C, $mp$ 9 °C, is prepared as shown in Figure 2. As described above, 2AT is used as the precursor to T2AA (Cl) and T2CA (Cl). It is important that the process used to synthesize 2AT yields the lowest amount of 3-acetylthiophene isomer.

**Thiophene Dicarboxylic Acids.** Adipic acid, or its dichlorinated derivative, provides a means of forming thiophene 2,5-dicarboxylic acid [37]. Thiophene aldehydes and other substituents exhibit pronounced fluorescence [38]:

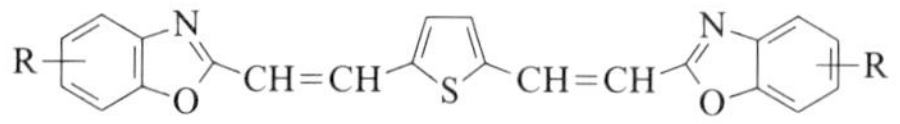

More complicated thiophene polybenzoxazoles are used as optical brighteners, and may be synthesized via the 2,5 carboxylic acid groups.

## 5.2. Bromothiophenes

The classical use of bromine as a leaving group has played an important part in the labor-atory and commercial synthesis of many thiophene derivatives and products.

**2-Bromothiophene** (2BT), $C_4H_3Br$, $M_r$ 78.17, $bp$ 153.5 °C, is easily produced to a good purity (see Fig. 2) [3].

Tipepidine (**15**) [39], an antitussive, is produced from 2BT.

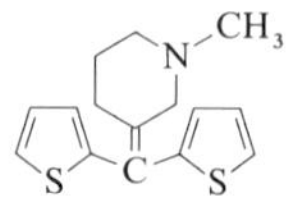

**15**

**3-Bromothiophene** (3BT), $C_4H_3Br$, $bp$ 160 °C. In order to obtain 3- (4-) substituted thiophenes, more elaborate chemistry is required than straight-forward electrophilic substitution. Successive bromination to 2,3,5-tribromothiophene can be followed by removal of the 2- and 5-Br atoms (see Fig. 2), but, a newer, simpler process is now patented [6] which is essentially free of 2-Br isomer.

3-Bromothiophene may be used to produce thiophene 3-malonic acid (TMA), the side-chain intermediate for the production of ticarcillin and temocillin from 6-aminopenicillinic acid (6APA). Timentin is a powerful antibiotic, produced from ticarcillin plus clavulanic acid [40]. An alternative route to TMA is via 3-methylthiophene (see below).

The growing interest in conducting polymers based on polythiophenes is extensively reported [41], [42]. In order to produce a workable polythiophene, a fairly long side-chain may be introduced at the 3-position of the thiophene ring. Choice of the structure of this side-chain is a compromise between the workability and the maintenance of required conductivity. It generally lies between $C_4$ and $C_{12}$ $n$-alkyl. Polymerization is via the 2- and 5-positions.

Removal of the 3-Br atom and replacement by an alkyl group is a means of obtaining the desired monomer.

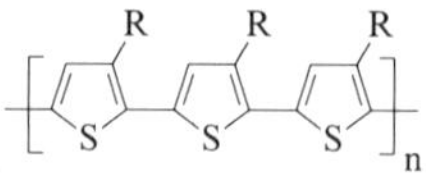

$$R = C_5 - C_{10} \text{ alkyl}$$

Conducting polythiophene

**Tetrahydrothiophene** (THT), $C_4H_8S$, $M_r$ 88.18, $bp$ 121.1 °C, $mp$ −96 °C, is easily confused in import/export statistics with thiophene itself, particularly since it may also be called thiophane; THT was certainly produced in commercial quantities before thiophene, and it remains a commercial product.

Manufacture is not from thiophene, but by VP replacement of the oxygen atom in tetrahydrofuran by sulfur from $H_2S$. The plant and catalyst is very similar to a thiophene unit. However, THT is used for its very strong odor, either on its own or, more commonly, within a cocktail, e.g. alkylsulfides, as a powerful gas odorant. The saturation of comparatively sweet-smelling thiophene thus produces vastly different odor characteristics.

**2-Methylthiophene** (2MT), $C_5H_6S$, *bp* 112.5 °C, *mp* 63.4 °C, is commercially produced to 98 wt % purity by VP cyclization (see Fig. 4).

It is a feasible raw material for the production of several of the important 2-substituted thiophenes. It may be used to manufacture 2-chloromethylthiophene (2-thenyl chloride), but this and further syntheses must be completed in situ, because of its instability.

**3-Methylthiophene** (3MT), *bp* 115 °C, is commercially produced by an analogous process to 2MT, and to the same purity.

Thenyldiamine (**16**) [43], an antihistamine, has been produced since the late 1940s. The synthesis requires a source of 3MT to prepare the 3-thenylbromide precursor.

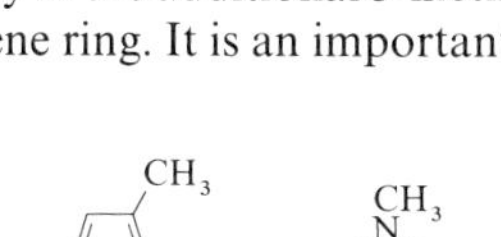

**16**

Morantel (**17**) [44] is a potent antihelmintic, differing from pyrantel ($\rightarrow$ Anthelmintics, **A 2**, p. 340), only in the additional 3-methyl group on the thiophene ring. It is an important market for 3MT.

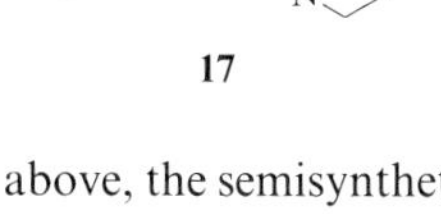

**17**

As stated above, the semisynthetic penicillins ticarcillin (**18**) [42] and timentin possess a 3-thienyl side-chain; 3MT, via free radical bromination of the methyl side-chain, is an alternative precursor to TMA.

The antifungal tioconazole (**19**) [45] is synthesized from 2-chloro-3-methylthiophene.

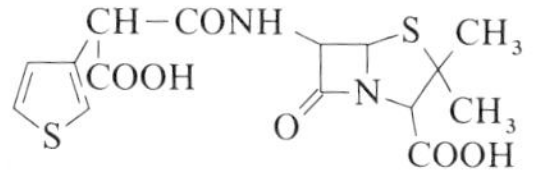

**18**

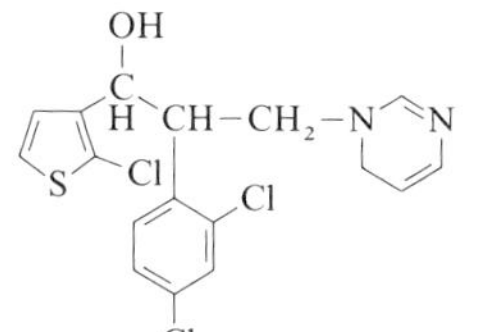

**19**

Cetiedil (**20**) [46] is a peripheral vasodilator which incorporates a thiophene ring in the 3-position; 3MT provides a means of synthesis.

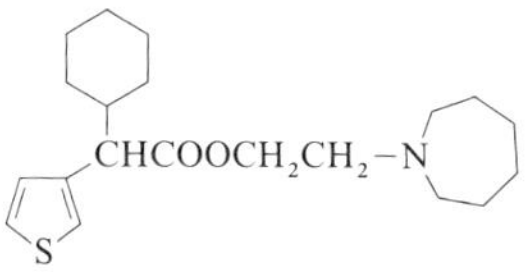

**20**

## 5.3. Thiophene

The use of thiophene mixed with other constituents to produce a denaturant has recently been developed [47]. Since thiophene is unknown in fermented alcohol, the addition of a marker containing thiophene facilitates the identification of alcohol denatured in this way. The thiophene is very difficult to remove.

Uses of thiophene itself have otherwise been restricted to analytical techniques, where a standard mass of sulfur is required for measurement control. Thiophene may also be used for quantitative deactivation of cracker catalysts.

## 5.4. Polysubstituted Derivatives

The commercial availability of a range of thiophenes substituted with an amino group, together with other ring substituents, is based on a modification of the Gewald reaction [48], for 2-$NH_2$ derivatives. Substitution of an amino group in the 3-position requires custom-designed chemistry [49], [50].

**Methyl 3-amino-4-methylthiophene 2-carboxylate** $C_7H_{10}NO_2S$. The 3-amino group is used to attach a side-chain to the thiophene ring for manufacture of the local anesthetic carticaine (**21**) [51]. This pharmaceutical is also produced with a 5-chloro group on the thiophene ring.

**21**

**Methyl 3-aminothiophene 2-carboxylate** $C_6H_7NO_2S$. This derivative is used to produce a sulfonylurea post-emergent herbicide, Harmony (M6316 DuPont) (**22**), which has extremely rapid biodegradeability [52]. This characteristic, together with the very low dose requirements, are important environmental safeguards.

The NSAI tenoxicam (**23**) [53] is one of a range of oxicams which happens to have a thiophene moiety and may be synthesized from this aminothiophene carboxylate.

**22**

**23**

The amino group, when in the 2-position of a substituted thiophene derivative, may be diazotized to a range of monoazo disperse dyes [54]. Coupling is with a pyrazolone or, e.g., phenol, and colors toward the yellow/green part of the spectrum are produced.

Substituted thiophene molecules are found in various components of animal and vegetable flavorings. The combination of $H_2S$ with unsaturated aldehydes may be the source of such thiophene derivatives in, e.g., cooked beef, pork liver, onion, and coffee [55], [56].

**Table 2.** Toxicological data for thiophene and derivatives

| | Thiophene | 2-Methyl-thiophene | 3-Methyl-thiophene | 2-Acetyl-thiophene | 2-Bromo-thiophene |
|---|---|---|---|---|---|
| **Acute Toxicity** | | | | | |
| Oral LD$_{50}$ | no data | 3200 mg/kg (mouse) | 2300 mg/kg (rat) | 50 mg/kg (mouse) | 200–250 mg/kg (rat) |
| Dermal LD$_{50}$ | no data | no data | 2000 mg/kg (guinea pig) | 370 mg/kg (rat) | 134 mg/kg (rat) |
| Inhalation LC$_{50}$ | no data | no data | 10 g/m$^3$, 2 h (rat) | 1460 mg/m$^3$, 1 h (rat) | 1.04 mg/L, 4 h (rat) |
| **Irritation** | | | | | |
| Skin | possibly mild | expected | yes | not reported | yes |
| Eye | possibly mild | expected | yes | predicted | yes |
| **Skin sensitization** | no data | no data | not found in tests | no data | no data |
| **Occupational exposure limits** | none specified | none specified | none specified | 2 ppm (provisional) | 1 ppm (provisional) |
| **Ecotoxicity** | possibly only slight harm to aquatic organisms | possibly only slight harm to aquatic organisms | possible only slight harm to aquatic organisms | harmful to aquatic organisms | expected only slight harm to aquatic organisms |
| **Persistence** | expect ultimate biodegradability | expect ultimate biodegradability | expect ultimate biodegradability | expect ultimate biodegradability | expect very slow biodegradability |
| **Stability** | | | | | |
| Materials to avoid | strong oxidizers | strong oxidizers | strong oxidizers | strong oxidizers | strong oxidizers |
| Conditions | stable under normal conditions | stable under normal conditions | stable under normal conditions | stable under normal conditions | slow decomposition unless cool/dark |
| Combustion products | toxic oxides of S + CO, CO$_2$ | toxic oxides of S + CO, CO$_2$ | toxic oxides of S + CO, CO$_2$ | toxic oxides of S + CO, CO$_2$ | HBr during storage (prolonged), CO, CO$_2$ |

As well as the fused ring structures illustrated above, which are valuable commercial products, others are worth mentioning, because of their potential markets.

Ticlopidine (**24**) [57] is a potent pharmaceutical, used to prevent blood platelet aggregation in the treatment of strokes. One of three possible isomeric structures of the thiophene analog of saccharin, is $10^3$ times sweeter than sucrose, and possesses all the characteristics of saccharin itself.

**24**

Many methods are available to synthesize fused thieno rings to a nitrogen heterocyclic [58], [59].


# 6. Toxicology

Toxicological data for thiophene and its derivatives are given in Table 2.


# 7. References

[1] L. S. Fuller, J. W. Pratt, F. S. Yates, *Manuf. Chem. Aerosol News* (1978) no. 5; *Inf. Chim.* (1979) no. 10, 19.
[2] A. Gilman, D. A. Shirley, *J. Am. Chem. Soc.* **71** (1949) 1870.
[3] I. Hirao, *Yakugaku Zasshi* **73** (1953) 1023; *Chem. Abstr.* **48** (1954) 10723.
[4] C. Troyanowsky, *Bull. Soc. Chim. Fr* **1955**, 424.
[5] S. Gronowitz, *Ark. Kemi* **7** (1954) 267; *Acta Chem. Scand.* **13** (1959) 1054.
[6] Shell Int. Res., EP 029586, 1989 (P. Grosvenor, L. Fuller).
[7] Y. Golfarb, Y. L. Danyusherskii, *Bull. Acad. Sci. USSR Div. Chem. Sci. (Engl. Transl.)* **1956**, 1385.
[8] E. Campaigne, W. M. Le Suer, *J. Am. Chem. Soc.* **70** (1948) 1555.
[9] N. I. Putskin, A. N. Sopokin, *Chem. Abstr.* **51** (1957) 16419.
[10] A. P. Terentev, G. M. Kaltski, *Chem. Abstr.* **46** (1952) 11178.
[11] A. Buzas, J. Teste, *Bull. Soc. Chim. Fr.* **1960**, 793.
[12] H. Hartough, I. Kosak, *J. Am. Chem. Soc.* **69** (1947) 3093.
[13] E. Campaigne, W. L. Archer, *J. Am. Chem. Soc.* **75** (1953) 989.
[14] Private Communication.
[15] *Ullmann*, 4th ed., **22**, 416.
[16] Pennsalt Chem. Corp., GB 963385, 1964 (T. E. Deger, B. Bucholtz, R. H. Goshorn).
[17] Midland York Tar Dist., GB 1345203, 1974 (N. R. Clark, W. E. Webster).
[18] Soc. Nat. Elf Aquitaine, GB 1585647, 1981.
[19] J. A. Clarke, O. Meth-Cohn, *Tetrahedron Lett.* **52** (1975) 4705–4708.
[20] Monsanto, US 2581868, 1952.
[21] Lenard, Solmssen, *J. Am. Chem. Soc.* **70** (1948) 2066.
[22] Sandoz, US 2717251, 1955; US 2757175, 1956.
[23] Pfizer, US 3502661, 1968/70.
[24] Boehringer Mannheim, DE 1815922.
[25] Sandoz, US 3524824, 1970.
[26] Syde, Malleray, *Bull. Soc. Chim. Fr.* **1963**, 1276.
[27] Standard Oil, US 2576031, 1951.
[28] Janssen, GB 1446239, 1975.
[29] CERPHA, DE 2048372, 1971.
[30] Pfizer, US 4556672, 1971.
[31] Nat. Res. Dev. Corp., GB 810196, 1959; US 3082155, 1963.
[32] Glaxo, FR 1384197, 1965.
[33] Lilly, BE 618663, 1962.
[34] Merck, DE 2129675, 191; DE 2203653, 1971.
[35] Sandoz, DE 2111071, 1971.
[36] Sandoz, US 3272826, 1966; BE 636717, 1964.
[37] *Tetrahedron Lett.* **42** (1970) 3719–3722.
[38] R. E. Atkinson, F. E. Hardy, *J. Chem. Soc. B* **1971**, 357–361.
[39] Tanabe, JP 17988, 1962.
[40] Beecham, US 3282926, 1966.
[41] *Eur. Chem. News*, April 9, 1991. R. D. McCullough, R. D. Lowe, *J. Chem. Soc., Chem. Commun.* **1992**, no. 1, 70–72.
[42] M. L. Blohm, P. C. Van Dort, J. E. Pickett, *CHEMTEC* **1992**, no. 2, 105–107. K.-H. Kocher, *Hoechst High Chem. Mag.* **1992**, no. 12, 39–42.
[43] E. Campaigne, W. M. Le Suer, *J. Am. Chem. Soc.* **71** (1949) 333.
[44] Pfizer, GB 1120587, 1968.
[45] Pfizer, US 4062966, 1977 (Gymer).
[46] Innothera, FR 1460571, 1966 (Pons, Robba).
[47] ACNA, EP-Appl. 424886, 1991.
[48] K. Gewald, *Chem. Ber.* **98** (1965) 3571. K. Gewald, *Chem. Ber.* **99** (1966) 94.
[49] H. Fiesselmann, *Chem. Abstr.* **55** (1961) 6497c/17651d.
[50] Shell Int. Res., EP 298542, 1989.
[51] Hoechst, US 3855243, 1974.
[52] Meeting of Weed Soc. of America, Seattle **25** (1980) 10.
[53] Hoffmann-LaRoche, DE 2537070, 1976.
[54] ICI, GB 1394365; GB 1394367; GB 1394368.
[55] M. Shankaranarayana, *CRC Crit. Rev. Food Technol.* **4** (1974) 395.
[56] R. Wilson, *J. Agric. Food. Chem.* **21** (1973) 873.
[57] Parcor (Sanofi), US 4127580, 1978 (E. Braye).
[58] Schneller, *Int. J. Sulfur Chem.* (1974) 309–316. J. M. Barker, *Adv. Heterocycl. Chem.* **21** (1977) 65.
[59] Croda Synthetic Chem., EP 0003645, 1982 (O. Meth-Cohn).


**Thiophenol → Thiols and Organic Sulfides**
**Thiophosphates → Phosphorus Compounds, Organic**
**Thiosulfates → Sulfites, Thiosulfates, and Dithionites**

# Thiourea and Thiourea Derivatives

BERND MERTSCHENK, FERDINAND BECK, SKW Trostberg AG, Trostberg, Federal Republic of Germany (Chap. 1)

WOLFGANG BAUER, Cassella AG, Frankfurt/Main, Federal Republic of Germany (Chap. 2)



Thiourea is a chemically interesting compound, which has three functional groups: amino, imino, and thiol. This results from tautomerism between thiourea and isothiourea:

$$
\begin{array}{ccc}
H_2N & & H_2N \\
\diagdown & & \diagdown \\
& C=S \rightleftharpoons & C-SH \\
\diagup & & \diagup \\
H_2N & & HN \\
\text{Thiourea} & & \text{Isothiourea}
\end{array}
$$

Because of this polyfunctionality and also because of its complex-forming properties, thiourea has been widely used for more than 30 years [1]–[5]. Its most important uses are as a starting material for nitrogen- and sulfur-containing heterocycles and formamidinesulfinic acid (thiourea dioxide), as a reaction partner for aldehydes, and as a component of addition compounds and complexes.

Thiourea derivatives are thermally stable, mostly crystalline compounds. They can be produced in good yields from readily available starting materials by simple syntheses. This class of compounds, described in detail in several review articles [6]–[11], has gained importance in the pharmaceutical sector, in plant protection, in various technical applications, and in the synthesis of heterocycles.

# 1. Thiourea

## 1.1. Physical Properties

Thiourea [62-56-6], 2-thiourea, $CH_4N_2S$, $M_r$ 76.12, $d_4^{20}$ 1.405, is a white, odorless solid, and crystallizes in a rhombic bipyramidal structure with four molecules per unit cell. It is the only known molecular crystal which is ferroelectric within a certain temperature range. Space group $D_{2h}^{16}-P_{nma}$.

Thiourea is soluble in protic and aprotic polar solvents (Table 1) and insoluble in nonpolar ones.

**Table 1.** Solubility of thiourea

| In $H_2O$ | | | | | | |
|---|---|---|---|---|---|---|
| Temperature, °C | 0 | 20 | 40 | 60 | 80 | 100 |
| Solubility, g/100 g | 4.9 | 13.7 | 30.7 | 70.9 | 138 | 233 |
| In MeOH | | | | | | |
| Temperature, °C | | 25 | 40.7 | 53.7 | 61.9 | |
| Solubility, g/100 g | | 11.9 | 16.4 | 22 | 24.6 | |
| In EtOH | | | | | | |
| Temperature, °C | 20 | 31.9 | 45 | 58 | 64.7 | |
| Solubility, g/100 g | 3.6 | 4.7 | 6.3 | 8.5 | 9.8 | |

Thiourea has no sharp melting point, as it rearranges relatively rapidly to ammonium rhodanide ($NH_4SCN$) from 135 °C. On rapid heating, a melting point of 180 °C has been determined. The heat of combustion at constant volume is 1482.2 kJ/mol (19.47 kJ/g), and at constant pressure 1485.6 kJ/mol (19.52 kJ/g).

## 1.2. Chemical Properties

Although it is known in only one form, thiourea can also react as the tautomeric iminothiol (see Section 2.2). Salts such as $(NH_2)_2CS \cdot HNO_3$ or $(NH_2)_2CS \cdot H_2SO_4$ are formed with strong acids [3]. Addition compounds and complexes can be synthesized with metal oxides or inorganic salts in acid solution, e.g., $[Cr\{(NH_2)_2CS\}_3]Cl_3$.

In alkaline solution, a large number of heavy metal ions can be precipitated quantitatively as sulfides with thiourea. These precipitates are readily filtered, so that the method can be used for analytical determination of metal salts [12]. The production of light-sensitive PbS layers [13] and the purification of mercury-containing wastewater [14] are based on this reaction.

Thiourea forms clathrates (inclusion compounds) with branched aliphatic and cyclic hydrocarbons [15]. This property can be used, e.g., for separating branched and straight-chain aliphatic hydrocarbons [16].

Thiourea can be alkylated and acylated, and reacts with aldehydes and ketones. With formaldehyde, for example, it forms $N$-(hydroxymethyl)thiourea ($H_2N-CS-NH-CH_2-OH$), which condenses to resin-like products on heating [17]. By condensation of thiourea with chloroketones, diketones, hydroxyketones, oxocarboxylic acids, and similar ketone derivatives, a large number of heterocycles can be produced. From 2-haloketones or 2-haloaldehydes (in the enol form) 2-aminothiazole derivatives are formed:

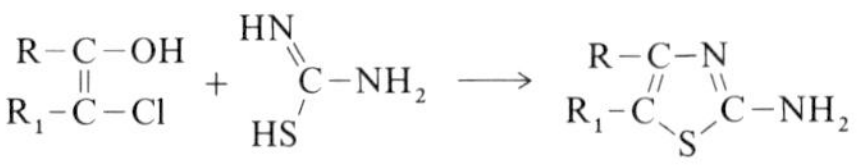

1,1,-Dihaloalkanes condense with thiourea to 1,3,5-dithiazines:

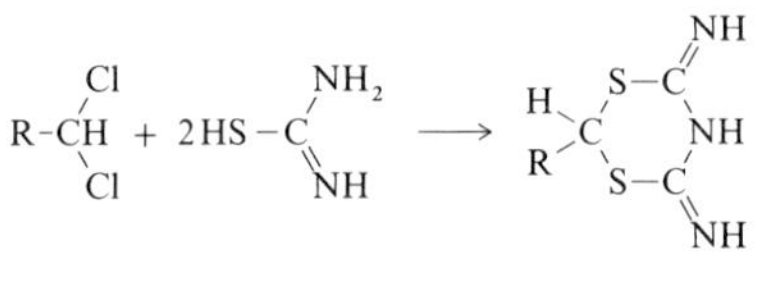

2-Hydroxyketones give 4-imidazoline-2-thiones:

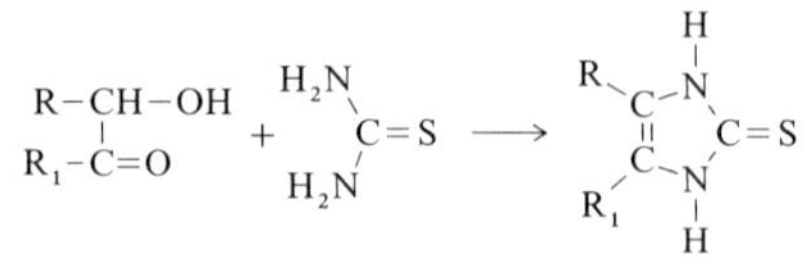

With 2-oxocarboxylic acids (e.g., pyruvic acid) 2,3-dihydro-2-thioxo-4$H$-imidazole-4-one derivatives are formed:

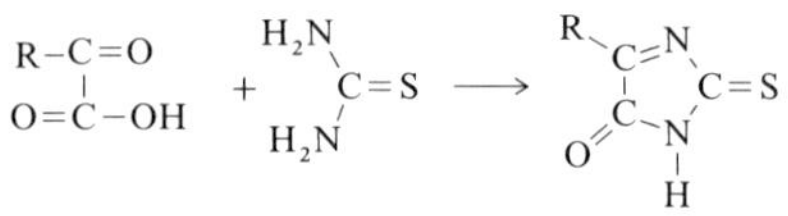

Malonic esters react with thiourea to thiobarbituric acid:

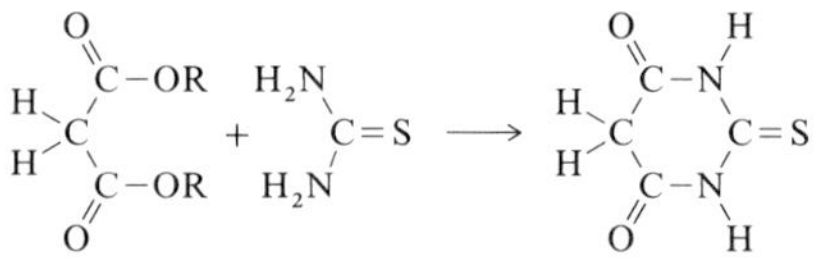

2-Oxocarboxylic acid esters condense with thiourea to thiouracils, and 1,3-diketones to mercaptopyrimidines:

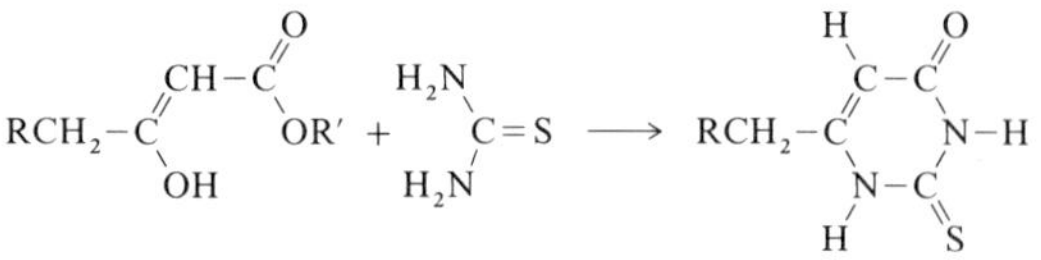

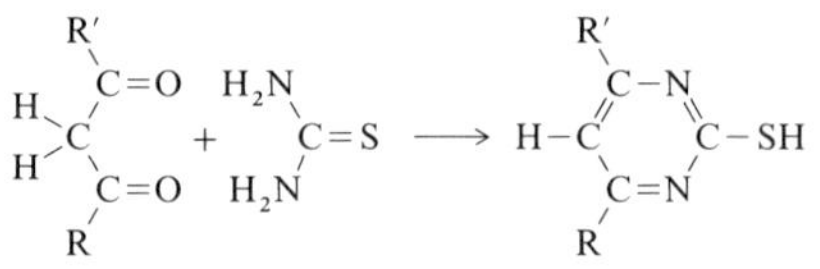

## 1.3. Production

Thiourea was first prepared by Reynolds by thermal rearrangement of ammonium rhodanide at ca. 150 °C:

$$NH_4SCN \rightleftharpoons (NH_2)_2CS$$

As this is an equilibrium reaction and separation of the two substances is quite difficult, better methods of preparation have been investigated.

The reaction between carbon disulfide and ammonia or ammonium carbonate under pressure at ca. 140 °C has not achieved industrial importance [18].

Production of thiourea is now carried out by treating technical-grade calcium cyanamide ($CaCN_2$) with hydrogen sulfide or one of its precursors, e.g., ammonium sulfide or calcium hydrogen sulfide [19]. The calcium cyanamide must contain at least 23 % nitrogen and must be free from calcium carbide ($CaC_2$), as otherwise explosive acetylene is liberated by water or hydrogen sulfide.

In Germany, thiourea is produced in a closed system by reaction of calcium cyanamide with hydrogen sulfide:

$$CaCN_2 + 3H_2S \longrightarrow Ca(SH)_2 + (NH_2)_2CS$$

$$2CaCN_2 + Ca(SH)_2 + 6H_2O \longrightarrow 2(NH_2)_2CS + 3Ca(OH)_2$$

$$Ca(OH)_2 + CO_2 \longrightarrow CaCO_3 + H_2O$$

A reaction vessel is charged continuously with an aqueous suspension of calcium cyanamide and a mixture of hydrogen sulfide and carbon dioxide, such that the above reactions go to completion. The converted reaction mixture is removed continuously from the reactor and excess $H_2S$ is removed by stripping. Waste-gas from the reactor contains hydrogen sulfide, which is purified in a scrubber connected to the reactor.

The stripped thiourea solution is filtered to remove the precipitated calcium carbonate formed as one of the reaction products. The filtrate is evaporated in vacuum and thiourea is crystallized by cooling. It is filtered, washed, and dried in an air stream.

## 1.4. Environmental Protection

In Germany, the scrubbed waste gases resulting from the production and filtration of thiourea are treated in an incinerator. All installations where thiourea dust is handled are equipped with air extraction. The extracted air is collected and, together with the waste air from the product drier, is purified in a scrubber.

Mother liquors containing thiourea are removed batchwise from the production process and are used in high-temperature incineration processes to remove $NO_x$. Side products of low solubility separated from the mother liquor are incinerated externally in a licensed plant.

The precipitated calcium carbonate arising in the production of thiourea is reprocessed (e.g., in cement factories or brick works). This is a contribution to the conservation of natural lime resources.

## 1.5. Quality Requirements

The German producer gives the following specification for the standard product:

| | |
|---|---|
| Thiourea | $\geq 99.0\%$ |
| Water | $\leq 0.30\%$ |
| Ash | $\leq 0.10\%$ |
| Rhodanide ($SCN^-$) | $\leq 0.15\%$ |
| Other N-containing compounds | $\leq 0.5\%$ |
| Iron | $\leq 10$ mg/kg |

Special qualities with higher purity are also produced.

## 1.6. Chemical Analysis

Tests of the content (purity) of thiourea can be carried out titrimetrically with Ce(IV) and potassium iodate solutions [20]. Since the commercial product is of relatively high quality, it is more efficient to determine the impurities as a means of quality control. The main impurity, dicyandiamide, is best determined by TLC. The incineration residue (ash) and the water content are also measured. A quantitative determination of sulfate, heavy metals, and rhodanide can also be carried out.

Traces of thiourea (e.g., in water) can be determined by HPLC ($C_{18}$ reverse-phase with water as the mobile phase and subsequent UV detection at 245 nm) [21].

Reaction of thiourea with sodium nitrite in acetic acid solution to give thiocyanate, which gives a red coloration with $Fe^{3+}$ ions, is also suitable for the determination of trace amounts. The red coloration is used to determine thiourea photometrically at 450 nm [22].

## 1.7. Storage and Transportation

Thiourea is packed in polyethylene bags ($\leq 50$ kg) or in folding containers ($\leq 1500$ kg). Bags and containers must be kept tightly closed, and in a cool, dry place. The contents must be protected from the action of light. Thiourea is not regarded as dangerous goods in transport regulations (RID/ADR, etc.).

## 1.8. Uses

Thiourea has a wide range of uses, e.g., for producing and modifying textile and dyeing auxiliaries [23], [24], in the production and modification of synthetic resins [25], in repro technology [26], in the production of pharmaceuticals (sulfathiazoles, tetramisole [27], and cephalosporins [28], in the production of industrial cleaning agents (e.g., for photographic tanks [29] and metal surfaces in general [30], [31]), for engraving metal surfaces [32], as an isomerization catalyst in the conversion of maleic to fumaric acid [33], in copper refining electrolysis [34], in electroplating (e.g., of copper) [35], and as an antioxidant (e.g., in biochemistry) [36].

Other uses are as an additive for slurry explosives [37], as a viscosity stabilizer for polymer solutions (e.g., in drilling muds [38]), and as a mobility buffer in petroleum extraction [39]. The removal of mercury from wastewater from chlorine–alkali electrolysis [14] and gold and silver extraction from minerals [40], [41] are also of economic importance.


## 1.9. Legal Aspects

Thiourea is included in international inventories of existing chemicals, e.g., EINECS (European Union), TSCA Inventory (United States), DSL (Canada), ACOIN and AICS (Australia), and MITI (Japan). In Germany and in the European Union, it is classified as a suspected carcinogen (MAK III B [42], [43]; EU cancer category 3). Similar evaluations have been made by the International Agency for Research on Cancer (Lyon, France) [44], and within the framework of the National Toxicology Program in the United States [45]. Within the European Union, thiourea must be labeled as a hazardous substance with hazard symbol Xn (St Andrew's cross–harmful), and R 22 (harmful if swallowed), R 40 (possible risk of irreversible effects), S 22 (do not breathe dust), S 24 (avoid contact with skin), and S 36/37 (wear suitable protective clothing and gloves). In the European Union Council Regulation on Existing Chemicals, thiourea is named in Annex I [46]. In Germany, the substance is classified in water hazard class 2 (hazardous for waters).


## 1.10. Economic Aspects

The quantity of thiourea produced and sold worldwide is ca. 10 000 t/a. The only producer in Western Europe is SKW Trostberg, Germany. Thiourea is also produced by Sakai Chemicals in Japan and at least seven Chinese companies.

According to estimations of SKW Trostberg, the largest producer, the quantity of thiourea produced worldwide is divided into the following main areas of use:

| | |
|---|---|
| Intermediate in the production of thiourea dioxide | ca. 30% |
| Ore leaching | 25% |
| Diazo papers | ca. 15% |
| Catalyst in fumaric acid synthesis | ca. 10% |

The remainder is divided between smaller areas of use (Section 1.8).

The United States EPA recently carried out a survey of the uses of thiourea in the United States. It is used, for example, in animal glue liquefiers, in silver and metal cleaners, in the production of flame-retardant resins, and as a vulcanization accelerator [45].


## 1.11. Toxicology [42]–[44], [47]

**Acute Toxicity and Local Tolerance.** The oral $LD_{50}$ for mice has been reported to be 1000 mg/kg [48], for rabbits 10 000 mg/kg [49], and for rats 1750 mg/kg [50] and 6200–6300 mg/kg [51]. After intraperitoneal administration $LD_{50}$ values varying between 4 and 1340 mg/kg were found for rats [52], [53]. Lung edema and pleural effusion have been described as symptoms. The dermal $LD_{50}$ is > 2800 mg/kg for rabbits [50] and > 6810 mg/kg for rats [51]. After inhalation of dust for 4 h (particle size 0.8–4.7 µm) $LC_{50}$ for rats was > 170 mg/m$^3$ [50], after aerosol inhalation (10% solution) > 195 mg/m$^3$ [50]. An alcoholic solution (droplet size < 10 µm) gave $LC_{50}$ > 900 mg/m$^3$ [51].

In skin tolerance tests (normal and abraded rabbit skin, 24 h application), slight to distinct erythema and slight edema were observed [50]. A 4 h application of 0.5 g gave no reaction [51]. Application of a 10% solution into the conjunctival sac of a rabbit's eye gave no reaction [50], but 100 mg of the solid substance gave a slight reddening and swelling [51]. Thiourea does not have a sensitizing effect in the guinea pig [51].

**Subacute and Chronic Toxicity.** In long-term application of thiourea, the effect on the thyroid gland stands out [54]. The substance disturbs incorporation of iodine in the synthesis of thyroxine and thus leads to a release of thyrotropic hormone from the pituitary gland, followed by severe stimulation of the thyroid gland, resulting in hyperplasia, adenomas, and, under certain circumstances, formation of thyroid carcinomas. Administration of thyroxine extensively prevents these effects.

Nonspecific effects of thiourea include hemosiderosis in the spleen, lymph nodes, and intestinal villi [55], weight loss of the liver, kidneys and spleen [54], anemia, leukopenia, increased erythropoiesis, and reduced bone growth.

Some examples of long-term tests are given below: rats received 0.625% thiourea in the diet for 14 days. This resulted in 52% growth inhibition [56]; 2.5 g per kilogram food (over 13 weeks) did not affect the body weight of mice [57]. Administration of 2.5 ppm in the drinking water of rats over 13 weeks gave no reaction [58]; after 27.5 mg kg$^{-1}$ d$^{-1}$ for 53 weeks in the drinking water normal growth was observed in rats [59]. Subcutaneous application in the mouse of 500 mg kg$^{-1}$ d$^{-1}$ for 10 d led to a slight decrease in the colloid content of the thyroid gland [60].

**Mutagenicity** [42], [43], [47]. In the Ames test with *Salmonella typhimurium* (TA 98, 100, 1535, 1537, 1538; 10–5000 µg per plate) thiourea was shown to be nonmutagenic, with and without activation (S9 mix) [51], [61]–[63]. A DNA repair test with isolated rat hepatocytes (0.3–30 mmol thiourea) also gave negative results [51]. A further test of this type gave a slightly positive result, as did a test with Chinese hamster cells V79 [64]. A CHO/HGPRT cell mutation test [65], a test for mitotic recombination (*Saccharomyces cerevisiae* D3) [66], and a micronucleus test (rat, 2 × 350 mg/kg, oral) [50] were also negative. A host-mediated assay (mouse) with TA 1530, TA 1538 (125 mg/kg thiourea, i.p.) showed a slight increase in the mutation rate, but the same test with TA 1535 and *Saccharomyces cerevisiae* (1000 mg/kg thiourea, i.p.) did not [48].

**Carcinogenicity.** A cell transformation test with Syrian golden hamster embryo cells (0–100 g/ml) was negative [67].

Experiments on mice with different test periods (4–10 months) and dose levels (0.1–0.5% in drinking water) showed a slight enlargement of the thyroid gland and slight thyroid hyperplasia [68]. Application of 2% thiourea in the diet over 1.5 years led to hyperplasia and cystic changes in the thyroid gland, but not to carcinomas [69]. In rats receiving 0.25% in drinking water for 5–23.5 months, adenomas and carcinomas in the thyroid gland were observed [70]; in other tests (0.2% in drinking water) tumors in the orbita, nose, and ear regions [71] were observed as well as in the liver (0.01–0.25% thiourea in the diet) [72].

Tumor formation is associated with excessive stimulation of the thyroid gland by thyrotropic hormone from the pituitary gland. The fact that thiourea has only a weak mutagenic potential also indicates an epigenetic mechanism for tumor formation. Since tumors are also found occasionally in organs other than the thyroid gland, thiourea is classified as a suspected carcinogen [42], [43].

**Embryotoxicity.** In mice and rats, maternally toxic oral doses of 1000 mg/kg body weight were found to be embryotoxic [73]. In rat fetuses, whose dams had received 2000 ppm in drinking water, maturation defects were apparent in the central nervous system, skeleton, and eyes [74]. These effects are attributed to the thyreostatic activity of thiourea. Therefore, at dosages which are not sufficient to inhibit thyroid function effects are not expected [42]. A single oral dose of 480 mg/kg was found to be neither maternally toxic nor teratogenic in rats [75].

**Absorption and Metabolism.** In humans, thiourea is rapidly and completely absorbed from the gastrointestinal tract following oral administration. Most is excreted unchanged in the urine [76]. Small amounts appear as sulfate and ether sulfate, as shown by tests with radioactive material in rats [77]. A whole-body autoradiography with $^{14}$C-thiourea showed enrichment and longer residence time in the thyroid gland compared with other tissues [78]. Following cutaneous application of 2 g/kg as an aqueous solution in rabbits, ca. 4% of the dose was detected in the urine, but only 0.1% after application of the solid [50].

**Effects on Humans.** Thiourea has been used temporarily in therapy because of its thyroid-depressant activity. After initial doses of 2–3 g/d, dosages of 100–200 mg/d were administered, and finally 25–70 mg was recommended as the daily dose. Doses of 10–15 mg/d were ineffective in 5 of 8 people [79]. Its use in therapy has been frequently associated with side effects, e.g., gastrointestinal disorders, pain in the muscles and joints, leukopenia, fever, and reddening of the skin. There are only a few reports on observations in the workplace [80], [81]. Apart from hypothyroidism, symptoms appeared which were similar to side effects known from therapeutic use.

In the use and processing of thiourea, contact dermatitis has occasionally been reported, e.g., on use in silver polish and diazo (blueprint) paper, sometimes after UV irradiation [82]–[86].

A Dutch expert committee recommended an occupational exposure limit of 0.5 mg/m$^3$ [47]. This value is also cited in the documentation of the German Research Association with reference to the thyroid gland [42], [43].


# 2. Thiourea Derivatives

## 2.1. Physical Properties

Melting points and solubilities of thiourea derivatives are given in [7], [9], [11].

The free energy of activation for rotation about the C–N bond determined by dynamic NMR spectroscopy is in the range $\Delta G^+ = 29$–$50$ kJ/mol [87].

Various physical methods, e.g., neutron diffraction [88] and IR spectroscopy [89], confirm a strong contribution of the polar, planar structure I to the ground state:

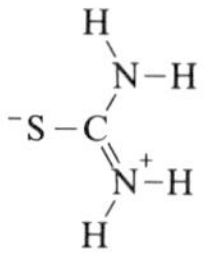

Monoalkyl-substituted thioureas exist as mixtures of $E$ and $Z$ isomers. In monoaryl substitution products, the $E$ isomer is preferred [90].

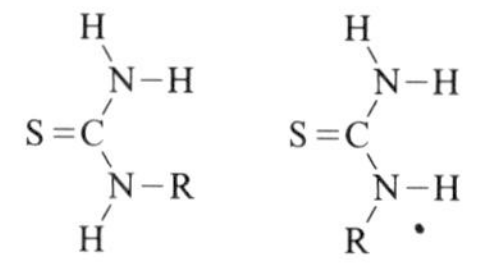

## 2.2. Chemical Properties

The chemistry of thiourea and its derivatives is dominated by two structural features:

1) The nucleophilicity of sulfur gives isothiuronium salts in good yields on protonation [91], alkylation, and arylation [9].
2) The polyfunctional character of the thiourea system can be used to synthesize 4-, 5-, 6-, and 7-membered heterocycles [9], [92].

Halogenation and oxidation reactions of monoaryl-substituted thioureas give 2-aminobenzothiazoles [93], important in the production of, e.g., polyester dyes [94].


## 2.3. Production

Methods for the production of thioureas can be divided into seven groups [11]:

1) Reaction of alkyl, aryl, and heteroaryl isothiocyanates with ammonia, or primary or secondary amines can be used universally, is simple to carry out on a preparative scale, and gives mono-, di-, and trisubstituted thioureas in good yields:

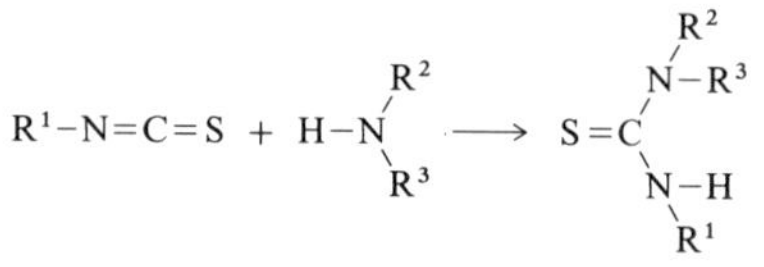

2) Reaction of primary amines with carbon disulfide is of fundamental importance for the synthesis of $N,N$-diaryl-substituted thioureas and results in quantitative yield with aromatic amines. Dithiocarbamates are formed as isolable intermediates ($\rightarrow$ Dithiocarbamic Acid and Derivatives, **A 9**, pp. 2–5), which are converted to thioureas with loss of hydrogen sulfide:

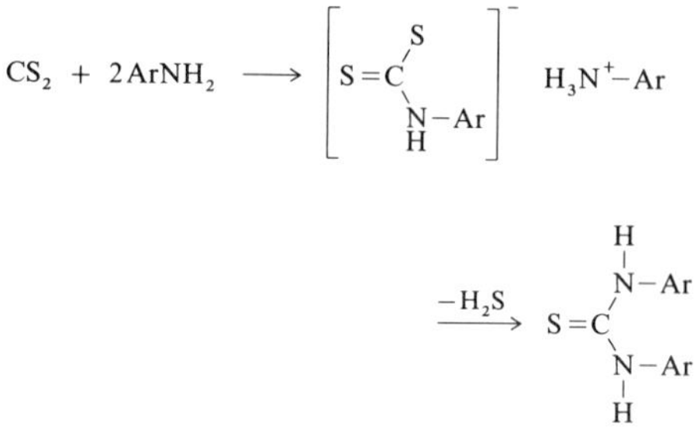

Nitroarenes can also be used as starting materials if the reaction is carried out in an alkaline medium [95].

3) Primary and secondary amines react with thiophosgene, forming thiocarbamoyl chlorides as intermediates, which are then converted to di- and tetrasubstituted thioureas:

$$
S=C\begin{smallmatrix}Cl\\Cl\end{smallmatrix} + H-N\begin{smallmatrix}R^1\\R^2\end{smallmatrix} \longrightarrow S=C\begin{smallmatrix}Cl\\N-R^2\\|\\R^1\end{smallmatrix} \longrightarrow S=C\begin{smallmatrix}N(R^3)R^4\\N-R^2\\|\\R^1\end{smallmatrix}
$$

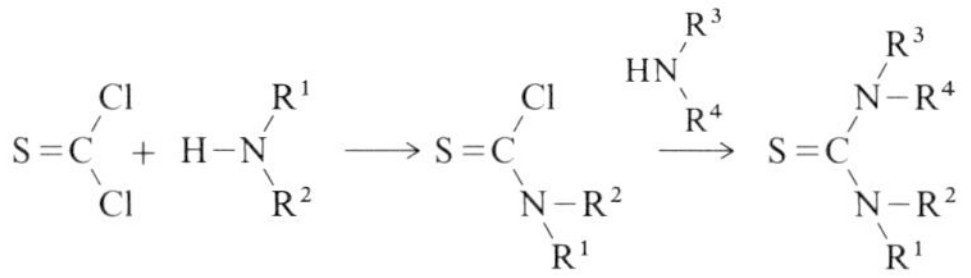

However, other thiocarbonic acid derivatives, e.g., thiocarbonylbisazoles or diesters of trithiocarbonic acid, react only with primary amines.

4) By analogy with the thiourea synthesis from ammonium thiocyanate, primary alkyl- and arylammonium thiocyanates are converted by thermolysis into monosubstituted thioureas in good yields:

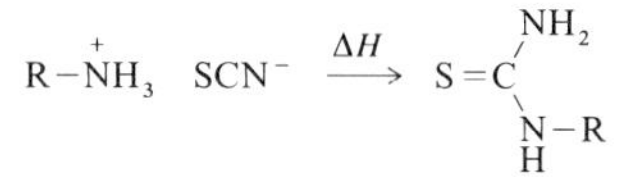

$$
R-\overset{+}{N}H_3 \quad SCN^- \xrightarrow{\;\Delta H\;} S=C\begin{smallmatrix}NH_2\\N-R\\|\\H\end{smallmatrix}
$$

5) Addition of hydrogen sulfide to alkyl- and aryl-substituted cyanamides offers an effective route to $N$-mono- and $N,N$-disubstituted thiourea derivatives by analogy with the industrial thiourea synthesis from calcium cyanamide:

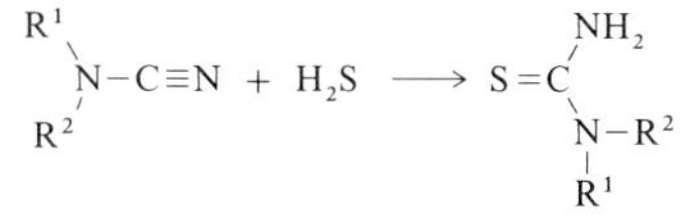

$$
\begin{smallmatrix}R^1\\R^2\end{smallmatrix}N-C\equiv N + H_2S \longrightarrow S=C\begin{smallmatrix}NH_2\\N-R^2\\|\\R^1\end{smallmatrix}
$$

6) For the production of sterically hindered, tetrasubstituted thioureas, sulfurization of the corresponding urea derivatives with phosphorus pentasulfide is suitable:

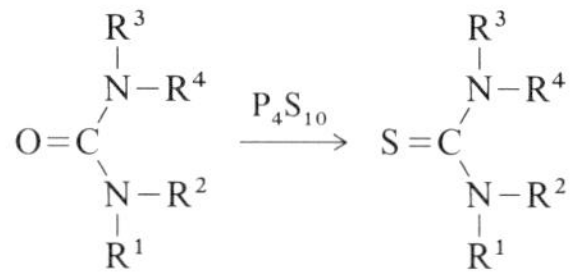

$$
O=C\begin{smallmatrix}N(R^3)-R^4\\N-R^2\\|\\R^1\end{smallmatrix} \xrightarrow{\;P_4S_{10}\;} S=C\begin{smallmatrix}N(R^3)-R^4\\N-R^2\\|\\R^1\end{smallmatrix}
$$

**Table 2.** Uses of thiourea derivatives as vulcanization accelerators [102]

| Compound | CAS No. | Polymer | Trade name | Producer |
|---|---|---|---|---|
| 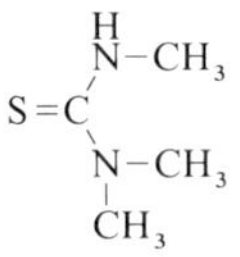 S=C, N-CH₃ (H), N-CH₃ / CH₃ | [2489-77-2] | polychloroprene | Thiate EF-2 | Vanderbilt |
| 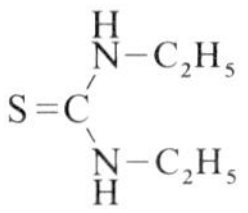 S=C, N-C₂H₅ (H), N-C₂H₅ (H) | [105-55-5] | polychloroprene EPDM * | Pennzone<br>Robac DETU<br>Sanceller EUR<br>Thiate H | Pennwalt<br>Robinson<br>Sanshin<br>Vanderbilt |
| 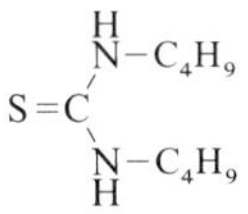 S=C, N-C₄H₉ (H), N-C₄H₉ (H) | [109-46-6] | polychloroprene EPDM * | Robac DBTU<br>Thiate U | Robinson<br>Vanderbilt |
| 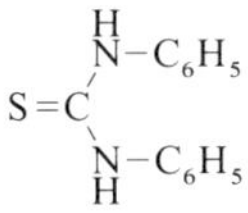 S=C, N-C₆H₅ (H), N-C₆H₅ (H) | [102-08-9] | polychloroprene EPDM * | A-1 Thiocarbanilid<br>Stabilizer C | Monsanto<br>Mobay |
| 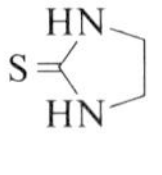 ethylenethiourea (HN–/HN–)C=S | [96-45-7] | polychloroprene | Akrochem ETU-22<br>Robac 22<br>Sanceller 22<br>Valcazit NPV/C | Akron Chemical<br>Robinson<br>Sanshin<br>Mobay |
| 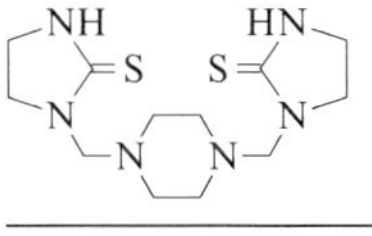 bis-ethylenethiourea derivative | [14764-02-4] | polychloroprene | Vulafor 322 | Vulnax |

* EPDM = ethylene–propylene–diene terpolymers.

**Table 3.** Pharmacologically active thiourea derivatives

| Compound | CAS No. | Use | Generic name | Trade name | Producer |
|---|---|---|---|---|---|
| 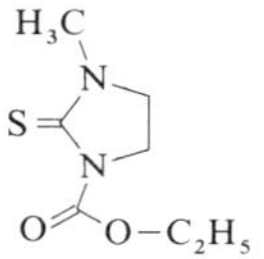 | *[22232-54-8]* | thyrotherapeutic agent (thyreostatic) | carbimazole | Atirozidina<br>Basolest<br>Neo-Carbimazole<br>Neo-Mercazole<br>Neo-Morphazole<br>Neo-Thyreostat<br>Neo-Tomizol<br>Neral<br>Tyrazol | Confas<br>Pharmachemie<br>Landerlan<br>Schering<br>Nicholas<br>Herbrand<br>Novag<br>Conradson<br>Neofarma |
| 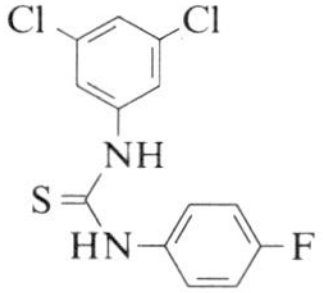 | *[790-69-2]* | fungicide, dermatological agent | loflucarban | Fluonilid | Continental Pharma |
| 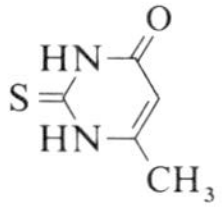 | *[56-04-2]* | thyrotherapeutic agent (thyreostatic) | methylthiouracil | Methiacil<br>Methiocil<br>MTU<br>Thimecil<br>Thyreostat<br>Thyrostabil<br>Tiouracil | Schwarz<br>Helvepharm<br>Philopharm<br>Physicians' Drug<br>Herbrand<br>Streuli<br>Lefa |
| 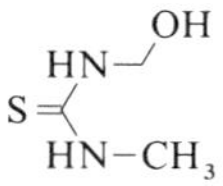 | *[15599-39-0]* | antiseptic | noxytiolin | Gynaflex<br>Noxyflex-S | Geistlich<br>Geistlich |
| 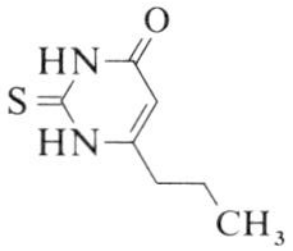 | *[51-52-5]* | thyrotherapeutic agent (thyreostatic) | propylthiouracil | Propycil<br>Propyl-Thiocil<br>Propyl-Thyracil<br>Prothiucil<br>Thyreostat II<br>Tiotil<br>Tirostaz | Kali-Chemie<br>Teva<br>Merck-Frosst<br>Donau-Pharmazie<br>Herbrand<br>Pharmacia<br>Yurtoglu |
| 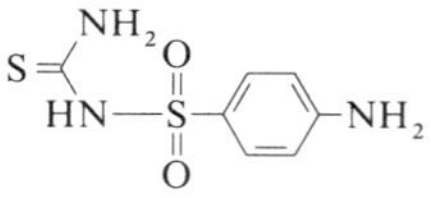 | *[515-49-1]* | antiinfective agent | sulfathiourea | Badional<br>Fontamide<br>Salvoseptyl | Bayer<br>Specia<br>Alkaloida |

**Table 3.** (continued)

| Compound | CAS No. | Use | Generic name | Trade name | Producer |
|---|---|---|---|---|---|
| 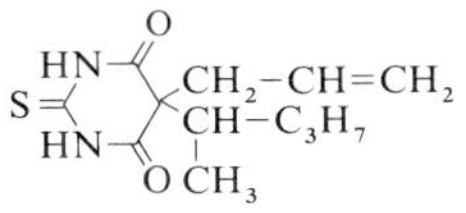 | *[77-27-0]*<br>*[337-47-3]* Na-salt | ultrashort narcotic, anesthetic | thiamylal | Bio-Tal<br>Bio-Tel<br>Surital<br>Thioseconal | Philips Roxane<br>Philips Roxane<br>Parke Davis<br>Lilly |
| 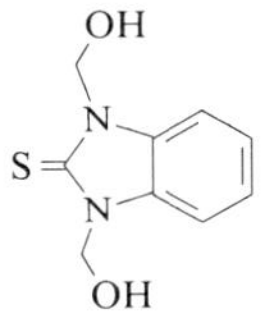 | *[6028-35-9]* | thyreostatic | thiabenzazolin | Thyreocordon | Donau-Pharmazie |
| 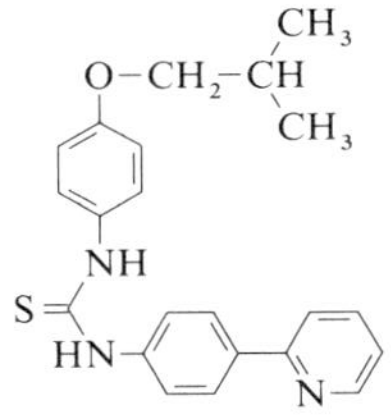 | *[92-97-7]* | tuberculostatic | thiocarbanidin | Thioban | Parke Davis |
| 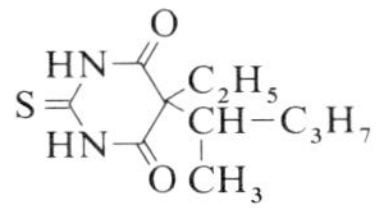 | *[76-75-5]*<br>*[71-73-8]* Na-salt | narcotic | thiopental | Farmotal<br>Intraval<br>Leopental<br>Nesdonal<br>Pentothal<br>Thio-Barbityral<br>Thionembutal<br>Tiobarbital<br>Trapanal | Farmitalia Carlo Erba<br>May & Baker<br>Leo<br>Specia<br>Abbott<br>Amino<br>Abbott<br>Miro<br>Byk Gulden |
| 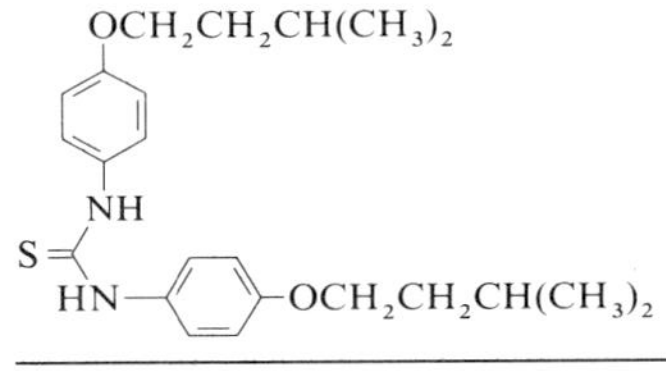 | *[910-86-1]* | tuberculostatic | thiocarlide | Amixyl<br>Disoxyl<br>Isoxyl | Inibsa<br>Ferrosan<br>Continental Pharma,<br>Inibsa |

**Table 3.** (continued)

| Compound | CAS No. | Use | Generic name | Trade name | Producer |
|---|---|---|---|---|---|
| (structure: $O=P(OC_2H_5)_2$ thiourea-linked tolylsulfonyl phenyl derivative) | [52406-01-6] | anthelmintic (veterinary) | uredofos | Sansalid | Röhm & Haas |

7) Cyclic thioureas can be obtained by the methods known for open-chain thioureas, and by the reaction of diamines with thiourea [96]:

$$\text{NH}_2 + \text{H}_2\text{N}{-}\text{C}{=}\text{S} \xrightarrow{\Delta} \text{cyclic } \text{N}{-}\text{C}{=}\text{S}$$

## 2.4. Uses

**Analysis.** For the quantitative determination of thiourea compounds, titration with phenyl iodosoacetate is suitable [97]. Chromatographic methods have been developed for separation [98].

Because of their complex-forming properties [99], thiourea derivatives can be used as analytical reagents [100] and for the extraction of noble metals [101].

**Industrial Applications.** The thiourea derivatives listed in Table 2 are used as accelerators for the vulcanization of polychloroprene and ethylene–propylene–diene terpolymers (EPDM) [102].

$N$-Substituted thiourea derivatives are used as corrosion inhibitors, as antioxidants in the rubber industry, in electroplating processes [8], and in the production of light-sensitive materials [103].

**Biological Activity.** Some thioureas are used as active substances in human and veterinary medicine [104]–[106], as insecticides [107]–[109], and as rodenticides [110]. The structures of thiourea derivatives which are important as pharmaceuticals and as plant protection agents and pesticides are given in Tables 3 and 4.

## 2.5. Toxicology

The toxicological properties of a selection of alkyl- and aryl-substituted thiourea derivatives are described in [111].

More recent studies have dealt with metabolism and the effect of steric factors on the toxicity of $N$-phenylthiourea derivatives [112], and with the mutagenic, teratogenic, and carcinogenic properties of imidazolidine-2-thione (ethylenethiourea) [96-45-7], a metabolite of ethylene-bisthiocarbamate fungicides [113].

**Table 4.** Thiourea derivatives as plant protection agents and pesticides [107]

| Thiourea derivative | CAS No. | Use | Generic name | Trade name | Producer |
|---|---|---|---|---|---|
| 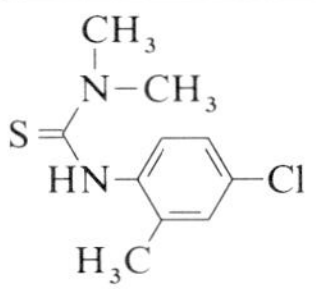 | *[28217-97-2]* | acaricide, anti-tick agent | chloromethiuron | Dipofene | Ciba-Geigy |
| 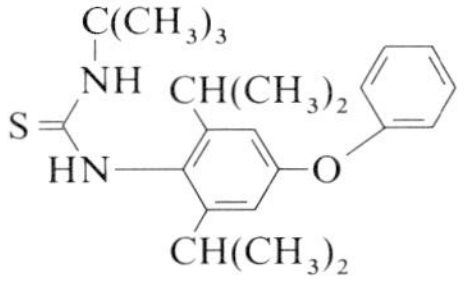 | *[80060-09-9]* | insecticide, acaricide | diafenthiuron | Pegasus<br>Polo | Ciba-Geigy<br>Ciba-Geigy |
| 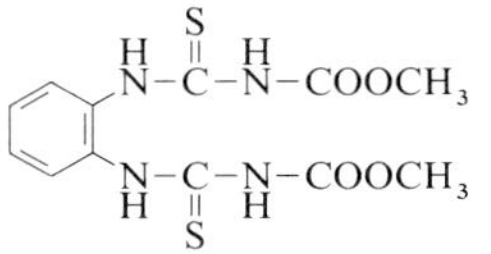 | *[23564-05-8]* | fungicide | thiophanate methyl | Topsin M<br>Cercobin M<br>Mildothane<br>Cycosin | Nippon Soda<br>Nippon Soda<br>Rhone-Poulenc<br>American Cyanamid |
| 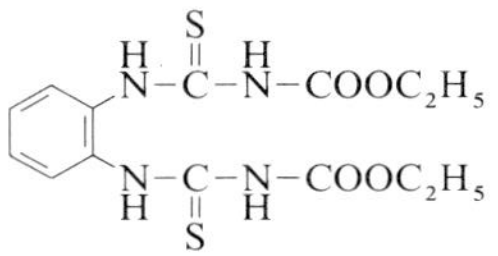 | *[23564-06-9]* | fungicide, anthelmintic | thiophanate | Topsin<br>Nemafax<br>Verdamax | Nippon Soda<br>May & Baker<br>May & Baker |

# 3. References

General References

[1] *Beilstein*, 4th ed., E IV 3, 342 ff. (1977).
[2] *Houben-Weyl*, **9**, 799–803, 884–899.
[3] *Gmelin*, 8th ed., (14), part D6, Kohlenstoff, 1978.
[4] SKW Trostberg AG, Produktstudie Thioharnstoff, Trostberg 1980.
[5] BUA, Stoffdossier Thioharnstoff, in press.
[6] D. C. Schröder, *Chem. Rev.* **55** (1955) 181–221.
[7] E. E. Reid: *Organic Chemistry of Bivalent Sulfur*, Chem. Publishing Co., New York 1963, pp. 11–65.
[8] V. V. Mozolis, S. P. Iokubaitite, *Russ. Chem. Rev. (Engl. Transl.)* **42** (1973) 587–595.
[9] F. Duus in D. Barton, W. D. Ollis (eds.): *Comprehensive Organic Chemistry*, vol. 3, Pergamon Press, Oxford 1979, pp. 452–486.
[10] D. H. Reid: *Organic Compounds of Sulphur, Selenium, and Tellurium*, The Chemical Society, London, vol. 1, 1970, pp. 209–216; vol. 2, 1973, pp. 236–243; vol. 3, 1975, pp. 267–279; vol. 4, 1977, pp. 142–153; vol. 5, 1979, pp. 139–151; vol. 6, 1981, pp. 165–174.
[11] *Houben-Weyl*, **E 4**, 484–521.

Specific References

[12] R. Bauer, I. Wehling, *Fresenius' Z. Anal. Chem.* **199** (1964) 171–183.
[13] F. Kicinsky, *Chem. Ind. (London)* **1** (1948) 54–57.
[14] D. Napoli, G. Ratti, G. Tubiello, *Quad. Ist. Ric. Acque* **45** (1979) 347–361.
[15] K. Gawalek: *Einschlußverbindungen*, VEB Deutscher Vlg. der Wissenschaften, Berlin 1969, pp. 226–236.
[16] W. Schlenk, *Justus Liebigs Ann. Chem.* **573** (1952) 142–162.
[17] BASF, DE-OS 1241612, 1962 (H. Scheuermann et al.).
[18] G. Inghilleri, *Gazz. Chim. Ital.* **39** (1909) 634–639.
[19] S. A. Miller, B. Bann, *J. Appl. Chem.* **6** (1956) 89–93.
[20] R. P. Agarwal, S. K. Ghosh, *Chim. Anal. (Paris)* **47** (1965) 407–488.
[21] SKW Trostberg AG, unpublished.
[22] K. Hayashi et al., *Bunseki Kagaku* **26** (1977) 514–515.
[23] Farbenfabriken Bayer, FR 1576094, 1969.
[24] Ciba Geigy, DE-OS 2059373, 1971 (J. Wegmann).
[25] SKW Trostberg, DE-OS 2027085, 1971 (H. Michaud et al.).
[26] Ricoh Co., Ltd., GB 1218933, 1971 (T. Iwaoka et al.).
[27] S. A. Janssen, US 3274209, 1960 (A. Raeymaekers).
[28] Lonza, EP 0045005, 1984 (A. Huwiler et al.).
[29] Degussa, DE-OS 2511433, 1975 (H. Knorre et al.).
[30] Trio Chemical Works, US 2907716, 1959 (R. J. Wassermann).
[31] Geigy, FR 1577582, 1968 (J. Casanova).
[32] IBM, US 4588471, 1986 (J. Griffith et al.).
[33] W. Schliesser, *Angew. Chem.* **74** (1962) 429–430.
[34] F. J. Kaltenböck, H. Wöbking, *Metall (Berlin)* **40** (1986) 1144–1146.
[35] A. Hung, *J. Electrochem. Soc.* **132** (1985) 1047–1049.
[36] E. Van Driesche et al., *Anal. Biochem.* **141** (1984) 184–188.
[37] ICI, DE-OS 2553055, 1976 (G. H. McCallum et al.).
[38] Philips Petroleum, US 4124073, 1978 (D. R. Wier).
[39] Shell Int. Res. Maatschappij B.V., DE-OS 2715026, 1977 (S. L. Wellington).
[40] T. Groenewald, *J. South Afr. Inst. Min. Metall.* **1977**, 217–223.
[41] SKW Trostberg AG, company brochure, Laugung und Gewinnung von Edelmetallen mit Thioharnstoff, Trostberg 1984.
[42] D. Henschler: *Gesundheitsschädliche Arbeitsstoffe. Toxikologisch-arbeitsmedizinische Begründung von MAK-Werten, Thioharnstoff*, Verlag Chemie, Weinheim 1988.
[43] DFG: *Occupational Toxicants. Critical Data Evaluation for MAK Values and Classification of Carcinogens*, vol. 1, Verlag Chemie, Weinheim 1991, p. 301.
[44] *IARC Monogr. Eval. Carcinog. Risk Chem. Man* **7** (1974) 95; Suppl. **4** (1982) 270.
[45] U.S. Department of Health and Human Services: *National Toxicology Program. Sixth Annual Report on Carcinogens*, National Institute of Environmental Health Services, Research Triangle Park, NC, 1991.
[46] Council of the European Communities, Council Regulation (EU) of March 23, 1993 on the Evaluation and Control of the Environmental Risks of Existing Substances, Official Journal of the European Communities, L 84, vol. 36, 5th April 1993.
[47] Dutch Expert Committee for Occupational Standards: *Health-Based Recommended Occupational Exposure Limits for Thiourea*, Directorate-General of Labour of Ministry of Social Affairs and Employment, Monograph RA II/90, 1990.
[48] V. F. Simmon, H. S. Rosenkranz, E. Zeiger, L. A. Poirier, *JNCI J. Natl. Cancer Inst.* **62** (1979) 911–918.
[49] F. B. Flinn, J. M. Geary, *Contr. Boyce Thompson Inst.* **11** (1940) 241–247.
[50] A. P. de Groot et al., Berichte von CIVO (Central Institute for Nutrition and Food Research), TNO, in Zeist (Holland) an SKW Trostberg AG 1975–1983.
[51] F. Korte, H. Greim: Überprüfung der Durchführbarkeit von Prüfungsvorschriften und der Aussagekraft der Grundprüfung des E. Chem. G. UFO-PLAN-Nr. 10704006/01, 1981.
[52] S. H. Dieke, C. P. Richter, *J. Pharmacol. Exp. Ther.* **83** (1945) 195–202.
[53] G. S. Wiberg, H. C. Grice, *Food. Cosmet. Toxicol.* **3** (1965) 597–603.
[54] E. Huf, F. Auffarth, *Naunyn-Schmiedebergs Arch. Exp. Pathol. Pharmakol.* **206** (1949) 394–415.
[55] L. Arvy, M. Gabe, *Compt. Rend. Soc. Biol.* **144** (1950) 487–488.
[56] C. C. Smith, *J. Pharmacol. Exp. Ther.* **100** (1950) 408–420.
[57] H. P. Morris, C. S. Dubnik, A. J. Dalton, *J. Natl. Cancer Inst.* **7** (1946) 159–169.
[58] M. R. Osheroff, Hazleton Laboratories America Inc., 13-Week Drinking Water Study in Rats with Thiourea, HLA Study No. 2319-119, Bericht an SKW Trostberg AG 1987.
[59] A. Hartzell, *Contr. Boyce Thompson Inst.* **12** (1942) 471–480.
[60] R. P. Jones, *J. Pathol. Bacteriol.* **58** (1946) 483–493.
[61] J. McCann, E. Choi, E. Yamasaki, B. N. Ames, *Proc. Natl. Acad. Sci. USA* **72** (1975) 5135–5139.
[62] H. S. Rosenkranz, L. A. Poirier, *JNCI J. Natl. Cancer Inst.* **62** (1979) 873–891.
[63] V. F. Simmon, *JNCI J. Natl. Cancer Inst.* **62** (1979) 893–899.

[64] K. Ziegler-Skylakakis, S. Rossberger, U. Andrae, *Arch. Toxicol.* **58** (1985) 5–9.

[65] M. L. Augustine, N. K. Poulsen, J. E. Heinze, *Environ. Mutagen.* **4** (1982) 389–390.

[66] V. F. Simmon, *JNCI J. Natl. Cancer Inst.* **62** (1979) 901–909.

[67] R. J. Pienta, J. A. Poiley, W. B. Lebberz, *Int. J. Cancer* **19** (1977) 642–655.

[68] E. Vazquez-Lopez, *Br. J. Cancer Res.* **3** (1949) 401–414.

[69] A. Gorbman, *Cancer Res.* **7** (1947) 746–758.

[70] H. D. Purves, W. E. Griesbach, *Br. J. Exp. Pathol.* **28** (1947) 46–53.

[71] A. Rosin, H. Ungar, *Cancer Res.* **17** (1957) 302–305.

[72] O. G. Fitzhugh, A. A. Nelson, *Science (Washington D.C., 1883)* **108** (1948) 626–628.

[73] S. Teramoto, M. Kaneda, H. Aoyama, Y. Shirasu, *Teratology* **23** (1981) 335–342.

[74] M. Kern, Z. Tatàr-Kiss, P. Kertai, I. Földes, *Acta Morphol. Acad. Sci. Hung.* **28** (1980) 259–267.

[75] J. A. Ruddick, W. H. Newsome, L. Nash, *Teratology* **13** (1976) 263–266.

[76] R. H. Williams, G. A. Kay, *Am. J. Physiol.* **143** (1945) 715–722.

[77] J. Schulman, R. P. Keating, *J. Biol. Chem.* **183** (1950) 215–221.

[78] P. Slanina, S. Ullberg, L. Hammarström, *Acta Pharmacol. Toxicol.* **32** (1973) 358–368.

[79] A. W. Winkler, E. B. Man, T. S. Danowski, *J. Clin. Invest.* **26** (1947) 446–452.

[80] A. G. Zaslavskaja, *Klin. Med. (Moscow)* **42** (1964) 129–132.

[81] Y. N. Talakin et al., *Gig. Tr. Prof. Zabol.* **11** (1990) 18–20.

[82] A. Dooms-Goossens et al., *Br. J. Dermatol.* **116** (1987) 573–579.

[83] A. Dooms-Goossens et al., *Contact Dermatitis* **19** (1988) 133–135.

[84] D. S. Nurse, *Contact Dermatitis* **6** (1980) 153–154.

[85] J. C. Van der Leun, *Arch. Dermatol.* **113** (1977) 1611.

[86] J. K. Kellett, M. H. Beck, G. Auckland, *Contact Dermatitis* **11** (1984) 124.

[87] D. R. Eaton, K. Zaw, *J. Inorg. Nucl. Chem.* **38** (1976) 1007.

[88] D. Mullen, G. Heger, W. Treutmann, *Z. Kristallogr.* **148** (1978) 95.

[89] Y. Mido, T. Yamanaka, R. Awata, *Bull. Chem. Soc. Jpn.* **50** (1977) 27.

[90] W. Walter, K. P. Rueß, *Justus Liebigs Ann. Chem.* **743** (1971) 167.

[91] J. T. Edward, I. Lantos, G. D. Derdall, S. C. Wong, *Can. J. Chem.* **55** (1977) 812, 2331.

[92] T. S. Griffin, T. S. Woods, D. L. Klayman in A. R. Kartritzky, A. J. Boulton (eds.): *Advances in Heterocyclic Chemistry*, vol. 18, Academic Press, New York 1975, pp. 100–145.

[93] Hoechst, EP 207416, 1986 (H. Rentel, T. Papenfuhs); EP 529600, 1992 (S. Dapperheld, H. Volk, K. Peter).

[94] M. A. Weaver, L. Shuttleworth, *Dyes Pigm.* **3** (1982) 81.

[95] Olin Mathieson Chem., US 3636104, 1972 (E. H. Kober, G. F. Ottmann).

[96] A. V. Bogatskii, N. G. Luk'yanenko, T. I. Kirichenko, *Chem. Heterocycl. Compd. (Engl. Transl.)* **19** (1983) 577.

[97] K. K. Verma, *Fresenius' Z. Anal. Chem.* **275** (1979) 287.

[98] H. Nakamura, Z. Tamura, *J. Chromatogr.* **96** (1974) 195.

[99] L. Beyer, J. Hartung, R. Köhler, *J. Prakt. Chem.* **333** (1991) 373.

[100] S. Singh, S. P. Mathur, R. S. Thakur, M. Katyal, *Acta Cien. Indica Chem.* **13** (1987) 40.

[101] S. Singh, S. P. Mathur, R. S. Thakur, L. Keemti, *Orient. J. Chem.* **3** (1987) 203.

[102] *Kirk-Othmer*, **20**, 337.

[103] Fuji Photo Film, JP 0504449, 1991 (H. Kawakami, K. Iwakura); *Chem. Abstr.* **118** (1993) 263906u.

[104] S. N. Pandeya, P. Ram, V. Shankar, *J. Sci. Ind. Res.* **40** (1981) 458.

[105] M. Negwer: *Organic-Chemical Drugs and their Synonyms*, 6th ed., Akademie Verlag, Berlin 1987.

[106] Swiss Pharmaceutical Society (ed.): *Index Nominum, International Drug Directory 1990/1991*, Medpharm Scientific Publ., Stuttgart 1990.

[107] C. R. Worthing, R. J. Hance (eds.): *The Pesticide Manual*, 9th ed., The British Crop Protection Council, Farnham 1991.

[108] K. H. Büchel: *Chemistry of Pesticides*, J. Wiley & Sons, New York 1983, p. 315.

[109] J. Drabek et al., *Spec. Publ. R. Soc. Chem.* **79** (1990) 170.

[110] R. Wegler: *Chemie der Pflanzenschutz- und Schädlingsbekämpfungsmittel*, Springer Verlag, Berlin 1970, pp. 605–606.

[111] *Ullmann*, 4th ed., **23**, pp. 172–173.

[112] A. E. Miller, J. J. Bischoff, K. Pae, *Chem. Res. Toxicol.* **1988**, 169.

[113] R. A. Frakes, *Regul. Toxicol. Pharmacol.* **8** (1988) 207.

**Thomas Phosphate → Phosphate Fertilizers**